Molecular

Cell

Biology

Harvey Lodish
Member of the Whitehead Institute for Biomedical Research;
Professor of Biology, Massachusetts Institute of Technology

David Baltimore
Ivan R. Cottrell Professor Molecular Biology and Immunology,
Massachusetts Institute of Technology

Arnold Berk
Professor, University of California at Los Angeles; Director of
the Molecular Biology Institute

S. Lawrence Zipursky
Professor of Biological Chemistry, University of California at
Los Angeles; Investigator, Howard Hughes Medical Institute

Paul Matsudaira
Member of the Whitehead Institute for Biomedical Research;
Associate Professor of Biology, Massachusetts Institute of
Technology

James Darnell
Vincent Astor Professor and Head of Molecular Cell Biology
Laboratory, Rockefeller University

MOLECULAR CELL

BIOLOGY

THIRD EDITION

Harvey Lodish

David Baltimore

Arnold Berk

S. Lawrence Zipursky

Paul Matsudaira

James Darnell

**SCIENTIFIC
AMERICAN
BOOKS**

An Imprint of W. H. Freeman and Company, New York

Cover illustration by Nenad Jakasevic

Library of Congress Cataloging-in-Publication Data

Molecular cell biology/James Darnell . . . [et al.].—3d ed.
     p.   cm.
   Second edition's main entry under the heading for Darnell.
   Includes bibliographical references and index.
   ISBN 0-7167-2380-8
   1. Cytology.   2. Molecular biology.   I. Darnell, James E.
QH581.2.D37   1995   94-22376
574.87′6042—dc20   CIP

Printed in the United States of America

Scientific American Books is a subsidiary of Scientific American, Inc.
Distributed by W. H. Freeman and Company, 41 Madison Avenue,
New York, New York 10010 and 20 Beaumont Street,
Oxford OX1 2NQ England

1 2 3 4 5 6 7 8 9 0   HAW   9 9 8 7 6 5

To our supportive
families for
their patience;
to our teachers
and students for
their inspiration;
and to the many
scientists who
generated the
knowledge that
made this
book possible.

Preface

➤ *Our Traditional Strengths*

In the preface to the first edition of *Molecular Cell Biology* we asserted that the new techniques of molecular biology would soon unify all experimental biology. The next nine years confirmed the validity of our position. Although once separate, the fields of biochemistry, genetics, cell biology, physiology, developmental biology, and now even much of neurobiology have become wedded by fundamental experimental approaches. Their common quests are threefold:

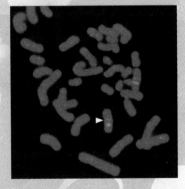

Figure 8-28a

• to understand how gene expression is controlled so that cells synthesize the right proteins at the right time in the right amounts

• to know the structure and function of proteins that carry out specific biologically important tasks: proteins

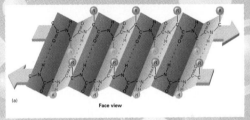

Figure 3-12a

that regulate ion flow, cause cell motility or contractility, catalyze a degradative or synthetic reaction, induce an embryonic structure, or send or receive a chemical signal

• to understand how the properties of cells, tissues, and the other organisms relate to the properties of the proteins and other molecules that they contain.

Thus our focus continues to be the central dogma of the new integrated science of cell biology: the hierarchy of genes ↔ proteins ↔ organelles ↔ cells ↔ tissues ↔ organs ↔ organisms. Our emphasis on the fundamental experimental tools and tech-

> ➤ *"At the top of this hierarchy of biological organization lie the genes. They not only specify the structure of proteins but ultimately they define the organization of cells, direct cells to form tissues and organs, and maintain the integration of function that gives the organism its unity."*

niques that we regard as the necessary foundation of this new integrated science of *Molecular Cell Biology* has been maintained.

Part I introduces students to the basic concepts and experimental methodologies of molecular cell biology. Here we discuss the basic chemical and physical principles of cellular activity, essential features of DNA replication, transcription, and translation, the structure and function of proteins—the working molecules of cells—and the genetic and molecular technology needed to study cells at the molecular level.

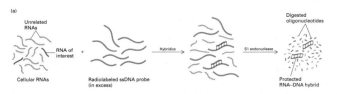

Figure 7-33a

In Part II the student is taught how genes work. Chapter 9 discusses the structure of genes and chromosomes and the genetic events that led to the evolution of present-day genomes. Chapter 10 explains the mechanisms of DNA replication, recombination, and repair. Chapter 11 focuses on the regulation of gene expression at the level of initiation of RNA synthesis, and Chapter 12 discusses post-transcriptional control of gene expression: the mechanism and control of RNA splicing, translation, and degradation. The regulation of gene expression during development of prokaryotes, eukaryotic microorganisms, and animals is the focus of Chapter 13.

In Part III we concentrate on the ways in which proteins work together to make a living cell. First, in Chapter 14, we describe

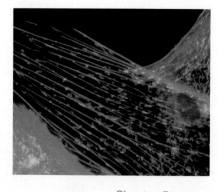

Chapter 5 opener

the structure of cell membranes and of membrane proteins. Chapter 15 examines the role of membrane proteins in the transport of small molecules into and out of the cell. Chapters 16 and 19 describe the assembly of the membranes and organelles that make up a cell, focusing on the targeting of proteins to their correct destinations. Two chapters detail the generation of ATP in the cytosol and mitochondria (Chapter 17) and by photosynthesis (Chapter 18).

Part IV emphasizes the interactions of cells with each other, both in normal and abnormal situations. Cell surface receptors, intracellular "second messengers," and the complex network of cell-to-cell communication required to coordinate differentiation, growth, metabolism, and behavior are the focuses of Chapter 20. Chapter 21 focuses on signaling in the nervous system and on the generation and propagation of action potentials. Chapters 22 and 23 turn to the complex system of fibers— the cytoskeleton— that is responsible for the cell's shape and motility. Chapter 24 focuses on the many proteins,

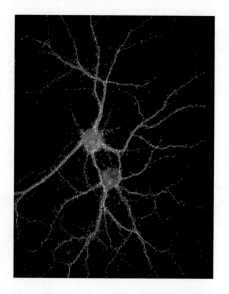

Figure 21-4

polysaccharides, proteoglycans, and other polymers that form the extracellular matrix surrounding animal cells and the plant cell wall and on how these molecules influence growth and differentiation of cells. Chapter 25, "Regulation of the Cell Cycle," emphasizes how cells integrate many internal and external signals in order to regulate their growth and division. In cancer these growth-regulatory circuits are impaired, and Chapter 26 explains how and why this happens. Chapter 27 integrates many of the themes of earlier chapters by showing how different cells function cooperatively to provide immunity from invading microorganisms.

➤ *Much is New*

New Chapters: Chapters 1 through 8, which deal with molecules, cells, proteins, and experimental techniques, have been reorganized and expanded to help students master the essential concepts and experimental techniques in molecular and cellular biology:

• A new Chapter 1 introduces the cell, defines the hierarchy of biological organization: small molecules ↔ genes ↔ proteins ↔ organelles ↔ cells ↔ tissues ↔ organs ↔ organisms, and explains why the study of cells is at the very center of modern biological research.

• Chapter 3, "Protein Structure and Function," presents a cohesive view of the structure and function of proteins, the principal classes of proteins—including enzymes and antibodies—and the important (and some very new) ways to isolate and characterize proteins. Chapter 3 further under-scores the importance of proteins in all cellular activities.

> ➤ *"The facts in biology are only as good as the tools available for investigation."*

Two new chapters focus the book on the most powerful techniques used by modern cell biologists to research a wide range of topics:

• Chapter 7, "Recombinant DNA Technology," contains much new material on the polymerase chain reaction and on the techniques for isolating, mutating, and analyzing genes that are employed throughout modern biology.

• Chapter 8, "Genetic Analysis in Cell Biology," illustrates how an analysis of mutant cells or organisms can lead to a deep understanding of the roles of individual proteins. It shows how genetics is used to identify genes and proteins that are important in a developmental pathway, and to analyze the steps in a biosynthetic or signaling pathway. We learn how genes causing human genetic diseases are iden-

tified and cloned, and how to introduce mutant genes into living plants or animals and thus study the roles of the encoded proteins.

• Chapter 11, "Regulation of Transcription Initiation," and Chapter 12, "Transcription Termination, RNA Processing and Posttranscriptional Control," allow greatly expanded coverage of these fast moving areas. Included is up-to-date coverage of the interactions of transcription initiation factors in regulating gene expression and newly uncovered mechanisms by which gene expression is controlled at the levels of RNA chain elongation, RNA splicing, and mRNA degradation. Examples range from the control of sex determination in *Drosophila* to the control of expression of the HIV genome, the virus that causes AIDS.

• Chapter 13, "Gene Control in Development," illustrates how differentiation and development in bacteria, yeasts, and animals is controlled by selective gene expression. The unifying theme is switches in gene expression: the decision to turn on or off a particular gene during differentiation. Carefully chosen examples include: the decision by bacteriophage λ to enter or not the lysogenic state, the control

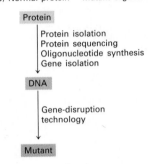

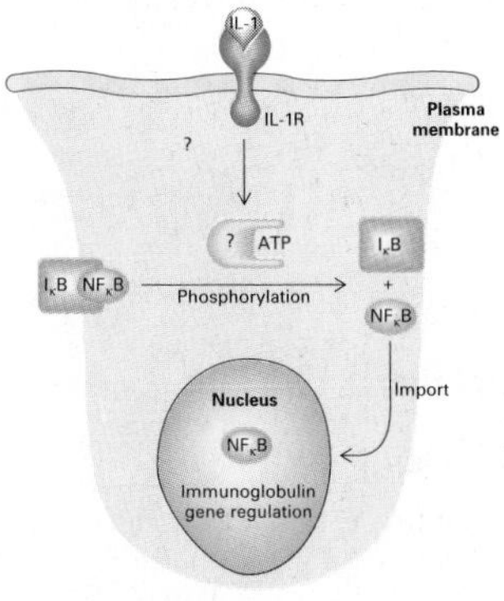

Figure 8-1

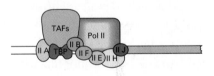

Figure 11-53

Figure 13-36

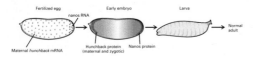

Figure 13-33

of gene expression during yeast mating, and the ways in which specific mRNAs are localized in the *Drosophila* egg and how this controls the patterning of the early embryo.

• **Chapter 20,** "Cell-to-Cell Signaling," contains new sections about the structure and function of cell receptors for growth factors, and how activation of a cell-surface receptor results in modification of specific transcription factors and induction of expression of specific genes.

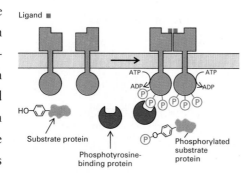

Figure 20-4g

• **Chapter 24,** "Multicellularity: Cell–Cell and Cell–Matrix Interactions," includes very recent developments that show how specific contacts between a cell and hormones bound to the extracellular matrix, or between proteins on the surface of two adjacent cells, trigger cell differentiation.

• The last several years have revealed much about how the cell cycle is regulated: how cells are triggered to begin DNA replication and how they regulate the start of mitosis. A totally new **Chapter 25** details these discoveries, and also explains how cells integrate signals from many internal and external molecules.

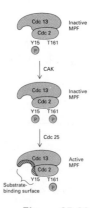

Figure 25-22

• Insights into the evolution of genes and genomes from the cloning and sequencing of many gene families (Chapter 9).

• New appreciation that some genetic diseases are due to alterations in mitochondrial DNA (Chapter 19).

• The proteins that cause intracellular vesicles to bud from and fuse with each other, and that sort and target proteins to their final destinations, and the importance of genetic analyses in identifying them (Chapters 16 and 19).

• The folding of newly made proteins, and the roles of chaperones in assisting this process (Chapters 3, 16, and 19).

• Newly uncovered oncogenes and tumor suppressor genes: their roles in regulating the cell cycle and cell proliferation (Chapters 20 and 25) and the mutations in them that lead to malignancy (Chapter 26).

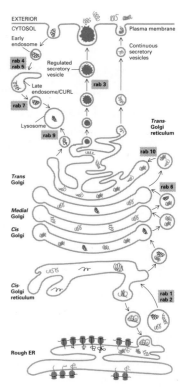

Figure 16-43b

• New types of membrane transport proteins, such as water channels that control water flow and cell volume, and ATP-powered pumps (Chapter 15); and a molecular understanding of the function of voltage-gated ion channels in conducting nerve impulses (Chapter 21).

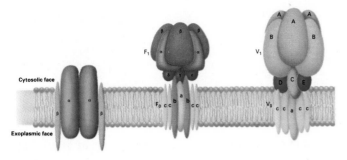

Figure 15-10

- The proteins that allow other proteins to cross membranes specifically and to insert in them (Chapters 16 and 19).

- New types of cell surface receptors, particularly in nerve cells, and their role in learning and memory (Chapter 21).

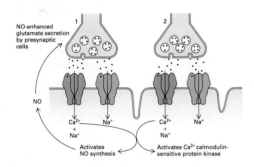

Figure 21-56

- The three-dimensional structure of actin and myosin, and the properties of newly identified proteins that regulate the structure and assembly of actin and other cytoskeletal proteins (Chapters 22 and 23).

Figure 22-21

New Authors: The exciting advances in every area of molecular cell biology during the four years since the second edition have driven these many changes in this third edition. The most obvious change, however, is the inclusion of three new authors, who are both outstanding teachers and experts in many of the fields that have been given increased coverage:

Arnold Berk has made major contributions to our understanding of proteins that control gene expression and also has developed some of the most widely used techniques for the study of messenger RNA.

Paul Matsudaira has employed a wide variety of experimental techniques to dissect the functions of cytoskeletal proteins and also has developed popular methods for sequencing proteins.

S. Lawrence Zipursky's studies on the fruit fly *Drosophila melanogaster* have led to new insights in how the developmental fate of cells is affected by their interactions with other cells.

All of us have worked together from the beginning to improve the overall structure of this edition and to better interrelate major concepts in each chapter. Reading and rewriting one another's chapters has been an education for all of us and, we hope, has given the book a greater cohesion and consistency.

▶ *New Study Tools*

In response to the advice of several dedicated teachers, Review Questions have been added to the third edition. We thank David Scicchitano of New York University, who so ably took on this task and created questions that review and reinforce the concepts and techniques to which students have been introduced. These end-of-chapter exercises will help students review and integrate the chapter's most important concepts and to integrate them with concepts learned in previous chapters. The text's cohesiveness is also improved by the new Chapter Introductions, which explain clearly how the material in that chapter relates to the text's overall hierarchy. For instance, the introduction to Chapter 8 explains *why* we need genetic tools to study cells. Similarly, short Summaries are now placed at critical points throughout each chapter, and Chapter-Ending Summaries have been rewritten to help students better focus on the chapter's major ideas and how they relate to the text's overall themes.

SUPPLEMENTS

We are pleased to announce that two outstanding supplements are available:

• The *Student Companion for Molecular Cell Biology*, first created to provide additional support for students using the second edition of *Molecular Cell Biology*, has been updated and revised in concert with the third edition. The revision, prepared by David Rintoul, Ruth Welti, Muriel Lederman, and Brian Storrie, once again is designed for use as an assigned text supplement or as a self-study guide for students taking molecular cell biology, reviewing and reinforcing the concepts, problems, and open questions of modern molecular cell biological research. Following a question and answer format (with all answers contained in an answer key), each chapter proceeds from simpler to more advanced problems that parallel a corresponding chapter in *Molecular Cell Biology*. Using this format, the *Companion* provides students of varying backgrounds a useful study resource.

• Full-color overhead transparencies of key figures from the text; available to qualified adopters.

For more information on the *Student Companion for Molecular Cell Biology* and the overhead transparencies, please contact Sales Support, W. H. Freeman and Company, 41 Madison Avenue, New York, NY 10010. Instructors in the Eastern United States and Canada may call 1-800-347-9411; those in the Western United States and Canada may call 1-800-347-9415.

ACKNOWLEDGMENTS

In updating, revising, and rewriting this book of well over a thousand pages, we were given invaluable help by many colleagues. We thank the following individuals who generously gave of their time and expertise by making major contributions to specific chapters in their areas of interest, or by reading and commenting on one or more chapters:

John N. Ableson	*California Institute of Technology*
Seth Alper	*Harvard Medical School*
Gregory Antipa	*San Francisco State University*
David J. Asai	*Purdue University*
Utpal Banerjee	*University of California at Los Angeles*
Rolf Benzinger	*University of Virginia at Charlottesville*
Doug Black	*University of California at Los Angeles*
Martin Blumenfeld	*University of Minnesota at St. Paul*
Bonnie Anderson Bray	*St. Luke's Roosevelt Hospital Center and Columbia University College of Physicians and Surgeons*
Chavela Carr	*Whitehead Institute and Massachusetts Institute of Technology*
Marco Colombini	*University of Maryland*
Riccardo Dalla-Favera	*Columbia University*
Ernest DuBrul	*University of Toledo*
Allen Ebens	*University of California at Los Angeles*
Larry A. Feig	*Tufts University School of Medicine*
Alice B. Fulton	*University of Iowa College of Medicine*
Paul Garrity	*University of California at Los Angeles*
Mark H. Ginsberg	*The Scripps Research Institute*
Jay Gralla	*University of California at Los Angeles*
Bruce Greenberg	*University of Waterloo*
Karen Greif	*Bryn Mawr College*
Michael Grunstein	*University of California at Los Angeles*
Robert Gunsalus	*University of California at Los Angeles*
Yoav Henis	*Tel Aviv University*
Deborah Hettinger	*Texas Lutheran College*
Andre Jagendorf	*Cornell University*
Daniel E. Johnson	*University of Pittsburgh*
Barbara Johnson-Wint	*Northern Illinois University*
Erik M. Jorgensen	*University of Utah*
Chris Kaiser	*Massachusetts Institute of Technology*
Paul R. Kerkof	*University of New Mexico*

Dan Kiehart	Duke University Medical Center	Arthur B. Pardee	Dana Farber Cancer Institute
Monty Krieger	Massachusetts Institute of Technology	Peter Pauw	Gonzaga University
Janis Kuby	San Francisco State University; University of California at Berkeley	Laura E. Pence	Massachusetts Institute of Technology
		Francesca Pignoni	University of California at Los Angeles
John Lammert	Gustavus Adolphus College	Deborah Robinson	Mount Allison University
Judith Lengyel	University of California at Los Angeles	Erkki Ruoslahti	La Jolla Cancer Institute
		Lori S. Rynd	Pacific University
Thomas J. Lindell	University of Arizona	Edmund Samuel	Southern Connecticut State University
Daniel Linzer	Northwestern University	Irvin Schmoyer	Muhlenberg College
Martin A. Lodish	The Hill School, Pottstown, PA	David Scicchitano	New York University
Carl Lundeen	University of North Carolina at Wilmington	James Sellers	National Institute of Health
George Mackie	University of Western Ontario	Neil Simister	Brandeis University
		Melvin J. Simon	California Institute of Technology
Eckhard and Eva Mandelkow	Max Planck Institute	Robert Simons	University of California at Los Angeles
Jim Manley	Columbia University		
Kenneth Marians	Memorial Sloan-Kettering Cancer Center	Roger D. Sloboda	Dartmouth College
		Gary S. Stein	University of Massachusetts
Andrea Mastro	Pennsylvania State University	Kathy Svoboda	Boston University School of Medicine
Douglas McAbee	University of Notre Dame		
William H. McClain	University of Wisconsin at Madison	Elizabeth Sztul	Princeton University
		Lothar Träger	Klinikum der Johann Wolfgang Goethe-Universität
Sara Moss McCowen	Virginia Commonwealth University		
Jeffrey H. Miller	University of California at Los Angeles	Bernard Trumpower	Dartmouth Medical School
		Judith Tsipis	Brandeis University
Dan Minor	Massachusetts Institute of Technology	George E. Veomett	University of Nebraska at Lincoln
Frank Monette	Boston University	David H. Vickers	University of Central Florida
Thomas S. Moore, Jr.	Louisiana State University		
E. N. Moudrianakis	Johns Hopkins University	Christopher Watters	Middlebury College
Jeanette Natzle	University of California at Davis	Gareth D. White	Massachusetts Institute of Technology
Stanley Nelson	University of California at Los Angeles	Steven White	San Jose State University
		Leslie Wilson	University of California at Santa Barbara
John Ninneman	Adams State College	Alan T. Wortman	University of Wisconsin at Lacrosse
Harry S. Noller	University of California at Santa Cruz		
Alma Moon Novotny	University of Houston at Clear Lake	Ana-Belen Ybarronda	The Scripps Research Institute
Thomas Owens	Cornell University	Kenton Zavitz	University of California at Los Angeles

It is often hard for non-authors to realize that production of a book such as this requires the talents and devoted efforts of many people over several years. We have been fortunate to have many exceptional people at W. H. Freeman and Company work with us on this and the two previous editions. Our particular thanks go to Linda Chaput, former president, for nurturing this book from its inception 14 years ago, and for helping us to plan this revision. We thank Randi Rossignol and Janet Tannenbaum, our development editors, and Mary Shuford, director of development, for providing us with exceptional guidance and support on the art program and text—from our first to our last drafts; we appreciate their devoted and highly competent efforts throughout the revision process. Shana Ederer, also in the development group, deserves our special thanks for her continuing and unstinting support work. Travis Amos and Larry Marcus were extremely energetic and resourceful in locating and developing sources for photographs and molecular models.

The attention of the entire project editorial team—Penny Hull, Philip McCaffrey, Laura Spagnoli, and Kay Ueno—to editorial questions on both text and art has been indispensable. Kay Ueno, our project editor, directed the efforts of an expert team of copy editors. We thank them—Gloria Hamilton, Elmarie Hutchinson, Bill O'Neal, and especially Ruth Steyn, for their excellent contributions in clarifying and organizing our occasionally muddled text and figures. Walter Hadler, our proofreader, provided that necessary additional reading. We greatly appreciate the efforts of Ellen Cash, production manager, who expertly coordinated all aspects of production. The new "look" of the book, and the expert blending of text and illustrations on each page, are due to the efforts of Blake Logan, our designer, and our layout artists Stan Hatzakis and Natasha Sylvester. Chris McAuliffe, Bill Page, and Denise Wiles ably coordinated the exceptional work of our illustrators Tomo Narashima, Ian Worpole, and Network Graphics.

Our own office staff, including Lois Cousseau, Carol Eng, Katherine Kanamori, Maureen Murray, and Cynthia Peterson, provided invaluable support, ranging from locating specific information and assembling reprints to helping with corrections in the text. The help of Udayakumara Liyanage in locating references for many of the chapters is also appreciated.

All these people and many others made the enormous task of writing this new edition, if not actually pleasant, at least tolerable.

Again to our families we say simply, thank you.

We believe that a comprehension of modern biology is needed both by those who use biological concepts professionally and by the general public, who increasingly will be faced with decisions about integrating new biological understanding into the fabric of their lives. We hope that this edition of *Molecular Cell Biology* will help both groups to better comprehend the revolution in understanding of living systems that is being generated by research laboratories around the world.

Harvey Lodish
David Baltimore
Arnold Berk
S. Lawrence Zipursky
Paul Matsudaira
James Darnell

January 1995

Contents in Brief

►**Part I**
Laying the Groundwork **1**

1 The Dynamic Cell 3

2 Chemical Foundations 15

3 Protein Structure and Function 51

4 Nucleic Acids, the Genetic Code, and Protein Synthesis 101

5 Cell Organization, Subcellular Structure, and Cell Division 141

6 Manipulating Cells and Viruses in Culture 189

7 Recombinant DNA Technology 221

8 Genetic Analysis in Cell Biology 263

►**Part II**
Control of Cellular Activity by the Nucleus **304**

9 The Molecular Anatomy of Genes and Chromosomes 307

10 DNA Replication, Repair, and Recombination 365

11 Regulation of Transcription Initiation 405

12 Transcription Termination, RNA Processing, and Posttranscriptional Control 485

13 Gene Control in Development 543

►**Part III**
Building and Fueling the Cell **592**

14 Membrane Structure: The Plasma Membrane 595

15 Transport Across Cell Membranes 633

16 Synthesis and Sorting of Plasma Membrane, Secretory, and Lysosomal Proteins 669

17 Cellular Genetics: Formation of ATP by Glycolysis and Oxidative Phosphorylation 739

18 Photosynthesis 779

19 Organelle Biogenesis: The Mitrochondrion,
    Chloroplast, Peroxisome, and Nucleus   809

▶ **Part IV**
**Integrative and Specialized
Cellular Events**   **850**

20 Cell-to-Cell Signaling   853

21 Nerve Cells   925

22 Microfilaments: Cell Motility and Control
    of Cell Shape   991

23 Microtubules and Intermediate Filaments   1051

24 Multicellularity: Cell-Cell and
    Cell-Matrix Interactions   1123

25 Regulation of the Eukaryotic Cell Cycle   1201

26 Cancer   1247

27 Immunity   1295

Contents

Chapter-Opening Illustrations · xiv

►Part I
Laying the Groundwork · 1

1 THE DYNAMIC CELL · 3

Evolution: Biology as a Historical Science · 4

The Construction of Cells · 5

Cells Are Surrounded by Water-Impermeable Membranes · 6

The Biological World Has Two Types of Cells · 7

Membranes Serve Functions Other Than Segregation · 8

Eukaryotic Cells Contain Organelles Thought to Have Evolved as Independent Organisms · 9

The Molecules of Life · 9

Genetics Allowed DNA Organization to Be Analyzed Before the Structure of DNA Was Discovered · 10

The Ultimate Triumph of Genetics Is the Human Genome Project · 12

RNA Is the Molecule That Is the Product of the Genome · 12

Cells Have Both a Fixed Identity and an Ability to Change · 12

Molecular Cell Biology · 14

2 CHEMICAL FOUNDATIONS · 15

Energy · 16

Covalent Bonds · 16

Each Atom Can Make a Defined Number of Covalent Bonds · 17

The Making or Breaking of Covalent Bonds Involves Large Energy Changes · 18

Covalent Bonds Exhibit Precise Orientations · 18

Polar Covalent Bonds Result from Unequal Sharing of Electrons · 19

Asymmetric Carbon Atoms and the Structure of Amino Acids and Carbohydrates · 20

The α Carbon in Amino Acids Is a Chiral Carbon · 20

Chiral Carbons Influence the Three-Dimensional Structure of Carbohydrates · 20

*Noncovalent Bonds and the Structures of
Biological Molecules*   21

The Hydrogen Bond Underlies Water's
Biological Properties   22

Ionic Interactions Are Attractions Between
Oppositely Charged Ions   23

Van der Waals Interactions Are Caused
by Transient Dipoles   24

Hydrophobic Bonds Cause Nonpolar
Molecules to Adhere to Each Other   25

Binding Specificity Can Be Conferred by
Multiple Weak Bonds   26

*Biomembranes: Hydrophobic Sheets
Separating Aqueous Compartments*   27

Phospholipids Are the Principal Components
of Biomembranes   27

Phospholipids Spontaneously Form Bilayers
in Aqueous Solutions   27

Chemical Equilibrium   28

*pH and the Concentration of Hydrogen
Ions*   30

Water Dissociates into Hydronium and
Hydroxyl Ions   30

Acids Release Hydrogen Ions, and Bases
Combine with Hydrogen Ions   31

Biological Molecules Can Have Both Acidic
and Basic Groups   31

The Henderson-Hasselbalch Equation
Describes the Relationships between
pH and Equilibrium Constants for
Acids and Bases   32

Buffers Maintain the pH of Cells and of
Extracellular Fluids   32

*Biochemical Energetics: Free Energy in
Biochemical Reactions*   34

The Change in Free Energy, ΔG, Determines
the Direction of a Chemical Reaction   34

The ΔG of a Reaction Depends on Changes
in Heat and Entropy   34

Temperature, Concentrations of Reactants,
and Other Parameters Affect the ΔG
of a Reaction   35

The Standard Free-Energy Change, $\Delta G^{\circ\prime}$, Can
Be Determined from Measurement of the
Equilibrium Constant, K_{eq} 36

The Generation of a Concentration Gradient
Requires an Expenditure of Energy   37

Many Cellular Processes Involve the Transfer
of Electrons in Oxidation-Reduction
Reactions   37

An Unfavorable Chemical Reaction Can
Proceed If It Is Coupled with an
Energetically Favorable Reaction   39

Hydrolysis of the Phosphoanhydride Bonds
in ATP Releases Substantial Free Energy   40

ATP Is Used to Fuel Many Cellular
Processes   42

Polymers of Glucose with Specific Glycosidic
Linkages Serve as Storage Reservoirs   43

Activation Energy and Reaction Rate   44

Energy Is Required to Initiate a Reaction   44

Enzymes Catalyze Biochemical Reactions   45

Summary   47

Review Questions   48

References   49

3   PROTEIN STRUCTURE AND FUNCTION   51

General Structure of Proteins   52

Form and Function Are Inseparable in Protein
Architecture   52

Amino Acids, the Building Blocks of Proteins,
Differ Only in Their Side Chains   52

Polypeptides Are Amino Acids Connected by
Peptide Bonds   56

Four Levels of Structure Determine the Shape
of Proteins   56

Polypeptides Can Be Chemically Analyzed
and Synthesized   59

Three-Dimensional Protein Structure Is
Determined through X-Ray Crystallography
and NMR Spectroscopy   59

Graphical Representations of Proteins
Highlight Internal Organization or
Surface Structures   61

Secondary Structures Are Crucial Elements in
Protein Architecture   63

Motifs Are Regular Combinations of
Secondary Structures   66

Structural and Functional Domains Are
Modules of Tertiary Structure   67

Sequence Homology Suggests Functional and
Evolutionary Relationships
among Proteins   68

Many Proteins Contain Tightly Bound
Prosthetic Groups   70

Chemical Modifications Alter the Biological Activity of Proteins   71

Proteolytic Processing and Protein Splicing Alter Protein Activity   72

A Protein Can Be Unfolded by Heat, Extreme pH, and Certain Chemicals   73

Many Denatured Proteins Can Refold into Their Native State In Vitro   73

Folding of Proteins In Vivo Is Promoted by Chaperones   74

Enzymes   75

The Active Site of an Enzyme Is a Cage of Amino Acids That Binds Substrates and Catalyzes Reactions   75

Proteases Degrade Proteins by Reducing Activation Energy for Peptide-Bond Hydrolysis   76

Coenzymes Are Essential for Certain Enzyme-Catalyzed Reactions   78

Activity of Some Enzymes Depends on a Conformational Change Induced by Substrate Binding   80

The Catalytic Activity of an Enzyme Can Be Characterized Mathematically   81

Enzymatic Activity Can Be Regulated by Various Mechanisms   83

Antibodies   86

Antibodies Are Multidomain, Multisubunit Proteins   86

Antigen-Binding Site Complements the Surface of the Antigen   87

Antibodies Are Valuable Tools for Identifying and Purifying Proteins   87

Antibodies Can Catalyze Chemical Reactions   87

Techniques for Purifying and Characterizing Proteins   88

Centrifugation Can Separate Particles and Molecules That Differ in Mass or Density   88

Electrophoresis Separates Molecules According to Their Charge: Mass Ratio   92

Liquid Chromatography Resolves Proteins by Mass, Charge, and Binding Affinity   94

Highly Specific Assays Detect Individual Proteins   96

Summary   97

Review Questions   99

References   100

4 NUCLEIC ACIDS, THE GENETIC CODE, AND PROTEIN SYNTHESIS   101

Nucleic Acids: Linear Polymers of Nucleotides   102

DNA   103

The Native State of DNA Is a Double Helix of Two Antiparallel Chains with Complementary Nucleotide Sequences   103

The Two Strands Can Separate, Causing DNA to Denature   108

Many DNA Molecules Are Circular   109

Linking Number, Twist, and Writhe Describe DNA Superstructure   110

RNA: The Basic Chemical Structure and Its Function in Gene Expression   111

Rules for the Synthesis of Proteins and Nucleic Acids and Macromolecular Carpentry   113

Nucleic Acid Synthesis   114

Nucleic Acid Polymerization Can Be Described by Four Rules   115

Organization of Genes in DNA Differs in Prokaryotes and Eukaryotes   116

Eukaryotic Primary RNA Transcripts Are Processed to Form Functional mRNAs   119

Protein Synthesis: The Three Roles of RNA in Translation   119

Messenger RNA Carries Information from DNA in a Three-Letter Genetic Code   120

Experiments with Synthetic mRNAs and Trinucleotides Break the Genetic Code   122

Folded Structure of tRNA Is Integral to Its Function   124

Aminoacyl-tRNA Synthetases Activate tRNA   126

Each tRNA Molecule Is Recognized by a Specific Aminoacyl-tRNA Synthetase   128

Ribosomes Are Protein-Synthesizing Machines   128

The Steps in Protein Synthesis   133

AUG Is the Initiation Signal in mRNA   133

Initiation Factors, tRNA, mRNA, and the Small Ribosomal Subunit Form an Initiation Complex   133

Ribosomes Provide Three tRNA-Binding (A, P, and E) during Protein Elongation   136

Polypeptide Termination Requires Protein Factors That Specifically Recognize UAA, UAG, and UGA   138

Summary   138

Review Questions   139

References   139

5   CELL ORGANIZATION, SUBCELLULAR STRUCTURE, AND CELL DIVISION   141

Prokaryotic and Eukaryotic Cells   142

Prokaryotes Have a Relatively Simple Structure   142

Eukaryotic Cells Have Complex Systems of Internal Membranes and Fibers   143

Prokaryotes and Eukaryotes Contain Similar Macromolecules   143

Prokaryotes and Eukaryotes Differ in the Amount of DNA per Cell   144

The Organization of DNA Differs in Prokaryotic and Eukaryotic Cells   147

Light Microscopy and Cell Architecture   148

The Resolution of Standard Light (Bright-Field) Microscopy Is Limited to about 0.2 μm   148

Immunofluorescence Microscopy Reveals Specific Proteins and Organelles within a Cell   150

Fluorescence Microscopy Can also Measure the Local Concentration of Ca^{2+} Ions and the Intracellular pH   153

The Confocal Scanning Microscope Produces Vastly Improved Fluorescent Images   154

Phase-Contrast and Nomarski Interference Microscopy Visualize Unstained Living Cells   156

Electron Microscopy   158

Transmission Electron Microscopy Depends on the Differential Scattering of a Beam of Electrons   158

Minute Details Can Be Visualized on Viruses and Subcellular Particles   159

Scanning Electron Microscopy Visualizes Details on the Surface of Cells or Particles   161

Sorting Cells and Their Parts   161

Flow Cytometry Is Used to Sort Cells Optically   161

Fractionation Methods Isolate Subcellular Structures   161

The Biomembranes and Organelles of the Eukaryotic Cell   166

The Plasma Membrane Has Many Varied and Essential Roles   167

The Eukaryotic Nucleus Is Bounded by a Double Membrane   167

The Nucleus Contains the Nucleolus, a Fibrous Matrix, and DNA-Protein Complexes   168

The Cytosol Contains Many Cytoskeletal Elements and Particles   168

The Endoplasmic Reticulum Is an Interconnected Network of Internal Membranes   170

Golgi Vesicles Process Secretory and Membrane Proteins and Sort Them to Their Proper Destinations   172

Lysosomes Are Acidic Organelles That Contain a Battery of Degradative Enzymes   173

Vacuoles in Plant Cells Store Small Molecules and Enable the Cell to Elongate Rapidly   174

Peroxisomes Produce and Degrade Hydrogen Peroxide   174

The Mitochondrion Is the Principal Site of ATP Production in Aerobic Cells   174

Chloroplasts Are the Sites of Photosynthesis   175

Cilia and Flagella Are Motile Extensions of the Eukaryotic Plasma Membrane   175

The Plasma Membrane Binds to the Cell Wall or the Extracellular Matrix   176

Cell Division and the Cell Cycle   177

In Prokaryotes DNA Replication Is Followed Immediately by Cell Division   177

In Eukaryotic Cells DNA Synthesis and Cell Division Occur in Special Phases of the Cell Cycle   177

Mitosis Is the Complex Process That Apportions the New Chromosomes Equally between Daughter Cells   178

Plant Cells Show Some Variations in Mitosis   180

Yeast Cells Have a Simplified Division   181

Meiosis Is the Form of Cell Division in Which Haploid Germ Cells Are Produced from Diploid Cells   181

Summary   185

Review Questions   186

References   187

6 MANIPULATING CELLS AND VIRUSES IN CULTURE   189

Growth of Microorganisms in Culture   190

Many Microorganisms Can Be Grown in Minimal Medium   190

Mutant Strains of Bacteria and Yeast Can Be Isolated by Replica Plating   190

Growth of Animal Cells in Culture   193

Rich Media Are Required for Culture of Animal Cells   193

Most Cultured Animal Cells Only Grow on Special Solid Surfaces   193

Primary Cell Cultures Have a Finite Life Span   194

Transformed Cells Can Grow Indefinitely in Culture   196

The Use of Hybrid Cells in Genetic Analysis of Animal Cells and Production of Monoclonal Antibody   198

Genes Can Be Mapped to Specific Chromosomes with Interspecific Hybrid Cells   199

Mutants in Purine- and Pyrimidine-Salvage Pathways Are Good Selective Markers   200

Hybridomas Are Fused Lymphoid Cells That Make Monoclonal Antibodies   201

Viruses: Structure and Function   202

Viral Capsids Are Regular Arrays of One or a Few Types of Proteins   202

Most Viral Host Ranges Are Narrow   204

Viruses Can Be Cloned and Counted in Plaque Assays   205

Viral Growth Cycles Are Classified as Lytic or Lysogenic   205

Four Types of Bacterial Viruses Are Widely Used in Biochemical and Genetic Research   206

Experiments with Plant Viruses Proved That RNA Can Act as a Genetic Material   208

Six Animal-Virus Classes Are Recognized Based on Genome Composition and Pathway of mRNA Synthesis   208

Radioisotopes: Indispensable Tools for Following Biological Activity   212

Several Factors Determine the Choice of a Radiolabel   213

Radiolabeled Molecules Can Be Detected by Visual and Quantitative Methods   214

Intracellular Precursor Pools Affect the Outcome of Pulse-Chase Experiments   215

Synthesis Time of Macromolecules Can Be Estimated from Labeling Experiments   216

The Dintzis Experiment Demonstrated That Proteins Are Synthesized from the Amino End to the Carboxyl End   216

Summary   218

Review Questions   219

References   220

7 RECOMBINANT DNA TECHNOLOGY   221

DNA Cloning with Plasmid Vectors   222

Plasmids Are Extrachromosomal Self-Replicating DNA Molecules   222

E. coli Plasmids Can Be Engineered for Use as Cloning Vectors   222

Plasmid Cloning Permits Isolation of DNA Fragments from Complex Mixtures   224

Production of Recombinant Plasmids   225

Restriction Enzymes Cut DNA Molecules at Specific Sequences   225

Many Restriction Enzymes Generate DNA Fragments with "Sticky" Ends   226

DNA Ligase Covalently Links Restriction Fragments   227

Restriction Fragments Are Readily Inserted into Plasmid Vectors   228

Formation and Uses of Synthetic DNA   228

λ-Phage Cloning Vectors and Construction of a Genomic Library   229

Bacteriophage λ Can Be Modified for Use as a Cloning Vector and Assembled In Vitro   230

Nearly Complete Genomic Libraries of Higher Organisms Can Be Prepared by λ Cloning   231

Larger DNA Fragments Can Be Cloned in Cosmids and Other Vectors   233

Construction of a cDNA Library   234

cDNAs Are Produced by Copying Isolated mRNAs with Reverse Transcriptase   235

cDNAs Can Be Enzymatically Converted to Double-Stranded DNA and Cloned   235

Identification of Specific Clones in a Genomic or cDNA Library   236

Membrane Hybridization Can Be Used to Screen a Library   237

Certain cDNAs and Synthetic Oligonucleotides Are Used as Probes   238

Expression Cloning Identifies Specific Clones Based on Properties of the Encoded Proteins   240

Analyzing and Sequencing Cloned DNA   240

Cleavage with an Appropriate Restriction Enzyme Separates a Cloned DNA from Its Vector   240

Gel Electrophoresis Resolves DNA Fragments of Different Size   242

Multiple Restriction Sites Can Be Mapped on a Cloned DNA Fragment   243

Pulsed-Field Gel Electrophoresis Separates Large DNA Molecules   244

Nucleotide Sequencing of Cloned DNA Fragments Paves the Way for Sequencing Entire Genomes   245

Analysis of Specific Nucleic Acids in Complex Mixtures   248

Southern Blotting Detects Specific DNA Fragments   248

Northern Blotting Detects Specific RNAs   249

Nuclease Protection Is Used to Quantitate Specific RNAs and Map the DNA Regions Encoding Them   249

Transcription Start Sites Can Be Mapped by S1 Protection and Primer Extension   251

Designing Expression Systems That Produce Abundant Amounts of Specific Proteins   252

Full-Length Proteins Encoded by Cloned Genes Can Be Produced in *E. coli* Expression Systems   252

Proteins with Post-Translational Modifications Can Be Produced in Eukaryotic Expression Systems   253

Proteins Encoded by Cloned Genes and cDNAs Can Be Expressed In Vitro   253

The Polymerase Chain Reaction: An Alternative to Cloning   254

From Protein to Gene and from Gene to Protein with Recombinant DNA Technology   256

Summary   257

Review Questions   258

References   260

8   GENETIC ANALYSIS IN CELL BIOLOGY   263

The Isolation and Characterization of Mutants   264

Mutations Are Recessive or Dominant   264

Mutations Involve Large or Small DNA Alterations   266

Mutations Occur Spontaneously and Can Be Induced   267

Some Human Diseases Are Caused by Spontaneous Mutations   268

Various Genetic Screens Are Used to Identify Mutants   269

Complementation Analysis Determines If Different Mutations Are in the Same Gene   274

Metabolic and Other Pathways Can Be Genetically Dissected   274

Suppressor Mutations Can Identify Genes Encoding Interacting Proteins   274

Genetic Mapping of Mutations   277

Segregation Patterns Indicate Whether Mutations Are on the Same or Different Chromosomes   277

Chromosomal Mapping Locates Mutations on Particular Chromosomes   277

Recombinational Analysis Can Map Genes Relative to Each Other on a Chromosome   279

DNA Polymorphisms Are Used to Map Human Mutations   279

Some Chromosomal Abnormalities Can Be Mapped by Banding Analysis   281

Molecular Cloning of Genes Defined by Mutations   284

Physical Maps of Human Chromosomes Y and 21 Have Been Constructed by Screening YAC Clones for Sequence-Tagged Sites   284

Physical and Genetic Maps Can Be Correlated   284

Physical Mapping of Selected Genomic Regions is the First Step in Cloning Many Genes   285

Mutation-Defined Genes Are Identified in Candidate Regions by Comparing Mutant and Wild-Type DNA Structure and mRNA Expression   287

Protein Structure Is Deduced from cDNA Sequences   289

Gene Replacement and Transgenic Animals   291

Specific Sites in Cloned Genes Can Be Altered In Vitro   291

DNA Can Be Transferred into Eukaryotic Cells in Various Ways   292

Normal Genes Can Be Replaced with Mutant Alleles in Yeast and Mice   292

Foreign Genes Can Be Introduced into Plants and Animals   296

Gene Therapy Involves Use of Transgenes to Treat Genetic Diseases   299

Summary   300

Review Questions   301

References   302

Part II
Control of Cellular Activity by the Nucleus   304

9 MOLECULAR ANATOMY OF GENES AND CHROMOSOMES   307

Molecular Definition of a Gene   308

Most Prokaryotic Genes Lack Introns and Those Encoding Related Proteins Form Operons, Which Produce Polycistronic mRNA   308

Most Eukaryotic Transcription Units Produce Monocistronic mRNAs   308

Simple Eukaryotic Transcription Units Give Rise to One mRNA   309

Complex Eukaryotic Transcription Units Give Rise to Alternative mRNAs   309

Some Genes Do Not Encode Protein   310

Organization of Genes on Chromosomes   311

Genomes of Higher Eukaryotes Contain Much "Nonfunctional" DNA   311

Cellular DNA Content Does Not Correlate with Phylogeny   312

About a Quarter to Half of All Eukaryotic Protein-Coding Genes Are Solitary   313

Gene Families Are Formed by Gene Duplication and Encode Homologous Proteins   313

Pseudogenes Are Duplicated Genes That Have Become Nonfunctional   314

rRNAs, tRNAs, and Histones Are Encoded by Tandemly Repeated Genes   315

Discovery of Repetitious DNA Fractions   316

Repeated DNA Reassociates More Rapidly Than Nonrepeated DNA   316

Reassociation Experiments Reveal Three Major Classes of Eukaryotic DNA   317

Simple-Sequence DNA   318

Higher Eukaryotes Contain Several Types of Simple-Sequence DNA   318

Most Simple-Sequence DNA Is Located in Specific Chromosomal Regions   319

Differences in Lengths of Simple-Sequence Tandem Arrays Permit DNA Fingerprinting   319

Immediate-Repeat DNA and Mobile DNA Elements   320

Movement of Bacterial Mobile Elements Is Mediated by DNA   323

Movement of Some Eukaryotic Mobile Elements Is Mediated by DNA   326

Two Major Categories of Retrotransposons Are Found in Eukaryotic Cells   328

Most Mobile Elements in Yeast Are Viral Retrotransposons   328

Copia Retrotransposons Are the Most Common *Drosophila* Mobile Elements   333

LINES and SINES, the Most Abundant Mobile Elements in Mammals, Are Nonviral Retrotransposons   334

Retrotransposed Copies of Cellular RNAs Are Present in Eukaryotic Chromosomes   336

Functional Rearrangements in Chromosomal DNA   338

Salmonella Flagellar Antigens Can Switch through Inversion of a Transcription-Control Region   338

Yeast Mating Types Can Switch by Gene Conversion   339

Trypanosome Surface Antigens Undergo Frequent Changes via Gene Conversion   341

Generalized DNA Amplification Produces Polytene Chromosomes   343

Localized DNA Amplification of rRNA and Other Genes Occurs in Some Eukaryotic Cells   343

Vertebrate Genes Encoding Antibodies
Are Assembled from Gene Segments by
Controlled Deletion of Intervening
DNA   344

*Organizing Cellular DNA into
Chromosomes*   344

Prokaryotic Chromosomes Contain Highly
Compacted Circular DNA Molecules with
a Single Replication Origin   344

Eukaryotic Nuclear DNA Associates with
Highly Conserved Histone Proteins to
Form Chromatin   346

Chromatin Exists in Extended and
Condensed Forms   347

*Morphology and Functional Characteristics
of Eukaryotic Chromosomes*   349

Chromosome Number and Shape Are Species
Specific   349

Nonhistone Proteins Provide a Structural
Scaffold for Long DNA Loops in
Chromosomes   349

Chromatin Contains Small Amounts of
DNA-Binding Proteins in Addition to
Histones and Scaffold Proteins   354

Each Chromosome Contains One Linear
DNA Molecule   354

Stained Chromosomes Have Characteristic
Banding Patterns   354

Heterochromatin Consists of Chromosome
Regions That Do Not Uncoil   355

Three Functional Elements Are Required for
Replication and Stable Inheritance of
Chromosomes   355

Yeast Artificial Chromosomes Can Be Used
to Clone Megabase DNA Fragments   359

Summary   359

Review Questions   360

References   361

10 DNA REPLICATION, REPAIR, AND
RECOMBINATION   365

*General Features of Chromosomal
Replication*   366

DNA Replication Is Semiconservative   366

Most DNA Replication Is Bidirectional   366

DNA Replication Begins at Specific
Chromosomal Sites   370

DNA Replication in E. coli   372

DnaA Protein Initiates Replication in
E. coli   372

DnaB Is a Helicase That Unwinds
Duplex DNA   372

Primase Catalyzes Formation of RNA
Primers for DNA Synthesis   374

At a Growing Fork One Strand Is
Synthesized Discontinuously from Multiple
Primers   374

DNA Polymerase III Synthesizes Both the
Leading and Lagging Strands   375

The Leading and Lagging Strands Are
Synthesized Concurrently   376

Interaction of Tus Protein with Termination
Sites Stops DNA Replication   378

Eukaryotic DNA Replication   378

Eukaryotic Proteins That Replicate SV40
DNA In Vitro Exhibit Similarities and
Differences with *E. coli* Replication
Proteins   378

Telomerase Prevents Progressive Shortening
of Lagging Strands during DNA
Replication   380

*Role of Topoisomerases in DNA
Replication*   381

Type I Topoisomerases Relax DNA by
Nicking and Closing One Strand of
Duplex DNA   381

Type II Topoisomerases Change DNA
Topology by Breaking and Rejoining
Double-Stranded DNA   382

Replicated Circular DNA Molecules Are
Decatenated by Type II Topoisomerases   383

Linear Daughter Chromatids Also Are
Separated by Type II Topoisomerases   383

Repair of DNA   385

Proofreading by DNA Polymerase Corrects
Copying Errors   385

Environmental DNA Damage Can Be
Repaired by Several Mechanisms   386

Excision Repair in *E. coli* Removes Bulky
Chemical Adducts Caused by UV Light
and Carcinogens   387

Genetic Studies in Eukaryotes Have
Identified DNA-Repair Genes   388

*Recombination between Homologous
DNA Sites*   389

Holliday Recombination Model Is Supported by Observation of Predicted Intermediate Structures   389

Recombination in *E. coli* Occurs by Three Similar Pathways That All Require RecA Protein   391

Site-Specific Integration of λ Phage Mimics a Homologous Recombination Event   395

Studies in Yeast Are Providing Insights into Meiotic Recombination   396

Gene Conversion Can Occur near the Crossover Point during Reciprocal Recombination   396

Summary   400

Review Questions   401

References   402

11 REGULATION OF TRANSCRIPTION INITIATION   405

Early Genetic Analysis of lac-*Operon Control in* E. coli   406

Enzymes Encoded at *lac* Operon Can Be Induced and Repressed   407

Mutations in *lacI* Cause Constitutive Expression of the *lac* Operon   407

Operator Constitutive Mutations Identify Binding Site for *lac* Repressor   408

Mutations in Promoter Prevent Expression of *lac* Operon   409

Regulation of *lac* Operon Depends on Cis-Acting DNA Sequences and Trans-Acting Proteins   409

Molecular Mechanisms of Transcription Initiation in Bacteria   411

Induction of the *lac* Operon Leads to Increased Synthesis of *lac* mRNA   411

E. coli RNA Polymerase Generally Initiates Transcription at a Unique Position on DNA Template   411

Protein-Binding Sites in *lac* Control Region Have Been Identified by Sequence Comparison and Footprinting Experiments   412

RNA Polymerase Interacts with Specific Promoter Sequences   413

σ^{70} Subunit of RNA Polymerase Functions as an Initiation Factor   414

α-Subunit Dimer Bends to rRNA Promoters in −40 to −60 Region   416

Binding of *lac* Repressor to the *lac* Operator Blocks Transcription Initiation by RNA Polymerase   416

Most Bacterial Repressors Are Homodimers Containing α Helices That Insert into Adjacent Major Grooves of Operator DNA   418

Positive Control of the *lac* Operon Is Exerted by cAMP-CAP   421

Cooperative Binding of cAMP-CAP and RNA Polymerase to *lac* Control Region Activates Transcription   423

Control of Transcription from All Bacterial Promoters Involves Similar But Distinct Mechanisms   423

Transcription from Some Promoters Is Initiated by Alternative Sigma (σ) Factors   424

RNA Polymerase Containing σ^{54} Is Regulated by Proteins That Bind at Enhancer Sites Distant from the Transcription Initiation Site   425

Eukaryotic Gene Control: Purposes and General Principles   426

Most Genes in Higher Eukaryotes Are Regulated by Controlling Their Transcription   427

DNA Regulatory Sites Often Are Located Many Kilobases from Eukaryotic Transcription Start Sites   428

Structure and Function of Eukaryotic Nuclear RNA Polymerases   430

Three Eukaryotic Polymerases Catalyze Formation of Different RNAs   430

Eukaryotic RNA Polymerases Have Complex Subunit Structure   430

The Largest Subunit in RNA Polymerase II Has an Essential Carboxyl-Terminal Repeat   432

RNA Polymerase II Initiates Transcription at DNA Sequences Corresponding to the 5′ Cap of mRNAs   433

Cis-Acting Regulatory Sequences in Eukaryotic DNA   435

The TATA Box Positions RNA Polymerase II for Transcription Initiation in Many Genes   435

Promoter-Proximal Elements Help Regulate Many Eukaryotic Genes   436

Transcription by RNA Polymerase II Often Is Stimulated by Distant Enhancer Sites   439

Most Eukaryotic Genes Are Regulated by Multiple Transcription-Control Elements   441

Eukaryotic Transcription Factors   442

Biochemical and Genetic Techniques Have Been Used to Identify Transcription Factors   442

Many Transcription Factors Are Modular Proteins Composed of Distinct Functional Domains   445

A Variety of Protein Structures Form the DNA-Binding Domains of Eukaryotic Transcription Factors   447

Heterodimeric Transcription Factors Increase Regulatory Diversity   452

A Diverse Group of Amino Acid Sequences Are Found in Activation Domains   452

RNA Polymerase II Transcription-Initiation Complex   453

Transcription-Initiation Complex Contains Many Proteins Assembled in a Specific Order   453

Transcription Activators Influence Assembly of Initiation Complex   456

Some Eukaryotic Regulatory Proteins Function as Repressors   456

Regulating the Activity of Eukaryotic Transcription Factors   456

Expression of Many Transcription Factors Is Restricted to Specific Cell Types   457

Some Transcription Factors Are Controlled by Lipid-Soluble Hormones   458

Polypeptide Hormones Signal Phosphorylation of Some Transcription Factors   462

Influence of Chromatin Structure on Eukaryotic Transcription Initiation   464

Association of Genes with Heterochromatin Can Lead to Their Repression   464

Transcriptionally Inactive DNA Regions Are Resistant to DNase I   466

Cytosine Methylation Is Associated with Inactive Genes in Vertebrates   468

Transcription by RNA Polymerase I   468

Pre-rRNA DNA from All Eukaryotes Is Similar   468

Only Essential Function of Polymerase I Is to Produce pre-rRNA   468

Species-Specific Initiation Factors Are Utilized by RNA Polymerase I   469

Transcription by RNA Polymerase III   470

tRNA Genes Bind Two Multisubunit Initiation Factors   471

5S-rRNA Gene Binds Three Initiation Factors   471

TBP Is Required for Transcription Initiation by All Three Eukaryotic RNA Polymerases   472

Other Transcription Systems   473

T7 and Related Bacteriophages Express Monomeric, Largely Unregulated RNA Polymerases   473

Mitochondrial DNA Transcription Exhibits Features Typical of Bacteriophage, Bacteria, and the Eukaryotic Nucleus   473

Chloroplasts Contain an RNA Polymerase Homologous to the *E. coli* Enzyme   474

Archaebacteria Have an RNA Polymerase and Putative General Transcription Factors Similar to Those of the Eukaryotic Nucleus   475

Summary   475

Review Questions   478

References   480

12 TRANSCRIPTION TERMINATION, RNA PROCESSING, AND POSTTRANSCRIPTIONAL CONTROL   485

Transcription Termination in Prokaryotes   486

Rho-Independent Termination Sites Have Characteristic Sequences   486

Attenuation Can Cause Premature Chain Termination   486

Transcription Termination at Some Sites Requires Rho Factor   488

Antitermination Can Prevent Premature Chain Termination   490

Eukaryotic Transcription-Termination Control   491

HIV Tat Protein Is an RNA-Binding Antitermination Protein   492

Premature Termination of c-*myc* Transcription Occurs in Nondividing Cells   492

RNA Polymerase II Pauses during
Transcription of *Drosophila* Heat-Shock
Genes under Normal Conditions   492

mRNA Processing in Higher Eukaryotes   494

mRNA Precursors Are Associated with
Abundant Nuclear Proteins Containing
Conserved RNA-Binding Domains   494

HnRNP Proteins May Have Multiple
Functions   496

Pre-mRNAs Are Cleaved at Specific
3′ Sites and Rapidly Polyadenylated
in Animal Cells   498

RNA-DNA Hybridization Reveals Spliced
Out Introns   498

Splice Sites in Pre-mRNAs Exhibit Short,
Conserved Sequences   500

Excision of Introns and Splicing of Exons
in Pre-mRNA Occur via Two
Transesterification Reactions   501

Small Nuclear Ribonucleoprotein Particles
Assist in Splicing   501

Portions of Two Different RNAs Are
Trans-Spliced in Some Organisms   508

Self-Splicing Group II Introns Provide Clues
to Evolution of snRNPs   508

Regulation of RNA Processing Controls
Expression of Some Proteins   509

*Subnuclear Organization and Transport of
Nuclear mRNA to the Cytoplasm*   513

Most Transcription and RNA Processing
Occurs in a Limited Number of Domains
in Mammalian Cell Nuclei   513

Messenger Ribonucleoproteins (mRNPs)
Exit the Nucleus through Nuclear
Pore Complexes   515

5′-Cap Structures Are Recognized by the
Nuclear Transport Mechanism   516

Pre-mRNAs Associated with Spliceosomes
Are Not Transported to the Cytoplasm   518

mRNA Remains Associated with Protein
in the Nucleus and Cytoplasm   518

Transport of mRNPs to the Cytoplasm Is
Regulated by Some Viral Proteins   518

*Regulation of mRNA Cytoplasmic
Localization, Stability, and Translation*   520

Some mRNAs Are Directed to Specific
Cytoplasmic Sites by Sequences in Their
3′ Untranslated Regions   521

Stability of Cytoplasmic mRNAs Varies
Widely   522

Degradation Rate of Some Eukaryotic
mRNAs Is Regulated   524

Translation of a Few mRNAs Is Regulated
by Specific RNA-Binding Proteins   525

Antisense RNA Regulates Translation of
Transposase mRNA in Bacteria   527

Processing of rRNAs and tRNAs   528

Pre-rRNA Binds Proteins, Then Is Cleaved
and Methylated in the Nucleolus   528

Pre-rRNA Genes Act as Nucleolar
Organizers   530

Self-Splicing Group I Introns in Some
Pre-rRNAs Were the First Examples of
Catalytic RNA   530

Processing of Pre-tRNA Involves Cleavage,
Modification of Bases, and Sometimes a
Unique Type of Splicing   532

RNA Editing   535

RNA Editing Regulates Protein Function
in Mammals   535

RNA Editing in Trypanosome
Mitochondria Drastically Alters mRNA
Sequences   536

Summary   537

Review Questions   539

References   540

13   GENE CONTROL IN DEVELOPMENT   543

*Lysogeny or Lysis in λ-Phage Infection of
E. coli*   544

Phage Mutants Unable to Undergo
Lysogeny Fall into Three Main
Complementation Groups   544

cI Protein Maintains Lysogeny by
Repressing and Activating Transcription
from Different Promoters   545

cII and cIII Proteins Are Critical to
Establishment of Lysogeny   547

Induction of Lytic Cycle Requires
Derepression of *cro* Gene   548

Cro and cI Have Similar DNA-Binding
Domains But Interact Differently with
λ Operators   549

Choice between Lysis and Lysogeny Involves
Regulatory Mechanisms Found in More
Complex Developmental Systems   550

xxviii   Contents

Cell-Type Specification and Mating-Type Conversion in Yeast   550

Cell Type–Specific Gene Expression in Yeast Is Regulated by Numerous DNA-Binding Proteins   551

Mating-Type Conversion Is Determined by Transcriptional Regulation of the *HO* Locus   554

Silencer Elements Repress Expression at *HML* and *HMR*   556

Myogenesis in Mammals   556

Embryonic Somites Give Rise to Myoblasts, the Precursors of Skeletal Muscle Cells   557

Certain Fibroblasts Can Be Converted into Muscle (Myotubes)   558

The *myoD* Gene Can Trigger Muscle Development   558

Myogenic Proteins Are HLH Transcription Factors   560

Myogenic Gene Activation Depends on Specific Amino Acids in MyoD   560

Id Protein Inhibits Activity of MyoD   560

Knockout Experiments Have Demonstrated Role of Myogenic Proteins In Vivo   560

Neurogenesis in Drosophila *and Mice*   561

Drosophila Sensory Hairs Arise from Proneural Clusters, Which Express Achaete and Scute Proteins   562

A Single Sensory Organ Precursor Develops from a Proneural Cluster in *Drosophila*   562

Drosophila Neurogenesis and Mammalian Myogenesis May Occur via Analogous Pathways Involving HLH Proteins   562

MASH1, a Homolog of Achaete and Scute Proteins, Regulates Neurogenesis in the Mouse   562

Specification of Other Cell Types Is Controlled by Different Classes of Transcription Factors   565

Regional Specification during Drosophila *Embryogenesis*   565

Drosophila Has Two Life Forms   565

Patterning Information Is Generated during Oogenesis and Early Embryogenesis   565

Morphogens Regulate Development as a Function of Their Concentration   568

Four Maternal Gene Systems Regulate Regionalization in the Early Embryo   568

Specification of the Anterior Region Depends on the Maternal *bicoid* Gene   568

Protein Encoded by Maternal *nanos* Gene Represses Translation of *hunchback* mRNA in Posterior Region   572

Hunchback Protein Regulates Expression of Several Gap Genes along the Anteroposterior Axis   573

Initial Patterning along Dorsoventral Axis Depends on Dorsal Protein   573

Maternal Terminal Genes Regulate Early Patterning of the Extreme Anterior and Posterior Ends of the Embryo   575

Subsequent Anteroposterior Patterning Is Regulated by a Cascade of Transcription Factors Expressed from Three Groups of Zygotic Genes   576

Selector Genes Control Regional Identity and Development of Adult Structures   581

Mammalian Homologs of Drosophila *ANT-C and BX-C*   584

Mammalian Hox Genes Are Colinear Homologs of *Drosophila* HOM-C Genes   584

Mutations in Hox Genes Result in Homeotic Transformations in the Developing Mouse   585

Summary   587

Review Questions   588

References   590

▶ **Part III**

Building and Fueling the Cell   592

14   MEMBRANE STRUCTURE AND THE PLASMA MEMBRANE   595

General Architecture of Lipid Membranes   596

All Membranes Contain Phospholipids and Proteins   596

The Phospholipid Bilayer Is the Basic Structural Unit of Biological Membranes   599

Phospholipid Bilayers Exhibit Two-Dimensional Fluidity That Depends on Temperature and Composition   599

Several Types of Evidence Point to the Universality of the Phospholipid Bilayer   602

Phospholipid Bilayers and Biological Membranes Form Closed Compartments   603

Membrane Proteins   604

  Proteins Interact with Membranes in
  Different Ways   604

  Transmembrane Proteins Contain Long
  Segments of Hydrophobic Amino Acids
  Embedded in the Phospholipid Bilayer   604

  Proteins Can Be Removed from Membranes
  by Detergents or High-Salt Solutions   604

  Many Integral Proteins Contain Multiple
  Transmembrane α Helices   606

  Porins Are Transmembrane Proteins
  Composed of Multiple β Strands   609

  Some Integral Proteins Are Bound to the
  Membrane by Covalently Attached
  Hydrocarbon Chains   610

  Interfacial Catalysis Involves Soluble
  Enzymes Acting at Membrane Surfaces   612

  The Orientation of Proteins in Membranes
  Can Be Experimentally Determined   612

Glycoproteins and Glycolipids   612

  Many Integral Proteins Contain Sugars
  Covalently Linked to Their Exoplasmic
  Domains   613

  Many Glycolipids Are Located in the Cell-
  Surface Membrane   614

Principles of Membrane Organization   615

  All Integral Proteins Bind Asymmetrically
  to the Lipid Bilayer   615

  The Two Membrane Leaflets Have
  Different Lipid Compositions   616

  Freeze-Fracture and Deep-Etching
  Techniques Reveal the Two Membrane
  Faces in Electron Microscopy   616

  Most Integral Proteins and Lipids Are
  Laterally Mobile in Biomembranes   616

  Some Membrane Proteins Interact with
  Cytoskeletal Components   616

  Erythrocytes Have an Unusual Plasma
  Membrane That Is Tightly Anchored
  to the Cytoskeleton   620

Specializations of the Plasma Membrane   623

  Plasma Membranes of Polarized Cells Are
  Divided into Two Regions with Different
  Compositions and Functions   623

  Tight Junctions Seal Off Body Cavities and
  Restrict Diffusion of Membrane
  Components   625

  Desmosomes and Gap Junctions
  Interconnect Cells and Control Passage of
  Molecules between Them   628

Summary   628

Review Questions   629

References   631

15 TRANSPORT ACROSS CELL MEMBRANES   633

*Major Types of Membrane Transport
Proteins*   634

*Diffusion of Small Molecules across Pure
Phospholipid Bilayers*   635

*Uniporter-Catalyzed Transport of Specific
Molecules*   636

  Three Main Features Distinguish Uniport
  Transport from Passive Diffusion   637

  Two General Models Have Been Proposed
  for Transporters   638

  Glucose Entry into Erythrocytes Is
  Mediated by a Uniporter   638

*Ion Channels, Intracellular Ion
Environment, and Membrane Electric
Potential*   640

  Ionic Gradients and Electric Potential
  Are Maintained across the Plasma
  Membrane   640

  Certain K^+ Channels Generate the
  Membrane Electric Potential   641

  Ion Concentration Gradients and Electric
  Potential Drive the Movement of
  Ions across Biological Membranes   643

Active Ion Transport and ATP Hydrolysis   644

  Ion Pumps Can Be Grouped into Three
  Classes (P, V, and F)   644

  Ca^{2+} ATPase Maintains Low Cytosolic
  Ca^{2+} Concentration   645

  Coupling of ATP Hydrolysis and Ion
  Pumping by P-Class ATPases Involves an
  Ordered Kinetic Mechanism   647

  Na^+/K^+ ATPase Maintains the Intracellular
  Concentrations of Na^+ and K^+ in Animal
  Cells   648

  V-Class H^+ ATPases Pump Protons across
  Lysosomal and Vacuolar Membranes   650

  The Multidrug-Transport Protein is an ATP-
  Powered Pump and ATP-Dependent Cl^-
  Channel   651

*Cotransport Catalyzed by Symporters and
Antiporters*   652

  Na^+-Linked Symporters Import Amino
  Acids and Glucose into Many Animal
  Cells   652

Na$^+$-Linked Antiporter Exports Ca^{2+} from Cells   654

Band 3 Is an Anion Antiporter That Exchanges Cl$^-$ and HCO$_3^-$ across the Erythrocyte Membrane   655

H$^+$/K$^+$ ATPase and Anion Antiporter Combine to Acidify the Stomach Contents While Maintaining Cytosolic pH Near Neutrality   657

Several Symporters and Antiporters Regulate Cytosolic pH   658

Plant and Prokaryotic Membrane Transport Proteins   659

H$^+$ Pumps and Anion Channels Establish Electric Potential and a Steep H$^+$ Concentration Gradient across the Plant-Vacuole Membrane   659

Proton Antiporters Enable Plant Vacuoles to Accumulate Metabolites and Ions   659

The Potential across the Plasma Membrane of Plant, Bacterial, and Fungal Cells Is Generated by Proton Pumping   660

Proton Symporters Import Many Nutrients into Bacteria   660

Osmosis, Water Channels, and the Regulation of Cell Volume   661

Osmotic Pressure Causes Water Movement across Membranes   662

Water Channels Are Necessary for Bulk Osmotic Flow of Water across Membranes   663

Some Animal Cells Regulate Their Volume by Modulating Their Internal Osmotic Strength   663

Changes in Intracellular Osmotic Pressure Cause Leaf Stomata to Open   664

Summary   665

Review Questions   666

References   667

16   SYNTHESIS AND SORTING OF PLASMA MEMBRANE, SECRETORY, AND LYSOSOMAL PROTEINS   669

The Synthesis of Membrane Lipids   671

Phospholipids Are Synthesized in Association with Membranes   671

Special Membrane Proteins Allow Phospholipids to Equilibrate in Both Membrane Leaflets   671

Phospholipids Move from the ER to Other Cellular Membranes   673

Sites of Synthesis of Organelle and Membrane Proteins   674

All Nuclear-Encoded Proteins Are Made by the Same Cytosolic Ribosomes   675

Membrane-Attached and Membrane-Unattached Ribosomes Synthesize Different Proteins   676

Overall Pathway for Synthesis of Secretory, Lysosomal, and Membrane Proteins   676

Newly Made Secretory Proteins Are Localized to the Lumen of the Rough ER   676

Many Organelles Participate in Protein Secretion   677

All Secretory Proteins Move from the Rough ER to Golgi Vesicles to Secretory Vesicles   677

The Steps in Protein Secretion Can Be Studied Genetically   678

Plasma Membrane Glycoproteins Follow the Same Maturation Pathway as Continuously Secreted Proteins   680

The Transport of Secretory and Membrane Proteins into or across the ER Membrane   681

A Signal Sequence on Nascent Secretory Proteins Targets Them to the ER and Is then Cleaved Off   682

Several Receptor Proteins Mediate the Interaction of Signal Sequences with the ER Membrane   683

Polypeptides Cross the ER Membrane in Protein-Lined Channels   686

ATP-Hydrolyzing Chaperone Proteins Prevent Protein Misfolding and Are Essential for Translocation of Secretory Proteins into the ER   687

Topogenic Sequences in Integral Membrane Proteins Allow Them to Achieve Their Proper Orientation in the ER Membrane   688

Post-Translational Modifications of Secretory and Membrane Proteins in the Rough ER   694

Disulfide Bonds Are Formed in the ER Lumen Soon after Synthesis   694

Chaperone Proteins Facilitate the Folding of Newly Made Proteins   695

The Formation of Oligomeric Proteins
Occurs in the ER   696

Quality Control in the ER   697

Only Properly Folded Proteins Are
Transported from the Rough ER to the
Golgi Complex   697

Unassembled or Misfolded Proteins Are
Often Degraded within the ER   698

ER-Specific Proteins Are Retained in the
Rough ER or Are Returned There from
the *Cis*-Golgi   698

*Protein Glycosylation: Discrete Steps in
the ER and Golgi Complex*   699

Different Structures Characterize *N*- and
O-Linked Oligosaccharides   699

Nucleotide Sugars Are the Precursors of
Oligosaccharides   700

O-Linked Oligosaccharides Are Formed by
the Sequential Addition of Sugars   703

The ER and Golgi Membranes Contain
Transporters for Nucleotide Sugars   703

The Diverse *N*-Linked Oligosaccharides
Share Certain Structural Features That
Reflect a Common Precursor   704

The Processing *N*-Linked Oligosaccharides
Involves the Sequential Removal and
Addition of Sugar Residues   704

Modifications to *N*-Linked Oligosaccharides
Are Completed in the Golgi Vesicles   706

The Movement of Proteins through the
Secretory Pathway Can Be Monitored
by Following the Processing of
N-Linked Oligosaccharides   707

N-Linked and *O*-Linked Oligosaccharides
May Stabilize Maturing Secretory and
Membrane Proteins   708

Phosphorylated Mannose Residues Target
Proteins to Lysosomes   709

Genetic Defects Have Elucidated the Role
of Mannose Phosphorylation   711

*The Mechanism and Regulation of Vesicular
Transport to and from the ER and the
Golgi Complex*   711

Two Types of Coated Vesicles Transport
Proteins from Organelle to Organelle   711

Clathrin Forms a Lattice Shell around
Coated Pits and Vesicles   711

A Chaperone Protein Catalyzes the
Depolymerization of Clathrin-Coated
Vesicles   713

A Type of Coated Vesicle without Clathrin
Mediates ER-to-Golgi Transport and
Transport within the Golgi   713

The Steps in Vesicular Transport Can Be
Studied Biochemically and Genetically   714

A Family of Small GTP-Binding Proteins
May Target Transport Vesicles to Their
Correct Destinations   716

*Golgi and Post-Golgi Sorting and Processing
of Membrane and Secretory Proteins*   718

Sequences in the Membrane-Spanning
Domain Cause the Retention of Proteins
in the Golgi   719

Different Vesicles Are Used for Continuous
and Regulated Protein Secretion   719

Secretory and Membrane Proteins Undergo
Several Proteolytic Cleavages During the
Late Maturation Stages   720

The Proteolytic Maturation of Insulin
Occurs in Acidic, Clathrin-Coated
Secretory Vesicles   722

*Sorting of Membrane Proteins Internalized
from the Cell Surface*   722

In Receptor-Mediated Endocytosis, Cell
Surface Receptors Are Internalized in
Clathrin-Coated Vesicles   722

The Low-Density Lipoprotein (LDL)
Receptor Binds and Internalizes
Cholesterol-Containing Particles   724

Mutant LDL Receptors Reveal a Signal
for Internalizing Receptors into
Clathrin-Coated Pits   724

Receptors and Ligands Dissociate in an
Acidic Late Endosome/CURL Organelle   726

Transferrin Delivers Iron to Cells by
Receptor-Mediated Endocytosis   727

Some Proteins Internalized by Endocytosis
Remain within the Cell, or Are
Transported across the Cell and
Secreted   728

Proteins Are Sorted in Several Different
Ways to Different Domains of the Plasma
Membrane   729

Viruses and Toxins Enter Cells by
Receptor-Mediated Endocytosis   731

Summary   734

Review Questions   735

References   737

17 CELLULAR ENERGETICS: FORMATION OF
    ATP BY GLYCOLYSIS AND OXIDATIVE
    PHOSPHORYLATION   739

    Energy Metabolism in the Cytosol   740

    In Clycolysis, ATP Is Generated by
    Substrate-Level Phosphorylation   742

    Some Eukaryotic and Prokaryotic Cells
    Metabolize Glucose Anaerobically   744

    Mictochondria and the Metabolism of
    Carbohydrates and Lipids   745

    The Outer and Inner Membranes of the
    Mitochondrion Are Structurally and
    Functionally Distinct   745

    Acetyl CoA Is a Key Intermediate in the
    Mitochondrial Metabolism of Pyruvate
    and Fatty Acids   748

    The Citric Acid Cycle Oxidizes the Acetyl
    Group of Acetyl CoA to CO_2 and
    Reduces NAD and FAD to NADH
    and $FADH_2$ 749

    Electrons Are Transferred from NADH and
    $FADH_2$ to Molecular O_2 by Electron-
    Carrier Proteins   750

    A Similar Electrochemical Protein Gradient
    Is Used to Generate ATP from ADP and
    P_i in Mitochondria, Bacteria, and
    Chloroplasts   751

    The Proton-Motive Force, ATP Generation,
    and Transport of Metabolites   752

    Closed Vesicles Are Required for the
    Generation of ATP   752

    The Proton-Motive Force Is Composed of a
    Proton Concentration Gradient and a
    Membrane Electric Potential   753

    The F_0F_1 Complex Couples ATP Synthesis
    to Proton Movement Down the
    Electrochemical Gradient   753

    Reconstitution of Close Membrane Vesicles
    Supports the Role of the Proton-Motive
    Force in ATP Synthesis   756

    Many Transporters in the Inner
    Mitochondrial Membrane Are Powered
    by the Proton-Motive Force   758

    Inner-Membrane Proteins Allow the Uptake
    of Electrons from Cytosolic NADH   759

    NADH, Electron Transport, and Proton
    Translocation   759

    Electron Transport in Mitochondria Is
    Coupled to Proton Translocation   759

    The Mitochondrial Electron Transport Chain
    Transfers Electrons from NADH to O_2 761

    Most Electron Carriers Are Oriented in the
    Transport Chain in the Order of Their
    Reduction Potentials   765

    Three Electron Transport Complexes Are
    Sites of Proton Translocation   766

    The Q Cycle Increases the Number of
    Protons Transported by the $CoQH_2$–
    Cytochrome C Reductase Complex   767

    The Cytochrome C Oxidase Complex
    Couples the Reduction of Oxygen to the
    Translocation of Protons   768

    Metabolic Regulation   770

    Respiration Is Controlled by the
    Production of ATP through the Proton-
    Motive Force   770

    An Endogenuous Uncoupler in Brown-Fat
    Mitochondria Converts H^+ Gradients
    to Heat   770

    The Rate of Glycolysis Depends on the
    Cell's Need for ATP and Is Controlled
    by Multiple Allosteric Effectors   771

    The Oxidation of Fatty Acids Occurs in
    Peroxisomes without Production
    of ATP   772

    Summary   773

    Review Questions   774

    References   776

18 PHOTOSYNTHESIS   779

    An Overview of Photosynthesis   780

    Photosynthesis Occurs on Thylakoid
    Membranes   780

    Photosynthesis Consists of Both "Light"
    and "Dark" Reactions   782

    The Light-Absorbing Step of
    Photosynthesis   783

    Each Photon of Light Has a Defined
    Amount of Energy   783

    Chlorophyll a Is the Primary Light-
    Absorbing Pigment   783

    The Absorption of Light by Reaction-Center
    Chlorophylls Causes a Charge Separation
    across the Thylakoid Membrane   784

    Molecular Analysis of Bacterial
    Photosynthesis   786

    Purple Photosynthetic Bacteria Utilize
    Only One Photosystem and Do Not
    Evolve O_2 786

Photoelectron Transport in the Photosynthetic Reaction Center of Purple Bacteria Results in a Charge Separation   788

Photosynthetic Bacteria also Carry Out Noncyclic Electron Transport   790

Molecular Analysis of Photosynthesis in Plants   790

Plants Utilize Two Photosystems, PSI and PSII, with Different Functions in Photosynthesis   790

Both PSI and PSII Are Essential for Photosynthesis in Chloroplasts   792

PSII Splits H_2O 793

Electrons Are Transported from PSII to PSI   794

PSI Forms NADPH   796

PSI Can Also Function in Cyclic Electron Flow   796

PSI and PSII Are Functionally Coupled   796

CO_2 Metabolism during Photosynthesis   797

CO_2 Fixation Is Catalyzed by Ribulose 1,5-Bisphosphate Carboxylase   797

CO_2 Fixation Is Activated in the Light   800

Photorespiration Liberates CO_2 and Consumes O_2 800

Peroxisomes Play a Role in Photorespiration   800

The C_4 Pathway for CO_2 Fixation Is Used by Several Tropical Plants   802

Sucrose Is Transported from Leaves through the Phloem to All Plant Tissues   803

Summary   805

Review Questions   806

References   807

19 ORGANELLE BIOGENESIS: THE MITOCHONDRION, CHLOROPLAST, PEROXISOME, AND NUCLEUS   809

An Overview of Organelle Biogenesis Outside the Secretory Pathway   810

Mitochondrial DNA: Structure, Expression, and Variability   812

Cytoplasmic Inheritance and DNA Sequencing Have Established the Existence of Mitochondrial Genes   812

The Size and Coding Capacity of mtDNA Varies in Different Organisms, Reflecting Evolutionary Movement of DNA between Mitochondrion and Nucleus   813

Proteins Encoded by Mitochondria DNA Are Synthesized on Mitochondrial Ribosomes   816

Mitochondrial Genetic Codes Are Different from the Standard Nuclear Code, and They Differ among Organisms   816

In Animals, Mitochondrial RNAs Undergo Extensive Processing   817

Mutations in Mitochondrial DNA Cause Several Genetic Diseases in Man   817

Synthesis and Localization of Mitochondrial Proteins   819

Most Mitochondrial Proteins Are Synthesized in the Cytosol as Precursors   819

Matrix-Targeting Sequences Direct Imported Proteins to the Mitochondrial Matrix   820

Mitochondrial Receptors Bind Matrix-Targeting Sequences   824

Intermediates in Translocation of Proteins into the Mitochondrion Can Be Accumulated and Studied   824

The Uptake of Mitochondrial Proteins Requires Energy   824

Matrix Chaperones Are Essential for the Import and Folding of Mitochondrial Proteins   826

Proteins Are Targeted to the Correct Submitochondrial Compartment by Multiple Signals and Several Pathways   827

Certain Mitochondrial Proteins Are Essential for Life   829

The Synthesis of Mitochondrial Proteins Is Coordinated   829

Chloroplast DNA and the Biogenesis of Chloroplasts and Other Plastids   830

Chloroplast DNA Contains over 120 Different Genes   830

Several Uptake-Targeting Sequences Direct Proteins Synthesized in the Cytosol to the Appropriate Chloroplast Compartment   832

Proplastids Can Differentiate into Chloroplasts or Other Plastids   835

Peroxisome Biosynthesis   837

All Peroxisomal Proteins Are Imported from the Cytosol   837

Genetic Diseases Have Helped to Elucidate the Process of Peroxisome Biogenesis   838

*Protein Traffic into and out of the
Nucleus*   840

Nuclear Proteins Are Selectively Imported
into Nuclei   840

Nuclear Pores Are the Portals for Protein
Transport   841

Multiple Types of Nuclear Localization
Sequences Direct Proteins and
Ribonucleoproteins to the Nucleus   842

Receptor Proteins in Nuclear Pores Bind
Nuclear Proteins for Import   844

Summary   844

Review Questions   845

References   847

▶ **Part IV**
**Integrative and Specialized
Cellular Events**   **850**

20   CELL-TO-CELL SIGNALING: HORMONES
AND RECEPTORS   853

Overview of Extracellular Signaling   854

Signaling Molecules Operate over Various
Distances in Animals   855

Receptor Proteins Exhibit Ligand-Binding
Specificity and Effector Specificity   856

Hormones Can Be Classified Based on Their
Solubility and Receptor Location   856

Effects of Many Hormones Are Mediated by
Second Messengers   857

Cell-Surface Receptors Can Be Categorized
into Four Major Classes   859

The Synthesis, Release, and Degradation of
Hormones Are Regulated   860

*Identification and Purification of Cell-Surface
Receptors*   865

Hormone Receptors Are Detected by Binding
Assays   865

K_D Values for Cell-Surface Hormone
Receptors Approximate the Concentrations
of Circulating Hormones   866

Affinity Techniques Permit Purification of
Receptor Proteins   866

Many Receptors Can Be Cloned without
Prior Purification   867

*Seven-spanning G Protein–Linked
Receptors*   869

Binding of Epinephrine to β- and
α-Adrenergic Receptors Induces Tissue-
Specific Responses Mediated by cAMP   870

Analogs Provide Information about
Essential Features of Hormone Structure
and Are Useful as Drugs   871

Studies with Mutant β-Adrenergic Receptors
Identify Residues That Interact with
Catecholamines   872

Trimeric Signal-Transducing G_S Protein
Links β-Adrenergic Receptors and Adenylate Cyclase   873

$G_{S\alpha}$ Belongs to GTPase Superfamily of
Intracellular Switch Proteins   876

Some Bacterial Toxins Irreversibly Modify G
Proteins   877

Adenylate Cyclase Is Stimulated and
Inhibited by Different Receptor-Ligand
Complexes   879

Analogous Regions in All Seven-spanning
Receptors Determine G Protein and
Ligand Specificity   881

Degradation of cAMP Also Is Regulated   881

*Role of cAMP in the Regulation of Cellular
Metabolism*   881

cAMP and Other Second Messengers
Activate Specific Protein Kinases   881

Epinephrine Stimulates Glycogenolysis
in Liver and Muscle Cells   882

cAMP-Dependent Protein Kinase Regulates
the Enzymes of Glycogen Metabolism   884

Kinase Cascade Permits Multienzyme
Regulation and Amplifies Hormone
Signal   885

Cellular Responses to cAMP Vary among
Different Cell Types   885

Receptor Tyrosine Kinases   886

SH2-Containing Proteins Bind to Specific
Phosphotyrosine Residues in Activated
RTKs   886

Ras Protein Is a Key Component of RTK
Signaling Pathways in Many Eukaryotes   887

Genetic Analysis of *Drosophila* Eye
Development Identified Three Proteins
That Link RTKs to a Kinase Cascade   891

GRB2 Is an Adapter Protein That Binds to
Activated RTKs   893

Sos Protein Is Localized to the Plasma
Membrane by Binding to the SH3
Domains in GRB2   893

A Highly Conserved Kinase Cascade
Transmits RTK-Mediated Signals
Downstream from Ras   894

Ras-Coupled RTKs Transduce Extracellular Signals by a Common Pathway   896

Yeast Mating-Factor Receptors Are Linked to G Proteins That Transmit Signals to MAP Kinase   897

Other Important Second Messengers   899

Cellular Effects of Ca^{2+} Depend on Its Cytosolic Level and Often Are Mediated by Calmodulin   899

Ca^{2+} Ions and cAMP Induce Hydrolysis of Muscle Glycogen   901

Inositol 1,4,5-Trisphosphate Causes the Release of Ca^{2+} Ions from the ER   901

Release of Intracellular Ca^{2+} Stores Also Is Mediated by Ryanodine Receptors in Muscle Cells and Neurons   904

1,2-Diacylglycerol Activates Protein Kinase C   904

Multiplex Signaling Pathways   905

Some Activated RTKs Stimulate Activity of Phospholipase C_γ 905

Multiple G Proteins Transduce Signals from Seven-Spanning Receptors to Different Effector Proteins   905

$G_{\beta\gamma}$ Acts Directly on Some Effectors in Mammalian Cells   906

The Insulin Receptor and Regulation of Blood Glucose   907

Insulin Has Short-Term Effects on Glucose Metabolism and Long-Term Growth-Promoting Effects   907

Insulin Signaling Pathway Involves a Soluble "Relay" Protein That Does Not Bind to the Receptor   910

Insulin and Glucagon Work Together to Maintain a Stable Blood Glucose Level   911

Regulation of Cell-Surface Receptors   912

Receptors for Many Peptide Hormones Are Down-Regulated by Endocytosis   912

Phosphorylation of Cell-Surface Receptors Modulates Their Activity   913

From Plasma Membrane to Nucleus   914

Activation of Some Transcription Factors Occurs via Several Signaling Pathways Coupled to G Protein–Linked Receptors and RTKs   914

STATs Are Transcription Factors Activated by Protein Tyrosine Kinases Associated with Cell-Surface Receptors   916

Summary   918

Review Questions   920

References   922

21 NERVE CELLS   925

Neurons, Synapses, and Nerve Circuits   926

Specialized Regions of Neurons Carry Out Different Functions   926

Synapses Are Specialized Sites Where Neurons Communicate with Other Cells   929

Neurons Are Organized into Circuits   931

The Action Potential and Conductance of Electric Impulses   932

The Resting Potential Is Generated Mainly by Open Potassium Channels   933

Opening and Closing Ion Channels Cause Specific, Predictable Changes in the Membrane Potential   935

Membrane Depolarizations Would Spread Only Short Distances without Voltage-Gated Cation Channels   935

Opening of Voltage-Gated Sodium Channels Depolarizes the Nerve Membrane during Conductance of an Action Potential   936

Voltage-Dependent Sodium Channel Proteins Propagate Action Potentials Unidirectionally without Diminution   938

Opening of Voltage-Gated Potassium Channels Causes Repolarization of the Plasma Membrane during an Action Potential   938

Movements of Only a Few Sodium and Potassium Ions Generate the Action Potential   939

Myelination Increases the Rate of Impulse Conduction   940

Action Potentials Are Generated in an All-or-Nothing Fashion by Summation of Electric Disturbances   943

Molecular Properties of Voltage-Gated Ion Channel Proteins   943

Patch Clamps Permit Measurement of Ion Movements through Single Sodium and Potassium Channels   944

All Voltage-Gated Ion Channels Have a Similar Molecular Structure   946

Shaker Mutants in *Drosophila melanogaster* Led to the Cloning of a Large Family of Voltage-Gated Potassium Channel Proteins   947

Study of Toxin-Resistant Mutants Led to the Identification of Amino Acids That Line the Ion-Conducting Pore of the Potassium Channel   948

A Complete *shaker* K+ Channel Is Assembled from Four Subunits   949

The S4 Segment Comprises the Voltage Sensor   949

The N-terminal Segment of the *shaker* Protein Causes Channel Inactivation   949

Potassium Channel Proteins Are Diverse   950

The Sodium Channel Protein Has Four Homologous Transmembrane Domains, Each Similar to a Potassium Channel Polypeptide   951

All Voltage-Gated Ion Channel Proteins Probably Evolved from a Common Ancestral Gene   951

Synapses and Impulse Transmission   952

Impulse Transmission across Electric Synapses Is Nearly Instantaneous   952

Chemical Synapses Can Be Fast or Slow, Excitatory or Inhibitory, and Can Exhibit Signal Amplification and Computation   953

Many Types of Receptors Bind the Same Neurotransmitter   955

Synaptic Transmission and the Nicotinic Acetylcholine Receptor   956

Acetylcholine Is Synthesized in the Cytosol and Stored in Synaptic Vesicles   956

Exocytosis of Synaptic Vesicles Is Triggered by Opening of Voltage-Gated Calcium Channels and a Rise in Cytosolic Calcium   958

Multiple Proteins Align Synaptic Vesicles with the Plasma Membrane and Participate in Vesicle Exocytosis and Endocytosis   960

The Nicotinic Acetylcholine Receptor Protein Is a Ligand-Gated Cation Channel   962

Spontaneous Exocytosis of Synaptic Vesicles Produces Small Depolarizations in the Postsynaptic Membrane   962

The Nicotinic Acetylcholine Receptor Contains Five Subunits, Each of Which Contributes to the Cation Channel   963

Hydrolysis of Acetylcholine Terminates the Depolarization Signal   965

Functions of Other Neurotransmitters, Their Receptors, and Their Transporters   965

GABA and Glycine Receptors Are Ligand-Gated Anion Channels Used at Many Inhibitory Synapses   966

Cardiac Muscarinic Acetylcholine Receptor Activates a G Protein and Open Potassium Channels   967

Different Catecholamine Receptors Affect Different Intracellular Second Messengers   967

A Serotonin Receptor Modulates Potassium Channel Function via the Activation of Adenylate Cyclase   968

Neurotransmitter Transporters Are the Proteins Affected by Drugs Such as Cocaine   970

Some Peptides Function as Both Neurotransmitters and Neurohormones   970

Endorphins and Enkephalins Are Neurohormones That Inhibit Transmission of Pain Impulses   971

Sensory Transduction: The Visual and Olfactory Systems   971

The Light-Triggered Closing of Sodium Channels Hyperpolarizes Rod Cells   972

Absorption of a Photon Triggers Isomerization of Retinal and Activation of Opsin   974

Cyclic GMP Is a Key Transducing Molecule   975

Rod Cells Adapt to Varying Levels of Ambient Light   976

Color Vision Utilizes Three Opsin Pigments   977

More Than a Thousand Different G-Protein–Coupled Receptors Detect Odors   978

Memory and Neurotransmitters   979

Mutations in *Drosophila* Affect Learning and Memory   979

Gill-Withdrawal Reflex in *Aplysia* Exhibits Three Elementary Forms of Learning   979

A Novel Glutamate Receptor Is the Coincidence Detector in Long-Term Potentiation Exhibited by Many Synapses in the Mammalian Brain   982

Retrograde Signaling by the Gas Nitric Oxide May Be a Part of Long-Term Potentiation   982

Mice Defective in the Hippocampal Ca^{2+}-Calmodulin-Activated Protein Kinase Are Impaired in Long-Term Potentiation and in Spatial Learning—the Beginnings of a Molecular Psychology   984

Summary   984

Review Questions   986

References   987

22   MICROFILAMENTS: CELL MOTILITY AND CONTROL OF CELL SHAPE   991

Actin Filaments   992

All Eukaryotic Cells Contain Abundant Amounts of Actin   994

The Actin Sequence Has Changed Little during Evolution   994

ATP Holds Together the Two Lobes of the Actin Monomer   994

G-Actin Assembles into Long F-Actin Polymers   995

F-Actin Is a Helical Polymer of Identical Subunits   995

F-Actin Has Structural and Functional Polarity   996

Actin Architectures   996

The Actin Cytoskeleton Is Organized into Bundles and Networks of Filaments   997

Actin Bundles and Networks Are Connected to the Membrane   998

Cortical Networks of Actin Filaments Stiffen Cell Membranes and Immobilize Integral Membrane Proteins   999

Dystrophin Anchors a Cortical Actin Network Directly to the Extracellular Matrix   1000

Actin Bundles Support Projecting Fingers of Membrane   1002

The Dynamics of Actin Assembly   1003

Actin Polymerization in Vitro Proceeds in Three Steps   1003

ATP Enhances Assembly from One End of a Filament   1003

Fungal Toxins Disrupt the Monomer-Polymer Equilibrium   1006

Actin-Binding Proteins Control the Lengths of Actin Filaments   1006

A Family of Actin-Severing Proteins Generates New Filament Ends by Breaking Actin Filaments   1007

Actin Filaments Are Stabilized by Actin-Capping Proteins   1009

Many Movements Are Driven by Actin Polymerization   1010

Myosin: A Cellular Engine that Powers Motility   1012

Myosin Is a Diverse Family of Proteins Characterized by Distinct Head, Neck, and Tail Domains   1014

The Myosin Tail Domain Regulates Binding to Membranes or the Assembly of Thick Filaments   1015

The Myosin Head Domain Is an Actin-Activated ATPase   1015

Myosin Heads Walk along Actin Filaments   1015

A Myosin Head Takes an 11-nm Step Each Time an ATP Molecule Is Hydrolyzed   1017

X-Ray Crystallography Reveals the Atomic Structure of the Motor Domain   1018

Conformational Changes in the Head Couple ATP Hydrolysis to Movement   1020

Muscle, A Specialized Contractile Machine   1021

Some Muscles Contract, Others Generate Tension   1022

Striated Muscles Contain a Regular Array of Actin and Myosin   1023

In Smooth Muscle, Thick and Thin Filaments Are Not in Regular Arrays   1025

Thick and Thin Filaments Slide Past Each Other During Contraction   1025

A Third Filament System of Long Proteins Organizes the Sarcomere   1026

Calcium from the Sarcoplasmic Reticulum Triggers Contraction   1026

Calcium Activation of Myosin Light Chains Regulates Contraction in Smooth Muscle and Invertebrate Muscle   1030

Actin and Myosin in Nonmuscle Cells   1032

Actin and Myosin II Are Arranged in Sarcomere-Like Structures   1032

Contractile Actin Bundles Are Attached to Specialized Sites at the Plasma Membrane   1035

Myosin II Stiffens Cortical Membranes   1036

Actin and Myosin II Have Essential Roles in
Cytokinesis   1036

Myosins I and V Move Membrane-Bounded
Cargoes along Actin Filaments   1038

Membrane-Bound Myosin in Vesicle
Movements   1039

Myosin I and Myosin II Are Not Essentially
Required for Cell Migration   1039

Cell Motility   1040

Movements of Fibroblasts Involve Controlled
Polymerization and Rearrangements of
Actin Filaments   1041

Ameboid Movement Involves Reversible
Gel-Sol Transitions of an Actin Network   1043

Cell Movements Are Coordinated by Various
Second Messengers and Signal
Transduction Pathways   1044

Summary   1046

Review Questions   1048

References   1049

23   MICROTUBULES AND INTERMEDIATE
FILAMENTS   1051

Microtubule Structures   1052

Tubulin Subunits Comprise the Wall of a
Microtubule   1052

Microtubules Form a Diverse Array of Both
Permanent and Transient Structures   1054

Microtubules Grow from Microtubule-
Organizing Centers   1055

The Microtubule-Organizing Center
Determines the Polarity of Cellular
Microtubules   1055

Multiple Tubulin Genes and Chemical
Modification Leads to Tubulin Diversity   1059

Microtubule Dynamics   1061

Microtubule Assembly and Disassembly
Occur by Preferential Addition and Loss
of $\alpha\beta$ Dimers at the (+) End   1061

Dynamic Instability Is an Intrinsic Property of
Microtubules   1064

Colchicine and Other Anti-Cancer Drugs
Poison Microtubule Assembly or
Disassembly   1066

Microtubule-Associated Proteins   1067

MAPs Organize Bundles of Microtubules   1067

MAPs Stabilize Microtubules   1070

*Kinesin, Dynein, and Intracellular
Transport*   1070

Fast Axonal Transport Occurs along
Microtubules   1070

Microtubules Provide Tracks for the
Movement of Pigment Granules   1072

Intracellular Membrane Vesicles Travel
Along Microtubules   1072

Microtubule Motor Proteins Promote Vesicle
Translocation along Microtubules   1075

Kinesin Is a (+) End–Directed Motor
Protein   1075

Dynein Is a (−) End–Directed Motor
Protein   1078

Multiple Motor Proteins Are Associated with
Membrane Vesicles   1078

Cilia and Flagella: Structure and Movement   1079

All Eukaryotic Cilia and Flagella Contain
Bundles of Doublet Microtubules   1080

Ciliary and Flagellar Beating Is Produced by
Controlled Sliding of Outer Doublet
Microtubules   1084

Dynein Arms Generate the Sliding Forces   1084

Axonemal Dyneins Are Multi-Headed Motor
Proteins   1086

Flagellar Beat Requires Conversion of Sliding
to Bending   1086

Genetic Studies Provide Information about the
Roles of the Central Microtubules and the
Radial Spokes   1087

Calcium Regulates the Direction of
Swimming   1087

Axonemes Assemble from Basal Bodies   1088

Basal Bodies Closely Resemble Centrioles   1089

*Microtubule Dynamics and Motor Proteins
during Mitosis*   1090

The Mitotic Apparatus Is a Microtubule
Machine for Separating Chromosomes   1091

The Kinetochore Is a Specialized Attachment
Site at the Chromosome Centromere   1091

Yeast Centromeres Bind a Single
Microtubule   1094

Centrosome Duplication and Migration
During Interphase and Prophase Initiate
the Assembly of the Mitotic Apparatus

During Prophase, Kinesin-Related Proteins and Cytoplasmic Dynein Participate in the Movements of Kinetochores and Centrosomes — 1097

Assembly of the Mitotic Apparatus Involves Dynamic Microtubules — 1098

At Metaphase Forces at the Kinetochore Move Chromosomes to the Equator of the Spindle — 1100

During Anaphase Chromosomes Separate and the Spindle Elongates — 1100

Astral Microtubules Determine Where Cytokinesis Takes Place — 1103

Plant Cells Build a New Cell Wall During Cell Division — 1105

Intermediate Filaments — 1106

Intermediate Filaments Are Classified into Five Types — 1106

All Subunit Proteins of Intermediate Filaments Have a Similar Structure — 1109

Intermediate Filaments Are Dynamic Polymers in the Cell — 1111

Phosphorylation of the N-Terminal Domain Regulates Polymerization of Intermediate Filaments during Mitosis — 1112

Intermediate Filament-Associated Proteins Cross-Link Intermediate Filaments to Membranes and Microtubules — 1113

Summary — 1116

Review Questions — 1118

References — 1119

24 MULTICELLULARITY: CELL-CELL AND CELL-MATRIX INTERACTIONS — 1123

The Extracellular Matrix: Primary Components and Functions — 1124

Collagen: A Class of Multifunctional Fibrous Proteins — 1126

The Basic Structural Unit of Collagen Is a Triple Helix — 1127

Most Exons in Fibrous Collagen Genes Encode Gly-X-Y Sequences — 1128

Collagen Fibrils Form by Lateral Interactions of Triple Helices — 1128

Denatured Collagen Polypeptides Cannot Renature to Form a Triple Helix — 1131

Procollagen Chains Assemble into Triple Helices in the Rough ER and Are Modified in the Golgi Complex — 1131

Collagen Is Assembled into Fibrils after Secretion — 1133

Mutations in Collagen Reveal Aspects of Its Structure and Biosynthesis — 1134

Collagens Form Diverse Structures — 1134

Type IV Collagen Forms the Two-Dimensional Reticulum of the Basal Lamina — 1135

Hyaluronan and Proteoglycans — 1136

Hyaluronan Is an Immensely Long, Negatively Charged Polysaccharide That Forms Hydrated Gels — 1136

Hyaluronan Inhibits Cell-Cell Adhesion and Facilitates Cell Migration — 1137

Proteoglycans Comprise a Diverse Family of Cell-surface and Extracellular-matrix Macromolecules — 1139

Proteoglycans Can Bind Many Growth Factors — 1142

Multiadhesive Matrix Proteins and Their Cell-Surface Receptors — 1143

Laminin and Nidogen Are Principal Structural Proteins of All Basal Laminae — 1143

Integrins Are Cell-Surface Receptors That Mediate Adhesion to the Extracellular Matrix and Cell-Cell Interactions — 1144

Fibronectins Bind Many Cells to Fibrous Collagens and Other Matrix Components — 1146

Fibronectins Promote Cell Adhesion to the Substratum — 1148

Fibronectins Promote Cell Migration — 1149

Cell-Cell Adhesion: Adhesive Proteins — 1150

Adhesive Proteins Mediate Cell-Cell Interactions — 1150

E-Cadherin Is a Key Adhesive Protein Expressed by Epithelial Cells — 1150

Cadherins Influence Morphogenesis and Differentiation — 1152

N-CAMS Mediate Ca^{2+}-Independent Adhesion of Cells in Nervous Tissue and Muscle — 1153

Movement of Leukocytes into Tissues Requires Sequential Interaction of Specific Adhesive Proteins — 1153

Cell-Cell Adhesion: Cell Junctions — 1155

Three Types of Desmosomes Impart Rigidity to Tissues — 1156

Intermediate Filaments Stabilize Epithelia by Connecting Spot Desmosomes — 1156

Hemidesmosomes Connect Epithelial Cells to the Basal Lamina   1157

Gap Junctions Allow Small Molecules to Pass between Adjacent Cells   1157

Connexin, a Transmembrane Protein, Forms Cylindrical Channels in Gap Junctions   1159

Dorsoventral Patterning During Embryogenesis   1159

Embryologic Development Is Directed by Induction   1160

Transforming Growth Factor β (TGFβ) Has Numerous Inductive Effects in Invertebrates and Vertebrates   1161

TGFβ Homolog Encoded by the decapentaplegic Gene Controls Dorsoventral Patterning in Drosophila Embryos   1162

Sequential Inductive Events Regulate Early Xenopus Development   1163

Formation of Internal Organs and Organization of Tissues   1167

The Basal Lamina Is Essential for Differentiation of Many Epithelial Cells   1167

Direct Cell-Cell Contact Regulates Kidney Induction   1168

Hedgehog Organizes Pattern in the Chick Limb and the Drosophila Wing   1169

Developmental Regulation by Direct Cell-Cell Contact   1172

Boss Is a Cell-Surface Inductive Ligand for the Sev Receptor   1172

Cell-Surface Notch and Delta Proteins Control Signaling between Many Different Types of Cells   1173

Regulation of Neuronal Outgrowth   1175

Individual Neurons Can Be Identified Reproducibly and Studied   1175

Growth Cones Guide the Migration and Elongation of Developing Axons   1176

Different Neurons Navigate along Different Outgrowth Pathways   1177

Different Extracellular-Matrix Components Are Permissive for Neuronal Outgrowth   1178

Three Genes Define Dorsoventral Outgrowth in C. elegans   1179

A Chemoattractant Related to Unc6 Is Produced in the Floor Plate of the Vertebrate Neural Tube   1180

Extracellular Signals Can Repel Growth Cones   1181

Different Growth Cones Navigate along Different Axons   1182

The Basal Lamina at the Neuromuscular Junction Directs Differentiation of Regenerating Nerve and Muscle   1185

Structure and Function of the Plant Cell Wall   1187

Cellulose Molecules Form Long, Rigid Microfibrils   1188

Other Polysaccharides Bind to Cellulose to Generate a Complex Wall Matrix   1188

Cell Walls Contain Lignin and an Extended Hydroxyproline-Rich Glycoprotein   1190

Plants Grow Primarily by Auxin-Induced Cell Enlargement   1190

The Orientation of Newly Made Cellulose Microfibrils Is Affected by the Microtubule Network   1192

Plasmodesmata Interconnect the Cytoplasm of Adjacent Cells in Higher Plants   1193

Summary   1194

Review Questions   1196

References   1197

25   REGULATION OF THE EUKARYOTIC CELL CYCLE   1201

Phases of the Cell Cycle   1202

Experimental Systems in Cell-Cycle Research   1203

Control of Entry into and Exit from Mitosis   1203

The Same Factor Promotes Oocyte Maturation and Mitosis in Somatic Cells   1203

Mitotic Cycling in Early Embryos Depends on Synthesis and Degradation of Cyclin B   1207

MPF-Catalyzed Phosphorylation of Nuclear Lamins and Other Proteins Induces Early Mitotic Events   1212

Protein Degradation and Dephosphorylation Trigger Late Mitotic Stages   1216

Biochemical Studies with Xenopus Egg Extracts Identified Central Role of MPF in Regulating Entry into Mitosis   1219

Regulation of MPF Activity   1219

MPF Catalytic Subunit Is Encoded by cdc2+ Gene in S. pombe   1221

MPF Catalytic Subunit Contains Activating
and Inhibitory Sites That Are
Phosphorylated   1222

Structure of Human Cyclin-Dependent
Kinase 2 Suggests How Phosphorylation
Regulates MPF Activity   1224

Entry into Mitosis Is Controlled by Multiple
Mechanisms That Regulate MPF
Activity   1224

Control of Entry into the S Phase   1226

S. cerevisiae Cdc28 Is Functionally
Equivalent to S. pombe Cdc2   1227

S Phase–Promoting Factor Consists of a
Catalytic Subunit and G_1 Cyclin   1228

Various Cdc28-Cyclin Heterodimers
Regulate Progress through the Cell
Cycle in S. cerevisiae   1230

Cdc28-G_1 Cyclin Complexes May Activate
Transcription Factors at START   1230

A Cdk-Cyclin Complex Regulates a DNA
Initiation Factor in Xenopus   1231

Cell-Cycle Control in Mammalian Cells   1232

Mammalian Restriction Point Is Analogous
to START in Yeast Cells   1232

Multiple Cdks and Cyclins Regulate
Passage of Mammalian Cells through
the Cell Cycle   1232

Growth Factor–Induced Expression of
Two Classes of Genes Returns G_0
Mammalian Cells to the Cell Cycle   1235

Activity of Transcription Factor E2F Is
Required for Entry into S Phase   1236

*Role of Checkpoints in Cell-Cycle
Regulation*   1237

Presence of Unreplicated DNA Prevents
Entry into Mitosis   1238

Defects in Assembly of the Mitotic Spindle
Prevent Exit from Mitosis   1239

DNA Damage Prevents Entry into S
Phase and Mitosis   1239

Summary   1240

Review Questions   1242

References   1243

26 CANCER   1247

Characteristics of Tumor Cells   1248

Malignant Tumor Cells Are Invasive and
Can Spread   1248

Alterations in Cell-to-Cell Interactions Are
Associated with Malignancy   1249

Tumor Cells Lack Normal Controls on
Cell Growth   1251

Use of Cell Cultures in Cancer Research   1251

Fibroblastic, Epithelial, and Nonadherent
Cells Grow Readily in Culture   1251

Some Cell Cultures Give Rise to Immortal
Cell Lines   1252

Certain Factors in Serum Are Required for
Long-Term Growth of Cultured Cells   1253

Malignant Transformation Leads to Many
Changes in Cultured Cells   1254

Transcription of Oncogenes Can Trigger
Transformation   1258

*Oncogenes and Their Proteins:
Classification and Characteristics*   1258

Oncogenes Were Initially Identified in
Viruses and Tumor Cell DNA   1258

Five Types of Proteins Participate in
Control of Cell Growth   1259

Oncoproteins Affect the Cell's Growth-
Control Systems in Various Ways   1260

Apoptosis, or Induced Cell Suicide, Is One
Mechanism of Protection Against
Cancer   1267

Oncoproteins Act Cooperatively in
Transformation and Tumor Induction   1267

Consistent Chromosomal Anomalies
Associated with Tumors Point to the
Presence of Oncogenes   1268

Inherited Human Propensities to Develop
Cancer Point to Tumor-Suppressor
Genes   1269

DNA Viruses as Transforming Agents   1270

DNA Viruses Can Transform
Nonpermissive Cells by Random
Integration of the Viral Genome into
the Host-Cell Genome   1270

Transformation by DNA Viruses
Requires Interaction of a Few
Independently Acting Viral Proteins   1271

*RNA-Containing Retroviruses as
Transforming Agents*   1273

Virion-Producing Infection Cycle of Retroviruses Requires Integration into the
Host-Cell Genome   1273

Oncogenic Transducing Retroviruses
Convert Cellular Proto-Oncogenes
into Oncogenes   1275

Nononcogenic Transducing Retroviruses Have Been Constructed Experimentally   1277

Slow-Acting Carcinogenic Retroviruses Can Activate Nearby Cellular Proto-Oncogenes after Integration into the Host-Cell Genome   1278

Human Tumor Viruses   1279

Chemical Carcinogens   1280

Most Chemical Carcinogens Must Undergo Metabolic Conversion to Become Active   1280

The Carcinogenic Effect of Chemicals Depends on Their Interaction with DNA   1284

The Role of Radiation and DNA Repair in Carcinogenesis   1284

Ineffective or Error-Prone Repair of Damaged DNA Perpetuates Mutations   1285

Some Defects in DNA-Repair Systems Are Associated with High Cancer Rates in Humans   1285

The Multicausal, Multistep Nature of Carcinogenesis   1286

Epigenetic Alterations May Occur in Teratocarcinomas   1286

Some Cancer-Inducing Chemicals Act Synergistically   1286

Natural Cancers Result from the Interaction of Multiple Events over Time   1287

Human Cancer   1287

Summary   1288

Review Questions   1289

References   1291

27 IMMUNITY   1295

Overview   1296

Antibodies Bind to Epitopes and Have Two Functional Domains   1297

Antibody Reaction with Antigen Is Reversible   1299

Antibodies Come in Many Classes   1300

Antibodies Are Made by B Lymphocytes   1302

The Immune System Has Extraordinary Versatility   1302

Clonal Selection Theory Underlies All Modern Immunology   1303

The Immune System Has a Memory   1304

Other Parts of the Immune Response Are Carried Out by T Lymphocytes   1306

B Cells and T Cells Have Identifying Surface Markers   1306

Macrophages Play a Central Role in Stimulating Immune Responses   1307

Cells Responsible for the Immune Response Circulate throughout the Body   1308

Tolerance Is a Central Concept of Immunology   1309

Immunopathology Is Disease Caused by the Immune System   1309

Antibodies, B Cells, and the Generation of Diversity   1310

Heavy-Chain Structure Differentiates the Classes of Antibodies   1310

Antibodies Have a Domain Structure   1310

The N-Terminal Domains of H and L Chains Have Highly Variable Structures That Constitute the Antigen-Binding Site   1311

Several Mechanisms Generate Antibody Diversity   1315

DNA Rearrangement Generates Antibody Diversity   1315

A Single Recombination Event Generates Diversity in L Chains   1315

Imprecision of Joining Makes an Important Contribution to Diversity   1318

Lambda Diversity Derives from Multiple Constant Regions   1319

H-Chain Variable Regions Derive from Three Libraries   1319

Recognition Sequences for All Joining Reactions Are Virtually Indistinguishable   1320

The Synthesis of Immunoglobulins Is Like That of Other Extracellular Proteins   1321

The Antigen-Independent Phase of B-Lymphocyte Maturation   1321

B-Lymphocytes Go through an Orderly Process of Gene Rearrangement   1321

The Antigen-Independent Phase Can Generate Cells with 10^{11} Different Specificities   1322

The Immune System Requires Allelic
  Exclusion   1323

Antibody Gene Rearrangement and
  Expression Are Controlled by
  Transcription Factors   1323

T Lymphocytes   1324

There Are Two T-Cell Receptor Molecules   1325

T-Cell Receptors Recognize a Foreign
  Antigen in Combination with a
  Self-Molecule   1325

T Cells Are Educated in the Thymus to
  React with Foreign Proteins but Not
  Self-Proteins   1329

T Cells Respond to Antigen by Either
  Killing Cells or Secreting Protein Factors   1330

*The Antigen-Dependent Phase of the
Immune Response*   1332

B-Cell Activation Involves B Cell–T$_H$ Cell
  Collaboration   1333

Activation Is the Result of Cellular
  Stimulation   1335

Antibody Secretion by Activated B Cells
  Entails Many Cellular Changes   1335

The Activation Process Produces Both
  Plasma Cells and Memory Cells   1335

The Change from Surface Antibody to
  Secreted Antibody Requires the
  Synthesis of an Altered H Chain   1335

Antibody Class Switching Also Requires an
  Altered H Chain   1337

Memory Cells Participate in the Secondary
  Immune Response   1338

Somatic Mutation of Variable Regions
  Follows from Activation   1338

Tolerance Is Achieved Partly by Making
  B Cells Unresponsive   1339

Summary   1340

Review Questions   1341

References   1342

Glossary   G-1

Index   I-1

Chapter Opening Illustrations

Chapter 1 An artist's rendition of the interior of a eukaryotic cell. Depicting a cell's interior is difficult because electron micrographs provide detailed pictures of only a thin slice of a cell, while the cell itself is a three-dimensional object with a very complex interior structure. Thus an artist can create a special sense of the cell's inner workings by using color and shading. Here the artist rendered the organelles inside the cell as he imagined them rather than as a faithful reconstruction from electron micrographs. The blue object is the cell's nucleus with the DNA visible inside as a coil. The red strands emerging from the nucleus are RNA molecules. In the rest of the cell is the cytoplasm, which contains many organelles like the red, kidney-shaped mitochondria and the sectioned orange vesicles. The green stack of flattened vesicles near the nucleus is the Golgi apparatus, and the other flat vesicles represent the cell's endoplasmic reticulum. All of these cellular elements are described in later chapters. [This picture, drawn by Tomo Narashima, originally appeared on the cover of the second edition of this book.]

Chapter 2 Three-dimensional model of an ATP molecule. The atoms are represented by spheres of the appropriate Van der Waals radius; carbon atoms are gray, nitrogens are blue, phosphorus are yellow, and oxygen is red. The model was based on the three-dimensional coordinates of the atoms in several nucleotide-protein complexes, derived from the crystalline structures of the molecules, in the computerized file of the Protein Data Bank. [Photograph courtesy Gareth White.]

Chapter 3 A two-dimensional crystalline array of 50S ribosomal subunits from *B. stearothermophilus*. Crystals were grown on phospholipid monolayers, rapidly frozen in vitreous ice, and imaged in a 100 kV cryoelectron microscope. In this 20 Å resolution projection map of the crystal, areas of high protein mass are displayed as bright colors. [Adapted from A. Avila-Sakar and W. Chiu, 1994, *J. Mol. Biol.* **239**:689–697. Photograph courtesy of A. Avila-Sakar and W. Chiu.]

Chapter 4 This image shows *E. coli* RNA polymerase (the bright, round object) bound to a 580 base pair fragment of bacteriophage λ DNA containing a promoter sequence that specifies a transcriptional start site. In binding to the DNA, the polymerase bends the linear DNA at a 90° angle. The picture was obtained by a modification of electron microscopy called scanning force microscopy. [From W. A. Rees et al. 1993, *Science* **260**:1646–1649; Photograph courtesy of R. W. Keller.]

Chapter 5 The actin filaments (red) and mitochondria (green) in a rat astrocyte. Astrocytes isolated from the cerebral cortex of a newborn rat were cultured for two weeks. They were fixed, permeabilized, and then stained with phalloidin, a protein that binds to actin filaments, that was tagged with rhodamine and that generates a red fluorescence. The cells were then stained with a fluorescein-tagged antibody (green fluorescence) specific to the mitochondrial enzyme cytochrome oxidase (antibody provided by Giampietro Schiavo and Cesare Montecucco). [Photomicrograph by Olaf Mundigl and Pietro DeCamilli.]

Chapter 6 Formation of syncytia in NIH 3T3 cells that express truncated Moloney murine leukemia virus envelope proteins. The nuclei are stained with Hoechst dye 33258 (blue) and the location of the envelope protein is indicated by the red rhodamine staining. Cultured NIH 3T3 cells were transfected by electroporation with DNA that encodes a truncated envelope protein that promotes cell-cell membrane fusion and formation of syncytia (multinucleated cells). The truncation is necessary for making the envelope protein competent for membrane fusion. After DNA transfection the cells were grown on glass cover slips and fixed 48 hours later with a 15-minute incubation with 4% paraformaldehyde in phosphate buffered saline. The cells were stained for 30 minutes with a rat monoclonal antibody, 83A25, directed against the gp70 envelope protein, and then with goat anti-rat immunoglobulin G antibodies coupled to rhodamine. The cells were then incubated for 2 minutes with Hoechst dye 33258 and the cover slips were mounted in Fluormount and viewed with a fluorescence microscope. [Photograph courtesy of David Sanders, Whitehead Institute for Biomedical Research.]

Chapter 7 Detection of HIV-1 nucleic acid in human lymphocytes by *in situ* PCR. Lymphocytes isolated from peripheral blood were fixed, permeabilized and subjected to PCR with HIV-1 specific primers. Amplified DNA (green) was detected by hybridization to a complementary oligonucleotide probe conjugated with 5-carboxyfluorescein. Nuclei were counterstained (red) with propidium iodide. Green fluorescent cells were isolated with a fluorescence activated cell sorter and visualized by confocal microscopy. [Photograph courtesy of Bruce Patterson, M.D.]

Chapter 8 Two sibling mice at about 10 weeks of age. The larger mouse carries a chimeric gene, which includes the structural gene for rat growth hormone under the control of the mouse metallothionein promoter. This gene was introduced into the mouse germline using transgenic technology. Mice carrying the transgene grow 2–3 times as fast as their normal siblings and can reach nearly twice their size. Since mice carrying the transgene are fertile and the gene has been incorporated into chromosomes of the germline the transgene can be passed on to subsequent generations. The introduction of transgenes is an important technique used widely in the study of mouse development, physiology, and behavior. [From R. D. Palmiter et al., 1982, *Nature* 300:611–615; Photograph courtesy of R. L. Brinster.]

Chapter 9 Localization of the gastrin/choleycystokinin-B receptor gene to the tip of the short arm of chromosome 11 by fluorescence in situ hybridization. The gastrin/cholecystokinin-B receptor cDNA was labeled with biotinylated-dUTP, hybridized to propidium iodide stained human metaphase chromosomes, and detected with FITC-conjugated avidin and anti-avidin antibodies. [See D. B. Zimonjic et al., 1994. *Cytogenet. Cell Genet.* 65:184; Photograph courtesy of T. Matsui.]

Chapter 10 During S phase of the cell cycle, DNA replication proteins assemble into large complexes or replication foci containing many replication forks (e.g., tens to hundreds) and thousands of replication proteins. During interphase these components are distributed uniformly throughout the nucleoplasm. In this figure antibodies to a DNA methylase were used to visualize replication foci in S-phase cells. Antibodies to other replication factors also show recruitment into these structures during S phase. [From H. Leohardt, A. Page, H.-U Weier and T. H. Bestor, 1992, *Cell* 71, 865–873.]

Chapter 11 An active region of transcription producing a "puff" in a *Drosophila* polytene chromosome. Chromosomes were stained with fluorescently labeled antibodies against the heat shock transcription factor (red) and RNA polymerase II (green). Regions of overlap appear yellow. The transcription factor is concentrated near the 5' end of the transcription unit comprising the puff and at additional positions along the polytene chromosomes. [Photograph courtesy of John R. Weeks and Arno L. Greenleaf.]

Chapter 12 The non-snRNP pre-mRNA splicing factor SC35 localizes in a speckled distribution in interphase nuclei (orange regions). HeLa cell SC35 was visualized by immunostaining with a fluorescently labeled antibody. An optical section of the immunostaining pattern is superimposed over a differential interference contrast image of the cells. [Photograph courtesy of David L. Spector, Cold Spring Harbor Laboratory.]

Chapter 13 The expression of the myogenin gene in this mouse embryo was followed using a reporter transgene in which the promoter of the myogenin gene was fused to the structural gene encoding *E. coli* β-galactosidase. The blue color indicates the pattern of transgene expression. By joining different regions of the myogenin promoter to β-galactosidase investigators have discovered distinct elements that control expression in the limb buds and myotomes. [From T.-C Cheng, M. C. Wallace, J. P. Merlie, and E. N. Olson, 1993, *Science* **261**:215–217; photograph courtesy of Eric N. Olson.]

Chapter 14 Model of a phospholipid bilayer derived from a molecular dynamics calculation. In cell membranes the fatty acid "tails" of the phospholipids are in a fluid-like state. To calculate a model for a bilayer in this state, the computer first generated energetically favored conformations of individual phospholipids, then allowed many to interact with each other in a model bilayer until a gel-like state of the hydrocarbon core was achieved. The carbon atoms in the fatty acids had the same ability to rotate about the C–C bonds as did carbon atoms in solution of pure hydrocarbon, such as $C_{16}H_{34}$. [Photograph courtesy of R. M. Venable and R. W. Pastor.]

Chapter 15 Immunofluorescence localization of the CHIP28/Aquaporin water channel protein (yellow-green fluorescence) in the outer medulla of rat kidney. The tissue section was stained with Evans Blue, which generates a red fluorescence. Aquaporin is localized to the apical (brush border) and basolateral plasma membranes of cells in the S3 segment of the proximal tubule, and is absent from other cells in this region of the kidney. [See I. Sabolic, et al., 1992, *Am. J. Physiol.* 263:C1225–C1233. Photograph courtesy of Dennis Brown.]

Chapter 16 Human skin fibroblasts were incubated at low temperature with a fluorescent (BODIPY) analog of the lipid sphingomyelin, labeling the plasma membrane of these cells. The cells were warmed to 37°C for 5 min to allow endocytosis to occur. The fluorescent lipid molecules remaining on the plasma membrane were then removed by incubating the cells with a protein solution, so that the hundreds of fluorescently labeled endosomal vesicles within the cell were readily visible. Each fluorescent dot is an endosome in the cytoplasm; the black hole in the center is the nucleus, which is unlabeled. Despite its appearance, this is not a photograph of a nebula from outer space. [Photomicrograph by Richard E. Pagano and Ona C. Martin, Carnegie Institution of Washington.]

Chapter 17 Two populations of mitochondria in a living mink fibroblast, treated with the lipophilic cation JC-1, fluoresce green and yellow. Formation of an aggregate of the dye within the mitochondria is caused by a high membrane potential and generates a red fluorescence; the red fluorescence combines with the green fluorescence from the unaggregated dye to generate a yellow fluorescence. Within single long mitochondria in this cell multiple regions with yellow fluorescence, separated by green fluorescence, are evident, indicating that within long continuous mitochondria heterogeneity in membrane potentials is possible. [See S. T. Smiley et al., 1991, *Proc. Nat'l. Acad. Sci. USA* **88**:3671–3675; Photograph courtesy Lan Bo Chen.]

Chapter 18 Abundant chloroplasts, each about 10 μm in diameter, are seen in this unstained section of a leaf from the pond weed *Elodea*. [Photograph by Dwight R. Kuhn.]

Chapter 19 Firefly luciferase, a peroxisomal matrix protein, is transported to peroxisomes of normal human fibroblasts, but remains cytoplasmic in cells from a Zellweger syndrome patient. The fibroblasts (on coverslips) were microinjected with mRNA encoding the luciferase. After overnight incubation in a humidified, CO_2 incubator at 37°C, the cells were fixed, permeabilized and labeled with appropriate primary (rabbit anti-luciferase) and secondary (FITC anti-rabbit) antibodies. The punctate immunofluorescence observed in normal human HS68 cells (*left*) is indicative of peroxisomal luciferase. The fibroblast cell line (GM6231) from the human patient (*right*) does not import luciferase into peroxisomes, but shows a cytoplasmic signal instead of the punctate signal. Magnification 165X. [See P. Walton et al., 1993, *Mol. Cell Biol.* **12**:531–541; photographs courtesy of Suresh Subramani.]

Chapter 20 PC12 cells overexpressing the epidermal growth factor (EGF) receptor respond to EGF by extending neurites. Activated forms of MAP kinase kinase injected into these cells lead to neurite extension also. In this experiment a dominant-interfering mutation was injected into a subpopulation of PC12 cells. Uninjected cells (red) extend neurites whereas injected cells (yellow/green) are inhibited. MAP kinase kinase is part of signal transduction cascade conserved from yeast to humans. [From S. Cowley and H. Paterson, P. Kemp and C. Marshall, 1994, *Cell* **77**:841–852; photograph courtesy of H. Paterson, Chester Beatty Laboratories.]

Chapter 21 Long-term (18 day) culture of rat brain cells from the cortex. The field contains several mature neurons with long axons that extend beyond the end of the field; these axons contain a nerve-specific intermediate filament protein that is detected by a yellow/green fluorescing antibody. The field also contains many astrocytes, a type of glial cell. These cells contain the glial-cell-specific intermediate filament protein GFAP (Glial Fibrillary Acidic Protein) that is detected by the red fluorescing antibody. [Photograph courtesy of Nancy L. Kedersha.]

Chapter 22 Three-dimensional reconstructions of a human polymorphonuclear leukocyte as it moves across a glass slide. Taken 40 seconds apart, the time lapse sequence shows a leukocyte arching up and then descending to contact the surface. The bulk of the cytoplasm then flows to fill the pseudopodium, and the cell, now ellipsoidal in shape, repeats the sequence when it lifts the pseudopodium off the surface. The three-dimensional shape of the cell was calculated by merging into a continuous stack outlines of the cell taken at different levels of focus. The process is analogous to constructing a three-dimensional model from a two-dimensional contour map. [Adapted from J. Murray et al., 1992, *Cell Motil. Cytoskeleton* **22**:211–223. This drawing is based upon a photograph by D. R. Soll.]

Chapter 23 An immunofluorescence micrograph showing the microtubule (light orange) and vimentin (purple) cytoskeletons in a human skin fibroblast. In many parts of the cell, the two systems of cytoskeletal fibers are co-linear (seen as red), indicating they may be directly connected; however, in other parts of the same cell, the fibers show different organizations. In Figure 23-53, the distribution of vimentin filaments, microtubules, and actin filaments in this same cell are shown in separate panels. [Photograph courtesy of V. Small.]

Chapter 24 Tight cell-cell contacts formed by cultured human squamous skin carcinoma cells. The permeabilized cells are stained with three fluorescent molecules. The green fluorescence results from a fluorescein-coupled antibody that detects a monoclonal antibody specific for an as-yet uncharacterized cell-cell adhesion molecule; the staining is at the sites of cell-cell contacts. The red fluorescence results from a Texas-red coupled antibody that detects a monoclonal antibody specific for a cell protein mainly associated with the Golgi complex, and the Hoechst dye fluoresces blue when bound to DNA; here it stains the cell nuclei. [Photograph courtesy Nancy L. Kedersha.]

Chapter 25 Metaphase in a cultured newt lung cell. Microtubules were visualized by indirect immunofluorescence. Chromosomes were stained with Hoechst 33342. [From J. C. Waters, R. W. Cole, and C. L. Reider, 1993, *J. Cell. Biol.* **122**:361–372; photograph courtesy of Conly L. Reider.]

Chapter 26 Structure of the p53 protein bound to DNA. The structure was derived by X-ray crystallography. A short segment of DNA containing a p53-binding site (dark blue) was crystallized with a fragment of the p53 protein known to have the DNA binding site (light blue). Six p53 residues which are frequently mutated in human tumors are highlighted in yellow. They all turn out to be in the region of DNA contact of the protein. A red zinc atom is also indicated. [Photograph courtesy of Nikola P. Pavletich.]

Chapter 27 Peptides bound into the grooves on MHC class I molecules. Two peptides are shown, one from vesicular stomatitis virus (*left*, MHC shown in gold) and one from Sendai virus (*right*, MHC shown in light blue). The amino termini of the peptides are at the bottom. Peptide oxygens and nitrogens are shown in red and blue, respectively. Some structured water molecules are shown as larger blue spheres. The picture illustrates how deeply the peptide is buried in the MHC molecule. It shows that the antigens recognized by T cell receptors, which are peptides bound to MHC molecules, consist of just the upper surfaces of the bound peptides, actually representing no more than 25 percent of the total peptide surface. [From M. Matsumara et al., 1992, *Science* **257**:927.]

Molecular

Cell

Biology

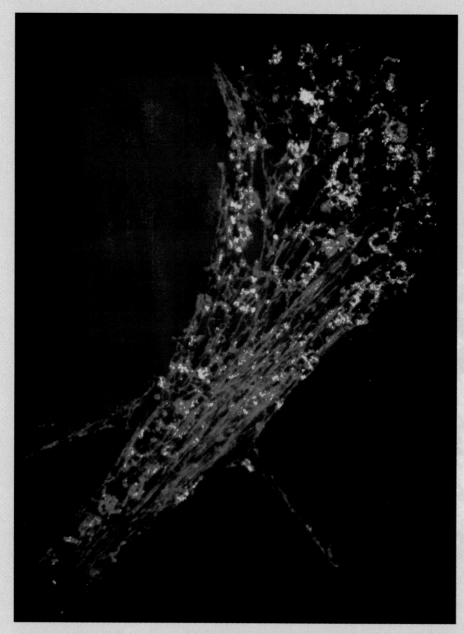

▲ Messenger RNA molecules are associated with the fibrous cytoskeleton in a cultured cortical nerve cell. Messenger RNA was detected by in situ hybridization of permeabilized cells using an oligo(dT) DNA probe specific for the poly (A) sequence at the 3′ end of virtually all mammalian mRNAs. In this electron micrograph, the gold particles, colored peach, detect the hybridized DNA probe, and the cytoskeleton is colored blue. Attachment of messenger RNAs to the cytoskeleton, composed primarily of microtubules and actin filaments, permits synthesis of proteins near their sites of function. [Photograph courtesy of G. Bassell and K. S. Kosick.]

Laying the
Groundwork

Molecular cell biology is a perspective on living systems that arose from the union of three other disciplines: biochemistry, genetics, and anatomical cell biology. In Part I, we present the scope of the book (Chapter 1) and begin to lay the groundwork for understanding the synthesis represented by molecular cell biology (Chapters 2–8). We present here both the elements of the more classical disciplines that relate to the new synthesis and some of the methodologies that underlie contemporary understanding in the field. We include laboratory procedures among the fundamentals because, as pointed out in Chapter 1, the facts of biology are only as good as the methods used to uncover them. Today's understanding of biological processes is sure to be modified as methodologies and new information accumulate. Furthermore, biology, more than physics or chemistry, is a historical science rooted in billions of years of evolution; as such, we depend primarily upon experimental information, with little help from theory, to evaluate living systems.

Chapter 2 begins with a discussion of the chemical foundations of cellular life. Cells are composed of many different molecules that interact in water and are compartmentalized by fatty membranes. Therefore the fundamentals of chemical bonding, weak and reversible molecular interactions, and the properties of water and of nonaqueous environments are keys to understanding how cells are made and how they function. Chapter 3 considers the properties of the most versatile molecules of cells, their proteins. Here we learn about the extraordinary variety of protein structure and function that has evolved from permutations of just 20 amino acid building blocks. In Chapter 4, we continue the exposition of biological chemistry by turning to the nucleic acids, molecules whose remarkable coding capacity puts them at the center of living systems. This chapter shows how just four nucleotide building blocks can be strung together in DNA to provide the information for specifying all of the hundreds of thousands of proteins in cells. It traces the flow of information from DNA to RNA to proteins and describes in outline the mechanisms that allow for this information flow.

In Chapter 5 we turn directly to cells and describe their organization. Here we learn about both the architecturally complex cells with a nucleus and the simpler cells that lack one. We begin here the description of methods, emphasizing the power of microscopy and of cell separation techniques. We also consider one of the central events of cell life, the division of a cell to form two progeny cells. Chapter 6 describes the central methodology of modern cell biology, the ability to grow cells in culture. It also introduces one of the most potent probes of cellular life, viruses, which grow in cells. Viruses have much less genetic information than cells and therefore provide a stripped-down version of certain cellular events that is often particularly amenable to investigation.

We describe in Chapter 7 the revolutionary ability to manipulate DNA, which developed starting in the 1970s and gets more powerful with each passing year. "Recombinant DNA technology" is an expanding set of techniques that allows investigators to find, modify, and functionally characterize genes at will. Coupled to rapid and precise methods of determining DNA sequence and mapping genes, it has become the method of choice for understanding how genes direct the formation and properties of cells, whether they be from plants, animals or microbes. It is also at the base of a new industry, biotechnology, and of new advances in medical treatment, particularly gene therapy. We end Part I with Chapter 8, an exposition of another approach to understanding biological organization: genetics. This formal science grew up before there was any molecular understanding of information flow in biological systems. It provides mutants in individual genes that are probes we use to understand the roles of particular proteins in cellular and organismal biology. Today we can even make predefined mutants in mice, revealing the complex biology of mammals with a previously unimaginable precision. These latter two chapters set the stage for the presentation of particular aspects of cellular biology in Parts II through IV of the book.

1

The Dynamic Cell

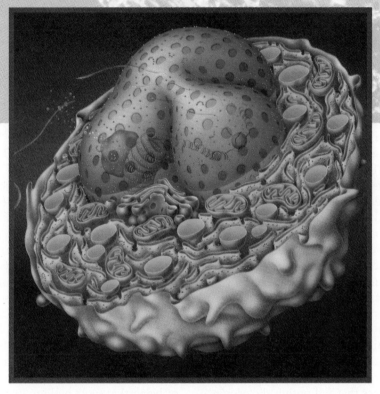

▲ An artist's rendering of a eukaryotic cell.

Once Earth cooled sufficiently, molecules that would later form parts of living beings appeared on our planet. Over millions of years, these molecules slowly joined into units able to reproduce themselves, and the first cells were born. Today there are millions of types of cells in the organisms on earth that evolved from those early beginnings. Scientists are trying to unravel the complexity they pose, hoping to understand how a surprisingly small constellation of molecules can produce such variety of structure and function. The next thousand plus pages contain the authors' attempt to systematically describe the current state of knowledge about cells. It may seem to you, our new reader, a daunting challenge. It is, however, a story of such overwhelming inventiveness and sheer beauty of construction that we trust that the joy of realization will repay the effort of concentration.

Leafing through this book, you will find biochemistry, metabolism, molecular biology and genetics, and you may wonder why these topics are in a cell biology book. To answer this question we need to develop some perspective on how biological systems are organized and how they evolved. In the end, we should see that cells, proteins and genes are so interconnected that no part of biology can be effectively studied without reference to analyses at both the molecular and cellular levels of organization. In this chapter we attempt to provide an overview of molecular cell biology and a framework for understanding the primacy of cells in the organization of biological systems. The chapter provides both a nontechnical summary of the book's material and a rationale for its organization and choice of subject matter.

Living systems, including the human body, consist of such closely interrelated elements that no single element can be fully appreciated in isolation from the others. Organisms contain organs; organs are composed of tissues; tissues consist of cells; and cells are formed from molecules, the most versatile of which are the proteins. At the top of this hierarchy of biological organization lie the genes. Genes not only specify the structure of proteins, but they define the organization of cells, direct cells to form tissues and organs, and maintain the integration of function that gives the organism its unity. Genes are nothing more than a coded representation of the elements of life, a code written in four chemical letters and displayed as a continually varying sequence in the DNA molecule. How that code is deciphered, how information is processed in living systems so that an integrated organism results, is one of the key topics of this book and of the science of molecular biology.

The unity of living systems is coordinated by many levels of interrelationships: molecules carry messages from organ to organ and cell to cell; tissues are delineated and integrated with other tissues by non-cellular membranes secreted by cells; cells gain identity from contact with other cells—generally all the levels into which we fragment biological systems interconnect. To learn about biological systems, however, we must take a segment at a time. A logical partition is to present the biology of cells as a unit, because an organism can be viewed as consisting of interacting cells, with cells being the closest thing to an autonomous biological unit we can define. Their autonomy is shown by our ability to grow dissociated cells in the laboratory in the absence of any of the normal organization in the living organism. The integration of cellular activity into tissues, the development of organisms by growth and specialization of cells, and the metabolic events fueling the dynamism of living systems are all topics on which we will touch, but they are all topics that fall within the province of other subdisciplines in the tapestry of biological science.

The notion of the cell as a central unit of biological organization is not just a didactic convenience. All organisms are made from cells and small organisms consist of single cells. Cells can live in the absence of the rest of the organism from which they were taken and thus are truly alive. Organisms only grow by the growth and division of cells. This was all realized first by Matthias Schleiden, working with plants in 1838, and a year later by Theodor Schwann, who was studying animals. Rudolf Virchow, a German pathologist enunciated the cell theory, in all its glory, in 1858:

Every animal appears as the sum of vital units, each of which bears in itself the complete characteristics of life.

Note that Virchow still clung to the idea inherited from the past that cells are "vital units," by which he meant that cells had a special property that endowed them with Life. Today we know that living systems are composed of individual chemical constituents that work closely together to allow cells to function, grow, and divide, and we no longer need think that Life is a property separate from the constituent parts of cells.

Cells were seen earlier than the nineteenth century, but it was not realized how fundamental they were. In the last half of the seventeenth century, the Dutch naturalist Antonie van Leeuwenhoek, using a simple magnifying lens, first saw that pond water contained a wide array of what he called "animalcules." (Some of Leeuwenhoek's original drawings, which include algal and other cells (Figure 1-1), have been recovered.) About the same time, Robert Hooke used the word "cells" to describe the units he saw in dead samples of cork (Figure 1-2), but it took almost two centuries of technical development of microscopes, tissue preservation, and tissue slicing before Schleiden and Schwann realized the generality of cellular organization.

▶ Evolution: Biology as a Historical Science

Biology is fundamentally different from physics and chemistry because it is a study of the results of historical events that played out over billions of years rather than the study of fundamental properties of matter. The properties of matter do limit how biological systems can work, but they do not dictate any particular system's form. The variety of physical shapes compatible with life, for instance, can be partly appreciated from the fossils of the strange creatures that came into being in the Cambrian Period, some 570 million years ago (Figure 1-3). It is the interplay of events over time that dictates today's biology, a historical process we call evolution. It was Charles Darwin's (Figure 1-4) great insight that organisms are all related in a great chain of being that gave biology its character as a historical science. The principle of interrelationship enunciated by Darwin, that organisms vary randomly and the fittest are then selected by the forces of their environment, guides biological thinking to this day. We now know that alterations in the structure and organization of genes provide the soil of random variation that nurtures evolutionary change.

Even for scientists brought up in the evolutionary tradition, it has come as a surprise to learn in recent years just how closely related are the genes of different species. Genes are so conserved that a gene appropriated from a human will sometimes function in a yeast cell and will often function in a fly cell. Clearly, one of the principles of evolution is to maintain unchanged many of the aspects of cellular life over billions of years even while great changes in external form and capability are occurring. At the molecular level of analysis, the changes fueling evolution are surprisingly subtle ones.

The creative part of the evolutionary process is adaptation to rapidly changing environments and the conquest of new environmental niches. Even as it selects alterations in cellular functions, however, evolution remains a conservative historical process in the sense that new structures are

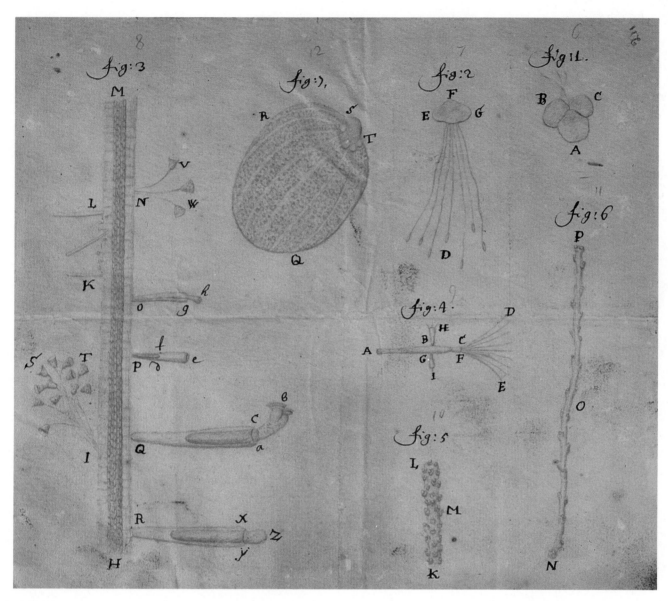

▲ FIGURE 1-1 Antonie van Leeuwenhoek's drawings represent the first look at microscopic creatures, revealing a world of microorganisms, some of which we now know are single celled. [By permission of the President and Council of the Royal Society.]

created rarely, but, more often, old structures are adapted to new circumstances. Change occurs more rapidly if previously evolved pieces can be rearranged or multiplied rather than waiting for a wholly new approach to emerge. The cellular organization of organisms plays a fundamental role in this process because it allows change to come about by small alterations in previously evolved cells, giving them new capabilities. One of the characteristics of genes that helps evolution work conservatively is that genes are fragmented into segments, each segment separated from the next by "meaningless" DNA sequence. Because these fragments can, over evolutionary time, become duplicated and rearranged into new patterns, evolution of new genes from old genetic material is facilitated by this organization.

▶ The Construction of Cells

Although generalizations in biology usually lack the theoretical underpinnings found in physics, there are very clear commonalities among living systems that give biology a unity. One is the style of cellular construction.

Cells Are Surrounded by Water-Impermeable Membranes

A cell, because it is a limited space, must have an outer border, and the construction of that border represents one of the most fundamental considerations in biological

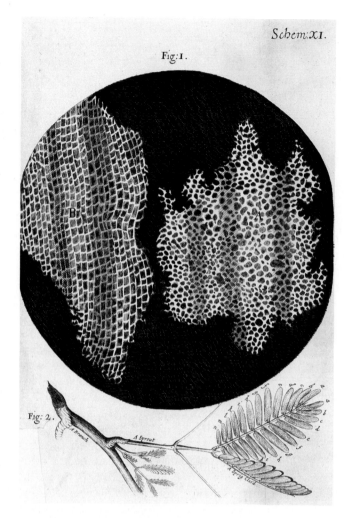

▲ FIGURE 1-2   In 1665, Robert Hooke drew the cut surface of cork that he observed through a light microscope; he called the spaces in the pattern "cells." [Courtesy of Houghton Library, Harvard University.]

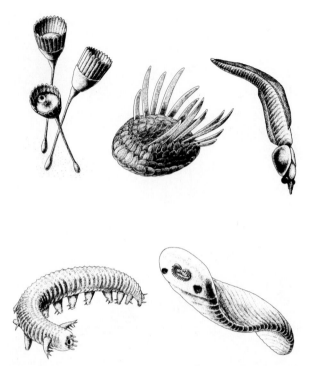

▲ FIGURE 1-3   Reconstruction of animals whose fossils were found in the Burgess Shale. The shale formed from mud in which creatures that existed 530 million years ago were embedded. All of those depicted have ceased to exist, and most of them have no obvious near relative on Earth today. They are evidence that in the Cambrian Period, starting 570 million years ago, there was the most remarkable burst of evolutionary diversification in world history. Many new types of organisms evolved, only a very few of which gave rise to the diversity of organisms on Earth today. [Reprinted from *Wonderful Life*: The Burgess Shale and the Nature of History by Stephen Jay Gould, with illustrations by Marianne Collins, with the permission of W. W. Norton & Company, Inc. Copyright © 1989 by Stephen Jay Gould.]

organization. The outer shell of cells, like any shell, is built to hold the interior contents from leaking out into the surrounding environment. The chemical processes of cellular life generally take place in a watery solution, and the intracellular constituents of cells are largely molecules that are easily dissolved in water. Similarly, the environment around cells is a watery one, the blood and other bodily fluids being solutions in water. Cells then, in order to maintain their integrity, need to be surrounded by an environment through which water cannot flow. A membrane composed of fatty molecules serves this purpose.

We all know from common experience that "oil and water don't mix." That maxim is all one needs to appreciate how a cell is constructed. When oil is poured on water, the oil spreads into a thin film, and that film is analogous to the film of fat that surrounds cells. We call the film that

limits a cell its plasma membrane. Biological membranes differ from a pure oil film in that the molecules that make the membrane have both oily and watery portions; they have long fatty chains, but they also have a head group that is water-soluble by virtue of being electrically charged. Thus membranes are formed because these bipartite molecules spontaneously orient themselves to form a double layer having a fatty interior with external surfaces bonded to the surrounding water by the charged head groups (Figure 1-5). The membrane is given rigidity by interspersion of cholesterol molecules, a molecule we have come to hate because of its association with heart disease, but one that is required to build the outer membrane of all of our cells. Hence from an understanding of the contrasting properties of watery solutions and oily layers, an understanding of cellular construction emerges.

In spite of the rigidity provided by cholesterol, membranes composed of fat are not very strong, so numerous mechanisms for strengthening the borders of cells have evolved. In plants the plasma membrane is surrounded by a rigid cell wall. Mammalian cells have proteins attached to their exterior surfaces that provide some stability, and cells are linked together through these proteins to provide integrity to tissues. Tissues and organs are often covered by strong networks of proteins and other molecules to provide rigidity and protection and to wall off the various compartments of the body. Single-celled organisms, like bacteria, have special outer coats to protect them.

The Biological World Has Two Types of Cells

There are actually two different types of cells in the biological world, cells with a nucleus and cells that lack one (Figure 1-6). The anuclear cells, called prokaryotic cells, have no internal compartments and therefore, in them, there is a seamless coupling of gene function with other events in the cell. Prokaryotic cells evolved before others and are the bacteria of today. Eukaryotic cells, those that have a nucleus, are later products of evolution, and they display much more complicated modes of gene regulation. The nucleus surrounds the cell's DNA with a specialized membrane, segregating the genes from the rest of the cell that constitutes the cytoplasm. This allows the functions of DNA to be uncoupled from events in the cytoplasm, which, in turn, allows the evolution of complex gene regulatory processes. All multicellular organisms are formed of

▲ FIGURE 1-4 Charles Darwin (1809–1882) four years after his epical voyage on *H.M.S. Beagle*. He had already begun private notebooks formulating his concept of evolution which would be published in *Origin of Species* (1859). [Courtesy of John Moss/Black Star.]

(a)

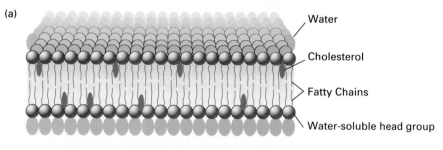

Water

Cholesterol

Fatty Chains

Water-soluble head group

(b)

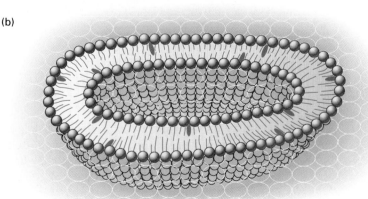

◄ FIGURE 1-5 Cells are formed with an outer fatty shell and an interior watery space. (a) The fatty molecules (lipids) align and orient themselves in a 2-layered sheet with the fat as an interior space. Cholesterol molecules provide some rigidity to the fatty layer. Each of the lipid molecules has a head group (*circular structure*) that combines easily with water. Thus both sides of the sheet are lined by water-seeking head groups, mainly charged phosphates. (b) A bilayer can fold to form an enclosed cell, with a watery space inside as well as surrounding the cell. In actuality, the interior space is much larger relative to the volume of the lipid.

(a)

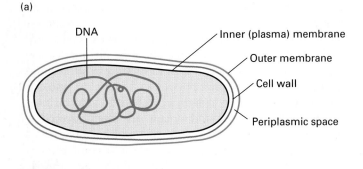

DNA

Inner (plasma) membrane

Outer membrane

Cell wall

Periplasmic space

(c)

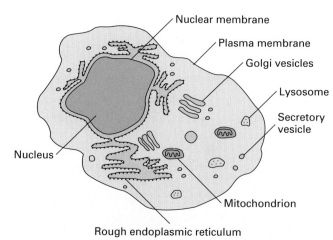

Nuclear membrane

Plasma membrane

Golgi vesicles

Lysosome

Secretory vesicle

Nucleus

Mitochondrion

Rough endoplasmic reticulum

(b)

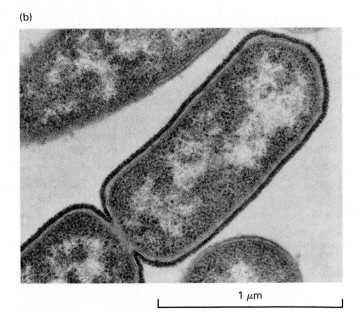

1 μm

(d)

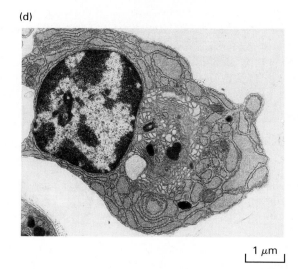

1 μm

▲ FIGURE 1-6   Comparison of the structure of prokaryotic and eukaryotic cells. (a) Diagram of a prokaryotic (bacterial) cell. (b) Photograph of a bacterial cell (*Bacteroides fragilis*) taken through an electron microscope. Around the cell there is a complex outer shell consisting of two lipid-based membranes and a cell wall. The membranes regulate entry into cells of substances needed for cell life and exit from the cell of substances produced inside it, as well as create a periplasmic space around the cell that is neither inside nor outside. The DNA inside the cell stays together in the center of the cell but is not surrounded by a membrane. (c) Diagram of a eukaryotic cell. (d) Electron micrograph of a eukaryotic cell (a plasma cell). Only a single membrane surrounds the cell, but the interior of the cell contains many compartments known as organelles. Some of their names are indicated in (c), and they will be discussed in later chapters. Notable is the nucleus where the cellular DNA is segregated within the nuclear membrane. [Part (b) courtesy of D. Chase; part (d) from P. C. Cross and K. L. Mercer, *Cell and Tissue Ultrastructure: A Functional Perspective*, W. H. Freeman and Company, New York, 1993.]

aggregated, often highly specialized, eukaryotic cells, but there are also many unicellular organisms with nuclei; yeast cells are an example.

Eukaryotic cells not only have a nucleus, but evolutionary processes have also endowed them with other intracellular compartments called organelles (see Figure 1-6c). For instance, they have lysosomes, which are membrane-limited compartments specialized for carrying out degradative processes. Each of the many types of organelles in eukaryotic cells is devoted to a particular set of functions.

Membranes Serve Functions Other Than Segregation

Although membranes are valuable as a way to segregate the watery interior of the cell from its environment, or to segregate intracellular events from one another, they have other important functions; for one thing, they make energy storage possible. Because membranes separate watery compartments from one another, if an ion or a molecule dissolved in water is moved through a membrane into a new cellular compartment, it will not be able to freely diffuse out of the compartment into which it was moved. It takes energy to move the molecule but, once moved, the molecule stores that energy by virtue of its entrapment. Formally, this storage of energy is just like the storage of energy in a battery; therefore, membranes not only delineate compartments, but also they are active participants in the cell's dynamism.

Another function of membranes is to act as anchors for molecules, primarily proteins. These proteins have functions that require membrane-binding. For instance, the passage of water-soluble molecules through membranes is carried out by transporters that are embedded in the membrane. Also, cells send information to each other by releasing signaling molecules. The outer membranes of cells have proteins, known appropriately as receptors, that bind the circulating signaling molecules. These signaling molecules allow the individual activities of the many cells in the body to be coordinated. The receipt of a signaling molecule by a receptor causes the transient organization of particular types of intracellular proteins, called signal transduction proteins, into an activated complex at the interior face of the cell's outer membrane, from which it directs alterations of events in the cell's cytoplasm or nucleus.

Eukaryotic Cells Contain Organelles Thought To Have Evolved as Independent Organisms

Another striking event that occurred during the evolution of nucleated cells was the import of bacteria to become their energy-producing mitochondria and, in plant cells, chloroplasts. In today's world these organelles of the cell are dependent on functions encoded in the nucleus. However, they retain some genetic functions of their own, so that nucleated cells today have multiple genetic systems in them: a predominant system controlled from the nucleus and secondary systems with their own DNA in the mitochondria and chloroplasts. In the organelles, the DNA is not surrounded by a nucleus, reminding us of its bacterial origin. It is quite remarkable that the energy generation mechanisms of all cells come directly from bacteria and were never separately evolved in higher cells.

► *The Molecules of Life*

The life of a cell is a complicated orchestration of many events. These include a multitude of specific chemical transformations, provision of sufficient energy, formation of organelles, movement of materials to their appointed place in the cell, and growth and division when new cells are needed. Many life-threatening diseases result from apparently small mistakes in these processes: phenylketonuria when a dietary constituent cannot be digested; cystic fibrosis when an ion channel made in the cytoplasm is unable to make its way to the cell surface; sickle cell disease

when hemoglobin precipitates inside red blood cells rather than being soluble; muscular dystrophy when a molecule that should help organize the interior of muscle cells is not functional. To keep all cells in the body in appropriate condition so that the whole organism can function requires complex and precise directions.

To appreciate how evolution has provided the directions that allow the cell to carry out its myriad activities, we need first to consider the implications of a central dichotomy of cellular life: there are two fundamentally different types of molecules in cells, small molecules and macromolecules, and they are formed in cells by two different types of synthetic processes. Small molecules are made and altered by individual steps of chemical transformation; these can be so extensive that a vegetarian diet can support the growth of an animal. Macromolecules, in contrast, are linear polymers made by linking a defined set of small molecules (monomers) together through repetitive use of a single chemical linkage step.

Small molecules and macromolecules play fundamentally different roles in the cell. The key function of small molecules is to be the substrates for making macromolecules, and the cell is careful to provide the mix of small molecules needed for macromolecular synthesis. Small molecules also do things in their own right; for instance, they store and distribute the energy for all cellular processes, and they are broken down to extract chemical energy, as when sugar is degraded to CO_2 and H_2O with the release of the energy bound up in the molecule. Small molecules can act as signals to cells; nerve cells communicate with each other by releasing and sensing small molecules, and the powerful effect on our body of a frightening event comes from the instantaneous flooding of the body with a small molecule hormone that mobilizes the "fight or flight response."

Macromolecules, though, are the most interesting and characteristic molecules of living systems; in a true sense the evolution of life as we know it is the evolution of macromolecular structures. The chief macromolecules in the cell, in terms of variety of function, are the proteins. Proteins are the do-ers of the cell; they catalyze small molecule transformations, allow cells to move and do work, and maintain internal cell rigidity. Moreover, proteins control the genes that determine cell constitution and function, move molecules across membranes, and even direct synthesis of themselves and other macromolecules. They come in many shapes and sizes (Figure 1-7), and the elucidation of their structure remains one of the most active areas of scientific investigation. Proteins are actually polymerized from only 20 types of monomers (amino acids). That such a limited set of building blocks can do so much is a continual marvel, even to those who work daily with proteins. They are the true glory of the biological world.

The macromolecule about which we most often read in the news is not protein but DNA. This molecule is extraordinary on many counts and thus easily captures the imagination of writers and even artists (Figure 1-8). Not only

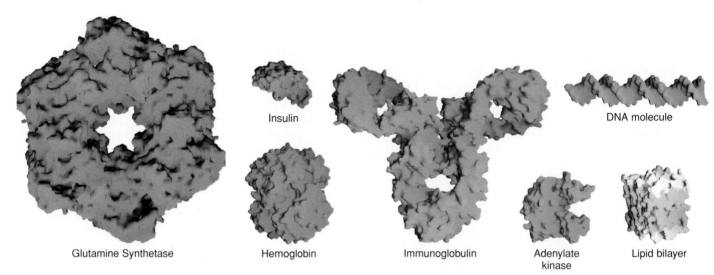

Insulin

DNA molecule

Glutamine Synthetase

Hemoglobin

Immunoglobulin

Adenylate kinase

Lipid bilayer

▲ FIGURE 1-7   Shapes and sizes of proteins. Shown is a representative set of proteins, drawn to a common scale and compared with the sizes of a lipid bilayer sheet and a DNA molecule. Each protein has a defined three-dimensional shape held together by numerous chemical bonds. The illustrated proteins include enzymes (adenylate kinase and glutamine synthetase), a hormone (insulin), an antibody (immunoglobulin), and the blood's oxygen carrier (hemoglobin). [Courtesy of Gareth White. Lipid bilayer: adapted from Rich-ard M. Venable, et al., 1993, *Science* **262**:223; Immunoglobulin G: J. Deisenhofer, 1981, *Biochemistry* **20**:2361, and H. D. Kratzin et al., (1989) Biol. Chem. Hoppe-Seyler **370**:263; glutamine synthetase: M. M. Yamashita, et al., 1989, Journal of Biological Chemistry **264**:17681; adenylate kinase: D. Dreusicke, et al., 1988, J. of Mol. Biol. **199**:359; insulin: E. N. Baker et al., 1988, *Philos. Trans. R. Soc. London* **319**:369; hemoglobin: B. Shaanan, 1983, J. Mol. Bio. **171**:31.]

does its double-helical structure make it one of nature's most magnificent constructions, but its functional properties also make it the cell's master molecule. It contains a coded representation of all of the cell's proteins; other molecules like sugars or fats are made by proteins, so their structures are indirectly coded in DNA. It also contains a coded set of instructions about when the proteins are to be made and in what quantities. DNA does all of this using only four types of monomers (nucleotides) arrayed in one to 50 or more strings called chromosomes, each containing many millions of nucleotide units.

Genetics Allowed DNA Organization to be Analyzed Before the Structure of DNA was Discovered

Since DNA is simply a long string of four different small molecules arrayed in a sequence that encodes the cell's information, extracting the information held in the molecule is a problem of daunting complexity. Even now that we can determine the sequence in long stretches of DNA, understanding the information in a piece of DNA is difficult because so many different events are encoded there. Some of DNA is devoted to encoding protein structure, but much of it is devoted to controlling when and how much of each protein is made. The fact that sequences encoding proteins are often broken up by non-coding sequences adds more complexity. Furthermore, the control instructions written in DNA can be very subtle, and we are far from under-

▲ FIGURE 1-8   An artist's rendition of a short piece of a DNA molecule. [Tom Otterness, *DNA Chain* 1989–90; courtesy of Brooke Alexander, New York. Photograph by D. James Dee.]

(a)

(b)

◄ FIGURE 1-9 (a) Gregor Mendel (1822–1884), the father of genetics. Although he published his analysis of heredity in peas in 1865, it was only rediscovered and credited as a revolutionary discovery in 1900. (b) Barbara McClintock (1902–1992) an early worker on corn genetics who provided the first microscopic evidence of chromosome breakage and reunion during recombination. Later McClintock showed that some chromosome segments are "mobile" in the genome, for which she received the Nobel Prize in 1983. [Part (a) Courtesy of the Moravian Museum, Brno; part (b) Courtesy of Nik Kleinberg/Picture Group.]

standing all of them even today. In the 1860s, however, a monk named Gregor Mendel (Figure 1-9a) gave the scientific world a tool for understanding the organization of DNA that is still used today. He realized that differences between organisms, like the different colors of pea flowers, are distributed among the offspring of a mating between two organisms in a regular way that can be understood only if the traits are determined by discrete entities, later called genes. It was learned early in the 20th century, mainly due to the work of Thomas Hunt Morgan and his school studying the fruit fly, *Drosophila melanogaster,* that genes are linked together in arrays on linear structures that would bind dyes and were therefore called chromosomes; we now know that within the chromosome is the long molecule of DNA that actually holds the genes. Later, Barbara McClintock (Figure 1-9b), working with corn, showed that genes on chromosomes can become rearranged and that individual genes can be mobile within the set of chromosomes. What Mendel defined as a gene is generally the region of DNA involved in the synthesis of one particular protein. Mendel's insight and the efforts of many others led to the science of genetics, allowing work to proceed on the organization and capability of DNA long before the discovery of the double helical structure of DNA by Watson and Crick in 1953.

Remarkably, the science of genetics developed through 50 years without any knowledge of the chemical structure of the gene. Equally remarkable, it was not until about 1940, through the efforts of George W. Beadle, Edward L. Tatum, and Boris Ephrussi working with both *Drosophila* and the fungus *Neurospora,* that the function of genes was established as the repository for the information to synthesize a protein. Actually, Archibald Garrod, in 1909, had realized that humans who inherit mutant genes have particular defects in defined enzymes (proteins that catalyze chemical reactions), but it was too early in the evolution of the field for the link between genes and enzymes to be recognized as a principle. Through the period when the structure of neither genes nor their products was understood, the science developed extraordinary depth and strength, giving testament to the remarkable powers of abstract thought of the early geneticists.

The Ultimate Triumph of Genetics Is the Human Genome Project

Today there is a productive interplay between genetics, which views genes as abstract units along a string, and molecular biology, which views genes as chemical entities. Nowhere is that interplay more evident than in the Human Genome Project, one of humankind's most ambitious and exciting scientific undertakings. The Human Genome Project is a commitment by the American scientific community, in partnership with the government, to determine the location and structure of all of the genes in the human chromosomes (We call the total nuclear DNA complement of an organism its genome). This project will lead, perhaps early in the twenty-first century, to a knowledge of the complete sequence of all of the DNA in the human genome. The American effort is part of an international effort coordinated by the Human Genome Organization. Because human genetics is particularly difficult, due to the obvious inability to control breeding, and because so much of our knowledge of even human biology comes from the study of non-human systems, the Human Genome Project includes extensive work on the DNA of many organisms of experimental interest. For each organism, scientists are making higher and higher resolution maps, locating defined pieces of DNA relative to one another. In 1994 the maps of the human chromosomes already had thousands of recognizable pieces of DNA mapped relative to one another (see Figure 1-10).

The goals of the Human Genome Project are multiple, and they illustrate how biology today has evolved into a science that is moving ahead both conceptually and practically. The cell biologist of tomorrow will work with a complete catalog of the genes in an organism of interest, allowing the set of genes making proteins relevant to a particular cellular event to be reliably predicted and their structure to be immediately known. Also, the Project already has moved from a focus on genes of experimental organisms and humans to mapping genes in organisms of commercial or health interest, like tomatoes or the bacterium that causes tuberculosis. Maps of human DNA allow us to locate genes of medical importance, those, for example, that cause specific diseases such as cystic fibrosis. Understanding the genomes of bacteria might facilitate identification of targets for which chemists could design new antibiotics, a particular hope with tuberculosis, a disease that is now often resistant to all known therapeutic agents. Maps of plants could help us locate genes that underlie important commercial traits, like fruit weight, that could then be incorporated into more productive plants. Soon the Human Genome Project should allow scientists to identify genes that are responsible for complicated human problems such as obesity or heart disease.

The Human Genome Project is a testament to the power that modern biology has achieved both as a theoretical and a practical science. Because the Project is focused on human genes, however, its success raises difficult issues for the scientific community and the world at large about how much we want to understand our own genetic constitution when that information might be used, for instance, to limit our access to insurance or employment. The Human Genome Project is going to show that each of us, no matter how healthy we seem, harbors deleterious genes, so that we all run the risk of discrimination. Realizing the potential of modern biology while defining and limiting the inappropriate use of the information is a challenge to both our science and our society.

RNA Is the Molecule That Is the Product of the Genome

The Human Genome Project will bring us detailed knowledge of human DNA and will provide a great resource for experimental biologists. However, it will also bring to biology a formidable challenge: to fit all of the human genes into a framework within which their particular functions can be understood. Genes actually do not specifically direct synthesis of proteins; they do it through an intermediate. Genes are the template for the synthesis of RNA, a molecule chemically very similar to DNA, but one that serves functions very different from those of DNA. Some RNA molecules are products of genes that do not encode proteins and directly serve roles similar to those of proteins: catalyzing chemical reactions or providing a structural scaffold for building multi-protein aggregates. RNA's most important role, however, is to act as a template for the synthesis of specific proteins, leading us to the so-called Central Dogma of biology: DNA specifies RNA, which specifies protein.

Using computer language, biologists often say that the synthesis of RNA represents a "read-out" of the genome. Each cell of the body reads out a subset of the information in the genome, providing itself with the set of RNAs and proteins it needs to fulfill its particular role in the body. Because our nervous system is our most complex part, it uses a larger fraction of the total information in the genome than does any other tissue type. Selective RNA synthesis is perhaps the most fundamental activity of the cell. So the Human Genome Project will give us all the genes, but until we understand cell-specific RNA synthesis, as well as the properties of the encoded proteins, we will not be able to appreciate how the DNA actually makes a cell.

► Cells Have Both a Fixed Identity and an Ability to Change

A cell in an adult organism can be viewed as a steady-state system. The DNA is constantly being read out into a particular constellation of RNAs, and these are specifying a particular set of proteins. As these proteins function, they

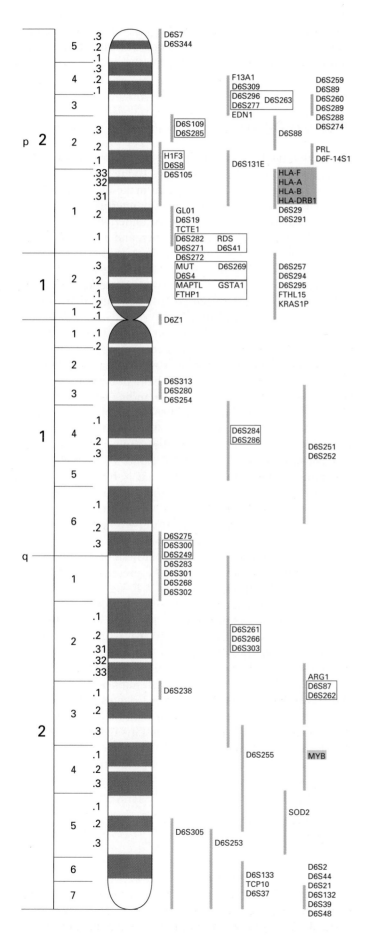

◄ FIGURE 1-10   Diagram of human chromosome 6. The chromosome itself is indicated on the left and segments where certain genes have been mapped are indicated by vertical lines (orange) at the right. The purple bands on the chromosome indicate portions that can be stained darkly by particular dyes. The code at the left of the chromosome is used to designate regions, with a *p* indicating the short arm and a *q* indicating the long arm of the chromosome. Thus the *HLA region* (green) is between 6p22.1 and 6p21.2, with 6p22.1 interpreted as chromosome 6, short arm (p), subregion 2, band 2.1. The codes to the right of the vertical lines indicate the location of two types of genes, those whose biological meaning is known and those that are simply pieces of DNA, parts of whose sequences have been determined (mostly designated with a *D* followed by a number). For instance, the HLA genes were defined by antibodies that react with the particular proteins encoded by genes in that region. Boxed items indicate those genes whose location, but not their exact order, is known. Many more proteins whose genes map to that region are known, as we will see when we get to Chapter 27, where the equivalent region of the mouse genome is shown at much higher resolution in Figure 27-33. Also indicated is MYB (yellow), a gene first recognized by its ability to cause cancer (see Chapter 26). [From Andreas Ziegler and Harry T. Orr, 1993, *Genome Digest* (HUGO Europe) **1**(2):3.]

are also being degraded and replaced by new ones, and the system is balanced so that the cell neither grows, shrinks, nor changes its functions. This static view of the cell, however, misses three important perspectives on cellular life. One is that cells do change their functions over time. For instance, when you eat a meal, the cells in your liver modify themselves to deal with the particular chemical constituents of the food you have ingested. When a pathway in the brain is strongly stimulated, those brain cells respond by making specific new RNAs encoding proteins that modify the cells, probably as part of the consolidation of a memory trace. Thus a full understanding of cell biology requires both an appreciation for how the genome's readout is set and for how it is altered by the environment.

A second aspect of cellular dynamics is that a nongrowing cell can begin to grow, preparatory to its division into two cells. Cell growth and division may be a response to an external regulatory signal, or it may come from relief of an internal cellular inhibitor. The growth process of a cell is organized into a cycle of events that has many check points used to determine if the various parts of the cells are duplicating in an orderly fashion. Whether a given cell will grow and divide is a highly regulated decision of the body, assuring that the adult replaces worn out cells or makes more cells in response to a new need. Examples are the growth of muscle in response to exercise or damage, and the proliferation of red blood cells when a person ascends to a higher altitude and needs more capacity to capture oxygen. There is, however, one major and devastating disease of cells, cancer, where cells multiply even though they are not needed by the body.

The third type of cellular change is most dramatic and complicated: the cellular growth and differentiation that occurs during the development of an organism from a fertilized egg. This is a process of extensive cell multiplication and cellular change. A mammal that starts as one cell becomes an organism with hundreds of diverse cell types such as muscle, nerve, skin and hair. Here we see at its most dramatic the power of DNA to control cellular behavior because development is an orchestrated set of easily tens of thousands of cellular changes that occurs virtually without fail. The almost perfect resemblance of 'identical' twins (Figure 1-11) is a testament to the power of DNA to reproducibly direct the development of a human being.

Nowhere is the variety of cellular activities and responses better illustrated than in the body's immune system. It is there that many cell types come together in organized tissues specifically designed to allow the body to distinguish its own cells from those of foreign invaders. Within the immune system we see specialized cells develop that can recognize invading cells; we see the formation of tissues from cells that originate in various parts of the body; we see cells actively surveying their environment with surface receptor proteins like antibodies, and we see cells change their properties as a consequence of their reaction to a foreign substance, allowing the body to rid itself of invaders.

► *Molecular Cell Biology*

Classically, the field of cell biology has been largely a matter of morphology and physiology. Today, with our deep understanding of the molecules that make up cells, and of how cells process information, we can interpret pictures of cells as the visual representation of underlying molecular events. We fuse biochemistry, genetics, and molecular biology with morphology and physiology to produce a dynamic understanding of cellular life, seeing the cell as a complex, changing entity shaped by the interaction of its DNA with both its external environment and that of the organism in which it resides.

Developing today's scientific understanding of cellular complexity and dynamism has been a long and difficult process, involving the work of many thousands of scientists over the last century and a half. Because biology is such a historical science, the understanding we have at any one time is largely a function of the tools of analysis that are available. Presenting the science of biology without a clear description of how the understanding was achieved robs the science of its essence and paints a false picture of the status of our knowledge. The facts in biology are only as good as the tools available for investigation, and as those tools become more effective, old concepts are sometimes totally reformulated; therefore, we need to present information in an experimental context.

The chapters that follow will put flesh on the bones of this introductory scan. At times the presentation will involve discussions of the chemistry of living systems or the physics of measuring devices. At times it will involve very detailed discussion of particular issues, with little reference to the wider issues of cell biology. Be assured that each topic in this book presents an important perspective on cell biology and that the whole tapestry will come into focus as the individual elements are mastered. In particular, the authors hope that the notion of knowledge as a product of experimental investigation will be continually reinforced so that the reader will see biology as a living science, one in which changing knowledge continually generates fresh perspectives and fresh opportunities for productive impacts on our society.

▲ FIGURE 1-11   Pair of identical twins. [© Bob Sacha]

2 Chemical Foundations

T he aim of molecular cell biology is an explanation of the structure and function of organisms and cells in terms of the properties of individual molecules, such as proteins and nucleic acids. This chapter is a review of many important chemical concepts required to comprehend the workings of a cell. All of the topics addressed here relate to important concepts and experiments presented later in the book.

We begin with a discussion of energy and its transformations. We then review covalent bonds, which connect individual atoms in a molecule, and the structures of carbohydrates, which illustrate the importance of relatively small differences in the arrangements of covalent bonds. Non-covalent bonds are important stabilizing forces between groups of atoms within larger molecules, and between different molecules; thus we discuss the most important ones: hydrogen and hydrophobic bonds, ionic interactions, and van der Waals interactions. Because water is the major constituent of cells and of the spaces

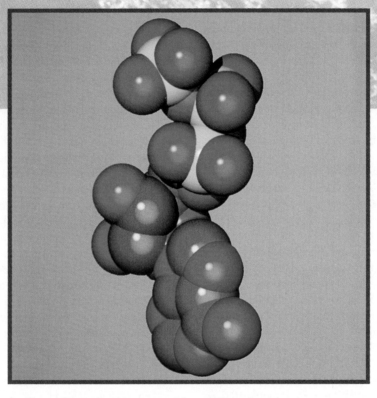

▲ Three-dimensional structure of an ATP molecule.

between cells, we discuss its structure and its interactions with various molecules. The aqueous environment found inside a cell is separated from that outside the cell by a membrane, built of a bilayer of phospholipids. Such membranes, which also define the perimeter of eukaryotic organelles, are described next. We then discuss chemical equilibrium and the reactions of chemicals dissolved in water, in particular, the pH of a solution, its buffering capacity, and the reactions of acids and bases. We then describe some important concepts for understanding the direction of reactions: free energy, entropy, and enthalpy. We describe how an unfavorable chemical reaction can proceed if it is coupled with an energetically favorable one, and the central role of ATP (adenosine triphosphate) in capturing and transferring energy in cellular metabolism. After discussing oxidation and reduction reactions and their associated free energy changes, we describe the factors that control the rates of chemical reactions, laying the groundwork for later chapters on cellular metabolism.

➤ *Energy*

The production of energy, its storage, and its use are as central to the economy of the cell as they are to the management of the world's resources. Cells require energy to do all of their work, including the synthesis of glucose from carbon dioxide and water in photosynthesis, the contraction of muscles, and the replication of DNA.

Energy may be defined as the ability to do work—a concept that is easy to grasp when it is applied to automobile engines and electric power plants. When we consider the energy associated with chemical bonds and chemical reactions within cells, however, the concept of work becomes less intuitive.

There are two principal forms of energy: kinetic and potential. *Kinetic energy* is the energy of movement—the motion of a car, for example, or the motion of molecules. The second form of energy, *potential energy,* or stored energy, is more important in the study of biological or chemical systems.

Heat, or *thermal energy,* is a form of kinetic energy—the energy of the motion of molecules. For heat to do work, it must flow from a region of higher temperature—where the average speed of molecular motion is greater—to one of lower temperature. Differences in temperature often exist between the internal and external environments of cells; however, cells generally cannot harness these heat differentials to do work. Even in warm-blooded animals that have evolved a mechanism for thermoregulation, the kinetic energy of molecules is used chiefly to maintain constant organismic temperatures.

Radiant energy is the kinetic energy of photons, or waves of light, and is critical to biology. Radiant energy can be converted to thermal energy, for instance when light is absorbed by molecules and the energy is converted to molecular motion. In the process of *photosynthesis,* light energy is absorbed by chlorophyll and is ultimately converted into other types of energy, such as that stored in chemical bonds.

One of the major forms of *electric energy* is also kinetic—the energy of moving electrons or other charged particles.

Several forms of *potential energy* are biologically significant. Central to biology is the potential energy stored in the bonds connecting atoms in molecules. Indeed, most of the biochemical reactions described in this book involve the making or breaking of at least one chemical bond. We recognize this energy when chemicals undergo energy-releasing reactions. The sugar glucose, for example, is high in potential energy. Cells degrade glucose continually, and the energy released when glucose is metabolized is harnessed to do many kinds of work.

A second biologically important form of potential energy, to which we shall refer often, is the energy in a *concentration gradient.* When the concentration of a substance on one side of a permeable barrier, such as a membrane, is different from that on the other side, the result is a concentration gradient. All cells form concentration gradients between their interior and the external fluids by selectively exchanging nutrients, waste products, and ions with their surroundings. Also, compartments within cells frequently contain different concentrations of ions and other molecules.

A third form of potential energy in cells is an *electric potential*—the energy of charge separation. For instance, there is an electric field of ~200,000 volts per cm across the outer, or "plasma" membrane of virtually all cells.

All forms of energy—both in inanimate objects and in the living world—are *interconvertible,* in accordance with the first law of thermodynamics, which states that *energy is neither created nor destroyed, but can be converted from one form to another.*[*] In photosynthesis, for example, as we have just seen, the radiant energy of light is transformed into the chemical potential energy of the bonds between the atoms in a sucrose or starch molecule. In muscles and nerves, such chemical potential energy is transformed, respectively, into kinetic and electric energy. In all cells, chemical potential energy, released by breakage of certain chemical bonds, is used to generate potential energy in the form of concentration and electric potential gradients. Similarly, energy stored in chemical concentration gradients or electric potential gradients is used to synthesize chemical bonds, or to transport other molecules "uphill" against a concentration gradient. This latter process occurs during the transport of nutrients such as glucose into certain cells and of many waste products out of cells.

Because all forms of energy are interconvertible, they can be expressed in the same units of measurement—namely, the calorie (cal) or the kilocalorie (1 kcal = 1000 cal).[†]

➤ *Covalent Bonds*

There are two kinds of associations between atoms and molecules: strong, *covalent bonds* and weaker, *noncovalent bonds.* Covalent bonds hold the atoms within an individual molecule together. Noncovalent bonds determine the three-dimensional architecture of large biological molecules and molecular complexes through cooperativity: al-

---

[*] The exceptions to this rule are nuclear reactions, in which mass is converted to energy.

[†] A *calorie* (cal) is defined as the amount of thermal energy required to heat 1 cm^3 of water by 1°C from 14°C to 15°C. Many biochemistry textbooks use the *joule* (J) instead: 1 cal = 4.184 J, so the two measures can be interconverted quite readily. The energy changes in chemical reactions, such as the making or breaking of chemical bonds, are measured in *kilocalories per mole* (kcal/mol) in this book. One mole of any substance is the amount that contains 6.02×10^{23} items of that substance, which is known as *Avogadro's number.* Thus, one can speak of a mole of photons, or 6.02×10^{23} photons. The weight of a mole of a substance in grams (g) is the same as its *molecular weight.* For example, the molecular weight of water is 18, so the weight of 1 mole of water is 18 g.

though no one noncovalent bond is strong, the effect of many weak bonds functioning together can be very powerful. Noncovalent bonds are more easily broken, making them the basis of many dynamic biological processes.

Each Atom Can Make a Defined Number of Covalent Bonds

Electrons move around the nucleus of an atom in clouds called *orbitals*, which lie in a series of concentric *shells*, or energy levels; electrons in outer shells have more energy than those in inner shells. Each shell has a maximum number of electrons that it can hold.

Electrons fill the innermost shells of an atom first; then the outer shells. The energy level of an atom is lowest when all of its orbitals are filled, and an atom's reactivity depends on how many electrons it needs to complete its outermost orbital. In most cases, in order to fill the outermost orbital, the electrons within it form *covalent bonds* with other atoms. In a covalent bond, two atoms are held close together because electrons in their outermost orbitals are shared by both atoms.

Most of the molecules in living systems contain only six different atoms: hydrogen, carbon, nitrogen, phosphorus, oxygen, and sulfur. The outermost orbital of each atom has a characteristic number of electrons:

$$ \dot{H} \qquad \cdot \dot{C} \cdot \qquad \cdot \ddot{N} \cdot \qquad \cdot \ddot{P} \cdot \qquad \cdot \ddot{O} \cdot \qquad \cdot \ddot{S} \cdot $$

These atoms readily form covalent bonds with other atoms and rarely exist as isolated entities. As a rule, each type of atom forms a characteristic number of covalent bonds with other atoms. A hydrogen atom, with one electron in its outer shell, forms only one bond, such that its outermost orbital becomes filled with two electrons. A carbon atom has four electrons in its outermost orbitals; it usually forms four bonds, as in methane (CH_4), in order to fill its outermost orbital with eight electrons:

$$ H : \overset{\displaystyle H}{\underset{\displaystyle H}{\ddot{C}}} : H \quad \text{or} \quad H - \overset{\displaystyle H}{\underset{\displaystyle H}{\overset{|}{\underset{|}{C}}}} - H $$

Nitrogen and phosphorus each have five electrons in their outer shells, which can hold up to eight electrons. Nitrogen atoms can form up to four covalent bonds. In ammonia (NH_3), the nitrogen atom forms three covalent bonds; one pair of electrons (the two dots on the right) around the atom are in an orbital not involved in a covalent bond:

$$ H : \overset{\displaystyle H}{\underset{\displaystyle H}{\ddot{N}}} : \quad \text{or} \quad H - \overset{\displaystyle H}{\underset{\displaystyle H}{\overset{|}{\underset{|}{N}}}} : $$

In the ammonium ion (NH_4^+), the nitrogen atom forms four covalent bonds, again filling the outermost orbital with eight electrons:

$$ H : \overset{\displaystyle H}{\underset{\displaystyle H}{\ddot{N}^+}} : H $$

Phosphorus can form up to five covalent bonds, as in phosphoric acid (H_3PO_4). The H_3PO_4 molecule is actually a "resonance hybrid," a structure between the two forms shown, in which the phosphorus has four or five covalent bonds with four oxygen atoms. Again, the nonbonding electrons are shown as pairs of dots:

$$ H - O - \overset{\displaystyle \overset{H}{|} \; \overset{:O:}{|}}{\underset{\displaystyle \underset{||}{O}}{P}} - O - H \quad \longleftrightarrow \quad H - O - \overset{\displaystyle \overset{H}{|} \; \overset{:O:}{|}}{\underset{\displaystyle \underset{|}{:O^-:}}{\overset{+}{P}}} - O - H $$

Esters of phosphoric acid form the backbone of nucleic acids, as discussed in Chapter 4; phosphates play key roles in cellular energetics, as discussed in Chapters 17 and 18, and in the regulation of cell function (Chapters 20 and 25).

The difference between the bonding patterns of nitrogen and phosphorus is primarily due to the relative sizes of the two atoms: the smaller nitrogen atom has only enough space for four bonding pairs of electrons to be arranged without creating destructive repulsions between them, while the larger sphere of the phosphorus atom allows more electron pairs to be arranged around it without the pairs being too close together.

Both oxygen and sulfur contain six electrons in their outermost orbitals. However, an atom of oxygen usually forms only two covalent bonds, as in molecular oxygen, O_2:

$$ \ddot{O} : : \ddot{O} \qquad \text{or} \qquad \ddot{O} = \ddot{O} $$

Primarily because its outermost orbital is larger than that of oxygen, sulfur can form as few as two covalent bonds, as in hydrogen sulfide (H_2S), or as many as six, as in sulfur trioxide (SO_3) or sulfuric acid (H_2SO_4):

$$ H - \ddot{S} : \qquad \ddot{O} = \overset{\displaystyle \overset{..O:}{}}{\underset{\displaystyle O:}{S}} \qquad H - O - \overset{\displaystyle \overset{O}{||}}{\underset{\displaystyle \underset{||}{O}}{S}} - O - H $$

Esters of sulfuric acid are important constituents of the proteoglycans (Chapter 24) that comprise part of the matrix surrounding most animal cells.

The Making or Breaking of Covalent Bonds Involves Large Energy Changes

Covalent bonds are very stable links between atoms; the energy required to break a covalent bond is much greater than the thermal energy available at room or body temperatures (25°C or 37°C). Thermal energy at 25°C is less than 1 kcal/mol (kilocalorie per mole) while, for instance, the energy required to break a C–C bond in ethane is about 83 kcal/mol:

$$H_3C : CH_3 \longrightarrow H_3C \cdot + \cdot CH_3 \qquad \Delta E = +83 \text{ kcal/mol}$$

where ΔE represents the difference between the total energy of all of the bonds in the reactants and in the products. The positive value indicates that an input of energy is needed to cause the reaction, and that the products contain more energy than the reactants. The high energy needed for breakage of the ethane bond means that, at room temperature (25°C), well under 1 in 10^{12} ethane molecules exists as a pair of $\cdot CH_3$ radicals. The covalent bonds in biological molecules (Table 2-1) have ΔE values similar to that of the C–C bond in ethane.

Chemical changes will tend to take place when the energy needed to break bonds is supplied by some other source, including energy released by the formation of other chemical bonds. In reactions where covalent bonds are made and broken, the differences in energy between reactants and products typically range from 1 to 20 kcal/mol (much lower than the energies listed in Table 2-1). In the course of a reaction, though, molecules usually adopt temporary states of much higher energy, called *transition states* or *activated complexes*, and the high energy requirement for achieving these states, often in the range of 100 kcal/mol, means that many chemical reactions, even though favorable in terms of the energy differential between reactants and products, will not occur over a measurable time scale. That is the reason why the covalent structures of molecules remain intact over very long periods of time. This high energy input needed before the bonds will break is called the *activation energy* and is discussed in detail near the end of this chapter. Specific biological catalysts—the enzymes—act by reducing the activation energy requirements of particular reactions and thus channel chemical reactions in specific directions.

Covalent Bonds Exhibit Precise Orientations

When two or more atoms form covalent bonds with another central atom, these bonds are oriented at precise angles to one another. The angles are determined by the mutual repulsion of the outer electron orbitals of the central atom. In methane, for example, the central carbon atom is bonded to four hydrogen atoms whose positions define the four points of a tetrahedron, so that the angle between any two bonds is 109.5° (Figure 2-1). Here, each bond is a *single bond*—a single pair of electrons shared between carbon and a hydrogen atom. When two atoms share two

TABLE 2-1 The Energy Required to Break Some Covalent Bonds Important in Biological Systems*			
Type of Bond	Energy (kcal/mol)	Type of Bond	Energy (kcal/mol)
SINGLE BOND		DOUBLE BOND	
O—H	110	C=O	170
H—H	104	C=N	147
P—O	100	C=C	146
C—H	99	P=O	120
C—O	84		
C—C	83	TRIPLE BOND	
S—H	81	C≡C	195
C—N	70		
C—S	62		
N—O	53		
S—S	51		

*Note that double and triple bonds are stronger than single bonds.

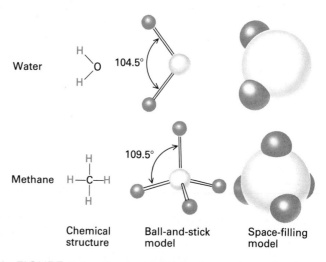

Water   104.5°

Methane   109.5°

Chemical structure   Ball-and-stick model   Space-filling model

▲ FIGURE 2-1   Bond orientations in a water molecule and in a methane molecule. Each molecule is represented in three ways. The atoms in the ball-and-stick models are smaller than they actually are in relation to bond length, to show the bond angles clearly. The sizes of the electron clouds in the space-filling models are more accurate.

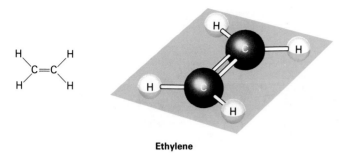

Ethylene

▲ FIGURE 2-2   In an ethylene molecule, the carbon atoms are connected by a double bond, causing all the atoms to lie in the same plane. (Atoms connected by a single bond can rotate freely about the bond axis; those connected by a double bond cannot.)

pairs of electrons—for example, when a carbon atom is linked to only three other atoms, the bond is a *double bond*:

$$\begin{array}{c} \diagdown \\ \diagup \end{array} C=$$

In this case, the carbon atom and all three atoms linked to it lie in the same plane (Figure 2-2). Atoms connected by a double bond cannot rotate freely about the bond axis. The rigid planarity imposed by double bonds has enormous significance for the shape of large biological molecules such as proteins and nucleic acids. (In *triple bonds,* two atoms share six electrons. These are rare in biological molecules.)

The outer electron orbitals not involved in covalent bond formation also contribute to molecular configurations. For example, the outer shell of the oxygen atom has two such pairs of electrons. In the water molecule, the orbitals of these nonbonding electrons have a specific spatial orientation and a high electron density that compresses the angle between the covalent H–O–H bonds to 104.5° (Figure 2-1).

Polar Covalent Bonds Result from Unequal Sharing of Electrons

In a covalent bond, one or more pairs of electrons are shared between two atoms. In certain cases, the bonded atoms exert different attractions for the electrons of the bond, resulting in unequal sharing of the electrons. The power of an atom in a molecule to attract electrons to itself, called *electronegativity,* is measured on a scale from 4.0 (for fluorine, the most electronegative atom) to a hypothetical zero (Table 2-2). In a covalent bond in which the atoms either are identical or have the same electronegativity, the bonding electrons are shared equally. Such a bond is said to be *nonpolar.* This is the case for C–C as well as for C–H bonds. However, if two atoms differ in electronegativity, the bond is said to be *polar;* one end will be slightly negatively charged (δ^-), and the other end will be slightly positively charged (δ^+). In a water molecule, for example, the oxygen atom, with an electronegativity of 3.5, attracts the bonded electrons more than do the hydrogen atoms, each of which has an electronegativity of 2.1; that is, the bonding electrons spend more time around the oxygen atom than around the hydrogens. Because both hydrogen atoms are on the same side of the oxygen atom, that side of the molecule has a slight positive charge, whereas the other side has a slight negative charge. A molecule such as water that has a net separation of positive and negative charges is said to have a *dipole moment.* (Figure 2-3). Some molecules, such as CO_2, have two polar bonds:

$$O^{\delta-}=C^{\delta+}=O^{\delta-}$$

TABLE 2-2   The Electronegativities of Some Atoms Important in Biological Systems

Element	Electronegativity
Fluorine (F)	4.0 (most electronegative)
Oxygen (O)	3.5
Chlorine (Cl)	3.0
Nitrogen (N)	3.0
Bromine (Br)	2.8
Sulfur (S)	2.5
Carbon (C)	2.5
Iodine (I)	2.5
Phosphorus (P)	2.1
Hydrogen (H)	2.1
Magnesium (Mg)	1.2
Calcium (Ca)	1.0
Lithium (Li)	1.0
Sodium (Na)	0.9
Potassium (K)	0.8 (least electronegative)

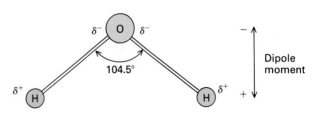

▲ FIGURE 2-3   The water molecule is polar and has a dipole moment. The symbol δ represents a partial charge (a weaker charge than the one on an electron or a proton) and thus the H–O bonds are polar. The size and direction of the partial separation of the positive and negative charges in the molecule is termed the dipole moment.

in which the polarities of the individual bonds cancel each other out, resulting in a molecule without a dipole moment.

► *Asymmetric Carbon Atoms and the Structure of Amino Acids and Carbohydrates*

A carbon (or any other) atom bonded to four dissimilar atoms is *asymmetric:* the bonds can be arranged in three-dimensional space in two different ways, producing molecules called *stereoisomers,* or *optical isomers,* that are mirror images of each other. One isomer is said to be right-handed and the other left-handed, a property called *chirality.* Most molecules in cells contain at least one asymmetric carbon atom, often called a *chiral carbon* atom. The different stereoisomers of a molecule usually have completely different biological activities.

The α Carbon in Amino Acids Is a Chiral Carbon

Almost all amino acids, the building blocks of the proteins (Chapter 3), have one asymmetric carbon atom, called the *α carbon,* or C_α, which is bonded to four different atoms or groups of atoms. In the amino acid alanine, for instance (Figure 2-4), this carbon atom is bonded to $-NH_2$, $-COOH$, $-H$, and $-CH_3$. By convention, the two mirror-image structures are called the D (*dextro*) and the L (*levo*) isomers of the amino acid (see Figure 2-4). The two isomers cannot be interconverted without breaking a chemical bond. With rare exceptions, only the L forms of amino acids are found in proteins.

Chiral Carbons Influence the Three-Dimensional Structure of Carbohydrates

The three-dimensional structures of carbohydrates provide excellent examples of the structural and biological importance of chiral carbon atoms, even in simple molecules. A carbohydrate is constructed of carbon (*carbo-*) and hydrogen and oxygen (*-hydrate,* or water). The formula for the simplest carbohydrates—the *monosaccharides,* or *simple sugars*—is $(CH_2O)_n$, where $n = 3, 4, 5, 6,$ or 7. All monosaccharides contain hydroxyl (–OH) groups and either an aldehyde or a keto group:

$$\underset{\textbf{Aldehyde}}{C-\overset{\overset{\textstyle O}{\|}}{C}-H} \qquad \underset{\textbf{Keto}}{C-\overset{\overset{\textstyle O}{\|}}{C}-C}$$

D-*Glucose* $(C_6H_{12}O_6)$, as we shall see later, is the principal source of energy for most cells in higher organisms.

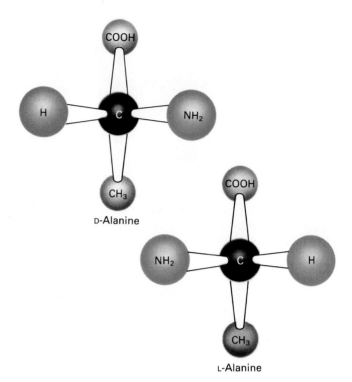

▲ FIGURE 2-4   Stereoisomers of the amino acid alanine. The α carbon is black.

Carbon atoms 2, 3, 4, and 5 in the linear form of D-glucose (Figure 2-5, top) are asymmetric, as each of these is bonded to four different atoms or groups. If the hydrogen atom and the hydroxyl group attached to carbon atom 2 (the *2 carbon*) were interchanged, the resulting molecule would be a different sugar, D-mannose, and could not be converted to glucose without breaking and making covalent bonds. Enzymes can distinguish between this single point of difference.

D-Glucose can exist in three different forms: a linear structure and two different hemiacetal ring structures (Figure 2-5). If the aldehyde group on the 1 carbon reacts with the hydroxyl group on the 5 carbon, the resulting hemiacetal, D-*glucopyranose,* contains a six-membered ring. Similarly, condensation of the hydroxyl group on the 4 carbon with the aldehyde results in the formation of D-*glucofuranose,* a hemiacetal containing a five-membered ring. Although all three forms of D-glucose exist in biological systems, the pyranose form is by far the most abundant.

When D-glucopyranose is formed from the linear molecule, carbon 1 becomes asymmetric, forming α-D-glucopyranose (shown in Figure 2-5) or the alternative β-D-glucopyranose. In the planar depiction of the pyranose ring (the *Haworth projection*) used in Figure 2-5, the –OH attached to carbon 1 "points" down (below the plane of projection) in α-D-glucopyranose, and points up (above the plane of projection) in the β form. Enzymes can also distinguish between this single point of difference, and thus the α and β forms have specific biological roles.

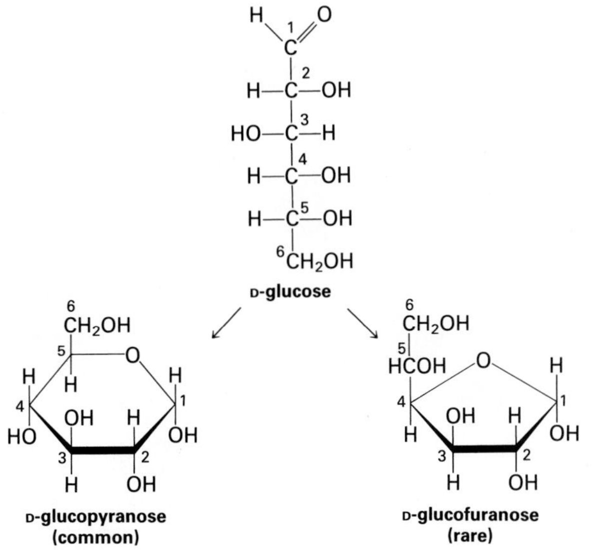

A FIGURE 2-5 Three alternative configurations of D-glucose. The ring forms are generated from the linear molecule by the reaction of the aldehyde at the 1 carbon with the hydroxyl on the 5 carbon or on the 4 carbon.

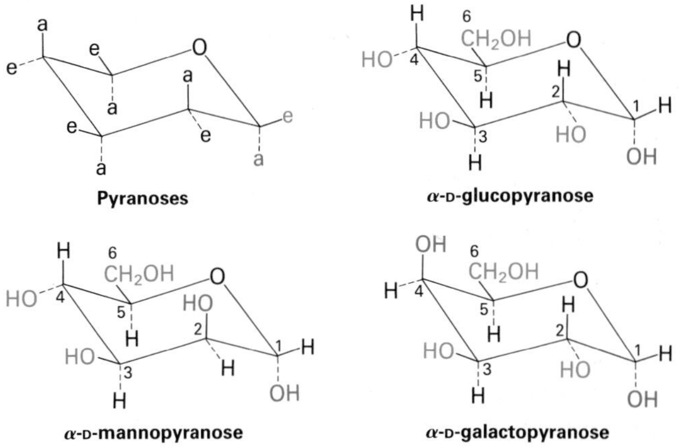

A FIGURE 2-7 Chair conformations of glucose, mannose, and galactose in their pyranose forms. The chair is the most stable conformation of a six-membered ring. (In an alternative form, called the *boat*, both carbon 1 and carbon 4 lie above the plane of the ring.) In the generalized pyranose ring at the top, a = axial atoms and e = equatorial atoms; the four bonds at each of the ring carbon atoms are tetrahedral. In α-D-glucopyranose, all of the hydroxyl groups except the one bonded to the 1 carbon are equatorial. In α-D-mannopyranose, the hydroxyl groups bonded to the 2 and 1 carbons are axial. In α-D-galactopyranose, the hydroxyl groups bonded to the 1 and 4 carbons are axial.

L-Glucose, the mirror image of D-glucose, is virtually unknown in biological systems; except for L-fucose, all sugars found in cells are of the D type. It is one of the unsolved mysteries of molecular evolution why only D forms of sugars were utilized, and not the chemically equivalent L forms.

Most biologically important sugars are six-carbon sugars, or *hexoses*, which are structurally related to D-glucose. *Mannose*, as noted, is identical to glucose except for the orientation of the substituents on the 2 carbon. If the pyranose forms of glucose and mannose are shown as Haworth projections (Figure 2-6), the hydroxyl group on the 2 carbon of glucose points downward whereas that on mannose points upward. Similarly, *galactose*, another hexose, dif-

fers from glucose only in the orientation of the hydroxyl group on the 4 carbon (see Figure 2-6).

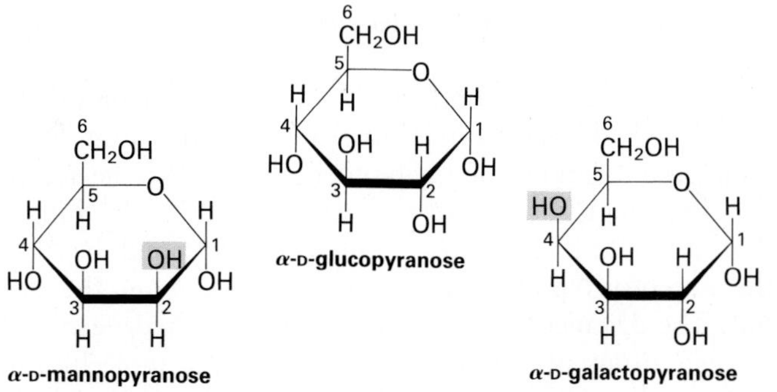

The Haworth projection is an oversimplification because the actual pyranose ring is not planar. Rather, the molecule adopts a conformation in which each of the ring carbons is at the center of a tetrahedron, just like the carbon in methane (see Figure 2-1). The preferred conformation of pyranose structures is the *chair* (Figure 2-7), in which the bonds going from a ring carbon to other atoms may take two directions: *axial* (perpendicular to the ring) and *equatorial* (in the plane of the ring).

➤ *Noncovalent Bonds and the Structures of Biological Molecules*

Carbohydrates illustrate the importance of subtle differences in covalent bonds in generating molecules with different biological activities. However, many of the forces that maintain the structures of large molecules, such as proteins and nucleic acids, are not covalent bonds. The forces that stabilize the three-dimensional architecture of individual large molecules and that bind one molecule to another are often much weaker, and thus they are sometimes called *interactions* rather than bonds. The energy released in the formation of noncovalent bonds is only

A FIGURE 2-6 Haworth projections of the structures of glucose, mannose, and galactose in their pyranose forms. The hydroxyl groups with different orientations from those of glucose are indicated in orange.

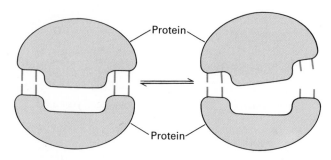

▲ FIGURE 2-8   The importance of multiple weak bonds in stabilizing an association between two large molecules. In the complex on the left, four noncovalent bonds bind two protein molecules together. Even if two of the bonds are broken, as in the complex on the right, the remaining two bonds will stabilize the structure, facilitating the re-formation of the broken bonds.

1–5 kcal/mol. Because the average kinetic energy of molecules at room temperature (25°C) is about 0.6 kcal/mol, many molecules will have enough energy to break these weak bonds. At physiological temperatures (25–37°C), weak bonds have a transient existence, but many can act together to produce highly stable structures (Figure 2-8).

Because noncovalent bonds play such crucial roles in biological structures, any student of the cell should be well acquainted with them. The four main types are the hydrogen bond, the ionic interaction, van der Waals interactions, and the hydrophobic bond. We shall consider each of these bonds in turn.

The Hydrogen Bond Underlies Water's Chemical and Biological Properties

Normally, a hydrogen atom only forms a covalent bond with one other atom. However, a hydrogen atom covalently bonded to a donor atom, D, may form an additional *hydrogen bond* as a weak association with an acceptor atom, A:

$$D^{\delta-}\!\!-\!H^{\delta+} + :A^{\delta-} \rightleftharpoons D^{\delta-}\!\!-\!H^{\delta+}\!\cdots\!A^{\delta-}$$

Hydrogen bond

Properties of Hydrogen Bonds In order for a hydrogen bond to form, the donor atom must be electronegative and thus the covalent D–H bond must be polar. The acceptor atom also must be electronegative and its outer shell must have at least one nonbonding pair of electrons that attracts the δ^+ charge of the hydrogen atom. In biological systems, both donors and acceptors are usually nitrogen or oxygen atoms, especially those atoms in amino (–NH₂) and hydroxyl (–OH) groups. Because all covalent N–H and O–H bonds are polar, their H atoms can participate in hydrogen bonds. By contrast, C–H bonds are nonpolar

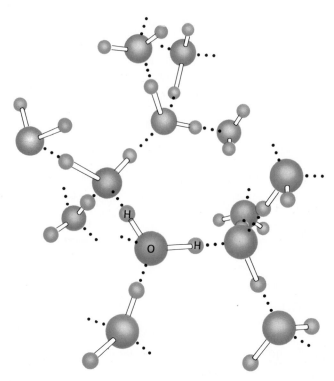

▲ FIGURE 2-9   In liquid water, each H_2O molecule apparently forms transient hydrogen bonds with several others, creating a fluid network of hydrogen-bonded molecules. The precise structure of liquid water is still not known with certainty.

and so these H atoms are never involved in a hydrogen bond.

Water molecules provide a classic example of hydrogen bonding. The hydrogen atom in one water molecule is attracted to a pair of electrons in the outer shell of an oxygen atom in an adjacent molecule (Figure 2-9). Not only do water molecules hydrogen-bond with one another, as in Figure 2-9, they also form hydrogen bonds with other kinds of molecules (Figure 2-10). The presence of hydroxyl or amino groups makes many molecules soluble in water. For instance, these groups in methanol (CH_3OH) and methylamine (CH_3NH_2) can form several hydrogen bonds with water, enabling the molecules to dissolve in it to high concentrations.

Most hydrogen bonds are 0.26–0.31 nm long (Table 2-3). The hydrogen atom is closer to the donor atom D, to which it remains covalently bonded, than it is to the acceptor. The length of the covalent D–H bond is a bit longer than it would be if there were no hydrogen bond, because the acceptor "pulls" the hydrogen away from the donor. The distance between the nuclei of the hydrogen and oxygen atoms of adjacent hydrogen-bonded molecules in water is approximately 0.27 nm, about twice the length of covalent H–O bonds in water.

Hydrogen Bonds as a Stabilizing Force An important feature of all hydrogen bonds is directionality. In the

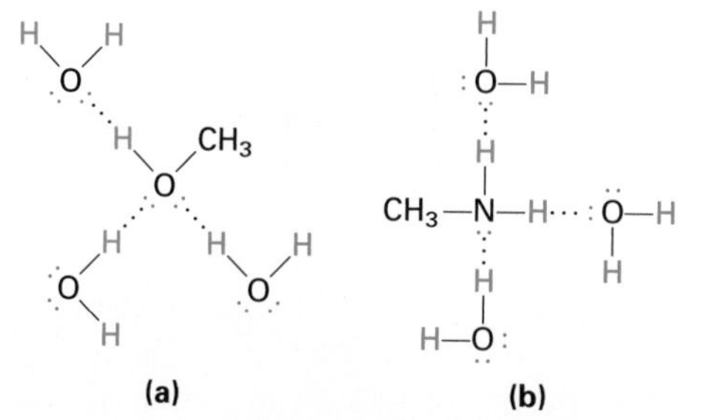

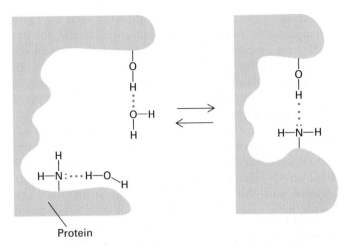

▲ FIGURE 2-10   Hydrogen bonds between (a) methanol and water and (b) methylamine and water. Each of the two pairs of nonbonding electrons in the outer shell of oxygen can accept a hydrogen atom in a hydrogen bond. Similarly, the single pair of unshared electrons in the outer shell of nitrogen is capable of becoming an acceptor in a hydrogen bond. The −OH oxygen and the −NH₂ nitrogen can also be the donor in hydrogen bonds to oxygen atoms in H₂O.

▲ FIGURE 2-11   Formation of a hydrogen bond (red dots) between a −OH and a −NH₂ group in a protein involves the disruption of existing hydrogen bonds (red dots) joining these groups to water.

strongest hydrogen bonds, the donor, the hydrogen, and the acceptor atoms all lie in a straight line. Nonlinear hydrogen bonds are weaker than linear ones; still, multiple nonlinear hydrogen bonds help to stabilize the three-dimensional structures of many proteins. Hydrogen bonds are important stabilizing forces in nucleic acids as well as in proteins, as will be detailed in Chapters 3 and 4. However, the strengths of these hydrogen bonds are only 1 or 2 kcal/mol, much weaker than the 5 kcal/mol characteristic of the hydrogen bonds between water molecules. The strength of a hydrogen bond in water, in turn, is much weaker than a covalent H–O bond. The strengths of other hydrogen bonds are similar to that of the hydrogen bond in water.

As depicted in Figure 2-11, the groups that will form a hydrogen bond in a protein will initially be hydrogen-bonded to water. The formation of the new hydrogen bond involves disruption of these hydrogen bonds with water,

and thus the net change in energy may not be much more than the 0.6 kcal/mol of available thermal energy. It is only because of the aggregate strength of multiple hydrogen bonds that they play a central role in the architecture of large biological molecules in aqueous solutions.

The Structures of Liquid Water and Ice   Hydrogen bonding between water molecules is of crucial importance because all life requires an aqueous environment and water comprises about 80 percent of the weight of most cells. The mutual attraction of its molecules causes water to have melting and boiling points at least 100°C higher than they would be if water were nonpolar; in the absence of these intermolecular attractions, water on earth would exist primarily as a gas.

Ordinary ice is a crystal. Each oxygen atom forms two covalent bonds with two hydrogen atoms; it also forms two hydrogen bonds with hydrogen atoms of adjacent molecules. The exact structure of liquid water is still unknown. It is believed to contain many icelike, maximally hydrogen-bonded networks that are presumably so transient and small that stable crystals do not form (see Figure 2-9). Most likely, water molecules are in rapid motion, constantly making and breaking hydrogen bonds with adjacent molecules. As the temperature of water increases toward 100°C, the kinetic energy of its molecules becomes greater than the energy of the hydrogen bonds connecting them, and the gaseous form of water appears.

Ionic Interactions Are Attractions between Oppositely Charged Ions

In some compounds, the bonded atoms are so different in electronegativity that the bonding electrons are never

TABLE 2-3 **Typical Hydrogen Bond Lengths***	
Bond	**Length (nm)**
OH···O⁻	0.26
OH···O	0.27
OH···N	0.28
⁺NH···O	0.29
NH···O	0.30
NH···N	0.31

*The values listed are the distances between the nuclei of the donor and the acceptor atom.

shared: these electrons are always found around the more electronegative atom. In sodium chloride (NaCl), for example, the bonding electron contributed by the sodium atom is completely transferred to the chlorine atom. Even in solid crystals of NaCl, the sodium and chlorine atoms are ionized, so it is more accurate to write the formula for the compound as Na^+Cl^-.

Because the electrons are not shared, the bonds in such compounds cannot be considered covalent. They are, rather, *ionic interactions* that result from the attraction of a positively charged ion—a *cation*—for a negatively charged ion—an *anion*. Unlike covalent or hydrogen bonds, ionic bonds do not have fixed or specific geometric orientations because the electrostatic field around an ion—its attraction for an opposite charge—is uniform in all directions. However, crystals of salts such as Na^+Cl^- do have very regular structures because that is the energetically most favorable way of packing together positive and negative ions. The force that stabilizes ionic crystals is called the *lattice energy*.

In aqueous solutions, simple ions of biological significance, such as Na^+, K^+, Ca^{2+}, Mg^{2+}, and Cl^-, do not exist as free, isolated entities. Instead, each is surrounded by a stable, tightly held shell of water molecules (Figure 2-12). An ionic interaction occurs between the ion and the oppositely charged end of the water dipole. K^+ provides one example:

$$K^+ \; {}^{\delta-}O \begin{matrix} H^{\delta+} \\ \\ H^{\delta+} \end{matrix}$$

Ions play an important biological role when they pass through narrow pores, or channels, in membranes. For example, ionic movements through membranes are essential for the conduction of nerve impulses and for the stimulation of muscle contraction. One way channels accommodate some ions and exclude others is by size, and an estimation of the size of an ion should include its shell of water molecules.

Most ionic compounds are quite soluble in water because a large amount of energy is released when ions tightly bind water molecules. This is known as the *energy of hydration*. Oppositely charged ions are shielded from one another by the water and tend not to recombine. Salts like Na^+Cl^- dissolve in water because the energy of hydration is greater than the lattice energy that stabilizes the crystal structure. In contrast, certain salts, such as $Ca_3(PO_4)_2$, are virtually insoluble in water; the large charges on the Ca^{2+} and PO_4^{3-} ions generate a formidable lattice energy that is greater than the energy of hydration.

Molecules with polar bonds also can attract water molecules, as can molecules that easily form hydrogen bonds. Such *polar molecules* can dissolve in water and are

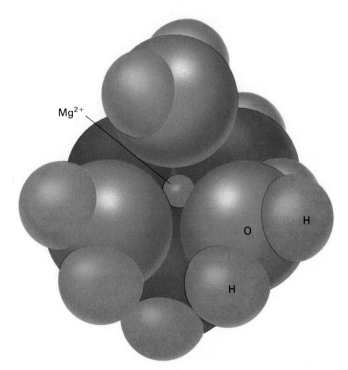

▲ FIGURE 2-12   In water, six H_2O molecules cluster around a magnesium ion (Mg^{2+}). These molecules are held tightly in place by electrostatic interactions between the positive Mg^{2+} and the partial negative charge on the oxygen of the water.

said to be *hydrophilic* (Greek, "water-loving"). Typical chemical groups that interact well with water are the hydroxyl (–OH) and amino (–NH$_2$) groups and the peptide and ester bonds:

Peptide   **Ester**

Van der Waals Interactions Are Caused by Transient Dipoles

When any two atoms approach one another closely, they create a weak, nonspecific attractive force that produces a van der Waals interaction, named for Dutch physicist Johannes Diderik van der Waals (1837–1923), who first described it. *Van der Waals interactions* are nonspecific and are caused by transient dipoles in all atoms. Momentary random fluctuations in the distribution of the electrons of any atom give rise to a transient unequal distribution of electrons—a *dipole*. If two noncovalently bonded atoms are close enough together, the transient dipole in one atom will perturb the electron cloud of the other. This perturbation generates a transient dipole in the second atom, and the two dipoles will attract each other weakly. Similarly, a polar covalent bond in one molecule will attract an oppositely oriented dipole in another.

Such interactions between atoms, carried out by either transient induced or permanent dipoles, and known collectively as van der Waals interactions, are seen in all types of molecules, both polar and nonpolar. In particular, van der Waals interactions are responsible for the cohesion among molecules of nonpolar liquids and solids, such as heptane, $CH_3-(CH_2)_5-CH_3$, that cannot form hydrogen bonds or ionic interactions with other molecules. When they are present, these stronger interactions will override most of the influence of a van der Waals interaction.

The van der Waals attraction decreases rapidly with increasing distance and is effective only when atoms are quite close to one another. However, if atoms get too close together, they become repelled by the negative charges in their outer electron shells. When the van der Waals attraction between two atoms exactly balances the repulsion between their two electron clouds, the atoms are said to be in *van der Waals contact* (Figure 2-13). Each type of atom has a *van der Waals radius* at which it is in van der Waals contact with other atoms. Two covalently bonded atoms are closer together than two atoms that are merely in van der Waals contact. Table 2-4 lists the van der Waals and covalent radii of some important atoms in biological systems.

The energy of the van der Waals interaction is about 1 kcal/mol, only slightly higher than the average thermal energy of molecules at 25°C. Thus the van der Waals interaction is even weaker than the hydrogen bond, which typi-

TABLE 2-4 Van der Waals Radii and Covalent (Single-Bond) Radii of Some Biologically Important Atoms*		
Atom	Van der Waals Radius (nm)	Covalent Radius for a Single Bond (nm)
H	0.10	0.030
O	0.14	0.074
F	0.14	0.071
N	0.15	0.073
C	0.17	0.077
S	0.18	0.103
Cl	0.18	0.099
Br	0.20	0.114
I	0.22	0.133

*The internuclear distance for a covalent bond or a van der Waals interaction is approximately the sum of the values for the two participating atoms. Note that the van der Waals radius is about twice as long as the covalent radius.

cally has an energy of 1–2 kcal/mol in aqueous solutions. The attraction between two large molecules can be appreciable, however, if they have precisely complementary shapes, so that they make many van der Waals contacts when they come into proximity. The van der Waals contacts are among the interactions that take place between an antibody molecule and its specific antigen and between many enzymes and their specifically bound substrates.

Hydrophobic Bonds Cause Nonpolar Molecules to Adhere to Each Other

Nonpolar molecules do not contain ions nor possess a dipole moment and do not become hydrated. Because they are insoluble or almost insoluble in water, they are said to be *hydrophobic* (Greek, "water-fearing"). The covalent bonds between two carbon atoms and between carbon and hydrogen atoms are the most common nonpolar bonds in biological systems. *Hydrocarbons*—molecules made up only of carbon and hydrogen—are virtually insoluble in water. The large hydrophobic molecule tristearin (Figure 2-14), a component of animal fat, is also insoluble in water, even though its six oxygen atoms participate in some slightly polar bonds between carbon and oxygen. When shaken in water, tristearin forms a separate phase similar to the separation of oil from water in an oil-and-vinegar salad dressing.

The force that causes hydrophobic molecules or nonpolar portions of molecules to aggregate together rather

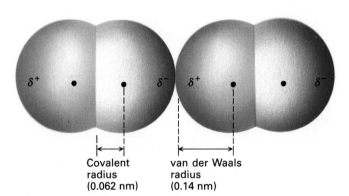

▲ FIGURE 2-13 Two oxygen molecules in van der Waals contact. They are attracted by transient dipoles in their electron clouds. Each type of atom has a van der Waals radius, at which van der Waals interactions with other atoms are optimal. Because atoms repel one another if they are close enough together for their outer electron shells to overlap, the van der Waals radius is a measure of the size of the electron cloud surrounding an atom. (Note that the covalent radius indicated here is for the double bond of O=O; the single-bond covalent radius of oxygen is slightly longer, as shown in Table 2-4.)

Covalent radius (0.062 nm)   van der Waals radius (0.14 nm)

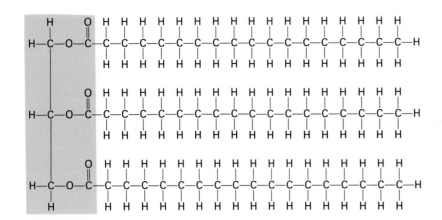

◄ FIGURE 2-14   The chemical structure of tristearin, or tristearoyl glycerol, a component of natural fats. It contains three molecules of the fatty acid stearic acid, $CH_3(CH_2)_{16}COOH$, esterified to one molecule of glycerol, $HOCH_2CH(OH)CH_2OH$. The atoms at the polar end of the molecule (blue) are hydrophilic; the rest of the molecule is highly hydrophobic.

than to dissolve in water is called the *hydrophobic bond.* This is not a separate bonding force; rather, it is the result of the energy required to insert a nonpolar molecule into water. A nonpolar molecule cannot form hydrogen bonds with water, so it distorts the usual water structure, forcing the water to make a cage of hydrogen bonds around it. Water molecules are normally in constant motion, and the formation of such cages restricts the motion of a number of water molecules; the effect is to increase the structural organization of water. This situation is energetically unfavorable; it decreases the *entropy,* or randomness, of the population of water molecules, as we will discuss further on page 34.

Water molecules are opposed to having their motion restricted by forming cages around individual hydrophobic molecules, or portions of hydrophobic molecules. This opposition is the major reason why molecules such as tristearin and heptane are essentially insoluble in water and form interactions mainly with other hydrophobic molecules. Nonpolar molecules can bond together, albeit weakly, through van der Waals interactions. The net result of the hydrophobic and van der Waals interactions is a very powerful tendency for hydrophobic molecules to interact with each other, and not with water.

Small hydrocarbons such as butane

$$CH_3—CH_2—CH_2—CH_3$$

are somewhat soluble in water [90 g/L (grams per liter) at 25°C and 1 atm (atmosphere) of pressure], because they can dissolve without disrupting the water lattice appreciably. Note, however, that 1-butanol

$$CH_3—CH_2—CH_2—CH_2OH$$

mixes completely with water in all proportions. The replacement of just one hydrogen atom with the polar –OH group allows the molecule to form hydrogen bonds with water and greatly increases its solubility.

Simply put, *like dissolves like.* Polar molecules dissolve in polar solvents such as water, while nonpolar molecules dissolve in nonpolar solvents such as hexane but not in water.

Binding Specificity Can Be Conferred by Multiple Weak Bonds

Besides contributing to the stability of large biological molecules, multiple noncovalent bonds can also confer *specificity* by determining how large molecules will fold or which regions of different molecules will bind together. All

▲ FIGURE 2-15   The binding of a hypothetical pair of proteins by two ionic bonds, one hydrogen bond, and one large combination of hydrophobic and van der Waals interactions. The structural complementarity of the surfaces of the two molecules gives rise to this particular combination of weak bonds and hence to the specificity of binding between the molecules.

weak bonds are effective only over a short range and require close contact between the reacting groups. For noncovalent bonds to form properly, there must be a complementarity between the sites on the two interacting surfaces. Figure 2-15 illustrates how several different weak bonds can bind two protein chains together. Almost any other arrangement of the same groups on the two surfaces would not allow the molecules to bind so tightly. Such multiple, specific interactions bind together the two chains of DNA (Chapter 4) and allow protein molecules to fold into a unique three-dimensional shape (Chapter 3).

► Biomembranes: Hydrophobic Sheets Separating Aqueous Compartments

Biomembranes provide another example of an important biological structure stabilized by multiple noncovalent bonds. All biological membranes have a hydrophobic core, and they all separate two aqueous solutions. The plasma membrane, for example, separates the interior of the cell from its surroundings. Similarly, the membranes that surround the organelles of eukaryotic cells separate one aqueous phase—the cell cytosol—from another—the interior of the organelle.

Although biological membranes contain proteins that give each type of membrane its unique properties, all biomembranes are built of the same kinds of phospholipid molecules linked together by multiple noncovalent interactions. The structure and function of membrane components are detailed in Chapter 14. Here, we briefly cover the chemical aspects of biomembrane phospholipids.

Phospholipids Are the Principal Components of Biomembranes

Key membrane components are *fatty acids,* molecules that contain a long hydrocarbon chain attached to a carboxyl group (–COOH). Fatty acids differ in length and in the extent and position of their double bonds. Most fatty acids in cells have 16, 18, or 20 carbon atoms and 0–3 double bonds. Fatty acids with no double bonds are said to be *saturated*; those with at least one double bond are *unsaturated*. Fatty acids are insoluble in water and salt solutions.

The lipids in membranes contain long-chain fatty acyl groups, but these are linked (usually by an ester bond) to small, highly hydrophilic groups. Consequently, membrane lipids, unlike tristearin, do not clump together in droplets but orient themselves in sheets to expose their hydrophilic ends to the aqueous environment. Molecules in which one end (the "head") interacts with water and the other end (the "tail") is hydrophobic are said to be *amphipathic* (Greek, "tolerant of both").

The tendency of amphipathic molecules to form organized structures spontaneously in water is the key to the

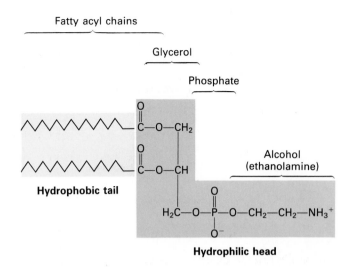

▲ FIGURE 2-16   The structure of phosphatidylethanolamine, a typical phospholipid. The hydrophobic and hydrophilic parts of the molecule are yellow and blue, respectively.

structure of cell membranes, which typically contain a large proportion of amphipathic lipids. The *phospholipids* are the most abundant amphipathic lipids. *Phosphoglycerides*, a principal class of phospholipids, contain fatty acyl side chains esterified to two of the three hydroxyl groups of a glycerol molecule; the third hydroxyl group is esterified to phosphate. The simplest phospholipid, *phosphatidic acid,* contains only these components. In most phospholipids, however, the phosphate group is also esterified to a hydroxyl group on a hydrophilic compound such as ethanolamine

$$HO—CH_2—CH_2—NH_3^+$$

or serine, choline, or glycerol, so that the hydrophilic head group is somewhat larger (Figure 2-16). Both of the fatty acyl side chains may be saturated or unsaturated, or one chain may be saturated and the other unsaturated. The negative charge on the phosphate as well as the charged groups or hydroxyl groups on the alcohol esterified to it interact strongly with water.

Phospholipids Spontaneously Form Bilayers in Aqueous Solutions

Phospholipids assume three different forms in aqueous solutions: micelles, bilayer sheets, and liposomes (Figure 2-17). The type of structure formed by a pure phospholipid or a mixture of phospholipids depends on the length of the fatty acyl chains and their degree of saturation, on the temperature, on the ionic composition of the aqueous medium, and on the mode of dispersal of the phospholipids in the solution. In all three forms, hydrophobic interactions cause the fatty acyl chains to aggregate together and exclude water molecules from the "core."

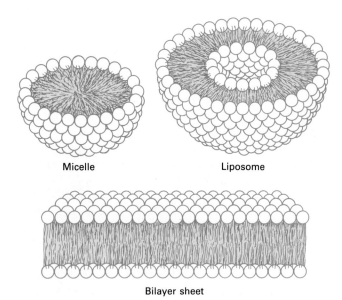

Micelle

Liposome

Bilayer sheet

▲ FIGURE 2-17   Cross-sectional views of the three structures that can be formed by phospholipids in aqueous solutions: a spherical micelle with a hydrophobic interior composed entirely of fatty acyl chains; a sheet of phospholipids in a bilayer; and a spherical liposome, larger than a micelle, comprising one phospholipid bilayer with an aqueous center.

Phospholipids of the composition present in cells spontaneously form symmetric sheetlike structures called *phospholipid bilayers*. Bilayers are two molecules thick. The hydrocarbon side chains of each layer minimize contact with water by aligning themselves tightly together in the center of the bilayer, forming a hydrophobic core that is about 3 nm thick. Their close packing is stabilized by van der Waals attractions between the hydrocarbon side chains of adjacent phospholipids. Ionic and hydrogen bonds stabilize the interaction of the polar head groups with water. A phospholipid bilayer can be of almost unlimited size—from micrometers to millimeters in length or width—and can contain tens of millions of phospholipid molecules. Because of their hydrophobic core, bilayers are impermeable to salts, sugars, and other small hydrophilic molecules.

➤ *Chemical Equilibrium*

The life of a cell involves several thousand different kinds of chemical reactions. Any chemical—whether or not it is in a cell—can, in principle, undergo various chemical reactions. The *rate* at which these reactions can take place under given conditions (concentration, temperature, pressure, and so on) and the *extent* to which they can proceed dictate which reactions actually occur. A *catalyst* is a substance that increases the rate of a reaction but is not itself permanently changed. It brings reactants together but does not end up among the products of the reaction—a function

aptly reflected in the Chinese term for catalyst, *tsoo mei*, which literally means "marriage broker."

Whether or not a catalyst is present, when reactants first come together—before any products have been formed—their rate of reaction is determined in part by their initial concentrations. As the reaction products accumulate, the concentration of each reactant decreases and so does the reaction rate. Meanwhile, some of the product molecules begin to participate in the reverse reaction, which re-forms the reactants. This reaction is slow at first but speeds up as the concentration of products increases. Eventually, the rates of the forward and reverse reactions become equal, so that the concentrations of reactants and products stop changing. The mixture is then said to be in *chemical equilibrium*.

The ratio of products to reactants at equilibrium depends on the temperature as well as on the nature and concentration of the compounds. The *pressure* affects the concentrations of reactants and products (particularly of gases), and thus also affects the equilibrium ratio. Under standard physical conditions (25°C and 1 atm pressure, for biological systems), the ratio of products to reactants at equilibrium, which is always the same for a given reaction, is termed the *equilibrium constant* (K_{eq}). For the simple reaction $A + B \rightleftharpoons X + Y$, the equilibrium constant is given by

$$K_{eq} = \frac{[X][Y]}{[A][B]} \qquad (1)$$

where brackets indicate equilibrium concentrations. In general, for a reaction

$$aA + bB + cC + \cdots \rightleftharpoons zZ + yY + xX + \cdots \qquad (2)$$

where capital letters represent particular molecules or atoms and lowercase letters represent the number of each in the reaction formula, the equilibrium constant is given by

$$K_{eq} = \frac{[X]^x[Y]^y[Z]^z}{[A]^a[B]^b[C]^c} \qquad (3)$$

Importantly, for any reaction the equilibrium constant is equal to

$$K_{eq} = \frac{\text{rate constant for forward reaction}}{\text{rate constant for reverse reaction}} \qquad (4)$$

In the generalized reaction (2) above, the rate of the forward (left to right) reaction will be

$$\text{Rate}_{forward} = k_f[A]^a[B]^b[C]^c$$

where k_f is the rate constant for the forward reaction. Similarly, the rate of the reverse (right to left) reaction is

$$\text{Rate}_{\text{reverse}} = k_r[X]^x[Y]^y[Z]^z$$

where k_r is the rate constant for the reverse reaction.

Since at equilibrium the rate of the forward reaction equals that of the reverse reaction, we can write

$$k_f[A]^a[B]^b[C]^c = k_r[X]^x[Y]^y[Z]^z$$

and thus, since

$$K_{eq} = \frac{[X]^x[Y]^y[Z]^z}{[A]^a[B]^b[C]^c}$$

therefore,

$$K_{eq} = k_f/k_r$$

Also importantly, catalysts do not alter the equilibrium constant: a catalyst accelerates the rates of the forward and reverse reactions by the same factor.

To illustrate several points concerning equilibrium, we shall use a fairly simple biochemical reaction: the interconversion of the compounds glyceraldehyde 3-phosphate (G3P) and dihydroxyacetone phosphate (DHAP). This reaction, which occurs during the breakdown of glucose, is catalyzed by the enzyme triosephosphate isomerase:

Glyceraldehyde 3-phosphate

Dihydroxyacetone phosphate

The equilibrium constant for this reaction under standard conditions is

$$K_{eq} = \frac{[\text{DHAP}]}{[\text{G3P}]} = 22.2$$

Thus the ratio of the concentrations of G3P and DHAP is 1:22.2 when the reaction reaches equilibrium. In practice, one measures the concentrations of reactants and products when a reaction is at equilibrium, and uses these values to calculate the equilibrium constant.

Whenever a reaction involves a single reactant and a single product, the ratio of product to reactant at equilibrium is equal to the equilibrium constant K_{eq} and is independent of the concentrations of product and reactant initially present. As in all reactions, this ratio is also independent of the reaction rate. In the presence of an enzyme or other catalyst, the reaction rate may increase, but the final ratio of product to reactant will always be the same. The magnitude of the equilibrium constant has no bearing on the rate of the reaction or on whether the reaction will take place at all under normal conditions. Despite the large equilibrium constant for the conversion of G3P to DHAP, for example, in an aqueous solution in which no enzyme catalyst is present, it requires so much energy to rearrange the bonds that no detectable reaction actually occurs.

When a reaction involves multiple reactants and/or products, the equilibrium concentration of a *particular* product or reactant depends on the initial concentrations of all reactants and products as well as on the equilibrium constant. Consider, for example, the hydrolysis (cleavage by addition of water) of a quantity of the dipeptide glycylalanine (GA) to glycine (G) and alanine (A):

Glycylalanine

Glycine   **Alanine**

Here,

$$K_{eq} = \frac{[G][A]}{[GA]}$$

(The concentration of water—55 M—does not change significantly during normal aqueous chemical reactions and, by convention, is not included in the calculation of equilibrium ratios.) The equilibrium is strongly in the direction of the formation of glycine and alanine. In other words, most of the glycylalanine is hydrolyzed at equilibrium. However, an excess of one of the products can drive the reaction in the reverse direction. For instance, suppose that the initial reaction mixture contains a small amount of glycylalanine and a large amount of alanine. As the reaction proceeds, the concentration of alanine [A] will always greatly exceed the concentration of glycine [G] produced by hydrolysis. This must reduce the equilibrium ratio of [G] to [GA], because K_{eq} remains constant. Thus, the reaction GA $\rightleftharpoons$ G + A can be driven to the left by an excess of the product alanine. More generally, in reactions involving more than one reactant or product, changes in the concentration of any one reactant will affect the concentrations at equilibrium of all of the reactants and products.

The above reactions exemplify the formation or cleavage of covalent bonds. Some important reactions involve, instead, the binding of one large molecule to another, such as the interaction of a protein P with a specific sequence D on a molecule of DNA, which frequently causes the expression of a gene to turn on or off. Just as in the reactions discussed above, in this case we can write

$$PD \rightleftharpoons P + D$$

where PD is the specific complex of DNA and protein, and

$$K_{eq} = \frac{[P][D]}{[PD]}$$

For typical reactions involving the binding of a protein to a specific DNA sequence, the value of K_{eq} is 10^{-10} M. We can relate the magnitude of this equilibrium constant to events within a cell: First, realize that a bacterial cell contains *one molecule* of DNA. Assume, for simplicity, that the cell also contains *one molecule* of the relevant protein. Further assuming the cell to be a cylinder 2 μm long and 1 μm in diameter, its volume will be 1.5×10^{-15} liters, and [P], the total concentration of the one molecule of P within the cell, will be

(1 molecule/1.5×10^{-15} L) $\div$ (6.02×10^{23} molecules/mol) = 1.1×10^{-9} M

Similarly, [D], the concentration of the one molecule of DNA, will also be 1.1×10^{-9} M. Using the equilibrium constant, we can calculate that at equilibrium the concentration of the protein-DNA complex will be

$$[PD] = 8.15 \times 10^{-10} \text{ M}$$

the concentration of free (unbound) P will be

$$[P] = 2.85 \times 10^{-10} \text{ M}$$

the concentration of DNA unbound to P will be

$$[D] = 2.85 \times 10^{-10} \text{ M}$$

and

$$K_{eq} = \frac{(2.85 \times 10^{-10} \text{ M})(2.85 \times 10^{-10} \text{ M})}{(8.15 \times 10^{-10} \text{ M})} = 10^{-10} \text{ M}$$

In other words, 74 percent (= 8.15×10^{-10} M $\div$ 1.1×10^{-9} M) of the time the protein molecule will be bound to the bacterial DNA and 26 percent of the time the DNA will not have the protein bound to it. Thus, if one knows the equilibrium constants for binding reactions and the numbers of the participating molecules within a cell, one can predict the concentration of the resulting complex.

▶ pH and the Concentration of Hydrogen Ions

The solvent inside cells and in all extracellular fluids is water. An important characteristic of any aqueous solution is the concentration of positively charged hydrogen ions (H^+) and negatively charged hydroxyl ions (OH^-). Because these ions are the dissociation products of H_2O, they are constituents of all living systems, and they are liberated by many reactions that take place between organic molecules within cells.

Water Dissociates into Hydronium and Hydroxyl Ions

When a water molecule dissociates, one of its polar H–O bonds breaks. The resulting hydrogen ion—a proton— has a short lifetime as a free particle and quickly combines with a water molecule to form a *hydronium ion* (H_3O^+). For convenience, however, we refer to the concentration of hydrogen ions in a solution, [H^+], even though we really mean the concentration of hydronium ions, [H_3O^+].

The dissociation of water is a reversible reaction,

$$H_2O \rightleftharpoons H^+ + OH^-$$

and at 25°C,

$$[H^+][OH^-] = 10^{-14} \text{ M}^2$$

where M symbolizes *molarity,* or moles per liter (mol/L). In pure water, [H^+] = [OH^-] = 10^{-7} M.

The concentration of hydrogen ions in a solution is expressed conventionally as its *pH:*

$$pH = -\log[H^+] = \log \frac{1}{[H^+]}$$

In pure water at 25°C, $[H^+] = 10^{-7}$ M, so

$$pH = -\log [10^{-7}] = 7.0$$

On the pH scale, 7.0 is considered neutral: pH values below 7.0 indicate acidic solutions and values above 7.0 indicate basic (alkaline) solutions (Table 2-5). In a 0.1 M solution of hydrogen chloride (HCl) in water, $[H^+] = 0.1$ M because virtually all of the HCl has dissociated into H^+ and Cl^- ions. For this solution

$$pH = -\log 0.1 = 1.0$$

In fact, pH values can be less than zero, since a 10 M solution of HCl will have a pH of -1.

Its pH is one of the most important properties of a biological fluid. The pH of the cytosol of cells is normally about 7.2. In certain organelles of eukaryotic cells, such as the lysosomes and vacuoles, the pH is much lower, about 5; this value of $[H^+]$ is more than 100 times higher than its value in the cytosol. Lysosomes contain many degradative enzymes that function optimally in an acidic environment, whereas their action is inhibited in the near-neutral environment of the cytosol. Maintenance of a specific pH is imperative for some cellular structures to function properly. On the other hand, dramatic shifts in cellular pH may play an important role in controlling cellular activity. For example, the pH of the cytosol of an unfertilized sea urchin egg is 6.6. Within 1 minute of fertilization, however, the pH rises to 7.2 (that is, $[H^+]$ decreases to about one-fourth its original value). The change in pH appears to trigger the growth and division of the egg.

Acids Release Hydrogen Ions, and Bases Combine with Hydrogen Ions

In general, any molecule or ion that tends to release a hydrogen ion is called an *acid* and any molecule or ion that readily combines with a hydrogen ion is called a *base*. Thus hydrogen chloride is an acid. The hydroxyl ion is a base, as is ammonia (NH_3), which readily picks up a hydrogen ion to become an ammonium ion (NH_4^+). Many organic molecules are acidic because they have a carboxyl group ($-COOH$) that tends to dissociate to form the negatively charged carboxylate ion ($-COO^-$):

$$X-C\underset{OH}{\overset{O}{\big\langle}} \rightleftharpoons X-C\underset{O^-}{\overset{O}{\big\langle}} + H^+$$

where X represents the rest of the molecule. The amino group ($-NH_2$), a part of many important biological molecules, is a base because, like ammonia, it can take up a hydrogen ion:

$$X-NH_2 + H^+ \rightleftharpoons X-NH_3^+$$

When acid is added to a solution, $[H^+]$ increases (the pH goes down). Consequently, $[OH^-]$ decreases because hydroxyl ions readily combine with the hydrogen ions to form water. Conversely, when a base is added to a solution, $[H^+]$ decreases (the pH goes up). Because $[H^+][OH^-] = 10^{-14}$ M², any increase in $[H^+]$ is coupled with a decrease in $[OH^-]$, and vice versa. No matter how acidic or alkaline a solution is, it always contains both ions: neither $[OH^-]$ nor $[H^+]$ is ever zero. For example, if $[H^+] = 0.1$ M (pH = 1.0), then $[OH^-] = 10^{-13}$ M.

Biological Molecules Can Have Both Acidic and Basic Groups

The degree to which a dissolved acid releases hydrogen ions or to which a base takes them up depends partly on

TABLE 2-5   The pH Scale

	Concentration of Hydrogen Ions (mol/l)	pH	Example
⇑ Increasing acidity	10^{-0}	0	
	10^{-1}	1	Gastric fluids
	10^{-2}	2	Lemon juice
	10^{-3}	3	Vinegar
	10^{-4}	4	Acid soil
	10^{-5}	5	Lysosomes
	10^{-6}	6	Cytoplasm of contracting muscle
Neutral	10^{-7}	7	Pure water and cytoplasm
	10^{-8}	8	Sea water
	10^{-9}	9	Very alkaline natural soil
	10^{-10}	10	Alkaline lakes
	10^{-11}	11	Household ammonia
⇓ Increasing alkalinity	10^{-12}	12	Lime (saturated solution)
	10^{-13}	13	
	10^{-14}	14	

the pH of the solution. Amino acids such as alanine (see Figure 2-4) provide an example. These molecules have the general formula

$$
\begin{array}{c}
NH_2 \\
| \\
H-C-COOH \\
| \\
R
\end{array}
$$

(where R represents the rest of the molecule), but in neutral solutions (pH = 7.0) they exist predominantly in the doubly ionized form

$$
\begin{array}{c}
NH_3^+ \\
| \\
H-C-COO^- \\
| \\
R
\end{array}
$$

Such a molecule, containing both a positive and a negative ion, is called a *zwitterion*. Zwitterions, having no net charge, are neutral.

In solutions at low pH, carboxylate ions (–COO$^-$) recombine with the abundant hydrogen ions, so that the predominant form of the amino acid molecule is

$$
\begin{array}{c}
NH_3^+ \\
| \\
H-C-COOH \\
| \\
R
\end{array}
$$

At high pH, the scarcity of hydrogen ions decreases the chance that an amino group or a carboxylate ion will pick up a hydrogen ion, so that the predominant form of an amino acid molecule at high pH is

$$
\begin{array}{c}
NH_2 \\
| \\
H-C-COO^- \\
| \\
R
\end{array}
$$

The Henderson-Hasselbalch Equation Describes the Relationships between pH and Equilibrium Constants for Acids and Bases

Many molecules used by cells have multiple acidic or basic groups, each of which can release or take up a proton. In the laboratory, it is often essential to know the precise state of dissociation of each of these groups at various pH values. The dissociation of an acid group HA, such as acetic acid (CH_3COOH), is described by

$$ HA \rightleftharpoons H^+ + A^- $$

The equilibrium constant, K_a, for this reaction is

$$ K_a = \frac{[H^+][A^-]}{[HA]} $$

By taking the logarithm of both sides and rearranging the result, we can derive the very useful relation between the equilibrium constant and pH known as the *Henderson-Hasselbalch equation*:

$$ \log K_a = \log \frac{[H^+][A^-]}{[HA]} = \log [H^+] + \log \frac{[A^-]}{[HA]} $$

$$ -\log [H^+] = -\log K_a + \log \frac{[A^-]}{[HA]} $$

Substituting pH for $-\log [H^+]$ and pK_a for $-\log K_a$, we have

$$ pH = pK_a + \log \frac{[A^-]}{[HA]} $$

The pK_a of any acid is equal to the pH at which one-half of the molecules are dissociated and one-half are neutral. This can be derived by observing that if $pK_a = pH$, then $\log ([A^-]/[HA]) = 0$, or $[A^-] = [HA]$. The Henderson-Hasselbalch equation allows us to calculate the degree of dissociation of an acid if both the pH of the solution and the pK_a of the acid are known. Experimentally, by measuring the concentration of A^- and of HA as a function of the solution's pH, one can calculate the pK_a of the acid and thus the equilibrium constant for the dissociation reaction.

Buffers Maintain the pH of Cells and of Extracellular Fluids

A growing cell must maintain a constant pH in the cytoplasm of about 7.4 despite the production, by metabolism, of many acids, such as lactic acid and CO_2 (which reacts with water to form carbonic acid, H_2CO_3). Cells have a reservoir of weak bases and weak acids, called *buffers*, which ensure that the cell's pH remains relatively constant. Buffers do this by "soaking up" H^+ or OH^- when these ions are added to the cell or are produced by metabolism.

The ability of a buffer to minimize changes in pH is related to its pK_a value. To understand this point, we also need to understand how the fraction of molecules in the undissociated form (HA) depends on pH. A *titration curve* (Figure 2-18) shows these relationships: at one pH unit below the pK_a of an acid, 91 percent of the molecules are in the HA form; at one pH unit above the pK_a, 91 percent are in the A^- form.

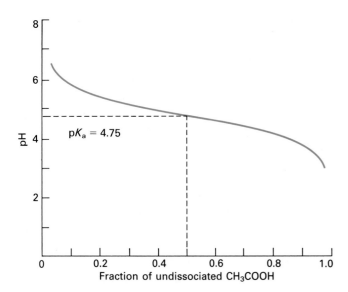

$$H_3PO_4 \rightleftharpoons H_2PO_4^- + H^+ \qquad pK_a = 2.1$$

$$H_2PO_4^- \rightleftharpoons HPO_4^{2-} + H^+ \qquad pK_a = 7.2$$

$$HPO_4^{2-} \rightleftharpoons PO_4^{3-} + H^+ \qquad pK_a = 12.7$$

Figure 2-19 shows the titration curve for phosphoric acid. The pK_a for the dissociation of the second proton (pH 7.2) is similar to the pH of the cytosol. Because pK_a = pH = 7.2, then, according to the second equation and the Henderson-Hasselbalch equation, about 50 percent of cellular phosphate is $H_2PO_4^-$ and 50 percent is HPO_4^{2-}. (Actually, 0.499973 is $H_2PO_4^-$, 0.499973 is HPO_4^{2-}, 0.0000039 is H_3PO_4, and 0.0000016 is PO_4^{3-}, as can be calculated by exact solution of the Henderson-Hasselbalch equations using all three dissociation constants.) Thus, phosphate is an excellent buffer at pH values around 7.2,

▲ FIGURE 2-18   The titration curve of acetic acid (CH_3COOH). The pK_a for the dissociation of acetic acid to H^+ and CH_3COO^- is 4.75. Because pH is measured on a logarithmic scale, the solution changes from 91 percent CH_3COOH at pH 3.75 to 9 percent CH_3COOH at pH 5.75. The acid has maximum buffering capacity in this pH range.

If we add additional acid (or base) to a solution of an acid (or a base) at its pK_a value (a 1:1 mixture of HA and A^-), the pH of the solution changes, but it changes less than it would if the original acid (or base) had not been present. This is because protons released by the added acid are taken up by the original A^- form of the acid, or the hydroxyl ions generated by the added base are neutralized by protons released by the original HA. This *buffering* capacity of acids and bases declines rapidly at more than one pH unit from the pK_a. In other words, the addition of the same number of moles of acid to a solution containing a mixture of HA and A^- that is at a pH near the pK_a of the acid will cause less of a pH change than it would if the HA and A^- were not present, or if the pH were far from the pK_a value.

All biological systems contain one or more buffers. Phosphoric acid (H_3PO_4) is a physiologically important buffer; phosphate ions are present in considerable quantities in cells and are an important factor in maintaining, or buffering, the pH of the cytosol. Phosphoric acid has three groups that are capable of dissociating:

$$\begin{array}{c} O \\ \| \\ HO—P—OH \\ | \\ OH \end{array}$$

but the three protons do not dissociate simultaneously; loss of each can be described by a discrete pK_a constant:

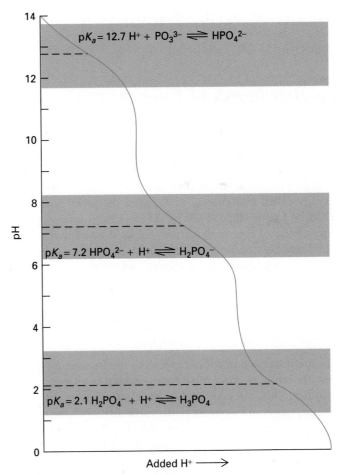

▲ FIGURE 2-19   The titration curve of phosphoric acid (H_3PO_4). This biologically ubiquitous molecule has three hydrogen atoms that dissociate at different pH values; thus, phosphoric acid has three pK_a values, as noted on the graph. The shaded areas denote the pH ranges—within one pH unit of the three pK_a values—where the buffering capacity of phosphoric acid is maximum; in these regions the addition of acid (or base) will cause the least change in the pH.

the approximate pH of the cytosol of cells, and at pH 7.4, the pH of human blood.

In nucleic acids, phosphate is found as a diester. It is linked to two carbon atoms of adjacent ribose sugars:

$$-\overset{|}{\underset{|}{C}}-O-\overset{O}{\underset{OH}{\overset{\parallel}{P}}}-O-CH_2-$$

The pK_a for the dissociation of the single –OH proton is about 3, which is similar to the pK_a for the dissociation of the first proton from phosphoric acid. Thus, at neutral pH, each phosphate residue in deoxyribonucleic acid (DNA) or ribonucleic acid (RNA) is actually negatively charged, which is why DNA and RNA are called nucleic *acids*:

$$-\overset{|}{\underset{|}{C}}-O-\overset{O}{\underset{O^-}{\overset{\parallel}{P}}}-O-CH_2-$$

► *Biochemical Energetics: Free Energy in Biochemical Reactions*

Because biological systems are generally held at constant temperature and pressure, it is possible to predict the direction of a chemical reaction by using a measure of potential energy called *free energy*, or *G*, after the great American chemist Josiah Willard Gibbs (1839–1903), a founder of the science of thermodynamics. Gibbs showed that under conditions of constant pressure and temperature, as generally found in biological systems, "all systems change in such a way that free energy is minimized."

The Change in Free Energy, ΔG, Determines the Direction of a Chemical Reaction

We are interested in what happens to the free energy when one molecule or molecular configuration is changed into another. Thus our concern is with relative, rather than absolute, values of free energy—in particular, with the difference between the values before and after the change, written ΔG, where, as usual, Δ stands for difference and

$$\Delta G = G_{products} - G_{reactants}$$

In mathematical terms, Gibbs's law—that systems change to minimize free energy—is a set of statements about ΔG:

If ΔG is negative for a chemical reaction or mechanical process, the forward reaction or process (from left to right as written) will tend to occur spontaneously.

If ΔG is positive, the reverse reaction (from right to left as written) will tend to occur.

If ΔG is zero, both forward and reverse reactions occur at equal rates; the reaction is at equilibrium.

The ΔG of a Reaction Depends on Changes in Heat and Entropy

At any constant temperature and pressure, two factors determine the ΔG of a reaction and thus whether the reaction will tend to occur: the change in bond energy between reactants and products and the change in the randomness of the system. Gibbs showed that free energy can be defined as

$$G = H - TS$$

where *H* is the bond energy (*enthalpy*) of the system, *T* is its temperature in degrees Kelvin (K), and *S* is its *entropy*, a measure of randomness. If temperature remains constant, a reaction proceeds spontaneously only if there is a decrease in free energy (a negative ΔG) in the equation

$$\Delta G = \Delta H - T\,\Delta S$$

The *enthalpy*, *H*, of reactants or of products is equal to their total bond energies; the overall change in enthalpy, ΔH, is equal to the overall change in bond energies. In an *exothermic* reaction, the products contain less bond energy than the reactants have, the liberated energy is converted to heat (the energy of molecular motion), and ΔH is negative. In an *endothermic* reaction, the products contain more bond energy than the reactants, heat is absorbed, and ΔH is positive. Reactions tend to proceed if they liberate energy (if $\Delta H < 0$), but this is only one of two important parameters of free energy to consider; the other is entropy.

Entropy, *S*, is a measure of the degree of randomness or disorder of a system. Entropy increases as a system becomes more disordered and decreases as it becomes more structured. Consider, for example, the diffusion of solutes from one solution into another one in which their concentration is lower. This important biological reaction is driven only by an increase in entropy; in such a process ΔH is near zero. To see this, suppose that a 0.1 M solution of NaCl is separated from a 0.01 M solution by a membrane through which Na^+ and Cl^- ions can diffuse (Figure 2-20). The ions from the 0.1 M solution can diffuse over a larger volume than they could before, as can the ions from the

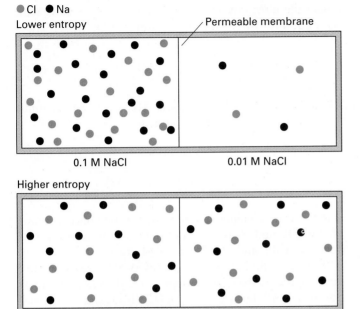

● Cl ● Na
Lower entropy

0.1 M NaCl   Permeable membrane   0.01 M NaCl

Higher entropy

0.055 M NaCl   0.055 M NaCl

▲ FIGURE 2-20   The entropy of a concentrated salt solution increases when the solution is allowed to mix with a dilute one. In this example, a membrane permeable to Na^+ and Cl^- ions is placed between the concentrated and dilute solutions; the Na^+ and Cl^- diffuse through the membrane until their concentration is equal on both sides. This gives the ions from both the concentrated and dilute solutions more room in which to move, with the result that the randomness, or entropy, of the system is increased.

more dilute solution; thus the disorder of the system increases. Maximum entropy is achieved when all ions can diffuse freely over the largest possible volume—that is, when the concentrations of all ions are the same on both sides of the membrane. If the two solutions are of equal volume and if the degree of hydration of Na^+ and Cl^- does not change significantly on dilution, ΔH will be approximately zero and the negative free energy of the diffusion reaction illustrated in Figure 2-20 will be due solely to the positive value of ΔS:

$$\Delta G = -T \, \Delta S$$

We saw on page 26 how the formation of hydrophobic bonds is driven primarily by a change in entropy. That is, if a long hydrophobic molecule, such as heptane or tristearin, is dissolved in water, the water molecules are forced to form a cage around it, restricting their free motion. This imposes a high degree of order on their arrangement and lowers the entropy of the system ($\Delta S < 0$). Because the entropy change is negative, hydrophobic molecules do not

dissolve well in aqueous solutions and tend to stay associated with each other.

To summarize the relationships between free energy, enthalpy, and entropy, an exothermic reaction ($\Delta H < 0$) that increases entropy ($\Delta S > 0$) occurs spontaneously ($\Delta G < 0$). An endothermic reaction ($\Delta H > 0$) can still occur spontaneously if ΔS is positive enough that the $T \, \Delta S$ term can overcome the positive ΔH. If there is no change in free energy with the conversion of reactants into products ($\Delta G = 0$), then the system will be at equilibrium: any conversion of reactants to products will be balanced by an equal conversion of products to reactants.

Many biological reactions lead to an increase in order, and thus a decrease in entropy ($\Delta S < 0$). An obvious example is the reaction that links amino acids together to form a protein. A solution of protein molecules has a lower entropy than does a solution of the same amino acids unlinked, because the free movement of any amino acid in a protein is restricted when it is bound in a long chain. For the linking reaction to proceed, a compensatory decrease in free energy must occur elsewhere in the system, as is discussed in Chapter 4.

Temperature, Concentrations of Reactants, and Other Parameters Affect the ΔG of a Reaction

The change in free energy of a reaction (ΔG) is influenced by temperature, pressure, and the initial concentrations of reactants and products. Most biological reactions—like others that take place in aqueous solutions—also are affected by the pH of the solution.

The *standard free-energy change* of a reaction, $\Delta G^{\circ\prime}$, is the value for the change in free energy under the conditions of 298 K (25°C), 1 atm pressure, pH 7.0 (as in pure water), and initial concentrations of 1 M for all reactants and products except protons, which are kept at pH 7.0. Table 2-6 gives values of $\Delta G^{\circ\prime}$ for some typical biochemical reactions. The sign of $\Delta G^{\circ\prime}$ depends on the direction in which the reaction is written. If the reaction A → B has a $\Delta G^{\circ\prime}$ of $-x$ kcal/mol, then the reverse reaction B → A will have a $\Delta G^{\circ\prime}$ value of $+x$ kcal/mol.

Most biological reactions differ from standard conditions, particularly in the concentrations of reactants. However, we can estimate free-energy changes for different temperatures and initial concentrations, using the following equation:

$$\Delta G = \Delta G^{\circ\prime} + RT \ln Q = \Delta G^{\circ\prime} + RT \ln [\text{products}]/[\text{reactants}]$$

where R is the gas constant of 1.987 cal/(degree · mol), T is the temperature (in degrees Kelvin), and Q is the initial ratio of products to reactants (constructed like the equilibrium constant, page 28). For example, again using as our

TABLE 2-6   Values of $\Delta G^{\circ\prime}$, the Standard Free-Energy Change, for Some Important Biochemical Reactions

Reaction	$\Delta G^{\circ\prime}$ (kcal/mol)
HYDROLYSIS	
Acid anhydrides:	
Acetic anhydride + H_2O $\longrightarrow$ 2 acetate	−21.8
PP_i + H_2O $\longrightarrow$ $2P_i$*	−8.0
ATP + H_2O $\longrightarrow$ ADP + P_i	−7.3
Esters:	
Ethylacetate + H_2O $\longrightarrow$ ethanol + acetate	−4.7
Glucose-6-phosphate + H_2O $\longrightarrow$ glucose + P_i	−3.3
Amides:	
Glutamine + H_2O $\longrightarrow$ glutamate + NH_4^+	−3.4
Glycylglycine + H_2O $\longrightarrow$ 2 glycine (a peptide bond)	−2.2
Glycosides:	
Sucrose + H_2O $\longrightarrow$ glucose + fructose	−7.0
Maltose + H_2O $\longrightarrow$ 2 glucose	−4.0
ESTERIFICATION	
Glucose + P_i $\longrightarrow$ glucose 6-phosphate + H_2O	+3.3
REARRANGEMENT[†]	
Glucose 1-phosphate $\longrightarrow$ glucose 6-phosphate	−1.7
Fructose 6-phosphate $\longrightarrow$ glucose 6-phosphate	−0.4
Glyceraldehyde 3-phosphate $\longrightarrow$ dihydroxyacetone phosphate	−1.8
ELIMINATION	
Malate $\longrightarrow$ fumarate + H_2O	+0.75
OXIDATION	
Glucose + $6O_2$ $\longrightarrow$ $6CO_2$ + $6H_2O$	−686
Palmitic acid + $23O_2$ $\longrightarrow$ $16CO_2$ + $16H_2O$	−2338
PHOTOSYNTHESIS	
$6CO_2$ + $6H_2O$ $\longrightarrow$ six-carbon sugars[‡] + $6O_2$	+686

* PP_i = pyrophosphate; P_i = phosphate.
[†] The three reactions shown occur during glycolysis, the first steps in conversion of glucose to CO_2.
[‡] The two principal products of photosynthesis are sucrose, a disaccharide, and starch, a long polymer of glucose.
SOURCE: Lehninger, A. L. 1975. *Biochemistry*, 2d ed. Worth, p. 397.

model the interconversion of glyceraldehyde 3-phosphate (G3P) and dihydroxyacetone phosphate (DHAP)

$$G3P \rightleftharpoons DHAP$$

we have Q = [DHAP]/[G3P] and $\Delta G^{\circ\prime}$ = −1840 cal/mol. The equation for ΔG then becomes

$$\Delta G = -1840 + 1.987\, T \ln \frac{[DHAP]}{[G3P]}$$

and we can calculate ΔG for any set of concentrations of DHAP and G3P. If the initial concentrations of both [DHAP] and [G3P] are 1 M, then $\Delta G = \Delta G^{\circ\prime}$ = −1840 cal/mol, because $RT \ln 1 = 0$. The reaction will tend to proceed from left to right, in the direction of formation of DHAP. If, however, [DHAP] begins as 0.1 M and [G3P] as 0.001 M, with other conditions being standard, then $Q = 0.1/0.001 = 100$, and

$$\begin{aligned} \Delta G &= -1840 + (1.987)(298) \ln (100) \\ &= -1840 + (1.987)(298)(4.605) \\ &= -1840 + 2727 = +887 \text{ cal/mol} \end{aligned}$$

Clearly, the reaction will now proceed in the direction of formation of G3P.

In a reaction $A + B \rightleftharpoons C$, in which two molecules combine to form a third, the equation for ΔG becomes

$$\Delta G = \Delta G^{\circ\prime} + RT \ln Q = \Delta G^{\circ\prime} + RT \ln \frac{[C]}{[A][B]}$$

The direction of the reaction will shift more toward the right (toward formation of C) if either [A] or [B] is increased.

The Standard Free-Energy Change, $\Delta G^{\circ\prime}$, Can Be Determined from Measurement of the Equilibrium Constant, K_{eq}

A chemical mixture at equilibrium is already in a state of minimal free energy: no free energy is being generated or released. Thus, for a system at equilibrium, we can write

$$0 = \Delta G = \Delta G^{\circ\prime} + RT \ln Q$$

At equilibrium the value of Q is the equilibrium constant K_{eq}, so that

$$\Delta G^{\circ\prime} = -RT \ln K_{eq}.$$

TABLE 2-7   Values of $\Delta G^{\circ\prime}$ for Some Values of K_{eq}

K_{eq}	$\Delta G^{\circ\prime}$ (cal/mol)*
0.001	+4086
0.01	+2724
0.1	+1362
1.0	0
10	−1362
100	−2724
1000	−4086

* Calculated from the formula $\Delta G^{\circ\prime} = -2.3\,RT \log K_{eq}$.

This means that by measuring the concentration of reactants and products at equilibrium one can calculate the value of $\Delta G^{\circ\prime}$. Using again the example of the interconversion of G3P and DHAP, $K_{eq} = 22.2$ and thus we can calculate that $\Delta G^{\circ\prime} = -1840$ cal/mol.

Expressed in terms of base 10 logarithms, the equation relating K_{eq} and $\Delta G^{\circ\prime}$ becomes

$$\Delta G^{\circ\prime} = -2.3RT \log K_{eq}$$

or

$$-\frac{\Delta G^{\circ\prime}}{2.3RT} = \log K_{eq}$$

or, by the definition of logarithms,

$$K_{eq} = 10^{-(\Delta G^{\circ\prime}/2.3RT)}$$

Note again that if $\Delta G^{\circ\prime}$ is negative, then $K_{eq} > 1$; that is, the formation of products from reactants is favored (Table 2-7).

The Generation of a Concentration Gradient Requires an Expenditure of Energy

A cell must often accumulate chemicals, such as glucose and K^+ ions, in greater concentrations than exist in its environment. Consequently, the cell must transport these chemicals against a concentration gradient. To find the amount of energy required to transfer one mole of a substance from outside the cell, where its concentration is C_1, to inside the cell, where its concentration is C_2, we use the equation

$$\Delta G = \Delta G^{\circ\prime} + RT \ln Q$$

where Q is the ratio of C_2 to C_1. For such a reaction, $\Delta G^{\circ\prime} = 0$ since no chemical bonds are formed or broken and no heat is taken up or released. Thus

$$\Delta G = RT \ln \frac{C_2}{C_1}$$

If the concentrations differ by a factor of 10, then at 25°C, $\Delta G = RT \ln 10 = 1.36$ kcal/mol of substance transported. (Such calculations assume that a molecule of a given substance inside a cell is identical to a molecule of that substance outside and that the substance is not sequestered, bound, or chemically changed by the transport.)

The "uphill" transport of molecules against a concentration gradient ($C_2 > C_1$) clearly does not take place spontaneously and so $\Delta G > 0$; it requires the input of cellular chemical energy. As discussed in Chapter 15, this energy is often released by the hydrolysis of ATP. Conversely, when a substance moves down its concentration gradient in crossing a membrane, ΔG has a negative value and the transport can be coupled to a reaction that has a positive ΔG, say the movement of another substance uphill across a membrane.

Many Cellular Processes Involve the Transfer of Electrons in Oxidation-Reduction Reactions

Many chemical reactions result in the transfer of electrons from one atom or molecule to another; this transfer may or may not accompany the formation of new chemical bonds. The loss of electrons from an atom or a molecule is called *oxidation*, and the gain of electrons by an atom or a molecule is called *reduction*. Because electrons are neither created nor destroyed in a chemical reaction, if one atom or molecule is oxidized, another must be reduced. For example, oxygen draws electrons from Fe^{2+} (ferrous) ions to form Fe^{3+} (ferric) ions, a reaction that occurs as part of the process by which carbohydrates are degraded in mitochondria. Each oxygen atom receives two electrons, one from each of two Fe^{2+} ions:

$$2Fe^{2+} + \tfrac{1}{2}O_2 \longrightarrow 2Fe^{3+} + O^{2-}$$

Thus Fe^{2+} is oxidized, and O_2 is reduced. Oxygen similarly accepts electrons in many oxidation reactions in aerobic cells.

In the transformation of succinate into fumarate, which also takes place during carbohydrate breakdown in mitochondria, succinate loses two hydrogen atoms (Figure 2-21), which is equivalent to a loss of two protons and two electrons. Thus succinate is said to be oxidized in its conversion into fumarate, and another molecule—flavin adenine dinucleotide (FAD), which accepts the electrons—is

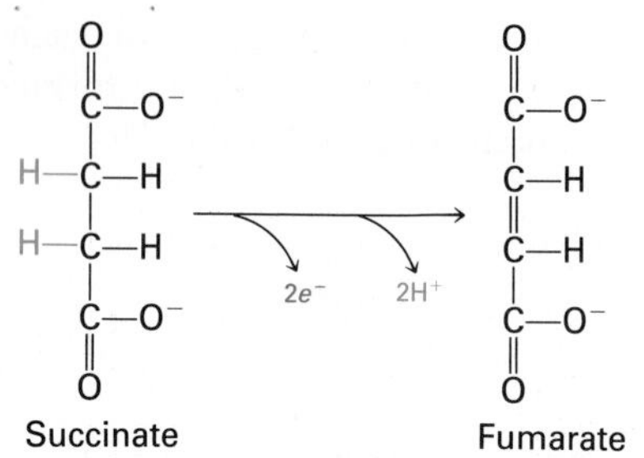

▲ FIGURE 2-21   The conversion of succinate to fumarate is an oxidation reaction: two electrons and two protons are released. This reaction occurs in mitochondria as part of the citric acid cycle, which functions in the final stages of the oxidation of glucose by oxygen to form carbon dioxide.

reduced (to $FADH_2$). Many biologically important oxidation and reduction reactions involve the removal or the addition of hydrogen atoms (protons plus electrons) rather than the transfer of isolated electrons.

Standard Reduction Potentials   To describe oxidation-reduction reactions, such as the reaction of ferrous ion (Fe^{2+}) and oxygen (O_2), it is easiest to divide them into two half-reactions:

Oxidation of Fe^{2+}: $2Fe^{2+} \longrightarrow 2Fe^{3+} + 2e^-$

Reduction of O_2: $2e^- + \frac{1}{2}O_2 \longrightarrow O^{2-}$

In this case, the reduced oxygen (O^{2-}) readily reacts with water to form two hydroxide ions:

$$O^{2-} + H_2O \rightleftharpoons 2\,OH^-$$

and thus we can also write the half-reaction for reduction of O_2 as

$$2e^- + \frac{1}{2}O_2 + H_2O \longrightarrow 2\,OH^-$$

or, since water dissociates into protons and hydroxide ions, the half-reaction can be written as

$$2e^- + \frac{1}{2}O_2 + 2H^+ \longrightarrow H_2O$$

The readiness with which an atom or a molecule takes up an electron is its *reduction potential, E*. Reduction potentials are measured in volts (V) from an arbitrary zero point set at the reduction potential of the following half-reaction under standard conditions (25°C, 1 atm, and reactants at 1 M):

$$H^+ + e^- \underset{\text{oxidation}}{\overset{\text{reduction}}{\rightleftharpoons}} \frac{1}{2}H_2$$

The value of E for a molecule or an atom under standard conditions is its *standard reduction potential, E'_0* (Table 2-8). Standard reduction potentials may differ somewhat from those found under the conditions in a cell, because the concentrations of reactants in a cell are not 1 M. A positive reduction potential means that a molecule or ion (say, Fe^{3+}) has a higher affinity for electrons than the H^+ ion does in the standard reaction. A negative reduction potential means that the substance—for example, acetate (CH_3COO^-) in its reduction to acetaldehyde (CH_3CHO)— has a lower affinity for electrons. In an oxidation-reduction reaction, electrons move spontaneously toward atoms or molecules having more positive reduction potentials. In other words, a compound having a more negative reduction potential can reduce—or transfer electrons to—one having a more positive potential.

The Relationship between Changes in Free Energy and Reduction Potentials   In an oxidation-reduction reaction, the total voltage change (change in electric potential), ΔE, is the sum of the voltage changes (reduction potentials) of the individual oxidation or reduction steps. Because all forms of energy are interconvertible, we can express ΔE as a change in chemical free energy (ΔG). The charge in one mole (6×10^{23}) of electrons is 96,500 coulombs (96,500 joules per volt)—a quantity known as the *faraday constant* ($\mathscr{F}$), after British physicist Michael Faraday (1791–1867). Thus we can write

$$\Delta G \text{ (cal/mol)} = -n\mathscr{F}\,\Delta E = -n(96{,}500/4.184)\,\Delta E \text{ (volts)}$$

where n is the number of electrons transferred and 4.184 is the factor used to convert joules into calories. Note that an oxidation-reduction reaction with a positive ΔE value will have a negative ΔG, and thus will tend to proceed from left to right.

The reduction potential is customarily used to describe the electric energy change that occurs when an atom or a molecule gains an electron. In an oxidation-reduction reaction, we also use the *oxidation potential*—the voltage change that takes place when an atom or molecule *loses* an electron—which is simply the negative of the reduction potential:

Reduction: $Cu^{2+} + e^- \longrightarrow Cu^+$ $E'_0 = +0.35$ V

Oxidation: $Cu^+ \longrightarrow Cu^{2+} + e^-$ $E'_0 = -0.35$ V

The voltage change in a complete oxidation-reduction reaction, in which one molecule is reduced and another is oxidized, is simply the sum of the oxidation and reduction potentials of the atoms or molecules in the partial reactions. Consider, for example, the change in electric poten-

TABLE 2-8   Values of the Standard Reduction Potential, E_0', and Standard Free Energy Values, $\Delta G^{\circ\prime}$, for Some Important Oxidation-Reduction Reactions (pH 7.0, 25°C)

Oxidant	Reductant	n^*	$E_0'(V)^\dagger$	$\Delta G^{\circ\prime}$ (kcal/mole)‡
Succinate + CO_2	α-Ketoglutarate	2	−0.67	+30.9
Acetate	Acetaldehyde	2	−0.60	+27.7
Ferredoxin (oxidized)	Ferredoxin (reduced)	1	−0.43	+ 9.9
$2H^+$	H_2	2	−0.42	+19.4
NAD^+	$NADH + H^+$	2	−0.32	+14.8
$NADP^+$	$NADPH + H^+$	2	−0.32	+14.8
Glutathione (oxidized)	Glutathione (reduced)	2	−0.23	+10.6
Acetaldehyde	Ethanol	2	−0.20	+ 9.2
Pyruvate	Lactate	2	−0.19	+ 8.7
Fumarate	Succinate	2	+0.03	− 1.4
Cytochrome c (+3)	Cytochrome c (+2)	1	+0.22	− 5.1
Fe^{3+}	Fe^{2+}	1	+0.77	−17.8
$\frac{1}{2}O_2 + 2H^+$	H_2O	2	+0.82	−37.8

$^*n$ is the number of electrons transferred.
$^\dagger E_0'$ refers to the partial reaction, written as Oxidant + e^- ⟶ reductant.
‡Calculated from the equation $\Delta G^{\circ\prime} = -n\mathcal{F}E_0'$.
source: Stryer, L. 1988. *Biochemistry*, 3d ed. W. H. Freeman and Company, p. 400.

tial (and, correspondingly, in standard free energy) when succinate is oxidized by oxygen:

$$\text{Succinate} + \tfrac{1}{2}O_2 \rightleftharpoons \text{fumarate} + H_2O$$

In this case, the partial reactions are:

Succinate $\rightleftharpoons$ fumarate + $2H^+ + 2e^-$	$E_0' = -0.03$ V $\Delta G^{\circ\prime} = +1.39$ kcal/mol ($n = 2$ electrons)
$\frac{1}{2}O_2 + 2e^- + 2H^+ \rightleftharpoons H_2O$	$E_0' = +0.82$ V $\Delta G^{\circ\prime} = -37.88$ kcal/mol ($n = 2$ electrons)
Sum: Succinate + $\frac{1}{2}O_2 \rightleftharpoons$ fumarate + H_2O	$\Delta E_0' = +0.79$ V $\Delta G^{\circ\prime} = -36.49$ kcal/mol

The overall reaction has a positive $\Delta E_0'$ or, equivalently, a negative $\Delta G^{\circ\prime}$ and thus, under standard conditions, will tend to occur from left to right.

An Unfavorable Chemical Reaction Can Proceed If It Is Coupled with an Energetically Favorable Reaction

Many chemical reactions in cells have a positive ΔG; they are energetically unfavorable and will not proceed spontaneously. One example is the synthesis of small peptides, such as glycylalanine, or proteins, from amino acids. How can such a reaction proceed? The cell's solution is to couple a reaction that has a positive ΔG to a reaction that has a negative ΔG of larger magnitude, so that the sum of the two reactions has a negative ΔG. Suppose that the reaction

$$A \rightleftharpoons B + X$$

has a $\Delta G^{\circ\prime}$ of +5 kcal/mol and that the reaction

$$X \rightleftharpoons Y + Z$$

has a $\Delta G^{\circ\prime}$ of −10 kcal/mol. In the absence of the second reaction, there would be much more A than B at equilibrium. The occurrence of the second process, by which X becomes Y + Z, changes that outcome: because it is such a favorable reaction, it will pull the first process toward the formation of B and the consumption of A.

The $\Delta G^{\circ\prime}$ of the overall reaction will be the sum of the $\Delta G^{\circ\prime}$ values of each of the two partial reactions:

$$
\begin{array}{lll}
A \rightleftharpoons B + X & \Delta G^{\circ\prime} = +5\ \text{kcal/mol} \\
\underline{X \rightleftharpoons Y + Z} & \underline{\Delta G^{\circ\prime} = -10\ \text{kcal/mol}} \\
Sum:\quad A \rightleftharpoons B + Y + Z & \Delta G^{\circ\prime} = -5\ \text{kcal/mol}
\end{array}
$$

The overall reaction releases energy. In cells, energetically unfavorable reactions of the type $A \rightleftharpoons B + X$ are often coupled to the hydrolysis of the compound adenosine triphosphate (ATP), a reaction with a negative change in free energy ($\Delta G^{\circ\prime} = -7.3$ kcal/mol), so that the overall reaction has a negative $\Delta G^{\circ\prime}$.

Hydrolysis of the Phosphoanhydride Bonds in ATP Releases Substantial Free Energy

Cells extract energy from foods through a series of reactions that exhibit negative free-energy changes. Much of the free energy released is not allowed to dissipate as heat but is captured in chemical bonds formed by other molecules for use throughout the cell. In almost all organisms, the most important molecule for capturing and transferring free energy is *adenosine triphosphate,* or *ATP* (Figure 2-22).

The useful free energy in an ATP molecule is contained in *high-energy phosphoanhydride bonds,* which are formed from the condensation of two molecules of phosphate by the loss of water:

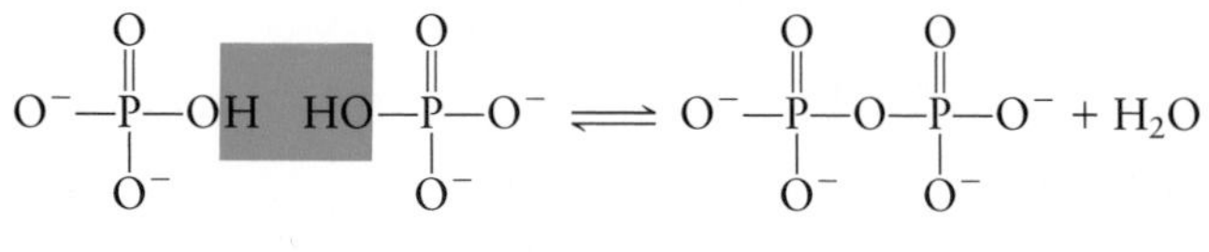

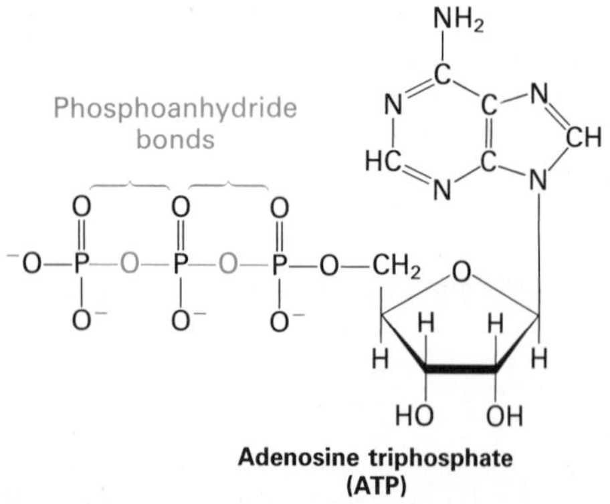

Adenosine triphosphate (ATP)

▲ FIGURE 2-22   The structure of adenosine triphosphate (ATP), in which two high-energy phosphoanhydride bonds (red) link the three phosphate groups.

An ATP molecule has two phosphoanhydride bonds and is often written

$$\text{Adenosine—p}\sim\text{p}\sim\text{p}$$

or simply

$$\text{Ap}\sim\text{p}\sim\text{p}$$

where p stands for a phosphate group and $\sim$ denotes a high-energy bond.

Breaking of a phosphoanhydride bond has a highly negative $\Delta G^{\circ\prime}$, as in hydrolysis:

$$\text{Ap}\sim\text{p}\sim\text{p} + H_2O \longrightarrow \text{Ap}\sim\text{p} + P_i + H^+$$

or

$$\text{Ap}\sim\text{p}\sim\text{p} + H_2O \longrightarrow \text{Ap} + PP_i + H^+$$

or

$$\text{Ap}\sim\text{p} + H_2O \longrightarrow \text{Ap} + P_i + H^+$$

where P_i stands for inorganic phosphate and PP_i for inorganic pyrophosphate, two phosphate groups linked by a phosphoanhydride bond. As the top two reactions show, the removal of a phosphate or a pyrophosphate group from ATP leaves adenosine diphosphate (ADP) or adenosine monophosphate (AMP), respectively.

The phosphoanhydride bond is an ordinary covalent bond, but it releases about 7.3 kcal/mol of free energy (under standard biochemical conditions) when it is broken. In contrast, hydrolysis of the phosphoester bond in AMP, forming inorganic phosphate and adenosine, releases only about 2 kcal/mol of free energy. This is why phosphoanhydride bonds are termed "high-energy" bonds, even though the $\Delta G^{\circ\prime}$ for the reaction of succinate with oxygen is much higher, -37 kcal/mol.

In cells, the free energy released by the hydrolysis of phosphoanhydride bonds is transferred to other molecules. This supplies them with enough free energy to undergo reactions that would otherwise be unfavorable. For example, if the reaction

$$B + C \longrightarrow D$$

is energetically unfavorable ($\Delta G > 0$), it can be made favorable by linking it to the hydrolysis of the terminal phosphoanhydride bond in ATP. Some of the energy in this phosphoanhydride bond is used to transfer a phosphate group to one of the reactants, forming a phosphorylated intermediate, B~p. The intermediate thus has enough free energy to react with C, forming D and free phosphate:

$$B + Ap{\sim}p{\sim}p \longrightarrow B{\sim}p + Ap{\sim}p$$

$$B{\sim}p + C \longrightarrow D + P_i$$

Thus, the overall reaction is

$$B + C + Ap{\sim}p{\sim}p \longrightarrow D + Ap{\sim}p + P_i$$

which is energetically favorable. Chapter 4 illustrates in detail how the hydrolysis of ATP is coupled to protein formation from amino acids; in the above example B and C would represent amino acids and D a dipeptide. Cells keep the ratio of ATP to ADP and AMP high, often as high as 10:1. Thus reactions in which the terminal phosphate group of ATP is transferred to another molecule will be driven even further along.

The $\Delta G^{\circ\prime}$ for hydrolysis of a phosphoanhydride bond in ATP (-7.3 kcal/mol) is twice the $\Delta G^{\circ\prime}$ for hydrolysis of a phosphoester bond, such as that in glucose 6-phosphate (-3.3 kcal/mol) (Table 2-9). Why is this so? A principal reason is that ATP and its hydrolysis products ADP and P_i are highly charged at neutral pH. Three of the four ionizable protons in ATP are fully dissociated at pH 7.0 and the fourth, with a pK_a of 6.95, is about 50 percent dissociated. The closely spaced negative charges in ATP repel each other strongly. When the terminal phosphoanhydride bond is hydrolyzed, some of this stress is removed by the separation of the hydrolysis products ADP^{3-} and HPO_4^{2-}; that is, the separated ADP^{3-} and HPO_4^{2-} will tend not to recombine to form ATP. In glucose-6-phosphate, by contrast, there is no charge repulsion between the phosphate group and the carbon atom to which it is attached. One of the hydrolysis products, glucose, is uncharged, and will not repel the negative HPO_4^{2-} ion, and thus there is less resistance to glucose and HPO_4^{2-} recombining to form glucose-6-phosphate.

Many other bonds—particularly those between a phosphate group and some other substance—have the same high-energy character as phosphoanhydride bonds. The phosphoanhydride bond of ATP is not the most or the least energetic of these bonds (Table 2-9). Why, then, is ATP the most important cellular molecule for capturing and transferring free energy? The free energy of hydrolysis of ATP is sufficiently great that reactions in which the terminal phosphate group is transferred to another molecule have a substantially negative $\Delta G^{\circ\prime}$. However, if the magnitude of the $\Delta G^{\circ\prime}$ for hydrolysis of the phosphoanhydride bond were much higher than it is, cells might require too much energy to form this bond in the first place.

TABLE 2-9   Values of $\Delta G^{\circ\prime}$ for the Hydrolysis of Various Biologically Important Phosphate Compounds*

Compound	$\Delta G^{\circ\prime}$ (kcal/mol)
PHOSPHOENOLPYRUVATE	-14.8
CREATINE PHOSPHATE	-10.3
PYROPHOSPHATE	-8.0
ATP (to ADP + P_i)	-7.3
ATP (to AMP + PP_i)	-7.3
GLUCOSE 1-PHOSPHATE	-5.0
GLUCOSE 6-PHOSPHATE	-3.3
GLYCEROL 3-PHOSPHATE	-2.2

* The bond that is cleaved is indicated by the wavy line.

ATP Is Used to Fuel Many Cellular Processes

If the terminal phosphoanhydride bond of ATP were to rupture by hydrolysis to produce ADP and P_i, energy would be released in the form of heat. Because the action of enzymes couples ATP hydrolysis to other reactions, however, much of this energy is converted to more useful forms, some of which are summarized in Figure 2-23. Energy from ATP is used to synthesize large cellular molecules—for example, to link amino acids to form proteins, or to connect single sugars (monosaccharides) to form polysaccharides. The energy is also used to synthesize many of the small molecules required by the cell. The hydrolysis of ATP supplies the energy needed to move individual cells from one location to another and to contract muscle cells. This energy also plays an important role in the transport of molecules into or out of the cell, usually against a concentration gradient. Gradients of ions, such as Na^+ and K^+, across a cellular membrane are produced by the action of membrane-embedded enzymes, called *ion pumps,* that couple the hydrolysis of ATP to the "uphill" movement of ions. The resulting ion concentration gradients are responsible for the generation of an electric potential across the membrane. This potential is the basis for the electric activity of cells and, in particular, for the conduction of impulses by nerves.

What energy sources are required to form high-energy bonds? Plants and microorganisms trap the energy in light through *photosynthesis:* in the chloroplasts of plant cells and in photosynthetic bacteria, chlorophyll pigments absorb the energy of light, which is then used to synthesize ATP from ADP and P_i. Our current understanding of the mechanism of these photosynthetic processes is described in Chapter 18. Much of the ATP produced in photosynthesis is used to help convert carbon dioxide to polysaccharides that are polymers of six-carbon sugars such as glucose:

$$6CO_2 + 6H_2O \xrightarrow{\text{ATP} \quad \text{ADP} + P_i} C_6H_{12}O_6 + 6O_2$$

Animals release free energy during the oxidation of food molecules in the process of *respiration*. All synthesis of ATP in animal cells and in nonphotosynthetic microorganisms results from the chemical transformation of energy-rich dietary or storage molecules.

One predominant source of energy in cells is the six-carbon sugar glucose. When one mole (180 g) of glucose reacts with oxygen under standard conditions according to the following reaction, 686 kcal of energy is released:

$$C_6H_{12}O_6 + 6O_2 \longrightarrow 6CO_2 + 6H_2O$$
$$\Delta G^{\circ\prime} = -686 \text{ kcal/mol}$$

If glucose is simply burned in air, all of this energy is released as heat. In the cell, through an elaborate set of enzyme-catalyzed reactions, the metabolism of one molecule of glucose is coupled to the synthesis of as many as 36 molecules of ATP from 36 molecules of ADP:

$$C_6H_{12}O_6 + 6O_2 + 36P_i + 36ADP \longrightarrow$$
$$6CO_2 + 6H_2O + 36ATP$$

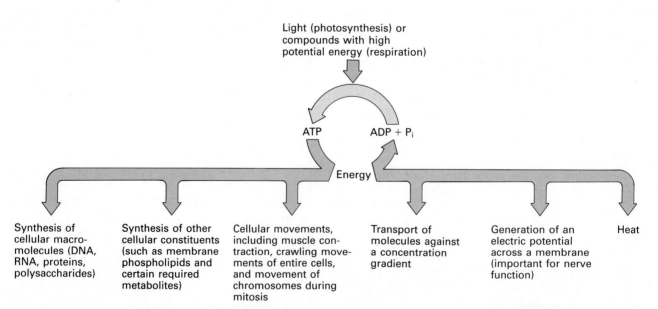

▲ FIGURE 2-23   The ATP cycle. ATP is formed from ADP and P_i by photosynthesis in plants and by the metabolism of energy-rich compounds in most cells. The hydrolysis of ATP to ADP and P_i is linked to many key cellular functions; the free energy released by the breaking of the phosphoanhydride bond is trapped as usable energy.

Because one high-energy phosphoanhydride bond in ATP represents 7.3 kcal/mol, about 263 kcal of energy is conserved in ATP per mole of glucose metabolized (an efficiency of 263/686, or about 38 percent). This type of cellular metabolism is termed *aerobic* because it is dependent on the oxygen in the air. Aerobic *catabolism* (degradation) of glucose is found in all higher plant and animal cells and in many bacterial cells. The overall result of glucose respiration,

$$C_6H_{12}O_6 + 6O_2 \longrightarrow 6CO_2 + 6H_2O$$

is an exact reversal of the photosynthetic reaction in which polymers of six-carbon sugars are formed:

$$6CO_2 + 6H_2O \longrightarrow C_6H_{12}O_6 + 6O_2$$

except that the energy of light is essential for the photosynthetic reaction. Respiration and photosynthesis are the two major processes constituting the *carbon cycle* in nature: sugars and oxygen produced by plants are the raw materials for respiration and the generation of ATP by plant and animal cells alike; the end products of respiration, CO_2 and H_2O, are the raw materials for the photosynthetic production of sugars and oxygen. The only net source of energy in this cycle is sunlight. Thus, directly or indirectly, photosynthesis is the source of chemical energy for almost all cells.

However, communities of organisms do exist in deep ocean vents where sunlight is completely absent. These unusual bacteria derive the energy for converting ADP and P_i into ATP from the oxidation of reduced inorganic compounds present in the dissolved vent gas that originates in the center of the earth. Unfortunately, little is yet known about the biology of these organisms.

Polymers of Glucose with Specific Glycosidic Linkages Serve as Storage Reservoirs

Because D-glucose is the principal source of cellular energy, it is important for cells to maintain a reservoir of it. Some sugars are stored as *disaccharides,* which consist of two monosaccharides linked together by a C–O–C bridge called a *glycosidic bond;* two examples are shown in Figure 2-24. The disaccharide lactose is the major sugar in milk; sucrose is a principal product of plant photosynthesis.

In the formation of any glycosidic bond, the 1 carbon atom of one sugar molecule reacts with a hydroxyl group of another. As in the formation of most biopolymers, the linkage is accompanied by the loss of water. In principle, a large number of different glycosidic bonds can be formed between two sugar residues. Glucose could be bonded to fructose, for example, by any of the following linkages: $\alpha(1 \rightarrow 1)$, $\alpha(1 \rightarrow 2)$, $\alpha(1 \rightarrow 3)$, $\alpha(1 \rightarrow 4)$, $\alpha(1 \rightarrow 6)$, $\beta(1 \rightarrow 1)$, $\beta(1 \rightarrow 2)$, $\beta(1 \rightarrow 3)$, $\beta(1 \rightarrow 4)$, or $\beta(1 \rightarrow 6)$, where α or β specifies the conformation at the 1 carbon in glucose and the number following the arrow indicates the fructose carbon to which the glucose is bound. Only the $\alpha(1 \rightarrow 2)$ linkage occurs in sucrose because of the specificity of the enzyme (the biological catalyst) for the linking reaction.

Glycosidic linkages also join chains of monosaccharides into longer polymers. *Cellulose* is a long-chained polymer of glucose linked together by $\beta(1 \rightarrow 4)$ glycosidic bonds. Cellulose is the major constituent of plant cell walls and is the most abundant organic chemical on earth. Its important structural and functional features are described in Chapter 24. The most common storage carbohydrate in animal cells is *glycogen,* a very long, branched polymer of glucose linked together by $\alpha(1 \rightarrow 4)$ glycosidic bonds. As much as 10 percent of the weight of liver can be glycogen. The synthesis and utilization of this polymer are described in Chapter 20.

◄ FIGURE 2-24
The formation of glycosidic linkages to generate the disaccharides lactose and sucrose. The lactose linkage is $\beta(1 \rightarrow 4)$; the sucrose linkage is $\alpha(1 \rightarrow 2)$. Sucrose is one of the major products of photosynthesis in higher plants, and lactose is an abundant nutrient in human and cow's milk.

➤ *Activation Energy and Reaction Rate*

Many chemical reactions that exhibit a negative $\Delta G^{\circ\prime}$ do not proceed unaided at a measurable rate. For example, in pure aqueous solutions, glyceraldehyde 3-phosphate (G3P), our recurrent model, is a fairly stable compound that reacts very slowly or not at all, yet in cells it can undergo several different reactions (Figure 2-25), each of which has a negative $\Delta G^{\circ\prime}$.

Similarly, a mixture of hydrogen and oxygen gases sealed in a flask at room temperature will remain quiescent indefinitely. If, however, the gases are exposed to an electric spark or a flame, the mixture explodes. The hydrogen gas burns vigorously, combining with the oxygen to form water:

$$2H_2 + O_2 \longrightarrow 2H_2O$$

This reaction releases a considerable amount of energy (57.8 kcal/mol of H_2O formed).

To understand this seemingly ambiguous behavior, consider the mechanism of the reaction of hydrogen with oxygen. Three molecules, two H_2 and one O_2, must come together in such a way that the bonds can rearrange to form the products. At the high temperature in a hydrogen flame, many molecules are moving so fast that when two of them collide, one hydrogen molecule breaks up into single atoms:

$$2H_2 \longrightarrow H_2 + 2H$$

This reaction requires an input of energy. Some of the kinetic energy of the two colliding molecules is used to break one of their covalent bonds; thus the products of the reaction do not have as much kinetic energy as the intact molecules did. The kinetic energy used to break the bond has been transformed into chemical potential energy; the single hydrogen atoms react very easily with oxygen molecules:

$$4H + O_2 \longrightarrow 2H_2O$$

This reaction releases energy—more energy, in fact, than was absorbed by the energy-requiring reaction in which the H_2 was broken apart. If we consider the two reactions as separate stages of an overall reaction, then the overall reaction has a negative change in enthalpy ($\Delta H < 0$) because the second stage releases more bond energy than the first absorbs.

Energy Is Required to Initiate a Reaction

The input of energy required to initiate a reaction is called the *activation energy*. In a burning flame, the kinetic energy of many molecules of H_2 or O_2 is great enough to generate the activation energy. At room or body temperature, however, the average kinetic energy of a typical gas such as H_2 is about 1.5 kcal/mol. Although many molecules will have more kinetic energy than this average, the chances are virtually zero that any one H_2 molecule will have enough kinetic energy to equal the activation energy (~100 kcal/mol) and thereby initiate the reaction. Thus a mixture of hydrogen and oxygen will not react until some of the molecules acquire enough kinetic energy—say, from the heat in a flame—to overcome the activation-energy barrier. The energetic relation between the initial reactants and the products of a reaction can usually be described by a diagram (Figure 2-26).

All biochemical reactions with a negative ΔG (such as the formation of H_2O from H_2 and O_2) have an activation

◄ FIGURE 2-25 Glyceraldehyde 3-phosphate, like most cellular molecules, can undergo any of several reactions with negative $\Delta G^{\circ\prime}$ values: oxidation (by O_2) to 3-phosphoglyceric acid, hydrolysis to glyceraldehyde and phosphate (P_i), and rearrangement to dihydroxyacetone phosphate. In the absence of enzymes or other catalysts, however, these reactions cannot occur in aqueous solution. Different enzymes catalyze each reaction; the presence of a specific enzyme is required for a particular reaction to proceed.

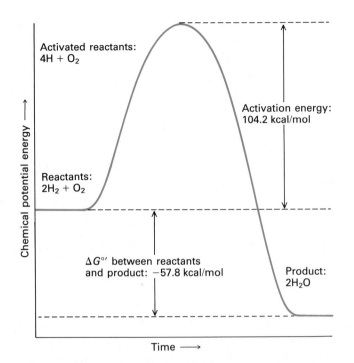

▲ FIGURE 2-26   The reaction of hydrogen with oxygen requires an initial input of 104.2 kcal/mol (the activation energy) even though the products have a much lower free energy than the reactants do. The energy content of the reactants is depicted at each stage.

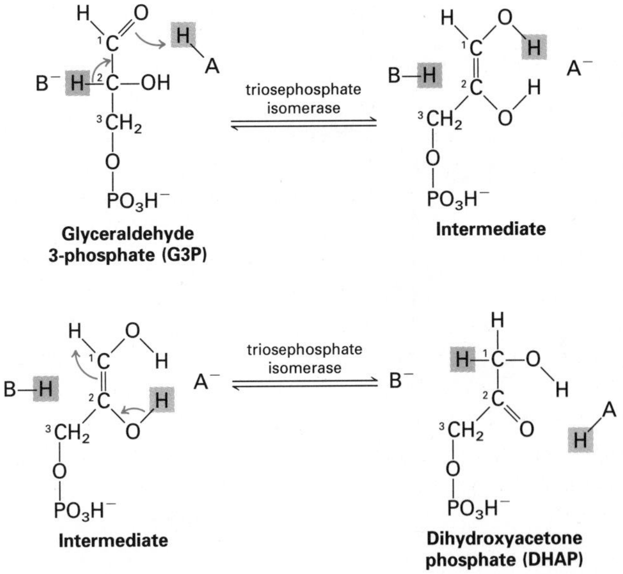

▲ FIGURE 2-27   The conversion of glyceraldehyde 3-phosphate (G3P) to dihydroxyacetone phosphate (DHAP) involves an intermediate. Two groups, a base B⁻ and an acid HA, are parts of triosephosphate isomerase, the enzyme that catalyzes this reaction. To form the intermediate, B⁻ abstracts a proton (blue) from the 2 carbon of G3P; HA adds a proton (red) to the keto-oxygen on the 1 carbon. To convert the intermediate to DHAP, BH donates its proton to the 1 carbon (regenerating the original B⁻) and A⁻ abstracts a proton from the −OH on the 2 carbon (regenerating HA). The curved arrows denote the movements of pairs of electrons that accompany the making and breaking of these bonds. [See D. Straus, et al., 1985, *Proc. Nat'l. Acad. Sci.* **82**:2272.]

energy; they proceed through one or more intermediate *transition states* whose content of free energy is greater than that of either the reactants or products. In some reactions, certain covalent bonds are moved to a strained position in the transition state, and an input of energy is essential for this to happen. In other reactions, to form the transition state electrons must be excited, requiring an input of energy; only then can they pair up in a covalent bond in the product. In still others, molecules need only enough energy to overcome the mutual repulsion of their electron clouds to get close enough to react. For example, the conversion of glyceraldehyde 3-phosphate (G3P) to dihydroxyacetone phosphate (DHAP) involves at least one intermediate (Figure 2-27). As the intermediate forms, the following changes take place simultaneously: a proton is removed from one carbon, another proton is donated to an oxygen, and pairs of electrons move from one bond to another. The activation energy required by each of these partial reactions contributes to the overall activation energy needed to form this reaction intermediate, which then rearranges to generate the final reaction product (bottom of Figure 2-27).

Thus each stage in a multistep reaction has its own activation energy (Figure 2-28). For the overall reaction to proceed, the highest activation energy must be achieved. Because biochemical reactions occur at moderate temperatures, the kinetic energy of colliding molecules is generally insufficient to provide the necessary activation energy; in most cases, enzymes reduce the activation energy requirement, allowing specific reactions to proceed.

Enzymes Catalyze Biochemical Reactions

Two significant rate-regulating factors for biological systems are the concentrations of the reactants and the pH. A reaction involving two or more different molecules proceeds faster at high concentrations because the molecules are more likely to encounter one another. The pH determines the dissociation state of the various acidic and basic groups on biological molecules; in general, only one of the

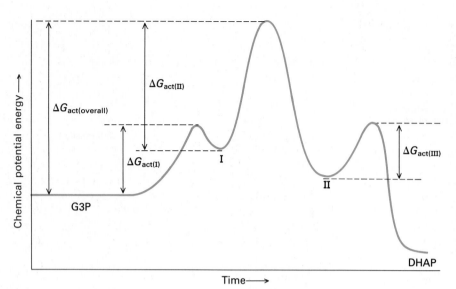

◄ FIGURE 2-28   Hypothetical energy changes in the conversion of glyceraldehyde 3-phosphate (G3P) to dihydroxyacetone phosphate (DHAP). The troughs in the curve represent stable intermediates in the reaction. The vertical distance from a trough to the succeeding crest represents the activation energy (ΔG_{act}) required for one intermediate to be converted to the next or the final product (for example, $\Delta G_{act(II)}$ represents the activation energy required for the conversion of intermediate I to intermediate II). The total activation energy $\Delta G_{act(overall)}$ is the difference between the free energy of the reactants and that represented by the highest crest along the pathway.

possible ionic forms (e.g., the zwitterion form of an amino acid) can undergo a particular reaction.

However, the most important determinants of biochemical reaction rates are *enzymes,* the proteins that act as catalysts (page 75). As we discussed earlier, enzymes, like all catalysts, cause reactions to reach equilibrium faster. A catalyst accelerates the rates of forward and reverse reactions by the same factor. It does not alter the change in free energy or the equilibrium constant; in fact, whether or not a catalyst is present, the equilibrium constant is equal to the ratio of the rate of the forward reaction to that of the reverse reaction.

Enzymes and all other catalysts act by reducing the activation energy required to make a reaction proceed (Figure 2-29). To achieve this, an enzyme binds either a single *substrate* (the substance on which the enzyme acts) or a set of similar substrates. Each enzyme catalyzes a single chemical reaction on the bound substrate. Triosephosphate isomerase, the enzyme that catalyzes the conversion of G3P to DHAP, binds G3P as its substrate. Once G3P is bound, the requisite movements of its hydrogen atoms, protons, and electrons are all facilitated by specific chemical groups on the parts of the enzyme adjacent to the bound substrate (see Figure 2-27).

Some enzymes bind a substrate in a way that strains certain of its bonds and makes it easy for these bonds in the substrate to undergo a reaction. That is, the enzyme stabilizes the transition state of the reaction by tightly binding a specific form of the substrate in which certain bonds are strained (Figure 2-29). Other enzymes form a covalent bond with the substrate that enables a different part of the substrate to undergo a reaction; after this happens the bond between the enzyme and substrate is broken. Still other enzymes bind multiple substrates in a way that brings them together so they can react readily with one another. In each case, the overall effect is to reduce the activation energy needed for formation of the reaction intermediate. Thus the presence or absence of particular enzymes in a cell or in extracellular fluids determines which of many possible chemical reactions will occur.

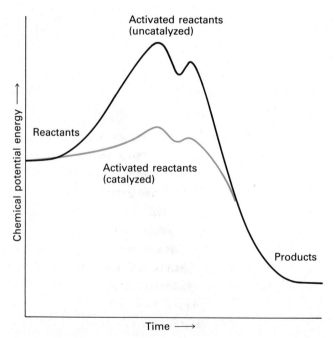

▲ FIGURE 2-29   A catalyst accelerates the rate of a reaction by reducing the activation energy requirement. Catalysts do not alter the free energy of reactants or products or affect their equilibrium concentrations.

SUMMARY

Atoms in a molecule are held together in a fixed orientation chiefly by covalent chemical bonds. Such bonds consist of pairs of electrons shared by two atoms, and they require relatively high energies to break them (50–200 kcal/mol). Many covalent bonds between unlike atoms are polar, meaning that the electrons spend more time around the more electronegative atom.

Carbohydrates (which serve as important energy reservoirs and structural components in cells) illustrate the significance of subtle differences in the arrangements of covalent bonds: As a linear molecule, D-glucose ($C_6H_{12}O_6$) contains four asymmetric carbon atoms; when D-glucose forms a pyranose ring, another of its carbons [carbon 1] becomes asymmetric, to yield α-D-glucopyranose or β-D-glucopyranose. Two other important sugars, mannose and galactose, differ from glucose only in the positions of the H and OH substituents on one carbon atom.

A number of interactions that are weaker than covalent bonds help to determine the shape of many large biological molecules and to stabilize complexes composed of two or more different molecules. Although any single noncovalent bond may be very weak (most release energy of 0.5–5 kcal/mol when they form), several such bonds between molecules or between the parts of one molecule can result in very stable structures. The attraction between two large molecules can be quite strong if they have complementary sites on their surfaces.

Noncovalent bonds in biological systems fall into four main types. In a hydrogen bond, a hydrogen atom covalently bonded to a donor atom associates with an acceptor atom whose nonbonding electrons attract the hydrogen. Hydrogen bonds among water molecules are largely responsible for the properties of both liquid water and the crystalline solid (ice) form. Ionic bonds result from the electrostatic attraction between the positive and negative charges of ions. In aqueous solutions, all cations and anions are surrounded by a tightly bound shell of water molecules. The weak and relatively nonspecific van der Waals interactions are created whenever any two atoms approach each other closely. Hydrophobic bonds occur among nonpolar molecules, such as hydrocarbons, in an aqueous environment. Hydrophobic bonds result from the van der Waals forces that arise between the nonpolar molecules and from the reduction in entropy of water that would otherwise occur if "cages" of water molecules surrounded individual hydrophobic molecules. Hydrophobic bonds are responsible for the organization of phospholipids into biomembranes, and the hydrophobic core of a biomembrane separates two aqueous solutions, such as the cytosol of a cell from its external environment.

The most useful measure of chemical reactions in biological systems is the change in free energy, ΔG. This value depends on the change in heat or enthalpy, ΔH (chiefly the overall change in bond energies), the change in entropy, ΔS (the randomness of molecular motion), and the temperature, T: $\Delta G = \Delta H - T\,\Delta S$. Chemical reactions tend to proceed in the direction for which ΔG is negative. The direction of a chemical reaction can be predicted if both its standard free energy change, $\Delta G^{\circ\prime}$, and the concentrations of reactants and products are known. The equilibrium constant, K_{eq}, is mathematically related to $\Delta G^{\circ\prime}$: $K_{eq} = 10^{-(\Delta G^{\circ\prime}/2.3RT)}$, and the value of $\Delta G^{\circ\prime}$ can be calculated from the experimentally determined concentrations of reactants and products at equilibrium. A chemical reaction having a positive ΔG can proceed spontaneously if it is coupled with a reaction having a negative ΔG of larger magnitude. Such coupled reactions are involved in the synthesis of proteins from amino acids, the synthesis of other molecules, cellular movement, and the transport of compounds into or out of a cell against a concentration gradient. All of these energetically unfavorable cellular reactions are fueled by hydrolysis of one or both of the two phosphoanhydride bonds in adenosine triphosphate (ATP). In cells, these bonds are a principal source of chemical potential energy.

In plant cells, ATP is generated from adenosine diphosphate (ADP) and inorganic phosphate (P_i) through the use of energy absorbed from light; much of the ATP is consumed in the synthesis of polymers of six-carbon sugars from CO_2 and H_2O. In animal cells and in nonphotosynthetic microorganisms, most of the ATP is generated during the oxidation (by O_2) of glucose to CO_2 and H_2O. Directly or indirectly, photosynthesis is the ultimate source of chemical energy for almost all cells.

Whether a reaction will actually proceed depends on how much activation energy it requires. Sometimes the kinetic energy of the reactant molecules is sufficient to overcome the activation-energy barrier. If the activation energy is too high to permit the reaction to occur at a measurable rate, catalysts can speed up the reaction by decreasing the activation energy requirement. A catalyst accelerates both the forward and the reverse reactions to the same extent; it does not change ΔG, $\Delta G^{\circ\prime}$, or K_{eq}. Enzymes are biological catalysts that generally facilitate only one of the many possible transformations that a molecule can undergo.

REVIEW QUESTIONS

1. Review the fundamental characteristics of covalent and noncovalent chemical bonds. Refer to the following chemical structures when answering the questions.

$CH_3CH_2CH_2CH_2CH_2CH_3$ $CH_2=CHCH_2CH_2CH_2CH_3$

_n_-Hexane   **1-Hexene**

CH_3CH_2OH CH_3NH_2 $CH_3\overset{\overset{\displaystyle O}{\|}}{C}CH_3$

Ethanol   **_n_-Methylamine**   **Acetone**

KCl $CH_3CH_2\overset{\overset{\displaystyle O}{\|}}{C}H$ $^+NH_3-CH_2-\overset{\overset{\displaystyle O}{\|}}{C}-O^-$

Potassium   **Propanal**   **Glycine**
chloride

$^+H_3N-\overset{\overset{\displaystyle COO^-}{|}}{\underset{\underset{\displaystyle CH_2OH}{|}}{C}}-H$ $^+NH_3-\overset{\overset{\displaystyle O}{}}{\underset{\underset{\displaystyle CH_3}{|}}{CH}}-\overset{\overset{\displaystyle O}{\|}}{C}-O^-$

Serine   **Alanine**

a. Which of these molecules would exhibit free rotation around all their carbon-carbon bonds? Which would exhibit hindered rotation around these bounds? Identify the location(s) of hindered rotation.

b. Which of the molecules would have a permanent dipole moment? Label the δ^+ and δ^- positions in those that would.

c. Which two of these are structural isomers of each other?

d. Which of these molecules have chiral centers? For those that do, identify the atom that confers chirality.

e. If an aqueous solution of each of these were prepared, which of the compounds would exhibit hydrogen bonding when dissolved? Which would interact in solution by van der Waal's forces? Which would interact via hydrophobic bonds? Which would have actual charges associated with them?

2. Review the concepts of equilibrium constants and strong and weak acids. Acetic acid, CH_3COOH, is a weak acid that partially dissociates according to the following:

$$CH_3COOH \rightleftharpoons CH_3COO^- + H^+$$

a. What would be the equation that would describe the K_{eq} for this process? Is this the same as the equation that yields K_a?

b. The pK_a for this reaction at 25°C and 1 atm is 4.7. What is the concentration of H^+ in this solution? What is the ratio of CH_3COO^- to CH_3COOH in a solution with a pH of 7.2?

c. Assume that you have 1 liter of a solution that is 0.499 M in CH_3COO and 0.001 M in CH_3COOH. What is the pH? If you add 0.050 moles of concentrated HCl, what would be the resulting pH? (Because you use concentrated HCl, assume that the volume change is negligible.) Suppose that you have 1 liter of a solution that is 0.250 M in CH_3COO^- and 0.250 M in CH_3COOH. What is the pH of this solution? What would be the resulting pH if you added 0.050 moles of concentrated HCl to it? Why is the change in pH dramatically different for each of these situations even though you added the same amount of HCl?

3. Cesium chloride (CsCl) is a salt that is often used in the laboratory in a process known as buoyant density gradient centrifugation, an experimental technique with which you will become familiar in later chapters. During the dissolution of CsCl at room temperature, the solution becomes cold to the touch.

a. What are the signs of ΔH, ΔS, and ΔG when CsCl goes into solution?

b. What is the ultimate driving force for the dissolution of CsCl in water? Explain your answer.

The following is an important biochemical reaction:

$$\text{Glutamate + oxaloacetate} \underset{k_2}{\overset{k_1}{\rightleftharpoons}} \alpha\text{-ketoglutarate + aspartate}$$

The $\Delta G^{\circ\prime}$ for this reaction is -1.15 kcal/mol. Assume that you have prepared a solution of glutamate and oxaloacetate at 25°C, 1 atm pressure, and you are going to monitor the formation of the products.

a. What would happen to K_{eq}, k_1, k_2, and the rate of the forward reaction if the initial concentrations of the reactants were increased? decreased?

b. This reaction is catalyzed by an enzyme called aspartate aminotransferase. What would happen to K_{eq}, k_1, k_2, and the rate of the forward reaction if this enzyme were added to the system?

c. What would happen to K_{eq}, k_1, k_2, and the rate of the forward reaction if the initial concentrations of the reactants were increased? decreased?

d. The sign of $\Delta G^{\circ\prime}$ suggests that this reaction is favorable. What does it tell you about the rate of the reaction? Explain your answer.

References

ALBERTY, R. A., and R. J. SILBEY. 1992. *Physical Chemistry*. Wiley. Chapters 1, 2, 3, 4, and 5.

ATKINS, P. W. 1993. *The Elements of Physical Chemistry*. W. H. Freeman and Company. Chapters 2, 3, 5, 7, 9.

ATKINS, P. W. 1992. *General Chemistry*. Scientific American Books.

BUTTLER, J. 1964. *Solubility and pH Calculations*. Addison-Wesley.

CANTOR, P. R., and C. R. SCHIMMEL. 1980. *Biophysical Chemistry*. W. H. Freeman and Company. Part 1, Chapter 5.

CASTELLAN, G. W. 1983. *Physical Chemistry*. Benjamin Cummings. Chapters 1, 5, 6, 7, 8, 9, and 10.

DAVENPORT, H. W. 1974. *ABC of Acid-Base Chemistry*, 6th ed. University of Chicago Press.

EDSALL, J. T., and J. WYMAN. 1958. *Biophysical Chemistry*. Academic Press. Vol. 1.

EISENBERG, D., and D. CROTHERS. 1979. *Physical Chemistry with Applications to the Life Sciences.* Benjamin-Cummings.

GENNIS, R. B. 1989. *Biomembranes: Molecular Structure and Function.* Springer-Verlag, New York, Inc. Chapters 1, 2, and 3.

GUYTON, A. C. 1991. *Textbook of Medical Physiology.* Saunders. Chapter 30.

HILL, T. J. 1977. *Free Energy Transduction in Biology.* Academic Press.

KEMP, D. S., and F. VELLACCIO. 1980. *Organic Chemistry.* Worth. Chapters 1 and 5.

KLOTZ, I. M. 1978. *Energy Changes in Biochemical Reactions.* Academic Press.

LEHNINGER, A. L., D. L. NELSON, and M. M. COX. 1993. *Principles of Biochemistry,* 2d ed. Worth. Chapters 1, 3, 4, and 13.

MURRAY, R. K., et al. 1990. *Harper's Biochemistry.* Lange. Chapters 3, 12, and 15.

NICHOLLS, D. G., and S. J. FERGUSON. 1992. *Bioenergetics 2.* Academic Press.

SHARON, N. 1980. Carbohydrates. *Sci. Am.* 243(5):90–116.

STRYER, L. 1995. *Biochemistry,* 4th ed. W. H. Freeman and Company. Chapters 1, 8, 17, and 21.

TANFORD, C. 1980. *The Hydrophobic Effect: Formation of Micelles and Biological Membranes,* 2d ed. Wiley.

WATSON, J. D., et al. 1988. *Molecular Biology of the Gene,* 4th ed. Benjamin-Cummings. Chapters 2 and 5.

WOOD, W. B., et al. 1981. *Biochemistry: A Problems Approach,* 2d ed. Benjamin-Cummings. Chapters 1, 5, and 9.

3 Protein Structure and Function

▲ 50S ribosomal subunits frozen into a 2D crystal.

Proteins, the working molecules of a cell, carry out the program of activities encoded by genes. This program requires the coordinated effort of many different types of proteins: some build the structure of the cell, while others process materials by catalyzing intracellular or extracellular chemical reactions. These vital functions are so entwined in the physiology of cells that it is difficult to think of any cell activity that does not involve proteins; perhaps there are none. Proteins first evolved as rudimentary molecules that facilitated a limited number of chemical reactions. Gradually, the structure of proteins was so adapted that they could catalyze an extraordinary range of chemical reactions—from creating or destroying bonds to transferring chemical groups from one protein to another. From the standpoint of evolution, proteins elevated cellular chemical reactions to a level of efficiency and speed that is nearly impossible to attain in a test tube. Now we realize that proteins are functional or active only when they are folded in a precisely dictated three-dimensional form.

In this chapter, we will study how the structure of a protein gives rise to its function. A repeating theme in biology is that large, complex structures are built from small, simple components. Proteins follow this principle. They are built from single chains of *amino acids,* the building blocks of proteins. Since 20 common amino acids are found in natural proteins, a 100-aa protein has 20^{100} (more than 10^{130}) possible linear combinations of amino acids. Only some combinations form stable and functional proteins; many do not, and such unstable proteins are rapidly degraded. The presence of certain sequences of neighboring amino acids that favor formation of discrete structures is thought to confer stability on a protein. Like letters of an alphabet, a sequence of amino acids forms "words" that denote particular structures. Some amino acid words specify a spiral *helix;* others stand for a straight *strand* or a meandering *random coil.* Portions of an entire protein assume the structures determined by its component amino acid sequences, thus building up the protein's overall three-dimensional structure from helices, strands, and coils.

The first section of this chapter examines protein architecture: the structure and chemistry of amino acids; how amino acids are linked together into a single chain; the forces that guide the chain to fold into simple helices and sheets; and how the internal packing of amino acids forms the skeleton and three-dimensional structures of proteins. In the next two sections, the functions of two classes of proteins, enzymes and antibodies, are described. Enzymes are protein catalysts and serve as general examples of how proteins bind molecules with a high degree of specificity, how they carry out chemical reactions, and how their activity is regulated by other molecules. Antibodies are proteins that recognize foreign molecules and help eliminate them from the body. Because the structure of antibodies must be adapted to binding molecules of widely different structures, they are good models for understanding how proteins mutually recognize each other. The last section of the chapter summarizes the three most commonly used techniques for isolating proteins and characterizing their structures: centrifugation, chromatography, and electrophoresis.

As discussed in the next chapter, the sequence of bases in DNA is translated into the sequence of amino acids in proteins. The deciphering of the genetic code signaled the beginning of modern molecular biology. Study of proteins has shown that the linear amino acid sequence of a protein specifies its spatial organization. Although the mechanism by which the linear sequence of proteins is synthesized is now understood in detail, the rules determining their three-dimensional structure remain to be uncovered.

► *General Structure of Proteins*

Proteins are designed to bind every conceivable molecule—from simple ions to large complex molecules like fats, sugars, nucleic acids, and other proteins. They catalyze an extraordinary range of chemical reactions, provide structural rigidity to the cell, control flow of material through membranes, regulate the concentrations of metabolites, act as sensors and switches, cause motion, and control gene function. The three-dimensional structure of proteins has evolved to carry out these functions efficiently and under precise control. The spatial organization of proteins is a key to understanding how they work. In a sense, we must think like cellular architects or protein engineers and constantly ask ourselves, "How would I design a protein machine to carry out analogous activities?"

Form and Function are Inseparable in Protein Architecture

A few protein structures are shown in Figure 3-1 to illustrate how the design and function of proteins are intimately linked. For instance, a barrel-like nuclear pore, a complex of several proteins, sits in the nuclear membrane and acts as a channel through which molecules traverse in or out of the nucleus (Figure 3-1a). Actin filaments of the cytoskeleton (Figure 3-1b) form a framework of girders and struts, which reinforce the cytoplasm like steel rods in concrete. Some proteins have grooves in their surface, which are logical binding sites for a variety of molecules, especially rod-shaped or filamentous ones. For example, reverse transcriptase, which copies RNA into DNA, is grooved so that RNA slides along the surface of the protein (Figure 3-1c). The two C-shaped β subunits of DNA polymerase III, clamp around double-stranded DNA to form a donut-shaped molecule (Figure 3-1d). This structure holds the polymerase complex firmly to DNA as the protein travels along the DNA strand. A delight in studying the structures of proteins is uncovering simple but ingenious ways that nature has built a protein to perform a particular function.

One of the major enigmas in biology today is how proteins, which are constructed from only 20 different amino acids, carry out the incredible array of diverse tasks that they do. Unlike the intricate branched structure of carbohydrates, proteins are single, unbranched chains of amino acid monomers. The unique shape of proteins arises from noncovalent interactions between regions in the linear sequence of amino acids. Only when a protein is in its correct three-dimensional conformation is it able to function efficiently. A key concept in understanding how a protein works is that *function is derived from three-dimensional structure, and three-dimensional structure is specified by amino acid sequence.*

Amino Acids, the Building Blocks of Proteins, Differ Only in Their Side Chains

Amino acids are the monomeric building blocks of proteins. The α carbon atom (C_α) of amino acids, which is adjacent to the carboxyl group, is asymmetric; that is, it is bonded to four different chemical groups (Figure 3-2). They are an amino (NH_2) group, a carboxyl (COOH) group, a hydrogen (H) atom, and one variable group, called a side chain or R group. Of the two stereoisomeric forms (Figure 2-4), almost all amino acids in proteins are the L form. Amino acids are also dipolar ions (two separated charged groups), or zwitterions, because at the pH of the cytoplasm, amino and carboxyl groups are ionized as $-NH_3^+$ and $-COO^-$. All 20 different amino acids have this same general structure, but their side chain groups vary in size, shape, charge, hydrophobicity, and reactivity.

The amino acids can be considered the alphabet in which linear proteins are "written." Students of biology must be familiar with the special properties of each letter of this alphabet, which are determined by the side chain. Amino acids can be classified into a few distinct categories based primarily on their solubility in water, which is influenced by the polarity of their side chains (Figure 3-3). Polar

(a)

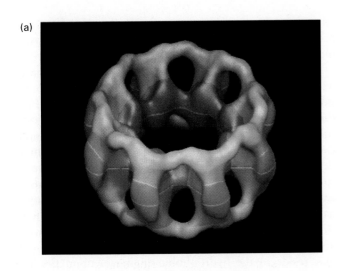

(c)

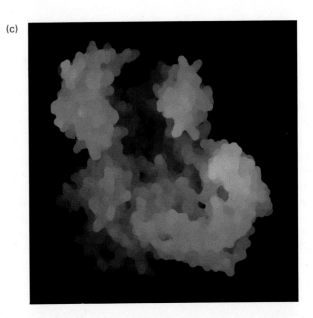

(b)

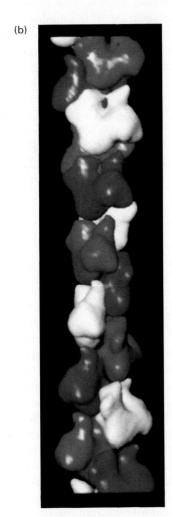

(d)

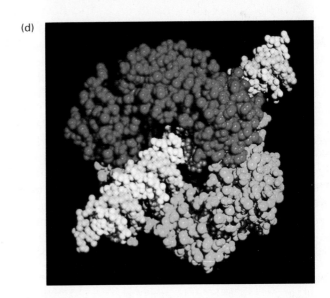

measures 5.5 nm on a side and 3.5 nm thick. This figure shows 14 subunits, alternately colored red, white, and blue. The cell cytoplasm is criss-crossed by actin filaments that are as long as 50–100 μm and have various functions in the cell (Chapter 22). (c) Reverse transcriptase, an enzyme present in RNA viruses that copies the viral RNA genome into DNA. Reverse transcriptase from the AIDS-causing virus HIV measures 11 × 3.0 × 4.5 nm. RNA to be copied lies in a groove on the surface of the enzyme. (d) Two β subunits of DNA polymerase (red and yellow) from *E. coli* clamped around a DNA molecule (light blue). The ring has an outer diameter of 8.0 nm and an inner diameter of 3.5 nm. DNA polymerase is required for DNA synthesis; this association of the β subunits and DNA assures that polymerase remains complexed with DNA during DNA replication (Chapter 10). [Part (a) from J. E. Hinshaw, B. O. Carragher, and R. A. Milligan, 1992, *Cell* **69:**1133; part (b) from M. F. Schmid et al., 1994, *J. Cell Biol.* **124:**341; part (c) from T. A. Steitz et al., 1992, *Science* **256:**1783; part (d) from Kuriyan et al., 1992, *Cell* **69:**425]

▲ FIGURE 3-1   Gallery of protein structural models. (a) The nuclear pore, a complex of proteins with a total molecular weight (MW) of 1.2×10^8, an outer diameter of 133 nm, a height of 70 nm, and a central hole 42 nm in diameter. Transport of molecules into and out of the nucleus occurs through such pores located in the nuclear membrane (Figure 5-33). (b) An actin filament, a polymer built by end-to-end association of identical actin subunits. Each subunit

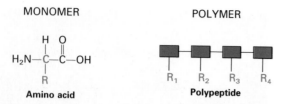

▲ FIGURE 3-2 Amino acids are the monomeric subunits that are linked together to form polypeptides and proteins. The α carbon (green) of each amino acid is bonded to four different chemical groups and thus is asymmetric. The side chain, or R group (red), is unique to each amino acid. Different linear combinations of the 20 naturally occurring amino acids generate the diversity of polypeptides. Thus the short polypeptide shown here, containing only four amino acids, has 20^4, or 160,000, possible structures.

amino acids tend to be on the surface of proteins and keep them soluble in water. In contrast, nonpolar amino acids avoid water and aggregate to form the water-insoluble core of proteins. The polarity of amino acid side chains thus is one of the forces responsible for shaping the final three-dimensional structure of proteins.

Hydrophilic, or water-soluble, amino acids have ionized or polar side chains. At neutral pH, *arginine* and *lysine* are positively charged; *aspartic acid* and *glutamic acid* are negatively charged and exist as aspartate and glutamate. These four amino acids are the prime contributors to the overall charge of a protein. A fifth amino acid, *histidine*, has an imidazole side chain, which has a pK_a of 6.8, the pH of the cytoplasm. As a result, small shifts of cellular pH will change the charge of histidine side chains:

pH 5.8   pH 7.8

At pH 5.8, 90 percent of histidine residues are protonated and thus carry a positive charge, whereas at pH 7.8, 90 percent are uncharged. The activities of many proteins are modulated by pH through protonation of histidine side chains. *Asparagine* and *glutamine* are uncharged but have polar amide groups with extensive hydrogen-bonding capacities. Similarly, *serine* and *threonine* are uncharged but have polar hydroxyl groups, which also participate in hydrogen bonds with other polar molecules. Because the charged and polar amino acids are hydrophilic, they are usually found at the surface of a water-soluble pro-

tein where they contribute not only to the solubility of the protein in water but also form binding sites for charged molecules.

Hydrophobic amino acids have aliphatic side chains, which are insoluble or only slightly soluble in water. The side chains of *alanine, valine, leucine, isoleucine,* and *methionine* consist entirely of hydrocarbons, except for the sulfur atom in methionine, and all are nonpolar. *Phenylalanine, tyrosine,* and *tryptophan* have large bulky aromatic side groups. As explained in Chapter 2, hydrophobic molecules avoid water by coalescing into an oily or waxy droplet. The same forces cause hydrophobic amino acids to pack in the interior of proteins away from the aqueous environment. In Chapter 14, we will see in detail how hydrophobic residues line the surface of membrane proteins that reside in the hydrophobic environment of the lipid bilayer.

Lastly, cysteine, glycine, and proline exhibit special roles in proteins because of the unique properties of their side chains. The side chain of *cysteine* contains a reactive sulfhydryl (SH) group, which can oxidize to form a disulfide (S–S) bond to a second cysteine:

Regions within a polypeptide or in separate polypeptides sometimes are cross-linked covalently through disulfide bonds. Although disulfide bonds are rare in intracellular proteins, they are commonly found in extracellular proteins, where they help maintain the native, folded structure. When the sulfhydryl group in cysteine remains reduced, this amino acid is very hydrophobic. The smallest amino acid, *glycine*, has a single hydrogen atom as its R group. Its small size allows it to fit into tight spaces. Unlike any of the other common amino acids, *proline* has a cyclic ring that is produced by formation of a covalent bond between its R group and the amino group on C_α. Proline is very rigid and its presence creates a fixed kink in a polypeptide chain. Proline and glycine are sometimes found at points on a protein's surface where the polypeptide chain loops back into the protein.

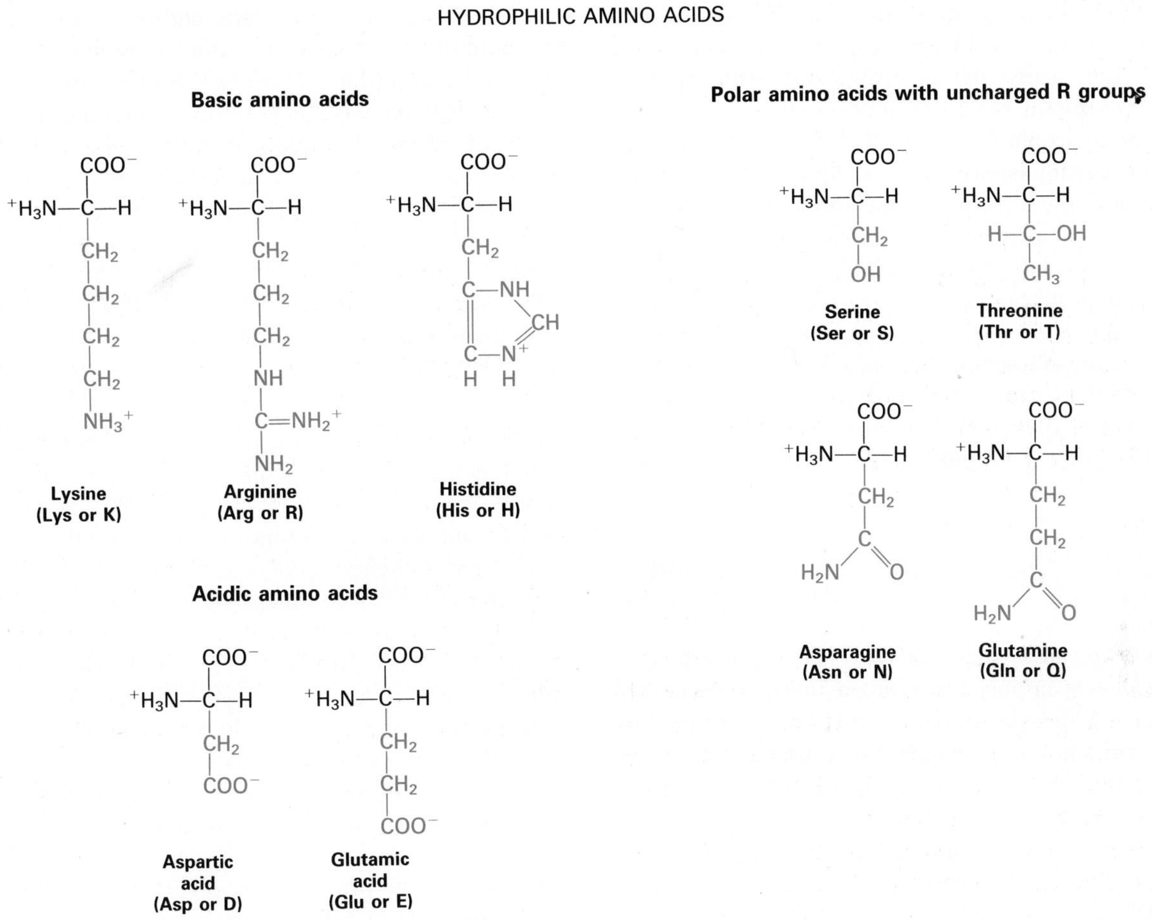

HYDROPHILIC AMINO ACIDS

Basic amino acids

Polar amino acids with uncharged R groups

Lysine
(Lys or K)

Arginine
(Arg or R)

Histidine
(His or H)

Serine
(Ser or S)

Threonine
(Thr or T)

Asparagine
(Asn or N)

Glutamine
(Gln or Q)

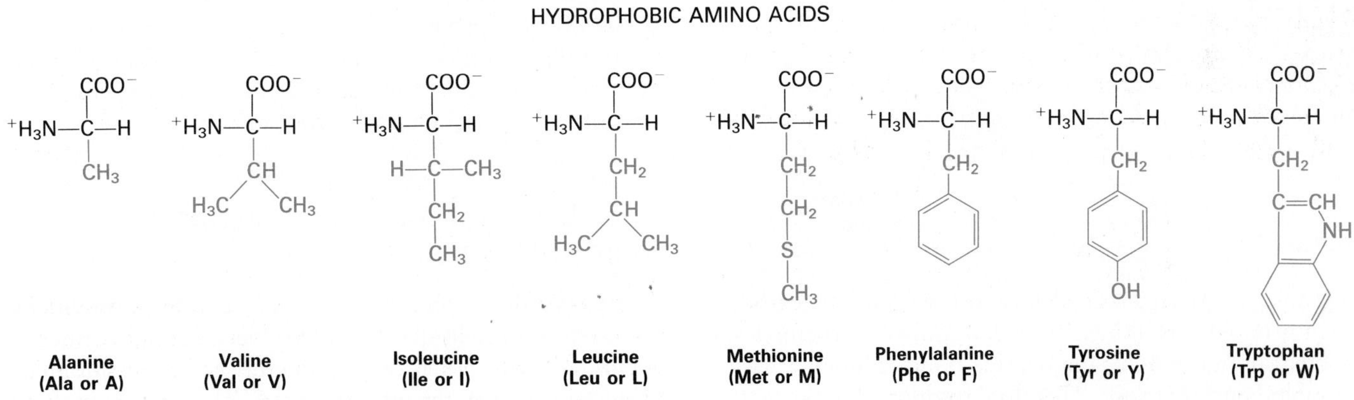

Acidic amino acids

Aspartic
acid
(Asp or D)

Glutamic
acid
(Glu or E)

HYDROPHOBIC AMINO ACIDS

Alanine
(Ala or A)

Valine
(Val or V)

Isoleucine
(Ile or I)

Leucine
(Leu or L)

Methionine
(Met or M)

Phenylalanine
(Phe or F)

Tyrosine
(Tyr or Y)

Tryptophan
(Trp or W)

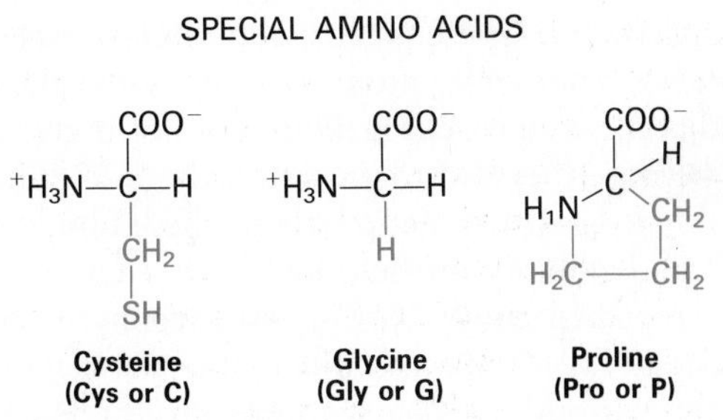

SPECIAL AMINO ACIDS

Cysteine
(Cys or C)

Glycine
(Gly or G)

Proline
(Pro or P)

▲ FIGURE 3-3   The structures of the 20 common amino acids grouped into three categories: hydrophilic, hydrophobic, and special amino acids. The side chain determines the characteristic properties of each amino acid. Shown are the zwitterion forms, which exist at the pH of the cytoplasm. In parentheses are the three-letter and one-letter abbreviations for each amino acid.

Some amino acids are more abundant in proteins than other amino acids. Cysteine, tryptophan, and methionine are rare amino acids; together they constitute approximately 5 percent of the total amino acids in a protein. Charged and hydrophobic amino acids such as lysine, arginine, glutamic acid, aspartic acid, alanine, valine, and leucine are among the most abundant amino acids; each can account for 5–10 percent of the total amino acids in a typical protein. Although amino acids vary in their molecular weights, a useful number to remember is 115, the average molecular weight (MW) of amino acids (taking their abundance in an average protein into account). This value allows one to estimate the number of residues from the molecular weight of a protein or vice versa; for example, a 50,000-MW protein would contain approximately 435 residues.

Polypeptides Are Amino Acids Connected by Peptide Bonds

Nature has evolved a single chemical linkage, the *peptide bond,* to connect amino acids into a linear, unbranched chain called a *polypeptide.* The peptide bond is formed by a condensation reaction between the amino group of one amino acid and the carboxyl group of another; a water molecule is liberated in the process:

Because of electron delocalization within the carbonyl group (C=O), as shown by the resonance structures depicted in Figure 3-4a, the peptide bond exhibits a partial double-bond character. Therefore rotation about a peptide bond is restricted compared with rotation about a typical C—N single bond. Consequently, the atoms in the peptide bond and its adjacent C_α atoms lie in the same plane; however, adjacent amino acid *residues* are not necessarily coplanar because they can rotate about the C—C_α and N—C_α bonds (Figure 3-4b). Rotation about these bonds contributes to the flexibility of the polypeptide chain.

The repeated amide N, C_α, and carbonyl C atoms of each residue form the *backbone* of the polypeptide from which the various side-chain groups project. As a consequence of the peptide linkage, the backbone has polarity, since all the amino groups lie to the same side of the C_α

atoms. This leaves at opposite ends of the polypeptide a free (unlinked) amino group (the *N-terminus*) and a free carboxyl group (the *C-terminus*). A polypeptide is conventionally depicted with its N-terminal amino acid on the left and its C-terminal amino acid on the right:

Many terms are used to denote polymeric chains of amino acids. A random sequence of amino acids linked by peptide bonds into a single chain, usually via polymerization of one or a few amino acids, is called a *polyamino acid.* A short chain of amino acids linked by peptide bonds and having a defined sequence is a *peptide;* longer peptides are referred to as *polypeptides.* Peptides generally contain fewer than 20–30 amino acid residues, whereas polypeptides contain as many as 4000 residues. We reserve the term *protein* for a polypeptide (or a complex of polypeptides) that has a three-dimensional structure. It is implied that proteins and peptides represent natural products of a cell, whereas a polyamino acid is an artificial substance that is not normally found in a cell.

The size of a protein or a polypeptide is reported as its mass in daltons (a dalton is one atomic mass unit) or as its molecular weight (a dimensionless number). For example, a 10,000-MW protein has a mass of 10,000 daltons (Da), or 10 kilodaltons (kDa). In the last section of this chapter, we will discuss different methods for measuring the sizes and other physical characteristics of proteins.

Four Levels of Structure Determine the Shape of Proteins

X-ray crystallographic analysis has revealed four hierarchical levels of protein structure. These levels are illustrated in Figure 3-5, which depicts the structure of hemagglutinin, a surface protein on the influenza virus. This protein binds to the surface of animal cells, including human cells, and is responsible for the infectivity of the flu virus.

The *primary* structure of a protein is the linear arrangement, or *sequence,* of amino acid residues that constitute the polypeptide chain. Sequencing of a protein determines the number and order of amino acid residues composing it, that is, its primary structure.

Secondary structure refers to the local organization of parts of a polypeptide chain, which can be arranged in several ways. A single polypeptide may exhibit all types of secondary structure. Without any stabilizing interactions, a polypeptide assumes a *random-coil* conformation. How-

(a)

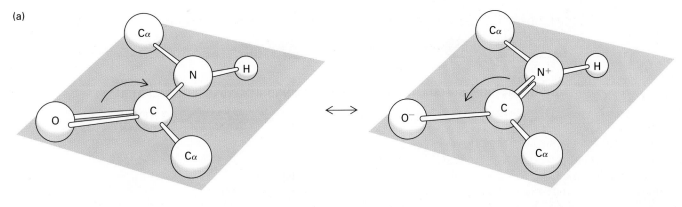

(b)

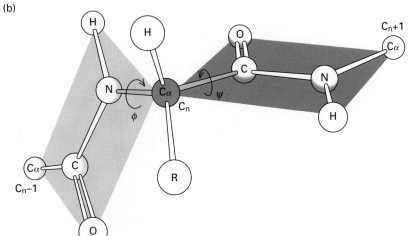

◄ **FIGURE 3-4** Flexibility of bonds in the polypeptide backbone. (a) The carbon-nitrogen linkage in a peptide bond has a partial double-bond character due to resonance. Thus rotation about a peptide bond is limited, making each peptide group in a polypeptide planar. (b) Rotation is possible about the two covalent single bonds (yellow) that connect each C_α to the two adjacent planar peptide units, although some restrictions apply to the values of ϕ and ψ for different amino acids. Therefore, despite the inflexibility of the peptide bond, the polypeptide backbone exhibits considerable flexibility.

ever, when stabilizing hydrogen bonds form between certain residues the backbone folds into one of two geometric arrangements: an *α helix* (a spiral polypeptide) or a *β strand* (an extended polypeptide). Lateral association of *β* strands forms *β sheets*. Finally, *turns* are U-shaped four-residue segments stabilized by hydrogen bonds between their arms. They are located at the surfaces of proteins and redirect the polypeptide chain toward the interior.

The next higher level of structure, *tertiary* structure, is the three-dimensional arrangement of amino acids residues. In contrast to secondary structure, which is stabilized by hydrogen bonds, tertiary structure results from hydrophobic interactions between the nonpolar side chains. The hydrophobic interactions hold the helices, strands, and random coils in a compact internal scaffold. A protein's size and shape is dependent not only on its sequence but also on the number, size, and arrangement of secondary structures. For proteins that consist of a single polypeptide chain, *monomeric proteins,* tertiary structure is the highest level of organization.

Multimeric proteins contain several protein subunits held together by noncovalent bonds. *Quaternary* structure describes the number (*stoichiometry*) and relative positions

of subunits in multimeric proteins. Hemagglutinin is a trimer of three identical subunits; other multimeric proteins can be composed of any number of identical or different subunits. Often, the arrangement of subunits is regular or symmetrical. The hemagglutinin trimer, for example, displays a three-fold rotational symmetry; its subunits are arranged like a pie divided into three pieces. Rotation by 120° about the center of the protein moves one subunit into the exact position of its neighbor. In addition to rotational symmetry, subunits in other proteins can be arranged as if they are mirror images of one another, an example of mirror symmetry. Your left and right hands are examples of mirror symmetrical objects.

In a fashion similar to the hierarchy of structures that make up a protein, proteins themselves are part of a hierarchy of cellular structures. Proteins can associate into larger structures termed *macromolecular assemblies.* A macromolecular assemblage of proteins is represented by the structure of the protein coat of a virus, a bundle of actin filaments, the nuclear pore complex, and other large microscopic objects. Macromolecular structures in turn combine with other cell biopolymers like lipids, carbohydrates, and nucleic acids to form complex cell organelles. Refer-

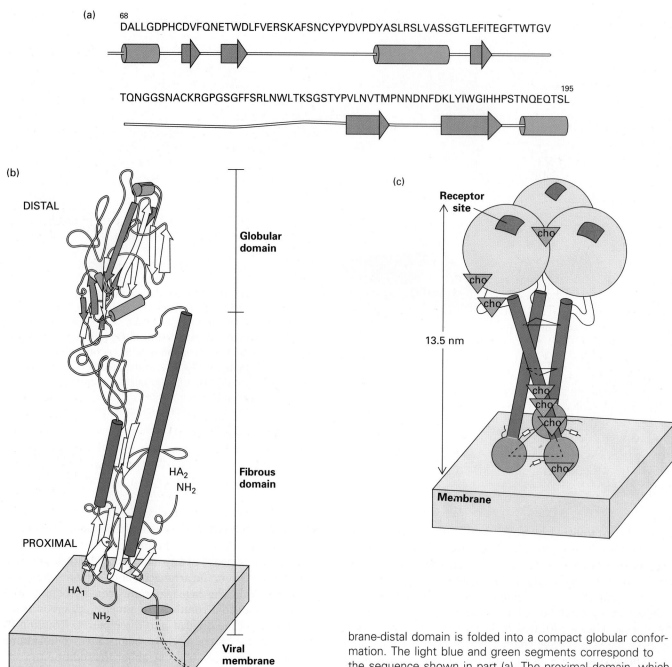

(a)
68
DALLGDPHCDVFQNETWDLFVERSKAFSNCYPYDVPDYASLRSLVASSGTLEFITEGFTWTGV

195
TQNGGSNACKRGPGSGFFSRLNWLTKSGSTYPVLNVTMPNNDNFDKLYIWGIHHPSTNQEQTSL

(b)

DISTAL

Globular
domain

HA₂
NH₂

Fibrous
domain

PROXIMAL

HA₁

NH₂

Viral
membrane

COOH

(c)

Receptor
site

cho

cho

cho

13.5 nm

cho
cho
cho

cho

Membrane

▲ FIGURE 3-5   Four levels of structure in hemagglutinin, which is a long multimeric molecule whose three identical subunits are composed of two chains, HA₁ and HA₂. (a) Primary structure is illustrated by the amino acid sequence of residues 68–195 of HA₁. This region is used by influenza virus to bind to animal cells. The one-letter amino acid code is used. Secondary structure is diagrammatically represented below the sequence showing regions of the polypeptide chain that are folded into α helices (light blue cylinders), β strands (light green arrows), and random coils (white lines). (b) Tertiary structure constitutes the folding of the helices and strands in each HA subunit into a compact structure that is 13.5 nm long and divided into two domains. The mem-

brane-distal domain is folded into a compact globular conformation. The light blue and green segments correspond to the sequence shown in part (a). The proximal domain, which lies adjacent to the viral membrane, has a stemlike conformation due to alignment of two long helices of HA₂ (dark blue) with β strands in HA₁. The longer of these helices in HA₂ consists of 49 residues and is 74 nm long; it is one of the longest helices found in proteins. The polypeptide backbone folds back on itself as short turns or longer loops, which usually lie at the surface of the molecule; these secondary structures connect the helices and strands in each chain. (c) Quaternary structure of HA is stabilized by lateral interactions among the long helices (dark blue) in the subunit stems, forming a triple-stranded coiled-coil stalk. Each of the distal globular domains in trimeric hemagglutinin has a site (red) for binding sialic acid molecules on the surface of target cells. Like many membrane proteins, HA has several covalently bound carbohydrate (CHO) chains. [Parts (b) and (c) adapted from A. Wilson, J. J. Skehel, and D. C. Wiley, 1981, *Nature* **289**:366.]

ring back to Chapter 1, we can now understand the place of proteins in the biology of cells, tissues, and organisms.

Polypeptides Can Be Chemically Analyzed and Synthesized

Before examining the details of secondary, tertiary, and quaternary structure, we briefly describe techniques for analyzing a protein's primary structure and creating synthetic peptides.

Determining the Amino Acid Composition and Sequence of Proteins The primary structure of a protein is characterized in two ways: by its overall amino acid composition and by its precise amino acid sequence. The amino acid composition of a protein gives the same information as an elemental analysis of a molecule—the types of amino acids present and their abundance but not their linear order. In contrast, the sequence of a protein is like a fingerprint, it uniquely establishes the identity of a protein—the linear order of amino acids. The composition of a protein is easily calculated from the sequence. Two proteins can differ in their sequence but nonetheless have identical amino acid compositions. Composition and sequence are determined by chemical methods based on the ability to cleave the peptide backbone at the peptide bond.

The first step in obtaining an amino acid composition is to degrade the protein completely into single amino acids. This is accomplished by hydrolyzing all the peptide bonds with a strong acid (Figure 3-6a). The mixture of free amino acids is passed through a column of silica (sand) beads, a technique known as liquid chromatography. Each amino acid will flow through the column at a characteristic rate. By monitoring the outflow (effluent) from the column, the amino acids present in the original protein can be identified and their abundance calculated.

The classic method for determining the amino acid sequence of a protein involves *Edman degradation,* an end-labeling procedure in which the amino group at the N-terminus of a polypeptide is chemically modified with phenylisothiocyanate (PITC). Treatment of the labeled polypeptide with acid frees the N-terminal residue as a PTH derivative, leaving the polypeptide with one less amino acid and a new N-terminus (Figure 3-6b). Subjecting a polypeptide to sequential cycles of coupling and acid cleavage thus shortens it one residue at a time; the PTH-amino acid at each position then is identified by high-pressure liquid chromatography.

Sophisticated physical methods such as mass spectrometry, which can measure the molecular mass of a large molecule to within 1 dalton, are now able to determine the sequences of peptides generally containing 20 or fewer amino acids. In this approach, peptide bonds are randomly cleaved by collision with helium gas, generating a family of truncated peptides. From analysis of the differences in mass among these peptides by mass spectrometry, the amino acid sequence of the original peptide can be deduced

(Figure 3-6c). Mass spectrometry is especially useful in identifying modified amino acids in a sequence.

Before about 1980, biologists commonly used the Edman chemical procedure for determining protein sequences. However, recombinant DNA methods developed in the 1970s and 1980s permit the sequence of a protein to be deduced from the sequence of its mRNA or gene. Because sequencing of mRNA and DNA generally is faster than chemical sequencing of proteins, this approach is now the most popular way to determine protein sequences, especially of large proteins.

Creating Synthetic Peptides Peptides of <50 residues can now be synthesized in a test tube from monomeric amino acids by techniques developed in the 1960s by Bruce Merrifield. All the techniques are based on recreating the linkages between amino acids by condensation reactions that form peptide bonds.

As outlined in Figure 3-7, synthesis begins from the C-terminus. The first step involves the covalent coupling of the first amino acid, via its carboxyl group, to an insoluble resin. The amino group of this residue then condenses with the carboxyl group of the second amino acid, forming a peptide bond; repetition of the condensation reaction adds subsequent amino acids. Thus, peptides are constructed sequentially by coupling the C-terminus of a monomeric amino acid with the N-terminus of the growing peptide. To prevent unwanted reactions involving the amino groups and carboxyl groups of the side chains during the coupling steps, a protecting (blocking) group is attached to the side chains. Without these protecting groups, branched peptides would be generated. In the last steps of synthesis, the side chain–protecting groups are removed and the peptide is cleaved from the resin.

Synthetic peptides, which are identical to the peptides synthesized in vivo, have become important tools in studies of proteins and cells. For example, short synthetic peptides of 10–15 residues can function as *antigens* to trigger production of *antibodies* in animals. A synthetic peptide can trick the animal into producing antibodies that bind the full-sized, natural protein antigen. As discussed later in this chapter and in other chapters, antibodies are useful in localizing proteins in cells and in isolating proteins from mixtures. Synthetic peptides also have been helpful in elucidating the rules that determine the secondary and tertiary structure of proteins. By systematically varying the sequence of synthetic peptides, researchers have studied the influence of various amino acids on protein conformation.

Three-dimensional Protein Structure Is Determined through X-ray Crystallography and NMR Spectroscopy

The detailed three-dimensional structures of some 800 proteins have been established by the painstaking efforts of many scientists, notably, Max Perutz and John Kendrew,

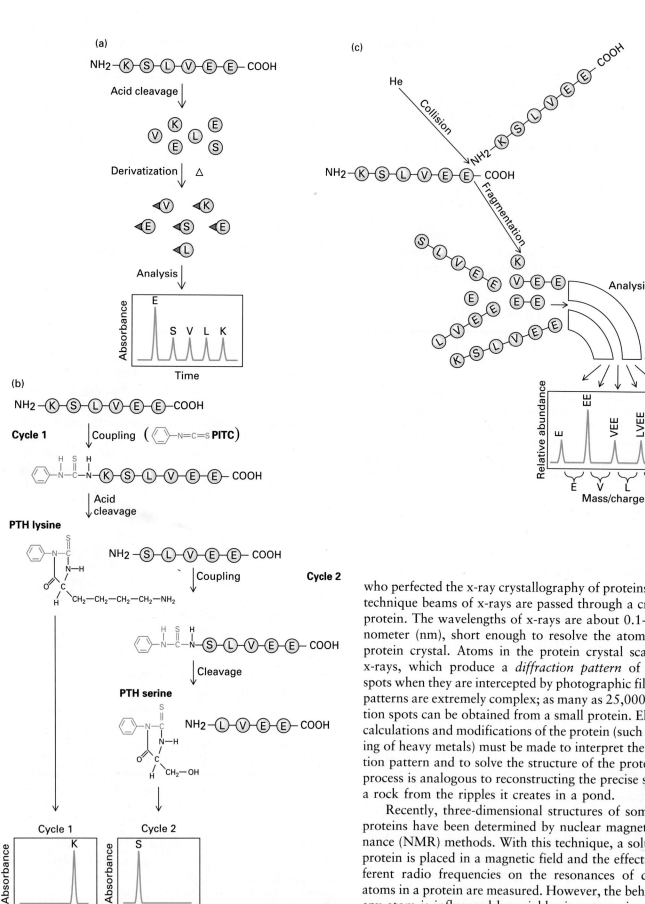

who perfected the x-ray crystallography of proteins. In this technique beams of x-rays are passed through a crystal of protein. The wavelengths of x-rays are about 0.1–0.2 nanometer (nm), short enough to resolve the atoms in the protein crystal. Atoms in the protein crystal scatter the x-rays, which produce a *diffraction pattern* of discrete spots when they are intercepted by photographic film. Such patterns are extremely complex; as many as 25,000 diffraction spots can be obtained from a small protein. Elaborate calculations and modifications of the protein (such as binding of heavy metals) must be made to interpret the diffraction pattern and to solve the structure of the protein. The process is analogous to reconstructing the precise shape of a rock from the ripples it creates in a pond.

Recently, three-dimensional structures of some small proteins have been determined by nuclear magnetic resonance (NMR) methods. With this technique, a solution of protein is placed in a magnetic field and the effects of different radio frequencies on the resonances of different atoms in a protein are measured. However, the behavior of any atom is influenced by neighboring atoms in adjacent residues; the closely spaced residues are more perturbed than distant residues. From the magnitude of the effect, the

◄ FIGURE 3-6 Determination of amino acid composition and sequence of a protein. (a) The first step in determining amino acid composition is acid cleavage (6 N HCl for 24 h at 110°C) to liberate the individual amino acids. The free amino acids are reacted with Ninhydrin, a reagent that turns violet when it covalently binds with amino groups. The derivatized amino acids are separated by ion-exchange chromatography (see Figure 3-39b). Each amino acid elutes from the column in a characteristic time and is detected by its absorbance at 450 nm, which is proportional to its concentration. (b) Edman degradation, the classic chemical method for determining amino acid sequence, involves a repetitive three-step procedure. In the first step, the polypeptide is reacted with phenylisothiocyanate (PITC), which forms a covalent bond with the terminal amino group. In the second step, the N-terminal amino acid is cleaved from the polypeptide by acid hydrolysis, yielding the cyclic phenylthiohydantoin (PTH) derivative of the N-terminal amino acid and a polypeptide that is shorter at its N-terminus by one residue. These two steps are then repeated with the shortened polypeptide. The PTH derivative formed in each cycle is identified by liquid chromatography. (c) Sequencing by mass spectrometry proceeds in two parts. First, a peptide (usually shorter than 20 residues) is introduced into a vacuum where it collides with a stream of gas, typically helium. The collision breaks the peptide at random peptide bonds, producing a mixture of shorter, ionized (charged) peptides. Within this mixture are two fragment ion series consisting of fragments that are one residue shorter, either at their C-terminus or N-terminus; the latter is depicted here. Second, the mass of the fragment ions are determined with a mass spectrometer. In this instrument, the fragment ions are accelerated through a curved magnet field. The amount they are deflected by the field is dependent on their charge and momentum (mass × velocity); because ions of the same charge have the same velocity, their deflection is a function of their mass. The mass of each fragment can be measured with an accuracy <0.1 of the mass of a hydrogen atom. The amino acid sequence is deduced from differences in mass between consecutive members of a series. With the exception of leucine and isoleucine, which have identical masses, the mass difference uniquely identifies the missing amino acid at that position.

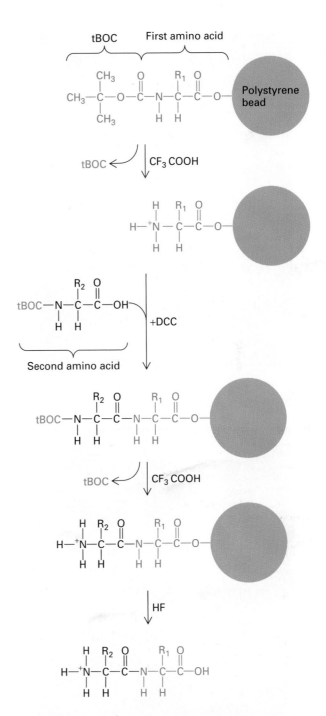

distances between residues can be calculated; these distances then are used to generate a model of the three-dimensional structure of the protein. Although the NMR approach does not require crystallization of a protein, a definite advantage, this technique is limited to relatively small proteins up to about 150 amino acids in length.

Graphical Representations of Proteins Highlight Internal Organization or Surface Features

Different ways of depicting proteins convey different types of information. The simplest way to represent three-dimensional structure is to trace the course of the backbone

▲ FIGURE 3-7 Solid-phase peptide synthesis. The first amino acid (blue) of the desired peptide is attached at its carboxyl end by esterification to a polystyrene bead. The amino group of the first amino acid in the peptide under construction is blocked by the attachment of a *tert*-butyloxycarbonyl (tBOC) group (red), which is removed by treatment with trifluoroacetic acid (CF_3COOH). The resulting free amino group forms a peptide bond with a second amino acid, which is presented with a reactive carboxyl group and a blocked amino group, together with the coupling agent dicyclohexylcarbodiimide (DCC). The process is repeated until the desired product is obtained; the peptide is then chemically cleaved from the bead with hydrofluoric acid (HF). [See R. B. Merrifield, L. D. Vizioli, and H. G. Boman, 1982, *Biochemistry* **21**:5020.]

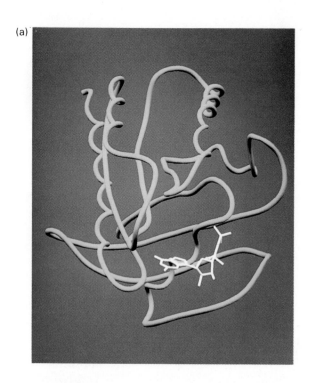

(a)

(b)

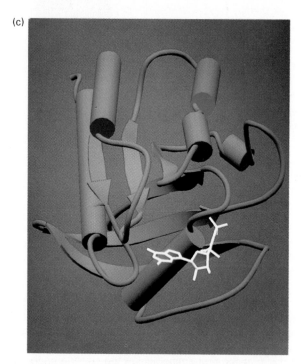

(c)

(d)

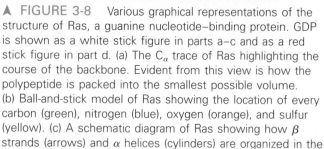

▲ FIGURE 3-8   Various graphical representations of the structure of Ras, a guanine nucleotide–binding protein. GDP is shown as a white stick figure in parts a–c and as a red stick figure in part d. (a) The C$_\alpha$ trace of Ras highlighting the course of the backbone. Evident from this view is how the polypeptide is packed into the smallest possible volume. (b) Ball-and-stick model of Ras showing the location of every carbon (green), nitrogen (blue), oxygen (orange), and sulfur (yellow). (c) A schematic diagram of Ras showing how β strands (arrows) and α helices (cylinders) are organized in the protein. Note the turns and loops connecting pairs of helices and strands. (d) The water-accessible surface of Ras. Painted on the surface are regions of positive charge (blue) and negative charge (red). Here we see that the surface of a protein is not smooth but has lumps, bumps, and crevices. The molecular basis for specific binding interactions lies in the uneven distribution of charge over the surface of the protein. [Courtesy of Gareth White. Adapted from L. Tong, et al., 1991, J. Mol. Biol. **217**:503.]

atoms with a solid line (Figure 3-8a); the most complex model shows the location of every atom (Figure 3-8b). The former shows the overall organization of the polypeptide chain without consideration of the amino acid side chains; the latter details the interactions among atoms that form the backbone and that stabilize the protein's conformation. Even though both views are useful, they do not show how different regions of the backbone are folded.

We use two shorthand symbols to represent secondary structure: cylinders for helices and arrows for strands. A polypeptide without any regular structure is shown as a string. This type of representation (Figure 3-8c) emphasizes the organization of the secondary structure of a protein, and various combinations of secondary structures are easily seen.

However, none of these three ways of representing protein structure conveys much information about the protein surface, which is of interest because this is where ligands bind to a protein. Computer analysis in which a water molecule is rolled around the surface of a protein can identify the atoms that are in contact with the watery environment. On this water-accessible surface, regions having common chemical (hydrophobicity or hydrophilicity) and electrical (basic or acidic) character can be mapped. Such models (Figure 3-8d) show the texture of the protein surface and the distribution of charge, both of which are important parameters of binding sites. This view represents a protein as seen by another molecule.

Secondary Structures Are Crucial Elements of Protein Architecture

In an average protein, 60 percent of the polypeptide chain exists as two regular secondary structures, α helices and β sheets; the remainder of molecule is in random coils and turns. Thus, α helices and β sheets are the major internal supportive elements in proteins. In this section, we explore the forces that favor formation of secondary structures. In later sections, we examine how these structures can pack into larger arrays.

The α Helix   In the early 1950s, Linus Pauling and Robert B. Corey came to realize that polypeptide segments composed of certain amino acids tend to assume a regular spiral, or helical, conformation, called the α helix. In this secondary structure, the carbonyl oxygen of each peptide bond is hydrogen bonded to the amide hydrogen of the amino acid four residues towards the C-terminus. This uniform arrangement of bonds confers a polarity on a helix because all the hydrogen-bond donors have the same orientation. The peptide backbone twists into a helix having 3.6 amino acids per turn (Figure 3-9). Each "step" up this spiral staircase of amino acids (*axial rise per residue*) represents a distance of about 0.15 nm along the axis. The in-

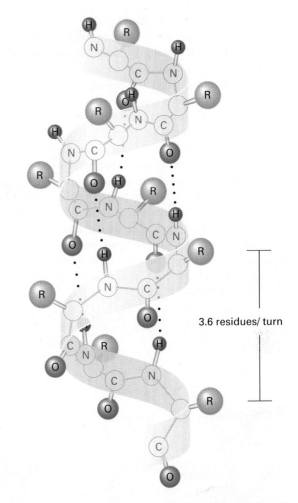

3.6 residues/ turn

▲ FIGURE 3-9   Model of the α helix. The polypeptide backbone is folded into a spiral that is held in place by hydrogen bonds (black dots) between backbone oxygen atoms and hydrogen atoms. Note that all the hydrogen bonds have the same polarity. The outer surface of the helix is covered by the side-chain R groups.

flexible, stable arrangement of amino acids in the α helix holds the backbone as a rodlike cylinder from which the side chains point outward. An important consequence of this arrangement is that the hydrophobic or hydrophilic quality of the helix is determined entirely by the side chains. The polar groups of the peptide backbone are already involved in hydrogen bonding in the helix and thus are unable to affect the hydrophobicity or hydrophilicity of the helix.

Certain amino acid sequences adopt α-helical conformations more readily than others. What determines this propensity is still debated, but some simple factors that destabilize an α helix are evident. For instance, proline is rarely found in α-helical regions because the fixed angles about its C_α bonds would "kink" or "break" the helix backbone. Also, proline in the backbone lacks an amide nitrogen to make a stabilizing hydrogen bond. For opposite reasons, glycine is also an infrequent participant. Its hydrogen side group permits too much flexibility around the C_α bonds, and the helix "bends" at a glycine residue in the backbone. In addition to constraints imposed by backbone flexibility, the side chains can inhibit helix formation. For instance, identically charged residues on the surface of a helix will repel each other and destabilize the helix.

As a rodlike element of protein structure, the α helix serves many functions. One function is purely structural. Oxygen-binding proteins like myoglobin and hemoglobin, for example, are based on a core of eight short α-helical rods connected by bends. Interactions between the helices are important in forming the core, and this pattern of α helices, called the *globin fold*, is found in other proteins

(Figure 3-10a). In some proteins, α helices function as binding sites, especially for other proteins and DNA. In many DNA-binding proteins, a cylindrical α helix fits snugly in the groove of double-stranded DNA, allowing strong interactions between the protein and DNA molecules (Figure 3-10b). Fibrous proteins illustrate a third function of the helix. These proteins often contain chains of α-helical segments, which can span long distances or form tough, sturdy structures when twisted about each other like a rope (Figure 3-10c).

Many α helices are *amphipathic*: they expose hydrophilic side chains on one side of the helix and hydrophobic side chains on the opposite side. This occurs when hydrophobic residues are regularly spaced three or positions apart in the linear sequence (Figure 3-11a). One way of visualizing this arrangement is to look down the center of an α helix and then project the amino acid residues onto

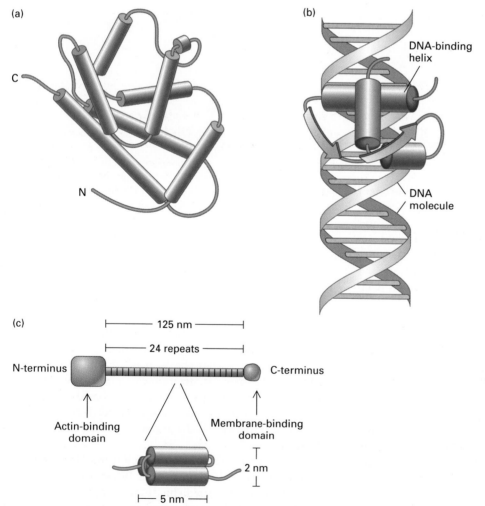

(a)

C

N

(b)

DNA-binding helix

DNA molecule

(c)

|← 125 nm →|

|← 24 repeats →|

N-terminus   C-terminus

Actin-binding domain   Membrane-binding domain

2 nm

|← 5 nm →|

◄ FIGURE 3-10 Three functional roles of α helices in proteins. (a) In oxygen-binding proteins, eight helices (blue) form the globin fold, which constitutes 70 percent of the subunit structure. The internal structure of most globular proteins is formed from individual or clusters of α helices. (b) In many DNA-binding proteins, an α helix fits in the major groove of the double-stranded DNA molecule, so that interaction is facilitated. Depicted in this figure is part of the glucocorticoid receptor, which activates some genes when bound to the hormone. (c) A trio of helices pack into a bundle. In dystrophin, the three helix bundle is repeated 24 times to form a long, flexible rod which separates two domains, one that binds actin filaments and the other that binds a membrane glycoprotein. The absence of either domain causes several forms of muscular dystrophy, a muscle-wasting disease of children and young adults. The three helix bundle is used in other structural proteins (Table 22-1) as a spacer between actin-binding domains and calcium-binding domains. [Part (a) after C. Branden and J. Tooze, 1991, *Introduction to Protein Structure*, Garland; part (b) after B. Luisi et al., 1991, *Nature* **352**:497; part (c) modified from Y. Yan, et al., 1993, *Science* **262**:2027.]

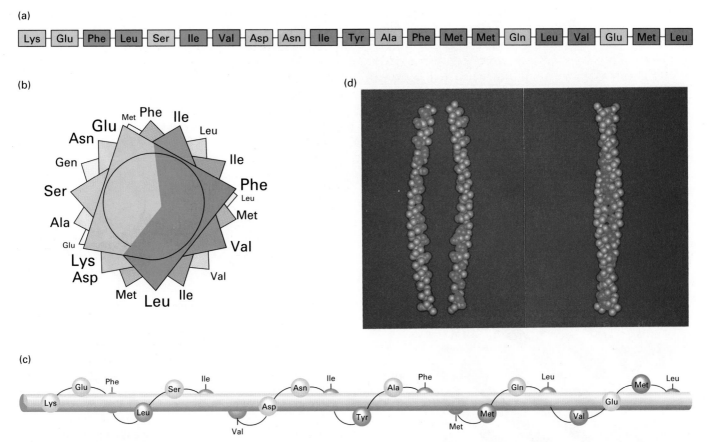

▲ FIGURE 3-11   Regions of an α helix may be amphipathic. (a) A polypeptide in which the residues appear to be randomly arranged. Note however that hydrophobic (red) residues are regularly spaced one or two residues apart. (b) By looking down the axis of this polypeptide in an α helix and then projecting the amino acid residues on the plane of the paper, we see that the residues are arranged in the form of a wheel with hydrophobic (red) and hydrophilic (yellow) residues clustered on opposite sides. The abbreviations correspond to the residues shown in (a). Because a helix makes one turn every 3.6 residues, the pattern is repeated every 18 residues; for example, residues Lys and Glu exactly overlap, as do residues Glu and Met. (c) In another representation of this amphipathic polypeptide, the hydrophobic residues form a stripe down one side and the hydrophilic residues form a stripe down the other side of the peptide backbone. (d) Two amphipathic helices (*left*) can interact to form a coiled-coil structure (*right*), which is common in rodlike fibers. The interaction between helices results from the "knobs-into-holes" packing of the hydrophobic residues (red). [Part (d) from C. Cohen and D. A. D. Parry, 1990, *Proteins: Struc. Funct. Genet.* **7**:1. Courtesy of C. Cohen.]

the plane of the paper. The residues will appear as a wheel, and in the case of an amphipathic helix, the hydrophobic residues all lie on one side of the wheel and the hydrophilic ones on the other side (Figure 3-11b). An amphipathic helix can also be represented as shown in Figure 3-11c.

As discussed in Chapter 14, amphipathic α helices are important structural elements in proteins that act as pores, or channels, in the cell membrane. The walls of a membrane pore are lined by a single layer of amphipathic α helices. Their hydrophilic surfaces face the outside of the pore, where water-soluble molecules and ions can pass through, and their hydrophobic surfaces face the lipid core of the membrane.

Amphipathic α helices also wrap around each other to form *coiled coils* (Figure 3-11d). The amino acid sequence of each such helix contains a characteristic seven-residue "heptad" repeat: hydrophobic residues predominate at positions 1 and 4 in each heptad. In a coiled coil, hydrophobic side chains project like "knobs" from one helix and interdigitate into the gaps, or "holes," between the hydrophobic side chains of the other helix along the contact surface. Nearly forty years ago Francis Crick proposed that such interactions, "knobs-into-holes" packing, would be the basis for binding between helices. Through coiled-coil interactions, protein subunits are able to form dimers or trimers or to polymerize into rodlike fibers.

The β Sheet   Another regular secondary structure, the β sheet, consists of laterally packed β strands. Each β strand is a short (5–8 residue), nearly fully extended polypeptide chain. The backbone atoms in adjacent β strands, either within the same or different polypeptide chains, can hydrogen bond, forming β sheets (Figure 3-12a). The planarity of the peptide bond forces the sheet to be pleated; hence,

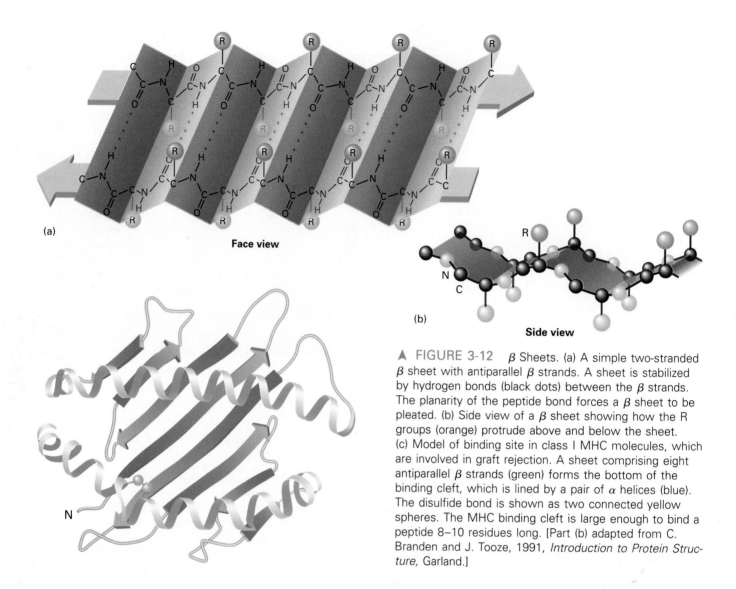

Face view

(a)

(b)

Side view

▲ **FIGURE 3-12** β Sheets. (a) A simple two-stranded β sheet with antiparallel β strands. A sheet is stabilized by hydrogen bonds (black dots) between the β strands. The planarity of the peptide bond forces a β sheet to be pleated. (b) Side view of a β sheet showing how the R groups (orange) protrude above and below the sheet. (c) Model of binding site in class I MHC molecules, which are involved in graft rejection. A sheet comprising eight antiparallel β strands (green) forms the bottom of the binding cleft, which is lined by a pair of α helices (blue). The disulfide bond is shown as two connected yellow spheres. The MHC binding cleft is large enough to bind a peptide 8–10 residues long. [Part (b) adapted from C. Branden and J. Tooze, 1991, *Introduction to Protein Structure,* Garland.]

this structure is also called a β pleated sheet, or simply pleated sheet. Like α helices, β strands have a polarity defined by the orientation of the peptide bond. Therefore, in a pleated sheet, adjacent β strands can be oriented antiparallel or parallel with respect to each other. In both arrangements of the backbone, the side chains project from both faces of the sheet (Figure 3-12b).

In some proteins, β sheets form the floor of a binding pocket (Figure 3-12c). In many structural proteins, multiple layers of pleated sheets provide toughness. Silk fibers, for example, consist almost entirely of stacks of antiparallel β sheets. The fibers are flexible because the stacks of β sheets can slip over each other. However, they are also resistant to breakage because the peptide backbone is aligned parallel with the fiber axis.

Turns Composed of three or four residues, turns are compact, U-shaped secondary structures stabilized by a hydro-

gen bond between their end residues. They are located on protein surfaces and form a sharp bend that redirects the polypeptide backbone back toward the interior. Glycine and proline are commonly present in turns. The lack of a large side chain in the case of glycine and the presence of a built-in bend in the case of proline allow the polypeptide backbone to fold into a tight U-shaped structure. Without turns, a protein would be large, extended, and loosely packed. A polypeptide backbone also may contain long bends, or *loops*. In contrast to turns, which exhibit a few defined structures, loops can be formed in many different ways.

Motifs Are Regular Combinations of Secondary Structures

Certain combinations, or *motifs,* of two or three secondary structures, usually α helices, β strands, and loops, are

(a) Helix–loop–helix motif

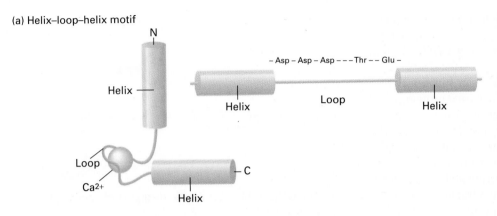

(b) Zinc-finger motif

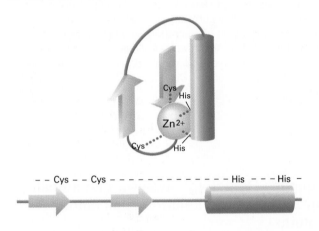

▲ FIGURE 3-13   Secondary structure motifs. (a) Helix-loop-helix motif is a characteristic feature of many calcium-binding proteins. Oxygen-containing R groups of residues in the loop form a ring around a Ca^{2+} ion. The 14-aa loop sequence *(right)*, which is rich in invariant hydrophilic residues, is found in various calcium-binding proteins. (b) Zinc-finger motif, which is present in many proteins that bind nucleic acids. A Zn^{2+} ion is held between a pair of β strands (green) and a single α helix (blue) by a pair of cysteine and histidine residues. In the 25-aa sequence of this motif *(right)*, the invariant cysteines usually occur at positions 3 and 6, and the invariant histidines at positions 20 and 24.

found in a variety of proteins. A motif has a characteristic sequence and usually is associated with a particular function. One example is the *helix-loop-helix* motif, which binds calcium ions in many calcium-binding proteins (Figure 3-13a). The presence of certain hydrophilic residues at invariant positions in the loop is a signature of these proteins. Oxygen atoms in the invariant residues bind a calcium ion through hydrogen bonds.

Another common motif is the *zinc finger* (Figure 3-13b), named for its zinc-binding capacity and fingerlike shape. The finger is a bundle of three secondary structures: an α helix and a pair of antiparallel β strands. A pair of histidine residues and a pair of cysteine residues located at opposite ends of the finger sequence form a cage, which is nestled between a loop connecting the strands and the base of the helix. A zinc ion holds the bundle together by binding the histidine and cysteine residues. Stabilized by zinc, the helix of the finger fits into the groove of a DNA molecule. This motif generally is present in proteins that bind RNA or DNA.

Additional secondary structure motifs will be examined in discussions of other proteins. The presence of the same motif in different proteins with similar functions

clearly indicates that nature reuses certain combinations of secondary structures rather than inventing new ones.

Structural and Functional Domains Are Modules of Tertiary Structure

The tertiary structure of large proteins is often subdivided into distinct globular or fibrous regions called *domains.* Structurally, a domain is a compactly folded region of polypeptide. For large proteins, domains can be recognized in structures determined by x-ray crystallography or in images captured by electron microscopy. These discrete globular or filamentous regions are well distinguished or physically separated from other parts of the protein, but connected by the polypeptide chain. Hemagglutinin, for example, contains a globular domain and a fibrous domain (see Figure 3-5c).

A domain consists of 100–300 residues in various combinations of α helices, β sheets, turns, and random coils. Often a domain is characterized by some interesting structural feature. Examples include an unusually abundant number of a particular amino acid (proline-rich domain, acidic domain, glycine-rich domain), sequences found in multiple proteins (SH3, or Src homology region 3), and a particular secondary structure motif ("kringle" domain, zinc-finger motif).

Domains sometimes are defined in functional terms based on observations that the activity of a protein is localized to a small region. For instance, a particular region or regions of a protein may be responsible for its catalytic activity (e.g., kinase domain) or binding ability (e.g., DNA-binding domain, membrane-binding domain). Functional domains often are identified experimentally by whittling down a protein to its smallest active fragment by cleaving the polypeptide backbone with proteases. Alternatively, the DNA encoding a protein can be subjected to mutagenesis, so that segments of the protein's backbone are removed (Chapter 7). The activity of the truncated protein product synthesized from the mutated gene is then monitored.

The functional definition of a domain is less rigorous than a structural definition; however, if the three-dimensional structure of a protein has not been determined, identification of functional domains can provide useful information about the protein. Because the activity of a protein usually depends on a proper three-dimensional structure, a functional domain consists of at least one and usually several structural domains.

The organization of tertiary structure into domains further illustrates the principle that complex molecules are built from simpler components. Like secondary structure motifs, tertiary structure domains are incorporated as modules into different proteins, thereby modifying their functional activities. The modular approach to protein architecture is particularly easy to recognize in large proteins, which tend to be a mosaic of different domains and thus can perform different functions simultaneously.

The epidermal growth factor (EGF) domain is one example of a module that is present in several proteins. EGF is a soluble peptide hormone that binds to cells in the skin and connective tissue, causing them to divide. It is generated by proteolytic cleavage between repeated EGF domains in the EGF precursor protein, which is anchored in the cell membrane by a membrane-spanning domain (Figure 3-14). Six conserved cysteine residues form three pairs of disulfide bonds that hold EGF in its native conformation. EGF also serves as a domain in various proteins including tissue plasminogen activator (TPA), a protease that is used to dissolve blood clots in heart attack victims; Neu protein, which is involved in embryonic differentiation; and Notch protein, a cell-adhesion molecule that glues cells together.

Besides the EGF domain, these proteins contain additional domains that also are found in other proteins. For example, TPA possesses a chymotryptic domain, a common feature in proteins that catalyze proteolysis. This domain is related to chymotrypsin, a secreted protease discussed in detail in a later section.

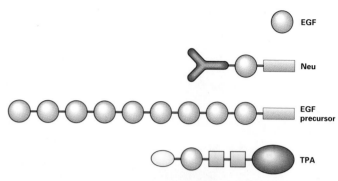

▲ FIGURE 3-14   Schematic diagrams of various proteins illustrating their modular nature. Epidermal growth factor (EGF) is generated by proteolytic cleavage of a precursor protein containing multiple EGF domains (orange). The EGF domain also occurs in Neu protein and in tissue plasminogen activator (TPA). Other domains, or modules, in these proteins include a chymotrypic domain (purple), immunoglobulin domain (green), fibronectin domain (yellow), membrane-spanning domain (pink), and kringle domain (blue). [Adapted from I. D. Campbell and P. Bork, *Curr. Opinions Struc. Biol.* **3**:385.]

Sequence Homology Suggests Functional and Evolutionary Relationships among Proteins

Since Fred Sanger determined the first primary sequence of a protein, the hormone insulin, in 1951, more than 100,000 protein sequences have been reported. Many of these are sequences of the same protein isolated from different organisms. Because of current efforts to sequence the genomes of various organisms, the number of sequenced proteins will expand exponentially in the coming years. As discussed in previous sections, however, the function of a protein depends on its three-dimensional structure. We now know that the secondary and tertiary structure of a protein is inherent in its amino acid sequence.

Recognition that the sequence determines the three-dimensional structure of a protein began to emerge in the 1960s when Max Perutz compared the structures of myoglobin and hemoglobin determined from x-ray crystallographic analysis (Figure 3-15a). He immediately noted that the subunits of hemoglobin (a tetramer of two α and two β subunits) resembled myoglobin (a monomer). Although the sequences of the two proteins were unknown at the time, Perutz proposed that the similar arrangement of α helices in the two proteins is a consequence of their having similar amino acid sequences. Later sequencing of myoglobin and hemoglobin revealed that identical or conserved (chemically similar) residues occur in identical positions throughout large stretches of both proteins (Figure 3-15b).

(a)

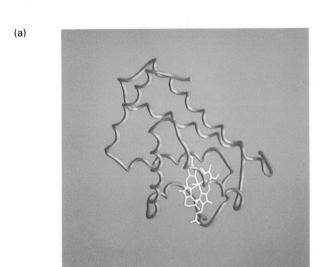

Myoglobin

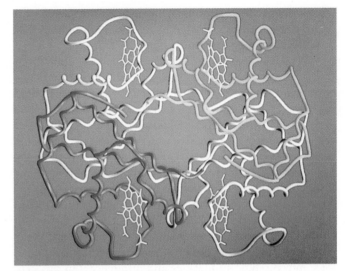

Hemoglobin

(b)

	1				5					10					15	
Myoglobin	Val	Leu	Ser	Glu	Gly	Glu	Trp	Gln	Leu	Val	Leu	His	Val	Trp	Ala	Lys
Hb α chain	Val	Leu	Ser	Pro	Ala	Asp	Lys	Thr	Asn	Val	Lys	Ala	Ala	Trp	Gly	Lys
Myo	Val	Glu	Ala	Asp	Val	Ala	Gly	His	Gly	Gln	Asp	Ile	Leu	Ile	Arg	Leu
Hb	Val	Gly	Ala	His	Ala	Gly	Glu	Tyr	Gly	Ala	Glu	Ala	Leu	Glu	Arg	Met
Myo	Phe	Lys	Ser	His	Pro	Glu	Thr	Leu	Glu	Lys	Phe	Asp	Arg	Phe	Lys	His
Hb	Phe	Leu	Ser	Phe	Pro	Thr	Thr	Lys	Thr	Tyr	Phe	Pro	His	Phe		Asp
Myo	Leu	Lys	Thr	Glu	Ala	Glu	Met	Lys	Ala	Ser	Glu	Asp	Leu	Lys	Lys	His
Hb	Leu	Ser	His						Gly	Ser	Ala	Gln	Val	Lys	Gly	His
Myo	Gly	Val	Thr	Val	Leu	Thr	Ala	Leu	Gly	Ala	Ile	Leu	Lys	Lys	Lys	Gly
Hb	Gly	Lys	Lys	Val	Ala	Asp	Ala	Leu	Thr	Asn	Ala	Val	Ala	His	Val	Asp
Myo	His	His	Glu	Ala	Glu	Leu	Lys	Pro	Leu	Ala	Gln	Ser	His	Ala	Thr	Lys
Hb	Asp	Met	Pro	Asn	Ala	Leu	Ser	Ala	Leu	Ser	Asp	Leu	His	Ala	His	Lys
Myo	His	Lys	Ile	Pro	Ile	Lys	Tyr	Leu	Glu	Phe	Ile	Ser	Glu	Ala	Ile	Ile
Hb	Leu	Arg	Val	Asp	Pro	Val	Asn	Phe	Lys	Leu	Leu	Ser	His	Cys	Leu	Leu
Myo	His	Val	Leu	His	Ser	Arg	His	Pro	Gly	Asp	Phe	Gly	Ala	Asp	Ala	Gln
Hb	Val	Thr	Leu	Ala	Ala	His	His	Pro	Ala	Glu	Phe	Thr	Pro	Ala	Val	His
Myo	Gly	Ala	Met	Asn	Lys	Ala	Leu	Glu	Leu	Phe	Arg	Lys	Asp	Ile	Ala	Ala
Hb	Ala	Ser	Leu	Asp	Lys	Phe	Leu	Ala	Ser	Val	Ser	Thr	Val	Leu	Thr	Ser
Myo	Lys	Tyr	Lys	Glu	Leu	Gly	Tyr	Gln	Gly-153							
Hb	Lys	Tyr	Arg-141													

▲ FIGURE 3-15   Protein homology. (a) Models of the tertiary structures of the oxygen-carrier proteins myoglobin and hemoglobin based on x-ray crystallographic analysis. Note the similarity in the tertiary structures of myoglobin and the two α subunits (dark and light blue) and two β subunits (pink shading) of hemoglobin. The planar white structure in the center of each polypeptide chain is the heme prosthetic group. (b) Amino acid sequences of sperm whale myoglobin and the α chain of normal adult human hemoglobin (Hb). The sequences are aligned to maximize the number of identical residues (red) in the two polypeptides. The gaps presumably represent insertions or deletions of nucleotides in the DNA sequences during the evolution of myoglobin and the hemoglobin α chain from a common ancestor. The β chain of hemoglobin is also highly homologous with myoglobin and the α chain. Most of the conserved residues hold the heme group in place or are responsible for maintaining the hydrophobic interior of the protein. [Parts (a) and (b) courtesy of Gareth White. Myoglobin adapted from S. E. V. Phillips, 1980, *J. Mol. Biol.* **142**:531; hemoglobin adapted from B. Shaanan, 1983, *J. Mol. Biol.* **171**:31.]

The two proteins also exhibit similar functions: myoglobin is the oxygen-carrier protein in muscle, and hemoglobin, the oxygen-carrier protein in blood.

As data concerning protein sequences and three-dimensional structures accumulated, the concept that similar sequences fold into similar secondary and tertiary structures was confirmed. The propensity of each amino acid to occur in the various types of secondary structures has been calculated from the amino acid sequence of secondary structures extracted from databases of the three-dimensional structures of proteins. This tabulation of the folding information inherent in the sequence is now being used in attempts to predict the three-dimensional structure of various proteins from their amino acid sequences.

In the classical taxonomy of the eighteenth and nineteenth centuries, organisms were classified according to their morphologic similarities and differences. In this century, the molecular revolution in biology has given birth to "molecular" taxonomy: the classification of proteins based on similarities and differences in their amino acid sequences. This new taxonomy provides much information about protein function and evolutionary relationships. If the similarity among proteins from different organisms is significant over their entire sequence, then the proteins are *homologs* of each other, and they probably carry out similar functions. Sequence similarity also suggests an evolutionary relationship among proteins; that is, they evolved from a common ancestor. We can therefore describe homologous proteins as belonging to the same "family" and can trace their lineage from comparisons of sequences. Closely related proteins have the most similar sequences; distantly related proteins have only faintly similar sequences.

The kinship among homologous proteins is most easily visualized from a tree diagram based on sequence analyses. For example, the amino acid sequences of hemoglobins from different species suggest that they evolved from an ancestral monomeric, oxygen-binding protein (Figure 3-16). Over time, this ancestral protein slowly changed, giving rise to myoglobin, which remained a monomeric protein, and to the α and β subunits, which evolved to associate into the tetrameric hemoglobin molecule. As the tree diagram in Figure 3-16 shows, evolution of the globin protein family parallels that of the vertebrates. In Chapter 7, we examine how mutations in genes generate protein families.

Many Proteins Contain Tightly Bound Prosthetic Groups

The activities of some proteins require the presence of a *prosthetic group*: a small nonpeptide molecule or metal that binds tightly to a protein, keeping the protein in a fixed conformation and participating in binding ligands. For example, each of the four hemoglobin chains binds and

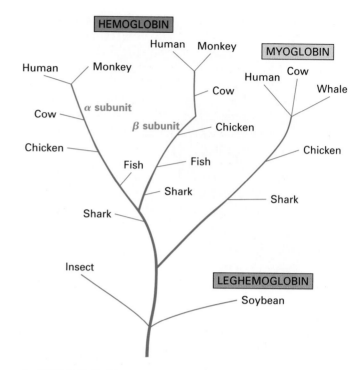

▲ FIGURE 3-16   Evolutionary tree showing how the globin protein family arose starting from the most primitive oxygen-binding proteins, leghemoglobins, in plants. Sequence comparisons have revealed that evolution of the globin proteins parallels the evolution of vertebrates. Major junctions occurred with the divergence of myoglobin from hemoglobin and the later divergence of hemoglobin in the α and β subunits. [Adapted from R. E. Dickerson and I. Geis, 1983, *Hemoglobin: Structure, Function, Evolution, and Pathology,* Benjamin-Cummings.]

enfolds a prosthetic group called *heme,* which consists of an iron atom held in a cage by protoporphyrin:

The heme groups are the oxygen-binding components of hemoglobin. Heme is also present in the cytochromes of the electron transport chain; in this case, it functions to bind electrons. Other electron transport proteins employ sulfur or flavin as prosthetic groups. In addition to acting as carriers of oxygen or electrons, prosthetic groups can act as antennae. For example, proteins involved in vision or photosynthesis contain retinal or chlorophyll, which absorb energy from sunlight. As we discuss in a later section, the activity of numerous enzymes also depends on the presence of a prosthetic group. Prosthetic groups can be linked to proteins noncovalently, as in hemoglobin, or covalently, as in cytochrome.

Chemical Modifications Alter the Biological Activity of Proteins

Nearly every protein in a cell is chemically altered after its synthesis on a ribosome. Such modifications may alter the activity, life span, or cellular location of proteins depending on the nature of the alteration. Protein alterations fall into two categories: chemical modification and processing. *Chemical modification* involves the linkage of a chemical group to the terminal amino or carboxyl groups of the backbone or to reactive groups in the side chains of internal residues. *Processing*, which involves the removal of peptide segments, is discussed in the next section. Both types of alteration are important steps in protein synthesis and are examined in detail in Chapters 4 and 16.

Two chemical modifications affect the terminal residues (Figure 3-17a). *Acetylation,* the addition of an acetyl group (CH_3CO) to the amino group of the N-terminal residue is the most common form of chemical modification,

involving an estimated 80 percent of all proteins. This modification may play an important role in controlling the life span of proteins within cells, as nonacetylated proteins are rapidly degraded by intracellular proteases. A chemically more complex modification is *fatty acid acylation,* the covalent attachment of a lipid fatty acid group to the N-terminus. Such a lipid "tail" functions to anchor a protein to the lipid bilayer of cellular membranes (see Figure 14-20). This modification is one mechanism used by cells to restrict certain proteins to membranes.

The internal residues in proteins can be modified by attachment of a variety of chemical groups to their side chains. Examples of some common modified residues are shown in Figure 3-17b. *Glycosylation,* the attachment of carbohydrate groups to the side chains of asparagine, serine, or threonine is important in the synthesis of many secreted and cell-surface *glycoproteins* (Chapter 16). Another common and key modification is *phosphorylation,* the substitution of a phosphate group for a hydroxyl group in serine, threonine, and tyrosine residues:

Phosphoserine

Phosphotyrosine

(a) Modified terminal residues

Acetylated N-terminus

Acylated N-terminus

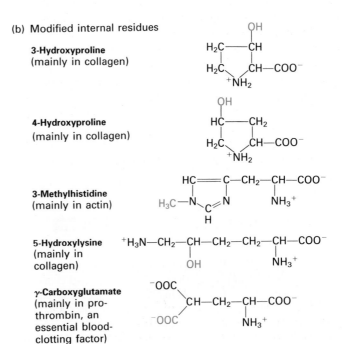

(b) Modified internal residues

3-Hydroxyproline (mainly in collagen)

4-Hydroxyproline (mainly in collagen)

3-Methylhistidine (mainly in actin)

5-Hydroxylysine (mainly in collagen)

γ-Carboxyglutamate (mainly in prothrombin, an essential blood-clotting factor)

▲ **FIGURE 3-17** Examples of common chemical modifications to amino acid residues in proteins. Such modifications occur after a polypeptide chain is synthesized. In each case, the substituted group (red) has replaced a hydrogen ion. (a) N-terminal residues may be modified by acetylation and fatty acid acylation. In the example shown of acylation, a myristate group has been added to the N-terminus. (b) Internal residues may be modified by hydroxylation, methylation, and carboxylation. Not shown are examples of glycosylated and phosphorylated residues, which also are common in proteins.

This easily reversible reaction serves as a mechanism for regulating the activity of many proteins. Addition of phosphate groups is carried out by protein *kinases;* their removal is catalyzed by protein *phosphatases.* In general, the chemical modification of any residue involves a pair of enzymes: one to add the group and another to remove the group. The counteracting activities of such modifying enzymes provide cells with a "switch" that can turn on or turn off the function of various proteins.

Proteolytic Processing and Protein Splicing Alter Protein Activity

In addition to chemical modification of residues, processing of proteins alters their activity. In the most common form of processing, residues are removed from the C- or N-terminus of a polypeptide by cleavage of the peptide bond in a reaction catalyzed by proteases. Proteolytic

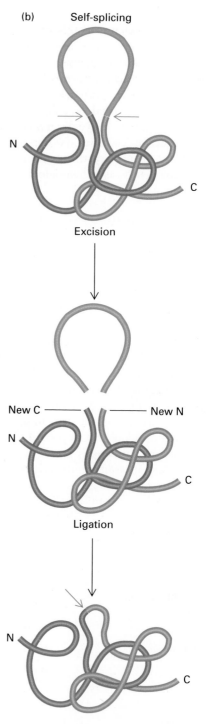

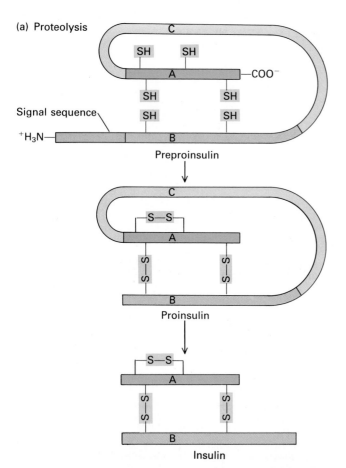

signal sequence from the amino end of the molecule. The remaining 84 amino acids constitute proinsulin, a molecule in which all the correct disulfide bridges are present. While the hormone is being packaged for secretion, the 33-residue C segment is removed via two proteolytic cleavages, yielding the A and B chains of insulin. (b) In self-splicing, a segment of a polypeptide (red) is removed by cleavage of two peptide bonds (red arrows); one new peptide bond then forms (blue arrow), regenerating a continuous polypeptide backbone. The process is autocatalytic and does not depend on enzymes. The excised segment is thought to be exposed at the surface of the folded protein.

▲ FIGURE 3-18   Protein processing by proteolysis and self-splicing. (a) Three successive enzyme-catalyzed cleavages produce insulin from preproinsulin. The first cleavage occurs immediately after synthesis of preproinsulin, a single chain of 108 amino acids. This cleavage removes the 24-aa

cleavage is a common mechanism of activation or inactivation, especially of enzymes involved in blood coagulation or digestion. Proteolysis also generates active peptide hormones from larger precursor polypeptides. The single polypeptide chain of preproinsulin, for example, undergoes several successive cleavages to produce insulin (Figure 3-18a). A more detailed description of the processing of membrane and secreted proteins is presented in Chapter 16.

A new type of processing, termed protein *self-splicing*, has been very recently discovered. Splicing refers to a process analogous to editing film: an internal segment of polypeptide is removed and the ends of the polypeptide are rejoined (Figure 3-18b). The splicing mechanism involves the cleavage of two peptide bonds and the subsequent formation of one new peptide bond. Unlike proteolytic processing, protein self-splicing is an autocatalytic process, which proceeds by itself without the involvement of enzymes. The mechanism by which the excised peptide extricates itself from the protein is poorly understood. As we will see in Chapter 12, processing of RNA molecules also involves an autocatalytic self-splicing mechanism, which is understood quite well and may serve as a model for understanding protein self-splicing.

A Protein Can Be Unfolded by Heat, Extreme pH, and Certain Chemicals

Any polypeptide chain containing n residues could, in principle, fold into 8^n conformations. The number 8^n is based on the fact that only *eight* bond angles are stereochemically allowed in the polypeptide backbone. In general, however, all molecules of any protein species adopt a single conformation, called the *native state*, which is the most stably folded form of the molecule. The native state is maintained through noncovalent bonds between the amino acids and any prosthetic groups and between the protein and its environment.

Proteins unfold, or *denature*, under various conditions that disrupt the weak bonds stabilizing their native conformation. Thermal energy from heat, for example, can break weak bonds. Extremes of pH can alter the charges on amino acid side chains, disrupting ionic and hydrogen

bonds. Chemicals such as urea or guanidine hydrochloride, when present at sufficient concentrations (usually 6–8 M), disrupt both hydrogen and hydrophobic bonds, causing proteins to unfold:

In the denatured state, a protein loses both its compact conformation and activity. Most denatured proteins precipitate in solution because hydrophobic groups, normally buried inside the molecules, interact with similar regions of other unfolded molecules, causing them to form an insoluble aggregate.

Many Denatured Proteins Can Refold into Their Native State in Vitro

Many proteins that are completely unfolded in 8 M urea and β-mercaptoethanol (which reduces disulfide bonds)

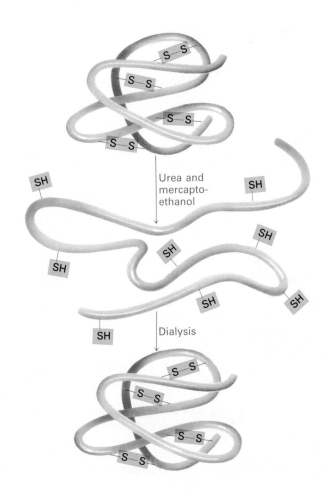

▶ FIGURE 3-19 Most polypeptides can be completely denatured by treatment with an 8 M urea solution containing mercaptoethanol (HSCH₂CH₂OH). The urea breaks intramolecular hydrogen and hydrophobic bonds, and the mercaptoethanol reduces each disulfide bridge to two –SH groups. When these chemicals are removed by dialysis, the –SH groups on the unfolded chain oxidize spontaneously to re-form disulfide bridges, and the polypeptide chain simultaneously refolds into its native conformation.

can *renature* (refold) into their native state when the denaturing reagents are removed by dialysis. During renaturation, all the disulfide, hydrogen, and hydrophobic bonds that stabilize the native conformation are re-formed. Thus, in this case proteins can be carried through a denaturation-renaturation cycle, which first destroys and then reestablishes their original structure and function (see Figure 3-19 on the opposite page).

The observation by Christian Anfinsen of such reversible denaturation and renaturation of ribonuclease, an enzyme that degrades RNA, provided the first clue that the information for folding a protein lies in its sequence. In general, small monomeric proteins or isolated protein domains can undergo multiple reversible denaturation-renaturation cycles. Large proteins are difficult to refold for reasons that are not fully understood. Instead of renaturing into soluble proteins, they often precipitate from solution in macroscopic insoluble aggregates.

The mechanism by which proteins refold in vitro has been elucidated in experiments in which the renaturing conditions are carefully adjusted and the refolding reaction is interrupted at various time intervals. Such studies indicate that proteins refold in stages characterized by formation of discrete, partially folded intermediates (Figure 3-20). In the initial stages of folding, hydrophobic regions of the molecule condense, secondary structures assemble, and after formation of long-range interactions and some rearrangements, the native tertiary conformation is reached. Because no other cofactors or proteins are required, at least in the test tube, protein folding is a self-assembly process.

Folding of Proteins in Vivo Is Promoted by Chaperones

Folding of proteins in vitro is inefficient; only a minority undergo complete folding within a few minutes. Clearly, proteins must fold correctly and efficiently in vivo, otherwise cells would waste much energy in the synthesis of

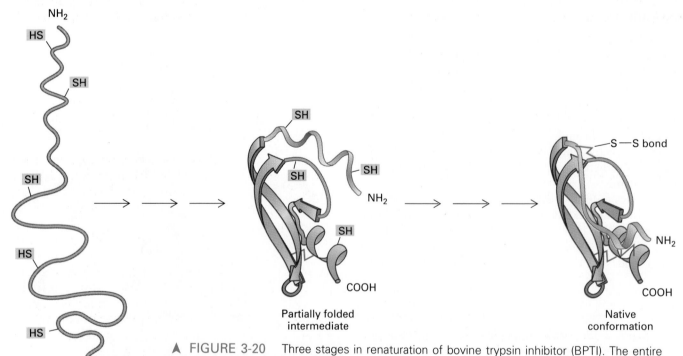

▲ FIGURE 3-20   Three stages in renaturation of bovine trypsin inhibitor (BPTI). The entire polypeptide chain of reduced and denatured BPTI assumes a random conformation. Under appropriate refolding conditions, the molecule assumes a compact, but nonnative, conformation. In this folding intermediate, the chain has packed into three β strands and an α-helical segment (green), but the N-terminal segment (orange) remains unfolded because four of the six sulfhydryl groups (yellow) are still reduced. After several steps during which additional disulfide bonds form (blue), BPTI assumes its native conformation. [Adapted from E. E. Creighton, 1993, *Proteins: Structure and Molecular Properties*, 2d ed., W. H. Freeman and Co.]

nonfunctional proteins and in the degradation of misfolded or unfolded proteins. More than 95 percent of the proteins present within cells have been shown to be in their native conformation, despite high protein concentrations ($\approx$ 100 mg/ml), which usually cause proteins to precipitate in vitro.

The explanation for the cell's remarkable efficiency in promoting protein folding probably lies in *chaperones,* a newly discovered family of proteins found in all organisms from bacteria to humans. Chaperones are located in every cellular compartment, bind a wide range of proteins, and may be part of a general protein-folding mechanism. Chaperones aid protein folding in two ways: first, by binding and stabilizing unfolded or partially folded proteins, thereby preventing these proteins from being degraded; and second, by directly facilitating their folding. The ability of chaperones to bind and stabilize their target proteins is specific and dependent on ATP hydrolysis. Binding of chaperones to partially folded proteins suggests that the folding process could be regulated at intermediate steps.

➤ *Enzymes*

A major premise of this chapter—that protein function depends on structure—is illustrated by the previous discussion of ribonuclease denaturation and renaturation. In this and the next section, we examine the relationship between the three-dimensional structure and the function of two classes of proteins, enzymes and antibodies, in some detail.

Almost every chemical reaction in a cell is catalyzed by a class of proteins called *enzymes.* As discussed in Chapter 2, catalysts increase the rates of reactions that are already energetically favorable by lowering the activation energy. In the test tube, catalysts such as charcoal and platinum facilitate reactions but often at high temperatures, extremes of high or low pH, or in organic solvents. In contrast to these harsh conditions, enzymes must catalyze chemical reactions in the mild conditions of a cell: 37°C, pH 6.5–7.5, and aqueous solvents.

The ability of enzymes to function as catalysts under conditions where nonbiological catalysts would be ineffectual is exemplified by two striking properties: their enormous catalytic power and their specificity. Quite often, the rate of an enzymatically catalyzed reaction is 10^6–10^{12} times that of an uncatalyzed reaction under otherwise similar conditions. Specificity denotes an enzyme's ability to act selectively on one substance or a small number of chemically similar substances. This level of discrimination is achieved because the site where the reactants bind has a precise size and shape that only fits the correct molecules. An example of specificity is provided by the enzymes that act on amino acids. As noted earlier, natural amino acids are L stereoisomers. Not surprisingly, enzyme-catalyzed reactions involving L-amino acids occur much more rapidly than do those involving D-amino acids, even though both stereoisomers of a given amino acid are the same size and possess the same R groups.

Chemicals that are changed in a reaction catalyzed by an enzyme are the *substrates* of that enzyme. The name of an enzyme usually indicates its substrate or function. The suffix *-ase* is commonly appended to the name of the type of molecule on which an enzyme acts. Thus prote**ases** degrade proteins, phosphat**ases** remove phosphate residues, and ribonucle**ases** cleave RNA molecules.

The number of different types of chemical reactions that occur in any one cell is very large: an animal cell, for example, normally contains 1000–4000 different types of enzymes, each of which catalyzes a single chemical reaction or set of closely related reactions. Certain enzymes are found in the majority of cells, because they catalyze synthesis of common cellular products (e.g., proteins, nucleic acids, and phospholipids) or are involved in the production of energy by the conversion of glucose and oxygen to carbon dioxide and water. Other enzymes are present only in a particular type of cell (e.g., a liver cell or a nerve cell), because they catalyze some chemical reaction unique to that cell type. Although most enzymes are located within cells, some are secreted and function in the blood, lumen of the digestive tract, or other extracellular space. Some microbial enzymes are secreted from and are active outside the organism.

The Active Site of an Enzyme Is a Cage of Amino Acids That Binds Substrates and Catalyzes Reactions

Certain amino acid side chains of an enzyme are important in determining its specificity and its ability to accelerate a chemical reaction. In the native conformation of an enzyme, these side chains are brought into proximity, forming the *active site.* Active sites thus consist of two important regions: one that recognizes and binds the substrate(s), and one that catalyzes the reaction once the substrate(s) has been bound. In some enzymes, the catalytic site is part of the substrate-binding site. The amino acids that make up the active site do not need to be adjacent in the linear polypeptide sequence; rather, folding of the molecule results in juxtaposition of these amino acids, forming a space in which the substrate sits.

Binding of a substrate to an enzyme usually involves the formation of multiple noncovalent ionic, hydrogen, and hydrophobic bonds, as well as van der Waals interactions (Figure 3-21). The array of chemical groups in the active site of the enzyme is precisely arranged in three-dimensional space, so that its specific substrate fits more

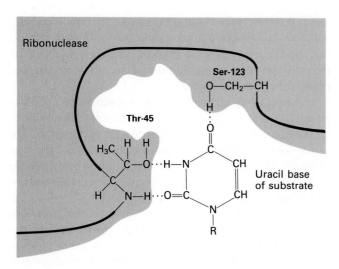

▲ FIGURE 3-21   The specific binding of a substrate to an enzyme involves the formation of multiple noncovalent bonds. Here, two amino acid residues of the enzyme ribonuclease bind uracil, part of its substrate, by three hydrogen bonds (dots). Substrates without the two C=O groups and one NH group in the appropriate positions would be unable to bind or would bind less tightly. Other regions of the enzyme, not depicted here, bind other parts of the RNA substrate by hydrogen bonds and van der Waals interactions.

snugly and therefore is more tightly bound than any other molecule (with the exception of some enzyme inhibitors). For this reason, the reaction can occur readily.

In the active site, covalent bonds between the enzyme and the substrate may be formed (and then broken), converting the substrate into a strained configuration called the *transition-state intermediate*. This intermediate either is converted into product or reverts back to substrate. Bonding between residues in the active site and the substrate help stabilize the transition state, thereby allowing more time for the reaction to be completed. As explained in Chapter 2, the *activation energy* is the energy required for formation of the transition state (see Figure 2-26). An enzyme, by virtue of its three-dimensional binding site, reduces the activation energy of a reaction compared with an uncatalyzed reaction involving the same reactants. The ability to bind transition-state intermediates is the one property that distinguishes enzymes from other proteins. If a protein cannot bind a transition-state intermediate, then it cannot catalyze a reaction.

Proteases Degrade Proteins by Reducing Activation Energy for Peptide-Bond Hydrolysis

Digestion of food is a disassembly process in which complex biopolymers are degraded to their simple monomer

building blocks by hydrolytic (bond-breaking) enzymes. To see how enzymes work, we will focus on the mechanism by which two well-characterized proteases—*trypsin* and *chymotrypsin*—cleave the peptide bond of proteins.

Because proteases are so destructive to proteins, it is crucial that their activity is controlled. Trypsin and chymotrypsin are synthesized in the pancreas and secreted into the small intestine as inactive precursors, or *zymogens*, called trypsinogen and chymotrypsinogen, respectively. In the protease-rich environment of the small intestine, the zymogens are converted to the active enzymes, which then begin to hydrolyze peptide bonds of ingested proteins. This delay in activation serves an important regulatory purpose by preventing the enzymes from digesting the pancreatic tissue in which they are made. Two irreversible proteolytic cleavages activate chymotrypsin (Figure 3-22). One cleav-

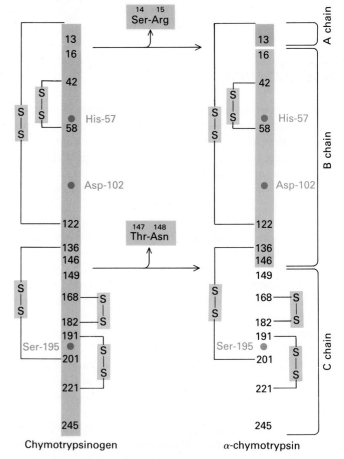

▲ FIGURE 3-22   A linear representation of the conversion of chymotrypsinogen into chymotrypsin by the excision of two dipeptides. These reactions yield three separate chains (A, B, and C), which are covalently linked by disulfide bonds (yellow) in the active enzyme. In the folded, native conformation of chymotrypsin, histidine 57, aspartate 102, and serine 195 are located in the active site.

age removes serine 14 (the serine at position 14) and arginine 15 from chymotrypsinogen; the other removes threonine 147 and asparagine 148. Removal of these two dipeptides activates the protease function of the enzyme.

The hydrolysis of peptide bonds is energetically favorable ($\Delta G^{\circ\prime} = -2$ kcal/mol). Nonetheless, the activation energy for uncatalyzed hydrolysis of peptide bonds in a protein in neutral aqueous solution at room temperature is so high that little or no hydrolysis occurs even after several months. For this reason, hydrolysis of proteins for amino acid analysis requires prolonged incubation (24 h) in a solution of 6 M hydrochloric acid at 110°C. Yet a molecule of trypsin or chymotrypsin can catalyze the hydrolysis of up to 100 peptide bonds per second at 37°C and neutral pH.

Furthermore, whereas the uncatalyzed hydrolysis of proteins results in the indiscriminate cleavage of all peptide bonds, proteases are selective in their action. Chymotrypsin, for example, only cleaves peptide bonds at the carboxyl ends of residues with large hydrophobic side chains, namely, phenylalanine, tyrosine, and tryptophan (Figure 3-23). Although the amino acid sequence of trypsin is similar to that of chymotrypsin, it exhibits a different specificity, cleaving peptide bonds on the C-terminal side of the basic residues lysine and arginine.

Substrate Binding by Chymotrypsin The reaction mechanism of chymotrypsin was deduced, in part, from its three-dimensional structure determined by x-ray crystallography (Figure 3-24). The enzyme contains three subunits, the A, B, and C chains, which have 13, 131, and 97 residues, respectively (see Figure 3-22). The quaternary structure of this extracellular protein is stabilized by disulfide bridges between the subunits. The molecule contains a *hydrophobic cleft,* a crevice in the surface of the protein that is bordered by the side chains of several hydrophobic amino acid residues. This cleft serves as the binding site for specific amino acid residues on the substrate. Because of its shape, the residues lining the cleft are in the right position to participate in hydrophobic interactions with the large hydrophobic side chains of phenylalanine, tyrosine, and tryptophan. Charged side chains or small hydrophobic residues on the substrate cannot make the noncovalent bonds necessary to fit into the cleft. As a result, chymotrypsin cleaves the peptide bonds after tyrosine, tryptophan, and phenylalanine residues but not other residues.

Mechanism of Catalysis by Chymotrypsin The catalytic activity of chymotrypsin depends on three amino acid residues: histidine 57, aspartate 102, and serine 195. These amino acids are distant from one another in the pri-

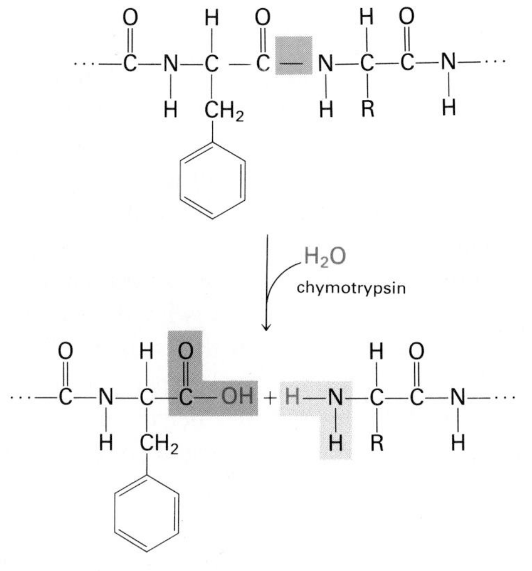

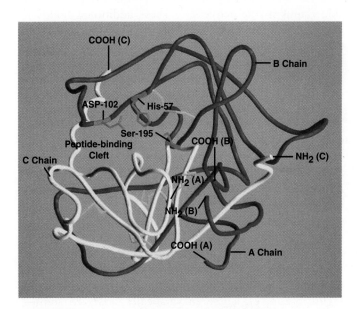

▲ FIGURE 3-24   A three-dimensional computer-generated model of chymotrypsin determined from x-ray crystallographic analysis. The N- and C-termini of the three chains are indicated with the chain designation in parentheses. The five disulfide bonds are shown in yellow. The three residues (orange) that are crucial for catalytic activity are brought into proximity in the native folded conformation, although they are widely separated in the sequence. [Courtesy of Gareth White. Adapted from H. Tsukada and D. M. Blow, 1985, J. Mol. Biol. **184**:703.]

▲ FIGURE 3-23   The hydrolysis of a peptide bond (blue) by chymotrypsin. This enzyme hydrolyzes the peptide bond following phenylalanine (shown), tyrosine, and tryptophan residues. As a result, these residues always lie at the C-terminus of one of the peptides (green) produced by cleavage.

mary structure of the enzyme but close together in the folded, native conformation (see Figure 3-24). When chymotrypsinogen is proteolytically activated, the polypeptide conformation is altered to bring these three residues into their final, correct alignment, forming the catalytic region of the active site in chymotrypsin.

From our previous discussion, recall that catalysis occurs when the activation energy for a chemical reaction is lowered. The limiting step in a reaction is formation of the transition-state intermediate, a short-lived form containing strained high-energy bonds. In the hydrolysis of a peptide bond, a tetrahedral transition-state intermediate is formed when a hydroxyl group forms a covalent bond with the carbonyl carbon in the peptide:

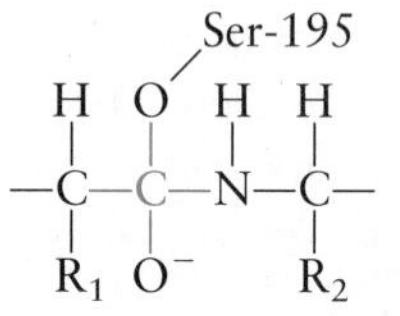

This intermediate is unstable in solution due to the unshared electrons contributed by the oxygen atoms; however, in chymotrypsin-catalyzed hydrolysis, the intermediate is stabilized by hydrogen bonds to amino acid side chains in the active site. These additional bonds in effect reduce the energy required to form the transition-state intermediate.

Hydrolysis of a polypeptide by chymotrypsin proceeds in two main steps. First, a peptide bond is broken and the carboxyl group is transferred to the hydroxyl residue of serine 195, forming a stable acylenzyme intermediate and releasing a peptide with a free amino group (R_1NH_2). Second, the acylenzyme intermediate is hydrolyzed, restoring the enzyme to its original state and releasing a peptide with a free carboxyl group (R_2COOH). The overall reaction, as illustrated in Figure 3-25, involves two tetrahedral transition-state intermediates.

Aspartate 102 and histidine 57 facilitate the acylation reaction by removing the proton from serine 195 and adding it to the nitrogen of the departing amino group (see Figure 3-25a, b). In a similar manner, aspartate 102 and histidine 57 facilitate the hydrolysis of the acylenzyme (see Figure 3-25d, e). These enzymatically catalyzed steps—transfer of a proton from the enzyme to the substrate, formation of a covalent acylenzyme intermediate, and hydrolysis of the acylenzyme—all drastically reduce the overall activation energy of the proteolysis reaction.

The hydroxyl group on serine 195 is unusually reactive. The concept of an "active" serine residue at the active site of chymotrypsin predated determination of the enzyme's crystal structure. It was already known, for example, that diisopropylfluorophosphate, a potent inhibitor of

chymotrypsin, reacts only with the hydroxyl on serine 195 to form a stable covalent enzyme derivative:

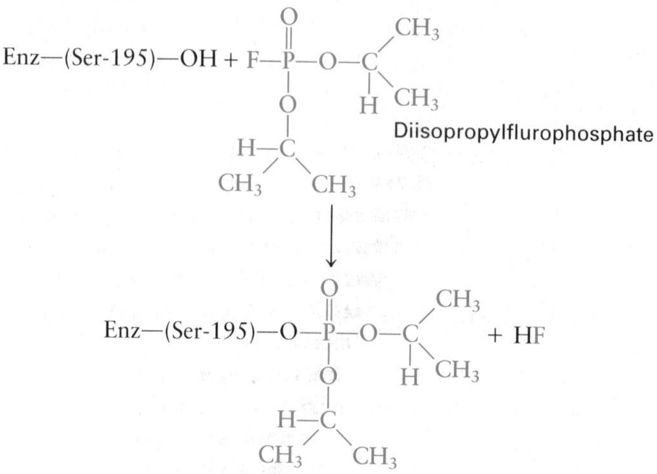

This chemical modification of serine 195 irreversibly inactivates chymotrypsin.

Comparison of Chymotrypsin and Trypsin About 40 percent of the amino acids in chymotrypsin and trypsin are the same; in particular the amino acid sequence in the vicinity of the key serine residue is identical in the two enzymes:

195
—Gly—Asp—Ser—Gly—Gly—Pro

The three-dimensional structures and catalytic mechanisms of the two enzymes also are quite similar, indicating that they evolved from a common ancestral polypeptide.

Nonetheless, as noted previously, trypsin and chymotrypsin have quite different specificities, the result of differences in the side chains of the amino acids that line the substrate-binding site. In fact, the only major structural difference between the two enzymes is found in this region. The binding site of chymotrypsin is lined by hydrophobic side chains, which interact with the hydrophobic residues of phenylalanine, tryptophan, and tyrosine. In contrast, the binding site of trypsin contains negatively charged side chains, which bind positively charged lysine and arginine residues in a substrate.

Coenzymes Are Essential for Certain Enzyme-Catalyzed Reactions

Many enzymes contain a *coenzyme*—a tightly bound small molecule or metal (prosthetic group) essential to enzymatic activity. Vitamins required in trace amounts in the diet are

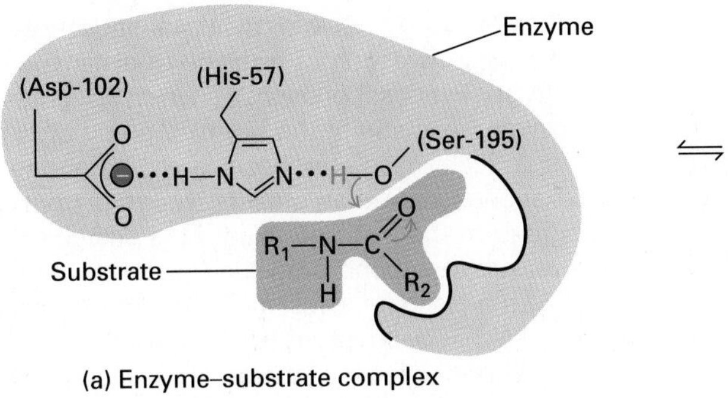

(a) Enzyme–substrate complex

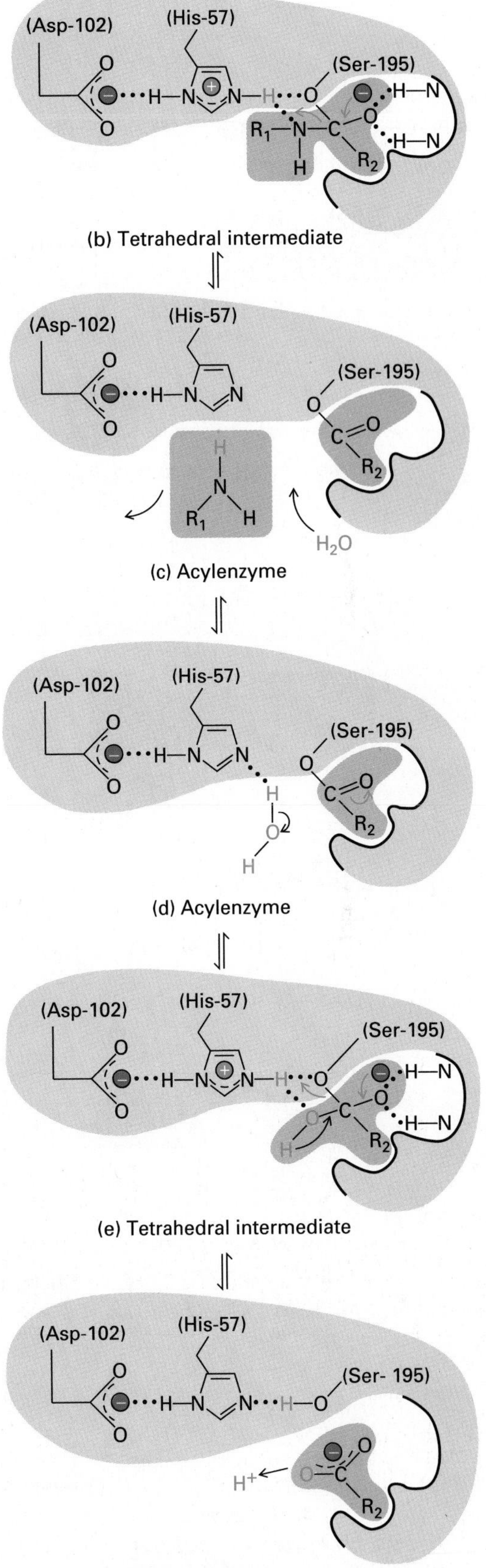

(b) Tetrahedral intermediate

(c) Acylenzyme

(d) Acylenzyme

(e) Tetrahedral intermediate

(f) Enzyme–product complex

▲ FIGURE 3-25   The mechanism of hydrolysis of a peptide bond catalyzed by chymotrypsin. Red curved arrows represent the movement of electrons in the charge-relay system that shifts negative charge (blue) in the course of the reaction. In the first step (a), substrate binds to the enzyme, so that the bond to be hydrolyzed is positioned near the hydroxyl group of Ser-195. Catalysis is initiated by a shift of bonds between active site residues and the peptide substrate. A new bond forms between the second nitrogen of His-57 and the hydroxyl hydrogen of Ser-195, which allows the oxygen of Ser-195 to bond with the carbonyl carbon of the peptide backbone. These rearrangements produce a tetrahedral intermediate (b) in which the carbonyl carbon temporarily has four single bonds and the oxygen has excess electrons. The hydrogen bound to the second nitrogen in His-57 then adds to the nitrogen of the substrate, cleaving the peptide bond to yield an acylenzyme intermediate and R_1NH_2, which is released from the enzyme (c). A similar charge-relay system then is induced in the acylenzyme as His-57 removes a proton (H) from a water molecule (d). The OH^- thus generated attacks the carboxyl carbon of the acylenzyme to form a second tetrahedral intermediate (e). In the final step (f), the bond between the tetrahedral carbon and the oxygen of Ser-195 is cleaved, yielding the original enzyme and R_2COO^-, which is released from the enzyme. [Adapted from R. M. Stroud et al., 1975, in *Proteases and Biological Control*, E. Reich et al., eds., Cold Spring Harbor Laboratory, p. 25.]

often converted to coenzymes. For instance, coenzyme A, is derived from the vitamin pantothenic acid, and the coenzyme pyridoxal phosphate is derived from vitamin B_6. To illustrate just one example of how coenzymes function, we consider the conversion of the amino acid histidine into histamine, a potent dilator of small blood vessels. In the first step of this reaction, the aldehyde group of pyridoxal phosphate forms a covalent bond with the α amino group of histidine, yielding a covalent intermediate called a *Schiff base* (Figure 3-26). Rearrangement of the bonds in the Schiff base weakens the bond between the C_α and the carboxylate group of the histidine, leading to release of the

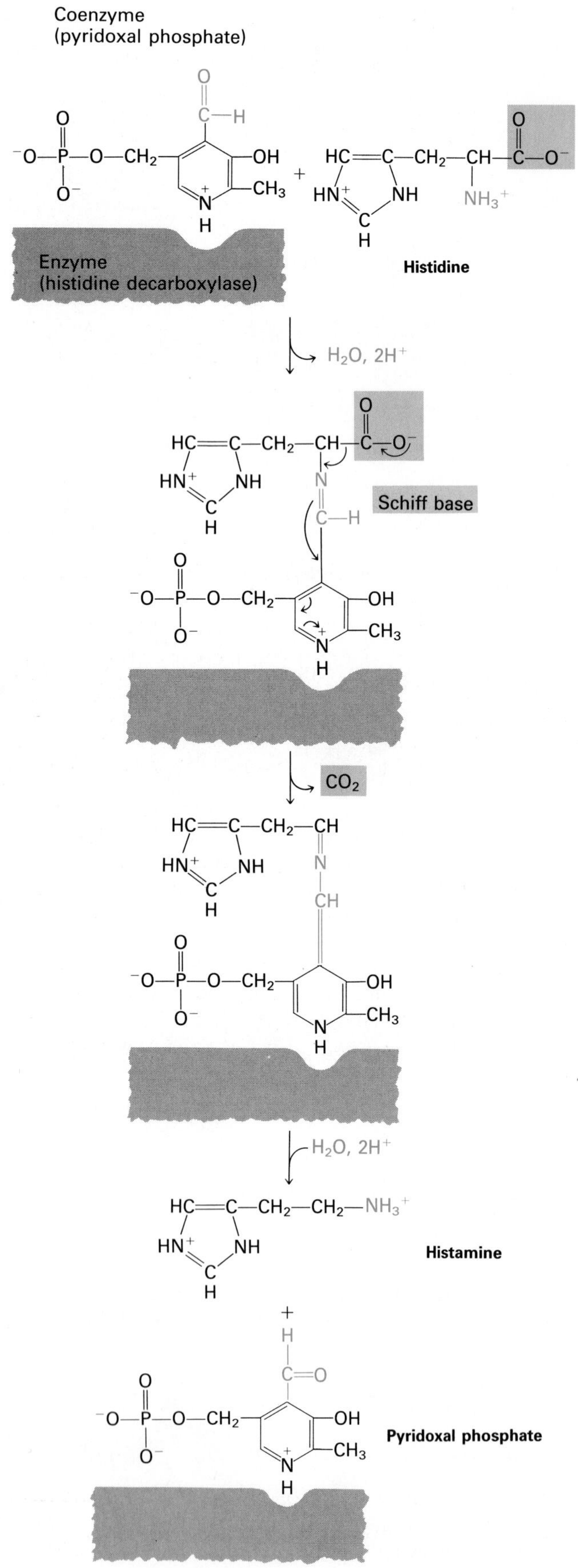

Coenzyme
(pyridoxal phosphate)

Enzyme
(histidine decarboxylase)

Histidine

H₂O, 2H⁺

Schiff base

CO₂

H₂O, 2H⁺

Histamine

Pyridoxal phosphate

◄ FIGURE 3-26   Role of pyridoxal phosphate, a coenzyme, in the conversion of histidine to histamine by histidine carboxylase. Pyridoxal phosphate, which is tightly bound to the enzyme, reacts with the α amino group of histidine, forming a Schiff base. The positive charge on the nitrogen of pyridoxal phosphate then attracts electrons from the carboxylate group of the histidine (blue) via a charge-relay system. This weakens the bond between C_α and the carboxylate group of the histidine, causing release of CO_2 and generating a second Schiff base intermediate. Hydrolysis of this intermediate yields histamine, the other reaction product, and regenerates the pyridoxal phosphate–enzyme complex. Pyridoxal phosphate is a coenzyme for many other enzymes that catalyze reactions involving amino acids.

latter as carbon dioxide. The involvement of pyridoxal phosphate thus lowers the activation energy for cleavage of the crucial bond in histidine. In the final step of the reaction, the Schiff base is hydrolyzed, regenerating the coenzyme, pyridoxal phosphate, and forming the product, histamine.

Activity of Some Enzymes Depends on a Conformational Change Induced by Substrate Binding

In the case of some enzymes, the active site and the substrate exhibit a complementary charge, shape, or both, so that the two molecules can interact to form a complex stabilized by a variety of noncovalent bonds. Such an interaction, which resembles the fitting of a key into a lock, occurs by what is called the *lock-and-key* mechanism (Figure 3-27a).

In other enzymes, the binding of substrate induces a conformational change in the enzyme that brings the catalytic residues close enough to the substrate-binding (or *recognition*) site so that they can attack the substrate. In effect, a functional active site is formed only when substrate binds to such enzymes, a mechanism termed *induced fit* (Figure 3-27b). Molecules that attach to the recognition site of these enzymes but that do not induce the necessary conformational change are not substrates.

Induced-fit substrate binding is displayed by hexokinase, which catalyzes the first step in the degradation of glucose by cells—the transfer of a phosphate residue from ATP to carbon 6 of glucose:

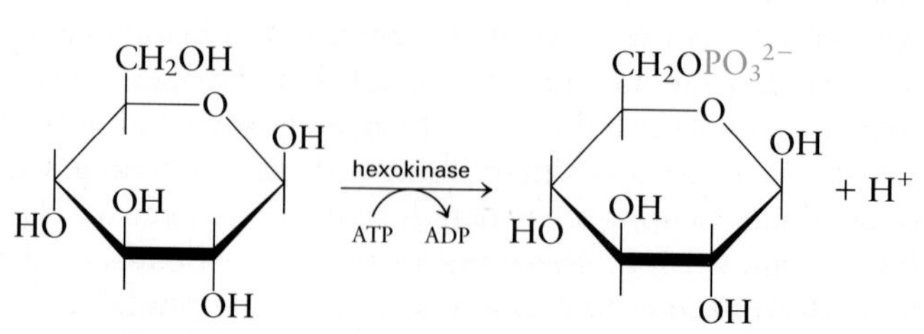

Glucose

hexokinase

ATP   ADP

Glucose 6-phosphate

+ H⁺

(a) LOCK-AND-KEY

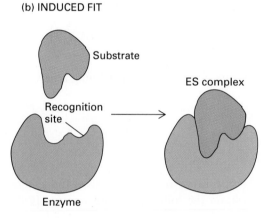

(b) INDUCED FIT

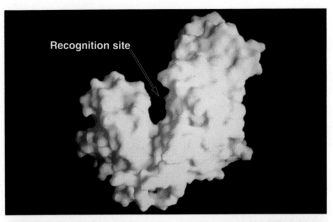

▲ FIGURE 3-27 Two mechanisms that position substrate for catalysis. (a) In enzymes that exhibit the lock-and-key mechanism, the active site contains residues involved in both the binding of substrate and catalysis. In this case, binding of substrate to the complementary active site positions it for catalysis. (b) In enzymes that exhibit the induced-fit mechanism, binding of substrate induces a conformational change in the enzyme that brings the residues involved in catalysis into proximity of the substrate.

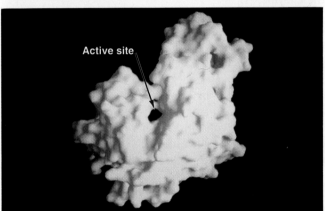

▲ FIGURE 3-28 Computer-generated images of hexokinase illustrating the induced-fit mechanism. (*top*) Hexokinase has an open conformation when glucose is not bound to it. ATP binds at the base of the deep cleft between the two domains. (*bottom*) Binding of glucose (not shown) induces a conformational change that brings the two domains closer together and positions the glucose molecule close enough to the bound ATP that the reaction can proceed. [Courtesy of Gareth White. Adapted from C. M. Anderson, R. E. Stenkamp, and T. A. Steitz, 1978, *J. Mol. Biol.* **123**:15.]

X-ray crystallography has shown that hexokinase consists of two domains. The binding of glucose in the cleft between the domains induces a major conformational change that brings these domains closer together (Figure 3-28); as a result, ATP, which sits at the base of the cleft, is brought into position to transfer a phosphate group to glucose. Only glucose and closely related compounds can induce this conformational change, ensuring that the enzyme is used to phosphorylate only the correct molecules. Glycerol, ribose, and even water can bind to hexokinase at the recognition site but cannot induce the requisite conformational change; hence they are not substrates of the enzyme.

The Catalytic Activity of an Enzyme Can Be Described Mathematically

Enzymatic specificity is usually quantified in relative terms: the reaction with a good substrate may occur, for example, 10,000 times faster than it occurs with a poor substrate.

The catalytic action of an enzyme on a given substrate can be described by two parameters: K_m (the Michaelis constant), which measures the affinity of an enzyme for its substrate, and V_{max}, which measures the maximal velocity of the reaction at saturating substrate concentrations. Equations for K_m and V_{max} are most easily derived by considering the simple reaction

$$\text{Substrate} \rightleftharpoons \text{product}$$

in which the rate of product formation (v) depends on the concentration of substrate, [S], and on the concentration of the enzyme, [E].

For an enzyme with a single catalytic site, Figure 3-29a shows how the rate of product formation depends on [S] when [E] is kept constant. At low concentrations of S, the reaction rate is proportional to [S]; as [S] is increased, the rate does not increase indefinitely in proportion to [S] but eventually reaches a maximum velocity, V_{max}, at which it becomes independent of [S]. V_{max} is proportional to [E] and to a catalytic constant k_{cat} that is an intrinsic property of the individual enzyme; halving [E] reduces the reaction rate at all values of [S] by one-half.

When interpreting curves such as those in Figure 3-29, bear in mind that all enzymatically catalyzed reactions include at least three steps: (1) the binding of a substrate (S) to an enzyme (E) to form an enzyme-substrate complex (ES); (2) the conversion of ES to the enzyme-product complex (EP); and (3) the release of the product (P) from EP, to yield free P:

$$E + S \underset{\text{binding}}{\overset{}{\rightleftharpoons}} ES \xrightarrow{\text{catalysis}} EP \xrightarrow{\text{release}} P + E$$

In the simplest case, when the release of P is very rapid, we can simplify the reaction equation as follows:

$$E + S \underset{k_2}{\overset{k_1}{\rightleftharpoons}} ES \xrightarrow{k_{cat}} E + P$$

In this case, the rate of product formation v is proportional to the concentration of ES and to the catalytic constant k_{cat} for the given enzyme:

$$v = k_{cat}[ES] \tag{1}$$

For a *steady-state* reaction in which the rate of formation of ES is equal to the rate of disappearance of ES, the following expression holds:

$$k_1[E][S] = k_2[ES] + k_{cat}[ES] \tag{2}$$
$$= (k_2 + k_{cat})[ES] \tag{3}$$

We can rewrite this last equation to give

$$[E] = \frac{(k_2 + k_{cat})[ES]}{k_1[S]} \tag{4}$$

If we designate the total amount of enzyme as E_{tot}, the sum of the free enzyme and bound enzyme is expressed as

$$[E_{tot}] = [E] + [ES] \tag{5}$$

By substituting equation (4) into (5), we get the following:

$$[E_{tot}] = \frac{(k_2 + k_{cat})}{k_1[S]}[ES] + [ES] \tag{6}$$
$$= [ES]\left[1 + \left(\frac{k_2 + k_{cat}}{k_1}\right)\left(\frac{1}{[S]}\right)\right] \tag{7}$$

We can simplify equation (7), by defining a new constant, K_m,

$$K_m = \frac{k_2 + k_{cat}}{k_1} \tag{8}$$

and then substituting it into (7); after rearranging the terms, we obtain the following expression for [ES]:

$$[ES] = \frac{[E_{tot}]}{1 + K_m/[S]} \tag{9}$$

If we now substitute equation (9) into equation (1), the velocity of the reactions is given by

$$v = k_{cat}\frac{[E_{tot}]}{1 + K_m/[S]} \tag{10}$$
$$= k_{cat}[E_{tot}]\frac{[S]}{[S] + K_m} \tag{11}$$

When the substrate is present in excess, all the substrate-binding sites on the enzyme are filled, and the reaction velocity is V_{max}, the maximum rate of product formation. Under this condition, [S] is large and much greater than K_m, so that

$$\frac{[S]}{[S] + K_m} \longrightarrow 1$$

Thus, $V_{max} = k_{cat}[E_{tot}]$. Substituting into equation (11) gives

$$v = V_{max}\frac{[S]}{[S] + K_m} \tag{12}$$

which fits the curves shown in Figure 3-29. By rearranging this equation, we can show that $K_m = [S]$ when v is half the maximal velocity. Thus both V_{max} and K_m for a particular enzyme and substrate can be determined from experimental curves of reaction velocity versus substrate concentration.

The slowest step in most enzymatic reactions is the conversion of the enzyme-substrate complex ES to the free enzyme E and product P. In such cases, k_{cat} is much less than k_2, and equation (8) becomes

$$K_m \cong \frac{k_2}{k_1} = K_{eq}$$

where K_{eq} is the equilibrium constant for binding of S to E. Thus the parameter K_m describes the affinity of an enzyme for its substrate. The smaller the value of K_m, the more avidly the enzyme can bind the substrate from a dilute solution and the smaller the concentration of substrate needed to reach half-maximal velocity (Figure 3-29b). The concentrations of the various small molecules in a cell vary widely, as do the K_m values for the different enzymes that act on them. Generally, the intracellular concentration of a substrate is approximately the same as or greater than the K_m value of the enzyme to which it binds.

Enzymatic Activity Can Be Regulated by Various Mechanisms

Most reactions in cells do not occur independently of one another or at a constant rate. Instead, the catalytic activity of enzymes is so regulated that the amount of reaction product is just sufficient to meet the needs of the cell. As a result, the steady-state concentrations of substrates and products will vary depending on cellular conditions. The flow of material in an enzymatic pathway is controlled by several mechanisms.

Feedback Inhibition by the Product of a Reaction Pathway Consider the series of reactions leading to biosynthesis of the amino acid isoleucine, which is used by cells primarily in the synthesis of proteins. The amount of isoleucine needed by a cell depends on its rate of protein synthesis. The first step in synthesis of isoleucine is elimination of the α amino group from threonine, yielding α-ketobutyrate. Threonine deaminase, the enzyme that catalyzes this reaction, plays a key role in regulating the level of isoleucine. In addition to its substrate-binding site for threonine, threonine deaminase contains a binding site for isoleucine. When isoleucine is bound there, the enzyme molecule undergoes a conformational change, so that it cannot function as efficiently. Thus isoleucine acts as an inhibitor of the reaction for the conversion of threonine. If the isoleucine concentration in the cell is high, the binding of isoleucine to threonine deaminase reduces the rate of isoleucine synthesis:

(a)

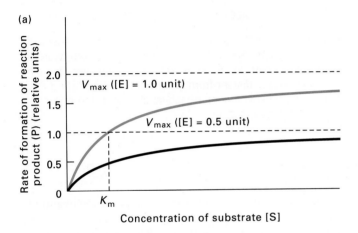

(b)

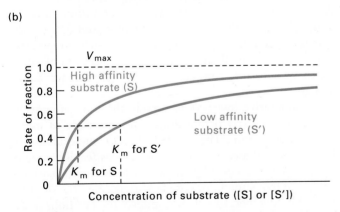

▲ FIGURE 3-29 Dependence of the velocity of an enzyme-catalyzed reaction on substrate concentration. (a) The rates of a hypothetical reaction S → P at two different concentrations of enzyme [E] as a function of substrate concentration [S]. The [S] that yields a half-maximal reaction rate is the K_m, a measure of the affinity of E for S. Doubling the concentration of enzyme causes a proportional increase in the reaction rate, so that the maximal velocity V_{max} is doubled; the K_m, however, is unaltered. (b) The rates of the reactions catalyzed by an enzyme with substrate S, for which the enzyme has a high affinity, and with substrate S′, for which the enzyme has a low affinity. Note that the V_{max} is the same with both substrates but that K_m is higher for S′, the low-affinity substrate.

This is an example of *feedback inhibition,* whereby an enzyme that catalyzes one of the reactions in a multistep pathway is inhibited by the ultimate product of the pathway.

In isoleucine synthesis, as in most cases of feedback inhibition, the final product of the reaction pathway inhibits the enzyme that catalyzes the first step. Thus, none of the intermediate products in the isoleucine pathway are formed. Feedback inhibition of enzyme function is reversible. If the concentration of free isoleucine is lowered, bound isoleucine dissociates from the enzyme, which then reverts to its active conformation. The binding of the feedback inhibitor isoleucine to threonine deaminase and its subsequent release can be described by the equilibrium binding constant K_i, which is similar to the Michaelis constant K_m used to describe substrate binding:

$$[E \cdot Ile]_{inactive} \underset{}{\overset{K_i}{\rightleftharpoons}} [Ile] + [E]_{active}$$

$$K_i = \frac{[Ile][E]_{active}}{[E \cdot Ile]_{inactive}}$$

Allosteric Regulation by Effector Molecules Some enzymes are regulated by *effectors,* small molecules that bind at a site distant from the active site but nevertheless can modulate catalytic activity. There are two types of effectors: *activator* molecules, which increase enzymatic activity, and *inhibitor* molecules, which decrease enzymatic activity. Effectors regulate enzymes by *allostery* (from the Greek for "another shape"). An allosteric effector induces a conformational change in the tertiary structure of an enzyme that is propagated to the substrate-binding and/or catalytic site of the enzyme, causing a change in its activity (Figure 3-30).

Enzymes may have multiple allosteric sites at which different effectors can bind. The presence of multiple sites permits complex and finely tuned control of an enzyme's activity. In some cases, several activators can enhance enzymatic activity; in others, some inhibitors but not others may be able to counteract the effect of an activator.

Cooperativity among Subunits of Multimeric Proteins Many enzymes and other proteins contain more than one polypeptide chain; that is, they are multimeric. Some enzymes are composed of identical subunits, each containing an active site and, possibly, a regulatory site, whereas others are composed of structurally different subunits. In the latter, active sites and regulatory sites may be located on different subunits.

In many cases, binding of an activator, inhibitor, or substrate to one subunit of a multimeric protein causes a conformational change, usually small, that triggers a change in quaternary structure. This quaternary rearrangement favors similar conformational changes in the other

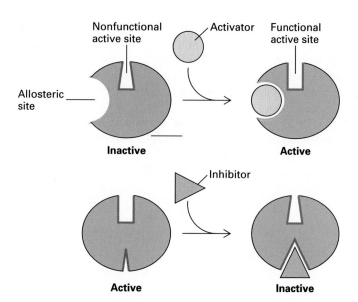

▲ FIGURE 3-30   Schematic diagrams showing allosteric activation *(top)* and inhibition *(bottom)* of an enzyme (blue). An effector molecule, either an activator or inhibitor, can bind at a site spatially distant from an enzyme's active site. Effector binding induces a conformational change in the enzyme that modulates its activity.

subunits, thereby increasing their affinity for the ligand initially bound. For enzymes that act in this fashion (i.e., several subunits interact *cooperatively*), a given increase or decrease in substrate or effector concentration causes a larger change in the reaction rate than would occur if the subunits acted independently. Because of such cooperative interactions, a small change in the concentration of an effector or substrate can lead to large changes in catalytic activity.

Cooperative interactions among the four subunits in hemoglobin clearly demonstrate one advantage of multimeric organization. The binding of an O_2 molecule to the heme group of any one of the four chains (each hemoglobin chain binds one O_2) induces a local conformational change in that subunit (Figure 3-31a). Because the subunits fit snugly together, this change induces the positions of the two α and two β chains in the tetramer to rearrange. In this new configuration of subunits, any local conformational changes that accompany O_2 binding can then occur more readily in the remaining subunits, increasing their affinity for oxygen. As a result, the binding of a second O_2 makes the quaternary structural change even more likely. The large scale effect of this subunit cooperativity is that hemoglobin is able to take up or lose four O_2 molecules over a much narrower range of oxygen pressures than it would otherwise. As a result, hemoglobin is almost completely oxygenated at the oxygen pressure in the lungs, and it is also largely deoxygenated at the oxygen pressure in the tissue capillaries.

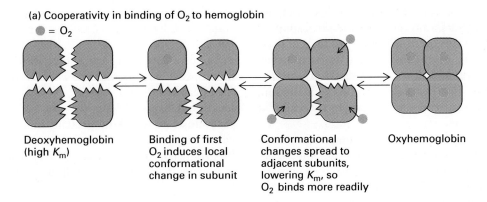

(a) Cooperativity in binding of O_2 to hemoglobin

● = O_2

Deoxyhemoglobin (high K_m)

Binding of first O_2 induces local conformational change in subunit

Conformational changes spread to adjacent subunits, lowering K_m, so O_2 binds more readily

Oxyhemoglobin

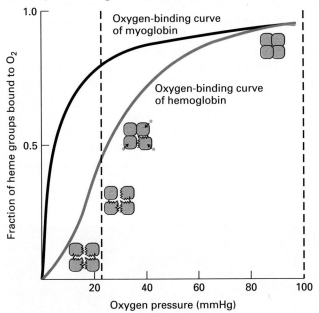

(b) Comparison of O_2 binding by myoglobin and hemoglobin

Oxygen-binding curve of myoglobin

Oxygen-binding curve of hemoglobin

▲ FIGURE 3-31   Effect of cooperativity among subunits on oxygen binding by hemoglobin. (a) Schematic diagrams illustrating conformational changes that occur in hemoglobin due to cooperative interactions among subunits. An almost all-or-none effect on binding is exhibited by the cooperative binding sites. (b) Comparison of oxygen binding to myoglobin (top curve) and hemoglobin (bottom curve) at different oxygen pressures. At the high oxygen pressures in the lung, hemoglobin in the arterial blood becomes saturated with oxygen. As the blood is delivered to capillaries in muscle or other tissues, the oxyhemoglobin is exposed to lower oxygen pressures. Over a narrow range of pressures (20–40 mm Hg), oxygen is released from hemoglobin but becomes bound to myoglobin because myoglobin has a higher affinity for oxygen (dashed line) than hemoglobin; therefore, oxygen is readily transferred from hemoglobin in the blood to myoglobin in muscle. The deoxyhemoglobin returns to the lung where it then rebinds oxygen.

The contrast between hemoglobin and myoglobin is revealing. Myoglobin is a single-chain, oxygen-binding protein found in muscle. The oxygen-binding curve of myoglobin has the characteristics of a simple equilibrium reaction:

$$E + O_2 \rightleftharpoons E—O_2$$

Myoglobin has a greater binding affinity for O_2 (a lower K_{O_2}) than hemoglobin at all oxygen pressures (Figure 3-31b). Thus at the oxygen pressure in capillaries, O_2 moves from hemoglobin into muscle cells, where it binds to

myoglobin, ensuring the efficient transfer of O_2 from blood to tissues.

In addition to enzymes and oxygen-carrying proteins, membrane-embedded receptor proteins are also likely to be multimeric. These membrane proteins must transmit a conformational signal from one side of a membrane to the other. They do so via a ligand-mediated rearrangement in quaternary structure or in the monomer-multimer equilibrium. Such signaling mechanisms are discussed in detail in Chapter 20.

Regulation of Enzymes at Other Levels   The activities of enzymes are extensively regulated in order that the numerous enzymes in a cell can work together harmoniously. All metabolic pathways are closely controlled at all times. Synthetic reactions occur when the products of these reactions are needed; degradative reactions occur when molecules must be broken down. The regulatory mechanisms that we have just discussed—feedback inhibition, allostery, and cooperativity—affect enzymes locally at their site of action.

Regulation of cellular processes, however, involves more than simply turning enzymes on and off. Some regulation is accomplished by keeping enzymes in compartments where the delivery of substrate or exit of product is controlled. In many cases, the compartments are organelles, such as the mitochondria, nuclei, or lysosomes. Compartmentation permits competing reactions to occur simultaneously in different parts of a cell. In addition to compartmentation, cellular processes are also regulated by enzyme synthesis and destruction. Often enzymes are synthesized at low rates when the cell has no need for their activities; however, upon increased demands by the cell (for instance, appearance of substrate), new enzyme is synthesized. Later, the pool of enzyme is lowered when levels of substrate decrease or the cell becomes inactive.

➤ *Antibodies*

Enzymes are not the only proteins that bind tightly and specifically to smaller compounds. The insulin receptor on the surface of a liver cell, for example, can bind to insulin so tightly that the receptors on a cell are half-saturated when the insulin concentration is only 10^{-9} M. This protein does not bind to most other compounds present in blood; it mediates the specific actions of insulin on liver cells. A molecule other than an enzyme substrate that can bind specifically to a macromolecule is often called a *ligand* of that macromolecule.

The capacity of proteins to distinguish among different molecules is developed even more highly in blood proteins called *antibodies*, or *immunoglobulins*, than in enzymes. Animals produce antibodies in response to the invasion of an infectious agent (e.g., a bacterium or a virus) or when they are exposed to any foreign chemical or material. The antibody-inducing agent is called an *antigen*. The presence of antigen causes an organism to make a large quantity of different antibody proteins, each of which may bind to a slightly different region of the antigen. The constellation of antibodies induced by a given antigen may differ from one member of a species to another. Antibodies can act as a signal for the elimination of infectious agents. For example, when antibodies bind to antigen on the surface of a bacterium, virus, or virus-infected cell, certain white blood cells (macrophages/monocytes) recognize the invading body as foreign and respond by destroying it.

Experimentally, animals produce antibodies in response to the injection of almost any foreign material; such antibodies bind specifically and tightly to the foreign substance but, like enzymes, do not bind to dissimilar molecules. The specificity of antibodies is exquisite: they can distinguish among proteins that differ by only a single amino acid and among the cells of different individual members of the same species. Antibodies are discussed at length in Chapter 27, but we introduce them here because they illustrate well how proteins bind other proteins. Because of their specificity and the ease with which they can be produced, antibodies are critical reagents in many experiments discussed in the following chapters.

Antibodies Are Multidomain, Multisubunit Proteins

Immunoglobulins are Y-shaped molecules formed from two types of polypeptides: heavy chains and light chains (Figure 3-32). The heavy chains run the length of the molecule; their C-terminal regions pair to form a stem. Visually we can distinguish three globular domains: two identical domains corresponding to each arm and the third composing the stem. Each arm of the antibody molecule contains a single light chain linked to a heavy chain by disulfide bonds. The N-terminal regions of both heavy and light chains lie at the tip of each arm and are distinguished by having highly *variable* sequences. The remaining portions

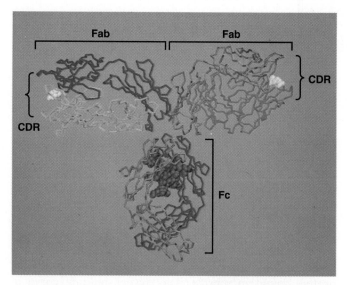

▲ FIGURE 3-32   Structure of an antibody molecule, which consists of two identical heavy chains (blue and orange) and two identical light chains (yellow and green). The Y-shaped molecule contains two identical Fab domains, forming the arms, and one Fc domain, forming the stem. Antigen molecules bind to the complementarity-determining regions (CDRs), which are highly variable regions located at the ends of each arm. Antibodies contain carbohydrate moieties (pink), and thus are glycoproteins. [From A. Levine, 1992, *Viruses*, W. H. Freeman and Company, p. 53.]

of the amino acid sequences in both chains are *constant* (i.e., nearly identical) among antibodies with different specificities. The arms are the business end of an antibody molecule, since the variable region at the end of each arm can bind a single antigen molecule. Because of its dimeric structure, an antibody can bind two antigen molecules.

Antigen-binding Site Complements the Surface of the Antigen

X-ray crystallographic analysis of antigen-antibody complexes has revealed that the antigenic specificity of an antibody is dependent on three highly variable regions near the end of each arm. Collectively called the *complementarity-determining region* (CDR), these regions form the antigen-binding site, which physically matches the antigen like a glove.

Most large antigens have multiple different sites, called *antigenic determinants* (or *epitopes*) that can induce production of specific antibodies; each of these antibodies binds to its own inducing epitope. For example, several antibodies against lysozyme, an enzyme that degrades the carbohydrate coat of bacteria, bind to different epitopes on lysozyme. Although the different epitopes on lysozyme differ greatly in their chemical properties, the interaction between lysozyme and antibody is complementary in all cases; that is, the surface of the antibody's antigen-binding site fits into that of the corresponding epitope as if they were molded together (Figure 3-33). The intimacy of contact between the two surfaces (determined by their hydrophilic, hydrophobic, and ionic bonds and van der Waals interactions) is responsible for the binding specificity exhibited by an antibody.

Antibodies Are Valuable Tools for Identifying and Purifying Proteins

Because they bind so selectively to proteins, antibodies can be used experimentally to isolate one protein from a complex mixture. In one commonly used technique, *antibody-affinity chromatography,* a pure antibody is chemically coupled to small agar beads, which are then packed in a short, narrow glass tube to form a column. When a protein solution is applied to the column, only the antigen adheres; proteins not bound by the antibody-coated beads pass through the column unimpeded (Figure 3-34). Afterwards, the adherent protein is eluted by adding a solution that disrupts the binding between the protein and the antibody. The use of antibody-affinity chromatography greatly speeds the purification of a protein by eliminating many wasteful steps in purification.

Antibodies have other uses. As described in Chapter 5, antibodies tagged with fluorescent dyes, gold particles, or enzymes are used to identify the presence and location of their antigens in cells or tissue sections by light or electron microscopy. Similarly, the presence and amount of antigen in cells can be determined by electrophoresing cellular extracts, transferring the separated components to nitrocellulose membranes, and then incubating the membranes with antibody. This technique is described at the end of this chapter.

Antibodies Can Catalyze Chemical Reactions

As we've seen, naturally occurring antibodies bind antigen molecules with exquisite specificity, but they exhibit no

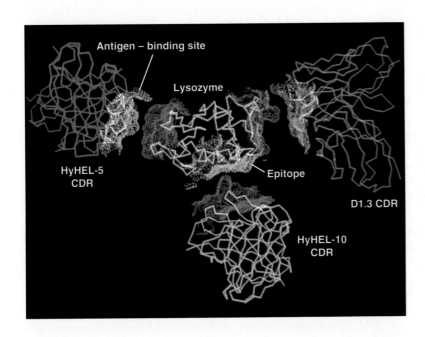

◄ FIGURE 3-33   The C_α backbone tracings of a single lysozyme molecule *(center)* and the CDRs of three different monoclonal antibodies, HyHEL-5, HyHEL-10, and D1.3, which are specific for three different epitopes on lysozyme. Note how the surface of the antigen-binding site on the different antibodies is complementary to the corresponding epitope on lysozyme. [From D. Davies and E. Padlan, 1990, *Ann. Rev. Biochem.* **59**:439.]

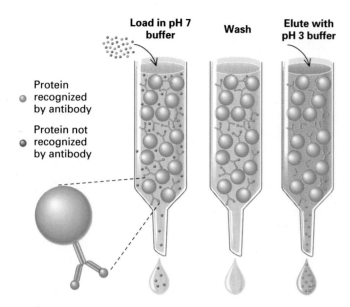

Load in pH 7 buffer   Wash   Elute with pH 3 buffer

Protein recognized by antibody

Protein not recognized by antibody

▲ FIGURE 3-34   Purification of a protein from a mixture by antibody-affinity chromatography. The mixture first is filtered through a column consisting of small beads coated with antibody molecules that are specific for the desired protein. Only that protein binds to the antibody matrix; any other proteins in the mixture pass through the column. To elute the bound protein, an acidic solution is added to disrupt the antigen-antibody complexes; the pure protein then is collected from the bottom of the column.

Techniques for Purifying and Characterizing Proteins

A protein must be purified before its structure and the mechanism of its action can be studied. However, because proteins vary in size, charge, and water solubility, no single method can be used to isolate all proteins. To isolate one particular protein from the estimated 10,000 different proteins in a cell is a daunting task that requires methods both for *separating* proteins and for *detecting* the presence of specific proteins.

Any molecule, whether protein, carbohydrate, or nucleic acid, can be separated from other molecules based on large differences in some physical characteristic. Although the sequence of amino acids in a protein uniquely determines its function, the most useful physical characteristic for separation of proteins is *size*, defined as either length or mass. Therefore, the size of protein molecules, and other polymers such as RNA or DNA, is one of the most frequent measurements in molecular cell biology. In this section, we briefly outline different techniques for separating proteins based on their size and/or other properties. The newer techniques are so simple and effective that they may not be appreciated as "physics in action." These techniques also apply to the separation of nucleic acids and other biomolecules. Finally, we consider general methods for detecting, or *assaying*, specific proteins.

catalytic activity. A key difference between enzymes and antibodies, as pointed out by Linus Pauling, is that enzymes bind the strained transition-state conformation of molecules, whereas antibodies bind the ground state, or normal conformation, of molecules. What happens if an antibody is tricked into binding the transition-state conformation of a molecule? As anticipated, the antibody now can act like an enzyme.

Such enzymes, called *catalytic antibodies,* have been produced by protein-engineering techniques (Chapter 7) and selection of antibodies that bind stable analogs of transition-state intermediates; such an analog resembles the normal transition-state intermediate of a reaction. By binding selectively to the transition state, catalytic antibodies, like enzymes, lower the free energy of the transition state, thereby increasing the rate of the reaction. In the earliest experiments, antibodies were generated against a phosphonate analog (Figure 3-35). This analog is a stable phosphorous-containing compound that resembles the tetrahedryl transition state of esters as they are hydrolyzed. Although made against phosphonates, these catalytic antibodies selectively hydrolysed esters at a rate several thousand-fold faster than control nonester molecules. Because antibodies against nearly any molecule can be generated, this special use of antibodies opens new opportunities in medicine and industry.

Centrifugation Can Separate Particles and Molecules That Differ in Mass or Density

The first step in a typical protein-purification scheme is centrifugation. The principle behind centrifugation is that two particles in suspension (cells, organelles, or molecules) having different masses or densities will settle to the bottom of a tube at different rates. Remember, mass is the weight of a sample (measured in *grams*), whereas density is the ratio of its weight to volume (*grams/liter*). Proteins vary greatly in mass but not in density. The average density of a protein is 1.37 g/l. Unless a protein has an attached lipid or carbohydrate, its density will not vary by more than 15 percent from this value. Table 3-1 lists the density and other physical characteristics of blood proteins. You can quickly sense the differences among proteins from this table. Heavier or more dense molecules sediment more quickly than lighter or less dense molecules; with time, a pellet of molecules forms at the bottom of the tube. The remaining liquid, the supernatant, contains the nonpelleted material.

A centrifuge is an instrument that speeds this process by subjecting the particles to centrifugal forces as great as 600,000 times the force of gravity g. The centrifugal force is proportional to the rotation rate of the rotor (measured in revolutions per minute, or rpm) and the distance of the tube from the center of the rotor. Modern ultracentrifuges

(a)

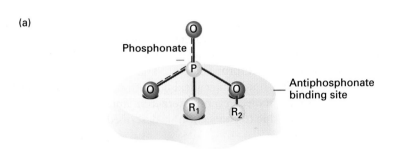

(b)

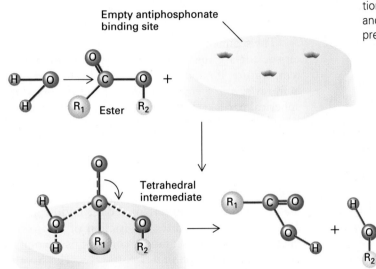

Carboxylic acid   Alcohol

◄ FIGURE 3-35 Schematic diagrams of catalysis by an antibody. (a) Binding site on antiphosphonate antibody binds phosphonate, a tetrahedral phosphorus compound. Note that the antiphosphonate binding site interacts with three atoms in the phosphonate molecule. (b) The tetrahedral transition-state intermediate in the hydrolysis of an ester, which resembles a phosphonate molecule, is bound by the antiphosphonate antibody. This binding stabilizes the transition state (i.e., lowers the activation energy), so that the reaction products, an acid and alcohol, are generated at a faster rate in the presence of the antibody than in its absence.

TABLE 3-1 Physical Characteristics of Selected Blood Proteins

Protein	Molecular Weight (kDa)	Sedimentation Constant (S)*	Density^{-1} (cm^3/g)	pI	Concentration in Plasma (mg/ml)
Immunoglobulin M	1000	18–20	0.723	5.1–7.8	0.8–0.9
α_2-Macroglobulin	820	19.6	0.735	5.4	2.65 (men) 3.35 (women)
High-density lipoprotein	435	5.5	1.093	—	0.37–1.17
Factor I (fibrinogen)	341	7.63	0.723	5.8	2–6
Factor XIII (fibrinase)	320	9.9	0.730	—	—
Immunoglobulin G	153	6.6–7.2	0.739	5.8–7.3	12–18
Ceruloplasmin	134	7.1	0.714	4.4	0.27–0.39
Plasminogen	81	4.2	0.715	5.6	0.48
Lactoferrin	77	4.93	0.716	8.2–9.2	0.004
Transferrin	76	4.9	0.725	5.2	2–4
Factor II (prothrombin)	70.2	5.11	0.719	3.8	0.11–0.23
Serum albumin	69	4.6	0.733	4.9	35–45
Transcortin	51.7	3.79	0.708	—	0.041
Orosomucoid	44.1	3.11	0.675	2.7	0.75–1.0
Retinol-binding protein	21	2.3	0.720	4.4–4.8	0.04–0.06
Lysozyme	15	2.19	0.721	10.5	0.005
β_2-Microglobulin	11.5	1.6	0.727	—	0.0013

* S = svedberg unit = 10^{-13} s.

reach speeds of 60,000 rpm or greater and generate forces sufficient to sediment particles with masses greater than 10,000 daltons. For a particle located 6 cm from the rotational axis of a typical ultracentrifuge, 60,000 rpm corresponds to a centrifugal force of 250,000 times gravity (250,000g). However, even at such tremendous forces, small particles with masses of 5000 Da or less diffuse too freely to settle uniformly through a centrifugal field. Centrifugation is used for two basic purposes: (1) as a preparative technique to separate one type of material from others and (2) as an analytical technique to measure physical properties (e.g., molecular weight, density, shape, and equilibrium binding constants) of macromolecules.

The Sedimentation Constant   The movement of a protein in a centrifugal field, which is determined by its density and shape, can be mathematically described by its sedimentation constant s. Because the centrifugal force and the density and viscosity of the medium can all be measured in a centrifuge, the radius and mass of a molecule can be calculated from the set of equations that define the sedimentation constant.

To understand how s is dependent on density and shape, let us think about an analogous example. Suppose you have two balls, both 6 cm in diameter, but one is made of styrofoam and the other of iron. If you place both on the surface of water, the styrofoam ball will float because it is less dense than water, whereas the iron ball will sink because it is more dense than water. Under the influence of gravity, objects will sink in solutions of lesser density, but their shape influences the rate of sinking. Consider a paper-thin iron sheet having the same mass as the 6-cm-diameter iron ball. The sheet will sink more slowly than the ball because it has more surface area and therefore will encounter more friction than the compactly shaped ball. Thus, the density and shape of these objects determine their rate of sedimentation.

The analogy holds true in the world of molecules. When a particle suspended in a liquid medium is subjected to centrifugal force, it will move if its density d is greater than the density of the surrounding medium d_0. The speed of movement v in a stationary medium is proportional to the gravitational acceleration g. In a centrifugal field, g is replaced by the centrifugal acceleration c, which is equal to $(2\pi\omega)^2 x$, where ω is the revolutions per unit of time and x is the distance of the particle from the axis of rotation.

As the particle moves, it encounters friction with the medium, which slows its movement. As the particle accelerates, not only does its velocity v increase, but so does friction. The frictional force ϕ is equal to fv, where f is a frictional coefficient related to the shape of the particle. For a spherical particle, $f = 6\pi\eta r$, where η is the viscosity of the medium and r is the radius of the particle. Note that f increases with the radius of the particle. Because an elongated molecule tumbles in solution, carving out a sphere of space that it occupies, it behaves as if it has a larger radius. Thus, an elongated molecule experiences greater friction

than a spherical molecule of the same mass and density. The velocity increases until the frictional force balances the centrifugal force P_c, after which time the particle continues to move at a uniform velocity. The motion of spherical particles in a fluid under ideal conditions (particles larger than solvent molecules; no interaction among particles; no disturbance due to convection) is described by *Stokes' law*:

$$v = 2cr^2\left(\frac{d - d_0}{9\eta}\right)$$

The sedimentation constant s equals v/c and is characteristic for a given particle in a given medium at a given temperature. If r and x are expressed in cm, g or c in cm/s^2, ω in rev/s, d in g/cm^3, η in g/cm/s, and the mass m of the spherical particle in g, then the sedimentation constant can be calculated from the expression

$$s = \frac{v}{c} = \frac{{}^4\!/_3\pi r^3(d - d_0)}{6\pi\eta r} = \frac{m[1 - (d_0/d)]}{f}$$

under standard conditions of sedimentation in water at 20°C, which standardizes the friction of the sedimenting particles. The sedimentation constant is commonly expressed in *svedbergs*(S); $1\,S = 10^{-13}$ s. The S values for various blood proteins are given in Table 3-1.

Differential Centrifugation   The most common use of centrifugation is the partial purification of proteins by separating soluble material from insoluble material (Figure 3-36a). The centrifugal force and duration of centrifugation are adjusted to ensure that the insoluble materials sediment into a pellet. After a starting mixture of a cell homogenate is poured into a tube and spun in a centrifuge, cell organelles such as nuclei collect into a pellet, but the soluble proteins remain in the supernatant. The supernatant fraction still contains a large mixture of proteins, which can be collected by decanting the supernatant and then subjected to further purification methods.

Rate-Zonal Centrifugation   Based on differences in their mass, proteins can be separated by centrifugation through a solution, usually containing sucrose (an inert sugar), of increasing density called a *density gradient*. When mixtures of proteins are layered on top of a sucrose gradient in a tube and subjected to centrifugation, they migrate down the tube at a rate controlled by the factors that affect the sedimentation constant. The proteins start from a thin zone at the top of the tube and separate into bands, or zones (actually disks), of proteins of different masses. This density-gradient separation technique is called *rate-zonal centrifugation* (Figure 3-36b). Samples are centrifuged just long enough to separate the molecules of interest. If they are centrifuged for too short a time, the molecules will not separate sufficiently. If they are centrifuged much longer than necessary, all of the molecules will end up in a pellet at the bottom of the tube. If the bottom

(a) DIFFERENTIAL CENTRIFUGATION

Sample is poured into tube

(b) RATE-ZONAL CENTRIFUGATION

Sample is layered on top of gradient

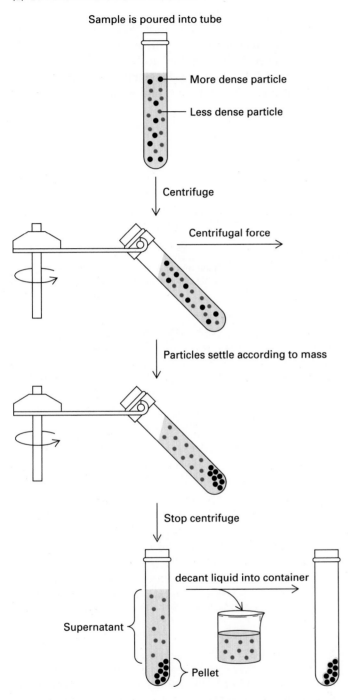

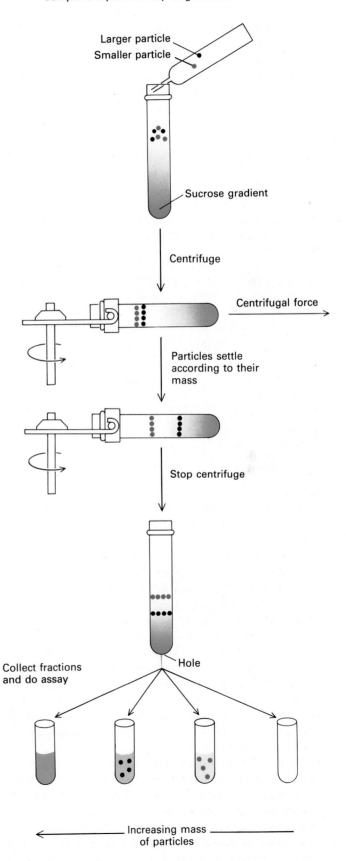

▲ FIGURE 3-36   Two common centrifugation techniques for separating particles. (a) Differential centrifugation separates a mixture of particles (macromolecules, cell organelles, and cells) that differ in mass or density. The most-dense particles (black) collect at the bottom of the tube as a pellet. The least-dense particles (green) remain in the liquid supernatant, which can be transferred to another tube. (b) Rate-zonal centrifugation separate particles or molecules that differ in mass but may be similar in shape and density (e.g., RNA molecules). Here two particles of different mass separate into two zones.

of the tube is punctured, the different zones containing proteins that differ in mass can be collected.

Although the sedimentation rate is strongly influenced by particle mass, rate-zonal centrifugation is seldom effective in determining *precise* molecular weights because variations in shape also affect sedimentation rate. The exact effects of shape are hard to assess, especially for proteins and single-stranded nucleic acid molecules that can assume many complex shapes. Nevertheless, rate-zonal centrifugation has proved to be the most practical method for separating many different types of polymers and particles. A second density-gradient technique called *equilibrium density-gradient centrifugation,* which is used mainly to separate DNA or organelles (see Figure 5-30).

Electrophoresis Separates Molecules According to Their Charge: Mass Ratio

Electrophoresis is a technique for separating, or *resolving,* molecules in a mixture under the influence of an applied electric field. Dissolved molecules in an electric field move, or migrate, at a speed determined by their charge: mass ratio. For example, if two molecules have the same mass and shape, the one with the greater net charge will move faster toward an electrode. The separation of small molecules, such as amino acids and nucleotides, is one example of the many uses of electrophoresis. In this case, a small drop of sample is deposited on a strip of filter paper or other porous substrate, which is then soaked with a conducting solution. When an electric field is applied at the ends of the strip, small molecules dissolved in the conducting solution move along the strip at a rate corresponding to the magnitude of their charge.

SDS-Polyacrylamide Gel Electrophoresis Because many proteins or nucleic acids that differ in size and shape have nearly identical charge: mass ratios, electrophoresis of these macromolecules in solution results in little or no separation of molecules of varying lengths. However, successful separation of proteins and nucleic acids can be accomplished by electrophoresis in various types of *gels* (semisolid suspensions in water), rather than in a liquid solution. Electrophoretic separation of proteins is most commonly performed in *polyacrylamide* gels. These gels are cast between a pair of glass plates by polymerizing a solution of acrylamide monomers into polyacrylamide chains and simultaneously cross-linking the chains into a semisolid matrix. The *pore size* of a gel can be varied by adjusting the concentrations of polyacrylamide and the cross-linking reagent.

When a mixture of proteins is applied to a gel and an electric current applied, smaller proteins migrate faster than larger proteins through the gel. The rate of movement is influenced by the gel's pore size and the strength of the electric field. The pores in a highly cross-linked polyacrylamine gel are quite small. Such a gel could resolve small proteins and peptides, but large proteins would not be able to move through it.

In what is probably the most powerful technique for resolving protein mixtures, proteins are exposed to SDS (sodium dodecylsulfate) before and during gel electrophoresis (Figure 3-37). SDS is a detergent and common cleaning agent found in toothpaste and shampoo; its chemical

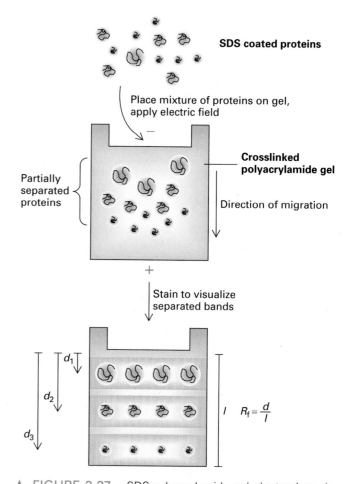

▲ FIGURE 3-37   SDS-polyacrylamide gel electrophoresis, a common technique for separating proteins. The protein mixture first is treated with SDS, a negatively charged detergent that binds to proteins. This binding dissociates multimeric proteins and forces all polypeptide chains into denatured conformations with nearly identical charge:mass ratios. During electrophoresis, the SDS-protein complexes migrate through the polyacrylamide gel. Small proteins are able to move through the pores more easily, and faster, than larger proteins. Thus the proteins separate into bands according to their size as they migrate through the gel. The separated protein bands are visualized by staining with a dye. The molecular weight of a separated protein can be calculated by comparing its R_f value, which equals the distance traveled in the gel d divided by the running length of the gel l, with the R_f values of protein standards of known molecular weight.

formula is $CH_3(CH_2)_{11}SO_4^-Na^+$. Approximately 1.4 g of SDS binds to each gram of protein. Because the negative charges on bound SDS molecules repel each other, multimeric proteins dissociate into their subunits and all polypeptide chains are forced into extended conformations with similar charge:mass ratios.

SDS treatment thus eliminates the effect of differences in shape, so that chain length, which reflects mass, is the sole determinant of the migration rate of proteins in SDS-polyacrylamide electrophoresis. Even chains that differ in molecular weight by less than 10 percent can be separated by this technique. Moreover, the molecular weight of a protein can be estimated by comparing the distance it migrates through a gel with the distances that proteins of known molecular weight migrate.

Two-Dimensional Gel Electrophoresis Electrophoresis of all cellular proteins through an SDS gel can separate proteins having relatively large differences in molecular weight but cannot resolve proteins having similar molecular weights (e.g., a 41-kDa protein from a 42-kDa protein). To separate proteins of similar mass, another physical characteristic must be exploited. Most commonly, this is electric charge, which is determined by the number of acidic and basic residues in a protein. Two unrelated proteins having similar masses are unlikely to have identical net charges because their sequences, and thus the number of acid and basic residues, are different.

In two-dimensional electrophoresis, proteins are separated in two sequential steps: first by their charge and then by their mass (Figure 3-38). In the first step, a cell extract is

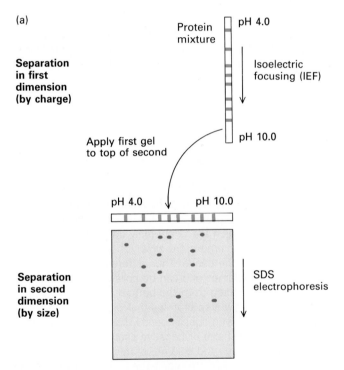

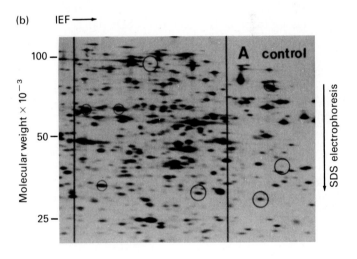

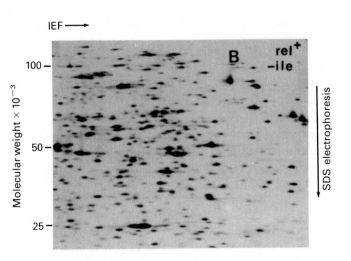

▲ FIGURE 3-38   Two-dimensional gel electrophoresis, a technique for separating proteins based on their charge and their mass. (a) Preparation of a two-dimensional protein gel by isoelectric focusing (IEF) followed by SDS electrophoresis. (b) Two-dimensional gels of protein extracts from cells growing in a medium containing isoleucine *(top)* and from cells placed for a brief time in a medium lacking isoleucine *(bottom)*. Each spot represents a single polypeptide; polypeptides can be detected by dyes, as here, or by other reagents. These patterns are reproducible, so that changes in individual proteins can be detected. In this example, certain spots (circled) in the top gel are absent from, or much fainter in, the bottom gel; these differences represent changes in protein synthesis by the cells in response to amino acid starvation. [Part (b) from P. H. O'Farrell, 1978, *Cell*, **14**:545.]

fully denatured by high concentrations (8 M) of urea and then layered on a glass tube filled with polyacrylamide that is saturated with a solution of *ampholytes,* a mixture of polyanionic and polycationic molecules. When placed in an electric field, the ampholytes will separate and form a continuous gradient based on their net charge. The most highly polyanionic ampholytes will collect at one end of the tube, and the most polycationic ampholytes will collect at the other end. This gradient of ampholytes establishes a pH gradient. Charged proteins will migrate through the gradient until they reach their *pI,* or *isoelectric point,* the pH at which the net charge of the protein is zero. This technique, called *isoelectric focusing* (IEF), can resolve proteins that differ by only one charge unit.

Proteins that have been separated on an IEF gel can then be separated in a second dimension based on their molecular weights. To accomplish this, the IEF gel is extruded from the tube and placed lengthwise on a second polyacrylamide gel, this time formed as a slab saturated with SDS. When an electric field is imposed, the proteins will migrate from the IEF gel into the SDS slab gel and then separate according to their mass. The sequential resolution of proteins by their charge and mass can achieve excellent separation of cellular proteins. For example, two-dimensional gels have been very useful in studying the expression of various genes in differentiated cells because as many as 1000 proteins can be resolved simultaneously.

Proteins separated in gels can be detected by various stains. For instance, Coomassie blue, a deep blue dye originally developed to color fabrics, commonly is used to stain proteins. The total amount of protein in a spot can be estimated from the intensity of the color. Radioisotopes also are used to detect proteins; this more sophisticated method is discussed in Chapter 6.

Liquid Chromatography Resolves Proteins by Mass, Charge, or Binding Affinity

Liquid chromatography, a third commonly used technique to separate mixtures of proteins, nucleic acids, and other molecules, is based on the principle that molecules dissolved in a solution will interact (bind and dissociate) with a solid surface. If the solution is allowed to flow across the surface, then molecules that interact frequently with the surface will spend more time bound to the surface and thus move more slowly than molecules that interact infrequently with the surface. Liquid chromatography is performed in a tube packed tightly with spherical beads. The nature of these beads determines whether separation of proteins depends on differences in mass, charge, or binding affinity.

Gel Filtration Chromatography In this type of chromatography, the column is composed of porous beads made from polyacrylamide, dextran (a bacterial polysaccharide), or agarose, which is derived from seaweeds. Un-

like electrophoresis in which proteins migrate through the pores in a solid gel, proteins flow around the spherical beads in gel filtration chromatography. However, the surface of the beads is punctured by large holes, and proteins will spend some time within these holes; as a result, their movement through a column is retarded.

Because smaller proteins can penetrate into the beads more easily than larger proteins, they travel through a gel filtration column more slowly than larger proteins (Figure 3-39a). Thus the total volume of liquid required to elute a protein from the column depends on its mass. By use of proteins of known mass, the elution volume can be used to estimate the mass of a protein in a mixture.

Ion-Exchange Chromatography In a second type of liquid chromatography, called ion-exchange chromatography, proteins are separated based on differences in their charge. This technique makes use of specially modified beads whose surfaces are covered by amino groups or carboxyl groups and thus carry either a positive (NH_3^+) or negative (COO^-) charge at neutral pH.

The proteins in a mixture carry various net charges at any given pH. When a solution of a protein mixture flows through a column of positively charged beads, only pro-

> FIGURE 3-39   Three commonly used liquid chromatographic techniques. (a) Gel filtration chromatography separates proteins that differ in size. A mixture of proteins is carefully layered on the top of a glass cylinder packed with porous beads. As the proteins are carried down the column by the flow of liquid, pores in the beads act as filters. Smaller proteins, which are temporarily trapped in the pores, travel through the column more slowly than larger proteins, which flow around the beads. Thus different proteins have different elution volumes and can be collected in separate liquid fractions collected from the bottom. (b) Ion-exchange chromatography separates proteins that differ in net charge in columns packed with special beads that carry either a positive charge (shown here) or negative charge. As a mixture of proteins flows through the column, proteins having the same net charge as the beads are repelled and do not bind to the beads, whereas proteins having the opposite charge bind to the beads. Bound proteins, in this case negatively charged, are eluted by passing a salt gradient (usually of NaCl or KCl) through the column. As the ions bind to the beads, they desorb the protein. Weakly charged proteins are eluted first followed by more highly charged proteins in a gradient of increasing salt concentration. (c) Affinity chromatography separates proteins based on affinity for a specific ligand, which is covalently attached to beads packed in a column. As a mixture of proteins passes through the column, only the protein(s) with high affinity for the ligand bind; all the nonbinding proteins flow through the column. The bound protein can be dislodged from the beads and eluted with a concentrated solution containing the ligand or by changing the pH or ionic strength of the buffer. In antibody-affinity chromatography, a specific antibody is used as the ligand.

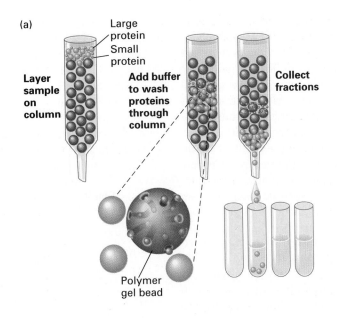

(a)

Large protein
Small protein

Layer sample on column

Add buffer to wash proteins through column

Collect fractions

Polymer gel bead

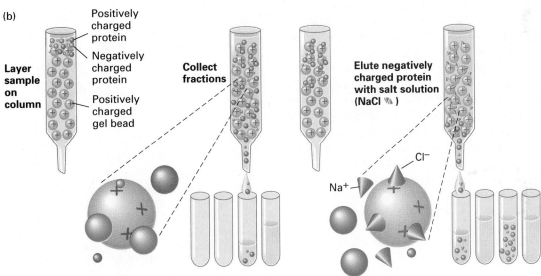

(b)

Positively charged protein
Negatively charged protein
Positively charged gel bead

Layer sample on column

Collect fractions

Elute negatively charged protein with salt solution (NaCl ▼)

Cl⁻

Na⁺

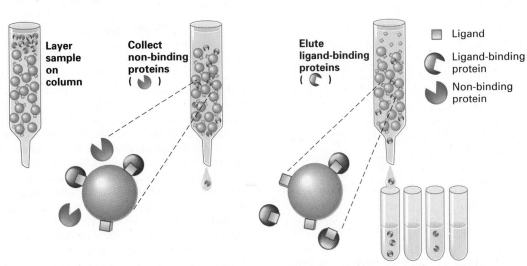

(c) Affinity chromatography

Layer sample on column

Collect non-binding proteins ( )

Elute ligand-binding proteins ( )

Ligand

Ligand-binding protein

Non-binding protein

teins with a net negative charge (acidic proteins) adhere to the beads; neutral and basic proteins flow unimpeded through the column (Figure 3-39b). The acidic proteins are then eluted selectively by passing a gradient of increasing concentrations of salt through the column. At low salt concentrations, protein molecules and beads are attracted by their opposite charges. At higher salt concentrations, negative salt ions bind to the positively charged beads, displacing the negatively charged proteins. In a gradient of increasing salt concentration, weakly charged proteins are eluted first and highly charged proteins are eluted last. Similarly, a negatively charged column can be used to retain and fractionate positively charged (basic) proteins.

Affinity Chromatography   A third form of chromatography, called affinity chromatography, relies on the ability of a protein to bind specifically to another molecule. As illustrated in Figure 3-34, an antibody can be used as an affinity reagent to isolate its corresponding antigenic protein from a mixture of proteins. Similarly, various small molecules, such as substrates for enzymes and ligands for other proteins, can be covalently attached to beads. Columns packed with such beads will retain only the proteins that bind the small molecule; the remaining proteins, regardless of their charge or mass, will pass through the column without binding to it (Figure 3-39c). The proteins bound to the column then are eluted by adding an excess of ligand, or by changing the salt concentration or pH.

By ingenious application of affinity chromatography, a wide variety of proteins can be purified in just a few steps from complex protein mixtures. In the example shown in Figure 3-40, the cytoskeletal proteins, actin and villin, form a complex which is specifically separated from an extract of intestinal cells containing hundreds of proteins by use of a DNase I affinity column. This affinity separation purifies two proteins from a cell extract in one step.

Highly Specific Assays Can Detect Individual Proteins

Purification of a protein, or any other molecule, requires a specific assay that can detect the molecule of interest in column fractions or gel bands. An assay capitalizes on some highly distinctive characteristic of a protein: the ability to bind a particular ligand, to catalyze a particular reaction, or to be recognized by a specific antibody. An assay must also be simple and fast in order to minimize errors and the possibility that the protein of interest is denatured or degraded while the assay is performed. The goal of any purification scheme is to isolate sufficient amounts of a given protein for study; thus a useful assay must also be sensitive enough that only a small proportion of the available material is consumed. Many common protein assays require just 10^{-9} to 10^{-12} g of material.

Chromogenic Enzyme Reactions   Many assays are tailored to detect some functional aspect of a protein. For

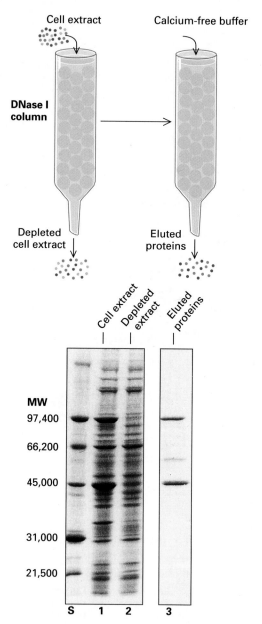

▲ FIGURE 3-40   Purification of actin and villin from an extract of intestinal cells by affinity chromatography. *(Top)* In the presence of calcium, proteins actin and villin form a complex that binds to an affinity column made from beads coated with DNase I. The bound proteins can be eluted with a calcium-free buffer. *(Bottom)* SDS gel electrophoresis demonstrates separation achieved with the DNase I column. The original cell extract (lane 1) is depleted of two proteins (with molecular weights of about 95,000 and 42,000) after passage through the column (lane 2). The fraction eluted from the column by calcium-free buffer contains two proteins, actin and villin (lane 3). Lane S is of protein standards with known molecular weights. [Photograph courtesy of Mark Chafel.]

example, enzyme assays are based on the ability to detect the loss of substrate or the formation of product. Many enzyme assays utilize *chromogenic* substrates, which change color during the course of the reaction. (Some substrates are naturally chromogenic; if they are not, they can be linked to a chromogenic molecule.) Because of the specificity of an enzyme for its substrate, only samples that contain the enzyme will change color in the presence of a chromogenic substrate and other required reaction components; the rate of the reaction provides a measure of the quantity of enzyme present.

Immunoblotting   One of the most powerful methods for detecting a particular protein in a complex mixture combines the superior resolving power of gel electrophoresis, the specificity of antibodies, and the sensitivity of enzyme assays. Called *immunoblotting*, or *Western blotting*, this

three-step procedure is commonly used to separate proteins and then identify a specific protein of interest (Figure 3-41).

The first step in an immunoblot assay is to separate a protein mixture on an SDS-polyacrylamide gel. Then, a paper-thin membrane made from nitrocellulose, which tenaciously binds most proteins, is applied to the face of the gel. When an electric field is applied, proteins are driven out of the gel and transferred to the membrane. This process is called blotting because the membrane picks up the proteins as if it were a blot impression of the gel. In the second step, the membrane is soaked in a solution of an antibody (Ab_1) specific for the protein of interest. Only the band containing this protein binds the antibody, forming a layer of antibody molecules. In the final step, the membrane is "developed" with an enzyme-linked antibody (Ab_2) to identify the band containing the protein of interest.

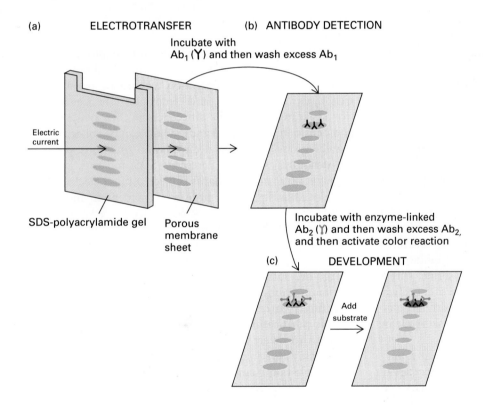

(a) **ELECTROTRANSFER**   (b) **ANTIBODY DETECTION**

Incubate with
Ab_1 (Y) and then wash excess Ab_1

Electric
current

SDS-polyacrylamide gel   Porous
membrane
sheet

Incubate with enzyme-linked
Ab_2 (Y) and then wash excess Ab_2,
and then activate color reaction

(c)   **DEVELOPMENT**

Add
substrate

◄ FIGURE 3-41   Immunoblotting, or Western blotting, is a superior three-step procedure for separating and detecting a protein in a mixture. (a) A protein mixture is electrophoresed through an SDS gel, and then from the gel onto a membrane. (b) The membrane is flooded with a solution of antibody (Ab_1) specific for the desired protein. Only the band containing this protein binds the antibody, forming a layer of antibody molecules. After time to allow for binding, the membrane is washed to remove unbound Ab_1. (c) In the development step, the membrane first is incubated with a second antibody (Ab_2) that binds to the Ab_1-coated protein forming a "sandwich" of antibody molecules. This second antibody is covalently linked to alkaline phosphatase, which catalyzes a chromogenic reaction. Finally, the substrate is added and a deep purple precipitate forms, marking the band containing the desired protein.

SUMMARY

Proteins play a pivotal role in the biology of cells. They carry out the activities encoded in the sequences of genes and build the structure of cells. Proteins are biopolymers made of 20 amino acid building blocks linked into a single chain by peptide bonds. Amino acids differ in their side chains and are classified according to their solubility in

water: soluble amino acids have polar or charged side chains; insoluble amino acids have nonpolar or hydrophobic side chains.

Proteins exhibit four levels of structural organization. The first and most simple level is the protein sequence—the linear order of amino acid residues in the chain. The second level involves folding of the polypeptide backbone into secondary structures including helices, strands, loops, and random coils. At the third level, secondary structures are packed into a globular or fibrous mass to form the tertiary structure, or overall three-dimensional shape, of a protein. Quatenary structure describes the stoichiometry and arrangement of protein subunits in multimeric proteins.

In some proteins, tertiary structure is subdivided into independently folded units called domains. In addition to hydrophobic bonds, tertiary and quatenary structure are maintained by disulfide bonds between cysteine residues and by prosthetic groups such as heme. The forces that hold proteins in their native conformation can be disrupted by heat or chemicals. In many cases, a denatured species can refold to form a fully active protein; however, protein folding sometimes requires assistance by chaperone proteins. In all cases the information for refolding resides in the amino acid sequence.

The structures of proteins are analyzed by chemical and physical methods. Protein sequence can be determined chemically by repeated cycles of the Edman reaction, but a more popular method for obtaining the sequence of a large protein is sequencing of the mRNA or DNA that encodes it. Secondary structures can be recreated by peptides synthesized from monomer amino acids with the Merrifield procedure. Three-dimensional structure is revealed by x-ray crystallographic analysis and NMR methods.

Numerous structural studies have shown that homologous proteins have similar sequences, shapes, and functions. Resemblance between proteins indicates they evolved from a common ancestor. The function of many proteins requires the presence of prosthetic groups and post-translational modification. One type of modification is the addition of chemical groups to certain amino acid residues: for example, phosphorylation of serine, threonine, and tyrosine side chains. A second type of modification is removal of a segment of the backbone from the N- or C-terminus by proteases or internally by a self-splicing mechanism.

Two important types of proteins, enzymes and antibodies, illustrate many of the characteristic properties of proteins. Enzymes are protein catalysts whose activity permits metabolic reactions to occur under physiologic conditions of temperature and pH. Catalysis occurs in the enzyme active site, which contains a substrate-binding site that matches the structure of substrate molecules, accounting for the specificity of enzyme activity. Once bound by an enzyme molecule, the substrate undergoes chemical changes that result in a strained conformation, the transition state. Stabilized by the enzyme, the transition state has a lowered activation energy, and the reaction proceeds to completion. In serine proteases, such as trypsin and chymotrypsin, the peptide bond is cleaved through a charge-relay mechanism involving hydrogen-bonded intermediates in the active site. Some enzymes require a coenzyme for catalytic activity. Enzymatic reactions may be regulated by feedback inhibition, allostery, and subunit cooperativity. The activity of enzymes also is controlled by their compartmentation and life span in cells.

Antibodies are defense proteins that bind antigens. The specificity and affinity of the interaction between an antibody and antigen are determined by complementarity between binding sites on the two molecules. Antibodies are useful reagents in cell biology. When coupled to small beads to form affinity columns, antibodies can separate antigen molecules quickly and in high yield from cellular extracts. Antibodies also can be engineered to act as enzymes.

Differences in size and charge can be exploited to isolate one protein species from a mixture of thousands of proteins. Three major separation techniques are centrifugation, chromatography, and electrophoresis. Centrifugation separates proteins by differences in their mass, shape, and density into discreet bands, which can be collected separately into test tubes. In electrophoretic methods, proteins are denatured with the detergent SDS and then migrate through a polymer gel that is subjected to an electric field. Proteins differing by as little as 1 percent in molecular weight can be resolved by SDS gel electrophoresis. Three forms of liquid chromatography are commonly used to separate proteins. In gel filtration chromatography, proteins are separated according to their size. Differences in charge are sufficient to resolve proteins by ion-exchange chromatography. Specific binding interactions are the basis for affinity chromatography.

No purification can be accomplished without an assay, a means to detect the presence of the protein of interest. Most assays for enzymes rely on a chromogenic substrate, a molecule that changes color when hydrolyzed by the enzyme. Other assays detect the presence of a protein by antibodies directed against the protein. Immunoblotting combines the resolution of electrophoresis, the specificity of antibodies, and the sensitivity of enzymatic reactions into a powerful method for detecting proteins.

REVIEW QUESTIONS

1. You have learned that a protein's function is "derived from three dimensional structure, and three-dimensional structure is specified by amino acid sequence." Review the hierarchical arrangement of protein structure, including the concepts of domains and motifs. The human hemoglobin β chain contains an α-helical structure known as the H helix. The linear arrangement of the amino acids in this region is

. . . Pro–Pro–Val–Glu–Ala–Ala–Tyr–Glu–Lys–Val–Val–Ala–Gly–Val–Ala–Asn–Ala–Leu–Ala–His–Lys–Tyr–His. . . .

   a. Draw this sequence in a schematic form using the example of the helical wheel in Figure 3-11b. How many turns are in this helix? What is its approximate length?

   b. Which of the amino acids in the H-helix are hydrophobic? Which are hydrophilic? Based on the linear sequence shown, what can you deduce about the nature of this α-helix?

   c. Based on the distribution of the polar and nonpolar amino acids, where would the faces of the H-helix be oriented?

   d. There are methods available to biologists that allow them to change specific amino acids in proteins in order to examine the effect on protein structure and function. (You will learn about these techniques in later chapters.) If the underlined tyrosine residue were changed to a proline, what would be the effect on the primary, secondary, and tertiary structures of the β chain of hemoglobin?

2. Review the features of enzyme active sites, including the lock-and-key and induced-fit models for enzyme action, as well as the two parameters that describe enzyme kinetics, K_M and V_{max}. The hydration of CO_2 to form carbonic acid, H_2CO_3, is important biologically and is represented as follows:

$$CO_2 + H_2O \rightleftharpoons H_2CO_3$$

The enzyme carbonic anhydrase catalyzes this reaction, converting 10^5 molecules of CO_2 into carbonic acid per molecule of enzyme in one second. The K_M for the reaction is approximately 8×10^{-3} M.

   a. What effect does carbonic anhydrase have on the equilibrium of this reaction? Does it change K_{eq}? Explain your answer. What happens to ΔG when carbonic anhydrase is added to the reaction?

   b. Assume that you are working under conditions where the enzyme is saturated with substrate. What would happen to K_M, V_{max}, ΔG, and K_{eq} for carbonic anhydrase under each of the following conditions?
     (1) The concentration of CO_2 is doubled.
     (2) The concentration of carbonic anhydrase is doubled.

   c. Refer to the following table for this question.

[CO$_2$] (M)	V (M/s)		
	No Inhibitor	Inhibitor A	Inhibitor B
0.0020	0.0012	0.0006	0.00086
0.0040	0.0020	0.0010	0.0015
0.0080	0.0030	0.0015	0.0024
0.016	0.0040	0.0020	0.0034
0.032	0.0048	0.0024	0.0044

These data for the reaction catalyzed by carbonic anhydrase were obtained using purified enzyme at a concentration of 10^{-8} M. What happens to K_M and V_{max} in the presence of each inhibitor of carbonic anhydrase relative to the reaction where no inhibitor is present? What are the values of K_M and V_{max} in each of the three cases? Speculate on the general mechanism by which these inhibitors exert their effects.

3. The purification of a specific protein from a heterogeneous mixture of proteins is of great importance to the biologist. Review some of the fundamental techniques described for purifying a protein to homogeneity. Refer to the table on the following page when answering the questions.

   a. You have a mixture containing only these five proteins. You have no antibodies available that recognize these proteins. Devise a scheme for separating them; justify your proposed method.

   b. How would you prove that you have successfully separated each specific protein?

   c. Once each of these proteins is pure, what might be done to permit rapid purification of each in future separations?

Protein	Molecular weight (kD)	Properties
Phosphorylase *a*	97.4	This enzyme is involved in glycogen degradation. Phosphorylase *a* is identical to phosphorylase *b*, but a serine residue has a phosphate group attached to it, making it the active form of the enzyme.
Phosphorylase *b*	97.4	Phosphorylase *b* is identical to phosphorylase *a*, but it has no phosphate group present on a serine residue; it is the inactive form of the enzyme.
Catalase	57.5	This protein is responsible for the decomposition of hydrogen peroxide.
Histone 2A	14.5	This protein is rich in basic amino acids and has a net positive charge at neutral pH; it has a strong affinity for DNA.
Lysozyme	14.3	This enzyme hydrolyses certain polysaccharides.

References

General References

STRYER, L. 1995. *Biochemistry*, 4th ed. W. H. Freeman and Company. Chapters 1–4, 7–9, 11, 12, and 31.

Proteins

ANFINSEN, C. B. 1973. Principles that govern the folding of protein chains. *Science* 181:223–230.

BRANDEN, C., and J. TOOZE. 1991. *Introduction to Protein Structure.* Garland.

CANTOR, C. R., and P. R. SCHIMMEL. 1980. *Biophysical Chemistry.* W. H. Freeman and Company. Parts 1, 2, and 3.

CHOTHIA, C. 1984. Principles that determine the structure of proteins. *Ann. Rev. Biochem.* 53:537–572.

CREIGHTON, T. E. 1993. *Proteins: Structures and Molecular Properties,* 2d ed. W. H. Freeman and Company.

DICKERSON, R. E., and I. GEIS. 1969. *The Structure and Action of Proteins.* Benjamin-Cummings.

DICKERSON, R. E., and I. GEIS. 1983. *Hemoglobin: Structure, Function, Evolution, and Pathology.* Benjamin-Cummings.

DOOLITTLE, R. 1985. Proteins. *Sci. Am.* 253(4):88–96.

FREIFELDER, D. 1982. *Physical Biochemistry: Applications to Biochemistry and Molecular Biology.* 2d ed. W. H. Freeman and Co.

HENDRICKSON, W. A., and K. WUTHRICH. 1991–1993. *Macromolecular Structures.* Current Biology Ltd. An annual series that compiles atomic structures of biological macromolecules reported in the previous year.

LESK, A. M. 1991. *Protein Architecture: A Practical Approach.* IRL Press.

MERRIFIELD, R. B. 1986. Solid-phase synthesis. *Science* 232:341–347.

PERUTZ, M. F. 1964. The hemoglobin molecule. *Sci. Am.* 211(5):64–76 (Offprint 196). This article and the next describe the three-dimensional structure of hemoglobin and the conformational changes that occur during oxygenation and deoxygenation.

PERUTZ, M. F. 1978. Hemoglobin structure and respiratory transport. *Sci. Am.* 239(6):92–125 (Offprint 1413).

PERUTZ, M. F. 1991. *Protein Structure and Function.* W. H. Freeman. This book describes for the intelligent layperson how protein structures are determined in a chapter entitled "Diffraction Without Tears". Later chapters use protein structures to illustrate important principles of structure and function.

RICHARDSON, J. S. 1981. The anatomy and taxonomy of protein structure. *Adv. Protein Chem.* 34:167–339.

ROSE, G. D., et al. 1985. Hydrophobicity of amino acid residues in globular proteins. *Science* 229:834–838.

ROSSMAN, M. G., and P. ARGOS. 1981. Protein folding. *Ann. Rev. Biochem.* 50:497–532.

SCHULZ, G. E., and R. H. SCHIRMER. 1979. *Principles of Protein Structure.* Springer-Verlag. A general treatment of protein structure.

Enzymes

DRESSLER, D. H., and H. POTTER. 1991. *Discovering Enzymes.* Scientific American Library.

FERSHT, A. 1985. *Enzyme Structure and Mechanism,* 2d ed. W. H. Freeman and Company.

WALSH, C. 1979. *Enzymatic Reaction Mechanisms.* W. H. Freeman and Company. A detailed discussion of the chemical bases of action of many types of enzymes.

Antibodies

TONEGAWA, S. 1985. The molecules of the immune system. *Sci. Am.* 253(4):122–130.

LERNER, R. A., BENKOVIC, S. J., and P. G. SCHULTZ. 1991. At the crossroads of chemistry and immunology: catalytic antibodies. *Science* 252:659–667.

4

Nucleic Acids, the Genetic Code, and the Synthesis of Macromolecules

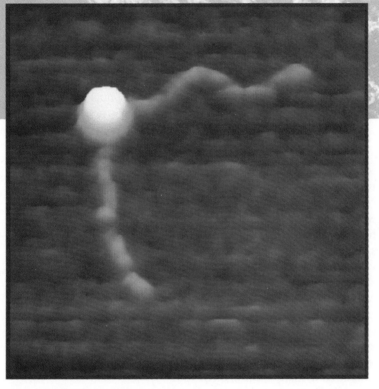

▲ *E. coli* RNA polymerase (the bright, round object) binds to DNA at a promoter sequence and bends the DNA.

The extraordinary versatility of proteins in catalyzing chemical reactions and building cellular structures was featured in the previous chapter. In this chapter we consider the nucleic acids, the molecules that (1) contain the information prescribing amino acid sequence in proteins and (2) serve as the structures on which amino acids are forged into proteins. *Deoxyribonucleic acid* (DNA) is the storehouse, or cellular library, that contains the information required to build a cell or organism. The exact duplication of this information in any species from generation to generation assures the genetic continuity of that species. The information is arranged in units identified by classical geneticists from Gregor Mendel through Thomas Hunt Morgan as *genes,* hereditary units controlling identifiable traits of an organism. In this chapter we explain how information storage in DNA is accomplished in a *genetic code* and how this information is copied, or *transcribed,* into *ribonucleic acid* (RNA) templates called *messenger RNA* (mRNA). Since this RNA carries the instructions from the DNA for the correct order of amino acids during protein synthesis, it must be accurately produced and its concentration must be carefully regulated. The remarkably accurate, stepwise assembly of amino acids into proteins occurs by a process of *translation* of messenger RNA. Each DNA code word, or *codon,* that is transcribed into mRNA is then interpreted by a second type of RNA called *transfer RNA;* with the aid of a third type of RNA, *ribosomal RNA,* and its associated proteins, the correct amino acids are linked by peptide bonds to make proteins.

The elucidation of the basic details of, first, the structure of DNA and then the synthesis of DNA, RNA, and protein is the monumental achievement of molecular biology from the discovery of the DNA structure in 1953 through the 1970s. To grasp the principles of macromolecular synthesis, we first consider the building blocks of the nucleic acids and the general structures of DNA and RNA. We introduce the basic mechanisms of DNA and RNA chain synthesis, including a discussion of gene structure, which reveals why molecular processing is required to make functional RNA molecules. The chapter closes with an outline of how the several RNA molecules participate in protein synthesis. Since the events of macromolecular synthesis are so central to all biological functions—growth control, differentiation, and the specialized chemical and physical properties of cells—they will arise again and again in later chapters, but an elementary grasp of the fundamentals of DNA, RNA, and protein synthesis is necessary to follow the subsequent discussions without difficulty.

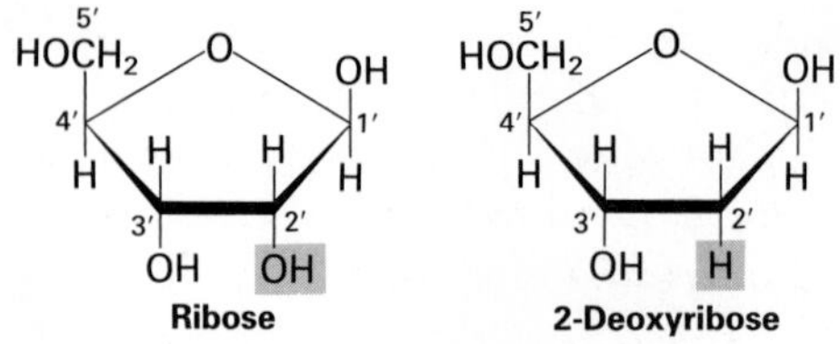

► *Nucleic Acids: Linear Polymers of Nucleotides*

The instructions about the types of protein and the quantity of each that a cell can synthesize come from *nucleic acids*—the molecules that store and transmit information in cells. As in many systems of communication, this information is passed as a code—in this case, a genetic code—that must be decoded at the site where it is used. In this chapter we examine first the chemical structures of DNA, the nucleic acid that stores the encoded information, and of RNA, which is the active agent in decoding the stored information.

Cells have two chemically similar information-carrying molecules: deoxyribonucleic acid (DNA) and ribonucleic acid (RNA). Both DNA and RNA in their *primary structures* are linear polymers composed of monomers (single chemical units) called *nucleotides*. Cellular RNAs range in length from less than one hundred to many thousands of units. The number of units in a cellular DNA molecule can exceed a hundred million.

DNA and RNA each consist of only four different nucleotides. A nucleotide has three parts: a phosphate group, a *pentose* (a five-carbon sugar molecule), and an organic *base* (Figure 4-1). In RNA, the pentose is always *ribose*; in

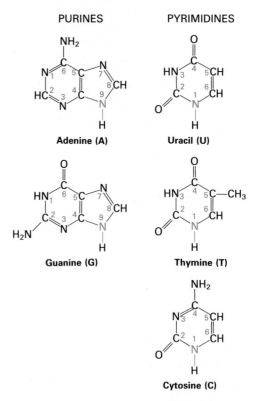

▲ FIGURE 4-2   Haworth projections of the structures of ribose and deoxyribose. By convention, the carbon atoms of the pentoses are numbered with primes. In nucleotides and nucleic acids, the 5′ carbon is linked in an ester bond to the phosphate and the 1′ carbon is linked to the base.

DNA, it is *deoxyribose* (Figure 4-2). The only other difference in the nucleotides of DNA and RNA is that one of the four nucleic acid bases differs between the two polymers.

The bases adenine, guanine, and cytosine (found in both DNA and RNA); thymine (found only in DNA); and uracil (found only in RNA) are often abbreviated A, G, C, T, and U, respectively. The base components of nucleic acids are either *purines* (A, G) or *pyrimidines* (C, T, or U)

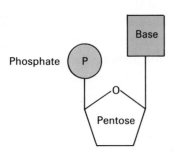

▲ FIGURE 4-1   A schematic diagram of a nucleotide.

▲ FIGURE 4-3   The chemical structures of the principal bases in nucleic acids. In nucleic acids and nucleotides, the 9 nitrogen atom of purines and the 1 nitrogen atom of pyrimidines (red) are bonded to the 1′ carbon of ribose or deoxyribose.

(Figure 4-3). A purine has a pair of fused rings; a pyrimidine has only one ring. Both purines and pyrimidines are *heterocyclic*: the rings are built of more than one kind of atom—in this case, nitrogen in addition to carbon. The presence of nitrogen atoms contributes to the chemical reactivity of the nucleic acid bases, but none is protonated (positively charged) at neutral pH. The acidic character of nucleotides is due to the presence of phosphate, which dissociates at the pH found inside cells, freeing hydrogen ions and leaving the phosphate negatively charged.

Atoms in the sugar component of a nucleotide provide the link between the base and the phosphate group. Whether the sugar is ribose or deoxyribose, its 1′ carbon atom is attached to the 9 nitrogen of a purine or to the 1 nitrogen of a pyrimidine. The hydroxyl group on the 5′ carbon atom of the sugar is replaced by an ester bond to the phosphate group (Figure 4-4).

Cells and extracellular fluids contain small concentrations of building blocks called *nucleosides*, combinations of a base and a sugar without a phosphate. The four *ribonucleosides* and *deoxyribonucleosides* are named in Table 4-1. Nucleotides, which have one, two, or three attached phosphate groups, are also referred to as *nucleoside phosphates*. The nucleoside phosphates that are used as building blocks are esterified at the 5′ hydroxyl (see Figure 4-4). *Nucleoside monophosphates* have a single esterified phosphate, *diphosphates* contain a pyrophosphate group

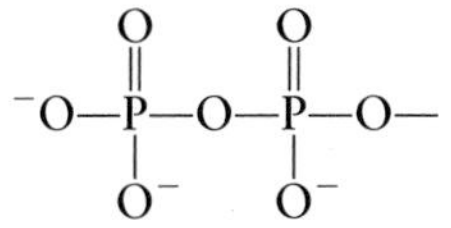

$$^-O-\overset{\overset{\displaystyle O}{\|}}{P}-O-\overset{\overset{\displaystyle O}{\|}}{P}-O-$$
$$\;\;\;\;\;\underset{O^-}{}\;\;\;\;\;\underset{O^-}{}$$

and *triphosphates* have a third phosphate. Table 4-1 names all the various forms of nucleoside phosphates. A supply of

▲ FIGURE 4-4   The chemical structure of a typical nucleotide, adenosine 5′-monophosphate (AMP). The C–N linkage between the sugar and the base shown here is a β linkage, meaning that the base lies above the plane of the ribose, as conventionally drawn. (In the α linkage, the base would lie below the plane.) Only the β linkage is found in cellular nucleotides. One additional phosphate (P–O–P) in ester linkage would constitute adenosine 5′-diphosphate (ADP), and two additional phosphates (P–O–P–O–P) adenosine 5′-triphosphate (ATP).

nucleoside triphosphates is necessary for the synthesis of nucleic acids. These compounds also serve many other functions in the cell; for example, the triphosphate ATP is the most widely used energy carrier in the cell.

When nucleotides polymerize to form nucleic acids, the hydroxyl group attached to the 3′ carbon of a sugar of

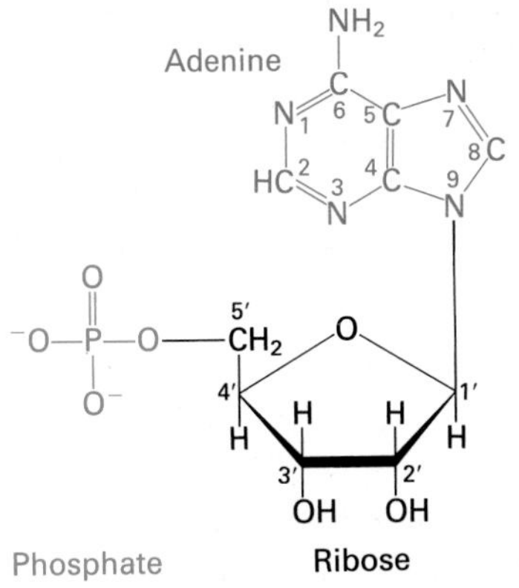

TABLE 4-1 Naming Nucleosides and Nucleotides				
	Bases			
	Purines		Pyrimidines	
	Adenine (A)	Guanine (G)	Cytosine (C)	Uracil (U) / Thymine [T]
Nucleosides { in RNA	Adenosine	Guanosine	Cytidine	Uridine
Nucleosides { in DNA	Deoxyadenosine	Deoxyguanosine	Deoxycytidine	Deoxythymidine
Nucleotides { in RNA	Adenylate	Guanylate	Cytidylate	Uridylate
Nucleotides { in DNA	Deoxyadenylate	Deoxyguanylate	Deoxycytidylate	Thymidylate
Nucleoside monophosphates	AMP	GMP	CMP	UMP
Nucleoside diphosphates	ADP	GDP	CDP	UDP
Nucleoside triphosphates	ATP	GTP	CTP	UTP
Deoxynucleoside mono-, di-, and triphosphates	dAMP, etc.			

one nucleotide forms an ester bond to the phosphate of another nucleotide, eliminating a molecule of water:

$$\text{(base)}_1 \qquad \text{O} \qquad \text{(base)}_2$$
$$\text{(sugar)}-\text{OH} + \text{HO}-\overset{\text{O}}{\underset{\text{O}^-}{\overset{\|}{\text{P}}}}-\text{O}-\text{(sugar)} \longrightarrow$$

$$\text{(base)}_1 \qquad \text{O} \qquad \text{(base)}_2$$
$$\text{(sugar)}-\text{O}-\overset{\text{O}}{\underset{\text{O}^-}{\overset{\|}{\text{P}}}}-\text{O}-\text{(sugar)} + \text{H}_2\text{O}$$

This condensation reaction is similar to that in which a peptide bond is formed. Thus a single nucleic acid strand is a phosphate-pentose polymer (a polyester) with purine and pyrimidine bases as side groups. The links between the nucleotides are called *phosphodiester bonds*. Like a polypeptide, a nucleic acid strand has a chemical orientation: the *3' end* has a free hydroxyl group at the 3' carbon of a sugar; the *5' end* has a free hydroxyl or phosphate group at the 5' carbon of a sugar (Figure 4-5). This directionality, plus the fact that (as we shall see) synthesis proceeds 5' to 3', has given rise to the convention that polynucleotide sequences are written and read in the $5' \rightarrow 3'$ direction (from left to right); for example, the sequence AUG is assumed to be (5')AUG(3'). (Although, strictly speaking, the letters A, G, C, T, and U stand for bases, they are also often used in diagrams to represent the whole nucleotides containing these bases.) The orientation of a nucleic acid strand is an extremely important property of the molecule.

➤ DNA

The modern era of molecular biology began in 1953, when James D. Watson and Francis H. C. Crick described the double-helical structure of DNA based on the analysis of x-ray diffraction patterns coupled with careful model building. A closer look at the "thread of life," as the DNA molecule is sometimes called, shows why the discovery of its basic structure suggests its function.

The Native State of DNA Is a Double Helix of Two Antiparallel Chains with Complementary Nucleotide Sequences

DNA consists of two associated polynucleotide strands that wind together through space in a helical fashion which is often described as a *double helix*. The two sugar-phosphate backbones are on the outside of the double helix, and the bases project into the interior. The adjoining bases in each strand stack on top of each other in parallel planes (Figure 4-6). The orientation of the two strands is antiparallel (their $5' \rightarrow 3'$ directions are opposite); they are held together by the cooperative energy of many hydrogen

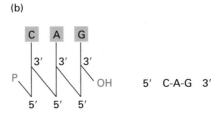

▲ FIGURE 4-5   Alternative ways of representing nucleic acid chains, in this case a single strand of DNA containing only three bases: cytosine (C), adenine (A), and guanine (G). (a) Chemical structure of the trinucleotide CAG. The nucleotide at the 3' end has a free hydroxyl group (i.e., unbonded to another nucleotide) on the 3' carbon of the deoxyribose. Similarly, the 5' end has a free 5' hydroxyl or, as shown here, a phosphate. (b) Two common simplified methods of representing polynucleotides. The "stick" diagram (*left*) shows the sugars as vertical lines and the phosphodiester bonds as slanting lines; the bases are denoted by their single-letter abbreviations. In the simplest representation (*right*), the bases are indicated by single letters; by convention, the chain is always written in the $5' \rightarrow 3'$ direction (left to right). Note that thymine (T) occurs only in DNA; it is replaced by uracil (U) in RNA.

bonds in addition to hydrophobic interactions. The opposite strands are held in precise register by a regular base pairing between the two strands: A is paired with T by two hydrogen bonds; G is paired with C by three hydrogen

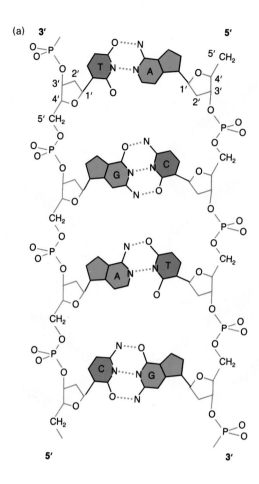

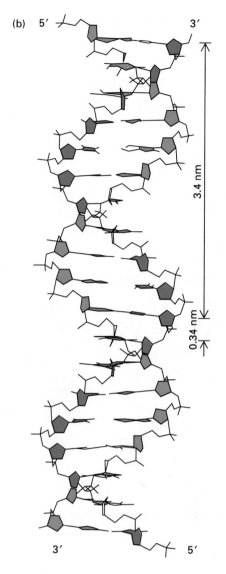

▲ FIGURE 4-6   Two representations of the DNA double helix. (a) Chemical structure of double-helical DNA, unraveled to show the sugar-phosphate backbones (black), base-paired bases (blue and red), and the hydrogen bonds between the bases (red dotted lines). The backbones run in opposite directions; the 5′ and 3′ ends are named for the orientation of the 5′ and 3′ carbon atoms of the sugar rings. Each base pair has one purine base—adenine (A) or guanine (G)—and one pyrimidine base—thymine (T) or cytosine (C)—connected by hydrogen bonds. (b) Skeletal model of double-helical DNA. The deoxyribose sugar rings and bases in one strand are shown in red, and in blue in the other strand; the phosphodiester bonds are black. The sugar-phosphate back-

bone is on the outside of the double helix, while the bases (shown on edge) project inward and are stacked in parallel "rungs," 0.34 nm apart. In the most common form of DNA (the B form), each strand makes a complete 360° turn every 3.4 nm. [Part (a) from R. E. Dickerson, 1983, *Scientific American,* **249**(6):94–111. Part (b) from L. Stryer, 1988, *Biochemistry*, 3d ed., W. H. Freeman and Company, p. 77.]

bonds (Figure 4-7). This *base-pair complementarity* is a consequence of the size, shape, and chemical composition of the bases. Hydrophobic and van der Waals interactions between the stacked adjacent base pairs also contribute significantly to the overall stability of the double helix.

To maintain the geometry of the double helix, a larger purine (A or G) must pair with a smaller pyrimidine (C or T); in natural DNA it is almost always A with T and G with C. However, in theory and in synthetic DNAs other interactions can occur. For example, a guanine (a purine)

could theoretically form hydrogen bonds with a thymine (a pyrimidine), causing only a minor distortion in the helix. The space available in the helix also would allow pairing between the two pyrimidines cytosine and thymine. While neither G-T nor C-T base pairs are normally found in DNA, G-U base pairs in helical regions of RNA are quite common.

Two polynucleotide strands can, in principle, form either a right-handed or a left-handed helix (Figure 4-8), but the geometry of the sugar-phosphate backbone is more

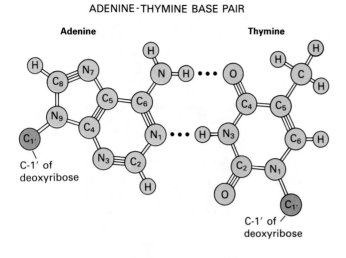

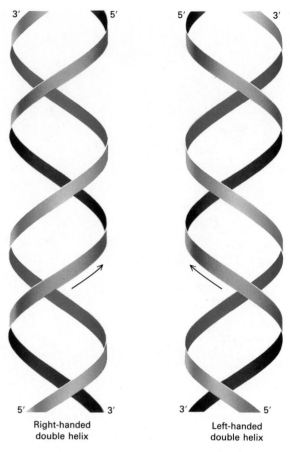

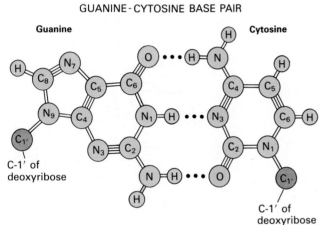

▲ FIGURE 4-7   Ball-and-stick models of A-T and G-C base pairs. Purines are shown in red; pyrimidines, in blue.

▲ FIGURE 4-8   Two possible helical forms of DNA are mirror images of each other. The geometry of the sugar-phosphate backbone of DNA causes natural DNA to be right-handed. (Right-handed and left-handed are defined by convention.)

compatible with the former, and therefore natural DNA is right-handed. The x-ray diffraction pattern of DNA indicates that the stacked bases are regularly spaced 0.34 nm apart along the helix axis (see Figure 4-6b). The helix makes a complete turn every 3.4 nm; thus there are about 10 pairs per turn. This is referred to as the *B form* of DNA. An alternative, more compact, *A form* has about 11 base pairs per turn, and one turn is about 2.3 nm. DNA in non-aqueous solutions can exist in the A form. On the outside of the B-form molecule, the spaces between intertwined strands form two helical grooves of different widths described as the *major* groove and the *minor* groove (Figure 4-9a). Consequently, part of each base is accessible from outside the helix to molecules that bind to the DNA by contacting chemical groups within the grooves.

In addition to normal B DNA, certain short DNA polymers can adopt an alternative left-handed configuration instead of the normal right-handed helix. This structure is called *Z DNA* because the bases seem to zigzag when viewed from the side (Figure 4-9b). Whether such

left-handed helices occur in nature is at present unknown. If they do occur in natural molecules, they could be a recognition signal in local regions of DNA.

Although the multitude of hydrogen and hydrophobic bonds between the polynucleotide strands provide stability to DNA, the double helix is somewhat flexible about its long axis because, unlike the α helix in proteins (see Figure 3-9), there are no hydrogen bonds between successive residues in a strand. This property allows DNA to bend, for example, when complexed with a DNA-binding protein (Figure 4-10a).

While the standard B-DNA model may describe the structure of most DNA, most of the time in cells, it is now believed that particular regions of DNA, especially that bound to protein, depart from the standard structure. In fact, sequence-dependent deformation of DNA about the long axis—creating *bent DNA*—is now recognized in naked DNA without proteins, as well as in cases in which DNA in combination with proteins is forced to bend. For example, the adenylate (A) residues, if all are in one strand,

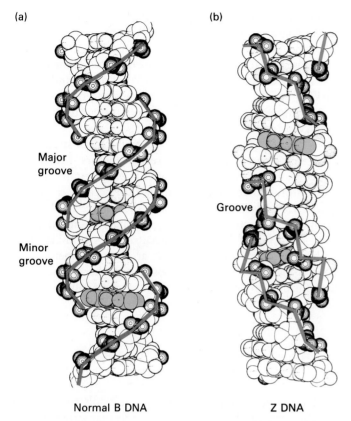

(a)   (b)

Major
groove

Minor
groove

Groove

Normal B DNA   Z DNA

▲ FIGURE 4-9   Space-filling models of (a) normal right-handed B DNA and (b) left-handed Z DNA. The red line in each model connects the phosphate residues along the chain; note the zig-zag irregularity of the Z-DNA backbone. Both B and Z DNA contain the same base pairs, which lie nearly perpendicular to the axis of the helix; two base pairs in each model are highlighted in blue. B DNA has two grooves (one major and one minor), whereas Z DNA has a single quite deep groove, which extends to the axis of the helix. [Courtesy of A. Rich.]

▼ FIGURE 4-10   Bending of DNA. (a) Protein binding may cause the inherently flexible DNA to bend. Here, a linear DNA (*left*) is shown binding a protein (*center*); the resulting bend in the DNA (*right*) is easily seen by comparison with the linear molecule. (b) Bending of DNA also is favored by the presence of several consecutive adenylate (A) residues in one strand, because they tend to stack at a greater angle to the helix axis than the complementary thymidylate (T) residues in the opposite strand. To visualize this effect, assume that linear DNA is unraveled (step 1), leaving a flat, ladder-like structure similar to that in Figure 4-6a. If the sequence consists of tracts of five A-T base pairs (red lines) interspersed with sections of five random base pairs (black lines), then the A-T base pairs will assume a characteristic tilt relative to the helix axis. The helix axis then reorients locally (step 2) to attain the most favorable base stacking at the junctions between the dissimilar regions. If the helix is then rewound (step 3), the overall curvature of the molecule is visible. [Part (a) from A. K. Aggarwal et al., 1988, *Science* **242**:899; courtesy of S. C. Harrison. Part (b) adapted from D. Crothers et al., 1990, *J. Biol. Chem.* **265**:7093.]

(a)

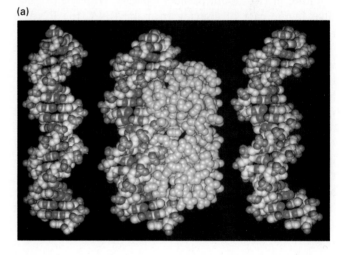

(b)

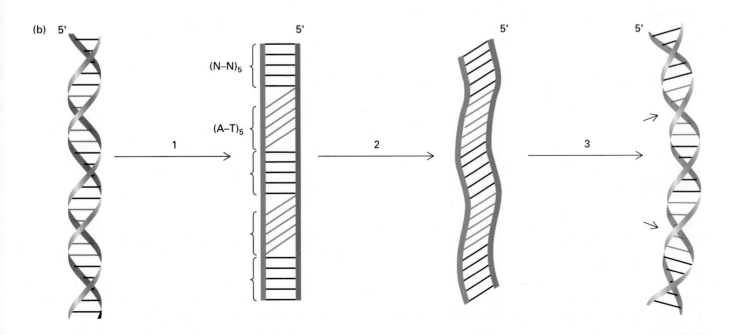

$(N{-}N)_5$

$(A{-}T)_5$

1   2   3

in A-T–rich DNA tend to stack at a greater angle to the helix axis than the thymidylate (T) residues in the opposite strand, leading to bending of the DNA even without attached proteins (Figure 4-10b). Naturally occurring bent DNA has been visualized in the electron microscope and can be detected by its physical properties. (It migrates faster through a porous substrate than does a linear DNA rod of the same number of residues.) Almost certainly, bent DNA is important because it accommodates the binding of special proteins and, by bending, brings together distant regions of DNA so that bound proteins can interact.

The Two Strands Can Separate, Causing the DNA to Denature

In DNA replication and in the copying of RNA from DNA, the strands of the helix must separate at least temporarily. As we discuss later, during DNA synthesis two new strands are made (one copied from each of the original strands), resulting in two double helices identical with the original one. In the case of copying to make RNA, the RNA is released and the two DNA strands can pair again.

The unwinding and separation of DNA strands can be induced experimentally: if a solution of DNA is heated, the thermal energy increases molecular motion to eventually break all the hydrogen bonds and other forces that stabilize the double helix, and the strands separate; the DNA is said to be *denatured* (Figure 4-11). The "melting" of DNA is revealed by measuring the absorption of ultraviolet (UV) light, which is routinely used to measure DNA concentration because of the high absorbance of UV light by nucleic acid bases. Native double-stranded DNA absorbs less than half as much light at 260 nm than does the equivalent

amount of single-stranded DNA (Figure 4-12a). Thus, as DNA denatures, its absorption of UV light increases. Near the denaturation temperature, a small increase in temperature causes an abrupt, near simultaneous, loss of the multiple, weak, cooperative interactions (hydrogen and base-stacking bonds) holding the two strands together, so that denaturation rapidly occurs throughout the entire length of the DNA.

The *melting temperature, T_m,* at which the strands of DNA will separate depends on several factors. Molecules that contain a greater number of G-C pairs require higher temperatures to denature (Figure 4-12b) because the three hydrogen bonds in G-C pairs make them more stable than are A-T pairs with two hydrogen bonds. In addition to heat, solutions of low ion concentration destabilize the double helix, causing it to melt at lower temperatures. DNA is also denatured by exposure to other agents that destabilize hydrogen bonds, such as alkaline solutions and concentrated solutions of formamide or urea:

$$\underset{\text{Formamide}}{\overset{\displaystyle O}{\underset{\displaystyle \|}{HC-NH_2}}} \qquad \underset{\text{Urea}}{\overset{\displaystyle O}{\underset{\displaystyle \|}{H_2N-C-NH_2}}}$$

The single-stranded molecules that result from denaturation form random coils without a regular structure. By lowering the temperature or increasing the ion concentration, the two complementary strands in time will renature into a perfect double helix (see Figure 4-11). Two DNA strands not related in sequence will remain as random coils and will not renature and, most important, will not greatly inhibit correct DNA partner strands from finding each other. As discussed in detail in Chapter 7, denaturation

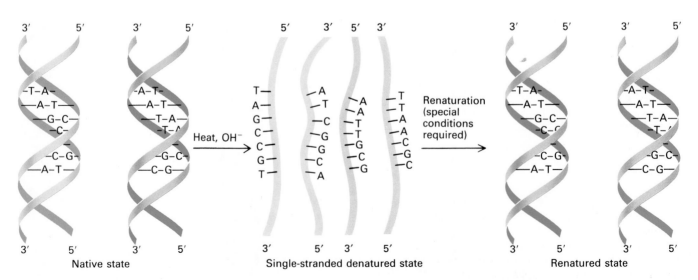

FIGURE 4-11   The denaturation and renaturation of double-stranded DNA molecules.

Native state   Single-stranded denatured state   Renatured state

(a)

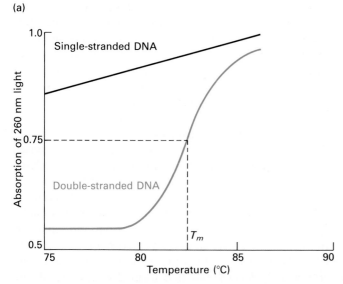

(b)

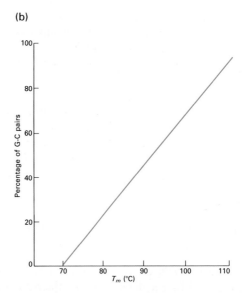

▲ FIGURE 4-12   (a) The absorption of ultraviolet light of 260-nm wavelength by solutions of single-stranded and double-stranded DNA. As regions of double-stranded DNA unpair, the absorption of light by those regions increases almost twofold. The temperature at which half the bases in a double-stranded DNA sample have denatured is denoted T_m

(for temperature of melting). Light absorption by single-stranded DNA changes much less as the temperature is increased. (b) The T_m is a function of the G + C content of the DNA; the higher the G + C percentage, the greater the T_m.

and renaturation of DNA are the basis of *nucleic acid hybridization*, a powerful technique in which reassociation of complementary DNA strands is used to study the relatedness of two DNA samples.

Many DNA Molecules Are Circular

All bacterial and many viral DNAs are circular molecules. Circular DNA molecules also occur in mitochondria that are present in almost all eukaryotic cells and chloroplasts that are present in plants and some single-cell eukaryotes.

Each of the two strands in a circular DNA molecule forms a closed structure without free ends. Just as is the case for linear DNA, elevated temperatures or alkaline pH destroy the hydrogen bonds and other interactions that stabilize double-helical circular DNA molecules. Unlike linear DNA, however, the two strands of circular DNA cannot unwind and separate, so they form an interlocked, tangled mass of single-stranded DNA under denaturing conditions (Figure 4-13a).

(a)

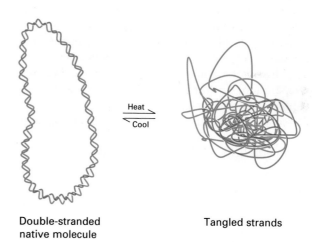

Double-stranded   Tangled strands
native molecule

(b)

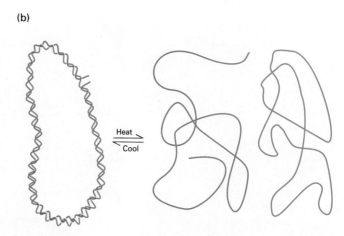

Nick in one strand   Strands completely separated

► FIGURE 4-13   Denaturation of circular DNA. (a) If both strands are closed circles, denaturation disrupts the double helix, but the two single strands become tangled about each other and cannot separate. (b) If one or both strands are nicked, however, the two strands will separate on thermal denaturation.

Only if the native circular DNA is *nicked* (i.e., one of the strands is cut), will the two strands unwind and separate when the molecule is denatured. One strand then remains circular, and the other becomes linear (Figure 4-13b). The nicking of circular DNA can occur naturally during the process of DNA replication or experimentally by cleaving a single phosphodiester bond in the circle with deoxyribonuclease (a DNA-degrading enzyme). The study of circular DNA molecules lacking free ends first uncovered the complicated geometric shape changes that the double-stranded DNA molecule must undergo when it is not free to denature.

Linking Number, Twist, and Writhe Describe DNA Superstructure

The topology of a DNA molecule can be described by three parameters. The first is determined by the helical turns of the two strands and is called *linking number*. It is an integer that is equal to the number of times within the boundaries being considered the two strands make a complete 360° turn . Such a turn has the effect of linking the strands if one tried to pull them apart without untwisting them. This linking is often illustrated and discussed in circular double-stranded molecules where it is easiest to see that the helical turns link the two strands and prevent separation. However many chromosomes are long linear molecules, sections or domains of which are physically separated from the remainder of the chromosome by tightly bound proteins at each end of a domain that prevent unwinding and strand separation. Such anchoring effectively creates the same topology as in a circle.

The second parameter that contributes to the topology of a DNA molecule is *twist*, which is related to the frequency, or periodicity, of the turns of the helix. In purified, isolated DNA at normal physiologic conditions of salt and temperature, an *average* of one right-handed helical turn (and one link) occurs every 10.6 bases. Twist can vary from segment to segment within a DNA molecule. The twist of DNA inside cells is not always uniform over short distances, because the DNA is associated with protein. For example, in the DNA wrapped around histone protein complexes called *nucleosomes* (Chapter 9), the twist is about 10.1 bases per helical turn.

A circular DNA molecule like that of the SV40 virus, which is about 5300 bases long, would be expected to have a linking number of 500 (5300 bases ÷ 10.6 bases/turn) under the salt and temperature conditions found in cells. If this were so, the molecule would form a relaxed circle, termed *form II*, when removed from virions and separated from its bound protein. Electron micrographs of isolated SV40 DNA, however, show that it is in a supercoiled state referred to as *form I* and that form II exists only if one strand is nicked (Figure 4-14).

The accepted explanation for the supercoiling of free SV40 DNA is that the helix is *underwound* by about 25 turns; that is, its linking number is about 475 (rather than the expected 500). This occurs because an untwisted region is left when the newly replicated DNA is separated from its associated protein (Figure 4-15a). Such an unwound region is energetically unfavorable, so the molecule attempts to adjust. If one chain were broken, the chains would spontaneously wind around each other another 25 turns—that is, the linking number would increase (Figure 4-15b). Without a chain break, however, either the average twist throughout the whole circle would have to decrease (to take up the "slack") or the molecule would have to coil on itself, producing supercoils (Figure 4-15c). Supercoils are the alternative chosen in nature when the temperature and salt concentrations are normal; neither the linking number

Form I

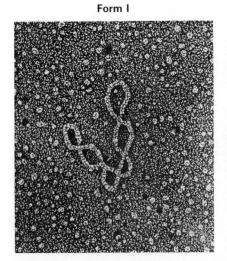

Form II

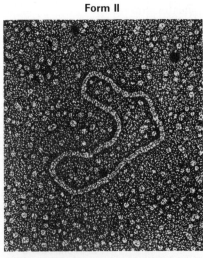

◄ FIGURE 4-14 Electron micrographs of isolated SV40 DNA in the supercoiled configuration (Form I) and relaxed-circle configuration (Form II). Form II occurs only when one strand is nicked. In the absence of a nick, this circular DNA undergoes supercoiling, as diagrammed in Figure 4-15. Only a few of the possible supercoils are visualized in one photograph.

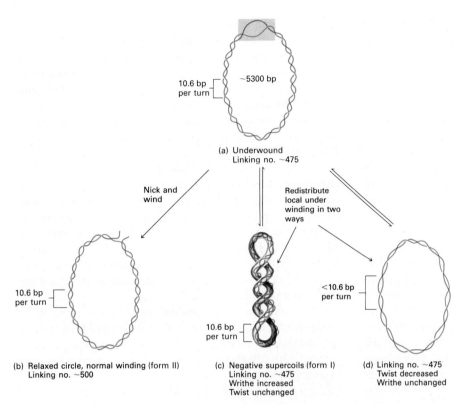

~5300 bp

10.6 bp per turn

(a) Underwound
Linking no. ~475

Nick and wind

Redistribute local underwinding in two ways

10.6 bp per turn

10.6 bp per turn

<10.6 bp per turn

(b) Relaxed circle, normal winding (form II)
Linking no. ~500

(c) Negative supercoils (form I)
Linking no. ~475
Writhe increased
Twist unchanged

(d) Linking no. ~475
Twist decreased
Writhe unchanged

◄ FIGURE 4-15   DNA supercoiling: interrelationship of twist (*T*), writhe (*W*), and linking number (*L*) of SV40 DNA. When isolated from virions and freed of protein, SV40 DNA is underwound (a) and has a linking number of about 475, rather than the 500 expected based on the number of nucleotides it contains (*L* = 5300 bp ÷ 10.6 bp/twist = 500). This underwinding is represented as a nonhelical region (yellow) in the DNA. The molecule could overcome this energetically unfavorable underwinding by breakage of one strand, followed by spontaneous winding of the helix until the normal *L* was attained, to give the relaxed form II (b). In the absence of a chain break, the underwound molecule could either (c) form supercoils (*W* increases; *T* unchanged) to give form I or (d) distribute the underwinding evenly throughout the molecule (*T* decreases; *W* unchanged). Under normal salt and temperature conditions, the supercoiling in (c) occurs, and supercoiled form I is observed (see Figure 4-14).

nor the average twist changes during the formation of supercoils. Because one less link between the strands would result in one supercoil, there would be 25 supercoils in the case just mentioned.

Because the right-hand winding of the strands in the DNA helix (see Figure 4-8) is regarded as *positive*, supercoils that form to compensate for the effects of underwinding in a helix (see Figure 4-15c) are said to be *negative* supercoils. This interconvertibility between supercoiling and underwinding is probably used in helping proteins bind to DNA. An *overwound* DNA helix—that is, one with more than 10.6 base pairs per turn and a linking number greater than expected—would tend to form *positive* supercoils in the opposite direction. (Such positive supercoils accumulate ahead of a growing fork in DNA replication and in front of RNA polymerase during transcription and must be relieved by enzymes called *topoisomerases* that break the DNA, dispel the tension, and reseal the strand; see Chapters 10 and 22.)

The supercoiling phenomenon, which is reciprocally related to twist when there is no change in linking number, is an expression of a third parameter—called *writhe*—that describes the pathway of the DNA backbone in space. While the linking number is a simple term (an integer, in fact), the exact definitions of twist and writhe are complex topological statements. However, in simplified terms $L = T + W$, where L is the linking number, T the twist, and W the writhe. As we discussed above, twist is an angular function related to the frequency of turns around the central axis of the helix within any given section of the DNA, whereas writhe is a property of the whole molecule that is driven by the energy inherent in assuming a certain twist. Unlike the linking number, which is a property of the whole molecule, twist and writhe can vary in different parts of the molecule, but they always change in a reciprocal relationship when L is constant.

► RNA: The Basic Chemical Structure and Its Function in Gene Expression

RNA is generally similar to DNA in chemical makeup; however, the sugar component (ribose) of RNA has an additional hydroxyl group at the 2′ position (see Figure 4-2), and thymine in DNA is replaced by uracil in RNA (see Figure 4-3). As in DNA the linkage between adjacent units is a 5′ → 3′ phosphodiester linkage (see Figure 4-5). The hydroxyl group on C-2 of ribose makes RNA more chemically labile than DNA. For example, RNA is cleaved into mononucleotides by alkaline solution, whereas DNA is not. Since RNA is also a long polyphosphodiester chain, it can take on the same configurations as DNA; it can be double-stranded or single-stranded, linear or circular. It can also participate in a hybrid helix composed of one

RNA strand and one DNA strand, with a slightly different structure from the common B-DNA structure. RNA can bind specific proteins to form ribonucleoprotein particles.

However, the most important discoveries about RNA structure do not concern its participation in strictly linear arrays. RNAs can also form regular three-dimensional structures that function in carrying out the task of genetic expression—getting the information in DNA converted to proteins. For example, small RNA molecules called *transfer RNA* (tRNA), adopt a well-defined three-dimensional architecture in solution (discussed later in this chapter), that is crucial in protein synthesis. Larger RNA molecules (especially, as we shall discuss, ribosomal RNA) have locally well defined three-dimensional structures, with more flexible links in between. In this sense RNA molecules are like proteins with structured domains connected by less structured, flexible stretches. The structured domains of an RNA are determined mainly by local base pairing that follows the same rules as in DNA. Looping of the RNA strand allows complementary sequences to form short double-helical regions, yielding *stem-loop* or *hairpin* structures (Figure 4-16a). The flexible, single-stranded loops may fold and interact, giving higher-order conformations such as the *pseudoknot* (Figure 4-16b).

As discussed in Chapter 12, experiments begun in the late 1970s have revealed that the folded domains of RNA molecules have catalytic capacities just as the folds (helices and β strands) of proteins do. Less is understood about the three-dimensional structure of catalytic RNA sites, but catalytic activities have definitely been identified. One of the most prominently studied is the ability of RNA to catalyze the cutting of an RNA chain. One example of such a *ribozyme* (RNA enzyme) is illustrated in Figure 4-17. The

substrate chain is base-paired with a site on the ribozyme, and the cleavage is induced at a precise nucleotide in the substrate. Other reactions, particularly the *phosphotransferase* reaction, which unites two distant segments of an RNA chain after cleavage has occurred and is called *splicing*, will be described fully later.

While research continues to reveal ever-increasing examples of catalytic function in *isolated* RNA, the bulk of RNA in cells is bound to proteins in complexes, the largest and most prominent of which are called *ribosomes*. These structures provide surfaces on which protein synthesis can occur. A large and still growing number of other, smaller ribonucleoprotein structures called *small nuclear ribonucleoproteins* ("snurps") play crucial roles in the processing of messenger RNA molecules and in other cell functions. There are also small cytoplasmic ribonucleoproteins such as the signal recognition particle that aids in protein secretion (Chapter 16).

The biological functions and the chemical properties of the large number of different RNA molecules therefore differ. Although individual RNAs were first recognized to perform the utterly crucial task of gene expression—the production of functional messenger RNA and its translation into protein—we now know that many other types of RNA exist and act as three-dimensional catalytic molecules. In this respect, RNA function presents some analogy to protein function. We discuss later in this chapter the direct roles of RNA in protein synthesis, and in Chapter 12 the possible catalytic role of RNA in messenger RNA formation. While these catalytic actions of RNA in modern-day cells may be limited, they may have been during early evolution the molecular forerunners of all biochemical processes.

(a) Secondary structure

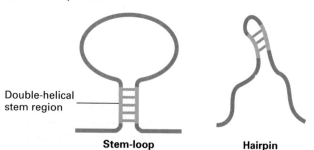

Double-helical stem region

Stem-loop   **Hairpin**

(b) Tertiary structure

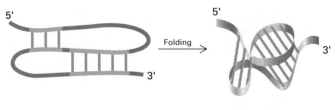

Pseudoknot

▲ FIGURE 4-16   RNA secondary and tertiary structures. (a) Base pairing between distant complementary segments of RNA molecules can form structured domains containing stem-loops, hairpins, and other secondary structures. In stem-loops, the single-stranded loop (dark red) between the base-paired helical stem (light red) may be hundreds or even thousands of nucleotides long, whereas in hairpins, the short turn may contain as few as 6–8 nucleotides. (b) Interactions between the flexible loops may result in three-dimensional folding to form tertiary structures such as the pseudoknot. This tertiary structure resembles a figure-eight knot, but the free ends do not pass through the loops, so no knot is actually formed. [Part (b) adapted from C. W. A. Pleij, K. Rietveld, and L. Bosch, 1985, *Nucl. Acids Res.* **13**:1717.]

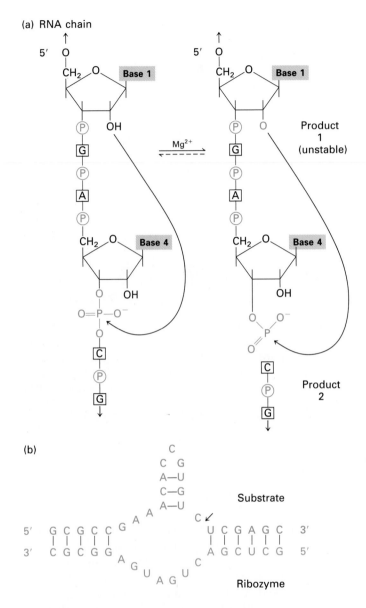

(a) RNA chain

(b)

Substrate

Ribozyme

◀ FIGURE 4-17
The transesterification reaction and RNA self-cleavage. (a) The transesterification reaction cuts an RNA chain (*left*) but leaves the energy of the internucleotide bond intact because the number of phosphodiester bonds (red) remains the same (product 1). In product 1 the free hydroxyl of nucleotide 1 has attacked the phosphate of nucleotide 4 and has become "transesterified." Product 1 is unstable and either joins another RNA chain in a splicing reaction (see Ch. 12) or is cleaved by H_2O. (b) Avocado sunblotch viroid, a small circular RNA molecule, can self-cleave at specific sites. The catalytic RNA, or ribozyme, base-pairs with the substrate RNA (red) and induces cleavage at a specific nucleotide (arrow). [Adapted from R. H. Symons, 1992, *Ann. Rev. Biochem.* **61**:641.]

the *central dogma* of molecular biology. Note however that proteins are required to facilitate the flow of information. A discussion of the mechanisms of the synthesis of nucleic acids or proteins could begin with any one of the three kinds of polymers, since the problems to be solved have much in common. In considering how the synthesis of any of these polymers begins, progresses, and ends, four general rules may be drawn from the experimental results:

1. Proteins and nucleic acids are made up of a limited number of different subunits. Although the number of amino acids is theoretically limitless, and several dozen have been identified as metabolic products in various organisms, only 20 different amino acids are used in making proteins. Likewise, only five nitrogenous bases are used to construct RNA and DNA in cells. Cell-free enzyme preparations and cells in culture can be "fooled" into incorporating chemical relatives of the 5 bases or 20 amino acids, but this almost never happens in nature.

2. The subunits are added one at a time. A priori, biological polymers could be built by aligning the subunits in the correct order on a template, or mold, and *simultaneously* fusing them all. But it does not happen that way. The assembly of proteins and nucleic acids is a step-by-step procedure in only one chemical direction: protein synthesis begins at the amino (NH_2–) terminus and continues through to the carboxyl (COOH–) terminus; nucleic acid synthesis begins at the 5' end and proceeds to the 3' end (Figure 4-18).

3. Each chain has a specific starting point, and growth proceeds in one direction to a fixed terminus; this requires start and stop signals that are well defined.

4. The primary synthetic product is usually modified. The functional form of a nucleic acid or a protein molecule is often not the same length as its initially synthesized form. The original chain is often inactive or incomplete. To make an active chain, specific enzymes act to change the length of the original chain: proteins are often shortened, and RNA chains almost always

▶ Rules for the Synthesis of Proteins and Nucleic Acids and Macromolecular Carpentry

The intricate relation in cells between the synthesis of DNA, RNA, and protein is circular and can be diagrammed as follows:

$$DNA \longrightarrow RNA \longrightarrow protein$$

DNA directs the synthesis of RNA, and RNA then directs the synthesis of protein; special proteins then can catalyze the synthesis of both RNA and DNA. This cyclic relationship occurs in all cells. The flow of information from nucleic acids to protein, but not the reverse, has been called

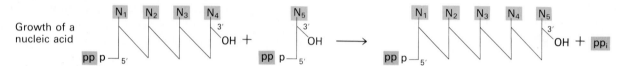

Growth of a polypeptide

Growth of a nucleic acid

▲ FIGURE 4-18   Chain elongation in the in vivo synthesis of both proteins and nucleic acids proceeds by sequential addition of monomeric units. In protein synthesis, one amino acid at a time is added to the carboxyl end of the growing chain; thus growth of a polypeptide chain proceeds from the N-terminus to the C-terminus. (R₁, R₂, etc., denote side chains of amino acids.) In the synthesis of nucleic acids, one nucleotide at a time is added to the 3'-hydroxyl group at the end of the growing chain; thus nucleic acids grow from the 5' end to the 3' end. With each addition, a pyrophosphate group (PP_i) is released, but the first nucleotide, at the 5' end, retains its triphosphate group. (N_1, N_2, etc., denote purine and pyrimidine bases.) The nucleic acid chain is represented by a stick diagram.

are; DNA chains are made in pieces and then linked together. Distant segments of long RNA chains may be spliced together, omitting the intervening sequence; protein chains may be cross-linked by covalent bonds (Figure 4-19). Primary chains can also undergo chemical additions, either during the formation of the chain or after its synthesis is complete. Methyl groups can be added to specific sites in DNA, RNA, and proteins; phosphate groups and a wide variety of oligosaccharides can be added to proteins. The details of this macromolecular carpentry are discussed in later chapters.

➤ *Nucleic Acid Synthesis*

The ordered assembly of the basic units in DNA (deoxyribonucleotides) or RNA (ribonucleotides) involves somewhat simpler cellular mechanisms than the correct assembly of the amino acids in a protein chain. As Watson and Crick remarked in their proposal of the double-helical DNA structure: "It has not escaped our notice that the specific pairing we have postulated immediately suggests a possible copying mechanism for the genetic material." Their prediction that the synthesis of new strands of DNA

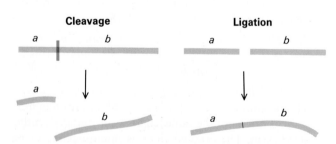

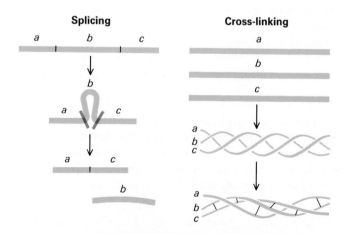

▲ FIGURE 4-19   Proteins and nucleic acids typically undergo one or more modifications after the basic polymeric chains are formed. Besides the modifications illustrated here, specific chemical groups may be added. Various types of postpolymerization changes are discussed in later chapters.

occurs by copying parental strands of DNA has proved to be correct. The phenomena associated with the duplication of DNA and the copying of DNA into RNA have given rise to two very broad and intensive areas of research in molecular biology: DNA replication and its role in cell growth and division, and the mechanism and the control of the synthesis of specific mRNAs during cell adaptation and differentiation.

Nucleic Acid Polymerization Can Be Described by Four Rules

We now consider a few general principles governing the synthesis of chains of DNA or RNA and briefly discuss some properties of the enzymes that carry out such synthesis. We will need to refer to these general principles as we examine the detailed relation of these crucial events to cell growth and differentiation in other chapters:

1. *Both DNA and RNA chains are produced in cells by copying a preexisting DNA strand according to the rules of Watson-Crick base pairing.* The DNA from which the new strand is copied is called a *template.* The information in the template is preserved: although the first copy has a complementary sequence, not an identical one, a copy of the copy produces the original (template) sequence again. In the replication of a double-helical DNA molecule, or *duplex*, both original DNA strands are copied, and when copying is finished, they separate from each other into two new duplexes, each consisting of one of the two original strands plus its copy. In some viruses, RNA molecules are produced by copying preexisting RNA molecules; in one class, the *retroviruses*, DNA is produced by copying RNA. However, the vast majority of cellular RNA and DNA synthesis in cells is from a preexisting DNA template.

2. *Nucleic acid strand growth is in one direction: $5' \rightarrow 3'$.* Because nucleic acids are phosphodiesters, each strand has a definite chemical orientation: a $5'$ (phosphate) end and a $3'$ (hydroxyl) end. All RNA and DNA synthesis, both cellular and viral, proceeds in the same chemical direction: from the $5'$ to the $3'$ end (see Figure 4-18). The nucleotides used in the construction of nucleic acid chains are $5'$ triphosphates of ribonucleosides or deoxyribonucleosides. Strand growth is energetically unfavorable but is driven by the energy available in the triphosphates. The α phosphate of the incoming nucleotide attaches to the $3'$ hydroxyl of the ribose (or deoxyribose) of the preceding residue to form a phosphodiester bond, releasing a pyrophosphate (PP_i). The equilibrium of the reaction is driven further toward chain elongation by pyrophosphatase, which catalyzes the cleavage of PP_i into two molecules of inorganic phosphate (P_i) (see Table 2.9).

3. *Special enzymes called polymerases make RNA or DNA.* The enzymes that copy DNA to make more DNA are *DNA polymerases;* those that produce RNA by copying DNA are *RNA polymerases.* Accurate synthesis of a nucleic acid by a polymerase always requires a template.

The copying of DNA by an RNA polymerase to make RNA is called *transcription.* Because the two DNA strands are complementary, rather than identical, they have different protein-coding potentials. It was believed for many years that only one strand of the DNA duplex gives rise to usable information when transcribed into RNA—and this is *almost* always true. In rare cases, however, this rule is violated: limited sections of DNA encode proteins on both strands.

RNA polymerases can initiate a new nucleic acid strand; DNA polymerases cannot (they require a primer). An RNA polymerase can find an appropriate initiation site on duplex DNA, bind the DNA, temporarily "melt," or separate, the two strands in that region, and begin generating a new RNA strand (Figure 4-20). Although not indicated in this figure, the location and regulated use of RNA start sites to produce mRNA requires perhaps a dozen or more proteins in eukaryotes and several proteins even in bacteria. This subject is covered in Chapter 11.

The nucleotide at the terminal $5'$ end of a growing RNA strand is chemically distinct from the nucleotides within the strand in that it retains all three phosphate groups. When an additional nucleotide is added to the $3'$ end of the growing strand, only the α phosphate is retained; the β and γ phosphates are lost (Figure 4-21).

DNA polymerases cannot initiate chain synthesis de novo; instead, DNA polymerases add nucleotides to the hydroxyl group at the $3'$ end of a preexisting RNA or DNA strand, called a *primer,* that is bound (hybridized) to the template strand. If RNA is the primer, the polynucleotide copied from the template is RNA at the $5'$ end and DNA at the $3'$ end. Cells have several different types of DNA polymerases; the physiologic role of these enzymes is described in Chapter 10.

4. *Duplex DNA synthesis requires a special growing fork.* Because duplex base-paired DNA consists of two intertwined strands, the base-pair copying of one strand requires the unwinding of the original duplex and a release or absorption of the resulting torsional force. This process is most likely accomplished by specific "unwinding proteins" and by topoisomerases, which nick the DNA, allow it to swivel, and then reseal it. Synthesis of daughter strands proceeds at or near the so-called *growing fork* between the unwound parental strands. Accurate initiation of DNA replication is a complex process that we examine in detail in Chapter 10.

All nucleic acid strands grow in the $5' \rightarrow 3'$ direction, but the two DNA strands of a duplex are antiparallel. On one parental strand, synthesis of a daughter strand, called the *leading strand,* proceeds continu-

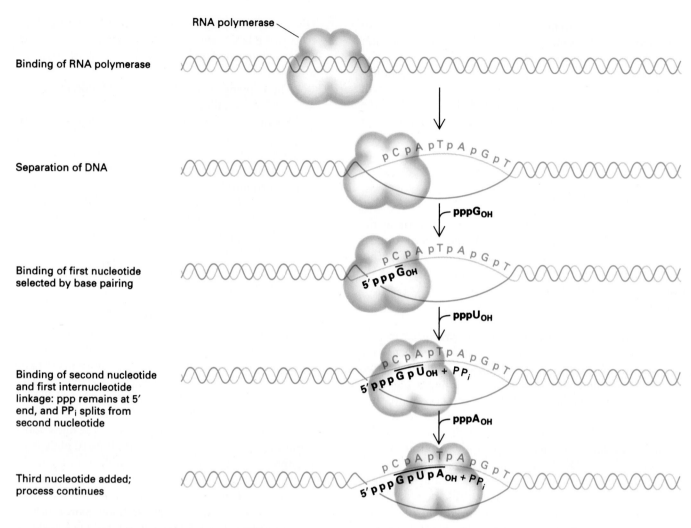

Binding of RNA polymerase

Separation of DNA

Binding of first nucleotide selected by base pairing

Binding of second nucleotide and first internucleotide linkage: ppp remains at 5′ end, and PPᵢ splits from second nucleotide

Third nucleotide added; process continues

▲ FIGURE 4-20   The transcription of a DNA strand into an RNA strand is catalyzed by an RNA polymerase, which can initiate the synthesis of strands de novo on DNA templates. The nucleotide at the 5′ end of an RNA strand is chemically distinct from the nucleotides within the strand in that it retains all three of its phosphate groups.

ously in the $5' \rightarrow 3'$ direction, in the direction of the growing fork (Figure 4-22). On the other parental strand, because $5' \rightarrow 3'$ strand growth is in the opposite direction from the movement of the growing fork, discontinuous segments of DNA are synthesized as new sections of the parental strand are exposed. This discontinuous daughter strand is called the *lagging strand*. The short, discontinuous segments, called *Okazaki fragments* after their discoverer Reiji Okazaki, are linked by *DNA ligase* to form a continuous strand. At least 20 different proteins participate in a DNA growing fork.

Organization of Genes in DNA Differs in Prokaryotes and Eukaryotes

With an appreciation of the biochemistry of stepwise assembly of polynucleotides we now want to focus briefly on the large-scale arrangement of information in DNA and

how this arrangement dictates the requirements for RNA manufacture so that information transfer goes smoothly.

Geneticists locate functional elements in the DNA by finding or purposely producing *mutant* organisms that differ from normal, or *wild-type*, organisms. As is discussed in Chapter 8, the production, detection and the analysis of mutant cells or organisms is one of the most important of all biological experimental approaches. The basis for each one of these heritable differences in organisms is a change in a *gene*, the hereditary unit carried in the DNA that has the information for each independent function in the organism. The number of genes in cells varies widely (Table 4-2). The simplest prokaryotic cells (e.g., *E. coli*), which have no nucleus, contain about 3000–5000 genes. Eukaryotic cells possess a nucleus; the simplest eukaryotes, such as baker's yeast (*Saccharomyces cerevisiae*), have ≈5000–10,000 genes per cell. Fruit flies (*Drosophila*) have perhaps 20,000 genes per cell, and mammals, including humans, have somewhere between 100,000 and 500,000 genes per

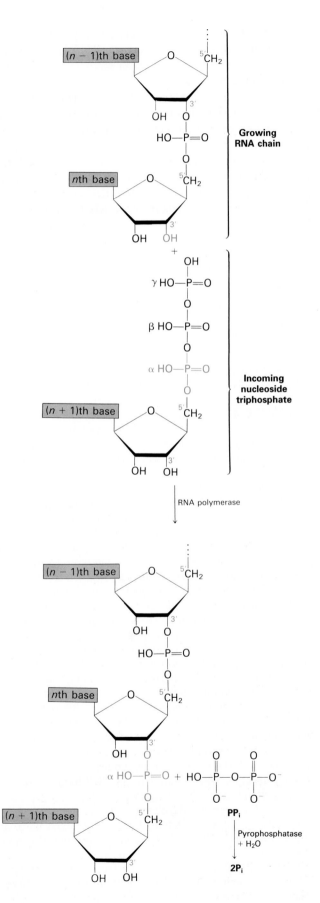

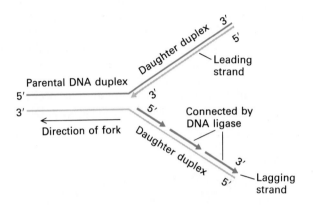

◄ FIGURE 4-21   During RNA chain growth, the α phosphate (red) of the incoming nucleoside triphosphate is linked to the 3' hydroxyl of the previous nucleotide added to the growing chain, thus extending the chain. The β and γ phosphates are released as pyrophosphate (PP$_i$), which then is cleaved to yield two molecules of inorganic phosphate (P$_i$).

▲ FIGURE 4-22   During replication of DNA, the duplex is sequentially unwound, exposing a single-stranded region (the growing fork) where complementary daughter strands are synthesized. Since nucleic acid chains can grow only in the 5' → 3' direction (indicated by arrowheads on daughter strands), only one new strand—the leading strand—is synthesized continuously. Synthesis of the other new strand—the lagging strand—proceeds discontinuously, initially forming fragments, because replication must begin at a new (downstream) start site on the parental strand as each new region of DNA is exposed at the growing fork. DNA ligase connects the lagging-strand fragments to form a continuous strand.

TABLE 4-2	**List of Estimated Gene Numbers**	
	Amount of DNA (bp)	**Estimated Number of Genes***
PROKARYOTES		
E. coli	4.5×10^6	4,000
EUKARYOTES		
Yeast (*S. cerevisiae*)	1.5×10^7	5,000 to 10,000
Caenorhabitis elegans	5.0×10^7	10,000
Drosophila melanogaster	1.5×10^8	20,000
Humans	4.0×10^9	100,000 to 500,000

* These estimates are based on partial sequencing of a small part of the DNA from these organisms but should be accurate within a factor of 2 or 3.

cell. The vast majority of genes carry information to build protein molecules, and the RNA copies of such genes are the messenger RNA molecules of cells.

In bacterial cells the most common arrangement of genes has a powerful and appealing logic: genes devoted to a single metabolic goal, say, the synthesis of the amino acid tryptophan, are found in a contiguous linear array in the bacterial DNA. This gene order makes it possible to produce a continuous strand of RNA that carries the message for a related series of enzymes devoted to making tryptophan (Figure 4-23). Each section of the mRNA represents the unit (or gene) that instructs the protein-synthesizing apparatus to make a particular protein. Such an arrangement of genes in a functional group is called an *operon*, because it can be operated as a unit by one decision to make mRNA from the entire operon. In bacterial DNA the

genes are closely packed with very few gaps, and the DNA is transcribed directly into colinear mRNA and then translated into protein even while later stretches of the mRNA are still being produced.

This economic clustering of genes devoted to a single metabolic function does not occur in eukaryotes, even simple ones like yeasts that metabolically are similar to bacteria. Rather, related eukaryotic genes generally are physically separated in the DNA and are transcribed from individual start sites, producing one mRNA from each gene. Moreover, when researchers first compared the informational regions of eukaryotic mRNAs with the DNAs encoding them, they were astounded to find that the protein-coding sequences of a mRNA were encoded in separate sections in the DNA; that is, *eukaryotic genes existed in pieces* separated by non-protein-coding segments (Figure

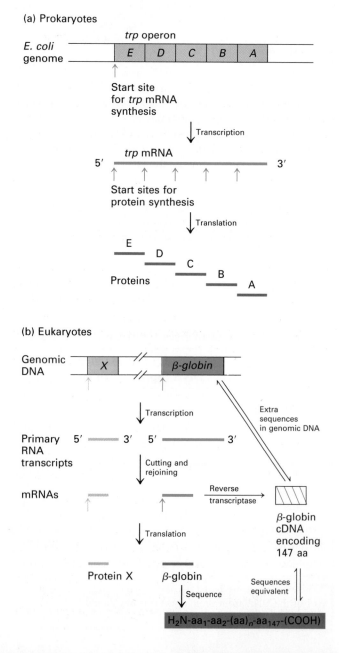

► FIGURE 4-23   Comparison of gene organization, transcription, and translation in prokaryotes and eukaryotes. (a) The tryptophan (*trp*) operon is a continuous segment of the *E. coli* chromosome containing five genes (blue) that encode the enzymes necessary for the stepwise synthesis of tryptophan. The entire operon is transcribed from one start site (blue arrow) into one long continuous *trp* mRNA(red). Translation of this mRNA begins at five different start sites, yielding five proteins (green). Proteins E and D associate to form the first enzyme in the tryptophan biosynthetic pathway; protein C catalyzes the intermediate step; and proteins A and B form tryptophan synthetase, the final enzyme. Thus the order of the genes in the bacterial genome parallels the sequential function of the encoded proteins in the tryptophan pathway. (b) Eukaryotic genes, even related ones, are physically separated in the genomic DNA. Each gene is transcribed from its own start site to yield a primary transcript, which is processed into a functional mRNA encoding a single protein. Among the earliest evidence indicating that eukaryotic genes contain noncoding segments came from studies with β-globin. By techniques described in Chapter 7, β-globin mRNA was copied into a DNA copy (called cDNA) by reverse transcriptase, a viral enzyme. Comparison of the nucleotide sequence of this cDNA with the known amino acid sequence of β-globin showed that the cDNA (and thus its mRNA template) correctly encoded the protein sequence. When the region of genomic DNA encoding β-globin was isolated and sequenced, it was unexpectedly found to be about twice as long as the cDNA. However, the primary β-globin RNA transcript, which is located in the nucleus, was shown to correspond with the β-globin gene. These findings suggest that eukaryotic primary transcripts contain both the coding and noncoding segments of the corresponding gene; cutting of the primary transcript and rejoining of the protein-coding segments yields a functional mRNA.

4-23b). This astonishing finding, first discovered in viruses that infect eukaryotic cells, implied that the initial RNA copy (or *transcript*) had to be clipped apart and carefully stitched back together in order to construct many of the mRNAs found in eukaryotic cells.

Eukaryotic Primary RNA Transcripts Are Processed to Form Functional mRNAs

In eukaryotic cells the primary RNA transcript produced by copying a protein-coding gene in the cell nucleus under-

(a)

(b)

goes several modifications collectively termed *RNA processing*. These changes yield a functional mRNA capable of being translated into protein.

Eukaryotic primary transcripts are modified at both ends, and these modifications are retained in mRNAs. To the initiating, or 5′, nucleotide of the primary transcript is added the *5′ cap* (Figure 4-24), which may serve to protect mRNA from enzymatic degradation. Processing at the 3′ end of the primary transcript involves cleavage by an endonuclease to yield a free 3′-hydroxyl group to which a string of adenylic acid residues are added by an enzyme called *poly-A polymerase*. The resulting poly-A tail may be 100–250 bases long, depending on the species. Poly-A polymerase does not require a template and can determine the approximately correct number of A residues to add.

The final step in the synthesis of many different eukaryotic mRNA molecules is dictated by the "genes in pieces" nature of most eukaryotic genes: internal cleavage of the RNA to excise noncoding segments, called *introns*, followed by splicing together of the coding segments, called *exons*. This extraordinary process is carried out within the nucleus with the aid of small ribonucleoprotein particles; the details of this process are discussed in Chapter 12. Many eukaryotic mRNAs also contain noncoding regions at each end; these are referred to as the 5′ and 3′ untranslated regions. Figure 4-25 summarizes the basic steps in RNA processing.

▶ *Protein Synthesis: The Three Roles of RNA in Translation*

Whereas DNA stores the information for protein synthesis, and RNA carries out the instructions encoded in DNA, most biological activities are carried out by proteins; their accurate synthesis is at the heart of cellular function.

Because the linear order of amino acids in each protein determines its function, the mechanism that maintains this order during protein synthesis is critical. In the design and function of the protein-synthesizing apparatus, three kinds of RNA molecules perform different but cooperative functions (Figure 4-26):

- *Messenger RNA* (mRNA) encodes the genetic information copied from DNA in the form of a sequence of bases that specifies a sequence of amino acids.

- *Transfer RNA* (tRNA) is the key to deciphering the code. The amino acids specified by the base sequence of an mRNA molecule are each attached to specific tRNAs, then carried to and deposited at the growing end of a polypeptide chain. The correct tRNA with its attached amino acid is selected at each step because each specific tRNA molecule can "read," by base pairing, its complementary base sequence in the mRNA.

▲ FIGURE 4-24   (a) The structure of the 5′ methylated cap of eukaryotic mRNA. The distinguishing chemical features are the 5′ → 5′ linkage of 7-methylguanylate to the initial nucleotide of the mRNA molecule and the methyl group at the 2′ hydroxyl of the ribose of the first nucleotide (Base 1). Both these features occur in all animal cells and in cells of higher plants; yeasts lack the methyl group on Base 1. The ribose of the second nucleotide (Base 2) also is methylated in vertebrates. (b) *S*-adenosylmethionine (*S*-Ado-Met) is the source of the methyl (CH_3) group for the two methyl transferase steps: the guanylate is methylated first, then the 2′ hydroxyl. [Part (a) see A. J. Shatkin, 1976, *Cell* **9**:645; part (b) see S. Venkatesan and B. Moss, 1982, *Proc. Nat'l. Acad. Sci. USA* **79**:304.]

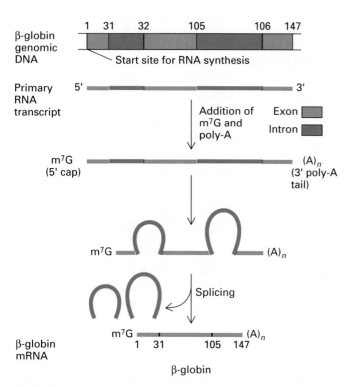

▲ **FIGURE 4-25**   Overview of RNA processing in eukaryotes using β-globin gene as an example. The β-globin gene contains three protein-coding segments, called exons, and two intervening noncoding segments, called introns. After addition of the 5′ cap and 3′ poly-A tail, the introns are removed during splicing of the exons. The sequence of the resulting mRNA corresponds to that of the protein β-globin. The small numbers refer to positions in the sequence of β-globin, which contains 147 amino acids.

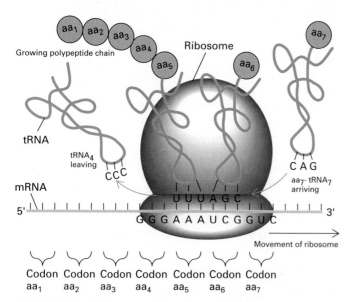

▲ **FIGURE 4-26**   The three roles of RNA in protein synthesis. Messenger RNA (mRNA) is translated into protein by the joint action of transfer RNA (tRNA) and the ribosome, which is a bipartite structure composed of numerous proteins and two major ribosomal RNA (rRNA) molecules. [Adapted from A. J. F. Griffiths, et al., 1993, An Introduction to Genetics Analysis, 5 ed., W. H. Freeman and Company.]

- *Ribosomal RNA* (rRNA) has important functions by itself, including attracting the mRNA and very likely catalyzing peptide-bond formation. In addition, rRNA binds a set of proteins to form *ribosomes*; rRNA plus the associated ribosomal proteins provide binding sites for all the accessory molecules necessary for protein synthesis. Ribosomes bearing bound tRNAs and special proteins can physically move along an mRNA molecule to translate its encoded genetic information into protein.

Translation refers to the whole process by which the base sequence of an mRNA is used to order and to join the amino acids in a protein. The three types of RNA participate in this essential protein-synthesizing pathway in all cells; in fact, the development of the three distinct functions of RNA was probably the molecular key to the origin of life.

In this section, we first explain the genetic code and the role of mRNA in carrying coded information. Next, we describe the structure of tRNA and its elementary biochemistry to show how the language of nucleic acids is converted to the language of proteins. Finally, we summarize the present understanding of how the ribosome serves in organizing the steps of protein synthesis. The biochemical events in protein synthesis and the required protein factors are discussed in the final section of this chapter.

Messenger RNA Carries Information from DNA in a Three-Letter Genetic Code

RNA contains ribonucleotides of adenine, cytidine, guanine, and uracil; DNA contains deoxyribonucleotides of adenine, cytidine, guanine, and thymine. Because four nucleotides cannot specify the linear arrangement of the 20 possible amino acids in a one-to-one manner, a *group* of nucleotides is required to symbolize each amino acid. The code employed must be capable of specifying at least 20 "words."

If two nucleotides were used to code for one amino acid, then only 16 (or 4^2) different code words could be formed, which would be an insufficient number. However, if a group of three nucleotides is used for each code word, then 64 (or 4^3) code words can be formed. Any code using groups of three or more nucleotides will have more than enough units to encode 20 amino acids. Many such coding systems are mathematically possible. However, the actual system used by cells is a *triplet code*, with a specified starting point in the mRNA. Each triplet is called a *codon*; 61 of the 64 possible codons specify individual amino acids (Table 4-3).

Because there are 61 codons for 20 amino acids in the general code, many amino acids have more than one codon; in fact, leucine, serine, and arginine each have six (Table 4-4). The different codons for a given amino acid

TABLE 4-3   The Genetic Code (RNA to Amino Acids)*

First Position (5' end)	Second Position				Third Position (3' end)
	U	C	A	G	
U	Phe	Ser	Tyr	Cys	U
	Phe	Ser	Tyr	Cys	C
	Leu	Ser	Stop (och)	Stop	A
	Leu	Ser	Stop (amb)	Trp	G
C	Leu	Pro	His	Arg	U
	Leu	Pro	His	Arg	C
	Leu	Pro	Gln	Arg	A
	Leu	Pro	Gln	Arg	G
A	Ile	Thr	Asn	Ser	U
	Ile	Thr	Asn	Ser	C
	Ile	Thr	Lys	Arg	A
	Met (start)	Thr	Lys	Arg	G
G	Val	Ala	Asp	Gly	U
	Val	Ala	Asp	Gly	C
	Val	Ala	Glu	Gly	A
	Val (Met)	Ala	Glu	Gly	G

* Stop (och) stands for the ochre termination triplet, and Stop (amb) for the amber, named after the bacterial strains in which they were identified. AUG is the most common initiator codon; GUG usually codes for valine, but, rarely, it can also code for methionine to initiate an mRNA chain.

TABLE 4-4   The Degeneracy of the Genetic Code

Number of Synonymous Codons	Amino Acid	Total Number of Codons
6	Leu, Ser, Arg	18
4	Gly, Pro, Ala, Val, Thr	20
3	Ile	3
2	Phe, Tyr, Cys, His, Gln, Glu, Asn, Asp, Lys	18
1	Met, Trp	2
Total number of codons for amino acids		61
Number of codons for termination		3
Total number of codons in genetic code		64

A surprising discovery was that some nucleotide sequences contain overlapping information, still in the form of a triplet code. This occurs if a one- or two-base shift in the reading frame, either to the right or left, coincides with a second reading frame starting from another AUG start site that is followed by a second (different) set of protein-coding triplets. In this case, translation of the mRNA can occur in two reading frames, at least for some distance, yielding different polypeptides (Figure 4-27). The vast majority of mRNAs can be read in only one frame because stop codons are encountered in the other two possible reading frames. Another unusual finding was that some mRNAs contain sites that cause *frame shifting*. At such sites the protein-synthesizing machinery may read four nucleotides as one amino acid and then continue reading triplets, or it may back up one base and read all succeeding

are said to be *synonymous*. The code itself is termed *degenerate*, which means that it contains redundancies.

The "start" (*initiator*) codon AUG specifies the amino acid methionine: all protein chains in prokaryotic and eukaryotic cells begin with this amino acid. GUG is used as the initiator codon for methionine in a few bacterial mRNAs (frequently in *Micrococcus luteus,* for example), but AUG predominates in bacteria. In eukaryotes CUG occasionally is used as an initiator codon for methionine. The three codons UAA, UGA, and UAG do not specify amino acids but constitute "stop" (*terminator*) signals at the ends of protein chains in almost all cells. The sequence of codons that runs from a specific start site to a terminating codon is called a *reading frame*. This precise linear array of ribonucleotides in groups of three in mRNA specifies the precise linear sequence of amino acids in a protein and also signals where synthesis of the protein chain starts and stops.

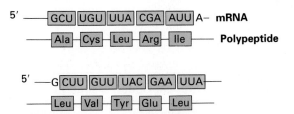

▲ FIGURE 4-27   Example of how the genetic code—an overlapping, commaless triplet code—can be read in two different frames. If translation of the mRNA sequence shown begins at two different upstream start sites (not shown), then two overlapping reading frames are possible; in this case, the codons are shifted one base to the right in the lower frame. As a result, different amino acids are encoded by the same nucleotide sequence. Many instances of such overlaps have been discovered in viral and cellular genes of prokaryotes and eukaryotes.

triplets in the new frame. These frame shifts are not common events, but a few dozen such instances are known.

The meaning of each codon is the same in most known organisms—a strong argument that life on earth evolved only once. Recently the genetic code has been found to differ for a few codons in mitochondria, in ciliated protozoans, and in *Acetabularia,* a single-celled plant (Table 4-5). A good many of these changes involve reading of normal stop codons as amino acids. For example, UGA is read as tryptophan in *Mycoplasma,* very small bacteria, and in nonplant mitochondria from protozoans, yeasts, worms, insects, and humans. In addition UAA and UAG are read as glutamine in some protozoans and in *Acetabularia,* and UGA is read as cysteine in *Euplotes.* In these cases there is not an exchange of one amino acid for another but an amino acid is read instead of a stop sign. It is now thought that these exceptions to the general code are later evolutionary developments; that is, at no single time was the code immutably fixed, although massive changes were not tolerated once a general code began to function early in evolution.

Experiments with Synthetic mRNAs and Trinucleotides Break the Genetic Code

The recognition of the existence of mRNA and how it functions led to the solution of the genetic code—one of the great triumphs of modern biochemistry. The underlying experimental work was largely carried out with cell-free extracts from bacteria. All the necessary components for protein synthesis except mRNA (tRNAs, ribosomes, amino acids, and the energy-rich nucleotides ATP and GTP) were present in these extracts. On the addition of chemically defined synthetic mRNAs, the extracts formed specific polypeptides. For example, synthetic mRNA composed only of U residues yielded polypeptides made up only of phenylalanine. Thus it was concluded that UUU codes for phenylalanine. Likewise, poly C and poly A coded for single amino acids (Figure 4-28). (Poly G did not work because it assumes an unusable stacked structure.)

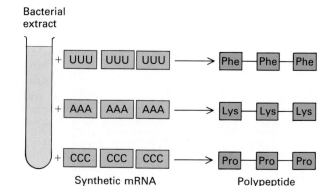

▲ FIGURE 4-28   The genetic code was worked out largely by using bacterial extracts that contained all the components necessary for protein synthesis except mRNA. When synthetic mRNAs consisting entirely of a single type of nucleotide were added to the extracts, polypeptides composed of a single type of amino acid formed as indicated. [See M. W. Nirenberg and J. H. Matthei, 1961, *Proc. Nat'l. Acad. Sci. USA* **47**:1588.]

Next, synthetic mRNAs with alternating bases were used. For example, the polypeptide made in response to the synthetic mRNA,

$$...A\,C\,A\,C\,A\,C\,A\,C\,A\,C\,A\,C...$$

contained alternating threonine and histidine residues. A further experiment was needed to determine whether threonine was encoded by ACA and histidine by CAC or vice versa. An mRNA made of repeated sequences of AAC,

$$...A\,A\,C\,A\,A\,C\,A\,A\,C\,A\,A\,C\,A...$$

was found to stimulate the synthesis of three kinds of polypeptide chains: all asparagine, all threonine, and all glutamine. Apparently, the decoding mechanism could start at any nucleotide and read the mRNA as one of three different repeated codons, depending on where it started: all

TABLE 4-5 Unusual Codon Usage in Nuclear and Mitochondrial Genes			
Codon	Universal Code	Unusual Code	Occurrence*
UGA	Stop	Trp	*Mycoplasma, Spiroplasma,* mitochondria of many species
CUG	Leu	Thr	Mitochondria in yeasts
UAA, UAG	Stop	Gln	*Acetabularia, Tetrahymena,* Paramecium, etc.
UGA	Stop	Cys	*Euplotes*

*Unusual code is used in nuclear genes of the listed organisms and in mitochondrial genes as indicated.
SOURCE: S. Osawa et al., 1992, *Microbiol. Rev.* **56**:229.

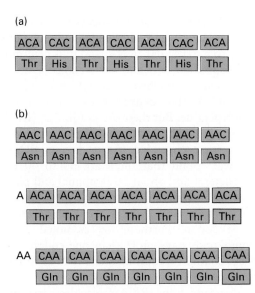

(a)

ACA	CAC	ACA	CAC	ACA	CAC	ACA
Thr	His	Thr	His	Thr	His	Thr

(b)

AAC	AAC	AAC	AAC	AAC	AAC	AAC
Asn	Asn	Asn	Asn	Asn	Asn	Asn

A | ACA | ACA | ACA | ACA | ACA | ACA | ACA |
  |-----|-----|-----|-----|-----|-----|-----|
  | Thr | Thr | Thr | Thr | Thr | Thr | Thr |

AA | CAA | CAA | CAA | CAA | CAA | CAA | CAA |
   |-----|-----|-----|-----|-----|-----|-----|
   | Gln | Gln | Gln | Gln | Gln | Gln | Gln |

◄ FIGURE 4-29   Assigning the codons for threonine and histidine. (a) When a synthetic mRNA with alternating A and C residues was added to a protein-synthesizing bacterial extract, the resulting polypeptide contained alternating threonine and histidine residues. (b) To determine that the codon assignment shown in (a) was correct (rather than the alternative ACA = His and CAC = Thr), an mRNA consisting of repeats of AAC was tested. This mRNA, which can be read in three frames, yielded three types of polypeptides. Since only the ACA codon was common to both experiments, it must encode threonine; thus CAC must encode histidine in (a). The assignments AAC = asparagine (Asn) and CAA = glutamine (Gln) were derived from additional experiments. [See H. G. Korana, 1968, reprinted in *Nobel Lectures: Physiology or Medicine (1963–1970)*, Elsevier (1973), p. 341.]

AAC, all ACA, or all CAA. The only codon in common between the two synthetic mRNAs just discussed was ACA, and the only amino acid in common in the polypeptide products was threonine. Therefore, ACA was assigned to threonine (Figure 4-29). Comparisons of the coding capacity of many such mixed polynucleotides revealed a substantial part of the genetic code.

In a second type of experiment with bacterial extracts, a specific RNA trinucleotide caused tRNA bearing a specific amino acid to bind to a ribosome. Thus it was possible to match every amino acid with at least one trinucleotide (Figure 4-30). In all, 61 of the 64 possible trinucleotides were assigned as codons for the 20 amino acids. Trinucleotides UAA, UGA, and UAG did not encode amino acids.

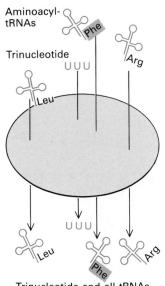

Trinucleotide and all tRNAs pass through filter

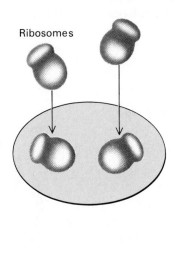

Ribosomes stick to filter

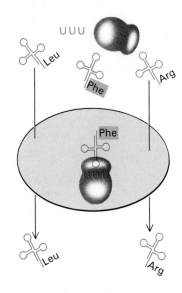

Complex of ribosome, UUU, and Phe-tRNA sticks to filter

▲ FIGURE 4-30   Marshall Nirenberg and his collaborators used extracts of *E. coli* to which they added chemically synthesized trinucleotides to decipher the entire genetic code. They prepared 20 bacterial extracts containing all 20 aminoacyl-tRNAs (tRNAs with an amino acid attached). In each extract sample, a different amino acid was radioactively labeled (green); the other 19 amino acids were present on tRNAs but remained unlabeled. Aminoacyl-tRNAs and trinucleotides passed through a nitrocellulose filter without binding (*left panel*); ribosomes, however, did bind to the filter (*center panel*). Each possible trinucleotide was tested separately for its ability to attract a specific tRNA by adding it with ribosomes to samples from each of the 20 aminoacyl-tRNA mixtures. The sample was then filtered. If the added trinucleotide caused the radiolabeled aminoacyl-tRNA to bind to the ribosome, then radioactivity would be detected on the filter (a positive test); otherwise, the label would pass through the filter (a negative test). By synthesizing and testing all possible trinucleotides, the researchers were able to match all 20 amino acids with one or more codons (e.g., phenylalanine with UUU as shown here). [See M. W. Nirenberg and P. Leder, 1964, *Science* **145**:1399.]

When synthetic mRNAs were used to direct in vitro protein synthesis, polypeptides formed much more inefficiently than when natural mRNAs were used, and the lengths of the newly made polypeptide chains were variable. The coding ability of the synthetic mRNA had produced reliable results in experiments designed to decipher the code, but true proteins were programmed by mRNA only when natural mRNAs were added to the bacterial extracts. The first successful synthesis of a specific protein occurred when the mRNA of bacteriophage F2 (a virus) was added to bacterial extracts and the coat, or capsid, protein (the "packaging" protein that covers the virus particle) was formed. With the use of real mRNAs, it was soon discovered that AUG encoded methionine at the start of almost all proteins and that the three trinucleotides UAA, UGA, and UAG that did not encode any amino acid were "stop" (terminator) codons.

Folded Structure of tRNA Is Integral to Its Function

The next advance in understanding how the code embodied in mRNA is translated into protein came from discoveries about the function and structure of transfer RNAs. All tRNAs have two functions: to chemically link to a particular amino acid and to recognize a codon in mRNA so that the amino acid can be added to a growing peptide chain. Each tRNA molecule is recognized by one and only

one of the 20 enzymes called *aminoacyl-tRNA synthetases*. Each of these enzymes can add one and only one of the 20 amino acids to a particular tRNA, and thus determines the first level of specificity in amino acid choice. Once its correct amino acid is attached, a tRNA then recognizes a codon in mRNA, thus delivering its amino acid to the growing polypeptide. But there are not just 20 tRNAs, one for each amino acid, nor are there as many as 61 total tRNAs, one for each codon. As studies on tRNAs proceeded, it was found that there are 30 to 40 different tRNAs in bacteria and about 50 in animal and plant cells. Thus there are several different tRNAs for most amino acids in most cells, and there is not a unique tRNA for every single codon. Furthermore, as noted previously, many amino acids have more than one codon—some as many as six codons—requiring some tRNAs to recognize more than one codon.

To understand how tRNA molecules carry out their functions, let us consider the structure of these small RNAs, which are 70–80 nucleotides long. In solution all tRNA molecules are folded into a similar three-dimensional structure, a *stem-loop* arrangement that resembles a cloverleaf when drawn in two dimensions. The four stems are short double helices stabilized by Watson-Crick base pairing; three of the four stems have loops containing seven or eight bases at their ends, while the remaining, unlooped stem contains the free 3′ and 5′ ends of the chain (Figure 4-31a). At the center of one loop, three nucleotides termed

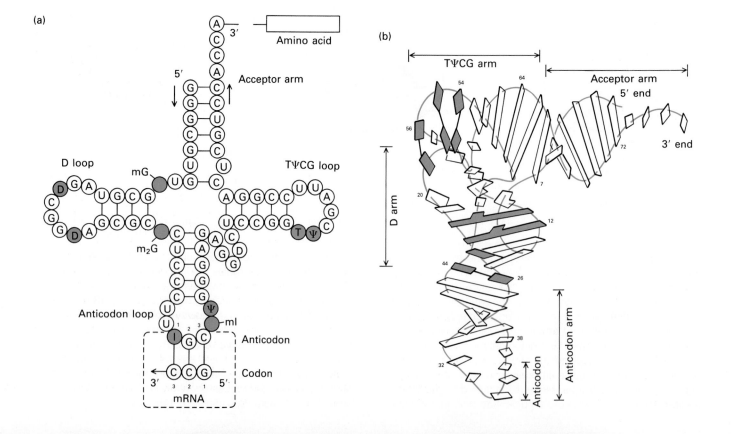

the *anticodon* can form base pairs with the three complementary nucleotides forming a codon in mRNA. As discussed later, an amino acid can be attached to the unlooped *amino acid acceptor stem*. Viewed in three dimensions, a tRNA molecule has an L shape with the anticodon loop and acceptor stem forming the ends of the two arms (Figure 4-31b).

After a tRNA molecule is synthesized, several of its nucleic acid bases typically are modified. For example, most tRNAs are synthesized with a four-base sequence of UUCG near the middle of the molecule. The first uridylate is methylated to become a thymidylate; the second is rearranged into a pseudouridylate (abbreviated Ψ), in which the ribose is attached to carbon 5 instead of to nitrogen 1. These modifications produce a characteristic TΨCG loop in an unpaired region at approximately the same position in nearly all tRNAs (see Figure 4-31a).

If perfect Watson-Crick base pairing were demanded between codon and anticodon, cells would have to contain 61 different tRNA species, one for each codon that specifies an amino acid. As noted above, however, cells contain fewer tRNAs. The explanation for this finding lies in the capability of a single tRNA anticodon to recognize more than one, but not every, codon corresponding to a given amino acid. This broader recognition can occur because of so-called *wobble* pairing between the third base in a codon and the first base in the corresponding anticodon (Figure 4-32). Although the first and second bases of a codon are paired with the third and second bases of the anticodon, respectively, in standard Watson-Crick fashion, a number of nonstandard interactions can occur between bases in the wobble position (Figure 4-33). Particularly important is

the G-U base pair, which structurally fits almost as well as the standard G-C pair. Thus, a given anticodon with G in the first (wobble) position can base-pair with the two corresponding codons that have either pyrimidine (C or U) in the third position (see Figure 4-32). For example, the phenylalanine codons UUU and UUC (5′ → 3′) are both recognized by the tRNA that has GAA (5′ → 3′) as the anticodon. In fact, all mRNA triplets of the type NNPyr (N = any base; Pyr = pyrimidine) encode a single amino acid and are decoded by a tRNA with G in the first anticodon position. Although A rarely is found in the anticodon wobble position, many tRNAs have an unusual modified base

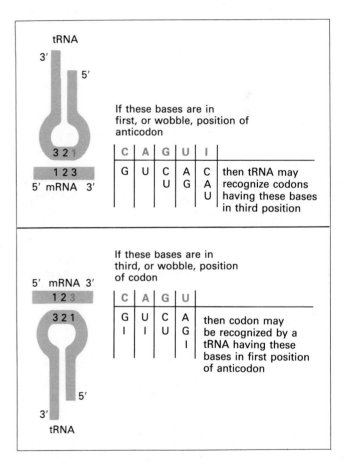

▲ **FIGURE 4-32** The first and second bases in an mRNA codon form Watson-Crick base pairs with the third and second bases, respectively, of a tRNA anticodon. However, the base in the third (or wobble) position of an mRNA codon often forms a nonstandard base pair with the base in the first (or wobble) position of a tRNA anticodon. Wobble pairing allows a tRNA to recognize more than one mRNA codon; conversely, it allows a codon to be recognized by more than one kind of tRNA, although each tRNA will bear the same amino acid. Note that a tRNA with I (inosine) in the wobble position can "read" (become paired with) three different codons (see Figure 4-33), and a tRNA with G or U in the wobble position can read two codons. Although A is theoretically possible in the wobble position of the anticodon, it is almost never found in nature.

◄ **FIGURE 4-31** (a) The primary structure of yeast alanine tRNA (tRNA^Ala), the first such sequence determined. This molecule is synthesized from the nucleotides A, C, G, and U, but some of the nucleotides, shown in red, are modified after synthesis: D = dihydrouridine, I = inosine, T = thymine, Ψ = pseudouridine, and m = methyl group. The primary structure can be diagrammed as a cloverleaf consisting of four base-paired stems and three loops: the D loop (for dihydrouridine, a virtually constant constituent of this loop); the anticodon loop; and the TΨCG loop (for thymidylate, pseudouridylate, cytidylate, and guanylate, which are almost always present in sequence in this loop). As discussed later, attachment of an amino acid to the acceptor arm yields an aminoacyl-tRNA. (b) A drawing of the three-dimensional structure of yeast tRNA^Phe with the standard Watson-Crick base pairs shown as rectangular bars connecting the segments of the folded molecule. A number of nonstandard molecular interactions (red) help to stabilize the L-shaped molecule. [Part (a), see R. W. Holly et al., 1965, *Science* **147**:1462. Part (b) prepared from x-ray crystallographic data; see J. L. Sussman and S. H. Kim, 1976, *Science* **192**:853.]

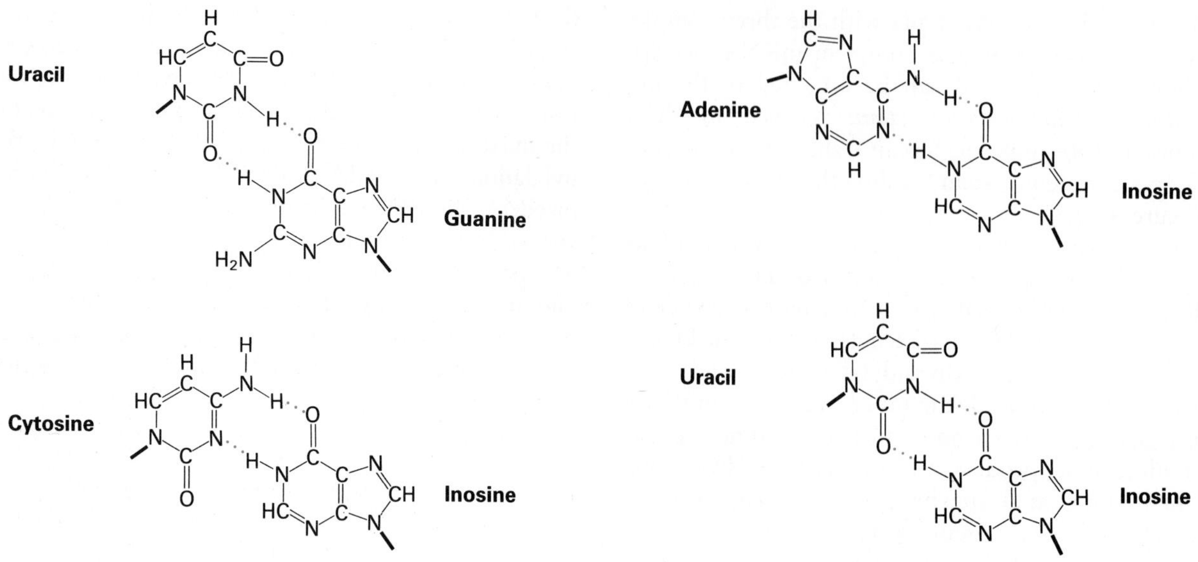

▲ FIGURE 4-33   The nonstandard, wobble base pairs U-G, C-I, A-I, and U-I. The heavy black lines indicate the bonds by which the nitrogenous bases attach to the 1' carbon of ribose.

in this position. One of the most important unusual wobble-position bases in plants and animals is *inosine,* a deaminated product of adenine, which can base-pair with A, C, and U (see Figure 4-33). A given transfer RNA with inosine in the wobble position thus can recognize the corresponding mRNA triplets with A, C, or U in the third codon position (see Figure 4-32). For this reason, inosine-containing tRNAs are heavily employed in translation of the synonymous codons that specify a single amino acid (e.g., the six codons for leucine; see Table 4-4).

Aminoacyl-tRNA Synthetases Activate tRNA

We now can see how a particular tRNA recognizes the codon(s) specifying a given amino acid, but how does the tRNA molecule become attached to the appropriate amino acid in the first place? As mentioned previously, this crucial process is catalyzed by the 20 different aminoacyl-tRNA synthetases (ARSs), each of which recognizes one amino acid and all its compatible, or *cognate,* tRNAs. These coupling enzymes link an amino acid to the free 2' or 3' hydroxyl of the ribose of the adenosine at the 3'-terminus of tRNA molecules. This two-step linkage reaction requires the cleavage of an ATP molecule:

$$\text{Enzyme} + \text{amino acid} + \text{ATP} \xrightarrow{\text{Mg}^{2+}}$$
$$\text{enzyme(aminoacyl-AMP)} + \text{PP}_i \qquad (1)$$

$$\text{tRNA} + \text{enzyme(aminoacyl-AMP)} \longrightarrow$$
$$\text{aminoacyl-tRNA} + \text{AMP} + \text{enzyme} \qquad (2)$$

In the first step, the enzyme, the amino acid, and ATP form a complex, releasing PP_i; in the second step, the aminoacyl group is transferred from the enzyme complex to the tRNA with release of AMP (Figure 4-34). About half the ARSs transfer the aminoacyl group to the 2' hydroxyl of the terminal adenosine (class I), and about half, to the 3' hydroxyl (class II). The resulting aminoacyl-tRNA retains the energy of the ATP, and the amino acid residue is said to be *activated.* The equilibrium of the reaction is driven further toward activation of the amino acid because pyrophosphatase then splits the high-energy phosphoanhydride bond in pyrophosphate. The overall reaction is

$$\text{Amino acid} + \text{ATP} + \text{tRNA} \xrightarrow{\text{enzyme}}$$
$$\text{aminoacyl-tRNA} + \text{AMP} + 2\text{P}_i$$

Recently the amino acid sequence of almost all the ARSs from *E. coli* and many from eukaryotes has been determined. These analyses have shown that representative enzymes in each of the two classes possess repeated sequence "motifs" characteristic of its class (Table 4-6). The three-dimensional structure of an ARS complexed with its cognate tRNA has been solved for GlnRS (a class I enzyme) and AspRS (a class II enzyme). As the models in Figure 4-35 show, GlnRS folds around its cognate tRNA differently than AspRS does. Thus during evolution two different types of aminoacyl-tRNA synthetases arose. The three-dimensional structures of other ARS-tRNA complexes are being determined.

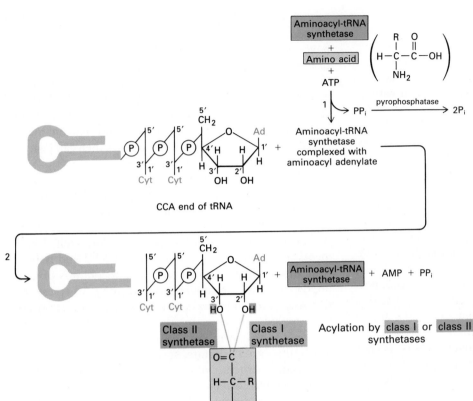

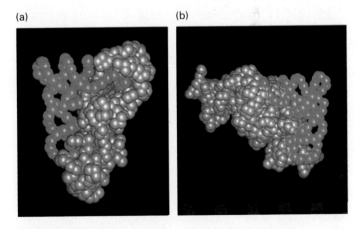

◄ FIGURE 4-34   Aminoacylation of tRNA. Amino acids are covalently linked to cognate tRNAs by amino-acyl-tRNA synthetases. Each of these enzymes recognizes one kind of amino acid and all the individual tRNAs that recognize codons for that amino acid. The two-step reaction requires energy from the hydrolysis of ATP. The equilibrium of the overall reaction favors the indicated products because the pyrophosphate (PP_i) released in step 1 is converted to inorganic phosphate (P_i) by a pyro-phosphatase. The 3' end of all tRNAs, to which the amino acid attaches, has the sequence CCA. Class I synthetases (purple) attach the amino acid to the 2' hydroxyl of the terminal adenylate in tRNA; class II synthetases (green) attach the amino acid to the 3' hydroxyl. (Ad = adenine; Cyt = cytosine).

TABLE 4-6   Classification of Aminoacyl-tRNA Synthetases (ARSs)*

Class I Synthetases (2'—OH)	Class II Synthetases (3'—OH)
Glu (α)	Gly ($\alpha_2\beta_2$)
Gln (α)	Ala (α_4)
Arg (α)	Pro (α_2)
Val (α)	Ser (α_2)
Ile (α)	Thr (α_2)
Leu (α)	Asp (α_2)??
Met (α_2)	Asn (α_2)
Tyr (α_2)??	His (α_2)
Trp (α_2)?	Lys (α_2)

*Class I ARSs add amino acids to the 2' hydroxyl at the 3' end of tRNAs; class II ARSs, to the 3' hydroxyl. The α or β in parentheses refer to the subunit composition of each enzyme, with α and β designating different polypeptides. The sequences of the α and β chains in the various enzymes are different.
SOURCE: G. Eriani et al., 1990, *Nature* 347:203.

(a)   (b)

▲ FIGURE 4-35   Computer-generated models of the complexes between glutamine and aspartate aminoacyl-tRNA synthetases (GlnRS and AspRS) and tRNA^{Gln} and tRNA^{Asp}, respectively. GlnRS is a class I enzyme; AspRS is a class II enzyme. Only the phosphates and C_α backbones are shown. The tRNA-synthetase interaction differs in the two complexes: (a) GlnRS contacts tRNA^{Gln} on the D loop and the minor groove of the acceptor stem, whereas (b) AspRS contacts tRNA^{Asp} on the anticodon loop and major groove of the acceptor stem (see Figure 4-31). [From M. Ruff et al., 1991, *Science* 252:1682.]

Each tRNA Molecule Is Recognized by a Specific Aminoacyl-tRNA Synthetase

The identification of a tRNA by its cognate ARS is just as important to the correct translation of the genetic code as codon-anticodon pairing. Once a tRNA is loaded with an amino acid, codon-anticodon pairing directs the tRNA into the proper ribosome site; if the wrong amino acid is attached to the tRNA, an error in protein synthesis results.

A classic experiment demonstrated this phenomenon. A cysteine residue already attached to a tRNA^{Cys} (a cysteine-specific tRNA) was chemically changed into alanine, so that the cysteinyl-tRNA^{Cys} became alanyl-tRNA^{Cys} (Figure 4-36). When used in the synthesis of a polypeptide, this aminoacyl-tRNA added alanine to the growing chain in response to a *cysteine*, not an *alanine*, codon. In the completed polypeptide, all the usual cysteine residues had been replaced with alanine, proving that only the anticodon of an aminoacyl-tRNA—and not the attached amino acid—is involved in the recognition step that causes the amino acid to be incorporated into a protein.

How each ARS identifies cognate tRNAs is not yet understood, but important recent progress has been achieved. One ARS can add the same amino acid to two (or more) different tRNAs with different anticodons encoding the same amino acid. Therefore each of these tRNAs must have a similar binding site that is recognized by the synthetase. By techniques described in Chapter 7, DNA molecules up to about 100 nucleotides long can be synthesized chemically. One approach for studying the amino acid *identity* of tRNAs is to produce synthetic genes for tRNAs with normal and various mutant sequences; the normal and mutant tRNAs produced by such synthetic genes then can be tested for their ability to bind purified ARSs.

Although very probably no single structure or sequence completely determines a specific tRNA identity, some important contributions to the identity of several *E. coli* tRNAs are known. Perhaps the most logical identity site in a tRNA molecule is the anticodon itself. Experiments in which the anticodons of tRNA^{Met} and tRNA^{Val} were interchanged showed that the anticodon is of major importance in determining the identity of these two tRNAs. More recent evidence is provided by the x-ray crystallographic analysis of the tRNA^{Gln}-GlnRS complex (see Figure 4-35a). In this complex, each of the anticodon bases neatly fits into a separate, specific "pocket" in the three-dimensional structure of GlnRS. Thus this ARS specifically recognizes the correct anticodon.

However, the anticodon may not be the principal identity element in other tRNAs. Figure 4-37 shows identity elements that have been identified in several different tRNAs. The simplest case is presented by tRNA^{Ala}: a single base pair (G3-U70) in the acceptor stem is necessary and sufficient for recognition of this tRNA by its cognate ARS. Furthermore, insertion of this single base pair into the acceptor stem of a minor species of tRNA^{Phe} converts its identity from phenylalanine to alanine. In several other tRNAs, combinations of nucleotides have been shown to be important for recognition by cognate ARSs. Solution of the three-dimensional structure of additional tRNAs with their cognate ARS should provide a clear molecular understanding of the rules governing this critical step in protein synthesis.

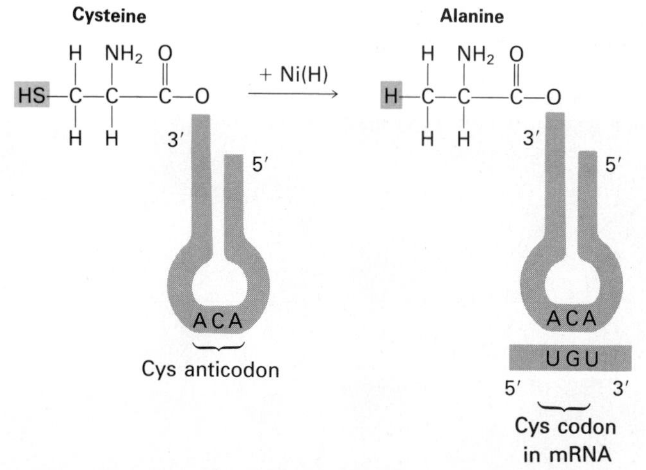

Ribosomes Are Protein-Synthesizing Machines

The highly specific chemical interactions involved in translating mRNA into protein do not take place in free solution inside a cell. This critical process, resulting in formation of more than one million peptide bonds each second in an average mammalian cell, requires the interaction of many chemical groups on RNAs and assisting proteins. Protein synthesis would be very inefficient if each of these participants had to react in free solution: simultaneous collisions between the necessary components of the reaction would be so rare that the rate of amino acid polymerization would be very slow indeed. Instead, the mRNA with its encoded information and the individual tRNAs loaded

▲ FIGURE 4-36 Experimental demonstration that codon-anticodon pairing directs addition of activated amino acids during protein synthesis. Cysteine was activated by attachment to tRNA^{Cys} and then chemically changed into alanine. When the altered aminoacyl-tRNA was added to an in vitro protein-synthesizing system, the resulting polypeptide contained alanine wherever the mRNA contained the UGU codon for cysteine. Thus, correct translation of the genetic code into protein depends on two tRNA recognition events: pairing with the proper mRNA codon and activation of the amino acid specified by that codon. [See F. Chappeville et al., 1962, *Proc. Nat'l. Acad. Sci. USA* **48**:1086.]

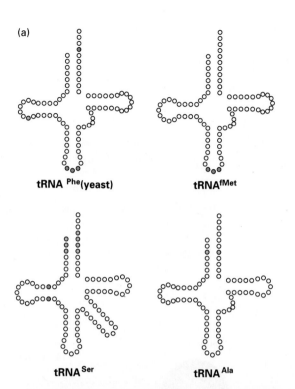

tRNA^{Phe}(yeast)   tRNA^{fMet}

tRNA^{Ser} tRNA^{Ala}

with their correct amino acids are brought together by their mutual binding to the most abundant RNA-protein complex in the cell—the ribosome. This two-part machine directs the elongation of a polypeptide at a rate of three to five amino acids added per second. Small proteins of 100–200 amino acids are therefore made in a minute or less. However the largest known protein, *titin*, found in muscle, has 30,000 amino acid residues and requires 2–3 h to make. Thus the machine that accomplishes this task must be precise and persistent.

With the aid of the electron microscope, ribosomes were first discovered as discrete, rounded structures prominent in animal tissues secreting large amounts of protein; initially, however, they were not known to play a role in protein synthesis. When reasonably pure preparations were achieved, ribosomes were seen to be very consistent in size. The role of ribosomes as the sites of protein synthesis was demonstrated by radiolabeling experiments, diagrammed in Figure 4-38, showing that labeled amino

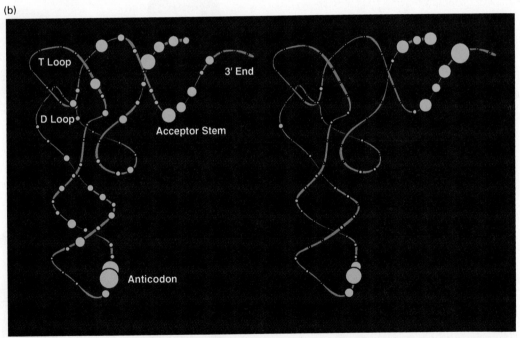

T Loop

D Loop

3' End

Acceptor Stem

Anticodon

▲ FIGURE 4-37   Identity elements in tRNA involved in recognition by aminoacyl-tRNA synthetases (ARSs). (a) Major identity elements in four tRNAs are indicated by colored circles denoting nucleotides that have been shown to be involved in interacting with the cognate ARS. Each nucleotide in the cloverleaf diagram of a tRNA molecule is represented by a circle. In each case, additional identity elements may yet be discovered. (b) The 67 known tRNA sequences in *E. coli* were compared by computer analysis; the conserved nucleotides in different tRNAs that recognize the same amino acid are shown as yellow circles in the left drawing with the tRNA chain in blue; increasing size indicates increasing conservation of a base at a given position. In the right drawing, nucleotides experimentally demonstrated to have a role in identity are shown as yellow circles. In this case, the circle size indicates the relative frequency that a given position acts as an identity element. [See L. D. H. Schulman and J. Abelson, 1988, *Science* **240**:1590. Part (a) adapted from W. H. McClain and H. B. Nicholas, Jr., 1987, *J. Mol. Biol.* **194**:635; part (b) from W. H. McClain, 1993, *J. Mol. Biol.* **234**:257.]

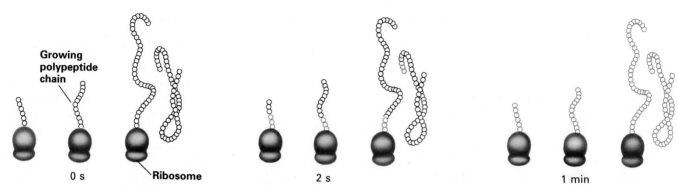

▲ FIGURE 4-38   Experimental demonstration that ribosomes are the sites of protein synthesis. When cells are exposed to radioactive amino acids (red) for very short times, the labeled amino acids are first incorporated into growing peptide chains associated with ribosomes. After longer exposure times, completed proteins, free of the ribosomes, contain more and more of the labeled amino acids. [See K. McQuillen, R. B. Roberts, and R. J. Britten, 1959, *Proc. Nat'l. Acad. Sci. USA* **45**:1437.]

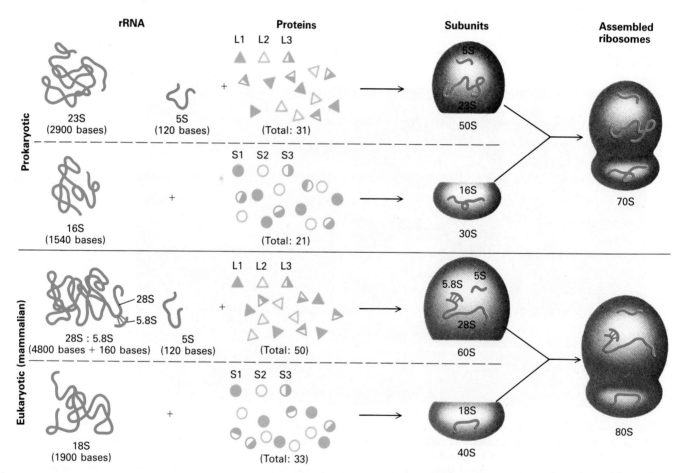

▲ FIGURE 4-39   In all cells, each ribosome consists of a large and a small subunit. The different subunits contain rRNAs of different lengths as well as a number of different proteins (indicated by different shadings). All ribosomes contain two major rRNA molecules. Prokaryotic ribosomes also have one small 5S rRNA about 120 bases long. Eukaryotic ribosomes have two small rRNAs: a 5S molecule similar to the prokaryotic 5S, and a 5.8S molecule 160 bases long. The proteins are named L1, L2, etc., and S1, S2, etc., depending on whether they are found in the large or the small subunit. Some cell organelles also have ribosomes: chloroplast ribosomes are similar to prokaryotic ribosomes; ribosomes in mitochondria have smaller RNAs and fewer proteins than prokaryotic ribosomes.

acids first became associated with growing polypeptide chains on ribosomes before the label appeared in finished chains.

The ribosome is a complex composed of individual ribosomal RNA (rRNA) molecules and more than 50 proteins, organized into a large subunit and a small subunit (Figure 4-39). The proteins in the two subunits differ, as do the molecules of rRNA. The large ribosomal subunit contains a larger rRNA molecule; the small subunit contains a small rRNA. The subunits and the rRNA molecules are commonly designated in terms of the Svedberg (S), a measure of the sedimentation rate of suspended particles centrifuged under standard conditions (Chapter 3). The lengths of the rRNA molecules, the quantity of proteins in each subunit, and consequently the sizes of the subunits differ in prokaryotic and eukaryotic cells. Perhaps of more interest, however, are the great structural and functional similarities among ribosomes from all species. This consistency is another reflection of the common evolutionary origin of the most basic constituents of living cells.

The sequences of the small and large ribosomal RNAs from several hundred organisms are now known. Although the primary nucleotide sequences of these rRNAs vary considerably, the same parts of each type of rRNA can be paired into stem-loops, giving each rRNA the same overall structural pattern in all organisms. Proof of such stem-loop structures in rRNA in solution was obtained by treating rRNA with chemical agents that cross-link paired bases and then digesting the samples with enzymes that destroy single-stranded rRNA, but not any cross-linked (base-paired) regions. Finally, the intact, cross-linked rRNA that remained was collected and sequenced, thus identifying the stem-loops in the original rRNA. Forty stem-loops have been located at similar positions along the sequences of prokaryotic and eukaryotic small rRNAs (Figure 4-40) and

(a)

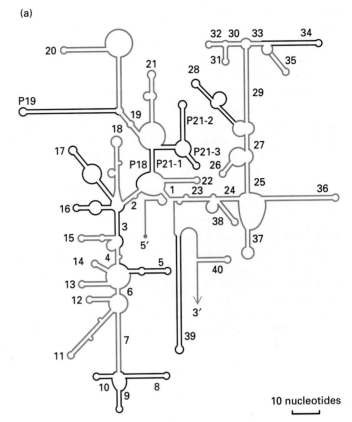

10 nucleotides

(b)

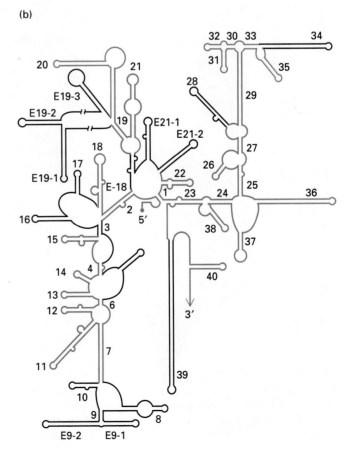

▶ FIGURE 4-40   Secondary structural maps of small ribosomal RNAs. Based on the known primary sequences of several hundred small rRNAs, the folded stem-loop structures shown here would have maximal base-pairing in (a) prokaryotic and (b) eukaryotic small rRNAs. The existence of these stem-loops in rRNA in solution has been demonstrated experimentally as discussed in the text. The stems, which are base-paired double helices, numbered 1–40 have been found in both prokaryotic and eukaryotic small rRNAs. In general, the length and position of the 40 helices are well conserved in all species, although the exact sequence varies from species to species. The most highly conserved regions are represented as red lines. Stem-loops found only in prokaryotes or only in eukaryotes are preceded by P or E, respectively. [Adapted from E. Huysmans and R. DeWachter, 1987, *Nucleic Acids Res.* **14**:73.]

an even larger number of regularly observed stem-loops exist in the large rRNAs. It seems clear that the fundamental protein-synthesizing machinery in all present-day cells arose only once and has been modified about a common plan during evolution.

Despite the complexity of the ribosome, great progress has been made in unraveling the overall structure of bacterial ribosomes and in identifying reactive sites in the rRNA that bind specific proteins, mRNA, and tRNA and participate in important steps in protein synthesis. Each constituent rRNA and protein has been purified and its sequence determined. Furthermore, the bacterial ribosomal particle can be reassembled when the isolated proteins and rRNA are mixed, and the reassembled ribosomes synthesize protein. The dimensions and overall structure of the ribosome, including prominent surface features, has been revealed by high-resolution electron-microscope studies, especially those using antibodies to identify the location of specific proteins (Figure 4-41).

More recent experiments have revealed which ribosomal proteins are neighbors and which segments of rRNA

are bound to which proteins. For example, treatment of ribosomes with ribonucleases, which digest RNA, releases RNA fragments plus attached proteins. Both the individual proteins and pieces of RNA can then be specifically identified. Perhaps the most powerful experiments have utilized chemical reagents to modify single-stranded RNA that is unprotected by binding either to protein or to another RNA. If the total sequence of the RNA is known, then the location of the modified nucleotides can be located within the molecule. (This technique, known as "footprinting," is also useful with DNA; it is described in Chapter 6.) Experiments of this type have been used to identify the specific bases in rRNA that are protected by tRNA as the two types of RNA cooperate in translation. The capacity of individual proteins or tRNAs to protect specific long regions and individual nucleotides in rRNA has allowed models of the large and small ribosomal subunits of *E. coli* to be constructed. One of the major advances from such studies has been the localization of a region of large ribosomal RNA where the aminoacylated ends of tRNA bind to participate in peptide synthesis.

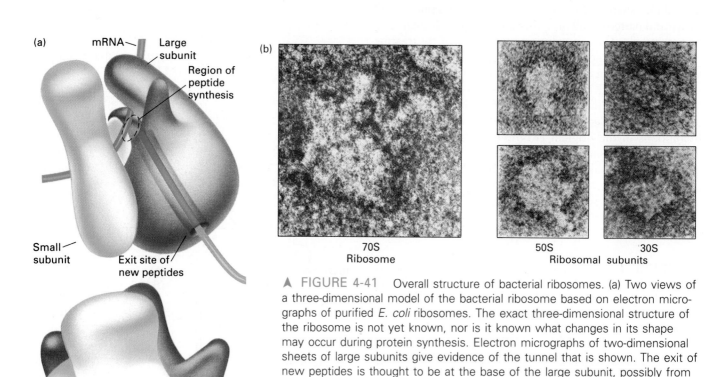

▲ FIGURE 4-41   Overall structure of bacterial ribosomes. (a) Two views of a three-dimensional model of the bacterial ribosome based on electron micrographs of purified *E. coli* ribosomes. The exact three-dimensional structure of the ribosome is not yet known, nor is it known what changes in its shape may occur during protein synthesis. Electron micrographs of two-dimensional sheets of large subunits give evidence of the tunnel that is shown. The exit of new peptides is thought to be at the base of the large subunit, possibly from the tunnel. The region of peptide synthesis has been inferred from the binding of aminoacyl- and peptidyl-tRNAs to specific rRNA regions and to the cross-linking to certain proteins that are located in this region. The path of mRNA is reasonable but is conjecture. (b) Electron micrographs of a 70S ribosome and of 50S and 30S ribosomal subunits. The subunits are viewed from different angles. [Drawings after J. A. Lake. Photographs from J. A. Lake, 1976, *J. Mol. Biol.* **105**:131. See A. Yonath, K. R. Leonard, and H. G. Wittman, 1987, *Science* **236**:813.]

➤ *The Steps in Protein Synthesis*

Protein synthesis usually is considered in three stages—initiation, elongation, and termination. The various components involved in each stage and the distinct biochemical events that occur are diagrammed in Figure 4-42.

AUG Is the Initiation Signal in mRNA

The first event of the initiation stage is attachment of a free molecule of methionine (Met) to the end of a $tRNA^{Met}$ by a specific aminoacyl-tRNA synthetase. There are at least two types of $tRNA^{Met}$: one designated $tRNA_i^{Met}$ that can initiate protein synthesis, and another that can incorporate methionine within the growing protein chain. The same enzyme, methionyl-tRNA synthetase (MetRS), can attach methionine to both tRNAs, but only methionyl-$tRNA_i^{Met}$ (i.e., activated methionine attached to $tRNA_i^{Met}$) can bind to the small ribosomal subunit to begin the process of protein synthesis. (In bacteria, the amino group of the methionine of methionyl-$tRNA_i^{Met}$ is modified by the addition of a formyl group and is sometimes designated N-formylmethionyl-$tRNA_i^{Met}$. However, methionyl-$tRNA_i^{Met}$ [abbreviated Met-$tRNA_i^{Met}$] is commonly used to designate the initiator tRNA in all cells.) The Met-$tRNA_i^{Met}$ (assisted by a protein-GTP complex) and a small ribosomal subunit bind to the mRNA at a specific site most often located quite near the AUG initiation codon (see Figure 4-42a).

In most bacteria, the small ribosomal subunit identifies initiation sites through the interaction of short nucleotide sequences in the small 16S rRNA and the mRNA. On the mRNA this *Shine-Dalgarno sequence,* named for its discoverers, is near a protein start site and complementary to a sequence at or very near the 3′ end of the 16S-rRNA molecule. Thus

mRNA 5′–UAAGGAGG–(5–10 nucleotides)–AUG
                | | | | | | | |
      HO–AUUCCUCC–(≈1400 nucleotides)–5′ rRNA

Every protein start signal does not match this rRNA sequence exactly, but on average six out of eight nucleotides match. Thus bacterial rRNA plays a direct role in recruiting a ribosome to a protein start site on the mRNA. Bacterial ribosomes can initiate protein synthesis at these sites in the middle of mRNAs thousands of nucleotides long (see Figure 4-23a).

In eukaryotic cells the mechanism by which a small ribosomal subunit finds start sites is not fully understood. Almost all eukaryotic mRNAs have a single start site at which synthesis of a single protein begins (see Figure 4-23b). The first signal thought to be recognized is the 5′ cap present on all eukaryotic mRNAs (Figure 4-24); however, some viral mRNAs, which are translated by the host

cell machinery in infected eukaryotic cells, lack a 5′ cap. In this case recognition occurs with the aid of additional protein factors discussed below. Usually, after cap recognition, the bound ribosomal subunit then is thought to slide along the mRNA to locate an AUG. Frequently the first AUG is used, but the presence of certain nucleotides surrounding the initiating AUG greatly increases the effectiveness of initiation. This sequence, referred to as a *Kozak sequence,* for Marilyn Kozak, who showed its importance, is

mRNA 5′–ACCAUGG–

In support of this model for end recognition are experiments showing that artificial circular mRNAs that lack ends are not translated. Also mRNAs without the 5′ cap are translated very poorly, and mutations introduced into Kozak sequences greatly decrease initiation frequency. However, even though viral mRNAs, such as those of poliovirus, lack the cap structure, eukaryotic ribosomes find their start site more than 600 nucleotides from the 5′ end of the viral mRNA. In addition, persuasive evidence recently has been found for a few cellular mRNAs that are translated from initiation sites located within the mRNA as much as 200–600 nucleotides from the 5′ end. Therefore, it is generally agreed that initiation of translation of most eukaryotic mRNAs involves recognition of the cap followed either by use of the first downstream AUG or by the locating of a 5′-proximal AUG with a consensus sequence surrounding the AUG codon. But there also seem to be other nucleotide structures within eukaryotic mRNA that can, at least in a few cases, direct ribosome binding and protein initiation. Whether these internal start sites interact directly with rRNA as does the Shine-Dalgarno sequence in bacteria is not known.

Initiation Factors, tRNA, mRNA, and the Small Ribosomal Subunit Form an Initiation Complex

A group of proteins called *initiation factors* (IF) help the small ribosomal subunit find the initiation site. Without these proteins, the complex of mRNA, Met-$tRNA_i^{Met}$, and the small ribosomal subunit poised at the AUG initiation codon does not form; the required presence of each of these proteins for initiation to occur has allowed biochemists to identify and characterize all of them. Three initiation factors have been characterized from prokaryotic cells, and at least five (some of which have several protein components) are known in eukaryotic cells (see Figure 4-42). In prokaryotes, IF_3 is critical in finding the AUG. In eukaryotes, the large eIF_4 complex helps to ensure that the 5′ end of the mRNA is single-stranded (i.e., the factor serves both to

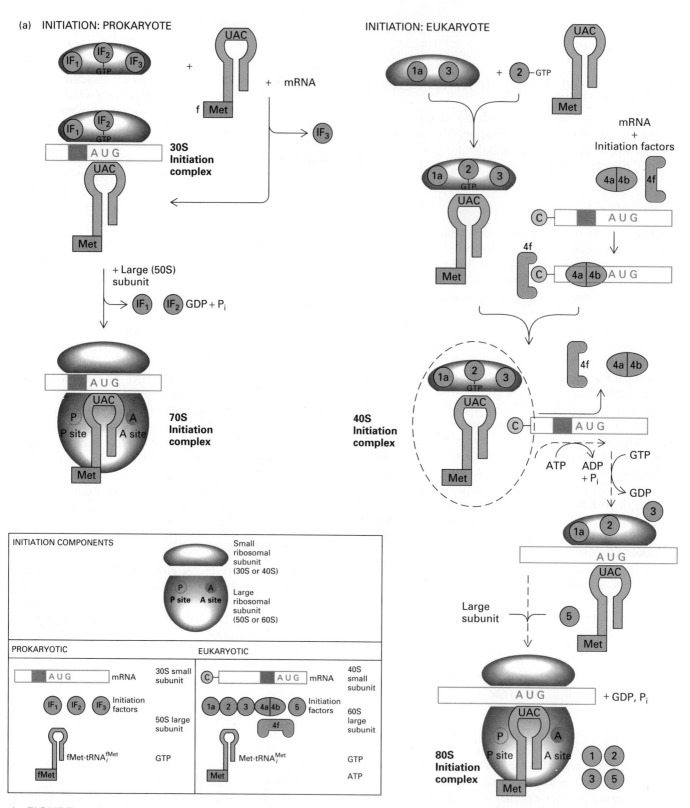

▲ FIGURE 4-42 The three stages of the translation of the genetic message from mRNA into protein. (a) *Initiation.* An initiation factor (IF₂ in prokaryotes and eIF₂ in eukaryotes) binds a molecule of GTP and a molecule of Met-tRNA$_i^{Met}$ to form a ternary complex. (The initiating methionine is formylated in prokaryotes.) This complex binds to other initiation factors (IF₃, eIF₃, etc.), to mRNA, and to the small ribosomal subunit to make the 30S initiation complex in prokaryotes and the 40S complex in eukaryotes. Once the Met-tRNA$_i^{Met}$ is positioned correctly at the AUG initiation codon, a large ribosomal subunit, using the ribosomal binding site on the mRNA (red box), joins to complete the 70S or

80S initiation complex. (During these steps, GTP is hydrolyzed in prokaryotes.) The Met-tRNA$_i^{Met}$ bearing the first amino acid is now bound to the ribosome at the P (peptidyl-tRNA) site. The initiation complex is ready to begin synthesis of the peptide chain. (b) *Elongation.* The growing polypeptide is always attached to the tRNA that brought in the last amino acid. A new aminoacyl-tRNA (Phe-tRNA^{Arg} here) binds to the ribosome at the A site. During elongation in prokaryotes, the protein complex Tu-Ts catalyzes the binding of each aminoacyl-tRNA to the ribosome. In eukaryotic cells, the similar proteins are called *elongation factors* (*EF₁* and *EF₁β*). An activated Tu-GTP complex binds to the TΨCG loop found

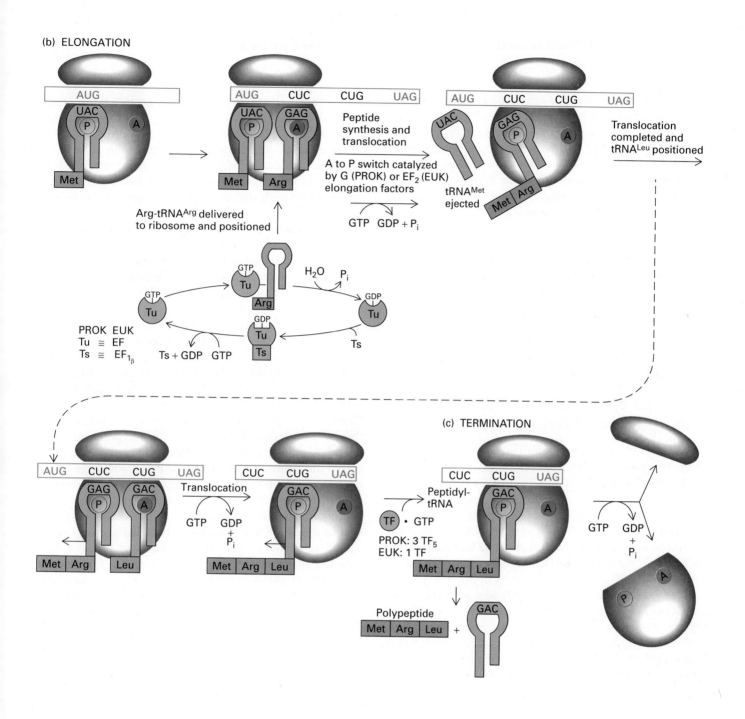

(b) ELONGATION

Arg-tRNA^{Arg} delivered to ribosome and positioned

Peptide synthesis and translocation

A to P switch catalyzed by G (PROK) or EF$_2$ (EUK) elongation factors

GTP → GDP + P$_i$

tRNA^{Met} ejected

Translocation completed and tRNA^{Leu} positioned

PROK EUK
Tu ≅ EF
Ts ≅ EF$_{1\beta}$

Ts + GDP GTP

Translocation

GTP → GDP + P$_i$

(c) TERMINATION

Peptidyl-tRNA

TF · GTP

PROK: 3 TF$_5$
EUK: 1 TF

GTP → GDP + P$_i$

Polypeptide
Met | Arg | Leu +

in all tRNAs and allows the tRNA to associate correctly with the ribosome at the A site. GTP is hydrolyzed, and the cycle is repeated when Ts aids in reactivating the Tu. After the incoming aminoacyl-tRNA is correctly placed in the A site— that is, once the codon-anticodon pairing is correct—the peptide chain (here, the first methionine) is transferred to the amino group of the newly arrived aminoacyl-tRNA, generating a peptidyl-tRNA that has acquired an additional amino acid. (In our example, the compound is methionyl-arginyl-tRNA^{Arg}.) At this stage, the peptidyl-tRNA is bound to the ribosome at the A site, and the ribosome moves one codon down the mRNA chain. (Here, the mRNA codons are illustrated with spaces separating them for convenience.) The translocation reaction is catalyzed (in bacteria by the elongation factor G

and in eukaryotes by the elongation factor EF$_2$) using energy from the hydrolysis of GTP. With this movement, the empty tRNA is released from the P site and the peptidyl-tRNA is shifted to the P site. This sequence of events is repeated for every amino acid added to the growing chain. Thus two molecules of GTP are used: one to position the tRNA and one in the addition of each amino acid during translocation. (c) *Termination*. When the ribosome arrives at the UAG codon, the translation is completed with the aid of termination factors (three proteins in bacteria, one in eukaryotes). Hydrolysis of the peptidyl-tRNA on the ribosome, with the release of the completed polypeptide and the last tRNA, is followed by dissociation of the two ribosomal subunits. This final step also requires GTP hydrolysis.

bind the cap and to unwind any secondary structure that may exist); it also ensures that the 5′ end is ready for the small subunit to locate the AUG. Only after the small subunit binds at the AUG is the large subunit added to the complex (see Figure 4-42a). GTP hydrolysis provides the energy required for many of the steps in protein synthesis. For example, the positioning of the large subunit in the bacterial initiation complex is aided by $IF_2 \cdot GTP$ and requires GTP hydrolysis.

In general, ribosomes and initiation factors from different bacterial species can be substituted for one another in protein synthesis in the test tube. The same is true for eukaryotic ribosomes and initiation factors, even when the mixture consists of, for example, a combined extract of human and yeast cells. But translation of prokaryotic mRNA by eukaryotic ribosomes (and vice versa) is poor.

Ribosomes Provide Three tRNA-binding Sites (A, P, and E) during Protein Elongation

There are two major tRNA-binding sites on the ribosome, one incoming site, called the A site (for aminoacyl-tRNA), and the P site (for peptidyl-tRNA), where new peptide bonds are formed. A third site, the E site, is transiently occupied by the deacylated end of the tRNA that has completed amino acid transfer and is about to be ejected. (This site is not shown in Figure 4-42 but is represented in Figure 4-43).

As the contributor of the first amino acid of the chain, the Met-tRNA$_i^{Met}$ enters the P position. In the hypothetical peptide shown in Figure 4-42b, the incoming arginyl-tRNA^{Arg} (the second amino acid in the chain is arginine) is then bound to the A position. A peptide bond is formed between the carboxyl group of the methionine and the amino group of the Arg-tRNA^{Arg} to make the dipeptide methionyl-arginyl-tRNA^{Arg}; in this process, the tRNA^{Arg} moves from the A site to the P site, displacing the now inactivated tRNA$_i^{Met}$. The hydrolysis of GTP furnishes the energy for this translocation of the peptidyl-tRNA. In our example, the third codon is (5′)CUG(3′); therefore, a leucinyl-tRNA^{Leu} with an anticodon of (3′)GAC(5′) or (3′)GAI(5′) is required. After the Leu-tRNA^{Leu} attaches to the A site, the leucine is incorporated into the new peptidyl chain and the translocation process repeats itself.

Recent experiments have added important refinements to our understanding of these events, particularly the movement of a tRNA molecule during the peptide-synthesis cycle. Recall that the L-shaped tRNA molecule has two major "business" ends: the anticodon loop at one end, which locates the correct codon on the mRNA, and the aminoacyl (acceptor) end, which donates an attached amino acid to the growing peptide (see Figure 4-31b). By testing which end of the tRNA molecule is associated with

which part of the ribosome (and thus protected from reacting with certain chemical reagents) at each step of the peptide-synthesis cycle, researchers have detected six tRNA binding states. In three of these states, both ends of the tRNA are associated with one of the major sites (A, P, or E), but in three "hybrid" states, tRNA assumes intermediate positions with one end in one site and one in another (Figure 4-43).

These studies have revealed that the anticodon loop is the first part of the incoming aminoacyl-tRNA to be bound, which is logical because only a tRNA carrying the correct amino acid is desired. This binding, which occurs between the anticodon and the 16S rRNA of the small ribosomal subunit, is designated the A/T state (see Figure 4-43b). Next the aminoacyl end of the incoming tRNA makes contact with specific nucleotides in the 23S rRNA of the large ribosomal subunit. This binding state, termed A/A, with both ends of the incoming tRNA bound to the ribosome corresponds to what was formerly called the A state. Now the aminoacyl end of the incoming tRNA "leans over" and makes contact with other nucleotides in the 23S rRNA (thereby displacing the acceptor end of the previous, outgoing tRNA); in this A/P state, the acceptor end of the incoming tRNA is in the P site, where the growing peptide chain can be transferred to the aminoacylated amino acid, lengthening the peptide by one residue. This peptidyltransferase reaction is the actual chain elongation step in which the new amino acid is added to the growing chain. As shown in the last two panels of Figure 4-43b, after transfer of the growing peptide to the acceptor end of the incoming tRNA in the P site, the anticodon end of this tRNA also moves to the P site (P/P state); concurrently the anticodon end of the outgoing tRNA (in the P/E state) is displaced, leaving this tRNA in the E state, bound only to the E site from which it is ejected. The A site is now unfilled and ready for the next aminoacyl tRNA. During these movements, the mutual orientation of the ribosomal subunits is thought to change in a way that moves the mRNA by one codon, thus bringing the next codon into position for decoding by the next tRNA.

A revolutionary piece of biochemistry has been achieved to explain the critical peptidyltransferase reaction, which actually effects the transfer of the growing peptide chain to the incoming activated amino acid (the acyl-amino acid) on the tRNA. Evidence suggests that in bacterial protein synthesis the 23S rRNA in the large ribosomal subunit itself may carry out the crucial peptidyltransferase function. When the vast majority of the protein is carefully removed from the large ribosomal subunit, and the remaining rRNA is mixed with analogs of aminoacylated-tRNA and peptidyl-tRNA, the peptidyltransferase reaction still occurs.

In summary then, 16S rRNA in the small ribosomal subunit attracts mRNA through RNA-RNA recognition; tRNAs bind to ribosomal sites, making direct contact with both 16S and 23S rRNA (the latter in the large subunit);

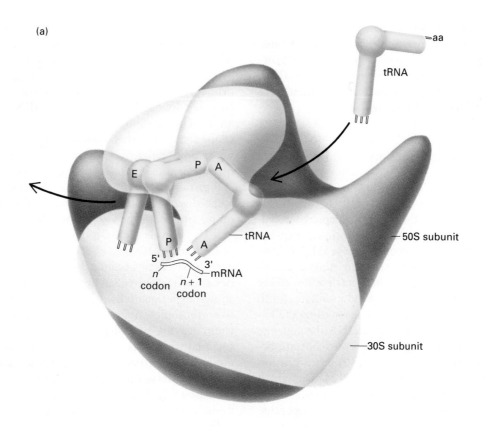

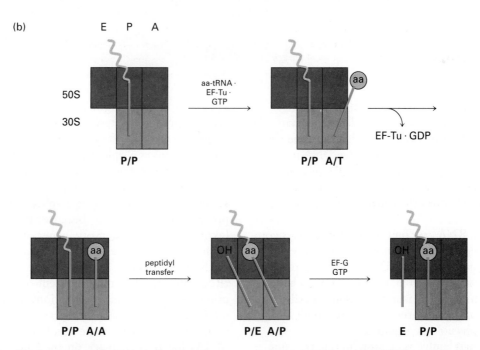

▲ FIGURE 4-43   Binding states of tRNA during translation of mRNA on a ribosome. (a) Orientation of a tRNA molecule relative to the large and small ribosomal subunits in the three main binding states designated A/A, P/P, and E. The blue A and P designate sites on the tRNA that bind to the 16S rRNA of the small ribosomal subunit; the red A, P, and E designate tRNA-binding sites on the tRNA that bind to the 23S rRNA of the large subunit. The highlighted area on the right indicates roughly where the complex of the incoming aminoacyl-tRNA and Tu·GTP interacts with the ribosome (see Figure 4-42b). The pathway of the tRNA through the ribosome is indicated by thick arrows. (b) Schematic representation of movement of tRNA on ribosome showing hybrid binding states—A/T, A/P, and P/E—in addition to major states illustrated in (a). Straight blue lines represent the tRNA molecule, and wavy green lines the growing polypeptide chain. See text for discussion. [Adapted from H. F. Noller, 1991, *Ann. Rev. Biochem.* **60**:191.]

and the actual peptide-synthesis reaction involves 23S rRNA. Clearly, many of the reactions necessary for peptide synthesis are *RNA mediated*. These findings provide powerful support for the increasingly prevalent belief that RNA chemistry holds the key to understanding the earliest steps in the evolution of biological systems.

Polypeptide Termination Requires Protein Factors that Specifically Recognize UAA, UAG and UGA

For polypeptides longer than the tripeptide depicted in Figure 4-42, the process of translocation continues step-by-step until all the amino acids encoded by the mRNA have been added. In each translocation step, the ribosome with its attached peptidyl-tRNA moves three nucleotides closer to the 3′ end of the mRNA. As illustrated in Figure 4-42b this movement is probably brought about by using the energy of GTP hydrolysis to propel the mRNA through the ribosome. Because some hydrogen bonds in rRNA are between distant nucleotides (see Figure 4-40), translocation may occur through a contraction-relaxation cycle in which the folding of the ribosome (and rRNA) changes.

In the mRNA encoding the hypothetical tripeptide shown in Figure 4-42, the three bases following the leucine

codon are UAG. This codon, like the codons UAA and UGA, signals the release of the peptidyl-tRNA complex when recognized by protein *termination factors* (*TFs*). There are at least two such factors in prokaryotes and probably also in eukaryotes. Almost simultaneously the complex divides into an uncharged tRNA molecule lacking an attached amino acid and a newly completed protein chain that can either assume its final shape or combine with additional protein subunits. (Actually, the interaction of two peptide chains can begin while a protein chain is still growing.) After releasing its peptidyl-tRNA, the ribosome disengages from the mRNA, divides into two subunits, and is ready to start the whole cycle again.

The peptide synthesis depicted in Figure 4-42 oversimplifies the release process; a tripeptide might, in fact, not even be released because the growing peptide chain on a ribosome is "buried" within the ribosome. This has been shown experimentally: a brief proteolytic digestion of active ribosomes destroys most of each growing peptide chain, leaving an undigested piece 35 amino acids long associated with each ribosome. The logical place to accommodate and protect about 35 amino acids is the tunnel recently observed in the large subunit (see Figure 4-41). At the base of the large subunit (near the presumed end of the tunnel), growing peptides can react with added antibodies, indicating that this is the exit site for the new polypeptide.

SUMMARY

DNA stores genetic information in all cells in the form of a three-letter code and is the chemical link from generation to generation. Messenger RNA copied from DNA retrieves information from DNA; mRNA, transfer RNA, and ribosomal RNA are the key elements in the protein-synthesizing machinery that decodes the genetic messages to make new proteins.

Both DNA and RNA are composed of nucleotides containing cytidine, adenine, and guanine acids plus thymidine acid in DNA and uridine acid in RNA. The two strands of the DNA helix are antiparallel (5′ → 3′/3′ → 5′). The standard form (B form) of DNA is not invariant in cells; bent DNA and a left-handed Z form of DNA are both known and are of likely physiologic significance.

While DNA acts only as a storehouse, the single-stranded RNA copied from it is a crucial, active participant in many cell reactions. In fact, RNA can assume three-dimensional shapes with chemically reactive surfaces and act at least in a limited fashion as proteins do in catalyzing biochemical reactions. By far the best understood role of RNA in present-day cells is in gene expression. The genes of all organisms are copied into mRNA. Messenger RNA in bacterial cells often contains information for several

proteins because bacterial genes devoted to individual biochemical tasks are often clustered into operons, transcription units for multiple genes. Eukaryotic mRNAs almost always encode a single protein, and the messenger RNA is very often a mosaic made of pieces of primary RNA transcripts that must be cut and spliced together. This RNA processing is required because the coding sequences (exons) in many eukaryotic genes are separated by intervening non-coding sequences (introns). The translation of mRNAs by tRNAs and ribosomes, which are made of rRNA plus more than 50 different proteins, shows the importance of RNA as a functioning catalytic molecule: mRNA, at least in bacteria, recruits ribosomes by actively binding the rRNA in the small ribosomal subunit; the tRNAs locate their place on the ribosome by binding to both the small and the large rRNAs; and the large ribosomal RNA likely plays a role in peptide synthesis.

Thus while proteins are the efficient and almost universal chemical catalysts for most reactions in present-day cells, the nucleic acids, because of their ability to be copied, are the replicating storehouses of information and the regulated conveyors of instructions within cells.

REVIEW QUESTIONS

1. Review the concept of DNA replication, paying particular attention to the direction in which new DNA is synthesized. For the following series of questions, refer to the sequence of double-stranded DNA at the end of this section.

   a. If the arrow under the sequence indicates the direction of replication fork progression, which strand would act as the template for leading strand synthesis? Which strand would act as the template for lagging strand synthesis? Label the 3' and 5' ends of the newly synthesized DNA.

   b. Assume that you have the strand marked W in the above sequence in a test tube as single-stranded DNA, and you wish to synthesize the C strand. What reagents and enzymes would you need to add to your reaction to do this?

   c. 2'-Deoxynucleoside triphosphate precursors are required for DNA synthesis. The phosphate groups in these nucleotides are referred to as α, β, and γ as indicated.

$$PO_3—PO_2—PO_2—\text{deoxyribose—base}$$
$$\gamma \qquad \beta \qquad \alpha$$

   If you wanted to make the C strand DNA from question 2 radioactive, you could use a nucleoside triphosphate containing a radioactive isotope of phosphorus, ^{32}P. Which position of the nucleotide would you need to have labeled with ^{32}P in order to obtain radioactive C strand? Explain your answer.

2. Review the concept of transcription and compare it to DNA replication.

   a. What are the differences and similarities between replication and transcription? Be certain to compare the nucleotide precursors and the requirements for successful initiation of nucleic acid synthesis by the polymerase involved.

   b. Refer to the DNA sequence in question 1. If this sequence belonged to a segment of DNA that was being transcribed, and the arrow indicated the direction of transcription through this segment of DNA, which strand would act as the template for RNA synthesis? What would be the sequence of this portion of the transcript? Be certain to label the 3' and 5' ends.

   c. Refer to Figure 4-17 in this chapter. As indicated in the last step of the figure, the addition of a base onto position n to make the message $n + 1$ bases in length requires a series of events, each of which contributes differently to the free energy of elongation. During transcription, RNA polymerase is bound in a complex that contains the RNA-DNA hybrid. The addition of a base during elongation can be broken down into two steps:
   (1) $NTP \rightarrow NMP + PP_i$; $\Delta G^{\circ\prime} \approx -7.2$ kcal/mole
   (2) $RNA(NMP_{(n)}) + NMP \rightarrow RNA(NMP_{(n+1)})$;
       $\Delta G^{\circ\prime} \approx +7.0$ kcal/mole
   The overall $\Delta G^{\circ\prime}$ for 1 and 2 is equal to the sum of these reactions and has a value of -0.2 kcal/mole. This is not enough free energy to drive transcription. What, ultimately, is the source of energy for RNA synthesis? Do you think a similar argument would apply to replication where dNTPs are used as the precursors?

3. Review the fundamentals of translation, including the steps involved in aminoacyl-tRNA synthesis and assembly of the ribosome-mRNA complex.

   a. Refer to your answer from question 2b above. Assume that this transcript represents a portion of an actual mRNA from *E. coli*. Identify the Shine-Dalgarno sequence. If this particular region were deleted from the transcript, what would be the effect on translation? Explain your answer.

   b. Identify a potential open reading frame from the transcript that would yield a peptide twelve amino acids in length; identify the AUG codon at the translation initiation site. Refer to the genetic code presented on page 121, and use this to translate the message as far as possible. Would the resulting peptide be soluble in an aqueous environment? Explain your answer.

   c. If the underlined base pair were changed such that the A became a T and the T became an A, what would happen to the protein encoded by this sequence?

W   5'–CTAGTAAGGAGGTGCGAATGCTATGAAAGCAACTAAACTAGTACTTGGGGCGGTGATTTAACC–3'
C   3'–GATCATTCCTCCACGCTTACGATACTTTCGTTGA*TTT*GATCATGAACCCCGCCACTAAATTGG–5'

$\longrightarrow$

References

General References

LEHNINGER, A. L., D. L. NELSON, and M. M. COX. 1993. *Principles of Biochemistry*, 2d ed. Worth.
The New Age of RNA. 1993. *FASEB J.*, vol. 7.
STRYER, L. 1995. *Biochemistry*, 4th ed. W. H. Freeman and Company.

Nucleic Acids

BAUER, W. R., F. H. C. CRICK, and J. H. WHITE. 1980. Supercoiled DNA. *Sci. Am.* 243(1):118–133 (Offprint 1474). A mathematical model for describing and analyzing DNA supercoiling.
CECH, T. 1986. RNA as an enzyme. *Sci. Am.* 255(5):64–75.
CECH, T. R., and B. L. BASS. 1986. Biological catalysis by RNA. *Ann. Rev. Biochem.* 55:599–629.

CROTHERS, D. M., T. E. HARAN, and J. G. NADEAU. 1990. Intrinsically bent DNA. *J. Biol. Chem.* **265**:7093–7099.

DICKERSON, R. E. 1983. The DNA helix and how it is read. *Sci. Am.* **249**(6):94–111.

FELSENFELD, G. 1985. DNA. *Sci. Am.* **253**(4):58–66.

KORNBERG, A., and T. A. BAKER. 1992. *DNA Replication*, 2d ed. W. H. Freeman and Company. Chapter 1. A good summary of the principles of DNA structure.

LONG, D. M., and O. C. UHLENBECK. 1993. Self-cleaving catalytic RNA. *FASEB J.* **7**:25–30.

MCCLURE, W. R. 1985. Mechanism and control of transcription initiation in prokaryotes. *Ann. Rev. Biochem.* **54**:171–204.

SAENGER, W. 1988. *Principles of Nucleic Acid Structure.* Springer-Verlag. A comprehensive treatise on the structures of RNA, DNA, and their constituents.

SWEETSER, D., M. NONET, and R. A. YOUNG. 1987. Prokaryotic and eukaryotic polymerases have homologous core subunits. *Proc. Nat'l. Acad. Sci. USA* **84**:1192–1196.

WANG, J. C. 1980. Superhelical DNA. *Trends Biochem. Sci.* **5**:219–221.

YOUNG, R. A. 1991. RNA polymerase II. *Ann. Rev. Biochem.* **60**:689–715.

The Genetic Code

JUKES, T. H. 1977. How many anticodons? *Science* **198**:319–320.

KHORANA, H. G. 1968. Nucleic acid synthesis in the study of the genetic code. Reprinted in *Nobel Lectures: Physiology or Medicine* (1963–1970). Elsevier (1973).

NIRENBERG, M. W., and P. LEDER. 1964. RNA codewords and protein synthesis. *Science* **145**:1399–1407.

OSAWA, S., et al. 1992. Recent evidence for evolution of the genetic code. *Microbiol. Rev.* **56**:229–264.

WOESE, C. B. 1967. *The Genetic Code.* Harper & Row.

Messenger RNA: Splicing and Initiation Signals

BRENNER, S., F. JACOB, and M. MESELSON. 1961. An unstable intermediate carrying information from genes to ribosomes for protein synthesis. *Nature* **190**:576–581.

GUTHRIE, C, and B. PATTERSON. 1988. Spliceosomal snRNAs. *Ann. Rev. Genet.* **23**:387–419.

JACKSON, R. J. 1991. Initiation without an end. *Nature* **353**:14–15.

KOZAK, M. 1991. An analysis of vertebrate mRNA sequences: intimations of translational control. *J. Cell Biol.* **115**:887–903.

MATTAJ, I. W., D. TOLLERVEY, and B. SERAPHIN. 1993. Small nuclear RNAs in messenger RNA and ribosomal RNA processing. *FASEB J.* **7**:47–53.

NIRENBERG, M. W., and J. H. MATTHEI. 1961. The dependence of cell-free protein synthesis in *E. coli* upon naturally occurring or synthetic polyribonucleotides. *Proc. Nat'l. Acad. Sci. USA* **47**:1588–1602.

SONENBERG, N. 1988. Cap-binding proteins of eukaryotic messenger RNA: functions in initiation and control of translation. *Prog. Nucl. Acid Res. Mol. Biol.* **35**:173–206.

SONENBERG, N. 1991. Picornavirus RNA translation continues to surprise. *Trends in Genetics* **7**:105–106.

STEITZ, J. A. 1988. "Snurps." *Sci. Am.* **258**(6):56–65.

WAHLE, E., and W. KELLER. 1992. The biochemistry of 3′-end cleavage and polyadenylation of messenger RNA precursors. *Ann. Rev. Biochem.* **61**:419–440.

Transfer RNA and Aminoacyl-tRNA Synthetase

BJORK, G. R., et al. 1987. Transfer RNA modification. *Ann. Rev. Biochem.* **56**:263–287.

CAVARELLI, J., and D. MORAS. 1993. Recognition of tRNAs by aminoacyl-tRNA synthetases. *FASEB J.* **7**:79–86.

HOAGLAND, M. B., et al. 1958. A soluble ribonucleic acid intermediate in protein synthesis. *J. Biol. Chem.* **231**:241–257.

HOLLEY, R. W., et al. 1965. Structure of a ribonucleic acid. *Science* **147**:1462–1465.

MCCLAIN, W. H. 1993. Transfer RNA identity. *FASEB J.* **7**:72–78.

RICH, A., and S.-H. KIM. 1978. The three-dimensional structure of transfer RNA. *Sci. Am.* **240**(1):52–62 (Offprint 1377).

ROULD, M. A., J. J. PERONA, and T. A. STEITZ. 1991. Structural basis of anticodon loop recognition by glutaminyl-tRNA synthetase. *Nature* **352**:213–218.

SCHULMAN, L. H. 1991. Recognition of tRNAs by aminoacyl-tRNA synthetases. *Prog. Nucl. Acid Res. Mol. Biol.* **41**:23–87.

SCHULMAN, L. H., and H. PELKA. 1988. Anticodon switching changes the identity of methionine and valine transfer RNAs. *Science* **242**:765–768.

Ribosomes

BERNABEU, C., and J. A. LAKE. 1982. Nascent polypeptide chains emerge from the exit domain of the large ribosomal subunit. *Proc. Nat'l. Acad. Sci. USA* **79**:3111–3115.

MOAZED, D., and H. F. NOLLER. 1991. Sites of interaction of the CCA end of peptidyl-tRNA with 23S rRNA. *Proc. Nat'l. Acad. Sci. USA* **88**:3725–3728.

NOLLER, H. F. 1993. tRNA-rRNA interactions and peptidyl transferase. *FASEB J.* **7**:87–89.

OAKES, M. I., et al. 1986. DNA hybridization electron microscopy: ribosomal RNA nucleotides 1392–1407 are exposed in the cleft of the small subunit. *Proc. Nat'l. Acad. Sci. USA* **83**:275–279.

OLSEN, G. J., and C. R. WOESE. 1993. Ribosomal RNA: a key to phylogeny. *FASEB J.* **7**:113–123.

STERN, S., B. WEISER, and H. F. NOLLER. 1988. Model for the three-dimensional folding of 16S ribosomal RNA. *J. Mol. Biol.* **204**:447–481.

YONATH, A., K. R. LEONARD, and H. G. WITTMAN. 1987. A tunnel in the large ribosomal subunit revealed by three-dimensional image reconstitution. *Science* **236**:813–816.

The Steps in Protein Synthesis

CASKEY, T. H. 1980. Peptide chain termination. *Trends Biochem. Sci.* **5**:234–237.

HERSHEY, J. W. B. 1991. Translational control in mammalian cells. *Ann. Rev. Biochem.* **60**:717–755.

JACOB, W. F., M. SANTER, and A. E. DAHLBERG. 1987. A single base change in the Shine-Dalgarno region of 16S rRNA of *E. coli* affects translation of many proteins. *Proc. Nat'l. Acad. Sci. USA* **84**:4757–4761.

KOZAK, M. 1986. Point mutations define a sequence flanking the AUG initiator codon that modulates transcription by eukaryotic ribosomes. *Cell* **44**:283–292.

MOAZED, D., and F. NOLLER. 1989. Interaction of tRNA with 23S rRNA in the ribosomal A, P, and E sites. *Cell* **57**:585–587.

NOLLER, H. F., V. HOFFARTH, and L. ZIMNIAK. 1992. Unusual resistance of peptidyl transferase to protein extraction procedures. *Science* **256**:1416–1419.

SAKS, M. E., J. R. SAMPSON, and N. N. ABELSON. 1994. The transfer RNA identity problem: a search for rules. *Science* **263**:191–197.

5

Cell Organization, Subcellular Structure, and Cell Division

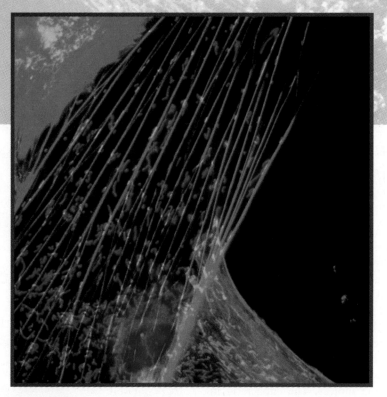

▲ The actin filaments (red) and mitochondria (green) in a rat astrocyte.

All the cells in all organisms have in common certain structural features, such as the architecture of their membranes. Many complicated metabolic events are also carried out in basically the same way in all organisms: the replication of DNA, the synthesis of proteins, and the production of chemical energy by the conversion of glucose to carbon dioxide. Before the introduction of the electron microscope, all cells were also generally assumed to share similar basic principles of organization. As noted in Chapter 1, bacterial cells are much smaller than typical animal, plant, or even fungal cells. Because the treatment of some bacteria with a Feulgen stain (which stains DNA) revealed central masses of nuclear material, it seemed possible that bacterial cells, like cells of higher organisms, possessed a defined nucleus and other internal compartments. But the limited resolution of the light microscope (only about 0.2 μm) does not provide an accurate image of the internal structure of bacterial cells. Not until the 1950s, when a variety of morphologic, biochemical, and genetic techniques became available, was it established that two types of cells—*eukaryotic cells* (literally, cells with a true nucleus) and *prokaryotic cells* (cells with no defined nucleus)—have persisted independently for perhaps a billion or more years of biological evolution. Although prokaryotes and eukaryotes differ radically in their organization, they apparently evolved from the same type of cell, but the nature of this ancestral cell is unknown.

This chapter begins with an overview of cell structure that focuses on the similarities and differences between prokaryotes and eukaryotes. It then describes the methods cell biologists use to study cell organization: the classic techniques of microscopy and cell fractionation used on fixed or killed cells, which provide only a static picture, and some newer developments, which reveal the movements and internal structures of living cells. We then examine the main components of eukaryotic cells—the internal organelles and fibers—and techniques for their isolation and purification. The extracellular substances that surround cells and give them shape and strength are also investigated. Finally, we discuss the most crucial aspect of cell division, the duplication of the cell in such a way that every daughter cell receives an identical copy of its genetic material. We touch on the marked difference between the way that prokaryotic and eukaryotic cells coordinate DNA synthesis and the subsequent equal partitioning of DNA during cell division. The following chapter will detail the ways in which different types of cells are cultured in the laboratory. The concepts and experimental techniques introduced in these two chapters form the basis of modern molecular cell biology.

➤ *Prokaryotic and Eukaryotic Cells*

Eukaryotes include all members of the protist, fungus, animal, and plant kingdoms—from the most primitive ferns to the most complex flowering plants, and from amebas and simple sponges to insects and mammals. Although the variety of organisms classified as eukaryotes is large, they all share certain structural features.

Prokaryotes include all bacteria, divided into two separate lineages: *eubacteria* and *archaebacteria*. Most bacteria studied in laboratories are eubacteria (Figure 5-1); in general, when bacterial structure or metabolism is discussed in this book, eubacteria are the subject. The eubacteria include the *photosynthetic organisms* formerly known as *blue-green algae* but better known today as *cyanobacteria*. Less is known about the archaebacteria, which grow in unusual environments. The *methanogens* live only in oxygen-free milieus such as swamps. These bacteria generate methane (CH_4), also known as "swamp gas," by the reduction of carbon dioxide. Other archaebacteria include the *halophiles*, which require high concentrations of salt to survive, and the *thermoacidophiles*, which grow in hot (80°C) sulfur springs, where a pH of less than 2 is common.

Interestingly, the archaebacteria are believed by many to represent a third cell lineage. Although they are prokaryotes, they have numerous eukaryotic features. Like eubacteria, archaebacteria have a circular genomic DNA, yet, like eukaryotes, they have introns in their tRNA genes. Their ribosomal subunits form functional ribosomes with eukaryotic ribosomal subunits and not with those from eubacteria. And, in contrast to eubacteria, they lack a peptidoglycan layer in their cell wall. (These prokaryotic and

eukaryotic characteristics will be discussed below; here, it is sufficient to note the differences between the archaebacteria and eubacteria.)

Prokaryotes Have a Relatively Simple Structure

The Plasma Membrane In general, only one type of membrane, the *plasma membrane*, forms the boundary of the cell proper in prokaryotes. Like all biological membranes, the structure of the plasma membrane is based on a

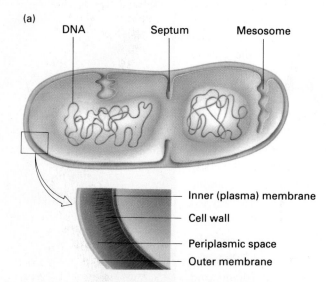

(a)

DNA   Septum   Mesosome

Inner (plasma) membrane
Cell wall
Periplasmic space
Outer membrane

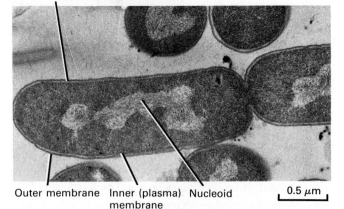

(b)   Periplasmic space and cell wall

Outer membrane   Inner (plasma)   Nucleoid   0.5 μm
                  membrane

▲ FIGURE 5-1   (a) Structure of a gram-negative eubacterial cell; note the periplasmic space between the inner and outer membranes. (b) Electron micrograph of a thin section of *E. coli*, a gram-negative bacterium. The micrograph shows the inner (plasma) membrane, the outer surface membrane that is part of the cell wall, and the nucleoid, the DNA-containing fibrous central region of the cell. [Part (b) courtesy of I. D. J. Burdett and R. G. E. Murray.]

phospholipid bilayer (see Figure 2-17), which is permeable to certain gases, such as oxygen and carbon dioxide, and to water—these substances can diffuse across it. However, it is virtually impermeable to most molecules that the cell must obtain from its environment, such as sugars, amino acids, and inorganic ions (for example, K^+ or Cl^-). The plasma membrane utilizes many membrane proteins called *channels,* or *transporters,* which allow these and other molecules to enter or leave the cell.

The Outer Membrane and Cell Wall   Eubacterial species can be divided into two classes. *Gram-negative bacteria* (those not stained by the Gram technique), such as the common intestinal bacterium *Escherichia coli* (*E. coli*), are surrounded by two surface membranes (Figure 5-1b). The inner membrane is the actual plasma membrane—the major permeability barrier of the cell. The outer membrane is unusual in that it is permeable to many chemicals having a molecular weight of 1000 or more; it contains proteins called *porins,* which line channels large enough to accommodate such molecules. Between the two membranes lie the *cell wall,* containing *peptidoglycan* (a complex of proteins and oligosaccharides that gives rigidity to the cell), and the *periplasm* (a space generally occupied by proteins secreted by the cell). *Gram-positive bacteria,* such as *Bacillus polymyxa,* have only a plasma membrane and a cell wall.

Eukaryotic Cells Have Complex Systems of Internal Membranes and Fibers

Both prokaryotic and eukaryotic cells are surrounded by a plasma membrane. However, unlike most prokaryotic cells, eukaryotic cells also contain extensive internal membranes that enclose specific regions, separating them from the rest of the *cytoplasm,* the region of the cell lying outside of the nucleus. These membranes define a collection of subcellular structures called *organelles.*

The Many Organelles of a Eukaryotic Cell   As noted in Chapter 1, the largest organelle in a cell is generally the *nucleus,* which contains most of the cellular DNA and is the site of synthesis of cellular RNAs. Most eukaryotic cells contain many other organelles (Figures 5-2 and 5-3): the *mitochondria,* in which the oxidation of small molecules generates most cellular ATP; the *rough* and *smooth endoplasmic reticula* (ERs), a network of membranes in which glycoproteins and lipids are synthesized; *Golgi vesicles,* which direct membrane constituents to appropriate places in the cell; *peroxisomes* in all eukaryotes and *glyoxisomes* in plant seeds, which metabolize hydrogen peroxide; and assorted smaller vesicles. Animal cells contain *lysosomes,* which degrade many proteins, nucleic acids, and lipids. Plant cells contain *chloroplasts,* the site of photosynthesis.

Both plant cells and certain eukaryotic microorganisms contain one or more *vacuoles* (see Figure 5-3a on page 6), large fluid-filled organelles that store many nutrient and waste molecules and also participate in the degradation of cellular proteins and other macromolecules. Each type of organelle plays a unique role in the growth and metabolism of the cell, and each contains a collection of specific enzymes that catalyze requisite chemical reactions. Some of this specificity resides in the organelle membranes, to which a number of the enzymes and other proteins are bound.

The Cytoskeleton   The *cytosol* is the part of the cytoplasm that is not contained in membrane-limited organelles. The cytosol of eukaryotic cells contains an array of fibrous proteins collectively called the *cytoskeleton.* Among these fibers are the *microfilaments,* built of the protein actin; the somewhat wider *microtubules,* built of tubulin; and the *intermediate filaments,* built of one or more rod-shaped protein subunits. Cytoskeletal fibers give the cell strength and rigidity. They also control movement within the cell; microtubules, for instance, are critical to chromosomal movement during cell division. Some cytoskeletal fibers may connect to organelles or provide tracks along which organelles move.

Eukaryotic Cell Walls   Plant cells are surrounded by a rigid cell wall containing cellulose and other polymers, which also contributes to the strength and rigidity of the cell. Fungi are also surrounded by a cell wall, but of different composition from the walls of bacterial or plant cells. During cell growth, a cell wall must expand; during cell division, a new wall must be laid down between the two daughter cells. Animal cells generally are not surrounded by walls.

Prokaryotes and Eukaryotes Contain Similar Macromolecules

The volume of a typical animal or plant cell is several hundred times that of a typical bacterial cell, yet the chemical composition of prokaryotic and eukaryotic cells is strikingly similar. By weight, about 70 percent of a typical cell is water. Other small molecules, including salts, lipids, amino acids, and nucleotides, account for another 7 percent. About 20 percent is protein, 2 percent is RNA, and less than 1 percent is DNA, the genetic material. These proportions are actually useful to know, as seen by the following examples.

Calculating the Abundance of Important Proteins   Consider a hepatocyte, the major cell type in the liver. It is roughly a cube 15 μm on a side, with a volume of $(0.0015 \text{ cm})^3$, or 3.4×10^{-9} ml. Assuming a cell density of ≈ 1.03 g/ml, the cell would weigh 3.5×10^{-9} g, 20 percent of which, or 6.9×10^{-10} g, would be protein. Since the molecular weight of a typical protein is 50,000 g/mol (roughly 400 amino acids), we can use the amount of protein per cell, together with Avogadro's number (6.02×10^{23} molecules/mol), to calculate the number of protein molecules per liver cell:

$$\frac{(6.9 \times 10^{-10} \text{ g}) \times (6.02 \times 10^{23} \text{ molecules/mol})}{5 \times 10^4 \text{ g/mol}}$$

or about 8.3×10^9 molecules of protein per cell. Liver cells contain about 10,000 different types of mRNA molecules, and thus about 10,000 different types of protein. Typical proteins therefore will be present at about 1,000,000 copies per liver cell, but many are much less abundant. Important hormone receptors, such as the cell-surface protein that binds insulin and sends a signal to the inside of the cell, are present at about 20,000 copies per cell. This may seem like a large number, but a simple calculation will indicate that the insulin receptor comprises about 0.0002 percent of total cell protein, making it a very rare protein and a correspondingly difficult one to study.

Determining Rates of Protein Synthesis   HeLa cells are a well-studied line of cells derived from a human cervical carcinoma that grow well in cell cultures. These cells have a composition typical of many kinds of mammalian cells. A HeLa cell contains about 4×10^6 ribosomes and about 7×10^5 mRNA molecules, roughly 150 times as many of each as are found in the much smaller *E. coli* cell (Table 5-1). A HeLa cell has about 5×10^9 protein molecules, consisting of 5000–10,000 different polypeptide species; the *E. coli* cell has about 3×10^6 protein molecules, consisting of 1000–2000 different species.

Since each ribosome takes about 30 seconds to synthesize an average protein of ≈ 400 amino acids, in 24 hours the 4 million ribosomes in a HeLa cell would synthesize about 1.1×10^{10} protein molecules. This is almost exactly what is required to allow the cell to divide every 24 hours, given that some proteins have very short lifetimes and need to be synthesized repeatedly during cell growth. This calculation indicates that the number of ribosomes within a cell is regulated so as to meet precisely the needs of a cell for new protein. The levels of all macromolecules within a cell are precisely controlled, as later chapters will emphasize repeatedly.

Prokaryotes and Eukaryotes Differ in the Amount of DNA per Cell

The differences in genetic organization between typical prokaryotic and eukaryotic cells become obvious when we consider the amount of DNA per cell (Table 5-2). The genome of *E. coli* contains 4.4×10^{-15} g, or 0.0044 picograms (pg), of DNA, an amount equal to 4×10^6 base pairs. Because three DNA bases encode each amino acid in

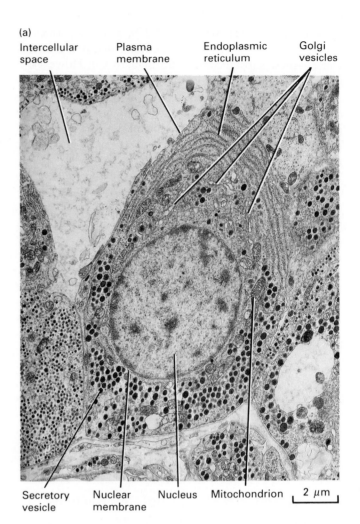

(a)

Intercellular space   Plasma membrane   Endoplasmic reticulum   Golgi vesicles

Secretory vesicle   Nuclear membrane   Nucleus   Mitochondrion   2 μm

▲ **FIGURE 5-2**   (a) Electron micrograph of a thin section of a hormone-secreting cell from the rat pituitary, showing the subcellular features typical of many animal cells. (b) Structure of a "typical" animal cell (*opposite page*). Not every animal cell contains all of the organelles, granules, and fibrous structures shown here, and other substructures can be present in some cells. Animal cells also differ considerably in shape and in the prominence of various organelles and substructures. [Part (a) courtesy of Biophoto Associates.]

a protein and an average protein contains about 400 amino acids, approximately 1200 DNA base pairs are used to encode each protein species. Thus *E. coli* DNA has a maximum coding capacity of about 3300 different proteins. Not all of the bacterial DNA encodes proteins, however, although a large part of it does. An *E. coli* cell may actually contain as many as 2000 different species of mRNAs and thus of proteins.

All eukaryotic cells contain more DNA than prokaryotic cells do. Yeast cells, which have some of the smallest genomes among eukaryotes, contain about three times as much DNA as *E. coli* does. The cells of higher plants and animals typically have 40–1000 times as much DNA as

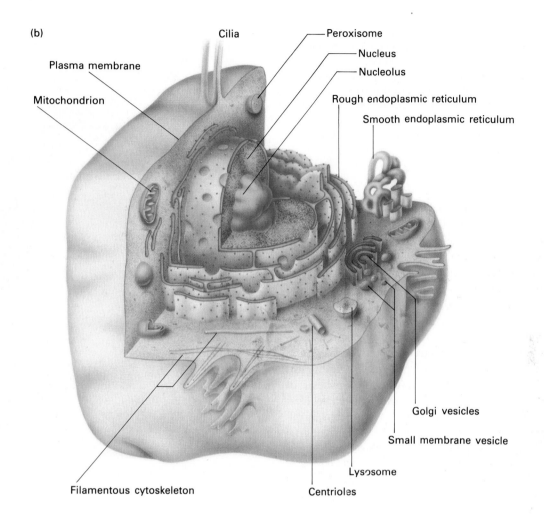

(b)

Cilia

Peroxisome

Nucleus

Nucleolus

Plasma membrane

Mitochondrion

Rough endoplasmic reticulum

Smooth endoplasmic reticulum

Golgi vesicles

Small membrane vesicle

Lysosome

Filamentous cytoskeleton

Centrioles

TABLE 5-1   The Main Macromolecular Components of *E. coli* and HeLa Cells

Component	Amount per HeLa Cell	Amount per *E. coli* Cell
Total dry weight	400 picograms (pg)	0.4 pg
Total DNA	15 pg*	0.017 pg†
Total RNA	30 pg	0.10 pg
Total protein	300 pg	0.2 pg
Cytoplasmic ribosomes	4×10^6	3×10^4
Cytoplasmic tRNA molecules	6×10^7	4×10^5
Cytoplasmic mRNA molecules	$7 \times 10^{5\ddagger}$	$4 \times 10^{3\ddagger}$

* HeLa cells are hypotetraploid: they contain about four copies of each chromosome. The normal diploid human DNA complement is about 5 pg per cell.
† A rapidly growing *E. coli* cell contains, on average, four DNA genomes. Each genomic DNA weighs 0.0044 pg.
‡ An average chain length of 1500 nucleotides is assumed.

➤ FIGURE 5-3   (a) Electron micrograph of a thin section of a leaf cell from *Phleum pratense,* showing a large internal vacuole, parts of five chloroplasts, and the cell wall. Although a nucleus is not evident in this micrograph, plant cells do contain a nucleus and other features of eukaryotic cells, as depicted in (b), structure of a "typical" plant cell (*opposite page*). Note the continuity of the plasma membrane at the plasmodesmata. [Part (a) courtesy of Biophoto Associates/Myron C. Ledbetter/Brookhaven National Laboratory.]

(a)

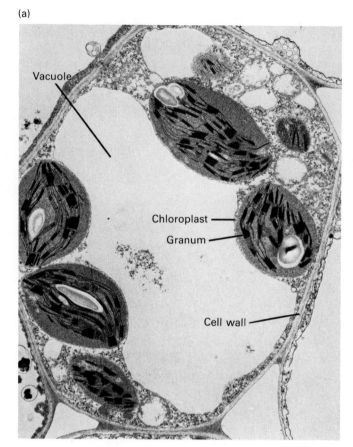

E. coli; the genomes of some amphibians are 40,000 times as large. If each segment of DNA were used to encode a single type of protein, typical animal cells would be able to encode as many as 3×10^6 different types of proteins. However, only a small fraction of the total DNA in a eukaryotic cell usually encodes proteins. It is widely believed that invertebrates can make about 20,000 proteins and that humans can make about 100,000 to 500,000—far fewer than the theoretical capacities of their respective genomes. However, this calculation assumes that each region of DNA encodes only one type of protein. In vertebrates a single immunoglobulin gene leads to the synthesis of several million different kinds of antibody molecules. As detailed in Chapter 27, this variety is generated by rearrangements of DNA segments that encode parts of immunoglobulins and by mutations in these rearranged genes.

TABLE 5-2   The DNA Content of Various Cells

| Organism | Size of DNA Genome | | Maximum Number of Proteins Encoded* | Number of Chromosomes (haploid)† |
	Number of Base Pairs	Total Length (mm)		
PROKARYOTIC				
Escherichia coli (bacterium)	4×10^6	1.36	3.3×10^3	1
EUKARYOTIC				
Saccharomyces cerevisiae (yeast)	1.35×10^7	4.60	1.125×10^4	17
Drosophila melanogaster (insect)	1.65×10^8	56	1.375×10^5	4
Homo sapiens (human)	2.9×10^9	990	2.42×10^6	23
Zea mays (corn)	5.0×10^9	1710	4.0×10^6	10

* Assuming 1200 base pairs per protein.
† Most insect and human cells are diploid, so they have twice the number of chromosomes shown.

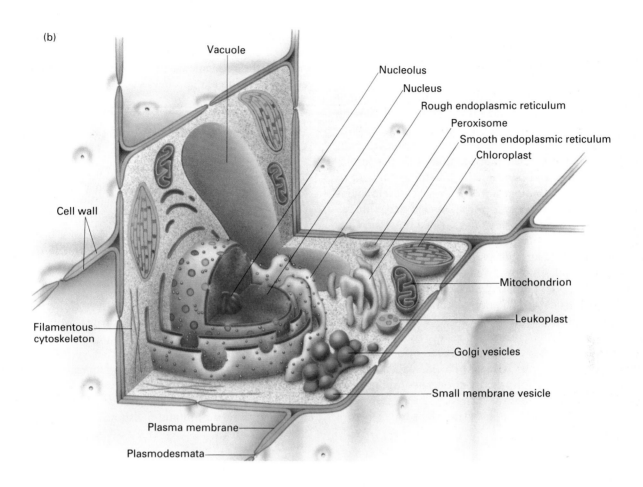

(b)

Vacuole

Nucleolus

Nucleus

Rough endoplasmic reticulum

Peroxisome

Smooth endoplasmic reticulum

Chloroplast

Cell wall

Mitochondrion

Leukoplast

Filamentous cytoskeleton

Golgi vesicles

Small membrane vesicle

Plasma membrane

Plasmodesmata

The Organization of DNA Differs in Prokaryotic and Eukaryotic Cells

In all prokaryotes studied to date, most or all cellular DNA is in the form of a single circular molecule. The cell is said to have a single *chromosome*, although the arrangement of DNA within this chromosome differs greatly from that within the chromosomes of eukaryotic cells. Rapidly growing cells may have up to four copies of this chromosome.

Prokaryotic cells lack, as we have said, a membrane-bound nucleus; most of the genomic DNA lies in the central region of the cell (see Figure 5-1). The DNA must be folded back on itself many times: one *E. coli* chromosomal DNA molecule stretched out to its full length would be over 1 millimeter (mm) long, or 1000 times as long as the cell itself.

The nuclear DNA of all eukaryotic cells, in contrast, is divided between two or more chromosomes which, except during cell division, are contained in a membrane-bound nucleus. The number and size of individual chromosomes vary widely among different eukaryotes (see Chapter 9). Yeasts, for example, have 12–18 chromosomes, each of which contains, on average, only 20 percent of the DNA in an *E. coli* chromosome. Human cells, at the other extreme, contain two sets of 23 chromosomes, each of which has about 30 times the amount of DNA present in an *E. coli* chromosome. Each eukaryotic chromosome is believed to contain a single, linear, double-stranded DNA molecule.

Despite their many differences, prokaryotic and eukaryotic cells have many biochemical pathways in common, and in most aspects the translation of mRNA into proteins is similar in all cells. Hence prokaryotes and eukaryotes are believed to be descended from the same primitive cell. Their divergence must have occurred before the separation of plant and animal cells. All extant prokaryotic and eukaryotic cells and organisms are the result of over one billion years of biological evolution. It is not surprising, then, that cells are so well adapted to their own environmental niches.

➤ *Light Microscopy and Cell Architecture*

There is no one "correct" view of a cell; thus it is essential to understand some of the details of the key cell-viewing techniques, the types of images they produce, and their limitations. Schleiden and Schwann, using a primitive light microscope, first described individual cells as the fundamental unit of life, and light microscopy has continued to play a major role in biological research. The development of electron microscopes has greatly extended the ability to resolve subcellular particles and has yielded much new information on the organization of plant and animal tissues. The nature of the images depends on the type of light or electron microscope employed and on the way in which the cell or tissue has been prepared. Each technique is designed to emphasize particular structural features of the cell; Figure 5-4 shows how a typical cell, a human leukocyte, appears when three different techniques are used.

Although the most common application of light and electron microscopy—to visualize fixed, killed cells—reveals much information, a critical question about such results is how true to life is the image of a biological specimen that has been fixed, stained, and dehydrated before examination? Thus the following two sections describe not only the classic methods used to view fixed cells but also some of the refinements that allow microscopy of unaltered or less altered specimens.

The Resolution of Standard Light (Bright-Field) Microscopy Is Limited to about 0.2 μm

The Compound Microscope   The *compound microscope* (Figure 5-5), the most common microscope in use today, contains several lenses that magnify the image of a specimen under study. The total *magnification* is a product of the magnification of the individual lenses: if the *objective lens* magnifies 100-fold (a 100× lens, the maximum usually employed) and the *eyepiece* magnifies 10-fold, the final magnification recorded by the human eye or on film will be 1000-fold.

Resolution   The most important property of any microscope is not its magnification but its power of *resolution*—its ability to distinguish between two very closely positioned objects. Merely enlarging the image of a specimen accomplishes nothing if the image is blurry.

The resolution of a microscope lens is numerically equivalent to D, the minimum distance between two distin-

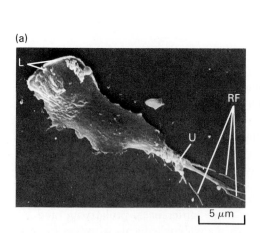

(a)

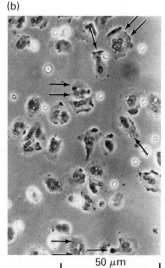

(b)

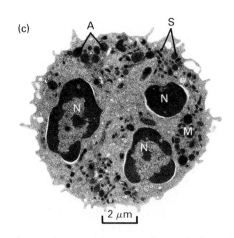

(c)

▲ FIGURE 5-4   Views of human leukocytes produced by three different microscopic techniques. (This white blood cell is polymorphonuclear, meaning its nucleus contains many irregularly shaped lobes.) (a) A scanning electron micrograph. The three-dimensional appearance of the cell surface is characteristic of images obtained by this technique. The shape of this cell indicates that it migrates: it has a wide, flattened projection (a lamellipodium, L) at its leading edge and a narrow tail (the uropod, U) ending in retraction fibers (RF). (b) A light micrograph (using phase-contrast optics) of polymorphonuclear leukocytes attached to a glass slide. Some cells have spread out and become firmly attached to the glass at many points on their surfaces; these stationary cells are indicated by double arrows. Other cells are migrating along the glass; their direction of movement is indicated by a single arrow. Note that only the nuclei and a few larger cytoplasmic particles are visible with this technique. (c) A transmission electron micrograph. This thin section of a cell shows numerous granules of two types: the larger azurophil granules (A) and the smaller specific granules (S). Also evident are three lobes of the single nucleus (N) and some mitochondria (M), which are not present in large numbers in leukocytes. [Courtesy of M. J. Karnofsky; cells prepared by J. M. Robinson, Department of Pathology, Harvard Medical School.]

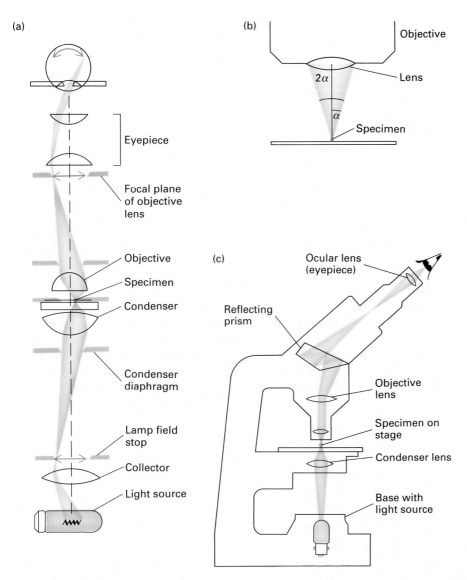

◀ FIGURE 5-5   The optical pathway in a compound optical microscope. (a) The specimen is usually mounted on a transparent glass slide and positioned on the movable specimen stage of the microscope. Light from a bright source is focused by the collector and condenser lenses onto the specimen. The objective lens picks up the light transmitted by the specimen and focuses it on the focal plane of the objective lens, creating a magnified image of the specimen. This image can be recorded directly. Usually, though, the image on the objective focal plane is magnified by the ocular lens, or eyepiece, which is focused on this objective focal plane; it picks up the light emanating from the already magnified image of the specimen and projects it onto the plane of the human eye or a piece of photographic film. The lamp field stop, the condenser diaphragm, and the other apertures restrict the amount of light entering or leaving a lens. (b) An important parameter is the half-angle, α, of the cone of light entering the objective lens from the specimen. The larger the value of α, the finer the resolution the objective lens can provide. (c) Diagram of a modern compound microscope. [Parts (a) and (b) adapted from B. Wilson, 1976, *The Science and Art of Basic Microscopy*, Figures 3-9 and 3-6.]

guishable objects; the smaller the value of D, the better the resolution. D depends on three parameters, all of which must be considered in order to achieve the best possible resolution: the *angular aperture, α,* or half-angle of the cone of light entering the lens from the specimen; the *refractive index, N,* of the air or fluid medium between the specimen and the objective lens; and the *wavelength, λ,* of incident light:

$$D = \frac{0.61 \, \lambda}{N \sin \alpha}$$

Decreasing the value of $λ$ or increasing either N or $α$ will decrease the value of D and thus improve the resolution.

The angular aperture depends on the width of the objective lens and its distance from the specimen (see Figure 5-5b). Moving the objective lens closer to the specimen will increase $\sin \alpha$ and reduce D. Intuitively, one can recognize

that increasing $α$ allows more light from the specimen to enter the objective lens.

The refractive index measures the degree to which a medium bends a light ray that passes through it; the refractive index of air is defined as $N = 1.0$. One useful way to decrease D and thereby improve the resolution is to use an *immersion oil* as the medium between the specimen and the objective lens. Since the refractive index of such oils is 1.5, the resolution will be improved 1.5-fold. An intuitive explanation for this improvement is that a medium with a higher refractive index than air, if placed between the specimen and the objective lens, will "bend" more of the light emanating from the specimen such that it goes into the lens.

Finally, the shorter the wavelength of incident light, the lower will be the value of D and the better the resolution.

However, the *limit of resolution* of a light microscope using visible light is about 0.2 μm; no matter how many times the image is magnified, the microscope can never re-solve objects less than ≈0.2 μm apart or reveal details smaller than ≈0.2 μm in size.

To calculate the limit of resolution, we need to know that the maximum angular aperture for the best objective lenses is 70° (sin 70° = 0.94). With the visible light of shortest wavelength (blue, λ = 450 nm) and with air (N = 1.0) above the sample, the finest resolution will be

$$D = \frac{0.61 \times 450 \text{ nm}}{1.0 \times 0.94} = 292 \text{ nm}$$

or about 0.3 μm. If immersion oil (N = 1.5) is introduced, then

$$D = \frac{0.61 \times 450 \text{ nm}}{1.5 \times 0.94} = 194 \text{ nm}$$

or about 0.2 μm.

Actually, the resolution of the light microscope can now be smaller than this. If we know the precise size and shape of an object—say, a 5-nm sphere of gold—and if we use a video camera to record the microscopic image as a digital image, then a computer can calculate the position of the *center* of the object to within a few nanometers. This technique has been used, to nanometer resolution, for tracking the movement of gold particles attached via anti-bodies to specific proteins on the surface of living cells.

Preparing Samples for Light Microscopy   Specimens for light microscopy are commonly fixed with a solution containing alcohol or formaldehyde, compounds that de-nature most proteins and nucleic acids. Formaldehyde also cross-links amino groups on adjacent molecules; these co-valent bonds stabilize protein–protein and protein–nucleic acid interactions and render the molecules insoluble and stable for subsequent procedures. Usually the sample is then embedded in paraffin or plastic and cut into thin sec-tions of one or a few micrometers thick (Figure 5-6). Alter-natively, the sample can be frozen without prior fixation and then sectioned; this avoids the denaturation of en-zymes by fixatives such as formaldehyde.

Since the resolution of the light microscope is ≈0.2 μm and mitochondria and chloroplasts are ≈1 μm long (about the size of eubacteria), theoretically one should be able to see these organelles. However, most cellular constituents are not colored and absorb about the same degree of visible light, so that they are hard to distinguish under a light microscope; often not even the nucleus of a cell can be seen. (See Figure 5-16a.) The final step in preparing a speci-men for observation, therefore, is to stain it in order to visualize the main structural features of a cell or tissue. Many chemical stains bind to molecules that have specific features: for example, *hematoxylin* and *eosin* bind to many different kinds of proteins and stain different cells differ-

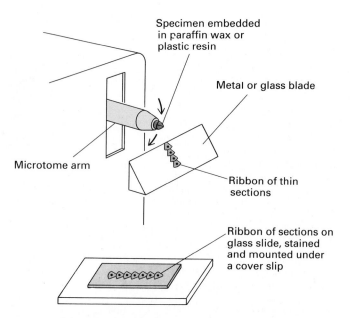

▲ FIGURE 5-6   Preparation of tissues for light micros-copy. A piece of fixed tissue is dehydrated by soaking it in alcohol-water solutions, in pure alcohol, and finally in a sol-vent such as xylene. The specimen is then placed in warm liquid paraffin, which is allowed to harden. A piece of the specimen is mounted on the arm of a microtome. The arm moves up and down over the metal or glass blade, cutting specimen sections a few micrometers thick.

ently (Figure 5-7); *benzidine* binds to heme-containing pro-teins and nucleic acids; and the *fuchsin* used in Feulgen staining binds to DNA.

If an enzyme catalyzes a reaction that produces a col-ored or otherwise visible precipitate from a colorless pre-cursor, the enzyme may be detected in cell sections by their colored reaction products. This last technique, called *cyto-chemical staining*, can be used with both light and electron microscopes (Figure 5-8).

Immunofluorescence Microscopy Reveals Specific Proteins and Organelles within a Cell

Perhaps the most powerful techniques for localizing pro-teins within a cell by light microscopy make use of the *fluorescence microscope* and *antibodies* specific for the desired protein. A chemical is said to be *fluorescent* if it absorbs light at one wavelength (the *excitation wave-length*) and emits light at a specific and longer wavelength within the visible spectrum. In modern fluorescence micro-scopes (Figure 5-9), only fluorescent light emitted by the sample is used to form an image; light of the exciting wave-length induces the fluorescence but is then absorbed by filters placed between the objective lens and the eye or camera.

> FIGURE 5-7   A light-microscopic view of a section of loose connective tissue supporting the lining of the small intestine, stained with the dyes hematoxylin and eosin. Note that at least seven different cell types are distinguishable by this staining protocol: fibroblasts (F) with their elongated nuclei, erythrocytes (Er) within small blood vessels, and several types of white blood cells: plasma cells (P), which synthesize antibodies, lymphocytes (L), which function in immune responses, eosinophils (Eo) and neutrophils (N), which participate in defense against parasites and other infectious agents, and macrophages (M), which ingest foreign particles and organisms. [From P. R. Wheater, H. G. Burkitt, and V. C. Daniels, 1987, *Functional Histology; A Text and Colour Atlas*, 2d ed., Churchill Livingstone, Figure 4.16, p. 62. Photo Researchers, Inc.]

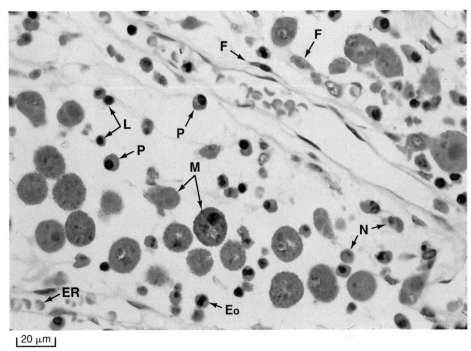

(a)

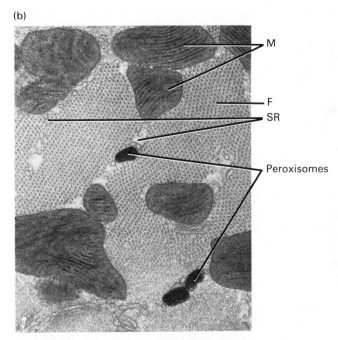

(b)

▲ FIGURE 5-8   Cytochemical staining, using light and electron microscopes. (a) Light micrograph of a cross section of human skeletal muscle stained for succinate dehydrogenase, an enzyme found only in mitochondria. At this magnification the stained mitochondria appear as purple dots; in skeletal muscle there are several different types of cells that differ in the number of mitochondria. (b) Electron micrograph of a thin section of rat heart muscle stained for catalase, an enzyme found in small organelles called peroxisomes. Catalase is detected by treating cell sections with hydrogen peroxide and a dye; the catalase-catalyzed oxidation of the dye produces an insoluble dense precipitate. Also visible (unstained) are mitochondria (M); vesicles of the sarcoplasmic reticulum (SR), a set of membranes that store Ca^{2+} ions; and (in cross section) actin and myosin contractible fibers (F). [Part (a) from P. R. Wheater, H. G. Burkitt, and V. C. Daniels. 1987. *Functional Histology; A Text and Colour Atlas*, 2d ed., Churchill Livingstone, Figure 1.23b, p. 25. Photo Researchers, Inc.; part (b) from H. D. Fahmi and S. Yokota, 1981, Schweiger, H. G., ed., *International Cell Biology 1980–1981*, Springer-Verlag, p. 640. Courtesy of H. D. Fahmi.]

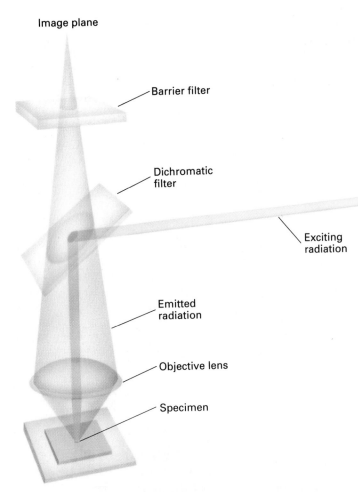

FIGURE 5-9   The epi-fluorescence microscope. Light from a multiwavelength source moves through an exciter filter, which allows only the desired wavelength of exciting radiation to pass. This radiation is reflected by the dichromatic filter and focused by the objective lens onto the sample; fluorescent molecules in the sample are then excited to emit light (fluoresce) at a specific and longer wavelength. This light is focused by the objective lens; most of it passes through the dichromatic filter and is not reflected. A final barrier filter blocks any residual light with the frequency of the exciting radiation.

Three very useful dyes for microscopy are *rhodamine*, Cy3, which emit red light, and *fluorescein*, which emits green light. Such dyes as fluorescein, Cy3, and rhodamine have a low, nonspecific affinity for biological molecules, but they can be chemically coupled to purified antibodies specific to almost any desired macromolecule: a fluorescent dye–antibody complex, when added to a permeabilized cell or tissue section, will bind to the chosen antigens, which then light up when illuminated by the exciting wavelength (Figure 5-10).

Fluorescence microscopy can also be applied to live cells. For example, purified actin may be chemically linked to a fluorescent dye. Careful biochemical studies have es-

tablished that this "tagged" molecule is indistinguishable in function from its normal counterpart: if the tagged protein is *microinjected* into a cultured cell, the cellular and the tagged actin copolymerize into normal long actin fibers. This technique can also be used to study individual microtubules within a cell.

In another technique, a chemically synthesized lipid is covalently linked to a fluorescent dye and added to a cell culture, where it is taken up into all cell membranes—particularly (for unknown reasons) into the endoplasmic reticulum (ER). The fluorescent image from such labeled live cells shows the lacelike ER network (Figure 5-11) that more conventional fluorescence microscopy can only portray in fixed cells.

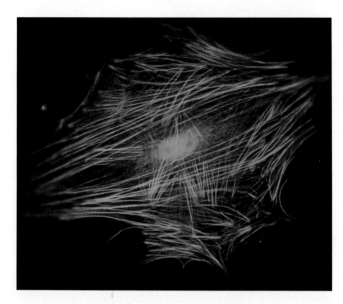

FIGURE 5-10   Fluorescence micrograph showing the distribution of long actin fibers in a cultured fibroblast cell. A fixed human skin fibroblast was permeabilized with a detergent and stained with a fluorescent anti-actin antibody before viewing. [Courtesy of E. Lazarides.]

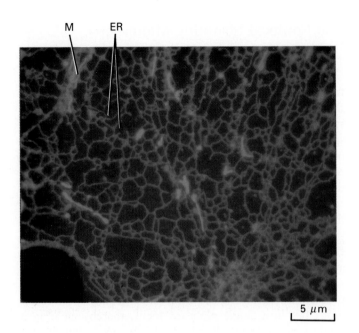

◄ FIGURE 5-11   Fluorescence micrograph of the endoplasmic reticulum (ER) in a flattened region of a living monkey kidney epithelial cell. The ER is a set of elongated membrane vesicles fibers that fuse with each other to form a network, or reticulum. The larger, nonreticular fluorescent structures are mitochondria (M). [See C. Lee and L. B. Chen, 1988. *Cell* **44**:37–46. Courtesy of L. B. Chen.]

Fluorescence Microscopy Can also Measure the Local Concentration of Ca^{2+} Ions and the Intracellular pH

Changes in the cytosolic concentration of Ca^{2+} ions or pH will frequently signal changes in cellular metabolism. The Ca^{2+} concentration in the cytosol of resting cells, for instance, is about 10^{-7} M. Many hormones or other stimuli frequently cause a rise in Ca^{2+} to 10^{-6} M; this, in turn, causes changes in cellular metabolism, such as contraction of muscle, as is described in Chapter 20. The fluorescent properties of certain dyes, such as *fura-2* (Figure 5-12),

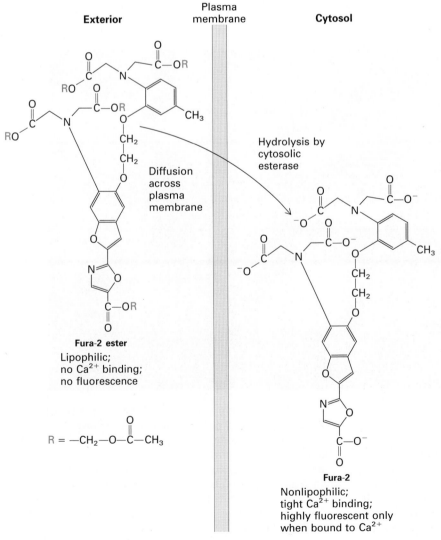

◄ FIGURE 5-12   The cytosolic concentration of Ca^{2+} can be monitored continuously by the fluorescence of Ca^{2+}–fura-2 complexes. When added to a medium, the lipophilic fura-2 ester (*left*) diffuses across the plasma membrane and is hydrolyzed to fura-2 by cytosolic esterases. Nonlipophilic fura-2 (*right*) cannot cross cellular membranes and remains in the cytosol. Fura-2 is not fluorescent unless Ca^{2+} is present, and the fluorescence of Ca^{2+}–fura-2 complexes is proportional to the concentration of Ca^{2+} ion in the cytosol.

facilitate measurement of the concentration of free Ca^{2+} in the cytosol. Fura-2 contains several carboxylate groups that tightly bind Ca^{2+} ions but no other cellular cations. The fluorescence of fura-2 is enhanced when Ca^{2+} is bound; over a certain range of Ca^{2+} concentration, the fluorescence is proportional to this concentration. By examining cells continuously in the fluorescence microscope, one can quantify rapid changes in fura-2 fluorescence and thus in the level of cytosolic Ca^{2+}. The fluorescence of other dyes is sensitive to pH, and can be used in a similar way to monitor the cytosolic pH of living cells.

In large cells, different Ca^{2+} concentrations can actually be detected in specific regions of the cytosol. In one example, Figure 5-13 shows the changes in Ca^{2+} levels that can be seen after fertilization of a sea urchin egg. When the sperm penetrates the egg, the level of Ca^{2+} ions rises in the adjacent region of the cytosol and then gradually increases throughout the egg. This spreading increase in cytosolic Ca^{2+} triggers the fusion of small vesicles with the plasma membrane, causing changes in the cell surface that prevent penetration by additional sperm.

The Confocal Scanning Microscope Produces Vastly Improved Fluorescent Images

Immunofluorescence microscopy has its limitations. The fixatives employed to preserve cell architecture often destroy the *antigenicity* of a protein—its ability to bind to its specific antibody. Also, the method is generally difficult to use on thin cell sections; the embedding media often fluoresce themselves, obscuring the specific signal from the antibody. In microscopy of whole cells, the fluorescent light comes from molecules above and below the plane of focus; thus the observer sees a superposition of fluorescent images from molecules at many depths in the cell, making it difficult to determine the actual three-dimensional molecular arrangement (see Figure 5-10).

The confocal scanning microscope avoids the last problem by permitting the observer to visualize fluorescent molecules in a single plane of focus, thereby creating a vastly sharper image (Figure 5-14). At any instant in this *confocal imaging*, only a single small part of a sample is

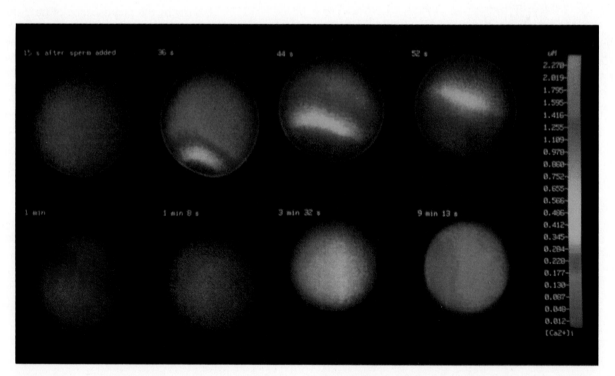

▲ FIGURE 5-13   Changes in the local concentration of Ca^{2+} ions in a sea urchin egg following fertilization. The Ca^{2+} throughout the cell was monitored at different times after fertilization using a microscope and fura-2 fluorescence (see Figure 5-12). For graphic purposes, the Ca^{2+} concentrations are expressed in a calibrated color scale (*right*) in units of micromolar Ca^{2+}. The Ca^{2+} concentration rises initially at the point of sperm entry (the lower left part of the cell); the elevated level then spreads like a wave. Eventually, the Ca^{2+} concentration becomes uniformly high throughout the cell and then falls uniformly to the resting state. [See R. Y. Tsien and M. Poenie, 1986, *Trends Biochem. Sci.* **11**:450–455. Images courtesy of J. Alderton, M. Poenie, R. A. Steinhardt, and R. Y. Tsien.]

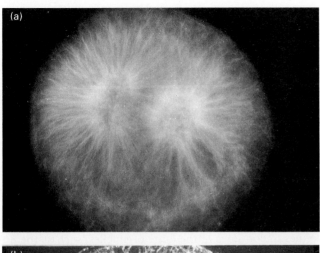

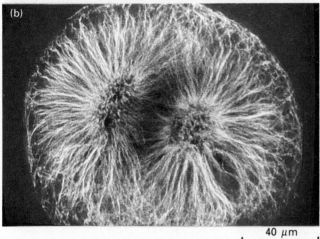

40 μm

▲ FIGURE 5-14   The advantage of confocal fluorescence microscopy. A mitotic fertilized egg from a sea urchin (*Psammechinus*) is lysed with a detergent, exposed to an anti-tubulin antibody, and then exposed to a fluorescein-tagged antibody that binds to the first antibody. (a) When viewed by conventional fluorescence microscopy, the mitotic spindle is blurred due to the background glow of fluorescence from tubulin above and below the plane of focus. (b) The confocal microscopic image is sharp, particularly in the center of the mitotic spindle; fluorescence is detected only from molecules in the focal plane. [From J. G. White, W. Amos, and M. Fordham, 1987, *J. Cell Biol.* **104**:41–48.]

illuminated with exciting light from a focused laser beam, which rapidly moves to different spots in the sample focal plane. Images from these spots are recorded by a video camera and stored in a computer, and the composite image is displayed on a computer screen. By a refinement known as *optical sectioning*, a computer records *serial sections*—fluorescent images of planes at different depths of the sample (Figure 5-15)—and combines the stack of images into one three-dimensional image (here, of the distribution of the protein tubulin within a cell).

(a)

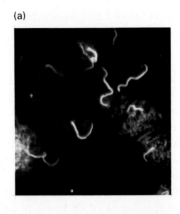

(b)   (c)

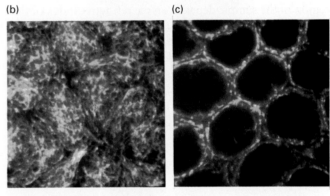

(d)

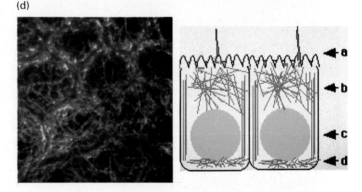

▲ FIGURE 5-15   Optical sectioning of the microtubule cytoskeleton of a cultured kidney epithelial cell (MDCKII), using the confocal microscope and indirect immunofluorescence with an antibody directed against β-tubulin. The epithelial cells were grown on permeable filter supports. Under these conditions the cells develop a functional polarity which is reflected in a polarized cytoskeleton. The images were taken at four different depths, about 4 μm apart at the locations indicated in the drawing. (a) shows the primary cilia which are present in interphase and extend from the apical membrane. (b) is at the level of the apical dense network of microtubules. (c) is at the level of the nuclei; the bright punctate staining are the longitudinal bundles of microtubules which extend from the apical domain to the basal part of the cell. (d) shows the basal network of microtubules which probably extend from the ends of the longitudinal microtubule bundles. [Courtesy of Sigrid Reinsch, Eric Karsenti, and Ernst Stelzer.]

Phase-Contrast and Nomarski Interference Microscopy Visualize Unstained Living Cells

Two optical techniques are notable for giving detailed views of transparent, live, unstained cells and tissues (Figure 5-16). Both convert small differences in refractive index and thickness between parts of the specimen (say, between the nucleus and cytosol) or between the specimen and the surrounding medium to differences of light and dark in the final image (Figure 5-17).

Phase-contrast microscopy (Figure 5-18) generates an image in which the degree of darkness or brightness of a region of the sample depends on the refractive index of that region.

Nomarski, or *differential, interference microscopy* generates an image that looks as if the specimen is casting a shadow to one side (Figure 5-16c and d); the "shadow" primarily represents a difference in refractive index rather than literal thickness. This technique utilizes plane-polarized light. In Nomarski interference micros-

copy, a prism splits the beam of incident light so that one part of the beam passes through one region of a specimen and the other part passes through a closely adjacent region; a second prism then reassembles the two beams. Minute differences in thickness or in the refractive index between adjacent parts of a sample are converted into a bright image (if the two beams are in phase when they recombine) or a dark one (if they are out of phase).

The phase-contrast microscope is especially useful in examining the structure and movement of the larger organelles, such as the nucleus and mitochondria, in live cultured cells. The greatest disadvantage of this technique is that it is suitable for observing only single cells or thin cell layers. The Nomarski interference technique, in contrast, only defines the outlines of the large organelles, such as the nucleus and vacuole (see Figure 5-16c). However, thick objects, such as the nuclei in a worm (Figure 5-16d), can be observed by combining this technique with optical sectioning.

Both phase-contrast and Nomarski interference microscopy can be used in *time-lapse microscopy,* in which

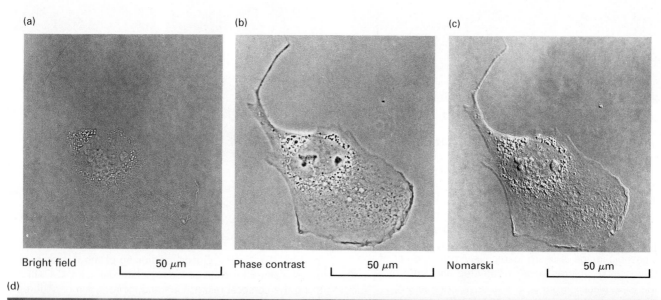

(a) Bright field   50 μm

(b) Phase contrast   50 μm

(c) Nomarski   50 μm

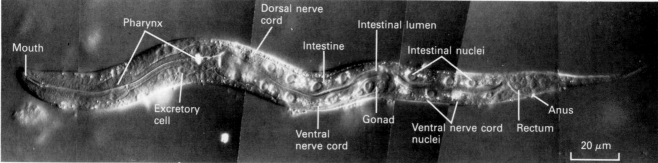

(d)

Mouth — Pharynx — Dorsal nerve cord — Intestine — Intestinal lumen — Intestinal nuclei — Excretory cell — Ventral nerve cord — Gonad — Ventral nerve cord nuclei — Rectum — Anus — 20 μm

▲ **FIGURE 5-16**   A live, cultured CV-1 fibroblast cell viewed by (a) bright field microscopy, where few subcellular structures are visible, (b) phase-contrast microscopy and (c) Nomarski interference (differential interference) microscopy. (d) A newly hatched larvae of the nematode *Caenor-* *habditis elegans,* viewed with Nomarski optics. The individual nuclei of many of the organism's 500 cells are visible. [Parts a, b, c courtesy of Matthew J. Footer; part (d) from J. E. Sulston and H. R. Horvitz, 1977, *Devel. Biol.* **56**:110.]

(a)

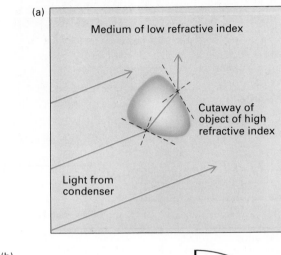

(b)

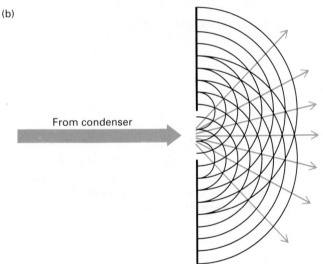

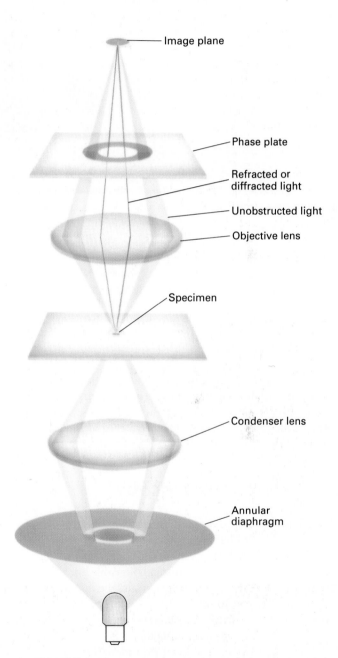

▲ FIGURE 5-17   Two ways in which a specimen can redirect light from a condenser lens. (a) Light moves more slowly in a medium of higher refractive index. Thus, a beam of light is refracted (bent) once as it passes from air into a transparent object and again when it departs. Consequently, part of a light wave that passes through a specimen will be refracted and will be out of phase (out of synchrony) with the part of the wave that does not pass through the specimen. How much their phases will differ depends on the difference in refractive index along the two paths and on the thickness of the specimen. If the two parts of the light wave are recombined, the resultant light will be brighter if they are in phase and less bright if they are out of phase. (b) Light waves impinging on a pinhole in an opaque object spread out in all directions. Overlapping waves emanating from different sides of the hole reinforce each other in some directions (straight arrows); to an observer in one of those directions, the pinhole will seem bright. This phenomenon is called diffraction. Similarly, when light impinges on an opaque object, the edges diffract the light waves, and the edges will seem bright when viewed in some directions (straight arrows) and dark in others.

▲ FIGURE 5-18   The optical pathway of the phase-contrast microscope. Incident light passes through an annular diaphragm, which focuses a circular annulus (ring) of light on the sample. Light that passes unobstructed through the specimen is focused by the objective lens onto the thick gray ring of the phase plate, which absorbs some of the direct light and alters its phase by one-quarter of a wavelength. If a specimen refracts or diffracts the light, the phase of some light waves is altered and the light waves are redirected through the thin, clear region of the phase plate. The refracted and unrefracted light are recombined at the image plane to form the image.

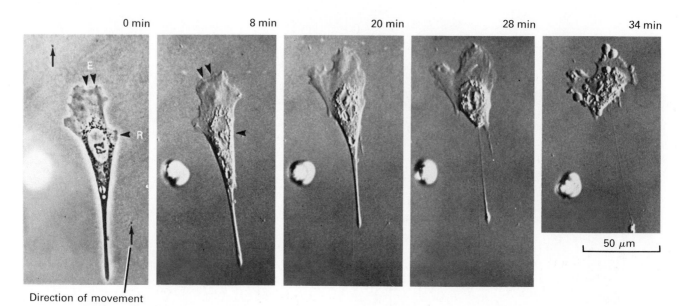

0 min   8 min   20 min   28 min   34 min

50 μm

Direction of movement

▲ FIGURE 5-19   Time-lapse micrographs show the movement of a cultured fibroblast cell along a glass surface. A bit of debris on the substratum serves as a reference point. The first image, at 0 min, was obtained by phase-contrast microscopy. Successive images of the same cell, obtained by Nomarski optics, show the lamella at the right of the cell retracting (R) and the lamellipodia at the leading edge of the cell extending (E). In the frame taken at 8 min, the leading edge has moved forward about 9 μm and the lamellipodia there form a thin flat sheet. By 28 min, the broad leading edge has spread and separated into two lamellae; the thin trailing edge of the cell has begun to retract into the cell body. By 34 min, retraction of the trailing edge is almost complete; only a thin thread of cytoplasm from the trail is left behind, anchored to the substratum. [From W.-T. Chen, 1981, *J. Cell Sci.* **49**:1.]

the same cell is photographed at regular intervals over periods of several hours. This procedure allows the observer to study cell movement, provided the microscope's stage can control the temperature of the specimen and the gas environment (Figure 5-19).

➤ *Electron Microscopy*

The fundamental principles of electron microscopy are similar to those of light microscopy: the major difference is that electromagnetic lenses, not optical lenses, focus a high-velocity electron beam instead of visible light. Because electrons are absorbed by atoms in air, the entire tube between the electron source and the viewing screen is maintained under an ultra-high vacuum.

Quantum physics teaches us that a beam of subatomic particles such as electrons can be considered either as a stream of discrete particles or as a series of waves. In typical electron microscopes, electrons have the properties of a wave with a wavelength of only 0.005 nm. Recall that the minimum distance D at which two objects can be distinguished is proportional to the wavelength λ of the light that illuminates the objects. Thus the limit of resolution for the electron microscope is theoretically 0.005 nm (less than the diameter of a single atom), or 40,000 times better than the resolution of the light microscope and 2 million times

better than that of the unaided human eye. The effective resolution of the electron microscope in the study of biological systems is considerably less than this ideal, however: under optimal conditions, a resolution of 0.10 nm can be obtained with transmission electron microscopes.

Transmission Electron Microscopy Depends on the Differential Scattering of a Beam of Electrons

The *transmission electron microscope* directs a beam of electrons through a specimen. Electrons are emitted by the tungsten *cathode* when it is electrically heated. The electric potential of the cathode is kept at 50,000–100,000 volts; that of the *anode*, near the top of the tube, is zero. This drop in voltage causes the electrons to accelerate as they move toward the anode. A *condenser lens* focuses the electron beam onto the sample; *objective* and *projector lenses* focus the electrons that pass through the sample and project them onto a viewing screen or a piece of photographic film (Figure 5-20).

Like the light microscope, the transmission electron microscope is used to view thin sections of a specimen, but the fixed sections must be much thinner for electron microscopy (only 50–100 nm, about 0.2 percent of the thickness of a single cell; Figure 5-21). Clearly, only a small piece of a cell can be observed in any one section. The

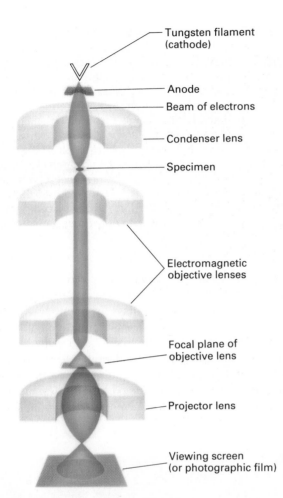

▲ FIGURE 5-20   The optical path in a transmission electron microscope. A beam of electrons emanating from a heated tungsten filament is focused onto the specimen plane by the magnetic condenser lens. The electrons passing through the specimen are focused by a series of objective and projector lenses to form a magnified image of the specimen on a fluorescent viewing screen or a piece of photographic film. The entire column, from the electron generator to the screen, is maintained at a very high vacuum.

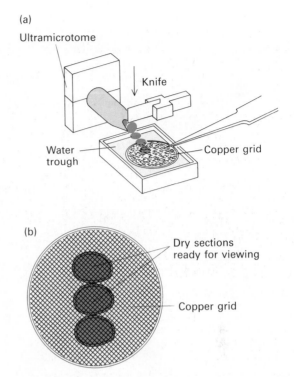

▲ FIGURE 5-21   Preparation of a sample of tissue for transmission electron microscopy. The tissue is dissected, cut into small cubes, and plunged into a fixing solution that cross-links and immobilizes proteins. (Glutaraldehyde is frequently used; osmium tetroxide, another fixing substance, stains intracellular membranes and certain macromolecules.) The sample is dehydrated by placing it in successively more concentrated solutions of alcohol or acetone; it is then immersed in a solution of plastic embedding medium and put in an oven. Heat causes the solution to polymerize into a hard plastic block, which is trimmed; sections less than 0.1 μm thick are then cut with an ultramicrotome, a fine slicing instrument with a diamond blade (a). The sections are floated off the blade edge onto the surface of water in a trough. A copper grid coated with carbon or some other material is used to pick up the sections, which are then dried (b).

image depends on variations in the scattering of the incident electrons by molecules in the preparation. Without staining, the beam of electrons passes through a cell or tissue sample uniformly: the entire sample appears uniformly bright and there is little differentiation of components. Staining techniques are therefore used to reveal the location and distribution of specific materials. Heavy metals, such as gold or osmium, appear dark on the micrograph because they scatter (diffract) most of the incident electrons; scattered electrons are not focused by the electromagnetic lenses and do not form the image. Osmium tetroxide preferentially stains certain cellular components, such as membranes, which appear black in the micrographs (the transmission electron micrographs in Figures 5-1 to 5-4 were taken of sections stained with this sub-

stance). Gold particles serve as electron-dense tags; when coated with proteins such as protein A that, in turn, bind antibody molecules, they can detect specific target proteins in thin sections (Figure 5-22).

Minute Details Can Be Visualized on Viruses and Subcellular Particles

The transmission electron microscope is also used to obtain information about the shapes of purified viruses, fibers, enzymes, and other subcellular particles. In one technique, called *metal shadowing*, a thin layer of evaporated metal, such as platinum, is laid at an angle on a biological specimen (Figure 5-23). An acid bath dissolves the biological material, leaving a metal *replica* of its surface which can

(a)

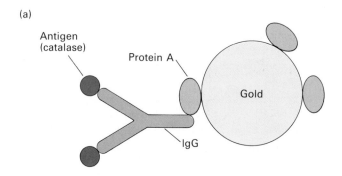

(b)

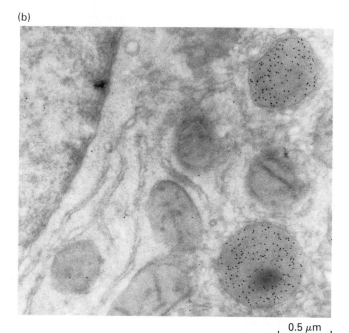

0.5 μm

◄ FIGURE 5-22   The use of antibodies to detect the subcellular location of a specific protein, the enzyme catalase, by electron microscopy. A slice of liver tissue is fixed with glutaraldehyde and sectioned. An IgG antibody to catalase is then added to the section. When complexed with gold particles 4 nm in diameter, protein A from the bacterium *Staphylococcus aureus* binds tightly to the common Fc domain (see Figure 3-32) of the antibody. Each gold particle makes the resulting immune complex (a) visible in the electron microscope. (b) The location of the gold particles (black dots) in the peroxisomes indicates the presence of the catalase. [From H. J. Geuze et al., 1981, *J. Cell Biol.* **89**:653. Reproduced from the *Journal of Cell Biology* by copyright permission of The Rockefeller University Press.]

then be examined in the transmission electron microscope (Figure 5-24). If the metal is deposited on only one side of the sample, the image seems to have "shadows," where the metal appears dark and the shadows appear light.

Electron microscopy cannot be used to study live cells because they are generally too vulnerable to the required conditions and preparatory techniques. In particular, the absence of water causes macromolecules to become denatured and nonfunctional. However, one technique allows

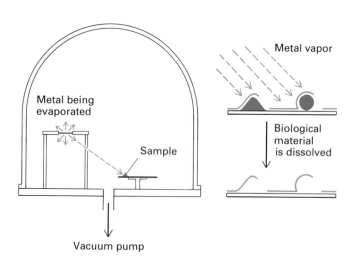

▲ FIGURE 5-23   Metal shadowing. This technique makes surface details on very small particles visible in the electron microscope. The sample on a grid is placed in a vacuum container. A filament of a heavy metal, such as platinum, is heated so that the metal evaporates and some of it falls over the sample grid in a very thin film. The biological material is then dissolved by acid, so that the observer views only the metal replica of the sample.

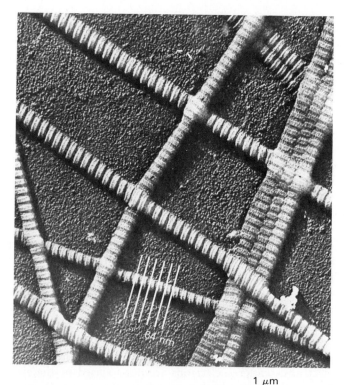

1 μm

▲ FIGURE 5-24   A platinum-shadowed replica of the substructural fibers of calfskin collagen, the major structural element of tendons, bone, and similar tissues. The fibers are about 200 nm thick; a characteristic 64-nm repeated pattern (white parallel lines) is visible along the length of each fiber. [Courtesy of R. Bruns.]

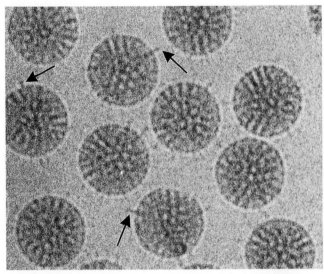

100 nm

▲ FIGURE 5-25   Electron micrograph of unstained rotavirus particles. A thin suspension of virus particles in water is applied to an electron microscopy grid and frozen. It is then visualized in a transmission electron microscope equipped with a sample stage cooled with liquid nitrogen. The low temperature prevents the ice surrounding the particles from evaporating in the vacuum. Because many biological specimens scatter more electrons than water does, the investigator can observe a very thin specimen without fixing, staining, or dehydrating it. Note the minute spikes (arrows) visible on some particles. [From B. V. Venkataram Prasad, et al., 1988, *J. Mol. Biol.* **199**:269–274.]

the observer to examine hydrated, unfixed, and unstained biological specimens directly in the electron microscope. An aqueous suspension of a sample is applied in an extremely thin film to a grid, is frozen in liquid nitrogen and maintained in this state by means of a special mount, and is then observed in the electron microscope. The very low temperature (−196°C) keeps the water from evaporating, even in a vacuum, and the sample can be observed in detail in its native, hydrated state without shadowing or fixing it (Figure 5-25).

Scanning Electron Microscopy Visualizes Details on the Surface of Cells or Particles

The *scanning electron microscope* allows the investigator to view the surfaces of unsectioned specimens, which cannot be viewed with transmission equipment. The sample is fixed and dried and coated with a thin layer of a heavy metal, such as platinum, by evaporation in a vacuum (see Figure 5-23); in this case, the sample is rotated so that the platinum is deposited uniformly on the surface. An intense electron beam inside the microscope scans rapidly over the sample. Molecules in the specimen are excited and release

secondary electrons that are focused onto a scintillation detector; the resulting signal is displayed on a cathode-ray tube. Because the number of secondary electrons produced by any one point on the sample depends on the angle of the electron beam in relation to the surface, the scanning electron micrograph has a three-dimensional appearance (see Figure 5-4a). The resolving power of scanning electron microscopes is only about 10 nm, much less than that of transmission instruments.

➤ *Sorting Cells and Their Parts*

Most animal and plant tissues contain a mixture of cell types. However, an investigator often wishes to study a pure population of one type of cell. In some cases, cells differ in some physical property that allows them to be separated. White blood cells (leukocytes) and red blood cells (erythrocytes), for instance, have very different densities because erythrocytes have no nucleus, and so these cells can be separated on the basis of density. Most cell types cannot be differentiated so conveniently, however, and other cell-separation techniques have had to be developed.

Flow Cytometry Is Used to Sort Cells Optically

A *flow cytometer* can identify different cells by measuring the light they scatter, or the fluorescence they emit, as they flow through a laser beam; thus it can sort out cells of a particular type from a mixture. Indeed, a *fluorescence-activated cell sorter,* or *FACS* (Figure 5-26), an instrument based on flow cytometry, can select one cell from thousands of other cells. For example, if an antibody specific to a certain cell-surface molecule is linked to a fluorescent dye, any cell bearing this molecule will bind the antibody and will then be separated from other cells when it fluoresces in the FACS. Once sorted from the other cells, the selected cell can then be grown in culture.

Other uses of flow cytometry include the measurement of a cell's DNA and RNA content and the determination of its general shape and size. The FACS can make simultaneous measurements of the size of a cell (from the amount of scattered light) and the amount of DNA it contains (from the amount of fluorescence from a DNA-binding dye).

Fractionation Methods Isolate Subcellular Structures

Although microscopy can localize a particular protein to a specific subcellular fiber or organelle, it is essential to isolate quantities of each of the major subcellular organelles to study their structures and metabolic functions in detail. Rat liver has been used in many classic studies because it is

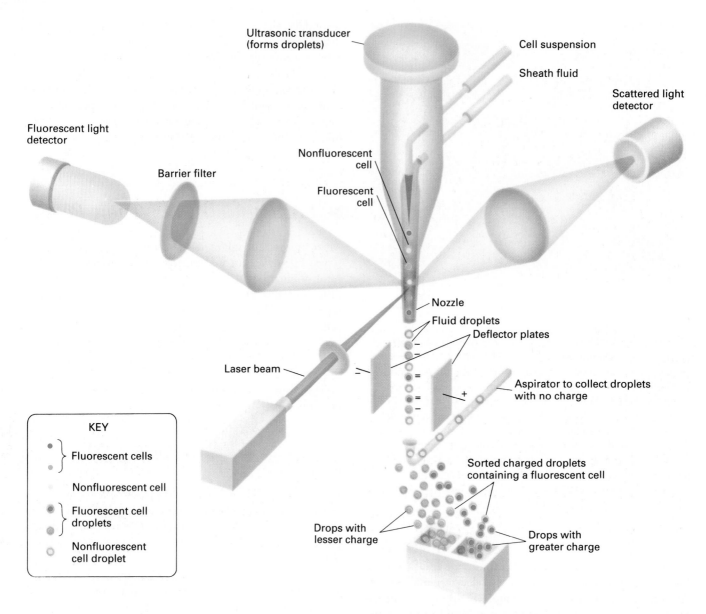

▲ FIGURE 5-26   The fluorescence-activated cell sorter (FACS). A concentrated suspension of cells is allowed to react with a fluorescent antibody or dye that binds to a particle or molecule such as DNA. The suspension is then mixed with a buffer (the sheath fluid), the cells are passed single-file through a laser light beam, and the fluorescent light emitted by each cell is measured. The light scattered by each cell can be measured simultaneously; from this the size and shape of the cell can be determined. The suspension is then forced through a nozzle, which forms tiny droplets containing at most a single cell. At the time of formation, each droplet is given an electric charge proportional to the amount of fluorescence of its cell. Droplets with no charge or different electric charges (due to different amounts of bound dye) are separated by an electric field and collected. It takes only milliseconds to sort each droplet, so up to 10 million cells per hour can pass through the machine. In this way, cells that have desired properties can be separated and then grown. [After D. R. Parks and L. A. Herzenberg, 1982, *Methods Cell Biol.* **26**:283.]

abundant in a single cell type (Figure 5-27). However, the same isolation principles apply to virtually all cells and tissues, and modification of these cell fractionation techniques can be used to separate and purify any desired components.

Disrupting the Cell to Release Its Contents   The initial step for purifying subcellular structures is to rupture the cell wall, the plasma membrane, or both. First, the cells are suspended in a solution of appropriate pH and salt content, usually isotonic sucrose (0.25 M) or a combination of salts similar in composition to those in the cell's interior. Many cells can then be broken by stirring the cell suspension in a high-speed blender or by exposing it to high-frequency sound (*sonication*). Plasma membranes can also be sheared by special pressurized tissue homogenizers in which the cells are forced through a very narrow space between the plunger and the vessel wall. Generally, the cell

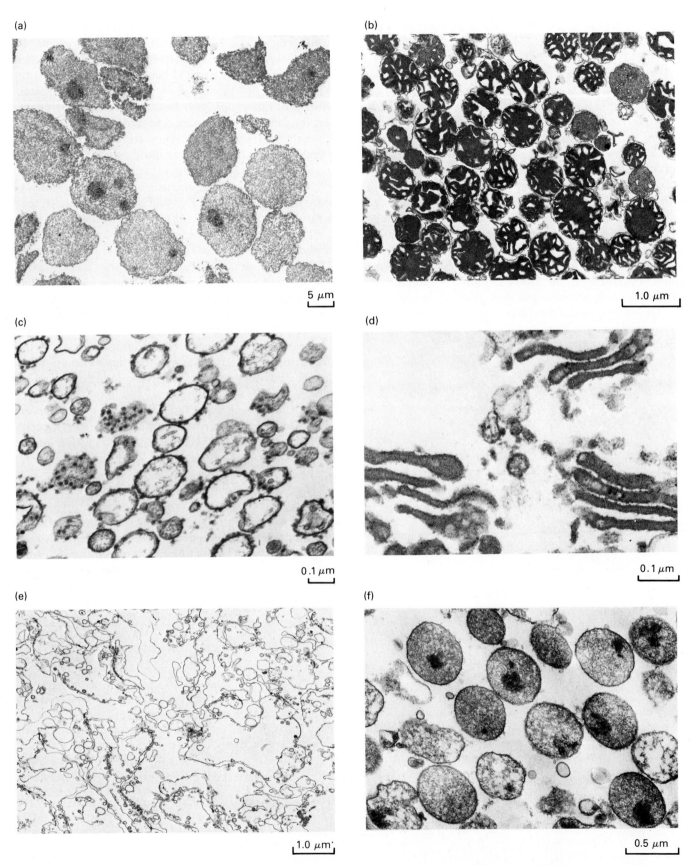

(a)

5 μm

(b)

1.0 μm

(c)

0.1 μm

(d)

0.1 μm

(e)

1.0 μm

(f)

0.5 μm

▲ FIGURE 5-27   Electron micrographs of purified rat liver organelles: (a) nuclei, (b) mitochondria, (c) rough endoplasmic reticulum, sheared into smaller vesicles termed microsomes, (d) Golgi vesicles, (e) plasma membranes, and (f) peroxisomes. [Parts (a)–(e) courtesy of S. Fleischer and B. Fleischer; part (f) courtesy of P. Lazarow.]

solution is kept at 0°C to best preserve enzymes and other constituents after their release from the stabilizing forces of the cell.

Sometimes *osmotic flow* (Figure 5-28) is enlisted to aid in rupturing the cells by placing them in a *hypotonic solution,* one with a lower concentration of salts than is in the cell's interior. This is done because the plasma membrane is freely permeable to water but poorly permeable to the salts and other small molecules—the solutes—within the cell. Recall that water flows across a semipermeable membrane in the direction that will yield the same concentration of water on both sides of the membrane; thus water will flow from a solution of high water (low solute) concentration to one of low water (high solute) concentration. Consequently, water flows from the low solute concentration (the *hypotonic solution*) outside the cell to the normal solute concentration (the *isotonic solution*) within the cell, causing the cells to swell and then more easily rupture.

Disrupting the cell produces a mix of suspended cellular components, the *homogenate,* from which the desired organelles must now be retrieved.

Centrifuging the Homogenate to Purify Organelles   In Chapter 3 we discussed the principles of centrifu-

(a) Isotonic medium

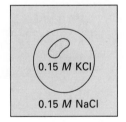

(b) Hypotonic medium

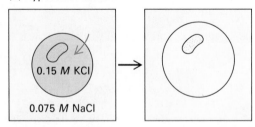

(c) Hypertonic medium

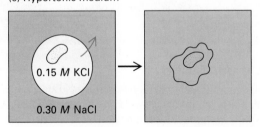

gation and the uses of centrifugation techniques for separating proteins and nucleic acids. Similar approaches are used for separating and purifying the various organelles, which differ in both size and density.

Most fractionation procedures begin with *differential-velocity,* or *rate-zonal, centrifugation* (Figure 5-29). Generally, the cellular homogenate is first filtered or centrifuged at relatively low speeds to remove unbroken cells. Then centrifuging at a slightly faster speed or for a longer duration will selectively pellet the nucleus—the largest organelle (usually 5–10 μm in diameter). A centrifugal force of 600 g (600 times the force of gravity) is necessary to sediment nuclei; this is generated by a typical centrifuge rotor operating at 500 revolutions per minute (rpm). The undeposited material is next centrifuged at a higher speed (15,000 g × 5 min), which deposits the mitochondria, chloroplasts, lysosomes, and peroxisomes. A subsequent centrifugation in the ultracentrifuge (100,000 g × 60 min) results in the deposition of the plasma membrane and fragments of the endoplasmic reticulum. A force of 100,000 g requires about 50,000 rpm in an ultracentrifuge; at this speed, the rotor chamber is kept in a high vacuum to reduce heating due to friction between air and the spinning rotor. The recovery of ribosomal subunits, small polyribosomes, and particles such as complexes of enzymes requires additional centrifugation at still higher speeds. Only the *cytosol*—the soluble aqueous portion of the cytoplasm—remains undeposited after centrifugation at 300,000 g × 2 h.

Rate-zonal centrifugation does not yield totally pure organelle fractions. One method for further purifying fractions is *equilibrium density-gradient centrifugation,* which separates cellular components according to their density. The impure organelle fraction is layered on top of a solution that contains a gradient of a dense nonionic substance, such as sucrose or glycerol. The tube is centrifuged at a high speed (about 40,000 rpm) for several hours, allowing

◄ FIGURE 5-28   Animal cells respond to the osmotic strength of the surrounding medium. Sodium, potassium, and chloride ions do not move freely across the cell membrane, but water does (arrows). (a) When the medium is isotonic, there is no net flux of water into or out of the cell. (b) When the medium is hypotonic, water flows into the cell until the ion concentration inside and outside the cell is the same. Here, the initial cytosolic ion concentration is twice the extracellular ion concentration, so the cell tends to swell to twice its original volume, at which point the internal and external ion concentrations are the same. (c) When the medium is hypertonic, water flows out of the cell until the ion concentration inside and outside the cell is the same. Here, the initial cytosolic ion concentration is one-half the extracellular ion concentration, so the cell is reduced to one-half its original volume.

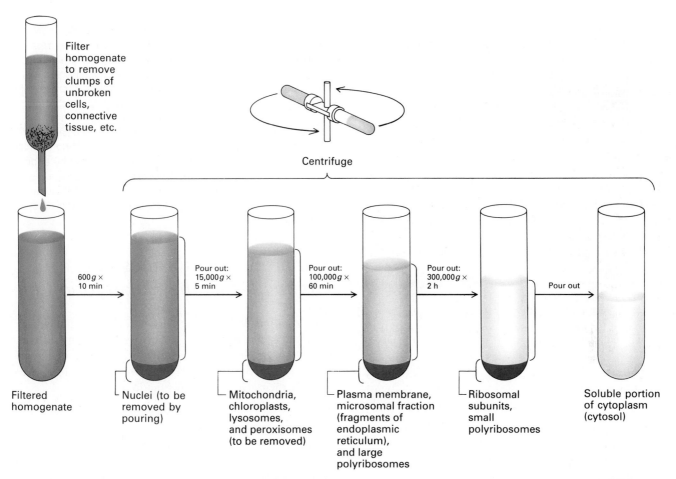

▲ FIGURE 5-29 Cell fractionation by rate-zonal centrifugation. The different sedimentation rates of various cellular components make it possible to separate them partially by centrifugation. Nuclei and viral particles can sometimes be purified completely by such a procedure. Each soluble fraction can be further separated by density-gradient centrifugation.

each particle to migrate to an equilibrium position where the density of the surrounding liquid is equal to the density of the particle. In typical preparations from animal cells, the rough endoplasmic reticulum (with a density of 1.20 g/cm^3) separates well from the Golgi vesicles (with a density of 1.14 g/cm^3) and from the plasma membrane (with a density of 1.12 g/cm^3). (The higher density of the rough endoplasmic reticulum is due largely to the ribosomes bound to it.) This method also works well for the resolution of lysosomes, mitochondria, and peroxisomes from the same initial fraction of a rate-zonal centrifugation (Figure 5-30).

Using Immunological Techniques to Obtain Pure Preparations of Organelles Many fractions are not pure even after rate-zonal and equilibrium density-gradient centrifugation. The next step is to take advantage of the fact that the membrane of each organelle contains unique enzymes and other proteins, allowing one to purify organelles by immunological techniques, using antibodies specific for these organelle-membrane proteins. One example is the purification of a particular class of cellular vesicles that contain the protein *clathrin* on their outer surface (Figure 5-31). An antibody to clathrin, bound to a bacterial carrier, selectively removes these vesicles from a crude preparation of membranes. A recent advance is the use of tiny metallic beads that are coated with specific antibodies. Organelles bound to these beads can be purified simply by adhesion to a small magnet on the side of the test tube.

Assaying for Purity The purity of organelle preparations can be assessed by examining morphology (through use of the electron microscope) or by quantifying marker molecules. For example, the protein cytochrome *c* is present only in mitochondria, so finding this protein in a fraction of lysosomes would indicate its contamination by mitochondria. Similarly, catalase is present only in peroxisomes; acid phosphatase, only in lysosomes; and ribosomes, only in the rough endoplasmic reticulum or the cytosol.

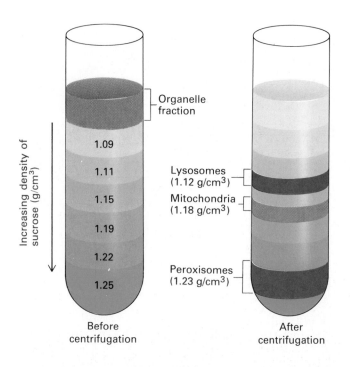

◄ FIGURE 5-30   Separation of organelles from rat liver by density-gradient centrifugation. Material in a pellet from a separation by rate-zonal centrifugation at 15,000 *g* (see Figure 5-29) is resuspended and layered on a density gradient composed of layers of increasingly more dense sucrose solutions in a centrifuge tube. Under centrifugation, each organelle migrates to its appropriate equilibrium density and remains there. Before cell disruption in this particular experiment, the liver is perfused with a solution containing a small amount of detergent, which is taken into the cells by endocytosis and transferred to the lysosomes but does not lyse them. Thus the lysosomes are less dense than they would normally be, affording a "clean" separation from the mitochondria.

➤ *The Biomembranes and Organelles of the Eukaryotic Cell*

Later chapters will examine the key roles played in eukaryotic cellular metabolism by the cytosol, the many different cellular organelles, and the membranes that define the boundaries of the organelles. Here we present a brief overview of their structures and functions.

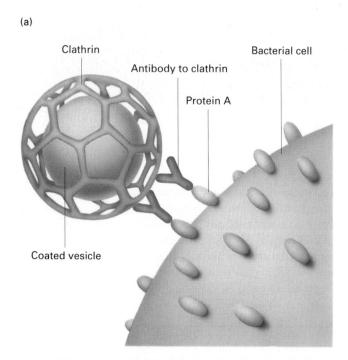

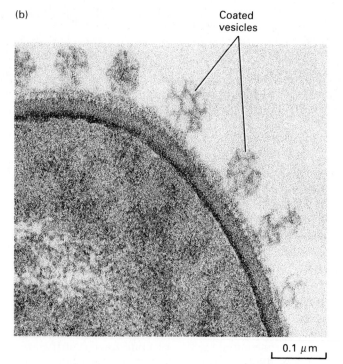

▲ FIGURE 5-31   Immunological purification of clathrin-coated vesicles. (a) A suspension of membranes from rat liver is incubated with an antibody to a clathrin antibody (a protein that coats the outer surface of certain cytoplasmic vesicles). The surface of the bacterium *Staphylococcus aureus* contains protein A, which binds antibodies nonspecifically (see Figure 5-22). By adding bacteria to the mixture of antibodies and membranes, and then recovering the bacteria by low-speed centrifugation, the clathrin-coated vesicles bound (by the antibody and protein A) to the bacterium are selectively purified. (b) A thin-section electron micrograph of clathrin-coated vesicles bound to these bacteria. [See E. Merisko, M. Farquahr, and G. Palade, 1982, *J. Cell Biol.* **93**:846–858; part (b) courtesy of G. Palade.]

Biomembranes (Chapter 14) separate two aqueous compartments. Separating the cell cytoplasm from the extracellular medium is the plasma membrane, the semipermeable barrier that defines the outer perimeter of the cell; other membranes similarly separate the interior of organelles from the cytosol. In Chapter 2 we learned that a phospholipid bilayer (see Figure 2-17) is the structural foundation for all biological membranes. Because of their 3-nm-thick hydrophobic core, bilayers are impermeable to salts, sugars, and other small hydrophilic molecules. Proteins are essential constituents of all biomembranes, and allow the various cellular membranes to function in specific ways. All specific functions of all cellular membranes are determined by the types of membrane proteins that are present. Membranes also provide anchoring points for many cytoskeletal proteins.

The Plasma Membrane Has Many Varied and Essential Roles

Some of the essential functions of the plasma membrane are to allow nutrients to enter the cell, to keep out unwanted materials in the extracellular milieu, to transport metabolic wastes out into the extracellular fluid, and to prevent needed metabolites and ions from leaving the cell. The plasma membrane also maintains the proper ionic composition and osmotic pressure of the cytosol. To serve these functions, the plasma membrane contains specific transport proteins that allow the passage of certain small molecules but not others.

Several major functions of the plasma membrane involve communications and interactions between cells. Few of the cells in a multicellular animal exist as isolated entities; rather, groups of cells with related specializations combine to form tissues, as discussed in detail in Chapter 24. Specialized areas of the plasma membrane contain proteins that make contact with other cells to strengthen tissues and to allow the exchange of metabolites between cells (Figure 5-32). Other proteins in the plasma membrane act as anchoring points for many of the fibrous structures that permeate the cytosol.

The Eukaryotic Nucleus Is Bounded by a Double Membrane

The eukaryotic nucleus, the largest organelle, is surrounded by two membranes, each one a phospholipid bilayer containing many different types of proteins. The inner nuclear membrane defines the nucleus itself. In many cells, the outer nuclear membrane is continuous with the rough endoplasmic reticulum (an extensive cytoplasmic membrane system to be discussed later), and the space between the inner and outer nuclear membranes is continu-

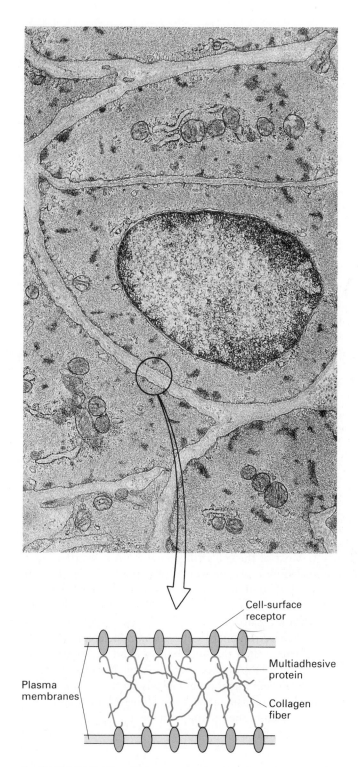

▲ FIGURE 5-32 Electron micrograph of smooth muscle in the wall of a small artery. The muscle cells are separated by relatively wide spaces that contain fibrous collagen and proteoglycans. Connections between plasma membrane receptor proteins and components of the extracellular matrix allow the cells to adhere to each other and give this tissue its strength and resistance to shear forces, as is depicted schematically in the diagram. [From D. W. Fawcett, 1981, *The Cell*, 2d ed., Saunders/Photo Researchers, Inc.]

ous with the lumen, or inner cavity, of the rough endoplasmic reticulum.

The two nuclear membranes appear to fuse at the nuclear pores (Figure 5-33). The distribution of nuclear pores is particularly vivid when the nucleus is viewed by the *freeze-fracture technique* (Figures 5-33b and 5-34). These ringlike pores are constructed of a specific set of membrane proteins, and they function as channels that regulate the movement of material between the nucleus and the cytosol.

The Nucleus Contains the Nucleolus, a Fibrous Matrix, and DNA-Protein Complexes

The major physiological function of the nucleus is to direct the synthesis of RNA. In a growing or differentiating cell, the nucleus is the site of vigorous metabolic activity. In other cells, such as resting mast cells (blood-borne cells that release histamine when triggered by allergens) and mature erythrocytes from nonmammalian vertebrates, the nucleus is inactive or dormant and minimal synthesis of DNA and RNA takes place.

Investigators still know little about the large-scale organization of the cell nucleus, although they have gained some knowledge of how DNA is packaged in the nucleus. In a nucleus that is not dividing, the chromosomes are dispersed and are only about 25 nm thick; they cannot be observed in the light microscope. However, a suborganelle of the nucleus, the *nucleolus*, is easily recognized under the light microscope. The *nucleolar organizer*, a region of one or more chromosomes in the nucleolus, contains many copies of the DNA that directs the synthesis of ribosomal RNA. Most of the cell's ribosomal RNA is synthesized in the nucleolus; some ribosomal proteins are added to ribosomal RNAs within the nucleolus as well. The finished or partly finished ribosomal subunit passes through a nuclear pore into the cytosol.

In the electron microscope, the nonnucleolar regions of the nucleus, called the *nucleoplasm*, can be seen to have areas of high DNA concentration, often closely associated with the nuclear membrane. Fibrous proteins called *lamins* form a two-dimensional network along the inner surface of the inner membrane, giving it shape and apparently binding DNA to it.

The Cytosol Contains Many Cytoskeletal Fibers and Particles

The cytosol is the fluid region of the cell cytoplasm that lies outside of the organelles. Initially, the cytosol was thought to be a fairly homogeneous "soup" in which all of the organelles floated. It is now known that the cytosol of eukaryotic cells contains a *cytoskeleton* composed of at least

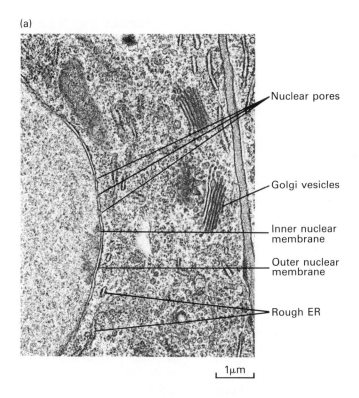

(a)

Nuclear pores

Golgi vesicles

Inner nuclear membrane

Outer nuclear membrane

Rough ER

1 μm

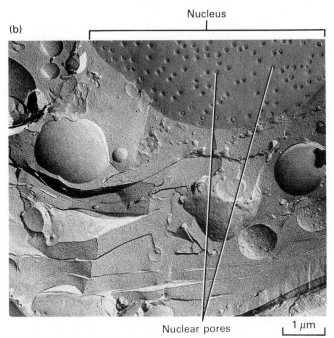

(b)

Nucleus

Nuclear pores

1 μm

▲ FIGURE 5-33   Views of the nuclear envelope. (a) Electron micrograph of a thin section of an *Arabidopsis* (plant) cell, showing the pores between the nuclear membranes. The rough endoplasmic reticulum and stacks of Golgi vesicles are also visible. (b) A freeze-fractured preparation of an onion root-tip cell, showing the nucleus and pores in the nuclear membrane, which appear to be filled with a granular substance. [Part (a) courtesy of Biophoto Associates/Myron C. Ledbetter/Brookhaven National Laboratory; part (b) courtesy of D. Branton.]

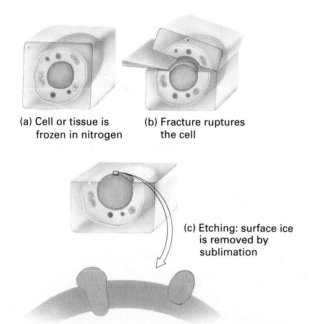

(a) Cell or tissue is frozen in nitrogen

(b) Fracture ruptures the cell

(c) Etching: surface ice is removed by sublimation

Exposed nuclear membrane with membrane proteins and particles

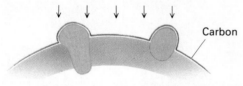

Carbon

(d) Carbon is added to form a continuous surface

Platinum

(e) Surface is shadowed with a thin layer of platinum

(f) Tissue is dissolved with acid; carbon-metal replica can be viewed under the electron microscope

▲ FIGURE 5-34   Production of a freeze-fractured, deep-etched image of a cell. (a) A preparation of cells or tissues is quickly frozen in liquid nitrogen at −196°C, which instantly immobilizes cell components. (b) The block of frozen cells is fractured with a sharp blow from a cold knife. The fracture plane is irregular, usually occurring along the plasma membrane or surfaces of organelles. (c) The specimen is then placed in a vacuum, where the surface ice is removed by sublimation; this technique is called deep-etching or freeze-etching. (d) A thin layer of carbon is evaporated vertically onto the surface to produce a carbon replica. After metal shadowing (see Figure 5-23) with platinum (e), the organic material is removed by acid (f), leaving a carbon-metal replica of the tissue ready for examination with the electron microscope. In prints of the electron micrograph the image is usually reversed: carbon-coated areas appear light and platinum-shadowed areas appear dark.

three classes of fibers—tubulin-containing microtubules (20 nm in diameter), actin microfilaments (7 nm in diameter), and intermediate filaments (10 nm in diameter). At least five types of intermediate filaments, each made of a different type of protein, have been identified in various animal cells. The cytoskeleton helps to maintain cell shape and mobility and provides anchoring points for other cellular structures.

Most cytosolic fibrous sytems are several microns in length, but they can be seen as long elements only in sections that by chance happen to be in the plane of the fiber bundles (Figure 5-35). This is because sections of cells or tissues for standard electron microscopy must be thinner than 0.1 μm. Serial sectioning of the tissue sample can compensate for these shortcomings by tracing a fiber from one image into the next to reconstruct its three-dimensional architecture. However, serial sectioning can be tedious, as 200 sections are needed to examine a cell 20 μm thick. High-voltage electron microscopes improve matters somewhat: they employ electric potential differences of one million volts and can be used to view sections up to 1 μm thick, so that fewer sections are needed for reconstruction of three-dimensional images.

Transmission electron microscopy can provide a striking view of cytosolic fibers in an unsectioned cell (Figure 5-36). The cell is first treated with nonionic detergents to dissolve the plasma membrane and the organelle membranes and to remove the cytosol. The insoluble cytoskeleton that remains is then cooled to the temperature of liquid helium (−269°C, or 4° above absolute zero) within milliseconds by the *quick-freeze technique,* which prevents the formation of ice crystals and distortions in cytoskeletal architecture. By this technique, the cytosol of cultured animal cells is resolved almost exclusively into a network of microfilaments, microtubules, and intermediate filaments, which crisscross each other in complex patterns so that different types of cytoskeletal fibers contact each other at many points. In cultured cells, actin microfilaments often occur in bundles of long fibers (see Figure 5-10) that appear to be connected by small fibrous proteins.

The cytosol of many cells also contains *inclusion bodies,* granules that are not bounded by a membrane. Some cells—specifically, muscle cells and hepatocytes (the principal cell type in the liver)—contain cytosolic granules of glycogen (see Figure 5-37), a glucose polymer that functions as a storage form of usable cellular energy. In well-fed animals, glycogen can account for as much as 10 percent of the wet weight of the liver. The cytosol of the specialized fat cells in adipose tissue contains large droplets of almost pure triacylglycerols, a storage form of fatty acids.

In addition, the cytosol is a major site of cellular metabolism and contains a large number of different enzymes. Its protein composition is high (about 20–30 percent of the cytosol is protein, and the cytosol contains 25–50 percent of the total protein within cells). Because of the high pro-

Cytosol

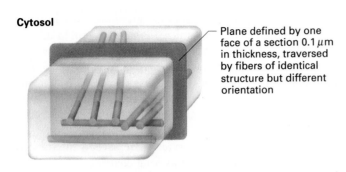

Plane defined by one
face of a section 0.1 µm
in thickness, traversed
by fibers of identical
structure but different
orientation

Section

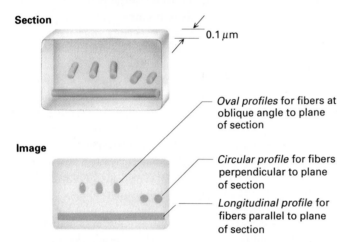

0.1 µm

Oval profiles for fibers at
oblique angle to plane
of section

Image

Circular profile for fibers
perpendicular to plane
of section

Longitudinal profile for
fibers parallel to plane
of section

▲ FIGURE 5-35   A section through a bundle of fibers
can generate very different images, depending on the angle
of the cut with respect to the plane of the fibers.

tein concentration, organized complexes of proteins can
form even if the energy that stabilizes them is weak. Many
investigators believe that the cytosol is highly organized,
with most proteins either bound to fibers or otherwise localized in specific regions.

➤ FIGURE 5-36   Electron micrograph of a platinum replica of a cytoskeleton prepared by quick-freezing and deep-
etching. A fibroblast cell is immersed in the detergent Triton
X-100 to remove soluble cytoplasmic proteins and membranes, and frozen rapidly to −269°C. While it is frozen,
water in the cytosol is removed ("etched away") by sublimation in a vacuum, resulting in exposure of the nonvolatile protein fibers. Once coated with a thin layer of platinum, these
fibers are visible in the ordinary transmission electron microscope. Prominent are bundles of actin microfilaments termed
stress fibers (SF), which are thought to connect segments of
the plasma membrane and anchor the cell to the substratum.
Also visible are two thicker microtubules (MT) and a more
diffuse meshwork of filaments studded with grape-like clusters, which are probably polyribosomes (R). [From J. E.
Heuser and M. Kirschner, 1980, *J. Cell Biol.* **86:**212. Reproduced from the *Journal of Cell Biology* by copyright permission of The Rockefeller University Press.]

The Endoplasmic Reticulum Is an Interconnected Network of Internal Membranes

Generally, the largest membrane in a eukaryotic cell is the
endoplasmic reticulum (ER)—a network of interconnected, closed membrane vesicles (see Figure 5-11). The
endoplasmic reticulum has a number of functions in the
cell but is notably important in the synthesis of many membrane lipids and proteins. The *smooth endoplasmic reticulum* is smooth because it lacks ribosomes; regions of the

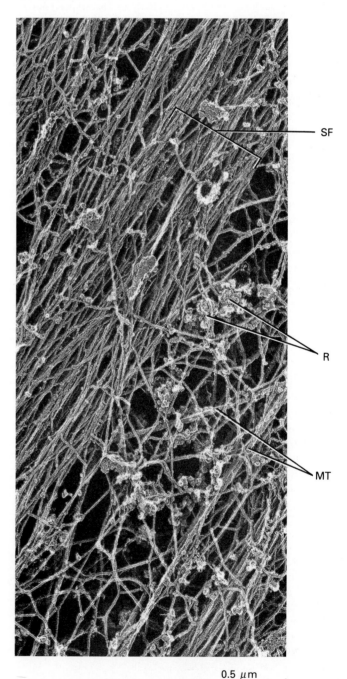

0.5 µm

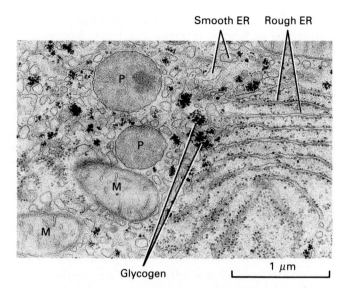

Smooth ER   Rough ER

P

P

M

M

Glycogen

1 μm

▲ FIGURE 5-37   Electron micrograph of a section of a rat hepatocyte, showing two mitochondria (M), two peroxisomes (P), rough and smooth endoplasmic reticula, and glycogen rosettes. [Courtesy of P. Lazarow.]

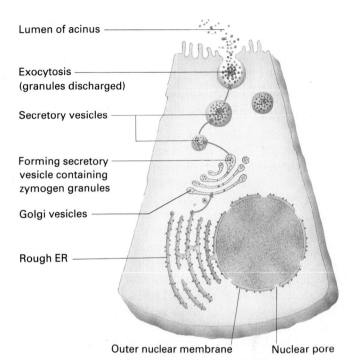

Lumen of acinus

Exocytosis (granules discharged)

Secretory vesicles

Forming secretory vesicle containing zymogen granules

Golgi vesicles

Rough ER

Outer nuclear membrane   Nuclear pore

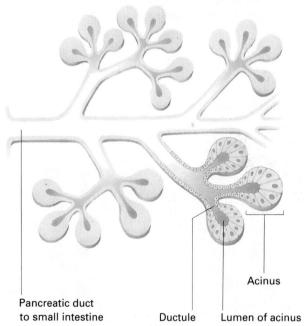

Pancreatic duct to small intestine

Ductule

Acinus

Lumen of acinus

rough endoplasmic reticulum are studded with ribosomes (Figure 5-37).

The Smooth Endoplasmic Reticulum   The smooth endoplasmic reticulum is the site of synthesis and metabolism of fatty acids and phospholipids. Many cells have very little smooth ER; in the hepatocyte, by contrast, smooth ER is abundant. Enzymes in the smooth ER of the liver modify or detoxify chemicals such as pesticides or carcinogens by chemically converting them into more water-soluble, conjugated products that can be secreted from the body. High doses of such compounds result in a large proliferation of the smooth ER in liver cells.

The Rough Endoplasmic Reticulum   Ribosomes bound to the rough endoplasmic reticulum synthesize certain membrane and organelle proteins and virtually all proteins to be secreted from the cell, as we shall see in Chapter 16. The ribosomes that fabricate secretory proteins are bound to the rough ER by the nascent polypeptide chain of the protein. As the growing secretory polypeptide emerges from the ribosome, it passes through the rough ER membrane, with the help of specific proteins in the membrane. The newly made secretory proteins accumulate in the lumen (inner cavity) of the rough ER before being transported to their next destination.

All eukaryotic cells contain a discernible amount of rough ER because it is needed for the synthesis of plasma membrane proteins and proteins of the extracellular matrix. Rough ER is particularly abundant in cells that are specialized to produce secreted proteins. For example, the pancreatic acinar cells synthesize digestive enzymes such as chymotrypsin (Figure 5-38), and plasma cells produce

▲ FIGURE 5-38   The pathway taken by newly synthesized secretory proteins in a rat pancreatic acinar cell. Immediately after synthesis, the secretory proteins are found in the lumen of the rough ER. Transfer vesicles transport them to the Golgi vesicles. Next they are concentrated in secretory vesicles containing granules of zymogens (pancreatic enzyme precursors, such as chymotrypsinogen). The cell is oriented so that the secretory vesicles form under its apical surface— the plasma membrane region that faces a ductule of the acinus. Fusion of the vesicles with the plasma membrane, triggered by hormones or nerve stimulation, releases the vesicles' contents into the ductule (exocytosis); from there, the inactive precursors move to the intestine, where they are proteolytically activated into digestive enzymes (see Figure 3-22).

serum antibodies; in both types of cells, a large part of the cytosol is filled with rough ER.

Golgi Vesicles Process Secretory and Membrane Proteins and Sort Them to Their Proper Destinations

Several minutes after their synthesis, most proteins leave the rough ER within small membrane-bounded transport vesicles that bud off from regions of the rough ER not coated with ribosomes. The vesicles carry the proteins to the luminal cavity of another membrane-limited organelle, the *Golgi complex* (see Figure 5-27d), a series of vesicles located near the nucleus in many cells.

Three-dimensional reconstructions from serial sections of a Golgi complex reveal a series of flattened membrane vesicles or sacs, surrounded by a number of more or less spherical membrane vesicles (Figure 5-39). The stack of flattened *Golgi vesicles* has three defined regions—the *cis*, the *medial*, and the *trans*. The transfer vesicles from the rough ER fuse with the *cis* region of the Golgi complex, where they deposit their proteins. These proteins then progress from the *cis* to the *medial* to the *trans* region.

Within each region are different enzymes that modify secretory and membrane proteins differently, depending on their structures and their final destinations.

After secretory proteins are modified in the Golgi vesicles, they are transported out of the complex by a second set of transport vesicles, which seem to bud off the *trans* side of the Golgi complex (see Figure 5-39). Some of these transport vesicles, termed *coated vesicles,* are surrounded by an outer protein cage composed primarily of the fibrous protein clathrin (see Figure 5-31). Some vesicles contain membrane proteins destined for the plasma membrane; others, proteins for the lysosome or for other organelles. How intracellular transport vesicles "know" which membranes to fuse with and where to deliver their contents is discussed in Chapter 16.

In many cells, the membranes of the secretory transport vesicles quickly fuse with the plasma membrane, releasing their secretory protein contents into the extracellular space—a process termed *exocytosis.* In other cells, the secretory transport vesicles fuse with similar vesicles, forming intracellular membrane-limited storage reservoirs for their secretory proteins (Figure 5-40). These reservoirs do not release their contents into the extracellular fluid until

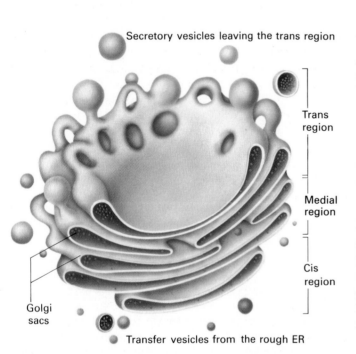

Secretory vesicles leaving the trans region

Trans region

Medial region

Cis region

Golgi sacs

Transfer vesicles from the rough ER

▲ FIGURE 5-39   Three-dimensional model of the complex of Golgi vesicles, built by analyzing micrographs of serial sections through a secretory cell. Transfer vesicles that have budded off from the rough ER fuse with the *cis* membranes of the Golgi complex. In acinar cells, the secretory vesicles that bud off the sacs on the *trans* membranes store secretory proteins, such as chymotrypsinogen, in concentrated form. [After a model by J. Kephart.]

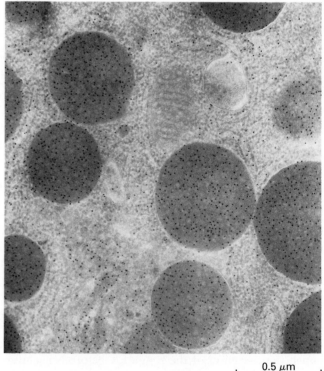

0.5 μm

▲ FIGURE 5-40   Electron micrograph showing secretory vesicles containing zymogen granules in a rat pancreatic acinar cell. To detect amylase, a protein typically found in zymogen granules, sections are treated with an antibody to amylase; 8-nm gold-particle complexes with protein A are then added to reveal the bound antibody (see Figure 5-22). [Courtesy of H. J. Geuze.]

an appropriate signal (such as a hormone) stimulates the cell to fuse its storage vesicles with the plasma membrane. Figure 5-40 shows secretory vesicles containing zymogen granules (particles composed of pancreatic enzyme precursors) just under the apical surface of a pancreatic acinar cell.

Lysosomes Are Acidic Organelles That Contain a Battery of Degradative Enzymes

No part of a cell is immortal; even in growing cells, where macromolecules and organelles are increasing, membranes, proteins, and other constituents are constantly being degraded. The *lysosomes,* membrane-limited organelles found in animal cells, degrade many membranes and organelles that have outlived their usefulness to the cell; they also degrade proteins and particles taken up by the cell.

Lysosomes vary in size and shape, and several hundred may be present in a typical cell. *Primary lysosomes* are roughly spherical and do not contain obvious particulate or membrane debris. *Secondary lysosomes* are larger and irregularly shaped and do contain particles or membranes that are being digested (Figure 5-41). Secondary lysosomes are so named because they appear to result from the fusion of primary lysosomes with other membrane organelles. How aged or defective organelles are marked for degradation and transferred to lysosomes is not known, but occasionally a bit of mitochondrion or other membrane can be seen within a lysosome (see Figure 5-41).

Lysosomes contain *acid hydrolases*—enzymes that degrade polymers into their monomeric subunits: phosphatases remove phosphate groups from mononucleotides and other compounds, such as phospholipids; nucleases degrade RNA and DNA into mononucleotide building blocks; proteases degrade a variety of proteins and peptides; still other enzymes degrade complex polysaccharides and lipids into smaller units.

The Degradative Lysosomal Enzymes Function at pH 5 All of the lysosomal enzymes work only at acid pH values. To enable the enzymes to function, the inside of a lysosome is maintained at about pH 4.8 by a hydrogen ion pump in the lysosomal membrane. The acid pH helps to denature proteins and make them accessible to the action of the lysosomal hydrolases, whose structures resist acid denaturation. The enzymes are inactive at the neutral pH values of cells and most extracellular fluids. Thus if a lysosome releases its contents into the cytosol, where the pH is between 7.0 and 7.3, no degradation of cytosolic components takes place.

Tay-Sachs Disease Is the Result of a Defective Lysosomal Enzyme The importance of lysosomes in the degradation of cellular membrane constituents is demonstrated by the existence of mutants that are defective in certain lysosomal hydrolases. *Tay-Sachs disease* is an inherited recessive disorder that results in mental retardation, derangement of the central nervous system, and death by the age of 5. In normal people, the ganglioside G_{M2}, a constituent of the plasma membranes of many mammalian cells (nerve cells in particular), is continually synthesized and degraded; the membranes of the brain cells in Tay-Sachs victims accumulate G_{M2} due to the absence of a specific lysosomal hydrolase, β-N-hexosaminidase A, a key enzyme in the normal turnover of G_{M2} (Figure 5-42). The excess G_{M2} is believed to cause all of the symptoms of Tay-Sachs disease. There are a large number of human lysosome-storage diseases in which, as in Tay-Sachs disease, a missing lysosomal enzyme causes the lysosomes to fill up with partly degraded cellular material.

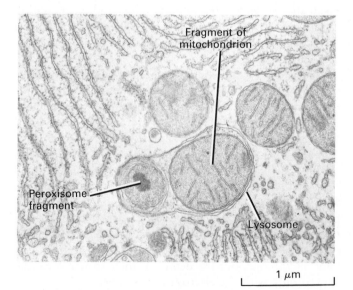

▲ FIGURE 5-41   The cytoplasm of a rat liver cell, illustrating an autophagic ("self-eating") vesicle; this secondary lysosome probably resulted from the fusion of one or more primary lysosomes with an aged or defective mitochondrion and a similarly disabled peroxisome; note that the vesicle contains bits of these organelles. [Courtesy of D. Friend.]

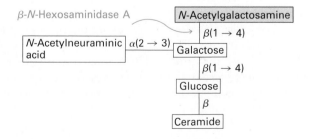

Ganglioside G_{M2}

▲ FIGURE 5-42   The first step in the degradation of the ganglioside G_{M2}. The red arrow indicates the activity site of β-N-hexosaminidase A, the degradative enzyme deficient in victims of Tay-Sachs disease.

Vacuoles in Plant Cells Store Small Molecules and Enable the Cell to Elongate Rapidly

Most plant cells (and also green algae and many microorganisms, such as yeasts) contain at least one membrane-limited internal vacuole (see Figure 5-3a). The number and size of vacuoles depend on both the type of cell and its stage of development; a single vacuole may occupy as much as 80 percent of a mature plant cell. Within the vacuoles, plant cells store water, ions, certain waste products, and such food materials as sucrose. In particular, vacuoles store highly valuable nitrogen-containing compounds, such as amino acids, derived from atmospheric nitrogen by the energy-requiring process known as *nitrogen fixation*.

The concentration of dissolved materials is much higher within the vacuoles than it is in the cytosol or extracellular fluids. The vacuolar membrane is semipermeable: it is permeable to water but is impermeable to the small molecules stored within it. Therefore water will tend to move by osmotic flow (Figure 5-28) into the vacuole from the cytosol. This causes the vacuole to expand, creating hydrostatic pressure, or *turgor*, inside the cell. This pressure is balanced by the mechanical resistance of the cellulose-containing cell wall that surrounds all plant cells. Most plant cells have a turgor of 5–20 atm; their cell walls must be strong enough to react to this pressure in a controlled way. Unlike animal cells, plant cells can elongate extremely rapidly—at rates of 20–75 μm/h. This elongation, which usually accompanies plant growth, occurs when the somewhat elastic cell wall stretches under the pressure created by water taken into the vacuole.

Vacuoles also act as receptacles for waste products and excess salts taken up by the plant and may function similarly to lysosomes in animal cells. Vacuoles have an acidic pH, maintained by a proton pump in the vacuole membrane, and contain a battery of degradative enzymes.

Peroxisomes Produce and Degrade Hydrogen Peroxide

The *peroxisomes* are a class of small, membrane-limited organelles found in the cytoplasm of all animal cells and many plant cells: *glyoxisomes* are similar organelles found in plant seeds that oxidize stored lipids as a source of carbon and energy for growth. For a long time these two organelles were believed to be lysosomes because their morphology resembles that of lysosomes. However, cell-fractionation studies have established that the enzymes in these organelles differ greatly, in both composition and function, from those in lysosomes. Peroxisomes and glyoxisomes contain enzymes that degrade fatty acids and amino acids; a product of these reactions is hydrogen peroxide (H_2O_2), a corrosive substance. Peroxisomes also contain copious amounts of the enzyme catalase, which degrades hydrogen peroxide:

$$2H_2O_2 \xrightarrow{\text{catalase}} 2H_2O + O_2$$

Many peroxisomes contain a crystalline array of catalase molecules, which is seen in the electron microscope as a densely stained "core" (see Figures 5-27f and 5-37). Chapter 17 discusses the role of peroxisomes in the metabolism of lipids; Chapter 18, the role of glyoxisomes in photosynthesis.

The Mitochondrion Is the Principal Site of ATP Production in Aerobic Cells

The high-energy bonds in ATP are the usual source of chemical energy for cellular growth and metabolism, and the principal sources of ATP in nonphotosynthetic cells are fatty acids and glucose. The complete aerobic degradation of glucose to CO_2 and H_2O is coupled to the synthesis of as many as 32 molecules of ATP:

$$C_6H_{12}O_6 + 6O_2 + 32P_i + 32ADP \longrightarrow$$
$$6CO_2 + 6H_2O + 32ATP + 32H_2O$$

In eukaryotic cells, the initial stages of glucose degradation occur in the cytosol, where two ATPs per glucose molecule are generated. The terminal stages, including those involving oxygen, occur in the mitochondria, where as many as 30 ATPs are generated (even this value is uncertain, since much of the energy released in mitochondrial oxidation can be used for other purposes, such as heat generation and the transport of molecules into or out of the mitochondrion, making less energy available for ATP synthesis). Similarly, during the oxidation of fatty acids to CO_2 virtually all of the ATP is formed in the mitochondrion. Thus the mitochondrion can be regarded as the "power plant" of the cell. The details of ATP formation in the cytosol and the mitochondria require a chapter of their own (Chapter 17); here, we describe some of the basic structural features of the organelle.

Most eukaryotic cells contain many mitochondria, which occupy up to 25 percent of the volume of the cytoplasm. Mitochondria are among the largest organelles in the cell (generally only the nucleus, vacuoles, and chloroplasts are larger). They are large enough to be seen under a light microscope, but the details of their structure can be viewed only with the electron microscope.

Mitochondria contain two very different membranes: an outer one and an inner one (Figure 5-43; see also Figure 5-27b). The outer membrane is composed of about half lipid and half protein. It contains proteins that render the

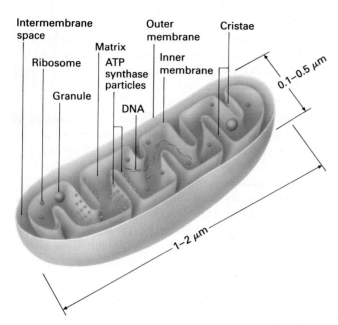

▲ FIGURE 5-43   Three-dimensional diagram of a mitochondrion cut longitudinally.

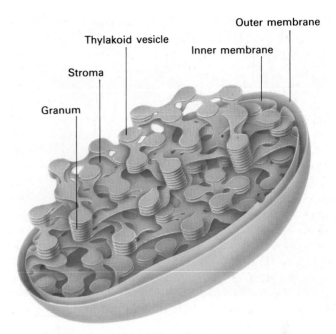

▲ FIGURE 5-44   Three-dimensional diagram of a chloroplast. The internal membrane (thylakoid) vesicles are fused together into stacks, or grana, which reside in a matrix (the stroma). All the chlorophyll in the chloroplast is contained in the membranes of the thylakoid vesicles.

membrane permeable to molecules having molecular weights as high as 10,000. In this respect, the outer membrane is similar to the outer membrane of gram-negative bacteria (Figure 5-1). The inner membrane is much less permeable, and it is about 20 percent lipid and 80 percent protein—a higher proportion of protein than occurs in other cellular membranes. The surface area of the inner membrane is greatly increased by a large number of infoldings, or *cristae,* that protrude into the *matrix,* or central space. The matrix and cristae are the sites of the enzymes that catalyze the final oxidation of sugars and lipids and the synthesis of ATP. Mitochondria have their own DNA, which is located in the matrix; its role in the synthesis of mitochondrial proteins is detailed in Chapter 19.

Chloroplasts Are the Sites of Photosynthesis

Except for vacuoles, the chloroplasts are the largest and most characteristic organelles in the cells of plants and green algae (see Figure 5-3a). Like mitochondria, chloroplasts do not have fixed positions and often migrate from place to place within cells. They can be as long as 10 μm and are typically 0.5–2 μm thick, but they vary in size and shape in different cells, especially among the algae. On a weight basis, 35 percent of the chloroplast is lipid, 5 percent is protein, and 7 percent is pigment. Predominant among the pigments is *chlorophyll,* the substance that absorbs light and gives the chloroplast its green color.

Like the mitochondrion, the chloroplast is surrounded by an outer and an inner membrane (Figure 5-44). Chloro-

plasts also contain an extensive internal membrane system made up of *thylakoid vesicles* (interconnected vesicles flattened to form disks), often grouped in stacks called *grana* (see Figure 5-3a). A space called the *stroma* surrounds the thylakoids. *Carbon dioxide fixation*—the conversion of CO_2 to intermediates during the synthesis of sugars— occurs in the stromal space. The thylakoid vesicles contain chlorophyll and other pigments and enzymes that absorb light and generate ATP during photosynthesis. Like ATP formation in mitochondria, photosynthesis also requires a chapter of its own (Chapter 18).

Chloroplasts, like mitochondria, have their own DNA; it encodes some of the key chloroplast proteins. Like mitochondrial proteins, most chloroplast proteins are made in the cytosol and incorporated in the organelle later, as described in Chapter 19.

Cilia and Flagella Are Motile Extensions of the Eukaryotic Plasma Membrane

The surfaces of animal cells and eukaryotic microorganisms often contain a number of protuberances and extensions that serve specific and important functions. *Cilia* and *flagella* are examples. These similar motile structures extend from the plasma membranes of unicellular organisms

▲ FIGURE 5-45 Scanning electron micrograph of the surface of hamster trachea cells, showing abundant clusters of cilia, which protrude into the lumen of the trachea. [Courtesy of E. R. Dirksen.]

and many animal cells (Figure 5-45). Bunches of microtubule filaments run the length of the central cores of cilia and flagella; they are structurally similar to the microtubules that compose the mitotic spindle, discussed later. Cilia beat backward and forward; flagella, which are longer than cilia, typically undulate in a whiplike manner. Both motions can propel a cell; the beating of the single flagellum, for example, is the only means of locomotion for sperm. Cilia are also present on many fixed epithelial cells, such as trachea cells; there, the cilia propel liquids and small particles along the surface of the sheet of cells that lines the airways.

The Plasma Membrane Binds to the Extracellular Matrix or the Cell Wall

Many essential eukaryotic cell entities lie outside the plasma membrane. A variety of proteins and polysaccharides attached to the outer surfaces of animal cells participate in the formation of specific contacts and junctions between cells; the plasma membrane also bears proteins that bind the cells to several matrix components. (Figure 5-32). Such interactions impart strength and rigidity to multicellular tissues.

Much of the volume of *connective tissue* in animals consists of the spaces between cells. Loose connective tissue forms the bedding on which most small glands or epithelial cells lie. It contains only a few cells and a number of

fibers of various diameters. The blood capillaries that bring oxygen and nutrients to the cells in all parts of the body are embedded in loose connective tissue. A principal function of the amorphous spaces between cells in this tissue is to allow these nutrients to diffuse freely into the cells of the epithelia and the glands.

The plasma membrane of plant cells is intimately involved in the assembly of cell walls; new layers of wall are laid down next to the plasma membrane as a cell matures (Figure 5-46). The walls are built primarily of *cellulose*, a rodlike polysaccharide formed from $\beta(1\rightarrow4)$-linked glucose monomers. As discussed in Chapter 24, the cellulose molecules aggregate, by hydrogen bonding, into bundles of fibers; other polysaccharides within the wall cross-link the

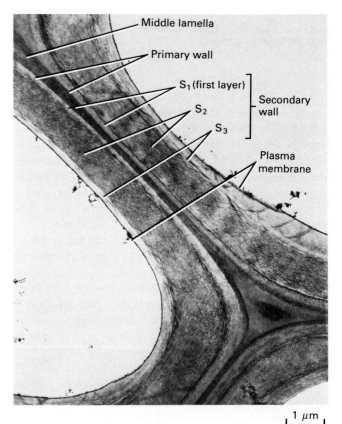

▲ FIGURE 5-46 Electron micrograph of a thin section showing parts of the cell walls separating three *Taxus canadensis* (plant) cells. The principal layers of each wall are evident: the middle lamella, the primary wall, and the three layers of secondary wall (S₁, S₂, and S₃). As the cell matures, the layers of cellulose fibers are laid down one by one in the sequence shown. The fibers in each layer run in a different direction from those in the preceding layer. The plasma membrane is adjacent to the S₃ layer, the youngest stratum of the cell wall. [Courtesy of Biophoto Associates/Myron C. Ledbetter/Brookhaven National Laboratory.]

cellulose fibers. In woody plants, a very different compound, *lignin* (a complex water-insoluble polymer of phenol and other aromatic monomers), imparts strength and rigidity to the cell walls. Other chemicals also are found in the walls of various plant cells; for example, waxes prevent plant tissues and proteins from drying out.

➤ Cell Division and the Cell Cycle

Before explaining in the next chapter how different kinds of cells are grown, studied, and manipulated in the laboratory, we need to discuss perhaps the most crucial aspect of cell division: the way a cell duplicates itself so that each daughter cell receives an identical copy of its genetic material. Prokaryotic and eukaryotic cells differ markedly in the way they coordinate DNA synthesis and in the subsequent equal partitioning of DNA during cell division. Therefore the two cell types require different methods of genetic analysis.

No specialized structure for division is visible in prokaryotes by light microscopy. Since the last century, however, biologists have recognized eukaryotic chromosomes in the microscope and understood their importance in cell division: *mitosis* in diploid somatic (body) cells ensures the genetic identity of equal daughter cells; *meiosis* produces haploid germ cells (the sperm and the egg in animals and plants) and ensures, in both unicellular and multicellular eukaryotes, genetic continuity between generations.

In Prokaryotes DNA Replication Is Followed Immediately by Cell Division

As noted earlier, the genome of a prokaryotic cell is a single circular molecule of DNA containing about 4×10^6 base pairs, about 1.4 mm in length (see Table 5-2). If growing bacteria are briefly exposed to radioactive thymidine, almost every cell is labeled, indicating that each organism is making DNA. Thus, of the time required for a bacterium to divide, most is spent synthesizing DNA. In a medium containing a high concentration of nutrients, *E. coli* divides every 30 min, and all but a minute or so of this time is used to complete one round of DNA synthesis.

During prokaryote cell division, a cell wall forms that divides the cell in two. It is likely that the two daughter DNA duplexes are linked at different points on the plasma membrane to ensure that one of the new DNA duplexes is delivered to each daughter cell (Figure 5-47). In certain bacteria, infolded plasma membrane regions termed *mesosomes* may provide anchors for DNA. Even in these organisms, however, there is no visible condensation and decondensation of the DNA, as there is in eukaryotic cells during mitosis.

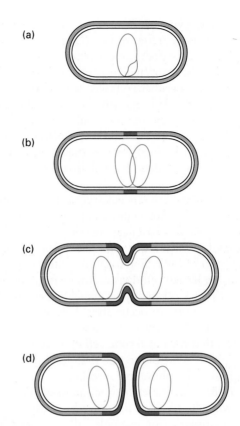

▲ FIGURE 5-47   DNA replication and cell division in a prokaryote. The DNA in prokaryotic cells is attached to the cell membrane and remains attached during cell division. (a) The circular chromosome (blue), which has already begun replication, is attached to the plasma membrane. (b) When DNA replication is complete, the new chromosome has an independent point of attachment to the membrane. New membrane and cell wall forms midway along the length of the cell. (c) As more sections of membrane and cell wall form, part of this growth gives rise to a septum dividing the cell. (d) Cell division is complete; the new chromosome is attached to the plasma membrane of each daughter cell.

In Eukaryotic Cells DNA Synthesis and Cell Division Occur in Special Phases of the Cell Cycle

Different eukaryotic cells can grow and divide at quite different rates. Yeast cells, for example, can divide every 120 min, and the first divisions of fertilized eggs in the embryonic cells of sea urchins and insects take only 15–30 min because one large preexisting cell (see Figure 5-13) is subdivided. However, most growing plant and animal cells take 10–20 h to double in number, and some duplicate at a much slower rate. Many cells in adults, such as nerve cells and striated muscle cells, do not divide at all; others, like the fibroblasts that assist in healing wounds, grow on demand but are otherwise quiescent.

Still, every eukaryotic cell that divides must be ready to donate equal genetic material to two daughter cells. DNA synthesis in eukaryotes does not occur throughout the cell division cycle but is restricted to a part of it before cell division. A simple experiment establishes this point: If root tips, in which cells divide frequently, are exposed to radio-active thymidine for a few minutes, only a fraction of the cells incorporate high levels of label into DNA; the remainder are not labeled at all. Moreover, the labeled cells are not in the process of dividing, but will divide several hours later.

The relationship between eukaryotic DNA synthesis and cell division was thoroughly analyzed in cultures of mammalian cells that were all capable of growth and division. In contrast to bacteria, it was found, eukaryotic cells spend only a part of their time in DNA synthesis, and it is completed hours before cell division (mitosis). Thus a gap of time occurs after DNA synthesis and before cell division; another gap was found to occur after division and before the next round of DNA synthesis. These experiments led to the conclusion that the eukaryotic *cell cycle* consists of an M (*mitotic*) phase, a G_1 phase (the first gap), the S (DNA *synthesis*) phase, a G_2 phase (the second gap), and back to M (Figure 5-48). The phases between mitoses (G_1, S, and G_2) are known collectively as the *interphase*. Many nondividing cells in tissues (for example, all quiescent fibroblasts) suspend the cycle after mitosis and just prior to DNA synthesis; such "resting" cells are said to have exited from the cell cycle and to be in the G_0 state.

As shown in Figure 5-49, it is possible to identify cells when they are in one of the three interphase stages of the cell cycle, by using a fluorescence-activated cell sorter (FACS; see Figure 5-26) to measure their relative DNA content: a cell that is in G_1 (before DNA synthesis) has a defined amount x of DNA; during S (DNA replication), it has between x and $2x$; and when in G_2 (or M), it has $2x$ of DNA.

The regulatory events that guide the cell from phase to phase are now well understood, and require a chapter of their own (Chapter 25).

Mitosis Is the Complex Process That Apportions the New Chromosomes Equally Between Daughter Cells

Mitosis has been recognized for a century as the mechanism in eukaryotes for partitioning the genetic material equally at cell division. Beginning with the union of egg and sperm, each somatic (body) cell of sexually reproducing organisms has two copies of each morphologically distinct chromosome, one inherited from the male parent and one from the female parent (see Table 5-2 footnote). Each somatic cell is thus said to be *diploid*, or $2n$, where n is the number of chromosome types. The process of mitosis ensures that when a somatic cell divides in two, each resulting cell is still $2n$.

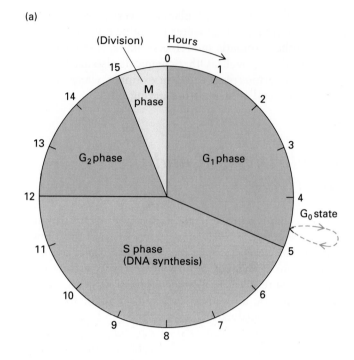

(a)

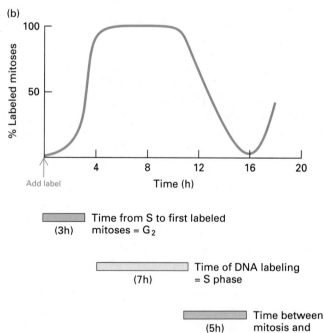

(b)

▲ FIGURE 5-48   The cell cycle in a mammalian cell having a generation time of 16 hr. (a) The three phases spanning the first 15 hr or so—the G_1 (first gap) phase, the S (DNA synthetic) phase, and the G_2 (second gap) phase—make up the interphases. DNA is synthesized in S, and other cellular macromolecules are synthesized throughout G_1, S, and G_2. The remaining hour is the M (mitotic) phase, during which the cell actually divides. (b) The phases of the cell cycle were determined by exposing a culture briefly to labeled thymidine, which is incorporated into DNA, and then observing the time of appearance of labeled mitotic cells.

(a)

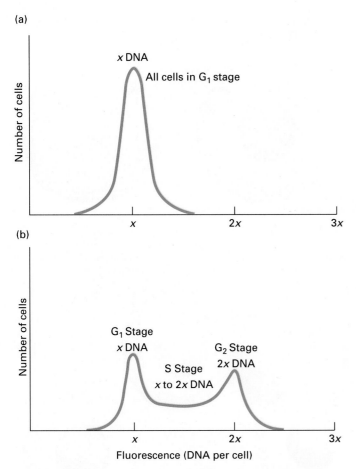

(b)

▲ FIGURE 5-49   Measuring DNA content to identify cell cycle phases. 3T3 mouse fibroblast cells were allowed to grow rapidly in culture or were removed from growth factors for 3 days. The cells were then reacted with a fluorescent dye that binds to DNA, and a fluorescence-activated cell sorter was used to measure the fluorescence level of each cell, which is proportional to the cell's content of DNA. Virtually all of the cells in the nongrowing population (a) had an amount of DNA, x, that indicated they had not yet begun DNA replication and were in the G_1 phase of the cell cycle. In the growing population (b), in contrast, while some of the cells were in G_1, other cells had twice this amount of DNA, $2x$, indicating that they were in G_2: they had replicated their DNA but had not yet divided. Still other cells in this group had an intermediate amount of DNA, indicating that they were in the S phase: in the process of replicating their DNA.

During *interphase*, the period between mitoses, chromosomes are not visible by light microscopy; DNA-protein complexes called *chromatin* are dispersed throughout the nucleus during this time. However, it is during the S (synthesis) stage of interphase that DNA replication takes place.

During *mitosis*, the period of cell division, a very characteristic series of microscopic events is observed. Although the events unfold continuously, they are conventionally divided into four substages: *prophase, metaphase, anaphase,* and *telophase* (Figure 5-50).

The beginning of a mitotic cell division is signaled by the appearance of the *chromosomes,* stainable as thin threads inside the nucleus (Figure 5-50b). During prophase and metaphase, a chromosome is visible as two identical coiled filaments, the *chromatids* (often called *sister chromatids*), held together by a constricted region, the *centromere.* By late prophase, the chromatids have become shorter and more densely packed (Figure 5-50c). Each chromatid contains one of the two new daughter DNA molecules produced in the preceding S phase of the cell cycle; thus each cell that enters mitosis is *4n*: it has four copies of each chromosomal DNA.

As we said in our discussion of the cytoskeleton, the *microtubules* are extremely important in directing chromosomal movement during cell division. Microtubules form the scaffold of the mitotic cell; they comprise the tracks along which the chromosomes move and they ensure that the chromosomes are apportioned appropriately to the daughter cells. The small cylindrical particles termed *centrioles* also play a key role, in organizing the network of microtubules during the prophase and metaphase stages of cell division. Centrioles are themselves constructed of microtubules, and they duplicate during interphase, forming daughter centrioles.

At the beginning of mitosis the two centrioles (each with its daughter centriole) begin to move apart, and microtubules radiate from each centriole in all directions, forming *asters.* The centrioles end up at opposite ends of the cell, and the surrounding regions, from which the microtubules radiate, are termed the *poles,* or polar regions, of the cell. These radiating microtubules, together with associated proteins, are called the *spindle.* Some of the microtubules form a network that links the centrioles as they move apart; other microtubules connect the centrioles with the *kinetochores.* These granular regions, visible in the centromeres of the chromatids, catalyze the movement of chromosomes along microtubules.

By the end of prophase, the nuclear membrane has become invisible in the light microscope. Electron microscopy of cells in late prophase reveals that the nuclear membrane breaks down into a large number of small, flattened membrane vesicles.

By *metaphase,* the condensed sister chromatids are connected to each pole of the cell by the microtubules attached to their kinetochores. During metaphase, the chromatids all migrate to the *equatorial plane* of the cell (Figure 5-50d). The microtubules attached to the kinetochores play a role in orienting the chromosomes in the center of the mitotic spindle.

Anaphase (Figure 5-50e) is marked by the separation of the two sister chromatids at their centromeres; as nearly as we know, the separation happens simultaneously to the entire set of chromosomes. The kinetochores attached to each member of a chromatid pair—now independent chromosomes—cause the chromosomes to migrate along mi-

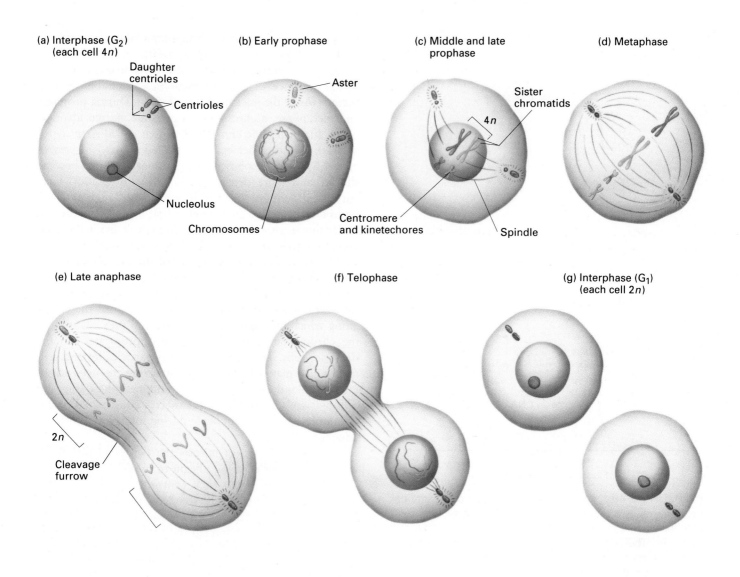

(a) Interphase (G_2) (each cell $4n$)
Daughter centrioles
Centrioles
Nucleolus
Chromosomes

(b) Early prophase
Aster

(c) Middle and late prophase
Sister chromatids
$4n$
Centromere and kinetechores
Spindle

(d) Metaphase

(e) Late anaphase
$2n$
Cleavage furrow

(f) Telophase

(g) Interphase (G_1) (each cell $2n$)

crotubules to opposite poles of the cell, which ensures that each new daughter cell receives an equal chromosome set. How the kinetochores cause the chromosomes to move to opposite poles is just becoming understood; current ideas and experiments relevant to chromosome movement are discussed in Chapter 23.

In *telophase* (Figure 5-50f), the chromosomes start to uncoil and become less condensed. Nuclear membrane vesicles are seen joining together to become two new nuclear membranes surrounding the two sets of daughter chromosomes. Simultaneously, there is division of the cell cytoplasm and separation, or *cytokinesis,* of the two daughter cells.

The Division of Mitochondria and Chloroplasts Is Uncoupled from the Cell Cycle   The growth and division of mitochondria and chloroplasts is not coupled to nuclear division, and the incorporation of proteins and lip-

ids into these organelles occurs continuously during the interphase period of the cell cycle. As these organelles increase in size, one or more daughters "pinch off" in a manner similar to the way in which bacteria grow and divide. Similarly, the replication of mitochondrial DNA occurs throughout interphase. At mitosis each daughter cell receives approximately the same quantity of mitochondria and mitochondrial DNA, but there is no system for apportioning exactly equal amounts of mitochondrial DNA to the daughter cells.

Plant Cells Show Some Variations in Mitosis   Most higher plant cells do not contain visible centrioles. An analogous region of the cell acts as a microtubule-organizing center, from which the spindle microtubules radiate. Because a plant cell is surrounded by a rigid cell wall, the shape of the cell does not change greatly in mitosis (Figure 5-51). During telophase, the new cell membrane and cell

◄ FIGURE 5-50   The stages of mitosis and cytokinesis in an animal cell. (a) *Interphase.* The G₂ stage of interphase immediately precedes the beginning of mitosis. Chromosomal DNA has been replicated and bound to protein during the S phase, but the chromosomes are not yet seen as distinct structures. The nucleolus is the only nuclear substructure that is visible under the light microscope. In a diploid cell before DNA replication there are two morphologic chromosomes of each type, and the cell is said to be 2n. In G₂, after DNA replication, the cell is 4n. There are four copies of each chromosomal DNA. Since the sister chromosomes have not yet separated from each other, they are called sister chromatids. (Morphologic types are distinguished by color in the figure.)

(b) *Early prophase.* The centrioles, each with a newly-formed daughter centriole, begin moving toward opposite poles of the cell; the chromosomes can be seen as long threads. The nuclear membrane begins to disaggregate into small vesicles.

(c) *Middle and late prophase.* Chromosome condensation is completed; each visible chromosome structure is composed of two chromatids held together at their centromeres. Each chromatid contains one of the two newly replicated daughter DNA molecules. The microtubular spindle begins to radiate from the regions just adjacent to the centrioles, which are moving closer to their poles. Some spindle fibers reach from pole to pole; most go to chromatids and attach at kinetochores.

(d) *Metaphase.* The chromosomes move toward the equator of the cell, where they become aligned in the equatorial plane. The sister chromatids have not yet separated. This is the phase in which morphologic studies of chromosomes are usually carried out.

(e) *Anaphase.* The two sister chromatids separate into independent chromosomes. Each contains a centromere that is linked by a spindle fiber to one pole, to which it moves. Thus one copy of each chromosome is donated to each daughter cell. Simultaneously, the cell elongates, as do the pole-to-pole spindles. Cytokinesis begins as the cleavage furrow starts to form.

(f) *Telophase.* New membranes form around the daughter nuclei; the chromosomes uncoil and become less distinct, the nucleolus becomes visible again, and the nuclear membrane forms around each daughter nucleus. Cytokinesis is nearly complete, and the spindle disappears as the microtubules and other fibers depolymerize. Throughout mitosis the "daughter" centriole at each pole grows until it is full-length. At telophase the duplication of each of the original centrioles is completed, and new daughter centrioles will be generated during the next interphase.

(g) *Interphase.* Upon the completion of cytokinesis, the cell enters the G₁ phase of the cell cycle and proceeds again around the cycle.

wall are formed from membrane vesicles that fuse together in a plane that is perpendicular to a line separating the two nuclei.

Yeast Cells Have a Simplified Division   The yeast *Saccharomyces cerevisiae* exemplifies the simplified cell-division apparatus in certain yeasts (Figure 5-52). After S phase, when division is about to begin, a bud—the future daughter cell—emerges from a thickened region of the cell wall called a *plaque.* The bud gradually enlarges. The chromosomes (although too small to be easily visible under the light microscope) are attached to a structure, the *spindle pole body* (SPB), which duplicates like the centriole in animal cells, signaling the beginning of nuclear division. Microtubules form to guide the chromosomes, and one spindle pole body enters the enlarging bud. The chromosomes then separate, and one set enters the bud to form the genetic complement of the new daughter cell. Finally, the bud, now complete, separates from the parent cell.

In yeasts, the cell derived from the bud is called the daughter cell, and the original cell is the mother. Initially, the daughter cell is smaller than the mother cell, an important distinction because the cells are not *biochemically* equal. However, both cells do have an identical set of chromosomes. Thus, the genetic consequences of mitosis are the same in yeasts as in plant and animal cells: the two postmitotic cells are genetically identical.

Meiosis Is the Form of Cell Division in which Haploid Germ Cells Are Produced from Diploid Cells

In all multicellular organisms and in single-celled organisms that are diploid at some phase of their life cycle, one important type of cell division departs from the plan of mitosis; this is *meiosis*, the division that gives rise to *gametes*—sperm and egg cells—in higher plants and animals. Like the body cells of most multicellular organisms, a premeiotic cell is diploid—it contains two of each morphologic type of chromosome. The two chromosomes of each type are descended from different parents, so they are *homologous* (i.e., their genes are similar but are usually not identical). In meiosis, *one* round of DNA replication, which makes the cell 4n, is followed by *two* separate cell divisions, yielding four *haploid* (1n) cells that contain only one chromosome of each homologous pair (Figure 5-53). There is an *interphase* between these two divisions that is unique in that it lacks a DNA synthesis phase.

The First and Second Meiotic Divisions   During the *first meiotic division*, each morphologic chromosome (actually two sister chromatids) aligns at the cell equator paired with its homologous partner. No division of centromeres occurs, however, so the sister chromatids remain together. One homolog (both sister chromatids) is randomly selected to travel to one daughter cell; the other homolog of that chromosome goes into the other cell. Because one of these homologous chromosomes came from an egg and the other from a sperm, parental characteristics are reassorted randomly into each new germ cell at the meiotic stage. In the *second meiotic division*, the centromeres do divide, as in mitosis, and one sister chromatid (now an independent chromosome) of each morphologic type is apportioned into each haploid daughter cell.

(a)

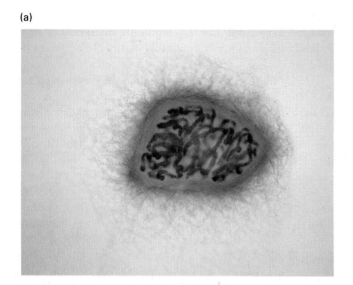

(b)

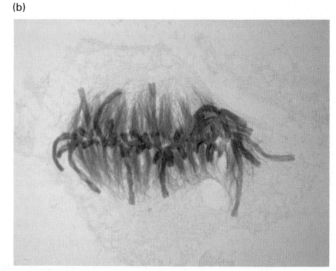

(c)

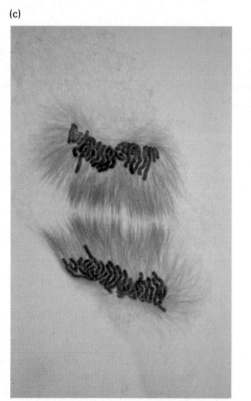

(d)

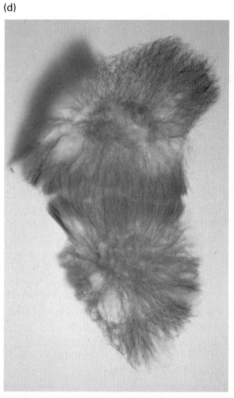

▲ FIGURE 5-51   Light micrographs of mitosis in a plant. Microtubules are stained red and chromosomes are counterstained blue. (a) *Prophase*. The chromosomes are beginning to condense. (b) *Metaphase*. The chromosomes are aligned on the equatorial plane. (c) *Anaphase*. The sister chromatids have separated and are moving toward opposite poles of the spindle. (d) *Telophase*. New membranes are forming around the daughter nuclei, and the chromosomes are uncoiling and becoming less distinct. [Courtesy of Andrew Bajer.]

(a)

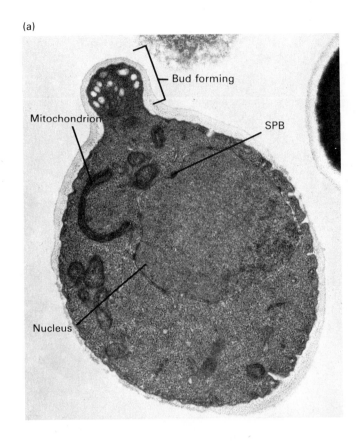

(b)

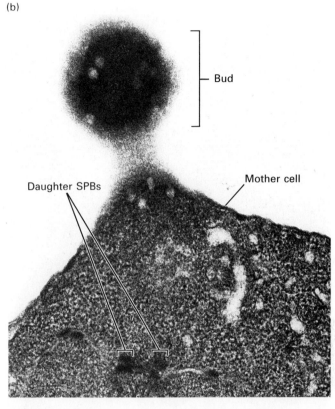

(c)

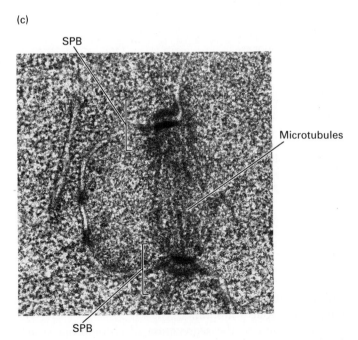

(d)

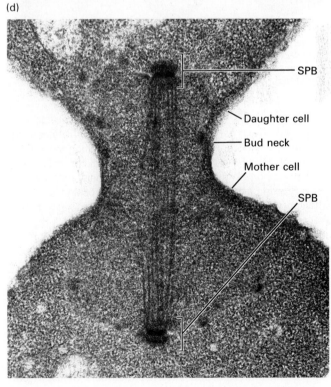

▲ FIGURE 5-52   Cell division in *Saccharomyces cerevisiae*. (a) Yeast cells divide by budding, not by classic mitosis. As in animal cells, DNA replication occurs during the S stage. In mitosis, the distribution of chromosomes to the new cells is determined by the spindle pole body (SPB) located at the nuclear periphery. (b) The beginning of nuclear division is signaled by the division of the SPB. (c) When the two daughter SPBs separate, the microtubules connecting them can be seen. (d) A portion of the nucleus containing one SPB then enters the bud. [Part (a) courtesy of B. Byers; Parts (b)–(d) courtesy of Y-H. Tu.]

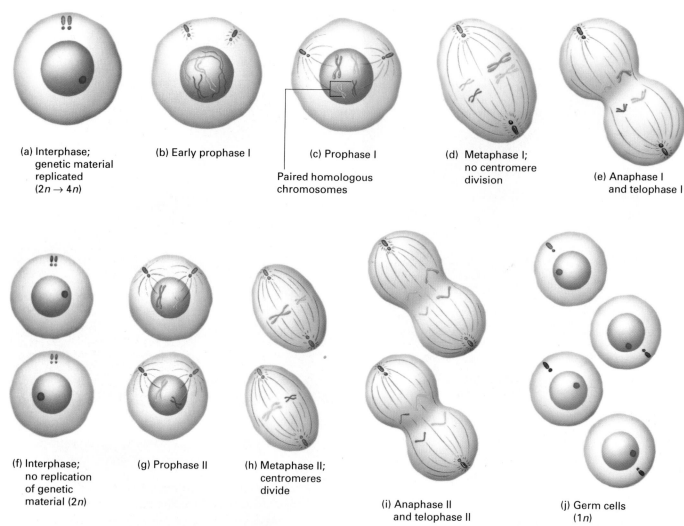

(a) Interphase; genetic material replicated ($2n \rightarrow 4n$)

(b) Early prophase I

(c) Prophase I

Paired homologous chromosomes

(d) Metaphase I; no centromere division

(e) Anaphase I and telophase I

(f) Interphase; no replication of genetic material ($2n$)

(g) Prophase II

(h) Metaphase II; centromeres divide

(i) Anaphase II and telophase II

(j) Germ cells ($1n$)

▲ FIGURE 5-53   The steps in meiosis. Steps (a) through (c) are carried out as in mitosis (see Figure 5-50). Metaphase (d) differs in two important ways. Each morphologic chromosome (actually two sister chromatids) aligns at the cell equator with its homologous partner during synapsis, the stage of chromosome pairing that allows exchange (crossing over) between the two. No division of centromeres occurs, however, so one homologue (both sister chromatids) of a morphologic type goes into one cell; the other homologue of that morphologic type goes into the other cell. Which homologue goes to which cell is chosen randomly. Thus the genetic material of each of the two chromosomes of the same morphologic type separates (segregates) independently. Because one of these chromosomes came from an egg and one originally from a sperm, parental characteristics are reassorted randomly into each new sperm and egg at the meiotic stage. In the second meiotic division, the centromeres do divide, and each cell receives one chromosome of each morphologic type, to become haploid ($1n$).

The Genetic Consequences of Meiosis The segregation of homologous chromosomes in meiosis is random; that is, the maternal and paternal members of each pair segregate independently. The example in Figure 5-54 shows only two morphologic types of chromosomes, a long (L) and a short (S); each one can be either maternal (m) or paternal (p) in heritage (Lm, Lp, Sm, Sp). Thus in this case, four types of gametes can be formed: LmSm, LpSp, LmSp, LpSm. The number of possible varieties of meiotic segregants is 2^n, where n is the haploid number of chromosomes. For humans, with 23 chromosomes, the number is about 8 million.

In addition to the segregation of chromosomes, another critical genetic event occurs in most (but not all) meiotic cells. Before the first meiotic division, the chromosomes of each homologous pair—that is, the maternal and paternal chromosomes—align with each other, an act called *synapsis*. At this time, *recombination* between chromatids can occur (Figure 5-55). This swapping of material between maternally and paternally derived chromosomes

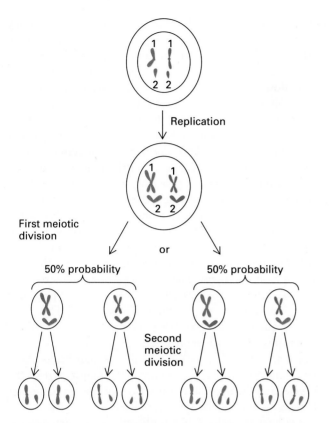

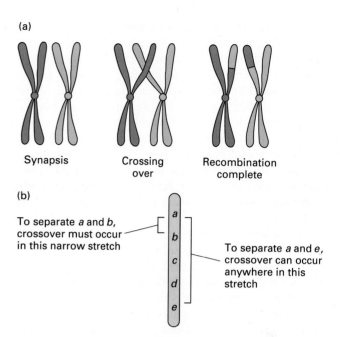

▲ FIGURE 5-55 The exchange of genetic material during synapsis. (a) Crossing over between homologous chromosomes during meiotic metaphase I. (b) The shorter the distance between two genes on a chromatid, the less likely they are to be separated by crossing over.

▲ FIGURE 5-54 The genetic consequences of meiosis are illustrated for an organism in which the haploid number of chromosomes is 2. Each diploid cell contains pairs of homologous chromosomes, one inherited from each parent. (Here, the maternal and paternal chromosomes are distinguished by color.) In the first meiotic division, each daughter cell receives either the maternal or paternal chromosome from each homologous pair. Which chromosome is incorporated into which daughter is random, and each homologous pair of chromosomes is segregated independently of the other pairs. Thus, in this simple case, there are two possible outcomes of the first division, each having an equal probability of 50 percent. In the second meiotic division, the chromatids separate to generate a total of four haploid daughter cells, each containing one chromosome of each type.

is called *crossing over*. Combined with random segregation, crossing over is the source of new combinations of genes in interbreeding populations. These two phenomena also underlie the classic methods of genetic mapping by breeding experiments, which will be considered in greater detail in Chapter 8. Genetic traits that almost always segregate together are said to be *linked* and are inferred to be controlled by nearby genes on the same chromosome; the more frequently recombination occurs between two genes on the same chromosome, the farther apart the genes are.

SUMMARY

Two types of cells exist in our biological universe: prokaryotic cells, with no nuclei (including bacteria of the archaebacterial and eubacterial lineages), and eukaryotic cells, with nuclei (including animals, plants, and many microorganisms). Prokaryotes and eukaryotes have a similar overall chemical composition and share many metabolic pathways, but prokaryotic cells are generally smaller and differ from eukaryotic cells in terms of structural elements and genetic processes.

All biomembranes are composed of a bilayer of phospholipid to which specific proteins are attached. In most respects, the internal architecture and central metabolic pathways are similar in the cells of all plants, animals, and unicellular eukaryotic organisms. Eukaryotic cells have intracellular membranes and organelles that are never found in prokaryotes: examples include the lysosomes, which degrade cellular membranes and other materials; the peroxisomes, which synthesize and degrade hydrogen peroxide; the endoplasmic reticulum, where secretory proteins and membrane proteins are synthesized; the Golgi vesicles, which sort and process proteins, particularly secretory and membrane proteins; the mitochondria, where the terminal stages of sugar and lipid oxidation and ATP synthesis take place; and the chloroplasts, where photosynthesis occurs in plants. The nucleus is surrounded by a nuclear membrane composed of an inner and an outer membrane; the outer membrane is usually continuous with the endoplasmic reticulum. Pores in the nuclear membrane allow particles to enter and leave the nucleus; lamins lining the inner surface of the inner membrane give it rigidity and bind some nuclear DNA.

Both plant and bacterial cells have cell walls, but these are very different in composition. Peptidoglycan, in particular, is unique to prokaryotes. Plant cell walls are built of cellulose and other polymers; the wall is the major determinant of cell shape and gives the cell rigidity. Animal cells lack a wall; they are surrounded by an extracellular matrix consisting of collagen, proteoglycans, and other components that give strength and rigidity to tissues and organs.

Vacuoles often fill much of the plant cell and generate turgor pressure that pushes the plasma membrane against the cell wall. Animal cells, in contrast, will shrink or swell, respectively, if placed in solutions of too high or too low osmotic strength.

The cytosol of all animal and plant cells contains a network of fibrous proteins, the cytoskeleton, which gives the cell structural stability and contributes to cell movement. Three principal types of protein filaments constitute the cytoskeleton: actin microfilaments, microtubules, and intermediate filaments.

The plasma membrane serves many functions: it selectively absorbs nutrients, expels wastes, and anchors the fibers of the intracellular cytoskeleton or the extracellular matrix or wall. The membrane contains proteins that serve as hormone receptors and that bind a cell to its neighbors. Cilia and flagella, extensions of the plasma membrane, can propel the cell or move liquids past it.

Various microscopic techniques generate different views of the cell with different resolutions. Standard (bright-field) light microscopy is best for stained or colored cells or tissue sections. Immunofluorescence microscopy allows specific proteins and organelles to be detected in fixed cells; the movements of microinjected fluorescent proteins can be followed in living cells. Confocal imaging, which allows the observer to view fluorescent molecules in a single plane of a specimen, permits optical sectioning of the sample and produces very sharp images. Phase-contrast and Nomarski optics enable scientists to view the details of live, unstained cells and to monitor cell movement.

Electron microscopy has a vastly higher resolution than light microscopy but generally requires the specimen to be fixed, sectioned, and dehydrated. After metal shadowing, the transmission electron microscope allows the investigator to visualize the structural details of such particles as viruses and collagen fibers. Unfixed, unstained specimens can be viewed in the electron microscope if they are frozen in hydrated form. The scanning electron microscope can be used to view unsectioned cells or tissues; it produces images that appear to be three-dimensional. The techniques of quick-freezing and deep-etching generate striking micrographs of cytoskeletal fibers.

Prokaryotic cells contain a single, circular molecule of DNA. Replication of DNA occupies the major part of the cell cycle, and during cell division each of the two daughter cells receives one of the daughter DNA molecules.

In eukaryotic cells, DNA replication is restricted to the S (synthesis) phase of the cell cycle. This period is preceded by the G_1 gap period, and is followed by the G_2 gap period. The M (mitosis, or cell division) phase follows G_2; it involves condensation of the duplicated chromosomes, their alignment at the metaphase plate, the separation of the sister chromatids, and their segregation into each of the two daughter cells. Mitosis ensures that each of the two daughter cells has the same component of DNA as the parent cell. During meiosis, a diploid cell undergoes one round of DNA replication followed by two rounds of cell division. The resulting haploid germ cells have only one copy of each homologous chromosome, and the maternal and paternal members of each chromosome pair segregate independently. Crossing over between homologous chromosomes during meiosis is the source of new combinations of genes in interbreeding populations.

REVIEW QUESTIONS

1. Data concerning the molecular composition of cells are often useful in estimating various biological parameters. In this chapter, this was demonstrated by using the average size of a hepatocyte to assess the number of molecules of protein in a cell. Review the logic behind these calculations.

Gluconeogenesis, a metabolic pathway responsible for the synthesis of glucose from smaller carbon-containing precursors, occurs in certain tissues within the body, particularly the liver. An early step in the process involves the conversion of pyruvate to oxaloacetate, a reaction that is catalyzed by the enzyme pyruvate carboxylase as follows:

$$\text{pyruvate} + CO_2 + ATP + H_2O \rightleftharpoons$$
$$\text{oxaloacetate} + ADP + P_i + 2H^+$$

The K_Ms for pyruvate and ATP in this reaction are 4×10^{-4} M and 6×10^{-5} M, respectively.

   a. There are approximately 10^{10} ATP molecules in a hepatocyte. By referring to the data presented in the text, estimate the concentration of ATP in a liver cell? Based on this value and the K_M for ATP in the above reaction, what can you infer about whether or not its concentration limits the conversion of pyruvate to oxaloacetate within the cell?

   b. If there were approximately 10^8 molecules of pyruvate in a hepatocyte, what would be the concentration of this molecule in the cell? How does this value relate to the K_M for pyruvate in this reaction when it occurs within the cell? Would increasing the concentration of pyruvate increase the rate of the reaction?

2. Review the features of the different microscopic techniques presented in this chapter. Be certain to compare and contrast the advantages and disadvantages of each. Likewise, review the importance of resolution in conjunction with magnification.

   As mentioned in this chapter, you can envision a liver cell as a cube measuring approximately 15 μm on each edge. Also, recall that a typical hepatocyte has approximately 20,000 receptors on its cell surface that are specific for the binding of the hormone insulin. Use these data to answer the following questions.

   a. What is the approximate surface area of a hepatocyte?
   b. If the insulin receptors were equally spaced on the surface of the cell, what would be the average distance between each?
   c. Assume you have a fluorescent antibody that binds to the insulin receptor, and that you are working with a fluorescence microscope under conditions where the resolution is 200 nm. Would the individual receptors be distinguishable from each other, assuming that one receptor would be visible, or would you expect to see a diffuse cloud of emitted light? Explain your answer.
   d. What other microscopic techniques might you use to improve your ability to discern the location of the individual insulin receptors?

3. Review the methods discussed in this chapter for fractionating subcellular organelles.

References

Many of the topics listed below are treated in more detail in later chapters of this book, where additional articles are cited.

Cells as the Basic Structural Units

MARGULIS, L., and K. V. SCHWARTZ. 1987. *Five Kingdoms: An Illustrated Guide to the Phyla of Life on Earth,* 2d ed. W. H. Freeman and Company. Electron micrographs and diagrams of representative cells from the principal groups of prokaryotes, eukaryotic microorganisms, plants, and animals.
DE DUVE, C. 1984. *A Guided Tour of the Living Cell.* Scientific American Library. Vols. 1 and 2. A simple idiosyncratic introduction to cell structure and function.

Plant and Microbial Cells

BROCK, T. D., and M. T. MADIGAN. 1991. *Biology of Microorganisms,* 6th ed. Prentice-Hall.

The following table summarizes several of the organelles that can be isolated from cells; it also lists the marker enzyme or protein that is characteristic of each.

Organelle	Marker
Mitochondria	Cytochrome c
Peroxisomes	Catalase
Lysosomes	Acid phosphatase
Plasma membrane	Amino acid permeases
Rough endoplasmic reticulum	Ribosomes

There are numerous examples of enzymes that catalyze identical reactions but are encoded by two different genes. In many of these situations, these enzymes are isolated in the cell within different compartments. An example of this involves aspartate aminotransferase which catalyzes the following reaction:

$$\text{glutamate} + \text{oxaloacetate} \rightleftharpoons \text{aspartate} + \alpha\text{-ketoglutarate}$$

There is a form of this enzyme in the mitochondria and another form in the cytosol in rodent hepatocytes.

   a. Devise a scheme for purifying each of the two forms of this enzyme from rat liver using an approach that involves the isolation of subcellular organelles coupled with techniques learned in previous chapters.
   b. How would you show that you have enriched the mitochondrial fraction? Likewise, how would you show that your cytosolic supernatant does not contain mitochondria?
   c. How might you demonstrate that you have purified each of the two forms of aspartate aminotransferase to homogeneity?

NEIDHARDT, F., INGRAHAM, J., and SCHAECHTER, M. 1990. *Physiology of the Bacterial Cell.* Sinauer Associates.
LEDBETTER, M. C., and K. R. PORTER. 1970. *Introduction to the Fine Structure of Plant Cells.* Springer-Verlag.
RAVEN, P. H., R. F. EVERT, and H. CURTIS. 1986. *Biology of Plants,* 4th ed. Worth. An introduction to the structures and functions of plants and their cells.

Mammalian Cells (Histology Texts)

FAWCETT, D. W. 1993. Bloom and Fawcett: *A Textbook of Histology,* 12th ed. Chapman & Hall. An excellent text containing detailed descriptions of the structures of mammalian organs, tissues, and cells.
CORMACK, D. H. 1992. *Essential Histology.* Lippincott. Another excellent histology text.
WEISS, L., and L. LANSING, eds. 1983. *Histology—Cell and Tissue Biology,* 5th ed. Elsevier Biomedical.
WHEATER, P. R., H. G. BURKITT, and V. C. DANIELS. 1987. *Functional Histology: A Text and Colour Atlas,* 2d ed. Churchill Livingstone.

Mammalian Cells (Atlases)

The following books are sources of detailed electron micrographs illustrating the structures of most mammalian cells and tissues.

CROSS, P. A., and K. L. MERCER. 1993. *Cell and Tissue Ultrastructure: A Functional Perspective.* W. H. Freeman and Company.
FAWCETT, D. W. 1981. *The Cell,* 2d ed. Saunders.
KESSEL, R. G., and R. H. KARDON. 1979. *Tissues and Organs: A Test Atlas of Scanning Electron Microscopy.* W. H. Freeman and Company.

Microscopy and the Internal Architecture of Cells

The following books and reviews cover all of the standard techniques in light and electron microscopy.

Light Microscopy

CONN, P. M., ed. 1990. *Quantitative and Qualitative Microscopy.* Academic Press. Methods in Neurosciences, Vol. 3.
FARKAS, D. L., et al. 1993. Multimode light microscopy and the dynamics of molecules, cells and tissues. *Ann. Rev. Physiol.* 55:785–817.
HERMAN, B., and J. J. LEMASTERS, eds. 1993. *Light Microscopy: Emerging Methods and Applications.* Academic Press.
MASON, W. T. ed. 1993. *Fluorescent and Luminescent Probes for Biological Activity.* Academic Press.
MATSUMOTO, B., ed. 1993. *Cell Biological Applications of Confocal Microscopy.* Academic Press. Methods in Cell Biology, Vol. 38.
SLAYTER, E. M. 1993. *Light and Electron Microscopy.* Cambridge University Press.
TAYLOR, D. L., and Y.-L. WANG, eds. 1990. *Fluorescence Microscopy of Living Cells in Culture.* Part B: *Quantitative Fluorescence Microscopy—Imaging and Spectroscopy.* Academic Press. Methods in Cell Biology, Vol. 30.
WANG, Y.-L., and D. L. TAYLOR, eds. 1989. *Fluorescence Microscopy of Living Cells in Culture.* Part A: *Fluorescent Analogs, Labeling Cells, and Basic Microscopy.* Academic Press. Methods in Cell Biology, Vol. 29.
WILLINGHAM, M. C., and I. PASTAN. 1985. *An Atlas of Immunofluorescence in Cultured Cells.* Academic Press.

Electron Microscopy

CHIU, W. 1993. What does electron cryomicroscopy provide that x-ray crystallography and NMR spectroscopy cannot? *Ann. Rev. Biophys. Biomol. Struct.* 22:233–235.
EVERHART, T. E., and T. L. HAYES. 1972. The scanning electron microscope. *Sci. Am.* 226(1):54–69.
HEUSER, J. 1981. Quick-freeze, deep-etch preparation of samples for 3-D electron microscopy. *Trends Biochem. Sci.* 6:64–68.
PEASE, D. C., and K. R. PORTER. 1981. Electron microscopy and ultramicrotomy. *J. Cell Biol.* 91:287s–292s.
WATT, I. M. 1985. *The Principles and Practice of Electron Microscopy.* Cambridge University Press.
WISCHNITZER, S. 1981. *Introduction to Electron Microscopy,* 3d ed. Pergamon Press.

Flow Cytometry

BATTYE, F. L., and K. SHORTMAN. 1991. Flow cytometry and cell-separation procedures. *Curr. Opin. Immunol.* 3:238–241.
CIVIN, C. I., et al. 1990. Positive stem cell selection—Basic science. *Prog. Clin. Biol. Res.* 333:387–402.
DARZYNKIEWICZ, Z., and H. A. CRISSMAN, eds. 1989. *Flow Cytometry.* Academic Press. Methods in Cell Biology, Vol. 33.
ORMEROD, M. G., ed. 1990, *Flow Cytometry: A Practical Approach.* IRL Press.

Cellular Membrane Systems and Organelles

These articles and books describe the properties of individual organelles and techniques for purifying them.

BAINTON, D. 1981. The discovery of lysosomes. *J. Cell Biol.* 91:66s–76s.
COURTOY, P. J., J. QUINTART, and P. BAUDHUIN. 1984. Shift of equilibrium density induced by 3,3'-diaminobenzidine cytochemistry: a new procedure for the analysis and purification of peroxidase-containing organelles. *J. Cell Biol.* 98:870–876.
DE DUVE, C. 1975. Exploring cells with a centrifuge. *Science* 189:186–194. The Nobel Prize lecture of a pioneer in the study of cellular organelles. (See also Palade, below.)
DE DUVE, C. 1983. Microbodies in the living cell. *Sci. Am.* 248(5), 74–84.
DE DUVE, C., and H. BEAUFAY. 1981. A short history of tissue fractionation. *J. Cell Biol.* 91:293s–299s.
HOLTZMAN, E. 1989. *Lysosomes.* Plenum Press.
HOWELL, K. E., E. DEVANEY, and J. GRUENBERG. 1989. Subcellular fractionation of tissue culture cells. *Trends Biochem. Sci.* 14:44–48.
LIPSKY, N. G., and R. E. PAGANO. 1985. A vital stain for the Golgi apparatus. *Science* 228:745–747.
LOUVARD, D., H. REGGIO, and G. WARREN. 1982. Antibodies to the Golgi complex and endoplasmic reticulum. *J. Cell Biol.* 92:92–107.
MERISKO, E., M. E. FARQUHAR, and G. PALADE. 1982. Coated vesicle isolation by immunoadsorption on *Staphylococcus aureus* cells. *J. Cell Biol.* 92:846–857.
NEUFELD, E. 1991. Lysosome storage diseases. *Ann. Rev. Biochem.* 60:257–280.
PALADE, G. 1975. Intracellular aspects of the process of protein synthesis. *Science* 189:347–358. The Nobel Prize lecture of a pioneer in the study of subcellular organelles. (See also de Duve, above.)
SCHEELER, P. 1981. *Centrifugation in Biology and Medical Science.* Wiley.
TOLBERT, N. E., and E. ESSNER. 1981. Microbodies: peroxisomes and glyoxysomes. *J. Cell Biol.* 91:271s–283s.

The Organization of the Cytoplasm

The following books and articles describe the structure of the cytoplasm, focusing on the cytoskeletal fibers and on cell motility.

BERSHADSKY, A., and J. VASILEV. 1988. *Cytoskeleton.* Plenum Press.
CARRAWAY, K. L., and C. A. C. CARRAWAY. 1992. *The Cytoskeleton: A Practical Approach.* IRL Press.
FULTON, A. B. 1980. How crowded is the cytoplasm? *Cell* 30:345–347.
SCHLIWA, M. 1986. *The Cytoskeleton: An Introductory Survey.* Springer-Verlag. Cell Biology Monographs, Vol. 13.

Cell Division and Cell Cycle

CROSS, F., J. ROBERTS, and H. WEINTRAUB. 1989. Simple and complex cell cycles. *Ann. Rev. Cell Biol.* 5:341–395.
Cold Spring Harbor Symposia on Quantitative Biology, 56. 1991. *The Cell Cycle.*
FORSBURG, S. L., and P. NURSE. 1991. Cell cycle regulation in the yeasts *Saccharomyces cerevisiae* and *Schizosaccharomyces pombe.* *Ann. Rev. Cell Biol.* 7:227–256.
STAIGER, C., and J. DOONAN. 1993. Cell division in plants. *Curr. Opin. Cell Biol.* 5(2):226–231.
KIRSCHNER, M. 1992. The cell cycle then and now. *Trends Biochem. Sci.* 17:281–285.
MURRAY, A. W., and M. KIRSCHNER. 1989. Dominoes and clocks: the union of two views of the cell cycle. *Science.* 246:614–621.
MURRAY, A. and T. HUNT. 1993. *The Cell Cycle, an Introduction.* Oxford University Press.
Frontiers in Biology: The Cell Cycle. 1991. *Science* 246:603–640. A series of papers by many authors.

6

Manipulating Cells and Viruses in Culture

Understanding current advances in molecular cell biology requires a familiarity with the most commonly used biological materials and knowledge of the latest experimental techniques. In this chapter, we describe the primary techniques used to culture and manipulate certain cells, viruses, and experimental organisms. Biologists choose to study these cells and organisms because they provide experimentally favorable examples of important molecular events or processes such as the control of gene activity; formation of organelles; secretion of proteins; and differentiation of cells and construction of organ systems.

In the first part of the chapter, we discuss the culture of microbial cells, animal cells, and hybrid cells in vitro and give some examples of their experimental uses. Many studies involve cells or organisms with mutations affecting a

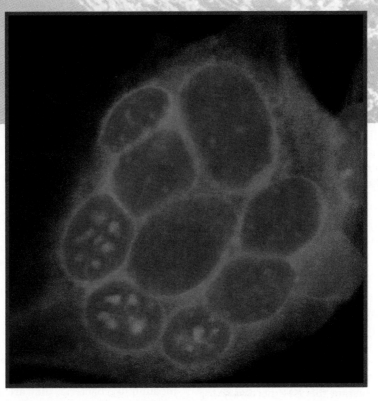

▲ Synctia of cultured 3T3 fibroblasts induced by expression of a mutant viral envelope glycoprotein.

specific biochemical or developmental pathway. In such studies, an important criterion for using a particular cell type or organism is the ease with which mutants can be isolated and characterized and the mutated genes identified and cloned (Chapter 8). Because viruses contain a small number of genes (4 to ≈100 in different types of viruses), viral mutants commonly are used to study the action of a limited set of genes designed to carry out restricted, specific molecular tasks. To provide a basis for understanding such studies, we consider the basic properties of viruses and their uses in various types of research.

In the final section of this chapter, we discuss the use of radioisotopes in detecting small numbers of biological molecules and in tracing molecular interconversions via biochemical pathways.

➤ Growth of Microorganisms in Culture

Biologists generally prefer to study cells in culture, rather than in intact organisms, for several reasons. First, most animal and plant tissues are composed of several different types of cells, whereas specific types of cells can be grown in culture; thus cultured cells are more homogeneous in their properties than in vivo cells in tissues. Second, experimental conditions can be controlled much more rigorously in culture than in an organism. By manipulating the growth conditions, for example, one can measure the effects of specific chemicals and growth factors on a particular cell type in culture. A third reason for preferring cultured cells is that in many cases a single cell can be readily grown into a colony, a process called *cell cloning,* or simply *cloning.* The resulting strain of cells, which is *genetically homogeneous,* is called a *clone.* This simple technique, which is commonly used with many bacteria, yeasts, and mammalian cell types makes it easy to isolate genetically distinct clones of cells.

Many Microorganisms Can be Grown in Minimal Medium

Among the advantages of using microorganisms such as the bacterium *Escherichia coli* and the yeast *Saccharomyces cerevisiae* are their rapid growth rate and simple nutritional requirements, which can be met with a minimal medium (Table 6-1). A minimal medium for such microorganisms can contain glucose as the sole source of carbon; metabolism of glucose to smaller molecules (e.g., CO_2, ethanol, or acetic acid) can generate the ATP necessary for energy-requiring activities of the cells. The sole nitrogen source in a minimal medium can be NH_4Cl, from which the cells can synthesize all the necessary amino acids and other nitrogen-containing metabolites. Salts are the only other component of a minimal medium.

Both prokaryotes (i.e., bacteria) and single-celled eukaryotes such as yeast, both of which grow in nature as single cells, are easily grown in culture dishes—usually on top of *agar,* a semisolid base of plant polysaccharides. The agar is first dissolved in a heated nutrient medium, and the solution is poured into petri dishes; as it cools, the solution solidifies. A dilute suspension of cells then is dispersed on top of the agar; in time each cell grows into a discrete *colony* (Figure 6-1a). Since the cells in a colony all derive from a single cell, they form a clone and have identical genomes (DNA). All the cells in a clone generally express the same set of genes and contain the same enzymes and other constituents in similar proportions. The division time for *E. coli* and similar microorganisms ranges from 20 min to 1 h. Thus a single *E. coli* cell, which divides approximately every 30 min, can grow into a colony containing 10^7–10^8 cells in 12 hours ($2^{24} = 1.7 \times 10^7$).

TABLE 6-1   Growth Media for Common Bacteria and Yeasts*

MINIMAL MEDIUM[†]

Carbon source: glucose or glycerol

Nitrogen source: NH_4^+ (e.g., $NaNH_4HPO_4$) or an organic compound such as histidine

Salts: Na^+, K^+, Mg^+, Ca^{2+}, SO_4^{2-}, Cl^-, and PO_4^{3-}

Trace elements

RICH MEDIUM

Partly hydrolyzed animal or plant tissue (rich in amino acids, short peptides, and lipids)

Yeast extract (rich in vitamins and enzyme cofactors, nucleic acid precursors, and amino acids)

Carbon source, nitrogen source, and salts as in minimal medium

---

* For more detailed information, see R. W. Davis, D. Botstein, and J. W. Roth, 1982, *A Manual for Genetic Engineering: Advanced Bacterial Genetics,* Cold Springs Harbor Laboratory.

[†] Typical for most bacteria and yeasts. Some photosynthetic bacteria (e.g., *Rhodospirillum rubrum,* cyanobacteria, and blue-green algae) require CO_2 as the carbon source. Some nitrogen-fixing bacteria (e.g., *Azotobacter*) require atmospheric N_2. Other organisms have special needs: for example, *Hemophilus* strains require factors found in whole blood.

Because yeasts, unlike most eukaryotic organisms, grow as single cells and can be grown in a simple defined medium like bacteria, they are a popular choice for studies of eukaryotic cell function. The entire life cycle of yeasts can be studied in culture, and colonies can be grown from a single vegetative cell (a growing cell) or from a single spore (a dormant cell). Studies with yeasts have provided valuable information on such subjects as the cellular mechanisms for controlling DNA synthesis; genetic recombination; and the gene products necessary to the orderly cycle of events leading to replication and segregation of chromosomes and ultimately cell division.

Mutant Strains of Bacteria and Yeast Can Be Isolated by Replica Plating

Random mutations in microorganisms cultured, or *plated,* on a petri dish containing a rich agar medium (see Table 6-1) may produce cells that are unable to synthesize a necessary metabolite (e.g., the amino acid arginine). If the rich medium supplies the metabolite whose synthesis is prevented by the mutation, then the mutant cell can survive and grow into a colony in which all the cells have the same mutation. Such a mutant colony, say an Arg⁻ colony, can be detected by *replica plating,* a technique developed by

(a)

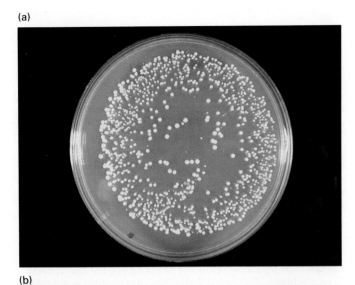

(b)

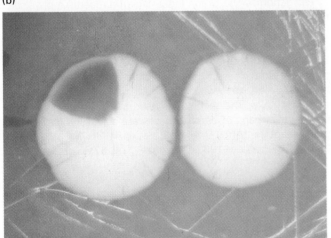

◄ FIGURE 6-1   (a) Colonies of the yeast *Saccharomyces cerevisiae* growing on a plate of agar containing only glucose, adenine, and salts. Each colony is a clone of cells. (b) Close-up view of two yeast colonies. As the cells in the colonies grew, a mutation affecting adenine biosynthesis occurred in some cells; as a result, an orange pigment derived from one of the biosynthetic intermediates accumulated. The orange sectors in these colonies are "subclones," the descendants of the original mutant cells. [See N. A. Levin, M-A. Bjornsti, and G. F. Fink, 1993, *Genetics* **133**:799. Courtesy of Dr. Nikki Levin.]

Joshua and Esther Lederberg in the 1940s (Figure 6-2). The initial plate of colonies growing on rich medium is called the *master plate*. A circular, velvet-covered stamp equal in size to the master plate is pressed onto it, so that the stamp picks up cells from each colony. The stamp is then used to deposit cells onto a second plate containing minimal agar medium without arginine (Arg⁻ medium) and also onto a third plate containing minimal agar medium with arginine (Arg⁺ medium). Cells that can synthesize arginine grow into colonies on both the Arg⁻ and Arg⁺ plates. However, a clone of cells on the master plate with a mutation in a gene required for arginine synthesis cannot grow on the Arg⁻ medium, resulting in a space on the Arg⁻ plate corresponding to the position of this clone on the master plate

➤ FIGURE 6-2   Replica plating is used to detect random mutations in bacterial and yeast cultures that produce cells differing in a single genetic trait from all other cells in the culture. A velvet-covered stamp equal in size to a petri dish is pressed onto the master plate used to culture the bacteria on rich medium. The bacteria picked up on the stamp are then deposited in the same arrangement on new (selective) plates. The media in the new plates can be chosen to test different genetic traits such as nutritional auxotrophy (the inability to grow in the absence of a specific nutrient). In the case shown, the selective plates can detect cells with a mutation in the pathway of arginine biosynthesis.

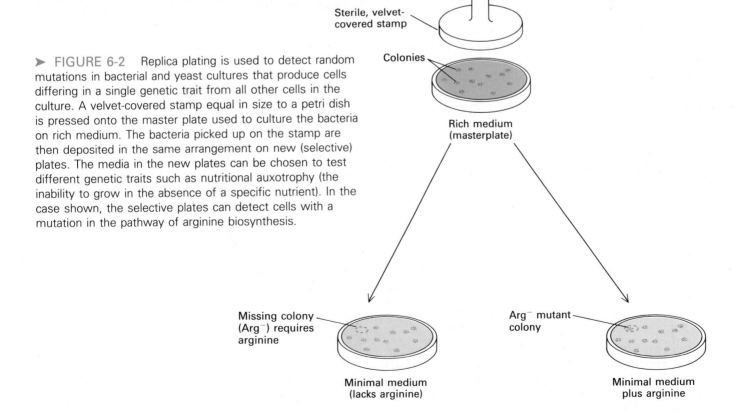

Sterile, velvet-covered stamp

Colonies

Rich medium (masterplate)

Missing colony (Arg⁻) requires arginine

Minimal medium (lacks arginine)

Arg⁻ mutant colony

Minimal medium plus arginine

and Arg⁺ plate. This clone of cells, thus identified as being defective in the synthesis of arginine, then can be isolated from the master plate.

Repeated replica plating of a master plate with paired minimal-medium plates containing or lacking different specific nutrients (e.g., amino acids, nucleic acid bases, vitamins) can identify clones of cells with mutations in the genes required for synthesis of different biological molecules. This approach has been used to isolate thousands of different bacterial and yeast *nutritional auxotrophs*; these are strains that are unable to grow in the absence of a specific nutrient.

Under standard culture conditions, random mutations occur in *E. coli* genes at the frequency of 1 in 10^6–10^8 base pairs per generation. Thus, by the time a single cell has divided to produce a few hundred or thousand cells, some

▼ FIGURE 6-3   Genetic and physical maps of the circular chromosome of *Escherichia coli*. The abbreviations on the genetic map (outside circle and enlarged segment) relate to the metabolic properties of the gene products; for example, *pyrE, asnA,* and *glnA* (blue) were each identified in a bacterial clone with a gene defect that made the addition of a pyrimidine, asparagine, or glutamine, respectively, necessary for growth. A strain with a genetic deficiency is said to be genetically marked, and the resulting genetic trait is termed a marker for that strain. A group of such markers can define a strain with great precision. Some of the regions of *E. coli* have been mapped so completely that every gene is known. An example is *ilv* (red), the locus for enzymes that catalyze the synthesis of isoleucine, leucine, and valine; each red capital letter represents a gene encoding a known purified enzyme. The numbers inside the genetic map indicate approximate percentages of the distance around the circular *E. coli* genome. The green inner circle represents a physical map of the *E. coli* DNA. The entire length of the genome is about 4×10^6 base pairs (bp), abbreviated 4000 kilobases (kb) or 4 megabases (Mb). Segment A is the largest (about 600 kb); V is the smallest (about 20 kb). The ends of each piece are sites cleaved by a restriction endonuclease (an enzyme) called *Not*I, which recognizes a particular sequence of 8 base pairs. The techniques used to construct physical and genetic maps of chromosomes are discussed in detail in other chapters. [After C. L. Smith et al., 1986, *Science* **236**:1448; M. Riley, 1993, *Microbiol. Rev.* **57**:862.]

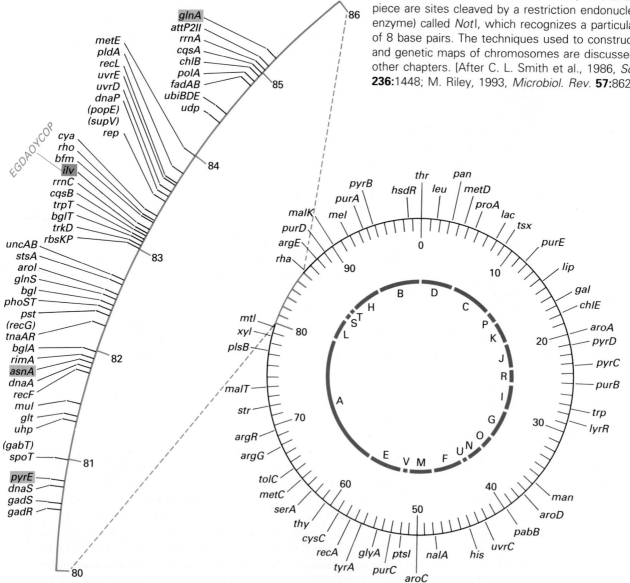

individual genes will differ among cells in the clone (Figure 6-1b). To maintain the genetic homogeneity of a clone, it must be recloned frequently under *selective* conditions that permit only cells with a particular set of desired characteristics to survive and grow. For example, to maintain an Arg⁻ clone, isolated as described above, the cells are replated on minimal medium plus arginine; only cells that can synthesize all the metabolites required for growth except arginine will grow on this selective medium.

The use of mutant clones both to map the position of genes on *E. coli* and yeast chromosomes and to unravel various biosynthetic pathways is discussed in Chapter 8. Almost all of the estimated 3000 genes in *E. coli* have been located on its circular genetic map, and the physical map of the single circular DNA molecule that constitutes the *E. coli* chromosome has been correlated with the genetic map (Figure 6-3). The techniques for constructing the physical map are described in Chapter 7. For practical purposes, every one of these genes can be assumed to be identical in every cell in a cloned colony of *E. coli* or in a liquid culture started from such a colony.

► Growth of Animal Cells in Culture

Animal cells are more difficult to culture than microorganisms because they require many more nutrients and typically grow only when attached to specially coated surfaces. Despite these difficulties, various types of animal cells, including both undifferentiated and differentiated ones, can be cultured successfully.

Rich Media Are Required for Culture of Animal Cells

Ten amino acids, referred to as the *essential amino acids,* cannot be synthesized by vertebrate animals and thus must be obtained from their diet (Table 6-2). Animal cells grown in culture also must be supplied with these ten essential amino acid, namely, arginine, histidine, isoleucine, leucine, lysine, methionine, phenylalanine, threonine, tryptophan, and valine. In addition, most cultured cells require cysteine, glutamine, and tyrosine. In the intact animal, these three amino acids are synthesized by specialized cells; for example, liver cells make tyrosine from phenylalanine, and both liver and kidney cells can make glutamine. Animal cells both within the organism and in culture can synthesize the ten remaining amino acids (see Figure 3-3); thus these amino acids need not be present in the diet or culture medium. The other essential components of a medium for culturing animal cells are vitamins, which the cells cannot make at all or in adequate amounts; various salts; glucose; and *serum,* the noncellular part of the blood (Table 6-3, *top*).

TABLE 6-2 Daily Requirement of Essential Amino Acids (for College-Age Males)*

Amino Acid	Grams
Arginine	0†
Histidine	unknown‡
Isoleucine	1.30
Leucine	2.02
Lysine	1.50
Methionine	2.02
Phenylalanine	2.02
Threonine	0.91
Tryptophan	0.46
Valine	1.50

* The amino acid requirements of men were established by feeding experiments. All of the listed amino acids are necessary to maintain body mass. However, arginine and histidine can be omitted from the diet for a short time without damaging body cells.
† Required by infants and growing children.
‡ Essential, but the precise requirement is not yet established.
SOURCE: W. C. Rose et al., 1955, *J. Biol. Chem.* **217**:987.

Serum, a mixture of hundreds of proteins, contains various factors needed for proliferation of cells in culture. For example, the hormone insulin is required for growth of many cultured vertebrate cells, and transferrin (an iron-transporting protein) is essential for incorporation of iron into cells in culture. Although many animal cells can grow in a serum-containing medium, such as Eagle's medium, certain cell types require specific protein factors that are not present in serum. Precursors of red blood cells, for instance, require the hormone erythropoietin, and T lymphocytes of the immune system require interleukin 2 (IL-2). Culture of such cells requires addition of the appropriate growth factor to the medium. A few mammalian cell types can be grown in a completely defined, serum-free medium supplemented with trace minerals, protein growth factors, and other components (Table 6-3, *bottom*).

Most Cultured Animal Cells Only Grow on Special Solid Surfaces

Within the tissues of intact animals, most cells tightly contact and interact specifically with other cells via various cellular junctions. The cells also contact a complex network of secreted proteins and carbohydrates called the *extracellular matrix,* which fills the spaces between cells. The matrix, whose constituents are secreted by cells themselves, helps bind the cells in tissues together; it also provides a lattice through which cells can move, particularly during the early stages of animal differentiation.

TABLE 6-3   Growth Media for Mammalian Cells

SERUM-CONTAINING MEDIUM (EAGLE'S MEDIUM)

Essential amino acids	Those listed in Table 6-2 plus cysteine, glutamine, and tyrosine (all at 10^{-4} to 10^{-5} M)
Vitamins	Choline, folic acid, nicotinamide, pantothenate, pyridoxal, and thiamine (1 mg/l); inositol (2 mg/l); riboflavin (0.1 mg/l)
Salts	Na^+, K^+, Ca^{2+}, Mg^{2+}, Cl^-, PO_4^{3-}, HCO_3^-
Glucose	0.9 g/l
Dialyzed serum*	5–10% of total volume

DEFINED (SERUM-FREE) MEDIUM

Amino acids	As above plus alanine and asparagine (10^{-4} M)
Vitamins, salts, glucose	As above
Other additions:	
Fatty acids	Linoleic acid, lipoic acid
N compounds	Hypoxanthine, thymidine, putrescine
Carbon source:	Pyruvate
Trace elements:	Cadmium (Cd), manganese (Mn), molybdenum (Mo), nickel (Ni), tin (Sn), vanadium (V)
Hormones and growth factors	Insulin, transferrin, hydrocortisone, fibroblast growth factor, epidermal growth factor

* Serum contains hundreds of proteins with a total protein concentration of 50–70 mg/ml. Albumin is the most plentiful serum protein (30–50 mg/ml). Various growth factors are present in serum at very low concentrations: for example, growth hormone (34 ng/ml) and insulin (0.2 ng/ml). Serum from fetal calves is used frequently because it contains higher concentrations of certain growth factors than serum from other species.

SOURCE: H. Eagle, 1959, *Science* **130**:432; S. E. Hutchings and G. H. Sato, 1978, *Proc. Nat'l, Acad. Sci. USA* **75**:901.

The extracellular matrices in various animal tissues all contain fibrous *collagen* proteins, *hyaluronic acid*, complexes of polysaccharides and proteins called *proteoglycans*, and glycoproteins (proteins linked to chains of carbohydrates). However, the exact composition of the matrix in different tissues varies, reflecting the specialized function of a tissue. In connective tissue, for example, the major protein of the extracellular matrix is a type of collagen that forms insoluble fibers with a very high tensile strength. Fibroblasts, the principal cell type in connective tissue, secrete this type of collagen as well as the other matrix components (Figure 6-4). Receptor proteins in the plasma membrane of cells bind various matrix elements, imparting strength and rigidity to tissues (see Figure 5-32).

The tendency of animal cells in vivo to interact with each other and with the surrounding extracellular matrix is mimicked in their growth in culture. Unlike bacterial and yeast cells, which can be grown in suspension, most cultured animal cells require a surface to grow on. Many types of cells can adhere to and grow on glass, or on specially treated plastics with negatively charged groups on the surface (e.g., SO_3^{2-}). The cultured cells secrete collagens and other matrix components; these bind to the culture surface

and function as a "bridge" between it and the cells. Cells cultured from single cells on a glass or a plastic dish form visible colonies in 10–14 days (Figure 6-5). Some tumor cells can be grown in suspension, a considerable experimental advantage because equivalent samples are easier to obtain from suspension cultures than from colonies grown in a dish.

Primary Cell Cultures Have a Finite Life Span

Primary cell cultures are derived from normal animal tissue. Commonly skin or whole embryos are used to establish primary cell cultures. To prepare tissue cells for culture (or to remove adherent cells from a culture dish for biochemical studies), trypsin or another protease is used to destroy the proteins in the junctions that normally interconnect cells. The cell type that usually predominates in primary cultures is called a *fibroblast* because it has the morphology and secretes the types of proteins characteristic of the fibroblasts that form fibrous connective tissue in animals. Cultured fibroblasts, however, are not as differentiated as tissue fibroblasts. Following appropriate stimula-

Fibroblast   Bundle of collagen fibers

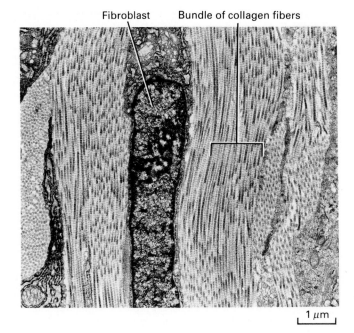

1 μm

▲ FIGURE 6-4   Electron micrograph of collagen fibers and the parent fibroblast cells that secreted them. The repeated pattern of staining along the collagen fibers is seen at higher magnification in the shadowed image in Figure 5-24. [From D. Eyre, 1980, *Science* **207**:1314.]

tion, these cultured cells can differentiate into many cell types (e.g., fat cells, connective tissue cells, muscle cells). In most studies, cultured "fibroblasts" are treated simply as convenient prototypical cells for study.

When first placed in culture, fibroblasts grow well and continue to double for 50–100 generations. Such a lineage of cells originating from one initial culture is called a *cell strain*. Eventually, however, the cells reach a "crisis," grow very slowly (if at all) for a few more generations, and finally cease growing altogether. This growth pattern is a reflection of the normal aging process of organisms. Cell death, often called *apoptosis*, is a normal feature of the differentiation and maturation of an organism, and the number of divisions of all types of cells is precisely controlled. For this reason, fibroblasts from newborns grow

➤ FIGURE 6-5   Cultured mammalian cells viewed at three magnifications. (a) A single mouse cell attached to a plastic petri dish, viewed through a scanning electron microscope. To separate attached cells so they can be plated individually, a cell culture is treated with a protease such as trypsin. (b) A photomicrograph of a single colony of human HeLa cells about 1 mm in diameter, produced from a single cell after growth for 2 weeks. (c) After staining cells in a 6-cm-diameter petri dish, individual colonies can easily be seen and counted. [See P. I. Marcus, S. J. Cieciura, and T. T. Puck, 1956, *J. Exp. Med.* **104**:615. Part (a) courtesy of N. K. Weller; parts (b) and (c) courtesy of T. T. Puck.]

(a)

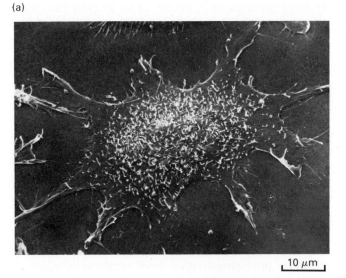

10 μm

(b)

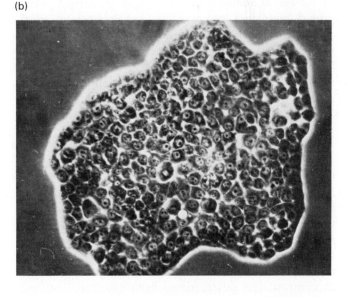

(c)

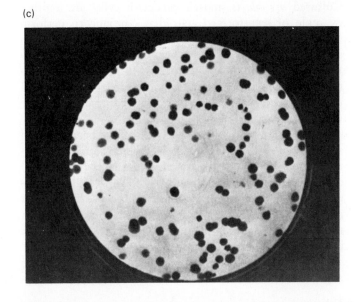

for more generations in culture than do cells from older individuals, and no nutritional regime or set of growth factors has been discovered that allows primary fibroblasts from adults to undergo additional divisions in culture.

Although primary cultures of cell strains are useful for many types of studies, their limited life span precludes certain types of experiments. Some cell types, however, can undergo a genetic change that permits them to grow indefinitely. Such *immortal* animal cells are used in a wide variety of studies.

Transformed Cells Can Grow Indefinitely in Culture

All mammalian cell lines capable of *indefinite* growth in culture are derived either from tumor cells taken directly from an animal or from cultured cells that have undergone *transformation*, a change that causes them to behave as tumor cells. In both cases, certain normal growth-regulating processes no longer operate, so the cells continue to grow indefinitely. Because tumor cells may arise by transformation of normal cells within an animal, they also are referred to as transformed cells. The term *cell line* is used for such continuously growing cells to distinguish them from cultured cell strains with a finite life span.

Differentiated Cell Lines   Although many transformed cells are relatively undifferentiated, some cell lines carry out many of the functions characteristic of the normal differentiated cells from which they are derived. One example is certain hepatoma cell lines (e.g., HepG2) that synthesize most of the serum proteins made by normal hepatocytes (the major cell type in the liver) from which they are derived (Table 6-4). These highly differentiated hepatoma cells are often studied as models of normal hepatocytes. In contrast, less differentiated hepatoma cell lines (e.g., PLC/PRF/5) synthesize far fewer liver-specific proteins. Cultured *myoblasts* (muscle precursor cells) are another example of transformed cells that continue to perform many functions of a specialized, differentiated cell. When grown in culture, transformed myoblasts can be induced to fuse to form myotubes, which resemble differentiated multinucleated muscle cells, and synthesize many, if not all, of the specialized proteins associated with contraction (Figure 6-6).

Certain lines of epithelial cells also have been cultured successfully. With few exceptions, all the internal and external body surfaces are covered with a layer of epithelial cells called an *epithelium* (Figure 6-7). These highly differentiated cells are said to be *polarized* because the plasma membrane is organized into at least two discrete regions. In the epithelial cells that line the intestine, for example, that portion of the plasma membrane facing the intestine, the *apical* surface, is specialized for absorption; the rest of the plasma membrane, the *basolateral* surface, mediates trans-

TABLE 6-4   Liver-Specific Proteins Synthesized by Two Human Hepatoma Cell Lines in Culture[*]

Protein	HepG2 Hepatoma	PLC/PRF/5 Hepatoma
Albumin	+	−
α_2-Macroglobulin	+	+
α_1-Antitrypsin	+	+
α_1-Antichymotrypsin	+	−
Transferrin	+	+
Haptoglobin	+	−
Gc-globulin	−	−
Complement C3	+	+
Complement C4	+	−

[*] All of the listed proteins are serum proteins synthesized and secreted by normal hepatocytes, the major cell type in the liver. A + indicates that the hepatoma cell line makes a particular protein; a − indicates that it does not. Since the HepG2 line makes nearly all liver-specific proteins, it is considered to be highly differentiated, whereas PLC/PRF/5, which makes fewer liver-specific proteins, is less differentiated. SOURCE: B. B. Knowles, C. C. Howe, and D. P. Aden, 1980, *Science* **209**:497.

Early stage of myotube

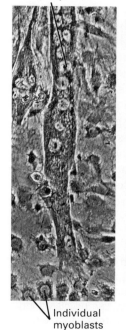 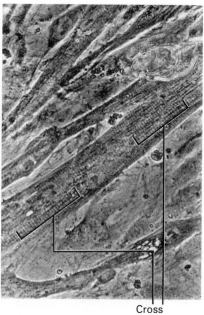

Individual myoblasts

Cross striations

▲ FIGURE 6-6   Cultured transformed line of rat myoblasts. *(Left)* This cell line grows indefinitely as single cells in culture. *(Right)* When growth of cultured myoblasts is stopped (e.g., by removing serum from the medium), the cells fuse to produce myotubes with the characteristic cross striations of differentiated muscle cells.

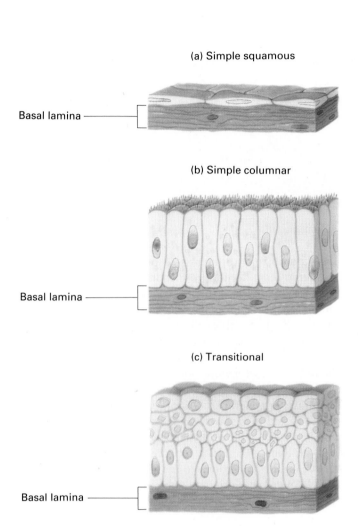

(a) Simple squamous

Basal lamina

(b) Simple columnar

Basal lamina

(c) Transitional

Basal lamina

(d) Stratified squamous
(nonkeratinized)

Basal lamina

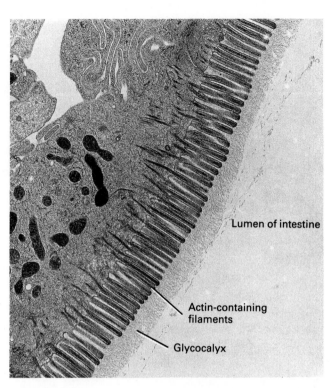

Lumen of intestine

Actin-containing
filaments

Glycocalyx

▲ FIGURE 6-8   Transmission electron micrograph of the apical surface of an intestinal epithelial cell. Note the large number of fingerlike projections (microvilli), which have actin filaments in their core, and the fibrous network of glycoproteins (the glycocalyx) overlaying the apical surface. The glycocalyx contains various digestive enzymes. [Courtesy of S. Ito.]

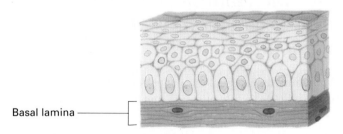

▲ FIGURE 6-7   Principal types of epithelium. (a) Simple squamous epithelia, composed of thin cells, line the blood vessels and many body cavities. (b) Simple columnar epithelia consist of elongated cells including mucus-secreting cells (in the lining of the stomach and cervical tract) and absorptive cells (in the lining of the small intestine). (c) Transitional epithelia, composed of several layers of cells with different shapes, line certain cavities subject to expansion and contraction (e.g., the urinary bladder). (d) Stratified squamous (nonkeratinized) epithelia line surfaces such as the mouth and vagina; these linings resist abrasion and generally do not participate in the absorption or excretion of materials into or out of the cavity.

port of nutrients from the cell to the blood and forms junctions with adjacent cells and the underlying extracellular matrix called the *basal lamina*. As shown in Figure 6-8, the apical plasma membrane of intestinal epithelial cells is folded into a large number of projections called *microvilli*, which have actin filaments in their core. This specialization, also called the *brush border*, greatly increases the surface area, and thus the absorption rate of the intestinal lining. Attached to the brush border is the *glycocalyx*, a fibrous network of glycoproteins containing enzymes that catalyze the final stages in the digestion of carbohydrates. A detailed discussion of these complex cells is presented in Chapter 14.

Despite the complexity of epithelial cells, certain epithelial-cell lines form polarized monolayers when grown in culture. One such line composed of Madin-Darby canine kidney (MDCK) cells forms a continuous polarized sheet, one cell thick, that exhibits many of the properties of the normal canine kidney epithelium from which it was derived (Figure 6-9). This type of preparation has proved valuable as a model for studying the functions of epithelial cells.

Undifferentiated Cell Lines   Many specialized animal cells cannot be cultured at all. This is not so surprising,

(a)

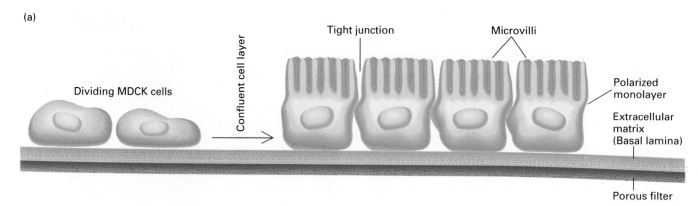

Dividing MDCK cells

Confluent cell layer

Tight junction

Microvilli

Polarized monolayer

Extracellular matrix (Basal lamina)

Porous filter

(b)

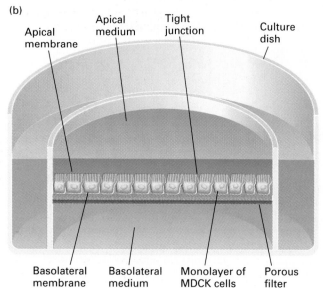

Apical membrane

Apical medium

Tight junction

Culture dish

Basolateral membrane

Basolateral medium

Monolayer of MDCK cells

Porous filter

(c)

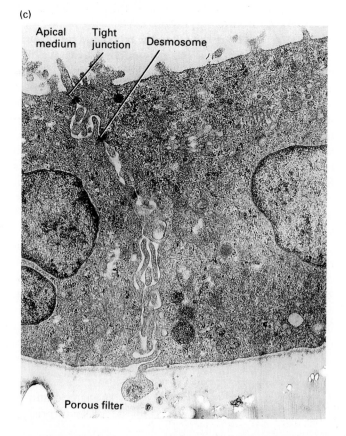

Apical medium

Tight junction

Desmosome

Porous filter

since certain fully differentiated normal cells (e.g., muscle and nerve cells) do not grow continuously in the body. In some cases, however, the less differentiated, dividing precursors of such cells can be cultured and will differentiate into nondividing cells exhibiting at least some of the properties of the corresponding differentiated cells. Cultures of myoblasts, mentioned previously, are an example of this phenomenon (see Figure 6-6).

Even poorly differentiated cell lines have been used to advantage in general studies on RNA, DNA, and protein synthesis and in research on the structural elements common to all cells. One of the most popular poorly differentiated cell lines is the HeLa cell (see Figure 6-5b,c), the first human cell to be grown continuously in the laboratory. The HeLa cell was originally obtained in 1952 from a malignant tumor (carcinoma) of the uterine cervix.

► *The Use of Hybrid Cells in Genetic Analysis of Animal Cells and Production of Monoclonal Antibody*

Spontaneous *fusion* of animal cells in culture occurs infrequently, but the rate of fusion increases greatly in the presence of certain viruses that have a lipoprotein envelope

◄ FIGURE 6-9   Culture of Madin-Darby canine kidney (MDCK) cells, a line of differentiated epithelial cells. (a) MDCK cells form a polarized epithelium when grown to confluence on a porous membrane filter coated on one side with collagen and other proteins of the basal lamina, the extracellular matrix that supports an epithelial layer. (b) Special culture dishes allow the cells to be bathed with an appropriate medium on each side of the filter; the apical surface faces the medium that bathes the upper side. (c) Electron micrograph of parts of two MDCK cells grown in tissue culture on a permeable filter. Tight junctions and desmosomes are specialized regions of the plasma membrane that connect adjacent cells. [Part (c) courtesy of R. Van Buskirk, J. Cook, J. Gabriels, and H. Eichelberger.]

similar to the plasma membrane of animal cells. A viral glycoprotein in this envelope promotes cell fusion; the mechanism of this effect is discussed in Chapter 16. Cell fusion also is promoted by polyethylene glycol, which causes the plasma membranes of adjacent cells to adhere to each other and to fuse (Figure 6-10). In most fused animal cells, the nuclei eventually also fuse, producing viable cells that contain chromosomes from both "parents." The fusion of two cells that are genetically different yields a hybrid cell called a *heterokaryon.*

Genes Can Be Mapped to Specific Chromosomes with Interspecific Hybrid Cells

Because some animal cells can be cultured from single cells in a well-defined medium, it is possible to select for genetically distinct *somatic* (body) *cells,* just as is done with bacterial and yeast cells. Moreover, during mitosis the chromosomes in a somatic cell are large and highly visible after staining, making it easy to distinguish individual chromosomes and to observe their arrangements (Chapter 9). In the branch of cell biology called *somatic-cell genetics,* fusion of cultured somatic cells has been used to map genes to chromosomes and to study gene function when nuclei are introduced into novel cellular environments.

Hybrids of cultured cells from different mammals— for example, human and rodent cells or cells of different rodents (rats, mice, and hamsters)—have been widely used in somatic-cell genetics. As hybrids of human and mouse cells grow and divide, they gradually lose human chromosomes in random order, but retain the mouse chromosomes. In a medium that can support growth of both mouse and human cells, the hybrids eventually lose all human chromosomes. However, in a medium in which human cells can grow but mouse cells lack one enzyme needed for growth, the one human chromosome that contains the gene encoding the needed enzyme will be retained, while all the other human chromosomes eventually will be lost.

By using various media in which mouse cells cannot grow, because they lack a particular enzyme, but human cells can, *panels* of hybrid cell lines have been established. Each cell line in a panel contains either a single human chromosome or a small number of human chromosomes, and a full set of mouse chromosomes. Because each chromosome can be identified visually under a light microscope, such hybrid cells provide a means for assigning, or "mapping," individual genes to specific chromosomes. For example, suppose a hybrid cell line is shown microscopically to contain a particular human chromosome. The cell line can then be tested biochemically for the presence of various human enzymes, exposed to specific antibodies to detect surface antigens, or subjected to DNA hybridization and cloning techniques (Chapter 7) to locate particular DNA sequences. The gene(s) encoding any protein(s) or containing DNA sequence(s) detected in such tests must be located on the particular human chromosome carried by the cell line. Panels of mouse-hamster hybrid cells also have been established; in these cells, the majority of mouse chromosomes are lost, allowing mouse genes to be mapped to specific mouse chromosomes.

(a)

(b)

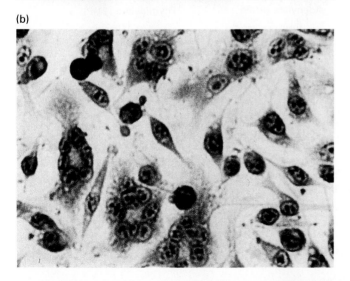

◄ FIGURE 6-10   Fusion of cultured animal cells. (a) Unfused growing mouse cells with a single nucleus per cell. (b) Fused mouse cells with 2–5 nuclei per cell. Fusion was induced by treatment with polyethylene glycol (45 percent) for 1 min. The number of nuclei per fused cell (heterokaryon) is determined by the polyethylene glycol concentration and the time of exposure. By adjustment of these factors, the number of heterokaryons containing only two nuclei can be maximized. [From R. L. Davidson and P. S. Gerald, 1976, *Som. Cell Genet.* **2**:165.]

Mutants in Purine- and Pyrimidine-Salvage Pathways Are Good Selective Markers

One metabolic pathway has been particularly useful in cell-fusion experiments. Most animal cells can synthesize the purine and pyrimidine nucleotides de novo from simpler carbon and nitrogen compounds, rather than from already formed purines and pyrimidines (Figure 6-11, *top*). The folic acid antagonists *amethopterin* and *aminopterin* interfere with the donation of methyl and formyl groups by tetrahydrofolic acid in the early stages of de novo synthesis of glycine, purines, nucleoside monophosphates, and thymidine monophosphate. These drugs are called *antifolates,* since they block reactions involving tetrahydrofolate, an active form of folic acid. Many cells, however, contain enzymes that synthesize the necessary nucleotides from pu-

rine bases and thymidine; these *salvage pathways* bypass the metabolic blocks imposed by antifolates (Figure 6-11, *bottom*).

A number of mutant cell lines have been isolated that lack the enzyme needed to carry out one of the steps in the salvage pathway. For example, cell lines lacking *thymidine kinase* (TK) have been selected because such cells are resistant to the otherwise toxic thymidine analog 5-bromodeoxyuridine. Cells containing TK convert 5-bromodeoxyuridine into 5-bromodeoxyuridine monophosphate. This nucleoside monophosphate is then converted into a nucleoside triphosphate by other enzymes and is incorporated by DNA polymerase into DNA where it exerts its toxic effects. This pathway is blocked in cells with a mutation in the *TK* gene preventing the production of functional TK. Similarly, cells lacking the HGPRT enzyme have been se-

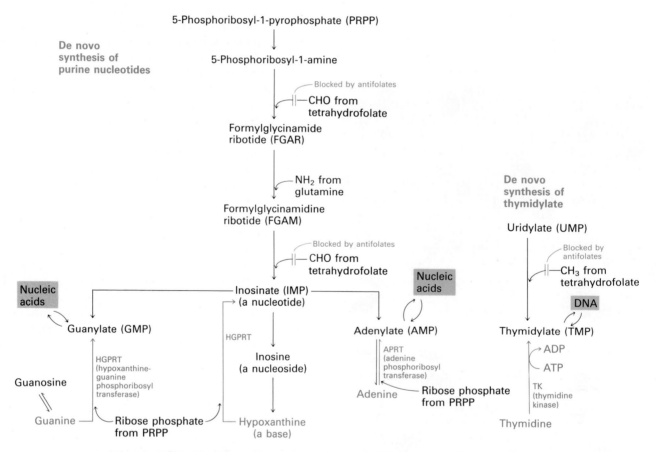

Salvage pathways to obtain purine nucleotides and thymidylate

▲ FIGURE 6-11   Under normal circumstances, cultured animal cells synthesize purine nucleotides and thymidylate by de novo pathways that require the transfer of a methyl or formyl group from an activated form of tetrahydrofolate (e.g., N^5,N^{10}-methylenetetrahydrofolate), shown in the upper portion of the diagram. Antifolates, such as aminopterin and amethopterin, block the reactivation of tetrahydrofolate, preventing purine and thymidylate synthesis. A second mecha-

nism for obtaining nucleotides, called a salvage pathway is shown in the bottom part of the diagram. The enzymes of the salvage pathways are HGPRT, APRT, and TK. These pathways are not blocked by antifolates. If the medium contains purine bases or nucleosides and thymidine, most mammalian cells can use them directly to make nucleotides by these pathways.

lected because they are resistant to the otherwise toxic guanine analog 6-thioguanine. HGPRT⁻ cells and TK⁻ cells are useful partners in cell fusions with one another or with cells that have salvage-pathway enzymes but that are differentiated and cannot grow in culture by themselves.

The selective medium most often used to culture such hybrid cells is called *HAT medium*, because it contains hypoxanthine (a purine), aminopterin, and thymidine. Normal cells can grow in HAT medium because even though aminopterin blocks de novo synthesis of purines and TMP, the thymidine in the media is transported into the cell and converted to TMP by TK and the hypoxanthine is transported and converted into usable purines by HGPRT. On the other hand, neither TK⁻ nor HGPRT⁻ cells can grow in HAT medium. However, hybrids formed by fusion of TK⁻ cells and HGPRT⁻ cells or from mutant cells and normal cells can grow in HAT medium because both functional salvage-pathway enzymes are present in the fused cells.

Hybridomas Are Fused Lymphoid Cells That Make Monoclonal Antibodies

As noted in Chapter 3, antibodies are synthesized by B lymphocytes. Each normal B lymphocyte in an animal is capable of producing a single type of antibody directed against a specific determinant, or epitope, on an antigen molecule. If an animal is injected with an antigen, B lymphocytes that make antibody recognizing the antigen are stimulated to grow and proliferate. Each antigen-activated B lymphocyte forms a clone of cells in the spleen or lymph nodes, with each cell of the clone producing identical antibody, termed *monoclonal antibody*. Because most natural antigens contain multiple epitopes, exposure of an animal to an antigen usually stimulates formation of several different B-lymphocyte clones, each producing a different antibody; a mixture of antibodies that recognize different epitopes on the same antigen is said to be *polyclonal*.

For many types of studies involving antibodies, monoclonal antibody is preferable to polyclonal antibody. However, biochemical purification of monoclonal antibody from serum is not feasible, in part because the concentration of any given antibody is quite low. For this reason, researchers looked to culture techniques in order to obtain usable quantities of monoclonal antibody. Because primary cultures of normal B lymphocytes do not grow indefinitely, such cultures have limited usefulness for production of monoclonal antibody. This limitation can be avoided by fusing normal B lymphocytes with cancerous lymphocytes called *myeloma cells*, which are immortal.

Fusion of myeloma cells with normal antibody-producing cells from a rat or mouse spleen yields *hybridomas*. Like myelomas, hybridomas are immortal, and each hybridoma produces the monoclonal antibody encoded by its B-lymphocyte partner. Many different myeloma-cell lines from mice and rats have been established; from these, HGPRT⁻ lines have been selected as described above. If

such mutant myeloma cells are fused with normal B lymphocytes, the resulting hybridomas can easily be separated from the unfused cells, since only the fused cells can grow in HAT medium (Figure 6-12). Each selected hybridoma

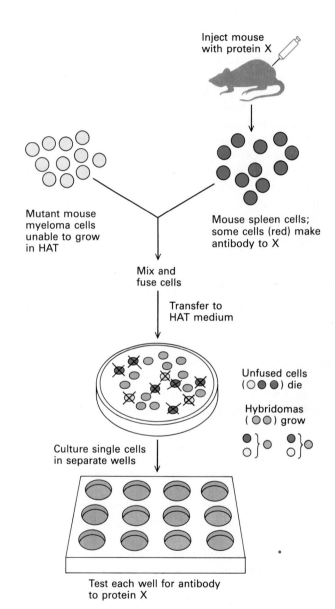

▲ FIGURE 6-12   Procedure for producing a monoclonal antibody to protein X. Immortal myeloma cells that lack HGPRT, an enzyme of the purine-salvage pathway (see Figure 6-11), are fused with normal antibody-producing spleen cells, which can make HGPRT. When plated on HAT medium, the unfused cells do not grow: the mutant myeloma cells because they cannot make purines via the salvage pathway, and the spleen cells because they have a limited life span in culture. Thus only fused cells, called hybridomas, formed from a myeloma cell and spleen cell, survive on HAT medium. Each hybridoma cell forms a clone producing a single antibody. Once a hybridoma clone that produces a desired antibody is identified, the clone can be cultured to yield large amounts of that antibody.

clone then is tested for production of the desired antibody; any clone producing that antibody then is grown in large cultures from which a substantial quantity of pure monoclonal antibody can be obtained.

Such pure antibodies are very valuable research reagents. For example, a monoclonal antibody that interacts with protein X can be used to label, and thus locate, protein X in specific cells of an organ or in specific cell fractions. Once identified, even very scarce proteins can be isolated by affinity chromatography in columns to which the monoclonal antibody is bound (see Figure 3-34). Monoclonal antibodies also have become important diagnostic and therapeutic tools in medicine. Monoclonal antibodies that bind to and inactivate toxic proteins (toxins) secreted by bacterial pathogens are used to treat diseases caused by these pathogens. Other monoclonal antibodies are specific for cell-surface proteins expressed by certain types of tumor cells; chemical complexes of such monoclonal antibodies with toxic drugs are being developed for cancer chemotherapy.

modify host-cell processes so as to maximize viral replication. For example, the roles of certain cellular factors in initiation of protein synthesis were revealed because viral proteins interrupt their action. Finally, when certain genes carried by cancer-causing viruses integrate into chromosomes of a normal animal cell, the normal cell can be converted to a cancer cell.

Since many viruses can infect all cells to which they are exposed, genetically modified viruses often are used to carry foreign DNA into a cell. For instance, one type of adenovirus efficiently infects cells lining the air passages in the lungs, causing a common type of cold. Recombinant DNA techniques discussed in Chapter 7 have been used to replace some of the disease-causing genes in this adenovirus with the *CFTR* gene, which is defective in individuals with cystic fibrosis. This recombinant adenovirus currently is being used to introduce a normal *CFTR* gene into the airway-lining cells of cystic fibrosis patients. This is one example of the growing list of "gene therapy" treatments.

Because of the extensive use of viruses in many cell-biology experiments, the basic aspects of the structure and function of viruses are described in this section.

➤ Viruses: Structure and Function

A virus is a small parasite that cannot reproduce by itself. Once it infects a susceptible cell, however, a virus can direct the cell machinery to produce more viruses. Each virus has either RNA or DNA as its genetic material; no known virus has both. The nucleic acid may be single- or double-stranded. The entire infectious virus particle, called a *virion,* consists of the nucleic acid and an outer shell of protein. The simplest viruses contain only enough RNA or DNA to encode four proteins. The most complex can encode 100–200 proteins.

Virus-infected animal cells are important model systems for the study of basic aspects of cell biology. In many cases, DNA viruses utilize cell enzymes for synthesis of their DNA genome, and most viruses utilize normal cellular enzymes for synthesis of their mRNAs and proteins. Most viruses dominate cellular macromolecular synthesis during the late phase of infection, synthesizing large amounts of a small number of viral mRNAs and proteins instead of the thousands of normal cellular macromolecules. For instance, animal cells infected by influenza or vesicular stomatitis virus synthesize only one or two types of glycoproteins, which are encoded by viral genes, whereas uninfected cells produce hundreds of glycoproteins. Such virus-infected cells have been used extensively in studies on synthesis of cell-surface glycoproteins. Similarly, much information about the mechanism of DNA replication has come from studies with bacterial cells and animal cells infected with simple DNA viruses, since these viruses depend almost entirely on cellular proteins to replicate their DNA. Viruses also often express proteins that

Viral Capsids Are Regular Arrays of One or a Few Types of Protein

The nucleic acid of a virion is enclosed within a protein coat, or *capsid,* composed of multiple copies of one protein or a few different proteins, each of which is encoded by a single viral gene. Because of this structure, a virus is able to encode all the information for making a relatively large capsid in a small number of genes. This efficient use of genetic information is important, since only a limited amount of RNA or DNA, and therefore a limited number of genes, can fit into a virion capsid. A capsid plus the enclosed nucleic acid is called a *nucleocapsid.*

Nature has found two basic ways of arranging the multiple capsid-protein subunits and the viral genome into a nucleocapsid. The simpler structure is a protein helix with the RNA or DNA protected within. Tobacco mosaic virus (TMV) is a classic example of the helical nucleocapsid. In TMV the protein subunits form broken disklike structures, like lock-washers, which form the helical shell of a long rodlike virus when stacked together (Figure 6-13a).

The other major structural class of viruses, called icosahedra, or quasi-spherical, viruses, is based on the *icosahedron,* a solid object built of 20 identical faces each of which is an equilateral triangle (Figure 6-13b). Each of 20 triangular faces is constructed of three identical capsid subunits, making a total of 60 subunits per capsid. At each of the 12 vertices, five subunits make contact symmetrically. Thus all protein subunits are in *equivalent* contact with each other. Tobacco satellite necrosis virus has such a sim-

(c) A large icosahedral virus

(a) Section of a helical virus

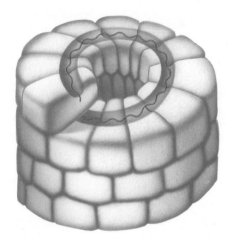

(b) A small icosahedral virus

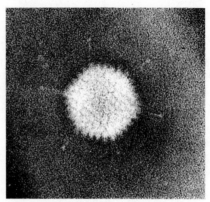

▲ FIGURE 6-13   Two basic geometric shapes of viruses. (a) In some viruses, the protein subunits form helical arrays around an RNA or DNA molecule (red), which runs in a helical groove within the enclosing protein tube. The electron micrograph is of tobacco mosaic virus, illustrating the rod shape of this type of virus. (b and c) In other viruses, the capsid proteins associate to form polyhedra with iscosahedral (20-sided) symmetry. The simplest and smallest of these quasi-spherical viruses have only fivefold symmetry at each vertex (b). In larger viruses of this type, the bonding between capsid-protein subunits not at the vertices is quasi-equivalent: the subunits on the vertices maintain fivefold symmetry, but those making up the surfaces in between exhibit sixfold symmetry. Although the actual shape of the protein subunits in these viruses is not a flat triangle as illustrated, the overall effect when the subunits are assembled is of a roughly spherical structure with triangular faces. The electron micrograph is of an adenovirus, showing surface projections at the vertices. [After S. E. Luria et al., 1978, *General Virology*, 3d ed., Wiley, pp. 39–40. Photograph of TMV courtesy of R. C. Valentine; photograph of adenovirus courtesy of Robley C. Williams, Univ. of California.]

ple icosahedral structure. However, most quasi-spherical viruses are larger and contain more proteins than can be accommodated on 20 identical faces. These proteins form shells whose subunits are in *quasi-equivalent* contact. Here, the proteins at the icosahedral vertices remain arranged in a fivefold symmetry, but additional subunits cover the surfaces between in a pattern of sixfold symmetry (Figure 6-13c).

The atomic structures of a number of icosahedral viruses have been determined by x-ray crystallography (Figure 6-14). The first three such viruses to be analyzed—tomato bushy stunt virus, poliovirus, and rhinovirus (the common cold virus)—exhibit a remarkably similar design, in terms of the rules of icosahedral symmetry as well as in the details of their surface proteins. In each virus, at atomic resolution, clefts ("canyons") can be observed near the points at which the viral surface proteins meet. These clefts are believed to interact with cell-surface receptors when the virus attaches to a host cell.

In some viruses, the symmetrically arranged nucleocapsid is covered by an external membrane, or *envelope*, which consists mainly of a phospholipid bilayer but also contains one or two types of virus-encoded glycoproteins (Figure 6-15). The phospholipids in the viral envelope are similar to those in the plasma membrane of an infected host cell. The viral envelope is, in fact, derived by budding from that membrane, but contains mainly viral glycoproteins.

The components of simple viruses such as TMV, which consists of a single RNA molecule and one protein species, undergo *self-assembly* if they are mixed in solution. More complex viruses containing a dozen or more protein species do not spontaneously assemble in vitro. The multiple components of such viruses assemble within infected cells in stages, first into subviral particles and then into completed virions. Such complex viruses produce some virus-encoded proteins that assist in the assembly of the virion; however, such assembly proteins are not themselves components of the completed virion.

(a)

Protein subunits

VP1

VP3

VP2

Picorna viruses

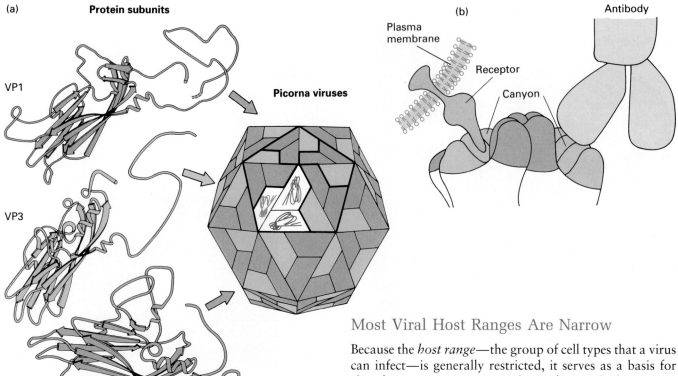

(b)

Antibody

Plasma membrane

Receptor

Canyon

▲ FIGURE 6-14   Structure of picornaviruses. These ico-sahedral viruses include poliovirus and the rhinoviruses, which cause the common cold. (a) The picornavirus capsid is composed of three proteins (VP1, VP2, and VP3) whose three-dimensional conformations differ only slightly. The model of a picornavirus based on x-ray crystallographic analyses shows that the vertices with fivefold symmetry contain five VP1 molecules; the vertices with sixfold symmetry contain three VP2 molecules and three VP3 molecules. (b) Many picornaviruses have an indentation, or "canyon," on their surface; this surface region may interact with a host-cell receptor, allowing the virus to bind to cells. Antibody to a viral surface protein could neutralize the virus either by preventing the receptor from entering the canyon or by obstructing entry of the virus-antibody complex into a host cell. [See M. S. Rossman and R. R. Rueckert, 1987, *Microbiol. Sci.* **4:**206.]

Most Viral Host Ranges Are Narrow

Because the *host range*—the group of cell types that a virus can infect—is generally restricted, it serves as a basis for classifying viruses. A virus that infects only bacteria is called a *bacteriophage*, or simply a *phage*. Viruses that infect animal or plant cells are referred to generally as *animal viruses* or *plant viruses*. A few viruses can grow in both plants and the insects which feed on the plants. The highly mobile insects serve as a *vector* for transferring these viruses between susceptible plant hosts; for example, potato yellow dwarf virus can grow in leafhoppers (insects that

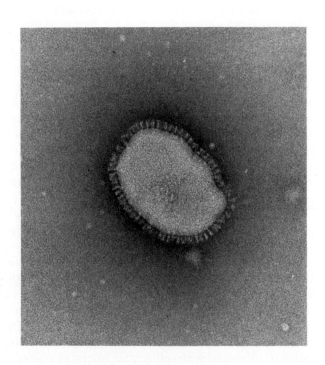

➤ FIGURE 6-15   Electron micrograph of a negatively stained particle of influenza virus. The virus particle is surrounded by a phospholipid bilayer; the large spikes protruding outward from the membrane are trimers of hemagglutinin protein and tetramers of neuraminidase protein. Inside is the nucleocapsid. [Courtesy of A. Helenius and J. White.]

feed on potato plant leaves) as well as in potato plants. Some strictly animal viruses also have wide host ranges; *vesicular stomatitis virus* grows in insects and in many different types of mammalian cells. Nevertheless, most animal viruses do not cross phyla, and some (e.g., poliovirus) infect only closely related species such as primates. The host-cell range of some animal viruses is further restricted to a limited number of cell types because only these cells have appropriate surface receptors to which the virions can attach.

Viruses Can Be Cloned and Counted in Plaque Assays

Virology is a quantitative science in which experimenters can accurately determine the number of infectious viral particles in a sample. Such an assay is performed by culturing a dilute sample of viral particles on a plate of host cells and then counting the number of local lesions, called *plaques,* that develop on the plate (Figure 6-16). A plaque develops on the plate wherever a single virion initially infected a single cell. The virus replicates in this initial host cell and then lyses the cell, releasing many progeny virions which infect the neighboring cells on the plate. After a few of these cycles of infection, enough cells are lysed to produce a visible plaque in the layer of remaining uninfected cells. Since all the progeny virions in a plaque are derived from a single parental virus, they constitute a virus clone. This type of plaque assay is in standard use for bacterial and animal viruses. Plant viruses can be assayed similarly by counting local lesions on plant leaves inoculated with viruses.

Viral Growth Cycles Are Classified as Lytic or Lysogenic

A protein on the surface of a virus binds, or *adsorbs,* specifically to a *receptor* protein on a host cell to begin an infection. This interaction determines the host range of a virus. Then, in one of various ways, the viral DNA or RNA crosses the plasma membrane into the cytoplasm. The entering genetic material may still be accompanied by inner viral proteins, but the capsid is typically left behind. The DNA genome of most DNA-containing viruses (with some

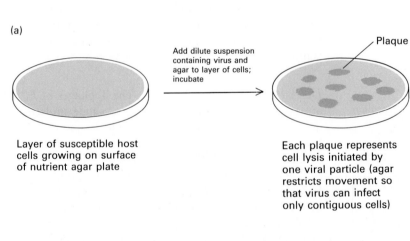

(a)

Layer of susceptible host cells growing on surface of nutrient agar plate

Add dilute suspension containing virus and agar to layer of cells; incubate

Plaque

Each plaque represents cell lysis initiated by one viral particle (agar restricts movement so that virus can infect only contiguous cells)

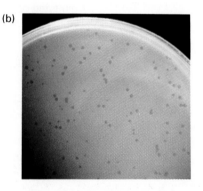

(b)

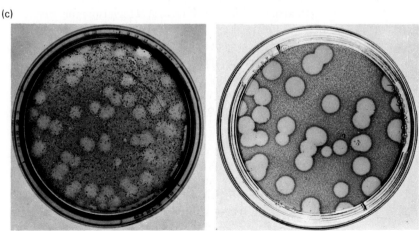

(c)

◄ FIGURE 6-16   Plaque assay for determining number of infectious particles in a viral suspension. (a) Each lesion, or plaque, that develops on a plate containing host cells results from growth of a single virion, forming new particles that infect neighboring cells. The virions in each plaque constitute a pure viral clone. (b) Plate illuminated from behind shows plaques formed by λ bacteriophage plated on *E. coli.* (c) Plates showing plaques produced by two animal viruses: *(left)* western equine encephalomyelitis virus plated on chicken embryo fibroblasts; *(right)* poliovirus plated on HeLa cells. [Part (b) courtesy of Barbara Morris; part (c, *left*) courtesy of R. Dulbecco; part (c, *right*) from S. E. Luria et al., 1978, *General Virology,* 3d ed., Wiley, p. 26.]

associated proteins) ends up in the cell nucleus, where the cellular DNA is, of course, also found. Once inside the cell, the viral DNA interacts with the host's machinery for transcribing mRNA, and the viral mRNA is translated into viral proteins by the host cell ribosomes, tRNA and translation factors. Most viral protein products fall into one of three categories: special enzymes needed for viral replication; inhibitory factors that stop host-cell DNA, RNA, and protein synthesis; and proteins used in the construction of new virions. The latter proteins generally are made in much larger amounts than the other two types. After the synthesis of hundreds to thousands of new virions has been completed, most infected bacterial cells and some infected plant and animal cells rupture, or lyse, releasing all the virions at once. In many plant and animal viral infections, however, no discrete lytic event occurs; rather, the dead host cell releases the virions as it gradually disintegrates.

These events—adsorption, penetration, replication, and release—describe the *lytic cycle* of viral replication

(Figure 6-17). The outcome is the production of a new round of viral particles and death of the cell. In the case of enveloped viruses, the virions "bud" from the host cell, thereby acquiring their outer phospholipid envelope; this envelope contains mostly viral glycoproteins (Figure 6-18).

In some cases, after a viral DNA molecule enters a host cell, it becomes integrated into the host chromosome, where it remains quiescent and is replicated as part of the cell's DNA from one generation to the next; this is called the *lysogenic cycle* (Figure 6-19). Under certain conditions, the integrated viral DNA is activated and enters the lytic cycle. Bacterial viruses of this type are called *temperate phages.* The genomes of a number of animal viruses also can integrate into the host-cell genome. Probably the most important are the *retroviruses,* described briefly later in this chapter and extensively in Chapter 26. A few phages and animal viruses can infect a cell and cause new virion production without killing the cell or becoming integrated.

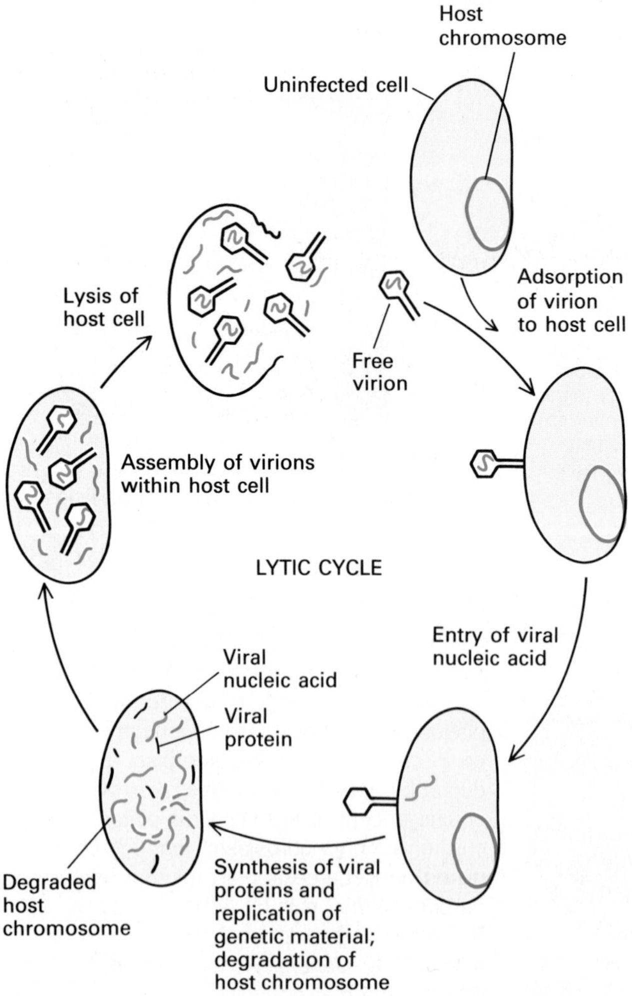

Four Types of Bacterial Viruses Are Widely Used in Biochemical and Genetic Research

Bacterial viruses have played a crucial role in the development of molecular cell biology. Thousands of different bacteriophages have been isolated; many of these are particularly well suited for studies of specific biochemical or genetic events. Here, we briefly describe four types of bacteriophages, all of which infect *E. coli,* that have been especially useful in molecular-biology research.

DNA Phages of the T Series   The T phages of *E. coli* are large lytic phages that contain a single molecule of double-stranded DNA. This molecule is about 2×10^5 base pairs long in T2, T4, and T6 viruses and about 4×10^4 base pairs long in T1, T3, T5, and T7 viruses. A T phage DNA enters an *E. coli* cell through the tail (see Figure 6-17). T phages have complex virions made of a helical protein "tail" attached to an icosohedral "head" filled with the viral DNA. The phage DNA then directs a program of events that produces approximately 100 new phage particles in about 20 min, at which time the infected cell lyses and releases the new phages. The discovery of messenger RNA was made by observing that the base composition (i.e., percentages of A, G, U, and C) of RNA synthesized after infection of *E. coli* with bacteriophage T2 did not resemble that of *E. coli* DNA, but was similar to the base composition of T2 DNA.

Temperate Phages   Typical of this class of phages is *E. coli* bacteriophage λ, which has one of the most studied genomes. On entering an *E. coli* cell, the double-stranded λ DNA can enter either the lytic cycle (like T phages) or the lysogenic cycle. In the latter case, the viral DNA forms a circle and approaches the circular host DNA at a specific site. Enzymes break both circular molecules of DNA and then rejoin the broken ends, so that the viral DNA becomes

▲ FIGURE 6-17   Steps in the lytic cycle are illustrated by infection of *E. coli* with bacteriophage T4.

> FIGURE 6-18   Maturation of an enveloped virus. Inside the host cell, virus-encoded glycoproteins are synthesized on the endoplasmic reticulum (ER) membrane and become inserted into it; by means of small transport vesicles the glycoproteins move through the Golgi complex to the host-cell plasma membrane where they project from its surface. The viral nucleocapsid, containing other viral proteins and nucleic acid, is initially free in the cytosol; it exits from the cell by binding to the viral glycoproteins in the plasma membrane and then budding through the membrane, acquiring a phospholipid envelope containing the inserted viral glycoproteins.

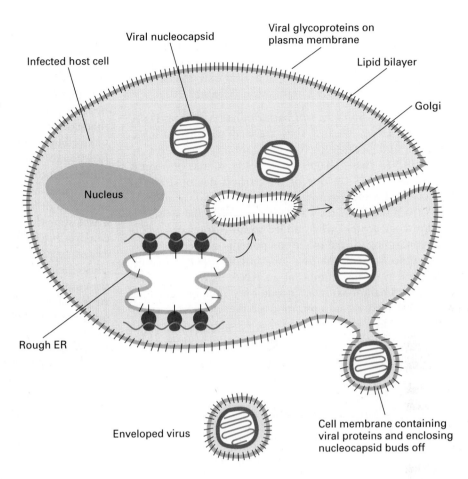

inserted into the host DNA (see Figure 6-19). The carefully controlled action of viral genes maintains λ DNA as part of the host chromosome by suppressing the lytic functions of the phage. Under appropriate stimulation, the chromosomal λ DNA is activated and enters the lytic cycle (Chapter 13).

Small DNA Phages   The genome of some bacteriophages encodes only 10–12 proteins, ≈5–10 percent of the number encoded by T phages. These small DNA phages are typified by the φX174 and the rodlike M13 phages. These were the first organisms in which the entire DNA sequence of the genome was determined, permitting extensive understanding of the viral life cycle. The viruses in this

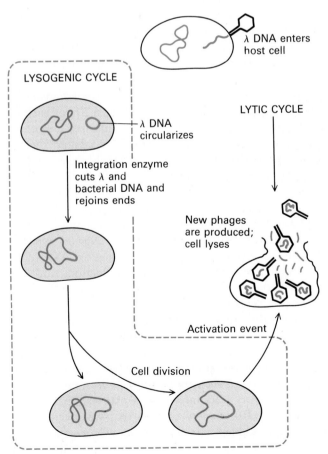

◄ FIGURE 6-19   Steps in the lysogenic cycle of bacteriophage λ in *E. coli*. After a λ phage infects a host cell, its DNA (red) may be integrated into the host-cell DNA (blue) and remain there as the host grows and divides. Under certain conditions (e.g., irradiation with ultraviolet light), the viral DNA is activated, separates from the host chromosome, and initiates the viral lytic cycle (see Figure 6-17), resulting in production of new virions.

group are so simple that they do not encode most of the proteins required for replication of their DNA but depend on cellular proteins for this purpose. For this reason, they have been useful in identifying and analysing the cellular proteins involved in DNA replication (Chapter 10).

RNA Phages   Some *E. coli* bacteriophages contain RNA instead of DNA. Because they are easy to grow in large amounts and because their RNA genomes also serve as their mRNA, these phages are a ready source of a pure species of mRNA. In one of the earliest demonstrations that cell-free protein synthesis can be mediated by mRNA, RNA from these phages was shown to direct the synthesis of viral coat protein when added to an extract of *E. coli* cells containing all the other components needed for protein synthesis. Also, the first long mRNA molecule to be sequenced was the genome of an RNA phage. These viruses, the smallest known, encode only four proteins: an RNA polymerase for replication of the viral RNA, two capsid proteins, and an enzyme that dissolves the bacterial cell wall and allows release of the intracellular virus particles into the medium.

Experiments with Plant Viruses Proved That RNA Can Act as a Genetic Material

The study of plant viruses inspired some of the first experiments in molecular biology. In 1935, Wendell Stanley purified and partly crystallized TMV; other plant viruses were crystallized soon thereafter. Pure proteins had been crystallized only a short time before Stanley's work, and it was at first thought that the TMV crystals were pure protein. Later studies showed that these crystalline preparations also contained RNA. In fact, plant viruses may contain either RNA or DNA, but for many years only plant viruses containing RNA were known.

Experiments with TMV were important in establishing nucleic acids as the informational molecules in viruses. For example, two groups, A. Gierer and G. Schramm in Germany and H. Fraenkel-Conrat and R. Williams in the United States, first demonstrated in 1956 that pure RNA from TMV could direct the production of infectious TMV particles. Further, it was shown that separate preparations of TMV protein and nucleic acid could reassemble into infectious virions. Finally, when protein from one TMV strain was mixed with RNA from another TMV strain, the reassembled hybrid virions also were infectious. Analysis of cells infected with such hybrid virions revealed that the source of the RNA determined which of the two viral strains was produced, further confirming RNA as the viral genetic material (Figure 6-20). These classic experiments, plus the study of self-assembly of the TMV protein and its RNA molecule, made plant viruses the most popular subjects of biophysical studies throughout the 1940s and 1950s.

Six Animal-Virus Classes Are Recognized Based on Genome Composition and Pathway of mRNA Synthesis

Animal viruses come in a variety of shapes, sizes, and genetic strategies. In this book, we are concerned with viruses that exhibit at least one of two features: they utilize important cellular pathways, closely mimicking a normal cellular function to form their molecules, or they can integrate their genomes into those of normal cells.

Viruses originally bore the names of the diseases they caused or of the animals or plants they infected. However, it was soon discovered that many different kinds of viruses can produce the same symptoms or the same apparent disease states; for example, several dozen different viruses can cause the red eyes, runny nose, and sneezing referred to as the common cold. Clearly, the original way of classifying viruses obscured many important differences in their structures and life cycles.

What *is* central to a viral life cycle are the types of nucleic acids formed during replication and the pathway by which mRNA is produced. The relation between the viral mRNA and the nucleic acid of the infectious particle is the basis of a simple means of classifying viruses. In this system, a viral mRNA is designated as a *plus strand* and its complementary sequence, which cannot function as an mRNA, is a *minus strand*. A strand of DNA complementary to a viral mRNA is also a *minus strand*. Production of a plus strand of mRNA requires that a minus strand of RNA or DNA be used as a template. Using this system, six classes of animal viruses are recognized. Bacteriophages and plant viruses also can be classified in this way, but the system has been used most widely in animal virology because representatives of all six classes have been identified. The composition of the viral genome and its relationship to the viral mRNA are illustrated in Figure 6-21 for each of the six classes of virus. Table 6-5 summarizes important properties of common viruses in each class and the research areas in which they have been widely used. Electron micrographs of several viruses are shown in Figure 6-22.

DNA Viruses (Classes I and II)   The genomes of both class I and class II viruses consist of DNA. Various types of DNA viruses are commonly used in studies on DNA replication, genome structure, mRNA production, and cell transformation.

Class I viruses contain a single molecule of double-stranded DNA (dsDNA). In the case of the most common type of class I animal virus, the DNA enters the cell nucleus where cellular enzymes are diverted to replication of the viral DNA and its transcription into viral mRNA. Examples of such viruses include the *adenoviruses*, which cause upper respiratory tract infections in many animals; *SV40* (simian virus 40), a monkey virus that was accidentally discovered in kidney-cell cultures from wild monkeys used

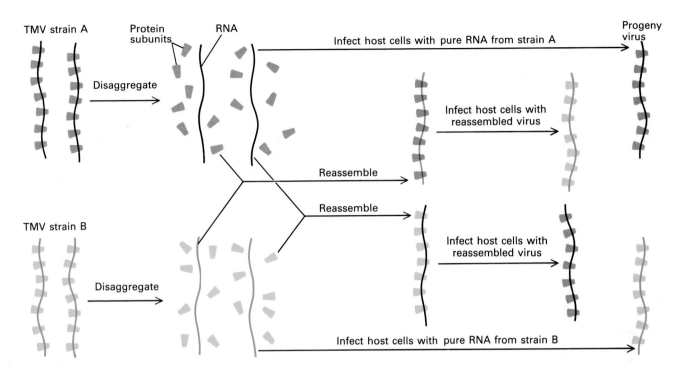

▲ FIGURE 6-20   Experimental demonstration that information for TMV protein is carried in TMV RNA. TMV particles were disaggregated, yielding protein and RNA; the isolated RNA but not the protein was shown to be infectious. Protein and RNA from two different viral strains were reassembled and host cells were infected with the reconstituted hybrid virions; the viral progeny formed in these cells was shown to correspond to the type of RNA in the hybrid, not to the type of protein. The infectivity of the reconstituted whole virus is much greater than that of the pure viral RNA. [See H. Fraenkel-Conrat, 1958, *The Harvey Lectures, 1957–1958*, Series 53, Academic Press, p. 56.]

in the production of poliovirus vaccines; and *herpesviruses*, which cause various inflammatory skin diseases including chickenpox. The second type of class I virus, the *poxviruses*, replicates in the host-cell cytoplasm. Typical of this class is *variola* (smallpox) and *vaccinia*, an attenuated (weakened) poxvirus used in vaccinations to induce immunity to smallpox. These very large, brick-shaped viruses ($0.1 \times 0.1 \times 0.2$ μm) carry their own enzymes for synthesizing viral mRNA and DNA in the cytoplasm.

Class II viruses, called *parvoviruses* (from Latin *parvo*, poor), are simple viruses that contain one molecule of single-stranded DNA (ssDNA). Some parvoviruses encapsidate (enclose) both plus and minus strands of DNA, but in separate virions; others encapsidate only the minus strand. In both cases, the ssDNA is copied inside the cell into dsDNA, which is then itself copied into mRNA.

RNA Viruses (Classes III–VI)   All the animal viruses belonging to classes III–VI have RNA genomes. A wide range of animals, from insects to human beings, are infected by viruses in each of these classes. These viruses have been particularly useful in studies on mRNA synthesis and

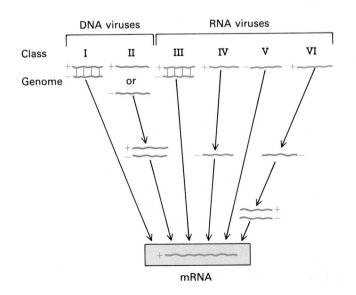

◄ FIGURE 6-21   Classification of animal viruses based on the composition of their genomes and pathway of mRNA formation. DNA is shown in blue; RNA, in red. The viral mRNA is designated as a plus strand, which is synthesized from a minus strand of DNA or RNA. See Table 6-5 for examples of viruses in each class.

TABLE 6-5   Animal Viruses Commonly Used in Molecular Biology

Class*/Virus	Known Hosts	Genome Size (kb)[†]	Envelope	Other Properties	Research Areas in Which Virus Is Used
CLASS I (DNA)					
Adenoviruses (class Ia)	Vertebrates	36	No	Replicates in host-cell nucleus; uses host enzymes for viral DNA and mRNA synthesis	mRNA synthesis and regulation; DNA replication; cell transformation
Herpesviruses (class Ia)	Vertebrates	150	Yes		
SV40 (class Ia)	Primates	5.2	No		
Vaccinia virus (class Ib)	Vertebrates	200	Yes	Replicates in host-cell cytoplasm using viral enzymes	Genome structure; mRNA synthesis by viral enzymes
CLASS II (DNA)					
Parvoviruses	Vertebrates	1–2	No		DNA replication
CLASS III (RNA)					
Reoviruses	Vertebrates	1.2–4.0[‡]	No	Has a genome of 10 dsRNA segments; uses viral enzymes to replicate	mRNA synthesis by viral enzymes; mRNA translation
CLASS IV (RNA)					
Poliovirus (class IVa)	Primates	7	No	Synthesizes a single mRNA, which is translated into a polyprotein that is cleaved to yield functional proteins	Viral RNA replication; interruption of host mRNA translation; polyprotein cleavage
Sindbis virus (class IVb)	Vertebrates, insects	10	Yes	Synthesizes at least two mRNAs each of which is translated into a polyprotein that is cleaved to yield functional proteins	Membrane formation; glycoprotein biosynthesis and intracellular transport
CLASS V (RNA)					
Vesicular stomatitis virus (class Va)	Vertebrates, insects	12	Yes	Has a virus-specific RNA polymerase that produces several mRNAs from its nonsegmented genome	Membrane formation; glycoprotein biosynthesis and intracellular transport
Influenza virus (class Vb)	Mammals and birds	1.0–3.3[‡]	Yes	Has a genome of 8 ssRNA segments; uses a virus-specific RNA polymerase to produce mRNAs	Membrane formation; glycoprotein biosynthesis and intracellular transport
CLASS VI (RNA)					
Retroviruses	Vertebrates, insects, yeasts	5–8	Yes	Contain reverse transcriptase, which copies RNA genome into DNA; the viral DNA integrates into host genome	Cell transformation; function of oncogenes; AIDS

* Class refers to the strategy for mRNA synthesis as illustrated in Figure 6-21.
[†] Size is given in kilobases (1 kb = 1000 nucleotides) for single-stranded nucleic acids and kilobase pairs for double-stranded nucleic acids.
[‡] Reoviruses and influenza have segmented RNA genomes; the length of each segment is in the range indicated.

Vaccinia (I)   Herpes (I)   Adenovirus (I)   Human wart virus (I)

Reovirus (III)   Sindbis virus (IV) Poliomyelitis virus (IV)   Influenza (V)

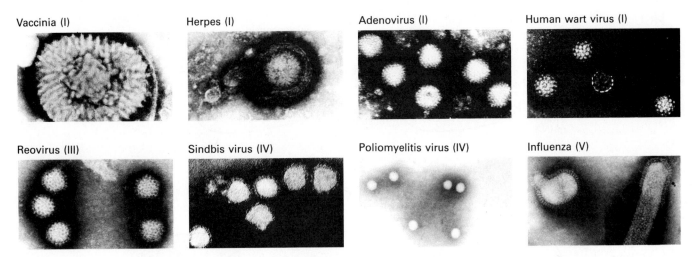

▲ FIGURE 6-22   Electron micrographs of representative animal viruses, all taken at the same magnification (×50,000) and shadowed with phosphotungstic acid. The class of each virus is in parentheses. [From P. Choppin, 1965, *Viral and Rickettsial Infections of Man,* Lippincott, frontispiece. Courtesy of P. Choppin.]

translation (class III); glycoprotein synthesis, membrane formation, and intracellular transport (classes IV and V); and cell transformation and oncogenes (class VI).

Class III viruses contain double-stranded genomic RNA (dsRNA). The minus RNA strand acts as a template for the synthesis of plus strands of mRNA. The virions of all class III viruses known to date have segmented genomes containing 10–12 double-stranded RNA segments, each of which encodes one or two polypeptides. In these viruses, the virion itself contains a complete set of enzymes that can utilize the minus strand of the genomic RNA as a template for synthesis of mRNA in the test tube as well as in the cell cytoplasm after infection. A number of important studies have used class III viruses as a source of pure mRNA.

Class IV viruses contain a single plus strand of genomic RNA, which is identical to the viral mRNA. Since the genomic RNA encodes proteins, it is infectious by itself. During replication of class IV viruses, the genomic RNA is copied into a minus strand, which then acts as a template for synthesis of more plus strands, or mRNA. Two types of class IV viruses are known. In class IVa viruses, typified by *poliovirus,* viral proteins are first synthesized, from a single mRNA species, as a long polypeptide chain, or *polyprotein,* which is then cleaved to yield the various functional proteins. Class IVb viruses, also called *togaviruses* (from Latin *toga,* cover) because the virions are surrounded by a lipid envelope, synthesize at least two species of mRNA in a host cell. One of these mRNAs is the same length as the virion's genomic RNA; the other corresponds to the 3′ third of the genomic RNA. Both mRNAs are translated into polyproteins. Included in class IVb are a large number of rare insect-borne viruses including *Sindbis virus.* Once called arboviruses (arthropod-borne viruses), class IVb vi-

ruses include those that cause yellow fever and viral encephalitis in human beings.

Class V viruses contain a single negative strand of genomic RNA, whose sequence is complementary to that of the viral mRNA. The genomic RNA in the virion acts as a template for synthesis of mRNA but does not itself encode proteins. Two types of class V viruses can be distinguished. The genome in class Va viruses, which include the viruses causing measles and mumps, is a single molecule of RNA. A virus-specific RNA polymerase present in the virion catalyzes synthesis of several mRNAs, each encoding a single protein, from the genomic template strand. Class Vb viruses, typified by *influenza virus,* have segmented genomes; each segment acts as a template for the synthesis of a different mRNA species. In most cases, each mRNA produced by a class Vb virus encodes a single protein; however, some mRNAs can be read in two different frames to yield two distinct proteins. As with class Va viruses, a class Vb virion contains a virus-specific polymerase that catalyzes synthesis of the viral mRNA. Thus the genomic RNA (a minus strand) in both types of class V viruses is not infectious in the absence of the virus-specific polymerase. The influenza RNA polymerase initiates synthesis of each mRNA by a unique mechanism. In the host-cell nucleus, the polymerase cuts off 12–15 nucleotides from the 5′ end of a cellular mRNA or mRNA precursor; this oligonucleotide acts as a "primer" that is elongated by the polymerase to form viral (+) mRNAs, using the genomic (−) RNA as a template.

Class VI viruses are also known as *retroviruses,* because the RNA of their genome (a single plus strand) directs the formation of a DNA molecule. The DNA molecule ultimately acts as the template for synthesis of viral mRNA (Figure 6-23). Initially, a viral enzyme called *re-*

➤ FIGURE 6-23
The life-cycle of a retrovirus. Viral RNA is shown in red; viral DNA, in blue; host-cell DNA, in light blue.

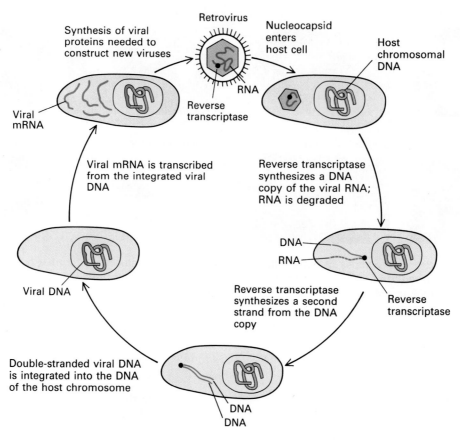

verse transcriptase copies the viral RNA genome into a single minus strand of DNA; the same enzyme then catalyzes synthesis of a complementary plus strand. The resulting dsDNA is integrated into the chromosomal DNA of the infected cell. Finally, the integrated viral DNA is transcribed by the cell's own machinery into RNA, which either acts as a viral mRNA or becomes enclosed in a virion, thereby completing the retroviral growth cycle. If a retrovirus contains cancer-causing genes (called *oncogenes*), any cell that it infects is transformed into a tumor cell, as detailed in Chapter 26. Among the known human retroviruses are *human T-cell lymphotrophic virus* (HTLV), which causes a form of leukemia, and *human immunodeficiency virus* (HIV), which causes acquired immune deficiency syndrome (AIDS). Both of these viruses can infect only specific cell types, primarily certain cells of the immune system and, in the case of HIV some brain and glial cells.

➤ Radioisotopes: Indispensable Tools for Following Biological Activity

The cells and viruses described in the previous sections are used in two main kinds of experiments: genetic studies in-

volving the isolation and characterization of mutants defective in a particular function, which are described in Chapter 8, and biochemical studies of cell structure and function. Because the latter studies often involve use of radioactively labeled molecules, we detail the basic properties of radioisotopes and common experimental procedures involving radiolabeled compounds in the remainder of this chapter.

Since World War II, when radioactive materials first became widely available as byproducts of work in nuclear physics, chemists and biologists have fashioned an almost limitless variety of radioactive chemicals. Today, radioactively labeled precursors of macromolecules greatly simplify many standard biochemical assays and significantly enhance the ability of researchers to follow biochemical events in whole cells as well as in cell extracts. Almost all experimental biology depends on the use of radioactive compounds.

At least one atom in a radiolabeled molecule is present in a radioactive form, called a *radioisotope* (Table 6-6). The presence of a radioisotope does not change the chemical properties of a molecule. For example, enzymes, both in vivo and in vitro, catalyze reactions involving labeled substrates just as readily as those involving nonlabeled substrates. Because radioisotopes emit easily detected particles, the fate of radiolabeled molecules can be traced in cells and cellular extracts.

TABLE 6-6 Commonly Used Radioisotopes

Radioisotope	Half-Life	Energy of Emitted Particle (MeV)*	Mean Path Length in Water (μm)	Specific Activity (mCi/mA)[†]	Specific Activities of Labeled Compounds (mCi/mmol)[‡]
Tritium (hydrogen-3)	12.35 yr	0.0186	0.47	2.92×10^4	10^2-10^5
Carbon-14	5730 yr	0.156	42	62.4	$1-10^2$
Sulfur-35	87.5 days	0.167	40	1.50×10^6	$1-10^6$
Phosphorus-33	25.5 days	0.248	—	5.32×10^6	$10-10^4$
Phosphorus-32	14.3 days	1.709	2710	9.2×10^6	$10-10^5$
Iodine-131	8.07 days	0.806	—	1.6×10^7	10^2-10^4
Iodine-125	60 days	0.035	—	2.2×10^6	10^2-10^4

* MeV = 10^6 electronvolts. The maximum energy for each emission is given. The particle emitted is a β particle, except in the case of ^{131}I and ^{125}I, which emit γ rays.

[†] The unit mCi (millicuries) is a measure of the number of disintegrations per time unit: 1 mCi = 2.2×10^9 disintegrations per minute. The unit mA (milliatoms) is the atomic weight of the element expressed in milligrams.

[‡] These values are for commercially available compounds of interest in biological research that contain the indicated radioisotope.

SOURCE: New England Nuclear Corporation, Boston.

Several Factors Determine the Choice of a Radiolabel

The choice of which labeled compound to use in a particular experiment involves several considerations. Some labeled compounds, for instance, are not suitable for studies with whole cells because they do not enter cells. One prominent example is ATP, as well as most other phosphorylated compounds (e.g., glucose 6-phosphate). Although ^{32}P-labeled ATP can contribute phosphorus-32 during RNA and DNA synthesis in a cell-free system, it cannot do so with whole cells, because it never gets into the cells. On the other hand, labeled orthophosphate (^{32}PO$_4^{3-}$) in the medium does enter both bacterial and animal cells and then is incorporated into nucleotides and eventually into cellular RNA and DNA. In cell-free extracts, however, orthophosphate is not efficiently incorporated into the nucleoside triphosphate precursors of RNA and DNA. For this reason, radioactive orthophosphate is not used to label nucleic acids in extracts.

Hundreds of biological compounds (e.g., amino acids, nucleosides, and numerous metabolic intermediates) la-

beled with various radioisotopes are commercially available. These preparations vary considerably in their *specific activity*, which is the amount of radioactivity per unit of material, measured in disintegrations per minute (dpm) per millimole. The specific activity of a labeled compound depends on the ratio of unstable potentially radioactive atoms to stable nonradioactive atoms. It also depends on the probability of decay of the radioisotope, indicated by its *half-life*, which is the time required for half of the atoms to undergo radioactive decay. In general, the shorter the half-life of a radioisotope, the higher its specific activity (see Table 6-6).

A labeled compound must have a high enough specific activity so that the radioactivity incorporated into cellular molecules is high enough to be accurately detected. For example, methionine and cysteine labeled with sulfur-35 (^{35}S) are widely used to label cellular proteins because preparations of these amino acids with high specific activities ($>10^{15}$ dpm/mmol) are available. Likewise, commercial preparations of ^{3}H-labeled nucleic acid precursors have much higher specific activities than the corresponding ^{14}C-labeled preparations. In most experiments, the former are preferable because they allow RNA or DNA to be adequately labeled after a shorter time of incorporation or require a smaller cell sample.

Various phosphate-containing compounds in which every phosphorus atom is the radioisotope phosphorus-32 are readily available. Because of their high specific activity, ^{32}P-labeled nucleotides are routinely used to label nucleic acids in cell-free systems; as discussed in the next section, however, phosphorus-32 is not suitable for certain types of incorporation experiments designed to localize labeled materials within the cell. The radioisotope iodine-125 (^{125}I), which also is available in almost pure form, can be covalently linked to a protein or nucleic acid to yield preparations with a high specific activity. Such attachment of iodine-125 can be achieved enzymatically or chemically and generally does not drastically affect the properties of a macromolecule.

Radiolabeled Molecules Can Be Detected by Visual and Quantitative Methods

Depending on the nature of an experiment, labeled compounds are detected by autoradiography, a semiquantitative visual assay, or their radioactivity is determined in an appropriate "counter," a highly quantitative assay that permits the concentration of a radiolabeled compound to be estimated. In some experiments, both types of detection are used.

Autoradiography In autoradiography, a cell or cell constituent is labeled with a radioactive compound and then overlaid with a photographic emulsion sensitive to radiation. Development of the emulsion yields small silver grains whose distribution corresponds to that of the radioactive material (Figure 6-24a).

Autoradiographic studies of whole cells have been crucial in determining the intracellular sites where various macromolecules are synthesized and their subsequent movements within cells. For example, when cells are incubated for a short time with [3H]thymidine, a unique DNA precursor, most of the radioactivity is localized to the nucleus, identifying the nucleus as the major site of DNA synthesis (Figure 6-24b). Even after cells are incubated for prolonged periods with [3H]thymidine, virtually all of the radioactivity remains in the nucleus, indicating that the DNA remains there. Similarly, the site of synthesis of RNA is revealed by incubating cells for 1 min with [3H]uridine, a unique RNA precursor, in this case, all the autoradiographic grains are found over the nucleus. After a longer period of incorporation, however, many autoradiographic grains are located over the cytoplasm, indicating that RNA, in contrast to DNA, is transported from the nucleus.

The same type of experiment with radiolabeled amino acids identifies the cytoplasm as the site of synthesis of all cellular proteins. As discussed in Chapter 16, the transport of secretory proteins from their site of synthesis on the rough endoplasmic reticulum to the cell surface where they

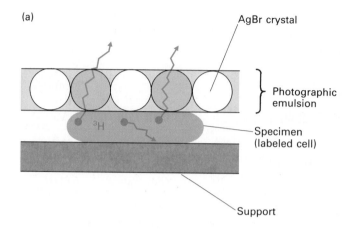

(a)

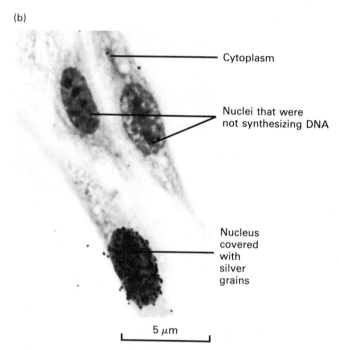

(b)

▲ FIGURE 6-24   The technique of autoradiography. (a) A radiation-sensitive photographic emulsion containing silver salts (AgBr) is placed over labeled cells attached to a glass slide (for the light microscope) or to a carbon-coated grid (for the electron microscope). The cell regions containing the labeled molecules emit radioactive particles, along the tracks of which silver is deposited. When the photographic emulsion is developed, the silver deposited appears as dark grains under the light microscope and as curly filaments in the electron microscope. (b) When cells are incubated with [3H]thymidine for a short period, any DNA synthesized during this labeling period incorporates the labeled precursor and can be localized by autoradiography. The fibroblasts shown here, from Chinese hamsters, were labeled for 1 h and then prepared for autoradiography. Small black grains almost entirely cover the nucleus of one cell, indicating that the cell synthesized DNA during the labeling period. The two other cells were not synthesizing DNA (the larger dark areas in their nuclei are nucleoli). The absence of black grains in the cytoplasm indicates that DNA is not synthesized there. [Part (a) adapted from E. D. P. DeRobertis and E. M. F. DeRobertis, 1979, *Cell and Molecular Biology*, Saunders, p. 62; part (b) courtesy of D. M. Prescott.]

are secreted was first documented by a technique called electron-microscope autoradiography, which allows each silver filament resulting from a radioactive disintegration to be observed.

In autoradiographic studies, the ability of the experimenter to localize the site at which the radioisotope is incorporated is affected by the energy of the particles emitted during radioactive disintegrations. For example, the β particles emitted by phosphorus-32 are so energetic that the streaks they make on a photographic emulsion can be as long as 1 mm, much longer than the diameter of individual cells. In contrast, the β particles emitted by tritium create tracks on a photographic emulsion that are only about 0.47 μm long; thus ^{3}H-labeled structures can be located within cells to an accuracy of about 0.5–1.0 μm, or about one-fifth the diameter of the nucleus of mammalian cells. Because tritium emits the least-energetic particles of all the common radioisotopes (see Table 6-6), it is highly preferred for locating labeled compounds or structures within cells.

Quantitative Determination of Radiolabeled Compounds   The amount of radioactivity in a labeled material can be measured with either of two different instruments: a *Geiger counter* or a *scintillation counter*. The former detects ions produced in a gas by the β particles or γ rays emitted from a radioisotope. In the latter, a radiolabeled sample is mixed with a liquid containing a fluorescent compound that emits a flash of light when it absorbs the energy of the β particles or γ rays released during decay of the radioisotope; a phototube in the instrument detects and counts these light flashes.

In the usual experimental design, a radioactive precursor is incubated with whole cells or cell-free extracts; the cellular constituents then are isolated and purified in various ways; finally, the radioactivity of the various fractions is measured with a counter. For example, to identify the site of RNA synthesis, cells can be incubated for a short period with [^{3}H]uridine and then subjected to a fractionation procedure to separate the various organelles (Chapter 5). The specific activity of the nuclear fraction (dpm/mg protein) is found to be much higher than that of any other organelle fraction, thus confirming the nucleus as the site of RNA synthesis.

A combination of labeling and biochemical techniques and of visual and quantitative detection methods are often employed. For instance, to identify the major proteins synthesized by a particular cell type, a sample of the cells is incubated with a radioactive amino acid (e.g., ^{35}S-labeled methionine) for a few minutes. The mixture of cellular proteins then is resolved by gel electrophoresis (Chapter 3), and the gel is subjected to autoradiography; radioactive bands correspond to newly synthesized proteins, which have incorporated the radiolabeled amino acid. Alternatively, the proteins can be resolved by liquid chromatography (Chapter 3), and the radioactivity in the eluted fractions determined quantitatively with a counter.

Intracellular Precursor Pools Affect the Outcome of Pulse-Chase Experiments

The purpose of some studies is to label a macromolecule and then trace the labeled compound as it moves through various cellular compartments or is incorporated into other molecules within cells. This can be accomplished with a *pulse-chase experiment* in which cells are incubated briefly with a radiolabeled precursor (the *pulse*) followed by removal and replacement of the labeled precursor by an excess of unlabeled precursor (the *chase*). The fate of the newly synthesized compounds that incorporate the label during the pulse is tracked by measuring the radioactivity of the cells or cell factions at various times thereafter.

In this type of experiment, the small labeled precursor added to the medium (e.g., an amino acid, nucleoside, or phosphate ion) first enters an intracellular *pool* of small molecules, which are free to diffuse throughout the cytoplasm and nucleus; subsequently, the precursor is incorporated into a macromolecule (e.g., protein or nucleic acid). Depending on the growth conditions of a cell culture, the quantities of compounds in the intracellular pool may vary. Likewise, the rates at which different precursors are absorbed, utilized, and secreted by cells varies. Both of these characteristics of small molecules in the soluble pool must be considered in the design and interpretation of pulse-chase experiments.

The clearest pulse-chase effect occurs when the labeled precursor is rapidly exchanged between the intracellular pool and external medium. In this case, the labeled precursor that accumulates in cells during the pulse is rapidly depleted during the chase; thus incorporation of label into macromolecules soon ceases, and the radioactivity is associated only with macromolecules synthesized during the pulse. As shown in Figure 6-25a, pulse-chase experiments with labeled amino acids exhibit this pattern, because rapid equilibration occurs between amino acids inside the cells and in the medium.

Ribonucleosides and deoxyribonucleosides, however, become phosphorylated soon after they enter the cell pool, and phosphorylated compounds do not generally leave the cell. Thus labeled nucleosides can enter the cell, but no equilibrium is established between the nucleic acid precursors in the medium and their phosphorylated counterparts in the cell. Nevertheless, a practically useful pulse-chase effect can be obtained in experiments with radioactive *deoxyribonucleosides*, because the deoxyribonucleotide content of the cell pool is sufficient for only a few minutes of DNA synthesis. Labeled thymidine, for example, can be satisfactorily chased even though it is phosphorylated to TMP and TTP, because an amount of TTP equal to that in the pool is utilized every few minutes for replication of DNA.

Labeled *ribonucleosides* behave differently, because it takes several hours for enough RNA synthesis to occur to consume the content of the ribonucleotide pool in animal

(a)

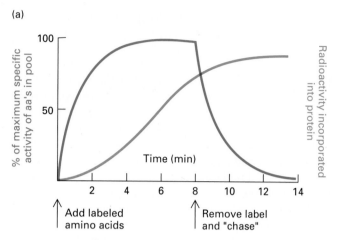

(b)

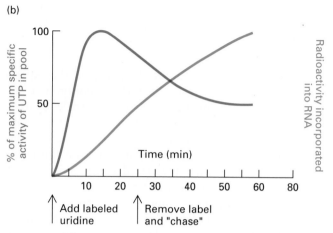

▲ FIGURE 6-25   Typical results from pulse-chase experiments in which the radiolabeled precursors are amino acids (a) or uridine (b). Growing cells are exposed to a pulse of labeled precursor for a short period (e.g., 5–10 min); during this time the label enters the intracellular pool of small, diffusible molecules (blue curves) and then is incorporated into newly synthesized macromolecules (red curves). The labeled precursor then is chased by placing the cells in a medium containing unlabeled precursor. In experiments with labeled amino acids (a), the chase rapidly reduces the radioactivity in the intracellular amino acid pool, because of rapid equilibration between amino acids inside cells and in the medium; thus a marked pulse-chase effect is seen. In experiments with labeled uridine (b), however, no marked pulse-chase effect is seen; that is, the incorporation of radioactivity into RNA continues for several hours after the chase. In this case, the labeled uridine in the pool, which has been phosphorylated, cannot leave the cell and equilibrate with unlabeled uridine in the medium during the chase. See text for further discussion.

cells. In most cultured animal cells, the pool does absorb a pulse of labeled ribonucleosides such as [3H]uridine within 10 min or so. However, a marked chase response (one that occurs within a few minutes) is not possible. Although the addition of unlabeled ribonucleosides to the exterior medium may further expand the ribonucleotide content of the cell pool, thus diluting the radioactive ribonucleotides within it and decreasing the rate of RNA labeling, the amount of incorporated label does not clearly level off until several hours after the chase begins (Figure 6-25b).

Synthesis Time of Macromolecules Can Be Estimated from Labeling Experiments

When a radioactive precursor first enters a cell, it can only label the macromolecules that are in the process of being constructed. For example, if a radioactive amino acid is added to a culture, the nascent (unfinished and still growing) protein chains are the first proteins to be labeled. As time passes, an increasing number of completed chains contain the radioactive label. The time required to form a specific type of macromolecule can be estimated by sampling a labeled cell culture at very short intervals and comparing the amount of radioactivity in the nascent macromolecules, which are attached to their templates, with the amount of free (completed) macromolecules (Figure 6-26).

Experiments of this type with cultured HeLa cells have shown that elongation of nascent protein chains occurs at a rate of about 10 amino acids per second. Thus synthesis of a typical protein containing 600 or so amino acids takes about 1 min.

The Dintzis Experiment Demonstrated That Proteins Are Synthesized from the Amino End to the Carboxyl End

As we've seen, researchers can use labeling experiments to locate the sites at which macromolecules are synthesized and to determine their rate of synthesis. In a classic labeling experiment (Figure 6-27), Howard Dintzis analyzed newly finished protein chains, revealing their step-by-step formation from the amino to the carboxyl terminus.

Dintzis used *reticulocytes* (the next-to-final stage in the formation of red blood cells in the bone marrow) because the α and β chains of hemoglobin constitute more than 90 percent of the total protein synthesized by these cells. After adding radioactive amino acids to a suspension of reticulocytes, Dintzis collected samples at short intervals. The labeled protein was separated from the reaction mixture, then the completed α and β chains were resolved. Each chain was digested with trypsin, which cleaves polypeptides on the carboxyl side of arginine and lysine residues, yielding a characteristic set of fragments. At the time of this experiment, the amino acid sequence of both the globin chains was known, as was the position of each tryptic fragment within the chains.

Nascent
macromolecules   +   Completed
attached to template   macromolecules

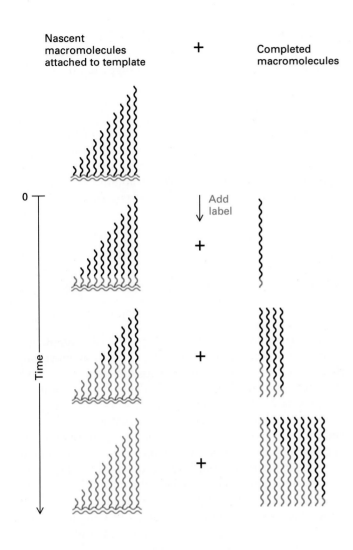

◀ FIGURE 6-26   Estimation of macromolecular synthesis time by radiolabel-incorporation experiments. When cells are incubated with radiolabeled precursors (e.g., amino acids or nucleosides), the label (red) first appears in nascent macromolecules in the process of synthesis on templates (blue). As time passes, molecules that contain more radioactive label are completed and leave the templates. At the end of an interval equivalent to the synthesis time of a single macromolecule, the total amounts of radioactivity in all finished and all unfinished molecules are equal. By taking periodic samples, separating the nascent chains from the completed macromolecules, and determining the radioactivity in each, researchers can estimate the synthesis time.

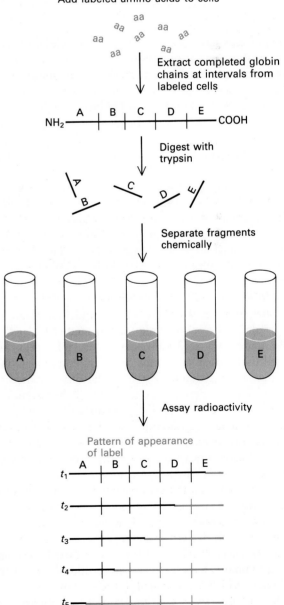

▶ FIGURE 6-27   Outline of the Dintzis experiment, which demonstrated that proteins are synthesized from the amino to the carboxyl end. A suspension of reticulocytes was incubated with radiolabeled amino acids; samples were periodically removed and the completed globin chains extracted. The chains were cleaved with trypsin into fragments (A–E), and the radioactivity in each fragment determined. Because the position of the tryptic fragments within the molecule was known, Dintzis could determine the relative time required for appearance of the label in various parts of the chain. As shown at the bottom, radioactivity (red) first appeared in the carboxyl-terminal E fragment (sample t_1). In the later samples (t_2–t_5), the label appeared successively in order in the other fragments, with the amino-terminal A fragment labeled last. [See H. Dintzis, 1961, *Proc. Nat'l. Acad. Sci. USA* **47**:247.]

Dintzis reasoned that the first completed chains to contain the radioactive label would be those that were almost complete when the label was added. Thus the *first* portion of the finished chains to contain label would be near the end at which chain synthesis finished and, the *last* portion of the finished globin chains to become labeled would lie at the end where chain synthesis started. As shown in Figure 6-27, the radioactive label always appeared in the carboxyl-terminal tryptic fragment first and in the amino-terminal fragment last, with intermediate fragments becoming consecutively labeled in the order in which they lay between the two termini. From these results, Dintzis concluded that synthesis begins at the amino terminus of each chain and moves in a step-by-step progression to the carboxyl end of the chain.

Similar experiments involving the labeling of nascent chains of RNA and DNA are described in Chapters 10 and 12. The logic of these studies parallels that of the Dintzis experiment: the shortest labeled molecules in a nascent set will be those whose sequence is near the start site; increasingly longer members will contain additional sequences progressively more remote from the start site.

SUMMARY

Cultured cells have played an important role in the development of molecular cell biology. By far the most popular bacterium for research purposes is *E. coli*; analysis of the many mutant *E. coli* strains available has made this the best-understood cell in the world of biology. The yeast species *S. cerevisiae* is the most frequently used single-cell eukaryote; yeast genetics also has reached a very advanced state. Studies with bacteria, bacterial viruses, and yeast established the basic principles of molecular biology.

Animal cells generally are more difficult to grow in vitro than bacterial cells. Culture of many animal cells requires complex media containing many nutrients and growth factors; many animal-cell types only grow when attached to a surface and cannot be cultured in suspension. A primary cell culture is derived from normal animal cells, which will not continue to grow indefinitely. Transformed cells, including tumor cells, will grow indefinitely in culture; the term cell line refers to an indefinitely growing culture derived from a single transformed cell. Even relatively undifferentiated cell lines (e.g., HeLa cells) have proved useful for many kinds of study. In addition, many cell lines exhibit characteristics of the differentiated cells from which they derive; examples include hepatomas that synthesize many liver-specific proteins and epithelial cells that form polarized monolayers. Specialized cells from bone marrow, liver, muscle, and other tissues often are isolated and used directly when they cannot be successfully cultured.

Hybrid cells produced by fusion of cell lines from one species with cells from another can be used to map genes to specific chromosomes. Fusion of normal, primary antibody-producing B lymphocytes with cultured lines of myeloma cells produces hybrid cells, called hybridomas, that are immortal; each hybridoma produces a single monoclonal antibody.

Viruses have been used to investigate many aspects of cell structure and function. These small parasites, which reproduce only within a host cell, consist of a simple genome, containing from a few to 200 genes at most, enclosed within a protein coat, the capsid. Some viruses have an outer envelope, which is partly derived from the host-cell plasma membrane. Viruses are a source of small, reproducible gene sequences; these, as well as viral mRNA and proteins, are easily purified. Viruses exhibit two types of growth cycles: the lytic cycle, which involves production of new viral particles and leads to death of the host cell, and the lysogenic cycle, in which the viral DNA is integrated into the host-cell genome and replicated as the cell grows and divides. An understanding of the basic properties of viruses and a familiarity with the experimental uses of different types of virus are indispensable to the cell biologist of today.

Several radioactive isotopes (e.g., ^{3}H, ^{14}C, and ^{32}P) are used to label biologically important molecules such as amino acids and nucleosides. These tracers are widely used in both cell-free and whole-cell systems to investigate various metabolic processes. Autoradiography provides a visual, semiquantitative measure of incorporation of a tracer into a cellular macromolecule; isolation of labeled macromolecules and measurement of their radioactivity in a counter provides a quantitative measure of incorporation. Important considerations in the use of radioisotopes include the specific activity of the labeled precursor; the energy of the particles emitted during radioactive decay; the speed at which different labeled precursors enter cells; and the extent and rate of exchange between small precursor molecules in the cell and in the medium. For example, tritiated (^{3}H) compounds give the best autoradiographic images because the low-energy β particles they emit produce better-defined images on a photographic emulsion than do higher-energy particles emitted by other radioisotopes. Compounds labeled with phosphorus-32 can be analyzed with great sensitivity because this radioisotope has a relatively short half-life and emits high-energy β particles, which are detected efficiently.

REVIEW QUESTIONS

1. You are keenly interested in studying the metabolism of aromatic amino acids in *Escherichia coli*, particularly the synthesis of phenylalanine. To initiate your investigations, you decide to isolate phenylalanine auxotrophs by exposing wild type *E. coli* to a mutagen. An auxotroph is a mutant that cannot grow on media lacking certain essential molecules that are normally synthesized by the wild type strain; hence, in this case, a phenylalanine auxotroph would require phenylalanine to be present in the medium in order to grow. A mutagen is a chemical or physical agent that interacts with DNA and causes a mutation to occur.

   a. Prior to exposing your wild type strain of *E. coli* to a potent mutagen, ethyl methanesulfonate (EMS), you isolate a well-defined, single colony from a culture dish containing minimal medium in the agar. Why is this step essential before you try to isolate auxotrophs requiring phenylalanine?

   b. You have prepared a large quantity of your wild type clone, and you expose it to EMS. How would you use replica plating to isolate several phenylalanine auxotrophs?

   c. You have successfully obtained three phenylalanine auxotrophs from your wild type *E. coli* strain. Your advisor points out that phenylalanine and tyrosine have very similar structures, suggesting that there might be overlap in the biosynthetic pathways for each. You and your advisor even devise schematic pathways with hypothetical precursors X and Y to show this. They are as follows:

   (1) $X \rightarrow Y \rightarrow phe \rightarrow tyr$

   (2) $X \rightarrow Y \rightarrow tyr \rightarrow phe$

   (3) $X \rightarrow Y \nearrow^{tyr}_{\searrow phe}$

   You are assigned the task of finding mutants of *E. coli* that are tyrosine auxotrophs. In addition, you must isolate mutants that are double auxotrophs in that they require both tyrosine and phenylalanine to grow. How would you do this? Could you make use of the master plates employed in the previous question? Explain your answer. You successfully isolate tyrosine auxotrophs and phenylalanine:tyrosine auxotrophs. Based on this, which of the above pathways presents a feasible model for the syntheses of these two aromatic amino acids? Explain your answer.

   d. Tyrosine is classified as a nonessential amino acid in animal cell culture. In reality, cells from vertebrates synthesize tyrosine from the essential amino acid phenylalanine by a reaction catalyzed by the enzyme phenylalanine hydroxylase. In comparison to your data obtained by studying *E. coli*, do you think that the biosynthetic pathway for tyrosine is the same in cells derived from vertebrates? What does this tell you about extrapolating data from prokaryotes to eukaryotes?

2. As you will learn in subsequent chapters, genetically modified forms of bacteriophage λ are often used in gene cloning. Bacteriophage λ stocks are frequently stored, and it is necessary to know the concentration of viable bacteriophage present.

   You have been given a stock of phage λ, and are assigned the task of determining the concentration of phage. To do this, you prepare a series of dilutions of the stock, add 100 μl of each to a culture of *E. coli*, mix this with agar, and pour it evenly over a plate containing a solidified layer of agar. You incubate the plates for twelves hours, and then count the individual plaques that are formed. You are also careful to include a control where no phage sample was added to the bacteria. The data are summarized in the table below.

Dilution of Stock	μl of Dilution Added to Agar	PFUs/Plate	
		Trial 1	Trial 2
	0	0	0
1:1	100	*	*
1:10	100	*	*
1:10²	100	*	*
1:10³	100	*	*
1:10⁴	100	310	360
1:10⁵	100	37	29
1:10⁶	100	4	5
1:10⁷	100	0	0

A * indicates that the plate had too many plaques to be counted accurately.

   a. What is the number of plaque forming units (PFUs) per ml of phage stock? Assuming that each PFU was generated from a single infected *E. coli*, what is the concentration of viable phage in the stock?

   b. If you were interested in using bacteriophage λ in an experiment, do you think it would be wise to choose a single plaque as a source of the virus? Why or why not?

3. Review the means for detecting radioactivity, particularly the quantitative details. What is meant by specific activity of a radiolabeled compound?

   a. You have a stock solution of [^{32}P]dATP with a specific activity of 5×10^{15} dpm/mmole at a concentration of 3.3 μM. You also have 1 ml of a stock solution of 5 mM dATP that is not radioactive. You add 5 μl of the radioactive dATP solution to the 1 ml of nonradioactive stock. What is the new specific activity? What is the total dATP concentration? How much did the actual dATP concentration change?

   b. If you added [^{32}P]dATP to cultured cells, would the cellular DNA incorporate radioactivity? Explain your answer. If it would not, what alternative method would you use to obtain radioactive DNA?

4. Review the classes of viruses that infect bacteria and animal cells.

   a. The small polyhedral phage ϕX174 has a chromosome consisting of single-stranded circular DNA that contains eleven genes. Theoretically, the combined molecular weights of the proteins encoded by these eleven genes exceed the amount of information in the phage chromosome.

   A segment of one mRNA produced by ϕX174 is shown.

   $$-\text{AUGAGU}-(\text{NNN})_n-\text{GUUUAUGGUACGCUG}-$$

   In this transcript, $(\text{NNN})_n$ indicates a series of amino acid-encoding triplets with no stop codons present. How many potential translation start sites can you identify? Do each of these read in the same frame, or do they yield different open reading frames? How might this account for the ability of the small chromosome to encode eleven proteins?

   b. Refer to Figure 6-21. For each of the six classes, discuss the enzymes and nucleic acid precursors that are essential for copying the viral genetic information into +mRNA suitable for translation.

References

Growth of Microorganisms in Culture

AUSUBEL, F., et al., eds. 1993. *Current Protocols in Molecular Biology.* Part 1: *E. coli,* plasmids, and bacteriophages. Current Protocols.

GUTHRIE, C., and G. F. FINK, eds. 1991. *Methods in Enzymology.* Vol. 194: Guide to Yeast Genetics and Molecular Biology. Academic Press.

INGRAHAM, J. L., O. MAALØE, and F. C. NEIDHARDT. 1983. *Growth of the Bacterial Cell.* Sinauer Associates.

MILLER, J. 1992. *A Short Course in Bacterial Genetics.* Cold Spring Harbor Laboratory Press.

NEIDHARDT, F. C., et al. 1987. *Escherichia Coli and Salmonella Typhimurium: Cellular and Molecular Biology.* American Society for Microbiology.

SAMBROOK, J., T. MANIATIS, and E. F. FRITSCH, eds. 1989. *Molecular Cloning* 2d ed. Cold Spring Harbor Laboratory Press.

Growth of Animal Cells in Culture

BARNES, D. W., D. A. SIRBASKY, and G. H. SATO, eds. 1984. *Cell Culture Methods for Molecular and Cell Biology.* Alan R. Liss.

BUTTLER, M. 1991. *Mammalian Cell Biotechnology.* IRL Press.

CONN, M. P. 1990. *Methods in Neuroscience.* Vol. 2: Cell Culture. Academic Press.

EVANS, M. J., and M. H. KAUFMAN. 1981. Establishment in culture of pluripotential cells from mouse embryos. *Nature* 292:154–156.

FOSKETT, K. J., and S. GRINSTEIN, eds. 1990. *Noninvasive Techniques in Cell Biology.* Wiley-Liss.

FRESHNEY, R. I. 1992. *Animal Cell Culture: A Practical Approach.* IRL Press.

FUCHS, F., and H. GREEN. 1981. Regulation of terminal differentiation of cultured human keratinocytes by vitamin A. *Cell* 25:617–625.

HOWLETT, A. R., and M. J. BISSELL. 1993. The influence of tissue microenvironment (stroma and extracellular matrix) on the development and function of mammary epithelium. *Epithelial Cell Biol.* 2:79–89.

LIN, C. Q., and M. J. BISSELL. 1993. Multi-faceted regulation of cell differentiation by extracellular matrix. *FASEB J.* 7:737–743.

TYSON, C. A., and J. A. FRAZIER, eds. 1993. *Methods in Toxicology.* Vol. I (Part A): *In Vitro* Biological Systems. Academic Press. This volume includes methods for growing many types of primary cells in culture.

WATSON, J. D., and H. J. McKENNA. 1992. Novel factors from stromal cells: bone marrow and thymus microenvironments. *Int. J. Cell Cloning* 10:144–252.

Cell Fusion

Somatic-Cell Genetics and Gene Mapping

DAVIDSON, R. L., and P. S. GERALD. 1976. Improved techniques for the induction of mammalian cell hybridization of polyethylene glycol. *Som. Cell Genet.* 2:165–176.

D'EUSTACHIO, P., and F. H. RUDDLE. 1983. Somatic-cell genetics and gene families. *Science* 220:919–924.

EGLITIS, M. A. 1991. Positive selectable markers for use with mammalian cells in culture. *Hum. Gene Ther.* 2:195–201.

McKUSICK, V. A., and J. S. AMBERGER. 1993. The morbid anatomy of the human genome: chromosomal location of mutations causing disease. *J. Med. Genet.* 30:1–26.

RUDDLE, F. H. 1982. A new era in mammalian gene mapping: somatic-cell genetics and recombinant DNA methodologies. *Nature* 294:115–119.

SCRIVER, C. R., A. L. BEAUDET, W. S. SLY, and D. VALLE, eds. 1989. *The Metabolic Basis of Inherited Disease.* 6th ed. Part 6: *The Disorders of Purine and Pyrimidine Metabolism.* McGraw-Hill.

YANG, N. S. 1992. Gene transfer into mammalian somatic cells in vivo. *Crit. Rev. Biotechnol.* 12:335–356.

Hybridoma Cells and Production of Monoclonal Antibodies

BORREBAECK, C., and I. HAGEN, eds. 1990. *Electromanipulations in Hybridoma Technology.* Stockton Press.

HARLOW, E., and D. LANE. 1988. *Antibodies: A Laboratory Manual.* Cold Spring Harbor Laboratory. Chapter 6: Monoclonal Antibodies and Chapter 7: Growing Hybridomas.

KOHLER, G., and C. MILSTEIN. 1975. Continuous cultures of fused cells secreting antibody of predefined specificity. *Nature* 256:495–497.

MILSTEIN, C. 1980. Monoclonal antibodies. *Sci. Am.* 243(4):66–74.

O'KENNEDY, R., and P. ROBEN. 1991. Antibody engineering: an overview. *Essays Biochem.* 26:59–75.

YELTON, D. E., and M. D. SCHARFF. 1981. Monoclonal antibodies: a powerful new tool in biology and medicine. *Ann. Rev. Biochem.* 50:657–680.

Viruses: Structure and Function

FIELDS, B. N., and D. M. KNIPE, eds. 1990. *Fundamental Virology.* Raven.

LURIA, S. E., et al. 1978. *General Virology,* 3d ed. Wiley.

ROSSMAN, M. G., and R. R. RUECKERT. 1987. What does the molecular structure of viruses tell us about viral functions? *Microbiol. Sci.* 4:206–214.

WEISS, R., et al., eds. 1984. *RNA Tumor Viruses.* Cold Spring Harbor Laboratory.

Use of Radioisotopes to Study Cell Metabolism

CONN, M. P. 1990. *Methods in Neuroscience.* Vol 3: Quantitative and Qualitative Microscopy. Academic Press. Section I contains a description of autoradiography techniques.

DINTZIS, H. 1961. Assembly of the peptide chains of hemoglobin. *Proc. Nat'l. Acad. Sci. USA* 47:247–261.

EAGLE, H., and K. A. PIEZ. 1958. The free amino acid pool of cultured human cells. *J. Biol. Chem.* 231:533–545.

FREEMAN, L. M., and M. D. BLAUFOX. 1975. *Radioimmunoassay.* Grune & Stratton.

PARDUE, M. L., and J. G. GALL. 1969. Molecular hybridization of radioactive DNA to the DNA of cytological preparations. *Proc. Nat'l. Acad. Sci. USA* 64:600–604.

PUCKETT, L., and J. E. DARNELL, JR. 1977. Essential factors in the kinetic analysis of RNA synthesis in HeLa cells. *J. Cell Phys.* 90:521–534.

ROGERS, A. W. 1979. *Techniques of Autoradiography,* 3d ed. Elsevier/North-Holland.

7 Recombinant DNA Technology

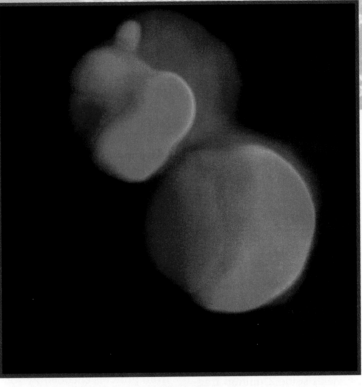

▲ Detection of human immunodeficiency virus (HIV-1) in two human lymphocytes by in situ polymerase chain reaction.

Once the structure of DNA and the genetic code were unraveled, it became clear that many deep biological secrets were locked up in the sequence of bases in DNA. But identifying the sequences of long regions of DNA, much less altering them at will, seemed a distant dream. An avalanche of technical discoveries in the 1970s drastically changed this perspective and has led to astounding advances in molecular cell biology in the past few years based on the analysis and manipulation of macromolecules, particularly DNA.

The discovery of two types of enzymes provided the impetus for these recent developments and permit *DNA cloning*. One type, called restriction enzymes, cut the DNA from any organism at specific sequences of a few nucleotides, generating a reproducible set of fragments. The other type, called DNA ligases, can insert DNA restriction fragments into replicating DNA molecules producing *recombinant DNA*. The recombinant DNA molecules then can be introduced into appropriate cells, most often bacterial cells; all the descendants from a single such cell, called a *clone,* carry the same recombinant DNA molecule. Once a clone of cells bearing a desired segment of DNA is isolated, unlimited quantities of this DNA can be prepared. In addition, DNA sequences up to ≈60 bases long can now be chemically synthesized by entirely automated procedures. Recombinant DNAs thus can be produced containing either natural DNA fragments resulting from restriction-enzyme cleavage or any desired chemically synthesized mutant sequences.

The availability of restriction enzymes also facilitated development of techniques for rapid *DNA sequencing* in the late 1970s. A long DNA molecule is first cleaved with restriction enzymes into a reproducible array of fragments, whose order in the original molecule is determined. Procedures also were developed for determining the sequence of bases in fragments up to 500 nucleotides long. Thus there was no longer any obstacle to obtaining the sequence of a DNA containing 10,000 or more nucleotides. Suddenly, any DNA could be isolated and sequenced. With the aid of computer-automated procedures for sequencing and for storing, comparing, and analyzing data, scientists will sequence the entire human genome in the next few years.

Any cloned DNA segment, whether natural, modified, or completely synthetic, can be reinserted into cells and tested for biological activity. Almost overnight, this group of techniques, often collectively referred to as *recombinant DNA technology*, became the dominant approach for studying many basic biological processes, as examples in the following chapters will illustrate. In this chapter, we describe the various techniques that compose recombinant DNA technology. The power and success of this new technology promise to bring many practical benefits, particularly in medicine and agriculture.

➤ *DNA Cloning with Plasmid Vectors*

The essence of cell chemistry is to isolate a particular cellular component and then analyze its chemical structure and activity. In the case of DNA, this is feasible for relatively short molecules such as the genomes of small viruses. But genomes of even the simplest cells are much too large to directly analyze in detail at the molecular level. The problem is compounded for complex organisms. The human genome, for example, contains about 6×10^9 base pairs (bp) in the 23 pairs of chromosomes. Cleavage of human DNA with restriction enzymes, which produce about one cut for every 3000 base pairs, yields some 2 million fragments, far too many to separate from each other directly. This obstacle to obtaining pure DNA samples from large genomes has been overcome by recombinant DNA technology. With these methods virtually any gene can be purified, its sequence determined, and the functional regions of the sequence explored by altering it in planned ways and reintroducing the DNA into cells and into whole organisms.

The essence of recombinant DNA technology is the preparation of large numbers of identical DNA molecules. A DNA fragment of interest is linked through standard $3' \rightarrow 5'$ phosphodiester bonds to a *vector* DNA molecule, which can replicate when introduced into a host cell. When a single *recombinant* DNA molecule composed of a vector plus an inserted DNA fragment is introduced into a host cell, the inserted DNA is reproduced along with the vector, producing large numbers of recombinant DNA molecules that include the fragment of DNA originally linked to the vector. In this section, the general procedure for cloning DNA fragments in *E. coli* plasmids is described. Further details and techniques are presented in later sections.

Plasmids Are Extrachromosomal Self-Replicating DNA Molecules

Plasmids are circular, double-stranded DNA (dsDNA) molecules that are separate from a cell's chromosomal DNA. These extrachromosomal DNAs occur naturally in bacteria and in the nuclei of yeast and some higher eukaryotic cells, existing in a parasitic or symbiotic relationship with their host cell. Plasmids range in size from a few thousand base pairs to more than 100 kilobases (kb). Like the host-cell chromosomal DNA, plasmid DNA is duplicated before every cell division. During cell division, at least one copy of the plasmid DNA is segregated to each daughter cell, assuring continued propagation of the plasmid through successive generations of the host cell.

Many naturally occurring plasmids contain genes that provide some benefit to the host cell, fulfilling the plasmid's portion of a symbiotic relationship. For example, some bacterial plasmids encode enzymes that inactivate antibiotics. Therefore, a bacterial cell containing such a plasmid is resistant to the antibiotic and can replicate in an environment containing the antibiotic, whereas the same type of bacterium lacking the drug-resistance plasmid is killed.

Such drug-resistance plasmids have become a major problem in the treatment of a number of common bacterial pathogens. As antibiotic use became widespread, plasmids containing several drug-resistance genes evolved, making their host cells resistant to a variety of different antibiotics simultaneously. Many of these plasmids also contain "transfer genes" encoding proteins that transfer a copy of the plasmid to other host cells of the same or related bacterial species. Plasmid transfer occurs through a complex macromolecular tube called a *pilus*, which is constructed from proteins encoded by some of the transfer genes. Such transfer can result in the rapid spread of drug-resistance plasmids, expanding the number of antibiotic-resistant bacteria in an environment such as a hospital. Coping with the spread of drug-resistance plasmids is an important challenge for modern medicine.

E. Coli Plasmids Can Be Engineered for Use as Cloning Vectors

The plasmids most commonly used in recombinant DNA technology replicate in *E. coli*. Generally, these plasmids have been engineered to optimize their use as vectors in DNA cloning. For instance, to simplify working with plasmids, their length is reduced; many plasmid vectors are only ≈3 kb in length, which is much less than that of naturally occurring *E. coli* plasmids. (The circumference of plasmids usually is referred to as their "length," even though plasmids are almost always circular DNA molecules.) Most plasmid vectors contain little more than the essential nucleotide sequences required for their use in

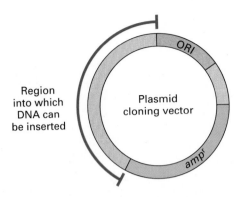

▲ FIGURE 7-1   Diagram of a simple cloning vector derived from a plasmid, a circular, double-stranded DNA molecule that can replicate within an *E. coli* cell. Plasmid vectors are ≈1.2–3 kb in length and contain a replication origin (ORI) sequence and a gene that permits selection usually by conferring resistance to a particular drug. Here the selective gene is *amp*^{r}; it encodes the enzyme β-lactamase, which inactivates ampicillin. Exogenous DNA can be inserted into the bracketed region without disturbing the ability of the plasmid to replicate or express the *amp*^{r} gene.

DNA cloning: a replication origin, a drug-resistance gene, and a region in which exogenous DNA fragments can be inserted (Figure 7-1).

Plasmid DNA Replication   The *replication origin* (ORI) is a specific DNA sequence of 50–100 base pairs that must be present in a plasmid for it to replicate. Host-cell enzymes bind to ORI, initiating replication of the circular plasmid. Once DNA replication is initiated at ORI, it continues around the circular plasmid regardless of its nucleotide sequence (Figure 7-2). Thus any DNA sequence inserted into such a plasmid is replicated along with the rest of the plasmid DNA; this property is the basis of molecular DNA cloning.

Selection of Transformed Cells   In 1944, O. T. Avery, C. M. Macleod, and M. McCarty first demonstrated gene transfer with isolated DNA obtained from *Streptococcus pneumoniae*. This process involved the genetic alteration of a bacterial cell by the uptake of DNA isolated from a genetically different bacterium and its recombination with the host-cell genome. Their experiments provided the first evidence that DNA is the genetic material. Later studies showed that such genetic alteration of a recipient cell can result from the uptake of exogenous extrachromosomal DNA (e.g., plasmids) that does not integrate into the host-cell chromosome. The term *transformation* is used to denote the genetic alteration of a cell caused by the uptake and expression of foreign DNA regardless of the mechanism involved. (Note that *transformation* has a second meaning defined in Chapter 6, namely the process by which normal cells with a finite life span in culture are converted into continuously growing cells similar to cancer cells.)

The phenomenon of transformation permits plasmid vectors to be introduced into and expressed by *E. coli* cells. In order to be useful in DNA cloning, however, a plasmid vector must contain a *drug-resistance gene* encoding an enzyme that inactivates a specific antibiotic. For example, the ampicillin-resistance gene (*amp*^{r}) encodes β-lactamase, which inactivates the antibiotic ampicillin. After plasmid vectors are incubated with *E. coli*, those cells that take up the plasmid can be easily selected from the larger number of cells that do not by growing them in an ampicillin-containing medium. The ability to select transformed cells is critical to DNA cloning by plasmid-vector technology because the transformation of *E. coli* with isolated plasmid DNA is inefficient.

Normal *E. coli* cells cannot take up plasmid DNA from the medium. Exposure of cells to high concentrations of certain divalent cations, however, makes a small fraction of cells permeable to foreign DNA by a mechanism

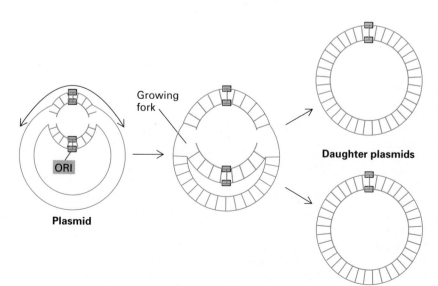

◀ FIGURE 7-2   Plasmid DNA replication. Newly synthesized daughter strands are shown in red. Once DNA replication is initiated at the origin (yellow), it continues in both directions around the circular molecule until the advancing growing forks merge and two daughter molecules are produced. The origin (ORI) is the only specific nucleotide sequence required for replication of the entire circular DNA molecule.

that is not understood. In a typical procedure, *E. coli* cells are treated with $CaCl_2$ and mixed with plasmid vectors; commonly, only 1 cell in about 10,000 or more cells becomes *competent* to take up the foreign DNA. Each competent cell incorporates a *single* plasmid DNA molecule, which carries an antibiotic-resistance gene. When the treated cells are plated on a petri dish of nutrient agar containing the antibiotic, only the rare transformed cells containing the antibiotic-resistance gene on the plasmid vector will survive. All the plasmids in such a colony of selected transformed cells are descended from the single plasmid taken up by the cell that established the colony.

Plasmid Cloning Permits Isolation of DNA Fragments from Complex Mixtures

As discussed in the next section, a DNA fragment of a few base pairs up to ≈20 kilobases (kb) can be inserted into a plasmid vector. When such a recombinant plasmid transforms an *E. coli* cell, all the antibiotic-resistant progeny cells that arise from the initial transformed cell will contain plasmids with the same inserted sequence of DNA (Figure 7-3). The inserted DNA is replicated along with the rest of the plasmid DNA and segregates to daughter cells as the colony grows. In this way, the initial fragment of DNA is replicated in the colony of cells into a large number of identical copies. Since all the cells in a colony arise from a single transformed parental cell, they constitute a *clone* of cells. The initial fragment of DNA inserted into the parental plasmid is referred to as *cloned DNA*, since it can be isolated from the clone of cells.

DNA cloning allows fragments of DNA with a particular nucleotide sequence to be isolated from a complex mixture of fragments with many different sequences. As a simple example, assume you have a solution containing four different types of DNA fragments, each with a unique sequence (Figure 7-4). Each fragment type is individually in-

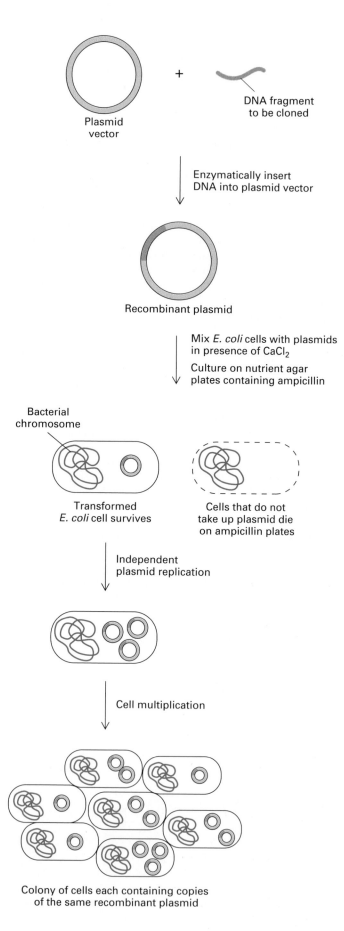

► FIGURE 7-3   General procedure for cloning a DNA fragment in a plasmid vector. Double-stranded DNA is represented by single lines here. Although not indicated by color, the plasmid contains a replication origin and ampicillin-resistance gene. Uptake of plasmids by *E. coli* cells is stimulated by high concentrations of $CaCl_2$. Even in the presence of $CaCl_2$, transformation occurs with a quite low frequency, and only a few cells are transformed by incorporation of a single plasmid molecule. Cells that are not transformed die on ampicillin-containing medium. Once incorporated into a host cell, a plasmid can replicate independently of the host-cell chromosome. As a transformed cell multiplies into a colony, at least one plasmid segregates to each daughter cell.

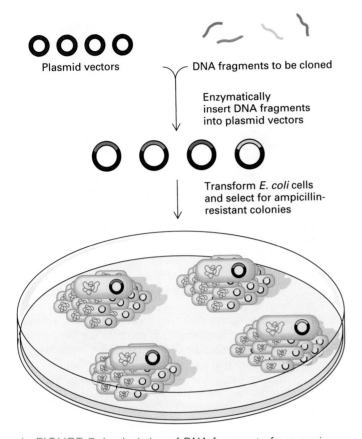

the initial mixture are isolated from each other in the separate bacterial colonies. DNA cloning thus is a powerful, yet simple method for purifying a particular DNA fragment from a complex mixture of fragments and producing large numbers of the fragment of interest.

➤ Production of Recombinant Plasmids

To clone specific DNA fragments in a plasmid vector, as just described, or in other vectors discussed in later sections, the fragments must be produced and then inserted into the vector DNA. As noted in the introduction, restriction enzymes and DNA ligases are utilized to produce such recombinant cloning vectors.

Restriction Enzymes Cut DNA Molecules at Specific Sequences

Restriction enzymes are bacterial enzymes that recognize specific 4- to 8-bp sequences, called *restriction sites,* and then cleave both DNA strands at this site. Since these enzymes cleave DNA within the molecule, they are also called *restriction endonucleases* to distinguish them from *exonucleases,* which digest nucleic acids from an end. Many restriction sites, like the *Eco*RI site shown in Figure 7-5a, are short inverted repeat sequences; that is, the restriction-site sequence is the same on each DNA strand when read in the 5′ → 3′ direction. Because the DNA isolated from an individual organism has a specific sequence, restriction enzymes cut the DNA into a reproducible set of fragments called *restriction fragments* (Figure 7-6).

The word *restriction* in the name of these enzymes refers to their function in the bacteria from which they are isolated: a restriction endonuclease destroys (restricts) incoming foreign DNA (e.g., bacteriophage DNA or DNA taken up during transformation) by cleaving it at all the restriction sites in the DNA. Another enzyme, called a *modification enzyme,* protects a bacterium's own DNA from cleavage by modifying it at or near each potential cleavage site. The modification enzyme adds a methyl group to one or two bases, usually within the restriction site. When a methyl group is present there, the restriction endonuclease is prevented from cutting the DNA (Figure 7-5b). Together with the restriction endonuclease, the methylating enzyme forms a *restriction-modification system* that protects the host DNA while it destroys foreign DNA. Restriction enzymes have been purified from several hundred different species of bacteria, allowing DNA molecules to be cut at a large number of different sequences corresponding to the recognition sites of these enzymes (Table 7-1).

▲ FIGURE 7-4 Isolation of DNA fragments from a mixture by cloning in a plasmid vector. Four distinct DNA fragments, represented in red, green, dark purple, and light purple, are inserted into plasmid cloning vectors, yielding a mixture of recombinant plasmids each containing a single DNA fragment. *E. coli* cells treated with CaCl₂ are incubated with the mixture of recombinant plasmids and then plated on nutrient agar containing ampicillin. Each colony of transformed, antibiotic-resistant cells that grows (represented by a group of cells) arises from a single cell that took up one or another of the recombinant plasmids; all the cells in a given colony thus carry the same DNA fragment. Overnight incubation of *E. coli* at 37°C produces visible colonies containing about a million cells. Since the colonies are isolated from each other on the culture plate, copies of the DNA fragments in the original mixture are separated in the individual colonies. Although not indicated visually, the transformed cells contain multiple copies of a given plasmid.

serted into a plasmid vector. The resulting mixture of recombinant plasmids is incubated with *E. coli* cells under conditions that facilitate transformation; the cells then are cultured on antibiotic selective plates. Since each colony that develops arose from a single cell that took up a *single* plasmid, all of the cells in a colony harbor the identical type of plasmid characterized by the DNA fragment inserted into it. As a result, copies of the DNA fragments in

(a)

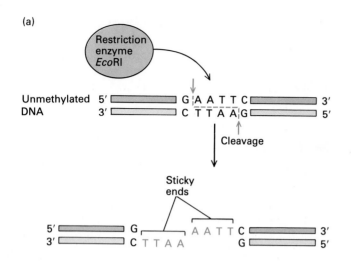

(b)

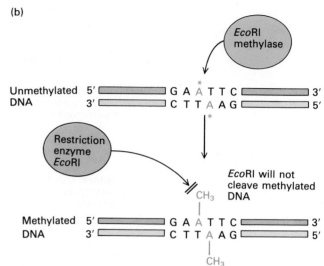

▲ FIGURE 7-5   Restriction-recognition sites are short DNA sequences recognized and cleaved by various restriction endonucleases. (a) *Eco*RI, a restriction enzyme from *E. coli*, makes staggered cuts at the specific 6-bp inverted repeat sequence shown. This cleavage yields fragments with single-stranded, complementary sticky ends. Many other restriction enzymes also produce fragments with sticky ends. The sticky ends on two fragments derived by cleavage of different DNAs with the same restriction enzyme readily base-pair, and the fragments can be enzymatically joined (ligated) to produce recombinant DNAs. (b) Bacterial cells with restriction endonucleases also contain corresponding modification enzymes that methylate bases in the restriction-recognition site. For example, *E. coli* cells containing the *Eco*RI restriction enzyme also contain the *Eco*RI methylase, a modification enzyme that catalyzes addition of a methyl group to two adenines in the *Eco*RI recognition sequence. The methylated restriction site is not cleaved by *Eco*RI, assuring that a cell making this restriction enzyme does not destroy its own DNA.

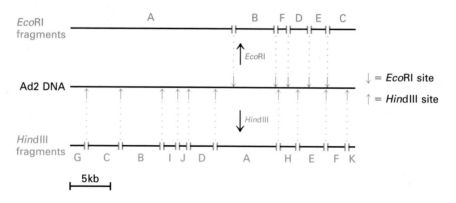

▲ FIGURE 7-6   Fragments produced by cleavage of the ≈36-kb DNA genome from adenovirus 2 (Ad2) by *Eco*RI and another restriction enzyme, *Hin*dIII from *Haemophilus influenzae*. Double-stranded DNA is represented by single black lines in this figure. Digestion of Ad2 DNA (*center*) with *Eco*RI generates six *Eco*RI fragments (*top*); these result from cleavage at each *Eco*RI restriction site (GAATTC) in the Ad2 sequence. Digestion with *Hin*dIII cleaves the Ad2 DNA at each *Hin*dIII site (AAGCTT), generating eleven specific fragments (*bottom*), all different from the *Eco*RI fragments. By convention, restriction fragments are labeled A–Z in order of decreasing size. By techniques described later, the order of fragments in the original DNA can be determined, thus mapping the restriction sites on the DNA (indicated by short arrows). Such a "restriction-site map" for various restriction enzymes is a unique characteristic of each DNA.

Many Restriction Enzymes Generate DNA Fragments with "Sticky" Ends

As illustrated in Figure 7-5a, *Eco*RI makes staggered cuts in the two DNA strands. Many other restriction enzymes make similar cuts, generating fragments that have a single-stranded "tail" at both ends. The tails on the fragments generated at a given restriction site are complementary to those on all other fragments generated by the same restriction enzyme. At room temperature, these single-stranded regions, often called "sticky ends," can transiently base-pair with those on other DNA fragments generated with the same restriction enzyme, regardless of the source of the DNA. This base pairing of sticky ends permits DNA from widely differing species to be ligated, forming chimeric molecules.

TABLE 7-1   Examples of the Actions of Restriction Endonucleases

Source Microorganism	Enzyme*	Recognition Site (↓)†	Number of Cuts (kb)‡			
			λ (50)	Ad2 (36)	SV40 (5.2)	pBR322 (4.3)
Arthrobacter luteus	*Alu*I	AG ↓ CT	143	158	34	14
Thermus aquaticus	*Taq*I	T ↓ CGA	121	50	1	13
Haemophilus parahaemolyticus	*Hph*I	GGTGA + 5	168	99	4	18
Haemophilus gallinarum	*Hga*I	GACGC + 8	102	87	0	12
Escherichia coli	*Eco*RI	G ↓ AATTC	5	5	1	1
Haemophilus influenzae	*Hin*dIII	A ↓ AGCTT	6	12	6	1
Nocardia otitiscaviaruns	*Not*I	GC ↓ GGCCGC	0	7	0	0
Streptomyces fimbriatus	*Sfi*I	GGCCN₄ ↓ NGGCC	0	3	1	0

*Enzymes are named with abbreviations of the bacterial strains from which they are isolated; the Roman numeral indicates the enzyme's priority of discovery in that strain (for example, *Alu*I was the first restriction enzyme to be isolated from *Arthrobacter luteus*).

†Recognition sequences are written $5' \rightarrow 3'$ (only one strand is given) with the cleavage site indicated by an arrow. For example, G ↓ GATCC is an abbreviation for

$$\downarrow$$
(5′)GGATCC(3′)
(3′)CCTAGG(5′)
$$\uparrow$$

The cleavage site for *Hph*I and *Hga*I occurs five or eight bases away from the recognition sequence; N indicates any base.

‡These columns list the number of cleavage sites recognized by specific endonucleases on the DNA of bacteriophage λ (λ), adenovirus type 2 (Ad2), simian virus 40 (SV40), and an *E. coli* plasmid (pBR322). The sizes of the DNAs in kilobases (kb) are in parentheses. Note that the actual number of cuts in these sequences deviates from the expected number in random sequences, which would be given by $L/4^n$, where n is the length of the site recognized by an enzyme and L is the length of the sequence.
SOURCE: R. J. Roberts, 1988, *Nuc. Acids Res.* **16**(supp):r271.

DNA Ligase Covalently Links Restriction Fragments

During in vivo DNA replication, DNA ligase catalyzes formation of $3' \rightarrow 5'$ phosphodiester bonds between the short fragments of the discontinuously synthesized DNA strand at a replication fork (see Figure 4-22). In recombinant DNA technology, purified DNA ligase is used to covalently join the ends of restriction fragments in vitro. This enzyme can catalyze the formation of a $3' \rightarrow 5'$ phosphodiester bond between the 3′-hydroxyl end of one restriction-fragment strand and the 5′-phosphate end of another restriction-fragment strand during the time that the sticky ends are transiently base-paired (Figure 7-7). When DNA ligase and ATP are added to a solution containing restriction fragments with sticky ends, the restriction fragments are covalently ligated together through the standard $3' \rightarrow 5'$ phosphodiester bonds of DNA.

Some restriction enzymes cleave both DNA strands exactly in the middle of the recognition site (e.g., *Alu*I in Table 7-1). These restriction enzymes generate DNA restriction fragments with "blunt" ("flush") ends in which all of the nucleotides at the fragment ends are base-paired to nucleotides in the complementary strand. In addition to

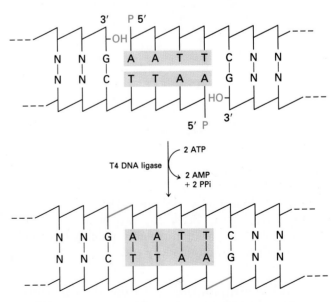

▲ FIGURE 7-7   Ligation of *Eco*RI sticky ends by DNA ligase from bacteriophage T4. After the complementary sticky ends (orange) on two fragments transiently base-pair, the adjacent 3′-hydroxyl and 5′-phosphate groups (red) are covalently joined (ligated). One ATP is consumed for each phosphodiester bond (red) formed.

(a) Sequence of polylinker

	*Sac*I		*Sma*I		*Xba*I		*Pst*I		*Hind*III

GAATTCGAGCTCGGTACCCGGGGATCCTCTAGAGTCGACCTGCAGGCATGCAAGCTT

| *Eco*RI | | *Kpn*I | | *Bam*HI | | *Sal*I | | *Sph*I | |

(b) Insertion of EcoRI restriction fragments

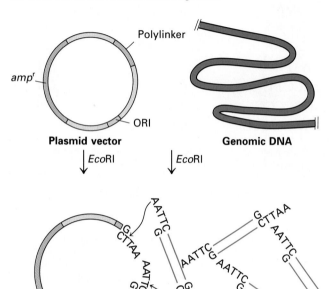

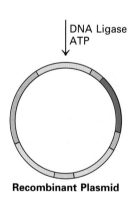

▲ FIGURE 7-8   Plasmid vectors containing a polylinker, or multiple cloning-site sequence, commonly are used to produce recombinant plasmids carrying exogenous DNA fragments. (a) Sequence of a polylinker that includes one copy of the recognition site, indicated by red brackets, for each of the 10 restriction enzymes indicated. Polylinkers are chemically synthesized and then are inserted into a plasmid vector. Only one strand is shown. (b) Insertion of genomic restriction fragments into the pUC19 plasmid vector, which contains the polylinker shown in (a). (The length of the polylinker in relation to the rest of the plasmid is greatly exaggerated here.) One of the restriction enzymes whose recognition site is in the polylinker is used to cut both the plasmid molecules and genomic DNA, generating singly-cut plasmids and restriction fragments with complementary sticky ends (letters at ends of green fragments). By use of appropriate reaction conditions, insertion of a single restriction fragment per plasmid can be maximized. Curved lines indicate base pairing between sticky ends prior to ligation. Note that the restriction sites are reconstituted in the recombinant plasmid. [See C. Yanisch-Perron, J. Vieira, and J. Messing, 1985, *Gene* **33**:103.]

constructed with a *polylinker*, or *multiple cloning-site sequence*, a synthetic sequence that contains one copy of several different restriction sites (Figure 7-8a). When such a vector is treated with a restriction enzyme that recognizes a recognition sequence in the polylinker, it is cut at that sequence, generating sticky ends. In the presence of DNA ligase, DNA fragments produced with the same restriction enzyme will be inserted into the plasmid (Figure 7-8b). The ratio of DNA fragments to be inserted to cut vectors and other reaction conditions are chosen to maximize the insertion of one restriction fragment per plasmid vector. The recombinant plasmids produced in in vitro ligation reactions then can be used to transform antibiotic-sensitive *E. coli* cells as shown in Figure 7-4. All of the cells in each antibiotic-resistant clone that remains after selection contain plasmids with the same inserted DNA fragment, but different clones carry different fragments.

ligating complementary sticky ends, the DNA ligase from bacteriophage T4 can ligate any two blunt DNA ends. However, blunt-end ligation requires a higher DNA concentration than ligation of sticky ends.

Restriction Fragments Are Readily Inserted into Plasmid Vectors

By use of restriction enzymes to create fragments with sticky ends and DNA ligase to covalently link them, foreign DNA can be inserted into plasmid vectors in vitro in a straightforward procedure. *E. coli* plasmid vectors can be

► Formation and Uses of Synthetic DNA

Advances in synthetic chemistry now permit the chemical synthesis of single-stranded DNA (ssDNA) molecules of any sequence up to about 60 nucleotides in length. Synthetic DNA has a number of applications in recombinant DNA technology. Complementary ssDNAs can be synthesized and hybridized to each other to form a dsDNA with sticky ends. Such completely synthetic dsDNAs can be cloned into plasmid vectors just like DNA restriction fragments prepared from living organisms. For example, the 57-bp polylinker sequence shown in Figure 7-8 was chemically synthesized and then inserted into plasmid vectors to facilitate the cloning of fragments generated by different restriction enzymes. This example illustrates the use of synthetic DNAs to add convenient restriction sites where they otherwise do not occur. As described later in the chapter, synthetic DNAs are used in sequencing DNA and as probes to identify clones of interest. Synthetic DNAs also can be substituted for natural DNA sequences in cloned DNA to study the effects of specific mutations; this topic is examined in the next chapter.

The technique for chemical synthesis of DNA oligonucleotides is outlined in Figure 7-9. Note that chains grow in the $3' \rightarrow 5'$ direction, opposite to the direction of DNA chain growth catalyzed by DNA polymerases. Once the chemistry for producing synthetic DNA was standardized, automated instruments were developed that allow researchers to program the synthesis of oligonucleotides of specific sequences up to about 60 nucleotides long.

➤ λ-Phage Cloning Vectors and Construction of a Genomic Library

Most DNA cloning is done with *E. coli* plasmid vectors because of the relative simplicity of this procedure. However, the number of individual clones that can be obtained by plasmid cloning is limited by the relatively low efficiency of *E. coli* transformation and the small number (only a few hundred) of individual transformed colonies

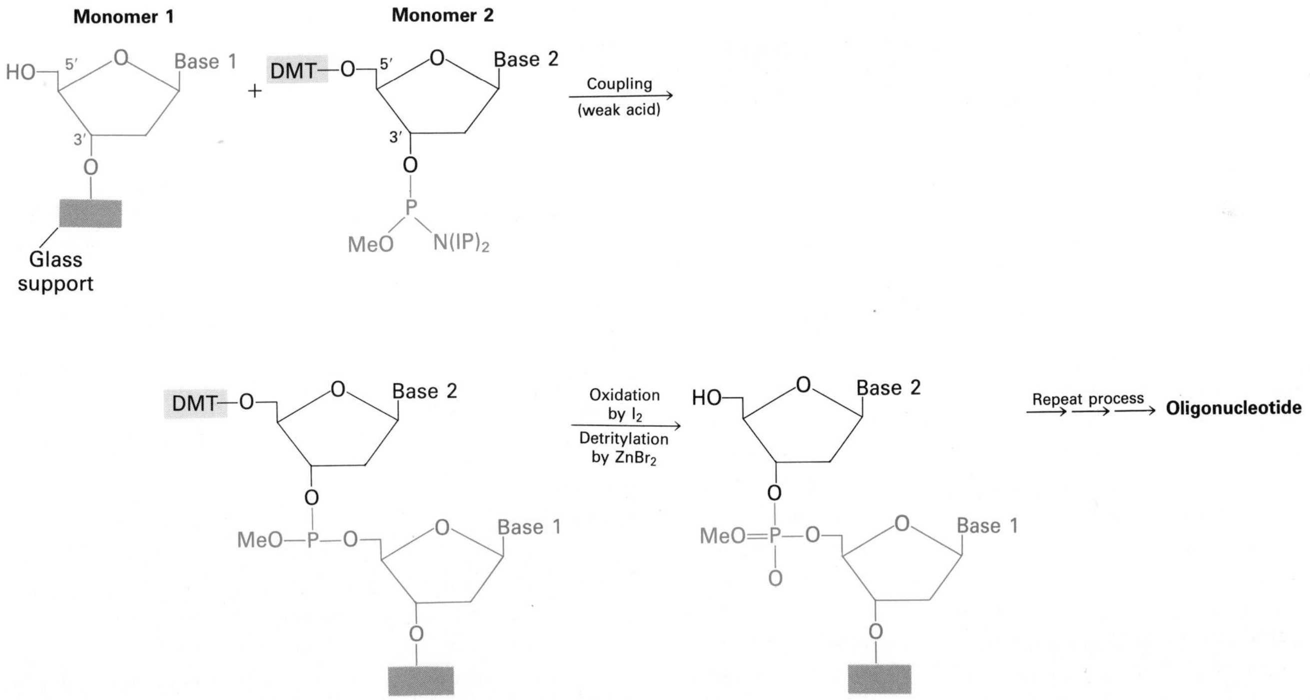

▲ FIGURE 7-9   Chemical synthesis of oligonucleotides by sequential addition of reactive nucleotide derivatives in the $3' \rightarrow 5'$ direction. The first nucleotide (monomer 1) is bound to a glass support by its 3' hydroxyl; its 5' hydroxyl is available. The next nucleotide in the sequence (monomer 2) is derivatized by addition of 4',4'-dimethoxytrityl (DMT) to its 5' hydroxyl, thus blocking this hydroxyl from reacting; in addition, a highly reactive methylated diisopropyl phosphoramidite group (red letters) is attached to the 3' hydroxyl. When the two monomers are mixed in the presence of a weak acid, they form a $5' \rightarrow 3'$ phosphodiester bond with the phosphorus in the trivalent state. Oxidation with iodine (I_2) increases the phosphorus valency to five; removal of the DMT group by detritylation with zinc bromide ($ZnBr_2$) frees the 5' hydroxyl, and the process is repeated. When synthesis is complete, all the methyl groups on the phosphates are removed at the same time at alkaline pH, and the bond linking monomer 1 to the glass support is cleaved. [See S. L. Beaucage and M. H. Caruthers, 1981, *Tetrahedron Lett.* **22**:1859.]

(a)

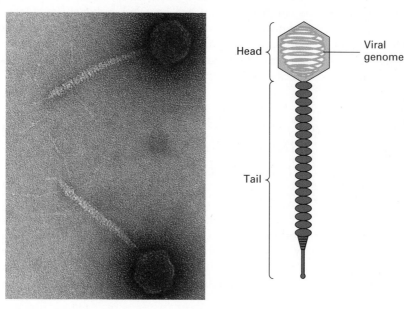

Head

Tail

Viral genome

◄ FIGURE 7-10   (a) Electron micrograph of bacteriophage λ virions and schematic diagram of one virion. (b) Simplified map of the λ-phage genome. Genes encoding proteins required for assembly of the head and tail map at the left end; those encoding additional proteins required for the lytic cycle map at the right end. Some regions of the genome can be replaced by exogenous DNA (diagonal lines) or deleted (dotted area) without affecting the ability of λ phage to infect host cells and assemble new virions, permitting insertion of up to ≈25 kb of DNA between the J and N genes. The λ genome has been mapped with some 60 different genes. Only a few individual genes are shown in this diagram. Small numbers indicate positions in kilobases (kb). [Photograph courtesy of R. Duda and R. Hendrix.]

(b) λ-Phage genome

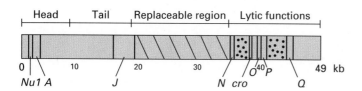

that can be detected on a typical petri dish. These limitations make plasmid cloning of all the genomic DNA of higher organisms impractical. For example, ≈1.5 × 10⁵ clones carrying 20-kb DNA fragments are required to represent the total human haploid genome, which contains ≈3 × 10⁹ base pairs. Fortunately, cloning vectors derived from bacteriophage λ have proved to be a practical means for obtaining the required number of clones to represent large genomes. Such a collection of λ clones that includes *all* the DNA sequences of a given species is called a *genomic library*. Once a genomic library is prepared, it can be screened for λ clones containing a sequence of interest.

Bacteriophage λ Can Be Modified for Use as a Cloning Vector and Assembled In Vitro

Bacteriophage λ is probably the most extensively studied bacterial virus, and a great deal is known about its molecular biology and genetics. A λ-phage virion has a head region, which contains the viral DNA genome, and a tail

region, which functions in infecting *E. coli* host cells (Figure 7-10a). When a λ virion infects a host cell, only the λ DNA enters the cell. As discussed in Chapter 6, the viral DNA then undergoes either lytic or lysogenic growth. In lytic growth, the viral DNA is replicated and assembled into more than 100 progeny virions in each infected cell, killing the cell in the process and releasing the replicated virions (see Figure 6-17). In lysogenic growth, the viral DNA inserts in the bacterial chromosome where it is passively replicated along with the host-cell chromosome as the cell grows and divides (see Figure 6-19).

As diagrammed in Figure 7-10b, the λ genes encoding the head and tail proteins as well as various proteins involved in the lytic and lysogenic growth pathways are clustered in discrete regions of the ≈50-kb viral genome. The genes involved in the lysogenic pathway and other viral genes not essential for the lytic pathway are irrelevant for use of bacteriophage λ as a cloning vector. These genes are removed from the viral DNA and replaced with other DNA sequences of interest. Up to ≈25 kb of foreign DNA can be inserted into the λ genome, resulting in a recombinant DNA that can be packaged to form virions capable of replicating and forming plaques on *E. coli* host cells.

The key to the high efficiency of λ-phage cloning is the ability to assemble λ virions in vitro. During the in vivo assembly of λ virions within infected host cells, viral heads and tails initially are assembled separately from multiple copies of the various proteins that compose these complex structures. Replication of λ DNA in a host cell generates long multimeric DNA molecules, call *concatomers*, that consist of multiple copies of the viral genome linked end to end and separated by specific nucleotide sequences called *COS sites*. Two λ proteins, designated Nu1 and A, bind to COS sites and direct insertion of the DNA between two adjacent COS sites into a preassembled head. This process results in the packaging of a single ≈50-kb λ genome from the multimeric concatomer into each preassembled head. Host-cell chromosomal DNA is not inserted into the λ heads because it does not contain any copies of the COS sequence. Once λ DNA is inserted into a preassembled λ head, the preassembled tail is attached producing a complete virion (Figure 7-11).

To prepare infectious λ virions carrying recombinant DNA, the phage-assembly process is carried out in vitro. In one method, *E. coli* cells are infected with a λ mutant defective in A protein, one of the two proteins required for packaging λ DNA into preassembled phage heads. These cells accumulate preassembled "empty" heads; since tails only attach to heads "filled" with DNA, preassembled tails also accumulate in these cells. After these cells are lysed experimentally, an extract containing high concentrations of heads and tails is prepared. When this extract is mixed with A protein and recombinant λ DNA, which contains a COS site, the DNA is packaged into the empty heads. The tails in the extract then combine with the filled heads, yielding complete virions carrying the recombinant λ DNA.

The recombinant virions produced by this method are fully infectious and can efficiently infect *E. coli* cells. Each virion particle binds to receptors on the surface of a host cell and injects its packaged recombinant DNA into the cell. This infection process is about a thousand times more efficient than transformation with plasmid vectors. For instance, ≈10^6 transformed colonies per microgram of recombinant plasmid DNA can be obtained routinely, whereas ≈10^9 plaques representing λ clones can be obtained per microgram of recombinant λ DNA.

Nearly Complete Genomic Libraries of Higher Organisms Can Be Prepared by λ Cloning

With the availability of λ-phage cloning vectors, preparation of genomic libraries for higher organisms, including

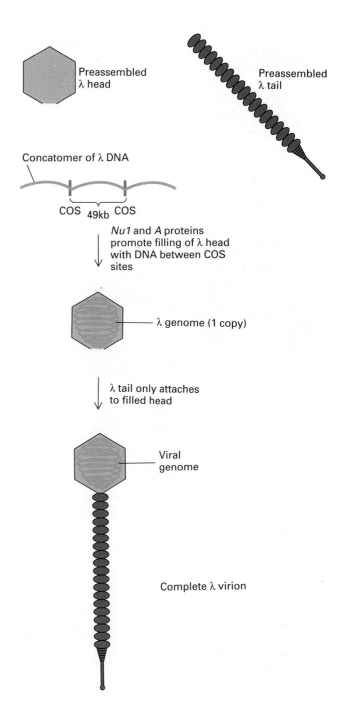

▲ FIGURE 7-11 Assembly of bacteriophage λ virions. Empty heads and tails are assembled from multiple copies of several different λ proteins. During the late stage of λ infection, long DNA molecules called concatomers are formed; these multimeric molecules consist of copies of the λ genome linked end to end and separated by COS sites (red), a protein-binding nucleotide sequence that occurs once in each copy of the λ genome. Binding of the λ proteins Nu1 and A to COS sites promotes insertion of the DNA between two adjacent COS sites into an empty head. After the heads are filled with DNA, preassembled λ tails are attached, producing complete λ virions capable of infecting *E. coli* cells.

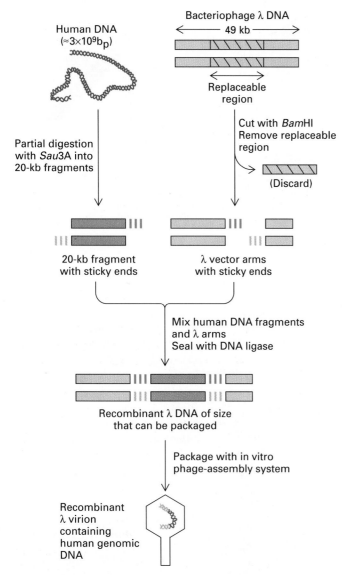

▲ FIGURE 7-12 Construction of a genomic library of human DNA in a bacteriophage λ vector. The λ DNA first is treated to remove the replaceable region (see Figure 7-10b); the non-essential regions (dotted area in Figure 7-10b) have been deleted from λ vectors to maximize the size of the exogenous DNA fragment that can be inserted. In this example, the replaceable region of the λ DNA is cut out with *Bam*HI, and the total DNA from human cells is partially digested with *Sau*3A. These two restriction enzymes produce fragments with complementary sticky ends (red lines). The λ vector arms and ≈20-kb genomic fragments are mixed, ligated, and packaged in vitro to produce recombinant λ-phage virions, which are plated on a lawn of *E. coli* cells. Multiple λ vectors have been constructed containing different restriction sites, so that restriction fragments generated by a variety of restriction enzymes can be cloned in λ vectors. A particular 20-kb region of the human genome would occur approximately once in every 1.5×10^5 recombinant λ virions. However, about 10^6 recombinant phages are required to assure that fragments encompassing the entire human genome have a 90–95 percent chance of being included in the library.

Figure labels:

Human DNA ($\approx 3 \times 10^9$ bp)

Bacteriophage λ DNA ← 49 kb →

Replaceable region

Cut with *Bam*HI Remove replaceable region

(Discard)

Partial digestion with *Sau*3A into 20-kb fragments

20-kb fragment with sticky ends

λ vector arms with sticky ends

Mix human DNA fragments and λ arms Seal with DNA ligase

Recombinant λ DNA of size that can be packaged

Package with in vitro phage-assembly system

Recombinant λ virion containing human genomic DNA

humans, now is feasible. A genomic library is a set of λ (or plasmid) clones that collectively contain every DNA sequence in the genome of a particular organism. Figure 7-12 summarizes the general procedure for constructing a λ genomic library. The λ DNA first is treated with a restriction endonuclease, producing fragments with sticky ends, called λ *vector arms,* that contain all the genes necessary for lytic growth. This step frees the nonessential region in the middle of the λ genome; this region is separated from the λ arms and discarded. Genomic DNA then is extracted from a cell type that contains all the genetic information of the organism under study. Sperm cells or cells of an early embryo often are used as sources of mammalian DNA. The extracted DNA then is cleaved by a restriction enzyme to produce ≈20-kb fragments with sticky ends complementary to the sticky ends on the λ vector arms being used.

The λ arms and the collection of genomic DNA fragments are mixed in about equal amounts. The complementary sticky ends on the fragments and λ arms hybridize and then are joined covalently by DNA ligase. Each of the resulting recombinant DNA molecules contains a foreign DNA fragment located between the two arms of the λ vector DNA. The ligated recombinant DNAs then are packaged into λ virions in vitro as described above. Only DNA molecules of the correct size can be packaged to produce fully infectious recombinant λ virions.

Finally, the recombinant λ virions are plated on a lawn of *E. coli* cells to generate a large number of recombinant λ plaques. Since each plaque arises from a single recombinant virion, all the progeny λ phages that develop are genetically identical and constitute a clone carrying a particular genomic DNA insert. The different plaques correspond to distinct phage clones, each carrying a different DNA insert, and collectively they constitute a λ genomic library.

In constructing a library of genomic DNA, the DNA commonly is cleaved with a restriction enzyme that recognizes a 4-bp restriction site (e.g., *Sau*3A as shown in Figure 7-12). A specific 4-bp sequence will occur on average once every $4^4 = 256$ base pairs. Complete digestion of the human haploid genome, which contains ≈3×10^9 base pairs, would yield somewhat more than 10^7 nonoverlapping different fragments. However, to increase the probability that all regions of the genome are successfully cloned and represented in the λ genomic library, the genomic DNA usually is only partially digested to yield overlapping restriction fragments of ≈20 kb (Figure 7-13). In a large λ library constructed from such overlapping restriction fragments, a specific sequence of genomic DNA may be contained in several "overlapping" clones.

The size of a genomic library for a given organism depends on the amount of DNA in that organism's haploid genome. If the human genome of about 3×10^9 base pairs is cleaved into 20-kb fragments for insertion into a λ vector, then roughly 1.5×10^5 different recombinant λ-phage virions would be required to constitute a complete library. Because the restriction fragments of human DNA are in-

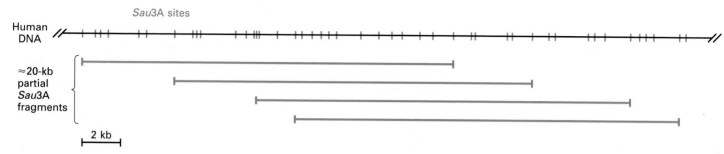

▲ FIGURE 7-13   Production of overlapping restriction fragments by partial digestion of human genomic DNA with Sau3A. This restriction endonuclease recognizes the 4-bp sequence GATC and produces fragments with single-stranded sticky ends of the same sequence on the 5′ end of each strand. A hypothetical region of human genomic DNA showing the Sau3A recognition sites (red) is shown at the top. Partial digestion of this region of DNA would yield a vari-ety of overlapping fragments (blue) ≈20 kb long. Use of such overlapping fragments increases the probability that all sequences in the genomic DNA will be represented in a λ library. In a large λ library, a specific sequence of genomic DNA (e.g., the gene encoding β-globin) often is contained in or extends over several "overlapping" clones, generated from overlapping restriction fragments.

corporated into phages randomly, about 10^6 recombinant phages are necessary to assure that each region of human DNA has a 90–95 percent chance of being included.

Each plaque produced by a recombinant bacterio-phage λ contains large numbers of recombinant virions and, consequently, large numbers of cloned DNA frag-ments. Hybridization methods for identifying recombinant λ clones of interest are described in a later section. Because these methods allow specific detection of very small plaques, as many as 5×10^4 plaques can be screened on a typical petri dish. Thus only 20–30 petri dishes, each con-taining about 5×10^4 λ plaques, are sufficient to represent the entire human genome. In contrast, to screen 10^6 recom-binant plasmids carrying the entire human genome would require about 5000 petri dishes, because only 200 or so transformed E. coli colonies can be detected on a typical petri dish.

Larger DNA Fragments Can Be Cloned in Cosmids and Other Vectors

Both λ-phage and E. coli plasmid vectors are useful for cloning only relatively small DNA fragments. The largest DNA fragment that can be cloned in a λ-phage vector is about 25 kb long; insertion of larger fragments produces recombinant λ DNA that is too large to be packaged into viral heads. Plasmid cloning is practical only with frag-ments up to about 20 kb long because the efficiency of E. coli transformation with plasmid DNA falls off rapidly for larger plasmids. Several other vectors, however, have been developed for cloning larger fragments of DNA (Table 7-2).

One common method for cloning larger fragments makes use of elements of both plasmid and λ-phage clon-ing. In this method, called *cosmid cloning*, recombinant plasmids containing inserted fragments up to 45 kb long can be efficiently introduced into E. coli cells. A cosmid vector is produced by inserting the COS sequence from λ-phage DNA into a small E. coli plasmid vector about 5 kb long. Like other plasmid vectors discussed earlier, cosmid vectors contain a replication origin (ORI), an anti-biotic-resistance gene (e.g., amp^r), and a polylinker se-quence containing numerous restriction-enzyme recogni-tion sites (Figure 7-14). Next, the cosmid vector is cut with a restriction enzyme and then ligated to 35- to 45-kb re-striction fragments of foreign DNA with complementary sticky ends. If the concentration of foreign DNA is high enough, the ligation reaction generates long DNA mole-cules containing multiple restriction fragments of the for-eign DNA separated by the 5-kb cosmid DNA. These li-gated DNA molecules, which resemble the concatomers that form during replication of λ phage in a host cell, can be packaged in vitro as described earlier.

TABLE 7-2   Maximum Size of DNA That Can Be Cloned in Vectors

Vector Type	Length of Cloned DNA (kb)
Plasmid	20
Bacteriophage λ	25
Cosmid	45
P1 vector	100
YAC (yeast artificial chromosome)	1000

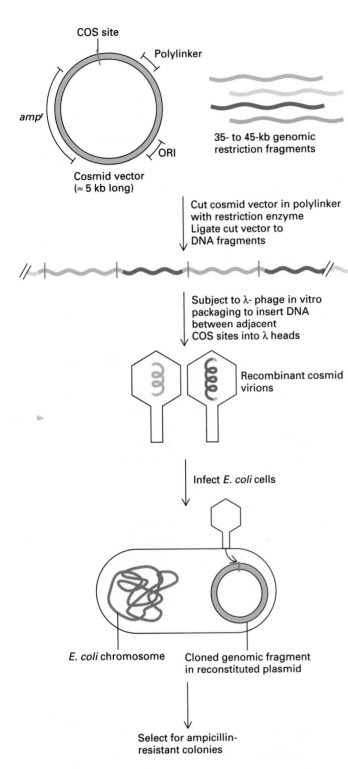

COS site

Polylinker

amp^r

ORI

Cosmid vector
(≈ 5 kb long)

35- to 45-kb genomic
restriction fragments

Cut cosmid vector in polylinker
with restriction enzyme
Ligate cut vector to
DNA fragments

Subject to λ- phage in vitro
packaging to insert DNA
between adjacent
COS sites into λ heads

Recombinant cosmid
virions

Infect E. coli cells

E. coli chromosome   Cloned genomic fragment
in reconstituted plasmid

Select for ampicillin-
resistant colonies

▲ FIGURE 7-14   General procedure for cloning DNA fragments in cosmid vectors. This procedure has the high efficiency associated with λ-phage cloning and permits cloning of restriction fragments up to ≈45 kb long. In this example, four different types of recombinant cosmid virions could be generated, each carrying one of the genomic fragments (green and orange). Plating of the recombinant virions on *E. coli* cells would yield four different types of colonies. Note that the lengths of vector DNA and genomic fragments are not to scale. See text for further discussion.

In the packaging reaction, the λ Nu1 and A proteins bind to COS sites in the ligated DNA and direct insertion of the DNA between two adjacent COS sites into empty phage heads. Packaging will occur as long as the distance between adjacent COS sites does not exceed about 50 kb (the approximate size of the λ genome). Phage tails then are attached to the filled heads, producing viral particles that contain a recombinant cosmid DNA molecule rather than the λ genome. When these virions are plated on a lawn of *E. coli* cells, they bind to phage receptors on the cell surface and inject the packaged DNA into the cells.

Since the injected DNA does not encode any λ proteins, no viral particles form in infected cells and no plaques develop on the plate. Rather, the injected DNA circularizes, forming in each host cell a large plasmid containing the cosmid vector and an inserted DNA fragment. This plasmid replicates and is segregated to daughter cells like other *E. coli* plasmids (see Figure 7-3), and the colonies that arise from transformed cells can be selected on antibiotic plates. The high efficiency of λ-phage infection of *E. coli* cells makes cosmid cloning a practical method of generating plasmid clones carrying DNA fragments up to 45 kb long. Since many genes of higher eukaryotes are on the order of 30–40 kb in length, cosmid cloning increases the chances of obtaining DNA clones containing the entire sequences of genes.

A recently developed approach similar to cosmid cloning makes use of larger *E. coli* viruses such as bacteriophage P1. Recombinant plasmids containing DNA fragments up to ≈100 kb long can be packaged in vitro with the P1 system. Still larger fragments of DNA can be cloned in yeast artificial chromosomes (YACs). To understand how YACs function requires an explanation of the basic elements of eukaryotic chromosomes; this topic is covered in Chapter 9.

▶ Construction of a cDNA Library

In higher eukaryotes, many genes are transcribed into mRNA only in specialized cell types. For example, mRNAs encoding globin proteins are found only in erythrocyte precursor cells, called reticulocytes. Likewise, the mRNA encoding albumin, the major protein in serum, is produced only in liver cells where albumin is synthesized. The specific DNA sequences expressed as mRNAs in a particular cell type can be cloned by synthesizing DNA copies of the mRNAs isolated from that type of cell, and then cloning the DNA copies in plasmid or bacteriophage λ vectors. DNA copies of mRNAs are called *complementary DNAs* (cDNAs); clones of such DNA copies of mRNAs are called *cDNA clones*. In addition to representing only the sequences expressed as mRNAs in a particular cell type, cDNA clones have the advantage that introns present in genomic DNA clones are absent from cDNA clones. Just as a large collection of clones containing fragments of genomic DNA representing the entire genome of a species is called a genomic library, a large collection of cDNA copies of all the mRNAs in a cell type is called a *cDNA library*.

cDNAs Are Produced by Copying Isolated mRNAs with Reverse Transcriptase

The first step in preparing a cDNA library is to isolate the total mRNA from the cell type or tissue of interest. Nature has greatly simplified the isolation of eukaryotic mRNAs: the 3′ end of nearly all eukaryotic mRNAs consists of a string of 50–250 adenylate residues, called a *poly-A tail*. Because of their poly-A tail, mRNAs can be easily separated from the much more prevalent rRNAs and tRNAs present in a cell extract by use of a column to which short strings of thymidylate (oligo-dTs) are linked to the matrix (Figure 7-15). When a cell extract is passed through an oligo-dT column, the mRNA poly-A tails base-pair with the oligo-dTs, binding the mRNAs to the column. Since rRNAs, tRNAs, and other molecules do not bind to the column, they can be washed away. The bound mRNAs are recovered by elution with a low-salt buffer.

The enzyme *reverse transcriptase*, which is found in retroviruses (see Figure 6-23) is then used to synthesize a strand of DNA complementary to each mRNA molecule (Figure 7-16). This enzyme can polymerize deoxynucleoside triphosphates into a complementary DNA strand using an RNA molecule as template. Like other DNA polymerases, this enzyme can only add nucleotides to the 3′ end of a preexisting primer base-paired to the template. Added free oligo-dT serves this function by hybridizing to the 3′ poly-A tail of each mRNA template.

cDNAs Can Be Enzymatically Converted to Double-Stranded DNA and Cloned

After cDNA copies of isolated mRNAs are synthesized, the mRNAs are removed by treatment with alkali, which hydrolyzes RNA but not DNA. The single-stranded cDNAs then are converted to double-stranded DNA molecules. To do this, the 3′ end of each cDNA strand is elongated by adding several residues of a single nucleotide (e.g., dG) through the action of *terminal transferase*, a unique DNA polymerase that does not require a template, but simply adds deoxynucleotides to free 3′ ends. A chemically synthesized oligo-dC primer then is hybridized to this 3′ oligo-dG. A DNA strand complementary to the original cDNA strand then is synthesized by a DNA polymerase, which uses the oligo-dC as a primer. These reactions produce a complete double-stranded DNA molecule corresponding to each of the mRNA molecules in the original preparation. Each double-stranded-DNA contains an oligo-dC–oligo-dG double-stranded region at one end and an oligo-dT–oligo-dA double-stranded region at the other end.

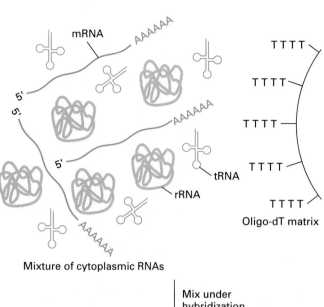

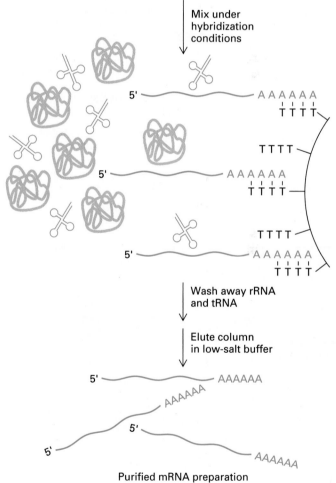

▲ FIGURE 7-15   Isolation of eukaryotic mRNA by oligo-dT column affinity chromatography. Isolated cytoplasmic RNA consists mostly of ribosomal RNAs (rRNAs) and transfer RNAs (tRNAs) shown in orange and blue. The much less abundant mRNAs (red) have 3′ poly-A tails, which hybridize to oligo-dT covalently coupled to the column matrix. After hybridization, the rRNAs and tRNAs are washed out of the column; then the mRNAs are eluted with a low-salt buffer. The resulting purified mRNA preparation contains many different mRNA molecules encoding different proteins.

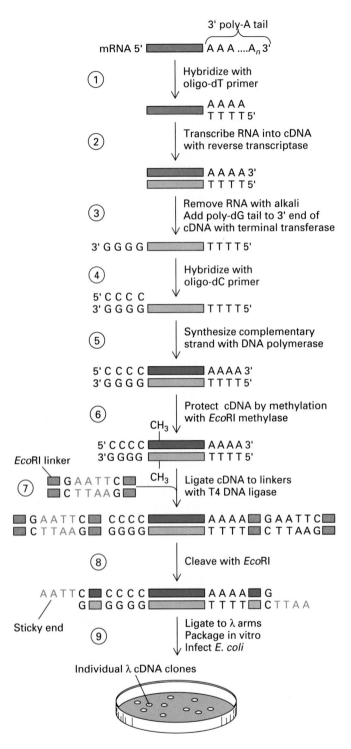

◄ FIGURE 7-16   Preparation of a bacteriophage λ cDNA library. A mixture of mRNAs, isolated as shown in Figure 7-15, is used to produce cDNAs corresponding to all the cellular mRNAs (steps 1–3). The single-stranded cDNAs (light green) are then converted into double-stranded cDNAs, which are treated with *Eco*RI methylase to prevent subsequent digestion by *Eco*RI (steps 4–6). The protected double-stranded cDNAs are ligated to a synthetic double-stranded *Eco*RI-site linker at both ends and then cleaved with the corresponding restriction enzyme, yielding cDNAs with sticky ends (red letters); these are incorporated into λ-phage cloning vectors, and the resulting recombinate λ virions are plated on a lawn of *E. coli* cells (steps 7–9). See text for further discussion.

less efficient than the ligation of DNA fragments with compatible sticky ends. Nonetheless, the ligation reaction can be driven to completion by using high concentrations of linkers. The resulting double-stranded cDNAs, which contain a restriction-site linker at each end, are treated with the restriction enzyme specific for the linker; this generates cDNA molecules with sticky ends at each end. To prevent digestion of any cDNAs that by chance have a recognition sequence for this restriction enzyme within the cDNA sequence, the mixture of double-stranded cDNAs is treated with the appropriate modification enzyme before addition of the linkers. This enzyme methylates specific bases within the restriction-site sequence, preventing the restriction enzyme from digesting the methylated sites (see Figure 7-5b).

The final step in construction of a cDNA library is ligation of the restriction-cleaved double-stranded cDNAs, which now have sticky ends, to plasmid or λ-phage vectors that have been cut to generate complementary sticky ends. The recombinant vectors then are plated on a lawn of *E. coli* cells, producing a library of plasmid or λ clones (see Figure 7-16). Each clone carries a cDNA derived from a single mRNA.

► Identification of Specific Clones in a Genomic or cDNA Library

Suppose you have isolated a particular protein and want to isolate the gene that encodes it. A complete genomic λ library from mammals contains at least a million different clones; a cDNA library must contain as many clones to include the sequences of scarce mRNAs. How are specific clones of interest identified in such large collections? The most common method involves screening a library by *hybridization* with radioactively labeled DNA or RNA *probes*. In an alternative method, called *expression cloning*, a specific clone in a library of cloned DNA is identified based on some property of its encoded protein. In this section, we first discuss the hybridization method and then describe expression cloning.

Next, short double-stranded DNA segments (usually ≈10–12 base pairs long) containing the recognition site for a particular restriction enzyme are ligated onto both ends of the double-stranded cDNAs. These short fragments, called restriction-site *linkers*, are prepared by hybridizing chemically synthesized complementary oligonucleotides. As noted earlier, the DNA ligase from bacteriophage T4 can ligate "blunt-ended" double-stranded DNA molecules that do not have sticky ends. Such blunt-end ligations are

Membrane Hybridization Can Be Used to Screen a Library

As discussed in Chapter 4, under the conditions of temperature and ion concentration found in cells, DNA is maintained as a duplex (two-stranded) structure by the hydrogen bonds that form between the A and T bases and G and C bases in each strand (see Figure 4-7). DNA duplexes can be denatured (melted) into single strands by heating them, usually in a dilute salt solution (e.g., 0.01 M NaCl), or by raising the pH above 11. If the temperature is lowered and the ion concentration in the solution is raised, or if the pH is lowered to neutrality, the A-T and G-C base pairs reform between complementary single strands (see Figure 4-11). This process goes by many names: renaturation, reassociation, hybridization, annealing. In a mixture of nucleic acids, only complementary single strands (or strands containing complementary regions) will reassociate; the extent of their reassociation is virtually unaffected by the presence of noncomplementary strands. Such *molecular hybridization* can take place between two complementary strands of either DNA or RNA, or between an RNA strand and a DNA strand.

To utilize molecular hybridization in the detection of specific DNA clones, single-stranded DNA from recombinant DNA molecules is attached to a solid support, commonly a nitrocellulose filter or treated nylon membrane (Figure 7-17). When a solution containing single-stranded nucleic acids is dried on such a membrane, the single strands become irreversibly bound to the solid support in a manner that leaves most of the bases available for hybridizing to a complementary strand. Although the chemistry of this irreversible binding is not well understood, the procedure is very useful. The membrane is then incubated in a solution containing a radioactively labeled single-stranded DNA (or RNA) that is complementary to some of the nucleic acid bound to the membrane. Under hybridization conditions (near neutral pH, 40–65°C, 0.3–0.6 M NaCl), this labeled probe hybridizes to the complementary nucleic acid bound to the membrane. Any excess probe that does not hybridize is washed away, and the labeled hybrids are detected by autoradiography of the filter.

The procedure for screening a λ library with this membrane-hybridization technique is outlined in Figure 7-18. The recombinant λ virions present in plaques on a lawn of *E. coli* are transferred to a nylon membrane by placing the membrane on the surface of the petri dish. Many of the viral particles in each plaque absorb to the surface of the membrane, but many virions remain in the plaques on the surface of the nutrient agar in the petri dish. In this way a *replica* of the petri dish containing a large number of individual λ clones is reproduced on the surface of the membrane. The original petri dish is refrigerated to store the collection of λ clones. The membrane is then incubated in an alkaline solution, which disrupts the virions, releasing and denaturing the encapsulated DNA. The membrane is

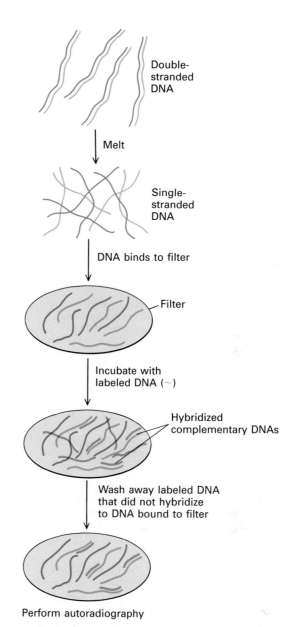

▲ FIGURE 7-17   Membrane-hybridization assay for detecting complementary regions of DNA (and RNA). This general assay, which is widely used in recombinant DNA technology, permits as little as 1 part in 10^6 of a particular DNA or RNA to be detected.

then dried, fixing the recombinant λ DNA to the membrane's surface. Next, the membrane is incubated with a radiolabeled probe under hybridization conditions. Unhybridized probe is washed away, and the filter is subjected to autoradiography.

The appearance of a spot on the autoradiogram indicates the presence of a recombinant λ clone containing DNA complementary to the probe. The position of the spot on the autoradiogram corresponds to the position on the original petri dish where that particular clone formed a

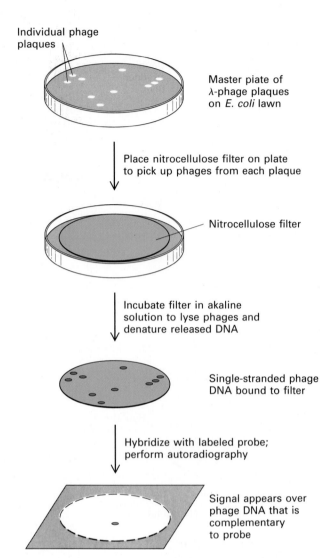

Individual phage plaques

Master plate of λ-phage plaques on *E. coli* lawn

Place nitrocellulose filter on plate to pick up phages from each plaque

Nitrocellulose filter

Incubate filter in akaline solution to lyse phages and denature released DNA

Single-stranded phage DNA bound to filter

Hybridize with labeled probe; perform autoradiography

Signal appears over phage DNA that is complementary to probe

▲ FIGURE 7-18   Identification of a specific clone from a λ-phage library by membrane hybridization to a radiolabeled probe. The position of the signal on the autoradiogram identifies the desired plaque on the plate. In practice, in the initial plating of a library the plaques are not allowed to develop to a visible size so that up to 50,000 recombinants can be analyzed on a single plate. Phage particles from the identified region of the plate are isolated and replated at low density so that the plaques are well separated. Then pure isolates can be obtained by repeating the plaque hybridization as shown in the figure.

plaque. Since the original petri dish still contains many infectious virions in each plaque, viral particles from the identified clone can be recovered for replating by aligning the autoradiogram and the petri dish and removing viral particles from the clone corresponding to the spot. A similar technique can be applied for screening a plasmid library in *E. coli* cells.

Certain cDNAs and Synthetic Oligonucleotides Are Used as Probes

Identification of specific clones by the membrane-hybridization technique depends on the availability of labeled probes that will hybridize with specific DNA sequences. The method used to prepare a particular probe depends on whether the gene of interest encodes a protein that is expressed at a high or low level.

cDNA Probes Prepared from Abundant mRNAs   In some cases, a particular mRNA is sufficiently abundant in a particular tissue or cell type that it can be readily isolated and used directly to prepare a probe. For example, over 90 percent of the mRNA isolated from reticulocytes encodes α or β globin. Specific probes for the α- and β-globin genes can be prepared by isolating the globin mRNAs from reticulocytes on an oligo-dT column (see Figure 7-15) and then synthesizing the cDNA with reverse transcriptase in the presence of radiolabeled nucleoside triphosphates. Because of the ease with which globin probes can be prepared, the genes encoding the α and β globins were among the first to be cloned. This same approach has been used to prepare probes directly from several other mRNAs expressed at high levels in certain tissues or cell types.

Synthetic Oligonucleotide Probes   In many cases, however, a gene of interest encodes a protein that is not expressed at a high level. For example, the enzyme β-N-hexosaminidase A, which is defective in the inherited disorder Tay-Sachs disease (see Figure 5-42) is present at a very low concentration in human cells. Likewise, the mRNA encoding this enzyme is so rare that it cannot be isolated directly and used to prepare a cDNA probe. Nonetheless, specific oligonucleotide probes for the genes encoding this and other low-abundance proteins can be chemically synthesized based on their amino acid sequence.

Because an oligonucleotide probe containing only ≈20 nucleotides is sufficient to screen a library, only a small portion (consisting of six or seven amino acids) of the total protein needs to be sequenced. To prepare a specific probe by this method, the protein of interest usually is purified by sequential column chromatography and SDS-polyacrylamide gel electrophoresis (Chapter 3). The purified protein is digested with one or more proteases (e.g., trypsin) into specific peptides (Figure 7-19). The N-terminal amino acid sequences of a few of these peptides is determined by sequential Edman degradation (see Figure 3-6). Based on the genetic code, oligonucleotide probes encoding the determined peptide sequences can be synthesized and radiolabeled (see Figure 7-9).

In choosing which peptide sequences to use in the design of oligonucleotide probes, the degeneracy of the genetic code must be considered (see Table 4-4). As a result of this degeneracy, many amino acids are encoded by multiple

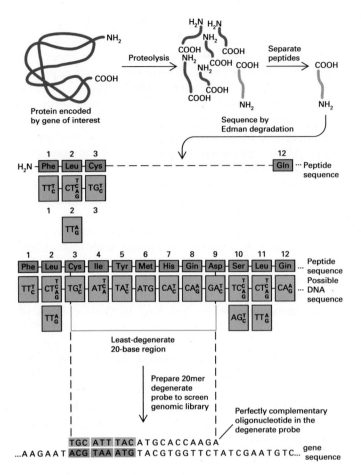

▲ FIGURE 7-19   Designing oligonucleotide probes based on protein sequence. An isolated protein is digested with a selective protease such as trypsin, which specifically cleaves peptide bonds on the carboxyl-terminal side of lysine and arginine residues. The resulting peptides are separated, and several are partially sequenced from their N-terminus by sequential Edman degradation (see Figure 3-6b). The determined sequences then are analyzed to identify the 6- or 7-amino acid region that can be encoded by the smallest number of possible DNA sequences. Because of the degeneracy of the genetic code, the 12-aa sequence (light green) shown here theoretically could be encoded by any of the DNA triplets below it, with the possible alternative bases at the same position indicated. For example, Phe-1 is encoded by TTT or TTC; Leu-2 is encoded by one of six possible triplets (CTT, CTC, CTA, CTG, TTA, or TTG). The region with the least degeneracy for a sequence of 20 bases (20mer) is indicated by the dark red bracket. There are 48 possible DNA sequences in this 20-base region that could encode the peptide sequence Cys-Ile-Tyr-Met-His-Gln-Asp. Since the actual sequence of the gene is unknown, a degenerate 20mer probe consisting of a mixture of all the possible 20-base oligonucleotides is prepared. If a genomic library is screened with this degenerate probe, the one oligonucleotide that is perfectly complementary to the actual gene sequence (blue) will hybridize to it.

codons. Since the specific codons used to encode particular amino acids in the protein of interest are unknown, oligonucleotides containing all possible combinations of codons must be synthesized to assure a perfect match with the gene. For this reason, peptides containing arginine, leucine, or serine are to be avoided if possible, since six different codons encode each of these amino acids. The best amino acids for making such probes are tryptophan and methionine (one codon each); the next best are phenylalanine, tyrosine, histidine, aspartic acid, glutamic acid, asparagine, cysteine, lysine, and glutamine (two codons each).

Once several peptides have been sequenced, the 6- or 7-aa stretch that can be encoded by the smallest number of possible DNA sequences is determined. This approach is illustrated in Figure 7-19 for a partial amino acid sequence. For example, the amino acids in the sequence extending from position 3 through 8 (Cys-Ile-Tyr-Met-His-Gln) can be encoded by 2, 3, 2, 1, 2, and 2 possible codons, respectively. Consequently, 48 (= $2 \times 3 \times 2 \times 1 \times 2 \times 2$) different 18-base DNA sequences could encode this one sequence of amino acids. The GA added at the 3′ end of these 18-base sequences must be complementary to the gene since the next amino acid in this peptide, Asp-9, is encoded by two codons that both start with GA. To be certain of obtaining a probe based on this amino acid sequence that hybridizes perfectly to the unique sequence present in the gene, all 48 of the 20mer probes must be synthesized. A mixture of 20mer probes based on any other portion of this peptide sequence would have to contain considerably more than 48 oligonucleotides, because of the presence of leucine and/or serine residues, each encoded by six different codons.

In practice, rather than separately synthesizing all possible oligonucleotides that can encode the selected portion of a peptide sequence, researchers usually synthesize a mixture of probes at one time. This can be done by adding more than one nucleotide precursor to the synthesis reaction at those points in the sequence that can be encoded by alternative bases. Such a mixture of oligonucleotides is often called a "degenerate" probe, because it is based on the degeneracy of the genetic code. The final step in preparing this type of probe is to radiolabel the oligonucleotides, usually by transferring a ^{32}P-labeled phosphate group from ATP to the 5′ end of each oligonucleotide using *polynucleotide kinase* (Figure 7-20).

In the most common approach, a radiolabeled degenerate probe is used to screen a λ cDNA library using the membrane-hybridization technique. This screening will identify clones that hybridize to the perfectly complementary oligonucleotide present in the probe mixture. Under the usual experimental conditions, oligonucleotides that differ from the cDNA sequence at one or two bases also will hybridize. The nucleotide sequence of the entire cDNA clone can then be determined by methods described in a later section. From the nucleotide sequence the complete

$$[\gamma\text{-}^{32}\text{P}]\text{ATP} \quad \text{5' end of oligonucleotide} \xrightarrow[\text{kinase}]{\text{polynucleotide}} \text{ADP} \quad 32\text{P-labeled oligonucleotide}$$

$$\underset{\alpha\ \beta\ \gamma}{\text{A p p p}} + \text{HONpNpN...} \xrightarrow{\hspace{2cm}} \text{App} + \text{pNpNpN...}$$

▲ **FIGURE 7-20** Radiolabeling of an oligonucleotide at the 5' end with phosphorus-32. The three phosphate groups in ATP are designated the α, β, and γ phosphates in order of their position away from the ribose ring. ATP containing the radioactive isotope ^{32}P in the γ-phosphate position is called $[\gamma\text{-}^{32}\text{P}]$ATP. Kinase is the general term for enzymes that transfer the γ-phosphate of ATP to specific substrates. Polynucleotide kinase can transfer the ^{32}P-labeled γ-phosphate of $[\gamma\text{-}^{32}\text{P}]$ATP to the 5' end of a polynucleotide chain (either DNA or RNA). This reaction is commonly used to radiolabel synthetic oligonucleotides.

amino acid sequence of the encoded protein can be predicted. The nucleotide sequence can be compared with the other peptide sequences initially determined from the peptide fragments to verify that the correct cDNA clone has been isolated.

Once a cDNA clone is obtained, radiolabeled cDNA can be prepared and used to probe a genomic library. This identifies cloned fragments of genomic DNA that contain the gene encoding the protein. Because the genes of higher eukaryotes, often contain large introns, the total length of the gene may be considerably greater than the corresponding cDNA. The genomic DNA also includes sequences outside of the protein-coding region that regulate the transcription of the gene and sequences of the primary RNA transcript required for RNA splicing and polyadenylation.

Expression Cloning Identifies Specific Clones Based on Properties of the Encoded Proteins

Genomic and cDNA libraries can also be screened for the properties of a specific protein encoded in the cloned DNA. This approach uses special cloning vectors, called λ *expression vectors,* in which the cloned DNA is transcribed into the complementary mRNA, which in turn is translated into the encoded protein. For example, λ-phage vectors have been constructed so that the junction of inserted DNA lies in a region of the vector that is transcribed and translated at a high rate. Cloned DNA inserted at this position is transcribed into mRNA in every cell infected by this type of vector. If the cloned DNA contains a protein-coding sequence inserted in the same reading frame as the vector protein, a *fusion protein* is synthesized with an amino terminus encoded by the vector and the remainder of the fusion protein encoded by the cloned DNA (Figure 7-21).

When replica nitrocellulose filters are prepared from a recombinant library constructed in a λ expression vector, fusion proteins expressed from each individual clone are bound to the nitrocellulose filter. The replica filter can be screened by procedures capable of detecting specific fusion proteins. For example, a monoclonal antibody specific for a protein of interest can be incubated with replica filters of a λ cDNA expression library. If one of the λ clones expresses a fusion protein that includes the region of the protein bound by that monoclonal antibody, antibody molecules will bind to the filter at the position of that specific clone. After washing the filter to remove unbound antibody, the position of the specific clone is detected by incubation with a second radioactively labeled antibody that recognizes the first antibody, followed by autoradiography of the filter.

In this method, termed *expression cloning,* any molecule that binds to a protein of interest with high affinity and specificity can be labeled and used as a probe to identify clones expressing the interacting protein. For instance, expression cloning has been useful in identifying cDNA clones encoding proteins that bind to specific DNA sequences; many such proteins are involved in controlling transcription. In this case, a labeled synthetic double-stranded DNA probe is incubated with replica filters prepared from a cDNA library cloned into a λ expression vector. Binding of the labeled DNA by fusion proteins locates the positions of desired clones on the original filter. As described in a later section, other types of expression vectors can be used to produce large amounts of a protein from a cloned gene.

▶ Analyzing and Sequencing Cloned DNA

Once a particular genomic DNA fragment or cDNA has been cloned, it can be separated from the vector DNA and analyzed. The most complete characterization of a cloned DNA requires determination of its nucleotide sequence; from this sequence, the amino acid sequence of the encoded protein can be deduced. The sequence of genomic DNA includes introns as well as exons and regions that control gene expression by determining the type of cell in which the encoded protein is expressed as well as the time in development and the amount of protein produced. Genomic DNAs also include replication origins and sequences important in determining how the DNA associates with proteins in chromosomes. In subsequent chapters, we consider how cells use DNA sequences for these functions. In this section, techniques for characterizing and finally sequencing cloned DNA are outlined.

Cleavage with an Appropriate Restriction Enzyme Separates a Cloned DNA from Its Vector

The first step in analyzing a cloned DNA is to separate it from the plasmid or λ vector carrying it. This can be done

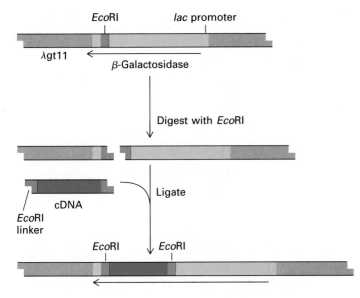

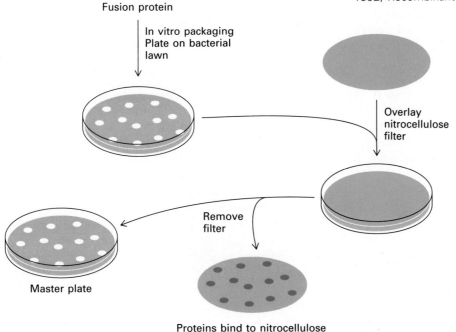

◄ FIGURE 7-21   Use of λ expression cloning to identify a cloned DNA based on binding of the encoded protein to a monoclonal antibody. The λgt11 vector was engineered to express the *E. coli* protein β-galactosidase at high level. The only *Eco*RI recognition site (red) in this vector lies near the 3' end of the β-galactosidase gene. If a cDNA (green), or protein-coding fragment of genomic DNA, is inserted into this *Eco*RI site in the correct orientation and proper reading frame, it will be expressed as a fusion protein in which most of the β-galactosidase sequence is at the N-terminal end and the protein sequence encoded by the inserted DNA is at the C-terminal end. Plaques resulting from infection with recombinant λgt11 will contain high concentrations of such fusion proteins. These proteins can be transferred and bound to a replica filter, which then is incubated with a monoclonal antibody (blue) that recognizes the protein of interest. Rinsing the filter washes away antibody molecules that are not bound to the specific fusion protein attached to the filter. Bound antibody usually is detected by incubating the filter with a second radiolabeled antibody (dark red) that binds to the first antibody. Any signals that appear on the autoradiogram are used to locate plaques on the master plate containing the gene of interest. [Adapted from J. D. Watson et al., 1992, *Recombinant DNA*, 2d ed., Scientific American Books.]

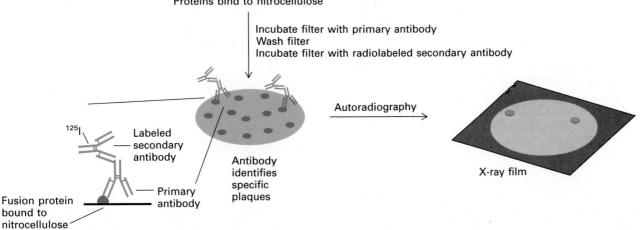

by cleaving the recombinant vector with the same restriction enzyme used to insert the DNA fragment initially. During ligation of a cut vector and DNA fragments generated with the same restriction enzyme, the restriction recognition sequence is regenerated between the DNA fragments and vector (see Figure 7-8). Subsequent treatment with the same restriction enzyme will cut the recombinant vector at the same sites, releasing the vector and cloned DNA, which then can be separated by gel electrophoresis.

Gel Electrophoresis Resolves DNA Fragments of Different Size

As described in Chapter 3, gel electrophoresis is a powerful method for separating proteins according to size. Gel electrophoresis also is used to separate DNA and RNA molecules by size. DNA and RNA molecules are highly charged near neutral pH because the phosphate group in each nucleotide contributes one negative charge. As a result, DNA and RNA molecules move toward the positive electrode during gel electrophoresis. Smaller molecules move through the gel matrix more readily than larger molecules, so that molecules of different length, such as restriction fragments, separate (Figure 7-22). Because the gel matrix restricts random diffusion of the molecules, molecules of different length separate into "bands" whose width equals that of the well into which the original protein mixture was placed. The resolving power of gel electrophoresis is so great that single-stranded DNA molecules up to about 500 nucleotides long can be separated if they differ in length by only one nucleotide. This high resolution is a critical aspect of the procedures for sequencing DNA described later.

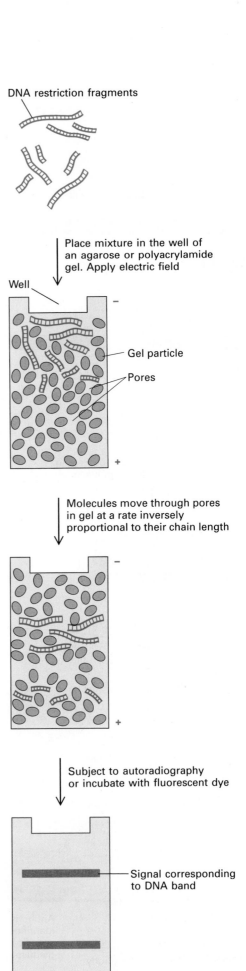

➤ FIGURE 7-22   Separation of DNA fragments of different lengths by gel electrophoresis. A gel is prepared by pouring a liquid containing either melted agarose or unpolymerized acrylamide between two glass plates a few millimeters apart. As the agarose solidifies or the acrylamide polymerizes into polyacrylamide, a gel matrix forms consisting of particles and interconnecting channels, or pores, whose size depends on the concentration of the gel material. Because the pores are larger in agarose gels than in polyacrylamide gels, large DNA fragments are separated on agarose gels and small DNA fragments on polyacrylamide gels. The mixture of DNA fragments to be separated is layered in a well at the top of the gel and an electric current is passed through the gel. DNA fragments move toward the positive pole at a rate inversely proportional to their length, forming bands that can be visualized by autoradiography (if the fragments are radiolabeled) or by treatment with a fluorescent dye. Agarose gels can be run in a horizontal direction; in this case, the melted agarose is allowed to harden on a single horizontal glass or plastic plate. This is not easily done with polyacrylamide gels because oxygen in the atmosphere inhibits polymerization of acrylamide.

FIGURE 7-23 Structural formula of ethidium, a fluorescent dye commonly used to detect DNA and RNA in agarose and polyacrylamide gels. The bromide salt of ethidium generally is used.

Two methods are common for visualizing separated DNA bands on a gel. If the DNA is not radiolabeled, the gel is incubated in a solution containing the fluorescent dye ethidium, a planar molecule that binds to DNA by intercalating between the base pairs (Figure 7-23). This binding concentrates the ethidium in the DNA and also increases its intrinsic fluorescence. As a result, when the gel is illuminated with ultraviolet light, the regions of the gel containing DNA fluoresce much more brightly than the regions of the gel without DNA (Figure 7-24a). Radioactively labeled DNA can be visualized by autoradiography of the gel. In this case, the gel is laid against a sheet of photographic film

in the dark, exposing the film at the positions where labeled DNA is present. When the film is developed, a photographic image of the DNA is observed (Figure 7-24b).

Multiple Restriction Sites Can Be Mapped on a Cloned DNA Fragment

In addition to separating restriction fragments of different lengths, gel electrophoresis provides a means for estimating the length of fragments. The distance that a restriction fragment migrates in a gel is inversely proportional to the logarithm of its length. Thus the length of a restriction fragment can be determined fairly accurately by comparison with restriction fragments of known length subjected to electrophoresis on the same gel (see Figure 7-24a).

The ability to determine the length of restriction fragments makes it possible to locate the positions of restriction sites relative to each other on a DNA molecule (e.g., a newly cloned DNA fragment). A diagram showing the positions of restriction sites on a DNA molecule is called a *restriction map,* and the process of determining these positions is called *restriction-site mapping.* Figure 7-25 illustrates the procedure for mapping two restriction sites relative to each other when only one copy of each site is present in a fragment. In this simple case, three fragment

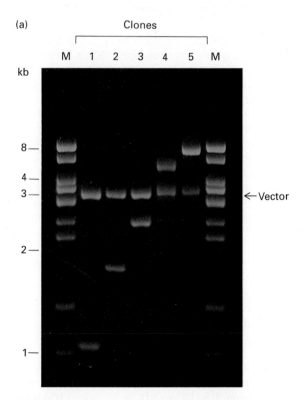

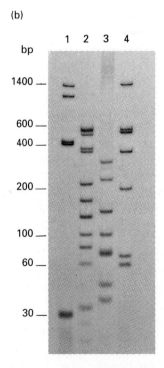

FIGURE 7-24 Separation of restriction fragments by gel electrophoresis. (a) Several different plasmid clones were digested with *Eco*RI and the digested DNA was subjected to agarose gel electrophoresis to separate the cloned fragments from the plasmid vector DNA. Each *Eco*RI-cut plasmid was layered on a separate well. As a "marker" for estimating the sizes of the cloned fragments, adenovirus 2 (Ad2) was

digested with *Hind*III (see Figure 7-6) and layered in the left well of the gel. The lengths of the Ad2 *Hind*III fragments were calculated from the Ad2 DNA sequence. (b) Autoradiogram of 32 p-labeled fragments separated by polyacrylamide gel electrophoresis. Lengths of fragments in the left lane are indicated in base pairs. Parts (a) and (b) courtesy of Carol Eng.

(a)

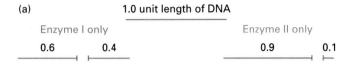

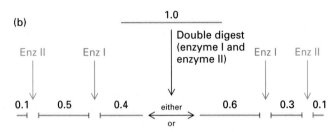

(b)

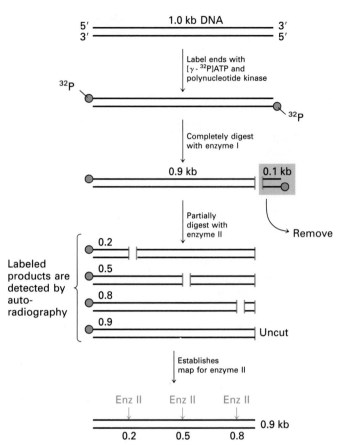

▲ FIGURE 7-25   Mapping the recognition sites for two restriction enzymes relative to each other in a DNA fragment containing one copy of each site. For simplicity, double-stranded DNA is represented by a single line. (a) The fragment is exposed separately to two restriction enzymes (I and II). Each enzyme cuts the fragment once, generating two subfragments, whose lengths are determined by gel electrophoresis. (b) The fragment also is digested with both enzymes simultaneously. Since the lengths of the resulting fragments will depend on the relative position of the two restriction sites, the sites can be mapped based on the lengths observed. By continuing this process with different pairs of enzymes, the investigator can construct a detailed map of restriction sites.

▲ FIGURE 7-26   Mapping the multiple copies of the recognition site for a given restriction enzyme in a DNA fragment. The fragment first is doubly end-labeled (red circles) and then is digested completely with an appropriate restriction enzyme (I) to produce fragments labeled at *one end* only. The larger fragment is partially digested with a second enzyme (II), so that no more than one cut is made in any fragment molecule. Complete digestion would generate only one labeled product (here, the 0.2-kb subfragment), whereas partial digestion generates a labeled product for each restriction site. From the lengths of the labeled pieces, the positions of the multiple recognition sites for enzyme II (red arrows) in the original DNA can be inferred. [See H. O. Smith and M. Birnstiel, 1976, *Nucl. Acids Res.* **3**:2387.]

samples are digested: one with enzyme I, one with enzyme II, and one with both enzymes.

When a DNA fragment contains multiple copies of the recognition site for one or more restriction enzymes, the mapping procedure is more complicated. In this case, the sites for each enzyme must be mapped before the sites for different enzymes can be mapped relative to each other. The first step in this procedure is radiolabeling both ends of the fragment with [γ-^{32}P]ATP and polynucleotide kinase (see Figure 7-20). As shown in Figure 7-26, the doubly end-labeled fragment is treated with a restriction enzyme that cuts the fragment just once, and the resulting singly end-labeled fragments are separated. These fragments then are partially digested with the enzyme whose multiple recognition sites are being mapped.

Multiple restriction sites in a cloned DNA fragment can be mapped by use of these two methods with multiple restriction enzymes. Each distinct DNA sequence has a characteristic restriction-site map (see Figure 7-6). Such maps can be used to align partially overlapping cloned DNA fragments. Also, specific small regions within a large cloned DNA fragment can be prepared by digesting the cloned fragment with various combinations of restriction enzymes; the smaller subfragments then can be isolated by gel electrophoresis.

Pulsed-Field Gel Electrophoresis Separates Large DNA Molecules

The gel electrophoretic techniques described so far can resolve DNA fragments up to ≈20 kb in length. Larger DNAs, ranging from 2×10^4 to 10^7 base pairs (20 kb to 10 megabases [Mb]) in length, can be separated by size with *pulsed-field gel electrophoresis*. This technique depends on the unique behavior of large DNAs in an electric field that is turned on and off (pulsed) at short intervals.

When an electric field is applied to large DNA molecules in a gel, the molecules migrate in the direction of the field and also stretch out lengthwise. If the current then is stopped, the molecules begin to "relax" into random coils. The time required for relaxation is directly proportional to the length of a molecule. The electric field then is reapplied at 90° or 180° to the first direction. Longer molecules relax less than shorter ones during the time the current is turned off; as a result, longer molecules start moving in the direction imposed by the new field more slowly than shorter ones. Repeated alternation of the field direction gradually forces large DNA molecules of different size farther and farther apart.

Pulsed-field gel electrophoresis is very important for purifying long DNA molecules up to $\approx 10^7$ base pairs in length (Figure 7-27). The technique is required for analyzing cellular chromosomes, which range in size from about 5×10^5 base pairs (smallest yeast chromosomes) to 2–3×10^8 base pairs (animal and plant chromosomes). Very large chromosomes must be digested into fragments of 10^7 base pairs or less before they can be analyzed. Such large restriction fragments can be generated with restriction enzymes that cut at rarely occurring 8-bp restriction sites.

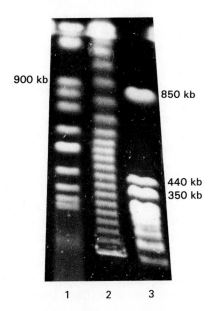

▲ FIGURE 7-27 Pulsed-field gel electrophoretic separation of large DNA molecules. Lane 1 shows individual DNA molecules; each band represents one chromosome from the yeast *S. cerevisiae*. Lane 3 shows *Not*I restriction fragments of the *E. coli* chromosome, which were used in mapping this genome. Lane 2 shows a "ladder" of concatemers of λ-phage DNA in which each unit is ≈48.5 kb long (see Figure 7-11). The band at the bottom is a single unit; the other bands are successively ≈48.5 kb longer. This ladder can be used to estimate the length of long DNA fragments electrophoresed in parallel. [See C. L. Smith et al., 1987, *Nucl. Acids Res.* **15**:4481. Photograph courtesy of C. L. Smith.]

Nucleotide Sequencing of Cloned DNA Fragments Paves the Way for Sequencing Entire Genomes

Virtually all the information required for the growth and development of an organism is encoded in the DNA of its genome. The availability of techniques to produce and separate DNA restriction fragments a few hundred nucleotides long led to development of two procedures for determining the exact nucleotide sequence of stretches of DNA up to ≈500 nucleotides long. These DNA sequencing methods together with the technology for constructing a library representing the entire genome of an organism make it possible to determine the exact sequence of the entire DNA of that organism.

The total genomes of many viruses and much of the *E. coli* genome have already been sequenced. Automation of the techniques for sequencing DNA and computerized storage of the sequence data are facilitating the current effort to determine the sequence of the entire human genome. Within the next decade, if not sooner, researchers also are likely to complete sequencing the entire genomes of several important experimental organisms including the yeast *Saccharomyces cerevisiae*, the round worm *Caenorhabditis elegans*, the fruit fly *Drosophila melanogaster*, and the mouse. Knowledge of these DNA sequences will undoubtedly revolutionize our understanding of how cells and organisms function.

Maxam-Gilbert Method In the late 1970s, A. M. Maxam and W. Gilbert devised the first method for sequencing DNA fragments containing up to ≈500 nucleotides. In this method, four samples of an end-labeled DNA restriction fragment are chemically cleaved at different specific nucleotides. The resulting subfragments are separated by gel electrophoresis, and the labeled fragments are detected by autoradiography. As illustrated in Figure 7-28, the sequence of the original end-labeled restriction fragment can be determined directly from parallel electrophoretograms of the four samples.

Sanger (Dideoxy) Method A few years later, F. Sanger and his colleagues developed a second method of DNA sequencing, which now is used much more frequently than the Maxam-Gilbert method. The Sanger method is also called *dideoxy sequencing* because it involves use of 2′,3′-dideoxynucleoside triphosphates (ddNTPs), which lack a 3′-hydroxyl group (Figure 7-29). In this method, the single-stranded DNA to be sequenced serves as the template strand for in vitro DNA synthesis; a synthetic 5′-end–labeled oligodeoxynucleotide is used as the primer. As shown in Figure 7-30, four separate polymerization reactions are performed, each with a low concentration of one of the four ddNTPs in addition to higher concentrations of the normal deoxynucleoside triphosphates (dNTPs). In each reaction, the ddNTP is randomly incorporated at the positions of the corresponding dNTP; such addition of a

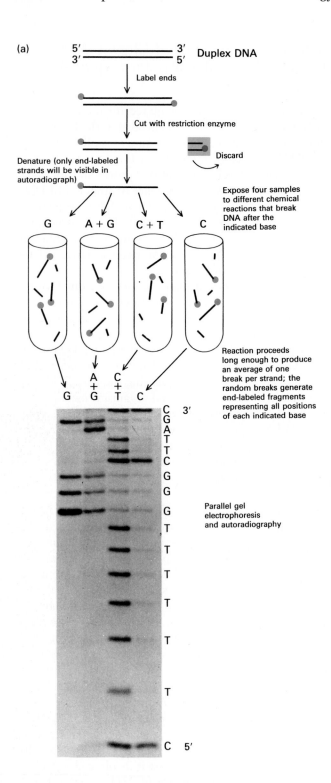

(a)

Duplex DNA

Label ends

Cut with restriction enzyme

Discard

Denature (only end-labeled strands will be visible in autoradiograph)

Expose four samples to different chemical reactions that break DNA after the indicated base

G   A + G   C + T   C

Reaction proceeds long enough to produce an average of one break per strand; the random breaks generate end-labeled fragments representing all positions of each indicated base

Parallel gel electrophoresis and autoradiography

(b)

Modification of a base by dimethylsulfate

Release of the modified base by heat

Backbone scission by alkali

Labeled fragment

▲ FIGURE 7-28   Maxam-Gilbert method for sequencing DNA fragments up to ≈500 nucleotides in length. (a) The double-stranded fragment to be sequenced is labeled at the 5′ ends with ^{32}P (see Figure 7-20). The label (red circle) is removed from one end, and the fragment then is denatured. Four identical samples of the prepared fragment are subjected to four different chemical reactions that selectively cut the DNA backbone at G, G + A, C + T, or C residues. The reactions are controlled so that each labeled chain is likely to be broken only once. An example of the reaction that cleaves at a G is shown in (b). The labeled subfragments created by all four reactions have the label at one end and the chemical cleavage point at the other. Gel electrophoresis and autoradiography of each separate mixture yield one radioactive band for each nucleotide in the original fragment. Bands appearing in the G and C lanes can be read directly. Bands in the A + G lane that are not duplicated in the G lane are read as A. Bands in the T + C lane that are not duplicated in the C lane are read as T. The sequence is read from the bottom of the gel up. [See A. Maxam and W. Gilbert, 1977, *Proc. Nat'l. Acad. Sci. USA* **74**:560. Photograph from L. Stryer, 1988, *Biochemistry*, 3d ed., W. H. Freeman and Company, p. 120; courtesy of Dr. David Dressler.]

ddNTP terminates polymerization because the absence of a 3′ hydroxyl prevents addition of the next nucleotide. The mixture of terminated fragments from each of the four reactions is subjected to gel electrophoresis in parallel; the separated fragments then are detected by autoradiography. The sequence of the original DNA template strand can be read directly from the resulting autoradiogram (see Figure 7-30c). Once the sequence for a particular cloned DNA fragment is determined, primers for overlapping fragments can be chemically synthesized based on that sequence. The sequence of a long continuous stretch of DNA thus can be determined by individually sequencing the overlapping cloned DNA fragments that compose it.

Informatics: The Storage, Distribution and Analysis of DNA Sequence Data   Vast amounts of DNA sequence have already been determined, and the pace at which new sequences are characterized is continuously accelerating. Computers are necessary to store and distribute this enormous volume of data. *Informatics* is the rapidly

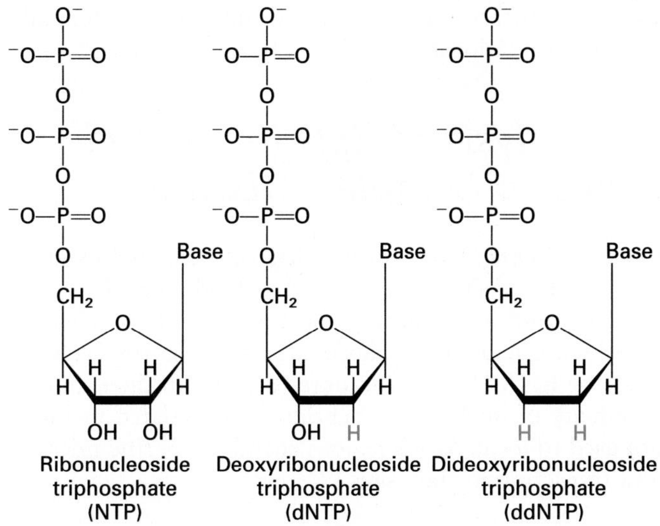

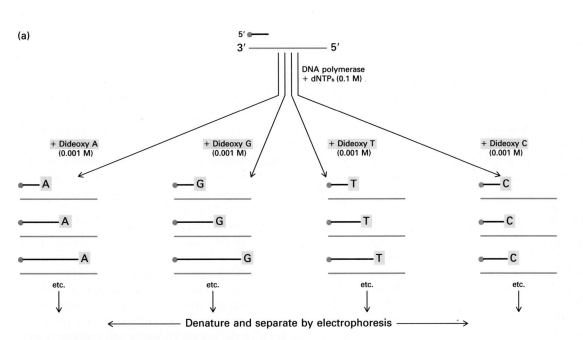

► FIGURE 7-29   Structures of ribonucleoside triphosphate (NTP), deoxyribonucleoside triphosphate (dNTP), and dideoxyribonucleoside triphosphate (ddNTP).

Ribonucleoside triphosphate (NTP)

Deoxyribonucleoside triphosphate (dNTP)

Dideoxyribonucleoside triphosphate (ddNTP)

(a)

5′
3′   5′

DNA polymerase
+ dNTPs (0.1 M)

+ Dideoxy A (0.001 M)   + Dideoxy G (0.001 M)   + Dideoxy T (0.001 M)   + Dideoxy C (0.001 M)

A   G   T   C

A   G   T   C

A   G   T   C

etc.   etc.   etc.   etc.

◄——— Denature and separate by electrophoresis ———►

▲ FIGURE 7-30   Sanger (dideoxy) method for sequencing DNA fragments. (a) A single strand of the DNA to be sequenced (blue line) is hybridized to a 5′-end–labeled synthetic deoxynucleotide primer. The primer is elongated in four separate reaction mixtures containing the four normal deoxynucleoside triphosphates (dNTPs) plus one of the four dideoxynucleoside triphosphates (ddNTPs) in a ratio of 100 to 1. A ddNTP molecule can add at the position of the corresponding normal dNTP, but when this occurs, chain elongation stops because the ddNTP lacks a 3′ hydroxyl. In time, each reaction mixture will contain a mixture of prematurely terminated chains ending at every occurrence of the ddNTP (yellow). (b) Three of the labeled chains that would be generated from the specific DNA sequence shown in the presence of ddGTP. (c) An actual audioradiogram of a polyacrylamide gel in which more than 300 bases can be read. Each reaction was carried out in duplicate using Sequenase™, a commercial preparation of the DNA polymerase from bacteriophage T7. [Part (c) courtesy of United States Biochemical Corporation.]

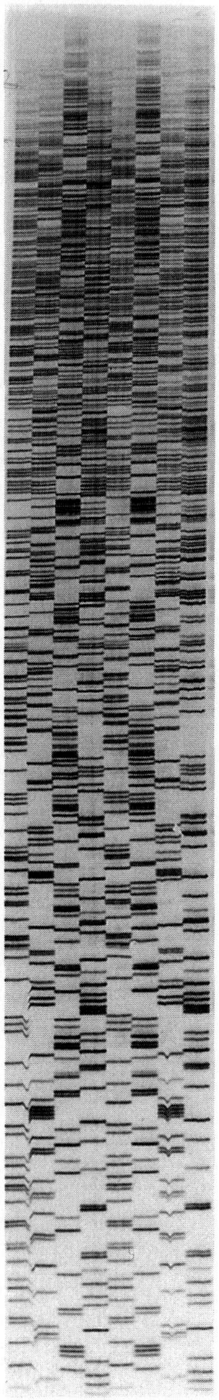

(b)

5′ ³²P-TAGCTGACTC 3′
3′   ATCGACTGAGTCAAGAGCTATTGGGCTTAA . . .

DNA polymerase
+dATP, dGTP, dCTP, dTTP
+ddGTP in low concentration

5′ ³²P-TAGCTGACTCAG 3′
3′   ATCGACTGAGTCAAGAGCTATTGGGCTTAA . . .
+
5′ ³²P-TAGCTGACTCAGTTCTCG 3′
3′   ATCGACTGAGTCAAGAGCTATTGGGCTTAA . . .
+
5′ ³²P-TAGCTGACTCAGTTCTCGATAACCCG 3′
3′   ATCGACTGAGTCAAGAGCTATTGGGCTTAA . . .

(c)

CAGTCAGT

developing area of computer science devoted to collecting, organizing, and analyzing DNA and protein sequences. The principal data banks where such sequences are stored are the GenBank at the Los Alamos National Laboratory, and the EMBL Data Library at the European Molecular Biology Laboratory in Heidelberg. These data bases continuously exchange newly reported sequences and make them available to molecular cell biologists throughout the world through an international computer network. Newly derived sequences can be compared with previously determined sequences to search for similarities, called *homologous sequences*. Protein-coding regions can be translated into amino acid sequences, which also can be compared. Because of degeneracy in the genetic code, related proteins often exhibit more homology than the genes encoding them.

As discussed in Chapter 3, proteins with similar functions often contain homologous sequences that correspond to important functional domains in the three-dimensional structure of the proteins. Discovery that a protein encoded by a newly cloned gene exhibits such homologies with proteins of known function can provide revealing insights into the function of the cloned gene. For example, the human gene *NF1* was identified and cloned by methods described in Chapter 8. This gene is associated with the inherited disease neurofibromatosis 1, which results in the development of multiple tumors in the peripheral nervous system. Before the *NF1* gene was cloned, there was little understanding of the molecular basis of the disease. After a cDNA clone of *NF1* was isolated and sequenced, the deduced sequence of the NF1 protein was checked against all other protein sequences in GenBank. A region of homology was identified with a protein called GAP, which interacts with the Ras protein. This latter protein is encoded by the *ras* gene, which is mutated in many human tumors. As we examine in detail in Chapters 20 and 26, the GAP and Ras proteins normally function to control cell growth and differentiation in response to signals from neighboring cells. Because of its sequence homology with GAP, researchers hypothesized that NF1 also would interact with Ras. The NF1 protein subsequently was expressed from the cloned gene by methods described later and shown to interact with Ras. This finding suggests that in individuals with neurofibromatosis, who have a defective *NF1* gene, a mutant NF1 protein, expressed in cells of the peripheral nervous system, interacts abnormally with Ras to signal inappropriate cell division.

This example illustrates how insight into the molecular basis of an inherited disease can be gained by identifying and cloning the associated mutant gene and then comparing the sequence of the encoded protein with the sequences of other proteins stored in data banks. This general approach for revealing the molecular role of various proteins will undoubtedly increase as the sequences of more proteins with known functions are determined and as new computer methods are devised for identifying potentially significant relationships among sequences.

▶ Analysis of Specific Nucleic Acids in Complex Mixtures

Once a specific DNA sequence has been isolated by cloning, the cloned DNA can be used as a probe to detect the presence, and in some cases the amounts, of complementary nucleic acids in complex mixtures such as total cellular DNA or RNA. These procedures depend on the exquisite specificity of nucleic acid hybridization. Related methods are used to locate DNA regions encoding specific mRNAs and transcription start sites.

Southern Blotting Detects Specific DNA Fragments

The technique of *Southern blotting*, named after its originator Edward Southern, can identify specific restriction fragments in a complex mixture of restriction fragments. The DNA to be analyzed, such as the total DNA of an organism, is digested to completion with a restriction enzyme. For an organism with a complex genome, this digestion may generate millions of specific restriction fragments. The complex mixture of fragments is subjected to gel electrophoresis to separate the fragments according to size. However, many different fragments are of exactly the same length, and these do not separate from each other.

Even though all the fragments are not resolved by gel electrophoresis, an individual fragment that is complementary to a specific DNA clone can be detected. The restriction fragments present in the gel are denatured with alkali and transferred onto a nitrocellulose filter or nylon membrane by blotting (Figure 7-31). This procedure preserves the distribution of the fragments in the gel, creating a replica of the gel on the filter, much like the replica filter produced from plaques of a λ library. The filter then is incubated under hybridization conditions with a specific radiolabeled DNA probe usually generated from a cloned restriction fragment. The DNA restriction fragment that is complementary to the probe hybridizes, and its location on the filter can be revealed by autoradiography.

Southern blotting permits a comparison between the restriction map of DNA isolated directly from an organism and the restriction map of cloned DNA. This is necessary to be certain that no rearrangements have occurred during the cloning procedure such as might happen if two restriction fragments that do not normally lie next to each other were inadvertently ligated together before ligation into a cloning vector. Southern blotting also is used to map restriction sites in genomic DNA next to the sequence of a cloned DNA fragment. This provides a rapid method of

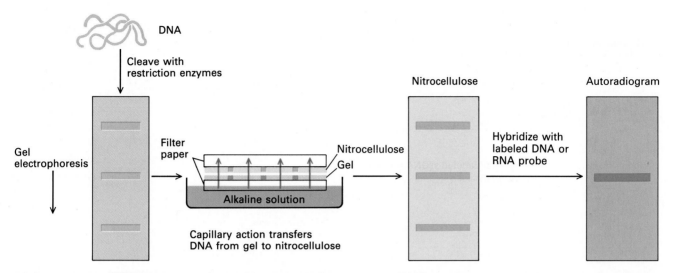

▲ **FIGURE 7-31** The Southern-blot technique for detecting the presence of specific DNA sequences following gel electrophoresis of a complex mixture of restriction fragments. The diagram depicts three restriction fragments in the gel, but the procedure can be applied to a mixture of millions of DNA fragments. [See E. M. Southern, 1975, *J. Mol. Biol.* **98**:508.]

comparing the restriction maps of different individual organisms in the region surrounding a cloned fragment. Deletion and insertion mutations are readily detected, as well as sequence differences in specific restriction sites.

Northern Blotting Detects Specific RNAs

Northern blotting, somewhat humorously named because it is patterned after Southern blotting, is used to detect a particular RNA in a mixture of RNAs. A RNA sample, often the total cellular RNA, is denatured by treatment with an agent (e.g., formaldehyde) that prevents hydrogen bonding between base pairs, ensuring that all the RNA molecules have an unfolded, linear conformation. The individual RNAs then are separated according to size by gel electrophoresis and transferred to a nitrocellulose filter to which the extended denatured RNAs adhere. The filter then is exposed to a labeled DNA probe and subjected to autoradiography. Because the amount of a specific RNA in a sample can be estimated from a Northern blot, the procedure is widely used to compare the amounts of a particular mRNA in cells under different conditions (Figure 7-32).

Nuclease Protection Is Used to Quantitate Specific RNAs and Map the DNA Regions Encoding Them

Another important method for detecting and quantitating specific RNA molecules employs endonucleases that digest single-stranded but not double-stranded nucleic acids. The method was originally designed using endonuclease S1, an enzyme from the mold *Aspergillus oryzae* that digests

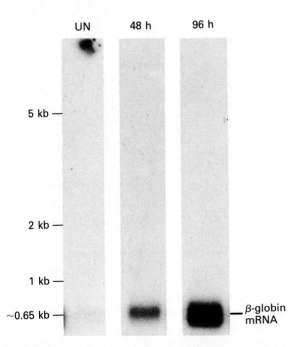

▲ **FIGURE 7-32** Northern blots of β-globin mRNA in extracts of erythroleukemia cells that are growing but uninduced (UN lane) and in cells that are induced to stop growing and allowed to differentiate for 48 h or 96 h. The density of a band is proportional to the amount of mRNA present. The β-globin mRNA is barely detectable in uninduced cells but increases more than 1000-fold after 96 h of differentiation. [Courtesy of L. Kole.]

(a)

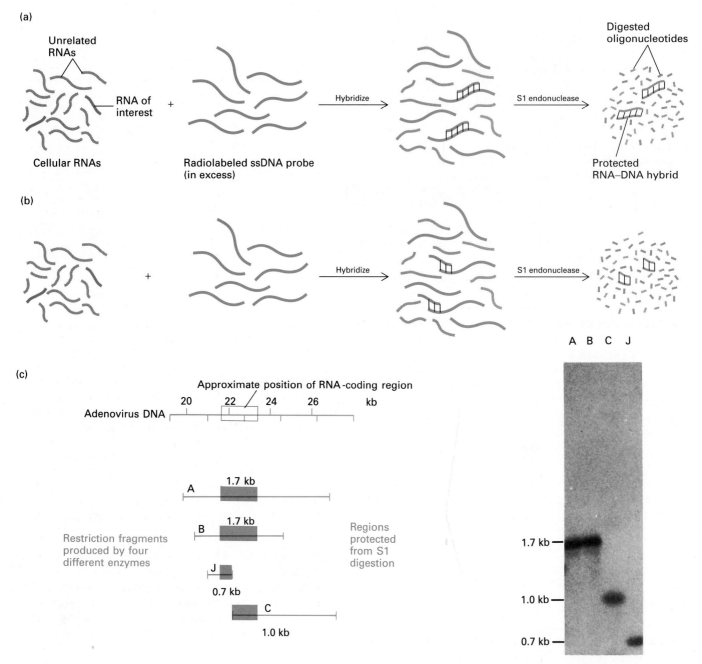

▲ FIGURE 7-33   Nuclease-protection method for quantitating specific RNAs in a mixture and mapping them. (a) A
radiolabeled, single-stranded DNA probe (red) is mixed with a
mixture of cellular RNAs; the probe hybridizes only to the
complementary RNA (blue), which is a small fraction of the
total RNA sample. In this example, the probe contains a
sequence complementary to the entire RNA of interest.
Digestion with S1 endonuclease degrades all the unprotected
(unhybridized) RNA and DNA sequences, leaving a double-
stranded RNA-DNA hybrid equal in length to the RNA. The
protected hybrid is detected by gel electrophoresis followed
by autoradiography. The density of the resulting band is
proportional to the amount of the hybridized RNA in the original mixture. (b) With a "partial" DNA probe, containing only a
portion of the DNA sequence complementary to the RNA,
the protected S1-digestion product is shorter than the RNA

and equal in length to the complementary region of the
probe. (c) An example of mapping an RNA on the genome of
adenovirus. A 1.7-kb RNA was approximately mapped to the
region between 21 and 24 from the left end of the 36-kb
viral genome. Four radiolabeled restriction-fragment probes
(A, B, C, and J) from this region of the viral DNA were prepared, hybridized with RNA from virus-infected cells, and
then treated with S1 endonuclease. An autoradiogram of the
S1-digestion products is shown at the right. Probes A and B
produced S1-digestion products of 1.7 kb, indicating the RNA
sequence maps entirely within these restriction fragments.
The results with the partial probes C and J map the RNA
sequence relative to the restriction site separating fragments
C and J. [Photograph in part (c) from A. J. Berk and P. A.
Sharp, 1977, *Cell* **12**:721; copyright M.I.T.]

single-stranded RNA and DNA but not double-stranded molecules. A labeled DNA strand, or probe, complementary to an RNA of interest is prepared from a cloned DNA. A source of RNA, such as the total polyadenylated RNA isolated from a particular tissue or type of cultured cell, is incubated with a high concentration of the labeled DNA probe under conditions where all the RNA complementary to the probe hybridizes to it (Figure 7-33a). The preparation then is treated with endonuclease S1, which digests all the unhybridized RNA and probe molecules. However, the double-stranded region in the RNA-DNA hybrids is protected from nuclease digestion, but the noncomplementary regions of the probe are trimmed to the length of the complementary RNA. The digested preparation is subjected to gel electrophoresis followed by autoradiography of the gel to detect the protected probe. The amount of radioactivity in the resulting band is a measure of the amount of RNA complementary to the probe in the initial sample of RNA. Nuclease protection also can be performed with a complementary labeled RNA and ribonuclease A, a single-strand–specific pancreatic ribonuclease.

The DNA region that encodes a particular RNA can be mapped with the nuclease-protection technique by use of restriction-fragment probes in which one end is complementary to only a portion of the RNA of interest. In this case, the RNA-DNA hybrid protected from S1 digestion is shorter than the RNA being probed; its length corresponds to that of the DNA region extending from one end of the coding region to a restriction site within it (Figure 7-33b). Comparison of the protected doubled-stranded fragments obtained with two or more such "partial" probes can map the RNA sequence relative to restriction sites in the complementary DNA (Figure 7-33c).

Transcription Start Sites Can Be Mapped by S1 Protection and Primer Extension

As discussed in Chapter 11, some of the DNA regulatory elements that control transcription of a protein-coding sequence (gene) into mRNA are located near the transcription *start site*. Mapping the start site for synthesis of a particular mRNA often helps in identifying the DNA regulatory sequences that control transcription of the corresponding gene. Two methods are used to map the 5′ end of a particular mRNA on a complementary DNA: *S1 protection* and *primer extension*. The first step in both methods is to identify the general region of a DNA that includes the start site of interest (Figure 7-34a); this can be done by Northern-blot analysis or nuclease protection using various cloned restriction fragments.

In the S1-protection method, the identified DNA region is treated with appropriate restriction enzymes to produce a single-stranded DNA fragment that will hybridize with the 5′ portion of the mRNA. This fragment is radiolabeled at the 5′ end, hybridized with the mRNA, and then

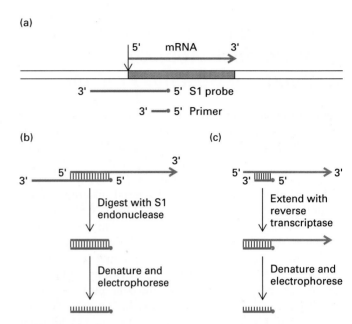

▲ **FIGURE 7-34** Two methods for mapping the start site for transcription of a particular gene in a region of DNA of known sequence. (a) Diagram of the DNA fragment containing the gene of interest (light blue) and the corresponding mRNA (red). The end-labeled (red dot) single-stranded DNA fragment used as a probe in the S1 mapping technique and the end-labeled oligonucleotide primer used in the primer-extension technique are shown below the position of their sequence in the DNA. (b) In the S1 mapping technique, the probe is hybridized with the mRNA, and unpaired nucleic acid is then digested with S1 endonuclease (see Figure 7-33). Denaturation leaves a labeled DNA fragment whose length accurately marks the distance of the starting nucleotide of the mRNA from the nucleotide that hybridized with the labeled DNA end. (c) In the primer-extension technique, a short (approximately 20 nucleotides) oligodeoxynucleotide is synthesized and end-labeled. After the primer is hybridized to the mRNA, it is extended by reverse transcriptase until it reaches the first nucleotide of the mRNA. The length of the primer-extension product, determined by gel electrophoresis, measures the distance from the 5′ end of the primer to the 5′ end of the mRNA.

trimmed with S1 endonuclease (Figure 7-34b). From the length of the labeled probe segment protected from digestion, the position of the start site in the original DNA can be located.

The primer-extension method uses a synthetic oligonucleotide that is complementary to an approximately 20-nucleotide stretch of the mRNA located 50–100 nucleotides from its 5′ end. This synthetic oligonucleotide is end-labeled at the 5′ end and then used to prime DNA synthesis by reverse transcriptase with the mRNA as the template (Figure 7-34c). The position of the start site can be mapped from the length of the resulting extension product.

➤ Designing Expression Systems That Produce Abundant Amounts of Specific Proteins

Many proteins with interesting or useful functions are normally expressed at very low concentrations. A case in point is *granulocyte colony-stimulating factor* (G-CSF). This human protein hormone stimulates the development and replication of granulocytes, the phagocytic white blood cells critical to defense against bacterial infections. Both granulocytes and the cells that produce G-CSF are very sensitive to chemotherapeutic agents used in the treatment of cancer. As a result, one of the most serious side effects of chemotherapy is a fall in the concentration of granulocytes, making patients prone to life-threatening infections. G-CSF is normally made in very low concentrations in the bone marrow where granulocytes differentiate and by some cultured human cell lines. Isolation of G-CSF from these sources is a tedious process yielding minuscule amounts of purified protein. By the techniques described in this section, G-CSF can now be produced in large enough amounts for therapeutic use in cancer patients to diminish the impact of chemotherapy on granulocyte production.

The first step in obtaining large amounts of low-abundance proteins such as G-CSF is to purify a sufficient quantity of the protein to allow determination of a portion of its amino acid sequence. (As discussed in Chapter 3, only very small amounts of a protein are needed for amino acid sequencing.) A synthetic oligonucleotide probe is designed based on the determined amino acid sequence; this probe is used to identify a cDNA encoding the protein of interest, as described previously. Once the desired cDNA is cloned, large amounts of the encoded protein often can be synthesized in engineered *E. coli* cells.

Previously we discussed a λ-phage expression vector that produces fusion proteins consisting of β-galactosidase joined to a protein fragment encoded by a DNA inserted into the vector; this type of expression system can be used to identify DNA clones encoding a specific protein (see Figure 7-21). Here, in contrast, we discuss expression vectors designed to produce full-length proteins at high levels.

Full-Length Proteins Encoded by Cloned Genes Can be Produced in *E. coli* Expression Systems

G-CSF and many other low-abundance proteins can be expressed at high levels in *E. coli* through use of specially designed expression vectors. Although many different expression vectors have been constructed, they all take advantage of the molecular mechanisms that control transcription and translation in *E. coli*.

Plasmid Expression Vectors Carrying a Strong, Regulated Promoter The most commonly used *E. coli* expression vectors are assembled by ligation of a basic plasmid vector, containing a replication origin (ORI) and selectable antibiotic-resistance gene, to a DNA sequence that functions as a strong, regulated *promoter*. A promoter is a DNA sequence where RNA polymerase initiates transcription. At a strong, regulated promoter, transcription is initiated many times per minute under specific environmental conditions.

One example of this type of expression vector contains a cloned fragment of the *E. coli* chromosome that includes the *lac* promoter and the *lacZ* gene encoding β-galactosidase. Transcription from the *lac* promoter only occurs when lactose, or a lactose analog such as isopropyl-thiogalactoside (IPTG), is added to the culture medium. IPTG generally is used because it cannot be metabolized, and therefore its concentration does not change as the cells grow. After addition of IPTG, the *lacZ* gene is transcribed into mRNA, which then is translated to yield many copies of the β-galactosidase protein (Figure 7-35a).

To modify this plasmid for production of G-CSF, the *lacZ* gene is replaced with a cDNA encoding G-CSF using restriction enzymes and DNA ligase. In this process, the *lac* promoter, which is required for efficient transcription, must be maintained just before the start site of the inserted cDNA. In *E. coli* cells transformed by the resulting plasmid, transcription of the G-CSF cDNA and expression of G-CSF protein occurs in the presence of IPTG (Figure 7-35b).

Plasmid Expression Vectors Carrying the T7 Late Promoter A more complicated expression system, involving two levels of amplification, can produce larger amounts of a desired protein than the system just described. This second-generation system depends on the regulated expression of *T7 RNA polymerase*, an extremely active enzyme that is encoded in the DNA of bacteriophage T7. The T7 RNA polymerase transcribes DNA beginning within a specific 23-bp promoter sequence called the *T7 late promoter*. Copies of the T7 late promoter are located at several sites on the T7 genome, but none is present in *E. coli* chromosomal DNA. As a result, in T7-infected cells, T7 RNA polymerase catalyzes transcription of viral genes but not of *E. coli* genes.

In this expression system, recombinant *E. coli* cells first are engineered to carry the gene encoding T7 RNA polymerase next to the *lac* promoter. In the presence of IPTG, these cells transcribe the T7 gene at a high rate and synthesize abundant amounts of T7 RNA polymerase. These cells then are transformed with plasmid vectors that carry a copy of the T7 late promoter and, adjacent to it, the cDNA encoding the desired protein (Figure 7-36). When IPTG is added to the culture medium containing these transformed, recombinant *E. coli* cells, large amounts of

(a)

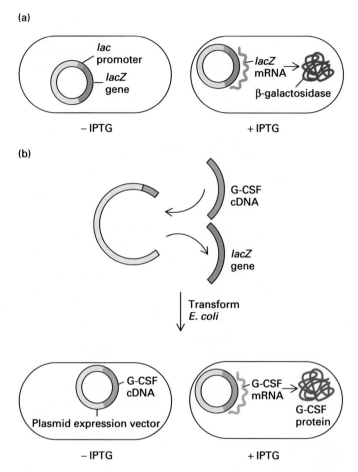

T7 RNA polymerase are produced. The polymerase then binds to the T7 late promoter on the plasmid expression vectors, catalyzing transcription of the inserted cDNA at a high rate. Since each *E. coli* cell contains many copies of the expression vector, prodigious amounts of mRNA corresponding to the cloned cDNA can be produced in this system.

Proteins with Post-Translational Modifications Can Be Produced in Eukaryotic Expression Systems

Most of the enzymes used in recombinant DNA technology (e.g., restriction enzymes, DNA polymerases, DNA ligases, polynucleotide kinase, and reverse transcriptase) are now produced commercially in *E. coli* expression systems. Large quantities of many eukaryotic proteins of interest, such as G-CSF, also can be produced in these systems. However, some eukaryotic proteins cannot be produced in active form in *E. coli* cells. These include proteins that are extensively modified during or following their synthesis, such as glycoproteins to which carbohydrate groups are added. *E. coli* lacks the enzymes that catalyze many of the post-translational modifications found on eukaryotic proteins.

To overcome this limitation of *E. coli* expression systems, researchers have developed *eukaryotic expression vectors* that permit addition of appropriate post-translation modifications to expressed proteins. Such vectors can be used in various types of eukaryotic cells to direct abundant synthesis of eukaryotic proteins from cloned genes.

Proteins Encoded by Cloned Genes and cDNAs Can Be Translated In Vitro

In the *E. coli* and eukaryotic expression systems just described, proteins are synthesized in vivo within living cells.

▲ FIGURE 7-35   A simple *E. coli* expression vector utilizing the *lac* promoter. (a) The expression vector plasmid contains a fragment of the *E. coli* chromosome containing the *lac* promoter and the neighboring *lacZ* gene. In the presence of the lactose analog IPTG, RNA polymerase normally transcribes the *lacZ* gene producing *lacZ* mRNA, which is translated into the encoded protein, β-galactosidase. (b) The *lacZ* gene can be cut out of the expression vector with restriction enzymes and replaced by the G-CSF cDNA. When the resulting plasmid is transformed into *E. coli* cells, addition of IPTG and subsequent transcription from the *lac* promoter produces G-CSF mRNA, which is translated into G-CSF protein.

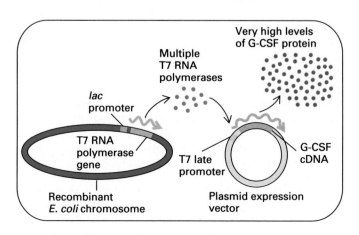

◄ FIGURE 7-36   Two-step expression vector system based on bacteriophage T7 RNA polymerase and T7 late promoter. The chromosome of a specially engineered *E. coli* cell contains a copy of the T7 RNA polymerase gene under the transcriptional control of the *lac* promoter. When transcription from the *lac* promoter is induced by addition of IPTG, the T7 RNA polymerase gene is transcribed, and the mRNA is translated into the enzyme. The T7 RNA polymerase molecules produced then initiate transcription at a very high rate from the T7 late promoter on the expression vector. Multiple copies of the expression vector are present in such cells, although only one copy is diagrammed here. The large quantity of mRNA transcribed from the cDNA cloned next to the T7 late promoter is translated into abundant protein product.

In vitro expression of proteins encoded by cloned DNA is possible with the T7 late-promoter expression vector diagrammed in Figure 7-36. In this system, the mRNA encoded by the cloned DNA is synthesized in vitro using the purified expression vector and purified T7 RNA polymerase. The mRNA then is translated in vitro to yield the desired protein. Although an in vitro expression system yields less protein than an in vivo system, it allows the desired protein to be radioactively labeled. In addition to the T7 system, plasmid vectors containing late promoters and the corresponding RNA polymerases from related bacteriophages (e.g., T3, T5, and SP6) also can be used for in vitro production of proteins from cloned DNA.

Cell extracts are used to translate the in vitro synthesized mRNA into protein. These extracts, which usually are prepared from rabbit reticulocytes or wheat germ, are rich in the components required for translation: ribosome subunits, tRNAs, amino acyl-tRNA synthases, and initiation, elongation, and termination factors (Chapter 4). Extracts first are treated with low concentrations of micrococcal nuclease to eliminate any endogenous mRNAs. At low levels, this enzyme, which is active only in the presence of Ca^{2+}, digests mRNA but not tRNA or rRNA. After digestion of mRNA is complete, the enzyme is inactivated by addition of EGTA, a chelating-agent that binds Ca^{2+} with a much higher affinity than the nuclease does. When an in vitro synthesized mRNA is added to a treated extract, only the corresponding protein is produced, since the endogenous mRNAs have been destroyed. If a radiolabeled amino acid such as [^{35}S]methionine is included in the translation reaction, the protein product will be labeled. The labeled protein can be used in binding experiments to test its interaction with other proteins and in various other experimental systems.

➤ *The Polymerase Chain Reaction: An Alternative to Cloning*

An alternative to cloning, called the *polymerase chain reaction* (PCR), can be used to directly amplify rare specific DNA sequences in a complex mixture when the ends of the sequence are known. This method of amplifying rare sequences from a mixture has vastly increased the sensitivity of procedures used in human genetics testing. For example, the β-globin gene in a small sample of DNA isolated from an individual can be specifically amplified by the PCR to determine if the person is a carrier of the mutant sickle-cell allele. Quantities of amplified DNA sufficient for sequencing can be prepared rapidly; subsequent sequencing reveals if the mutant allele is present in the sample.

A typical PCR is outlined in Figure 7-37. Genomic DNA is digested into large fragments using a restriction enzyme and then is heat-denatured into single strands. Two synthetic oligonucleotides complementary to the 3' ends of the DNA segment of interest are added in great excess to the denatured DNA, and the temperature is lowered to 50–60°C. The genomic DNA remains denatured, because the complementary strands are at too low a concentration to encounter each other during the period of incubation, but the specific oligonucleotides, which are at a very high concentration, hybridize with their complementary sequences in the genomic DNA. The hybridized oligonucleotides then serve as primers for DNA chain synthesis, which begins upon addition of a supply of deoxynucleotides and a temperature-resistant DNA polymerase such as that from *Thermus aquaticus* (a bacterium that lives in hot springs). This enzyme, called *Taq polymerase*, can extend the primers at temperatures up to 72°C. When synthesis is complete, the whole mixture is heated further (to 95°C) to melt the newly formed DNA duplexes. When the temperature is lowered again, another round of synthesis takes place because excess primer is still present. Repeated cycles of synthesis (cooling) and melting (heating) quickly amplify the sequence of interest. At each round, the number of copies of the sequence between the primer sites is doubled; therefore the desired sequence increases exponentially.

The PCR is so effective at amplifying specific DNA sequences that DNA isolated from a single human cell can be analyzed for mutations associated with various genetic diseases. In one reported case, this approach was used to screen in vitro fertilized human embryos prepared from sperm and ova from a couple who both are carriers of the genetic disorder cystic fibrosis. This disease results from mutation in the *CFTR* gene, which is located on chromosome 7. The DNA isolated from a single embryonic cell

➤ FIGURE 7-37   The polymerase chain reaction (PCR). The top left of the figure indicates a region of DNA to be amplified. One strand is shown in dark blue and the other in light blue. Oligonucleotide primers are synthesized corresponding to ≈20 nucleotides from each end of the region to be amplified. Arrowheads indicate the 3' ends of primers and extended strands. An excess of the two primers together with the heat-resistant DNA polyerase Taq polymerase and dNTPs are added to a preparation of DNA that includes the sequence to be amplified. The solution is heated to 95°C to melt the DNA into single strands and then is cooled to 60°C; at the lower temperature, the primers anneal (hybridize) to the complementary strands. Continued incubation at 60°C allows Taq polymerase to extend the primers, resulting in the synthesis of two partial doubled-stranded DNA molecules that include the sequence between the primers. After a second cycle of heating and cooling, four partial double-stranded DNA molecules are synthesized; each includes the sequence between the primers. After the third cycle, two of the eight DNA molecules formed are completely double stranded and extend from the 5' end of primer 1 to the 5' end of primer 2. Subsequent cycles of heating and cooling amplify these DNA molecules by a factor of two each cycle. In twenty cycles, the DNA sequence between the primers is amplified more than a millionfold.

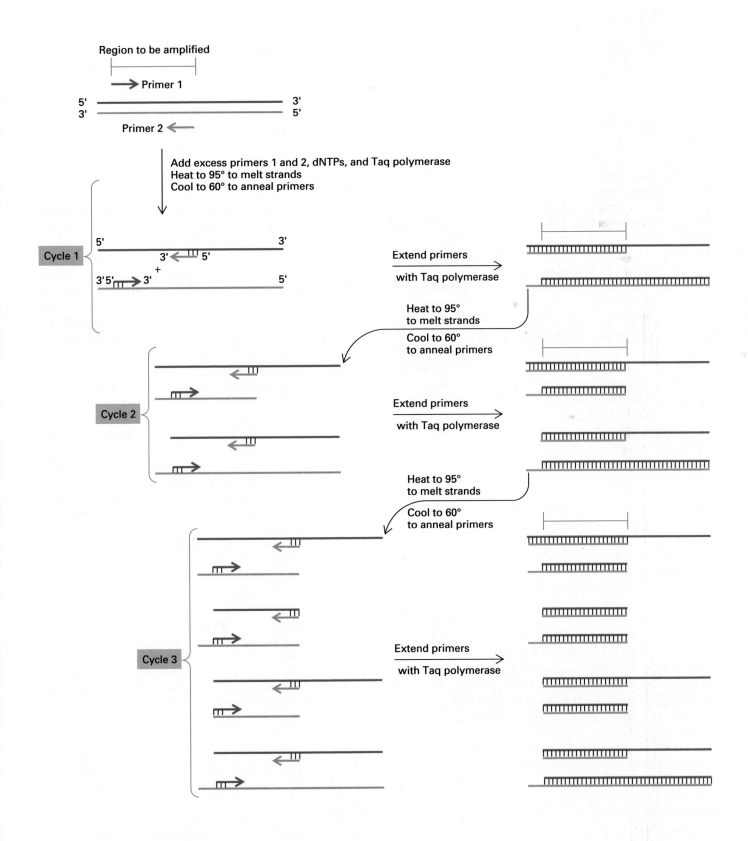

was subjected to PCR amplification and then analyzed for mutations identified in one of the two copies of chromosome 7 in each parent. In this way embryos that had inherited the wild-type chromosome from at least one parent were identified and then transferred to the mother's uterus. (Removal of a single cell from an in vitro fertilized human embryo has no apparent effect on subsequent development of the embryo after it is implanted in a receptive uterus.) By use of this procedure, carrier couples can be assured of having children that will not be at risk for cystic fibrosis. Another medical application of the PCR is early detection of infection with HIV, the virus that causes acquired immunodeficiency syndrome (AIDS). The PCR is so sensitive that it can detect HIV at very early stages in the disease (before symptoms appear) when only a few thousand blood cells in a patient are infected with the virus.

In basic research, the PCR also has numerous applications. For example, this procedure allows the recovery and rapid amplification of entire DNA sequences between any two ends whose sequences are known; the amplified sequences then can be ligated into standard cloning vectors. Only fragments of ≈ 2 kb or less can be amplified efficiently; however, further development of this technique will likely extend this upper limit.

The PCR also provides a powerful approach to cloning a cDNA based on the partial amino acid sequence of a purified protein. The amino acid sequence of two peptides isolated from the protein are used to design two degenerate oligonucleotide mixtures containing all possible DNA sequences encoding the two peptides. Rather than using these oligonucleotides as probes for direct screening of a cDNA library, as described previously, they are used as primers in a PCR. First, cDNA is synthesized from total cellular mRNA using reverse transcriptase. The cDNA is then used as the template for a PCR performed with the two degenerate oligonucleotide primers. This reaction amplifies the region of the cDNA between the sequences encoding the peptides used to design the degenerate primers. The PCR procedure effectively selects the correct oligonucleotides for priming DNA synthesis from the degenerate oligonucleotide mixtures, because only DNA synthesized from the correct cDNA template will hybridize to oligonucleotides present in both degenerate primer mixtures. For exponential amplification to take place, priming must occur from both ends of a fragment. Even if an oligonucleotide in one of the degenerate mixtures hybridizes to an incorrect cDNA and primes DNA synthesis, the DNA strand that is synthesized will not be amplified, because it will not contain a sequence complementary to one of the oligonucleotides in the second degenerate primer mixture. The cDNA sequence amplified by this procedure contains the unique sequence of the naturally occurring mRNA encoding the region between the two peptides originally sequenced. This unique DNA sequence can then be radioactively labeled and used as a probe for screening a cDNA or genomic library.

▶ From Protein to Gene and From Gene to Protein with Recombinant DNA Technology

In the past, researchers have had two basic approaches for unraveling the molecular mechanisms underlying various cellular processes. One approach involves the biochemical purification and analysis of a protein based on its functional characteristics by techniques described in Chapter 3. The other approach involves the characterization and mapping of genes defined by mutations using the classical genetic analyses described in Chapter 8. Recombinant DNA technology and methods for chemical synthesis of DNA provide the link between protein and gene. With the methods discussed in this chapter, today's molecular cell biologists can begin with an isolated protein and clone the gene that encodes it. By adding the techniques described in Chapter 8, they can begin with the concept of a gene identified by the characteristics of a mutant organism and isolate a cDNA clone containing the gene. Ultimately, the encoded protein can be produced in sufficient quantities for detailed study. The marriage of biochemical and genetic approaches by recombinant DNA technology provides an enormously powerful strategy for studying the role of particular proteins in complex processes.

For example, suppose that a small amount of a protein with an interesting function is isolated and purified (Figure 7-38 *left*). Portions of its amino acid sequence can be determined and used to design oligonucleotide probes to isolate a cDNA encoding the protein. The cDNA can then be incorporated into an expression vector to produce much larger amounts of the protein, greatly simplifying further studies. The cDNA also can be used as a probe to isolate the gene encoding the protein from a genomic library.

Once the gene encoding a particular protein has been isolated and cloned, it can be studied in various ways. For example, techniques described in Chapter 11 permit direct analysis of the genomic DNA sequences that regulate activity of a gene, and expression of the encoded protein, during development or in response to changes in the cellular environment. Moreover, as discussed in Chapter 8, a normal gene can be replaced by a mutant form of the gene in some organisms. Such gene replacement allows researchers to directly test the function of the gene in the physiology and development of the organism. Portions of the natural gene sequence also can be replaced with synthetic DNA, thus introducing specific mutations in the encoded protein. With this technique, investigators can analyze the effect of specific changes in a protein's structure on its function. These are powerful approaches for studying how a particular protein functions in complex cellular processes. They depend on isolation of the DNA encoding a specific protein through the use of data on its amino acid sequence or generation of specific antibody against the protein.

FROM PROTEIN TO GENE	FROM GENE TO PROTEIN
Isolate protein on the basis of its molecular function (e.g., enzymatic or hormonal activity)	Isolate genomic clone corresponding to an altered trait in mutants (e.g., nutritional auxotrophy, inherited disease, developmental defect)
↓	↓
Determine partial amino acid sequence of the protein	Use genomic DNA to isolate a cDNA for the mRNA encoded by the gene
↓	↓
Synthesize oligonucleotides that correspond to portions of the amino acid sequence	Sequence the cDNA to deduce amino acid sequence of the encoded protein
↓	↓
Use oligonucleotides as probes to select cDNA or genomic clone encoding the protein from library	Compare deduced amino acid sequence with that of known proteins to gain insight into function of the protein
↓	↓
Sequence isolated gene	Use expression vector to produce the encoded protein

◄ FIGURE 7-38   Overview of alternative strategies for studying the roles of specific proteins in complex cellular processes. This chapter highlights the protein-to-gene strategy; Chapter 8 describes the gene-to-protein strategy. In both cases, recombinant DNA technology can link a gene and its encoded protein, so that traditional biochemical and genetic approaches are combined in powerful ways.

Alternatively, suppose that a gene encoding an unknown protein is identified by classical genetic analyses of mutants exhibiting an altered trait (Figure 7-38 *right*). Genomic clones carrying the gene can be isolated, as described in Chapter 8; the cloned gene then can be used to probe a cDNA library to identify a cDNA clone complementary to the mRNA corresponding to the gene. The complete amino acid sequence of the encoded protein can be deduced by sequencing the cDNA (even though the protein has never actually been isolated). Comparison of this amino acid sequence with the sequences of previously studied proteins reveals if it is related to other known proteins and may provide clues about the function of the protein. By use of an expression vector, sufficient amounts of the protein can be produced to test hypotheses about its function.

As we will see in later chapters, this approach has been crucial in identifying and isolating proteins involved in the development of *Drosophila* and mice starting with mutants exhibiting defects in development. For example, genes that are critical early in the development of *Drosophila* and mice have been shown to encode proteins closely related to a particular class of transcription factors, which are proteins that control the transcription of multiple genes (Chapters 11 and 13). Other genes with important functions later in *Drosophila* development have been shown to encode proteins homologous to polypeptide-hormone receptors characterized from mammals. These proteins span the plasma membrane of the cell, communicating information about the extracellular environment to the interior of the cell (Chapters 13 and 20). It is easy to imagine how these two classes of proteins could have a vital role in the development of a multicellular organism and to understand why mutations in the genes encoding them have disastrous consequences.

SUMMARY

Recombinant DNA technology has reshaped the way biological research is carried out today. The technology depends on the use of restriction enzymes, which cut DNA at specific 4- to 8-bp sequences, generating a reproducible set of restriction fragments from the genome of any organism. The sequences left at the termini of many restriction fragments can be covalently joined with the enzyme DNA ligase, allowing insertion of restriction fragments into plasmid vectors, circular DNA molecules that replicate in *E. coli* cells. If such recombinant plasmids carry an antibiotic-resistance gene, they can transform *E. coli* cells to antibiotic resistance. Selection of the rare cells that take up the plasmid is done on plates containing the appropriate antibiotic; on such plates only these transformed cells grow into colonies. Since each colony arises from a single transformed cell, it constitutes a clone of cells all carrying the same recombinant plasmid. In this way, a complex mixture of restriction fragments carried in plasmid vectors can be separated into discrete *E. coli* clones each composed of cells harboring a particular restriction fragment.

Restriction fragments also can be ligated to the arms of the bacteriophage λ genome. Such recombinant λ DNA containing inserted fragments of up to ≈25 kb can be efficiently packaged in vitro into infectious λ virions. When the resulting mixture of recombinant λ virions is used to generate plaques on a lawn of *E. coli*, each plaque is a recombinant λ clone containing a distinct inserted restriction fragment. Cloning in λ vectors is so efficient that nearly all of the genomic DNA of complex multicellular organisms can be represented in a λ genomic library of a few million independent λ clones. Cosmids are plasmid vectors that contain the λ-phage COS sequence, which allows them to be packaged in vitro into λ virions. By use of cosmid vectors, DNA fragments up to ≈45 kb in length can be introduced into *E. coli* cells. A similar strategy using bacteriophage P1 permits cloning of DNA fragments up to ≈100 kb long.

The ability to manipulate recombinant DNA was vastly extended by the development of techniques for chemically synthesizing single-stranded DNA molecules of any sequence containing up to about 60 nucleotides. Such synthetic oligonucleotides have many uses.

Messenger RNAs can be converted into complementary *cDNAs* with reverse transcriptase. Since almost all eukaryotic mRNAs have a 3′ poly-A tail, a mixture of cDNAs representing nearly all the mRNAs extracted from a particular tissue or cell type can be synthesized by priming reverse transcription with oligo-dT. This mixture of cDNAs can be converted into double-stranded DNA and inserted into plasmid or λ vectors to generate a cDNA library.

Two general methods are used to identify specific clones encoding known proteins within a genomic or cDNA library. One method involves membrane hybridization with radiolabeled, single-stranded cDNA or synthetic oligonucleotide probes. The other method utilizes λ expression vectors to synthesize the protein encoded by the cloned DNA. In this case, a specific clone is identified based on the ability of the encoded protein to specifically bind a monoclonal antibody or other ligand.

Cloned genomic restriction fragments and cDNAs can be separated from the vectors carrying them by cleavage with the restriction enzyme used to prepare the recombinant cloning vector. The cloned DNA then is isolated from the vector DNA by polyacrylamide or agarose gel electrophoresis, which separates single-stranded DNA molecules according to chain length. Restriction sites for multiple different restriction enzymes can be mapped on a cloned fragment, generating a restriction map that is characteristic for each distinct DNA sequence.

The Maxam-Gilbert and Sanger (dideoxy) methods are used to sequence cloned genomic restriction fragments and cDNAs. Both methods depend on the ability of polyacrylamide gel electrophoresis to separate single-stranded DNA molecules differing in length by only one nucleotide. From the nucleotide sequence of a cloned gene, the complete amino acid sequence of the encoded protein is easily deduced. If this sequence exhibits homology with sequences of other previously studied proteins, the protein is likely to have similar biochemical activities. A protein identified in this way can be produced and isolated by use of an expression vector carrying the cDNA encoding it. As discussed in the next chapter, cloning of genes identified by genetic analysis of mutants and sequencing of these genes can provide valuable information about cellular processes disrupted by specific mutations.

The availability of cloned restriction fragments led to development of two very useful methods to detect a specific DNA or RNA in a complex mixture: Southern blotting, which detects specific DNA fragments, and Northern blotting, which detects specific RNAs. Both methods involve gel electrophoresis of a nucleic acid mixture, transfer (blotting) of the separated DNA or RNA molecules to a filter, and exposure of the filter to a specific, radiolabeled DNA probe, which hybridizes to complementary molecules.

The polymerase chain reaction (PCR), which is used to amplify a particular DNA sequence, has numerous applications in basic research and medicine. With this technique, the DNA from a single human cell can be amplified enough to allow sequencing of a particular region, providing an exquisitely sensitive way to detect mutations associated with inherited human diseases.

This chapter has focussed on the experimental strategy of starting with a protein identified and purified by biochemical techniques and eventually isolating the gene encoding it. In the next chapter, we examine the opposite strategy of starting with a gene identified by genetic analysis and eventually isolating the protein it encodes. Both strategies employ many of the same recombinant DNA techniques and are potent tools for analyzing how macromolecules and cells function.

REVIEW QUESTIONS

1. Review the action of restriction endonucleases, paying particular attention to the properties of the sites in DNA which they cleave. Also review the concept of a polylinker as shown in Figure 7-8. What is the advantage of having such a region in a plasmid used for cloning?

For the following question, use the information in the table and in the polylinker sequence below.

Restriction Endonuclease	Sequence of Cleaved DNA*
AgeI	A ↓ CCGGT
BamHI	G ↓ GATCC
Cfr10I	Pu ↓ CCGGPy
HpaII	C ↓ CGG
HindIII	A ↓ AGCTT
KpnI	GGTAC ↓ C
NaeI	GCC ↓ GGC
PstI	CTGCA ↓ G
SacI	GAGCT ↓ C
SalI	G ↓ TCGAC
SmaI	CCC ↓ GGG
SphI	GCATG ↓ C
XbaI	T ↓ CTAGA
XmaI	C ↓ CCGGG

* An arrow indicates the cleavage site.

Suppose that you are going to use a plasmid to clone a region of DNA. See Box 1a for the polylinker region present in your vector; each of the six base pair restriction sites found in this region is unique in that it is not found elsewhere in the plasmid.

a. You cut the plasmid with AgeI. Where will it cut the polylinker region? Will it produce sticky or blunt ends?

b. You could potentially use this same enzyme to digest the DNA you wish to insert into this plasmid, but you notice that one other enzyme cleaves DNA at the same sequence as AgeI. Which is it? Would you be able to use it to digest the DNA you wish to insert into the vector? Could any of the other restriction enzymes be used to produce DNA fragments that can be inserted into the plasmid vector cut with AgeI?

c. A colleague of yours suggests that you digest the sample of DNA you wish to insert by using XmaI. Do you think this is feasible? If recombinants were formed when you used XmaI, which of the listed restriction enzymes would cut the new plasmids at the junction between the insert and the vector polylinker?

2. Review the techniques for preparing cDNA and genomic DNA libraries, as well as the means of selecting clones of interest. There are several mutagenic chemical agents that induce changes in DNA by forming methyl groups that react with guanine to form O^6-methylguanine (mG). This particular lesion can instruct DNA polymerase to place a T opposite the modified G as opposed to a C, thus producing base changes in DNA. During evolution, cells have devised a way of removing mG from DNA by a reaction involving a protein that is used only once in the reaction and is not regenerated; it is called O^6-alkylguanine–DNA alkyltrans-

ferase (AGT), and the reaction is described as follows:

$$O^6\text{-methylguanine} + R\text{-S-H} \longrightarrow \text{guanine} + R\text{-S-CH}_{3'}$$

where R–S–H is the enzyme. In this reaction, the methyl group is transferred to a cysteine residue, leaving guanine intact in the DNA and resulting in an inactive enzyme. AGT from humans has been characterized extensively. In this enzyme, the amino acid sequence surrounding the cysteine that accepts the methyl group is

-Pro-Ile-Leu-Ile-Pro-Cys-His-Arg-Val-

The underlined Cys represents the methyl-accepting group.

a. You wish to clone the gene for AGT from humans. Would you prepare a cDNA library or a genomic library to do this? What would be the advantage or disadvantage of each? Would you need to consider the conservation of the peptide sequence referred to above? What is the relevance of expression of the gene encoding AGT to your decision? Refer to your answer to these questions to defend your choice. If you chose to prepare both types of libraries, describe how you would obtain them.

b. Assume that you have prepared a cDNA library from the cells you are using to clone AGT, and you employed bacteriophage λ as the vector. How would you select plaques harboring the cDNA of interest? If you used the amino acid sequence given above, how many degenerate probes could exist, and how would you prepare them?

3. Compare and contrast the methods of Southern blotting and Northern blotting. What is each used for? Pay particular attention to the different conditions required for nucleic acid transfer in each case. Can RNA be transferred in an alkaline solution?

O^6-Alkylguanine–DNA alkyltransferase (AGT) activity (see question 2) is not found in all cells. You have a series of human cell lines in your laboratory, some of which are actually derived from tumors.

a. You are interested in using the cloned AGT DNA from the previous problem to detect mRNA corresponding to AGT in each of the cell lines. How would you do this? The results you obtain are shown in the following table, along with those showing the actual protein activity as measured by a specific assay for AGT.

Cell Line	Relative AGT mRNA	Relative AGT Activity
A	++++	++++
B	++++	−
C	−	−
D	−	−
E	+	+

Box 1a
5'-GAATTCGAGCTCGGTACCCGGGGATCCACCGGTGTCGACCTGCAGGCATGCAAGCTT-3'

Relate these results to expression of the *AGT* gene and translation of the message.

b. Suppose that you wanted to prepare large quantities of the AGT protein. Devise a method for doing this.

c. You wish to determine if the *AGT* gene is present in the cell lines lacking AGT mRNA by using the technique of Southern blotting. What restriction enzyme would you choose for this purpose, or would you choose several? Suppose that you choose the restriction enzymes *Eco*RI and *Hin*dIII to digest the DNA from each of your cell lines. You then resolve the DNA by electrophoresis, transfer it to a membrane, and use your clone to probe the filters. The results are in the following table.

Cell Line	Sizes of Bands (base pairs)		
	*Eco*RI	*Hin*dIII	*Eco*RI/*Hin*dIII
A	25,000	18,000	14,000
		12,000	11,000
B	25,000	30,000	25,000
C	25,000	18,000	14,000
		12,000	11,000
D	—*	—	—
E	25,000	18,000	14,000
		12,000	11,000

*A dash (—) denotes that no bands were detected.

Construct a map showing the relative positions of the restriction sites for *Hin*dIII and *Eco*RI at the genetic locus that encodes AGT. Is this map unique, or is there more than one possible solution? If you wanted to distinguish among several possible restriction maps for these results, what additional data would you need?

d. Based on data from the Southern, Northern, and AGT activity measurements, explain further the pattern of AGT activity found in each cell line.

References

General Methods of Recombinant DNA Technology

AUSUBEL, F. M., et al. 1987. *Current Protocols in Molecular Biology.* Wiley.

SAMBROOK, J., E. F. FRITSCH, and T. MANIATIS. 1989. *Molecular Cloning.* Cold Spring Harbor Laboratory.

DNA Cloning with Plasmid Vectors

BERG, P. 1981. Dissections and reconstructions of genes and chromosomes. *Science* 213:296–303.

BOLIVAR, F., et al. 1977. Construction and characterization of new cloning vehicles. II: A multipurpose cloning system. *Gene* 2:95–107.

COHEN, S. N., et al. 1973. Construction of biologically functional bacterial plasmids in vitro. *Proc. Nat'l. Acad. Sci. USA* 70:3240–3244.

OLD, R. W., and S. B. PRIMROSE. 1985. *Principles of Gene Manipulation: An Introduction to Genetic Engineering.* Blackwell Scientific Publications.

WATSON, J. D., et al., 1992, *Recombinant DNA,* 2d ed. Scientific American Books.

Bacteriophage λ as a Cloning Vehicle

COLLINS, J., and B. HOHN. 1978. Cosmids: a type of plasmid gene-cloning vector that is packageable in vitro in bacteriophage λ heads. *Proc. Nat'l. Acad. Sci. USA* 75:4242–4246.

DUNN, I. S., and F. R. BLATTNER. 1987. Charons 36 to 40: Multi enzyme, high-capacity, recombination-deficient replacement vectors with polylinkers and polystuffers. *Nucl. Acids Res.* 15:2677–2701.

HOHN, B. 1979. In vitro packaging of λ and cosmid DNA. *Methods Enzymol.* 68:299–308.

FRISCHAUF, A.-M., et al. 1983. Lambda replacement vectors carrying polylinker sequences. *J. Mol. Biol.* 170:827–848.

MURRAY, N. E., and K. MURRAY. 1974. Manipulation of restriction targets in phage λ to form receptor chromosomes for DNA fragments. *Nature* 251:476–483.

RAMBACH, A., and P. TIOLLAIS. 1974. Bacteriophage λ having *Eco*RI endonuclease sites only in the nonessential region of the genome. *Proc. Nat'l. Acad. Sci. USA* 71:3927–3931.

THOMAS, M., J. R. CAMERON, and R. W. DAVIS. 1974. Viable molecular hybrids of bacteriophage lambda and eukaryotic DNA. *Proc. Nat'l. Acad. Sci. USA* 71:4579–4583.

Chemical Synthesis of DNA

CARUTHERS, M. H. 1985. Gene synthesis machines: DNA chemistry and its uses. *Science* 230:281–285.

ITAKURA, K., and A. D. RIGGS. 1980. Chemical DNA synthesis and recombinant DNA studies. *Science* 209:1401–1405.

ITAKURA, K., J. J. ROSSI, and R. B. WALLACE. 1984. Synthesis and use of synthetic oligonucleotides. *Ann. Rev. Biochem.* 53:323–356.

cDNA Cloning

GUBLER, U., and B. J. HOFFMAN. 1983. A simple and very efficient method for generating cDNA libraries. *Gene* 25:263–289.

HAN, J. H., C. STRATOWA, and W. J. RUTTER. 1987. Isolation of full-length putative rat lysophospholipase cDNA using improved methods for mRNA isolation and cDNA cloning. *Biochemistry* 26:1617–1632.

JENDRISAK, J., R. A. YOUNG, and J. D. ENGEL. 1987. Cloning cDNA into λgt10 and λgt11. *Methods Enzymol.* 152:359–385.

OKAYAMA, H., and P. BERG. 1982. High-efficiency cloning of full-length cDNA. *Mol. Cell. Biol.* 2:161–170.

Identifying Clones of Interest

BENTON, W. D., and R. W. DAVIS. 1977. Screening λgt recombinant clones by hybridization to single plaques in situ. *Science* 196:180–183.

GRUNSTEIN, M., and D. S. HOGNESS. 1975. Colony hybridization: a method for the isolation of cloned DNAs that contain a specific gene. *Proc. Nat'l. Acad. Sci. USA* 72:3961–3965.

HALL, B. D., and S. SPIEGELMAN. 1961. Sequence complementarity of T2 DNA and T2-specific RNA. *Proc. Nat'l. Acad. Sci. USA* 47:137–146.

SINGH, H., et al. 1988. Molecular cloning of an enhancer binding protein: isolation by screening of an expression library with a recognition site DNA. *Cell* 52:415–423.

WALLACE, R. B., et al. 1981. The use of synthetic oligonucleotides as hybridization probes. II: Hybridization of oligonucleotides of mixed sequence to rabbit β-globin DNA. *Nucl. Acids Res.* 9:879–887.

WETMUR, J. G., and N. DAVIDSON. 1968. Kinetics of renaturation of DNA. *J. Mol. Biol.* 31:349–370.

YOUNG, R. A., and R. W. DAVIS. 1983. Efficient isolation of genes by using antibody probes. *Proc. Nat'l. Acad. Sci. USA* 80:1194–1198.

Restriction Enzymes

DAVIES, K. E., ed. 1988. *Genome Analysis: A Practical Approach.* Oxford: IRL Press.

MERTZ, J. E., and R. W. DAVIS. 1972. Cleavage of DNA by RI restriction endonuclease generates cohesive ends. *Proc. Nat'l. Acad. Sci. USA* 69:3370–3374.

NATHANS, D., and H. O. SMITH. 1975. Restriction endonucleases in the analysis and restructuring of DNA molecules. *Ann. Rev. Biochem.* 44:273–293.

NELSON, M., and M. MCCLELLAND. 1992. The effect of site-specific methylation on restriction-modification enzymes. *Nucl. Acids Res.* 20 (Suppl.):2145–2157.

ROBERTS, R. J. 1992. Restriction enzymes and their isoschizomers. *Nucl. Acids Res.* 20 (Suppl.):2167–2180.

SMITH, H. O. 1970. Nucleotide sequence specificity of restriction endonucleases. *Science* 205:455–462.

Gel Electrophoresis of DNA Fragments

AIJ, C., and P. BORST. 1972. The gel electrophoresis of DNA. *Biochem. Biophys. Acta* 269:192–200.

ANDREWS, A. T. 1986. *Electrophoresis,* 2d ed. Oxford University Press.

CANTOR, C. R., C. L. SMITH, and M. K. MATTHEW. 1988. Pulsed-field gel electrophoresis of very large DNA molecules. *Ann. Rev. Biophys. Biophys. Chem.* 17:41–72.

CARLE, G. F., M. FRANK, and M. V. OLSON. 1986. Electrophoretic separations of large DNA molecules by periodic inversion of the electric field. *Science* 232:65–70.

RICKWOOD, D., and B. D. HAMES, eds. 1982. *Gel Electrophoresis of Nucleic Acids.* London: IRL Press Ltd.

SHARP, P. A., B. SUGDEN, and J. SAMBROOK. 1973. Detection of two restriction endonuclease activities in *Hemophilus parainfluenzae* using analytical agarose-ethidium bromide electrophoresis. *Biochemistry* 12:3055–3062.

SMITH, C. L., et al. 1987. A physical map of the *Escherichia coli* K12 genome. *Science* 236:1448–1453.

SMITH, H. O., and M. L. BERNSTIEL. 1976. A simple method for DNA restriction-site mapping. *Nucl. Acids Res.* 3:2387–2398.

DNA Sequencing

MAXAM, A. M., and W. GILBERT. 1977. A new method for sequencing DNA. *Proc. Nat'l. Acad. Sci. USA* 74:560–564.

MAXAM, A. M., and W. GILBERT. 1980. Sequencing end-labeled DNA with base-specific chemical-cleavages. *Methods Enzymol.* 65:499–560.

SANGER, F. 1981. Determination of nucleotide sequences in DNA. *Science* 214:1205–1210.

Analysis of Specific Nucleic Acids in Complex Mixtures

ALWINE, J. C., D. J. KEMP, and G. R. STARK. 1977. A method for detection of specific RNAs in agarose gels by transfer to diazoben-zyloxymethyl-paper and hybridization with DNA probes. *Proc. Nat'l. Acad. Sci. USA* 74:5350–5354.

BERK, A. J., and P. A. SHARP. 1977. Sizing and mapping of early adenovirus mRNAs by gel electrophoresis of S1 endonuclease digested hybrids. *Cell* 12:721–732.

SOUTHERN, E. M. 1975. Detection of specific sequences among DNA fragments separated by gel electrophoresis. *J. Mol. Biol.* 98:503–515.

THOMAS, P. S. 1980. Hybridization of denatured RNA and small DNA fragments transferred to nitrocellulose. *Proc. Nat'l. Acad. Sci. USA* 77:5201–5205.

Expression Vectors

ELROY-STEIN, O., T. R. FUERST, and B. MOSS. 1989. Cap-independent translation of mRNA conferred by encephelomyocarditis virus 5' sequence improves the performance of the vaccinia virus/bacteriophage T7 hybrid expression system. *Proc. Nat'l. Acad. Sci. USA* 86:6126–6130.

LUCKOW, V. A., and M. D. MILLER. 1989. High level expression of non-fused foreign proteins with *Autographa california* nuclear polyhedrosis virus expression vectors. *Virology* 170:31–39.

NAGATA, S., et al. 1986. Molecular cloning and expression of cDNA for human granulocyte colony-stimulating factor. *Nature* 319:415–418.

PELHAM, H. R. B., and R. J. JACKSON. 1976. An efficient mRNA dependent translation system from reticulocyte lysates. *Eur. J. Biochem.* 67:247–256.

TABOR, S., and C. C. RICHARDSON. 1985. A bacteriophage T7 RNA polymerase promoter system for controlled exclusive expression of specific genes. *Proc. Nat'l. Acad. Sci. USA* 82:1074–1078.

SOUZA, L. M., et al. 1986. Recombinant human granulocyte-colony stimulating factor: effects on normal and leukemic myeloid cells. *Science* 232:61–65.

STUDIER, F. W., and B. A. MOFFATT. 1986. Use of bacteriophage T7 RNA polymerase to direct selective high-level expression of cloned genes. *J. Mol. Biol.* 189:113–134.

Polymerase Chain Reaction

ERLICH, H., ed. 1992. *PCR Technology: Principle and Applications for DNA Amplification.* W. H. Freeman and Company.

ERLICH, H. E., R. A. GIBBS, and H. H. KAZAZIAN JR., eds. 1989. *Polymerase Chain Reaction.* Cold Spring Harbor Laboratory.

HANDYSIDE, A. H., et al. 1992. Birth of a normal girl after in vitro fertilization and preimplantation diagnostic testing for cystic fibrosis. *New Engl. J. Med.* 327:905–909.

INNIS, M. A., et al. 1990. *PCR Protocols: A Guide to Methods and Applications.* Academic Press.

SAIKI, R. K., et al. 1988. Primer-directed enzymatic amplification of DNA with a thermostable DNA polymerase. *Science* 239:487–491.

8 Genetic Analysis in Cell Biology

▲ Mice carrying the rat growth hormone transgene grow to nearly twice the adult size of their siblings lacking the transgene.

In previous chapters, we learned how proteins are isolated, how their structures are determined, and how the genes encoding known proteins are cloned. Our primary concern, however, is what a protein does in the organism. In principle, the in vivo function of a protein can be deduced by seeing what effect removal of the protein has on the cell or, in the case of a multicellular organism, on the whole organism. In practice, removing a protein is done indirectly by identifying organisms in which the nucleotide sequence of the gene encoding the protein is altered or deleted. Such changes in the DNA sequence, called *mutations,* can lead to loss of the encoded protein or to a change in its structure. The affected organisms, called *mutants,* are identified by virtue of differences in their appearance, physiology, behavior, or growth properties compared with normal, wild-type (nonmutant) organisms. By comparing specific DNA sequences from mutant and normal organisms, researchers can correlate the abnormal features of the mutant organism with differences in the expression or structure of specific proteins. Alternatively, specific mutations can be introduced into cloned genes and the mutated genes then introduced into intact organisms (e.g., yeast and mice); again, comparison of the normal and mutant organisms provides clues about the in vivo functioning of the encoded protein.

In this chapter we consider some basic concepts in genetics and various genetic techniques that are useful in studying how proteins carry out specific cellular processes and the order in which specific proteins function. In the first section, we discuss the isolation of mutants and their characterization using classical genetic analyses. In the second section, we describe the various steps in the genetic mapping of mutations. The use of recombinant DNA techniques presented in Chapter 7 to isolate and clone genes defined by mutations is explained next. And in the final section of the chapter, we consider molecular techniques

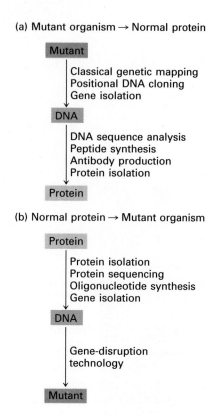

(a) Mutant organism → Normal protein

Mutant

Classical genetic mapping
Positional DNA cloning
Gene isolation

DNA

DNA sequence analysis
Peptide synthesis
Antibody production
Protein isolation

Protein

(b) Normal protein → Mutant organism

Protein

Protein isolation
Protein sequencing
Oligonucleotide synthesis
Gene isolation

DNA

Gene-disruption
technology

Mutant

◄ FIGURE 8-1   Flow chart summarizing two different approaches towards understanding the role of specific proteins in normal cellular and organismal function. (a) Mutations causing changes in the structure or amount of proteins can lead to defects in normal organismal function and development. The procedures for isolating and cloning genes defined by mutations and then identifying the encoded proteins are described in this chapter. (b) In other cases, the techniques described in previous chapters are used to isolate proteins and then clone genes encoding specific proteins. In the last section of this chapter, we consider in vitro mutagenesis of cloned genes and their introduction into organisms. This approach produces a mutant organism in which the function of the specific protein encoded by the mutated gene is disrupted.

for introducing cloned genes, including deliberately mutated genes, into the genome of eukaryotes. As the discussion in this and the previous chapter illustrates, the marriage of two genetic disciplines, classical genetics and recombinant DNA technology, forms a powerful approach for understanding biological function (Figure 8-1). Chapter 7 focused on the protein-to-gene strategy, whereas this chapter describes the gene-to-protein strategy.

➤ The Isolation and Characterization of Mutants

The development and function of an organism is in large part controlled by genes. Mutations can lead to changes in the structure of an encoded protein or to a decrease or complete loss in its expression. Because a change in the DNA sequence results in all copies of the encoded protein being affected, mutations can be particularly damaging to a cell or organism. In contrast, any alterations in the sequences of RNA or protein molecules that occur during their synthesis are less serious because many copies of each RNA and protein are synthesized.

Many different types of mutants have been identified in organisms ranging from bacteria to humans. Mutants can differ from their normal counterparts in a variety of ways. Some mutations cause only subtle changes; for ex-

ample, certain mutations in the fruit fly, *Drosophila melanogaster,* result in failure of a single type of neuronal cell to develop, but mutant flies otherwise are normal. Other mutations lead to significant changes in development, cellular function, appearance, and/or behavior of an individual. Several *Drosophila* mutants are illustrated in Figures 8-2 and 8-3. None of these mutations are lethal; some mutations, however, result in organismal death.

Geneticists often distinguish between the *genotype* and *phenotype* of an organism. The function and physical appearance of an individual is referred to as its phenotype, whereas the entire set of genes carried by an individual is its genotype. However, the two terms commonly are used in a more restricted sense: that is, genotype usually denotes whether an individual carries mutations in a single gene (or a small number of genes), and phenotype denotes the physical and/or functional consequence of that genotype. Figure 8-4 illustrates the phenotypic traits, or "characters," studied by Gregor Mendel in his classical experiments, which provided the foundation for modern genetics.

Isolation and characterization of mutants in a variety of experimental organisms have been used to investigate many different fundamental biological processes. Genetic analyses of mutants defective in a particular process can reveal (a) the number of genes required for phenotypic expression; (b) the order in which gene products act in the process; and (c) whether the proteins encoded by different genes interact with each other. Genetic studies of this type have helped unravel various metabolic pathways, regulatory mechanisms, and developmental processes.

Mutations Are Recessive or Dominant

A fundamental genetic difference among organisms is whether their cells carry a single set of chromosomes or two copies of each chromosome. The former are referred to as *haploid*; the latter, as *diploid*. Many simple unicellular organisms are haploid, whereas complex multicellular organisms (e.g., fruit flies, mice, humans) are diploid.

(a)

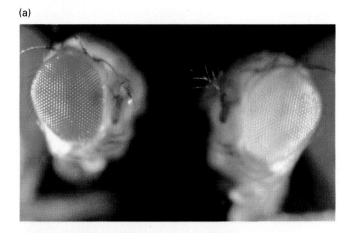

(c)

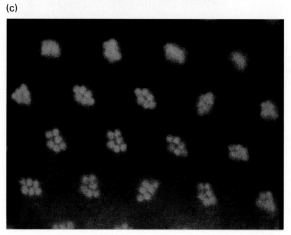

(b)

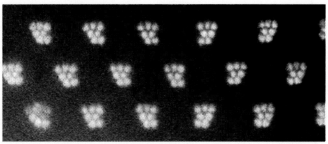

▲ FIGURE 8-2   Eye mutants in *Drosophila*. (a) White-eyed mutants lack the pigment that is present in the normal bright red eyes of fruit flies. The white mutant, the first known *Drosophila* mutant, was identified in the early part of the 20th century. (b, c) A special optical technique permits visualization of the different types of photoreceptor cells in each of the 800 modules (ommatidia) that compose the *Drosophila* compound eye. With this technique, seven of the eight photoreceptor cells are easily seen in a wild-type eye (b), whereas only six are visible in the *sevenless* mutant eye (c). A single cell, the R7 photoreceptor cell, is missing from each eye module in this mutant. Developmental, genetic, and molecular analysis of the sevenless gene has provided insights into the basic mechanisms by which cells signal each other during development. [Part (a) from A. J. E. Griffiths et al., 1993, *An Introduction to Genetic Analysis*, 5th ed., W. H. Freeman and Company, p. 61; parts (b) and (c) courtesy of U. Banerjee.]

(a)

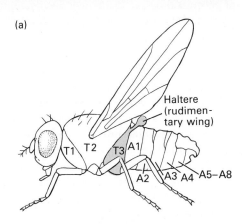

Haltere (rudimentary wing)

T1   T2   T3   A1

A2   A3   A4   A5–A8

(b)

(c)

▲ FIGURE 8-3   (a) Drawing of a normal fruit fly. The third thoracic segment (T3) has a rudimentary wing called a haltere. (b) A normal fly with a single pair of wings, which arise from the T2 segment. (c) A mutant fly in which the ultrabithorax gene is inactivated. In this mutant, the T3 segment, which normally does not give rise to wings, is transformed into a segment that does. Study of this locus has led to remarkable progress in understanding the mechanisms that pattern not only the body of the fruit fly but also the body of mammals. [From E. B. Lewis, 1978, *Nature* **276:**565; photographs courtesy of E. B. Lewis. Reprinted by permission from *Nature*; copyright 1978 Macmillan Journals Limited.]

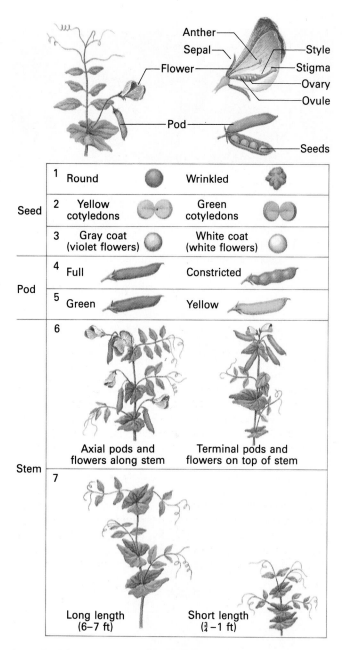

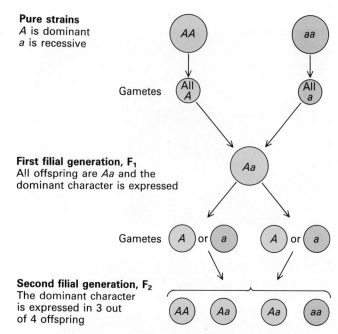

Pure strains
A is dominant
a is recessive

Gametes

First filial generation, F₁
All offspring are *Aa* and the
dominant character is expressed

Gametes

Second filial generation, F₂
The dominant character
is expressed in 3 out
of 4 offspring

▲ **FIGURE 8-5**   Mendel's experiments revealed that an individual organism has two alternative hereditary units for a given phenotypic trait: a dominant unit (*A*) and a recessive unit (*a*). In crossing pure strains of peas, he discovered that all F₁ plants expressed the dominant unit; however, the recessive unit apparently was still present because it emerged in some of the F₂ plants. Mendel concluded that the two hereditary units, now called alleles, do not "blend" in the parent plants but remain discrete and segregate in the formation of pollen and ovules (gametes). Each gamete receives only one of the two parental alleles. This conclusion is known as Mendel's first law.

▲ **FIGURE 8-4**   The seven "characters" of garden peas observed and described by Mendel. To cross two strains, the flower of one strain is opened before self-pollination occurs, the anthers are removed, and the ovules are dusted with pollen from the other strain. The offspring of such a cross are called the F₁ generation. These easily noted differences in phenotype facilitated identification of strains that differed in only one trait. [From M. W. Strickberger, 1976, *Genetics*, 2d ed., Macmillan, p. 115.]

type to be observed; that is, the individual must be homozygous for the mutant allele. In contrast, the phenotypic consequences of a *dominant* mutation are observed in a heterozygous individual carrying one mutant and one normal allele. The genetic studies of Mendel in the nineteenth century revealed that genes can exist in alternative allelic forms and demonstrated the difference between dominant and recessive alleles (Figure 8-5).

Some alleles can be associated with both a recessive and a dominant phenotype. For instance, fruit flies heterozygous for the mutant *stubble* (*Sb*) allele have short and stubby body hairs rather than the normal long, slender hairs; the mutant allele is dominant in this case. In contrast, flies homozygous for this allele die during development. Thus the recessive phenotype associated with this allele is lethal, whereas the dominant phenotype is not.

Mutations Involve Large or Small DNA Alterations

Mutations involving a change in a single base pair, often called a *point mutation*, or a deletion of a few base pairs

Different forms of a gene (e.g., normal and mutant) are referred to as *alleles*. Since diploid organisms carry two copies of each gene, they may carry identical alleles (i.e., be *homozygous* for a gene) or different alleles (i.e., be *heterozygous* for a gene). A *recessive* mutation is one in which both alleles must be mutant in order for the mutant pheno-

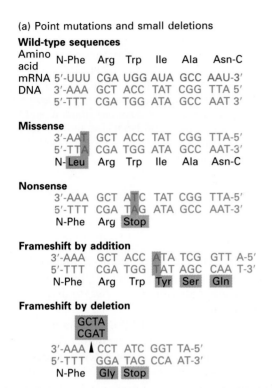

(a) Point mutations and small deletions

Wild-type sequences

Amino acid	N-Phe	Arg	Trp	Ile	Ala	Asn-C
mRNA	5'-UUU	CGA	UGG	AUA	GCC	AAU-3'
DNA	3'-AAA	GCT	ACC	TAT	CGG	TTA 5'
	5'-TTT	CGA	TGG	ATA	GCC	AAT 3'

Missense

3'-AAT GCT ACC TAT CGG TTA-5'
5'-TTA CGA TGG ATA GCC AAT-3'
N-Leu Arg Trp Ile Ala Asn-C

Nonsense

3'-AAA GCT ATC TAT CGG TTA-5'
5'-TTT CGA TAG ATA GCC AAT-3'
N-Phe Arg Stop

Frameshift by addition

3'-AAA GCT ACC ATA TCG GTT A-5'
5'-TTT CGA TGG TAT AGC CAA T-3'
N-Phe Arg Trp Tyr Ser Gln

Frameshift by deletion

GCTA
CGAT

3'-AAA ▲ CCT ATC GGT TA-5'
5'-TTT GGA TAG CCA AT-3'
N-Phe Gly Stop

(b) Chromosomal abnormalities

Inversion

Deletion

Balanced translocation

Insertion

▲ **FIGURE 8-6** Different types of mutations. (a) Point mutations, which involve alteration in a single base pair, and small deletions generally directly affect the function of only one gene. A wild-type peptide sequence and the mRNA and DNA encoding it are shown at the top. Altered nucleotides and amino acid residues are highlighted in green. Missense mutations lead to a change in a single amino acid in the encoded protein. In a nonsense mutation, a nucleotide base change leads to the formation of a stop codon (purple). This results in premature termination of translation, thereby generating a truncated protein. Frameshift mutations involve the addition or deletion of any number of nucleotides that is not a multiple of three, causing a change in the reading frame.

Consequently, completely unrelated amino acid residues are incorporated into the protein prior to encountering a stop codon. (b) Chromosomal abnormalities involve alterations in large segments of DNA. Presumably these abnormalities arise due to errors in the mechanisms for repairing double-stranded breaks in DNA. Chromosomes (I or II) are shown as single thick lines with the regions involved in a particular abnormality highlighted in green or purple. Inversions occur when a break is rejoined to the correct chromosome but in an incorrect orientation; deletions, when a segment of DNA is lost; translocations, when breaks are rejoined to the wrong chromosomes; and insertions, when a segment from one chromosome is inserted into another chromosome.

generally affect the function of a single gene (Figure 8-6a). Changes in a single base pair may produce one of three types of mutation: a *missense* mutation, which results in a protein in which one amino acid is substituted for another; a *nonsense* mutation, in which a stop codon replaces an amino acid codon, leading to premature termination of translation; or a *frameshift* mutation, which leads to introduction of unrelated amino acids and generally stop codons. Small deletions have similar effects as frameshift mutations, although one third of these will be in-frame and result in removal of a small number of contiguous amino acids.

The second major type of mutation involves large-scale changes in chromosome structure and can affect the functioning of numerous genes, resulting in major phenotypic consequences. Such *chromosomal mutations,* or abnormalities, can involve deletion of several contiguous genes, in-

version of genes on a chromosome, or the exchange of large segments of DNA between nonhomologous chromosomes (Figure 8-6b).

Mutations Occur Spontaneously and Can Be Induced

Mutations arise spontaneously at low frequencies due to the chemical instability of purine and pyrimidine bases and to errors during DNA replication. Natural exposure of an organism to certain environmental factors, such as ultraviolet light and chemical carcinogens (e.g., aflatoxin B1), also can cause mutations.

A common cause of spontaneous point mutations is the deamination of cytosine to uracil in the DNA double helix. Subsequent replication leads to a mutant daughter cell in which a T-A base pair replaces the wild-type C-G

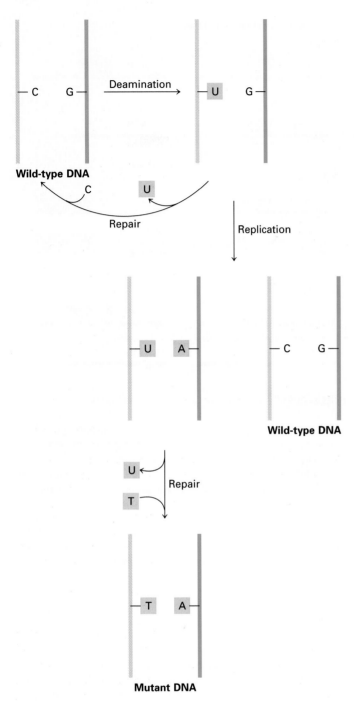

Wild-type DNA

Wild-type DNA

Mutant DNA

◄ FIGURE 8-7   Formation of a spontaneous point mutation as a result of deamination of cytosine (C) to form uracil (U). The resulting U-G base pair is recognized as abnormal and may be repaired to restore the wild-type C-G pair. Alternatively, the mutation can be fixed in the DNA through replication. After one round of DNA replication, one daughter DNA molecule will have a U-A base pair (mutant) and the other will have a C-G pair (wild type). The uracil is removed and replaced by thymine, generating a mutant DNA in which a T-A pair replaces a C-G pair.

treatments are referred to as *induced* mutations. Generally, chemical mutagens induce point mutations, whereas ionizing radiation gives rise to large chromosomal abnormalities.

Ethylmethane sulfonate (EMS), a commonly used mutagen, alkylates guanine in DNA, forming O^6-ethylguanine (Figure 8-9a). During subsequent DNA replication, O^6-ethylguanine directs incorporation of thymine, not cytosine, resulting in formation of mutant cells in which a G-C base pair is replaced with an A-T base pair (Figure 8-9b).

Some Human Diseases Are Caused by Spontaneous Mutations

Many common human diseases, often devastating in their effects, are due to mutations in single genes. Genetic diseases arise by spontaneous mutations in germ cells (egg and sperm), which are transmitted to future generations. For example, *sickle cell disease,* which affects 1 in 500 individuals of African descent, is caused by a single missense mutation at codon 6 of the β-globin gene; as a result of this mutation, the glutamic acid at position 6 in the normal protein is changed to a valine in the mutant protein. This alteration has a profound effect on hemoglobin, the oxygen-carrier protein of erythrocytes, which consists of two α-globin and two β-globin subunits (see Figure 3-15a). The deoxygenated form of the mutant protein is insoluble in erythrocytes and forms crystalline-like arrays. The erythrocytes of affected individuals become rigid and their transit through capillaries is blocked, causing severe pain and tissue damage. Because the erythrocytes of heterozygous individuals are resistant to the parasite causing malaria, which is endemic in Africa, the mutant allele has been maintained. It is not that individuals of African descent are more likely than others to acquire a mutation causing the sickle cell defect, but rather the mutation has been maintained in this population by interbreeding.

Spontaneous mutation in somatic cells (i.e., non-germline body cells) also is an important mechanism in certain human diseases including retinoblastoma, which is associated with retinal tumors in children. The hereditary form of retinoblastoma, for example, results from a germ-line

base pair (Figure 8-7). Although DNA replication generally is carried out with high fidelity, errors occasionally occur. The mechanism by which one type of error occurs during synthesis of DNA is illustrated in Figure 8-8. In the example shown, the mutant DNA contains nine additional base pairs.

In order to increase the frequency of mutation in experimental organisms, researchers often treat them with high doses of chemical mutagens or expose them to ionizing radiation. Mutations arising in response to such

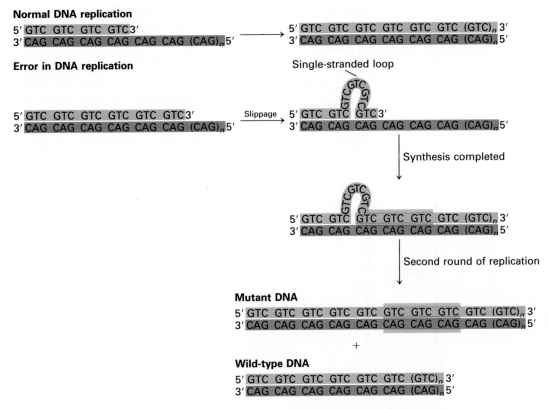

▲ FIGURE 8-8   One mechanism by which errors in DNA replication produce spontaneous mutations. The replication of only one strand is shown; the other strand is replicated normally, as shown at the top. A replication error may arise in regions of DNA containing tandemly repeated sequences (in this case, GTC) when a portion of the newly synthesized strand (light blue) loops out into a single-stranded form. This slippage displaces the newly synthesized strand back along the template strand (dark blue), with its 3′ end still paired with the template. As a result, the DNA-synthesizing enzymes copy a region of the template strand a second time, leading to an increase in length of nine nucleotides (yellow) in this example. A subsequent round of DNA replication results in the production of one normal duplex DNA molecule and one mutant duplex containing the additional nucleotides.

mutation in one *Rb* allele and a second somatically occurring mutation in the other *Rb* allele (Figure 8-10a). When an *Rb* heterozygous retinal cell undergoes somatic mutation, it is left with no normal allele; as a result, the cell proliferates in an uncontrolled manner, giving rise to a retinal tumor. A second form of this disease, called sporadic retinoblastoma, results from two independent mutations disrupting both *Rb* alleles (Figure 8-10b). Since only one somatic mutation is required for tumor development in children with hereditary retinoblastoma, it occurs at a much higher frequency than the sporadic form, which requires acquisition of two independently occurring somatic mutations.

Various Genetic Screens Are Used to Identify Mutants

The procedures used to identify and isolate mutants depend on whether the mutation is recessive or dominant and on the experimental organism. These procedures, referred to as *genetic screens*, generally are designed to identify and isolate recessive mutations induced by treatment with a mutagen. In haploid organisms (e.g., prokaryotes and yeast), the defects caused by induced mutations are seen immediately in the progeny of the mutagenized population. In diploid organisms, however, phenotypes resulting from recessive mutations can be observed only in individuals homozygous for the mutant alleles.

Genes that encode proteins essential for life are among the most interesting and important ones to study. Phenotypic expression of mutations in essential genes leads to death of the individual. So how does one isolate organisms with a lethal mutation and maintain them from one generation to the next? In prokaryotes and haploid eukaryotes such as yeast, essential genes can be studied through the use of *conditional* mutations. For instance, a mutant protein may be fully functional at 30°C but completely inactive at 37°C, whereas the normal protein would be fully

(a)

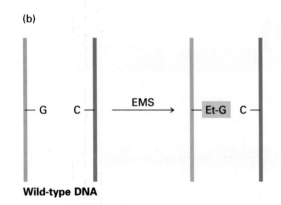

EMS   Guanine O^6-Ethylguanine   Thymine

◄ FIGURE 8-9   Induction of point mutations by ethylmethane sulfonate (EMS), a
commonly used mutagen. (a) EMS alkylates
guanine at the O^6 position, forming O^6-ethyl-
guanine (Et-G), which base pairs with thymine.
(b) Two rounds of DNA replication of a strand
containing Et-G yields a mutant DNA in which
a G-C base pair is replaced with an A-T pair.
Cells also have repair enzymes that can
remove the ethyl group from Et-G.

(b)

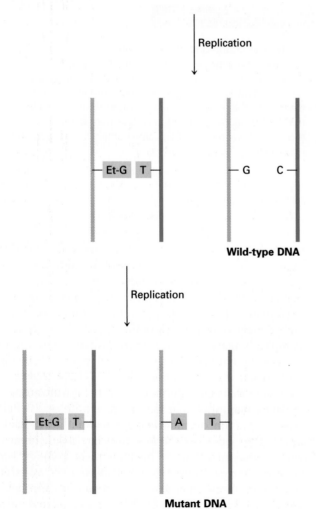

Wild-type DNA

Replication

Wild-type DNA

Replication

Mutant DNA

(a) **Hereditary retinoblastoma**

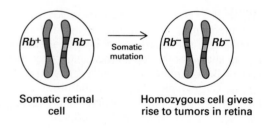

Somatic retinal   Homozygous cell gives
cell   rise to tumors in retina

(b) **Sporadic retinoblastoma**

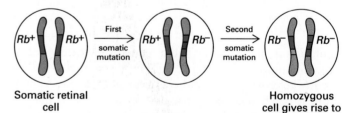

Somatic retinal   Homozygous
cell   cell gives rise to
                                             tumors in retina

▲ FIGURE 8-10   Role of spontaneous somatic mutation
in retinoblastoma, a childhood disease marked by retinal
tumors. Tumors arise from retinal cells that carry two mutant
Rb^- alleles. (a) In hereditary retinoblastoma, a child receives
a normal Rb^+ allele from one parent and a mutant Rb^- allele
from the other parent. A single mutagenic event in a heterozygous somatic retinal cell that inactivates the normal allele
will result in a cell homozygous for two mutant Rb^- alleles.
(b) In sporadic retinoblastoma, a child receives two normal
Rb^+ alleles. Two separate somatic mutations, inactivating
both alleles in a particular cell, are required to produce a
homozygous Rb^- Rb^- retinal cell.

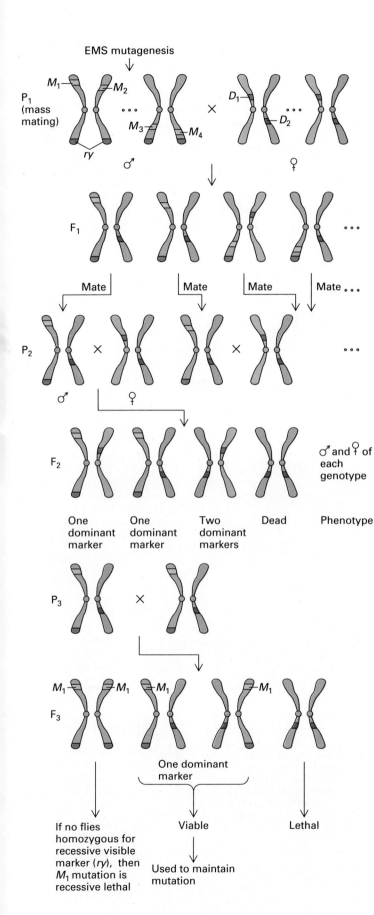

EMS mutagenesis

P_1 (mass mating)

M_1 M_2 $\times$ D_1 D_2

M_3 M_4

ry

♂   ♀

F_1 •••

Mate   Mate   Mate   Mate •••

P_2 $\times$ $\times$ •••

♂   ♀

F_2 ♂ and ♀ of each genotype

One dominant marker | One dominant marker | Two dominant markers | Dead | Phenotype

P_3 $\times$

M_1 M_1 M_1 M_1
F_3

One dominant marker

If no flies homozygous for recessive visible marker (ry), then M_1 mutation is recessive lethal

Viable

Used to maintain mutation

Lethal

◄ FIGURE 8-11   Procedure used to identify and maintain recessive lethal mutations on chromosome 3 (an autosome) in *Drosophila*, a diploid organism. This approach requires three sequential crosses. First, many males are treated with a mutagen (e.g., EMS), producing flies carrying various mutations (M_1, M_2, etc.) in their germ-line cells (sperm). These males also carry a nonlethal recessive mutation that gives rise to a visible phenotype in homozygotes; the marker in this example is called *rosy* (*ry*) and confers an altered eye color. In the first cross (P_1), the mutagenized males are mass-mated to a large number of females, traditionally in pint-size milk bottles. The females carry dominant visible mutations (D_1 and D_2) on chromosome 3; these are nonlethal in heterozygotes but are lethal in homozygotes. In the second cross (P_2), individual heterozygous F_1 males carrying mutagenized chromosome 3 are mated individually to non-mutagenized females in small culture vials. The F_2 progeny homozygous for either dominant marker will die; those heterozygous for both markers are easily identified and excluded. The remaining F_2 progeny include males and females that have one mutagenized chromosome carrying the *ry* marker and one nonmutagenized chromosome carrying a single dominant visible marker. These heterozygous brothers and sisters are mated individually in the third cross (P_3). The absence of flies with rosy-colored eyes in the F_3 progeny indicates induction of a lethal mutation (m_1 in this example) on chromosome 3.

functional at both temperatures. A temperature at which the mutant phenotype is observed is called *nonpermissive*; a *permissive temperature* is one at which the phenotype is not observed. Mutant strains can be maintained at a permissive temperature; then, for analysis, a subculture can be set up at a nonpermissive temperature.

Figure 8-11 outlines a procedure for inducing, identifying, and maintaining recessive lethal mutations in *Drosophila*, a diploid organism. Male fruit flies are treated with a mutagen and then mated with females, yielding F_1 progeny that are heterozygous for any induced mutations. Because these mutations are recessive, the mutant phenotype is not observed in the F_1 generation, and two additional crosses are needed to reveal the mutant phenotype. By using fly strains carrying known mutations (called *markers*) that give rise to visible phenotypes, researchers can distinguish heterozygous F_2 progeny carrying one mutagenized chromosome and one normal chromosome from siblings with other genotypes. Mating of these F_2 heterozygous siblings produces an F_3 generation in which one-fourth of the flies will be homozygous for any mutation induced on the mutagenized chromosome, and if the mutation is in a gene essential for viability, they will not survive; one-fourth will be homozygous for the normal allele; and one-half will be heterozygous. The effects of the mutation can then be assessed in the homozygous class that does not survive, and the mutation can be maintained in the flies that are heterozygous for the mutation.

We describe here two genetic screens that have been particularly useful in cell biology research: one for mutations affecting regulation of the cell cycle in yeast; the other for mutations affecting embryogenesis in *D. melanogaster*. The use of these mutants in revealing fundamental molecular mechanisms is discussed in later chapters.

Cell-Cycle Mutants in Yeast In the late 1960s and early 1970s, L. H. Hartwell and colleagues set out to identify genes important in regulation of the cell cycle in the yeast *S. cerevisiae*. Cell division in this yeast occurs through a budding process, and the size of the bud, which is easily visualized by light microscopy, is an indication of the cell's position in the cell cycle (see Figure 5-52).

The screen for cell-cycle mutants involved two steps. First, mutagenized yeast cells temperature-sensitive for growth were identified (Figure 8-12a). The identified mutants then were analyzed by video microscopy at the nonpermissive temperature for cell-division defects (see Figure

(a)

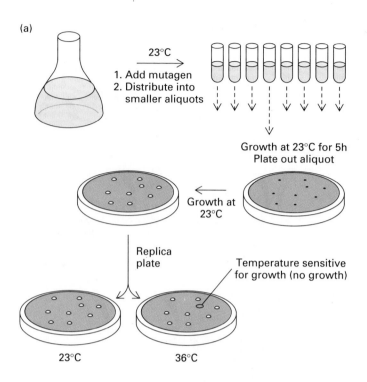

(b)

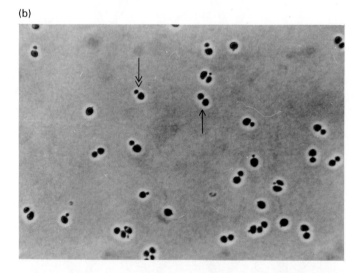

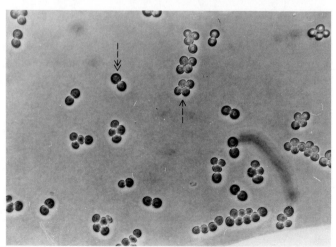

➤ FIGURE 8-12   Two-step genetic screen used to identify cell-cycle mutants in yeast. (a) Yeast cells were grown in a large liquid culture, treated with a chemical mutagen, and then subcultured into smaller aliquots. After a 5-h growth period at 23°C, aliquots from each tube were separately plated onto agar-containing petri dishes and incubated at 23°C. Colonies that developed on these plates were replica plated (see Figure 6-2) onto two plates: one was incubated at the permissive temperature (23°C); the other at the nonpermissive temperature (36°C). The temperature-sensitive colonies that grew at 23°C but not at 36°C were assessed to determine whether they were blocked at specific stages in the cell cycle. (b) Time-lapse photography of temperature-sensitive mutants identified in (a) permitted detection of yeast mutants with cell-cycle defects. The top photograph shows *cdc13* mutant cells growing at the permissive temperature just prior to being shifted to the nonpermissive temperature. The bottom photograph shows exactly the same field after a 6-h incubation at the nonpermissive temperature 36°C. By comparing the size of the buds in the top photograph with the morphology of the corresponding dividing cells in the bottom, the stage in the cell cycle in which the mutated gene is required can be determined. For example, when the bud is large (single-headed arrow), the cells will divide at the nonpermissive temperature but then are blocked prior to the next cell division; they thus appear in the bottom photograph as a cluster of four cells of equivalent size. In contrast, cells with very small buds (double-headed arrow) fail to divide at the nonpermissive temperature and appear in the bottom photograph as two cells of equivalent size. [Part (a) see L. H. Hartwell, 1967, *J. Bacteriol.* **93**:1662; part (b) from J. Culotti and L. H. Hartwell, 1971, *Exp. Cell Res.* **67**:391.]

8-12b). These yeast mutants were not simply slow growing as they might be if they carried a mutation affecting general cellular metabolism; rather, they grew normally but showed a stage-specific block in growth at the nonpermissive temperature. The cell-cycle stage at which cell growth was arrested at the nonpermissive temperature indicated when the protein encoded by the mutated gene was required.

One particularly important cell-division cycle (cdc) mutant identified in these screens was designated cdc28. When cdc28 cells that had just divided at the permissive temperature were placed at the nonpermissive temperature, they did not proceed through an additional division. However, when cdc28 cells were allowed to begin a cell cycle at the permissive temperature and then were shifted to the higher temperature at later times, the cells proceeded through the cell cycle and divided, but they failed to undergo an additional round of cell division. Recent work described in detail in Chapter 25 has shown that the CDC28 gene encodes a protein kinase that phosphorylates

particular serine and threonine residues. This enzyme is required for cells to proceed from the G_1 phase to the S phase of the cell cycle (see Figure 5-48). The CDC28 gene appears to be a critical regulator of the cell cycle in most, if not all, eukaryotic organisms.

Embryonic Mutants in Fruit Flies Current understanding of the molecular mechanisms regulating development of multicellular organisms is based, in large part, on genetic screens in *Drosophila*. C. Nüsslein-Volhard, E. Wiechaus, and their colleagues systematically screened for recessive lethal mutations affecting embryogenesis in *Drosophila* using a scheme similar to that shown in Figure 8-11. Dead homozygous embryos carrying lethal recessive mutations identified by this screen were analyzed for specific defects in the cuticular structures on the embryo surface (Figure 8-13). A detailed picture of embryonic development has emerged from the characterization of these defects and the analysis of both the structure of the encoded proteins and their patterns of expression during

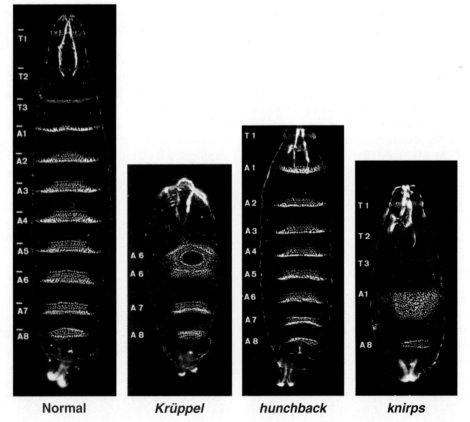

| Normal | Krüppel | hunchback | knirps |

▲ FIGURE 8-13   Comparison of normal *Drosophila* embryo and three dead embryos carrying recessive embryonic lethal mutations identified by a genetic screen similar to that outlined in Figure 8-11. Mutations of this type have played a key role in studying establishment of the pattern of the *Drosophila* embryo along the anterior-posterior axis (i.e., from head to tail). Defects in embryonic pattern are assessed by analyzing the cuticular structures on the surface of dead embryos. The anterior end of the embryo is towards the top. Thoracic and abdominal segments are designated T2–T3 and A1–A8, respectively. [From C. Nusslein-Volhard and E. Wiechaus, 1980, *Nature* **287**:795.]

embryogenesis. We will discuss some of the fundamental discoveries based on these genetic studies in Chapter 13.

Complementation Analysis Determines If Different Mutations Are in the Same Gene

In many genetic studies, different recessive mutations associated with the same phenotype are studied to determine whether such mutations are in the same gene or in different genes. If two mutations, A and B, are in the same gene, then a diploid organism heterozygous for both mutations (i.e., carrying one A allele and one B allele) will exhibit the mutant phenotype. In contrast, if mutation A and B are in separate genes, then diploid heterozygotes carrying a single copy of each mutant allele will exhibit the wild-type (normal) phenotype. In this case, the mutations are said to complement each other. *Complementation analysis* of a set of mutants exhibiting the same phenotype can distinguish the individual genes in a set of functionally related genes, all of which must function to produce a given phenotypic trait.

Complementation analysis has been used in many organisms including *S. cerevisiae*. Haploid yeast cells exist in one of two different mating types, **a** or *α*. Mating of **a** cells with *α* cells yields **a**/*α* diploids, which can be subjected to complementation analysis like other diploid organisms. For example, the yeast genome encodes four enzymes required for growth on galactose (Figure 8-14a). If any one of these enzymes is absent or defective, yeast cells cannot grow on galactose. Figure 8-14b illustrates complementation analysis of Gal⁻ yeast strains defective for growth on galactose. When Gal⁻ strains with mutations in different *GAL* genes are mated, the resulting diploid cells will grow on galactose, because the wild-type gene in each strain will compensate for the genetic defect in the other. In contrast, diploids formed from Gal⁻ strains that are mutated in the same gene will not grow on galactose.

Metabolic and Other Pathways Can Be Genetically Dissected

Various types of analysis can order the genes involved in biochemical pathways and other cellular processes. A fairly straightforward example involves the genetic dissection of the biochemical pathway for synthesis of arginine in the bread mold *Neurospora crassa*. Four different mutant strains that are unable to synthesize arginine and require arginine for growth (called arginine *auxotrophs*) were identified years ago. Each of the steps in biosynthesis of arginine is catalyzed by an enzyme encoded by a separate gene. The order of action of the different genes, hence the order of the biochemical reactions in the pathway, were determined by assessing which mutants could grow on different intermediates (Figure 8-15). Numerous biochemical pathways have been dissected by this type of study.

Other types of cellular processes also are amenable to genetic analysis. For example, the maturation pathway for secretory proteins in yeast has been dissected and ordered by analysis of a set of conditional temperature-sensitive secretion-defective (*sec*) mutants. In these mutant strains, the secretion of all proteins is blocked at the higher (non-permissive) temperature but is normal at the lower (permissive) temperature. At the higher temperature, *sec* mutants accumulate proteins in the rough endoplasmic reticulum (ER), Golgi complex, or secretory vesicles (Figure 8-16a). At least 60 gene products are required to complete the maturation pathway as defined by the number of genes in which mutations give rise to a secretion defect. Studies with double-mutant *sec* strains carrying mutations in two different genes have shown that the pathway must be ordered in the following sequence: rough ER → Golgi → secretory vesicles (Figure 8-16b). This maturation pathway is believed to apply to all secretory proteins in all eukaryotic organisms, including plants.

Suppressor Mutations Can Identify Genes Encoding Interacting Proteins

The phenomenon of *genetic suppression* can be used to identify proteins that specifically interact with one another in the living cell. The underlying logic is as follows. Point mutations may lead to structural changes in protein A that disrupt its ability to associate with another protein (protein B) involved in the same cellular process. Similarly, mutations in protein B might lead to small structural changes that would inhibit its ability to interact with protein A. In rare cases small structural changes in protein A may be suppressed by compensatory changes in protein B. In these rare cases, strains carrying a specific mutant allele of protein A or B would be mutant, but strains carrying both would be normal. This is analogous to changes made in a lock and key.

Identification of such suppressor mutations has been elegantly applied in studies of the cytoskeletal protein actin in yeast. A strain of yeast that was temperature sensitive for growth and carried a mutant actin allele called *act1-1* was plated at the nonpermissive temperature. A few cells were capable of growth at this temperature; these *revertants* were shown to have a second mutation in another gene, called *SAC6*, that allowed the *act1-1* mutants to grow. This *sac6* mutation acted as a dominant suppressor of the *act1-1* mutation, so that the double mutants (*act1-1 sac6*) exhibited the wild-type phenotype. This suppression was found to be allele specific: that is, the *sac6* mutation suppressed the *act1-1* mutation but not other *act1* mutations. Single mutants carrying any one of several different *sac6* mutations were, like *act1-1* mutants, temperature sensitive for growth. Remarkably, some *act1* mutations were found to be dominant suppressors of the recessive temperature-sensitive lethality of various *sac6* mutations.

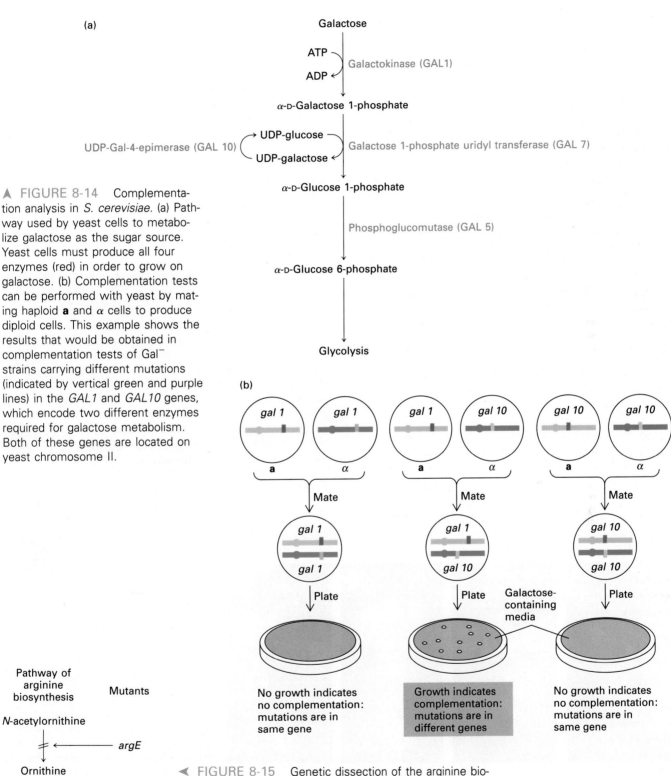

(a)

Galactose

ATP
ADP

Galactokinase (GAL1)

α-D-Galactose 1-phosphate

UDP-Gal-4-epimerase (GAL 10)   UDP-glucose / UDP-galactose

Galactose 1-phosphate uridyl transferase (GAL 7)

α-D-Glucose 1-phosphate

Phosphoglucomutase (GAL 5)

α-D-Glucose 6-phosphate

Glycolysis

▲ FIGURE 8-14 Complementation analysis in *S. cerevisiae*. (a) Pathway used by yeast cells to metabolize galactose as the sugar source. Yeast cells must produce all four enzymes (red) in order to grow on galactose. (b) Complementation tests can be performed with yeast by mating haploid **a** and α cells to produce diploid cells. This example shows the results that would be obtained in complementation tests of Gal⁻ strains carrying different mutations (indicated by vertical green and purple lines) in the *GAL1* and *GAL10* genes, which encode two different enzymes required for galactose metabolism. Both of these genes are located on yeast chromosome II.

(b)

gal 1   *gal 1*   *gal 1*   *gal 10*   *gal 10*   *gal 10*

a   α   **a**   α   **a**   α

Mate   Mate   Mate

gal 1 / *gal 1*   *gal 1* / *gal 10*   *gal 10* / *gal 10*

Plate   Plate   Plate

Galactose-containing media

No growth indicates no complementation: mutations are in same gene

Growth indicates complementation: mutations are in different genes

No growth indicates no complementation: mutations are in same gene

Pathway of arginine biosynthesis   Mutants

N-acetylornithine

argE

Ornithine

argF

Citrulline

argG

Arginosuccinate

argH

Arginine

◄ FIGURE 8-15 Genetic dissection of the arginine biosynthetic pathway. Mutants in the bread mold *N. crassa* that required arginine for growth were identified many years ago; analysis of these mutants provided the first genetic dissection of a biochemical pathway. Four separate genes were shown to be required for arginine biosynthesis by complementation analysis. These were ordered into a pathway by assessing the ability of different mutants to grow on different biosynthetic intermediates. For instance, *argE* mutants can grow on ornithine (and other intermediates later in the pathway), whereas *argH* mutants can grow only on arginine.

(a) Single mutants

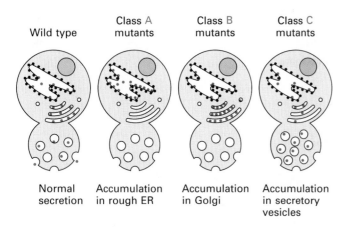

Wild type	Class A mutants	Class B mutants	Class C mutants
Normal secretion	Accumulation in rough ER	Accumulation in Golgi	Accumulation in secretory vesicles

Possible interpretations

Rough ER $\xrightarrow{A}$ Vesicles $\xrightarrow{C}$ Golgi $\xrightarrow{B}$ $\Big\{$ Exocytosis

or

Rough ER $\xrightarrow{A}$ Golgi $\xrightarrow{B}$ Vesicles $\xrightarrow{C}$ $\Big\{$ Exocytosis

(b) Double mutants: Class B and C

Accumulation in Golgi

Conclusion

Rough ER $\longrightarrow$ Golgi $\longrightarrow$ Vesicles $\longrightarrow$ $\Big\{$ Exocytosis

▲ **FIGURE 8-16** Genetic dissection of maturation pathway of secretory proteins in yeast. (a) Temperature-sensitive mutants carrying a single *sec* mutation accumulate secretory proteins (red dots) in various cellular compartments when grown at the higher (nonpermissive) temperature. The observed accumulation patterns are consistent with the two possible maturation sequences shown. (b) Double-mutant yeast cells carrying defects in both the class B and class C secretory processes accumulate proteins in the Golgi complex, not in secretory vesicles. Thus class B mutations act an earlier point in the maturation pathway than class C mutations; the correct sequence is shown at the bottom. [See P. Novick et al., 1981, *Cell* **25**:461.]

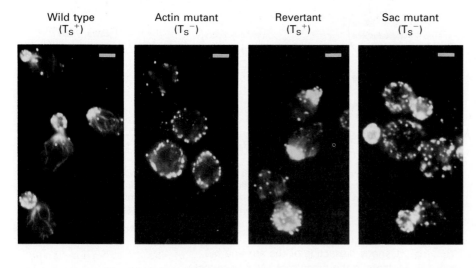

Wild type (Ts⁺)	Actin mutant (Ts⁻)	Revertant (Ts⁺)	Sac mutant (Ts⁻)

Genotype

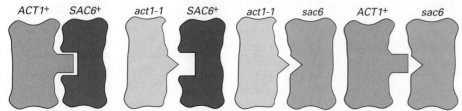

$ACT1^+$ $SAC6^+$	$act1\text{-}1$ $SAC6^+$	$act1\text{-}1$ $sac6$	$ACT1^+$ $sac6$

◄ **FIGURE 8-17** Genetic suppression in yeast involving temperature-sensitive (Ts⁻) mutations in the actin gene (*ACT1*) and in the *SAC6* gene, which encodes an actin-binding protein. (*Top*) Immunofluorescence micrographs of wild-type yeast, two Ts⁻ single-mutant strains, and a double-mutant revertant. In wild-type yeast, actin is distributed asymmetrically, whereas in the actin mutant (genotype = *act1-1 SAC6⁺*) and the Sac mutant (*ACT1⁺ sac6*), it is distributed randomly. The double-mutant revertant (*act1-1 sac6*) grows at high temperature and shows the wild-type actin distribution. (*Bottom*) Schematic diagrams interpreting these results in terms of compensatory changes in the structure of actin (orange) and the SAC protein (green). The inability of one mutant and one wild-type protein to interact leads to the mutant phenotype. However, the productive mutant-mutant interaction suppresses the mutant phenotype. [Photograph from A. E. M. Adams et al., 1989, *Science* **243**:231.]

In summary, then, mutations in either the *SAC6* or *ACT1* gene confer the same recessive lethal phenotype, and specific lethal alleles of each gene can act as dominant suppressors of specific lethal alleles of the other gene, resulting in a viable organism. This reciprocal suppression argues strongly for a direct interaction in vivo between the proteins encoded by the two genes (Figure 8-17). Indeed, biochemical studies have shown that these two proteins, ACT1 and SAC6, do indeed interact (Chapter 22), and immunolocalization studies indicate that the two proteins are present in the same part of the cell.

▶ *Genetic Mapping of Mutations*

Although genetic analysis can provide important insights into cellular pathways, its most profound impact on molecular cell biology has been to facilitate identification of the proteins that actually carry out these processes. *Genetic mapping* of a mutation-defined gene determines the precise position of the gene along the length of a particular chromosome relative to other genes. Mapping is an important step towards cloning of a gene; once a gene has been cloned, identification of its encoded protein is relatively easy.

Segregation Patterns Indicate Whether Mutations Are on the Same or Different Chromosomes

As discussed in Chapter 5, during meiosis each chromosome segregates independently. Therefore, traits controlled by genes on separate chromosomes also segregate independently (Figure 8-18). The observation that two mutant traits segregate independently indicates that the mutations are located on different chromosomes. See Figure 5-55 and below for discussion of meiotic recombination. Conversely, mutant gene loci that segregate together with a high frequency are on the same chromosome; such loci are referred to as *linked*.

Chromosomal Mapping Locates Mutations on Particular Chromosomes

The localization of a mutation to a particular chromosome is the first step in genetic mapping of a mutation. A simple example involves recessive mutations on the X chromosome in *Drosophila*. (In fruit flies, as in humans, males are genotypically XY and females are XX.) Such *X-linked recessive* mutations exhibit a distinctive *sex-linked* segregation pattern in various crosses (Figure 8-19). Mutant males mated to normal, homozygous females produce no phenotypically affected progeny. All the sons from a female homozygous for the mutation will have the mutant phenotype, and all the daughters will be heterozygous and thus unaffected. Although heterozygous females are phenotypically normal, they act as *carriers*, transmitting the mutant

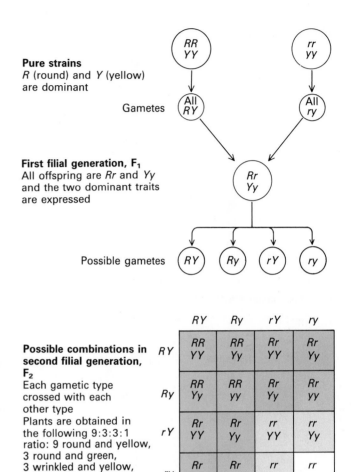

▲ FIGURE 8-18   When Mendel crossed strains of peas that were pure (homozygous) for each of two traits—say, round (*RR*) and yellow (*YY*) peas with wrinkled (*rr*) and green (*yy*) peas—the plants of the F₂ generation exhibited four different combinations of the two traits, which always appeared in the ratio 9:3:3:1. Mendel accounted for this ratio by proposing that the two traits (form and color) were assorted among the offspring independently of each other. We now know that such independently segregating traits are controlled by genes on separate chromosomes. This conclusion is known as Mendel's second law.

allele to 50 percent of their male progeny. Thus any recessive mutation for which this sex-linked segregation pattern is observed can be mapped to the X chromosome.

Human genetic diseases exhibit several inheritance patterns depending on the nature of the mutations that cause them. For example, Duchenne muscular dystrophy (DMD), a muscle degenerative disease that specifically affects males, is caused by a recessive mutation on the X chromosome. This clinically important mutation exhibits the typical sex-linked segregation pattern (Figure 8-20a).

In contrast to DMD, cystic fibrosis results from a recessive mutation in an autosome. Such *autosomal recessive* mutations exhibit a quite different segregation pattern.

(a) Mutant male × wild type homozygous female

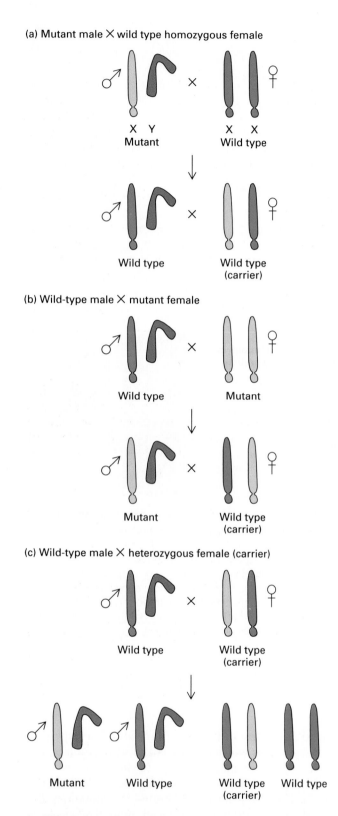

(b) Wild-type male × mutant female

(c) Wild-type male × heterozygous female (carrier)

▲ FIGURE 8-19 Unique segregation pattern of X-linked recessive traits. In this example, segregation of mutant X chromosomes (yellow) and of wild-type X chromosomes (blue) is shown. Observation of this segregation pattern for a mutant phenotypic trait maps the gene affected by the mutation to the X chromosome.

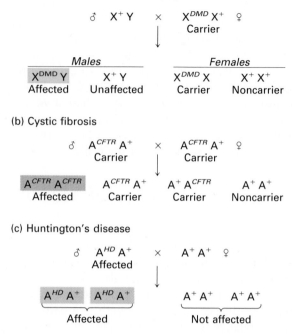

(a) Duchenne muscular dystrophy

$$\delta \quad X^+ Y \quad \times \quad X^{DMD} X^+ \quad \mathcal{Q}$$
Carrier

Males		Females	
$X^{DMD} Y$	$X^+ Y$	$X^{DMD} X$	$X^+ X^+$
Affected	Unaffected	Carrier	Noncarrier

(b) Cystic fibrosis

$$\delta \quad A^{CFTR} A^+ \quad \times \quad A^{CFTR} A^+ \quad \mathcal{Q}$$
Carrier Carrier

$A^{CFTR} A^{CFTR}$	$A^{CFTR} A^+$	$A^+ A^{CFTR}$	$A^+ A^+$
Affected	Carrier	Carrier	Noncarrier

(c) Huntington's disease

$$\delta \quad A^{HD} A^+ \quad \times \quad A^+ A^+ \quad \mathcal{Q}$$
Affected

$A^{HD} A^+$	$A^{HD} A^+$	$A^+ A^+$	$A^+ A^+$
Affected		Not affected	

▲ FIGURE 8-20 Segregation patterns for three human genetic diseases. Wild-type chromosomes are indicated by superscript plus sign. (a) Duchenne muscular dystrophy is caused by a recessive mutation in the *DMD* gene on the X chromosome, which exhibits the typical sex-linked segregation pattern. Males born to mothers heterozygous for a *DMD* mutation have a 50 percent chance of inheriting the mutant allele and being affected. Females born to heterozygous mothers have a 50 percent chance of being carriers. (b) Cystic fibrosis results from a recessive mutation on an autosome (A). Both males and females can be affected but must carry two mutant alleles of the *CFTR* gene to show the mutant phenotype. Both parents must be heterozygous for their children to be at risk of being affected or being carriers. (c) Huntington's disease is caused by an autosomal dominant mutation. Both males and females can be affected, and only one mutant allele is needed to confer the mutant phenotype. If either parent is heterozygous for the mutant *HD* allele, their children have a 50 percent chance of inheriting the mutant allele and getting the disease.

First, both males and females can be affected with cystic fibrosis. Second, both parents must be heterozygous carriers of a mutant *CFTR* allele in order for their children to be at risk of being affected with the disease. Each child of heterozygous parents has a 25 percent chance of receiving both mutant *CFTR* alleles and thus being affected; a 50 percent chance of receiving one normal and one mutant allele and thus being a carrier, and a 25 percent chance of receiving two normal alleles (Figure 8-20b).

Autosomal dominant mutations are associated with a third segregation pattern. Huntington's disease, a neural degenerative disease that generally strikes in mid to late life is caused by this type of mutation. If either parent carries a mutant *HD* allele, each of their children (regardless of sex) has a 50 percent chance of inheriting the mutant allele and being affected (Figure 8-20c).

Mapping of a gene to a specific autosome is more complicated. Once one mutation has been mapped to a particular chromosome, it can be used as a marker in linkage studies to identify other mutations located on that chromosome. In humans, small differences, or polymorphisms, in DNA sequences among individuals serve as molecular markers; this phenomenon is discussed in a later section.

Recombinational Analysis Can Map Genes Relative to Each Other on a Chromosome

The position of a particular gene relative to other genes on a chromosome can be assessed in many organisms by recombination mapping. This technique was devised one night in 1913 by A. Sturtevant while he was an undergraduate working in the laboratory of T. H. Morgan at Columbia University. Originally used in studies on *Drosophila,* recombination mapping is still used today to assess the distance between two genetic loci on the same chromosome in many experimental organisms.

The basis for Sturtevant's gene-mapping technique is the exchange, or *recombination,* of DNA sequences between homologous chromosomes that occurs naturally during meiosis (see Figure 5-55). As noted earlier, genetic traits that segregate together during meiosis more frequently than expected from random segregation are said to be linked and are inferred to be controlled by genes on the same chromosome. *The more frequently recombination occurs between two genes on the same chromosome, the farther apart they are.* The frequency of exchange between two points along the length of a chromosome, called the *recombination frequency,* is proportional to the distance in base pairs separating the two points.

Although the molecular mechanisms underlying meiotic recombination are not well understood, a critical first step in this process is the introduction of breaks in the DNA; the reciprocal exchange of the chromosomal segments occurs at the position of the break. These breaks are thought to occur at random along the length of the DNA; the farther apart two genes are, the more likely that a DNA break will occur between them at meiosis. By using many different genetic traits, or *markers,* distributed along the length of a chromosome, the position of a mutation can be determined by assessing its segregation with respect to these marker genes during meiosis. Thus the more markers that are available, the more precisely a mutation can be mapped.

As more and more mutations are mapped, the linear order of genes along the length of a chromosome, referred to as a *genetic map,* can be constructed. The more genes that are localized along a chromosome, the more detailed the map. (See Figure 6-3 for a genetic map of *E. coli,* whose single circular chromosome has been mapped in great detail.) By convention, one *genetic-map unit* is defined as the distance between two positions along a chromosome that results in one recombinant individual in 100 progeny. The distance corresponding to this 1 percent recombination frequency is called a *centimorgan* (in honor of Sturtevant's mentor, Morgan). Determination of the genetic-map distance between two loci in *Drosophila* by recombinational analysis is illustrated in Figure 8-21.

Comparison of the physical distances between known genes, determined by molecular analysis, with their recombination frequency indicates that in *Drosophila* a 1 percent recombination frequency represents a distance of about 400 kilobases. The relationship between recombination frequency (i.e., genetic-map distance) and physical distance differs among organisms.

DNA Polymorphisms Are Used to Map Human Mutations

In the experimental organisms commonly used in genetic studies, many phenotypic markers are readily available for genetic mapping of mutations. This is not the case for map-

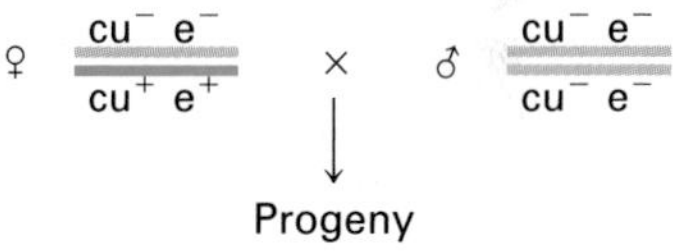

Phenotype	Genotype From ♀	From ♂	Total
Parental: curled wings, dark body	$cu^- e^-$	$cu^- e^-$	389
normal wings, normal body	$cu^+ e^+$	$cu^- e^-$	414
Recombinant: curled wings, normal body	$cu^- e^+$	$cu^- e^-$	104
normal wings, dark body	$cu^+ e^-$	$cu^- e^-$	93
			1000

$$\text{Genetic-map distance} = \frac{\text{Recombinants}}{\text{Recombinants} + \text{Parentals}} \times 100 = \frac{104 + 93}{1000} \times 100 = 19.7 \text{ centimorgans}$$

▲ FIGURE 8-21 Determination of genetic-map distance between two *Drosophila* loci by recombinational analysis. In *Drosophila,* meiotic recombination occurs only in females. Here, the distance between two recessive mutations causing curled wings (*cu*) and darkly pigmented, or ebony, bodies (*e*) is determined. Wild-type alleles are indicated with a superscript plus sign; mutant alleles with superscript minus sign. Mutant chromosomal regions are diagrammed as light green lines; wild-type regions as dark blue lines.

ping loci associated with genetically transmitted diseases in humans. However, through recombinant DNA technology many molecular DNA markers are now available. One type of commonly used DNA marker is referred to as a *restriction fragment length polymorphism* (RFLP). RFLPs are variations among individuals in the length of restriction fragments from identical regions of the genome.

In all organisms natural DNA sequence variations occur throughout the genome. Because most of the human genome does not code for protein (Chapter 9), a large amount of sequence variation is acceptable in humans. Indeed, it has been estimated that in humans base changes between different individuals can be detected every 200 nucleotides or so. These variations in DNA sequence, referred to as *DNA polymorphisms,* may create or destroy restriction-enzyme recognition sites. As a consequence, the pattern of restriction fragment lengths from a region of the genome may differ between two individuals (Figure 8-22a). Loss of a site will result in the appearance of a larger fragment and the disappearance of two smaller fragments. Formation of a new site will result in the loss of a larger frag-

ment and its replacement with two smaller fragments. These changes in DNA structure are used as molecular markers in a fashion analogous to phenotypic markers used in mapping by meiotic recombination. Figure 8-22b illustrates how RFLP analysis of a family can detect allelic forms within a DNA region.

How do polymorphisms arise? As we discussed in an earlier section, a finite number of mutations occur spontaneously in the DNA sequence. Some of these will occur in coding sequences, leading to production of defective protein sequences. Others will occur in noncoding sequences; in most cases, these will have no effect on function. For the purposes of genetic markers, the effects on organismal function of these changes in DNA sequence are unimportant. In humans, polymorphisms often appear to be associated with single-base changes; CpG dinucleotide sites are "hot spots" for single-base changes. Consequently restriction-enzyme cleavage sites that contain this dinucleotide are often polymorphic. The length of DNA between two restriction sites bracketing short sequences of DNA that are repeated in a head-to-tail fashion are often highly vari-

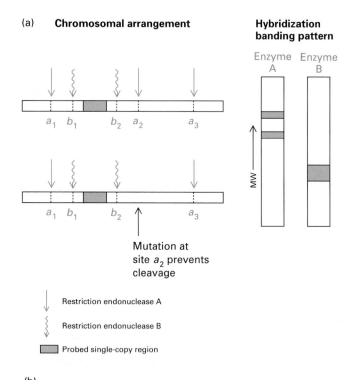

(a)   Chromosomal arrangement

Hybridization banding pattern

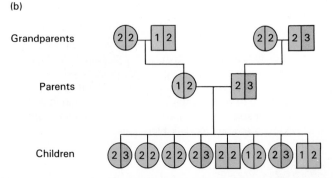

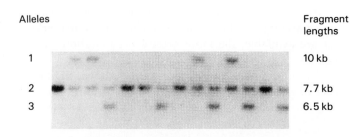

◀ FIGURE 8-22   Analysis of restriction fragment length polymorphisms (RFLPs). (a) In the example shown, DNA is treated with two different restriction enzymes, which cut DNA at different sequences. The resulting fragments are subjected to Southern blot analysis; that is, they are separated by electrophoresis, transferred to nitrocellulose, and detected by hybridization with a radioactive probe, which binds to the indicated DNA region (green). Since no differences between the two chromosomes occur in the sequences recognized by the B enzyme, only one fragment is produced, as indicated by a single hybridization band. However, treatment with enzyme A produces fragments of two different lengths (two bands are seen), indicating that a mutation has caused the loss of one of the *a* sites. (b) RFLP analysis of the DNA from eight children, their parents, and grandparents detected the presence of three alleles for a region known to be present on chromosome 5. The DNA samples were cut with the restriction enzyme *Taq*I and analyzed by the Southern blot procedure. In this family, this region exists in three allelic forms characterized by *Taq*I sites spaced 10, 7.7, or 6.5 kb apart. Each individual has two alleles; some contain allele 2 (7.7 kb) on both chromosomes, and others are heterozygous at this site. Circles indicate females; squares indicate males. [After H. Donis-Keller et al., 1987, *Cell* **51**:319.]

able. These variable repeat regions are on the order of 14–70 base pairs long; they are found as single copies or in arrays containing up to 100 repeat units. Recently, even simpler shorter repeat sequences (e.g. CA repeats) have been used. Since these repeats are generally less than 200 base pairs they can be amplified easily by PCR using unique flanking DNA sequences as primers. It is estimated that these repeats are found on the order of once every 40 Kb. Presumably arrays of repeats are formed by recombination or a slippage mechanism of either the template or newly synthesized strands during DNA replication (see Figure 8-8). Polymorphic restriction sites are limited in that there are generally only 2 alleles at any given locus. In contrast, polymorphic repeat sequences will give rise to many different alleles and hence will be multiallelic and of more general use in mapping disease loci.

In experimental organisms such as *Drosophila*, recombination mapping is facilitated not only by the availability of readily visible markers but by the ability to carry out specifically desired genetic crosses and to obtain large numbers of progeny from a given cross. For humans, of course, this is not the case, necessitating an alternative approach. Since many families can be identified in which individuals are at risk for a genetic disease (i.e., both parents are heterozygous for an autosomal recessive mutation associated with a particular disease), the frequency with which specific polymorphic markers segregate with the mutant allele causing the disease is a measure of their proximity to the locus. The more families afflicted with a particular disease that are available for study and the more DNA polymorphisms identified, the more precisely the disease-associated gene can be mapped.

In most family studies, however, it is highly unlikely that more than the products of 100 meioses can be obtained limiting the localization to no less than 1 cM. An alternative strategy and one that will become more widely used is called *linkage disequilibrium*. In this approach one assumes that a genetic disease commonly found in a particular community was likely to be caused by a single mutation many generations ago. This ancestral chromosome will carry closely linked markers that will be conserved through many generations. Markers which are further away on the chromosome will tend to become separated from the disease gene by recombination. This allows human geneticists to take all the affected individuals in a population and to assess the distribution of specific markers; DNA markers tightly associated with the disease can be used to localize the disease locus to a relatively small region. This approach was used to localize the region for the gene causing diastrophic dysplasia in a large Finnish population to 60 kilobases.

Some Chromosomal Abnormalities Can Be Mapped by Banding Analysis

Many mutations involve large changes in DNA structure that can be mapped by direct light-microscope observation of chromosomes. Normal human metaphase chromosomes exhibit characteristic banding patterns when stained with various dyes (Figure 8-23). The nonuniform staining of

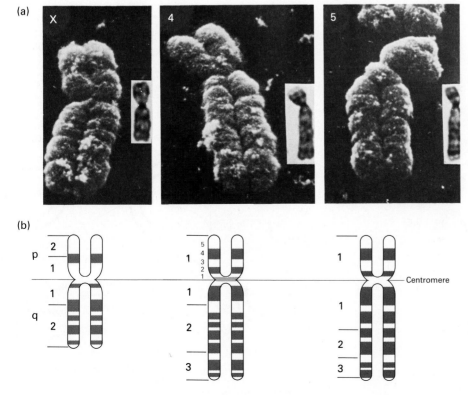

◄ FIGURE 8-23 Human male chromosomes X, 4, and 5 after G-banding. The chromosomes were subjected to brief proteolytic treatment and then stained with Giemsa reagent, producing distinctive bands at characteristic places. (a) Scanning electron micrographs show constrictions at the sites where bands appear in light micrographs (*insets*). (b) Standard diagrams of G bands (purple). Regions of variable length (green) are most common in the centromere region (e.g., on chromosome 4). The numbering system to indicate positions on the chromosomes is illustrated: p = short arm, q = long arm. Each arm is divided into major sections (1, 2, etc.), and subsections. The short arm of chromosome 4, for example, has five subsections. DNA located in the fourth section would be said to be in p14. [Part (a) from C. J. Harrison et al., 1981, *Exp. Cell Res.* **134**:141; courtesy of C. J. Harrison.]

(a)

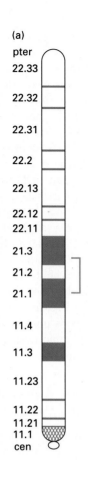

(b)

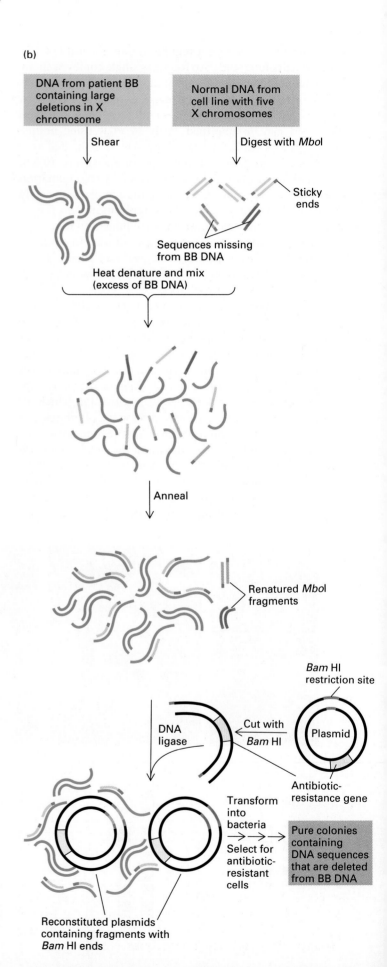

(c)

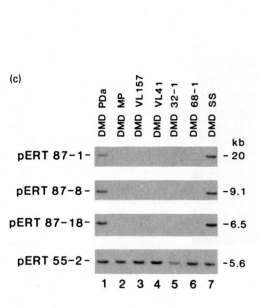

◄ FIGURE 8-24 Cytological mapping and cloning of gene associated with Duchenne muscular dystrophy. (a) Schematic diagram of banding pattern of human X chromosome; the red bracket indicates the position of the cytologically visible deletion in patient BB with muscular dystrophy. In addition to muscular dystrophy, this patient had several other genetic diseases due to loss of other genes adjacent to the *DMD* gene. (b) Isolation of probes mapping close to *DMD* gene using DNA from patient BB. Shearing of DNA from BB yielded fragments of various lengths with heterogeneous ends, which generally were blunt. Normal DNA was cleaved with the restriction endonuclease *Mbo*I, which produces small fragments with sticky ends (red) complementary to the sticky ends on fragments resulting from *Bam*HI cleavage. (To increase the proportion of normal X-chromosome DNA in this sample, a cell line carrying five X chromosomes was used as the source of the normal DNA.) The sheared BB DNA and the *Mbo*I-cut normal DNA were each heat denatured and then mixed. Since a large excess of BB DNA was used, most of the *Mbo*I fragments hybridized with it; the only *Mbo*I fragments that renatured with each other were those within the region containing the sequences deleted in the BB DNA (orange and green). These renatured fragments were selectively cloned in a DNA vector containing a *Bam*HI cloning site. Using this procedure, L. M. Kunkel and colleagues produced 125 clones, four of which mapped to the region deleted in the BB DNA. One of these specific clones, designated pERT87, was used to identify overlapping clones, and small regions of these overlapping clones (spanning about 38 kb) were used to probe the DNA of other muscular dystrophy patients. (c) Southern blots of DNA from several patients with muscular dystrophy probed with the pERT87 probe. These patients did not have the other diseases exhibited by patient BB, and their DNA showed no cytologically visible mutations. However, as the blot shows, the DNA from five patients (patients MP, VL157, VL41, 32-1, and 68-1) lack the sequence corresponding to the pERT87 probe. Since the *DMD* gene is very large (about 2 megabases), some patients with defects in the *DMD* gene are not identified with the pERT87 probe but are identified with other probes. [Parts (a) and (c) from A. P. Monaco et al., 1985, *Nature* **316:**842; part (b) see U. Franke et al., 1985, *Am. J. Hum. Genet.* **37:**250.]

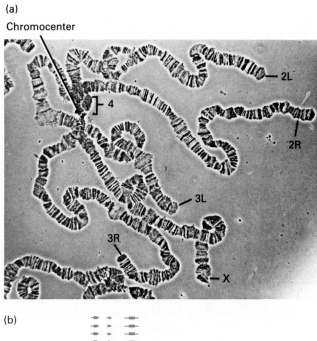

(a)

Chromocenter

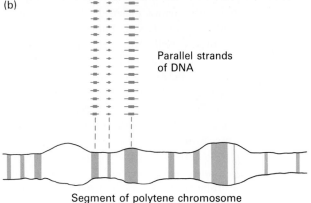

(b)

Parallel strands of DNA

Segment of polytene chromosome

▲ FIGURE 8-25 (a) Light micrograph of polytene salivary gland chromosomes from *Drosophila melanogaster* stained to reveal the very reproducible banding pattern. In these four chromosomes (X, 2, 3, and 4), a total of approximately 5000 bands can be distinguished. The centromeres of all four chromosomes often appear fused at the *chromocenter;* the smallest chromosome (4), a dot chromosome, is also associated with the chromocenter. The tips of the metacentric 2 and 3 chromosomes are labeled (L = left arm; R = right arm). The tip of the acrocentric X chromosome is also labeled. (b) The DNA in salivary polytene chromosomes is repeated about 1000 times; the duplicated DNA fibers are thought to remain parallel. Therefore, any staining property in a chromosome with one DNA duplex would be amplified 1000 times to produce a transverse band (light blue). [Part (a) courtesy of J. Gall.]

chromosomes with these dyes presumably reflects local differences in chromosome structure discussed in Chapter 9.

Differences in the banding patterns of normal and mutant chromosomes can help map the mutations causing certain human diseases. For instance, the X chromosome from some patients with Duchenne muscular dystrophy has an altered banding pattern reflecting a deletion of part of the chromosome that is large enough to be cytologically visible (Figure 8-24a). Analysis of the DNA from one such patient (BB) has been useful in localizing the chromosomal position of the affected *DMD* gene and cloning it (Figure 8-24b). Clones produced from this analysis can be used to probe the DNA from other DMD patients who do not have cytologically visible mutations (Figure 8-24c).

In *Drosophila*, the light-microscope study of chromosomal abnormalities is facilitated by the very large chromosomes of the larval salivary gland (Figure 8-25). These chromosomes have undergone many rounds of DNA replication without cell division, resulting in the colinear arrangement of 1000–2000 copies of each chromosome. The large size of these chromosomes, referred to as *polytene chromosomes*, has permitted compilation of very detailed maps of stained regions (bands) and unstained regions (interbands). Comparison of the banding patterns from normal flies and mutant flies can localize some mutations to quite specific chromosomal regions.

➤ *Molecular Cloning of Genes Defined by Mutations*

In the first two sections of this chapter, we described how to isolate and characterize mutants and to map mutations. In this section, we discuss the use of recombinant DNA techniques to isolate mutation-defined genes. It is this approach, often referred to as *positional cloning*, that has revolutionized the study of developmental biology and human genetics.

In *E. coli* and yeast a specific gene defined by mutations can be isolated by transforming cells exhibiting the mutant phenotype with various plasmids that collectively represent the total genomic DNA. Those plasmids that carry the gene of interest will restore the mutants to the wild-type phenotype, which can be identified easily. Plasmids can rescue mutants because they carry a normal copy of genes. This *DNA-rescue* approach is feasible in *E. coli* and yeast because collections of plasmids representing the entire genome are available and can be effectively transferred into mutants. Although individual cloned genes can be introduced into genetically tractable higher organisms (e.g., *Drosophila, Caenorhabditis elegans,* and mice), these methods are not efficient enough to be used for screening plasmid libraries representing the entire genome.

The cloning of genes defined by mutations in higher organisms, including humans, is feasible only if DNA clones representing a limited region of the genome that includes the gene of interest can be identified. To obtain these specific clones, two steps are required: construction of the *physical map* of the genome, which is based on the actual distances between different nucleotide stretches, and correlation of the physical map with the genetic map in which distances are measured in recombination frequencies. Efforts to establish physical maps for a number of experimentally important genetically tractable higher organisms and for humans are currently under way.

The general strategy for isolating genes defined by mutations in higher organisms involves three steps:

• Determine the genetic-map position of the mutation

• Obtain DNA clones for the region of the chromosome roughly corresponding to the genetic-map position

• Search for differences in the structure of mutant and wild-type DNA or in the expression of specific mRNAs to identify the gene whose function has been disrupted

Physical Maps of Human Chromosomes Y and 21 Have Been Constructed by Screening YAC Clones for Sequence-Tagged Sites

Physical maps of both the human Y chromosome and the long arm of chromosome 21 have been completed using essentially the same approach in both cases. As discussed in Chapter 7, long segments of human DNA can be cloned in yeast artificial chromosomes (YACs). These clones can be ordered by using the polymerase chain reaction (PCR) to screen them for *sequence-tagged sites* (STSs). These are randomly spaced, unique sequences of DNA known to map to particular chromosomes or parts of chromosomes. Figure 8-26 illustrates this procedure in the simple case of three contiguous YAC clones containing six STSs randomly distributed over the region spanned by the clones.

Detecting and ordering a set of YACs for an entire chromosome is done in a similar way. However, instead of subjecting individual YACs to PCR analysis, which is far too laborious, pools of YACs are screened. Those pools in which PCR products are detected then are divided into smaller pools, and the PCR analysis is repeated. This process is continued until a set of YACs is identified each of which contains at least one STS and that collectively contain all the STSs. The detail of a map constructed with this procedure depends on the number of STSs used in the PCR analyses and the number of DNA clones available for analysis. In the mapping of human chromosome 21, for example, 120,000 clones from two separate YAC libraries were screened in pools; in addition 14,000 YACs isolated from a library prepared specifically from chromosome 21 were screened individually. By use of 198 STSs, researchers identified 810 positive clones and ordered them into a contiguous map.

Physical and Genetic Maps Can Be Correlated

In order for a physical map to be helpful in identifying a segment of DNA corresponding to a specific mutation, the cloned DNA must contain specific landmarks that can be identified by genetic linkage. Such landmarks include known genetic loci and RFLPs. The availability of extensive physical and genetic maps of the genome of the genetically tractable worm *C. elegans* has greatly facilitated the cloning of genes defined by mutation. In this organism, the genetic-map position of a mutation immediately defines a physical interval of DNA within which the gene is located.

One strategy for correlating physical and genetic maps is outlined in Figure 8-27 for a defined region of the *Drosophila* genome. In this example, the objective was to localize the *shibire* (*shi*) locus, which is on the X chromosome, in order to clone it. This locus had already been mapped by recombinational analysis with visible phenotypic markers spread along the entire length of the X chromosome. From a collection of *Drosophila* YAC clones, a 275-kb clone

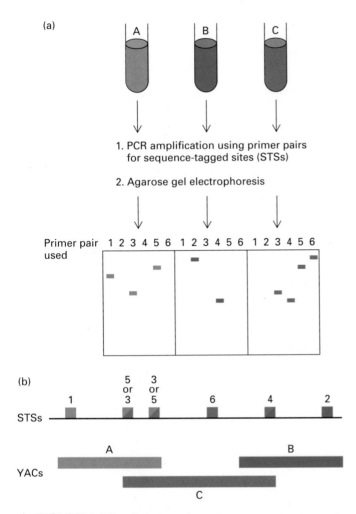

(a)

A   B   C

1. PCR amplification using primer pairs
   for sequence-tagged sites (STSs)

2. Agarose gel electrophoresis

Primer pair   1 2 3 4 5 6   1 2 3 4 5 6   1 2 3 4 5 6
used

(b)

STSs

         1   5   3   6   4   2
                  or   or
                  3   5

YACs

                    A   B

                              C

▲ FIGURE 8-26   Ordering of contiguous overlapping YAC clones. (a) Aliquots of DNA prepared from each YAC clone (A, B, and C) are subjected to PCR amplification (see Figure 7-37) using primer pairs (1–6) corresponding to the ends of various sequence-tagged sites (STSs). Only those clones containing STSs with ends complementary to particular primers will be amplified. Electrophoretic analysis then shows which YAC clones contain which STSs. For example, clone A contains STS 1, 3, and 5. (b) Based on the gel patterns in this example, the three YAC clones can be ordered as shown. However, the positions of STS 3 and 5 relative to each other cannot be determined from this analysis. This general approach has been used to generate a complete physical map of the human Y chromosome and the long arm of human chromosome 21.

viduals, the greater the probability of obtaining a recombination event between an RFLP and the locus of interest. Once a region of DNA containing a gene of interest has been identified, the specific location of the gene must be determined, since the gene itself may correspond to only a small segment of the region defined by RFLP analysis. We discuss below the approaches that are used to identify a gene of interest within a larger segment of genomic DNA.

Physical Mapping of Selected Genomic Regions Is the First Step in Cloning Many Genes

Although complete physical maps of the genomes of many experimental organisms and humans will soon be completed, many genes are now cloned by constructing a physical map of a limited region of the genome containing the gene of interest. One of the first steps in isolating a particular gene is to identify cloned DNA segments that map in close proximity to that gene. These clones can be identified by assessing the gene's segregation patterns either with respect to DNA clones that recognize different polymorphic markers or to specific genetic loci that have been cloned previously. In most genetically tractable organisms, many DNA clones mapping throughout the genome are already available. As discussed in the previous section, mapping of a large number of polymorphic sites (detected by cloned DNA) relative to a gene of interest increases the likelihood that a clone mapping close to that gene will be identified.

In Situ Hybridization   The chromosomal location of a cloned DNA segment can be determined by hybridizing it to metaphase chromosomes in mouse and humans and to interphase polytenized chromosomes in *Drosophila*. In this technique, referred to as *in situ hybridization*, cloned DNA is labeled in vitro using nucleoside triphosphates that are conjugated to biotin (a naturally occurring prosthetic group that can be used as a molecular tag) or contain a fluorescent nucleoside derivative. The labeled DNA is placed on a microscope slide to which chromosomes are attached; the sample then is treated to allow visualization by light microscopy of regions where the labeled DNA probe hybridizes to the chromosomes (Figure 8-28). This technique has been particularly useful in chromosomal mapping of DNA clones in *Drosophila*, because the cytological map of the polytene chromosomes of this organism is known at high resolution.

Specific DNA clones also can be mapped to particular regions of human chromosomes by screening for their presence, with the Southern-blot technique (see Figure 7-31), in hybrid cells formed between cultured human and rodent cells. As discussed in Chapter 6, the human chromosomes in these hybrids typically are unstable, but stable hybrids containing one or a few human chromosomes can be produced. Furthermore, if human cells carrying chromosomal translocations are used, stable hybrid lines can be estab-

containing the genetic-map position of *shi* was identified. To locate the region within this YAC clone containing *shi*, recombinational analysis was carried out using 10 RFLPs and a closely linked visible phenotypic marker.

The length of DNA to which a specific mutation can be localized depends upon the density of cloned loci or RFLPs in the region of interest. In humans, it also depends on the number of individuals that can be studied: the more indi-

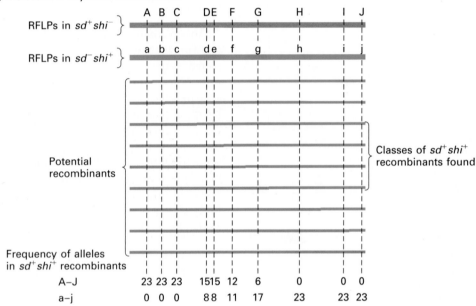

(a) Isolation of recombinants between scalloped (*sd*) and shibire (*shi*) genes

Identified 23 *sd⁺ shi⁺* ♂ recombinants

(b) Molecular analysis of recombinants

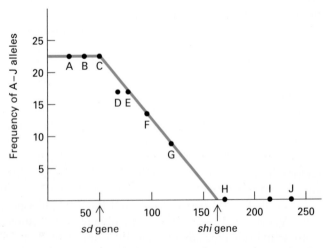

(c) Frequency of A–J alleles from *sd⁺ shi⁻* chromosome in *sd⁺ shi⁺* recombinants

Position of RFLPs in YAC clone

▲ FIGURE 8-27   Correlation of the physical and genetic maps of the region of the *Drosophila* X chromosome containing the shibire (*shi*) locus. The genetic-map position of *shi* was known from recombinational analysis with visible phenotypic markers spread along the length of the X chromosome. In addition, a 275-kb YAC clone containing the *shi* locus had been identified. In order to clone *shi*, its specific location within this YAC clone had to be determined. (a) As a first step, 23 male recombinants carrying both the *shibire* locus and a closely linked, visible phenotypic marker called *scalloped* (*sd*) were isolated. (b) These recombinants then were analyzed for the presence of 10 RFLPs that differed in *shi⁺* chromosomes (a–j) and *shi⁻* chromosomes (A–J). This analysis showed, for example, that none of the *sd⁺ shi⁺* recombinants found contained just A, whereas all contained A–C and 15 of the 23 recombinants contained A–D. (c) The frequency of the A–J alleles in these recombinants was plotted against the known positions of the RFLPs within the 275-kb YAC clone. The intersection of the curve with the X axis predicts the position of the *shi* gene. This analysis led to the successful localization and cloning of the *shi* gene. [See A. M. van der Bliek and E. M. Meyerowitz, 1991, *Nature* **351**:411.]

(a)

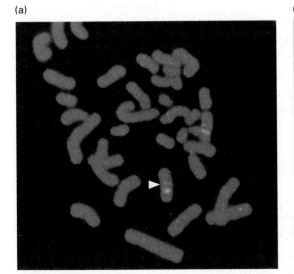

(b)

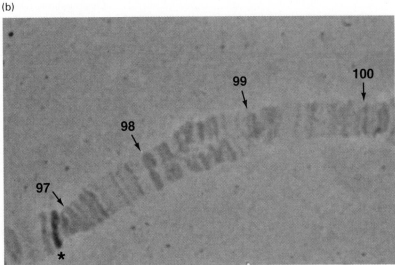

▲ FIGURE 8-28   Chromosomal mapping of cloned DNA segments by in situ hybridization. In this procedure cloned DNA is labeled by one of several possible methods and then hybridized to chromosomes attached to microscope slides. (a) A human metaphase chromosome hybridized with a fluorescent-labeled probe for the glutamate dehydrogenase (*GD*) gene and with a probe specific for the centromere of chromosome 10 (arrowhead). The thick yellow hybridization band below the arrowhead locates the *GD* gene on chromosome 10. Chromosomes were visualized by epifluorescence microscopy. (b) A *Drosophila* salivary gland chromosome hybridized with a DNA probe labeled with biotin-derivitized nucleotides. Hybridization is detected with a biotin-binding protein called avidin that is covalently bound to an enzyme such as alkaline phosphatase. On addition of a soluble substrate, the enzyme catalyzes a reaction that results in formation of an insoluble colored precipitate at the site of hybridization (asterisk). The numbers indicate the band divisions. Note the local separation of the two homologs between bands 98–99. [Part (a) from P. Deloukas et al., 1993, *Genomics* **17**:676; courtesy of W. Holick and D. van Loon, F. Hoffmann-LaRoche Inc. Part (b) courtesy of F. Piguoni.]

lished in which only part of a human chromosome is represented. Southern-blot screening of such lines can localize human DNA clones to small chromosomal regions.

Chromosome (DNA) Walking   A molecular technique called *chromosome*, or *DNA*, *walking* can be used to isolate contiguous regions of genomic DNA beginning with a previously cloned DNA fragment that maps near a gene of interest. In this reiterative process, overlapping DNA clones are isolated; the process is repeated until a clone containing the desired gene is obtained.

The first step in a DNA walk is radiolabeling of the starting clone and hybridizing it to a genomic DNA library (Figure 8-29). After clones that hybridize are identified, their DNA is isolated and mapped by restriction-endonuclease cleavage. Typically, only part of the cloned DNA fragments hybridize with the probe; the nonhybridizing parts extend into contiguous DNA regions. Small regions of these overlapping cloned fragments that extend the farthest from the probe in each direction are isolated, radiolabeled, and used to probe the genomic DNA library once again. In this way, DNA that extends farther away in both directions from the starting point can be identified. The larger the steps taken in the walk, the faster large segments

of contiguous DNA can be isolated. Thus DNA vectors that are used to clone large pieces of DNA (e.g., cosmids and YACs) are particularly valuable in chromosome walking. Although the starting clone may be considered close to a gene of interest by genetic recombination, it may be a considerable distance from the gene at the molecular level. In *Drosophila*, for instance, a genetic-map distance of 1 centimorgan (i.e., a 1 percent recombination frequency) is equal to about 400,000 base pairs.

Mutation-Defined Genes Are Identified in Candidate Regions by Comparing Mutant and Wild-Type DNA Structure and mRNA Expression

Once a set of overlapping DNA clones encompassing a gene of interest is obtained, how is that gene localized to a particular clone? As a first step, genomic DNA from mutants is subjected to Southern-blot analysis with DNA probes representing different regions of the chromosomal walk of the normal genome. Generally, Southern blotting can successfully identify chromosomal rearrangements and small deletions in the mutant DNA but is not sensitive

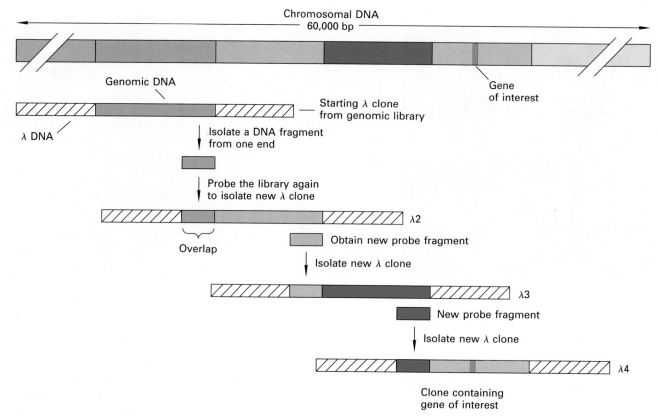

▲ FIGURE 8-29   Chromosome (DNA) walking. This technique can be used to isolate overlapping DNA fragments starting with a previously cloned DNA fragment that maps near a gene of interest (dark red). The walk is continued until a clone containing the desired gene is identified. In this example, the chromosomal DNA fragments are cloned in λ phage. The starting clone (green) is used as a probe to screen a genomic DNA library for overlapping sequences; usually a walk is conducted in both directions, since it generally is not known how far the gene of interest is from the starting clone or whether it is to the left or right. For simplicity, a one-way walk is illustrated here. The clones isolated in the walk are used as probes in Southern-blot analysis of genomic DNA from mutants; in this way, the DNA clone containing the region corresponding to the gene of interest can be identified. Normally, a walk would involve more than the three steps shown here. In some cases, part of the desired gene may already be cloned, and DNA walking may be necessary to isolate the entire gene. By initiating walks from several different starting clones in parallel (these walks will eventually link up), the process of isolating overlapping clones for long contiguous stretches of DNA is accelerated. [From J. D. Watson et al., 1992, *Recombinant DNA*, 2d ed., W. H. Freeman and Company, p. 128.]

enough to identify point mutations. Compared with normal DNA, Southern blots of mutant DNA carrying large deletions will often not contain sequences homologous to the probe, whereas blots of mutant DNA that have rearrangements or smaller deletions will show fragments of different sizes. Point mutations are not detected by Southern blotting unless they either destroy or create a restriction-enzyme recognition site, creating a RFLP. As shown in Figure 8-22a, creation of a new recognition site will result in the loss of one fragment and its replacement with two smaller ones. Conversely, loss of a site will lead to the loss of two smaller fragments and the appearance of a larger one.

In general, Southern-blot analysis will define a region of the genome containing the gene of interest. In many cases the interval defined may not contain the entire gene, or it may contain other genes as well. Screening for point mutations and the identification of sequences transcribed into mRNAs in the region are further steps to isolation and characterization of the gene.

Analysis of mRNA Expression   To further narrow the region containing the desired gene, the mRNA encoding its gene product must be identified. Phenotypic analysis of a mutant may suggest tissues in which the mRNA is likely to be expressed. For instance, a mutation that phenotypically affects muscle, but no other tissue, might be in a gene that is expressed only in muscle tissue. The expression of mRNA in both normal and mutant tissues generally is determined by Northern blotting (see Figure 7-32) or in situ hybridization of labeled DNA or RNA to tissue sections. Northern blots permit comparison of both the level of expression and size of mRNAs in mutant and wild-type tissues. Although the sensitivity of in situ hybridization is lower than that of Northern-blot analysis, it can be very helpful in identifying an mRNA that is expressed at low levels in a given tissue but at very high levels in a subclass of cells within that tissue (Figure 8-30). An mRNA that is altered or missing in different mutants compared with wild-type individuals is an excellent candidate for encoding the protein whose function is disrupted in the mutants.

Identification of Point Mutations In many cases, point mutations may result in no detectable change in the level of expression or electrophoretic mobility of mRNAs.

(a)

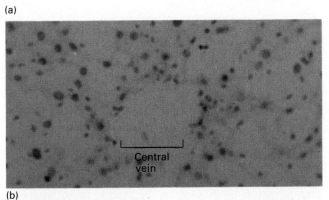

(b)

▲ FIGURE 8-30 Autoradiograph showing in situ hybridization. A mouse liver section was exposed to labeled RNA complementary to glutamine synthetase mRNA. The label was allowed time to hybridize; then, unhybridized labeled RNA was washed away. (a) A light microscopic view of the autoradiograph showing cords of liver cells (hepatocytes) around a central vein. The barely visible dark grains around the central vein are in the first layer of hepatocytes. (b) The second view is a dark-field picture, which shows the grains (white dots) with much greater contrast. [See F. Kuo et al., 1988, *Mol. Cell Biol.* **8**:4966. Photographs courtesy of F. Kuo.]

Thus if comparison of mRNA expression in mutants and normal individuals reveals no detectable differences in the candidate mRNAs, a search for point mutations in the DNA regions encoding the mRNAs is undertaken. The overall strategy is to scan for regions of DNA that carry point mutations and then to sequence that region of the mutant allele. Of several different methods devised to screen for single-base changes, two are discussed here: one is applicable to DNA fragments as large as 2 kb; the other to fragments less than about 400 base pairs.

In the first method, a wild-type, single-stranded DNA fragment is radiolabeled at on one end and then is hybridized to unlabeled, single-stranded DNA from the mutant. The resulting heteroduplex DNA is treated with a reagent, such as hydroxylamine or osmium tetroxide, that selectively modifies C or both C and T, respectively, in regions of single-stranded DNA corresponding to the position of the mismatches between the mutant and wild-type DNAs. The DNA backbone at the position of the derivitized DNA then is selectively cleaved using piperidine. Gel electrophoresis followed by autoradiography localizes the position of the mismatch relative to the labeled end (Figure 8-31a).

The second method for detecting point mutations is used with smaller fragments and depends on the property of single-stranded DNA molecules of the same length to assume different conformations depending upon their nucleotide sequence. Single-base differences can lead to altered conformations—referred to as *single-stranded conformation polymorphisms* (SSCPs)—that can be detected by electrophoresis on nondenaturing polyacrylamide gels (Figure 8-31b). One disadvantage of this method is that the mutant chromosome is likely to carry naturally occurring polymorphisms unrelated to the gene of interest; these can lead to misidentification of the DNA fragment carrying the gene of interest. For this reason, the more mutant alleles available for analysis, the more likely that a gene will be correctly identified.

Protein Structure Is Deduced from cDNA Sequence

Once the mRNA encoded by a gene defined by mutation has been identified, the corresponding cDNA can be cloned (see Figure 7-16) using specific genomic fragments which recognize the mRNA as a probe. Sequencing of the cDNA permits the sequence of its encoded protein to be deduced. This is an important step in linking genetics to the cell and molecular biology of development and function, since proteins with similar structures often have similar functions.

As described in Chapter 7, two large data bases of DNA and protein sequences are maintained and available to biologists throughout the world. Thus one can efficiently search for sequence similarities between a new sequence and all those previously determined. The sequence of a protein deduced from the cDNA of a newly isolated

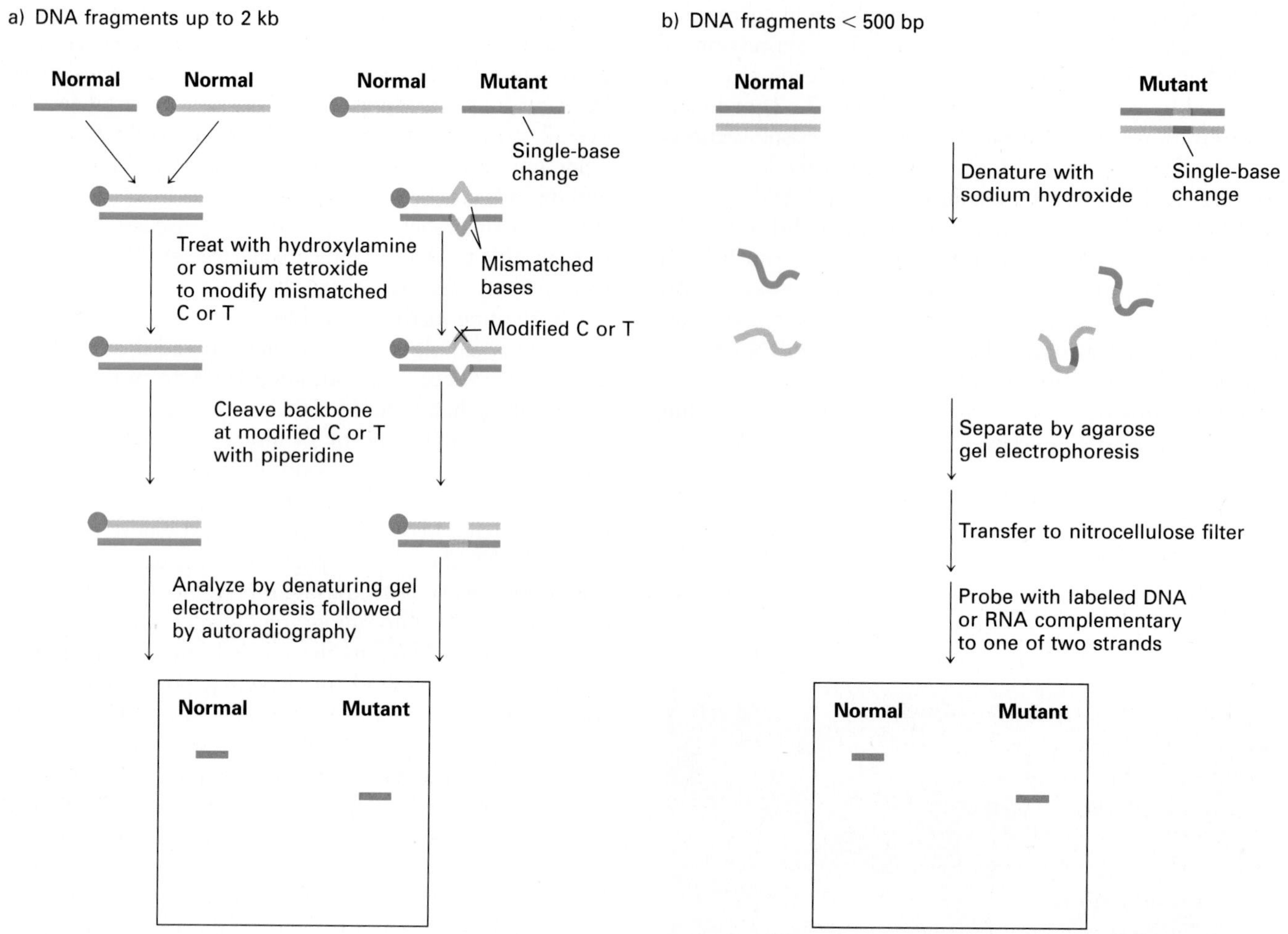

a) DNA fragments up to 2 kb

b) DNA fragments < 500 bp

▲ FIGURE 8-31   Detection of point mutations. (a) In DNA fragments up to 2 kb in length, single-base changes (green) can be detected by chemical cleavage at the mismatched bases in mutant-normal heteroduplexes. One strand of the normal DNA is radiolabeled at one end (red dot) and then is hybridized with the complementary normal or mutant strand. Hydroxylamine or osmium tetroxide modifies any mismatched C or T; the modified backbone is susceptible to cleavage by piperidine. The shortened labeled fragment is detected by gel electrophoresis and autoradiography in comparison with normal DNA. (b) Single-stranded conformation polymorphisms (SSCPs) can be used to detect single-base changes (dark green) in DNA restriction fragments less than about 500 bp. Single-stranded DNA molecules that differ by only one base frequently show different electrophoretic mobilities in nondenaturing gels. Differences between normal and mutant DNA are revealed by hybridization with labeled probes.

gene is likely to show similarities with the structures of previously determined proteins. In those cases where the biochemical function of a structurally similar class of proteins is known, it is possible to generate biochemically testable hypotheses. For example, if a newly identified protein is similar to proteins known to regulate the transcription of genes, the mutant phenotype may result from lack of expression of a particular gene or set of genes in the affected tissue.

Even if a newly determined sequence does not show similarities with proteins of known function, certain properties of the protein (e.g., whether it is a secreted from the cell or spans the cell membrane) may be predicted. As discussed in Chapter 3, various biochemical and immunologic techniques can be used to generate antibodies to proteins predicted from DNA sequence analysis. These antibodies then can be used to determine which cells produce a specific protein and its subcellular distribution (e.g., nuclear or cytoplasmic). Again this information can provide a critical link between the phenotype caused by a mutation and the underlying cellular and molecular mechanisms. Many examples of the central importance of this approach to cell and molecular biology are discussed in subsequent chapters.

➤ Gene Replacement and Transgenic Animals

In the preceding sections, we discussed the isolation of genes defined by mutations affecting particular processes; this approach can provide valuable information about the molecular mechanisms underlying various cellular processes and the in vivo function of genes. Many genes, however, have been identified based on the biochemical properties of their encoded protein, the sequence similarity of the encoded protein with proteins of known function, or their interesting patterns of expression in development. In the absence of mutant forms of such genes, their in vivo functions may be unclear. By mutating a specific gene and then replacing a normal copy with a mutant form, scientists can assess its in vivo function. This technique, referred to a *gene knockout*, is in essence the reverse of the approach described in the previous sections.

Other techniques permit the introduction of foreign genes or altered forms of an endogenous gene into an organism. For the most part, these techniques do not result in replacement of the endogenous gene, but rather the integration of additional copies of it. Such introduced genes are called *transgenes*; the organisms carrying them are referred to as *transgenics*. Transgenes can be used to study organismal function and development in a variety of different ways. For instance, genes that are normally expressed at specific times and places during development can be genetically engineered in vitro to be expressed in different tissues at different times and then reintroduced into the animal to assess the cellular and/or organismal consequences. For example, the antennapedia (*Antp*) gene in *Drosophila* normally controls leg development, but misexpression of this gene in the developing antenna transforms it into a leg.

The production of both gene-targeted knockout and transgenic animals makes use of techniques for mutagenizing cloned genes in vitro and then transferring them into eukaryotic cells. We briefly describe these procedures first, then discuss the production and uses of knockout and transgenic organisms.

Specific Sites in Cloned Genes Can Be Altered in Vitro

Specific sequences in cloned genes can be altered in vitro and then introduced into experimental organisms. This approach has been exploited primarily to study two questions. First, what is the relationship between the structure of a particular protein and its biological function? And second, what are the specific DNA sequences required to determine the expression pattern of a particular gene?

A variety of enzymatic and chemical methods are available for producing site-specific mutations in vitro. In recent years, however, the most common methods use specific oligonucleotides as mutagens. Because oligonucleotides of any desired sequence can be chemically synthesized (see Figure 7-9), oligonucleotide-based mutagenesis can generate precisely designed deletions, insertions, and point mutations in a DNA sequence. Figure 8-32 illustrates the use of this strategy to produce a deletion.

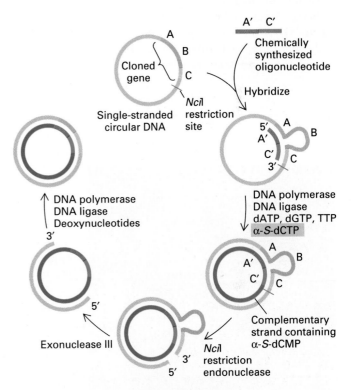

▲ FIGURE 8-32   In vitro mutagenesis of cloned genes with chemically synthesized oligonucleotides. In this example, a cloned gene that normally contains three segments (A, B, and C) is mutagenized by deletion of segment B. An oligonucleotide consisting of A' and C' segments, which are complementary to A and C, is chemically synthesized and then is hybridized to the complementary single-stranded DNA containing the gene to be mutagenized. In the hybrid molecule, segment B forms a loop. The oligonucleotide serves as a primer for synthesis of a complementary strand; one of the nucleotide triphosphates used is substituted with sulfur (instead of one of the oxygens) at the α position. The heteroduplex DNA is treated with the restriction endonuclease *Nci*I; because this enzyme cannot cleave the sulfur-substituted strand (green), a single-strand nick is created in the original template strand (purple). The nicked molecule then is treated with exonuclease III, which cleaves nucleotides in the 3' to 5' direction, thereby removing the B segment. The resulting gap in the template strand is filled in by DNA polymerase and closed by DNA ligase.

DNA Can Be Transferred into Eukaryotic Cells in Various Ways

Production of both knockout and transgenic organisms requires the transfer of DNA into eukaryotic cells. Many cells readily take up DNA from the medium. Yeast cells, for instance, can be treated with enzymes to remove their thick outer walls; the resulting *spheroplasts* will take up DNA added to the medium. Plant cells also can be converted to spheroplasts, which will take up DNA from the medium. Cultured mammalian cells take up DNA directly, particularly if it is first converted to a fine precipitate by treatment with calcium ions. DNA also can be injected directly into the nuclei of both cultured cells and developing embryos (Figure 8-33). Another popular method for introducing DNA into yeast, plant, and animal cells is called *electroporation*. Cells subjected to a brief electric shock of several thousand volts become transiently permeable to

DNA. Presumably the shock briefly opens holes in the cell membrane allowing the DNA to enter the cells before the holes reseal.

Once the foreign DNA is inside the host cell, enzymes that probably function normally in DNA repair and recombination join the fragments of foreign DNA with the host cell's chromosomes. Since only a relatively small fraction of cells take up DNA, a selective technique must be available to identify the transgenic cells. In most cases the exogenous DNA includes a gene encoding a marker such as drug resistance. The introduced DNA can insert into the host genome in a highly variable fashion showing no site specificity, can replace an endogenous gene by homologous recombination, or it can remain as an independent extrachromosomal DNA molecule referred to as an *episome*.

Normal Genes Can Be Replaced with Mutant Alleles in Yeast and Mice

Gene knockout is a technique for selectively inactivating a gene by replacing it with a mutant allele in an otherwise normal organism. This technique of disrupting gene function, which has been widely used in yeast and mice, is a powerful tool for unraveling the mechanisms by which basic cellular processes occur.

Gene Knockout in Yeast After foreign, or exogenous, yeast DNA is taken up by diploid yeast cells, recombination generally occurs between the introduced DNA and the homologous chromosomal site in the recipient cell. Because of this specific, targeted recombination of identical stretches of DNA, called *homologous recombination*, any gene in yeast chromosomes can be replaced with a mutant allele (Figure 8-34). The resulting yeast cells, carrying one mutant allele and one wild-type allele, generally grow normally. To determine whether the knockout gene controls an obligatory function, recombinants containing the mutant allele on one chromosome are treated to induce meiosis and sporulation; each diploid cell produces four haploid spores, which are tested for viability. One of the first genes tested in this way was the one encoding actin, a prominent cytoskeletal protein in higher organisms. Haploid yeast spores without a normal actin gene cannot grow (Figure 8-35).

This technique also is useful in assessing the role of proteins identified solely on the basis of DNA sequence. For instance, the entire sequence of the genome of a number of organisms will be determined during the next 5–10 years. Stretches of DNA that exhibit long open reading frames or homology to genes encoding known proteins are likely to be transcribed and translated into as-yet unidentified proteins. In yeast, gene knockouts could be used to determine whether such regions are important for specific cellular functions that are phenotypically detectable. This

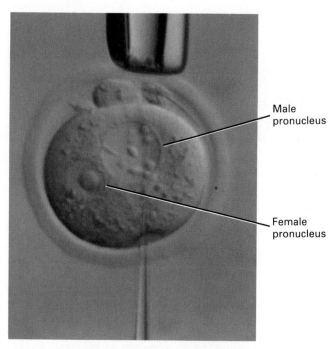

Male pronucleus

Female pronucleus

▲ FIGURE 8-33   DNA can be directly injected into individual cells under the microscope through fine-tipped glass needles, or pipettes, which control pumping of small volumes (≈10^{-10} ml). In this fertilized mouse egg, the two parental nuclei have not yet fused. While the egg is held by a blunt pipette, a DNA solution is injected through a fine-tipped pipette inserted into the male pronucleus. Such injected eggs are viable and can be transplanted into the uterus of a primed mouse. The implanted eggs will develop into mice that contain the injected DNA in every cell. [Courtesy of R. L. Brinster.]

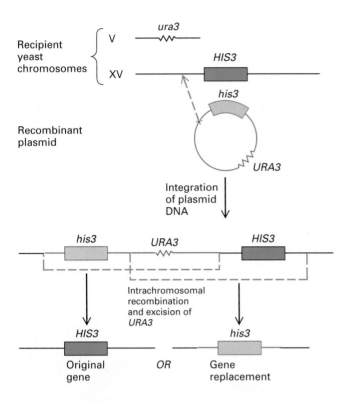

◄ FIGURE 8-34   Replacement of the normal *HIS3* gene by homologous recombination with mutant *his3* gene in the yeast *S. cerevisiae*. Recombinant DNA technology is used to prepare a plasmid containing two yeast genes: *his3* (yellow), a deletion mutation that encodes only part of the sequence of one enzyme in the histidine biosynthetic pathway; and *URA3*, which encodes an enzyme in the uracil biosynthetic pathway. The *URA3* gene is included simply as a selective marker. The recipient cell (a uracil-requiring strain) takes up the plasmid, which can integrate into the *ura3* gene (not of interest in this case) or into the *HIS3* gene. Recombinant cells are selected by their ability to grow in the absence of uracil. Haploid *URA3* yeast cells in which the normal *HIS3* gene is replaced (knocked out) by the *his3* mutant gene are detected by replica plating in the presence and absence of histidine. [See S. Scherer and R. W. Davis, 1979, *Proc. Nat'l. Acad. Sci. USA* **76**:4951.]

that this chromosome contains 182 open reading frames of sufficient length to encode proteins longer than 100 amino acids (considered in this analysis to be the minimum length of a naturally occurring protein). The sequences of the proteins that could be encoded by 116 of these putative protein-coding regions exhibited no obvious homology to any known proteins. The in vivo functions of 55 of these regions were tested by gene knockouts: 3 regions were required for viability; 42 nonessential genes were tested, of which 14 showed a mutant phenotype and 28 did not. The large number of putative genes with no detectable phenotype is quite surprising. In some cases the lack of a phenotype could indicate the existence of backup or compensatory pathways in the cell. Alternatively, the mutations may give rise to only subtle defects that would require more in-depth phenotypic analysis to uncover.

Gene Knockout in Mice   Gene-targeted knockout mice are a powerful experimental system for studying embryogenesis and behavior; they also may be useful model systems for studying certain human genetic diseases. The

technique thus provides a potentially powerful approach to identifying and studying new genes and the proteins that they encode.

This gene-knockout approach already has been used to analyze yeast chromosome III, whose complete DNA sequence is known. Analysis of the DNA sequence indicated

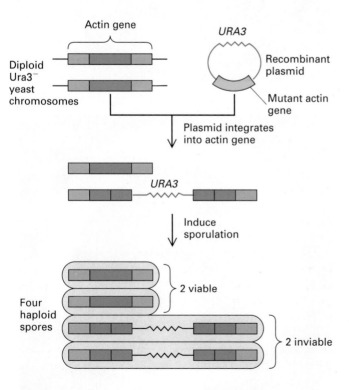

◄ FIGURE 8-35   Demonstration that actin gene is required for yeast viability by gene-targeted knockout. A recombinant plasmid containing the *URA3* gene and a mutant actin gene (yellow) is introduced into uracil-requiring (Ura3⁻) diploid yeast cells. Cells that integrate the plasmid are selected by their growth in the absence of uracil. Since the recombinant cells contain one normal and one disrupted actin gene, they can synthesize actin. To determine whether the actin gene is essential, meiosis and sporulation are induced by starving the cells; each diploid cell produces four haploid spores. The wild-type spores, which are Ura3⁻, can grow, but those with the disrupted actin gene do not. [Adapted from D. Shortle et al., 1982, *Science* **217**:371.]

procedure for producing gene-targeted knockout mice involves the following steps:

- Mutant alleles are introduced by homologous recombination into *embryonic stem (ES) cells.*

- ES cells containing a knockout mutation in one allele of the gene being studied are introduced into early mouse embryos. The resultant mice will be *chimeras* containing tissues derived from both the transplanted ES cells and host cells. These cells can contribute to both the germ-cell and somatic-cell populations.

- Chimeric mice are mated to assess whether the mutation is incorporated into the germ line.

- Chimeric mice each heterozygous for the knockout mutation are mated to produce homozygous knockout mice.

The isolation and culture of embryonic stem cells, which are derived from the blastocyst, is illustrated in Figure 8-36. These cells can be grown in culture through many generations. Exogenous DNA containing a mutant allele of the gene being studied is introduced into ES cells by transfection. The introduced DNA recombines with chromosomal sequences in about 1 cell out of 100 (i.e., 1 percent recombination frequency). In some cells, the added DNA recombines with the homologous chromosomal site, but recombination at other chromosomal sites (i.e., nonhomologous recombination) occurs 10^3–10^4 times more frequently. The small fraction of cells in which homologous recombination takes place can be identified by a combination of positive and negative selection: positive selection to identify cells in which any recombination occurs and negative selection to remove cells in which recombination takes place at nonhomologous sites.

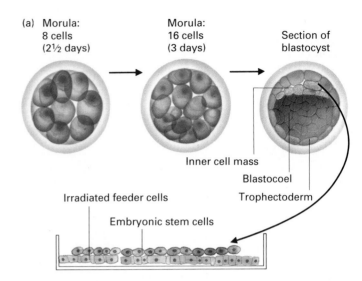

(a)

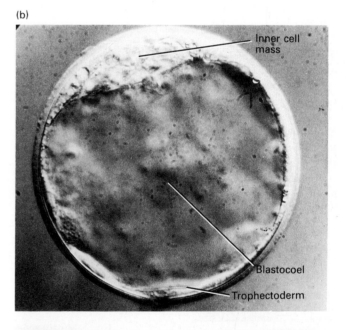

(b)

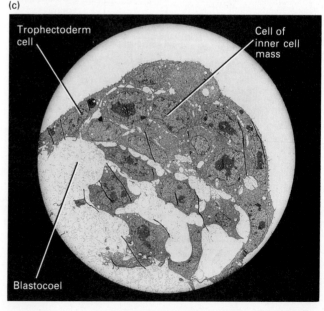

(c)

➤ FIGURE 8-36   Preparation of embryonic stem (ES) cells. (a) Fertilized mouse eggs divide slowly at first; after 4½ days, they form the blastocyst, a hollow structure composed of about 100 cells surrounding an inner cavity called the blastocoel. Only ES cells, which constitute the inner cell mass, actually form the embryo. Other cells form the trophectoderm, which gives rise to the membranes (amnion and placenta) by which the embryo is attached to the uterine wall. Embryonic stem cells can be removed from the blastocyst and grown on lethally irradiated "feeder cells." To produce knockout mice, exogenous DNA carrying a mutant allele is introduced into the ES cells; cells in which homologous recombination occurs are selected (see Figure 8-37) and reintroduced into blastocysts (see Figure 8-38). (b) A light micrograph of a blastocyst. (c) An electron micrograph of a section through a blastocyst. [See E. Robertson et al., 1986, *Nature* **323**:445; photographs courtesy of R. Calarco.]

For this selection scheme to work, the DNA constructs introduced into ES cells need to include, in addition to sequences used to knock out the gene of interest, two selective genes (Figure 8-37). One of these additional genes (neo^R) confers neomycin resistance; it permits positive selection of cells in which either homologous (specific) or nonhomologous (random) recombination has occurred. The second selective gene, the thymidine kinase gene from herpes simplex virus (tk^{HSV}) confers sensitivity to gancylovir, a cytotoxic nucleotide analog; this gene permits negative selection of ES cells in which nonhomologous recombination has occurred. Only ES cells that undergo homologous recombination (i.e., gene-targeted specific insertion of the DNA construct) can survive this selection scheme.

Once ES cells heterozygous for a knockout mutation in the gene of interest are obtained, they are injected into a recipient mouse blastocyst, which subsequently is transferred into a surrogate pseudopregnant mouse (Figure 8-38). If the ES cells also are homozygous for a visible marker trait (e.g., coat color), then chimeric progeny carrying the knockout mutation can be identified easily. These are then mated with mice homozygous for another allele of the marker trait to determine if the knockout mutation is incorporated into the germ line. Finally, mating mice, each heterozygous for the knockout allele, will produce progeny homozygous for the knockout mutation.

Use of Knockout Mice to Study Human Genetic Diseases Gene knockout has promise as a way to produce model systems for studying inherited human diseases. Such model systems would be powerful tools for investi-gating the nature of genetic diseases and the efficacy of different types of treatment, and for developing effective gene therapies to cure these often devastating diseases.

Recent studies on cystic fibrosis illustrate this use of the knockout technique. Cystic fibrosis, which afflicts about 1 in 2000 Caucasians, is caused by an autosomal

a) Formation of ES cells carrying a knockout mutation

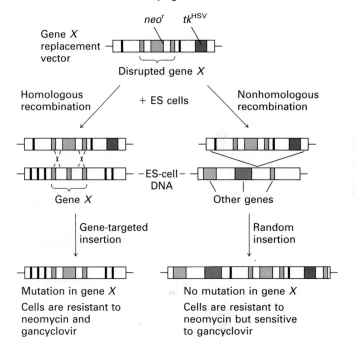

b) Positive and negative selection of recombinant ES cells

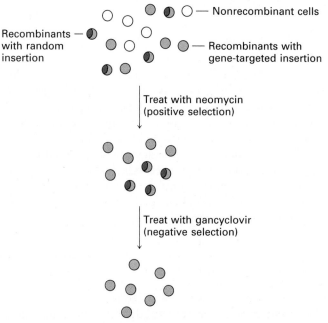

➤ FIGURE 8-37   Isolation of mouse ES cells with a gene-targeted disruption by positive and negative selection. (a) When exogenous DNA is introduced into ES cells, random insertion via nonhomologous recombination occurs much more frequently than gene-targeted insertion via homologous recombination. Recombinant cells in which one copy of the gene *X* (orange) is disrupted can be obtained by using a recombinant vector that carries gene *X* disrupted with *neo*^{r} (pink), a neomycin-resistance gene, and, outside of the region of homology, *tk*^{HSV} (purple), the thymidine kinase gene from herpes simplex virus. The viral thymidine kinase, unlike the endogenous mouse enzyme, can convert the nucleotide analog gancylovir into the monophosphate form; this is then modified to the triphosphate form, which inhibits cellular DNA replication in ES cells. Thus gancylovir is cytotoxic for recombinant ES cells carrying the *tk*^{HSV} gene. (b) Recombinant cells are selected by treatment with neomycin, since cells that fail to pick up DNA or integrate it into their genome are neomycin sensitive. The surviving recombinant cells are treated with gancylovir. Only cells with a targeted disruption in gene *X* will survive. [See S. L. Mansour et al., 1988, *Nature* **336**:348.]

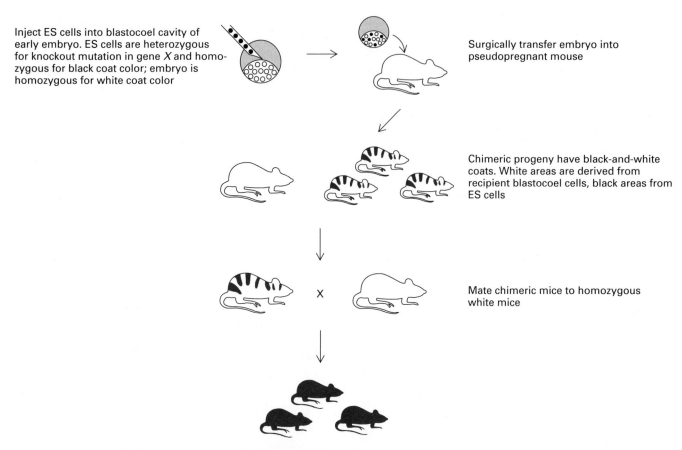

Inject ES cells into blastocoel cavity of early embryo. ES cells are heterozygous for knockout mutation in gene *X* and homozygous for black coat color; embryo is homozygous for white coat color

Surgically transfer embryo into pseudopregnant mouse

Chimeric progeny have black-and-white coats. White areas are derived from recipient blastocoel cells, black areas from ES cells

Mate chimeric mice to homozygous white mice

Black progeny develop from germ-line cells derived from ES cells and are heterozygous for disrupted gene *X*

▲ FIGURE 8-38   General procedure for producing gene-targeted knockout mice. Embryonic stem (ES) cells heterozygous for a knockout mutation in a gene of interest (*X*) and homozygous for a marker gene (e.g., black coat color) are transplanted into the blastocoel cavity of 4.5-day embryos that are homozygous for an alternate marker (e.g., white coat color). The early embryos then are implanted into a pseudo-pregnant female. Some of the resulting progeny are chimeras, indicated by their black and white coats. Chimeric mice then are backcrossed to white mice; black progeny from this mating have ES-derived cells in their germ line. Intercrossing of these black mice produces individuals homozygous for the disrupted allele, that is, knockout mice. [Adapted from M. R. Capecchi, 1989, *Trends Genet.* **5**:70.]

recessive mutation in the *CFTR* gene (see Figure 8-20b). This gene was cloned by positional cloning strategies, and the biochemical function of its encoded protein studied. Using the human gene, researchers isolated the homologous mouse gene and subsequently introduced mutations in it. The gene-knockout technique was then used to produce homozygous mutant mice, which showed symptoms (or phenotype) similar to those of humans with cystic fibrosis. Thus these knockout mice may provide a model system for studying this genetic disease.

Foreign Genes Can Be Introduced into Plants and Animals

In the previous section we discussed techniques for replacing one form of a gene with another through homologous recombination. In this section we discuss methods for producing transgenic mice, fruit flies, and plants. Transgenic organisms carry cloned genes that have integrated randomly into the host genome.

Transgenic technology has numerous experimental applications and potential agricultural and therapeutic value. For instance, dominantly acting alleles of genes causing tumors can be used to produce transgenic mice, thus providing an animal model for studying cancer. In *Drosophila*, transgenes often are used to determine whether a cloned segment of DNA corresponds to a gene defined by mutation. If the cloned DNA is indeed the gene in question, then introducing it as a transgene into a mutant will transform the mutant into a phenotypically normal individual. Transgenic plants may be commercially valuable in agriculture. Plant scientists, for example, have developed transgenic tomatoes that exhibit reduced production of ethylene, which promotes fruit ripening. The ripening process is delayed in these transgenic tomatoes, thus prolonging their shelf life. Finally, transgenic technology is a critical component in the burgeoning field of gene therapy for human genetic diseases.

Transgenic Mice   As noted in the discussion of knockout mice, specific integration of exogenous DNA into the genome of mouse cells by homologous recombination occurs at a very low frequency. In contrast, the frequency of random integration of exogenous DNA into the mouse genome at nonhomologous sites is very high. Because of this phenomenon, the production of transgenic mice is a highly efficient and straightforward process.

As outlined in Figure 8-39, foreign DNA containing a gene of interest is injected into one of the two pronuclei (the male and female haploid nuclei contributed by the parents) of a fertilized mouse egg before they fuse. The injected DNA has a good likelihood of being randomly integrated into the chromosomes of the diploid zygote. Injected eggs then are transferred to foster mothers in which normal cell growth and differentiation occurs. About 10–30 percent of the progeny will contain the foreign DNA in equal amounts (up to 100 copies per cell) in all tissues, including germ cells. Immediate breeding and backcrossing (parent-offspring mating) of the 10–20 percent of these mice that breed normally can produce pure transgenic strains homozygous for the transgene.

Numerous examples of the use of transgenic mice for studying various aspects of normal mammalian biology are presented in other chapters. They also provide a model system for studying disease processes. For example, many forms of cancer are promoted by normal cellular genes acting in a dominant fashion due to their misregulated activity. Although transgenic mice carrying one of these genes, called *myc* develop normally, tumors form at a high frequency. The observation that only a small number of cells expressing the transgene develop tumors supports a model in which additional genetic changes are necessary for tumors to form. These mice may provide an important tool for identifying those changes.

Transgenic Fruit Flies   Foreign DNA can be incorporated into the *Drosophila* germ-line genome by the technique of P-element transformation (Figure 8-40). This technique makes use of a segment of the P element, a highly mobile DNA element, which can transpose (jump) from an extrachromosomal element into a chromosome. (Mobile DNA elements are discussed in detail in Chapter 9.) Generally, this procedure results in incorporation of a single copy of the transgene into the *Drosophila* genome. In contrast, transgenic mice carry multiple copies of the transgene incorporated into their chromosomes. In both organisms, however, the chromosomal insertion site is highly variable.

Flies that develop from injected embryos will carry some germ cells that have incorporated the transgene; some of their progeny will carry the transgene in all somatic and germ-line cells, giving rise to pure transgenic lines. Individuals carrying the transgene are recognized by expression of a marker gene (e.g., one affecting eye color) that is also present on the donor DNA. Although the transgenes in *Drosophila* and mice insert in chromosomal sites different from the position of the corresponding endogenous gene, they usually are expressed in the right tissue and right time during development. Examples of the importance of this technology for studying development are discussed in Chapter 13.

Transgenic Plants   In nature, plant cells often live in close association with certain bacteria, which may provide a convenient vehicle for introducing cloned DNA into plants. *Agrobacterium tumefaciens,* for example, attaches to the cells of dicotyledonous plants and causes the formation of plant tumors known as *galls.* (Plants with two leaflets from each seed are called dicotyledons, or *dicots;* plants with one leaflet are called *monocots.*) This bacterium introduces a circular DNA molecule called the *Ti* (tumor-inducing) *plasmid* into the plant cell in a manner

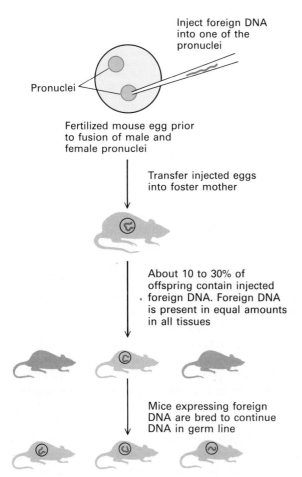

Inject foreign DNA into one of the pronuclei

Pronuclei

Fertilized mouse egg prior to fusion of male and female pronuclei

Transfer injected eggs into foster mother

About 10 to 30% of offspring contain injected foreign DNA. Foreign DNA is present in equal amounts in all tissues

Mice expressing foreign DNA are bred to continue DNA in germ line

▲ FIGURE 8-39   General procedure for producing transgenic mice. See text for discussion. [See R. L. Brinster et al., 1981, *Cell* **27**:223.]

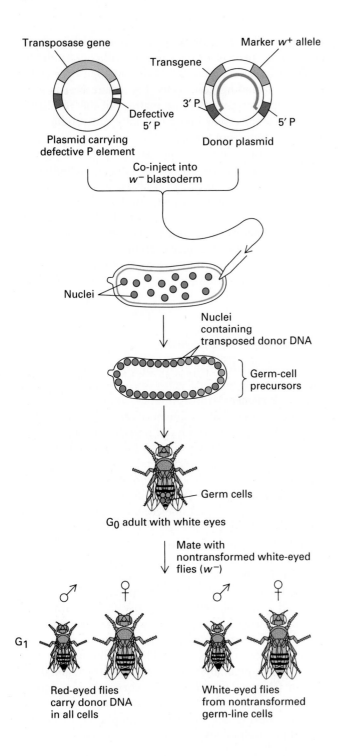

Transposase gene

Marker *w+* allele

Transgene

Defective
5′ P

3′ P

5′ P

Plasmid carrying
defective P element

Donor plasmid

Co-inject into
w− blastoderm

Nuclei

Nuclei
containing
transposed donor DNA

Germ-cell
precursors

Germ cells

G₀ adult with white eyes

Mate with
nontransformed white-eyed
flies (*w−*)

♂   ♀   ♂   ♀

G₁

Red-eyed flies
carry donor DNA
in all cells

White-eyed flies
from nontransformed
germ-line cells

◄ FIGURE 8-40   Generation of transgenic fruit flies by P-element transformation. The P element is a mobile genetic element that can move from one place in the genome to another. This movement (transposition) is catalyzed by transposase, which is encoded by the P element; the 3′ and 5′ ends of the P element are recognized by transposase and are required for transposition to occur. To produce transgenic fruit flies by this method, the functionally different regions of the P element are incorporated into two different bacterial plasmids. The donor plasmid contains three necessary elements: the transgene (orange); a marker gene (pink) used to indicate flies in which the plasmid DNA is transposed to a recipient chromosome; and both ends of the P element (dark purple)—3′ P and 5′ P—flanking the other two genes. It does not contain transposase. In this example, the marker is the dominant *w+* allele, which confers red eye color. The red bracket indicates the segment of the donor plasmid that can transpose into the fly genome. The other plasmid carries the P element (encoding transposase) with mutations in one end, which prevent it from transposing. The two plasmids are co-injected into blastoderm embryos homozygous for the recessive *w−* allele, which confers white eye color. Transposase synthesized from the gene on the P-element plasmid catalyzes transposition of the donor plasmid DNA into the fly genome. Because transposition occurs only in germ-line cells (not in somatic cells), all the G₀ adults that develop from injected embryos have white eyes. Mating of these flies with white-eyed flies will yield some G₁ red-eyed progeny carrying the transgene and the marker allele in all cells.

An especially useful characteristic of plants for transgenic studies is the ability of cultured plant cells to give rise to mature plants. Meristematic (growing) cells from dissected plant tissue or cells within excised parts of a plant will grow in culture to form *callus tissue,* an undifferentiated lump of cells. Under the influence of plant growth hormones, different plant parts (roots, stems, and leaves) develop from the callus and eventually grow into whole, fertile plants. When an agrobacterium containing a recombinant Ti plasmid infects a cultured plant cell, the newly incorporated foreign gene is carried into the plant genome. As noted above *A. tumefaciens* readily infects dicots (petunia, tobacco, carrot) but not monocots; reliable techniques for introducing genes into monocots are still being developed. Direct introduction of DNA by electroporation has been successful in rice plants, (which are monocots), and the future looks bright for the manipulation of other commercially important monocotyledonous crop plants. Also available for gene transfer experiments are cells of a tiny, rapidly growing member of the mustard family called *Arabidopsis thaliana.* This plant appears to be well suited to genetic analysis of a variety of developmental and physiologic processes. It takes up little space, is easy to grow, and has a small genome, and genes defined by mutations can be cloned by positional cloning strategies.

similar to bacterial conjugation. The plasmid DNA then recombines with the plant DNA. Since the Ti plasmid has been isolated, new genes can be inserted into it using recombinant DNA techniques and the Ti genes causing tumors can be disrupted. The resulting recombinant plasmid can then transfer desired genes into plant cells (Figure 8-41).

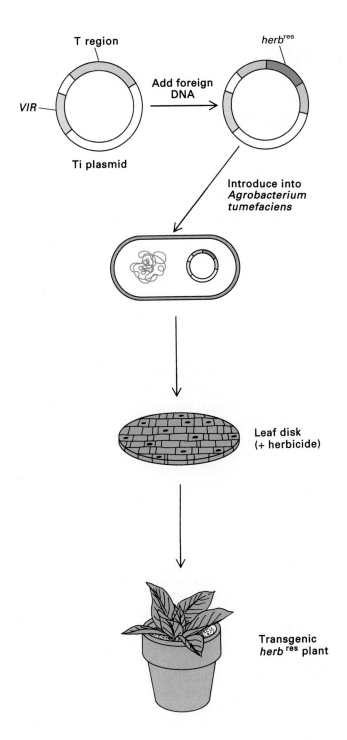

T region

herb^res

VIR

Ti plasmid

Add foreign
DNA

Introduce into
*Agrobacterium
tumefaciens*

Leaf disk
(+ herbicide)

Transgenic
herb^res plant

◄ FIGURE 8-41 Production of transgenic plants with recombinant Ti plasmids. In nature, the Ti (tumor-inducing) plasmid in *A. tumefaciens* gains entry into a plant and is integrated into the plant DNA due to the action of the *VIR* (virulent) region of the plasmid. Tumor-like growths (galls) result from action of the T region of the gene. By recombinant DNA techniques, a foreign gene is introduced within the T region of a Ti plasmid, thus destroying its tumor-inducing ability. An agrobacterium containing such a recombinant Ti plasmid then is used to introduce the foreign gene into plant cells. When a selectable transgene is used—here one conferring resistance to herbicide (*herb^res*)—recombinant plants can be selected. [See R. T. Fraley et al., 1983, *Proc. Nat'l. Acad. Sci. USA* **80**:4803.]

to marshall this same technology to cure disease by introducing normal genes (i.e., transgenes) into affected individuals. Indeed, some genetic disorders in experimental organisms have already been cured by gene therapy. Numerous technical issues, however, need to be resolved before gene therapy can become widely used in humans.

One critical issue is to determine how to reliably and safely introduce various genes into human cells. Since many of the genes associated with genetic diseases are expressed only in certain tissues or cell types, methods are needed to introduce genes selectively into specific cell types, such as blood cells (β-globin gene), lung tissue (cystic fibrosis gene), and muscle (Duchenne muscular dystrophy gene). In addition, many human genes are very large, requiring design of methods to introduce large DNA molecules; alternatively, truncated versions of large genes that still function normally might be engineered. Beyond such technical difficulties, human gene therapy raises various ethical issues. For example, although most gene therapy will likely involve somatic cells, the technology may permit introduction of genes into the germ line. In this case, the transgenes would be transmitted to future generations!

Nonetheless, the age of gene therapy is upon us. The first gene therapy experiment was begun in 1990 for treatment of a young girl with adenosine deaminase deficiency (ADA). Patients with this disorder have a severe immunodeficiency and are prone to infection; they usually die in the early years of childhood. A retroviral vector carrying the normal *ADA* gene was used to introduce the gene into T lymphocytes isolated from the patient. These genetically altered cells then were returned to the patient. Because T lymphocytes have a limited life span, the procedure had to be repeated every few months. This treatment led to a dramatic improvement in the young girl's health, allowing her to resist infection and participate in the same activities as other healthy children.

Gene Therapy Involves the Use of Transgenes to Treat Genetic Diseases

We saw in previous sections how recombinant DNA technology can be used to identify mutations giving rise to human genetic disorders and to isolate the corresponding gene from normal individuals. The hoped-for next step is

SUMMARY

The analysis of mutants provides a critical link between organismal function and development and the biochemical properties and functions of proteins. Although many mutations occur spontaneously, the frequency of mutations can be markedly enhanced by inducing them with chemical or radiation treatment. Spontaneously arising mutations have considerable medical importance. They also have proven to be important in uncovering fundamental biological processes. Mutations act in a dominant or a recessive fashion. In diploid organisms, the phenotype of a dominant mutation is observed when only one of the two alleles is mutant, whereas the phenotype of a recessive mutation is seen only in those individuals in which both alleles are mutant.

Since many processes are required for organismal survival, techniques have been devised to isolate and study mutations in essential genes. In bacteria and yeast, mutations inactivating essential genes can be identified as conditionally lethal temperature-sensitive mutations. The mutant phenotype can be observed at the nonpermissive temperature, and the mutant strain can be maintained and propagated at the permissive temperature. In diploid organisms, chromosomes carrying recessive lethal mutations can be maintained as heterozygotes, and the consequences of the mutations can be assessed in homozygotes.

Different types of genetic analyses reveal cell function. The complementation test can determine if mutations are in the same or different genes. Metabolic pathways and other multistep cellular processes can be dissected and ordered by analyzing in detail the phenotypes of single-gene disruptions, as well as the phenotypes of double mutants. The physical interactions between specific gene products can sometimes be inferred through the study of allele-specific extragenic suppressor mutations.

A critical step in the molecular analysis of a gene defined by mutations is its mapping to a particular location of the genome. This is typically done in several steps. First, the mutation is mapped to one of the chromosomes. Then, by assessing the frequency of meiotic recombination between specific genetic markers and the mutation of interest, the approximate position of the mutation along the length of the chromosome can be determined. In experimental organisms such as Drosophila, many markers with easily assessed phenotypes are available. Because such genetic markers are not available in humans, molecular markers in the form of restriction fragment length polymorphisms are used. In some cases the chromosomal location of specific mutations can be determined by light-microscope observation of chromosomes for evidence of deletions or rearrangements.

Knowing the precise map position of a mutation is a basis for designing molecular cloning strategies to identify the corresponding gene. In some organisms, such as C. elegans, extensive physical maps covering virtually the entire genome can be correlated with the genetic map; a genetic-map position in this organism corresponds to a set of DNA clones. Extensive physical maps of the human genome are becoming available and is anticipated that these maps will lead to the molecular characterization of many genetic diseases. In most cases, however, the first step to cloning a gene defined by a mutation in higher organisms is to isolate a DNA clone in the general region of the mutation. With this clone as the starting point, DNA walking is used to identify long stretches of contiguous DNA. The positions of mutations within the walk are determined using several different techniques to analyze DNA structure and mRNA expression. Once the region of the genomic DNA corresponding to the mutation is identified, then cDNAs corresponding to it are identified and sequenced. The sequence of the encoded protein can be deduced from the cDNA sequence. In many cases, the deduced protein sequence reveals similarities with proteins of known function and, hence, may provide clues to its function. With the primary sequence available, various strategies are available for generating antibodies to the predicted protein product. These antibodies facilitate the tissue and cellular localization of the protein and its eventual isolation.

Many genes of interest have been cloned not because of the effects of mutations in them, but because of their time of expression or the properties of their encoded proteins. A variety of different techniques are available to mutagenize such genes in vitro; the mutant genes can then be introduced into yeast and mice, replacing the endogenous wild-type gene. Such gene-targeted knockout is a powerful approach to assessing the biological function of any cloned yeast or mouse gene. Foreign genes can also be introduced into both unicellular and multicellular organisms and stably passed from one generation to another in a transgenic line. In mice, DNA is injected into one of the two pronuclei of a fertilized mouse egg and multiple copies of tandemly arrayed genes become incorporated in the genome. In Drosophila, the gene of interest is inserted into a transposable element on a plasmid, which then is injected into the early embryo. DNA transposes from the plasmid to the genome and inserts in an essentially random manner. Transgenic plants can be created using recombinant Ti plasmid DNA introduced by transfer from the bacterium Agrobacterium tumefaciens. Transgenic organisms are important experimental tools for investigation of basic biological questions, for generating disease models in mice, and for improving the quality of plant crops.

REVIEW QUESTIONS

1. In this chapter you learned about a powerful tool that is useful in many fields of biology—mutagenesis. Review some of the basic concepts behind the process. What are the different types of mutations that can be identified? What are the potential consequences of each in terms of protein structure and function?

You are working in a laboratory that uses the yeast *Saccharomyces cerevisiae* as a model organism for dissecting amino acid biosynthesis. You are particularly interested in investigating threonine metabolism. Your first task is to isolate a variety of mutants that require the presence of threonine in order to grow. You choose to use two mutagens to do this, ethyl methanesulfonate (EMS) and x-rays. You expose some yeast independently to each of these agents. The results are presented in the following table:

Mutagen	Mutant	Mating Type
EMS	thr⁻A	α
EMS	thr⁻B	α
X-rays	thr⁻C	a

a. You are interested in performing complementation studies using these mutants. Which combinations could you use? Explain your answer. Following these studies, you conclude that thr⁻A and thr⁻C complement each other, whereas other combinations do not. What could you conclude from this?

b. You attempt to obtain revertants of each of the mutants. To do this, you expose each to EMS. You manage to obtain revertants for thr⁻A and thr⁻B, but thr⁻C yields none. Based on your knowledge of each of the mutagens used to obtain the original mutants, speculate as to why this might be the case.

2. Understanding the patterns by which genetic diseases are inherited is extremely important. Review the concepts of homozygous and heterozygous alleles, as well as dominant and recessive mutations.

a. Huntington's disease is a debilitating condition characterized by uncontrollable convulsive movements; it is inherited in a dominant fashion. If a heterozygous woman carrying the Huntington's disease gene has children with a man that does not have the mutant allele, what is the chance that the offspring will develop the disease? Will the mother develop the condition? Explain your answers.

b. Phenylketonuria (PKU) is a condition in which the amino acid phenylalanine is not metabolized properly to tyrosine due to a deficiency in the enzyme phenylalanine hydroxylase. It is a recessive disorder, resulting in mental retardation and early death if left untreated. If a homozygous woman and a heterozygous man have children, what percentage of the offspring will show the symptoms of PKU? What percentage will be heterozygous and thus carry the mutant allele? If both parents are heterozygous for the condition, what chance is there that their offspring will develop PKU? If a woman who is homozygous for the normal allele had children with a man who is homozygous for the disease allele, what percent of the offspring would develop the condition?

c. One serious form of hemophilia involves a recessive X-linked condition. If a woman who is heterozygous for the disease allele has children with a male that is not a hemophiliac, what is the chance that their sons will be hemophiliacs? What is the chance that the daughters will be hemophiliacs? What percentage of the sons and daughters would carry the hemophilia conferring allele? If a woman who does not carry a gene for the condition had children with a man exhibiting X-linked hemophilia, what percentage of their sons and daughters would have the disease? In general, do you think men or women are at a greater risk for exhibiting X-linked hemophilia? Explain your answer.

3. The notion of recombination between homologous regions of chromosomes is extremely important. Review the concepts of genetic linkage and map units.

A   B Z   C

a   b z   c

a. Suppose that you want to determine the genetic distance of Z relative to A, B, and C on the *Drosophila* X-chromosome. How might you achieve this goal?

b. If B and Z were separated by less than 1 cM, what would be the approximate physical distance between these genes?

c. The B gene has been cloned, and DNA from this region is available to you. In addition, a contiguous set of YACs has been assembled for the X-chromosome. Both the B and Z genes give a readily visible phenotype when mutant. Describe an approach you would use to identify the region of cloned DNA encoding Z.

4. The ability to manipulate the mouse genome, either to produce transgenic mice or animals with gene knockouts, is very important. Review the principles for doing this, paying particular attention to Figures 8-36 through 8-39.

You have learned that the chemical agent EMS induces point mutations in DNA. Its methyl analog, methyl methanesulfonate, also induces point mutations by a similar mechanism involving the formation of O^6-methylguanine. (The system that repairs this lesion in DNA was described in problems 2 and 3 in Chapter 7.)

You have a theory that mice lacking this gene would be signifi-cantly more prone to tumor formation at low doses of methylat-ing carcinogens than the wild type mice.

a. You wish to prepare knockout mice lacking O^6-alkylgua-nine–DNA alkyltransferase (AGT), the enzyme that repairs O^6-methylguanine. Assuming that the gene has been cloned from mice, design a vector for performing an AGT knock-out in embryonic stem (ES) cells.

b. Once a suitable vector has been prepared, how would you introduce the DNA into ES cells? Describe the selection cri-terion for obtaining knockout cells as opposed to those that have integrated the vector via non-homologous recombina-tion. How would you then use these cells to obtain mice exhibiting the desired AGT knockout? How would you prove that the mice are now deficient in terms of a viable AGT gene?

c. Assume that you have obtained mice exhibiting the desired knockout for AGT. You perform experiments whereby the mice are exposed to methylating agents, and you discover that tumors are formed to a much higher extent in these mice relative to the wild type animal. Suppose that you wanted to develop a transgenic mouse containing a mutant form of the AGT protein rather than a knockout. How might you do this? Are there techniques available by which you can alter the AGT gene in a precise way?

References

General References in Genetics

AYALA, F. J., and J. A. KIGER, JR. 1984. *Modern Genetics*, 2d ed. Ben-jamin/Cummings Publishing Co.

GRIFFITHS, A. J. F., et al. 1993. *An Introduction to Genetic Analysis*, 5th ed. W. H. Freeman and Company.

WATSON, J. D., et al. 1992. *Recombinant DNA*, 2d ed. W. H. Freeman and Company.

Isolation and Characterization of Mutants

ADAMS, A. E. M., D. BOTSTEIN, and D. G. DRUBIN. 1989. A yeast actin-binding protein is encoded by sac6, a gene found by suppression of an actin mutation. *Science* 243:231.

BEADLE, G. W., and E. L. TATUM. 1941. Genetic control of biochemical reactions in *Neurospora*. *Proc. Nat'l. Acad. Sci. USA* 27:499.

BENZER, S. 1962. The fine structure of a gene. *Sci. Am.* 206(1):70.

COULSON, A., et al. 1986. Toward a physical map of the genome of the nematode *Caenorhabditis elegans*. *Proc. Nat'l. Acad. Sci. USA* 83:7821–7825.

FRANCKE, U., et al. 1985. Minor Xp21 chromosome deletion in a male associated with expression of Duchenne muscular dystrophy, chronic granulomatous disease, retinitis pigmentosa, and McLeod syndrome. *Am. J. Hum. Genet.* 37:250–267.

GELEHTER, T. D., and F. S. COLLINS. 1990. *Principles of Medical Genet-ics*. Williams and Wilkins.

HARTWELL, L. H. 1967. Macromolecular synthesis of temperature-sensi-tive mutants of yeast. *J. Bacteriol.* 93:1662.

HARTWELL, L. H., et al. 1974. Genetic control of the cell division cycle in yeast. *Science* 183:46.

JARVIK, J., and D. BOTSTEIN. 1975. Conditional-lethal mutations that suppress genetic defects in morphogenesis by altering structural proteins. *Proc. Nat'l. Acad. Sci. USA* 72:2738.

JONES, P. A., et al. 1992. Methylation, mutation and cancer. *BioEssays* 14:33–36.

KUNKEL, L. M., et al. 1985. Specific cloning of DNA fragments absent from the DNA of a male patient with an X-chromosome dele-tion. *Proc. Nat'l. Acad. Sci. USA* 82:4778.

MORGAN, T. H. 1936. *The Theory of the Gene*. Yale University Press.

NÜSSLEIN-VOLHARD, C., and E. WIECHAUS. 1980. Mutations affecting seg-ment number and polarity in *Drosophila*. *Nature* 287:795–801.

PAINTER, T. S. 1934. Salivary chromosomes and the attack on the gene. J. Hered. 25:465–476.

PETERS, J. A., ed. 1959. *Classic Papers in Genetics*. Prentice-Hall.

STURTEVANT, A. H. 1913. The linear arrangement of six sex-linked fac-tors in *Drosophila*, as shown by their mode of association. *J. Exper. Zool.* 14:43.

WHITE, R., and J-M. LALOUEL. 1988. Sets of linked genetic markers for human chromosomes. *Ann. Rev. Genet.* 22:259–279.

Molecular Cloning of Genes Defined by Mutations

BENDER, W., P. SPIERER, and D. HOGNESS. 1983. Chromosomal walking and jumping to isolate DNA from the *Ace* and *rosy* loci and the Bithorax complex in *Drosophila melanogaster*. *J. Mol. Biol.* 168:17–33.

BOTSTEIN, D., et al. 1980. Construction of a genetic linkage map in man using restriction fragment length polymorphisms. *Am. J. Genet.* 32:314–331.

CHUMAKOV, I., et al. 1992. Continuum of overlapping clones spanning the entire human chromosome 21q. *Nature* 359:380.

COOPER, D. N., and M. KRAWCZAK. 1989. Cytosine methylation and the fate of CpG dinucleotides in vertebrate genomes. *Hum. Genet.* 83:181–188.

COTTON, R. G. H., N. R. RODRIGUES, and R. D. CAMPBELL. 1988. Reactiv-ity of cytosine and thymine in single-base-pair mismatches with hydroxyamine and osmium tetroxide and its application to the study of mutations. *Proc. Nat'l. Acad. USA* 85:4397–4401.

FOOTE, S., et al. 1992. The Y chromosome: overlapping DNA clones spanning the euchromatic region. *Science* 258:60.

OLIVER, S. G., et al. 1992. The complete DNA sequence of yeast chro-mosome III. *Nature* 357:38–46.

OLSON, M., et al. 1989. A common language for physical mapping of the human genome. *Science* 245:1434.

ORITA, M., et al. 1989. Rapid and sensitive detection of point muta-tions and DNA polymorphisms using the polymerase chain reac-tion. *Genomics* 5:874.

ORITA, M., et al. 1989. Detection of polymorphisms of human DNA by gel electrophoresis as single-stranded conformation polymor-phisms. *Proc. Nat'l. Acad. Sci. USA* 86:2766.

ROSSITER, B. J. F., and C. T. CASKEY. 1990. Molecular scanning methods of mutation detection. *J. Biol. Chem.* 265:12753.

VOLRATH, D., et al. 1992. The human Y chromosome: a 43-interval map based on naturally recurring mutation. *Science* 258:52.

Gene Replacement and Transgenic Animals

CAPECCHI, M. R. 1989. Altering the genome by homologous recombina-tion. *Science* 244:1288–1292.

DAVEY, M. R., E. L. RECH, and B. J. MULLIGAN. 1989. Direct DNA trans-fer to plant cells. *Plant Mol. Biol.* 13:273–285.

EVANS, M. J., and M. H. KAUFMAN. 1981. Establishment in culture of pluripotential cells from mouse embryos. *Nature* 292:154–156.

FUNG-LEUNG, W-P., and T. W. MAK. 1992. Embryonic stem cells and homologous recombination. *Curr. Opin. Immunol.* 4:189–194.

HASTY, P., et al. 1991. Introduction of a subtle mutation into the *Hox-2.6* locus in embryonic stem cells. *Nature* 350:243–246.

HERMES, J. D., et al. 1989. A reliable method for random mutagenesis: the generation of mutant libraries using spiked oligonucleotide primers. *Gene* 84:143–151.

HORSCH, R. B., et al. 1988. Leaf disc transformation. *Plant Mol. Biol. Manual* A5:1–9.

MANSOUR, S. L., K. R. THOMAS, and M. R. CAPECCHI. 1988. Disruption of the proto-oncogene *int-2* in mouse embryo-derived stem cells: a general strategy for targeting mutations to non-selectable genes. *Nature* 336:348–352.

MEYEROWITZ, E. M. 1989. Arabidopsis, a useful weed. *Cell* 56:263–269.

OELLER, P. W., et al. 1991. Reversible inhibition of tomato fruit senescence by antisense RNA. *Science* **254**:437.

RUBIN, G. M., and A. C. SPRADLING. 1982. Genetic transformation of *Drosophila* with transposable element vectors. *Science* **218**:348.

SHESELY, E. G., et al. 1991. Correction of a human β^S-globin gene by gene targeting. *Proc. Nat'l. Acad. Sci. USA* **88**:4294–4298.

SPRADLING, A. C., and G. M. RUBIN. 1982. Transposition of cloned P elements into *Drosophila* germ line chromosomes. *Science* **218**:341.

TAYLOR, J., J. OTT, and F. ECKSTEIN. 1985. The rapid generation of oligonucleotide-directed mutations at high frequency using phosphorothioate-modified DNA. *Nucl. Acids Res.* **13**:8765–8785.

ZIMMER, A. 1992. Manipulating the genome by homologous recombination in embryonic stem cells. *Ann. Rev. Neurosci.* **15**:115–137.

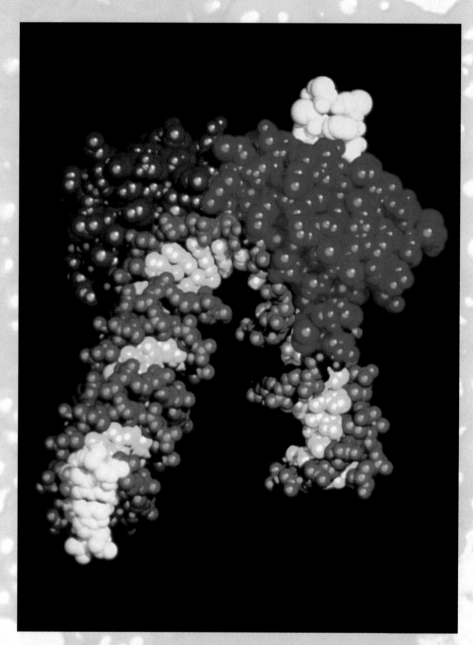

▲ The TATA-box binding polypeptide (TBP) bound to promoter DNA. Sites of transcription initiation in the nuclei of eukaryotic cells are selected when TBP binds to a DNA sequence called a promoter. Atoms shown in red and blue represent the two roughly symmetrical halves of the TBP DNA binding domain, which is highly conserved among all eukaryotes. The amino terminal domain of TBP from the plant *Arabidopsis thaliana* is shown in white. TBP binds to the minor groove of DNA, shown in green atoms for the sugar-phosphate backbone and yellow atoms for the bases, bending the DNA into an 80° angle. After more than forty additional polypeptides assemble around the TBP-promoter DNA complex, transcription initiates in the region represented by white RNA atoms at the lower left, three turns of the DNA helix from the TBP binding site. *Courtesy of Joseph L. Kim and Stephen K. Burley.*

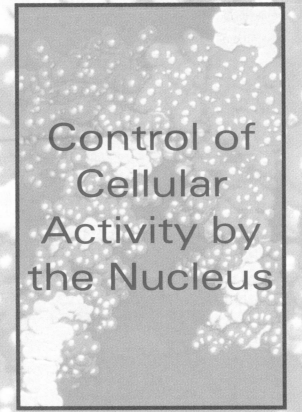

Control of Cellular Activity by the Nucleus

Having established the molecular constituents of living systems, we now turn to the logic by which they are combined to produce a living cell. That logic is embedded in the central dogma of biology: DNA makes RNA, which specifies the sequence of amino acids in protein. Although this dogma leaves out many of the important roles of RNA—and it is even violated by reverse transcription—it is the organizing principle behind cellular anatomy, physiology, and development. It puts DNA in the initiating position, the repository of the heredity of a cell, an organism, or a species. It puts RNA in the role of carrier, the working set of information diagrams for an individual cell at a certain time. It gives protein the role of output: proteins are the cells' workhorses and make one cell different from another. Part II concentrates on the informational molecules, describing how DNA is organized and how its enormous fund of information, enough for all the body's behavior, can be selectively provided in appropriate packets of RNA molecules for individual cells.

Chapter 9 begins with a description of the anatomy of DNA and how it is packaged into the cell's nucleus. We look at the molecular definition of a gene and build on that to show how genes are organized in DNA. This chapter describes the vast regions of mammalian DNA that lack genes and harbor highly repetitive sequences of questionable function. Chapter 9 then deals with the daunting problem of packaging lengths of DNA 100,000 times the length of a mammalian cell into a compact but accessible form inside the cell's nucleus. From this chapter will come an appreciation of the remarkable task that experimental biology has set itself for the end of this century and the beginning of the next: the Human Genome Project, which aims to define at the molecular level the entire DNA of human beings.

For the logic of the central dogma to succeed, it is crucial that each cell of the body—and even of the species—have approximately the same set of genetic instructions to work from. Thus as Chapter 10 describes, complex machinery has evolved to allow precise duplication of DNA each time a cell divides. Further machinery is described that is involved in perpetually scanning DNA for accidents and repairing them faithfully to maintain the integrity of DNA's information. A few alterations in DNA sequence do occur over time and they provide the necessary genetic variation for evolutionary selection. The chapter ends by describing how DNA is broken and recombined as germ cells form so that new members of the species receive new combinations of genetic information.

In Chapters 11 and 12 we turn to perhaps the most complicated issue in all of molecular biology—and probably the most central one—how genes are differentially expressed in particular cells, the issue of *gene control.* Chapter 11 describes a single type of event in cellular life, the initiation of RNA synthesis from a gene. Evolution has made this the central choice point in the use of DNA's information: whether initiation occurs or not determines whether a gene is expressed or not in a particular cell. Thus hundreds or perhaps thousands of transcription factor proteins are involved in this decision in a mammalian cell. Chapter 12 describes a series of further choice points in the use of cellular information that occur after an RNA molecule has been made. The most complex of these is the splicing process by which an initial transcript from DNA is processed to remove many internal regions, thus shaping the information content of an mRNA molecule. Further choice points come when the mRNA is translated into protein and when it is finally degraded.

Chapter 13 considers how all of this molecular machinery is brought together to allow an organism to develop. Here there are yet many open questions, as well as a daunting wealth of detail, so that only certain events are described that provide models for understanding the full range of biological potential in DNA. These events include choice points in the life of bacterial viruses and yeast that illustrate how differentiation of cellular functions comes about. Chapter 13 also describes how genetics has been a key tool in uncovering developmental strategies in multi-cellular organisms, with two organisms having received particular attention: the nematode worm, *C. elegans,* and the fruit fly, *Drosophila.*

9 The Molecular Anatomy of Genes and Chromosomes

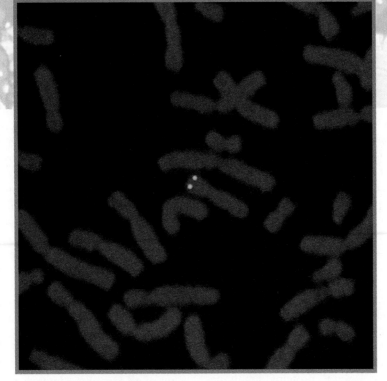

▲ Localization of a single gene (yellow dots) on metaphase chromosomes by fluorescence in situ hybridization.

Sequencing and other molecular analyses have yielded startling revelations about genomes. First of all, sequence analysis has suggested that a very large fraction of all vertebrate genomes, perhaps well over 90 percent, does not encode precursors to mRNAs or any other RNAs. No function for the majority of this "extra" DNA* has yet been found. In multicellular organisms, this noncoding DNA contains many regions that are "repetitious" (similar but not identical). Variations within the repetitious DNA stretches are so great that each single human, for example, can be distinguished by a DNA "fingerprint" based on these variations in repetitive sequences. Moreover, some repetitious DNA sequences are not found in constant positions in the DNA of individuals of the same species. Such "mobile" DNA segments, which are present in both prokaryotic and eukaryotic organisms, cause mutations when they move to new sites in the genome, and therefore may play an important role in evolution, even though they generally have no role in the life cycle of an individual organism.

In higher eukaryotes, the genes, regions of DNA sequence that encode proteins, lie amidst this sea of unstable, probably nonfunctional DNA. Sequencing studies on hundreds of genes have confirmed that noncoding introns are common in genes of multicellular plants and animals; less common, but sometimes present, in single-celled eukaryotes; and very rare in bacteria. Sequencing of the same protein-coding gene in a variety of eukaryotic species has shown that evolutionary pressure selects for maintenance of relatively similar sequences in exons. In contrast, wide sequence variation, even including total loss, occurs among introns, suggesting that most of the sequence of introns is nonfunctional. Cloning and sequencing have also confirmed the widespread existence of "families" of similar genes encoding proteins with related, but distinct functions. Members of a gene family arose during evolution

---

* Occasionally referred to as "junk" DNA in the literature, we do not use this term.

when an ancestral gene was duplicated, and the two copies of the gene evolved into distinct but related sequences. In some cases, however, gene duplicates have accumulated mutations that render them useless for encoding protein, although they are still maintained in the genome as sequences called "pseudogenes."

The sheer length of cellular DNA is a significant problem with which cells must contend. The length of a typical bacterial cell chromosome is about 10^3 times longer than the length of the cell. Mechanisms had to evolve to fold and organize this great length of DNA within the cell cytoplasm. The problem is even greater for higher organisms. The total length of DNA in the nucleus of a human cell is about 10^5 times the diameter of a typical cell. Eukaryotes have evolved specialized proteins that fold the nuclear DNA and organize it into the structures of DNA and protein visualized as individual chromosomes during mitosis.

In this chapter we first present a molecular definition of genes that is consistent with the detailed knowledge of DNA sequence now available. We also consider the large-scale structure of chromosomes, how DNA is packaged into compact complexes of DNA and proteins called chromatin, and describe the functional elements required for chromosome duplication and segregation.

➤ Molecular Definition of a Gene

Chapter 8 presented the definition of a gene based on classical genetic techniques such as complementation analysis. Most genes defined by classical genetic analysis of mutant phenotypes affect the function of a single protein. However, some mutations affect several proteins simultaneously. Understanding how this occurs requires knowledge of the molecular structure of genes.

In molecular terms, a gene can be defined as *the entire nucleic acid sequence that is necessary for the synthesis of a functional polypeptide or RNA molecule.* According to this definition, a gene includes more than the nucleotides encoding the amino acid sequence of a protein, referred to as the *coding region,* or a functional RNA such as an rRNA or tRNA. A gene also includes all the DNA sequences required to get a particular RNA transcript made. In some prokaryotic genes, DNA sequences controlling the initiation of transcription by RNA polymerase can lie thousands of base pairs from the coding region. In eukaryotic genes, such *transcription-control regions* can lie 50 kb or more from the coding region. Other critical noncoding regions in eukaryotic genes are the DNA sequences that specify signals for 3′ cleavage and polyadenylation, and for splicing of primary RNA transcripts; mutations in these RNA-processing signals prevent expression of a functional mRNA and thus of the encoded polypeptide.

Most Prokaryotic Genes Lack Introns, and Those Encoding Proteins with Related Functions Form Operons, Which Produce Polycistronic mRNAs

Sequencing of thousands of genes from eubacteria has revealed that the vast majority contain uninterrupted protein-coding regions. However, a few exceptions have been identified, including the gene for the enzyme thymidine synthetase encoded by bacteriophage T4. Similarly, genes encoding rRNAs and tRNAs generally are uninterrupted in eubacteria. In archaebacteria, however, a number of tRNA genes have been found to contain introns.

As discussed in Chapter 4, genes encoding enzymes involved in related functions generally are located next to each other in bacterial chromosomes. For example, the five genes encoding the enzymes required to synthesize the amino acid tryptophan from simple precursor molecules map in one contiguous stretch of ≈7 kb on the *E. coli* genome (see Figure 4-23a). The genes in this cluster comprise a single *transcription unit* referred to as an *operon*. The full set of genes is transcribed from the first gene in the cluster, *trpE,* to the last gene in the cluster, *trpA,* producing a single ≈7-kb mRNA molecule. Ribosomes initiate translation at the beginning of each of the genes in this mRNA producing the five polypeptides required for tryptophan synthesis. Since the *trp* mRNA encodes several polypeptides, it is said to be a *polycistronic* mRNA.

One consequence of the arrangement of bacterial genes into operons is that a single mutation can influence the expression of several proteins (Figure 9-1a). For example, if a single point mutation (e.g., a base-pair change) in the transcription-control region of the *trp* operon prevents initiation of transcription by RNA polymerase, then expression of all *five* of the polypeptides required for tryptophan synthesis is eliminated.

Most Eukaryotic Transcription Units Produce Monocistronic mRNAs

In contrast to polycistronic mRNAs, which are common in prokaryotes, most eukaryotic transcription units produce mRNAs that encode only one protein. This distinction correlates with a fundamental difference in mRNA translation in prokaryotes and eukaryotes. In prokaryotes, the small ribosomal subunit and initiation factors recognize and bind to specific related mRNA sequences called *ribosome-binding sites* (or Shine-Delgarno sequences after their discoverers) at the 5′ end of protein-coding regions (see Figure 4-43). Additional initiation factors and the large ribosomal subunit then join the small subunit and translation begins (see Figure 4-42). As a consequence, translation initiation can begin at internal sites in a polycistronic mRNA molecule.

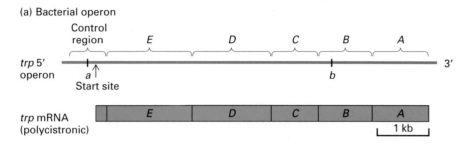

(a) Bacterial operon

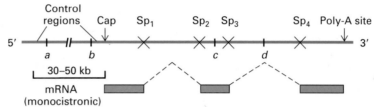

(b) Simple eukaryotic transcription unit

▲ FIGURE 9-1   Effect of various mutations on expression of bacterial operons and simple eukaryotic transcription units. (a) The *trp* operon in the *E. coli* genome contains five genes (*A–E*) encoding enzymes required for synthesis of tryptophan. Transcription of the entire operon yields an ≈7-kb polycistronic mRNA and is regulated by the transcription-control region, consisting of promoter/operator/attenuator sequences (Chapters 11 and 12), located just 5′ of the start site. A mutation within the transcription-control region (*a*) can prevent expression of all the proteins encoded by the *trp* operon. In contrast, a mutation within the *trpB* gene (*b*) affects TrpB protein only. (b) A simple eukaryotic transcription unit includes a region that encodes one protein extending from the 5′ cap site to the 3′ poly-A site and control regions. The sequences between splice sites (Sp) 1 and 2 and between 3 and 4 are introns, which are removed during processing of the primary transcripts and thus do not occur in the functional mRNA. (Dashed lines indicate spliced-out introns.) One transcription-control region, where transcription is initiated, is located just 5′ (upstream) of the cap site; a second is located 30–50 kb upstream of the cap site. Mutations in a transcription-control region (*a, b*) may reduce or prevent transcription, thus reducing or eliminating synthesis of the encoded protein. A mutation within an exon (*c*) may result in an abnormal protein with diminished activity. A mutation within an intron (*d*) that introduces a new splice site results in an abnormally spliced mRNA encoding a nonfunctional protein.

In contrast, in eukaryotes, the small ribosomal subunit and initiation factors generally first recognize the 5′-cap structure on an mRNA molecule and then interact with the large ribosomal subunit to initiate translation at the closest AUG codon. Consequently, in most cases only the sequence following the first AUG in an mRNA is translated; therefore, only a single polypeptide is produced. In other words, most eukaryotic mRNAs are *monocistronic*. Primary transcripts of eukaryotic protein-coding genes generally are processed into mRNAs that are translated to give a single type of polypeptide (see Figure 4-25).

Simple Eukaryotic Transcription Units Give Rise to One mRNA

Many eukaryotic genes (e.g., human β-globin gene) encode a single protein; such *simple transcription units* give rise to a single type of mRNA. Mutations in exons, introns, and transcription-control regions all may influence the expression of genes that constitute such simple transcription units (Figure 9-1b). However, mutations in simple eukaryotic transcription units, unlike those in bacterial operons, affect only one protein.

Because a simple eukaryotic transcription unit encodes only one mRNA and hence only one protein, mutations in such a unit are scored as individual cistrons in a complementation test (see Figure 8-14). In this case, then a "gene" defined by mutation and a transcription unit are directly related and refer to the same DNA sequence.

Complex Eukaryotic Transcription Units Give Rise to Alternative mRNAs

Although many transcription units in eukaryotes, especially in yeasts, are simple, *complex transcription units* are quite common in both invertebrates and vertebrates. The primary RNA transcript encoded by complex transcription units may contain more than one poly-A site and can be spliced in more than one way. Because of the different processing possibilities, the exons in a single complex transcription unit can be linked in alternative ways, yielding different mRNAs that encode distinct proteins (Figure 9-2).

(a) *Overlapping exons:* No complementation between any mutants; mutation *c* affects only mRNA₁

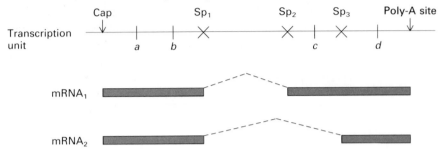

(b) *Multiple poly-A sites and nonoverlapping exons:* No complementation with *a* mutant; *b* and *c* mutants complement

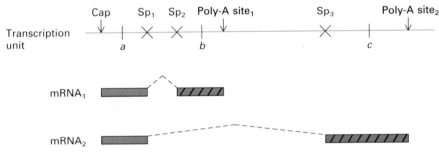

(c) *One poly-A site and nonoverlapping exons:* No complementation between *a, b,* or *e* mutants and any others; *c* and *d* mutants complement

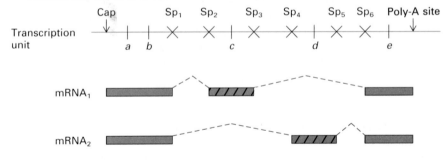

◄ FIGURE 9-2   Various types of complex eukaryotic transcription units and the effect of mutations on expression of the encoded proteins. Complex transcription units contain multiple splice sites (Sp) and may contain more than one poly-A site; their primary transcripts can be processed in alternative ways to yield two or more functional mRNAs. (Dashed lines indicate spliced-out introns.) Mutations within exons shared by all the alternative mRNAs (solid red) affect all the encoded proteins and therefore cannot complement. Mutations at sites within overlapping but nonidentical exons (e.g., mutations *c* and *d* in part [a]) also cannot complement each other, although mutation *c* affects only one of the encoded proteins. Mutations in nonoverlapping exons (red with black diagonal lines) affect only the protein containing the exon sequence and can complement (e.g., mutations *b* and *c* in part [b], and mutations *c* and *d* in part [c]). Although not shown here, mutations in the transcription-control regions upstream of a complex transcription unit can reduce or eliminate transcription of all the encoded proteins.

The same kinds of mutations that affect expression of proteins encoded in simple eukaryotic transcription units also affect proteins encoded in complex units. However, if a complex transcription unit is considered to contain as many genes as the different mRNAs that it encodes, then demonstrating the existence of each gene by complementation tests may be difficult. For example, a mutation in an exon that is shared by all the alternative mRNAs (e.g., exons shown in solid red in Figure 9-2) will affect all the alternative proteins expressed from a given complex transcription unit. In this case, no complementation can occur between two chromosomes carrying individual mutations in the shared exons; thus the individual protein-coding genes cannot be distinguished by genetic analysis. Mutations in the transcription-control regions will have the same affect. On the other hand, mutations in an exon present in only one of the alternative mRNAs encoded by a complex transcription unit (e.g., exons shown in red with diagonal lines in Figure 9-2b,c) will affect only the protein encoded by that mRNA. In this case, a chromosome carrying a mutation in one nonoverlapping exon can complement a chromosome carrying a mutation in a different nonoverlapping exon. In summary, a complex eukaryotic transcription unit is not coextensive with a gene defined by mutation (i.e., a potential complementation unit).

Some Genes Do Not Encode Protein

Although most genes are transcribed into mRNAs that encode proteins, clearly some RNAs, including tRNAs and rRNAs, do not encode proteins. However, because the DNA that encodes tRNAs and rRNAs can be mutated and detected in genetic experiments, we speak of tRNA genes and rRNA genes, even though the final products of these genes are RNA molecules and not proteins. Many other RNA molecules described in later chapters also do not encode proteins and therefore are transcribed from "genes" that do not encode proteins.

➤ *Organization of Genes on Chromosomes*

Having reviewed the relationship between transcription units and genes in prokaryotes and eukaryotes, we now consider the organization of protein-coding genes on chromosomes.

Genomes of Higher Eukaryotes Contain Much "Nonfunctional" DNA

Figure 9-3 diagrams the protein-coding regions in an 80-kb stretch of DNA from the yeast *S. cerevisiae* and in the β-globin gene cluster of humans, also about 80 kb long. Note that in the single-celled yeast, protein-coding regions are closely spaced along the DNA sequence, whereas only a small fraction of the human DNA encodes protein. Apparently, a considerable amount of DNA that does not encode protein is present in higher eukaryotes.

The origin of this remarkable difference between microorganisms and multicellular organisms, though not entirely understood, may lie in different selective pressures during evolution. For example, microorganisms must compete for limited amounts of nutrients in their environment, and metabolic economy thus is a critical characteristic.

Since synthesis of nonfunctional (i.e., non-protein-coding) DNA requires time and energy, presumably there was selective pressure to lose nonfunctional DNA during evolution of microorganisms. On the other hand, natural selection in vertebrates depends largely on their behavior. The energy invested in DNA synthesis is trivial compared with the metabolic energy required for the muscle movement associated with various behaviors; thus there was little selective pressure to eliminate nonfunctional DNA in vertebrates.

▼ FIGURE 9-3   Diagrams of ≈80-kb region from chromosome III of the yeast *S. cerevisiae* and the β-globin gene cluster on human chromosome 11. (a) In the yeast DNA, blue boxes indicate open reading frames; it is not clear whether all these potential protein-coding sequences are functional genes. (b) In the human DNA, the blue boxes represent transcribed regions that encode the indicated globin-type proteins. The arrangement of exons and introns is quite similar for all the globin-type genes; it is shown enlarged for the β-globin gene, with the small numbers indicating base pairs in each region of the gene. The human β-globin gene cluster contains two pseudogenes (diagonal lines); these regions are related to the functional globin-type genes but are not transcribed. Red arrows indicate the locations of an ≈300-bp sequence, called *Alu*, that is found at more than 500,000 positions in the human genome. This region of the human genome evolved from duplications of an ancestral globin gene and is likely to have a higher density of functional genes than most regions of human DNA. [Part (a) see S. G. Oliver et al., 1992, *Nature* **357**:28; part (b) see F. S. Collins and S. M. Weissman, 1984, *Prog. Nucl. Acid Res. Mol. Biol.* **31**:315.]

(a) *S. cerevisiae* (chromosome III)

tRNA gene   Open reading frame

(b) Human β-globin gene cluster (chromosome 11)

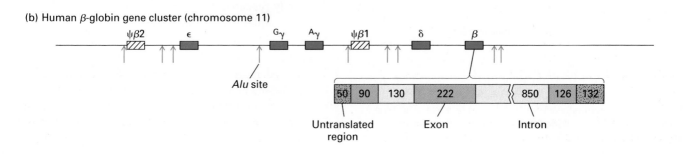

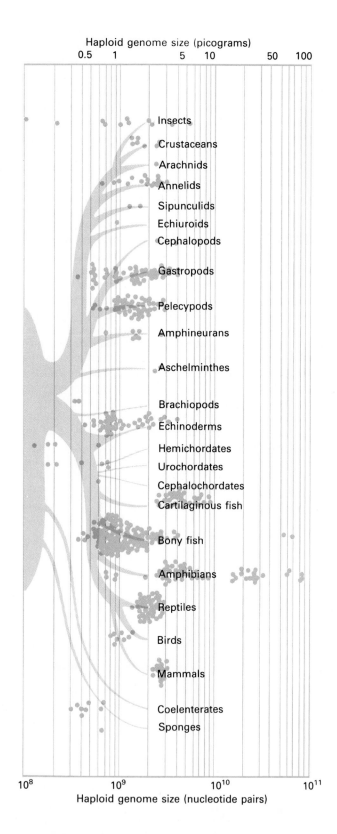

Haploid genome size (picograms)

Cellular DNA Content Does Not Correlate with Phylogeny

The total amount of chromosomal DNA in different animals and plants does not vary in a consistent manner with the apparent complexity of the organisms. Yeasts, fruit flies, chickens, and humans have successively larger amounts of DNA in their haploid chromosome sets (0.015, 0.15, 1.3, and 3.2 picograms, respectively), in keeping with what we perceive to be the increasing complexity of these organisms. Yet the vertebrates with the greatest amount of DNA per cell are amphibians (Figure 9-4), which are surely less complex than humans in their structure and behavior. Many plant species also have considerably more DNA per cell than humans have. For example, the DNA content per cell of wheat, broad beans, and garden onions (7.0, 14.6, and 16.8 picograms, respectively) ranges from about two to more than five times that of humans, and tulips have ten times as much DNA per cell as humans.

There is considerable intragroup variation in DNA content. All insects or all amphibians would appear to be similarly complex, but the amount of haploid DNA within each of these phyla varies by a factor of 100. The same variation in DNA content per cell is common within groups of plants that have similar structures and life cycles. For example, the broad bean contains about three to four times as much DNA per cell as the kidney bean.

These facts further suggest that much of the DNA in certain organisms is "extra" or expendable—that is, it does not encode RNA or have any regulatory or structural function. The total amount of DNA per haploid cell in an organism is referred to as the *C value*; the failure of C values to correspond to phylogenetic complexity is called the *C-value paradox*. This perplexing variation in genome size occurs mainly because eukaryotic chromosomes contain variable amounts of DNA with no demonstrable func-

▲ FIGURE 9-4   The amount of DNA per haploid chromosome set in a variety of animals. Variation within certain phyla (e.g., insects and amphibians) is quite wide. The blue "tree" represents evolutionary relationships between the indicated animals. Mammals are more closely related to birds and reptiles than to coelenterates and insects. [See R. J. Britten and E. Davidson, 1971, *Quart. Rev. Biol.* **46**:111.]

TABLE 9-1   Classification of Eukaryotic DNA

Protein-coding DNA
 Solitary genes
 Duplicated and diverged genes (functional gene families and nonfunctional pseudogenes)
Tandemly repeated genes encoding rRNA, 5S rRNA, tRNA, and some histone genes
Repetitious DNA
 Simple-sequence DNA
 Intermediate-repeat DNA (mobile DNA elements)
  Short interspersed elements (nonviral retrotransposons)
  Long interspersed elements (viral and nonviral retrotransposons)
Unclassified spacer DNA

tion. As discussed later, this includes repeated DNA stretches, some of which are never transcribed and most all of which are likely dispensable. The different classes of eukaryotic DNA sequences discussed in the following sections are summarized in Table 9-1.

About a Quarter to Half of All Eukaryotic Protein-Coding Genes Are Solitary

In multicellular organisms, roughly 25–50 percent of the protein-coding genes are represented only once in the haploid genome and thus are termed *solitary* genes. The remaining protein-coding genes belong to families of two or more similar genes.

A well-studied solitary protein-coding gene encodes chicken lysozyme. An enzyme that cleaves the polysaccharides composing bacterial cell walls, lysozyme is found in human tears and is an abundant component of chicken egg-white protein. Its activity helps to keep the surface of the eye and the chicken egg sterile. The 15-kb DNA sequence encoding chicken lysozyme constitutes a simple transcription unit (i.e., a single gene) containing four exons and three introns (Figure 9-5). The flanking regions, extending for about 20 kb upsteam and downstream from the transcription unit, do not encode any detectable mRNAs.

Gene Families Are Formed by Gene Duplication and Encode Homologous Proteins

Frequently, the DNA that lies within 5–10 kb of a particular gene contains sequences that are close but inexact copies of the gene. Such sequences, which are thought to have arisen by duplication of an ancestral gene, are referred to as *duplicated* protein-coding genes; duplicated genes probably constitute half of the protein-coding DNA in vertebrate genomes. A set of duplicated genes that encode proteins with similar but nonidentical amino acid sequences is called a *gene family*; the encoded closely related, homologous proteins constitute a *protein family*. Many protein

TABLE 9-2 Protein Families in Vertebrates and Invertebrates	
Family	Number of Proteins in Family
COMMON PROTEINS	
Actins	5–30
70 K Heat-shock proteins	3
Keratins	>20
Myosin, heavy chain	5–10
Protein kinases	10–100s
Transcription factors	10–100s (?)
Tubulins, α and β	3–15
INSECT PROTEINS	
Eggshell proteins (silk moth and fruit fly)	50
VERTEBRATE PROTEINS	
Globins (many species)	
α-globin	1–3
β-like globins	5
Immunoglobins, variable regions (many species)	500
Ovalbumin (chicken)	3
Transplantation antigens (mouse and human)	50–100
Visual pigment protein (human)	4
Vitellogenin (frog, chicken)	5

families contain from several to as many as 20 members; a few families contain hundreds of members (Table 9-2).

The genes encoding the β-like globins are a good example of a gene family. Two identical β-globin polypeptides combine with two identical α-globin polypeptides (encoded by another gene family) and with four small heme groups to form a hemoglobin molecule (see Figure 3-15). As shown in Figure 9-3b, the β-like globin gene family contains five functional genes designated β, δ, $^A\gamma$, $^G\gamma$, and ϵ; the encoded polypeptides are similarly designated. All the hemoglobins formed from the different β-like globins carry oxygen in the blood, but they exhibit somewhat different properties that are suited to specific roles in human physiology. For example, hemoglobins containing either $^G\gamma$ or $^A\gamma$ are expressed only during fetal life. Because these fetal hemoglobins have a higher affinity for oxygen than adult hemoglobins, they can effectively extract oxygen from the maternal circulation in the placenta. The lower oxygen affinity of adult hemoglobins, which are expressed after birth, permits better release of oxygen to the

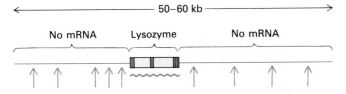

▲ FIGURE 9-5   The chicken lysozyme gene and its surrounding regions. This 15-kb simple transcription unit contains four exons (blue) and three introns (tan). The positions indicated by red arrows are repetitive sequences found at many sites elsewhere in the genome. [See P. Balducci et al., 1981, *Nucleic Acids Res.* **9**:3575.]

tissues, especially muscles, which have a high demand for oxygen during exercise.

The different β-globin genes probably arose by duplication of an ancestral gene, most likely as the result of an "unequal crossover" during recombination in a germ-cell (egg or sperm) precursor (Figure 9-6). Over evolutionary time the two copies of the gene that resulted accumulated random mutations; beneficial mutations that conferred some refinement in the basic oxygen-carrying function of hemoglobin were retained by natural selection. Repetitions of this process are thought to have resulted in the evolution of the contemporary globin-like genes observed in humans and other complex species today.

The various proteins that make up the cytoskeleton, which are present in almost all cells in varying amounts, form several different protein families. In vertebrates, these cytoskeletal proteins include the actins, tubulins, and intermediate filament proteins like the keratins. Although the physiologic rational for these gene families is not as obvious as it is for the globins, the different members of a family probably have similar but subtly different functions suited to the particular type of cell in which they are expressed.

The different genes composing a gene family often can be identified and isolated by molecular cloning. For example, clones of the α-globin and β-globin genes in a λ-phage library of human genomic DNA were identified by plaque hybridization with a cDNA probe prepared from mRNA isolated from reticulocytes, the precursors of mature erythrocytes (see Figure 7-18). However, because the embryonic- and fetal-globin genes are closely related in sequence to the adult-globin genes, they also hybridize to the adult-globin probe under the standard conditions of plaque hybridization. Consequently, the screening of the λ-phage

genomic library led to identification of clones not only of the adult-globin genes but also of the embryonic- and fetal-globin genes. Once the various globin-gene clones had been identified and isolated, they were distinguished by restriction site mapping and DNA sequencing.

Pseudogenes Are Duplicated Genes That Have Become Nonfunctional

Two regions in the human β-like globin gene cluster contain nonfunctional sequences similar to those of the functional β-like globin genes. Because no known proteins correspond to these regions, they are called *pseudogenes* (see Figure 9-3b). Sequence analysis shows that these copies retain the same apparent exon-intron structure as the functional β-like genes, suggesting that they also arose by duplication of the same ancestral gene. However, *sequence drift* during evolution resulted in the accumulation of sequences that either terminate translation or block mRNA processing, rendering such regions nonfunctional even if they were transcribed into RNA.

The human δ-globin gene may represent an intermediate in this process of drift. This gene produces very little mRNA, and future sequence drift that completely halts the activity of this infrequently used gene duplicate might well be tolerated by the organism. Such a "silencing" genetic event apparently occurred in the δ gene of gibbons some 5–10 million years ago. The initial mutation was probably in the transcription-control region of the δ gene, thus reducing its activity. Enough mutations subsequently accumulated in this gene to render it incapable of directing production of a protein; it thus became a pseudogene.

▼ FIGURE 9-6   Gene duplication resulting from unequal crossing over. Each parental chromosome (*top*) contains one ancestral globin gene (blue) containing three exons and two introns (see Figure 9-3b). Homologous repetitious *Alu* sequences lie 5′ and 3′ of the globin gene. The parental chromosomes are shown displaced relative to each other, so that the *Alu* sequences are aligned. Recombination between *Alu* sequences as shown would generate one recombinant chromosome with two copies of the globin gene and one chromosome with a deletion of the globin gene. Subsequent independent mutations in the duplicated genes could lead to slight changes in sequence that might result in slightly different functional properties of the encoded proteins. Unequal crossing over also can result from rare recombinations between unrelated sequences.

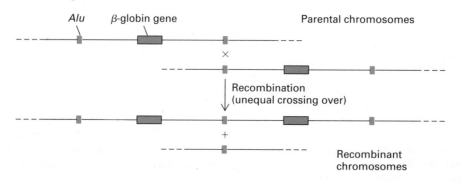

Present-day gibbons survive perfectly well with one adult β-like globin gene.

Pseudogenes have been identified in various other gene families, including the tubulin and actin gene families. In addition to the complete but nonfunctional gene copies that constitute pseudogenes, partial copies of some genes have been identified. For example, sequences corresponding to fragments of the 5′ and 3′ ends of the tubulin genes are quite common in human DNA. These presumably arose by unequal crossovers within the tubulin genes, rather than in adjacent regions as diagrammed in Figure 9-6. As discussed in a later section, other nonfunctional gene copies can arise by reverse transcription of mRNA into cDNA and integration of this intron-less DNA into a chromosome.

rRNAs, tRNAs, and Histones Are Encoded by Tandemly Repeated Genes

The genes for the rRNAs, each type of tRNA, and one family of proteins, the histones, which package nuclear DNA into chromatin, occur in invertebrates and some vertebrates as *tandemly repeated arrays*. These are distin-guished from the duplicated genes of gene families in that the multiple tandemly repeated genes encode identical or nearly identical proteins or functional RNAs. Most often copies of a sequence appear one after the other, in a head-to-tail fashion, over a long stretch of DNA. Within a tan-dem array of rRNA or tRNA genes, each copy is exactly, or almost exactly, like all the others. Although the tran-scribed portions of rRNA genes are the same in a given individual, the nontranscribed spacer regions between the transcribed regions can vary. Arrays of tandemly repeated histone DNA are somewhat more complex; however, each histone gene, too, has multiple identical copies.

The tandemly repeated rRNA, tRNA, and histone genes are needed to meet the great cellular demand for their transcripts. Most of the RNA in a cell consists of rRNA and tRNA. Assuming RNA polymerase molecules move at a fixed speed, there must be a limit to the number of RNA copies that transcription of a single gene can pro-vide during one cell generation, even if it is fully loaded with polymerase molecules. If more RNA is required than can be transcribed from one gene, multiple copies of the gene are necessary, as illustrated in Figure 9-7 for the syn-thesis of pre-RNA, which is processed into 18S, 5.8S, and

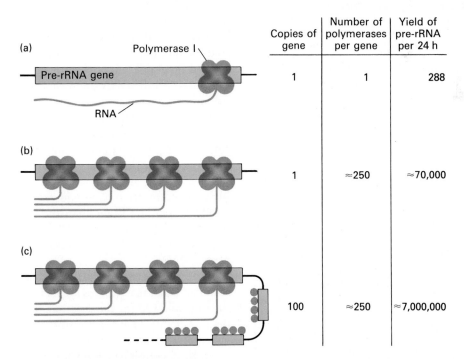

	Copies of gene	Number of polymerases per gene	Yield of pre-rRNA per 24 h
(a)	1	1	288
(b)	1	≈250	≈70,000
(c)	100	≈250	≈7,000,000

▲ FIGURE 9-7   Effect of copy number and loading with RNA polymerase I on rate of synthesis of pre-rRNA in human cells. Genes encoding pre-rRNAs, which are processed into the 18S, 5.8S and 28S rRNAs, are transcribed by the enzyme RNA polymerase I. Transcription of the pre-rRNA gene by a single molecule of RNA polymerase I takes about 5 min. (a) If a cell contained one copy of the pre-rRNA gene, which was transcribed by one polymerase at a time, it could pro-duce a maximum of 288 copies per 24 h. (b) The yield of pre-rRNA from a single copy of the pre-rRNA gene would increase substantially if the gene was maximally loaded with ≈250 polymerase molecules. (c) The highest rate of pre-rRNA synthesis is possible when a cell contains multiple copies of the pre-rRNA and these are transcribed by many polymerase molecules at one time. (Duplicate genes are indi-cated by small blue rectangles and polymerases by red cir-cles). In order to generate enough rRNA to divide every 24 h, human embryonic cells must have at least 100 copies of the rRNA gene and these must be near maximally loaded with RNA polymerase I.

28S rRNA. For example, during early embryonic development in humans, many embryonic cells have a doubling time of ≈ 24 h and contain 5–10 million ribosomes. To produce enough rRNA to form this many ribosomes, an embryonic human cell needs at least 100 copies of the pre-rRNA gene, and most of these must be close to maximally active for the cell to divide every 24 h (see Figure 9-7c). The importance of repeated rRNA genes is illustrated by *Drosophila* mutants called "bobbed" (because they have stubby wings), which lack a full complement of the tandemly repeated rRNA genes. A bobbed mutation that reduces the number of rRNA genes to less than ≈ 50 is a recessive lethal mutation.

Genes encoding many functional RNAs other than mRNA exist in multiple copies in eukaryotic cells (Table 9-3). All species, including yeasts, contain 100 or more copies of the genes encoding 5S rRNA and pre-rRNA. More than 20,000 copies of the 5S rRNA gene are present in frogs. The copy number for individual tRNA genes ranges from 10 to 100. The multiple copies of all the rRNA genes occur in tandem arrays.

▶ Discovery of Repetitious DNA Fractions

Besides the duplicated protein-coding genes and the tandemly repeated genes encoding rRNAs, tRNAs, and histones discussed in the previous section, eukaryotic cells contain multiple copies of other DNA sequences in the genome. These are generally referred to as *repetitious DNA* (see Table 9-1). Some of these sequences are quite short and occur as *tandem* repeats; others are much longer and are *interspersed* at many places in the genome. The existence of these repeated sequences was first recognized in reassociation experiments in which denatured eukaryotic DNA was observed to renature nonuniformly; that is, some of it reassociated much more rapidly than the bulk of cellular DNA. Here we briefly review the experimental evidence that led to discovery of the two major classes of repetitious DNA; later, we discuss each class in more detail.

Repeated DNA Reassociates More Rapidly Than Nonrepeated DNA

Suppose that the total DNA of an organism is broken into fragments with an average length of about 1000 base pairs. The DNA is then melted into single strands and placed under conditions that allow strand reassociation to occur (e.g., a favorable ion concentration and a favorable temperature). All the DNA fragments would re-form duplexes at about the same speed if none contained sequences that were repeated in the genome. However, a segment containing a sequence repeated many times in the genome would find a complementary partner more quickly than a segment with a sequence that occurs only once per haploid genome, because the repeated sequence would be present at a much higher concentration. Consequently the repeated sequence would reassociate faster than the fragment of unique sequence. For this reason, the DNA encoding pre-rRNA and that encoding 5S rRNA reassociate faster than does nonrepeated DNA.

The parameters that affect the degree to which single-stranded DNA reassociates are its initial concentration and the time allowed for the reaction. The C_0t of a reaction is the product of the concentration of the DNA measured in moles of nucleotide per liter C_0 and the reaction time t in seconds. A convenient term for comparing the reassociation rates of different DNA fractions is the $C_0t_{1/2}$ value—the C_0t at which one-half of a given fraction renatures. The lower the value of $C_0t_{1/2}$, the higher the reassociation rate. By comparing the $C_0t_{1/2}$ value of any particular DNA fraction with that of a "standard" nucleic acid (e.g., a viral or bacterial DNA of known length, both of which have either no or very few repetitive sequences), the approximate frequency of repeats within the fraction of interest can be determined.

TABLE 9-3   Copy Number of Tandemly Repeated Genes Encoding Structural RNAs in Several Eukaryotes*

Species	Number of Copies		
	Pre-rRNA Gene	5S-rRNA Gene	tRNA Genes[†]
Saccharomyces cerevisiae	140	140	250
Dictyostelium discoideum	180	180	?
Tetrahymena pyriformis			
Micronucleus[‡]	1	300	800
Macronucleus	200	300	800
Drosophila melanogaster			
X chromosome	250	165	860
Y chromosome	150	165	860
Xenopus laevis	450	24,000	1150
Human	≈ 250	2000	1300

*The copy numbers in this table were estimated by hybridizing saturating amounts of labeled RNA to DNA.
[†]The tRNA numbers include all tRNA sites and therefore represent more than 50 different tRNA genes in some organisms. Copy numbers for individual tRNAs range from 10–100.
[‡]The micronucleus is inactive in synthesis of pre-rRNA.
SOURCE: B. Lewin, 1980, *Gene Expression*, Vol. 2, Wiley, p. 876.

Reassociation Experiments Reveal Three Major Classes of Eukaryotic DNA

As the data in Figure 9-8 illustrates, the total DNA of the calf genome can be divided into three fractions based on their reassociation rate. Similar results have been obtained with the DNA from the other mammalian species tested. Although the same three DNA classes are found in all eukaryotes, their relative proportions may vary significantly from those observed in mammals; the yeast *S. cerevisiae*, for example, has very little simple-sequence DNA.

Rapid Reassociation Rate (Simple-Sequence DNA)
In typical reassociation experiments, about 10–15 percent of mammalian DNA reassociates almost immediately and has a $C_0 t_{1/2}$ value of 0.01 or less. This rapidly reassociating fraction, referred to as *simple-sequence DNA*, has proved

to be the simplest DNA to analyze. Simple-sequence DNA is composed largely of several different sets of short (5- to 10-bp) oligonucleotides repeated in long tandem arrays.

Intermediate Reassociation Rate (Intermediate-Repeat DNA) Another 25–40 percent of mammalian DNA reassociates at an intermediate rate over a broad range of $C_0 t_{1/2}$ values from 0.01 to about 10. Sequence analyses of cloned samples of this fraction—termed *intermediate-repeat DNA*, or *moderately repeated sequences*—from many different animals (e.g., fruit flies, frogs, rodents, and humans) have revealed that it is composed of a very large number of copies of a relatively few sequence families.

Intermediate-repeat DNA is interspersed throughout the mammalian genome and is classified into two types: *short interspersed elements (SINES)*, which contain 150–300 base pairs, and *long interspersed elements (LINES)*, which can be 5000–7000 base pairs long. The multiple copies of both SINES and LINES may not be *exact* repeats; neither SINES nor the vast majority of LINES code for protein. One major class of short intermediate repeats, or SINES, and perhaps about ten classes of long intermediate repeats, or LINES, have been found in mammalian genomes; each class may be present in thousands of copies.

Intermediate-repeat sequences are repeated less frequently in *Drosophila* and yeast DNA than in mammalian DNA; indeed, the similarity among these elements in higher and lower eukaryotes was recognized only in the 1980s. As discussed later, many intermediate-repeat sequences can move, or *transpose*, to new sites in cellular DNA. For this reason, they now are commonly referred to as *transposable elements*, or *mobile DNA elements*.

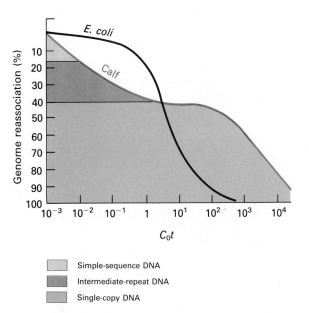

Simple-sequence DNA

Intermediate-repeat DNA

Single-copy DNA

▲ **FIGURE 9-8** Reassociation curves of calf DNA and of *E. coli* DNA. DNA from calf thymus or *E. coli* was broken into fragments of about 1000 base pairs, denatured into single strands, and reassociated for various periods of time and at various DNA concentrations. The extent of reassociation (%) was measured by a chromatographic technique and plotted against $C_0 t$ (molar concentration × time in seconds). The most rapidly reassociating calf DNA (yellow) renatures almost immediately. Two other broad fractions are noted: one that reassociates at an intermediate rate (green) and one that reassociates quite slowly (blue). The *E. coli* DNA reassociates over a fairly narrow range of $C_0 t$ values, which indicates that the *E. coli* fragments all tend to reassociate at the same speed. The calf DNA that reassociates at intermediate and rapid rates is repetitious DNA. The slowly reassociating fraction of calf DNA, which renatures about 500–1000 times more slowly than *E. coli* DNA, represents DNA sequences present only once per haploid genome. [See R. J. Britten and D. E. Kohne, 1968, *Science* **161**:529].

Slow Reassociation Rate (Single-Copy DNA) As shown in Figure 9-8, about 50–60 percent of mammalian DNA reassociates rather slowly, with $C_0 t_{1/2}$ values ranging from about 100 to 10,000. The most slowly reassociating sequences in mammalian DNA reanneal about 500 times more slowly than those in *E. coli* DNA. Because the amount of haploid DNA in mammalian cells is about 700 times that in *E. coli* and because almost all *E. coli* DNA is present in a single copy only, the slowly reassociating fraction of mammalian DNA is assumed to be *single-copy DNA*.

According to Mendelian genetics, only one copy of each gene is contained in the haploid DNA set; thus the single-copy DNA fraction is expected to contain most of the genes encoding mRNA. The reverse is not true: every single-copy DNA sequence does not necessarily perform a genetic function. It appears that only a small fraction of the total DNA in humans, on the order of 5 percent, actually encodes proteins or functional RNA molecules. The remainder of the single-copy DNA, which currently has no known function other than to separate functional DNA sequences, is referred to as *spacer DNA*.

➤ Simple-Sequence DNA

As noted previously, much of the simple-sequence DNA of higher organisms—the most rapidly reassociating DNA fraction—is composed of tandemly repeated, 5- to 10-bp oligonucleotide sequences. However, many instances of tandem repeats containing 20–200 nucleotides are now known to occur also in vertebrate and plant genomes (Table 9-4).

Higher Eukaryotes Contain Several Types of Simple-Sequence DNA

Simple-sequence DNA generally occurs in very long stretches up to 10^5 base pairs in length. Such DNA can be separated from the single-copy cellular DNA by equilibrium density-gradient centrifugation in an ultracentrifuge, because its average base composition, and hence buoyant density, differs from that of the single-copy cellular DNA. For example, centrifugation of *Drosophila virilis* DNA yields three density classes in addition to the main component of cellular DNA (Figure 9-9a). These long stretches of simple-sequence DNA are often referred to as *satellite* DNA, because these distinct density classes of DNA appeared to form "satellites" around the main density band of DNA in such equilibrium density gradients. However, not all simple-sequence DNA separates from main-band DNA to form satellite bands; therefore, the term simple-sequence DNA is preferred to satellite DNA.

Each animal or plant cell has several types of simple-sequence DNA characterized by different repeat units. In some species, the repeating oligomeric units of simple-sequence DNA have very similar sequences, which suggests an ancestral relationship between two or more different simple-sequence types. In *Drosophila virilis* cells, for example, the 7-bp repeats constituting satellite bands II and

---

TABLE 9-4   Examples of Simple-Sequence DNAs in Eukaryotes*

Organism	Base Pairs Per Repeat	Sequence of One Repeat Unit	Location
Drosophila melanogaster (fruit fly)	5	AGAAG (polypurine) ATAAT	Arms of Y chromosome; centromeric heterochromatin of chromosome 2; long arm of 2, near telomere
	7	ATATAAT	
	10	AATAACATAG AGAGAAGAAG	Centromeric heterochromatin of all chromosomes; tip of long arm of 2, near telomere
Drosophila virilis	7	ACAAACT (band I) ATAAACT (band II) ACAAATT (band III)	Centromeric heterochromatin
Cancer borealis (marine crab)	2	AT	?
Pagurus pollicaris (hermit crab)	4	ATCC	?
	3	CTG	?
Cavia poriella (guinea pig)	6	CCCTAA	Centromeric heterochromatin
Dipodomys ordii (kangaroo rat)	10	ACACAGCGGG	Centromeric heterochromatin
Cercopithecus aethiops (African green monkey; α sequences)*	172	—	Throughout chromosomes
Homo sapiens (human; alphoid sequences)*	171	—	Throughout chromosomes

* All eukaryotic species have more than one type of simple-sequence DNA, characterized by the sequence of the repeat unit; the table includes only selected examples. Many repeats are 10 bp or less in length, but several longer repeats, such as the primate α and alphoid sequences and human minisatellites are now known. The α and alphoid sequences are not shown because of their length.

SOURCE: K. Tartoff, 1975, *Ann. Rev. Genet.* 9:355; and A. J. Jeffreys et al., 1985, *Nature* 314:69.

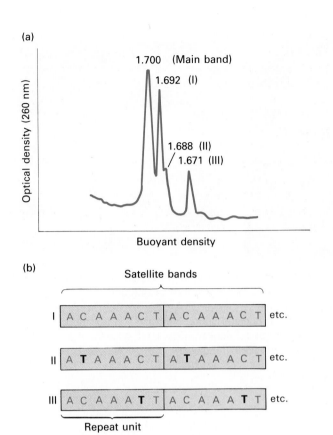

(a)

Optical density (260 nm)

1.700 (Main band)

1.692 (I)

1.688 (II)

1.671 (III)

Buoyant density

(b)

Satellite bands

I | A C A A A C T | A C A A A C T | etc.

II | A T A A A C T | A T A A A C T | etc.

III | A C A A A T T | A C A A A T T | etc.

Repeat unit

▲ **FIGURE 9-9**   Satellite (simple-sequence) DNA in *Drosophila virilis*. (a) The DNA from embryonic tissue was extracted and subjected to equilibrium density-gradient sedimentation in cesium chloride to separate DNAs differing in buoyant density. The DNA content of different zones in the gradient was monitored by measuring the absorption of ultraviolet light at 260 nm (i.e., the optical density). The main band of the DNA—that is, the greatest part of the DNA—has a density of 1.700. The three satellite bands, I, II, and III, are less dense. (b) DNA sequence analyses showed that each of the satellite bands is composed of a DNA that is a long tandem array of a seven-base repeat sequence. The repeat sequences in bands II and III differ from the band I sequence by only one base (bold black). (See J. G. Gall, E. H. Cohen, and D. D. Atherton, 1973, *Cold Spring Harbor Symp. Quant. Biol.* **38**:417.]

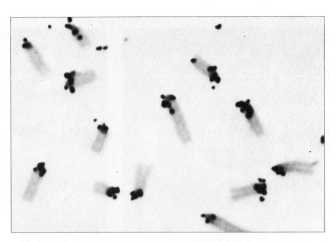

▲ **FIGURE 9-10**   Localization of simple-sequence DNA in mouse genome by in situ hybridization. Purified simple-sequence DNA from mouse cells was randomly transcribed by *E. coli* RNA polymerase to make labeled RNA. Chromosomes from cultured mouse cells were fixed and denatured on a microscope slide, and then the chromosomal DNA was hybridized in situ to the labeled RNA. Autoradiography of the telocentric mouse chromosomes (i.e., chromosomes in which the centromeres are located near one end) shows that most of the complementary simple-sequence DNA (silver grains) lies close to the centromere. [From M. L. Pardue and J. G. Gall, 1970, *Science* **168**:1356; courtesy of J. G. Gall.]

III each differ from the repeat of band I by a single base change (Figure 9-9b). In humans, at least 10 types of simple-sequence DNA exist. A single type can account for as much as 0.5–1 percent of the total human genome, an amount equivalent to $\approx 10^7$ base pairs, or three times the size of the *E. coli* genome.

Most Simple-Sequence DNA Is Located in Specific Chromosomal Regions

The location of simple-sequence DNA within chromosomes has been studied in several species by in situ hybridization experiments. In the chromosomes of mice and per-

haps most other mammals, much, but not all, of the simple-sequence DNA lies near centromeres, discrete chromosomal regions discussed in a later section (Figure 9-10). In the chromosomes of *Drosophila melanogaster*, simple-sequence DNA is likewise concentrated in the centromere region, but some also is located within the arms and end regions, called telomeres (see Table 9-4).

As DNA cloning has progressed, variations in simple-sequence DNA have been found that can be used to localize specific chromosomal sites. For instance, a particular simple sequence in the human genome is present only in the middle of the long arm of chromosome 16 (Figure 9-11). Thus this repetitive sequence is a useful molecular marker of this site in the human genome.

Differences in Lengths of Simple-Sequence Tandem Arrays Permit DNA Fingerprinting

Long stretches of simple-sequence DNA within a species exhibit a remarkable conservation in nucleotide sequence. In contrast, differences in the number of repeats, and thus in the length, of simple-sequence tandem arrays containing the same repeat unit are quite common. These differences in length result from unequal crossing over within regions of simple-sequence DNA during development of sperm and oocyte precursors and during meiosis (Figure 9-12a). As a consequence of this unequal crossing over, the simple-

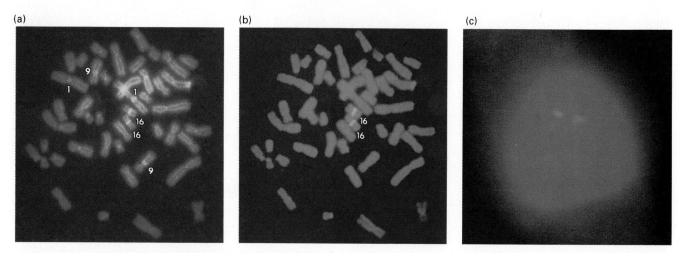

▲ FIGURE 9-11   Use of simple-sequence DNA as a chromosomal marker. (a) Human metaphase chromosomes stained with a fluorescent DNA dye. (b) The same chromosome spread hybridized in situ with a particular simple-sequence DNA that is labeled with a fluorescent biotin derivative. When the chromosomes are viewed under a different wavelength of light, the hybridized simple-sequence DNA appears as a yellow band on chromosome 16, thus locating this particular simple sequence at one site in the genome. (c) The simple-sequence DNA sites are still visible as yellow bands in the nucleus during interphase when the chromosomes are not condensed. [See R. K. Moyzis et al., 1987, *Chromosoma* **95**:378; courtesy of R. K. Moyzis.]

sequence tandem arrays in different individuals differ in length, a characteristic that provides the basis for *DNA fingerprinting*, a powerful technique for identifying individual DNA samples.

In humans and other mammals, some of the simple-sequence DNA exists in relatively short 1- to 5-kb regions made up of 20–50 repeat units each containing 15 to about 100 base pairs. These regions are called *minisatellites* to distinguish them from the more common regions of tandemly repeated simple-sequence DNA, which are 10^5–10^6 base pairs long. The sequences of the repeat unit in two human minisatellites are shown in Figure 9-12b. Even slight differences in the total lengths of various minisatellites from different individuals can be detected by conventional Southern-blot analysis.

To obtain the "DNA fingerprint" of an individual, a sample of DNA is digested with restriction enzymes that do not cut within minisatellite sequences; the fragments then are subjected to Southern-blot analysis with radioactive minisatellite probes (see Figure 7-31). Duplicate analyses with two to four different minisatellite probes is sufficient to provide a DNA fingerprint that is unique for each individual (Figure 9-13). The amount of DNA required for such an analysis can be extracted from the cells in a drop of blood obtained by finger puncture. Also, minute amounts of isolated DNA can be amplified by PCR (see Figure 7-37) for analysis of minisatellite tandem repeat lengths. DNA fingerprinting, which is superior to other methods (e.g., conventional fingerprinting) for identification of individuals, has been widely adapted by law enforcement agencies.

➤ *Intermediate-Repeat DNA and Mobile DNA Elements*

In the reassociation experiments discussed previously, 25–40 percent of mammalian DNA renatures at an intermediate rate (see Figure 9-8). The tandemly repeated genes encoding rRNA, tRNA, and histones and the closely related genes forming various gene families make up a small portion of this fraction. The term *intermediate-repeat DNA*, however, generally refers only to the bulk of this fraction, which consists of moderately repetitive DNA sequences interspersed throughout the genome and capable of moving, or transposing, to new sites in the genome.

The mobile DNA elements (or simply mobile elements) that constitute intermediate-repeat DNA are essentially molecular parasites that appear to have no specific function in the biology of their host organisms, but exist only to maintain themselves. For this reason, Francis Crick referred to these mobile DNA elements as "selfish DNA." These mobile DNA elements seem to have slowly accumulated over evolutionary time until the rate at which they transpose to new sites is balanced by the rate at which they are eliminated from the genomes of modern-day organisms. They are eliminated at a very slow rate by the deletion of segments of DNA containing them and by the accumulation of mutations until they can no longer be recognized to be related to the original mobile DNA element. Since mobile elements are eliminated from eukaryotic genomes so slowly, they have accumulated during evo-

(a) Generation of unequal tandem arrays

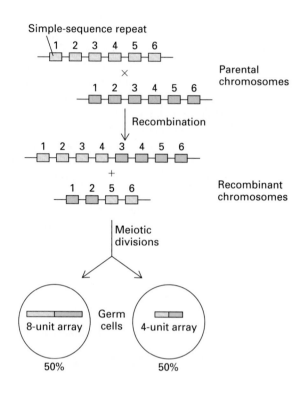

◄ FIGURE 9-12   Simple-sequence DNA is characterized by variations in the length of tandem arrays and constancy in sequence. (a) Differences in lengths of simple-sequence DNA tandem arrays can be generated by unequal crossing over during meiosis. In this example, unequal recombination within a stretch of DNA containing six copies (1–6) of a particular simple-sequence repeat unit yields germ cells containing either an 8-unit or 4-unit tandem array.
(b) Consensus sequences of repeat unit of human minisatellites λ33.1 and λ33.5 were determined from more than ten sets of repeats in each case. Red letters indicate positions in which base differences have been detected; red solid dot indicates a deletion. The 62-bp repeat unit of λ33.1 is much more highly conserved than the 17-bp unit of λ33.5. The total length of minisatellites ranges from 1 to 5 kb. Variations in length generated as shown in part (a) is the basis of DNA fingerprinting. [Part (b) see A. J. Jeffreys et al., 1985, *Nature* **314**:67.]

(b) Minisatellite consensus sequences

λ **33.1 minisatellite**   AAGGGTGGGCAGGAAGTGGAGTGTGTGCCTGCTTCCCTTCCCTGTCTTGTCCTGGAAACTCA

λ **33.5 minisatellite**   C GGGCAGG•AGGGGGAGG
                            T

lution to the point where they account for a large fraction of eukaryotic genomes. In humans mobile DNA elements account for ≈30 percent of the total genomic DNA.

Although mobile elements appear to have no direct function other than to maintain their own existence, their presence may have had a significant impact on evolution. As noted in Chapter 8, one form of spontaneous mutation results from the insertion of a mobile DNA element into or near a transcription unit. Of perhaps still greater significance in evolution, homologous recombination between

mobile DNA elements dispersed throughout ancestral genomes may have been important in generating gene duplications and other DNA rearrangements during evolution (see Figure 9-6). Such duplications and DNA rearrangements contributed greatly to the evolution of new genes. As discussed in an earlier section, gene duplication probably preceded the evolution of a new member of a gene family, which subsequently acquired distinct, beneficial functions. There is also evidence that during the evolution of higher eukaryotes, recombination between introns of distinct

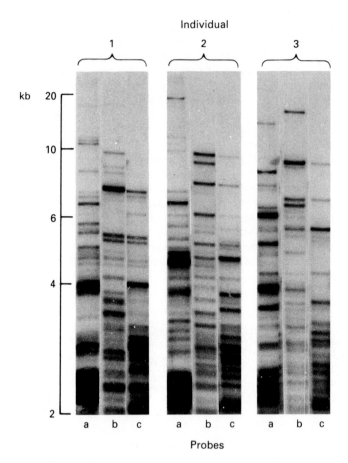

▲ FIGURE 9-13 Human DNA fingerprints. DNA samples from three individuals (1, 2, and 3) were subjected to Southern-blot analysis using the restriction enzyme *Hinf*1 and three labeled minisatellites as probes (λ33.6, 33.15, and 33.5; lanes a, b, and c, respectively). DNA from each individual produced a unique band pattern with each probe. Conditions of electrophoresis can be adjusted so that at least 50 bands can be resolved for each person with this restriction enzyme. The nonidentity of these three samples is easily distinguished. [From A. J. Jeffreys, V. Wilson, and S. L. Thein, 1985, *Nature* **316**:76; courtesy of A. J. Jeffreys.]

sequences, and the molecular basis of their movement, or *transposition* was deciphered.

When bacterial mobile elements were first discovered, researchers did not initially link them to intermediate-repeat DNA, which had been previously identified in eukaryotes. However, as the wide variations in the amounts and chromosomal positions of eukaryotic intermediate repeats were documented, their similarity with bacterial mobile elements was recognized. Thus the study of intermediate-repeat DNA in eukaryotes converged with research on mobile DNA elements in bacteria, although at first there was no apparent connection between the two classes of DNA.

Mobile elements are now known to transpose by one of two mechanisms: (1) directly as DNA or (2) via an RNA intermediate transcribed from the mobile element by an RNA polymerase and then converted back into double-stranded DNA by a reverse transcriptase (Figure 9-14).

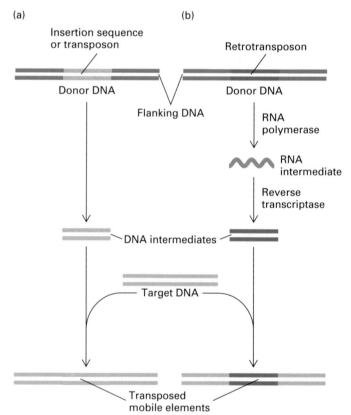

▲ FIGURE 9-14 Classification of mobile elements into two major classes based on transposition mechanism. (a) Bacterial insertion sequences and eukaryotic transposons (orange) move via a DNA intermediate. This mechanism is often referred to as transposition. (b) Eukaryotic mobile elements called retrotransposons (green) are first transcribed into an RNA molecule, which then is reversed-transcribed into double-stranded DNA. This mechanism is referred to as retrotransposition. In both cases, the double-stranded DNA intermediate is integrated into the target-site DNA to complete movement.

genes occurred, generating new genes made from novel combinations of preexisting exons. The term *exon shuffling* has been coined to refer to this type of evolutionary process. Consequently, although mobile DNA elements appear to be completely selfish molecular parasites, they may have contributed indirectly to the evolution of higher organisms by serving as sites of recombination leading to the evolution of new genes.

Barbara McClintock (Figure 1-9b) discovered the first *mobile elements* while doing classical genetic experiments in maize (corn) during the 1940s. She characterized genetic entities that could move into and back out of genes, changing the phenotype of corn kernals. Her theories were very controversial until similar mobile elements were discovered in bacteria, where they were characterized as specific DNA

Mobile elements that transpose through a DNA intermediate are generally referred to as *transposons*. (As discussed below, this term has a more specific meaning in reference to bacterial mobile elements.) The second mechanism of transposition is analogous to the infectious process of retroviruses described in Chapter 6. Consequently, mobile elements that transpose to new site in the genome via an RNA intermediate are called *retrotransposons*. Indeed, retroviruses can be thought of as retrotransposons that evolved genes encoding viral coats, thus allowing them to transpose between cells.

Movement of Bacterial Mobile Elements Is Mediated by DNA

Because the earliest molecular understanding of mobile DNA elements came from studies with *E. coli* mutants, we first discuss these bacterial elements and how they move and then consider their eukaryotic counterparts in later sections.

Bacterial Insertion Sequences The first molecular understanding of mobile elements came from study of *E. coli* mutants in which DNA sequences ≈1–2 kb in length were inserted into a gene, thus disrupting its function. These inserted stretches of DNA—called *insertion sequences*, or *IS elements*—have been visualized in electron-microscope studies. In one such study, DNA cloned from the galactose operon of an *E. coli gal3* mutant, which is unable to grow on galactose, was denatured and hybridized to DNA from the wild-type galactose operon (*gal*). Some of the resulting renatured double-stranded DNA molecules contained one strand of the mutant galactose operon hybridized to the complementary strand of the wild-type galactose operon. This type of molecule in which partially complementary DNA strands are hybridized to each other is called a *heteroduplex*. In the heteroduplex visualized in Figure 9-15, the mutant strand contains an ≈1.5-kb DNA sequence that is absent from the wild-type strand. This additional length of DNA, designated IS2, forms a loop of single-stranded DNA extending from the otherwise double-stranded heteroduplex molecule as shown in the diagram below the electron micrograph.

The insertion of IS2 into the galactose operon disrupts the coding region of a gene for one of the enzymes required to metabolize galactose, resulting in the phenotype produced by the *gal3* mutation. Similar mutations in other *E. coli* genes result from insertion of the same DNA sequence (i.e., IS2). More than 20 different IS elements have been found in *E. coli* and other bacteria.

Transposition of IS elements is a very rare event, occurring in only one in 10^5–10^7 cells per generation, depending on the IS element. Higher rates of transposition would probably result in too great a mutation rate for the host cell. At a very low rate of transposition, most host cells survive and therefore propagate the parasitic IS element.

(a)

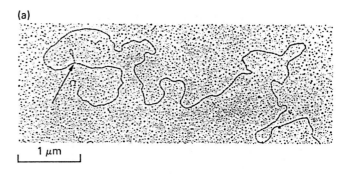

1 μm

(b)

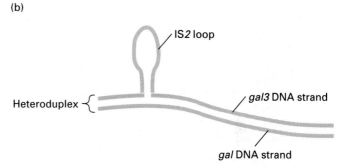

▲ FIGURE 9-15 Bacterial insertion sequence (IS element). (a) Electron micrograph of a heteroduplex formed from DNA extracted from two recombinant λ-phage vectors, one carrying the wild-type *E. coli* galactose operon (*gal*) and the other carrying a mutant form of the operon (*gal3*). The loop (arrow) corresponds to the insertion sequence (IS2) that causes this mutation. (b) Diagram of the region of the heteroduplex containing IS2 element. Because the IS2 integrated into the *gal3* DNA strand has no complement in the *gal* DNA strand, it cannot hybridize and forms a single-stranded loop extending from the double-stranded region of the heteroduplex. [Part (a) from A. Ahmed and D. Scraba, 1975, *Mol. Gen. Genet.* **136**:233; courtesy of A. Ahmed. See also H. Ohtsubo and E. Ohtsubo, 1978, *Proc. Nat'l. Acad. Sci. USA* **75**:615.]

Even though many of the rare transpositions that do occur inactivate essential genes, killing the host cell and the IS elements it carries, other host cells survive. Since IS elements transpose into approximately random sites, some transposed sequences enter nonessential regions of the genome (e.g., regions between genes), thereby expanding the number of IS elements in a cell. IS elements also can insert into plasmids or lysogenic viruses, which can be transferred to other cells. When this happens, IS elements can transpose into the chromosomes of virgin cells.

Structure of IS Elements and Mechanisms of Transposition The general structure of IS elements is diagrammed in Figure 9-16. DNA sequencing has revealed that an ≈50-bp *inverted repeat* invariably is present at each end of an insertion sequence. Between the inverted repeats is a protein-coding region, which encodes enzymes required for transposition of an IS element to a new site.

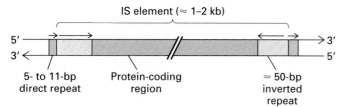

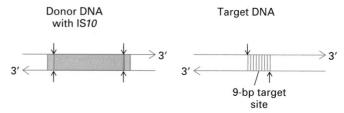

Donor DNA with IS*10*

Target DNA

▲ FIGURE 9-16   General structure of bacterial IS elements. These mobile elements contain a central region (dark orange), which encodes enzymes required for transposition, and inverted repeats at each end. The sequences of the inverted repeats may not be exactly identical but are characteristic of a particular IS element. Flanking each IS element are short direct repeats (light blue), which are generated from the target-site DNA during insertion of a mobile element. Thus the direct-repeat sequence depends on the site of insertion and is not characteristic of the IS element; however, the length of the direct repeats is constant for a given IS element. Arrows indicate sequence orientation. Note that the regions in this diagram are not to scale; the coding region between the inverted repeats makes up most of the length of an IS element.

(a) Transposase makes blunt-ended cuts in donor DNA and staggered cuts in target DNA

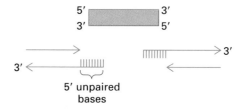

5′ unpaired bases

(b) Transposase ligates IS*10* to 5′ single-stranded ends of target DNA

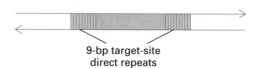

(c) Cellular DNA polymerase extends 3′ cut ends and ligase joins extended 3′ ends to IS*10* 5′ ends

9-bp target-site direct repeats

Some IS elements (e.g., IS*10*) encode a single enzyme, whereas others encode two enzymes; as discussed below, this difference relates to two different transposition mechanisms. In either case, IS-encoded proteins are expressed at a very low rate, accounting for the very low frequence of transposition. An important hallmark of IS elements is the presence of short *direct repeats*, containing 5–11 base pairs, immediately adjacent to both ends of the inserted element. The *length* of the direct repeat is characteristic of each type of IS element, but its *sequence* depends on the target site where a particular copy of the IS element is inserted. When the sequence of a mutated gene containing an IS element is compared to the sequence of the wild-type gene before insertion, only one copy of the short direct-repeat sequence is found in the wild-type gene. Duplication of this target-site sequence to create the second direct repeat adjacent to an IS element occurs during the insertion process.

Two classes of IS elements, differing in the transposition mechanism, have been identified. Some IS elements excise from one location and insert at a new position in the bacterial chromosome; this mechanism is termed *nonreplicative transposition*. Other IS elements insert a *copy* of themselves at a new position, but the original copy is retained in its original location; this mechanism is termed *replicative transposition*.

Duplication of target-site sequences during insertion is most easily understood for those IS elements that utilize nonreplicative transposition (Figure 9-17). This transposition process is catalyzed by a *transposase* enzyme encoded between the inverted repeats of the IS element. Transposase molecules bind to the inverted-repeat sequences present at

▲ FIGURE 9-17   Model for nonreplicative transposition of the bacterial insertion sequence IS*10*. Molecules of transposase, which is encoded by the IS element, bind to the ends of the inverted repeats, in the region of the terminal direct repeats (light blue) of the donor DNA, and cleave both strands precisely at the ends of the inverted repeats (arrows), which are not indicated by color. Other transposase molecules bind to an approximately random DNA target site and make staggered cuts in the two strands at the arrows. In this example, the target site is a 9-bp sequence, indicated by vertical red lines. Transposase then ligates the 3′ ends of the excised IS*10* to the 5′ single-stranded ends at the cleaved target site. The gap left in the resulting intermediate is filled in by cellular DNA polymerase; finally cellular DNA ligase forms the 3′ → 5′ phosphodiester bonds between the 3′ ends of the extended target DNA strands and the 5′ ends of the IS*10* strands. This process results in duplication of the target-site sequence on either side of the inserted IS element. Both target-site direct repeats flanking an IS element are in the same orientation. Their sequence is determined by the site of insertion. Note that the length of the target site and IS*10* are not to scale. [See H. W. Benjamin and N. Kleckner, 1989, *Cell* **59**:373 and, 1992, *Proc. Nat'l. Acad. Sci. USA* **89**:4648.]

Circular plasmid
with IS element

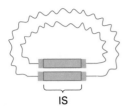

IS

Target DNA
(bacterial chromosome)

5-bp target site

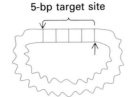

(a) Staggered cuts are made in
the target site and single-
stranded cuts are made in
the donor DNA.

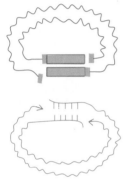

(b) Strand transfer occurs
between 3′ ends of IS
and 5′ ends of target
DNA

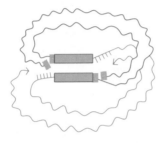

(c) Copying of IS begins at free
3′ ends of target DNA,
beginning with duplication
of target site

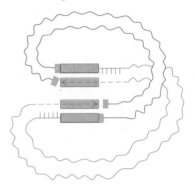

(d) Completion of replication
yields cointegrate
containing plasmid
DNA, target DNA,
and two copies
of IS element.
Cointegrate is
resolved by
site-specific
recombination.

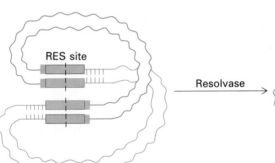

RES site

Resolvase →

Circular plasmid
with IS and
original direct
repeats

Target DNA with copy
of IS inserted between
duplicated target sites

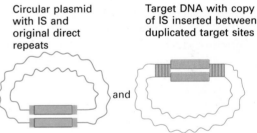

and

▲ FIGURE 9-18   Model for replicative transposition of certain bacterial insertion sequences, such as γδ. Transposition is depicted between an insertion sequence in a donor plasmid (yellow with direct repeats in blue) and a 5-bp target site in a recipient bacterial chromosome. Staggered cuts in the donor and target DNA are made at positions indicated by small black arrows. Arrowheads on cut target DNA indicate the 3′ ends; the small blue rectangles on ends of the cut IS element are the original direct repeats at the 5′ ends. This transposition mechanism results in two copies of the IS element: a newly synthesized copy inserted at the target site of the recipient chromosome with the 5-bp target-site duplicated and a regenerated copy in the donor plasmid including the original direct repeats. [See J. Shapiro, 1979, *Proc. Nat'l. Acad. Sci. USA* **76**:1933; K. Mizuuchi, 1983, *Cell* **35**:785.]

each end of the IS element in the donor DNA and cleave the DNA, precisely excising the element. Transposase molecules also bind to a target DNA molecule and make a staggered cut in the two DNA strands generating ends with 5′ single-stranded tails. This remarkable transposase, which has multiple activities, then ligates the 5′ nucleotides generated at the target site to the 3′ ends of the IS element strands cleaved out of the donor site. A DNA polymerase

encoded by the host cell then extends the 3′ ends of the target site, filling in the single-stranded gaps and generating a short repeat of the target-site sequence at either end of the newly inserted IS element. This is the origin of the short direct repeats that flank IS elements.

Figure 9-18 depicts the other transposition mechanism, replicative transposition, for movement of an IS element from a plasmid into a bacterial chromosome. IS elements using this mechanism encode a transposase that makes staggered cuts in the target DNA but makes only single-strand cuts in the donor DNA at the junctions with the IS element. As with the other mechanism of transposition, the 5' ends of the target DNA are ligated to the 3' end of the donor DNA; in this mechanism, however, when host-cell DNA polymerase extends the 3' ends of the target DNA, it continues to replicate throughout the entire length of the IS element, replicating each strand of the element. As a consequence of this replicative transposition, a new copy of the IS element is generated, and the entire plasmid containing the donor DNA becomes fused to the recipient chromosome.

The large circular intermediate in replicative transposition is called a *cointegrate*. A second enzyme encoded by this type of IS element, called *resolvase*, completes the transposition process. Resolvases are *site-specific recombinases*, which bind to a specific sequence (called the RES site) in each of the two copies of the IS element in the cointegrate. After binding these sequences and bringing them together through interactions of resolvase molecules bound to each RES site, the resolvases break and religate the four strands so that the donor and target DNAs are regenerated. The net result is a replicative transposition of the IS element to the target DNA and generation of a short direct repeat flanking the IS element at the site of insertion.

Mutations due to both types of IS elements revert to the wild-type phenotype at a low frequency; approximately one cell in 10^7 mutants regains the wild-type phenotype. Sequencing of cloned revertant DNA has shown that the original wild-type sequence is precisely regenerated. Reversion is thought to result from homologous recombination between the short direct repeats generated at the target site when IS elements insert. Recombination between these sequences regenerates the wild-type sequence in the bacterial chromosome and excises the IS element as a circular DNA molecule. Since the excised circle (unlike a plasmid) does not contain an origin of DNA replication, it does not replicate and is not passed on to successive generations of daughter cells.

Bacterial Transposons Transposons are composite mobile genetic elements that are larger than IS elements and contain one or more protein-coding genes in addition to those required for transposition. Many transposons are composed of an antibiotic-resistance gene flanked by two copies of the same type of IS element (Figure 9-19). Insertion of a transposon into plasmid or chromosomal DNA is readily detectable because of the acquired resistance to an antibiotic. Transposition occurs through the action of the IS-element transposase acting on the left end of the leftward IS element and the right end of the rightward IS element. In transposons, the internal inverted repeats of the flanking IS elements have accumulated mutations so that they are not recognized by the transposase. Transposition produces a short direct repeat of the target site on either

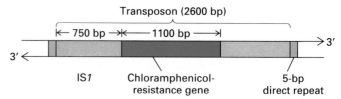

▲ FIGURE 9-19 General structure of bacterial transposons, such as Tn9 of *E. coli*. This transposon consists of a chloramphenical-resistance gene (dark blue) flanked by two copies of IS*1* (yellow), one of the smallest IS elements. Other copies of IS*1*, without the drug-resistance gene, are located elsewhere in the *E. coli* chromosome. The internal inverted repeats of IS*1* abutting the resistance gene are so mutated that transposase does not recognize them. During transposition, cuts are made at the positions indicated by small arrows, so the entire transposon is moved. Movement of this transposon from donor DNA (e.g., a plasmid) occurs via the replicative mechanism shown in Figure 9-18. The target-site sequence at the point of insertion becomes duplicated on either side of the transposon during transposition. Note that the 9-bp target site direct repeat (light blue) is not to scale.

side of the newly integrated transposon, just as for IS elements.

Transposons are very valuable tools for the bacterial geneticist. They can be introduced into cells on plasmids or viral genomes. Once transferred into a cell, transposons can act as mutagens that affect only a single cellular gene. Although transposition is a rare event, mutagenized cells are readily isolated because of their newly acquired antibiotic resistance gene. The site of the transposon-generated mutation can be determined readily by restriction-enzyme mapping, which reveals the insertion of the large transposon DNA. The precise sequence of bacterial DNA at the site of insertion can then be determined by dideoxy DNA sequencing (see Figure 7-30) using a primer complementary to the known sequence of the inverted repeats at the ends of the transposon.

Movement of Some Eukaryotic Mobile Elements Is Mediated by DNA

Some mobile elements in eukaryotes also transpose by the direct transfer of DNA intermediates. Included among these are the *P element* of *Drosophila*, which can be used in constructing transgenic *Drosophila* (see Figure 8-40), and the *Ac* and *Ds elements* present in maize, the first mobile elements to be discovered.

Drosophila P Element Approximately half of all the spontaneous mutations (i.e., mutations that occur without treating an organism with mutagens) observed in *Drosophila* are due to the insertion of mobile elements (Table 9-5). Although most of the mobile elements in *Drosophila*

TABLE 9-5   Classes of Intermediate-Repeat DNA in *Drosophila*

Class	Length (kb)	Characteristic Feature	Mobility Type
Copia	5.1	Long terminal repeats	Viral retro-transposon
dm297	7.0		
dm412	7.6		Viral
dm17.6	7.4	*Copia*-like	retro-
Opus	8.0		transposon
(≈12 others)			
F	Variable	3′ poly A tails	Nonviral retro-transposon
G			
FB (foldback)	Variable	Terminal inverted repeats	?
P	2.9	31-bp terminal inverted repeats	Transposon

SOURCE: D. J. Finnegan and D. H. Fawcett, 1986, *Oxford Surveys on Eukaryotic Genes* 3:1.

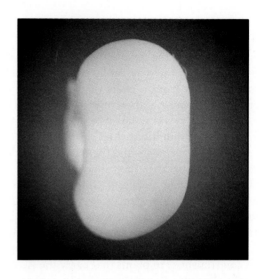

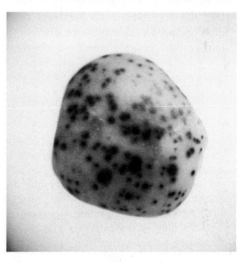

▲ FIGURE 9-20   Corn kernel phenotypes resulting from a mobile element in an anthocyanin gene. Corn kernels are purple if they express genes that make a pigment called anthocyanin. These kernels came from a plant with a mobile element in the *a* gene of the anthocyanin pathway. If the mobile element remains stationary, the *a* gene in all cells is defective and the kernels are all white (*top*); if the element is excised early during the development of the plant, a functional *a* gene is generated and the kernels are purple (not shown). If the mobile element is excised in some of the cells that proliferate to form the aleurone (pigment-forming layer of the kernel), only clones of cells descending from those in which the mobile element was excised synthesize anthocyanin. This gives rise to kernels with spots of purple pigment (*bottom*), each spot resulting from a clone of cells derived from a parental cell in which the mobile element was excised, generating a functional *a* gene. Large spots result from excision of the mobile element early in the development of the aleurone, so that several cell divisions occur following excision, resulting in a large number of cells with a functional *a* gene. Small spots result from excision late in the development of the aleurone, so that fewer cell divisions occur following excision of the element, giving rise to fewer cells with a functional *a* gene. [See J. A. Ranks, P. Mason, and N. Federoff, 1988, *Genes & Devel.* **2**:1364; courtesy of N. Fedoroff.]

function as retrotransposons, at least one—the P element—functions as a transposon and probably transposes by a mechanism similar to that used by bacterial insertion sequences, although the exact mechanism is not known. Like IS elements, the P element has an inverted repeat at each end, and both copies of this sequence are required for transposition. An 8-bp direct duplication of the target-site sequence is generated at either end of inserted P elements. As discussed earlier, similar short direct repeats result from the insertion of bacterial IS elements. The central portion of the P element encodes a protein required for transposition, which appears to be equivalent to the transposase of bacterial IS elements. One additional observation argues strongly that the P element transposes as a DNA molecule rather than via an RNA intermediate: the gene encoding the P-element transposase contains introns. Transposition via an RNA intermediate would be expected to remove the introns by RNA splicing, as occurs in yeast Ty elements discussed later.

Ac and Ds Elements in Corn   Barbara McClintock's original discovery of mobile elements came from observation of unusual revertible mutations in corn. Among the best-studied of these mutations are those that affect the production of any of the several enzymes required to make anthocyanin, a purple pigment that affects kernel color (Figure 9-20). Many different mutations in this pathway causing loss of kernel color have been found. Some of these are revertible at high frequency producing spotted kernels. The spots occur when the mutation reverts in cells that are proliferating to form the pigment layer of the kernel. Each of the spots results from a clone of cells derived from a single cell in which the mutation reverted. A second class of mutations does not revert unless they occur in the presence of the first class of mutations. McClintock called the first class *activator* (Ac) elements and the second class *dissociation* (Ds) elements because they also tended to be associated with chromosome breaks.

The agents responsible for these mutations have now been cloned and sequenced. The Ac element is about 4600 nucleotides long; the Ds elements are Ac elements with deletions (Figure 9-21). A Ds element cannot transpose by itself but can transpose in plants that carry the Ac element. Thus the Ac element appears to encode a transposase that functions like the transposase of bacterial insertion sequences. The Ac and Ds elements are also similar to IS elements in that they contain terminal inverted repeats, which are required for transposition. The terminal inverted repeats are thought to be recognized by the Ac transposase in a manner analogous to the recognition of the terminal inverted repeats of IS elements by IS-element transposases. Since Ds elements are simply Ac elements with internal deletions, they are also transposed by the action of the Ac transposase. Ds elements cannot transpose if an Ac element is not in the same cell because the transposase encoded by Ds elements in defective. Transposition of Ac and Ds elements involves duplication of target-site DNA, generating 8-bp direct repeats flanking the inserted element.

The mechanism of Ac transposition probably is similar to the transposition of bacterial IS elements that move without replication (see Figure 9-17). Because the site from which an Ac or Ds element transposes can be repaired to generate a functional gene, mutations due to the insertion of these elements are unstable.

Two Major Categories of Retrotransposons Are Found in Eukaryotic Cells

All eukaryotes studied from yeast to humans contain retrotransposons, mobile DNA elements that transpose through an RNA intermediate utilizing reverse transcriptase (see Figure 9-14). In humans, retrotransposons are present in hundreds of thousands of sites scattered throughout the genome, and the positions of many of these mobile elements vary from one individual to another. Although yeast and *Drosophila* DNA contain fewer copies of retrotransposons than human DNA, variation in retrotransposon content and position occurs among strains of the same species.

Retrotransposons are divided into two major categories. Members of one category are called *viral retrotransposons*, because they exhibit many similarities to the genomes of retroviruses. In particular, they contain ≈250- to 600-bp direct terminal repeats, also called *long terminal repeats*, or *LTRs*. Members of the second category lack LTRs and consequently are called *nonviral retrotransposons*. In mammals, the most abundant of these resemble cellular RNA molecules that are synthesized by RNA polymerase III. (The three types of eukaryotic nuclear RNA polymerases are described in Chapter 11.)

Most Mobile Elements in Yeast Are Viral Retrotransposons

Many strains of *S. cerevisiae* contain 50–100 copies of related ≈5-kb DNA sequences called *Ty elements*. Figure 9-22 shows a Southern blot of DNA from three strains of yeast digested with the restriction enzyme *Eco*RI and hybridized to radioactively labeled cloned Ty DNA. Each band corresponds to an *Eco*RI fragment containing the Ty-

▼ FIGURE 9-21   Structure of Ac and Ds mobile elements in maize. The protein-coding region of the activator (Ac) element contains five exons (dark orange) and is flanked by a short inverted repeat (light orange), whose sequence is shown. Dissociation (Ds) elements have the same general structure as the Ac element except that a portion of the coding sequence is deleted (indicated by brackets). Ds elements can transpose only in plants that also contain the Ac element, implying that Ac encodes the transposase required for mobility. As with other mobile elements, short direct repeats of the target-site sequence (light blue) are generated during transposition. The inverted repeats and direct repeats are not to scale. [Adapted from N. V. Fedoroff, 1989, *Cell* **56**:181.]

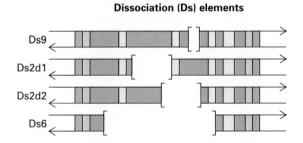

Three strains

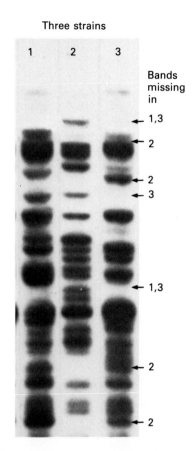

Bands missing in

← 1,3
← 2
← 2
← 3

← 1,3

← 2

← 2

▲ FIGURE 9-22   Total cellular DNA was isolated from three strains of *S. cerevisiae,* digested with the restriction enzyme *Eco*RI, and subjected to electrophoresis through an agarose gel. The gel was then subjected to the Southern-blot procedure (Figure 7-31) by transferring the separated DNA fragments to nitrocellulose filter paper, denaturing the bound DNA, and then hybridizing it to radioactively labeled cloned Ty DNA to reveal all the *Eco*RI fragments containing Ty sequence. The three strains shared some—but not all—DNA fragments that hybridized to the probe, indicating that Ty elements have inserted into distinct regions of the genome in each strain. [See J. R. Cameron, E. Y. Loh, and R. W. Davis, 1979, *Cell* **16**:739; courtesy of R. W. Davis.]

(a) Viral retrotransposons

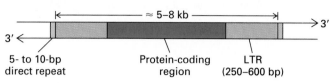

5- to 10-bp direct repeat

Protein-coding region

LTR (250–600 bp)

(b) Genes in retroviral DNA and viral retrotransposons
Retroviral DNA

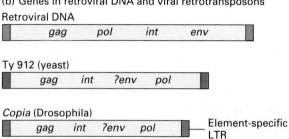

gag   pol   int   env

Ty 912 (yeast)

gag   int   ?env   pol

Copia (Drosophila)

gag   int   ?env   pol

Element-specific LTR

▲ FIGURE 9-23   (a) General structure of eukaryotic viral retrotransposons. These mobile elements consist of a central protein-coding region (dark green) flanked by two long terminal repeats (LTRs), which are direct repeats (light green). Like other mobile elements, integrated retrotransposons have short target-site direct repeats (light blue) at their 3′ and 5′ ends. Note that the different regions are not drawn to scale; the protein-coding region constitutes 80 percent or more of a retrotransposon. (b) Component genes of integrated retroviral DNA and two eukaryotic retrotransposons, yeast Ty elements and *copia* elements from *Drosophila.* The central region of retroviral DNA contains four genes—*gag, pol, int,* and *env;* the functions of the encoded proteins are described in the text. The central coding regions in eukaryotic viral retrotransposons encode analogous proteins. The colored regions at the ends of these elements represent the LTRs; although the LTRs differ from element to element, they are direct repeats containing 250–600 base pairs in all cases. Retroviral DNA is about 5–7 kb long; *copia* and Ty elements are about 5 kb long.

element sequence. The three strains contain many bands in common, resulting from copies of Ty elements at equivalent positions in the DNA of each strain. In addition, each strain contains unique Ty-containing *Eco*RI fragments resulting from Ty transpositions. Early evidence that Ty elements transpose came from an experiment in which five separate cultures were derived from a single clone of cells and grown for one month. A single colony was isolated from each of the five cultures and DNA from each colony was isolated. Southern blots like the one shown in Figure

9-22 showed that two of the five cultures acquired new Ty-containing *Eco*RI fragments, indicating that a Ty element had transposed to a new site during the one-month period.

Sequencing of Ty elements demonstrated that they have features in common with integrated retroviral DNA (Figure 9-23a). In particular, both have long terminal repeats (LTRs). In retroviruses, LTRs are generated from the 5′ and 3′ sequences of the viral genomic RNA by the complex series of steps diagrammed in Figure 9-24. Although

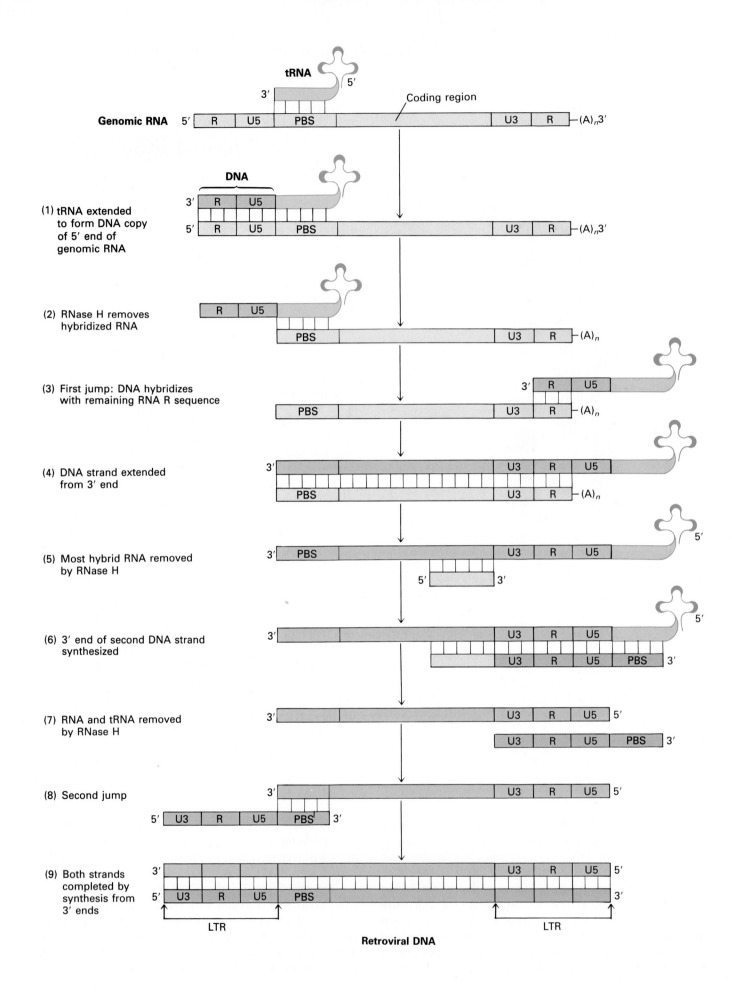

Genomic RNA — 5′ [R | U5 | PBS | Coding region | U3 | R] —(A)$_n$3′

tRNA

(1) tRNA extended to form DNA copy of 5′ end of genomic RNA

(2) RNase H removes hybridized RNA

(3) First jump: DNA hybridizes with remaining RNA R sequence

(4) DNA strand extended from 3′ end

(5) Most hybrid RNA removed by RNase H

(6) 3′ end of second DNA strand synthesized

(7) RNA and tRNA removed by RNase H

(8) Second jump

(9) Both strands completed by synthesis from 3′ ends

LTR

LTR

Retroviral DNA

◄ FIGURE 9-24   Generation of LTRs during reverse transcription of retroviral genomic RNA. A complicated series of nine events generates a double-stranded DNA copy of the single-stranded RNA genome of a retrovirus (*top*). The genomic RNA is packaged in the virion with a retrovirus-specific cellular tRNA hybridized to a complementary sequence near its 5′ end called the primer-binding site (PBS). The retroviral RNA has a short direct-repeat terminal sequence (R) at each end. The overall reaction is catalyzed by reverse transcriptase, which has both a deoxynucleotide-polymerizing activity and a RNase H activity, which digests the RNA strand in a DNA-RNA hybrid. The entire process yields a double-stranded DNA molecule that is longer than the template RNA and has a long terminal repeat (LTR) at each end. The different regions are not shown to scale. The PBS and R region are actually much shorter than the U5 and U3 regions, and the central coding region is very much longer (≈7500 nucleotides) than the other regions. [See E. Gilboa et al., 1979, *Cell* **18**:93.]

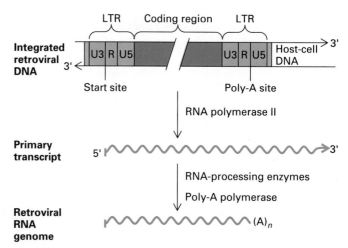

▲ FIGURE 9-25   Generation of retroviral genomic RNA from integrated retroviral DNA. The short direct repeat sequence (light blue) of target-site DNA is generated during integration of the retroviral DNA into the host-cell genome. The left LTR directs cellular RNA polymerase II to initiate transcription at the first nucleotide of the left R region. The primary transcript generated extends beyond the right LTR. The right LTR, now present in the RNA primary transcript, directs cellular enzymes to cleave the primary transcript at the last nucleotide of the right R region and to add a poly-A tail, yielding a retroviral RNA genome with the structure shown at the top of Figure 9-24.

the two LTRs of integrated retroviral DNA are identical in sequence, they perform distinct critical functions in the life cycle of a retrovirus. The leftward LTR functions as a promoter that directs host-cell RNA polymerase II to initiate transcription at the 5′ nucleotide of the R sequence (Figure 9-25). After the entire retroviral DNA has been transcribed, the RNA sequence corresponding to the rightward LTR directs host-cell RNA-processing enzymes to cleave the primary transcript and add a poly-A tail at the 3′ end of the R sequence, thus generating the retroviral RNA genome shown at the top of Figure 9-24. Consequently the LTRs, and the process of reverse transcription that generates them, are critical to the successful propagation of a retroviral RNA genome through successive cycles of infection. Reverse transcription generates a DNA sequence with LTR termini that direct the synthesis and processing of viral RNA with the precise terminal sequences to permit a subsequent round of reverse transcription. The finding that Ty elements contained LTRs immediately suggested that they transpose in a manner similar to the reverse transcription of retroviral RNA.

In addition to the *pol* gene encoding reverse transcriptase, the retroviral genome contains three other protein-coding genes: *gag*, which encodes a precursor protein that is processed by proteolytic cleavage into internal virion proteins; *env*, which encodes a precursor protein processed into the viral coat proteins; and *int*, which encodes the enzyme *integrase*. This enzyme functions analogously to the transposase encoded in bacterial insertion sequences, inserting the double-stranded DNA formed from viral RNA by reverse transcription into a target DNA site and, in the process, generating short direct repeats of the target-site sequence at either end of the inserted viral DNA se-

quence. Ty elements contain coding regions for proteins that are homologous to those of retroviruses (Figure 9-23b). In addition, Ty elements generate short direct repeats at the site of insertion, a characteristic shared by other mobile elements, as we have seen repeatedly.

Although these sequence comparisons provided strong evidence that Ty elements transpose through an RNA intermediate, the experiments depicted in Figure 9-26 provided even stronger, functional evidence for this conclusion. Ty elements normally transpose at a very low rate. As with bacterial IS elements, this low transposition rate assures that the frequency of Ty-insertion mutations, which would otherwise kill cells, also remains low. In experiment 1 shown in Figure 9-26, yeast cells were transformed with yeast plasmids containing a Ty element cloned next to a galactose-sensitive promoter. As expected from this arrangement, the production of Ty mRNA was much higher in the presence of galactose than in its absence. In addition, Ty-element transposition was stimulated by addition of galactose to the medium. This increased Ty transposition resulted from an increase in the amount of Ty mRNA, which could function as a template for reverse transcription, and in the amount of reverse transcriptase and integrase encoded by the Ty element. In addition, these cells

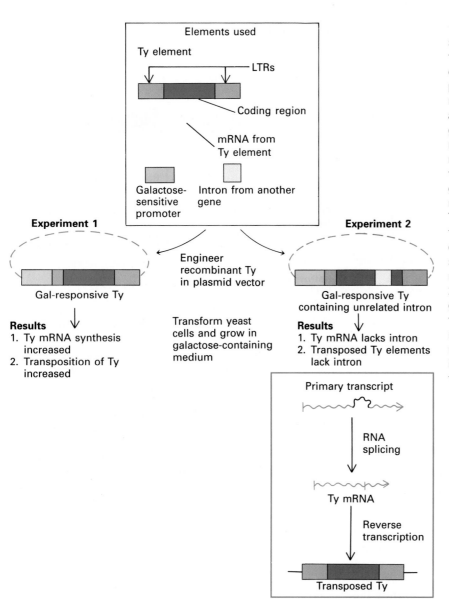

Elements used

Ty element

LTRs

Coding region

mRNA from
Ty element

Galactose-
sensitive
promoter

Intron from another
gene

Experiment 1

Gal-responsive Ty

Results
1. Ty mRNA synthesis
   increased
2. Transposition of Ty
   increased

Engineer
recombinant Ty
in plasmid vector

Transform yeast
cells and grow in
galactose-containing
medium

Experiment 2

Gal-responsive Ty
containing unrelated intron

Results
1. Ty mRNA lacks intron
2. Transposed Ty elements
   lack intron

Primary transcript

RNA
splicing

Ty mRNA

Reverse
transcription

Transposed Ty

◄ FIGURE 9-26 Experimental demonstration that yeast Ty element moves through an RNA intermediate. When yeast cells are transformed with a Ty-containing plasmid, the Ty element can transpose to new sites, although normally this occurs at a low rate. Using the elements diagrammed at the top, researchers engineered two different recombinant Ty elements in plasmid vectors. These were transformed into yeast cells, which were grown in galactose-containing medium or nongalactose medium. In experiment 1, the Ty element contained a promoter (yellow) that increases transcription in the presence of galactose. The results suggest that transcription into an mRNA intermediate is required for Ty transposition. In experiment 2, an intron (tan) from an unrelated yeast gene was inserted into the putative protein-coding region (dark green) of the recombinant galactose-responsive Ty element. The observed absence of the intron in transposed Ty elements is interpreted in the box at the bottom. [See J. Boeke et al., 1985, *Cell* **40**:491.]

accumulated particles that have the appearance of retroviral virions (Figure 9-27). However, these particles do not appear to move from cell to cell as infectious retroviruses do.

An even more revealing result was observed in experiment 2 in which an unrelated intron was inserted into the Ty DNA sequence. As in the first experiment, addition of galactose stimulated Ty transposition. Significantly, the resulting newly integrated Ty elements all lacked the inserted intron. The interpretation of this result is that the intron was spliced out of the Ty mRNA before it was reverse-transcribed into a double-stranded DNA copy, which subsequently inserted into the host-cell genome. The ob-

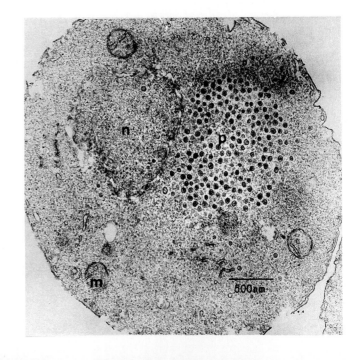

➤ FIGURE 9-27 Electron micrograph of intracellular virus-like particles in a yeast cell induced to express Ty RNA at a high level. **n** = nucleus; **m** = mitochondrion; **p** = particles. [See D. F. Garfinkel, J. D. Boeke, and G. R. Fink, 1989, *Cell* **42**:507; courtesy of J. D. Boeke.]

served removal of the intron from transposed Ty elements strongly implies that transposition occurs by reverse transcription of mRNA produced by transcription of Ty DNA. Recall that the presence of introns in *Drosophila* P elements is an argument that P elements do not transpose by reverse transcription of an RNA intermediate.

Copia Retrotransposons Are the Most Common *Drosophila* Mobile Elements

Ty elements are the primary class of mobile elements in yeast. In contrast, *Drosophila* contains several classes of mobile elements (see Table 9-5). As discussed previously, one of these, the P element, acts as a DNA transposon. The F and G elements have a 3′-terminal poly-A tail and appear similar to the nonviral retrotransposons found in mammals, which are described below.

By far the most common *Drosophila* intermediate-repeat elements, called *copia* (Latin for abundance), are ≈5–8 kb long and occur at many chromosomal sites. The presence of *copia* elements at multiple locations has been

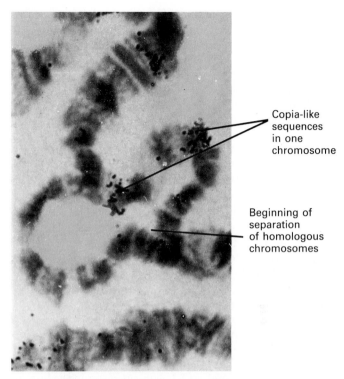

Copia-like sequences in one chromosome

Beginning of separation of homologous chromosomes

▲ FIGURE 9-29   Demonstration that repetitive DNA elements in *Drosophila* species are mobile. Labeled cloned *copia*-like sequences were hybridized to salivary gland chromosomes of a hybrid fly whose parents were from sibling species. Over most of their length, both polytene chromosomes of a homologous pair are aligned in register, but in a few regions they separate, as shown in this electron micrograph. At two sites in one of the chromosomes *copia* is detected by autoradiography, but it is lacking in the other chromosome. Similar experiments with many strains and closely related species have demonstrated that a particular repeat element has no fixed chromosomal locations in all strains. (See M. W. Young, 1979, *Proc. Nat'l. Acad. Sci. USA* **76**:6274; courtesy of M. W. Young.]

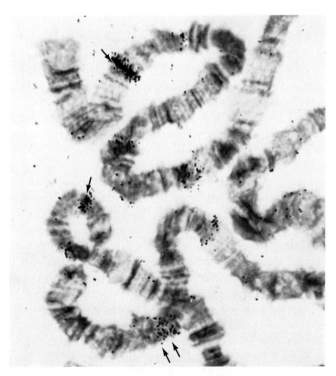

▲ FIGURE 9-28   Localization of *copia* elements, the most common type of intermediate-repeat DNA in *Drosophila*, by in situ hybridization to polytene chromosomes. Labeled cloned *copia* DNA was hybridized to fixed, denatured polytene chromosomes. Autoradiography revealed the presence of *copia* DNA at many sites in the genome (arrows at silver grains). Not all sites of hybridization are indicated and only a portion of the polytene chromosomes are shown. [See M. W. Young, 1979, *Proc. Nat'l. Acad. Sci. USA* **76**:6274; courtesy of M. W. Young.]

demonstrated by in situ hybridization experiments with polytene chromosomes (Figure 9-28). These unique chromosomes, found in the salivary gland of *Drosophila*, contain multiple associated copies of each chromosome and can be observed in the light microscope (see Figure 8-25).

The mobility of *copia*-like sequences was demonstrated in experiments in which two strains of *Drosophila melanogaster* that came from different parts of the world were mated. In the resulting hybrid progeny, the polytene chromosomes derived from each parent align with each other, but occasionally separate over part of their length. In some of these regions of separation, in situ hybridization to *copia*-like sequences clearly occurred in the chromosome derived from only one of the parents, indicating that one strain contains a copia insertion at that position in the chromosome and the other does not (Figure 9-29). Copia elements have sequences similar to those of retroviruses and Ty elements (see Figure 9-24), arguing strongly that *copia* elements also act as viral retrotransposons.

LINES and SINES, the Most Abundant Mobile Elements in Mammals, Are Nonviral Retrotransposons

The most abundant intermediate-repeat DNA sequences in mammals are divided into two classes mentioned previously: long interspersed elements (LINES) and short interspersed elements (SINES). In humans, full-length LINES are ≈6–7 kb long, while SINES are ≈300 bp long. Although these DNA sequences do not contain LTRs, evidence indicates that they transpose through reverse transcription of an RNA intermediate. Repeated sequences with characteristics of LINES have been observed in protozoa, insects, and plants, but for unknown reasons they are particularly abundant in the genomes of mammals. SINES also are found primarily in mammalian DNA.

Structure and Movement of L1 LINE Elements   The most common LINE elements constitute the L1 LINE family. Some 50,000 copies of L1 elements occur in the human genome, accounting for ≈5 percent of total human DNA. The general structure of L1 elements, based on a "consensus" (average) sequence, is diagrammed in Figure 9-30. Although the full-length L1 sequence is ≈6 kb long, variable amounts of the left end are absent at over 90 percent of the sites where this mobile element is found. L1 elements usually are flanked by direct repeats, the hallmark of mobile elements. The consensus sequence contains two long open reading frames (ORFs) ≈1 kb and ≈4 kb long. The function of the protein encoded by ORF1 is not known, but the protein encoded by ORF2 is similar in sequence to the reverse transcriptases of retroviruses and viral retrotransposons.

Evidence for the mobility of L1 elements has come from recent analyses of DNA from two male patients with genetic diseases resulting from mutations in the X chromosome. One patient had hemophilia resulting from mutation of the gene encoding factor VIII, one of the blood-clotting factors; the other patient had Duchenne muscular dystrophy, which results from mutation in the dystrophin gene. Cloning and sequencing of the affected genes from these two patients revealed that the mutations were due to insertions of a L1 element. Although the single X chromosome of males is derived from one of the maternal X chromosomes, the L1 insertions in these two patients were not present in either copy of their mothers' factor VIII or dystrophin genes. This finding indicates that new insertions occurred in the DNA of the oocytes (eggs) from which these patients developed. As is found with other transposable elements, the L1 insertions produced short direct repeats of the target-site sequence.

Experiments similar to those described earlier with yeast Ty elements (see Figure 9-26) demonstrated that L1 elements transpose through an RNA intermediate. In these experiments, an intron was introduced into a cloned mouse L1 element by recombinant DNA techniques. The recombinant L1 element was stably transformed into cultured hamster cells. The transformed DNA was detected in the hamster cells with the polymerase chain reaction (PCR) using primers located about 200 base pairs upstream and downstream from the region where the intron was inserted. The initial PCR-amplified fragment detected corresponded to the recombinant L1 sequence, including the intron, that was between the primers. After several cell doublings, however, a second PCR-amplified fragment was detected that corresponded to the L1 element from which the intron had been removed. This finding strongly suggests that over time the recombinant L1 element containing the inserted intron had transposed to new sites in the hamster genome through an RNA intermediate that underwent RNA splicing to remove the intron.

Since L1 elements do not contain LTRs, their mechanism of transposition through an RNA intermediate must differ from viral retrotransposition, in which the LTRs play a crucial role (see Figures 9-24 and 9-25). Recent in vitro transcription studies indicate that L1 DNA can be transcribed by RNA polymerase III. As discussed in Chapter 11, promoter sequences for this polymerase often lie just within the 5′ ends of the genes it transcribes. Thus, the function of the left LTR in viral retrotransposons, which

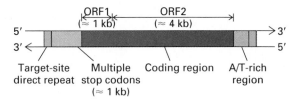

▲ FIGURE 9-30   Diagram of the general structure of an L1 LINE element, a common type of eukaryotic nonviral retrotransposon. An L1 element is flanked by direct repeats that are distinct for different copies of the element and appear to be generated from the target-site sequence during insertion. An ≈1-kb open reading frame (ORF1) occurs near the left (5′) end of the top strand following a region with multiple stop codons in all possible reading frames. The function of the protein encoded by ORF1 is unknown. The ≈4-kb ORF2 in the top strand encodes a protein that is similar to the reverse transcriptases of retroviruses and viral retrotransposons. The right end of an L1 element is rich in A residues in the 3′ end of the top strand and T residues in the 5′ end of the complementary strand. Unlike viral retrotransposons, L1 elements do not contain LTRs.

directs transcription by RNA polymerase II, probably is mimicked by an RNA polymerase III promoter sequence near the left end of L1 elements. The right LTR in viral retrotransposons signal where the poly-A tail is added to the RNA transcript; as shown in Figure 9-24, the R sequence preceding the poly-A tail is crucial in reverse transcription. L1 elements, however, contain an A/T-rich region near their right end, which generates a stretch of A residues in their RNA transcripts.

Based on these considerations, the model of L1 retroposition shown in Figure 9-31 has been proposed. RNA polymerase III terminates transcription after it encounters a string of T residues. Transcription of a full-length L1 element is thought to stop at the 3′ end of the first T-rich region in the flanking DNA downstream from an insertion site. The resulting RNA transcript terminates in a stretch of U residues containing a 3′ hydroxyl. Folding of the transcript and hybridization of the run of U residues to the run of A residues synthesized from the A/T-rich region at the right end of the L1 sequence generates a reverse transcriptase template. Addition of deoxyribonucleotides to the 3′ end of the RNA template yields a single-stranded DNA copy of the original L1 sequence. Subsequent synthesis of a complementary DNA strand and integration of the entire sequence into a target site in cellular DNA completes transposition of the element. Initiation of synthesis of the complementary DNA strand at variable internal positions in the L1 sequence would account for the deletion of variable amounts of the left end of the full-length L1 sequence observed at most sites.

In the L1 element diagrammed in Figure 9-30, the length of the open reading frames has been maximized. The vast majority of L1 sequences, however, contain stop codons and frameshift mutations in ORF1 and ORF2; these mutations probably have accumulated in most L1 sequences over evolutionary time. Maintenance of L1 transposition requires that only one L1 sequence in the genome maintain intact open reading frames encoding the reverse transcriptase and other proteins required for L1 retrotransposition. Mutations could accumulate in other L1 elements without interfering with their retrotransposition, which can be directed by the enzymes expressed from only one or a few intact L1 elements in the cell.

SINES and Alu *Sequences* The second major class of mammalian intermediate-repeat DNA can be visualized by electron-microscope analysis. When human DNA is mechanically sheared into fragments, reannealed briefly, and examined in the electron microscope, short double-stranded regions are observed; the remainder of the single-stranded fragments do not hybridize because they are not complementary (Figure 9-32). The double-stranded regions result from reannealing of related ≈300-bp sequences, which occur at least once in most fragments 5 kb or longer.. These short repeated sequences, designated

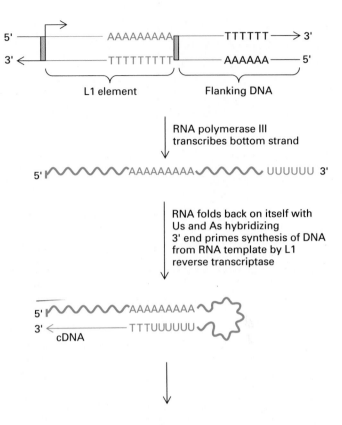

▲ FIGURE 9-31   Proposed mechanism for movement of L1 element via nonviral retrotransposition. According to this model, the L1 strand containing a 3′ A/T-rich region is transcribed by RNA polymerase III, beginning at the first 5′ nucleotide of the element and terminating at the first T/A-rich sequence in the flanking DNA. The string of U residues at the 3′ end of the transcript then base-pairs with the A-rich region at the 3′ end of the L1 sequence. The L1 reverse transcriptase then synthesizes a cDNA copy of the L1 sequence starting from the 3′ U in the transcript. The largely double-stranded copy of the L1 sequence formed then would be integrated into a cellular DNA target site by an integrase activity, which presumably would remove the RNA loop at the right end and generate the target-site direct repeats (light blue) at both ends. [Adapted from P. Jagadeeswaran, B. G. Forget, and S. M. Weissman, 1981, *Cell* **26**:141; S. W. Van Arsdell et al., 1981, *Cell* **26**:11.]

SINES, are present at 500,000 or more sites in the human genome, accounting for ≈5 percent of human DNA. SINES with related sequences also occur with similar high frequency in the genomes of other mammals.

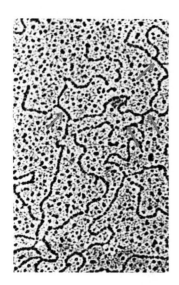
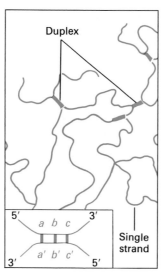

▲ FIGURE 9-32   Electron micrograph and corresponding diagram of human DNA fragments renatured under conditions that permit reassociation only of repetitive sequences. The reassociated duplex regions (arrows) appear thicker than the still-denatured unpaired regions. The duplex regions are about 300 bp long and occur on the average of once every 3000–5000 bp. Such repetitive sequences are called short interspersed elements (SINES). The inset illustrates base pairing between complementary strands of SINES inserted at different locations in the genome. [Photograph from P. L. Deininger and C. W. Schmid, 1976, *J. Mol. Biol.* **106**:773; courtesy of C. W. Schmid.]

Several hundred examples of SINES from various mammals have been cloned and sequenced. Although no two copies of these intermediate repeats are identical, their general sequence similarity shows that an ancestral relationship exists between these elements both within and among species. The sequence conservation is about 80 percent within a species, but falls to only about 50–60 percent among different mammalian species. Because many of these repetitive sequences in human DNA were found to contain a recognition site for the restriction enzyme *Alu*I, they were collectively called the *Alu* family. However, since these short interspersed elements are not precisely identical, many lack the *Alu* site; nonetheless, the name, though somewhat misleading, is convenient and widely used to refer to the most abundant type of human SINE. In addition to the hundreds of thousands of full-length *Alu* sequences in the human genome, many partial *Alu*-like sequences, clearly related to the *Alu* family but as short as 10 base pairs, have been found scattered between genes and within introns.

As shown in Figure 9-33, *Alu* elements have a remarkable sequence similarity with 7SL RNA, a 294-nucleotide cytoplasmic RNA. This small cellular RNA is part of a cytoplasmic ribonucleoprotein particle, the *signal recognition particle*, that aids in the secretion of newly formed polypeptides through the membranes of the endoplasmic reticulum (Chapter 16). Perhaps because the 7SL sequence is a functional RNA sequence, it is highly conserved even in species as diverse as *Drosophila*, mouse, and man. Recently, a small (≈100-nucleotide) RNA of *E. coli* has been sequenced and found to be similar to eukaryotic 7SL RNA. Thus this molecule has existed since early in evolution. However, neither *Drosophila* nor single-celled organisms have any *Alu*-type intermediate repeats (at least in large numbers). These findings suggest that 7SL RNA genes existed before *Alu* sequences and that the *Alu* sequences somehow arose fairly late in evolution from the 7SL sequences.

As described earlier for LINES, evidence for the mobility of SINES has come from analysis of DNA from a patient with a genetic disorder, in this case neurofibromatosis, which results from mutation in the *NF1* gene. DNA from this patient was shown to contain a new *Alu* sequence in one of the two *NF1* alleles; this insertion inactivates the allele and leads to occurrence of multiple neuronal tumors called neurofibromas. (Like hereditary retinoblastoma [see Figure 8-10], both *NF1* alleles must carry a mutation for neurofibromas to occur. In this patient, one *NF1* allele was inactivated by insertion of the *Alu* sequence; inactivating mutations in the other *NF1* allele in peripheral neurons led to the development of neurofibromas.) Several inherited recessive mutations causing disease in humans also have been found to result from insertion of *Alu* sequences in exons, thereby disrupting protein-coding regions.

Like all other mobile elements, *Alu* sequences usually are flanked by direct repeats. Unlike L1 elements, *Alu* sequences do not encode proteins; similar to L1 elements, *Alu* sequences are transcribed by RNA polymerase III and contain an A/T-rich region at one end. Consequently, they are thought to be retrotransposed by a mechanism similar to that proposed for L1 elements (see Figure 9-31), possibly by the reverse transcriptase and other required proteins expressed from functional L1 elements. *Alu* sequences appear to have retrotransposed widely through the human genome and are tolerated in sites where they do not disrupt gene function: between genes, and in both possible orientations within introns and the regions transcribed into the 5′ and 3′ untranslated regions of mRNAs (Figure 9-34). Although once postulated to function in controlling gene expression, *Alu* sequences are now thought to have no function, like other mobile elements, despite their widespread occurrence in mammalian genomes.

Retrotransposed Copies of Cellular RNAs Are Present in Eukaryotic Chromosomes

In addition to SINES and LINES, which constitute the bulk of intermediate-repeat DNA in mammals, other repetitive

```
              1                    15 16                   30 31                   45
7SL      •GCCGGGCGCGGTGG   CGCGTGCCTGTAGTC   CCAGCTACT•CGGGAG
Alu-cons  GGCCGGGCGCGGTGG   CTCACGCCTGTAATC   CCAGC•ACTTTGGGAG

              46                   60 61                   75 76                   90
7SL       GCTGAGGCTGGAGGA   TCGCTTGAGTCCAGG   AGTTC••••CCAGCC
Alu-cons  GCCGAGGCGGGCGGA   TCACCTGAGGTCAGG   AGTTCGAGACCAGCC

              91                  105 106                 120 121                 135
7SL       TGGGCAACATAGCGA   GACCCCGTCTCT•••   ••••••••••••••••
Alu-cons  TGGCCAACATGGTGA   AACCCCGTCTCTACT   AAAAATACAAAAATT

              136                 150 151                 165 166                 180
7SL      •GCCGGGCGCGGTGG   CGCGTGCCTGTAGTC   CCAGCTACTCGGGAG
Alu-cons  AGCCGGGCGTGGTGG   CGCGCGCCTGTAATC   CCAGCTACTCGGGAG

              181                 195 196                 210 211                 225
7SL       GCTGAGGCTGGAGGA   TCGCTTGAGTCCAGG   AGTTCTGGGCTGTAG
Alu-cons  GCTGAGGCAGGAGAA   TCGCTTGAACCCGGG   AGGCGGAGGTTGCAG

              226                 240 241                 255 256                 270
7SL       TGCGCCTGTGA••••G   CCAGTGCACTCCAGC   CTGGGCAACATAGCG
Alu-cons  TGAGCC•GAGATCGCG   CCAGTGCACTCCAGC   CTGGGCGACAGAGCG

              271                 285
7SL       AGACCCCGTCTCT
Alu-cons  AGACTCCGTCTCAAA   AAAAA
```

▲ FIGURE 9-33   Comparison of *Alu* sequences and 7SL RNA. The *Alu* consensus sequence (*Alu*-cons) was determined from the sequence of 125 separate *Alu* elements from human DNA. The consensus sequence is aligned with the portions of the 7SL RNA sequence that are most homologous with *Alu*. Bases in the *Alu* sequences that differ from the corresponding ones in the 7SL sequence are in red type; solid red dots indicate positions added to maximize homology. Both the *Alu* sequence and the 7SL sequence have two sections that are nearly identical (1–117 and 136–285). These two sections are separated in the *Alu* sequence by an A-rich region from positions 118 through 135 (yellow). *Alu* elements also contain a 3′ A-rich sequence, which is not shown here. [Adapted from J. Jurka and T. Smith, 1988, *Proc. Nat'l. Acad. Sci. USA* **85**:4775.]

sequences have been identified in this DNA fraction. The existence of these sequences came to light when researchers attempted to isolate human and mouse genomic DNA encoding the many known small nuclear RNAs (snRNAs),

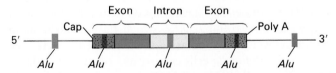

▲ FIGURE 9-34   Positions of *Alu* sequences in and around a typical protein-coding transcription unit. There is no fixed or regular location of *Alu* sequences within transcription units. The repetitive sequences may be contained within introns (tan) or within the noncoding regions at the 5′ and 3′ ends of transcription units (stippled blue), but, of course, not in the translated exons (solid blue). In many cases, *Alu*-type sequences have been found flanking transcription units, but they do not occur in similar positions in the homologous genes of related organisms, arguing that they probably do not function in regulating gene expression.

which are involved in RNA processing (Chapter 12), and the tRNAs, which function in translation. These efforts uncovered only a few true genes corresponding to the functional RNA sequences. However, a horde of partial or mutant copies of these genes were isolated and cloned from genomic DNA. Some of these DNA sequences were found to differ only slightly from the corresponding functional genes, but others represent truncated partial copies corresponding to either the 5′ or the 3′ end of the functional RNAs.

Because many of these nonfunctional copies of small RNAs are flanked by short direct repeats, they are thought to result from rare retrotransposition events that have accumulated through the course of evolution. It is postulated that at some point in the past, these RNAs were reverse-transcribed and then inserted into the genomic DNA of a germ cell; the individual that developed from that germ cell would carry the retrotransposed RNA and pass it on to future generations. Once inserted into the genome, retrotransposed RNAs cannot move to another site; thus

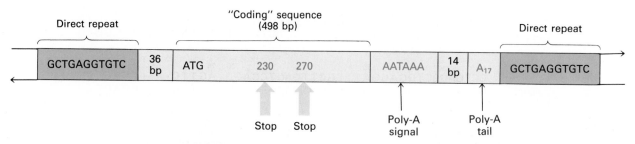

▲ FIGURE 9-35   A human β-tubulin processed pseudo-gene that probably arose by retrotransposition of functional β-tubulin mRNA. The genomic DNA region diagrammed here was detected because of its complementarity to a β-tubulin cDNA. The 498-bp "coding" region corresponds to the known amino acid sequence of β-tubulin except for base changes at nucleotides 230 and 270, which have introduced translation stop codons. The absence of introns, several of which are present in the true β-tubulin gene, and the pres-ence of a poly-A signal and a poly-A tail, which are character-istic of processed mRNA, suggest that this pseudogene arose by reverse transcription of β-tubulin mRNA and inser-tion of the resulting DNA into the chromosome. The transla-tion stop codons, which render an RNA product from the pseudogene nonfunctional, represent accumulated mutations. Also shown are the terminal direct repeats that are charac-teristic of mobile elements. [See C. D. Wilde et al., 1982, *Nature* **297**:83.]

they are not true mobile elements. However, the en-zymes involved in the rare retrotransposition of cellular RNAs most likely are encoded by a LINE or viral retrotransposon.

Mutated DNA copies of a wide variety of mRNAs also have been discovered integrated into chromosomal DNA. These are not duplicates of whole genes that have drifted into nonfunctionality (i.e., the pseudogenes discussed ear-lier in this chapter) because they lack introns and do not have flanking sequences similar to those of the functional gene copies. Instead, these DNA segments appear to be retrotransposed copies of spliced and polyadenylated (pro-cessed) mRNA. Compared with normal genes encoding mRNAs, these inserted segments generally contain multi-ple mutations, which are thought to have accumulated since they were first reverse-transcribed and randomly inte-grated into the genome. These nonfunctional genomic cop-ies of mRNAs are referred to as *processed pseudogenes*. As illustrated in Figure 9-35, most processed pseudogenes are flanked by short direct repeats, supporting the hypothesis that they are generated by rare retrotransposition events involving cellular mRNAs.

▶ *Functional Rearrangements in Chromosomal DNA*

In the previous section, we discussed the apparently ran-dom movement of mobile elements in genomic DNA. Al-though mobile DNA elements may well have played a role in evolution, their transposition appears to serve no direct, immediate function for the organism. In contrast to this are several types of functional rearrangements of DNA re-gions, which have been identified in both prokaryotes and eukaryotes. The mechanisms involved in these functional rearrangements include *inversion* of DNA segments, *gene conversion, DNA amplification,* and *deletion* of segments. Examples of each of these mechanisms, which play a role in the regulation of a few selected genes, are described in this section. However, by far the most frequent mechanism of gene control during normal development, differentia-tion, and reproduction involves the regulation of transcrip-tion, discussed in Chapters 11 and 13.

Salmonella Flagellar Antigens Can Switch through Inversion of a Transcription-Control Region

Salmonella typhimurium, a type of bacterium closely re-lated to *E. coli,* produces the nausea and diarrhea of com-mon "food poisoning" when ingested. The protein from which the *Salmonella* flagellum is constructed is one of the major *Salmonella* antigens to which the human immune system responds in eliminating this pathogenic bacterium. However, *Salmonella* cells can express two types of flagel-lar proteins called H1 and H2, which are encoded at dis-tant sites on the *Salmonella* chromosome. Any one *Salmo-nella* cell expresses only one type of flagellar protein, but as a clone of cells grows, some progeny spontaneously switch to expression of the other flagellar protein, a process known as *phase variation.* As a result, when individuals respond by making antibody against the major flagellar protein expressed by the *Salmonella* cells that have infected them, a small fraction of the bacterial cells are resistant to the antibody because they express the alternative type of flagellar protein.

The mechanism of phase variation has been studied by cloning and sequencing the genes involved, by analyzing mutants defective in the switching mechanism, and finally by developing an in vitro switching system using purified proteins that direct the process. The mechanism, outlined in Figure 9-36, involves inversion of a DNA segment that functions as a promoter, located adjacent to the *H2* operon. The *H2* operon contains two genes, one encoding the H2 flagellar protein and one encoding rH1, a specific repressor that inhibits transcription of the *H1* gene, which is located in a different region of the *Salmonella* genome. When the promoter region adjacent to the *H2* operon is in one 5′ → 3′ orientation, the *Salmonella* RNA polymerase binds to the promoter and initiates transcription of the *H2* operon. In this situation, or "phase," H2 and rH1 are both expressed; the later protein shuts off expression of H1, the other flagellar protein. Approximately once every thousand cell divisions, the segment containing the *H2* promoter is inverted. In this orientation of the *H2* promoter, RNA polymerase directs transcription away from the *H2* operon; as a result, neither H2 nor rH1 is expressed. In the absence of the repressor rH1, RNA polymerase can bind to the promoter of the *H1* gene and initiate transcription of that gene. In this phase, then, H1 is expressed and H2 is not.

The protein that catalyzes this inversion is a *site-specific recombinase* encoded by the *Hin* gene, which is completely contained within the inverting segment of DNA. The Hin protein is expressed only very rarely, so that inversion and the resulting phase variation is an infrequent phenomenon. Nonetheless, this process is important to the viability of the *Salmonella* species, because it extends the period of infection and, consequently, increases the number of new hosts to which *Salmonella* is exposed.

Yeast Mating Types Can Switch by Gene Conversion

Saccharomyces cerevisiae has a rather complicated life cycle and can grow as haploid or diploid cells. As discussed in Chapter 5, growing haploid and diploid yeast cells reproduce asexually by budding (see Figure 5-52). Each haploid *S. cerevisiae* cell exhibits one of two possible *mating types*, called a and α. Two haploid cells can "mate," or fuse, to form a diploid cell if they are of the opposite mating type (Figure 9-37).

A normal haploid cell switches its mating type each generation. This remarkable phenomenon has been traced to changes in DNA structure on chromosome III of *S. cerevisiae*. Three genetic loci directly involved in mating-type switching have been located (Figure 9-38). The central locus is termed *MAT*, the mating-type locus. The *MAT* locus is actively transcribed into mRNA, and the protein encoded by this mRNA acts to regulate genes not directly linked to (i.e., distant from) the *MAT* locus. These unlinked genes encode the proteins that give the cell its a or α phenotype. Two additional "silent" (nontranscribed) copies of the genetic information found at *MAT* are stored at loci termed *HML* and *HMR* (homothallic copies to the left and right of *MAT*, respectively, as the genetic map is conventionally drawn). Depending on the yeast strain, the silent *HML* locus contains sequences required for the α or a mating type; the *HMR* locus contains sequences for the opposite mating type. In the discussion here, *HML* carries α sequences and *HMR* carries a sequences. These sequences are transferred alternately from *HMLα* or *HMRa* into the *MAT* locus once during each cell generation. Once

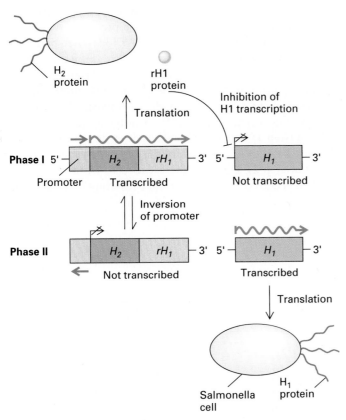

▲ FIGURE 9-36 Control of expression of *Salmonella* flagellar proteins (H1 and H2) by DNA inversion. The *H2* operon contains two genes: one encoding H2 and one encoding rH1, which represses transcription of the *H1* gene located in a different region of the *Salmonella* chromosome. When the upstream region (yellow) containing the *H2* promoter is in one 5′ → 3′ orientation (indicated by blue arrows), the *H2* operon is transcribed. In this phase (I), the cell's flagella contain H2 protein, and rH1 prevents transcription of the *H1* gene. Approximately once every 1000 cell doublings, the promoter-containing region upstream of *H2* is inverted by a site-specific recombination between the sequences at each end of the region. As a result, the *H2* operon is not transcribed; in the absence of rH1 protein, the *H1* gene is transcribed. In this phase (II), the cell's flagella contain H1 protein. See text for further discussion.

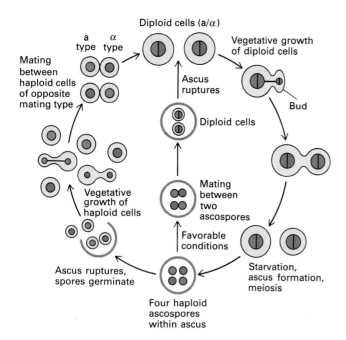

▲ FIGURE 9-37   Life cycle of *S. cerevisiae*. Two haploid cells that differ in mating type, called **a** and α, can mate to form an **a**/α diploid cell, which multiplies by budding. Under starvation conditions, diploid cells undergo meiosis, forming haploid ascospores. Rupture of the ascus releases haploid spores of one or the other mating type. In the presence of nutrients, haploid spores germinate into haploid cells, which also can grow vegetatively by budding.

this transfer occurs, the previously silent sequences can be transcribed.

The DNA from the three mating-type loci in yeast has been cloned and sequenced. Although *HML*α and *HMR*a contain some similar sequences, other sequences are distinct in the two loci. The most crucial sequences are called Y**a** and Yα. The presence of one or the other of these sequences at the *MAT* locus determines whether a cell makes **a** or α mRNAs, and hence whether it expresses an **a** or α phenotype (see Figure 9-38).

Switching of the sequence content at the *MAT* locus is accomplished by directed gene conversion, a process in which the sequence of one DNA region is changed (converted) to that of another region. The first step in this process is cleavage of both DNA strands between the Z and Y sequences in the *MAT* locus. This is accomplished by HO nuclease, a site-specific endonuclease present in wild-type HO$^+$ strains. The cut ends at the *MAT* locus then pair with the DNA of either *HMR*a or *HMR*α. However, since switching of the information at *MAT* is the rule in HO$^+$ strains, pairing of the cut ends is somehow preferentially directed to the silent locus with the opposite Y sequence.

As illustrated in Figure 9-39, after a cut is made at *MAT*a (step 1), the Y**a** sequence is degraded forming a gap in the *MAT* DNA. The ends of the gap in *MAT* then invade *HML*α (step 2). DNA synthesis proceeds from the free 3′-hydroxyl ends of the *MAT* strands, resulting in the copying

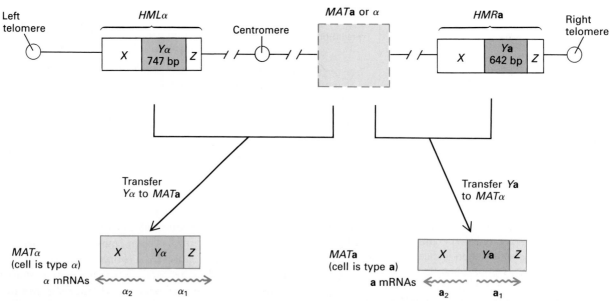

▲ FIGURE 9-38   Structure of the region on *S. cerevisiae* chromosome III that determines whether a yeast cell is the **a** or α mating type. The yeast chromosome (*top*) contains two silent (nontranscribed) loci, *HML*α and *HMR*a, and a central locus, termed *MAT,* which can exist in two forms containing either the Yα or Y**a** sequence (*bottom*). Transfer of the Yα or Y**a** sequence from one of the silent loci to *MAT* with the other sequence switches the mating type. The positions of the loci with respect to the centromere and telomeres of chromosome III are indicated.

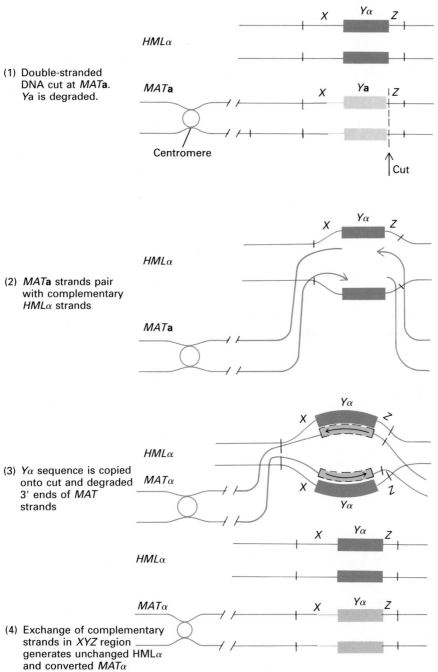

(1) Double-stranded DNA cut at *MAT***a**. Y**a** is degraded.

*HML*α

*MAT***a**

Centromere

Cut

(2) *MAT***a** strands pair with complementary *HML*α strands

*HML*α

*MAT***a**

(3) Y*α* sequence is copied onto cut and degraded 3′ ends of *MAT* strands

*HML*α

*MAT*α

*HML*α

*MAT*α

(4) Exchange of complementary strands in *XYZ* region generates unchanged HMLα and converted MATα

◄ FIGURE 9-39   Switching of **a** mating type to α mating type by gene conversion at the *MAT* locus on yeast chromosome III. Arrowheads indicate the 3′ direction. The Y**a** locus and part of *X* degraded in step (1) is shown in gray. In step (2), the cut ends of the *MAT* locus invade the *HML*α locus. Addition of nucleotides to the 3′ end of each cut strand results in duplication of the Y*α* information shown in light blue (3). Strand exchange yields a *MAT* locus containing the newly synthesized Y*α* locus and regenerates the original *HML*α locus (4). Thus the mating type has been converted from **a** to α. By an analogous process involving *HML***a**, the *MAT*α locus can be switched to *MAT***a** in the next generation. [See J. N. Strathern et al., 1982, *Cell* **31**:183.]

of the Y*α* sequence at *HML*α onto the cut ends of the *MAT* strands (step 3). An exchange of strands between the newly copied region of *HML*α and the remaining portion of *MAT* leads to regeneration of the *MAT* locus with the Y*α* sequence substituted for the original Y**a** sequence (step 4). Since *HML*α is regenerated, it is available to switch again in a later generation. During the next cell division, *MAT*α is switched to *MAT***a** by exactly the same process involving *HMR***a**. The regulation of mating-type switching is discussed in Chapter 13 (see Figure 13-14).

Trypanosome Surface Antigens Undergo Frequent Changes Via Gene Conversion

Trypanosomes are protozoans that can cause severe illness in animals and humans. One well-studied species, *Trypanosoma brucei*, contains highly *variable surface glycoproteins* (*VSGs*) that are antigenic. *T. brucei* has a mechanism for frequently changing one VSG to a similar but distinct antigen, thus enabling the protozoan to elude host immune responses that would inactivate it. Although this mechanism differs from the phase variation of *Salmonella* flagellar proteins, it serves the same purpose for the organism.

Although the mechanism of VSG switching is not understood fully, analysis of VSG DNA has provided some insights. Partially purified mRNAs encoding several different VSGs were used to make cDNAs, which then were hybridized with *T. brucei* genomic DNA. Because the VSG cDNAs hybridized to clusters of genomic DNA sequences called *basic copies* (BCs), the BC sequences were thought to encode VSGs. However, sequencing studies showed that the 3′-end sequence of all genomic BCs analyzed differs from that of the mRNAs, which have a common sequence at their 3′ end. In addition, the 5′-end sequence was found to differ in genomic BCs and VSG mRNAs. These findings suggest that VSG mRNAs are not synthesized directly from BCs.

Southern-blot analysis of trypanosome genomic DNA, with the labeled 3′ end of VSG cDNAs as a probe, showed that the transcription site for synthesis of VSG mRNA was located elsewhere in the genome. This active VSG gene was found to contain a transposed and duplicated copy, termed the *expression-linked copy* (ELC), of a BC sequence adjacent to a 3′-end sequence corresponding to that found in the VSG mRNAs. Upstream of the ELC are multiple transcription units encoding a 140-nucleotide RNA. As shown in Figure 9-40, a 35-nucleotide portion of this RNA is transpliced onto the 5′ end of the primary transcript synthesized from the ELC region. This 35-nucleotide, untranslated "leader" sequence has been found in many other trypanosome mRNAs, in addition to the VSG mRNAs; thus transplicing is a common event in these cells. This type of transplicing is discussed further in Chapter 12.

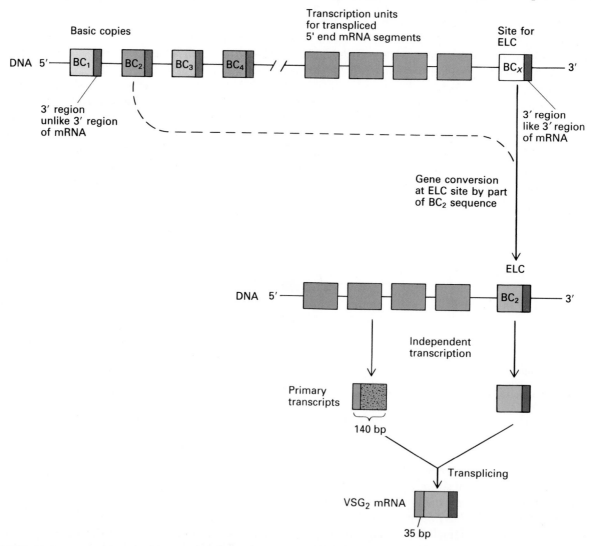

▲ FIGURE 9-40   General structure and arrangement of genomic region that encodes variable surface glycoproteins (VSGs) in trypanosomes. Active transcription in this region occurs only at the ELC (expression-linked copy) site, which has a constant 3′ end sequence (purple) and a variable region (white) corresponding to a transposed and duplicated copy of one of the basic copies (BCs). Gene conversion is the probable mechanism by which one BC coding region replaces another at the ELC site. A VSG mRNA is formed by transplicing of the transcript from the ELC site with a 35-nucleotide leader sequence contained in a 140-nucleotide RNA. [See P. Borst and G. M. Cross, 1982, *Cell* **29**:291.]

The chromosomal rearrangements giving rise to different VSG mRNAs in trypanosomes are random events that occur in about 1 in 10^6 cells. Thus these rearrangements differ from yeast mating-type switching, which is a programmed change occurring before each cell division. However, in VSG switching one DNA region is maintained and one region is changed, similar to yeasting mating-type switching. Thus, the same mechanism of gene conversion probably is responsible for both types of rearrangement. A similar DNA rearrangement may occur in *Paramecium*, another protozoan that changes its surface proteins often.

Generalized DNA Amplification Produces Polytene Chromosomes

All of the DNA rearrangements discussed so far—both functional and nonfunctional—involve changes in the position of sequences within the genome. Another type of rearrangement involves generalized amplification of DNA sequences, or *polytenization.*

The salivary glands of *Drosophila* species contain enlarged interphase chromosomes. When fixed and stained, these chromosomes are characterized by a large number of well-demarcated bands (see Figure 8-25). The enlargement of chromosomes in the salivary glands, and in other tissues as well, occurs when the DNA repeatedly replicates but the daughter chromosomes do not separate. The result is a *polytene* chromosome composed of many parallel copies of itself. Although most of a chromosome participates in polytenization, certain sequences, such as the simple-sequence DNA near the centromere, are not amplified. Furthermore, the ribosomal genes tend to be amplified less than other sequences during polytenization (Figure 9-41).

The molecular basis for the varying extent of replication along presumably linear chromosomal DNA molecules remains unknown, as does the reason for differences in the degree of polytenization in different cell types. For example, *Drosophila* salivary gland chromosomes appear to contain about 1000 copies of most DNA regions, whereas the polytene chromosomes in some intestinal cells contain only about 100 copies.

Localized DNA Amplification of rRNA and Other Genes Occurs in Some Eukaryotic Cells

The localized amplification of specific genes was first demonstrated in the rRNA genes of frog oocytes. Several other examples of this phenomenon have been described, although gene-specific amplification probably is not as widespread as polytenization. The amplified gene copies may remain in tandem at the site where the increase takes place, or they may be released as free-floating extra copies.

Frog oocytes are 2–3 mm in diameter and contain many stored ribosomes awaiting the burst in translation that occurs after a mature oocyte is fertilized. The rRNA

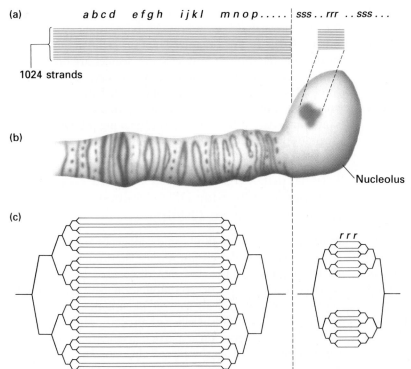

(a)   *a b c d   e f g h   i j k l   m n o p . . . . . | s s s . . r r r . . s s s . . .*

1024 strands

(b)

Nucleolus

(c)

r r r

◄ **FIGURE 9-41**   Model of DNA amplification to produce polytene chromosomes. (a) Sequences along the left arm of the *Drosophila hydei* X chromosome are represented by the letters *a* through *p*. Simple-sequence DNA (*sss*) and ribosomal genes (*rrr*) occur on the short right arm of the chromosome. In the salivary glands, all the DNA in the left arm of the chromosome is replicated in parallel many times; ten duplications would give $2^{10} = 1024$ copies. The simple-sequence DNA does not appear to increase at all; the ribosomal genes are replicated, but not as many times as the DNA in the left arm. (b) Morphology of the stained polytene X chromosome as it appears in the light microscope. The banding pattern results from reproducible packing of DNA and protein within each amplified site along the chromosome. (c) The proposed pattern of replication in the polytene *D. hydei* X chromosome. [See C. D. Laird et al., 1973, *Cold Spring Harbor Symp. Quant. Biol.* **38**:311.]

(a)

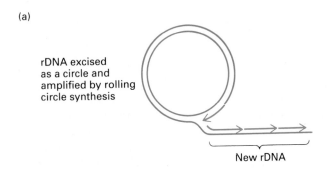

rDNA excised
as a circle and
amplified by rolling
circle synthesis

New rDNA

(b)

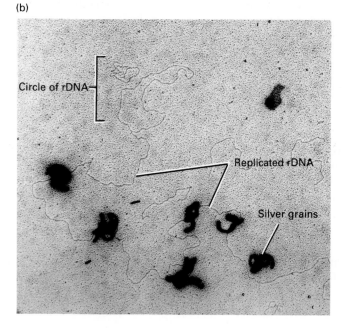

Circle of rDNA

Replicated rDNA

Silver grains

◀ FIGURE 9-42   Amplification of rRNA genes in frog oocytes. Replication of excised rDNA (blue) takes place by a mechanism called rolling circle replication. Newly synthesized daughter strands are shown in red. (b) Electron microscope autoradiograph of DNA extracted from frog oocytes exposed to radioactive thymidine. A circle of excised replicating rDNA is visible. That the long strand extending from the circle is newly synthesized DNA is indicated by the presence of silver grains, which correspond to sites of labeled thymidine incorporation. [See J. G. Gall, and M. L. Pardue, 1969, *Proc. Nat'l. Acad. Sci. USA* **63**:378. Photograph from J.-D. Rochaix, A. Bird, and A. Bakken, 1974, *J. Mol. Biol.* **87**:473.]

for these ribosomes is produced by transcription of a large pool of rDNA, which is greatly amplified during the development of the oocyte. This 2000-fold amplification occurs by excision of rDNA from the genome in the form of circular molecules, which then are copied by *rolling circle replication* (Figure 9-42). Although localized amplification of rRNA genes was first discovered in frog oocytes, it probably also occurs in many other animals with large egg cells.

Vertebrate Genes Encoding Antibodies Are Assembled from Gene Segments by Controlled Deletion of Intervening DNA

The great diversity of antibody molecules produced by vertebrates results in part by assembly of functional antibody genes from a relatively small number of distinct gene segments encoding various parts of the antibody molecule that interacts with antigen. A similar process is involved in the assembly of functional genes encoding the receptors on T cells, which also interact with antigen and have a critical function in the immune response. In both cases, DNA rearrangements permit a limited number of gene segments to be combined in a very large number of possible combina-

tions, each encoding a different antibody or T-cell receptor (TCR) capable of interacting with a specific antigen.

Antibody or TCR gene segments are assembled into functional continuous transcription units by deletion of up to hundreds or thousands of base pairs between the segments. This process, which is catalyzed by site-specific recombination enzymes, is discussed in detail in Chapter 27. We saw above how pathogenic organisms like *Salmonella* and *T. brucei* use directed DNA rearrangements to generate antigenic diversity in order to defend themselves from the immune response of vertebrates. In turn, related mechanisms of directed DNA rearrangements are used by vertebrates to generate the antibody diversity required to defend themselves from these and other pathogens.

▶ Organizing Cellular DNA into Chromosomes

Now that we have seen how protein-coding genes and other sequences are organized within cellular DNA, we turn to the question of how DNA molecules are organized within cells into the structures we observe as chromosomes. Because the total length of cellular DNA in prokaryotes and eukaryotes is a thousand up to a hundred thousand times the cell's length, the packing of DNA into chromosomes is crucial to cell architecture.

Prokaryotic Chromosomes Contain Highly Compacted Circular DNA Molecules with a Single Replication Origin

In bacterial cells without plasmids, all genes are located on a single large, circular chromosome. The circular nature of bacterial chromosomes was first discovered by analyzing the frequency of genetic recombination between mutant genes that produced easily assayed phenotypes, such as the inability to grow in the absence of a specific amino acid or the inability to grow on a particular sugar. As multiple genes were mapped on the *E. coli* chromosome by recom-

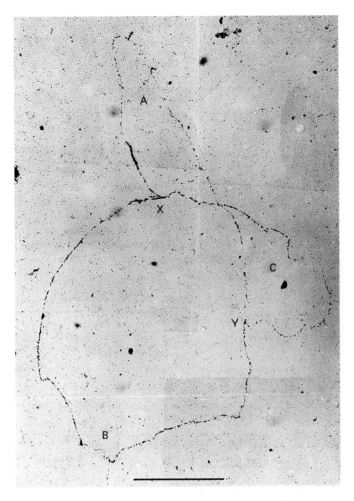

the long thread of chromosomal DNA. Autoradiography of the filter revealed ^{3}H-labeled circular DNA molecules with a total length of 1 mm, equivalent to the 4×10^6 base pairs composing the entire *E. coli* genome. The autoradiograph shown in Figure 9-43 shows a partially replicated *E. coli* chromosomal DNA molecule. The structure of the molecule demonstrates that the chromosome replicates from a single origin, as diagrammed for a small circular plasmid DNA molecule in Figure 7-2.

The 1-mm long DNA molecule of the *E. coli* chromosome is contained within cells that are only about 2 μm long and about 0.5–1 μm wide. A free DNA molecule of this size would form a random coil about 1000 times the volume of an *E. coli* cell. However, several mechanisms operate to compact *E. coli* chromosomal DNA sufficiently to fit inside the bacterial cell. For example, the large volume filled by the free DNA is due largely to charge repulsion between the negatively charged phosphate groups. In the cell, this effect is counteracted by association of the DNA with positively charged polyamines, such as spermine and spermidine (Figure 9-44), which shield the negative charges of the DNA phosphate groups. In addition, numerous small protein molecules associate with the chromo-

▲ FIGURE 9-43 Autoradiograph of the *E. coli* chromosome labeled with [^{3}H]-thymine. Points X and Y indicate the positions of DNA replication forks. Region B is the unreplicated parental chromosome. Regions A and C are the newly replicated nascent daughter chromosomes containing one parental strand and one newly replicated daughter strand. [From J. Cairns, 1964, *Cold Spring Harbor Symp. Quant. Biol.* **28**:43.]

binational analysis, no ends were found in the single linkage map that developed. Rather, every gene had other identified genes that could be mapped on either side of it. The whole linkage map generated a circle as shown in Figure 6-3.

In 1963 the circular structure of the *E. coli* chromosome was observed directly. *E. coli* cells were grown in the presence of ^{3}H-labeled thymine, which is incorporated only into DNA. The cells were then gently lysed by the addition of enzymes that degrade cell walls. Proteins were dissociated from the DNA by addition of detergents that denature protein, and the DNA was gently spread on the surface of a membrane filter. Special care was taken to avoid mixing the solution so as not to mechanically shear

```
                    +NH₃
                     |
                    CH₂
                     |
                    CH₂            +NH₃
                     |              |
                    CH₂            CH₂
                     |              |
                    +NH₂           CH₂
                     |              |
                    CH₂            CH₂
                     |              |
                    CH₂            CH₂
                     |              |
                    CH₂            +NH₂
                     |              |
                    CH₂            CH₂
                     |              |
                    +NH₂           CH₂
                     |              |
                    CH₂            CH₂
                     |              |
                    CH₂            +NH₃
                     |
                    CH₂         Spermidine
                     |
                    CH₂
                     |
                    +NH₃
                 Spermine
```

▲ FIGURE 9-44 Structures of spermine and spermidine.

somal DNA, causing it to fold into a more compact structure. The most abundant of these proteins, H-NS, is a dimer of a 15.6-kDa polypeptide. H-NS binds DNA tightly and compacts it considerably, as measured by an increased rate of sedimentation during centrifugation and decreased viscosity. There are about 20,000 H-NS molecules per cell, enough for one H-NS dimer per ≈400 base pairs of DNA.

Finally, *E. coli* chromosomal DNA is tightly supercoiled—that is, twisted upon itself like the circular SV40 DNA shown in Figure 4-14. As discussed in Chapter 10, an *E. coli* enzyme called DNA gyrase can introduce negative supercoils into DNA. Supercoiling contributes to the compaction necessary to fit chromosomal DNA into the bacterial cell. Figure 9-45 is an electron micrograph of an isolated, highly supercoiled *E. coli* chromosome attached to a fragment of cell membrane.

Eukaryotic Nuclear DNA Associates with Highly Conserved Histone Proteins to Form Chromatin

The problem of compacting cellular DNA is also significant for eukaryotic cells. When the DNA from eukaryotic nuclei is isolated in isotonic buffers (i.e., buffers with the same salt concentration found in cells, ≈0.15 M KCl), it is found associated with an equal mass of protein in a highly compacted complex called *chromatin*. The general structure of chromatin has been found to be remarkably similar in all eukaryotic cells.

The most abundant proteins associated with eukaryotic DNA are *histones,* a family of basic proteins found in all eukaryotic nuclei. The five major types of histone proteins—termed H1, H2A, H2B, H3, and H4—are easily separated by gel electrophoresis (Figure 9-46). The histone proteins are rich in basic amino acids, which contact negatively charged phosphate groups in DNA. In a fraction of the histone proteins of most cells, some of the basic amino acid side chains are modified by post-translational addition of methyl, acetyl (CH_3CO-), or phosphate groups, neutralizing the positive charge of the side chain or converting it to a negative charge.

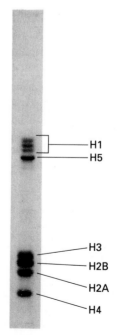

▲ FIGURE 9-45 Electron micrograph of an isolated folded *E. coli* chromosome. The highly supercoiled DNA is attached to a fragment of the cell membrane appearing as the most darkly staining material in the micrograph. Although the highly supercoiled nature of the *E. coli* chromosome is illustrated by this electron micrograph, the chromosome actually decondensed considerably during isolation. Within the cell, the chromosome has a diameter of <1 μm. The bar corresponds to 2 μm. [From H. Delius and A. Worcel, 1974, *J. Mol. Biol.* **82**:107.]

◄ FIGURE 9-46 Gel electrophoretic separation of histone proteins extracted from chicken blood cells. The major histone species—H2A, H2B, H3, and H4—are present in about equal amount. The other major histones are H1, which is found in white blood cells and most other vertebrate cells, and H5, which is similar to H1 and replaces it in the red blood cells of birds. The separation of H1 into three bands results from differences in the extent of phosphorylation of residues in the protein. [Courtesy of V. Allfrey.]

(a)

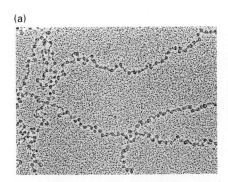

(b)

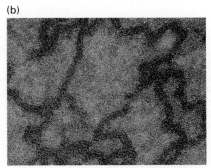

▲ FIGURE 9-47   Electron micrographs of extracted chromatin in extended and condensed forms. (a) Chromatin isolated in low ionic strength buffer has an extended "beads-on-a-string" appearance. The "beads" are nucleosomes (10-nm diameter) and the "string" is connecting DNA. (b) Chromatin isolated in buffer with a physiologic ionic strength (0.15 M KCl) appears as a condensed fiber 30 nm in diameter. [Left micrograph courtesy of S. McKnight and O. Miller, Jr.; right micrograph courtesy of B. Hamkalo and J. B. Rattner.]

The amino acid sequences of four histones (H2A, H2B, H3, and H4) from a wide variety of organisms are remarkably similar among distantly related species. For example, the sequences of histone H3 from sea urchin tissue and of H3 from calf thymus are identical except for a single amino acid, and only four amino acids are different in H3 from the garden pea and that from calf thymus. Minor histone variants encoded by genes that differ from the highly conserved major types also exist, particularly in vertebrates.

The amino acid sequence of H1 varies more from organism to organism than do the sequences of the other major histones. In certain tissues, H1 is replaced by special histones. For example, in the nucleated red blood cells of birds, a histone termed H5 is present instead of H1 (see Figure 9-46). Despite minor variations, the similarity in the amino acid sequences of the major histones among all eukaryotes is most impressive.

Chromatin Exists in Extended and Condensed Forms

When chromatin is extracted from nuclei and examined in the electron microscope, its appearance depends on the salt concentration to which it is exposed. At low salt concentration, isolated chromatin resembles "beads on a string" (Figure 9-47a). In this extended form, the string is a thin filament of DNA connecting the beadlike structures termed *nucleosomes*. Composed of DNA and histones, nucleosomes are about 10 nm in diameter and are the primary structural units of chromatin. If chromatin is isolated at physiologic salt concentration ($\approx$0.15 M KCl), it assumes a more condensed fiber-like form 30 nm in diameter (Figure 9-47b).

Structure of Nucleosomes Individual nucleosomes can be isolated by nuclease digestion of extracted chromatin, because the DNA component of nucleosomes is much less susceptible to digestion than is the linker DNA connecting nucleosomes. Partial nuclease treatment first releases groups of nucleosomes by digestion of the linker DNA between some of the nucleosomes. More extensive digestion produces nucleosome tetramers, trimers, and dimers. Eventually, nuclease treatment digests all the DNA between individual nucleosomes, so that all the nucleosomes are released. The DNA content of a single nucleosome plus the DNA linking neighboring nucleosomes varies between 160 and 200 base pairs in different organisms. After digestion of all the linker DNA, nucleosomes from all eukaryotes contain close to 146 base pairs of DNA.

A nucleosome is composed of a protein core with DNA wound around its surface like thread around a spool. The core is an octamer containing two copies each of histones H2A, H2B, H3, and H4. X-ray crystallography has shown that the octameric histone core is disk shaped (Figure 9-48). About 146 base pairs of DNA are wrapped slightly less than two turns around the core to form the nucleosome.

Assembly of Nucleosomes Newly replicated DNA quickly associates with already formed histone octamers. A model of nucleosome assembly has been proposed based on studies with rapidly dividing fertilized frog oocytes. Analysis of protein complexes isolated from early frog embryos revealed two acidic nonhistone proteins associated with the basic histone proteins that were not assembled into nucleosomes. One of these nonhistone proteins, called *nucleoplasmin*, was found bound to H2A and H2B; the other, called *N1 protein*, to H3 and H4. When partially purified preparations of these two complexes were mixed in the presence of DNA, nucleosomes were formed with release of free nucleoplasmin and N1 (Figure 9-49). Proteins resembling nucleoplasmin and N1 have been identified in other cell types. Thus, the proposed pathway of nucleosome assembly may operate in most cells.

(a)

(b)

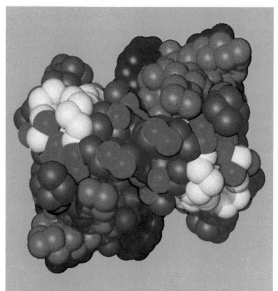

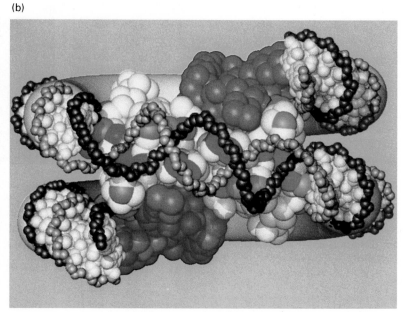

▲ FIGURE 9-48   Structure of the histone octamer and the nucleosome. (a) Model of octameric histone core based on a 3.1 Å resolution structure determined by x-ray crystallography. The histone core contains two copies each of H2A (light blue), H2B (dark blue), H3 (green), and H4 (white). The spheres represent amino acid residues, not atoms. The positively charged arginine and lysine residues are red. The amino termini of the histone proteins are not visualized by x-ray crystallography, but they are thought to extend outward from the top and bottom of this view of the histone octameric core. (b) Model of the nucleosome in which the octameric core is represented by one centrally located (H3/H4)$_2$ tetramer (white) flanked by two H2A/H2B dimers (blue); 146 base pairs of DNA (gray) are wrapped 1.75 supercoil turns around the histone core. In the central area of the picture, the DNA bases (light gray) have been stripped away and the path of the phosphodiester backbones is represented by medium and dark gray spheres; these are "undersized" in order to visualize the matching of the pattern of the positively charged residues (red spheres) on the surface of the histone octamer with the negatively charged DNA backbone. The α-helix dipoles are indicated by orange. [See G. Arents and E. N. Moudrianakis, 1993, *Proc. Nat'l. Acad. Sci.* **90**:10489; courtesy of E. N. Moudrianakis.]

Solenoid Structure of Condensed Chromatin   In its condensed form, chromatin appears as fibers ≈30 nm in diameter (see Figure 9-47b). A model for the structure of these thick fibers is shown in Figure 9-50. In this model, nucleosomes are packed into a spiral or solenoid arrangement with six nucleosomes per turn. A fifth histone, H1, is bound to the DNA on the inside of the solenoid, with one H1 molecule associated with each nucleosome. The unit of one nucleosome plus one bound H1 is referred to as a *chromatosome*. Under various conditions, condensed chromatin is further folded into giant supercoiled loops.

As noted earlier, when chromatin is extracted at the physiologic salt concentration, condensed 30-nm solenoid fibers are obtained. However, when extraction is done at a low salt concentration, H1 is released, yielding the extended beads-on-a-string form. Thus, depending on the extraction conditions, two forms of chromatin can be observed experimentally in vitro. As discussed in the next section, the chromatin in chromosomal regions that are not being transcribed exists predominantly in the condensed form, whereas that in regions being transcribed probably assumes the extended form.

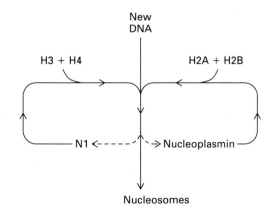

◀ FIGURE 9-49   Proposed pathway of nucleosome assembly in frog eggs. Both N1 and nucleoplasmin are acidic proteins that have been shown to associate with histones as indicated. [Adapted from S. M. Dilworth et al., 1987, *Cell* **51**:1009.]

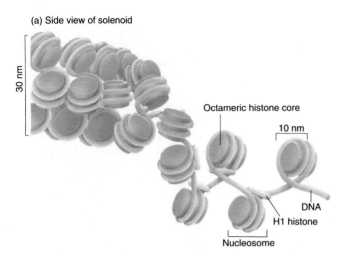

(a) Side view of solenoid

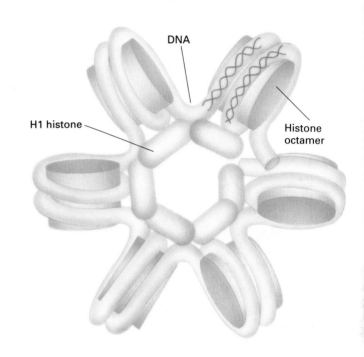

(b) View along axis of one turn of solenoid

▲ FIGURE 9-50   Two views of a model of the condensed chromatin fiber. The octameric histone core (see Figure 9-48a) is shown as a disk. Each nucleosome associates with one H1 molecule, and the fiber coils into a solenoid structure with a diameter of 30 nm. [Part (a) adapted from M. Grunstein, 1992, *Sci. Am.* **267**:68; part (b) adapted from A. Klug, 1985, *Proc. R. W. Welch Fdn. Conf. Chem. Res.* **39**:133.]

➤ *Morphology and Functional Characteristics of Eukaryotic Chromosomes*

Chromatin is further organized into large units hundreds to thousands of kilobases in length called *chromosomes*. Microscopic observations on the number and size of chromosomes and their staining patterns led to the discovery of many important general characteristics of chromosome structure.

Chromosome Number and Shape Are Species Specific

In nondividing cells the chromosomes are not visible, even with the aid of histologic stains for DNA (e.g., Feulgen or Giemsa stains) or electron microscopy. During mitosis and meiosis, however, the chromosomes condense and become visible in the light microscope. Therefore, almost all cytogenetic work (i.e., studies of chromosome morphology) has been done with condensed *metaphase* chromosomes obtained from dividing cells—either somatic cells in mitosis or dividing gametes during meiosis.

The condensation of metaphase chromosomes probably results from several orders of supercoiling of the 30-nm solenoid chromatin fibers. Recall from Chapter 5 that at the time of mitosis, cells have already progressed

through the S phase of the cell cycle and have replicated their DNA, so that the chromosomes which become visible during metaphase are duplicated structures (see Figure 5-50). Each metaphase chromosome consists of two sister *chromatids*, which are attached at the *centromere*. The number, sizes, and shapes of the metaphase chromosomes constitute the *karyotype*, which is distinctive for each species (Figure 9-51). In most organisms, all cells have the same karyotype. However, species that appear quite similar can have very different karyotypes, indicating that similar genetic potential can be organized on chromosomes in very different ways (Figure 9-52; Table 9-6).

Nonhistone Proteins Provide a Structural Scaffold for Long DNA Loops in Chromosomes

Although histones are the predominant proteins in chromosomes, nonhistone proteins are also involved in organizing the structure of chromosomes. Electron micrographs of histone-depleted chromosomes from HeLa cells reveal long loops of DNA containing 10–90 kb anchored to a *chromosome scaffold* composed of nonhistone proteins (Figure 9-53). This scaffold has the shape of the metaphase chromosome and persists even when the DNA is digested by nucleases. Figure 9-54 shows a model for the association of the 30-nm solenoid chromatin fiber with the

(b)

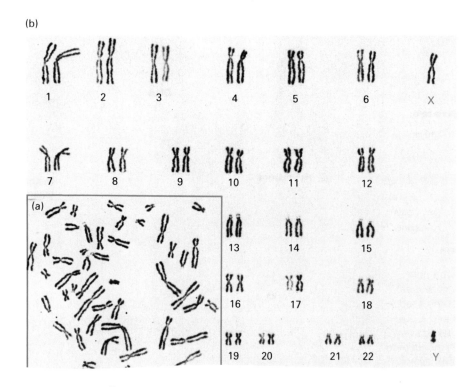

◄ FIGURE 9-51   Karyotype of human male metaphase chromosomes. (a) The chromosomes from one mitotic cell, a lymphocyte, were stained with orcein (a dye that reacts with DNA), and the cell was "squashed" on a microscope slide. (b) After identification of each pair of homologues, photographs of the chromosomes were mounted in sequence by size. Each number indicates a homologous pair, and the entire set is the karyotype. The unnumbered, unpaired X and Y chromosomes are the sex chromosomes. A metaphase chromosome consists of two identical sister chromatids; their point of association is the centromere. According to the position of the centromere, the chromosome is termed metacentric (centromere in the middle; e.g., chromosome 1), acrocentric (asymmetrical centromere; e.g., chromosome 4), or telocentric (centromere at an end; e.g., chromosome 14). [Courtesy of J. German.]

➤ FIGURE 9-52   Karyotypes of the Reeves muntjac (a) and the Indian muntjac (b), two species of small deer that are quite similar but do not interbreed. Despite the difference in the number of chromosomes in these animals, the two genomes contain about the same total amount of DNA. (The chromosomes for both deer are shown at the same magnification.) Similar plants can also have strikingly different karyotypes. [Part (a) (left) courtesy of K. W. Fink/Photo Researchers, Inc.; part (b) (left) courtesy of J. P. Ferrero/Jacana/Photo Researchers, Inc.; both karyotype photographs courtesy of R. Church.]

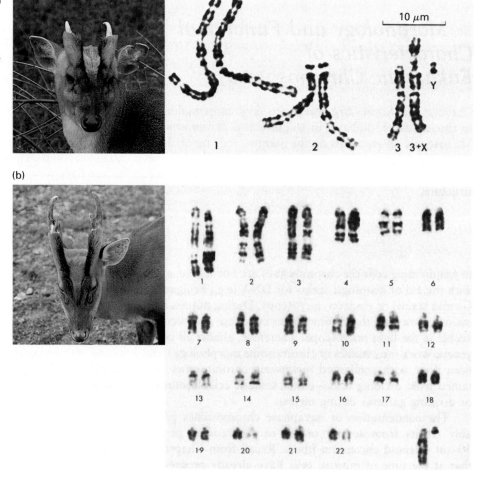

TABLE 9-6   Chromosome Numbers of Various Species*

Common Name	Species	Diploid Number	Common Name	Species	Diploid Number
ANIMALS (2n)			PLANTS AND FUNGI (2n)		
Human	*Homo sapiens*	46	Yeast	*Saccharomyces cerevisiae*	36 ±
Rhesus monkey	*Macaca mulatta*	42	Green algae	*Acetabularia mediterranea*	20 ±
Cattle	*Bos taurus*	60	Garden onion	*Allium cepa*	16
Dog	*Canis familiaris*	78	Barley	*Hordeum vulgare*	14
Cat	*Felis domesticus*	38	Wheat	*Triticum monoccum*	24
Horse	*Equus calibus*	64	Corn	*Zea mays*	20
House mouse	*Mus musculus*	40	Snapdragon	*Antirrhinum majus*	16
Rat	*Rattus norvegicus*	42	Tomato	*Lycopersicon esculentum*	24
Golden hamster	*Mesocricetus auratus*	44	Tobacco	*Nicotinana tabacum*	48
Guinea pig	*Cavia cobaya*	64	Kidney bean	*Phaseolus vulgaris*	22
Rabbit	*Oryctolagus cuniculus*	44	Pine	*Pinus* species	24
Chicken	*Gallus domesticus*	78 ±	Garden pea	*Pisum sativum*	14
Alligator	*Alligator mississipiensis*	32	Potato	*Solanum tuberosum*	48
Frog	*Rana pipiens*	26	Broad bean	*Vicia faba*	12
Carp	*Cyprinus carpio*	104			**Haploid number**
Silkworm	*Bombyx mori*	56			
House fly	*Musca domestica*	12	PLANTS AND FUNGI (1n)		
Fruit fly	*Drosophila melanogaster*	8	Slime mold	*Dictyostelium discoideum*	7
Flatworm	*Planaria torva*	16	Mold (fungus)	*Aspergillus nidulans*	8
Freshwater hydra	*Hydra vulgaris attenuata*	32	Pink bread mold	*Neurospora crassa*	7
			Penicillin mold	*Penicillium* species	4
Nematode	*Caenorhabditis elegans*	11/12 (male/female)	Green algae	*Chlamydomonas reinhardtii*	16

*In most organisms the chromosome number does not vary, but in some species (those with numbers listed as ±) either there is a natural variation in strains or the correct number has not been determined. In a few organisms the sexes differ; in nematodes, the male is haploid for the sex chromosome (X,O), and the female is diploid (X,X).
SOURCE: M. Strickberger, 1985, *Genetics*, 3d ed., Macmillan; and R. B. Flavell et al., 1974, *Biochem. Genet.* **12**:257.

chromosome scaffold in a metaphase chromosome. During interphase (when cells are not in mitosis or meiosis), the compaction of chromosomes is reduced, as depicted at the bottom of the figure.

In the purest preparations of metaphase chromosomes, only a few major scaffold proteins are found. One of these proteins has been identified as the enzyme *topoisomerase II* (topo II). As an enzyme, topo II can cleave double-stranded DNA, pass an uncut portion of the DNA between the cut ends, and then reseal the cut. As discussed in Chapter 10, this enzyme is critical in DNA replication, because it can unknot tangles of DNA that would otherwise form as the long parental strands unwind and daughter strands are

synthesized. Staining of metaphase chromosomes with fluorescent antibodies raised against highly purified topo II demonstrates that this enzyme is associated with the chromosome scaffold. Observation of such stained chromosomes at different levels of focus through the use of a confocal microscope ("optical sectioning") reveals that the scaffold has a regular helical pattern (Figure 9-55).

Even in interphase chromosomes, which are not as condensed as metaphase chromosomes, the DNA remains associated with topo II and hence with the chromosome scaffold. This association has been demonstrated by treating interphase cells with topoisomerase inhibitors. During interphase, when DNA replication occurs, topo II normally

(a)

Loops of DNA

Protein scaffold

(b)

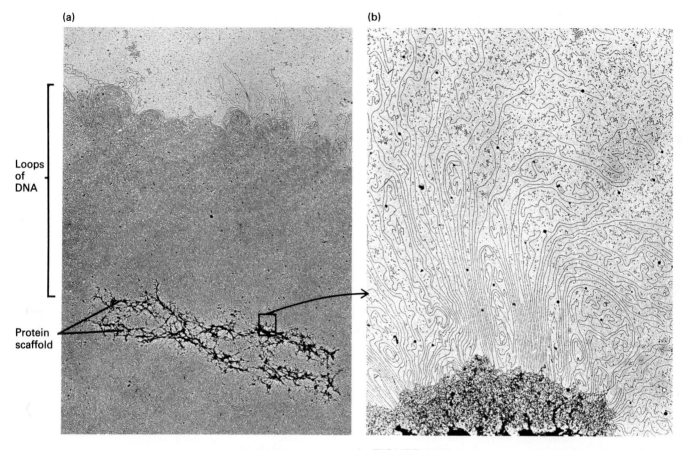

▲ FIGURE 9-53   Electron micrographs of a histone-depleted metaphase chromosome prepared from HeLa cells by treatment with a mild detergent. (a) At lower magnification, nonhistone protein scaffolding (dark structure), from which long loops of DNA protrude, is visible (50,000X). (b) A higher magnification of a section of a micrograph like (a) shows the apparent attachment of the DNA loops to the protein scaffold (150,000X). [From J. R. Paulson and U. K. Laemmli, 1977, *Cell* **12**:817. Copyright 1977 M.I.T.]

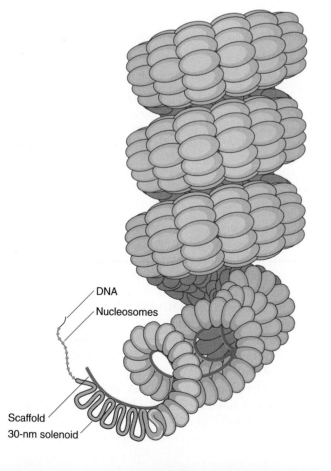

DNA

Nucleosomes

Scaffold

30-nm solenoid

◄ FIGURE 9-54   A model for the association of the 30-nm solenoid chromatin fiber with the chromosome scaffold in a metaphase chromosome. DNA and nucleosomes in the beads-on-a-string form of chromatin are shown in approximate relative scale to the 30-nm fiber. Loops of 30-nm fiber are associated with the chromosomal scaffold at the bases of the loops. These loops contain the 10–90 kb of DNA seen in Figure 9-53. In this model, the chromosome scaffold coils into a helix which is further coiled to produce the final compaction of DNA observed in a metaphase chromosome. [From A. J. F. Griffiths et al., *An Introduction to Genetic Analysis*, Fifth Edition, W. H. Freeman and Company, New York, 1993.]

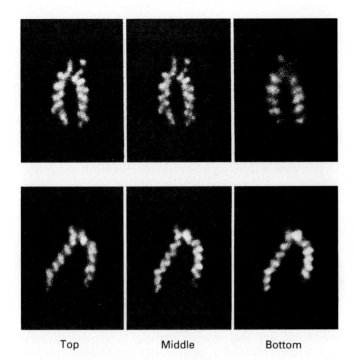

Top   Middle   Bottom

▲ **FIGURE 9-55** Two, partially uncoiled human metaphase chromosomes stained with fluorescent antibodies to topoisomerase II. The chromosomal scaffold is visualized at three depths of focus (top, middle, and bottom). Note that the tilt of the bright bands differs by 90° in the top and bottom focal positions, indicating a helical twist to the chromosome backbone. The helical coils visualized may correspond to the highest order coils of the model shown in Figure 9-54.

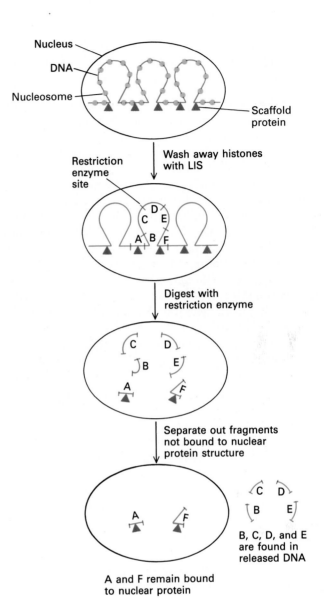

▲ **FIGURE 9-56** Mapping of sites at which organized loops of chromatin bind to scaffold-associated regions (SARs). Nuclei are washed free of all histones by treating them with the detergent lithium diiodosalicylate (LIS), which exposes the DNA to cleavage by restriction enzymes. The nuclei then are treated with a known restriction enzyme to yield fragments (A to F), which are analyzed by the Southern-blot technique. In the example shown, four of the six restriction enzyme fragments (B, C, D, and E) are not associated with scaffold proteins; these fragments are recovered outside of the nucleus. Extraction of the DNA that remains bound to the residual nonhistone nuclear proteins yields the two fragments A and F. (See J. Mirkovitch, M.-E. Mirault, and U. K. Laemmli, 1984, *Cell* **39**:223.]

acts to cut, untangle, and reseal the newly synthesized DNA molecules. During cleavage by topo II, the free 5′ phosphates on the DNA strands become covalently linked to tyrosine side chains of the enzyme. If interphase cells are treated with a drug that inhibits the resealing step of topo II activity, the cells accumulate ≈50-kb DNA fragments that have topo II attached to the ends. Studies of this type and others with the DNA-binding compound ethidium bromide, which can be used to measure the extent of DNA supercoiling, leave little doubt that proteins, including (if not only) topo II, are bound during interphase to fixed sites in mammalian DNA that are 30–90 kb apart.

The binding sites for topo II are called *scaffold-associated regions* (SARs). The procedure for mapping SARs is illustrated in Figure 9-56. Such mapping, which has been done with a number of individual genes, indicates that SARs occur *between*, not within, transcription units. Although the physiologic role of SARs has not yet been conclusively demonstrated, they may hold specific chromosome domains in such a way as to facilitate transcription or replication. There is evidence that in *Drosophila* SARs can insulate transcription units from each other, so that proteins regulating transcription of one gene do not influence

the transcription of a neighboring gene separated by a SAR. When chromatin is being transcribed by an RNA polymerase, it is thought to uncoil in the transcribed region between SARs into the 10-nm beads-on-a-string form.

Chromatin Contains Small Amounts of Other DNA-binding Proteins in Addition to Histones and Scaffold Proteins

The total mass of the histones associated with a DNA molecule is about equal to that of the DNA. Interphase chromatin and metaphase chromosomes also contain small amounts of other proteins. For instance, a growing list of DNA-binding *transcription factors,* nonhistone proteins that help regulate transcription, have been identified associated with interphase chromatin. The structure and function of these very rare, but critical, proteins are examined in Chapter 11. A few other nonhistone DNA-binding proteins are present in much larger amounts than the transcription factors. Some of these proteins exhibit high mobility during electrophoretic separation and thus have been designated HMG (high-mobility group). So far, no specific function has been demonstrated for HMGs. They are, however, necessary for cell viability as indicated by the observation that yeast mutants that cannot make HMGs are nonviable.

Each Chromosome Contains One Linear DNA Molecule

Nearly all DNA viruses seem to possess one molecule of DNA, and bacterial cells contain a single chromosomal DNA molecule. The general belief is that all eukaryotic chromosomes also contain a single long DNA molecule. Because the longest DNA molecules in human chromosomes are almost 10 cm long ($2–3 \times 10^8$ base pairs), they are difficult to handle experimentally without breaking. However, in lower eukaryotes, the sizes of the largest DNA molecules that can be extracted are consistent with the hypothesis that each chromosome contains a single DNA molecule. For example, physical analysis of the largest DNA molecules extracted from several genetically different *Drosophila* species and strains shows that they are from 6×10^7 to 1×10^8 base pairs long. These sizes match the DNA content of single stained metaphase chromosomes of *Drosophila melanogaster*, as measured by the amount of DNA-specific stain absorbed. Therefore, each chromosome probably contains a single DNA molecule.

The correspondence between the number of DNA molecules per cell and the number of chromosomes has been conclusively demonstrated in yeast cells. The DNA from each *S. cerevisiae* chromosome can be separated and indi-

vidually identified by pulsed-field gel electrophoresis (see Figure 7-27). The number of separated DNA molecules equals the number of genetic linkage groups (i.e., chromosomes) in yeast. The length of yeast chromosomal DNA ranges from about 1.5×10^5 to 10^6 base pairs.

Stained Chromosomes Have Characteristic Banding Patterns

Certain dyes selectively stain certain regions of metaphase chromosomes more intensely than other regions, producing banding patterns that are specific for individual chromosomes. The molecular basis for the regularity of chromosomal bands remains unknown, but they are very useful. When stained chromosomes are viewed microscopically, these bands serve as landmarks along the length of each chromosome and also help to distinguish chromosomes of similar size and shape.

Banding in Metaphase Chromosomes Quinacrine, a fluorescent dye that inserts between base pairs (*intercalates*) in the DNA helix, produces *Q bands.* However, because Q bands fade with time, other staining techniques are generally preferred in the laboratory. For example, chromosomes can be subjected briefly to mild heat or proteolysis and then stained with Giemsa reagent, a permanent DNA dye, to produce a pattern of *G bands* (see Figure 8-23). Treatment of chromosomes with a hot alkaline solution before staining with Giemsa reagent produces *R bands* in a pattern that is approximately the reverse of the G-band pattern. The distinctiveness of these banding patterns permits cytologists to identify specific parts of a chromosome and to locate the sites of chromosomal breaks and rearrangements (Figure 9-57). In addition, cloned DNA probes that have hybridized to specific sequences in the chromosomes can be located in particular bands.

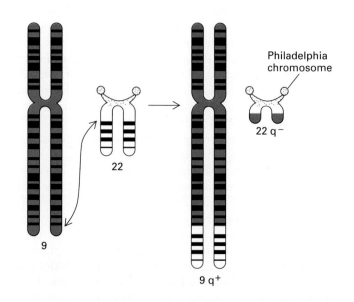

> FIGURE 9-57   The unique banding pattern associated with each metaphase chromosome permits analysis of many chromosomal abnormalities. For example, in all patients with chronic myelogenous leukemia, the leukemic cells contain the Philadelphia chromosome and an abnormal chromosome 9 (*right*). These result from a translocation between normal chromosomes 9 and 22 (*left*). [From J. Kuby, 1992, *Immunology*, W. H. Freeman and Co., p. 508.]

Banding in Interphase Polytene Chromosomes For over 50 years, cytologists have observed stainable bands in interphase polytene chromosomes in the salivary glands of *Drosophila melanogaster* and other dipteran insects (see Figure 8-25). Although the bands in human chromosomes probably represent very long folded or compacted stretches of DNA containing about 10^7 base pairs, the bands in *Drosophila* chromosomes represent much shorter stretches of only 50,000–100,000 base pairs.

The detailed banding of insect salivary gland chromosomes occurs because of polytenization (see Figure 9-41), which results in thick bundles of parallel interphase chromosomes that all have the same banding pattern across the width of the bundle. Molecular cloning experiments and mRNA mapping studies have suggested that each band contains a limited number of transcription units, perhaps only one unit in some cases. Since transcription units can also be found in interband stretches of DNA, however, the relationship between banding and function is still unclear. Nonetheless, the reproducible pattern of bands seen in *Drosophila* salivary gland chromosomes provides an extremely powerful method for locating specific DNA sequences along the lengths of the chromosomes in this species.

Heterochromatin Consists of Chromosome Regions That Do Not Uncoil

As cells exit from mitosis and the condensed chromosomes uncoil, certain sections of the chromosomes remain dark staining. The dark-staining areas, termed *heterochromatin,* are regions of condensed chromatin. Heterochromatin appears most frequently—but not exclusively—at the centromere and telomeres (ends) of the chromosome. The light-staining, less condensed portions are called *euchromatin.* The distinctions between heterochromatin and euchromatin are most clearly visualized in polytene chromosomes such as those of *Drosophila* salivary gland cells. Because heterochromatic regions apparently remain condensed throughout the life cycle of the cell, they have long been regarded as sites of inactive genes. In recent years, molecular studies have shown that most of the DNA in heterochromatin is highly repeated DNA that is never (or very seldom) transcribed. However, not all inactive genes and nontranscribed regions of DNA are visible as heterochromatin; and it is not necessarily true that no transcription occurs within heterochromatin.

Three Functional Elements Are Required for Replication and Stable Inheritance of Chromosomes

So far we have seen that the eukaryotic chromosome is a linear structure composed of a single DNA molecule that exists in a condensed or extended form and is associated with histones, scaffold proteins, and certain other less abundant proteins. Although chromosomes differ in length and number among species, cytogenetic studies long ago showed that they all behave similarly at the time of cell division. During mitosis, sister chromatids become attached to microtubules making up the mitotic spindle apparatus and align on the metaphase plate; they then separate at the centromeres, and one chromatid of each metaphase chromosome is distributed to each daughter cell (see Figure 5-50). More recently, recombinant DNA research with yeast cells has identified all of the chromosomal elements that are necessary for equal segregation of sister chromatids to occur during mitosis. The culmination of this work has been construction of *yeast artificial chromosomes (YACs).* In order to duplicate and segregate correctly, chromosomes must contain three functional elements: (1) special sequences (origins) involved in the initiation of DNA replication; (2) the centromere; and (3) the two ends, or telomeres.

Autonomously Replicating Sequences (Yeast Origins) If yeast cells lack a particular gene (e.g., one of the genes that encode an enzyme for synthesis of the amino acid leucine), they can be transformed with cloned plasmids containing the missing DNA (in this case, a *LEU* gene). Leu⁺ transformants, which can grow on media lacking leucine, are obtained much more frequently if the plasmid contains one of many sequences of approximately 100 base pairs called *autonomously replicating sequences* (Figure 9-58a). An autonomously replicating sequence (ARS) acts as an origin of replication, allowing the transformed circular plasmid to replicate in the yeast cell nucleus. The function of ARSs as origins of DNA replication is discussed in Chapter 10.

Centromeres In any culture of leucine-requiring yeast cells transformed with a simple circular *LEU/ARS*-containing plasmid, only about 5–20 percent of progeny cells contain the plasmid because mitotic segregation of the plasmids is faulty; most of the time plasmid DNA does not enter the bud. However, if restriction fragments of yeast genomic DNA are cloned into such circular plasmids, a small fraction of fragments are found to confer equal segregation of the plasmids to both mother and daughter cells following mitosis (Figure 9-58b). Because the fragments with this effect were found to be derived from the centromeres of yeast chromosomes, they are called CEN sequences. Such cloning experiments have led to isolation of sequences that improve mitotic segregation to the extent that >90 percent of the progeny of transformed Leu⁻ yeast cells contain the *LEU* plasmid. These cells and almost all of their descendants therefore grow well in a medium lacking leucine.

Once the yeast centromere regions that confer mitotic segregation were cloned, their sequences could be determined and compared. Comparison of different yeast centromeres has revealed three regions (I, II, and III) that are

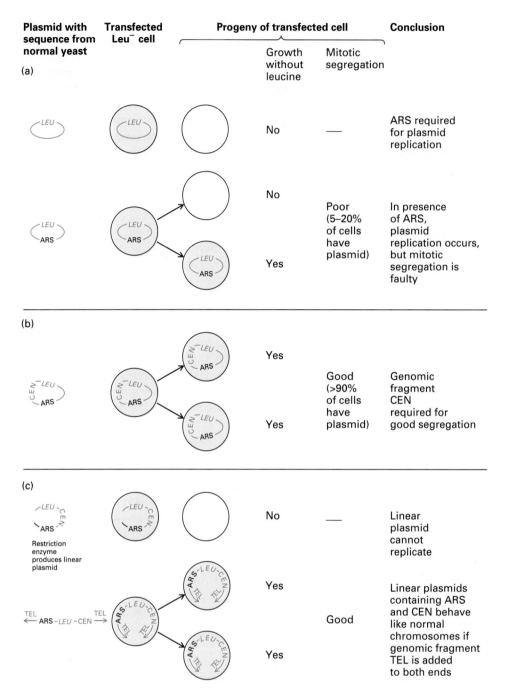

▲ FIGURE 9-58   Demonstration of functional chromosomal elements by transfection experiments with yeast cells that lack an enzyme necessary for leucine synthesis (Leu⁻ cells). In these experiments, plasmids containing the *LEU* gene from normal yeast cells are constructed and introduced into Leu⁻ cells by transfection. The plasmids then replicate along with the nuclear DNA of the cells; only cells containing a *LEU* plasmid will grow in a medium that is not supplemented with leucine. (a) The plasmid must possess sequences that allow autonomous replication (ARS) in order to replicate in the cell. However, even plasmids with ARS exhibit poor segregation during mitosis, and therefore do not appear in each of the daughter cells. (b) When randomly broken pieces of genomic yeast DNA are inserted into plasmids containing ARS and *LEU*, some of the transfected cells produce large colonies, indicating that a high rate of mitotic segregation among their plasmids is facilitating the continuous growth of daughter cells. The DNA recovered from plasmids in these large colonies contains yeast centromere (CEN) sequences. (c) When Leu⁻ yeast cells are transfected with linearized plasmids containing *LEU*, ARS, and CEN, no colonies grow. Addition of telomere (TEL) sequences gives the linearized plasmids the ability to replicate as new chromosomes that behave very much like a normal chromosome in both mitosis and meiosis. [See A. W. Murray and J. W. Szostak, 1983, *Nature* **305**:89.]

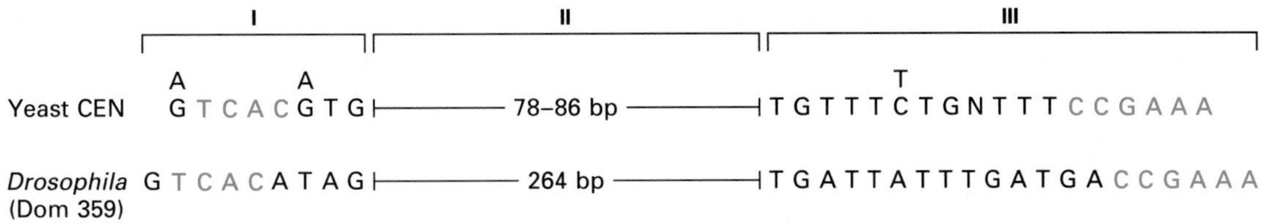
I
II
III
A A
T
Yeast CEN G T C A C G T G 78–86 bp T G T T T C T G N T T T C C G A A A
Drosophila (Dom 359) G T C A C A T A G 264 bp T G A T T A T T T G A T G A C C G A A A

▲ FIGURE 9-59 (*Top*) Consensus yeast CEN sequence based on analysis of ten yeast centromeres. All three distinctive regions are required for a centromere to function. Region II, although variable in sequence, is fairly constant in length and is rich in A and T residues. (*Bottom*) One *Drosophila* simple-sequence DNA that is located near the centromere has a repeat unit with some homology to the yeast consensus CEN, including two identical 4-bp and 6-bp stretches (red). [See L. Clarke and J. Carbon, 1985, *Ann. Rev. Genet.* **19:**29.]

necessary for a centromere to function (Figure 9-59). Short, fairly well conserved nucleotide sequences are present in regions I and III. Although region II seems to have a fairly constant length (78–86 base pairs), it contains no definite consensus sequence; however, it is rich in A and T residues. One *Drosophila* simple-sequence DNA, which comes from a centromeric region, has a repeat unit that bears some similarity to yeast CEN regions I and III, suggesting that similar mechanisms may control segregation in yeast and higher eukaryotes.

In yeast cells in the prophase stage of mitosis, the centromeric DNA is bound to a complex of proteins that recognize region III of the CEN sequence. As illustrated in Figure 9-60, this protein complex, designated CBF (centromere-binding factor), simultaneously binds to the centromere and to a microtubule extending from a spindle pole body (SPB), effectively tethering each chromatid to the microtubule spindle apparatus. The SPB, the functional analog of the centriole in animal cells, is embedded in the yeast nuclear envelope, which does not break down into multiple small vesicles during mitosis in yeast as it does in higher eukaryotes. The SPB is replicated at the beginning of mitosis, and the two structures move to opposite poles of the nucleus. When budding occurs, the microtubules extending between the two spindle pole bodies lengthen, pushing the spindle pole bodies apart until one enters the bud (see Figure 5-52).

The CBF acts like a "molecular motor," using the energy released by ATP hydrolysis to move an attached centromere along a microtubule in the direction of the spindle pole body. In this way, one sister chromatid of each metaphase chromosome is moved toward the SPB in the emerging daughter bud and the other toward the SPB remaining in the mother cell. The function of microtubules and the role of centromere-binding proteins are discussed in greater detail in Chapters 24 and 25.

Telomeres If circular plasmids containing an ARS and CEN sequence are cut once with a restriction enzyme, they become linear. Such linear plasmids do not replicate in

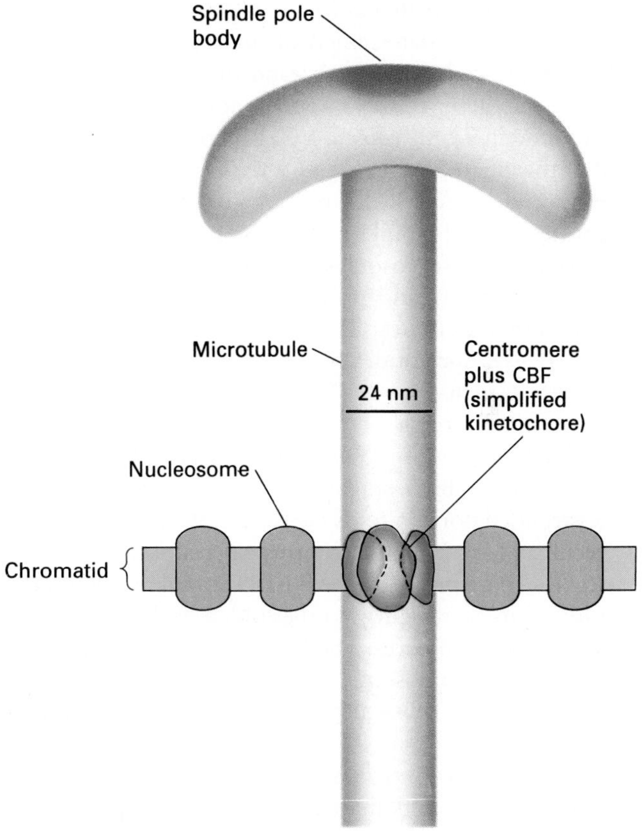
Spindle pole body
Microtubule
24 nm
Centromere plus CBF (simplified kinetochore)
Nucleosome
Chromatid

▲ FIGURE 9-60 A model of the binding of the yeast centromere region to a microtubule extending from the spindle polebody. The centromere-binding factor (CBF), a complex of three proteins, binds to both region III of CEN in chromatid DNA and to the microtubule. The CBF functions as a molecular motor that can move along the microtubule in the direction of the spindle pole body. This movement is thought to be responsible for segregation of sister chromatids to the mother and daughter cells during anaphase. Nucleosomes immediately adjacent to the centromere are diagrammed.

yeast cells unless they contain special telomeric (TEL) sequences ligated to their ends (Figure 9-58c). The first successful experiments involving transfection of yeast cells with linear plasmids were achieved by using the ends of a DNA molecule that was known to replicate as a linear molecule in the ciliated protozoan *Tetrahymena*. During part of the life cycle of *Tetrahymena*, much of the nuclear DNA is repeatedly copied in short pieces to form a so-called macronucleus. One of these repeated fragments was identified as a dimer of ribosomal DNA, the ends of which contained a repeated sequence $(G_4T_2)n$. When a section of this repeated TEL sequence was attached to the ends of linear yeast plasmids containing ARS and CEN, replication and good segregation of the plasmids occurred.

Considerable research has revealed that telomeric DNA has a characteristic type of sequence and is added to the ends of DNA molecules by a special enzyme termed *telomere terminal transferase*, or *telomerase*. The sequences of telomeres in a dozen or so organisms, including humans, have been determined; all are repetitive oligomers with a high G content in the strand that runs $5' \rightarrow 3'$ toward the telomere. For example, in addition to the G_4T_2 repeat found in *Tetrahymena*, a G_4T_4 repeat has been identified in the ciliated protozoan *Oxytricha*; a $G_{1-3}T$ repeat, in yeast; and an AGGGTT repeat, in humans and other vertebrates. These simple sequences are repeated at the very termini of chromosomes for a total of a few hundred base pairs in yeasts and protozoans and a few thousand base pairs in vertebrates. The 3' end of the G-rich strand extends 12–16 nucleotides beyond the 5' end of the complementary C-rich strand. This would normally suggest a single-stranded stretch of DNA in an otherwise duplex molecule. However, the telomere ends are not a simple duplex followed by a single-stranded region. Rather the single-stranded portion folds back on itself and, because of its peculiar G-rich nature, forms special (non-Watson-Crick) G–G base pairs (Figure 9-61). Certain proteins have been found to bind tightly to the folded-back structure at the very end of telomeres. These are thought to protect the ends of linear chromosomes from attack by exonucleases, which might otherwise degrade the free chromosomal ends.

Telomerases from several different ciliated protozoans have been isolated and characterized. Ciliated protozoans are a good source for isolation of this enzyme because they synthesize large amounts of it during the phase of their development when the macronucleus forms and large numbers of telomeres must be added to the ends of the multiple short DNA fragments. Telomerase from the protozoan *Tetrahymena* was discovered to add the G_4T_2 sequence found on *Tetrahymena* telomeres to the ends of a variety of telomeric primers from different organisms having different sequences. On the other hand, telomerase from the protozoan *Oxytricha* adds the G_4T_4 sequence found on *Oxytricha* telomeres. Thus the source of the enzyme and

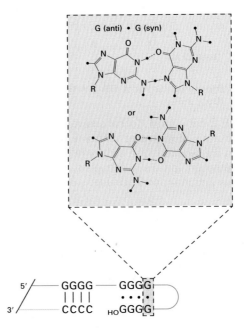

▲ FIGURE 9-61   Model for DNA structure in telomeres. The 3' end of the G-rich strand is folded back on itself and stabilized by unusual G-G base pairing in which one G residue is in the syn conformation rather than the anti conformation found in Watson-Crick base pairs. The mechanism by which telomerase lengthens telomeric sequences is described in Chapter 10.

not the sequence of the telomeric DNA primer determines the sequence of the DNA nucleotides added to the primer.

The mechanism underlying this specificity became evident when telomerases were discovered to be complexes of both protein and RNA. The sequence of the associated RNA serves as the template for the DNA sequence added to the ends of telomeres. This was proven by transforming *Tetrahymena* with a mutated form of the gene encoding the telomerase-associated RNA. The resulting telomerase added a DNA sequence complementary to the mutated RNA sequence to the ends of telomeric primers. Thus telomerase is a specialized form of a reverse transcriptase that carries its own internal RNA template to direct DNA synthesis.

This elaborate machinery for generating sequences at the end of chromosomes provides a mechanism for maintaining the length of linear chromosomes during replication. As discussed briefly in Chapter 4, DNA replication occurs at the growing fork where the two strands of the DNA helix have unwound. Because the polymerases that copy DNA can add nucleotides only in the $5' \rightarrow 3'$ direction and require a RNA primer, only one strand—the leading strand—can be synthesized continuously and in its entirety. The other strand—the lagging strand—is synthesized in short fragments that are ligated together after removal of the RNA primers (see Figure 4-22). Moreover, the lagging strand cannot be completed in its entirety all the way to the end of a linear duplex chromosome because RNA would have to prime synthesis at the very end of the lagging-strand template. As a result, successive rounds of replication would shorten chromosomes from their ends. By a process described in detail in Chapter 10, telomerase recognizes the ends of chromosomes and adds repetitive telomeric sequences, thereby providing a mechanism for complete synthesis of the lagging strand.

Yeast Artificial Chromosomes Can Be Used to Clone Megabase DNA Fragments

The research on circular and linear plasmids in yeast identified all the basic components of a yeast artificial chromosome (YAC). To construct YACs, TEL sequences from yeast cells or from the protozoan *Tetrahymena* are combined with yeast CEN and ARS sequences; to these are added DNA with selectable yeast genes and enough DNA from any source to make a total of more than 50 kb. (Smaller DNA segments do not work as well.) Such artificial chromosomes replicate in yeast cells and segregate almost perfectly, with only 1 daughter cell in 1000 to 10,000 failing to receive an artificial chromosome. During meiosis, the two sister chromatids of the artificial chromosome separate correctly to produce haploid spores.

Studies such as those depicted in Figure 9-58 strongly support the conclusion that yeast chromosomes, and probably all eukaryotic chromosomes, are linear, double-stranded DNA molecules with special regions—including centromere (CEN), telomere (TEL), and autonomously replicating sequences (ARSs)—that ensure replication and proper segregation. A technical point of considerable importance is that YACs can be used to clone very long chromosomal pieces from other species. For instance, YACs have been used extensively for cloning fragments of human DNA up to 1000 kb (1 megabase) in length, permitting the isolation of overlapping YAC clones that collectively encompass nearly the entire length of individual human chromosomes.

SUMMARY

In recent years, a wealth of new information about the DNA in the chromosomes of many different organisms has become available. In earlier times, genetic complementation analysis led to the definition of a gene as a unit of DNA that encodes one polypeptide (i.e., a cistron). As more and more was learned about DNA regulatory elements that control gene function in prokaryotes, it became clear that the expression of individual "genes" often depends on sequences that lie at some distance from them. Once modern molecular genetic techniques demonstrated multiple polyadenylation sites, multiple splicing sites, and other complexities in eukaryotic genes, the inadequacy of the definition of a gene by classical genetic tests (e.g., complementation) became clear.

The concept of the gene as a biological entity—that is, an inheritable DNA unit whose function is detected by observing the effect of a mutation—is still valid. However, according to the current molecular definition, a gene consists of *all* the DNA sequences necessary to produce a single protein or RNA product. Thus, the gene is no longer thought of as a single, contiguous protein-encoding stretch of DNA.

Bacterial genomes have little wasted space. Virtually all the DNA in a bacterial cell is packaged by abundant, small, nonspecific DNA-binding proteins and polyamines into a single compact, supercoiled ("folded") circular chromosome. Many of the protein-coding genes are arranged in operons, which are transcribed into polycistronic mRNAs. Little nonfunctional DNA is contained either within or between operons. The only repetitive sequences in prokaryotes are tandemly repeated ribosomal genes, insertion sequences (IS elements), and transposons. The only *E. coli* chromosomal DNA sequences that may serve no useful function are IS elements and some of the transposons.

Eukaryotic single-celled organisms such as yeast have only three to five times as much DNA per cell as prokaryotes. Probably only a small fraction of the genome of these lower eukaryotes is expendable. In contrast, the protein-coding genes of multicelled animals and plants make up a small fraction—in some cases as low as 1 percent—of the total cellular DNA. The eukaryotic transcription units encoding many proteins are islands separated by vast stretches of apparently nonfunctional DNA. Eukaryotic genomes also contain functional tandemly repeated genes encoding rRNA, tRNA, and histones.

The "extra," nonfunctional DNA in eukaryotes consists of introns, pseudogenes, simple-sequence DNA, mobile DNA elements (intermediate-repeat DNA), and unclassified spacer DNA between transcription units. The amount of simple-sequence DNA varies greatly among genomes that are otherwise comparable in size; this variation suggests that even if some simple-sequence DNA serves an important function, there may be vastly more than is required. Sequence analysis of numerous pseudo-genes has shown that these are derived from normal genes that have lost critical nucleotide sequences. Introns are present in some transcription units from all eukaryotic cell types; they abound in vertebrate genes and may constitute 80–90 percent of all the DNA within vertebrate transcription units. Introns perform no known function, and, like other extra DNA, they appear to be dispensable.

Mobile DNA elements are present in many eukaryotic cells. Some of these are similar in structure to bacterial transposons and like them move by a DNA-mediated mechanism. Most eukaryotic mobile elements, however, are retrotransposons, which move in the genome via RNA intermediates. Although most if not all mobile DNA elements have no useful function in the life cycle of organisms, they may affect evolution, as they introduce changes in genes and can serve as sites for homologous recombination leading to gene duplications and exon shuffling.

Functional DNA rearrangements in both prokaryotes and eukaryotes can occur by several mechanisms: inversion and deletion of DNA segments, gene conversion, and DNA amplification. Although these rearrangements play a role in controlling expression of a few genes, expression of the vast majority of genes is regulated by mechanisms that control the frequency of transcription initiation (Chapter 11).

Within eukaryotic cells, the nuclear DNA associates with histones, the most abundant proteins in eukaryotic nuclei, and small amounts of other proteins to form chromatin. Histones are basic proteins and have very similar amino acid sequences in all eukaryotes. The interaction of DNA and histones results in the compaction and spatial organization of the very large DNA molecule, whose length exceeds the cell diameter by a factor up to 10^5. Two copies each of histones H2A, H2B, H3, and H4 assemble into a histone octameric core around which a helical DNA molecule is wrapped, forming a nucleosome. In extended, beads-on-a-string chromatin, nucleosomes are separated by a stretch of linker DNA. Under certain conditions, histone H1 also associates with the DNA wrapped around nucleosome cores; the presence of H1 permits packing of chromatin into a condensed, solenoid form containing six nucleosomes per turn of the solenoid. Although not conclusively demonstrated, DNA that is not being transcribed probably exists in the solenoid form, and some structural change in chromatin packing (e.g., removal of H1 to gener-

ate the extended form) occurs when a region is transcribed. Finally, certain DNA sequences bind to a nonhistone protein scaffold, thereby dividing chromatin into long loops that extend outward from the scaffold. During mitosis, these long loops of chromosomal DNA, still attached at their base to the underlying protein scaffold, are condensed into a visible form.

Eukaryotic chromatin is further organized into chromosomes, hundreds to thousands of kilobases in length. Chromosomes are visible in the light microscope during mitosis and meiosis. The karyotype—the collective term for chromosome number, size, and shape—is species specific. Organisms with similar structures and degrees of complexity can differ greatly in their karyotypes and in the total amount of DNA in their genomes. Differences in the amount of DNA present in closely related organisms are due principally to variations in the amount of spacer and repetitious DNA. The staining of chromosomes with dyes after various treatments produces light and dark bands at characteristic places along their length. These bands, which can be resolved in the light microscope, are the landmarks by which cytogeneticists prepared early physical maps of the chromosomes of many species. Much more detailed physical maps can now be constructed based on the location of individual genes observed by in situ hybridization of cloned DNA probes. Together with extensive DNA sequencing these approaches will ultimately lead to complete mapping and sequencing of the human genome.

Each eukaryotic chromosome contains a linear DNA molecule ranging in length from 5×10^5 bp (smallest yeast chromosome) to 2×10^8 bp (human chromosome 1). Studies on plasmid transfection into yeast cells has shown that replication of linear chromosomes requires three DNA elements: autonomously replicating sequences (ARSs), which function as replication origins; a centromeric (CEN) sequence, to which the spindle attaches at nuclear division (mitosis or meiosis); and telomeric (TEL) sequences at each end, which allow replication of a linear DNA molecule in its entirety. Yeast artificial chromosomes (YACs) containing these three elements plus selectable markers replicate and segregate like normal chromosomes. Because YACs containing large DNA fragments up to 10^6 base pairs (a megabase) in length can be cloned, they are a powerful tool for studying chromosomes of higher eukaryotes.

REVIEW QUESTIONS

1. Review the concept of a karyotype. Are karyotypes based only on the number of chromosomes in a species, or are chromosomal shapes and sizes also important parameters? What is meant by a C value and the C-value paradox?

a. You have obtained the karyotype from a species of fruit fly, *Drosophila grimshawi*. This particular organism has five chromosomes. You must also obtain the karyotype of a related species, *Drosophila melanogaster*. Prior to obtaining a chromosomal spread for *D. melanogaster,* would you predict that the *D. grimshawi* karyotype would be similar or identical to it? Explain your answer.

b. Molecular biology has revolutionized the studies of phylogeny and evolutionary biology in that comparisons of chromosomal organization and structure at the DNA sequence level can be made among species. Therefore, not only can gross structural features—wing shape or body size, for example—and karyotypes be used to classify organisms, but the actual organization and DNA sequence of the genomes of different species can be employed to categorize them.

A group of genes in each *Drosophila* species encodes yolk proteins, also known as vitellogenins. How would you compare the similarity or differences between the vitellogenin genes in *D. grimshawi* and *D. melanogaster?* Be specific about the approaches you might take, and discuss the advantages and limitations of each in assessing the interspecies comparisons you wish to make. In terms of genes that encode proteins, do you think that you would need to be concerned about exon versus intron comparisons among species?

2. As indicated in this chapter, only a small percentage of the DNA in higher eukaryotes is actually transcribed or necessary for replication and other chromosomal functions; much of the genome has no apparent purpose. Prepare a list of the types of DNA that are actually transcribed. Review the nature of solitary protein-coding genes, gene families, and tandemly repeated genes. What are the advantages of arranging transcribed genes in these three different ways? Are all the members in each class capable of being transcribed, or are some silent?

Compare and contrast simple sequence DNA, SINES, and LINES. What are the functions of each? Are these types of DNA transcribed?

a. The use of the drug methotrexate is important in cancer therapy. This particular compound inhibits the enzyme dihydrofolate reductase (dhfr), which is essential in the biochemical pathway that ultimately produces the thymine necessary for DNA synthesis, resulting in cell death. After continued exposure to methotrexate, resistance to the drug can develop in cells, and they no longer die. It has been shown that the *dhfr* gene becomes amplified in some cells that become resistant to the drug, meaning that many copies of the *dhfr* gene are somehow made. Often, these copies are integrated into a chromosome and are arranged in tandem in a repeated pattern. How do you think that this might enable a cell to become resistant to methotrexate? Figure 9-7 may help answer this question.

b. Referring to Figure 9-6, suggest a mechanism that could account for amplification of the *dhfr* gene.

3. Review the features of mobile elements, including the mechanisms by which they move. Be certain to understand which organism is associated with each mobile element presented in the chapter.

Two features that are often found on mobile elements are the short direct repeats and the longer inverted repeats.

a. Refer to Figure 9-16 in this chapter that illustrates an IS element, including the inverted repeats and short direct repeats. Are these inverted repeats 100 percent homologous, or are they nearly homologous regions? Concentrate on only the upper strand of DNA in the portion of the illustration showing the inverted repeats. Do you think that the region containing the inverted repeat within the upper strand could base pair with the inverted repeat in the same strand if the DNA located between them could loop out? Would you still have the requisite antiparallel structure in this new base-paired region? Could such a structure form in the opposite strand?

b. Refer again to Figure 9-16. Are the short direct repeats 100% homologous or are there sequence differences? Based on the proposed mechanism accounting for the formation of the short direct repeats, do you think homology would always occur? Explain your answer.

4. Review the different types of transcription units discussed in this chapter. What are the differences between polycistronic and monocistronic mRNAs? In what general classes of organisms are each of these found? What is meant by simple versus complex transcription units in eukaryotes?

a. As you know, P elements are mobile elements found in *Drosophila*. They consist of inverted repeats having a segment of DNA located between them that encodes a transposase. Assume you are studying a P element having four exons present in this region. A primary transcript is made from this DNA in most *Drosophila* cells. In the germ cells, an active transposase results, but in somatic cells, a shorter, inactive protein is made that possesses no activity. What mechanisms might account for the ability of one primary transcript ultimately to encode two different protein products?

b. Based on the above information, when do you think transposition events occur in *Drosophila* carrying this P element? That is, would they occur in the somatic cells of the adult fly, or would they occur in the germ cells? What does this tell you about the ability to detect the effects of transposition in the fly genome having this particular P element?

References

Protein-Coding Genes and Gene Families

CLEVELAND, D. W., and K. F. SULLIVAN. 1985. Molecular biology and genetics of tubulin. *Ann. Rev. Biochem.* **54**:331–365.

EFSTRATIADIS, A., et al. 1980. The structure and evolution of the human β-globin gene family. *Cell* **21**:653–668.

GORIN, M. B., et al. 1981. The evolution of α-fetoprotein and albumin. I: A comparison of the primary amino acid sequences of mammalian α-fetoprotein and albumin. *J. Biol. Chem.* **256**:1954–1959.

HENTSCHEL, C. C., and M. L. BIRNSTIEL. 1981. The organization and expression of histone gene families. *Cell* **25**:301–313.

KAFATOS, F. C. 1983. Structure, evolution and developmental expression of the chorion multigene families in silk moths and *Drosophila.* In S. Subtelny and F. C. Kafatos, eds. *Gene Structure and Regulation in Development,* Alan R. Liss.

MIRKOVITCH, J., P. SPIERER, and U. K. LAEMMLI. 1986. Genes and loops in 320,000 base-pairs of the *Drosophila melanogaster* chromosome. *J. Mol. Biol.* **190**:255–258.

NATHANS, J. 1987. Molecular biology of visual pigments. *Ann. Rev. Neurosci.* **10**:163–194.

OHTA, T. 1983. On the evolution of multigene families. *Theor. Popul. Biol.* **23**:216–240.

PIATIGORSKY, J. 1984. Lens crystallins and their gene families. *Cell* **38**:620–621.

ROYAL, A., et al. 1979. The ovalbumin gene region: common features in the organization of three genes expressed in chicken oviduct under hormonal control. *Nature* **279**:125–132.

STEINEET, P. M., A. C. STEVEN, and D. R. ROOP. 1985. The molecular biology of intermediate filaments. *Cell* **42**:411–419.

Tandemly Repeated Genes Encoding rRNA

FEDEROFF, N. 1979. On spacers. *Cell* **16**:697–710.

WELLAUER, P. K., and I. B. DAWID. 1979. Isolation and sequence organization of human ribosomal DNA. *J. Mol. Biol.* **128**:289–303.

Repetitious DNA

BRITTEN, R. J., and D. E. KOHNE. 1968. Repeated sequences in DNA. *Science* **161**:529–540.

DEININGER, P. L. 1989. SINES: Short interspersed repeated DNA elements in higher eukaryotes. In Berg, D. E. and M. M. Howe, eds. *Mobile DNA*, American Society for Microbiology pp. 593–618.

DOOLITTLE, W. F., and C. SAPIENZA. 1980. Selfish genes, the phenotype paradigm and genome evolution. *Nature* **284**:601–603.

GALL, J. G. 1981. Chromosome structure and the C-value paradox. *J. Cell Biol.* **91**:3s–14s.

JEFFREYS, A. J., et al. 1988. Spontaneous mutation rates to new length alleles at tandem-repetitive hypervariable loci in human DNA. *Nature* **332**:278–280.

JEFFREYS, A. J., V. WILSON, and S. C. THEIN. 1985. Individual-specific fingerprints of human DNA. *Nature* **316**:76–78.

KORNBERG, J. R., and M. C. RYKOWSKI. 1988. Human genome organization: *Alu*, LINES, and the molecular structure of metaphase chromosome bands. *Cell* **53**:391–400.

MIKLOS, G. L. G., et al. 1988. Microcloning reveals a high frequency of repetitive sequences characteristic of chromosome 4 in the β-heterochromatin of *Drosophila*. *Proc. Nat'l. Acad. Sci. USA* **85**:2051–2055.

MOYZIS, R. K., et al. 1987. Human chromosome-specific repetitive DNA sequences: novel markers for genetic analysis. *Chromosoma* **95**:375–386.

SINGER, M. F. 1982. SINES and LINES: highly repeated short and long interspersed sequences in mammalian genomes. *Cell* **28**:433–434.

SINGER, M. F., V. KREK, J. P. MCMILLAN, G. D. SWERGOLD, and R. E. THAYER. 1993. LINE1: a human transposable element. *Gene* **135**:183–188.

Transposition and Transposable Elements

BENJAMIN, H. W., and N. KLECKNER. 1989. Intramolecular transposition by Tn10. *Cell* **59**:373–383.

BENJAMIN, H. W., and N. KLECKNER. 1992. Excision of Tn10 from the donor site during transposition occurs by flush double-strand cleavages at the transposon termini. *Proc. Nat'l. Acad. Sci. USA* **89**:4648–4652.

BERG, D. E., and M. M. HOWE, eds. 1989. *Mobile DNA*. American Society for Microbiology.

CALOS, M. P., and J. H. MILLER. 1980. Transposable elements. *Cell* **20**:579–595.

CAMERON, J. R., E. Y. LOH, and R. W. DAVIS. 1979. Evidence for transposition of dispersed *Repetitive* DNA families in yeast. *Cell* **16**:739–751.

ENGELS, W. R. 1989. P elements in *Drosophila melanogaster*. In D. E. Berg and M. M. Howe, eds., *Mobile DNA*, American Society for Microbiology, pp. 437–484.

FEDEROFF, N. V. 1989. About maize transposable elements and development. *Cell* **56**:181–191.

———. 1989. Maize transposable elements. In D. E. Berg and M. M. Howe, eds., *Mobile DNA*, American Society for Microbiology, pp. 375–412.

FINNEGAN, D. J., and D. H. FAWCETT. 1986. Transposable elements in *Drosophila melanogaster*. *Oxford Surveys on Eukaryotic Genes* **3**:1–62.

GALAS, D. J., and M. CHANDLER. 1989. Bacterial insertion sequences. In D. E. Berg and M. M. Howe, eds., *Mobile DNA*, American Society for Microbiology, pp. 109–162.

KLECKNER, N. 1981. Transposable elements in prokaryotes. *Ann Rev. Genet.* **15**:341–404.

MCCLINTOCK, B. 1956. Controlling elements and the gene. *Cold Spring Harbor Symp. Quant. Biol.* **21**:197–216.

O'HARE, K., and G. RUBIN. 1983. Structures of P transposable elements and their sites of insertion and excision in the *Drosophila melanogaster* genome. *Cell* **34**:25–35.

RUBIN, G. M., and A. C. SPRADLING. 1982. Genetic transformation of *Drosophila* with transposable element vectors. *Science* **218**:348–353.

SHAPIRO, J. A. 1979. Molecular model for the transposition and replication of bacteriophage Mu and other transposable elements. *Proc. Nat'l. Acad. Sci. USA* **76**:1933–1937.

SHAPIRO, J. A., ed. 1983. *Mobile Genetic Elements.* Academic Press.

SIMON, M., and I. HERSKOWITZ, eds. 1985. Genome rearrangement. In *UCLA Symposia on Molecular and Cellular Biology*, Vol. 20, Alan R. Liss.

SPRADLING, A. C., and G. M. RUBIN. 1982. Transposition of cloned P elements into *Drosophila* germ line chromosomes. *Science* **218**:341–347.

Retrotransposons

BINGHAM, P. M., and Z. ZACHAR. 1989. Retrotransposons and the FB transposon from *Drosophila melanogaster*. In D. E. Berg and M. M. Howe, eds., *Mobile DNA*, American Society for Microbiology, pp. 485–502.

BOEKE, J. D., et al. 1985. Ty elements transpose through an RNA intermediate. *Cell* **40**:491–500.

EVANS, J. P., and R. D. PALMITER. 1991. Retrotransposition of a mouse L1 element. *Proc. Nat'l. Acad. Sci. USA* **88**:8792–8796.

FANNING, T. G., and M. F. SINGER. 1987. LINE-1: a mammalian transposable element. *Biochem. Biophys. Acta* **910**:203–212.

GARFINKEL, D., J. BOEKE, and G. R. FINK. 1985. Ty element transposition: reverse transcriptase and virus-like particles. *Cell* **42**:507–517.

HAYNES, S. R., and W. R. JELINEK. 1981. Low molecular weight RNAs transcribed *in vitro* by RNA polymerase III from *Alu*-type dispersed repeats in Chinese hamster DNA are also found *in vivo*. *Proc. Nat'l. Acad. Sci. USA* **78**:6130–6134.

JAGADEESWARAN, P., B. G. FORGET, and S. M. WEISSMAN. 1981. Short interspersed repetitive DNA elements in eucaryotes: transposable DNA elements generated by reverse transcription of RNA Pol III transcripts? *Cell* **26**:141–142.

JELINEK, W. R., and C. W. SCHMID. 1982. Repetitive sequences in eukaryotic DNA and their expression. *Ann. Rev. Biochem.* **51**:813–844.

JURKA, J., and T. SMITH. 1988. A fundamental division in the *Alu* family of repeated sequences. *Proc. Nat'l. Acad. Sci. USA* **85**:4775–4778.

KINGSMAN, A. J., and S. M. KINGSMAN. 1988. Ty: a retroelement moving forward. *Cell* **53**:333–335.

MINAKAMI, R., et al. 1992. Identification of an internal *cis*-element essential for the human L1 transcription and nuclear factor(s) binding to the element. *Nucl. Acids Res.* **20**:3139–3145.

MOUNT, S. M., and G. M. RUBIN. 1985. Complete nucleotide sequence of the *Drosophila* transposable element *copia*: homology between *copia* and retroviral proteins. *Mol. Cell Biol.* **5**:1630–1638.

NARITA, N., et al. 1993. Insertion of a 5′ truncated L1 element into the 3′ end of exon 44 of the dystrophin gene resulted in skipping of the exon during splicing in a case of Duchenne muscular dystrophy. *J. Clin. Invest.* **91**:1862–1867.

VAN ARSDELL, S. W., et al. 1981. Direct repeats flank three small nuclear RNA pseudogenes in the human genome. *Cell* **26**:11–17.

WALLACE, M. R., et al. 1991. A de novo *Alu* insertion results in neurofibromatosis type 1. *Nature* **353**:864–868.

WILLIAMSON, V. M., E. T. YOUNG, and M. CIRIACY. 1981. Transposable elements associated with constitutive expression of yeast alcohol dehydrogenase II. *Cell* **23**:605–614.

YOUNG, M. W. 1982. Differing levels of dispersed repetitive DNA among closely-related species of *Drosophila*. *Proc. Nat'l. Acad. Sci. USA* **79**:4570–4574.

Functional Genomic Rearrangements

BORST, P., and D. R. GREAVES. 1987. Programmed gene rearrangements altering gene expression. *Science* **235**:658–667.

GOLDEN, J. W., S. J. ROBINSON, and R. HASELKORN. 1985. Rearrangement of nitrogen fixation genes during heterocyst differentiation in the cyanobacterium *Anabaena*. *Nature* **314**:419–423.

Salmonella Phase Variation

HEICHMAN, K. A., and R. C. JOHNSON. 1990. The Hin invertasome: protein-mediated joining of distant recombination sites at an enhancer. *Science* **249**:511–517.

HUGHES, K. T., P. YOUDERIAN, and M. I. SIMON. 1988. Phase variation in *Salmonella*: analysis of Hin recombinase and *hix* recombination site interaction in vivo. *Genes & Develop.* **2**:937–948.

SILVERMAN, M., and M. SIMON. 1980. Phase variation: genetic analysis of switching mutants. *Cell* **19**:845–854.

ZIEG, J., et al. 1980. Recombination switch for gene expression. *Science* **196**:170–172.

Yeast Mating-Type Switching

HERSKOWITZ, I., and O. OSHIMA. 1982. Control of cell type in *S. cerevisiae*: mating type and mating type interconversions. In J. N. Strathern, E. W. Jones, and J. R. Broach, eds., *Molecular Biology of the Yeast* Saccharomyces, Cold Spring Harbor Laboratory.

KOSTRIKEN, R., et al. 1983. A site-specific endonuclease essential for mating-type switching in *Saccharomyces cerevisiae*. *Cell* **35**:167–174.

NASMYTH, K. A. 1983. Molecular analysis of a cell lineage. *Nature* **302**:670–676.

NASMYTH, K. A. 1987. The determination of mother cell–specific mating type switching in yeast by a specific regulator of *HO* transcription. *EMBO J.* **6**:243–248.

STILLMAN, D. J., et al. 1988. Characterization of a transcription factor involved in mother cell specific transcription of yeast *HO* gene. *EMBO J.* **7**:485–494.

STRATHERN, J. N., E. W. JONES, and J. R. BROACH, eds. 1982. *Molecular Biology of the Yeast* Saccharomyces. Cold Spring Harbor Laboratory.

STRUHL, K. 1983. The new yeast genetics. *Nature* **305**:391–397.

Antigenic Variation in Trypanosomes

BORST, P. 1986. Discontinuous transcription and antigenic variation in trypanosomes. *Ann. Rev. Biochem.* **55**:701–732.

SHEA, C. M., G.-S. LEE, and L. H. T. VAN DER PLOEG. 1987. VSG gene 118 is transcribed from a cotransposed pol I–like promoter. *Cell* **50**:603–612.

VAN DER PLOEG, L. H. T. 1986. Discontinuous transcription and splicing in trypanosomes. *Cell* **47**:479–480.

Chromosomal Proteins: Bacterial Chromatin Proteins, Histones, and Nucleosomes

ARENTS, G., and E. N. MOUDRIANAKIS. 1993. Topography of the histone octamer surface: repeating structural motifs utilized in the docking of nucleosomal DNA. *Proc. Nat'l. Acad. Sci. USA* **90**:10489–10493.

DE BERNARDIN, W., T. KOLLER, and J. M. SOGO. 1986. Structure of *in vivo* transcribing chromatin as studied in simian virus 40 minichromosomes. *J. Mol. Biol.* **191**:469–482.

DILWORTH, S. M., S. J. BLACK, and R. A. LASKEY. 1987. Two complexes that contain histones are required for nucleosome assembly in vitro: role of nucleoplasmin and N1 in *Xenopus* egg extracts. *Cell* **51**:1009–1018.

EARNSHAW, W. C., and M. M. S. HECK. 1985. Localization of topoisomerase II in mitotic chromosomes. *J. Cell Biol.* **100**:1716–1725.

EARNSHAW, W. C., et al. 1985. Topoisomerase II is a structural component of mitotic chromosome scaffolds. *J. Cell Biol.* **100**:1706–1715.

FELSENFELD, G., and J. D. MCGHEE. 1986. Structure of the 30 nm chromatin fiber. *Cell* **44**:375–377.

GASSER, S. M., and U. K. LAEMMLI. 1987. A glimpse at chromosomal order. *Trends Genet.* **3**:16–22.

GOODWIN, G. H., J. M. WALKER, and E. W. JOHNS. 1978. The high mobility group (HMG) non-histone chromosomal proteins. In H. Busch, ed., *The Cell Nucleus*, Vol. 6, Academic Press.

GRAZIANO, V., S. E. GERCHMAN, D. K. SCHNEIDER, and V. RAMAKRISHNAN. 1994. Histone H1 is located in the interior of the chromatin 30-nm filament. *Nature* **368**:351-354.

HAGGREN, W., and D. KOLODRUBETZ. 1988. The *Saccharomyces cerevisiae* ACP2 gene encodes an essential HMG1-like protein. *Mol. Cell Biol.* **8**:1282–1289.

HEBBES, T. R., A. W. THORNE, and C. CRANE-ROBINSON. 1988. A direct link between core histone acetylation and transcriptionally active chromatin. *EMBO J.* **7**:1395–1402.

PHI-VAN, L., and W. H. STRATLING. 1988. The matrix attachment regions of the chicken lysozyme gene co-map with the boundaries of the chromatin domain. *EMBO J.* **7**:655–664.

RICHARD, T. J., et al. 1984. Structure of the nucleosome core particle at 7Å resolution. *Nature* **311**:532–537.

RICHMOND, T. J., M. A. SEARLES, and R. T. SIMPSON. 1988. Crystals of a nucleosome core particle containing defined sequence DNA. *J. Mol. Biol.* **199**:161–170.

SCHMID, M. B. 1990. More than just "histone-like" proteins. *Cell* **63**:451–453.

SOLOMON, M. J., P. L. LARSEN, and A. VARSHAVSKY. 1988. Mapping protein-DNA interactions in vivo with formaldehyde: evidence that histone H4 is retained on a highly transcribed gene. *Cell* **53**:937–947.

TRAVERS, A. A., and A. KLUG. 1987. The bending of DNA in nucleosomes and its wider implications. *Phil. Trans. Roy. Soc. (Lond.)* **B317**:537–561.

WIDOM, J., and A. KLUG. 1985. Structure of the 300 Å chromatin filament: x-ray diffraction from oriented samples. *Cell* **43**:207–213.

Chromosomal DNA Content, Scaffolding, Loops, and Banding Patterns

BOY DE LA TOUR, E., and U. K. LAEMMLI. 1988. The metaphase scaffold is helically folded: sister chromatids have predominantly opposite helical handedness. *Cell* **55**:937–944.

GALL, J. G. 1981. Chromosome structure and the C-value paradox. *J. Cell Biol.* **91**:3s–14s.

KAVENOFF, R., L. C. KLOTZ, and B. H. ZIMM. 1974. On the nature of chromosome-sized DNA molecules. *Cold Spring Harbor Symp. Quant. Biol.* **38**:1–8.

KELLUM, R., and P. SCHEDL. 1991. A position-effect assay for boundaries of higher order chromosomal domains. *Cell* **64**:941–950.

LUDERUS, M. E. E., et al. 1992. Binding of matrix attachment regions to lamin B_1. *Cell* **70**:949–959.

MACGREGOR, H. C., and J. M. VARLEY. 1983. *Working with Animal Chromosomes*. Wiley.

MIRKOVITCH, J., M.-E. MIRAULT, and U. K. LAEMMLI. 1984. Organization of the higher-order chromatin loop: specific DNA attachment sites on nuclear scaffold. *Cell* **39**:223–232.

MIRKOVITCH, J., P. SPIERER, and U. K. LAEMMLI. 1986. Genes and loops in 320,000 base-pairs of the *Drosophila melanogaster* chromosome. *J. Mol. Biol.* **190**:255–258.

SAWYER, J. R., and J. C. HOZIER. 1986. High resolution of mouse chromosomes: banding conservation between man and mouse. *Science* **232**:1632–1635.

SPIERER, P., et al. 1983. Molecular mapping of genetic and chrommometric units in *Drosophila melanogaster*. *J. Mol. Biol.* **168**:35–50.

Autonomously Replicating Sequences, Centromeres, and Telomeres

BLACKBURN, E. H. 1991. Structure and function of telomeres. *Nature* **350**:569–773.

BLACKBURN, E. H., and J. W. SZOSTAK. 1984. The molecular structure of centromeres and telomeres. *Ann. Rev. Biochem.* **53**:163–194.

CARBON, J., and L. CLARK. 1990. Centromere structure and function in budding and fission yeasts. *New Biologist* **2**:10–19.

CLARKE, L., and J. CARBON. 1985. The structure and function of yeast centromeres. *Ann. Rev. Genet.* **19**:29–56.

EARNSHAW, W. C., and J. E. TOMKIEL. 1992. Centromere and kinetochore structure. *Curr. Opin. Cell Biol.* **4**:86–93.

GREIDER, C. W., and E. H. BLACKBURN. 1987. The telomere terminal transferase of *Tetrahymena* is a ribonucleoprotein enzyme with two kinds of primer specificity. *Cell* **51**:887–898.

HEGEMANN, J. H., et al. 1988. Mutational analysis of centromere DNA from chromosome VI of *Saccharomyces cerevisiae*. *Mol. Cell Biol.* **8**:2523–2535.

HENDERSON, E., et al. 1987. Telomeric DNA oligonucleotides form novel intramolecular structures containing guanine-guanine base pairs. *Cell* **51**:899–908.

HYMAN, A A., et al. 1992. Microtubule-motor activity of a yeast centromere-binding protein complex. *Nature* **359**:533–536.

LECHNER, J., and J. CARBON. 1991. A 240-kd multisubunit protein complex, CBF3, is a major component of the budding yeast centromere. *Cell* **64**:717–725.

LUNDBLAD, U., and J. W. SZOSTAK. 1989. A mutant with a defect in telomere elongation leads to senescence in yeast. *Cell* **57**:633–643.

MORIN, G. B. 1991. The human telomere terminal transferase enzymes is a ribonucleoprotein that synthesizes TTAGGG repeats. *Cell* **59**:521–529.

MURRAY, A. W., and J. W. SZOSTAK. 1983. Construction of artificial chromosomes of yeast. *Nature* **305**:189–193.

SHAMPAY, J., and E. H. BLACKBURN. 1988. Generation of telomere-length heterogeneity in *Saccharomyces cerevisiae*. *Proc. Nat'l. Acad. Sci. USA* **85**:534–538.

YU, G.-L., et al. 1990. *In vivo* alteration of telomere sequences and senescence caused by mutated *Tetrahymena* telomerase RNAs. *Nature* **344**:126–132.

ZAKIAN, V. A. 1989. Structure and function of telomeres. *Ann. Rev. Genet.* **23**:579–604.

Yeast Artificial Chromosomes as Cloning Vectors

BURKE, D. T., G. F. CARLE, and M. V. OLSON. 1987. Cloning of large exogenous DNA into yeast by means of artificial chromosome vectors. *Science* **236**:806–812.

COHEN, D., I. CHUMAKOV, and J. WEISSENBACH. 1993. A first-generation physical map of the human genome. *Nature* **366**:698–701.

LARIN, A., A. P. MONACO, and H. LEHRACH. 1991. Yeast artificial chromosome libraries containing large inserts from mouse and human DNA. *Proc. Nat'l. Acad. Sci. USA* **88**:4123–4127.

10 DNA Replication, Repair, and Recombination

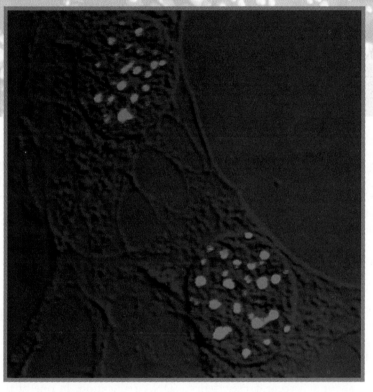

▲ DNA methyltransferase localizes to sites of DNA replication in 3T3 fibroblasts.

Before Watson and Crick discovered the duplex structure of DNA, one of the most mysterious aspects of biology was how genetic material was duplicated from one generation to the next. Recognition of the base-paired nature of the DNA duplex immediately gave rise to the notion that a template was involved in the transfer of information between generations, but a host of structural and biochemical questions soon followed. When replication has been completed, are old strands paired with new, or old with old and new with new? How does replication begin, and how does it progress along the chromosome? What mechanisms ensure only one round of replication before cell division? What enzymes take part in DNA synthesis, and what are their functions? How does duplication of the long helical duplex occur without the strands becoming tangled?

The events that allow assembly of new DNA chains are described collectively as DNA *synthesis*. DNA *replication* (or chromosomal replication), a more comprehensive term, encompasses not only DNA chain synthesis but also its initiation and termination. Studies of synthesis focus on enzymes and accessory factors at the point of chain growth. Studies of replication are in addition concerned with how DNA synthesis starts and stops in such a way that each chromosome is duplicated exactly. Another crucial issue in replication is the separation of the two new chromosomes. We discuss these processes in this chapter. The events of mitosis (or meiosis) during which the chromosomes are distributed to daughter cells are introduced in Chapter 5 and discussed in detail in Chapters 23 and 25.

For DNA to serve as the genetic link between generations, not only must the base sequence be copied correctly during replication, but the integrity of the sequence must be maintained continuously for accurate protein synthesis in all cells. As a consequence, it is not too surprising that cells possess enzymatic "repair" functions to keep DNA sequences accurate. Some of the enzymes that cut and repair damaged DNA also participate in genetic recombination, a mechanism by which exchange of duplex DNA sequences occur. The mechanisms of DNA repair and recombination are discussed in the later sections of this chapter.

► General Features of Chromosomal Replication

In this section, we consider several general features of DNA replication including its semiconservative nature, the bidirectional growth of new DNA strands from a common point of origin, and the common properties of DNA replication origins.

DNA Replication Is Semiconservative

The base-pairing principle inherent in the Watson-Crick model suggested that the two new DNA chains were copied from the two old chains. Although this mechanism provided for exact copying of genetic information, it raised a new question: Is replication a *conservative* process in which the two new strands form a new duplex and the old duplex remains intact? Or is replication a *semiconservative* process in which each old strand becomes paired with a new strand copied from it. This question was settled first for *E. coli* with proof that each newly formed duplex contains one old and one new DNA chain.

When *E. coli* cells are grown in ammonium salts containing ^{15}N instead of ^{14}N, the "heavy" atoms are incorporated into deoxyribonucleotide precursors and then into DNA. The resulting DNA is about 1 percent denser than the DNA containing only the normal isotope. Such "heavy" DNA and normal ("light") DNA form discrete bands during equilibrium density-gradient (isopyncnic) centrifugation in solutions of CsCl. Bacteria grown for several generations in a heavy medium will contain almost completely substituted (i.e., heavy) DNA. If cells containing all heavy DNA are then transferred to a normal (light) medium and allowed to replicate once, the resulting DNA has a density that is intermediate between the densities of heavy and light DNAs. After the cells undergo another doubling, their DNA consists of half heavy-light DNA and half completely light DNA (Figure 10-1).

This experiment was refined further by denaturing and then separating the long strands of heavy-light DNA into single strands. If these single strands are subjected to equilibrium centrifugation, two distinct bands are formed, showing that each of the long strands is either all heavy (H) or all light (L). These experiments showed that copying of the two original heavy strands in light medium yielded: two H-L duplexes; copying of this mixed duplex in light

medium led to H-L and L-L duplexes. This is *semiconservative* replication.

Experiments of a different design first demonstrated semiconservative DNA replication in eukaryotic chromosomes. Cultured cells were labeled during DNA synthesis with [^{3}H]thymidine. The *mitotic* chromosomes then were examined by autoradiography beginning at the first division after labeling, which occurred within a few hours. The autoradiographs revealed that all mitotic chromosomes were labeled in *both* chromatids (Figure 10-2). Presumably each chromatid represented one double-stranded DNA molecule, with one DNA strand in each chromatid being new (labeled) and one being old.

This interpretation was greatly strengthened by autoradiographs made after one further round of cell division in the absence of labeled thymidine. At this point one chromatid in each mitotic pair was labeled and one was largely unlabeled—a result consistent with the semiconservative model; the labeled chromatid contained the old labeled strand, and the unlabeled chromatid, the old unlabeled strand (see Figure 10-2b). Thus each labeled strand retained its label and was paired after a second replication with a new unlabeled strand. Evidence of semiconservative replication has been obtained with both plant and animal chromosomes. Apparently all cellular DNA in both prokaryotic and eukaryotic cells is replicated by a semiconservative mechanism.

Most DNA Replication Is Bidirectional

How does the DNA replication machinery proceed along a duplex? Several possible molecular mechanisms of DNA synthesis (chain growth) would allow semiconservative DNA replication. In one of the simplest possibilities, one new strand derives from one origin and the other new strand derives from another origin (Figure 10-3a). Only one strand of the duplex grows at each growing point. This mechanism does, in fact, occur in linear DNA viruses such as adenovirus. In such circumstances, the ends of the DNA molecules serve as fixed sites for the initiation and termination of replication. A second possibility envisions one origin and one *growing fork* (the fork of synthesis), which moves along the DNA in one direction with both strands of DNA being copied (Figure 10-3b). A third possibility is that synthesis might start at a single site and proceed in both directions, so that both strands are copied at each of *two* growing forks (Figure 10-3c). The available evidence suggests that the third alternative is the most common: DNA replication proceeds in both directions—that is, *bidirectionally*—from a given starting site, with both strands being copied at each fork. Thus two growing forks emerge from a single origin site.

In a circular chromosome (the form found in bacteria, plasmids, and some viruses), one origin often suffices, and

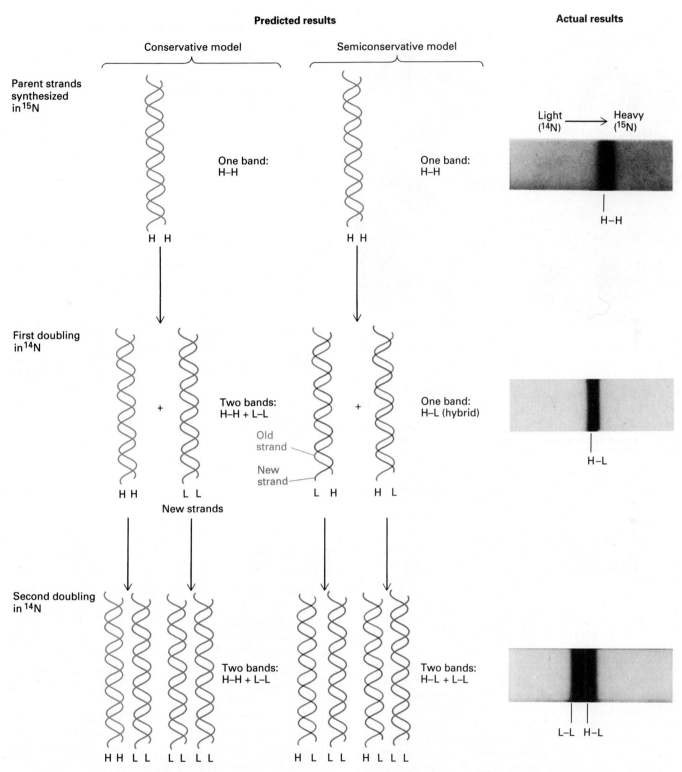

▲ FIGURE 10-1   The Meselson-Stahl experiment showed that DNA replication is semiconservative (i.e., each daughter duplex contains one new and one old strand) not conservative (i.e., one daughter duplex contains two old strands and the other duplex contains two new strands). *E. coli* cells were grown in a medium containing (^{15}N)ammonium salts until all the cellular DNA contained the "heavy" isotope. The cells were then transferred to a medium containing the normal "light" isotope (^{14}N). Samples were removed from the cultures periodically and analyzed by density-gradient equilibrium centrifugation in CsCl to separate heavy-heavy (H-H), light-light (L-L), and heavy-light (H-L) duplexes into distinct bands. The actual banding patterns were consistent with the semiconservative mechanism. [See M. Meselson and W. F. Stahl, 1958, *Proc. Nat'l. Acad. Sci. USA* **44**:671; photographs courtesy of M. Meselson.]

(a) Metaphase chromosomes after pulse-labeling

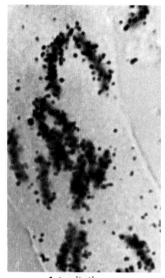
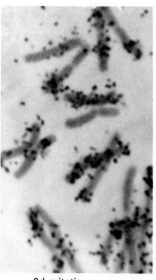

1st mitotic
division (≈8 h)

2d mitotic
division (≈24 h)

(b) Expected results with semiconservative replication

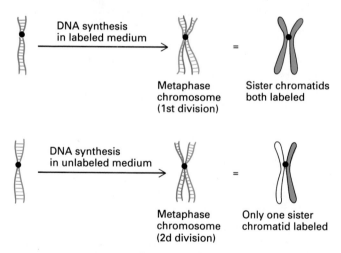

DNA synthesis
in labeled medium →

Metaphase
chromosome
(1st division)

Sister chromatids
both labeled

DNA synthesis
in unlabeled medium →

Metaphase
chromosome
(2d division)

Only one sister
chromatid labeled

▲ FIGURE 10-2   Experimental demonstration of semiconservative DNA replication in root cells of the lily (*Bellavalia
romana*). Growing root cells were pulse-labeled with
[³H]thymidine; samples then were taken periodically, stained,
and autoradiographed. Mitotic cells with metaphase chromosomes were first evident about 8 h after the labeling period.
(a) Autoradiograph of metaphase chromosomes during the
first division after labeling shows that *both* sister chromatids
are labeled, whereas during the second mitotic division, only
one sister chromatid of each pair is heavily labeled. Because
of sister chromatid exchange of whole segments of DNA
during the second mitosis, some otherwise unlabeled chromatids have patches of silver grains. (b) Diagram illustrates
how semiconservative replication would produce the
observed results. [See J. H. Taylor, P. S. Woods, and W. L.
Hughes, 1957, *Proc. Nat'l. Acad. Sci. USA* **43**:122; photographs courtesy of J. H. Taylor.]

(a) Unidirectional growth of single strands from two origins

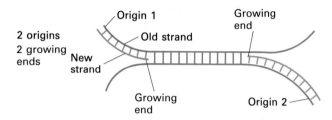

Origin 1

Growing
end

2 origins
2 growing
ends

Old strand

New
strand

Growing
end

Origin 2

(b) Unidirectional growth of both strands from one origin

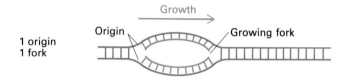

Growth

Origin

Growing fork

1 origin
1 fork

(c) Bidirectional growth of both strands from one origin

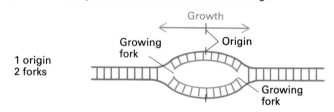

Growth

Growing
fork

Origin

1 origin
2 forks

Growing
fork

▲ FIGURE 10-3   Three mechanisms of DNA-strand
growth that are consistent with semiconservative replication.
The third mechanism—bidirectional growth of both strands
from a single origin—appears to be the most common in
both eukaryotes and prokaryotes.

the two resulting growing forks merge on the opposite side
of the circle to complete replication (see Figure 7-2). The
long linear chromosomes of eukaryotes contain multiple
origins; the two growing forks from a particular origin
continue to advance until they meet the advancing growing
forks from neighboring origins. Each region served by one
DNA origin is called a *replicon*.

Bidirectional replication was first detected by fiber autoradiography of labeled DNA molecules from cultured
mammalian cells (Figure 10-4). When cells are exposed alternately to high and low levels of [³H]thymidine, the labeled DNA tracks are first very dark and then lighter. Such
studies have revealed clusters of active replicons with
growing forks moving away from a common origin, thus
providing unambiguous evidence of bidirectional growth.
Most cellular DNA and many viral DNA molecules replicate bidirectionally. Such viruses serve as excellent models
for the study of cellular DNA replication.

Replication "Bubbles"   If DNA from replicating eukaryotic cells is extracted and examined by electron
microscopy, so-called replication "bubbles," or "eyes,"
extending from multiple origins of replication are visible

(a) Predicted fiber autoradiographic pattern

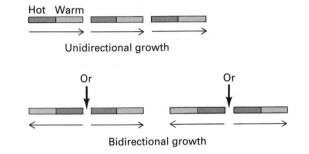

Unidirectional growth

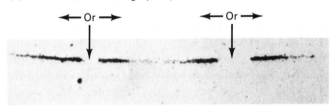

Bidirectional growth

(b) Actual fiber autoradiographic pattern

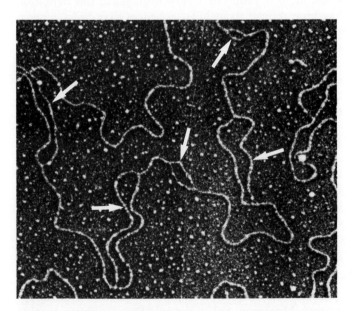

◄ FIGURE 10-4   Demonstration of bidirectional growth of cellular DNA chains. If cultured replicating cells are exposed alternately to high and low levels of [³H]thymidine, the resulting DNA will be heavily labeled ("hot") near replication origins (Or) and lightly labeled ("warm") farther away. When such labeled DNA is dried on a microscope slide as long linear molecules (fibers) and then exposed to a radiation-sensitive emulsion, autoradiographic signals should be produced corresponding to the hot-warm DNA regions. (a) Predicted patterns of autoradiographic bands for uni- and bidirectional DNA synthesis. (b) Actual fiber autoradiograph of DNA from cultured mammalian cells shows autoradiographic signals consistent with bidirectional synthesis. [See J. A. Huberman and A. D. Riggs, 1968, *J. Mol. Biol.*, **32**:327; J. A. Huberman and A. Tsai, 1973, *J. Mol. Biol.* **75**:5; Photographs courtesy of J. A. Huberman.]

(Figure 10-5). Although such micrographs do not constitute conclusive evidence for unidirectional or bidirectional fork movement, electron-microscope studies of bubbles in viral DNA have provided evidence for bidirectional replication. If circular viral DNA molecules at different stages of replication are cut with a restriction endonuclease that

recognizes a single site, the positions at the center of the replication bubble with respect to the restriction site can be determined (Figure 10-6). The most common result from such analyses is a series of ever-larger bubbles whose centers map to the same site, indicating bidirectional replication of both DNA strands from that site. Thus both fiber autoradiography and electron microscopy have indicated that bidirectional DNA replication is the general rule.

Number of Growing Forks and Their Rate of Movement Fiber autoradiography of DNA from cells labeled for various times allows the rate of growing-fork movement to be estimated. In *E. coli* cells one round of DNA replication takes about 42 min. Since the single circular *E. coli* chromosome of 4.4×10^6 base pairs is duplicated from one origin by two growing forks, the rate of fork movement is about 1000 bp/s per fork; the rate determined from fiber-labeling experiments on *E. coli* agrees with this value, indicating that the fiber-labeling technique can provide a reasonable estimate of the rate of growing-fork movement in vivo.

The rate of fork movement in human cells, based on fiber-labeling experiments, is only about 100 bp/s per fork. Since the entire human genome of 3×10^9 base pairs replicates in 8 h, it must contain at least 1000 growing forks. If there were only 1000 growing forks, each would be required to replicate about 10^6 base pairs; however, fiber autoradiography and electron microscopy (see Figures 10-4 and 10-5) indicate that growing forks are not spaced that far apart. A more likely estimate is that the human genome contains 10,000–100,000 replicons each of which functions for only part of the 8 h required for replication of the entire genome.

▲ FIGURE 10-5   Electron micrograph of DNA extracted from rapidly dividing nuclei of early *D. melanogaster* embryos. The arrows mark replication bubbles; the diameters of the DNA chain in both arms of these bubbles indicate that they are double-stranded. [See A. B. Blumenthal, H. J. Kreigstein, and D. S. Hogness, 1973, *Cold Spring Harbor Symp. Quant. Biol.* **38**:205; courtesy of D. S. Hogness.]

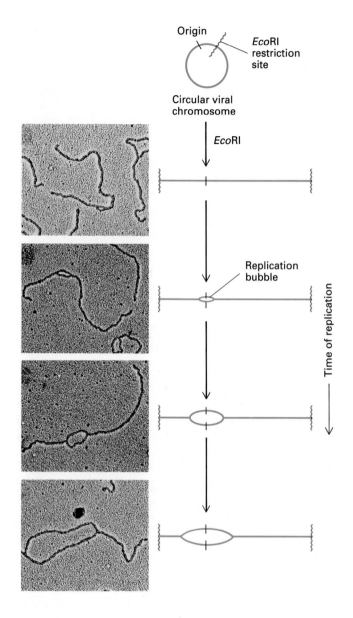

Demonstration of bidirectional chain growth from a single origin in viral DNA. The replicating DNA from SV40-infected cells was cut by the restriction enzyme *Eco*RI, which recognizes a single site, and examined by electron microscopy. The electron micrographs and corresponding diagrams show a collection of increasingly longer replication bubbles, the centers of which are a constant distance from each end of the cut molecules, thus indicating that chain growth occurs in two directions from a common origin. [See G. C. Fareed, C. F. Garon, and N. P. Salzman, 1972, *J. Virol.* **10**:484; photographs courtesy of N. P. Salzman.]

earliest synthesized DNA at the beginning of a round of synthesis was density-labeled with a heavy isotope. The cells then were transferred to medium containing the normal isotope to complete the round of synthesis and allowed to grow for several generations. The cells again were synchronized and then briefly labeled with a radioisotope. Analysis of the newly replicated radiolabeled DNA revealed that it also contained the heavy isotope, indicating that the same regions of DNA were copied each time at the beginning of replication. However, none of these experiments provided sufficient resolution to determine which specific nucleotide sequences in bacterial, yeast, and many viral DNAs function as origins of DNA replication.

A *replication origin* is defined experimentally as a stretch of DNA that is necessary and sufficient for replication of a circular DNA molecule, usually a plasmid or virus, in an appropriate host cell. In yeast, this definition has been refined to include sequences that direct replication once per S phase, the period of the cell cycle in which chromosomal duplication takes place. This important characteristic of DNA replication in eukaryotic cells will be considered in Chapter 25. We discuss three types of replication origin to illustrate some general conclusions about their nature: the *E. coli oriC*, the simian virus 40 (SV40) origin, and yeast autonomously replicating sequences. The *E. coli oriC* and the SV40 origin have been used extensively as substrates for studying replication in vitro. The detailed knowledge now available about the proteins required to start replication at the *E. coli* origin and the accumulating information about other origins and their use in vitro all suggest that most cellular DNA replication begins at specific sequences, possibly using similar mechanisms.

E. coli Replication Origin   It is now known that plasmids containing *E. coli oriC*, a DNA segment of about 240 nucleotides, are capable of independent and controlled replication. Moreover, *oriC* is the only *E. coli* segment capable of conferring independent replication. Important conserved sequence features in bacterial origins have been deduced from phylogenetic comparisons of *oriC* with the origins of five other bacterial species including the distant species *Vibrio harveyi*, a marine bacterium (Figure 10-7). Repetitive 9-bp and 13-bp sequences (these repeat se-

DNA Replication Begins at Specific Chromosomal Sites

Before investigators could tackle the problem of how DNA replication is initiated in vitro, they had to determine whether there are specific sites on chromosomes at which DNA replication always begins in vivo. Animal viruses were shown by electron-microscope studies to have replication bubbles whose centers were always in the same approximate site. The same also was demonstrated for circular bacterial and plant viruses and for bacterial, yeast, and mammalian plasmids. More than 20 years ago, double-labeling experiments with synchronized bacterial cells and synchronized animal cells also suggested that there is at least regional specificity for DNA-replication origins in both bacterial and animal cells. In these experiments, the

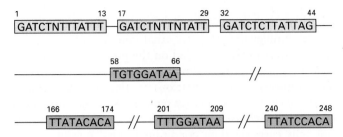

▲ FIGURE 10-7 Consensus sequence of the minimal bacterial replication origin based on analyses of chromosomes from six species. The 13-bp repetitive sequences (yellow) are rich in adenine and thymine residues; the 9-bp repetitive sequences (green) exist in both orientations. These sequences are referred to as 13-mers and 9-mers, respectively. Indicated nucleotide position numbers are arbitrary. This consensus sequence was derived from comparison of the following species: E. coli, Salmonella typhimurium, Enterobacter aerogenes, Klebsiella pneumoniae, Erwinia carotovora, and Vibrio harveyi. In the E. coli genome, an AT-rich region just to the left of oriC facilitates local unwinding of the duplex DNA. [See J. Zyskind et al., 1983, Proc. Nat'l. Acad. Sci. USA **80**:1164.]

quences are referred to as *9-mers* and *13-mers*, respectively) are characeristic of every bacterial origin sequence. As we will see later, these are important binding sites for the DnaA protein that initiates replication. In addition, in E. coli a segment of DNA with a relatively high AT content is located adjacent to *oriC*; this sequence appears to be important in facilitating local unwinding of the duplex to reveal the two single-stranded templates to be copied.

Yeast Autonomously Replicating Sequences The yeast genome, like all eukaryotic genomes, has multiple origins of replication. Cloning experiments indicate that about 400 origins exist in the 17 chromosomes of *S. cerevisiae;* more than a dozen of these have been characterized in detail. Each yeast origin sequence, called an *autonomously replicating sequence* (ARS), confers on a plasmid the ability to replicate in yeast and is a required element in yeast artificial chromosomes (see Figure 9-58). Detailed mutational analysis of one ≈ 180-bp ARS called ARS1 revealed only one essential element, a 15-bp segment designated element A. Three other short segments—the B_1, B_2, and B_3 elements—are required for efficient functioning of ARS1 (Figure 10-8). An 11-bp ARS consensus sequence was deduced by comparing the sequences required for functioning of many different DNA segments that act as ARSs. Element A in ARS1 was found to be identical at all 11 positions of the consensus sequence, and element B_2 at 9 of 11.

The DNA footprinting method, described in Chapter 11, has been used to show that a complex of proteins called ORC (origin-recognition complex) binds specifically to element A in ARS1 in an ATP-dependent manner. This complex also binds specifically to other ARSs tested. Discovery of exactly how an ARS and ORC function in initiating DNA replication awaits development of an in vitro system in which yeast proteins mediate replication of an ARS-containing plasmid.

SV40 Replication Origin A 65-bp region in the SV40 chromosome is sufficient to promote DNA replication both in animal cells and in vitro. Three segments of the

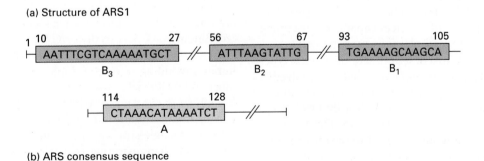

▲ FIGURE 10-8 (a) Diagram of one yeast autonomously replicating sequence, designated ARS1, showing regions required for its efficient functioning as a replication origin (only one strand is shown). In vitro mutagenesis of ARS1 generated a series of mutants, which were tested for their ability to transform yeast with high efficiency. This assay identified element A (orange) as an essential region for origin function. In a more sensitive assay, three other regions, B_1, B_2, and B_3, were shown to increase the efficiency of origin function. (b) The 11-bp consensus sequence deduced from comparisons of the essential regions in many different ARSs (both strands are shown). Element A of ARS1 contains this consensus sequence in its entirety, and element B matches it at 9/11 positions. [See Y. Marahrens and B. Stillman, 1992, *Science* **255**:817.]

SV40 origins are required for activity, as demonstrated by testing of origins containing specific mutations. As discussed below, researchers have used mammalian proteins and plasmids carrying the SV40 replication origin to study the molecular mechanisms of mammalian DNA replication.

Three Common Features of Replication Origins

Although the specific nucleotide sequences of replication origins from *E. coli*, yeast, and SV40 are very different, they share several properties. First, replication origins are unique DNA segments that contain multiple short repeated sequences. Second, these short repeat units are recognized by multimeric origin-binding proteins (not yet isolated for yeast), which play a key role in assembling the replication enzymes at the origin site. And third, the origin region usually is flanked by DNA sequences rich in adenine and thymine, which facilitates unwinding of duplex DNA. Unwinding is necessary to generate single-stranded regions that can serve as templates for DNA synthesis.

➤ *DNA Replication in E. coli*

The studies that we have reviewed so far demonstrate that DNA replication starts at unique chromosomal origins, proceeds bidirectionally, and is semiconservative. The biochemical mechanisms responsible for these critical cellular events were uncovered first in bacterial systems; more recently, they have been studied with proteins isolated from cultured eukaryotic cells. Because the biochemical mechanisms involved in *E. coli* DNA replication are now understood in considerable detail, our discussion focuses on these. Unraveling of these mechanisms depended on prior events: identification of the replication origin (*oriC*) in *E. coli* and development of a system capable of sustaining in vitro replication of small plasmids carrying *oriC*.

Before we plunge into a description of the several types of enzymes and numerous other factors that are involved in DNA replication, it is important to point out certain elementary problems in the copying of DNA by DNA polymerases. First, DNA polymerases involved in DNA replication are unable to unwind duplex DNA to separate the two strands that are to be copied. Second, the DNA duplex is an antiparallel helix: that is, the two strands are opposite ($5' \rightarrow 3'$ and $3' \rightarrow 5'$) in chemical polarity. However, all DNA polymerases catalyze nucleotide addition to the 3'-hydroxyl end of a growing chain and thus direct growth *only* in the $5' \rightarrow 3'$ direction (see Figure 4-22). And third, all DNA polymerses so far discovered only elongate a pre-existing primer strand of DNA or RNA; they cannot initiate chains. These last two properties of DNA polymerases, directionality and primer requirement, pose distinct problems in the copying of DNA. In this section, we describe the enzymological solutions to the unwinding, priming, and directionality problems.

DnaA Protein Initiates Replication in *E. coli*

Genetic studies first suggested that initiation of replication at *oriC* most likely depended on the protein encoded by a gene designated *dnaA*. Initially mutant strains carrying temperature-sensitive mutations in *dnaA* were isolated; these cells grew at permissive temperatures (e.g., 30°C) but not at nonpermissive temperatures (39–42°C). When *E. coli* cells carrying such conditional lethal mutations had begun DNA replication at the permissive temperature and then were shifted to the higher temperature, they completed the round of DNA synthesis already under way; however, they did not start another round of replication at the nonpermissive temperature. Nonconditional missense or deletion mutants in *dnaA* can be isolated only if a foreign replication origin (from a plasmid or bacteriophage) that is not dependent on *dnaA* is inserted into the *E. coli* chromosome. Such recombinant *dnaA* mutants, however, require an exogenous protein that recognizes the second origin. These findings pinpointed the DnaA protein as a prime candidate for interaction with *oriC*.

By cloning the *dnaA* gene in a bacteriophage, investigators obtained large amounts of pure DnaA protein, which then was biochemically analyzed. Studies in vitro showed that the protein binds to the four 9-mers in *oriC*, forming a multimeric complex that contains 10–20 protein subunits (Figure 10-9). The binding of DnaA to *oriC* requires ATP. Furthermore, although DnaA can bind to duplex *E. coli* DNA in the relaxed-circle form, it can initiate replication only if the DNA is negatively supercoiled (see Figure 4-14). The reason for this specificity is that DNA molecules with negative supercoils are easier to unwind locally (thus providing a single-stranded template region) than are DNA molecules without supercoils (see Figure 4-15). Supercoiling of DNA and the enzymes, called topoisomerases, that control the degree of DNA supercoiling are discussed in detail in a later section.

Binding of DnaA to the *oriC* 9-mers, which are rich in adenine and thymine and thus easily melted, facilitates the initial unwinding of *E. coli* duplex DNA. This process requires ATP and yields a so-called "open" complex (see Figure 10-9). When a mixture of *E. coli* DNA and DnaA protein is treated with an endonuclease that specifically recognizes single-stranded DNA, the DNA is cut in the origin region, demonstrating that it is unwound.

DnaB Is a Helicase That Unwinds Duplex DNA

Further unwinding of the two strands of the *E. coli* chromosome to generate unpaired template strands is mediated by the protein product of the *dnaB* locus, which is essential for DNA replication. In contrast to the behavior of temperature-sensitive *dnaA* mutants, DNA synthesis stops immediately in *dnaB* temperature-sensitive mutants upon shifting them to a nonpermissive temperature. This finding is consistent with the biochemical activity of DnaB as a *heli-*

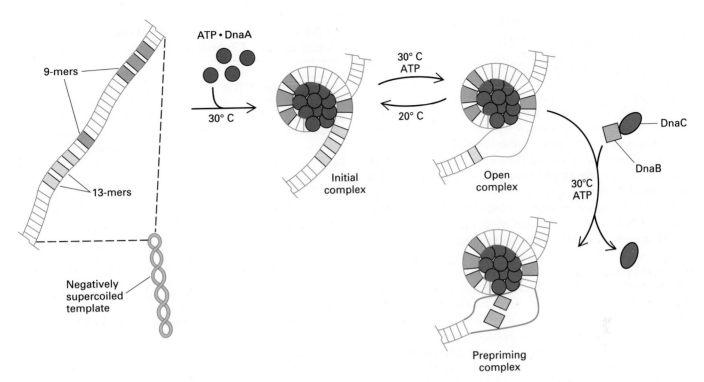

▲ FIGURE 10-9   Model of initiation of replication at *E. coli oriC*. The 9-mers and 13-mers are the repetitive sequences shown in Figure 10-7. DnaA is required only for initiation of replication, whereas DnaB and DnaC are required for initiation and chain elongation. The sole function of DnaC is to deliver DnaB, which has helicase activity, to the template. One helicase molecule binds to each DNA strand at *oriC*; the two molecules then proceed in opposite directions away from the origin. See text for discussion. [Adapted from C. Bramhill and A. Kornberg, 1988, *Cell* **52**:743.]

case. DnaB binds to both single strands in the open complex formed between DnaA and *oriC*. This binding requires ATP and the protein product encoded by the *dnaC* locus, and yields the prepriming complex (see Figure 10-9).

Helicases constitute a class of enzymes that can move along a DNA duplex utilizing the energy of ATP hydrolysis to separate the strands (Figure 10-10). The separated strands are inhibited from subsequently reannealing by the *E. coli single-strand–binding protein* (Ssb protein), which binds to both separated strands. When temperature-sensitive *dnaB* mutants are shifted to nonpermissive temperatures, unwinding ceases; as a result, DNA synthesis stops for lack of single-stranded templates.

▶ FIGURE 10-10   Experimental demonstration of helicase activity of *E. coli* DnaB protein. In the presence of ATP and single-strand–binding (Ssb) protein, purified DnaB can unwind a gapped DNA duplex. Unwinding has been found to occur with a tenfold preference in one direction (to the left in this diagram). Based on this observation, DnaB is thought to move along the strand to which it binds in the 5′ → 3′ direction. As discussed later, this strand acts as the template for synthesis of the lagging strand. The Ssb protein binds to unpaired DNA strands and prevents them from reannealing. [From A. Kornberg and T. Baker, 1992, *DNA Replication*, 2d ed., W. H. Freeman and Company.]

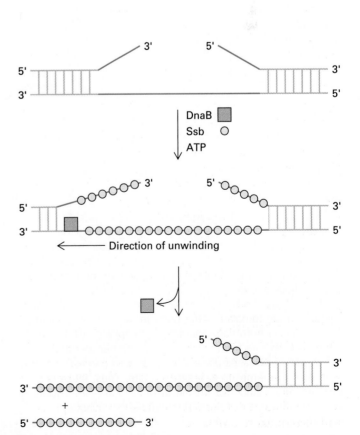

Primase Catalyzes Formation of RNA Primers for DNA Synthesis

An *E. coli* enzyme called *primase* catalyzes synthesis of short RNA molecules that function as primers for nucleotide addition by DNA polymerase (Table 10-1). *E. coli* strains with temperature-sensitive mutations in *dnaG*, which encodes primase, cannot replicate their DNA at the nonpermissive temperature, thereby establishing the essential role of primase. Several different mechanisms for binding of primase to the DNA template have been identified in vitro. In the simplest case, primase recognizes a specific nucleotide sequence in single-stranded DNA and synthesizes an RNA primer at that site. More commonly, priming on a single-stranded DNA requires binding of DnaB, which then promotes the binding of primase; after primase binds to the templae, it catalyzes synthesis of a primer (Figure 10-11). In the case of some bacteriophages, primase, DnaB, and several accessory proteins form a complex referred to as the *primosome,* which can move along DNA strands covered with Ssb protein and synthesize very short RNA primers at many different sites. The term primosome, however, is now more generally used to denote the complex between primase and helicase. In *E. coli* cells, primase binds to DnaB loaded at *oriC* and forms a primosome. The

TABLE 10-1 RNA Primers (Primase Products) Used in DNA Chain Initiation	
Replicating DNA	**RNA Oligonucleotide***
Bacteriophage T4	pppAC (N)$_3$
Bacteriophage T7	pppACCA
	pppACCC
E. coli	pppAC (N)$_{7-10}$
Yeast	pppA (N)$_{8-12}$
Drosophila	pppA (N)$_{7-9}$
Lymphoblastoid cells	pppA (N)$_8$
	pppG (N)$_8$

*N = any ribonucleotide. The indicated primer lengths for the eukaroytic cells are averages. Yeast and *Drosophila* RNA primers are rich in adenine, which accounts for about 75 percent of the bases.

primase within the primosome provides RNA primers for synthesis of both strands of duplex DNA.

At a Growing Fork One Strand Is Synthesized Discontinuously from Multiple Primers

We have seen how the activities of helicase and primase solve two of the problems inherent to DNA replication—unwinding of the duplex template and the requirement of DNA polymerases for a primer. Remember, though, that both strands of the DNA template are copied as the replication bubble enlarges. Each end of the bubble represents a growing fork where both new strands are synthesized (see Figure 10-3c). At each fork, one strand called the *leading strand* is synthesized *continuously* from a single primer on the leading-strand template and grows in the 5′ → 3′ direction (Figure 10-12 *top*). Growth of the leading strand proceeds in the same direction as movement of the growing fork.

Synthesis of the *lagging strand* is more complicated: the overall direction of growth must be 3′ → 5′ (complementary to its template strand), but DNA polymerase adds nucleotides only in the 5′ → 3′ direction. In both prokaryotes and eukaryotes these apparently incompatible requirements are met by a process involving *discontinuous* copying of the lagging strand from multiple primers (see Figure 10-12 *middle*). As synthesis of the leading strand progresses, sites uncovered on the lagging-strand template are copied into short RNA primers (<15 nucleotides) by primase (see Table 10-1). These primers then are elongated in the 5′ → 3′ direction (opposite to the overall direction of

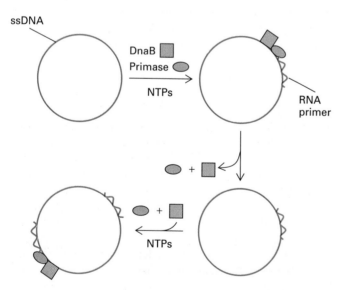

▲ FIGURE 10-11   General in vitro system for synthesis of short RNA primers from a single-stranded DNA template. DnaB is thought to bind to the template first and then attract primase to the template. After primase synthesizes a short RNA molecule from ribonucleotide triphosphates (NTPs), both proteins dissociate from the template. Another round of primer synthesis takes place after DnaB and primase reassociate with the template at a different site. Thus, these proteins act in a distributive manner. Even though the helicase activity of DnaB is not needed with a ssDNA substrate, it increases primer synthesis by promoting binding of primase.

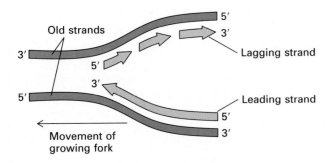

Old strands

3′

5′

5′

3′

Lagging strand

3′

5′

Leading strand

3′

Movement of
growing fork

Lagging-strand synthesis

(a) RNA oligonucleotides
(primer)
copied from
DNA

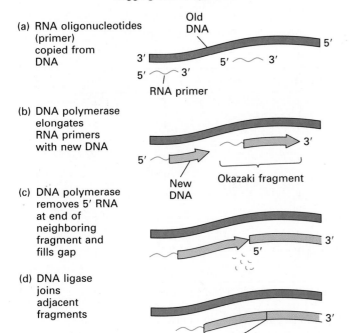

Old
DNA

5′

3′

5′   3′

5′   3′

RNA primer

(b) DNA polymerase
elongates
RNA primers
with new DNA

5′

3′

New
DNA

Okazaki fragment

(c) DNA polymerase
removes 5′ RNA
at end of
neighboring
fragment and
fills gap

5′

3′

(d) DNA ligase
joins
adjacent
fragments

3′

Ligation

Ligase reaction

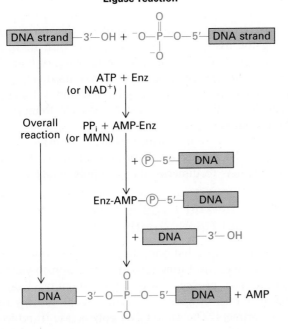

◄ **FIGURE 10-12**   The overall structure of a growing fork (*top*) and steps in synthesis of the lagging strand. Synthesis of the leading strand, catalyzed by DNA polymerase III, occurs by sequential addition of deoxyribonucleotides with release of PP$_i$ (see Figure 4-18). The more complicated, discontinuous synthesis of the lagging strand involves several distinct steps (*middle*, a–d). The reaction catalyzed by DNA ligase (*bottom*) joins the 3′-hydroxyl end of one Okazaki fragment to the 5′ phosphate of the adjacent fragment. During this reaction, ligase transiently attaches covalently to the 5′ phosphate, thus activating the phosphate group. *E. coli* DNA ligase uses NAD$^+$ as cofactor, generating NMN and AMP, whereas bacteriophage T4, commonly used in DNA cloning, uses ATP, generating PP$_i$ and AMP.

growing-fork movement) by DNA polymerase III. The resulting short fragments, containing RNA and DNA, are called *Okazaki fragments*, after their discoverer Reiji Okazaki. In bacteria and bacteriophages, Okazaki fragments contain 1000–2000 nucleotides, but in eukaryotic cells they are much shorter (100–200 nucleotides).

As each newly formed segment of the lagging strand approaches the 5′ end of the adjacent Okazaki fragment (the one just completed), DNA polymerase I takes over. The 5′ → 3′ *exonuclease activity* of this enzyme removes the RNA primer of the adjacent fragment; the polymerization activity of the enzyme simultaneously fills in the gap between the fragments by addition of deoxynucleotides. Finally, another critical enzyme, *DNA ligase*, joins adjacent completed fragments (see Figure 10-12 *bottom*).

DNA Polymerase III Synthesizes Both the Leading and Lagging Strands

Three DNA polymerases (I, II, and III) have been purified from *E. coli* (Table 10-2). As discussed in a later section, DNA polymerase I probably is the most important enzyme for gap filling during DNA repair. DNA polymerase II is encoded by one of the many genes activated in response to DNA damage that halts movement of the growing fork, the so-called SOS response. This polymerase also fills gaps and appears to facilitate DNA synthesis directed by damaged templates. This latter activity of DNA polymerase II tends to perpetuate mutations. Our discussion here focuses on DNA polymerase III. The observation that *E. coli* strains with a temperature-sensitive mutation in this enzyme stopped making DNA when shifted to a nonpermissive temperature indicated that it was critical. Studies with DNA replication in vitro subsequently demonstrated that DNA polymerase III is the functional enzyme at the growing fork in *E. coli*.

The DNA polymerase III holoenzyme is a very large (>600kDa), highly complex protein composed of 10 different polypeptides. Overall, the enzyme has an asymmet-

TABLE 10-2   Properties of DNA Polymerases

E. coli	I	II	III		
Polymerization: $5' \rightarrow 3'$	+	+	+		
Exonuclease activity: $3' \rightarrow 5'$	+	+	+		
$5' \rightarrow 3'$	+	−	−		
Synthesis from:					
Intact DNA	−	−	−		
Primed single strands	+	−	−		
Primed single strands plus single-strand-binding protein	+	−	+		
In vitro chain elongation rate (nucleotides per minute)	600	?	30,000		
Molecules present per cell	400	?	10–20		
Mutation lethal?	+	−	+		
Mammalian Cells*	α	$\beta^{\dagger}$	γ	δ	ϵ
Polymerization: $5' \rightarrow 3'$	+	+	+	+	+
Exonuclease activity‡: $3' \rightarrow 5'$ (editing function)	−	−	+	+	+
Synthesis from:					
RNA primer	+	−	−	+	?
DNA primer	+	+	+	+	+
Associated DNA primase	+	−	−	−	−
Sensitive to aphidicolin (inhibitor of cell DNA synthesis)?	+	−	−	+	+
Cell location:					
Nuclei	+	+	−	+	+
Mitochondria	−	−	+	−	−

* Yeast DNA polymerase I, II, and III are equivalent to polymerase α, β, and δ, respectively. I and III are essential for cell viability.

† Polymerase β is most active on DNA molecules with gaps of about 20 nucleotides and is thought to play a role in DNA repair.

‡ Eukaryotic enzymes undoubtedly have $5' \rightarrow 3'$ exonuclease activities to remove primers, but such reactions with strictly purified enzymes have not been carried out, except in the case of the herpes DNA polymerase, which has such an activity.

ric dimeric structure; it contains two copies of most subunits and two catalytic sites for nucleotide addition. Detailed biochemical studies have begun to reveal the role of the different subunits in polymerase function. The so-called *core polymerase*, composed of the α, ξ, and θ subunits, contains the active site for nucleotide addition. The central role of the remaining subunits is to convert the core polymerase from a *distributive* enzyme, which falls off the template strand after forming relatively short stretches of DNA containing 10–50 nucleotides, to a *processive* enzyme, which can form stretches of DNA with up to 5×10^5 nucleotides without being released from the template. This latter activity is necessary for efficient leading-strand synthesis.

The key to the processive nature of DNA polymerase III is the ability of the β subunit to form a donut-shaped dimer around duplex DNA and then associate with and hold the catalytic core polymerase at the primer-template terminus (Figure 10-13). Once tightly associated with the DNA, the β-subunit dimer functions like a "clamp," which

can move freely along the DNA, like a ring on a string, carrying the associated core polymerase with it. In this way, the active sites remain near the growing fork and the processivity of the core polymerase maximized. The directionality of movement of the β-subunit clamp is thought to be provided by the polymerase activity itself. Remarkably, of the six remaining subunits, five (γ, δ, δ', χ, and ψ) form the so-called γ complex, which mediates transfer of the β subunit to the duplex DNA–primer substrate. The final subunit (τ) acts to dimerize the core polymerase.

The Leading and Lagging Strands Are Synthesized Concurrently

So far we have seen that once the prepriming complex and an RNA primer are formed, chain elongation to yield the leading strand proceeds with little difficulty. Lagging-strand synthesis, however, proceeds discontinuously from multiple primers. The same DNA polymerase III molecule adds nucleotides to the growing points of both the leading

(a)

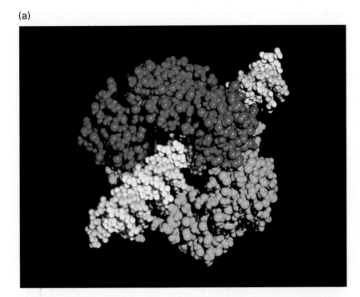

(b)

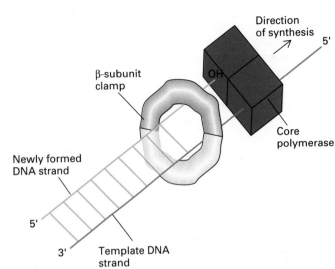

▲ FIGURE 10-13   Role of β subunits in increasing the processivity of *E. coli* DNA polymerase III. (a) Space-filling model based on x-ray crystallographic studies of dimeric β subunit binding to a DNA duplex. Two β subunits (red and yellow) form a donut-like clamp that remains tightly bound to a closed circular DNA molecule, but readily slides off the ends of a linear DNA molecule. (b) Schematic diagram of proposed association of the core polymerase (green) with the β-subunit clamp at the primer-template terminus; this

interaction is thought to keep the core from "falling off" the template. In the absence of the β subunit, the core polymerase synthesizes only short DNA fragments containing ≈20 nucleotides before it dissociates from the template; the enzyme can reinitiate synthesis from another primer-template terminus. In the presence of the β subunit, the processivity of the core polymerase increases more than 1000-fold. [Part (a) from X-P. Kong et al., 1992, *Cell* **69**:425.]

and lagging strands; moreover, nucleotide addition is now known to occur concurrently at both growing ends.

As noted earlier, dimeric DNA polymerase III contains two catalytic sites for nucleotide addition. It seems likely, but has not yet been proven, that the lagging-strand template wraps around one of the catalytic subunits, thus inverting the physical direction (but not the biochemical direction) of the growing lagging strand (Figure 10-14). This

inversion would place the 3′ growing ends of both the leading and lagging strands close to a catalytic site in the polymerase III molecule, so that deoxynucleotides can be added to the leading strand and lagging strand (or its RNA primer) at the same time.

➤ FIGURE 10-14   Proposed mechanism of concurrent synthesis of leading and lagging strands during DNA replication in *E. coli*. Looping of the lagging-strand template around one arm of dimeric DNA polymerase III brings the 3′ end of the growing lagging strand close to the catalytic site in one arm of the polymerase molecule. The leading strand is produced at the catalytic site in the other arm. As replication proceeds, more lagging-strand primers are formed (e.g., primer 3); these in turn are looped close to a catalytic site for elongation by the DNA polymerase molecule. The polymerase complex acts as an asymmetric dimer. The core polymerase associated with a dimeric β subunit remains attached to the leading-strand template and functions processively (see Figure 10-13). The core polymerase and helicase remain attached to the lagging-strand template, whereas the β subunit and the primase components act in a distributive fashion (i.e., dissociate from the growing fork and reassociate farther along the template). [Adapted from A. Kornberg, 1988, *J. Biol. Chem.* **263**:1.]

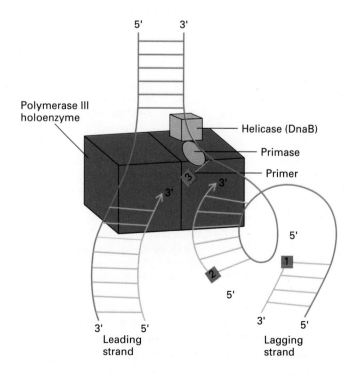

One consequence of this mechanism is that, at any given point of addition, the bases being copied from the parental template strands are located at different positions in the original duplex. In other words, the point in the template at which the lagging strand is being copied is displaced from the point in the template at which leading-strand copying is occurring. In addition, as each lagging-strand segment is completed, it must be released from the active site, which then is transferred to the next RNA primer.

Biochemical studies indicate that the core polymerase releases a completed Okazaki fragment but does *not* dissociate from the growing fork. In contrast, the β subunit and primase dissociate from the growing fork and then reassociate farther along the lagging-strand template; that is, these components act in a distributive manner. All these events must be closely coordinated in order for the growing

fork to move 500–1000 nucleotides per second while both strands are being replicated. The subunit composition of DNA polymerase III at the growing fork is not known yet. However, since all the polymerase subunits have been cloned recently and can be produced in quantity, the precise subunit composition at the growing fork now can be determined. Figure 10-15 summarizes the various proteins that function at the growing fork during replication of the *E. coli* chromosome.

Interaction of Tus Protein with Termination Sites Stops DNA Replication

Although much research attention has been focused on replication origins, *E. coli* DNA also has been shown to contain termination (TER) sites, which bind a specific protein called Tus. This protein may act to stop replication by preventing helicase (DnaB) from unwinding duplex DNA, thereby interrupting the function of the growing fork. The details of the mechanism that arrests the growing fork are not yet clear. Replication of the circular *E. coli* chromosome produces two interlocking daughter chromosomes, which must be separated. This process is discussed in the section on topoisomerases.

► *Eukaryotic DNA Replication*

Can we generalize the enzymatic mechanisms that replicate the single chromosome in simple prokaryotes such as *E. coli* to eukaryotes including higher mammals? As in *E. coli*, investigation of eukaryotic DNA-replication mechanisms initially was concentrated on characterization of different DNA polymerases from eukaryotic cells (see Table 10-2). This work was followed by development of in vitro systems for copying small chromosomes from animal viruses; viral DNA replication is dependent almost entirely on host-cell proteins. The primary effort has focused on replication of the SV40 chromosome; these studies have progressed so rapidly in recent years that the SV40 chromosome now can be replicated in vitro using only eight purified components from mammalian cells. The specific functions of these proteins are highly reminiscent of proteins required for replication of plasmids carrying *oriC*. Thus, the mechanistic problems involved in DNA replication, which are similar in all organisms, have been solved in most cases by use of similar types of proteins.

Eukaryotic Proteins That Replicate SV40 DNA in Vitro Exhibit Similarities and Differences with *E. coli* Replication Proteins

The mechanism of replication of the SV40 chromosome is schematically depicted in Figure 10-16. Similar to *E. coli*, replication is initiated at a unique location on the SV40

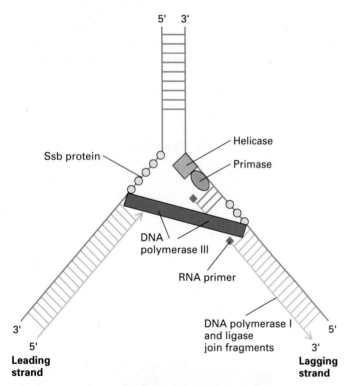

▲ FIGURE 10-15   Diagrammatic summary of major proteins that function at the *E. coli* growing fork. Helicase, encoded by the *dnaB* gene, binds to and moves along the lagging-strand template in the 5′ → 3′ direction, unwinding the duplex as it progresses. Primase, encoded by the *dnaG* gene, associates with helicase and synthesizes RNA primers distributively. DNA polymerase III elongates the lagging-strand primers, which subsequently are removed by the 5′ → 3′ exonuclease activity of DNA polymerase I. This enzyme also fills in the gaps between the lagging-strand fragments, which then are ligated together by DNA ligase. Since the core polymerase remains attached to the template, leading-strand synthesis can occur continuously.

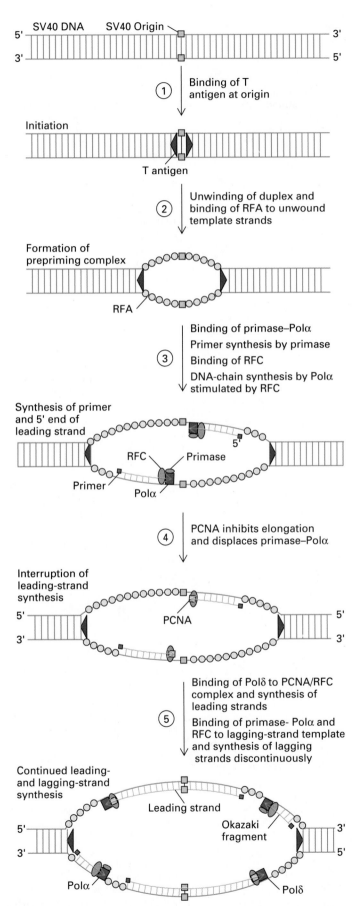

FIGURE 10-16 Model of in vitro replication of SV40 DNA by eukaryotic enzymes. In contrast to replication in *E. coli*, two different DNA polymerases catalyze elongation of the leading and lagging strands. Polymerase α (Pol α), which is tightly associated with a primase, forms the 5′ ends of the leading strands and then is displaced from the template. Association of polymerase δ (Pol δ) with PCNA increases the enzyme's processivity, so that it can synthesize the remainder of the leading strands. Lagging-strand synthesis downstream from the leading-strand primers is thought to be carried out by the combined action of primase and Pol α. RFC stimulates the activity of Pol α. See text for details. RFA = replication factor A; RFC = replication factor C; PCNA = proliferating cell nuclear antigen. [See T. Tsurimoto et al., 1990, *Nature* **346**:534.]

DNA by interaction of a virus-encoded, site-specific DNA-binding protein called *T antigen* (step 1). This multifunctional protein locally unwinds duplex DNA through its helicase activity. Opening of the duplex at the SV40 origin also requires ATP and replication factor A (RFA), a host-cell single-strand–binding protein with a function similar to that of Ssb in *E. coli* cells (step 2). As in *E. coli*, eukaryotic DNA replication occurs bidirectionally from RNA primers made by a primase; synthesis of the leading strand is continuous, while synthesis of the lagging strand is discontinuous. In contrast to the situation in *E. coli*, however, two distinct polymerases, α and δ, appear to function at the eukaryotic growing fork. Polymerase δ (Pol δ) is largely responsible for leading-strand synthesis; polymerase α (Pol α), which is tightly associated with a primase, is thought to synthesize the lagging strand.

Interestingly, initial leading-strand synthesis may be carried out by Pol α and Pol δ acting in a sequential fashion (see Figure 10-16). After the primase–Pol α complex binds to the unwound template strands, primase forms RNA primers; these are elongated for a short stretch by Pol α whose activity is stimulated by replication factor C (step 3). Binding of PCNA (proliferating cell nuclear antigen) at the primer-template terminus then displaces Pol α, thus interrupting leading-strand synthesis (step 4). Pol δ then binds to PCNA at the 3′ ends of the growing strands. The association of Pol δ with PCNA increases the processivity of the enzyme, so that it can continue synthesis of the leading strands. The function of PCNA thus appears to be highly analogous to that of the β subunit of *E. coli* polymerase III. Although these proteins are separated in evolutionary time by billions of years, they contain distinctly recognizable stretches of conserved amino acids. As unwinding of the duplex DNA progresses farther away from the origin, the primase–Pol α complex associates with the unwound template strands downstream from the leading-strand primers. Synthesis of the lagging strand then is carried out by combined action of the two enzymes and RFC, while leading-strand synthesis on the other side of the origin also proceeds (step 5). Finally, as in *E. coli*, eukaryotic topoisomerases probably play an important role in relieving torsional stress induced by growing-fork movement and in separating the two daughter chromosomes, as discussed in the next section.

Much has been learned about the eukaryotic proteins that function in DNA replication from in vitro studies of SV40 chromosome replication. However, studies on replication of *eukaryotic* DNA in vitro have been hampered for two reasons: the lack of eukaryotic experimental systems that can sustain in vitro replication initiated at cellular origins and by the lack of in vitro replication systems prepared from extracts of genetically tractable organisms such as yeast. The recent identification of a protein complex that binds to yeast chromosomal origins may be an important first step towards detailed research on cellular DNA replication using a combined genetic and biochemical strategy, which has proven so profitable in *E. coli*.

Telomerase Prevents Progressive Shortening of Lagging Strands during Eukaryotic DNA Replication

Unlike bacterial chromosomes, which are circular, eukaryotic chromosomes are linear and carry specialized ends called *telomeres*. As discussed in Chapter 9, telomeres consist of repetitive oligomeric sequences; for example, the yeast telomeric repeat sequence is $(5')$-$G_{1-3}T$-$(3')$. The need for a specialized region at the ends of eukaryotic chromosomes is apparent when we consider that all known DNA polymerases extend DNA chains in the $5' \rightarrow 3'$ direction and require a primer. As the growing fork approaches the end of a linear chromosome, synthesis of the leading strand continues to the end of the DNA template strand; the resulting completely replicated daughter chromatid then is released. However, because the lagging-strand template is copied in a discontinuous fashion, it cannot be replicated in its entirety. Without some special mechanism, the daughter chromatid resulting from lagging-strand synthesis would be shortened at each cell division.

The enzyme that prevents this progressive shortening of the lagging strand is a modified reverse transcriptase (see Chapter 7), called *telomerase*, which can elongate the lagging-strand template from its 3'-hydroxyl end. This unusual enzyme contains a catalytic site that polymerizes deoxyribonucleotides directed by a RNA template as well as the RNA molecule that functions as that template (Figure 10-17). The repetitive sequence added by telomerase is determined by the RNA associated with the enzyme. Once the 3' end of the lagging-strand template is sufficiently elongated, synthesis of the lagging strand can take place,

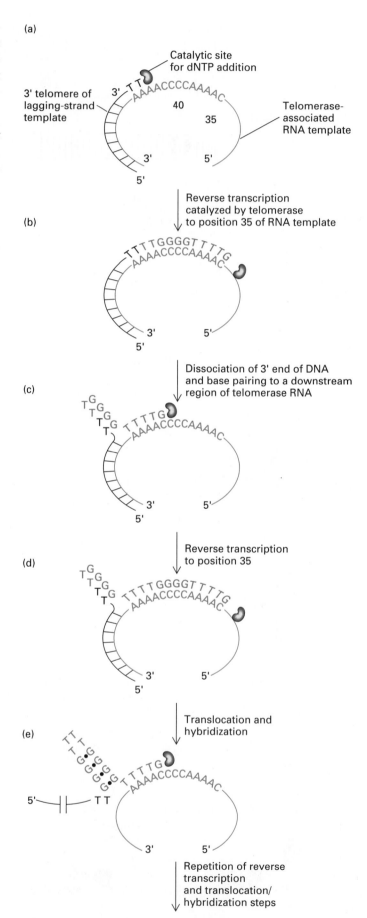

◄ FIGURE 10-17   Model of action of telomerase, a ribonucleoprotein complex that functions as a modified reverse transcriptase to elongate the 3′ telomeric end of the lagging-strand DNA template by a reiterative mechanism. This figure illustrates the action of the telomerase from *Oxytricha*, which adds a T_4G_4 repeat unit; other telomerases add slightly different sequences. (a) The 3′ end of the lagging-strand template (black) base-pairs with a unique region of the telomerase-associated RNA (red). Hybridization is facilitated by the match between the sequence at the end of the telomere and the sequence at the 3′ end of the RNA. (b) The telomerase catalytic site (green) then adds deoxyribonucleotides (blue) using the RNA molecule as a template; this reverse transcription proceeds to position 35 of the RNA template. (c) The strands of the resulting DNA-RNA duplex are thought to slip relative to one another, leading to displacement of a single-stranded region of the telomeric DNA strand and to uncovering of part of the RNA template sequence. (d,e) The lagging-strand telomeric sequence is again extended to position 35 by telomerase, and the DNA-RNA duplex undergoes translocation and hybridization as before. The slippage mechanism is thought to be facilitated by the non-Watson-Crick base pairing (black dots) between the displaced G residues (see Figure 9-61). Telomerases can add very long stretches of repeats by repetition of steps (d) and (e). [Adapted from D. Shippen-Lentz and E. H. Blackburn, 1990, *Nature* **247**:550.]

presumably from additional primers. The precise mechanism involved in synthesizing the lagging-strand 3′ telomere is not known.

► *Role of Topoisomerases in DNA Replication*

DNA is not a static, completely double-stranded structure; rather, it coils and bends in space, leading to changes in topology including formation of supercoils and local unwinding. The three parameters that describe DNA topology—linking number, twist, and writhe—are defined in Chapter 4 and illustrated in Figure 4-15. The enzymes that control the topology of DNA are critical to DNA replication in both prokaroyotic and eukaryotic cells. These enzymes function at several different steps in replication. First, replication of many *E. coli* phages and plasmids, including plasmids bearing *oriC*, requires that the template be *negatively supercoiled*. Second, movement of the growing fork induces formation of *positive supercoils*, ahead of the fork, which must be removed. And third, topoisomerases are required to separate the two daughter duplexes resulting from a single round of DNA replication. In this section we describe the two different classes of topoisomerases and the evidence that they are important in the process of DNA replication.

Type I Topoisomerases Relax DNA by Nicking and Closing One Strand of Duplex DNA

The first topoisomerase to be discovered, the *omega protein* in *E. coli*, can remove negative supercoils without leaving nicks in the DNA molecule. After the enzyme binds to a DNA molecule and cuts one strand, the free 5′ phosphate on the DNA is covalently attached to a tyrosine residue in the enzyme. The DNA strand that has not been cleaved is then passed through the single-stranded break. The cleaved strand is then re-sealed (Figure 10-18). By this mechanism, the enzyme removes one negative supercoil at a time,

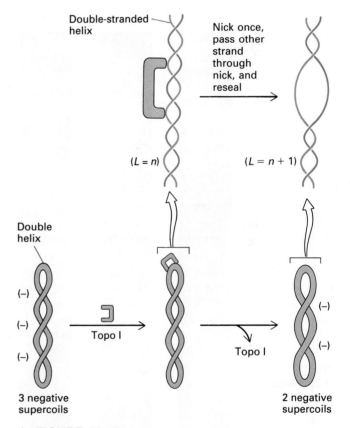

▲ FIGURE 10-18   Action of *E. coli* type I topoisomerase. This enzyme cleaves one strand of DNA and attaches covalently to the free 5′ phosphate; the uncut strand then is passed through the nick; finally the cut strand is resealed. *E. coli* topo I preferentially removes negative supercoils, whereas eukaryotic topo I enzymes can remove both positive and negative supercoils. During each round of nicking and resealing catalyzed by the *E. coli* enzyme, the linking number L is increased by 1 and one negative supercoil is removed. (The assignment of sign to supercoils is by convention with the helix stood on its end; in a negative supercoil the "front" strand falls from right to left as it passes over the back strand [as here]; in a positive supercoil, the front strand falls from left to right.)

thereby increasing the linking number by 1. An enzyme with this activity is classified as a *type I topoisomerase* (topo I).

The topo I from *E. coli* acts on negatively, but not positively, supercoiled molecules. In contrast, topo I enzymes from eukaryotic cells can remove both positive and negative supercoils. Because the relaxation (removal) of DNA supercoils by topo I is energetically favorable, the reaction proceeds without an energy requirement. If fully supercoiled SV40 DNA is treated with *E. coli* topo I in vitro for various time periods, numerous *topoisomers,* each containing a different number of supercoils, are produced (Figure 10-19). Thus the sequential action of topo I can remove all the supercoils in a DNA molecule.

Studies with gene-targeted knockout strains of *E. coli* have established that topo I is essential for viability. The enzyme is thought to help maintain the proper superhelical density of the *E. coli* chromosome by removing negative supercoils formed by action of type II topoisomerase (topo II), which is discussed below. Since *E. coli* topo I cannot remove positive supercoils, it is unlikely to play a role in growing-fork progression. In yeast, topo I is not essential for viability: mutations in the gene encoding this enzyme affect the growth rate of yeast cells but are not lethal. During in vitro replication of SV40 DNA, eukaryotic topo I can support growing-fork movement, since it is capable of removing both positive and negative supercoils. However, eukaryotic (yeast) topo II also can relax positive and negative supercoils, so the function of topo I during eukaryotic DNA replication in vivo is unclear. In both yeast and *E. coli,* topo I enzymes may be important in regulating aspects of transcription, as well as DNA damage and repair.

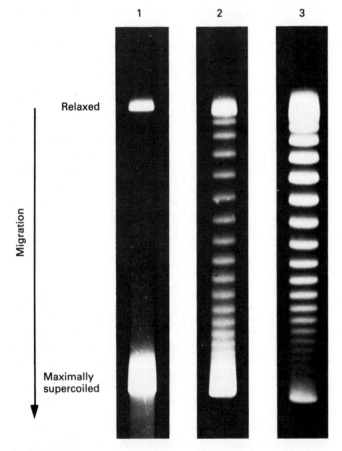

▲ FIGURE 10-19   Separation of SV40 DNA topoisomers containing different numbers of supercoils by gel electrophoresis. DNA was extracted from SV40 virions under conditions that ensure the maximal number of supercoils (lane 1). Aliquots were treated with *E. coli* type I topoisomerase for 3 min (lane 2) or 30 min (lane 3). About 25 bands, equal to the number of possible topoisomers, are visible in the electrophoretograms after topo I treatment, including the fully relaxed form. [From W. Keller and I. Wendel, 1974, *Cold Spring Harbor Symp. Quant. Biol.* **39**:199; courtesy of W. Keller.]

Type II Topoisomerases Change DNA Topology by Breaking and Rejoining Double-Stranded DNA

The first *type II topoisomerase* (topo II) to be described was isolated from *E. coli.* Termed *DNA gyrase,* this enzyme has the ability to cut a double-stranded DNA molecule, pass another portion of the duplex through the cut, and reseal the cut (Figure 10-20a). Because this maneuver has the effect of changing a positive supercoil into a negative supercoil, it changes the linking number of the DNA by 2. Hydrolysis of ATP is not required to induce supercoils but is required for the enzyme to turn over and introduce additional negative supercoils. The topo II enzymes from mammalian cells cannot, like *E. coli* DNA gyrase, increase the superhelical density at the expense of ATP; presumably no such activity is required in eukaryotes since binding of histones increases the potential superhelicity. All type II topoisomerases catalyze *catenation* and *decatenation,* that is, the linking and unlinking, of two different DNA duplexes (Figure 10-20b).

Temperature-sensitive gyrase mutants are unable to grow at nonpermissive temperatures, thus demonstrating that gyrase is necessary for DNA replication in *E. coli* cells. The enzyme functions to introduce negative supercoils near *oriC* in the template; as noted earlier, DnaA can initiate replication only on a negatively supercoiled template. Measurements of the degree of DNA supercoiling in *E. coli* suggest that there is one negative supercoil for each 15–20 turns of the DNA helix. A second crucial function of gyrase is to remove positive supercoils ahead of the growing fork during elongation of the growing strands (Figure 10-21).

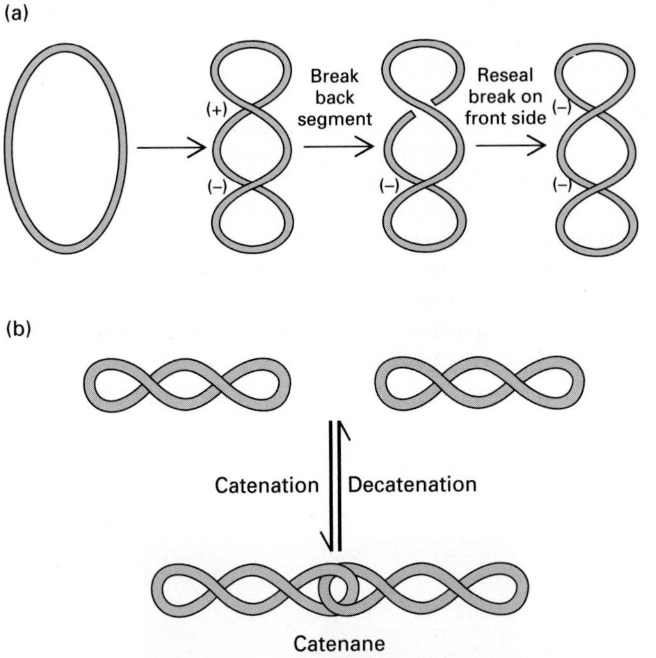

▲ FIGURE 10-20 Action of *E. coli* DNA gyrase, a type II topoisomerase. (a) Introduction of negative supercoils. The initial folding introduces no stable change, but the subsequent activity of gyrase produces a stable structure with two negative supercoils. Since one positive supercoil is replaced with one negative supercoil, the linking number *L* changes by 2. Eukaryotic topo II enzymes cannot introduce supercoils. (b) Catenation and decatenation of two different DNA duplexes. Both prokaryotic and eukaryotic topo II enzymes can catalyze this reaction. [See N. R. Cozzarelli, 1980, *Science* **207**:953.]

Replicated Circular DNA Molecules Are Separated by Type II Topoisomerases

A topo II activity apparently is needed to complete replication of circular DNA molecules. During replication the parental strands remain intact and retain their superhelicity. As the two growing forks approach each other, how does the as-yet-unreplicated parental duplex region become unpaired to allow replication to finish? This situation, which must also arise in the meeting of two growing forks in a eukaryotic chromosome, is illustrated in Figure 10-22. The last few helical turns in the parental DNA could be removed by changing the topology of the already replicated regions leaving the two newly complete daughter helices linked together as *catenanes,* covalently linked but not yet completely finished circles. Replication then could be completed before or after decatenation, leaving catenanes as the final replication product.

In *E. coli,* decatenation is catalyzed by DNA gyrase and a second type II enzyme, called *topoisomerase IV,* which genetic studies suggest is responsible for separating newly replicated molecules in vivo. For example, at nonpermissive temperatures, temperature-sensitive topo IV mutants die, and the mutant cells contain an accumulation of enlarged chromosome-protein complexes, called nucleoids, which presumably are intertwined DNA molecules. Furthermore, electron microscopy of such topo IV mutants carrying plasmids reveals accumulated interlocked plasmids—that is, catenanes of plasmid DNA (Figure 10-23).

To separate DNA catenanes, topo IV presumably binds to the interlocked duplexes and makes a double-stranded break in one molecule; while remaining attached to the substrate, it then passes the other molecule through the break and finally reseals the break in the cut molecule. Interestingly, although DNA gyrase can carry out decatenation in vitro, it cannot fully substitute for topo IV in vivo, as demonstrated by the lethal effects of topo IV mutations.

Linear Daughter Chromatids Also Are Separated by Type II Topoisomerases

Not only are yeast cells with a temperature-sensitive mutation in topo II nonviable at nonpermissive temperatures, but light-microscopic studies have revealed the participation of topo II in separating their linear chromosomes. Although individual yeast chromosomes are too small to be

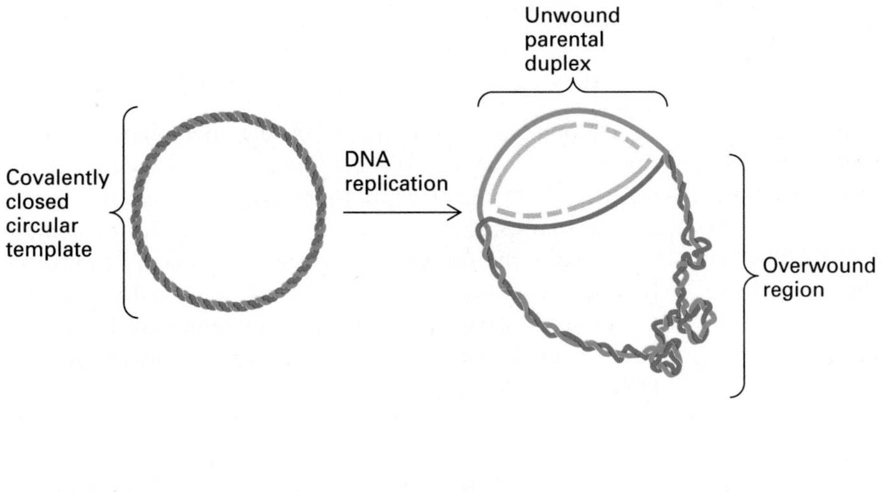

◄ FIGURE 10-21 Movement of the growing fork during DNA replication induces formation of positive supercoils in the duplex DNA ahead of the fork. In order for extensive DNA synthesis to proceed, the positive supercoils must be removed (relaxed). This can be accomplished by *E. coli* DNA gyrase and by eukaryotic type I and type II topoisomerases. [Adapted from A. Kornberg and T. Baker, 1992, *DNA Replication,* 2d ed., W. H. Freeman and Company, p. 380.]

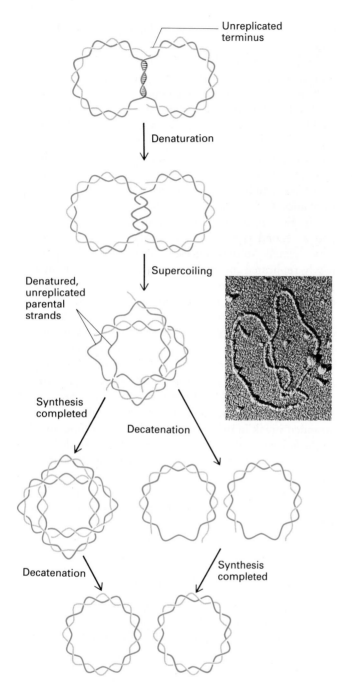

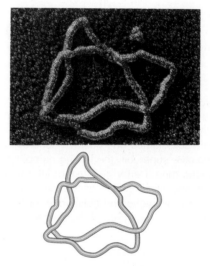

◄ FIGURE 10-22   Completion of replication of circular DNA molecules. Denaturation of the unreplicated terminus followed by supercoiling overcomes the steric and topological constraints of copying the terminus. At least with SV40 DNA, the final two steps (synthesis and decatenation) can occur in either order depending on experimental conditions. Parental strands are in dark colors; daughter strands in light colors. (*Inset*) Electron micrograph of two fully replicated SV40 DNA molecules interlocked twice. This structure would result if synthesis was completed before decatenation. Topo II can catalyze decatenation of such interlocked circles in vitro. [Drawing adapted from S. Wasserman and N. Cozzarelli, 1986, *Science* **232**:951. Micrograph from O. Sundin and A. Varshavsky, 1981, *Cell* **25**:659; courtesy of A. Varshavsky.]

▲ FIGURE 10-23   Electron micrograph of a plasmid catenane from *E. coli* cells with temperature-sensitive mutations in topoisomerase IV. Below the electron micrograph is a tracing of the two interlocked DNA molecules. Accumulation of catenanes in these mutants indicates that topo IV probably is the prokaryotic topoisomerase that catalyzes decatenation of newly replicated daughter chromosomes in vivo. [Photograph from D. E. Adams et al., 1992, *Cell* **71**:271.]

visualized by light microscopy, a structure called the nuclear body, which contains all the chromosomes clumped together, can be seen. When temperature-sensitive mutants are shifted to a nonpermissive temperature, the nuclear body, which usually divides at the junction of the mother and daughter cell, appears to get stuck in the passageway between the two cells (Figure 10-24).

As noted in Chapter 9, topoisomerase II also is a primary component of the nonhistone protein scaffolding to

which long DNA loops are attached in metaphase chromosomes (see Figure 9-53). This finding and the genetic studies in yeast strongly suggest that eukaryotic topo II resolves tangles that exist in newly replicated linear chromosomes. And finally, the similarity between the phenotypes of topo II mutants in yeast and topo IV mutants in bacteria argues that control of topological domains may be analogous in eukaryotic chromosomes and in small circular DNA molecules.

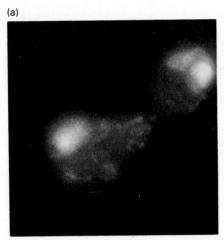

▲ FIGURE 10-24   Fluorescence micrographs of yeast cells with a temperature-sensitive mutation in topoisomerase II at 80 min after the beginning of the cell cycle. (a) At a permissive temperature (26°C), the DNA (bright stain) is divided between the mother cell on left and smaller daughter cell. (b) At a nonpermissive temperature (35°C), the DNA is caught at the junction between the mother and daughter cells. [From C. Holm et al., 1985, *Cell* **41**:554; courtesy of C. Holm.]

➤ *Repair of DNA*

Errors in DNA sequence can be induced by environmental factors such as radiation, mutagenic chemicals, and thermal decomposition of nucleosides; so-called copying errors also are occasionally introduced by DNA polymerases during replication. If these errors were left totally uncorrected, both growing and nongrowing somatic cells might accumulate so much genetic damage that they could no longer function. In addition, the DNA in germ cells might incur far too many mutations for viable offspring to be formed. Thus the correction of DNA sequence errors in all types of cells is important for survival.

Proofreading by DNA Polymerase Corrects Copying Errors

The enzymatic basis for maintaining the correct base sequence during DNA replication is most completely understood in *E. coli*. DNA polymerase III has been shown to introduce about 1 incorrect base in 10^4 internucleotide linkages during replication in vitro. Since an average *E. coli* gene is about 10^3 bases long, an error frequency of 1 in 10^4 base pairs would cause a potentially harmful mutation in every 10th gene during each replication, or 10^{-1} mutations per gene per generation. However, the measured mutation rate in bacterial cells is much less, about 10^{-5} to 10^{-6} mutations per gene per generation.

The mystery of this increased accuracy in vivo was, in part, cleared up when the proofreading function of DNA polymerases was discovered first in DNA polymerase I and then in DNA polymerase III. In this work, a synthetic poly-dA template–primer complex containing a mismatched base at the 3′-hydroxyl end of the primer was synthesized. When the mismatched complex was supplied to *E. coli* DNA polymerase I, the incorrectly hydrogen-bonded base was removed by a $3' \rightarrow 5'$ exonuclease activity (Figure 10-25). In DNA polymerase III, this function resides in the Θ subunit of the core polymerase. It is likely that if an incorrect base were accidentally incorporated during DNA synthesis, the polymerase would pause, excise it, and then recopy the region correctly. Proofreading is a property of almost all bacterial DNA polymerases. Both the δ and ϵ, but not the α, DNA polymerases of animal cells also have proofreading activity. It seems likely that this function is indispensable for all cells to avoid excessive genetic damage.

Genetic studies in *E. coli* have shown that proofreading does, indeed, play a critical role in maintaining sequence fidelity during replication. Mutations in the gene encoding the Θ subunit of DNA polymerase III inactivate the proofreading function and lead to a massive increase in the rate of spontaneous mutations by some 10^5 fold. *E. coli* possesses an additional mechanism for checking the fidelity of DNA replication by identifying mispaired bases in newly replicated DNA. This enzymatic repair machinery determines which strand is to be repaired by distinguishing the newly replicated strand (the one in which an error occurred during replication) from the template. This is accomplished by assessing the *methylation* state of each strand; *E. coli* DNA is methylated on both strands at A residues in the sequence GATC through a postreplication mechanism; a newly synthesized strand remains unmethylated for a short while after synthesis.

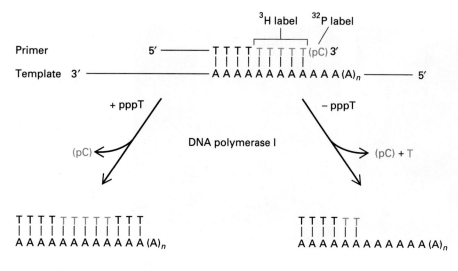

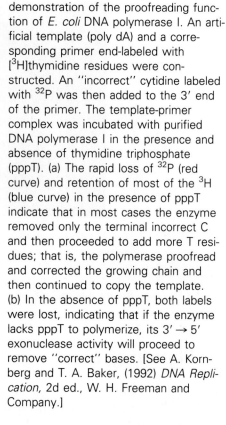

▲ FIGURE 10-25   Experimental demonstration of the proofreading function of *E. coli* DNA polymerase I. An artificial template (poly dA) and a corresponding primer end-labeled with [³H]thymidine residues were constructed. An "incorrect" cytidine labeled with ³²P was then added to the 3′ end of the primer. The template-primer complex was incubated with purified DNA polymerase I in the presence and absence of thymidine triphosphate (pppT). (a) The rapid loss of ³²P (red curve) and retention of most of the ³H (blue curve) in the presence of pppT indicate that in most cases the enzyme removed only the terminal incorrect C and then proceeded to add more T residues; that is, the polymerase proofread and corrected the growing chain and then continued to copy the template. (b) In the absence of pppT, both labels were lost, indicating that if the enzyme lacks pppT to polymerize, its 3′ → 5′ exonuclease activity will proceed to remove "correct" bases. [See A. Kornberg and T. A. Baker, (1992) *DNA Replication,* 2d ed., W. H. Freeman and Company.]

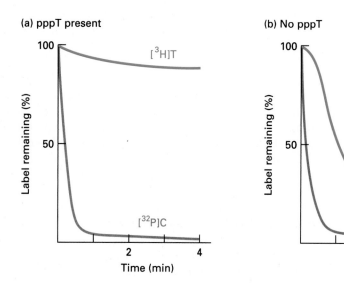

Environmental DNA Damage Can Be Repaired by Several Mechanisms

Many cells that divide very slowly or not at all (e.g., liver and brain cells) must use the information in their DNA for weeks, months, or even years. However, the DNA in all cells, even nongrowing ones, is subject to physical and chemical damage. Uncorrected base changes in the DNA of nongrowing cells would result in the production of faulty proteins at an unacceptable rate. The evolutionary response to this problem has been the development of DNA-repair systems. Table 10-3 lists the general types of DNA damage and their causes.

DNA-repair mechanisms have been studied most extensively in *E. coli* using a combination of genetic and bio-

chemical approaches. These studies reveal a remarkably diverse collection of enzymatic repair mechanisms and cellular responses to DNA damage with some 100 genes participating in these processes. Repair mechanisms can be divided into two broad categories: in one, the damaged DNA is directly repaired; in the other, more common, mechanism, the damaged region is removed (excised) by specialized nuclease systems and then DNA synthesis fills the gap. Excision-repair processes involve the following steps: (a) the damaged DNA segment must be recognized as distinct from undamaged DNA; (b) the damaged strand is cleaved by an endonuclease; (c) an exonuclease removes a portion of the damaged strand, leaving a single-stranded gap; and finally (d) a DNA polymerase and ligase collaborate in repairing the gap using the undamaged strand as a template.

TABLE 10-3 DNA Lesions That Require Repair

DNA Lesion	Example/Cause
Missing base	Removal of purines by acid and heat. (Under physiological conditions $\approx 10^4$ purines/day/cell in a mammalian genome); removal of altered bases (e.g., uracil) by DNA glycosylases
Altered base	Ionizing radiation; alkylating agents (e.g., ethylmethanesulfonate)
Incorrect base	Mutations affecting $3' \rightarrow 5'$ exonuclease proofreading of incorrectly incorporated bases
Bulge due to deletion or insertion of a nucleotide	Intercalating agents (e.g., acridines) cause addition or loss of a nucleotide during recombination or replication
Linked pyrimidines	Cyclobutyl dimers (usually thymine dimers) resulting from UV irradiation
Strand breaks	Breakage of phosphodiester bonds by ionizing radiation or chemical agents (e.g., bleomycin)
Cross-linked strands	Covalent linkage of two strands by bifunctional alkylating agents (e.g., mitoomycin)
$3'$-deoxyribose fragments	Disruption of deoxyribose structure by free radicals leading to strand breaks

SOURCE: Adapted from A. Kornberg and T. Baker, 1992, *DNA Replication*, 2nd ed., W.H. Freeman and Company, pp. 771–773.

Excision Repair in *E. coli* Removes Bulky Chemical Adducts Caused by UV Light and Carcinogens

In this section, we describe the *E. coli* excision-repair mechanism called *UvrABC repair,* which is used to repair DNA regions containing "bulky" chemical adducts. For example, this mechanism repairs *thymine-thymine dimers,* which are caused by UV light (Figure 10-26), and other altered bases attached to large chemical groups (e.g., carcinogens such as methylcholanthrene). Cells carrying mutations in the *uvrA, B,* or *C* locus are very sensitive to agents causing this sort of DNA damage. The availability of strains for producing large quantities of the UvrA, B, and C proteins has facilitated their purification and the characterization of their roles in this process.

As illustrated in Figure 10-27, the first step in this mechanism involves binding of a complex containing two molecules of UvrA and one molecule of UvrB to DNA. The formation and binding of this complex requires ATP. It

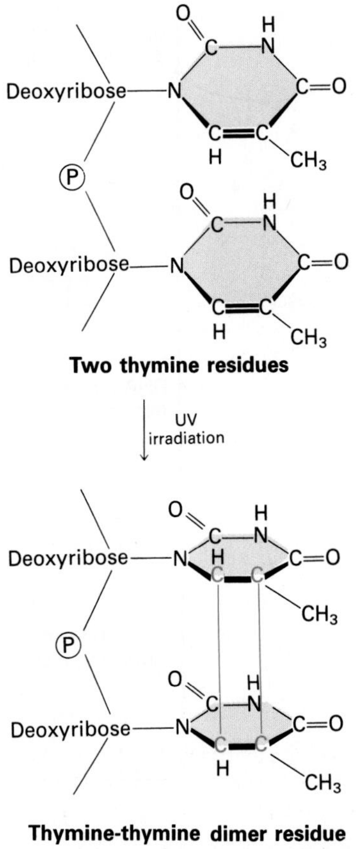

Two thymine residues

UV irradiation

Thymine-thymine dimer residue

▲ FIGURE 10-26   UV irradiation can cause adjacent thymine residues in the same DNA strand to become covalently attached. The resulting thymine-thymine dimer (cyclobutylthymine) may be repaired by an excision-repair mechanism.

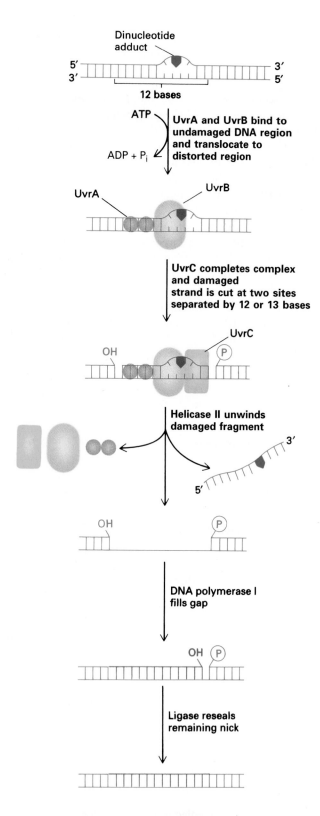

Dinucleotide adduct

12 bases

ATP

ADP + P_i

UvrA and UvrB bind to undamaged DNA region and translocate to distorted region

UvrA   UvrB

UvrC completes complex and damaged strand is cut at two sites separated by 12 or 13 bases

UvrC

OH   P

Helicase II unwinds damaged fragment

3'

5'

OH   P

DNA polymerase I fills gap

OH   P

Ligase reseals remaining nick

▲ FIGURE 10-27   Excision repair of DNA by *E. coli* UvrABC mechanism. In the example shown, the lesion results from a dinucleotide adduct to the DNA. This mechanism can repair various lesions, including thymine-thymine dimers, that produce distortions in the normal shape of DNA in a region. See text for discussion. [Adapted from A. Sancar and G. B. Sancar, 1988, *Ann. Rev. Biochem.* **57**:29.]

seems likely that the UvrA-UvrB complex initially binds to an undamaged segment and translocates along the DNA helix until a distortion caused by a DNA adduct is recognized; this translocation along the helix also requires ATP. Studies with the DNA footprinting method discussed in Chapter 11 have shown that UvrA alone can bind selectively to damaged DNA over a 33-bp region. This interaction is considerably weaker than that between damaged DNA and the UvrA-UvrB complex, which occurs over 19 base pairs. In the presence of ATP, a region of the damaged strand bound to the UvrA-UvrB complex becomes more susceptible to endonuclease cleavage.

DNA cleavage requires the UvrC protein, which binds to the damaged DNA/UvrA-UvrB complex. The precise position of the cleavage sites is determined by the nature of the DNA damage. In the case of thymine dimers, for instance, the endonuclease cleaves two phosphodiester bonds: one is located eight nucleotides 5' to the lesion, and one is located four or five nucleotides 3' to the lesion. After cleavage, the damaged strand is removed by a helicase; the resultant gap then is repaired by the combined actions of DNA polymerase I and DNA ligase.

Genetic Studies in Eukaryotes Have Identified DNA-Repair Genes

Several types of genetic studies in eukaryotes have identified DNA repair genes that encode proteins involved in DNA repair. The basic strategy is to search for mutants that exhibit increased sensitivity to UV light or other agents that damage DNA. Such mutants presumably are deficient in the wild-type excision-repair mechanisms that normally can repair damage caused by such agents. In the yeast *S. cerevisiae*, numerous UV-sensitive mutants and 10 radiation-sensitive (*RAD*) genes have been identified. Of these genes, six are probably involved in the initial events of recognizing the damaged DNA. Techniques of somatic-cell genetics have been used to generate mutant Chinese hamster ovary (CHO) cell lines that are abnormally sensitive to UV light or to DNA damage caused by mitomycin C. These mutations fall into eight complementation groups, indicating that at least eight different genes are involved in excision repair in hamsters. Finally, mutations mapping to seven complementation groups in humans all give rise to a hereditary disease called *xeroderma pigmentosa*. Patients with this disease are highly sensitive to UV light and have an increased incidence of melanomas and carcinomas. Several other human genetic diseases also are associated with increased sensitivity to DNA-damaging agents and cancer susceptibility (Table 10-4); the mutations causing all these diseases quite likely are in DNA-repair genes.

Both DNA sequencing of DNA-repair genes and functional studies suggest that *E. coli*, yeast, hamsters, and humans all employ quite similar molecular mechanisms for excision repair of DNA. For example, the predicted protein

TABLE 10-4 Human Hereditary Diseases Associated with DNA-repair Defects

Disease*	Sensitivity	Cancer Susceptibility	Symptoms
Ataxia telangiectasia	γ irradiation	Lymphomas	Ataxia, dilation of blood vessels in skin and eyes, chromosome aberrations, immune dysfunction
Bloom's syndrome	Mild alkylating agents	Carcinomas, leukemias, lymphomas	Photosensitivity, facial telangiectases, chromosome alterations
Cockayne's syndrome	UV irradiation		Dwarfism, retinal atrophy, photosensitivity, progeria, deafness, trisomy 10
Fanconi's anemia	Cross-linking agents	Leukemias	Hypoplastic pancytopenia, congenital anomalies
Xeroderma pigmentosum	UV irradition, chemical mutagens	Skin carcinomas and melanomas	Skin and eye photosensitivity, keratoses

* Other human hereditary disorders that may be related to DNA-repair defects include dyskeratosis congenita (Zinsser-Cole-Engman syndrome), progeria (Hutchinson-Gilford syndrome), and trichothiodystrophy.
SOURCE: From A. Kornberg and T. Baker, 1992, *DNA Replication*, 2d ed., W. H. Freeman and Company, p. 788.

product encoded by the yeast *RAD14* gene has considerable homology with the protein encoded by one of the genes mutated in xeroderma pigmentosa patients. In addition, DNA transfection experiments have revealed that certain human genes can rescue some UV-sensitive CHO cell mutants. Two of the human genes identified in this way have been shown to be related to the yeast *RAD3* and *RAD10* genes. The human protein with partial homology to the RAD10 protein also contains a region that is similar to part of *E. coli* UvrC.

▶ Recombination between Homologous DNA Sites

In the previous sections of this chapter, we discussed the enzymatic mechanisms by which the genome is faithfully reproduced from one generation to another through the process of DNA replication, and the phenomenon of DNA repair necessary to maintain the correct DNA sequence. In this section, we examine the mechanisms of recombination by which the genome can change from one generation to another.

Soon after Mendel's rules of independent gene segregation were rediscovered and the segregation of linked groups of genes on individual chromosomes was widely recognized, another great genetic discovery was made in *D. melanogaster*: blocks of genes from homologous chromosomes could be exchanged by the process of crossing over, or *recombination*. Recombination, which takes place

during meiosis in sexually reproducing organisms, provides a mechanism for generating genetic diversity beyond that achieved by the independent assortment of chromosomes. Genetic exchange by recombination occurs not only in animals and plants but also in prokaryotes, viruses, plasmids, and even in the DNA of cell organelles such as mitochondria.

The events in a *reciprocal recombination* are equivalent to the breakage of two duplex DNA molecules representing homologous but genetically distinguishable chromosomes, an exchange of *both* strands at the break, and a resolution of the two duplexes so that no tangles remain. The frequency of recombination between two sites is proportional to the distance between the sites. (As discussed in Chapter 8, this phenomenon is the basis of genetic mapping of genes defined by mutations.) In the remainder of this chapter, we describe various proteins involved in recombination and models of how they carry out this process.

Holliday Recombination Model Is Supported by Observation of Predicted Intermediate Structures

A proposal by Robin Holliday in 1964 is the basis of the most popular current models for the molecular events of recombination, which are illustrated in Figure 10-28. In step 1, a nick is made in one strand of each of the two homologous chromosomes that are going to recombine. Strand exchange then occurs at the site of the nicks, and

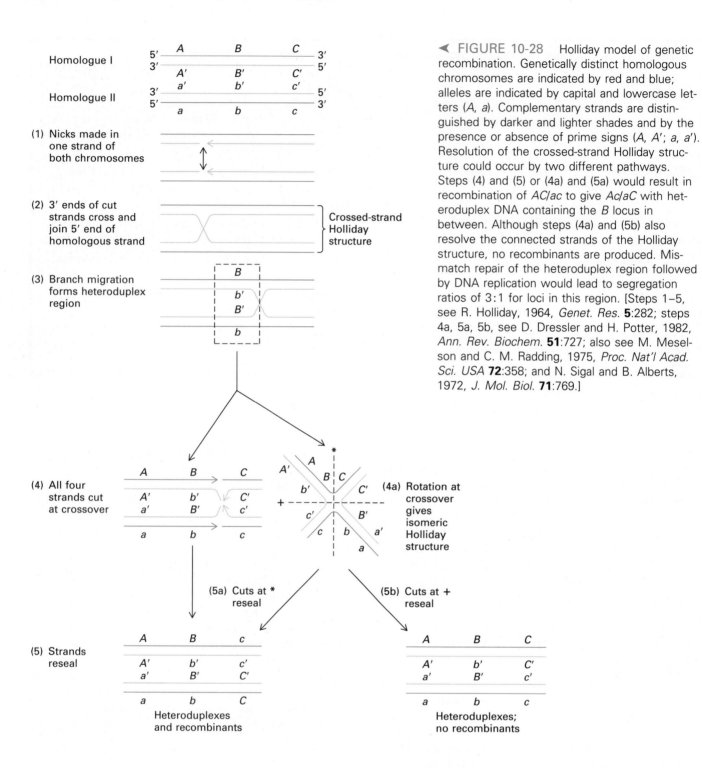

◄ FIGURE 10-28   Holliday model of genetic recombination. Genetically distinct homologous chromosomes are indicated by red and blue; alleles are indicated by capital and lowercase letters (*A, a*). Complementary strands are distinguished by darker and lighter shades and by the presence or absence of prime signs (*A, A'; a, a'*). Resolution of the crossed-strand Holliday structure could occur by two different pathways. Steps (4) and (5) or (4a) and (5a) would result in recombination of *AC/ac* to give *Ac/aC* with heteroduplex DNA containing the *B* locus in between. Although steps (4a) and (5b) also resolve the connected strands of the Holliday structure, no recombinants are produced. Mismatch repair of the heteroduplex region followed by DNA replication would lead to segregation ratios of 3:1 for loci in this region. [Steps 1–5, see R. Holliday, 1964, *Genet. Res.* **5**:282; steps 4a, 5a, 5b, see D. Dressler and H. Potter, 1982, *Ann. Rev. Biochem.* **51**:727; also see M. Meselson and C. M. Radding, 1975, *Proc. Nat'l Acad. Sci. USA* **72**:358; and N. Sigal and B. Alberts, 1972, *J. Mol. Biol.* **71**:769.]

the cut 3′ ends are joined to the 5′ ends of the homologous strand, producing a *crossed-strand Holliday structure* (step 2). The branch point then migrates, creating a *heteroduplex* region containing one strand from each parental chromosome (step 3).

Two mechanisms have been proposed for separation, or *resolution*, of the connected duplexes. According to the original proposal, all four strands are cut at the crossover site and the left side of chromosome I joins the right side of

chromosome II, and vice versa (steps 4 and 5 in Figure 10-28). Both strands in each of the resulting duplexes are recombinant; that is, all markers to the left and right of the crossover site have undergone reciprocal recombination. A later proposal simplifies the enzymatic cutting that is necessary to resolve the crossed-strand intermediate. Rotation of the Holliday structure at the crossover site forms a rotational isomer, or isomeric *Holliday structure* (step 4a). The two connected duplexes of this structure could be resolved

Homologue I

3′ ──────────────────────── 5′
5′ ──────────────────────── 3′

Homologue II

5′ ──────────────────────── 3′
3′ ──────────────────────── 5′

Nick and
strand displacement
by synthesis
at 3′ end

Uptake of
displaced single
strand, leaving
unpaired loop

Loop excised,
resulting in
single-strand
invasion

Rotation of
both chromosomes
produces two
crossover strands

Branch migration
creates
heteroduplex
on each
chromosome

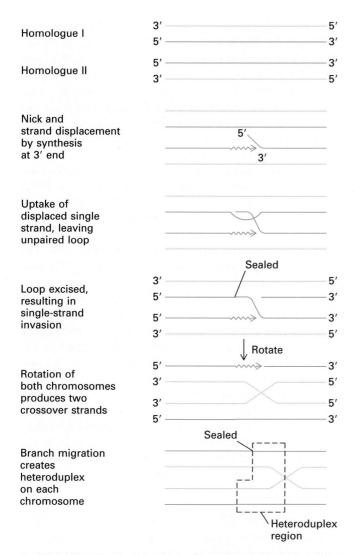

▲ FIGURE 10-29   Meselson-Radding model of genetic recombination. Genetically distinct homologous chromosomes are indicated by red and blue; complementary strands are distinguished by darker and lighter shades. In this modification of the Holliday model, formation of the crossed-strand Holliday structure begins with a nick in one of the chromosomes. After rotation at the crossed-strand site and branch migration of the crossed-strand intermediate, resolution would occur as shown in Figure 10-28. [See M. Meselson and C. Radding, 1975, *Proc. Nat'l. Acad. Sci. USA* **72**:358.]

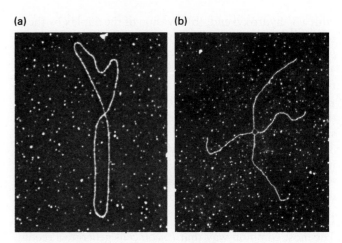

▲ FIGURE 10-30   Electron micrographs of plasmid DNA in the process of recombination. (a) Circular plasmid DNA in crossed-strand Holliday structure. (b) More highly magnified view reveals single-stranded ring in center of isomeric Holliday structure that results from rotation about the crossover point. [See H. Potter and D. Dressler, 1978, *Cold Spring Harbor Symp. Quant. Biol.* **43**:969; courtesy of D. Dressler.]

tion model, Matthew Meselson and Charles Radding suggested another mechanism for creating a crossed-strand Holliday structure; this mechanism requires a single-strand cut in only one chromosome (Figure 10-29). DNA synthesis from the nick leads to the formation of a 5′-phosphate end, which then invades the homologous duplex DNA. After branch migration of the crossed-strand region, rotation and resolution produce results similar to those described in Figure 10-28. As we will discuss in the next section, this model is supported by the finding that a single strand can be used to initiate a recombination event using a single *E. coli* protein in vitro.

Do Holliday intermediates actually exist? Viral and plasmid DNA molecules in the act of recombining can be extracted from both bacterial and animal cells. Electron micrographs of such molecules have revealed structures similar to the crossed-strand and isomeric Holliday structures (Figure 10-30). Thus, regardless of the mechanism initiating recombination, the final connection between the unresolved chromosomes seems to involve branch migration and chromosomal rotation.

Recombination in *E. coli* Occurs by Three Similar Pathways That All Require RecA Protein

Genetic analysis has revealed three different pathways of recombination in *E. coli*. Not only is the RecA protein required in all three pathways, but they also share the following steps: (a) generation of a single-stranded DNA segment

by cutting and rejoining of only two strands. If this involves the two strands that were not cut to generate the original Holliday intermediate, then recombinant duplex chromosomes containing a heteroduplex region are produced (step 5a). However, if resolution involves cutting of the two strands that were originally cut, the resulting duplex chromosomes contain a heteroduplex region but are not recombinants (step 5b).

In a further modification of the Holliday recombina-

with a 3'-hydroxyl end; (b) invasion of the duplex by the 3' *recombinogenic* end of the single-stranded DNA and complexing with regions of homology; (c) formation of a Holliday structure, which can undergo branch migration; and (d) endonuclease cleavage followed by ligation to yield recombinants. Recombination in *E. coli,* as in the Meselson-Radding model, is initiated by a single-strand cut; however, in *E. coli* the recombinogenic end is a 3'-hydroxyl end, whereas in the Meselson-Radding model it is a 5'-phosphate end (see Figure 10-29). In this section, we describe the primary *E. coli* recombination pathway, which is initiated by an enzyme complex called RecBCD.

Initiation of Recombination (RecBCD Enzyme)

The most common way that *E. coli* cells generate a recom-

binogenic single-stranded region of DNA probably is by action of the RecBCD enzyme. This enzyme complex, which is composed of proteins encoded by the *recB, C,* and *D* genes, specifically recognizes double-strand breaks. Such breaks occur naturally during bacterial conjugation, a process in which chromosomal DNA is transferred from one bacterium to another through direct cell contact. Double-stranded breaks also can be generated by exposure to x-rays and certain chemicals.

The mechanism of action of RecBCD was worked out in studies with bacteriophage λ, which has a linear DNA genome with free blunt ends equivalent to double-strand breaks. Certain regions of λ-phage DNA undergo recombination at higher frequencies than other regions in normal *E. coli* host cells but not in *recBCD* mutant host cells. The sites of increased recombination were named *CHI sites* because the Greek letter chi looks like a crossover point. Experiments with purified RecBCD enzyme and λ DNA indicate that the protein recognizes and binds to a free end of the λ-phage chromosome. The enzyme then functions as a helicase, moving along and remaining attached to the duplex DNA as it unwinds it (Figure 10-31). Because RecBCD enzyme unwinds the DNA faster than it is rewound, single-stranded loops are created as the protein progresses. After a CHI site has been passed, a specific nuclease activity of RecBCD cuts one of the exposed strands, leaving a free 3'-hydroxyl end just downstream from the CHI site. This end can now participate in the process of strand invasion, which is catalyzed by RecA protein.

Strand Invasion, Homologous Pairing, and Formation of Holliday-Type Structure (RecA Protein)

Studies with single-stranded DNA (ssDNA) showed that the protein encoded by the *recA* gene can bind to ssDNA. In the presence of target duplex DNA, the RecA-ssDNA complex can find and bind to a target-DNA region homologous to the ssDNA. RecA then can insert the ssDNA into the target DNA, displacing one of the preexisting strands and forming a heteroduplex Holliday-type structure (Figure 10-32).

In subsequent studies, RecA was shown to bind to ssDNA regions generated by action of the RecBCD enzyme. *E. coli* Ssb protein stimulates this reaction by binding to the single-stranded region and preventing intrastrand base pairing, which would inhibit binding of RecA. In the presence of ATP, RecA coats the single-stranded region and polymerizes, forming a filament that wraps around the entire length of the ssDNA. Because the polymerization of RecA occurs in the 5' → 3' direction along the DNA, coating takes place in a discontinuous fashion as a region of the duplex is unwound. X-ray crystallographic analysis of RecA suggests that each molecule has two DNA-binding sites, both of which may lie within the core of the filament.

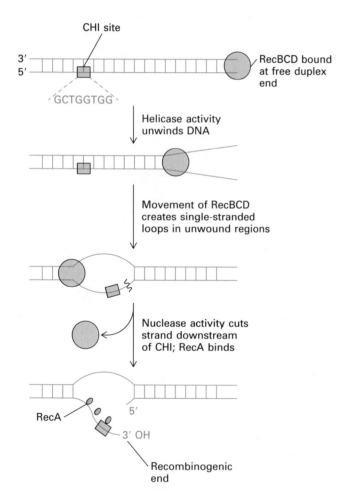

▲ **FIGURE 10-31**   Initiation of recombination by *E. coli* RecBCD enzyme. This enzyme, which has both helicase and nuclease activities, produces a single-stranded DNA region with a free 3'-hydroxyl end. This recombinogenic end binds RecA, which catalyzes strand invasion and formation of a Holliday structure. [See A. F. Taylor et al., 1985, *Cell* **41**:153.]

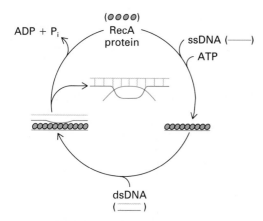

▲ FIGURE 10-32   Formation of Holliday-type structure by *E. coli* RecA. In the presence of ATP, RecA binds single-stranded DNA (ssDNA) and promotes insertion of the bound strand at a homologous region of double-stranded DNA (dsDNA), yielding a crossed-strand Holliday-type structure. The insertion reaction requires the ATPase activity of RecA. [See S. S. Flory et al., 1984, *Cold Spring Harbor Symp. Quant. Biol.* **49**:513.]

RecA effectively catalyzes the in vitro formation of Holliday structures between a double-stranded circular DNA containing a short single-stranded gap and a fully double-stranded linear DNA. Although formation of Holliday structures depends on RecA, their maintenance does not. Removal of RecA leaves stable Holliday structures, which have been used as substrates in studies on branch migration and resolution.

RecA protein also is important in DNA repair, in part, through its central role in regulating the *SOS response* of *E. coli* to UV irradiation. When *E. coli* cells are irradiated with UV light, a whole series of *SOS* genes that enable the bacterium to survive is activated, including the *recA* gene itself. In irradiated cells, the RecA protein promotes the proteolytic breakdown of various DNA-binding proteins. For instance, breakdown of the lambda cI repressor is dependent upon *recA* function; this inactivation of the repressor releases the dormant phage in irradiated cells. RecA protein does not appear to be a protease itself, but rather promotes the DNA-binding proteins to degrade themselves in a reaction requiring ATP and single-stranded DNA (in vivo the single-stranded DNA is probably formed by DNA breakdown induced by irradiation).

RecA protein also promotes the cleavage of LexA protein. This protein, the product of the regulatory gene *lexA*, is a repressor of a series of other genes, many of which are involved in DNA repair. For instance, both the *recA* gene itself and the genes for the UvrABC nuclease are expressed when the LexA protein is digested. The elaborate SOS response sheds light on why UV irradiation not only induces repair synthesis but also, because of the increased RecA production, increases recombination.

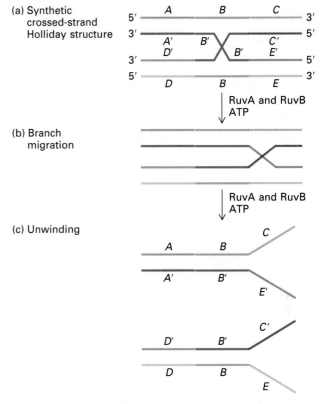

▲ FIGURE 10-33   Experimental demonstration of branch migration catalyzed by *E. coli* RuvA and RuvB proteins. A synthetic Holliday structure was produced by annealing synthetic single-stranded oligonucleotides in which only the center *BB′* region was homologous. Complementary strands are indicated by darker and lighter shades and the presence or absence of prime signs (*A,A′;B,B′*); segments with different sequences are indicated by color. (a) In the synthetic Holliday structure, the crossover involves only the homologous region (green). (b,c) Treatment of the Holliday structure with RuvA and RuvB in the presence of ATP leads to branch migration followed by unwinding to yield cruciform structures with nonhomologous single-stranded ends. Branch migration in the other direction (towards *A* and *D*) yields similar cruciform structures. [Adapted from H. Iwasaki et al., 1992, *Genes & Dev.* **6**:2214.]

Branch Migration and Resolution of Holliday Structures (Ruv Proteins)   Strains of *E. coli* with mutations in the *ruvA*, *ruvB*, or *ruvC* gene exhibit defective recombination, indicating that the proteins encoded by these genes play a role in recombination. Migration of the crossover point in Holliday structures is efficiently catalyzed by RuvA and RuvB. RuvA protein specifically recognizes the Holliday structure, whereas RuvB protein has the helicase activity necessary for promoting branch migration. Studies with synthetic oligonucleotide substrates have clarified the role of these two proteins (Figure 10-33).

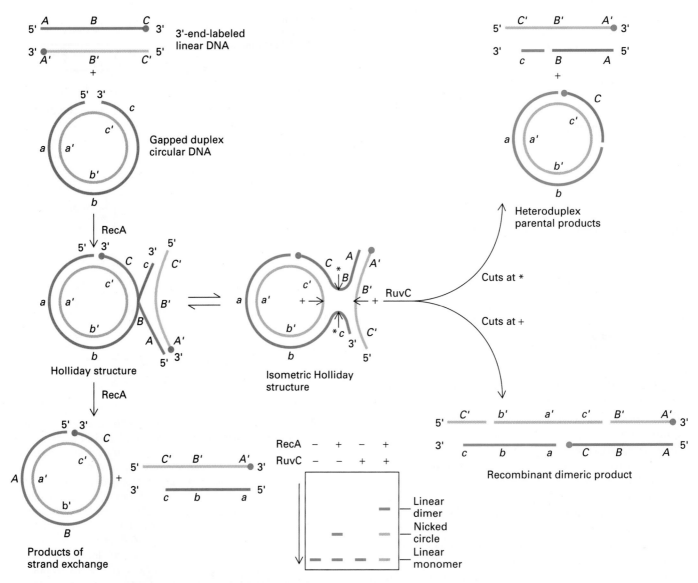

▲ FIGURE 10-34   Demonstration that *E. coli* RuvC protein can resolve Holliday structures. Genetically distinct DNA duplexes are indicated by red and blue; alleles are indicated by capital and lowercase letters (*A,a*). Complementary strands are distinguished by darker and lighter shades and the presence or absence of primes (*A, A'*). A linear DNA molecule end-labeled with ^{32}P (red dot) and a homologous circular DNA molecule containing a short region of single-stranded DNA, called gapped DNA, were incubated with RecA and ATP to produce Holliday structures. In the absence of RuvC, strand exchange catalyzed by RecA yielded two heteroduplex labeled products: a linear monomeric DNA and a nicked circular DNA, corresponding to the parental configuration (i.e., a nicked circle and linear monomer). When RuvC

was also present in the reaction mixture, it catalyzed specific cleavage of the Holliday intermediate. Cleavage at + sites would yield a recombinant linear dimer, whereas cleavage at * sites would yield heteroduplex DNAs in the parental configuration. (*Inset*) Diagram of an autoradiograph obtained following gel electrophoresis of the reaction products with a fixed amount of RecA in the presence and absence of RuvC. Note that in the presence of RecA and absence of RuvC, the radiolabel appeared in a linear monomeric DNA and nicked circular DNA. In the presence of RuvC, the primary product detected was a linear dimeric DNA, indicating that RuvC cleaves Holliday structures preferentially at the + sites. [Adapted from H. J. Dunderdale et al., 1991, *Nature* **354**:506.]

Holliday structures generated by RecA are efficiently resolved by the nuclease action of the RuvC protein, which binds specifically to these recombination intermediates

(Figure 10-34). RuvC only binds to crossed-strand regions that exhibit homology, suggesting that specific base pairing occurs within the crossover region. RuvC is not inhibited

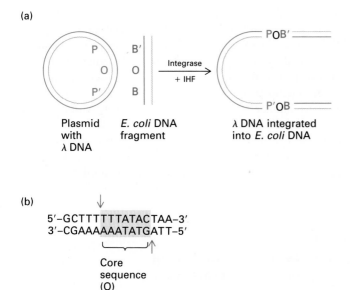

(a)

Plasmid with λ DNA | E. coli DNA fragment | λ DNA integrated into E. coli DNA

(b)

5′–GCTTTTTTATACTAA–3′
3′–CGAAAAAATATGATT–5′

Core
sequence
(O)

▲ FIGURE 10-35   Integration of phage λ into E. coli by recombination at the core O sequence in the λ attachment site (POP′) and E. coli attachment site (BOB′). (a) In vitro incubation of supercoiled plasmid containing the phage BOB′ segment and a linear fragment of the E. coli genome containing BOB′ with purified λ integrase and integration host factor (IHF) yields a linear DNA molecule in which the phage DNA is integrated to produce the order B′OP-plasmid-P′OB (right). Complementary strands are distinguished by darker and lighter shades. (b) The sequence of the 15-bp region that is similar in the bacterial and phage attachment sites is shown. The core region is identical. The arrows indicate the sites at which staggered cuts are made around the core sequence (O). (c) Steps in the integration of λ into E. coli, which is equivalent to genetic recombination between allelic markers (e.g., P and B) on homologous chromosomes. [See P. A. Kitts and H. A. Nash, 1987, Nature 329:346.]

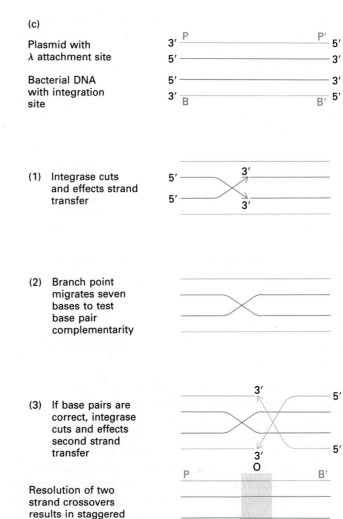

(c)

Plasmid with λ attachment site

Bacterial DNA with integration site

(1) Integrase cuts and effects strand transfer

(2) Branch point migrates seven bases to test base pair complementarity

(3) If base pairs are correct, integrase cuts and effects second strand transfer

Resolution of two strand crossovers results in staggered recombination

by the presence of RecA, nor does it require interaction with RecA, as demonstrated by its ability to resolve Holliday structures lacking RecA.

Site-Specific Integration of λ Phages Mimics a Homologous Recombination Event

Several examples of *site-specific* recombination have been discovered in both prokaryotic and eukaryotic cells. Site-specific recombination requires the recognition of unique nucleotide sequences in both DNA molecules by enzymes called *recombinases*, which then catalyze the joining of the two molecules. One well-studied example is the integration of bacteriophage λ into a particular site in the E. coli chromosome. Integration can be carried out in vitro with the

viral enzyme *integrase,* which in a stepwise fashion makes and then resolves a Holliday junction. Many, but not all, site-specific recombination systems proceed through a Holliday intermediate.

The site-specific integration of λ-phage DNA into the E. coli chromosome is thought to be quite similar to generalized homologous recombination. The genome of λ phage contains a 15-bp region, the attachment site, which contains an 7-bp core sequence identical to the integration (or attachment) site in the host-cell DNA. The in vitro integration system consists of a plasmid DNA that contains the phage attachment site (POP′), a linear fragment of the bacterial DNA containing its integration site (BOB′), purified integrase, and a bacterial host protein termed integration host factor (Figure 10-35 a,b). By using DNA molecules with mutations or single-strand breaks in the 15-bp homologous region, researchers have been able to stop the reaction at several stages and to collect intermediates in the reaction leading to integration.

The interpretation of all the experiments with various mutants is that phage integrase (as a dimer) makes staggered cuts in and catalyzes strand transfer between bacterial and phage DNA at the core sequence of the attachment sites (Figure 10-35c, step 1). As in the case of topoisomerases, the cut ends of the DNA are covalently bound to integrase during this reaction. The resulting strand-transfer structure is identical to the crossed-strand Holliday structure. If the neighboring sequences in the POP′ and BOB′ sites are correct, then a branch migration of seven bases occurs and the duplex remains perfectly paired (step 2). A second strand exchange (step 3) catalyzed by integrase then resolves the Holliday structure, resulting in the integration of the phage DNA into the bacterial chromosome.

Studies in Yeast Are Providing Insights into Meiotic Recombination

A central interest in exploring the mechanisms of homologous recombination in eukaryotes is its role in meiosis. Studies in yeast reveal that defects in meiotic recombination lead to defects in meiosis itself. It is unclear whether these meiotic defects reflect a requirement for meiotic recombination or whether intermediates that accumulate in mutant strains (e.g., double-strand breaks) lead to meiotic arrest. Nonetheless, it seems likely that an understanding of meiotic recombination will be an essential part of our understanding of the complex process of correctly pairing homologous chromosomes during meiosis.

Four different approaches have been taken to studying meiotic recombination in yeast. First, genetic screens have identified one panel of genes required for meiotic recombination and another for both meiotic and mitotic recombination. Second, four proteins have been purified that exhibit a strand-exchange activity similar to that of E. coli RecA. The roles of these yeast proteins in recombination remain unclear but can be addressed by gene knockout experiments. Third, analysis of DNA regions undergoing recombination have revealed recombination intermediates containing double-strand breaks with 5′ recessed ends (i.e., 3′ single-stranded tails). Based on this and other observations, a model for yeast recombination has been proposed. And fourth, using a differential hybridization scheme, investigators have identified genes that are selectively expressed during meiosis.

Genetic analysis reveals that DMC1 (disrupted meiosis, cDNA 1), one of the genes identified by differential hybridization, is required for meiotic recombination in yeast. Yeast cells with mutations in DMC1 accumulate intermediates that have double-strand breaks and 3′ single-

stranded tails. In vitro, E. coli RecA protein will bind to this type of intermediate; moreover, RecA and the yeast DMC1 protein exhibit sequence homology. Biochemical studies will be necessary to directly assess whether DMC1 functions in a manner similar to RecA. Interestingly, the protein encoded by the yeast RAD51 gene has an amino sequence similar to both RecA and DMC1. Mutations in RAD51 result in defects in both meiotic and mitotic recombination. Based on these very early observations, it is tempting to propose that both prokaryotic and eukaryotic recombination mechanisms may be quite similar.

Gene Conversion Can Occur near the Crossover Point during Reciprocal Recombination

Although markers at some distance from the crossover point are exchanged in a reciprocal fashion during recombination, an apparent nonreciprocal event sometimes occurs at or near the crossover point. This phenomenon is most easily studied in yeast in which each meiotic product can be scored in the haploid progeny spores. In a cross of multiply marked yeast strains that undergo recombination, most allelic markers segregate 2:2, but a few show 3:1 or 1:3 segregation (Figure 10-36). Such a nonreciprocal event is called *gene conversion* because one allele is apparently "converted" into another. It is now known that in gene conversion, the exact base sequence is represented at the converted site, obviously suggesting that exact copying of a DNA strand is involved. This property is referred to as *fidelity* of gene conversion. The occurrence of 1:3 and 3:1 events at equal frequencies is called *parity* of gene conversion.

The double-strand break model of yeast meiotic recombination shown in Figure 10-37 can account for the observed phenomenon of gene conversion. In this model, an intermediate with two crossovers and two heteroduplex regions is generated. The distortion in the heteroduplex regions caused by mispairing of the different alleles (e.g., D and d) can be recognized by a mismatch repair system, which removes a single-stranded region containing the mismatch and then fills in the gap substituting correctly paired base(s) for the mispaired one(s).

There are three possible results of mismatch repair at the heteroduplex region indicated as d/D in Figure 10-37. Conversion of the d allele to D results in a 3:1 ratio of D:d in the spores; conversion of D to d results in a 2:2 ratio; and no repair results in a 5:3 segregation ratio of D:d (Figure 10-38). In the latter case, the heteroduplex remains in the spore; after the first cell division, two daughter cells of different genotypes (reflecting the alleles in the heteroduplex) are generated, giving rise eventually to a sectored colony in which one-half of the colony is phenotypically D and the other *d*. If we view each DNA strand in the heteroduplex as giving rise to distinct progeny, then there are eight potential genotypes; this would lead to segregation of *D* and *d* in a 5:3 ratio in the event that repair does not occur. If the initial double-strand cut is on the chromosome carrying the *D* allele, rather than *d*, then the following segregation patterns would be observed: conversion of *d* to *D* results in 2:2 segregation; conversion of *D* to *d* results in 1:3 segregation; and no repair results in a 3:5 ratio of *D* to *d*.

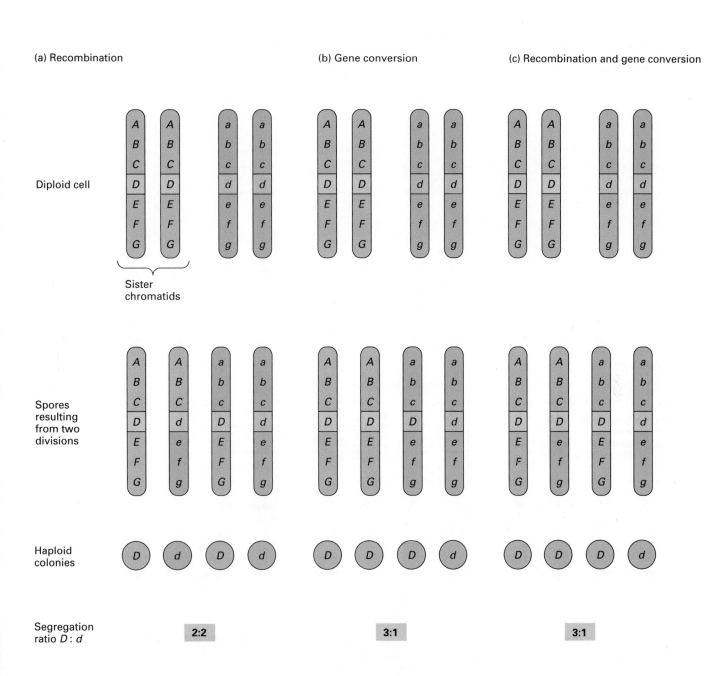

▲ FIGURE 10-36 Gene conversion during meiosis of multiply marked diploid yeast cells. Allelic markers are indicated by capital and lowercase letters. (a) Normal homologous recombination between chromatids yields haploid recombinant progeny spores in which markers segregate in a 2:2 ratio. (b) Gene conversion, a nonreciprocal event, results in a 3:1 (or 1:3) segregation ratio. In effect, one allele is converted into another; in this case, *d* to *D*. (c) Typically, gene conversion is observed for markers close to the crossover point in recombinants. See Figures 10-37 and 10-38 for molecular mechanisms whereby the different genotypes are generated.

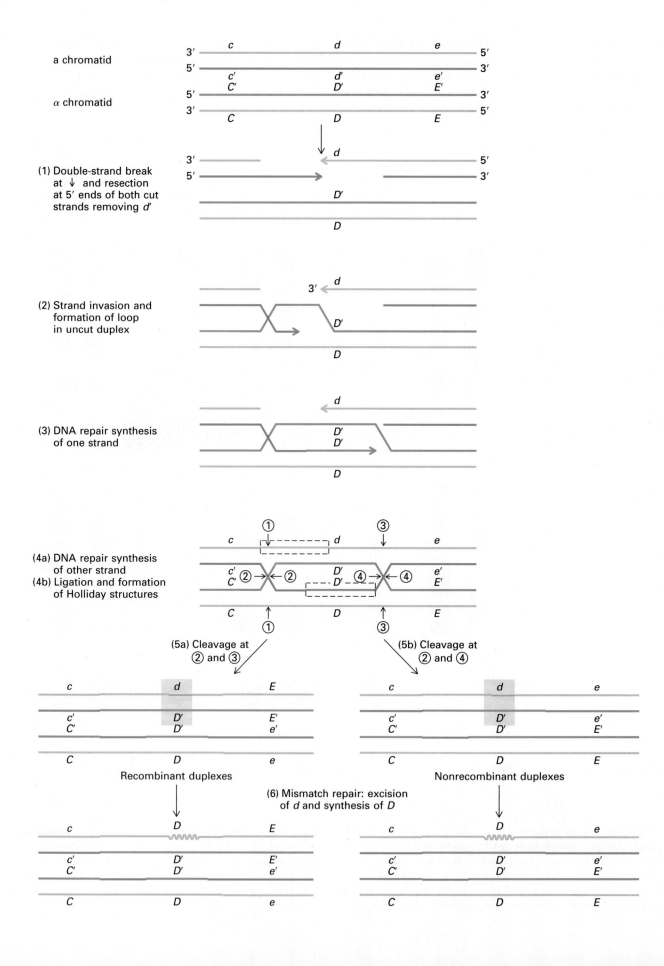

◄ FIGURE 10-37   Double-strand break model of meiotic recombination in the yeast *S. cerevisiae.* Only one a (blue) and one α (red) chromatid are shown (see Figure 10-36); alleles are indicated by capital and lowercase letters (*D*, *d*). Complementary DNA strands are indicated by darker and lighter shades and by the presence or absence of prime signs (*D*, *D'*). In the example shown, the initial double-strand break and resection of 5' ends occurs on the maternal chromatid, removing the *d'* marker. This is followed by strand invasion and DNA synthesis with the paternal *D* strand as the template. Repair synthesis of the other strand and ligation results in formation of a Holliday structure with two crossovers. (Repaired regions are marked by black dashed lines.) Resolution of this crossed-strand intermediate can occur in two ways, yielding either recombinant duplexes (5a) or nonrecombinant ones (5b). In both cases, one duplex contains a complementary *DD'* region, but the other contains a heteroduplex mismatched *dD'* region (yellow). Cells contain enzymes that can repair such mismatched segments by excising a single-strand segment containing the mismatch and using the other strand as a template for synthesis of a matching strand. In this example, *d* is removed and *D* is synthesized (jagged red segment), thus "converting" *d* to *D*. Whether recombination of distant markers is associated with gene conversion is determined by the position of the cuts that resolve the Holliday structure. (See Figure 10-38 for the possible segregation patterns resulting from mismatch repair.)

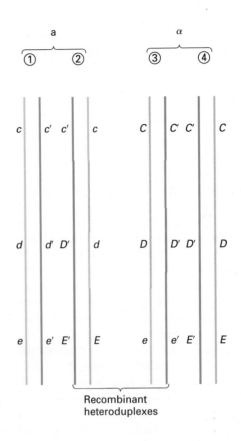

Recombinant heteroduplexes

► FIGURE 10-38   Segregation patterns arising from meiotic recombination in yeast depend on which parental chromatid undergoes a double-strand break; whether recombinant or nonrecombinant heteroduplexes are formed; whether mismatch repair of heteroduplexes occurs, and if so in which direction. Shown here are the expected segregation patterns resulting from meiosis in which the recombinant *dD'* heteroduplex diagrammed in Figure 10-37 is generated. (*Top*) The four chromatids in the diploid cell after DNA replication and recombination but prior to meiotic division. Each duplex is numbered for convenience. Note the *dD'* heteroduplex region in ②. (a,b) Spores (and haploid-cell colonies) exhibit different *d:D* segregation ratios depending on direction of mismatch repair. (c) If no mismatch repair occurs, one spore will contain the mismatched heteroduplex ②. The haploid daughter cells resulting from the first mitotic division will carry *d* or *D*, giving rise to a sectored colony. Each of the other spores divides to give two identical daughter cells. Thus of the eight daughter cells, three are *d* and five are *D*. (See J. W. Szostak, et al., 1983, *Cell* **33:**25, and H. Sun, et al., 1991, *Cell* **64:**1155.)

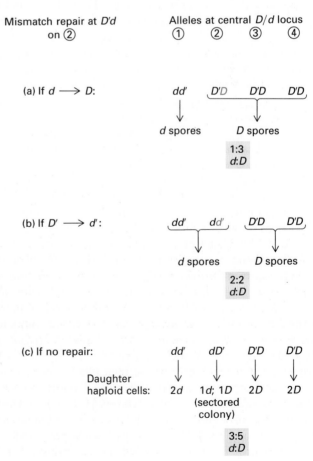

SUMMARY

The general principles of DNA replication seem to apply, with little modification, to all cells. Viewed at the level of whole chromosomes, DNA synthesis is initiated at special regions called origins; a bacterial chromosome has just one origin, whereas a eukaryotic chromosome has many. Synthesis usually proceeds bidirectionally away from an origin via two growing forks moving in opposite directions; this movement produces a replication bubble. The copying of the DNA duplex at the growing fork is semiconservative: that is, each daughter duplex contains one old strand and its newly made complementary partner.

The enzymes and other protein factors that perform the necessary functions at a growing fork have been studied most extensively in E. coli. Many similar proteins are known to function in eukaryotic cells, suggesting that the biochemical mechanism of DNA replication is similar in all cells. Most of the enzymatic events at the growing fork are a consequence of two properties of the DNA double helix and one property of DNA polymerases, the enzymes that copy DNA. First, the DNA helix contains antiparallel strands: that is, the $5' \rightarrow 3'$ direction of one strand is opposite to the $5' \rightarrow 3'$ direction of the other. Second, the strands of the duplex are interwound and cannot simply be pulled apart along their entire lengths all at once. Third, DNA polymerases require a nucleic acid primer—that is, the 3'-hydroxyl end of a DNA or an RNA molecule—to begin synthesis. All DNA-chain growth consequently proceeds in the $5' \rightarrow 3'$ direction. Thus, in order for both strands to be copied at a growing fork, one of the new strands must be synthesized in a discontinuous fashion.

The strand synthesized continuously in the direction of movement of the growing fork is called the leading strand. It extends the 3' end of an RNA primer base-paired to a template strand. Synthesis of the other strand, the lagging strand, occurs in the direction opposite to the overall direction of growing-fork movement. Lagging-strand synthesis occurs discontinuously from a series of short RNA primers formed by primase at various sites on the second template strand; the resulting segments of RNA plus DNA are called Okazaki fragments. A nuclease activity removes the RNA primers, and a DNA polymerase then fills the remaining gap by addition of deoxyribonucleotides. Finally, DNA ligase joins adjacent completed Okazaki fragments. Thus the growing fork moves along in one direction, and the principle of $5' \rightarrow 3'$ DNA-chain synthesis is preserved.

Many proteins assist DNA polymerase, primase, and ligase in the construction of a new chromosome. These include helicases, which can unwind DNA, and topoisomerases. The latter enzymes allow the strands of a duplex to swivel about each other so that the duplex can be unwound; these enzymes also can pass one DNA duplex through another to introduce or remove supercoils. Topoisomerases are important both in growing-fork movement and in resolving (untangling) finished chromosomes after DNA duplication. In addition, telomerase prevents shortening of lagging strands during replication of linear DNA molecules such as those of eukaryotic chromosomes.

Copying errors that occur during DNA replication are corrected by two mechanisms. First, DNA polymerases have a proofreading function by which they can recognize incorrect (mispaired) bases at the 3' end of the growing strand and then remove them by a $3' \rightarrow 5'$ exonuclease activity. Second, mispaired bases in the newly synthesized strand can be identified by a scanning mechanism, specifically removed, and replaced by DNA synthesis directed by the correct template. Errors in the DNA sequence also result from exposure to UV light and various chemicals. Such damage must be repaired so that stable mutations do not occur at an intolerable rate. In the excision-repair mechanism, a battery of enzymes recognizes damaged or mismatched base pairs, excises them, and generally widens the resulting single-stranded gap; DNA polymerases then copy the single strand and ligase joins the free ends.

During homologous recombination, two duplex DNA molecules are broken and strands are exchanged. This occurs in all organisms and plays an important role in generating genetic diversity. Probably at least two events can initiate recombination: (1) invasion of a duplex from the cut end of one strand of a homologous DNA as seen in E. coli and (2) invasion of one strand from the end of the duplex in which both strands have been cut as exemplified by studies in yeast. In both cases, however, it appears that the invading strand initiating recombination contains a 3' hydroxyl tail. These initial events are followed by rotation of the interconnected DNA molecules to create crossed-strand Holliday structures. Enzymatic creation and resolution of Holliday structures has been achieved in a few cases in the test tube. In both prokaryotes and eukaryotes strand invasion leads to regions of heteroduplex DNA. Repair of these regions gives rise to aberrant segregation patterns of closely linked markers. This phenomenon is called gene conversion.

1. A site at which replication of the genome begins in a particular organism is referred to as a replication origin. What are some of the features of this region in *E. coli*? What are the proteins that bind to the *E. coli* origin of replication, and how do they assist in the initiation of the process?

DNA polymerases cannot initiate the synthesis of DNA de novo, that is without a segment of nucleic acid to build onto. Contrast this to what you have learned about RNA polymerases, where a primer is not necessary to initiate RNA synthesis. Does this offer an explanation as to why primase is an RNA polymerase? What is the direction in which a DNA polymerase adds nucleotides during synthesis?

In *E. coli,* the genome is a closed circle of double-stranded DNA that is replicated in a bidirectional fashion from *oriC.* What happens when the replication forks meet? What is meant by the process of decatenation, and what enzyme is involved in the process?

The ends of linear chromosomes that are characteristic of eukaryotes are referred to as telomeres. Review the features of these structures and the processes that are required to replicate these regions.

a. You are investigating replication in human cells. If DNA synthesis occurs in a bidirectional fashion from a particular replication origin, would the leading and lagging strands be the same in each of the diverging forks? Explain your answer.

b. Refer to Figure 10-14. This drawing illustrates an important—albeit unproven—model for replication forks in *E. coli*. In this particular scenario, the lagging strand contains a loop. Contrast this diagram with that presented in Figure 10-15. Which of these probably portrays a more accurate picture of replication? Explain your answer.

c. Table 10-2 highlights the properties of DNA polymerases from *E. coli* and mammalian cells. In mammalian cells, which of the DNA polymerases are responsible for synthesizing the leading and lagging strands during DNA replication? Note that the role of DNA polymerase ϵ has not yet been clarified. Contrast the properties of DNA polymerases α and ϵ. What feature of DNA polymerase α might lead you to question its involvement in replication?

d. Although no $5' \rightarrow 3'$ exonuclease activity is reported for the mammalian DNA polymerases, a note in Table 10-2 indicates that they undoubtedly have such an activity to remove primers. What does this refer to? Why would such an activity be critical for successful replication of the genome?

2. Topoisomerases are vital for maintaining DNA structure. Review the properties of type I and type II topoisomerases. What are the principal differences between them? Where is topoisomerase II located within eukaryotic cells? Is topoisomerase II part of a special structure? Review the concepts of linking number (L), twist (T), and writhe (W) presented in Chapter 4. Recall that $L = T + W$.

a. If a closed circular piece of DNA were 6,102 base pairs in size and the values of L and T were 500 and 565 respectively, what would be the value of W? What would be the average number of base pairs per turn of the helix?

b. Suppose that *E. coli* topoisomerase I were used to reduce the number of supercoils to -60. If T remained unchanged, what would be the new value of L? How many times would topoisomerase I need to nick the original supercoiled molecule to change the linking number as described?

3. DNA repair can best be thought of as the collected biochemical systems within a cell that maintain the genome. Review the process in *E. coli* involving the *uvrA*, *uvrB*, and *uvrC* gene products that remove bulky adducts from DNA. What is known about the clearance of bulky DNA damage from the genome in eukaryotes?

a. Table 10-4 lists a variety of agents, both chemical and physical, that damage DNA, resulting in the formation of DNA lesions. Speculate on the possible effects that each of these lesions would have on replication and transcription if they remained unrepaired.

b. Figure 10-26 shows the chemical structure of a lesion produced in DNA following exposure to ultraviolet radiation. This particular moiety strongly blocks the progression of replication forks and transcription bubbles. It has been documented that the removal of these lesions from DNA in both *E. coli* and human cells does not occur at equal rates over the entire genome. The transcribed strand in some expressed genetic loci is cleared of dimers much faster than the nontranscribed strand or other quiescent domains of the genome. What advantages might such a transcription-dependent repair system offer the cell in terms of clearing damage that impedes DNA and RNA synthesis?

c. Some enzymes associated with removing abnormal adducts from DNA are regulated in a cell cycle dependent fashion in that they increase several fold towards the end of G_1. Do you think the evolution of such a system in cells confers an advantage in maintaining an intact genome?

4. Recombination is extremely important mechanism in cells that has a variety of functions, including the generation of genetic diversity. What is meant by reciprocal recombination? Pay particular attention to Figure 10-27; be sure you understand how to resolve Holliday junctions. Be aware that the resolution of a Holliday junction can result in the formation of heteroduplex DNA without concomitant formation of a recombinant duplex.

What is meant by a gene conversion event? When and how do they occur in yeast?

a. You are studying the process of recombination in yeast, and you have a particular interest in gene conversion events. The following sequence of DNA belongs to a region which you are characterizing in terms of the aberrant segregation patterns that are characteristic of gene conversions.

```
D   5'-ATATGCGCTAGGGCCCTACGCCGTGTAT-3'
D'  3'-TATACGCGATCCCGGGATGCGGCACATA-5'


d   5'-ATATGCGCTAGGCCCCTACGCCGTGTAT-3'
d'  3'-TATACGCGATCCGGGGATGCGGCACATA-5'
```

Are these regions 100% homologous? Assume that a recombinational event occurred such that D and d' were paired and D' and d were paired. Locate the site of heteroduplex formation that would result.

b. The restriction endonuclease *Apa*I cuts DNA at GGGCC*C at the position indicated by the asterisk. Assume that normal segregation of diploid yeast carrying the D and d chromosomes occurs. What would be the ratio of the offspring whose DNA would be cut with *Apa*I in this region to those who would not?

c. If a gene conversion event occurred in this region, could you use *Apa*I to determine in which direction the event occurred? That is, did it result in D → d or d → D? Describe the cutting patterns that would result in each case.

References

General References

KORNBERG, A., and T. A. BAKER. 1992. *DNA Replication*, 2d ed. W.H. Freeman and Company.

MARIANS, K. J. 1992. Prokaryotic DNA replication. *Ann. Rev. Biochem.* **61**:673–719.

MCMACKEN, R., and T. J. KELLEY, eds. 1987. DNA replication and recombination. In *UCLA Symp. Mol. Cell Biol., New Ser.*, Vol. 47. A. R. Liss.

STILLMAN, B. 1989. Initiation of eukaryotic DNA replication in vitro. *Ann. Rev. Cell Biol.* **5**:197–245.

WEST, C. F. 1992. Enzymes and molecular mechanisms of genetic recombinations. *Ann. Rev. Biochem.* **61**:603–640.

DNA Replication

BREWER, B. J. 1988. When polymerases collide: replication and the transcriptional organization of the *E. coli* chromosome. *Cell* **53**:679–686.

HUBERMAN, J. A., and A. D. RIGGS. 1968. On the mechanism of DNA replication in mammalian chromosomes. *J. Mol. Biol.* **32**:327–341.

HUBERMAN, J. A., and A. TSAI. 1973. Direction of DNA replication in mammalian cells. *J. Mol. Biol.* **75**:5–12.

MCCARROLL, T. M., and W. L. FANGMAN. 1988. Time of replication of yeast centromeres and telomeres. *Cell* **54**:505–513.

MESELSON, M., and F. STAHL. 1958. The replication of DNA in *E. coli*. *Proc. Nat'l. Acad. Sci. USA* **44**:671–782.

RAO, P. N., and R. T. JOHNSON. 1970. Mammalian cell fusion: studies on the regulation of DNA synthesis and mitosis. *Nature* **225**:159–164.

SUNDIN, O., and A. VARSHAVSKY. 1980. Terminal stages of SV40 DNA replication proceed via multiple intertwined catenated dimers. *Cell* **21**:103–114.

SUNDIN, O., and A. VARSHAVSKY. 1981. Arrest of segregation leads to accumulation of highly intertwined catenated dimers: discussion of the final stages of SV40 DNA replication. *Cell* **25**:659–669.

TAYLOR, J. H. 1958. The duplication of chromosomes. *Sci. Am.* **198**(6):36–42.

TSENG, B. Y., J. M. ERICKSON, and M. GOULIAN. 1979. Initiator RNA of nascent DNA from animal cells. *J. Mol. Biol.* **129**:531–545.

Properties of DNA Polymerases and DNA Synthesis in Vitro

BAKER, T. A., and A. KORNBERG. 1988. Transcriptional activation of initiation of replication from the *E. coli* chromosomal origin: an RNA-DNA hybrid near oriC. *Cell* **55**:113–123.

BONNER, C., et al. 1988. Purification and characterization of an inducible *Escherichia coli* DNA polymerase capable of insertion and bypass at abasic lesions in DNA. *J. Biol. Chem.* **263**:18946–18952.

BRAMHILL, D., and A. KORNBERG. 1988. Duplex opening by dnaA protein at novel sequences in initiation of replication at the origin of the *E. coli* chromosome. *Cell* **52**:743–755.

COTTERILL, S. M., et al. 1987. A cryptic proofreading 3'5' exonuclease associated with the polymerase subunit of the DNA polymerase-primase from *Drosophila melanogaster*. *Proc. Nat'l. Acad. Sci. USA* **84**:5635–5639.

DEAN, F. B., et al. 1987. Simian virus 40 large tumor antigen requires three core replication origin domains for DNA unwinding and replication in vitro. *Proc. Nat'l. Acad. Sci. USA* **84**:8267–8271.

DEAN, F. B., et al. 1989. SV40 replication in vitro. In C. C. Richardson and I. R. Lehman, eds., *Molecular Mechanisms in DNA Replication and Recombination*, A. R. Liss.

HÜBSCHER, U., and P. THÖMMES. 1992. DNA polymerase ε: in search of a function. *Trends Biochem. Sci.* **17**:55–58.

KONG, X.-P., et al. 1992. Three-dimensional structure of the β subunit of the *E. coli* DNA polymerase III holoenzyme: a sliding DNA clamp. *Cell* **69**:425–437.

NAGATA, K., R. A. GUGGENHEIMER, and J. HURWITZ. 1983. Adenovirus DNA replication in vitro: synthesis of full-length DNA with purified proteins. *Proc. Nat'l. Acad. Sci. USA* **80**:4266–4270.

OGAWA, T., et al. 1985. Initiation of enzymatic replication at the origin of the *E. coli* chromosome: contributions of RNA polymerase and primase. *Proc. Nat'l. Acad. Sci. USA* **82**:3562–3566.

PRELICH, G., and B. STILLMAN. 1988. Coordinated leading and lagging strand synthesis during SV40 DNA replication in vitro requires PCNA. *Cell* **53**:117–126.

ROBERTS, J. M., and G. D'URSO. 1988. An origin unwinding activity regulates initiation of DNA replication during mammalian cell cycle. *Science* **241**:1486–1489.

SHIPPEN-LENTZ, D., and E. H. BLACKBURN. 1990. Functional evidence for an RNA template in telomerase. *Science* **247**:546–552.

WIDES, R. J., et al. 1987. Adenovirus origin of DNA replication: sequence requirements for replication in vitro. *Mol. Cell Biol.* **7**:864–874.

TSURIMOTO, T., T. MELENDY, and B. STILLMAN. 1990. Sequential initiation of lagging and leading strand synthesis by two different polymerase complexes at the SV40 DNA replication origin. *Nature* **346**:534–539.

TSURIMOTO, T., and B. STILLMAN. 1991. Replication factors required for SV40 DNA replication in vitro. I. DNA structure-specific recognition of a primer template junction by eukaryotic DNA polymerases and their accessory proteins. *J. Biol. Chem.* **266**:1950–1960.

TSURIMOTO, T., and B. STILLMAN. 1991. Replication factors required for SV40 DNA replication in vitro. II. Switching of DNA polymerase α and δ during initiation of leading and lagging strand synthesis. *J. Biol. Chem.* **266**:1961–1968.

Origins of DNA Replication

BELL, S. P., and B. STILLMAN. 1992. ATP-dependent recognition of eukaryotic origins of DNA replication by a multiprotein complex. *Nature* **375**:128–134.

BOROWIEC, J. A., et al. 1990. Binding and unwinding—how T antigen engages the SV40 origin of DNA replication. *Cell* **60**:181–184.

BRAMHILL, D., and A. KORNBERG. 1988. A model for initiation at origins of DNA replication. *Cell* **54**:915–918.

BREWER, B. J., and W. L. FANGMAN. 1987. The localization of replication origins on ARS plasmids in *S. cerevisiae*. *Cell* **51**:463–471.

DE PAMPHILIS, M. L. 1993. Eukaryotic DNA replication: Anatomy of an origin. *Ann. Rev. Biochem.* **62**:29–63.

DIFFLEY, J. F. X., and B. STILLMAN. 1990. The initiation of chromosomal DNA replication in eukaryotes. *Trends Genet.* **6**:427–432.

HUBERMAN, J. A., et al. 1987. The in vivo replication origin of the yeast 2μm plasmid. *Cell* **51**:473–481.

MARAHRENS, Y., and B. STILLMAN. 1992. A yeast chromosomal origin of DNA replication defined by multiple functional elements. *Science* **255**:817–823.

SUGIMOTO, K., et al. 1979. Nucleotide sequence of *Escherichia coli* K-12 replication origin. *Proc. Nat'l. Acad. Sci. USA* **76**:575–579.

ZYSKIND, J. W., et al. 1983. Chromosomal replication origin from the murine bacterium *Vibrio harveyi* functions in *E. coli*: *oriC* consensus sequence. *Proc. Nat'l. Acad. Sci. USA* **80**:1164–1168.

Topoisomerases

ADAMS, D. E., et al. 1992. The role of topoisomerase IV in partitioning bacterial replicons and the structure of catenated intermediates in DNA replication. *Cell* **71**:271–288.

COZZARELLI, N. R. 1980. DNA gyrase and the supercoiling of DNA. *Science* **207**:953–960.

HOLM, C., T. STEARNS, and D. BOTSTEIN. 1989. DNA topoisomerase II must act at mitosis to prevent nondisjunction and chromosome breakage. *Mol. Cell Biol.* **9**:159–168.

LIU, L. F. 1989. DNA topoisomerase poisons as anti-tumor drugs. *Ann. Rev. Biochem.* **58**:351–375.

LIU, L. F., C.-C. LIU, and B. M. ALBERTS. 1980. Type II DNA topoisomerases: enzymes that can unknot a topologically knotted DNA molecule via a reversible double-strand break. *Cell* **19**:697–707.

OSHEROFF, N., ZECHIEDRICH, E. L., and GALE, K. C. 1991. Catalytic function of topoisomerase II. *Bioessays* **13**:269–273.

WANG, J. C. 1985. DNA topoisomerases. *Ann. Rev. Biochem.* **54**:665–698.

YANG, L., et al. 1987. Role of DNA topoisomerase in simian virus 40 DNA replication in vitro. *Proc. Nat'l. Acad. Sci. USA* **84**:950–954.

Repair of DNA

CLARK, A. J. 1991. *rec* genes and homologus recombination proteins in *Escherichia coli*. *Biochimie* **73**:523–532.

FRIEDBERG, E. C. 1991. Eukaryotic DNA repair: glimpses through the yeast *Saccharomyces cerevisiae*. *BioEssays* **13**:295–301.

GLAZER, P. M., et al. 1987. DNA mismatch repair detected in human cell extracts. *Mol. Cell Biol.* **7**:218–224.

GROSSMAN, L., et al. 1988. Repair of DNA-containing pyrimidine dimers. *FASEB J.* **2**:2696–2701.

MELLON, I., et al. 1986. Preferential DNA repair of an active gene in human cells. *Proc. Nat'l. Acad. Sci. USA* **83**:8878–8882.

MODRICH, P. 1991. Mechanisms and biological effects of mismatch repair. *Ann. Rev. Genet.* **25**:229–253.

MURLI, S., and WALKER, G. C. 1993. SOS mutagenesis. *Curr. Opin. Genet. Dev.* **3**:719–725.

OH, E. Y., et al. 1989. ATPase activity of the UvrA and UvrAB protein complexes of the *Escherichia coli* UvrABC endonuclease. *Nucl. Acids Res.* **17**:4145–4159.

SANCAR, A., and G. B. SANCAR. 1988. DNA repair enzymes. *Ann. Rev. Biochem.* **57**:29–68.

SEELEY, T. W., and L. GROSSMAN. 1990. The role of *Escherichia coli* UvrB in nucleotide excision repair. *J. Biol. Chem.* **265**:7158–7165.

VAN DUIN, M., et al. 1988. Evolution and mutagenesis of the mammalian excision repair gene *ERCC-1*. *Nucl. Acids Res.* **16**:5305–5322.

WILLIS, A. E., and T. LINDAHL. 1987. DNA ligase deficiency in Bloom's syndrome. *Nature* **325**:355–357.

Recombination

BISHOP, D. K., et al. 1992. DMC-1: a meiosis-specific yeast homolog of *E. coli* Rec A required for recombination, synaptonemal complex formation and cell cycle progression. *Cell* **69**:439–456.

COX, M. M., and I. R. LEHMAN. 1987. Enzymes of general recombination. *Ann. Rev. Biochem.* **56**:229–262.

DIXON, D. A., and S. C. KOWALCZYKOWSKI. 1991. Homologous pairing in vitro stimulated by the recombination hotspot, chi. *Cell* **66**:361–371.

DUCKETT, D. R., et al. 1988. The structure of the Holliday junction, and its resolution. *Cell* **55**:79–89.

DUNDERDALE, H. J., et al. 1991. Formation and resolution of recombination intermediates by *E. coli* RecA and RuvC proteins. *Nature* **354**:506–510.

GONDA, D. K., and C. M. RADDING. 1983. By searching processively RecA protein pairs DNA molecules that share a limited stretch of homology. *Cell* **34**:647–654.

HABER, J. E. 1992. Exploring the pathways of homologous recombination. *Curr. Opin. Cell Biol.* **4**:401–412.

HOLLIDAY, R. 1964. A mechanism for gene conversion in fungi. *Genet. Res.* **5**:282–304.

IWASAKI, H., et al. 1992. *Escherichia coli* RuvA and RuvB proteins specifically interact with Holliday junctions and promote branch migration. *Genes & Dev.* **6**:2214–2220.

LANDY, A. 1989. Dynamic, structural, and regulations of lambda site-specific recombinations. *Ann. Rev. Biochem.* **58**:913–949.

MESELSON, M., and C. M. RADDING. 1975. A general model for genetic recombination. *Proc. Nat'l. Acad. Sci. USA* **72**:358–361.

MÜLLER, B., et al. 1990. Enzymatic formation and resolution of Holliday junctions in vitro. *Cell* **60**:329–336.

NASH, H. A., and C. A. ROBERTSON. 1989. Heteroduplex substrates for bacteriophage lambda site-specific recombination: cleavage and strand transfer products. *EMBO J.* **8**:3523–3533.

RADDING, C. M. 1991. Helical interactions in homologous pairing and strand exchange driven by RecA protein. *J. Biol. Chem.* **266**:5355–5358.

SCHIESTL, R. H., and S. PRAKASH. 1988. *RAD1*, an excision repair gene of *Saccharomyces cerevisiae*, is also involved in recombination. *Mol. Cell Biol.* **8**:3619–3626.

SHINOHARA, A., H. OGAWA, and T. OGAWA 1992. Rad51 protein involved in repair and recombination in *S. cerevisiae* is a RecA-like protein. *Cell* **69**:457–470.

SMITH, G. R. 1983. Chi, hot spots of generalized recombinations. *Cell* **34**:709–710.

SMITH, G. R. 1989. Homologous recombination in *E. coli*: multiple pathways for multiple reasons. *Cell* **58**:807–809.

STORY, R. M., I. T. WEBER, and T. A. STEITZ. 1992. The structure of the *E. coli* RecA protein monomer and polymer. *Nature* **355**:318–325.

SUN, H., D. TRECO, and J. W. SZOSTAK. 1991. Extensive 3'-overhanging, single-stranded DNA associated with the meiosis-specific double-strand breaks at the *ARG4* recombination initiation site. *Cell* **64**:1155–1161.

SZOSTAK, J. W., et al. 1983. The double-strand-break repair model for recombination. *Cell* **33**:25–35.

TAYLOR, A. F. 1992. Movement and resolution of Holliday junctions by enzymes from *E. coli. Cell* **69**:1063–1065.

TAYLOR, J. H. 1958. Sister chromatid exchanges in tritium-labeled chromosomes. *Genetics* **43**:515–529.

TSANEVA, I. R., B. MÜLLER, and S. C. WEST. 1992. ATP-dependent branch migration of Holliday junctions promoted by the RuvA and RuvB proteins of *E. coli. Cell* **69**:1171–1180.

VON WETTSTEIN, D. S., W. RASMUSSEN, and P. B. HOLM. 1984. The synaptonemal complex in genetic segregation. *Ann. Rev. Genet.* **18**:331–414.

11

Regulation of Transcription Initiation

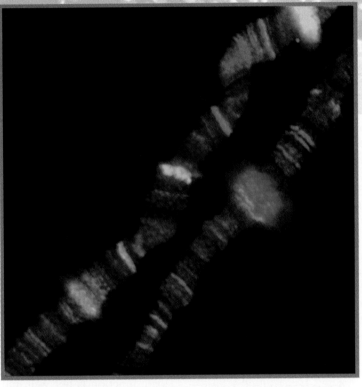

▲ An active region of transcription producing a "puff" in a *Drosophila* polytene chromosome.

One of the underlying principles of molecular cell biology is that *the actions and properties of each cell type are determined by the proteins it contains.* But what determines the types and amounts of the various proteins that characterize a particular cell type? Or that allow a single-celled organism to respond to changes in its environment? The determining factors are the concentration of each protein's corresponding mRNA, the frequency at which the mRNA is translated, and the stability of the protein itself. The concentration of various mRNAs is, in turn, determined largely by which genes are transcribed and their rate of transcription in a particular cell type. Thus the differential transcription of different genes largely determines the actions and properties of cells.

The term *gene expression* commonly refers to the entire process whereby the information encoded in a particular gene is decoded into a particular protein. Theoretically, regulation at any one of the various steps in this process could lead to differential gene expression. Synthesis of mRNA requires that an RNA polymerase initiate transcription, polymerize ribonucleoside triphosphates complementary to the DNA coding strand, and then terminate transcription. In prokaryotes, ribosomes and initiation factors have immediate access to newly formed RNA transcripts, which function as mRNA without further modification. In eukaryotes, the initial RNA transcript is processed by addition of a poly-A tail and splicing, which removes noncoding introns, yielding a functional mRNA. The mRNA then is transported from its site of synthesis in the nucleus to the cytoplasm where translation occurs. Finally, the stability of an mRNA affects its concentration in both prokaryotic and eukaryotic cells.

The various steps in gene expression have been analyzed for hundreds of specific genes in prokaryotes and eukaryotes. Although examples of regulation of each of these steps have been found, the decision to initiate transcription—the first step in gene expression—is the most important control point in determining whether or not most genes are expressed and how much of the encoded mRNAs, and consequently proteins, are produced. In this chapter, we review current understanding of the molecular events that determine when transcription is initiated. Chapter 12 considers other regulatory mechanisms that control the subsequent steps in gene expression. Chapter 13 presents an overview of how these multiple levels of gene regulation contribute to the development of specific types of cells in multicellular organisms.

In considering the mechanisms that regulate gene expression in different types of cells, it is useful to keep in mind an important difference between the purposes of gene regulation in single-celled versus multicellular organisms. In bacteria, gene control serves mainly to allow the single cell to adjust to changes in its nutritional environment so that its growth and division can be optimized. Although gene control in multicellular organisms in a few cases involves response to environmental changes, its most characteristic and biologically far-reaching purpose is the regulation of a genetic program that underlies embryologic development and tissue differentiation. We begin our discussion by considering the simpler bacterial cell because many of the principles of transcription control that operate in bacteria apply to the more complex transcription process in eukaryotic cells. In most of this chapter, we use the term "bacteria" to refer to eubacteria. The other major and very distinct group of prokaryotes, called *archaebacteria,* is considered at the end of the chapter.

➤ *Early Genetic Analysis of* lac-*Operon Control in* E. coli

The molecular details of how gene expression is controlled were first discovered in bacteria through a combination of genetic and biochemical experiments. One of the first examples to be well understood was the *lac* operon of *E. coli,* largely through the work of Francois Jacob, Jacques Monod, and their colleagues in the 1960s (Figure 11-1).

▲ FIGURE 11-1   Francois Jacob (1920—, *left*) and Jacques Monod (1919–1976, *center*) of the Pasteur Institute in Paris, together with Andre Lwoff (1902—, *right*) with whom they shared the Nobel Prize in 1965. [Photograph by Rene Saint-Paul, Paris; courtesy of F. Jacob.]

The *lac* operon includes two genes, *Z* and *Y*, that are required for metabolism of lactose (the major sugar in milk) and a third gene, *A* (Figure 11-2). The *lacY* gene encodes *lactose permease,* which spans the *E. coli* cell membrane and uses the energy available from the electrochemical gradient across the membrane to pump lactose into the cell (Chapter 15). The *lacZ* gene encodes *β-galactosidase,* which splits the disaccharide lactose into the monosaccharides glucose and galactose (Figure 11-3a); these sugars are further metabolized through the action of enzymes encoded in other operons. The *lacA* gene encodes *thiogalactoside transacetylase,* an enzyme whose physiologic function is not well understood.

The enzymes *β*-galactosidase and lactose permease are easily assayed. In particular, the colorless compound 5-bromo-4-chloro-3-indolyl-*β*-D-galactoside (Figure 11-3b), commonly called *X-gal,* is hydrolyzed by *β*-galactosidase into an insoluble, intensely blue product, providing a simple colorimetric assay for the enzyme in living cells. To assay lactose permease, radiolabeled lactose can be added to the medium in which cells are growing and its rate of entry into cells determined.

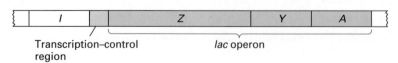

▲ FIGURE 11-2   The *lac* operon includes three genes: *lacZ,* which encodes β-galactosidase; *lacY,* which encodes lactose permease; and *lacA,* which encodes thiogalactoside transacetylase. Binding of regulatory proteins to sites in the control region immediately upstream of the *lacz* gene regulate its transcription. The *lacl* gene, which encodes the lac repressor, maps adjacent to the *lac* control region.

(a)

(b)

X-gal

(c)

Isopropyl-β-D-thiogalactoside
(IPTG)

◄ FIGURE 11-3   (a) β-Galactosidase catalyzes the hydrolysis of lactose into glucose and galactose. (b) X-gal (5-bromo-4-chloro-3-indolyl-β-D-galactoside) is a colorless analog of lactose. Hydrolysis of X-gal by β-galactosidase yields an intensely blue product, providing a sensitive colorimetric assay for the enzyme. (c) IPTG, another lactose analog, is a strong inducer of *lac* transcription but is not digested by β-galactosidase. When added to culture media, its concentration remains constant for long periods, resulting in continuous induction of *lac* transcription.

Enzymes Encoded at *lac* Operon Can Be Induced and Repressed

E. coli cells grown in a medium containing glucose as the sole source of carbon and energy have very low levels of β-galactosidase and lactose permease activity. When these cells are switched to a medium containing lactose, the activities of both β-galactosidase and lactose permease increase a thousandfold within about 20 min. Such large changes in enzyme activity are a common phenomenon in bacteria. Bacteria rapidly adapt to changes in their environment by *inducing* the activity of the enzymes needed to metabolize the available nutrients. Early studies showed that the increase in β-galactosidase activity resulted from the synthesis of new β-galactosidase enzyme molecules. Synthesis starts within minutes of switching from a glucose to lactose medium. Furthermore, synthesis of β-galactosidase protein is shut off, or *repressed*, within minutes of switching from a lactose-containing medium back to a medium containing glucose. The synthesis of lactose permease and thiogalactoside transacetylase also is rapidly induced when cells are placed in a medium containing lactose as the only carbon source and repressed when the cells are switched to a medium without lactose. Thus all three genes of the *lac* operon are *coordinately regulated*.

Some molecules similar in structure to lactose can induce expression of the *lac*-operon genes even though they cannot be hydrolyzed by β-galactosidase. Such small molecules (i.e., smaller than proteins) are called *inducers*. One

of these, isopropyl-β-D-thiogalactoside (Figure 11-3c), abbreviated IPTG, is particularly useful in genetic studies of the *lac* operon, because it can diffuse into cells and, since it is not metabolized, its concentration remains constant throughout an experiment.

Insight into the mechanisms controlling synthesis of β-galactosidase and lactose permease first came from the study of mutants in which control of β-galactosidase expression was abnormal. The isolation of such mutants was greatly aided by the ease with which β-galactosidase can be colorimetrically assayed. For example, X-gal incorporated into nutrient agar plates is transported by lactose permease into cells where it is hydrolyzed by β-galactosidase, producing the blue product without interfering with cell growth. Colonies of cells producing β-galactosidase appear blue; colonies that do not produce β-galactosidase are white, and the two are easily distinguished. Simple colorimetric plate assays such as this made it possible to rapidly identify and characterize mutant cells with abnormal control of *lac*-operon expression. These studies have revealed that trans-acting proteins and cis-acting DNA sequences regulate expression of the *lac* operon.

Mutations in *lacI* Cause Constitutive Expression of *lac* Operon

When wild-type *E. coli* cells are plated on media containing X-gal plus lactose as the major carbon source, all the colonies that grow appear blue. When the cells are plated

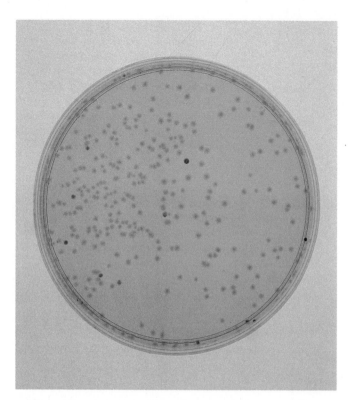

▲ FIGURE 11-4   Growth of mutagenized *E. coli* cells on nutrient agar containing X-gal and glucose as the sole carbon source. Under these conditions, wild-type cells do not produce β-galactosidase and thus cannot hydrolyze X-gal. A small number of mutant cells can hydrolyze X-gal, forming blue colonies. These constitutive mutants are now known to be *lacI⁻* and do not produce functional *lac* repressor.

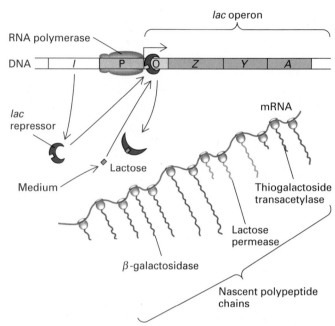

▲ FIGURE 11-5   Jacob and Monod model of transcriptional regulation of the *lac* operon by *lac* repressor. When *lac* repressor binds to a DNA sequence called the operator (light yellow), which lies just upstream of the *lacZ* gene, transcription of the operon by RNA polymerase is blocked. Binding of lactose to the repressor causes a conformational change, so that the repressor no longer binds to the operator. RNA polymerase then is free to transcribe the *lac* genes, and the resulting polycistronic mRNA is translated into the encoded proteins. The polymerase initially binds to a region termed the promoter (dark yellow). [Adapted from A. J. F. Griffiths et al., 1993, *An Introduction to Genetic Analysis*, 5th ed., W. H. Freeman and Co.]

on media containing X-gal plus glucose as the carbon source, the resulting colonies appear white; in this case, β-galactosidase synthesis is repressed and there is not sufficient β-galactosidase in the cells to hydrolyze the X-gal to its colored product. However, when the cells are exposed to chemical mutagens before plating on X-gal/glucose plates, rare blue colonies appear (Figure 11-4). In most cases, when cells from these blue colonies are recovered and grown in media containing glucose, they are found to express all the genes of the *lac* operon at much higher levels than wild-type cells in the same medium. Such mutants are called *constitutive* because they fail to repress the *lac* operon in media lacking lactose and instead continuously, or constitutively, express the enzymes. These mutations were mapped by recombination analysis (Chapter 8) to a region on the *E. coli* chromosome to the left of the *lacZ* gene, a region called the *lacI* gene (see Figure 11-2).

Jacob and Monod reasoned that such constitutive mutants probably had a defect in a protein that normally repressed expression of the *lac* operon in the absence of lactose. Hence they called the protein encoded by the *lacI* gene the *lac repressor* and proposed that it binds to a site

on the *E. coli* genome where transcription of the *lac* operon is initiated, thereby blocking transcription. They further hypothesized that when lactose is present in the cell, it binds to the *lac* repressor, decreasing its affinity for the repressor-binding site on the DNA. As a result, the repressor falls off the DNA and transcription of the *lac* operon is initiated, leading to synthesis of β-galactosidase, lactose permease, and thiogalactoside transacetylase (Figure 11-5).

Operator Constitutive Mutations Identify Binding Site for *lac* Repressor

The model proposed by Jacob and Monod predicted that a specific DNA sequence near the transcription start site of the *lac* operon is a binding site for *lac* repressor. They reasoned that mutations in this sequence, which they termed the *operator* (O), would prevent the repressor from binding, thus yielding constitutive mutants that could be identified on indicator plates like X-gal/glucose plates. However, most constitutive mutants recovered on indicator plates have mutations in *lacI*, as discussed above. This is because the *lacI* gene is much larger than the operator and consequently is disrupted by random mutations more frequently than the operator is.

To distinguish between mutations in the *lacI* gene, which inactivate the repressor, and mutations in the operator, which prevent repressor binding, Jacob and Monod mutagenized cells carrying two copies of the wild-type *lacI* gene, one on the bacterial chromosome and one on a plasmid. Because the probability of separate mutations inactivating both copies of *lacI* in a given cell is vanishingly small, only the rare cells with mutations in the operator sequence would be recovered on indicator plates. Using this approach, Jacob and Monod isolated mutants that expressed the *lac* operon constitutively even when two copies of the wild-type *lacI* gene encoding the *lac* repressor were present in the same cell. These mutations, designated O^c for *operator constitutive*, mapped to one end of the *lac* operon, as the model predicted (see Figure 11-5).

Mutations in Promoter Prevent Expression of the *lac* Operon

Another type of mutation revealed an additional important DNA sequence required for normal regulation of *lac*-operon expression. Most mutations that prevent expression of β-galactosidase in cells exposed to an inducer such as IPTG map in *lacZ*. These mutations result in amino acid substitutions that inactivate β-galactosidase enzymatic ac-

tivity or introduce a premature stop codon in *lacZ* resulting in a truncated β-galactosidase protein without enzymatic activity. But a rare class of β-galactosidase mutations map to a region between *lacI* and the operator, in a region termed the *promoter* (P). Cells carrying these mutations also cannot induce expression of the *lacY* and *lacA* genes; that is, these mutations prevent expression of the entire *lac* operon. According to the Jacob and Monod model, such promoter mutations block initiation of transcription by RNA polymerase (see Figure 11-5). Consequently, no *lac* mRNA and therefore no *lac* proteins are synthesized, even when *lac* repressor binds IPTG and comes off the *lac* operator. As described below, direct biochemical experiments with purified RNA polymerase showed that these promoter mutations block binding of RNA polymerase to the promoter region.

Regulation of *lac* Operon Depends on Cis-acting DNA Sequences and Trans-acting Proteins

Analyses of the effects of mutations in cells containing two copies of *lac* DNA, one on the bacterial chromosome and one on a plasmid known as an F-factor, provided further insight into regulation of *lac*-operon expression. Table 11-1 shows the results of assays for β-galactosidase and lactose permease in cells with one or two copies of the *lac* operon. The genotype of *lac* DNA on the F-factor plasmid is shown in red type.

TABLE 11-1   Effect of *E. coli* Genotype on Expression of β-Galactosidase and Lactose Permease in the Presence and Absence of Inducer (IPTG)

Strain	Genotype*	β-Galactosidase −IPTG	β-Galactosidase +IPTG	Permease −IPTG	Permease +IPTG
1	$I^+P^+O^+Z^+Y^+$	−	+	−	+
2	$I^+P^+O^cZ^+Y^+$	+	+	+	+
3	$I^+P^+O^cZ^+Y^+ / I^+P^+O^+Z^+Y^+$	+	+	+	+
4	$I^+P^+O^cZ^-Y^+$	−	−	+	+
5	$I^+P^+O^cZ^-Y^+ / I^+P^+O^+Z^+Y^+$	−	+	+	+
6	$I^+P^+O^cZ^+Y^-$	+	+	−	−
7	$I^+P^+O^cZ^+Y^- / I^+P^+O^+Z^+Y^+$	+	+	−	+
8	$I^-P^+O^+Z^+Y^+$	+	+	+	+
9	$I^-P^+O^+Z^-Y^+ / I^+P^+O^+Z^+Y^+$	−	+	−	+
10	$I^-P^+O^+Z^+Y^+ / I^+P^+O^+Z^-Y^+$	−	+	−	+
11	$I^-P^+O^+Z^+Y^+ / I^+P^+O^+Z^+Y^-$	−	+	−	+
12	$I^+P^-O^+Z^+Y^+ / I^+P^+O^+Z^-Y^-$	−	−	−	−

*Genotype of *lac* region on bacterial chromosome is shown in black type. Some strains contain a second copy of *lac* DNA carried on an F-factor plasmid; the genotype of the plasmid DNA is shown in red type.

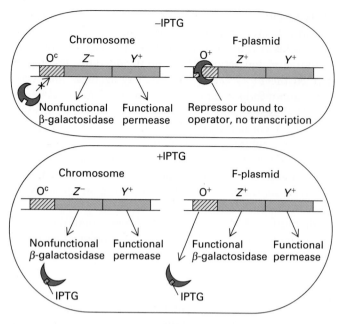

▲ **FIGURE 11-6**   Experimental demonstration that O^c mutations are cis-acting. *E. coli* cells containing two copies of the *lac* operon are diagrammed. In these cells (strain 5 in Table 11-1), the *lac* operon on the bacterial chromosome has an O^c mutation in the operator and a mutation in the *lacZ* gene (*lacZ*⁻), which inactivates the β-galactosidase enzyme. The *lac* operon on the F-plasmid is wild-type. (*Top*) In the absence of IPTG, the chromosomal *lac* operon is transcribed because the O^c mutation prevents binding and repression by *lac* repressor. A functional lactose permease is expressed, but the β-galactosidase protein produced from the *lacZ*⁻ gene is defective, so no β-galactosidase activity is observed. (*Bottom*) In the presence of IPTG, both *lac* operons in the cell are transcribed. Since the *lac* operon on the plasmid has a wild-type *lacZ* gene, β-galactosidase activity is detected.

Strain 2 has the phenotype expected for a cell containing one copy of the *lac* operon with an O^c mutation: the cells express β-galactosidase and lactose permease in both the presence and absence of the inducer IPTG. Note that the same phenotype is observed even when cells contain a copy of the wild-type *lac* operon on an F-factor plasmid (strain 3), demonstrating that the O^c mutation is dominant over O⁺. Significantly, the results for strains 5 and 7 indicate that the O^c mutation only affects expression of *lac* genes on the same DNA molecule. This result is consistent with the Jacob and Monod model. Strain 5 expresses lactose permease in the absence of inducer because of the O^c mutation on the bacterial chromosome (Figure 11-6). However, β-galactosidase activity is not expressed because mRNA transcribed from the mutant *lacZ* gene on the bacterial chromosome is translated into a nonfunctional protein. The O^c mutation does not cause expression of the wild-type *lacZ* gene on the F-factor plasmid. Consequently, no β-galactosidase activity is detected in strain 5 cells in the absence of IPTG. Comparable results are seen for lactose permease expression in strain 7. The O^c mutation is said to be *cis-acting* because it only influences expression of genes on the same DNA molecule in which the O^c mutation occurs (i.e., genes in cis to the mutation).

Mutations in *lacI* produce different effects (see Table 11-1). When cells with a single *lac* operon have a *lacI*⁻ mutation, they express β-galactosidase and lactose permease constitutively because no functional repressor is made (strain 8). Unlike the O^c mutation described above, the *lacI*⁻ mutation is recessive to the wild-type *lac*⁺ gene. For example, when a wild-type copy of the *lac* operon is introduced on an F-factor plasmid (strain 9), normal control over *lac* expression is observed; that is, the wild-type *lacI*⁺ gene is dominant over the mutant *lacI*⁻ gene. Furthermore,

► **FIGURE 11-7**   Experimental demonstration that the *lacI*⁺ gene is trans-acting. (*Top*) Cells carrying a single *lacI*⁻ gene produce an inactive repressor; as a result, they express β-galactosidase and lactose permease constitutively. (*Bottom*) When a wild-type *lacI*⁺ gene is introduced into *lacI*⁻ cells on a F-factor plasmid, the transformed cells produce functional repressor, which can bind to both *lac* operators. As a result, these cells do not express β-galactosidase or lactose permease in the absence of inducer.

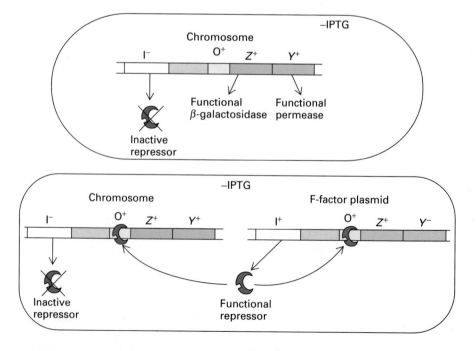

the results with strains 10 and 11 show that the wild-type *lacI*[+] gene is not cis-acting like O[c] mutations. In these strains, the wild-type *lacI*[+] gene on the F-factor plasmid exerts control over the *lacZ* and *lacY* genes on the bacterial chromosome. This finding is easy to understand since the *lacI*[+] gene encodes a protein, which is free to diffuse through the cell and bind to any *lac* operator in the cell (Figure 11-7). The *lacI*[+] gene is said to be *trans-acting* because it can control expression of genes on DNA molecules other than the one containing the *lacI*[+] gene (i.e., genes in trans to *lacI*[+]).

In general, cis-acting mutations are in DNA sequences that function as binding sites for proteins that control the expression of nearby genes. For example, the cis-acting O[c] mutations prevent binding of the *lac* repressor to the operator. Similarly, mutations in the *lac* promoter are cis-acting (see Table 11-1, strain 12), since they alter the binding site for RNA polymerase. When RNA polymerase cannot initiate transcription of the *lac* operon, none of the genes in the operon can be expressed irrespective of the function of the repressor. In general, genes that act in trans to regulate expression of genes on other DNA molecules encode diffusable products. In most cases these are proteins, but in some cases RNA molecules can act in trans to regulate gene expression.

➤ Molecular Mechanisms of Transcription Initiation in Bacteria

In this section, we describe various types of studies that have provided support for the model of transcriptional control proposed by Jacob and Monod. These experiments have led to a detailed understanding of how regulatory proteins, including repressors and activators, and *E. coli* RNA polymerase work together to regulate transcription initiation at bacterial promoters. Although we frequently describe experimental results with the *lac* operon to illustrate various points in this discussion, many of the principles apply generally to transcriptional control in bacteria.

Induction of the *lac* Operon Leads to Increased Synthesis of *lac* mRNA

The Jacob and Monod model based on genetic analyses predicts that growth of bacteria in the presence of lactose induces increased production of *lac* mRNA by derepressing transcription of the *lac* operon. RNA-DNA hybridization experiments with recombinant bacteriophage DNA containing the *lac* operon demonstrated that inducer does indeed regulate the synthesis of *lac* mRNA. In these experiments, radiolabeled RNA was isolated from *E. coli* cells and hybridized to an excess of bacteriophage DNA containing the *lac* operon so that all complementary mRNA

was hybridized. Unhybridized RNA was digested with ribonuclease A, which hydrolyzes single-stranded RNA to small oligonucleotides. When labeled RNA was isolated from cells grown in a medium containing the inducer IPTG, far more labeled *lac* mRNA was recovered than when labeled mRNA was isolated from cells grown in repressing conditions in a glucose-containing medium. These biochemical experiments directly demonstrated that induction of the *lac* operon results in increased synthesis of *lac* mRNA.

E. coli RNA Polymerase Generally Initiates Transcription at a Unique Position on DNA Template

Bacteria have a single type of RNA polymerase that catalyzes synthesis of mRNAs, rRNAs, and tRNAs encoded in the bacterial chromosome. In *E. coli* RNA polymerase is one of the largest proteins in the cell. The major form of this enzyme is composed of five subunits (Figure 11-8): two large subunits called β' (156 kDa) and β (151 kDa), two copies of a smaller subunit called α (37 kDa), and one copy of a subunit called σ^{70} (70 kDa). Of the ≈3000 RNA polymerase molecules in an *E. coli* cell, about half are actively engaged in transcribing DNA into RNA at any one time in rapidly growing cells. Most of the remainder are loosely associated with DNA, moving randomly along the DNA until they encounter a promoter sequence.

Even before development of recombinant DNA technology, researchers used purified RNA polymerase and viral DNA to show that a polymerase molecule generally initiates transcription at a unique position (a single base) in the template DNA lying upstream of the coding sequence.

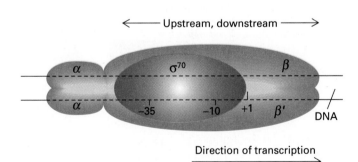

▲ FIGURE 11-8   Schematic representation of the major form of *E. coli* RNA polymerase bound to DNA. By convention, the transcription initiation site is generally numbered +1. Base pairs extending in the direction of transcription are said to be downstream of the start site; those extending in the opposite direction are upstream. As discussed later, the σ^{70} subunit binds to specific sequences near the −10 and −35 positions in the promoter. The α subunits lie close to the DNA in the upstream direction. The β and β' subunits associate with the start site.

(For some promoters, the polymerase initiates at two or three alternative neighboring bases.) The ribonucleoside triphosphate used by RNA polymerase to initiate transcription in an in vitro reaction can be recognized because it retains its 5′ triphosphate, whereas all other nucleotides in the RNA chain contain a single phosphate in ester linkage to the preceding nucleotide. The base pair where transcription initiates is called the transcription initiation site, or *start site*. By convention, the transcription initiation site in the DNA sequence of a transcription unit is usually numbered +1. Base pairs extending in the direction of transcription are assigned positive numbers and those extending in the opposite direction ("upstream" from the direction of transcription) are assigned negative numbers.

Protein-Binding Sites in *lac* Control Region Have Been Identified by Analysis of Mutant Control Sequences and Footprinting Experiments

The region of the *lac* operon containing the operator and promoter, called the *control region,* was one of the first regions of DNA to be sequenced. By comparing the sequence of the control region from wild-type cells with the sequences from operative constitutive (O^c) and promoter mutants, investigators identified these critical cis-acting DNA control sites (Figure 11-9).

Further analysis of the sites within the *lac* control region that bind protein was accomplished with a technique called DNase I footprinting. In this technique, which is illustrated in Figure 11-10, a region of DNA that is bound to a protein is protected from digestion with DNase I. When DNA samples digested in the presence and absence of a DNA-binding protein are electrophoresed, the region protected by the bound protein appears as a gap, or "footprint," in the array of bands resulting from digestion in the absence of protein. The products of DNA sequencing reactions of the DNA sample being footprinted generally are analyzed on the same gel in order to determine the precise sequence protected from DNase I digestion by the bound protein. Footprint experiments have shown that the *lac* control region contains binding sites for three proteins (see Figure 11-9): RNA polymerase, *lac* repressor, and a protein termed CAP complexed to cAMP (cAMP-CAP), which is described later.

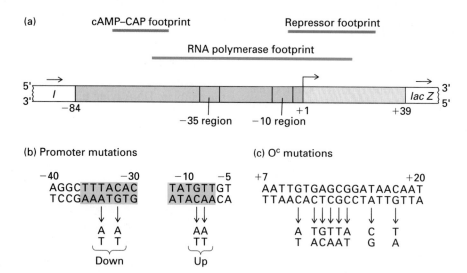

▲ FIGURE 11-9   Structure of the *lac*-operon transcription-control region. (a) Diagram showing important segments of the control region. The region coding the N-terminus of β-galactosidase is to the right, and the region coding the C-terminus of *lac* repressor is to the left. Regions protected from DNase I digestion (footprints) by cAMP-CAP, RNA polymerase, and *lac* repressor are indicated by colored bars at the top. The sequence of the entire control region has been determined. Mutational analyses identified three critical cis-acting regions: two in the promoter (dark yellow) near −35 and −10 and one between +7 and +20 in the operator (light yellow). (b) Sequences of −35 and −10 regions of the *lac* promoter with four single-base mutations indicated. Promoter mutations can cause either a decrease (Down) or increase

(Up) in expression of the *lac* operon. (c) Sequence of *lac* operator with single-base O^c mutations indicated. The mutations lead to constitutive expression of the *lac* operon in the absence of inducer. [See R. C. Dickson et al., 1975, *Science* **187**:27 (sequence of control region and promoter mutations). W. Gilbert, A. Maxam, and A. Mirzabekov, 1976, in N. O. Kjeldgaard and O. Mallow, eds., *Control of Ribosome Synthesis,* Academic Press (sequence of O^c mutations). W. S. Reznikoff and J. N. Abelson, 1978, in J. H. Miller and W. S. Reznikoff, eds., *The Operon,* Cold Spring Harbor Laboratory Press; J. Majors, 1977, PhD. thesis, Harvard University; and W. Gilbert, 1976, in R. Losick and M. Chamberlain, eds., *RNA Polymerase,* Cold Spring Harbor Laboratory Press (sequence of promoter mutants).

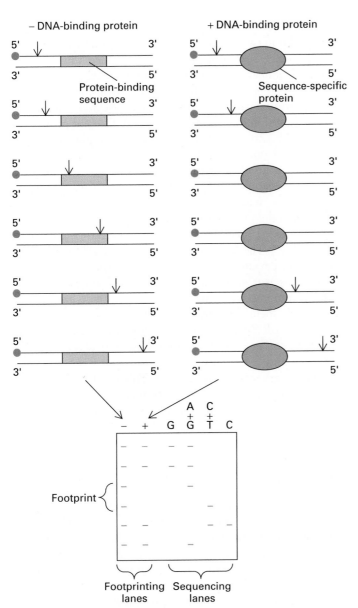

― DNA-binding protein

+ DNA-binding protein

Protein-binding sequence

Sequence-specific protein

A C
+ +
G G T C

― +

Footprint

Footprinting lanes   Sequencing lanes

◄ FIGURE 11-10   DNase I footprinting. (*Top*) A DNA fragment is labeled at one end with ^{32}P (red dot) as in the Maxam-Gilbert sequencing method (see Figure 7-28). Portions of the sample then are digested with DNase I in the presence and absence of a protein that binds to a specific sequence in the fragment. DNase I hydrolyzes the phosphodiester bonds of DNA between the 3' oxygen on the deoxyribose of one nucleotide and the 5' phosphate of the next nucleotide. A low concentration of DNase I is used so that on average each DNA molecule is cleaved just once (vertical arrows). In the absence of a DNA-binding protein, the sample is cleaved at all possible positions between the labeled and unlabeled ends of the original fragment. The two samples of DNA then are separated from protein, denatured to separate the strands, and electrophoresed. The resulting gel is analyzed by autoradiography, which detects only labeled strands and reveals fragments extending from the labeled end to the site of cleavage by DNase I. (*Bottom*) Diagram of hypothetical autoradiogram of the gel for the minus protein sample above reveals bands corresponding to all possible fragments produced by DNase I cleavage (―lane). In the sample digested in the presence of a DNA-binding protein, two bands are missing (+lane); these correspond to the DNA region protected from digestion by bound protein and are referred to as the footprint of that protein. This protected region can be precisely aligned with the DNA sequence if sequencing reactions are performed on the original end-labeled DNA and the products electrophoresed on the same gel. In this example, the products of four Maxam-Gilbert sequencing reactions are shown.

RNA Polymerase Interacts with Specific Promoter Sequences

The original experimental data of the footprint of *E. coli* RNA polymerase on the *lac* control region is shown in Figure 11-11 (lane 4). In a specific DNA sequence, DNase I cleaves some phosphodiester bonds more frequently than others. Consequently, the distribution of fragments in the absence of protein (lane 3) is not as uniform in an actual experiment as diagrammed in Figure 11-10. RNA polymerase produced a clear footprint over a region of ≈70 base pairs. Some sites on each strand within this 70-bp region were not protected (note spaces between the brackets in Figure 11-11). This finding indicates that a polymerase molecule lies along one surface of the DNA double helix in the region between ≈−20 and ≈−50 leaving the phosphodiester bonds on the other side of the helix exposed to DNase I.

The part of the *lac* control region largely protected by RNA polymerase is indicated in Figure 11-9. This protein-binding region includes the sequences that are mutated in *lac*-promoter mutants that exhibit decreased *lac*-operon expression. Studies with these promoter-mutant DNA sequences and purified RNA polymerase showed that the enzyme has greatly reduced affinity for the mutant DNA. Thus it is believed that the polymerase contacts bases in the wild-type *lac* DNA sequence when it binds to the promoter.

By now, the nucleotide sequences of the promoters adjacent to the start sites of hundreds of *E. coli* operons have been determined. These promoters can be classified according to their "strength"; that is, the relative frequency of transcription initiation (times per minute) at each promoter. Thus, RNA polymerase initiates transcription at a high frequency at "strong" promoters and at a low frequency at "weak" promoters.

The sequences of several particularly strong *E. coli*

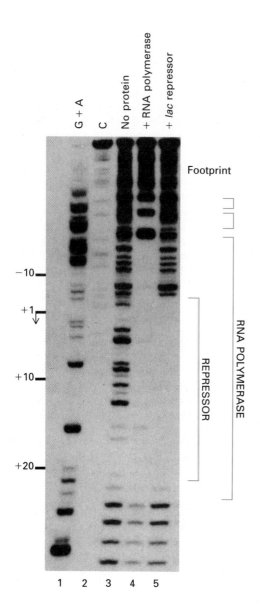

▲ FIGURE 11-11   Footprints of RNA polymerase and *lac* repressor on *lac* control-region DNA. Lane 3 shows the fragments produced by digestion of the free DNA with DNase I. The variation in density of the bands indicates that the enzyme cleaves some phosphodiester bonds more readily than others. Lanes 4 and 5 show the digestion products formed in the presence of RNA polymerase and *lac* repressor, respectively. The brackets on the right indicate the DNA regions protected from DNase I digestion by bound RNA polymerase or repressor. Lanes 1 and 2 show the products of two Maxam-Gilbert sequencing reactions: from these lanes, the gel bands can be correlated with the nucleotide sequence of the *lac* control region. The positions in the sequence are noted on the left; the arrowhead indicates the direction of transcription. This footprinting experiment was performed with the 5' label on the right end of the bottom strand as shown in Figure 11-9. [From A. Schmitz and D. J. Galas, 1979, *Nucl. Acids Res.* **6**:111.]

promoters are shown in Figure 11-12a. These sequences are similar in two regions, one near -10 and the other near -35. (The -10 region sometimes is called the *Pribnow box* after an early discoverer.) The consensus sequences of the two regions also are shown, along with single base-pair mutations that cause a significant decrease in the frequency of transcription initiation from several different promoters (Figure 11-12b). All of these damaging mutations cause deviations from the -10 and -35 consensus sequences. In addition, a number of promoter mutations that increase gene expression have been identified. Such up mutations increase the match to the -10 and -35 consensus sequences in the promoters where they occur. Two up mutations in the -10 region of the *lac* promoter are indicated in Figure 11-9. The *lac* promoter, which is a relatively weak promoter, deviates at several positions from the -35 and -10 consensus sequences. Footprinting studies with mutated promoters carrying any of these single-base mutations have identified these bases as particularly important for binding RNA polymerase. It should be noted that different *E. coli* promoters do not have identical sequences in the -10 and -35 regions, suggesting that RNA polymerase has a rather broad specificity. This feature of -10 and -35 promoter regions contrasts with restriction-enzyme cut sites; all copies of a particular restriction site have the same sequence.

σ^{70} Subunit of RNA Polymerase Functions as an Initiation Factor

Three types of evidence indicate that σ^{70} is the RNA polymerase subunit that binds to the -10 and -35 regions of most *E. coli* promoters. One line of evidence depends on the ability of UV irradiation to cause atoms in DNA to form covalent bonds with closely situated amino acid residues. When RNA polymerase–promoter complexes are irradiated with UV light, covalent bonds form between bases in the -10 and -35 promoter regions and the amino acids of σ^{70}.

Identification of *second-site mutations* in *E. coli* strains carrying mutations in the promoters controlling various operons provides a second line of evidence. All such second-site mutations that increased expression of genes from the mutant promoters (i.e., *suppressed* the effect of the initial promoter mutations) were mapped to the gene encoding the σ^{70} subunit. These results strongly suggest that the altered amino acids in σ^{70} in the *second-site revertants* make direct contact with the altered bases in the mutant promoters, thereby overcoming the effect of the promoter mutations. Presumably, the alterations in σ^{70} improve its interaction with the altered bases in the -10 and -35 promoter regions. By analogy with suppressor mutations discussed in Chapter 8, the ability of second-site mutations in the σ^{70} gene to suppress promoter mutations

(a) Strong *E.coli* promoters

```
tyr tRNA    TCTCAACGTAACACTTTACAGCGGCG··CGTCATTTGATATGATGC·GCCCCGCTTCCCGATAAGGG
rrn D1      GATCAAAAAAATACTTGTGCAAAAAA··TTGGGATCCCTATAATGCGCCTCCGTTGAGACGACAACG
rrn X1      ATGCATTTTTCCGCTTGTCTTCCTGA··GCCGACTCCCTATAATGCGCCTCCATCGACACGGCGGAT
rrn (DXE)₂  CCTGAAATTCAGGGTTGACTCTGAAA··GAGGAAAGCGTAATATAC·GCCACCTCGCGACAGTGAGC
rrn E1      CTGCAATTTTTCTATTGCGGCCTGCG··GAGAACTCCCTATAATGCGCCTCCATCGACACGGCGGAT
rrn A1      TTTTAAATTTCCTCTTGTCAGGCCGG··AATAACTCCCTATAATGCGCCACCACTGACACGGAACAA
rrn A2      GCAAAAATAAATGCTTGACTCTGTAG··CGGGAAGGCGTATTATGC·ACACCCCGCGCCGCTGAGAA
λ Pᴿ       TAACACCGTGCGTGTTGACTATTTTA·CCTCTGGCGGTGATAATGG··TTGCATGTACTAAGGAGGT
λ Pᴸ       TATCTCTGGCGGTGTTGACATAAATA·CCACTGGCGGTGATACTGA··GCACATCAGCAGGACGCAC
T7 A3       GTGAAACAAAACGGTTGACAACATGA·AGTAAACACGGTACGATGT·ACCACATGAAACGACAGTGA
T7 A1       TATCAAAAAGAGTATTGACTTAAAGT·CTAACCTATAGGATACTTA·CAGCCATCGAGAGGGACACG
T7 A2       ACGAAAAACAGGTATTGACAACATGAAGTAACCATGCAGTAAGATAC·AAATCGCTAGGTAACACTAG
fd VIII     GATACAAATCTCCGTTGTACTTTGTT··TCGCGCTTGGTATAATCG·CTGGGGGTCAAAGATGAGTG
                          -35                        -10           +1  ⌁
```

(b) Consensus sequences of σ⁷⁰ promoters

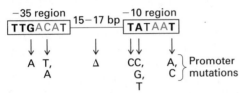

FIGURE 11-12   Strong promoters recognized by *E. coli* RNA polymerase containing σ⁷⁰. (a) Sequences of some strong promoters with spaces (dots) introduced to maximize homology in −35 region and −10 region. These sequences correspond to the top strand of the promoter with transcription proceeding to the right (see Figure 11-9). Bases that match the −35 and −10 consensus sequences are highlighted in yellow. Each sequence shown controls transcription from a particular gene or operon. The six rrn sequences control genes encoding rRNA. The λ, T7, and fd sequences, which are on viral genomes, direct transcription by the RNA polymerase of an infected *E. coli* host cell. (b) Consensus sequences of −35 and −10 regions, which are separated by 15–17 base pairs. Mutations known to significantly decrease the frequency of transcription from a number of different promoters are indicated. In the consensus sequences, the frequency with which the indicated base occurs at each position in different σ⁷⁰ promoters is indicated as follows: red letters, >75 percent; boldface black letters, 50–75 percent; black letter, 40–50 percent. The sequences of the −10 and −35 regions in the *lac* promoter, a weak promoter, deviates from the consensus sequences at several positions (see Figure 11-9b). [See W. R. McClure, 1985, *Ann. Rev. Biochem.* **54**:171 (consensus sequences); W. Siebenlist, R. B. Simpson, and W. Gilbert, 1980, *Cell* **20**:269 (promoter mutations).]

indicates that the σ⁷⁰ subunit interacts with promoter DNA.

The third type of experiment demonstrating that σ⁷⁰ interacts directly with promoter DNA solved a curious paradox. Although the results just discussed indicate that σ⁷⁰ is the subunit of RNA polymerase that interacts with the −10 and −35 promoter regions, isolated σ⁷⁰ does not bind to promoters. The solution to this paradox came from analyzing DNA binding by deletion mutants of σ⁷⁰ constructed using recombinant DNA technology. Mutant σ⁷⁰ subunits in which the N-terminal half of the polypeptide is deleted were found to bind specifically to promoter DNA. Presumably, the N-terminus of wild-type σ⁷⁰ prevents the polypeptide from interacting with promoters until it has associated with the other RNA polymerase subunits.

The strength of a promoter is determined largely by the affinity of the promoter sequence for RNA polymerase. Although other factors contribute to how rapidly a bound polymerase will initiate, most strong promoters have a good match to the −35 and −10 consensus sequences, as noted previously. Addition of ribonucleoside triphosphates to a polymerase-promoter complex results in transcription initiation and chain elongation. After the polymerase transcribes approximately 10 base pairs, the σ⁷⁰ subunit is released. Consequently the σ⁷⁰ subunit acts as an *initiation factor* required for transcription initiation, but not for RNA-strand elongation once initiation has taken place. The form of the polymerase that continues transcribing, consisting of the β′, β, and two α subunits, is called the *core polymerase*. The form of the polymerase that also contains the σ⁷⁰ subunit and can initiate transcription is called the *holoenzyme*.

The initial complex formed when RNA polymerase binds to a promoter region is called a *closed complex* because the DNA strands near the transcription start site still are base-paired. Before polymerization can proceed, the polymerase must separate the hydrogen-bonded base pairs in the region of the start site. Following this step, the polymerase can mediate base pairing between the template strand and the incoming ribonucleotide triphosphates to be added to the 3' end of the growing RNA strand.

The length of the unpaired, separated region in this open complex was determined by exposing open complexes to dimethyl sulfate or potassium permanganate. Dimethyl sulfate methylates the N_1 of adenine in single-stranded DNA but not in double-stranded DNA in which this nitrogen is hydrogen bonded to thymine and buried on the inside of the double helix (see Figure 4-7). Similarly, permanganate interacts with the C_5–C_6 bond of thymine in single-stranded DNA but not in double-stranded DNA where the bond is not accessible to attack because it is buried within the double-helical structure. Treatment of open complexes with these reagents have revealed reactive bases at positions −9 to +2 or +3, indicating that the separated region extends for about 12 base pairs.

α-Subunit Dimer Binds to rRNA Promoters in −40 to −60 Region

E. coli cells contain six copies of the pre-rRNA gene, and the promoters that control their transcription are among the strongest in the cell (see Figure 11-12). Together they account for more than half of all transcription in rapidly growing cells. A-T rich sequences located −40 to −60 base pairs from the transcription initiation site are required for this very high transcription rate. The α subunits of RNA polymerase can be isolated from the β and β' subunits in the form of an α-subunit dimer. Recent work has shown that these isolated α-subunit dimers bind to the −40 to −60 region of the strong rRNA promoters. RNA polymerase containing a mutated α subunit that does not bind the −40 to −60 region of a rRNA promoter does not transcribe the rRNA promoter at a high rate, but does transcribe other promoters normally. These results make it clear that one function of the α subunits is to contribute to the high activity of rRNA promoters by interacting with the −40 to −60 region.

Binding of lac Repressor to the lac Operator Blocks Transcription Initiation by RNA Polymerase

The Jacob and Monod model predicts that lac repressor will block transcription initiation from the lac promoter by binding to the lac operator. Proving this point and analyzing how the blocking occurs required experimentation with the purified lac repressor protein. Lac repressor was originally identified and purified on the basis of its ability to tightly bind lactose and lactose analogs such as IPTG, which are potent inducers of lac transcription. DNase I footprinting experiments with the purified protein showed

that it binds to a specific sequence in the lac control region (Figure 11-11, lane 5). The region of DNA protected from DNase I digestion by lac repressor corresponds with the location of O^c mutations (see Figure 11-9). Purified lac repressor was found to have a greatly decreased affinity for lac control-region DNA containing these O^c mutations. These results indicate that interactions between the lac repressor and bases in the wild-type operator sequence are responsible for the tight binding of lac repressor to lac-operator DNA. When these bases are mutated in O^c mutants, some of the interactions between lac repressor and the operator DNA are eliminated resulting in decreased affinity of the repressor for the mutant operator DNA.

DNA-binding experiments with purified lac repressor and isolated lac-operator DNA demonstrated that addition of lactose or IPTG to the binding reaction vastly decreased the affinity of the repressor for operator DNA, directly confirming a major element of the Jacob and Monod model. This effect of IPTG on binding of lac repressor to the operator can be observed directly in living E. coli cells through the technique of in vivo footprinting. In this technique, living cells are treated with the strong methylating agent dimethyl sulfate (DMS), which can diffuse through the cell membrane. Dimethyl sulfate methylates the N_7 of guanine bases in double-stranded DNA; this nitrogen is exposed in the major groove of B-form DNA. (As noted above, N_1 of guanine is protected in double-stranded DNA.) Sufficient DMS is added to methylate about one in every few hundred guanines. Sites in DNA where lac repressor interacts closely with the N_7 of a guanine are protected from methylation. This is similar to a bound protein protecting DNA from digestion by DNase I.

To demonstrate the effect of inducer on binding of lac repressor in vivo, E. coli cells were grown in the presence and absence of IPTG and then treated with DMS. The isolated DNA from each sample was treated with the restriction enzyme HaeIII, which cuts in the lac control region upstream and downstream from the start site. The resulting digested DNA was denatured to separate the strands and treated with piperidine, which cleaves DNA at methylated guanine (mG) residues, yielding a series of successively shorter fragments all ending at the position of a 5' mG (Figure 11-13a). Electrophoresis then revealed two sites that were protected in cells grown in the absence of IPTG: the +5 and +7 guanines of the noncoding strand within the lac operator (lane 1). These two guanines were not protected in lacI⁻ cells, demonstrating that it is the lac repressor that makes close contact with these guanines in vivo. These guanines were methylated in wild-type cells grown in the presence of IPTG (lane 2), showing that the inducer causes lac repressor to dissociate from the lac operator in living cells.

The lac-operator sequence contains numerous guanine residues that are not protected from methylation. Because DMS is a small molecule that can diffuse into the DNA-repressor complex, only those guanines in which N_7 is very closely apposed to protein atoms are protected. In contrast,

(a) In vivo footprinting of *lac* operator

(1) Grow *E. coli* cells in presence or absence of IPTG (inducer)
(2) Treat cells with dimethyl sulfate at a concentration that methylates one in every ≈200 guanine residues
(3) Isolate DNA, digest with *Hae*III.
(4) Denature to separate strands
(5) Treat with piperidine

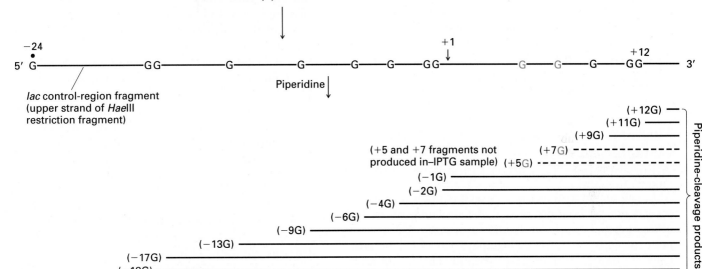

(b) *lac*-operator regions protected by bound repressor

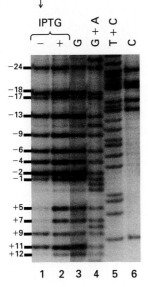

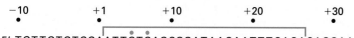

— Protected from DNase I digestion
• Protected from methylation

▲ **FIGURE 11-13** In vivo footprinting of *lac* operator region. (a) The general procedure outlined at the top yields a population of short fragments of the *lac* control region resulting from piperidine cleavage at all the N₇-methylated guanine (mG) residues resulting from dimethyl sulfate (DMS) methylation of GS that are not protected by binding of *lac* repressor. The positions of the guanines in the upper strand of the *Hae*III restriction fragment and the corresponding piperidine-cleavage products are shown in the middle. Protected residues are in red type. DNA was isolated from DMS treated cells, digested with HacIII, denatured and cleaved at all mGs with piperdine. The cleaved DNA was then separated by gel electrophoresis and DNA from the *lac* control region was detected by the Southern blotting procedure (see Figure 7-31). An autoradiograph of the piperidine-cleavage products from an actual in vivo footprinting experiment is shown at the bottom. Lanes 1 and 2 are from cells grown in the absence and presence of IPTG, respectively. Note that bands at the +5 and +7 positions are missing in the −IPTG sample. Lanes 3–6 show the products of Maxam-Gilbert sequencing reactions on *lac* operator DNA from untreated cells. These lanes permit assignment of the positions of the 5′ mGs that produced the cleavage products within the DNA sequence, as shown to the left of the gel. The region of the autoradiogram reproduced shows the bases from −24 to +12. (b) Binding of *lac* repressor protects only two guanines in the upper strand (red dots) from methylation. In contrast, ≈25-base sequences on each strand are protected from DNase I digestion by bound repressor. This difference is due to the difference in size of dimethyl sulfate and DNase I. See text for discussion. [Photograph in part (a) from H. Nick and W. Gilbert, 1985, *Nature* **313**:795.]

the repressor protects an ≈25-base sequence of each strand from digestion by DNase I, which is itself a large protein molecule (Figure 11-13b).

DNase I footprinting experiments show that the binding sites for RNA polymerase and *lac* repressor in the *lac*-operon control region overlap between about −5 and +20 (see Figures 11-9 and 11-11). Consequently when *lac* repressor is bound to the *lac* operator, it is in a position to prevent interactions between RNA polymerase and DNA at the +1 position required for transcription initiation. As predicted by the Jacob and Monod model, binding of *lac* repressor to *lac* DNA does prevent RNA polymerase from initiating transcription in in vitro reactions.

Most Bacterial Repressors Are Homodimers Containing α Helices That Insert into Adjacent Major Grooves of Operator DNA

Although the three-dimensional structure of the *lac* repressor has not yet been determined, the structures of several other bacterial repressors bound to operator DNA have been solved by x-ray crystallography. Most of these proteins are dimers of a single polypeptide chain, and each monomer of the repressor binds to one-half of the operator DNA. Not surprisingly, most operator sequences are short inverted repeats, as illustrated for the *lac* operator in Figure 11-14. Each half of the inverted repeat sequence in an operator is called a *half-site*. Consequently, an operator can be thought of as two half-sites, each of which binds one monomer of a dimeric repressor protein.

For many dimeric bacterial repressors, one α helix in each subunit, termed the *recognition helix,* fits into the major groove of one operator half-site (Figure 11-15a). Atoms in the side chains of polar residues in the protein form specific hydrogen bonds with atoms in the bases of the major groove; these bonds plus van der Waals interactions determine the sequence specificity of repressor binding. In addition, charge interactions between positively charged lysines and arginines in the repressor and nega-

tively charged phosphate groups in the DNA, as well as hydrogen bonds involving phosphate oxygens, contribute to the strength of the repressor-operator interaction.

In general, monomeric repressor molecules dimerize in such a way that the two recognition helices are positioned precisely to enter the major grooves in the two symmetrical halves of the operator DNA (Figure 11-15b; also see Figure 13-8). Thus, each subunit of a dimeric repressor can make similar contacts with bases in each symmetrically disposed half-site of the operator. Interactions also may occur between atoms of the repressor and atoms in the minor groove of the operator DNA (see Figure 13-4). In general, a single repressor monomer makes too few interactions to bind to DNA stably. However, a repressor dimer makes twice as many interactions as the monomer, a sufficient number for repressors to bind to their operator sites with high affinity ($K_d = \approx 10^{-8}$–10^{-11} M). The recognition helix that protrudes from the surface of bacterial repressors to enter the DNA major groove and "read" the DNA sequence is usually supported in the protein structure in part by hydrophobic interactions with a second α helix just N-terminal to it (e.g., the α_2 helix in Figure 11-15b). This recurring structural element in many bacterial repressors is called a *helix-turn-helix* motif.

Although most bacterial repressors have this dimeric structure with each monomer inserting an α helix into the major groove of an operator half-site, some repressors exhibit alternative protein structures. For example, the Arc repressor of bacteriophage P22 is a tetramer (Figure 11-16). This repressor makes contacts with the bases of its operator half-site through the side chains of two antiparallel β strands that fit into the DNA major groove.

Ligand-Induced Conformational Change in the trp *Repressor* Like the *lac* repressor, the affinity of many bacterial repressors for DNA is altered when the protein binds small molecules. The *allosteric* conformational change responsible for this alteration in DNA binding affinity has been observed for the *trp* repressor by x-ray crystallography of the protein in the absence and presence of tryptophan.

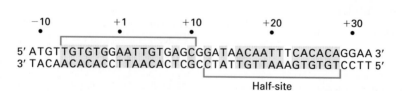

▲ FIGURE 11-14   The *lac* operator sequence is a nearly perfect inverted repeat centered around the GC base pair at position +11. The 17-bp sequence of the top strand beginning at −7 is nearly identical in the 5′ → 3′ direction to the 17-bp sequence of the bottom strand in the 5′ → 3′ direction beginning at +28. Identical base pairs in the two halves of the inverted repeat are highlighted in yellow. Each half of the inverted-repeat sequence is called an operator half-site (brackets).

(a) Bacteriophage 434
repressor

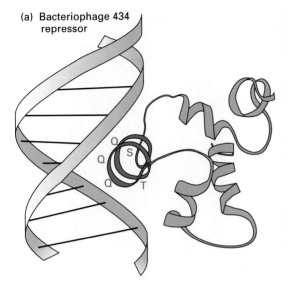

(b) Cro protein

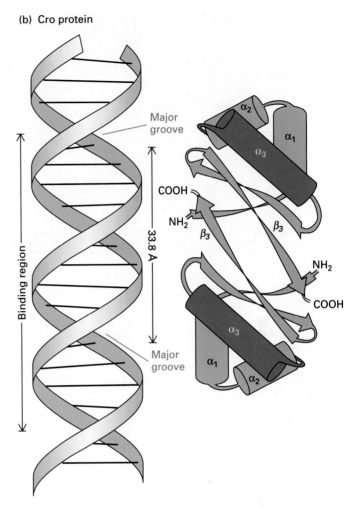

▲ FIGURE 11-15   Many bacterial repressors interact with operators by inserting an α helix of the protein (the recognition or sequence-reading helix) into a major groove of the DNA. Shown here are two bacteriophage repressors, which bind to operator sequences in the viral genome, thereby preventing transcription by host-cell enzymes. (a) One monomer of the dimeric bacteriophage 434 repressor (*right*), represented as a ribbon diagram, interacting with DNA (*left*). Three glutamine (Q) residues, one serine (S), and one threonine (T) in the recognition helix (dark green) form hydrogen bonds with specific bases in a major groove of the operator DNA. (b) Dimeric Cro protein, a repressor from bacteriophage λ, with α helices represented as cylinders and β strands as arrows. The two monomers dimerize in such a way that the distance between the α3 recognition helices (dark green) matches the distance between adjacent major grooves along one side of the operator DNA helix. The α2 helices help position the recognition helices so they can insert into the DNA major grooves. See Figures 13-4 and 13-8 for additional representations of repressor-operator interactions. [Part (a) adapted from S. C. Harrison, 1991, *Nature* **353**:715; part (b) adapted from G. Felsenfeld, 1985, "DNA," *Readings from Scientific American: The Molecules of Life*, Scientific American, Inc., pp. 13–23.]

Trp repressor regulates the expression of genes required for the synthesis of the amino acid tryptophan (see Figure 9-1a). Whereas *lac* repressor binds operator DNA in the absence of its ligand (lactose), *trp* repressor binds DNA only when it has bound its ligand, tryptophan. In this way *trp* repressor blocks production of the several enzymes required for tryptophan synthesis when sources of tryptophan in the environment are high. In the absence of tryptophan, the protein is referred to as the *trp aporepressor*. The recognition helices of the *trp* aporepressor are too close together to fit into the major grooves of the *trp*-operator half-sites (Figure 11-17). Binding of tryptophan to the aporepressor causes it to undergo a conformational change that moves these α helices apart, so that they are separated by the precise distance required to fit into the adjacent major grooves of the *trp* operator half-sites.

Electrophoretic Mobility Shift Assay   DNA-binding proteins can be measured quantitatively with the electrophoretic mobility shift assay (EMSA), which also is referred to as the gel-shift or band-shift assay. In this assay, the electrophoretic mobility of a radiolabeled DNA fragment is determined in the presence and absence of a sequence-specific DNA-binding protein. Protein binding generally reduces the mobility of a DNA fragment, causing a shift in the location of the fragment band detected by autoradiography.

(a)

(b)

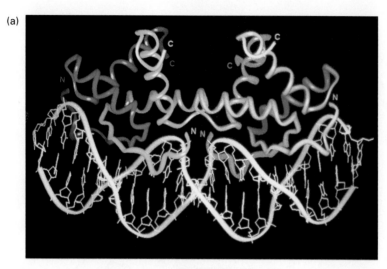

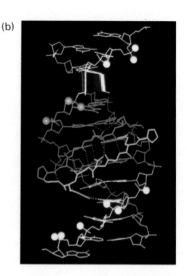

▲ FIGURE 11-16 Arc protein, a repressor from bacteriophage P22, forms a tetramer containing antiparallel β strands that contact atoms in the major groove of its operator half-sites. (a) Structure of the Arc tetramer–operator complex. The operator DNA is shown in white, and each Arc monomer is shown as a ribbon diagram of a different color. (b) Diagram of the interactions of Arc side chains with atoms in the major groove of an operator half-site. Only the Arc residues in the β strands that enter the DNA major groove are shown. The DNA is shown in blue and white. Phosphate atoms contacted by the Arc repressor are shown as spheres. Red and yellow stick diagrams show the amino acid residues of the two antiparallel β strands that contact DNA atoms in the major groove. Dashed lines represent hydrogen bonds between side chains and between side chains and bases. [From B. E. Raumann, et al., 1994, *Nature* **367**:754.]

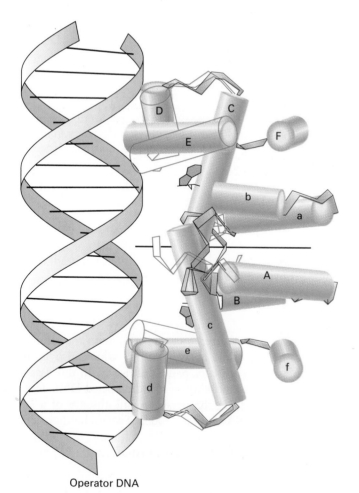

Operator DNA

◄ FIGURE 11-17 Conformational change in the *trp* aporepressor caused by the binding of tryptophan. The α helices in the dimeric protein are shown as cylinders identified by upper-case letters in one monomer and lower-case letters in the other. The recognition helices (E and e) in the aporepressor (transparent with red holding line) are too close together to fit into adjacent major grooves in the operator DNA. When the aporepressor binds tryptophan (light red), the N-terminal ends of helices E and e move apart by 8 Å. As a result, helices E and e in the repressor (orange) fit neatly into the operator DNA. [Adapted from R.-G. Zhang et al., 1987, *Nature* **327**:591.]

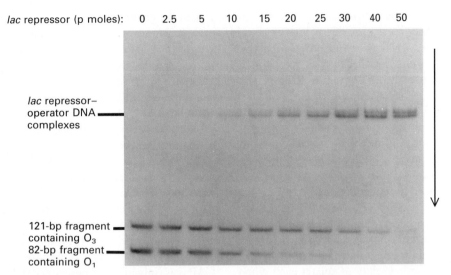

lac repressor (p moles):   0   2.5   5   10   15   20   25   30   40   50

lac repressor–
operator DNA
complexes

121-bp fragment
containing O₃
82-bp fragment
containing O₁

▲ FIGURE 11-18   Electrophoretic mobility shift assay for DNA-binding proteins. In this example, increasing amounts of *lac* repressor were incubated with a mixture of two ^{32}P-labeled *lac* operator fragments: an 82-bp fragment containing O_1, a high-affinity repressor-binding site, and a 121-bp fragment containing O_3, a low-affinity repressor-binding site. The samples then were electrophoresed on a polyacrylamide gel. An autoradiogram of the gel is shown. The two repressor-operator DNA complexes comigrate. However, the O_1-containing fragment is completely bound at a lower concentration of repressor than is the O_3 fragment. This finding shows that *lac* repressor binds to O_1 with a greater affinity than to O_3. [From M. Fried and D. M. Crothers, 1981, *Nucl. Acids Res.* **9**:6505.]

One advantage of this technique over footprinting assays is that binding can be detected when only a small fraction of the labeled DNA fragment is bound. This simplifies quantitative analysis of DNA binding reactions. For example, in the analysis depicted in Figure 11-18, the DNA fragment containing the high-affinity O_1 site in the *lac* operator is bound by a lower concentration of *lac* repressor than another operator fragment containing O_3, a low-affinity site. EMSA is widely used to assay sequence-specific DNA-binding proteins during their purification.

Positive Control of the *lac* Operon Is Exerted by cAMP-CAP

E. coli prefers metabolizing glucose above all other sugars, probably because glucose is a central molecule in carbohydrate metabolism. When it enters an *E. coli* cell, glucose is utilized directly without the induction of any new enzymes. If glucose is present in a medium to which lactose, arabinose, or any of a number of other sugars is added, the enzymes needed to metabolize the other sugar are not induced to high levels until the glucose is used up. When these observations were first made, it seemed likely that a breakdown product of glucose (a *catabolite*) prevented synthesis of the mRNAs encoding a wide variety of sugar-metabolizing enzymes. Consequently, this phenomenon is referred to as *catabolite repression*. However, further studies have shown that, in fact, this apparent repression does not result from the action of a transcription repressor.

When *E. coli* is starved for glucose, it synthesizes an unusual nucleotide named cyclic adenosine-3′,5′-monophosphate, generally referred to as cyclic AMP, or cAMP (Figure 11-19). This molecule was originally discovered by enzymologists working with vertebrate enzymes whose activity differed in the presence or absence of cAMP. In bacteria, an increase in the cAMP level seems to be an "alert" signal indicating a low glucose level. For instance, when bacteria are grown in a medium containing both lactose and glucose, the intracellular cAMP level initially is low, and the enzymes encoded by the *lac* operon are ex-

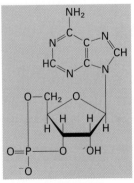

Cyclic AMP

▲ FIGURE 11-19   Structure of cyclic AMP (cAMP).

pressed only at low levels. Once most of the glucose has been metabolized, cAMP levels in the cells rise, and β-galactosidase is induced to high levels. When dibutyryl cyclic AMP, an analog of cAMP that can pass through the cell membrane, is added to media containing lactose and glucose, β-galactosidase synthesis occurs at a high level before all the glucose has been metabolized. Similar results for enzymes required to metabolize other sugars suggested that cAMP activates transcription of the various operons encoding these enzymes.

Once the effects of cAMP on transcription were suspected, it was hypothesized that one or more proteins must mediate, or "interpret," the connection between high cAMP levels (glucose starvation) and the need to induce other sugar-metabolizing enzymes. To test this theory, a search was made for mutant cells that could not grow in the presence of any of a wide variety of sugars even when glucose was absent from the medium. Individual mutant *E. coli* colonies were found that had simultaneously lost the ability to metabolize lactose, galactose, maltose, or arabinose.

These mutants fell into two classes. One class did not make cAMP in the absence of glucose. These were defec-

tive in the enzyme adenyl cyclase, which converts ATP into cAMP. The other half could make cAMP but did not increase their synthesis of the enzymes for metabolizing any of the sugars mentioned above in the presence of those sugars. This finding suggested that the second group of mutants contained a defective protein that could not respond to cAMP. Subsequently, a protein that could bind to cAMP was isolated from normal cells and was found to be absent in the second group of mutants. This protein now is generally called *catabolite activator protein* (CAP), but sometimes is referred to as cAMP receptor protein (CRP). In vitro transcription experiments provided conclusive evidence that the *lac* operon is controlled by cAMP and CAP. When cloned DNA carrying the *lac* operon was incubated with protein extracts prepared from *E. coli* cells, the rate of synthesis of *lac* mRNA in the absence of either cAMP or CAP was only about 5 percent of the rate in the presence of both. Thus, maximum transcription from the *lac* operon requires the presence of the cAMP-CAP complex.

DNase I footprinting experiments revealed that purified cAMP-CAP binds to a specific sequence in the *lac* control region called the CAP site. The CAP site is just upstream from the RNA polymerase–binding site (Figure

(a) Glucose present (cAMP low); no lactose: No *lac* mRNA

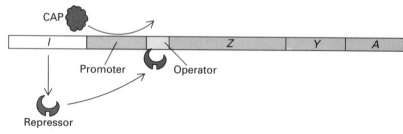

(b) Glucose present (cAMP low); lactose present

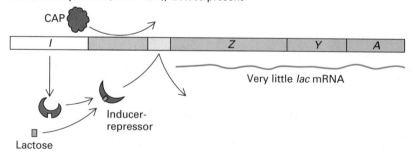

(c) No glucose present (cAMP high); lactose present

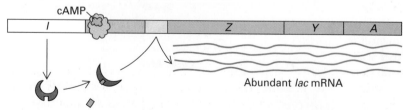

◄ FIGURE 11-20   Negative and positive control of *lac*-operon transcription by the *lac* repressor and cAMP-CAP, respectively. (a) In the absence of lactose, no *lac* mRNA is formed because repressor bound to the *lac* operator prevents transcription. (b) In the presence of glucose and lactose, the *lac* repressor binds lactose and undergoes a conformational change, so that it does not bind to the *lac* operator. However, cAMP is low, because glucose is present, and thus cAMP-CAP does not bind to the CAP site in the operator. As a result, RNA polymerase does not bind efficiently to the *lac* promoter and little *lac* mRNA is synthesized. (c) In the presence of lactose and the absence of glucose, maximal transcription of the *lac* operon occurs: the *lac* repressor does not bind to the *lac* operator, the concentration of cAMP increases, and the resulting cAMP-CAP complex forms and binds at the CAP site, stimulating binding and initiation by RNA polymerase.

11-9). Mutations in the CAP site that prevent in vitro binding of cAMP-CAP also prevent high-level expression of the *lac* operon in vivo. These findings indicate that the cAMP-CAP complex must bind to this site to stimulate transcription. Thus, bound cAMP-CAP *activates* transcription (positive control), whereas bound *lac* repressor inhibits transcription (negative control), as summarized in Figure 11-20.

Cooperative Binding of cAMP-CAP and RNA Polymerase to *lac* Control Region Activates Transcription

Several observations indicate that cAMP-CAP activates transcription of the *lac* operon by directly contacting an α subunit of an RNA polymerase bound at the *lac* promoter. The structure of the complex between cAMP-CAP and DNA has been determined by X-ray crystallography (Figure 11-21). Like most bacterial repressors, CAP is a dimer containing two helices that insert into adjacent major grooves of an approximate inverted repeat sequence. Unlike many repressors, binding of cAMP-CAP bends the flexible DNA molecule around itself, so that the DNA is curved through an angle of approximately 90 degrees.

In solution, cAMP-CAP and RNA polymerase bind to each other weakly in the absence of DNA ($K_d \approx 3 \times 10^{-7}$ M). By itself RNA polymerase binds fairly weakly to the *lac* promoter, which deviates at several positions from the -10 and -35 consensus sequences. Similarly, cAMP-CAP does not have high affinity for the CAP site in the *lac* control region compared with its affinity for an optimal DNA sequence. However, when cAMP-CAP and RNA polymerase bind to the *lac* control region simultaneously, they stimulate each other's binding. This phenomenon, referred to as *cooperative binding,* occurs for the same reason that a repressor dimer has higher DNA-binding affinity than a repressor monomer. When cAMP-CAP and RNA polymerase are bound to their adjacent binding sites, they are at a very high relative concentration. Consequently, even though they have relatively low affinity for each other, they remain bound forming a cAMP-CAP/RNA polymerase complex, which makes more contacts with the DNA than either cAMP-CAP or RNA polymerase alone.

Strong evidence that direct contact between cAMP-CAP and RNA polymerase is required for activation came from strains producing a mutant CAP. The cAMP-CAP complex in these mutants binds and bends DNA to the same extent as wild-type cAMP-CAP but fails to activate *lac* transcription. These mutations block the weak interaction between cAMP-CAP and RNA polymerase and occur in a bulge on the surface of CAP (see Figure 11-21). The residues identified by these mutations are thought to be the ones that contact the surface of RNA polymerase. When these residues are changed, the protein-protein interaction between cAMP-CAP and RNA polymerase no longer oc-

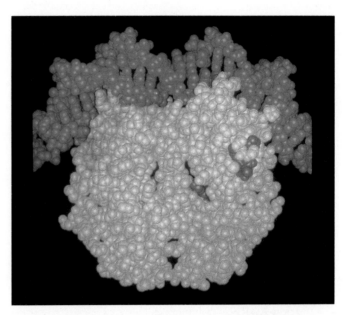

▲ FIGURE 11-21   Structure of the cAMP-CAP DNA complex. CAP is shown in light blue, and DNA and cAMP are shown in red. CAP protein bends the DNA double helix around its surface. Amino acid residues where mutations eliminate activation of the *lac* promoter are highlighted in dark blue. The DNA sequence near the right edge of the figure corresponds to the region (near -50) where the C-terminus of an α subunit of RNA polymerase is thought to interact with the highlighted residues in CAP. [See S. Schultz, G. Shields, and T. Steitz, 1991, *Science* **253**:1001; Y. Zhou, X. Zhang, and R. H. Ebright, 1993, *Proc. Nat'l. Acad. Sci. USA* **90**:6081. Photograph courtesy of R. H. Ebright.]

curs, and the mutant cAMP-CAP complex fails to activate *lac* transcription.

The region of RNA polymerase that interacts with CAP was identified in experiments with a mutant form of RNA polymerase that has a deletion of the C-terminus of the α subunits. This mutant polymerase does not respond to cAMP-CAP activation of the *lac* promoter. Consequently, the surface of CAP highlighted in Figure 11-21 is thought to bind to the C-terminus of one of the polymerase α subunits. The resulting stabilization of polymerase binding to the *lac* promoter is thought to activate *lac* transcription. The cAMP-CAP/RNA polymerase interaction also may stimulate formation of the open complex and other steps in transcription initiation subsequent to polymerase binding.

Control of Transcription from All Bacterial Promoters Involves Similar But Distinct Mechanisms

The various studies on control of transcription initiation from the *lac* promoter have revealed several general principles that are applicable to other bacterial promoters:

- Transcription is initiated from a single region of the template DNA by RNA polymerase, whose σ^{70} subunit binds tightly to promoter sequences.

- Once bound to a promoter, RNA polymerase separates the DNA strands, forming an open complex; as a result, ribonucleoside triphosphates can base-pair to the coding strand.

- RNA polymerase then moves the transcription "bubble" of separated DNA strands down the template as it polymerizes ribonucleoside triphosphates complementary to the DNA template strand.

- After ≈ 10 ribonucleotides have been polymerized, the σ^{70} subunit is released and the core polymerase ($\alpha_2\beta\beta'$) continues transcription of the template.

- Binding of repressors to operator sequences, which overlap promoters, interfere with transcription initiation by RNA polymerase.

- Activators bind upstream from promoters or on the opposite side of the DNA helix from RNA polymerase; contact between an activator and RNA polymerase stimulates transcription initiation.

Although these principles are thought to apply to most *E. coli* promoters, studies with more than 100 different promoters have shown that regulation of transcription from each promoter has its unique aspects. For example, cAMP-CAP activates transcription from both the *lac* operon and *gal* operon, which encodes enzymes required for metabolism of galactose. However, to activate *gal*-operon transcription, cAMP-CAP must interact with a different surface on RNA polymerase than it binds to in activating *lac*-operon transcription.

Likewise, some repressor proteins regulate transcription of several different operons located at separate sites on the *E. coli* chromosome. The *trp* repressor discussed earlier is a well-studied example. In addition to the *trp* operon, this repressor represses transcription of *aroH*, a single-gene operon that encodes an enzyme required for the synthesis of all aromatic amino acids, and *trpR*, another single-gene operon encoding the *trp* repressor itself. Such coordinately regulated operons constitute a *regulon*.

Although all three operons in this regulon are repressed by the *trp* repressor, the extent of repression varies greatly. The *trp* operon is repressed seventyfold; *aroH*, only about twofold; and *trpR*, threefold. These differences result from differences in the affinity of *trp* repressor for the specific operator in each of these operons, differences in the position of the operators relative to the -10 and -35 sequences of each promoter, and differences in the strengths of the three promoters. Thus the specific nucleotide sequences of the promoters and operators in a regulon allow the same repressor to differentially regulate transcription of the component operons. It is worth noting that none of the three operators in the *trp*-repressor regulon have the optimal sequence for binding *trp* repressor. Rather, they evolved operator sequences that result in regulation appropriate for the metabolic functions of the operons they control.

Transcription from Some Promoters Is Initiated by Alternative Sigma (σ) Factors

The major form of RNA polymerase in *E. coli* contains σ^{70}, and most promoters interact with this form of the enzyme. *E. coli* also contains several specialized forms of RNA polymerase, which contain alternative sigma factors that recognize different consensus promoter sequences. In addition to σ^{70}, four *E. coli* sigma factors have been identified: σ^{28}, σ^{32}, σ^{38}, and σ^{54} (Table 11-2). The promoters recog-

---

TABLE 11-2   Sigma Factors of *E. coli*

Sigma Factor	Promoters Recognized	Promoter Consensus	
		-35 Region	-10 Region
σ^{70}	Most genes	TTGACA	TATAAT
σ^{32}	Genes induced by heat shock	TCTCNCCCTTGAA	CCCCATNTA
σ^{28}	Genes for motility and chemotaxis	CTAAA	CCGATAT
σ^{38}	Genes for stationary phase	?	?
		-24 Region	-12 Region
σ^{54}	Genes for nitrogen metabolism and other functions	CTGGNA	TTGCA

SOURCES: C. A. Gross, M. Lonetto, and R. Losick, 1992, in *Transcriptional Regulation*, S. L. McKnight and K. R. Yamamoto, eds. Cold Spring Harbor Laboratory Press; D. N. Arnosti and M. J. Chamberlin, 1989, *Proc. Nat'l. Acad. Sci. USA* **86**:830; K. Tanaka, et al., 1993, *Proc. Nat'l. Acad. Sci., USA* **90**:3511.

nized by each of these alternative sigma factors are associated with specific groups of genes. Both σ^{32} and σ^{28} have regions of amino acid sequence similarity to σ^{70}. Like σ^{70}, they both recognize DNA sequences about -10 and -35 base pairs from the transcription-initiation site. The recently discovered σ^{38} is also similar in sequence to σ^{70} and recognizes many of the same promoters recognized by σ^{70}; consequently, it probably also recognizes sequence at -10 and -35. In contrast, σ^{54} has little sequence similarity with the other sigma factors and recognizes sequences approximately -12 and -24 base pairs from the initiation site.

The gram-positive soil bacterium *Bacillus subtilis* encodes several additional sigma factors, which are expressed in a specific order when these cells develop into spores in response to nutrient deprivation. These sigma factors recognize the promoters associated with sets of genes required at distinct times in the development of the spore. Some complex bacteriophages also encode specialized sigma factors expressed in a specific temporal order following infection. These redirect the cellular core polymerase to transcribe phage genes rather than cellular genes, in an order determined by the time at which the different phage-encoded sigma factors are expressed.

RNA Polymerase Containing σ^{54} Is Regulated by Proteins That Bind at Enhancer Sites Distant from the Transcription Initiation Site

Operons transcribed by RNA polymerases containing, σ^{70}, σ^{32}, or σ^{28} are regulated by repressors and activators that bind to DNA near the region where polymerase binds. In almost all cases, repressors bind between $+30$ and -50 and activators between -30 and -65 in the control region. On the other hand, operons transcribed by RNA polymerases containing σ^{54} are regulated solely by activators that generally bind between -80 and -160 from the initiation site. The σ^{54}-activators can activate transcription even when their binding sites, called *enhancers*, are moved more than a kilobase upstream from the start site. Analyzing the way they accomplish this led to the surprising realization that proteins bound at distant sites in DNA can interact with each other while the DNA between their binding sites forms a large loop.

The best-characterized σ^{54}-activator—the NTRC protein (nitrogen regulatory protein C)—stimulates transcription from the promoter of the *glnA* gene. This gene encodes the enzyme glutamine synthetase, which synthesizes the amino acid glutamine from glutamic acid and ammonia. In vivo footprinting showed that the σ^{54}-containing RNA polymerase binds to the *glnA* promoter, forming a closed complex, before being activated by NTRC. The poised polymerase, however, cannot form an open complex and initiate transcription until it receives a signal from NTRC. NTRC, in turn, is regulated by a protein kinase (an enzyme that adds a phosphate group to a protein) called NTRB. In

response to low levels of glutamine, NTRB phosphorylates dimers of NTRC, which then bind to two enhancer sites upstream of the *glnA* promoter (Figure 11-22). Phosphorylated NTRC bound at the enhancers then stimulates the σ^{54}-polymerase bound at the promoter to separate the DNA strands and initiate transcription. This process requires ATP hydrolysis by phosphorylated NTRC, as shown by the finding that mutants defective in ATP hydrolysis are

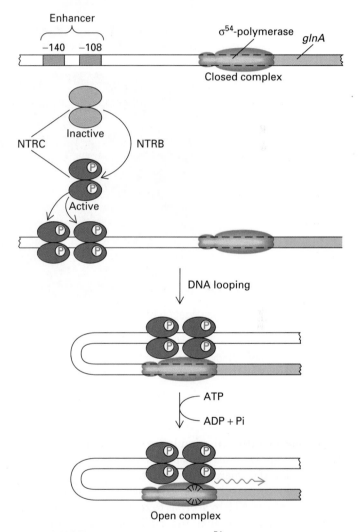

▲ **FIGURE 11-22** Activation of σ^{54}-containing RNA polymerase at *glnA* promoter by NTRC. The polymerase binds to the *glnA* promoter, forming a closed complex, before being activated. In response to a low concentration of organic nitrogen, a protein kinase called NTRB phosphorylates dimeric NTRC (purple), which then binds to two enhancers (orange) centered at -108 and -140 from the transcription start site. The bound phosphorylated NTRC dimers interact with the bound σ^{54}-polymerase, causing the intervening DNA to form a loop. The ATPase activity of NTRC then stimulates the polymerase to unwind the template strands at the start site, forming an open complex. Transcription of the *glnA* gene can then begin.

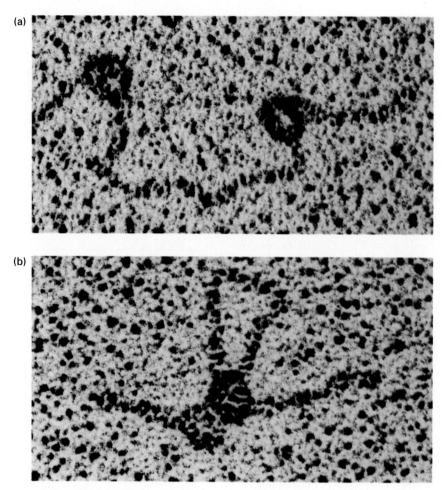

▲ FIGURE 11-23   Visualization of DNA looping and interaction of bound NTRC and σ^{54}-polymerase. (a) Electron micrograph of DNA restriction fragment with two phosphorylated NTRC dimers binding to the enhancer region near one end and σ^{54}-polymerase bound to the *glnA* promoter near the other end. (b) Electron micrograph of the same fragment preparation showing NTRC dimers and σ^{54}-polymerase binding to each other with the intervening DNA forming a loop between them. [From W. Su et al., 1990, *Proc. Nat'l. Acad. Sci. USA* **87**:5505; courtesy of S. Kustu.]

invariably defective in stimulating open-complex formation by σ^{54}-polymerase. It is postulated that ATP hydrolysis supplies the energy required for melting the DNA strands. In contrast, as we saw earlier, σ^{70}-polymerase does not require ATP hydrolysis to separate the strands at a start site, indicating that the molecular mechanisms of open-complex formation by the two forms of polymerase differ significantly.

Electron-microscope studies have shown that NTRC dimers bound at enhancers and σ^{54}-polymerase bound at the promoter directly interact, forming a loop in the flexible DNA backbone between the binding sites (Figure 11-23). Thus bound phosphorylated NTRC stimulates open-complex formation by directly contacting a σ^{54}-polymerase bound at the promoter. As discussed later, this activation mechanism is somewhat similar to transcriptional activation in eukaryotes.

► Eukaryotic Gene Control: Purposes and General Principles

The primary function of gene control in prokaryotes is to adjust the enzymatic machinery of the cell to its immediate nutritional and physical environment. Within a *single* bacterial cell, genes are reversibly induced and repressed by transcriptional control in order to allow growth and division, the raison d'être of bacterial existence. Single-celled eukaryotes, such as yeast cells, also seem to be designed only, or mainly, for the purpose of replicating; yeasts also possess many genes that are controlled in response to environmental variables (e.g., nutritional status, oxygen tension, and temperature). Even in the organs of higher animals—for example, the mammalian liver—some genes can respond reversibly to external stimuli such as noxious chemicals. In general, however, metazoan cells are protected from immediate outside influences; that is, most cells in metazoans experience a fairly constant environment. Perhaps for this reason, genes that respond to environmental changes constitute a much smaller fraction of the total number of genes in multicellular organisms than in single-celled organisms.

The most characteristic and exacting requirement of gene control in multicellular organisms is the execution of precise developmental decisions so that the right gene is activated in the right cell at the right time during the development of the many different cell types that collectively form a multicellular organism. In most cases, once a developmental step has been taken by a cell, it is not reversed. Thus these decisions are fundamentally different from bacterial induction and repression. Many differentiated cells (e.g., skin cells, red blood cells, lens cells of the eye, and antibody-producing cells) march down a pathway to final cell death in carrying out their genetic programs, and they leave no progeny behind. These fixed patterns of gene control leading to differentiation serve the need of the whole organism and not the survival of an individual cell.

Most Genes in Higher Eukaryotes Are Regulated by Controlling Their Transcription

Although examples of regulation at each step in eukaryotic gene expression, have been discovered, regulation of transcription initiation is by far the most widespread form of gene control in eukaryotes, as it is in prokaryotes. This has been demonstrated by direct measurements of the transcription rates of multiple genes. Such measurements show, for example, that genes expressed specifically in mammalian liver are transcribed in liver cells and are not transcribed in other cells of the body.

The simplest and most direct method of measuring transcription rates is to expose cells for a brief time (e.g., 5 min or less) to a labeled RNA precursor and then determine the amount of labeled nuclear RNA formed by its hybridization to a cloned DNA. This method can be used to determine transcription rates in cultured cells but is not practical in whole animals because the amount of labeled RNA obtained is insufficient for accurate measurements.

Even with cultured cells, a second method—called *nascent-chain* or ("run-on") *analysis*—often is preferred. In this method, nuclei are isolated from cells and allowed to incorporate ^{32}P from labeled ribonucleoside triphosphates directly into nascent (growing) RNA chains to produce highly labeled RNA preparations (Figure 11-24). The reactions are run for a brief period so that the RNA polymerase adds 300–500 nucleotides to nascent RNA chains. By hy-

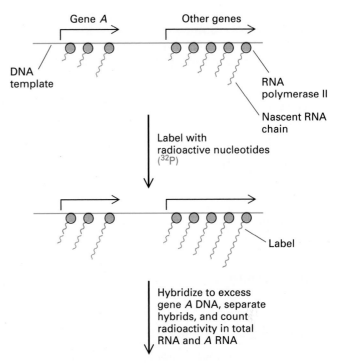

▲ FIGURE 11-24   Nascent-chain (run-on) assay for transcription rate of a gene. Labeled RNA is prepared in isolated nuclei by allowing extension of already initiated RNA chains. The average polymerase only moves a few hundred nucleotides and very little new initiation occurs. By hybridizing the labeled RNA to the cloned DNA for a specific gene (*A* in this case), the fraction of total RNA transcribed from that gene can be measured. [See J. Weber, W. Jelinek, and J. E. Darnell, 1977, *Cell* **10**:611.]

bridizing the labeled RNA to cloned DNA from a specific gene, the fraction of the total RNA copied from a particular gene (i.e., its relative transcription rate) can be determined. Relative transcription rates determined either by labeling of cultured cells or of isolated nuclei are nearly the same (Table 11-3), indicating that all of the polymerase molecules that are active at the time nuclei are taken from intact cells continue to function during incubation of isolated nuclei.

The run-on analysis of isolated nuclei is very widely used as a direct assay of relative transcription rates. For example, hepatocytes, the major type of cell in the mammalian liver, produce dozens of proteins that are not made in most other organs. These include both enzymes and numerous secreted proteins that constitute the major protein components of serum. The run-on experiment illustrated in Figure 11-25 shows that transcription of many genes encoding proteins expressed specifically in hepatocytes is readily detected in nuclei prepared from liver, but

TABLE 11-3   Comparison of Two Methods of Labeling Nuclear RNA for Use in Hybridization Experiments to Assay Transcription Rates*

DNA Sequences Used in Hybridization	Percent of Total Labeled RNA Hybridized	
	Whole Cells	Isolated Nuclei
Adenovirus genome:		
Cells early in infection	0.75	0.58
Cells late in infection	16.6	14.4
β-globin	0.01[†]	0.01[†]
Chinese hamster cDNAs	0.0001–0.001	0.0001–0.001
Ovalbumin cDNA	0.00018[‡]	0.00024[‡]
Conalbumin cDNA	0.00015[‡]	0.00022[‡]

*Nuclear RNA was labeled either by exposing whole cells to [^{3}H]uridine or by exposing isolated nuclei to [^{32}P]UTP. After brief incubation, the RNA was hybridized to the various DNAs indicated in the table and the percentage of the total label in the RNA:DNA hybrid was determined. This assay thus measures the relative transcription rate of the DNA used for hybridization.

[†]Erythroleukemia cells were used in assays of β-globin.

[‡]Cultured minced chicken oviducts were used in assays of ovalbumin and conalbumin genes.

SOURCE: See J. E. Darnell Jr., 1982, *Nature* **297**:365; and G. S. McKnight and R. D. Palmiter, 1979, *J. Biol. Chem.* **254**:9050.

not in nuclei from brain or kidney. Thus the expression of a wide variety of liver-specific proteins is regulated by controlling the transcription of their genes.

DNA Regulatory Sites Often Are Located Many Kilobases from Eukaryotic Transcription Start Sites

The basic principles shown to control transcription in prokaryotes apply to eukaryotic organisms as well. In most cases transcription initiates at a specific base pair in template DNA or at alternative sites within a few base pairs. Transcription is controlled by trans-acting proteins, called *transcription factors*, binding at cis-acting regulatory DNA sequences. However, as might be expected from the increased complexity of eukaryotic organisms, the processes regulating transcription initiation in eukaryotes are more complex than in prokaryotes. Most *E. coli* operons are transcribed by RNA polymerase containing σ^{70} and are controlled by repressors and activators that bind to DNA sequences within 60 base pairs of the transcription initiation site. Relatively few operons are transcribed by RNA

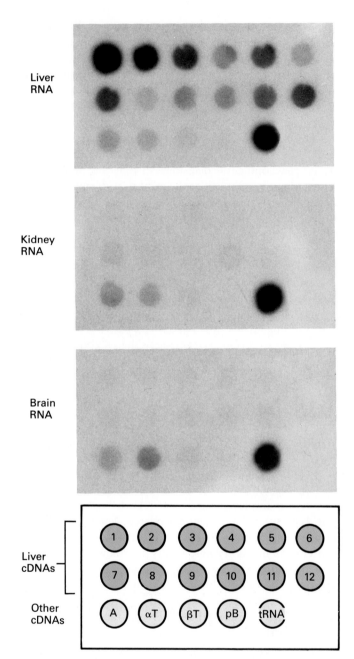

▲ FIGURE 11-25   Experimental demonstration of differential synthesis of 12 mRNAs encoding liver-specific proteins. Nuclei from mouse liver, kidney, and brain cells were exposed to [^{32}P]UTP, and the resulting labeled RNA was hybridized to various cDNAs fixed to nitrocellulose. After removal of unhybridized RNAs, the hybrids were revealed by autoradiography. The cDNAs labeled 1–12 encode proteins synthesized actively in liver (e.g., 4 = albumin; 3 = α_1-antitrypsin; 6 = transferrin) but not in most other tissues. The other cDNAs tested were actin (A) and α- and β-tubulin (αT, βT), which are proteins found in almost all cell types. Methionine tRNA and the plasmid DNA (pB) in which the cDNAs were cloned were included as controls. The pattern of spots in the three tissues, which represents the synthesis of specific mRNAs, clearly indicates that transcriptional control is the primary means for regulating production of these liver-specific proteins. [See E. Derman et al., 1981, *Cell* **23**:731; and D. J. Powell et al., 1984, *J. Mol. Biol.* **197**:21.]

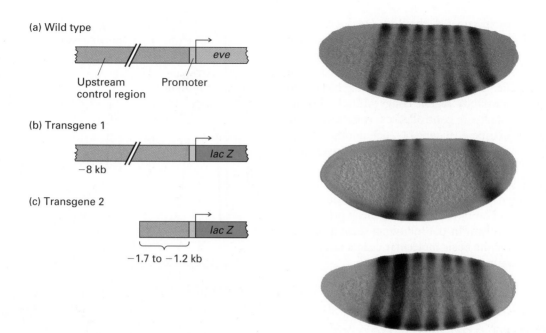

(a) Wild type

Upstream
control region

Promoter

eve

(b) Transgene 1

−8 kb

lac Z

(c) Transgene 2

−1.7 to −1.2 kb

lac Z

▲ FIGURE 11-26 Regulation of *even-skipped* (*eve*) expression in early *Drosophila* embryos. (a) A normal *Drosophila* embryo stained with antibody specific for the Even-skipped (Eve) protein. This protein is expressed in seven stripes (brown) each three or four cells wide on the surface of the embryo. (b) Embryo resulting from mating of transgenic flies carrying an engineered transgene consisting of the *E. coli lacZ* gene ligated to an 8-kb DNA fragment (orange/yellow) that lies upstream of the *eve* coding region and includes the *eve* promoter. Staining of the embryo with X-gal revealed that β-galactosidase (blue) was expressed in regions corresponding to Eve stripes 2, 3, and 7. This result demonstrated that expression of *eve* in these stripes is con-

trolled by elements in the 8-kb upstream region. (c) Embryo resulting from mating of transgenic flies carrying a DNA construct consisting of *lacZ*, an ≈140-bp segment including the *eve* promoter, and the region from −1.7 to −1.2 upstream of the *eve* coding region. Staining with X-gal and antibody to Eve revealed β-galactosidase (blue) in stripe 2, showing that expression of Eve stripe 2 is controlled by a specific ≈0.5-kb region upstream of the *eve* promoter. [See K. Harding et al., 1989, *EMBO J.* **8**:1205; T. Goto, P. Mac-Donald, and T. Maniatis, 1989, *Cell* **57**:413; S. Small, A. Blair, and M. Levine, 1992, *EMBO J.* **11**:4047. Photographs courtesy of S. Small and M. Levine.]

polymerase containing σ^{54} controlled by activators binding at sites more than 100 base pairs from the start site. This situation is well suited to the close spacing of genes on the *E. coli* chromosome.

The situation is reversed in eukaryotes. In higher eukaryotes, protein-coding genes are generally well separated in the genome. Regulatory transcription factors often bind at sites distant from the transcription start site, in some cases tens of thousands of base pairs either upstream or downstream from the initiation site. As a result of this arrangement, multiple transcription factors may control transcription from a single promoter, permitting complex control of transcription during development.

Numerous examples of such complex transcriptional control are discussed in Chapter 13. Here we describe one to illustrate this phenomenon. During early development of the *Drosophila* embryo, the protein encoded by the *even-skipped* (*eve*) gene is expressed in seven stripes, three or four cells wide, along the length of the embryo (Figure 11-26a). This anatomically complex pattern of *eve*-gene expression is regulated by multiple cis-acting DNA control

sequences spread over many kilobases in the region of the *Drosophila* genome containing the *eve* gene.

To study how this pattern of expression is determined, researchers cloned DNA from this region of the *Drosophila* genome using methods described in Chapters 7 and 8. *Drosophila* DNA encoding the N-terminus of the Even-skipped (Eve) protein and extending 8 kb upstream was ligated to the *lacZ* gene from *E. coli*. In this case, *lacZ* functions as a *reporter gene* whose expression is easily measured by the X-gal assay for β-galactosidase. This DNA construct was introduced into the *Drosophila* genome by P-element transformation (see Figure 8-40). Embryos from the transgenic flies expressed β-galactosidase in regions corresponding to Even-skipped stripes 2, 3, and 7 (Figure 11-26b). This result indicates that the cis-acting DNA sequences controlling expression of Even-skipped protein in these stripes lie within 8 kb of the first protein-coding exon of the *eve* gene. The DNA sequences controlling expression of Even-skipped in stripes 1, 4, 5, and 6 lie outside of this 8-kb region, either farther upstream or possibly downstream.

In subsequent experiments, the 8-kb *Drosophila* control region was cleaved into several restriction fragments, each of which was ligated to an ≈140-bp fragment containing the *eve* promoter, which in turn was ligated to the *lacZ* reporter gene. Experiments with these constructs demonstrated that a DNA region lying ≈1.2–1.7 kb upstream of the *eve* start site controlled expression of even-skipped protein in stripe 2 (Figure 11-26c); another region lying 3.0–3.9 upstream controlled expression equivalent to stripe 3.

These results illustrate the use of multiple transcription-control regions, scattered over many kilobases of DNA, to produce a complex pattern of gene expression in a multicellular organism. In the following sections of this chapter, we describe the basic molecular events underlying this type of eukaryotic transcription control.

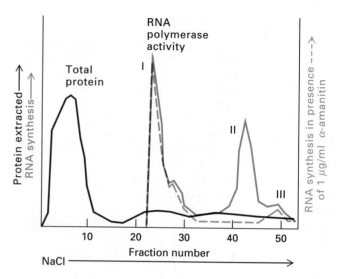

▲ FIGURE 11-27   The separation and identification of the three eukaryotic RNA polymerases by chromatographic analysis. A protein extract from the nuclei of cultured frog cells was passed through a DEAE Sephadex column to which charged proteins adsorb differentially, and adsorbed proteins were eluted (black curve) with NaCl solutions of increasing concentration. Successive fractions of the eluted proteins were assayed for the ability to transcribe DNA (red curve) in the presence of the four nucleoside triphosphates (including radioactive UTP). Most of the proteins did not bind to the column, but the enzymes did. The synthesis of RNA by each fraction in the presence of 1 μg/ml of α-amanitin also was measured (blue curve). Polymerases I and III are insensitive to the compound at that concentration, whereas polymerase II is sensitive, that is, it ceases RNA synthesis. (Polymerase III is sensitive to 10 μg/ml of α-amanitin, however; polymerase I is unaffected even by this higher concentration.) [See R. G. Roeder, 1974, *J. Biol. Chem.* **249**:241.]

➤ Structure and Function of Eukaryotic Nuclear RNA Polymerases

The nuclei of all eukaryotic cells examined so far (e.g., vertebrate, *Drosophila*, yeast, and plant cells) contain three different RNA polymerases, designated I, II, and III. These enzymes initially were recognized as distinct proteins when chromatographic purification of enzymes for RNA synthesis resulted in three fractions eluting at different salt concentrations during ion-exchange chromatography (Figure 11-27). The eukaryotic RNA polymerases also differ in their sensitivity to α-amanitin, a poisonous cyclic octapeptide produced by some mushrooms. Polymerase I is very insensitive to α-amanitin; polymerase II is very sensitive; and polymerase III has intermediate sensitivity.

Three Eukaryotic Polymerases Catalyze Formation of Different RNAs

Each eukaryotic RNA polymerase catalyzes transcription of genes encoding different classes of RNA. Polymerase I is located in the nucleolus and is responsible for synthesis of precursor rRNA (pre-rRNA), which is processed into the 28S, 5.8S, and 18S rRNAs. Polymerase III functions outside the nucleolus and transcribes the genes encoding tRNAs, 5S rRNA, and a whole array of small, stable RNAs. These latter RNAs include one involved in RNA splicing and the 7S RNA of the signal recognition particle involved in the transport of proteins into the endoplasmic reticulum (Chapter 16), but the functions of many are unknown at present. Polymerase II catalyzes transcription of all protein-coding genes; that is, it functions in production of mRNAs. RNA polymerase II also produces four small RNAs that take part in RNA splicing.

Eukaryotic RNA Polymerases Have a Complex Subunit Structure

Each of the three eukaryotic RNA polymerases is considerably more complex than *E. coli* RNA polymerase. All three contain two large subunits and 12–15 smaller subunits, some of which are present in two or all three of the polymerases. The best-characterized eukaryotic RNA polymerases are from the yeast *S. cerevisiae*; most of the yeast genes encoding the polymerase subunits have been cloned and sequenced. The subunits themselves have been separated by SDS polyacrylamide gel electrophoresis (Table 11-4). The three nuclear RNA polymerases from all eukaryotes so far examined are very similar to those of yeast, and where genes have been cloned and sequenced, the sequences are very homologous.

All three eukaryotic RNA polymerases contain core subunits with some sequence homology to the *E. coli* core polymerase ($\alpha_2\beta\beta'$). The largest and second largest sub-

TABLE 11-4 Subunits of *S. cerevisiae* Nuclear RNA Polymerases I, II, and III*

	I	II	III
	CORE SUBUNITS		
β'-like:	A190	B220	C160
β-like:	A135	B150	C128
α-like:	AC40	B44$_{(2)}$	AC40
	AC19		AC19
	COMMON SUBUNITS		
Same as II	ABC27$_{(2)}$ ABC23 ABC14.5 ABC10α ABC10β	Same as II	
	SPECIFIC SUBUNITS		
	A49	B32	C82
	A43	B16	C53
	A34.5	B12.6	C37
	A14	B12.5	C34
	A12.2		C31
			C25
			C11

*Subunits are named according to their apparent molecular weights in kilodaltons based on mobility in SDS polyacrylamide gel electrophoresis preceded by A, B, or C for subunits of RNA polymerases I, II, or III, respectively. Subunits shared by all three polymerases are designated ABC. Those shared by RNA polymerase I and III are designated AC. (2) indicates two molecules of the indicated subunit per polymerase.
SOURCE: A. Sentenac et al., 1992, in S. L. McKnight and K. R. Yamamoto, eds., *Transcriptional Regulation*, Cold Spring Harbor Laboratory Press, pp. 27–54.

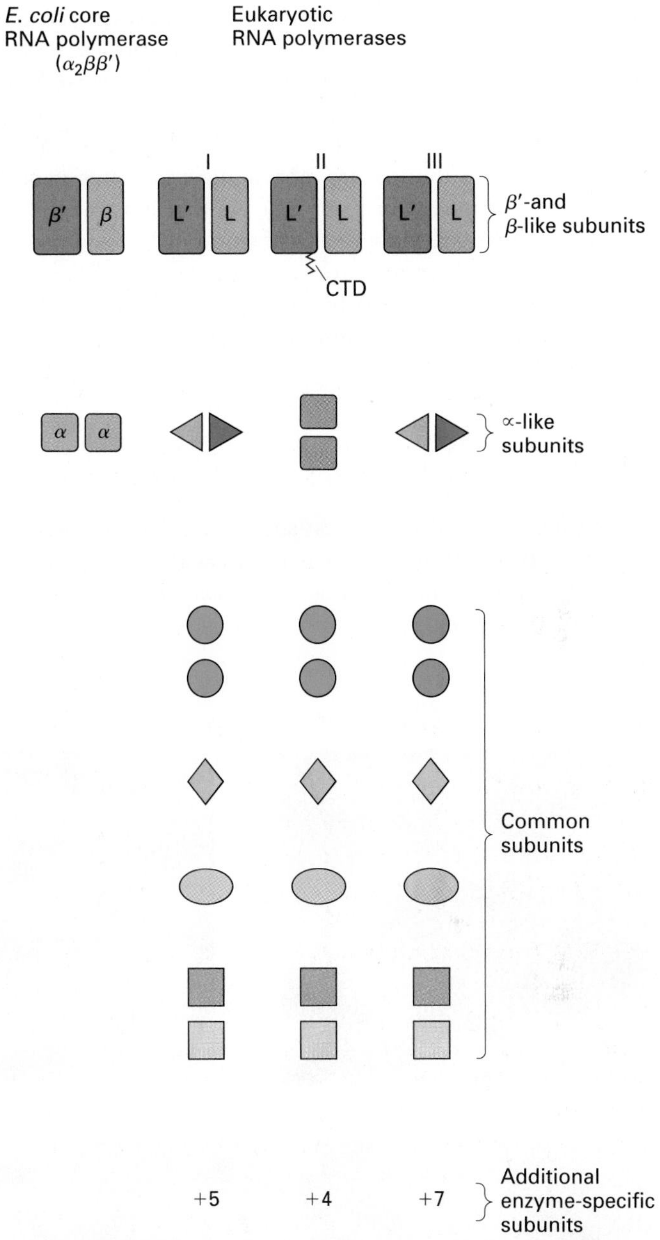

▲ FIGURE 11-28   Schematic representation of the subunit structure of yeast nuclear RNA polymerases. The largest (L') subunits of each of the three types of yeast RNA polymerases are related in sequence and are also related to the largest (β') subunit of *E. coli* RNA polymerase. Similarly, the second largest subunits of each of the three yeast RNA polymerases (L) are related to each other and to the second largest (β) subunit of *E. coli* RNA polymerase. The largest subunit of RNA polymerase II also contains an essential C-terminal domain (CTD) consisting of a heptapeptide repeat. RNA polymerase II contains two identical smaller subunits distantly related in sequence to the *E. coli* RNA polymerase α subunit. RNA polymerases I and II contain the same two nonidentical subunits with regions of sequence homology to the *E. coli* α subunit. All three polymerases share five other common subunits (two copies of the largest of these). In addition, each yeast polymerase contains four to seven unique smaller subunits.

units of each eukaryotic polymerase are related though distinct from each other and are similar to the *E. coli* β' and β subunits, respectively (Figure 11-28). Both yeast RNA polymerase I and III contain two subunits (AC40 and AC19) that have regions of homology with the *E. coli* α subunit. Yeast polymerase II contains two copies of a different subunit (B44) that exhibits a somewhat more distant sequence similarity to the *E. coli* α subunit. This extensive homology in the amino acid sequences of the core subunits in RNA polymerases from various sources indicates that this enzyme arose early in evolution and was largely conserved. This seems logical for an enzyme catalyzing a process so basic as copying RNA from DNA.

In addition to the core subunits related to the *E. coli* α, β, and β' subunits, all three yeast RNA polymerases contain five small common subunits; two copies of one of these and a single copy of each of the others is present in an enzyme molecule. Finally, each eukaryotic polymerase has four to seven enzyme-specific subunits that are not present in the other two polymerases. The functions of these multiple polymerase subunits are not understood, but gene-knockout experiments in yeast indicate that most of them are essential for cell viability. Disruption of the few polymerase subunit genes that are not absolutely essential for viability results in very poorly growing cells. Thus, it seems likely that every one of the multiple subunits must be present for eukaryotic RNA polymerases to function normally.

The Largest Subunit in RNA Polymerase II Has an Essential Carboxyl-Terminal Repeat

The carboxyl end of the largest subunit of RNA polymerase II contains a stretch of seven amino acids that is nearly precisely repeated multiple times. Neither polymerase I nor III contains these repeating units. This heptapeptide repeat, with a consensus sequence of Tyr-Ser-Pro-Thr-Ser-Pro-Ser, is known as the *carboxyl-terminal domain* (CTD). Yeast polymerase II contains 26 or more repeats; the mammalian enzyme has 52 repeats; and an intermediate number of repeats occurs in polymerase II from nearly all other eukaryotes. The CTD is critical for viability, and at least 10 copies of the repeat must be present for yeast to survive.

In rapidly growing cells, phosphate groups are attached to the serines and some tyrosines in the CTD. In vitro experiments with model promoters show that RNA polymerase II with an unphosphorylated CTD is used to initiate transcription, but that the CTD becomes phosphorylated as the polymerase transcribes away from the promoter. Although further research is needed to determine if this occurs at all types of promoters, analysis of polytene chromosomes from *Drosophila* larvae salivary glands are consistent with this idea. These polytene chromosomes consist of approximately one thousand aligned individual

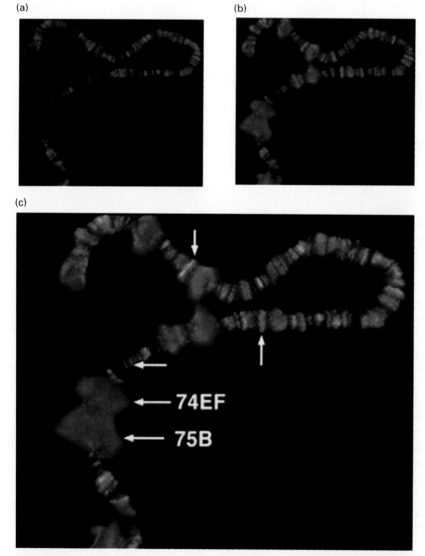

(a)

(b)

(c)

◀ FIGURE 11-29 Experimental demonstration that carboxyl-terminal domain (CTD) of RNA polymerase II is phosphorylated during transcription. Salivary gland polytene chromosomes were prepared from *Drosophila* larvae just before molting. The preparation was treated with a rabbit antibody specific for phosphorylated CTD and with a goat antibody specific for unphosphorylated CTD. The preparation then was stained with fluorescein-labeled anti-goat antibody and rhodamine-labeled anti-rabbit antibody. When the stained sample was viewed through a fluorescence microscope, regions containing unphosphorylated CTD appeared green (a) and those containing the phosphorylated form appeared red (b). Part (c) is an overlay of the micrographs in (a) and (b), showing phosphorylated and unphosphorylated CTD simultaneously. The molting hormone ecdysone induces very high rates of transcription in the puffed regions labeled 74EF and 75B; note that only phosphorylated CTD is present in these regions. Smaller, puffed regions transcribed at high rates also are visible. Nonpuffed sites that stain red (up arrow) or green (horizontal arrow) also are indicated, as is a site staining both red and green, producing a yellow color (down arrow). [From J. R. Weeks, 1993, *Genes & Dev.* **7**:2329; courtesy of J. R. Weeks and A. L. Greenfield.]

chromosomes, making it possible to visualize them in the light microscope. The polytene chromosome shown in Figure 11-29 was prepared just before molting of the larva. The large chromosomal "puffs" induced at this time in development are regions where the genome is very actively transcribed. Staining with antibodies specific for the phosphorylated or unphosphorylated CTD demonstrated that the highly transcribed puffed regions contain RNA polymerase II with a phosphorylated CTD.

RNA Polymerase II Initiates Transcription at DNA Sequences Corresponding to the 5′ Cap of mRNAs

As discussed earlier, bacterial RNA polymerases generally initiate transcription at a specific base within promoters lying just upstream of coding regions in the genome. Moreover, binding of repressors and activators to nearby sites are largely responsible for the transcriptional control of bacterial operons. Consequently, once eukaryotic genes began to be cloned and sequenced, determining the site of transcription initiation was a prerequisite for studying the mechanisms of transcriptional control in eukaryotes. Here we describe some of the methods used to map the start site for RNA polymerase II.

Analysis of Nascent Transcripts   We saw earlier how nascent-chain (run-on) assays can be used to determine relative transcription rates of various genes (see Figure 11-24). Recall that if cultured cells are exposed to a radioactive label for very brief times, only growing (nascent) RNA chains are labeled. Analysis of these labeled nascent transcripts permits the start site for a transcription unit to be located, as outlined in Figure 11-30. Figure 11-31 shows an example of this type of in vivo analysis for mapping the promoter of the adenovirus major late transcription unit to ≈6 kb from the left end of the viral genome.

In Vitro "Run-off" Assays   To more precisely locate initiation start sites, the in vitro run-off transcription assay can be used. We will illustrate use of this assay to determine the specific base pair at which transcription of the adenovirus genome begins during the late phase of infection. Restriction fragments from the promoter region of the adenovirus genome were incubated with radiolabeled ribonucleoside triphosphates and a protein extract prepared from isolated HeLa cell nuclei. When RNA polymerase II in the extract reaches the end of a fragment, it runs off the template and terminates the RNA chain (Figure 11-32). Since the positions of the restriction sites in the DNA template were known, the lengths of the run-off transcripts indicate the location of the transcription start site. Analyses of the 5′ end of the resulting RNA transcripts showed that they all contained a RNA cap structure identical to that present at the 5′ end of all eukaryotic mRNAs. The 5′

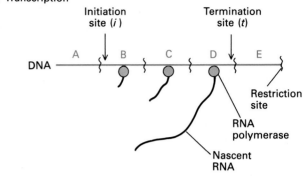

(1) Transcription

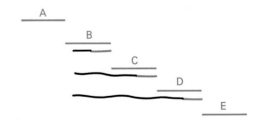

(3) Hybridization of labeled nascent RNA with purified DNA fragments A–E

Results: −   +   +   +   −

▲ **FIGURE 11-30**   Mapping of transcription initiation site by analysis of nascent transcripts synthesized in vivo. (1) A section of DNA with transcription initiation and termination sites is illustrated in the act of being transcribed. Restriction sites within this DNA are indicated. (2) Cells are exposed for a brief time, producing labeled nascent RNA molecules. (3) After the cells are disrupted, the isolated RNA is hybridized to restriction fragments A–E shown in (a). The shortest labeled RNA transcripts hybridize to the restriction fragment (B) that contains the initiation site; successively longer RNAs hybridize to fragments downstream from the initiation site. The longest labeled RNAs hybridize to the fragment (D) containing the termination site.

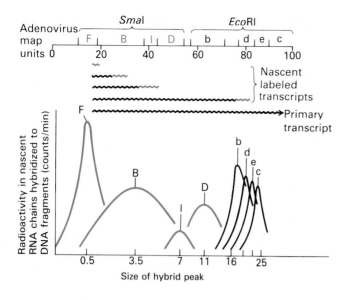

▲ FIGURE 11-31   Mapping of initiation site for adenovirus major late transcription unit by nascent-transcript analy-

sis. A restriction map of the adenovirus genome locating the
positions of SmaI and EcoRI fragments is shown at the top.
One map unit on the adenovirus genome equals ≈360 base
pairs. Adenovirus-infected cells were labeled for 5 min; the
RNA was isolated and separated according to size by centrifugation. Each size fraction then was hybridized to the restriction fragments shown at the top; unhybridized RNA was
removed by digestion with ribonuclease A. The radioactivity
hybridizing to each restriction fragment was plotted against
the length of the nascent RNA. Blue curves represent label
hybridized to SmaI fragments; black curves, label hybridized
to EcoRI fragments. The SmaI fragment F hybridized to the
shortest nascent RNA; fragments mapping to the right all the
way to the end of the genome hybridized to successively
longer nascent RNAs. Therefore, the adenovirus genome
includes a single long transcription unit with a promoter
within the SmaI F fragment. Since the 2.5-kb F fragment,
from 11 to 18 map units, hybridized to the shortest RNAs
with an average length of ≈0.5 kb, the initiation site for RNA
synthesis lies about 0.5 kb to the left of the right end of the
F fragment, at ≈16 map units. [See J. Weber et al., 1977,
Cell **10**:611; R. M. Evans et al., 1977, *Cell* **12**:733.]

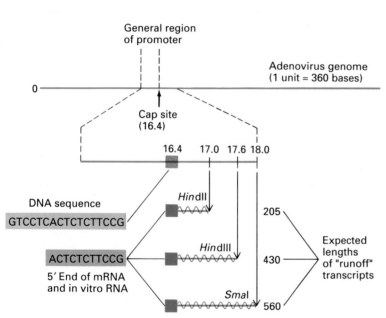

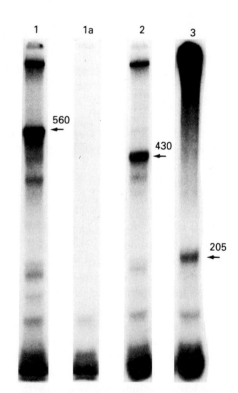

▲ FIGURE 11-32   Precise mapping of initiation site of
adenovirus major late transcription unit by in vitro run-off
analysis. Three samples of purified adenovirus DNA that
encodes the cap site (located at approximately 16.4 map
units) were digested with one of three restriction enzymes
that cut at 17.0, 17.6, and 18.0. The cut DNA templates
were mixed and incubated with RNA polymerase II, extracts
of HeLa cells, and labeled ribonucleoside triphosphates.
When the RNA polymerase reaches a cut end, it "runs off"
the template (*left*). If the start site is the cap site, the labeled
RNA products should stretch from the cap site to the cut
end; this prediction was verified by subjecting the labeled
runoff transcripts to gel electrophoresis and autoradiography
(*right*). Lanes 1, 2, and 3 show RNA made from DNA cut at
18.0, 17.6, and 17.0, respectively. The sample analyzed in
lane 1a is the same as that in lane 1, except that α-amanitin,
an inhibitor of RNA polymerase II, was included in the transcription mixture. The green boxes in the diagram indicate an
11-nucleotide sequence that is present at the capped end of
all the late adenovirus mRNAs and of the in vitro RNA transcripts as well. Thus the starting point for in vitro transcription by RNA polymerase II corresponds to the cap site in
mRNA. Under the conditions of this experiment, RNA polymerase II also makes end-to-end transcripts of the restriction
fragments, forming the bands at the top of the gel. Low-
molecular-weight HeLa RNAs also are labeled, forming the
bands at the bottom of the gel. [See R. M. Evans and E. Ziff,
1978, *Cell* **15**:1463; P. A. Weil et al., 1979, *Cell* **18**:469. Photograph courtesy of R. G. Roeder.]

cap was added by enzymes in the nuclear extract. The nucleotide to which the cap was added must have been used to initiate the RNA chain because the mechanism of cap addition requires that the RNA have a 5' tri- or diphosphate (see Figure 4-24). Only an initiating nucleotide retains its 5' triphosphate. A 5' end generated by cleavage of a longer RNA has only a single 5' phosphate and cannot serve as the substrate for cap addition. The sequence at the 5' end of the RNA transcripts produced in vitro was the same as that at the 5' end of late adenovirus mRNAs synthesized in vivo. Transcription thus starts at the corresponding sequence on the coding strand of the adenovirus DNA that corresponds to the capped 5' end of late adenovirus mRNAs.

Similar in vitro run-off assays soon were conducted with other cloned eukaryotic genes including those encoding β-globin, ovalbumin, and various histones. In each case, the transcription start site was accurately located and found to be equivalent to the 5' sequence of the corresponding mRNA. Thus, synthesis of eukaryotic mRNAs by RNA polymerase II begins at the DNA sequence encoding the 5' end of the mRNA. Today, the start site for synthesis of a newly characterized mRNA generally is determined simply by identifying the DNA sequence corresponding to the 5' end of the mRNA. This is most often done with the primer-extension and nuclease-protection assays (see Figure 7-34).

► Cis-acting Regulatory Sequences in Eukaryotic DNA

Once transcription start sites in eukaryotic DNA had been identified, analysis of the DNA sequences controlling initiation of transcription could begin. In this section, the identification and properties of various transcription-control regions in eukaryotic DNA are described.

The TATA Box Positions RNA Polymerase II for Transcription Initiation in Many Genes

The first eukaryotic genes to be sequenced and studied in in vitro transcription systems were viral genes and cellular genes that are very actively transcribed either at particular times of the cell cycle (e.g., the histone genes) or in specific differentiated cell types (e.g., globins, immunoglobulins, ovalbumin, silk fibroin). In all of these rapidly transcribed genes, a highly conserved sequence called the *TATA box* was found ≈25–35 base pairs upstream of the start site (Figure 11-33). As we will see, most in vitro transcription studies have used genes containing a TATA box. A single-base change in this nucleotide sequence drastically decreases in vitro transcription of TATA-containing promoters (Figure 11-34). In most cases, sequence changes between the TATA box and start site do not significantly affect the transcription rate. If base pairs are deleted between the TATA box and the normal transcription start site, transcription of the altered, shortened template initiates at a new site ≈25 base pairs downstream from the TATA box. Consequently, the TATA box acts similarly to an *E. coli* promoter to position RNA polymerase II for transcription initiation. The nucleotide at which RNA synthesis begins is an adenine in about 50 percent of the genes sequenced (see Figure 11-33).

Genes with an Initiator-containing Promoter Instead of a TATA box, some eukaryotic genes contain an alternative promoter element called an *initiator*. Most naturally occurring initiator elements have a C at position −1 and an A at the transcription start site (+1). Directed mutagenesis of mammalian genes with an initiator-containing promoter has revealed that the nucleotide sequence immediately surrounding the start site determines the strength of such promoters (Figure 11-35). Based on this type of study, the highly degenerate consensus initiator sequence shown at the bottom of Figure 11-35 has been proposed.

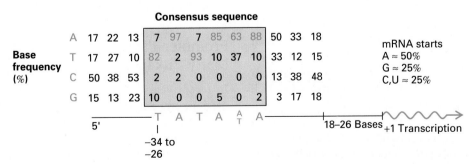

▲ FIGURE 11-33   Comparison of nucleotide sequences upstream of start site in 60 different vertebrate protein-coding genes. Each sequence was aligned to maximize homology in the region from −35 to −20. The tabulated numbers are the percentage frequency of each base at each position. Maximum homology occurs over a six-base region, referred to as the TATA box, whose consensus sequence is shown at the bottom. The initial base in mRNAs encoded by genes containing a TATA box most frequently is an A. [See R. Breathnach and P. Chambon, 1981, *Ann. Rev. Biochem.* **50**:349; P. Bucher, 1990, *J. Mol. Biol.* **212**:563.

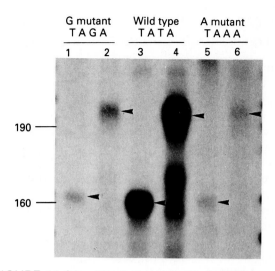

▲ FIGURE 11-34   Effect of substitution in TATA box on transcription by RNA polymerase II. The second 5′ T in the wild-type TATA box of the conalbumin gene was replaced, by in vitro mutagenesis, with a G or an A. The wild-type and mutated DNAs were cut with two different restriction enzymes and then incubated in an in vitro run-off transcription mixture with labeled ribonucleoside triphosphates. Gel electrophoresis and autoradiography of the RNA products from the wild-type gene revealed two run-off transcripts containing about 160 (lane 3) and 190 (lane 4) nucleotides; synthesis of these products from the two mutated DNAs was greatly diminished [See B. Wasylyk and P. Chambon, 1981, *Nucl. Acids Res.* **9**:1813; courtesy of P. Chambon.]

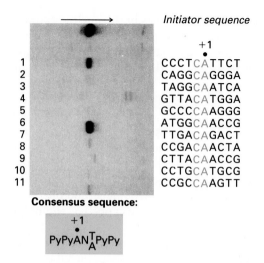

▲ FIGURE 11-35   Determination of consensus sequence of the initiator promoter element by mutagenesis. Most naturally occurring initiator elements have a C at position −1 and A at +1 (transcription start site). In this experiment, a wild-type transcription-control region containing an initiator promoter element was mutagenized so as to contain random DNA sequences at the positions just upstream and downstream of the CA dinucleotide at −1/+1. These sequences then were tested for their ability to function as promoters during in vitro transcription using an extract of nuclei isolated from cultured HeLa cells. An autoradiogram of the RNA products is shown. The large band in lane 1 represents the RNA initiated at the terminal transferase gene initiator promoter element. Lanes 2–11 represent the RNA transcripts produced from the mutated initiator elements whose sequences (in the top strand) are shown on the right. Clearly, these sequences vary greatly in their promoter activity. Based on the analysis of these and additional synthetic sequences, the highly degenerate initiator consensus sequence shown at the bottom was proposed. Py represents a pyrimidine (C or T) and N represents any of the four bases. [From R. Javahery, et al., 1994, *Mol. Cel. Biol.* **14**:116.]

Genes without TATA Boxes or Initiator Elements

Transcription of genes with promoters containing a TATA box or initiator element begins at a well-defined initiation site. However, transcription of many genes has been shown to begin at any one of multiple possible sites over an extended region, often 20–200 base pairs in length. As a result, such genes give rise to mRNAs with multiple alternative 5′ ends. These genes, which generally are transcribed at low rates (e.g., genes encoding the enzymes of intermediary metabolism), do not contain a TATA box or an initiator.

Most genes of this type contain a CG-rich stretch of 20–50 nucleotides within the first 100–200 bases upstream of the start site region. As we discuss later, a transcription factor called SP1 recognizes these CG-rich sequences. The dinucleotide CG is statistically underrepresented in vertebrate DNAs, and the presence of CG-rich regions just upstream from transcription-initiation sites is a distinctly nonrandom distribution. These CG-rich regions often are called "CpG islands" because they occur in a "sea" of DNA sequences low in CG residues. These CpG islands can be identified by their susceptibility to restriction enzymes (e.g., *Hpa*II) that have CG in their recognition sequences. The presence of a CpG island in a newly

cloned DNA fragment suggests that it may contain a transcription-initiation region.

Promoter-Proximal Elements Help Regulate Many Eukaryotic Genes

In recent years, recombinant DNA techniques have been used to systematically mutate the nucleotide sequences upstream of the start sites of various eukaryotic genes in order to identify transcription-control regions. By now, hundreds of eukaryotic genes have been analyzed and scores of related transcription-control regions have been identified. Such control regions lying within 100–200 base pairs upstream of the promoter often are referred to as *promoter-proximal elements*. In some cases, these regulatory elements are cell-type specific.

Analysis of 5'-Deletion Series One approach frequently taken to determine the upstream border of a transcription-control region for a mammalian gene involves constructing a set of 5' *deletions*. In this approach, DNA restriction fragments that include the cap site of the gene being analyzed and varying amounts of upstream DNA are cloned in a bacterial plasmid vector next to an easily assayed reporter gene. Commonly used reporter genes include the *E. coli lacZ* gene encoding β-galactosidase, an *E. coli* gene encoding the enzyme chloramphenicol acetyl transferase (CAT), and the firefly gene encoding luciferase. Each of these reporter genes express enzymes that are easily assayed in cell extracts. The engineered plasmids are separately transfected into cultured cells, or used to prepare transgenic organisms, to determine if the inserted DNA stimulates expression of the reporter gene over the very low level observed with the bacterial vector DNA alone. If a fragment does stimulate reporter-gene expression, it contains transcription-control sequences. Construction and analysis of a 5'-deletion series permits identification of the 5' borders of upstream control regions. In the example shown in Figure 11-36, two control elements are identified: one maps between deletions 2 and 3, and the other maps between deletions 4 and 5.

Analysis of Linker Scanning Mutations Once the 5' border of a transcription-control region is determined, analysis of *linker scanning mutations* can pinpoint the sequences with regulatory functions that lie between the border and the transcription start site. In this approach, a systemic set of constructs with contiguous overlapping mutations are assayed for their expression of a reporter gene (Figure 11-37). Use of this technique to identify promoter-proximal elements of the thymidine kinase (*tk*) gene

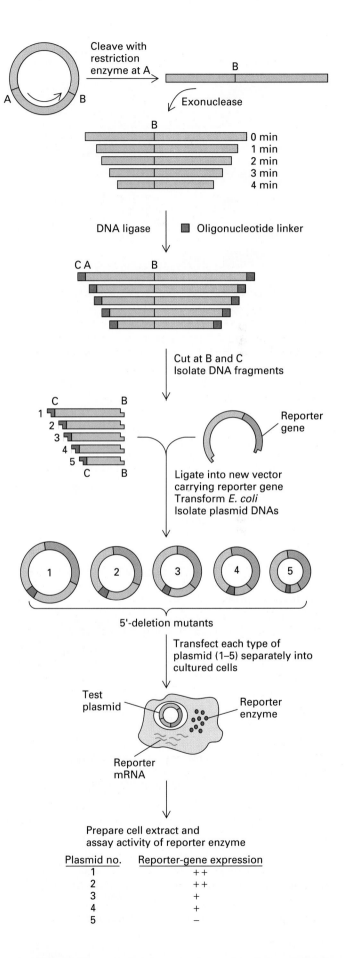

► FIGURE 11-36   Construction and analysis of a 5'-deletion series. A DNA fragment (orange) containing a transcription start site is cloned into a plasmid vector (*upper left*). The plasmid is linearized by digestion with a restriction enzyme (A) that cleaves at the upstream end of the fragment being analyzed. The linearized DNA is then digested with an exonuclease for different periods of time so that increasing lengths of DNA are removed from each end. After addition of a synthetic oligonucleotide linker and digestion with restriction enzymes B and C, the deleted fragments are cloned into a plasmid vector with an easily assayed reporter gene (light blue). Plasmids with deletions of various lengths 5' to the transcription start site are then transfected into cultured cells (or used to prepare transgenic organisms) and expression of the reporter gene is assayed. The results of this hypothetical example (*bottom*) indicate that the test fragment contains two control elements. The 5' end of one lies between deletions 2 and 3; the 5' end of the other lies between deletions 4 and 5.

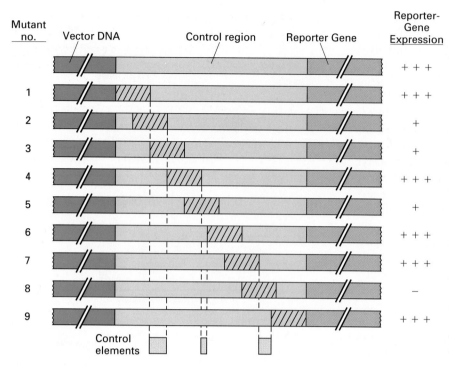

Reporter-Gene Expression

Analysis of linker scanning mutations to identify transcription-control elements. A region of eukaryotic DNA (yellow) that supports high level expression of a reporter gene (blue) is cloned in a plasmid vector as diagrammed at the top. Overlapping linker scanning (LS) mutations (crosshatch) are introduced from one end of the region being analyzed to the other. These mutations result from scrambling the nucleotide sequence in a short stretch of the DNA. The mutant plasmids are transfected separately into cultured cells, and the activity of the reporter-gene product is assayed. In the hypothetical example shown here, LS mutations 1, 4, 6, 7, and 9 have little or no effect on expression of the reporter gene, indicating that the regions altered in these mutants contain no control elements. Reporter-gene expression is significantly reduced in mutants 2, 3, 5, and 8, indicating that control elements (light orange) lie in the intervals shown at the bottom.

(a)

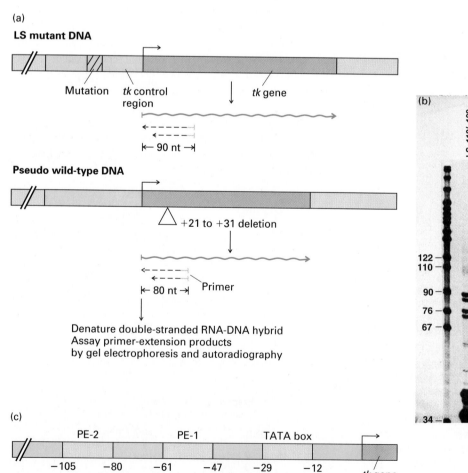

(c)

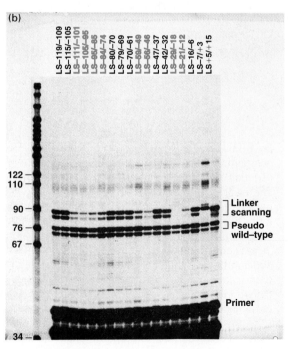

from herpes simplex virus (HSV) is illustrated in Figure 11-38. In this case, DNA constructs corresponding to the HSV *tk* control region with linker scanning mutations were microinjected into frog oocytes, which initiate transcription from this promoter at a high rate. The amount of HSV *tk* mRNA expressed from the mutant templates was determined by the primer-extension method (see Figure 7-34c). The results demonstrate that the DNA region upstream of the HIV *tk* gene contains three separate transcription-control sequences: a TATA box in the interval from -29 to -12, and two other control elements further upstream (from -105 to -80 and from -61 to -47).

Flexibility in the Spacing between Promoter-Proximal Elements In *E. coli* operons transcribed by σ^{70}-polymerase, repressor-binding and activator-binding sites are located within about 50 base pairs of the -10 and -35 promoter sequences. In the case of the *lac* control region, insertion of 5 base pairs between the CAP site and the promoter eliminates activation by the cAMP-CAP complex, because the bound complex is not aligned properly to interact with a polymerase molecule bound to the promoter (see Figure 11-9).

◀ FIGURE 11-38 Identification of promoter-proximal elements controlling the thymidine kinase (*tk*) gene of herpes simplex virus (HSV) by analysis of linker scanning mutations. (a) A series of DNA constructs containing linker scanning mutations in the *tk* control region was prepared. Each construct was coinjected with a pseudo wild-type *tk* gene into *Xenopus laevis* oocytes in which the HSV *tk* gene is transcribed at a high rate. After 24 h, RNA was isolated and assayed by the primer-extension method using a labeled RNA primer (light red) complementary to a short section of *tk* mRNA (see Figure 7-34c). The psuedo wild-type gene, which has a 10-bp deletion mutation, served as an internal control on the transcriptional activity of injected oocytes and the recovery of RNA. (b) The primer-extension products were analyzed by gel electrophoresis and autoradiography. Assay of the RNA transcribed from the LS mutants yielded two main labeled products: one contained 90 nucleotides (nt) corresponding to extension all the way to the 5′ end of the RNA; the other was slightly shorter due to incomplete extension. Assay of the RNA transcribed from the pseudo wild-type gene likewise yielded two products: one 80 nucleotides long, and the other slightly shorter. The labeled bands from each linker scanning mutant are compared with those from the psuedo wild-type gene in the same injected oocytes. (Weak background bands seen in multiple lanes are ignored.) The LS mutants are named by the upstream interval in which the wild-type *tk* sequence is mutated. Note the decreased density of bands, indicating decreased *tk* transcription, for the LS mutant DNAs labeled in red type. (c) Based on these results, promoter-proximal elements (orange) in the control region of the *tk* gene were mapped. [Photograph from S. L. McKnight and R. Kingsbury, 1982, *Science* **217**:316.]

To test the spacing constraints on control elements in the HSV *tk* promoter identified by analysis of linker scanning mutations, researchers prepared and assayed constructs containing small deletions and insertions between the elements (Figure 11-39). The results showed that in this promoter some variation in the distance between the TATA box and the promoter-proximal elements (PE-1 and PE-2) is tolerated before transcription is diminished significantly, but large insertions decrease transcription. Similar results have been observed with several other eukaryotic promoter regions analyzed by these methods.

Transcription by RNA Polymerase II Often Is Stimulated by Distant Enhancer Sites

As noted earlier, transcription from many eukaryotic promoters can be stimulated by control elements, called *enhancers,* that are located thousands of base pairs away from the start site. Such long-distance transcriptional control is relatively rare in *E. coli* occurring principally in the case of operons transcribed by σ^{54}-polymerase.

In eukaryotic transcription systems, the first enhancer to be discovered was in the simian virus 40 (SV40) genome. In this work, two types of plasmids were constructed: one containing the rabbit β-globin gene and the other containing the β-globin gene plus a 366-bp fragment of SV40 DNA that included the SV40 early promoter. These were transfected into cultured cells derived from a tissue that does not express β-globin. As the results in Figure 11-40 show, cells transfected with the β-globin gene alone produced little β-globin mRNA, whereas those transfected with the β-globin gene plus the SV40 fragment synthesized considerable β-globin mRNA.

Further analysis of this region of SV40 DNA revealed that the sequence responsible for this dramatic enhancement of transcription of the β-globin gene is an ≈100-bp sequence lying ≈100 base pairs upstream of the SV40 early transcription start site. The sequence was called the SV40 enhancer because it enhances transcription from promoters present on the same plasmid. In SV40 it functions to stimulate transcription from viral promoters. The SV40 enhancer stimulates transcription from all mammalian promoters that have been tested when it is inserted in either orientation anywhere on a plasmid carrying the test promoter, even when it is thousands of base pairs from the transcription start site.

An extensive linker scanning mutational analysis of the SV40 enhancer showed that no single mutation completely eliminated enhancer function (Figure 11-41). Mutations with the greatest effect diminished enhancer stimulation about 80 percent compared with wild-type activity; such mutant enhancers still exerted considerable transcriptional stimulation. These results indicate that the SV40 enhancer is composed of multiple elements each of which contrib-

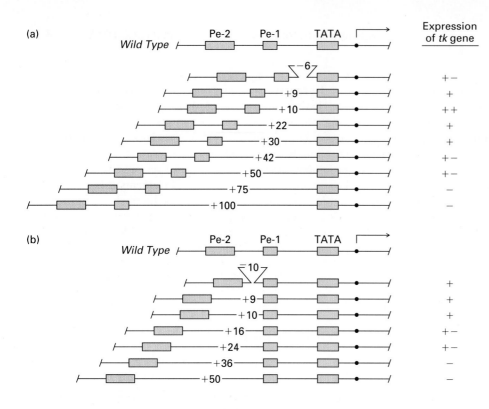

(a)

	Expression of *tk* gene
Wild Type — Pe-2 Pe-1 TATA	
−6	+ −
+9	+
+10	+ +
+22	+
+30	+
+42	+ −
+50	+ −
+75	−
+100	−

(b)

Wild Type — Pe-2 Pe-1 TATA	
−10	+
+9	+
+10	+
+16	+ −
+24	+ −
+36	−
+50	−

◄ FIGURE 11-39 Experimental demonstration that variations in spacing between promoter-proximal elements is tolerated in eukaryotic genes. Constructs containing various deletions and insertions between the promoter-proximal elements of the HSV *tk* control region were prepared. These were coinjected along with pseudo wild-type DNA into *Xenopus* oocytes and production of *tk* mRNA was assayed as shown in Figure 11-38. The results summarized at the right indicate that small deletions and insertions between the TATA box and PE-1 (a) or between PE-1 and PE-2 (b) have only a slight effect on transcription, whereas large insertions reduce transcription significantly. [Adapted from S. L. McKnight, 1982, *Cell* **31**:355.]

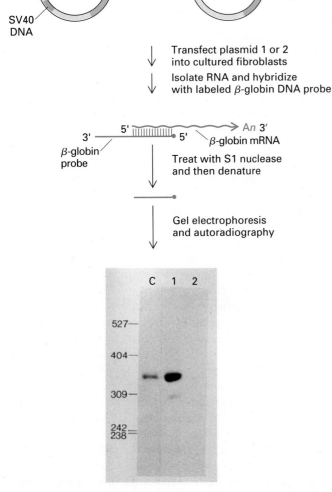

SV40 DNA

β-globin gene

↓ Transfect plasmid 1 or 2 into cultured fibroblasts
↓ Isolate RNA and hybridize with labeled β-globin DNA probe

3′ 5′ An 3′
β-globin probe 5′ β-globin mRNA

↓ Treat with S1 nuclease and then denature

↓ Gel electrophoresis and autoradiography

C 1 2

527—
404—
309—
242—
238—

◄ FIGURE 11-40 Identification of SV40 enhancer region. (*top*) Plasmids containing the β-globin gene with or without a 366-bp fragment of SV40 DNA (orange) were constructed. Each plasmid was transfected separately into cultured fibroblasts, which do not normally express β-globin. The amount of β-globin mRNA synthesized by transfected cells was assayed by the S1 nuclease-protection method (see Figure 7-34b). The probe used in this assay was a restriction fragment, generated from a β-globin cDNA clone, that was complementary to the 5′ end of β-globin mRNA. The 5′ end of the probe was labeled with ^{32}P (red dot). When β-globin mRNA hybridized to the probe, an ≈340-nucleotide fragment of the probe was protected from digestion by S1 nuclease. Autoradiography of electrophoresed S1-protected fragments (*bottom*) revealed that cells transfected with plasmid 1 (lane 1) produced much more β-globin mRNA than those transfected with plasmid 2 (lane 2). Lane C is a control assay of β-globin mRNA isolated from reticulocytes, which actively synthesize β-globin. These results show that the SV40 DNA fragment contains an element, called an enhancer, that greatly stimulates synthesis of β-globin mRNA. [Adapted from J. Banerji, S. Rusconi, and W. Schaffner, 1981, *Cell* **27**:299–308.]

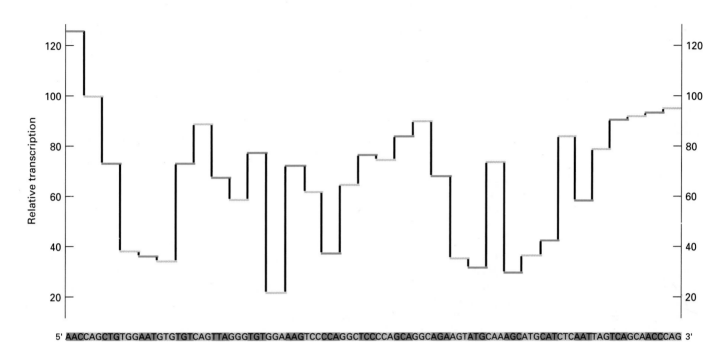

5' AACCAGCTGTGGAATGTGTGTCAGTTAGGGTGTGGAAAGTCCCCAGGCTCCCCAGCAGGCAGAAGTATGCAAAGCATGCATCTCAATTAGTCAGCAACCCAG 3'

▲ FIGURE 11-41   Analysis of linker scanning mutations in enhancer region of SV40 DNA. The wild-type sequence of the SV40 enhancer region is shown at the bottom. Successive 3-bp segments of the enhancer were mutated, and the resulting series of mutant enhancers were assayed for their ability to stimulate expression of the β-globin gene in a transfection assay similar to that illustrated in Figure 11-40. The graph shows the transcription rate of each mutant relative to that with the wild-type SV40 enhancer. The results indicate that the enhancer region contains multiple control sequences. [Adapted from M. Zenke et al., 1986, *EMBO J.* **5**:387.]

utes to the total activity of the enhancer. As discussed later, each of these regulatory elements is a protein-binding site.

Soon after discovery of the SV40 enhancer, enhancers were identified in other viral genomes and in eukaryotic cellular DNA. In some cases, they were found 50 or more kilobases from the promoter they controlled. Analyses of many different eukaryotic cellular enhancers have shown that they can occur upstream from a promoter, downstream from a promoter, within an intron, or even downstream from the final exon of a gene. Enhancers often are cell-type specific. For example, the genes encoding antibodies (immunoglobulins) contain an enhancer within the second intron that can stimulate transcription for all promoters tested, but only in B lymphocytes, the type of cells that normally express antibodies. Analyses of the effects of deletions and linker scanning mutations in cellular enhancers have shown that, like the SV40 enhancer, they generally are composed of multiple elements, which contribute to the overall activity of the enhancer.

Most Eukaryotic Genes Are Regulated by Multiple Transcription-Control Elements

Initially, enhancers and promoter-proximal elements were thought to be distinct types of transcription-control elements. However, as more enhancers and promoter-proximal elements were analyzed, the distinctions between them became less clear. For example, both types of element generally can stimulate transcription even when inverted, and both types can be cell-type specific.

The general consensus now is that a spectrum of control regions regulate transcription by RNA polymerase II. Enhancers that can stimulate transcription from a promoter tens of thousands of base pairs away (e.g., the SV40 enhancer) are at one extreme. Promoter-proximal elements, such as the upstream elements controlling the HSV *tk* gene, which lose their influence when moved an additional 15–20 base pairs farther from the promoter, are at the other extreme. Researchers have identified a large number of transcription-control regions that can stimulate transcription from distances between these two extremes.

Figure 11-42a summarizes the locations of transcription-control sequences for a hypothetical mammalian gene. Transcription initiates at the cap site encoding the first nucleotide of the first exon of an mRNA. For many genes, especially those encoding abundantly expressed proteins, a TATA box located 25–30 base pairs upstream from the cap site directs RNA polymerase II to the start site. Promoter-proximal elements within the first ≈200 base pairs upstream of the cap site stimulate transcription. Enhancers, which strongly activate transcription, frequently in a specific differentiated cell type, usually are 100–200 base

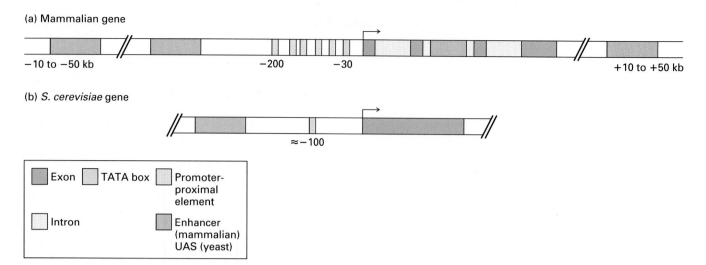

(a) Mammalian gene

−10 to −50 kb   −200   −30   +10 to +50 kb

(b) *S. cerevisiae* gene

≈−100

Exon   TATA box   Promoter-proximal element

Intron   Enhancer (mammalian) UAS (yeast)

▲ FIGURE 11-42   *General pattern of regulatory sequences that control gene expression in yeast and multicellular organisms (invertebrates, vertebrates, and plants). (a) Genes of multicellular organisms contain both promoter-proximal elements and enhancers and often a TATA box. The latter not only contributes to the level of transcription but also positions RNA polymerase II to initiate transcription ≈30* base pairs downstream. Enhancers may be either upstream or downstream and as far away as 50 kb from the transcription start site. In some cases, promoter-proximal elements occur downstream from the start site as well. (b) Most yeast genes contain only one regulatory region, called an upstream activating sequence (UAS), and a TATA box, which is ≈100 base pairs upstream from the start site.

pairs long. Although enhancers often lie within a few kilobases of the cap site, in some cases they lie much farther upstream or downstream from the cap site or within an intron. Some genes are controlled by more than one enhancer region, as in the case of the *Drosophila even-skipped* gene (see Figure 11-26).

The *S. cerevisiae* genome contains regulatory elements called *upstream activating sequences* (UASs), which function similarly to enhancers and promoter-proximal elements in higher eukaryotes. Most yeast genes contain only one UAS, which generally lies within a few hundred base pairs of the cap site. In addition, yeast genes contain a TATA box ≈100 base pairs upstream from the transcription start site (Figure 11-42b).

▶ Eukaryotic Transcription Factors

As in *E. coli*, the various eukaryotic transcription-control elements described in the previous section are binding sites for regulatory proteins. Binding of these proteins, called *transcription factors*, generally stimulates transcription, although eukaryotic regulatory proteins that repress transcription also have been identified. In this section, we discuss the identification, purification, and structures of proteins that function as transcription activators.

Biochemical and Genetic Techniques Have Been Used to Identify Transcription Factors

In yeast, *Drosophila*, and other genetically tractable eukaryotes, classical genetic studies have identified genes en-coding transcription factors. However, in mammals and vertebrates, which are not amenable to such genetic analysis, most transcription factors have been identified by biochemical purification.

Biochemical Isolation of Transcription Factors

Once a DNA regulatory element has been identified by the kinds of mutational analyses described in the previous section, proteins in an extract of cell nuclei can be assayed by their ability to bind specifically to the identified sequence. In this approach, DNA fragments containing the identified regulatory sequence are used in an electrophoretic mobility shift assay (see Figure 11-18) or DNase I footprinting assay (see Figure 11-10) to identify a putative transcription factor as it is purified by column chromatography. A particular type of affinity chromatography (see Figure 3-39c), called *sequence-specific DNA affinity chromatography*, is a powerful technique for the final purification step. Long DNA strands containing multiple copies of the transcription factor–binding site are synthesized and hybridized. These synthetic molecules then are coupled to a solid support to create a sequence-specific affinity column. A partially purified extract containing the desired transcription factor is applied to the column in a low-salt buffer (100 mM KCl). Proteins that do not bind to the specific binding site are washed off the column with additional low-salt buffer. Proteins with low affinity for the binding site are eluted with buffer containing 300 mM KCl. Finally, highly purified transcription factor is eluted with buffer containing 1 M KCl. The entire procedure for identifying DNA regulatory elements and isolating transcription factors that bind to them is summarized in Figure 11-43. As a final test, the ability of the isolated protein to stimulate transcription of a template containing the corre-

(a) Identifying DNA regulatory sequences

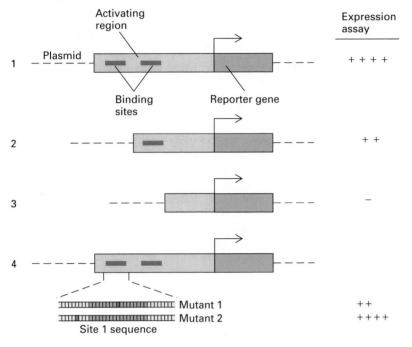

	Expression assay
1	+ + + +
2	+ +
3	−
4	
Mutant 1	+ +
Mutant 2	+ + + +

(b) Identifying cognate protein and locating precise binding sequence

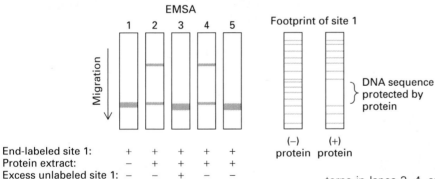

End-labeled site 1: + + + + +
Protein extract: − + + + +
Excess unlabeled site 1: − − + − −
Excess unlabeled mut. 1: − − − + −
Excess unlabeled mut. 2: − − − − +

(c) Purifying cognate protein for specific DNA binding sequence

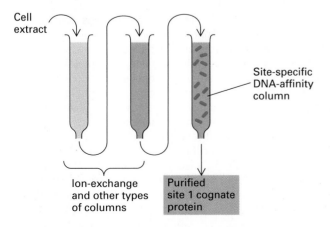

▼ FIGURE 11-43 General procedure for detecting and isolating DNA regulatory elements and the transcription factors (cognate proteins) that bind to them. (a) A DNA fragment containing potential regulatory sequences (yellow), such as the region upstream from a specifically regulated gene, is cloned next to a reporter gene (blue) encoding an easily assayed protein (1). After the cloned DNA is transfected into cultured cells, expression of the reporter gene is assayed by measuring the activity of its encoded protein; alternatively, synthesis of mRNA from the correct promoter site can be assayed either by the primer-extension or nuclease-protection method. Analysis of deleted forms of the cloned DNA (2 and 3) can identify segments whose absence causes a decrease in transcription. The precise location of the regulatory sequence within such an activating segment can be identified by testing the effects of linker scanning mutations and finally point mutations on expression of the reporter gene (4). For example, a mutation (red) within site 1 would destroy its activating ability, whereas a mutation outside the site would not. (b) To identify a cognate protein that binds to an identified regulatory element, electrophoretic mobility shift assays (EMSA) of nuclear extracts are conducted using end-labeled DNA fragments containing the identified element (in this case site 1). The results with various combinations of assay reactants are illustrated in the gel diagrams. The "shifted" band in lane 2 results from binding of a protein in the extract to the labeled fragment. That this protein binds specifically to site 1 is demonstrated by the band patterns in lanes 3, 4, and 5. In lane 3, the excess unlabeled wild-type DNA binds all the protein, so none is available for binding to the labeled fragment; thus no shifted band is observed. In lane 4, the mutant 1 DNA cannot bind the protein because of the point mutation in site 1; thus the protein is free to bind to the labeled fragment, generating the shifted band. In lane 5, excess unlabeled mutant 2, which carries a mutation outside of site 1, binds all the protein because site 1 is intact; thus no protein is available to bind the labeled fragment, and no shifted band is observed. The specific protein-binding sequence within an identified regulatory element can be pinpointed by footprinting experiments of the end-labeled fragment used in the EMSA (see Figure 11-10). (c) Any DNA-binding protein identified in (b) can be purified from nuclear extracts by column chromatography using the EMSA to identify column fractions containing the protein. Several chromatographic steps are used to partially purify the protein; final purification is obtained by affinity chromatography on a column containing the double-stranded DNA sequence (blue) to which the protein binds specifically as determined in the footprinting experiment. [For sequence-specific DNA-affinity chromatography see J. T. Kadonaga and R. Tjian, 1986, *Proc. Nat'l. Acad. Sci. USA* **83**:5889.]

(a) SP1-binding sites in SV40 genome

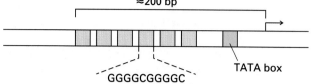

GGGGCGGGGC

TATA box

(b) SP1 transcription-activating assay

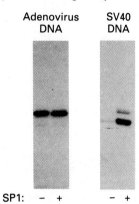

Adenovirus DNA   SV40 DNA

SP1:   −   +   −   +

◄ FIGURE 11-44   In vitro transcription-activating activity of SP1, a DNA-binding protein that recognizes 10-bp GC-rich sequences. (a) The SV40 genome contains six copies of a GC-rich promoter-proximal element upstream of the early promoter. The optimal SP1-binding sequence is shown for one of these sites, but SP1 also binds to related GC-rich sequences with high affinity. SP1 was isolated based on its ability to bind to this region of the SV40 genome and purified as illustrated in Figure 11-43c. (b) The activating activity of purified SP1 can be assayed in an in vitro transcription system. In this assay, template DNA and a partially purified nuclear extract from cultured HeLa cells containing RNA polymerase II and associated general transcription factors are incubated with labeled ribonucleoside triphosphates. The labeled RNA products are subjected to electrophoresis and autoradiography. Shown here are electrophoretograms from assays with adenovirus and SV40 DNA in the absence (−) and presence (+) of SP1. SP1 had no significant effect on transcription from the adenovirus promoter, which contains no SP1-binding sites. In contrast, SP1 stimulated transcription from the SV40 promoter about tenfold. [Adapted from M. R. Briggs et al., 1986, Science 234:47.]

sponding binding sites is assayed in an in vitro transcription reaction (Figure 11-44).

Once a transcription factor is isolated, its partial amino acid sequence can be determined and used to clone the gene or cDNA encoding it, as outlined in Chapter 7. The isolated gene can then be used to test the ability of the encoded protein to stimulate transcription in an in vivo transfection assay (Figure 11-45).

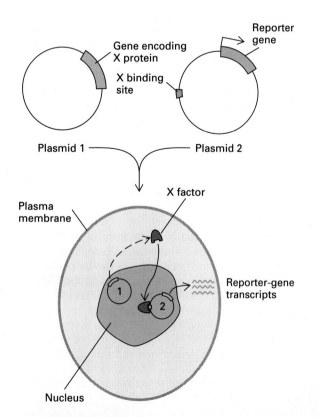

◄ FIGURE 11-45   In vivo assay for transcription-factor activity. The assay system requires two plasmids. One plasmid contains the gene (light purple) encoding the putative transcription factor (X protein). The second plasmid contains a reporter gene (blue) and a binding site for X protein (orange). Both plasmids are simultaneously introduced into host cells that lack the gene encoding X protein and the reporter gene. The production of reporter-gene RNA transcripts is measured; alternatively, the activity of the encoded protein can be assayed. If reporter-gene transcription is greater in the presence of the X-encoding plasmid, then the protein is an activator; if transcription is less, then it is a repressor. By use of plasmids encoding a mutated or rearranged transcription factor, important domains of the protein can be identified.

Genetic Identification of Genes Encoding Transcription Factors In yeast, genes encoding transcription factors were first identified through classical genetic analysis. For example, one of the yeast genes required for growth on galactose is called *GAL4*. Incubation of wild-type yeast cells in galactose media results in more than a thousandfold increase in the concentration of mRNAs encoding the enzymes catalyzing galactose metabolism. This activation of mRNA expression is not observed in *gal4* mutants. (In *S. cerevisiae* wild-type genes are designated with capital letters in italics, while recessive mutant alleles of the gene are indicated with lowercase letters in italics. The encoded protein is designated by the name of the gene in Roman type, with the first letter capitalized, e.g., Gal4.) Directed mutagenesis studies like those described in the previous section identified UASs for the induced genes. Each of these UASs was found to contain one or more copies of a related 17-bp sequence called UAS$_{GAL}$. When a copy of UAS$_{GAL}$ was cloned upstream of a TATA box followed by a *lacZ* reporter gene, expression of *lacZ* was activated in galactose media in wild-type cells, but not in *gal4* mutants. This indicated that UAS$_{GAL}$ is a transcription-control element activated by the GAl4 protein in galactose media.

The *GAL4* gene was isolated by complementation of a *gal4* mutant with a library of wild-type yeast DNA (Chapter 8). Using recombinant DNA techniques, the Gal4 protein was expressed in *E. coli* and found to bind to UAS$_{GAL}$. Thus, the Gal4 protein binds to UAS$_{GAL}$ sequences and activates transcription from a nearby promoter when cells are placed in galactose media.

Classical genetic studies in a number of other organisms including *Drosophila*, the nematode *C. elegans*, and higher plants have uncovered several genes encoding transcription factors. For example, many mutations that interfere with normal *Drosophila* development have been identified. One of these inactivates the ultrabithorax (*Ubx*) gene, causing an extra pair of wings to develop from the third thoracic segment (see Figure 8-3). The wild-type *Ubx* gene was cloned and sequenced, and the encoded protein expressed in large amounts using recombinant DNA techniques. Transcription assays showed that the Ubx protein functions as a transcription factor. The remarkable change in phenotype observed in *Ubx⁻* mutants indicates that Ubx protein influences transcription of a large number of *Drosophila* genes.

Many Transcription Factors Are Modular Proteins Composed of Distinct Functional Domains

A remarkable set of experiments with the yeast Gal4 protein demonstrated that this transcription factor is composed of separable functional domains: a *DNA-binding domain*, which interacts with specific DNA sequences, and an *activation domain*, which interacts with other proteins to stimulate transcription from a nearby promoter. In these experiments, a series of *gal4* deletion mutants were tested for their ability to activate transcription of a reporter gene (*lacZ*) in yeast cells (Figure 11-46a); binding of the encoded mutant Gal4 proteins to the UAS$_{GAL}$ sequence also was assayed. Transcription activation was measured in cells lacking the GAL4 gene, so that wild-type Gal4 protein would not interfere with the analysis. A small deletion from the N-terminus of Gal4 eliminated its ability to bind, whereas a series of mutant proteins with deletions from the C-terminus extending all the way to amino acid 74 retained the ability to bind specifically to UAS$_{GAL}$ (Figure 11-46b). These results demonstrate that a domain present in the N-terminal 74 amino acids of Gal4 is capable of binding to this DNA sequence. A similar analysis of the yeast transcription factor Gcn4, which regulates genes required for the synthesis of many amino acids, showed that it contains a different DNA-binding domain within its C-terminal 60 amino acids.

As expected, deletion of the N-terminal region of Gal4 eliminated its ability to activate expression of the reporter gene, because it could not bind to the UAS$_{GAL}$ sequence. Deletion of ≈125 or more amino acids from the C-terminus of Gal4 was required to completely eliminate its activation ability; these deletions did not affect its DNA-binding activity. Surprisingly, when the N-terminal DNA-binding domain of Gal4 was fused directly to various C-terminal fragments, the resulting truncated proteins, which lacked most of the internal portion of the protein, retained the ability to stimulate expression of the reporter gene (see Figure 11-46b, *bottom*). Thus the C-terminal ≈100-aa region of Gal4 contains an activation domain that, when fused to the N-terminal DNA-binding domain, is capable of stimulating transcription from the reporter gene.

Similar experiments with Gcn4 indicated that it contains an ≈20-aa activation domain near the middle of its sequence. The wild-type Gcn4 protein contains a region between this activation domain and its C-terminal DNA-binding domain that is extremely sensitive to digestion by proteases. Such highly protease-sensitive regions generally are relatively unstructured, flexible stretches of polypeptide chain, which can readily bend into a conformation that fits easily into the active site of a protease.

Further evidence for the existence of distinct activation domains in Gal4 and Gcn4 came from experiments in which their activation domains were fused to a DNA-binding domain from an entirely different protein, the *E. coli* LexA repressor. LexA has an N-terminal DNA-binding domain that binds specifically to *lexA* operator sequences. In these studies, a reporter gene with an upstream *lexA* operator was introduced into yeast cells. Expression of the LexA DNA-binding domain in these yeast cells did not activate expression of the reporter gene. But introduction of DNA constructs containing the coding sequence for

(a) Reporter-gene construct

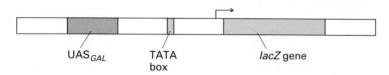

(b) Wild-type and mutant Gal4 proteins

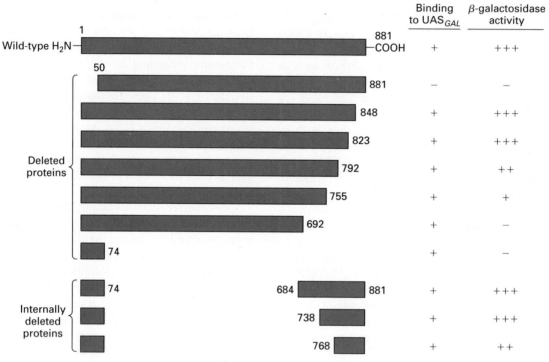

▲ FIGURE 11-46   Experimental demonstration of separate functional domains in yeast Gal4 protein, a transcription factor that binds to the UAS_GAL regulatory element in DNA. (a) Diagram of DNA construct containing a *lacZ* reporter gene with an added TATA box ligated to UAS_GAL, which contains several Gal4-binding sites. The reporter-gene construct and DNA encoding wild-type or mutant Gal4 were simultaneously introduced into mutant (*gal4*) yeast cells and the β-galactosidase activity was assayed. Activity will be high if the introduced *GAL4* DNA encodes a functional protein. (b) Schematic diagrams of wild-type Gal4 and various mutant forms. Small numbers refer to positions in the wild-type sequence. Deletion of 50 amino acids from the N-terminal end destroyed the ability to bind to UAS_GAL and to stimulate expression of β-galactosidase from the reporter gene. Proteins with extensive deletions from the C-terminal end still bound to UAS_GAL. These results localize the DNA-binding domain to the N-terminal end of Gal4. Proteins with a deletion of 58 amino acids from the C-terminal end had no decrease in the ability to stimulate expression of β-galactosidase; those with a deletion of 89 amino acids exhibited a small decrease. The ability to activate β-galactosidase expression was not entirely eliminated unless 126–189 or more amino acids were deleted. Thus, the activation domain lies in the C-terminal region of Gal4. Fusion proteins containing the N-terminal and C-terminal segments also were able to stimulate expression of β-galactosidase, indicating that the central region of Gal4 is not crucial for its function in this assay. [See J. Ma and M. Ptashne, 1987, *Cell* **48**:847; I. A. Hope and K. Struhl, 1986, *Cell* **46**:885; R. Brent and M. Ptashne, 1985, *Cell* **43**:729.]

the LexA DNA-binding domain ligated to the coding sequence for either the Gal4 or Gcn4 activation domain led to expression of the reporter gene in yeast cells. In this case, a fusion protein consisting of the DNA-binding domain from one transcription factor and the activation domain from a different factor was expressed in vivo and activated transcription. Thus, entirely novel transcription factors composed of prokaryotic and eukaryotic elements can be constructed.

Studies such as these have now been carried out with many eukaryotic transcription factors. Activation domains in mammalian transcription factors are frequently assayed by fusing them to the Gal4 DNA-binding domain since mammalian cells do not contain an endogenous transcrip-

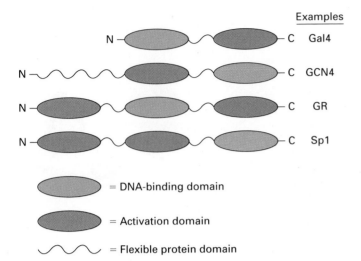

Examples

N — Gal4

N — GCN4

N — GR

N — Sp1

= DNA-binding domain

= Activation domain

= Flexible protein domain

▲ **FIGURE 11-47** Schematic diagrams illustrating the modular structure of eukaryotic transcription activators. These transcription factors may contain more than one activation domain but rarely contain more than one DNA-binding domain. Gal4 and Gcn4 are yeast transcription activators. The glucocorticoid receptor (GR), which also contains a hormone-binding domain (not shown), activates transcription of target genes in the presence of certain hormones. Sp1 binds to regulatory elements in the SV40 genome, activating transcription by RNA polymerase II. Diagrams are not to scale.

tion factor that binds to the UAS_{GAL} sequence. The structural model of eukaryotic activators that has emerged from these studies is a modular one in which one or more activation domains is connected to a sequence-specific DNA-binding domain through relatively flexible protein domains (Figure 11-47). In some cases, amino acids included in the DNA-binding domain also contribute to transcriptional activation. As discussed below, activation domains are thought to function through protein-protein interactions with transcription factors bound at the promoter. The flexible protein domains in activators, which connect the DNA-binding domains to activation domains, may explain why alterations in the spacing between control elements is so well tolerated in eukaryotic control regions (see Figure 11-39). When the DNA-binding domains of neighboring transcription factors are shifted in their relative positions on the DNA, their activation domains may still be able to interact because they are attached to their DNA-binding domains through flexible protein regions.

A Variety of Protein Structures Form the DNA-binding Domains of Eukaryotic Transcription Factors

Eukaryotic transcription factors contain a variety of structural motifs that interact with specific DNA sequences. As with most bacterial activators and repressors, α helices in the DNA-binding domain of eukaryotic transcription fac-

tors are oriented so that they lie in the major groove of DNA where protein atoms make specific hydrogen bonds and van der Waals interactions with atoms in the DNA. Interactions with sugar-phosphate backbone atoms and, in some cases, with atoms in the DNA minor groove contribute to binding. X-ray crystallographic analyses of several complexes between specific protein-binding sites in DNA and isolated transcription factor DNA-binding domains have revealed a number of structural motifs that can present an α helix to the major groove.

Transcription factors often are classified according to the type of DNA-binding domain they contain. Most of the structural classes of DNA-binding domains have characteristic consensus amino acid sequences. Consequently, newly characterized transcription factors frequently can be classified once the corresponding genes or cDNAs are cloned and sequenced. A few of the more common classes of DNA-binding domains whose three-dimensional structures have been determined are described here and illustrated in Figure 11-48. Many additional classes are recognized, and new classes are still being characterized. The genomes of higher eukaryotes may encode dozens of classes of DNA-binding domains and literally hundreds of transcription factors.

Homeodomain Proteins The structure of the DNA-binding domain from the *Drosophila* Engrailed protein is diagrammed in Figure 11-48a. Transcription factors with this type of DNA-binding domain are called *homeodomain* proteins, a name derived from a group of *Drosophila* genes in which the conserved sequence encoding this structural motif was first noted. As discussed in detail in Chapter 13, mutations in these genes, called *homeotic genes,* result in the transformation of one body part into another during development. Two of the most-studied of these genes are designated *antennapedia (Antp)* and *ultrabithorax (Ubx).* When *Antp* and *Ubx* cDNAs were first cloned and sequenced, their encoded proteins were found to share a highly conserved 60-aa region; this same conserved region was identified in the proteins encoded by other homeotic genes as they were cloned and sequenced. Because this conserved region was often diagrammed in a box when the sequences of these related proteins were compared, it came to be known as the "homeobox." The conserved sequence was later found in vertebrate genes, including human genes, recently shown to have critical functions in development. Related sequences also are present in the yeast proteins encoded at the *MATa1* and *MATα2* loci, which regulate the mating type of the yeast cell, a process analogous to control of cell type in multicellular organisms (Chapter 13). X-ray crystallography has shown that the α2 protein folds into a structure very similar to that of homeodomain transcription factors.

Zinc-Finger Proteins A number of different proteins have regions that fold around a central Zn^{2+} ion, producing a compact domain from a relatively short length of the

(a)

(b)

(c)

(d)

(e)

(f)

(g)

(h)

(i)

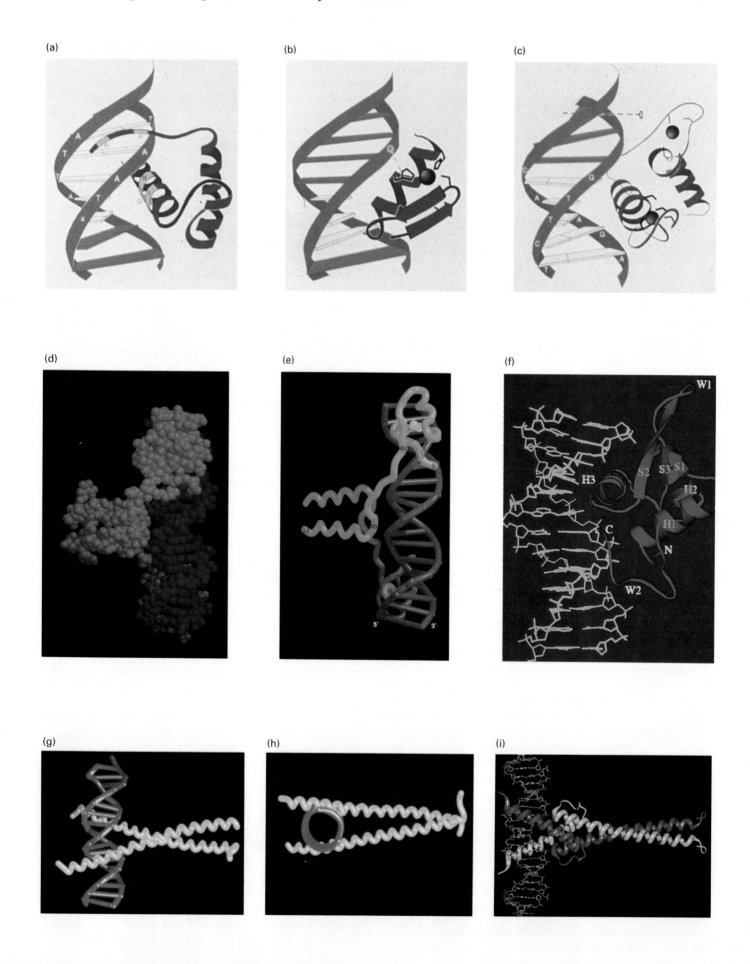

◄ FIGURE 11-48 Structural models of the DNA-binding domains of several classes of eukaryotic transcription factors and their interactions with DNA. All models are based on x-ray crystallographic analysis of protein-DNA complexes. (a) Homeodomain from Engrailed protein, a transcription factor that is expressed during *Drosophila* embryogenesis. (b) C_2H_2 zinc-finger domain from Zif268 protein, which contains three such domains. (c) C_4 zinc-finger domain from glucocorticoid receptor, a transcription factor in the steroid hormone receptor superfamily. Only one DNA-binding domain from this dimeric protein is shown. In (a), (b), and (c), the polypeptide backbone (blue) is represented as a ribbon diagram. DNA is in red with base pairs in the recognition site directly contacted by the protein structure in white type. Purple spheres represent sulfur atoms. (d) C_6 zinc-finger domain of Gal4 protein shown as a space-filling model. (e) C_6 zinc-finger domain of Gal4 as a trace diagram, with the polypeptide backbone in blue and sulfur atoms in yellow. (f) Winged-helix (forkhead) domain from HNF3, a hepatocyte-specific transcription factor. (g) Leucine-zipper domain from yeast Gcn4 with the two polypeptide backbones of the subunit C-termini of the dimeric protein in yellow. (h) The Gcn4-DNA complex viewed down the DNA axis. The dimerization domain in Gcn4 contains precisely spaced leucine residues. Some DNA-binding proteins with this same general motif contain other hydrophobic amino acids in these positions; hence, this structural motif is generally called a basic zipper. (i) A helix-loop-helix domain followed immediately by a leucine-zipper domain from the dimeric Max transcription factor. The helix-loop-helix motif extends from the DNA-binding helices on the left to approximately where the chains first cross; this is followed immediately by a leucine-zipper coiled-coil region in the dimeric protein. [Parts (a), (b), and (c) from S. C. Harrison, 1991, *Nature* **353**:715; parts (d) and (e) from R. Marmorstein et al., 1992, *Nature* **356**:408; part (f) provided by K. L. Clark and S. K. Burley. See 1993, *Nature* **364**:412; parts (g) and (h) from T. E. Ellenberger et al., 1992, *Cell* **71**:1223, part (i) provided by A. R. Ferre-D Amare and S. K. Burley; see 1993, *Nature* **363**:38.]

polypeptide chain. Termed a *zinc finger*, this structural motif was first recognized in DNA-binding domains but now is known to occur in proteins that do not bind to DNA. To date, three classes of zinc-finger proteins have been identified.

Figure 11-48b shows the structure of one type of zinc-finger DNA-binding domain, termed the C_2H_2 (or class I) *zinc finger*. The name is derived from the sequence of a repeating unit initially identified in the DNA-binding domain of transcription factor IIIA, which is required for transcription of 5S rRNA genes by RNA polymerase III. Each repeating unit has the consensus sequence (Tyr/Phe) X Cys X_{2-4} Cys X_3 (Phe/Tyr) X_5 Leu X_2 His X_{3-4} His, where X is any amino acid (Figure 11-49). Each repeating unit binds one zinc ion through the two cysteine (C) and two histidine (H) side chains. The name "zinc finger" was coined because a two-dimensional diagram of the structure resembles a finger. When the three-dimensional structure was solved, it became clear that the binding of the zinc ion by the two cysteines and two histidines folds the relatively short polypeptide sequence into a compact domain, which can insert its α helix into the major groove of DNA.

The C_2H_2 zinc finger is one of the most common DNA-binding motifs in eukaryotic transcription factors. More than a thousand of these consensus sequences are in the current protein sequence data base. The repeating units in these proteins can interact with successive groups of base pairs, primarily within the major groove, as the protein wraps around the DNA double helix (Figure 11-50).

A second type of zinc-finger structure is found in more than 100 transcription factors in the steroid hormone receptor superfamily. As discussed later, the function of many transcription factors in this family is regulated by lipid-soluble hormones, which can diffuse through cell membranes. Deletion-mutation analysis of these proteins showed that regions with the consensus sequence Cys X_2 Cys X_{13} Cys X_2 Cys X_{14-15} Cys X_5 Cys X_9 Cys X_2 Cys formed the DNA-binding domain. The two groups of four

► FIGURE 11-49 Amino acid sequence of the DNA-binding domain of TFIIIA from the frog *Xenopus laevis*. The single-letter amino acid code is used, and the sequence is arranged to maximize alignment of the repeated amino acids tyrosine (Y) or phenylalanine (F), leucine (L), and the invariant pairs of cysteines (C) and histidines (H). [Adapted from J. Miller, A. D. McLachlan, and A. Klug, 1985, *EMBO J.* **4**:1609.]

```
                 1            10             20            30
                 |            |              |             |
1                Y I C S F A D C G A A Y N K N W K L Q * A H L C * K H 37
2 T G E K * P F P C K E E G C E K G F T S L H H L T * R H S L * T H 67
3 T G E K * N F T C D S D G C D L R F T T K A N M K * K H F N R F H 98
4 N I K I C V Y V C H F E N C G K A F K K H N Q L K * V H Q F * S H 129
5 T Q Q L * P Y E C P H E G C D K R F S L P S R L K * R H E K * V H 159
6 A G - - * - Y P C K K D D S C S F V G K T W T L Y L K H V A E C H 188
7 Q D - - * L A V C - - D V C N R K F R H K D Y L R * D H Q K * T H 214
8 E K E R T V Y L C P R D G C D R S Y T T A F N L R * S H I Q S F H 246
9 E E Q R * P F V C E H A G C G K C F A M K K S L E * R H S V * V H 276
```

Repeat number

Amino acid position

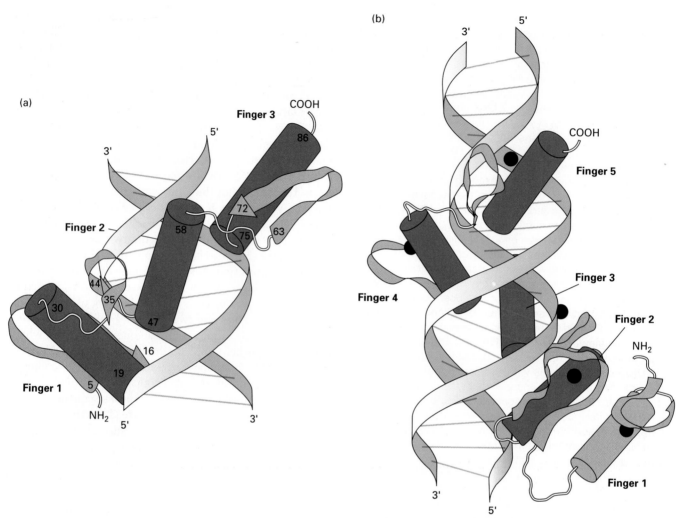

▲ FIGURE 11-50   Model of complexes formed between DNA and DNA-binding domains in two C_2H_2 zinc-finger proteins. Helical regions are represented by cylinders, and β strands by broad arrows. Helices that interact with DNA (blue) are shown in dark shade. (a) A three-finger protein named Zif268, which is expressed by G_0-arrested mouse fibroblast cells immediately after treatment with serum growth factors. All three fingers interact with DNA. Small numbers indicate positions in the amino acid sequence. (b) A five-finger protein called GL1, which is encoded by a gene that is amplified in a number of human tumors. Note that finger 1 of GL1 does not interact with DNA. Small black circles indicate Zn^{2+} ions. Part (a) is shown at a larger scale than part (b). [Part (a) adapted from N. P. Pavletich and C. O. Pabo, 1991, *Science* **252**:809; part (b) adapted from N. P. Pavletich and C. O. Pabo, 1993, *Science* **261**:1701.]

critical cysteines in this region each binds a Zn^{2+} ion. Consequently, this motif was named the C_4 (or class 2) *zinc finger* by analogy with the C_2H_2 zinc-finger structure. However, when the three-dimensional structures of these DNA-binding domains were solved, the C_4 motif was found to be quite distinct from the C_2H_2 motif. A particularly important difference between the two is that C_2H_2 zinc-finger proteins generally contain three or more repeating units and bind as monomers, whereas C_4 zinc-finger proteins generally contain one unit of the type shown in Figure 11-48c and bind as homodimers or heterodimers composed of two different C_4 zinc-finger proteins.

Like bacterial repressors, which often are described as helix-turn-helix proteins (see Figure 11-15), C_4 zinc-finger homodimers have twofold rotational symmetry (Figure 11-51). Also like bacterial systems, the consensus DNA sequences that bind C_4 homodimers are inverted repeats. However, heterodimers of these proteins bind to two direct repeats of the nucleotide sequence recognized by the DNA-binding domain of one protein. This must mean that the two protein monomers are associated differently in heterodimers than in the symmetrical homodimers. As of yet, the three-dimensional structure of heterodimers of C_4 zinc-finger proteins has not been solved. As discussed below, the ability of transcription factors to form heterodimers has important implications for the diversity of DNA sites that can function as regulatory elements.

The DNA-binding domain in the yeast Gal4 protein exhibits the third type of zinc-finger motif, which is known as the C_6 (or class 3) *zinc finger*. Proteins of this class have the consensus sequence Cys X_2 Cys X_6 Cys X_{5-6} Cys X_2 Cys X_6 Cys in which the six cysteines bind two zinc ions, folding the region into a compact globular domain (see Figure 11-48d and e). The Gal4 protein binds DNA as a dimer in which the monomers associate through hydrophobic interactions along one face of their α-helical regions. This type of interaction between α helices, to form a *coiled coil*, also occurs in dimeric leucine-zipper proteins and is discussed in more detail below.

Winged-Helix (Forkhead) Proteins The DNA-binding domains in histone H5 and several transcription factors that function during early development of *Drosophila* and mammals have the *winged-helix* motif, also called the *forkhead* motif (see Figure 11-48f). Like C_2H_2 zinc-finger proteins, winged-helix proteins generally bind to DNA as monomers.

Leucine-Zipper Proteins Another structural motif present in a large class of transcription factors is exemplified by the DNA-binding domain of yeast Gcn4. The first transcription factors recognized in this class contained the hydrophobic amino acid leucine at every seventh position in the C-terminal portion of their DNA-binding domains. These proteins bind to DNA as dimers, and mutagenesis of the leucines showed that they were required for dimerization. Consequently, the name *leucine zipper* was coined to denote this structural motif.

X-ray crystallographic analysis of complexes between DNA and the Gcn4 DNA-binding domain has shown that the dimeric protein contains two extended α helices that "grip" the DNA molecule, much like a pair of scissors, at two adjacent major grooves separated by about half a turn of the double helix (see Figure 11-48g and h). The portions of the α helices contacting the DNA include basic residues that interact with phosphates in the DNA backbone and additional residues that interact with specific bases in the major groove.

Gcn4 forms dimers via hydrophobic interactions be-

tween the C-terminal regions of the α helices, forming a coiled-coil structure. This structure is common in proteins containing amphipathic α helices in which hydrophobic amino acid residues are regularly spaced alternately three or four positions apart in the sequence. As a result of this characteristic spacing, the hydrophobic side chains form a stripe down one side of the α helix (see Figure 3-11). These hydrophobic stripes make up the interacting surfaces between the α-helical monomers in a coiled-coil dimer.

As noted above, the first transcription factors in this class to be analyzed contained leucine residues at every seventh position in the dimerization region and thus were named leucine-zipper proteins. However, additional DNA-binding proteins containing other hydrophobic amino acids in these positions subsequently were identified. Like

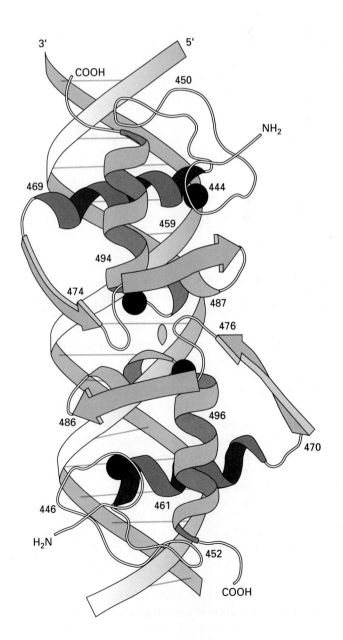

► FIGURE 11-51 Structure of the DNA-binding domains of the dimeric glucocorticoid receptor interacting with DNA (blue). In this ribbon diagram, α-helical regions are shown as spirals, and β strands as broad arrows. Two α helices (dark purple) interact with the DNA. Small numbers indicate positions in the amino acid sequence. Like all C_4 zinc-finger homodimers, this transcription factor has twofold rotational symmetry; the center of symmetry is shown by the yellow elipse. Black circles indicate Zn^{2+} ions. [Adapted from B. F. Luisi et al., 1991, *Nature* **352**:497.]

leucine-zipper proteins, they form dimers containing a C-terminal coiled-coil region and N-terminal DNA-binding domain. The term *basic zipper* (bZip) now is frequently used to refer to all proteins with these common structural features. Many basic-zipper transcription factors are heterodimers of two different polypeptide chains, each containing one basic-zipper region.

Helix-Loop-Helix Proteins   The DNA-binding domain of another class of dimeric transcription factors contains a structural motif very similar to the basic-zipper motif except that a nonhelical loop of the polypeptide chain separates the two α-helical regions in each monomer (see Figure 11-48i). Termed a *helix-loop-helix*, this motif was predicted from the amino acid sequences of these proteins, which contain an N-terminal α helix with basic residues, a middle loop region, and a C-terminal region with hydrophobic amino acids spaced at intervals characteristic of an amphipathic α helix. As with basic-zipper proteins, different helix-loop-helix proteins can form heterodimers.

Heterodimeric Transcription Factors Increase Regulatory Diversity

Three types of DNA-binding proteins discussed in the previous section can form heterodimers: C_4 zinc-finger proteins, basic-zipper proteins, and helix-loop-helix proteins. Other classes of transcription factors whose structures have not yet been determined also form heterodimeric proteins. One consequence of heterodimer formation is an expansion of the number of potential DNA sequences that a family of factors can bind. Heterodimer formation also allows different combinations of activation domains to be brought together at regulatory sequences. In addition, there are examples of basic-zipper and helix-loop-helix proteins that block DNA binding when they dimerize with another polypeptide otherwise capable of binding DNA. When these inhibitory factors are expressed, they repress transcriptional activation by the factors with which they interact.

The rules governing the interactions of members of a transcription-factor class are complex. In the example shown in Figure 11-52, factors A, B, and C can interact with each other, but the inhibitory factor interacts only with factor A. This combinatorial complexity expands both the number of DNA sites from which these factors can activate transcription and the ways in which they can be regulated. This is not possible for transcription factors that bind only as monomers or homodimers.

A Diverse Group of Amino Acid Sequences Are Found in Activation Domains

Although the three-dimensional structures of the DNA-binding domains from numerous eukaryotic transcription factors have been determined, the structure of not one acti-

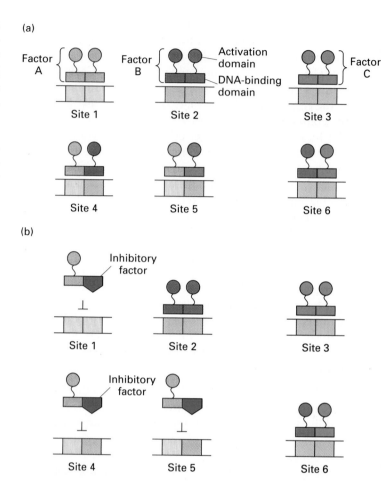

▲ FIGURE 11-52 C_4 zinc-finger, basic-zipper, and helix-loop-helix proteins can form heterodimers with other members of the same class. (a) In the example shown, factors A, B, and C can each interact with each other, permitting the three transcription factors to bind to six different DNA sequences (sites 1–6) and creating six combinations of activation domains. (Each binding site is divided into two half-sites). Four different factor monomers could combine to make 12 homo- and heterodimeric factors; five monomers could make 20 dimeric factors; and so forth. (b) When an inhibitory factor (green) is expressed that interacts only with factor A, binding to sites 1, 4, and 5 is inhibited, but binding to sites 2, 3, and 6 is unaffected.

vation domain has yet been solved. Nonetheless, activation domains defined by mutation analysis exhibit common amino acid sequence features in some cases.

For examples, Gal4, Gcn4, and most other yeast transcription factors analyzed so far are rich in acidic amino acids (aspartic and glutamic acids). Deletion analyses of numerous transcription factors from mammals and *Drosophila* have identified several classes of activation domains. Some are glutamine rich, some are proline rich, and some are rich in the closely related amino acids serine and threonine, both of which have hydroxyl groups. However, some strong activation domains that are not particularly rich in any specific amino acid also have been identified.

Most activation domains characterized in yeast transcription factors also stimulate transcription in mammalian cells, whereas a number of mammalian activation domains do not stimulate transcription when tested in yeast. Thus, while some activation mechanisms function in all eukaryotic cells, some mechanisms may have evolved since the divergence of yeast and animals. Before we examine what has been learned about these activation mechanisms, we first consider the proteins that function at the transcription start site.

➤ RNA Polymerase II Transcription-Initiation Complex

When the purified *E. coli* core polymerase is associated with the σ^{70} subunit, it can initiate transcription in vitro from strong promoters. Thus, as discussed earlier, σ^{70} acts like an initiation factor, which is released from the template after polymerization of the initial 10 or so ribonucleotide triphosphates. In contrast, none of the purified eukaryotic RNA nuclear polymerases can initiate transcription in vitro from promoter regions known to be transcribed in vivo. In vitro transcription catalyzed by purified RNA polymerase II initially was achieved only when the reaction mixtures included extracts prepared from isolated nuclei. This finding indicated that initiation factors in the nuclear extract are required for RNA polymerase II to recognize transcription-initiation sites. Purification of these factors by column chromatography revealed that several proteins assemble at the promoter region along with RNA polymerase II into a *transcription-initiation complex,* which is nearly as large as a ribosome.

The polymerase II initiation factors are called *general transcription factors,* because they are thought to be required for transcription of all genes that are transcribed by RNA polymerase II. (In contrast, the transcription factors discussed in the previous section bind to specific sites in particular genes.) The currently characterized general transcription factors required for initiation by human RNA polymerase II are listed in Table 11-5. These factors are designated *TFIIA, TFIIB,* etc., for transcription factor for RNA polymerase II plus an identifying letter. General transcription factors with similar activities have been isolated from rat liver, *Drosophila* embryos, and yeast. The genes encoding a number of these proteins have been cloned and sequenced. In all cases, equivalent general transcription factors from different eukaryotes are highly conserved.

Transcription-Initiation Complex Contains Many Proteins Assembled in a Specific Order

Most studies of initiation-complex assembly have used promoters with consensus TATA boxes. The major late

TABLE 11-5 Subunit Masses of Human General Transcription Factors Required for Transcription Initiation by RNA Polymerase II*

Factor	Native Mass (kDa)	Subunit Mass (kDa)
TFIID:	≈750	
TBP		38
TAFs		230 and smaller (>8)
TFIIA	≈100	34
		19
		14
TFIIB	33	33
TFIIF	210	30 (2)
		74 (2)
TFIIE	180	34 (2)
		56 (2)
TFIIH	230	90
		62
		43
		41
		35
TFIIJ	—	—

* Most general transcription factors are multimeric proteins. TFIID consists of a single TATA box–binding protein (TBP) and more than eight TBP-associated factors (TAFs). TFIIJ is not yet well characterized, and the native mass of TFIIA has not been accurately determined. TFIIC is not listed because purification of this factor revealed it to be a previously characterized protein that affected the in vitro transcription assay for trivial reasons. The term TFII I is reserved for a factor that stimulates transcription from a restricted set of promoters. Numbers in parentheses indicate the number of copies of a subunit present in the native factor.

promoter of adenovirus is frequently used because it is one of the strongest promoters for transcription in vitro. Footprinting and electrophoretic mobility shift assays have revealed that most of the general factors assemble in a specific order on the adenovirus major late promoter (Figure 11-53).

TFIID is the first general transcription factor to bind to a TATA-box promoter. The best-characterized subunit of TFIID, called the TATA-box binding protein, or TBP, binds the TATA-box sequence specifically. All eukaryotic TBPs analyzed have very similar C-terminal domains of 180 residues. The sequence of this region is 80 percent identical in the yeast and human proteins, and most differences are conservative substitutions. This conserved C-terminal domain functions as well as the full-length protein in in vitro transcription. (The function of the N-terminal domain of TBP, which varies greatly in sequence and length

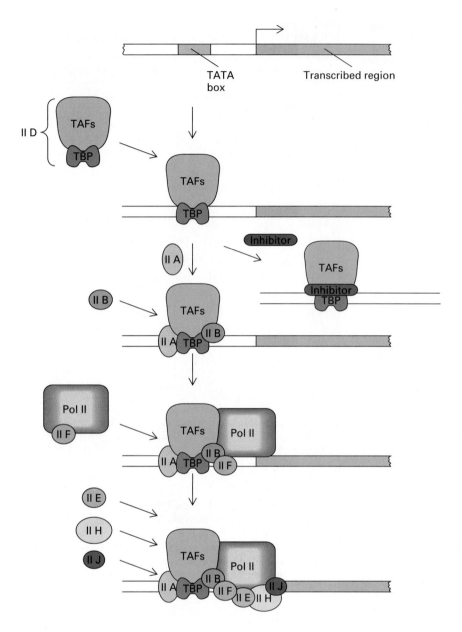

▲ FIGURE 11-53   Model for assembly of the RNA polymerase II initiation complex on a TATA-containing promoter. Assembly begins with the binding of TFIID to the TATA box. TFIID is composed of one TATA box–binding subunit called TBP (dark blue) and more than eight other subunits (TAFs), represented by one large symbol (light blue). Inhibitors (green) can bind to the TFIID-promoter complex, blocking the binding of other general transcription factors. Binding of TFIIA to the TFIID-promoter complex prevents inhibitor binding, forming the TFIID-TFIIA-promoter complex (D-A complex). TFIIB then adds to the D-A complex, followed by binding of a preformed complex between TFIIF and RNA polymerase II (red). However, transcription is not initiated by the bound polymerase until TFIIE, TFIIH, and TFIIJ have added to the complex in that order.

among different eukaryotes, is not currently understood.) The structure of the conserved C-terminal domain has been solved by x-ray crystallography, as has the structure of TBP complexed to a TATA-box promoter (Figure 11-54). TBP is a monomer that folds into a saddle-shape structure; the two halves of the molecule exhibit an overall dyad symmetry but are not identical, unlike dimeric transcription factors. TBP binds DNA quite differently than any other DNA-binding protein yet characterized, interacting with the minor groove and bending the DNA molecule considerably.

Nuclear extracts also have been shown to contain inhibitors that bind to the TFIID-promoter complex, thereby blocking further assembly. However, binding of TFIIA to the TFIID-promoter complex prevents binding of these inhibitors, allowing the assembly process to continue. A complex of RNA polymerase II and TFIIF associates with the complex about halfway through the assembly process, but the polymerase cannot initiate transcription until the remaining factors TFIIE and H bind (see Figure 11-53). Initiation by RNA polymerase II requires hydrolysis of the $\beta-\gamma$ bonds of ATPs. One of the last factors to add to the complex, TFIIH, which has DNA helicase activity, can use the energy from hydrolysis of ATP to separate the strands of the duplex template DNA. This protein is suspected to mediate unwinding of the strands at the start site, allowing the polymerase to initiate transcription. TFIIH also has a protein kinase activity, which can transfer the γ-phosphates of ATPs to multiple serines in the C-terminal repeat domain of the largest RNA polymerase II subunit.

(a)

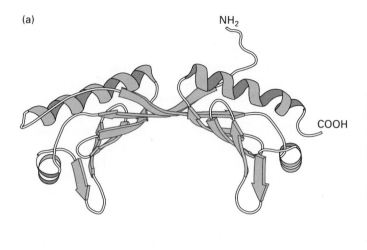

(b)

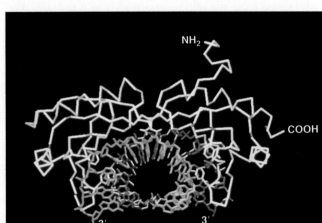

(c)

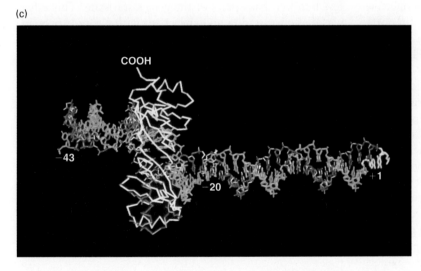

(d)

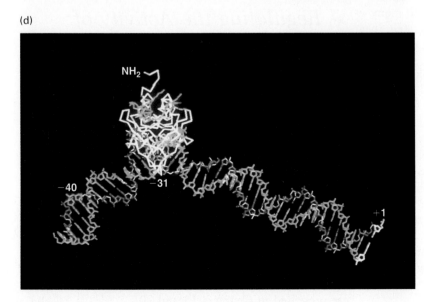

▲ FIGURE 11-54   Structure of the TATA box-binding protein (TBP) and the complex formed between TBP and the adenovirus major late promoter. (a) A ribbon diagram of the conserved C-terminal domain of TBP from the plant *Arabidopsis thaliana*. Spirals represent α-helical regions, and broad arrows represent β strands. The folded protein exhibits a dyad symmetry, but the distribution of amino acid side chains (not shown) on the surface of the protein is not symmetrical; thus the two halves of the protein are clearly distinguishable. (b) A view of the TBP-promoter complex with the protein in the same orientation as in (a). The protein is represented by a line connecting the α carbons of adjacent amino acid residues. The underside of the saddle-shaped TBP molecule contacts the sugar-phosphate backbone and the minor groove of the DNA (strands in red and light blue). (c) A view of the TBP-promoter complex viewed from the top of the TBP molecule as shown in (a). Note the distortion in the DNA double helix in the binding region and the two bends in the DNA backbone that result as the minor groove binds to TBP. Lysine side chains required to bind TFIIA are shown in dark blue on the lower surface of the TBP molecule. The transcription start site (+1) and positions −20 and −43 in the promoter are indicated. (d) A view of the TBP-promoter complex from the left side of TBP as shown in (a). The view further illustrates the bend in promoter DNA produced by binding of TBP. As in (c), lysine side chains required for binding of TFIIA are shown in dark blue on upper surface. [Part (a) adapted from D. B. Nikolov et al., 1992, *Nature* **360:**40; parts (b–d) from J. L. Kim, B. D. Nikolov, and S. K. Burley, 1993, *Nature* **365:**520. Photographs courtesy of S. K. Burley.]

This reaction is thought to be important for releasing the polymerase from the initiation complex as it initiates transcription. The initiation complex, including the polymerase, contains at least forty polypeptides and has a total molecular weight of greater than two million daltons! This is about 70 percent the size of a prokaryotic ribosome.

Transcription Activators Influence Assembly of Initiation Complex

Now that we have some understanding of the proteins that act at the promoter to assist in transcription initiation by RNA polymerase II, we can return to the question: How do transcription activators stimulate transcription? The working hypothesis is that transcription factors bound at promoter-proximal elements and enhancers act to control the assembly of the initiation complex and the rate at which the poised RNA polymerase in this complex initiates transcription. Although little is yet known about the mechanism of transcriptional activation, some evidence supports this general hypothesis. For example, an initiation complex can be assembled in vitro with an incomplete TFIID lacking the TAF subunits, but transcription from this complex is not stimulated by transcription activators bound to promoter-proximal elements. This result demonstrates that the TAF subunits of TFIID are required for transcription control. Some activation domains bind the TFIID complex by contacting TBP or specific TAF subunits. Other activation domains bind TFIIB. Likewise, transcription factors bound to distant enhancers are thought to interact with the initiation complex, forming a DNA loop between the enhancer and the promoter as has been demonstrated for the interaction between *E. coli* NTRC bound at the *glnA* enhancer and σ^{54}-polymerase bound at the *glnA* promoter (see Figure 11-23). The molecular mechanisms by which these interactions stimulate transcription are not understood. Presently, it is widely assumed that with so many polypeptides in the initiation complex, there are likely to be many specific mechanisms by which activation domains stimulate transcription.

Some Eukaryotic Regulatory Proteins Function As Repressors

Most eukaryotic transcription factors that have been studied extensively are activators, which stimulate transcription. However, proteins that repress transcription also have been identified in eukaryotes. Some repressor proteins function by binding to DNA sequences that overlap activator-binding sites. Other repressors function by binding to sequences that overlap a transcription start site, much like prokaryotic repressors. In both cases, binding of a repressor molecule to a specific DNA site blocks binding of proteins required to initiate transcription.

In many cases, however, eukaryotic repressors inhibit transcription without interfering with the binding of an activator or general transcription factor. One important example is the protein encoded by the *Wilms' tumor* (*WT1*) gene, which is expressed preferentially in the developing kidney. Children who inherit mutations in both the maternal and paternal *WT1* genes, so that they produce no functional WT1 protein, invariably develop kidney tumors early in life. The WT1 protein has a C_2H_2 zinc-finger DNA-binding domain, and binding sites for the protein were discovered in the control region of the gene encoding a transcription activator called EGR-1. The experiment outlined in Figure 11-55 demonstrated that WT1 protein repressed transcription of a reporter gene linked to the *EGR-1* promoter region.

Eukaryotic transcription repressors like WT1 appear to be the functional converse of activators. They can inhibit transcription from a gene they do not normally regulate when their cognate binding sites are placed within a few hundred base pairs of the gene's start site. This effect was demonstrated in an experiment with Krüpple protein, which represses transcription of several genes during embryonic development of *Drosophila*. Like activators, many eukaryotic repressors have two functional domains: a DNA-binding domain and a repression domain. When the Krüpple repression domain was fused to the DNA-binding domain of the *E. coli lac* repressor, the resulting fusion protein inhibited transcription of a reporter gene linked to upstream *lac* operator sites.

As discussed previously for activation domains, a variety of amino acid sequences can function as repression domains. Little information is yet available on the mechanism by which eukaryotic repressors inhibit transcription.

▶ *Regulating the Activity of Eukaryotic Transcription Factors*

We have seen in the preceding sections that transcription of eukaryotic genes is regulated by combinations of activators and repressors that bind to specific DNA regulatory sequences. Whether or not a specific gene in a multicellular organism is expressed in a particular cell at a particular time is largely a consequence of the binding and activity of the transcription factors that interact with the regulatory sequences of that gene. Clearly, since different proteins are expressed in different cells at different times in development, the activity of transcription factors must be controlled. The isolation of transcription factors and characterization of the genes that encode them have allowed some success in determining how they are regulated. A few principles of their control are discussed here; other examples are presented in Chapters 13 and 20.

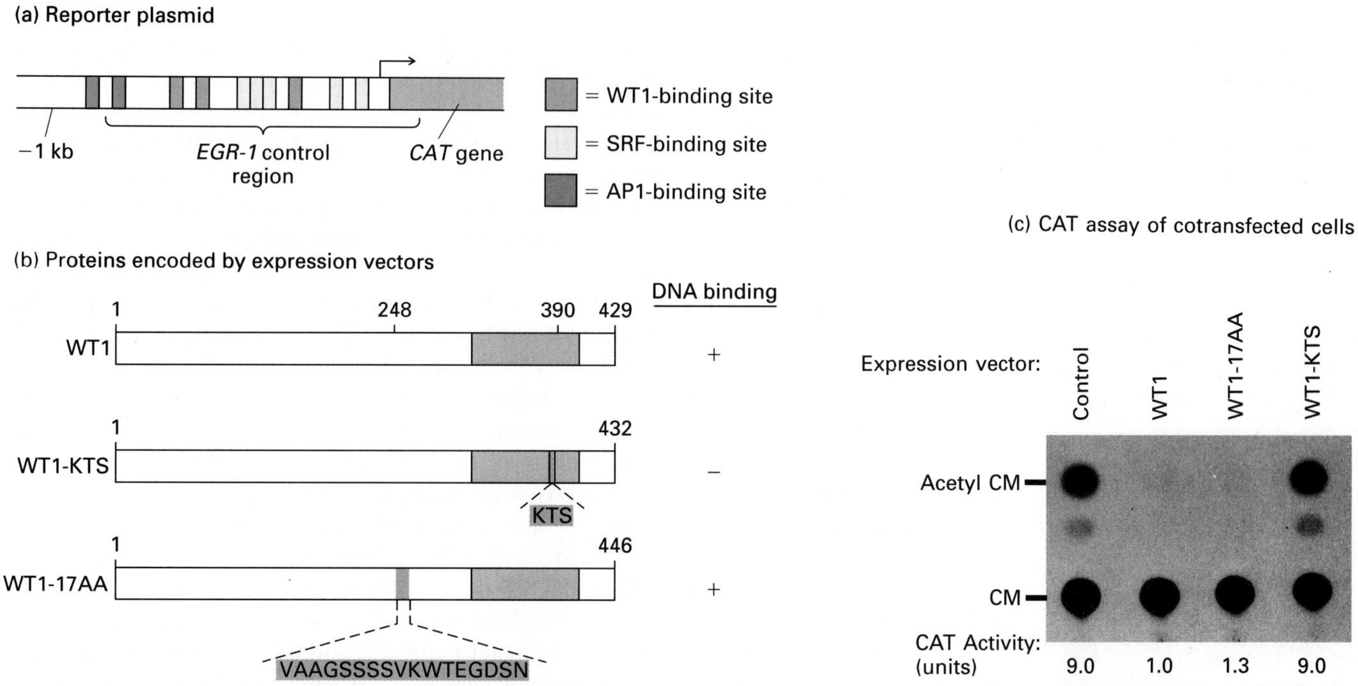

(a) Reporter plasmid

−1 kb   *EGR-1* control region   *CAT* gene

= WT1-binding site
= SRF-binding site
= AP1-binding site

(c) CAT assay of cotransfected cells

(b) Proteins encoded by expression vectors

DNA binding

WT1   1   248   390   429   +

WT1-KTS   1   432   −
           KTS

WT1-17AA   1   446   +
           VAAGSSSSSVKWTEGDSN

Expression vector: Control   WT1   WT1-17AA   WT1-KTS

Acetyl CM

CM

CAT Activity: (units)   9.0   1.0   1.3   9.0

▲ FIGURE 11-55   Experimental demonstration that the protein encoded by the *Wilms' tumor* (*WT1*) gene represses transcription. The in vivo assay system illustrated in Figure 11-45 was used to measure the activity of WT1 and two alternative forms of the protein. (a) The reporter plasmid in these experiments carried the *EGR-1* promoter region ligated to the *E. coli* gene that encodes chloramphenicol acetyl transferase (CAT). This enzyme, which is not present in untransfected mouse fibroblasts, transfers an acetyl group (–CH₃COOH) from acetyl coenzyme A (CoA) to chloramphenicol (CM), producing acetyl CM. Expression of CAT from the reporter gene was assayed by incubating extracts prepared from transfected cells with ¹⁴C-labeled CM and acetyl CoA, separating the reaction products by thin layer chromatography, and visualizing them by autoradiography. The *EGR-1* promoter region contains binding sites for WT1 and two ubiquitous activators, SRF and AP1. (b) The reporter plasmid was cotransfected with one of three plasmids expressing the three forms of WT1 resulting from alternative splicing of *WT1* mRNA. The C-terminal DNA-binding region of WT1 (orange) contains four C₂H₂ zinc fingers; WT1-KTS has an insertion of three amino acids in the DNA-binding region of WT1; WT1-17AA has a 17-aa sequence inserted into the WT1 sequence outside of the DNA-binding region. The inserted amino acids are indicated in the one-letter code. DNA-binding tests demonstrated that both WT1 and WT1-17AA bind to the *EGR-1* promoter DNA, whereas WT1-KTS does not. (c) Autoradiogram of reaction products from CAT assay of cotransfected cells shows that both WT1 and WT1-17AA repressed expression of the *CAT* reporter gene, compared with a control vector that did not express WT1, but WT1-KTS did not. These results indicate that WT1 must bind to the *EGR-1* promoter region in order to exert its repression activity. [Adapted from S. L. Madden et al., 1991, *Science* **253**:1550; photograph courtesy of Frank Rauscher.]

Expression of Many Transcription Factors is Restricted to Specific Cell Types

An obvious critical control point for cells is transcription of the genes encoding transcription factors themselves. For example, let us consider the expression of specific genes in mammalian liver cells (hepatocytes). Transcription-control regions of more than a half dozen genes whose products are formed only or mainly in hepatocytes have been dissected by the techniques of deletion and site-specific mutagenesis. Several generalizations can be drawn from these studies. First, liver-specific genes carried on plasmids can be expressed in hepatoma cells, which are derived from liver tumors (normal hepatocytes cannot be cultured for long periods easily), but not in other types of cultured cells. Second, multiple protein-binding sites, some close to and others distant from the transcription start sites, are required for maximal expression of liver-specific genes. Finally, at least four transcription factors important for hepatocyte-specific gene expression are present in hepatocytes and a few other cell types but are not present in most cells.

A brief description of the protein-binding sites in one gene expressed mainly in hepatocytes and the proteins that bind these required sites will illustrate these generalizations. Transthyretin (TTR)—a serum protein that binds

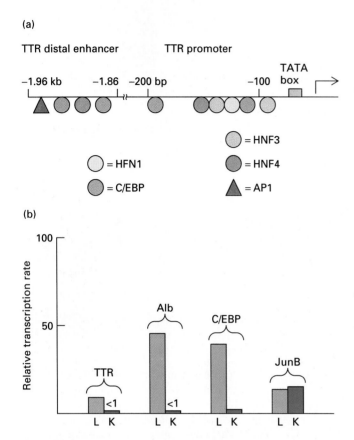

▲ FIGURE 11-56 (a) Binding sites for regulatory proteins that control transcription of mouse transthyretin (TTR) gene, which is expressed primarily in liver cells. HNF = hepatocyte nuclear factor. (b) Bar graph showing relative transcription of genes encoding TTR, albumin (Alb), C/EBP (one of the TTR transcription factors), and JunB (a widely distributed transcription factor that is a component of AP1) in liver (light blue) and kidney cells (dark blue). Synthesis of methionyl tRNA was used as a standard (100 units) in each cell type. The liver-specific transcription of TTR, albumin, and C/EPB is obvious. (See R. Costa, D. R. Grayson, and J. E. Darnell, 1989, *Mol. Cell Biol.* **9**:1415; K. Xanthopoulus et al., 1989, *Proc. Nat'l. Acad. Sci. USA* **86**:4117.]

thyroid hormone—is made in adults only in hepatocytes and choroid plexus cells, which secrete cerebrospinal fluid and its constituent proteins. There are ten definitely identified protein-binding sites in the regulatory regions of the TTR gene (Figure 11-56a). A family of leucine-zipper transcription factors called AP1, which are found in most cells, bind to one site. The other nine sites bind four proteins, all of which are present in liver-cell extracts at much higher concentrations than in any other cell type. Different genes that encode prominent hepatocyte-specific proteins, such as serum albumin or α_1-antitrypsin, have different arrangements of protein-binding sites and use overlapping but not identical sets of factors. Thus there is no single arrangement of sites that dictates liver-specific gene expression.

One of the transcription factors that acts on several liver specific genes is C/EBP, the first leucine-zipper protein

to be cloned and sequenced. Others include hepatocyte nuclear factor 1 (HNF1), which contains a homeobox domain; HNF4, a C_2H_2 zinc-finger protein; and HNF3, whose DNA-binding domain is the prototype of the winged-helix group (see Figure 11-48f). By use of cloned probes for the genes encoding these transcription factors, their mRNAs were shown to be limited in cell distribution, being prominent in hepatocytes and a few other cell types. In addition to hepatocytes, C/EBP is expressed in fat, intestinal, and some brain cells but not in kidney cells (Figure 11-56b), whereas HNF4 is expressed in intestinal and kidney cells but not brain cells. HNF1 is present mainly in hepatocytes among adult cells. The distribution of these transcription factors is controlled by regulating transcription of the genes that encode them.

Exactly how many cell-enriched or cell-limited transcription factors exist cannot be settled at present, but judging from results available with liver cells, red blood cells, and a few other cell types, transcription factors with a limited distribution could number in the hundreds. How these proteins act in groups and why particular factors act on a particular gene in a particular cell are not yet known. It is undoubtedly true that different combinations of transcription factors operate in different tissues. For example, although transthyretin (TTR) is expressed in liver cells and in choroid plexus cells, most other liver-specific genes are *not* transcribed in the choroid plexus. Thus, liver and choroid plexus cells may share some but not all factors. In fact, HNF1, HNF3, and C/EBP are not present in choroid plexus cells, although HNF4 is. Since transthyretin mRNA is formed at a very high rate in choroid plexus cells, there must be other factors in these cells that help activate transcription of the TTR gene in that cell type. The differential cellular distribution of transcription factors in adult cells results largely from differential transcription of the genes encoding the factors. The question of how particular sets of transcription-factor genes are activated in specific cell types is a critical area of developmental biology discussed in Chapter 13.

Some Transcription Factors Are Controlled by Lipid-Soluble Hormones

The activities of many transcription factors are regulated. For example, the transcription of many eukaryotic genes is controlled by extracellular signals called *hormones*, which are secreted from one cell type and travel through extracellular fluids to affect the function of cells at a different location in the organism. One class of hormones comprises small, lipid-soluble molecules, which can diffuse through plasma and nuclear membranes to interact with intracellular transcription factors directly (Figure 11-57). These lipid-soluble hormones, including many different types of steroid hormones, retinoids, and thyroid hormones bind to and regulate specific members of a large superfamily of related transcription factors. The first members of this superfamily were identified as specific intracellular high-

Cortisol

Retinoic acid

Thyroxine

▲ FIGURE 11-57   Examples of lipid-soluble hormones that bind to members of the steroid hormone receptor superfamily of transcription factors. Cortisol is a steroid hormone that binds to the glucocorticoid receptor. Like other steroid hormones, it is synthesized from cholesterol. Retinoic acid is produced from vitamin A and is the ligand for the retinoic acid receptor. Thyroxine is synthesized from tyrosine residues in the protein thyroglobulin in the thyroid gland. It is a ligand for the thyroid hormone receptor.

affinity binding proteins, or "receptors," for steroid hormones, leading to the name *steroid hormone receptor superfamily* for this class of transcription factors.

Domain Structure of Steroid Hormone Receptors
The first steroid hormone receptor to be highly purified and shown to be a DNA-binding protein was specific for glucocorticoid hormones (e.g., cortisol), which are produced mainly in the adrenal gland. The DNA-binding ability of this receptor initially was studied with DNA from mouse mammary tumor virus (MMTV). A 1.2-kb fragment of MMTV DNA whose transcription is increased in the presence of glucocorticoids was used in these experiments (Figure 11-58). Binding of the glucocorticoid receptor–hormone complexes to the fragment was visualized by electron microscopy. Footprinting experiments showed that receptor-hormone complexes bind at four to six sites in the DNA fragment in a region stretching from 84–305 base pairs upstream from the transcription start site.

Purification of the glucocorticoid receptor allowed cloning and sequencing of a cDNA and then genomic clones encoding the receptor. Cloning of the genes encoding other hormone receptors soon followed, including those for estrogens, progesterone, thyroid hormone, vita-

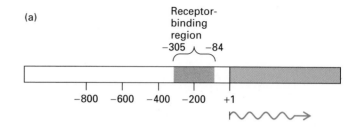

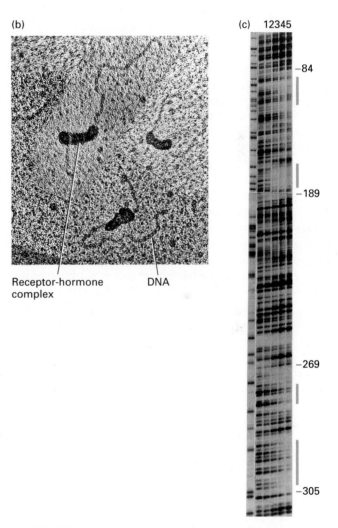

▲ FIGURE 11-58   Binding of glucocorticoid receptor to transcription-control region of mouse mammary tumor virus (MMTV) DNA. (a) The region of the MMTV genome containing the binding sites for the glucocorticoid receptor and the start site (+1) for hormone-dependent RNA synthesis is shown. (b) Electron micrographs of glucocorticoid receptor-hormone complexes bound to the DNA diagrammed in (a). (c) A DNase footprint of the promoter region of the MMTV genome in the presence and absence of glucocorticoid receptor protein, which protects four regions (bars) from digestion. Increasing amounts of glucocorticoid receptor were added in lanes 2–5; none was present in lane 1. The numbers indicate positions of nucleotides upstream from the start site. [See F. Payvar et al., 1983, *Cell* **35**:391; photographs courtesy of K. Yamamoto.]

GRE   5'AGAACA(N)₃ TGTTCT 3'
        3'TCTTGT(N)₃ ACAAGA 5'

ERE   5'AGGTCA(N)₃ TGACCT 3'
        3'TCCAGT(N)₃ ACTGGA 5'

VDRE   5'AGGTCA(N)₃ AGGTCA 3'
        3'TCCAGT(N)₃ TCCAGT 5'

RARE   5'AGGTCA(N)₅ AGGTCA 3'
        3'TCCAGT(N)₅ TCCAGT 5'

▲ FIGURE 11-59   Consensus sequences of the DNA sites (response elements) that bind the glucocorticoid receptor (GRE), estrogen receptor (ERE), vitamin D₃ receptor (VDRE) and retinoic acid receptor (RARE). The inverted repeats in GRE and ERE and direct repeats in VDRE and RARE are indicated by red arrows. [See K. Umesono et al., 1991, *Cell* **65**:1255; A. M. Naar et al., 1991, *Cell* **65**:1267.]

min D, and retinoic acid, a metabolic derivative of vitamin A that has powerful effects on limb bud development in embryos and on skin renewal in adult mammals. The characteristic nucleotide sequences of the DNA sites, called *response elements*, that bind several major hormone receptors have been determined (Figure 11-59). The sequences of the consensus response elements for the glucocorticoid and estrogen receptors are 6-bp inverted repeats separated

by any 3 base pairs. This finding suggested that these steroid hormone receptors would bind to DNA as symmetrical dimers, as was later shown from the x-ray crystallographic analysis of the glucocorticoid receptor's C₄ zinc-finger DNA-binding domain (see Figure 11-51). Some response elements, such as those for the vitamin D₃ and retinoic acid receptors, are direct repeats of the same sequence recognized by the estrogen receptor, separated by 3–5 base pairs. The receptors that bind to such direct-repeat response elements also do so as dimers. In this case, the monomers must interact with each other to form nonsymmetrical dimeric receptors, but none of their structures has yet been determined.

Comparison of the amino acid sequences of various hormone receptors reveals a remarkable conservation in both their amino acid sequences and the different functional regions of the proteins (Figure 11-60). All the hormone receptors have a unique amino-terminal region of variable length (100–500 amino acids) containing regions that function as activation domains. The DNA-binding domain maps near the center of the primary sequence and has the C₄ zinc-finger motif (see Figure 11-48c). The hormone-binding domain lies near the carboxyl-terminal end of these receptors.

Direct binding of hormone molecules has been demonstrated for most steroid-receptor proteins, and studies with recombinant proteins have provided decisive evidence about the functions of the various domains. For example, if the DNA-binding domain of the human estrogen receptor is replaced with the similar region of the glucocorticoid receptor, the recombinant protein binds to glucocorticoid

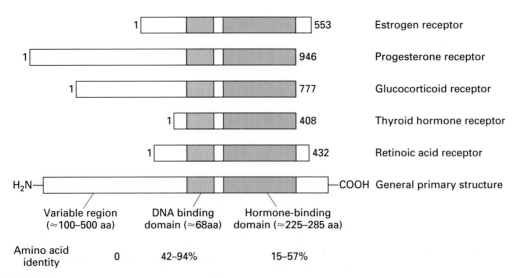

▲ FIGURE 11-60   General design of human transcription factors in the steroid hormone receptor superfamily. The centrally located DNA-binding domain exhibits considerable sequence homology among different receptors, and the C-terminal hormone-binding domain somewhat lower homology. The N-terminal regions in various receptors vary in length, have unique sequences, and contain one or more activation domains. This general pattern has been found in the estrogen receptor (553 amino acids [aa]), progesterone receptor (946 aa), glucocorticoid receptor (777 aa), thryoid hormone receptor (408 aa), and retinoic acid receptor (432 aa). [See R. M. Evans, 1988, *Science* **240**:889.]

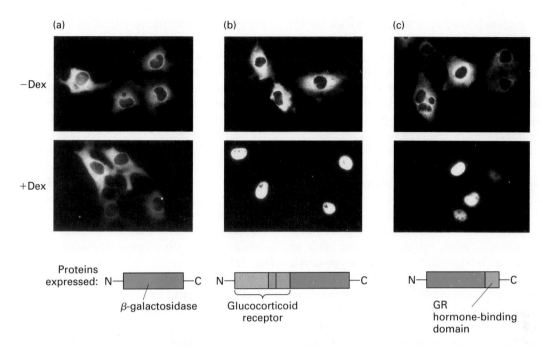

Proteins
expressed: N—[ ]—C   N—[ ]—C   N—[ ]—C
         β-galactosidase   Glucocorticoid   GR
                             receptor   hormone-binding
                                                domain

▲ FIGURE 11-61   Experimental demonstration that hormone-binding domain of the glucocorticoid receptor mediates translocation to the nucleus in the presence of hormone. Cultured animal cells were transfected with expression vectors encoding the proteins diagrammed at the bottom. (a) Expression of β-galactosidase was detected by immunofluorescence with a labeled antibody specific for β-galactosidase. The enzyme was localized to the cytoplasm in the presence and absence of the glucocorticoid hormone dexamethasone (Dex). (b) When a fusion protein consisting of β-galactosidase and the entire 794-aa rat glucocorticoid receptor (GR) was expressed in the cultured cells, it was present in the cytoplasm in the absence of hormone but was transported to the nucleus in the presence of hormone. (c) Fusion of the 382-aa receptor hormone-binding domain (green) alone also resulted in hormone-dependent transport of the fusion protein to the nucleus. [Photographs from D. Picard and K. R. Yamamoto, 1987, *EMBO J.* **6**:3333.]

response elements in DNA. However, stimulation of transcription by the recombinant receptor depends on estrogen treatment of the cells in which the recombinant receptor protein is made.

Mechanism of Hormonal Control of Activity Experiments with deletion mutants of various steroid hormone receptors demonstrated that in the absence of hormone, the hormone-binding domain inhibits transcriptional activation. This inhibition is relieved when hormone is bound. Deletion of the hormone-binding domain from the glucocorticoid receptor resulted in constitutive activity; that is, the deleted receptor stimulated transcription from genes with glucocorticoid response elements in the absence of hormone. Remarkably, when cultured animal cells were transfected with recombinant DNA containing the β-galactosidase gene fused to the hormone-binding region of the glucocorticoid receptor, the cellular location of the expressed fusion protein was influenced by the presence of hormone (Figure 11-61). In the absence of glucocorticoid hormone, the β-galactosidase fusion protein was found exclusively in the cytoplasm; however, in cells treated with hormone, the fusion protein was located in the nucleus.

These startling results are compatible with the model shown in Figure 11-62. In the absence of hormone, the glucocorticoid receptor is anchored in the cytoplasm as a

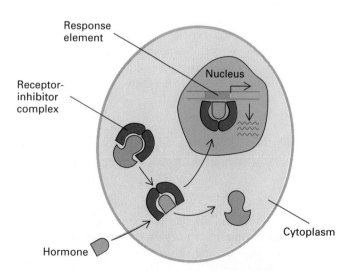

▲ FIGURE 11-62   Model of hormone-dependent gene activation by the glucocorticoid receptor. In the absence of hormone, the receptor (dark purple) is bound to an inhibitor (green) in the cytoplasm. When hormone is added, it binds to the receptor, releasing the receptor from the inhibitor. The receptor-hormone complex can translocate to the nucleus, where the DNA-binding domain interacts with the corresponding response element, thereby activating transcription of the target gene.

large protein aggregate perhaps complexed with an inhibitor protein. In this situation, the receptor cannot interact with target genes; hence, no transcriptional activation occurs. Binding of hormone releases the glucocorticoid receptor from the cytoplasmic anchor, allowing it to enter the nucleus where it can bind to response elements associated with target genes. Other members of the steroid receptor superfamily interact with inhibitors in the nucleus from which they are released when they bind their specific hormone.

The thyroid hormone receptor differs functionally from the glucocorticoid receptor in two important respects: it binds to its DNA response elements in the absence of hormone, and the bound protein represses transcription rather than activating it. When thyroid hormone binds to the thyroid hormone receptor, the protein is converted from a repressor to an activator.

It is not yet known how binding of hormone to these receptors changes the protein structure such that the glucocorticoid receptor is released from its cytoplasmic anchor in one case, and the function of the activation domain of the thyroid hormone receptor is altered in the other. Nonetheless, these two steroid hormone receptors demonstrate that the activity of transcription factors can be regulated either by controlling their DNA-binding activity or by altering the function of their activation domain.

Orphan Receptors The ligands for the hormone-binding domains in many members of the steroid receptor superfamily are as-yet unknown. These DNA-binding proteins, referred to as *orphan receptors,* were discovered by screening cDNA libraries with probes specific for the nucleotide sequence encoding the highly conserved DNA-binding domain characteristic of the steroid hormone receptors. The precise role of orphan receptors, and the unknown hormones that presumably regulate their activity, is an important subject of current research.

Polypeptide Hormones Signal Phosphorylation of Some Transcription Factors

Although lipid-soluble hormones can diffuse through the plasma membrane and interact directly with transcription factors in the cytoplasm or nucleus, the second major class of hormones, peptide and protein hormones, cannot. Instead, these hormones function by binding to specific cell-surface receptors, which then pass the signal that they have bound hormone to proteins within the cell, a process called *signal transduction.*

In many cases, the mechanism by which a hormonal signal is transduced into an activating signal for transcription factors involves phosphorylation. A simple example is

provided by *interferon* γ (INFγ), a hormone released by antigen-stimulated T helper lymphocytes, which are critical in the immune response (Chapter 27). When INFγ binds to a specific receptor protein that is present on the surface of most cells, it induces expression of a number of genes, producing an antiviral state that decreases the susceptibility of the cells to infection by a broad variety of viruses. Interferon γ also is very important in stimulating the function of various cells that participate in the immune response. To analyze how this hormone causes induction of a specific set of genes, researchers isolated a cDNA clone encoding a transcription factor required for INFγ-mediated induction using the procedure outlined in Figure 11-43. The encoded protein is $\approx$91 kDa and is called Stat1α, for signal transducer and activator of transcription.

After cultured cells are treated with INFγ, the DNA-binding activity of Stat1α increases rapidly, in parallel with the rapid rise of transcription of inducible genes. This induction of Stat1α DNA-binding activity by INFγ occurs even in cells treated with cycloheximide, an inhibitor of protein synthesis. Thus, some type of posttranslational modification of preexisting Stat1α activates its DNA-binding activity. By staining cells with fluorescein-labeled anti-Stat1α antibody, researchers demonstrated that Stat1α translocates from the cytoplasm to the nucleus following INFγ treatment, with kinetics similar to that of gene induction (Figure 11-63a). Analysis of Stat1α from INFγ-treated cells showed that hormone treatment leads to phosphorylation of a specific tyrosine residue in the protein. Furthermore, the phosphorylated protein was found to form a homodimer, whereas the unphosphorylated protein is a monomer. When the critical tyrosine was changed to a phenylalanine by site-specific mutagenesis, the mutant Stat1α failed to activate target genes in a transfection experiment, and failed to translocate to the nucleus (Figure 11-63b). The model for INFγ-mediated activation of Stat1α suggested by these results is illustrated in Figure 11-64. Clearly, INFγ-induced phosphorylation of Stat1α is critical to this protein's ability to stimulate transcription. Recent cloning experiments have identified several other proteins with structural homology to Stat1α; further research may show that they activate other sets of genes in response to different polypeptide hormones.

Phosphorylation of specific serine or threonine residues has been shown to regulate the activity of a number of other transcription factors. In some cases, phosphorylation of the free transcription factor either enhances or decreases its DNA-binding activity. In other cases, the inactive, nonphosphorylated transcription factor binds to its DNA recognition sequence. Phosphorylation then alters the functioning of the activation domain, so that the protein can stimulate transcription, much like hormone binding affects the thyroid hormone receptor bound to DNA.

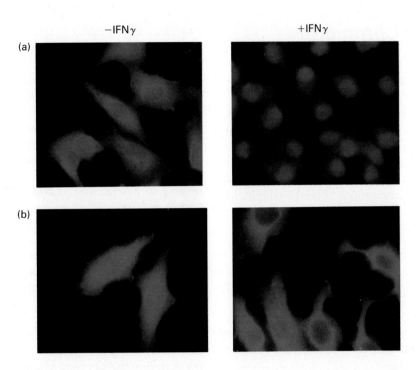

▲ FIGURE 11-63   Experimental demonstration that INFγ-dependent translocation of the transcriptional activator Stat91 to the nucleus depends on a critical tyrosine residue. U3A cells, which do not express Stat91, were transformed with Stat91 expression vectors and then stained with a fluorescein-labeled anti-Stat91 antibody. (a) When transformed cells expressing wild-type Stat91 were treated with IFNγ for 30 min, the protein became localized in the nucleus. (b) IFNγ-mediated nuclear translocation did not occur in transformed cells expressing a mutant Stat91 in which tyrosine at position 701 was replaced by phenylalanine. [From K. Shuai et al., 1993, *Science* **261**:1744; courtesy of J. E. Darnell.]

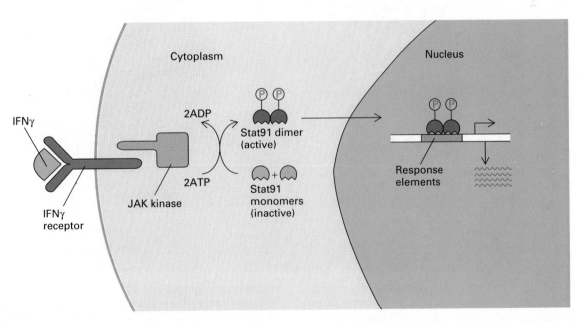

▲ FIGURE 11-64   Model for IFNγ-mediated gene activation by phosphorylation and dimerization of Stat91. The phosphorylated dimer translocates to the nucleus where it binds to corresponding response elements. JAK is a protein kinase that phosphorylates Stat91 in response to binding of IFNγ by cell-surface receptors. [See K. Shuai et al., 1992, *Science* **258**:1808.]

➤ Influence of Chromatin Structure on Eukaryotic Transcription Initiation

We have not yet considered the fact that eukaryotic DNA associates with histones to form nucleosomes and can be folded into higher-order chromatin structures (see Figure 9-50 and 9-54). There is considerable evidence that tightly folded, or condensed, chromatin—called heterochromatin—can inhibit transcription. Several examples of this phenomenon and the effect of cytosine methylation on transcription are discussed in this section.

Association of Genes with Heterochromatin Can Lead to Their Repression

The clearest examples of repression of genes due to their association with heterochromatin come from *Drosophila* and yeast, but similar processes likely occur in many other eukaryotes including humans.

Position Effect Variegation in Drosophila   The striking eye-color phenotype shown in Figure 11-65 results from a chromosomal rearrangement in which the wild type w^+ locus, which encodes an enzyme required for synthesis of the normal red eye pigment, is transposed next to centromeric heterochromatin. At this chromosomal location, repression of w^+ occurs in some cells but not others. When repression does occur, it is passed on to daughter cells, so that clones of eye cells in the adult fly appear white; other clones of cells expressing the normal pigment appear red, leading to patches of red and white eye color. This phenomenon is referred to as *position effect variegation*, because the variegating (randomly changing) phenotype results from the position of the w^+ gene near a centromere. The repressed state is said to be *epigenetic* since it is inherited but does not depend on a particular DNA sequence. (e.g., w^+ is expressed in other cells with the identical genomic sequence).

Position effect variegation is thought to result from inclusion of the w^+ gene in a chromosomal region where the tightly packed chromatin structure prevents access of transcription factors that would normally bind to the w^+

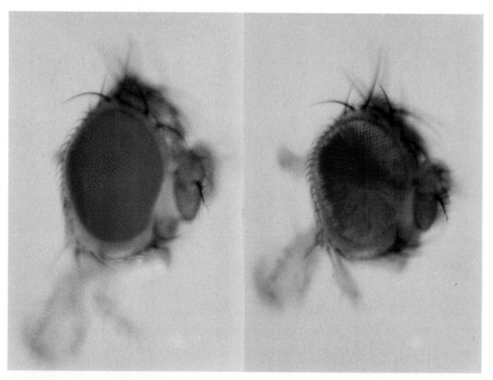

▲ FIGURE 11-65   Position effect variegation influencing eye color in *Drosophila*. The red eye color of a normal *Drosophila* adult is changed to a patchy, variegated phenotype when the promoter for the w^+ gene is brought within ≈25 kb of centromeric heterochromatin on the X chromosome. This rearrangement results from an inversion of the chromosomal DNA by recombination between two moderately repeated sequences (Chapter 9). Additional DNA rearrangements in progeny of affected flies that move w^+ away from centromeric heterochromatin restores normal eye color. This finding indicates that the variegated eye color results from placement of the w^+ promoter near heterochromatin, not from another type of mutation in the w^+ transcription-control region. [From K. D. Tartoff et al., 1989, *Dev. Genet.* **10**:162; courtesy of K. D. Tartoff.]

transcription-control sequences and stimulate expression. One such region is the centromere. The precise chromosomal domain of repression is thought to be somewhat variable from cell to cell at an early time in eye development but then is stably inherited by subsequent daughter cells.

Silent Mating-Type Loci in Yeast A similar phenomenon occurs at the silent mating-type loci of yeast in which the repression process has been easier to study. As discussed in Chapter 9, yeast chromosome III contains three mating-type loci. Only the central *MAT* locus, which contains either a or α sequences, is transcribed, thereby determining whether a cell has the a or α phenotype (see Figure 9-38). Two other loci, *HML* and *HMR*, which contain copies of the a or α sequences, are located near the telomeres. The mechanism that represses, or "silences," the gene copies at *HML* and *HMR* is essential to the haploid yeast cell, because expression of both the α and a proteins in the same cell is incompatible with mating.

In any one haploid yeast cell, the active *MAT* locus contains two α genes *or* two a genes. (As we will see in Chapter 13, the proteins encoded by these genes function as transcription factors that either activate or repress mating type–specific genes.) The two genes at *MAT* are transcribed in opposite directions from two promoters near the center of the locus. These two promoters are controlled by a UAS located between them. Precisely the same promoter and UAS are present in the silent *HML* and *HMR* loci. Consequently, the function of the transcription factors that interact with these sequences is somehow blocked at HML and HMR.

Repression of the silent mating-type loci requires sequences, called *silencers*, located near both ends of each silent locus, 3′ to the two genes. Any gene placed in the vicinity of these silencers by recombinant DNA techniques is repressed, even a tRNA gene transcribed by RNA polymerase III, which uses a different set of general transcription factors from RNA polymerase II. Consequently, the repression mechanism is not specific for the transcription factors that recognize the UAS associated with the a and α genes.

Several lines of evidence indicate that repression of the *HML* and *HMR* loci results from a condensed chromatin structure that sterically blocks transcription factors from interacting with the DNA. In one telling experiment, the gene encoding a DNA methylase of *E. coli* was introduced into yeast cells under the control of a yeast promoter so that the enzyme was expressed. This enzyme methylates adenine residues in the sequence GATC. Methylation at this sequence can be assayed easily using restriction enzymes that digest either the methylated or unmethylated sequence. Using these methods, researchers demonstrated that the *E. coli* methylase expressed in yeast cells was able to methylate GATC sequences within the *MAT* locus and

other regions of the yeast genome, but not within the *HML* and *HMR* loci.

Another indication that a specific chromatin structure is required for silencing at these loci came from studies of histone mutants. Yeast cells contain only two copies of the genes encoding each of the four histones that make up the nucleosome core. Consequently it has been possible to substitute mutant genes for the wild-type genes in order to test the effects of specific histone mutations. Mutations in the N-terminal region of histone H4 derepress *HML* and allow the *E. coli* DNA methylase to gain access to GATC sequences in *HML*. This result suggests that a specific interaction of the H4 N-terminus is required for formation of a fully repressing chromatin structure.

In addition to the histone H4 gene, several other genes have been discovered that are necessary for repression of the silent loci. Mutations in these genes also allow the *E. coli* DNA methylase to gain access to sites in *HML* and *HMR*. Mutation of one specific gene called *SIR1* does not completely derepress the silent loci in all cells. Interestingly, in this mutant, those cells in which the silent loci are repressed pass on the repressed state to their progeny for many generations. This is very similar to the position effect variegation observed in *Drosophila* where the w^+ locus is repressed in some clones of cells and not others.

Inactive X Chromosome in Mammalian Females One important case of heterochromatinization that correlates with gene inactivation in mammals is the random inactivation and condensation of one of the two female sex chromosomes (the X chromosomes) in virtually all the diploid cells of females. The inactive X, which appears as heterochromatin throughout the cell cycle, is visible during interphase as a dark-staining, peripheral nuclear structure called the *Barr body* after its discoverer, the Canadian cytologist M. L. Barr. The process of X inactivation, often termed *lyonization* after the British cytogeneticist Mary Lyons, who first recognized the phenomenon, can lead to important genetic consequences in individuals who have a defective gene on one of their two X chromosomes.

Each cell in a female has two X chromosomes, one of maternal origin (X_m) and one of paternal origin (X_p). Early during embryologic development, inactivation of either the X_m or the X_p chromosome occurs in each cell. The female embryo thus becomes a mosaic of cells: about half the cells have an inactive X_m, and the other half have an inactive X_p. All subsequent daughter cells maintain the same inactive X chromosomes as their parent cells. As a result, the adult female is a mosaic of clones, some expressing the genes on the X_m chromosome and the rest expressing the genes on the X_p.

The effects of X inactivation are evident in the expression of coat color in cats. Male cats, which normally have only one X chromosome, can be black or yellow, but they are almost never a mixture of both. In contrast, certain females, called *calicos,* have coats that are patches of black and yellow. The two colors have been traced to a gene on the X chromosome. The males have either the black (Bl^+) or the yellow (Bl^-) allele of this gene on their single X chromosome, whereas the calicos are females that are heterozygous (Bl^+/Bl^-). The black patches on the heterozygotes are produced by clones in which the active X chromosome carries the Bl^+ allele, and the yellow patches mark clones in which the active X has the Bl^- allele. The rare calico male has a sex chromosome constitution of XXY, in which one X is also inactivated.

Lyonization occurs early in development. The choice of which X chromosome to inactivate in each cell is random and does not depend on a particular DNA sequence. That is, X inactivation is an epigenetic event, since it affects gene expression but is not determined solely by the DNA sequence. The mechanism of X inactivation, which currently is being studied, may involve repression by heterochromatin somewhat like position effect variegation in *Drosophila*; however, DNA methylation also may be involved, as discussed below.

Transcriptionally Inactive DNA Regions Are Resistant to DNase I

In higher eukaryotes, transcriptionally inactive genes generally are found in less accessible chromatin structures than are transcriptionally active genes. This phenomenon can be demonstrated by digesting isolated nuclei with DNase I. Following digestion, the DNA is completely separated from chromatin protein, digested to completion with a restriction enzyme, and analyzed by Southern blotting (see Figure 7-31). When a gene is cleaved at random sites by DNase I, the Southern-blot band corresponding to that gene is lost. This method has been used to show that the inactive β-globin gene in nonerythroid cells is much more resistant to DNase I than is the active β-globin gene in erythroid precursor cells (Figure 11-66).

Observations such as these suggest that early in the activation of a gene in higher eukaryotes, the chromatin in the chromosomal region that includes the gene is decondensed, permitting transcription factors to gain access to the control sequences. At present, however, no convincing evidence supports such a mechanism of gene control. It may be that the decondensation of chromatin structure reflected by increased sensitivity to DNase I digestion is simply a consequence of transcriptional activity. In the specialized centromeric and telomeric heterochromatin regions, the chromatin structure clearly prevents transcription factors from interacting with transcription-control sequences, as discussed in the previous section. Whether this occurs more generally in the genome of higher eukaryotes is not yet clear.

Although the functional significance of DNase I–sensitive chromatin still is not clear, large loops of DNA in

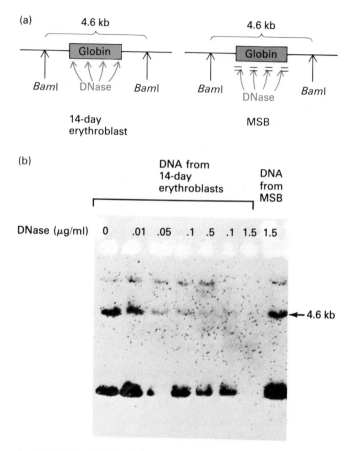

▲ **FIGURE 11-66**   Demonstration that transcriptionally active genes are more susceptible than inactive genes to DNase I digestion. Chick embryo erythroblasts at 14 days actively synthesize globin, whereas cultured undifferentiated MSB cells do not. Nuclei from each type of cell were isolated and exposed to increasing concentrations of DNase. The nuclear DNA was then extracted and treated with the restriction enzyme *Bam*I, which cleaves the DNA around the globin sequence and normally releases a 4.6-kb globin fragment (a). The DNase- and *Bam*I-digested DNA was subjected to Southern blot analysis with a labeled cloned adult globin DNA used as a probe to detect the 4.6-kb fragment. As shown in (b) the nuclear DNA from the 14-day globin-synthesizing cells was sensitive to DNase digestion, but the DNA from the MSB cells was not. [See J. Stalder et al., 1980, *Cell* **19**:973; photograph courtesy of H. Weintraub.]

which transcription is occurring can be visualized projecting out from denser underlying tightly packed chromatin. An example is the hormone-induced "puffs" of highly transcribed genes visible in *Drosophila* salivary gland polytene chromosomes (see Figure 11-29). Another good example of this phenomenon is found in the gnatlike fly *Chironomous tentans.* The salivary glands of this insect also contain easily visualized polytene chromosomes, which contain Balbiani rings analogous to *Drosophila* chromosomal puffs. When the chromatin in such regions is

(a)

(b)

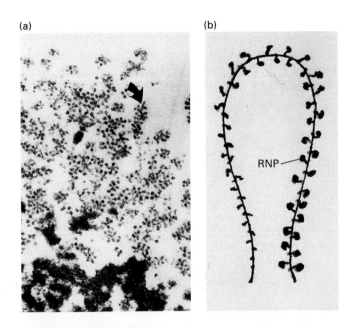

RNP

(c)

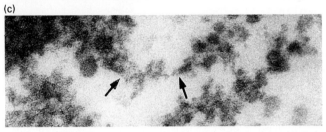

▲ FIGURE 11-67 Electron micrographs of transcription loops in the Balbiani rings of *Chironomous tentans* salivary glands. (a) Active transcription loops of 75S RNA with characteristic granules of ribonucleoprotein (arrow) extend upward from the condensed, dark-staining, nontranscribed chromatin at the bottom. (b) Reconstruction of a single DNA transcription loop from serial thin sections reveals a gradual increase in size of the associated ribonucleoprotein (RNP) particles that reflects the increasing length of the RNA transcripts being synthesized. (c) A single high-power view of one thin section shows the thin chromatin thread (arrows) near the origin of transcription, where no large RNP particles are visible. No return of the distal end of a loop is nearby, indicating that the beginning and end of a transcription loop join the condensed chromatin at different sites separated by some distance. [From C. Erisson et al., 1989, *Cell* **56**:631; photographs courtesy of B. Daneholt.]

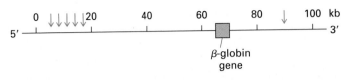

Human DNA introduced	Expression of β-globin in transgenic mice	
	Frequency	Level
β-globin gene (64–70)	1/3	Low
β-globin gene and hypersensitive sites (0–22, 64–70, 82–94)	7/7	High

▲ FIGURE 11-68 Effect of DNase I-hypersensitive sites on expression of human β-globin gene in transgenic mice. Hypersensitive sites (red arrows) are located ≈50 kb 5′ and ≈20 kb 3′ to the β-globin gene in developing human erythrocytes. A transgene must include these sites along with the β-globin gene itself and its promoter-proximal elements in order for transgenic mice to express human β-globin mRNA at high levels and a high frequency. This finding indicates that such hypersensitive sites play a role in transcription control. [Adapted from F. Grosveld et al., 1987, *Cell* **51**:975.]

sectioned and examined by serial reconstruction, loops of DNA being transcribed are visible (Figure 11-67).

DNase I–Hypersensitive Sites Although chromosomal DNA in the region of a transcribed gene is more sensitive to digestion by DNase I than DNA in a transcriptionally silent region, it is still partially protected from DNase I digestion compared to naked DNA. This is because most of the DNA is bound to the surface of histone octamers in nucleosomes. However, within transcriptionally active regions of chromatin, some sites are nearly as sensitive to DNase I digestion as naked DNA. These *DNase I–hypersensitive sites* occur in regions where transcription factors are bound, and probably result from digestion at sites immediately adjacent to the bound factors. DNase I–hypersensitive sites can be mapped by Southern blotting and may be useful in identifying transcription factor–binding sites for a gene of interest.

For example, DNase I–hypersensitive sites have been mapped ≈50 kb 5′ and ≈20 kb 3′ of the human β-globin gene in developing red blood cells (Figure 11-68). When a restriction fragment containing the β-globin gene and adjacent sequences that did not contain the DNase I–hypersensitive sites was introduced into the germ line of mice, only a small fraction of the resulting transgenic mice expressed the human transgene, and those that did so synthesized extremely low levels of human β-globin mRNA. However, when the 5′ and 3′ hypersensitive sites were included in the construct introduced into the germ line, all the transgenic mice expressed the transgene at levels comparable to the endogenous mouse adult β-globins. These results indicate that the DNase I–hypersensitive sites are important transcription-control regions required for normal β-globin expression during erythrocyte development.

General Repression by Nucleosomes Compared with "naked" DNA unassociated with protein, nucleosomal DNA exhibits a generalized repression of *basal* transcription in vitro (i.e., transcription in the absence of activators). The effects of decreasing the density of nucleosomes on in vivo transcription has been studied in yeast cells. The idea behind these experiments was to reduce nucleosome density by reducing the production of histones. This was accomplished by engineering yeast cells in which the histone genes are expressed from the GAL1 promoter and a UAS$_{GAL}$. In these cells, expression of the histone genes is stimulated in media containing galactose and is repressed in media containing glucose. When the engineered cells were switched from galactose medium to glucose medium, histone synthesis ceased. After the cells completed a round of DNA replication in glucose medium, the nucleosome density in daughter cells fell to about half that in normal cells. In these cells, uninduced transcription from all regulated promoters increased, indicating that one function of nucleosomes in vivo is to repress transcription from uninduced promoters. A current area of active research is aimed at understanding how transcription activators overcome the repressive effect of nucleosomes.

Cytosine Methylation Is Associated with Inactive Genes in Vertebrates

Considerable evidence shows that in vertebrates inactive genes often contain the modified cytidine residue 5-methylcytidine (mC) followed immediately by a G in the DNA sequence. For example, regions of DNA that are actively engaged in transcription lack 5'-methylcytidine residues. If plasmids are methylated at CG sites before transfection, the plasmid genes often are not transcribed in recipient cells. Likewise, genes that become methylated after transfection also are inactive. Furthermore, soon after replication of a methylated DNA region, methyl groups are added to the newly synthesized daughter DNA strands. The parental strand containing mCG dinucleotides retains its methyl groups after replication, and these may function as signals for methylation of CG sites in the daughter strand as follows:

Parental strand: (5') mC G (3')
Daughter strand: (3')   G C (5')

$$+CH_3 \downarrow \text{DNA methylase}$$

Parental strand: (5') mC G (3')
Daughter strand: (3')   GCm (5')

Various types of studies have demonstrated conclusively that a close *correlation* exists between methylation and transcriptional inactivation. In some cases, methylation at particular sites prevents binding of transcription factors. Normal embryologic development in mammals clearly depends on cytosine methylation. For example, gene-targeted knockout of the major DNA methylase in the mouse results in embryonic death in mice homozygous

for the mutation. Although little is yet known about how active genes are protected from being methylated, many workers conjecture that during replication of transcriptionally active DNA, associated proteins shield it from DNA methylases.

► Transcription by RNA Polymerase I

Recall that RNA polymerase I, one of the three RNA polymerases present in eukaryotes, is dedicated to the synthesis of only one type of RNA molecule, called *pre-rRNA*. The primary pre-rRNA transcript is processed into the 18S, 5.8S, and 28S rRNAs found in vertebrate ribosomes or their functional equivalents in other eukaryotes.

Pre-rRNA DNA from All Eukaryotes Is Similar

Cloning and sequencing of the DNA encoding pre-rRNA from many species has shown that this DNA shares several properties in all eukaryotes (Figure 11-69). First, the pre-rRNA transcription units are arranged in long tandem arrays. Second, the genomic regions corresponding to the three finished rRNAs are always arranged in the same 5' → 3' order: 18S, 5.8S, and 28S. Third, in all eukaryotic cells (and even in bacteria), the pre-rRNA transcription unit is longer than the sum of the three finished rRNA molecules (Table 11-6). For example, in human cells, the primary 45S pre-rRNA transcript is about 13.7 kb long, whereas 28S rRNA is 5.1 kb, 5.8S RNA is 160 bases, and 18S RNA is 1.9 kb. Thus only about half of the primary transcript appears in the final rRNA products. The other half, called *transcribed spacer RNA*, is degraded rapidly and has no known function.

The transcribed spacer regions in pre-rRNAs from different species vary in length but always are located at both ends and between the 18S, 5.8S, and 28S regions. Stepwise digestion of the spacer sequences by nucleolytic enzymes eventually generates the finished rRNAs (see Figure 12-50). The 5.8S and 28S rRNAs are hydrogen-bonded to each other in the large ribosomal subunit, whereas the 18S rRNA is a component of the small ribosomal subunit.

Only Essential Function of Polymerase I Is to Produce pre-rRNA

For many years, pre-rRNA was recognized as the major product synthesized by RNA polymerase I, but it was impossible to determine whether other minor rRNAs might be produced by this enzyme. A recent experiment has now demonstrated that the only activity of RNA polymerase I required for viability in yeast is synthesis of pre-rRNA.

(a) RIBOSOMAL TRANSCRIPTION UNITS

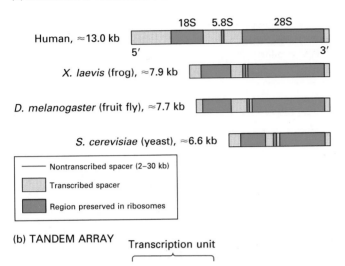

▲ FIGURE 11-69   General structure of pre-rRNA transcription units. (a) In all eukaryotes, pre-rRNA transcription units contain three regions (red) that appear in ribosomes as 18S, 5.8S, and 26 to 28S. The order of these regions in the genome is always 5′ → 3′ as shown in the four examples here. Variations in the lengths of the transcribed spacer regions (tan) account for the major differences in the lengths of pre-rRNA transcription units from different organisms. (b) The genomes of all animals contain multiple tandem copies of the pre-rRNA transcription unit. The nontranscribed spacer regions between these copies can vary greatly among different species from ≈2 kb in frogs to ≈30 kb in humans.

A temperature-sensitive mutation in the large subunit of polymerase I was identified in yeast. Cells with this mutation did not grow at the nonpermissive temperature. However, when the mutant cells were transformed with a plasmid containing the pre-rRNA gene under the control of the *GAL1* promoter and UAS, the cells were able to form colonies at the nonpermissive temperature, provided they were plated on galactose medium to induce transcription from the *GAL1* promoter. Although the transformed cells did not contain functional RNA polymerase I at the nonpermissive temperature, they could grow because the pre-rRNA gene, normally transcribed by polymerase I, could now be transcribed by polymerase II. This finding shows that any other RNAs that might be produced by RNA polymerase I are not required for viability.

Species-Specific Initiation Factors Are Utilized by RNA Polymerase I

Analysis of linker scanning mutations of the promoter region of the human pre-rRNA gene has identified two transcription-control regions. The *core* element is essential for transcription and includes the transcription-initiation site. An ≈50-bp long *upstream control element* (UCE), beginning at about −100 from the start site, stimulates in vitro transcription ten- to a hundredfold.

Protein purification led to the discovery of two DNA-binding proteins that assist RNA polymerase I to correctly initiate and transcribe pre-rRNA genes (Figure 11-70). One of these, called *upstream binding factor* (UBF), was purified on the basis of its specific binding to the UCE, but was found to bind to both the UCE and the upstream por-

TABLE 11-6	Lengths of Primary rRNA Transcripts and Cytoplasmic rRNAs in Various Species				
	Primary transcript*		Ribosomal RNA Length*		
	S Value	Length (kb)	26S–28S (kb)	16S–18S (kb)	Percentage of Precursor Preserved
Escherichia coli (prokaryote)	30	6.0	3.0	1.5	75
Saccharomyces cerevisiae (yeast)	37	6.6	3.8	1.7	77
Dictyostelium discoideum (slime mold)	37	7.4	4.1	1.8	80
Drosophila melanogaster (fruit fly)	34	7.7	4.1	1.8	76
Xenopus laevis (frog)	40	7.9	4.5	1.9	81
Gallus domesticus (chicken)	45	11.2	4.6	1.8	57
Mus musculus (mouse)	45	13.7	5.1	1.9	51
Homo sapiens (human)	45	13.7	5.1	1.9	51

* The lengths of the various RNA molecules are estimates based on gel electrophoresis and direct measurements of electron micrographs. The size is of the first major product with definite 5′ and 3′ ends.
SOURCE: B. Lewin, 1980, *Gene Expression*, vol. 3, Wiley, p. 867.

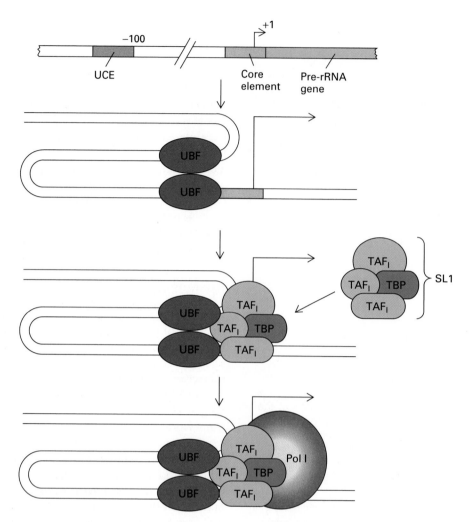

▲ FIGURE 11-70   Model for the assembly of the RNA polymerase I initiation complex. The control region of pre-rRNA transcription units contain a core promoter element, which overlaps the start site, and an upstream control element (UCE) located ≈100 base pairs upstream. Upstream binding factor (UBF) binds to both the UCE and core element, and the two bound molecules are thought to make protein-protein interactions causing the intervening DNA to loop out. Selectivity factor 1 (SL1) then binds to the UBF-DNA complex and the remaining free segment of the core element. SL1 is a multimeric protein composed of TBP and three TBP-associated factors (TAF$_I$) with molecular weights of 110, 63, and 48 kDa. Finally, RNA polymerase I (red) binds, completing assembly of the initiation complex.

tion of the core element, even though these sites have little apparent sequence similarity. The other protein, named *selectivity factor 1 (SL1)*, then binds and stabilizes the complex. After SL1 binding, RNA polymerase I binds and initiates transcription. In contrast to initiation by RNA polymerase II, transcription initiation by RNA polymerase I does not require hydrolysis of the β–γ phosphate of ATP.

Surprisingly, purification of SL1 revealed that the TATA-binding protein (TBP) central to transcription initiation by RNA polymerase II is one of the SL1 subunits. The remaining subunits of human SL1 are called TAF$_I$s for TBP-associated factors required for RNA polymerase I. They appear to be distinct from the TFIID-TAFs (now called TAF$_{II}$s) involved in transcription catalyzed by RNA polymerase II.

Transcription by RNA Polymerase III

Transcription by RNA polymerase III produces small, stable RNAs including tRNAs, the 5S rRNA associated with the large ribosomal subunit, one of the small nuclear RNAs required for pre-mRNA splicing, and the 7S RNA associated with the signal recognition particle involved in secretion of proteins and the insertion of membrane-spanning proteins into cellular membranes (Chapter 16). The functions of many other small RNAs produced by RNA polymerase III are as yet undiscovered. Here we review initiation of transcription of tRNA and 5S-rRNA genes by RNA polymerase III.

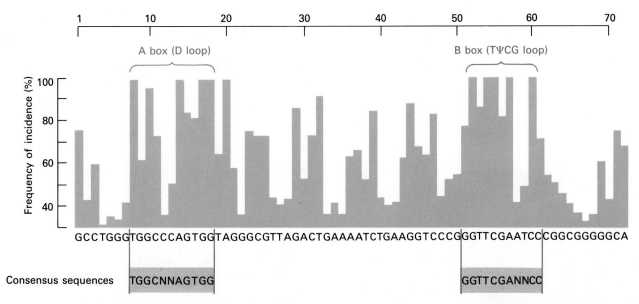

▲ **FIGURE 11-71** Frequency of occurrence (%) of the most common nucleotides in 80 different eukaryotic tRNA genes. Two consensus sequences, called the A box and the B box, are present. The A box encodes the D loop and the B box encodes the TΨCG loop in tRNA; both loops are con-stant features in different tRNA molecules. The A and B boxes also function as promoter elements in tRNA genes. [See D. H. Gauss and M. Sprinzl, 1981, *Nucl. Acids Res.* **9**:r1; G. Galli, H. Hofstetter, and M. L. Birnstiel, 1981, *Nature* **294**:626.]

tRNA Genes Bind Two Multisubunit Initiation Factors

As described in the next chapter, the initial transcripts produced from tRNA genes are precursor molecules, which are processed into mature tRNAs; in this regard, tRNA genes are similar to pre-rRNA genes. An early surprising discovery was that the transcription-control regions of tRNA genes lie entirely within the transcribed sequence. Two such *internal control regions* are required. These DNA sequences encode the most highly conserved portions of eukaryotic tRNAs, the D-loop and the TΨCG loop (see Figure 4-31). Consequently, evolution has used these two highly conserved sequences for two distinct purposes: (1) to encode essential regions of tRNA molecules required for protein synthesis, and (2) to bind proteins necessary for transcription initiation by RNA polymerase III. In discussing these sequences as promoter elements, they are referred to as the *A box* and *B box* (Figure 11-71).

The DNA-binding factors required for RNA polymerase III to initiate transcription have been best characterized in *S. cerevisiae*. Two complex initiation factors, called *TFIIIC* and *TFIIIB*, have been purified from yeast. TFIIIC is composed of six polypeptides with a total molecular weight of ≈600 kDa. Assembly of the initiation complex on tRNA genes begins by binding of TFIIIC to the A box and B box in the promoter (Figure 11-72a). TFIIIB then binds ≈50 base pairs upstream of the A box. TFIIIB can bind to many different sequences at this position, indicating that it is directed to bind to DNA by its interaction with

TFIIIC rather than by sequence-specific interactions with DNA. TFIIIB is made up of three subunits. One is TBP, which we can now see is a subunit of a general initiation factor for all three nuclear RNA polymerases. The second, called *BRF* (for TFIIB-related factor) is similar in sequence to TFIIB, and may perform a similar function in initiation by RNA polymerase III as TFIIB does for RNA polymerase II. The third subunit of TFIIIB is a 90-kDa polypeptide called B″. Once TFIIIB has bound, then RNA polymerase III can bind and initiate transcription in the presence of ribonucleoside triphosphates. Initiation by RNA polymerase III does not require hydrolysis of an ATP β–γ bond similar to polymerase I. Once TFIIIB binds, TFIIIC can be removed without affecting initiation by RNA polymerase III. Thus, TFIIIC can be thought of as an assembly factor for the critical initiation factor, TFIIIB.

5S-rRNA Gene Binds Three Initiation Factors

In addition to TFIIIC and TFIIIB, transcription of 5S-rRNA genes requires a third initiation factor called *TFIIIA*. Binding of this factor, which is a monomer, to the 5S-rRNA gene begins assembly of the initiation complex. The binding site for TFIIIA, called the C box, was identified by deletion mapping of the 5S-rRNA promoter region. Located 81–99 base pairs downstream from the start site, it is an internal control region like the A and B boxes of tRNA genes.

(a) tRNA gene

(b) 5S-rRNA gene

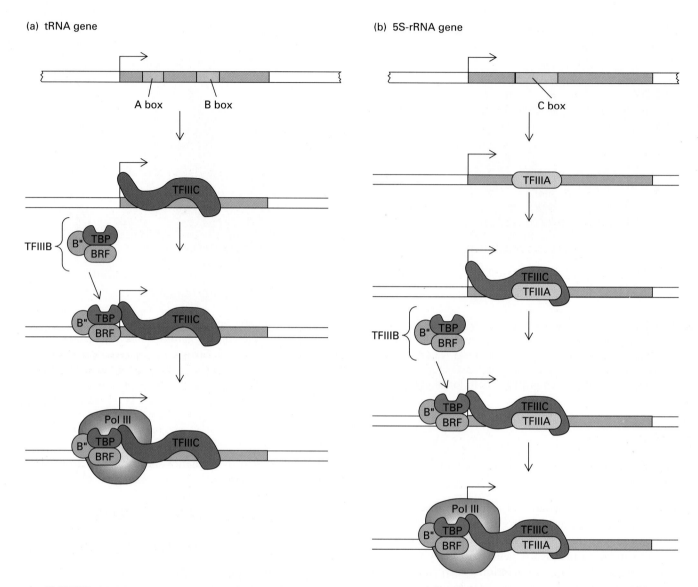

▲ FIGURE 11-72   Models for assembly of RNA polymerase III initiation complexes at tRNA and 5S-rRNA promoters in yeast. (a) In transcription of tRNA genes, TFIIIC, a large multisubunit protein, first binds with high affinity to the B box promoter element and with lower affinity to the A box. TFIIIC acts as an assembly factor for binding the trimeric TFIIIB to any DNA sequence upstream of the tRNA gene. TBP (dark blue) is one of the subunits of TFIIIB. After binding of TFIIIB, RNA polymerase III (red) binds and initiates transcription. (b) In transcription of 5S-rRNA genes, TFIIIA initially binds to the C box promoter element. TFIIIC then binds to TFIIIA and is held in place via protein-protein interactions in the same position relative to the start site as in tRNA genes. TFIIIB binds next, followed by RNA polymerase III as in transcription of tRNA genes.

Synthesis of 5S rRNA is initiated by binding of TFIIIA to the C box (Figure 11-72b). Once TFIIIA has bound, TFIIIC binds to the gene at a similar position relative to the start site as when TFIIIC binds to a tRNA gene. TFIIIB then binds, interacting analogously with TFIIIC as it does in a tRNA gene. Once TFIIIB has bound, RNA polymerase III can bind and initiate transcription. TFIIIA thus acts as an assembly factor for binding of TFIIIC, which then performs the same function that it serves on tRNA genes as an assembly factor for TFIIIB.

TBP Is Required for Transcription Initiation by All Three Eukaryotic RNA Polymerases

As we saw earlier, binding of the TBP subunit of TFIID begins assembly of the initiation complex on TATA-containing promoters transcribed by RNA polymerase II (see Figure 11-53). The finding that TBP is a subunit of critical initiation factors for RNA polymerase I and III was surprising, since the promoters recognized by these enzymes do not contain TATA boxes. Furthermore, a mutant yeast TBP with a reduced ability to bind to the TATA box has been isolated and shown to have an alteration in the TATA-binding surface of the protein. The mutant TBP inhibits transcription initiation by RNA polymerase II, but not by RNA polymerase I and III, in an in vitro system. These results suggest that the TBP-DNA interaction at RNA polymerase I and III promoters differs from the interaction at RNA polymerase II promoters. This hypothesis makes the involvement of TBP in initiation by all three polymerases even more puzzling. Perhaps TBP interacts with one or more of the subunits shared by all three polymerases. Future research should reveal how similar the function of TBP is in each of these initiation reactions.

► *Other Transcription Systems*

At this point we have presented an overview of transcription systems in eubacteria such as *E. coli* and *Bacillus subtilis* and the nuclear RNA polymerases of eukaryotes. Other RNA polymerases are involved in transcribing specific genes expressed during the middle and late phases of infection by certain bacteriophage, the genomes of eukaryotic mitochondria and chloroplasts, and the genome in archaebacterial cells. Although less thoroughly understood in some cases, these systems have interesting parallels and relationships to the more extensively studied systems described earlier in the chapter.

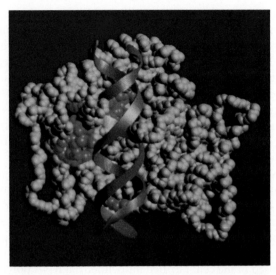

▲ **FIGURE 11-73**   Structural model of T7 RNA polymerase bound to DNA based on x-ray crystallographic analysis. Note that the DNA double helix fits into a deep cleft in the surface of the protein. The amino acid at position 748, which determines the specificity of the enzyme for the T7 promoter versus the T3 promoter, is highlighted in red at the bottom of the figure. Other residues that are critical for the enzyme's polymerizing activity also are highlighted in red. To date, this is the only RNA polymerase whose three-dimensional structure has been determined to atomic resolution. [From R. Sousa et al., 1993, *Science* **364**:593.]

T7 and Related Bacteriophages Express Monomeric, Largely Unregulated RNA Polymerases

Bacteriophage T7 completely dominates macromolecular synthesis in its host *E. coli* cell within a few minutes after infection. All ribosomes in the cell rapidly become committed to the synthesis of virus-encoded proteins exclusively. The virus achieves this by initially directing the synthesis of a virus-encoded RNA polymerase, which does not recognize *E. coli* promoters, and a set of proteins that inactivate the *E. coli* RNA polymerase. Consequently, when the host RNA polymerase is inactivated, transcription of the host genome stops; because *E. coli* mRNAs are very short-lived, the synthesis of most *E. coli* proteins soon ceases. The T7 RNA polymerase initiates transcription from promoter sequences on the viral DNA that are completely different from *E. coli* promoters. The enzyme binds to a 23-bp sequence that overlaps the transcription initiation site.

The T7 RNA polymerase—a single polypeptide chain of 98 kDa—is the simplest RNA polymerase known. The enzyme is probably close to the simplest possible protein molecule capable of carrying out the functions of an RNA polymerase: specific promoter binding, initiation, elongation, and termination. Because the virus is committed to expressing the proteins that make up the virion coat at maximal level, T7 RNA polymerase is not regulated signif-

icantly. An enormously active, unregulated enzyme, T7 RNA polymerase, and viral polymerases from the related bacteriophages such as T3 and SP6, is extremely useful for in vitro RNA synthesis and in bacterial expression systems (Chapter 7).

T7 RNA polymerase is the first RNA polymerase whose three-dimensional structure has been solved to atomic resolution by x-ray crystallography. The molecule contains a large cleft with the dimensions necessary to fit a DNA double helix. The location of mutations known to affect catalytic function fit nicely with a model for how the enzyme binds to DNA (Figure 11-73). *E. coli* DNA polymerase I probably binds DNA similarly, since it contains a similar large cleft. High-resolution electron-microscope studies of *E. coli* RNA polymerase and yeast RNA polymerase II show similar clefts. Consequently, this type of structure appears to be a feature of all RNA (and probably DNA) polymerases.

Mitochondrial DNA Transcription Exhibits Features Typical of Bacteriophage, Bacteria, and the Eukaryotic Nucleus

Mitochondria contain a distinct DNA genome and protein-synthesis system within the inner mitochondrial membrane, in the compartment known as the mitochondrial matrix. This DNA encodes a subset of the proteins essential to the mitochondrion's principle function: ATP synthesis by oxidative phosphorylation. In vertebrates, mitochondrial DNA is an ≈16-kb circle; it encodes 13 integral membrane proteins, which assemble into the inner mitochondrial membrane, as well as the rRNAs and tRNAs required for translation of the mitochondrial mRNAs. There is strong evidence that mitochondria evolved from early eubacterial cells that developed a symbiotic relationship with ancient eukaryotic cells.

Mitochondrial RNA polymerases are far simpler than the nuclear RNA polymerases or even the RNA polymerases of modern-day bacteria. The mitochondrial RNA polymerase from *S. cerevisiae* consists of a 145-kDa subunit with RNA polymerizing activity and a 43-kDa specificity factor essential for initiating transcription at the start sites used in the cell. Purified mitochondrial RNA polymerase from the frog *Xenopus laevis* has a similar structure. The yeast genes encoding the mitochondrial RNA polymerase subunits have been cloned, and their sequences reveal an interesting relationship with bacterial polymerases. The large subunit clearly is related to the monomeric RNA polymerases of bacteriophage T7 and related bacteriophages. However, the mitochondrial enzyme is functionally distinct from the bacteriophage enzyme in its dependence on the small subunit for transcription from the proper start sites. This small subunit is related to bacterial σ factors, which determine the start sites for the bacterial core polymerase composed of the β', β and two α subunits. Thus mitochondrial RNA polymerase appears to be a hybrid of the simple bacteriophage RNA polymerases and the bacterial RNA polymerases of intermediate complexity.

Similar to bacteriophage RNA polymerases, the promoter sequences recognized by mitochondrial RNA polymerases include the transcription start site. A-rich promoter sequences have been characterized in yeast, plants, and animals. The circular, human mitochondrial genome contains two related 15-bp promoter sequences, one for

the transcription of each strand (Figure 11-74). Each strand is transcribed in its entirety; the long primary transcripts are then processed to yield mitochondrial mRNAs, rRNAs, and tRNAs, as discussed in Chapter 19. A 22-kDa basic protein called *mtTF1*, which binds immediately upstream from the two mitochondrial promoters, greatly stimulates transcription. A homologous protein found in yeast mitochondria is required for maintenance of mitochondrial DNA and probably performs a similar function. Interestingly, the DNA-binding domain of mtTF1 is related to the DNA-binding domain of the nuclear RNA polymerase I transcription factor UBF. Also, the binding sites for mtTF1 are in the opposite orientation relative to the direction of transcription for the two mitochondrial promoters. Therefore, it appears that mtTF1 can bind upstream from the mitochondrial RNA polymerase in either orientation and still activate transcription. This is a characteristic more typical of eukaryotic activators than prokaryotic activators. Thus the mitochondrial transcription system appears to combine mechanisms of transcription control from bacteriophage, bacteria, and the eukaryotic nucleus.

Chloroplasts Contain an RNA Polymerase Homologous to the *E. coli* Enzyme

Plant chloroplasts are also thought to have evolved from an ancient eubacterial cell, in this case one capable of photosynthesis. This ancestor of modern chloroplasts likely established a symbiotic relationship with the eukaryotic ancestor of modern plants. In contrast to mitochondrial DNA, chloroplast DNA has remained fairly large, 120–160 kb in different plants. Sequencing of chloroplast DNAs revealed genes with considerable homology to the genes encoding *E. coli* RNA polymerase, α, β, and β' subunits; in chloroplast DNA, however, the gene encoding the β' homolog is split into two pieces, encoding proteins homologous to the N- and C-terminal portions of the *E. coli* protein. The polypeptides encoded by these chloroplasts

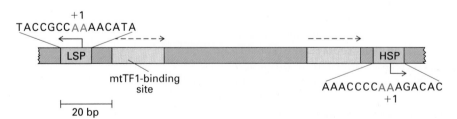

▲ FIGURE 11-74   Diagram of the promoter region in human mitochondrial DNA, a 16-kb circular molecule that contains two 15-bp promoter sequences (yellow). The two DNA strands are referred to as the light strand and heavy strand based on their density in CsCl gradients. The full length of the light strand is transcribed from the light-strand promoter (LSP), and the heavy strand is transcribed in the opposite direction from the heavy-strand promoter (HSP). The nonsymmetrical binding sites (orange) for mitochondrial transcription factor (mtTF1) lie just upstream of the two promoters in opposite 5′ → 3′ orientations (dotted arrows) relative to the promoters. The related but nonidentical sequences of LSP and HSP are shown. Mitochondrial RNA polymerase initiates transcription at the A residues in red.

genes later were found to be subunits of an RNA polymerase purified from chloroplasts; as yet, no gene encoding an equivalent to an *E. coli* σ factor has been identified.

Several chloroplast promoters have been characterized through the construction of 5'-deletion mutants and analysis by in vitro transcription, and two have been analyzed in vivo by fusing 5' deletions to a reporter gene and transfecting the constructs into chloroplasts. These studies have shown that some chloroplast promoters are quite reminiscent of the *E. coli* σ^{70}-promoter, with similar sequences in the −10 and −35 regions. Transcription from one promoter analyzed in vivo, however, depended on sequences from about −20 to +60, quite different from most *E. coli* promoters. This promoter may be recognized by a second RNA polymerase, which most likely is encoded in the nuclear genome and imported into the organelle. Analysis of chloroplast transcription is still in its infancy, but at this point it is clear that at least one transcription system is highly homologous to transcription in *E. coli* and other bacteria.

Archaebacteria Have an RNA Polymerase and Putative General Transcription Factors Similar to Those of the Eukaryotic Nucleus

Archaebacteria, or *Archaea*, constitute a third group of organisms along with eubacteria, such as *E. coli*, and eukaryotes. This group, which was first recognized in the early 1970s, includes single-celled organisms that are found in extreme environments and are difficult to culture. Recent studies of these organisms reveal that their transcription enzymes and possibly factors are much more similar to those of the eukaryotic nucleus than those of eubacteria. This, along with similarities in the macromolecules involved in translation (rRNAs, tRNAs, and initiation factors), argue strongly that the eubacterial evolutionary line split off from a common ancestor before the split between archaebacteria and eukaryotes, making modern eukaryotes more closely related to archaebacteria than to eubacteria.

Like eubacteria, archaebacteria have a single RNA polymerase, but it contains 13 distinct subunits (14 in some), a complexity equivalent to that of eukaryotic nuclear RNA polymerases. Examples of genes encoding nine of these subunits have been cloned and sequenced. The deduced amino acid sequences of seven subunits show significant homology to eukaryotic subunits, including the two largest, which are common to all three of the nuclear RNA polymerases. However, two of the archaebacterial subunits show little detectable similarity in sequence to either eubacterial or eukaryotic RNA polymerase subunits.

Like eubacteria, archaebacteria transcribe operons into polycistronic mRNAs. However, archaebacterial promoters are similar to eukaryotic promoters. An A/T-rich consensus sequence—TTTA(A/T)T—occurs 26 nucleotides upstream from the transcription start site. Mutations in this sequence eliminate in vitro transcription. As in eukaryotic TATA-containing promoters, insertions or deletions following this sequence lead to a shift in the transcription start site to maintain the spacing downstream from the promoter element. In addition, genes with sequence similarity to those encoding eukaryotic TFIIB and TBP have been cloned from archaebacterial DNA. Although the function of the encoded proteins has not yet been tested, their discovery raises the possibility that the mechanisms of transcription initiation in archaebacteria is similar to those in eukaryotes. So far, little is known concerning how transcription is regulated in this fascinating group of organisms.

SUMMARY

While gene expression can be regulated at many levels, in most cases it is the *initiation of transcription* that is controlled. In single-celled organisms, most gene regulation adapts cellular metabolism to the supply of nutrients in the changing environment. In multicellular eukaryotes, gene regulation permits the differentiation of various cell types, resulting in the development of tissues and organs that function cooperatively in a single organism.

In bacteria, a single transcription unit, or *operon*, can contain several genes. Transcription begins at a specific base pair (or sometimes, alternative neighboring base pairs) determined by a DNA sequence called a promoter, which is a binding site for RNA polymerase. RNA polymerase is a large protein composed of β', β, and two α subunits (β'βα$_2$) and one of a few alternative initiation factors called σ subunits; the σ subunit is released from the core β'βα$_2$ polymerase after initiation. Initiation from most *E. coli* promoters requires the σ^{70} initiation factor, which binds to bases near −10 and −35 from the initiation site at +1. Polymerase first binds to form a closed complex in which the template DNA strands are not yet separated. Then it separates the strands over a distance of 12–13 base pairs surrounding the transcription start site, forming an open complex, and initiates polymerization of RNA complementary to the coding strand.

The "strength" of a promoter refers to how frequently RNA polymerase initiates transcription. In the absence of regulatory proteins, promoter strength is determined by the affinity of the σ^{70} subunit of RNA polymerase for the -10 and -35 sequences, the ability of the polymerase to form an open complex, and the ability of the polymerase to transcribe away from the promoter, all of which are affected by the DNA sequence of the promoter region. Transcription initiation can be further regulated by the action of repressor and activator proteins. Repressors bind to DNA sequences called operators which overlap the region contacted by polymerase bound to the promoter. Repressors either sterically block polymerase binding, inhibit open-complex formation, or prevent transcription away from the promoter region. Activators of σ^{70}-polymerase generally bind to DNA on the opposite side of the double helix from the polymerase in the region from -20 to -50, or just upstream of the polymerase near -60. In one of the best-studied cases of activation, activation of *lac*-operon transcription by CAP protein, the CAP activator stimulates transcription by binding to a surface of polymerase and increasing the affinity of polymerase for the promoter. Activators also may stimulate open-complex formation and transcription away from the promoter region.

Mutations in an operator sequence that decrease repressor binding result in constitutive (unregulated) transcription. Operator and promoter mutations are cis-acting; that is, they only affect the expression of genes on the same DNA molecule in which the mutation occurs. In contrast, genes encoding repressors and activators are trans-acting, since the proteins they encode can diffuse through the cell to act at control regions on different DNA molecules than the DNA molecules encoding the repressor or activator.

Many bacterial repressors and activators contain an α helix on their surface, which inserts into the major groove of a binding site on DNA. High-affinity binding results from multiple hydrogen bonds and van der Waals interactions between atoms in the protein and atoms in the DNA; therefore, high-affinity binding of a given protein only occurs with a restricted set of DNA sequences. Repressors and activators are generally dimeric, sometimes tetrameric proteins. For dimers, binding is stabilized by contacts between the two symmetrically disposed DNA-recognition α helices of the dimer and two symmetrically disposed half-sites in the cognate DNA site.

Promoters transcribed by RNA polymerase containing a σ^{54} subunit are regulated quite differently from those transcribed by σ^{70}-polymerase. Activators bound to sites ≈ 100 base pairs upstream from the polymerase transiently bind to the polymerase by "looping out" the intervening DNA and stimulate open-complex formation by a process that requires ATP hydrolysis. The polymerase then initiates transcription.

Transcription initiation in eukaryotes is more complex than in prokaryotes, but it employs many of the same principles. Eukaryotes use three types of RNA polymerases to transcribe different classes of RNAs. The three polymerases each contain related core subunits, which are homologous to the β', β, and α subunits of the bacterial core polymerase, but they also contain 10–13 additional subunits, including six shared by all three polymerases. RNA polymerase I synthesizes pre-rRNA, which is processed into 28S, 18S, and 5.8S rRNAs. RNA polymerase II transcribes all the protein-coding genes and the genes encoding four small RNAs required for splicing mRNA precursors. RNA polymerase III produces 5S rRNA, tRNAs, one of the small RNAs required for splicing mRNA precursors, and some other small, stable RNAs, most with unknown functions.

Strong promoters for protein-coding genes transcribed by RNA polymerase II initiate transcription from a single region containing either a TATA box (at ≈ -30 to -25 in higher plants and animals, or ≈ -100 in *S. cerevisiae*) or an initiator, which spans the start site. Like the core bacterial polymerase, the eukaryotic RNA polymerases require initiation factors. However rather than a single σ subunit acting as an initiation factor, multiple proteins are required; some of these bind to promoter sequences first and then direct the polymerase to the transcription start site. For RNA polymerase II, initiation is best understood for promoters with a TATA box. Here, the multisubunit TFIID protein binds to the TATA-box sequence through its TBP subunit. An extremely large initiation complex (>2000 kDa) then assembles through the sequential binding of TFIIA, TFIIB, a complex of TFIIF with RNA polymerase II, TFIIE, TFIIH and TFIIJ.

As in bacterial systems, transcription initiation by eukaryotic RNA polymerase II is stimulated by activators and inhibited by repressors. The process is similar to regulation of *E. coli* σ^{54}-polymerase in that the regulatory proteins bind to DNA distant from the initiation complex and ATP hydrolysis is required for open-complex formation and initiation. Activators can bind tens to tens of thousands of base pairs from the initiation complex and still greatly stimulate transcription. Similar to the direct interaction between *E. coli* σ^{54}-polymerase and its activators, many eukaryotic activators are thought to bind directly to proteins in the initiation complex with the flexible intervening DNA being "looped out." Activator-binding sites within ≈ 200 base pairs of a promoter are called promoter-proximal elements. Activator-binding sites located at greater distances upstream or downstream within introns or 3' of the transcription unit are called enhancers. Generally, multiple activator proteins bind within an enhancer, which is usually one hundred or more base pairs in length. Transcription from most eukaryotic promoters in higher eukaryotes is thus regulated by combinations of several activators and repressors bound at promoter-proximal elements and enhancers.

Many activators are modular proteins containing distinct, separable DNA-binding domains and activation domains. Functional synthetic activators, composed of a DNA-binding domain from one activator or bacterial repressor plus an activation domain from another activator, are readily constructed with recombinant DNA techniques. Protein regions linking DNA-binding domains and activation domains are thought to be flexible. The DNA-binding domains of eukaryotic activators include a variety of different protein structures most of which insert an α helix into the major groove of DNA. Many classes of eukaryotic activators can form heterodimeric proteins with other members of the same class. This increases the variety of DNA sequences specifically bound by the proteins and also increases the types of activation domains that can be present in one activator protein.

The activation domains in some eukaryotic activators interact with the TBP-associated factors (TAFs), which along with TBP form TFIID, the first protein to bind during initiation-complex assembly. Activation domains also can contact TBP and TFIIB directly, but how any of these interactions stimulate transcription is not yet understood. Currently, eukaryotic repressors are not as thoroughly studied as activators, but many appear to be analogous to activators in that they repress transcription from distant binding sites and are modular proteins containing separable DNA-binding and repression domains.

Since activators and repressors regulate transcription, the genes expressed by a specific type of cell in a multicellular organism are determined by which activators and repressors are expressed in that cell. Genes expressed specifically in liver hepatocytes have control regions with binding sites for activators expressed in hepatocytes as well as more ubiquitously expressed activators. How combinations of activators, some cell type specific and some ubiquitous, result in cell type–specific transcription is not currently understood. The regulatory transcription factors expressed in a specific cell type are determined by the developmental history of the cell, a subject discussed in Chapter 13.

The expression of some genes is regulated by extracellular signals such as hormones. Lipid-soluble hormones diffuse through the plasma membrane and nuclear membrane where they bind and activate transcription factors. Most transcription factors of this type are in the steroid receptor superfamily and have related structures. In the absence of hormone, these transcription factors are bound by inhibitors. When hormone binds to a specific domain of the receptor, the inhibitors are released. In some cases (e.g., the glucocorticoid receptor), the inhibitor prevents binding of the factor to DNA. In other cases (e.g., the thyroid receptor), the factor is bound to its cognate DNA sites and

functions as a repressor in the absence of hormone; binding of hormone converts it to an activator.

In contrast to lipid-soluble hormones, peptide and polypeptide hormones cannot diffuse through the plasma membrane. Instead, they interact with specific cell-surface receptors, which transduce the signal into the cytoplasm. These cell-surface receptors often have protein kinase activity or are indirectly linked to protein kinases. Gene activation by interferon γ involves a simple signal-transduction pathway in which a protein kinase linked directly to the cell-surface receptor phosphorylates and activates a transcription factor in the cytoplasm. The activated transcription factor then is transported into the nucleus where it interacts with its specific binding sites, or response elements, and activates transcription. In other more complex signal transduction pathways, discussed in Chapter 20, multiple steps may intercede between the binding of a hormone to a cell-surface receptor and phosphorylation of transcription factors in the nucleus. Phosphorylation can regulate the DNA-binding activity of an activator or the function of its activation domain. Regulation of either process results in transcriptional regulation.

The template for transcription in eukaryotes is not free DNA but chromatin, DNA associated with histones and other proteins. Nucleosomes repress the low level of transcription that is observed in the absence of activators. In *Drosophila* and yeast, genes sometimes are packaged into condensed chromatin structures in which transcription factors are sterically blocked from interacting with their cognate DNA sites. DNA sequences that organize such condensed chromatin structures are called silencers in yeast. Similar processes may be responsible for lyonization, the repression of most of the genes on one X chromosome in female mammals. In vertebrates, many of the nontranscribed genes in a specific cell type and many of the inactive genes on the repressed X chromosome of females have a 5-methyl group on the C residue in CG dinucleotides in transcription-control regions. This methylation contributes to repression.

Transcription initiation by RNA polymerases I and III also requires initiation factors, which must bind to DNA control regions before the polymerase can bind and initiate. Initiation factors bound close to the transcription start site for each of the three eukaryotic nuclear RNA polymerases share a common subunit, the TATA box–binding protein (TBP), even though the rRNA and tRNA promoters do not contain a TATA box.

Mitochondria and chloroplasts contain relatively small genomes encoding some of the proteins essential for the function of those organelles. Chloroplasts contain an RNA polymerase that is quite similar to the bacterial core polymerase, reflecting the likely evolutionary origin of chloroplasts from a bacterial partner in a symbiotic relationship with the eukaryotic ancestor of modern-day plants. The mitochondrial genome is transcribed by a nuclear-encoded RNA polymerase composed of only two polypeptide chains. The largest subunit has sequence homology with the simplest of all known RNA polymerases, those present in a group of related bacteriophages. These bacteriophage RNA polymerases are composed of a single polypeptide chain of approximately 100 kDa, probably the simplest protein that can carry out all the basic steps of transcription: promoter binding, separation of the template strands, polymerization of ribonucleoside triphosphates complementary to the template strand, and termination of transcription. Archaebacteria have a multisubunit RNA polymerase and promoters with an essential A/T-rich sequence located 26 base pairs upstream from the transcription initiation site. Genes encoding proteins homologous to TFIIB and TBP have been discovered in these organisms. Consequently, transcription in archaebacteria may be similar in many ways to eukaryotic nuclear transcription.

REVIEW QUESTIONS

1. The process of transcription in all organisms can be described in terms of three fundamental stages: initiation, elongation, and termination. In the first section of this chapter, the process of initiation of transcription in prokaryotes was emphasized using *E. coli* RNA polymerase as a paradigm. Review the properties of this RNA polymerase. What are the subunits of the core enzyme? What is the principal difference between the core enzyme and the holoenzyme?

Review the concepts of *cis*-acting and *trans*-acting regulatory elements. In general, what is meant by each?

Refer to Figure 11-5 depicting the *lac* operon. You have grown wild type *E. coli* in a minimal medium containing glucose as the sole carbon source. You then divide the medium containing the bacteria into two portions labeled "A" and "B." To "A," you add nothing, and use it as a control; to "B," you add the potent mutagen ethyl methanesulfonate (EMS).

a. You plate *E. coli* from either "A" or "B" onto dishes containing X-gal and glucose, but no lactose or IPTG. The colonies resulting from the control are all white, but a small number of colonies on the plate containing bacteria that were exposed to EMS are blue. What is the source of the blue color? Prior to characterizing the mutations that result in the formation of the blue colonies, you are told to prepare a list of possible genotypes that could explain the resulting phenotype. What would you place on the list? Explain your answer.

b. You carefully isolate one of the blue colonies, and use it to inoculate medium containing glucose; you also isolate one of the wild type colonies from the plate containing *E. coli* not exposed to EMS. You then sequence the regulatory region *O* in each. How would you do this? When you compare the sequence for *O* from the mutant colony to that from the wild type *E. coli*, they are identical. Do you think the mutation is in a *trans*-acting or a *cis*-acting element? Explain your answer.

c. You decide to isolate another blue colony from the plate described in 1 above. You sequence the *O* region, and discover that a GC → AT transition has occurred at position +17 within the region. Based on these data, would you propose that the mutation is in a *cis*-acting or *trans*-acting element? Explain your answer.

d. You take the same colony isolated in (c) above, grow it in minimal medium, divide the bacteria into two portions, and expose one to EMS. You plate bacteria from each of the two portions onto dishes containing X-gal and glucose, but no IPTG or galactose. The bacteria that were not exposed to EMS still yield only colonies that are blue; however, the bacteria that were exposed to EMS produce several white colonies. Why would this happen? You isolate one of the white colonies, inoculate medium, grow bacteria, and use these as a source of DNA to sequence the *O* region from the *lac* operon. You discover that it is still the same as the mutant *O* region from (c) above. You then plate these bacteria onto a dish containing minimal medium, X-gal, and IPTG; blue colonies result. What does this tell you? Propose a genotype for this revertant that is compatible with the above data.

2. Recall that transcription in the nucleus of eukaryotes is mediated by three RNA polymerases. Where is each polymerase located in the nucleus, and what transcripts are synthesized by each? Review the general transcription factors involved in initiation by RNA polymerases I, II, and III. Use Figures 11-53, 11-70, and 11-71 as guides to understanding how the multiprotein initiation complex is formed for each RNA polymerase. Which factor(s) is (are) common for initiation by each polymerase?

Specific transcription factors are responsible for regulating gene expression mediated by RNA polymerase II. What are the general classes of transcription control proteins that are bound to DNA.

3. You are investigating a gene called *das* that you have isolated from Chinese hamster ovary cells. You do not know the function of the Das protein, but you have cloned the *das* gene, including a 300 base pair region located upstream from the +1 start site for transcription. You are very interested in dissecting the 300 base

pair region to determine the location of the promoter sequence and to characterize it. You decide to do this by performing a controlled exonuclease digestion as described in Figure 11-36; you obtain the following:

Region   I. $\underline{-300 \quad -250 \quad -200 \quad -150 \quad -100 \quad -50 \quad +1}$

Region II. $\underline{-300 \quad -250 \quad -200 \quad -150 \quad -100}$

Region III. $\underline{-300 \quad -250 \quad -200}$

Region IV. $\underline{-300 \quad -250}$

Region V. $\underline{-300}$

You add linkers to each of the above, and prepare a series of plasmids with the deleted upstream regions fused to the gene encoding firefly luciferase with an upstream TATA box. (Firefly luciferase is the enzyme that metabolizes the molecule luciferin, a process that results in the generation of light that is associated with fireflies on warm summer nights.) You transfect Chinese hamster ovary cells with the plasmids, prepare cell extracts, and, using luciferin as a substrate, assay for firefly luciferase activity. The results are presented below.

Upstream Region	Relative Firefly Luciferase Activity
I	++++
II	++++
III	++
IV	−
V	−

a. Based on these data, where would you place the promoter-proximal elements in the upstream region?
b. You are interested in trying to purify the protein(s) that interact(s) with the most upstream promoter-proximal element. Devise a strategy for doing this. What method would you use to assay for the presence of the protein?
c. You manage to purify a protein that recognizes this promoter-proximal element for the *das* gene. You have a mutant hamster cell line that is *das⁻* in the sense that you cannot detect any mRNA encoded by the *das* gene. You have used Southern blots to demonstrate that the gene for encoding the Das protein is present, and you have devised the hypothesis that the promoter region in the mutant is defective. You decide to use the polymerase chain reaction to test this notion. How would you do this? Much to your disappointment, you discover that the *das* gene promoter region in the *das⁻* mutant is identical to that found in normal cells. You know that transcription of the vast majority of genes in the mutant is normal. You then propose that the transcription factor you isolated from the normal cells is defective in

the mutant. How would you test this? If this factor really were responsible for the mutant phenotype, could this be reflected by an inability of it to bind to its consensus sequence? How would you test this? If you found that the mutant factor recognizes the proper consensus sequence, does that rule out the possibility that it is responsible for the observed phenotype? Explain your answer, and propose experiments to test your ideas.

4. You are investigating a transcription factor that binds directly to DNA. You know that it recognizes the following five sequences. (Only one strand of each is shown, but double-stranded DNA is required for binding.)

1. 5'-ATTGCATAGGCTAATAGCTCTGTT-3'
2. 5'-CAGTTAAGCCTGTAGCTT-3'
3. 5'-GCTGTGTACTAAAGCTGATAGCAGGGCTTA-3'
4. 5'-CTATGCTCGTAGCAC-3'
5. 5'-CGAGTGTTTTGCTCGTCGCAAAATGTGCCCTAGGC-3'

a. You are studying the interaction of the purified transcription factor with these sequences. You label sequence 1 at the 5' end using T4 polynucleotide kinase and [³²P]ATP. Which of the phosphates in the ATP molecule would need to contain the radioactive isotope in order for the DNA to become labeled? You now use radiolabeled DNA sequence 1 along with the purified protein in an EMSA. You see the expected shift in labeled DNA as the amount of protein is increased. You perform the assay again in an identical manner, but you add a 100-fold excess of unlabeled sequence 3 to the mixture. What do you think would happen to the intensity of the shifted bands in this case? Explain your answer.
b. You perform an experiment like that in (a) above, but this time, you add a 100-fold excess of an unlabeled DNA sequence that is also 25 base pairs in length but is not recognized by the transcription factor. What effect would this have on the intensities of the shifted bands?
c. Based on the five sequences above, propose a consensus sequence that is recognized by the protein. Do you think this sequence would be protected or digested in a DNase footprint assay? Would you be able to use the DNase footprinting method to pinpoint the protected regions for each of these sequences? How would you do this?

5. Control of gene expression in eukaryotic cells is extraordinarily complex. One level of control that can inhibit transcription occurs when a gene is present in condensed, tightly packed DNA known as heterochromatin. In contrast, euchromatin contains many of the actively expressed genes. An excellent example of this concept in this chapter is the repression of the silent HML and HMR loci in yeast. Review this material. What is meant by diminished access to tightly packed DNA due to steric hindrance? What other examples support the notion that condensed chromatin structures can prevent transcription, possibly due to steric factors?

a. You have two yeast cell lines. In cell line A, a gene encoding firefly luciferase (see Question 3) under the control of a yeast RNA polymerase II promoter is stably integrated into a chromosome near the silent HML locus. In cell line B, the same gene and promoter are present in a region near the expressed MAT locus. You homogenize cells from each line, and assay for luciferase activity. In cell line A, no luciferase activity is observed, but cell line B yields a strong luciferase signal. Explain these results.

b. Recall that ionizing radiation induces breaks in DNA. Likewise, ethyl methanesulfonate (EMS) damages DNA by placing ethyl groups on bases and phosphate residues, pro-

ducing numerous lesions in DNA, including breaks. You expose cell line A and B to equal, low doses of ionizing radiation or EMS. You then isolate DNA from each, cut with a restriction endonuclease, and use Southern analysis to determine the amount of full-length luciferase gene in each cell line. You discover that the amount of full-length luciferase gene diminishes at equal rates in both cell lines with increasing doses of ionizing radiation. When EMS is used, however, the full-length luciferase gene is damaged to a much higher degree in cell line B when compared to cell line A. Devise a viable explanation for these results.

References

Prokaryotic Gene Control

General References

BECKWITH, J. R., and D. ZIPSER, eds. 1970. *The Lactose Operon.* Cold Spring Harbor Laboratory.

GRALLA, J. D. 1991. Transcription control—lessons from an *E. coli* promoter data base. *Cell* 66:415–418.

MCCLURE, W. R. 1985. Mechanism and control of transcription initiation in prokaryotes. *Ann. Rev. Biochem.* 54:171–204.

MCKNIGHT, S. L., and K. R. YAMAMOTO, eds. 1992. *Transcriptional Regulation.* Cold Spring Harbor Laboratory Press.

NEIDHARDT, F., ed. 1987. *Escherichia coli and Salmonella typhimurium.* American Society of Microbiology.

PABO, C. O., and R. T. SAUER. 1992. Transcription factors: structural families and principles of DNA recognition. *Ann. Rev. Biochem.* 61:1053–1095.

E. coli *RNA Polymerase*

LOSICK, R., and M. J. CHAMBERLIN, eds. 1976. *RNA Polymerase.* Cold Spring Harbor Laboratory.

ROSS, W., et al. 1993. A third recognition element in prokaryotic promoters: site-specific DNA binding by the α subunit of RNA polymerase. *Science* 262:1407–1413.

Sigma (σ) Factors

BURGESS, R. R., et al. 1969. Factor stimulating transcription by RNA polymerase. *Nature* 221:43–46.

GROSS, C. A., M. LONETTO, and R. LOSICK. 1992. Bacterial sigma factors. In S. L. McKnight and K. R. Yamamoto, eds. *Transcriptional Regulation.* Cold Spring Harbor Laboratory Press, pp. 129–178.

HELMANN, J. D., and M. J. CHAMBERLAN. 1988. Structure and function of bacterial sigma factors. *Ann. Rev. Biochem.* 57:839–872.

KUSTU, S., A. K. NORTH, and D. S. WEISS. 1991. Prokaryotic transcriptional enhancers and enhancer-binding proteins. *Trends Biochem. Sci.* 16:397–402.

LONETTO, M., M. GRIBSKOV, and C. A. GROSS. 1992. The σ^{70} family: sequence conservation and evolutionary relationships. *J. Bact.* 174:3843–3849.

TANAKA, K., et al. 1993. Heterogeneity of the principal sigma factor in Escherichia coli: the rpoS gene product, σ^{38}, is a second principal sigma factor of RNA polymerase in stationary-phase Escherichia coli. *Proc. Nat'l. Acad. Sci. USA.* 90:3511–3515.

Bacterial Promoters

MCCLURE, W. R. 1985. Mechanism and control of transcription initiation in prokaryotes. *Ann. Rev. Biochem.* 54:171–204.

PRIBNOW, D. 1975. Bacteriophage T7 early promoters: nucleotide sequences of two RNA polymerase binding sites. *J. Mol. Biol.* 99:419–443.

SIEBENLIST, U., R. B. SIMPSON, and W. GILBERT. 1980. *E. coli* RNA polymerase interacts homologously with two different promoters. *Cell* 20:269–281.

Negative Control of Transcription in Bacteria

BECKWITH, L. J. 1987. Genetics at the Pasteur Institute: substance and style. *ASM News* 53:551–555.

GILBERT, W., and B. MÜLLER-HILL. 1967. Isolation of the *lac* repressor. *Proc. Nat'l. Acad. Sci. USA* 58:2415–2421.

JACOB, F., and J. MONOD. 1961. Genetic regulatory mechanisms in the synthesis of proteins. *J. Mol. Biol.* 3:318–356.

MARTIN, K., L. HUO, and R. F. SCHLIEF. 1986. The DNA loop model for *ara* repression: AraC protein occupies the proposed loop sites *in vivo* and repression negative mutations lie in these same sites. *Proc. Nat'l. Acad. Sci. USA* 83:3654–3658.

OEHLER, S., et al. 1990. The three operators of the *lac* operon cooperate in repression. *EMBO J.* 9:973–979.

PARDEE, A. B., F. JACOB, and J. MONOD. 1959. The genetic control and cytoplasmic expression of inducibility in the synthesis of β-galactosidase by *E. coli J. Mol. Biol.* 1:165–178.

Positive Control of Transcription in Bacteria

DE CROMBRUGGHE, B., and I. PASTON. 1978. Cyclic AMP, the cyclic AMP receptor protein, and their dual control of the galactose operon. In J. H. Miller and W. S. Reznikoff, eds., *The Operon.* Cold Spring Harbor Laboratory.

LOSICK, R., and P. STRAGIER. 1992. Crisscross regulation of cell-type specific gene expression during development of *B. subtilis. Nature* 355:601–604.

MAGASANIK, B. 1989. Gene regulation from sites near and far. *New Biol.* 1:247–251.

NORTH, A. K., et al. 1993. Prokaryotic enhancer-binding proteins reflect eukaryotic-like modularity: the puzzle of nitrogen regulatory protein C. *J. Bact.* 175:67–73.

RAIBAUD, O. 1989. Nucleoprotein structures at positively regulated bacterial promoters: homology with replication origins and some hypotheses on the quaternary structure of the activator proteins in the complexes. *Mol. Microbiol.* 3:455–458.

REITZER, L. J., and B. MAGASANIK. 1986. Transcription in glnA in *E. coli* is stimulated by activator bound to sites far from the promoter. *Cell* 45:785–792.

REZNIKOFF, W. S. 1992. Catabolite gene activator protein activation of *lac* transcription. *J. Bact.* 174:655–658.

SCHLEIF, R. 1992. DNA looping. *Ann. Rev. Biochem.* 61:199–223.

ZUBAY, G., D. SCHWARTZ, and J. BECKWITH. 1970. Mechanism of activation of catabolic-sensitive genes: a positive control system. *Proc. Nat'l. Acad. Sci. USA* 66:104–110.

Compound Control of Bacterial Transcription

CASHEL, M., and K. E. RUDD. 1987. The stringent response. In F. Neidhardt, ed., *Escherichia coli and Salmonella typhimurium,* American Society of Microbiology.

GOTTESMAN, S. 1984. Bacterial regulation: global regulatory networks. *Ann. Rev. Genet.* **18**:415–442.

Levels of Eukaryotic Gene Control, Eukaryotic Transcription Units, and Transcription Rates

BLAU, H. M. 1988. Hierarchies of regulatory genes may specify mammalian development. *Cell* **53**:673–674.

DARNELL, J. E., JR. 1982. Variety in the level of gene control in eukaryotic cells. *Nature* **297**:365–371.

DERMAN, E., S. GOLDBERG, and J. E. DARNELL. 1976. hnRNA in HeLa cells: distribution of transcription sizes estimated from nascent molecule profile. *Cell* **9**:465–472.

DERMAN, E., et al. 1981. Transcriptional control in the production of liver-specific mRNAs. *Cell* **23**:731–739.

MCKNIGHT, G. S., and R. D. PALMITER. 1979. Transcriptional regulation of the ovalbumin and conalbumin genes by steroid hormones in chick oviduct. *J. Biol. Chem.* **254**:9050–9058.

NEVINS, J. R. 1982. Adenovirus gene expression: control at multiple steps of mRNA biogenesis. *Cell* **28**:1–2.

Eukaryotic Nuclear RNA Polymerases

DARST, S. A., et al. 1991. Three-dimensional structure of yeast RNA polymerase II at 16 Å resolution. *Cell* **66**:121–128.

KOLODZIEJ, P. A., et al. 1990. RNA polymerase II subunit composition, stoichiometry, and phosphorylation. *Mol. Cell. Biol.* **10**:1915–1920.

ROEDER, R. G. 1975. Multiple forms of deoxyribonucleic acid-dependent ribonucleic acid polymerase in *Xenopus laevis. J. Biol. Chem.* **249**:241–248.

TREICH, I., et al. 1992. RPC10 encodes a new mini subunit shared by yeast nuclear RNA polymerases. *Gene Exp.* **2**:31–37.

YOUNG, R. A. 1991. RNA polymerase II. *Ann. Rev. Biochem.* **60**:689–715.

Promoters for RNA Polymerase II

BAKER, C. C., and E. B. ZIFF. 1980. Biogenesis, structures, and sites of encoding of the 5′ termini of adenovirus-2 mRNAs. *Cold Spring Harbor Symp. Quant. Biol.* **44**:415–428.

BUCHER, P. 1990. Weight matrix descriptions of four eukaryotic RNA polymerase II promoter elements from 502 unrelated promoter sequences. *J. Mol. Biol.* **212**:563–578.

CONTRERAS, R., and W. FIERS. 1981. Initiation of transcription by RNA polymerase(s)II in permeable SV40-infected or noninfected CV1 cells: evidence for multiple promoters of SV40 late transcription. *Nucl. Acids Res.* **9**:215–236.

CORDEN, J., et al. 1980. Promoter sequences of eukaryotic protein-coding genes. *Science* **209**:1406–1414.

DYNAN, W. S. 1986. Promoters for housekeeping genes. *Trends Genet.* **2**:196–197.

GROSSCHEDL, R., et al. 1981. Point mutation in the TATA box curtails expression of sea urchin H2A histone gene in vivo. *Nature* **294**:178–180.

KOLLMAR, R., and P. J. FARNHAM. 1993. Site-specific initiation of transcription by RNA polymerase II. *Proc. Soc. Exp. Biol. Med.* **203**:127–139.

SMALE, S. T. 1994. Core promoter architecture for eukaryotic protein-coding genes. In R. C. Conoway and J. W. Conoway, eds., *Transcription: Mechanisms and Regulation,* Raven Press, pp. 63–81.

SMALE, S. T., and D. BALTIMORE. 1989. The "initiator" as a transcription control element. *Cell* **57**:103–113.

Transcription-Control Elements for RNA Polymerase II

BANERJI, J., S. RUSCONI, and W. SCHAFFNER. 1981. Expression of β-globin gene is enhanced by a remote SV40 DNA sequence. *Cell* **27**:299–308.

BOULET, A. M., C. R. ERWIN, and W. J. RUTTER. 1986. Cell-specific enhancers in the rat exocrine pancreas. *Proc. Nat'l. Acad. USA* **83**:3599–3603.

FROMM, M., and P. BERG. 1982. Deletion mapping of DNA regions required for SV40 late and early region promoter function *in vivo. J. Mol. Appl. Genet.* **1**:457–481.

GROSSCHEDL, R., and M. L. BIRNSTIEL. 1980. Spacer DNA sequences upstream of the T-A-T-A-A-A-T-A sequence essential for promotion of H2A histone gene transcription in vivo. *Proc. Nat'l. Acad. Sci. USA* **77**:7102–7106.

GUARENTE, L. 1988. UASs and enhancers: common mechanism of transcriptional activation in yeast and mammals. *Cell* **52**:303–305.

MCKNIGHT, S. L. 1982. Functional relationships between transcription control signals of the thymidine kinase gene of herpes simplex virus. *Cell* **31**:355–365.

MCKNIGHT, S. L., and R. KINGSBURY. 1982. Transcription control signals of a eukaryotic protein-coding gene. *Science* **217**:316–324.

NUSSINOV, R. 1991. Signals in DNA sequences and their potential properties. *Computer Appl. Biosciences* **7**:295–299.

PINKERT, C. A., et al. 1987. An albumin enhancer located 10 kb upstream functions along with its promoter to direct efficient, liver-specific expression in transgenic mice. *Genes & Dev.* **1**:268–276.

SCHLEIF, R. 1987. Why should DNA loop? *Nature* **327**:369–370.

SERFLING, E., M. JASIN, and W. SCHAFFNER. 1985. Enhancers and eukaryotic gene transcription. *Trends Genet.* **1**:224–230.

ZENKE, M., et al. 1986. Multiple sequence motifs are involved in SV40 enhancer function. *EMBO J.* **5**:387–397.

Eukaryotic Transcription Factors

BRIGGS, M. R., et al. 1986. Purification and biochemical characterization of the promoter-specific transcription factor Sp1. *Nature* **234**:47–52.

GUARENTE, L. 1987. Regulatory proteins in yeast. *Ann. Rev. Genet.* **21**:425–452.

HARRISON, S. C. 1991. A structural taxonomy of DNA-binding domains. *Nature* **353**:715–719.

HOPE, I. A., and K. STRUHL. 1986. Functional dissection of a eukaryotic transcriptional activator protein, GCN4 of yeast. *Cell* **46**:885–894.

JOHNSON, P. F., and S. L. MCKNIGHT. 1989. Eukaryotic transcriptional regulatory proteins. *Ann. Rev. Biochem.* **58**:799–839.

JOHNSTON, S. A., and J. E. HOPPER. 1982. Isolation of the yeast regulatory gene *GAL4* and analysis of its dosage effects on the galactose/melibiose regulon. *Proc. Nat'l. Acad. Sci. USA* **79**:6971–6975.

JONES, N. C., P. W. J. RIGBY, and E. B. ZIFF. 1988. Trans-acting protein factors and the regulation of eukaryotic transcription: lessons from studies on DNA tumor viruses. *Genes & Dev.* **3**:267–281.

LAMB, P., and S. L. MCKNIGHT. 1991. Diversity and specificity in transcriptional regulation: the benefits of heterotypic dimerization. *Trends Biochem. Sci.* **16**:417–422.

MA, J., and M. PTASHNE. 1987. Deletion analysis of GAL4 defines two transcriptional activating segments. *Cell* **48**:847–853.

MA, J., and M. PTASHNE. 1987. A new class of yeast transcriptional activators. *Cell* **51**:113–119.

MITCHELL, P. J., and R. TJIAN. 1989. Transcriptional regulation in mammalian cells by sequence-specific DNA-binding proteins. *Science* **245**:371–378.

MORIMOTO, R. I. 1992. Transcription factors: positive and negative regulators of cell growth and disease. *Curr. Opin. Cell Biol.* **4**:480–487.

PABO, C. O., and R. T. SAUER. 1992. Transcription factors: structural families and principles of DNA recognition. *Ann. Rev. Biochem.* **61**:1053–1095.

STRUHL, K. 1987. Promoters, activator proteins, and the mechanism of transcriptional initiation in yeast. *Cell* **49**:295–297.

VARMUS, H. E. 1987. Oncogenes and transcriptional control. *Science* **238**:1337–1339.

Initiation of Transcription by Eukaryotic RNA Polymerases

Transcription of Protein-Coding Genes by RNA Polymerase II

BURATOWSKI, S., et al. 1989. Five intermediate complexes in transcription initiation by RNA polymerase II. *Cell* 56:549–561.

DAVISON, B. L., et al. 1983. Formation of stable preinitiation complexes between eukaryotic class B transcription factors and promoter sequences. *Nature* 301:680–686.

DYNLACHT, B. D., T. HOEY, and R. TJIAN. 1991. Isolation of coactivators associated with the TATA-binding protein that mediate transcriptional activation. *Cell* 66:563–576.

FLORES, O., et al. 1991. The small subunit of TFIIF recruits RNA polymerase II into the preinitiation complex. *Proc. Nat'l. Acad. Sci. USA* 88:9999–10003.

GILEADI, W., W. J. FEAVER, and R. D. KORNBERG. 1992. Cloning of a subunit of yeast RNA polymerase II transcription factor B and CTD kinase. *Science* 257:1389–1392.

HAHN, S., et al. 1989. Isolation of the gene encoding the yeast TATA-binding protein TFIID: a gene identical to the SPT15 suppressor of Ty element insertions. *Cell* 58:1173–1181.

HERNANDEZ, N. 1993. TBP, a universal eukaryotic transcription factor? *Genes & Dev.* 7:1291–1308.

HORIKOSHI, M., et al. 1989. Cloning and structure of a yeast gene encoding a general transcription initiation factor TFIID that binds to the TATA box. *Nature* 341:299–303.

KIM, J. L., D. B. NIKOLOV, and S. K. BURLEY. 1993. Co-crystal structure of TBP recognizing the minor groove of a TATA element. *Nature* 365:520–527.

KIM, Y., et al. 1993. Crystal structure of a yeast TBP/TATA-box complex. *Nature* 365:512–520.

LU, H., et al. 1992. Human general transcription factor IIH phosphorylates the C-terminal domain of RNA polymerase II. *Nature* 358:641–645.

NIKOLOV, D. B., et al. 1992. Crystal structure of TFIID TATA-box binding protein. *Nature* 360:40–46.

SAWADOGO, M., and A. SENTENAC. 1990. RNA polymerase B (II) and general transcription factors. *Ann. Rev. Biochem.* 59:711–754.

SAYRE, M. H., H. TSCHOCHNER, and R. D. KORNBERG. 1992. Reconstitution of transcription with five purified transcription factors and RNA polymerase II from *Saccharomyces cerevisiae. J. Biol. Chem.* 267:23376–23382.

SCHAEFFER, L., et al. 1993. DNA repair helicase: a component of BTF2 (TFIIH) basic transcription factor. *Science* 260:58–63.

WEIL, P. A., et al. 1979. Selective and accurate initiation of transcription of the Ad2 major late promoter in a soluble system dependent on purified RNA polymerase II and DNA. *Cell* 18:469–484.

ZAWEL, L., and D. REINBERG. 1993. Initiation of transcription by RNA polymerase II: a multistep process. *Prog. Nucl. Acid Res. Mol. Biol.* 44:67–108.

ZHOU, Q., et al. 1992. Holo-TFIID supports transcriptional stimulation by diverse activators and from a TATA-less promoter. *Genes & Dev.* 6:1964–1974.

Transcription of pre-rRNA Genes by RNA Polymerase I

BELL, S. P., et al. 1988. Functional cooperativity between transcription factors UBF1 and SL1 mediates human rRNA synthesis. *Science* 241:11192–11197.

COMAI, L., N. TANESE, and R. TJIAN, 1992, The TATA-binding protein and associated factors are integral components of the RNA polymerase I transcription factor, SL1. *Cell* 68:965–976.

DUBE, C., et al. 1988. Human ribosomal RNA genes: orientation of the tandem array and conservation of the 5′ end. *Science* 239:64–68.

ELION, E. A., and J. R. WARNER. 1986. An RNA polymerase I enhancer in *Saccharomyces cerevisiae. Mol. Cell Biol.* 6:2089–2097.

MILLER, O. L. 1981. The nucleolus, chromosome, and visualization of genetic activity. *J. Mol. Biol.* 91:15s–27s.

NAGI, Y., and M. NOMURA. 1991. Synthesis of large rRNAs by RNA polymerase II in mutants of *Saccharomyces cerevisiae* defective in RNA polymerase I. *Proc. Nat'l. Acad. Sci. USA* 88:3962–3966.

PAULE, M. R. 1994. Transcription of ribosomal RNA by eukaryotic RNA polymerase I. In R. C. Conaway and J. W. Conaway, eds., *Transcription: Mechanisms and Regulation,* Raven Press, pp. 83–106.

REEDER, R. H., 1992. Regulation of transcription by RNA polymerase I. In S. McKnight, and K. R. Yamamoto, eds., *Transcriptional Regulation,* Monograph 22, Cold Spring Harbor Laboratory Press, pp. 315–348.

SCHNAPP, G., et al. 1994. The HMG box-containing nucleolar transcription factor UBF interacts with a specific subunit of RNA polymerase I. *EMBO J.* 13:190–199.

Transcription of tRNA and 5S-rRNA Genes by RNA Polymerase III

BROWN, D. D., and J. B. GURDON. 1978. Cloned single repeating units of 5S DNA direct accurate transcription of 5S RNA when injected into *Xenopus* oocytes. *Proc. Nat'l. Acad. Sci. USA* 75:2849–2953.

GABRIELSEN, O. S., and A. SENTENAC. 1991. RNA polymerase III(C) and its transcription factors. *Trends Biochem. Sci.* 16:412–416.

GEIDUSCHEK, E. P., and G. A. KASSAVETIS. 1992. RNA polymerase III transcription complexes. In S. McKnight, and K. R. Yamamoto, eds., *Transcriptional Regulation,* Cold Spring Harbor Laboratory Press, pp. 247–280.

KASSAVETIS, G. A., et al. 1990. *S. cerevisiae* TFIIIB is the transcription initiation factor proper of RNA polymerase III, while TFIIIA and TFIIIC are assembly factors. *Cell* 60:235–245.

LASSAR, A. B., P. L. MARTIN, and R. G. ROEDER. 1983. Transcription of class III genes: formation of preinitiation complexes. *Science* 222:740–748.

SAKONJU, S., D. F. BOGENHAGEN, and D. D. BROWN. 1980. A control region in the center of the 5S RNA gene directs specific initiation of transcription. I: The 5′ border of the region. *Cell* 19:13–25.

Mechanisms of Transcriptional Activation in Eukaryotes

BERGER, S. L., et al. Genetic isolation of ADA2: a potential transcriptional adapter required for function of certain acidic activation domains. *Cell* 70:251–265.

BOYER, T. G., and A. J. BERK. 1993. Functional interaction of adenovirus E1A with holo-TFIID. *Genes & Dev.* 7:1810–1823.

CAREY, M., et al. 1990. A mechanism for synergistic activation of a mammalian gene by GAL4 derivatives. *Nature* 345:361–364.

CHOY, B., and M. R. GREEN. 1993. Eukaryotic activators function during multiple steps of preinitiation complex assembly. *Nature* 366:531–536.

DRAPKIN, R., A. MERINO, and D. REINBERG. 1993. Regulation of RNA polymerase II transcription. *Curr. Opin. Cell. Biol.* 5:469–476.

HOEY, T., et al. 1993. Molecular cloning and functional analysis of *Drosophila* TAF110 reveal properties expected of coactivators. *Cell* 72:247–260.

HORIKOSHI, M., et al. 1988. Mechanism of action of a yeast activator: direct effect of GAL4 derivatives on mammalian TFIID-promoter interactions. *Cell* 54:665–669.

KOLESKE, A. J., and R. A. YOUNG. 1994. An RNA polymerase II holoenzyme responsive to activators. *Nature* 368:466–469.

LIEBERMAN, P. M., and A. J. BERK. 1994. A mechanism for TAFs in transcriptional activation: activation domain enhancement of TFIID-TFIIA-promoter DNA complex formation. *Genes & Dev.* 8:995–1006.

MEISTERERNST, W., et al. 1991. Activation of class II gene transcription by regulatory factors is potentiated by a novel activity. *Cell* 66:981–993.

PTASHNE, M. 1988. How eukaryotic transcriptional activators work. *Nature* 335:683–689.

ROBERTS, S. G., et al. 1993. Interaction between an acidic activator and transcription factor TFIIB is a requirement for transcriptional activation. *Nature* 363:741–744.

ROEDER, R. G. 1991. The complexities of eukaryotic transcription initiation: regulation of preinitiation complex assembly. *Trends Biochem. Sci.* 16:402–408.

STRINGER, K. F., C. J. INGLES, and J. GREENBLATT. 1990. Direct and selective binding of an acidic transcriptional activation domain to the TATA-box factor TFIID. *Nature* 345:783–786.

TJIAN, R., and T. MANIATIS. 1994. Transcriptional activation: a complex puzzle with few easy pieces. *Cell* 77:5–8.

WANG, W., J. D. GRALLA, and M. CAREY. 1992. The acidic activator of GAL4-AH can stimulate polymerase II transcription by promoting assembly of a closed complex requiring TFIID and TFIIA. *Genes & Dev.* 6:1716–1727.

Regulating Activity of Transcription Factors

BEATO, M. 1989. Gene regulation by steroid hormones. *Cell* 56:335–344.

DARNELL, J. E., JR., I. M. KERR, and G. R. STARK. 1994. JAK-STAT pathways and transcriptional activation in response to interferons and other extracellular signaling proteins. *Science.* 264:1415–21

EVANS, R. M. 1989. Molecular characterization of the glucocorticoid receptor. *Recent Prog. Hormone Res.* 45:1–22.

FALVEY, E., and U. SCHIBLER. 1991. How are the regulators regulated? *Faseb J.* 5:309–314.

GREEN, S., and P. CHAMBON. 1988. Nuclear receptors enhance our understanding of transcription regulation. *Trends Genet.* 4:309–314.

LAI, E., and J. E. DARNELL, JR. 1991. Transcriptional control in hepatocytes: a window on development. *Trends Biochem. Sci.* 16:427–430.

HUNTER, T., and M. KARIN. 1992. The regulation of transcription by phosphorylation. *Cell* 70:375–387.

LOHNES, D., et al. 1992. Retinoid receptors and binding proteins. *J. Cell Sci.* (Suppl.) 16:69–76.

MANGELSDORF, D. J., et al. 1993. Retinoid receptors. *Recent Prog. Hormone Res.* 48:99–121.

YAMAMOTO, K. R., et al. 1992. Combinatorial regulation at a mammalian composite response element. In S. L. McKnight and K. R. Yamamoto, eds., *Transcriptional Regulation*, Cold Spring Harbor Laboratory Press, pp. 1169–1192.

Eukaryotic Repressors

BANIAHMAD, A., A. C. KOHNE, and R. RENKAWITZ. 1992. A transferable silencing domain is present in the thyroid hormone receptor, in the v-erbA oncogene product and in the retinoic acid receptor. *EMBO J.* 11:1015–1023.

HAN, K., and MANLEY, J. L. 1993. Transcriptional repression by the *Drosophila* even-skipped protein: definition of a minimal repression domain. *Genes & Dev.* 7:491–503.

HERSCHBACH, B. M., and A. D. JOHNSON. 1993. Transcriptional repression in eukaryotes. *Ann. Rev. Cell Biol.* 9:479–509.

JOHNSON, A. D., and I. HERSKOWITZ. 1985. A repressor (MAT α2 product) and its operator control expression of a set of cell-type specific genes in yeast. *Cell* 42:237–247.

LEVINE, M., and J. L. MANLEY. 1989. Transcriptional repression of eukaryotic promoters. *Cell* 59:405–408.

LICHT, J. D., et al. 1990. *Drosophila* Krüppel protein is a transcriptional repressor. *Nature* 346:76–79.

RENKAWITZ, R. 1990. Transcription repression in eukaryotes. *Trends Genet.* 6:192–197.

SAHA, S., et al. 1993. New eukaryotic transcriptional repressors. *Nature* 363:648–652.

Effect of Chromatin Structure and Methylation on Gene Activity

ADAMS, C. C., and J. L. WORKMAN. 1993. Nucleosome displacement in transcription. *Cell* 72:305–308.

ANTEQUERA F., and A. BIRD. 1993. CpG islands. In J. P. Jost and H. P. Saluz, eds., *DNA Methylation: Molecular Biology and Biological Significance*, Birkhauser Verlag, pp. 169–185.

BIRD, A. P. 1993. Genomic imprinting: imprinting on islands. *Curr. Biol.* 3:275–277.

CROSTON, G. E., and J. T. KADONAGA. 1993. Role of chromatin structure in the regulation of transcription by RNA polymerase II. *Curr. Opin. Cell Biol.* 5:417–423.

DOERFLER, W. 1993. Patterns of de novo DNA methylation and promoter inhibition: studies on the adenovirus and human genomes. In J. P. Jost and H. P. Saluz, eds., *DNA Methylation: Molecular Biology and Biological Significance*, Birkhauser Verlag, pp. 262–299.

FELSENFELD, G. 1992. Chromatin as an essential part of the transcriptional mechanism. *Nature* 355:219–224.

GRAESSMANN, M., and A. GRAESSMANN. 1993. DNA methylation, chromatin structure, and the regulation of gene expression. In J. P. Jost and H. P. Saluz, eds., *DNA Methylation: Molecular Biology and Biological Significance*, Birkhauser Verlag, pp. 404–424.

KORNBERG, R. D., and LORCH, Y. 1992. Chromatin structure and transcription. *Ann. Rev. Cell Biol.* 13:187–205.

LAURENSON, P., and J. RINE. 1992. Silencers, silencing, and heritable transcriptional states. *Microbiol. Rev.* 56:543–560.

LEE, D. Y., et al. 1993. A positive role for histone acetylation in transcription factor access to nucleosomal DNA. *Cell* 72:73–84.

LI, E., T. H. BESTOR, and R. JAENISCH. 1992. Targeted mutation of the DNA methyltransferase gene results in embryologic lethality. *Cell* 69:915–926.

RAZIN, A., and H. CEDAR. 1991. DNA methylation and gene expression. *Microbiol. Rev.* 55:451–458.

SINGER-SAM, J., and A. D. RIGGS. 1993. X chromosome inactivation and DNA methylation. In J. P. Jost and H. P. Saluz, eds., *DNA Methylation: Molecular Biology and Biological Significance*, Birkhauser Verlag, pp. 358–384.

SINGH, J., and A. J. S. KLAR. 1992. Active genes in budding yeast display enhanced in vivo accessibility to foreign DNA methylases: a novel in vivo probe for chromatin structure of yeast. *Genes & Dev.* 6:186–196.

STOGER, R., et al. 1993. Maternal-specific methylation of the imprinted mouse *Igf2r* locus identifies the expressed locus as carrying the imprinting signal. *Cell* 73:61–71.

TARTOFF, K. D., et al. 1989. Towards an understanding of position effect variegation. *Dev. Genet.* 10:162–176.

TAYLOR, I. C., et al. 1991. Facilitated binding of GAL4 and heat shock factor to nucleosomal templates: differential function of DNA binding domains. *Genes & Dev.* 5:1285–1298.

THOMPSON, J. S., A. HECHT, and M. GRUNSTEIN. 1993. Histones and the regulation of heterochromatin in yeast. *Cold Spring Harbor Symp. Quant. Biol.* 58:247–256.

ZLATNOVA, J. S., and K. E. VAN HOLDE. 1992. Chromatin loops and transcriptional regulation. *Crit. Rev. Euk. Gene Exp.* 2:211–224.

Other Transcription Systems

Bacteriophage RNA Polymerases

RASKIN, C. A., et al. 1992. Substitution of a single bacteriophage T3 residue in bacteriophage T7 RNA polymerase at position 748 results in a switch in promoter specificity. *J. Mol. Biol.* 228:506–515.

SOUSA, R., et al. 1993. Crystal structure of bacteriophage T7 RNA polymerase at 3.3 Å resolution. *Nature* 364:593–599.

Mitochondrial Transcription

CLAYON, D. A. 1991. Replication and transcription of vertebrate mitochondrial DNA. *Ann. Rev. Cell Biol.* 7:453–478.

CONSTANZA, M. C., and T. D. FOX. 1990. Control of mitochondrial gene expression in *Saccharomyces cerevisiae Ann. Rev. Genet.* 24:91–113.

JAEHNING, J. A. 1993. Mitochondrial transcription: is a pattern emerging? *Mol. Microbiol.* 8:1–4.

Chloroplast Transcription

HESS, W. R., et al. 1993. Chloroplast *rps*15 and the *rpo*B/C1/C2 gene cluster are strongly transcribed in ribosome-deficient plastids: evidence for a functioning non-chloroplast-encoded RNA polymerase. *EMBO J.* **12**:563–571.

HU, J., and L. BOGORAD. 1990. The 180- 120-, and 38-kilodalton polypeptides are encoded in chloroplast genes. *Proc. Nat'l. Acad. Sci. USA* **87**:1531–1535.

HU, J., R. F. TROXLER, and L. BOGORAD. 1991. Maize chloroplast RNA polymerase: the 78-kilodalton polypeptide is encoded by the plastid *rpo*C1 gene. *Nucl. Acids Res.* **19**:3431–3434.

KLEIN, U., J. D. DE CAMP, and L. BOGORAD. 1992. Two types of chloroplast gene promoters in *Chlamydomonas reinhardtii*. *Proc. Nat'l. Acad. Sci. USA* **89**:3453–3457.

Archaebacterial Transcription

OUZOUNIS, C., and C. SANDER. 1992. TFIIB, an evolutionary link between the transcription machineries of Archaebacteria and eukaryotes. *Cell* **71**:189–190.

ZILLIG, W., et al. 1992. RNA polymerases and transcription in archaebacteria. *Biochem. Soc. Symp.* **58**:79–88.

12

Transcription Termination, RNA Processing, and Posttranscriptional Control

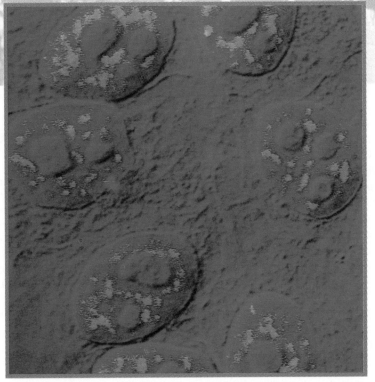

▲ Localization of an RNA splicing factor to discrete regions of interphase nuclei. Courtesy of David L. Spector.

In the previous chapter, we saw that regulation of most genes occurs at the first step in gene expression: initiation of transcription. However, once transcription of a gene has been initiated, synthesis of the encoded RNA requires that RNA polymerase transcribe the entire transcription unit and not terminate prematurely. The initial 5'-capped RNA transcripts produced from eukaryotic protein-coding genes are not functional and undergo processing during which the 3' end is cleaved and polyadenylated and introns are removed during splicing of exons. Finally, the mature, functional mRNA is transported from the nucleus to the cytoplasm. An additional type of processing, referred to as RNA editing, has recently been discovered. This process, which is distinct from splicing, involves modification of the sequence of an mRNA after its synthesis. As with protein-coding genes, the initial transcripts produced from tRNA genes and most rRNA genes also undergo processing to yield the corresponding functional RNAs.

The multiple steps in the synthesis of RNAs provide opportunities for additional levels of gene control beyond the regulation of transcription initiation. In the case of a protein-coding gene, the amount of protein expressed also can be regulated by controlling the stability of the corresponding mRNA and the frequency of translation initiation. In addition, the cellular locations of some mRNAs are regulated, so that newly synthesized protein is concentrated where it is needed. All of the regulatory mechanisms that control gene expression following transcription are referred to as *posttranscriptional control*. In this chapter, we consider the various steps in the synthesis of mRNA, tRNA, and rRNA following transcription initiation and describe examples of how regulation at these steps contributes to the control of gene expression.

➤ Transcription Termination in Prokaryotes

Because protein-coding regions are closely spaced in the genomes of microorganisms, independent control of neighboring genes is possible only if a transcription *termination site* lies between them. The absence of termination sites between the genes in a single operon (e.g., *E. coli trp* operon) allows the genes to be coordinately regulated. However, genes in the operon downstream are independently regulated because transcription by *E. coli* RNA polymerase terminates following the last gene of the *trp* operon. As we discuss in this section, *E. coli* possesses two principal mechanisms of transcription termination. One requires no protein components other than RNA polymerase itself; this is the only mechanism observed in in vitro systems with purified *E. coli* RNA polymerase. The other mechanism requires an additional protein, the transcription-termination factor called *Rho*.

Termination also can occur within a few hundred base pairs of a transcription-initiation site, preventing transcription of downstream genes. Generally, such early, or premature, termination is regulated. Two termination-control mechanisms—called *attenuation* and *antitermination*—have been identified in *E. coli*. One leads to premature chain termination; the other prevents it, thereby permitting expression of downstream genes. Thus, in addition to regulation of transcription initiation, termination control can provide secondary regulation of bacterial gene expression in some cases.

Rho-Independent Termination Sites Have Characteristic Sequences

DNA sequences where purified *E. coli* RNA polymerase terminates transcription without the aid of the Rho factor are called Rho-independent termination sites. These sites have two characteristic features: a series of T residues and preceding these a GC-rich self-complementary region with several intervening nucleotides. The resulting RNA transcript thus has a series of U residues at the 3' end preceded by a GC-rich self-complementary sequence, such as (5')CCCACTNNNNAGTGGG(3'). When such a self-complementary region of a growing RNA chain is synthesized, the complementary sequences base-pair with one another, forming a *stem-loop* structure (Figure 12-1).

GC-rich stem-loop RNA structures interact with the surface of *E. coli* RNA polymerase, causing it to pause during the process of transcription. The rU-dA base pairs between the 3' end of the nascent RNA chain and the template strand of the DNA are extremely unstable compared to other types of Watson-Crick base pairs. This short $(rU)_n$-$(dA)_n$ hybrid region is thought to melt during the time that the polymerase pauses, releasing the RNA chain from the

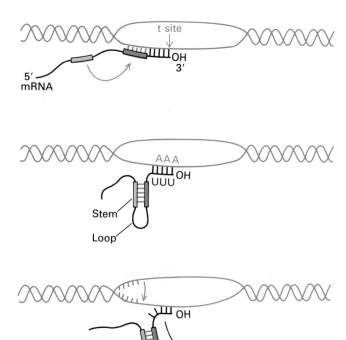

▲ FIGURE 12-1   Rho-independent termination of transcription in *E. coli*. The sequence of mRNA synthesized near a termination site (t) contains a string of U residues preceded by a GC-rich region with dyad symmetry (red boxes). The bases of the symmetrical region of the mRNA form a stem-loop by base pairing. This structure interacts with RNA polymerase and causes it to pause during elongation. This pausing coupled with the weak base pairing of rU-dA base pairs at the termination site displaces the mRNA chain and signals the polymerase to release from the template. [See T. Platt, 1981, *Cell* **24**:10.]

transcription complex. Mutations that weaken the dyad symmetry or that result in fewer consecutive U residues decrease termination at these sites.

An example of such a Rho-independent termination site occurs at the end of the *trp* operon. Synthesis of *trp* mRNA ceases at a site 36 bases beyond the stop codon of the *trpA* gene. At this site there are four consecutive U residues preceded by a 22-nucleotide sequence high in G and C that can fold back on itself and base-pair to form a stem-loop structure (Figure 12-2).

Attenuation Can Cause Premature Chain Termination

We have seen how binding of tryptophan to the *trp* repressor causes the protein to undergo a conformational change so that it binds tightly to the *trp* operator, inhibiting transcription initiation (see Figure 11-17). This inhibition,

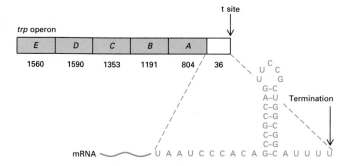

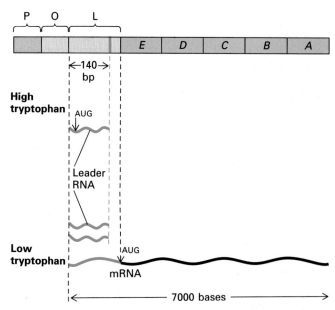

▲ FIGURE 12-2 Sequence of *trp* termination (t) site, a Rho-independent site. The *trp* operon, composed of five genes (blue), is followed by a 36-bp sequence at the end of which termination occurs. The 3' end of the corresponding mRNA has a GC-rich stem-loop structure preceding the final four U residues, a characteristic feature of mRNAs produced from genes with Rho-independent termination sites. [See T. Platt, 1981, *Cell* **24**:10.]

▲ FIGURE 12-3 Attenuation provides a secondary mechanism for controlling expression of the *trp* operon. The leader sequence (L), which lies between the operator (O) and the first structural gene (*E*), contains an attenuator site (red band) at which transcription is terminated depending on the concentration of tryptophan in the medium. The promoter-operator-leader region, which is ≈200 bp long, is not drawn to scale with the structural genes (see Figure 12-2). AUG indicates the start codons for translation of the leader sequence (red) and *trpE* mRNA (black). [See T. Platt, 1978, in J. H. Miller and W. S. Reznikoff, eds., *The Operon*, Cold Spring Harbor Laboratory.]

however, is not complete, and a second regulatory mechanism, called *attenuation,* is used to further inhibit expression of the *trp* operon when tryptophan is plentiful.

Attenuation was discovered during analysis of *E. coli* mutants that manufactured greater-than-normal amounts of tryptophan-synthesizing enzymes. These mutants make a normal *trp* repressor and have a normal *trp* operator and promoter; however, their repressed level of *trp* enzymes is higher than normal, and they respond to the absence of tryptophan with an even higher rate of transcription. Comparison of the tryptophan genes from this group of mutants with those from wild-type cells showed that short stretches of DNA containing fewer than 50 base pairs were deleted in the mutants. The deletions occurred just after the promoter-operator region in the normal *trp* operon and before the coding sequences for the *trp* enzymes.

How could the deletion of a short region of DNA downstream from the initiation site for *trp* mRNA cause increased synthesis of this mRNA? In the hope of answering this question, experimenters sequenced the beginning of the *trp* mRNA molecule and the DNA region near the *trp* promoter in both normal and mutant *E. coli* strains. A molecule of *trp* mRNA contains 162 nucleotides upstream (i.e., toward the 5' end of the molecule) from the AUG codon that constitutes the translation initiation site for the first of the five *trp*-encoded enzymes (Figure 12-3). This stretch of mRNA is called the *leader sequence* (L). In normal *E. coli* cells, only small amounts of *trp* mRNA sequences are made when tryptophan is abundant because *trp* repressor inhibits transcription initiation. The small number of *trp* transcripts that are synthesized consist of leader RNA only; the remainder of the ≈7-kb *trp* operon is not transcribed. When tryptophan is scarce, however, the entire *trp* operon is transcribed, including the leader sequence and all the coding sequences for the *trp*-encoded enzymes.

Even when the entire *trp* mRNA sequence is formed in large amounts, more leader sequences than whole mRNA molecules are made. These observations revealed that the leader sequence contains an *attenuator,* a site where a choice is made between elongation of the growing *trp* transcript by RNA polymerase and termination. When tryptophan is scarce, about 25–50 percent of the RNA polymerase molecules continue transcribing past the attenuator. When tryptophan is abundant, little initiation of transcription takes place, but virtually all of the transcripts that are initiated terminate at the attenuator. Mutant cells from which the attenuator region has been deleted produce more *trp* mRNA under all conditions than do normal cells. Thus the *trp* operon is controlled not only by the repressor-operator mechanism discussed in Chapter 11 but also by termination of RNA synthesis at a site 140 nucleotides from the start site. The number of RNA polymerase molecules that pass the attenuator depends on the concentration of tryptophan in the medium. Therefore, regulation of the *trp* operon is not an all-or-none affair; rather, secondary control by attenuation (i.e., termination at the attenuator site) permits a cell to finely balance the amount of the tryptophan-synthesizing enzymes formed with the cell's need for tryptophan.

Attenuation depends on the formation of a particular stem-loop structure in the mRNA leader sequence. Formation of this structure depends in turn on the rate of ribosomal translation of the leader sequence, which is engaged by a ribosome soon after it is synthesized. Whether the ribosome efficiently translates the leader sequence depends on the supply of charged tRNAs for the amino acids encoded by the sequence—particularly tryptophan.

The mechanism controlling attenuation depends on the ability of the same mRNA leader sequence to form alternative secondary structures, one of which causes termination. Four regions of leader sequence critical to the mechanism are diagrammed in Figure 12-4a,b. The sequence of region 2 can base-pair with either region 1 or with region 3. Similarly, region 3 can base-pair with either region 4 or region 2. As RNA polymerase transcribes the leader DNA, newly synthesized region 2 base-pairs immediately with region 1 in the nascent RNA. The resulting stem-loop secondary structure interacts with RNA polymerase causing it to pause during elongation, but termination does not occur because the stem-loop is not followed by a series of U residues; thus after pausing, the polymerase continues to elongate the leader transcript. The pause provides sufficient time to assure that a ribosome will initiate translation at a start codon near the 5′ end of the leader.

Region 1 of the leader contains two successive UGG codons, for tryptophan. When tryptophan is present in sufficient quantity, the ribosome translates the leader transcript rapidly, melting the base pairing between regions 1 and 2. When regions 3 and 4 of the RNA are synthesized, they pair forming a stem-loop followed by a series of U residues, so that transcription is terminated by the Rho-independent mechanism (Figure 12-4c, *left*). On the other hand, when the supply of tryptophan is low, the ribosome pauses at each tryptophan codon in region 1 of the leader. Since the ribosome is paused over region 1, region 1 cannot base-pair with region 2, and region 2 is free to base-pair with region 3 as soon as it is synthesized (Figure 12-4c, *right*). The polymerase does not terminate after formation of the 2-3 stem-loop because it is not followed closely by a string of U residues. In this situation, region 3 is sequestered in the 2-3 hybrid region and thus cannot base-pair with region 4 when it is synthesized. Consequently, the terminating RNA structure does not form and RNA polymerase continues transcription. This mechanism maximizes attenuation when tryptophan is present in abundance and minimizes attenuation when tryptophan is scarce. The small leader peptide translated from the leader sequence is rapidly degraded.

A similar attenuation mechanism occurs in several operons encoding enzymes that synthesize amino acids.

➤ **FIGURE 12-4**   Mechanism of attenuation of *trp*-operon transcription. (a) Diagram of the 140-nucleotide *trp* leader RNA. Colored regions are critical to the control of attenuation. (b) Different secondary structures that occur in the leader sequence of *trp* mRNA as the result of alternative base pairing. Colored regions correspond to those shown in part (a). The 5′ region is not shown. (c) Translation of the *trp* leader sequence begins from the 5′ end soon after it is synthesized, while synthesis of the rest of the *trp* mRNA molecule continues. At high tryptophan concentrations, formation of the 3-4 stem-loop followed by a series of 3′ U residues causes termination by the Rho-independent mechanism. At low tryptophan concentrations, region 3 is sequestered in the 2-3 stem-loop and cannot base-pair with region 4. In the absence of the stem-loop structure required for termination, transcription of the *trp* coding sequences continues. See text for discussion. [Part (b) adapted from C. Yanofsky, 1981, *Nature* **289**:751; part (c) adapted from T. Platt, 1981, *Cell* **24**:10.]

For example, the mRNAs encoding the enzymes for synthesis of phenylalanine and histidine contain leaders with 7 of 14 and 7 of 16 codons for phenylalanine and histidine, respectively, preceding a protein stop codon in the first 50 bases of their sequence. In the *ilv-GMEDA* operon, which encodes enzymes required for the synthesis of isoleucine, leucine, and valine, the leader sequence contains codons for all three amino acids and a deficiency in any one of them reduces attenuation.

Attenuation also occurs in operons that do not encode amino acid biosynthetic enzymes. In the *E. coli bgl* operon, which codes for enzymes that metabolize glucose-containing polysaccharides, attenuation is controlled by a sequence-specific RNA-binding protein that binds to and stabilizes an RNA secondary structure, thereby precluding formation of a terminating stem-loop structure. The RNA-binding activity of this protein is controlled by its phosphorylation in response to glucose-containing sugars in the environment. When glucose is present, it binds to and activates an enzyme the removes an inhibitory phosphate from this RNA-binding protein. The active protein then binds to a specific sequence near the 5′ end of nascent *bgl* transcripts and prevents attenuation. In this way, a sequence-specific RNA-binding protein can control transcription in *E. coli* by controlling attenuation, just as sequence-specific DNA-binding proteins control transcription initiation by acting as repressors or activators.

Transcription Termination at Some Sites Requires Rho Factor

About half the termination sites in *E. coli* require the Rho factor. Rho-dependent termination sites were first recognized during in vitro studies of transcription of λ-phage

(a) *trp* leader RNA

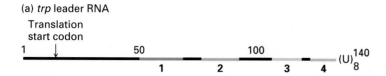

(b) Alternative secondary structures

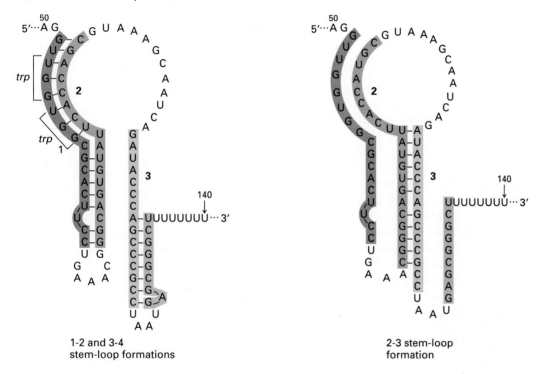

1-2 and 3-4
stem-loop formations

2-3 stem-loop
formation

(c) Translation of *trp* leader

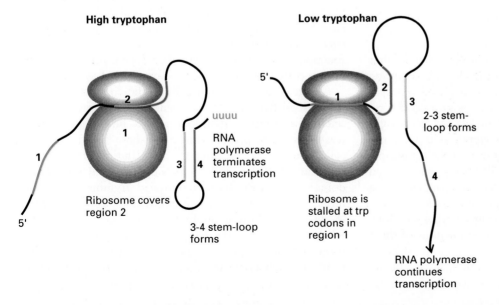

High tryptophan

Ribosome covers
region 2

RNA
polymerase
terminates
transcription

3-4 stem-loop
forms

Low tryptophan

2-3 stem-
loop forms

Ribosome is
stalled at trp
codons in
region 1

RNA polymerase
continues
transcription

DNA by purified *E. coli* RNA polymerase. In this in vitro system, transcription of the λ DNA begins at two promoter sites, P_L and P_R; these same promoters function in vivo immediately after a virion infects an *E. coli* cell. Transcription of λ DNA by *E. coli* RNA polymerase proceeds in two directions: left from P_L and right from P_R. The λ RNA transcripts formed in vitro with pure RNA polymerase are several thousand nucleotides long, much longer than those present inside cells during the early stages of λ infection. However, when extracts of uninfected cells are added to the transcription reaction with pure RNA polymerase, two discrete products are formed identical to the RNA transcripts found inside infected cells immediately after infection—one containing ≈500 nucleotides and one containing ≈1000 nucleotides. These transcripts extend from P_L and P_R to specific termination sites. In other words, in the absence of some factor in the uninfected cell extract, transcription of λ DNA in vitro does not terminate at the same sites as it does in vivo. The protein in the uninfected cell extract responsible for termination at these sites in λ DNA was purified and is now called Rho.

Many Rho-dependent chain-terminating sites have been identified in λ and *E. coli* DNA. Comparisons of the sequences of these sites have revealed no obvious similarities. Although it is not known precisely how the Rho factor causes termination, it is known that this ≈50 kDa protein forms a hexamer and that a 70- to 80-base segment of the growing RNA transcript wraps around the outside of the hexamer. This activates an ATPase activity of Rho that is associated with its movement along the RNA in the 3′ direction until it eventually unwinds the RNA-DNA hybrid at the active site of RNA polymerase. Whether transcription is terminated or not seems to depend on whether Rho moves sufficiently fast to "catch up" with the polymerase. Consequently, polymerase pausing during elongation is thought to be an important component of Rho-dependent termination.

Antitermination Can Prevent Premature Chain Termination

The timing of λ-phage gene expression immediately after infection is regulated by proteins that control transcription termination. As mentioned previously, in the early stages of λ-phage infection, transcription begins in a central region of the λ DNA and proceeds to the left and to the right of two promoters, P_L and P_R. Transcription initially terminates at t_L and t_R, yielding relatively short RNA transcripts corresponding to about 1000 and 500 bases from P_L and P_R, respectively (Figure 12-5a). The P_L transcript encodes a viral protein called N, which prevents termination of transcription at t_L and t_R and thus permits synthesis of long RNA transcripts corresponding to other portions of the λ genome (Figure 12-5b). Studies with λ mutants demonstrated that for N to prevent termination, a specific se-

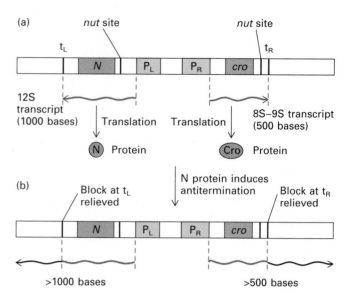

▲ **FIGURE 12-5** Transcription from P_L and P_R during the early stages of λ-phage infection. (a) Immediately after infection, two relatively short transcripts are synthesized from P_L and P_R; transcription terminates at the Rho-dependent terminators t_L and t_R. These transcripts encode Cro, a DNA-binding protein that induces the lytic pathway, and N protein, which acts with other proteins to prevent termination at t_L and t_R. (b) Later in infection, when N accumulates, antitermination complexes assemble at the *nut* sites in the RNA, and much longer transcripts are produced.

quence called *nut* (for N-utilization) must occur between the promoter and the termination site.

The isolation of *E. coli* mutants that do not support antitermination by N, and are consequently resistant to infection by λ, led to the identification of cellular proteins required for N function called NusA (for N utilization substance A), NusB, NusE, and NusG. Of course, these proteins have other functions in uninfected cells. NusA is an *elongation factor*, and NusE is one of the proteins of the small ribosomal subunit (S10). These cellular proteins function together to prevent Rho-dependent termination during transcription of the ribosomal RNA genes.

Protein binding studies have shown that the RNA transcribed from the *nut* sequence contains two protein-binding regions: a stem-loop called box A, which interacts with N protein, and a 12-base linear sequence called box B, which interacts with the cellular proteins NusB and S10 (Figure 12-6a). N contains a RNA-binding domain rich in the basic amino acid arginine that recognizes box A in *nut* RNA. This interaction has been assessed in an electrophoretic mobility shift assay with labeled RNA, analogous to the assay used to study the binding of proteins to specific DNA sequences (see Figure 11-18). The results of such assays indicate that the 5-bp stem in box A

(a) Sequence of *nut* site

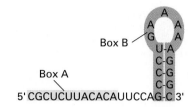

(b) Formation of antitermination complex

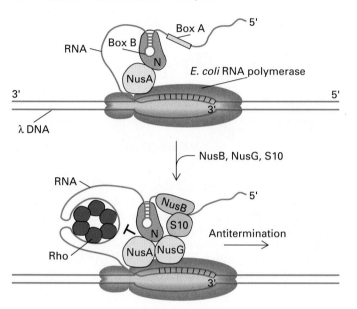

▲ **FIGURE 12-6** Antitermination by λ-phage N protein and cellular proteins. (a) The *nut* site in λ-phage RNA transcript contains a stem-loop (box A) that binds N protein and a linear region (box B) that interacts with the cellular proteins NusB and S10. (b) After N protein interacts with the NusA-polymerase complex, rapid binding of NusB, NusG, and S10 produces a stable antitermination complex, which can transcribe many kilobases despite the occurrence of termination signals. Inhibition of the terminating action of hexameric Rho factor is diagrammed. See text for discussion. [Adapted from J. Greenblatt, J. R. Nodwell, and S. W. Mason, 1993, *Nature* **364**:401.]

can be any sequence so long as it is fully base-paired, but the loop has to be (5′)GAAAA(3′) for N to bind to the RNA with high affinity. Interactions among the proteins involved in N-mediated antitermination have been studied by linking one protein to a solid support such as cellulose and determining whether another protein will bind to the resulting matrix.

Based on these binding studies and mutant analyses, the model of N-mediated antitermination shown in Figure 12-6b has been proposed. N protein binds to the *nut* box A in the nascent transcript and then interacts with NusA complexed with RNA polymerase. Although the interaction between N and NusA is quite weak, the strong binding

of N to the *nut* RNA sequence increases its localized concentration sufficiently for N and NusA to interact at the N protein concentration found in cells. Once N and NusA interact, NusB, NusG, and S10 rapidly bind, yielding an antitermination complex that is stabilized by multiple protein-protein contacts. As the polymerase moves along the template DNA away from the *nut* site, the antitermination complex remains associated with the enzyme and the *nut* RNA sequence, so a RNA loop of increasing size forms. By an unknown mechanism, the antitermination complex can block termination at both Rho-dependent and Rho-independent termination sites.

Thus, antitermination at t_L and t_R in the λ-phage genome results from a complex of proteins, some of which bind to a specific sequence in the nascent RNA and some of which bind to RNA polymerase. Why does bacteriophage λ employ this elaborate mechanism to prevent termination at sites that could simply be deleted? There are several reasons that all concern the ability of λ to undergo either a lytic or a lysogenic mode of infection (see Figure 6-19). In the molecular mechanism that determines which mode of infection an infecting λ genome will undergo, it is important for Cro protein to be expressed before proteins encoded to the right of t_R. Once lysogeny is established, it is critical that the genes to the right of t_R and the left of t_L remain off, otherwise the lytic cycle would be induced at an inappropriate time. Transcription from P_R and P_L is inhibited by λ repressor, which is continuously expressed in lysogenized cells (see Chapter 13). Repression, however, is not absolute. The low level of transcription from these promoters that does occur in a lysogenized cell terminates at t_R and t_L because insufficient N protein is produced to antiterminate. Once again, we see that control of termination early in a transcription unit provides another level of regulation in addition to the regulation of transcription initiation.

▶ *Eukaryotic Transcription-Termination Control*

Much less is understood about the mechanism(s) of transcription termination in eukaryotes. Purified RNA polymerase III terminates after polymerizing a series of U residues, without a requirement for an upstream stem-loop secondary structure. Termination by RNA polymerase I requires a polymerase-specific termination factor that binds a specific DNA sequence at the end of the rRNA transcription unit. Termination by RNA polymerase II is poorly understood. In vitro nuclear run-on experiments (see Figure 11-24) indicate that RNA polymerase II terminates at multiple possible sites over a distance of 0.5–2 kb beyond the poly-A addition site in most mammalian transcription units. It is not yet clear how RNA polymerase II is prevented from terminating before the poly-A site, but is allowed to terminate following the poly-A site. This is currently an intense area of investigation because recent studies have revealed that regulation of transcription termination controls the genes of human immunodeficiency virus (HIV), the causative agent of AIDS, in infected eukaryotic cells. Regulation of termination also controls expression of *c-myc*, a cellular gene involved in cell-cycle control that becomes deregulated in Burkitt's lymphoma (Chapter 26).

HIV Tat Protein Is an RNA-Binding Antitermination Protein

Currently, HIV provides the best-understood example of regulated transcription termination in eukaryotes. Analysis of HIV mutants has revealed that efficient expression of HIV genes requires a small viral protein encoded at the *tat* locus. Nuclear run-on transcription assays have demonstrated that cells infected with *tat⁻* mutants produce viral transcripts that hybridize to promoter-proximal HIV DNA restriction fragments but not to restriction fragments farther downstream from the promoter. In contrast, cells infected with wild-type HIV synthesize long viral transcripts that hybridize to restriction fragments throughout the single HIV transcription unit. Thus Tat protein causes RNA polymerase II to read through a transcriptional block, much the same as the λ-phage N protein.

Like N, Tat is a sequence-specific RNA-binding protein. It binds to a sequence called TAR, which is located near the 5' end of the HIV transcript (Figure 12-7a). In vitro studies with purified Tat have shown that binding of Tat protein to TAR depends on the overall secondary structure of the RNA and an unpaired UCU bulge; this is analogous to the structural requirement for binding of λ-phage N protein to *nut* RNA (see Figure 12-6a). Moreover, the RNA-binding domain of Tat contains an arginine-rich region whose amino acid composition is similar to the arginine-rich region of N that interacts with *nut* RNA. A 4-base loop region of TAR also has been shown to interact with certain cellular proteins, much like the interaction of *E. coli* NusB and S10 with box B in λ-phage *nut* RNA. These similarities between Tat and N protein raise the possibility that Tat may interact with cellular proteins and RNA polymerase II by looping of the growing RNA transcript, thereby preventing premature termination of transcription of the HIV genome (Figure 12-7b).

Premature Termination of c-*myc* Transcription Occurs in Nondividing Cells

Control of transcription termination also is important in regulating transcription of *c-myc*. This gene encodes one subunit of a heterodimeric transcription factor that functions in the control of cellular proliferation. Sometimes *c-myc* is translocated to a position in the genome where it is close to the immunoglobulin-gene enhancer (see Chapter 26). This translocation leads to constitutive, unregu-

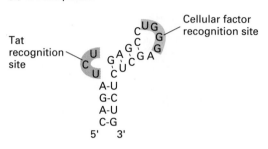

(a) TAR sequence

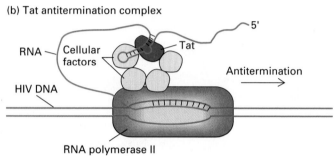

(b) Tat antitermination complex

▲ FIGURE 12-7   Antitermination by HIV Tat protein. (a) Sequence of TAR element in HIV transcript showing secondary structure required for antitermination and sequences recognized by Tat and cellular factors. (b) Proposed model of Tat antitermination complex. See text for discussion. [Adapted from J. Greenblatt, J. R. Nodwell, and S. W. Mason, 1993, *Nature* **364**:401.]

lated expression of the *c-myc* gene, which is thought to contribute to the transformation of normal cells into cancer cells. For this reason, *c-myc* is classified as a *proto-oncogene*.

In normal nondividing cells, *c-myc* is not expressed; but when cellular proliferation is stimulated by growth factors, *c-myc* is rapidly induced. Nuclear run-on experiments have shown that *c-myc* transcripts in nondividing cells hybridize only to promoter-proximal DNA restriction fragments from the *c-myc* transcription unit. After addition of appropriate growth factors, transcripts hybridize with restriction fragments from throughout the *c-myc* transcription unit. These results, similar to those with the *tat⁻* and wild-type HIV transcription unit, indicate that expression of *c-myc* also is regulated by controlling transcription termination. Currently little is known about how this regulation of *c-myc* is achieved.

RNA Polymerase II Pauses during Transcription of *Drosophila* Heat-Shock Genes under Normal Conditions

A different regulatory mechanism controls elongation during transcription of *heat-shock* genes in *Drosophila*. Both prokaryotic and eukaryotic cells respond to elevated temperatures and other agents that denature proteins by mo-

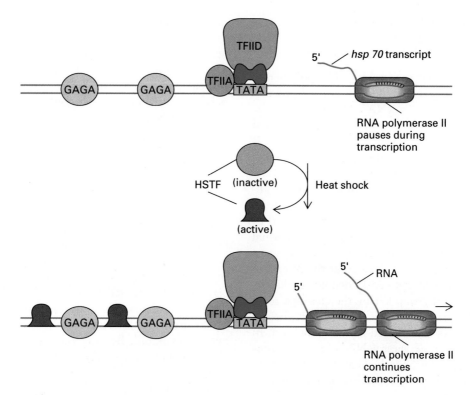

▲ FIGURE 12-8   Control of chain elongation during transcription of *Drosophila* *hsp70* gene by RNA polymerase II. In the absence of heat shock (*top*), RNA polymerase II pauses after synthesizing ≈25 nucleotides of the *hsp70* transcript. Transcription requires GAGA, a transcription factor that binds to promoter-proximal elements, and general transcription factors that bind in the TATA-box region. Although TFIID and TFIIA are shown here, the precise nature of these bound proteins is unclear. After heat shock (*bottom*), the heat-shock transcription factor (HSTF) is converted from an inactive to an active DNA-binding form. Binding of activated HSTF to specific promoter-proximal elements releases the paused polymerase by an unknown mechanism, leading to rapid transcription of the rest of the *hsp70* gene. Bound HSTF also stimulates rapid transcription initiation by additional RNA polymerase II molecules. [See H-s. Lee et al., 1992, *Genes & Dev.* **6:**284.]

mentarily suspending translation of most mRNAs and transcription of most genes and turning to the transcription of a limited set of genes encoding proteins that increase cellular survival. Little is actually known of the function of most of these heat-shock proteins. However, the amino acid sequence conservation among species for some of these proteins is the highest for any known proteins. For example, Hsp70 (70-kDa heat-shock protein) from *E. coli* is 50 percent identical with the human protein, and the *Drosophila* protein is 85 percent identical with that from humans. Hsp70 binds to hydrophobic amino acids in unfolded proteins and prevents their precipitation in the cell. Binding to unfolded proteins activates an intrinsic ATPase activity in Hsp70, and the energy released is thought to be used in some way to refold the denatured protein.

Transcription of *hsp70* and other heat-shock genes in *Drosophila* is induced from a low basal level to extremely high levels within seconds of heat shock. Part of the reason for this extremely rapid induction is that in the absence of heat shock, RNA polymerase II initiates transcription of *hsp70* but in most cases transcribes only ≈25 nucleotides and then pauses (Figure 12-8, *top*). Following heat shock, the heat-shock transcription factor (HSTF) is rapidly activated and binds to specific heat-shock response elements upstream of the *hsp70* promoter. HSTF binding appears to release the paused polymerase, allowing rapid transcription of the gene as well as frequent reinitiation of transcription by new RNA polymerase II molecules (Figure 12-8b). At present the mechanism that causes RNA polymerase to pause after initiating transcription at the heat-shock promoter is not well understood, although a ubiquitous *Drosophila* transcription factor, called the GAGA factor (after the sequence of its binding site) has been implicated in the mechanism. This type of regulated elongation is well adapted to extremely rapid induction and may be used in controlling expression of other eukaryotic genes in addition to heat-shock genes.

➤ *mRNA Processing in Higher Eukaryotes*

The initial RNA product synthesized by RNA polymerase II, called a *primary transcript*, undergoes several processing steps before a functional mRNA is produced (Figure 12-9). RNA polymerase initiates transcription at the first nucleotide of the first exon of a gene (see Figure 11-32). It is now known that shortly after transcription begins, the 5′ end of the nascent RNA is capped with 7-methyl-guanylate (see Figure 4-24). As noted in the previous section, transcription by RNA polymerase II terminates at any one of multiple sites ≈0.5–2 kb downstream from the 3′ end of the last exon in the transcript. The 3′ end of a functional mRNA then is generated by endonucleolytic cleavage at a specific sequence, the *poly-A site*, located at the 3′ end of the final exon. A stretch of 100–250 adenine residues is added to the 3′-hydroxyl group left by the cleavage reaction. Finally, introns are removed by RNA splicing before the completed mRNA is transported to the cytoplasm. In this section, we discuss how these RNA processing steps occur and give examples of how they can be regulated to control expression of proteins encoded by complex eukaryotic transcription units.

mRNA Precursors Are Associated with Abundant Nuclear Proteins Containing Conserved RNA-Binding Domains

Nascent RNA and mRNA intermediates in the nuclei of eukaryotic cells do not exist as free RNA molecules. From the time nascent transcripts first emerge from RNA polymerase II until they are transported into the cytoplasm, they are associated with an abundant set of nuclear proteins, as numerous in growing eukaryotic cells as histones. These proteins were first characterized as being the major protein components of heterogeneous ribonucleoprotein particles (hnRNPs), which contain *heterogeneous nuclear RNA* (hnRNA), a collective term referring to mRNA precursors (pre-mRNA) and other nuclear RNAs of various sizes. The proteins in these ribonucleoprotein particles can be dramatically visualized with fluorescent-labeled monoclonal antibodies (Figure 12-10).

High-dose UV irradiation causes covalent cross-links to form between RNA bases and closely associated proteins. This technique has been used to identify a large number of hnRNP proteins. In this work, cells were treated with UV radiation and their nuclei were isolated. Extracts of these nuclei then were chromatographed on an oligo-dT cellulose column (see Figure 7-15) under conditions that denature proteins. The only proteins recovered from this procedure were those covalently linked to RNAs containing a poly-A tail. Because these proteins must have been closely associated with the nuclear RNA in living cells, they could be identified as hnRNP proteins. The major hnRNP proteins identified by this cross-linking technique were used to prepare monoclonal antibodies. Treatment of cell extracts from unirradiated cells with these antibodies identified a complex set of abundant hnRNP proteins in human cells, designated A1–U and ranging in size from 34 to 120 kDa (Figure 12-11). Characterization of the mRNAs encoding these proteins has shown that some of them (e.g., A2 and B1) are related proteins derived by alternative splicing of exons from the same transcription unit.

The hnRNP proteins bind to RNA with extremely high

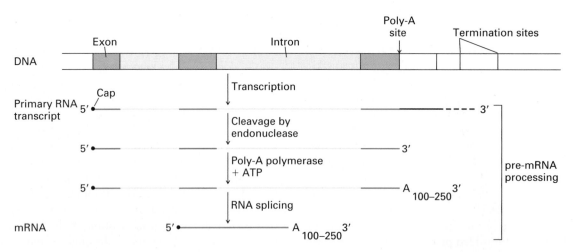

▲ **FIGURE 12-9**   Steps in pre-mRNA processing in eukaryotes. This processing occurs in the nucleus, yielding a functional, mature mRNA, which is transported to the cytoplasm. Transcription by RNA polymerase II produces a primary RNA transcript extending from the 5′ capped nucleotide, which is retained in the mature RNA, to one of several alternative 3′ nucleotides located ≈0.5–2 kb downstream from the poly-A site. The primary transcript is cleaved at the poly-A site, a string of A residues is added, and the introns are spliced out. The poly-A tail contains ≈250 A residues in mammals, ≈150 in insects, and ≈100 in yeasts.

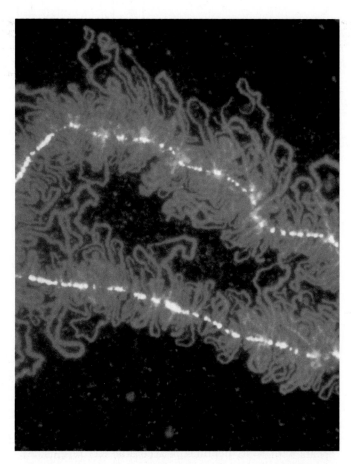

▲ FIGURE 12-10   Visualization of hnRNP protein associated with nascent transcripts in an oocyte of the newt *Nophthalmus viridescens*. A portion of a "lampbrush" chromosome is shown. The chromosome axis fluoresces white after staining with the DNA-specific dye DAPI. The hnRNP protein fluoresces red after staining with a monoclonal antibody against a specific hnRNP protein. [Courtesy of M. Roth and J. Gall.]

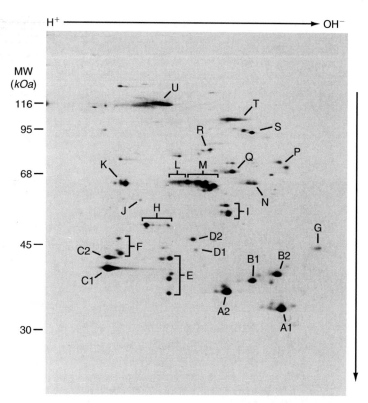

▲ FIGURE 12-11   Separation of hnRNP proteins from human Hela cells. The hnRNPs were isolated from the nucleoplasm prepared from [35S]methionine-labeled HeLa cells by immunoprecipitation with a monoclonal antibody that binds to the C proteins. Proteins were separated by nonequilibrium pH gradient gel electrophoresis in the first dimension (left to right) followed by SDS-polyacrylamide gel electrophoresis in the second dimension (top to bottom). An autoradiogram of the gel is shown with molecular weights indicated at the left. [From G. Dreyfuss et al., 1993, *Ann. Rev. Biochem.* **62**:289.]

affinity but differing specificity. In studies with homopolymeric RNAs, different groups of hnRNP proteins were found to bind to poly-A, poly-C, poly-G, and poly-U, suggesting that different proteins might associate with different regions of an mRNA precursor depending on the base composition and sequence of the region. For example, the hnRNP proteins A1, C, and D bind preferentially to the pyrimidine-rich sequences at the 3′ ends of introns, discussed in a later section. Like transcription factors, most hnRNP proteins have a modular structure. They contain one or more RNA-binding domains and at least one other domain that is thought to interact with other proteins. Several different types of RNA-binding domains, called *RNA-binding motifs,* have been identified by constructing deletions of hnRNP proteins and testing their ability to bind RNA.

RNP Motif   The most common RNA-binding motif in hnRNP proteins, called the RNP motif, is also found in many other RNA-binding proteins. This motif contains about 90 amino acids and includes one highly conserved 8-aa sequence (RNP1) towards the C-terminal end and a region of hydrophobic amino acids (RNP2) towards the N-terminal end (Figure 12-12a). The structure of the RNP motif present in the hnRNP C protein was shown by NMR techniques to consist of four β strands and two α helices (Figure 12-13). RNA is thought to associate with the surface of the protein formed by the β strands and N- and C-termini.

RGG Box   Deletion analysis of hnRNP U protein identified a 26-residue sequence containing five RGG (Arg-Gly-Gly) repeats with several interspersed aromatic amino acids (Figure 12-12b). Several other RNA-binding proteins contain a similar sequence. The consensus sequence of this motif is called the RGG box, although the structure of this domain has not yet been determined. These arginine-rich RNA-binding domains are similar to the RNA-binding domains of the λ-phage N and HIV Tat proteins.

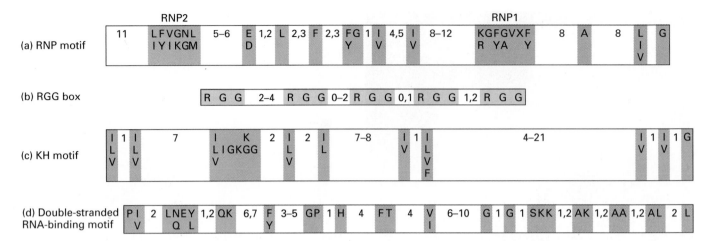

	RNP2											RNP1				

(a) RNP motif

| 11 | L F V G N L / I Y I K G M | 5–6 | E D | 1,2 | L | 2,3 | F | 2,3 | F G Y | 1 | I V | 4,5 | I V | 8–12 | K G F G V X F / R YA Y | 8 | A | 8 | L I V | G |

(b) RGG box

| R G G | 2–4 | R G G | 0–2 | R G G | 0,1 | R G G | 1,2 | R G G |

(c) KH motif

| I L V | 1 | I L V | 7 | | I K / L I G K G G V | 2 | I L V | 2 | I L | 7–8 | | I V | 1 | I L V F | 4–21 | | I V | 1 | I V | 1 | G |

(d) Double-stranded RNA-binding motif

| P I V | 2 | L N E Y Q L | 1,2 | Q K | 6,7 | F Y | 3–5 | G P | 1 | H | 4 | F T | 4 | V I | 6–10 | G | 1 | G | 1 | S K K | 1,2 | A K | 1,2 | A A | 1,2 | A L | 2 | L |

▲ FIGURE 12-12   Consensus sequences of four RNA-binding motifs found in hnRNP proteins and some other RNA-binding proteins. Conserved amino acid residues, represented by the one-letter code, are highlighted in color; vertically aligned letters indicate alternative amino acids at a given sequence position that are present in different proteins. Small numbers indicate the number of nonconserved residues separating the conserved sequences. The N-terminal end of each motif is to the left; the C-terminal end, to the right. [Adapted from C. G. Burd and G. Dreyfuss, 1994, *Science* **265**:615.]

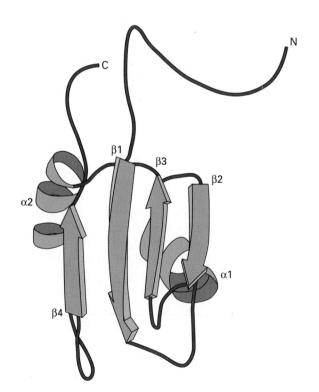

▲ FIGURE 12-13   Ribbon diagram of three-dimensional structure of the RNP motif in hnRNP C protein. RNA is thought to interact with the surface facing out from the page. The two α helices are in blue; the four β strands in green. [From G. Dreyfuss et al., 1993, *Ann. Rev. Biochem.* **62**:289.]

Other RNA-Binding Motifs

Other RNA-Binding Motifs   The hnRNP K protein contains an RNA-binding domain called the KH motif, for K homology (Figure 12-12c). To date, this motif has been identified in a small number of RNA-binding proteins, including the protein encoded by the fragile-X gene (*FMR1*). Mutations in *FMR1* are associated with the most common form of heritable mental retardation. The KH motif is about 45 residues long; it is generally found in two or more repeats with RGG peptides between the KH repeats. Another RNA-binding motif has been observed in a number of proteins that bind double-stranded RNA but not double-stranded DNA (Figure 12-12d). Although some RNA-binding proteins contain domains with the zinc-finger motif (see Figure 11-48b), zinc-finger structures have not yet been described in an hnRNP protein.

HnRNP Proteins May Have Multiple Functions

The association of mRNA precursors with hnRNP proteins may prevent formation of short secondary structures dependent on base-pairing of complementary regions, thereby making the pre-RNA accessible for interaction with other macromolecules. This function would be analogous to that of single-strand–binding (Ssb) protein at a DNA replication fork (Chapter 10). Support for this idea comes from the finding that hnRNP proteins accelerate hybridization between complementary RNAs (Figure 12-14), just as Ssb protein accelerates hybridization between complementary DNA strands. However, the diversity of hnRNP proteins suggests that they have other functions as well. For example, various hnRNP proteins may interact with different portions of a pre-mRNA and contribute to the structure recognized by RNA-processing factors.

Cell-fusion experiments have shown that some hnRNP proteins cycle in and out of the cytoplasm, whereas others remain localized in the nucleus (Figure 12-15). This finding

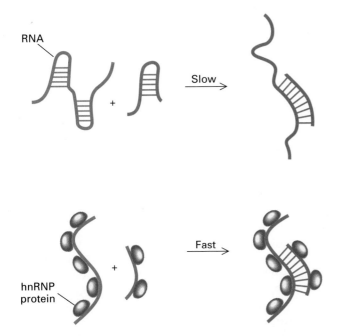

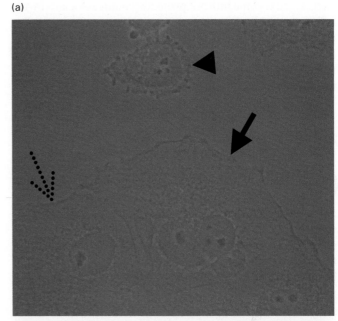

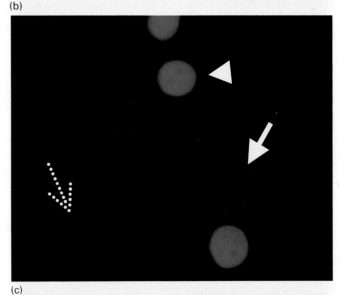

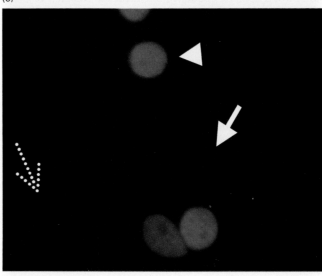

▲ FIGURE 12-14 Hybridization of RNA molecules in vitro is stimulated by hnRNP proteins. This finding suggests that hnRNP proteins prevent formation of RNA secondary structures, thereby facilitating base pairing between different complementary molecules. They may have a similar function in vivo. [Adapted from D. S. Portman and G. Dreyfuss, 1994, *EMBO J.* **13**:213.]

➤ FIGURE 12-15 Experimental demonstration that human hnRNP A1 protein can cycle in and out of cytoplasm, but human hnRNP C protein cannot. Heterokaryons were prepared by treating HeLa cells and cultured *Xenopus* cells with polyethylene glycol. The cells were treated with cycloheximide immediately after fusion to prevent protein synthesis. After 2 h, the cells were fixed and stained with fluorescent-labeled antibodies specific for hnRNP C and hnRNP A1. These antibodies do not bind to the homologous *Xenopus* proteins. (a) A fixed preparation viewed by phase-contrast microscopy includes unfused Hela cells (arrowhead) and *Xenopus* cells (dotted arrow), as well as fused heterokaryons (solid arrow). In the heterokaryon in this micrograph, the round Hela-cell nucleus is to the right of the oval-shaped *Xenopus* nucleus. (b,c) When the same preparation was viewed by fluorescence microscopy, the stained hnRNP C protein was green and the stained hnRNP A1 protein was red. Note that the unfused *Xenopus* cell on the left is unstained, confirming that the antibodies are specific for the human proteins. In the heterokaryon, hnRNP C only appears in Hela-cell nuclei (b), whereas hnRNP A1 appears in both nuclei (c). Since protein synthesis was blocked after cell fusion, some of the labeled hnRNP A1 must have left the Hela-cell nucleus, moved through the cytoplasm, and entered the *Xenopus* nucleus in the heterokaryon. [See S. Pinol-Roma and G. Dreyfuss, 1992, *Nature* **355**:730; courtesy of G. Dreyfuss.]

suggests that some hnRNP proteins remain associated with fully processed mRNA when it is transported from the nucleus to the cytoplasm, exchange with cytoplasmic proteins, and then are transported back to the nucleus. The role of hnRNP proteins in transport of mRNA is discussed in more detail in a later section.

Pre-mRNAs Are Cleaved at Specific 3′ Sites and Rapidly Polyadenylated in Animal Cells

In animal cells, all mRNAs, except histone mRNAs, have a 3′ poly-A tail. Early studies of pulse-labeled adenovirus and SV40 RNA demonstrated that the viral primary transcripts extend beyond the poly-A sequence in the viral mRNAs. These results suggested that A residues are added to a 3′ hydroxyl generated by endonucleolytic cleavage, but the predicted downstream RNA fragments are degraded so rapidly in vivo that they cannot be detected. The cleavage mechanism was firmly established using in vitro processing reactions performed with extracts of HeLa-cell nuclei where both predicted cleavage products were observed.

Early sequencing of cDNA clones from animal cells showed that nearly all mRNAs contain the sequence AAUAAA 10–35 nucleotides *upstream* from the poly-A tail. Most of the rare exceptions contain the alternative hexanucleotide sequence AUUAA. Polyadenylation of RNA transcripts from transfected genes is virtually eliminated when the template DNA encoding the AAUAAA sequence is mutated to any other sequence except one encoding AUUAAA. The unprocessed RNA transcripts produced from such mutant templates do not accumulate in nuclei, but are rapidly degraded. Further mutagenesis of sequences within a few hundred bases of poly-A sites revealed that a second signal *downstream* from the cleavage site is required for efficient cleavage and polyadenylation of most pre-mRNAs in animal cells. This downstream poly-A signal is not a specific sequence but rather a GU-rich or simply a U-rich region within ≈50 nucleotides of the cleavage site.

The mechanism of pre-mRNA cleavage and polyadenylation currently is being uncovered by purification of the required proteins from HeLa-cell nuclear extracts. According to the current model (Figure 12-16), a 350-kDa protein complex called *cleavage and polyadenylation specificity factor* (CPSF), composed of three or four different polypeptides, first forms an unstable complex with the upstream AU-rich poly-A signal. Then at least three additional proteins—a 200-kDa heterotrimer called cleavage stimulatory factor (CStF) and two as-yet poorly characterized cleavage factors (CFI and II)—bind to the CPSF-RNA complex. Interaction between CstF and the GU- or U-rich downstream poly-A signal stabilizes the multiprotein complex. Finally, a poly-A polymerase (PAP) binds to the complex before cleavage can occur. This requirement for PAP binding links cleavage and polyadenylation, so that the free 3′ ends generated are rapidly polyadenylated. Assembly of this large, multiprotein cleavage-polyadenylation complex around the AU-rich poly-A signal in an RNA transcript is analogous in many ways to formation of the transcription-initiation complex at the AT-rich TATA box of a template DNA molecule (see Figure 11-53).

Following cleavage at the poly-A site, polyadenylation proceeds in two phases. Addition of the first ≈12 A residues occurs slowly, followed by rapid addition of up to 200–250 more A residues. The rapid phase requires the binding of multiple copies of a poly-A–binding protein containing the RNP motif. This protein is designated *PABII* to distinguish it from the first poly-A–binding protein characterized, which binds to the poly-A tail of cytoplasmic mRNAs. PABII binds to the short A tail initially added by PAP, stimulating polymerization of additional A residues by PAP (see Figure 12-16). By an unknown mechanism, PABII slows polymerization greatly when the poly-A tail reaches a length of 200–250 residues.

RNA-DNA Hybridization Reveals Spliced Out Introns

RNA splicing was discovered during analysis of adenovirus mRNA synthesis. In these studies, the abundant viral mRNA encoding the major virion capsid protein, called hexon, was isolated by gel electrophoresis of cytoplasmic polyadenylated RNA. To map the region of the viral DNA coding for hexon mRNA, researchers hybridized the isolated mRNA to the coding strand and the RNA-DNA hybrid was visualized in the electron microscope (Figure 12-17). Three loops of single-stranded DNA (A, B, and C) were observed; these correspond to the three introns in the hexon gene. Since these intron sequences in the viral genomic DNA are not present in mature hexon mRNA, they loop out between the exon sequences that hybridize to their complementary sequences in the mRNA.

Similar analyses of hybrids between RNA isolated from the nuclei of infected cells and viral DNA revealed RNAs that were colinear with the viral DNA (primary transcripts) and RNAs with one or two of the introns removed (processing intermediates). These results, together with the findings that the 5′ cap and 3′ poly-A tail of mRNA precursors are retained in mature cytoplasmic mRNAs, led to the realization that introns are removed from primary transcripts as exons are spliced together. For short transcription units, RNA splicing usually follows cleavage and polyadenylation of the 3′ end of the primary transcript. But for long transcription units containing multiple exons, splicing of exons in the nascent RNA sometimes begins before transcription of the gene is complete.

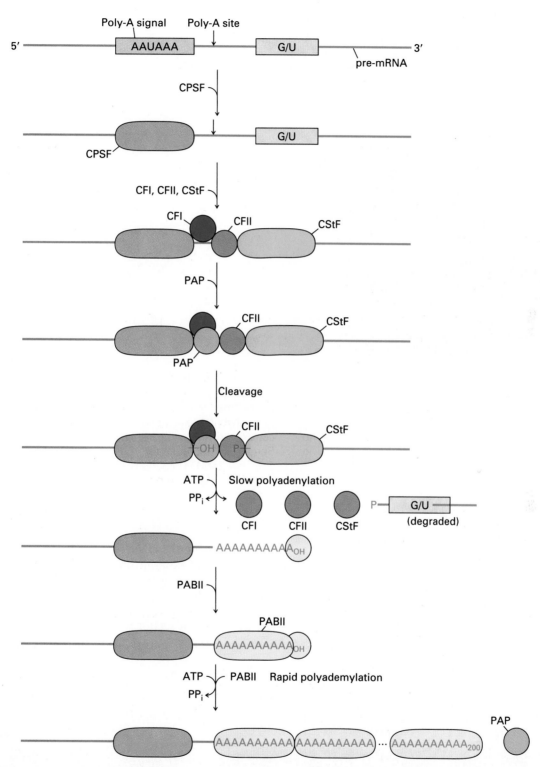

▲ FIGURE 12-16   Model for cleavage and polyadenylation of pre-mRNAs in mammalian cells. Cleavage-and-polyadenylation specificity factor (CPSF) binds to an upstream AU-rich polyadenylation (poly-A) signal. CStF interacts with a downstream GU- or U-rich sequence (yellow), stabilizing the complex that also includes CFI and CFII. Binding of poly-A polymerase (PAP) then stimulates cleavage at a poly-A site, which usually is 10–35 nucleotides 3′ of the upstream poly-A signal. The cleavage factors are released, as is the downstream RNA cleavage product, which is rapidly degraded. Bound PAP adds ≈12 A residues at a slow rate to the 3′-hydroxyl group generated by the cleavage reaction. Binding of poly-A–binding protein II (PABII) to the initial short poly-A tail accelerates the rate of addition by PAP. After 200–250 A residues have been added, PABII signals PAP to stop polymerization.

(a)

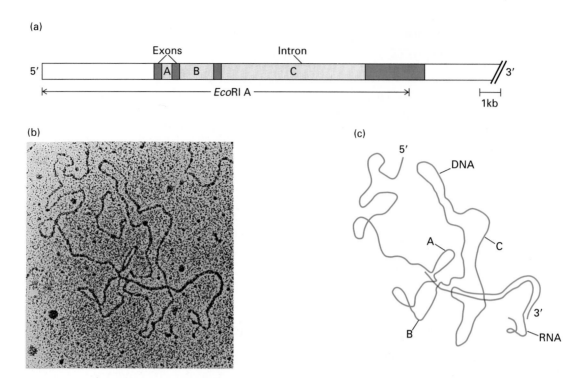

▲ **FIGURE 12-17** Demonstration that introns are spliced out by electron microscopy of RNA-DNA hybrid between adenovirus DNA and the mRNA encoding hexon, a major viral protein. (a) Diagram of the *Eco*RI A fragment of adenovirus DNA, which extends from the left end of the genome to just before the end of the final exon of the hexon gene. The gene consists of three short exons and one long (≈3.5-kb)

exon separated by three introns of ≈1, 2.5, and 9 kb. (b) Electron micrograph of hybrid between *Eco*RI A fragment diagrammed in (a) and hexon mRNA. (c) Diagram of structure of the hybrid molecule in (b). The loops marked A, B, and C correspond to the introns indicated in (a). [Part (b) from S. M. Berget, C. Moore, and P. A. Sharp, 1977, *Proc. Nat'l. Acad. Sci. USA* **74**:3171; courtesy of P. A. Sharp.]

Splice Sites in Pre-mRNAs Exhibit Short, Conserved Sequences

The location of splice sites in a pre-mRNA can be determined by comparing the sequence of genomic DNA with that of the cDNA prepared from the corresponding mRNA. Sequences that are present in the genomic DNA

but absent from the cDNA represent introns and indicate the positions of exon-intron boundaries. Such analysis of a large number of different mRNAs revealed moderately conserved, short consensus sequences at intron-exon boundaries in eukaryotic pre-mRNA; in higher organisms, a pyrimidine-rich region just upstream of the 3' splice site also is common (Figure 12-18). The only universally con-

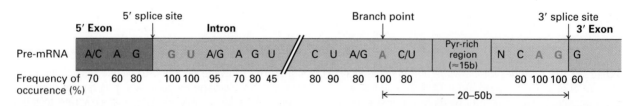

▲ **FIGURE 12-18** Consensus sequences around 5' and 3' splice sites in vertebrate pre-mRNAs. The only invariant bases are the (5')GU and (3')AG of the intron, although the flanking bases indicated are found at frequencies higher than expected based on a random distribution. A pyrimidine-rich region (light blue) near the 3' end of the intron is found in most cases. The branch-point **A,** also invariant, usually is 20–

50 bases from the 3' splice site. The central region of the intron, which may range from 40 bases to 50 kilobases in length, generally is unnecessary for splicing to occur. [See R. A. Padgett et al., 1986, *Ann. Rev. Biochem.* **55**:1119; E. B. Keller and W. A. Noon, 1984, *Proc. Nat'l. Acad. Sci. USA* **81**:7417.]

served nucleotides are the (5′)GU and (3′)AG in the intron. Deletion analyses of the center portion of introns in various pre-mRNAs have shown that generally only 30–40 nucleotides at each end of an intron are necessary for splicing to occur at normal rates.

Recombinant DNAs containing the 5′ exon-intron junction of one transcription unit (e.g., SV40 late region) and the 3′ intron-exon junction of another (e.g., mouse β-globin gene) have been prepared and introduced into cultured cells. Spliced mRNA molecules are formed in which the two exon sequences are joined and the chimeric intron is deleted precisely. The formation of correctly spliced mRNAs in such experiments indicates that the cell's splicing machinery can recognize and correctly join heterologous 5′ and 3′ splice sites.

Excision of Introns and Splicing of Exons in Pre-mRNA Occur via Two Transesterification Reactions

Experiments with cell extracts that accurately excise introns and join exons at the correct sites in a pre-mRNA molecule were critical for understanding the mechanism of RNA splicing. Analysis of the intermediates formed during such in vitro splicing reactions led to the conclusion that introns are not cut out as linear molecules (Figure 12-19). Surprisingly, an intron is removed as a *lariat* structure in which the 5′ G of the intron is joined in an unusual 2′,5′-

phosphodiester bond to an adenosine near the 3′ end of the intron. This adenosine is called the *branch point* because it forms an RNA branch in the lariat structure.

The finding that excised introns have a branched lariat structure led to the discovery that splicing of exons proceeds via two sequential *transesterification reactions* as illustrated in Figure 12-20. In each reaction, one phosphate-ester bond is exchanged for another. Since the number of phosphate-ester bonds in the molecule is not changed in either reaction, no energy is consumed. The net result of these two transesterification reactions is that two exons are ligated and the intervening intron is released as a branched lariat structure.

Small Nuclear Ribonucleoprotein Particles Assist in Splicing

Six small U-rich RNAs are abundant in the nuclei of mammalian cells. Designated U1 through U6, these small nuclear RNAs (*snRNAs*) range in length from 107 to 210 nucleotides (Figure 12-21a). Even before splicing was accomplished in vitro, several observations led to the suggestion that snRNAs assisted in the splicing reaction. First, the short consensus sequence at the 5′ end of introns (CAG|GUAAGU) was found to be complementary to a sequence near the 5′ end of the snRNA called U1. Second, snRNAs were found associated with hnRNPs in nuclear extracts.

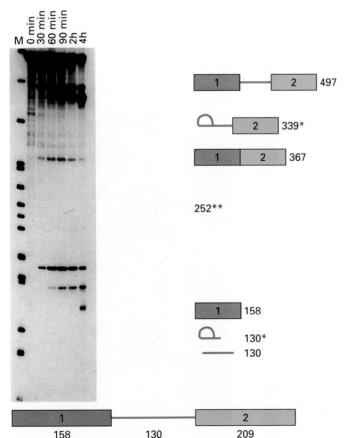

◄ FIGURE 12-19   Analysis of RNA products formed in an in vitro splicing reaction. A nuclear extract from HeLa cells was incubated with 497-nucleotide radiolabeled RNA (*bottom*) that contained portions of two exons (red) from human β-globin mRNA separated by a 130-nucleotide intron (blue). After incubation for various times, the RNA was purified and subjected to electrophoresis and autoradiography, along with RNA markers (lane M). Small numbers indicate the number of nucleotides in the various species. Much of the slower-migrating starting RNA (497) was correctly spliced, yielding a 367-nucleotide product. The excised intron (130*) migrated slower than expected based on its molecular weight, indicating that it is not a linear molecule. Likewise, one of the reaction intermediates (339*) exhibited an anomolously slow electrophoretic mobility. Additional analysis indicated that in both cases the intron had a lariat structure resulting in the slow mobility. The 252** band, an aberrant product of the in vitro reaction, is greatly reduced in reactions in which the RNA is capped. [See R. A. Padgett et al., 1984, *Science* **225**:898; from B. Ruskin et al., 1984, *Cell* **38**:317; Photograph courtesy of Michael R. Green.]

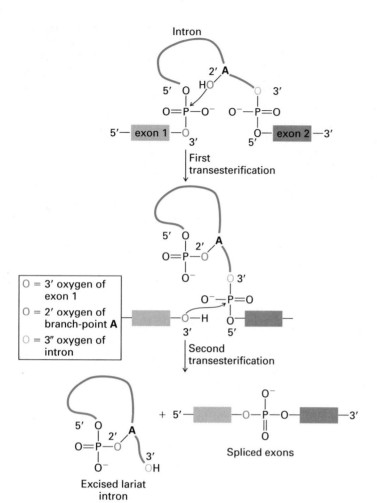

◄ FIGURE 12-20   Splicing of exons in pre-mRNA occurs via two transesterification reactions. In the first reaction, the ester bond between the 5' phosphorus of the intron and the 3' oxygen (red) of exon 1 is exchanged for an ester bond with the 2' oxygen (dark blue) of the branch-site **A** residue. In the second reaction, the ester bond between the 5' phosphorus of exon 2 and the 3' oxygen (light blue) of the intron is exchanged for an ester bond with the 3' oxygen of exon 1, releasing the intron as a lariat structure and joining the two exons. Arrows show where the activated hydroxyl oxygens react with phosphorus atoms.

➤ FIGURE 12-21   (a) Gel electrophoretic separation of ^{32}P-labeled small RNAs isolated from nuclei of cultured Chinese hamster cells. All the labeled RNAs have been sequenced. The U1–U6 species are rich in U residues, and all except U3 are involved in RNA splicing. U3 is found in the nucleolus where it is involved in processing of rRNA precursors. The 5.8S and 5S RNAs are ribosomal RNAs. The 7S RNA is a component of the signal recognition particle involved in transport of proteins into the lumen of the endoplasmic reticulum. The 5.8S, 5S, and 7S RNAs are present largely in the cytoplasm, whereas the U RNAs are located primarily in the nucleus. (b) Cell stained with anti-Sm antibody, which is specific for snRNP proteins that bind to U1, U2, U4, and U5 snRNA. The stained particles are yellow. The nucleolus is not stained because U3 snRNA does not bind Sm protein. [Part (a) courtesy of W. Jelinek and S. Haynes; part (b) courtesy of M. Lerner.]

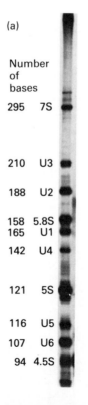

(a)

Number of bases

295	7S
210	U3
188	U2
158	5.8S
165	U1
142	U4
121	5S
116	U5
107	U6
94	4.5S

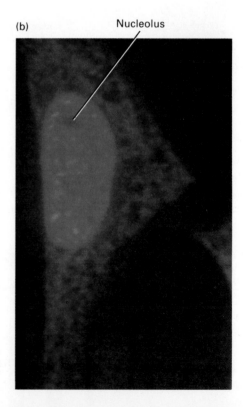

(b)   Nucleolus

The snRNAs associate in the nucleus with six to ten proteins to form small nuclear ribonucleoprotein particles (*snRNPs*). Some of these proteins are common to all snRNPs, and some are specific for individual snRNPs. For unknown reasons, antisera from patients with the autoimmune disease systemic lupus erythematosus (SLE) contain antibodies (called *anti-Sm antibody*) directed against a protein that is common to the U1, U2, U4, and U5 snRNPs (Figure 12-21b). Antisera from some SLE patients were found to have greater specificity for one or another of the individual snRNPs. These specific antisera have been widely used in characterizing components of the splicing reaction. For example, when antiserum that is specific for U1 snRNP is added to an in vitro splicing mixture, the splicing reaction is interrupted, confirming the importance of the U1 snRNP.

In subsequent studies, a synthetic oligonucleotide that is complementary to and thus hybridizes with the 5'-end region of U1 snRNA was found to block splicing, indicating that this is the region that binds to pre-mRNA during the splicing reaction. Further evidence for the importance of base pairing between the 5' end of U1 snRNA and the conserved 5' splice-site sequence of pre-mRNAs came from experiments with genes that were mutated in the 5' splice-site consensus sequence of an intron. When genes containing these mutations were transfected into cells, splicing of the corresponding pre-mRNAs was blocked. However, when a mutant gene was cotransfected with a mutant U1 snRNA gene containing a compensating sequence change that restored base pairing with the mutant 5' splice site, splicing was restored (Figure 12-22). This result argued strongly that base pairing between the 5' splice site of a

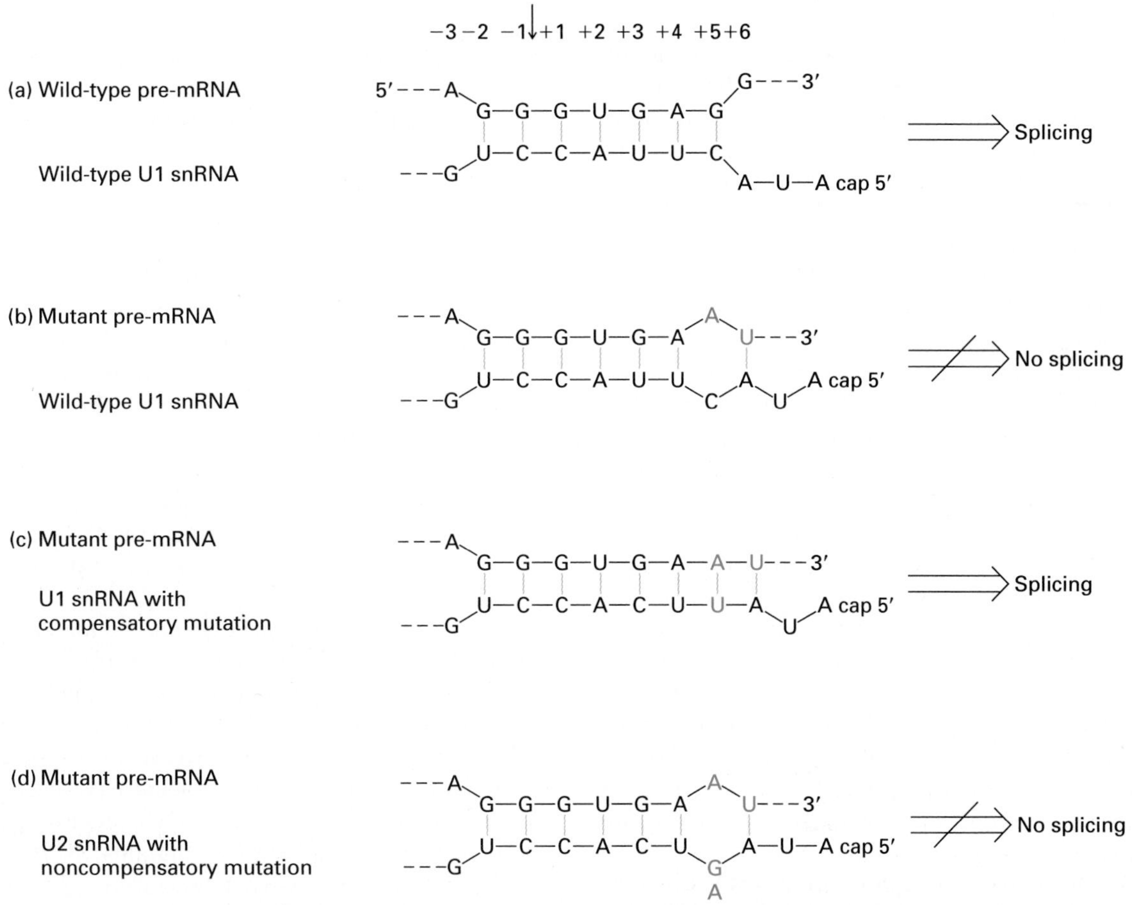

▲ FIGURE 12-22   Effect of mutations in the 5' splice site of a pre-mRNA and the 5' region of U1 snRNA on splicing. An adenovirus gene containing the wild-type or a mutant 5' splice-site sequence was cotransfected with a wild-type or mutant U1 snRNA into cultured cells. Subsequent analysis of the RNA in transfected cells revealed whether splicing of the adenovirus exons occurred. Mutant bases are in red. (a) The wild-type pre-mRNA and U1 snRNA base-pair from positions −2 to +5 around the splice site (↓); splicing occurred. (b) Mutations at +5 and +6 in the pre-

mRNA prevented base pairing at position +5 and blocked splicing. (c) A mutation in U1 snRNA in which C was changed to U at the fourth position restored base pairing and splicing. (d) When the fourth position in U1 snRNA was changed to a G or A, splicing was not restored. The finding that only a compensatory mutation in U1 snRNA restored splicing indicates that the regions diagrammed normally base-pair during RNA splicing. [See Y. Zhuang and A. M. Weiner, 1986, *Cell* **46**:827.]

pre-mRNA and the 5′ region of U1 snRNA is required for RNA splicing.

After discovery of the lariat structure of excised introns, a consensus sequence was recognized in the region flanking the branch point in pre-mRNAs (see Figure 12-18). In the yeast *S. cerevisiae*, virtually all introns have the sequence UACUAAC in the region of the branch-point A (in boldface). Except for the branch-point A, this yeast sequence is complementary to an internal sequence in U2 snRNA. Compensatory mutation experiments, similar to those just described with U1 snRNA and 5′ splice sites, demonstrated that base pairing between U2 snRNA and the branch-site sequence in pre-mRNA is critical to splicing. Significantly, the branch-point A itself, which is not base-paired to U2 snRNA, "bulges out," allowing its 2′ hydroxyl to participate in the first transesterification reaction of RNA splicing (see Figure 12-20).

These studies with U1 and U2 snRNAs indicate that during splicing they base-pair with pre-mRNA as shown in Figure 12-23a. Additional compensatory mutation experiments demonstrated that other RNA-RNA interactions also occur during splicing (Figure 12-23b,c). Based on the results of these experiments, identification of reaction intermediates, and other biochemical analyses, the spliceosomal splicing model depicted in Figure 12-24 was proposed. According to this model, U1 and U2 snRNAs, as part of the U1 and U2 snRNPs, base-pair with the 5′ splice site and branch-point regions of an intron, respectively. In yeast, base pairing between U2 snRNA and the branch site occurs over six bases (see Figure 12-23a). In higher eukaryotes where the branch-site sequence is less highly conserved, the association of U2 snRNP with pre-mRNA is assisted by a protein called U2AF, which binds to the pyrimidine-rich region near the 3′ splice site. U2AF interacts with RNA through an RNP motif and is thought to interact with other proteins required for splicing through a domain containing repeats of the dipeptide serine-arginine (the SR motif). The snRNAs in the U4 and U6 snRNPs base-pair over an extended complementary region (see Figure 12-23b); this complex then associates with U5 snRNP. The U4/U6/U5 complex then associates, presumably via protein-protein interactions, with the previously formed complex consisting of pre-mRNA base-paired to U1 and U2 snRNPs. The resulting high-molecular-weight (60S) ribonucleoprotein complex is called a *spliceosome* (Figure 12-25).

After formation of the spliceosome, extensive rearrangements occur in the pairing of snRNAs and the pre-mRNA. The U4 and U6 snRNAs dissociate from each other; U6 snRNA then base-pairs to a sequence in U2 snRNA that is just 5′ to the sequence that interacts with the branch site in pre-mRNA. U1 is thought to dissociate from the 5′ splice site in the pre-mRNA, after which U5 base-pairs with exon sequences flanking the splite sites. These rearrangements yield the interactions shown in Figure 12-23c.

▶ FIGURE 12-23   Base pairing between snRNAs and pre-mRNA splice-site sequences during RNA splicing. Secondary structures in the U snRNAs that are not altered during splicing are shown in diagrammatic line form. The purple rectangles represent sequences that bind snRNP proteins recognized by anti-Sm antibody. (a) The 5′ region of U1 snRNA initially base-pairs with nucleotides at the 5′ end of the intron (dark blue) and 3′ end of the 5′ exon (dark red) of the pre-mRNA; U2 snRNA base-pairs with a sequence that includes the branch-point **A**, although this residue is not base-paired. The yeast branch-site sequence is shown here. (b) An internal region of U6 snRNA (light blue) initially base-pairs with the 5′ end of U4 snRNA (light red). (c) Rearrangements later in the splicing process result in U6 snRNA (light blue) base pairing with U2 snRNA (dark red), and U5 snRNA (black) interacting with four nucleotides in the exons, displacing U1 snRNA from the pre-mRNA. U2 snRNA remains base-paired to the intron near the pyrimidine-rich region (blue rectangle). The interactions between U5 snRNA and the exons were detected by formation of covalent cross-links between the exon bases and base analogs incorporated into substrate RNA. [See E. J. Sontheimer and J. A. Steitz, 1993, *Science* **262**:1989; adapted from M. J. Moore, C. C. Query, and P. A. Sharp, 1993, in R. Gesteland and J. Atkins, eds., *The RNA World,* Cold Spring Harbor Press, pp. 303–357.]

The rearranged spliceosome then catalyzes the two transesterification reactions that result in RNA splicing. After the second transesterification reaction, the ligated exons are released from the spliceosome while the lariat intron remains associated with the snRNPs. This final intron-snRNP complex is unstable and dissociates. The individual snRNPs released can participate in a new cycle of splicing. The excised intron is rapidly degraded by a "debranching enzyme," which hydrolyzes the 5′,2′-phosphodiester bond at the branch point, and other nuclear RNases.

It is estimated that at least one hundred proteins are involved in RNA splicing, making this process comparable in complexity to protein synthesis and initiation of transcription. Some of these splicing factors are associated with snRNPs, but others are not. In yeast, various proteins that are required for pre-mRNA splicing have been identified by analysis of *PRP* (precursor RNA processing) *mutants*. In these temperature-sensitive mutants, RNA splicing is blocked at nonpermissive temperatures. In mammalian systems, purification of nuclear extracts has led to identification of proteins required for specific steps in the spliceosomal splicing cycle. Many of the yeast *PRP* genes have been cloned, as have some genes encoding mammalian splicing factors. Sequencing of these genes has revealed that the encoded proteins contain domains with the RNP motif and the SR motif; some proteins also exhibit sequence homologies to known RNA helicases. As noted earlier, RNP domains are involved in binding RNA; SR do-

(a)

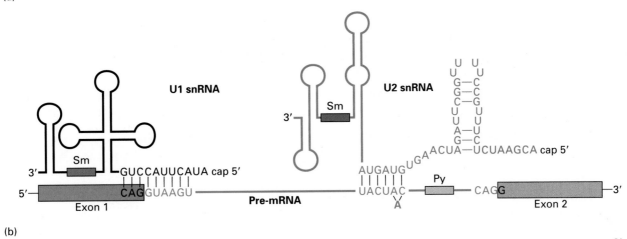

(b)

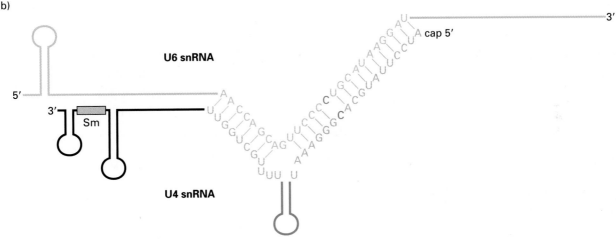

(c)

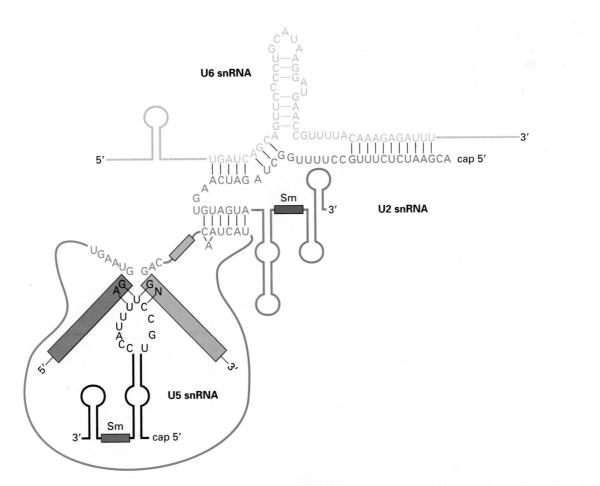

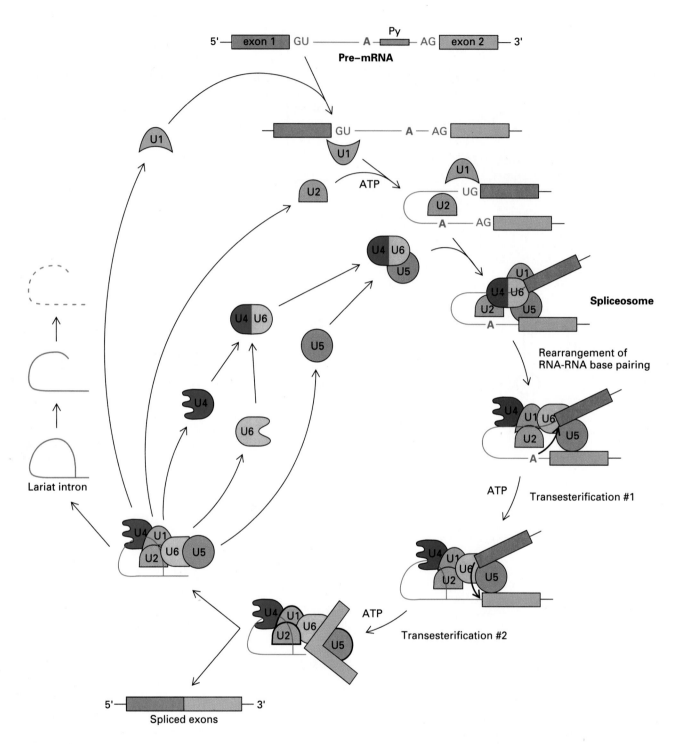

▲ FIGURE 12-24   The spliceosomal splicing cycle. The splicing snRNPs (U1, U2, U4, U5, and U6) associate with the pre-mRNA and with each other in an ordered sequence to form the spliceosome. This large ribonucleoprotein complex then catalyzes the two transesterification reactions that result in splicing of the exons (red) and excision of the intron (dark blue) as a lariat structure. Although ATP hydrolysis is not required for the transesterification reactions, it is thought to provide the energy necessary for rearrangements of the spliceosome structure that occur during the cycle. Note that the snRNP proteins in the spliceosome are distinct from the hnRNP proteins discussed earlier. Some hnRNP proteins bind to the pyrimidine-rich region near the 3' end of introns; others bind to other regions of the pre-mRNA. See text for discussion and Figure 12-23 for base-pairing interactions between the snRNAs. [See S. W. Ruby and J. Abelson, 1991, *Trends Genet.* **7**:79; adapted from M. J. Moore, C. C. Query, and P. A. Sharp, 1993, in *The RNA World,* Cold Spring Harbor Press, pp. 303–357.]

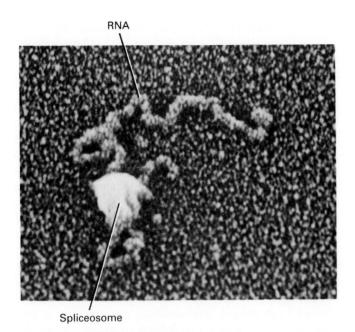

RNA

Spliceosome

▲ FIGURE 12-25 Electron micrograph of a spliceosome. Extracts of HeLa cells were mixed with a β-globin pre-mRNA; the reaction was interrupted before splicing was completed, so that the spliceosomes, containing snRNPs and the pre-mRNA substrate, could be purified. [From R. Reid, J. Griffith, and T. Maniatis, 1988, *Cell* **53**:949; courtesy of J. Griffith.]

mains are involved in protein-protein interactions and also may contribute to binding RNA. RNA helicases may be necessary for the base-pairing rearrangements that occur in snRNAs during the spliceosomal splicing cycle, particularly the dissociation of U4 from U6 and of U1 from the 5′ splice site.

Portions of Two Different RNAs Are Trans-Spliced in Some Organisms

Virtually all functional mRNAs in vertebrate and insect cells are derived from a single pre-mRNA by removal of internal introns and splicing of exons. In contrast, in trypanosome and euglenoid protozoans, mRNAs are constructed by splicing together separate RNA molecules. This process, referred to as *trans-splicing*, is also used in the synthesis of 10–15 percent of the mRNAs in the round worm *Caenorhabditis elegans*, an important model organism for studying embryonic development.

In trypanosomes, a single type of leader RNA of 140 nucleotides is produced in abundance from tandemly repeated transcription units. In a two-step reaction analogous to spliceosomal pre-mRNA splicing, a 39-nucleotide portion of the leader RNA, termed a *mini-exon* is spliced to the 5′ end of protein-coding exons in primary transcripts, which lack internal introns (Figure 12-26). The 5′ mini-exon, present in all trypanosome mRNAs, is thought to assist in initiation of translation. Many protein-coding

➤ FIGURE 12-26 Trans-splicing of leader RNA to a protein-coding exon in a polycistronic primary transcript in *Trypanosoma* occurs by two transesterification reactions. First, a branch-point **A** just upstream of a pyrimidine-rich region (light blue) 5′ of the coding exon (light red) in the transcript is joined with the 3′ region of the leader RNA, freeing the mini-exon (orange). Second, the 3′ end of the free mini-exon joins the 5′ end of the coding exon, excising a branched product containing the 3′ region of the leader RNA and the region of the primary transcript 5′ of the coding region. These steps are equivalent to the first and second transesterification reactions in pre-mRNA splicing in higher eukaryotes (see Figure 12-20). Splicing signals specific enzymes to cleave and polyadenylate the 3′ end of the coding exon, yielding a functional mRNA. [Adapted from L. Bonen, 1993, *FASEB J.* **7**:40.]

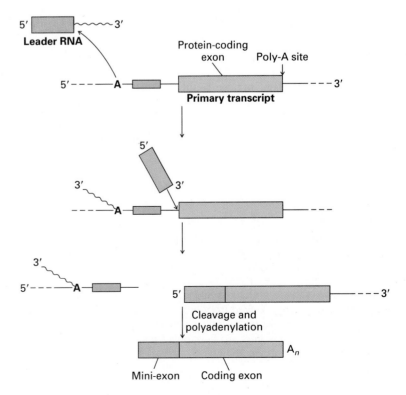

transcription units in trypanosomes are polycistronic, resembling bacterial operons. Individual mRNAs are produced from the polycistronic precursors by trans-splicing. Splicing of a 5′ mini-exon to a coding region triggers cleavage and polyadenylation at the 3′ end of the exon. Consequently, trypanosomes use trans-splicing and linked cleavage and polyadenylation to combine the operon organization of polycistronic transcription units characteristic of bacteria with the monocistronic organization of mRNAs characteristic of eukaryotes.

Self-Splicing Group II Introns Provide Clues to the Evolution of snRNPs

A remarkable class of introns has been discovered in protein-coding genes and some rRNA and tRNA genes of chloroplasts and mitochondria from plant and fungal cells. Under conditions of elevated temperature and Mg^{2+} (concentrations far higher than present in vivo), pure preparations of RNA transcripts containing these introns slowly splice out the introns. Since no protein is involved in these in vitro reactions, it was realized that this type of intron is a *self-splicing intron*.

The first self-splicing introns recognized were discovered in nuclear rRNA genes of protozoans. Now called *group I introns*, these are discussed in a later section. The self-splicing introns from mitochondria and chloroplasts described in this section were discovered later and consequently were called *group II introns*. The discovery of self-splicing introns led to the realization that RNA can function as a catalyst for chemical reactions that change covalent bonds. This insight revolutionized the way RNA function is viewed. As discussed in Chapter 4, RNA is now thought to catalyze peptide-bond formation during protein synthesis in ribosomes. Other examples of RNA catalysis are discussed later in this chapter.

Even though their precise sequences are not highly conserved, all group II introns fold into a conserved, complex secondary structure containing numerous stem-loops (Figure 12-27a). Self-splicing by a group II intron occurs via two transesterification reactions, involving intermediates and products analogous to those found in nuclear pre-mRNA splicing. The mechanistic similarities between group II intron self-splicing and spliceosomal splicing has led to the hypothesis that snRNAs function analogously to the stem-loops in the secondary structure of group II introns.

An extension of this hypothesis is that introns in present-day nuclear pre-mRNAs evolved from ancient group II self-splicing introns through the progressive loss of internal RNA structures, which concurrently evolved into trans-acting snRNAs that perform the same functions. According to this hypothesis, snRNAs interact with 5′ and 3′ splice sites of pre-mRNAs and with each other to produce an RNA structure functionally analogous to that of group II self-splicing introns (Figure 12-27b). In support of this kind of evolutionary model, group II intron mutants have been constructed in which domain V and part of domain I are deleted. Such mutants are defective in self-splicing, but when RNA molecules equivalent to the deleted regions are added to the in vitro reaction, self-splicing occurs. This finding demonstrates that these domains in group II introns can be trans-acting, like snRNAs.

The similarity in the mechanisms of group II intron self-splicing and spliceosomal splicing of pre-mRNAs also leads to the hypothesis that the splicing reaction is catalyzed by the snRNA, not the protein, components of spliceosomes. As mentioned above, group II introns self-splice only under nonphysiological conditions. Proteins called *maturases*, which bind to group II intron RNA, are required for rapid splicing under in vivo conditions. Maturases are encoded by group II introns themselves, and are thought to stabilize the precise three-dimensional interactions of the intron RNA required to catalyze the two splicing transesterification reactions. The snRNP proteins may function analogously in spliceosomes to stabilize a precise geometry of snRNAs and intron nucleotides required to catalyze pre-mRNA splicing.

The evolution of snRNAs may have been an important step in the rapid evolution of higher eukaryotes. As inter-

▶ FIGURE 12-27 Schematic diagrams comparing the secondary structures of group II self-splicing introns (a) and U snRNAs present in the spliceosome (b). The first transesterification reaction is indicated by black arrows; the second reaction, by blue arrows. The branch-point A is boldfaced. The similarity in these structures suggests that the spliceosomal snRNAs evolved from group II introns, with the trans-acting snRNAs being functionally analogous to the corresponding domains in group II introns. [Adapted from P. A. Sharp, 1991, *Science* **254**:663.]

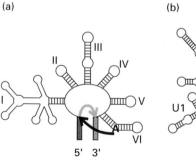

(a)   Group II intron

(b)   U snRNAs in spliceosome

Pre-mRNA

nal intron sequences were lost and their functions in RNA splicing supplanted by trans-acting snRNAs, the remaining intron sequences would be free to diverge. This in turn likely facilitated the evolution of new genes through exon shuffling (Chapter 9). It also permitted the increase in protein diversity that results from alternative RNA splicing and an additional level of gene control resulting from regulated RNA splicing.

Before we consider the regulation of RNA processing, one more remarkable property of group II introns deserves mention, namely, their ability to behave as mobile DNA elements in the genome. The maturases that increase the rate of self-splicing of these introns also contain a domain that is homologous to reverse transcriptases. Thus group II introns can move in the genome like other nonviral retrotransposons discussed in Chapter 9. As is generally true for mobile DNA elements, transposition of group II introns is rare. However, when a group II intron does transpose, it does not inactivate the gene into which it inserts, because the inserted intron is spliced out of the transcript produced from the target gene by self-splicing!

Regulation of RNA Processing Controls Expression of Some Proteins

As explained in the previous sections, conversion of a 5' capped RNA transcript into a functional mRNA involves two primary steps: (1) cleavage and polyadenylation at the 3' end and (2) ligation of exons with the concomitant excision of introns, or RNA splicing. To understand how regulation of RNA processing can control gene expression, we need to recall that higher eukaryotes contain both simple and complex transcription units. The primary transcripts produced from the former contain one poly-A site and exhibit only one pattern of RNA splicing, even if multiple introns are present; thus simple transcription units encode a single mRNA (see Figure 9-1b). In contrast, the primary transcripts produced from complex transcription units can be processed in alternative ways to yield different mRNAs (see Figure 9-2).

In theory, the expression of both simple and complex transcription units could be controlled by on-off regulation of the cleavage and polyadenylation of their pre-mRNAs. Although this step probably occurs constitutively (i.e., is not regulated) in the case of most genes, a few examples of this type of posttranscriptional regulation have been discovered. In the case of pre-mRNAs produced from complex transcription units, cleavage and polyadenylation at different poly-A sites and/or splicing of different exons yields mRNAs encoding distinct proteins. Such alternative processing pathways of complex pre-mRNAs usually are regulated, often in a cell type–specific manner. That is, one possible mRNA is expressed in one type of cell or tissue, while another is expressed in different cells or tissues. Since about 5 percent of all transcription units in higher eukary-otes are complex units, this type of posttranscriptional regulation is a significant gene-control mechanism in such organisms. In this section, we describe one example of on-off control of cleavage/polyadenylation and two examples of differential RNA splicing to yield only one mRNA in a specific cell type or tissue.

U1A Autoregulation The best-understood example of on-off regulation of the cleavage/polyadenylation step in RNA processing involves U1A protein, one of the several proteins in U1 snRNP, which plays a critical role in RNA splicing (see Figure 12-24). U1A protein binds to a 7-base sequence in the snRNA component of U1 snRNP. The pre-mRNA encoding U1A itself also contains two copies of this 7-base sequence (one with a single-base change) just upstream of its polyadenylation signal. When a U1A molecule is bound to each copy of this 7-base sequence in a U1A pre-mRNA in an in vitro system containing poly-A polymerase and all the factors necessary for cleavage and polyadenylation, cleavage occurs but addition of the poly-A tail does not (Figure 12-28). This blockage of polyadenylation requires interaction between the C-terminal end of the bound U1A molecules and poly-A polymerase.

The finding that U1A blocks polyadenylation of its own pre-mRNA in vitro can account for the autoregulation of U1A observed in vivo. When U1A exceeds the level of U1 snRNA, the excess U1A protein binds to its two binding sites in U1A pre-mRNAs, thereby blocking poly-adenylation (see Figure 12-28). Since U1A binding does not prevent cleavage of the pre-mRNAs, free 3' ends are generated; these are thought to be rapidly degraded by an exonuclease, thus eliminating further processing and production of U1A mRNA. As a result, synthesis of U1A protein is decreased until all the excess is used in formation of new U1 snRNPs. Once this occurs, no free U1A protein is available to bind to newly transcribed U1A pre-mRNAs, which then can be processed normally. This on-off mechanism of posttranscriptional regulation thus coordinates expression of U1A protein with U1 snRNP assembly.

Cell Type–Specific Expression of Alternative Fibronectins The fibronectin gene provides a good example of gene control by tissue-specific RNA splicing (Figure 12-29). The fibronectin primary transcript contains multiple exons; these are spliced in different ways to produce different forms of this protein, which is secreted by both fibroblasts and hepatocytes (as well as other types of cells). The fibronectin produced by fibroblasts is a secreted, fibrous protein that interacts with components of the extracellular matrix as well as cell-surface receptors, thereby binding cells to the extracellular matrix. Splicing of fibronectin pre-mRNA in fibroblasts yields an mRNA containing two exons, designated EIIIA and EIIIB, which encode protein domains that interact with cell-surface receptors in many cell types, thus making fibroblast fibronectin adher-

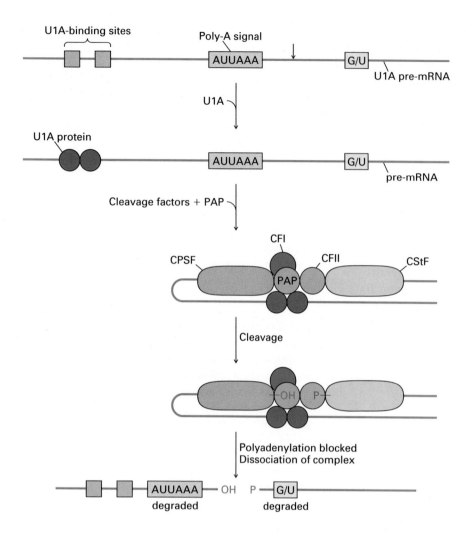

◄ FIGURE 12-28   On-off regulation of polyadenylation of U1A pre-mRNA. U1A is a sequence-specific RNA-binding protein that is a component of the U1 snRNP. Two U1A-binding sites are located upstream from the unusual AUUAAA poly-A signal (↓) in U1A pre-mRNA. When excess U1A protein accumulates, it binds to these sites in the nascent U1A pre-mRNA. Following assembly of the cleavage-and-polyadenylation complex on the pre-mRNA, cleavage occurs, but the bound U1A molecules interact with poly-A polymerase (PAP), preventing it from adding A residues to the free 3'-hydroxyl group. In the absence of polyadenylation, both cleavage products are degraded, and thus no U1A protein is produced.

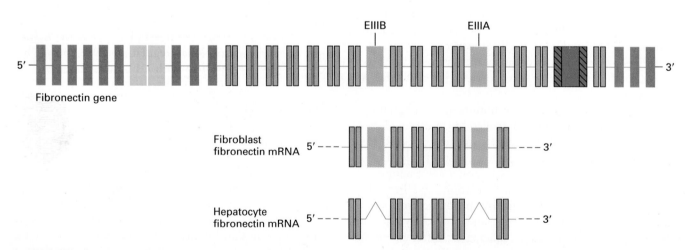

▲ FIGURE 12-29   Cell type-specific splicing of fibronectin pre-mRNA in fibroblasts and hepatocytes. The ≈75-kb fibronectin gene (*top*) contains multiple exons that likely evolved from an ancestral gene through a number duplications of exons. Exons with homologous sequences are shown in the same color. In this diagram, introns (dark blue lines) are not drawn to scale; most of them are much longer than any of the exons. The fibronectin mRNA produced in fibroblasts includes the EIIIA and EIIIB exons, whereas these exons are spliced out of fibronectin mRNA in hepatocytes.

ent to cell surfaces. In contrast, splicing of fibronectin pre-mRNA in hepatocytes "skips" these two exons, yielding an mRNA that does not encode the corresponding domains. As a result, the fibronectin secreted by hepatocytes does not adhere strongly to cell surfaces, allowing it to circulate in the serum. Numerous other examples of regulated splicing of alternative exons have been discovered. One of the most thoroughly studied involves sexual differentiation in *Drosophila*.

Drosophila *Sex Determination*   The isolation and characterization of genes required for normal sex determination in *Drosophila* showed that three of these genes operate in a cascade of gene regulation by controlling the splicing of specific primary transcripts. The first of these genes to be expressed, called *sex-lethal* (*sxl*), is transcribed from an early promoter (P_E) that is active only in very early female embryos (Figure 12-30a). The primary *sxl* transcript is spliced into an mRNA with two exons, resulting in

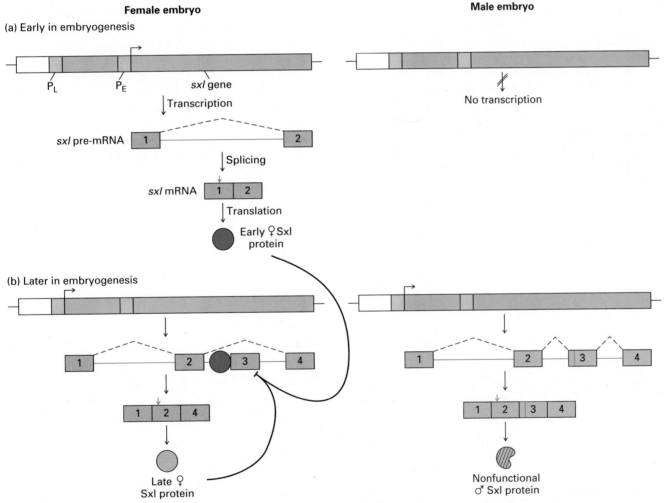

▲ FIGURE 12-30   Expression of Sex-lethal (Sxl) protein during *Drosophila* embryogenesis. In pre-mRNAs, exons are shown as red boxes and introns by blue line; splicing is indicated by dashed lines. Red vertical arrow indicates AUG start codon for translation in mRNAs. (a) Early in development, *sxl* pre-mRNA is synthesized from P_E, which is active only in female embryos. (b) Later in development, a different *sxl* pre-mRNA is synthesized from P_L in both males and females. In males, this transcript is spliced to yield an mRNA with four exons. Because exon 3 contains an in-frame stop codon (dark red), no functional Sxl protein is produced in males. In females, binding of the early Sxl protein to the late *sxl* pre-mRNA prevents splicing of exons 2 and 3. The resulting mRNA, containing three exons, is translated into functional late Sxl protein, which also binds to the late *sxl* pre-mRNA, ensuring its continued production in females. See text for further discussion.

expression of Sex-lethal (Sxl) protein in early female embryos. Sex-lethal protein is a sequence-specific RNA-binding protein that binds to RNA through an RNP motif at the C-terminal end.

As development of the *Drosophila* embryo proceeds, transcription of the *sxl* gene from P_E is repressed, while transcription from an upstream late promoter (P_L) is induced (Figure 12-30b). Transcription from P_L occurs in both males and females and continues from this time in development onward. The *sxl* transcript produced from P_L contains four exons; the first one is located upstream and the second one downstream of the first exon in the early *sxl* transcript produced from P_E. In both males and females, the first exon of the P_L-derived late transcript is spliced to the second exon, at a splice site that is not used in processing the early transcript. Subsequent splicing of P_L transcripts differs in males and females due to the presence of Sex-lethal protein resulting from transcription from P_E in females.

In males, the second exon of the late transcript is spliced to a third exon, which in turn is spliced to a fourth exon encoding the RNA-binding RNP motif. The third exon of the resulting mRNA contains a stop codon (UGA). Consequently, translation of this mRNA yields a Sxl protein that lacks the RNA-binding domain and is nonfunctional. In females, the Sxl protein produced from the early P_E transcript binds to specific sequences in the late *sxl* pre-

mRNA, preventing the splicing of exon 2 to exon 3. Instead, exon 2 is spliced to exon 4, producing an mRNA in which exon 3 of the male-specific *sxl* mRNA containing the stop codon is "skipped." This late female-specific mRNA is translated into a protein containing the RNP motif that binds specifically to the *sxl* transcript. As a result, the late female Sxl protein, which differs from the early female Sxl protein at the N-terminus, but has the identical RNP motif at its C-terminus, also regulates splicing of the late *sxl* primary transcript in females. In this way, the late female Sxl protein autoregulates its own production, ensuring continued expression of a functional Sxl protein after repression of the early promoter and activation of the late promoter.

Sxl protein also regulates expression of the *transformer* (*tra*) gene, the second gene to be expressed in the regulatory cascade leading to sexual differentiation in *Drosophila*. This control involves binding of Sxl protein to the 3′ pyrimidine-rich region in the intron between exons 1 and 2 in *tra* pre-mRNA (Figure 12-31). The bound Sxl blocks binding of U2AF and U2 snRNPs in this region, preventing splicing to the 5′ end of exon 2 and favoring splicing to an alternative splice site at the 5′ end of exon 3. As a result, female embryos express functional Transformer (Tra) protein. As in the case of the *sxl* gene, male embryos produce *tra* mRNAs containing a stop codon; these are translated into a nonfunctional protein. Although

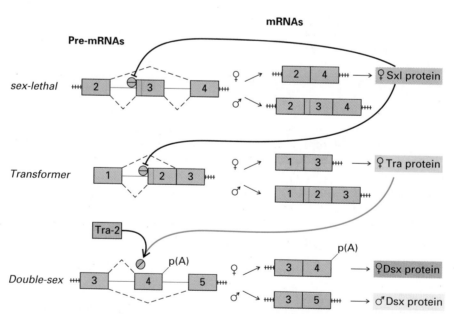

▲ FIGURE 12-31 Cascade of splicing regulation in expression of *sex-lethal, transformer,* and *double-sex* genes in *Drosophila* embryos. For clarity, only the exons (light red boxes) and introns (dark blue lines) where regulated splicing occurs are shown. Splicing is indicated by dashed lines above (female) and below (male). Dark red bands contain in-frame stop codons, which prevent synthesis of functional protein. Thicker lines indicate repression (black) or activation (red) of splicing by binding of the indicated RNA-binding pro-

teins at the sites shown. As a result of this cascade of regulated splicing, distinct Double-sex (Dsx) proteins are produced in female and male embryos. These repress transcription of genes required for sexual differentiation of the opposite sex. See text for further discussion. [Adapted from M. J. Moore, C. C. Query, and P. A. Sharp, 1993, in R. Gesteland and J. Atkins, eds., *The RNA World*, Cold Spring Harbor Press, pp. 303–357.]

Tra protein does not contain an RNA-binding domain, it forms a complex with another protein, Transformer 2 (Tra2), that does. The Tra-Tra2 complex binds to a repeated 13-base sequence in the pre-mRNA synthesized from the *double-sex* (*dsx*) gene. In this case, the bound proteins *activate* an upstream 3' splice site that deviates considerably from the consensus 3' splice-site sequence (see Figure 12-18). This splice site is not used in males, because they produce no functional Tra protein and therefore cannot activate the splice site. Both Tra and Tra2 proteins contain serine-arginine (SR) repeats similar to those in U2AF protein; as noted earlier, U2AF has been shown to interact with the pyrimidine-rich region near 3' splice sites and with several proteins required for RNA splicing. The SR repeats in the Tra-Tra2 complex bound to the *dsx* pre-mRNA bind to SR repeats in one of the general splicing factors. This interaction promotes assembly of a complex of splicing proteins that activates the upstream 3' splice site, leading to production of a female-specific *dsx* mRNA, which is polyadenylated at the end of exon 4 (see Figure 12-31). This mRNA is translated into a female-specific Double-sex protein that represses transcription of genes required for male sexual development. In male embryos, which lack Tra protein, the third exon of the *dsx* transcript is spliced to an alternative exon (exon 5); this is spliced to additional downstream exons, producing a male-specific *dsx* mRNA. This mRNA encodes an alternative transcription repressor that inhibits female sexual development.

As mentioned earlier, about 5 percent of all pre-mRNAs in higher eukaryotes are subject to this kind of regulated splicing. Regulated splicing can lead to the production of different proteins (e.g., male- and female-specific Double-sex proteins) or can control expression of functional proteins by including or excluding a stop codon as in the *Drosophila sex-lethal* and *transformer* genes. It is likely that in many other examples of regulated RNA splicing, sequence-specific RNA-binding proteins either inhibit or activate splice sites. During in vitro splicing of pre-mRNA from complex transcription units, differences in the relative concentrations of hnRNP proteins and essential splicing factors also can influence the selection of alternative splice sites. Consequently, it has been suggested that differences in the concentrations of general splicing factors and hnRNP proteins between different cell types also may contribute to cell type–specific splicing of pre-mRNAs from complex transcription units.

► *Subnuclear Organization and Transport of Nuclear mRNA to the Cytoplasm*

After pre-mRNAs have been processed in the nucleus, each of the resulting mature mRNAs is transported into the cytoplasm through one of the ≈4000 nuclear pores of a typical mammalian cell. The mechanisms by which nuclear mRNAs are transported from their sites of synthesis and processing to nuclear pores is not presently understood. But recent results suggest that pre-mRNAs are bound to subnuclear structures where RNA processing takes place and that movement of mature mRNAs to nuclear pores may not occur by random diffusion.

Most Transcription and RNA Processing Occurs in a Limited Number of Domains in Mammalian Cell Nuclei

Microscopic studies with fluorescent-labeled nucleic acid probes and monoclonal antibodies indicate that the bulk of transcription and RNA processing takes place in a limited number of domains (foci) in mammalian cell nuclei. Digital imaging microscopy, for example, localized total polyadenylated RNA (including unspliced and partially spliced pre-mRNA and nuclear mRNA) in a 1-μm-thick section of a human fibroblast nucleus. A protein splicing factor called SC-35, which is required for spliceosome assembly, was localized to the center of these same loci (Figure 12-32a). Most of the nuclear polyadenylated RNA was found to lie between dense regions of chromatin (Figure 12-32b). Other studies using anti-Sm antibody to stain snRNPs have shown that most of the cellular spliceosomal snRNPs are also localized to these discrete foci.

To analyze the relationship of a specific nuclear RNA to these foci, total nuclear polyadenylated RNA and the abundant, large (≈75-kb) fibronectin primary transcript were observed simultaneously. Most of the fibronectin nuclear RNA was associated with one or two of the foci (Figure 12-33a). The fibronectin gene, in turn, was closely associated with the fibronectin transcripts (Figure 12-33b). These results indicate that transcription and RNA processing do not occur randomly throughout the nucleus; rather, the nucleus is organized into discreet domains (≈20–100 in human fibroblasts) where the bulk of transcription and RNA processing occurs.

This highly organized view of the nucleus implies that there is an underlying nuclear substructure. It has been known for many years that when mammalian cells are treated with a mild nonionic detergent, digested extensively with DNase I to remove most of the DNA, and extracted with high concentrations of salt, a fibrillar network of protein and RNA remains in the region of the nucleus (Figure 12-34). This protein network has been called the *nuclear matrix*, or *nuclear skeleton*. The multiple protein components of the nuclear matrix prepared in this way have not been fully characterized. However, snRNPs remain associated with the nuclear matrix prepared from detergent-extracted, DNase I–treated cells. Moreover, when the nuclear matrix is prepared with a low concentration of salt, pre-mRNAs associated with the matrix un-

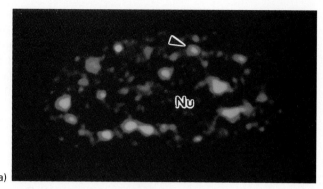

(a)

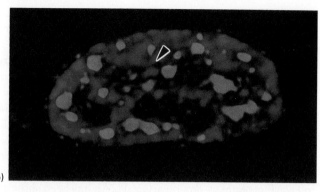

(b)

▲ FIGURE 12-32 Foci of polyadenylated RNA and RNA splicing factors in the nucleus of a mammalian fibroblast. Digital imaging microscopy was used to reconstruct a 1-μm-thick section of a stained human fibroblast nucleus. (a) Section stained with red rhodamine-labeled poly-dT to detect polyadenylated RNA and with a green fluorescein-labeled monoclonal antibody to detect the essential RNA-splicing protein SC-35. Regions where the stains overlap appear yel-

low. The arrow points to one focus. Nu = nucleolus. (b) The same section shown in (a) stained to detect polyadenylated RNA (red) and DNA (blue), which was visualized with the DNA-specific dye DAPI. The arrow points to a region that contains low levels of DNA but no detectable polyadenylated RNA, demonstrating that polyadenylated RNA does not occur in all areas between regions of chromatin. [From K. C. Carter et al., 1993, *Science* **259**:1330.]

(a)

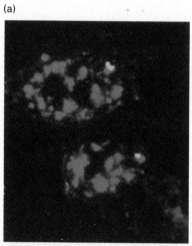

(b)

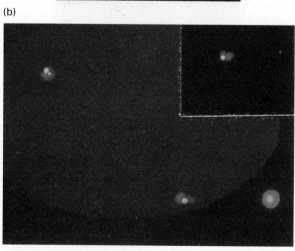

◄ FIGURE 12-33 Localization of polyadenylated RNA, fibronectin RNA transcripts, and the fibronectin gene in a fibroblast nucleus by immunofluorescence microscopy. (a) Two nuclei stained with red rhodamine-labeled poly-dT to detect polyadenylated RNA and with green fluorescein-labeled fibronectin cDNA to detect fibronectin transcripts. The fibronectin signal, localized to a few loci, appears yellow because it overlaps the red polyadenylated RNA. (b) A diploid fibroblast nucleus in which the DNA sequence just upstream of the fibronectin gene was detected by hybridization to digoxigenin-labeled DNA followed by treatment with red rhodamine-labeled anti-digoxigenin antibody. Fibronectin RNA transcripts were detected by hybridization to biotinylated DNA from within the transcribed region of the fibronectin gene followed by treatment with green fluorescein-labeled anti-biotin antibody. Note that the two fibronectin gene regions (red and yellow) are closely associated with the RNA transcripts (green). The inset shows another example of a region of hybridization in a fibroblast nucleus at higher magnification. The red dot in the lower right corner is a fluorescent bead marker. [From Y. Xing et al., 1993, *Science* **259**:1326.]

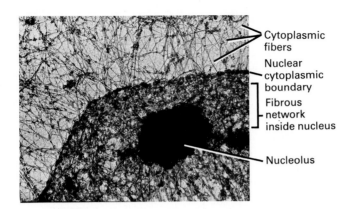

Cytoplasmic fibers

Nuclear cytoplasmic boundary

Fibrous network inside nucleus

Nucleolus

◄ FIGURE 12-34   Transmission electron micrograph showing the nuclear matrix (skeleton) of a HeLa cell. Cells were treated with a nonionic detergent to remove membranes; digested with DNase to remove most of the DNA, histones, and chromatin-associated protein; and then extracted with 0.25 M ammonium sulfate. A whole mount of the remaining material was prepared. [From S. Penman et al., 1982, *Cold Spring Harbor Symp. Quant. Biol.* **46**:1013.]

dergo splicing when ATP is added. These results suggest that the RNA-processing foci observed microscopically may be associated with specific regions of the nuclear matrix.

Messenger Ribonucleoproteins (mRNPs) Exit the Nucleus through Nuclear Pore Complexes

It is not yet clear how fully processed mRNAs, associated with hnRNP proteins (as *nuclear mRNPs*), are transported from the foci where RNA splicing takes place to the nuclear pores through which they pass to the cytoplasm. The double-membrane nuclear envelope is a specialized extension of the rough endoplasmic reticulum (see Figure 5-33). Nuclear pore complexes, which span the double membrane, are well visualized in electron micrographs of the nuclear envelope microdissected from the large nuclei of amphibian oocytes (Figure 12-35). Nuclear pore complexes are composed of dozens of proteins with a total molecular mass of ≈120 million daltons. They are immense compared with the eukaryotic ribosome of ≈4 mil-

(a)

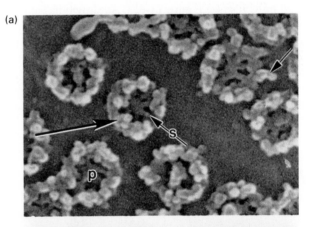

(b)

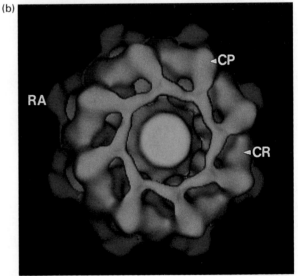

➤ FIGURE 12-35   Structure of the nuclear pore complex. (a) Scanning electron micrograph of nuclear envelopes from a *Xenopus* oocyte viewed from the cytoplasmic face. p = plugs; s-labeled arrow = spokes; unlabeled arrows point to globular and rod-shaped particles associated with nuclear pore structures. (b) Model of nuclear pore complex viewed from the cytoplasm prepared by computer-assisted processing of cryoelectron micrographs. CP = cytoplasmic particle; CR = cytoplasmic ring; RA = radial arm. (c) Schematic diagram of cross-section of nuclear pore complex embedded in the nuclear envelope. [Part (a) from M. W. Goldberg and T. D. Allen, 1993, *J. Cell Sci.* **106**:261; part (b) from C. W. Akey and M. Radermacher, 1993, *J. Cell Biol.* **122**:1.]

(c)

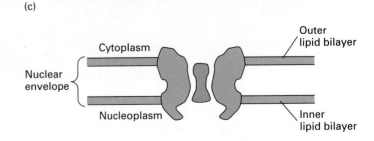

Cytoplasm

Outer lipid bilayer

Nuclear envelope

Nucleoplasm

Inner lipid bilayer

lion daltons. The nuclear pore complex functions as a gated channel through which ribonucleoproteins are selectively transported to the cytoplasm by a mechanism that requires ATP hydrolysis.

In one revealing analysis, mRNPs containing an unusually large, abundantly expressed mRNA were visualized by electron microscopy as they passed through nuclear pores in the salivary glands of larvae of the insect *Chironomous tentans*. The ≈30-kb mRNAs of these mRNPs encode secreted proteins that glue the insect larvae to a solid support, such as a twig, in preparation for metamorphosis. They are expressed from genes in large chromosomal puffs called Balbiani rings, where nascent pre-mRNAs associate with hnRNP proteins to form 50-nm coiled hnRNPs (see Figure 11-67). As shown in Figure 12-36, the nuclear mRNPs also are coiled but appear to uncoil during their passage through nuclear pores; as they enter the cytoplasm, the mRNPs are bound by ribosomes. The observation that mRNPs become associated with ribosomes during transport indicates that the 5′ end leads the way through the nuclear pore complex. Even though ribo-somes engage the 5′ end of an mRNP as it enters the cytoplasm, it is unlikely that the process of protein synthesis "pulls" the mRNA into the cytoplasm, since inhibitors of protein synthesis do not immediately block the transport of newly synthesized mRNA.

5′-Cap Structures Are Recognized by the Nuclear Transport Mechanism

Studies of snRNA transport have demonstrated that the 5′ cap of RNAs is somehow recognized by the nuclear transport mechanism, supporting the notion that mRNPs move through the nuclear pore complex in the 5′ → 3′ direction. U1, U2, U4, and U5 snRNAs, which are synthesized by RNA polymerase II, initially are produced with a m^7G cap identical to that of pre-mRNAs at their 5′ ends (see Figure 4-24). Following their synthesis, these snRNAs are transported to the cytoplasm where they assemble with Sm protein and other proteins to form the corresponding snRNPs. Before the assembled snRNPs are transported

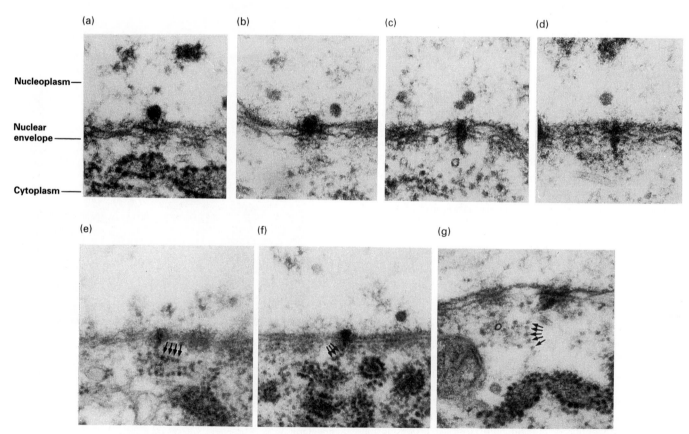

Nucleoplasm —

Nuclear envelope —

Cytoplasm —

▲ FIGURE 12-36   Electron micrographs of Balbiani ring mRNPs passing through nuclear pore complexes in salivary glands of a *Chironomous tentans* larva. (a–d) The mRNPs appear to uncoil as they pass through a nuclear pore. (e–g) As they enter the cytoplasm, the mRNPs appear to associate with ribosomes (arrows). [From H. Mehlin, B. Daneholt, and U. Skoglund, 1992, *Cell* **69**:605.]

back to the nucleus where they participate in RNA splicing, the 5′ cap is further methylated to yield $m^{2,2,7}G$. In contrast, U6 snRNA, which is synthesized by RNA polymerase III, has a nucleotide triphosphate at its 5′ end. U6 snRNA is not transported into the cytoplasm following its synthesis but is retained in the nucleus. Here it associates with the U6 snRNP–specific proteins, which are transported into the nucleus from their site of synthesis in the cytoplasm.

The discovery that capped snRNAs initially are transported into the cytoplasm, but that uncapped U6 snRNA is not, suggested that the cap structure might be required for transport to the cytoplasm. To test this idea, researchers constructed a U1 snRNA gene in which the U6 promoter, recognized by RNA polymerase III (Pol III), replaced the normal U2 promoter, recognized by RNA polymerase II (Pol II). Unlike most promoters recognized by Pol III, the entire U6 promoter lies upstream of the transcribed region. The wild-type U1 snRNA gene and the engineered U1 snRNA gene containing the Pol III promoter were separately microinjected into *Xenopus* oocyte nuclei along with radiolabeled GTP. The results of this experiment, described in Figure 12-37, showed that U1 snRNA must have the usual 5′ cap to be transported to the cytoplasm.

(a)

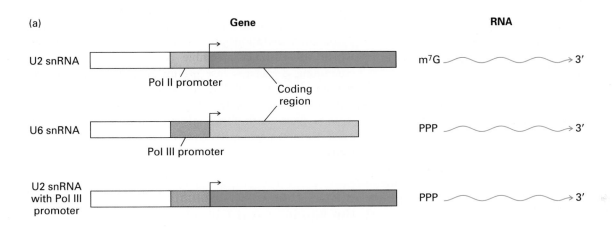

(b)

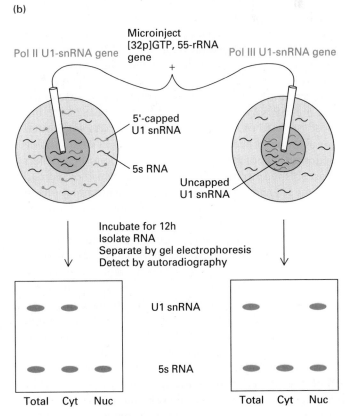

▲ FIGURE 12-37 Experimental demonstration that transport of U1 snRNA from the nucleus to the cytoplasm requires a 5′ cap structure. (a) Diagrams of genes encoding U1 snRNA and U6 snRNA, which are transcribed by RNA polymerase II and III, respectively, and of an engineered U1 gene with a Pol III promoter. Newly transcribed wild-type (Pol II) U1 snRNA has a 5′ cap (m^7G), whereas U6 snRNA and Pol III U1 snRNA produced from the engineered gene do not. (b) Diagram of experimental protocol and results. Multiple copies of the wild-type U1-snRNA gene and of the engineered Pol III U1-snRNA gene were microinjected separately into *Xenopus* oocyte nuclei. [^{32}P]GTP was coinjected to label the RNA synthesized; the 5S-rRNA gene served as an internal control for RNA synthesis. Both injected snRNA genes had a mutation that prevented any U2 snRNA that entered the cytoplasm from assembling into snRNPs and reentering the nucleus. After incubation, RNA from unfractionated oocytes (Tot) and from the cytoplasm (Cyt) and nucleus (Nuc) of other oocytes was isolated and analyzed. The resulting autoradiograms diagrammed at the bottom show that almost all the capped U1 snRNA was transported to the cytoplasm, whereas all the uncapped U1 snRNA was retained in the nucleus. Synthesis and transport of the control 5S rRNA was similar in cells injected with either type of U1-snRNA gene. [Adapted from J. Hamm and I. W. Mattaj, 1990, *Cell* **63**:109.]

In other experiments, U1 snRNAs synthesized in vitro were microinjected into *Xenopus* oocyte nuclei. Only U1 snRNA with the usual 5′ monomethylated cap was transported into the cytoplasm; if it lacked a cap or had the trimethylated cap normally synthesized in the cytoplasm, the synthetic snRNA remained in the nucleus. Moreover, transport of U1 snRNA containing the usual 5′ cap was inhibited if a high concentration of a cap analog was coinjected with the snRNA. This cap analog presumably competes with capped U1 snRNA for binding to an unidentified protein that recognizes cap structures and is required for transport of a large class of RNPs to the cytoplasm. It is not known whether this protein is a component of the nuclear pore complex or a separate nuclear protein that binds to cap structures and then interacts with the nuclear pore complex.

Pre-mRNAs Associated with Spliceosomes Are Not Transported to the Cytoplasm

It is critical that only fully processed mRNAs be transported from the nucleus to the cytoplasm, because translation of incompletely processed pre-mRNAs containing introns would produce defective proteins, which might interfere with the function of the corresponding normal proteins. We describe two types of experiments that indicate the association of incompletely processed pre-mRNAs with spliceosomes prevents their transport to the cytoplasm.

In one type of experiment, a gene encoding a pre-mRNA with a single intron that is efficiently spliced out was mutated to introduce deviations from the consensus splice-site sequences. Mutation of the invariant GT at the 5′ end of the intron or of the invariant AG at the 3′ end resulted in pre-mRNAs that were retained in the nucleus. However, mutation of both the 5′ and 3′ splice sites in the same pre-mRNA resulted in efficient transport of the unspliced pre-mRNA. When mutant pre-mRNAs with mutations in a single splice site were analyzed in in vitro splicing reactions, the mutant pre-mRNAs were bound by snRNPs to form a spliceosome, but RNA splicing was blocked and the RNA was not released. Pre-mRNAs with mutations in both the 5′ and 3′ splice sites were not efficiently bound by snRNPs.

A similar phenomenon is observed in patients with thalassemia, an inherited disease that results in abnormally low levels of globin proteins. Many instances of this disease are due to mutations in globin-gene splice sites that decrease the efficiency of splicing. The resulting unspliced globin pre-mRNAs are retained in reticulocyte nuclei and rapidly degraded. A low rate of RNA splicing may occur at *cryptic splice sites*, sequences in the pre-mRNA that are inefficiently recognized by snRNPs and do not serve as splice sites unless an authentic splice site is mutated. The corresponding abnormally spliced globin RNAs are found at low levels in the cytoplasm of thalassemic reticulocytes.

Genetic experiments in yeast also have been used to assess the role of spliceosomes in retaining associated pre-mRNAs in the nucleus. In this work, a synthetic intron was inserted into a β-galactosidase gene in such a way that only the unspliced pre-mRNA could be translated to produce a functional enzyme; RNA splicing of the pre-mRNA from the engineered gene caused a frameshift in the mRNA sequence that prevented synthesis of a functional enzyme. Wild-type cells transformed with the synthetic gene did not produce β-galactosidase from it because the unspliced pre-mRNA was not transported to the cytoplasm. Yeast mutants capable of transporting this unspliced pre-mRNA were identified because they expressed higher levels of β-galactosidase than wild-type cells. These mutant strains were found to contain mutations either in the genes encoding the PRP proteins required for formation of spliceosomes or in the DNA sequence encoding the bases in U1 snRNA that base-pair with pre-mRNAs (see Figure 12-23a). These mutations are thought to decrease the rate at which spliceosomes assemble on the synthetic intron-containing β-galactosidase pre-mRNA, allowing some of the unspliced pre-mRNA to be transported to the cytoplasm, where it is translated into functional enzyme. Results from both types of experiments indicate that association of a pre-mRNA with a spliceosome blocks its transport to the cytoplasm.

mRNA Remains Associated with Protein in the Nucleus and Cytoplasm

Our discussion so far indicates that pre-mRNA and mRNA is associated with various proteins in ribonucleoprotein complexes. We can summarize the protein components involved in processing of pre-mRNAs and transport of mRNPs to the cytoplasm as shown in Figure 12-38. A nascent mRNA transcript associates with hnRNP proteins while transcription is in process. Likewise, in many cases, spliceosomes assemble at splice sites in a nascent transcript, so that splicing at some sites may occur before cleavage and polyadenylation at the 3′ end; the latter can only occur after transcription is complete. After spliceosomes and the 3′-processing complex dissociate from a mature mRNA, some of the hnRNP proteins are removed as the resulting nuclear mRNP is transported through nuclear pores, with the 5′ end going first. In the cytoplasm, the hnRNPs that remain associated with the mature mRNA dissociate and return to the nucleus. Cytoplasmic proteins associate with the mRNA, forming cytoplasmic mRNPs. One of these, the poly-A–binding protein (PABP), covers the 3′ poly-A tail of the mRNA.

Transport of mRNPs to the Cytoplasm Is Regulated by Some Viral Proteins

Clearly, transport of mRNPs containing mature, functional mRNAs, from the nucleus to the cytoplasm, involves

Although the mechanism underlying the preferential transport of adenovirus mRNAs is not understood, one intriguing hypothesis is that an hnRNP protein required for transport becomes limiting late in infection, when prodigious amounts of viral mRNAs are synthesized. According to this hypothesis, the viral E1B-E4 complex, which is localized to nuclear foci of viral replication and transcription, concentrates the postulated limiting hnRNP transport protein near these viral replication/transcription centers, so that it can preferentially bind to RNA transcripts from the viral DNA.

HIV Rev Protein A retrovirus, HIV integrates a DNA copy of its RNA genome into the host-cell DNA. The integrated viral DNA, or provirus, contains a single transcription unit. The single primary transcript can be spliced in alternative ways to yield three classes of mRNAs: a 9-kb unspliced mRNA; ≈4-kb mRNAs formed by removal of one intron; and ≈2-kb mRNAs formed by removal of two or more introns (Figure 12-39a). After their synthesis in the host-cell nucleus, all three classes of HIV mRNAs are transported to the cytoplasm and translated into viral proteins; some of the 9-kb unspliced RNA is used as the viral genome in new viral particles.

Studies with HIV mutants showed that a virus-encoded protein called *Rev* is required for transport of unspliced and singly spliced viral mRNAs from the nucleus to the cytoplasm of infected cells (Figure 12-39b). Subsequent biochemical experiments demonstrated that Rev binds to a specific *Rev-response element* (RRE) present in HIV RNA. In cells infected with HIV mutants lacking the RRE, unspliced and singly spliced viral mRNAs remain in the nucleus, demonstrating that RRE is required for Rev-mediated stimulation of transport.

Subsequent experiments, based on those described earlier with genes mutated at the 5′ and/or 3′ splice sites, have given insight into how binding of Rev protein to RRE stimulates transport of certain HIV mRNAs. Recall that pre-mRNAs with a mutation in the 5′ or 3′ splice site are assembled into spliceosomes but cannot complete the splicing reaction; because the mutant pre-mRNAs are not released from spliceosomes, they are not transported into the cytoplasm. When a RRE is engineered into a pre-mRNA with a mutation in the 5′ or 3′ splice site, the unspliced pre-mRNA is transported into the cytoplasm in cells expressing Rev. This finding indicates that binding of Rev to a pre-mRNA somehow permits the spliceosome-associated RNA to be transported to the cytoplasm.

In cells infected with *rev*-HIV mutants, only multiple spliced viral mRNAs are released from spliceosomes and transported to the cytoplasm. However, in cells infected with wild-type HIV, Rev binds to the RRE in the viral primary transcript, allowing the spliceosome-associated unspliced and singly spliced mRNAs to be transported (see Figure 12-39a). How Rev accomplishes this is the subject of intense investigation, since development of a drug that

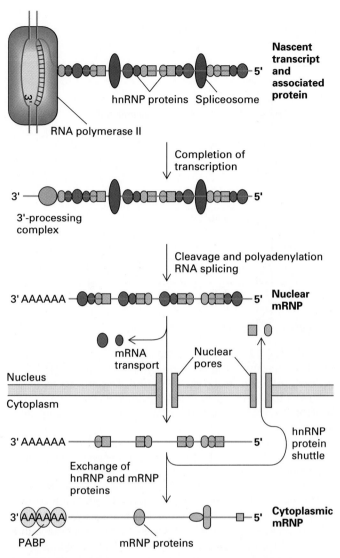

▲ FIGURE 12-38 Summary of protein components involved in synthesis and transport of mature mRNA. PABP = poly-A–binding protein. See text for discussion.

a complex mechanism that is crucial to gene expression. In theory, regulation of this transport step could provide another means of gene control. To date, however, only a few examples of such regulation, all involving viral mRNAs, have been discovered.

Preferential Transport of Adenovirus mRNAs
During the late phase of adenovirus infection, two viral proteins designated E1B and E4 form a complex that stimulates the transport of mature viral mRNAs and inhibits the transport of most cellular mRNAs. Remarkably, mRNAs produced from any gene introduced into the genome of a recombinant adenovirus are preferentially transported over cellular mRNAs, indicating that no specific sequence in viral mRNAs is required for this preferential transport.

(a)

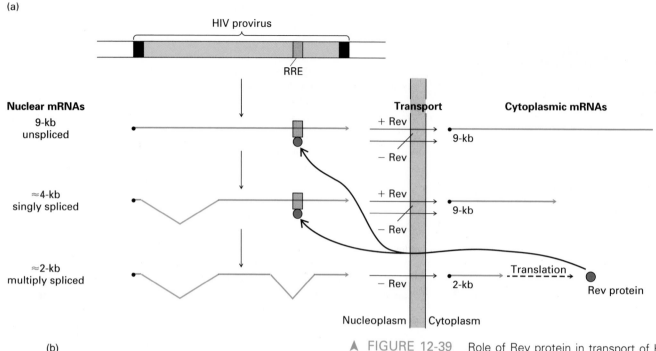

(b)

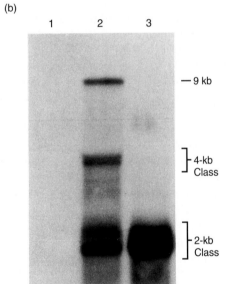

▲ FIGURE 12-39 Role of Rev protein in transport of HIV mRNAs from the nucleus to the cytoplasm. (a) The HIV genome, which contains overlapping coding regions, is transcribed into a single 9-kb primary transcript. Several ≈4-kb mRNAs result from splicing out of one intron (blue angled line), and several ≈2-kb mRNAs from splicing out of two or more introns. After transport to the cytoplasm, each RNA species is translated into different viral proteins. Rev protein, encoded by a 2-kb mRNA, interacts with the Rev-response element (RRE) in the unspliced and singly spliced mRNAs, stimulating their transport to the cytoplasm. (b) Northern blot of cytoplasmic HIV RNAs from uninfected cells (lane 1), cells infected with wild-type HIV (lane 2), and cells infected with a rev^- HIV mutant (lane 3). [Adapted from B. R. Cullen and M. H. Malim, 1991, *Trends Biochem. Sci.* **16**:346.]

could enter infected cells and block Rev function would prevent HIV replication. Such a drug might be an effective therapy to prevent development of AIDS in HIV-infected individuals.

► Regulation of mRNA Cytoplasmic Localization, Stability, and Translation

Like nuclear pre-mRNAs and mature mRNAs before transport, cytoplasmic mRNAs are associated with protein in ribonucleoprotein complexes. However, the ratio of protein to RNA is lower in cytoplasmic mRNPs than in nuclear mRNPs (see Figure 12-38). The most abundant cytoplasmic mRNP protein is the cytoplasmic poly-A–binding protein (PABP), which is tightly associated with the poly-A tails of mRNAs. In mammalian cells, PABP is distinct from the nuclear poly-A–binding protein, PABII, involved in the synthesis of the poly-A tail (see Figure 12-16).

The translation of many mRNAs in the cytoplasm does not take place free in solution. For example, polyribosomes can be intimately associated with membranes of the rough endoplasmic reticulum. The proteins synthesized at these sites are exported from the cell or become components of new membranes. Some mRNAs encoding proteins

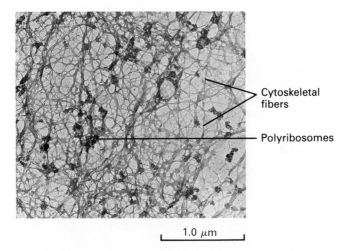

▲ FIGURE 12-40 A transmission electron micrograph of a whole mount of a human diploid fibroblast. The cell was prepared by removing lipids and most proteins with mild detergents. The remaining cytoskeletal structure then was fixed with glutaraldehyde. The area of the cytoplasm shown contains many fibers and dark groups of objects whose approximate diameters are equal to those of ribosomes. [From S. Penman et al., 1982, *Cold Spring Harbor Symp. Quant. Biol.* **46**:1013.]

destined to associate with the cytoskeleton, a cytoplasmic network of diverse fibers, are also not freely diffusable.

The association of some mRNAs with the cytoskeleton has been demonstrated by microscopic examination of individual cells extracted with a mild nonionic detergent, which dissolves lipid membranes but does not denature proteins. If cultured cells are treated briefly with such a detergent, the cell membrane becomes porous and many proteins (about 75 percent of the total) and other materials (e.g., more than 90 percent of the tRNAs) wash out of the cells. Many polyribosomes, however, are retained in such preparations (Figure 12-40). Digestion of these preparations with a combination of ribonuclease A and ribonuclease T1 suggests that poly-A tails are associated with the cytoskeleton. Ribonuclease A hydrolyzes phosphodiester bonds that are 3′ of Cs and Us, and ribonuclease T1 hydrolyzes bonds 3′ of Gs. Consequently, digestion of RNA with a combination of these two enzymes degrades most of the RNA, but leaves poly-A stretches undigested. When detergent-extracted cells are digested with ribonucleases A and T1, most of the resistant poly-A remains associated with the cytoskeletal preparation. Treatment of cells with cytochalasin D, which disrupts actin microfilaments (Chapter 22), causes most of the poly-A to be released when the cells are subsequently extracted with a mild detergent. These observations have led to the suggestion that the poly-A–binding protein associated with mRNA poly-A tails may associate with the portion of the cytoskeleton made of actin microfilaments.

Some mRNAs Are Directed to Specific Cytoplasmic Sites by Sequences in Their 3′ Untranslated Region

Among the many mRNAs synthesized within a cell, some have the unusual property of being localized to specific regions of the cell cytoplasm. In all cases examined thus far, this localization is specified by sequences in the 3′ untranslated region of the mRNA. Such mRNA localization can be visualized in mammalian myoblasts (muscle precursor cells) as they differentiate into myotubes, the fused, multinucleated cells that make up muscle fibers. Myoblasts are motile cells that extend cytoplasmic regions, called lamellipodia, from the leading edge (see Figure 5-19). Extension of lamellipodia during cell movement requires polymerization of β-actin. Sensibly, β-actin mRNA is concentrated in the leading edges of myoblasts, the region of the cell cytoplasm where the encoded protein is needed for motility.

When myoblasts fuse into syncytial myotubes, β-actin expression is repressed and the muscle-specific α-actin is induced. In contrast to β-actin mRNA, α-actin mRNA is restricted to the perinuclear regions of myotubes. Both mRNAs can be observed in cultured myoblast cells that are in the process of differentiating. In these cells, β-actin mRNA is localized at the periphery of the cell and α-actin mRNA is restricted to the perinuclear region (Figure 12-41). To test the ability of actin mRNA sequences to direct the cytoplasmic localization of an mRNA, fragments of α- and β-actin cDNAs were inserted into a plasmid vector that expresses β-galactosidase from a strong viral promoter (Figure 12-42a). The expression vector was constructed so that the actin sequences would be incorporated into the 3′ untranslated region of the β-galactosidase mRNAs expressed from the vectors. Differentiating myoblasts were transfected with the recombinant expression vectors and later were treated with X-gal to detect β-galactosidase activity. When hydrolyzed by β-galactosidase, X-gal produces an insoluble blue compound. The results showed that inclusion of the 3′ untranslated end of α- or β-actin cDNAs directed localization of the expressed β-galactosidase, whereas the 5′ untranslated and coding regions did not (Figure 12-42b,c).

How the 3′ untranslated region of actin mRNAs induce specific cytoplasmic localization is not yet understood. Treatment of cultured myoblasts with cytochalasin D, which disrupts actin microfilaments, leads to a rapid delocalization of actin mRNA, showing that cytoskeletal actin microfilaments are required. Disruption of other cytoskeletal components, however, does not alter the localization of actin mRNAs. Presumably proteins bind to specific sequences in the 3′ untranslated regions of actin mRNAs and interact with specific components of microfilaments, possibly including motor proteins that translocate along the length of microfilaments (Chapter 22).

The most elaborate examples of mRNA localization involve certain mRNAs in the cytoplasm of *Drosophila*

(a)

(b)

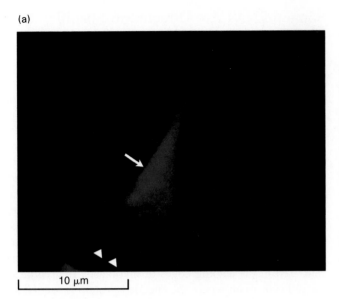

10 μm

▲ FIGURE 12-41 Localization of α- and β-actin mRNAs in a differentiating myoblast. The mRNAs were detected by in situ hybridization with a fluorescent-labeled oligonucleotide probe specific for α-actin mRNA (green) or β-actin mRNA (red). Some of the β-actin mRNA is concentrated near the leading edge of the cell (arrowheads), whereas the α-actin mRNA in the same cell is confined to the perinuclear region (arrow). [From E. H. Kislauskis et al., 1993, *J. Cell. Biol.* **123**:165.]

oocytes. Messenger RNAs encoded by the *bicoid* and *nanos* genes are synthesized in nurse cells, which surround the developing oocyte, and then move through cytoplasmic bridges into the oocyte. The *bicoid* and *nanos* mRNAs become localized in small regions at the anterior and posterior poles of the oocyte, respectively. Localization of these mRNAs is critical to establishing Bicoid and Nanos protein gradients in the early embryo, which are essential for normal embryonic development (Chapter 13). Analysis of *Drosophila* mutants defective in early embryonic development has identified three genes required for localization of *bicoid* mRNA. Mutants in these genes alter *bicoid* mRNA distribution at particular points during the localization process. One of these genes, *staufen,* encodes an RNA-binding protein that is believed to bind to sequences in the 3′ untranslated region of *bicoid* mRNA required for its localization. Localization of *nanos* mRNA may be even more complicated, since eight genes have been identified that are required for its localization.

Stability of Cytoplasmic mRNAs Varies Widely

The concentration of an mRNA is a function of both its rate of synthesis and its rate of degradation. The same rate of transcription produces a much higher steady-state concentration of an mRNA with a longer lifetime than of an mRNA that is degraded rapidly once it reaches the cytoplasm. The stability of an mRNA also determines how rapidly the synthesis of the encoded protein can be shut down. For a stable mRNA, synthesis of the encoded protein persists long after transcription of the gene is repressed. Most bacterial mRNAs are unstable; they decay exponentially with a half-life typically of a few minutes (Table 12-1). This allows the cell to rapidly readjust the synthesis of proteins to accommodate changes in the cellular environment. Most cells in multicellular organisms, on the other hand, exist in a fairly constant environment and carry out a specific set of functions over periods of days to months or even the lifetime of the organism (nerve cells, for example). Accordingly, most mRNAs of higher eukaryotes have half-lives of many hours.

However, some proteins in eukaryotic cells are required only for short periods of time and must be expressed in bursts. For example, hormones called lymphokines, which coordinate cell-cell interactions between the cells involved in the immune response of mammals (Chapter 27) are synthesized and secreted in short bursts. Similarly, many of the transcription factors that regulate the onset of the S phase of the cell cycle (Chapter 25) are synthesized for brief periods only. The mRNAs encoding these proteins have unusually short half-lives, and their expression is controlled by rapidly turning gene transcription on and off. Since the mRNAs are short lived, this results in the required burst of protein synthesis. Many of these short-

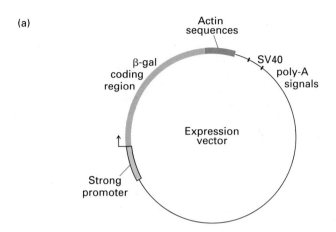

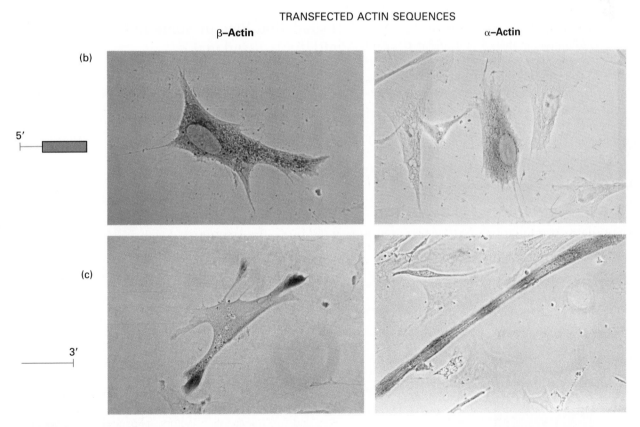

▲ FIGURE 12-42 Experimental demonstration that 3′ untranslated region of α- and β-actin mRNAs direct localization of a β-galactosidase reporter gene. (a) Diagram of β-galactosidase expression vector. A cDNA (red) corresponding to the 5′ untranslated and coding sequences *or* 3′ untranslated region of α- or β-actin mRNA was linked 3′ of the β-galactosidase coding region (blue). The vector also carried a strong promoter (yellow) from Rous sarcoma virus and two poly-A signals from SV40. The resulting plasmids were transfected separately into differentiating myoblasts. Subse- quently, the cells were stained with X-gal to detect β-galactosidase. (b) Transfection with plasmids containing the 5′ untranslated and coding sequences of actin cDNAs resulted in no localization of β-galactosidase, indicated by diffuse blue color resulting from hydrolysis of X-gal. (c) In contrast, transfection with 3′ untranslated regions led to localization of β-galactosidase to lamellipodia (β-actin) or peri- nuclear regions (α-actin). [Photographs from E. H. Kislaukis et al., 1993, *J. Cell. Biol.* **123**:165.]

TABLE 12-1   Half-Lives of Messenger RNAs

Cell	Cell Generation Time	mRNA Half-Lives[*]	
		Average	Range Known for Individual Cases
Escherichia coli	20–60 min	3–5 min	2–10 min
Saccharomyces cerevisiae (yeast)	3 h	22 min	4–40 min
Cultured human or rodent cells	16–24 h	10 h	30 min or less (histone and *c-myc* mRNAs) 0.3–24 h (specific mRNAs of cultured cells)

[*]For information on specific mRNA half-lives for *E. coli*, see A. Hirashima, G. Childs, and M. Inouye, 1973, *J. Mol. Biol.* **119**:373; for yeast, see L.-L. Chia and C. McLaughlin, 1979, *Mol. Gen. Genet.* **170**:137; and for mammalian cells, see M. M. Harpold, M. Wilson, and J. E. Darnell, 1981, *Mol. Cell Biol.* **1**:188.

lived mRNAs contain multiple copies of the sequence AUUUA, often overlapping, in their 3' untranslated region (Figure 12-43a). When these AU-rich sequences are inserted into the 3' untranslated region of genes encoding stable mRNAs, such as the globin genes, the resulting recombinant mRNAs are unstable (Figure 12-43b). The mechanism by which these sequences destabilize mRNAs is not yet understood.

Degradation Rate of Some Eukaryotic mRNAs Is Regulated

In several cases, the degradation rate of specific eukaryotic mRNAs is regulated. For example, when rat mammary tissue is cultured in the presence of the hormone prolactin, the concentration of the mRNA encoding the milk protein casein is about 30,000 molecules per cell. When the me-

(a) 3' Sequences of unstable mRNAs

GMCSF   UAAUAUUUAUAUAUUUAUAUUUUUAAAAUAUUUAUUUAUUUAUUUAUUUAA

IFN-β   UUUUGAAAUUUUUAUUAAAUUAUGAGUUAUUUUUAUUUAUUUAAAUUUUAUUUU

IL-1   UUAUUUUUUAAUUAUUAUUUAUAUAUGUAUUUAUAAAUAUAUUUAAGAUAAU

TNF   AUUAUUUAUUAUUUAUUUAUUAUUUAUUUAUUUA

c-Fos   GUUUUUAAUUUAUUUAUUAAGAUGGAUUCUCAGAUAUUUAUAUUUUUAUUUU

c-Myc   UAAUUUUUUUUAUUUAAGUACAUUUUGCUUUUUAAAGUUGAUUUUUUUCU

(b)   Activity of AUUUA repeats

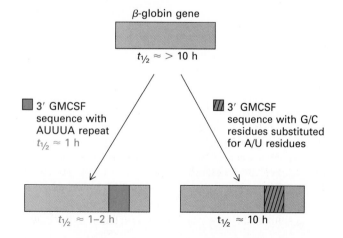

β-globin gene

$t_{1/2} \approx > 10$ h

3' GMCSF sequence with AUUUA repeat
$t_{1/2} \approx 1$ h

3' GMCSF sequence with G/C residues substituted for A/U residues
$t_{1/2} \approx 10$ h

$t_{1/2} \approx 1$–2 h $t_{1/2} \approx 10$ h

▲ FIGURE 12-43   (a) Sequences of 3' untranslated regions of several unstable mRNAs encoding four lymphokines and two transcription factors, c-Fos and c-Myc, which regulate entry into the S phase of the cell cycle. Note the characteristic AUUUA sequences (red), some of which are overlapping. Lymphokines: GMCSF = granulocyte-macrophage colony-stimulating factor; IFN-β = interferon β; IL-1 = interleukin 1; and TNF = tumor necrosis factor. (b) Destabilizing effect of AUUUA sequences on mRNA half-life ($t_{1/2}$). Cultured cells were transfected separately with expression vectors containing the diagrammed β-globin sequences and the half-life of the encoded mRNA was determined. Insertion of AUUUA sequences (red) into the β-globin gene dramatically reduced the stability of β-globin mRNA. [See G. Shaw and R. Kamen, 1986, *Cell* **46**:659.]

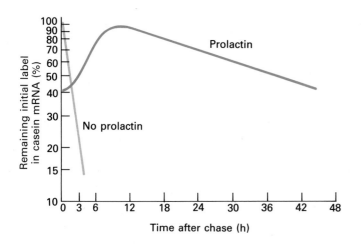

◄ FIGURE 12-44   Stabilization of casein mRNA by pro-lactin. Breast tissue from lactating rats was placed in a culture medium lacking prolactin for 4 h. One sample was then labeled with [³H]uridine in the presence of prolactin; another was labeled in its absence. After 3 h, the label was removed; a chase of unlabeled nucleosides was added; and samples were taken at intervals for determination of the remaining labeled casein mRNA by hybridization. The culture with pro-lactin accumulated labeled mRNA for several hours because of the ineffective chase of RNA precursors. The eventual decay of labeled prolactin mRNA in the presence of prolactin clearly was much less rapid than in the absence of prolactin. [See W. A. Guyette, R. J. Matusik, and J. M. Rosen, 1979, *Cell* **17**:1013.]

dium lacks prolactin, the level of casein mRNA falls a hundredfold to 300 molecules per cell. In vitro run-on analysis has shown that prolactin treatment of cultured breast tissue causes only a threefold increase in transcription of the casein gene. Thus most of the prolactin-induced increase in casein mRNA concentration results from an enormous increase in the stability of casein mRNA (Figure 12-44).

The mechanism regulating mRNA stability in cells of higher eukaryotes is best understood in the case of the mRNA encoding the transferrin receptor. In vertebrates, ingested iron is carried through the circulation bound to a protein called transferrin. Without transferrin, the solubility of iron ions is too low to provide for cellular requirements. The transferrin-iron complex is brought into cells by interacting with the specific cell-surface transferrin receptor. Intracellular concentrations of iron are regulated with great precision because excess iron produces hydroxyl radicals, which are extremely reactive and have a variety of toxic effects on cells. On the other hand, sufficient iron is required by multiple enzymes that use iron as a cofactor. When iron stores in the cell are sufficient, the import of the transferrin-iron complex is reduced by increasing the degradation rate of transferrin-receptor (TfR) mRNA, which quickly leads to a decrease in the level of transferrin receptor. When iron stores in the cell fall, TfR mRNA is stabilized, leading to increased synthesis of the receptor protein.

Analysis of deletion mutations in the TfR gene has revealed that this regulation of TfR mRNA stability depends on certain repeated sequences, called *iron-response elements* (IREs), in the 3′ untranslated region of the mRNA. Each IRE is ≈30 bases long and can form a stem-loop structure in which the loop contains five specific bases. The stems in the IREs in TfR mRNA contain AU-rich sequences similar to the 3′ untranslated AUUUA sequences that destabilize lymphokine mRNAs. When iron concentrations are adequate, these AU-rich sequences are thought to promote degradation of TfR mRNA by the same mechanism that leads to rapid degradation of lymphokine mRNAs.

When the iron concentration falls slightly, a cytoplasmic protein called *IRE-binding protein* (IRE-BP) binds to the IREs, recognizing the sequence in the loop and the secondary structure of the stem (Figure 12-45). The presence of IRE-BP bound to the IREs is thought to block recognition of the destabilizing AU-rich sequences by the proteins that would otherwise degrade TfR mRNA. These findings for TfR mRNA suggest that other mRNAs whose stability is known to be regulated also may contain response elements that interact with specific proteins, causing a decrease in their degradation rate.

Translation of a Few mRNAs Is Regulated by Specific RNA-binding Proteins

Regulation of the final step in gene expression, mRNA translation, has been demonstrated in a number of cases. One mechanism, involving a specific protein-mRNA interaction, regulates translation of mRNAs encoding some ribosomal proteins in bacteria and mRNAs encoding iron-metabolizing proteins in eukaryotes.

Translational Control of Ribosomal Protein Synthesis in Bacteria   One example of such translational regulation in bacteria involves ribosomal protein S8 (i.e., protein number 8 associated with the small ribosomal subunit). This ribosomal protein inhibits in vitro translation of the mRNA encoding ribosomal protein L5 (i.e., protein number 5 associated with the large subunit). A comparison of the sequences of a stem structure in 16S RNA and of the 5′ end of the mRNA encoding L5 protein suggests a physical basis for the translational arrest caused by excess S8 protein. During the assembly of bacterial ribosomes, S8 protein binds tightly to the stem portion of one particular stem-loop in 16S rRNA. The sequence of L5 mRNA in the region of the start codon can form a stem-loop whose sequence is similar to that of the S8-binding site in 16S rRNA

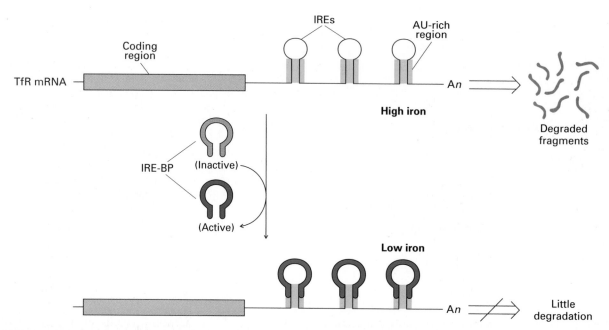

▲ FIGURE 12-45   Model for iron-dependent regulation of stability of transferrin-receptor (TfR) mRNA. The 3′ iron-response elements (IREs) in this mRNA have a stem-loop structure containing AU-rich sequences (yellow) that promote mRNA degradation. At low intracellular iron concentrations, the conformation of IRE-BP (dark green) is such that it binds to the IREs, thereby inhibiting degradation. As a result, the level of transferrin receptor increases, so that more iron can be brought into the cell.

(Figure 12-46). S8 protein can bind to this sequence in L5 mRNA, but has a higher affinity for its binding site in 16S rRNA. Consequently, when 16S rRNA is available, S8 protein binds to it preferentially and is assembled into ribosomes. But when more S8 protein than 16S rRNA is present, the excess S8 protein binds to L5 mRNA, inhibiting the initiation of translation by sterically blocking the interaction of the 30S ribosomal subunit with the initiation codon and Shine-Delgarno sequence in L5 mRNA. When subsequent transcription of rRNA genes produces more rRNA, S8 protein binds to its higher-affinity binding site in 16S RNA, releasing L5 mRNA and allowing further synthesis of L5 protein.

Control of Eukaryotic Translation by IRE-BP   At present, the only known examples of translational control in eukaryotes mediated by a specific RNA-binding protein

➤ FIGURE 12-46   The primary and secondary structures of portions of 16S rRNA and the mRNA encoding ribosomal protein L5 in bacteria. The red box in 16S rRNA indicates the binding site for S8 ribosomal protein. Green shading highlights the nucleotides that are identical in the two RNAs. The striking similarity in the sequences and stem-loop structures in these RNAs accounts for the ability of S8 protein to also bind to L5 mRNA, leading to arrest of its translation. [See P. O. Olins and M. Nomura, 1981, *Nucl. Acids Res.* **9**:1757.]

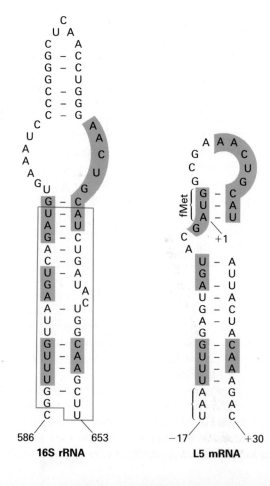

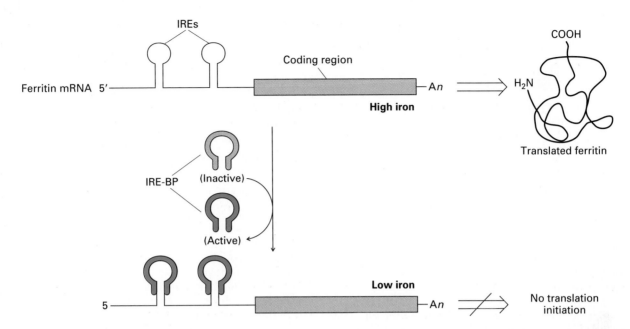

▲ FIGURE 12-47 Regulation of translation of ferritin mRNA by IRE-BP. At low iron concentrations, binding of active IRE-BP blocks translation initiation. The same mechanism controls translation of the mRNA encoding ALA syn-thase. In these two mRNAs, the IREs are located at the 5' end and do not contain the AU-rich sequences that promote mRNA degradation as they do in transferrin receptor mRNA (see Figure 12-45).

involves the mRNAs encoding several proteins that function in iron metabolism. IRE-BP, the protein that regulates degradation of TfR mRNA, also regulates the translation of the mRNAs encoding the two chains of ferritin and the enzyme 5-aminolevulinate (ALA) synthase in erythrocytes. Ferritin is an intracellular protein that binds iron ions, thereby preventing accumulation of toxic levels of free ions. When intracellular stores of iron are low, translation of ferritin mRNAs is repressed, so that the iron transported into the cell by the transferrin receptor is available for iron-requiring enzymes; when iron is in excess, ferritin synthesis is derepressed so that free iron ions are bound by newly synthesized ferritin. ALA synthase catalyzes the first step in synthesis of heme, the iron-containing prosthetic group in hemoglobin. When iron concentrations are low, production of ALA synthase is repressed; as a result, iron is shunted away from heme synthesis and is made available for iron-requiring enzymes essential for cell viability. When iron concentrations are adequate, production of ALA synthase is derepressed, permitting heme synthesis.

Translation of the ferritin and ALA synthase mRNAs is regulated by binding of IRE-BP to IREs (iron-response elements) at the 5' ends of the mRNAs (Figure 12-47). Unlike the 3' IREs in TfR mRNAs (see Figure 12-45), the 5' IREs in ferritin and ALA synthase mRNAs do not have AU-rich stems and do not promote mRNA degradation. When iron concentrations are low, IRE-BP is active and binds to the IREs, inhibiting translation initiation by blocking the ability of the 40S ribosomal subunit to scan for the first AUG from the 5' cap where it initially binds (see Figure 4-42). When iron concentrations are high, IRE-BP is inactive and does not bind to the 5' IREs, so that translation initiation can proceed.

Antisense RNA Regulates Translation of Transposase mRNA in Bacteria

An additional mechanism of translational control, called *antisense control*, occurs in bacterial cells. This type of regulation is mediated by antisense RNA, which contains sequences complementary to the region of an mRNA containing an initiation codon. Hybridization of the complementary antisense RNA blocks recognition of the initiation codon and binding of the 30S ribosomal subunit to the Shine-Delgarno sequence, thereby preventing initiation of translation.

Expression of the transposase encoded by the bacterial insertion sequence IS*10* is regulated by the antisense translation-control mechanism. As discussed in Chapter 9, transposase catalyzes transposition of this mobile DNA element (see Figure 9-17). If too much transposase were expressed, so many mutations would result from IS*10* transposition that the host cell would not survive. Normally, this does not occur because of antisense control. IS-*10* contains two promoters: one called P_{IN} directs transcription of the strand coding for transposase; the other called P_{OUT} lies within the transposase gene and directs transcription of the noncoding strand, producing an antisense RNA complementary to the 5' end of transposase mRNA (Figure 12-48). Because P_{OUT} is a much stronger promoter than P_{IN}, antisense mRNA is produced in greater abundance than transposase mRNA. Hybridization of the antisense RNA to most of the much rarer transposase

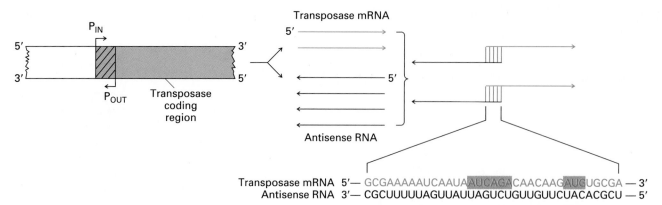

Transposase mRNA 5'— GCGAAAAAUCAAUAAUCAGACAACAAGAUGUGCGA — 3'
Antisense RNA 3'— CGCUUUUUAGUUAUUAGUCUGUUGUUCUACACGCU — 5'

▲ FIGURE 12-48   Antisense control of expression of IS 10 transposase. The 5' end of transposase mRNA (red) transcribed from P_{IN} is complementary to the 5' end of antisense RNA (black) transcribed from P_{OUT}. Because P_{OUT} is a much stronger promoter than P_{IN}, more antisense RNA than transposase mRNA is produced, so that nearly all of the transposase mRNA hybridizes to the more abundant antisense RNA. Since the AUG start codon and the ribosome-binding Shine-Delgarno sequence (green shading) are in the hybridized region, initiation of translation of transposase mRNA is blocked. Very rarely, translation of a transposase mRNA is initiated before hybridization to an antisense RNA, leading to a low rate of IS10 transposition.

mRNA prevents translation, thereby assuring that the rate of synthesis of transposase and, in turn, the frequency of transposition are compatible with survival of the host cells.

The strategy of antisense control has been adapted by researchers to inhibit gene expression experimentally in eukaryotic cells. In this approach, vectors are constructed that express high levels of a RNA complementary to the RNA transcript of a target gene. In some cases, introduction of such an antisense vector into cells has inactivated the target gene. This repression of gene expression, however, is not due to direct inhibition of translation as in bacterial systems. Rather, hybridization of the antisense RNA to the target-gene transcript interferes with RNA processing in the nucleus. In other cases, antisense expression vectors have failed to completely eliminate the activity of proteins encoded by some target genes, perhaps because even low levels of some proteins, particularly enzymes, may be sufficient to perform their required functions in the cell. Also, detailed analysis of antisense control in bacteria has revealed that the secondary structures of the complementary RNAs greatly influence the rate at which they hybridize, and thus the efficiency of antisense control. The RNAs transcribed from P_{OUT} and P_{IN} IS10 have been optimized by natural selection to hybridize at extremely high rates at physiological temperature and salt concentrations, so transposase expression is inhibited very effectively.

▶ Processing of rRNA and tRNA

Approximately 80 percent of the total RNA in rapidly growing mammalian cells (e.g., cultured HeLa cells) is rRNA, and 15 percent is tRNA; protein-coding mRNA thus constitutes only a small portion of the total RNA. The primary transcripts produced from most rRNA genes and from tRNA genes, like pre-mRNAs, are extensively processed to yield the mature, functional forms of these RNAs.

The 28S and 5.8S rRNAs associated with the large (60S) ribosomal subunit and the 18S rRNA associated with the small (40S) ribosomal subunit in higher eukaryotes are encoded by a single *pre-rRNA* transcription unit (see Figure 11-69). Transcription by RNA polymerase I yields a 45S (13.7-kb) primary transcript (pre-rRNA), which is processed into the mature 28S, 18S, and 5.8S rRNAs found in cytoplasmic ribosomes. Discovery of this processing was the first indication that mature cytoplasmic RNAs are derived from larger precursor RNAs synthesized in the nucleus. Although cytoplasmic tRNAs are much smaller than the rRNAs derived from pre-rRNA, they also are produced from larger precursors (pre-tRNAs), synthesized by RNA polymerase III, which undergo various types of processing. RNA polymerase III also catalyzes synthesis of 5S rRNA, the other RNA present in the large ribosomal subunit in eukaryotes. However, unlike pre-tRNAs and pre-rRNAs, the primary transcripts produced from 5S-rRNA genes do not undergo extensive processing. The synthesis and processing of pre-rRNA occurs in the nucleolus, whereas tRNAs and 5S RNA are produced in the nucleoplasm outside of the nucleolus.

Pre-rRNA Binds Proteins, Then Is Cleaved and Methylated in the Nucleolus

Eukaryotic cells contain multiple pre-rRNA genes arranged in tandem. These are transcribed in the nucleolus, and the nascent pre-rRNA transcripts are immediately

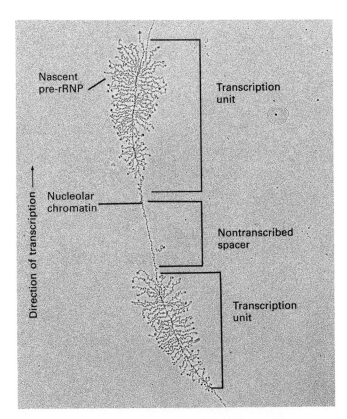

Direction of transcription →

Nascent pre-rRNP

Nucleolar chromatin

Transcription unit

Nontranscribed spacer

Transcription unit

▲ FIGURE 12-49   Electron micrograph of pre-rRNA transcription units from nucleolus of a frog oocyte. Each "feather" represents pre-rRNA molecule associated with protein in a pre-rRNP emerging from a transcription unit. Pre-rRNA transcription units are arranged in tandem, separated by spacer regions of nucleolar chromatin. [Courtesy of Y. Osheim and O. J. Miller, Jr.]

bound by proteins, forming pre-ribonucleoprotein particles, or pre-rRNPs (Figure 12-49). Several ribonucleoprotein particles of different sizes have been extracted from mammalian nucleoli. The largest of these (80S) contains an intact 45S pre-rRNA molecule. This RNA molecule is cut in a series of cleavage steps that ultimately yield the mature rRNAs found in ribosomes. The small (40S) ribosomal subunit containing 18S rRNA is finished and delivered to the cytoplasm much faster than is the large (60S) ribosomal subunit (Figure 12-50). Because of this difference in processing times, the pool of nuclear precursors for large ribosomal subunits is larger than that for small ribosomal subunits. In fact, the most prominent RNA component in HeLa-cell nucleoli is 32S pre-rRNA (equivalent to yeast 27S rRNA), the immediate precursor of 28S rRNA (equivalent to 25S rRNA in yeast). In growing HeLa cells, for example, there are about five times as many 32S molecules as intact 45S primary transcripts. Some of the proteins in the pre-rRNPs found in nucleoli remain associated with the mature ribosomal subunits; other of these proteins are re-

stricted to the nucleolus and assist in assembly of the subunits (Figure 12-51).

The cleavages that lead to the mature rRNAs occur at specific sequences in the pre-rRNA sequence and may be catalyzed by small nucleolar RNAs, or *snoRNAs*. These nucleolar RNAs, which range in size from 87 to 275 bases, include the U3 RNA mentioned earlier, as well as several other uridine-rich RNAs. Like snRNAs, the snoRNAs are associated with proteins in ribonucleoprotein complexes called *snoRNPs*. These snoRNPs have been shown to associate with pre-rRNA and the various processing intermediates in the nucleolus. Just as several of the snRNPs involved in pre-mRNA processing include Sm protein, several snoRNPs contain a 34-kDa protein called fibrillarin. Biochemical studies in vertebrates and analyses of yeast mutants indicate that some snoRNAs are required for specific cleavages in pre-rRNA. These findings suggest that all the cleavages in pre-rRNA may be catalyzed by snoRNAs, just as pre-mRNA splicing is thought to be catalyzed by snRNAs. Remarkably, a snoRNP designated MRP not only is required for pre-rRNA processing but also functions in mitochondria to process RNA primers for mitochondrial DNA synthesis. This intriguing discovery has led to the suggestion that MRP might help to coordinate mitochondrial replication with cell growth and division.

In addition to cleavage steps that shorten pre-rRNA (see Figure 12-50b), processing also involves methylation of pre-rRNA with methionine acting as the methyl donor. In human cells, more than 100 methyl groups are added to specific bases and to the riboses of specific ribonucleotides; most of these methyl groups can be detected even in nascent pre-rRNA molecules. In addition, a few methyl groups are added to 45S pre-rRNA molecules after transcription is completed, and four methyl groups are added to 18S rRNA in the cytoplasm. All of the methyl groups added to the pre-rRNA primary transcript are preserved during processing (i.e., they are found in 28S and 18S rRNA in the cytoplasm). The positions of these methyl groups are highly conserved in vertebrate cells (frogs and humans); many of the same sites also are present in yeast. If cells are deprived of methionine, processing of pre-rRNA is interrupted, suggesting that methyl groups have some role in processing.

As noted earlier, 5S-rRNA genes are transcribed by RNA polymerase III in the nucleoplasm outside of the nucleolus. Without further processing, 5S RNA migrates to the nucleolus, where it assembles with the 28S and 5.8S rRNAs and proteins into large ribosomal subunits (see Figure 12-50a). When assembly of ribosomal subunits in the nucleolus is complete, they move—by mechanisms as-yet unknown—to the cytoplasm, where they appear first as free subunits. Since the lumina of the nuclear pores are only about 10–20 nm in diameter and the large ribosomal subunit is about the same size, a conformational change in one or both of these structures may occur to permit passage of the subunit to the cytoplasm.

(a)

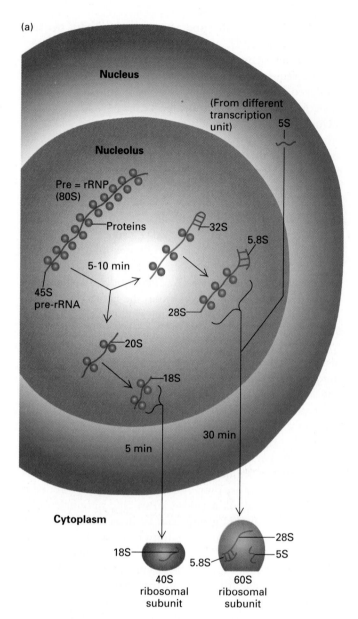

(b)

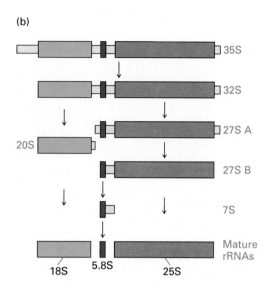

▲ **FIGURE 12-50**   Processing of pre-rRNA and assembly of ribosomes in eukaryotes. (a) Major intermediates and times required for various steps in pre-rRNA processing in higher eukaryotes. Ribosomal and nucleolar proteins associate with 45S pre-rRNA soon after its synthesis, forming an 80S pre-rRNP. Synthesis of 5S rRNA occurs outside of the nucleolus. The extensive secondary structure of rRNAs (see Figure 4-40) is not represented here. Note that RNA constitutes about two-thirds of the mass of the ribosomal subunits, and protein about one-third. (b) Pathway for processing of 6.6-kb (35S) pre-rRNA primary transcript in *S. cerevisiae*. The transcribed spacer regions (tan), which are discarded during processing, separate the regions corresponding to the mature 18S, 5.8S, and 25S rRNAs. All of the intermediates diagrammed have been identified; their sizes are indicated in red type. [Part (b) adapted from S. Chu et al., 1994, *Proc. Nat'l. Acad. Sci. USA* **91**:659.]

Pre-rRNA Genes Act as Nucleolar Organizers

When pre-rRNA genes initially were identified in the nucleolus by in situ hybridization, it was not known whether any other DNA was required to form the nucleolus. Subsequent experiments with transgenic *Drosophila* strains demonstrated that a single complete pre-rRNA transcription unit induces formation of a small nucleolus (Figure 12-52). Thus a single pre-rRNA gene is sufficient to be a *nucleolar organizer*, and all the other components of the ribosome diffuse to the newly formed pre-rRNA. The structure of the induced nucleolus appears, at least by light microscopy, to be the same as, except smaller than, a normal nucleolus containing 200 or so pre-rRNA genes.

Self-Splicing Group I Introns in Some Pre-rRNAs Were the First Examples of Catalytic RNA

The first indication that RNA can catalyze reactions, like enzymes, came during the study of ribosomal RNA biogenesis in the protozoan *Tetrahymena thermophila*. Researchers found that DNA encoding pre-rRNA in this organism contains an intervening intron in the region that encodes the large rRNA molecule. Careful searches for even one gene without the extra sequence failed. So it appeared that splicing is required to make ribosomal RNA in these organisms.

To further study this phenomenon, researchers prepared recombinant *Tetrahymena* DNA containing the two

(a)

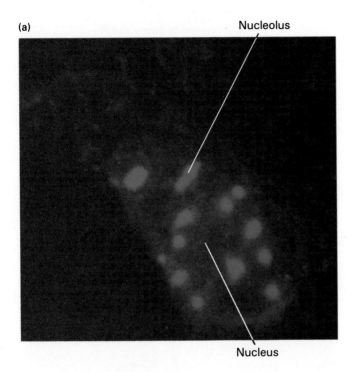

Nucleolus

Nucleus

◄ FIGURE 12-51   Antibody staining of nucleolar and ribosomal proteins. (a) Cell stained with serum from a patient with systemic lupus erythematosis (SLE). This serum reacts with antigens present specifically in nucleoli. (b) Cell stained with rabbit antisera raised against ribosomal proteins. Note that both the nucleolus and cytoplasmic ribosomes are brightly stained. [Courtesy of M. R. Lerner.]

rRNA exons and intervening 408-base intron. In vitro transcription of this DNA yielded a transcript that was spliced at the correct sites when incubated by itself, without assistance from any protein. The first step in removal of the intron requires guanosine; although bound to a specific site in the intron, this guanosine is not part of the RNA chain and functions as a cofactor in the splicing reaction (Figure 12-53, *left*). The free 3′ hydroxyl of this guanosine is linked to the 5′ phosphate of the nucleotide at the 5′ end of the intron. This transesterification reaction, which requires no energy, cuts the transcript, leaving a free 3′ hydroxyl on the uridine at the end of the 5′ exon. Cleavage at the 3′ end of the intron occurs during a second transesterification that links the exons in a 5′,3′-phosphodiester bond.

Following the discovery of self-splicing in *Tetrahymena* pre-rRNA, a whole raft of self-splicing sequences were found in pre-rRNAs from other single-celled organisms, in mitochondrial and chloroplast pre-rRNAs, in several pre-mRNAs from certain *E. coli* bacteriophages, and in some tRNA primary transcripts from eubacteria. The

(b)

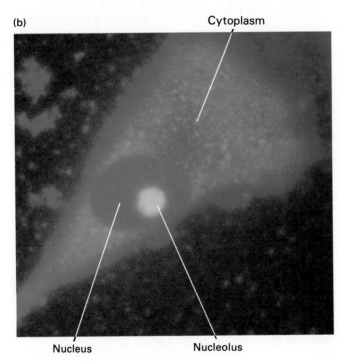

Cytoplasm

Nucleus

Nucleolus

➤ FIGURE 12-52   Micrograph of a polytene chromosome preparation from a fruit fly carrying a single pre-rRNA transgene. The cells were pulse-labeled with [³H]uridine before subjecting them to Giemsa staining and autoradiography; the silver grains (black dots) represent newly synthesized RNA. The large pale structure in the center is a normal nucleolus. The smaller one (arrow) is a new nucleolus formed at the end of the transgene. [See G. H. Karpen, J. E. Schafer, and C. D. Laird, 1988, *Genes & Devel.* **2**:1745; courtesy of G. H. Karpen.]

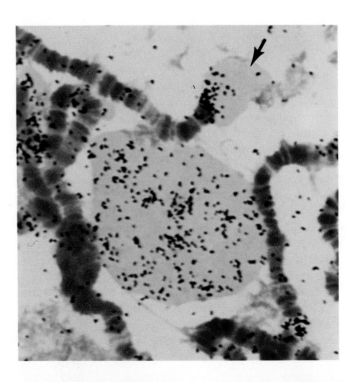

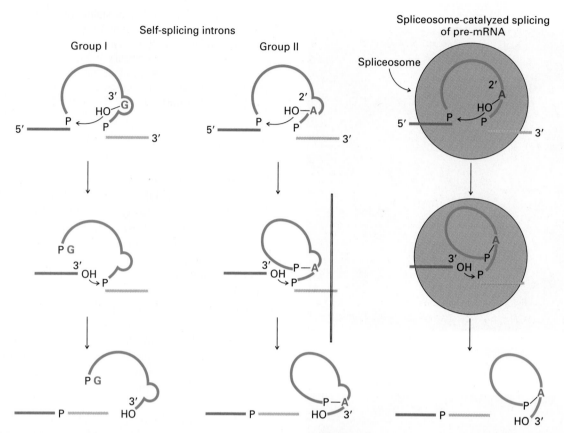

▲ FIGURE 12-53   Splicing mechanisms in group I and group II self-splicing introns and spliceosome-catalyzed splicing of pre-mRNA. The intron is shown in blue; the exons to be joined in red. In group I introns (*left*), a guanosine cofactor (**G**) associates with the active site. The 3′-hydroxyl group of this guanosine participates in a transesterification reaction with the phosphate at the 5′ end of the intron; this reaction is analogous to that involving the 2′-hydroxyl groups of the branch-site **A** in group II introns and pre-mRNA introns spliced in spliceosomes. The subsequent transesterification that links the 5′ and 3′ exons is similar in all three splicing mechanisms. Note that spliced-out group I introns are linear structures, unlike the branched intron products in the other two cases. [Adapted from P. A. Sharp, 1987, *Science* **235**:769.]

self-splicing sequences in all these precursors, referred to as group I introns, use guanosine as a cofactor and can fold by internal base pairing to juxtapose closely the two exons that must be joined. As discussed earlier, certain mitochondrial and chloroplast pre-mRNAs and tRNAs contain a second type of self-splicing intron, designated group II. Like group II introns, some group I introns also are mobile DNA elements. As illustrated in Figure 12-53, the splicing mechanisms used by group I introns, group II introns, and spliceosomes are generally similar, involving two transesterification reactions. Clearly, in the self-splicing introns, RNA functions as a *ribozyme,* an RNA sequence with catalytic ability.

It should be noted that the group I intron within the 28S region of *Tetrahymena* and certain other pre-rRNA transcription units is unrelated to the transcribed spacer sequences that separate the 18S, 5.8S, and 28S regions in the majority of organisms (see Figure 11-69). In particular, the self-splicing mechanism that removes group I introns differs from the cleavage mechanism by which spacer sequences are removed during processing of pre-rRNA discussed in the previous section (see Figure 12-50b).

Processing of pre-tRNA Involves Cleavage, Modification of Bases, and Sometimes a Unique Type of Splicing

The primary transcripts produced from tRNA genes by RNA polymerase III are considerably longer than the tRNAs found in the cytoplasm, which average 75–80 nucleotides in length. Cytoplasmic tRNAs also contain numerous modified bases that are not present in tRNA primary transcripts. Precursor tRNA (pre-tRNA) was initially recognized in pulse-labeling experiments with cultured HeLa cells. Gel electrophoresis of the RNA synthesized by cells labeled for a short time with [³H]uridine revealed the presence of products that were up to 40 bases longer than tRNA but contained some of the modified bases found in tRNA. When actinomycin D, which prevents RNA synthesis, was added to the cells after a brief period of labeling with [³H]uridine, the amount of these longer products decreased, while the amount of tRNA increased, indicating that the initial longer transcripts are precursors of tRNA.

Many genes encoding tRNAs have been cloned and sequenced. Comparison of the sequences of such genes with their corresponding tRNAs has shown that some eukaryotic nuclear tRNA genes contain introns. For example, the pre-tRNA transcribed from the yeast tyrosine tRNA (tRNA^Tyr) gene contains a 14-base intron that is not present in mature tRNA^Tyr (Figure 12-54). Some archaebacterial tRNA genes also contain introns. The introns in nuclear pre-tRNAs are shorter than those in pre-mRNAs, and they do not contain the splice-site consensus sequences

found in pre-mRNAs. Pre-tRNA introns also are clearly distinct from the much longer self-splicing group I and group II introns found in chloroplast and mitochondrial pre-rRNAs.

Unlike the RNA-splicing mechanisms discussed previously, splicing of pre-tRNA does *not* involve transesterification reactions. Rather, as outlined in Figure 12-55, pre-tRNA first is cleaved at two places, on each side of the intron, thereby excising the intron; second, the exon ends are ligated. These two distinct steps can be carried out separately by partially purified protein fractions. This mechanism differs from the splicing mechanisms illustrated in Figure 12-53 in which the 5′ end of the intron is released in the first step and the 3′ end is released in the second step at the same time the 5′ and 3′ exons are ligated.

Certain mutations in pre-tRNA that change its secondary structure prevent the splicing reaction, indicating that pre-tRNA molecules must be folded into a particular secondary structure for intron excision to occur. Since introns always are found in the anticodon loop of pre-tRNAs, pre-

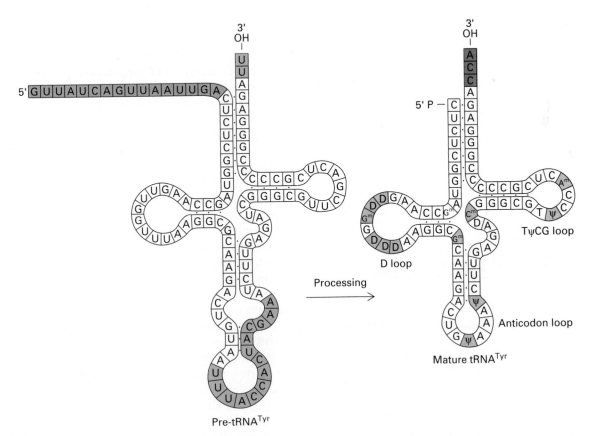

▲ FIGURE 12-54 Processing of tyrosine pre-tRNA involves four types of changes. A 14-nucleotide intron (blue) in the anticodon loop is removed by splicing. A 16-nucleotide sequence (green) is cleaved by RNase P from the 5′ end. U residues at the 3′ end are replaced by the CCA sequence (red) found in all mature tRNAs. Numerous bases in the

stem-loops are converted to characteristic modified bases (yellow). Not all pre-tRNAs contain introns that are spliced out during processing, but they all undergo the other types of changes shown here. D = dihydrouridine; Ψ = pseudouridine. See Figure 12-56 for structures of modified bases.

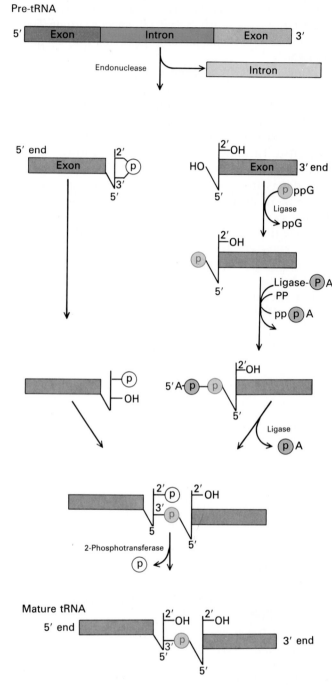

Pre-tRNA

Mature tRNA

▲ FIGURE 12-55   Mechanism of splicing in pre-tRNA. Several features distinguish this splicing mechanism from the mechanisms utilized by self-splicing introns and spliceosomes (see Figure 12-53): the excision of the intron in one step; requirement for GTP and ATP in ligation of exons; formation of a 2′,3′-cyclic monophosphate on the cleaved end of the 5′ exon; and catalysis by proteins (enzymes) rather than RNA. The phosphate in the 3′ → 5′ linkage joining the exons originates from GTP. ATP activates the ligase through formation of a covalently bonded ligase-AMP intermediate. The 2′-phosphate on the 5′ exon is removed in the final step. [See E. M. Phizicky and C. L. Greer, 1993, *Trends Biochem. Sci.* **18**:31.]

tRNAs most likely are folded similarly to mature tRNAs, thereby bringing the two intron-exon junctions into proximity (see Figure 12-54). Following excision of the intron, a 2′,3′-cyclic phosphomonoester forms on the cleaved end of the 5′ exon. The multistep reaction joining the two exons requires two nucleoside triphosphates: a GTP, which contributes the phosphate group for the 3′ → 5′ linkage in the finished tRNA molecule; and an ATP, which forms an activated ligase-AMP intermediate. The 2′-phosphate on the 5′ exon is removed in the final step (see Figure 12-55).

Whereas not all pre-tRNAs contain introns, they all contain a 5′ sequence of variable length that is absent from mature tRNAs (see Figure 12-54). These extra 5′ nucleotides are removed by the endonuclease *ribonuclease P* (RNase P), a ribonucleoprotein enzyme. *E. coli* RNase P consists of a 14-kDa polypeptide and a 377-nucleotide RNA called M1 RNA, which was one of the earliest catalytic RNA molecules to be recognized. At high Mg^{2+} concentrations, isolated M1 RNA recognizes and cleaves *E. coli* pre-tRNAs. The RNase P polypeptide increases the rate of cleavage by M1 RNA, allowing it to proceed at physiological Mg^{2+} concentrations.

In addition to splicing and 5′ cleavage, about 10 percent of the bases in pre-tRNAs are modified enzymatically. Three types of base modifications occur (see Figure 12-54): replacement of U residues at the 3′ end of pre-tRNA with a CCA sequence, which is found at the 3′ end of all tRNAs; addition of methyl and isopentenyl groups to purine residues and 2′ methylation of ribose; and conversion of specific uridines to dihydrouridine, pseudouridine, or ribothymidine residues (Figure 12-56). When intron-containing yeast tRNA genes are microinjected in *Xenopus* oocyte nuclei, correctly processed tRNAs are produced, indicating that enzymatic systems for processing tRNAs have been conserved over a wide evolutionary range.

Experiments with nucleated and enucleated *Xenopus* oocytes have demonstrated that processing of pre-tRNA occurs only in the nucleus. Fractionation of cultured mammalian cells in nonaqueous solvents, which prevent leakage of nuclear contents, also has shown that pre-tRNAs are present only in the nucleus and completed mature tRNAs only in the cytoplasm. Like all macromolecules transported from the nucleus to the cytoplasm, mature tRNAs are transported through nuclear pore complexes. Interestingly, U6 snRNA, another RNA synthesized by RNA polymerase III, remains in the nucleus, whereas tRNAs are efficiently transported to the cytoplasm. This difference raises the possibility that in the nucleus mature tRNAs, like mature mRNAs and rRNAs, are bound by specific proteins that facilitate their transport through nuclear pores. Once in the cytoplasm, tRNAs are passed between aminoacyl-tRNA synthetases (see Figure 4-35), elongation factors, and ribosomes (see Figure 4-42). Thus the three major types of RNA, mRNA, rRNA, and tRNA, are largely associated with proteins and spend little time free in the cell.

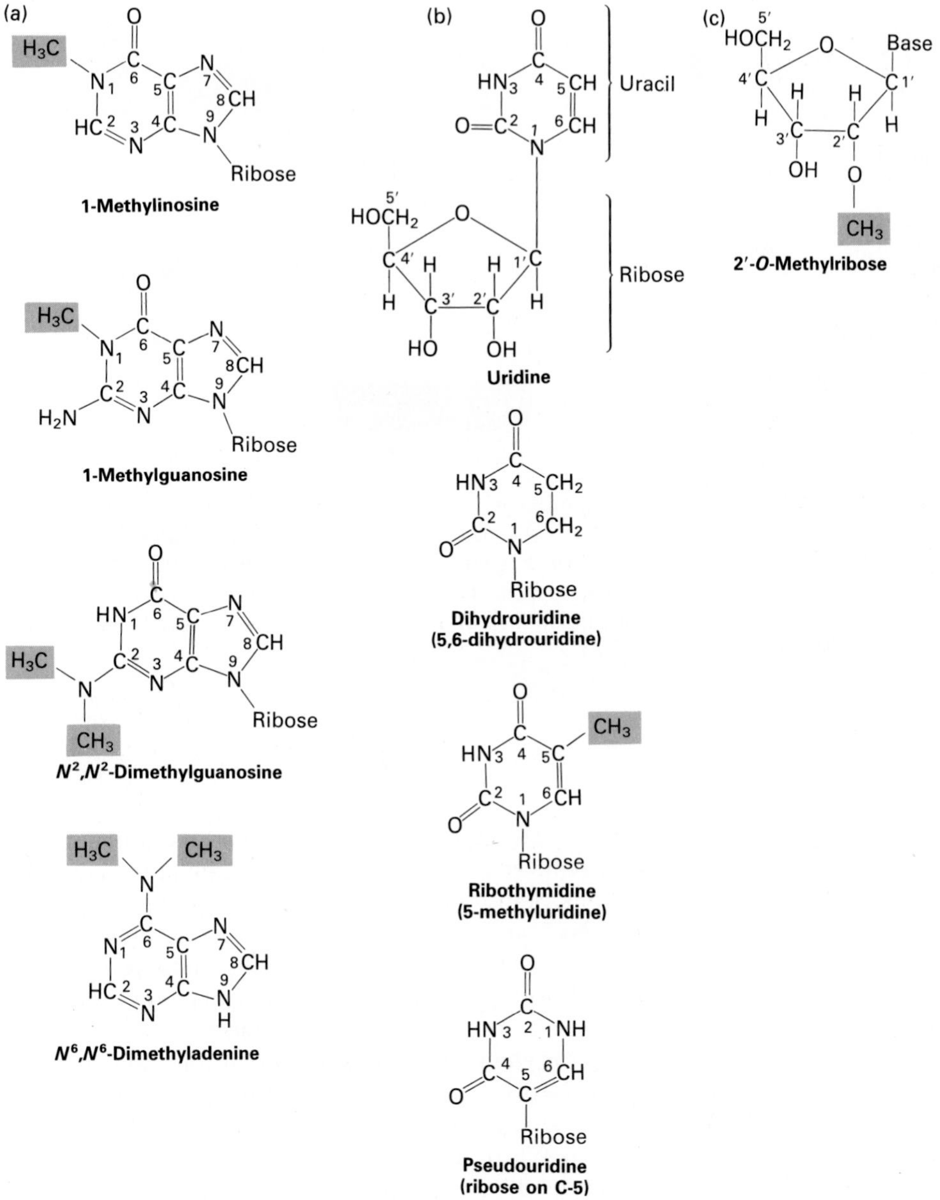

▲ FIGURE 12-56 Structures of some modified bases commonly found in mature tRNAs. (a) Methylated purine nucleosides. (b) The common nucleoside uridine and its common derivatives. (c) 2'-O-Methylribose, which can be linked to any base in tRNAs.

(a)
1-Methylinosine
1-Methylguanosine
N²,N²-Dimethylguanosine
N⁶,N⁶-Dimethyladenine

(b)
Uracil
Ribose
Uridine
Dihydrouridine (5,6-dihydrouridine)
Ribothymidine (5-methyluridine)
Pseudouridine (ribose on C-5)

(c)
2'-O-Methylribose

➤ RNA Editing

Sequencing of numerous cDNA clones and of the corresponding genomic DNAs from multiple organisms led in the mid-1980s to the unexpected discovery of a previously unrecognized type of pre-mRNA processing. In this type of processing, called *RNA editing,* the sequence of a pre-mRNA is altered; as a result, the sequence of the corresponding mature mRNA differs from the exons encoding it in genomic DNA. RNA editing is widespread in the mitochondria of protozoans and plants and in chloroplasts; in these organelles, more than half the sequence of some mRNAs is altered from the sequence of the corresponding

primary transcripts! In higher eukaryotes, RNA editing is much rarer, and thus far only single-base changes have been observed. Such minor editing, however, turns out to have important functional consequences in some cases.

RNA Editing Regulates Protein Function in Mammals

In mammals, the *apo-B* gene encodes two alternative forms of apolipoprotein B (Apo-B), a serum protein. The cell type–specific expression of these two forms in hepatocytes and intestinal epithelial cells is regulated by mRNA editing. In hepatocytes, an ≈500-kDa protein called Apo-B100 is

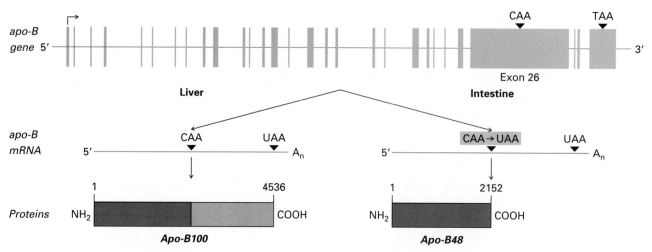

▲ FIGURE 12-57   RNA editing of *apo-B* mRNA. The *apo-B* mRNA produced in the liver has the same sequence as the exons in the primary transcript. This mRNA is translated into Apo-B100, which has two functional domains: a N-terminal domain (green) that associates with lipids and a C-terminal domain (orange) that binds to LDL receptors on cell membranes. In the *apo-B* mRNA produced in the intestine, the CAA codon in exon 26 is edited to a UAA stop codon. As a result, intestinal cells produce Apo-B48, which corresponds to the N-terminal domain of Apo-B100. [Adapted from P. Hodges and J. Scott, 1992, *Trends Biochem. Sci.* **17**:77.]

expressed, whereas Apo-B48 is expressed in intestinal epithelial cells. The ≈240-kDa Apo-B48 corresponds to the N-terminal region of Apo-B100. Both Apo-B proteins transport lipids in the serum in the form of large lipoprotein complexes. However, only the large protein made in the liver delivers cholesterol to body tissues by binding to the low-density lipoprotein (LDL) receptor present on all cells.

Comparison of Apo-B mRNA from hepatocytes and intestinal cells showed that nucleotide 6666 of the mRNA was changed from the C encoded in the DNA of both cell types to a U in the mRNA from intestinal cells. This change, which converts a CAA codon for glutamine to a UAA stop codon, results in synthesis of the shorter Apo-B48 in intestinal cells (Figure 12-57). The enzyme that performs the posttranscriptional conversion of C_{6666} to U has been partially purified and used to show that an RNA as short as 26 nucleotides with the sequence surrounding nucleotide 6666 can be recognized and edited. Although the *apo-B* gene is expressed only in the liver and intestine, editing-enzyme activity can be detected in many tissues, raising the possibility that the same type of editing occurs in as-yet unrecognized mRNAs.

A second example of mRNA editing in mammals involves the pre-mRNAs encoding the subunits of receptors for the neurotransmitter glutamate. Glutamate receptors, which span the cell membrane of neurons, have important functions in memory and learning. In response to release of the neurotransmitter glutamate, these transmembrane proteins open ion channels initiating nerve impulses

(Chapter 21). Several different glutamate receptor subtypes are assembled from a family of related glutamate receptor subunits. Some of the glutamate receptor channels conduct both Na^+ and Ca^{2+} ions, whereas others only conduct Na^+ ions. This difference, which has important consequences for nerve function, is controlled by an RNA-editing enzyme. This enzyme changes a specific CAG codon for glutamine to a CGG codon for arginine in the pre-mRNAs for a specific group of receptor subunits. The edited codon encodes an amino acid in the wall of the ion channel. When arginine is present at this position in the receptor, Ca^{2+} ions cannot pass through the channel. In mRNA preparations from brain, some of the glutamate receptor subunit mRNAs are only partially edited, suggesting that the RNA-editing activity may be higher in some neurons than others. If so, regulation of the editing enzyme may contribute to memory and learning.

RNA Editing in Trypanosome Mitochondria Drastically Alters mRNA Sequences

The most extreme form of RNA editing uncovered so far occurs in the mitochondria of trypanosomes. These parasitic protozoa have a single large mitochondrion called a *kinetoplast*. Many of the mRNA precursors produced from kinetoplast DNA are extensively edited by the insertion and deletion of U residues. Insertions predominate and in some cases are so extensive that the length of the transcript is almost doubled. The mechanism of this extensive editing

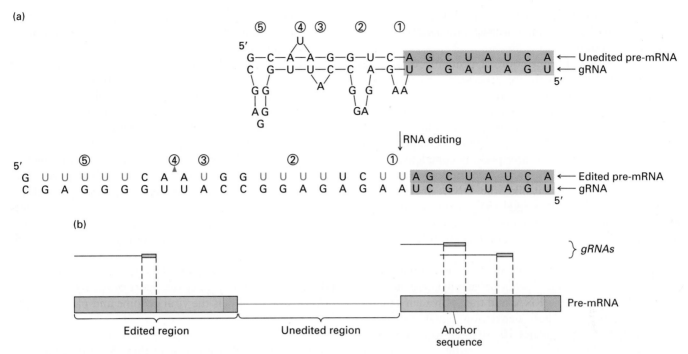

▲ FIGURE 12-58   Mechanism of RNA editing in kineto-plast mRNAs of trypanosomes. (a) An anchor sequence (blue) in the unedited pre-mRNA hybridizes to the 5′ anchor sequence (yellow) of a guide RNA (gRNA), which then directs the addition or deletion of U residues. Only a small region of the pre-mRNA is diagrammed. Added Us are shown in red; Δ = deleted U. Circled numbers identify sites where editing occurs. (b) Many gRNAs only hybridize to sequences in previously edited regions of an mRNA. Three such gRNAs are shown aligned with the corresponding anchor sequences (blue) in two extensively edited regions of a primary transcript. [Adapted from B. Bass, 1993, in R. Gesteland and J. Atkins, eds., *The RNA World*, Cold Spring Harbor Press, pp. 383–418.]

is currently being analyzed, but the favored hypotheses involve RNA-catalyzed transesterification reactions similar to the reaction that adds a G to the 5′ end of group I introns (see Figure 12-53). It is already clear that small *guide RNAs* (gRNAs) encoded at distant regions of the kinetoplast genome serve as templates for the editing process. The 5′ end of a gRNA hybridizes to a short region of unedited pre-mRNA, called an anchor sequence, while its 3′ end functions as a template for the editing process (Figure 12-58a). Many of the gRNAs that participate in the editing of extensively edited regions do not hybridize to anchor sequences in the primary transcript, but rather to sequences in partially edited intermediates (Figure 12-58b).

It is not clear why trypanosome kinetoplasts utilize such an elaborate mechanism to produce mRNAs. However, a recent analysis of RNA editing in different groups of trypanosomes revealed that editing is most extensive in the earliest trypanosomes to have evolved. This finding led the authors to suggest that RNA editing in trypanosome kinetoplasts may be a "molecular fossil" of the mechanism of RNA synthesis during an early stage in the evolution of cells.

SUMMARY

The expression of most genes is regulated at the initial step in mRNA synthesis by controlling the frequency of transcription initiation by an RNA polymerase. In the case of some genes, transcript elongation following initiation is controlled by various mechanisms that either cause or prevent premature termination before the full-length transcript is synthesized. In many bacterial genes, this type of regulation is accomplished by an attenuation mechanism in which the regulated formation of a stem-loop in the nascent RNA transcript directly signals RNA polymerase to terminate. Bacteriophage λ genes are regulated by an antitermination mechanism in which a viral protein binds to a specific sequence in a nascent RNA transcript and then, in combination with several host-cell proteins, interacts with the RNA polymerase synthesizing the transcript and prevents it from terminating. A similar process appears to regulate elongation during transcription of HIV genes in infected cells. Very rapid induction of heat-shock genes is possible because RNA polymerase II initiates transcription and then pauses; under the appropriate conditions, the heat-shock transcription factor is quickly activated and binds to heat-shock genes, releasing polymerase to continue transcription.

In eukaryotes, transcription of protein-coding genes yields pre-mRNAs, which are processed into functional mature mRNAs. This provides additional opportunities for gene control beyond the regulatory mechanisms influencing transcription initiation and termination. In eukaryotic cells, nascent pre-mRNAs associate with a complex set of hnRNP proteins before transcription is completed. These RNA-binding proteins probably have multiple functions. They may act somewhat like the DNA single-strand–binding (Ssb) protein; they likely help form the structures recognized by RNA-processing factors; and some appear to play a role in transport of nuclear mRNAs to the cytoplasm.

The 3′ ends of animal-cell mRNAs are formed when several proteins associate with a polyadenylation signal in a pre-mRNA, cleave the RNA, and then polyadenylate the resulting 3′ end. The polyadenylation signal usually includes an AAUAAA sequence 10–30 nucleotides upstream from the cleavage site and a downstream GU- or U-rich sequence. This step is rarely, but sometimes regulated.

Most pre-mRNAs in multicellular organisms are spliced to remove introns. Splicing involves two transesterification reactions that probably are catalyzed by snRNAs associated with specific proteins in snRNPs. The snRNPs associate with splice sites to form a 60S particle called a spliceosome in which the splicing reactions occur. The snRNAs associate with consensus sequences at 5′ and 3′ splice sites that almost always contain a GU and AG as the first two and last two nucleotides of an intron, respectively. Something like a hundred proteins are required for spliceosomal splicing of pre-mRNA introns, making this process similar in complexity to transcription initiation and translation.

Self-splicing introns present in some precursor RNAs can splice in the complete absence of proteins, proving that these splicing reactions are catalyzed by the pre-RNA molecules themselves. Group II self-splicing introns all contain a similar set of RNA secondary structures and splice by a mechanism analogous to spliceosomal splicing of pre-mRNA precursors. This finding suggests that snRNAs evolved from the secondary structure domains of ancestral self-splicing introns. Group I self-splicing introns fold into an alternative secondary structure and splice by a different but analogous splicing mechanism that requires a guanosine nucleoside cofactor.

About one in every twenty transcription units in higher eukaryotes is complex, that is alternative exons in the pre-mRNA can be included in the final mature mRNA. Alternative splicing increases the diversity of proteins that can be expressed from a single transcription unit. Alternative splicing usually is regulated so that some of the possible splicing pathways are used in some cell types and alternative splicing pathways are used in other cell types. In a few cases, sequence-specific RNA-binding proteins have been shown to bind near specific splice sites to either inhibit or activate splicing to the nearby site.

Microscopic immunolocalization experiments revealed that most RNA splicing occurs in ≈20–100 nuclear foci in mammalian cells. These foci may be organized through interactions with a nuclear skeleton, a system of protein fibers in the nucleus that may be analogous to the cytoskeletal framework in the cytoplasm. Fully processed mRNAs, in association with proteins as mRNPs, exit the nucleus through large, complex, symmetrical nuclear pore structures, which span the double-membrane nuclear envelope. The association of pre-mRNAs with spliceosomes prevents their translocation through nuclear pores, insuring that mRNAs are not transported until splicing is complete. A 5′ cap is required for transport of many RNAs. Transport of viral mRNPs is regulated in cells infected by some viruses. HIV Rev protein promotes transport of the unspliced pre-mRNA that is used as the viral genome and packaged into HIV virions.

Some mRNAs are directed to specific locations in the cytoplasm through sequences in their 3′ untranslated regions. Localization requires components of the cytoskeleton. The majority of mRNAs may be associated with cytoskeletal components through their poly-A tails, which are bound by the cytoplasmic poly-A–binding protein. Most eukaryotic mRNAs are fairly stable, with half-lives of many hours. But some eukaryotic mRNAs and almost all bacterial mRNAs are very unstable, allowing expression of the encoded protein to be rapidly shut off when transcription of the gene is repressed. Unstable eukaryotic mRNAs often, but not invariably, have AU-rich sequences in their 3′ untranslated regions. The stability of some eukaryotic mRNAs is regulated and this has a dramatic affect on the amount of the encoded protein synthesized. The stability of the mRNA encoding transferrin receptor is regulated by a protein that binds to specific sequences in the 3′ untranslated region stabilizing the mRNA. The same RNA-binding protein represses translation of other mRNAs that have a protein-binding site near their 5′ ends.

Pre-rRNAs and pre-tRNAs are extensively processed by cleavage and modification of nucleotides. Synthesis and processing of pre-rRNAs and assembly of ribosome subunits occur in nucleoli, which assemble around rRNA genes. Some pre-tRNAs contain short introns in their anticodon loops. These are spliced out by a completely different mechanism from pre-mRNA splicing and self-splicing introns.

RNA editing is a recently discovered type of processing that alters the sequences of RNAs after they are formed. Two instances of RNA editing have been shown to affect mammalian mRNAs by changing specific bases. One is clearly developmentally regulated, and the second probably is regulated; both have a profound influence on the function of the encoded proteins. RNA editing is common in chloroplast RNAs and mitochondrial RNAs of fungi and protozoans. The most extreme cases occur in trypanosome mitochondrial RNAs; nearly half the length of some mature mRNAs results from addition of U residues following transcription.

Even though gene expression is most commonly regulated by controlling transcription initiation, the numerous examples presented in this chapter show that for specific genes mechanisms have evolved to regulate nearly every imagineable aspect of mRNA synthesis and function. In particular, proteins that bind to specific RNA sequences can regulate transcription termination, polyadenylation, RNA splicing, transport from the nucleus to cytoplasm, localization in the cytoplasm, mRNA stability, and translation initiation.

REVIEW QUESTIONS

1. In *E. coli*, there are two fundamental ways to stop transcription elongation: Rho-dependent termination and Rho-independent termination. What is Rho? What, if any, are the characteristic features of the stop sites for Rho-dependent termination? What are the features of the termination signal for genes whose expression are controlled by Rho-independent termination?

What is meant by the process of attenuation? Be certain you understand the *trp* operon as a paradigm for attenuation. How is the availability of tryptophan involved in the attenuation process?

Bacteriophage λ uses a process called antitermination to control viral genes expressed early following infection of *E. coli*. What proteins are involved in bacteriophage λ antitermination, and how do they prevent termination by Rho?

In eukaryotes, each of the three RNA polymerases has different termination signals. What is known about each of them for RNA polymerases I, II, and III?

a. You are investigating the expression of a gene in *E. coli* that is regulated by Rho-independent chain termination, and you have obtained the sequence of the mRNA encoded by it. A portion of the mRNA is

5′–AUGUCGCAAUCGUCUAAGAUUGCGAUGUG–3′.

Draw a stem-loop structure for this segment of the mRNA.

b. You isolate a mutant that has significantly diminished termination when this gene is expressed. You sequence the region in the mutant that encompasses the locus identified in 1a above, and the mutant mRNA corresponding to it is

5′–AUGUCGCAAUCGUCUAAGACGAUGUGCCC–3′.

Compare this sequence to that for the wild-type sequence in 1a. Why would termination be decreased in this mutant? What type of mutation occurred?

2. The processing of RNA in eukaryotes is extremely important. Be certain you understand the temporal and spatial aspects of the process of post-transcriptional modification. What is meant by capping? Are all RNA molecules that are synthesized in eukaryotes capped? Which are or are not? How is the polyA-tail that is characteristic of mRNA added? Are all mRNA molecules polyadenylated?

Once RNA molecules are synthesized in the nucleus, proteins begin to associate with them. Describe some of the general features of these proteins. What is the RNP motif? What is the RGG box?

Following the addition of the 5′ cap and the poly-A tail, introns are removed from the message prior to transport from the nucleus to the cytoplasm. The splicing of exons is coupled to the removal of the introns. What factors facilitate the splicing reactions? What is a transesterification reaction? What is meant by a lariat structure, and how is one formed? What constitutes a spliceosome?

One of the most important breakthroughs concerning the processing of RNA was the discovery of self-splicing introns. Distinguish between group I and group II introns. How is each self-splicing? What is maturase, and how is it involved in self-splicing?

a. You have sequenced a region of a gene from *Saccharomyces cerevisiae* that contains an exon/intron/exon junction. A portion of the result is shown in Box 2a below. The region marked I contains the exon/intron boundary, and the region marked II contains the intron/exon boundary. The dashed line indicates the remaining bases within the intron. The arrow above the region shows the direction of transcription.

What would be the pre-mRNA encoded by this portion of the gene? Mark the exon/intron boundary in region I and the intron/exon boundary in region II using a slash (/). What would be the sequence of the mRNA from this locus following removal of the intron?

Box 2a

5′–GCTGTGGCCAGGTAAGTG-----TACTAACCTCGTGCTGTATGTAGTCTCCAGGTGT–3′
3′–CGACACCGGTCCATTCAC-----ATGATTGGAGCACGACATACATCAGAGGTCCACA–5′
           I   II

Box 2b Mutant
5'–GCTGTGGCCAGGTAAGTG-----TACTAGCCTCGTGCTGTATGTAGTCTCCAGCTGT–3'
3'–CGACACCGGTCCATTCAC-----ATGATCGGAGCACGACATACATCAGAGGTCGACA–5'

---

b. You have isolated a mutant that has the sequence in Box 2b above. The mRNA corresponding to this region is synthesized, but the intron is not removed. Why might this be the case?

3. The transport of messenger RNA from the nucleus to the cytoplasm is of critical importance to the cell since translation occurs in the cytoplasm. How do the mRNPs gain access to the cytoplasm from the nucleus? What are some of the structural requirements of the mRNPs for them to be transported? What prevents the transport of hnRNPs that have not been processed to mRNPs?

a. You have prepared two recombinant plasmids containing the gene that encodes firefly luciferase. In plasmid A, the gene is under control of the U1 promoter. In plasmid B, the luciferase gene is located downstream from the U6 promoter. Each of these plasmids was injected into frog oocytes, and, following an appropriate period of time, the total RNA, nuclear RNA, or cytoplasmic RNA was isolated. Each sample of RNA was resolved by electrophoresis, transferred to a membrane by Northern blotting, and hybridized with labeled probe for the firefly luciferase RNA. The results indicate the presence (+) or absence (−) of firefly luciferase RNA are shown in the table below.

	Firefly Luciferase RNA	
	Plasmid A	Plasmid B
Total RNA	+	+
Nuclear RNA	−	+
Cytoplasmic RNA	+	−

Explain these results.

4. Another important post-transcriptional modification is that which involves RNA editing. To what does this refer? In what organisms and organelles does it occur? What are some of the potential effects of RNA editing on the ultimate expression of a protein?

a. You have isolated and sequenced the cDNA for a protein you are interested in; a portion of the region is shown as triplets that encode amino acids. The arrow indicates the direction of transcription.

$\longrightarrow$
5'--- AAG TCA TTT TTG GGA CCC AAG CAT ---3'
3'--- TTC AGT AAA AAC CCT GGG TTC GTA ---5'

What would be the transcript for this region? What would be the resulting amino acid sequence?

b. You clone the actual gene for this region and discover that the region above is present in an exon, but there is a single change at the position underlined in red.

$\longrightarrow$
5'--- AAG TCA TTT TAG GGA CCC AAG CAT ---3'
3'--- TTC AGT AAA ATC CCT GGG TTC GTA ---5'

If this region were transcribed, what would be the resulting transcript? What would be the corresponding amino acid sequence? How would RNA editing explain your observation that the cDNA predicts a different sequence than that encoded within the actual genome?

5. Ultimately, the expression of a gene to yield a protein can be controlled at a variety of levels. Such precise regulation is vitally critical for maintaining the integrity of a cell. The levels of control are obviously not identical for all genes. Furthermore, not all the types of regulatory mechanisms apply to all genes.

Recall that the central dogma of molecular biology dictates that nucleic acids encode proteins, but that the reverse is not true. In other words, the flow of information within a cell is unidirectional.

Consider each of the following four questions first for prokaryotes and then for eukaryotes. Prepare a list of the control points requested.

a. Consider transcription of genes that encode proteins. In general, what are the levels that govern the actual production of full-length transcript?

b. Once a transcript has been made, what are the post-transcriptional modifications that can regulate RNA expression? What are some of the mechanisms that govern the stability of a transcript?

c. Consider translation of mRNA. How can this process be regulated?

d. Following translation of a message, the protein itself can be structurally modified as part of post-translational modifications. List several of these features that can affect a protein's function.

References

Transcription Termination and Its Regulation

BOGENHAGEN, D. F., and D. D. BROWN. 1981. Nucleotide sequences in *Xenopus* 5S DNA required for transcription termination. *Cell* **24**:261–270.

CULLEN, B. R. 1990. The HIV-1 Tat protein: an RNA sequence-specific processivity factor? *Cell* **63**:655–657.

DAS, A. 1993. Control of transcription termination by RNA-binding proteins. *Ann. Rev. Biochem.* **62**:893–930.

GREENBLATT, J., J. R. NODWELL, and S. W. MASON. 1993. Transcription antitermination. *Nature* **364**:401–406.

LANDICK, R., and C. L. TURNBOUGH, JR. 1992. Transcriptional attenuation. In S. L. McKnight and K. R. Yamamoto, eds., *Transcrip-*

tional Regulation, Cold Spring Harbor Laboratory Press, pp. 407–446.

LEE, H-S., et al. 1992. DNA sequence requirements for generating paused polymerase at the start of *hsp70. Genes and Dev.* 6:284–295.

MARTIN, F. H., and I. TINOCO, JR. 1980. DNA-RNA hybrid duplexes containing oligo (dA:rU) sequences are exceptionally unstable and may facilitate termination of transcription. *Nucl. Acids Res.* 8:2295–2299.

PLATT, T. 1986. Transcription termination and the regulation of gene expression. *Ann. Rev. Biochem.* 55:339–372.

RICHARDSON, J. P. 1991. Preventing the synthesis of unused transcripts by rho factor. *Cell* 64:1047–1049.

SPENCER, C. A., and M. GROUDINE. 1990. Transcription elongation and eukaryotic gene regulation. *Oncogene* 5:777–785.

YAGER, T. D., and R. S. VON HIPPEL. 1987. Transcription elongation and termination in *Escherichia coli.* In F. C. Neidhardt et al., eds., *Escherichia coli and Salmonella typhimurium: Cellular and Molecular Biology,* Vol. 1, American Society of Microbiology, pp. 1241–1275.

YANOFSKY, C. 1988. Transcription attenuation. *J. Biol. Chem.* 263:609–612.

Processing of mRNAs

hnRNP Proteins

DREYFUSS, G., et al. 1993. hnRNP proteins and the biogenesis of mRNA. *Ann. Rev. Biochem.* 62:289–321.

PINOL-ROMA, S., and G. DREYFUSS. 1992. Shuttling of pre-mRNA binding proteins between nucleus and cytoplasm. *Nature* 355:730–732.

Cleavage and Polyadenylation in Animal Cells

BIENROTH, S., W. KELLER, and E. WAHLE. 1993. Assembly of a processive messenger RNA polyadenylation complex. *EMBO J.* 12:585–594.

BIENROTH, S., et al. 1991. Purification of the cleavage and polyadenylation factor involved in 3′ processing of messenger RNA precursors. *J. Biol. Chem.* 266:19768–19769.

FITZGERALD, M., and T. SHENK. 1981. The sequence 5′-AAUAAA-3′ forms part of the recognition site for polyadenylation of late SV40 mRNAs. *Cell* 24:251–260.

FORD, J. P., and M.-T. HSU. 1978. Transcription pattern of in vivo-labeled late simian virus 40 RNA: equimolar transcription beyond the mRNA 3′ terminus. *J. Virol.* 28:795–801.

GILMARTIN, G. M., and J. R. NEVINS. 1991. Molecular analysis of 2 poly(A) site-processing factors that determine the recognition and efficiency of cleavage of pre-mRNA. *Mol. Cell. Biol.* 11:2432–2438.

GUNDERSON, S. I., et al. 1994. The human U1A snRNP protein regulates polyadenylation via a direct interaction with poly(A) polymerase. *Cell* 76:531–541.

MOORE, C. L., and P. A. SHARP. 1985. Accurate cleavage and polyadenylation of exogenous RNA substrate. *Cell* 41:845–855.

MOORE, C. L., and P. A. SHARP. 1986. Analysis of RNA cleavage at the adenovirus-2 L3 polyadynlation site. *EMBO J.* 8:1929–1938.

MURTHY, K. G. K., and J. L. MANLEY. 1992. Characterization of the multisubunit cleavage-polyadenylation specificity factor from calf thymus. *J. Biol. Chem.* 267:14804–14811.

NEVINS, J. R., and J. E. DARNELL, JR. 1978. Steps in the processing of Ad2 mRNA: poly(A)+ nuclear sequences are conserved and poly(A) addition precedes splicing. *Cell* 15:1477–1493.

PROUDFOOT, N.J., and G. G. BROWNLEE. 1976. 3′ noncoding region sequences in eukaryotic messenger RNA. *Nature* 263:211–214.

RAABE, T., F. J. BOLLUM, and J. L. MANLEY. 1991. Primary structure and expression of bovine poly(A) polymerase. *Nature* 353:229–234.

SHEETS, M. D., and M. WICKENS. 1989. Two phases in the addition of a poly(A) tail. *Genes & Dev.* 3:1401–1412.

TAKAGAKI, Y., et al. 1990. A multisubunit factor, CstF, is required for polyadenylation of mammalian pre-mRNAs. *Genes & Dev.* 4:112–120.

TAKAGAKI, Y., et al. 1992. The human 64-kDa polyadenylation factor contains a ribonucleoprotein-type RNA-binding domain and unusual auxiliary motifs. *Proc. Nat'l. Acad. Sci. USA* 89:1403–1407.

WAHLE, E. 1991. Purification and characterization of a mammalian polyadenylate polymerase involved in the 3′ end processing of messenger RNA precursors. *J. Biol. Chem.* 266:3131–3139.

WAHLE, E. 1991. A novel poly(A)-binding protein acts as a specificity factor in the second phase of messenger RNA polyadenylation. *Cell* 66:759–768.

WAHLE, E., and W. KELLER. 1992. The biochemistry of 3′-end cleavage and polyadenylation of messenger RNA precursors. *Ann. Rev. Biochem.* 61:419–440.

WAHLE, E., et al. 1991. Isolation and expression of cDNA clones encoding mammalian poly(A) polymerase. *EMBO J.* 10:4251–4257.

ZARKOWER, D., et al. 1986. The AAUAAA sequence is required both for cleavage and polyadenylation of simian virus 40 pre-mRNA in vitro. *Mol. Cell. Biol.* 6:2317–2323.

Messenger RNA Splicing

AGABIAN, N. 1990. *Trans*-splicing of nuclear pre-mRNAs. *Cell* 61:1157–1160.

AMREIN, H., M. L. HEDLEY, and T. MANIATIS. 1994. The role of specific protein-RNA and protein-protein interactions in positive and negative control of pre-mRNA splicing by Transformer 2. *Cell* 76:735–746.

GREEN, M. R. 1991. Biochemical mechanisms of constitutive and regulated pre-mRNA splicing. *Ann. Rev. Cell Biol.* 7:559–599.

GUTHRIE, C. 1991. Messenger RNA splicing in yeast: clues to why the spliceosome is a ribonucleoprotein. *Science* 253:157–163.

HOROWITZ, D. S., and A. R. KRAINER. 1994. Mechanisms for selecting 5′ splice sites in mammalian pre-mRNA splicing. *Trends Genet.* 10:100–105.

KANDELS-LEWIS, S., and B. SERAPHIN. 1993. Role of U6 snRNA in 5′ splice site selection. *Science* 262:2035–2039.

LESSER, C. F., and C. GUTHRIE. 1993. Mutations in U6 snRNA that alter splice site specificity: implications for the active site. *Science* 262:1982–1988.

MANIATIS, T. 1991. Mechanisms of alternative pre-mRNA splicing. *Science* 251:33–34.

MATTHEW, K. R., C. TSCHUDI, and E. ULLU. 1994. A common pyrimidine-rich motif governs *trans*-splicing and polyadenylation of tubulin polycistronic pre-mRNA in trypanosomes. *Genes & Dev.* 8:491–501.

MCKEOWN, M. 1992. Alternative mRNA splicing. *Ann. Rev. Cell Biol.* 8:113–155.

MCKEOWN, M. 1993. The role of small nuclear RNAs in RNA splicing. *Curr. Opin. Cell Biol.* 5:448–454.

MOORE, M. J., C. C. QUERY, and P. A. SHARP. 1993. Splicing of precursors to messenger RNAs by the spliceosome. In R. Gesteland and J. Atkins, eds., *The RNA World,* Cold Spring Harbor Press, pp. 303–357.

MOORE, M. J., and P. A. SHARP. 1993. Evidence for two active sites in spliceosome provided by stereochemistry of pre-mRNA splicing. *Nature* 365:364–368.

NILSEN, T. W. 1994. RNA–RNA interactions in the spliceosome: unraveling the ties that bind. *Cell* 78:1–4.

RIO, D. C. 1992. RNA binding proteins, splice site selection, and alternative pre-mRNA splicing. *Gene Expr.* 2:1–5.

RUBY, S. W., and J. ABELSON. 1991. Pre-mRNA splicing in yeast. *Trends Genet.* 7:79–86.

RYMOND, B. C., and M. ROSBASH. 1992. Yeast pre-mRNA splicing. In J. R. Broach, J. R. Pringle, and E. W. Jones, eds., *The Molecular and Cellular Biology of the Yeast Saccharomyces: Gene Expression,* Cold Spring Harbor Press, pp. 143–192.

SALDANHA, R., et al. 1993. Group I and group II introns. *FASEB J.* 7:15–24.

SHARP, P. A. 1994. Nobel lecture: Split genes and RNA splicing. *Cell* 77:805–815.

SMITH, C. W., J. G. PATTON, and B. NADAL-GINARD. 1989. Alternative splicing in the control of gene expression. *Ann. Rev. Genet.* 23:527–577.

SONTHEIMER, E. J., and J. A. STEITZ. 1993. The U5 and U6 small nuclear RNAs as active site components of the spliceosome. *Science* 262:1989–1996.

Nuclear Substructure and mRNA Transport through Nuclear Pores

CARTER, K. C., et al. 1993. A three-dimensional view of precursor messenger RNA metabolism within the mammalian nucleus. *Science* 259:1330–1335.

CHANG, D. D., and P. A. SHARP. 1989. Regulation by HIV Rev depends upon recognition of splice sites. *Cell* 59:789–795.

CULLEN, B. R., and M. H. MALIM. 1991. The HIV-1 Rev protein: prototype of a novel class of eukaryotic post-transcriptional regulators. *Trends Biochem. Sci.* 16:346–350.

DARGEMONT, C., and L. C. KUHN. 1992. Export of mRNA from microinjected nuclei of *Xenopus laevis* oocytes. *J. Cell Biol.* 118:1–9.

HAMM, J., and I. W. MATTAJ. 1990. Monomethylated cap structures facilitate RNA export from the nucleus. *Cell* 63:109–118.

HINSHAW, J. E., B. O. CARRAGHER, and R. A. MILLIGAN. 1992. Architecture and design of the nuclear pore complex. *Cell* 69:1133–1142.

KRUG, R. M. 1993. The regulation of export of mRNA from nucleus to cytoplasm. *Curr. Opin. Cell Biol.* 5:944–949.

LEGRAIN, P., and M. ROSBASH. 1989. Some *cis*- and *trans*-acting mutants for splicing target pre-mRNA to the cytoplasm. *Cell* 57:573–583.

MEHLIN, H., B. DANEHOLT, and U. SKOGLUND. 1992. Translocation of a specific premessenger ribonucleoprotein particle through the nuclear pore studied with electron microscope tomography. *Cell* 69:605–613.

NEWMEYER, D. D. 1993. The nuclear pore complex and nucleocytoplasmic transport. *Curr. Opin. Cell. Biol.* 5:395–407.

NYMAN, U., et al. 1986. Intranuclear localization of snRNP antigens. *J. Cell Biol.* 102:137–144.

ROSBASH, M., and R. H. SINGER. 1993. RNA travel: tracks from DNA to cytoplasm. *Cell* 75:399–401.

SPECTOR, D. L. 1993. Nuclear organization and pre-mRNA processing. *Curr. Opin. Cell Biol.* 5:442–448.

SPECTOR, D. L., X.-D. FU, and T. MANIATIS. 1991. Associations between distinct pre-mRNA splicing components and the cell nucleus. *EMBO J.* 10:3467–3481.

XING, Y., et al. 1993. Higher level organization of individual gene transcription and RNA splicing. *Science* 259:1326–1330.

ZEITLIN, S., R. C. WILSON, and A. EFSTRATIADIS. 1989. Autonomous splicing and complementation of in vivo-assembled spliceosomes. *J. Cell. Biol.* 108:765–777.

Regulation of mRNA Cytoplasmic Localization, Stability, and Translation

ALTMAN, M., and M. TRACHSEL. 1993. Regulation of translation initiation and modulation of cellular physiology. *Trends Biochem. Sci.* 18:429–432.

GAVIS, E. R., and R. LEHMANN. 1992. Localization of *nanos* RNA controls embryonic polarity. *Cell* 71:301–313.

GORLACH, M., C. G. BURD, and G. DREYFUSS. 1994. The mRNA poly(A)-binding protein: localization, abundance, and RNA-binding specificity. *Exp. Cell Res.* 211:400–407.

HENTZE, M. W. 1991. Determinants and regulation of cytoplasmic mRNA stability in eukaryotic cells. *Biochem. Biophys. Acta* 1090:281–292.

KISLAUKIS, E. H., et al. 1993. Isoform-specific 3'-untranslated sequences sort α-cardiac and β-cytoplasmic actin messenger RNAs to different cytoplasmic compartments. *J. Cell Biol.* 123:165–172.

KLAUSNER, R. D., et al. 1993. Regulating the fate of mRNA: the control of cellular iron metabolism. *Cell* 72:19–28.

LENK, R., et al. 1977. A cytoskeleton structure with associated polyribosomes from HeLa cells. *Cell* 10:67–78.

MACDONALD, P. M., et al. 1993. RNA regulatory element BLE1 directs the early steps of *bicoid* mRNA localization. *Development* 118:1233–1243.

MELEFORS, O., and M. W. HENTZE. 1993. Translational regulation by mRNA/protein interactions in eukaryotic cells: ferritin and beyond. *Bioessays* 15:85–90.

SACHS, A. B. 1993. Messenger RNA degradation in eukaryotes. *Cell* 74:413–421.

SHAW, G., and R. KAMEN. 1986. A conserved AU sequence from the 3' untranslated region of GM-CSF mediates selective mRNA degradation. *Cell* 46:659–667.

SIMONS, R. W., and N. KLECKNER. 1988. Biological regulation by antisense RNA in prokaryotes. *Ann. Rev. Genet.* 22:567–600.

SUNDEL, C. L, and R. H. SINGER. 1991. Requirement of microfilaments in sorting of actin messenger RNA. *Science* 253:1275–1277.

TANEJA, K. L., et al. 1992. Poly(A) RNA codistribution with microfilaments: evaluation by in situ hybridization and quantitative digital imaging microscopy. *J. Cell Biol.* 119:1245–1260

Processing of rRNAs and tRNAs

ALTMAN, S., L. KIRSEBOM, and S. TALBOT. 1993. Recent studies of ribonuclease P. *FASEB J.* 7:7–14.

CECH, T. R. 1990. Self-splicing of group I introns. *Ann. Rev. Biochem.* 59:543–568.

CECH, T. R. 1993. Catalytic RNA: structure and mechanism. *Biochem. Soc. Trans.* 21:229–234.

DARR, S. C., J. W. BROWN, and N. R. PACE. 1992. The varieties of ribonuclease P. *Trends Biochem. Sci.* 17:178–182.

FOURNIER, M. J., and E. S. MAXWELL. 1993. The nucleolar snRNAs: catching up with the spliceosomal snRNAs. *Trends Biochem. Sci.* 18:131–135.

GUERRIER-TAKADA, C., and S. ALTMAN. 1984. Catalytic activity of an RNA molecule prepared by in vitro transcription. *Science* 223:285–286.

KRUGER, K., et al. 1982. Self-splicing RNA: autoexcision and autocyclization of the ribosomal RNA intervening sequence of *Tetrahymena*. *Cell* 31:147–157.

MATTAJ, I. W., D. TOLLERVEY, and B. SERAPHIN. 1993. Small nuclear RNAs in messenger RNA and ribosomal RNA processing. *FASEB J.* 7:47–53.

PHIZICKY, E. M., and C. L. GREER. 1993. Pre-tRNA splicing: variation on a theme or exception to the rule? *Trends Biochem. Sci.* 18:31–34.

WOOLFORD, J. L., JR., and J. R. WARNER. 1991. The ribosome and its synthesis. In J. R. Broach, J. R. Pringle, and E. W. Jones, eds., *The Molecular and Cellular Biology of the Yeast Saccharomyces*, Vol. 1, Cold Spring Harbor Press, pp. 587–626.

ZIMMERMANN, R. A., and A. E. DAHLBERG, eds. 1993. *Ribosomal RNA: Structure, Evolution, Processing, and Function in Protein Synthesis*, Telford, Caldwell, NJ.

RNA Editing

BASS, B. L. 1993. RNA editing: new uses for old players in the RNA world. In *The RNA World*, R. Gesteland and J. Atkins, eds, Cold Spring Harbor Press, pp. 383–418.

BENNE, R., et al. 1986. Major transcripts of the frameshifted *coxII* gene from trypanosome mitochondria contains four nucleotides that are not encoded in the DNA. *Cell* 46:819–826.

BLUM, B., N. BAKALARA, and L. SIMPSON. 1990. A model for RNA editing in kinetoplastic mitochondria: "guide" RNA molecules transcribed from maxicircle DNA provide the edited information. *Cell* 60:189–198.

BLUM, B., et al. 1991. Chimeric gRNA-mRNA molecules with oligo(U) tails covalently linked at sites of RNA editing suggest that U addition occurs by transesterification. *Cell* 65:543–554.

CHEN, S.-H., et al. 1987. Apolipoprotein B-48 is the product of a messenger RNA with an organ-specific in-frame stop codon. *Science* 238:363–366.

HODGES, P., and J. SCOTT. 1992. Apolipoprotein B mRNA editing: a new tier for the control of gene expression. *Trends Biochem. Sci.* 17:77–81.

MASLOV, D. A., et al. 1994. Evolution of RNA editing in kinetoplastid protozoa. *Nature* 368:345–348.

POWELL, L. M., et al. 1987. A novel form of tissue-specific RNA processing produces apolipoprotein B-48 in intestine. *Cell* 50:831–840.

SIMPSON, L. D., A. MASLOV, and B. BLUM. 1993. RNA editing in *Leishmania* mitochondria. In R. Benne, ed., *RNA Editing: The alteration of Protein Coding Sequences of RNA*, Ellis Horwood Limited, Sussex, England, pp. 53–85.

SOMMER, B., et al. 1991. RNA editing in brain controls a determinant of ion flow in glutamate-gated channels. *Cell* 67:11–19.

13 Gene Control in Development

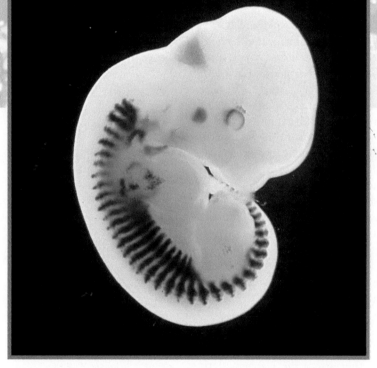

▲ The myogenin gene, required for the development of skeletal muscle, is expressed in the limb buds and myotomes in the 11.5 day mouse embryo.

The molecular basis for gene control in eukaryotic cells is one of the most actively studied areas in all of biology. Classical and molecular genetics discover how genes participate in controlled programs that result in the development of plants and animals. The ultimate goal is to unravel the mystery of the conversion of a fertilized egg into a complex multicellular organism comprising a vast spectrum of different cell types. Over the past decade, developmental biologists have made remarkable progress towards this goal using the general approach described in Chapter 8. That is, they have identified mutations affecting development with classical genetic techniques and then identified the corresponding gene with recombinant DNA technology (see Figure 8-1). This approach has been used to study the developmental decisions involved in λ-phage infection of *E. coli* and in specification of the three different cell types in the yeast *Saccharomyces cerevisiae*. The same general approach applies to multicellular organisms such as the fruit fly *Drosophila melanogaster* and the simple soil worm *Caenorhabditis elegans*. These studies have revealed that many developmentally important genes encode transcription factors, suggesting that regulation of transcription is a primary strategy for controlling development in all organisms.

In this chapter we consider the regulation of gene expression during development in several model systems. These studies highlight the diverse molecular mechanisms used to control gene expression as well as the conservation of basic strategies to generate different developmental states. We discuss mechanisms controlling developmental switches between alternative pathways, mechanisms for generating specialized cell types (e.g., muscles and neurons), and finally the role of transcription factors in regulating the spatial organization of the embryo. In Chapters 20 and 24 we will discuss the mechanisms by which development is also regulated by extracellular signals.

➤ *Lysogeny or Lysis in λ-Phage Infection of E. coli*

Upon entering an *E. coli* cell, a λ-phage virion can pursue two very different pathways (Figure 13-1a). Activation of one set of phage genes results in the *lytic pathway* in which the viral DNA is replicated many times and then is packaged into viral coats, causing lysis of the host cell and release of up to 100 λ virions (see Figure 6-17). Activation of another set of phage genes results in the *lysogenic pathway* in which the viral DNA is integrated through site-specific recombination into the host-cell genome (see Figure 10-35) where it remains in a dormant state. An integrated bacteriophage is called a *prophage;* a host cell carrying a prophage is referred to as a *lysogen* or lysogenized cell. Although the lysogenic state is very stable, certain physiologic conditions (e.g., DNA damage induced by UV light) can destabilize the prophage, resulting in its excision from the host-cell genome and its entry into the lytic pathway. In this section we describe the genetic and molecular analyses that have elucidated the complex transcription-control program regulating the choice between the lytic and lysogenic developmental pathways. As discussed in later sections, developmental decisions in more complex biological systems share common themes with this developmental choice in λ phage.

Phage Mutants Unable to Undergo Lysogeny Fall into Three Main Complementation Groups

Identification of mutations that affect the lysogenic/lytic decision was a critical first step towards uncovering the molecular mechanisms regulating it. When wild-type bacteriophage λ is plated on *E. coli*, most of the bacteria are lysed by the virus, but in a small proportion of the cells the virus enters the lysogenic pathway. If a single phage virion lysogenizes the first cell that it infects, no plaque will form.

(a)

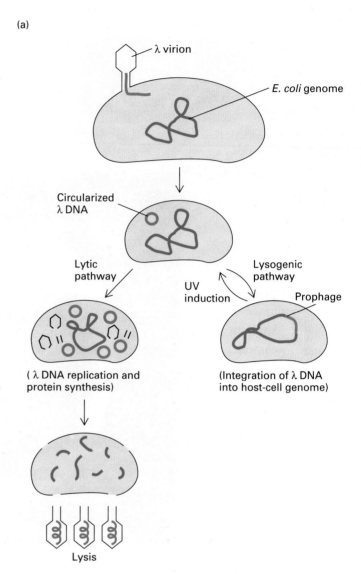

(b)

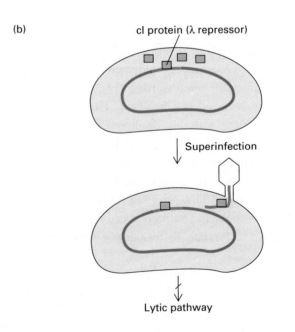

➤ FIGURE 13-1   Alternative λ-phage developmental pathways. (a) After a λ virion infects an *E. coli* cell, it enters either the lytic or lysogenic pathway depending on which set of viral genes is activated. Treatment of a lysogenized cell with UV light activates the integrated prophage to excise from the cellular genome, thereby inducing the lytic pathway. (b) In a lysogenized cell, the only viral protein expressed is cI (λ repressor), which is necessary to maintain the lysogenic state. The cI protein represses transcription of lytic genes in a superinfecting λ virion, thereby protecting the cell from lysis. [See M. Ptashne, 1992, *A Genetic Switch,* Blackwell Scientific Publications and Cell Press.]

If, however, after the first round of lytic growth, a fraction of the newly infected cells are lysogenized (about 1 in 20 for wild-type λ under normal growth conditions), then the plaque that is formed will be turbid. The turbidity is due to the growth of lysogenized cells, which are resistant to reinfection (often referred to as superinfection) because a DNA-binding protein expressed from the prophage represses expression of lytic genes in the superinfecting bacteriophage (Figure 13-1b).

Mutants unable to undergo the lysogenic program of development give rise to clear plaques because all cells infected by λ are lysed. These mutants were found to fall primarily into three different complementation groups referred to as *cI, cII,* and *cIII*. The cI protein represses the lytic pathway and is required both for the establishment and maintenance of the lysogenic state. In contrast, both cII and cIII proteins regulate expression of the *cI* locus during establishment of the lysogenic state but are not required for its maintenance. That cI protein is required for maintenance of the lysogenic state, not only for its establishment, was demonstrated with λ phage carrying a temperature-sensitive *cI* mutation. At permissive temperatures, these mutant phage virions could enter the lysogenic state; when host cells carrying the mutant prophage were shifted to a nonpermissive temperature, however, the phage entered the lytic pathway.

We can view the lysogenic and lytic states as two different developmental states, which must be established and then maintained. For simplicity, we begin with a discussion of how lysogeny is maintained; this is followed by a brief description of the factors that regulate the choice in an infected cell between the lytic and lysogenic pathways.

cI Protein Maintains Lysogeny by Repressing and Activating Transcription from Different Promoters

The finding that cI, or λ *repressor,* is the only protein needed to maintain lysogeny once it is established raises several questions:

- How does cI repress the lytic pathway?
- How is expression of *cI* maintained to ensure stability of the prophage through many cell divisions?
- How is the level of cI controlled within a range that allows rapid derepression and induction of the lytic pathway in response to environmental factors such as UV light?

The answers to these questions involve the interactions of cI with a short stretch of λ DNA containing three contiguous cI-binding sites, or *operators,* termed O_R1, O_R2, and O_R3 (Figure 13-2a). Binding of cI to O_R1 and O_R2 represses transcription from P_R, a promoter at the right-hand side of the operator region. P_R controls transcription of *cro,* which encodes a protein (Cro) that represses lysogeny, thereby promoting lytic growth as discussed below. In addition, binding of cI to O_R1 and O_R2 stimulates transcription in the left-hand direction from another promoter. Because this promoter controls transcription of the *cI* gene itself, it is called P_{RM} (promoter of repressor maintenance). Finally, when cI is bound to all three operators, it represses transcription from both P_R and P_{RM} ensuring that the repressor is not overproduced. This feedback mechanism allows the level of cI to be rapidly reduced in response to environmental conditions, leading to induction of the lytic state.

Cooperative Binding of cI to λ Operators   The interactions between cI protein and the three operators—O_R1, O_R2, and O_R3—are responsible for maintaining λ DNA in the lysogenic state. All three operators contain 17 base pairs and have similar but not identical sequences. The cI protein, which has two distinct globular domains, spontaneously forms dimers. The N-terminal domain binds to the DNA operator, and the C-terminal domain mediates dimerization. The affinity of dimeric cI for the operators is in the order $O_R1 > O_R2 > O_R3$. Binding of cI to isolated O_R2 is about tenfold weaker than binding of cI to O_R1. However, prior binding of cI to O_R1 facilitates binding of a second cI molecule to O_R2 due to interactions between the dimerization domains of the two molecules (see Figure 13-2b). This cooperative binding ensures that O_R1 and O_R2 are each simultaneously bound to cI dimers. Because of the orientation of dimeric cI molecules bound at O_R2 and O_R3, no interaction is possible between their dimerization domains (see Figure 13-2c); thus cooperative binding does not increase the affinity of cI for O_R3. At the concentrations of cI present in lysogenized cells, O_R1 and O_R2 are largely saturated, whereas O_R3 is largely unoccupied. In this situation, RNA polymerase cannot bind to P_R, and transcription of *cro* is repressed.

Cooperative binding of protein to DNA is an important regulatory mechanism in both simple bacterial systems and in more complex developmental systems. It provides an effective switching mechanism for turning on or off very different programs of gene expression in response to small differences in protein concentration. The importance of cooperative binding in maintaining the stable yet dynamic potential of a λ prophage can be appreciated by examining Figure 13-3. A two- to threefold decrease in the cI concentration at the levels expressed in lysogens does not switch off the lysogenic state, whereas a slightly larger decrease in the cI level does, thereby inducing the lytic pathway (red curve). At this concentration of cI, the *cro* gene is transcribed at 50 percent of its maximal rate, providing enough

(a)

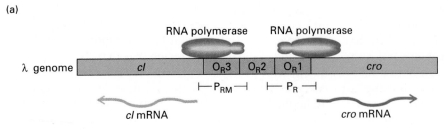

(b)

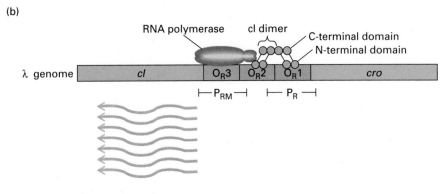

(c)

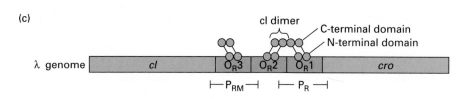

◄ FIGURE 13-2   Maintenance of lysogeny by cI protein. (a) In the absence of cI protein, the *cro* and *cl* genes in the λ genome are transcribed. Cro protein represses lysogeny. (b) Once lysogeny is established, *cl* is transcribed from the λ DNA integrated into the host-cell genome. Binding of dimeric cI protein (purple) to O_R1 and O_R2 blocks transcription of *cro* from P_R and stimulates transcription of *cl* from P_{RM}. Thus cI regulates its own expression (autoregulation). The sequences of P_R and P_{RM} overlap O_R1 and O_R3, respectively. The C-terminal domain of cI promotes dimerization, and the N-terminal domains in the dimer bind to the DNA operators. Interaction between the C-terminal domains of two cI dimers leads to cooperative binding of cI to O_R1 and O_R2. (c) At high cI concentrations, the dimer also binds to O_R3, thereby preventing transcription from P_{RM}. Because of the orientation of the molecules, interaction is not possible and cooperative binding does not increase the affinity of cI for O_R3. Thus, at the usual concentration of cI in lysogenized cells, O_R1 and O_R2 are largely saturated, but O_R3 is mostly unbound. [Adapted from M. Ptashne, 1992, *A Genetic Switch,* Blackwell Scientific Publications and Cell Press.]

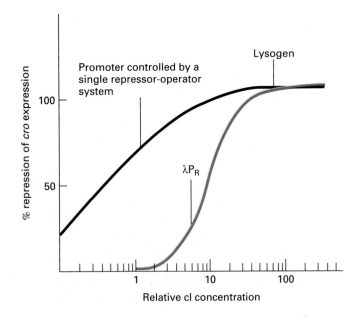

◄ FIGURE 13-3   Role of cooperative binding in regulatory switching. The red curve shows repression of *cro* transcription from P_R as a function of cI concentration (note the log scale). A two- to threefold decrease in cI level from that present in a lysogenized cell causes a minimal change in Cro expression. In the middle part of the curve, a relatively small change in cI level leads to a substantial increase in Cro expression (i.e., relief of repression at P_R by binding of cI to O_R1 and O_R2). As shown by the black curve, if P_R were controlled by only one operator and cI did not bind cooperatively, then a much greater reduction in cI concentration would be required to achieve comparable derepression of Cro expression. The arrow indicates the approximate cI concentration in the lysogenic state. [Adapted from M. Ptashne, 1992, *A Genetic Switch,* Blackwell Scientific Publications and Cell Press.]

Cro protein to drive the lytic cycle. If just one operator controlled the *cro* promoter and no cooperative binding occurred, comparable derepression would require a hundredfold decrease in the cI level (black curve). The ability of a prophage to quickly switch to the lytic pathway in response to UV-induced damage to DNA permits excision of the prophage, production of virions, and lysis before death of the host cell, thus ensuring survival of λ. In more complex developmental systems, transcription-control on/off switches may respond to even smaller changes in the concentration of DNA-binding proteins.

Autoregulation of Transcription by cI Protein   As noted above, cI dimers bound to O_R1 and O_R2 physically block binding of RNA polymerase to P_R, thereby repressing transcription of the *cro* gene. In contrast, transcription of the *cI* gene from P_{RM} is stimulated significantly by binding of cI dimers to O_R1 and O_R2. The mechanism of this stimulation, although not fully worked out, probably involves protein-protein interactions, similar to cAMP-CAP activation of the *lac* promoter (Figures 11-20 and 11-21). When mutations in the λ genome affect certain acidic residues in cI, the mutant protein cannot stimulate transcription from P_{RM}. However, certain "compensatory" mutations in *E. coli* RNA polymerase, which restore the ability of the mutant cI protein to act as a transcription activator, have been identified. Discovery of such suppressor mutations (see Figure 8-17) is strong evidence that these acidic residues in cI interact with RNA polymerase, thereby promoting transcription from P_{RM} (Figure 13-4). Hence, cI stimulates transcription of the gene encoding it; this *autoregulation* is an efficient mechanism for maintaining the lysogenic state. As discussed in later sections, autoregulation is a common mechanism for maintaining particular developmental states in multicellular organisms.

cII and cIII Proteins Are Critical to Establishment of Lysogeny

Having discussed how the lysogenic state is maintained, we now consider the molecular events that lead to production of sufficient cI to establish lysogeny. When a λ virion first enters an *E. coli* cell, transcription of certain viral genes by host-cell RNA polymerase begins from P_R and P_L; this latter promoter lies to the left of the *cI* gene. Even though bacterial transcripts often are polycistronic, initial transcription of the λ genome from P_R and P_L does not extend past *cro* and *N*, terminating at DNA sites designated T_L and T_R. As a result, two mRNAs are produced: one encodes Cro protein; the other encodes N protein (Figure 13-5).

The expressed Cro protein then binds to O_R3, thereby directly repressing transcription of the *cI* gene from P_{RM} and indirectly promoting the lytic pathway. As the level of N protein increases, it binds to a DNA sequence called nut

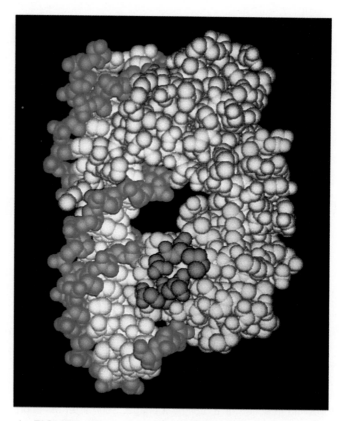

▲ FIGURE 13-4   Computer model of a dimeric cI molecule bound to λ operator DNA based on crystallographic analysis of cocrystallized cI and λ DNA. In this model, the two cI monomers are shown in tan and light green; the two DNA strands, in dark and light blue. As discussed later, each monomer makes contact with bases within adjacent major grooves of the DNA; the "arm" of the tan monomer wraps around the major groove facing the reader. The residues in red are acidic amino acids required for positive autoregulation of the *cI* gene by the protein and probably represent contact points between cI protein and RNA polymerase. [See S. R. Jordan and C. O. Pabo, 1988, *Science* **242**:893. Computer model and photograph courtesy of R. Bushman and M. Ptashne.]

(N utilization), which is located upstream of both T_L and T_R. Interaction of RNA polymerase with N protein bound to nut allows transcription past T_L and T_R, resulting in production of cII and cIII proteins, both of which are critical for establishing the lysogenic state. The cII protein is a positive regulator of *cI* transcription, not from P_{RM} but from another strong promoter designated P_{RE} (promoter of repressor establishment). Transcription of *cI* from P_{RE} is very efficient, leading to a burst in synthesis of cI protein (see Figure 13-5). Subsequent binding of cI to O_R1 and O_R2 then represses transcription of *cro* from P_R and activates transcription of *cI* from P_{RM}, as discussed in the previous section (see Figure 13-2). In addition, cII promotes

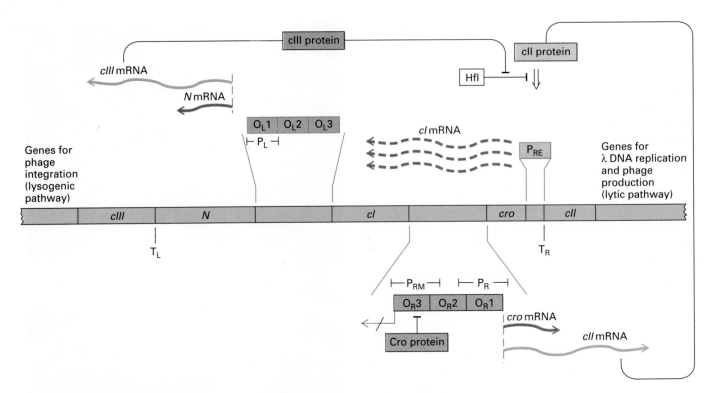

▲ FIGURE 13-5   Establishment of the lysogenic state following infection of *E. coli* by λ phage. Soon after infection, transcription from P_R and P_L leads to production of two short, monocistronic mRNAs (dark red) encoding Cro and N protein, respectively. Binding of Cro protein to O_R3 represses transcription of the *cl* gene from P_RM. Binding of N protein to nut sites (not shown) allows transcription past T_R and T_L, generating two longer polycistronic mRNAs (light red), which encode Cro plus cII or N plus cIII. Transcription of *cl* from P_RE is stimulated by cII, causing a burst in production of *cl* mRNA (broken red lines). Binding of cl to O_R1 and O_R2 and to O_L1 and O_L2 then represses expression of Cro and N protein, respectively, thereby inhibiting the lytic pathway. A host-cell protease, called Hfl, can degrade cII protein, thereby reducing its stimulation of *cl* transcription; cIII protein protects cII from protease degradation. The light red arrow from P_RM indicates that it is a weak promoter that will work efficiently only with cl bound to O_R2.

establishment of lysogeny by activating transcription of the *int* gene, which is located elsewhere in the λ genome. This gene encodes a site-specific recombinase (integrase) that catalyzes integration of λ DNA into the host-cell chromosome.

Since entry of a λ virion into a cell sets into motion molecular events leading toward both lysogeny and lysis, what factors determine which pathway predominates? The cellular factors that determine the choice between the two pathways in an individual cell are not clear; however, several conditions are known to favor production of cI, and hence the lysogenic state, in cell populations. For instance, when λ-infected *E. coli* cells are grown in a rich medium, lysis occurs preferentially; in contrast, lysogeny is more common in starved cells, perhaps because such cells are unable to supply the metabolic requirements for the lytic pathway.

This physiologic regulation of expression of specific λ genes is mediated by a host-cell protease encoded by the *Hfl* (*high frequency of lysogeny*) locus. The level of this protease, which degrades cII protein, is considerably higher in cells grown in a rich medium than in those grown in a poor medium. Moreover, cIII protein protects cII from degradation by Hfl protease, providing an additional level of regulatory complexity. These observations indicate that the lytic/lysogenic decision is determined by the stability of the transcription regulator cII, which itself depends on the physiologic state of the cell. From these considerations, a powerful conclusion can be drawn: although genetic programs exist for alternative pathways of development, the pathway choice can be influenced by extracellular stimuli (Figure 13-6).

Induction of Lytic Cycle Requires Derepression of *cro* Gene

As mentioned earlier, DNA-damaging events can release the prophage in a lysogenized cell, thereby inducing the lytic pathway. For example, when a lysogen is subjected to UV irradiation, it undergoes a number of biochemical changes that allow it to repair DNA. One of these changes is proteolysis of a number of proteins, including cI. The cleaved repressor no longer dimerizes and thus loses its

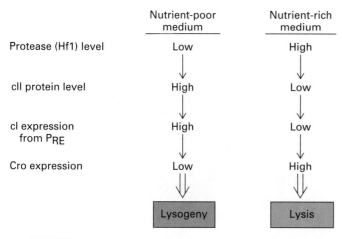

	Nutrient-poor medium	Nutrient-rich medium
Protease (Hf1) level	Low	High
cII protein level	High	Low
cI expression from P$_{RE}$	High	Low
Cro expression	Low	High
	Lysogeny	Lysis

▲ FIGURE 13-6 The choice between lytic and lysogenic pathways depends in part on the nutritional state of the host cells. This physiologic regulation of gene expression is mediated by Hfl protease. At high levels of Hfl, associated with growth in a rich medium, lysis occurs preferentially; low levels of Hfl, associated with a poor medium, allow establishment of lysogeny. Once established, lysogeny is maintained largely by cI protein, which represses transcription of *cro* from P$_R$ and activates transcription of *cI* from P$_{RM}$ (see Figure 13-2b).

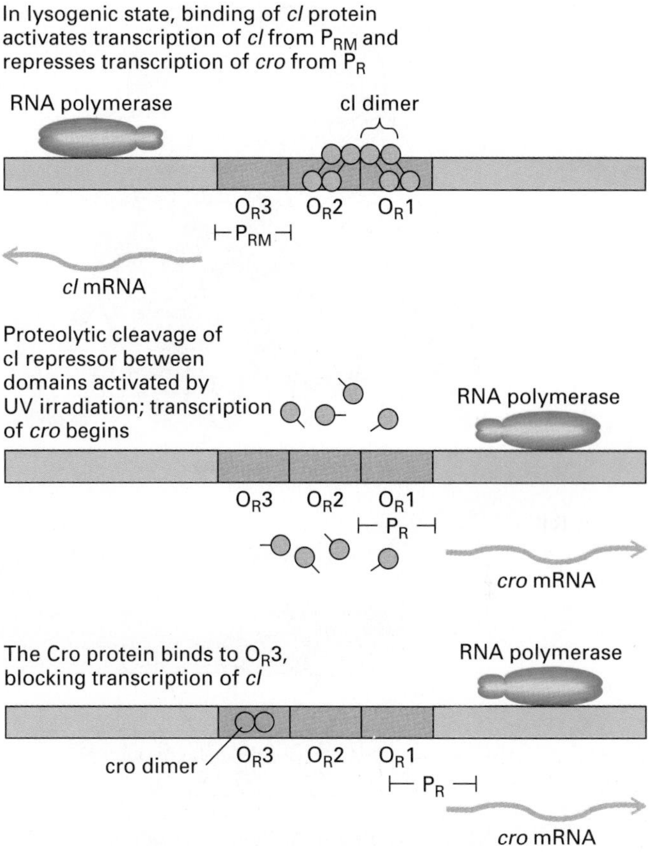

▲ FIGURE 13-7 Maintenance and release from lysogeny by binding of cI protein (λ repressor) and Cro protein to operator regions of the λ genome. The base sequences in the three O$_R$ sites are similar but not identical. Dimeric cI binds in the order O$_R$1 > 2 > 3, whereas Cro dimers bind in the order O$_R$3 > 2 > 1. Because cI has about a 10 times higher affinity than Cro for DNA, it is preferentially bound even when present at only about 10 to 20 percent the level of Cro. [See M. Ptashne et al., 1980, *Cell* **19**:1; A. D. Johnson et al., 1981, *Nature* **294**:217.]

ability to bind to DNA; bound cI dimers also are cleaved and fall off the DNA. As a result, repression of the *cro* gene is relieved, and transcription of the gene from P$_R$ is initiated (Figure 13-7).

Like cI, Cro protein forms dimers, which bind to the right-hand λ operator sites. However, the affinity of dimeric Cro for the operators is in the order O$_R$3 > O$_R$2 > O$_R$1, which is opposite to the cI-binding order. Binding of Cro to O$_R$3 shuts down transcription of the *cI* gene from P$_{RM}$. As the concentration of cI decreases, its repression of P$_L$ is relieved, so that transcription of the *N* gene can begin (see Figure 13-5). As discussed earlier, N protein permits transcription beyond T$_R$, leading to synthesis of mRNAs encoding proteins in the lytic pathway. In the absence of a sufficient amount of cI, the lytic cycle ensues.

Cro and cI Have Similar DNA-binding Domains But Interact Differently with λ Operators

The DNA-binding domains of dimeric Cro and cI have similar three-dimensional structures (Figure 13-8a). The side chains of certain amino acid residues in two of these helices—the *recognition helices*—penetrate adjacent major grooves of λ DNA and interact with specific base pairs in the operator site. Differences in these amino acid–base pair contacts for Cro and cI largely account for the differences

in their operator specificity noted above (Figure 13-8b). Although the binding specificity of Cro and cI is determined primarily by specific interactions involving residues in the recognition helices, the strength of the binding depends largely on interactions of specific regions in each protein with defined subsets of phosphate residues in the DNA backbone. Interestingly, in multicellular organisms a single DNA site may be recognized by several different, but structurally related, transcription factors. In some cases, differences in the affinity of these factors for a given DNA site may play an important role in determining the physiologic consequences of these protein-DNA interactions.

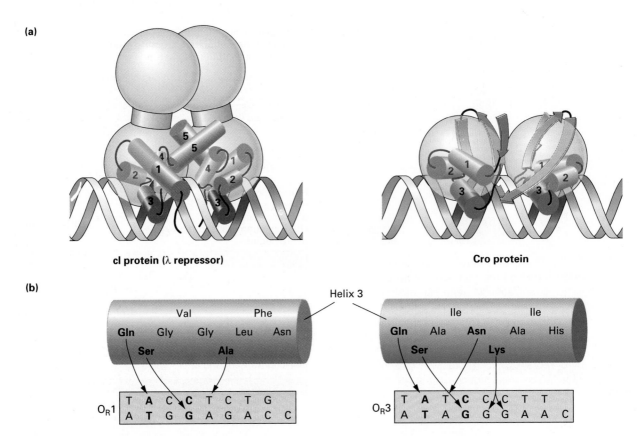

(a)

cl protein (λ repressor)

Cro protein

(b)

Helix 3

▲ FIGURE 13-8   (a) Structural models of dimeric cl and Cro proteins based on x-ray crystallographic analyses. In both proteins, the helices numbered 1, 2, and 3 have similar orientations. Helix 3 (recognition helix) inserts into the DNA major groove, establishing amino acid–base interactions. Helix 2 helps position helix 3 in the major groove. Because Cro lacks large N-terminal domains that could facilitate interactions between neighboring dimers, it does not exhibit cooperative binding as does cl. (b) Amino acid–base interactions between cl and O_R1 (*left*) and between Cro and O_R3

(*right*). The 17-bp λ operators (only half is shown) have similar but not identical sequences; the amino acid sequences of the cl and Cro recognition helices also differ. The side chains of the amino acids in boldface point toward the DNA and make contact with specific nucleotides as indicated by the arrows. Differences in these and other interactions not depicted account for the different affinities of cl and Cro for the operator sites. Cro binds most strongly to O_R3; cl, to O_R1. [Adapted from M. Ptashne, 1992, *A Genetic Switch*, Blackwell Scientific Publications and Cell Press.]

Choice between Lysis and Lysogeny Involves Regulatory Mechanisms Found in More Complex Developmental Systems

In summary, our examination of the choice between the lytic and lysogenic pathway has revealed the following important concepts:

- Different developmental pathways are defined by transcription of different sets of genes.

- Distinct mechanisms exist for the establishment and maintenance of different developmental pathways. Maintenance of a developmental pathway often occurs through an autoregulatory transcription-control mechanism.

- Transcription regulators often bind cooperatively to regulatory sites on DNA providing an efficient and stable on/off switch.

- Different transcription regulators can bind to the same DNA site(s) with different affinities and different physiologic consequences.

- The same DNA-binding protein may act as both a repressor and activator of transcription.

In the following sections, we will see that many of these same mechanisms play a role in regulating more complex developmental systems.

▶ Cell-Type Specification and Mating-Type Conversion in Yeast

Multicellular organisms contain numerous distinct cell types. Although cell types often are distinguished morphologically, the fundamental property characterizing a cell type is its particular pattern of gene expression. Certain genes are expressed in all cells; others in just a small population of cells; yet others are expressed only under certain physiologic conditions. Detailed studies of cell-type specification in the yeast *S. cerevisiae* have provided insights into the various mechanisms by which such cell-specific patterns of gene expression are generated. Also, the control of mating-type switching, which was discussed in Chapter 9, involves transcription-control mechanisms that probably operate in other developmental systems.

Cell Type–Specific Gene Expression in Yeast Is Regulated by Numerous DNA-binding Proteins

There are three different cell types in yeast: haploid **a** and α cells, and diploid **a**/α cells. Each of the three cell types expresses a unique set of genes. All haploid cells express certain haploid-specific genes; in addition, **a** cells express **a**-specific genes, and α cells express α-specific genes. In **a**/α diploid cells, diploid-specific genes are expressed, whereas haploid-specific, **a**-specific, and α-specific genes are not. As illustrated in Figure 13-9, regulation of this cell type–specific transcription is mediated by cell type–specific proteins (α1, α2, and **a**1) encoded at the *MAT* locus in combination with a transcription factor called *MCM1*, which is expressed in both haploid and diploid cells.

MCM1, a multifunctional dimeric transcription factor, exhibits different activity in haploid **a** and α cells (Figure 13-10). In **a** cells, MCM1 binds to the P site in **a**-specific upstream regulatory sequences (URSs), thereby stimulating transcription of **a**-specific genes. In α cells, however, the activity of MCM1 is modulated by its association with α1 or α2. Binding of a MCM1-α1 complex to an α-specific URS stimulates transcription of α-specific genes, whereas binding of MCM1 and α2 to an **a**-specific URS blocks transcription. In addition, association of MCM1 with pheromone-activated STE12 protein promotes transcription of certain **a**- and α-specific genes.

*Gene Repression by α2-MCM1 and α2-**a**1 Complexes* Each **a**-specific gene has a 31-bp URS containing binding sites for both MCM1 and α2. Binding of an MCM1 dimer to the P site in this URS activates transcription, but the simultaneous binding of an α2 dimer represses transcription of the **a**-specific gene (see Figure 13-10). Although both MCM1 and α2 can bind independently to an **a**-specific URS, binding of these two proteins is highly cooperative due to physical interaction between them. This highly cooperative binding of MCM1 and α2 ensures that transcription of **a**-specific genes is repressed in α cells (and diploid cells).

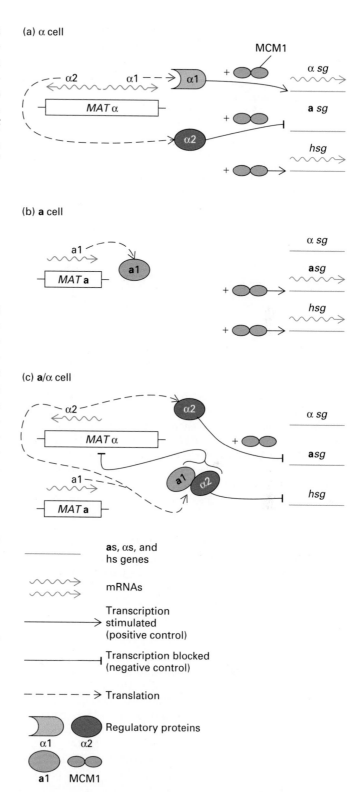

(a) α cell

(b) **a** cell

(c) **a**/α cell

as, αs, and hs genes

mRNAs

Transcription stimulated (positive control)

Transcription blocked (negative control)

Translation

Regulatory proteins

α1 α2

a1 MCM1

▲ FIGURE 13-9 Regulation of cell type–specific genes in yeast by regulatory proteins encoded at the *MAT* locus together with MCM1, a constitutive transcription factor produced by all three cell types. As a result of this regulation, each cell type exhibits a distinctive pattern of gene expression: **a**s = **a**-specific genes/mRNAs; αs = α-specific genes/mRNAs; hs = haploid-specific genes/mRNAs.

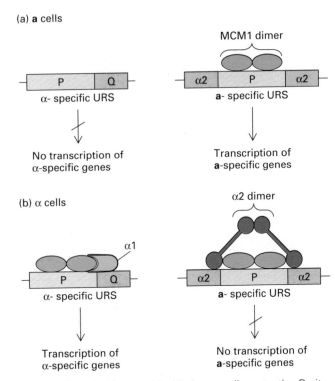

(a) **a** cells

MCM1 dimer

α- specific URS

a- specific URS

No transcription of
α-specific genes

Transcription of
a-specific genes

(b) α cells

α2 dimer

α1

α- specific URS

a- specific URS

Transcription of
α-specific genes

No transcription of
a-specific genes

▲ FIGURE 13-10   MCM1 binds as a dimer to the P site
in α-specific and **a**-specific upstream regulatory sequences
(URSs). The consensus sequences of the P box in α- and **a**-
specific genes differ slightly. (a) In **a** cells, MCM1 stimulates
transcription of **a**-specific genes. MCM1 does not bind efficiently to the P site in the α-specific URS in the absence of
α protein. (b) In α cells, the activity of this multifunctional
transcription factor is modified by its association with α1 or
α2. The α1-MCM1 complex stimulates transcription of α-
specific genes, whereas the α2-MCM1 complex blocks transcription of **a**-specific genes. The α2-MCM1 complex also is
produced in diploid cells, where it has the same blocking
effect on transcription of **a**-specific genes (see Figure 13-9c).

➤ FIGURE 13-11   Relative binding affinities of MCM1,
α2, and MCM1-α2 complex to wild-type and mutant **a**-
specific upstream regulatory sequences (URSs). MCM1 and
α2 each bind alone to **a**-specific URSs with relatively low
affinity, but the cooperative binding of the MCM1-α2 complex exhibits high affinity. (a,b) Insertion of three base pairs
(blue) on either side of the P site, which binds MCM1, does
not affect the binding affinity of either MCM1 or α2. (c) The
high-affinity, cooperative binding of these two dimeric proteins requires correct spacing of the P site and α2-binding
site, as demonstrated by the much lower affinity of the
MCM1-α2 complex for the mutant URS. Since the α2-binding
sequence occurs frequently within the genome, the dependence of the highly cooperative MCM1-α2 binding on the
spacing and orientation of **a**-specific URSs increases the
specificity of α2 function. [See D. L. Smith and A. D. Johnson, 1992, *Cell* **68**:133.]

MCM1 promotes binding of α2 to an **a**-specific URS
by orienting the two DNA-binding domains of the α2
dimer to the α2-binding sequences flanking the P site in
this regulatory sequence (Figure 13-11). In an **a**-specific
URS, the two α2-binding sites are separated by about 25
base pairs, which is equivalent to 2.5 turns of the DNA
helix. Since a dimeric α2 molecule binds to both sites, each
DNA site is referred to as a *half-site*. The relative position
of both half-sites and their orientation are highly conserved
among different **a**-specific URSs. Nevertheless, changing
the orientation or spacing of the half-sites has little effect
on the binding affinity of isolated α2 dimers in the absence
of MCM1, suggesting that the two monomeric α2 subunits
have considerable flexibility. However, the highly cooperative, high-affinity binding of α2 to an **a**-specific URS in the
presence of MCM1 requires a precise spacing and orientation of the α2 half-sites. A short amino acid sequence just
N-terminal of the homeodomain in α2 has been shown to
interact specifically with MCM1. Presumably, this interaction constrains the flexibility of the α2 dimer, so that it
binds with high affinity only to uniquely oriented and
spaced α2 half-sites. Thus the affinity of α2 for sites in an
a-specific URS is influenced by its association with MCM1.

The "relaxed" specificity of α2 protein may expand
the range of genes that it can regulate. For instance, in **a**/α
diploid cells, α2 forms a heterodimer with **a**1 that acts to
repress both haploid-specific genes and the gene encoding
α1 (see Figure 13-9c). The example of α2 suggests that
relaxed specificity may be a general strategy for increasing
the regulatory range of a particular protein.

The mechanism by which α2 represses transcription is
different from the mechanism of repression by cI in λ
phage. As described previously, cI physically blocks bind-

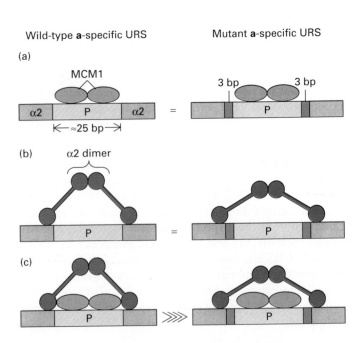

Wild-type **a**-specific URS

Mutant **a**-specific URS

(a)

MCM1

3 bp   3 bp

α2   P   α2

≈25 bp

(b)

α2 dimer

P

P

(c)

P

P

ing of RNA polymerase to P_L and P_R. In contrast, the MCM1-α2 or a1-α2 complex in yeast interacts with two proteins designated Tup1 and Ssn6, which do not themselves bind to DNA, to form a large complex that represses transcription of many genes. The precise mechanism by which these complexes mediate transcriptional repression is not known.

Gene Activation by α1-MCM1 Complex Transcription of α-specific genes is stimulated by the simultaneous binding of α1 and MCM1 to the URSs associated with these genes (see Figure 13-10b). These regulatory sequences contain two adjacent DNA sequences, the so-called P box and Q box. Although MCM1 protein binds efficiently to a perfect palindromic sequence, it does not bind to the P box in α-specific URSs, which is an imperfect palindromic sequence. Similarly, α1 does not bind alone to the Q box in an α-specific URS. The simultaneous binding of these proteins, however, occurs with high affinity and turns on transcription from the PQ site.

Gene Activation by MCM1 Plus Pheromone-Activated STE12 Protein An important feature of the yeast life cycle is the ability of haploid a and α cells to mate, that is, attach and fuse giving rise to a diploid a/α cell. Each haploid cell type secretes a different mating factor, a small polypeptide called a *pheromone,* and expresses a cell-surface receptor that recognizes the pheromone secreted by cells of the other type. Thus a and α cells both secrete and respond to pheromones. Binding of the mating factors to their receptors leads to expression of a set of genes that direct arrest of the cell cycle in G_1 and attachment/fusion of haploid cells to form diploid cells. In the presence of sufficient nutrients, the diploid cells will continue to grow. Starvation, however, induces diploid cells to progress through meiosis, each yielding four haploid spores (Figure 13-12). If the environmental conditions become conducive to vegetative growth, the spores will germinate and undergo mitotic division.

Studies with yeast mutants have provided insights into how the a and α pheromones activate cell-surface receptors, leading to a cascade of biochemical steps that ultimately change gene expression, thereby directing development toward mating. Surprising similarities have been uncovered between the mechanisms by which yeast respond to mating factors and higher eukaryotes respond to various extracellular factors that promote growth and differentiation (Table 22-1).

Haploid yeast cells carrying mutations in the sterile 12 (*STE12*) locus cannot respond to pheromones and do not mate. The *STE12* gene encodes a DNA-binding protein that binds to a certain DNA sequence referred to as the *pheromone-responsive element* (PRE). This element is present in many different a- and α-specific URSs. Binding of mating factors induces a cascade of signaling events,

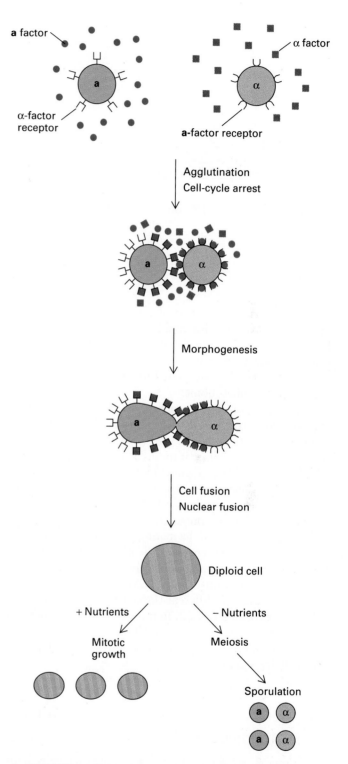

▲ FIGURE 13-12   Haploid yeast cells produce pheromones, or mating factors, and pheromone receptors. The α cells produce α factor and a receptor; the a cells produce a factor and α receptor. Binding of the mating factors to their cognate receptors on cells of the opposite type leads to gene activation, resulting in mating and production of diploid cells. In the presence of sufficient nutrients, these cells will grow as diploids. Without sufficient nutrients, cells will undergo meiosis and form four haploid spores.

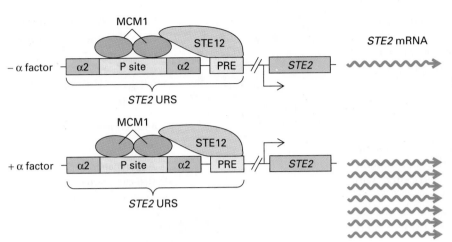

◄ FIGURE 13-13   The *STE2* URS, an **a**-specific URS, contains an MCM1-binding P site and a nearby STE12-binding site, which is a pheromone-response element (PRE). In the presence of the α mating factor, transcription of *STE2*, an **a**-specific gene, increases substantially. STE12 protein cannot bind to a single PRE in the absence of MCM1 bound to an adjacent P site. See text for discussion.

resulting in phosphorylation of various proteins including the STE12 protein. This rapid phosphorylation is correlated with an increase in the ability of STE12 to stimulate transcription. It is not yet known, however, whether STE12 must be phosphorylated to stimulate transcription in response to pheromone.

Interaction of STE12 protein with DNA has been studied most extensively at the URS controlling transcription of the *STE2* gene, an **a**-specific gene encoding the receptor for the α pheromone. Adjacent to this **a**-specific URS is a PRE, which binds STE12 (Figure 13-13). When **a** cells are treated with α pheromone, transcription of *STE2* increases in a process that requires STE12 protein. Presumably, pheromone-induced upregulation of the α receptor increases the efficiency of the mating process. STE12 protein has been found to bind most efficiently to PRE in the *STE2* URS when MCM1 is simultaneously bound to the adjacent P site. We saw previously that MCM1 can act as an activator or a repressor at different URSs depending on whether it complexes with α1 or α2. In this case, MCM1 functions as an activator at yet another URS by interacting with STE12, a transcription factor whose activity is modified by extracellular signals.

Mating-Type Conversion Is Determined by Transcriptional Regulation of the *HO* Locus

In Chapter 9, we discussed the mechanism whereby haploid yeast cells switch their mating type. Recall that two silent (nontranscribed) copies of the *MAT* locus—designated *HML* and *HMR*—are located on yeast chromosome 3 in addition to the active (transcribed) *MAT* locus (see Figure 9-38). The phenotype of haploid yeast cells is determined by the mating-type sequence (**a** or α) that they carry in the central *MAT* locus. The mechanism of gene conversion, by which the information of the *MAT* locus is switched (**a** → α or α → **a**), begins with a site-specific cleavage at *MAT* by the HO endonuclease (see Figure

9-38). It is now known that mating-type conversion in haploid yeast cells can occur only at a specific stage of the cell cycle and only in one of the two mitotic products, the so-called mother cell (Figure 13-14). This asymmetry in mating-type conversion can be demonstrated by testing individual mother and daughter cells for mating type; this

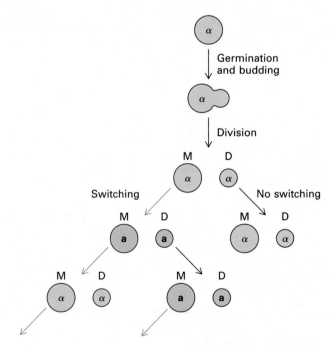

▲ FIGURE 13-14   Specificity of mating-type conversion. Under appropriate conditions, haploid yeast spores germinate and undergo mitotic division by budding. The mother cell (M), which is larger than the daughter cell (D) can, before the next DNA duplication (i.e., during G$_1$ phase), undergo a switch in the DNA sequences at the *MAT* locus. The first step in this gene conversion is catalyzed by HO endonuclease. Switching can occur in both directions (**a** → α and α → **a**) but only in haploid mother cells in the G$_1$ phase of the cell cycle.

can be done with the α and **a** mating factors, or pheromones (Figure 13-15).

A similar asymmetry in the developmental potential of two mitotic products is a common occurrence in multicellular organisms. For example, many differentiated cells are generated from *stem cells*, which can divide asymmetrically to yield a stem cell plus a sibling cell that is more specialized and has lost some of its developmental potential (Figure 13-16). The mechanisms that have been shown to control mating-type conversion in yeast may provide insight into stem-cell development more generally.

As noted already, mating-type switching exhibits three types of specificity: it occurs only in haploid cells, during the G_1 phase of the cell cycle, and in one mitotic product (the mother cell). The secret of this threefold specificity lies in regulation of transcription of the *HO* locus, which encodes the endonuclease that initiates the conversion process. This regulation occurs through two adjacent regulatory sequences—referred to as *URS1* and *URS2*—that lie ≈ 110 base pairs upstream of the *HO* locus (Figure 13-17).

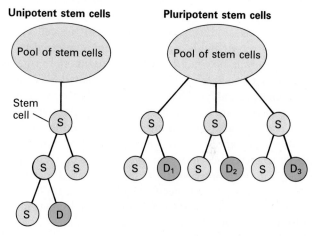

▲ FIGURE 13-16 The production of differentiated cells (D) from stem cells (S). Unipotent stem cells produce a single type of differentiated cell, whereas pluripotent stem cells may produce two or more types of differentiated cells.

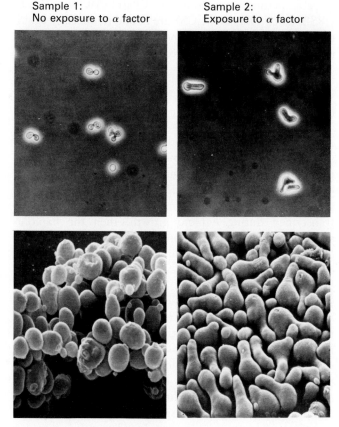

Sample 1:
No exposure to α factor

Sample 2:
Exposure to α factor

▲ FIGURE 13-15 Detection of mating-type phenotype (**a** or α) in yeast. Cells of unknown mating type were divided into two samples: one was exposed to the α factor (sample 2) and the other was not (sample 1). Sample 1 shows normal budding cells and daughter cells in a phase-contrast light micrograph (*top*) and a scanning electron micrograph (*bottom*). The corresponding micrographs of sample 2 show that cell division is inhibited in these cells. Since each mating factor inhibits cell division in cells of the opposite mating type, the sample cells are **a**. [See J. Thorner, 1980, in *The Molecular Genetics of Development*, T. A. Leighton and W. A. Loomis, eds., Academic Press. Courtesy of J. Thorner, R. Kunisawa, and M. Davis.]

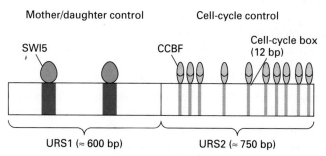

▲ FIGURE 13-17 Upstream regulatory region that controls transcription of the *HO* locus. This gene encodes a site-specific endonuclease that initiates mating-type switching. This region consists of two adjacent sequences, URS1 and URS2, which contain protein-binding sites (dark green and orange). Binding of SWI5 and CCBF (cell-cycle box factor) as shown is required for expression of HO endonuclease. SWI5 is active only in the mother cell despite being synthesized during the previous cell cycle; thus URS1 confers mother-cell specificity. URS2 contains multiple binding sites for CCBF, a complex containing SWI4 and SWI6. The activity and/or expression of CCBF is restricted to G_1; thus URS2 confers cell-cycle specificity. Not shown are multiple binding sites for the **a**1/α2, which is produced by diploid cells and represses transcription of *HO*, thus conferring haploid-cell specificity. [See B. J. Andrews and I. Herskowitz, 1989, *Cell*, **57**:21; G. Tebb et al., 1993, *Genes & Dev.* **7**:517.]

Haploid-Cell Specificity The *HO* locus is a haploid-specific gene whose transcription is repressed in diploid yeast cells. This repression is mediated by a heterodimeric complex comprising α2 and a1, which are encoded by *MAT*α and *MAT*a, respectively (see Figure 13-9c). This complex binds to multiple sites within URS1 and URS2, thereby preventing transcription of *HO* in diploid cells.

Cell-Cycle Specificity URS2, which is ≈750 base pairs long, contains 10 repeats of a 12-bp sequence, termed the cell-cycle box, that functions in stimulating expression of HO endonuclease during the G_1 phase of the cell cycle. This 12-bp repeat binds a protein complex referred to as CCBF (cell-cycle box factor), which contains the proteins encoded by two *switch* (*SWI*) *genes* designated *SWI4* and *SWI6*. Genetic analysis has shown that both SWI4 and SWI6 proteins are required for mating-type switching. These proteins also are important in regulating the cell cycle–specific expression of other genes. SWI4 binds specifically to the cell-cycle box in the absence of SWI6, suggesting that SWI6 is necessary for the activating function of CCBF but not for its site-specific DNA-binding ability. The activity and/or expression of CCBF is now known to be responsive to CDC28 kinase, a central component of the cell-cycle machinery that is highly active in late G_1 (Chapter 25).

Mother-Cell Specificity The key protein in restricting mating-type conversion to mother cells is SWI5, which binds to two short sequences within URS1 and is required for transcription of the *HO* locus. Although *HO* transcription occurs only in late G_1, the SWI5 protein is synthesized in all stages of the cell cycle except G_1. Presumably, SWI5 synthesized in the previous cell cycle is selectively functional in mother cells following division. The central role of SWI5 in mating-type switching is underscored by experiments with recombinant yeast in which SWI5 is expressed during G_1 (under control of a heterologous promoter); in such recombinants, switching occurs in both mother and daughter cells.

The mechanism by which SWI5 is rendered functional only in mother cells is not known. One straightforward possibility is that SWI5 is asymmetrically distributed at mitosis: that is, mother cells but not daughter cells inherit SWI5. However this does not appear to be the case since both mother and daughter cells stain with antibodies to SWI5. A molecular understanding of the mechanism by which SWI5 is selectively active in mother cells may provide insight into the more general question of how developmental potential is inherited differently by the two products of a cell division.

Other Inhibitors and Activators of HO *Transcription* In addition to the inhibitors and activators of *HO* transcription described above, several other proteins play roles in controlling expression of HO endonuclease. Among these are the SWI1, 2, and 3 proteins, which like SWI4–6 are required for mating-type conversion. A clue to the function of these proteins came from identification of yeast cells carrying mutations in the *SIN* genes. These *switch-independent* mutations lead to inappropriate transcription of the *HO* locus. Molecular characterization of several *SIN* genes has shown that they encode components of chromatin, suggesting that the *SIN*-gene products may repress *HO* transcription by maintaining the *HO* regulatory region in a configuration that prevents binding of the positively acting SWI proteins. Presumably, then, the *SWI* genes promote *HO* transcription by antagonizing the function of the *SIN* genes.

The molecular mechanisms by which these different levels of regulation converge on the *HO* locus to precisely control its transcription are not known. One model proposes that binding of SWI5 to URS1 promotes the activity of SWI1, 2, and 3. These proteins in turn somehow counteract the effect of the proteins encoded by the *SIN* genes, thereby permitting binding of CCBF (SWI4/SWI6) to URS2. Once SWI5 and CCBF are bound to the *HO* regulatory region in G1, expression of the HO endonuclease and mating-type switching proceed.

Silencer Elements Repress Expression at *HML* and *HMR*

As noted earlier, the *HML* and *HMR* loci on yeast chromosome 3 contain "extra" silent (nontranscribed) copies of the α and a sequences (see Figure 9-38). The location of the extra a and α sequence in *HML* and *HMR*, respectively, varies in different yeast strains. If these extra copies were transcribed during haploid growth, then the haploid-specific genes would be repressed and the haploid cells could not mate (i.e., the haploid cells would be phenotypically like diploid cells).

As discussed in Chapter 11, *silencer elements* are responsible for specific repression of the a and α sequences associated with *HML* and *HMR*. Recent studies suggest that an important transcription-control mechanism may involve assembly of silencer-associated regions into higher-order chromatin structures inaccessible to the transcriptional machinery. Since many eukaryotic genes are silent in different cell types, the ability to shut off transcription completely probably is widespread among eukaryotes.

► *Myogenesis in Mammals*

Animals exhibit a remarkable diversity of cell types. These different cells carry out a wide variety of different functions and exhibit an enormous spectrum of different morphologies. These differences are due to both qualitative and quantitative differences in the expression of specific genes. Cell biologists do not yet understand for any unique cell type in a multicellular organism the complete set of regulatory molecules that makes it different from other cells. However, considerable progress has been made towards identifying DNA-binding proteins that play critical roles in controlling the transcription of cell type–specific genes (Chapter 11). It appears as though cell type–specific patterns of gene expression reflect the activity of specific transcription factors or unique combinations of them. In this and the next section, we consider development of muscle cells in mammals and nerve cells in *Drosophila*, two well-studied examples of cell-type specification in higher eukaryotes.

The development of mammalian skeletal muscle (a subclass of muscle that mediates most voluntary movement) is a favorable system for investigating the role of

transcription factors in controlling cell-type specification. Skeletal muscle cells are elongated multinucleate fibers. They synthesize a large number of specific proteins that are necessary for generating their unique structure and highly specialized functions (Chapter 22). The development of mammalian muscle can be studied both in the intact organism and in in vitro systems in which muscle precursor cells can be induced to differentiate into muscle.

Embryonic Somites Give Rise to Myoblasts, the Precursors of Skeletal Muscle Cells

Mammalian myogenesis, summarized in Figure 13-18, can be divided into three stages: myoblast determination, migration, and differentiation into muscle. Precursor skeletal muscle cells, called *myoblasts*, arise from *somites*, which are blocks of mesodermal cells found lateral to the neural tube in the embryo (Figure 13-19). Myoblasts are commit-

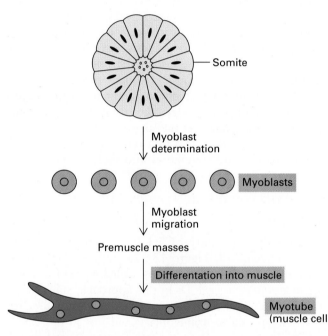

▲ FIGURE 13-18 Schematic diagram of three stages in development of skeletal muscle in mammals. Somites are collections of embryonic mesodermal cells, some of which become determined as myoblasts. Myoblasts, which are precursor skeletal muscle cells, are distinct from other somite-derived precursor cells. Myoblasts migrate to form premuscle masses in the limbs and elsewhere, where they differentiate into multinucleate skeletal muscle cells, called myotubes.

➤ FIGURE 13-19 Embryonic determination of myoblasts in mammals. (a) Skeletal muscle is derived from embryonic structures called somites, which are blocks of mesodermal cells. (b) After formation of the neural tube, each somite differentiates into a dermomyotome, which gives rise to skin and muscle, and a sclerotome, which develops into skeletal structures. Myoblasts form at each edge of the dermomyotome. Lateral myoblasts migrate to the limb bud. Axial myoblasts form the myotome. (c) The dermotome gives rise to skin elements (dermis), and the myotome to axial muscle. [See B. A. Williams and C. P. Ordahl, 1994, *Development* **120**:785–796. Adapted from M. Buckingham, 1992, *Trends Genet.* **8**:144.]

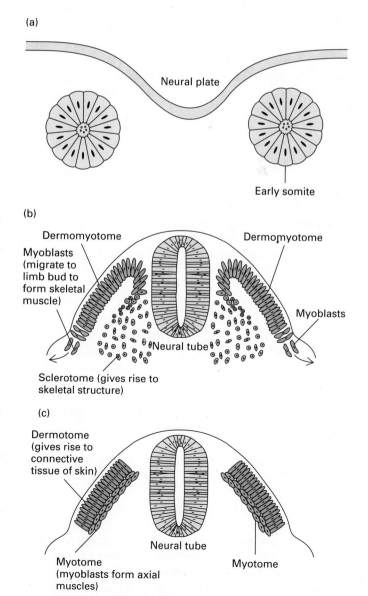

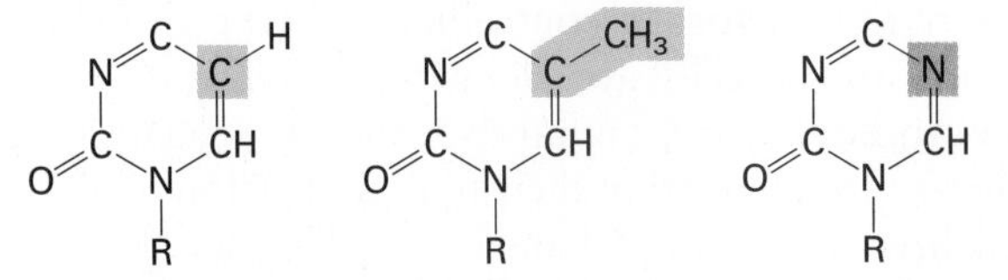

Cytidine   5-methylcytidine   5-azacytidine

▲ FIGURE 13-20   Methylation of cytidine residues in the genome to yield 5-methylcytidine in some cases prevents gene transcription. When cultured cells are grown in the presence of 5-azacytidine, which cannot be methylated, it replaces cytidine residues and may lead to activation of previously repressed genes. R = phosphorylated deoxyribose.

ted to become muscle but have not yet differentiated; hence, they are referred to as *determined*. Myoblasts migrate from their site of origin in somites, termed the *dermomyotome*, to regions where skeletal muscles will form (e.g., developing limbs). Here the cells align, stop dividing, fuse to form a *syncytium* (a cell containing many nuclei but sharing a common cytoplasm), and differentiate into muscle. We refer to this multinucleate skeletal muscle cell as a *myotube*. Concomitant with cellular fusion there is a dramatic rise in the expression of genes necessary for muscle development and function.

Certain Fibroblasts Can Be Converted into Muscle (Myotubes)

Studies in the late 1970s demonstrated that a fibroblast cell line called C3H 10T1/2 could be induced to differentiate into myotubes by incubation in the presence of 5-azacytidine (Figure 13-20). Upon entry into cells, 5-azacytidine is converted to 5-azacytidine triphosphate and then is incor-

porated into DNA in place of cytidine. In contrast to cytidine, 5-azacytidine cannot be methylated due to the presence of a nitrogen at position 5 in the pyrimidine ring. As discussed in Chapter 11, methylation of cytidine residues in DNA can prevent transcription of genes. When cells are grown in 5-azacytidine, so that cytidine residues are replaced with a derivative that cannot be methylated, genes previously inactivated by methylation can become activated. The high frequency of C3H 10T1/2 cells converted into myotubes suggested to early workers that reactivation of one or a small number of closely linked genes is sufficient to drive a program of muscle development.

If this conclusion were correct, the researchers reasoned that DNA isolated from C3H 10T1/2 cells grown in the presence of 5-azacytidine, so-called *azamyoblasts*, should be able to drive untreated cells to become muscle (Figure 13-21). When this hypothesis was tested, 1 in 10^4 cells transfected with DNA isolated from azamyoblasts was converted into myotubes, a frequency consistent with the earlier conclusion that one or a small set of closely linked genes is responsible for converting fibroblasts into myotubes. These studies set the stage for the isolation and characterization of the genes that regulate the entire program of myoblast differentiation into skeletal muscle.

The *myoD* Gene Can Trigger Muscle Development

Four genes that can convert C3H 10T1/2 cells into muscle have been identified. Figure 13-22a outlines the screen used to identify one of these genes, called *myoD* for *myogenic determination gene D*. In this approach, subtractive hybridization was used to isolate cDNAs corresponding to mRNAs expressed in azamyoblasts but not in untreated C3H 10T1/2 cells. The isolated azamyoblast-specific cDNAs then were used as probes to screen an azamyoblast

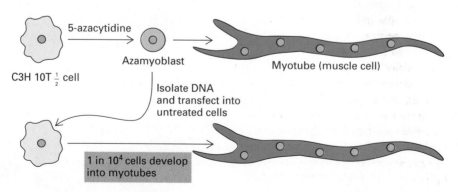

C3H 10T $\frac{1}{2}$ cell   Azamyoblast   Myotube (muscle cell)

Isolate DNA and transfect into untreated cells

1 in 10^4 cells develop into myotubes

▲ FIGURE 13-21   Experimental system for studying mammalian myogenesis. A fibroblastic cell line called C3H 10T1/2 can be converted into muscle cells by incubating them with 5-azacytidine. Under appropriate conditions, intermediate precursor cells, termed azamyoblasts, accumulate. DNA iso-

lated from azamyoblasts can drive conversion of untreated C3H 10T1/2 cells into muscle cells. [See S. M. Taylor and P. A. Jones, 1979, *Cell* **17**:771; A. B. Lasser et al., 1986, *Cell* **47**:649.]

(a) Screen for azamyoblast-specific genes

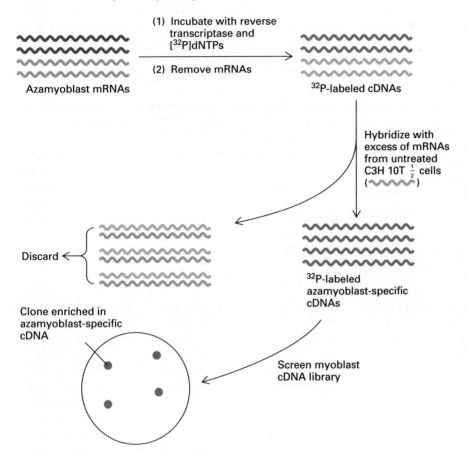

(1) Incubate with reverse transcriptase and [^{32}P]dNTPs

(2) Remove mRNAs

Azamyoblast mRNAs

^{32}P-labeled cDNAs

Hybridize with excess of mRNAs from untreated C3H 10T $\frac{1}{2}$ cells (⁓⁓⁓)

Discard

^{32}P-labeled azamyoblast-specific cDNAs

Clone enriched in azamyoblast-specific cDNA

Screen myoblast cDNA library

(b) Assay for myogenic activity of *myoD* cDNA

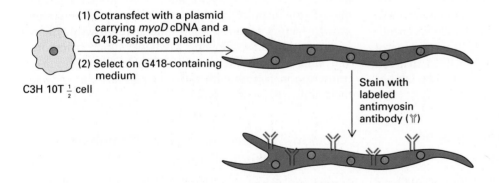

(1) Cotransfect with a plasmid carrying *myoD* cDNA and a G418-resistance plasmid

(2) Select on G418-containing medium

C3H 10T $\frac{1}{2}$ cell

Stain with labeled antimyosin antibody (Y)

◄ **FIGURE 13-22** Identification and assay of genes that drive myogenesis. (a) Azamyoblast mRNAs were isolated from cell extracts on an oligo-dT column (see Figure 7-15). Incubation with reverse transcriptase and [^{32}P]dNTPs yielded radiolabeled cDNAs. The cDNAs were mixed with mRNAs from *untreated* C3H 10T½ cells; only cDNAs derived from mRNAs (light red) produced by both azamyoblasts and untreated cells hybridized. This technique of subtractive hybridization yielded labeled azomyoblast-specific cDNAs (dark blue), at least some of which correspond to genes required for myogenesis. These cDNAs then were used as probes to screen an azomyoblast cDNA library. (b) A cDNA isolated as shown in part (a), designated *myoD,* was shown to drive conversion of C3H 10T½ cells into muscle cells, identified by their binding of labeled antibodies against myosin, a muscle-specific protein. G418 is an antibiotic.

cDNA library to identify clones corresponding to mRNAs enriched in azamyoblasts. Three such clones then were tested for their ability to trigger muscle development in C3H 10T½ cells. First, three recombinant plasmids, each carrying one of the azamyoblast-specific cDNAs linked to a strong promoter, were constructed. C3H 10T½ cells then were transfected with one of these plasmids plus a second plasmid carrying a gene conferring resistance to G418, an antibiotic. Transfected cells selected on a medium containing G418 were stained with an antibody to myosin, a muscle-specific protein, to determine their ability

to express muscle-specific genes (Figure 13-22b). One of the tested cDNA clones, designated *myoD*, converted C3H 10T½ cells into muscle at a high frequency, whereas the other two did not. Colonies of *myoD*-transfected cells were indistinguishable from C3H 10T½ cells treated with 5-azacytidine, and both types of cells exhibited myotube-like properties. The *myoD* cDNA also was found to convert a number of other cell lines into muscle.

Based on these findings, the *myoD* gene was proposed to play a key role in muscle development. A similar approach has identified three other genes—the *myogenin*, *myf5,* and *mrf4* genes—that also function in muscle development. As discussed in a later section, knockout experiments have demonstrated the importance of three of these genes in muscle development in the intact mouse.

Myogenic Proteins Are HLH Transcription Factors

The four myogenic proteins—MyoD, Myf5, myogenin, and Mrf4—are all members of the *helix-loop-helix* (*HLH*) family of DNA-binding transcription factors (see Figure 11-48i). HLH proteins form homo- and heterodimers that bind to a common DNA site with the consensus sequence CANNTG. Referred to as the *E box,* this sequence is present in multiple copies in most muscle-specific enhancers. Although MyoD can bind to a single E box, it exerts an activating effect only when bound to two or more sites. Binding to multiple sites occurs cooperatively. This cooperative binding may provide a mechanism for an all-or-none response, as discussed previously for binding of cI protein to λ-phage DNA, and also increase the specificity of MyoD. Because of its minimal sequence requirement, the E-box consensus sequence is likely to be present at many locations within the genome (i.e., on a purely random basis the E-box sequence will be found every 256 nucleotides).

The affinity of MyoD for DNA is tenfold greater when it binds as a heterodimer complexed with the E2A protein than when it binds as a homodimer. Although E2A is an HLH protein that is expressed in many cell types, it also is required for myogenesis. For example, inhibition of E2A protein function in C3H 10T½ cells antagonizes myogenic conversion by azacytidine. This effect was demonstrated by transfecting C3H 10T½ cells with plasmid DNA encoding RNA complementary to E2A mRNA. Transcription of transfected plasmid templates produces an RNA that can hybridize to the endogenous E2A mRNA, thereby inhibiting its translation into E2A protein.

Myogenic Gene Activation Depends on Specific Amino Acids in MyoD

The DNA-binding domains of both E2A and MyoD have similar but not identical amino acid sequences, and both proteins recognize the E-box sequence in DNA. These findings raise two questions: What restricts the activation of muscle-specific genes to muscle cells? Why are genes that encode nonmuscle proteins and have CANNTG sites in their enhancers not activated in muscle cells? Some insight into these questions has come from in vitro mutagenesis studies in which various E2A variants were produced.

Wild-type E2A protein cannot by itself drive C3H 10T½ cells to a myogenic fate, although it can bind to MyoD-recognition sites (E box) controlling muscle-specific genes. However, an E2A variant with three amino acid substitutions corresponding to residues present in MyoD was found to convert C3H 10T½ cells to myotubes.

Two of these substitutions are in the basic DNA-binding region of the protein, and one is just adjacent to this region. The specificity of activation of muscle-specific genes thus may depend not on binding of transcription factors per se but on the quality of binding. For example, the unique DNA-protein interactions specified by MyoD (or the E2A variant) may promote a specific conformational change that allows MyoD to mediate transcriptional activation. Similarly, binding of MyoD to regulatory regions controlling genes not expressed in muscle cells may not result in a MyoD conformation compatible with activation. Finally, the specificity of activation of different subsets of E-box containing genes may reflect the spectrum of HLH proteins in a cell that can function as dimerization partners, as well as other proteins bound at adjacent sites within transcription-control regions.

Id Protein Inhibits Activity of MyoD

Screens for genes related to *myoD* led to identification of another HLH protein that lacks a basic region and hence is unable to bind to DNA. However, this protein can form dimers with MyoD and E2A, thereby inhibiting formation of E2A-MyoD dimers and hence their binding to DNA. Accordingly, this protein is referred to as *Id* for *inhibitor of DNA binding.* Interestingly, in proliferating azamyoblasts, which express MyoD, E2A, and Id, the MyoD-binding (or E2A-binding) site in the promoter of the muscle creatine kinase (MCK) gene is not occupied. This presumably reflects the formation of inactive MyoD-Id or E2A-Id complexes. When these cells are induced to differentiate into muscle (for instance by the removal of serum-containing growth factors required for proliferative growth), the Id concentration falls, allowing MyoD-E12 dimers to form and bind to the MCK promoter. Hence, dimerization of transcription factors with different partners can regulate their ability to bind to DNA.

Knockout Experiments Have Demonstrated Role of Myogenic Proteins in Vivo

Expression of any one of the four myogenic proteins in C3H 10T½ cells can induce the cells to differentiate into muscle in vitro. The functions of these proteins in the intact animal during normal myogenesis have been studied in gene-knockout experiments. In these studies, mice were prepared with gene-targeted knockout mutations in the genes encoding MyoD, Myf5, or myogenin (see Figure 8-38).

Mice with either the *myoD* or *myf5* gene knocked out have normal muscle, whereas those with the *myogenin* gene knocked out are missing the vast majority of skeletal muscle (Table 13-1). In the myogenin knockout mice, myoblasts accumulate at sites normally occupied by skeletal muscle, indicating that myogenin is not required for formation of myoblasts but is required for their differentiation

TABLE 13-1   Effect of Knockout of Myogenic Genes in Mice

| Gene Knocked Out | Phenotype* | | | Role of Myogenic Protein |
	Viable	Myoblasts	Muscle	
myoD	Yes	+	+	?
myf5	Yes	+	+	?
myoD; myf5	No	−	−	Required for myoblast formation or survival
myogenin	No	+	−	Required for myoblast differentiation into muscle

* + sign indicates that myoblasts or mature muscle cells are found at normal sites; − sign indicates that they are not.
SOURCE: T. G. Braun et al., 1992, *Cell* **71**:369; P. Hasty et al., 1993, *Nature* **364**:501; M. A. Rudnicki et al., 1992, *Cell* **71**:383; M. A. Rudnicki et al., 1993, *Cell* **75**:1351.

into myotubes. The simple, but erroneous, conclusion from these findings is that Myf5 and MyoD are not required for muscle development. However, since either protein can drive a myogenic program in cell culture, the loss of one gene may be compensated by the function of the other. Indeed, mice homozygous for mutations in both *myf5* and *myoD* die shortly after birth and lack skeletal muscle. In contrast to the myogenin mutants, myoblasts do not accumulate in the *myf5; myoD* double mutants, suggesting that the Myf5 and MyoD proteins are required for the formation or survival of myoblasts.

The results of these gene-knockout experiments are consistent with the observation that azacytidine-treated C3H 10T½ cells express Myf5 and MyoD prior to fusion but express myogenin only as they fuse to form a syncytium, which then differentiates to form a myotube. As the model in Figure 13-23 illustrates, MyoD and Myf5 are thought to have similar but overlapping functions in selecting cells from developing somites to become myoblasts; that is, they are required for myoblast determination during normal myogenesis. Myogenin, then, promotes the differentiation of myoblasts into myotubes. The fourth myogenic protein, Mrf4, is expressed later in development and may play a role in the maintenance of muscle cells.

► Neurogenesis in Drosophila and Mice

The *Drosophila* genome contains an ≈100-kb gene complex, termed the *achaete-scute complex* (AS-C), that controls neurogenesis. This stretch of genomic DNA contains four genes designated *achaete* (*ac*), *scute* (*sc*), *lethal of scute* (*l'sc*), and *asense* (*a*), all of which encode HLH proteins that are remarkably similar to the myogenic HLH proteins discussed in the previous section. The role of *Drosophila* AS-C has been studied most thoroughly in certain sensory organs that are innervated by a single neuron; development of these sensory organs is regulated by the products of AS-C. Genes with functions analogous to *ac* and *sc* also have been identified in mice.

Two other genes encoding HLH proteins play critical roles in development of *Drosophila* sensory organs: the *daughterless* (*da*) gene promotes their development, whereas the *extramachrochaete* (*emc*) gene inhibits their development. Thus mutations in *da* lead to deletion of sensory structures; mutations in *emc* lead to development of ectopic sensory structures. The *da* and *emc* genes encode proteins that are analogous in structure and function to mammalian E2A and Id, respectively. For example, hetero-

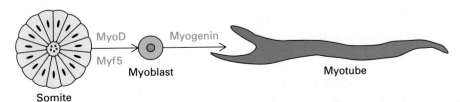

▲ FIGURE 13-23   Model of genetic control of mammalian skeletal muscle in vivo based on gene-targeted knockout experiments. According to this model, MyoD and Myf5 serve a redundant function in myoblast determination, while myogenin is necessary for differentiation of myoblasts into myotubes.

dimeric complexes of Da protein with Ac or Sc protein bind to DNA better than the homodimeric forms of Ac and Sc. And like Id, Emc is an HLH protein that lacks the basic domain and binds to Ac and Sc proteins, thus inhibiting their association with Da and binding to DNA.

Drosophila Sensory Hairs Arise from Proneural Clusters, Which Express Achaete and Scute Proteins

To illustrate the functions of the *Drosophila* neurogenic genes, we examine the development of the sensory hairs located on the second thoracic segment of the adult fly. The epidermis of the second thoracic segment and the associated wing are derived from a monolayer of cells (a columnar epithelium) called the *wing imaginal disc*. The distribution of sensory hairs that arise from the imaginal disc is highly reproducible. Each hair is part of a sensory organ that contains four cells: a hair that protrudes from the epidermis; the socket into which the hair is inserted; a neuron that transmits the sensory information; and finally a cell associated with the neuron referred to as a support cell. These cells are derived from a single cell by two sequential divisions. This "grandmother cell" is referred to as a *sensory organ precursor* cell (SOP), or more generally as a neural precursor cell. The pattern of SOPs in the developing imaginal disc presage the pattern of hairs in the adult. These cells do not migrate from another location but rather arise in distinct positions within the imaginal disc. Each SOP emerges from a cluster of cells, the *proneural cluster,* that express the Achaete and Scute proteins.

A Single Sensory Organ Precursor Develops from a Proneural Cluster in *Drosophila*

All the cells within a proneural cluster probably have the potential to become an SOP and develop into a sensory-hair organ. Selection of a single cell to develop as an SOP represents a choice between two pathways of cellular development: epidermal or neural. This choice depends on interactions between cells within a proneural cluster mediated in part by two cell-surface proteins encoded by the *Notch* and *Delta* loci (Figure 13-24a). Mutations at these two loci result in the appearance of multiple hairs arising from a single proneural cluster. A combination of biochemical and genetic evidence indicates that these two proteins directly interact: Notch acts as a cell-surface receptor transmitting an inhibitory signal to the nucleus, and Delta acts as the ligand. Although the mechanism whereby this cellular interaction leads to selection of a single SOP is not known, a likely hypothesis is that interaction promotes inactivation of *achaete* and *scute* in all but one cell in a proneural cluster (Figure 13-24b). The selected SOP then begins to express *asense*, which encodes another HLH protein, as well as other genes whose products determine the type of neuron that develops, a process termed *neuronal specification*. As production of Asense protein increases in an SOP, synthesis of Achaete and Scute proteins decreases.

Genetic studies have shown that the Extramachrochaete (Emc) protein plays a role in limiting the function of *ac* and *sc*. Loss-of-function mutations in *emc* lead to formation of multiple sensory hairs from a single proneural cluster, whereas gain-of-function mutations suppress SOP formation. During normal development, the region of a proneural cluster from which the SOP will arise is marked by lower expression of Emc protein than found in regions giving rise to epidermal structures.

Drosophila Neurogenesis and Mammalian Myogenesis May Occur via Analogous Pathways Involving HLH Proteins

Although many details concerning myogenesis in mammals and neurogenesis in *Drosophila* are still not understood, the two developmental pathways appear to be generally analogous. Both pathways involve *determination* of precursor cells (myoblasts or SOPs) from particular embryonic structures (somites or wing imaginal discs) and *differentiation* of these precursor cells into either muscle cells or nerve cells. (Unlike myogenesis, however, neurogenesis in *Drosophila* does not involve migration of the precursor cells to another site before differentiation.) Regulation of both pathways depends on several HLH proteins that interact with each other and with DNA to control gene transcription (Figure 13-25). Furthermore, flies that carry mutations in either *achaete* or *scute* lack different classes of sensory hairs, whereas *ac-sc* double mutants lack all sensory hairs. In a sense, these effects are similar to the failure of myoblasts to form in *myf5; myoD* double-mutant mice.

MASH1, a Homolog of Achaete and Scute Proteins, Regulates Neurogenesis in the Mouse

A mammalian homolog of Achaete and Scute, called *MASH1* (for *mammalian achaete-scute homolog 1*) is expressed widely in the developing central nervous system and within regions of the autonomic nervous system. *MASH1* cDNA originally was isolated from a rat cell line derived from a progenitor cell that gives rise to part of the autonomic nervous system, the branch of the nervous system innervating structures such as the heart, skin, and glands. Screening of a cDNA library prepared from this cell

(a)

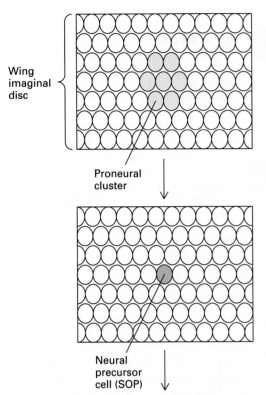

Wing imaginal disc

Expression of proneural genes (e.g., *achaete* and *scute*) in proneural cluster

Proneural cluster

Selection of one cell in cluster mediated by cellular interactions involving Notch and Delta proteins

Expression of neural precursor genes (e.g., *asense*) by SOP

Neural precursor cell (SOP)

Neuronal specification

▲ FIGURE 13-24   Neurogenesis in *Drosophila*.
(a) A single cell in a proneural cluster is selected to become a neural precursor cell, or SOP, in a process requiring *achaete* and *scute*. Expression of *asense* and other neural precursor genes by an SOP mediate differentiation into a specific type of neuron.
(b) Expression of Achaete protein over time is visualized in this series of photomicrographs of proneural clusters (PSA, PPA, and DC) stained with anti-Achaete antibody. DC is an overlap of two proneural clusters each of which gives rise to an SOP (ADC and PDC). After the SOP is selected in a proneural cluster, expression of Achaete is turned off; thus the SOP is no longer visualized with anti-Achaete. Each proneural cluster has a dynamic and distinctive pattern of Achaete expression. [Part (b) from J. B. Skeath and S. B. Carroll, 1991, *Genes & Dev.* **5**:984.]

(b)

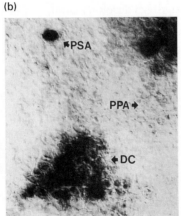
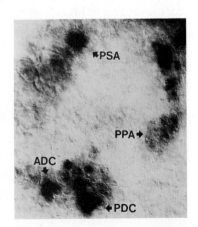
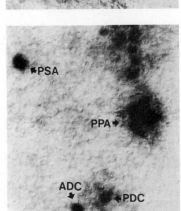
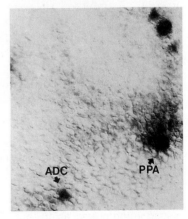

Fly neurogenesis

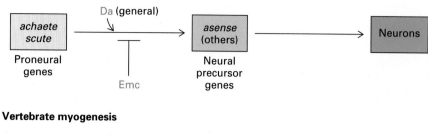

Vertebrate myogenesis

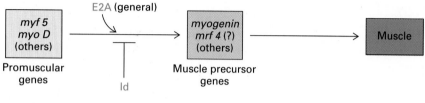

Precursor determination   Differentiation

◄ FIGURE 13-25   Comparison of genes that regulate *Drosophila* neurogenesis and mammalian myogenesis. All these genes encode HLH transcription factors and exert analogous functions in both neurogenesis and myogenesis. In both cases, functions of the earliest-acting genes (*left*) are under both positive and negative control by other HLH proteins (red type). The products of these genes are required for precursor determination and subsequent differentiation or specification into mature muscle cells and neurons. See text for discussion. [Adapted from Y. N. Jan and L. Y. Jan, 1993, *Cell* **75**:827.]

line with an *achaete* or *scute* probe identified *MASH1* cDNA, which has considerable sequence homology with the *Drosophila* probes.

Mice with knockout mutations in *MASH1* die in the first 24 h after birth, perhaps due to a physiologic defect (e.g., in feeding or breathing) resulting from developmental abnormalities observed in the autonomic nervous system of these mutants. For instance, mutant *MASH1* mice show a striking loss of neurons in the autonomic nervous system (Figure 13-26) and in the olfactory system, but the number of non-neuronal cells (e.g., glial cells) is unaffected. This phenotype could result from a defect in the determination

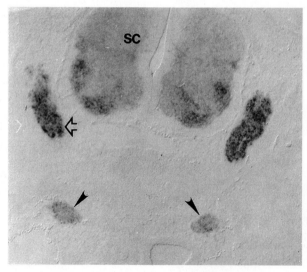

Normal

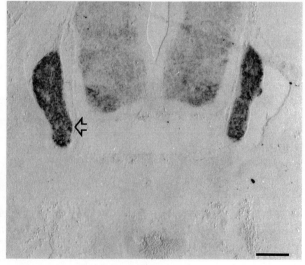

MASH1 mutant

▲ FIGURE 13-26   Photomicrographs of spinal cord sections from a normal mouse and one carrying a mutation in the *MASH1* gene, a homolog of *achaete-scute* in *Drosophila*. These sections were stained with a molecular marker specific for neurons. Sympathetic ganglia (solid arrowheads) are stained in the normal animal but lacking in the *MASH1* mutant; these ganglia contain neurons that are part of the autonomic nervous system. In contrast, dorsal root ganglia, composed of sensory neurons, are stained in both animals (open arrowheads). [From F. Guillemot et al., 1993, *Cell* **75**:463.]

or differentiation of neuronal precursor cells, or in their ability to survive. Further analyses of these mutant mice showed that the olfactory epithelium lacks neuronal precursor cells; in contrast, the autonomic nervous system contains neuronal precursor cells but no fully differentiated neurons. As we saw earlier, the *achaete* and *scute* genes in *Drosophila* function in determination of sensory organ precursors (see Figure 13-24a); in development of some sensory organs, these genes also play a role in differentiation. Thus the mammalian *MASH1* gene and fly *achaete* and *scute* genes may be functional analogs as well as sequence homologs.

Specification of Other Cell Types Is Controlled by Different Classes of Transcription Factors

Numerous experimental results strongly suggest that HLH transcription factors play an important role in controlling cell-type specification of skeletal muscle and certain neurons in multicellular organisms. It is unlikely that all cell types will be specified by HLH proteins. For instance, biochemical studies indicate that genes expressed specifically in erythroid cells, which give rise to erythrocytes, are controlled by a zinc-finger transcription factor called GATA protein. Mice with a knockout mutation in the *GATA* gene are severely anemic due to the failure of erythroid cells to differentiate properly. Another example involves hepatic nuclear factor (HNF) α, β, and γ, a newly recognized class of DNA-binding proteins that control liver-specific genes and are expressed in liver precursor cells. HNFs are highly homologous to a *Drosophila* protein, called Forkhead protein, that functions in differentiation of endodermal cells, the same type of embryonic cells that give rise to the mammalian liver.

► *Regional Specification during* Drosophila *Embryogenesis*

In previous sections of this chapter, we discussed the transcription-control mechanisms that specify different cell types in yeast, *Drosophila*, and mammals. Each cell type expresses specific subsets of genes that determine its biochemical and morphological characteristics. In addition to different cell types, multicellular organisms often exhibit striking regional differences in their cellular organization. For instance, the tissue in hands and feet are composed of the same cells organized in very different ways. What mechanisms determine how cells are organized in different parts of an organism? What are the mechanisms that specify one end of the developing embryo will become a head and the other the tail? How are the number, order, and form of digits established? What controls the size, position, and interrelationship between different organs?

In the remainder of this chapter, we discuss how the embryo is subdivided into distinct spatial domains, a process called *regionalization*. Regionalization requires precise orchestration of cellular interactions, cell movements, and changes in gene expression. Our understanding of the molecular and cellular basis of regionalization has been revolutionized by genetic analysis of early *Drosophila* development. Remarkably, mammalian genes homologous to the fly early-patterning genes appear to have similar functions, suggesting that at least in part, the basic blueprint of a body plan has been conserved through evolution. In this section, we first review some basic features of *Drosophila* development and then consider the role played by different transcription factors in regionalization. In the final section of this chapter, we describe the role of homologous genes in mammalian development. The contribution of cellular interactions to development is examined in Chapter 24.

Drosophila Has Two Life Forms

The entire life cycle of *Drosophila* occurs within only 9–10 days (Figure 13-27a). The organism has two forms, a worm-like *larval* form and the adult fly form, separated by a period of metamorphosis called *pupation*. Within 1 day the fertilized egg develops into a larva; three subsequent larval stages, or *instars* as they are called, require about 4 more days. The larva is actually a separate animal from the adult. During early embryogenesis about a dozen groups of cells, termed *imaginal discs*, are set aside and are carried in the abdomen of the larva (Figure 13-27b). These groups of cells give rise to the adult epidermal structures (wings, legs, etc.). Other groups of precursor cells give rise to adult internal organs such as portions of the gut, the vast majority of the musculature, and the central nervous system. Some larval cells also are conserved in the adult. After the last larval stage, an outer shell is formed. The larval cells are broken down and nutrients derived therefrom are used in the growth and development of the cells that give rise to the different body parts of the adult fly. Pupation takes another 4 days or so. At the end of pupation the shell splits and an adult fly emerges.

Patterning Information Is Generated during Oogenesis and Early Embryogenesis

The blueprint for constructing a fruit fly, including critical spatial information, is in large part laid down in the egg before fertilization. Production of an egg (*oogenesis*) oc-

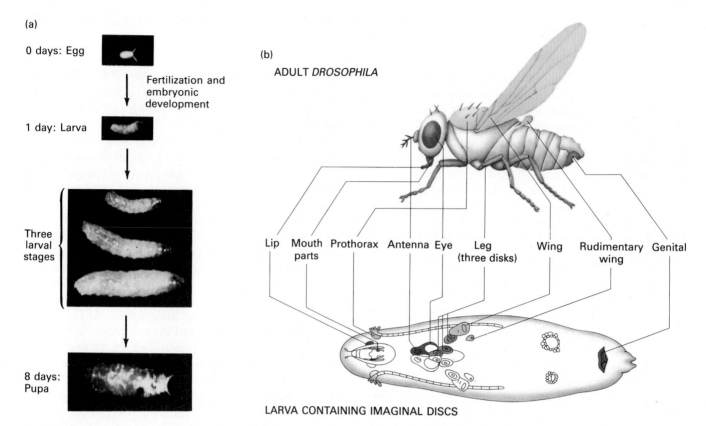

(a)

0 days: Egg

Fertilization and embryonic development

1 day: Larva

Three larval stages

8 days: Pupa

(b)

ADULT *DROSOPHILA*

Lip   Mouth parts   Prothorax   Antenna   Eye   Leg (three disks)   Wing   Rudimentary wing   Genital

LARVA CONTAINING IMAGINAL DISCS

▲ FIGURE 13-27   The development of *D. melanogaster* takes 9–10 days. (a) The fertilized egg develops into a blastoderm and undergoes cellularization in a few hours. The larva, a segmented form, appears in about 1 day and passes through three stages (instars), over a 4-day period, developing into a prepupa. Pupation takes ≈4–5 days ending with the emergence of the adult fly from the pupal case. (b) Groups of ectodermal cells called imaginal discs are set aside at specific sites in the larval body cavity. From these the various body parts indicated develop during pupation. Other precursor cells give rise to adult muscle, the nervous system, and other internal structures. [Part (a) from M. W. Strickberger, 1985, *Genetics*, 3d ed., Macmillan, p. 38; reprinted with permission of Macmillan Publishing Company. Part (b) adapted from same source and J. W. Fristrom et al., 1969, in E. W. Hanly, ed., *Park City Symposium on Problems in Biology*, University of Utah Press, p. 381.]

curs in *ovarioles*, which collectively form the fly ovary (Figure 13-28a). At the distal end of an ovariole, a stem cell divides asymmetrically generating a single germ cell, which divides four times to generate 16 cells. One of these cells completes meiosis, becoming an *oocyte*; the other 15 cells become *nurse cells*. These nurse cells synthesize protein and RNA, which is transported by a series of cytoplasmic bridges (a link between adjacent cells that allows passage of proteins and RNA) into the oocyte. These molecules are necessary for maturation of the oocyte and early stages of embryogenesis. Each group of 16 cells is surrounded by a single layer of *follicle cells*, which form the egg shell. As an oocyte matures within an ovariole, new germ cells are produced from a stem cell, displacing the previously generated

► FIGURE 13-28   Oocyte development in *Drosophila*, during which regionalization of the embryo begins. (a) General structure of the ovaries and internal genitalia. Each ovary in a female fly comprises a cluster of ovarioles, which function like assembly lines. At the distal tip of an ovariole, a stem cell divides to generate a precursor germ cell; as this cell progresses through the ovariole, it divides and matures. The mature oocyte is released into the oviduct, where it is fertilized by sperm from a previous mating, which are stored in the seminal vesicle. The fertilized egg then is laid through the vulva. (b) Structure of the developing oocyte at three stages in its maturation. Each developing unit, or follicle, consists of a developing oocyte, nurse cells, and a layer of somatic cells called follicle cells. (*Stage 1*) Early in oogenesis, the oocyte is about the same size as the neighboring nurse cells. (*Stage 2*) The nurse cells begin to synthesize mRNAs and proteins necessary for oocyte maturation, and the follicle cells begin to form the egg shell. Midway through oogenesis, the oocyte has increased in size considerably. (*Stage 3*) The mature egg is surrounded by the vitelline coat and chorion, which compose the egg shell. The nurse cells and follicle cells have been discarded, but some of the mRNAs synthesized by nurse cells, which become localized in discrete spatial domains of the oocyte, function in early patterning of the embryo. Polar granules are distinct cytoplasmic structures located in the posterior region of the egg. This is the region in which germ cells arise. [Adapted from A. J. F. Griffiths et al., 1993, *An Introduction to Genetic Analysis*, 5th ed., W. H. Freeman and Company, p. 643.]

oocyte, nurse cells, and surrounding follicle cells. The egg is released into the oviduct, where it is fertilized.

The polarity of the embryo is presaged in the mature oocyte (Figure 13-28b). Some of the mRNAs produced by nurse cells become localized in very discrete spatial domains of the oocyte. Interactions between the follicle cells and the oocyte also are important in generating spatial information that is carried over to the embryo.

Embryogenesis is activated by fertilization. The first 13 nuclear divisions of the fertilized ovum are synchronous and occur remarkably rapidly, each division occurring about every 10 min. Because nuclear division is not accompanied by cell division, a syncytium forms. As the nuclei divide, they begin to migrate outward toward the plasma membrane of the embryo. By 3 h after fertilization, the nuclei have reached the surface of the embryo, which at this stage is referred to as the *syncytial blastoderm* (Figure 13-29). Cell membranes then form around the nuclei, generating the *blastula*. Early-patterning events in *Drosophila* occur before formation of the blastula, that is, before cellularization of the embryo. The mechanisms regulating spatial organization in the embryo rely in large part on the diffusion of developmentally important proteins within the developing syncytium. This is very different from the strategies of many other organisms such as mammals in which early-patterning events occur between cells and hence must rely on intercellular communication (see Chapter 24).

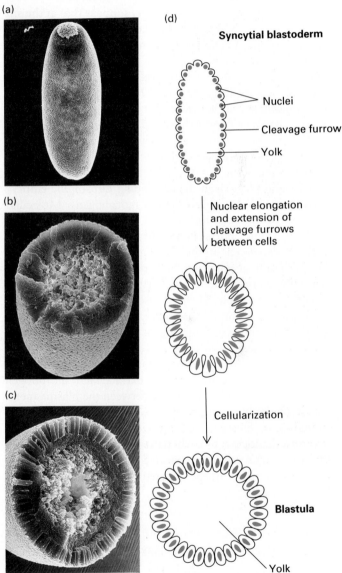

▲ FIGURE 13-29   Formation of blastula during early embryogenesis in *Drosophila*. Nuclear division is not accompanied by cell division until about 2000–4000 nuclei have formed. Electron micrographs of embryos before cellularization show surface bulges overlying individual nuclei (a) and absence of cell membranes (b), which are evident after cellularization (c). Note separation of the nuclei of so-called pole cells, which give rise to germ cells, at the posterior end (*top*) of the embryo in (a). Change of syncytial blastoderm into blastula is illustrated in corresponding diagrams (d), in which pole cells are not shown. [See R. R. Turner and A. P. Mahowald, 1976, *Dev. Biol.* **50**:95; photographs courtesy of A. P. Mahowald.]

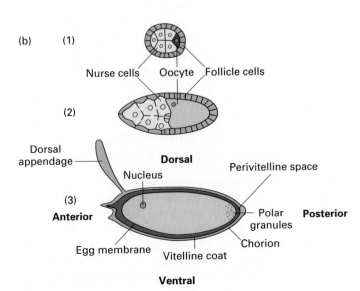

Morphogens Regulate Development as a Function of Their Concentration

A central concept in developmental biology is that of a *morphogen,* a substance that specifies cell identity as a function of its concentration. A continuous gradient of morphogen concentration can elicit a set of unique cellular responses at a finite number of *threshold* concentrations: above the threshold, one response is illicited; below it, cells respond differently. As an example, suppose that at the site in the embryo where a particular morphogen is synthesized, its concentration is high enough to establish fate A for cells in the immediate vicinity. As the distance from this site increases, the concentration of the morphogen decreases. At some distance (e.g., 10 cell diameters), a threshold concentration is reached; cells that experience morphogen levels below the threshold are consigned to cell fate B. Still farther away, the morphogen concentration may reach another threshold below which cell fate C is established, and so on.

Several proteins in the developing *Drosophila* embryo have been shown to function as morphogens. Molecular analysis of mutants, for instance, has revealed that gradients of transcription factors translated from maternal mRNA deposited in the developing oocyte control the transcription of genes in the early embryo, or *zygote,* in spatially restricted domains. The products of these zygotic genes, in turn, control the transcription of other zygotic genes, further refining specific spatial domains of gene activity in the embryo. Morphogens controlling development along the anterior-posterior axis diffuse within the blastoderm. In contrast, patterning along the dorsoventral axis occurs in response to a graded extracellular morphogen located in the perivitelline space between the plasma membrane of the blastoderm and the vitelline membrane surrounding it. This external type of pattern regulation may be most analogous to early patterning in mammalian embryos in which cells are separated by membranes and morphogens are likely to be secreted (see Chapter 24).

Four Maternal Gene Systems Regulate Regionalization in the Early Embryo

The systematic approach to identifying mutations in *Drosophila* affecting early development has led to the characterization of four different maternal gene systems that regulate axis determination in the embryo. Three systems affect development along the anteroposterior axis; one, along the dorsoventral axis. Mutations affecting early-pattern formation were identified and classified on the basis of their disruption of the outer (cuticular) structures of the embryo, which occur in a highly reproducible pattern in wild-type embryos. As illustrated in Figure 13-30, four classes of mutations were found each of which affects different regions of the embryo: anterior regions that give rise to the head and thorax; posterior regions that give rise to the

abdominal segments; the extreme anterior and posterior regions that give rise to the termini; and regions that determine patterning along the dorsoventral axis.

In the rest of this section, we consider three important molecular mechanisms for controlling embryonic regionalization (patterning) that have been unraveled by analysis of various *Drosophila* developmental mutants:

- Spatial localization of specific mRNAs that the mother deposits in the egg

- Conversion of gradients of specific molecules in the embryo into region-specific transcription of patterning genes

- Specification of cell fate in response to graded expression of an extracellular signal

We describe examples of these patterning mechanisms involving the anterior, posterior, and dorsoventral systems. The terminal system, which relies heavily on a highly conserved intercellular communication pathway (see Chapter 20) is considered briefly.

In order to decipher the molecular and cellular basis of these patterning mechanisms, investigators had to (1) determine the pattern of mRNA expression and the distribution of the encoded proteins in different spatial domains of the embryo and (2) assess the effects of mutations on patterning and on the expression of other patterning genes. As noted already, the effect of mutations on patterning can be assessed visually by examining the patterns of cuticular structures on the surface of dead embryos (see Figure 13-30). These cuticular structures are formed by the cells directly beneath them. Early gene products can be visualized micropically by in situ hybridization to detect specific mRNAs and staining with antibody to detect specific proteins. The use of these techniques to localize several important early gene products is illustrated in Figure 13-31. Gene expression in fly embryos also can be detected by use of a reporter construct in transgenic flies. In this method, the *E. coli lacZ* gene is fused to a promoter element that normally controls transcription of a *Drosophila* gene of interest. Expression of the *lacZ* gene, which encodes β-galactosidase, thus serves as a "stand-in" or "reporter" for expression of the fly gene. This technique is described in Chapter 11.

Specification of the Anterior Region Depends on the Maternal *bicoid* Gene

The first morphogen to be described at the molecular level was the protein encoded by the *bicoid* locus. We first describe the mechanism by which the Bicoid protein gradient is established in the early embryo and then discuss the mechanisms by which specific cellular responses are induced by different concentrations of Bicoid protein. It is important to emphasize, however, that the establishment

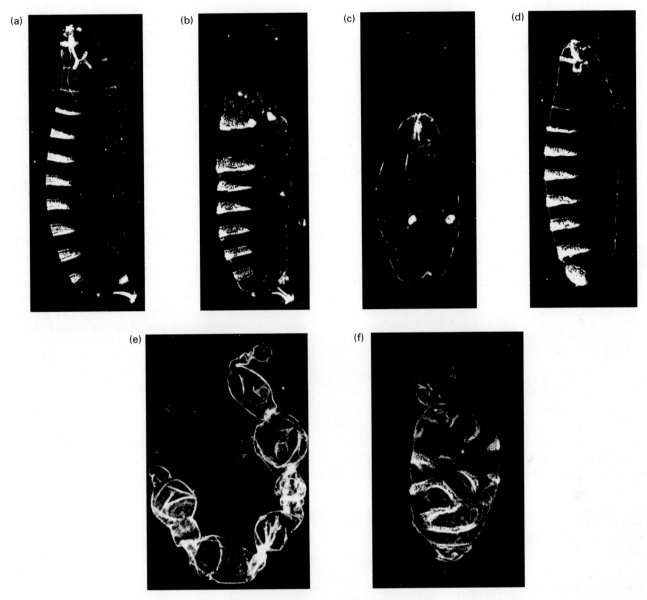

▲ FIGURE 13-30   Abnormal patterns in the outer (cuticular) structures of the *Drosophila* embryo result from mutations in four different maternal gene systems that regulate formation of the anterior/posterior and dorsoventral axes during early embryogenesis. Epidermal cells form different cuticular structures in a highly reproducible fashion; these structures serve as indicators, or markers, of regional identity along the axes in the fly embryo. The earliest steps in axis determination depend on synthesis of certain mRNAs by the mother and deposition of these mRNAs onto the developing oocyte. Embryos derived from mothers homozygous for mutations in these early-patterning genes exhibit various types of abnormal cuticular patterns. In all the embryo cuticular preparations shown here, anterior is toward the top and ventral is toward the left. (a) The wild-type pattern. (b) Mutations in the *bicoid* locus (anterior system) disrupt development of the anterior abdominal segments, the thorax, and regions of the head. (c) Mutations in the posterior system lead to loss of abdominal segments. An *oskar* mutant is shown. (d) Mutations in the terminal system affect development at both ends of the embryo. A *torsolike* mutant is shown here. (e,f) Mutations in the dorsoventral system can lead to dorsalization or ventralization. In a *dorsal* mutant (e), the entire embryo forms dorsal structures; in a *cactus* mutant (f), the entire embryo becomes ventralized. [From D. St. Johnston and C. Nüsslein-Volhard, 1992, *Cell* **68**:201.]

of regional identity by Bicoid is likely to differ from the mechanism in most developing systems because, as noted earlier, the early fly embryo is a syncytium rather than a collection of separate cells. Thus Bicoid can diffuse from one region of the early fly embryo through a common cytoplasm shared by all the nuclei. In contrast, in many other embryos, including those of vertebrates, the cells have membranes isolating them from diffusible cytoplasmic signals. Establishing gradients in such systems requires the use of secreted molecules.

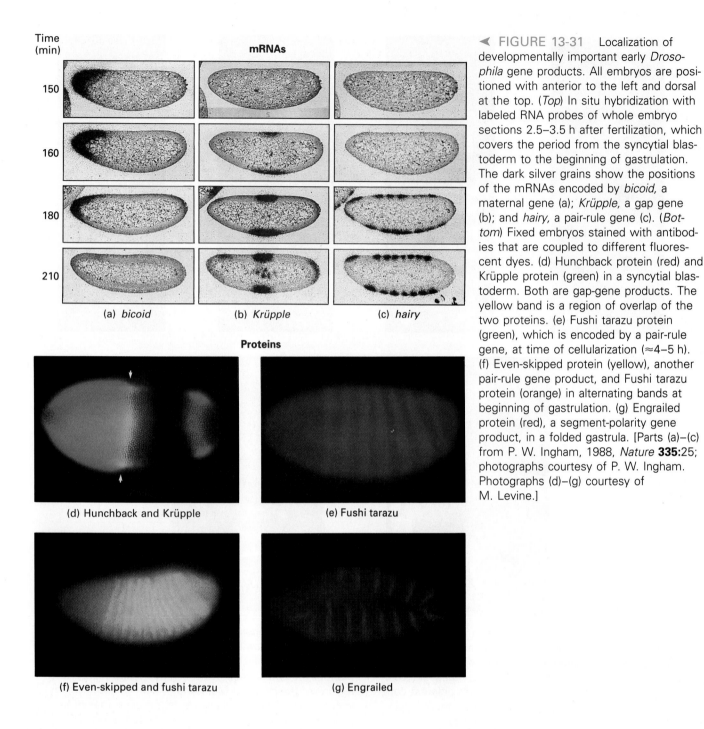

Time
(min)

mRNAs

150

160

180

210

(a) *bicoid*   (b) *Krüpple*   (c) *hairy*

Proteins

(d) Hunchback and Krüpple   (e) Fushi tarazu

(f) Even-skipped and fushi tarazu   (g) Engrailed

◄ FIGURE 13-31 Localization of developmentally important early *Drosophila* gene products. All embryos are positioned with anterior to the left and dorsal at the top. (*Top*) In situ hybridization with labeled RNA probes of whole embryo sections 2.5–3.5 h after fertilization, which covers the period from the syncytial blastoderm to the beginning of gastrulation. The dark silver grains show the positions of the mRNAs encoded by *bicoid*, a maternal gene (a); *Krüpple*, a gap gene (b); and *hairy*, a pair-rule gene (c). (*Bottom*) Fixed embryos stained with antibodies that are coupled to different fluorescent dyes. (d) Hunchback protein (red) and Krüpple protein (green) in a syncytial blastoderm. Both are gap-gene products. The yellow band is a region of overlap of the two proteins. (e) Fushi tarazu protein (green), which is encoded by a pair-rule gene, at time of cellularization ($\approx$4–5 h). (f) Even-skipped protein (yellow), another pair-rule gene product, and Fushi tarazu protein (orange) in alternating bands at beginning of gastrulation. (g) Engrailed protein (red), a segment-polarity gene product, in a folded gastrula. [Parts (a)–(c) from P. W. Ingham, 1988, *Nature* **335**:25; photographs courtesy of P. W. Ingham. Photographs (d)–(g) courtesy of M. Levine.]

The *bicoid* mRNA, which is synthesized in mothers and deposited in the oocyte, is localized to the most anterior region, or anterior pole, of the early fly embryo (see Figure 13-31a). Embryos produced by mothers that are homozygous for *bicoid* mutations lack head and thoracic tissue (see Figure 13-30a). Other maternal mRNAs either are unlocalized or are localized to different regions of the embryo (e.g., *gurken* mRNA at the anteriodorsal edge and *nanos* mRNA posteriorly). During oogenesis, maternal mRNAs are synthesized in nurse cells and then transported into the maturing oocyte (see Figure 13-28b). The precise

molecular mechanism by which *bicoid* mRNA is localized to the anterior pole is not known. However, localization has been shown to depend on the 3' untranslated end of *bicoid* mRNA and the products of three other maternal genes. Mutations that result in failure to localize *bicoid* mRNA produce a phenotype similar to, though less severe than, the phenotype associated with mutations in the *bicoid* gene itself.

The *bicoid* gene encodes a DNA-binding protein, whose binding region is a homeobox. This structural motif (see Figure 11-48a), is found in many different

▶ FIGURE 13-32 Experimental demonstration that transcription of genes directly regulated by Bicoid protein reflects the concentration of Bicoid and its affinity for protein-binding sites in the promoter. (a–c) Increasing the number of *bicoid* genes in the mother changed the Bicoid gradient in the early embryo, leading to a corresponding change in the gradient of Hunchback protein (red) expressed from the zygotic gene. (d,e) The *hunchback* promoter has been shown to contain three high-affinity and three low-affinity Bicoid-binding sites. Transgenic flies carrying a reporter gene linked to a synthetic promoter containing either four high-affinity sites (d) or four low-affinity sites (e) were prepared. Expression of the reporter-gene product was dependent on the affinity of the Bicoid-binding sites in the promoter. [Adapted from D. St. Johnston and C. Nüsslein-Volhard, 1992, *Cell* **68**:201.]

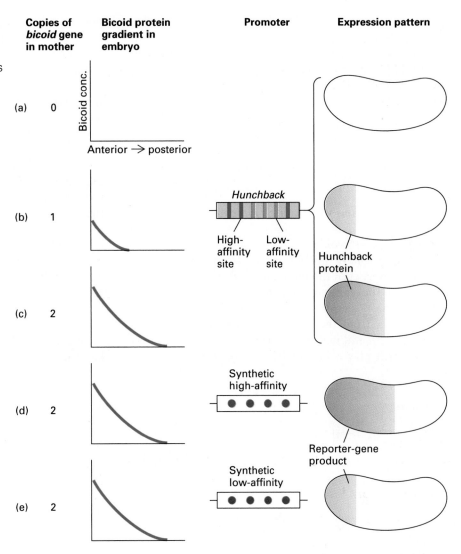

proteins that play developmentally important roles in a variety of organisms. Once Bicoid protein is synthesized at the anterior end of the embryo, it diffuses posteriorly forming a protein gradient along the anteroposterior axis. Evidence that the Bicoid protein gradient determines anterior structures was obtained through injection of synthetic *bicoid* mRNA at different locations in the embryo. This treatment led to formation of anterior structures at the site of injection with progressively more posterior structures forming at increasing distances from the injection site.

The graded concentration of Bicoid protein along the length of the embryo determines the region in which the *hunchback* gene is transcribed in the zygote. Transcription of *hunchback* requires Bicoid protein, and *hunchback* mutations affect many of the same anterior structures as *bicoid* mutations. The *bicoid* mRNA concentration at the anterior pole of the embryo can be raised experimentally by increasing the number of *bicoid* genes in the mother, since *bicoid* mRNA is synthesized by the mother and de-

posited in the egg. Embryos produced from such mothers exhibit an abnormal Bicoid gradient in which higher concentrations of Bicoid extend more posteriorly than in normal embryos. As a consequence, the posterior boundary of *hunchback* transcription moves posteriorly.

The experimental results depicted in Figure 13-32 provide strong evidence that Bicoid protein directly regulates transcription of *hunchback*. For example, the spatial distribution of expressed Hunchback protein parallels that of the Bicoid protein gradient (see Figure 13-32a–c). Moreover, analysis of the *hunchback* promoter just upstream of the transcription-start site has shown that it contains three low-affinity and three high-affinity binding sites for Bicoid protein. Studies with transgenic flies carrying reporter genes driven by synthetic promoters containing either all high-affinity or low-affinity Bicoid-binding sites have demonstrated that the affinity of the site determines the threshold concentration of Bicoid at which gene transcription is activated. That is, in response to the same Bicoid protein gradient in the embryo, expression of a reporter gene controlled by a promoter carrying high-affinity Bicoid-binding sites occurs in more posterior regions than does transcription of a reporter gene carrying low-affinity sites (see Figure 13-32d,e). In addition, the number of Bicoid-binding sites occupied at a given concentration has been shown to determine the amplitude, or level, of the response.

Protein Encoded by Maternal *nanos* Gene Represses Translation of *hunchback* mRNA in Posterior Region of the Embryo

In addition to anteriorly localized *hunchback* mRNA produced by zygotic transcription of the *hunchback* gene under control of Bicoid protein, the early embryo contains maternal *hunchback* mRNA, which is uniformly distributed. However, even though *hunchback* mRNA is present throughout the embryo, Hunchback protein is not observed in the posterior region. This exclusion of Hunchback protein from the posterior region depends on the products of several maternal genes, the posterior group of genes, which play a key role in determining the antero-posterior axis, in particular the development of the abdomen. One of the posterior genes, called *nanos*, encodes a morphogen that functions in repressing translation of maternal *hunchback* mRNA. Like the *bicoid* gene, *nanos* is transcribed in nurse cells during oogenesis, and its mRNA is transported to the oocyte. In contrast to *bicoid* mRNA, *nanos* mRNA is localized to the posterior pole of the early embryo. Nanos protein acts in conjunction with Pumilio protein, which is encoded by a uniformly distributed maternal mRNA, to repress translation of maternal *hunchback* mRNA. Since Nanos is localized in the posterior region of the embryo, this repression occurs preferentially in the posterior region. The remaining posterior group of maternal genes are necessary for the specific synthesis and localization of *nanos* mRNA to the posterior pole of the embryo. Mutations in these genes (e.g., *oskar*) as well as in *nanos* itself lead to loss of abdominal segments (see Figure 13-30c). In short, the sole function of Nanos protein is to repress translation of maternal *hunchback* mRNA in the posterior region of the embryo. Indeed, normal development can take place in the absence of *nanos* if the embryo lacks maternal *hunchback* mRNA (Figure 13-33).

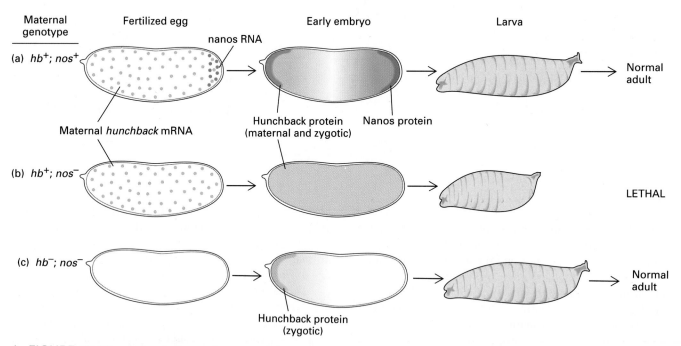

▲ FIGURE 13-33   Role of maternally derived Nanos protein in excluding Hunchback protein from posterior region of *Drosophila* embryo. mRNAs are indicated by colored dots. There are two sources of *hunchback* (*hb*) mRNA in the early embryo: that derived from the mother, which is uniformly distributed, and that derived from zygotic transcription of *hunchback*, which is localized anteriorly (see Figure 13-31d). (a) In embryos produced by wild-type female flies, maternal *nanos* mRNA is localized posteriorly. Once this mRNA is translated, the resulting Nanos protein inhibits translation of maternal *hunchback* mRNA in the posterior region of the embryo. (b) In the absence of Nanos, translation of maternal *hunchback* mRNA in the posterior region leads to a failure of the posterior structures to form normally and the embryo dies. (c) Since *hunchback* is also required zygotically, viable flies homozygous for both *hunchback* and *nanos* cannot be produced. However, pole cells (germ line precursors) that are homozygous for both mutations can be transplanted from an early embryo into a surrogate embryo. Germ line cells that carry mutations in both the *hunchback* and *nanos* genes produce embryos that develop normally. This finding shows that Nanos functions solely to prevent translation of maternal *hunchback* mRNA in the posterior region. [Adapted from P. Lawrence, 1992, *The Making of a Fly: The Genetics of Animal Design*, Blackwell Scientific Publications.]

In addition to Nanos protein, inhibition of *hunchback* mRNA translation requires specific sequences in the 3′ untranslated region of the *hunchback* mRNA. These sequences, called Nanos-response elements, are located downstream of the protein-coding sequences in the mRNA. Inhibition of *hunchback* mRNA translation is dependent upon the concentration of Nanos protein and the number and specific sequence of the Nanos-response elements. The molecular mechanisms of this inhibition of translation is not yet known.

Hunchback Protein Regulates Expression of Several Gap Genes along the Anteroposterior Axis

The *hunchback* gene as well as several other zygotic genes involved in patterning of the abdominal region are collectively called *gap genes*. These genes are expressed within 2 h following fertilization and just before cellularization of the embryo. The first gap gene to be expressed is *hunchback*; as discussed earlier, its transcription is regulated by maternally derived Bicoid protein.

Hunchback protein is a morphogen that functions as a transcription activator or repressor of several other genes, including *Krüpple*, *knirps*, and *giant*. These genes also encode transcription factors, which control subsequent patterning events in localized domains of the embryo. Localization of these proteins in the early embryo is shown in Figure 13-34. Hunchback protein controls both the anterior and posterior boundaries of *Krüpple* expression. At high concentrations, Hunchback protein represses *Krüpple*. Below a critical threshold concentration, Hunchback protein no longer represses *Krüpple*, but instead acts as a transcriptional activator. The second threshold sets a posterior boundary below which Hunchback protein no longer activates *Krüpple*. The Knirps and Giant proteins are each located in two domains. Expression in their anterior domains is dependent on Bicoid protein. Expression of *Knirps* and *Giant* is repressed by Hunchback at distinct threshold concentrations, thereby setting the anterior boundaries of their more posterior domains. The posterior boundary of the posterior Knirps and Giant domains are determined by the product of another gap gene, *tailless*.

Initial Patterning along Dorsoventral Axis Depends on Dorsal Protein

Pattern formation along the dorsoventral axis of the *Drosophila* embryo provides another example of how a gradient of developmental information gives rise to distinct cell fates. As noted above, patterning along the anterior/posterior axis depends on morphogens that diffuse through the common cytoplasm of the syncytial blastoderm. In contrast, patterning along the dorsoventral axis involves diffusible secreted cues acting in a graded way through a cell-

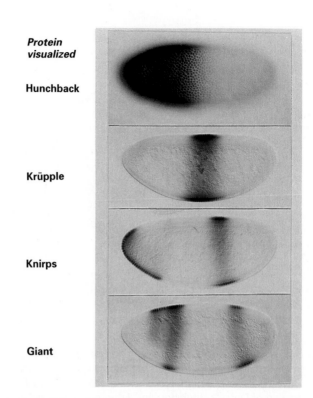

▲ FIGURE 13-34   Localization of gap-gene products in early *Drosophila* embryos visualized by staining with specific antibodies against Hunchback, Krüpple, Knirps, and Giant proteins. Anterior is to the left. Hunchback protein, which functions as both a transcription activator and repressor, is a morphogen that regulates transcription of the *Krüpple, knirps,* and *giant* genes. See text for discussion. [Adapted from G. Struhl, P. Johnston, and P. Lawrence, 1992, *Cell* **69**:237.]

surface receptor. This type of mechanism also may operate in embryos of other species that are not syncytia (i.e., whose cells are surrounded by a plasma membrane).

Through a series of complex steps that are only now being identified a specific signal is generated in the most ventral region of the perivitelline space separating the plasma membrane of the embryonic syncytium from the egg shell. Although it has not been directly shown, it seems likely that the graded distribution of this signal along the dorsoventral axis leads to graded stimulation of Toll protein, a membrane-bound receptor encoded by maternal *Toll* mRNA and uniformly distributed along the surface of the embryo. That is, the ligand of Toll protein probably functions as a morphogen in this system. High levels of Toll activation trigger ventral-cell fates, whereas lower levels trigger more dorsal-cell fates.

Signal-Dependent Nuclear Localization of Dorsal Protein   What effect does graded activation of the Toll receptor have on gene expression in the fly embryo? Various genetic studies focused attention on the Dorsal protein

(a)

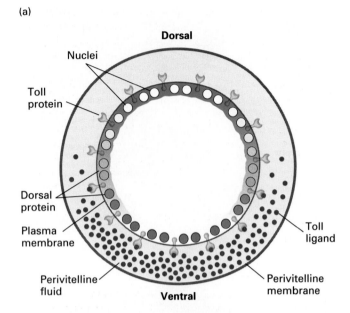

(b)

Wild type

Spätzle mutant

▲ FIGURE 13-35   Nuclear localization of maternally derived Dorsal protein. (a) Schematic diagram of cross section of early *Drosophila* embryo showing gradient of nuclear localization of Dorsal protein (green). Although Dorsal initially is uniformly distributed in the cytoplasm, activation of Toll protein, a cell-surface receptor, leads to its nuclear localization. It is thought that the dorsoventral gradient of the Toll ligand leads to differential activation of Toll protein, which in turn establishes the Dorsal gradient, with nuclear localization greatest on the ventral side and least on the dorsal side. The

Toll ligand probably is preferentially generated on the ventral side of the embryo in the perivitelline fluid, which is located between the vitelline and plasma membranes. (b) Photomicrographs of early *Drosophila* embryos from a wild-type female and from a female homozygous for the *spätzle* mutation. *Spätzle* is likely to encode the Toll ligand. Disruption of activation of Toll protein in the *spätzle* mutant disrupts nuclear localization of Dorsal protein, which is visible in the wild-type embryo. [Part (b) from C. A. Rushlow et al., 1989, *Cell* **59**:1165.]

and its activity. Maternal *dorsal* mRNA is distributed uniformly in the cytoplasm of the early embryo. Although Dorsal protein initially also is distributed uniformly in normal embryos, the protein is redistributed into a gradient of *nuclear localization* (Figure 13-35a). In the most ventral parts of the embryo, Dorsal protein is localized almost exclusively in nuclei with very little in the cytoplasm. In lateral regions, the protein appears to be approximately evenly distributed between nuclei and the cytoplasm. Finally, in the most dorsal regions of the embryo, the protein is largely cytoplasmic.

Female flies homozygous for a *dorsal* mutation produce embryos with no ventral structures, indicating that Dorsal protein is required for formation of ventral structures. In the absence of Dorsal protein, dorsal structures form over the entire embryo (see Figure 13-30e). In mutants in which the Toll receptor is constitutively active throughout the embryo, the Dorsal protein is located in nuclei throughout the dorsoventral axis and cells located dorsally exhibit ventral-cell properties. Thus, nuclear localization of Dorsal appears crucial to establishing ventral-cell fates. Once in the nucleus, Dorsal regulates expression of various genes, as discussed below. Thus, in contrast to anterior/posterior patterning, which reflects the presence

or absence of a transcription factor, dorsoventral patterning reflects access of a transcription factor to the nucleus, and hence to its target genes.

In embryos carrying the *spätzle* mutation, activation of the Toll receptor is disrupted and no nuclear localization of Dorsal protein is observed (Figure 13-35b). This finding indicates that Dorsal is translocated to nuclei in response to activation of Toll. In this respect, Dorsal is analogous to NF$_\kappa$B, a mammalian transcription factor that is translocated to the nucleus of immune-system cells in response to extracellular signals (Figure 13-36). Not only are Dorsal and NF$_\kappa$B functionally analogous, the two proteins also exhibit sequence homology.

Transcriptional Activation and Repression by Dorsal Protein   Dorsal protein has been shown to function as both a transcription repressor and activator. For instance, it activates the *twist* gene and represses the *zen* gene in the ventral regions of the embryo. The Dorsal-binding sites in the regulatory regions controlling *twist* activation and *zen* repression have been identified. To test whether the differences between these binding sites determine whether the target gene is activated or repressed, Dorsal-binding sites in the *twist* gene were replaced by those from *zen* and vice

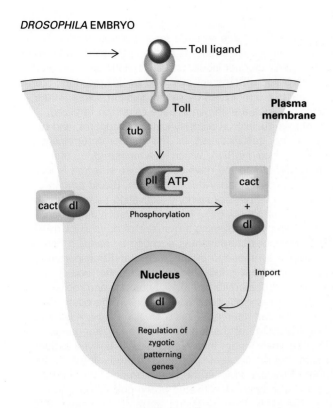

DROSOPHILA EMBRYO

MAMMALIAN LYMPHOCYTE

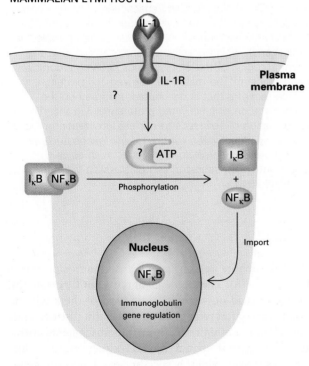

◄ **FIGURE 13-36** Signal-dependent nuclear translocation of transcription factors occurs in both *Drosophila* embryos and mammalian lymphocytes. Dorsal protein is homologous to NF$_\kappa$B, a transcription factor that regulates transcription of immunoglobulin genes in mammals. Both proteins are localized to the nucleus in response to a specific extracellular signal. Cactus protein (cact) and I$_\kappa$B bind to and inhibit nuclear localization of Dorsal and NF$_\kappa$B, respectively. Pelle protein (pll) is a protein kinase required for translocation of Dorsal; a mammalian homolog of Pelle has not yet been identified. Mutations in the *tube* gene disrupts nuclear localization of Dorsal, but the function of the Tube protein is not known. Whether phosphorylation by Pelle regulates Dorsal and/or Cactus in vivo is not known. [Adapted from C. A. Shelton and S. A. Wasserman, 1993, *Cell* **72**:515.]

Inhibition of Dorsal Function Three cytoplasmic proteins encoded by the *pelle*, *cactus*, and *tube* genes are important in regulating the nuclear localization of Dorsal (see Figure 13-36). The sequence of Cactus protein is similar to that of I$_\kappa$B, a protein that binds to NF$_\kappa$B (remember NF$_\kappa$B is similar in structure to Dorsal) and inhibits its translocation to the nucleus. Mutations in *cactus* lead to constitutive nuclear localization of the Dorsal protein and subsequent ventralization of the entire embryo (see Figure 13-30f). Hence, it is likely that Cactus binds to Dorsal, thereby inhibiting its nuclear localization. The function of Tube is not known. Pelle, however, has been shown to be a protein kinase. In principle, then, Pelle could affect the activity of the Dorsal, Tube, or Cactus proteins. Biochemical experiments will be necessary to determine the mechanisms by which Pelle and Tube promote the dissociation of Cactus from Dorsal and the subsequent nuclear localization and transcriptional activation or repression of genes by Dorsal.

Maternal Terminal Genes Regulate Early Patterning of the Extreme Anterior and Posterior Ends of the Embryo

Although the mechanisms involved in early patterning of the head and tail regions of the *Drosophila* embryo are not well understood, an extracellular signal clearly is involved. This feature is similar to patterning along the dorsoventral axis. Genetic analysis has identified several maternal genes, comprising the terminal gene system, that regulate development of both anterior and posterior structures in the *Drosophila* embryo. Mutations in these genes lead to defects at the termini of the embryo (see Figure 13-30d).

Several terminal genes have been cloned and the functions of their encoded proteins currently are being studied. One of these genes, called *torso*, encodes a transmembrane receptor protein that has tyrosine kinase activity (see Chapter 20). The Torso protein, encoded by a maternal mRNA, is distributed uniformly along the surface of the blastoderm. The maternally-encoded Torsolike protein is

versa. Nevertheless, Dorsal continued to activate the altered *twist* gene and repressed the altered *zen* gene. These findings argue for the importance of the context (i.e., other DNA sequences and other proteins bound to neighboring sites) in which specific DNA regulatory sites are located in determining the pattern of gene expression driven by specific DNA-binding proteins.

produced by anterior and posterior follicle cells and secreted into the perivitelline space. Whether this protein directly binds and activates the Torso protein or regulates production of a ligand is not known. In either case, Torso is selectively activated in the anterior and posterior regions, even though it is distributed uniformly on the blastoderm surface.

Activation of the Torso receptor triggers an intracellular signaling cascade that stimulates expression of various zygotic genes including the gap gene *tailless*, which encodes a steroid receptor–like transcription factor. This signal-transduction cascade, which can be triggered by various cell-surface signals, has been remarkably conserved through evolution; it is discussed in detail in Chapter 20.

Subsequent Anterior/Posterior Patterning Is Regulated by a Cascade of Transcription Factors Expressed from Three Groups of Zygotic Genes

In the previous several sections, we have examined early patterning of the *Drosophila* embryo, which initially specifies the anteroposterior and dorsoventral axes. This early regionalization depends on the products encoded by maternal mRNAs, which function in a region-specific fashion to regulate zygotic transcription of early-patterning genes. Figure 13-37 summarizes the spatial distribution of some of the maternal and zygotic gene products involved in axis determination in the early embryo.

Subsequent to these early events, additional patterning occurs as embryonic development progresses. In this section, we briefly discuss subsequent patterning events affecting the anteroposterior axis; these are mediated by proteins encoded by two groups of zygotic genes, termed *pair-rule* and *segment-polarity* genes. The mechanisms that confer additional patterning along the dorsoventral axis are discussed in Chapter 24.

Many of the regionalization events discussed in this and earlier sections occur before cellularization of the syncytial blastoderm and utilize transcription factors that diffuse within the common cytoplasm of the syncytium. After plasma membranes are fully formed and cellularization is completed, interactions among cells refine the pattern. Most of the genes that function in axis determination (i.e., the maternal, gap, pair-rule, and segment-polarity genes) are expressed transiently and act to establish spatial domains where *selector genes* are expressed. Selector genes, which are expressed into the adult, are required to specify and maintain regional identity along the anterior/posterior axis.

Before continuing, we need to define *segment* and *parasegment*, two particularly confusing terms convention-

▶ **FIGURE 13-37**   Summary of contributions of four maternal gene systems required for axis determination in *Drosophila* embryos. Anterior is to the left; ventral at bottom. Maternal gene product labels are red; zygotic gene product labels are black; mRNAs are indicated by colored dots; protein products are indicated by varied gradient colors. Three embryonic stages are shown in each case: the fertilized ovum before nuclear division; syncytial blastoderm; and blastula soon after cellularization. (a) The anterior determinant is encoded by *bicoid* mRNA, which is synthesized in nurse cells and transported into the developing oocyte where it is localized to the anterior end. After fertilization, Bicoid protein is produced and diffuses posteriorly, activating transcription of the zygotic *hunchback* gene in a concentration-dependent fashion, thereby establishing a gradient of *hunchback* mRNA. (b) The posterior determinant is encoded by *nanos* mRNA, which also is synthesized by nurse cells and transported to the oocyte where it is localized to the posterior end. The Nanos protein presumably diffuses anteriorly and represses translation of maternal *hunchback* mRNA in the posterior region of the early embryo. This repression is necessary for normal expression of the zygotic *knirps* and *giant* genes. (c) The terminal system involves maternally derived extracellular signaling molecules (e.g., Torsolike protein), which are thought to be laid down in the vitelline membrane during oogenesis and become activated at the poles following fertilization. The signal then binds and activates a membrane-bound receptor encoded by the maternal *torso* gene. Activation of Torso protein stimulates transcription of *tailless* and other zygotic genes. (d) The dorsoventral system also involves a maternally derived extracellular signaling molecule (e.g., Toll ligand), which is activated and/or released in the most ventral region of the embryo, thus defining a dorsoventral ligand gradient. The maternally derived Toll protein, a cell-surface receptor that is distributed uniformly on the embryo surface, thus is activated preferentially in the ventral region. Activation of Toll leads to nuclear localization of Dorsal protein, which activates ventral-specific genes such as *twist* and represses dorsal-specific genes such as *zen*. [Adapted from D. St. Johnston and C. Nüsslein-Volhard, 1992, *Cell* **68**:201.]

ally used to denote regions along the anterior/posterior axis (Figure 13-38). The segments of the adult fly originally were named based on visual examination of the adult fly; thus each segment corresponds to a visually distinct unit, not a developmental unit. In contrast, parasegments correspond to the actual spatial domain along the anterior/posterior axis over which a specific set of selector genes exert their patterning function. Imagine trying to divide any repetitive structure aligned head to tail; regardless of the specific borders chosen the pattern will remain repetitive (with the exception of the two ends). During early development, the embryo is divided into parasegments with the anterior border of each parasegment demarcated by a

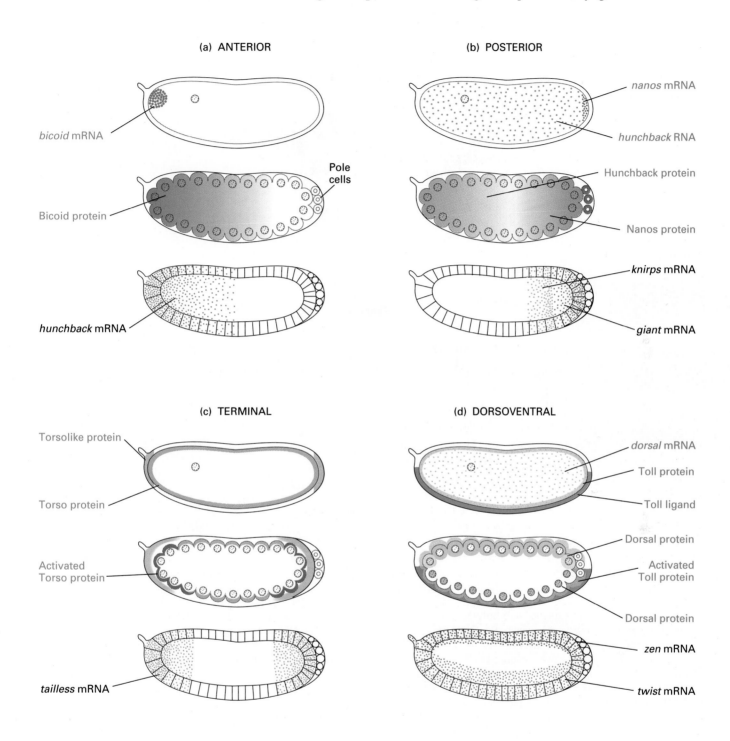

(a) ANTERIOR

bicoid mRNA

Bicoid protein

Pole cells

hunchback mRNA

(b) POSTERIOR

nanos mRNA

hunchback RNA

Hunchback protein

Nanos protein

knirps mRNA

giant mRNA

(c) TERMINAL

Torsolike protein

Torso protein

Activated Torso protein

tailless mRNA

(d) DORSOVENTRAL

dorsal mRNA

Toll protein

Toll ligand

Dorsal protein

Activated Toll protein

Dorsal protein

zen mRNA

twist mRNA

sharp band of cells expressing a particular patterning gene called *engrailed*. Each parasegment roughly contains the posterior portion of one segment and the anterior portion of the segment located just posterior to it. A segment, then, is roughly divided in half by expression of the *Engrailed* protein. This boundary defines a developmental unit within each segment called a *compartment*: cells posterior to this boundary will become part of the so-called posterior compartment and those anterior to it part of the anterior compartment. Once the border between these compartments is established, there is no mixing of cells across it.

As discussed earlier, the products of certain maternal genes (e.g., *bicoid, nanos,* and *torso*) control expression of zygotic genes such as *hunchback, Krüpple, knirps,* and *tailless*), which function in patterning of the anterior/posterior axis. Because mutations in these zygotic genes result in the loss of contiguous segments (i.e., there are gaps in the segmentation pattern), they are collectively called gap genes.

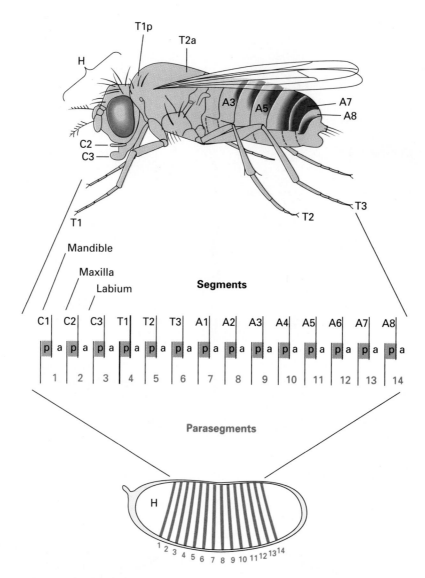

▲ FIGURE 13-38 The relationship between segments in the adult fly and parasegments, which are developmental units corresponding to the domains of activity of selector genes. The head segments are designated C1–C3; thoracic segments, T1–T3; and abdominal segments, A1–A8. The anterior border of each parasegment is marked by a sharp band of cells expressing Engrailed protein (red). Each seg- ment is divided into an anterior (a) and posterior (p) compart- ment. Engrailed expression marks cells in the posterior compartment. There is no mixing of cells across the ante- rior/posterior compartment boundary of each segment. [From P. A. Lawrence, 1992, *The Making of a Fly: The Genetics of Animal Design*, Blackwell Scientific Publications.]

Proteins encoded by maternal and gap genes control tran- scription of the pair-rule genes, the next set of genes to be expressed in the developing embryo.

Pair-Rule Genes The products of the pair-rule genes, which include *fushi tarazu, hairy,* and *even-skipped,* are located in 14 stripes of cells, or parasegments, covering the central part of the embryo (see Figure 13-31e,f). Mutations in a single pair-rule gene remove seven parasegments, ei- ther the even or odd ones. Gene expression in each stripe appears to be controlled independently by the action of different transcription factors encoded by gap and mater- nal genes. For example, expression of Even-skipped stripe 2 is controlled by the maternally derived Bicoid protein and the gap proteins Hunchback, Krüpple, and Giant. These proteins exert their effect by binding to a clustered set of regulatory sites located upstream of *eve* (Figure 13- 39a). Hunchback and Bicoid activate transcription of *eve,* whereas Krüpple and Giant repress its transcription. The segment of DNA containing these regulatory sites will

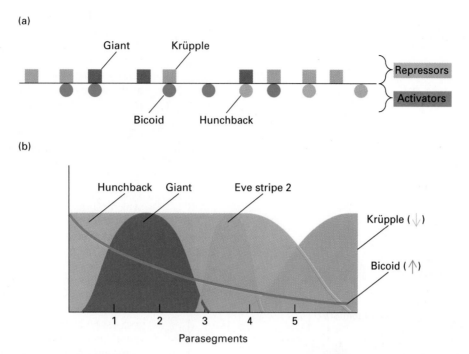

▲ FIGURE 13-39   Expression of the Even-skipped (Eve) stripe 2 in the *Drosophila* embryo. (a) Diagram of 815-bp regulatory region controlling transcription of the pair-rule gene *eve* in stripe 2; this segment begins 909 bp upstream of the transcription start site. This region contains binding sites for Bicoid and Hunchback proteins, which activate transcription of *eve*, and for Giant and Krüpple proteins, which repress transcription. (b) Concentration gradients of the four proteins that regulate expression of Eve stripe 2. The coordinated effect of the two repressors (↓) and two activators (↑) determine the precise boundaries of the second anterior Eve stripe. Expression of other stripes is regulated independently by other combinations of transcription factors encoded by maternal and gap genes. [See S. Small et al., 1991, *Genes & Dev.* **8**:827.]

drive expression of a *lacZ* reporter gene specifically in stripe 2. The coordinated effect of these proteins, each of which has a unique concentration gradient along the anterior/posterior axis, precisely regulates the boundaries of stripe 2 expression (Figure 13-39b).

Segment-Polarity Genes   Next, each parasegment is further subdivided through the action of segment-polarity genes such as *wingless* and *engrailed*. By this stage in embryogenesis, cellularization is complete, and all cells are surrounded by a plasma membrane; thus further patterning depends on intracellular signaling. Several segment-polarity genes encode secreted proteins that signal developmental events in neighboring cells. Segment-polarity genes are responsible for generating the patterns of cells within each segment. In the embryonic cuticle, pattern is defined by cuticular protrusions (denticles) and regions devoid of them, which are smooth. Each cuticular pattern reflects a different combination of pair-rule and segment-polarity genes. These genes also are frequently responsible for determining the polarity of the pattern within a parasegment; the orientation of the denticles also can be altered by the segment-polarity genes. The *engrailed* gene is expressed in the most anterior band of cells of each of the well-defined parasegments (see Figure 13-38).

Selector Genes   The next genes to be transcribed in the regulatory hierarchy that controls regionalization of the Drosophila embryo are the selector genes. The products of these genes regulate development within parasegmental domains. By this stage in embryogenesis, each cell along the anterior/posterior axis expresses a unique combination of transcription factors, which control subsequent cell development. It is thought that the patterns of selector-gene expression are largely determined by various gap-gene products. These expression patterns are determined in the early embryo and maintained into the adult fly. Continuous expression of selector genes is required to specify the organization of structures along the anterior/posterior axis. Two of the best-studied selector genes are *Antennapedia* (*Antp*), which dominates development of the fourth parasegment, and *Ultrabithorax* (*Ubx*), which largely controls the sixth parasegment. The products of the *Antp* and *Ubx* genes are first expressed during the third hour of embryogenesis. Figure 13-40 summarizes the cascade of transcription factors that determines their patterns of expression. The function of selector genes is examined in more detail in the next section.

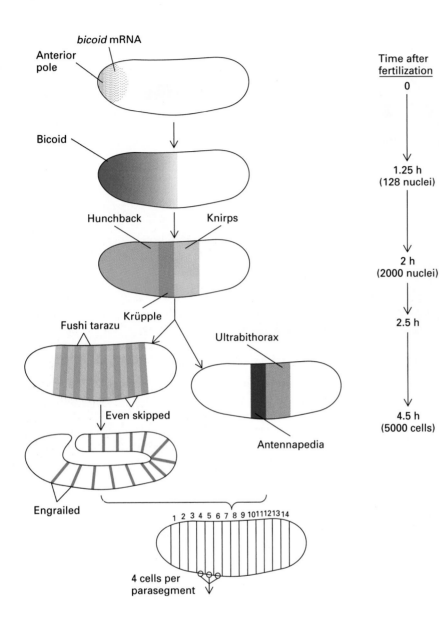

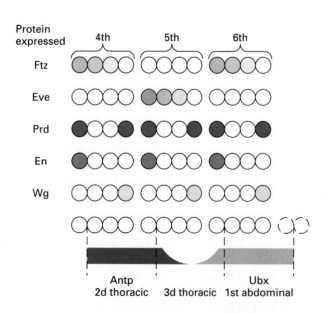

Time after fertilization

0

1.25 h (128 nuclei)

2 h (2000 nuclei)

2.5 h

4.5 h (5000 cells)

◄ FIGURE 13-40   Summary of sequential expression of various genes during early development of the *Drosophila* embryo and localization of their gene products within the embryo. Maternal *bicoid* mRNA is localized at the anterior pole of the egg, but Bicoid protein, which is synthesized soon after fertilization, diffuses to form a gradient. In most cases, an mRNA and its corresponding protein are present in the same regions of the embryo. The expression domains of three gap genes is shown: *hunchback*, *Krüpple*, and *knirps*. Not shown is Nanos protein, which represses translation of *hunchback* mRNA in the posterior region. Expression of each of the stripes corresponding to the proteins encoded by the pair-rule genes *fushi tarazu* (*ftz*), *even-skipped* (*eve*), and *paired* (*prd*) is determined by specific combinations of Bicoid and various gap-gene products. The segment-polarity gene *engrailed* (*en*) is expressed at the anterior end of each parasegment; it plays an important role along with other segment-polarity genes such as *wingless*(*wg*) in patterning of each parasegment. Cellularization occurs after 2.5 h, and gastrulation (including folding of the embryo) occurs at about 4.5 h. By this time, each parasegment consists of four bands of cells, represented by four circles at the bottom. Within each parasegment, each band is characterized by expression of a unique set of pair-rule and segment-polarity genes, several of which are illustrated. These expression patterns act as positional values distinguishing each cell band in a parasegment. The gap-gene products largely determine the expression domains of selector genes such as *Antennapedia* (*Antp*) and *Ultrabithorax* (*Ubx*). The proteins encoded by selector genes specify the organization of adult structures within the context of the positional identity of cells within each parasegment.

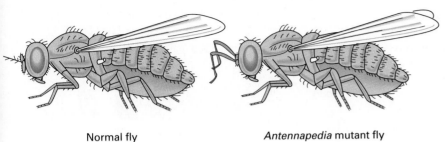

◄ FIGURE 13-41   Misexpression of Antp protein in the developing antenna, due to a regulatory mutation or induced expression of an *Antp* transgene, leads to its transformation into a leg, a structure whose development normally is controlled by Antp in thoracic segments. [From W. McGinnis and M. Kuziora, 1994, *Sci. Am.* **270**(2):58.]

Normal fly   *Antennapedia* mutant fly

Selector Genes Control Regional Identity and Development of Adult Structures

Selector genes are required to determine the structures of the various body parts. The transformation of one body part into another is termed *homeosis,* and mutations giving rise to such transformations are called *homeotic.* Mutations in selector genes often (but not always) are homeotic, providing particularly vivid evidence of the role of genes in regional specification. For instance, misexpression of the *Antennapedia* gene in the primordium of the antenna results in its development into a leg rather than an antenna (Figure 13-41). Both structures are appendages covered with sensory structures although the type and distribution of these structures are quite different.

Selector genes act in specification of epidermal structures (the external surface), the musculature, neural tissue, and gut tissue along the axis. Here we discuss only the role of selector genes in regulating development of the epidermis, because this function is understood best. Two clusters of selector genes—the antennapedia complex (ANT-C) and the bithorax complex (BX-C)—play a central role in controlling regionalization in *Drosophila* and have been studied in great detail. Both of these gene clusters are located on chromosome III. Recent studies, briefly described in the next section, indicate that similar genes play analogous roles in regulating axis determination in mammals.

Organization of Genes in BX-C and ANT-C   The BX-C contains three structural genes each encoding a homeobox-containing transcription factor. These three genes are called *Ultrabithorax (Ubx), abdominal A (abdA),* and *Abdominal B (AbdB).* The coding sequences of these genes make up only a small part of the some 300 kb of DNA encompassed by the complex (Figure 13-42a). Mutational analysis has shown that noncoding regions of these

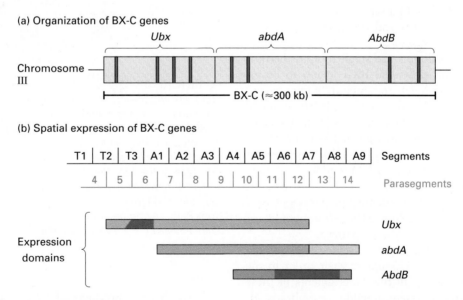

(a) Organization of BX-C genes

(b) Spatial expression of BX-C genes

▲ FIGURE 13-42   (a) Organization of genes within the bithorax complex (BX-C) on *Drosophila* chromosome III. Transcription of all three genes is from right to left. The exons (blue) make up a relatively small part of the BX-C. The large introns (tan) play an important role in regulating the specific spatial and temporal patterns of transcription of the BX-C genes along the anterior/posterior axis. The intron/exon structure of AbdB is uncertain. (b) Expression patterns of the products (mRNAs and/or proteins) of the *Ultrabithorax (Ubx), abdominal A (abdA),* and *Abdominal B (AbdB)* genes in parasegments 4–14. Darker levels of shading indicate higher concentrations of gene products. Ubx protein is expressed between the anterior border of parasegment (PS) 5 and the posterior border of PS12; AbdA protein is expressed between the anterior border of PS7 and the posterior border of PS14; expression of AbdB protein begins at the anterior border of PS10 and terminates within PS14. [Part (a) adapted from M. Peifer, et al., 1987, *Genes Dev.* **1**:891–898; Part (b) adapted from T. A. Kaufman et al., 1990, *Adv. Genet.* **27**:309.]

genes play a key role in regulating the specific pattern of their expression of these genes within different parasegments. Remarkably, the order of the genes within the complex (*Ubx, abdA, AbdB*) corresponds to their linear order of expression along the body axis (anterior → posterior) as illustrated in Figure 13-42b.

The ANT-C is separated from the BX-C by some 10 cM on *Drosophila* chromosome III. It contains five protein-coding genes including *Deformed* (*Dfd*), *Sex combs reduced* (*Scr*), and *Antennapedia* (*Antp*). These genes control development of parasegments 0 and 1 (*Dfd*), 2 and 3 (*Scr*), and 4 and 5 (*Antp*). As with genes in the BX-C, the linear order of genes in the ANT-C corresponds to their pattern of expression along the body axis. This organization of both the ANT-C and BX-C is not likely to be serendipitous, as it is observed in homologous gene clusters in higher organisms. The functional significance of this organization is not yet understood.

Determination of Parasegment Identity by BX-C

Specification of the identity of parasegments 5–14 in the fly embryo requires the BX-C. Removal of the entire BX-C leads to embryonic death. However, by analyzing the cuticle of dead larvae, investigators have assessed the role of the BX-C gene products in specifying various parasegments along the anterior/posterior axis. For example, deletion of the entire BX-C causes transformation of parasegments 5–12 into parasegment 4, as well as changes in parasegments 13 and 14 (Figure 13-43a). In a sense, then, the BX-C represses parasegment 4 identity and allows the more posterior parasegments to be specified.

Analysis of various double and single mutants permits the contribution of the individual BX-C genes to be assessed. In effect, this approach amounts to "adding back" one or two genes to flies that lack all three. If *Ubx* is added back (i.e., *abdA⁻ AbdB⁻* double mutants), parasegments 4–6 form normally, whereas parasegments 7–13 are transformed into parasegment 6 (Figure 13-43b); parasegment 14 also is abnormal. If in addition *abdA* is added back (i.e., *AbdB⁻* single mutants), parasegments 4–9 are normal, but parasegments 10–13 have the identity of parasegment 9 (Figure 13-43c). Finally, adding back *AbdB* would, of course, restore the normal wild-type pattern.

The order in which the genes are added back, as depicted in Figure 13-43, corresponds to their order in the BX-C. Furthermore, nearly all the parasegments in these mutants exhibit patterns present in wild-type embryos, although some parasegments in the mutants exhibit patterns characteristic of other parasegments in the wild type. In contrast are mutants that exhibit abnormal, not simply transformed, parasegmental patterns. An example is the single mutant with the genotype *Ubx⁺ abdA⁻ AbdB⁺*. In embryos with this genotype, parasegments 10–13 are abnormally patterned; that is, they do not correspond morphologically to any wild-type parasegment. This finding

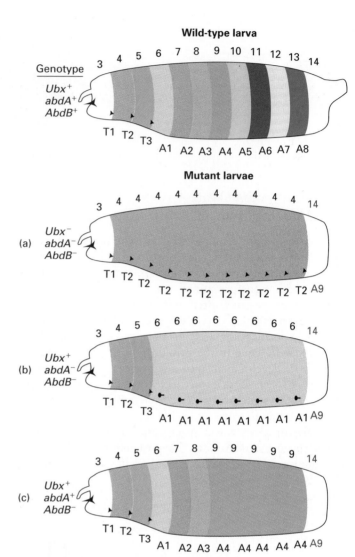

▲ FIGURE 13-43   Contribution of BX-C genes—*Ubx, abdA,* and *AbdB*—to determination of parasegment identity. The numbers above each larva indicate the parasegments; those below, the corresponding segments. The cuticular pattern of larvae is used to assign an identity to each parasegment (PS), which is indicated by color, as depicted in the wild type at the top. Red PS and segment labels indicate abnormal patterns that do not correspond exactly to any found in wild-type larvae. See text for discussion. [Adapted from P. A. Lawrence, 1992, *The Making of a Fly: Genetics of Animal Design,* Blackwell Scientific Publications.]

suggests that the products of *Ubx* and *AbdB*, in the absence of the *abdA* gene product, do not provide recognizable patterning information. Presumably, during normal development of the epidermis, *Ubx* and *AdbB* are never expressed without *abdA*. Indeed, analysis of other "out-of-order" mutants indicates that the BC-X genes must be expressed along the body axis in the order *Ubx, abdA,* and *AbdB*—their order in the genome—for normal patterning information to be generated.

(a) Wild type (BX-C⁺)

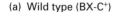

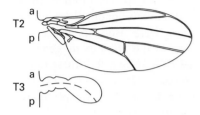

T2 a / p

T3 a / p

(b) *bithorax* mutant

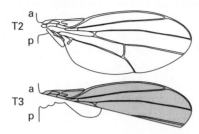

T2 a / p

T3 a / p

(c) *postbithorax* mutant

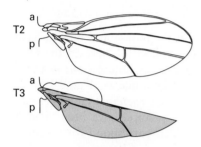

T2 a / p

T3 a / p

(d) *Haltere* mimic mutant

T2 a / p

T3 a / p

◀ FIGURE 13-44 Mutations affecting the expression domain of Ubx protein lead to abnormal development of the wing and haltere, which normally are found in the T2 and T3 segments, respectively. (a) Schematic diagram of wing and haltere in normal (BX-C⁺) fly. (b) The *bithorax* mutation leads to loss of Ubx expression in the anterior compartment of T3; as a result, part of the haltere is transformed into the anterior part (T2a) of the wing (yellow). (c) In *postbithorax* mutants, Ubx is not expressed in the T3p compartment, leading to development of this compartment into the posterior part (T2p) of the wing (yellow). (d) The *Haltere mimic* mutation causes expression of Ubx in T2, leading to transformation of a wing into haltere (yellow). [Adapted from P. A. Lawrence, 1992, *The Making of a Fly: The Genetics of Animal Design*, Blackwell Scientific Publications.]

The cuticular patterns illustrated in Figure 13-43 result from loss-of-function mutations in BX-C genes. Researchers also have assessed the effect of gain-of-function mutations in transgenic flies. For instance, transgenic embryos carrying *Ubx* under control of a heat-shock promoter express the protein encoded by *Ubx* uniformly along the anterior/posterior axis (see Figure 13-42b for normal distribution of Ubx protein). In these transgenic embryos, parasegments 6–14 form normally, but parasegments 1–5 are transformed into parasegment 6. During normal development, the identity of these anterior parasegments is controlled by the ANT-C. Some Ubx protein is expressed in parasegment 5, although at a lower concentration than in parasegment 6. Thus the level of the BX-C gene products, not simply their presence or absence, plays a role in specifying parasegment identity.

Control of Morphogenesis by BX-C Genes
Homeotic genes not only determine parasegmental identity in the embryo, they also regulate development of adult structures. For instance, as noted previously, mutations in the *Antp* gene transform an antenna into a leg (see Figure 13-41). Another striking example is provided by mutations in the BX-C that affect development of the wing and haltere (a rudimentary wing that serves as a balancing organ). Normally, the wing develops from both the anterior and posterior compartments of the second thoracic segment, designated T2a and T2p, and the haltere develops from both compartments of the third thoracic segment, T3a and T3p (Figure 13-44a). In *bithorax* mutants, which do not express Ubx protein in T3a, the anterior portion of a wing develops in the T3a compartment; in *postbithorax* mu-

tants, which do not express Ubx in T3p, the posterior portion of a wing develops in the T2p compartment (Figure 13-44b,c). Not surprisingly, mutations that result in no expression of Ubx in T3 cause transformation of the haltere into a wing (see Figure 8-3). Clearly, the Ubx protein acts to suppress wing development in T3. The converse transformation, from a wing to a haltere, is seen in flies carrying the *Haltere mimic* mutation (Figure 13-44d). This mutation leads to high expression of Ubx protein throughout T2, where it normally is expressed only at low levels.

Mechanism of Action of Selector Genes Although we have described just a few examples, extensive analyses of gain-of-function and loss-of-function mutations in the ANT-C and BX-C have led to several conclusions:

- Parasegment identity is determined by specific combinations and concentrations of the proteins encoded by selector genes.

- Abnormal combinations of selector-gene products give rise to abnormal patterning.

- Loss of expression of a selector gene leads to a posterior → anterior transformation in which posterior regions take on morphologic characteristics of more anterior regions.

- Ectopic expression of a selector gene (gain of function) in a region anterior to where it normally is expressed causes suppression of selector genes normally expressed more anteriorly. This phenomenon, called *phenotypic suppression*, results in an anterior → posterior transformation in morphology.

Finally, what do selector genes actually control? It is relatively easy to envision how a transcription factor generally can regulate formation of a specific cell type by controlling the expression of cell type–specific genes, such as muscle creatine kinase in developing muscle. But how do selector genes regulate the difference between a leg and an antenna or between a wing and a haltere? It seems highly unlikely that these genes activate unique sets of structural genes, since the cell types comprising these different tissues are often indistinguishable. It is thought that selector genes act upon the same sets of genes in different ways (perhaps controlling the timing of proliferation and differentiation). A major challenge to developmental biologists is to understand the mechanisms by which different selector genes activate different pathways of morphogenesis.

➤ *Mammalian Homologs of* Drosophila *ANT-C and BX-C*

A milestone in developmental biology was the discovery that a region of homology existed in all the structural genes within the major *Drosophila* homeotic complexes, the ANT-C and BX-C, which collectively are referred to as HOM-C. This homologous region encodes about 60 contiguous amino acids in each of the proteins encoded by the

HOM-C genes. This amino acid sequence forms the DNA-binding domain, which has the homeobox motif in these proteins. Shortly after this discovery, mammalian homologs of the HOM-C genes were isolated using the homeobox-containing regions of the *Drosophila* genes as DNA probes. Recent studies have shown that these mammalian genes, like the homologous *Drosophila* genes, play a critical role in regulating the development of specific regions along the anterior/posterior axis.

Mammalian Hox Genes Are Colinear Homologs of *Drosophila* HOM-C Genes

Molecular cloning studies in the mouse and human have identified four gene clusters, collectively termed the *Hox complex* (Hox-C), containing genes homologous to those in the *Drosophila* ANT and BX complexes (Figure 13-45). These mammalian gene clusters, designated HoxA–HoxD in humans and Hoxa–Hoxd in mice, have been divided into 13 regions, some of which correspond to identified genes. Each cluster is located on a different chromosome and contains a unique set of genes. Homologous genes located in different Hox clusters are referred to as *paralogs*. Whether genes (paralogs) corresponding to all the Hox regions shown in Figure 13-45 eventually will be identified or even exist is not known.

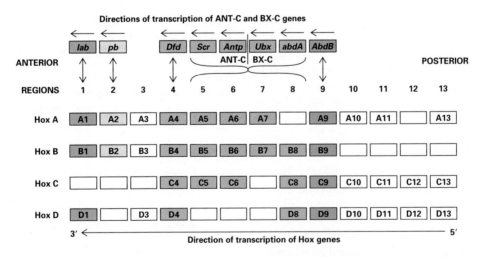

▲ FIGURE 13-45   Comparison of *Drosophila* ANT-C and BX-C genes and the four human Hox complexes. For purposes of alignment, the ANT-C and BX-C (*top*) are shown adjacent to each other, although they are separated on chromosome III in the fly genome. Each Hox complex (HoxA–HoxD) can be divided into 13 regions numbered 1–13. Genes in different Hox complexes with the same number are homologs called paralogs (e.g., *HoxA1* and *HoxB1*). Empty boxes indicate that the corresponding genes have not been identified. Similarities in expression patterns and amino acid sequence strongly suggest that the fly *labial (lab), proboscipedia (pb), Deformed (Dfd),* and *Abdominal B (AbdB)* genes are analogous to regions 1, 2, 3, and 9. The set of fly genes including *Sex combs reduced (Scr), Antennapedia (Antp), Ultrabithorax (Ubx),* and *abdominal A (abdA)* (red) is similar to regions 5–8, but precise correlations between the fly genes and Hox regions are not possible at present. Horizontal arrows indicate direction of transcription. Anterior and posterior refer to the order of expression of these genes along the body axis. [Adapted from M. P. Scott, 1992, *Cell* **71**:551.]

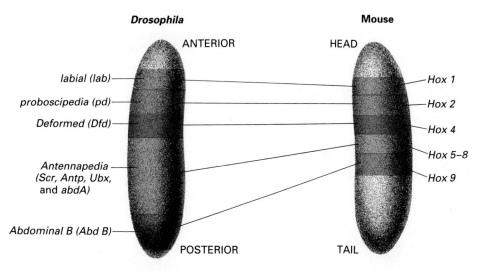

▲ FIGURE 13-46   Schematic diagram depicting expression domains of the indicated HOM-C genes and Hox genes in *Drosophila* and mouse embryos, respectively. Note that in the domain labeled Antennapedia, four genes are expressed: *Sex combs reduced* (*Scr*), *Antennapedia* (*Antp*), *Ultrabithorax* (*Ubx*), and *abdominal A* (*abdA*), which are analogous to Hox5–Hox8. Red lines indicate corresponding regions between *Drosophila* and mouse. The observed conservation in the order of expression along the axis suggests that a common mechanism regulates development of specific regions along the anterior-posterior axis. [Adapted from M. McGinnis and M. Kuziora, 1994, *Sci. Am.* **270**(2):58; drawing by Tomo Narashima, © 1994, Scientific American, Inc. All rights reserved.]

As discussed earlier, the genomic order of the BX-C and ANT-C genes is colinear with the order of expression of these genes along the anterior/posterior axis of *Drosophila*. This same feature of genomic organization has been demonstrated for the mouse Hox genes (Figure 13-46). Conservation of this pattern of gene expression suggests that a common mechanism regulates development of specific regions along the anterior/posterior axis in both *Drosophila* and mammals.

Mutations in Hox Genes Result in Homeotic Transformations in the Developing Mouse

The genes composing the *Drosophila* HOM-C were isolated based on developmental defects observed in flies carrying mutations within the complex. In contrast, the mammalian Hox genes have been isolated based on their homology with HOM-C genes not on their function. If Hox genes are functionally equivalent to *Drosophila* HOM-C genes, then mutations in the mammalian genes would be expected to produce homeotic transformations along the body axis. Gene-knockout and transgenic technology have been used to assess the functional role of several Hox genes in controlling regional identity in the mouse. The results of these studies support the view that the Hox genes play qualitatively similar roles in controlling

regional identity along the anterior/posterior axis in mammals as the HOM-C genes do in flies.

Gain-of-Function Mutations in Hoxd-4 Gene   The Hoxd-4 protein is a homolog of the *Drosophila* Deformed (Dfd) protein, which is encoded within the ANT-C and plays a role in regulating development in ectodermal parasegments 0 and 1. The anterior border of Hoxd-4 expression in the mouse embryo includes precursor cells that normally give rise to cervical vertebrae. Ectopic expression of Hoxd-4 in a domain anterior to this region was achieved by fusing *Hoxd-4* cDNA to the regulatory sequences that control transcription of *Hoxa-1*, which normally is expressed more anteriorly (see Figure 13-46). In newborn transgenic mice carrying this construct, which has the effect of a gain-of-function mutation, anterior (occipital) vertebrae are morphologically similar to more posterior (cervical) vertebrae (Figure 13-47). This homeotic transformation is similar to those resulting from ectopic expression of *Drosophila* selector genes discussed previously; that is, ectopic expression of a selector gene product results in an anterior → posterior transformation.

Loss-of-Function Mutations in Hoxb-4 Gene   The *Hoxb-4* gene, located in the Hoxb cluster, is a paralog of *Hoxd-4*. Although both genes are homologs of *Deformed* (*Dfd*) and occupy the same position within their respective clusters (see Figure 13-44), their expression patterns differ,

Wild type   **Ectopic *Hoxd-4*
mutant**

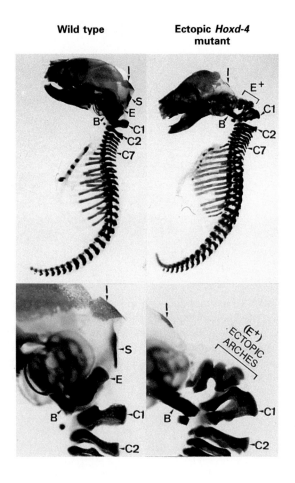

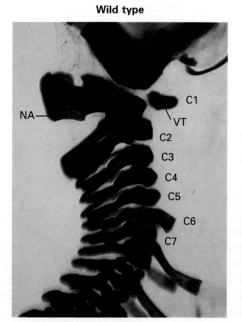

◄ FIGURE 13-47   Effect of ectopic expression of Hoxd-4 protein anterior to its usual expression domain on development of anterior structures in mouse embryos. The photographs show skeletal preparations of a normal newborn mouse and a transgenic newborn that misexpresses Hoxd-4; the lower photographs are higher magnifications of the upper ones. The transgenic mouse exhibits a reduction or loss of the supraoccipital (S) and extraoccipital (E) bones of the skull and the appearance of bony structures called ectopic archers (E$^+$), which are fused to the basioccipital (B) bone; these structures appear similar to neural arches characteristic of more posterior vertebrae. The cervical vertebrae are indicated by C followed by a number. [From T. Lufkin et al., 1992, *Nature* **359**:835.]

as do the amino acid sequences of the protein each encodes. In normal mice, the anterior border of Hoxb-4 expression gives rise to the anterior edge of the second cervical vertebra (C2). In mice homozygous for a knock-out mutation in *Hoxb-4*, C2 is morphologically like a first cervical vertebra (Figure 13-48). Thus, as in *Drosophila*, loss-of-function mutations in mouse *Hox* genes results in a posterior → anterior transformation.

In the mouse, paralogs in all four Hox clusters are expressed in the embryonic region giving rise to C2, although the functions of the four genes probably are not completely redundant. For instance, knockout of *Hoxb-4* does not produce complete transformation of C2 into C1, suggesting that different paralogs may contribute distinct aspects of pattern to a particular region.

➤ FIGURE 13-48   Effect of loss-of-function mutation in *Hoxb-4* on skeletal development in mouse embryos. The cervical (C) vertebrae are indicated by numerals. In the wild-type skeleton, C2 lacks the neural arch (NA) and ventral tubercle (VT) characteristic of C1. In the mutant skeleton, C2 is transformed into a structure similar to C1 by addition of a neural arch (open arrow) and ventral tubercle (closed arrow). [From R. Ramirez-Solis et al., 1993, *Cell* **73**:279.]

Wild type   ***Hoxb-4* mutant**

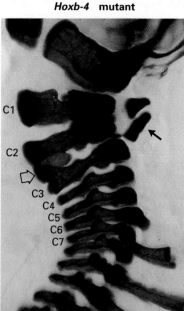

SUMMARY

Extensive research of several model developmental systems has shown that differential gene expression plays a central role in guiding development in eukaryotes. Developmental biologists have identified a variety of mechanisms used to control gene expression, yet noted that certain basic control strategies are conserved in diverse organisms.

A classic system for investigating control of gene expression involves the choice between the lytic or lysogenic developmental pathway following infection of *E. coli* by bacteriophage λ. The cI protein (λ repressor), which is required to establish and maintain the lysogenic state, plays a key role in regulating expression of various λ genes. This λ-encoded transcription factor binds to three tandemly arranged operators in the λ genome with the following order of affinity: $O_R1 > O_R2 > O_R3$. Simultaneous, cooperative binding of cI to O_R1 and O_R2 has two effects. It represses expression of Cro protein (which promotes the lytic pathway) by directly blocking binding of RNA polymerase to the *cro* promoter. It concurrently activates another promoter that controls transcription of the *cI* gene in the opposite direction. Another viral protein critical in the decision between the lytic or lysogenic pathway is cII, which is expressed at an early stage of λ infection and stimulates transcription of the *cI* gene. The stability of the cII protein, and hence its activating effect, depends on the nutritional state of the host cell. Cro protein also binds to O_R1, O_R2, and O_R3 but with a reversed order of affinity than cI. Binding of Cro to O_R3 inhibits cI expression, thereby inducing the lytic pathway. Studies of λ gene expression following infection of a host cell revealed the complexity of gene regulatory networks in even the simplest systems and identified autoregulation and cooperative binding as important mechanisms in establishing stable developmental states.

Studies of mating-type control in yeast have provided insight into more complex regulatory networks. Mating type is controlled by the *MAT* locus, which encodes transcription factors regulating cell type–specific gene expression. Genetic and molecular studies have provided a detailed picture of the combinatorial control of gene expression that leads to the specification of three yeast cell types: **a** and α haploid cells and the **a**/α diploid. These studies have revealed that a single DNA-binding protein can combine with other DNA-binding proteins to provide different patterns of gene expression. For instance, the MCM1 protein can act alone to activate **a**-specific genes in **a** cells and it can form complexes with the α1 protein to activate α-specific genes in α cells. The MCM1 protein also can form complexes with Ste12, a transcription factor whose activity is affected by extracellular signals.

Silent (nontranscribed) copies of the mating-type sequences are present in the *HML* and *HMR* loci, which are located on the same chromosome as MAT. By the mechanism of gene conversion (Chapter 9), an inactive mating-type sequence can be substituted for the opposite sequence in the *MAT* locus, thereby switching the mating type. The switching process is regulated by multiple controls, which converge on the expression of HO endonuclease, the enzyme catalyzing the first step in gene conversion. This enzyme is expressed only in haploid mother cells during the G_1 phase of the cell cycle. Each level of control is mediated by a distinct set of transcription repressors or activators that act on regulatory sites upstream of the *HO* gene. In addition, a complex interplay among more generalized factors promotes or antagonizes *HO* transcription.

Multicellular organisms consist of many different cell types exhibiting diverse functions and morphologies. Cell type specification depends on qualitative and quantitative differences in the expression of specific sets of genes. The analysis of muscle development (myogenesis) in mammals and neuron development (neurogenesis) in *Drosophila* has revealed the role of HLH-type transcription factors in specifying both cell types. Development of precursor muscle cells (myoblasts) depends on muscle-specific MyoD and Myf5, and development of neuronal cells depends on Achaete and Scute. Expression of these HLH proteins is controlled by general positive and negative regulatory proteins, which also are HLH transcription factors. Further differentiation of precursor cells requires other HLH proteins (e.g., myogenin in the case of muscle and Asense protein in the case of neuronal cells). Mutant studies have shown that a mammalian homolog of Achaete and Scute, called MASH1, is crucial for development of a subclass of neurons in the mouse. Whether specification of other cell types is controlled by HLH proteins is not known.

An important feature of animal development is the process of regionalization. Genetic and molecular analysis of gene expression during *Drosophila* embryogenesis has led to a detailed description of the control of gene expression in different spatial domains within the developing fly. Early development in *Drosophila* is controlled by four different sets of genes: three regulate patterning along the anterior/posterior axis (the anterior, posterior and terminal systems), and one regulates patterning along the dorsoventral axis. The anterior system controls the development of the head and thorax. The establishment of differential cell states in the anterior region of the fly embryo is critically dependent upon the transcription factor Bicoid. Maternal *bicoid* mRNA is localized to the extreme anterior of the early embryo; following translation, Bicoid protein diffuses toward the posterior, producing a Bicoid gradient. The Bicoid concentration gradient determines the precise spatial domain where the zygotic gap gene *hunchback* is transcribed. The resulting gradient of Hunchback protein determines the expression domains of several other gap genes (e.g., *Krüpple, knirps,* and *giant*). Maternally derived Nanos protein, which is localized to the posterior region of the early embryo, functions as the posterior determinant. Nanos inhibits the translation of maternal *hunchback* mRNA, which is uniformly distributed, thereby excluding Hunchback protein from the posterior region. The combined action of the proteins encoded by the maternal *bicoid* gene and zygotic gap genes control transcription of the pair-rule genes; the differential expression of these genes divides the embryo into 14 developmental units called parasegments. The segment-polarity genes then regulate patterning of cells within each parasegment. Finally, selector genes act within parasegments to generate the organization of cells within different body regions. The terminal system controls development of the most anterior and posterior structures in the embryo. Uniformly distributed Torso receptor is specifically activated at the two poles, inducing expression of zygotic gap genes such as *tailless.*

The determination of different cellular fates along the dorsoventral axis utilizes yet another means of regulating gene expression. It is postulated that a gradient of an extracellular signal is present in the space between the egg shell and the plasma membrane of the early embryo. The receptor for this signal, encoded by maternal *Toll* mRNA, is distributed uniformly on the embryo's surface. Because of the signal gradient, activation of Toll also is graded along the dorsoventral axis, with the greatest activation occurring on the ventral side. Activation of Toll leads to graded nuclear localization of the maternally derived Dorsal protein: in ventral regions of the embryo, Dorsal is largely present in nuclei; whereas in dorsal regions, it is located mostly in the cytoplasm. Nuclear localization of Dorsal is required for subsequent development of ventral structures. Dorsal is highly homologous to $NF_\kappa B$, a mammalian transcription factor that regulates expression of immunoglobulin genes. Like Dorsal, $NF_\kappa B$ is translocated to the nucleus following stimulation of cells by an extracellular signal.

The role of selector genes in determining regional identity (e.g., location of adult structures) along the anterior/posterior axis has been studied extensively in *Drosophila* and more recently in mammals. These genes, which occur in complexes or clusters, encode homeobox-containing transcription factors. In *Drosophila,* HOM-C comprises two different gene clusters referred to as the Antennapedia complex (ANT-C) and bithorax complex (BX-C). Although these two complexes are separated by a large chromosomal segment in *Drosophila,* the homologous mammalian genes are contiguous, forming a cluster called Hox. Mammals have four Hox clusters located on four different chromosomes. Remarkably, the regional expression of selector genes along the anterior/posterior axis of the embryo reflects their order in the genome; the significance of this conserved organization is not known. In both the mouse and fly, loss-of-function mutations in HOM-C or Hox genes generally result in transformation (homeosis) of the affected body region into one anterior to it (i.e., a posterior → anterior transformation). Conversely, ectopic expression (equivalent to gain-of-function mutations) of a selector gene in a region anterior to its normal expression domain usually leads the affected region to develop into a structure characteristic of a more posterior region (i.e., an anterior → posterior transformation). The conservation of the structure and function of selector-gene complexes demonstrated to date strongly suggests that the basic mechanisms for determining regional identity along the body axis have been maintained throughout evolution.

REVIEW QUESTIONS

1. The differential mechanisms that dictate whether bacteriophage λ will infect *E. coli* via a lytic or lysogenic pathway yields an important model system for understanding complex genetic interactions. Review the factors that regulate the choice between lysogeny and lytic development, paying particular attention to cI, cII, cIII, N, Cro and integrase proteins. Be sure that you understand the events that occur upon infection of *E. coli* by bacteriophage λ.

a. If the O_R1 site were mutated such that cI could no longer bind, what would happen to the level of Cro protein in infected *E. coli*? Would you expect the lytic or lysogenic response to predominate for this particular bacteriophage λ mutant?

b. Suppose that a mutant bacteriophage λ carries a gene that produces a defective Cro protein. If this particular Cro molecule could not bind to the O_R3 binding site, what would happen to the expression of the *cI* gene? Would this favor a lytic or lysogenic response?

c. Assume that the N protein of a particular phage lambda mutant could not bind to the Nut site. What would happen to the levels of cII protein and cIII protein? What would be the subsequent effect on *cI* gene expression? Would this favor a lytic or lysogenic response?

d. In Chapter 7, you learned that bacteriophage λ could be adapted for use as a vector in gene cloning experiments. If you were to use phage λ for such endeavors, would you want the lytic or lysogenic pathway to predominate? Refer to Figure 7-10. Notice that one of the genes present in the replaceable region of wild type phage λ encodes cIII protein. Why would the absence of the *cIII* gene make the adapted phage λ more suitable as a cloning vector? Explain your answer.

2. The ability of yeast to switch mating types is an extremely important example of how discrete regions of the genome can actually be changed in a programmed fashion, in this case by a gene conversion event.

a. How are the **a**-specific genes inhibited in a yeast cell with an α mating type? Likewise, how are the α-specific genes inhibited in a yeast cell with an **a** mating type? Be sure to include MCM1 and other factors in your discussion.

b. You have obtained purified **a** factor pheromone, and you add it to a yeast culture to determine if they are **a** or α mating type. What would you expect to see if they were **a** mating type? What would you observe if they were α mating type?

c. You are working with a well characterized strain of yeast that has a mutation in the **a** pheromone receptor. The mutant phenotype results in the receptor having no affinity for the pheromone. The actual mutant receptor is expressed and found on the cell surface. What mating type would these cells be? What would be the effect of adding **a** or α pheromone to these cells?

3. Place the following genes into the table below, including the general class of gene and the type of protein it encodes: (a) *bicoid* (b) *nanos* (c) *hunchback* (d) *knirps* (e) *fushi tarazu* (f) *cactus* (g) *engrailed* (h) *Toll* (i) *pelle* (j) *Ultrabithorax*

	Transcription Factor	Receptor Protein	Other
Maternal-effect genes			
Gap gene			
Pair rule gene			
Segment polarity gene			
Selector gene			

4. *Drosophila melanogaster* has become an extraordinary model species for gaining an understanding of the complexities of development. Beginning with oogenesis, review the general stages of fly development. Why have investigators chosen this organism for study?

a. What is a morphogen? What is meant by a morphogen gradient? If a particular morphogen of interest were a transcription factor, how might a gradient of it be affective in a *D. melanogaster* blastoderm in regulating development? Would such a mechanism hold true in the early development of mammalian embryos?

b. Describe three methods for visualizing the expression of genes microscopically.

c. What are *bicoid* and *nanos*? Are these mRNA products localized in the oocyte? Where are these mRNAs made? What type of protein is encoded by *bicoid* mRNA? Describe the role of Nanos protein and Bicoid protein in the expression of Hunchback protein. Be certain to mention the sources of *hunchback* mRNA. If *nanos* mRNA were present to a lower extent at the posterior pole in a mutant, what would be the effect on the boundary of Hunchback protein expression? Would it move posteriorly or anteriorly?

d. What general type of factor is the Dorsal protein? What is the source of *dorsal* mRNA? How is *dorsal* mRNA distributed in the embryo during development? Refer to Figure 13-35. You have an agent that can block the Toll receptor and prevent binding of the natural ligand. What would happen to the localization of Dorsal protein? Suppose you were studying a mutant in which the normal interaction between Cactus and Dorsal proteins were abolished. What would be the pattern of dorsal protein localization? Would the blocking agent have an effect in this mutant?

e. What is the general function of the Torso protein? Where is it found?

f. What are the pair-rule genes, segment-polarity genes, and selector genes? Discuss their interrelationships.

5. You are interested in studying the expression of the *Krüpple* gene. You decide to use P-element mediated transformation to prepare transgenic flies having the promoter for Krüpple placed upstream from the *E. coli lacZ* gene. (See Chapters 8 and 9; refer to Question 4 in Chapter 9 as well.)

a. What experiments would you need to do to demonstrate that you have actually obtained a transgenic line that you desire?

b. The enzyme produced by the *lacZ* gene is β-galactosidase. Suppose that you used antibodies against β-galactosidase to determine its location during development of the embryo. At what point following fertilization would you expect to see staining of the protein by the antibody? Explain your answer.

References

Cell-Type Specification in Bacteriophage λ and Yeast

ANDREWS, B. J., and I. HERSKOWITZ. 1989. Identification of a DNA-binding factor involved in cell-cycle control of the yeast *HO* gene. *Cell* 57:21–29.

ANDREWS, B. J., and I. HERSKOWITZ. 1989. The yeast SWI4 protein contains a motif present in developmental regulators and is part of a complex involved in cell-cycle–dependent transcription. *Nature* 342:830–833.

DOLAN, J. W., and S. FIELDS. 1990. Overproduction of the yeast STE12 protein leads to constitutive transcriptional induction. *Genes & Dev.* 4:492–502.

ELBLE, R., and B-K. TYE. 1991. Both activation and repression of a-mating-type-specific genes in yeast require transcription factor MCM1. *Proc. Nat'l. Acad. Sci. USA* 88:10966–10970.

HERSKOWITZ, I. 1989. A regulatory hierarchy for cell specialization in yeast. *Nature* 342:749–757.

KELEHER, C. A., et al. 1992. Ssn6-Tup1 is a general repressor of transcription in yeast. *Cell* 68:709–719.

LAURENSON, P., and J. RINE. 1992. Silencers, silencing, and heritable transcriptional states. *Microbiol. Rev.* 56:543–560.

LOWNDES, N. F., et al. 1992. SWI6 protein is required for transcription of the periodically expressed DNA synthesis genes in budding yeast. *Nature* 357:505–508.

LYDALL, D., G. AMMERER, and K. NASMYTH. 1991. A new role for MCM1 in yeast: cell cycle regulation of *SWI5* transcription. *Genes & Dev.* 5:2405–2419.

MA, H. 1994. The unfolding drama of flower development: recent results from genetic and molecular analyses. *Genes & Dev.* 8:745–746.

NASMYTH, K., et al. 1990. The identification of a second cell cycle control on the *HO* promoter in yeast: cell cycle regulation of SWI5 nuclear entry. *Cell* 62:631–647.

PETERSON, C. L., and I. HERSKOWITZ. 1992. Characterization of the yeast *SWI1, SWI2,* and *SWI3* genes, which encode a global activator of transcription. *Cell* 68:573–583.

PTASHNE, M. 1992. *A Genetic Switch.* Blackwell Scientific Publications and Cell Press.

SMITH, D. L., and A. D. JOHNSON. 1992. A molecular mechanism for combinatorial control in yeast: MCM1 protein sets the spacing and orientation of the homeodomains of an α2 dimer. *Cell* 68:133–142.

SONG, O-K., et al. 1991. Pheromone-dependent phosphorylation of the yeast STE12 protein correlates with transcriptional activation. *Genes & Dev.* 5:741–750.

TEBB, G., et al. 1993. SWI5 instability may be necessary but is not sufficient for asymmetric *HO* expression in yeast. *Genes & Dev.* 7:517–528.

VERSHON, A. K., and A. D. JOHNSON. 1993. A short, disordered protein region mediates interactions between the homeodomain of the yeast α2 protein and the MCM1 protein. *Cell* 72:105–112.

WOLBERGER, C., et al. 1991. Crystal structure of a MATα2 homeodomain-operator complex suggests a general model for homeodomain-DNA interactions. *Cell* 67:517–528.

Muscle and Neural Specification

BENEZRA, R., et al. 1990. The protein Id: a negative regulator of helix-loop-helix DNA-binding proteins. *Cell* 61:49–59.

BRAUN, T., et al. 1992. Targeted inactivation of the muscle regulatory gene *Myf-5* results in abnormal rib development and perinatal death. *Cell* 71:369–382.

CAMPUZANO, S., and J. MODOLELL. 1992. Patterning of the *Drosophila* nervous system: the *achaete-scute* gene complex. *Trends Genet.* 8:202–208.

CUBAS, P., and J. MODOLELL. 1992. The *extramacrochaetae* gene provides information for sensory organ patterning. *EMBO J.* 11:3385–3393.

DAVIS, R. L., H. WEINTRAUB, and A. B. LASSAR. 1987. Expression of a single transfected cDNA converts fibroblasts to myoblasts. *Cell* 51:987–1000.

DAVIS, R. L., et al. 1990. The myoD-binding domain contains a recognition code for muscle-specific gene activation. *Cell* 60:733–746.

ELLIS, H. M., D. R. SPANN, and J. W. POSAKONY. 1990. The *extramacrochaetae* gene, a negative regulator of sensory organ development in *Drosophila*, defines a new class of helix-loop-helix proteins. *Cell* 61:27–38.

GUILLEMOT, F., et al. 1993. Mammalian *achaete-scute* homolog 1 is required for the early development of olfactory and autonomic neurons. *Cell* 75:463–476.

HASTY, P., et al. 1993. Muscle deficiency and neonatal death in mice with a targeted mutation in the *myogenin* gene. *Nature* 364:501–506.

NABESHIMA, Y., et al. 1993. *Myogenin* gene disruption results in perinatal lethality because of severe muscle defect. *Nature* 364:532–535.

RUDNICKI, M. A., et al. 1992. Inactivation of *MyoD* in mice leads to up-regulation of the myogenic HLH gene *Myf-5* and results in apparently normal muscle development. *Cell* 71:383–390.

RUDNICKI, M. A., et al. 1993. MyoD or Myf-5 is required for the formation of skeletal muscle. *Cell* 75:1351–1359.

VAN DOREN, M., H. M. ELLIS, and J. W. POSAKONY. 1991. The *Drosophila extramacrochaetae* protein antagonizes sequence-specific DNA binding by *daughterless/achaete-scute* protein complexes. *Development* 113:245–255.

WEINTRAUB, H., et al. 1991. The *myoD* gene family: nodal point during specification of the muscle cell lineage. *Science* 251:761–766.

Regionalization of the Embryo

GAVIS, E. R., and R. LEHMANN. 1992. Localization of *nanos* RNA controls embryonic polarity. *Cell* 71:301–313.

GEISLER, R., et al. 1992. The gene *cactus*, which is involved in dorsoventral pattern formation of *Drosophila*, is related to the $I_\kappa B$ gene family of vertebrates. *Cell* 71:613–621.

INGHAM, P. W., and A. M. ARIAS. 1992. Boundaries and fields in early embryos. *Cell* 68:221–235.

IP, Y. T., et al. 1991. The *dorsal* morphogen is a sequence-specific DNA-binding protein that interacts with a long-range repression element in *Drosophila*. *Cell* 64:439–446.

IP, Y. T., et al. 1992. *dorsal-twist* interactions establish *snail* expression in the presumptive mesoderm of the *Drosophila* embryo. *Genes & Dev.* 6:1518–1530.

JIANG, J., et al. 1992. Individual *dorsal* morphogen binding sites mediate activation and repression in the *Drosophila* embryo. *EMBO J.* 11:3147–3154.

LAWRENCE, P. A. 1992. *The Making of a Fly: The Genetics of Animal Design.* Blackwell Scientific Publications.

LUFKIN, T., et al. 1992. Homeotic transformation of the occipital bones of the skull by ectopic expression of a homeobox gene in transgenic mice. *Nature* 359:835–841.

MCGINNIS, W., and R. KRUMLAUF. 1992. Homeobox genes and axial patterning. *Cell* 68:283–302.

MCGINNIS, W., and M. KUZIORA. 1994. The molecular architects of body design. *Sci. Am.* 270:58–66.

NÜSSLEIN-VOLHARD, C., and E. WIECHAUS. 1980. Mutations affecting segment number and polarity in *Drosophila*. *Nature* 287:795–801.

PAN, D., and A. J. COUREY. 1992. The same *dorsal* binding site mediates both activation and repression in a context-dependent manner. *EMBO J.* 11:1837–1842.

In Part III we turn to the elements of cells that allow them to function appropriately in the overall economy of the organism. Our focus now changes from the nucleus, where information handling occurs, to the cytoplasm, where most of a cell's activities take place. Although each cell type carries out specialized activities, this Part will be primarily concerned with the principles by which cells are constructed and organized. Considering the extraordinary diversity of specialized cellular activities—contraction by muscle cells, sugar metabolism by liver cells, electrical conduction by nerves, antibody secretion by lymphocytes—the underlying similarity of cells is remarkable. Among organisms, the major difference in cellular design is between the simple cells of prokaryotes, which lack both a nucleus and other membrane-limited internal organelles, and the more complicated cells of eukaryotes, which have both a nucleus and extensive compartmentalization in their cytoplasm. Our concern here will be mainly with eukaryotic cells.

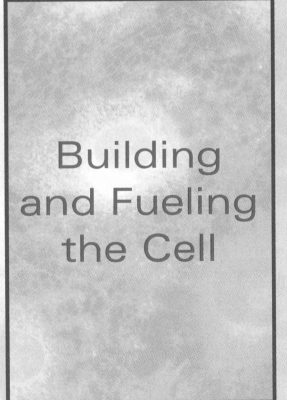

Building and Fueling the Cell

The organelles of the eukaryotic cell cytoplasm are surrounded by membranes, and the whole cell is delimited by its plasma membrane, so membrane biogenesis will be a leitmotif of Chapters 14 to 19. Nothing takes place in a cell without an expenditure of energy, making energy generation, storage, and channeling another major concern. A key issue is how proteins are targeted to particular places in the cell, or even secreted out of cells. These three topics—membranes, energy, and protein targeting—are fundamental considerations of cell architecture and activity. As cells have evolved, there has been a continual specialization of their use of membranes, as in the tough but easily deformed outer membrane of the red blood cell; repeated channeling of energy into new uses, as in the conduction of impulses by nerve cells; and targeting of new proteins to extend the range of cellular functions.

Chapter 14, which opens Part III, discusses the structural principles of membranes. It focuses first on the particular properties of phospholipids that allow them to form membrane sheets to surround cells and organelles. Then the chapter turns to the key membrane of the cell, the plasma membrane, which defines its outer limits of an individual cell. It is the surface through which the cell interacts with the other cells of the body and ultimately with the outside world. Membranes are impermeable to charged or polar compounds, which is why they can delimit cells and organelles, but these types of molecules must be able to get across membranes in a controlled manner if the cell is to be any more than a bag of water. Thus Chapter 15 describes the all-important activity of transport across cell membranes. Membranes are specialized by what molecules they will transport and what molecules they hold back. Furthermore, organelles and cells can act as batteries that store energy because they selectively hold charged molecules and release them on demand.

Proteins are made in the cytoplasm either by ribosomes that are freely soluble or ones that are tethered to membranes. But proteins must find many specific locations in cells. This problem of sorting requires that proteins contain signals for localization and that the cell have the machinery necessary to recognize the signals and move the proteins to particular locations. Chapter 16 describes one level of protein sorting, how the cell distinguishes between proteins destined to be soluble constituents of the cytoplasm, ones targeted to the plasma membranes, ones that are to be secreted from the cell and ones designed to be taken up by lysosomes.

Every activity of a cell, from the synthesis of a protein to the contraction of a muscle, requires energy, and thus cellular energetics is a major consideration of cell design. Every cell has energy generation equipment, especially the mitochondria in eukaryotic cells, which extract energy from sugars, storing it in the energy currency of cells, ATP. Chapter 17 describes the biochemistry of energy generation, focusing on the remarkable way that cells couple the energy that can be generated, by virtue of charge separation across membranes, with the energy stored in ATP. Chapter 18 turns to the equally remarkable and related process by which plant as well as prokaryotic cells fix the energy of light by coupling it to the synthesis of sugars. In plants, this process of photosynthesis, which also liberates oxygen and therefore replenishes the air's supply, occurs in chloroplasts. Finally, in Chapter 19 we consider how mitochondria, peroxisomes, and chloroplasts are formed by describing the selective uptake of proteins into these and other organelles. We end the chapter by describing the unique process by which nuclei selectively import proteins and ribonucleoproteins from the cytosol.

14 Membrane Structure: The Plasma Membrane

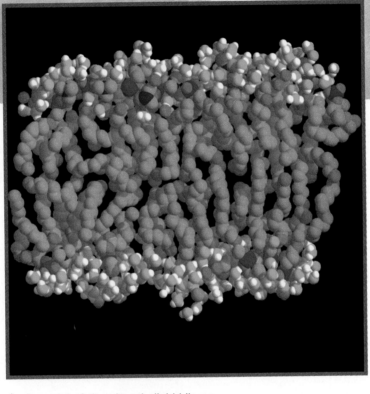

A model of the phospholipid bilayer.

Every cellular membrane carries out highly specialized functions. Although every membrane has the same basic phospholipid bilayer structure, a different set of membrane proteins enables each cellular membrane to carry out its distinctive activities (Figure 14-1). The plasma membrane that surrounds every cell, for example, contains a set of specific proteins that affect many aspects of the cell's behavior: Certain proteins provide anchors for cytoskeletal fibers or for components of the extracellular matrix that give the cell its shape. Still others bind signaling molecules, provide a passageway across the membrane for certain molecules, or regulate the fusion of the membrane with other membranes in the cell. Enzymes bound to the plasma membrane catalyze reactions that would occur with difficulty in an aqueous environment.

A multitude of internal membranes in each eukaryotic cell enclose separate compartments that perform specialized tasks: photosynthesis in the chloroplast, oxidative phosphorylation in the mitochondrion, degradation of macromolecules in the lysosome, and so on. Each organelle membrane contains a unique set of proteins that are essential for its proper functioning. Prokaryotes generally lack internal membranes; however, internal membrane vesicles in photosynthetic bacteria contain the proteins that absorb light and catalyze other initial steps of photosynthesis. Far from being a mere bag of soluble components, the cell is now viewed as a highly organized entity with many functional subcompartments each limited by one or more membranes.

Here, we discuss the basic principles that govern the organization of proteins and phospholipids in all biological membranes and describe the structure and function of several characteristic membrane proteins. To illustrate the

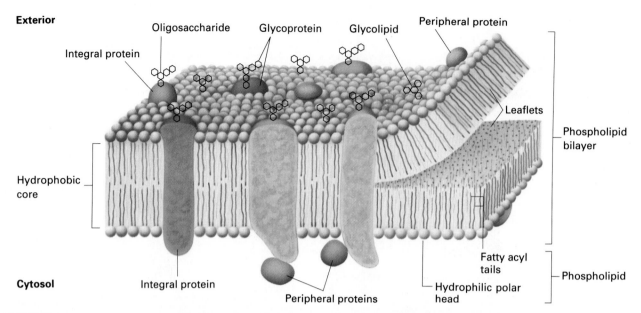

▲ FIGURE 14-1   Schematic diagram of a typical biological membrane. The phospholipid bilayer, the basic structure of all cellular membranes, consists of two leaflets of phospholipid molecules whose fatty acyl tails form the hydrophobic interior of the bilayer; their polar, hydrophilic head groups line both surfaces. Integral proteins have one or more regions embedded in the lipid bilayer; most span the bilayer as shown. Peripheral proteins are primarily associated with the membrane by specific protein-protein interactions. Oligosaccharides bind mainly to membrane proteins; however, some bind to lipids, forming glycolipids.

general principles of membrane organization, we look at the plasma membranes in several typical mammalian cells. The relative simplicity of the erythrocyte (red blood cell) membrane allows us to examine how it interacts with a submembranous cytoskeleton to give the cell its shape and flexibility. The membranes of pancreatic acinar cells and intestinal epithelial cells demonstrate how different regions of the same plasma membrane can be specialized to perform different tasks.

➤ *General Architecture of Lipid Membranes*

Subcellular fractionation techniques (see Figures 5-30 and 5-31) can partially separate and purify several important biological membranes, including the plasma and mitochondrial membranes, from many kinds of cells. These preparations are often contaminated with membranes from other organelles. However, the plasma membrane of human erythrocytes can be isolated in near purity because these cells contain no internal membranes.

All Membranes Contain Phospholipids and Proteins

All membranes, regardless of their source, contain proteins and lipids (Table 14-1). The protein:lipid ratio

varies greatly: the inner mitochondrial membrane is 76 percent protein; the myelin membrane is only 18 percent. Because of its high phospholipid content, myelin can electrically insulate the nerve cell from its environment.

The lipid composition varies among different mem-

TABLE 14-1   Chemical Composition of Some Purified Membranes

Membrane	Percentage by Weight		
	Protein	Lipid	Carbo-hydrate
Myelin	18	79	3
Plasma membrane:			
Human erythrocyte	49	43	8
Mouse liver	44	52	4
Ameba	54	42	4
Chloroplast spinach lamellae	70	30	0
Halobacterium purple membrane	75	25	0
Mitochondrial inner membrane	76	24	0

SOURCE: G. Guidotti, 1972, *Ann. Rev. Biochem.* **41**:731.

branes. All membranes contain a substantial proportion of phospholipids, predominantly *phosphoglycerides*, which have a glycerol backbone (Figure 14-2). All membrane phospholipids are amphipathic, having both hydrophilic and hydrophobic portions. *Sphingomyelin*, a phospholipid that lacks a glycerol backbone, also is commonly found in plasma membranes. Instead of a glycerol backbone, it contains *sphingosine*, an amino alcohol with a long unsaturated hydrocarbon chain. A fatty acyl side chain is linked to the amino group of sphingosine by an amide bond to form a *ceramide* (Figure 14-3). The terminal hydroxyl group of sphingosine is esterified to phosphocholine; thus the hydrophilic head of sphingomyelin is similar to that of phosphatidylcholine.

Nearly all fatty acyl chains found in the membranes of eukaryotic cells have an even number of carbon atoms (usually 16, 18, or 20). Unsaturated fatty acyl chains normally have one double bond, but some have two, three, or four (Table 14-2). In general, all such double bonds are of the cis configuration:

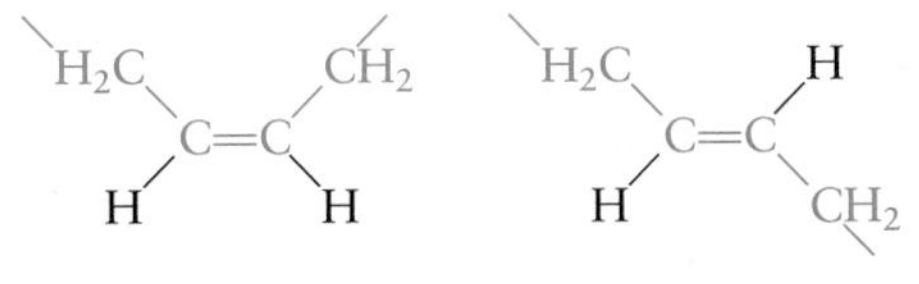

A cis double bond introduces a rigid kink in the otherwise flexible straight chain of a fatty acid (Figure 14-4).

A major difference among phospholipids concerns the charge carried by the polar head groups at neutral pH. Some phosphoglycerides (e.g., phosphatidylcholine, phosphatidylethanolamine) have no net electric charge; others (e.g., phosphatidylglycerol, cardiolipin, phosphatidylserine) have a net negative charge. A few rare phospholipids carry a net positive charge at neutral pH. Nonetheless, the polar head groups in all phospholipids can pack together into the characteristic bilayer structure. Sphingomyelin and glycolipids are similar in shape to phosphoglycerides and can form mixed bilayers with them.

Cholesterol and its derivatives constitute another important class of membrane lipids, the *steroids*. The basic structure of steroids is the four-ring hydrocarbon shown in

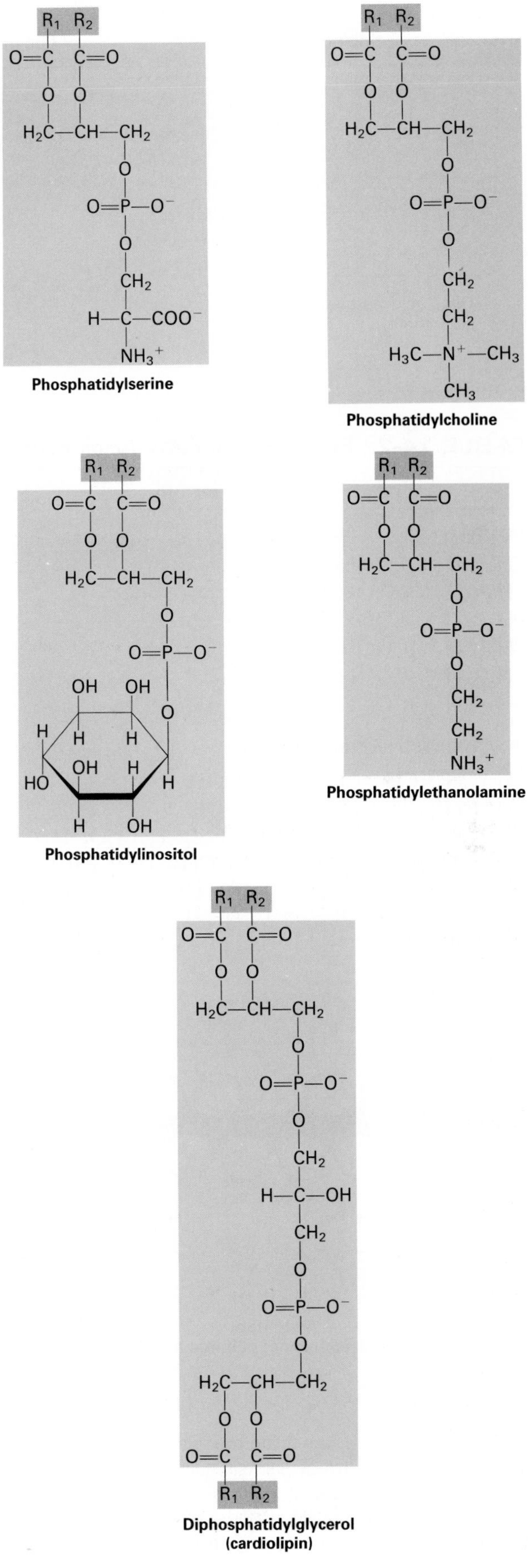

➤ FIGURE 14-2   The structures of some common phosphoglycerides found in cellular membranes. The fatty acyl side chains (R_1 and R_2) can be saturated, or they can contain one or more double bonds. Note that diphosphatidylglycerol contains four fatty acids: it is composed of two molecules of phosphatidic acid linked to carbons 1 and 3 of a central glycerol. Hydrophobic portions of the molecules are shown in red; hydrophilic, in blue.

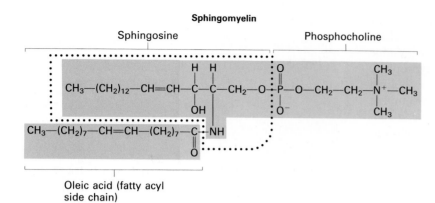

Sphingomyelin

Sphingosine

Phosphocholine

Oleic acid (fatty acyl side chain)

◄ FIGURE 14-3   The structure of a typical sphingomyelin molecule. A group of phospholipids that lack a glycerol backbone, sphingomyelin may contain a different fatty acyl side chain than oleic acid (shown here). Linkage of sphingosine (outlined by black dots) to a fatty acid via an amide bond forms a ceramide. The hydrophobic portion of the molecule is shown in red; the hydrophilic, in blue.

TABLE 14-2   Some Typical Fatty Acids Found in Cells

Chemical Formula	Systematic Name	Common Name
SATURATED FATTY ACIDS		
$CH_3(CH_2)_{10}COOH$	*n*-Dodecanoic	Lauric
$CH_3(CH_2)_{12}COOH$	*n*-Tetradecanoic	Myristic
$CH_3(CH_2)_{14}COOH$	*n*-Hexadecanoic	Palmitic
$CH_3(CH_2)_{16}COOH$	*n*-Octadecanoic	Stearic
$CH_3(CH_2)_{18}COOH$	*n*-Eicosanoic	Arachidic
$CH_3(CH_2)_{22}COOH$	*n*-Tetracosanoic	Lignoceric
UNSATURATED FATTY ACIDS		
$CH_3(CH_2)_5CH{=}CH(CH_2)_7COOH$		Palmitoleic
$CH_3(CH_2)_7CH{=}CH(CH_2)_7COOH$		Oleic
$CH_3(CH_2)_4CH{=}CHCH_2CH{=}CH(CH_2)_7COOH$		Linoleic
$CH_3CH_2CH{=}CHCH_2CH{=}CHCH_2CH{=}CH(CH_2)_7COOH$		Linolenic
$CH_3(CH_2)_4(CH{=}CHCH_2)_3CH{=}CH(CH_2)_3COOH$		Arachidonic

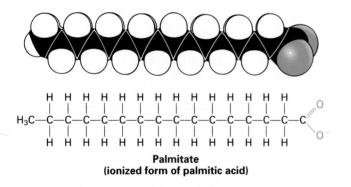

Palmitate
(ionized form of palmitic acid)

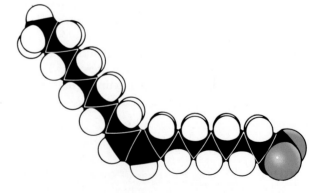

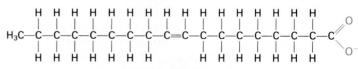

Oleate
(ionized form of oleic acid)

▲ FIGURE 14-4   Space-filling models and chemical structures of the ionized form of palmitic acid, a saturated fatty acid, and oleic acid, an unsaturated one. In saturated fatty acids, the hydrocarbon chain is linear; the cis double bond in oleate creates a kink in the hydrocarbon chain.

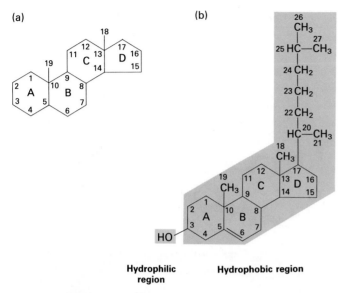

(a)

(b)

Hydrophilic   Hydrophobic region
region

▲ FIGURE 14-5   (a) The general structure of a steroid. All steroids contain the same four hydrocarbon rings, conventionally labeled A, B, C, and D, with the carbons numbered as shown. (b) The structure of cholesterol.

The Phospholipid Bilayer Is the Basic Structural Unit of Biological Membranes

Despite the variable compositions of biological membranes, the basic structural unit of virtually all biomembranes is the *phospholipid bilayer*: a sheetlike structure composed of two layers of phospholipid molecules whose polar head groups face the surrounding water and whose fatty acyl chains form a continuous hydrophobic interior about 3 nm thick (Figure 14-7). Each phospholipid layer in this lamellar structure is called a *leaflet*. The major driving force for the formation of phospholipid bilayers is, as discussed in Chapter 2, hydrophobic interactions between the fatty acyl chains of glycolipid and phospholipid molecules. Van der Waals interactions among the hydrocarbon chains favor close packing of these hydrophobic "tails." Hydrogen bonding and electrostatic interactions between the polar head groups and water molecules also stabilize the bilayer. Micelles (see Figure 2-17) are generally not formed by phospholipids in aqueous solutions, since the fatty acyl chains in sphingomyelins, glycolipids, and all phosphoglycerides are too large to fit into the interior of a micelle.

Figure 14-5a. Cholesterol (Figure 14-5b) is the major steroidal constituent of animal tissues; other steroids play more important roles in plants. Although cholesterol is almost entirely hydrocarbon in composition, it is amphipathic because it contains a hydroxyl group that interacts with water. Cholesterol is especially abundant in the plasma membrane of mammalian cells but is absent from most prokaryotic cells. As much as 30–50 percent of the lipids in plant plasma membranes are steroids (cholesterol and other steroids unique to plants).

Some lipids are found in a restricted range of membranes. For example, *galactolipids* (Figure 14-6) constitute as much as 70 percent of the lipids in chloroplast thylakoid membranes, and the phospholipid diphosphatidylglycerol, also called *cardiolipin* (see Figure 14-2), is highly enriched in the inner mitochondrial membrane. Neither of these lipids is generally present in other cellular membranes.

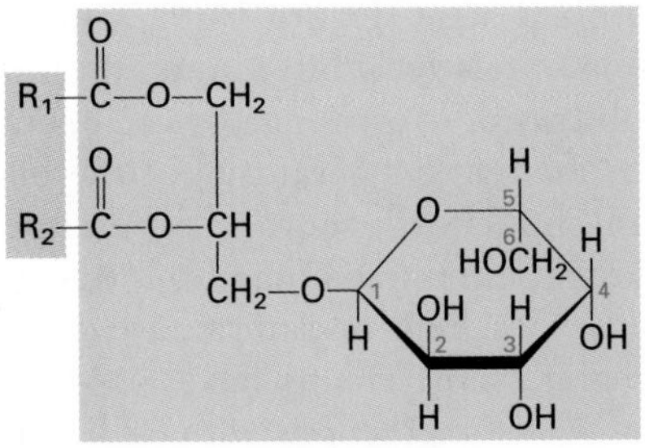

Phospholipid Bilayers Exhibit Two-Dimensional Fluidity That Depends on Temperature and Composition

In both pure phospholipid bilayers and in natural membranes, thermal motion permits phospholipid and glycolipid molecules to rotate freely around their long axes and to diffuse laterally within the membrane leaflet. Because such movements are lateral or rotational, the fatty acyl chains remain in the hydrophobic interior of the membrane. In both natural and artificial membranes, a typical lipid molecule exchanges places with its neighbors in a leaflet about 10^7 times per second and diffuses several micrometers per second at 37°C. Thus a lipid could diffuse the 1-μm length of a bacterial cell in only 1 s and of an animal cell in about 20 s.

In pure phospholipid bilayers, phospholipids do not migrate, or flip-flop, from one leaflet of the membrane to the other. In some natural membranes, however, they occasionally do so, catalyzed somehow by one or more membrane proteins. Energetically, such movements are extremely unfavorable, because the polar head of a phospholipid must be transported through the hydrophobic interior of the membrane.

The mobility of membrane lipids can be measured by *electron-spin resonance (ESR) spectroscopy*. In this technique, synthetic phospholipids containing a nitroxide group attached to a fatty acyl chain are introduced into otherwise normal phospholipid membranes. ESR measures the energy absorbed by the unpaired electron of the nitroxide group. ESR studies show that the hydrophobic interior of both synthetic and natural membranes generally

Galactolipid

▲ FIGURE 14-6   The structure of a galactolipid. Unlike phospholipids, galactolipids are uncharged. Nonetheless, they are amphipathic, having distinct hydrophobic (red) and hydrophilic (blue) regions.

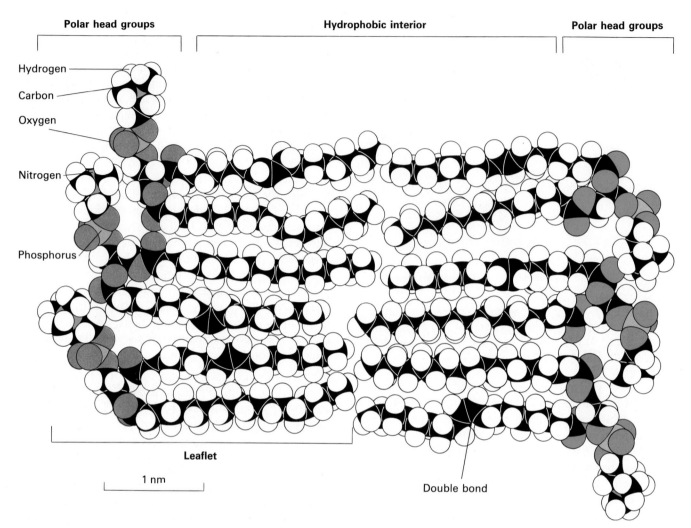

▲ FIGURE 14-7   A space-filling model of a typical phospholipid bilayer. The hydrophobic interior is generated by the fatty acyl side chains. Some of these chains have bends caused by double bonds. The different polar head groups all lie on the outer, aqueous surface of the bilayer. [From L. Stryer, 1988, *Biochemistry,* 3d ed., W. H. Freeman and Company, p. 289. Courtesy of L. Stryer.]

have a low viscosity and a fluidlike, rather than gel-like, consistency.

Two systems of pure phospholipid bilayers—liposomes and planar bilayers—have proved to be especially useful in studying the motions of lipid molecules within membranes (Figure 14-8). *Liposomes* are spherical vesicles up to 1 μm in diameter consisting of a phospholipid bilayer that encloses a central aqueous compartment; they are formed by mechanically dispersing phospholipids in water. *Planar bilayers* are formed across a hole in a partition that separates two aqueous solutions.

When a suspension of liposomes or a planar bilayer composed of a single type of phospholipid is heated, it undergoes an abrupt change in physical properties over a very narrow temperature range. This *phase transition* is due to

increased motion about the C–C bonds of the fatty acyl chains, which pass from a highly ordered, gel-like state to a more mobile fluid state (Figure 14-9). During the gel-to-fluid transition, a relatively large amount of heat (thermal energy) is absorbed over a narrow temperature range, which is the "melting temperature" of the bilayer.

In general, lipids with short or unsaturated fatty acyl chains undergo phase transition at lower temperatures than do lipids with long or saturated chains. Compared with long chains, short chains have less surface area to form van der Waals interactions with each other. Since the gel state is stabilized by these interactions, short-chain lipids melt at lower temperatures than long-chain lipids. Likewise, the kinks in unsaturated fatty acyl chains result in their forming less stable van der Waals interactions with

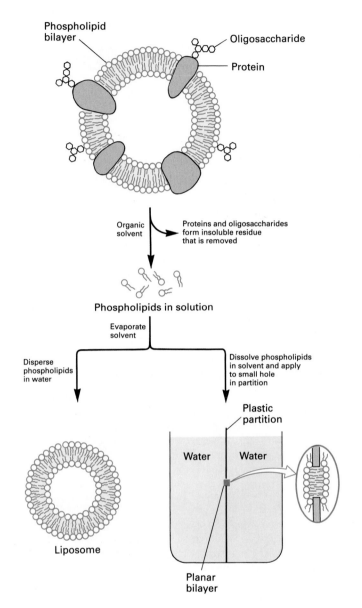

Experimental formation of pure phospholipid bilayers. A preparation of biological membranes is treated with an organic solvent, such as a mixture of chloroform and methanol (3:1), which selectively solubilizes the phospholipids and cholesterol. Proteins and carbohydrates remain in an insoluble residue. The solvent is removed by evaporation. *(Bottom left)* If the lipids are mechanically dispersed in water, they spontaneously form a liposome with an internal aqueous compartment. *(Bottom right)* A planar bilayer can form over a small hole in a partition separating two aqueous phases; such bilayers are often termed "black lipid membranes" because of their appearance.

tact with the aqueous solution near the polar head groups of the phospholipids; the steroid ring interacts with and tends to immobilize their fatty acyl chains. The net effect of cholesterol on membrane fluidity varies, depending on the lipid composition. Cholesterol restricts the random movement of the part of the fatty acyl chains lying closest to the outer surfaces of the leaflets, but it separates and disperses the tails of the fatty acyls and causes the inner regions of the bilayer to become slightly more fluid. At the high concentrations found in eukaryotic plasma membranes, cholesterol tends to make the membrane less fluid at growth temperatures near 37°C. At temperatures below the phase transition, cholesterol keeps the membrane in a fluid state by preventing the hydrocarbon fatty acyl chains of the membrane lipids from binding to each other, thereby offsetting the drastic reduction in fluidity that would otherwise occur at low temperatures.

By synthesizing a diverse array of phospholipids, as well as steroids such as cholesterol, cells maintain an appropriate fluidity of their plasma membranes as well as of all internal membranes.

other lipids than do saturated chains. As a result, unsaturated lipids maintain a more random, fluid state at lower temperatures than lipids with saturated fatty acyl chains.

Maintenance of bilayer fluidity appears to be essential for normal cell growth and reproduction. All cell membranes contain a mixture of different fatty acyl chains and are fluid at the temperature at which the cell is grown. Animal and bacterial cells adapt to a decrease in growth temperature by increasing the proportion of unsaturated to saturated fatty acids in the membrane, which tends to maintain a fluid bilayer at the reduced temperature.

Membrane cholesterol is another major determinant of bilayer fluidity. Cholesterol is too hydrophobic to form a sheet structure on its own, but it intercalates (is inserted) among phospholipids. Its polar hydroxyl group is in con-

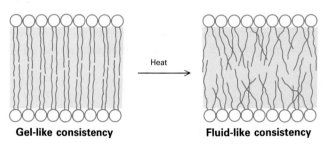

Gel-like consistency   **Fluid-like consistency**

▲ **FIGURE 14-9** Alternative forms of the phospholipid bilayer. Heat induces transition from a gel to a fluid over a temperature range of only a few degrees. The fluid phase is favored by the presence of short fatty acyl chains and by a double bond in the chains; thus these structural features reduce the melting temperature of bilayers.

Several Types of Evidence Point to the Universality of the Phospholipid Bilayer

One piece of evidence that the phospholipid bilayer structure is common to all biomembranes is that, as detailed above, many of the physical properties of pure phospholipid bilayers are similar to those of natural cellular membranes. Another is that either a single species of phospholipid, or a mixture of phospholipids with a composition approximating that found in natural membranes, spontaneously forms either planar bilayers or liposomes when dispersed in aqueous solutions (see Figure 14-8).

Some of the earliest direct experimental evidence that certain cellular membranes are built of phospholipid bilayers was obtained by E. Gorter and F. Grendel in 1925. These researchers extracted the lipids from human erythrocytes and floated them on the surface of a water solution. It was already known that under such conditions, phospholipids form a unimolecular film (monolayer) with the hydrophilic head groups facing the water and the hydrophobic tails pointing up into the air (Figure 14-10). Gorter and Grendel found that the area of the monolayer formed from extracted erythrocyte lipids was approximately twice that of the surface area of the original cells. Since erythrocytes contain no internal membranes, Gorter and Grendel concluded that the lipids are arranged in the membrane as a continuous bilayer. Interestingly, these investigators actually extracted only a portion of the erythrocyte lipids and underestimated the surface area of the red blood cell; because these two errors canceled each other out, they reached the correct conclusion!

Perhaps the best evidence for the bilayer structure comes from low-angle, x-ray diffraction analysis of the multimembrane *myelin sheath,* which is elaborated by Schwann cells and covers and insulates many mammalian nerve cells (Figure 14-11a). The myelin sheath, which is a series of stacked membranes, is the major membrane component of such nerves and can be separated from other cellular membranes in a pure state, permitting direct physical and chemical analyses. X-ray diffraction analysis of these stacked plasma membranes has revealed a regular variation in density that is consistent with a bilayer organization of each membrane unit (Figure 14-11b). In this organization, protein is located on either side of the mem-

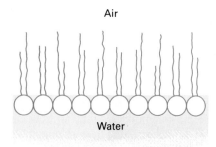

▲ FIGURE 14-10   A monolayer is formed when phospholipids are floated on the surface of an aqueous solution.

(a)

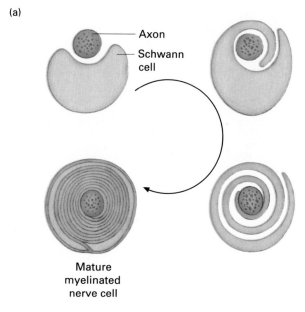

(b)

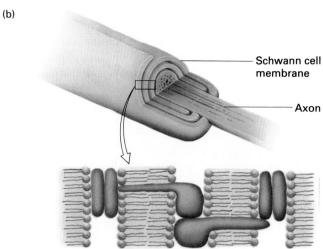

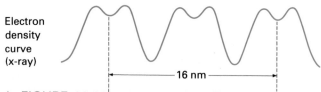

▲ FIGURE 14-11   Low-angle x-ray diffraction analysis of myelin membranes. This technique measures the density of matter and can be used to determine the distribution of lipid and protein in biomembranes. (a) During development of the nervous system, a large Schwann cell envelops the axon of a neuron. The continuous growth of the Schwann cell membrane into its own cytoplasm, together with rotation of the nerve axon, results in a laminated spiral of double plasma membranes around the axon. Mature myelin, a stack of plasma membranes of the Schwann cell, is relatively rich in phospholipids. (b) The profile of electron density—and thus of matter—obtained by x-ray diffraction studies on fresh nerve, and the relation of this profile to the protein and lipid components of the myelin membranes. [Adapted from W. T. Norton, 1981, in *Basic Neurochemistry*, 3d ed., G. J. Siegel et al., eds., Little, Brown, p. 68.]

brane, which has hydrophilic external faces and a central region of almost pure low-density hydrocarbon. Although some polypeptide segments pass through the lipid bilayer, these make up less than 10 percent of the inner mass of the membrane and are not detected in this type of analysis.

Electron microscopy has provided the most direct evidence for the universality of the bilayer structure. A thin section of a membrane is treated with an electron-opaque stain such as osmium tetroxide, which binds to the polar head groups of the phospholipids. When viewed in an electron microscope, a cross section of such an osmium tetroxide–stained cellular membrane looks like a railroad track: two thin dark lines (the stain–head group complexes) with a uniform light space of about 2 nm (hydrophobic tails) between them (Figure 14-12). Although some osmium tetroxide may bind to the double bonds of fatty acyl chains, most of it binds to the polar head groups.

Deviations from the basic bilayer structure occasionally occur, especially in membrane regions containing large amounts of specialized proteins; some membrane regions may lack lipid entirely. However, these are exceptions to the general pattern. One notable exception to the premise that the phospholipid bilayer forms the basic structural unit of biomembranes may be the thylakoid membranes of chloroplasts, whose major lipid components are galactolipids (see Figure 14-6). In aqueous solutions, purified galactolipids form inverted micelles, spherical structures composed of a single lipid layer. Unlike most lipid micelles, however, the polar head groups in galactolipid micelles face inward, rather than outward. The thylakoid membrane is 60–70 percent protein, and the bilayer may result from interactions among galactolipids, phospholipids, and membrane proteins.

Phospholipid Bilayers and Biological Membranes Form Closed Compartments

Perhaps the most significant biophysical property of phospholipids relevant to an understanding of biological membranes is that pure phospholipid membranes spontaneously seal to form closed structures that separate two aqueous compartments (see Figure 14-8). A sheet of phospholipid bilayer would be an unstable structure if it had a free edge, where the hydrophobic region of the bilayer would be in contact with water. Similarly, all cellular membranes surround closed compartments of the cell; all bilayers have an *internal face* (the side oriented toward the interior of the compartment) and an *external face* (the side presented to the environment). Because most organelles are surrounded by a single bilayer membrane, it is also useful to speak of the *cytosolic* and *exoplasmic faces* of the membrane, the *cytosol* being the part of the cytoplasm outside of organelles. The exoplasmic face of the plasma membrane, which is directed away from the cytosol toward the extracellular space, defines the outer limit of the cell (Figure 14-13). Some organelles, such as the nucleus and mito-

chondrion, are surrounded by two membranes; in these cases, the exoplasmic surface faces the lumen between the two membranes.

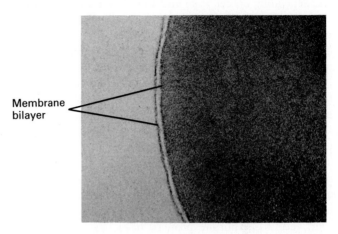

▲ **FIGURE 14-12**   Electron micrograph of thin section of an osmium-stained erythrocyte membrane. Note the "railroad track" appearance of the phospholipid bilayer at the cell surface. [Courtesy of J. D. Robertson.]

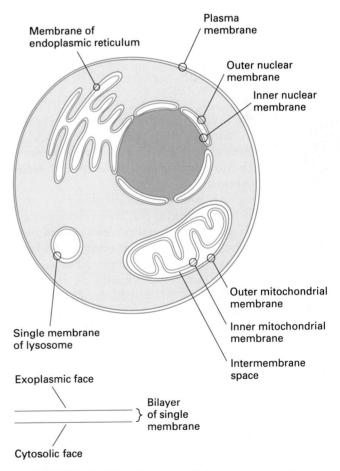

▲ **FIGURE 14-13**   Faces of cellular membranes. For organelles enclosed in two phospholipid membranes (e.g., the nucleus, chloroplast, mitochondrion), the exoplasmic faces (red) border the space between the inner and outer membranes.

➤ *Membrane Proteins*

The complement of proteins attached to a membrane varies, depending on cell type and subcellular location. Mitochondrial membrane proteins differ markedly from plasma membrane proteins; the plasma membrane components of a liver cell and an intestinal cell are not the same. We examine how specific proteins are deposited in specific membranes in Chapters 16 and 19. Here, we focus on the structures and functions of a few well-understood membrane proteins.

Some proteins are bound only to the membrane surface; others have one region buried within the membrane and domains on one or both sides of it. It is difficult to obtain enough pure protein to determine the sequences, let alone the interactions, of many important membrane proteins because they are present in such tiny amounts. However, certain membrane proteins have been purified and their orientation within the membrane determined experimentally. The sequences of many membrane proteins are now being determined from the sequences of their cloned cDNA. As discussed in later chapters, the membrane interactions of certain proteins can be predicted from the amino acid sequences.

Proteins Interact with Membranes in Different Ways

Membrane proteins can be classified into two broad categories—integral (intrinsic) and peripheral (extrinsic)—based on the nature of the membrane-protein interactions (see Figure 14-1). Most biomembranes contain both types of membrane proteins.

Integral membrane proteins, also called *intrinsic proteins,* have one or more segments that are embedded in the phospholipid bilayer. Most integral proteins contain residues with hydrophobic side chains that interact with fatty acyl groups of the membrane phospholipids, thus anchoring the protein to the membrane. Virtually all are *transmembrane* integral proteins, which span the phospholipid bilayer. They contain one or more membrane-spanning domains as well as domains, from four to several hundred residues long, extending into the aqueous medium on each side of the bilayer. In contrast, some integral proteins are anchored in only one of the two leaflets; these proteins contain covalently bound fatty acids that function to anchor them in the bilayer.

Peripheral membrane proteins, or *extrinsic proteins,* do not interact with the hydrophobic core of the phospholipid bilayer. Instead they are usually bound to the membrane indirectly by interactions with integral membrane proteins or directly by interactions with lipid polar head groups. The cytoskeletal proteins *spectrin* and *actin,* which are bound to the cytosolic face of the erythrocyte plasma membrane, and the enzyme *protein kinase* C, which shut-tles between the cytosol and the cytosolic face of the plasma membrane, are examples of peripheral proteins. Other peripheral proteins, including certain proteins of the extracellular matrix, are localized to the outer (exoplasmic) surface of the plasma membrane.

Transmembrane Proteins Contain Long Segments of Hydrophobic Amino Acids Embedded in the Phospholipid Bilayer

Two types of interactions keep transmembrane integral proteins embedded in membranes: ionic interactions with the polar head groups of the lipids and hydrophobic interactions with the lipid interior of the bilayer. *Glycophorin,* a major erythrocyte membrane protein, exhibits both types of interaction. As shown in Figure 14-14, the region of glycophorin that spans the membrane and interacts with the hydrophobic core of the bilayer consists of 34 residues (numbers 62–95) including one 23-aa sequence (73–95) composed entirely of hydrophobic (or uncharged) amino acids. This membrane-embedded region forms an α helix, with each amino acid residue adding 0.15 nm (1.5 Å) to the length of the helix (see Figure 3-9). Thus a helix containing 25 residues would be 3.75 nm long, just sufficient to span the hydrocarbon core of a phospholipid bilayer. The hydrophobic side chains protrude outward from the helix axis to form hydrophobic bonds with the fatty acyl chains. The polar carbonyl (C=O) and imino (NH) groups in each peptide bond are on the inside of the α helix, and every such group participates in a hydrogen bond to the helix axis. Thus the polar backbone groups in a transmembrane α helix are shielded from the surrounding hydrophobic fatty acyl chains.

In glycophorin, as in many other transmembrane proteins, positively charged amino acids (lysine and arginine) border the hydrophobic segment on the cytosolic surface and may interact with negatively charged phospholipid head groups predominant in the cytosolic leaflet of the erythrocyte membrane to hold the hydrophobic segment firmly in place.

Proteins Can Be Removed from Membranes by Detergents or High-Salt Solutions

To study the structure and function of membrane proteins, researchers must first isolate and purify them like other proteins. Integral membrane proteins present a special problem. In general, due to a preponderance of exposed hydrophilic groups, water-soluble globular proteins maintain their native conformation and remain individually suspended in an aqueous medium. In contrast, when integral proteins are separated from membranes and their hydrophobic regions exposed, the protein molecules aggregate and precipitate from aqueous solutions. Such proteins can be solubilized by detergents, which have affinity both for hydrophobic groups and for water.

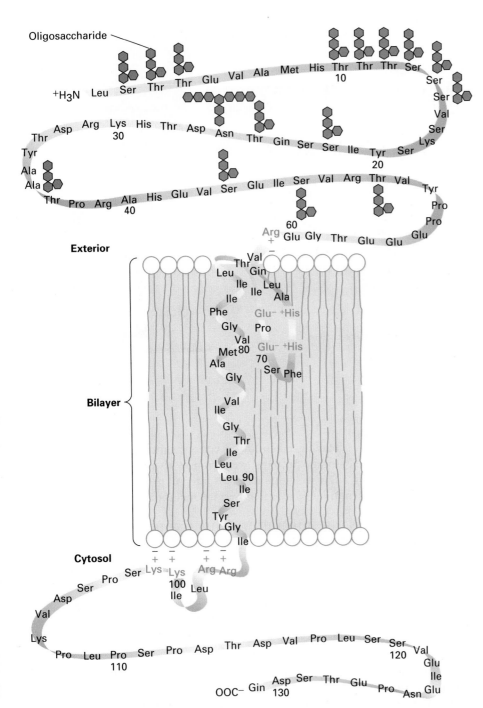

◀ FIGURE 14-14 Amino acid sequence and transmembrane disposition of glycophorin A from the erythrocyte membrane. This protein is a homodimer, but only one of its polypeptide chains is shown. Residues 62–95 are buried in the membrane, with the sequence from position 73 through 95 forming an α helix. Some evidence suggests that the negatively charged glutamic acid residues at positions 70 and 72 are ionically linked to the positively charged histidine residues at positions 66 and 67, allowing residues 62–72 to insert into the hydrophobic core of the bilayer. The ionic interactions shown between positively charged arginine and lysine residues and negatively charged phospholipid head groups in the cytosolic and exoplasmic faces of the membrane are hypothetical. Both the amino-terminal segment of the molecule, located outside the cell, and the carboxyl-terminal segment, located inside the cell, are rich in charged residues and polar uncharged residues, making these domains water-soluble. The exoplasmic domain has 16 carbohydrate residues attached to it; 15 are linked to serine or threonine chains, and one longer one is linked to an asparagine. Many other membrane proteins also contain covalently linked oligosaccharides, which are rich in negatively charged sialic acid groups. [See V. T. Marchesi, H. Furthmayr, and M. Tomita, 1976, *Ann. Rev. Biochem.* **45**:667; A. H. Ross et al., 1982, *J. Biol. Chem.* **257**:4152.]

Detergents are amphipathic molecules that disrupt membranes by intercalating into phospholipid bilayers and solubilizing lipids and proteins. Some detergents are natural products, but most are synthetic molecules developed for cleaning and for dispersing mixtures of oil and water (Figure 14-15). Ionic detergents, such as sodium deoxycholate and sodium dodecylsulfate (SDS), contain a charged group; nonionic detergents, such as Triton X-100 and octylglucoside, lack a charged group.

The hydrophobic part of a detergent molecule is attracted to hydrocarbons and mingles with them readily; the hydrophilic part is strongly attracted to water. Thus, in a mixture of oil and water, the detergent forms a monomolecular film at the boundary between the two substances.

Agitation of oil-water mixtures breaks large volumes of oil into smaller droplets. In the absence of detergent, these droplets recoalesce into their original aggregate. In the presence of detergent, however, each droplet is surrounded by a single layer of detergent molecules with outward-pointing hydrophilic ends, which makes the droplet act as a hydrophilic object that can be suspended in water and rinsed away.

At very low concentrations in pure water, detergents are dissolved as isolated molecules. As the concentration increases, the molecules begin to form micelles. These are small, spherical aggregates in which hydrophilic parts of the molecules face outward and the hydrophobic parts cluster in the center (see Figure 2-17). The *critical micelle*

➤ FIGURE 14-15   Structures of five common detergents. The bile salt sodium deoxycholate is a natural product; the others are synthetic ones. The hydrophobic portion of each molecule is shown in red; the hydrophilic portion, in blue.

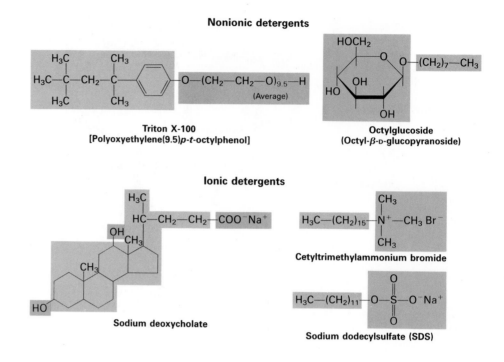

Nonionic detergents

Triton X-100
[Polyoxyethylene(9.5)*p-t*-octylphenol]

Octylglucoside
(Octyl-β-D-glucopyranoside)

Ionic detergents

Sodium deoxycholate

Cetyltrimethylammonium bromide

Sodium dodecylsulfate (SDS)

concentration (CMC) at which micelles form is characteristic of each detergent and is a function of the structures of its hydrophobic and hydrophilic parts.

Ionic detergents bind to the hydrophobic regions of membrane proteins as well as to the hydrophobic segments that form the core of water-soluble proteins. Because of their charge, these detergents also disrupt ionic and hydrogen bonds. At high concentrations, for example, sodium dodecylsulfate completely denatures proteins by binding to every side chain. Thus the mobility of SDS-solubilized proteins in gel electrophoresis is a good measure of their molecular weight (Figure 3-37).

Nonionic detergents act in different ways at different concentrations. At high concentrations (above the CMC), they solubilize biological membranes by forming mixed micelles of detergent, phospholipid, and integral membrane proteins (Figure 14-16). At low concentrations (below the CMC), these detergents may bind to the hydrophobic regions of most membrane proteins, making them soluble in aqueous solution. In this case, although mixed micelles are not formed, the solubilized protein will not aggregate during subsequent purification steps.

Most peripheral proteins are bound to specific integral membrane proteins by ionic or other weak interactions. Generally they can be removed from the membrane by solutions of high ionic strength (high salt concentrations), which disrupt ionic bonds, or by chemicals that bind divalent cations such as Mg^{2+}. Unlike integral proteins most peripheral proteins are soluble in aqueous solution and are not solubilized by nonionic detergents.

Having taken an overview of integral and peripheral membrane proteins, we turn now to a more detailed study of a few well-characterized integral proteins that are representative of many important membrane proteins.

Many Integral Proteins Contain Multiple Transmembrane α Helices

As noted previously, membrane-embedded α helices anchor glycophorin and other integral proteins to the phospholipid bilayer. It is now known that many transmembrane proteins contain multiple membrane-spanning α helices. Three examples of such proteins are discussed in this section.

Glycophorin   Although Figure 14-14 depicts glycophorin as a monomer with a single α helix spanning the bilayer, this protein is present in erythrocyte membranes as a dimer of two identical polypeptide chains. Even after denaturation with SDS, glycophorin migrates during polyacrylamide gel with the molecular weight of a dimer, rather than a monomer.

A simple experiment demonstrated that the two chains of glycophorin are held together primarily by specific interactions between their α-helical membrane-spanning domains. In this experiment, a peptide with the sequence of the transmembrane domain was synthesized chemically and labeled with a radioactive amino acid. The synthetic radiolabeled peptide was dissolved in a nonionic detergent and then added to a preparation of erythrocyte membranes; finally, the erythrocyte membrane proteins were

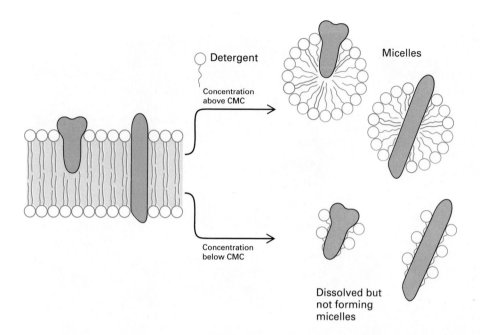

Solubilization of integral membrane proteins by nonionic detergents. At a concentration higher than its critical micelle concentration (CMC), a detergent solubilizes lipids and integral membrane proteins, forming mixed micelles containing detergent, protein, and lipid molecules. At concentrations below the CMC, many detergents (e.g., octylglucoside) can dissolve membrane proteins without forming micelles by coating the membrane-spanning regions. Since octylglucoside has a high CMC, it is particularly effective in solubilizing integral membrane proteins without denaturing them or forming mixed micelles.

solubilized, treated with SDS, and subjected to gel electrophoresis. The gel pattern of this preparation indicated that much of the glycophorin migrated with a molecular weight slightly larger than that of the monomer, and each glycophorin monomer had attached to it one molecule of the radiolabeled peptide. Thus the peptide had bound specifically to the membrane-spanning region of glycophorin within the erythrocyte membrane, disrupted the glycophorin dimers, and co-purified with glycophorin after the membrane was dissolved in detergents. Other synthetic peptides, of similar amino acid composition but with a different sequence, did not bind, indicating that the membrane-spanning segment of glycophorin could bind specifically only to another polypeptide segment of the same sequence. The two membrane-spanning α helixes of glycophorin are thought to form a coiled-coil structure (see Figure 3-10c) stabilized by specific interactions between the amino acid side chains at the interface of the two helices.

Bacterial Photosynthetic Reaction Center Successful x-ray crystallographic analysis of an integral membrane protein was not achieved until 1985, although by that year the molecular structures of more than 200 water-soluble proteins had already been determined. This initial structural determination to atomic resolution showed that the membrane-embedded *photosynthetic reaction center* (PRC) of the bacterium *Rhosopseudomonas viridis* comprises four polypeptides and several prosthetic groups, including four chlorophyll molecules. In this 1187-aa protein, three of the four polypeptides, designated L, M, and H, span the membrane (Figure 14-17a). The L and M subunits compose the bulk of the structure within the membrane: each forms a crescent consisting of a wall of five

transmembrane α helices. The H subunit, which is primarily globular, is anchored to the cytosolic surface of the membrane by a single transmembrane α helix. The fourth subunit, the *cytochrome,* is a peripheral protein bound to the exoplasmic segments of the other three polypeptides.

All the amino acid residues in the 11 transmembrane α helices that face the exterior of the molecule are hydrophobic (or nonpolar) and anchor the PRC to the fatty acyl groups of the membrane. Many side chains in the α helices that face the interior of the molecule are also hydrophobic and bind adjacent helices by van der Waals forces. However, some residues in these transmembrane α helices are polar. Several histidine and glutamate residues, for instance, bind iron atoms (Fe^{2+} or Fe^{3+}) present in the center of the transmembrane region (Figure 14-17b). These charged residues face the interior of the molecule.

Comparison of photosynthetic reaction centers from different bacterial species has shown that the overall structure of the PRC is highly conserved, although the identities of many of the amino acids are not. In particular, the side chains that protrude outward and anchor the PRC in the hydrocarbon core of membrane are always hydrophobic, but their identity is not conserved; it seems that any hydrophobic side chain can anchor the PRC in the lipid bilayer. In contrast, side chains that face inward and anchor helices together are highly conserved; here, van der Waals interactions demand specific side-chain interactions.

Recall that in water-soluble proteins, hydrophilic side chains are preponderant on the surface and hydrophobic side chains tend to be confined to the interior of the molecule. In the transmembrane region of multispanning integral proteins, such as the photosynthetic reaction center, the situation is reversed: hydrophobic side chains lie on the

(a)

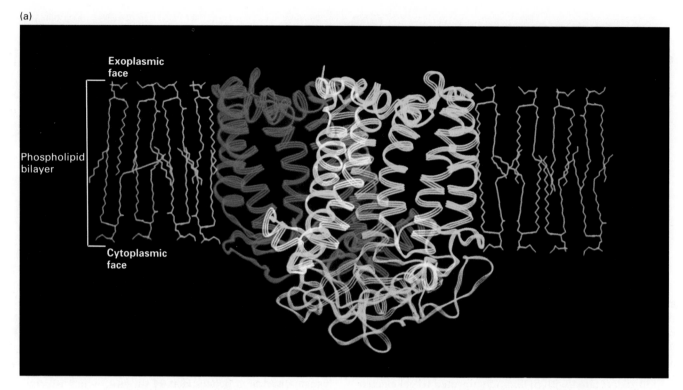

(b)

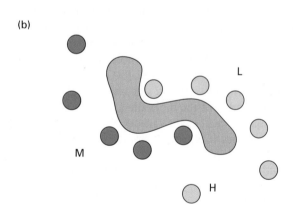

▲ FIGURE 14-17   Structure of the bacterial photosynthetic reaction center (PRC) as determined by x-ray crystallography. (a) Model showing the polypeptide backbone of the three membrane-spanning subunits of PRC: the L subunit (yellow) and the M subunit (purple) each form five transmembrane α helices and have a very similar structure overall; the H subunit (light blue) is anchored to the membrane only by a single transmembrane α helix. The fourth subunit (not shown) is a peripheral protein that binds to the exoplasmic sequences of the other subunits. (b) Schematic diagram of a cross-sectional view through the middle of the membrane showing the locations of the 11 transmembrane α helices in the PRC. The center of the protein (tan) is occupied by prosthetic groups such as free (nonheme) iron and the chlorophylls. [Part (a) courtesy of D. Rees; part (b) after R. Henderson, 1985, *Nature* **318**:598.]

outside of the protein molecule, exposed to the hydrocarbon interior of the membrane; hydrophilic side chains are found in the interior of the molecule. These membrane-embedded proteins are "inside-out" compared to water-soluble proteins.

Bacteriorhodopsin and Other Seven-Spanning Membrane Proteins   More than 150 integral proteins with seven membrane-spanning α helices have been identified. This class of "seven-spanning" membrane proteins is typified by *bacteriorhodopsin,* a protein found in a photosynthetic bacterium (Figure 14-18). Absorption of light by the retinal group attached to bacteriorhodopsin causes a conformational change in the protein that results in pump-

ing of protons from the cytosol across the bacterial membrane to the extracellular space. The proton concentration gradient thus generated across the membrane is used to synthesize ATP, as discussed in Chapter 18. Both the overall arrangement of the seven α helices in bacteriorhodopsin and the identity of most of the amino acids can be resolved by computer analysis of micrographs of two-dimensional crystals of the membrane-embedded protein taken at various angles to the electron beam.

Other seven-spanning membrane proteins include the *opsins* (eye proteins that absorb light), cell-surface receptors for many hormones, and receptors for odorous molecules. Amino acid sequence analysis of these proteins has shown that no amino acids are found in the same position

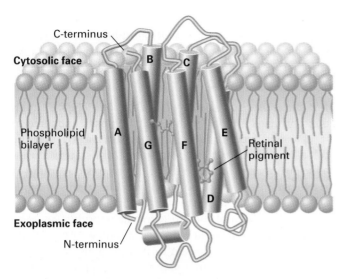

▲ FIGURE 14-18 Overall structure of bacteriorhodopsin as deduced from electron diffraction analyses of two-dimensional crystals of the protein in the bacterial membrane. The seven membrane-spanning α helices are labeled A–G. The retinal pigment is covalently attached to lysine 216 in helix G. The approximate position of the protein in the phospholipid bilayer is indicated. [Adapted from R. Henderson et al., 1990, *J. Mol. Biol.* **213**:899.]

in all of them, and only a few residues are conserved in even a substantial number of them. Nonetheless, each of these proteins contains seven stretches of amino acids long enough (>22 amino acids) and hydrophobic enough to span the phospholipid bilayer. Though direct evidence is lacking, it is thought that all of these proteins adopt a conformation in the membrane similar to that of bacteriorhodopsin. This is one of several examples of how investigators can predict the orientation of proteins in a membrane from the amino acid sequence alone.

Porins Are Transmembrane Proteins Composed of Multiple β Strands

The porins are a class of transmembrane proteins whose structure differs radically from that of other integral proteins. Several types of porin are found in the outer membrane of gram-negative bacteria such as *E. coli* (see Figure 5-1). The outer membrane protects an intestinal bacterium from harmful agents (e.g., antibiotics, bile salts, and proteases) but permits the uptake and disposal of small hydrophilic molecules including nutrients and waste products. The porins in the outer membrane of an *E. coli* cell provide channels for passage of disaccharides, phosphate, and similar molecules.

The amino acid sequences of porins are predominantly polar and contain no long hydrophobic segments typical of the integral proteins with α-helical membrane-spanning domains described previously. X-ray crystallography has revealed that porins are trimers of identical subunits, each containing 16 β strands that are arranged to form a barrel-shaped structure with a pore in the center (Figure 14-19). As noted in Chapter 3, half of the amino acid side chains of a β strand point in one direction, and the other half point in the opposite direction (Figure 3-12). In a porin monomer, the outward-facing side chains on each of the 16 β strands are hydrophobic and thus can interact with the fatty acyl groups of the membrane lipids or with other porin monomers. The side chains facing the inside of a porin monomer are predominantly hydrophilic; these side chains line the pore that allows small water-soluble molecules to cross the membrane.

All the transmembrane proteins examined to date contain either α-helical membrane-spanning domains or multiple β strands; perhaps other types of structures are yet to be uncovered among this important type of protein.

(a)

(b)

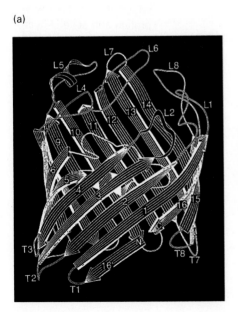
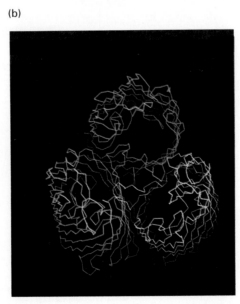

◄ FIGURE 14-19 Two views of the three-dimensional structure of the OmpF porin, a trimeric protein found in the *E. coli* outer membrane. (a) Side view of the monomer, with its 16 antiparallel β strands (numbered), which form a barrel. A pore is in the center of each monomer. Loops L1–L7 connect the β strands and lie at the outer surface of the outer membrane; loops T1–T8 face the periplasmic space on the "inside" of the bacterial outer membrane. (b) Top view of the trimer showing the transmembrane pore in the center of each subunit. [From S. W. Cowan et al., 1992, *Nature* **358**:727. Courtesy of J. Rosenbusch.]

Some Integral Proteins Are Bound to the Membrane by Covalently Attached Hydrocarbon Chains

In eukaryotic cells, as noted earlier, some integral membrane proteins do not span the entire bilayer but are anchored in only one leaflet. These proteins contain covalently attached hydrocarbon chains that bind the protein molecule to the plasma membrane. The lipid-bound proteins are grouped into three classes defined by the type of attached hydrocarbon chain (Figure 14-20).

Glycosylphosphatidylinositol-anchored Proteins Some cell-surface proteins are anchored to the exoplasmic face of the plasma membrane by a complex glycosylated phospholipid. This anchor contains two fatty acyl groups and several sugars, usually including *N*-acetylglucosamine, mannose, and inositol. Several enzymes, including alkaline phosphatase, fall into this class. Various experiments have shown that the phospholipid anchor is both necessary and sufficient for binding these cell-surface proteins to the membrane. For instance, the enzyme phospholipase C cleaves the phosphate-glycerol bond in phospholipids as

well as in glycosylphosphatidylinositol anchors (Figure 14-21). Treatment of cells with phospholipase C releases glycosylphosphatidylinositol-anchored proteins such as the Thy-1 protein and alkaline phosphatase from the cell surface.

Myristate-anchored Proteins A second class of lipid-bound proteins is found within the cytoplasm, anchored to the cytosolic face of the plasma membrane by myristic acid, a 14-carbon saturated fatty acid. (Myristate is always bound by an amide linkage to the glycine residue found at the N-terminus of such proteins.) Included in this group is p60$^{v\text{-}src}$ (or v-Src), a mutant form of the normal cellular protein c-Src. Both of these proteins have protein-tyrosine kinase activity; that is, they catalyze addition of phosphate groups to tyrosine residues in certain proteins. When synthesized in certain cultured mammalian cells, v-Src causes the cells to become transformed, or cancerous; it is therefore called a *transforming protein*. It is not known what protein or proteins must be phosphorylated by v-Src to cause a cell to become cancerous, but it is known that v-Src must be bound to the cytosolic face of the plasma membrane to transform cells. This has been shown experimen-

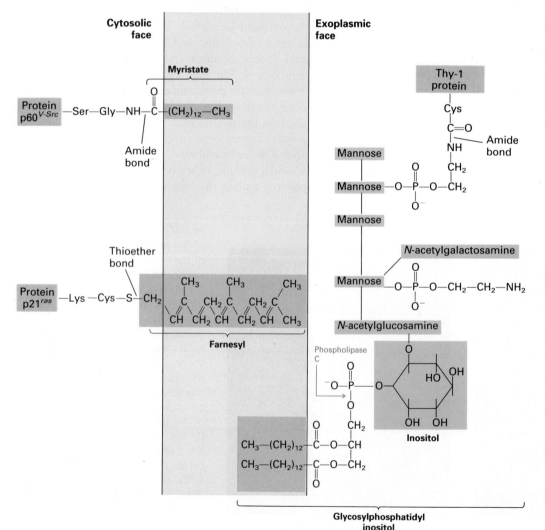

◄ **FIGURE 14-20** Three types of lipid anchors by which some integral proteins (orange) are attached to the plasma membrane. Myristate- and farnesyl-anchored proteins, which include various proteins involved in regulation of cell growth, are found only on the cytosolic face. Glycosylphosphatidylinositol-anchored proteins, which include the Thy-1 protein and several hydrolytic enzymes, are found only on the exoplasmic face. The membrane-embedded hydrocarbon chains are highlighted in red.

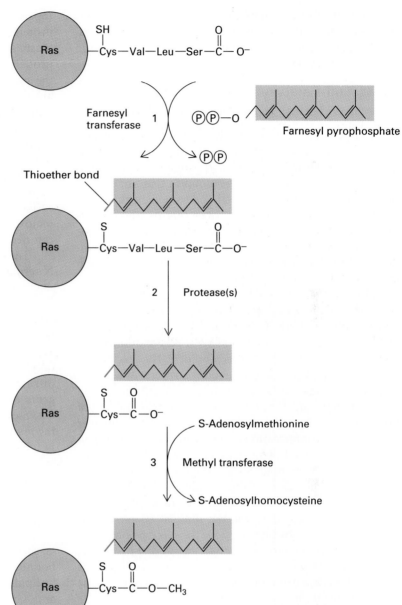

▲ FIGURE 14-21 Specificity of cleavage of phospholipids by phospholipases A_1, A_2, C, and D. Susceptible bonds are shown in color. R denotes the polar group attached to the phosphate, such as choline in phosphatidylcholine (see Figure 14-2) or inositol in glycosylphosphatidylinositol anchors (see Figure 14-20).

▶ FIGURE 14-22 Enzymatic steps by which Ras protein becomes covalently linked to a 15-carbon unsaturated hydrocarbon. Farnesyl pyrophosphate embedded in the membrane reacts with the –SH group of a cysteine residue four amino acids from the C-terminus of the protein; this reaction forms a thioether bond (red) and releases pyrophosphate. Then the three C-terminal residues are cleaved, and a methyl group is added to the carboxyl group of the cysteine residue. [After J. B. Gibbs, 1991, *Cell* **65**:1.]

tally with a mutant form of v-Src in which alanine is substituted for the normal N-terminal glycine. The mutant protein has normal kinase activity but is not linked to myristate, does not bind to the plasma membrane, and cannot transform cultured cells.

The myristate anchor is necessary, but not sufficient, for binding of c-Src to cell membranes. This protein also binds to the cytosolic domains of many cell-surface receptors, including receptors for growth hormones. Binding of the appropriate hormone to the exoplasmic face of the receptor protein stimulates c-Src activity. Thus these hormone receptors are also "c-Src receptors." Likewise, Src can be classified as a peripheral protein, because it binds to these cell-surface receptors, and as an integral protein, because of its myristate anchor.

Farnesyl-anchored Proteins The third class of lipid-

bound proteins is anchored to the cytosolic face of the plasma membrane by a 15-carbon farnesyl moiety or by a 20-carbon geranylgeranyl moiety, both of which are polyunsaturated hydrocarbons. The lipid is covalently linked by a thioether bond to a cysteine residue that is always located near the C-terminus. An example is p21^{ras} (or Ras); this 21-kDa protein encoded by the *ras* gene must be bound to the plasma membrane to function as a regulator of cell growth and division. As we detail in Chapter 26, certain mutated Ras proteins can transform cells (like v-Src) and are a major cause of human cancer.

Figure 14-22 illustrates the series of reactions by which the C-terminus of a Ras protein is modified by addition of a farnesyl moiety. All of these reactions are essential for proper functioning of the protein, as can be shown by adding an inhibitor of the first reaction to a culture of transformed (tumor) cells containing a mutant Ras protein. In the presence of such an inhibitor, the mutant Ras protein is nonfunctional, so the cell reverts to a normal, nontransformed phenotype. Such inhibitors are now being used in clinical trials against several human cancers.

Interfacial Catalysis Involves Soluble Enzymes Acting at Membrane Surfaces

Thus far we have considered the interactions of integral membrane proteins with the hydrophobic core of phospholipid bilayers. However, many water-soluble enzymes bind to the fatty acyl groups of phospholipids that are embedded in bilayers. One class of well-understood examples are the *phospholipases,* soluble enzymes that hydrolyze various bonds in the head groups of phospholipids and are important for degradation of damaged or aged cell membranes. The specificity of various phospholipases is shown in Figure 14-21.

How can water-soluble enzymes act at the interface between an aqueous solution and a lipid surface, given the tight packing of both the fatty acyl chains and the polar head groups in a phospholipid bilayer? How can such proteins reversibly interact with membranes? The mechanism of phospholipase A_2 provides some answers. When this enzyme is in aqueous solution, it has a conformation in which the active site is buried inside the protein. Binding of phospholipase A_2 to a phospholipid micelle or bilayer causes a small conformational change that fixes the protein to the phospholipid head groups and opens a hydrophobic channel next to the lipid surface (Figure 14-23). A phospholipid molecule can then slide toward the interface, with its fatty acyl chains interacting with the hydrophobic residues of the enzyme that line the internal channel. This movement of the substrate positions the specific bond to be cleaved close enough to the catalytic site so that hydrolysis can proceed. As this example illustrates, enzymes need not be embedded in a membrane in order to catalyze reactions involving membrane components.

The Orientation of Proteins in Membranes Can Be Experimentally Determined

Bacteriorhodopsin and the photosynthetic reaction center are the only integral proteins whose complete three-dimensional structures are known. As noted above, frequently it is sufficient to know which segments of a protein span the bilayer and which are localized to the cytosolic and exoplasmic faces. Sometimes, as in the case of seven-spanning proteins related to bacteriorhodopsin, this can be deduced from the amino acid sequence alone. Usually, though, a protein's orientation in the membrane must be determined directly.

One procedure for doing so is to label cells with a covalent reagent that cannot penetrate the membrane and reacts only with exposed regions of the integral protein on the exoplasmic face. For example, in the presence of hydrogen peroxide and radioactive iodine (^{125}I), the enzyme lac-

toperoxidase iodinates tyrosine and histidine side chains of proteins. Because lactoperoxidase cannot penetrate the plasma membrane, it reacts only with the exposed parts of proteins on the cell surface. To determine which parts of the protein contain the radioactive label, the membranes are dissolved in a detergent. The protein is purified (e.g., with an antibody-affinity column; see Figure 3-34) and then digested with a protease such as trypsin. The resulting peptides are separated and examined for their content of radioactive iodine.

Another procedure involves digestion of intact cells with proteases. In this case, only those regions of cell-surface proteins exposed to the extracellular medium are removed. By identifying the cleaved segments, researchers can deduce how a membrane protein is oriented.

These types of experiments have shown that the N-terminus of glycophorin and all the attached sugar residues are on the exoplasmic surface (see Figure 14-14). Only this exoplasmic segment is digested by proteases added to the cell exterior and is reactive with extracellular lactoperoxidase. The C-terminal segment (residues 96–131) faces the cytoplasm and reacts with such reagents only if the permeability barrier of the plasma membrane is disrupted. For example, addition of low amounts of detergents causes holes to form in the membrane and allows enzymes such as lactoperoxidase to enter the cytosol from the extracellular medium. The hydrophobic transmembrane segment does not react with any aqueous enzymes or agents. Base on these findings, experimenters concluded that glycophorin is a transmembrane protein.

▶ *Glycoproteins and Glycolipids*

Carbohydrates are found in many membranes, covalently bound either to proteins as constituents of *glycoproteins* or to lipids as constituents of *glycolipids*. Both glycoproteins and glycolipids are especially abundant in the plasma membrane of eukaryotic cells but are absent from the inner mitochondrial membrane, the chloroplast lamellae, and several other intracellular membranes. Bound carbohydrates increase the hydrophilic character of lipids and proteins and help to stabilize the conformations of many membrane proteins. Many secreted proteins also are glycoproteins.

In many glycoproteins and glycolipids, the carbohydrate components are oligosaccharides. Commonly present in these sugar chains are galactose, mannose, and the four sugars shown in Figure 14-24. Both *N*-acetylglucosamine and *N*-acetylgalactosamine contain an acetamide group

$$-NH-\overset{\overset{\displaystyle O}{\displaystyle \|}}{C}-CH_3$$

in place of the usual hydroxyl group on carbon 2. The substituent attached to carbon 5 of fucose is a methyl

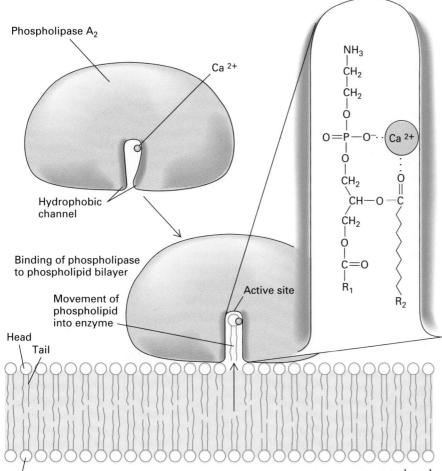

Phospholipase A₂

Ca²⁺

Hydrophobic channel

Binding of phospholipase to phospholipid bilayer

Movement of phospholipid into enzyme

Active site

Head

Tail

Phospholipid

> ◀ FIGURE 14-23 Mechanism of action of phospholipase A₂, an example of interfacial catalysis. When the soluble enzyme binds to a phospholipid bilayer, it undergoes a small conformational change, opening a channel lined with hydrophobic amino acids (red) that leads to the catalytic site. As a phospholipid molecule moves from the bilayer into the channel, an enzyme-bound Ca²⁺ (blue) binds to and localizes the phosphate in the head group and also binds to the oxygen of the carboxylate (see inset). This movement positions the ester bond (red) to be cleaved next to the catalytic site. [After D. L. Scott et al., 1990, *Science* **250**:1541; D. Blow, 1991, *Nature*, **351**:444.]

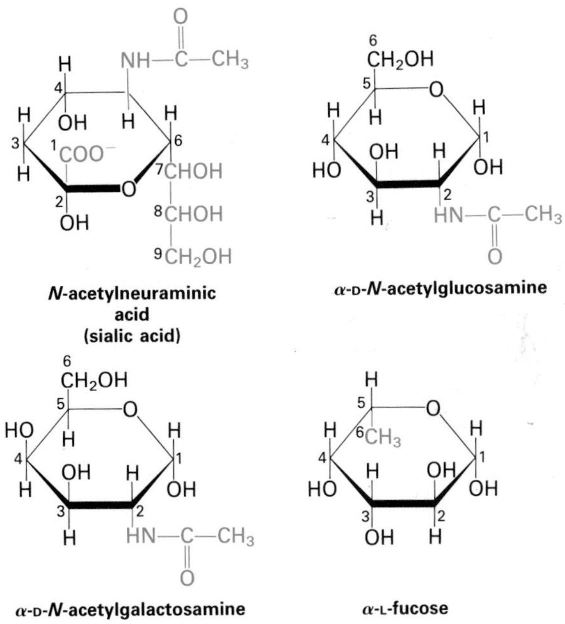

N-acetylneuraminic acid (sialic acid)

α-D-N-acetylglucosamine

α-D-N-acetylgalactosamine

α-L-fucose

▲ FIGURE 14-24 The structures of four sugars commonly found in glycoproteins and glycolipids. The substituents in red replace either –OH or –CH₂OH groups in the corresponding unsubstituted sugars.

group rather than the CH₂OH group present in most other sugars. N-acetylneuraminic acid, also called sialic acid, contains an acetamide group, as well as a three-carbon moiety and a carboxyl group, which gives the molecule a negative charge. Because of the presence of sialic acid residues, many glycoproteins and glycolipids have a net negative surface charge.

Many Integral Proteins Contain Sugars Covalently Linked to Their Exoplasmic Domains

The carbohydrate components of integral membrane proteins are almost always linked to the exoplasmic domains of these proteins. Such carbohydrates generally play no role in the catalytic function of membrane-associated enzymes; in fact, several modified glycoproteins that lack their carbohydrates have been shown to function normally. However, because of their multiple hydroxyl groups, bound carbohydrates increase the solubility of glycoproteins in water and are important for proper folding of the exoplasmic domains of many membrane glycoproteins (Chapter 16).

Sugar residues in glycoproteins are commonly linked to two different classes of amino acid residues. Sugars are classified as O-linked if they are bonded to the hydroxyl oxygen of serine, threonine, or (in collagen) hydroxylysine; they are classified as *N*-linked if they are bonded to the amide nitrogen of asparagine. The structures of *O*- and *N*-linked oligosaccharides are very different and usually do not contain the same sugars. *O*-linked oligosaccharides are generally shorter and more variable than *N*-linked ones and often contain only a few sugar residues. Figure 14-25 shows the structures of two common *O*-linked sugars.

N-linked oligosaccharides, which often have complex structures, are found in many membrane and secreted proteins (Figure 14-26). These structures contain *N*-acetylglucosamine, mannose, fucose, galactose, and *N*-acetylneuraminic acid. The proteins differ primarily in the number of branches (two, three, or four), in the number of *N*-acetylneuraminic acid residues (zero, one, two, or three), and in the chemical nature of the linkage of *N*-acetylneuraminic acid to galactose. Even oligosaccharides found at the same site in a single protein species are often heterogeneous.

Many Glycolipids Are Located in the Cell-Surface Membrane

The simplest glycolipid, *glucosylcerebroside* (Figure 14-27), contains a single glucose unit attached to a ceramide. One or more of the sugar residues in many glycolipids is *N*-acetylneuraminic acid; glycolipids containing this sugar are called *gangliosides* and are abundant in the plasma membrane of nerve cells and other mammalian cells. Glycolipids are always found in the exoplasmic leaflet of membranes and are situated mainly, but not exclusively, on the surface membrane of cells, with their polar carbohydrate chains facing outward toward the environment and away from the cell.

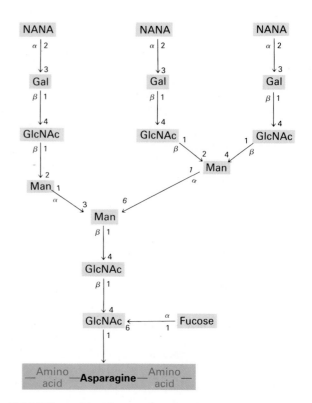

▲ FIGURE 14-26   The structure of a typical *N*-linked oligosaccharide, which is linked to the amide nitrogen of asparagine. These often complex oligosaccharides are attached to many membrane proteins and to secreted proteins (e.g., antibodies). NANA = *N*-acetylneuraminic acid; Gal = galactose; GlcNAc = *N*-acetylglucosamine; Man = mannose.

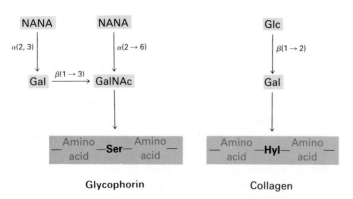

▲ FIGURE 14-25   The structures of two common *O*-linked oligosaccharides. Glycophorin (Figure 14-14) is an erythrocyte membrane glycoprotein; collagen, a major constituent of connective tissue, is synthesized and secreted by fibroblasts as well as certain epithelial cells. NANA = *N*-acetylneuraminic acid (sialic acid); GalNAc = *N*-acetylgalactosamine; Glc = glucose; Gal = galactose; Ser = serine; Hyl = hydroxylysine.

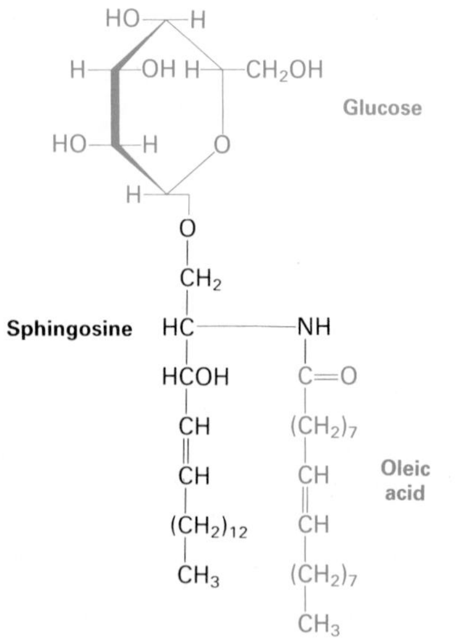

▲ FIGURE 14-27   The structure of glucosylcerebroside, one of the simplest glycolipids, consists of sphingosine, an oleic acid side chain, and a single glucose residue. This glycolipid is abundant in the myelin sheath.

The *blood-group antigens,* which can trigger harmful immune reactions, are glycoproteins or glycolipids present on the surface of erythrocytes and many other types of cells. The carbohydrate portion of these molecules, which constitute the antigenic determinant, are genetically determined in humans and other animals. The carbohydrates of the human blood groups (A, B, and O) have been studied in great detail. The A, B, and O antigens are structurally related oligosaccharides, which may be linked either to lipids or to proteins (Figure 14-28). The O antigen is a chain of fucose, galactose, N-acetylglucosamine, and glucose, with a ceramide lipid (or a hydroxyl protein side chain) linked to the glucose. The A antigen is similar to O, except that the A antigen contains an N-acetylgalactosamine attached to the outer galactose residue; the B antigen is also similar to O, except for an extra galactose residue attached to the outer galactose. All people have the enzymes that synthesize the O antigen. People with type A blood also have the enzyme that adds the extra N-acetylgalactosamine; those with type B blood have the enzyme that adds the extra galactose. People with type AB blood synthesize A and B antigens; those with type O make only the O antigen.

As shown in Table 14-3, people who cannot synthesize the A and/or B antigen normally have antibodies against the antigen(s) in their serum. Thus, when type B or AB blood is transfused into a person with blood type A or O, the anti-B antibodies of the recipient bind to the transfused erythrocytes and trigger their destruction by phagocytic cells. Similarly, type A or AB blood cannot be safely transfused into people with blood type B or O. People with blood type AB, who lack both anti-A and anti-B antibodies, can receive blood of any ABO type. People with type O blood can only be transfused with type O blood, since they have both anti-A and anti-B antibodies; however, type O people are universal donors, because their blood lacks both

TABLE 14-3   ABO Blood Groups

Blood-group Type	Antigens on RBCs	Serum Antibodies	Can Receive Blood Types
A	A	Anti-B	A and O
B	B	Anti-A	B and O
AB	A and B	None	All
O	None	Anti-A and anti-B	O

the A and B antigen and can thus be transfused into individuals of any ABO type. The ability of antibodies to distinguish subtle differences in macromolecular structure underscores the role that cell-surface oligosaccharides play in establishing the uniqueness of cells and organisms.

Principles of Membrane Organization

Thus far we have discussed primarily the structures of pure phospholipid vesicles and the properties of several glycolipids and integral membrane proteins. We now turn to more complex questions about the architecture of lipids and proteins in cellular membranes, particularly the plasma membrane. How are these lipids and proteins organized? How mobile are membrane proteins? First, we consider their asymmetry, which is essential to the function of all biological membranes.

All Integral Proteins Bind Asymmetrically to the Lipid Bilayer

Each type of integral membrane protein has a single, specific orientation with respect to the cytosolic and exoplasmic faces of a cellular membrane. All molecules of any particular integral membrane protein, such as glycophorin or the photosynthetic reaction center, lie in the same direction. This absolute asymmetry in protein orientation gives the two membrane faces their different characteristics. In contrast to phospholipids, proteins have never been observed to flip-flop across a membrane. Such movement would be energetically unfavorable: it would require a transient movement of hydrophilic amino acid and sugar

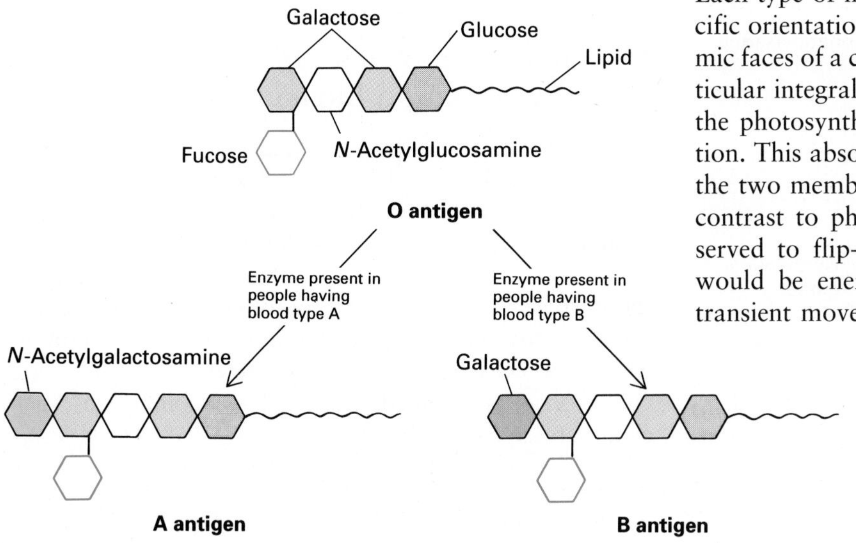

FIGURE 14-28   The structures of the human blood-group antigens, and some key enzymes that catalyze their synthesis. [See T. Feizi, 1990, *Trends Biochem Sci.* **15**:330 for a review.]

residues through the hydrophobic interior of the membrane. Accordingly, the asymmetry of membrane proteins is established during their biosynthesis (detailed in Chapters 16 and 19) and maintained throughout the protein's lifetime.

Membrane asymmetry is most obvious in the case of membrane glycoproteins and glycolipids. In the plasma membrane, all the O- and N-linked oligosaccharides in glycoproteins and all of the oligosaccharides in glycolipids are on the exoplasmic surface, exposed to the exterior of the cell. In the endoplasmic reticulum, they are found on the interior, or *lumenal,* membrane surface (which is analogous to the exoplasmic surface of the plasma membrane). Although this is the most common orientation of membrane glycoproteins, an exception is a small but important class of glycoproteins located in the nuclear matrix and nuclear membrane. In these proteins, which include transcription factors and components of the nuclear pore complex, single O-linked N-acetylglucosamine residues are attached to serine residues on the cytosolic face of the protein.

The Two Membrane Leaflets Have Different Lipid Compositions

The lipid compositions of the two membrane leaflets are quite different in all the plasma membranes analyzed thus far. In the human erythrocyte and in a line of canine kidney cells grown in culture, all of the glycolipids, and almost all of the sphingomyelin and phosphatidylcholine with its positively charged head group, are found in the exoplasmic

leaflet. In contrast, lipids with neutral or negative polar head groups, such as phosphatidylethanolamine and phosphatidylserine, are preferentially located in the cytosolic leaflet (Figure 14-29).

In contrast to glycolipids, which are always found exclusively in the exoplasmic leaflet, most kinds of phospholipid, as well as cholesterol, are generally present in both membrane leaflets, although they are often more abundant in one or the other. The relative abundance of a particular phospholipid in the two leaflets of a plasma membrane can be determined based on its susceptibility to hydrolysis by phospholipases added to the cell exterior. Phospholipids in the cytosolic leaflet are resistant to such hydrolysis because the phospholipases cannot penetrate to the cytosolic face of the plasma membrane. It is not clear how these differences in lipid composition arise in the two leaflets; certain lipids may bind to specific protein domains that occur preferentially in one membrane leaflet.

Freeze-fracture and Deep-etching Techniques Reveal the Two Membrane Faces in Electron Microscopy

The two faces of any cellular membrane can be separated and prepared for electron microscopy by the freeze-fracture and deep-etching techniques (see Figure 5-34). When a frozen specimen is fractured by a sharp blow, the fracture line frequently runs through the hydrophobic interior of the membrane, separating the two phospholipid leaflets. The fractured surfaces of many membranes reveal numerous protuberances, most of which are membrane proteins (Figures 14-30 and 14-31). In deep-etching studies, the cytoplasmic face of a membrane is customarily called the P (protoplasmic) face and the exoplasmic face is the E face. It is not unusual for most or all protuberances to be on one of the two surfaces and their mirror images, in the form of pits or holes, to be on the other. This may occur because the integral proteins are bound more tightly to the lipids in one leaflet than to those in the other.

Most Integral Proteins and Lipids Are Laterally Mobile in Biomembranes

As discussed earlier, phospholipids are laterally mobile within one leaflet of a planar phospholipid bilayer. Various experiments have shown that many integral membrane proteins also float quite freely within the plane of the membrane. In one such study, two different cells are fused for the purpose of observing the movement of their distinct surface proteins. At various times after incubation at 37 °C, the cells (say, mouse and human fibroblasts) are chilled on ice and immediately fixed with compounds, such as glutaraldehyde

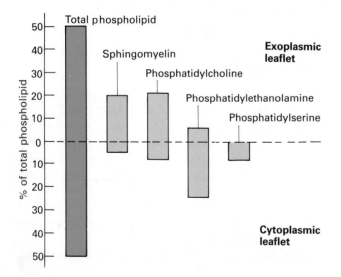

▲ FIGURE 14-29   Distribution of phospholipids in the two leaflets of the erythrocyte plasma membrane. Values are expressed as a percentage of total membrane lipids. Although 50 percent of the total phospholipid is found in each face, individual phospholipids are distributed asymmetrically. [See J. E. Rothman and J. Lenard, 1977, *Science* **195**:743.]

$$\underset{H}{\overset{O}{\underset{\|}{C}}} - (CH_2)_3 - \underset{H}{\overset{O}{\underset{\|}{C}}}$$

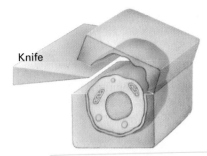

Knife

Fracture splits the plasma membrane

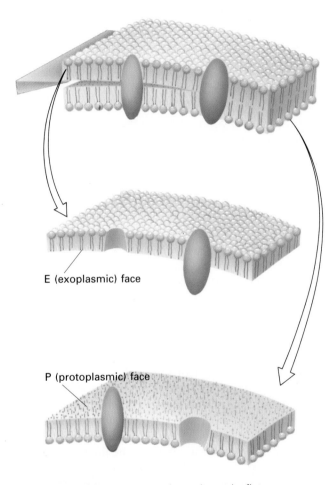

E (exoplasmic) face

P (protoplasmic) face

Detail of the two exposed membrane leaflets

◄ FIGURE 14-30   The freeze-fracture technique can separate the two phospholipid leaflets that form every cellular membrane. Membrane proteins and particles remain bound to one leaflet or the other.

Such experiments suggest that many integral proteins are free to diffuse in a sea of lipid in the two-dimensional space of the membrane. According to this concept, currently popularized as the *fluid mosaic model*, the membrane is viewed as a two-dimensional mosaic of laterally mobil phospholipid and protein molecules (see Figure 14-1). Integral membrane proteins consisting of two or more noncovalently linked polypeptide chains (e.g., dimeric glycophorin) float as a unit in the lipid.

Lateral protein diffusion occurs in intracellular membranes as well. Freeze fracturing reveals that the inner mitochondrial membrane contains stable aggregates of in-

(a)

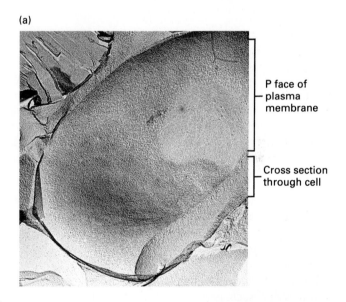

P face of plasma membrane

Cross section through cell

(b)   Intramembrane particles (band 3)

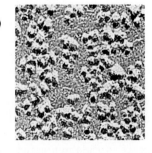

▲ FIGURE 14-31   Freeze-fracture, deep-etching image of an erythrocyte plasma membrane. (a) The P face of the plasma membrane and a cross section through the cell. The E face, or outer leaflet, has been fractured off, leaving just the P face. The particles in the cross section are hemoglobin. (b) The intramembrane particles at higher magnification. These particles are composed mainly of band 3, the major intramembrane protein. [Photographs courtesy of D. Branton.]

that cross-link lysine side chains and prevent further protein movement. The plasma membrane protein H-2 (found on most mouse cells) or the related HLA protein (found on most human cells) can be detected using specific H-2 or HLA antibodies coupled to fluorescent dyes. Immediately after fusion, the mouse and human antigens are grouped in separate areas, but they quickly diffuse throughout the cell. Soon both halves of the fused cells are equally fluorescent, demonstrating that most of the surface H-2 and HLA proteins were not rigidly held in place on the original mouse and human cells (Figure 14-32).

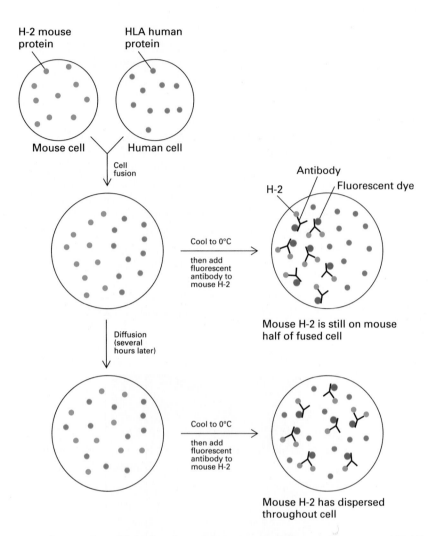

H-2 mouse protein

HLA human protein

Mouse cell

Human cell

Cell fusion

Antibody

H-2

Fluorescent dye

Cool to 0°C

then add fluorescent antibody to mouse H-2

Mouse H-2 is still on mouse half of fused cell

Diffusion (several hours later)

Cool to 0°C

then add fluorescent antibody to mouse H-2

Mouse H-2 has dispersed throughout cell

◄ FIGURE 14-32   Experimental demonstration that cell-surface proteins are laterally mobile. Immediately after human and mouse cells are fused (see Figure 6-10), surface antigens of the two cell types remain localized in their respective halves of the fused cell; they can be detected by fluorescent antibodies (in this case specific for mouse H-2 protein). After several hours of incubation, the mouse and human surface proteins are evenly distributed throughout the membrane of the fused cell. [See L. D. Frye and M. Edidin, 1970, *J. Cell Sci.* **7**:319.]

tegral proteins. If isolated vesicles prepared from this membrane are placed in a strong electric field, all the intramembrane particles, which bear a net negative electric charge, move to one end of each vesicle (Figure 14-33). When the field is turned off, the particles rapidly return to their original random distribution. This experiment demonstrates that such particles, which are complexes of as many as 15 membrane proteins, are laterally mobile.

A more quantitative estimate of the rate and extent of lateral movements of surface proteins and lipids can be obtained from *fluorescence recovery after photobleaching* (FRAP) studies where cells are treated with a fluorescent reagent that binds to a specific surface protein or lipid, which is uniformly distributed on the surface. A laser light is then focused on a small area of the surface, irreversibly bleaching the bound reagent and thus reducing the fluorescence in the illuminated area. In time, the fluorescence of the bleached area increases because unbleached fluorescent surface molecules diffuse into it and bleached ones diffuse outward (Figure 14-34). The extent of recovery of fluorescent molecules in the patch is proportional to the fraction of labeled molecules that are mobile in the membrane, and the rate of recovery of fluorescence can be used to calculate the *diffusion coefficient* (the rate at which the molecules diffuse).

FRAP studies with phospholipid-labeled fluorescent dyes have shown that in fibroblast plasma membranes, all the phospholipids are freely mobile over distances of about 0.5 μm; however, most of the lipids cannot diffuse over much longer distances. These findings suggest that protein-rich regions of the plasma membrane, about 1 μm in diameter, separate lipid-rich regions containing the bulk of the membrane phospholipid. Phospholipids are free to diffuse within such a region but not from one lipid-rich region to an adjacent one. Furthermore, the rate of lateral diffusion of lipids in the plasma membrane is nearly an order of magnitude slower than in pure phospholipid bilayers: diffusion constants of 10^{-8} cm^2/s and 10^{-7} cm^2/s are characteristic of the plasma membrane and a lipid bilayer, respectively. This difference suggests that lipids may be tightly but not irreversibly bound to certain integral proteins in some membranes.

Some Membrane Proteins Interact with Cytoskeletal Components

Numerous experiments similar to those discussed in the previous section have shown that depending on the cell type, 30–90 percent of all integral proteins in the plasma

(a)

(b)

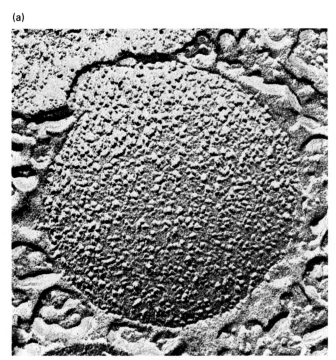

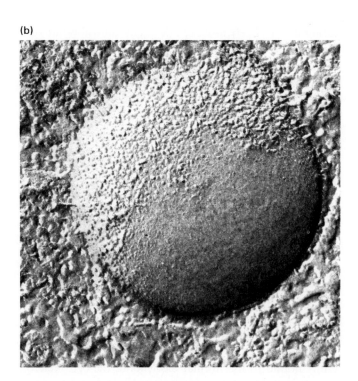

▲ FIGURE 14-33   Electron micrographs revealing mobility of protein particles in freeze-fractured vesicles prepared from the inner mitochondrial membrane. (a) Integral protein particles are visible as protuberances that are randomly distributed in the surface of a free-fractured vesicle. (b) After the vesicle was subjected to a strong electric field and then rapidly frozen, all the particles clustered at one end, showing that the particles can move laterally within the membrane plane in response to a voltage gradient. Their rate of movement is similar to that of many other integral membrane proteins. [Part (b) from A. E. Sowers and C. R. Hackenbrook, 1981, *Proc. Nat'l. Acad. Sci. USA.* **78**:6246. Part (a) courtesy of A. E. Sowers; part (b) courtesy of A. E. Sowers and C. R. Hackenbrook.]

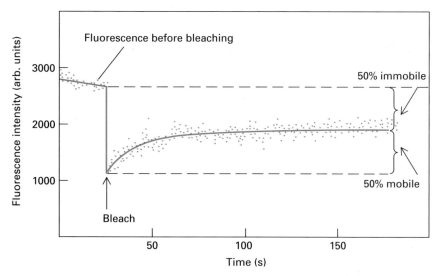

▲ FIGURE 14-34   Fluorescence recovery after photobleaching (FRAP) experiment with human hepatoma cells growing attached to a glass microscope slide. The cells were treated with a fluorescent monovalent fragment (containing a single binding site) of an antibody specific for the asialoglycoprotein receptor expressed on the surface of hepatoma cells. A small patch of the cell surface then was briefly exposed to laser light, which irreversibly bleached the fluorescence. The cells were incubated at 12°C (to prevent endocytosis of the bound antibody), and the fluorescence was monitored. In the experiment shown here, 50 percent of the fluorescence returned to the bleached area, indicating that 50 percent of the asialoglycoprotein receptor molecules in the illuminated patch of membrane were mobile, and 50 percent were immobile. The rate of fluorescence recovery is proportional to the rate of protein diffusion into the membrane region and thus to the diffusion coefficient. [See Y. I. Henis et al., 1990, *J. Cell Biol.* **111**:1409.]

membrane are freely mobile. The lateral diffusion rate of a protein in an intact membrane is generally 10–30 times lower than that of the same protein embedded in synthetic liposomes. These findings suggest that the mobility of integral proteins is restricted due to interactions with the rigid submembrane cytoskeleton (Chapter 5). Such interactions would need to be broken and remade as a protein diffuses laterally.

More direct evidence that interactions with the cytoskeleton normally restrict the movement of membrane proteins comes from study of "blebs" of the plasma membrane of fibroblasts and certain other cells. These occasional protrusions of the cell lack any cytoskeletal proteins. More than 90 percent of the integral proteins in blebs are mobile, and their rate of diffusion is five- to tenfold higher than that in more normal segments of the plasma membrane. Another striking piece of evidence is that treatment of intact cells with a nonionic detergent releases all of the lipids but only a fraction of the integral membrane proteins, even though these proteins, once purified, are freely soluble in the detergent. Many of these integral proteins must be tightly bound to the fibrillar elements of the cytoskeleton (e.g., actin and intermediate filaments), which resist extraction with detergents.

Erythrocytes Have an Unusual Plasma Membrane That Is Tightly Anchored to the Cytoskeleton

Perhaps the most-studied plasma membrane is that of the mammalian erythrocyte, which transports oxygen from the lungs to the tissues and returns carbon dioxide back to the lungs. The erythrocyte has no nucleus and no internal membranes; it is essentially a "bag" of hemoglobin containing relatively few other intracellular proteins. Although the erythrocyte normally adopts the shape of a biconcave disk 7 μm in diameter, it is very flexible and can squeeze through capillaries with much smaller diameters (Figure 14-35a).

In contrast to most other cellular membranes, the integral proteins of the erythrocyte plasma membrane cannot diffuse in the phospholipid bilayer and appear to be more-or-less uniformly distributed in the plane of the membrane, without large specialized patches. The erythrocyte cytoskeleton forms a dense fibrillar shell that underlies the entire plasma membrane (Figure 14-35b). This cytoskeletal network is attached to integral membrane proteins at many points, giving the erythrocyte plasma membrane its great strength and flexibility. This structure differs from that found in most other mammalian cells whose cytoskeleton typically courses throughout the cytoplasm and is anchored to the plasma membrane at relatively few points. Extensive study of erythrocytes has revealed how their integral membrane proteins are anchored to the underlying

(a)

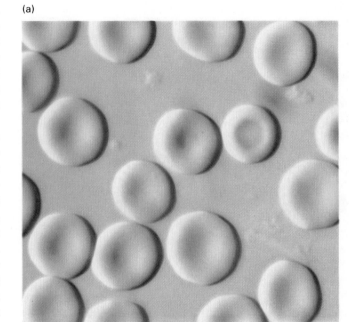

(b)

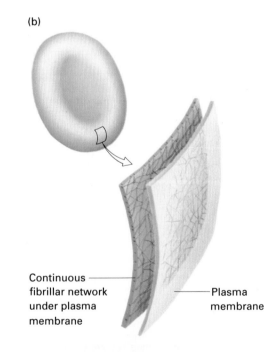

Continuous fibrillar network under plasma membrane

Plasma membrane

▲ FIGURE 14-35   (a) Normal human erythrocytes, viewed by differential interference light microscopy, are disk shaped; the nonvisible surface also is concave. (b) The erythrocyte cytoskeleton is a dense fibrillar shell located just under the plasma membrane. [Part (a) courtesy of M. Murayama, Biological Photo Service.]

cytoskeleton, how these interactions immobilize the membrane proteins, and how they generate the specific shape of a cell.

The analysis of erythrocyte membrane proteins is relatively straightforward. When the cells are placed in distilled water, the influx of water by osmosis causes them to swell. Eventually, the plasma membrane ruptures, releasing hemoglobin and other internal proteins. Because a hemoglobin-depleted erythrocyte retains the overall size and shape of the intact cell but is translucent, it is called a *ghost*. Under appropriate conditions, leaky membranes will reseal to form either right-side-out vesicles or inside-out vesicles (Figure 14-36). In the latter, the original cytoplasmic face of the plasma membrane now faces outward; these vesicles have proved useful in studies of the cytoplasmic surface of the erythrocyte membrane.

The two predominant erythrocyte integral membrane proteins, glycophorin and band 3, are glycoproteins. *Band 3 protein* is a dimer of two identical chains, each consisting of 929 amino acids (Figure 14-37). The C-terminal segment of each chain is embedded in the lipid membrane and makes multiple passages through it, forming probably 12 or 14 transmembrane α helices; these catalyze anion exchange across the plasma membrane (see Figures 15-19 and 15-20). The N-terminal segment of the molecule folds into discrete water-soluble domains that protrude into the cytosol and anchor the cytoskeleton to the membrane (Figure 14-38).

Erythrocyte cytoskeletons can be isolated by treating ghosts with a nonionic detergent, which solubilizes the membrane lipids and most of the membrane proteins, leaving the cytoskeleton as the insoluble fraction (Figure 14-39). The primary component of the cytoskeleton is *spectrin*, a long, fibrous protein. Two dimeric subunits of spectrin, each composed of an α and β polypeptide chain, associate to form head-to-head tetramers that are 200 nm in length. The ends of several spectrin tetramers are cross-linked at *junctional complexes*, which contain short actin filaments and other actin- and spectrin-binding proteins such as tropomyosin and adducin.

As illustrated in Figure 14-38, the cytoskeleton is attached to the plasma membrane by two proteins. One of these, called *band 4.1 protein*, is part of the junctional complex; it binds to spectrin and to the cytosolic domain of glycophorin, thus anchoring the cytoskeleton to the plasma membrane. The protein *ankyrin* also attaches the fibrous cytoskeletal shell to the plasma membrane. Ankyrin has two domains: one binds tightly and specifically to a site on the β chain of spectrin near the center of the tetramer; the other binds tightly to a region on band 3 protein that protrudes into the cytosol (see Figure 14-37). In erythrocyte ghosts, as in intact erythrocytes, most band 3 molecules are immobile and cannot diffuse laterally in the membrane plane because of these interactions with the cytoskeleton. (Glycophorin is probably also immobile, but this has not

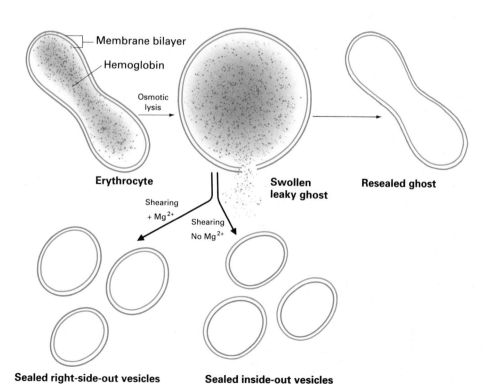

◄ FIGURE 14-36
Osmotic lysis of erythrocytes generates leaky membrane ghosts, which can reseal to form ghosts with the same size and shape as the original cells. If leaky ghosts are sheared by homogenization, smaller, nonleaky vesicles form. Depending on the presence or absence of Mg^{2+}, the vesicles will be predominantly inside out or right-side out.

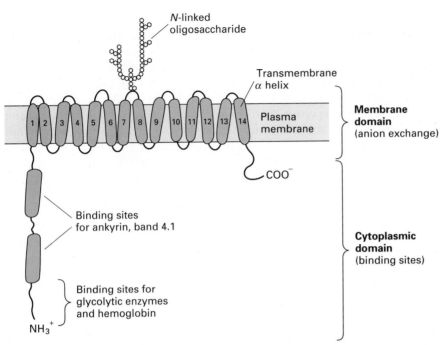

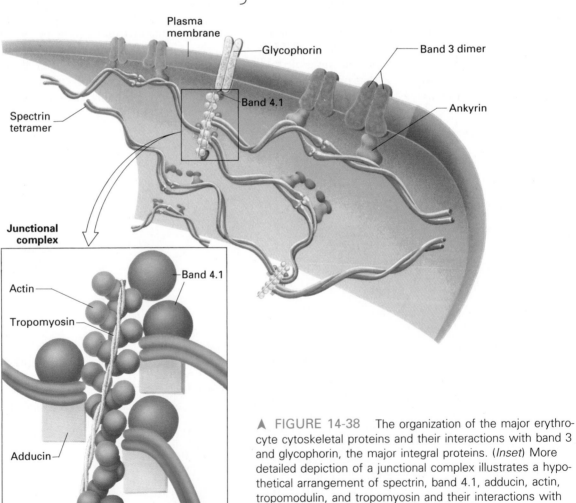

◄ FIGURE 14-37   Schematic drawing of band 3 protein (green) within the erythrocyte plasma membrane. Band 3 is a homodimer, but only one of the polypeptide chains is depicted here. Besides binding to the cytoskeletal protein ankyrin, band 3 binds to several glycolytic enzymes. [See R. Kopito and H. F. Lodish, 1985, *Nature* **316**:234; G. Jay and L. Cantley, 1986, *Ann. Rev. Biochem.* **55**:511. Drawing courtesy of Sam Lux.]

▲ FIGURE 14-38   The organization of the major erythrocyte cytoskeletal proteins and their interactions with band 3 and glycophorin, the major integral proteins. (*Inset*) More detailed depiction of a junctional complex illustrates a hypothetical arrangement of spectrin, band 4.1, adducin, actin, tropomodulin, and tropomyosin and their interactions with the ends of spectrin filaments. [Adapted from S. E. Lux, 1979, *Nature* **281**:426; E. J. Luna and A. L. Hitt, 1992, *Science* **258**:955.]

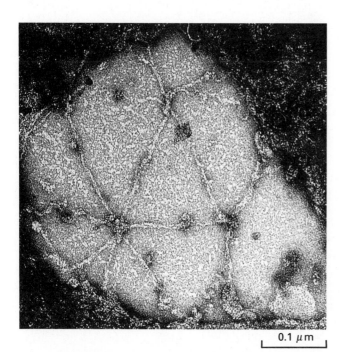

0.1 μm

▲ FIGURE 14-39   Electron micrograph of human erythrocyte cytoskeleton spread on a microscope grid. The long fibers are composed mainly of spectrin and can be seen to intersect at several junctional complexes (see Figure 14-38). [From T. J. Byers and D. Branton, 1985, *Proc. Nat'l. Acad. Sci. USA* **82**:6153. Courtesy of D. Branton.]

➤ Specializations of the Plasma Membrane

Unlike blood cells, which are independent, free-floating units, most animal and plant cells are organized into multicellular arrays that form solid tissues. In animals, cells in sheets or other aggregates can carry out their designated functions only if the plasma membrane is organized into at least two discrete regions, each specialized for very different tasks. Such cells are said to be *polarized*. Moreover, for an animal tissue to have strength, its cells must be able to form tight, strong contacts with their neighbors. Here, we explore the importance and variety of the plasma membrane's specialized regions and the nature of its cell-cell junctions.

Plasma Membranes of Polarized Cells Are Divided into Two Regions with Different Compositions and Functions

To illustrate the specializations of the plasma membrane found in polarized cells, we examine two important cell types: the epithelial cell lining the lumen of the small intestine and the pancreatic acinar cell, which secretes a number of digestive enzymes. The plasma membrane of both these cell types have two domains: an *apical* region facing a lumen that connects to the outside of the body and a *basolateral* region facing capillary-containing tissue.

Intestinal Epithelial Cells   With few exceptions, all internal and external body surfaces are covered with a polarized layer of epithelial cells. Different epithelia have specialized functions: for example, the epithelium lining the cervical tract and stomach contains mucus-secreting cells, and the stratified epithelium lining the mouth resists abrasion. A well-studied, highly polarized epithelial cell lines the lumen of the small intestine (see Figure 6-7b). Its two major functions are to absorb from the lumen of the small intestine small nutrient molecules that result from digestion of food and to transfer them across the single cell layer into the blood.

The lumenal (apical) surface of intestinal epithelial cells is highly specialized for absorption. This region, often called the *brush border* because of its appearance, consists of large numbers of fingerlike projections (100 nm in diameter) called *microvilli* (singular: *microvillus;* Figure 14-40a). These extensions of the cell surface greatly increase the membrane area and enhance the rate of absorption into the cells. A bundle of actin filaments that runs down the center of each microvillus gives rigidity to the microvilli.

Before being absorbed in the intestine, ingested proteins and carbohydrates are degraded in the intestine by

yet been studied.) When erythrocyte ghosts are placed in a solution of low ionic strength, the principal cytoskeleton proteins are removed. As a result, the ghosts lose their rigid shape and band 3 acquires lateral mobility.

The importance of spectrin is illustrated by certain mutant strains of mice that synthesize defective spectrin or no spectrin at all. In these mutants, the erythrocytes are spherical, rather than concave, and the rate of diffusion of band 3 protein within the membrane increases 50-fold. *Hereditary spherocytosis* and *elliptocytosis* are human diseases caused by similar genetic defects. In some cases, the mutant spectrin molecules fail to form head-to-head tetramers or bind defectively to band 4.1 protein or ankyrin. In other cases, ankyrin is defective or absent. In any case, the consequences are an unstable cytoskeleton and an abnormally shaped cell, with fragments of membrane occasionally budding from the surface. Because these abnormal erythrocytes are degraded by the spleen more rapidly than normal ones are, affected persons have fewer circulating erythrocytes and are said to be anemic. These hereditary defects further indicate that the spectrin-containing cytoskeleton is the major determinant of the rigidity of the erythrocyte membrane and acts to restrict the lateral motion of band 3 protein.

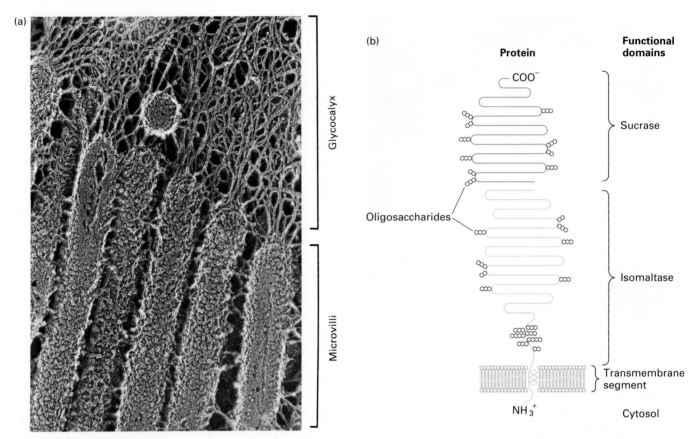

A FIGURE 14-40 Lumenal surface of intestinal epithelial cells. (a) Micrograph of microvilli, obtained by the deep-etching technique. The surface of each microvillus is covered with a series of bumps believed to be integral membrane proteins. The glycocalyx, which covers the apices (tips) of the microvilli, is composed of a network of glycoproteins and digestive enzymes including sucrase-isomaltase. (b) Overall structure and general orientation of sucrase-maltase in the plasma membrane of intestinal epithelial cells. [Part (a) from N. Hirokawa and J. E. Heuser, 1981, *J. Cell Biol.* **91**:399; part (b) see W. Hunziker et al., 1986, *Cell* **46**:227; G. M. Cowell et al., 1986, *Biochem J.* **237**:455. Photograph courtesy of N. Hirokawa and J. E. Heuser.]

pancreatic enzymes. First small peptides and oligosaccharides are produced; these are further broken down into monosaccharides and amino acids by peptidases and glycosidases that are bound to the exoplasmic face of the plasma membrane of the microvilli, extending into the intestinal lumen. Figure 14-40b depicts the membrane orientation of sucrase-isomaltase, a rodlike enzyme that catalyzes the hydrolysis of sucrose, a disaccharide, into the two monosaccharides glucose and fructose (see Figure 2-24). Other degradative enzymes in the microvillar plasma membrane have similar structures. These hydrolytic enzymes are components of the *glycocalyx,* a loose network covering the outer membrane surface. Composed of the oligosaccharide side chains of integral membrane glycoproteins, glycolipids, and peripheral glycoproteins including various enzymes, the glycocalyx (or cell coat) appears as a "fuzz" in electron micrographs of microvilli (see Figure 14-40a). The microvillar plasma membrane also contains transport proteins that allow monosaccharides, amino acids, and other compounds to pass from the intestinal lumen into the cell.

Once nutrients have been absorbed across the apical surface of intestinal epithelial cells, they move through the cells and are transported across the basolateral membrane into the blood. Transport proteins located in the basolateral membrane, which differ from those in the apical membrane, mediate movement of nutrients out of the cells. As discussed later, the lateral membrane also contains specialized regions involved in cell-cell adhesion, and the basal membrane contains proteins that anchor the cell to the basal lamina.

Pancreatic Acinar Cells   A pancreatic acinus is a more-or-less spherical aggregate of about a dozen cells (Figure 14-41). The lumen (central cavity) of an acinus is connected to a ductule that merges with ductules from the lumina of other acini to form a larger duct; this duct eventually leads into one of several main pancreatic ducts, which empty into the lumen of the small intestine (see Figure 5-38).

Like intestinal epithelial cells, pancreatic acinar cells have a plasma membrane divided into two functional regions. This similarity is not coincidental: acinar cells develop as outcroppings of the epithelial cells that line the fetal gut. Acinar cells synthesize degradative enzymes (e.g., amylases, proteases, ribonucleases) and store them as inactive precursors (zymogens) in membrane-lined secretory vesicles, which cluster under the apical region of the plasma membrane adjacent to the ductule. The secretory vesicles can fuse only with this part of the plasma membrane, ensuring that the digestive enzymes are released into the ductule, not into the surrounding tissue.

The basolateral membrane covers the sides of an acinar cell below the apical (lumen-facing) surface and extends along the base of the cell (see Figure 14-41). Nutrients in the blood in the surrounding vessels are transported through this region of the plasma membrane into the cell. The basolateral membrane also houses receptors for various peptide hormones that are synthesized and released into the blood by epithelial cells of the stomach and intestine when food is present. Binding of these hormones by their receptors in the pancreas triggers the secretion of digestive enzymes by acinar cells.

Tight Junctions Seal Off Body Cavities and Restrict Diffusion of Membrane Components

For polarized epithelial cells to function, extracellular fluids surrounding their apical and basolateral membranes must be kept separate. This is accomplished by *tight junctions,* which connect adjacent epithelial cells. These specialized regions of the plasma membrane form a barrier that seals off body cavities such as the intestinal lumen and pancreatic ductules from the blood. Tight junctions also prevent diffusion of membrane proteins and glycolipids between the apical and basolateral regions of the plasma membrane, insuring that these regions contain different membrane components.

Tight junctions are composed of thin bands of plasma membrane proteins that completely encircle a polarized cell and are in contact with similar thin bands on adjacent cells. In intestinal epithelial cells, tight junctions are usually located just below the apical microvillar surface (Figure 14-42). When thin sections of cells are viewed in an electron microscope, the plasma membranes of adjacent cells appear to touch each other at intervals and even to fuse (Figure 14-43). Because tight junctions prevent diffusion of small molecules directly from the intestinal lumen into the blood, intestinal epithelial cells must transport nutrients through the cells as previously described. In pancreatic acini, tight junctions likewise prevent leakage of secreted pancreatic proteins, including digestive enzymes, from the central cavities and ductules into the blood.

(a)

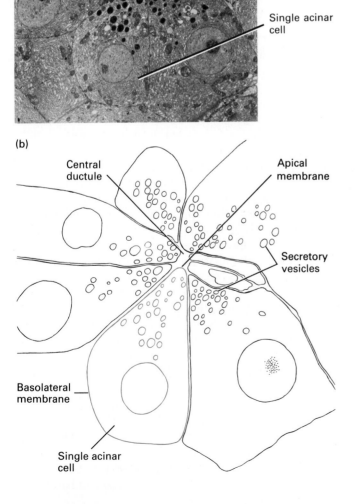

(b)

◄ FIGURE 14-41   (a) Low-magnification (×4000) electron micrograph of a rat pancreatic acinus showing the overall arrangement of the cells surrounding the central ductule. The nuclei are close to the base of the cells; the secretory vesicles are in the apical region, which borders the lumen. (b) Fusion of the membrane of a secretory vesicle with the apical plasma membrane causes exocytosis of the contents into the lumen. [(a) courtesy of Biophoto Associates.]

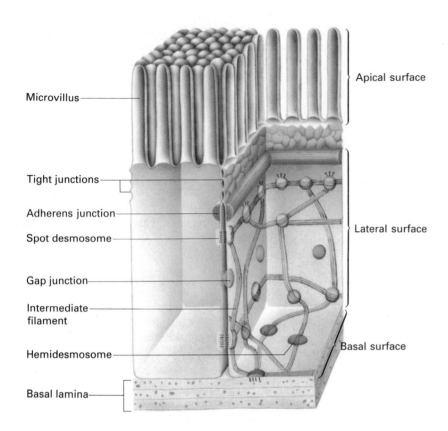

Microvillus

Tight junctions

Adherens junction

Spot desmosome

Gap junction

Intermediate filament

Hemidesmosome

Basal lamina

Apical surface

Lateral surface

Basal surface

◄ FIGURE 14-42 Schematic diagram of intestinal epithelial cells and principal types of cell junctions that connect them. The basal surface of the cells rests on a fibrous network of collagen and proteoglycans (basal lamina), which supports the epithelial cell layer. The apical surface faces the intestinal lumen.

That tight junctions are impermeable to most water-soluble substances can be demonstrated in an experiment in which lanthanum hydroxide (an electron-dense colloid of high molecular weight) is injected into the pancreatic blood vessel of an experimental animal; a few minutes later the pancreatic acinar cells are fixed and prepared for microscopy. As shown in Figure 14-44, the lanthanum hydroxide diffuses from the blood into the space that separates the basolateral surfaces of adjacent acinar cells, but cannot penetrate past the outermost tight junction. Other studies have shown that tight junctions also are impermeable to salts. For instance, when MDCK cells are grown in a medium containing very low concentrations of Ca^{2+}, they form a monolayer in which the cells are not connected by tight junctions; as a result, fluids flow freely across the cell layer. When Ca^{2+} is added to such a monolayer, tight junctions form within an hour (see Figure 6-9), and the cell layer becomes impermeable to fluids and salts.

Freeze-fracture electron microscopy affords a different view of the tight junction. The microvillar tight junction shown in Figure 14-45a appears to comprise an interlocking network of ridges in the plasma membrane. More specifically, there appear to be ridges on the cytosolic face of the plasma membrane of each of the two contacting cells. (Corresponding grooves not shown here are found on the exoplasmic face.) High magnification reveals that these ridges are made up of particles, believed to be protein, 3–4 nm in diameter. The tight junction is formed by a double row of these particles, one row donated by each cell (Figure 14-45b). The molecular structure of these junctions is not known, but protein particles on the two cells probably form extremely tight links with each other, essentially fusing the two plasma membranes and creating an impenetrable seal. Treatment of an epithelium with the protease trypsin destroys the tight junctions, implicating proteins as essential structural components of these junctions.

The ability of tight junctions to prevent diffusion of membrane components between the apical and basolateral

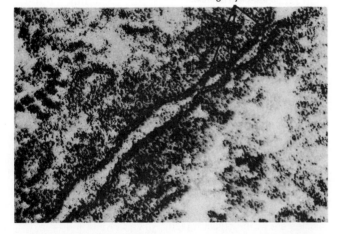

Contact points of tight junction

▲ FIGURE 14-43 Electron micrograph of a tight junction between two hepatocytes (liver cells), seen as three points of contact between the plasma membranes of adjacent cells. [Courtesy of D. Goodenough.]

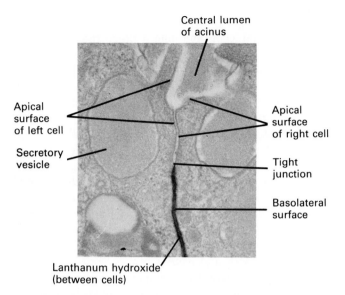

▲ FIGURE 14-44 Experimental demonstration that tight junctions prevent passage of water-soluble substances. Pancreatic acinar tissue is fixed and prepared for microscopy a few minutes after electron-opaque lanthanum hydroxide is injected into the blood of an experimental animal. As shown in this electron micrograph of adjacent acinar cells, the lanthanum hydroxide can penetrate between the cells but is arrested at the level of the tight junction. [Courtesy of D. Friend.]

membrane regions also can be demonstrated with monolayers of MDCK cells. Under appropriate culture conditions, adjacent MDCK cells are connected by tight junctions, which separate the medium into two sections: one bathing the apical surface of the MDCK monolayer; the other bathing the basolateral surface (see Figure 6-9). If liposomes containing a fluorescent-tagged glycoprotein are added to the apical medium, some of the liposomes eventually fuse with the apical surface, incorporating some fluorescent glycoprotein into the apical region of the monolayer. As long as the tight junctions between adjacent cells remain intact, no fluorescent-tagged protein is detectable in the basolateral surface of the cells. However, if the tight junctions are destroyed by removing Ca^{2+} from the medium, the fluorescent protein is soon detectable in the basolateral surface, indicating that it can diffuse from the apical to basolateral regions of the membrane.

Lipids in the cytosolic leaflets of apical and basolateral membranes have the same composition and apparently can diffuse from one region of the membrane to the other. In contrast, the lipid compositions of the exoplasmic leaflets of the apical and basolateral membrane regions are very different, and membrane lipids in the exoplasmic leaflets cannot diffuse through tight junctions. All the glycolipid in MDCK cells, for instance, is present in the exoplasmic face of the apical membrane, as are all proteins anchored

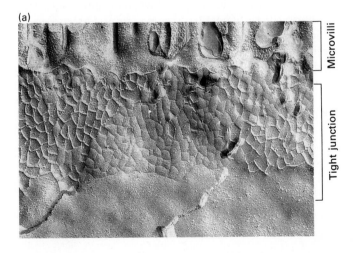

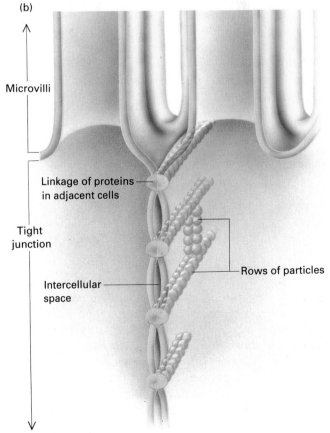

▲ FIGURE 14-45   (a) Freeze-fracture electron micrograph of a tight junction between two intestinal epithelial cells. The fracture plane passes through the plasma membrane of one of the two adjacent cells. The honeycomb-like network of ridges of particles below the microvilli forms the tight junction. (b) A model showing how the junction might be formed by linkage of the rows of particles in adjacent cells. [Part (a) courtesy of L. A. Staehelin; part (b) after L. A. Staehelin and B. E. Hull, 1978, Sci. Am. **238**(5):140.]

to the membrane by fatty acids linked to a glycosyl-phosphatidylinositol group (see Figure 14-20). In fact, the only lipids in the outer leaflet of the apical plasma membrane are glycolipids, proteins linked to glycosyl-phosphatidylinositol anchors, and cholesterol. Phosphatidylcholine, conversely, is present almost exclusively in the exoplasmic face of the basolateral plasma membrane.

Lipids in the exoplasmic leaflets of the apical or basolateral regions of the membrane of one cell cannot diffuse into the membrane of an adjacent cell connected to it by a tight junction. For example, one variant line of MDCK cells (strain I) contains a particular glycolipid called the *Forssman antigen* in the exoplasmic leaflet of the apical membrane, but another (strain II) does not. When grown together in culture, the two cell types form normal tight junctions with each other, but the Forssman glycolipid does not move from the type I to the type II cells; thus lipids do not diffuse past the tight junction from cell to cell.

Desmosomes and Gap Junctions Interconnect Cells and Control Passage of Molecules between Them

In order to function in an integrated manner, the aggregates of cells that compose organized tissues must adhere to each other and to the surrounding extracellular matrix and also control the movement of ions and small molecules between them. Two principal types of junctions, the desmosome and the gap junction, perform these functions; like the tight junction, these are specialized regions of the plasma membrane (see Figure 14-42).

In many organs, cells are bound tightly together by *desmosomes;* these interconnections confer mechanical strength on tissues. *Adherens junctions,* which are found primarily in epithelial cells, form a belt of cell-cell adhesion just under the tight junctions. *Spot desmosomes* are found in all epithelial cells and many other tissues, such as smooth muscle. They are buttonlike points of contact between cells, often thought of as a "spot-weld" between adjacent plasma membranes. *Hemidesmosomes,* similar in structure to spot desmosomes, anchor the plasma membrane to regions of the extracellular matrix. Bundles of intermediate filaments course through the cell, interconnecting spot desmosomes and hemidesmosomes.

Finally, *gap junctions* are distributed along the lateral surfaces of adjacent cells and allow them to exchange small molecules. Gap junctions help to integrate the metabolic activities of all cells in a tissue by regulating the passage of ions and small molecules from cell to cell. Among these are intracellular signaling molecules (e.g., cyclic AMP) and precursors of DNA and RNA.

Electron micrographs of animal tissue sections have shown that a space of about 20 nm ordinarily is present between the nonjunctional regions of plasma membranes of adjacent cells. This space contains integral membrane and extracellular surface glycoproteins that are important to intercellular adhesion.

In higher plants, individual cells are separated from each other by cell walls, and tight junctions and desmosomes are not present. However, many plant cells are interconnected by *plasmodesmata;* like gap junctions in animals, these structures allow low-molecular-weight metabolites and signaling molecules to move between cells.

An understanding of the structure and function of desmosomes requires knowledge about actin microfilaments and intermediate filaments; likewise, an understanding of gap junctions and plasmodesmata depends on knowledge of cellular metabolism and signaling. Therefore, we defer detailed discussion of these junctions until later chapters when these related topics are examined.

SUMMARY

The basic structure of all biological membranes is the closed phospholipid bilayer. The lipid composition of the bilayer varies among the diverse cellular membranes: glycolipids and cholesterol are abundant in the plasma membrane; large quantities of cardiolipin are found in the inner mitochondrial membrane. The phospholipid composition of the two leaflets in a single membrane may also differ, as in the erythrocyte plasma membrane, but such variations do not explain the unique properties of different biological membranes. Each type of membrane—or even patch of membrane—owes much of its individuality to the distinctive properties of its protein components.

Membrane proteins can be classified into two broad groups. An integral protein interacts directly with the phospholipid bilayer and usually contains one or more long α-helical sequences of hydrophobic amino acids that interact within the hydrophobic middle of the bilayer. Examples of such integral proteins are the major erythrocyte glycoproteins glycophorin and band 3, bacteriorhodopsin, and the bacterial photosynthetic reaction center, whose structure is known to atomic detail. Specific interactions among the membrane-spanning α helices stabilize the three-dimensional architecture of these proteins. A quite different architecture characterizes bacterial porins, whose monomeric subunits consist of antiparallel β strands arranged in a barrel shape; the outer surfaces of these proteins contain hydrophobic side chains that interact with the hydrophobic middle of the bilayer. Other integral proteins are bound to the bilayer by covalently attached fatty acids, by farnesyl and related hydrophobic molecules, or by lipid-containing glycosylphosphatidylinositol anchors. The second group, peripheral proteins, are bound to the membrane primarily, if not exclusively, by protein-protein (rather than protein-lipid) interactions. Examples of peripheral proteins are ankyrin, band 4.1, and other erythrocyte cytoskeletal proteins.

Biological membranes are highly asymmetric. All integral membrane proteins bind asymmetrically to the lipid bilayer, and all molecules of any one kind lie in the same direction. In addition, all the molecules of a particular membrane-bound enzyme face the same surface of the membrane; virtually all membrane oligosaccharides in glycolipids and glycoproteins face the exoplasmic surface of the bilayer. This asymmetry is related to all aspects of membrane function.

Various studies involving fluorescence recovery after photobleaching (FRAP) and other techniques have shown that all lipids and most integral proteins are laterally mobile in a membrane. Some membrane proteins, however, are immobile, probably because they are anchored to cytoskeletal components or to proteins of the extracellular matrix.

The plasma membrane of the mammalian erythrocyte differs from membranes of most tissue cells in that it contains a homogeneous array of surface proteins that cannot diffuse laterally. No cytoskeleton courses through the erythrocyte cell. Just under the plasma membrane, a lace-like cytoskeletal network constructed of the fibrous protein spectrin, actin, and accessory proteins is bound to two of the major integral membrane glycoproteins, glycophorin and band 3, by many highly specific protein-protein interactions. This submembrane spectrin network appears to be responsible for the shape of the erythrocyte.

Cells in organized tissues are connected by several types of cell junctions, which are specialized structures in the plasma membrane composed primarily of proteins. Some integral proteins form tight junctions between cells, sealing off fluids on different sides of a cell layer. In epithelial cells, such as those that form the pancreatic acinus or intestinal epithelium, tight junctions also define two domains of the plasma membrane: the apical and basolateral regions. Each region contains unique lipids and proteins that enable it to perform specialized functions, such as binding hormones or fusing with intracellular vesicles that contain secretory proteins. Tight junctions prevent diffusion of proteins and lipids between the apical and basolateral regions of the exoplasmic leaflet; however, the lipid components can diffuse laterally in the cytosolic leaflet. Adherens junctions and spot desmosomes bind the plasma membranes of adjacent cells in a way that gives strength and rigidity to the entire tissue; hemidesmosomes join surface cells to the underlying extracellular matrix. Gap junctions interconnect the cytoplasm of two adjacent cells, allowing small molecules to pass between them.

REVIEW QUESTIONS

1. Biological membranes must maintain cellular compartmentalization, permit communication between the interior and exterior of the cell, and allow for communication between the cytosol and internal organelles. The two essential components of a membrane are lipids and protein. Review the fundamental nature of the lipid environment in a membrane. Be certain that you can distinguish among the general structures of the lipids, including phosphoglycerides, sphingolipids, glycolipids, and steroids.

What is the difference between a saturated and unsaturated fatty acid? How does the degree of saturation affect the bending of the fatty acid chain? Are most of the double bonds in naturally occurring fatty acids in a *trans* or *cis* configuration? What is the effect on membrane fluidity of increasing the amount of unsaturated fatty acids found in the hydrophobic side chains of phosphoglycerides? What net charges are associated with the polar heads of membrane lipids?

When phospholipids are placed in an aqueous solution, liposomes can be formed by sonicating the mixture. (Sonication refers to the process of using sound waves to stir the solution vigorously.) Refer to the following table listing several derivatives of phosphatidylcholine having different R_1 and R_2 groups. The percent composition of the liposomes is also given. Using the sonication technique, you prepare liposomes from these mixtures at room temperature.

Preparation	R1 = palmitate R2 = palmitate	R1 = oleate R2 = palmitate	R1 = linoleate R2 = palmitate
A	80%	10%	10%
B	10%	20%	70%
C	40%	40%	20%

a. Rank the liposomes that would result in each preparation from the lowest to the highest membrane fluidity. Explain why you placed them in that order.

b. If band 3 protein were purified from erythrocytes and embedded in these liposomes, in which would it have the fastest lateral mobility? Explain your answer.

c. What would happen to the lateral mobility of band 3 protein in these liposomes if the temperature were lowered to 4°C? Explain your answer.

2. Recall that integral membrane proteins span the hydrophobic core of a membrane, often by having an α-helical domain that consists of numerous hydrophobic amino acids. Refer to Figure 14-14 for an illustration of this concept using glycophorin A as a model membrane-spanning protein. Notice the hypothetical interactions between positively charged *arg* and *lys* residues with the head groups of the lipid bilayer.

a. You have cloned the gene for glycophorin A, and you intend to delete the region that corresponds to amino acids 74 through 95. How would you do this? You replace the region with DNA encoding an α-helical domain that spans a biological membrane, but it is part of a different protein. You prepare the recombinant in a suitable expression vector, and transfect mammalian cells. You have two antibodies against glycophorin A: The first recognizes the N-terminal domain and the second recognizes the C-terminal domain. The antibody against the N-terminal domain binds to the surface of transfected cells but not to cells lacking the expression vector; the antibody against the C-terminal domain does not bind to either set of cells. Explain these results based on your knowledge of proteins that span the membrane with a single α helix.

b. Assume that the interaction between the positively charged amino acids and the lipid bilayer is real. If this protein were placed in liposomes containing only phosphatidylglycerol, do you think the interactions between the *arg* and *lys* residues and the bilayer would occur? What would happen to the interaction if the lipid composition of the liposome were changed to 90 percent phosphatidylcholine/10 percent phosphatidylglycerol? Do you think that the lipid environment has the potential to affect the three-dimensional protein structure and influence its function or activity? If so, what precautions would you need to take if you were investigating protein function by using artificial liposomes?

3. The rate (area/time) of a molecule's lateral movement in a lipid bilayer is often described by the following:

$$\text{Rate} = (2kT)/(3\pi r\eta)$$

where k is the Boltzmann constant (3.3×10^{-27} kcal/K), T is the absolute temperature, r is the radius of the molecule and is related to its size, and η is the viscosity which has a value of 2.4×10^{-11} kcal · s/cm^3 for a lipid bilayer. The viscosity of the lipid bilayer is much greater than that of the surrounding medium or the cytoplasm; therefore, it is the bilayer that dictates the rate of lateral mobility of integral membrane proteins. A rough estimate of the radius of an α helix—a structure that commonly spans the membrane in integral membrane proteins (see Figure 3-9)—is 0.25 nm.

a. You are interested in the rate at which proteins diffuse within the lipid membrane. You decide to investigate the lateral mobility of glycophorin A and a derivative of it which has a 200 amino acid protein fused to the N-terminus. You discover that both proteins have the same lateral mobility in a liposome. Explain your result. Does it contradict the above equation?

b. You decide to continue to characterize the mobility of proteins in membranes by comparing the lateral rate of movement of glycophorin A to that of purified bacteriorhodopsin (refer to Figure 14-18 for its structure) reconstituted in identical liposome environments. You discover that the mobility of bacteriorhodopsin is fourfold slower than that of glycophorin A. Explain this result.

c. If the "radius" of phosphatidylinositol were estimated at 0.5 nm, what would be its rate of lateral diffusion at 37°C? Suppose you have anchored the extracellular portion of the glycophorin A molecule to the inositol group of phosphatidylinositol. What would happen to the lateral mobility of the molecule?

4. Review the details of the cytoskeletal network of the erythrocyte. Be sure to understand the interactions of spectrin, adducin, actin, tropomyosin, ankyrin, band 3, and band 4.1; refer to Figure 14-36. Why is the erythrocyte a particularly useful system for studying membrane-protein structural and functional interactions? Which of the integral membrane proteins are constrained in terms of their lateral mobility in the lipid bilayer? How are they constrained?

In this chapter, reference was made to certain hereditary diseases associated with mutant spectrin binding defectively to protein 4.1 or ankyrin. If a red cell were deficient in spectrin, what effect would this have on the lateral mobility of band 3?

The two principal classes of membrane proteins are extrinsic and integral. What methods are used for isolating each class of membrane protein? What are the advantages of nonionic versus ionic detergents in purifying membrane proteins?

a. You want to investigate the integral membrane proteins that are embedded in the membranes of lysosomes in hepatocytes. You are aware that the internal pH of lysosomes is acidic, and you suspect that a transporter of some sort must assist the lysosome in maintaining the low pH relative to that found in the surrounding cytoplasm. Before you investigate the transporter itself, you decide to characterize the integral protein composition of the lysosomal membrane. Propose a scheme for doing this.

b. Suppose that you successfully purified the transporter protein. How could you confirm that it is found in the lysosome membrane and is not merely a contaminant from other parts of the cell?

References

Properties of Phospholipid Bilayers

DEY, I., et al. 1993. Molecular and structural compositions of phospholipid membranes in livers of marine and freshwater fish in relation to temperature. *Proc. Nat'l. Acad. Sci. USA* **90**:7498–7502.

GLASER, M. 1993. Lipid domains in biological membranes. *Curr. Biol.* **3**:475–481.

HAKOMORI, S. 1986. Glycosphingolipids. *Sci. Am.* **254**(5):44–53.

SINGER, S. J., and G. L. NICOLSON. 1972. The fluid mosaic model of the structure of cell membranes. *Science* **175**:720–731.

TANFORD, C. 1980. *The Hydrophobic Effect,* 2d ed. Wiley. Includes a good discussion of the interactions of proteins and membranes.

WARNOCK, D. E., et al. 1993. Determination of plasma membrane lipid mass and composition in cultured chinese hamster ovary cells using high gradient magnetic affinity chromatography. *J. Biol. Chem.* **268**:10145–10153.

WENDOLOSKI, J. J., et al. 1989. Molecular dynamics simulation of a phospholipid micelle. *Science* **243**:636–638.

YEAGLE, P. L. 1993. *The membranes of cells,* 2d ed. Academic Press.

Lipid Anchors of Membrane Proteins

ENGLUND, P. T. 1993. The structure and biosynthesis of glycosyl phosphatidylinositol protein anchors. *Ann. Rev. Biochem.* **62**:121–138.

GIBBS, J. B. 1991. Ras C-terminal processing enzymes—new drug targets? *Cell* **65**:1–4.

HOMANS, S. W., et al. 1988. Complete structure of the glycosyl-phosphatidylinositol membrane anchor of rat brain Thy-1 glycoprotein. *Nature* **333**:269–272.

LOW, M. G. 1989. Glycosyl-phosphatidylinositol: a versatile anchor for cell surface proteins. *FASEB J.* **3**:1600–1608.

SCHULTZ, A. M., L. E. HENDERSON, and S. OROSZLAN. 1988. Fatty acylation of proteins. *Ann. Rev. Cell Biol.* **4**:611–647.

SEABRA, M. C., et al. 1992. Rab geranylgeranyl transferase. *J. Biol. Chem.* **267**:14497–14507.

Structure of Integral Membrane Proteins

ALTENBACH, C., et al. 1990. Transmembrane protein structure: spin labeling of bacteriorhodopsin mutants. *Science* **248**:1088–92.

BLOW, D. 1991. Lipases reach the surface. *Nature* **351**:444–445.

COWAN, S. W. 1993. Bacterial porins: lessons from three high-resolution structures. *Curr. Biol.* **3**:501–507.

CRAMER, W. A., et al. 1992. Forces involved in the assembly and stabilization of membrane proteins. *FASEB J.* **6**:3397–3402.

DEISENHOFER, J., et al. 1985. Structure of the protein subunits in the photosynthetic reaction center of *Rhodopseudomonas viridis* at 3Å resolution. *Nature* **318**:618–624.

HENDERSON, R., et al. 1990. Model for the structure of bacteriorhodopsin based on high-resolution electron cryomicroscopy. *J. Mol. Biol.* **213**(4):899–929.

GOODSELL, D. S. 1991. Inside a living cell. *Trends Biochem. Sci.* **16**:203–206.

HENDERSON, R., and G. F. SCHERTLER. 1990. The structure of bacteriorhodopsin and its relevance to the visual opsins and other seven helix G-protein couple receptors. *Philos. Trans. R. Soc. Lond.* [*Biol*] **326**(1236):379–389.

HENNESSEY, E. S., and J. K. BROOME-SMITH. 1993. Gene fusion techniques for determining membrane-protein topology. *Curr. Biol.* **3**:524–531.

JENNINGS, M. L. 1989. Topography of membrane proteins. *Ann. Rev. Biochem.* **58**:999–1027.

LODISH, H. F. 1988. Multi-spanning membrane proteins: how accurate are the models? *Trends Biochem. Sci.* **13**:332–334.

POPOT, J-L. 1993. Integral membrane protein structure: transmembrane alpha helices as autonomous folding domains. *Curr. Biol.* **3**:532–540.

REES, D. C., L. DE ANTONIO, and D. EISENBERG. 1989. Hydrophobic organization of membrane proteins. *Science* **245**:510–513.

REES, D. C., et al. 1989. The bacterial photosynthetic reaction center as a model for membrane proteins. *Ann. Rev. Biochem.* **58**:607–633.

SCOOT, D. L., et al. 1990. Interfacial catalysis: the mechanism of phospholipase A_2. *Science* **250**:1541–1546.

Detergents

HELENIUS, A., and K. SIMONS. 1975. Solubilization of membranes by detergents. *Biochim. Biophys. Acta* **415**:29–79.

SILVIUS, J. R. 1992. Solubilization and functional reconstitution of biomembrane components. *Ann. Rev. Biophys. Biomol. Struct.* **21**:323–348.

Membrane Asymmetry

BRANTON, D. 1966. Fracture faces of frozen membranes. *Proc. Nat'l. Acad. Sci. USA* **55**:1048–1056.

BRETSCHER, M. S., and S. MUNRO. 1993. Cholesterol and Golgi apparatus. *Science* **261**:1280–1281.

DEVAUX, P. F. 1992. Protein involvement in transmembrane lipid asymmetry. *Ann. Rev. Biophys. Biomol. Struct.* **21**:417–439.

DEVAUX, P. F. 1993. Lipid transmembrane asymmetry and flip-flop in biological membranes and in lipid bilayers. *Curr. Biol.* **3**:489–494.

ROTHMAN, J., and J. LENARD. 1977. Membrane asymmetry. *Science* **195**:743–753.

Mobility of Membrane Proteins and Lipids

EDIDIN, M., S. C. KUO, and M. P. SHEETZ. 1991. Lateral movement of membrane glycoproteins restricted by dynamic cytoplasmic barriers. *Science* **254**:1379–1382.

FRYE, L. D., and M. EDIDIN. 1970. The rapid intermixing of cell-surface antigens after formation of mouse-human heterokaryons. *J. Cell Sci.* **7**:319–335.

HACKENBROCK, C. R. 1981. Lateral diffusion and electron transfer in the mitochondrial inner membrane. *Trends Biochem. Sci.* **6**:151–154.

JACOBSON, K., A. ISHIHARA, and R. INMAN. 1987. Lateral diffusion of proteins in membranes. *Ann. Rev. Physiol.* **49**:163–175.

LEE, G. M., A. ISHIHARA, and K. JACOBSON. 1991. Direct observation of brownian motion of lipids in a membrane. *Proc. Nat'l. Acad. Sci. USA* **88**:6274–6278.

SHEETZ, M. P. 1993. Glycoprotein motility and dynamic domains in fluid plasma membranes. *Ann. Rev. Biophys. Biomol. Struct.* **22**:417–431.

WEIR, M., and M. EDIDIN. 1988. Constraint of the translational diffusion of a membrane glycoprotein by its external domains. *Science* **242**:412–414.

ZHANG, F., G. M. LEE, and K. JACOBSON. 1993. Protein lateral mobility as a reflection of membrane microstructure. *Bioessays.* **15**:579–588.

Cytoskeletal Attachment of the Erythrocyte Membrane

BENNETT, V. 1992. Ankyrins: adaptors between diverse plasma membrane proteins and the cytoplasm. *J. Biol. Chem.* **267**:8703–8706.

BENNETT, V., and D. M. GILLIGAN. 1993. The spectrin-based membrane skeleton and micron-scale organization of the plasma membrane. *Ann. Rev. Cell Biol.* 9:27–66.

BYERS, T. M., and D. BRANTON. 1985. Visualization of the protein associations in the erythrocyte membrane skeleton. *Proc. Nat'l. Acad. Sci. USA* 82:6153–6157.

CHASIS, J. A., and N. MOHANDAS. 1992. Red blood cell glycophorins. *Blood* 80:1869–1879.

CONBOY, J. G. 1993. Structure, function, and molecular genetics of erythroid membrane skeletal protein 4.1 in normal and abnormal red blood cells. *Semin. Hematol.* 30:58–73.

ELGSAETER, A., et al. 1986. The molecular basis of erythrocyte shape. *Science* 234:1217–1223.

LIU, S. C., and L. H. DERICK. 1992. Molecular anatomy of the red blood cell membrane skeleton: structure-function relationships. *Semin. Hematol.* 29:231–243.

LUNA, E. J., and A. L. HITT. 1992. Cytoskeleton-plasma membrane interactions. *Science* 258:955–964.

MATAYOSHI, E. D., and T. M. JOVIN. 1991. Rotational diffusion of band 3 in erythrocyte membranes. Comparison of ghosts and intact cells. *Biochemistry* 30:3527–3538.

PALEK, J. 1993. Introduction: red blood cell membrane proteins, their genes and mutations. *Semin. Hematol.* 30:1–3.

PALEK, J., and K. E. SAHR. 1992. Mutations of the red blood cell membrane proteins: from clinical evaluation to detection of the underlying genetic defect. *Blood* 80:308–330.

PETERS, L. L., and S. E. LUX. 1993. Ankyrins: structure and function in normal cells and hereditary spherocytes. *Semin. Hematol.* 30:85–118.

Cell Polarity and Specialized Regions of the Plasma Membrane

BARTLES, J. R., L. BRAITERMAN, and A. HUBBARD. 1985. Endogenous and exogenous domain markers of the rat hepatocyte plasma membrane. *J. Cell Biol.* 100:1126–1138.

BOCK, G., and S. CLARK. 1987. Junctional complexes of epithelial cells. *Ciba Foundation Symposium 125.* Wiley.

GUMBINER, B. M. 1992. Epithelial morphogenesis. *Cell* 69:385–387.

HANZEL, D., et al., 1991. New techniques lead to advances in epithelial cell polarity. *Semin. Cell Biol.* 2:341–53.

KOBAYASHI, T., et al. 1992. A functional barrier to movement of lipids in polarized neurons. *Nature* 359:647–650.

LISANTI, M. P., et al. 1988. Polarized apical distribution of glycosyl-phosphatidylinositol-anchored proteins in a renal epithelial cell line. *Proc. Nat'l. Acad. Sci. USA* 85:9557–9561.

NELSON, W. J. 1991. Cytoskeleton functions in membrane traffic in polarized epithelial cells. *Semin. Cell Biol.* 2:375–385.

NELSON, W. J. 1992. Regulation of cell surface polarity from bacteria to mammals. *Science* 258:948–955.

RODRIGUEZ-BOULAN, E., and S. K. POWELL. 1992. Polarity of epithelial and neuronal cells. *Ann. Rev. Cell Biol.* 8:395–427.

SEMENZA, G. 1986. Anchoring and biosynthesis of stalked brush-border membrane proteins: glycosidases and peptidases of enterocytes and renal tubuli. *Ann. Rev. Cell Biol.* 2:255–313.

VAN MEER, G. 1993. Transport and sorting of membrane lipids. *Curr. Opin. Cell Biol.* 5:661–673.

Tight Junctions

ANDERSON, J. M., M. S. BALDA, and A. S. FANNING. 1993. The structure and regulation of tight junctions. *Curr Opin Cell Biol.* 5:772–8.

CITI, S. 1993. The molecular organization of tight junctions. *J. Cell Biol.* 121:485–9.

GUMBINER, B. M. 1993. Breaking through the tight junction barrier. *J. Cell Biol.* 123:1631–1633.

GUMBINER, B. M. 1993. Proteins associated with the cytoplasmic surface of adhesion molecules. *Neuron.* 11:551–64.

RUBIN, L. L. 1992. Endothelial cells: adhesion and tight junctions. *Curr. Opin. Cell Biol.* 4:830–833.

VAN MEER, G., B. GUMBINER, and K. SIMONS. 1986. The tight junction does not allow lipid molecules to diffuse from one epithelial cell to the next. *Nature* 322:639–641.

VAN MEER, G., and K. SIMONS. 1986. The function of tight junctions in maintaining differences in lipid composition between the apical and the basolateral cell-surface domains of MDCK cells. *EMBO J.* 5:1455–1464.

15 Transport across Cell Membranes

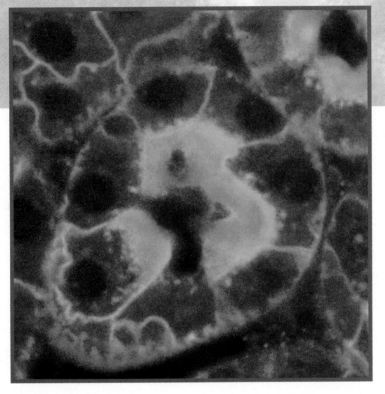

▲ The CHIP28/Aquaporin water channel in kidney tubules.

The plasma membrane is a selectively permeable barrier between the cell and the extracellular environment. Its permeability properties ensure that essential molecules such as glucose, amino acids, and lipids readily enter the cell, metabolic intermediates remain in the cell, and waste compounds leave the cell. In short, the selective permeability of the plasma membrane allows the cell to maintain a constant internal environment. Similarly, organelles within the cell often have a different internal environment from that of the surrounding cytosol, and organelle membranes maintain this difference. For example, within the animal-cell lysosome or the plant-cell vacuole—the organelles involved in digestive and degradative processes—the concentration of protons (H^+) is 100–1000 times that of the cytosol; this gradient is maintained by proteins in the organelle membrane.

An artificial membrane composed of pure phospholipid or of phospholipid and cholesterol is permeable to gases, such as O_2 and CO_2, and small, relatively hydrophobic molecules, such as ethanol (Figure 15-1). Such molecules can cross cellular membranes unaided by transport proteins. No metabolic energy is expended because movement is from a high to a low concentration of the molecule, down its chemical concentration gradient. As noted in Chapter 2, such transport reactions have a positive ΔS value (increase in entropy) and a negative ΔG (decrease in free energy).

In contrast, a pure phospholipid membrane is only slightly permeable to water and is essentially impermeable to most water-soluble molecules, such as glucose, nucleosides, and amino acids, and to ions such as hydrogen, sodium, calcium, and potassium. Proteins are required to transport such molecules and ions across all cellular membranes. Because different cell types require different mixtures of these low-molecular-weight compounds, the plasma membrane of each cell type contains a specific set

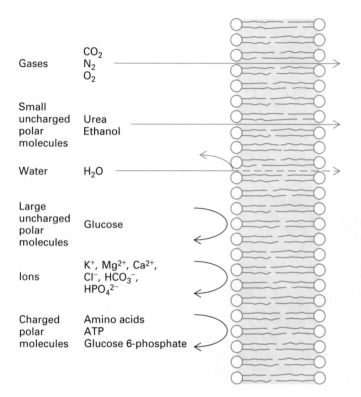

A pure artificial phospholipid bilayer is permeable to small hydrophobic molecules and small uncharged polar molecules. It is slightly permeable to water and impermeable to ions and to large uncharged polar molecules. When a small phospholipid bilayer separates two aqueous compartments (see Figure 14-8), membrane permeability can be easily determined by adding a small amount of radioactive material to one compartment and measuring its rate of appearance in the other compartment.

ATP-powered pumps (or simply *pumps*) are ATPases that use the energy of ATP hydrolysis to move ions across a membrane *against* a chemical concentration gradient or electric potential. This type of ion movement, referred to as *active transport*, is an example of a coupled chemical reaction (Chapter 2) in which an energetically unfavorable reaction—transport of ions "uphill" against a concentration

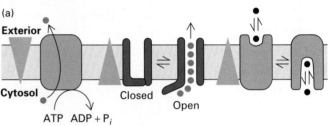

(a)

ATP-powered pump   Ion channel   Transporter
($10^0 - 10^3$ ions/s)   ($10^7 - 10^8$ ions/s)   ($10^2 - 10^4$ molecules/s)

of transport proteins that allow only certain ions or molecules to cross, as does the membrane surrounding each type of subcellular organelle.

In this chapter we present examples of each of the many types of transport proteins and discuss what is known about their structure and mechanisms of action. We also explain how different combinations of transport proteins in different subcellular membranes enable cells to carry out essential physiologic processes, such as the maintenance of cytosolic pH, the transport of glucose across the absorptive intestinal epithelium, the accumulation of sucrose and salts in plant-cell vacuoles, and the directed flow of water in both plants and animals. Often the same type of transport protein is involved in quite different physiologic processes.

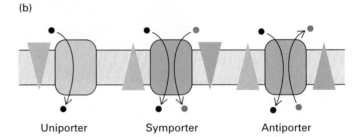

(b)

Uniporter   Symporter   Antiporter

▲ FIGURE 15-2   Schematic diagrams illustrating action of membrane transport proteins. (a) The three major types of transport proteins. Pumps utilize the energy of hydrolysis of a phosphoanhydride bond in ATP to power movement of specific ions (red circles) against their electrochemical gradient. (Gradients are indicated by triangles with the tip pointing towards lower concentration and/or electrical potentials.) Channels catalyze movement of specific ions (or water) down their electrochemical gradient. Transporters, which fall into three groups, facilitate movement of specific small molecules, such as glucose and amino acids (black circles), or of ions, such as Na^+ and H^+. (b) The three groups of transporters. Uniporters (also shown in part [a]) transport a single type of molecule down its concentration gradient. The movement of one molecule against its concentration gradient (black circles), driven by movement of one or more ions down an electrochemical gradient (red circles) is catalyzed by symporters and transporters. These cotransport proteins differ in the relative direction of movement of the transported molecule and cotransported ion.

► *Major Types of Membrane Transport Proteins*

Figure 15-2a illustrates the three major types of membrane transport proteins. One type couples ATP hydrolysis with the energetically unfavorable ("uphill") movement of ions. The other two types of membrane transport proteins are not ATPases, and their ability to transport ions or molecules is not dependent on ATP hydrolysis. Some of these proteins catalyze only energetically favorable ("downhill") movement of substances, whereas others couple the downhill movement of one substance to the uphill movement of a different substance.

gradient or membrane potential—is coupled to an energetically favorable reaction—the hydrolysis of ATP to ADP and P_i. The overall reaction—ATP hydrolysis and the "uphill" movement of ions—is energetically favorable. Such pumps maintain the low calcium and sodium ion concentrations inside virtually all animal cells relative to that in the medium, and generate the low pH inside animal-cell lysosomes, plant-cell vacuoles, and the lumen of the stomach.

Channel proteins transport water or specific types of ions down their concentration (or electric potential) gradients, energetically favorable reactions. They form a protein-lined passageway across the membrane through which multiple water molecules or ions move simultaneously, single file at a very rapid rate—up to 10^8 per second. For example, the plasma membrane of all animal cells contains potassium channel proteins that allow K^+, and only this ion, to move across the membrane down its concentration gradient. We will show how movement of K^+ through these always-open channels generates an electric potential across the plasma membrane. Many other types of channel proteins are usually closed, and open only in response to specific signals. Because these types of ion channels play a fundamental role in the functioning of nerve cells, they will be discussed in detail in Chapter 21.

The third class of membrane-transport proteins, called *transporters,* move a wide variety of ions and molecules across membranes. Transporters, in contrast to channel proteins, bind only one (or a few) substrate molecules at a time; after binding substrate molecules, the transporter undergoes a conformational change such that the bound substrate molecules, and only these molecules, are transported across the membrane. Because movement of each substrate molecule (or small number of molecules) requires a conformational change in the transporter, transporters move only about 10^2–10^4 molecules per second, a lower rate than that associated with channel proteins.

Three types of transporters have been identified (Figure 15-2b). *Uniporters* transport one molecule at a time down a concentration gradient. This type of transporter, for example, moves glucose or amino acids across the plasma membrane into mammalian cells. In contrast, antiporters and symporters catalyze movement of one type of ion or molecule *against* its concentration gradient coupled to movement of a different ion or molecule down its concentration gradient. Like ATP pumps, antiporters and symporters mediate coupled reactions in which an energetically unfavorable reaction is coupled to an energetically favorable reaction. Because symporters and antiporters catalyze "uphill" movement of certain molecules, they are often referred to as "active transporters" but, unlike pumps, no hydrolysis of ATP (or any other molecule) is involved.

We begin with a discussion of the mechanistically simplest type of transport: diffusion through a pure phospholipid bilayer. This is the mechanism by which gases and some small molecules, such as ethanol, enter and leave cells. In subsequent sections, we will see how transport cat-

alyzed by various membrane-transport proteins differs from simple diffusion.

▶ Diffusion of Small Molecules across Pure Phospholipid Bilayers

In the transport process called *passive diffusion,* a small molecule in aqueous solution dissolves into the phospholipid bilayer, crosses it, and then dissolves into the aqueous solution on the opposite side. No proteins are involved. The relative diffusion rate of a substance across the bilayer is proportional to its concentration gradient across the layer (Figure 15-3) and to its hydrophobicity. There is little specificity to the process, in that any small hydrophobic molecule will be transported.

The first step in passive diffusion is movement of a molecule from the aqueous solution into the hydrophobic

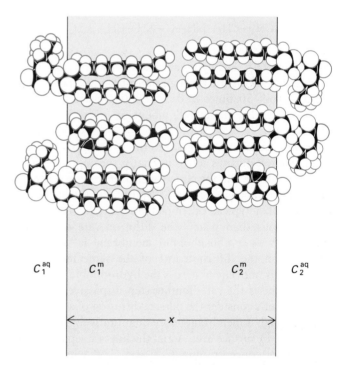

▲ FIGURE 15-3   A simple model for passive diffusion—the movement of small hydrophobic molecules directly across the phospholipid bilayer of a biological membrane. The upper half of the figure is a space-filling model of the lipid bilayer, showing phospholipid and cholesterol molecules. The lower half represents the hydrocarbon barrier (of thickness x) to diffusion. C_1^{aq} and C_2^{aq} are the concentrations of two solutions on sides 1 and 2 of the membrane; C_1^m and C_2^m are the corresponding membrane concentrations just within the hydrocarbon barrier.

interior of the phospholipid bilayer. The hydrophobicity of a substance is measured by its *partition coefficient K*, the equilibrium constant for its partition between oil and water. Since the composition of the interior of the phospholipid bilayer resembles that of oil, the partition coefficient of a substance moving across a bilayer equals the ratio of its concentration just inside the hydrophobic core of the bilayer C^m to its concentration in the aqueous solution C^{aq}:

$$K = \frac{C^m}{C^{aq}} \qquad 15\text{-}1$$

The partition coefficient is a measure of the relative affinity of a substance for lipid versus water. For example, urea

$$\underset{\displaystyle NH_2-\overset{\textstyle O}{\overset{\|}{C}}-NH_2}{}$$

has a $K = 0.0002$, whereas diethylurea (with two ethyl groups)

$$CH_3-CH_2-NH-\overset{\textstyle O}{\overset{\|}{C}}-NH-CH_2-CH_3$$

has a $K = 0.01$. Diethylurea, which is more hydrophobic than urea, will diffuse through phospholipid bilayer membranes about 50 times $(0.01 \div 0.0002)$ faster than urea. Diethylurea also enters cells about 50 times faster than urea.

Once a molecule moves into the hydrophobic interior of a bilayer, it diffuses across it; finally, the molecule moves from the bilayer into the aqueous medium on the other side of the membrane (see Figure 15-3). Because the hydrophobic core of a typical cell membrane is 100–1000 times more viscous than water, the diffusion rate of all substances across a phospholipid membrane is very much slower than the diffusion rate of the same molecule in water. Thus, movement across the hydrophobic portion of a membrane is the rate-limiting step in passive diffusion.

Now let's consider the passive diffusion of small molecules through a membrane more quantitatively. Suppose a membrane of surface area A and thickness x separates two solutions of concentrations C_1^{aq} and C_2^{aq}, where $C_1^{aq} > C_2^{aq}$ (see Figure 15-3). In this case, the diffusion rate is given by a modification of *Fick's law*, which states that the diffusion rate across the membrane dn/dt (in mol/s) is directly proportional to the *permeability coefficient P*, to the difference in solution concentrations $C_1^{aq} - C_2^{aq}$, and to the area A, or

$$\frac{dn}{dt} = PA(C_1^{aq} - C_2^{aq}) \qquad (15\text{-}2)$$

P, and thus the rate of passive diffusion, is proportional to the partition coefficient K and to the diffusion coefficient within the membrane D and is inversely proportional to membrane thickness x. To see the important point that the value of P for any substance is proportional to its partition coefficient K we can write the equation for flow of material *within* the membrane (which must equal flux of material across the membrane) as

$$\frac{dn}{dt} = \frac{D}{x} A(C_1^m - C_2^m) \qquad (15\text{-}3)$$

where C_1^m and C_2^m are the concentrations just inside the hydrophobic region of the membrane. Since K equals C_1^m/C_1^{aq} and, equivalently, C_2^m/C_2^{aq}, equation 15-3 becomes

$$\frac{dn}{dt} = \frac{KD}{x} A(C_1^{aq} - C_2^{aq}) \qquad (15\text{-}4)$$

Comparing this to equation 15-2, we see that

$$P = \frac{KD}{x} \qquad (15\text{-}5)$$

Thus, the rate of diffusion of any substance through the membrane will be proportional to its particular permeability coefficient P. Since D and x are the same for most substances, the rate of diffusion of any substance is thus proportional to its partition coefficient K. Quantitatively, the rate of passive diffusion of a water-soluble molecule is proportional to its hydrophobicity.

Fick's law does not apply to charged molecules. The diffusion of charged molecules across a membrane permeable to the ion is determined not only by the concentration gradient but also by any electric potential gradient that might exist across the membrane.

▶ Uniporter-Catalyzed Transport of Specific Molecules

Very few molecules enter or leave cells, or cross organelle membranes, unaided by proteins. Even transport of molecules, such as water and urea, that can diffuse across pure phospholipid bilayers is frequently accelerated by transport proteins. Thus we need to understand the properties of the different kinds of membrane-transport proteins and their many roles in cell and organism physiology. Many studies on the function of membrane transport proteins involve extraction and purification of a specific protein and its reincorporation into pure phospholipid bilayer membranes, such as liposomes (Figure 15-4). In this sys-

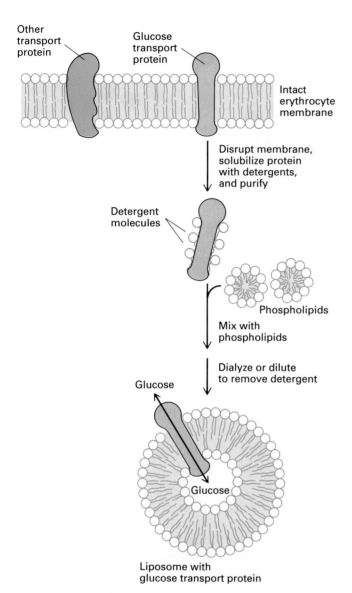

▲ FIGURE 15-4 Liposomes containing a single type of transport protein can be used to investigate properties of the transport process. Here, all the integral proteins of the erythrocyte membrane are solubilized by a nonionic detergent, such as octylglucoside. The glucose transport protein, a uniporter, can be purified by chromatography on a column containing a specific monoclonal antibody and then incorporated into liposomes made of pure phospholipids.

tem, the functional properties of the various membrane proteins can be examined without ambiguity.

We begin our discussion with the simplest membrane transport proteins—uniporters, which catalyze movement of one molecule at a time down a concentration gradient. In subsequent sections, we examine the other types of transport proteins shown in Figure 15-2 and describe how multiple transport proteins sometimes function together to accomplish a particular cellular process (e.g., movement of nutrients from the intestinal lumen to the blood).

Three Main Features Distinguish Uniport Transport from Passive Diffusion

Uniporters facilitate movement of specific small molecules across membranes down their concentration gradient. Similar to enzymes, uniporters accelerate a reaction that is already thermodynamically favored, and the movement of a substance across a membrane down its concentration gradient will have the same negative ΔG value whether or not a protein transporter is involved. This type of movement sometimes is referred to as *facilitated transport* (or facilitated diffusion). We stressed in Chapter 2 that many chemical reactions that are thermodynamically favored will not occur unless an appropriate enzyme is present; such is also the case with movement of hydrophilic molecules across biological membranes.

Three properties of uniporter-catalyzed movement of glucose and other small hydrophilic molecules across a membrane distinguish this type of transport from passive diffusion:

1. The rate of uniport transport is far higher than predicted by Fick's equation describing passive diffusion (Figure 15-5), because the transported molecules never

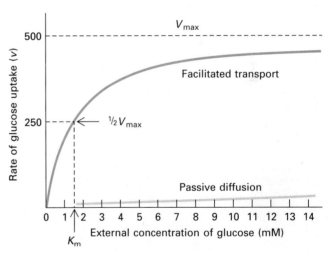

▲ FIGURE 15-5 Comparison of the observed uptake rate of glucose by erythrocytes (dark red curve) with the calculated rate if glucose were to enter solely by passive diffusion through the phospholipid bilayer (light red curve). The rate of glucose uptake (measured as micromoles per milliliter of cells per hour) in the first few seconds is plotted against the glucose concentration in the extracellular medium. In this experiment the initial concentration of glucose in the erythrocyte is zero, so that the concentration gradient of glucose across the membrane is the same as the external concentration. The glucose transporter in the erythrocyte membrane clearly increases the rate of glucose transport, compared with that associated with passive diffusion, at all glucose concentrations. Like enzymes, the transporter-catalyzed uptake of glucose exhibits a maximum transport rate V_{max} and is said to be saturable. The K_m is the concentration at which the rate of glucose uptake is half-maximal.

enter the hydrophobic core of the phospholipid bilayer; thus the partition coefficient K is irrelevant.

2. Uniport transport is specific. Each transporter protein catalyzes movement of only a single species of molecule or a single group of closely related molecules.

3. Uniport transport occurs via a limited number of transporter proteins, rather than throughout the phospholipid bilayer. Consequently, there is a maximum transport rate, V_{max}, that is achieved only when the concentration gradient across the membrane is very large, and each transporter is working at its maximal rate.

Figure 15-4 shows the rate of glucose uptake by erythrocytes at different external glucose concentrations. The initial intracellular glucose concentration is zero, so V_{max} is achieved at high external glucose concentrations. Since the concentration of glucose is usually higher in the extracellular medium than in the cell, the plasma membrane glucose transporters usually catalyze net movement of glucose in one direction: from the medium into the cell. However, if the concentration gradient is reversed, the glucose transporter, like all uniporters, is equally able to catalyze net movement in the reverse direction: from the cell into the medium. Such a situation occurs in liver cells during periods of starvation, when these cells synthesize glucose (from fatty acids, amino acids, and other small molecules) and release it into the blood.

Two General Models Have Been Proposed for Transporters

For many years, transporter proteins were pictured as shuttles that move from one side of the membrane to the other, or rotate an active site from one membrane face to the other. According to this *carrier model,* the transporter protein binds the molecule to be moved at one face, moves through the membrane, and releases the molecule at the other face. Carrier models do explain the properties of certain small peptide antibiotics that accelerate the movement of certain ions across membranes. For example, the antibiotic valinomycin increases transport of K^+ ions across biological membranes by forming a sphere around each K^+ ion. The hydrophobic amino acid side chains of the antibiotic lie on the outer surface, and six or eight oxygen atoms on the inside coordinately bind to the K^+ (Figure 15-6). The hydrophobic exterior makes the K^+-carrier complex soluble in the lipid interior of the membranes and thus facilitates movement of the ion through the interior of the membrane.

The carrier model, however, is implausible for protein-mediated transport through biological membranes, because shuttle movement of large proteins through the distinct hydrophobic and hydrophilic regions of cellular membranes would be energetically expensive. As discussed below, investigators now believe that membrane trans-

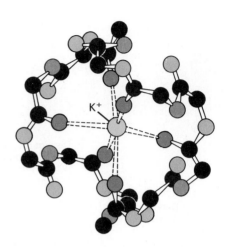

▲ FIGURE 15-6   Model of valinomycin complexed with one K^+ ion. The peptide folds around a single K^+ ion and six carboxyl oxygens (dark red) on interior residues bind to it. The periphery of the complex is composed of the hydrophobic side chains of the amino acids.

porters undergo conformational changes that permit bound ions or molecules to pass through the membrane.

Glucose Entry into Erythrocytes Is Mediated by a Uniporter

One of the best-characterized membrane-transport proteins is the erythrocyte glucose transporter, which catalyzes the uniport movement of glucose down its concentration gradient. We discuss the glucose transporter in some detail as an example of the uniport type of transport protein.

Alternative Conformational States of the Glucose Transporter   The glucose transporter alternates between two conformational states: in one, a glucose-binding site faces the outside of the membrane; in the other, a glucose-binding site faces the inside (Figure 15-7). Unidirectional transport of glucose from the exterior inwards occurs when the transporter, with glucose bound to the outward-facing site, undergoes a conformational change such that the outward-facing site is inactivated and the bound glucose moves through the protein and becomes attached to the newly-formed inward-facing site. After the glucose, bound to the inward-facing site, is released into the cell interior, the transporter (without a bound glucose) undergoes the reverse conformational change, inactivating the inward-facing glucose binding site, and recreating the outward-facing one. As noted above, the glucose transporter can also catalyze the net movement of glucose from the cytosol outwards; this reaction proceeds by reversal of steps a–d shown in Figure 15-7.

Kinetics of Transporter-Catalyzed Movement of Glucose   As noted previously, a plot of the entry rate of glucose into erythrocytes versus external glucose concen-

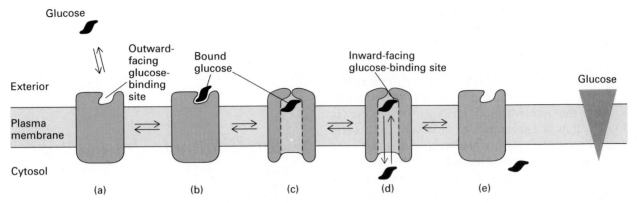

▲ FIGURE 15-7   Model of the mechanism of action of the glucose transporter, which is believed to shuttle between two conformational states. In one conformation (a, b, and e), the glucose-binding site faces outward; in the other (c, d), the binding site faces inward. Binding of glucose to the outward-facing binding site (a to b) triggers a conformational change in the transporter (b to c) moving the bound glucose through the protein such that it is now bound to the inward-facing binding site. Glucose can then be released to the inside of the cell (d); the transporter then undergoes the reverse conformational change (d to e), inactivating the inward-facing glucose binding site and regenerating the outward-facing one. If the concentration of glucose is higher inside the cell than outside, the cycle will work in reverse (e to a) catalyzing net movement of glucose from inside to out. This occurs in intestinal epithelial cells during transport of glucose from the intestine to the blood.

tration is not linear; rather, it is a curve that levels off at V_{max} at high external glucose concentrations (see Figure 15-5). The kinetics of the unidirectional transport of glucose (and other small molecules) from the outside of a cell inwards, catalyzed by a uniporter, can be described by the same type of equation used to describe a simple enzymatically catalyzed chemical reaction. For simplicity, let's assume that a substance S (say, glucose) is present initially only on the outside of the membrane. In this case, we can write

$$S_{out} + transporter \overset{K_m}{\rightleftharpoons}$$

$$S \cdot transporter\ complex \overset{V_{max}}{\rightleftharpoons} S_{in} + transporter$$

where K_m is the substance-transporter binding constant and V_{max} is the maximum transport rate of S into the cell. If C is the concentration of S_{out} (initially, the concentration of $S_{in} = 0$), then, by exactly the same derivation used for the Michaelis-Menton equation (Chapter 3) we can write

$$v = \frac{V_{max}}{1 + K_m/C} \qquad (15\text{-}6)$$

where v is the transport rate of the species into the cell; V_{max} is the rate of transport if all molecules of the transporter contain a bound S, which occurs at high S concentrations; and K_m is the substrate concentration at which half-maximal transport occurs across the membrane. (The lower the value of K_m, the more tightly the substrate binds to the transporter, and the greater the transport rate.) For the erythrocyte transporter, the K_m for glucose transport is 1.5 millimolar (mM); at this concentration roughly half of the transporters with outward-facing binding sites would have a bound glucose (see Figure 15-7). Blood glucose is normally 5 mM, or 0.9 g/l. At this concentration, the erythrocyte glucose transporter is functioning at 77 percent of the maximal rate V_{max}, as can be seen from Figure 15-5 or from equation 15-6, which describes this curve.

The kinetics of glucose transport are more complex and more revealing than this simple analysis suggests, because the slow (rate-determining) step in out → in transport of glucose is the conformational change in the transporter from one with an inward-facing glucose-binding site, unoccupied by glucose, to one with an unoccupied outward-facing glucose-binding site, as shown in Figure 15-7d,e. This conformational change is accelerated severalfold when glucose is bound to the inward-facing site (see Figure 15-7c). This phenomenon has strange consequences for the kinetics of glucose transport. For instance, if [14C]glucose is added to a suspension of cells whose intracellular glucose concentration is zero, the labeled glucose is transported inward at a particular initial rate. This *initial* rate of [14C]glucose transport is *accelerated* severalfold if unlabeled glucose is present inside the cells before addition of the labeled glucose. This unexpected experimental observation not only supported the conformational model of glucose-transporter function depicted in Figure 15-7, it also indicated that the rate-limiting step in out → in transport of glucose is the conformational change that converts an inward-facing site to an outward-facing site, a change that is accelerated by binding of glucose to the inward-facing site.

Specificity and Structure of the Glucose Transporter   The data in Table 15-1 show that the erythrocyte glucose transporter is highly specific. For example, the K_m for the nonbiological L-isomer of glucose is >3000 mM; thus at concentrations at which D-glucose is readily transported into the erythrocyte, L-glucose does not enter at a measurable rate. Although D-mannose and D-galactose, which differ from D-glucose in the configuration at one carbon atom (see Figure 2-6), are transported into erythrocytes at measurable rates, higher concentrations of these substrates are needed to half-saturate the transport reaction. Thus the glucose transporter has a lower affinity (indicated by higher K_m values) for these substrates than for the normal substrate D-glucose.

After glucose is transported into the erythrocyte, it is rapidly phosphorylated, forming glucose 6-phosphate, which cannot leave the cell. Because of this reaction, the first step in the metabolism of glucose (see Figure 17-2), the intracellular concentration of free glucose does not increase as glucose is taken up by the cell. Consequently, the glucose concentration gradient across the membrane is maintained, as is the rate of glucose entry into the cell.

The erythrocyte glucose-transport protein has been purified, cloned, and sequenced. An integral, transmembrane protein with a molecular weight (MW) of 45,000, it accounts for 2 percent of the protein in the plasma membrane of erythrocytes. When the purified glucose transporter is inserted into artificial liposomes (see Figure 15-4), it dramatically increases their permeability to D-glucose. All the properties of glucose entry into erythrocytes are retained in this artificial system: in particular, D-glucose, D-mannose, and D-galactose are taken up, but L-glucose is not.

Amino acid sequence and biophysical studies on the glucose transporter indicate that it contains 12 α helices that span the phospholipid bilayer. Although these transmembrane α helices contain predominantly hydrophobic amino acid side chains, several helices bear amino acid resi-

dues (e.g., serine, threonine, asparagine, and glutamine) whose side chains can form hydrogen bonds with the hydroxyl groups on glucose. These residues are thought to form the inward-facing and outward-facing glucose-binding sites in the interior of the protein.

▶ *Ion Channels, Intracellular Ion Environment, and Membrane Electric Potential*

Plasma membranes usually contain several uniporters that enable molecules such as amino acids, nucleosides, and sugars to enter or leave the cell down their chemical concentration gradients. The movement of ions across the plasma membrane and organelle membranes also is mediated by transport proteins: symporters and certain antiporters cotransport ions together with specific small molecules, whereas ion channels, ion pumps, and some antiporters transport only ions. In all cases, the rate and extent of ion transport across membranes is influenced not only by the ion concentrations on the two sides of the membrane but also by an electric potential that exists across the membrane. Thus, we turn now to the origin of the electric potential across the membrane, and its relationship to ion channels within the membrane.

Ionic Gradients and an Electric Potential Are Maintained across the Plasma Membrane

The specific ionic composition of the *cytosol* (the liquid component of the cytoplasm) usually differs greatly from that of the surrounding fluid. In virtually all cells—including microbial, plant, and animal cells—the cytosolic pH is kept near 7 and the cytosolic concentration of potassium ion (K^+) is much higher than that of sodium ion (Na^+). In particular, in both invertebrates and vertebrates, the concentration of K^+ is 20–40 times higher in cells than in the blood, while the concentration of Na^+ is lower in cells than in the blood (Table 15-2). The concentration of calcium ion (Ca^{2+}) free in the cytosol is generally less than 1 micromolar (10^{-6} M), a thousand or more times lower than that in the blood. Plant cells and many microorganisms maintain similarly high cytosolic concentrations of K^+ and low concentrations of Ca^{2+} and Na^+ even if the cells are cultured in very dilute salt solutions. The ATP-driven ion pumps (discussed later in this chapter) generate and maintain these ionic gradients.

The plasma membrane contains channel proteins that allow the principal cellular ions (Na^+, K^+, Ca^{2+}, and Cl^-) to move through it at different rates down their concentration gradients. Ion concentration gradients and selective movements of ions through channels create a difference in

---

TABLE 15-1   Specificity of the Erythrocyte Glucose Transporter

Substrate	K_m(mM)*	Substrate	K_m(mM)*
D-Glucose	1.5	D-Mannose	20
L-Glucose	>3000	D-Galactose	30

* Concentration in millimolar required for half-maximal rate of transport.
SOURCE: P. G. Lefebvre, 1961, *Pharmacol. Review* **13**:39.

TABLE 15-2 Typical Ion Concentrations in Invertebrates and Vertebrates

	Cell (mM)	Blood (mM)
SQUID AXON*		
K^+	400	20
Na^+	50	440
Cl^-	40–150	560
Ca^{2+}	0.0003	10
$X^{-\dagger}$	300–400	
MAMMALIAN CELL		
K^+	139	4
Na^+	12	145
Cl^-	4	116
HCO_3^-	12	29
X^-	138	9
Mg^{2+}	0.8	1.5
Ca^{2+}	<0.0002	1.8

* The large nerve axon of the squid is chosen as an example of an invertebrate cell because it has been used widely in studies of the mechanism of conduction of electric impulses.
$\dagger X^-$ represents proteins, which have a net negative charge at the neutral pH of blood and cells.

electric potential between the inside and the outside of the cell. The magnitude of this potential—$\approx$70 millivolts (mV) with the inside of the cell negative with respect to the outside—does not seem like much until we realize that the plasma membrane is only about 3.5 nm thick. Thus the voltage gradient across the plasma membrane is 0.07 V per 3.5×10^{-7} cm, or 200,000 volts per centimeter! (To appreciate what this means, consider that high-voltage transmission lines for electricity utilize gradients of about 200,000 volts per kilometer!) The plasma membrane, like all biological membranes, is an electrical device called a *capacitor*—a thin sheet of nonconducting material (the hydrophobic interior) surrounded on both sides by electrically conducting material (the polar head groups and the ions in the surrounding aqueous medium)—that can store an electric charge across it.

The ionic gradients and electric potential across the plasma membrane drive many biological processes. Opening and closing of Na^+, K^+, and Ca^{2+} channels are essential to the conduction of an electric impulse down the axon of a nerve cell. In many animal cells, the concentration gradient of Na^+ ions and the membrane electric potential power the uptake of other molecules against their concen-

tration gradient; amino acids frequently enter cells in this manner. In most cells, a rise in the concentration of Ca^{2+} ions in the cytosol is an important regulatory signal. In muscle cells, for instance, it initiates contraction; in the exocrine cells of the pancreas, it triggers secretion of digestive enzymes.

After describing here how ionic gradients and specific ion channel proteins generate a membrane electric potential, we examine the ATP-powered ion pumps that generate ion concentration gradients in the following section.

Certain K^+ Channels Generate the Membrane Electric Potential

In the experimental system outlined in Figure 15-8a, the distribution of K^+, Na^+, and Cl^- ions is similar to that between an animal cell and its aqueous environment. A membrane separates a 15-mM KCl/150-mM NaCl solution on the right side (representing the "outside" of the cell) from a 150-mM KCl/15-mM NaCl solution on the left side (the "inside"). A potentiometer (voltmeter) is connected to the solution on each side to measure any difference in electric potential across the membrane. If the membrane is impermeable to all ions, no ions will flow across it and there will be no electric potential difference across it.

Now suppose that the membrane contains Na^+ channel proteins that accommodate Na^+ ions but exclude K^+ and Cl^- ions. Na^+ ions then tend to move down their concentration gradient from the right side to the left, leaving an excess of negative Cl^- ions compared with Na^+ ions on the right side and generating an excess of positive Na^+ ions compared with Cl^- ions on the left side. The excess Na^+ on the left and Cl^- on the right remain near the respective surfaces of the membrane. There is now a separation of charge across the membrane, which a potentiometer can measure as an electric potential, or voltage. The right side is negative with respect to the left (Figure 15-8b). As more and more Na^+ ions move through channels across the membrane, the magnitude of this charge difference increases. However, continued movement of the Na^+ ions eventually is inhibited by the excess of positive charges accumulated on the left side of the membrane and by the attraction of Na^+ ions to the excess negative charge built up on the right side. The system soon reaches an equilibrium point at which the two opposing factors that determine the movement of Na^+ ions—the membrane electric potential and the ion concentration gradient—balance each other out. At equilibrium, no net movement of Na^+ ions occurs across the membrane. Thus the excess negative (Cl^-) charges bound to the right surface of the membrane are separated from and attracted to the excess positive (Na^+) ones on the left. In this way, the phospholipid membrane acts as a capacitor, and stores the charge across it exactly as does a capacitor in an electric circuit.

The magnitude of the resulting sodium equilibrium potential in volts (the electric potential across a membrane

(a) Membrane impermeable to Na⁺, K⁺, and Cl⁻

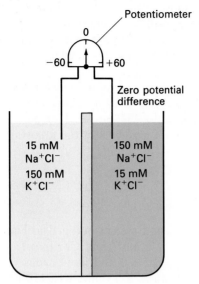

Potentiometer

Zero potential difference

15 mM Na⁺Cl⁻
150 mM K⁺Cl⁻

150 mM Na⁺Cl⁻
15 mM K⁺Cl⁻

(b) Membrane permeable to Na⁺ only

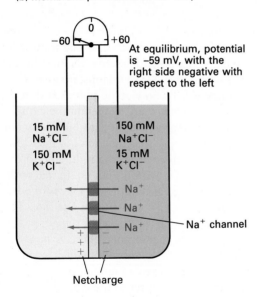

At equilibrium, potential is −59 mV, with the right side negative with respect to the left

15 mM Na⁺Cl⁻
150 mM K⁺Cl⁻

150 mM Na⁺Cl⁻
15 mM K⁺Cl⁻

Na⁺
Na⁺
Na⁺ — Na⁺ channel

Netcharge

(c) Membrane permeable to K⁺ only

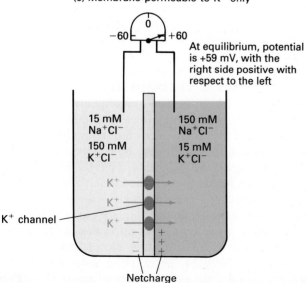

At equilibrium, potential is +59 mV, with the right side positive with respect to the left

15 mM Na⁺Cl⁻
150 mM K⁺Cl⁻

150 mM Na⁺Cl⁻
15 mM K⁺Cl⁻

K⁺ channel

K⁺
K⁺
K⁺

Netcharge

◄ **FIGURE 15-8** Experimental system for generating a transmembrane voltage potential. (a) An impermeable membrane separates a 150 mM KCl/15 mM NaCl solution from a 15 mM KCl/150 mM NaCl solution. No ions move across the membrane, and no difference in electric potential is registered on the potentiometer connecting the two solutions. (b) If the membrane is selectively permeable only to Na⁺, then Na⁺ ions diffuse from right to left, through Na⁺ channels (dark purple) down the Na⁺ concentration gradient. The Cl⁻ and K⁺ ions cannot cross the membrane, so a net positive charge builds up on the left side and a negative charge builds up on the right side. At equilibrium, the membrane potential caused by the charge separation becomes equal to the Nernst potential E_{Na} registered on the potentiometer, and the movement of Na⁺ ions in the two directions becomes equal. (c) If the membrane is selectively permeable only to K⁺, then K⁺ ions diffuse from left to right through K⁺ channels (light purple) down the K⁺ concentration gradient. The Na⁺ and Cl⁻ ions cannot cross the membrane, creating a net negative charge on the left side and a net positive charge on the right side. At equilibrium, the membrane electric potential is equal to E_K.

permeable only to Na⁺ ions) is given by the *Nernst equation*, which is derived from basic principles of physical chemistry:

$$E_{Na} = \frac{RT}{ZF} \ln \frac{[Na_l]}{[Na_r]} \qquad (15\text{-}7)$$

where R (the gas constant) = 1.987 cal/(degree · mol), or 8.28 joules/(degree · mol), T (the absolute temperature) = 293 K at 20°C, Z (the valency) = +1, F (the Faraday constant) = 23,062 cal/(mol · V), or 96,000 coulombs/(mol · V), and $[Na_l]$ and $[Na_r]$ are the Na⁺ concentrations on the left and right sides, respectively, at equilibrium. The Nernst equation is similar to the equations used to calculate the voltage change associated with oxidation or reduction reactions (Chapter 2), which also involve movement of electric charges. At 20°C, equation 15-7 reduces to

$$E_{Na} = 0.059 \log_{10} \frac{[Na_l]}{[Na_r]} \qquad (15\text{-}8)$$

In Figure 15-8b, $[Na_l]/[Na_r] = 0.1$, and $E_{Na} = -0.059$ V (−59 mV) with the right side negative with respect to the left.

If the membrane is permeable only to K⁺ ions and not to Na⁺ or Cl⁻ ions, the calculation is the same:

$$E_K = \frac{RT}{ZF} \ln \frac{[K_l]}{[K_r]} \qquad (15\text{-}9)$$

The *magnitude* of the membrane electric potential is the same (59 mV), except that the right side is now positive with respect to the left (Figure 15-8c). This is precisely the opposite polarity to that obtained with selective Na^+ permeability.

If the membrane is permeable to Na^+ and K^+ ions to the same degree, then each moves down its concentration gradient: Na^+ from the right side to the left, and K^+ from left to right. In this case, no membrane electric potential is expected and none is observed. An intermediate situation between these two extremes can also occur. When the membrane is permeable to both Na^+ and K^+ ions but more permeable to K^+ ions, the right side initially has a positive potential relative to the left, but the magnitude of the potential is somewhat less than $E_K = +59$ mV. Eventually, due to the diffusion of Na^+ and K^+ ions, there will be an equal concentration of both ions on both sides of the membrane and no membrane electric potential.

As noted earlier, the membrane potential across the plasma membrane of animal cells is about -70 mV; that is, the cytosolic face is negative with respect to the exoplasmic (outside) face. These membranes contain many open K^+ channels but few open Na^+ or Ca^{2+} channels. As a result, the major ionic movement across the plasma membrane is that of K^+ from the inside outward. This leaves an excess of negative charge on the inside, and an excess of positive on the outside, and is the major determinant of the inside-negative membrane potential. Quantitatively, the usual potential of -70 mV is close to that of the potassium equilibrium potential calculated from the Nernst equation.

The Na^+-K^+ ATPase, an ion pump discussed below, pumps K^+ ions into the cytosol from the extracellular medium, generating the K^+ concentration gradient. Movement of K^+ ions through K^+ channels from the cytosol outward, down their concentration gradient, generates the inside-negative membrane potential. These always-open K^+ channels, the so-called *resting K^+ channels,* have just been cloned and sequenced, but as yet we know little of their molecular structure. However, much is known of the structure and function of other types of K^+ channels that are found mainly in nerve cells and that open or close in response to changes in membrane potential or to other signals. These are detailed in Chapter 21 as are the important Na^+ and Ca^{2+} channels that are essential for the function of nerve cells.

Ion Concentration Gradients and Electric Potential Drive the Movement of Ions across Biological Membranes

Two forces govern the movement of such ions as K^+, Cl^-, and Na^+ across selectively permeable membranes: the membrane electric potential and the ion concentration gradient. These forces may act in the same direction or in opposite directions. The free-energy change ΔG corresponding to the transport of 1 mole of Na^+ ions from the

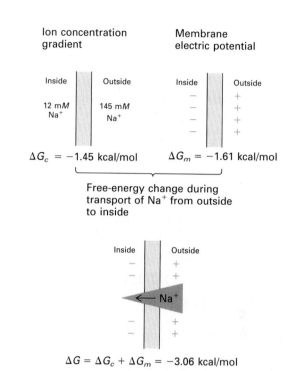

▲ FIGURE 15-9   Transmembrane forces acting on Na^+ ions. As with all ions, the movement of Na^+ ions across the plasma membrane is governed by the sum of two separate forces generated by the membrane electric potential and the ion concentration gradient. In the case of Na^+ ions, these forces usually act in the same direction.

outside (exterior) to the inside (cytosol) of a typical mammalian cell is about -3 kcal/mol (Figure 15-9). Since ΔG is <0, this reaction is thermodynamically favored. About half of this ΔG value is contributed by the membrane electric potential and half is contributed by the Na^+ concentration gradient. It is important to understand these forces in some detail, since the inward movement of Na^+ ions is used to power the uphill movement of several ions and small molecules into or out of animal cells catalyzed by symport and antiport proteins.

The free-energy change generated from the Na^+ concentration gradient is

$$\Delta G_c = RT \ln \frac{[Na_{in}]}{[Na_{out}]} \tag{15-10}$$

At the concentrations of Na_{in} and Na_{out} shown in Figure 15-9, which are typical for many cells, $\Delta G_c = -1.45$ kcal/mol, the change in free energy for the thermodynamically favored transport of 1 mol of Na^+ ions from outside to inside the cell if there were no membrane electric potential. The free-energy change generated from the membrane electric potential is

$$\Delta G_m = FE \tag{15-11}$$

where F = the Faraday constant and E = the membrane electric potential. If $E = -70$ mV, then $\Delta G_m = -1.6$ kcal mol, the change in free energy for the thermodynamically favored transport of 1 mol of Na^+ ions from outside to inside the cell if there were no Na^+ concentration gradient. Given both forces acting on Na^+ ions, the total ΔG will be the sum of the two partial values:

$$\Delta G = \Delta G_c + \Delta G_m = (-1.45) + (-1.61) = -3.06 \text{ kcal/mol}$$

In this typical example, the Na^+ concentration gradient and the membrane electric potential contribute almost equally to total ΔG for transport of Na^+ ions.

➤ Active Ion Transport and ATP Hydrolysis

We turn now to the ATP-powered pumps that transport ions against their concentration gradients. The Na^+-K^+ ATPase, for instance, pumps K^+ into the cell and Na^+ outwards, thus establishing the high cytosolic concentration of K^+ that is essential for generation of the cytosol-negative potential across the plasma membrane. The Ca^{2+} ATPases pump Ca^{2+} out of the cytosol into the extracellular medium or into intracellular organelles, thereby maintaining the concentration of Ca^{2+} in the cytosol much lower than that in the extracellular medium (see Table 15-2).

Early evidence for the existence of these pumps came from studies in which aerobic production of adenosine triphosphate (ATP) in a cell was inhibited by 2,4-dinitrophenol. The ion concentration inside the cell gradually approached that of the exterior environment as the ions moved through plasma membrane channels down their electric and concentration gradients. Eventually the cell died: partly because protein synthesis requires a high concentration of K^+ ions and partly because in the absence of a Na^+ gradient across the cell membrane, a cell cannot import certain nutrients such as amino acids. A significant fraction of available energy in every cell is required to maintain the concentration gradients of such ions as Na^+, K^+, and Ca^{2+} across the plasma and intracellular membranes. In nerve and kidney cells, for example, up to 25 percent of the ATP produced by the cell is used for ion transport, and in human erythrocytes, up to 50 percent of the energy stored in ATP molecules is used for this purpose. Thus a central issue of cellular metabolism is how permeation systems use energy.

Ion Pumps Can Be Grouped into Three Classes (P, V, and F)

Figure 15-10 illustrates the general structure of the three principal classes of ATP-powered ion pumps, called P, V,

and F, and Table 15-3 on page 646 summarizes their properties. (A fourth, less common class is described later.) Ion pumps in the P class, the simplest in structure, are composed of four transmembrane subunits: two α and two β polypeptides. The larger α subunit is phosphorylated during the transport process, and the transported ions are thought to move through this subunit. Included in this class are the Na^+-K^+ ATPase in the plasma membrane and several Ca^+ ATPases including one in the plasma membrane, which transports Ca^{2+} ions out of the cell, and another in the membrane of the sarcoplasmic reticulum (SR) of muscle cells, which transports Ca^{2+} ions from the cytosol to the SR lumen. A third member of the P class, found in certain acid-secreting cells, transports protons (H^+ ions) out of and K^+ ions into the cell.

Ion pumps of the V and F classes are similar in structure to each other but unrelated to P-class pumps. All known members of these two classes transport only protons. F-class pumps contain at least three kinds of transmembrane proteins and V-class pumps contain at least two kinds; both classes contain at least five kinds of extrinsic polypeptides that form the cytosolic domain (see Figure 15-10). Two of the transmembrane subunits and two of the extrinsic subunits in F-class pumps exhibit sequence homology with those in V-class pumps.

V-class ATP-powered pumps maintain the low pH of plant vacuoles and of lysosomes and other vesicles in animal cells by using the energy released by ATP hydrolysis to pump protons from the cytosolic to the exoplasmic face of the membrane up the proton electrochemical gradient. F-class proton pumps are found in bacterial plasma membranes and in mitochondria and chloroplasts. In contrast to V pumps, they generally function to power the synthesis of ATP from ADP and P_i by movement of protons from the exoplasmic to the cytoplasmic face of the membrane down the proton electrochemical gradient. Because of their importance in ATP synthesis in chloroplasts and mitochondria, F-class proton pumps are treated separately in Chapters 17 and 18.

All three classes of ATPases have one or more binding sites for ATP on the cytosolic face of the membrane. Although these proteins are often called ATPases, the system is tightly coupled so that the energy stored in the phosphoanhydride bond in ATP is not dissipated: ATP is not hydrolyzed into ADP and P_i unless ions are transported. The protein collects the free energy released during ATP hydrolysis and uses it to move ions uphill against a potential or concentration gradient. Here we discuss in some detail examples of P and V ion pumps, as well as a more recently discovered fourth class of ATP-driven pump. In the next section, we examine symporters and antiporters, which couple the movement of one substance (Na^+, for example) down its electrochemical gradient to the movement of another substance (say glucose) up its concentration gradient.

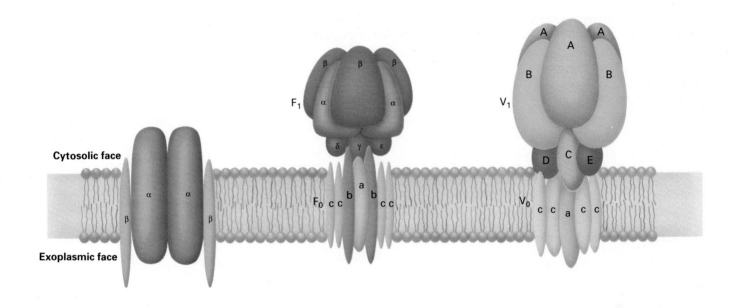

Subunit	Molecular weight (kD)	Subunit	Molecular weight (kD)	Subunit	Molecular weight (kD)
α	≈ 120	α	55	B	57
β	≈ 50	β	≈ 50	A	≈ 70
		γ	31	C	44
		δ	19	D	30
		ϵ	15	E	26
		a	30	a	20
		b	17	c	16
		c	8		

▲ FIGURE 15-10   The three classes of ATP-powered ion pumps. The P-class pumps are tetramers composed of two different polypeptides (α and β). All P-class pumps become phosphorylated as part of the transport cycle; the sequence around the phosphorylated residue, which is in the larger α subunit, is homologous among different pumps. V- and F-class pumps do not form phosphoprotein intermediates, and none of their subunits are related to those of P-class pumps. However, two of the cytosolic subunits and two of the transmembrane subunits in V- and F-class pumps exhibit homology (indicated by different shades of the same color). Each pair of homologous subunits is thought to have evolved from a common polypeptide. [Adapted from N. Nelson, 1992, *Curr. Opin. Cell Biol.* **4**:654.]

Ca^{2+} ATPase Maintains a Low Cytosolic Ca^{2+} Concentration

A prototype P-class ion pump is the Ca^{2+} ATPase, often called the calcium pump. It maintains the concentration of free Ca^{2+} ions in the cytosol at about 0.1–0.2 μM (1–2 $\times$ 10^{-7} M). (Not all Ca^{2+} in the cytosol is "free;" some is bound to the negative charges on phosphate, oxalate, or ATP molecules. In general, it is the concentration of unbound Ca^{2+} that is important.) In most eukaryotic cells, a Ca^{2+} ATPase located in the plasma membrane transports Ca^{2+} out of cell against its concentration gradient. Muscle cells contain a second, different Ca^{2+} ATPase that transports Ca^{2+} from the cytosol to the lumen of the sarcoplasmic reticulum, an internal organelle that concentrates and stores Ca^{2+} ions. As discussed in Chapter 22 the SR and its calcium pump (referred to as the *muscle calcium pump*) are critical in muscle contraction and relaxation: release of Ca^{2+} ions from the SR into the muscle cytosol causes contraction, and the rapid removal of Ca^{2+} ions from the cytosol by the muscle calcium pump induces relaxation.

Muscle Calcium Pump   Because the muscle calcium pump constitutes more than 80 percent of the integral protein in SR membranes, it is easily purified and character-

TABLE 15-3   Comparison of Major Classes of ATP-Powered Ion Pumps

Property	Class P	Class F	Class V
Number of different types of subunits	2	8 (minimum)	7 (minimum)
Ions transported	H^+, Na^+, K^+, Ca^{2+}	H^+ only	H^+ only
Characteristic functional features	Large α subunit is phosphorylated	Generally functions to synthesize ATP, powered by movement of H^+ down electrochemical gradient	Generally functions to hydrolyze ATP and to generate a transmembrane H^+ electrochemical gradient
Location of specific pumps	Plasma membrane of plants, fungi, bacteria (H^+ pump) Plasma membrane of higher eukaryotes (Na^+-K^+ pump) Plasma membrane of mammalian stomach cells (H^+-K^+ pump) Plasma membrane of all eukaryotic cells (Ca^{2+} pump) Sarcoplasmic reticulum membrane in muscle cells (Ca^{2+} pump)	Bacterial plasma membranes Inner mitochondrial membrane Thylakoid membrane of chloroplast	Vacuolar membranes in plants, yeast, other fungi Endosomal and lysosomal membrane in animal cells Plasma membrane of certain acid-secreting animal cells (e.g., osteoclasts and some kidney-tubule cells)

ized. A single transmembrane 100,000-MW α polypeptide contains the Ca^{2+} ATPase activity, transports two Ca^{2+} ions per ATP hydrolyzed, and requires Mg^{2+} to complex the ATP. [The function, and even the existence, of a β subunit is controversial.] The very high affinity of the cytosolic surface of this enzyme for the Ca^{2+} ion ($K_m = 10^{-7}$ M) allows it to transport Ca^{2+} very efficiently from the cytosol, where the free-Ca^{2+} concentration ranges from 10^{-7} M (resting cells) to more than 10^{-6} M (contracting cells), to the SR lumen, where the *total*-Ca^{2+} concentration can be as high as 10^{-2} M. The activity of the muscle Ca^{2+} ATPase is so regulated that if the free-Ca^{2+} concentration in the cytosol becomes too high, the rate of calcium pumping increases until the concentration is reduced to less than 1 μM.

The concentration of free Ca^{2+} within the sarcoplasmic reticulum is much less than the total concentration of 10^{-2} M. Two soluble proteins in the lumen of SR vesicles bind Ca^{2+}. One, *calsequestrin* (44,000 MW), is extremely acidic: 37 percent of its residues are aspartic and glutamic acids. Each molecule of calsequestrin binds 43 Ca^{2+} ions with K_m values ≈ 1 mM, amply justifying its name. The second protein, known as the *high-affinity Ca^{2+}-binding protein*, has a somewhat lower valency for Ca^{2+} but a higher affinity ($K_m = 3$–4 μM) than calsequestrin. Such proteins serve as a reservoir for intracellular Ca^{2+}, thereby reducing the concentration of free Ca^{2+} ions in the SR vesicles, and consequently decreasing the energy needed to pump Ca^{2+} ions into them from the cytosol.

Plasma-Membrane Calcium Pumps   The plasma membranes of animal, yeast, and probably plant cells contain Ca^{2+} ATPases that export Ca^{2+} ions from the cytosol into the extracellular medium. In many cases, the activity of these enzymes is stimulated by a rise in cytosolic free Ca^{2+} triggered, for instance, by hormone stimulation. The Ca^{2+}-binding regulatory protein *calmodulin* is an essential subunit of the erythrocyte and other plasma-membrane Ca^{2+} ATPases. A rise in cytosolic Ca^{2+} induces the binding of Ca^{2+} ions to calmodulin, which triggers an allosteric activation of the Ca^{2+} ATPase; as a result, the export of Ca^{2+} ions from the cell accelerates, and the original low cytosolic concentration of free Ca^{2+} (≈ 0.1 μM) is restored.

Coupling of ATP Hydrolysis and Ion Pumping by P-Class ATPases Involves an Ordered Kinetic Mechanism

The mechanism of action of the Ca^{2+} ATPase in the SR membrane is now known in some detail. It involves several steps that must occur in a defined order in order to couple ATP hydrolysis with ion pumping (Figure 15-11). When the protein is in one conformation, termed E1, first one and then a second Ca^{2+} ion binds with high affinity to sites on its cytosolic surface (step 1). Then an ATP binds to its site on the cytosolic surface; in a reaction that requires a Mg^{2+} ion to be tightly complexed to the ATP, the bound ATP is hydrolyzed to ADP and the liberated phosphate is transferred to a specific aspartate residue in the protein, forming a high-energy acyl phosphate bond, denoted by E1 ~ P (step 2). The protein then changes its conformation to E2 − P, propelling the two Ca^{2+} ions through it until they become weakly bound to Ca^{2+}-binding sites that are simultaneously formed on the exoplasmic surface of the protein, which faces the SR lumen (step 3). The Ca^{2+} ions then dissociate from the exoplasmic surface of the protein (step 4). Following this, the aspartyl-phosphate bond is hydrolyzed, causing E2 to revert to E1, inactivating the exoplasmic-facing Ca^{2+}-binding sites and regenerating the cytosolic-facing Ca^{2+}-binding sites (step 5).

Thus phosphorylation of the muscle calcium pump by ATP favors conversion of E1 to E2, and dephosphorylation favors the conversion of E2 to E1. While only E2 − P, not E1 ~ P, is actually hydrolyzed, the free energy of hydrolysis of the aspartyl phosphate bond in E1 ~ P is greater than that for E2 − P. The reduction in free energy of the as-partyl-phosphate bond in E2 − P, relative to E ~ P, can be said to power the E1 → E2 conformational change. The affinity of Ca^{2+} for the cytosolic-facing binding sites in E1 is a thousandfold greater than the affinity of Ca^{2+} for the exoplasmic-facing sites in E2; this difference enables the protein to transport Ca^{2+} unidirectionally from the cytosol outward.

Much evidence supports the model depicted in Figure 15-11. For instance, the ATPase has been isolated with phosphate linked to the aspartate residue, and spectroscopic studies have detected slight alterations in protein conformation during the E1 → E2 conversion.

Based on the amino acid sequence of the muscle calcium pump and various biochemical studies, investigators proposed the structural model for the catalytic α subunit shown in Figure 15-12. The protein contains 10 transmembrane α helices, at least four of which contain residues that bind Ca^{2+}; the membrane-spanning α helices are thought to form the passageway through which Ca^+ ions move. The bulk of the protein consists of three cytosolic globular domains, which have been shown to be involved in ATP binding, phosphorylation of aspartate, and energy transduction. These domains are connected by "stalks" to the membrane-embedded domain.

As noted previously, all P-class ATPases, regardless of which ion they pump, are phosphorylated during the transport process. The sequence data in Table 15-4 shows that the amino acid sequences around the phosphorylated aspartate in the catalytic α subunit are highly conserved in several enzymes of this type. Thus the mechanistic model in Figure 15-11 probably is generally applicable to all these ATP-powered ion pumps. In addition, the α subunits of all the P-class ion pumps examined to date have a similar molecular weight and, as deduced from their amino acid sequences derived from cDNA clones, have a similar arrangement of transmembrane α helices (see Figure 15-12).

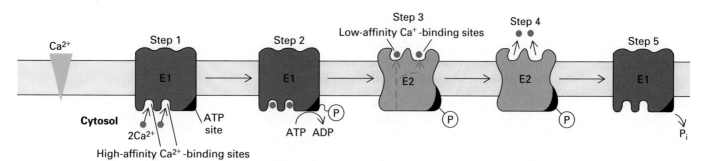

▲ FIGURE 15-11   Model of the mechanism of action of muscle Ca^{2+} ATPase which is located in the sarcoplasmic reticulum (SR) membrane. Only one α polypeptide of this P-class pump is shown. E1 and E2 are alternate conformational forms of the protein in which the Ca^{2+}-binding sites are on the cytosolic and exoplasmic faces, respectively. An ordered sequence of steps, as diagrammed here, is essential for coupling ATP hydrolysis and the transport of Ca^{2+} ions (blue circles) across the membrane. ~P indicates a high-energy acyl phosphate bond; —P indicates a low-energy phosphoester bond. See text for details. [Adapted from W. P. Jencks, 1980, *Adv. Emzomol.* **51**:75; W. P. Jencks, 1989, *J. Biol. Chem.* **264**:18855.]

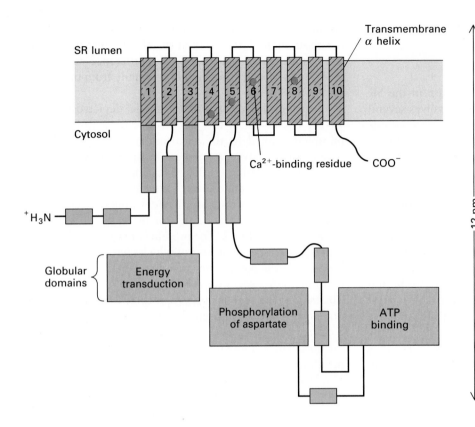

◄ FIGURE 15-12   Model for the structure of the catalytic subunit of muscle Ca^{2+} ATPase. The 10 transmembrane α helices are thought to form a channel through which Ca^{2+} ions move. Site-specific mutagenesis studies have identified four residues (red dots), located in four of the transmembrane helices (4, 5, 6, and 8), that participate in Ca^{2+} binding. Trypsin digestion releases three cytosolic globular domains, which constitute the bulk of the protein. One cytosolic domain functions in ATP binding; a second bears the aspartate that is phosphorylated/dephosphorylated; and the third is involved in energy transduction. [After D. H. MacLennan et al., 1985, *Nature* **316**:696; T. Toyofuku et al., 1992, *J. Biol. Chem.* **267**:14490.]

These findings strongly suggest that all these proteins evolved from a common precursor, although they now transport different ions.

Na⁺/K⁺ ATPase Maintains the Intracellular Concentrations of Na⁺ and K⁺ Ions in Animal Cells

A second P-class ion pump that has been studied in considerable detail is the Na⁺/K⁺ ATPase present in the plasma membrane. It has been solubilized and purified from the plasma membranes of several types of animal cells, including those in the mammalian kidney and the electric organ of eels (tissues very rich in this enzyme). This ion pump is a tetramer of subunit composition $\alpha_2\beta_2$. The β polypeptide is a 50,000-MW transmembrane glycoprotein, which is required for newly synthesized α subunits to fold properly but apparently is not involved directly in ion pumping. The α subunit is a 120,000-MW nonglycosylated polypeptide whose amino acid sequence and predicted membrane structure are very similar to those of the SR (muscle) Ca^{2+} ATPase. In particular, the Na⁺/K⁺ ATPase has a stalk on

---

TABLE 15-4   Amino Acid Sequence around the Phosphorylated Aspartate Residue in the Catalytic α Subunit of Several P-class Ion-translocating ATPases

P-class ATPase	Amino Acid Sequence	Subunit Composition
Rat stomach H⁺/K⁺ ATPase	TLGSTSVICSDKTGTLT	$\alpha_2\beta_2$
Sheep kidney Na⁺/K⁺ ATPase	TLGSTSTICSDKTGTLT	$\alpha_2\beta_2$
Rat brain α (+)isoform of the Na⁺/K⁺ ATPase	TLGSTSTICSDKTGTLT	$\alpha_2\beta_2$
Cardiac Muscle ATPase	TLGCTSVICSDKTGTLT	α

*The phosphorylated aspartate residue is in red type. The single-letter amino acid abbreviations are listed in Figure 3-3.
SOURCE: G. E. Shull and J. E. Lingrel, 1986, *J. Biol. Chem.* 261:16788; G. E. Shull et al., 1986, *Biochem.* 25:8125; D. H. MacLennan et al., 1985, *Nature* 316:696.

(a)

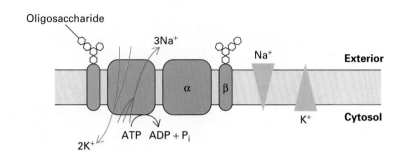

(b)

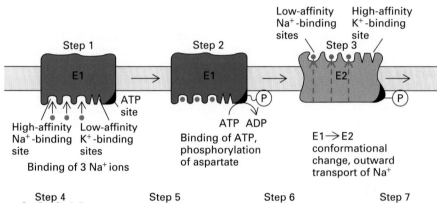

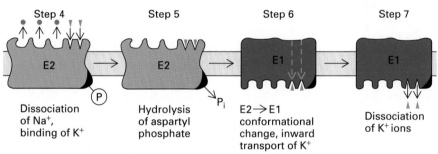

◄ FIGURE 15-13   Models for the structure and function of the Na$^+$/K$^+$ ATPase in the plasma membrane. (a) This P-class pump is a tetramer composed of two copies each of a small glycosylated β subunit and a large α subunit, which performs ion transport. Hydrolysis of one molecule of ATP to ADP and Pi is coupled to export of three Na$^+$ ions and import of two K$^+$ ions against their concentration gradients (gray triangles). It is not known whether only one α subunit, or both, in a single ATPase molecule transports ions. (b) Model of the mechanism of action of the Na$^+$/K$^+$ ATPase. For simplicity only a single α subunit is shown. This mechanism, like that for the Ca^{2+} ATPase, involves a high-energy acyl phosphate intermediate (E1 $\sim$ P) and conformational changes. In this case, hydrolysis of the E2-P intermediate powers transport of a second ion (K$^+$) inward. Na$^+$ ions are indicated by blue circles; K$^+$ ions, by red triangles. See text for details. [Adapted from P. Läuger, 1991, *Electrogenic Ion Pumps*, Sinauer Associates (Sunderland, MA), p. 178.]

the cytosolic face to which are bound domains that contain the ATP-binding site and the phosphorylated aspartate. The overall process of transport moves three Na$^+$ ions out of and two K$^+$ ions into the cell per ATP molecule split (Figure 15-13a).

Several lines of evidence indicate that the Na$^+$/K$^+$ ATPase is responsible for the coupled movement of K$^+$ and Na$^+$ into and out of the cell, respectively. For example, a strong correlation between the flux of Na$^+$ and K$^+$ ions across the plasma membrane and the activity of the Na$^+$/K$^+$ ATPase has been observed in a number of different tissues. In addition, the drug *ouabain*, which binds to a specific region on the exoplasmic surface of the protein, specifically inhibits Na$^+$/K$^+$ ATPase and also prevents cells from maintaining their Na$^+$/K$^+$ balance. Mutant cells that are resistant to ouabain have a point mutation in the Na$^+$/K$^+$ ATPase that prevents ouabain binding without affecting ion transport. Any doubt that the Na$^+$/K$^+$

ATPase is responsible for ion movement was dispelled by the demonstration that the enzyme, when purified from the membrane and inserted into liposomes, propels K$^+$ and Na$^+$ transport in the presence of ATP.

The mechanism of action of the Na$^+$/K$^+$ ATPase is similar to that of the Ca^{2+} ATPase. In the E1 conformational state the cytosolic-facing surface of the protein has three high-affinity sites for binding Na$^+$ ions (Figure 15-13b, step 1). The binding constant K_m for Na$^+$ to these cytosolic sites is 0.6 mM, a value considerably lower than the intracellular Na$^+$ ion concentration of $\approx$12 mM (see Table 15-2), indicating that normally Na$^+$ ions will fill these sites. Following binding of ATP, the terminal phosphate group is transferred to an aspartate residue, generating E1$\sim$P (step 2). The protein then undergoes a conformational change to E2, during which the three bound Na$^+$ ions move through the protein and become attached to low-affinity sites on the exoplasmic surface (step 3). The

Na$^+$ ions dissociate into the extracellular medium, and two K$^+$ ions bind to high-affinity sites on the exoplasmic face (step 4). The K_m for K$^+$ binding to these sites is about 0.2 mM, a lower value than the extracellular K$^+$ ion concentration of $\approx$4 mM (see Table 15-2), indicating that these sites will fill with K$^+$ ions. Then the aspartyl-phosphate bond is hydrolyzed (step 5), and the two bound K$^+$ ions move through the protein and become attached to low-affinity sites on the cytosolic surface (step 6). After these ions dissociate into the cytosol (step 7), the protein is ready to begin another cycle.

V-Class H$^+$ ATPases Pump Protons across Lysosomal and Vacuolar Membranes

Several ATP-powered proton pumps have been identified (see Figure 15-10). The H$^+$/K$^+$ ATPase, a P-class pump found in certain epithelial cells lining the stomach, is discussed later. Much more common are the V-class H$^+$ pumps in the membranes of lysosomes, endosomes, and plant vacuoles. These pumps function to acidify the lumen of these organelles.

The pH of the lysosomal lumen in growing cultured animal cells is usually $\approx$4.5–5.0. This can be measured precisely in living cells, which phagocytose fluorescent particles and transfer them to the lysosomes. The ability of different wavelengths of visible light to excite fluorescence is highly dependent on pH, and the lysosomal pH can be calculated from the spectrum of the fluorescence emitted. The cytosolic pH, in contrast, is $\approx$7.0; maintenance of this proton gradient of more than 100-fold between the lysosomal lumen and the cytosol depends on ATP production by the cell.

Isolated lysosomal and vacuolar membranes contain ATP-dependent proton pumps, which have been purified and incorporated into liposomes. These V-class H$^+$ ATPases differ greatly in structure and mechanism from the P class of ATPases in that they are composed of seven different polypeptide chains and are not phosphorylated/dephosphorylated during proton transport. As illustrated in Figure 15-10, V-class ATPases contain two discrete domains: a cytosolic-facing hydrophilic domain (V$_1$) composed of five different polypeptides and a transmembrane domain (V$_0$) containing several copies of proteolipid **c** and one copy of polypeptide **a**. The subunit composition of the cytosolic domain is A$_3$B$_3$CDE; the A and/or B subunits contain the site(s) where ATP binding and hydrolysis occur. Each transmembrane **c** subunit is thought to span the membrane several times; the **c** and **a** subunits together form the proton-conducting channel.

Similar V-class ATPases are found in the plasma membrane of certain acid-secreting cells, such as osteoclasts (bone-resorbing macrophage-like cells), which dissolve the calcium phosphate crystals that give bone its rigidity and strength. Osteoclasts bind to bone and seal off a small seg-

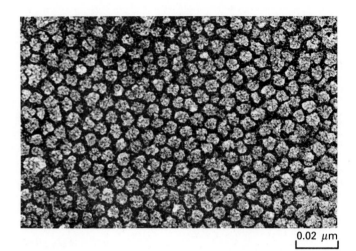

0.02 μm

▲ FIGURE 15-14   The plasma membrane of certain acid-secreting cells contains an almost crystalline array of V-class H$^+$ ATPases. This electron micrograph is of a platinum replica of the cytosolic surface of the apical plasma membrane of a toad bladder epithelial cell. Each stud is a single V-class H$^+$ ATPase ($\approx$600,000 MW) composed of several polypeptide subunits surrounding a central channel. [From D. Brown, S. Gluck, and J. Hartwig, 1987, *J. Cell Biol.* **105**:1637.]

ment of extracellular space between the plasma membrane and the surface of the bone. Secretion of HCl into this space by osteoclasts dissolves the bone. Another example is found in the mitochondria-rich epithelial cells lining the toad bladder; the apical plasma membrane of these cells contain many V-class H$^+$ ATPases, which function to acidify the urine (Figure 15-14). In addition to a typical V-class H$^+$ ATPase, the membrane of plant vacuoles contains a distinctive ion pump that utilizes the energy released by hydrolysis of inorganic pyrophosphate (PP$_i$) to pump protons into the vacuole. This PP$_i$-hydrolyzing proton pump, believed to be unique to plants, has an amino acid sequence different from any other ion-transporting proteins.

If the only transport protein in the membrane of a lysosome or vacuole were an ATP-powered proton pump, the lumen of the organelle would never become acidic. Addition of ATP to the cytosolic side of such a vesicle would initiate pumping of protons to the inside, but this would cause a buildup of positive charge on the inside of the vesicle membrane and a corresponding buildup of negative charges on the outside (the cytosolic face). In other words, the pump would generate an electric potential across the membrane, exoplasmic face positive, but not a proton concentration gradient. That is, the electric potential across the membrane would oppose further movement of protons from the cytosol into the vesicle lumen (or extracellular space). In order for an organelle lumen or an extracellular space (e.g., the outside of an osteoclast) to become acidic, movement of H^+ up its concentration gradient, powered by a proton pump, must be accompanied by (1) movement of an equal number of anions in the same direction or (2) movement of equal numbers of a different cation in the opposite direction. As discussed in detail later, the first process occurs in the membrane of plant vacuoles. The second occurs in the lining of the stomach, which contains a H^+/K^+ ATPase that pumps one H^+ outward and one K^+ inward.

The Multidrug-Transport Protein Is an ATP-Powered Pump and an ATP-Dependent Cl^- Channel

A series of rather unexpected observations led to recognition of a fourth class of ATP-powered transport protein in 1986. Oncologists, for instance, noted that tumor cells often became simultaneously resistant to several chemotherapeutic drugs with unrelated chemical structures; similarly, cell biologists observed that cultured cells selected for resistance to one toxic substance (e.g., colchicine, a microtubule inhibitor) frequently became resistant to several other drugs, including the anticancer drugs vinblastine and adriamycin. This resistance is due to a *multidrug-transport protein* known as P170, a 170,000-MW plasma membrane glycoprotein (Figure 15-15). This protein uses the energy derived from ATP hydrolysis to *export* a large variety of drugs from the cytosol to the extracellular medium. The P170 gene is frequently amplified in multidrug-resistant cells, resulting in a large overproduction of P170.

Most drugs transported by P170 are hydrophobic and diffuse from the culture medium across the plasma membrane into the cell; their active export from the cytosol by P170 increases the extracellular concentration of these drugs required to kill cells. Membrane vesicles containing P170 can carry out ATP-dependent transport of a number of drugs against their concentration gradients; as is the case with other transporters, transported drugs compete with

one another for transport by P170. P170 has been purified and incorporated into liposomes; different drugs have been shown to enhance the ATPase activity of these liposomes in a dose-dependent manner corresponding to their ability to be transported by P170. Thus, P170 is certainly an ATP-powered pump of certain small molecules.

What might be the normal function of P170? It is expressed in abundance in the liver, intestines, and kidney—sites from which natural toxic products are removed from the body. The P170 in these cells may move natural and metabolic toxins into the bile, intestinal lumen, or forming urine. Thus P170 evolved to transport a variety of toxic natural products and coincidentally acquired the ability to transport drugs whose structures are similar to those of these toxins. Tumors derived from these cell types, such as hepatomas (liver cancers), frequently are resistant to virtually all chemotherapeutic agents and thus difficult to treat, presumably because the tumors exhibit increased expression of the normal P170 protein.

Perhaps surprisingly, P170 is also a Cl^- channel. Expression of recombinant P170 protein in a cultured fibroblast, for instance, causes the appearance of a cell-volume–regulated Cl^- channel with properties similar to those of Cl^- channels in epithelial cells in which P170 is normally expressed. Activity of this channel is dependent on the presence of ATP, but since the Cl^- ions move down their concentration gradient, hydrolysis of ATP is unlikely to power transport of the Cl^- ions.

The *cystic fibrosis transmembrane regulator* (CFTR) protein exhibits some similarities to P170. Mutation in the *CFTR* gene encoding this protein causes cystic fibrosis (CF), the most common lethal autosomal recessive genetic disease of Caucasians. The disease is caused by defective Cl^- channels in the apical plasma membranes of epithelial cells in the lung, sweat glands, pancreas, and other tissues. In particular, increases of the intracellular second messenger cyclic AMP cause an increase in Cl^- transport by cells from normal, but not from CF, individuals. When the

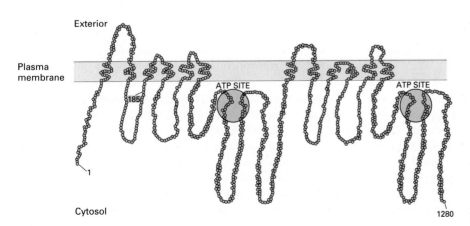

◄ FIGURE 15-15 Model of the human multidrug-transport protein. The two halves of this 1280-aa molecule have similar amino acid sequences, and there are two ATP-binding sites facing the cytosol. Since a point mutation at amino acid 185 changes the drug specificity of the transporter, this residue is believed to form part of the drug-binding site. [See M. M. Gottesman and I. Pastan, 1988, *J. Biol. Chem.* **263**:12163; G. Ferro-Luzzi Ames and H. Legar, 1992, *FASEB J.* **6**, 2660.]

CFTR gene was isolated by positional cloning (Chapter 8), the investigators were surprised to find that the sequence and predicted structure of the encoded CFTR protein were very similar to those of P170 (see Figure 15-15). CFTR protein clearly is a Cl^- channel whose activity is enhanced (in an unknown manner) by the presence of ATP. The purified protein, when incorporated into liposomes, forms Cl^- channels with properties similar to those in normal epithelial cells. And when the wild-type CFTR protein is expressed by recombinant techniques in cultured epithelial cells from CF patients, the cells recover normal chloride-channel activity. This latter result suggests that gene therapy might reverse the course of the disease. Since CFTR protein is similar to P170 in structure and its ATP-dependent chloride-channel activity, it may, like P170, also function as an ATP-powered pump of some as-yet unidentified molecule. In any case, much remains to be learned about this fascinating class of membrane proteins.

➤ Cotransport Catalyzed by Symporters and Antiporters

In addition to the substances transported by the multidrug-transport protein, Na^+, K^+, Ca^{2+}, and H^+ are the only known substances whose transport into or out of a cell against a concentration gradient is directly coupled to ATP hydrolysis. Yet cells often must import other molecules, such as glucose and amino acids, against a concentration gradient. To do this, a cell utilizes the energy stored in the transmembrane gradient of Na^+ or H^+ ions. For instance, the energetically favored movement of a Na^+ ion (the "cotransported" ion) into the cell across the plasma membrane, driven both by its concentration gradient and by the membrane electric potential, can be coupled obligatorily to movement of the "transported" molecule (e.g., glucose) against its concentration gradient. The transported molecule and cotransported ion can move in the same direction, in which case the process is called *symport:* when they move in opposite directions, the process is called *antiport* (Figure 15-2b).

Na$^+$-Linked Symporters Import Amino Acids and Glucose into Many Animal Cells

Importing glucose and amino acids from the lumen of the small intestine against their concentration gradients is a function of the microvilli of the intestinal epithelial cells (see Figure 14-40). Glucose and amino acids are transported from the intestinal lumen into these epithelial cells, which then export these substances to the blood (Figure 15-16a). The two stages of this transcellular transport process involve two sets of transporters: one in the microvilli on the apical surface of the plasma membrane; the other on the basolateral surface. A necessary condition for the trans-

cellular transport of glucose and amino acids is that the epithelial cell be polarized, with different sets of transport proteins localized in the basolateral and apical surfaces.

Glucose is transported against its concentration gradient from the intestinal lumen across the apical surface of the epithelial cells by the *sodium-glucose symporter* located in the microvillar membranes. This protein couples transmembrane movement of one glucose molecule to the transport of two Na^+ ions (Figure 15-16b):

$$2Na^+_{out} + glucose_{out} \rightleftharpoons 2Na^+_{in} + glucose_{in}$$

Movement of Na^+ from the intestinal lumen into the cell is driven by two forces: by the Na^+ ion concentration gradient (the Na^+ ion concentration is lower inside the cell than in the lumen) and by the inside negative membrane electric potential (see Figure 15-9). Quantitatively, the free-energy change for the symport transport of two Na^+ ions and one glucose molecule can be written

$$\Delta G = RT \ln \frac{[glucose_{in}][Na^+_{in}]^2}{[glucose_{out}][Na^+_{out}]^2} + 2FE$$

where E is the electric potential across the apical membrane and the other parameters are as defined previously. Although the values of the intracellular Na^+ concentration and the membrane electric potential are not known with certainty, we estimate from our earlier calculations that each contributes about -1.5 kcal/mol of Na^+ transported inward, or a total of about 3 kcal/mol (see Figure 15-9). Thus the change in free energy ΔG for transport of two moles of Na^+ inward is about -6 kcal. By substituting this value into the partial equation for glucose transport

$$\Delta G = RT \ln \frac{[glucose_{in}]}{[glucose_{out}]}$$

we can calculate that a ΔG of -6 kcal/mol can generate an equilibrium concentration of glucose inside the cell that is $\approx 30,000$ times greater than the exterior (luminal) concentration. Thus glucose from the intestinal lumen is concentrated in the cell by the action of the sodium-glucose symporter. By transporting two Na^+ ions per glucose, the intestinal sodium-glucose symport protein—called a *two-sodium/one-glucose symporter*—can accumulate glucose against a much steeper concentration gradient than if only one Na^+ ion were transported per glucose.

In the steady state, all the Na^+ ions transported from the intestinal lumen into the cell during Na^+-glucose symport, or the similar process of Na^+-amino acid symport, are pumped out across the basolateral membrane, often called the *serosal (blood-facing) membrane*. The Na^+/K^+ ATPase that accomplishes this is found in these cells exclu-

(a)

Basal lamina

Absorptive epithelial cells

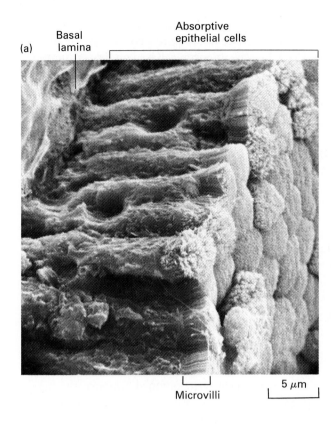

Microvilli

5 μm

(b)

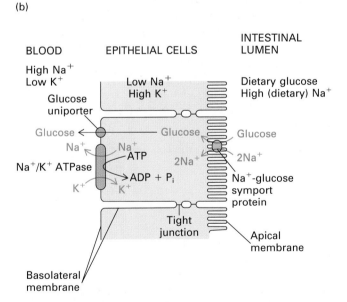

BLOOD

High Na⁺
Low K⁺

Glucose
uniporter

Glucose

Na⁺

Na⁺/K⁺ ATPase

K⁺

EPITHELIAL CELLS

Low Na⁺
High K⁺

Glucose

Na⁺

ATP

ADP + Pᵢ

K⁺

Tight
junction

Basolateral
membrane

INTESTINAL
LUMEN

Dietary glucose
High (dietary) Na⁺

Glucose

2Na⁺

2Na⁺

Na⁺-glucose
symport
protein

Apical
membrane

(c)

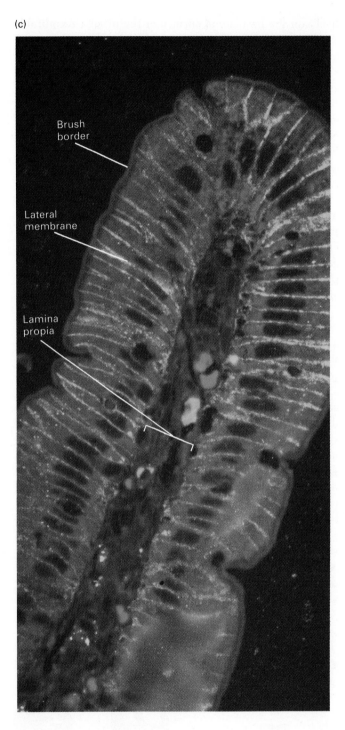

Brush
border

Lateral
membrane

Lamina
propia

▲ FIGURE 15-16   Transport of glucose from intestinal lumen into the blood. (a) Scanning electron micrograph of the absorptive epithelial cells of the small intestine. The cells reside on a basal lamina layer and contain abundant microvilli on their apical surface. (b) Entry of glucose from the intestinal lumen into the epithelial cells is catalyzed by a Na⁺-glucose symport protein (blue) located in the apical surface membrane. The Na⁺/K⁺ ATPase (green) in the basolateral surface membrane generates the Na⁺ gradient, which provides the energy for glucose uptake by pumping out the Na⁺ ions entering the cell by a Na⁺-glucose symporter. Glucose leaves the cell via a facilitated-diffusion uniporter (orange) in the basolateral membrane. (c) Immuno-microscopic localization of glucose transporters (uniporters) to the basal and lateral sides of rat intestine epithelial cells. The tissue section was stained with Evans blue, which generates a red fluorescence. The transporter, detected with a yellow-green-fluorescing antibody, is absent from the brush border. The lamina propria consists of vascularized connective tissue. [Part (a) from R. Kessel and R. Kardon, 1979, *Tissues and Organs: A Text-Atlas of Scanning Electron Microscopy,* W. H. Freeman and Company, p. 176. Part (c) see B. Thorens et al., 1990, *Am. J. Physiol.* **259:**C279; courtesy of B. Thorens.]

sively on the basolateral surface of the plasma membrane (see Figure 15-16b). Glucose and amino acids concentrated inside the cell by symport move outward through the basolateral membrane via uniport-transport proteins, such as the glucose transporter discussed earlier (see Figure 15-7), which also are localized to this region of the plasma membrane (Figure 15-16c). The net result is movement of Na^+ ions, amino acids, and glucose from the intestinal lumen across the intestinal epithelium into the blood. Tight junctions between the epithelial cells prevent these molecules from diffusing back into the intestinal lumen.

The epithelial cells lining kidney tubules also contain sodium-glucose symporters. These cells, which have an architecture similar to that of intestinal epithelial cells (see Figure 15-16), reabsorb glucose from the blood filtrate that is the forming urine back into the blood. In the first part of a kidney tubule, the epithelial cells transport glucose against a relatively small glucose concentration gradient. These cells utilize a second type of sodium-glucose symport protein—a *one-sodium/one-glucose symporter,* which has a high transport rate but cannot transport glucose against a steep concentration gradient. At the intracellular Na^+ concentration and membrane potential depicted in Figure 15-9, a one-sodium/one-glucose symporter could generate an intracellular glucose concentration ≈ 100 times that of the extracellular medium (here the forming urine). In the latter part of a kidney tubule, however, the epithelial cells take up the remaining glucose against a more than hundredfold glucose concentration gradient. To accomplish this, these cells contain in their apical membrane the same two-sodium/one-glucose symporter found in intestinal epithelial cells.

The two types of sodium-glucose symport proteins are similar in amino acid sequence, predicted structure, and mechanism but have evolved to transport glucose under different conditions. The mechanism of action of sodium-glucose symporters has not been elucidated, but confor-

mational changes similar to those that occur in uniport transporters, which do not require a cotransported ion (see Figure 15-7), may be important. Based on this assumption, an analogous model has been proposed for the sodium-glucose symporter (Figure 15-17).

In summary, the transepithelial movement of glucose and amino acids is driven by the cellular hydrolysis of ATP. The coupling of ATP hydrolysis to the entry of glucose (or amino acids) and Na^+ is indirect. ATP hydrolysis is used directly to generate a Na^+ and a K^+ concentration gradient and the K^+ gradient generates an inside-negative membrane potential. Both the Na^+ concentration gradient and the membrane potential are used as a source of energy to drive the uptake of glucose and amino acids against their concentration gradient from the lumen of the intestine or kidney tubules. These substances then move from the epithelial cells into the blood down their concentration gradient by uniporter-catalyzed transport.

Na^+-Linked Antiporter Exports Ca^{2+} from Cells

In the previous section, we examined symport, the movement of one substance (usually Na^+) down its concentration gradient coupled to the inward movement of a different substance (e.g., glucose or an amino acid) against its concentration gradient. Antiport is a similar process (see Figure 15-2), except that the inward movement of an ion such as Na^+ is coupled obligatorily to the outward movement of a different molecule:

$$Na^+_{out} + X_{in} \rightleftharpoons Na^+_{in} + X_{out}$$

where X is exported from the cell against a concentration gradient.

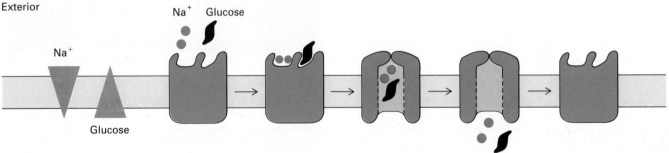

Exterior

Na^+ Glucose

Na^+

Glucose

Cytosol

▲ FIGURE 15-17   Proposed model for operation of a Na^+-glucose symport protein (blue), which has binding sites for two Na^+ ions and one glucose ≈ 3.5 nm apart on its exoplasmic surface. The simultaneous binding of Na^+ and glucose to these sites induces a conformational change, generating a transmembrane pore or tunnel that allows both Na^+ and glucose to pass into the cytosol. After this passage, the protein reverts to its original conformation. [See E. Wright, K. Hager, and E. Turk, 1992, *Curr. Opin. Cell Biol.* **4**:696 for details on the structure and function of this and related transporters.]

In cardiac muscle cells an Na^+/Ca^{2+} antiporter, rather than a plasma membrane Ca^{2+} ATPase, is the principal protein that reduces the concentration of Ca^{2+} in the cytosol. Its reaction can be written:

$$3Na^+_{out} + Ca^{2+}_{in} \rightleftharpoons 3Na^+_{in} + Ca^{2+}_{out}$$

Note that the movement of three Na^+ ions is required to power the export of one Ca^{2+} ion against a greater than 10000-fold concentration gradient (0.1 μM versus 2 mM). As in other muscle cells, a rise in the intracellular Ca^{2+} concentration in cardiac muscle triggers contraction. Thus the operation of the sodium-calcium antiporter lowers the cytosolic concentration of Ca^{2+} and reduces the strength of heart-muscle contraction. The Na^+/K^+ ATPase in the plasma membrane of cardiac cells, as in other body cells, creates the Na^+ concentration gradient used to power export of Ca^{2+} ions. Both the Na^+/K^+ ATPase and the sodium-calcium antiporter appear to be distributed throughout the plasma membrane of cardiac muscle cells, unlike the localized distribution of the Na^+/K^+ ATPase in polarized epithelial cells.

Drugs such as ouabain and digoxin are of great clinical significance: they increase the force of heart-muscle contraction and are widely used in the treatment of congestive heart failure. The primary effect of these drugs is to inhibit the Na^+/K^+ ATPase, thereby raising the intracellular Na^+ concentration. Because the sodium-calcium antiporter functions less efficiently with a lower Na^+ concentration gradient, fewer Ca^{2+} ions are exported and the intracellular Ca^{2+} concentration increases. This increase causes the muscle to contract more strongly.

Band 3 Is an Anion Antiporter That Exchanges Cl^- and HCO_3^- across the Erythrocyte Membrane

Not all antiport transport involves cations (positive ions) such as Na^+ and Ca^{2+}. Band 3, the predominant integral protein of the mammalian erythrocyte, is an *anion antiporter* that catalyzes the one-for-one exchange of singly charged anions (negative ions), such as Cl^- and HCO_3^-, across the plasma membrane. Since one negative ion is exchanged for another, there is no net movement of electric charge and the reaction is not affected by the membrane potential. The direction of the reaction is dependent only on the concentration gradients of the transported ions.

Transmembrane anion exchange is essential to an important function of the erythrocyte—the transport of waste carbon dioxide (CO_2), which is generated in peripheral tissues, to the lungs for excretion by respiratory exhalation (Figure 15-18). Waste CO_2 released from cells into the capillary blood diffuses across the erythrocyte membrane. In its gaseous form, CO_2 dissolves poorly in aqueous solutions, such as blood plasma, but the potent enzyme

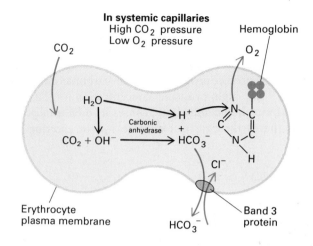

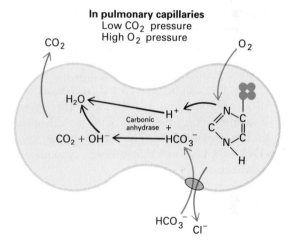

▲ FIGURE 15-18   Schematic drawings showing anion transport across the erythrocyte membrane in systemic and pulmonary capillaries. Band 3 protein (purple)—an anion antiporter—catalyzes the reversible exchange of Cl^- and HCO_3^- ions across the membrane and works in conjunction with carbonic anhydrase. In systemic capillaries, the overall reaction causes HCO_3^- to be released from the cell, which is essential for CO_2 transport from the tissues to the lungs. In the lungs, the overall reaction is reversed.

carbonic anhydrase inside the erythrocyte converts CO_2 to the water-soluble bicarbonate (HCO_3^-) anion:

$$OH^- + CO_2 \rightleftharpoons HCO_3^-$$

Since

$$H^+ + OH^- \rightleftharpoons H_2O$$

we can write the overall reaction for carbonic anhydrase as

$$H_2O + CO_2 \rightleftharpoons HCO_3^- + H^+$$

This process occurs while the hemoglobin in the erythrocyte is releasing oxygen into the blood plasma. The removal of oxygen from hemoglobin induces a change in its conformation that enables a histidine of a globin side chain to bind the proton produced by the carbonic anhydrase reaction. The high cytosolic concentration of HCO_3^-, formed by carbonic anhydrase, causes it to be transported out of the erythrocyte in exchange for an entering Cl^- anion:

$$HCO_3^-{}_{in} + Cl^-{}_{out} \rightleftharpoons HCO_3^-{}_{out} + Cl^-{}_{in}$$

(see Figure 15-18). The entire anion-exchange process is completed within 50 milliseconds (ms), during which time 5×10^9 HCO_3^- ions are exported from the cell down its concentration gradient. If anion exchange did not occur, HCO_3^- would accumulate inside the erythrocyte to toxic levels during periods such as exercise, when much CO_2 is generated. About 80 percent of the CO_2 in blood is transported as HCO_3^- generated inside erythrocytes; anion exchange allows about two-thirds of this HCO_3^- to be transported by blood plasma external to the cells, increasing the amount of CO_2 that can be transported from tissues to the lungs. Also, without anion exchange, the increased HCO_3^- concentration in the erythrocyte would cause the cytosol to become alkaline. The exchange of HCO_3^- (equal to $OH^- + CO_2$) for Cl^- causes the cytosolic pH to remain near neutrality.

The overall direction of this anion-exchange process is reversed in the lungs. CO_2 diffuses out of the erythrocyte and is eventually expelled in breathing. The lowered concentration of CO_2 within the cytosol drives the carbonic anhydrase reaction, as written above, from right to left: HCO_3^- reacts to yield CO_2 and OH^-. Oxygen binding to hemoglobin causes a proton to be released from hemoglobin; the proton combines with the OH^- to form H_2O. The lowered intracellular HCO_3^- concentration causes HCO_3^- to enter the erythrocyte in exchange for Cl^-.

Band 3, which has been studied extensively, catalyzes the precise one-for-one sequential exchange of anions on opposite sides of the membrane required to preserve electroneutrality in the cell; only once every 10,000 or so transport cycles does an anion move unidirectionally from one side of the membrane to the other. Band 3 also participates in other functions of the erythrocyte membrane, such as the binding of the cytoskeletal protein ankyrin (see Figure 14-38). As discussed in Chapter 14, band 3 is composed of a membrane-embedded domain, folded into at least 12 transmembrane α helices, and a cytosolic domain (see Figure 14-37). Experiments with protease-generated fragments of band 3 have shown that the membrane-bound domain in the protein catalyzes anion transport. The precise mechanism of anion transport is not known; as with other transport proteins, however, conformational changes that alternate a cytosolic-facing and an exoplasmic-facing anion-binding site appear to be important (Figure 15-19).

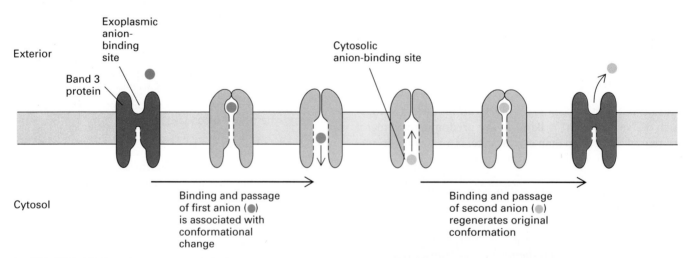

▲ FIGURE 15-19   Proposed model of anion cotransport by band 3 protein. The protein catalyzes a one-for-one exchange of anions, probably shuttling between two conformations. In one form (dark purple), it binds an anion (dark blue) on its exoplasmic surface, causing a conformational change that generates a tunnel through which the anion crosses the membrane and also creates an anion-binding site on the cytosolic surface (light purple). Binding of the second anion (light blue) to the cytosolic site and its passage across the membrane in the opposite direction regenerate the original conformation with the binding site on the outer surface. Experiments supporting this model have shown that band 3 can have an anion-binding site on the cytosolic or exoplasmic surface, but not on both surfaces simultaneously. [See H. Passow, 1986, *Rev. Physiol. Biochem. Pharmacol.* **103**:62, and S. L. Alper, 1991, *Annu. Rev. Physiol.* **53**:549.]

H$^+$/K$^+$ ATPase and Anion Antiporter Combine to Acidify the Stomach Contents While Maintaining Cytosolic pH Near Neutrality

The mammalian stomach contains a 0.1-M solution of hydrochloric acid (H$^+$Cl$^-$). This strongly acidic medium denatures many ingested proteins before they are degraded by proteolytic enzymes (e.g., pepsin) that function at acidic pH. Hydrochloric acid is secreted into the stomach by *parietal cells* (also known as oxyntic cells) in the gastric lining

(Figure 15-20a). These cells contain a *H$^+$/K$^+$ ATPase* in their apical membrane, which faces the stomach lumen. This enzyme is a P-class ATPase (see Figure 15-10), similar in structure and function to the Na$^+$/K$^+$ ATPase discussed earlier (see Figure 15-13). Unlike the Na$^+$/K$^+$ ATPase, which pumps one positively-charged ion outward [3Na$^+$ outward *versus*-2K$^+$ inward] during each cycle of ATP hydrolysis, the action of the H$^+$/K$^+$ ATPase is electroneutral. It exports one H$^+$ ion and imports one K$^+$ ion for each ATP hydrolyzed; thus there is no *net* movement of electric charge. This proton-potassium pump generates a concen-

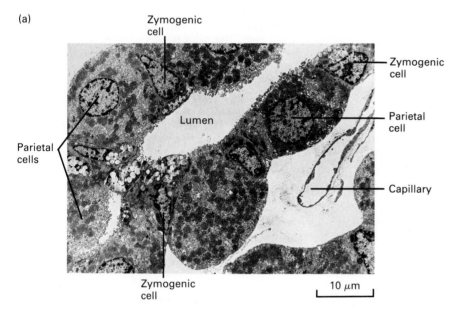

(a)

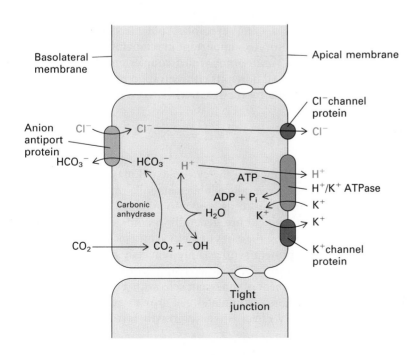

◄ FIGURE 15-20 Acidification of stomach lumen by parietal cells in the gastric lining. (a) Electron micrograph of acid-secreting parietal cells (gastric epithelial cells) as well as zymogenic cells that secrete the protease pepsinogen. Parietal cells contain microvilli on the luminal face; the basal surface rests on a basement membrane that, in turn, is fed by a small blood capillary. Parietal cells contain abundant mitochondria that provide energy for the transcellular transport of H$^+$ and Cl$^-$ ions for production of hydrochloric acid. (b) Schematic representation of four different transport proteins whose combined operation acidifies the stomach lumen while maintaining the pH and electroneutrality of the cytosol. The apical membrane of parietal cells contains a H$^+$/K$^+$ ATPase (a P-class pump) as well as Cl$^-$ and K$^+$ channel proteins. The basolateral membrane contains a band 3–like anion antiporter that exchanges HCO$_3^-$ and Cl$^-$ ions. See text for details. [Part (a) from R. Kessel and R. Kardon, 1979, *Tissues and Organs: A Text-Atlas of Scanning Electron Microscopy,* W. H. Freeman and Company, p. 170.]

tration of H^+ ions 10^6 times greater in the stomach lumen than in the cell cytosol (pH = 1.0 versus pH = 7.0).

Since $[H^+] \times [OH^-] = 10^{-14} M^2$, export of H^+ ions from the parietal-cell cytosol would be expected to result in a rise in the concentration of OH^- ions and thus a marked rise in cytosolic pH. However, during HCl pumping the pH of the cytosol remains neutral; how does this happen? The answer, again, is anion exchange (Figure 15-20b). The excess cytosolic OH^- combines with CO_2 that diffuses into the cell from the blood, forming HCO_3^- in a reaction catalyzed by cytosolic carbonic anhydrase. The HCO_3^- ion is transported across the basolateral membrane of the cell into the blood in exchange for an incoming Cl^- ion by means of an anion-exchange protein, which is similar in structure and function to the erythrocyte band 3 protein. Thus the H^+ ion that is generated by carbonic anhydrase is pumped outward across the apical membrane in exchange for a K^+ ion; likewise, the HCO_3^- ion generated by carbonic anhydrase is transported across the basolateral membrane in exchange for a Cl^- ion. The Cl^- ions thus imported into the cell exit through Cl^- channels in the apical membrane, entering the stomach lumen. To preserve electron neutrality, each Cl^- ion that moves outward across the apical membrane is accompanied by a K^+ ion that moves outward through a separate K^+ channel (see Figure 15-20b); these are the same K^+ ions that were pumped inward by the H^+/K^+ ATPase. The net result is accumulation of both H^+ and Cl^- ions (i.e., HCl) in the stomach lumen.

In summary, the coordinated actions of H^+/K^+ ATPases, anion antiporters, K^+ channels, and Cl^- channels, all localized to specific domains of the plasma membrane of parietal cells, are essential for net transport of H^+ and Cl^- ions into the stomach lumen, and thus its acidification, without any change in cytosolic pH. Tight junctions connect all the epithelial cells, preventing H^+ or Cl^- in the lumen from diffusing into fluids surrounding the basolateral membrane and HCO_3^- from diffusing into the lumen.

Several Symporters and Antiporters Regulate Cytosolic pH

In order for all cells to grow and divide, the pH of the cytosol must be maintained within very narrow limits (7.2–7.4). This task is compounded by the fact that most metabolic products of glucose are acidic, even though glucose is not ionized. The anaerobic metabolism of glucose yields lactic acid, and aerobic metabolism yields CO_2, which is hydrated by carbonic anhydrase to carbonic acid (H_2CO_3). These weak acids dissociate, yielding protons. If these protons were not exported from the cell, the cytosolic pH would drop precipitously.

Two types of transporters are employed to remove some of the "excess" protons generated during metabolism of animal cells (Figure 15-21). One, a $Cl^-/Na^+HCO_3^-$ trans-

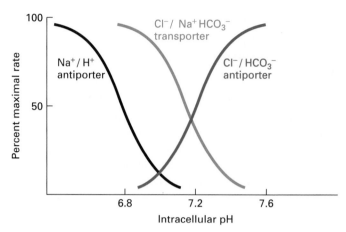

▲ **FIGURE 15-21**   Effect of intracellular pH on activity of membrane transport proteins that regulate the cytosolic pH of mammalian cells. The cytosolic pH of most growing cells is ≈7.2. Increases in the cytosolic pH above this value activate allosterically a Cl^-/HCO_3^- antiporter (blue curve) that exports HCO_3^- in exchange for Cl^-; this enzyme operates to lower the cytosolic pH towards the optimal value. A small reduction in pH activates a $Cl^-/Na^+HCO_3^-$ transporter (red curve) that imports HCO_3^- and Na^+ in exchange for Cl^-, thus raising the cytosolic pH. At even lower cytosolic pH values, a Na^+/H^+ antiporter is activated (black curve); this enzyme exports H^+ in exchange for Na^+, causing an even faster rise in the pH of the cytosol. [After S. L. Alper, 1991, *Annu. Rev. Physiol.* **53**:549.]

porter, imports one Na^+ ion down its concentration gradient, together with one HCO_3^-, in exchange for export of one Cl^- ion. The imported HCO_3^- ions combine with protons generated by metabolism to produce CO_2, which diffuses out of the cell. Thus the overall action of this transporter raises the cytosolic pH (reduces the H^+ concentration). Also important in removing excess protons is a *Na^+/H^+ antiporter*, which couples entry of one Na^+ ion into the cell down its concentration gradient to export of one H^+ ion by the reaction

$$Na^+_{out} + H^+_{in} \rightleftharpoons Na^+_{in} + H^+_{out}$$

Small changes in the cytosolic pH may have profound effects on the overall cellular metabolic rate. For instance, primary fibroblast cells grown to maximal density (confluence) in tissue culture generally become quiescent: DNA synthesis stops; the rates of RNA synthesis, glucose catabolism, and protein synthesis are reduced; and the cytosolic pH drops from the characteristic 7.4 of growing cells to ≈7.2. Treatment of quiescent cells with a mixture of serum growth factors restimulates cell growth and DNA synthesis. An early effect of these growth factors is a marked increase in the cytosolic pH to 7.4; this dramatic change is caused in part by stimulation of the Na^+/H^+ antiport, which expels protons into the medium. The rise in cytosolic pH is believed to help activate the metabolic pathways required for cell growth and division.

The plasma membranes of most cultured animal cells contain a band 3–like Cl^--HCO_3^- antiporter similar to the parietal-cell anion antiporter discussed previously. A rise in the cytosolic pH above 7 ($[OH^-] > 10^{-7}$ M) stimulates this anion-exchange protein, which then functions to lower the pH. Recall that a HCO_3^- ion can be viewed as a complex of OH^- and CO_2, so export of HCO_3^- lowers the cytosolic pH. Exchange of cytosolic HCO_3^- for extracellular Cl^- is powered by the out → in Cl^- concentration gradient:

$$Cl^-_{out} + HCO_3^-{}_{in} \rightleftharpoons Cl^-_{in} + HCO_3^-{}_{out}$$

The activity of the Cl^-/HCO_3^- antiporter is increased as the cytosolic pH rises, resulting in more rapid export of HCO_3^- and a lowering of the cytosolic pH (Figure 15-21). Similarly, the Na^+/H^+ antiporter and the $Cl^-/Na^+HCO_3^-$ transporter are activated when the pH of the cytosol is lowered, resulting in a rise in the cytosolic pH. In this manner the cytosolic pH of growing cells is maintained very close to pH 7.4.

▶ Plant and Prokaryotic Membrane Transport Proteins

Plants, fungi, and prokaryotes contain the same general types of transport proteins as animal cells (see Figure 15-2). However, they often use a different type of protein than animal cells to accomplish a given task (e.g., generation of membrane potential) and also face their own transport needs (e.g., concentrating sucrose in plant vacuoles). In this section, we describe a few important transport proteins found in plants, fungi, and prokaryotes.

H^+ Pumps and Anion Channels Establish an Electric Potential and a Steep H^+ Concentration Gradient across the Plant-Vacuole Membrane

As noted earlier, the membrane of plant vacuoles contains a V-class ATPase (see Figure 15-10) and a unique PP_i-powered pump, both of which function to pump H^+ ions into the vacuolar lumen against a concentration gradient. If these H^+ pumps functioned by themselves, they would generate an inside-positive electric potential but not a proton concentration gradient across the membrane. As illustrated in Figure 15-22, the vacuolar membrane also con-

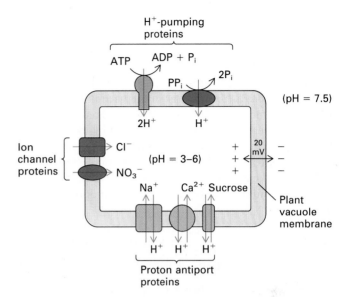

▲ FIGURE 15-22   Concentration of ions and sucrose by the plant vacuole. The vacuole membrane contains two types of proton pumps: a V-class H^+ ATPase (light green) and a unique pyrophosphate-hydrolyzing proton pump (dark green). These pumps generate a lowered luminal pH as well as an inside-positive electric potential across the vacuolar membrane due to the inward pumping of H^+ ions. The inside-positive potential powers the movement of Cl^- and NO_3^- from the cytosol through separate channel proteins (dark purple). Proton antiporters (light purple), powered by the H^+ gradient, accumulate Na^+, Ca^{2+}, and sucrose inside the vacuole. [After P. Rea and D. Sanders; 1987, *Physiol. Plant.* **71**:131; J. M. Maathuis and D. Saunders, 1992, *Curr. Opin. Cell Biol.* **4**:661; P. A. Rea et al., 1992, *Trends Biochem. Sci.* **17**:348.]

tains Cl^- and NO_3^- channels that transport these anions into the vacuole; anion entry against a concentration gradient is driven by the inside-positive potential generated by the H^+ pumps. Operation of both types of proton pumps in conjunction with these anion channels produces an inside-positive electric potential of about 20 mV across the vacuolar membrane and a H^+ concentration within the vacuole 30–30,000 times greater than that of the cytosol. Thus, the lumen of plant vacuoles is much more acidic (pH 3–6) than is the cytosol (pH 7.5).

Proton Antiporters Enable Plant Vacuoles to Accumulate Metabolites and Ions

The proton gradient and electric potential across the plant-vacuole membrane are used in much the same way as the Na^+ gradient and electric potential across the animal-cell plasma membrane: to power the selective uptake or extrusion of ions and small molecules. In the leaf, for example, excess sucrose generated during photosynthesis in the day is stored in the vacuole; during the night the stored sucrose moves into the cytoplasm and is metabolized to CO_2 and H_2O with concomitant generation of ATP from ADP and P_i.

Accumulation of sucrose in plant vacuoles is catalyzed by a *proton-sucrose antiporter* in the vacuolar membrane. The inward movement of sucrose is powered by the outward movement of H^+, which is favored by its concentration gradient (lumen > cytosol) and by the outward-negative potential across the vacuolar membrane (see Figure 15-22). Uptake of Ca^{2+} and Na^+ into the vacuole from the cytosol is similarly catalyzed by proton antiports.

The Potential Across the Plasma Membrane of Plant, Bacterial, and Fungal Cells Is Generated by Proton Pumping

The concentration gradients of Na^+ and K^+ across the plasma membrane of mammalian cells are established by the Na^+/K^+ ATPase, and the resting membrane potential of about −70 mV is generated mainly by outward movement of K^+ through potassium channels. Plant cells, as well as fungal and bacterial cells, lack a Na^+/K^+ ATPase. In these cells, a P-class H^+ ATPase in the plasma membrane pumps protons out of the cell, thus generating an inside-negative membrane potential. Much as in the vacuolar membrane, this membrane potential is used to power uptake of certain nutrients and ions. For instance, the unicellular freshwater algae *Chlorella* contains a proton-hexose symport that is utilized for uptake of sugars against their concentration gradients.

Many aerobic bacteria derive their energy from the oxidation of glucose to CO_2, during which electrons are transferred from key metabolic intermediates (pyruvate and reduced nicotinamide adenine dinucleotide, or NADH) to oxygen, the ultimate electron acceptor. The electron transporters are integral proteins in the bacterial plasma membrane. As electrons move along the transport chain, protons are actively transported out of the cell, which can result in a membrane electric potential of up to 100 mV (inside negative), as well as a 100-fold increase in the H^+ concentration in the surrounding medium relative to the cytosol. (The extracellular pH can become 5.0–5.5, while cytosolic pH remains at ~7.3.) As in mitochondria (Chapter 17), the energy stored in this proton gradient is principally used to generate ATP from ADP and P_i, using an F-class H^+ ATPase, and to import nutrients from the medium.

Proton Symporters Import Many Nutrients into Bacteria

Nutrient uptake by bacterial cells is a more formidable task than nutrient uptake by mammalian cells for two main reasons. First, concentrations of glucose, amino acids, and other nutrients are finely regulated in the blood that bathes mammalian cells, whereas bacteria are often subjected to

media of widely differing compositions. The common intestinal bacterium *E. coli*, for instance, also grows in soil and in freshwater lakes—environments with very different nutrient compositions. Second, bacteria often must concentrate nutrients such as sugars, amino acids, and vitamins against very high concentration gradients. Bacteria have evolved systems for nutrient uptake that can function at greater than 100-fold concentration gradients. Generally, these transporters are inducible: the quantity of a transport protein in the cell membrane is regulated according to both the concentration of the nutrient in the medium and the metabolic needs of the cell.

Bacteria use two very different systems for nutrient uptake. In one, not discussed here, phosphorylation of a sugar is part of the uptake mechanism; this system is believed to be unique to bacteria. In the other, a H^+ concentration gradient across the bacterial membrane powers nutrient uptake; this process is analogous to the use of the H^+ or Na^+ concentration gradients, respectively, across the plant-vacuole membrane or the plasma membrane of mammalian cells.

In *E. coli*, the sugar lactose is concentrated from the medium by the lactose transporter, more properly called the *proton-lactose symporter* (Figure 15-23). This 30,000-MW integral membrane protein is encoded by the Y gene of the *lac* operon. The cotransport of lactose can be demonstrated by poisoning *E. coli* cells with an inhibitor of

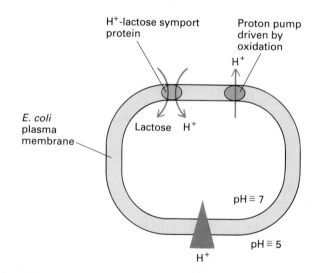

▲ FIGURE 15-23   The import of lactose into *E. coli* is fueled by the energy stored in a proton gradient across the plasma membrane. Oxidation of pyruvate to CO_2 is coupled to the pumping of protons out of the cytosol, generating as much as a 100-fold H^+ concentration gradient across the membrane. Lactose is imported by an H^+-lactose symport protein (encoded by the *lacY* gene). The inward movement of protons is obligatorily coupled to the inward movement of lactose.

oxidative phosphorylation (e.g., cyanide). Since the electron transport chain is blocked in poisoned cells, no protons are pumped across the plasma membrane and the cells do not develop a proton gradient (outside > cytosol). Experimentally, the medium pH and cytosolic pH become the same, and the cells do not concentrate lactose from the medium. When the medium surrounding the poisoned cells is acidified by the addition of HCl, the cells again concentrate lactose, using the energy of this artificially imposed proton gradient.

The sequence of the *E. coli* proton-lactose symporter has been determined. Although the structure of the protein at atomic resolution is not yet available, its sequence is compatible with 12 transmembrane α helices (Figure 15-24). It is thought that a conformational change in the protein associated with proton transport (down the H+ gradient) allows simultaneous transport of a bound lactose molecule up its concentration gradient. According to one model, histidine 322 and glutamate 325, both located in the same α helix, function as a relay system for proton transport. That is, histidine 322 receives a proton from the outside (exoplasmic) solution and transfers it to glutamate 325; from there the proton is passed into the inside (cytosolic) solution. This process is accompanied by a reversible conformational change in the protein during which a lactose molecule, bound to a site on the exoplasmic surface is transported through the protein, becomes bound to a site on the cytosolic face, and then is released into the cytosol.

Studies with *E. coli* mutants have supported this model. For example, mutation of the histidine at position 322 to an arginine abolishes proton-lactose symport activity and remarkably converts the protein to a lactose uniporter (facilitated-diffusion transporter), which can transport lactose only down its concentration gradient (similar to the glucose transporter). In addition, *E. coli* strains in which alanine 177 or tyrosine 236 is mutated to

certain other amino acids have an enhanced ability to transport maltose, a disaccharide similar to lactose. These findings indicate that alanine 177 and tyrosine 236 are involved in binding and transporting lactose (see Figure 15-24).

➤ Osmosis, Water Channels, and the Regulation of Cell Volume

In this section, we examine two types of phenomena that, at first glance, may seem unrelated: the regulation of cell volume in both plant and animal cells, and the *bulk flow* of water (the movement of water containing dissolved solutes) across one or more layers of cells. In humans, for example, water moves from the blood filtrate that will form urine across a layer of epithelial cells lining the kidney tubules and into the blood, thus concentrating the urine. (If this did not happen, one would excrete several liters of urine a day!) In higher plants, water and minerals are absorbed by the roots and move up the plant through conducting tubes (the *xylem*); water is lost from the plant mainly by evaporation from the leaves. What these processes have in common is *osmosis*—the movement of water from a lower to a higher solute concentration (Figure 15-25). We begin with a consideration of some basic facts about osmosis, and then show how they explain many physiologic properties of animals and plants.

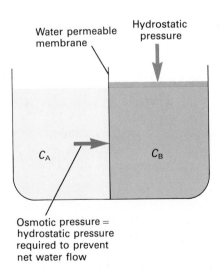

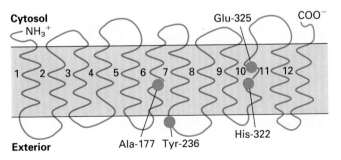

▲ FIGURE 15-24 Schematic diagram of secondary structure of *E. coli* proton-lactose symporter. Studies with mutants indicate that His-322 and Glu-325 are involved in proton transport, while Ala-177 and Tyr-236 are important in binding and transporting lactose. [See H. R. Kabak, 1987, *Biochemistry* **26**:2071; R. J. Brooker and T. H. Wilson, 1985, *Proc. Nat'l. Acad. Sci. USA* **82**:3959.]

▲ FIGURE 15-25 Experimental system for demonstrating osmotic pressure. Solutions A and B are separated by a membrane that is permeable to water but impermeable to all solutes. If C_B (the total concentration of solutes in solution B) is greater than C_A, water will tend to flow across the membrane from solution A to solution B. The osmotic pressure π between the solutions is the hydrostatic pressure that would have to be applied to solution B to prevent this water flow. From the van't Hoff equation, $\pi = RT(C_B - C_A)$.

Osmotic Pressure Causes Water Movement across Membranes

Most biological membranes are relatively impermeable to ions and other solutes, but, like all phospholipid bilayers, somewhat permeable to water. Permeability to water is increased by water channel proteins discussed below. Water tends to move across a membrane from a solution of low solute concentration to one of high. Or, in other words, since solutions with a high amount of dissolved solute have a lower concentration of water, water will move from a solution of high water concentration to one of lower. This process is known as *osmotic flow*.

Osmotic pressure is defined as the hydrostatic pressure required to stop the net flow of water across a membrane separating solutions of different compositions. In this context, the "membrane" may be a layer of cells or a plasma membrane. The osmotic pressure of a dilute solution is approximated by the *van't Hoff equation*:

$$\pi = RT(C_1 + C_2 + C_3 + \cdots + C_n)$$

where π is the osmotic pressure in atmospheres (atm) or millimeters of mercury, R is the gas constant, T is the absolute temperature, and C_1–C_n are the molar concentrations of all the solutes (ions or molecules). The total number of solute molecules is important. For example, a 0.5 M NaCl solution is actually 0.5 M Na^+ ions and 0.5 M Cl^- ions and has approximately the same osmotic pressure as a 1 M solution of glucose or lactose.

If the membrane is permeable to water but not to solutes, the osmotic pressure across the membrane is simply

$$\pi = RT \, \Delta C$$

where ΔC is the difference in total solute concentration between the two sides (see Figure 15-25). The water flow across a semipermeable membrane produced by a concentration gradient of 10 mM sucrose or 5 mM NaCl is balanced by a hydrostatic pressure of 0.22 atm, or 167 mm Hg.

Osmotic Control in Animal Cells If animal cells are placed in a *hypotonic solution*, in which the concentration of dissolved molecules, such as salts, is lower than it is inside the cells, the cells will swell. Some cells (e.g., erythrocytes) will actually burst as water enters them by osmotic flow (see Figure 14-36). Rupture of the plasma membrane by a flow of water into the cytosol is termed *osmotic lysis*. Immersion of all animal cells in a *hypertonic solution* (one with a *higher* than normal concentration of solutes) causes them to shrink as water leaves them by osmotic flow. Consequently, it is essential that animal cells be maintained in an isotonic medium, where the concentration of solutes is close to that of the cell cytosol (Figure 5-28).

Osmotic Control in Protozoan Cells Because plant, algal, fungal, and bacterial cells are surrounded by rigid cell walls, they do not absorb extra fluid and swell when placed in a hypotonic solution (even pure water). Although most protozoans (like animal cells) do not have a rigid cell wall, many have a mechanism for avoiding osmotic lysis. A *contractile vacuole*, present in many protozoans, takes up water from the cytosol (Figure 15-26). Unlike the plant vacuole, which also takes up water, the contractile vacuole periodically discharges its contents through fusion with the plasma membrane. Thus, even though water continuously enters the cell by osmotic flow, the contractile vacuole prevents too much water from accumulating in the cell and swelling it to the bursting point.

(a)

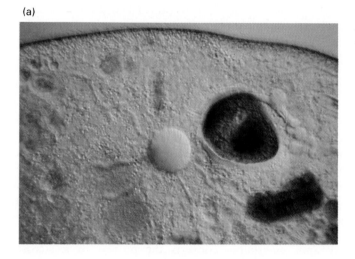

(b)

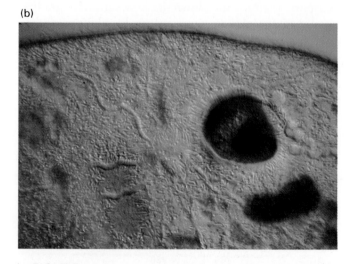

▲ FIGURE 15-26   The contractile vacuole in *Paramecium caudatum*, a typical ciliated protozoan. The vacuole is filled by radiating canals that collect fluid from the cytosol. When the vacuole is full, it fuses for a brief period with the plasma membrane and expels its contents. (a) A full vacuole and system of radiating canals. (b) A nearly empty vacuole; the radiating canals are collecting more fluid from the cytoplasm to refill it.

Water Channels Are Necessary for Bulk Osmotic Flow of Water across Membranes

Pure phospholipid bilayers are slightly permeable to water (see Figure 15-1), as are many cellular membranes. Frog oocytes and eggs, for instance, have a salt concentration of ≈150 mM, yet they do not swell when placed in pond water with a very low osmotic strength. Erythrocytes in contrast, swell or shrink in response to small changes in extracellular osmotic strength. For this reason, investigators long suspected that the plasma membranes of erythrocytes and other cell types contain water channel proteins that accelerate the osmotic flow of water.

Studies with a recently cloned 28-kDa erythrocyte membrane protein, designated Aquaporin, or CHIP, provide support for this hypothesis. The sequence of this protein suggests that it may contain six transmembrane α helices, but its sequence is unrelated to that of any other known membrane protein. When mRNA encoding the aquaporin protein was injected into frog oocytes, the protein was synthesized, and the oocytes simultaneously acquired an increased permeability to water, as evidenced by swelling of the oocytes when placed in a solution of low osmotic strength (Figure 15-27). This finding shows that aquaporin is a water channel. Aquaporin, or proteins related to it in amino acid sequence, are expressed in abundance in erythrocytes and in other cells (e.g., the kidney cells that resorb water from the urine) that exhibit high permeability for water. The functional form of this water channel protein is a tetramer of identical 28-kDa subunits, but precisely how it transports water is not known.

Some Animal Cells Regulate Their Volume by Modulating Their Internal Osmotic Strength

When certain animal cells are stressed by being placed in media of higher (hypertonic) or lower (hypotonic) salt concentrations than the cytosol, they can modulate their internal osmotic strength to keep their volume constant. For example, when a lymphocyte shrinks as a result of a hyperosmotic environment, the resulting drop in cytosolic pH

0.5 min

1.5 min

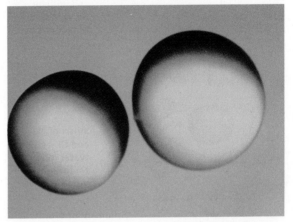

2.5 min

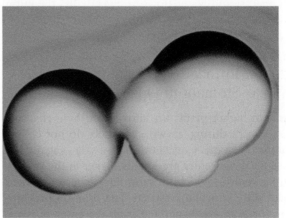

3.5 min

➤ FIGURE 15-27 Increased water permeability of frog oocytes expressing the erythrocyte water channel protein aquaporin. Control oocytes (left image of each panel) and oocytes that had been microinjected with mRNA encoding Aquaporin CHIP (right image of each panel) were kept in a solution of 0.1 mM salt, and then transferred to a hypotonic salt solution (0.035 M) for the time periods indicated. The volume of the control oocytes was unchanged, because they are poorly permeable to water, whereas the ones expressing the water channel protein swelled due to an osmotic influx of water. [Courtesy of Gregory M. Preston and Peter Agre, Johns Hopkins University School of Medicine.]

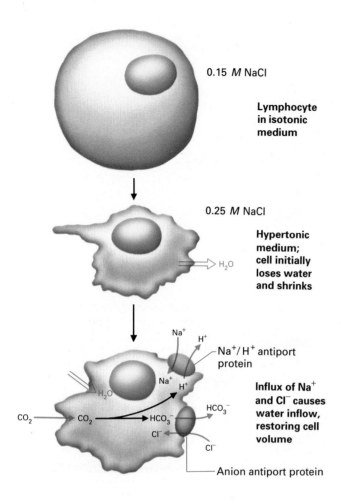

0.15 *M* NaCl

**Lymphocyte
in isotonic
medium**

0.25 *M* NaCl

H_2O

**Hypertonic
medium;
cell initially
loses water
and shrinks**

Na^+ H^+

Na^+

H^+

**Na^+/H^+ antiport
protein**

H_2O

**Influx of Na^+
and Cl^- causes
water inflow,
restoring cell
volume**

CO_2

CO_2

HCO_3^-

HCO_3^-

Cl^-

Cl^-

Anion antiport protein

◄ FIGURE 15-28   *Volume regulation in lymphocytes by modulation of intracellular salt concentration. When placed in a hypertonic solution of 0.25 M NaCl, the cell initially shrinks. Within minutes, however, the volume begins to increase due to the influx of Na^+ and Cl^- ions, catalyzed by two antiporters, and a resultant osmotic influx of water. Eventually, the original volume of the cell is restored, but at the expense of maintaining the normal intracellular salt concentration: the total intracellular salt concentration increases nearly to that of the medium. In a hypotonic solution, the antiporters operate in the opposite direction, so that the intracellular salt concentration decreases below normal. [See S. Grinstein et al., 1985, Fed. Proc. **44**:2508.]*

activates two membrane transport proteins: a *Na^+/H^+ antiporter*, which catalyzes the import of Na^+ in exchange for H^+, and an *anion antiporter*, which catalyzes import of Cl^- in exchange for HCO_3^- (Figure 15-28). The H^+ and HCO_3^- are formed from gaseous CO_2 by cytosolic carbonic anhydrase. In other types of cells, shrinking activates a *Na^+/K^+-$2Cl^-$ symporter* that catalyzes the simultaneous import of one Na^+, one K^+, and two Cl^- ions. In either case, an increase in cellular salt concentration and osmotic strength results, and water flows in. Eventually, the initial volume of the cell is restored, but the cytosolic salt concentration becomes much higher than normal.

Changes in Intracellular Osmotic Pressure Cause Leaf Stomata to Open

Most cells in higher plants are surrounded by a rigid cell wall and, except during growth, generally do not change their volume or shape. Concentrations of salts, sugars, and other dissolved molecules are usually higher in the vacuole than in the cytosol, which, in turn, has higher concentrations than the extracellular space. The osmotic pressure, called *turgor pressure*, generated from the entry of water

into the vacuole, pushes the cytosol and the plasma membrane against the resistant cell wall. Cell elongation is caused by the localized loosening of a region of the cell wall, followed by an influx of water into the vacuole.

The opening and closing of *stomata*—the pores through which CO_2 enters a leaf—is one important example of how plant cells can reverse their shape due to the osmotic movement of water (Figure 15-29). The external epidermal cells of a leaf are covered by a waxy cuticle that is largely impenetrable to water and to CO_2, a gas required for photosynthesis by the chlorophyll-laden mesophyll cells in the leaf interior. As CO_2 enters a leaf, water vapor is simultaneously lost—a process that can be injurious to the plant. Thus it is essential that the stomata open only during periods of light, when photosynthesis occurs; even then, they must close if too much water vapor is lost. Each stomate is surrounded by two *guard cells*, whose changes in shape (caused by changes in turgor pressure) open and close the pores.

Stomatal opening is caused by an increase in the concentration of ions or other solutes within the guard cells due to an opening of K^+ and Cl^- channels and an influx of K^+ and Cl^- ions from the environment (see Figure 15-29c), the metabolism of stored sucrose to four smaller compounds, or a combination of these two processes. Water enters the guard cells osmotically, increasing the turgor pressure. Since the guard cells are connected to each other only at their ends, the turgor pressure causes the cells to bulge outward, opening the stomatal pore between them. Stomatal closing is caused by the reverse process—a decrease in solute concentration and turgor pressure within the guard cells.

Stomatal opening is under tight physiologic control. A drop in CO_2 within the leaf, resulting from active photosynthesis, causes the stomata to open. When more water exits the leaf then enters it from the roots, the mesophyll cells produce the hormone abscissic acid, which causes K^+ efflux from the guard cells; water exits the cells osmotically, and the stomata close, protecting the leaf from further dehydration.

(a)

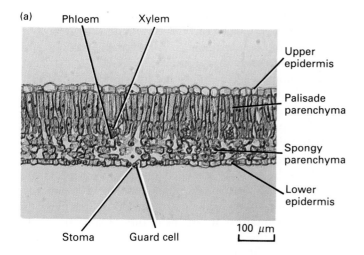

Phloem   Xylem

Upper
epidermis

Palisade
parenchyma

Spongy
parenchyma

Lower
epidermis

Stoma   Guard cell

100 μm

(b)

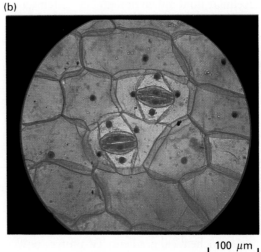

100 μm

(c)

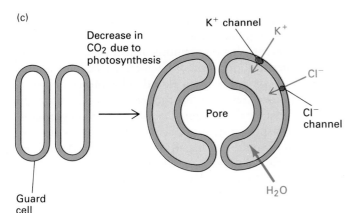

Decrease in
CO_2 due to
photosynthesis

K^+ channel

K^+

Cl^-

Cl^-
channel

Pore

H_2O

Guard
cell

▲ FIGURE 15-29   The opening and closing of stomata. (a) Cross section of a leaf from a lilac (*Syringa* sp) showing the surface epidermal cells and the internal photosynthetic parenchyma, or mesophyll, cells. Guard cells line the stomatal pore and, by changing their shape, open and close the pore. (b) Light micrograph of a stomata in a leaf of a wandering Jew (*Tradescantia* sp) plant. (c) Opening of stretch-activated K^+ and Cl^- channels in the plasma membrane of the guard cells is followed by an influx of K^+ and Cl^-. The increase in cytosolic KCl triggers the osmotic influx of water, causing the cells to bulge and opening the stomatal pore. [See D. J. Cosgrove and R. Hedrich, 1991, *Planta* **186**:143. Parts (a) and (b) courtesy Runk/Schoenberger, from Grant Heilman.]

SUMMARY

The plasma membrane regulates the traffic of molecules into and out of the cell. Gases and small hydrophobic molecules diffuse directly across the phospholipid bilayer at a rate proportional to their ability to dissolve in a liquid hydrocarbon, and pure phospholipid bilayers are slightly permeable to water. However, ions, sugars, amino acids, and sometimes water cannot diffuse across the phospholipid bilayer at sufficient rates to meet the cell's needs and must be transported by a group of integral membrane proteins including channels, transporters, and ATP-powered ion pumps.

The protein-catalyzed uptake of some molecules into the cell is driven only by a concentration gradient. For example, the glucose transporter (a uniport protein) in the erythrocyte plasma membrane forms a transmembrane passage that allows only bound glucose, and closely related molecules, to cross the bilayer down its concentration gradient.

An inside-negative electric potential of 50–70 mV exists across the plasma membrane of all cells. In animal cells, this potential is generated primarily by movement of cytosolic K^+ ions through K^+ channels to the external medium. In plants and fungi, in contrast, the membrane potential is maintained by the ATP-driven pumping of protons from the cytosol across the membrane.

Transport of some ions and molecules against a concentration gradient is catalyzed by ATP-powered pumps. These membrane transport proteins couple the energy-releasing hydrolysis of ATP with the energy-requiring transport of substances against their concentration gradient. The Na^+/K^+ ATPase, for example, pumps three Na^+ ions out of and two K^+ ions into the cell per ATP hydrolyzed. The Ca^{2+} ATPase pumps two Ca^{2+} ions out of the cell or, in muscle, into the sarcoplasmic reticulum per ATP hydrolyzed. These pumps create an intracellular ion milieu of high K^+, low Ca^{2+}, and low Na^+ that is very different from the extracellular fluid milieu of high Na^+, high Ca^{2+}, and low K^+. An ATP-powered H^+ pump in animal lysosomal and endosomal membranes and plant-vacuole membranes is responsible for maintaining a lower pH inside the organelles than in the surrounding cytosol. The various ATP-powered ion pumps can be grouped into three classes. In P-class pumps, which include the Na^+/K^+ ATPase and Ca^{2+} ATPase, phosphorylation of the α subunit and an alternation of conformational states are essential for coupled ion transport. The multisubunit V- and F-class ATPases pump protons exclusively, and a phosphorylated protein is not an intermediate in transport.

A small molecule or ion may be imported or exported against its concentration gradient by coupling its movement to that of another molecule or ion, usually H^+ or Na^+, down its electrochemical gradient. Two forces power the movement of H^+ or Na^+: the membrane potential and the ion concentration gradient. Entry of glucose and amino acids into absorptive cells against their concentration gradient is coupled by symport proteins to the energetically favorable entry of Na^+. In cardiac muscle cells, the export of Ca^{2+} ions is coupled to the import of Na^+ ions by an antiport protein. In other cases, the uptake of nutrients is driven by a proton gradient across the cell membrane: one example is the uptake of lactose by E. coli, catalyzed by a proton-lactose symporter; another is the uptake of sucrose, Na^+, Ca^{2+}, and other substances into plant vacuoles catalyzed by various proton antiport proteins.

The plasma membranes of many cells are differentiated so that specific membrane segments contain different transport proteins and carry out quite different transport processes. In the intestinal epithelial cell, the Na^+-glucose and Na^+-amino acid symporters are in the apical membrane region facing the intestinal lumen and the Na^+/K^+ ATPase and glucose and amino acid uniport proteins are in the basolateral membrane region facing the blood capillaries. This combination allows the cells to transport amino acids and glucose from the lumen to the blood. Acid-secreting cells in the stomach have ATP-driven H^+/K^+ pumps, K^+ channels, and Cl^- channels on the apical membrane and HCO_3^-/Cl^- antiporters on the basolateral membrane; the combined action of these proteins allows the cytosolic pH to be maintained at near neutrality, despite the active export of protons from the cells into the stomach lumen, causing its acidification.

Water channel proteins increase the water permeability of membranes. Most biological membranes are more permeable to water than to ions or other solutes, and water will move across them by osmosis from a solution of lower solute concentration to one of higher. This causes animal cells to swell or shrink when placed in hypotonic or hypertonic solutions, respectively, though some cells have the ability to restore their original volume by opening or closing ion transport proteins. Changes in the cytosolic concentration of solutes such as K^+ and Cl^- and the resultant osmotic efflux or influx of water control the respective closing and opening of stomata in leaves.

REVIEW QUESTIONS

1. Recall that two of the fundamental questions presented in Chapter 2 concerned the feasibility of any given chemical reaction: Will it occur, and how fast will it happen? These same ideas can be applied to molecules crossing membranes. The value of ΔG, which dictates if a reaction will occur as written, is calculated from the following:

$$\Delta G = \Delta G^{\circ\prime} + RT\ln(Q) + nF\,\Delta E.$$

In the case of problems related to the diffusion of molecules, $\Delta G^{\circ\prime}$ is always 0, and in the case of uncharged molecules, n is 0, and the equation reduces to

$$\Delta G = RT\ln(Q).$$

a. Assume that you have two 1 ml chambers separated by an impermeable partition having a hole in it. Such a device is illustrated in Figure 14-8. The hole, which is 1 mm in diameter, is filled with a lipid bilayer membrane. Each chamber contains both glucose and urea at certain concentrations indicated in the table below.

	Chamber I	Chamber II
[glucose]	.001 M	0.0001 M
[urea]	.001 M	0.0001 M

What would ΔG be for each of the following processes?

$$\text{glucose}_{(\text{Chamber I})} \rightleftharpoons \text{glucose}_{(\text{Chamber II})}$$

$$\text{urea}_{(\text{Chamber I})} \rightleftharpoons \text{urea}_{(\text{Chamber II})}$$

Are the values of ΔG equal, or are they different? What do they tell you about diffusion of each of these molecules? At what point would the values of ΔG become 0?

b. Review the derivation of Equation 15-2 that describes the rate at which diffusion through lipid bilayer membranes will occur. The permeability coefficients for glucose and urea are 5×10^{-8} cm/sec and 3×10^{-6} cm/sec, respectively. What would be the initial rates of diffusion for each of these molecules?

c. Do the values for ΔG for these processes contradict the rate data? Explain your answer.

d. How does a cell permit water soluble molecules to enter it? Are the mechanisms that have developed selective for specific molecules? What are the kinetics of such systems like?

2. Review the three different types of transport molecules discussed in this chapter: uniport, symport, and antiport proteins. Be sure you understand the differences in their properties. Give an example of each. Do all these transporters rely directly on ATP hydrolysis as a source of energy for transport? If not, do they rely indirectly on ATP hydrolysis for energy to transport molecules? Explain your answer.

a. You have isolated an integral membrane protein from the plasma membrane of hamster hepatocytes that is responsible for transporting Ca^{+2}. You are not certain about some of its transport features, and you decide to place it into liposomes to study it. How would you do this? Is there any reason to assume that the transporter will orient itself in only one direction in all the liposomes?

b. Assume that an average of two transport molecules is in each liposome. To one set of liposomes, you add the radioactive isotope $^{45}Ca^{+2}$; to another set, you add $^{45}Ca^{+2}$ and ATP. You then isolate the liposomes by gel filtration. No $^{45}Ca^{+2}$ is present in the liposomes incubated in the absence of ATP, but $^{45}Ca^{+2}$ is found inside liposomes incubated with ATP. Explain these findings. Given the fact that the transporter molecules are randomly oriented in the liposomes, are the $^{45}Ca^{+2}$ data reasonable?

3. An important concept in the transport of any molecule against its concentration gradient is the source of energy for performing the process. Review the plasma membrane antiport system responsible for extruding Ca^{+2} from the cytosol into the extracellular milieu.

a. Refer to Table 15-2 for the typical ion concentrations for Na^+ and Ca^{+2} within the cell and in the blood of mammalian cells. Calculate the free energy for the movement of one Na^+ ion into a cell. Assume that the resting membrane potential is -70 mV.

b. Calculate the free energy for the movement of one Ca^{+2} ion out of the cell.

c. Is there sufficient energy stored in the sodium gradient such that the free energy associated with one Na^+ entering the cell could be used to transport one Ca^{+2} out of the cell? What is the minimal number of sodium ions that would be needed to supply sufficient energy for the process? What is the actual stoichiometry for the Na^+-Ca^{+2} transporter?

4. Review the principle of osmosis. This concept is very important in biological systems. How do animal cells regulate their volume and osmotic strength?

a. Which of the solutions in the following table have the same osmolarity?

	[NaCl], M	[KCl], M	[urea], M
Solution A	0.100	0.050	—*
Solution B	0.050	0.050	—
Solution C	0.200	0.050	—
Solution D	0.200	—	0.100

* A dash (—) indicates none added

b. You place erythrocytes into these solutions and observe the results under a microscope. Nothing happens to the cells in solution A; the cells in solution B have lysed, and the red color due to hemoglobin is visible; the cells in solution C shrink and remain that way; and the cells in solution D initially shrink, but they eventually swell again to normal size. Explain these results.

References

Uniport Transport: The Glucose Transporter

BELL, G., et al. 1993. Structure and function of mammalian facilitative sugar transporters. *J. Biol. Chem.* **268**:19161–19164.

HENDERSON, P. J. 1993. The 12-transmembrane helix transporters. *Curr. Opin. Cell Biol.* **5**:708–721.

THORENS, B., M. J. CHARRON, and H. F. LODISH. 1990. Molecular physiology of glucose transporters. *Diabetes Care* **13**:209–218.

THORENS, B. 1993. Facilitated glucose transporters in epithelial cells. *Annu. Rev. Physiol.* **55**:591–608.

Ion Channels, Intracellular Ion Environment, and Membrane Electric Potential

GRAVES, J. S., ed. 1985. *Regulation and Development of Membrane Transport Processes.* Society of General Physiology/Wiley.

HILLE, B. 1991. *Ion Channels of Excitable Membranes.* 2d ed. Sinauer Associates (Sunderland, MA).

OFFNER, F. F. 1991. Ion flow through membranes and the resting potential of cells. *J. Membr Biol.* **123**:171–82.

RACKER, E. 1985. *Reconstitutions of Transporters, Receptors, and Pathological States.* Academic Press.

STEIN, W. D., and W. R. LIEB. 1986. *Transport and Diffusion across Cell Membranes.* Academic Press.

WOOLLEY, G. A., and B. A. WALLACE. 1992. Model ion channels: gramicidin and alamethicin. *J. Membr. Biol.* **129**:109–136.

ATP-Powered Transport

Coupling of ATP Hydrolysis and Transport

JENCKS, W. P. 1988. Energy-rich compounds and work. Fritsch Lipmann Memorial Symposium (1987). In Kleinkauf, van Dohren, and Jaenicke, eds., *The Roots of Modern Biochemistry,* Walter de Gruyter (Berlin), pp. 569–580.

JENCKS, W. P. 1989. How does a calcium pump pump calcium? *J. Biol. Chem.* **264**:18855–18858.

PEDERSEN, P. L., and E. CARAFOLI. 1987. Ion motive ATPases: I. Ubiquity, properties, and significance to cell function. *Trends Biochem. Sci.* **12**:146–150; II. Energy coupling and work output. *Trends Biochem. Sci* **12**:186–189.

SACHS, G., and K. MUNSON. 1991. Mammalian phosphorylating ion-motive ATPases. *Curr. Opin. Cell Biol.* **3**:685–694.

Na^+/K^+ ATPase

GLYNN, I. M. 1993. *All hands to the sodium pump. J. Physiol.* (*Lond*). **462**:1–30.

LINGREL, J. B., et al. 1994. Structure-function studies of the Na, K-ATPase. *Kidney Int. Suppl.* **44**:32–39.

SKOU, J. C., and M. ESMANN. 1992. The Na,K-ATPase. *J. Bioenerg. Biomembr.* **24**:249–261.

Ca²⁺ ATPase

CARAFOLI, E. 1992. The Ca²⁺ pump of the plasma membrane. *J. Biol. Chem.* **267**:2115–2118.

CARAFOLI, E., et. al. 1992. The molecular basis of the modulation of the plasma membrane calcium pump by calmodulin. *Ann. NY Acad. Sci.* **671**:58–68.

GROVER, A. K., and I. KHAN. 1992. Calcium pump isoforms: diversity, selectivity and plasticity. *Cell Calcium* **13**:9–17.

JENCKS, W. P. 1992. On the mechanism of ATP-driven Ca²⁺ transport by the calcium ATPase of sarcoplasmic reticulum. *Ann. NY Acad. Sci.* **671**:49–56.

MACLENNAN, L. D., T. TOYOFUKU, and J. LYTTON. 1992. Structure-function relationships in sarcoplasmic or endoplasmic reticulum type Ca²⁺ pumps. *Ann. NY Acad. Sci.* **671**:1–10.

Proton Pumps

BOWMAN, B. J., L. N. VAZQUEZ, and E. J. BOWMAN. 1992. The vacuolar ATPase of *Neurospora crassa. J. Bioenerg. Biomembr.* **24**:361–370.

CAPPELL, M. S. 1992. Omeprazole in the treatment of peptic ulcers and gastroesophageal reflux disease. *N J Med.* **89**:762–764. Omeprazole is a potent and specific inhibitor of the gastric H,K-ATPase.

FORGAC, M. 1992. Structure, function and regulation of the coated vesicle V-ATPase. *J. Exp. Biol.* **172**:155–169.

GLUCK, S. L. 1992. The structure and biochemistry of the vacuolar H⁺ ATPase in proximal and distal urinary acidification. *J. Bioenerg. Biomembr.* **24**:351–359.

NELSON, N. 1992. Organellar proton-ATPases. *Curr. Opin. Cell Biol.* **4**:654–660.

REA, P. A., et al. 1992. Vacuolar H⁺-translocating pyrophosphatases: a new category of ion translocase. *Trends Biochem. Sci.* **17**:348–353.

SACHS, G., et al. 1992. Structural aspects of the gastric H,K-ATPase. *J. Bioenerg. Biomembr.* **24**:301–308.

SERRANO, R. 1993. Structure, function and regulation of plasma membrane H⁺-ATPase. *FEBS Lett.* **325**:108–111.

SZE, H., J. M. WARD, and S. LAI. 1992. Vacuolar H⁺-translocating ATPases from plants: structure, function, and isoforms. *J. Bioenerg. Biomembr.* **24**:371–381.

Multidrug-Transport Protein and CFTR Protein

BEAR, C. E., et al. 1992. Purification and functional reconstitution of the cystic fibrosis transmembrane conductance regulator (CFTR). *Cell* **68**:809–818.

BEAR, C., and V. LING. 1993. Multidrug resistance and cystic fibrosis genes: complementarity of function? *Trends Genet.* **9**:67–68.

CHIN, K. V., I. PASTAN, and M. M. GOTTESMAN. 1993. Function and regulation of the human multidrug resistance gene. *Adv. Cancer Res.* **60**:157–180.

COLLINS, F. S. 1992. Cystic fibrosis: molecular biology and therapeutic implications. *Science* **256**:774–779.

DOIGE, C. A., and G. F. AMES. 1993. ATP-dependent transport systems in bacteria and humans: relevance to cystic fibrosis and multidrug resistance. *Annu. Rev. Microbiol.* **47**:291–319.

GOTTESMAN, M. M., and I. PASTAN. 1993. Biochemistry of multidrug resistance mediated by the multidrug transporter. *Annu. Rev. Biochem.* **62**:385–427.

HIGGINS, C. F. 1992. ABC transporters: from microorganisms to man. *Annu. Rev. Cell Biol.* **8**:67–113.

LING, V. 1992. P-glycoprotein and resistance to anticancer drugs. *Cancer* **69**:2603–2609. Charles F. Kettering Prize.

WELSH, M. J., et al. 1992. Cystic fibrosis transmembrane conductance regulator: a chloride channel with novel regulation. *Neuron* **8**:821–829.

WELSH, M. J., and A. E. SMITH. 1993. Molecular mechanisms of CFTR chloride channel dysfunction in cystic fibrosis. *Cell* **73**:1251–1254.

Symport and Antiport Transport

Cotransporters of Amino Acids, Glucose, and Ca²⁺

BALDWIN, S. A. 1993. Mammalian passive glucose transporters: members of an ubiquitous family of active and passive transport proteins. *Biochim. Biophys. Acta* **1154**:17–49.

CRESPO, L. M., C. J. GRANTHAM, and M. B. CANNELL. 1990. Kinetics, stoichiometry and role of the Na-Ca exchange mechanism in isolated cardiac myocytes. *Nature* **345**:618–621.

LANGER, G. A. 1992. Calcium and the heart: exchange at the tissue, cell, and organelle levels. *FASEB J.* **6**:893–902.

MAATHUIS, F. J., and D. SANDERS. 1992. Plant membrane transport. *Curr. Opin. Cell Biol.* **4**:661–669.

PHILIPSON, K. D., and D. A. NICOLL. 1992. Sodium-calcium exchange. *Curr. Opin. Cell Biol.* **4**:678–683.

PHILIPSON, K. D., and D. A. NICOLL. 1993. Molecular and kinetic aspects of sodium-calcium exchange. *Int. Rev. Cytol.* **137C**:199–227.

WRIGHT, E. M. 1993. The intestinal Na⁺/glucose cotransporter. *Annu. Rev. Physiol.* **55**:575–589.

WRIGHT, E. M., K. M. HAGER, and E. TURK. 1992. Sodium cotransport proteins. *Curr. Opin. Cell Biol.* **4**:696–702.

Anion-Exchange Proteins

ALPER, S. L. 1991. The band 3-related anion exchanger (AE) gene family. *Annu. Rev. Physiol.* **53**:549–564.

PASSOW, H. 1986. Molecular aspects of band 3 protein-mediated anion transport across the red cell membrane. *Rev. Physiol. Biochem. Pharmacol.* **103**:62–203.

RUETZ, S., A. E. LINDSEY, and R. R. KOPITO. 1993. Function and biosynthesis of erythroid and nonerythroid anion exchangers. *Soc. Gen. Physiol. Ser.* **48**:193–200.

TANNER, M. J. 1993. Molecular and cellular biology of the erythrocyte anion exchanger (AE1). *Semin. Hematol.* **30**:34–57.

Control of Cytosolic pH

BORON, W. F. 1986. Intracellular pH regulation in epithelial cells. *Annu. Rev. Physiol.* **48**:377–388.

COUNILLON, L., and J. POUYSSEGUR. 1993. Molecular biology and hormonal regulation of vertebrate Na⁺/H⁺ exchanger isoforms. *Soc. Gen. Physiol. Ser.* **48**:169–185.

TSE, M., et al. 1993. Structure/function studies of the epithelial isoforms of the mammalian Na⁺/H⁺ exchanger gene family. *J. Membr. Biol.* **135**:93–108.

Transport into Prokaryotic Cells

HENDERSON, P. J., et al. 1992. Sugar-cation symport systems in bacteria. *Int. Rev. Cytol.* **137**:149–208.

KABACK, H. R. 1992. In and out and up and down with lac permease. *Int. Rev. Cytol.* **137**:97–125.

KABACK, H. R. 1992. The lactose permease of *Escherichia coli*: a paradigm for membrane transport proteins. *Biochim. Biophys. Acta* **1101**:210–213.

SAHIN-TOTH, M., et al. 1992. Functional interactions between putative intramembrane charged residues in the lactose permease of *Escherichia coli. Proc. Nat'l. Acad. Sci. USA* **89**:10547–10551.

Osmosis, Water Channels, and Regulation of Cell Volume

AGRE, P., et al. 1993. Aquaporin CHIP: the archetypal molecular water channel. *Am. J. Physiol.* **265**:F463–F476.

GRINSTEIN, S., et al. 1992. Activation of the Na⁺/H⁺ antiporter during cell volume regulation. Evidence for a phosphorylation-independent mechanism. *J. Biol. Chem.* **267**:23823–23828.

SARKADI, B., and J. C. PARKER. 1991. Activation of ion transport pathways by changes in cell volume. *Biochim. Biophys. Acta* **1071**:407–427.

SCHROEDER, J. I., and R. HEDRICH. 1989. Involvement of ion channels and active transport in osmoregulation and signaling of higher plant cells. *Trends Biochem. Sci.* **14**:187–192.

VERKMAN, A. S. 1992. Water channels in cell membranes. *Annu. Rev. Physiol.* **54**:97–108.

16

Synthesis and Sorting of Plasma Membrane, Secretory, and Lysosomal Proteins

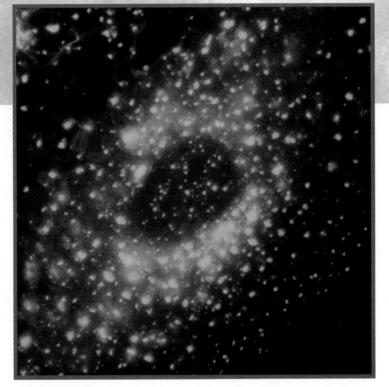

▲ Endosomes containing a fluorescent lipid.

In a typical mammalian cell we might find up to 10,000 to 20,000 different kinds of proteins, and in a yeast cell about 5000. Each of these proteins must be incorporated into the correct aqueous space (e.g., the lumen of the endoplasmic reticulum or the cytosol), membrane, or organelle in order to function properly, and each of the thirty or so membranes (e.g., the plasma membrane; the inner and outer mitochondrial membranes) and aqueous spaces in a cell must contain a discrete set of proteins in order to carry out its unique functions. Specific hormone receptor proteins, for example, must be delivered to the plasma membrane if the cell is to recognize hormones and specific ion-channel and transporter proteins are needed if the cell is to be able to import or export specific ions or other small molecules. Enzymes such as RNA and DNA polymerases must be targeted to the nucleus; still others, such as proteolytic enzymes or catalase, must go to the lysosome or peroxisome

respectively. Thus a critical factor in the "building" of a cell from its constituent proteins, lipids, nucleic acids, and other components is *protein targeting,* or *protein sorting*—directing each newly made polypeptide to a particular destination. How proteins are sorted to their correct destinations is the major subject of this chapter and Chapter 19.

Protein sorting occurs at several levels. Some proteins are encoded by the DNA of mitochondria and chloroplasts, are synthesized on ribosomes in these organelles, and are incorporated directly into compartments within these organelles. These proteins are discussed in Chapter 19. However, most proteins in the mitochondria and chloroplasts, and *all* of the proteins in the other organelles, particles, and membranes of a eukaryotic cell, are encoded by nuclear DNA and are synthesized on ribosomes in the cytosol. The sorting of each of these proteins to its correct destination requires multiple sorting signals and multiple sorting events.

For proteins synthesized on cytosolic ribosomes, the first level of sorting divides them into two broad categories: proteins synthesized on membrane-bound ribosomes of the *rough endoplasmic reticulum* (rough ER; the roughness is due to the ribosomes bound to it) and those synthesized on cytosolic ribosomes that are not attached to membranes. The latter proteins are released into the cytosol; some remain there, while others are specifically incorporated into the nucleus, mitochondrion, chloroplast, or peroxisome, and are subsequently sorted further to reach their correct destinations within these organelles. Chapter 19 will focus on the uptake and intraorganellar sorting of these cytosolic proteins.

This chapter focuses on the large class of proteins that are synthesized on ribosomes bound to the rough ER and enter the *secretory pathway* (Figure 16-1). Included in this class are proteins that cross the membrane of the ER *co-translationally*—that is, during synthesis—and become soluble in the ER lumen. Some of these soluble proteins will be secreted from the cell; after their synthesis, small transport vesicles carry them sequentially through the Golgi complex and then to secretory vesicles. During *exocytosis*, the final stage of their secretion, the secretory-vesicle membranes fuse with the plasma membrane, releasing the protein to the cell exterior. Other proteins that are soluble in the ER lumen remain there after synthesis, as resident ER proteins. Still others move on to the Golgi vesicles; some remain resident there, while others are targeted to lysosomes. Thus, a hierarchy of sorting events ensures that each protein reaches its proper destination. The membranes and lumina of the rough ER, the Golgi complex, and the secretory vesicles each contain certain resident enzymes that make specific chemical modifications on the proteins within. Further, each protein bears one or more targeting signals that direct it to its destination.

Another important class of proteins that are synthesized on the rough ER and thus enter the secretory pathway are *integral membrane proteins*. Each such protein has a unique orientation with respect to the membrane's phospholipid bilayer (Chapter 14). Sequences of up to 25 amino acids within the protein, called *topogenic sequences*, ensure that each protein acquires its proper orientation dur-

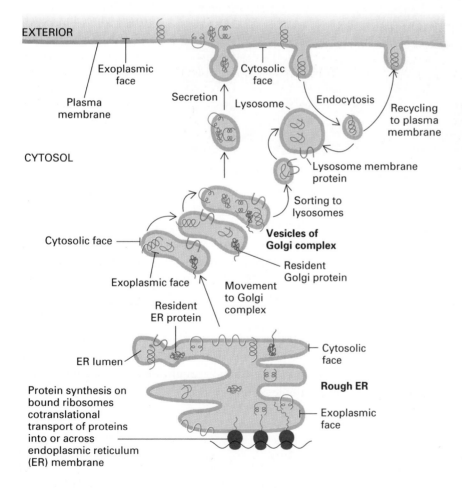

◀ FIGURE 16-1   The secretory pathway of protein synthesis and sorting. Proteins synthesized on ribosomes bound to the rough endoplasmic reticulum (ER) enter the secretory pathway. This class includes both integral membrane proteins that become inserted in the ER membrane and proteins that become soluble in the lumen of the ER. The proteins are subsequently sorted to their different destinations. Some (rough ER enzymes or structural proteins) remain resident in the ER. The remainder move via transport vesicles to the three classes of Golgi vesicles, where certain of them remain. The rest of the proteins move, via secretory vesicles, either to the plasma membrane—where soluble proteins are secreted—or to lysosomes. Plasma membrane proteins also move *via* transport vesicles from the rough ER to the Golgi to the plasma membrane; frequently they are internalized by endocytosis and then either sorted to lysosomes for degradation or recycled back to the plasma membrane. Each integral membrane protein (blue) retains its membrane orientation during all the sorting steps: some segments always face the cytosol (gray); others always face the exoplasmic space (yellow; i.e., the lumen of the ER, the Golgi vesicles, and other transport vesicles and the outside of the cell).

ing its insertion into the ER membrane. These proteins move, via transport vesicles, from the ER to their final destinations, such as the plasma membrane or the membranes of the lysosome or Golgi vesicles. During this transport, the membrane proteins always retain the same orientation—the same segments of the protein, for instance, always face the cytosol (Figure 16-1).

Only some of the integral plasma membrane proteins remain in that membrane. Others, particularly cell surface receptors, frequently are internalized into the cell by the process known as endocytosis, and are then sorted to lysosomes for degradation, to other intracellular vesicles, or back to the plasma membrane. Thus, an individual plasma membrane protein can undergo multiple sorting events during its transport to the plasma membrane and also after it arrives there.

Each of these processes will be detailed in this chapter, as will several important sequential modifications, such as the addition and modification of covalently bound carbohydrate, that occur to proteins as they move through the secretory pathway.

➤ The Synthesis of Membrane Lipids

Even though such structures as the inner mitochondrial membrane, the plasma membrane, and the rough and smooth ER membranes carry out very different functions, they are all phospholipid bilayers in which different classes of proteins are embedded. Cells synthesize new membranes only by expansion of existing membranes; phospholipids are synthesized in association with existing membranes and are immediately incorporated into them; similarly, proteins are inserted into existing membranes. As a result, organelles containing one or more membranes, such as the ER, mitochondria, and chloroplasts, grow by expansion of the existing organelle, which eventually divides into two or more "daughters." Accordingly, our study of membrane and organelle biogenesis begins with the biosynthesis of phospholipids and their incorporation into membranes.

Phospholipids Are Synthesized in Association with Membranes

All phospholipids and other membrane lipids, such as sphingomyelin and glycolipids, are amphipathic molecules. Because of their extremely hydrophobic tail regions, and despite their polar head groups (Figures 14-2 and 14-3), they dissolve very poorly in aqueous solutions. In fact, depending on their concentration and the ionic composition, they spontaneously form either micelles (spherical vesicles built of a phospholipid monolayer with all of the fatty acyl chains on the inside) or phospholipid bilayers (Figure 2-17). These physicochemical properties of phospholipids have profound implications for the biosynthesis of cellular membranes and membrane-containing organelles.

The extremely low solubility of phospholipids in aqueous solution makes the assembly of a new phospholipid bilayer from soluble components energetically difficult. Phospholipids are synthesized by enzymes bound to preexisting cellular membranes and incorporated into such membranes immediately after synthesis. As evidence for this, when cells are briefly exposed to radioactive phosphate ($^{32}PO_4^{3-}$) or to radiolabeled fatty acids or sugars, all phospholipids and glycolipids incorporating these precursor substances are seen to be associated with intracellular membranes; none are found free in the cytosol. Thus, phospholipids are incorporated into membranes during their synthesis.

In all cells, phospholipid synthesis occurs at the interface between a membrane and the cytosol. In bacterial cells, the synthesis takes place at the cytosolic face of the plasma membrane; in animal and plant cells, synthesis of most phospholipids is associated with the ER membrane, usually the smooth ER. Most cells can synthesize fatty acids from acetate (CH_3COO^-); many mammalian cells obtain fatty acids by hydrolysis of triglycerides. In the pathway for the synthesis of phosphatidylethanolamine (a typical phospholipid) in animal cells (Figure 16-2), one substrate—a CoA ester of a fatty acid (formed by reacting a free fatty acid with CoA using energy supplied by hydrolysis of a phosphoanhydride bond of ATP)—is an amphipathic molecule. The fatty acyl side chain is embedded in the cytosolic leaflet of the ER membrane, and the CoA portion protrudes into the cytosol. All other substrates and reaction products, such as ATP, ethanolamine, and cytidine triphosphate (CTP), are soluble constituents of the cytosol. Most of the enzymes that catalyze these reactions have one segment bound to or inserted into the ER membrane; the other protrudes into the cytosol.

Special Membrane Proteins Allow Phospholipids To Equilibrate in Both Membrane Leaflets

When newly made, a phospholipid is localized to the cytosolic leaflet of the ER membrane, but soon thereafter most phospholipids are found in both the cytosolic and the exoplasmic leaflets (see Figure 14-13), as they are in most cellular membranes. Phospholipids do not flip-flop spontaneously across a pure phospholipid bilayer. In many formed membranes, such as the erythrocyte membrane or the plasma membrane of nucleated cells, a phospholipid takes several hours to move from one membrane face to the other. How then does a newly made phospholipid move rapidly from the cytosolic to the exoplasmic leaflet of the ER?

It appears that ER membranes (and plasma membranes of bacteria) contain one or more proteins, known as *flippase* proteins, that can catalyze such a flip-flop process (Figure 16-3). The half-time for this movement—the time required for half of the newly-made phospholipid to move to the exoplasmic leaflet of the ER membrane—is only a

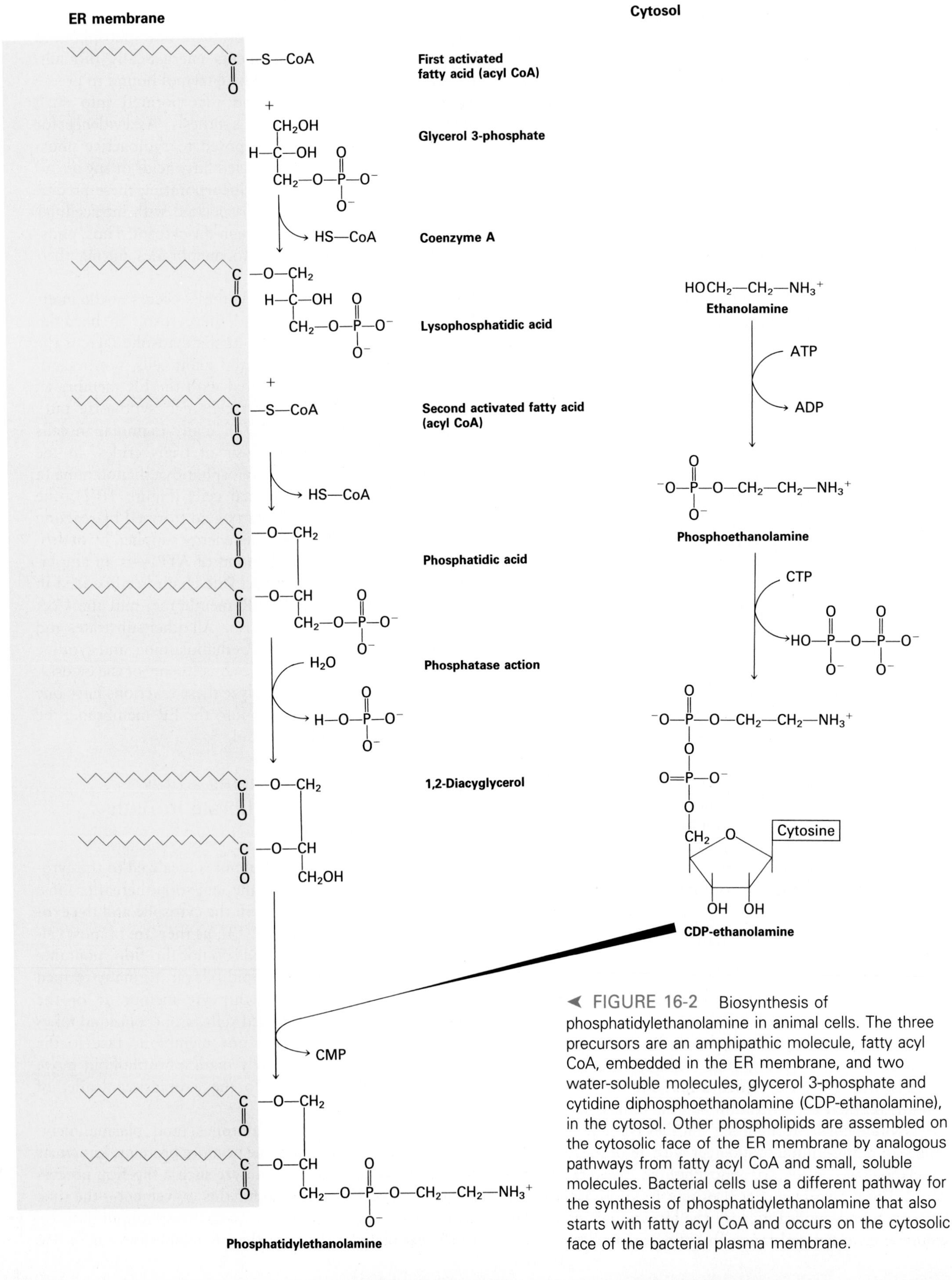

ER membrane

Cytosol

◄ FIGURE 16-2   Biosynthesis of phosphatidylethanolamine in animal cells. The three precursors are an amphipathic molecule, fatty acyl CoA, embedded in the ER membrane, and two water-soluble molecules, glycerol 3-phosphate and cytidine diphosphoethanolamine (CDP-ethanolamine), in the cytosol. Other phospholipids are assembled on the cytosolic face of the ER membrane by analogous pathways from fatty acyl CoA and small, soluble molecules. Bacterial cells use a different pathway for the synthesis of phosphatidylethanolamine that also starts with fatty acyl CoA and occurs on the cytosolic face of the bacterial plasma membrane.

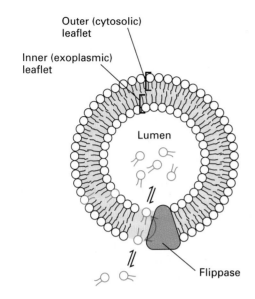

CH₃—CH₂—CH₂—C(=O)—O—CH₂
CH₃—CH₂—CH₂—C(=O)—O—CH
CH₂—O—P(=O)(O⁻)—O—CH₂—CH₂—N⁺(CH₃)₃

Dibutyrylphosphatidylcholine (⬭)

▲ FIGURE 16-3 Assay for a phospholipid flippase. The phospholipid dibutyrylphosphatidylcholine is water-soluble because it contains short, four-carbon fatty acyl chains. When added to a suspension of pure phospholipid vesicles (liposomes), the phospholipid spontaneously inserts itself into the outer leaflet, but because liposomes lack a flippase protein, it does not flip to the inner leaflet. A flippase protein, such as is found in ER membranes, catalyzes the movement of this small phospholipid to the inner (exoplasmic) leaflet; from there, it can spontaneously move into the aqueous lumen of the vesicle. Thus a flippase can be assayed as a protein that allows dibutyrylphosphatidylcholine added to the exterior of a vesicle to accumulate in the vesicle lumen. [After W. R. Bishop and R. M. Bell. 1985. *Cell* **42**:50–60; Y. Kawashima and R. M. Bell. 1987. *J. Biol. Chem.* **262**:16495–16502.]

few minutes. The erythrocyte membrane contains low amounts of a different kind of flippase protein, which catalyzes a slow movement of phospholipids from one leaflet to the other.

The two leaflets of a membrane often have different phospholipid compositions. The exoplasmic leaflet frequently is rich in phosphatidylcholine; and the cytosolic leaflet, in phosphatidylserine (Figure 14-29). How this lipid asymmetry is achieved is not known, but the mechanism in part may involve different affinities of phospholipids for the flip-flop protein or for specific regions of integral membrane proteins localized to the two faces.

Phospholipids Move from the ER to Other Cellular Membranes

Although phospholipids are synthesized in the ER, they are found in the membranes of all organelles. How do they move there? Membrane vesicles bud off the ER membrane and fuse with membranes of the Golgi complex; other vesicles bud from the Golgi complex and fuse with the membranes of other organelles. Such vesicles contain phospholipids and other membrane lipids as well as transported proteins (Figure 16-1). Certain phospholipids can be incorporated selectively into different transport vesicles, explaining the different phospholipid compositions of different organelle membranes. In animal cells, the addition of terminal sugars such as galactose and *N*-acetylneuraminic (sialic) acid to glycolipids occurs in the Golgi complex. Such glycolipids are found only in the exoplasmic leaflet (on the luminal side) of the Golgi membrane. They are moved to the plasma membrane via transport vesicles—probably the same vesicles that move newly made secretory and integral membrane proteins to the plasma membrane.

Not all movement of phospholipids is via vesicular transport. Phospholipid movement also involves *phospholipid exchange proteins*—water-soluble proteins that can remove phospholipids from the cytosolic membrane face of one organelle (say, the ER) and release them into the cytosolic face of the membrane of another organelle (Figure 16-4). Such proteins have been identified in the cytosol of

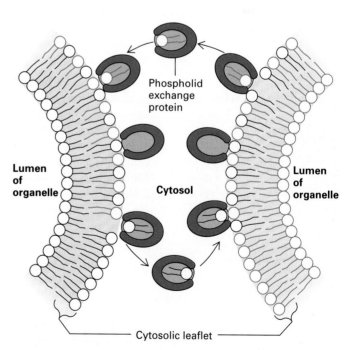

▲ FIGURE 16-4 The action of phospholipid exchange proteins. These proteins catalyze the movement of specific phospholipids between the cytosolic faces of different membranes. No phospholipids in the exoplasmic face are exchanged. [After A. Cleves, T. McGee, and V. Bankaitis. 1991. *Trends Cell Biol.* **1**:30–34.]

all types of eukaryotic cells in which they have been sought. Each exchange protein generally binds a single type of phospholipid.

When added to a mixture of isolated membranes, an exchange protein will equilibrate phospholipids among all membranes present. In cells, in contrast, it is thought that these proteins can account for the abundance of certain lipids in specific organelle membranes. In yeast cells, for example, it is thought that one particular exchange protein generates the high ratio of phosphatidylinositol to phosphatidylcholine that the Golgi membranes require to function properly. When yeast cells have a mutation that inactivates this exchange protein, they contain defective Golgi membranes with an abnormal phospholipid composition.

Mitochondria and chloroplasts, perhaps not surprisingly, synthesize some of their own membrane lipids and import others. Some phospholipids that are synthesized in the ER, such as phosphatidylethanolamine and phosphatidylcholine, are incorporated into mitochondria as well as into the plasma membrane and Golgi membranes. Since there are no vesicles that bud from the ER and fuse with mitochondria, the movement of these phospholipids from the ER to mitochondria is thought to involve phospholipid

exchange proteins. Certain phospholipids, particularly cardiolipin, are found only in the mitochondrial inner membrane; cardiolipin is synthesized on the matrix side of the inner mitochondrial membrane from imported precursors. In photosynthetic tissues, the chloroplast is the site for the synthesis of all chloroplast lipids, including glycolipids; at least in spinach leaves, the chloroplast is the site of fatty acid synthesis for the entire cell.

▶ Sites of Synthesis of Organelle and Membrane Proteins

Each subcellular organelle and each membrane contains a unique constellation of proteins. As examples, the ATP-ADP transporter (Chapter 17) is unique to the inner mitochondrial membrane, and certain hydrolytic enzymes are abundant in lysosomes. One possible method of directing proteins to their appropriate destinations in the cell might be via functionally different classes of ribosomes, each occupying a specific intracellular site and each translating only certain classes of messenger RNAs (mRNAs). Mito-

TABLE 16-1   Proteins Synthesized by Different Classes of Ribosomes

Location of Ribosome	Class of Protein Synthesized
Mitochondrion	All proteins encoded by mitochondrial DNA, mainly certain integral proteins of the inner membrane
Chloroplast	All proteins encoded by chloroplast DNA
Cytosolic ribosomes unbound to membranes	Soluble cytosolic proteins Cytoskeletal proteins (including those, such as spectrin and ankyrin, that are frequently bound to the cytosolic surface of integral membrane proteins) Proteins attached to membranes by covalently attached myristate or polyisoprenoid lipids (e.g., Ras, Src proteins) Chloroplast proteins encoded by nuclear DNA Mitochondrial proteins encoded by nuclear DNA Peroxisome proteins Glyoxisome proteins Nuclear proteins (histones, lamins, etc.)

Cytosolic ribosomes bound to ER membranes	Integral membrane glycoproteins
	    Plasma membrane
	    Nuclear membrane
	    Rough ER membrane
	    Golgi membrane
	    Lysosome membrane
	    Endosome membrane
	Secreted proteins (including extracellular matrix proteins, such as fibronectin and collagen, that are frequently bound to integral plasma membrane proteins)
	Lysosomal enzymes
	Rough ER enzymes
	Golgi complex enzymes

chondria and chloroplasts do contain unique populations of ribosomes, and all proteins encoded by mitochondrial DNA or chloroplast DNA are translated on ribosomes in the respective organelle (Table 16-1). However, specialized cytosolic ribosomes for the synthesis of specific proteins have never been found.

All Nuclear-Encoded Proteins Are Made by the Same Cytosolic Ribosomes

Abundant evidence indicates that all ribosomes in the cytosol of eukaryotic cells are functionally equivalent. Some ribosomes are always found tightly bound to the rough ER (it is the presence of these *membrane-bound ribosomes* that distinguishes the rough from the smooth ER), whereas others, termed *membrane-unattached ribosomes* appear to be free in the cytosol (Figures 16-5 and 16-6). (Many of these "free" ribosomes are actually bound to cytoskeletal fibers.) However, these two classes of ribosomes have the same protein and ribosomal RNA (rRNA) compositions. In cell-free protein-synthesizing systems containing a variety of added mRNAs, both classes of isolated ribosomes translate equally all nuclear-encoded mRNAs. Apparently, then, no functionally unique class of cytosolic ribosome is found only at a certain intracellular site to translate a specific class of mRNAs. Current evidence indicates that all information for intracellular protein distribution is located in the amino acid sequence of the newly synthesized protein itself.

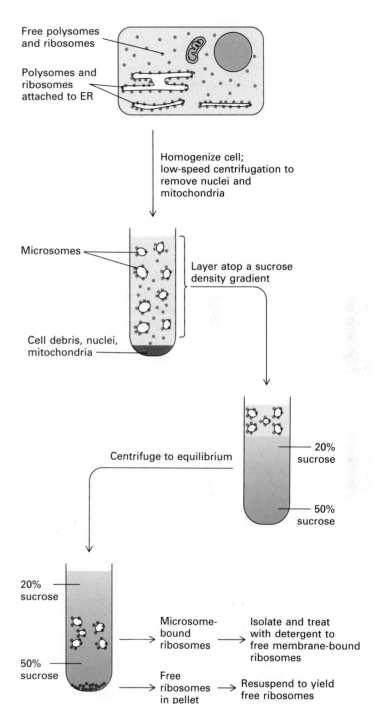

▲ FIGURE 16-6   Purification of membrane-unattached ribosomes from membrane-attached ribosomes and polysomes (clusters of ribosomes linked by mRNA). Cells are first homogenized to shear the rough endoplasmic reticulum into small vesicles termed microsomes. The homogenate is freed of nuclei and mitochondria by low-speed centrifugation and is then layered on a surcose density gradient. After prolonged centrifugation, membrane-unattached ribosomes and polysomes pellet to the bottom of the tube because they are more dense than the 50-percent sucrose solution. At equilibrium the microsomes form a band in the gradient, due to the low buoyant density of the phospholipids. Ribosomes bound to the microsomes can be freed of membranes by treatment with nonionic detergents.

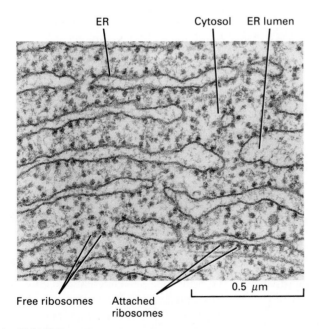

▲ FIGURE 16-5   Electron micrograph of ribosomes attached to the rough ER in a pancreatic exocrine cell. Most of the proteins synthesized by this cell are to be secreted and are formed on membrane-attached ribosomes. A few membrane-unattached (free) ribosomes are evident; presumably, these are synthesizing cytosolic or other nonsecretory proteins. [Courtesy of G. Palade.]

Membrane-Attached and Membrane-Unattached Ribosomes Synthesize Different Proteins

Although, when isolated, the two classes of cytosolic ribosomes are functionally equivalent, in the cell, membrane-attached and membrane-unattached ribosomes do translate different classes of nuclear-encoded mRNAs (Table 16-1). Proteins synthesized on membrane-attached ribosomes are those that enter the secretory pathway during their synthesis (Figure 16-1). This class includes not only secretory proteins, but also lysosomal enzymes and proteins localized to the lumen of the ER, to Golgi vesicles, and to secretory vesicles, as well as integral ER, Golgi, lysosomal, and plasma-membrane proteins.

Membrane-unattached ribosomes synthesize a large variety of proteins. These include soluble cytosolic proteins such as glycolytic enzymes, and cytoskeletal proteins including those, such as ankyrin (Figure 14-38), that are tightly bound to integral membrane proteins. These proteins are released into the cytosol, where they remain or bind to other cytosolic proteins (as for example, in the polymerization of actin or tubulin) or to membrane proteins. Some proteins, such as the Ras and Src cytosolic signaling proteins, acquire covalently bound fatty acids that aid their binding to specific cellular membranes (Figure 14-20).

We now consider some mechanisms of protein targeting and organelle biogenesis in more detail. We begin with a general discussion of the pathway of protein secretion, followed by a detailed consideration of the way in which nascent (still growing) secretory proteins are targeted to and cross the rough ER membrane.

► *Overall Pathway for Synthesis of Secretory, Lysosomal, and Membrane Proteins*

Here we present an overview of the function of the many organelles in the secretory pathway. Many of the experiments described here take advantage of cells that are specialized for the secretion of specific proteins (Table 16-2) and therefore contain organelles such as the rough ER and Golgi vesicles in abundance. For example, of the total protein made by hepatocytes (the principal cells of the liver), about 70 percent consists of serum proteins, such as albumin, that are secreted into the blood. Most of the proteins synthesized by the pancreatic acinar cells are digestive enzymes, such as chymotrypsinogen (the precursor of chymotrypsin; see Figure 3-22), ribonuclease, and amylase, that are packaged into zymogen vesicles and are secreted into the intestine. Most of the cytoplasm in these cells is filled with rough ER (Figure 16-5). Importantly, all cells secrete *some* proteins—extracellular matrix proteins such as collagens, proteoglycans, and fibronectin constitute about 5

percent of the protein made by most cultured cells—and all eukaryotic cells use essentially the same secretory pathway.

Newly Made Secretory Proteins Are Localized to the Lumen of the Rough ER

The *rough ER* is an extensive interconnected series of flattened sacs, generally lying in layers (see Figure 16-5). The *lumen*, or cisterna, of the rough ER defines a space topologically distinct from the cytosol. When cells are homogenized, the rough ER breaks up into small closed vesicles, termed *rough microsomes*, with the same orientation (ribosomes on the outside) as that found in the intact cell (see Figure 16-6). A simple experiment shows that immediately after their synthesis secretory proteins are localized in the lumen of ER vesicles, though they have been synthesized on cytosolic ribosomes (Figure 16-7): When purified mi-

TABLE 16-2   Classes of Secretory Proteins in Vertebrates

Protein Type	Example	Site of Synthesis
CONTINUOUSLY SECRETED PROTEINS		
Serum proteins	Albumin	Liver (hepatocyte)
	Transferrin (Fe transporter)	Liver
	Lipoproteins	Liver, intestine
	Immunoglobulins	Lymphocytes
Extracellular matrix proteins	Collagen	Fibroblasts, others
	Fibronectin	Fibroblasts, liver
	Proteoglycans	Fibroblasts, others
REGULATED SECRETORY PROTEINS		
Peptide hormones	Insulin	Pancreatic β-islet cells
	Glucagon	Pancreatic α-islet cells
	Endorphins	Neurosecretory cells
	Enkephalins	Neurosecretory cells
	ACTH	Anterior pituitary lobe
Digestive enzymes	Trypsin	Pancreatic acini
	Chymotrypsin	Pancreatic acini
	Amylase	Pancreatic acini, salivary glands
	Ribonuclease	Pancreatic acini
	Deoxyribonuclease	Pancreatic acini
Milk proteins	Casein	Mammary gland
	Lactalbumin	Mammary gland

crosomes are incubated with cytosol (which provides enzymes necessary for protein synthesis) and radioactive amino acids, the newly synthesized secretory proteins will be radioactive. The proteins remain associated with the microsomes, and since they are digested by added proteases only if the permeability barrier of the microsomal membrane is destroyed, it was concluded that they are inside the vesicles.

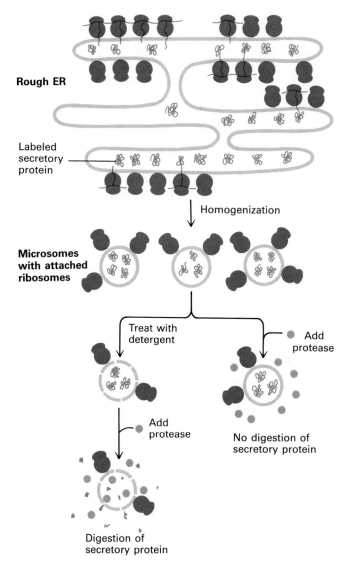

Rough ER

Labeled secretory protein

Homogenization

Microsomes with attached ribosomes

Treat with detergent

Add protease

Add protease

No digestion of secretory protein

Digestion of secretory protein

▲ **FIGURE 16-7**   Location of secretory proteins just after synthesis. Secretory proteins are sequestered in the lumen of the rough ER, as experiments similar to the one depicted here have proved. Newly synthesized secretory proteins associated with a microsome are not digested by an added protease, which remains outside the microsome. A detergent that makes the microsome membrane permeable allows some luminal proteins to leak out and the protease to enter. Under these conditions, newly made proteins are destroyed by the protease.

Many Organelles Participate in Protein Secretion

After a secretory protein has been transported across the ER membrane to the ER lumen, small transport vesicles containing the protein form from the ER and move to the membrane stacks known as the cis-*Golgi reticulum,* on the *cis* face of the Golgi complex (Figure 16-8). The protein moves through the Golgi complex to the *trans* face and then into a complex network of vesicles termed the trans-*Golgi reticulum.* From there, the protein is sorted to one of two types of secretory vesicles, which fuse with the plasma membrane and release the protein to the cell exterior.

In all cell types, at least some of the secretory proteins are secreted continuously. Examples of *continuous secretion* include collagen secretion by fibroblasts and secretion of serum proteins by hepatocytes (see Table 16-2). These proteins are sorted in the *trans*-Golgi reticulum into vesicles that immediately fuse with the plasma membrane, releasing their contents (Figure 16-8).

In certain cells, the secretion of a specific set of proteins is not continuous; these proteins are sorted in the *trans*-Golgi reticulum into vesicles that are stored inside the cell awaiting a stimulus for exocytosis (see Figures 16-8 and 5-40). Examples of this *regulated secretion* include the exocrine cells of the pancreas, which secrete precursors of digestive enzymes, and hormone-secreting endocrine cells such as those of the pancreas, which synthesize insulin and other hormones and store them in vesicles (see Table 16-2). The release of each of these proteins is initiated by different neural and hormonal stimuli. In most cases of regulated secretion studied so far, exocytosis is triggered by a rise in intracellular Ca^{2+} and consists of the fusion of the membrane of the secretory vesicle with the plasma membrane.

A large number of enzymes have been identified that act sequentially to modify secretory proteins during their maturation. Each enzyme is localized to a specific organelle and modifies the proteins as they pass through the organelle. Amino acid side chains can be altered, saccharide residues can be added and the oligosaccharides remodeled, specific proteolytic cleavages may take place, disulfide bonds can form, polypeptide chains may assemble into multiprotein complexes. Not all of these reactions occur to every secreted protein. However, collagen is an example of a secreted protein that is modified in all of these ways.

All Secretory Proteins Move from the Rough ER to Golgi Vesicles to Secretory Vesicles

Protein transport was initially established by studies on newly synthesized secretory proteins in pancreatic acinar cells, using electron microscope autoradiography after pulse-chase studies. When thin slices of pancreas in culture are treated with the radiolabeled amino acid leucine ([³H]leucine) for a brief period of time (the "pulse" label;

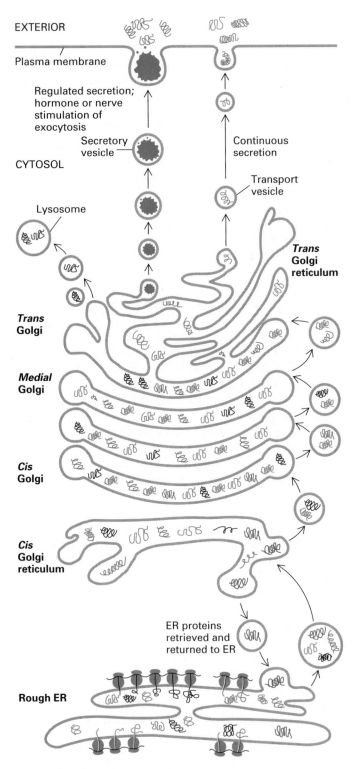

EXTERIOR

Plasma membrane

Regulated secretion; hormone or nerve stimulation of exocytosis

Secretory vesicle

Continuous secretion

CYTOSOL

Transport vesicle

Lysosome

Trans Golgi reticulum

Trans Golgi

Medial Golgi

Cis Golgi

Cis Golgi reticulum

ER proteins retrieved and returned to ER

Rough ER

◄ **FIGURE 16-8**   Maturation of secretory proteins. After their synthesis, secretory proteins are localized to the lumen of the rough ER. Always surrounded by membrane-bounded vesicles, they migrate first to the interconnected vesicles known as the *cis*-Golgi reticulum on the *cis* face of the Golgi membrane complex, where modifications mainly to carbohydrate chains occur. The proteins then undergo additional modifications as they migrate through the Golgi vesicles to the *trans* face and the *trans*-Golgi reticulum. Proteins are shuttled between the Golgi vesicles by small transport vesicles. In some cells, such as hepatocytes, the matured proteins are secreted continuously (constitutive secretion). In other cells, such as exocrine pancreatic cells, some proteins are stored in secretory vesicles, where they await a signal for secretion (regulated secretion). Proteins are also sorted in the *trans*-Golgi vesicles or the *trans*-Golgi reticulum for transport to lysosomes. Many proteins that function while soluble in the ER lumen migrate after synthesis from the ER to the *cis*-Golgi reticulum, whereupon they are incorporated into transport vesicles and retrieved to the ER.

Though not shown here, membrane proteins follow much the same maturation pathway as secretory proteins. Plasma membrane glycoproteins go through the same stages of maturation that continuously secreted proteins do. ER membrane proteins stop maturing in the ER, while membrane proteins destined to remain in the Golgi complex are formed in the rough ER and transported to the appropriate Golgi vesicle where they remain.

➤ FIGURE 16-9   The synthesis and movement of guinea-pig pancreatic secretory protein as revealed by electron microscope autoradiography. (a) At the end of a 3-min labeling period with [³H]leucine, the tissue is fixed, sectioned for electron microscopy, and subjected to autoradiography. Most of the labeled proteins (the autoradiographic grains) are over the rough ER. (b) Following a 7-min chase period with unlabeled leucine, most of the labeled proteins have moved to the Golgi vesicles. (c) After a 37-min chase, most of the proteins are over immature secretory vesicles. (d) After a 117-min chase, the majority of the proteins are over mature zymogen granules. [Courtesy of J. Jamieson and G. Palade.]

proteins are localized to the mature secretory vesicles (zymogen granules). Apparently, immature secretory vesicles fuse together to become mature secretory vesicles; during this time there is also concentration of their contents. Electron-microscope studies have been confirmed by fractionating cells after various pulse and chase times. The results clearly indicate that secretory proteins are always sequestered in membrane vesicles.

The Steps in Protein Secretion Can Be Studied Genetically

Yeasts secrete few proteins into the growth medium, but they do secrete continuously a number of enzymes that

see Figure 6-25), essentially all the incorporated radioactivity in the acini is seen to be located in the rough ER (Figure 16-9). The incorporation of radioactivity is then blocked by incubating the tissue with abundant unlabeled leucine (the "chase" period). After a 7-min. chase, most labeled proteins can be found in the Golgi vesicles. At later times, the radioactivity is located in immature secretory vesicles adjacent to the Golgi vesicles; at still later times, the labeled

(a)

(b)

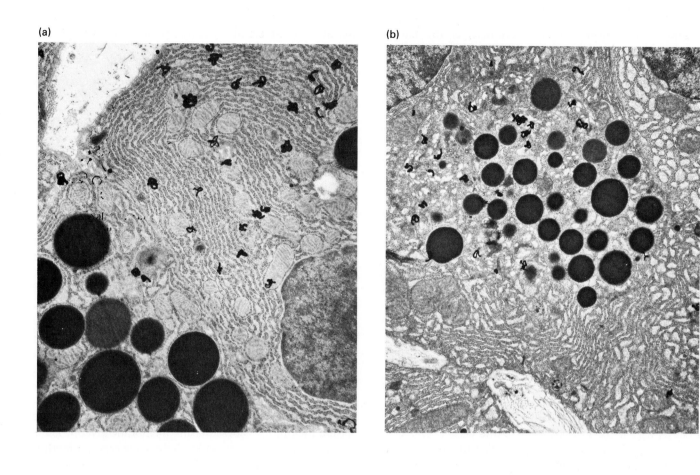

(c)

(d)

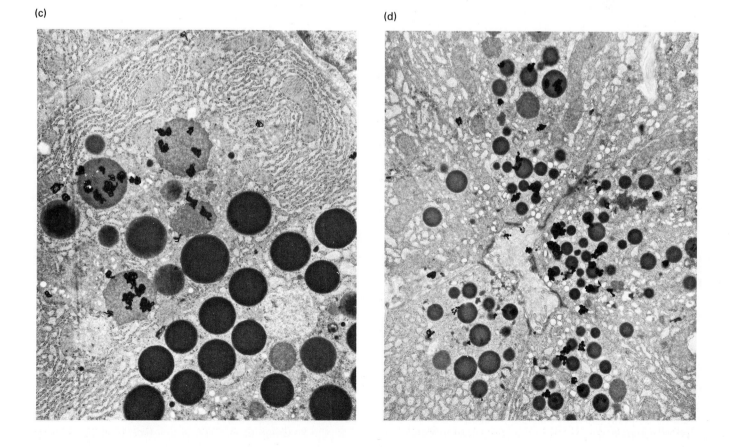

remain localized in the narrow space between the plasma membrane and the cell wall. The best-studied of these, invertase, hydrolyzes the disaccharide sucrose to glucose and fructose. A genetic analysis of protein secretion in a series of temperature-sensitive yeast mutants has defined at least five discrete steps in the maturation pathway for secretory proteins (Figure 16-10; see also Figure 8-16). A large set of temperature-sensitive mutant yeast strains was identified in which the secretion of all proteins is blocked at the higher ("nonpermissive") temperature (where the cells cannot grow) but is normal at the lower temperature (where the cells grow normally). When transferred from the lower to the higher temperature, each class of these so-called *sec* mutants accumulates secretory proteins in one organelle, because the secretory pathway is blocked at that point: either in the cytosol, in the rough ER, in small vesicles taking proteins from the ER to the Golgi complex, in the Golgi vesicles, or in the secretory vesicles. At least 30 gene products are required to complete the secretory pathway, and each of the five steps requires the action of multiple proteins, as we detail later in this chapter.

Studies with double *sec* mutants confirm that the order of the pathway is cytosol → rough ER → ER-to-Golgi transport vesicles → Golgi vesicles → secretory vesicles → exocytosis. For instance, when yeast cells contain double mutants with defects in both class B and class D secretory transport processes (see Figure 16-10), proteins accumulate in the rough ER, not in the Golgi vesicles; hence class B

mutations must act at an earlier point in the maturation pathway than class D mutations do.

The pathway for the maturation of secretory proteins elucidated by studies in yeasts and in pancreatic acinar cells is thought to function in all eukaryotic cells.

Plasma Membrane Glycoproteins Follow the Same Maturation Pathway as Continuously Secreted Proteins

The maturation pathway taken by continuously secreted proteins is also followed by plasma membrane glycoproteins. Well-studied examples include viral glycoproteins destined for the plasma membranes of infected cells, such as the hemagglutinin (HA) glycoprotein of the influenza virus (see Figure 3-5), glycophorin (see Figure 14-14), the plasma membrane $Na^+ - K^+$ ATPase (see Figure 15-13), and enzymes in plant plasma membranes that synthesize such cell wall components as cellulose. Pulse-labeling studies using radioactive amino acids, followed by subcellular fractionation and immunoprecipitation to detect radiolabeled proteins, have established that the newly made glycoproteins are inserted into the rough ER membrane and subsequently move through the Golgi vesicles en route to the plasma membrane. These plasma membrane glycoproteins also have been shown to undergo the same types of modifications in the same ER and Golgi compartments that secre-

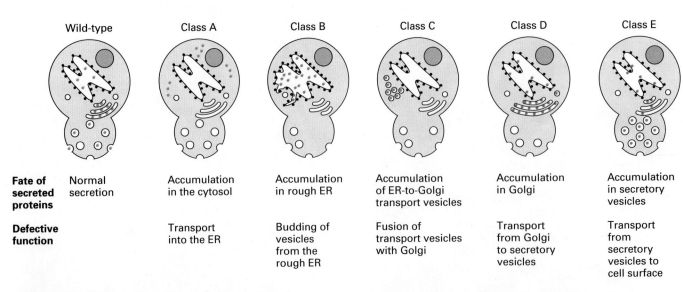

	Wild-type	Class A	Class B	Class C	Class D	Class E
Fate of secreted proteins	Normal secretion	Accumulation in the cytosol	Accumulation in rough ER	Accumulation of ER-to-Golgi transport vesicles	Accumulation in Golgi	Accumulation in secretory vesicles
Defective function		Transport into the ER	Budding of vesicles from the rough ER	Fusion of transport vesicles with Golgi	Transport from Golgi to secretory vesicles	Transport from secretory vesicles to cell surface

▲ FIGURE 16-10 Five steps in the maturation of secretory proteins as defined by yeast *sec* mutants that are temperature-sensitive for protein secretion. The collection of about 30 different *sec* genes can be grouped into six classes, corresponding to the site where newly made secre-

tory proteins (the red dots) accumulate when cells are shifted from the growing (permissive) temperature to the higher nonpermissive one. [See P. Novick et al. 1981. *Cell* **25**:461; C. A. Kaiser and R. Schekman. 1990. *Cell* **61**:723–733; and N. Green et al. 1992. *J. Cell Biol.* **116**:597.]

tory proteins do. For instance, the same *N*- and *O*-linked oligosaccharides (see Figures 14-25 and 14-26) are found on both classes of proteins.

➤ *The Transport of Secretory and Membrane Proteins into or across the ER Membrane*

Here we focus on the initial steps during protein biosynthesis: the transport of secretory proteins across the ER membrane or the insertion of membrane proteins into the ER membrane. Ribosomes that are synthesizing secretory pro-

teins and integral ER, Golgi, lysosomal, and plasma membrane proteins are tightly bound to the ER membrane; during protein synthesis, the secretory and lysosomal proteins cross the ER membrane and the proteins destined for the various membranes are inserted into the ER membrane. However, the initial stages of protein synthesis occur while the ribosome is in the cytosol, unbound to the membranes. This raises three key questions that we discuss in turn: How are certain proteins targeted to the secretory pathway? Since pure phospholipid membranes are impermeable to proteins, how do the secretory proteins actually cross the membrane? And what is the source of energy for the transport of proteins across biological membranes? Figure 16-11 summarizes the pathway by which secretory

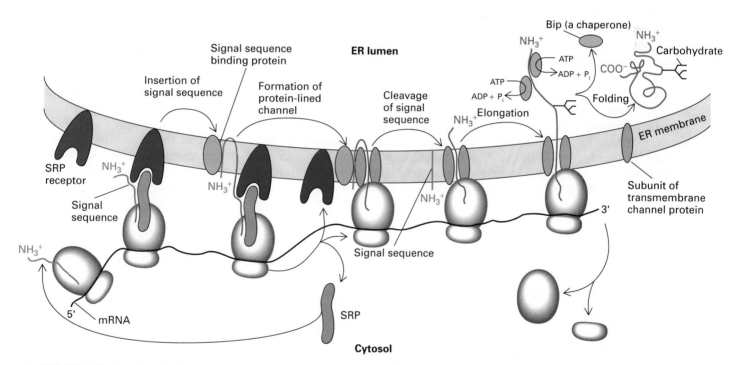

▲ FIGURE 16-11   Synthesis of secretory proteins on the ER. The N-terminal signal sequence emerges from the ribosome only when the polypeptide is about 70 amino acids long, because about 30 amino acids remain buried in the ribosome. An elongated signal recognition particle (SRP) then binds to the signal sequence, and the complex of SRP, nascent polypeptide, and ribosome binds to the ER membrane through the SRP receptor. The signal sequence inserts itself into the ER membrane by binding to an ER membrane protein, the signal sequence binding protein (purple). Several ER membrane proteins are recruited together to form a transmembrane channel through which the growing polypeptide crosses the phospholipid bilayer. During or just after this insertion step, SRP dissociates from the ribosome-peptide complex and is released into the cytosol. The SRP receptor is also freed to initiate the insertion of another secretory protein; the hydrolysis of GTP (not depicted here) by SRP and the SRP receptor is essential for the dissociation of SRP from the signal sequence and from the SRP receptor. The signal sequence is cleaved off in the ER lumen by signal

peptidase and the signal sequence is quickly degraded. The peptide chain continues to elongate and is extruded into the ER lumen through the transmembrane channel. The chaperone protein Bip binds to the growing chain on the luminal surface and then, powered by the hydrolysis of ATP, releases the nascent chain; this cycle of binding and release of the nascent chain by Bip is essential for translocation to continue, and it also facilitates the eventual folding of many secreted polypeptides.

Enzymes on the luminal surface add carbohydrates to the chain at asparagine, serine, and threonine residues. After synthesis is complete and the ribosomes are released, the remaining C-terminus of the secreted protein is drawn into the ER lumen, the transmembrane channel is disassembled, and the secreted protein assumes its final conformation.
[See T. A. Rapoport. 1991. *FASEB J.* **5**:2792–2798; P. J. Rapiejko and R. Gilmore. 1992. *J. Biol. Chem.* **117**:493–502; S. L. Sanders et al. 1992. *Cell* **69**:353–365; G. Migliaccio, C. V. Nicchitta, and G. Blobel. 1992. *J. Cell Biol.* **117**:15–25; and D. Görlich et al. 1992. *Nature* **357**:47–52.]

proteins cross the ER membrane, and we now discuss the steps in this process in some detail.

A Signal Sequence on Nascent Secretory Proteins Targets Them to the ER and Is then Cleaved Off

To begin, a sequence of 16–30 amino acid residues on a newly made secretory protein, called the *signal sequence,* directs the ribosome to the ER membrane and initiates the transport of the growing polypeptide across the ER membrane. Characteristically, a signal sequence is found at the N-terminus of the protein and it contains one or more positively charged amino acids followed by a continuous stretch of 6–12 hydrophobic residues (Table 16-3); otherwise, the signal sequences of various secretory proteins have little identity to each other. Signal sequences are not normally found on complete polypeptides made in cells, implying that the signal sequence is cleaved from the protein while it is still growing on the ribosome. The enzyme *signal peptidase,* which normally cleaves off the signal sequences, is localized to the ER lumen.

If we wish to detect the signal sequence, the mRNA must be translated in a cell-free system containing ribosomes, tRNAs, ATP and GTP, and cytosolic enzymes but with no ER membranes. In this case, the protein, with its signal sequence attached, is released into the cytosol (Figure 16-12a). Important information has been derived from

the use of microsome membranes that have been stripped of their own ribosomes by treatment with a chemical that chelates Mg^{2+} ions. If these membranes are present during the cell-free synthesis of a secretory protein, the protein is found in the ER lumen with the signal sequence removed (Figure 16-12b). If, however, these membranes are added to the reaction mixture after the secretory protein is completely synthesized, the protein generally is not incorporated into the ER lumen and its signal sequence remains attached (Figure 16-12a). Thus, a newly made secretory protein can cross the ER membrane and have its signal sequence removed only if the membrane has been present during protein synthesis.

In an extension of the experiment described in Figure 16-12, microsomes were added at different times after synthesis of a secretory protein began. These experiments showed that microsome membranes must be added before the first 70 or so amino acids are polymerized in order for the subsequently completed secretory protein to be localized in the ER lumen. At this point, about 40 amino acids protrude from the ribosomes, including the signal sequence that later will be cleaved off, and about 30 amino acids are buried in a channel or tunnel in the ribosome. Thus the transport of most secretory proteins into the ER lumen—particularly the transport of those with more than 100 amino acids—must occur during translation. However, a smaller secretory protein, such as the α mating factor of yeast (≈70 amino acids), can cross into the ER lumen after

---

TABLE 16-3   Amino Acid Sequences of Signal Peptides of Several Eukaryotic Secretory and Membrane Proteins

Secretory or Membrane Protein	Amino Acid Sequence*
Preproalbumin	Met-Lys-Trp-Val-Thr-**Phe-Leu-Leu-Leu-Leu-Phe-Ile-Ser-Gly-Ser-Ala-Phe**-Ser ↓ Arg . . .
Pre–IgG light chain	Met-Asp-Met-Arg-Ala-Pro-Ala-Gln-**Ile-Phe-Gly-Phe-Leu-Leu-Leu-Leu-Phe**-Pro-Gly-Thr-Arg-Cys ↓ Asp . . .
Prelysozyme	Met-Arg-Ser-**Leu-Leu-Ile-Leu-Val-Leu-Cys-Phe-Leu**-Pro-Leu-Ala-Ala-Leu-Gly ↓ Lys . . .
Preprolactin	Met-Asn-Ser-Gln-Val-Ser-Ala-Arg-Lys-Ala-Gly-Thr-**Leu-Leu-Leu-Leu-Met-Met**-Ser-Asn-Leu ↓ Leu . . .
VSV glycoprotein precursor	Met-Lys-Cys-**Leu-Leu-Tyr-Leu-Ala-Phe-Leu-Phe-Ile**-His-Val-Asn-Cys ↓ Lys . . .
Rat proinsulin-1	Met-Ala-**Leu-Trp-Met**-Arg-**Phe-Leu**-Pro-**Leu-Leu-Ala-Leu-Leu-Val-Leu-Trp**-Glu-Pro-Lys-Pro-Ala-Gln-Ala ↓ Phe . . .
Acetylcholine receptor γ subunit precursor	Met-**Val-Leu-Thr-Leu-Leu-Leu-Ile-Ile-Cys-Leu-Ala-Leu**-Glu-Val-Arg-Ser ↓ Glu . . .

source: D. P. Leader. 1979. *Trends Biochem. Sci.* 4:205 and T. A. Rapoport. 1985. *Curr. Top. Membr. Transport* 24:1–63.
*Hydrophobic residues are in **boldface;** arrows (↓) indicate the site of cleavage by signal peptidase.

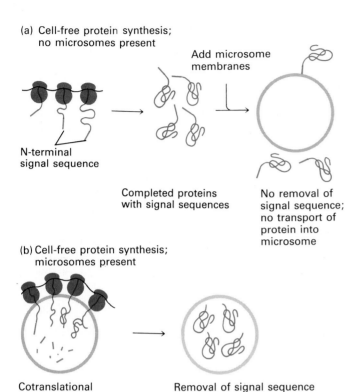

(a) Cell-free protein synthesis;
no microsomes present

Add microsome
membranes

N-terminal
signal sequence

Completed proteins
with signal sequences

No removal of
signal sequence;
no transport of
protein into
microsome

(b) Cell-free protein synthesis;
microsomes present

Cotranslational
transport of protein
into microsome

Removal of signal sequence

▲ **FIGURE 16-12** The cotranslational insertion of secretory proteins into microsomes (i.e., ER vesicles). Synthesis is carried out in a cell-free system (an extract of wheat germ is commonly used, since it contains functional ribosomes but no functional mRNA). (a) A secretory protein synthesized in a system without microsomes retains the N-terminal signal sequence. If microsomes are subsequently added, the protein is not transported across the vesicle membrane and the signal sequence is not removed. (b) By contrast, if microsomes are present during protein synthesis, the signal sequence is removed from the nascent chain and the protein is transported into the lumen of the vesicles.

its synthesis has been completed and it has been released from the ribosome. In the cytosol the α factor becomes bound to a chaperone protein (discussed later) that keeps it in an unfolded state; this allows it subsequently to cross the ER membrane.

Many experiments have shown that a signal sequence is all that is necessary to direct a protein from the cytosol to the lumen of the ER. One such study uses recombinant DNA techniques to produce a chimeric protein that has the signal sequence of a secretory protein attached to the N-terminus of a cytosolic protein that normally is not secreted, such as globin. (A *chimera* is a mythological monster that has the head of a lion, a goat's body, and a serpent's tail.) The globin part of the chimeric protein is

transported to the ER lumen, where the signal peptide is cleaved off exactly as it would be from a "normal" secretory protein.

All signal sequences on secretory proteins have a "core" of hydrophobic amino acids (see Table 16-3) that are essential for signal-sequence function; the specific deletion of several of these amino acids from the signal sequence, or the mutation of one of them to a charged amino acid, abolishes the ability of the protein to cross the ER membrane into the lumen. Other experiments indicate that any random N-terminal amino acid sequence, provided it is sufficiently long and hydrophobic, will cause the protein to be translocated to the ER lumen. These hydrophobic residues form a binding site that is critical for the interaction of signal sequences with the ER membrane.

Several Receptor Proteins Mediate the Interaction of Signal Sequences with the ER Membrane

Since secretory proteins are synthesized in association with the ER membrane but not with any other cellular membrane, some signal-sequence recognition system must target them there. A simple experiment identified the key element in this targeting process. A preparation of microsomes, stripped of their ribosomes, was exposed to a solution of 0.5 M NaCl, so that several proteins were removed from the membranes. When these treated microsome vesicles were recovered by centrifugation and added to a cell-free protein synthesis reaction, they were unable to support the transmembrane movement of newly synthesized secretory proteins. However, when the proteins removed by the NaCl treatment were returned to the preparation, secretory proteins were incorporated into the lumen of the vesicles. A single active component—the *signal recognition particle* (SRP)—was purified from the mixture of proteins extracted by NaCl. SRP contains six discrete polypeptides and a 300-nucleotide RNA and is an essential component for protein translocation across the ER membrane (Figure 16-13; see also Figure 16-11).

Some functions of SRP have been clarified by studies in which mRNAs for secretory proteins, such as the pituitary hormone precursor preprolactin, are translated in a cell-free protein synthesis system (Figure 16-14). When cell-free translation is carried out in the absence of SRP, full-length protein with its signal sequence is made. If SRP is added, but no microsomes, protein synthesis is arrested at about 70 amino acids. If microsome membranes are then added, the elongation of preprolactin chains resumes, the ribosomes bind to the membranes, the signal sequence is cleaved, and the prolactin protein is sequestered in the lumen of the microsomes. Thus SRP prevents the synthesis of a complete secretory protein in the absence of rough ER membranes. Without this *quality control* step, synthesis of

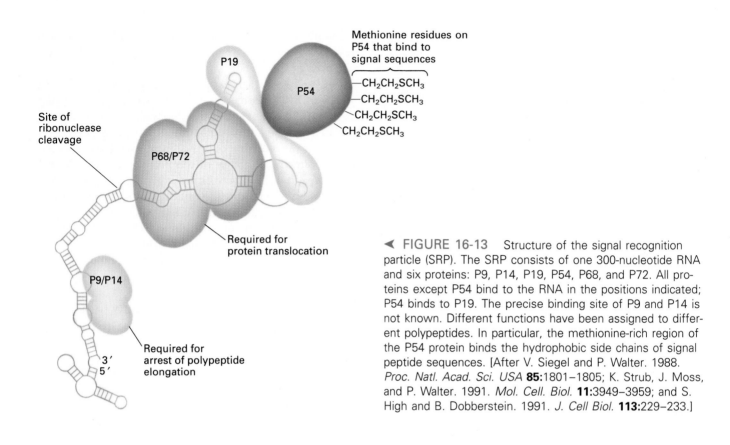

◀ FIGURE 16-13   Structure of the signal recognition particle (SRP). The SRP consists of one 300-nucleotide RNA and six proteins: P9, P14, P19, P54, P68, and P72. All proteins except P54 bind to the RNA in the positions indicated; P54 binds to P19. The precise binding site of P9 and P14 is not known. Different functions have been assigned to different polypeptides. In particular, the methionine-rich region of the P54 protein binds the hydrophobic side chains of signal peptide sequences. [After V. Siegel and P. Walter. 1988. *Proc. Natl. Acad. Sci. USA* **85**:1801–1805; K. Strub, J. Moss, and P. Walter. 1991. *Mol. Cell. Biol.* **11**:3949–3959; and S. High and B. Dobberstein. 1991. *J. Cell Biol.* **113**:229–233.]

the complete secretory protein would occur in the cytosol, and the protein might misfold or aggregate; this would prevent the protein from ever being transported across the ER membrane.

Many functions of SRP can be assigned to specific component proteins (see Figure 16-13). For instance, in the SRP-ribosome complex in which protein elongation is blocked (see Figure 16-14b), the 54,000-MW SRP protein (P54) can be chemically cross-linked to the signal sequence, evidence that this SRP protein is the one that binds to the signal sequence. P54 contains a segment with a large number of clustered methionine residues whose side chains are thought to protrude outward and bind to the hydrophobic side chains that form the "core" of a signal sequence. SRP can be cleaved by a ribonuclease into a smaller fragment containing part of the RNA bound to P9 and P14 and a larger fragment containing the other four proteins. The larger fragment cannot block the elongation of growing polypeptides but is completely normal in its ability to bind to signal sequences and to catalyze the insertion of nascent secretory proteins into the rough ER. This result establishes that the arrest of polypeptide elongation by SRP is not an essential aspect of polypeptide translocation to ER membranes and that elongation arrest is dependent on the SRP components P9 and P14.

An *SRP receptor* in the rough ER membrane binds SRP, and initiates the binding of the ribosome and nascent chain to the ER membrane (see Figure 16-11). This receptor contains two polypeptide subunits: integral proteins of about 640 and 300 amino acids, respectively. The larger subunit is anchored to the ER membrane by a hydrophobic sequence of amino acids near its N-terminus. The treatment of ER membranes with tiny amounts of protease cleaves the larger protein subunit very near its site of attachment to the membrane and releases a soluble form of the SRP receptor. This soluble fragment can relieve the block in chain elongation that is imposed by SRP in cell-free systems (see Figure 16-14c). The SRP receptor thus must bind to the SRP and possibly also to the ribosome.

The SRP and its receptor only initiate the transfer of the nascent chain across the ER membrane; then they dissociate from it, using the energy of GTP hydrolysis. GTP hydrolysis is also required to insert the signal sequence into the membrane. Both the P54 subunit of SRP and the larger subunit of the SRP receptor can bind GTP. The energy released by the hydrolysis of one or more molecules of GTP causes the SRP receptor to detach from SRP and SRP to detach from the ribosome-nascent chain complex once the signal sequence has been completely inserted into the ER membrane. After thus dissociating, the SRP and its recep-

(a) No SRP, no SRP receptor, no microsomes

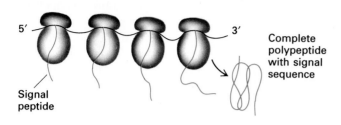

Complete polypeptide with signal sequence

(b) Plus SRP, no SRP receptor, no microsomes

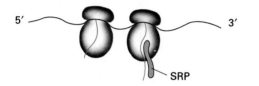

Elongation blocked at 70-100 amino acids

(c) Plus SRP, plus soluble SRP receptor, no microsomes

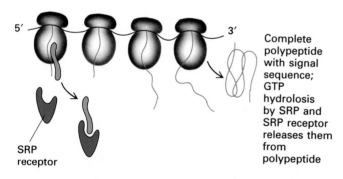

Complete polypeptide with signal sequence; GTP hydrolosis by SRP and SRP receptor releases them from polypeptide

(d) Plus SRP, plus microsomes with SRP receptor

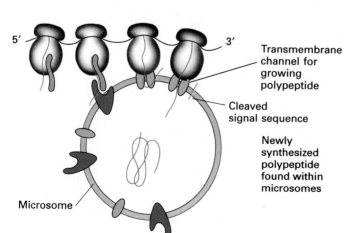

Transmembrane channel for growing polypeptide

Cleaved signal sequence

Newly synthesized polypeptide found within microsomes

◄ FIGURE 16-14   The properties of SRP and the SRP receptor studied in cell-free protein synthesis reactions. The mRNA that encodes preprolactin, a typical secretory protein made by the pituitary, is readily translated in a wheat-germ cell-free system without any microsomes. This system contains all the necessary factors for in vitro protein synthesis. In one series of experiments, SRP, SRP receptor, and microsomes were added to the system and the effects on protein synthesis were examined. (a) When SRP, the SRP receptor, and microsomes are all absent, the complete protein with its signal sequence is synthesized. (b) The addition of only SRP causes elongation to be blocked at 70–100 amino acids. At this stage, the 54,000-MW SRP protein can be chemically cross-linked to the signal sequence (direct evidence that SRP binds to the N-terminal residues on the growing chain). (c) The addition of a soluble SRP receptor removes this block in polypeptide chain elongation. The SRP receptor binds to SRP. Hydrolysis of GTP by SRP and the SRP receptor causes their release from the growing chain. (d) If SRP and microsomes containing the SRP receptor are added, a transmembrane channel forms and the newly made protein is translocated across the vesicle membrane. [See P. Walter and G. Blobel. 1981. *J. Cell Biol.* **91**:557; V. C. Krieg, P. Walter, and A. E. Johnson. 1986. *Proc. Natl. Acad. Sci. USA* **83**:8604–8608, and D. Zopf, H. Bernstein, and P. Walter. 1993. *J. Cell Biol.* **120**:1113–1121].

tor recycle to direct the insertion of additional proteins.

Several types of experiments indicate that SRP and the SRP receptor are essential to initiate the cotranslational (during translation) transport of a secretory protein across the mammalian ER membrane. For example, microsomal proteins can be solubilized in detergents and then reconstituted with phospholipid micelles, yielding vesicles that are fully competent in protein translocation. If the SRP receptor is removed (by immunoaffinity chromatography, Figure 3-34) from the solubilized microsomal proteins, vesicles reconstituted from the remaining proteins are nonfunctional; they regain their function if purified SRP receptor is added back to the reconstitution mixture. This is strong evidence for the essential role of the SRP receptor in protein translocation across the mammalian ER membrane. Other evidence for the role of the SRP receptor comes from studies of mutant yeasts that lack a functional SRP. These mutants grow very slowly and show defective secretion of most proteins. However, the secretion of a few proteins is normal, indicating that yeasts, and probably other eukaryotes, have (unknown) mechanisms for targeting certain secretory proteins to the ER membrane that do not involve SRP.

Other integral ER membrane proteins are thought to bind the signal sequence after its release and facilitate its insertion into the ER membrane. One of these proteins is the TRAM protein discussed later, that may function as the signal sequence binding protein (see Figure 16-11).

Polypeptides Cross the ER Membrane in Protein-Lined Channels

Once the signal sequence initiates translocation of the growing protein chain across the ER membrane, the protein chain passes directly from the large ribosomal subunit into the membrane. The growing chain is never exposed to the cytosol and does not fold until it reaches the ER lumen. Thus the protein crosses the membrane in an unfolded state. Later in this chapter and in Chapter 19 we shall see other evidence of an important generality: proteins must be unfolded before they can insert themselves into or cross membranes. If newly made secretory proteins were allowed to fold in the cytosol, they would be unable to cross the ER membrane.

The nascent chain traverses the ER membrane through a protein-lined channel in the membrane (Figure 16-15), in much the same way that ions pass through membrane channels (Figure 15-2). The transport channel for the nascent polypeptide may also participate in binding the 60S subunit of the ribosome to the ER membrane.

Three types of experiments have demonstrated the identities of the channel proteins. In one, the mRNA for a mutant secretory protein lacking a chain termination codon was translated in a cell-free system containing microsomes. The ribosomes translated the entire length of the mRNA but could not release the completed polypeptide; under this circumstance the nascent polypeptide becomes "stuck" crossing the ER membrane. When microsomes from mammalian cells were used, chemical cross-linking agents (Figure 20-8b) could covalently link the stuck nascent chain primarily to two abundant membrane proteins, which evidently form part of the mammalian transport channel. One of these proteins, named TRAM (*translocating chain-associated membrane protein*), also binds and can be chemically cross-linked to the signal sequence, and may be the signal sequence binding protein depicted in Figure 16-11. TRAM spans the ER membrane at least 8 times, and, as discussed below, indeed is essential for protein translocation.

The second of the required mammalian proteins is similar in amino acid sequence to the protein product of the yeast *sec61* gene. Genetic studies in yeast provided evidence for the role of the sec61 protein (sec61p) in ER protein translocation since *sec 61*, like the sec 62, and sec 63 genes, produce Class A *sec* mutants (see Figure 16-10), de-

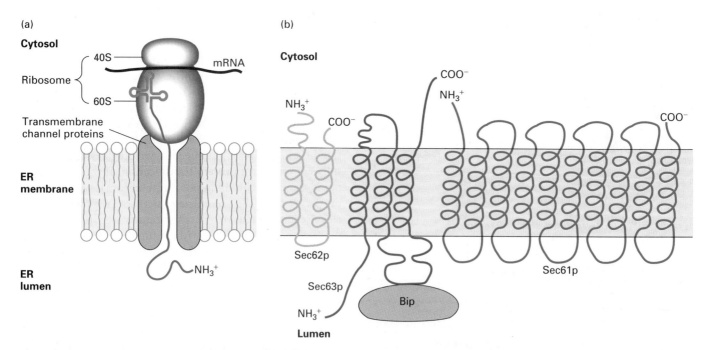

▲ **FIGURE 16-15**   Proteins involved in the transport of nascent secretory polypeptides across the ER membrane. (a) A set of proteins form a transmembrane channel through which the nascent chain passes; they also may anchor the 60S ribosomal subunit to the membrane. (b) In yeast, three ER membrane proteins, Sec61p (the product of the *sec61* gene), Sec62p, and Sec63p, are essential for the transport of nascent chains into the ER lumen and probably form part of the transmembrane channel. Sec61p binds to the large ribo-somal subunit and Sec63p binds to Bip, a chaperone in the ER lumen that is also essential for the translocation of nascent chains. [Part (b) after D. Feldheim, J. Rothblatt, and R. Schekman. 1992. *Mol. Cell. Biol.* **12**:3288–3296 and T. A. Rapoport. 1992. *Science* **258**:931–936. See also A. Müsch, M. Wiedmann, and T. A. Rapoport. 1992. *Cell* **69**:343–362; P. Walter. 1992. *Nature* **357**:22–23; and D. Feldheim et al. 1993. *Mol. Biol. Cell* **4**:931–939.]

fective in the translocation of secretory proteins into the ER lumen. The proteins coded by these yeast genes have multiple membrane-spanning α-helixes and at least the sec61p protein forms part of the channel in yeast ER membranes through which the nascent chain moves (Figure 16-15b). The *sec61* protein also binds tightly to ribosomes, and may serve to bind to the ER membrane the 60S subunit of those ribosomes that are synthesizing secretory proteins (see Figure 16-15a).

Finally, a biochemical experiment established that the only integral mammalian ER proteins required for translocation of nascent secretory proteins are the SRP receptor, TRAM, and a complex of proteins containing the protein homologous in sequence to the yeast sec61 protein, since phospholipid vesicles reconstituted in vitro and containing only these proteins are functional in translocating nascent secretory proteins. Thus, in mammalian cells the SRP receptor, TRAM, (and possibly an additional signal sequence binding protein) function to orient and insert the signal peptide (Figures 16-11 and 16-15). Proteins similar in sequence to the yeast sec61, sec62, and sec63 proteins form the channel through which the growing polypeptide moves.

ATP-Hydrolyzing Chaperone Proteins Prevent Protein Misfolding and Are Essential for Translocation of Secretory Proteins into the ER

As soon as a segment of a nascent secretory or membrane protein crosses into the ER lumen, it is bound by an abundant protein termed *Bip* (Figure 16-16; see also Figure 16-11). [The name is shorthand for heavy-chain *b*inding *p*rotein, since Bip was discovered by its interaction with the heavy chain of IgG antibodies (see Figure 3-32) in the ER lumen.] In the transport of proteins across the ER membrane, Bip serves two essential functions.

First, Bip binds to short hydrophobic sequences of amino acids of the sort that are normally found in the interior of a water-soluble protein. Preferentially, Bip binds to segments of a protein that have large hydrophobic amino acids such as tryptophan, phenylalanine, and leucine in every second position (Figure 16-16). By binding to such sequences on unfolded proteins that have just been translocated into the ER lumen, Bip prevents denaturation or nonspecific aggregation of the polypeptide. Bip then binds and hydrolyzes ATP, causing release of the bound polypeptide. In many cases the released polypeptide will quickly fold correctly (either as a monomeric protein or as part of an oligomer) with the hydrophobic segment properly buried in the protein interior; such folded proteins will not rebind to Bip. However, if the hydrophobic segment of the protein remains unfolded, Bip immediately rebinds to it. Thus, Bip uses the energy released by ATP hydrolysis to keep proteins unfolded and to prevent misfolding.

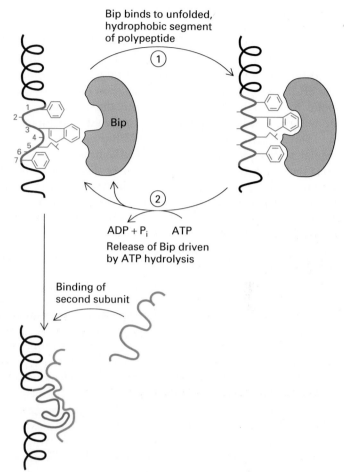

▲ FIGURE 16-16   Binding of the ER luminal protein Bip to nascent proteins. (1) Bip binds to exposed hydrophobic segments of polypeptides (colored blue), preventing protein denaturation or aggregation. Preferentially, Bip binds to segments of a protein in which large hydrophobic amino acids are in every second position, here numbered 1, 3, 5, and 7. (2) After binding and hydrolyzing ATP, Bip releases the bound polypeptide. If the polypeptide remains in the unfolded state, with its hydrophobic segment exposed, it rebinds to Bip. In contrast, if the released protein folds properly it will not rebind to Bip. Folding can involve a single polypeptide chain or, as illustrated here, can require binding of the chain to a second protein subunit. [After G. C. Flynn et al. 1991. *Nature* **353**:726–730 and S. Blond-Elguindi et al. 1993. *Cell* **75**:1–20. See D. Palleros et al. 1993. *Nature* **365**:664–666 for a model of ATP binding and hydrolysis by Bip.]

Second, many newly made proteins bind to Bip as soon as they are translocated into the ER lumen, and Bip is also required for the translocation of many secretory proteins into the ER. To show this, recombinant DNA–generated mutants of Bip protein from yeast were constructed in which Bip function was temperature-sensitive. When the temperature of growing cells was increased, all Bip functions ceased, and over the next hour, although overall protein synthesis was normal, the transport of proteins into the ER stopped. It is thought that in the absence of Bip, parts of certain nascent secretory proteins aggregate within the ER lumen, inhibiting further elongation of the nascent chain through the ER membrane and thus blocking the protein-translocation channels within the ER membrane.

The sequence of the 70,000-MW Bip protein is very similar to that of a class of 70,000-MW heat-shock proteins that bind to the unfolded regions of thermally denatured proteins and prevent their precipitation in the cell; this resembles the function of Bip in binding nascent secretory proteins. Bip and the other 70,000-MW heat-shock proteins are examples of proteins termed *chaperones* that prevent misfolding and thus assist the folding of many proteins (pages 54 and 695).

Topogenic Sequences in Integral Membrane Proteins Allow Them to Achieve Their Proper Orientation in the ER Membrane

In Chapter 14, we were introduced to several of the vast array of integral proteins that occur in the plasma membrane and other cellular membranes. All of these glycoproteins are synthesized on the ER and move to the plasma membrane via the Golgi complex along the same pathway followed by continuously secreted proteins. This brings us

to an important question: how are these proteins inserted into the ER membrane with their proper orientations?

The problem is complex, since integral proteins interact with membranes in many different ways, as exemplified by plasma membrane glycoproteins that have been studied in some detail (Figure 16-17). Many such proteins contain a single membrane-spanning segment: a sequence of 20–25 hydrophobic amino acids that forms a transmembrane α helix, anchoring the protein in the phospholipid bilayer. Some of these proteins have their hydrophilic N-terminal segment on the exoplasmic face and the hydrophilic C-terminal segment on the cytosolic face; other plasma membrane glycoproteins have the reverse orientation. Many plasma membrane proteins, such as the glucose transport protein, have multiple membrane-spanning α-helical segments.

Specific sequences of amino acids in these proteins, called *topogenic sequences,* function in different ways to determine the orientation of the protein in the membrane. We can begin our discussion of this function with the large class of integral membrane proteins that have a single α helix that spans the membrane once, the C-terminus in the cytosol, and the N-terminus exoplasmic. Examples include the red blood cell protein glycophorin, the insulin receptor, and proteins that form the surface spikes of some viruses, including the vesicular stomatitis virus (the VSV G protein) and the influenza virus (the HA protein) (Figure 16-17). Immediately after synthesis, each of these proteins spans the ER membrane with the same orientation that it

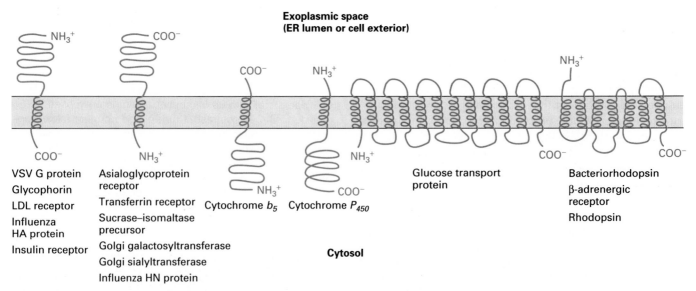

▲ FIGURE 16-17   Topologies of some integral membrane proteins synthesized on the rough ER. Segments of the protein chain in the ER membrane bilayer are shown as transmembrane α helixes; the portions outside the membrane are folded. Topogenic sequences of amino acids in the protein act during biosynthesis to ensure proper transmembrane orientation. [From W. Wickner and H. F. Lodish. 1985. *Science* **230**:400–407; E. Hartmann, T. A. Rapoport, and H. F. Lodish. 1989. *Proc. Natl. Acad. Sci.* **86**:5786–5790, and C. A. Brown and S. D. Black. 1989. *J. Biol. Chem.* **264**:4442.]

will have when it appears on the plasma membrane; as with all integral membrane proteins, the topology of these proteins is preserved as they are transported from the ER to the cell surface.

Two types of topogenic sequences function to orient this class of proteins in the ER membrane: an *N-terminal cleaved signal sequence* that is similar to the previously described signal sequence on secreted proteins, and a *stop-transfer membrane-anchor sequence* in the middle of the protein that becomes the membrane-spanning α helix. The insulin receptor, a typical member of this class of proteins, provides information about the latter.

Stop-Transfer Membrane-Anchor Sequences The signal sequence on a nascent insulin receptor, much like that of a secretory protein, is cleaved while the chain is still growing, and the new N-terminus of the growing polypeptide is extruded across the ER membrane into the lumen. However, unlike the case with secretory proteins, a sequence of about 22 hydrophobic amino acids in the middle of the insulin receptor *stops* the *transfer* of the protein through the membrane, and this same sequence remains *anchored* in the *membrane*—hence it is called a *stop-transfer membrane-anchor* sequence (Figure 16-18). The

C-terminus of the nascent chain remains in the cytosol and is not transferred across the ER membrane.

Support for such a model has come from studies using cloned cDNAs that encode the insulin receptor protein. Intact cloned cDNA, in an appropriate cellular expression vector, directs the synthesis of full-length insulin receptor protein, which is transported normally to the plasma membrane. Several types of mutant insulin receptor genes have been constructed: some are missing most or all of the DNA segment that encodes the membrane-spanning α-helical region of the receptor protein; others change two or more hydrophobic amino acids in this region to polar ones. In both these cases, the entire mutant protein is transported into the ER lumen and is eventually secreted from the cell. This establishes that the membrane-spanning α helix of the insulin receptor (and of other proteins in its class) functions as a stop-transfer sequence that prevents the C-terminus of the protein from crossing the ER membrane.

Uncleaved Internal Signal-Anchor Sequences As mentioned earlier, other proteins also span the ER membrane once, but with their N-terminus on the cytosolic face

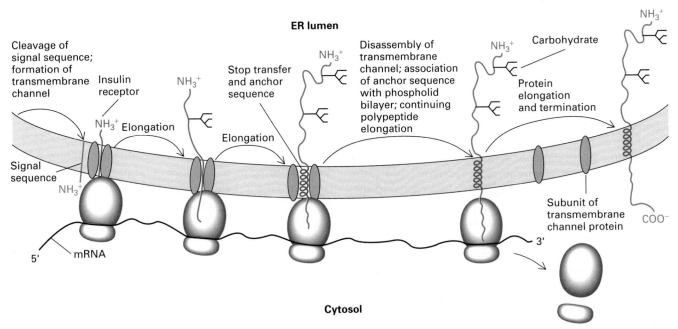

▲ **FIGURE 16-18** The synthesis and membrane insertion of the insulin receptor. Until the point at which the C-terminus of the protein is synthesized, this model is the same as the one for secretory proteins (see Figure 16-11). The hydrophobic stop-transfer membrane-anchor sequence of about 22 amino acids prevents the nascent chain from extruding further into the ER membrane, probably by causing disassembly of the transmembrane channel through which the nascent

chain travels; thus, this sequence anchors the insulin receptor to the phospholipid bilayer of the ER membrane. The ribosome is released from the membrane and completes the synthesis of the protein as a membrane-unbound ribosome in the cytosol. Other integral membrane proteins with similar structures (see Figure 16-17) are inserted in a similar manner. [After W. Wickner and H. F. Lodish. 1985. *Science* **230**:400–407.]

and their C-terminus on the exoplasmic (luminal) face (see Figure 16-17). Such proteins are also anchored to the membrane by a hydrophobic sequence of about 22 amino acids in the middle of the protein. The asialoglycoprotein receptor (see Figure 16-49) is perhaps the most intensively studied member of this group, but the biosynthesis of the others is similar. These proteins do not utilize a cleaved N-terminal signal sequence; rather, the hydrophobic stop-transfer membrane-anchor sequence also functions as an internal signal sequence (Figure 16-19a). As with cleaved signal sequences (Figure 16-11), the N-terminus of the internal uncleaved signal sequence faces the cytosol, and thus the N-terminus of the protein is cytosolic, while the C-terminus of the growing chain, like that of secretory proteins, is exoplasmic: it extrudes into the ER lumen.

Other proteins, such as those of the class known as cytochrome P450, utilize an internal uncleaved signal-anchor sequence, yet they assume an orientation in the ER membrane "backwards" from that of the asialoglycoprotein receptor class: the N-terminus of the internal signal-anchor sequence is in the ER lumen and the C-terminus (containing the catalytically active part of the protein) is in the cytosol (see Figure 16-17). Once the signal-anchor sequence inserts into the ER membrane, the rest of the protein is synthesized without the ribosome becoming attached to the ER membrane, and the C-terminus remains in the cytosol (Figure 16-19b).

Why does the orientation of the internal signal–membrane anchor sequence in this class of proteins (Figure 16-19b) differ from that in proteins like the asialoglycoprotein receptor (Figure 16-19a)? The answer resides not in the amino acid sequence of the hydrophobic signal-anchor itself, but in the nature of the hydrophilic amino acids flanking it. As a general rule, the flanking segment with the most numerous positively charged and the fewest negatively charged side chains remains on the cytosolic face of the membrane. A striking example is provided by the influenza neuraminidase protein, which is similar in orientation to the asialoglycoprotein receptor (see Figure 16-17), with three arginine residues just N-terminal to the signal-anchor sequence: mutation of these three to glutamate residues causes the neuraminidase protein to acquire the reverse orientation, with the N-terminus in the ER lumen and the C-terminus cytosolic. Other proteins, too, can be made to "flip" their orientation in the ER by mutating charged residues that flank the internal signal-anchor segment. Thus, the topogenic sequences for these proteins are the internal hydrophobic signal-anchor sequences together with the flanking hydrophilic residues.

Multiple Membrane-Spanning α-Helical Segments and Multiple Topogenic Sequences Many important proteins, such as ion pumps, ion channels, and trans-

▶ FIGURE 16-19   Internal signal-anchor sequences. (a) The synthesis and insertion of the asialoglycoprotein receptor and proteins of similar orientation. The internal signal sequence (a transmembrane α helix) functions exactly like the signal sequence at the N-terminus of a secretory protein: both direct the insertion of the growing chain into the ER membrane. The internal signal is uncleaved and is sufficiently long and hydrophobic to anchor the growing chain to the membrane. The C-terminus extrudes into the ER membrane, as in the case of a secretory protein. Thus the same topogenic sequence functions as an internal, uncleaved signal and as a membrane anchor; it causes the protein to assume an orientation with its N-terminus facing the cytosol and its C-terminus facing the ER lumen. (b) The synthesis and insertion of the cytochrome P450 polypeptide and others of similar membrane orientation. This protein also utilizes an internal signal sequence, but it inserts into the ER membrane with its N-terminus facing the ER lumen and its C-terminus remaining in the cytosol. No transmembrane channel forms around the growing polypeptide, and the C-terminus of the protein is synthesized without attachment of the ribosome to the ER membrane. Note that in both (a) and (b), charged amino acid residues flanking the internal signal sequence are thought to determine the orientation of the sequence in the membrane, with the more positively charged residues remaining on the cytosolic face. [Part (a) after M. Spiess and H. F. Lodish. 1986. *Cell* **44**:177–185; A. S. Shaw, P. Rottier, and J. K. Rose. 1988. *Proc. Natl. Acad. Sci. USA* **85**:7592–7596. Part (b) after E. Hartmann, T. A. Rapoport, and H. F. Lodish. 1989. *Proc. Natl. Acad. Sci.* **86**:5786–5790; J. Beltzer et al. 1991. *J. Biol. Chem.* **266**:973–978; M. Sakaguchi et al. 1992. *Proc. Natl. Acad. Sci.* **89**:16–19; and G. Parks and R. A. Lamb. 1993. *J. Biol. Chem.* **268**:19101–19109.]

porters span the membrane multiple times (Figure 16-17). Experiments showed that each membrane-spanning α helix in these proteins is a topogenic sequence (Figure 16-20).

The first α-helical segment is an internal uncleaved signal-anchor sequence similar to that of the asialoglycoprotein receptor. This segment initiates the insertion of the growing chain; as before (see Figure 16-19a), the SRP and the SRP receptor are involved in this step. In fact, if the first membrane-spanning α helix of the glucose transporter is placed (by recombinant DNA techniques) in the middle of an otherwise cytosolic protein, the chimera will insert into the ER membrane with an orientation similar to that of the asialoglycoprotein receptor: the N-terminus cytosolic and the C-terminus exoplasmic (Figure 16-19a). The second hydrophobic α helix in multispanning proteins is a stop-transfer membrane-anchor sequence. Thus the first two transmembrane α helixes insert into the membrane as a hairpin. The nascent protein chain continues to grow into the cytosol. According to this model, the third α helix

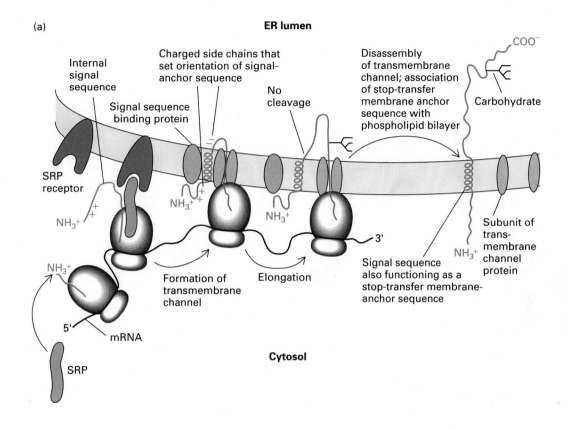

(a)

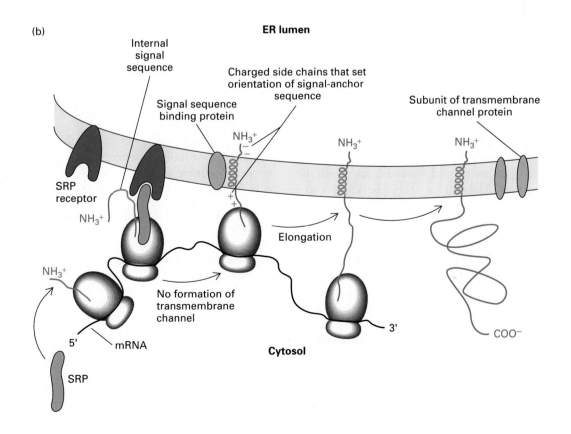

(b)

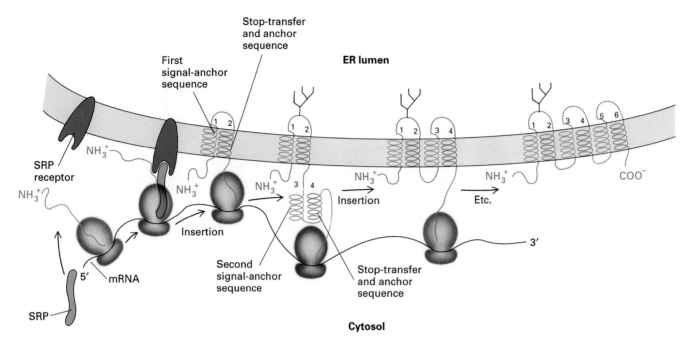

▲ **FIGURE 16-20**   The transmembrane insertion of a nascent glucose transporter. This protein, like many integral proteins, is believed to traverse the membrane as a series of α-helical segments. An internal, uncleaved signal-anchor sequence (red) is known to direct the binding of the nascent polypeptide chain to the rough ER membrane and to initiate cotranslational insertion; both SRP and the SRP receptor are involved in this step. In this model, the internal, uncleaved signal-anchor sequence and the sequence of nascent chain following it, which contains a stop-transfer membrane-anchor sequence, insert into the ER membrane as an α-helical hairpin. The nascent chain continues to grow in the cytosol. Subsequent α-helical hairpins could insert similarly, although SRP is required only for the insertion of the first signal-anchor sequence. Each helix is numbered in the figure. Although only six transmembrane α-helixes are depicted here, this transporter and proteins of similar structure have twelve or more. [See H. P. Wessels and M. Spiess. 1988. *Cell* **55**:61–70.]

would be another internal uncleaved signal-anchor sequence, and the fourth another stop-transfer membrane-anchor sequence. Thus helixes 3 and 4 would insert into the membrane as a hairpin, just as helixes 1 and 2 do. However, neither SRP nor the SRP receptor is required at this step. All transmembrane α helixes could insert into the ER bilayer in this way. The number of topogenic sequences would be equal to the number of transmembrane α helixes.

Attachment of a Glycosylphosphatidylinositol Membrane Anchor   Many cell-surface proteins are anchored to the phospholipid bilayer not by a sequence of hydrophobic amino acids, but by a covalently attached lipid (see Figure 14-20). The biosynthesis of an important class of these, those proteins anchored by a glycosylphosphatidylinositol (GPI) membrane anchor, is shown in Figure 16-21, and illustrates the only other known type of topogenic sequence in a protein. These proteins are synthesized and anchored to the ER membrane exactly like the insulin receptor is (Figure 16-18), except that a short sequence of amino acids in the exoplasmic domain adjacent to the membrane-spanning domain is recognized by an endoprotease that simultaneously cleaves off the original membrane-anchor sequence and adds the preformed lipid (GPI) anchor instead.

Why change one type of membrane anchor for another? For one thing the cytosol-facing hydrophilic domain is cut off during attachment of the GPI anchor. Proteins with GPI anchors can diffuse in the plane of the phospho-

▶ **FIGURE 16-21**   Addition of a glycosylphosphatidylinositol (GPI) membrane anchor to a membrane protein in the ER. A preformed GPI anchor is added to certain proteins originally anchored in the ER membrane by a segment of hydrophobic amino acids. A sequence (red) in the protein in the exoplasmic-facing domain, near the membrane-spanning region, is the signal for a proteolytic cleavage; simultaneously the ethanolamine in the anchor is added to the carboxyl group of the new C-terminus. Other sugar residues found in the anchors of different proteins (Figure 14-20) are added to this basic framework, probably in the Golgi complex. [After C. Abeijon and C. B. Hirschberg. 1992. *Trends Biochem. Sci.* **17**:32–36. See also I. W. Caras. 1991. *J. Cell Biol.* **113**:77–85 and K. Kodukula et. al. 1992. *Proc. Natl. Acad. Sci. USA* **89**:4982–4985.]

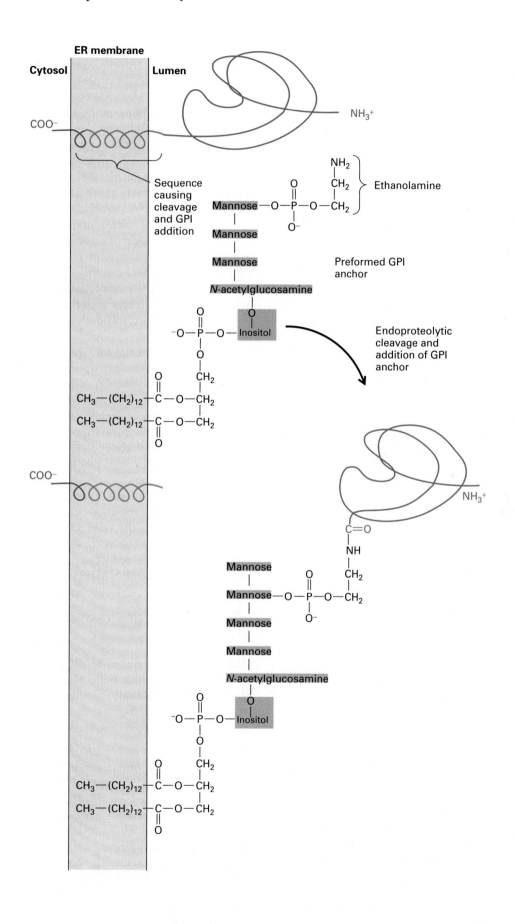

lipid bilayer membrane, whereas proteins anchored by membrane-spanning α helixes are frequently attached to the cytoskeleton by their cytosol-facing segments. Also, in polarized epithelial cells (see Figure 14-42), the GPI anchor targets the attached protein to the apical domain of the plasma membrane, as we discuss later in this chapter.

➤ *Post-Translational Modifications of Secretory and Membrane Proteins in the Rough ER*

Once their synthesis is complete, newly introduced polypeptides in the membrane and lumen of the ER must be folded, sorted, and transported. Secretory and membrane proteins undergo five principal modifications during their transit to the cell surface: (1) formation of disulfide bonds; (2) proper folding of the polypeptide, (3) addition and modification of carbohydrates; (4) specific proteolytic cleavages, and (5) formation of multichain proteins. Each modification takes place in a specific organelle through which these proteins pass. The first two and the fifth of these reactions take place exclusively in the ER, and addition of some carbohydrates and some proteolytic cleavages also occur in this organelle.

The modifications that occur in the ER are essential for the protein to reach its proper cellular location. Only properly folded and assembled proteins are transported from the ER to the Golgi complex and to the cell surface. As will be discussed in a later section, unfolded, misfolded, or partly folded and assembled proteins are selectively retained in the ER, or are retrieved to the ER from the *cis*-Golgi reticulum (see Figure 16-8). Frequently, misfolded proteins and unassembled subunits of multiple-polypeptide complexes are degraded within the ER. Such "quality control" in the transport of proteins to the Golgi vesicles ensures that misfolded proteins do not reach the cell surface, where they might be recognized by the immune system as a "foreign" protein, inducing an immunological response that can lead to the destruction of the cell.

We begin by discussing two modifications generally localized to the rough ER—the formation of disulfide bonds and the formation of multichain proteins. The addition and modification of carbohydrate chains requires a section of its own, since some of the reactions occur in the ER and others in the three types of Golgi vesicles.

Disulfide Bonds Are Formed in the ER Lumen soon after Synthesis

In eukaryotic cells, the formation of disulfide bonds (Cys—S—S—Cys) occurs in the lumen of the rough ER, never in the cytosol. Disulfide bonds are confined to secretory proteins and to the exoplasmic domains of membrane proteins; such bonds are important stabilizing forces in the tertiary structure of these proteins.

The tripeptide *glutathione* (Figure 16-22a) is the major thiol- (SH-) containing molecule in eukaryotic cells, and serves two important functions: to prevent the formation of disulfide bonds in the cytosol and to catalyze their formation in the ER. Glutathione shuttles between the reduced form, abbreviated GSH, and the oxidized form, a disulfide-linked dimer, abbreviated GSSG. The formation of a disulfide bond in a protein involves two thiol-disulfide exchange reactions with glutathione (Figure 16-22b). In the first, an —SH group of a cysteine residue reacts with one of the two sulfur atoms in a G—S—S—G disulfide bond, forming a Cys—S—S—G disulfide bond and one molecule of reduced GSH. In the second reaction, another cysteine residue in the protein reacts with the Cys—S—S—G bond, forming a Cys—S—S—Cys disulfide bond and a second molecule of GSH.

In the lumen of the ER, the ratio of GSH to GSSG is maintained at about 5:1, optimum for the formation of protein disulfide bonds. Consistent with this, experiments in which reduced and denatured secretory proteins are refolded in cell-free systems (see Figure 3-19) showed that disulfide bond formation is indeed maximal at a GSH:GSSG ratio of about 5:1. This observation suggests that the redox environment of the ER is optimal for disulfide bond formation.

In the cytosol of eukaryotic cells, glutathione is also present at high concentration, about 0.01 M. In contrast to the ratio in the ER, the GSH:GSSG ratio in the cytosol is over 50:1. Oxidized GSSG in the cytosol is reduced by the enzyme glutathione reductase, using electrons from the potent reducing agent NADPH (see Figure 17-3):

$$2NADPH + 2H^+ + GSSG \longrightarrow 2NAD^+ + 2GSH$$

Thus cytosolic proteins in bacterial and eukaryotic cells do not utilize the disulfide bond as a stabilizing force because the high GSH/GSSG ratio would drive the reactions in Figure 16-22b to the left—that is, in the direction of Cys—SH and away from Cys—S—S—Cys.

Rearrangements of Disulfide Bonds In proteins that contain more than one disulfide bond, the proper pairing of cysteine residues is essential for normal structure and activity. In some proteins, disulfide bonds are formed while the polypeptide is still growing on the ribosome, as in the case of the immunoglobulin (Ig) light chain, whose two disulfide bonds stabilize the two separate domains in this protein (see Figure 3-32). The disulfide bonds form sequentially: the first and second cysteines form a disulfide bond before the third cysteine has even been translated, automatically ensuring the correct pairing of cysteines. Similarly, the third cysteine pairs with the fourth to create the second disulfide bond.

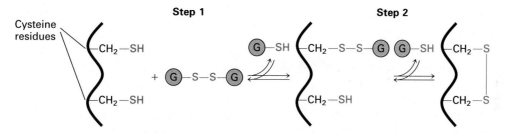

▲ FIGURE 16-22 Structure and function of glutathione. (a) The tripeptide glutathione is the major determinant of the oxidation-reduction state of the cytosol and the ER. In the ER the ratio of reduced glutathione (GSH) to the oxidized form (GSSG) is about 5:1, whereas in the cytosol it is ≈50:1. (b) Disulfide bonds in proteins form within the ER lumen by a thiol-disulfide exchange reaction with oxidized glutathione. [See C. Hwang, A. J. Sinskey, and H. F. Lodish. 1992. *Science* **257**:1496–1502.]

Disulfide-bond formation does not always occur sequentially. For example, proinsulin (see Figure 3-18) has three disulfide bonds that are not sequential—pairing is between cysteines 1 and 4, 2 and 6, and 3 and 5. In this case, the first disulfide bonds that form spontaneously by oxidation of the Cys–SH groups may not be characteristic of the proper folded conformation of the protein. Disulfide bonds often break and reform, eventually creating the proper S–S bonds and stabilizing the proper protein conformation, but studies of the spontaneous refolding of purified proteins with many S–S bonds showed that in cell-free systems this process is very slow, requiring minutes to hours. In the ER, the rearrangement process is accelerated by the enzyme *protein disulfide isomerase* (PDI), which is found in abundance in the ER of secretory tissues in such organs as the liver and pancreas (Figure 16-23). PDI acts by catalyzing the rearrangement of disulfide bonds in a broad range of protein substrates, allowing them to generate their thermodynamically most stable configurations.

Biotechnology and the Importance of Disulfide Bonds Failure to understand the process of formation of disulfide bonds has cost several biotechnology companies millions of dollars when they tried to synthesize functional therapeutic proteins in bacterial cells by recombinant DNA techniques. Most proteins developed for therapeutic use in humans or animals, such as blood proteins and hormones, are stabilized by disulfide bonds. When recombinant DNA techniques are used to synthesize mammalian secretory proteins in bacterial cells, the proteins generally are not secreted (even when a bacterial signal sequence replaces the normal one). Rather, they accumulate in the cytosol where they often denature and precipitate, in part because S–S bonds do not form. Indeed, biotechnologists must use sophisticated chemical methods to refold the denatured bacterially produced proteins. Thus bacterial cells are not suitable hosts for the synthesis of mammalian proteins that are normally stabilized by disulfide bonds. Biotechnologists must use cultured animal cells to produce such proteins (e.g., monoclonal antibodies and tissue plasminogen activator, an anticlotting agent) in quantity.

Chaperone Proteins Facilitate the Folding of Newly Made Proteins

As discussed in Chapter 3, many reduced, denatured proteins can spontaneously refold into their native state. In

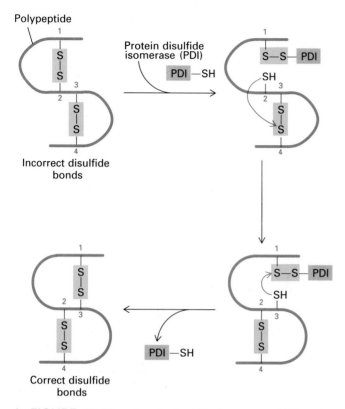

▲ FIGURE 16-23   Protein disulfide isomerase (PDI) and the rearrangement of disulfide bonds. PDI catalyzes the breakage and reformation of disulfide bonds and, in so doing, accelerates the folding of proteins containing multiple disulfide bonds. In the oxidizing environment of the ER lumen, disulfide bonds form in newly made secretory proteins, but the bonds are often incorrect. PDI contains an active-site cysteine residue with a free reduced sulfhydryl (SH) group, which reacts with disulfide (S–S) bonds on nascent or newly completed proteins to form an S–S bond between PDI and the protein. This bond, in turn, can react with a free SH on the protein to form a new S–S bond. In this way, the disulfide bonds on a protein can rearrange themselves until the most stable configuration for the protein is achieved. [After R. B. Freedman. 1984. *Trends Biochem. Sci.* **9**:438–441; R. Noiva and W. Lennarz. 1992. *J. Biol. Chem.* **267**:3553–3556.]

most cases such refolding requires hours to reach completion, yet new secretory proteins generally fold in the ER lumen within minutes after their synthesis. The ER contains several *chaperone proteins* that facilitate the folding of newly synthesized proteins within the ER lumen. PDI is one such chaperone, or folding catalyst. Bip, as already noted (Figure 16-16), is another: it transiently binds to exposed hydrophobic segments of proteins and prevents them from misfolding or forming aggregates, and thus enhances their ability to fold into the proper conformation.

Other examples of chaperones are peptidyl-prolyl isomerases, a family of enzymes that accelerate the rotation about peptidyl-prolyl bonds in unfolded segments of a polypeptide; such isomerizations are frequently rate-limiting for the folding of protein domains:

Most peptidyl-prolyl isomerases will catalyze the rotation of any exposed peptidyl-prolyl bond. Others have very specific substrates. NinaA, for example, an ER peptidyl-prolyl isomerase expressed in certain eye cells of *Drosophila*, is required exclusively for the folding of opsin (see Figure 21-52), the membrane protein that absorbs light and triggers the visual response and that is a major protein synthesized in these eye cells. Organisms that have mutations in the *ninaA* gene lack the proper visual responses yet are otherwise quite normal, which substantiates the premise that the principal, if not the only, function of the NinaA peptidyl-prolyl isomerase is to fold the visual protein opsin. This example illustrates an important point about chaperones: they can be remarkably specific in the proteins they help to fold; the folding of certain polypeptides in the ER can be absolutely dependent on a particular chaperone, while folding of other proteins may not require that catalyst.

The Formation of Oligomeric Proteins Occurs in the ER

Many important secretory and membrane proteins are *oligomers*: they are built of two or more polypeptides. In all cases, these oligomeric proteins are assembled in the ER. One important example is provided by the immunoglobulins, which contain two heavy (H) and two light (L) chains, all linked by S–S bonds (see Figure 3-32). Figure 16-24 shows the assembly of a second well-studied example, the influenza HA (hemagglutinin) protein, which forms the spikes protruding from the surface of the influenza virus particle (the spike structure is illustrated in Figures 3-5 and 16-54, and the virus structure in Figure 6-22). Each spike is formed within the ER from three copies of a precursor protein, termed HA_0, which has a single membrane-spanning α helix. In the Golgi complex, each of the three HA_0 proteins is cleaved to form two polypeptides, HA_1 and HA_2; thus, each spike in the virus particle contains three copies of HA_1 and three of HA_2 noncovalently bonded together. The trimer is stabilized by interactions between the exoplasmic domains of the constituent polypeptides as well as by interactions between the three cytosolic and membrane-spanning domains.

Each newly made HA_0 polypeptide requires approximately 7 min to fold and be incorporated into a trimer.

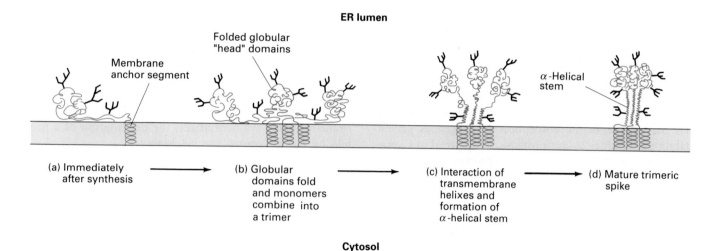

ER lumen

(a) Immediately after synthesis → (b) Globular domains fold and monomers combine into a trimer → (c) Interaction of transmembrane helixes and formation of α-helical stem → (d) Mature trimeric spike

Cytosol

▲ FIGURE 16-24   The folding of the hemagglutinin (HA) precursor polypeptide HA$_0$ and the formation of a trimer within the ER. (a) Immediately after its synthesis, the bulk of the HA$_0$ polypeptide is on the luminal side of the ER membrane, anchored by a hydrophobic α helix near its C-terminus. (b) First, the globular domains that will form the head of the spike (see Figure 16-54a) fold, stabilized by disulfide bonds. (c) Then three chains interact with each other, initially via their transmembrane α helices; this activity apparently triggers the formation of a long stem containing one α-helix from the luminal part of each HA$_0$ polypeptide. (d) Finally, interactions between the three globular heads occur, generating the mature trimeric spike. Not shown here is later proteolytic cleavage of HA$_0$ to HA$_1$ and HA$_2$ in the Golgi complex, well after the trimer has been formed; the molecular structure of the trimer after cleavage to HA$_1$ and HA$_2$ is depicted in Figures 16-54 and 3-5. [After M-J. Gething, K. McCammon, and J. Sambrook. 1986. *Cell* **46**:939–950; I. Braakman et al. 1991. *J. Cell Biol.* **114**:401–411.]

Several approaches have been used in influenza-infected cells to detect intermediates during folding of the monomers and their subsequent assembly into trimers. In a typical experiment, two monoclonal antibodies are used. One binds only to a region of the HA$_0$ polypeptide that is exposed on the monomer surface but is folded inside the rigid trimer. The other binds only to a region on the surface of the correctly folded trimer and thus it immunoprecipitates trimers, not monomers. In this experiment, virus-infected cells are pulse-labeled with a radioactive amino acid, and at various times during the subsequent chase period membranes are solubilized by detergent. Immediately after the pulse, all of the radioactive HA$_0$ protein is immunoprecipitated by the monomer-specific monoclonal antibody. During the chase period, increasing amounts of radioactive HA$_0$ protein react only with the trimer-specific monoclonal antibody, and decreasing amounts react with the other antibody. With this pulse-chase method, one can follow the monomer-to-trimer conversion of newly made HA$_0$ protein within a living cell.

➤ Quality Control in the ER

After their synthesis, both secretory and membrane proteins are packaged into small transport vesicles that carry them to the Golgi vesicles. Only properly folded proteins are generally allowed to enter these transport vesicles, so that incompletely folded proteins do not reach the plasma membrane or are not secreted. Some of these imperfect polypeptides accumulate in the ER; others are degraded there. Here we first discuss some examples of this *quality control* process for ensuring that only properly folded proteins reach the plasma membrane or are secreted; then we consider the molecular mechanisms involved.

Only Properly Folded Proteins Are Transported from the Rough ER to the Golgi Complex

Almost any mutation in a secretory or membrane protein that prevents it from folding properly will also prevent the polypeptide from moving out of the ER. Many mutations created in vitro in the cloned HA gene prevent the protein from folding properly and generating a trimer. Some involve small deletions or alterations of the membrane-spanning segment, or changes in the large, exoplasmic-facing N-terminus. Others are temperature-sensitive mutations in which the HA$_0$ protein folds properly into a trimer at a lower temperature but not at higher, nonpermissive temperatures. In all cases, the improperly folded HA$_0$ polypeptides remain in the membrane of the rough ER. In most cases, the exoplasmic domains of the mutant proteins are permanently bound to Bip (Figure 16-16). Thus Bip performs two related functions in the ER—assisting the folding of normal proteins by preventing their aggregation, and binding to irreversibly misfolded proteins.

Another example is a mutation in the secretory protein α_1-antiprotease, which is secreted by hepatocytes and macrophages; this protein binds to and inhibits trypsin and also the blood protease elastase. In the absence of α_1-antiprotease, elastase degrades the fine tissue in the lung that participates in the absorption of oxygen. A genetic inability to produce α_1-antiprotease, widespread in Caucasians, is the major genetic cause of emphysema (destruction of lung tissue by unchecked elastase). The defect is due to a single mutation in α_1-antiprotease in which glutamate 342 is replaced by lysine. The mutant α_1-antiprotease is produced normally in the rough ER but forms an almost crystalline aggregate that is not exported beyond the ER. The secretion of liver proteins also becomes impaired because the rough ER in the hepatocytes becomes filled with aggregated α_1-antiprotease.

Our final example concerns the most common mutation that causes cystic fibrosis: a deletion of amino acid 508 (ΔF508) in the CFTR protein (see page 651). The mutant protein does not move from the ER to the plasma membrane at normal body temperature (37°C), although it will at lower temperatures (25°C). Thus the ΔF508 mutation induces a temperature-sensitive defect in protein maturation: apparently, it causes the mutant protein to assume a slightly incorrect structure at 37°C but a normal structure at 25°C.

Unassembled or Misfolded Proteins Are Often Degraded within the ER

In addition to the retention of misfolded proteins, another form of quality control within the ER is the degradation of misfolded polypeptides, or unassembled subunits of oligomeric proteins. Two well-worked-out examples involve the α subunit of the acetylcholine receptor (see Figure 21-39), and the α subunit of the T cell receptor (see Figure 27-30). Both of these receptors are normally assembled into multisubunit oligomeric complexes while they are in the ER, and only a completely assembled oligomer is allowed to exit the ER en route to the Golgi complex.

If only one subunit of the oligomer is expressed (by recombinant DNA techniques) in a cell, but not the other subunits, the expressed polypeptide is retained in the ER and is selectively degraded there within one to two hours of its synthesis. Thus the ER contains a set of proteolytic enzymes that can somehow recognize unassembled polypeptides and degrade them.

At least some of this recognition is directed toward the membrane-spanning region of these polypeptides. For example, the α subunit of the T cell receptor has a single membrane-spanning segment with two charged amino acids that are normally bound by ionic linkage (within the phospholipid bilayer) to residues of opposite charge in two

other subunits of the T cell receptor. If these other subunits are absent, the charged residues in the α subunit cause the entire polypeptide to assume an abnormal conformation and trigger its degradation within the ER. Mutation of the charged residues to neutral ones prevents degradation of the α subunit in the ER when the other T cell receptor subunits are absent—evidence indicating that the membrane-spanning segment affects the degradation of a protein by the ER proteolytic machinery.

ER-Specific Proteins Are Retained in the Rough ER or Are Returned There from the *Cis*-Golgi

A related aspect of ER quality control is the retention in the ER lumen of resident proteins such as Bip (Figure 16-16) and PDI (Figure 16-23), which catalyze the folding of newly made proteins. Bip and PDI are soluble in the ER lumen. How are they retained in the ER to carry out their work, and prevented from being secreted?

The answers lie in a specific sequence in these resident ER proteins and in a receptor that recognizes that sequence. As in several other ER luminal proteins, the C-terminal sequence in both the PDI and Bip proteins is Lys-Asp-Glu-Leu (KDEL in the one-letter code shown in Figure 3-3). Several experiments demonstrated that this *KDEL sequence* is both necessary and sufficient for retention in the ER. For instance, when a mutant PDI protein lacking these four residues is synthesized in a fibroblast, the protein is entirely secreted. If a normally secreted protein is altered so that it contains these four amino acids at its C-terminus, the protein is retained in the ER.

The KDEL sequence binds to a receptor protein that causes the sequence to be retained in the ER. The *KDEL receptor* has been purified and cloned, and two surprising results emerged from these studies.

First, the number of KDEL receptors is only about one-tenth or less that of ER proteins with KDEL sequences, insufficient to bind all of these proteins to the ER membrane and thus retain them there. Apparently the KDEL receptor acts mainly to retrieve any proteins with the KDEL recognition sequence that have escaped to the *cis*-Golgi reticulum and return them to the ER. One piece of supporting evidence for this concept is that while some of the KDEL receptor protein is localized to ER membranes, most is found in the small transport vesicles that are in between the ER and the *cis* Golgi, and in the vesicles termed the *cis*-Golgi reticulum (Figure 16-25).

Second, the ER-localized proteins that carry the KDEL recognition sequence have oligosaccharide chains with modifications that are made only by enzymes that are localized to the *cis*-Golgi or *cis*-Golgi reticulum; thus, at some time the protein, with its KDEL recognition se-

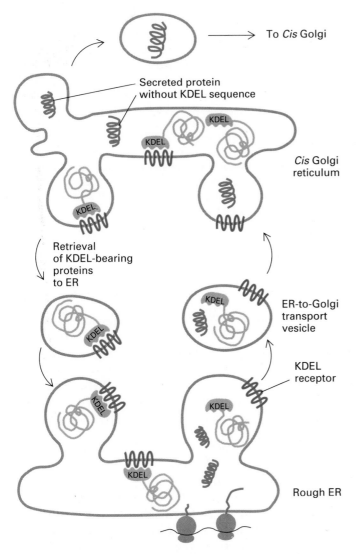

To *Cis* Golgi

Secreted protein
without KDEL sequence

KDEL

KDEL

KDEL

Cis Golgi
reticulum

KDEL

Retrieval
of KDEL-bearing
proteins
to ER

KDEL

KDEL

ER-to-Golgi
transport
vesicle

KDEL receptor

KDEL

KDEL

KDEL

KDEL

Rough ER

▲ FIGURE 16-25   Role of the KDEL receptor in the retrieval of ER-resident proteins. Many ER luminal proteins bear a C-terminal KDEL (Lys-Asp-Glu-Leu) sequence that localizes them to the ER. The KDEL receptor is located mainly in the *cis*-Golgi reticulum and in ER-to-Golgi transport vesicles; its chief function is to bind proteins with the KDEL recognition sequence and return them to the ER. [After J. Semenza et al. 1990. *Cell* **61**:1349–1357.]

quence, must have been transported at least to the *cis*-Golgi reticulum.

Thus, ER-specific proteins that carry the KDEL sequence, such as Bip and PDI, are able to leave the ER only to be retrieved by the KDEL receptor, but we do not yet know how the movements of the KDEL receptor are controlled.

Clearly, the transport of newly made proteins from the rough ER to the Golgi vesicles is a highly selective and regulated process; let us briefly review its stages: Certain

resident ER proteins are retained in the ER or are returned there from the *cis*-Golgi reticulum. Abnormally folded proteins are selectively retained in the ER, either because they form aggregates or because they are permanently bound to Bip or other ER chaperones. Some improperly folded proteins are selectively degraded in the ER. Only properly folded proteins are transported from the ER to the Golgi complex of vesicles.

The selective entry of proteins into membrane-bounded-transport vesicles is an important feature of protein targeting—one we will encounter several times in our study of the subsequent stages in the maturation of secretory and membrane proteins.

▶ Protein Glycosylation: Discrete Steps in the ER and Golgi Complex

Recall that following their synthesis in the rough ER, most proteins move in small transport vesicles to the *Golgi complex*—the series of organelles composed of both flattened and spherical vesicles (Figure 16-26). Although biologists used to refer to the Golgi complex as one organelle, the *cis, medial,* and *trans* vesicles of the Golgi contain different sets of enzymes that introduce different modifications to secretory and membrane proteins, and thus each region should be considered as a separate organelle.

Most plasma membrane and secretory proteins contain one or more carbohydrate chains, and glycosylation is the principal chemical modification to most such proteins. Some glycosylation reactions occur in the lumen of the ER; others, in the lumina of the *cis, medial,* and *trans*-Golgi vesicles. Certain proteins, in order to fold properly, must have carbohydrates added to them in the ER. Also, lysosomal enzymes, in order to be delivered to lysosomes, must have their oligosaccharide chains modified in the *cis* Golgi. As we shall see, glycosylation reactions can provide us with useful markers for following the movement of proteins as they travel from the ER and through the Golgi vesicles. Thus an understanding of the synthesis and function of the oligosaccharide chains attached to proteins is important, and requires some knowledge of their structures.

Different Structures Characterize *N*- and *O*-Linked Oligosaccharides

Recall that the structures of *N*- and *O*-linked oligosaccharides are very different, and that different sugar residues are usually found in each type (Figure 16-27; see also Figures 14-25 and 14-26). For instance, in *O*-linked sugars, *N*-acetylgalactosamine is invariably linked to serine or

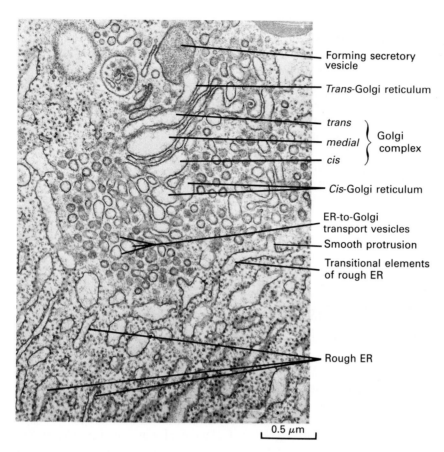

Forming secretory vesicle

Trans-Golgi reticulum

trans ⎫
medial ⎬ Golgi complex
cis ⎭

Cis-Golgi reticulum

ER-to-Golgi transport vesicles

Smooth protrusion

Transitional elements of rough ER

Rough ER

0.5 μm

◄ **FIGURE 16-26** Electron micrograph of the Golgi complex and ER in an exocrine pancreatic cell. The stacked vesicles of the Golgi complex and a forming secretory vesicle are evident. The rough ER contains transitional elements from which smooth protrusions appear to be budding. These buds form the vesicles that transport secretory and membrane proteins from the rough ER to the Golgi complex. Other vesicles seen at the periphery of the *cis* side of the Golgi complex form the *cis*-Golgi reticulum, which is an intermediate between the ER and the *cis* Golgi. [Courtesy G. Palade.]

threonine and, in collagens, galactose is linked to hydroxylysine. In *all* N-linked oligosaccharides, N-acetylglucosamine is linked to asparagine. O-linked oligosaccharides are generally short, often containing only one to four sugar residues (Figure 16-27). However, some O-linked oligosaccharides, such as those bearing the ABO blood group antigens, can be very long (see Figure 14-28). The N-linked oligosaccharides, in contrast, have a minimum of five sugars and always contain mannose as well as N-acetylglucosamine (see Figure 16-27). Although most cytosolic and nuclear proteins are not glycosylated, when they are, a single N-acetylglucosamine residue is linked to the serine or threonine hydroxyl group. These exceptions include a protein localized to the nuclear-pore complex, which we will encounter later, and several transcription factors.

Here we restrict our discussion to the biosynthesis of oligosaccharides in the lumina of the ER and the Golgi complex. The different structures of N- and O-linked oligosaccharides reflect differences in their biosynthesis. O-linked sugars are added one at a time, and each sugar transfer is catalyzed by a different enzyme. In contrast, for N-linked oligosaccharides a large preformed oligosaccharide containing 14 sugar residues is added, in the ER, to the asparagine residue; then, in the ER and in the Golgi, cer-

tain sugar residues are removed and others are added, one at a time, in a defined order with each reaction catalyzed by a different enzyme. We begin with the addition of single sugar residues, a process common to the biosynthesis of both N- and O-linked oligosaccharides.

Nucleotide Sugars Are the Precursors of Oligosaccharides

The incorporation of sugars into such polymers as glycoproteins requires an input of energy. The intermediates used in the biosynthesis of oligosaccharides are nucleoside diphosphate or monophosphate sugars (Figure 16-28). The ester bond between the phosphate residue and the carbon atom in the sugar is a high-energy bond, and the transfer of the sugar residue to a hydroxyl group on another sugar residue, as depicted in Figure 16-29, or to a serine or threonine residue, is energetically favored. All known glycosyltransferase enzymes that act on secretory proteins are integral membrane proteins with active sites facing the lumen of the organelle.

During the formation of O-linked glycoproteins and the modification of N-linked ones, each sugar transfer is catalyzed by a different type of glycosyltransferase. Each

(a)

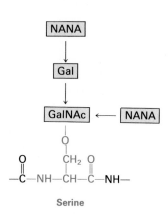

O-Linked oligosaccharide

◄ FIGURE 16-27   Structure of typical *N*-linked and *O*-linked oligosaccharides. (a) Structure of the *O*-linked oligosaccharide, linked to serine and threonine hydroxyl groups, in proteins such as glycophorin and the LDL receptor. As shown here, one negatively charged *N*-acetylneuraminic acid (sialic acid) is attached to the galactose and one is attached to the *N*-acetylgalactosamine, although only one of these is present in some cases. (b) Structures of typical asparagine-linked oligosaccharides, found in a variety of mammalian serum glycoproteins such as immunoglobulins. The five residues always found in *N*-linked oligosaccharides are in green. High-mannose oligosaccharides can have as few as three mannose residues and, in protozoans and yeast, as many as 60. Complex oligosaccharides have two or more branches, each containing at least one *N*-acetylglucosamine, galactose, and sialic acid residue. Hybrid oligosaccharides contain one branch that has the complex structure, and one or more "high-mannose" branches. [After R. Kornfeld and S. Kornfeld. 1985. *Ann. Rev. Biochem.* **45**:631–664.]

(b)

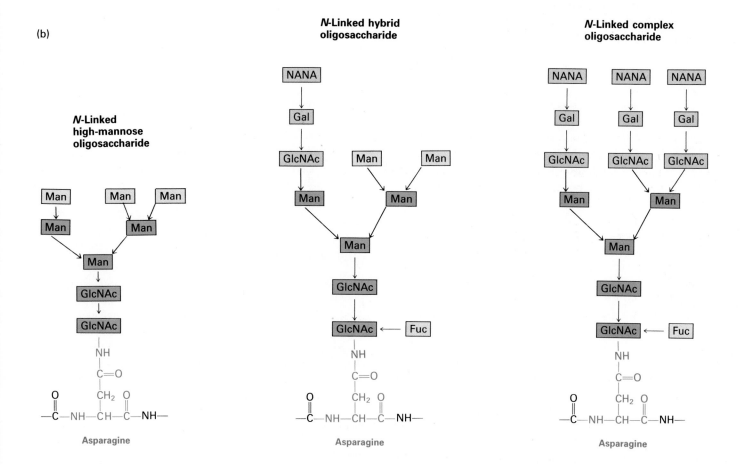

NANA = *N*-Acetylneuraminic acid (sialic acid)

Gal = Galactose

GlcNAc = *N*-Acetylglucosamine

GalNAc = *N*-Acetylgalactosamine

Man = Mannose

Fuc = Fucose

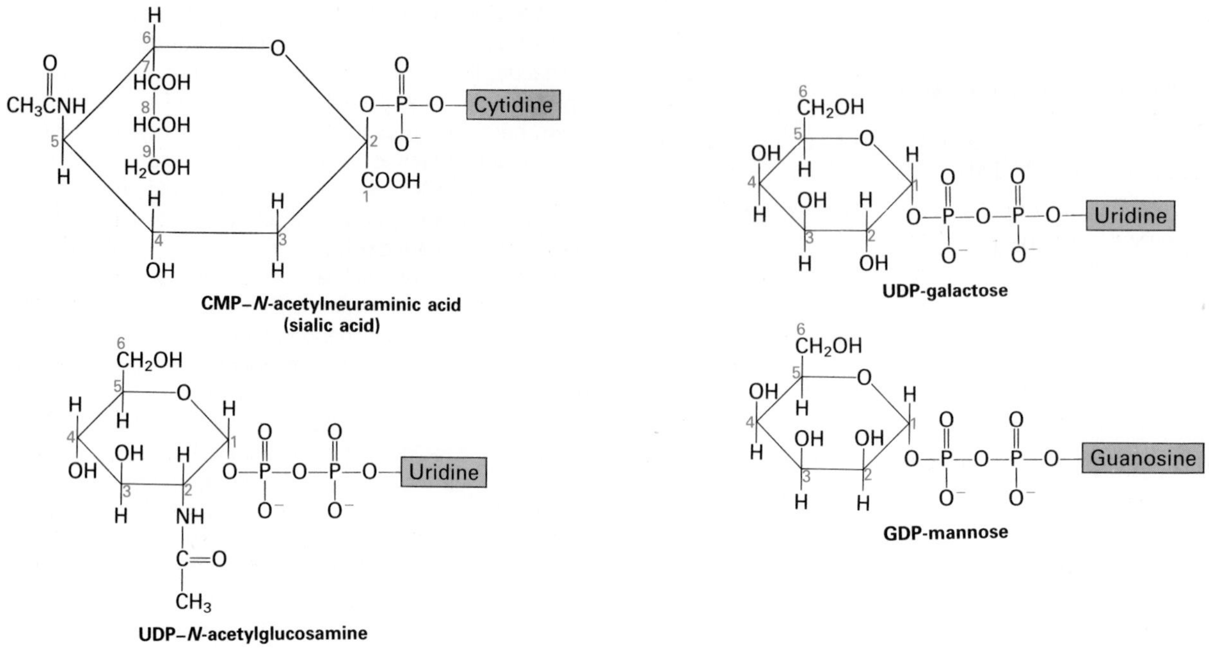

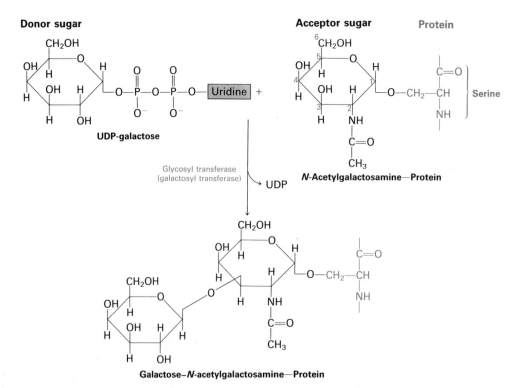

▲ FIGURE 16-28   Structures of some sugar nucleotides, precursors of the saccharide residues in glycoproteins.

▲ FIGURE 16-29   The addition of a galactose residue from UDP-galactose to carbon atom 3 of N-acetylgalactosamine attached to a protein, a step in the elongation of typical O-linked oligosaccharides. Glycosyltransferases are specific both for the nucleoside sugar donor and for the carbon atom of the acceptor sugar to which it is transferred.

glycosyltransferase enzyme is specific for both the donor sugar nucleotide and the acceptor molecule. The galactosyltransferase depicted in Figure 16-29, for instance, only transfers a galactose residue (from UDP-galactose), and only to the 3 carbon atom of an acceptor *N*-acetylgalactosamine residue. A different enzyme will transfer the galactose to the 4 carbon of *N*-acetylgalactosamine, and yet another to the 4 carbon of galactose.

O-Linked Oligosaccharides Are Formed by the Sequential Addition of Sugars

In the biosynthesis of the typical *O*-linked oligosaccharide depicted in Figure 16-27a, the *N*-acetylgalactosamine residue is transferred from UDP–*N*-acetylgalactosamine to the hydroxyl group of the serine or threonine residue by an enzyme (an *N*-acetylgalactosaminyltransferase) localized to the ER or the *cis*-Golgi reticulum. After the protein has moved to the *trans*-Golgi vesicles, the galactose residue is added to the *N*-acetylgalactosamine by a specific *trans*-Golgi galactosyltransferase (see Figure 16-29). The last steps in the biosynthesis of *O*-linked (and *N*-linked) sugars are the addition of two *N*-acetylneuraminic acid (also called sialic acid) residues; these steps also occur in the *trans* Golgi or the *trans*-Golgi reticulum.

Similarly, saccharide residues in glycolipids are added in the Golgi vesicles. These reactions also utilize glycosyltransferases with active sites that face the lumen. Since several oligosaccharides, such as the ABO antigens, are found attached to both glycoproteins and glycolipids, the same glycosyltransferase may be utilized to synthesize both.

The ER and Golgi Membranes Contain Transporters for Nucleotide Sugars

All nucleotide sugar substrates for oligosaccharide synthesis are fabricated in the cytosol, from nucleoside triphosphates and sugar phosphates (Figure 16-30), but the transfer of sugar residues occurs in the lumina of the rough ER and the Golgi. How do sugar nucleotides enter these organelles?

The ER membrane contains a specific transporter for UDP–*N*-acetylgalactosamine, and Golgi membranes contain specific transporters for several sugar nucleotides, including CMP–*N*-acetylneuraminic acid and UDP-galactose (Figure 16-31). These transporters are antiporters: when sugar nucleotides are imported, nucleotides such as UMP and CMP must be exported from the Golgi vesicles. First, UDP formed in a glycosyltransferase reaction, is hydrolyzed to UMP and P_i, and then UMP is transported out of the Golgi vesicles in a one-for-one exchange for UDP-galactose, UDP–*N*-acetylglucosamine, or UDP–*N*-acetylgalactosamine. Other antiporters catalyze a one-

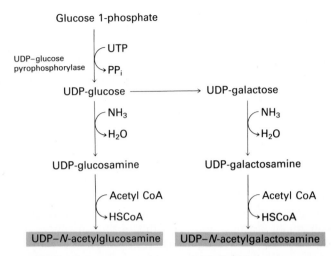

▲ **FIGURE 16-30**   Reactions in the synthesis of some common sugar nucleotides. These reactions occur in the cytosol, and the sugar nucleotides are then transported into the ER or Golgi.

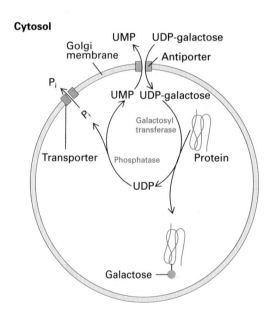

▲ **FIGURE 16-31**   The antiport uptake of nucleotide sugars into Golgi vesicles. UDP-galactose enters from the cytosol in an exchange for UMP that is mediated by an antiporter located in the Golgi membrane. UMP is produced by phosphatase action on UDP, a product of the galactosyltransferase reaction. A different transporter allows the inorganic phosphate formed from UDP to exit the Golgi vesicle. Other known antiporters allow CMP–*N*-acetylneuraminic acid to enter in exchange for CMP, and UDP–*N*-acetylglucosamine to enter in exchange for UMP. [See C. B. Hirschberg and M. D. Snider. 1987. *Ann. Rev. Biochem.* **56**:63–68 and C. Abeijon and C. B. Hirschberg. 1992. *Trends Biochem. Sci.* **17**:32–36.]

for-one exchange of CMP–*N*-acetylneuraminic acid for CMP and of GDP-mannose for GMP. In catalyzing such exchanges, these antiporters allow the concentration of nucleotide sugars in the Golgi vesicles or the ER to be maintained at a constant level—a requirement for oligosaccharide synthesis.

The Diverse *N*-Linked Oligosaccharides Share Certain Structural Features That Reflect a Common Precursor

All the *N*-linked oligosaccharides from a wide variety of proteins made by different organisms have been found to contain an identical unit consisting of three mannose residues and two *N*-acetylglucosamine residues in exactly the same configuration (see Figure 16-27). This reflects the fact that all *N*-linked oligosaccharides are formed from a common precursor, as we shall detail. However, it is also important to understand that *N*-linked oligosaccharides on different proteins have very different structures, which result from the sequential removal and addition of specific saccharide residues after the precursor oligosaccharide is transferred to the polypeptide. Numerous enzymes localized to the rough ER and several Golgi subcompartments participate in these multistep saccharide changes.

The *N*-linked oligosaccharides fall into several structural classes (see Figure 16-27). One class comprises the N-*linked complex oligosaccharides*, found on many serum proteins and viral glycoproteins. These sugars contain *N*-acetylglucosamine, mannose, fucose, galactose, and *N*-acetylneuraminic acid. The structural differences among the complex glycoproteins involve the number of branches (ranging from two to four), the number of *N*-acetylneura-

minic acid residues, and the nature of the chemical linkage between *N*-acetylneuraminic acid and galactose. A second class, the N-*linked high-mannose oligosaccharides*, contain only *N*-acetylglucosamine and mannose. Members of this class differ primarily in the number of mannose residues attached. In a third class are the *hybrid* N-*linked oligosaccharides*, in which one of the branches has a typical complex structure while the other two contain only mannose.

The Processing of *N*-Linked Oligosaccharides Involves the Sequential Removal and Addition of Sugar Residues

The biosynthesis of all *N*-linked oligosaccharides begins in the ER with a large *precursor oligosaccharide*. The structure of this precursor (Figure 16-32) is the same in plants, animals, and single-celled eukaryotes—a branched oligosaccharide containing three glucose, nine mannose, and two *N*-acetylglucosamine molecules; its structure can therefore be written as $(Glc)_3(Man)_9(GlcNAc)_2$. Five of its residues—three mannose and two *N*-acetylglucosamine (highlighted in Figure 16-32)—are the ones that are conserved in the structures of all *N*-linked oligosaccharides, as can be seen by comparing Figures 16-32 and 16-27b.

This precursor oligosaccharide is linked by a pyrophosphoryl residue to *dolichol*, a long-chain (75–95 carbon atoms) polyisoprenoid lipid (see Figure 16-32) that acts as a carrier for the oligosaccharide. The dolichol pyrophosphoryl oligosaccharide is formed in the ER in a complex set of reactions utilizing membrane-attached enzymes of the rough ER, in which *N*-acetylglucosamine, mannose, and glucose residues are added one at a time to dolichol phosphate (Figure 16-33). The final dolichol pyrophos-

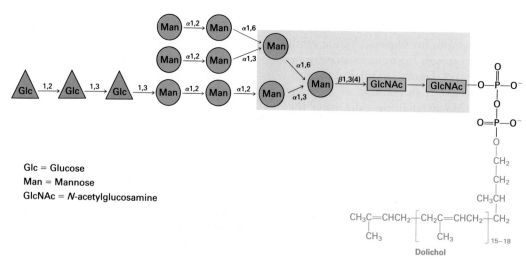

Glc = Glucose
Man = Mannose
GlcNAc = *N*-acetylglucosamine

▲ FIGURE 16-32 Structure of the dolichol pyrophosphoryl oligosaccharide precursor of *N*-linked oligosaccharides. Dolichol is strongly hydrophobic and long enough to span a phospholipid bilayer membrane four or five times. This oligosaccharide lipid is located on the rough ER, and the sugar residues face the lumen of the organelle. The five residues in the shaded area—three of mannose and two of *N*-acetylglucosamine—are conserved in the structures of all *N*-linked oligosaccharides (compare this figure with Figure 16-27b).

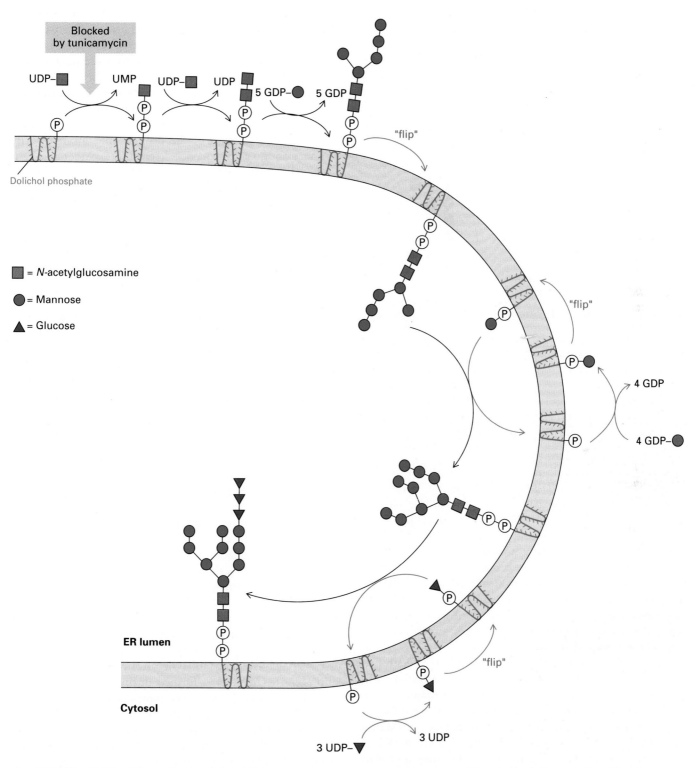

▲ FIGURE 16-33   Biosynthesis of the dolichol pyrophos-phoryl oligosaccharide precursor of *N*-linked oligosaccharides. Two *N*-acetylglucosamine and five mannose residues from UDP sugars are added one at a time to a dolichol phosphate on the cytosolic face of the ER. (Tunicamycin blocks the first enzyme in this pathway, and thus inhibits the synthesis of all *N*-linked oligosaccharides in cells.) Then the dolichol pyro-phosphoryl oligosaccharide is flipped to the luminal face (blue arrow) where the remaining four mannose and all three glu-cose residues are added. In the latter reactions the mannose or glucose is first transferred from a nucleotide sugar to a carrier dolichol phosphate on the cytosolic face of the ER; the carrier is then flipped to the luminal face where the glu-cose or mannose is transferred to the growing oligosaccha-ride, and the carrier is flipped back again to the cytosolic face. [After C. Abeijon and C. B. Hirschberg. 1992. *Trends Biochem. Sci.* **17**:32–36.]

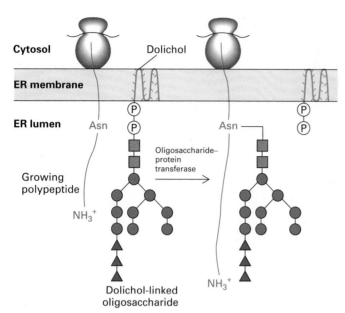

▲ FIGURE 16-34   The catalyzed transfer of an oligosaccharide from the dolichol carrier to a susceptible asparagine residue on a nascent protein. The oligosaccharide is transferred as a unit as soon as the asparagine crosses to the luminal side of the ER.

phoryl oligosaccharide is oriented so that the dolichol portion is firmly embedded in the ER membrane and the oligosaccharide portion faces the ER lumen.

The oligosaccharide is transferred en bloc by an ER enzyme, *oligosaccharide-protein transferase,* from the dolichol carrier to an asparagine residue on the nascent polypeptide (Figure 16-34). The asparagine residue must be in the tripeptide recognition sequence Asn-X-Ser or Asn-X-Thr (where X is any amino acid except proline). The transferase contains three subunits. Two are *ribophorins,* abundant integral ER transmembrane proteins whose cytosol-facing domains bind tightly to the larger subunit of the ribosome and thus localize the third transferase subunit, the actual enzyme, within the ER lumen near the growing polypeptide chain.

While the nascent polypeptide is still in the rough ER, immediately after the oligosaccharide is transferred to it, all three glucose residues and one particular mannose residue of the oligosaccharide are removed by three different enzymes (Figure 16-35, top left). The three glucose residues were the last residues to be added to the oligosaccharide as it was being formed (see Figure 16-33); thus they appear to act as a signal that the oligosaccharide is complete and ready to be transferred to a protein.

Modifications to *N*-Linked Oligosaccharides Are Completed in the Golgi Vesicles

The final modifications to the *N*-linked oligosaccharides take place in the Golgi complex, where three different ver-

sions of a processing pathway lead respectively to complex, high-mannose, and hybrid oligosaccharides. This pathway involves a coordinated, sequential set of reactions that add and remove specific sugar residues from the oligosaccharide. Different enzymes are localized to the *cis-*, the *medial-*, and the *trans-*Golgi vesicles, and act sequentially as the protein moves through these organelles en route to the cell's exterior.

The conformation of the protein and, ultimately, its primary amino acid sequence determines whether an oligosaccharide is attached to a particular Asn-X-Ser/Thr sequence in the ER, since not all such sequences are utilized. For instance, the rapid folding of a segment of a protein containing an Asn-X-Ser/Thr sequence may prevent the oligosaccharide transferase enzyme (Figure 16-34) from transferring an oligosaccharide to it. Similarly, the conformation of a segment of a protein may determine whether, in the Golgi, a particular *N*-linked oligosaccharide becomes complex or hybrid or high-mannose. In some cases the oligosaccharides attached to different asparagine residues in the same polypeptide are modified differently in the Golgi. Each specific cell type in an organism contains its own specific processing enzymes: thus the same protein produced by individual cell types may have differently processed carbohydrates. For example, certain asparagine residues in the HA glycoprotein of the influenza virus have complex oligosaccharides when the virus is grown in one cell type but high-mannose oligosaccharides when the virus is grown in another cell type.

The production of *complex N-linked oligosaccharides* is the net result of a stepwise, coordinated set of reactions in which five mannose residues are removed in addition to the one cut off in the ER; the three that ultimately remain are the conserved ones. Also, three *N*-acetylglucosamine residues, three galactose residues, one to three *N*-acetylneuraminic acid residues, and one fucose residue are added, one at a time, to each oligosaccharide chain (see Figure 16-35).

The *high-mannose oligosaccharides* found on the mature form of many glycoproteins (see Figure 16-27) have the same structures as the intermediates—$(Man)_8(GlcNAc)_2$, $(Man)_7(GlcNAc)_2$, $(Man)_6(GlcNAc)_2$, and $(Man)_5(GlcNAc)_2$—formed during the processing of complex oligosaccharides. The eight-mannose oligosaccharide $(Man)_8(GlcNAc)_2$ is formed if the *cis* Golgi mannose-cleaving enzyme α-mannosidase I, which catalyzes reaction 5 in Figure 16-35, cannot act on the particular *N*-linked oligosaccharide—perhaps because part of the oligosaccharide is buried in a crevasse in the protein surface. The oligosaccharide will be $(Man)_5(GlcNAc)_2$ if reaction 6 (mediated by *N*-acetylglucosaminyltransferase I) cannot occur, since the only substrate for the next enzyme in the pathway—the one catalyzing reaction 7—is the oligosaccharide $(GlcNAc)(Man)_5(GlcNAc)_2$, which is the product of reaction 6 and thus, if reaction 6 does not occur, no further processing of the oligosaccharide takes place.

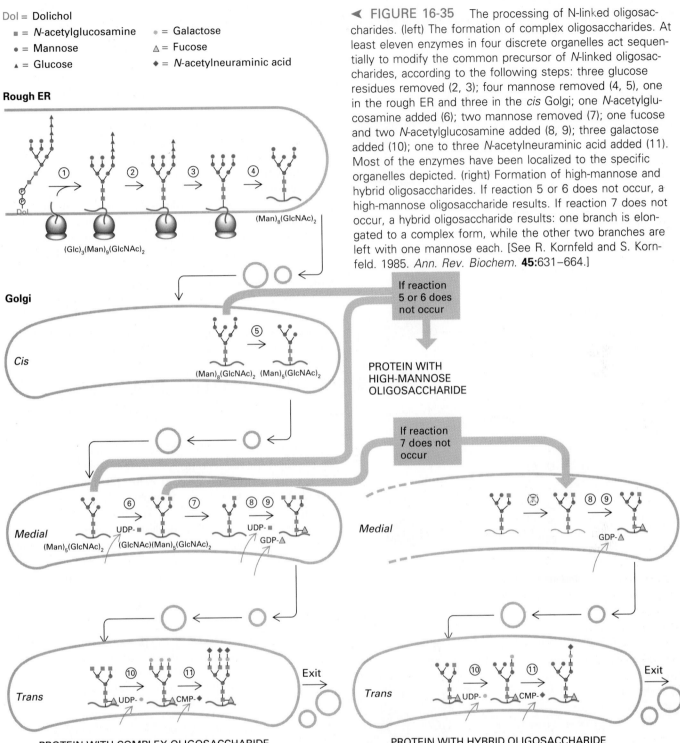

Dol = Dolichol

■ = *N*-acetylglucosamine ● = Galactose
● = Mannose △ = Fucose
▲ = Glucose ◆ = *N*-acetylneuraminic acid

Rough ER

(Glc)₃(Man)₉(GlcNAc)₂

(Man)₈(GlcNAc)₂

Golgi

Cis

(Man)₈(GlcNAc)₂ (Man)₅(GlcNAc)₂

If reaction 5 or 6 does not occur

PROTEIN WITH HIGH-MANNOSE OLIGOSACCHARIDE

If reaction 7 does not occur

Medial

(Man)₅(GlcNAc)₂ (GlcNAc)(Man)₅(GlcNAc)₂
UDP-■ UDP-■
GDP-△

Medial

GDP-△

Trans

UDP-● CMP-◆

Exit

PROTEIN WITH COMPLEX OLIGOSACCHARIDE

Trans

UDP-● CMP-◆

Exit

PROTEIN WITH HYBRID OLIGOSACCHARIDE

◄ FIGURE 16-35 The processing of N-linked oligosaccharides. (left) The formation of complex oligosaccharides. At least eleven enzymes in four discrete organelles act sequentially to modify the common precursor of N-linked oligosaccharides, according to the following steps: three glucose residues removed (2, 3); four mannose removed (4, 5), one in the rough ER and three in the *cis* Golgi; one *N*-acetylglucosamine added (6); two mannose removed (7); one fucose and two *N*-acetylglucosamine added (8, 9); three galactose added (10); one to three *N*-acetylneuraminic acid added (11). Most of the enzymes have been localized to the specific organelles depicted. (right) Formation of high-mannose and hybrid oligosaccharides. If reaction 5 or 6 does not occur, a high-mannose oligosaccharide results. If reaction 7 does not occur, a hybrid oligosaccharide results: one branch is elongated to a complex form, while the other two branches are left with one mannose each. [See R. Kornfeld and S. Kornfeld. 1985. *Ann. Rev. Biochem.* **45**:631–664.]

Finally, *hybrid oligosaccharides* (see Figure 16-27) result if the enzyme that catalyzes reaction 7 cannot act on the particular protein-attached oligosaccharide. Only the branch with an *N*-acetylglucosamine residue can be elongated to the complex form by reactions 8 through 11 (see Figure 16-35, right), while the other two branches remain unmodified.

Thus the *N*-linked oligosaccharides, as they travel through the ER and the Golgi complex, are modified in much the same way that an automobile is built on an assembly line. The protein is acted on sequentially by a large set of enzymes that add and remove sugar residues. The reaction product of one enzyme is the substrate of the next,

and groups of modifying enzymes are localized in different organelles. Even if reactions 5 through 11 do not occur, the attached protein will still be secreted normally.

The Movement of Proteins through the Secretory Pathway Can Be Monitored by Following the Processing of *N*-Linked Oligosaccharides

Subcellular fractionation experiments clearly show that the different oligosaccharide-processing enzymes are localized to different organelles. Galactosyltransferase (reaction 10

(a)

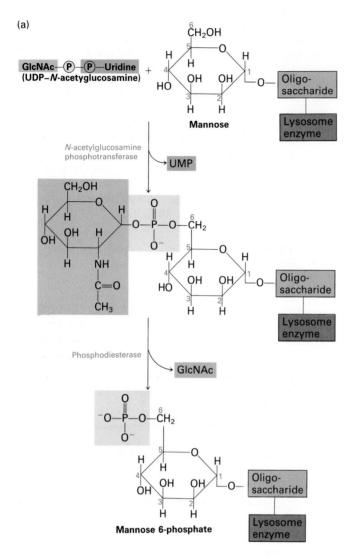

in Figure 16-35) is localized to the *trans*-most Golgi vesicles, and sialyltransferase (reaction 11) is found in the *trans* Golgi and the *trans*-Golgi reticulum. In fact, in subcellular fractionation experiments (see Figures 5-29 and 5-30), galactosyltransferase is frequently used as a marker enzyme for the *trans*-Golgi vesicles.

One can also use electron microscope autoradiography to determine the organelles in which specific sugars are added to protein-linked oligosaccharides. For instance, when cells are exposed briefly to [³H]mannose, and then fixed, sectioned, and developed with a radiation-sensitive photographic emulsion (Figure 6-24), virtually all incorporated radioactivity (observed as autoradiographic grains) is in the ER. This indicates that mannose residues are added to proteins in the ER, as depicted in Figure 16-35. Radioactivity from newly incorporated [³H]galactose is abundant over the *trans*-Golgi region, indicating that this sugar is added to protein in that organelle.

N-Linked and O-Linked Oligosaccharides May Stabilize Maturing Secretory and Membrane Proteins

In many cases, *N*- and *O*-linked oligosaccharides appear to be required for the secretion of proteins or the movement of plasma membrane glycoproteins to the cell surface. When the first stage in the formation of the oligosaccharide-lipid donor (see Figure 16-33) is blocked by the antibiotic tunicamycin, the polypeptide is synthesized, but it contains no *N*-linked sugar chains. Many proteins require these *N*-linked oligosaccharides in order to fold properly in the ER. The HA glycoprotein, for instance, cannot form a

(b)

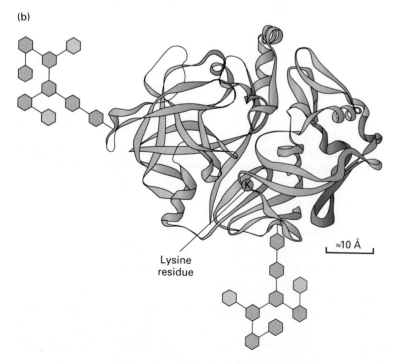

◄ FIGURE 16-36   The phosphorylation of mannose residues on lysosomal enzymes. (a) The two stages of mannose phosphorylation are catalyzed by separate enzymes. The first enzyme, *N*-acetylglucosamine phosphotransferase, transfers an *N*-acetylglucosamine phosphate group to carbon atom 6 of one or more mannose residues. The second enzyme, a phosphodiesterase, removes the *N*-acetylglucosamine group, leaving the phosphate attached to the 6 carbon of the mannose residue. (b) *N*-acetylglucosamine phosphotransferase recognizes only lysosomal enzymes as substrate. In this model of the structure of cathepsin D, a typical lysosomal enzyme, amino acid residues critical for phosphotransferase recognition are indicated in red. The sugars (yellow) on the two high-mannose oligosaccharides are those modified by the phosphotransferase, and are far from the phosphotransferase recognition site. [Part (b) adapted from A. B. Cantor, T. J. Baranski, and S. Kornfeld. 1992. *J. Biol. Chem.* **267**:23349–23356. See also S. Kornfeld. 1987. *FASEB J.* **1**:462–468.]

normal trimer (see Figure 16-24) if the addition of N-linked oligosaccharides is prevented by tunicamycin: the protein remains, misfolded, in the rough ER. Moreover, in the HA protein even the mutation of one asparagine to a glutamine residue, thereby preventing the addition of an N-linked oligosaccharide to that site, will cause the protein to accumulate in the ER in an unfolded state.

By contrast, other proteins fold properly and are secreted normally even if the addition of all N-linked oligosaccharides is blocked. For instance, whether glycosylated or unglycosylated, fibronectin (a constituent of the extracellular matrix) is secreted at the same rate and to the same extent by fibroblasts. However, unglycosylated fibronectin is more susceptible to degradation by tissue proteases than is normal glycosylated fibronectin. Similarly, recombinant erythropoietin, when synthesized without its normal N-linked oligosaccharides, is as potent as the normal hormone in stimulating the growth of red blood cell precursors in culture. However, when injected into humans, the hormone without its oligosaccharides is much less potent than the normal protein, presumably because it is degraded much faster than normal. These results establish that added carbohydrates confer stability on many extracellular proteins.

Phosphorylated Mannose Residues Target Proteins to Lysosomes

So far, we have seen that oligosaccharides often play an important role in ensuring the correct conformation and hence the stability of maturing proteins. Here we discuss a case in which a particular glycosylation pattern is the signal that targets lysosomal enzymes to lysosomes, and prevents their secretion.

Lysosomal enzymes, like secretory proteins, are cotranslationally inserted into the ER lumen, where they receive the same N-linked oligosaccharide, as do membrane and secretory proteins, $(Glc)_3(Man)_9(GlcNAc)_2$. In the cis Golgi, one or more mannose residues of these oligosaccharides become phosphorylated on carbon atom 6, and the phosphorylated mannose residues are the chemical signal that targets the protein to lysosomes. The phosphorylation of mannose residues is a two-step procedure involving two separate enzymes (Figure 16-36a).

N-Acetylglucosamine phosphotransferase (the first enzyme) uses only lysosomal proteins as substrates and does not catalyze reactions with other glycoproteins. Deglycosylated lysosomal enzymes are potent inhibitors of the phosphorylation of intact lysosomal enzymes by N-acetylglucosamine phosphotransferase, indicating that N-acetylglucosamine phosphotransferase recognizes the protein part of a lysosomal glycoprotein, not the oligosaccharide (which is the same as on secretory proteins which do not become phosphorylated). The N-acetylglucosamine phosphotransferase binds tightly to a recognition domain consisting of several amino acids, including one invariant lysine residue that is present on one part of the surface of

all lysosomal enzymes (Figure 16-36b). After binding to this site, it adds an N-acetylglucosamine phosphate residue to the oligosaccharide.

In order for the phosphorylated lysosomal enzyme to be diverted to lysosomes, it must bind to a specific receptor protein, the *mannose 6-phosphate receptor,* a transmembrane protein whose exoplasmic domain contains a region that binds mannose 6-phosphate very tightly and specifically. Most mannose 6-phosphate receptors are located in the *trans*-Golgi reticulum, where they bind the mannose 6-phosphate residue on the lysosomal protein (Figure 16-37). The mannose 6-phosphate receptor with its attached lysosomal enzyme becomes localized to a small region of membrane that is coated on its cytosolic face by the fibrous protein *clathrin.* As we discuss later (page 722), clathrin causes these membrane regions to pinch off to become specialized transport vesicles which then lose their clathrin coats (see Figure 16-37). These transport vesicles fuse with a sorting vesicle, an organelle often termed *CURL* (for *c*ompartment of *u*ncoupling of *r*eceptor and *l*igand) or the *late endosome,* which has an internal pH of ≈ 5.5. The low pH causes mannose 6-phosphate receptors to release their bound lysosomal enzymes because the receptor can bind mannose 6-phosphate at the slightly acidic pH ($\approx 6.5–7$) of the *trans*-Golgi reticulum but not at a pH less than ≈ 6. Furthermore, in the vesicle that transports the protein to a lysosome a phosphatase generally removes the phosphate from lysosomal enzymes, preventing any rebinding of the enzyme to the receptor.

Two types of vesicles bud from the late endosome/CURL. One type contains the lysosomal enzyme but not the mannose 6-phosphate receptor; after these vesicles bud from the sorting vesicles, they fuse with lysosomes, delivering the enzyme to its destination. The other type of vesicle recycles the mannose 6-phosphate receptor back to the *trans*-Golgi reticulum or, as discussed below, to the cell surface.

The lysosomal enzymes we have talked about so far are actually precursors, *proenzymes.* These catalytically inactive proenzymes are the ones sorted by the mannose 6-phosphate receptor. Late in its maturation a proenzyme undergoes a proteolytic cleavage which causes a conformational change in the protein, forming a smaller but enzymatically active polypeptide. This cleavage occurs in either the acidic late endosomal vesicle or the lysosome. Delaying the activation of lysosomal proenzymes until they reach the lysosome prevents them from digesting macromolecules in earlier compartments of the secretory pathway.

Finally, some mannose 6-phosphate receptors are found on the cell surface, where they bind the lysosomal proteins that occasionally are secreted that contain mannose 6-phosphate. Vesicles containing these plasma membrane receptors bound to lysosomal enzymes bud into the cytosol, and eventually fuse with late endosome/CURL organelles. Exactly as with lysosomal enzymes transported from the *trans* Golgi, these internalized lysosomal enzymes are delivered to lysosomes, while the mannose 6-phosphate

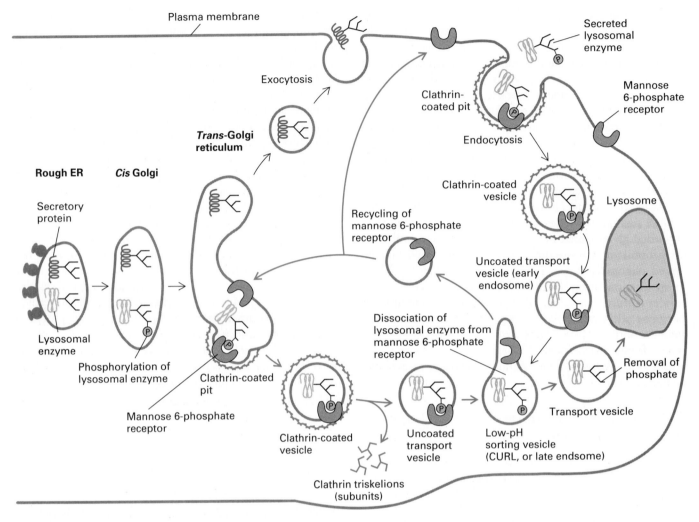

▲ **FIGURE 16-37**   The targeting of lysosomal enzymes to lysosomes. During biosynthesis, lysosomal enzymes migrate from the rough ER (left) to the *cis* Golgi, in which one or more mannose residues are phosphorylated. In the *trans*-Golgi reticulum, the enzymes bind to the membrane-attached, highly specific mannose 6-phosphate receptor, which directs the enzymes into vesicles coated with the fibrous protein clathrin. The clathrin surrounding these vesicles is rapidly depolymerized to its clathrin triskelion subunits, and the uncoated transport vesicle fuses with a CURL vesicle (also called a late endosome). The low pH of the CURL vesicle (≈5.5) causes the phosphorylated enzyme to dissociate from its receptor. The receptor recycles back to the Golgi, and the phosphorylated enzyme is incorporated into a different transport vesicle that buds from the CURL. In this vesicle the protein loses its phosphate group, and the vesicle soon fuses with a lysosome. The sorting of lysosomal enzymes from secretory proteins (left) thus occurs in the *trans*-Golgi reticulum, and these two classes of proteins are incorporated into different vesicles, which take different routes after they bud from the Golgi.

The mannose 6-phosphate receptor is also found on the plasma membrane (upper right), where it binds extracellular phosphorylated lysosomal enzymes that are occasionally secreted. These receptor-ligand complexes are internalized from the cell surface in clathrin-coated vesicles which also lose their clathrin coats and fuse with the CURL organelle. [See G. Griffiths et al. 1988. *Cell* **52**:329–341; S. Kornfeld. 1992 Ann. Rev. Biochem. **61**:307–330; and G. Griffiths and J. Gruenberg. 1991. *Trends Cell Biol.* **1**:5–9.]

receptors recycle either to the plasma membrane or to the *trans* Golgi reticulum (see Figure 16-37). This process is called *receptor-mediated endocytosis* and is used by many cell-surface receptors to bring bound proteins or particles into the cell, as we discuss in more detail on page 722.

The mechanism by which lysosomal enzymes are targeted to lysosomes illustrates several important general principles about protein sorting. First, membrane-imbedded receptors with their bound ligands diffuse into discrete regions of the membrane of an organelle—in this case the

trans-Golgi reticulum—where they are specifically incorporated into budding transport vesicles. Second, these transport vesicles fuse only with one specific organelle, here the late endosome/CURL. Third, cellular transport receptors recycle; in this case the low pH of the sorting vesicle causes the receptors to dissociate from the bound ligand and to be incorporated in yet other transport vesicles that recycle them back to the *trans*-Golgi reticulum or to the cell surface.

Genetic Defects Have Elucidated the Role of Mannose Phosphorylation

The discovery of the mannose 6-phosphate pathway began with a study of patients with *I-cell disease,* a severe genetic abnormality resulting in a deficiency of multiple lysosomal enzymes in fibroblasts and macrophages. Fibroblasts and macrophages from these patients contain large intracellular vesicles filled with glycolipids and extracellular components that would normally be degraded by lysosomal enzymes. Cells from affected persons lack the first enzyme required to phosphorylate mannose residues (see Figure 16-36a), and lysosomal enzymes are secreted rather than remaining sequestered in lysosomes. When the diseased fibroblasts are grown in a medium containing lysosomal enzymes bearing phosphorylated mannose, the enzymes are internalized by receptor-mediated endocytosis, using cell surface mannose 6-phosphate receptors, and the cellular content of these enzymes becomes almost normal. Hepatocytes, the predominant type of liver cells, do not use the mannose 6-phosphate pathway for sorting lysosomal enzymes to lysosomes. As a result, lysosomes in the hepatocytes from patients with I-cell disease do contain a normal complement of lysosomal enzymes, even though these cells are defective in mannose phosphorylation. How hepatocytes target newly made enzymes to lysosomes is unknown.

▶ *The Mechanism and Regulation of Vesicular Transport to and from the ER and the Golgi Complex*

Proteins move from one organelle to another in the secretory pathway by means of small transport vesicles. Here we discuss three crucial issues concerning vesicular traffic within cells. First, how are transport vesicles formed, and why are certain integral membrane proteins incorporated selectively into these vesicles and others not? Second, what is the molecular nature of the signal on a particular transport vesicle that causes it to bind only to a particular type of organellar membrane? How, for instance, does a vesicle containing secretory proteins from the rough ER "know" to move to and fuse with *cis*-Golgi and not with *trans*-Golgi membranes? How does a vesicle bearing lysosomal enzymes from the *trans*-Golgi reticulum "know" to fuse with the late endosome/CURL rather than with the plasma membrane? Third, what is the mechanism by which the membranes of the transport vesicle and the destination organelle fuse with each other?

Two Types of Coated Vesicles Transport Proteins from Organelle to Organelle

All eukaryotic cells contain a plethora of small membrane-limited vesicles (see Figure 16-26). Some of these vesicles have a protein coat on their cytosolic surface. At one time, it was thought that all coated vesicles contained the fibrous protein clathrin, and clathrin-coated vesicles were believed to transport plasma membrane glycoproteins and secretory proteins from the rough ER through the Golgi vesicles to the cell surface. However, such a general role for clathrin-coated vesicles seemed improbable, since a yeast mutant with its clathrin gene deleted still secretes proteins quite normally. Indeed, there are two types of coated vesicles. One type, which has a coat of clathrin on its cytosolic surface, forms from the plasma membrane or the *trans*-Golgi reticulum. The other type, which has a coat made of a different set of proteins, transports proteins from the ER to the Golgi and between different types of Golgi vesicles. Vesicles with the second type of coat also transport continuously secreted proteins from the *trans*-Golgi reticulum to the cell surface.

We begin our study of the mechanism of vesicular transport with a discussion of the formation and structure of clathrin-coated vesicles, followed by a discussion of the nonclathrin-coated vesicles. Both types of coats serve two functions—polymerization of the coat subunit proteins around the cytosolic face of a budding vesicle helps the vesicle to pinch off from the parent organelle, and the coat proteins help to select which membrane proteins will enter the transport vesicles.

Clathrin Forms a Lattice Shell around Coated Pits and Vesicles

Typical clathrin-coated vesicles have a soccer-ball appearance (Figure 16-38a). They are 50–100 nm in diameter, with a membrane-bounded vesicle inside a coat that is composed primarily of the fibrous protein clathrin.

Purified clathrin has the form of a three-limbed triskelion (from the Greek for three-legged; see Figure 16-38b and c). Each limb contains one clathrin heavy chain (180,000 MW) and one clathrin light chain ($\approx$35,000–40,000 MW). There are two types of light chains, α and β; their amino acid sequences are 60 percent identical, and their functional differences are not known. Even in the absence of membrane vesicles, triskelions combine, or polymerize, to form the cagelike structure that is found around a coated vesicle.

Clathrin polymerizes into a lattice along the cytosolic face of a region of membrane; this is believed to cause endocytosis, in which the region expands inward, first forming a *clathrin-coated pit* (Figure 16-39a), and ultimately pinches off from the membrane. After an actual vesicle forms, the clathrin "cage" is completed, producing a coated vesicle.

Proteins other than clathrin are required for the formation of coated vesicles. One, *dynamin,* is a $\approx$900-amino-acid cytosolic protein that binds and then hydrolyzes GTP. As evidence for its importance in endocytosis, the cellular expression of mutant dynamins that cannot bind GTP will

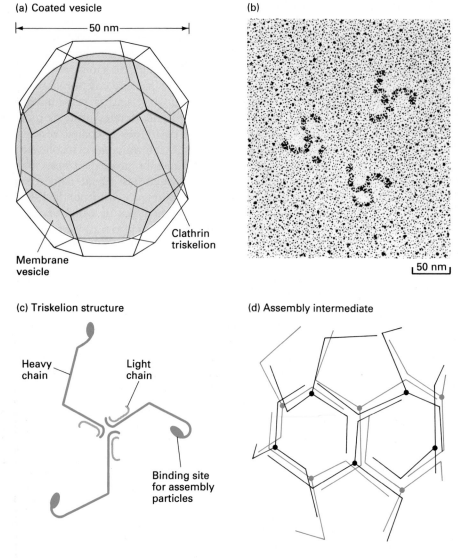

(a) Coated vesicle

50 nm

Membrane vesicle

Clathrin triskelion

(b)

50 nm

(c) Triskelion structure

Heavy chain

Light chain

Binding site for assembly particles

(d) Assembly intermediate

◄ FIGURE 16-38   Structure and assembly of a clathrin-coated vesicle. (a) A typical clathrin-coated vesicle contains a membrane vesicle (tan) about 40 nm in diameter surrounded by a fibrous network of 12 pentagons and 8 hexagons. The fibrous coat is constructed of 36 clathrin triskelions, one of which is shown here in red. One clathrin triskelion is centered on each of the 36 vertices of the coat. Coated vesicles having other sizes and shapes are believed to be constructed similarly: each vesicle contains 12 pentagons but a variable number of hexagons. (b) Electron micrograph of purified clathrin triskelions. (c) Detail of a clathrin triskelion. Each of three clathrin heavy chains has a specific bent structure. A clathrin light chain is attached to each heavy chain near the center; a globular domain is at each distal (outer) tip. Although it is not obvious in (a) or (b), each triskelion has an intrinsic curvature; when triskelions polymerize, they form a curved (not flat) structure. (d) An intermediate in the assembly of a clathrin-coated vesicle, containing 10 of the final 36 triskelions, illustrates the intrinsic curvature and the packing of the clathrin triskelions. [Part (a) see B. M. F. Pearse, 1987. *EMBO J.* **6**:2507–2512. Part (b) courtesy of D. Branton, from E. Ungewickell and D. Branton. 1981. *Nature* **289**:420. Reprinted by permission from Nature. Copyright 1981, Macmillan Journals Limited. Part (c) see F. M. Brodsky. 1988. *Science* **242**:1396–1402; J. E. Heuser and J. Keen. 1988. *J. Cell Biol.* **107**:877–886.]

block the formation of coated vesicles. However, how dynamin functions in this process is not known.

Between the clathrin coat and the vesicle membrane lies a 20-nm space containing *assembly particles*. Each particle (340,000 MW) contains one copy each of four polypeptides of 105,000, ≈100,000, ≈50,000, and ≈20,000 MW. Assembly particles are required for the clathrin cage to form around a vesicle: they bind to the globular domain at the end of each clathrin heavy chain in a triskelion (see Figure 16-38c and d) and promote the polymerization of clathrin triskelions into cages. By also binding to the cytosolic face of membrane proteins, assembly particles determine which proteins are specifically included in (or excluded from) the budding transport vesicle (Figure 16-39b).

Most important, there are two types of assembly particles, AP-1 and AP-2, that contain different, though related,

polypeptides. AP-2 assembly particles bind to the cytosol-facing domains of plasma membrane receptors that become internalized in clathrin-coated pits by the process of receptor-mediated endocytosis, such as the LDL receptor protein (see Figure 16-48), but they do not bind to plasma membrane proteins that are not internalized in these pits. AP-1 assembly particles, in contrast, bind to the cytosolic domains of proteins that bud from the *trans*-Golgi reticulum. The mannose 6-phosphate receptor is found in clathrin-coated pits that bud from the plasma membrane or the *trans*-Golgi reticulum (see Figure 16-37); the receptor's cytoplasmic tail binds to both AP-1 and AP-2 assembly particles. Thus assembly particles mediate the binding of clathrin to proteins in coated pits and distinguish between integral membrane proteins that are to be included in these transport vesicles and those that are to be left behind (Figure 16-39b).

(a)

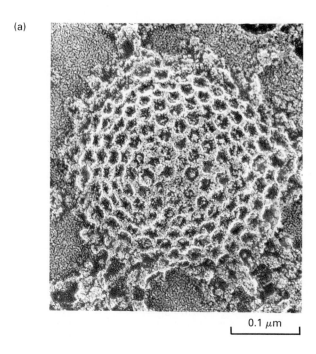

0.1 μm

(b)

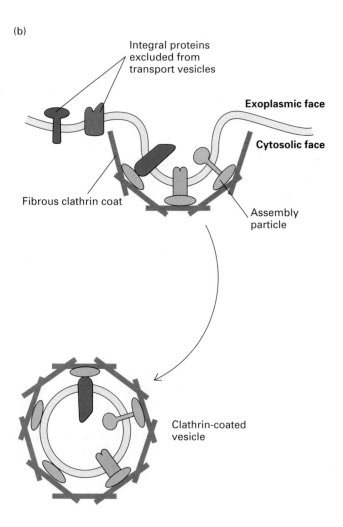

Integral proteins
excluded from
transport vesicles

Exoplasmic face

Cytosolic face

Fibrous clathrin coat

Assembly
particle

Clathrin-coated
vesicle

◀ FIGURE 16-39 Formation of a clathrin-coated vesicle. (a) A clathrin-coated pit on the cytoplasmic face of the plasma membrane of a fibroblast, showing the polygonal network of the fibrous clathrin. The cell was rapidly frozen in liquid helium, freeze-fractured, and then treated by the deep-etching technique. (b) A model for the selective incorporation of "passenger" integral membrane proteins into clathrin-coated vesicles. The cytosolic domains of certain membrane proteins bind specifically to assembly particles that, in turn, bind to clathrin as it polymerizes spontaneously over a region of membrane. Proteins that do not bind to assembly particles are excluded from these vesicles. [Photograph in (a) courtesy of J. Heuser.]

A Chaperone Protein Catalyzes the Depolymerization of Clathrin-Coated Vesicles

Clathrin-coated vesicles are stable at the pH and ionic composition of the cell cytosol, yet coated vesicles normally lose their clathrin coat and the assembly particles just after their formation (Figure 16-37). How does this happen?

A liver enzyme has been purified that depolymerizes the vesicles' coat to clathrin triskelions. This enzyme is the *chaperone protein hsc70*, a cytosolic protein present in all eukaryotic cells. It is related in amino acid sequence to Bip (Figure 16-16), a chaperone found in the ER, as well as to others in the mitochondrion and chloroplast (Figure 19-7). The hydrolysis of ATP to ADP and P_i by hsc70 is required to disrupt the multiple clathrin-clathrin interactions that stabilize the coated vesicle, causing the depolymerization of the clathrin coat into triskelions. The hsc70 protein apparently binds to the clathrin light chains (Figure 16-38c), since the purified light chains stimulate ATP hydrolysis by hsc70.

Both the formation of coated vesicles and the depolymerization of the coat must be highly regulated in the cell, as both processes occur simultaneously. But much remains to be learned about the molecular details of these processes.

A Type of Coated Vesicle Without Clathrin Mediates ER-to-Golgi Transport and Transport Within the Golgi

Nonclathrin-coated vesicles were first seen when isolated Golgi fractions were incubated in a solution containing ATP and cytosol: the fractions formed a large number of buds and vesicles (Figure 16-40). The vesicles were found to contain a protein coat on their cytosolic face, but it was composed not of clathrin but of αCOP, a protein similar in size to the clathrin heavy chain (see Figure 16-38c). These nonclathrin-coated vesicles transport proteins from the ER to the Golgi, and from one set of membranes in the Golgi to another.

(a)

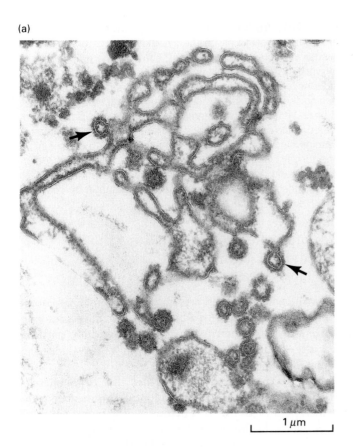

1 μm

(b)

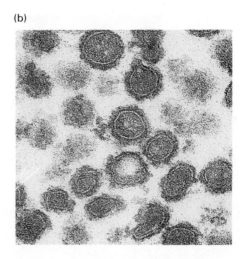

◀ FIGURE 16-40   Nonclathrin-coated vesicles. (a) Incubation of a Golgi fraction with cytosol and ATP causes buds and vesicles to form. These Golgi vesicles from rat hepatocytes have a distinct outer coat (arrows) but do not contain the protein clathrin. (b) In this purified preparation of nonclathrin-coated vesicles, the distinct coat associated with the cytosolic face of the membrane is easily visible. Such coated vesicles transport proteins from the ER to the Golgi and from one Golgi vesicle to another. [Electron micrographs courtesy of L. Orci; part (a) from P. Melancon et al. 1987. *Cell* **51**:1053–1062.]

αCOP forms the fibrous outer shell, normally surrounding a membrane vesicle. This outer coat, like clathrin, may cause a membrane to form budding vesicles. Between the αCOP coat and the vesicle's membrane is a set of proteins that may serve the same functions as the assembly particles in clathrin-coated vesicles—acting as a bridge between the cytosolic tails of membrane proteins and the fibrous cage that surrounds the vesicles, and mediating the specific incorporation of proteins into these coated vesicles (Figure 16-39b). One subunit, termed βCOP (≈110,000 MW), has amino acid sequences that are distantly related to one subunit of the assembly particles found in clathrin-coated vesicles. βCOP, together with αCOP and five other proteins that coat nonclathrin-coated vesicles, are also found in the cytosol of eukaryotic cells as a complex, called a *coatomer*. Coatomers dissociate from the nonclathrin-coated vesicles soon after they form.

The Steps in Vesicular Transport Can Be Studied Biochemically and Genetically

The goal of much current research is to identify all of the proteins involved in the formation of transport vesicles, all proteins that target these transport vesicles to their appropriate organelle, and all that catalyze membrane fusion. Later in the chapter, when we discuss receptor-mediated endocytosis, we will learn more about the formation and function of clathrin-coated vesicles. Here we discuss the two complementary experimental approaches that have revealed much about transport utilizing nonclathrin-coated vesicles—the biochemical analysis of cell-free systems that carry out transport, using non-clathrin-coated vesicles, from one organelle to another, and the molecular analysis of yeast mutants in which vesicular transport is blocked at a specific stage (see Figure 16-10). Many of the same proteins have now been identified by both approaches, and additional studies should eventually elucidate all proteins required for vesicle budding, targeting, and fusion.

A Biochemical Analysis of Vesicle Budding and Fusion

Figure 16-41 illustrates how a cell-free system can be used to demonstrate the specific movement of plasma membrane glycoproteins from a *cis*-Golgi vesicle (the *donor membrane*) to the *medial* Golgi (the *acceptor membrane*) by means of small transport vesicles. These experiments make use of cultured cells infected by the vesicular stomatitis virus (VSV), chosen because VSV synthesizes only *one* predominant glycoprotein, the VSV G protein, which accumulates in the plasma membrane. (Like the influenza HA protein, G protein forms the surface spikes of the virus.)

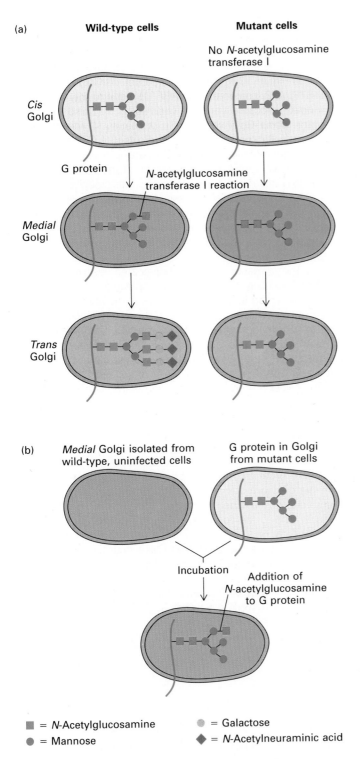

(a)

Wild-type cells   **Mutant cells**

Cis Golgi

No *N*-acetylglucosamine transferase I

G protein

N-acetylglucosamine transferase I reaction

Medial Golgi

Trans Golgi

(b)

Medial Golgi isolated from wild-type, uninfected cells

G protein in Golgi from mutant cells

Incubation

Addition of *N*-acetylglucosamine to G protein

■ = *N*-Acetylglucosamine   ● = Galactose
● = Mannose   ◆ = *N*-Acetylneuraminic acid

▲ FIGURE 16-41   A cell-free system exhibiting movement of newly made VSV glycoprotein (G protein) from donor *cis*-Golgi vesicles to acceptor *medial*-Golgi vesicles. Part (a) shows the properties of a mutant line of cultured fibroblasts that are essential for these experiments. The cells lack the enzyme *N*-acetylglucosamine transferase I (step 6 in Figure 16-35). In wild-type cells, this enzyme is localized to the *medial* Golgi and modifies asparagine-linked oligosaccharides by the addition of one *N*-acetylglucosamine. Thus, after VSV infection of wild-type cells, the oligosaccharide on the viral G protein is modified to a typical complex oligosaccharide, as shown in the *trans*-Golgi vesicle. In mutant cells, however, the G protein reaches the cell surface with a simpler high-mannose oligosaccharide containing only two *N*-acetylglucosamine and five mannose residues. (b) To detect movement of the G protein from one Golgi vesicle to another in cell-free systems, the mutant cells are infected with VSV. When Golgi vesicles are isolated from such infected mutant cells, all their G protein contains the high-mannose oligosaccharide lacking *N*-acetylglucosamine. The incubation of these donor Golgi vesicles with Golgi vesicles from normal, uninfected cells—the acceptor vesicles—results in the generation of G protein containing the additional *N*-acetylglucosamine. This G protein has traveled in transport vesicles from mutant *cis*-Golgi vesicles to wild-type *medial*-Golgi vesicles to receive the *N*-acetylglucosamine residue. [See W. E. Balch et al. 1984. *Cell* **39**:405; W. A. Braell et al. 1984. *Cell* **39**:511; W. E. Balch et al. 1984. *Cell* **39**:525–536.]

VSV is used to infect a mutant cell line that is missing one of the enzymes that modify *N*-linked oligosaccharides. Donor Golgi vesicles are purified from these mutant cells and are incubated with cytosol and acceptor Golgi membranes from normal cells. After incubation, the oligosaccharide on the G protein is seen to have acquired the modification that the missing enzyme makes in normal acceptor Golgi vesicles. Thus the G protein has moved from the

mutant donor vesicles to the acceptor Golgi. It does so by means of nonclathrin-coated transport vesicles that bud from the donor Golgi vesicles and then fuse with the acceptor Golgi. Details of the vesicular transport are shown in Figure 16-42. Similar assays also have been used to document the movement of proteins in cell-free systems from the rough ER to the *cis* Golgi and from the *medial* Golgi to the *trans* Golgi.

Fractionation of the cytosol has shown that several proteins are required for the formation of the transport vesicles (Figure 16-42b). One is the GTP-binding protein known as ARF: In the cytosol, ARF contains a bound GDP. The initiation of vesicle formation occurs when a Golgi-attached enzyme displaces the GDP and allows GTP to bind to ARF, enabling the ARF-GTP complex to bind to ARF receptors on the Golgi membrane. Next coatomer—the preformed complex of αCOP, βCOP, γCOP, and four other proteins—binds to ARF and other proteins on the cytosolic face of the Golgi membrane, inducing budding of the transport vesicle. The final fission that creates the completed transport vesicle requires fatty acyl CoA, but how this molecule functions is unknown. The critical experiment that demonstrated this model of vesicle formation was similar to the one described in Figure 16-40 except that purified proteins were used: purified Golgi vesicles were incubated with coatomer, ARF, fatty acyl CoA, and a derivative of GTP that could not be hydrolyzed, and the formation and release of nonclathrin-coated vesicles were observed.

Fusion of the transport vesicles with the acceptor Golgi membranes is a complex process that involves a number of proteins (Figure 16-42c). Two *integral membrane proteins* have been identified: a VAMP-like protein in the transport vesicles and a syntaxin-like protein in the acceptor Golgi vesicles are known to bind to each other in the first step in fusion. [VAMP itself is a protein in the secretory vesicles of nerve cells and syntaxin is a protein in the plasma membrane of these cells (see Figure 21-37). These proteins similarly bind to each other in the first stage of regulated exocytosis (see Figure 16-8), suggesting that proteins of related structure may carry out similar functions in many types of vesicle fusion events.] Several *cytosolic proteins* have been purified that catalyze different steps in vesicle fusion: one, NSF, is a tetramer of identical subunits that mediates the fusion of the transport vesicle membrane with the membrane of the acceptor *medial*-Golgi vesicle. Other proteins, called α-, β-, and γ-SNAPs (soluble NSF attachment proteins) are required for NSF to bind to Golgi membranes, and are also essential for the vesicle fusion reaction.

Genetic Evidence for the Role of Coatomer and SNAPs An analysis of the various classes of temperature-sensitive yeast *sec* mutants (Figure 16-10) has confirmed the involvement of several Sec proteins in the processes of budding and fusion of transport vesicles. For instance, the amino acid sequence of the γCOP coatomer subunit is similar to that of Sec21p, the product of the yeast *sec21* gene, a class B mutant. Functional Sec21p is required for the formation of ER-to-Golgi transport vesicles. Similarly, the mammalian NSF protein and the yeast Sec18 protein are similar in sequence, as are the mammalian α-SNAP and the yeast Sec17. Mutants in the yeast *sec17* or *sec18* genes are class C: cells accumulate ER-to-Golgi transport vesicles at the high, nonpermissive temperatures, and these vesicles fuse with the *cis* Golgi when the cells are placed at the lower, permissive temperature. Thus, the Sec17 and Sec18 proteins are required for vesicle fusion, in agreement with the biochemical demonstration that their mammalian counterparts, NSF and α-SNAP, are required for the fusion of transport vesicles with Golgi membranes.

NSF itself, or some protein very similar to it, also participates in the fusion of other intracellular vesicles. Indeed, NSF and the SNAPs may catalyze all membrane fusion reactions in cells. Then how do the various transport vesicles recognize only their correct acceptor vesicles? Presumably, such recognition involves the binding of specific molecules on the surface of the transport vesicles (such as VAMP-like proteins) with molecules on the acceptor vesicles (such as syntaxin-like proteins); many such "targeting proteins" are now being identified.

► FIGURE 16-42   Details of vesicular transport within the Golgi complex.

(a) Steps in the transport between *cis* and *medial* Golgi vesicles, as determined by the study of cell-free extracts shown in Figure 16-41. The incubation of *cis*-Golgi vesicles with GTP, ATP, coatomer, and ARF (1) leads to the formation of nonclathrin-coated transport vesicles (2) that bind to acceptor *medial*-Golgi vesicles (3). Uncoating (4) requires a cytoplasmic Rab protein and the hydrolysis of GTP to GDP and P$_i$; experimentally adding a GTP analog that cannot be hydrolyzed will inhibit this step. Fusion of the two membranes (5) requires ATP hydrolysis and four cytoplasmic proteins: NSF and three SNAPs (soluble NSF attachment proteins). This step can be blocked selectively by adding the chemical *N*-ethylmaleimide (NEM), which reacts with an essential SH group on NSF (hence the name, *N*EM-sensitive factor).

(b) The formation of nonclathrin-coated vesicles from the *cis* Golgi. Budding is initiated when molecules of ARF protein are caused (in an unknown manner by a membrane-bound enzyme) to exchange their bound GDP for GTP, and then bind to ARF receptors on *cis*-Golgi vesicles. Coatomer then binds to the cytosolic face of the *cis*-Golgi vesicle; its polymerization into the fibrous coat induces vesicle budding. Fatty acyl CoA is essential for the final separation of the transport vesicle from the donor membrane, but how it functions is not known.

(c) The attachment of a transport vesicle to the acceptor *medial*-Golgi vesicle just prior to membrane fusion (step 4 in Part a). The attachment of the two membranes is initiated by a Rab protein, and by binding of the cytosolic domain of a VAMP-like protein in the membrane of the transport vesicle with the cytosolic domain of a syntaxin-like protein in the membrane of the acceptor vesicle. Then the three SNAP proteins bind to the complex, followed by binding of NSF. Subsequently (step 5, not shown here), ATP hydrolysis triggers the fusion of the Golgi membrane with the transport vesicle membrane, which releases NSF and the SNAPs into the cytosol. [Part (a) after D. W. Wilson et al. 1991. *Trends Biochem. Sci.* **16**:333–337; Part (b) after J. Ostermann et al. 1993. *Cell* **75**:1015; Part (c) after S. W. Whiteheart et al. 1993. *Nature* **362**:353–355 and M. Bennett and R. Scheller. 1993. *Proc. Natl. Acad. Sci. USA* **90**:2559–2563.]

A Family of Small GTP-Binding Proteins May Target Transport Vesicles to Their Correct Destinations

A family of proteins that bind GTP appear to participate in the control of vesicular traffic in eukaryotic cells. These Rab proteins all contain ≈200 amino acids and have a similar overall structure (Figure 16-43a), yet each is localized to a different organelle. Purified Rab proteins bind and hydrolyze GTP, and it is thought that the cycle of GTP binding and hydrolysis ensures the fusion of the correct vesicles.

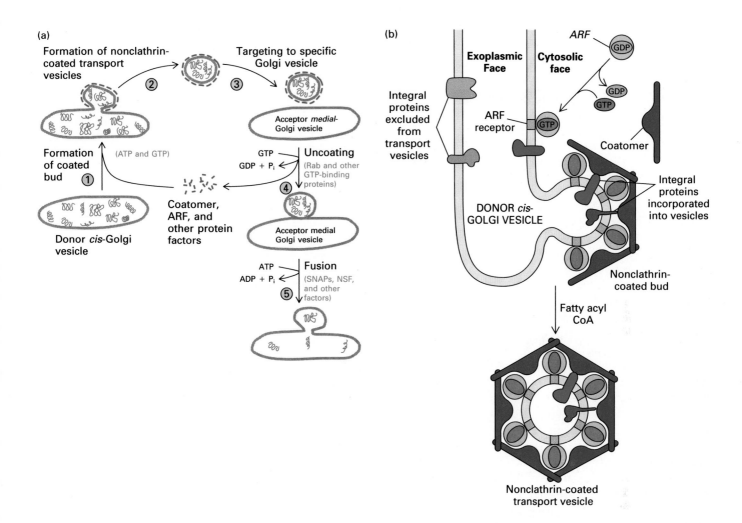

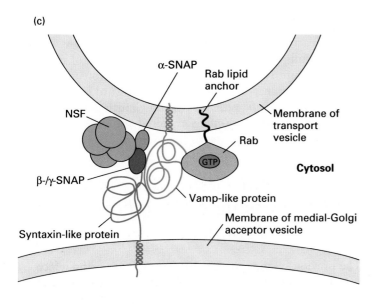

Specifically, a cytosolic Rab protein would exchange its bound GDP for a GTP and undergo a conformational change during the time it binds to a surface protein on a particular transport vesicle just as it is budding from the donor vesicle (see Figure 16-42b). The Rab-GTP complex would then facilitate the binding of the transport vesicle to its proper acceptor organelle and initiate the vesicle fusion process. Once vesicle fusion occurs, the GTP bound to the Rab protein would be hydrolyzed to GDP, triggering the release of the Rab protein from the membrane. The released Rab protein, bound to a cytosolic protein named GDI, exchanges its GDP for GTP only when it binds to another forming transport vesicle, completing the Rab "cycle."

Several lines of evidence support the involvement of specific Rab proteins in specific vesicle fusion events. For instance, Rab5 is localized to the *early endosomes*— organelles that form from clathrin-coated vesicles just after they bud from the plasma membrane (Figure 16-43b; see also Figure 16-37). The fusion of early endosomes with each other in cell-free systems, reflecting a process that normally occurs after receptor-mediated endocytosis (page 722), requires the presence of Rab5, but not of any other

(a)

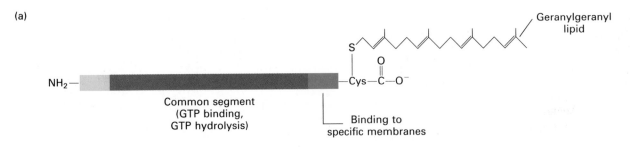

(b)

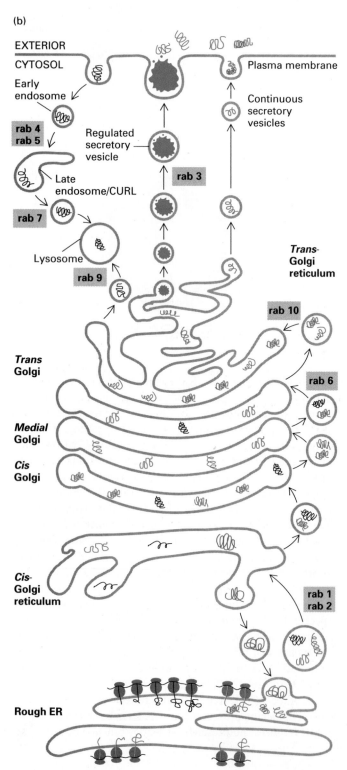

▲ **FIGURE 16-43**   Rab proteins. (a) Generalized structure of Rab proteins. Each contains a central section, highly similar in sequence in all Rab proteins, that binds GTP and then hydrolyzes it to GDP. A segment near the C-terminus, ≈35 amino acids long (blue), is very different in all Rab proteins and determines the membrane to which it binds. The C-terminal cysteine residue, like that in Ras protein (Figure 14-20) is covalently bound to a polyisoprenoid lipid, in this case a 20-carbon lipid termed geranylgeranyl. The N-terminal–most ≈20 amino acids are also very different in all Rab proteins. (b) Subcellular location of Rab proteins in mammalian cells. Each is found in a discrete transport vesicle, and is thought to participate in directing that transport vesicle to the appropriate acceptor organelle. [Part (a) adapted from P. Chavrier et al. 1991. *Nature* **353**:769–772; part (b) adapted from P. Chavrier et al. 1990. *Cell* **62**:317–329.]

Rab protein. Similarly, Rab1 is essential for ER-to-Golgi transport reactions to occur in cell-free extracts, while other Rab proteins, such as Rab5 or Rab7, are not. The yeast *sec4* gene is a member of the Rab gene family, and yeast cells expressing mutant sec4 proteins accumulate secretory vesicles that are unable to fuse with the plasma membrane (class E mutants in Figure 16-10). Thus, individual Rab proteins are clearly essential for specific vesicle fusion reactions to occur, but we do not yet know why specific Rab proteins bind only to specific types of membranes (Figure 16-43b), or whether the Rab proteins actually help the transport vesicles to move toward or bind to their proper acceptor vesicles. Certainly much information about the important proteins required for these vesicle-targeting reactions remains to be discovered.

▶ *Golgi and Post-Golgi Sorting and Processing of Membrane and Secretory Proteins*

The previous section dealt with the mechanisms by which transport vesicles bud off from donor organelles and bind to and fuse with appropriate acceptor membranes. Now we return to a fundamental question of protein sorting:

what are the signals on membrane and secretory proteins themselves that target them to their correct destinations? We shall consider in this section the sorting of enzymes to the correct Golgi membrane; the sorting of secretory proteins to regulated and continuously exocytosed vesicles (Figure 16-8); and the specific proteolytic cleavages that occur in these vesicles. The next section considers the sorting of proteins to the apical or basolateral domain of polarized epithelial cells and the sorting of proteins following receptor-mediated endocytosis from the cell surface. Our focus will be on determining the sequences in a particular protein that direct it to its destination and keep it there. We have already seen how a phosphorylated carbohydrate acts as a sorting sequence for lysosomal enzymes (Figure 16-37), and, in general, how sequences of amino acids in the cytosolic domains of membrane proteins cause them to be incorporated into coated vesicles (Figure 16-39b). We shall see that other signal sequences that target proteins are often relatively short stretches of amino acids ($\approx$10–20) that can be found in different parts of the protein, depending on the type of sorting event.

Sequences in the Membrane-Spanning Domain Cause the Retention of Proteins in the Golgi

All known Golgi enzymes are inserted into the ER membrane and move by transport vesicles to the Golgi, where they remain. Why do these enzymes stay in the Golgi and not move on to the plasma membrane as do the plasma membrane proteins? The single membrane-spanning α helix in each of these proteins is the *signal sequence* that causes its retention in the Golgi.

All Golgi-localized carbohydrate-modifying enzymes, such as galactosyltransferase and sialyltransferase, have a similar structure: a short N-terminal domain that faces the cytosol, a single transmembrane α helix, and a large C-terminal domain that faces the Golgi lumen and that contains the catalytic site (see Figure 16-17). There are no amino acid sequences common to the transmembrane segments of these Golgi enzymes. Nonetheless, several experiments have shown that the transmembrane α helixes of these proteins are both necessary and sufficient to cause them to remain in the Golgi. Thus, if the membrane-spanning α helix of the transferrin receptor (a plasma membrane protein) is replaced by that of the Golgi enzyme galactosyltransferase, the resultant chimeric protein, when expressed in cultured cells, is localized to the Golgi. Conversely, if the membrane-spanning α helix of galactosyltransferase is replaced by that of the transferrin receptor, the resultant chimeric protein is found on the plasma membrane, not in the Golgi.

How the membrane-spanning α helix itself causes Golgi localization is not known. One popular theory is that the membrane-spanning α helixes of several Golgi enzymes bind specifically together within the phospholipid bilayer, in much the same way that the 11 membrane-spanning α helices of the photosynthetic reaction center form very specific and stable interactions with each other (see Figure 14-17). The resulting complex of many Golgi enzymes might bind to cytoskeletal proteins, such as actin or tubulin, or in some other way be prevented from diffusing into the transport vesicles that would otherwise take them to the cell surface.

Different Vesicles Are Used for Continuous and Regulated Protein Secretion

All eukaryotic cells transport newly made plasma membrane glycoproteins and certain secretory proteins from the *trans*-Golgi, via nonclathrin-coated vesicles, to the plasma membrane, where they are continuously exocytosed. In contrast, certain specialized secretory cells store some secretory proteins in vesicles and secrete them only when triggered by a specific stimulus, the process known as *regulated secretion*. As one example, the pancreatic β-islet cells store newly made insulin in special secretory vesicles and secrete insulin only in response to a specific external signal—in this case an elevation in blood glucose. The exocrine pancreatic acinar cells similarly store a set of newly made digestive enzymes, such as ribonuclease, amylase, and trypsinogen (see Figure 5-40), and secrete them only in response to stimulation by specific hormones. These and many other cells simultaneously utilize two different classes of secretory vesicles, one for regulated protein secretion and the other for continuous secretion (see Figure 16-8). How are the two different types of secretory proteins, all of which are soluble in the lumen of the *trans*-Golgi reticulum, sorted to the correct type of secretory vesicle?

Experiments show that different proteins sorted to regulated secretory vesicles in different types of cells do share a common sorting mechanism. Consider proteins as diverse as ACTH (adrenocorticotropic hormone, normally made in pituitary cells), insulin, and trypsinogen. When recombinant DNA techniques are used to induce the synthesis of insulin and trypsinogen in pituitary tumor cells already synthesizing ACTH, all three proteins segregate in the same regulated secretory vesicles and are secreted together when a hormone binds to a receptor on the pituitary cells and causes a rise in cytosolic Ca^{2+}. Although these three proteins have no identical amino acid sequences that might serve as a sorting sequence, they obviously do share some feature that provides a sorting signal for their incorporation into the regulated secretory vesicles.

Morphologic evidence suggests that sorting into the regulated pathway is controlled by selective protein aggregation. Immature vesicles of this pathway—those with a clathrin coat that have just budded from the *trans*-Golgi reticulum—contain a core, visible in the electron micro-

(a)

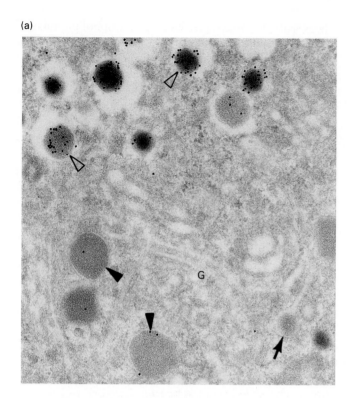

(b)

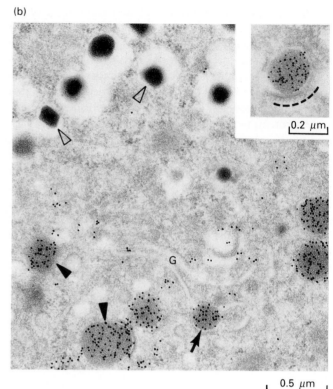

0.5 μm

▲ FIGURE 16-44   Intravesicular cleavage of proinsulin to form insulin. Proinsulin is packaged into clathrin-coated secretory vesicles before it is cleaved to insulin. Serial sections of the Golgi region of an insulin-secreting cell are stained with (a) a monoclonal antibody that detects insulin (and not proinsulin), and (b) an antibody that recognizes proinsulin (and not insulin). The antibodies are bound to electron-opaque gold particles and appear as dark dots in electron micrographs. Mature vesicles (open arrowheads) have a dense core and stain intensely with the antiinsulin antibody (a) but not with antiproinsulin (b). In contrast, immature secretory vesicles (closed arrowheads) and vesicles budding from the *trans* Golgi (arrows) contain less dense cores and stain with the proinsulin antibody (b) but not with antiinsulin (a). Also, the Golgi complex (G) is stained with antiproinsulin but not with antiinsulin. The inset in (b) shows the clathrin coat (dashed line) on a proinsulin-rich secretory vesicle. Since immature secretory vesicles contain proinsulin (not insulin), the proteolytic conversion of proinsulin to insulin must take place after the proinsulin is transported from the *trans*-Golgi reticulum to these vesicles. Note that immature vesicles budding from the *trans* Golgi (arrows) contain aggregates that include proinsulin and other regulated secretory proteins. [From L. Orci et al. 1987. *Cell* **49**:865–868. Electron micrographs courtesy of L. Orci.]

scope (see the arrows in Figure 16-44b), that consists of an aggregate of secretory proteins. It is significant that these aggregates are found in vesicles that are in the process of budding from the *trans*-Golgi reticulum, since it indicates that proteins destined for regulated secretory vesicles selectively aggregate together *prior* to their incorporation into the vesicles. Indeed, regulated secretory vesicles from many mammalian cells contain two proteins, chromogranin B and secretogranin II, that together form aggregates when incubated at the ionic conditions—pH ≈6.5 and 1 mM Ca^{2+}—that are thought to occur in the *trans*-Golgi reticulum; such aggregates will not form at the neutral pH of the ER. In cells, these aggregates include ACTH, insulin, or any other proteins that, depending on the cell type, are incorporated into regulated secretory vesicles, and exclude proteins that are normally incorporated into continuous,

not regulated, secretory vesicles. Thus, the selective aggregation of regulated secretory proteins together with chromogranin B or secretogranin II could be the basis of sorting these proteins into regulated secretory vesicles; those not incorporated into the aggregates would be secreted continuously by default.

Secretory and Membrane Proteins Undergo Several Proteolytic Cleavages During the Late Maturation Stages

For some secretory proteins—such as growth hormone—and certain viral membrane proteins, such as the VSV glycoprotein, removal of the N-terminal signal sequence of amino acids from the nascent chain is the only known proteolytic cleavage required to convert the polypeptide directly to the mature, active species. However, some plasma membrane and most secretory proteins go through an additional, intermediate stage—a relatively long-lived intracellular form, termed the *proprotein* (or, for future hormones, the *prohormone*), that requires further proteolytic processing to mature. Serum proteins such as albumin, hormones such as insulin and glucagon (Figure 16-45) and the yeast α mating factor (Figure 20-39), and membrane proteins such as the HA glycoprotein (see Figure 16-24) are synthesized as longer, inactive precursors, from which specific polypeptides are cleaved to generate the mature, active molecule.

In general, the proteolytic conversion of the proprotein to a mature molecule occurs in secretory vesicles, and involves an endoproteolytic cut at a site C-terminal to (i.e., after) dibasic amino acid recognition sequences such as Arg-Arg or Lys-Arg. Additional amino acids can be cleaved at the N-terminus or at both ends of the proprotein (see Figure 16-45). In proinsulin, the extra amino acids, collectively termed the *C peptide*, are located internally in the polypeptide; the N-terminal B chain and the C-terminal A chain of mature insulin are linked by disulfide bonds and remain attached when the C peptide is removed (see Figure 3-18).

The identification and purification of the proteases responsible for this maturational processing proved impossible for many years. The breakthrough came from the analysis of a yeast mutant, termed kex2, that synthesized the precursor of the α mating factor but could not process it to the functional hormone, so that the cells were defective in mating (see Figure 20-39). The product of the wild-type *KEX2* gene is the endoprotease that cleaves the α factor's precursor at a site C-terminal to Arg-Arg and Lys-Arg residues. Using the *KEX2* gene as a DNA probe, workers were able to clone genes of related sequence from mammalian cells. The mammalian genes turned out to encode a family

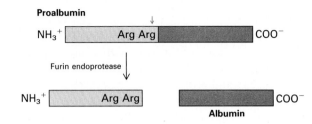
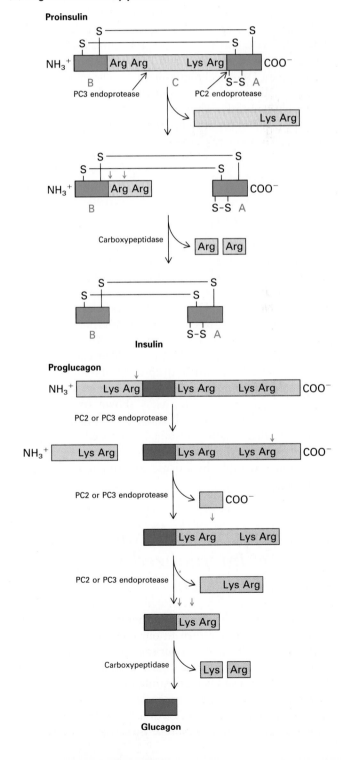

▶ FIGURE 16-45   Proteolytic processing of typical proproteins secreted in the continuous or regulated pathways. (a) The endoprotease furin, found in continuously secreted vesicles, converts proalbumin to albumin by a single cleavage on the C-terminal side of a sequence of two consecutive basic amino acids. (b) Two endoproteases, PC2 and PC3, both of which cleave after two basic amino acids, are localized to regulated secretory vesicles. Two cleavages of proinsulin, one by PC2 and the other by PC3, release the C peptide. The B chain is processed further by the action of a carboxypeptidase that sequentially removes the two arginines at the C-terminus. The processing of proglucagon to glucagon involves three successive cleavages by PC2 or PC3 endoproteases. Finally, a carboxypeptidase removes the two C-terminal basic amino acids from the last intermediate to form mature glucagon. [See J. T. Potts et al. 1980. *Ann. NY Acad. Sci.* **343**:38; L. C. Lopez et al. 1983. *Proc. Natl. Acad. Sci. USA* **80**:5485; P. J. Barr, 1991. *Cell* **66**:1–3; and D. Steiner et al. 1992. *J. Biol. Chem.* **267**:23435–23438.]

of endoproteases, all of which cut after an Arg-Arg or Lys-Arg sequence (Figure 16-45). One, called *furin,* is found in all mammalian cells; it processes proteins such as albumin that are secreted by the continuous pathway. Two, termed PC2 and PC3, are found only in cells that have the regulated secretory pathway: they are localized to regulated secretory vesicles and proteolytically cleave the pro- forms of many hormones at specific sites.

Despite their similar cleavage sites the various endoproteases are not equivalent in their action. For instance, PC2 and PC3 catalyze different cleavages in the conversion of proinsulin to insulin (Figure 16-45). Furthermore, if the gene encoding proinsulin is expressed in cultured fibroblasts that contain furin, then proinsulin, not insulin, is secreted. However, if PC2 and PC3 are also expressed by recombinant DNA techniques in these cells, then fully processed insulin is secreted, even though the protein is secreted continuously. Thus PC2 and PC3 are essential for the processing of proproteins normally secreted by the regulated pathway.

The Proteolytic Maturation of Insulin Occurs in Acidic, Clathrin-Coated Secretory Vesicles

Normally, in the forming of mature secretory vesicles several immature ones fuse together; the proteolytic cleavage of the propeptides occurs during this maturation process. That the conversion of proinsulin to insulin indeed occurs in newly formed regulated secretory vesicles is shown by the electron micrographs in Figure 16-44. An antibody specific for proinsulin (and not for insulin) detects proinsulin in the Golgi vesicles and in the cores of clathrin-coated secretory vesicles that are in the process of budding or have just budded from the *trans* Golgi. These immature vesicles contain no mature insulin, as is shown by an insulin-specific monoclonal antibody that does not react with proinsulin. Rather, mature insulin is localized to the more mature secretory vesicles, which can be distinguished in the electron micrographs by a dense core of almost crystalline insulin.

► Sorting of Membrane Proteins Internalized from the Cell Surface

Having followed the sorting and maturation of secretory proteins, let us now return to our consideration of membrane proteins. Thus far in our saga of the secretory pathway we have seen how newly made plasma membrane proteins are transported from the rough ER through the Golgi complex and then, via continuously secreted transport vesicles, to the cell surface. Here we continue our discussion of the sorting of plasma membrane proteins, focusing on the different fates of proteins following receptor-mediated endocytosis from the cell surface and on the sorting of proteins to the apical or basolateral domains (see Figure 14-42) of polarized epithelial cells. Clathrin-coated vesicles play key roles in these processes, and thus we include some recent information concerning the ways that different membrane proteins are selected for incorporation into different types of coated vesicles.

In Receptor-Mediated Endocytosis, Cell Surface Receptors Are Internalized in Clathrin-Coated Vesicles

Endocytosis is the process in which a small region of the plasma membrane invaginates to form a new intracellular membrane-limited vesicle about 0.05 to 0.1 μm in diameter. It is similar to the processes, discussed above, whereby transport vesicles bud off from the ER or Golgi.

There are two types of endocytosis. In *pinocytosis,* endocytic vesicles nonspecifically take up small droplets of extracellular fluid and any material dissolved in it. In *receptor-mediated endocytosis,* a specific receptor on the surface of the membrane binds tightly to the extracellular macromolecule that it recognizes, called the *ligand;* the plasma membrane region containing the receptor-ligand complex then undergoes endocytosis, becoming a transport vesicle. In this process, the rate of endocytosis of a ligand is limited by the amount of its corresponding cell surface receptor, and receptor-ligand complexes are selectively incorporated into the intracellular transport vesicles; most other plasma membrane proteins are excluded. Vertebrate cells bear many types of receptors on the cell surface that bind specific ligands tightly and with a high degree of specificity (Table 16-4). One example that we have already noted is the uptake of lysosomal enzymes by cell surface mannose 6-phosphate receptors (Figure 16-37).

Receptor-mediated endocytosis occurs via *clathrin-coated pits* and vesicles (Figure 16-46). We have already noted that clathrin is visible in electron micrographs as a proteinaceous layer coating the cytosolic side of pits that form from the *trans* Golgi or from the plasma membrane (Figure 16-46; also see Figures 16-38 and 16-39). Clathrin-coated pits make up about 2 percent of the surface of cells such as hepatocytes and fibroblasts.

Some receptors are clustered over clathrin-coated pits even in the absence of ligand. Other receptors diffuse freely in the plane of the plasma membrane but undergo a conformational change when binding to ligand, so that when the receptor-ligand complex diffuses into a clathrin-coated pit, it is retained there. Two or more types of receptor-bound ligands, such as LDL and transferrin, which are discussed below, can be seen in the same coated pit or vesicle.

As noted above, the spontaneous polymerization of clathrin triskelions is thought to cause the pits to expand and eventually to form clathrin-coated vesicles. These vesicles lose their coats after endocytosis, forming smooth-

TABLE 16-4   Proteins and Particles Taken Up by Receptor-Mediated Endocytosis in Animal Cells

Ligand	Function of Receptor-Ligand Complex	Cell Type
Low-density lipoproteins (LDLs)	Supply cholesterol	Most
Transferrin	Supplies iron	Most
Glucose- or mannose-terminal glycoproteins	Remove injurious agents from circulation	Macrophage
Galactose-terminal glycoproteins	Remove injurious agents from circulation	Hepatocyte
IgG Immuno-globulins	Transfer immunity to fetus	Fetal yolk sac; intestinal epithelial cells of neonatal animals
Phosphovitel-logenins	Supplies protein to embryo	Developing oocyte
Fibrin	Removes injurious agents	Epithelial cells
Insulin, other peptide hormones	Alter cellular metabolism; ligand and often receptor are degraded after endocytosis	Most

(a)

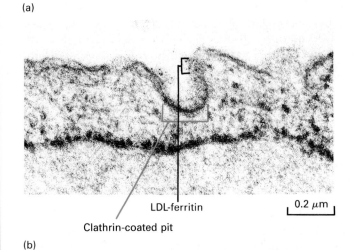

LDL-ferritin

Clathrin-coated pit

0.2 μm

(b)

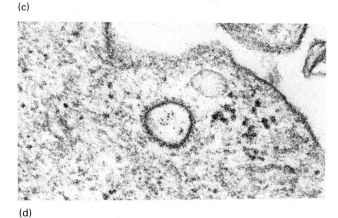

LDL-ferritin

(c)

(d)

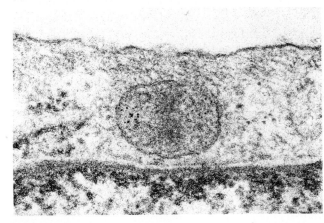

➤ FIGURE 16-46   The initial stages of receptor-mediated endocytosis of low-density lipoprotein (LDL) particles by cultured human fibroblasts, revealed by electron microscopy. (a) A coated pit, showing the clathrin coat on the inner (cytosolic) surface of the pit. The small dots over the pit are LDL particles covalently linked to the iron-containing protein ferritin. Each small iron particle in ferritin is visible under the electron microscope. (b) A pit containing LDL apparently closing on itself to form a coated vesicle. (c) A coated vesicle containing ferritin-tagged LDL particles. (d) Ferritin-tagged LDL particles in a smooth-surfaced early endosome 6 min after being added to cells. [Photographs courtesy of R. Anderson. Reprinted by permission from J. Goldstein, R. Anderson, and M. S. Brown. *Nature* **279**:679. Copyright 1979, Macmillan Journals Limited. See also M. S. Brown and J. Goldstein. 1986. *Science* **232**:34.]

surfaced vesicles called *early endosomes*. As with the mannose 6-phosphate receptor (Figure 16-37), endocytosed receptors are usually recycled intact to their point of origin, be it the plasma membrane or the *trans*-Golgi reticulum. Receptor endocytosis and recycling which is well understood (at least in outline), involves a number of types of intracellular vesicles (see Figure 16-37) which we will now discuss.

The Low-Density Lipoprotein (LDL) Receptor Binds and Internalizes Cholesterol-Containing Particles

Many endocytosed ligands have been observed in clathrin-coated pits and vesicles, and researchers believe that these membrane structures function as intermediates in the endocytosis of most (though not all) ligands bound to cell surface receptors. In most cases the receptor and ligand dissociate after endocytosis; the receptor returns to the cell surface and the ligand is transported to lysosomes where it is degraded. A typical and intensively studied example of this process involves the receptor for *low-density lipoprotein* (LDL). LDL is one of a variety of complexes that carry cholesterol through the bloodstream. Most cells manufacture receptors that specifically bind LDL, and utilize the cholesterol from endocytosed LDL particles for incorporation into membranes. In adrenal cortical cells cholesterol is also converted to steroid hormones and, in hepatocytes, to bile acids.

Cholesterol is insoluble in body fluids: whether it is ingested in foodstuffs or synthesized in the liver, it must be transported by a water-soluble carrier. The LDL particle (Figure 16-47) is a sphere 20–25 nm in diameter. Its outer surface is a monolayer membrane of phospholipids and cholesterol, in which one molecule of a very large (4563-amino-acids-long), protein called *apo-B* is embedded (see Figure 12-57). Inside is an extremely nonpolar core of cholesterol, all of which is esterified through the single hydroxyl group to a long-chain fatty acid. After endocytosis, the LDL particles are transported to lysosomes (Figure 16-48), where the apo-B protein is degraded to amino acids and the cholesterol esters are hydrolyzed to cholesterol and fatty acids. The cholesterol is incorporated directly into cell membranes or is re-esterified and stored as lipid droplets in the cell for later use; the fatty acids are used to make new phospholipids or triglycerides.

The LDL receptor is a single-chain glycoprotein of 839 amino acids. A sequence of 22 hydrophobic amino acids spans the plasma membrane once, presumably as an α helix. About 50 amino acids at the C-terminus face the cytosol and some of these are involved in binding the receptor to the coated pits. Within the 767 amino acids on the exoplasmic face, the most striking feature is an *N*-terminal segment of about 320 residues that is extremely rich in disulfide-bonded cysteine residues; it includes a sevenfold repeat of a sequence of 40 amino acids that contains the LDL binding site.

Mutant LDL Receptors Reveal a Signal for Internalizing Receptors into Clathrin-Coated Pits

The human LDL system has proved invaluable for studying receptor-mediated endocytosis because mutant forms of the receptor protein are available from persons who have the inherited disorder *familial hypercholesterolemia* (FH), which is characterized by high levels of cholesterol in the blood. Persons homozygous for these mutant alleles often die at an early age from heart attacks caused by atherosclerosis, a buildup of cholesterol deposits that ultimately block the arteries.

The genetic defects specifically affect the LDL receptor. In some persons with this disorder, the LDL receptor is simply not produced; in others, it binds LDL poorly or not at all. In one especially instructive case, the mutant receptor binds LDL normally but the LDL-receptor complex cannot be internalized by the cell and is distributed evenly over the cell surface rather than confined to clathrin-coated pits. Plasma membrane receptors for other ligands are internalized normally in clathrin-coated pits, but the mutant LDL receptor apparently cannot bind to the AP-2 assembly particles (see Figure 16-39b) in the coated pit. The mutant receptor has a single tyrosine-to-cysteine change in its cytosolic domain.

This and other mutant forms of the LDL receptor have been experimentally generated and expressed in fibroblasts; analysis has shown that a sequence of four amino acid residues surrounding this tyrosine in the cytosolic domain—Asp-Pro-X-Tyr (where X can be any amino acid)—is crucial for internalization. A mutation in any of the three crucial residues in this sequence reduces or abolishes the ability of the receptor to be incorporated into clathrin-coated pits. In particular, the fourth amino acid in this sequence must be tyrosine or another aromatic amino acid (phenylalanine, tryptophan) in order for the receptor to be internalized.

Many other plasma membrane receptors that are internalized into clathrin-coated pits, such as the transferrin receptor, contain a similar amino acid sequence in their cytosolic domains, and mutagenesis studies on these receptors have confirmed that these four amino acids form a general recognition signal for binding to clathrin-coated pits, presumably via an AP-2 assembly particle. A peptide consisting of these four residues adopts, in solution, a reverse β-turn conformation, and the AP-2 particle may bind to the tyrosine or other aromatic amino acid in such exposed β-turns. As further evidence for the importance of this sequence, a plasma membrane protein that is not

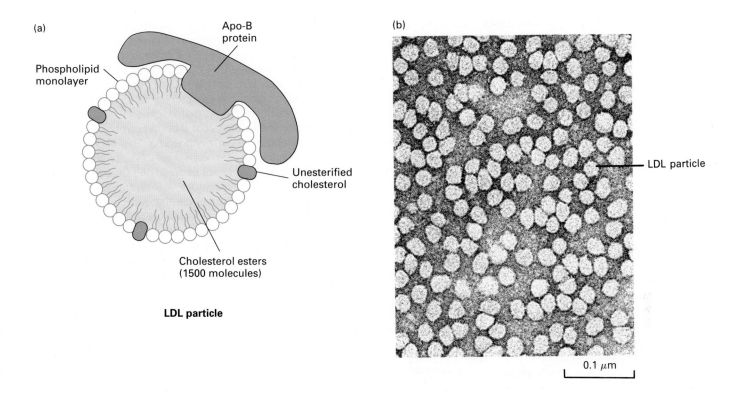

▲ FIGURE 16-47   Structure of an LDL particle. (a) The surface of the LDL particle is a monolayer of phospholipid and unesterified cholesterol. The hydrophobic core of the particle is rich in fatty acid esters of cholesterol. One copy of the hydrophobic apo-B protein is embedded in the membrane. (b) Electron micrograph of a negatively stained preparation of LDL particles. (c) Esters of cholesterol with fatty acids, mainly linoleic acid, are found in the interior of LDL particles. The lysosomal enzyme cholesterol esterase hydrolyzes the ester bond, forming a free fatty acid and cholesterol. [See R. Anderson. 1979. *Nature* **279**:679. Part (b) courtesy of R. Anderson. Reprinted by permission from Nature. Copyright 1979, Macmillan Journals Limited.]

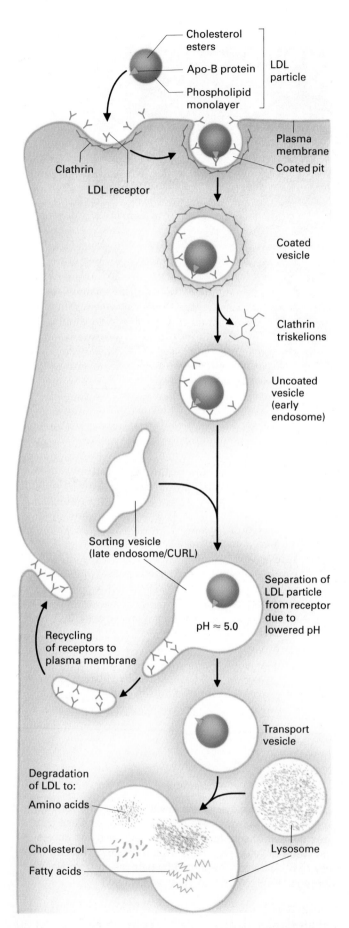

◄ FIGURE 16-48   Fate of LDL and its receptor after endocytosis. The same pathway is followed by other ligands, such as insulin and other protein hormones, that are internalized by receptor-mediated endocytosis and degraded in the lysosome. After an LDL particle binds to an LDL receptor on the plasma membrane, the receptor-ligand complex is internalized in a clathrin-coated pit that pinches off to become a coated vesicle. The clathrin coat then depolymerizes to triskelions, resulting in an early endosome. This endosome fuses with a sorting vesicle, known as CURL or late endosome, where the low pH (≈5) causes the LDL particles to dissociate from the LDL receptors. A receptor-rich region buds off to form a separate vesicle that recycles the LDL receptors back to the plasma membrane. A vesicle containing an LDL particle may fuse with another late endosome but ultimately fuses with a lysosome to form a larger lysosome. There, the apo-B protein of the LDL particle is degraded to amino acids and the cholesterol esters are hydrolyzed to fatty acids and cholesterol. Cholesterol is incorporated into cell membranes. Abundant imported cholesterol inhibits synthesis by the cell of both cholesterol and LDL receptor protein.

normally internalized into clathrin-coated pits can be made to internalize if these four amino acids are added to its cytosolic domain. The influenza HA protein (Figure 16-24) is one such protein; a mutant HA protein, genetically engineered to contain such a four-amino-acid recognition sequence in its cytosolic domain, is internalized into clathrin-coated pits.

Receptors and Ligands Dissociate in an Acidic Late Endosome/CURL Organelle

The overall rate of endocytotic internalization of the plasma membrane is quite high; cultured fibroblasts regularly internalize 50 percent of their cell surface proteins and phospholipids each hour. Most cell surface receptors that undergo endocytosis will repeatedly deposit their ligands within the cell and then recycle to the plasma membrane, once again to mediate the internalization of ligand molecules. For instance, the LDL receptor makes one round trip into and out of the cell every 10–20 min, for a total of several hundred trips in its 20-hour life span. How, then, does the internalized receptor release its ligand and get recycled to the cell surface?

Like the mannose 6-phosphate receptor (see Figure 16-37), the LDL and other cell surface receptors dissociate from their ligands within the acidic late endosome/CURL organelles (see Figure 16-48). These are spherical vesicles with tubular branching membranes and are found a few micrometers from the cell surface. The original experiments that defined this organelle (Figure 16-49) utilized the *asialoglycoprotein receptor*, a liver-specific protein that mediates the binding and internalization of abnormal gly-

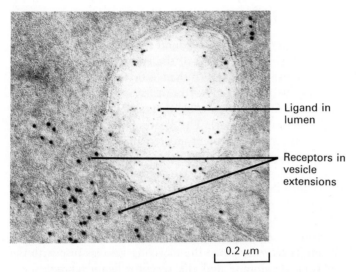

Ligand in
lumen

Receptors in
vesicle
extensions

0.2 µm

▲ FIGURE 16-49   An electron micrograph of the sorting
vesicle (late endosome/CURL organelle). An asialoglycoprotein receptor and its ligand, a galactose-terminal glycoprotein,
are both localized simultaneously in a hepatocyte CURL after
receptor-mediated endocytosis. The larger dark grains localize
the receptor in the tubular extensions budding off from the
vesicle. The smaller dark grains localize the asialoglycoprotein
ligand in the vesicle lumen. In this experiment, liver cells
were perfused with a galactose-terminal glycoprotein and
then were fixed and sectioned for electron microscopy. To
localize the receptor, the sections were stained with antibodies specific for the receptor, which were tagged with gold
particles 8 nm in diameter (the large electron-opaque grains
in the micrograph). Localization of the ligand was detected
by using galactose-terminal glycoprotein antibody linked to
gold particles 5 nm in diameter. [Courtesy of H. J. Geuze.
Copyright 1983, M.I.T. See Geuze, H. J., et al. 1983. *Cell*
32:277.]

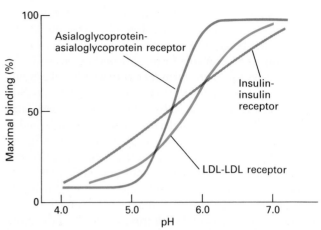

▲ FIGURE 16-50   Effect of pH on the binding of proteins
to their respective receptors. For all three of these receptor-
ligand complexes, the internalized ligands are transported to
lysosomes, where they are degraded, and the receptors are
recycled. Dissociation of receptor and ligand is believed to be
triggered by the low pH in the late endosome/CURL vesicle.
[See A. Dautry-Varsat, A. Ciechanover, and H. F. Lodish.
1983. *Proc. Natl. Acad. Sci. USA* **80**:2258.]

coproteins whose oligosaccharides terminate in galactose
rather than the normal sialic acid (hence the name
*asialo*glycoprotein; see Figure 16-27).

Between 5 and 10 min after internalization both the
receptors and their ligands accumulate in late endosome/
CURL vesicles. Electron microscopy localizes the ligand to
the vesicle lumen (Figure 16-49), where presumably it is no
longer bound to its receptor. Indeed, the spherical part of
these vesicles contains little receptor at all. In contrast, the
tubular membranes attached to the spherical vesicle are
rich in receptor and rarely contain ligand. Thus these membranes contain receptor that has dissociated from ligand,
and the late endosome/CURL must be the organelle in
which receptors uncouple from their ligand.

Beginning 15 min after internalization, ligands are
transferred to lysosomes, but the intact receptors themselves are never found in these organelles. The receptor-
rich elongated membrane vesicles that bud from the CURL
vesicle mediate the recycling of receptors back to the cell
surface. The spherical part of the CURL vesicles eventually
buds off transport vesicles that, with their cargo of ligand,
soon fuse with lysosomes. The LDL receptor, for example,
is never directed to a lysosome—to be degraded by potent
lysosome proteases—until it becomes damaged in some
way.

Why do receptors release their ligands in the late endo-
some/CURL? Late endosome/CURL vesicles have an internal pH of 5.0 to 5.5, compared to the lysosomal pH of
≈4.5–5.0. Endosomes have been partially purified and
contain a V-class ATP-dependent proton pump (see Figure
15-10). Clathrin-coated vesicles contain a similar proton
pump, and thus ligands encounter a progressively lower
pH as they move from coated vesicles through the various
early and late endosomes to lysosomes. Late endosome/
CURLs and coated vesicles also contain a Cl⁻ channel, allowing the proton pump to generate a concentration gradient of H⁺ ions, rather than the transmembrane electric potential that would form if only protons were transferred
from the cytosol to the vesicle lumen (page 650). The acidity in the late endosomes explains how receptors dissociate
from their tightly bound ligands: most receptors, including
the asialoglycoprotein, insulin, and LDL receptors, bind
their ligands tightly at neutral pH but release their ligands
if the pH is lowered to below 5.0, the condition found in
late endosomes (Figure 16-50).

Transferrin Delivers Iron to Cells by
Receptor-Mediated Endocytosis

Perhaps the best example of how changes in pH mediate
the sorting of receptors and ligands is the behavior of

transferrin, a major serum glycoprotein that transports iron to all tissue cells from the liver (the main site of iron storage in the body) and from the intestine (the site of iron absorption). The iron-free form, *apotransferrin*, binds two Fe^{3+} ions very tightly to form *ferrotransferrin*. All growing cells contain surface *transferrin receptors* that avidly bind ferrotransferrin ($K_m = 6 \times 10^{-9}$ M) at neutral pH, after which the receptor-bound ferrotransferrin is subjected to endocytosis. Like the components of LDL, the two bound Fe^{3+} atoms remain in the cell, but there the similarity with the fate of other endocytosed ligands, including LDL, ends: the apotransferrin part of the ligand is secreted from the cell within minutes, carried in the bloodstream to the liver or intestine, and reloaded with iron. Figure 16-51 summarizes the pathway.

What properties of this receptor-ligand complex cause the cell to retain the iron portion of the ligand and secrete the apotransferrin portion? The answer again lies in changes in pH—more specifically, in the unique ability of apotransferrin to remain bound to the transferrin receptor at the low pH (5.0–5.5) of the endocytic vesicles (Figure 16-51). At a pH of less than 6.0, the two bound Fe^{3+} atoms dissociate from ferrotransferrin and are transported from the late endosome/CURL vesicle into the cytosol (in an unknown manner). The apotransferrin formed by the dissociation of the iron atoms remains bound to the transferrin receptor at the pH of these vesicles. The transferrin receptor is recycled back to the surface (like the LDL receptor) with the apotransferrin still bound to it. Remarkably, although apotransferrin binds tightly to its receptor ($K_m = 6 \times 10^{-9}$ M) at a pH of 5.0 or 6.0, it does not bind measurably at neutral pH. Hence the apotransferrin dissociates from its receptor when the recycling vesicles fuse with the plasma membrane and the receptor-ligand complex encounters the neutral pH of the extracellular interstitial fluid or growth medium. The surface receptor is then free to bind another molecule of ferrotransferrin.

Some Proteins Internalized by Endocytosis Remain within the Cell, or Are Transported across the Cell and Secreted

In several receptor-ligand systems, found mainly in oocytes (egg cells), endocytosed material simply remains in the cells and is minimally processed. Developing insect and avian oocytes, for example, internalize yolk proteins and other proteins from the blood or surrounding cells. (Coated pits were first discovered in insect eggs, where they occupy a large portion of the plasma membrane.) A hen's egg is a single cell containing several grams of protein, virtually all of which is imported from the bloodstream by endocytosis. Vitellogenin, a precursor of yolk proteins (principally of lipovitellin and phosvitin), is synthesized by the liver and secreted into the bloodstream, from which it is endocytosed into the developing egg. Yolk proteins remain in storage granules within the egg and are used after fertilization as a source of amino acids and energy by the developing embryo. Egg-white proteins, such as ovalbumin, lyso-

Transferrin receptor — Ferrotransferrin

Apotransferrin dissociates from receptor at neutral pH

Exterior (pH 7.0)

Clathrin

pH ~ 6.0

Apotransferrin is recycled to cell surface

pH in vesicle is decreased by H^+ ATPase

pH ~ 5.0

Apotransferrin

Fe^{3+} Fe^{3+} ← Fe^{3+} Fe^{3+}
Fe^{3+} Fe^{3+} ← Fe^{3+} Fe^{3+}

CURL

To cytosol

Low pH causes release of Fe^{3+} from ligand; ligand remains bound to receptor

◄ FIGURE 16-51 The transferrin cycle that operates in all growing mammalian cells. After endocytosis, iron is released from the transferrin-receptor complex in the acidic late endosome/CURL compartment. The iron dissociates and remains in the cell. The transferrin protein remains bound to its receptor at this pH, and they cycle to the cell surface together. When the receptor-transferrin complex encounters the neutral pH of the exterior medium, the iron-free transferrin is released to gather its next load of iron. [See A. Ciechanover et al. 1983. *J. Biol. Chem.* **258**:9681.]

zyme, and conalbumin, are secreted by cells lining the hen oviduct, and are also endocytosed by the egg cell.

In *transcytosis*, or *transcellular transport*, some endocytosed material passes all the way through the cells and is exocytosed (secreted) from the plasma membrane at the opposite side. Transcytosis occurs mainly in sheets of polarized epithelial cells (see Figure 14-42). An example of transcytosis is the movement of maternal immunoglobulins across mammalian yolk-sac cells into the fetus and across the intestinal epithelial cells of the newborn mouse (Figure 16-52). The F_c receptor that mediates the transcytosis of immunoglobulins has the property of binding to its ligand at an acidic pH of 6 but not at neutral pH. Figure 16-52 shows how a difference in the pH of the extracellular media on the two sides of intestinal epithelial cells in new-

born mice allows immunoglobulins to move in one direction—from the lumen to the blood.

Proteins Are Sorted in Several Different Ways to Different Domains of the Plasma Membrane

In polarized epithelial cells the plasma membrane is divided into two surfaces, apical and basolateral, separated by tight junctions (see Figures 14-42 and 14-43); the surfaces contain different species of proteins. Transcytosis is only one of several processes by which membrane proteins are sorted to either the apical or basolateral domains. Multiple mechanisms ensure that all plasma membrane proteins will be localized correctly to the apical or basolateral domain; any one protein may be targeted by more than one mechanism. We understand in outline the mechanisms by which plasma membrane proteins are sorted in polarized cells, but the molecular mechanisms underlying the selective movements of these membrane proteins are not yet known.

Direct Sorting from the Golgi Complex to the Apical or Basolateral Plasma Membrane One mechanism for targeting proteins to the appropriate surface of the plasma membrane involves sorting in the *trans*-Golgi reticulum. A variety of microscopic and cell-fractionation studies all indicate that proteins destined for the apical and basolateral membranes are found together in the same *trans*-Golgi reticulum membranes. However, two types of vesicles, containing different membrane proteins, bud from the *trans*-Golgi reticulum; one moves to and fuses with the apical plasma membrane; the other goes to the basolateral. These vesicles contain distinct Rab proteins that may target them to the appropriate plasma membrane region.

Cultured MDCK epithelial cells, which maintain distinct apical and basolateral membranes (see Figure 6-9), have been useful in investigating this aspect of protein transport. When MDCK cells are infected with the influenza virus, a membrane-enveloped virus, progeny viruses bud only from the apical membrane; on the other hand, in cells infected with VSV, another enveloped virus, the virus buds only from the basolateral membrane. This is because the HA glycoprotein of the influenza virus (Figure 16-24) is transported from the Golgi complex exclusively to the apical membrane, and the VSV glycoprotein (G protein) is transported only to the basolateral membrane (Figure 16-53). Furthermore, when uninfected cells express (by recombinant DNA techniques) only the HA viral protein, all the HA accumulates only in the apical membrane, indicating that the targeting sequence resides in the HA glycoprotein itself and not in other viral proteins produced during viral infection. Despite much effort, workers have not been able to identify a unique sorting sequence in the HA and other apical plasma membrane proteins; it may be that

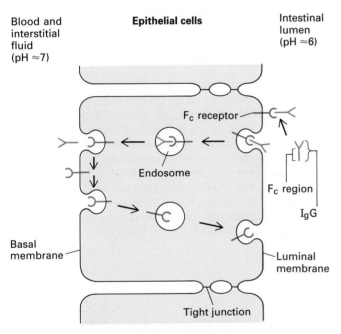

▲ FIGURE 16-52 Transcytosis. The movement of maternal immunoglobulins across the intestinal epithelial cells of newborn mice—transcytosis—involves both endocytosis and exocytosis. In newborn mice, the pH is ≈6 in the intestinal lumen and ≈7 on the opposite (blood-facing) side of the cell sheet. The particular F_c receptors on these cells bind to the F_c segment of IgG molecules only at pH values of 6 or lower, not at a pH of 7.0. Vesicles (endosomes) containing the F_c receptor–IgG complex form at the luminal surface, move across the cell, and fuse with the basal membrane, where they release the IgG. Unloaded receptors are recycled by transcytosis in the opposite direction: endosomes form from the basal membrane, move across the cell, and fuse with the luminal membrane. In other cells, proteins move by transcytosis from the basal to the apical surface of polarized cells.

Labels in figure:
Blood and interstitial fluid (pH ≈7)
Epithelial cells
Intestinal lumen (pH ≈6)
F_c receptor
Endosome
F_c region
IgG
Basal membrane
Luminal membrane
Tight junction

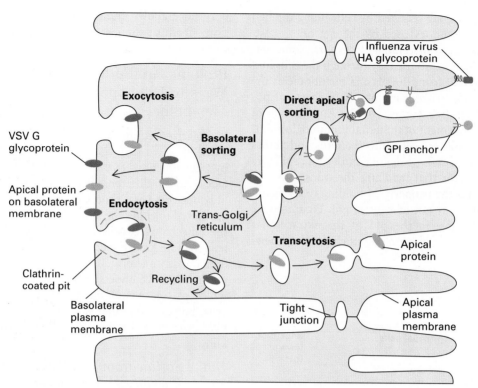

▲ FIGURE 16-53   The sorting of proteins destined for the apical and basolateral plasma membranes of epithelial cells. When a cultured line of kidney epithelial MDCK cells (see Figure 6-9) is infected simultaneously with VSV and influenza virus, the VSV glycoprotein (G protein) is found only on the basolateral membrane, whereas the HA glycoprotein of the influenza virus is found only on the apical membrane. Many proteins, such as aminopeptidase N and all proteins with a glycosylphosphatidylinositol (GPI) membrane anchor, are localized to the apical plasma membrane surface; other integral proteins such as the $Na^+K^+ATPase$ are sorted directly to the basolateral. Subcellular fractionation and immunoelectronmicroscopy show that all plasma membrane proteins are in the same rough ER and Golgi vesicles, but apical- and basolateral-directed proteins are in different post-Golgi vesicles; thus sorting to specific transport vesicles occurs within the *trans*-Golgi reticulum. In certain other polarized cells some apical and basolateral proteins are transported together to the basolateral surface, and then apical proteins move selectively, by endocytosis and transcytosis, to the apical membrane. [After K. Simons and A. Wandinger-Ness. 1990. *Cell* **62**:207–210 and K. Mostov et al. 1992. *J. Cell Biol.* **116**:577–583.]

each protein contains multiple sorting sequences, any one of which can target it to the appropriate plasma membrane domain. It is also not known how the sorting of integral plasma membrane proteins occurs in the *trans*-Golgi reticulum.

Membrane Lipids and Protein Sorting   The tight junction in polarized epithelial cells acts as a barrier against the diffusion of proteins and even lipids from the apical to the basolateral membrane. The exoplasmic leaflets of the apical and the basolateral membranes have very different lipid compositions, indicating that lipids in these leaflets cannot diffuse through tight junctions. All the glycolipid in MDCK cells, for instance, is present in the apical membrane; in fact, the exoplasmic leaflet of the apical plasma membrane contains only glycolipid and cholesterol. Phosphatidylcholine, in contrast, is present almost exclusively in the basolateral plasma membrane.

As we have seen, the addition of carbohydrate to glycolipids, like that to glycoproteins, is completed in the Golgi. It is thought that glycolipids in membranes of the *trans*-Golgi reticulum aggregate tightly with each other, probably through the hydroxyl group on the sphingosine lipid (see Figure 14-3). The resulting two-dimensional network of glycolipids is specifically incorporated into vesicles that carry proteins destined for the apical membrane. Importantly, all proteins anchored to the membrane by an attached glycosylphosphatidylinositol (GPI) molecule are also sorted into these apically targeted vesicles (Figure 16-53). Such proteins receive their GPI anchor in the ER (Figure 16-21), move to the Golgi, and then are directed to the apical plasma membrane. If any recombinant secretory protein is engineered so that it bears a GPI membrane anchor, it, too, is targeted directly to the apical plasma membrane, a result showing that the GPI anchor itself is an apical targeting signal. Thus interactions between lipids, or between lipids and proteins, rather than interactions only between proteins, may target certain proteins as well as glycolipids to the apical membrane.

Transcytosis and Protein Sorting Apical and basolateral proteins are sorted differently in hepatocytes, where the basolateral membranes face the blood as in intestinal epithelial cells, and the apical membranes form the bile canaliculus. In the hepatocytes, all newly made apical and basolateral proteins are first delivered together from the *trans*-Golgi reticulum to the basolateral membrane. From there, both basolateral and apical proteins are endocytosed in the same vesicles. Within the late endosome/CURL vesicles, their paths diverge. The endocytosed basolateral proteins are sorted into transport vesicles that recycle them to the basolateral membrane. In contrast, the apically destined endocytosed proteins are sorted into a class of transport vesicles that are transcytosed and fuse with the apical membrane (Figure 16-53).

Even in epithelial cells, such as MDCK cells, in which apical-basolateral protein sorting occurs in the Golgi, transcytosis may provide a "fail-safe" sorting mechanism: an apical protein sorted incorrectly to the basolateral membrane would be subjected to endocytosis and delivered to the apical membrane.

Cytoskeletal Attachment and Protein Sorting The attachment of integral membrane proteins to the cytoskeleton may also be involved in protein sorting, in that attachment serves as a *retention signal*. A fibrous cytoskeleton containing ankyrin and the spectrin-like protein fodrin (see Figure 22-6) is found on the cytosolic face of the basolateral membrane, but not of the apical membrane, in polarized epithelial cells. Several integral proteins that are selectively localized to the basolateral membrane, including band 3 (Figure 14-37) and the Na$^+$-K$^+$ ATPase (Figure 15-13), can bind tightly to ankyrin. Once these integral proteins are transported from the Golgi vesicles to the basolateral membrane, they become locked into the ankyrin-fodrin cytoskeleton and cannot undergo endocytosis and subsequent transcytosis to the apical membrane. However, basolateral proteins that are not locked to the cytoskeleton will be endocytosed and recycled back to the basolateral surface (Figure 16-53).

Viruses and Toxins Enter Cells by Receptor-Mediated Endocytosis

Many viruses and toxins exploit the specialized system of receptor-mediated endocytosis to enter the cytoplasm of animal cells. Most viruses and bacteria will appropriate, for use as receptors, cell surface proteins that the cell or organism surely never intended for the entry of pathogens. For instance, poliovirus binds to a protein whose real function is to bind cells to the extracellular matrix, and the receptor for certain retroviruses is a specific transporter for amino acids that is found in most animal cells. Often, the outcome is the death of the cell and, in viral and bacterial infections, the multiplication and spread of viral particles (or bacteria). A study of viral entry into cells can give us insight into the mechanism of membrane fusion.

Endocytosis and Infection by Membrane-Enveloped Viruses In many animal viruses, an outer phospholipid bilayer membrane surrounds the viral genetic material and protein coat. The virus is usually formed by budding from the plasma membrane of the host cell, and the viral membrane phospholipids are derived from the host's plasma membrane (Figure 6-18). The viral membrane also contains one or more virus-specific glycoproteins, which bind to a host's cell surface molecules to initiate infection. For example, the predominent glycoprotein of the influenza virus (which infects many types of mammalian cells) is the *HA (hemagglutinin) protein* that forms the larger spikes on the surface of the virus; the HA spikes bind to sialic acid, a sugar linked to cell surface proteins and lipids (Figure 16-27), initiating virus infection.

After it binds to a cell surface protein, the virus particle is endocytosed into an acidic endosome. At this stage, the viral and endosomal membranes fuse, releasing the genetic material from the inside of the virus into the cytosol of the host cell and initiating replication of the virus. How the membrane of the influenza virus fuses with the host-cell endosomal membrane has been studied in some detail and is likely to be the first membrane fusion system that is completely understood.

There is considerable evidence that the low pH within the enclosing endosome triggers the fusion of its membrane with the viral membrane. For instance, viral infection is inhibited by the addition of lipid-soluble bases, such as ammonia or trimethylamine, which raise the pH of normally acidic endosomes. Also, a conformational change in the HA spike occurs over a very narrow change in pH (5.0–5.5) and is critical for infectivity.

Fusion of Membranes Catalyzed by the Influenza HA Proteins The molecular events of the membrane fusion process are becoming understood. HA is synthesized (see Figure 16-24) as a precursor, HA$_0$; that is cleaved, probably by the endoprotease furin, into an HA$_1$ and an HA$_2$ subunit in the Golgi. Thus each HA spike protein consists of three HA$_1$ and three HA$_2$ subunits. The N-terminus of HA$_2$, generated by the proteolytic cleavage, constitutes the *fusion peptide*—a strongly hydrophobic amino acid sequence, Glu-Leu-Phe-Gly-Ala-Ile-Ala-Gly-Phe-Ile-Glu. At a pH of 7.0 the N-terminus of HA$_2$ is tucked into a crevice in the spike (Figure 16-54a). A pH of 5.0 induces a conformational change that disrupts the interactions between the fusion peptide and the rest of the spike. This induces several other conformational changes. First, the three globular HA$_1$ domains at the tip of the spike (blue in Figure 16-54) separate from one another. Second,

(a)

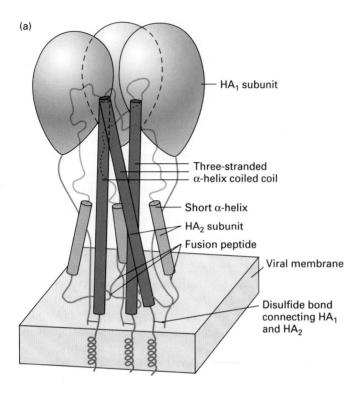

HA₁ subunit

Three-stranded α-helix coiled coil

Short α-helix

HA₂ subunit

Fusion peptide

Viral membrane

Disulfide bond connecting HA₁ and HA₂

◄ FIGURE 16-54   Membrane fusion directed by the influenza HA protein. A major conformational change of the influenza virus hemagglutinin HA spike occurs at pH ≈5.0, as shown by these schematic representations of the structure of the molecule (a) at pH = 7 and (b) at pH = 5. (a) Each globular domain (blue) at the tip of the native spike is composed of part of one HA₁ chain, and each HA₁ domain is linked to one HA₂ subunit by a disulfide bond at the base of the molecule near the viral membrane. Each HA₂ subunit contains a fusion (purple) peptide at its N-terminus, followed by a short (≈16-aa) α helix (light green), a ≈25-aa nonhelical loop (orange), and a long (≈47-aa) α-helix (dark green). The long α helices from the three HA₂ subunits form a three-stranded coiled coil. (b) At pH 5.0 the binding of the fusion peptide to other segments of HA₂ is disrupted. This induces major structural rearrangements in the protein: The three HA₁ domains separate from each other but remain tethered to the HA₂ subunits by the disulfide bonds at the base of the molecule. Each ≈25-aa loop (orange) in HA₂ rearranges into an α helix and combines with the short and long α helices of HA₂ to form a continuous 88-aa α helix. The three 88-aa helices form a three-stranded coiled coil 13.5 nm long that protrudes outward from the viral membrane. The fusion peptides, at the tip of the coiled coil, insert into the endocytic vesicle membrane; this triggers fusion of the viral and endosomal membranes. [Courtesy of Chavela Carr. Adapted from C. M. Carr and P. Kim. 1993. *Cell* **73**:823–832 and J. M. White and I. A. Wilson. 1987. *J. Cell Biol.* **105**:2887– 2896.]

(b)

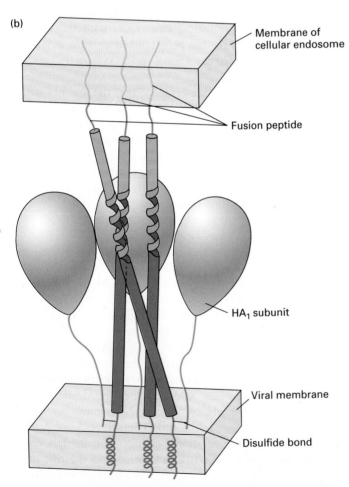

Membrane of cellular endosome

Fusion peptide

HA₁ subunit

Viral membrane

Disulfide bond

three segments comprising each HA₂ subunit rearrange to form a single α helix, 88 amino acids long. [In the native (pH 7.0) structure, the HA₂ subunit is composed of two α helixes (light green and dark green in Figure 16-54) connected by a long nonhelical loop (orange). During the pH 5-induced conformational change the connecting loop (orange) also rearranges into an α helix, thus forming the single, long (88-aa) α helix.] The 88-aa α helixes from the three HA₂ subunits then twist together into a three-stranded coiled-coil rod that protrudes more than 13 nm outward from the viral membrane. The fusion peptides are at the tip of the rod; they insert into the endocytic vesicle membrane, triggering fusion of the viral and endosomal membranes.

Multiple pH 5-activated HA spikes are essential for membrane fusion to occur. Figure 16-55 suggests one way by which the protein scaffold formed by many HA spikes, possibly together with other cellular proteins, could link together the viral and endosomal membranes and induce their fusion. This figure also illustrates some likely but still hypothetical intermediates in the fusion process.

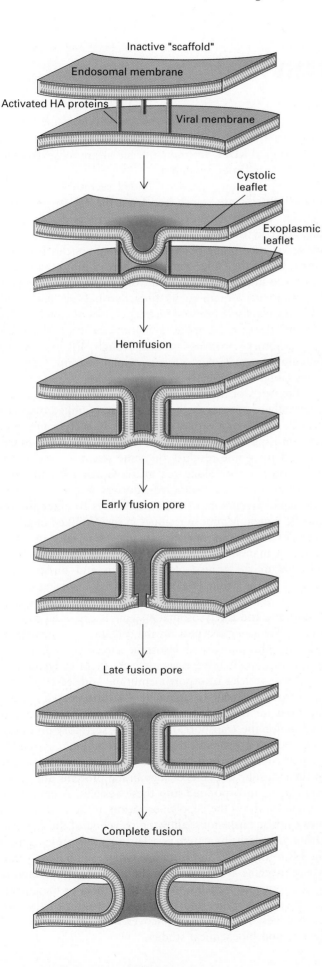

Inactive "scaffold"

Endosomal membrane

Activated HA proteins

Viral membrane

Cystolic leaflet

Exoplasmic leaflet

Hemifusion

Early fusion pore

Late fusion pore

Complete fusion

◄ FIGURE 16-55   Model of membrane fusion directed by the HA protein. A number of pH 5–activated HA proteins, possibly in concert with cellular membrane proteins, form a scaffold that connects a small region of the viral membrane and the endosomal membrane. By unknown mechanisms, first the exoplasmic leaflets of the two membranes fuse and form a continuous phospholipid leaflet, shown as the hemifusion intermediate. Then the cytosolic leaflets of the two membranes fuse, forming an early fusion pore; the pore then widens until complete fusion of the two membranes is achieved. [Adapted from J. R. Monck and J. M. Fernandez. 1992. *J. Cell Biol.* **119**:1395–1404.]

SUMMARY

Membranes grow by expansion of existing membranes. Phospholipids are synthesized on the cytosolic face of the smooth ER membrane or of the bacterial plasma membrane. In a reaction catalyzed by certain membrane proteins, newly made phospholipids equilibrate with the exoplasmic membrane face. The transport of phospholipids and membrane proteins between organelles is mediated by small vesicles that bud from one organelle and fuse with another, or for phospholipids, by phospholipid exchange proteins. Membrane-containing organelles, such as chloroplasts and mitochondria, grow by the expansion of existing organelles. Proteins and lipids are added to existing organelles, and eventually the organelle divides in two.

Although all cytosolic ribosomes are functionally equivalent, membrane-attached and membrane-unattached ribosomes synthesize different classes of proteins, depending on a signal sequence in the protein itself. An important class of proteins synthesized on ribosomes bound to the rough ER includes secretory proteins, enzymes destined for the ER, Golgi complex, and lysosomes, and integral plasma membrane proteins. Synthesis of this class of proteins is initiated on membrane-unattached ribosomes. A sequence of hydrophobic amino acids, the signal sequence, is recognized and bound by a signal recognition particle (SRP), which in turn is bound by an SRP receptor on the rough ER membrane. The SRP directs both the binding of the ribosome to the ER membrane and the insertion of the nascent protein into or across the membrane. Generally, such signal sequences are located at the N-terminus and are cleaved from the protein in the rough ER.

Proteins cross the ER membrane in an unfolded state through a protein-lined channel; this translocation requires an ATP-hydrolyzing chaperone protein in the ER lumen that also assists in protein folding. Topogenic sequences allow newly made membrane proteins to assume their appropriate orientations in the ER membrane. Topogenic sequences include signal sequences, stop-transfer membrane-anchor sequences, internal, uncleaved signal-anchor sequences, and sequences that replace a hydrophilic membrane anchor with a glycosylphosphatidylinositol membrane anchor.

Within the lumen of the rough ER, disulfide bonds are formed, facilitated by a higher ratio of oxidized to reduced glutathione than is found in the cytosol. Individual polypeptides, in turn, bind together, forming oligomeric proteins. Chaperone proteins such as Bip, peptidyl-prolyl isomerase, and protein disulfide isomerase facilitate steps in protein folding in the ER. The attachment of the ubiquitous high-mannose oligosaccharide, containing mannose, glucose, and N-acetylglucosamine, to specific asparagine residues of proteins also occurs in the lumen of the rough

ER. Certain resident ER proteins are retained in the ER by a C-terminal KDEL sorting sequence or are retrieved there from the cis-Golgi reticulum.

Generally, only properly folded secretory and membrane proteins are transferred from the ER to the Golgi vesicles. Unassembled or misfolded polypeptides are often degraded in the ER. Membrane and secretory proteins move via membrane-bounded transport vesicles from the rough ER to the Golgi complex. There, a number of additional modifications occur, some of which are important in targeting the protein to its final destination. The Golgi complex consists of several distinct classes of vesicles, each with its distinct enzymatic activities; the transport route for secretory proteins is thus rough ER → cis-Golgi reticulum → cis Golgi → medial Golgi → trans Golgi → trans-Golgi reticulum → secretory vesicles.

Enzymes localized in the ER and Golgi vesicles act sequentially to process the high-mannose N-linked oligosaccharide to one of a variety of complex N-acetylneuraminic acid–bearing forms. Also, oligosaccharides attached to serine and threonine residues on some secretory and membrane proteins are elongated in the Golgi. ER and Golgi membranes contain specific transporters that allow nucleotide-sugar precursors of oligosaccharides to enter the organelle lumen. Mannose residues in the N-linked oligosaccharides of lysosome enzymes are phosphorylated in the cis Golgi. A mannose 6-phosphate receptor then binds these proteins in the trans-Golgi reticulum and directs their transfer to the acidic late endosome/CURL sorting organelle, where the receptor is recycled to the Golgi or plasma membrane and the lysosomal enzyme is delivered to lysosomes. The same receptor on the cell surface binds extracellular, phosphorylated lysosome enzymes and, by receptor-mediated endocytosis, delivers them to lysosomes.

Small vesicles transport membrane and secretory proteins from organelle to organelle. Clathrin-coated vesicles are used for the endocytosis of cell surface receptors, as well as for the transport of regulated secretory and lysosome-targeted proteins from the trans Golgi. Assembly particles in the clathrin-coated vesicles recognize sequences in the cytosolic domains of membrane proteins and cause them to be incorporated into these vesicles. A chaperone protein catalyzes the depolymerization of the clathrin coat into clathrin triskelions, allowing recycling of the clathrin. Other, nonclathrin-coated, vesicles transport proteins from the ER to the Golgi and from one Golgi region to the next. Many proteins are involved in the processes of formation of the transport vesicles, in targeting them to the correct acceptor membrane, and in triggering fusion of the two membranes. They are being uncovered by a combination of genetic and biochemical studies.

The Golgi complex plays a key role in sorting newly made secretory and membrane proteins; proper sorting is thought to be directed by specific amino acid sequences on the proteins themselves. Some enzymes remain in the Golgi, retained there by sequences within their membrane-spanning domain. Some secretory proteins and most membrane proteins are targeted to vesicles that continuously fuse with the plasma membrane and exocytose their contents. Other secretory proteins are directed to precursors of regulated secretory vesicles, apparently by coprecipitation with secretogranins in the *trans*-Golgi reticulum. The regulated secretory proteins are concentrated and stored in secretory vesicles to await a neural or hormonal signal for exocytosis. Often additional proteolytic cleavages (generally after a sequence of two basic amino acids) occur within the newly made, usually acidic secretory vesicles, by a set of proteases unique to the regulated secretory pathway. Many hormones, such as insulin and ACTH, are synthesized as inactive precursor proteins and require such proteolytic cleavage to generate the active hormone. In the *trans* Golgi or *trans*-Golgi reticulum of many polarized epithelial cells, membrane proteins are sorted into vesicles destined for different regions of the plasma membrane; signals for apical targeting include a glycosylphosphatidylinositol anchor. In other polarized cells, particularly hepatocytes, all plasma membrane proteins are directed first to the basolateral membrane; apically destined ones are then delivered to the apical membrane by endocytosis and transcytosis.

Proteins that are internalized by endocytosis from the cell surface undergo various fates. Many receptors, such as the LDL receptor, release their ligand in the acidic milieu of the late endosome/CURL; the receptors are sorted into vesicles that recycle them to the plasma membrane, while the ligands, in contrast, are sorted into vesicles that fuse with lysosomes. The iron-carrier transferrin is endocytosed bound to the transferrin receptor; it releases its two bound iron atoms in the late endosome/CURL and is then recycled back to the plasma membrane, still bound to its receptor, and is secreted into the medium. Many viruses and toxins gain entry into cells by binding to specific cell surface receptors and undergoing endocytosis. The HA glycoprotein on the surface of the influenza virus undergoes a conformational change in the acidic endosome; this triggers the fusion of the viral and endosomal membranes, releasing the viral genome into the cytosol.

REVIEW QUESTIONS

1. Table 16-1 presents a summary of the different classes of proteins that are synthesized in a cell. What subcellular compartments are responsible for sorting proteins and directing them to the correct location?

The membranes of the organelles within cells have different features that are related to their function. As you learned in Chapter 14, the lipid composition of these membranes is often very dissimilar. Likewise, the composition of the lipids contained in an individual leaflet of a membrane is quite heterogeneous. Where are lipids synthesized relative to a given organelle membrane in which they are ultimately incorporated? Lipid flip-flop between the layers of a membrane is a rare event, and yet asymmetry is apparent. What is known about mechanisms that might assist in setting up this asymmetry and maintaining it?

Cytosolic ribosomes are the sites of synthesis of most protein in cells; however, in and of themselves, these structures are not specialized to synthesize different classes of proteins. What general types of proteins are made on free ribosomes? What proteins are synthesized on the rough endoplasmic reticulum?

a. Human growth hormone (hGH) is synthesized in the pituitary gland; it is a peptide with a molecular weight of 23 kD. You are investigating the synthesis of hGH in a cell culture system, and you want to determine if the newly made protein is found in the rough endoplasmic reticulum. How would you isolate the endoplasmic reticulum for this purpose?

b. In order to radiolabel the proteins in the cells you are investigating, you add [^{35}S]methionine to the culture medium.

You then isolate microsomes, and divide the sample into three portions. To sample A, you add detergent; to sample B, you add a protease; and to sample C, you add detergent and then a protease. These procedures are summarized in the table below. You then add an inhibitor that binds irreversibly to the protease to inactivate it. Following this, you disrupt the remaining microsomes. You separate hGH from other proteins by adding a highly specific antibody prepared against it to your mixtures, resulting in the generation of an antibody-antigen complex that forms a precipitate, which you isolate as a pellet by centrifugation. (This technique is referred to as immunoprecipitation.) You carefully wash the antibody-antigen complex in the pellet, resuspend it in a buffer containing SDS, and analyze the samples by electrophoresis and autoradiography. How many bands should appear on the x-ray film in lanes where hGH is present? The results are presented in the table, where a plus (+) indicates a band of molecular weight corresponding to hGH, and a minus (−) represents no detected protein. Explain these results.

	Detergent	Protease	hGH Protein
Sample A	+	−	+
Sample B	−	+	+
Sample C	+	+	−

2. The various pathways that proteins can take during the sorting process that targets them to different organelles are summarized in Figure 16-8; be certain you understand the general features illustrated.

The order in which proteins pass through the organelles of a cell, ultimately ending in secretion into the extracellular environment, was elucidated to a great extent using yeast, particularly by employing *sec* mutants. What are these mutants? How can they be used to discern the protein secretion pathway? Specifically, how do double mutants enable the geneticist to unravel the order of such complex processes?

Directing a protein into the lumen of the endoplasmic reticulum relies upon the presence of several factors, including the signal recognition particle (SRP), and SRP receptor. Review the functions of each of these.

a. You are characterizing the synthesis of prelysozyme in an in vitro translation system. You have isolated mRNA that encodes the entire prelysozyme molecule, including the signal peptide. You have also isolated mRNA from a mutant that lacks a large portion of the signal peptide (see Table 16-3). The molecular weight of prelysozyme protein without the signal peptide attached is 15.6 kD, and the size with the signal peptide attached is 17.8 kD. You prepare the in vitro translation system for each of the mRNA molecules, being certain to add the proper buffer and amino acids, including [^{35}S]methionine to radiolabel the proteins. Finally, you add SRP, SRP receptor, and microsomes as described in the table below. You incubate the mixtures, stop the reaction, immunoprecipitate the prelysozyme protein, and analyze the precipitate by electrophoresis and autoradiography (see Question 1). The results are presented as the size of the protein you obtain. Explain them. Refer to Figure 16-14 as a guide.

	SRP	SRP Receptor	Micro-somes	Size of Product Formed (kD)
	−	−	−	17.8
mRNA (wild type)	+	−	−	8.0
	+	+	−	17.8
	+	−	+	15.6
	−	−	−	15.6
mRNA (−signal peptide)	+	−	−	15.6
	+	+	−	15.6
	+	−	+	15.6

b. You wish to characterize further the in vitro translation products synthesized in the presence of SRP and microsomes. After incubation of each of the mRNA molecules with these two components, you add a protease prior to the immunoprecipitation step and electrophoretic analysis. Under these conditions, the 15.6 kD protein from the wild type mRNA is present, but no protein is found for the mutant mRNA. Explain the result.

3. The addition of sugar residues to proteins is quite important, not only for their proper function, but as a signal for their correct compartmentalization. What are the differences between *O*-linked and *N*-linked oligosaccharides? How are the sugars activated for synthesis of these oligosaccharides? Which of the carbohydrate groups is generally shorter? How are the oligosaccharides added to proteins? What are the different categories of *N*-linked oligosaccharides? In what compartments are the sugars added, and where and when do modifications occur?

The targeting of essential enzymes to the lysosome involves recognition of mannose 6-phosphate by a receptor. What are the remaining steps in the pathway that target the enzymes to the lysosome?

a. You are studying cells in culture that were obtained from a patient having I-cell disease? What are some of the molecular characteristics of this disorder?

b. The internalization of LDL receptors has much in common with the targeting of mannose 6-phosphate glycoproteins to lysosomes. Review the pathway for internalization of LDL receptors and their subsequent recycling to the cell surface.

c. You are studying a mutant cell line that secretes large quantities of mannose 6-phosphate glycoproteins. You discover that the CURLs in these cells have an internal pH of approximately 6.8 instead of 5.0. Furthermore, you know that the elevated pH is due to a faulty proton pump that is present in the CURL membrane. The mannose 6-phosphate receptors that are present are normal. Why would the mannose 6-phosphate proteins be secreted? What would happen to LDL–LDL receptor complexes in these cells?

4. The Golgi complex is very important for proper sorting of proteins destined for secretory vesicles, the plasma membrane, and other parts of the cell. Review the properties of the different Golgi vesicles.

What is clathrin? What are clathrin coated vesicles? How are they used for transport of secretory proteins from the Golgi? What is their function in receptor-mediated endocytosis? What are the signals that are used to direct proteins to different compartments or to remain in the Golgi apparatus?

a. You are studying a protein from the surface of hepatocytes that appears to be a receptor for LDL. You have isolated the cDNA for the receptor, and sequenced it. You enter the sequence data for this region into a computer and, using a sophisticated program, identify a relatively small cytosolic domain that is 70 amino acids in length. You are told to pay particular attention to *pro* residues near the N-terminus of this region and use them to search for the sequence asp-pro-X-tyr, where X is any amino acid. Why would this sequence be of interest in this situation?

b. You have found the sequence asp-pro-gln-tyr, which satisfies the above conditions. During the course of your research, you discover a patient whose DNA for the LDL receptor encodes *leu* at the position where *pro* should be in the sequence you found. You decide to perform site-directed mutagenesis to change the *pro* in the wild type receptor to a *leu* residue to explore the consequences of this change. How would you do this? What affect would this change have on the function of the LDL receptor protein? Why do you think proline is absolutely required in this sequence?

References

Synthesis of Membrane Lipids

BISHOP, W. R., and R. M. BELL. 1988. Assembly of phospholipids into cellular membranes: Biosynthesis, transmembrane movement and intracellular translocation. *Ann. Rev. Cell Biol.* 4:579–610.

DOWHAN, W. 1991. Phospholipid-transfer proteins. *Curr. Opin. Cell Biol.* 3:621–625.

FUNG, M. K., H. B. SKINNER, and V. A. BANKAITIS. 1992. Mechanistic insights relevant to protein secretion in yeast. *Curr. Opin. Genet. Dev.* 2:775–779.

HJELMSTAD, R. H., and R. M. BELL. 1991. Molecular insights into enzymes of membrane bilayer assembly. *Biochemistry* 30:1731–1740.

PAGANO, R. E. 1990. Lipid traffic in eukaryotic cells: mechanisms for intracellular transport and organelle-specific enrichment of lipids. *Curr. Opin. Cell Biol.* 2:652–663.

ROSENWALD, A. G. and R. E. PAGANO. 1993. Intracellular transport of ceramide and its metabolites at the Golgi complex: insights from short-chain analogs. *Adv. Lipid Res.* 26:101–118.

VAN ECHTEN, G., and K. SANDHOFF. 1993. Ganglioside metabolism: Enzymology, topology, and regulation. *J. Biol. Chem.* 268:5341–5344.

Transport of Proteins across the ER Membrane

DALBEY, R. E., and G. VON HEIJNE. 1992. Signal peptidases in prokaryotes and eukaryotes—a new protease family. *Trends Biochem. Sci.* 17:474–478.

GILMORE, R., 1993. Protein translocation across the endoplasmic reticulum membrane. *Cell* 75:589–593.

GÖRLICH, D. and T. A. RAPOPORT. 1993. Protein translocation into proteoliposomes reconstituted from purified components of the endoplasmic reticulum membrane. *Cell* 75:615–630.

LANDRY, S. J., and L. M. GIERASCH. 1991. Recognition of nascent polypeptides for targeting and folding. *Trends Biochem. Sci.* 16:159–163.

NUNNARI, J., and P. WALTER. 1992. Protein targeting to and translocation across the membrane of the endoplasmic reticulum. *Curr. Opin. Cell Biol.* 4:573–580.

RAPOPORT, T. A. 1992. Transport of proteins across the endoplasmic reticulum membrane. *Science.* 258:931–936.

SANDERS, S. L., and R. SCHEKMAN. 1992. Polypeptide translocation across the endoplasmic reticulum membrane. *J. Biol. Chem.* 267:13791–13794.

VON HEIJNE, G. 1990. The signal peptide. *J. Membr. Biol.* 115:195–201.

Topogenic Sequences in Integral Membrane Proteins

CARAS, I. W. 1991. Probing the signal for glycosylphosphatidylinositol anchor attachment using decay accelerating factor as a model system. *Cell Biol. Int. Rep.* 15:815–826.

CRAMER, W. A., et al. 1992. Forces involved in the assembly and stabilization of membrane proteins. *FASEB J.* 6:3397–3402.

ENGLUND, P. T. 1993. The structure and biosynthesis of glycosylphosphatidylinositol protein anchors. *Ann. Rev. Biochem.* 62:121–138.

HIGH, S., and B. DOBBERSTEIN. 1992. Mechanisms that determine the transmembrane disposition of proteins. *Curr. Opin. Cell Biol.* 4:581–586.

WESSELS, H. P., and M. SPIESS. 1988. Insertion of a multispanning membrane protein occurs sequentially and requires only one signal sequence. *Cell* 55:61–70.

Post-Translational Modifications of Proteins in the Rough ER

CRAIG, E. A., B. D. GAMBILL, and R. J. NELSON. 1993. Heat shock proteins: molecular chaperones of protein biogenesis. *Microbiol. Rev.* 57:402–414.

DOMS, R. W., et al. 1993. Folding and assembly of viral membrane proteins. *Virology* 193:545–562.

FREEDMAN, R. B. 1989. Protein disulfide isomerase: multiple roles in the modification of nascent secretory proteins. *Cell* 57:1069–1072.

GETHING, M. J., and J. SAMBROOK. 1992. Protein folding in the cell. *Nature* 355:33–45.

HURTLEY, S. M., and A. HELENIUS. 1989. Protein oligomerization in the endoplasmic reticulum. *Ann. Rev. Cell Biol.* 5:277–307.

NOIVA, R., and W. J. LENNARZ. 1992. Protein disulfide isomerase. A multifunctional protein resident in the lumen of the endoplasmic reticulum. *J. Biol. Chem.* 267:3553–3556.

OLSON, T. S., and M. D. LANE. 1989. A common mechanism for post-translational activation of plasma-membrane receptors? *FASEB J.* 3:1618–1624.

Quality Control in the ER: Retention and Degradation

BERGERON, J., M. BRENNER, D. THOMAS, and D. WILLIAMS. 1994. Calnexin: a new membrane-bound chaperone of the endoplasmic reticulum. *Trends Biochem. Sci.* 19:124–128.

BONIFACINO, J. S., and J. LIPPINCOTT-SCHWARTZ. 1991. Degradation of proteins within the endoplasmic reticulum. *Curr. Opin. Cell Biol.* 3:592–600.

KLAUSNER, R. D., and R. SITIA. 1990. Protein degradation in the endoplasmic reticulum. *Cell* 62:611–614.

LODISH, H. F. 1988. Transport of secretory and membrane glycoproteins from the rough endoplasmic reticulum to the Golgi. *J. Biol. Chem.* 263:2107–2110.

PELHAM, H. R. 1989. Control of protein exit from the endoplasmic reticulum. *Ann. Rev. Cell Biol.* 5:1–23.

ROSE, J. K., and R. W. DOMS. 1988. Regulation of protein export from the endoplasmic reticulum. *Ann. Rev. Cell Biol.* 4:257–288.

SITIA, R., and J. MELDOLESI. 1992. Endoplasmic reticulum: a dynamic patchwork of specialized subregions. *Mol. Biol. Cell.* 3:1067–1072.

Protein Glycosylation

ABEIJON, C., and C. B. HIRSCHBERG. 1992. Topography of glycosylation reactions in the endoplasmic reticulum. *Trends Biochem. Sci.* 17:32–36.

DRIOUICH, A., L. FAYE, and L. A. STAEHELIN. 1993. The plant Golgi apparatus: a factory for complex polysaccharides and glycoproteins. *Trends Biochem. Sci.* 18:210–214.

KORNFELD, R., and S. KORNFELD. 1985. Assembly of asparagine-linked oligosaccharides. *Ann. Rev. Biochem.* 45:631–664.

PAULSON, J. C. 1989. Glycoproteins: what are the sugar chains for? *Trends Biochem. Sci.* 14:272–276.

PAULSON, J. C., and K. J. COLLEY. 1989. Glycosyltransferases. Structure, localization, and control of cell type-specific glycosylation. *J. Biol. Chem.* 264:17615–17618.

ROTH, J. 1991. Localization of glycosylation sites in the Golgi apparatus using immunolabeling and cytochemistry. *J. Electron Microsc. Tech.* 17:121–131.

STANLEY, P. 1992. Glycosylation engineering. *Glycobiology* 2:99–107.

Synthesis of Lysosome and Vacuole Proteins

KORNFELD, S. 1992. Structure and function of the mannose 6-phosphate/insulinlike growth factor II receptors. *Ann. Rev. Biochem.* 61:307–330.

KORNFELD, S., and I. MELLMAN. 1989. The biogenesis of lysosomes. *Ann. Rev. Cell Biol.* 5:483–525.

NEUFELD, E. F. 1991. Lysosomal storage diseases. *Ann. Rev. Biochem.* 60:257–280.

VON FIGURA, K. 1991. Molecular recognition and targeting of lysosomal proteins. *Curr. Opin. Cell Biol.* 3:642–646.

The Mechanism and Regulation of Vesicular Transport to and from the ER and the Golgi Complex

BALCH, W. E. 1989. Biochemistry of interorganelle transport. A new frontier in enzymology emerges from versatile in vitro model systems. *J. Biol. Chem.* 264:16965–16968.

BENNETT, M. K., and R. H. SCHELLER. 1993. The molecular machinery for secretion is conserved from yeast to neurons. *Proc. Natl. Acad. Sci. USA* 90:2559–2563.

KREIS, T. E. 1992. Regulation of vesicular and tubular membrane traffic of the Golgi complex by coat proteins. *Curr. Opin. Cell Biol.* 4:609–615.

MELLMAN, I., and K. SIMONS. 1992. The Golgi complex: in vitro veritas? *Cell* 68:829–840.

NOVICK, P. J., and P. BRENNWALD. 1993. Friends and family: the role of the Rab GTPases in vesicular traffic. *Cell* 75:597–601.

NOVICK, P., and M. GARRETT. 1994. Vesicular transport: No exchange without receipt. *Nature* 369:18–19.

PELHAM, H. R. B., and S. MUNRO. 1993. Sorting of membrane proteins in the secretory pathway. *Cell* 75:603–605.

PRYER, N. K., L. J. WUESTEHUBE, and R. SCHEKMAN. 1992. Vesicle-mediated protein sorting. *Ann. Rev. Biochem.* 61:471–516.

ROTHMAN, J. E., and L. ORCI. 1992. Molecular dissection of the secretory pathway. *Nature* 355:409–415.

SIMONS, K., and M. ZERIAL. 1993. Rab proteins and the road maps for intracellular transport. *Neuron* 11:789–799.

WHITE, J. M. 1992. Membrane fusion. *Science* 258:917–924.

Golgi and Post-Golgi Sorting and Processing of Membrane and Secretory Proteins

BAUERFEIND, R., and W. B. HUTTNER. 1993. Biogenesis of constitutive secretory vesicles, secretory granules and synaptic vesicles. *Curr. Opin. Cell Biol.* 5:628–635.

HURTLEY, S. M. 1992. Golgi localization signals. *Trends Biochem. Sci.* 17:2–3.

HUTTNER, W. B., and S. A. TOOZE. 1989. Biosynthetic protein transport in the secretory pathway. *Curr. Opin. Cell Biol.* 1:648–654.

KLIONSKY, D. J., P. K. HERMAN, and S. D. EMR. 1990. The fungal vacuole: composition, function, and biogenesis. *Microbiol. Rev.* 54:266–292.

RAMBOURG, A., and Y. CLERMONT. 1990. Three-dimensional electron microscopy: structure of the Golgi apparatus. *Eur. J. Cell Biol.* 51:189–200.

TOOZE, S. A. 1991. Biogenesis of secretory granules. Implications arising from the immature secretory granule in the regulated pathway of secretion. *FEBS Lett.* 285:220–224.

Regulated Secretion

ALMERS, W. 1990. Exocytosis. *Ann. Rev. Physiol.* 52:607–624.

BURGESS, T. L., and R. B. KELLY. 1987. Constitutive and regulated secretion of proteins. *Ann. Rev. Cell Biol.* 3:243–293.

MILLER, S. G., and H. P. MOORE. 1990. Regulated secretion. *Curr. Opin. Cell Biol.* 2:642–647.

MONCK, J. R., and J. FERNANDEZ. 1994. The exocytic fusion pore and neurotransmitter release. *Neuron* 12:707–716.

Proteolytic Processing of Secretory and Membrane Proteins

DOUGLASS, J., O. CIVELLI, and E. HERBERT. 1984. Polyprotein gene expression: generation of diversity of neuroendocrine peptides. *Ann. Rev. Biochem.* 53:665–715.

FULLER, R. S., R. E. STERNE, and J. THORNER. 1988. Enzymes required for yeast prohormone processing. *Ann. Rev. Physiol.* 50:345–362.

JUNG, L. J., T. KREINER, and R. H. SCHELLER. 1993. Prohormone structure governs proteolytic processing and sorting in the Golgi complex. *Recent Prog. Horm. Res.* 48:415–436.

SOSSIN, W. S., and R. H. SCHELLER. 1991. Biosynthesis and sorting of neuropeptides. *Curr. Opin. Neurobiol.* 1:79–83.

STEINER, D. F., et al. 1992. The new enzymology of precursor processing endoproteases. *J. Biol. Chem.* 267:23435–23438.

The Internalization of Macromolecules and Particles—Pinocytosis

ANDERSON, R. G., et al. 1992. Potocytosis: sequestration and transport of small molecules by caveolae. *Science* 255:410–411.

MELLMAN, I. S., et al. 1980. Selective iodination and polypeptide composition of pinocytic vesicles. *J. Cell Biol.* 86:712–722.

Receptor-Mediated Endocytosis

BAKER, E. N., S. V. RUMBALL, and B. F. ANDERSON. 1987. Transferrins: insights into structure and function from studies on lactoferrin. *Trends Biochem. Sci.* 12:350–353.

DAUTRY-VARSAT, A., A. CIECHANOVER, and H. F. LODISH. 1983. pH and the recycling of transferrin during receptor-mediated endocytosis. *Proc. Natl. Acad. Sci. USA* 80:2258–2262.

GRUENBERG, J., and K. E. HOWELL. 1989. Membrane traffic in endocytosis: insights from cell-free assays. *Ann. Rev. Cell Biol.* 5:453–481.

HUNZIKER, W., and I. MELLMAN. 1991. Relationships between sorting in the exocytic and endocytic pathways of MDCK cells. *Semin. Cell Biol.* 2:397–410.

MOSTOV, K. E., and N. E. SIMISTER. 1985. Transcytosis. *Cell* 43:389–390.

RIEZMAN, H., Y. CHVATCHKO, and V. DULIC. 1986. Endocytosis in yeast. *Trends Biochem. Sci.* 11:325–328.

SMYTHE, E., and G. WARREN. 1991. The mechanism of receptor-mediated endocytosis. *Eur. J. Biochem.* 202:689–699.

Clathrin and Assembly of Coated Vesicles

BRODSKY, F. M. 1988. Living with clathrin: its role in intracellular membrane traffic. *Science* 242:1396–1402.

BRODSKY, F. M., et al. 1991. Clathrin light chains: arrays of protein motifs that regulate coated-vesicle dynamics. *Trends Biochem. Sci.* 16:208–213.

KEEN, J. H. 1990. Clathrin and associated assembly and disassembly proteins. *Ann. Rev. Biochem.* 59:415–438.

LAZAROVITS, J., and M. ROTH. 1988. A single amino acid change in the cytoplasmic domain allows the influenza virus hemagglutinin to be endocytosed through coated pits. *Cell* 53:743–752.

PEARSE, B. M. F. 1987. Clathrin and coated vesicles. *EMBO J.* 6:2507–2512.

SCHMID, S. L., and J. E. ROTHMAN. 1985. Enzymatic dissociation of clathrin in a two-stage process. *J. Biol. Chem.* 260:10044–10049.

Low-Density Lipoprotein Receptor and LDL Uptake

BROWN, M. S., and J. L. GOLDSTEIN. 1984. How LDL receptors influence cholesterol and atherosclerosis. *Sci. Am.* 251(5):58.

BROWN, M. S., and J. L. GOLDSTEIN. 1986. Receptor-mediated pathway for cholesterol homeostasis. *Science* 232:34–47 (Nobel Prize Lecture).

HOBBS, H. H., et al. 1990. The LDL receptor locus in familial hypercholesterolemia: mutational analysis of a membrane protein. *Ann. Rev. Genet.* 24:133–170.

Synthesis of Plasma-Membrane Proteins in Polarized Cells

APODACA, G., et al. 1991. The polymeric immunoglobulin receptor. A model protein to study transcytosis. *J. Clin. Invest.* 87:1877–1882.

BARTLES, J. R., and A. L. HUBBARD. 1988. Plasma-membrane protein sorting in epithelial cells: do secretory pathways hold the key? *Trends Biochem. Sci.* 13:181–184.

HUBBARD, A. L. 1991. Targeting of membrane and secretory proteins to the apical domain in epithelial cells. *Semin. Cell Biol.* 2:365–374.

MOSTOV, K., et al. 1992. Plasma membrane protein sorting in polarized epithelial cells. *J. Cell Biol.* 116:577–583.

RODRIGUEZ-BOULAN, E., and S. K. POWELL. 1992. Polarity of epithelial and neuronal cells. *Ann. Rev. Cell Biol.* 8:395–427.

SIMONS, K., and A. WANDINGER-NESS. 1990. Polarized sorting in epithelia. *Cell* 62:207–210.

Penetration of Viruses into Cells

HELENIUS, A. 1992. Unpacking the incoming influenza virus. *Cell* 69:577–578.

MARSH, M., and A. HELENIUS. 1989. Virus entry into animal cells. *Adv. Virus Res.* 36:107–151.

WEBSTER, R. G., and R. ROTT. 1987. Influenza virus A pathogenicity: the pivotal role of hemagglutinin. *Cell* 50:665–666.

WEIS, W., et al. 1988. Structure of the influenza virus hemagglutinin complexed with its receptor, sialic acid. *Nature* 333:426–431.

WILEY, D. C., and J. J. SKEHEL. 1987. Structure and function of the hemagglutinin membrane glycoprotein of the influenza virus. *Ann. Rev. Biochem.* 56:365–394.

17

Cellular Energetics: Formation of ATP by Glycolysis and Oxidative Phosphorylation

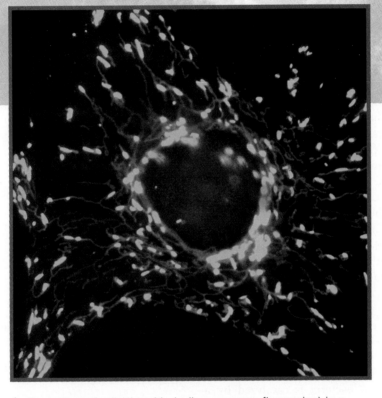

▲ Two types of mitochondria (yellow or green fluorescing) in a living cultured mink fibroblast.

All processes involved in the growth of cells require energy. In most cases, free energy is supplied by the hydrolysis of one of the high-energy phosphoanhydride bonds in adenosine triphosphate (ATP) by the reaction

$$ATP^{4-} + H_2O \rightleftharpoons ADP^{3-} + P_i^{2-} + H^+$$

where the change in free energy under standard conditions $\Delta G^{o\prime}$ is -7.3 kcal/mol, and where ADP = adenosine diphosphate and P_i = inorganic phosphate, HPO_4^{2-}.

The energy released by the hydrolysis of phosphoanhydride bonds powers many otherwise energetically unfavorable events, such as the transport of molecules against a concentration gradient, exemplified by the Na^+-K^+ ATPase, the movement (beating) of cilia, and the contraction of muscle. ATP is needed in a multitude of pathways, including the synthesis of nucleic acids and proteins from amino acids and nucleotides. ATP is a universal "currency" of chemical energy, found in all types of organisms. ATP must have occurred in the earliest life forms.

This chapter and the next focus on how cells generate the high-energy phosphoanhydride bond of ATP by the endergonic reaction (one requiring the input of free energy, G, in order to proceed): $P_i^{2-} + H^+ + ADP^{3-} \rightleftharpoons ATP^{4-} + H_2O$ where the change in free energy, $\Delta G^{o\prime}$, is $+7.3$ kcal/mol.

We shall concentrate on two of the most important processes. In Chapter 18, we discuss photosynthesis, detailing how light energy is converted to the chemical energy of a phosphoanhydride bond and how CO_2 and H_2O are converted to six-carbon sugars. Here, we examine the metabolic pathways by which glucose and fatty acids (the principal sources of energy in animal and most other nonphotosynthetic cells, including plant cells such as roots) are metabolized to CO_2. The complete aerobic degradation of glucose to CO_2 and H_2O is coupled to the synthesis of as many as 32 molecules of ATP:

$$C_6H_{12}O_6 + 6O_2 + 32P_i^{2-} + 32ADP^{3-} + 32H^+ \longrightarrow 6CO_2 + 32ATP^{4-} + 38H_2O$$

In eukaryotic cells and in bacteria, the initial enzymatically catalyzed chemical reactions in the pathway of glucose degradation occur in the cytosol. In eukaryotes, the final steps occur in the mitochondria, together with generation of most of the ATP, while in bacteria, which lack mitochondria, many of the final steps occur on the plasma membrane. Similarly, the final stages of metabolism of fatty acids can also occur in mitochondria, though in most cells metabolism of fatty acids occurs in peroxisomes without production of ATP. Localizing different metabolic pathways to different compartments of the cell prevents the pathways from interfering with each other. An important focus of this chapter is the way in which mitochondria convert the energy released by the oxidation of metabolic products of glucose and lipids to ATP phosphoanhydride bonds.

At first glance, photosynthesis and aerobic oxidation appear to have little similarity. However, a revolutionary finding in cell biology is that bacteria, mitochondria, and chloroplasts all use the same (or very nearly the same) process, called *chemiosmosis* (Figure 17-1), to generate ATP from ADP and P_i. The immediate energy sources that power this reaction are the proton concentration gradient and the membrane electric potential, collectively termed the *proton-motive force*. In photosynthesis, formation of this gradient is generated by the energy absorbed from light. In mitochondria, energy from the metabolism of sugars, fatty acids, and other substances, culminating in their oxidation by O_2, is used to pump protons across a mitochondrial membrane, generating a proton-motive force.

We have already seen that the proton-motive force is used in many ways. Proton concentration gradients supply energy for the transport of small molecules across a membrane against a concentration gradient; for example, the uptake of lactose by certain bacteria is catalyzed by a proton-sugar symport (see Figure 15-23), as is the concentration of ions and sucrose by plant vacuoles (see Figure 15-22). The rotation of bacterial flagella is also powered by the transmembrane proton concentration gradient (in contrast to the beating of eukaryotic cilia, which is powered by ATP hydrolysis). Conversely, ATP-powered proton pumps (see Figure 15-10) utilize the energy released by the hydrolysis of a phosphoanhydride bond to transport protons against the proton concentration gradient. The *chemiosmotic theory* is based on the principle introduced previously in the discussion of active transport: *the membrane potential, the concentration gradients of protons (and other ions) across a membrane, and the phosphoanhydride bonds in ATP are equivalent and interconvertible forms of chemical potential energy.*

Because of the central importance of the cytosol in energy metabolism, we begin our discussion of cellular energetics with the initial steps in glucose metabolism, which take place in the cytosol. We then discuss mitochondria and some details of the chemiosmotic process.

► *Energy Metabolism in the Cytosol*

Glycolysis Results in the Net Production of Two ATP Molecules from the Conversion of One Glucose Molecule to Two Pyruvate Molecules

In the initial stage of glucose metabolism, *glycolysis*, each glucose molecule is converted to two molecules of the three-carbon compound pyruvate $C_3H_3O_3^-$ (Figure

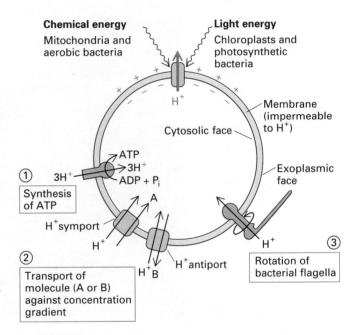

▲ FIGURE 17-1   The chemiosmotic process. Chemiosmosis requires sealed, closed membrane vesicles that are relatively impermeable to H^+. In photosynthesis, energy absorbed from light is used to move protons across the membrane, generating a transmembrane proton (a gradient of proton concentration and/or -motive force electric potential across the membrane). In mitochondria and aerobic bacteria, energy liberated by the oxidation of carbon compounds is used to move protons across the membrane. In all cases, the protons are moved across the membrane to the exoplasmic face. (In mitochondria and chloroplasts, the cytosolic face contacts the matrix or stromal space respectively. The exoplasmic face contacts the intermembrane space of mitochondria and the thylakoid lumen of chloroplasts; see Figure 14-13). The energy stored in the H^+ gradient can be used (1) to move protons from the exoplasmic face down their concentration gradient across the membrane, coupled to the synthesis of ATP from ADP and P_i, which occurs on the cytosolic membrane face in mitochondria, chloroplasts, and bacteria; (2) to pump metabolites across the membrane against their concentration gradient; and (3) to power the rotation of flagella in bacteria.

17-2). These chemical reactions, called the *glycolytic pathway*, take place in the cytosol and do not involve molecular oxygen. Glycolysis is highly regulated: just enough glucose to meet the cell's need for ATP is transported into the cell (Figure 15-5) or is generated by the hydrolysis of polymers such as glycogen (Figure 20-24).

All of the metabolic intermediates between the initial carbohydrate and the final product, pyruvate, are phosphorylated compounds. Four molecules of ATP are formed from ADP in glycolysis: two in the step catalyzed by phos-

phoglycerate kinase, when two molecules of 1,3-bisphosphoglycerate are converted to 3-phosphoglycerate, and two in the step catalyzed by pyruvate kinase (see Figure 17-2). However, two ATP molecules are consumed during earlier steps of this pathway: the first by the addition of a phosphate residue to glucose in the reaction catalyzed by hexokinase, and the second by the addition of a second phosphate to fructose 6-phosphate in the reaction catalyzed by phosphofructokinase-1. Thus there is a net gain of two ATP molecules. The balanced chemical equa-

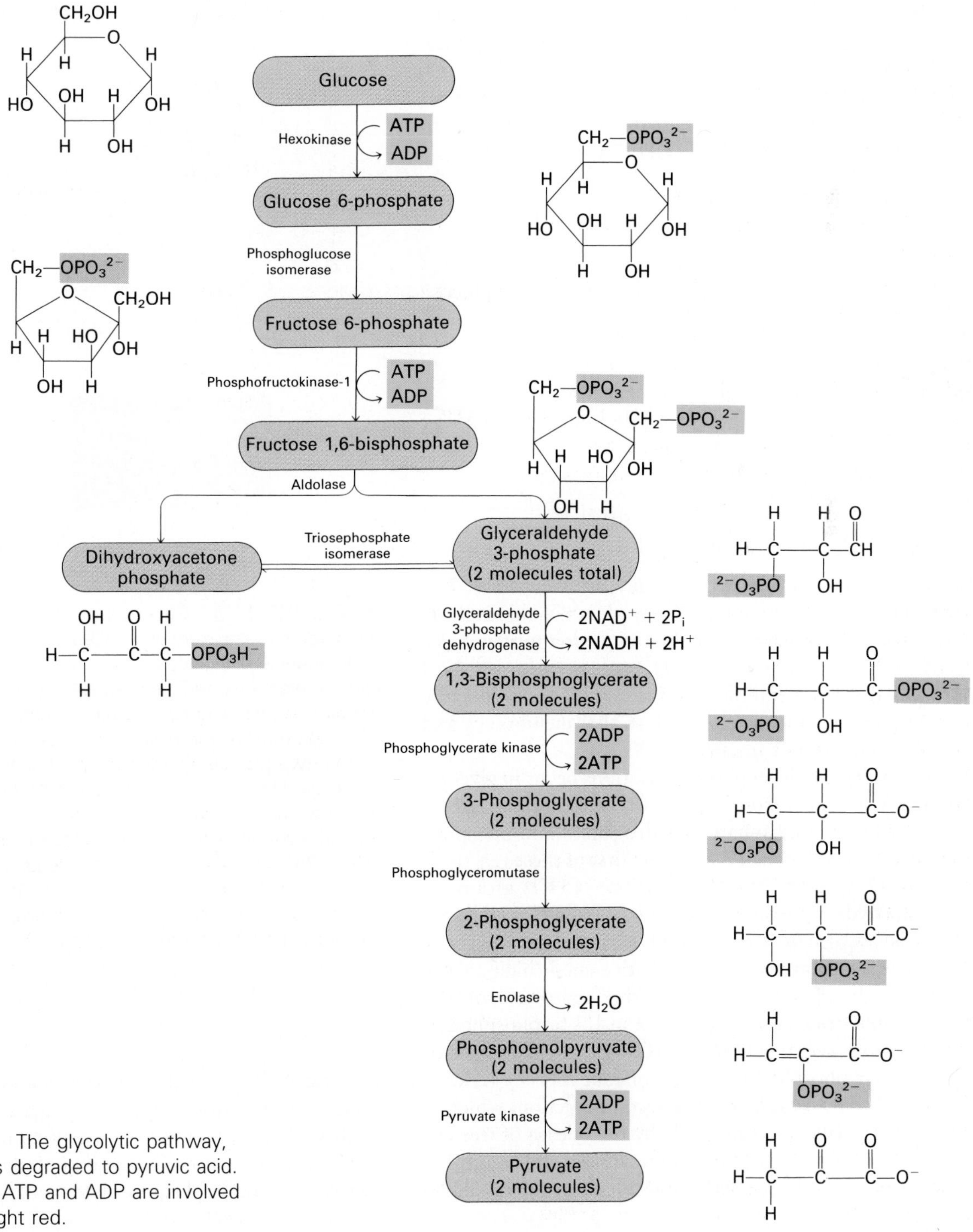

▲ FIGURE 17-2   The glycolytic pathway, by which glucose is degraded to pyruvic acid. Reactions in which ATP and ADP are involved are highlighted in light red.

tion for this series of reactions shows that four hydrogen atoms (four protons and four electrons) are also formed:

$$C_6H_{12}O_6 \longrightarrow 2C_3H_4O_3 + 4H$$

or

$$C_6H_{12}O_6 \longrightarrow 2C_3H_4O_3 + 4H^+ + 4e^-$$

The reaction that generates these H atoms is catalyzed by the enzyme glyceraldehyde 3-phosphate dehydrogenase. All four electrons and two of the four protons are transferred to two molecules of the oxidized form of the electron carrier nicotinamide adenine dinucleotide, NAD^+, to produce the reduced form, NADH (Figure 17-3):

$$2H^+ + 4e^- + 2NAD^+ \longrightarrow 2NADH$$

Thus the overall reaction for this first stage of glucose metabolism is

$$C_6H_{12}O_6 + 2NAD^+ + 2ADP^{3-} + 2P_i^{2-} \longrightarrow$$
$$2C_3H_4O_3 + 2NADH + 2ATP^{4-}$$

In Glycolysis, ATP Is Generated by Substrate-Level Phosphorylation

Cells synthesize ATP in two basically different ways. The immediate energy source for ATP synthesis in chloroplasts and mitochondria is provided by the proton-motive force across a membrane. In the other process, *substrate-level phosphorylation*, metabolites in the cytosol are chemically transformed by aqueous enzymes, so that membranes and ion gradients are not involved.

Two substrate-level phosphorylations occur in glycolysis. The first results from a pair of reactions catalyzed by glyceraldehyde 3-phosphate dehydrogenase and phosphoglycerate kinase (Figure 17-4). In the first of these reactions (Figures 17-4 and 17-5), the aldehyde (CHO) group on glyceraldehyde 3-phosphate is oxidized by NAD^+; the oxidation is coupled to the addition of a phosphate group, forming 1,3-bisphosphoglycerate with a single high-energy phosphate bond. The $\Delta G^{\circ\prime}$ for the hydrolysis of the phosphoanhydride bond to carbon 1 in 1,3-bisphosphoglycerate is more negative than the $\Delta G^{\circ\prime}$ for the hydrolysis of the terminal phosphoanhydride bond in ATP (-12 kcal/mol versus -7.3 kcal/mol). The second reaction, which is strongly *exergonic* (accompanied by the release of free energy), catalyzed by phosphoglycerate kinase (see Figure 17-4), transfers this high-energy phosphate group to ADP,

$$NAD^+ + H^+ + 2e^- \rightleftharpoons NADH$$
$$NADP^+ + H^+ + 2e^- \rightleftharpoons NADPH$$

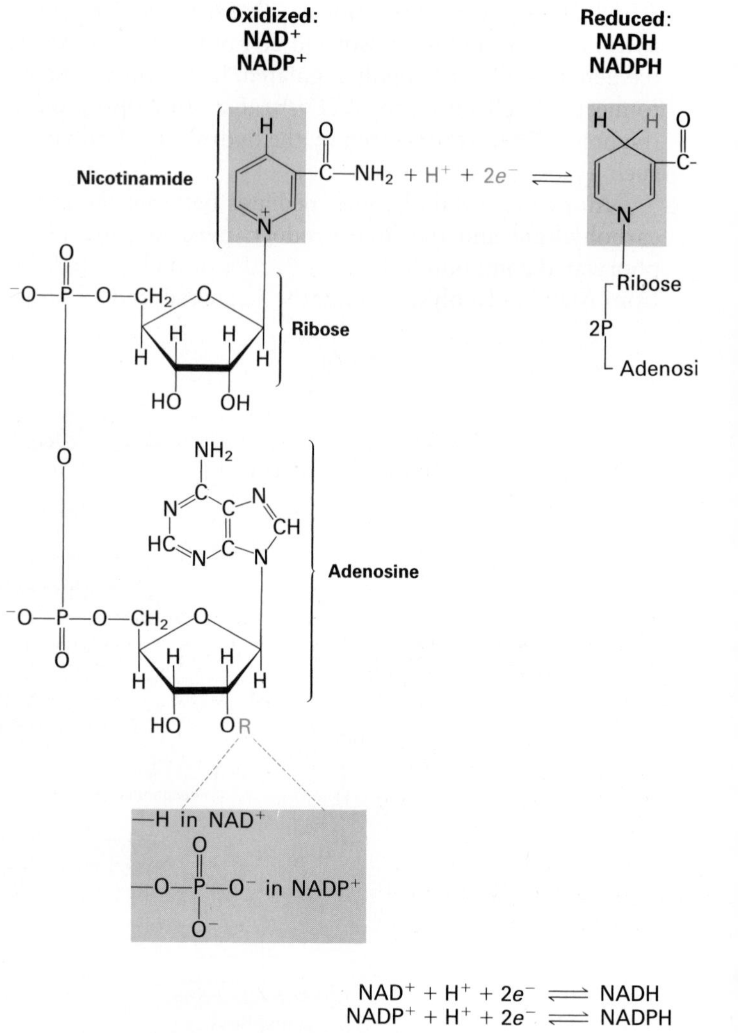

▲ FIGURE 17-3   Structure of NAD^+ and NADH. Nicotinamide adenine dinucleotide (NAD^+) and the related nicotinamide adenine dinucleotide phosphate ($NADP^+$) accept only pairs of electrons; reduction to NADH or NADPH involves the transfer of two electrons simultaneously. In most oxidation-reduction reactions in biological systems, a pair of hydrogen atoms (two protons and two electrons) are removed from a molecule. One of the protons and both electrons are transferred to NAD^+; the other proton is released into solution. Thus the overall reaction is sometimes written $NAD^+ + 2H^+ + 2e^- \rightleftharpoons NADH + H^+$. NADP is identical in structure to NAD except for the presence of an additional phosphate group. However, NAD and NADP participate in different sets of enzymatically catalyzed reactions.

forming ATP. The net change in standard free energy of these two reactions is negative ($\Delta G^{\circ\prime} = -4.5 + 1.5 = -3.0$ kcal/mol), so the two reactions overall are strongly exergonic. In the preceding stages of glycolysis, each molecule of fructose 1,6-bisphosphate generated two molecules

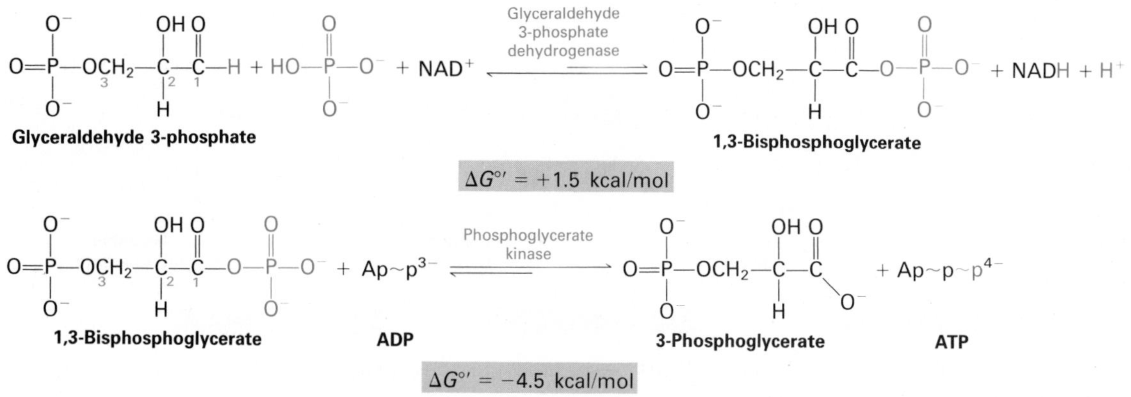

▲ FIGURE 17-4 The first of two substrate-level phosphorylation reactions in glycolysis. Glyceraldehyde 3-phosphate dehydrogenase and phosphoglycerate kinase are water-soluble enzymes that catalyze ATP synthesis in the cytosol. The three C atoms of the glycerol are numbered.

of glyceraldehyde 3-phosphate (see Figure 17-2); therefore, the catabolism of one glucose molecule has now generated two ATPs.

The second high-energy phosphate bond is formed when the product of the phosphoglycerate kinase reaction, 3-phosphoglycerate, is converted in a series of reactions to pyruvate (Figure 17-6). In the first of these reactions, catalyzed by phosphoglyceromutase, 3-phosphoglycerate is converted to 2-phosphoglycerate. In the next reaction, catalyzed by enolase, loss of water converts the phosphate-carbon bond in 2-phosphoglycerate from a low-energy bond ($\Delta G^{\circ\prime}$ of hydrolysis = -2 kcal/mol) to the high-energy bond ($\Delta G^{\circ\prime}$ of hydrolysis = -15 kcal/mol) in phosphoenolpyruvate. In the third reaction, catalyzed by pyruvate kinase, the phosphate in phosphoenolpyruvate is transferred to ADP—a strongly exergonic reaction ($\Delta G^{\circ\prime} = -7.5$ kcal/mol)—and pyruvate is produced.

These examples illustrate how interconversions of soluble chemicals by soluble enzymes are coupled to generate ATP. However, only two of the 32 ATP molecules gener-

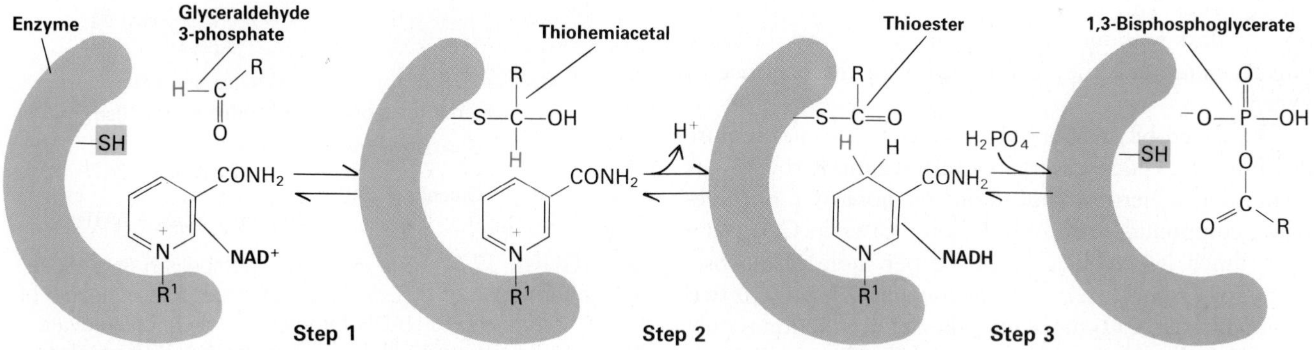

▲ FIGURE 17-5 The mechanism of action of glyceraldehyde 3-phosphate dehydrogenase. A thioester

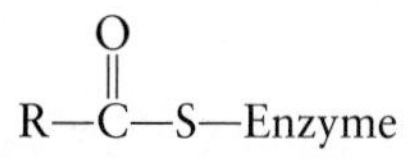

$$R-\overset{\overset{\displaystyle O}{\|}}{C}-S-Enzyme$$

is an energy-rich intermediate in the reaction. The sulfhydryl (SH) group is the side chain of cysteine at the active site; R symbolizes the rest of the glyceraldehyde 3-phosphate molecule. R′ is the rest of the NAD molecule. Step 1: The enzyme has bound NAD$^+$, and the −SH group on the enzyme reacts with glyceraldehyde 3-phosphate to form a thiohemiacetal. Step 2: A hydrogen atom (red) and two electrons are transferred to NAD$^+$, forming the reduced form NADH, and a proton from the O atom of the thiohemiacetal is simultaneously lost to the medium; the products are a thioester and NADH + H$^+$. Step 3: The thioester reacts with phosphate to produce 1,3-bisphosphoglycerate; NADH is freed from the enzyme surface, and the free enzyme with its −SH group is regenerated.

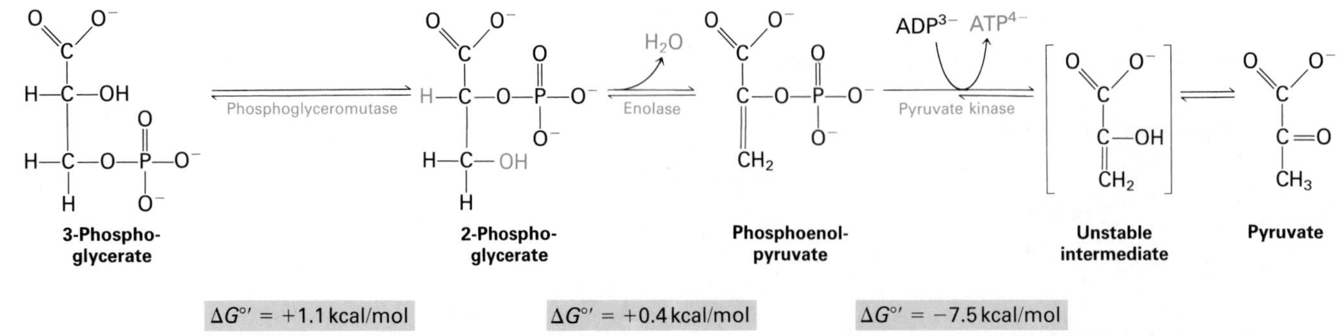

$\Delta G^{\circ\prime} = +1.1\,\text{kcal/mol}$ $\Delta G^{\circ\prime} = +0.4\,\text{kcal/mol}$ $\Delta G^{\circ\prime} = -7.5\,\text{kcal/mol}$

▲ **FIGURE 17-6**   Formation of the second pair of ATP molecules during glycolysis. This reaction, catalyzed by pyruvate kinase, is the second of the two substrate-level phosphorylations in glycolysis.

ated during the complete oxidation of glucose to CO_2 and H_2O are created during the conversion of glucose to pyruvate. The remaining 30 are synthesized in the mitochondria by a fundamentally different type of process that involves the generation and utilization of proton concentration gradients across the inner mitochondrial membrane.

Some Eukaryotic and Prokaryotic Cells Metabolize Glucose Anaerobically

Most eukaryotic cells are *obligate aerobes:* they grow only in the presence of oxygen and they metabolize glucose (or related sugars) completely to CO_2, with the concomitant production of a large amount of ATP. Most of these cells also generate some ATP by anaerobic metabolism. A few eukaryotes, including certain yeasts, are *facultative anaerobes:* they grow in either the presence or absence of oxygen. Annelids and mollusks, as examples, can live and grow for days without oxygen. Many prokaryotic cells are *obligate anaerobes:* they cannot grow in the presence of oxygen.

The anaerobic metabolism of glucose does not require mitochondria. Glucose is not converted entirely to CO_2 (as it is in obligate aerobes) but to one or more two- or three-carbon compounds, and only in some cases to CO_2. As a result, much less ATP is produced per mole of glucose. Yeasts, for example, ferment glucose anaerobically to two ethanol and two CO_2 molecules; the net production is only two ATP molecules per glucose molecule (Figure 17-7; see also Figure 17-2). Recall that during the initial stages of glycolysis (that is, during the conversion of glucose to pyruvate), two NAD molecules are reduced to NADH; as the pyruvate is converted into ethanol, the NADH from the initial stages is reoxidized to NAD. This *anaerobic fermentation* is the basis of the beer and wine industry.

During the prolonged contraction of mammalian skeletal muscle cells, oxygen becomes limited and glucose cannot be oxidized completely to CO_2 and H_2O. The cells ferment glucose to two molecules of lactic acid—again, with the net production of only two molecules of ATP per

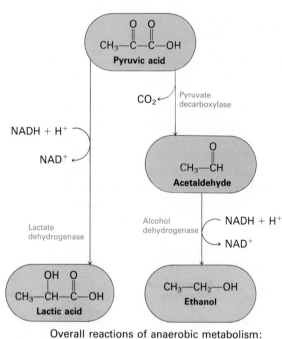

▲ **FIGURE 17-7**   The anaerobic metabolism of glucose. In the formation of pyruvate from glucose, one molecule of NAD^+ is reduced to NADH for each molecule of pyruvate formed (see Figure 17-2). To regenerate NAD^+, two electrons are transferred from each NADH molecule to an acceptor molecule. When oxygen supplies are low in muscle cells, the acceptor is pyruvic acid, and lactic acid is formed. In yeasts, acetaldehyde is the acceptor, and ethanol is formed.

glucose molecule (see Figures 17-2 and 17-7). The lactic acid causes muscle and joint aches. It is largely secreted into the blood; some passes into the liver, where it is reoxidized to pyruvate and either further metabolized to CO_2 or converted to glucose. Much lactate is metabolized to CO_2 by the heart, which is highly perfused by blood and can

(a)

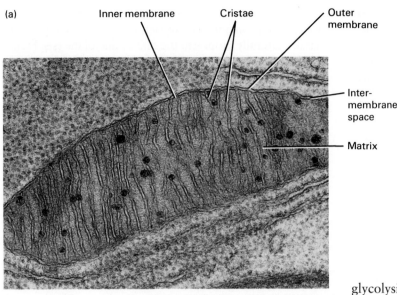

Inner membrane   Cristae   Outer membrane

Inter-membrane space

Matrix

(b)

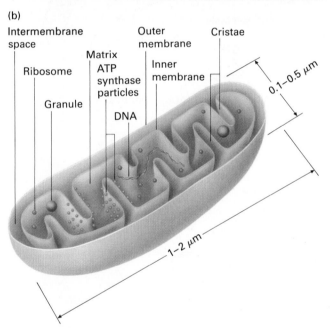

Intermembrane space

Ribosome

Granule

Matrix
ATP synthase particles

DNA

Outer membrane

Inner membrane

Cristae

0.1–0.5 μm

1–2 μm

◄ FIGURE 17-8   The structure of the mitochondrion. (a) Electron micrograph of a mitochondrion from a bat pancreas, showing the inner membrane with extensive cristae, the outer membrane, the intermembrane space, and small granules in the matrix. (b) A three-dimensional diagram of a mitochondrion cut longitudinally. The matrix contains the mitochondrial DNA and ribosomes. [Part (a) from D. W. Fawcett, *The Cell,* 2d ed., Saunders, 1981, p. 421. Courtesy of Don Fawcett.]

glycolysis is transported to the mitochondria, where it is oxidized by O_2 to CO_2:

$$CH_3-\overset{O}{\underset{}{C}}-\overset{O}{\underset{}{C}}-OH + 2\tfrac{1}{2}O_2 \longrightarrow 3CO_2 + 2H_2O$$

These oxidation reactions in the mitochondria generate the bulk of the ATP produced from the conversion of glucose to CO_2. Some energy released in mitochondrial oxidation is used for other purposes, such as heat generation and the transport of ADP into the mitochondrion and ATP out of the organelle into the cytosol.

The mitochondrion is really the "power plant" of the cell. The details of the reactions by which pyruvate is oxidized and ATP is formed depend on understanding the structure of the mitochondrion, which we now address.

The Outer and Inner Membranes of the Mitochondrion Are Structurally and Functionally Distinct

Most eukaryotic cells contain many mitochondria. They are among the larger organelles in the cell—each one is about the size of an *E. coli* bacterium—and they occupy as much as 25 percent of the volume of the cytoplasm. They are large enough to be seen under a light microscope, but the details of their structure can be viewed only with the electron microscope (Figure 17-8).

Mitochondria contain two very different membranes—the *outer membrane* and the *inner membrane*—which define two submitochondrial compartments: the *intermembrane space* between the two membranes, and the *matrix* or central compartment. The fractionation and purification of these membranes and compartments has made it possible to determine their protein and phospholipid compositions and to localize each reaction to a specific membrane or space.

continue aerobic metabolism at times when exercising skeletal muscles secrete lactate. Lactic acid bacteria (the organisms that "spoil" milk) and other prokaryotes also generate ATP by the fermentation of glucose to lactate.

➤ Mitochondria and the Metabolism of Carbohydrates and Lipids

In aerobic cells, glucose is oxidized completely to CO_2 by O_2. In the latter stages of this process, pyruvate formed in

The outer membrane defines the smooth outer perimeter of the mitochondrion. Transmembrane channels formed by the protein *mitochondrial porin* make this membrane freely permeable to most small molecules (<4,000 to 5,000 MW), including protons. While flow of metabolites across the outer membrane may limit their rate of mitochondrial oxidation, the inner membrane is the major permeability barrier between the cytosol and the mitochondrial matrix.

Freeze-fracture studies indicate that the inner membrane contains many protein-rich intramembrane particles that are laterally mobile in the membrane plane (see Figure 14-33). Some of these particles function in electron transport from NADH or reduced flavin adenine dinucleotide (FADH$_2$) to O$_2$ (Figure 17-9) and in ATP synthesis. Some particles are transporters that allow otherwise impermeable molecules, such as ADP and phosphate, to pass from the cytosol to the matrix, and other molecules, such as

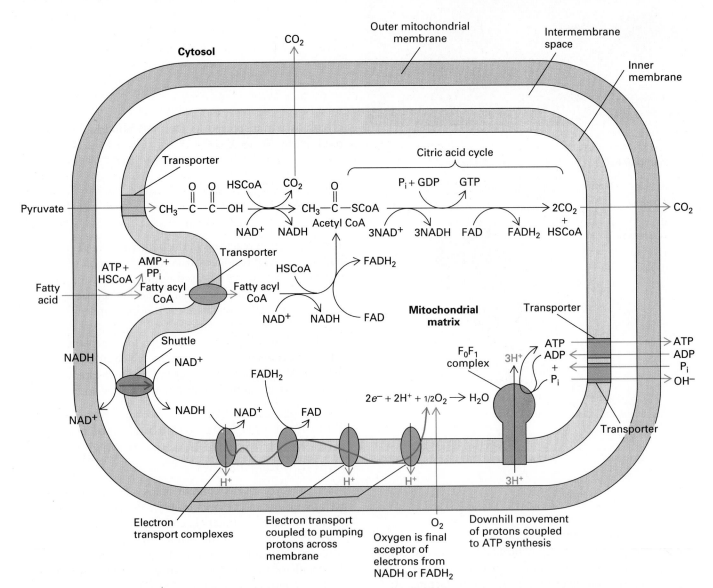

▲ FIGURE 17-9   Outline of the major metabolic reactions in mitochondria. The substrates of oxidative phosphorylation—pyruvate, fatty acids, ADP, and P$_i$—are transported to the matrix from the cytosol by transporters; O$_2$ diffuses into the matrix. NADH, which is generated in the cytosol during glycolysis, is not transported directly to the matrix because the inner membrane is impermeable to NAD$^+$ or NADH; instead, a shuttle system (see Figure 17-21) transports electrons from cytosolic NADH to the electron transport chain. ATP is transported to the cytosol in exchange for ADP and P$_i$, CO$_2$ diffuses into the cytosol across the mitochondrial membranes. HSCoA denotes free coenzyme A (CoA), and SCoA denotes CoA when it is esterified (see Figure 17-11). Fatty acids are linked to CoA on the outer mitochondrial membrane. Subsequently, the fatty-acyl group is removed from the CoA, linked to a carnitine carrier that transports it across the inner membrane, and then the fatty acid is reattached to a CoA on the matrix side of the inner membrane. The blue ovals depict the electron carrier protein complexes that transfer electrons from NADH and FADH$_2$ to O$_2$.

ATP, to move from the matrix into the cytosol. Protein constitutes 76 percent of the total inner membrane weight—a higher fraction than in any other cellular membrane. Cardiolipin (diphosphatidylglycerol, see Figure 14-2), a lipid concentrated in the inner membrane, reduces the permeability of the phospholipid bilayer to protons and thus enables a proton-motive force to be established across the inner membrane.

The inner membrane and the matrix are the sites of most reactions involving the oxidation of pyruvate and fatty acids to CO_2 and H_2O and the coupled synthesis of ATP from ADP and P_i. These complex processes involve many steps but can be subdivided into three groups of reactions, each of which occurs in a discrete membrane or space in the mitochondrion (see Figure 17-9):

1. Oxidation of pyruvate or fatty acids to CO_2, coupled to the reduction of the electron carriers NAD^+ and FAD (Figure 17-10) to NADH and $FADH_2$, respectively. These reactions occur in the matrix or on inner-membrane proteins facing it.

2. Electron transfer from NADH and $FADH_2$ to O_2. These reactions occur in the inner membrane and are coupled to the generation of a proton-motive force across it.

3. Harnessing of the energy stored in the electrochemical proton gradient for ATP synthesis by the F_0F_1 ATPase complex in the inner membrane.

The last two groups of reactions involve multisubunit proteins that are asymmetrically oriented in the inner mitochondrial membrane.

The highly convoluted foldings, or *cristae*, of the inner membrane greatly expand its surface area, enhancing its ability to generate ATP. In typical liver mitochondria, for example, the area of the inner membrane is about five times that of the outer membrane. In fact, the total area of all mitochondrial inner membranes in liver cells is about 17 times that of the plasma membrane. In heart and skeletal muscles, mitochondria contain three times as many cristae as are found in typical liver mitochondria—presumably reflecting the greater demand for ATP by muscle cells.

In plants, mitochondria are essential for ATP production in photosynthetic cells during dark periods when photosynthesis is not possible, and all of the time in roots and other nonphotosynthetic tissues. Stored carbohydrates, mostly in the form of starch, are hydrolyzed to glucose and then metabolized to pyruvate; as in animal mitochondria, pyruvate is oxidized to CO_2, with concomitant generation of ATP.

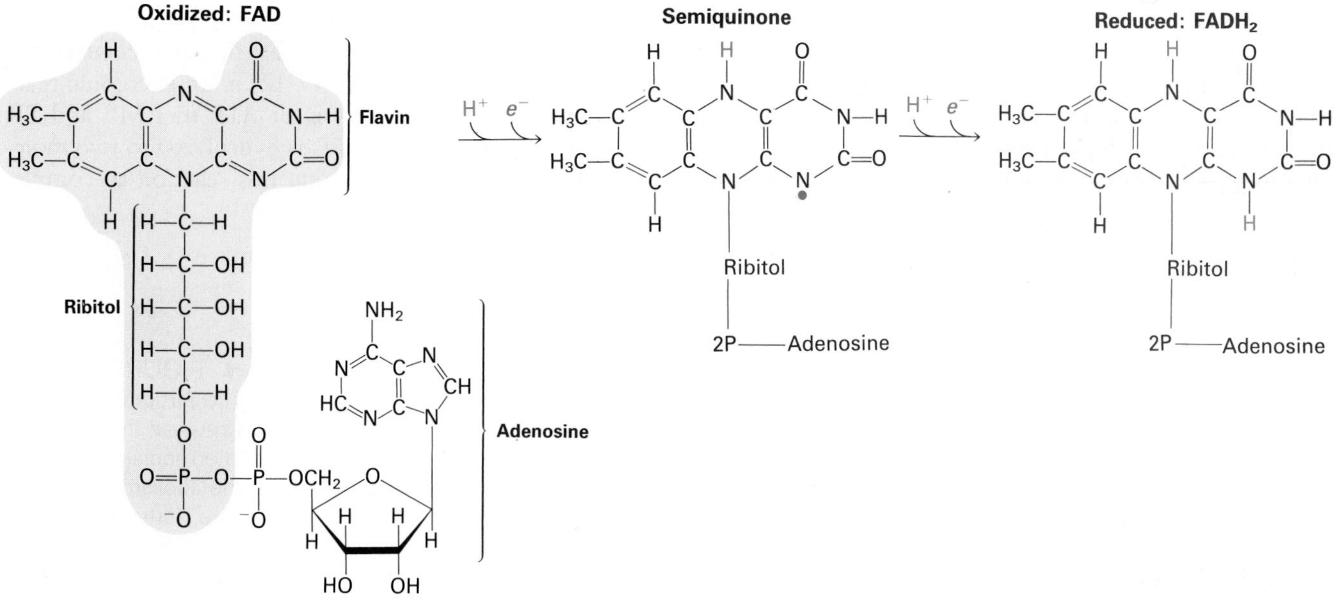

▲ FIGURE 17-10   Reduction of FAD to $FADH_2$. The cofactor flavin adenine dinucleotide (FAD) contains a three-ring flavin component that can accept one or two hydrogen atoms. The addition of one electron together with a proton (i.e., a hydrogen atom) generates a semiquinone intermediate. The semiquinone is a free radical because it contains an unpaired electron (denoted here by a blue dot), which is delocalized by resonance to all of the flavin ring atoms. The addition of a second electron and proton (i.e., a second hydrogen atom) generates the reduced form, $FADH_2$. Flavin mononucleotide (FMN) is a cofactor related to FAD that contains only the flavin–ribitol phosphate part of FAD (shown enclosed in the color screen).

Acetyl CoA Is a Key Intermediate in the Mitochondrial Metabolism of Pyruvate and Fatty Acids

Pyruvate, which is generated in the cytosol during glycolysis, is transported across the mitochondrial membranes to the matrix. The complete oxidation of pyruvate to form CO_2 and H_2O occurs in the mitochondrion and utilizes O_2 as the final electron acceptor (oxidizer).

Figure 17-9 traces the metabolism of pyruvate in the mitochondrion. Immediately on entering the matrix, pyruvate reacts with coenzyme A (HSCoA) to form CO_2 and the intermediate acetyl CoA (Figure 17-11)—a reaction catalyzed by the enzyme *pyruvate dehydrogenase*, a component of the matrix:

$$CH_3-\overset{O}{\underset{\|}{C}}-\overset{O}{\underset{\|}{C}}-O^- + HSCoA + NAD^+ \longrightarrow$$

$$CH_3-\overset{O}{\underset{\|}{C}}-SCoA + CO_2 + NADH$$

This reaction is highly exergonic ($\Delta G^{\circ\prime} = -8.0$ kcal/mol) and essentially irreversible. Pyruvate dehydrogenase is one of the most complex enzymes known. It is a giant molecule 30 nm in diameter (4.6 million MW), even larger than a ribosome, and contains 60 subunits composed of three different enzymes, several regulatory polypeptides, and five different coenzymes.

Fatty acids are also oxidized in the mitochondrion to produce acetyl CoA; the energy released is used to synthesize ATP from ADP and P_i. In most mammalian cells, in yeasts and in plants, and probably in most eukaryotes, fatty acids are degraded chiefly in peroxisomes; fatty acids containing over ≈ 20 CH_2 groups are degraded only in

these organelles. In peroxisomes, fatty acids are converted to acetyl CoA, but since H_2O_2 is generated, not ATP, we discuss peroxisomal oxidation at the end of this chapter.

Fatty acids are stored as triacylglycerols, primarily as droplets in adipose (fat-storing) cells. In response to hormones such as adrenaline, triacylglycerols are hydrolyzed in the cytosol to free fatty acids and glycerol:

$$CH_3-(CH_2)_n-\overset{O}{\underset{\|}{C}}-O-CH_2$$
$$CH_3-(CH_2)_n-\overset{O}{\underset{\|}{C}}-O-CH + 3H_2O \longrightarrow$$
$$CH_3-(CH_2)_n-\overset{O}{\underset{\|}{C}}-O-CH_2$$

$$3CH_3-(CH_2)_n-\overset{O}{\underset{\|}{C}}-OH + \begin{matrix} HO-CH_2 \\ HO-CH \\ HO-CH_2 \end{matrix}$$

Fatty acid   Glycerol

Fatty acids are released into the blood, from which they are taken up and oxidized by most cells. They are the major energy source for many tissues, particularly for heart muscle. In humans, the oxidation of fats is quantitatively more important than the oxidation of glucose as a source of ATP. In part, this is because the oxidation of 1 g of triacylglycerol to CO_2 generates about six times as much ATP as does the oxidation of 1 g of hydrated glycogen.

In the cytosol, free fatty acids are linked to coenzyme A to form an acyl CoA (Figure 17-12) in an exergonic reaction coupled to the hydrolysis of ATP to AMP and PP_i (inorganic pyrophosphate). PP_i is hydrolyzed to two molecules of phosphate (P_i), drawing this reaction to comple-

◄ FIGURE 17-11
The structure of acetyl CoA—an important intermediate in the metabolism of pyruvate, fatty acids, and many amino acids.

Coenzyme A (CoA)

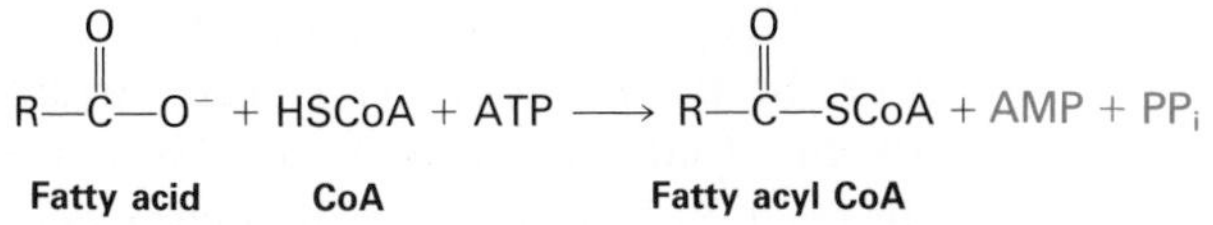

$$R-\overset{\overset{\displaystyle O}{\|}}{C}-O^- + HSCoA + ATP \longrightarrow R-\overset{\overset{\displaystyle O}{\|}}{C}-SCoA + AMP + PP_i$$

Fatty acid   **CoA**   **Fatty acyl CoA**

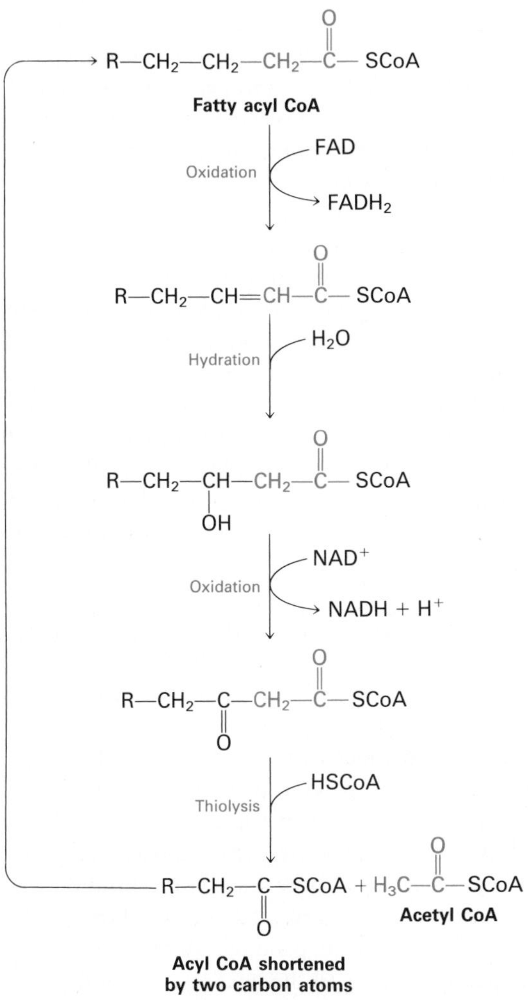

$$\longrightarrow R-CH_2-CH_2-CH_2-\overset{\overset{\displaystyle O}{\|}}{C}-SCoA$$

Fatty acyl CoA

Oxidation ⟨ FAD → FADH$_2$ ⟩

$$R-CH_2-CH=CH-\overset{\overset{\displaystyle O}{\|}}{C}-SCoA$$

Hydration ⟨ H$_2$O ⟩

$$R-CH_2-\underset{\underset{\displaystyle OH}{|}}{CH}-CH_2-\overset{\overset{\displaystyle O}{\|}}{C}-SCoA$$

Oxidation ⟨ NAD$^+$ → NADH + H$^+$ ⟩

$$R-CH_2-\overset{\overset{\displaystyle O}{\|}}{C}-CH_2-\overset{\overset{\displaystyle O}{\|}}{C}-SCoA$$

Thiolysis ⟨ HSCoA ⟩

$$R-CH_2-\overset{\overset{\displaystyle O}{\|}}{C}-SCoA + H_3C-\overset{\overset{\displaystyle O}{\|}}{C}-SCoA$$

Acetyl CoA

**Acyl CoA shortened
by two carbon atoms**

▲ FIGURE 17-12   Oxidation of fatty acids in mitochondria. Four enzymatically catalyzed reactions convert a fatty acyl CoA molecule to acetyl CoA and a fatty acyl CoA shortened by two carbon atoms. Concomitantly, one NAD$^+$ molecule is reduced to NADH and one FAD molecule is reduced to FADH$_2$. The cycle is repeated on the shortened acyl CoA until fatty acids with an even number of carbon atoms are completely converted to acetyl CoA. Fatty acids with an odd number of C atoms are rare; they are metabolized to one molecule of propionyl CoA and multiple acetyl CoAs.

tion. Then the fatty acyl group is transported across the inner mitochondrial membrane by a transporter protein and is reattached to another CoA molecule on the matrix side. Each molecule of acyl CoA in the mitochondrion is then oxidized to form one molecule of acetyl CoA and an acyl CoA shortened by two carbon atoms (see Figure 17-12). Concomitantly, one molecule apiece of NAD$^+$ and

FAD are reduced, respectively, to NADH and FADH$_2$. This set of reactions is repeated on the shortened acyl CoA until all C atoms are converted to acetyl CoA. For stearoyl CoA, the overall reaction is

$$CH_3-(CH_2)_{16}-\overset{\overset{\displaystyle O}{\|}}{C}-SCoA$$
$$+ 8HSCoA + 8FAD + 8NAD^+ + 8H_2O \longrightarrow$$
$$9CH_3-\overset{\overset{\displaystyle O}{\|}}{C}-SCoA + 8FADH_2 + 8NADH + 8H^+$$

In addition to its role in the oxidation of fatty acids and carbohydrates, acetyl CoA occupies a central position in the mitochondrial oxidation of many amino acids. It is also an intermediate in many biosynthetic reactions, such as the transfer of an acetyl group to lysine residues in histone proteins and to the N-termini of many mammalian proteins. Acetyl CoA is also a biosynthetic precursor of cholesterol and other steroids, and also of the isoprenyl groups that anchor proteins such as Ras to membranes (see Figure 14-20). In respiring mitochondria, however, the fate of the acetyl group of acetyl CoA is almost always oxidation to CO$_2$.

The Citric Acid Cycle Oxidizes the Acetyl Group of Acetyl CoA to CO$_2$ and Reduces NAD and FAD to NADH and FADH$_2$

The final stage of the oxidation of carbohydrates and lipids—the *citric acid cycle* (also called the *tricarboxylic acid cycle* and the *Krebs cycle*)—is a complex set of nine reactions (Figure 17-13). First consider the net reaction:

$$CH_3-\overset{\overset{\displaystyle O}{\|}}{C}-SCoA$$
$$+ 3NAD^+ + FAD + GDP^{3-} + P_i^{2-} + 2H_2O \longrightarrow$$
$$2CO_2 + 3NADH + FADH_2 + GTP^{4-} + 2H^+ + HSCoA$$

Note that there is no involvement of molecular O$_2$ in this cycle and that only one high-energy phosphate bond is synthesized by substrate-level phosphorylation (in GTP). The two carbon atoms in acetyl CoA are oxidized to two molecules of CO$_2$. Concomitantly, the released electrons are transferred to NAD$^+$ and FAD to form the reduced molecules NADH and FADH$_2$.

Now let us look at the cycle in detail. Note that Figure 17-13 depicts the oxidation of a single molecule of acetyl CoA, but we must keep in mind that the oxidation of one glucose molecule will have generated two molecules of acetyl CoA, and the oxidation of one molecule of stearic acid generates nine of acetyl CoA.

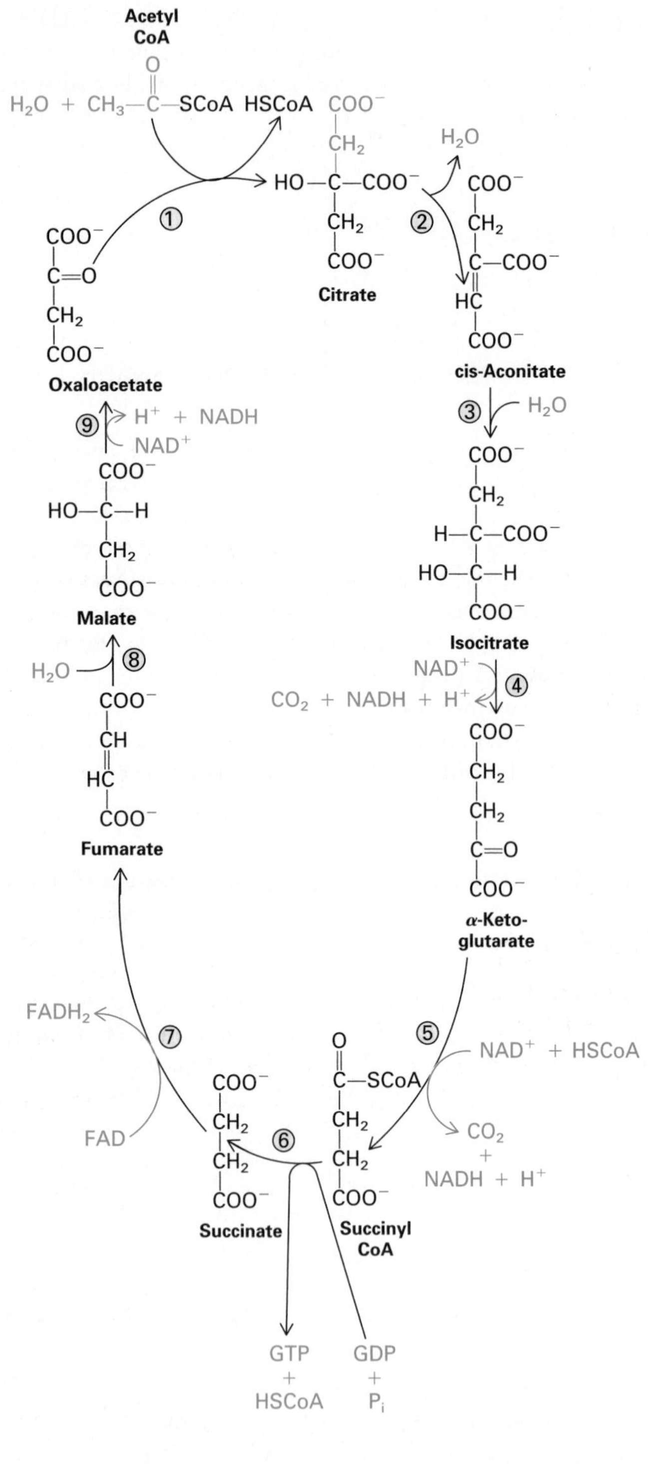

Acetyl
CoA

$H_2O + CH_3-\overset{\overset{\text{O}}{\|}}{C}-SCoA$ HSCoA COO^-

... (citric acid cycle diagram with the following labeled intermediates and reactions)

Oxaloacetate, Citrate, cis-Aconitate, Isocitrate, α-Ketoglutarate, Succinyl CoA, Succinate, Fumarate, Malate

Reaction 9: $H^+ + NADH$ / NAD^+
Reaction 2: H_2O
Reaction 3: H_2O
Reaction 4: NAD^+ ; $CO_2 + NADH + H^+$
Reaction 5: $NAD^+ + HSCoA$; $CO_2 + NADH + H^+$
Reaction 6: $GTP + HSCoA$ / $GDP + P_i$
Reaction 7: $FADH_2$ / FAD
Reaction 8: H_2O

As shown in Figure 17-13, the cycle begins with the condensation of the two-carbon acetyl group from acetyl CoA with the four-carbon molecule oxaloacetate. The product of reaction 1 is the six-carbon citric acid, for which the cycle is named. In reactions 2 and 3, citrate is isomerized to the six-carbon molecule isocitrate by the single enzyme aconitase. In reaction 4, isocitrate is oxidized to the five-carbon α-ketoglutarate, generating one CO_2 molecule and reducing one molecule of NAD^+ to NADH. In reaction 5, the α-ketoglutarate is oxidized to the four-carbon molecule succinyl CoA, generating the second CO_2 molecule formed during each turn of the cycle and reducing another NAD^+ molecule to NADH. In reactions 6–9, succinyl CoA is oxidized to oxaloacetate, regenerating the molecule that was initially condensed with acetyl CoA. Concomitantly, one FAD molecule is reduced to $FADH_2$ and one NAD^+ molecule to NADH. The conversion of succinyl CoA to succinate (reaction 6) is coupled to the synthesis of one GTP molecule (from GDP and P_i); this reaction is slightly exergonic ($\Delta G^{\circ\prime} = -0.8$ kcal/mol).

Most enzymes and small molecules involved in the citric acid cycle are soluble in aqueous solution and are localized to the matrix of the mitochondrion. This includes the water-soluble molecules CoA, acetyl CoA, and succinyl CoA, as well as NAD^+ and NADH. Succinate dehydrogenase (reaction 7) together with the $FAD/FADH_2$ and α-ketoglutarate dehydrogenase (reaction 5) are localized to the inner membrane.

The protein concentration of the mitochondrial matrix is 500 mg/ml (a 50 percent protein solution), and the matrix must have a viscous, gel-like consistency. When mitochondria are disrupted by gentle ultrasonic vibration or osmotic lysis, the six non–membrane-bound enzymes in the citric acid cycle are released as a very large multiprotein complex. The reaction product of one enzyme, it is believed, passes directly to the next enzyme without diffusing through the solution. However, much work is needed to determine the structure of the enzyme complex: biochemists generally study the properties of enzymes in dilute aqueous solutions of less than 1 mg/ml, and weak interactions between enzymes are often difficult to detect.

Electrons Are Transferred from NADH and $FADH_2$ to Molecular O_2 by Electron-Carrier Proteins

In summary, the reactions in the glycolytic pathway and citric acid cycle result in the conversion of one glucose molecule to six CO_2 molecules and the concomitant reduction of 10 NAD^+ to 10 NADH molecules and of two FAD to two $FADH_2$ molecules (Table 17-1). The reduced coenzymes are reoxidized by molecular O_2 in a multistep process. NADH and $FADH_2$ first transfer electrons to acceptor molecules in the inner mitochondrial membrane; the loss of electrons regenerates the oxidized forms of NAD^+ and FAD as well as the reduced form of the acceptor. The

▲ FIGURE 17-13   The citric acid cycle. First, a two-carbon acetyl residue from acetyl CoA condenses with the four-carbon molecule oxaloacetate (reaction 1) to form the six-carbon molecule citrate. Through a sequence of enzymatically catalyzed reactions (2–9), each molecule of citrate is eventually converted to oxaloacetate, losing two CO_2 molecules in the process. In four of the reactions, four pairs of electrons are removed from the carbon atoms: three pairs are transferred to three molecules of NAD^+ to form $3NADH + 3H^+$; one pair is transferred to the acceptor FAD to form $FADH_2$.

TABLE 17-1   Net Result of the Glycolytic Pathway and the Citric Acid Cycle

Reaction	CO_2 Molecules Produced	NADH Molecules Produced	$FADH_2$ Molecules Produced
1 Glucose molecule to 2 pyruvates	0	2	0
2 Pyruvates to 2 acetyl CoAs	2	2	0
2 Acetyl CoAs to 4 CO_2s	4	6	2
Total	6	10	2

electrons released from NADH and $FADH_2$ are transferred along the electron transport chain, a group of electron carriers all but one of which are integral components of the inner membrane. Eventually, they are transferred to O_2, forming H_2O.

The following overall reactions summarize these steps:

$$NADH + H^+ + \tfrac{1}{2}O_2 \longrightarrow NAD^+ + H_2O$$

$$\Delta G^{\circ\prime} = -52.6 \text{ kcal/mol}$$

$$FADH_2 + \tfrac{1}{2}O_2 \longrightarrow FAD + H_2O$$

$$\Delta G^{\circ\prime} = -43.4 \text{ kcal/mol}$$

As the negative $\Delta G^{\circ\prime}$ values indicate, the oxidation of these reduced coenzymes by O_2 are strongly exergonic reactions. More importantly, most of the free energy released during the oxidation of glucose to CO_2 is retained in the reduced coenzymes generated during glycolysis and the citric acid cycle. To see this, recall the large total change in standard free energy for the oxidation of glucose to CO_2 ($\Delta G^{\circ\prime} = -680$ kcal/mol). The oxidation of the 10 NADH and 2 $FADH_2$ molecules yields an almost equivalent change of $\Delta G^{\circ\prime} = 10 (-52.6) + 2 (-43.4) = -613$ kcal/mol of glucose. Thus over 90 percent of the potential free energy present in the glucose bonds that become oxidized is conserved in the reduced coenzymes. The reoxidation of these coenzymes by O_2 generates the vast majority of ATP phosphoanhydride bonds.

The free energy released during the oxidation of a single NADH or $FADH_2$ molecule by O_2 is sufficient to drive the synthesis of several molecules of ATP from ADP and P_i ($\Delta G^{\circ\prime} = +7.3$ kcal/mol for the reaction ADP + $P_i \rightarrow$ ATP). Thus it is not surprising that NADH oxidation and ATP synthesis do not occur in a single reaction. Rather, a step-by-step transfer of electrons from NADH to O_2, via the electron transport chain, allows the free energy to be released in small increments. At several sites during electron transport from NADH to O_2, protons are transported

across the inner mitochondrial membrane and a proton concentration gradient forms across it. Because the outer membrane is freely permeable to protons, the pH of the mitochondrial matrix is higher (i.e., the proton concentration is lower) than that of the cytosol and intermembrane space (see Figure 17-9). An electric potential across the inner membrane also results from the pumping of positively charged protons outward from the matrix, which becomes negative with respect to the intermembrane space. Thus free energy released during the oxidation of NADH or $FADH_2$ is stored both as an electric potential and a proton concentration gradient—collectively, the proton-motive force—across the inner membrane. The movement of protons back across the inner membrane, driven by this force, is coupled to the synthesis of ATP from ADP and P_i.

ATP synthesis is catalyzed by the F_0F_1 ATPase complex, also called the ATP synthase, a member of the F class of ATP-powered proton pumps (see Figure 15-10). Such complexes form the knob-like particles that protrude from the inner mitochondrial membrane (see Figure 17-9).

A Similar Electrochemical Proton Gradient Is Used to Generate ATP from ADP and P_i in Mitochondria, Bacteria, and Chloroplasts

Bacteria lack mitochondria, yet aerobic bacteria carry out the same processes of oxidative phosphorylation that occur in eukaryotic mitochondria. Bacterial enzymes that catalyze the reactions of both the glycolytic pathway and the citric acid cycle are localized to the bacterial cytosol; enzymes that oxidize NADH to NAD^+ and transfer the electrons to the ultimate acceptor O_2 are localized to the bacterial plasma membrane. The movement of electrons through these membrane carriers is coupled to the pumping of protons out of the cell (Figure 17-14). The movement of protons back into the cell, down their concentration gradient, is coupled to the synthesis of ATP; the membrane proteins involved in ATP synthesis are essentially identical in structure and function to those of the

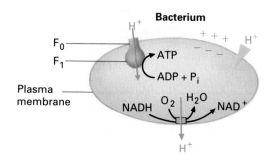

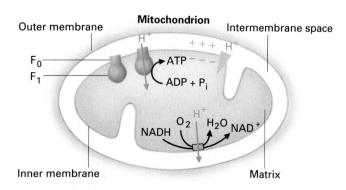

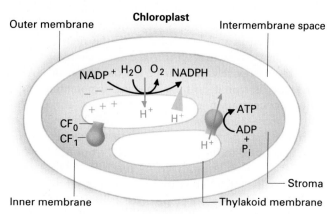

▲ FIGURE 17-14   Membrane orientation and the direction of proton movement in bacteria, mitochondria, and chloroplasts. The membrane surface facing a shaded area is a cytosolic face; the surface facing an unshaded area is an exoplasmic face. In bacteria, mitochondria, and chloroplasts, the F_0F_1 complexes always face the cytosolic face of the membrane. During electron transport, protons are always moved from the cytosolic face to the exoplasmic face, creating a proton concentration gradient (exoplasmic face > cytosolic face) and an electric potential (negative cytosolic face) across the membrane. During the generation of ATP, protons flow in the reverse direction (down their electrochemical proton gradient) through the F_0F_1 complexes.

Recall that every cellular membrane has a cytosolic face and an exoplasmic face. In chloroplasts, mitochondria, and bacteria, the F_0F_1 ATPase complex is always positioned so that the globular F_1 segment is on the cytosolic face. The F_1 segment catalyzes ATP synthesis from ADP and P_i. Thus in all cases, ATP is formed on the cytosolic face of the membrane. In the mitochondrion, for instance, ATP is synthesized on the matrix side of the inner membrane. (Note that the cytosolic face is toward the matrix or stroma, respectively, in the inner mitochondrial membrane and the thylakoid membrane.)

Protons always flow through the F_0F_1 ATPase complex across the membrane from the exoplasmic to the cytosolic face, driven by a combination of the proton concentration gradient (exoplasmic face > cytosolic face) and the membrane electric potential (the cytosolic face is negative with respect to the exoplasmic face). The polarity of this proton gradient and membrane potential is established by electron transport. During electron transport in mitochondria, protons are translocated from the cytosolic face to the exoplasmic face (see Figure 17-14), since the direction of proton movement is from the matrix to the intermembrane space. Similarly, in aerobic bacteria, the oxidation of carbohydrates is coupled to electron transport and to translocation of protons from the cytosol into the external medium.

During oxidative phosphorylation in both bacteria and mitochondria, the movement of electrons through the membrane is coupled to the generation of a proton-motive force, making the processes of electron transport, proton pumping, and ATP formation interdependent. The generation of this electrochemical proton gradient and its simultaneous dissipation in ATP formation normally occur and the processes are closely coupled.

Through the combined efforts of biochemists, biophysicists, microscopists, and geneticists, much has been learned about the process of oxidative phosphorylation. The following sections describe the way in which a proton concentration gradient is used to generate ATP in oxidative phosphorylation and the ways in which electron transport is coupled to the generation of a proton-motive force.

► The Proton-Motive Force, ATP Generation, and Transport of Metabolites

Closed Vesicles Are Required for the Generation of ATP

Much evidence shows that in mitochondria the coupling between the oxidation of NADH and $FADH_2$ by O_2 and the synthesis of ATP from ADP and P_i occurs only via the electrochemical proton gradient across the inner membrane. In the laboratory, adding oxygen and an oxidizable substrate such as pyruvate or succinate to isolated intact

mitochondrial F_0F_1 ATPase complex, which we examine in some detail later in this chapter. The proton-motive force across the bacterial plasma membrane is also used to power the uptake of nutrients such as sugars (see Figure 15-23). Chloroplasts also utilize an F_0F_1 complex to synthesize ATP; these particles are localized to the thylakoid membranes (see Figure 17-14).

mitochondria results in a net synthesis of ATP. However, ATP production is absolutely dependent on the integrity of the inner mitochondrial membrane. In the presence of minute amounts of detergents that make the membrane leaky, the oxidation of these metabolites by O_2 still occurs, but no ATP is made. Under these conditions, no transmembrane proton concentration gradient or membrane electric potential can be maintained.

The Proton-Motive Force Is Composed of a Proton Concentration Gradient and a Membrane Electric Potential

The proton-motive force that propels protons down their electrochemical gradient is a combination, or sum, of the proton concentration gradient and the membrane electric potential. A similar combination of forces drives the movement of Na^+ ions into a cell (see Figure 15-9).

Since protons are positively charged, moving them across the membrane generates an electric potential only if the membrane is poorly permeable to other cations and to anions. A proton concentration (pH) gradient can develop only if the membrane is permeable to a major anion, such as Cl^- (chloride), or if the protons are exchanged for another cation, such as K^+ (potassium). In the latter case, proton movement does not lead to a difference in electric potential across the membrane because there is always an equal concentration of positive and negative ions on each side of the membrane. However, proton movement does produce a pH gradient across the membrane, making the proton concentration gradient the major component of the proton-motive force in such cases. This occurs in the chloroplast thylakoid membrane during photosynthesis (Chapter 18). By contrast, the mitochondrial inner membrane is relatively impermeable to both anions and cations: a greater proportion of energy is stored as a membrane electric potential, and the actual pH gradient is smaller. In respiring mitochondria, the electric potential across the inner membrane is about 200 mV (the matrix being negative with respect to the intermembrane space), so that the membrane electric potential is the more significant component of the proton-motive force.

Since a difference of one pH unit represents a tenfold difference in H^+ concentration, a pH gradient of one unit across a membrane is equivalent to an electric potential of 59 mV (at 20° C). Thus we can define the proton-motive force, pmf, as

$$pmf = \Psi - \left(\frac{RT}{\mathscr{F}}\right) \cdot \Delta pH = \Psi - 59\Delta pH$$

where R is the gas constant of 1.987 cal/(degree-mol); T is the temperature (in degrees Kelvin); $\mathscr{F}$ is the Faraday constant (23,062 cal · V^{-1} · mol^{-1}); and Ψ is the transmembrane electric potential; Ψ and pmf are measured in milli-

volts. In respiring mitochondria, pmf $\simeq$ 220 mV; $\Psi \simeq$ 160 mV, and a pH of one unit (equivalent to ~60 mV) accounts for the remaining pmf.

Mitochondria and chloroplasts are much too small to be impaled with electrodes, so how can the electric potential and the pH gradient across the inner mitochondrial membrane be determined? The inner membrane is normally impermeable to potassium ions, but the antibiotic valinomycin (see Figure 15-6) is a potassium ionophore: a lipid-soluble peptide that selectively binds one K^+ ion in its hydrophilic interior, thus allowing K^+ ions to be transported across the otherwise impermeable phospholipid membrane (see Figure 17-28c). At equilibrium, the concentration of K^+ ions on both sides of the membrane is determined by the membrane electric potential, E (in mV), according to the Nernst equation (page 642):

$$E = -\frac{RT}{\mathscr{F}} \ln \frac{K_{in}}{K_{out}} = -59 \log \frac{K_{in}}{K_{out}}$$

When trace amounts of valinomycin and radioactive potassium ($^{42}K^+$) ions are added to a suspension of respiring mitochondria, oxidative phosphorylation proceeds and is largely unaffected. The $^{42}K^+$ ions accumulate inside the mitochondria in a K_{matrix}:$K_{cytosol}$ ratio of about 2500. By the Nernst equation

$$E = -59 \log \frac{K_{in}}{K_{out}} = -59 \log 2500 = -200 \text{ mV}$$

Thus the electric potential across the inner membrane is ~200 mV (negative inside matrix).

The fluorescent properties of a number of dyes are dependent on pH (Chapter 5). By trapping such dyes inside inner-membrane vesicles of mitochondria, we can measure the inside pH during oxidative phosphorylation; as noted, the matrix pH is typically one unit higher than that of the cytosol.

The F_0F_1 Complex Couples ATP Synthesis to Proton Movement down the Electrochemical Gradient

The F_0F_1 ATPase complex (Figure 17-15) couples proton movement down its electrochemical gradient with the synthesis of ATP from ADP and P_i. The enzyme complex has two principal components: F_0 and F_1.

The integral membrane complex F_0 contains three types of subunits, a, b, and c. In bacteria, these have the composition $a_1b_2c_{10}$. Each a subunit is thought to span the membrane eight times, each b once, and each c twice. In mitochondria, each F_0 complex also contains, depending on the species, two to five additional peptides of unknown function (Table 17-2).

(a)

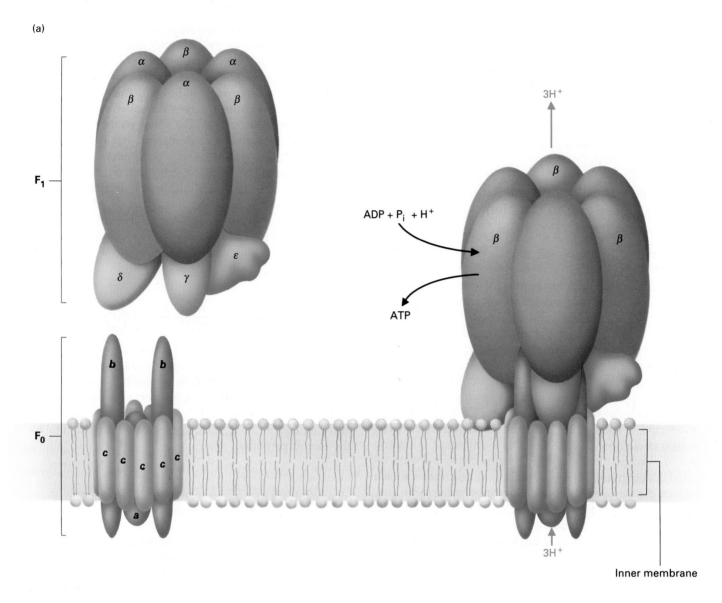

(b)

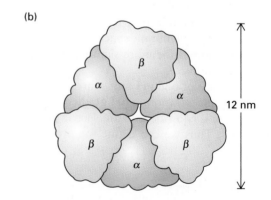

▲ FIGURE 17-15   Structure of ATP synthase (the F_0F_1 ATPase complex). (a) The ATP synthase of mitochondria, bacteria, and chloroplasts. This bipartite enzyme is composed of two oligomeric complexes, F_0 and F_1. The F_0 portion is an integral *membrane protein;* in bacteria, it consists of three subunits *(a, b,* and *c)* that form a proton conduction channel. The F_1 portion contains three copies each of subunits α and β and is bound to F_0 via subunits γ, δ, and ε. (b) Arrangement, viewed from the "top," of the α and β subunits of F_1, as determined by x-ray crystallography.

The synthesis of one ATP molecule from ADP and P_i occurs spontaneously at the catalytic site on a β subunit of F_1, due to tight binding of ATP at this site. Proton movement through F_0, driven by the proton-motive force, promotes the catalytic synthesis of ATP by causing the bound ATP to be released; this frees up the site for the binding of ADP and P_i, which, in turn, spontaneously combine to form another tightly bound ATP. Catalysis alternates between the three copies of the β subunit active site, as shown in Figure 17-17. [Part (a) courtesy of B. Trumpower; part (b) from M. Bianchet et al., 1991, *J. Biol. Chem.* **266:**21197–21201.]

TABLE 17-2 Protein Composition of Several Enzyme Complexes Involved in Electron Transport and ATP Synthesis

Bacteria	In Mitochondria
F_0	
3 subunits	*5–8 subunits*
a: proton conduction	*a*: proton conduction
b: proton conduction	*b*: proton conduction
c: proteolipid	*c*: proteolipid
	2–5 accessory proteins
CoQH$_2$–CYTOCHROME C REDUCTASE	
3 subunits	*9–11 subunits*
cytochrome *b*	cytochrome *b*
cytochrome c_1	cytochrome c_1
FeS protein	FeS protein
	6–8 accessory proteins
CYTOCHROME C OXIDASE	
3 subunits	*11–13 subunits*
I (contains hemes *a* and a_3 and Cu_b^+)	I (contains hemes *a* and a_3 and Cu_b^+)
II (contains Cu_a^+)	II (contains Cu_a^+)
III	III
	8–11 others, depending on species

F_0 contains a transmembrane channel through which protons flow to F_1. When F_0 or its purified proteolipid subunit *c* is experimentally incorporated into phospholipid vesicles (liposomes), the permeability of the vesicles to H$^+$ is greatly stimulated, indicating that the 10 copies of the proteolipid subunit *c* form a proton channel. Each subunit *c* contains two membrane-spanning α helixes, and an aspartate residue in one of these helixes is thought to participate in proton movement, since mutation of this aspartate, or chemical modification of this aspartate by the chemical dicyclohexylcarbodiimide, specifically blocks H$^+$ translocation.

Attached to F_0 is F_1, a water-soluble complex of five distinct polypeptides (α, β, γ, δ, and ϵ) with the composition $\alpha_3\beta_3\gamma\delta\epsilon$. F_1 forms the knobs that protrude on the matrix side of the inner membrane (see Figures 17-14 and 17-15a); it can be detached from the inner membrane by mechanical agitation. When F_1 is physically separated from the membrane, it is capable only of catalyzing ATP hy-

drolysis. Hence, it has been called the F_1 ATPase. Its natural function, however, is synthesis of ATP. Submitochondrial vesicles from which F_1 is removed cannot catalyze ATP synthesis; when F_1 particles reassociate with these vesicles, they once again become fully active in ATP synthesis (Figure 17-16).

F_1 can be dissociated into its component polypeptides, and the functions of several of these subunits are known, at least in outline. Each of the three copies of subunit β binds ATP and ADP and contains a catalytic site for the synthesis of ATP from ADP and P$_i$; subunits γ and δ and possibly ϵ bind to F_0 (see Figure 17-15a). ATP or ADP also binds to regulatory or allosteric sites on the three α subunits; this binding modifies the rate of ATP synthesis according to the level of ATP and ADP in the matrix.

Although it is not an integral membrane protein, F_1 probably forms the barrier that makes the inner membrane impermeable to protons, since removal of F_1 from the inner mitochondrial membrane makes it highly permeable to protons through the F_0 channel.

Exactly how the movement of protons from the intermembrane space through the F_0F_1 complex into the matrix space is coupled to ATP generation is the subject of intense debate and experimentation. The coupling between proton flow and ATP synthesis must be indirect, since the ATP, P$_i$, and ADP binding sites on the β subunits of the F_1 complex, where ATP synthesis occurs, are 9–10 nm from the surface of the mitochondrial membrane.

Concerning ATP formation, most evidence suggests that ADP and P$_i$, after binding to a β subunit, spontaneously form ATP, which remains tightly bound to F_1; experiments show that $\Delta G^{\circ\prime}$ is near 0 for the reaction in which ATP is formed from ADP and P$_i$ bound to isolated F_1 particles. Presumably, dissociation of the bound ATP from the protein is induced by a conformational change in the F_1 complex that, in turn, is caused by the flow of protons through it. In one hypothetical mechanism for ATP synthesis, called the *binding-change mechanism* (Figure 17-17), proton flux in essence causes a 120° rotation of the $\alpha_3\beta_3$ part of F_1, so that each β subunit is reoriented relative to subunits γ, δ, and ϵ and changes its binding affinities for ATP, ADP, and P$_i$.

The release of the bound ATP frees the catalytic site for further ADP and P$_i$ binding, which again spontaneously forms tightly bound ATP. The catalysis alternates between the three copies of the β subunit, which undergo sequential conformational changes.

Most experiments indicate that the passage of three protons through the F_0F_1 complex is coupled to the synthesis of one high-energy phosphate bond in ATP. A simple calculation indicates that the passage of more than one proton is required to synthesize one molecule of ATP. For example, for the reaction

$$H^+ + ADP^{3-} + P_i^{2-} \rightleftharpoons ATP^{4-} + H_2O$$

(a)

(b)

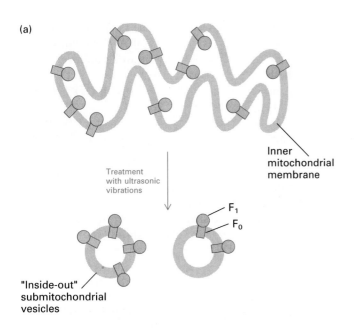

Treatment with ultrasonic vibrations

Inner mitochondrial membrane

F_1
F_0

"Inside-out" submitochondrial vesicles

"Inside-out" submitochondrial vesicles capable of both electron transport and ATP synthesis

Mechanical agitation

F_1 particles are incapable of electron transport or ATP synthesis but have ATPase activity

Membrane vesicles capable of electron transport, not of ATP synthesis

Reconstituted vesicles capable of both electron transport and ATP synthesis

▲ FIGURE 17-16   Demonstration that mitochondrial F_1 particles are required for ATP synthesis, not for electron transport. (a) "Inside-out" submitochondrial vesicles can be obtained by exposing the inner membrane to ultrasonic vibration, so that the vesicles are disrupted and reseal with the F_1 particles facing outside. These vesicles can transfer electrons from added NADH to O_2 with the concomitant synthesis of ATP from ADP and P_i. (b) The mechanical agitation of inside-out vesicles causes the F_1 particles to dissociate from the inner membrane. The membrane vesicles can still transport electrons from NADH to O_2 but cannot synthesize ATP. The subsequent addition of F_1 particles reconstitutes the native membrane structure, restoring the capacity for ATP synthesis.

at standard concentrations of ATP, ADP, and P_i

$$\Delta G^{\circ \prime} = +7.3 \text{ kcal/mol}$$

At the concentrations of reactants in the mitochondrion, however, ΔG is probably higher (+10 to +12 kcal/mol). The amount of free energy released by the passage of 1 mol of protons down an electrochemical gradient of 220 mV (0.22 V) can be calculated from the Nernst equation, setting $n = 1$ and measuring ΔE in volts:

$$\Delta G \text{ (cal/mol)} = -n\mathscr{F}\Delta E = -(23{,}062 \text{ cal} \cdot V^{-1} \cdot \text{mol}^{-1})\Delta E$$
$$= (23{,}062 \text{ cal} \cdot V^{-1} \cdot \text{mol}^{-1})(0.22 \text{ V})$$
$$= -5{,}080 \text{ cal/mol, or } -5.1 \text{ kcal/mol}$$

Since just over 5 kcal/mol of free energy is made available, the passage of at least two, and more likely three, protons is essential to the synthesis of each molecule of ATP from ADP and P_i.

Reconstitution of Closed Membrane Vesicles Supports the Role of the Proton-Motive Force in ATP Synthesis

The view that a proton-motive force is the immediate source of energy for ATP synthesis, introduced in 1961 by Peter Mitchell, was initially opposed by virtually all researchers working in oxidative phosphorylation and photosynthesis. They favored a mechanism similar to the well-elucidated substrate-level phosphorylation in glycolysis, in which electron transport is directly coupled to ATP synthesis (see Figures 17-4, 17-5, and 17-6). Electron transport through the membranes of chloroplasts or mitochondria was believed to generate a high-energy chemical bond directly (a phosphate linked to an enzyme by an ester bond, for example), which was then used to convert P_i and ADP

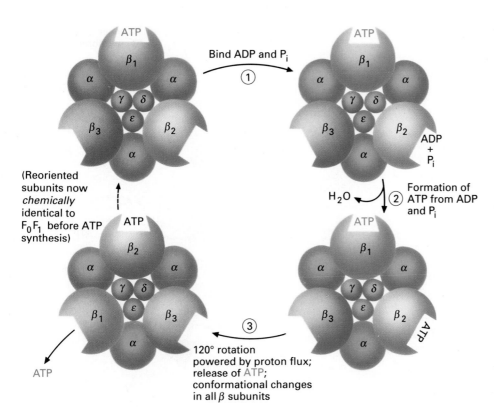

◄ FIGURE 17-17   Rotational model for the synthesis of ATP from ADP and P_i by F_0F_1 ATPase. The three β subunits (see Figure 17-15) alternate between three conformational states that differ in their binding affinities for ATP, ADP, and P_i. In reaction 1, ADP and P_i bind to one of the three β subunits (here, arbitrarily designated β_2) to form ATP spontaneously (reaction 2). Proton flux (reaction 3) powers a 120° rotation of the $\alpha_3\beta_3$ part of the F_1 complex relative to subunits γ, δ, and ε, causing ATP to be released from one β subunit (here, β_1) with concomitant conformational changes in all three β subunits. [After R. L. Cross, D. Cunningham, and J. K. Tamura, 1984, *Curr. Top. Cell Reg.* **24**:336; P. Boyer, 1989, *FASEB J.* **3**:2164–2178.]

to ATP. Despite intense efforts by a large number of investigators, however, no such intermediate could ever be identified.

But clear-cut evidence to support the role of the proton-motive force in ATP synthesis was not easy to obtain until the development of techniques to purify and reconstitute organelle membranes and membrane proteins. Then Mitchell's chemiosmotic mechanism became accepted, although many of its details are still unclear. Here, it is appropriate to summarize some key experiments that established the basic features of the proton-motive force.

One experiment directly demonstrated that an artificially imposed pH gradient can result in ATP synthesis (Figure 17-18). Chloroplast thylakoid vesicles containing F_0F_1 particles were equilibrated in the dark with a buffered solution at pH 4.0. When the pH in the thylakoid lumen was also 4.0, the vesicles were rapidly mixed with a solution at pH 8.0 containing ADP and P_i. A burst of ATP synthesis accompanied the transmembrane movement of protons driven by the 10,000-fold (10^{-4} M versus 10^{-8} M) concentration gradient. In reciprocal experiments on "inside-out" preparations of submitochondrial vesicles, an artificially generated membrane electric potential also resulted in ATP synthesis.

Particularly dramatic results were achieved in an experiment employing bacteriorhodopsin. This protein (see Figure 14-18) was asymmetrically incorporated into liposomes that also contained purified mitochondrial F_0F_1 complexes. Illumination caused the bacteriorhodopsin to pump protons into the vesicular lumen—the mechanism

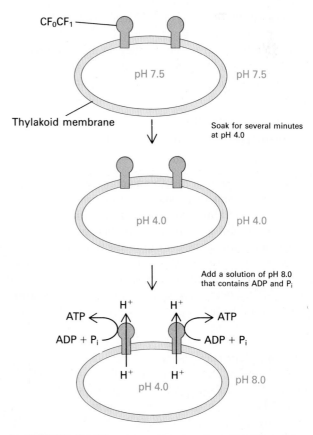

▲ FIGURE 17-18   Demonstration that ATP synthesis from ADP and P_i in chloroplast thylakoid membranes results from an artificially imposed pH gradient.

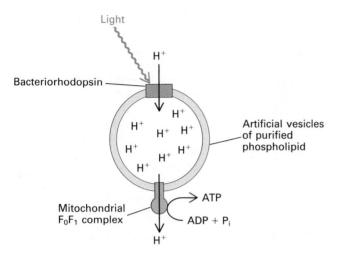

▲ FIGURE 17-19   Demonstration that the mitochondrial F_0F_1 complex is the site of ATP synthesis. Purified mitochondrial F_0F_1 complexes and bacteriorhodopsin are incorporated into the same artificial phospholipid vesicles (liposomes). Bacteriorhodopsin pumps protons into the lumen following illumination, which results in the synthesis of ATP from ADP and P_i.

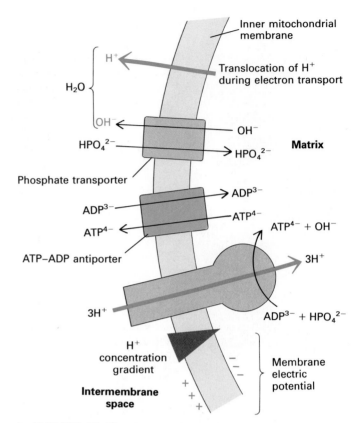

▲ FIGURE 17-20   The phosphate and ATP-ADP transport system in the inner mitochondrial membrane. The phosphate transporter couples the uptake of one HPO_4^{2-} to the outward movement of one OH^- anion. Similarly, the ATP-ADP antiporter exchanges one incoming ADP^{3-} for one ATP^{4-} exported from the matrix. The exported OH^- combines with a proton translocated outward by respiration, so that the net result is the uptake of one ADP^{3-} and one HPO_4^{2-} in exchange for one ATP^{4-}, powered by the outward translocation of one proton during electron transport. For every four protons translocated outward, three are used to synthesize one ATP molecule and one is used to export ATP in exchange for ADP and P_i. The outer membrane is not shown here because it is permeable to molecules smaller than 5,000 MW.

by which photosynthetic archaebacteria synthesizing bacteriorhodopsin generate a proton-motive force during photosynthesis. In the experiment, the resultant proton-motive force powered the synthesis of ATP from ADP and P_i; if no F_0F_1 was present, no ATP was made (Figure 17-19).

These findings left no doubt that the F_0F_1 complex is the ATP-generating enzyme and that ATP generation is dependent on an electrochemical proton gradient.

Many Transporters in the Inner Mitochondrial Membrane Are Powered by the Proton-Motive Force

The inner mitochondrial membrane contains a number of proteins that transport various metabolites into and out of the organelle, including pyruvate, malate, and the amino acids aspartate and glutamate. Two such proteins transport ADP and P_i from the cytosol to the mitochondrion in exchange for ATP formed by oxidative phosphorylation inside the mitochondrion and transported to the cytosol (Figure 17-20). Energy stored in the proton concentration gradient is used to power this exchange of ATP for ADP and P_i. The two components of this system are the *phosphate transporter* and the *ATP-ADP antiport protein*.

The phosphate transporter is an antiporter: it catalyzes the transmembrane exchange of one HPO_4^{2-} for one OH^- (Figure 17-20). ADP entry and ATP exit are similarly coupled: the ATP-ADP antiporter allows one molecule of ADP to enter only if one molecule of ATP exits simultaneously. The ATP-ADP antiporter, a dimer of two 30,000-MW subunits, makes up 10–15 percent of the protein in the inner

membrane, so it is one of the more abundant mitochondrial proteins. The overall reaction for the two antiporters functioning together is:

$$ADP^{3-}_{out} + HPO_4^{2-}_{out} + OH^-_{in} + ATP^{4-}_{in} \rightleftharpoons$$
$$ADP^{3-}_{in} + HPO_4^{2-}_{in} + OH^-_{out} + ATP^{4-}_{out}$$

Each OH^- transported outward combines with a proton, translocated during electron transport to the intermembrane space, to form H_2O. This drives the overall reaction—written above—to the right, in the direction of ATP export and ADP and P_i import. Because some of the protons translocated out of the mitochondrion during electron transport provide the power (by combining with the exported OH^-) for the ATP-ADP exchange, fewer protons are available for ATP synthesis. For every four protons translocated out, three are used to synthesize one ATP molecule and one is used to power the export of ATP from the mitochondrion in exchange for ADP and P_i. This expenditure of energy from the proton concentration gradient to export ATP from the mitochondrion in exchange for ADP and P_i ensures a high ratio of ATP to ADP in the cytosol, where the phosphoanhydride-bond energy of ATP is utilized to power many energy-requiring reactions.

Inner-Membrane Proteins Allow the Uptake of Electrons from Cytosolic NADH

As stated in our earlier discussion of glucose metabolism, two molecules of NAD^+ are reduced to NADH during the conversion of glucose to pyruvate in the cytosol. The electrons from NADH are ultimately transferred to O_2, concomitant with the generation of ATP. However, the inner mitochondrial membrane is impermeable to NADH, so how do the electrons from cytosolic NADH enter the mitochondrial electron transport system?

Several *electron shuttles* are employed. In the most widespread—the *malate shuttle* (Figure 17-21)—cytosolic NADH reduces the four-carbon oxaloacetate to malate. Malate crosses the inner mitochondrial membrane (in exchange for α-ketoglutarate) and reduces NAD^+, forming NADH and oxaloacetate in the matrix. But oxaloacetate does not pass directly back to the cytosol: it is first converted to the amino acid aspartate in order to cross in exchange for glutamate; once in the cytosol, the aspartate is reconverted to oxaloacetate. To allow the cycle to proceed, α-ketoglutarate is converted in the cytosol to the amino acid glutamate, which crosses to the matrix where it is reconverted to α-ketoglutarate. The net effect of this complex cycle is the oxidation of cytosolic NADH to NAD^+, together with the reduction of matrix NAD^+ to NADH.

► NADH, Electron Transport, and Proton Translocation

In the previous section we described studies that showed how a proton concentration gradient and a membrane electric potential can generate ATP and power the import and export of molecules. We now turn to the experiments which revealed that the source of the proton-motive force is the movement of electrons from the coenzymes NADH and $FADH_2$ (reduced from NAD^+ and FAD during glycolysis, the citric acid cycle, and the oxidation of fatty acids) to the final electron acceptor, O_2. Electron movement from NADH and $FADH_2$ to O_2 occurs in a number of steps that are catalyzed by a series of electron carriers in four multiprotein complexes in the inner mitochondrial membrane (Figure 17-22).

Electron Transport in Mitochondria Is Coupled to Proton Translocation

As is demonstrated by the experiment in Figure 17-23, isolated mitochondria maintained without O_2 do not oxidize NADH (or $FADH_2$ or succinate) because there is no ultimate electron acceptor. (Nor do they generate ATP.) The addition of a small amount of O_2 to such anaerobic mitochondria results in the oxidation of NADH:

$$2NADH + 2H^+ + O_2 \longrightarrow 2NAD^+ + 2H_2O$$

Each NADH molecule releases two electrons to the electron transport chain. These two electrons ultimately reduce one oxygen atom (one-half of an O_2 oxygen molecule), forming one molecule of water:

$$NADH \longrightarrow NAD^+ + H^+ + 2e^-$$

and

$$2e^- + 2H^+ + \tfrac{1}{2}O_2 \longrightarrow H_2O$$

Electron transport from NADH (or $FADH_2$) to O_2 is coupled to proton transport across the membrane, as the same experiment (Figure 17-23) also shows. In this experiment, 10 protons are transported out of the matrix for every electron pair transferred from NADH to O_2. As soon as O_2 is added to the suspension of mitochondria, the medium outside the mitochondria becomes acidic. During electron transport from NADH to O_2, protons translocate from the matrix to the intermembrane space; since the outer membrane is freely permeable to protons, the pH of the outside medium is lowered briefly.

Succinate is oxidized to fumarate (by the enzyme succinate dehydrogenase, a part of the succinate CoQ reductase complex) as part of the citric acid cycle (see Figure 17-13), and this causes the reduction of the electron carrier FAD to $FADH_2$. If we were to add succinate in place of NADH in the experiment depicted in Figure 17-23, the medium outside the mitochondria would again become acidic. However, only six protons would be transported outward from the matrix for every electron pair transferred from succinate to O_2: $FADH_2$ transfers electrons to the electron transport chain, but at a later point than NADH does (see Figure 17-22), and as a consequence fewer protons are transported from the matrix.

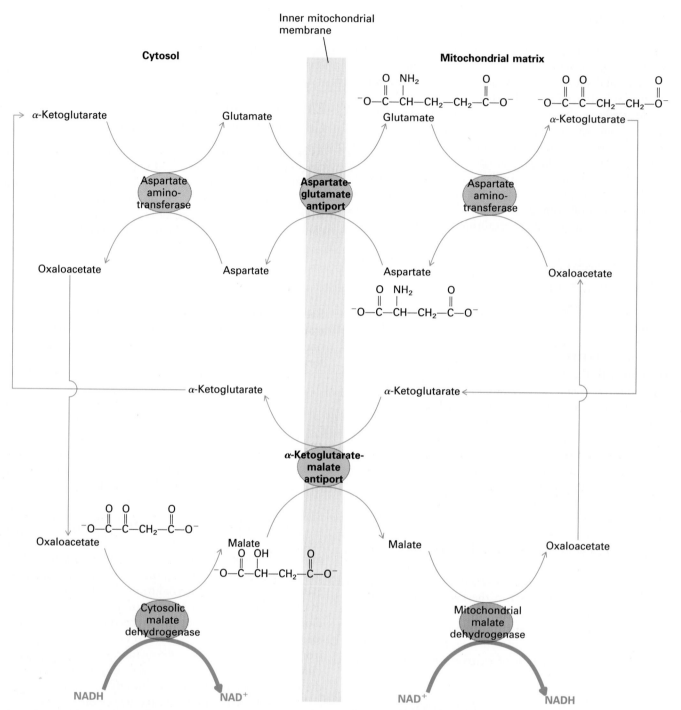

▲ FIGURE 17-21   The malate shuttle. This shuttle transports electrons from cytosolic NADH across the inner mitochondrial membrane. Cytosolic NADH reduces cytosolic oxaloacetate, forming NAD and malate *(lower left).* Malate enters the matrix in exchange for one molecule of α-ketoglutarate, catalyzed by the α-ketoglutarate–malate antiport. In the matrix, malate in turn reduces matrix NAD^+, forming NADH and oxaloacetate. The oxaloacetate is converted into the amino acid aspartate *(upper right),* which exits the matrix in exchange for glutamate catalyzed by the aspartate-glutamate antiporter. Once in the cytosol, aspartate is reconverted to oxaloacetate *(upper left),* completing the cycle. One counterclockwise turn of the entire cycle can be summarized as

$$\text{NADH}_{\text{cytosol}} + \text{NAD}^+{}_{\text{matrix}} \rightleftharpoons \text{NAD}^+{}_{\text{cytosol}} + \text{NADH}_{\text{matrix}}$$

[Courtesy of B. Trumpower.]

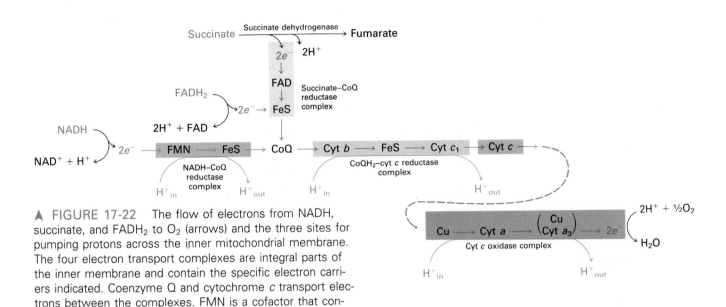

▲ FIGURE 17-22   The flow of electrons from NADH, succinate, and FADH$_2$ to O$_2$ (arrows) and the three sites for pumping protons across the inner mitochondrial membrane. The four electron transport complexes are integral parts of the inner membrane and contain the specific electron carriers indicated. Coenzyme Q and cytochrome c transport electrons between the complexes. FMN is a cofactor that consists of the flavin–ribitol phosphate part of FAD (see Figure 17-10).

The Mitochondrial Electron Transport Chain Transfers Electrons from NADH to O$_2$

When one molecule of NADH is oxidized to NAD$^+$, two electrons and one proton are released. Recall that protons are soluble in aqueous solution as hydronium ions (H$_3$O$^+$). Free electrons, by contrast, cannot exist in aqueous solutions. The electrons released from NADH (or FADH$_2$) are passed to O$_2$ along a chain of electron carriers.

Electron transport from NADH to O$_2$ is a thermodynamically downhill process: as electrons move from NADH to O$_2$, their potential declines by 1.14 V, which corresponds to 26.2 kcal/mol of electrons transferred, or ≈53 kcal/mol for a pair of electrons. Much of this energy is conserved at three stages of electron transport by the movement of protons across the inner mitochondrial membrane from the matrix to the intermembrane space.

Most electron carriers are prosthetic groups such as flavins, heme, iron-sulfur clusters, and copper, bound to protein particles on the inner membrane. Ubiquinone, often called *Coenzyme Q*, or simply *CoQ* is the only electron carrier that is not a protein-bound prosthetic group. The 12 or more electron carriers are grouped into four multiprotein complexes (Figure 17-24; see also Figure 17-22), each of which spans the inner mitochondrial membrane. Table 17-3 lists the prosthetic groups in each multiprotein complex.

We first take a detailed look at some individual carriers in the electron transport chain; we then consider the multiprotein complexes.

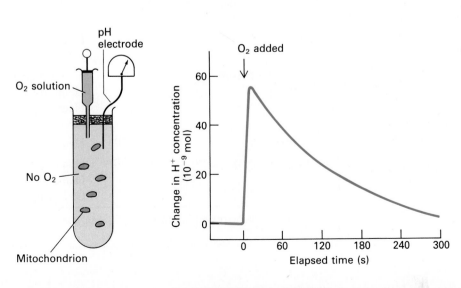

◄ FIGURE 17-23   Demonstration that electron transport from NADH or FADH$_2$ to O$_2$ is coupled to proton transport across the membrane. If a source of electrons for respiration, such as NADH, is added to a suspension of mitochondria depleted of O$_2$, no NADH is oxidized. When a small amount of O$_2$ is added to the system (arrow), the pH of the surrounding medium drops sharply—a change that corresponds to an *increase* in protons outside the mitochondria. Thus the oxidation of NADH by O$_2$ is coupled to the movement of protons out of the matrix. Once the O$_2$ is depleted, the excess protons slowly move back into the mitochondria (powering the synthesis of ATP) and the pH of the extracellular medium returns to its initial value.

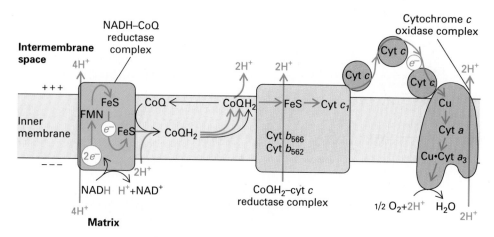

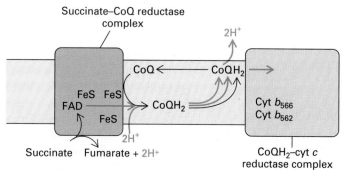

▲ FIGURE 17-24   The pathway of electron transport (blue) and proton transport (red) in the inner mitochondrial membrane.

The transport of a pair of electrons from one NADH to one CoQ molecule results in the transport of four protons from the matrix to the intermembrane space by the NADH-CoQ reductase complex. For every electron transported from $CoQH_2$ through the $CoQH_2$–cytochrome c reductase complex to cytochrome c, two protons are moved from the matrix into the intermembrane space, or four protons per electron pair. This involves the Q cycle (see Figure 17-29). Finally, the peripheral protein cytochrome c diffuses in the intermembrane space, transporting electrons (one at a time) from the $CoQH_2$–cytochrome c reductase complex to the cytochrome c oxidase complex. The transport of one electron through the cytochrome c oxidase complex is associated with the translocation of one proton; thus the transport of a pair of electrons, needed to reduce one oxygen atom, is accompanied by the translocation of two protons. The protons released into the matrix space by the oxidation of NADH are consumed in the formation of water from O_2, resulting in no net proton translocation from these reactions. Thus a total of 10 protons are translocated per pair of electrons moved from NADH to O_2.

The succinate-CoQ reductase complex (bottom) oxidizes succinate to fumarate, reducing one CoQ to $CoQH_2$; the $CoQH_2$ donates its electrons to the $CoQH_2$–cytochrome c reductase complex. No protons are translocated by the succinate CoQ reductase complex. For every pair of electrons transported from succinate to O_2 through the $CoQH_2$–cytochrome c reductase and cytochrome c oxidase complexes, six protons are translocated from the matrix into the intermembrane space.

Carriers in the Electron Transport Chain   The *iron-sulfur clusters Fe_2S_2 and Fe_4S_4* (Figure 17-25) are nonheme prosthetic groups and are important electron carriers. These clusters consist of Fe atoms bonded both to inorganic S atoms and to four S atoms on cysteine residues on the protein (see Figure 17-25). Some Fe atoms in the cluster bear a +2 charge while others have a +3 charge. However, the net charge of each Fe atom is actually between +2 and +3, because electrons in the outermost orbits are dispersed among the Fe atoms and move rapidly from one atom to another. Iron-sulfur clusters accept and release electrons one at a time; the additional electron is also dispersed over all the Fe atoms in the cluster.

TABLE 17-3 Components of the Mitochondrial Electron Transport Chain		
Enzyme Complex	**Mass (in Daltons)**	**Prosthetic Groups**
NADH-CoQ reductase	85,000	FMN FeS
Succinate-CoQ reductase	97,000	FAD FeS
$CoQH_2$–cytochrome c reductase	280,000	Heme b_{566} Heme b_{562} Heme c_1 FeS
Cytochrome c	13,000	Heme c
Cytochrome c oxidase	200,000	Heme a Heme a_3 Cu_a^+, Cu_b^+

SOURCE: J. W. De Pierre and L. Ernster, 1977, *Ann. Rev. Biochem.* 46:201.

(a)

(b)

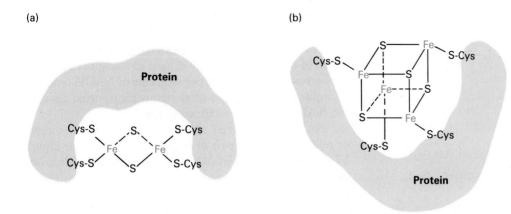

Protein

Cys-S　　　S　　　S-Cys

Fe　　Fe

Cys-S　　　S　　　S-Cys

Cys-S　　　S —— Fe　S-Cys

Fe　　S

Fe —— S

S　　Fe

Cys-S　　S-Cys

Protein

◄ FIGURE 17-25   Three-dimensional structures of some iron-sulfur clusters in electron-transporting proteins: (a) a dimeric (Fe_2S_2) cluster; (b) a tetrameric (Fe_4S_4) cluster. In both types of clusters, each Fe atom is bonded to four S atoms: some S atoms are molecular sulfur; others occur in the cysteine side chains of a protein. Such FeS clusters accept and release one electron at a time. [See W. H. Orme-Johnson, 1973, *Ann. Rev. Biochem.* **42**:159.]

$H_2C-CH_2-CH_2-\overset{\displaystyle CH_3}{CH}-CH_2-CH_2-CH_2-\overset{\displaystyle CH_3}{CH}-CH_2-CH_2-CH_2-\overset{\displaystyle CH_3}{CH}-CH_3$

HCOH　　　　CH_3

H_3C　　　　　　　$CH=CH_2$

N　　　N

Fe

N　　　N

OHC　　　　CH_3

CH_2　　CH_2

CH_2　　CH_2

CO_2^-　　CO_2^-

a-Type heme

Found in cytochromes *a* and a_3 of cytochrome *c* oxidase

◄ FIGURE 17-26   Heme prosthetic groups of respiratory-chain cytochromes in mitochondria. Note the differences in substituents on the porphyrin rings. Hemes accept and release one electron at a time.

$H_2C=CH$　　CH_3

H_3C　　　　　　$CH=CH_2$

N　　N

Fe

N　　N

H_3C　　　　　　CH_3

CH_2　　CH_2

$^-O_2C-CH_2$　　$CH_2-CO_2^-$

b-Type heme

Found in cytochromes b_{562} and b_{566} of $CoQH_2-$cytochrome *c* reductase complex

CH_3

(Protein) RS—CH　　CH_3

H_3C　　　　　　CH_3　　CH

N　　N

Fe　　　　SR (Protein)

N　　N

H_3C　　　　　　CH_3

CH_2　　CH_2

$^-O_2C-CH_2$　　$CH_2-CO_2^-$

c-Type heme

Found in cytochromes *c* and c_1

Cytochromes are proteins that contain a heme prosthetic group (Figure 17-26) similar to that in hemoglobin or myoglobin; the Fe in the center of the heme is the electron transporter. Electron transport occurs by oxidation and reduction:

$$Fe^{3+}_{ox} + e^- \rightleftharpoons Fe^{2+}_{red}$$

The cytochromes in the electron transport chain, in order of their entry (see Figure 17-22), are b_{566}, b_{562}, c_1, *c*, *a*, and a_3. The different cytochromes have slightly different heme structures and axial ligands of the Fe atom (see Figure 17-26); these generate different environments for Fe. Therefore, each cytochrome has a different reduction potential, or tendency to accept an electron—an important characteristic dictating unidirectional electron flow along the chain. Because the heme ring in cytochromes consists of alternating double- and single-bonded atoms, a large number of resonance forms exist, and the extra electron is delocalized to the heme carbon and nitrogen atoms as well as to the Fe ion.

Ubiquinone (Figure 17-27) (*coenzyme Q* or *CoQ*) is a carrier of hydrogen atoms (protons plus electrons). The oxidized quinone can accept a single electron to form a semiquinone, and then a second electron and two protons to form the fully reduced hydroquinone form, dihydroubiquinone. Both CoQ and the reduced form $CoQH_2$ diffuse freely in the mitochondrial inner membrane. CoQ is the collection point for electrons released by the NADH-CoQ

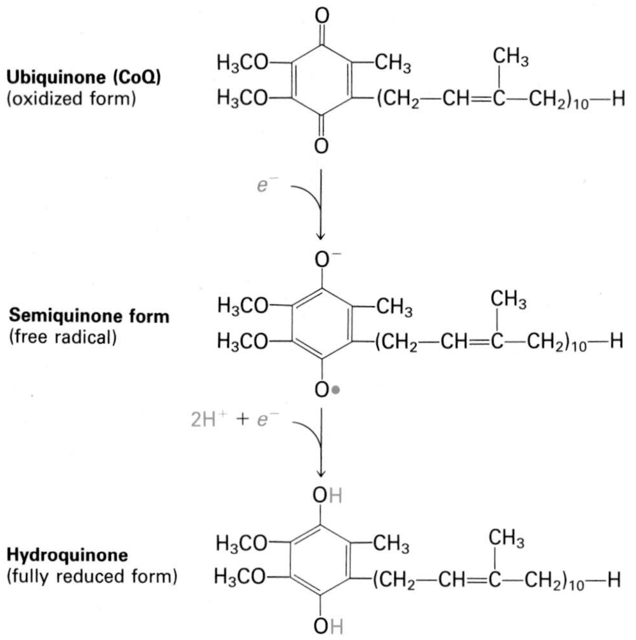

Ubiquinone (CoQ)
(oxidized form)

Semiquinone form
(free radical)

Hydroquinone
(fully reduced form)

▲ FIGURE 17-27   The structure of coenzyme Q (CoQ, also called ubiquinone) illustrating its ability to carry two protons and two electrons. Found in bacterial and mitochondrial membranes, CoQ is the only carrier in the electron transport system that is not tightly bound or covalently bonded to a protein. Because of its long hydrocarbon "tail" of isoprene units, CoQ is soluble in the hydrophobic core of phospholipid bilayers and is very mobile. The addition of one electron to oxidized CoQ results in a half-reduced (semiquinone) form. The semiquinone is a free radical; the unpaired electron (blue dot) is delocalized by resonance over the benzene ring and attached oxygen atoms.

reductase complex and the succinate-CoQ reductase complex, as well as for electrons from $FADH_2$. $CoQH_2$ donates its electrons to the $CoQH_2$–cytochrome c reductase complex.

The NADH-CoQ Reductase Complex   This multiprotein complex carries electrons from NADH to CoQ (see Figures 17-22 and 17-24). In this part of the electron transport chain NADH is oxidized to NAD^+:

$$NADH \longrightarrow NAD^+ + H^+ + 2e^-$$

and the two released electrons move through a series of carriers until CoQ is ultimately reduced to $CoQH_2$:

$$CoQ + 2e^- + 2H^+ \longrightarrow CoQH_2$$

As in the other stages of electron transport, the overall reaction

$$NADH + CoQ + 2H^+ \longrightarrow NAD^+ + H^+ + CoQH_2$$

is thermodynamically favorable. Each transported electron undergoes a drop in potential of ~360 mV, equivalent to a $\Delta G^{\circ\prime}$ of −16.6 kcal/mol for the two electrons transported. Much of this released energy is used to transport four protons across the membrane per molecule of NADH oxidized or, equivalently, per pair of electrons transported through the NADH-CoQ reductase complex.

NAD^+ is exclusively a two-electron carrier: it accepts or releases only one pair of electrons at a time (see Figure 17-3). In the NADH-CoQ reductase complex, electrons first flow from NADH to FMN (flavin mononucleotide), a cofactor related to FAD, and then to an iron-sulfur protein. FMN, like FAD, can accept two electrons, but does so one electron at a time (see Figure 17-10).

The Succinate-CoQ Reductase Complex   This complex contains succinate dehydrogenase, the enzyme that oxidizes a molecule of succinate to fumarate in the citric acid cycle (see Figure 17-13, step 7). The two electrons released in this reaction are transferred first to FAD, then to an iron-sulfur protein, and finally to CoQ, forming the reduced $CoQH_2$. The overall reaction catalyzed by this complex is

$$Succinate + CoQ \longrightarrow fumarate + CoQH_2$$

No protons are translocated across the membrane by the succinate-CoQ reductase complex.

The $CoQH_2$–Cytochrome c Reductase Complex   When $CoQH_2$ donates its electrons to the $CoQH_2$–cytochrome c reductase complex, this regenerates oxidized CoQ:

$$CoQH_2 \longrightarrow CoQ + 2H^+ + 2e^-$$

Within this complex the released electrons are transferred to an iron-sulfur protein and to two b-type cytochromes, then to cytochrome c_1. Finally, the two electrons are transferred to two molecules of the oxidized form of cytochrome c (a water-soluble intermembrane-space protein), forming reduced cytochrome c:

$$Cyt\ c^{3+}\ (oxidized) + e^- \longrightarrow Cyt\ c^{2+}\ (reduced)$$

Thus the overall reaction catalyzed by the $CoQH_2$–cytochrome c reductase complex is

$$CoQH_2 + 2\ Cyt\ c^{3+} \longrightarrow CoQ + 2H^+ + 2\ Cyt\ c^{2+}$$

As each pair of electrons pass through the $CoQH_2$–cytochrome c reductase complex, four protons are translocated from the matrix space through the inner mitochondrial membrane.

The Cytochrome c Oxidase Complex Cytochrome c, after being reduced by the CoQH$_2$–cytochrome c reductase complex, transports electrons, one at a time, to the cytochrome c oxidase complex. This electron transfer regenerates oxidized cytochrome c:

$$\text{Cyt } c^{2+} \text{ (reduced)} \longrightarrow \text{Cyt } c^{3+} \text{ (oxidized)} + e^-$$

Within this complex, electrons are transferred, again one at a time, first to a Cu^{2+} ion, then to cytochrome a, then to a complex of a second copper ion and cytochrome a_3, and finally to O_2, the ultimate electron acceptor, yielding H_2O:

$$2e^- + 2H^+ + \tfrac{1}{2}O_2 \longrightarrow H_2O$$

The overall reaction catalyzed by the cytochrome c oxidase complex is

$$2 \text{ Cyt } c^{2+} \text{ (reduced)} + 2H^+ + \tfrac{1}{2}O_2 \longrightarrow$$
$$2 \text{ Cyt } c^{3+} \text{ (oxidized)} + H_2O$$

During transport of each pair of electrons through the cytochrome c oxidase complex, two protons are translocated through the membrane (see Figure 17-24).

Lateral Mobility of Electron Carriers Each of the electron transport complexes are laterally mobile in the mitochondrial membrane, as was shown in the experiment depicted in Figure 14-33. The complexes are present in nonequal amounts: for each NADH-CoQ reductase complex, there are about three CoQH$_2$–cytochrome c reductase complexes and seven cytochrome c oxidase complexes. Furthermore, there do not appear to be stable contacts between any two complexes: electron transport from one complex to another only occurs by diffusion of electron shuttles. The lipid-soluble coenzyme Q picks up electrons from the NADH-CoQ reductase and succinate-CoQ reductase complexes and transfers them to the CoQH$_2$–cytochrome c reductase complex. Cytochrome c interacts with a specific site on the CoQH$_2$–cytochrome c reductase complex, where it picks up an electron. The reduced cytochrome c diffuses in the intermembrane space until it encounters a cytochrome c oxidase complex, to which it donates an electron. Mitochondrial electron flow, in summary, does not resemble an electric current through a wire, with each electron following the last. Rather, electrons are picked up by a carrier, one or two at a time, and then passed along to the next carrier in the pathway.

Most Electron Carriers Are Oriented in the Transport Chain in the Order of Their Reduction Potentials

As we saw in Chapter 2, the *reduction potential*, E, for a partial reduction reaction

$$\text{oxidized molecule} + e^- \rightleftharpoons \text{reduced molecule}$$

is a measure of the equilibrium constant of that partial reaction. For instance, for the partial reaction

$$NAD^+ + H^+ + 2e^- \rightleftharpoons NADH$$

the value of the standard reduction potential is negative:

$$E_0' = -0.32 \text{ V}$$

showing that this partial reaction tends to proceed toward the left; that is, toward the oxidation of NADH to NAD$^+$. By contrast, the standard reduction potential for the partial reaction

$$\text{Cyt } c_{ox}(Fe^{3+}) + e^- \rightleftharpoons \text{Cyt } c_{red}(Fe^{2+})$$

is positive:

$$E_0' = +0.22 \text{ V}$$

showing that this partial reaction tends to proceed toward the right; that is, toward the reduction of cytochrome c (Fe^{3+}) to c (Fe^{2+}). The final stage of the chain, the reduction of O_2 to H_2O

$$2H^+ + \tfrac{1}{2}O_2 + 2e^- \longrightarrow H_2O$$

has the most positive reduction potential:

$$E_0' = +0.816 \text{ V}$$

With the exception of the b cytochromes in the CoQH$_2$–cytochrome c reductase complex, the reduction potential of the electron carriers in mitochondria increases steadily from NADH to O_2; thus electron transport is thermodynamically favorable at every stage except one. We can think of the electrons released by NADH as having a high potential energy, with a fraction of this energy lost at each step as the electrons move from NADH to O_2, the ultimate electron acceptor. To emphasize this point, we can calculate the $\Delta G^{\circ\prime}$ value for cytochrome c oxidase, the reaction of the last complex in the electron transport chain (see Figures 17-22 and 17-24):

$$2 \text{ Cyt } c \text{ } (Fe^{2+}) + 2H^+ + \tfrac{1}{2}O_2 \rightleftharpoons 2 \text{ Cyt } c \text{ } (Fe^{3+}) + H_2O$$

The half-reactions are

$\text{Cyt } c \text{ } (Fe^{3+}) + e^- \rightleftharpoons \text{Cyt } c \text{ } (Fe^{2+})$	$E_0' = +0.26 \text{ V}$
$2H^+ + \tfrac{1}{2}O_2 + 2e^- \rightleftharpoons H_2O$	$E_0' = +0.82 \text{ V}$

The change in voltage for the total reaction is

$$\Delta E_0' = +0.82 - 0.26 = +0.56 \text{ V}$$

From Chapter 2

$$\Delta G^{\circ\prime} = -n\mathscr{F}\,\Delta E_0'$$

where n is the number of electrons involved—here, 2. Thus

$$\Delta G^{\circ\prime} = (-2)(23{,}062 \text{ cal} \cdot \text{V}^{-1} \cdot \text{mol}^{-1})(0.56 \text{ V})$$

$$= -25{,}829 \text{ cal/mol}$$

$$= -25.8 \text{ kcal/mol}$$

The reaction is strongly exergonic. The transfer of a pair of electrons from cytochrome c to O_2 releases a significant amount of energy that can be made to do useful work—in this case, to move protons from the matrix across the inner mitochondrial membrane to the intermembrane space. An understanding of how this coupling occurs requires a somewhat more detailed look at the structure of the electron transport chain.

Three Electron Transport Complexes Are Sites of Proton Translocation

Except for the succinate-CoQ reductase complex, each multiprotein complex is a site for proton transport across the inner mitochondrial membrane during electron movement. By selectively extracting mitochondrial membranes with detergents, we can isolate each of the complexes in near purity and then incorporate them into phospholipid vesicles. Then adding an appropriate electron donor and electron acceptor causes the movement of protons across the membrane. For example, the cytochrome c oxidase complex can be incorporated into phospholipid vesicles so that the binding site for the cytochrome c polypeptide is on the outside (Figure 17-28); the addition of reduced cytochrome c results in electron transfer through the oxidase complex to O_2. Direct measurements indicate that two protons are transported out of the vesicles for every electron pair transported (or, equivalently, for every two molecules of cytochrome c oxidized). Similar studies indicate that the NADH-CoQ reductase complex translocates four protons per pair of electrons transported (or, equivalently, four protons per NADH molecule oxidized). Note that protons are generated in the matrix by the oxidation of NADH to NAD$^+$ and H$^+$. These excess protons are consumed by the cytochrome c oxidase complex during the formation of H$_2$O (see Figure 17-24), completing a "proton loop" initiated by the NADH-CoQ reductase complex, but resulting in no net movement of protons.

(a)

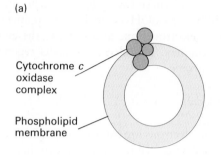

(b)

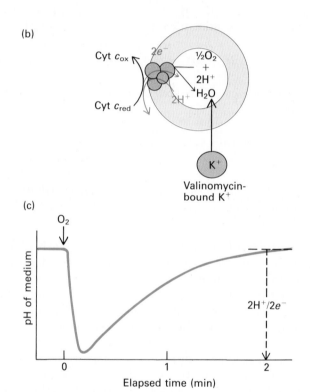

▲ **FIGURE 17-28**   Proton transport catalyzed by the cytochrome c oxidase complex. Segments of the mitochondrial electron transport system can be isolated and assembled into phospholipid membrane vesicles. If such a system is supplied with both a donor and an acceptor of electrons, protons are transported across the membrane. (a) When the cytochrome c oxidase complex is incorporated into vesicles, the cytochrome c binding site is positioned on the outer surface. (b) If O_2 and reduced cytochrome c are added, electrons are transferred to O_2 to form H$_2$O, and protons are transported from the inside to the outside of the vesicles. (c) This change is measured by a drop in the pH of the medium following the addition of O_2. Valinomycin and K$^+$ are added to the system to dissipate the electric membrane potential generated by the translocation of H$^+$, which would reduce the number of protons moved across the membrane. As the reduced cytochrome c becomes fully oxidized, protons leak back into the vesicles and the pH of the medium returns to its initial value. Studies show that two protons are transported per O atom reduced. Since two electrons are needed to reduce one O atom, one proton is transported per electron released by cytochrome c. [After B. Reynafarje, L. Costa, and A. Lehninger, 1986, *J. Biol. Chem.* **261**:8254–8262.]

While relatively little is known about the mechanism of proton translocation by the NADH-CoQ reductase complex, the mechanism of proton translocation by the two cytochrome complexes is now well understood.

The Q Cycle Increases the Number of Protons Transported by the CoQH₂– Cytochrome c Reductase Complex

Two protons are translocated across the membrane per electron (four per electron pair) transported through the CoQH₂–cytochrome c reductase complex (Figure 17-29). Because CoQ plays a key role, this process is known as the *proton-motive Q cycle*, or *Q cycle*. During this process, CoQ cycles between its reduced and oxidized states by accepting and releasing two protons and two electrons together:

$$CoQ + 2H^+ + 2e^- \rightleftharpoons CoQH_2$$

One molecule of CoQH₂, reduced on the matrix side of the inner membrane by the NADH-CoQ reductase or succi-nate-CoQ reductase complex, diffuses across the membrane to a site on the intermembrane side of the CoQH₂–cytochrome c reductase complex, where it releases its two protons into the intermembrane space; these represent two of the four protons translocated from the matrix to the intermembrane space per pair of electrons transported. Simultaneously, one of the two electrons from CoQH₂ is transported, via an iron-sulfur protein and cytochrome c_1, directly to cytochrome c. The other electron released from the CoQH₂ is a cycling electron. It moves through cytochromes b_{566} and b_{562} to the matrix surface where it reduces an oxidized CoQ molecule (reaction 3a, Figure 17-29), forming the partially reduced CoQ semiquinone anion, denoted by CoQ⁻· (the dot indicates that it is a free radical and the minus indicates that it is ionized; see Figure 17-27). When a second cycling electron, released from a second CoQH₂, is similarly transported through cytochromes b_{566} and b_{562}, it reduces the CoQ⁻·, forming, together with two protons picked up from the matrix space, CoQH₂. This CoQH₂ molecule diffuses to the binding site on the intermembrane surface of the CoQH₂–cytochrome c reductase complex, where it, too, releases its two protons into the intermembrane space, as well as one electron di-

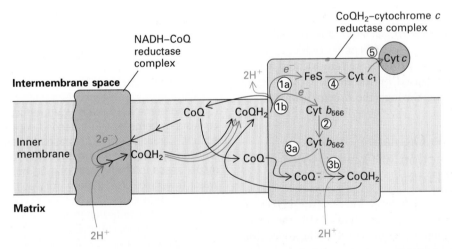

▲ FIGURE 17-29   The proton-motive Q cycle, by which four protons are translocated for each pair of electrons transported through the CoQH₂–cytochrome c reductase complex to cytochrome c.

One molecule of CoQH₂, reduced by the NADH-CoQ reductase complex, diffuses to a site on the intermembrane surface of the CoQH₂–cytochrome c reductase complex, where it simultaneously releases its two protons into the intermembrane space and its two electrons into the CoQH₂– cytochrome c reductase complex. (This represents two of the four protons translocated per pair of electrons transported.) One of the two electrons from CoQH₂ (reaction 1a) is picked up by the iron-sulfur complex and is transported first to cytochrome c_1 (reaction 4) and then to cytochrome c (reaction 5). The second electron released from the CoQH₂ (reaction 1b) is a cycling electron. It moves, via cytochromes b_{566} and b_{562} (reaction 2), to the matrix face of the complex where it is added to an oxidized CoQ molecule (reaction 3a), forming the semiquinone CoQ⁻·, a partially reduced, ionized free radical. The next time that a CoQH₂ molecule sends a cycling electron to cytochromes b_{566} and b_{562}, this electron is transferred to the CoQ⁻· (reaction 3b) and two protons are picked up from the matrix space, converting the CoQ⁻· into a new molecule of CoQH₂. This new CoQH₂ acts exactly like the original CoQH₂ from the NADH-CoQ reductase: it diffuses to the intermembrane surface, where it releases its two protons and its two electrons (reaction 1), starting the Q cycle again. As demonstrated by the calculation in the text, for every pair of electrons transferred through the CoQH₂– cytochrome c reductase complex to cytochrome c, four protons are translocated to the intermembrane space. [Adapted from B. Trumpower, 1990, *J. Biol. Chem.* **265**:11409–11412.]

rectly to cytochromes c_1 and c, and one electron to cycle through the cytochromes b_{566} and b_{562}.

The number of protons transported from the matrix space to the intermembrane space per $CoQH_2$ oxidized and thus per pair of electrons transported through the $CoQH_2$-cytochrome c reductase complex can be calculated as follows. Imagine 100 molecules of $CoQH_2$, reduced by the NADH-CoQ reductase complex, interacting with $CoQH_2$–cytochrome c reductase complexes. In the first passage through the cycle, 200 electrons will be released; 100 electrons will be transported to cytochrome c_1, 200 protons will be transported across the membrane into the intermembrane space, and 100 electrons will cycle through the b cytochromes and generate 50 new molecules of reduced $CoQH_2$. In the second cycle, these 50 molecules of $CoQH_2$ will transport 100 protons into the intermembrane space and generate 50 cycling electrons that will in turn generate 25 new molecules of reduced $CoQH_2$. In the third cycle, these 25 molecules of $CoQH_2$ will transport 50 protons and generate 25 cycling electrons that will generate 12.5 new molecules of reduced $CoQH_2$. Continuing this *ad infinitum*, the number of protons transported as a result of oxidation of the original 100 molecules of $CoQH_2$ by 200 electrons will be

$$200 + 100 + 50 + 25 + 12.5 + 6.25 + 3.125 + \ldots = 400$$

In summary, for every *pair of electrons* transported from $CoQH_2$ through the $CoQH_2$–cytochrome c reductase complex to cytochrome c, *four protons* are translocated across the membrane—two released by the initial $CoQH_2$, and two by electrons cycling through the b cytochromes.

The Cytochrome c Oxidase Complex Couples the Reduction of Oxygen to the Translocation of Protons

Cytochrome c, after its reduction by the $CoQH_2$–cytochrome c reductase complex, is reoxidized by the cytochrome c oxidase complex (Figure 17-30). The oxidation of four cytochrome c molecules is coupled to the reduction of one molecule of O_2, forming two molecules of water. The key to oxygen reduction by cytochrome c oxidase is an oxygen reduction center (see Figure 17-30a) consisting of one molecule of heme a_3 and one copper ion, called Cu_b^+, bound to subunit I of the cytochrome c oxidase complex (see Table 17-2).

Four molecules of reduced cytochrome c bind, one at a time, to a site on the oxidase that includes subunit II. An electron is transferred from the heme of each cytochrome c, first to a Cu^{++} ion, called Cu_a^{++}, bound to subunit II, then to the heme a bound to subunit I, and finally to the Cu_b^{++} and heme a_3 in the oxygen reduction center. The cyclic oxidation and reduction of the iron and copper in the reduction center, together with the uptake of four protons

from the matrix space, is coupled to the transfer of the four electrons to oxygen and the formation of water (Figure 17-30b). Intermediates in oxygen reduction include the peroxide anion (O_2^{2-}) and probably the hydroxyl radical ($OH^{\bullet}$). These intermediates would be harmful if they escaped from the reaction center, but they do so only rarely.

For every four electrons transferred from reduced cytochrome c through the cytochrome c oxidase complex (i.e., for every molecule of O_2 reduced to two H_2O molecules), four additional protons are translocated from the matrix space to the intermembrane space. All four protons move during steps 5–7 of the cycle depicted in Figure 17-30b, but the mechanism by which these protons are translocated is not known.

Current evidence thus suggests that 10 protons are transported from the matrix space across the inner mitochondrial membrane for every electron pair that is transferred from NADH to O_2: four protons are transported by

> FIGURE 17-30   Electron transport through the cytochrome c oxidase complex.

(a) Structure of the cytochrome c oxidase complex in the inner mitochondrial membrane. Subunits I, II, and III are considered to be the catalytic core of the enzyme and are the only three subunits in bacterial cytochrome c oxidases; the function of the 10 additional subunits in the mitochondrial enzyme is not known.

Reduced cytochrome c binds to a site on the oxidase complex that includes subunit II. From the heme of reduced cytochrome c, one electron is transferred, via a copper ion (Cu_a) and heme a, to the oxygen reduction center, which consists of heme a_3 and a second copper ion (Cu_b) bound to subunit I. Four electrons, released by four molecules of reduced cytochrome c, together with four protons from the matrix space, combine with one O_2 molecule to form two water molecules. Additionally, for each electron transferred from cytochrome c to oxygen, one proton is transported from the matrix to the intermembrane space, or a total of four for each molecule of O_2 reduced to two of H_2O. Ψ = membrane potential.

(b) Steps in the reduction of oxygen to water. Four electrons, released sequentially from four molecules of reduced cytochrome c, are transferred sequentially to the heme a_3 – Cu_b oxygen reduction center. In reactions 1 and 2, two electrons are added, reducing the Fe^{3+} in the heme to Fe^{2+}, and the Cu^2 to Cu^+. Then an O_2 molecule binds to the oxygen reduction center (reaction 3), immediately followed (reaction 4) by the transfer of two electrons, one from Fe^{2+} and one from Cu^+, forming the peroxide anion O_2^{2-} and regenerating Fe^{3+} and Cu^{2+}. Next (reaction 5) one electron and two protons are added, forming an unusual $Fe^{4+}=O^{2-}$ intermediate. The addition of the fourth electron (reaction 6) and two protons (reaction 7) forms two molecules of water and regenerates the initial Cu^{2+} Fe^{3+} oxygen reduction center. The translocation of four protons from the matrix space to the intermembrane space occurs during reactions 5 to 7.

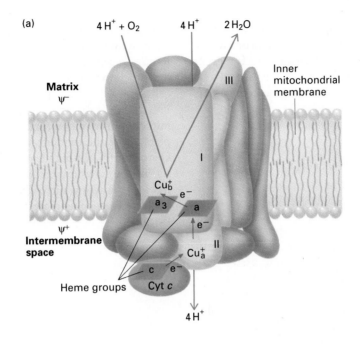

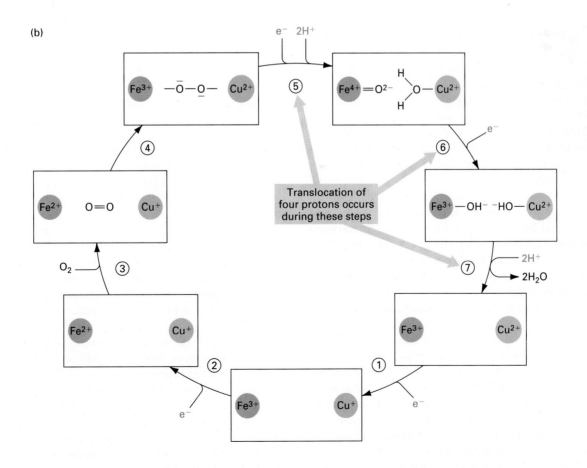

the NADH-CoQ reductase complex, four by the Q cycle in the CoQH$_2$–cytochrome c reductase complex, and two by the cytochrome c oxidase complex. Since the succinate CoQ reductase complex does not transport protons, six protons are transported across the membrane for every electron pair that is transferred from succinate (or FADH$_2$) to O$_2$, four by the Q cycle and two by the cytochrome c oxidase complex.

► *Metabolic Regulation*

All enzymatically catalyzed reactions and metabolic pathways are regulated. The mitochondrion synthesizes ATP only to meet cellular requirements. The import of metabolites to the mitochondrion and the export of ATP from it are also coordinated with ATP synthesis and, as we shall see, with the generation of body heat. Similarly, the conversion of glucose, fatty acids, and other metabolites to acetyl CoA is tightly regulated to produce only the amount of this substrate required for the citric acid cycle. Here, we consider how this regulation is achieved. We also discuss the important role of peroxisomes in the oxidation of fatty acids.

Respiration Is Controlled by the Production of ATP through the Proton-Motive Force

If intact isolated mitochondria are provided with NADH (or FADH$_2$), O$_2$, and P$_i$, but not with ADP, the oxidation of NADH and the reduction of O$_2$ rapidly cease as the amount of endogenous ADP is depleted by ATP formation. If ADP is readded, the oxidation of NADH is rapidly restored. Thus mitochondria can oxidize FADH$_2$ and NADH only as long as there is a source of ADP and P$_i$ to generate ATP. This phenomenon, termed *respiratory control*, illustrates how one key reactant can limit the rate of a complex set of interrelated reactions.

Intact cells and tissues also employ respiratory control. Cells oxidize only enough glucose to synthesize the amount of ATP required for their metabolic activities. Stimulation of a metabolic activity that utilizes ATP, such as muscle contraction, results in an increased level of cellular ADP; this, in turn, increases the oxidation rate of metabolic products in the mitochondrion.

The metabolic nature of respiratory control is now well understood. Recall that the oxidation of NADH, succinate, or FADH$_2$ is *obligatorily* coupled to proton transport across the inner mitochondrial membrane. If the resulting proton-motive force is not dissipated when protons are used to synthesize ATP from ADP and P$_i$ (or for some other purpose), both the transmembrane proton concentration gradient and the membrane electric potential will increase to very high levels. NADH oxidation will eventually cease, because it will require too much energy to move additional protons across the inner membrane against the

existing proton-motive force. Observers feel that the availability of ADP for ATP synthesis—respiratory control—is only one way in which mitochondrial oxidation is regulated in intact cells: a rise in cytosolic Ca^{2+} ions, as occurs in muscle cells during contraction, also triggers an increase in mitochondrial oxidation and ATP production in many cells.

Certain poisons, called *uncouplers,* render the inner mitochondrial membrane permeable to protons. Uncouplers allow the oxidation of NADH and the reduction of O$_2$ to continue at high levels but do not permit ATP synthesis. In the uncoupler 2,4-dinitrophenol (DNP), two electron-withdrawing nitro (NO$_2$) groups stabilize the negatively charged phenolate form:

Both the neutral and negatively charged forms of DNP are soluble in phospholipid membranes and in aqueous solution, so DNP can act as a proton shuttle. By transporting protons across the inner membrane, DNP short-circuits both the transmembrane proton concentration gradient and the membrane electric potential. Uncouplers such as DNP abolish ATP synthesis and dispense with any requirement for ADP in NADH oxidation or electron transport. The energy released by the oxidation of NADH in the presence of DNP is converted to heat.

An Endogenous Uncoupler in Brown-Fat Mitochondria Converts H$^+$ Gradients to Heat

Brown-fat tissue is specialized for the generation of heat. In contrast to *white-fat tissue,* which is specialized for the storage of fat, brown-fat tissue contains abundant mitochondria, which impart their dark brown color to the tissue.

In the inner membrane of brown-fat mitochondria, a 33,000-MW inner-membrane protein called *thermogenin* functions as a natural uncoupler of oxidative phosphorylation. Thermogenin does not form a proton channel, but is, rather, a proton transporter. Its amino acid sequence is similar to that of other mitochondrial transporters, such as the ATP-ADP antiporter (see Figure 17-20), and it functions at a rate (~100 protons per second) that is characteristic of transporters, but is one-million-fold slower than typical channel proteins (see Figure 15-2). Like other uncouplers, it short-circuits the proton concentration gradient and the electric potential across the inner mitochondrial membrane, converting energy released by NADH oxidation to heat. This protein is regulated to generate heat and maintain body temperature under different environmental conditions. For instance, during the adaptation of rats to cold, the ability of their tissues to generate heat (*thermogenesis*) is increased by the induction of synthesis of this inner-membrane uncoupler protein. In cold-adapted animals, thermogenin may make up 15 percent of the mitochondrial membrane protein.

Adult humans have little brown fat, but human infants have a great deal. In the newborn, thermogenesis by brown-fat mitochondria is vital to survival, as it also is in hibernating mammals. In fur seals and other animals naturally acclimated to the cold, muscle-cell mitochondria contain thermogenin, which permits a great deal of energy in the proton concentration gradient to be converted to heat and used to maintain body temperature.

The Rate of Glycolysis Depends on the Cell's Need for ATP and Is Controlled by Multiple Allosteric Effectors

The activity of the glycolytic pathway is also continuously regulated, so that the production of ATP and pyruvate is adjusted to meet the needs of the cell. *Phosphofructokinase-1*, which catalyzes the third reaction in the conversion of glucose to pyruvate (Figure 17-31), is the principal rate-limiting enzyme of the entire pathway (see Figure 17-2). The allosteric inhibition of phosphofructokinase-1 by citrate allows the activities of the glycolytic pathway to be coordinated with those of the citric acid cycle. If citrate—the product of the first step of the citric acid cycle—accumulates, its feedback inhibition of phosphofructokinase-1 reduces the generation of pyruvate and acetyl CoA, so that less citrate is formed. Intermediates in the citric acid cycle are also used in biosynthesis of amino acids; a buildup of citrate indicates that these intermediates are plentiful and that glucose need not be degraded for this purpose.

Phosphofructokinase-1 is allosterically *activated* by ADP and allosterically *inhibited* by ATP. This arrangement makes the rate of glycolysis very sensitive to intracellular levels of ATP and ADP. The allosteric inhibition of phosphofructokinase by ATP may seem unusual, since ATP is also a substrate of this enzyme. But the affinity of the substrate-binding site for ATP is much higher (has a lower K_m) than that of the allosteric site. Thus at low concentrations, ATP binds to the catalytic but not to the inhibitory allosteric site and enzymatic catalysis proceeds at near maximal rates. At high concentrations, ATP binds to the allosteric site, inducing a conformational change that inhibits phosphofructokinase-1 and reduces the overall rate of glycolysis.

The metabolite *fructose 2,6-bisphosphate* is another important allosteric activator of phosphofructokinase-1 (see Figure 17-31). Fructose 2,6-bisphosphate is formed from the glycolytic intermediate fructose 6-phosphate; the catalyst is phosphofructokinase-2, an enzyme different from phosphofructokinase-1. Fructose 6-phosphate accel-

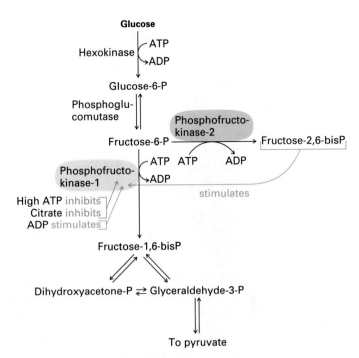

▲ FIGURE 17-31 Enzymatic control of the glycolytic pathway. Phosphofructokinase-1 is the main control point in the regulation of the glycolytic pathway. It is allosterically inhibited by ATP and citrate, allowing an excess of ATP or of intermediates from the citric acid cycle to slow the rate of glucose metabolism. Phosphofructokinase-1 is allosterically activated by fructose 2,6-bisphosphate, which is formed by phosphofructokinase-2 from fructose 6-phosphate. Thus an excess of fructose 6-phosphate indirectly activates the enzyme that converts it into fructose 1,6-bisphosphate, the next intermediate in glycolysis.

erates the formation of fructose 2,6-bisphosphate, which, in turn, activates phosphofructokinase-1. This type of control, by analogy with feedback control, is known as *feedforward activation*, in which the abundance of a metabolite (here, fructose 6-phosphate) induces an acceleration in its metabolism. Fructose 2,6-bisphosphate allosterically activates phosphofructokinase-1 in liver cells by decreasing the inhibitory effect of ATP and by increasing the affinity of phosphofructokinase-1 for one of its substrates, fructose 6-phosphate.

In addition to phosphofructokinase, two other enzymes in the glycolytic pathway are also subject to allosteric control. *Hexokinase* is allosterically inhibited by its reaction product, glucose 6-phosphate. *Pyruvate kinase* is allosterically inhibited by one of its reaction products, ATP; glycolysis slows down if too much ATP is present. The three glycolytic enzymes that are regulated by allosteric molecules catalyze reactions with the most negative $\Delta G°'$ values—reactions that are essentially irreversible under ordinary conditions. These enzymes are particularly suitable for regulating the entire glycolytic pathway. Additional control is exerted by glyceraldehyde 3-phosphate dehydrogenase, which catalyzes the reduction of NAD^+ to NADH. If cytosolic NADH builds up due to a slowdown in oxidative phosphorylation, this step in glycolysis will be slowed by mass action.

Thus the reactions of the glycolytic pathway and the citric acid cycle, oxidative phosphorylation, and thermogenesis are tightly controlled to produce the appropriate amount of ATP required by the cell. Glucose metabolism is controlled differently in various mammalian tissues to meet the metabolic needs of the organism as a whole. During periods of carbohydrate starvation, for instance, glycogen in the liver is converted to glucose 6-phosphate. Under these conditions, however, phosphofructokinase-1 is inhibited and glucose 6-phosphate is not metabolized to pyruvate; rather, it is converted to glucose and released into the blood to nourish the brain and muscles, which then oxidize the bulk of the available glucose. (Chapter 20 contains a more detailed discussion of the control of glucose metabolism in the liver and muscles.) In all cases, the activity of these enzymes is regulated by the level of small-molecule metabolites, generally by allosteric interactions or by phosphorylation.

The Oxidation of Fatty Acids Occurs in Peroxisomes without Production of ATP

Peroxisomes, also called *microbodies*, are small fatty acid–oxidizing organelles, ≈ 0.2–1 μm in diameter, that are lined by a single membrane (see Figure 5-37). They are present in all mammalian cells other than erythrocytes and are also found in plants, yeasts, and probably in most eukaryotic cells. Peroxisomes contain several oxidases—enzymes that use oxygen as an electron acceptor to oxidize organic substances, in the process forming hydrogen peroxide (H_2O_2), which is then degraded by catalase (Figure 17-32).

Oxidases $\quad O_2 \xrightarrow{\text{oxidases}} H_2O_2$
$\qquad\qquad\quad RH_2 \qquad R$

$2 H_2O_2 \xrightarrow{\text{catalase}} 2 H_2O + O_2$

Peroxidation $\quad H_2O_2 + R'H_2 \xrightarrow{\text{catalase}} 2 H_2O + R'$

▲ FIGURE 17-32   Peroxisomal respiration. Substrates (RH_2) for the peroxisomal oxidases *(upper reaction)* include not only the L- but also the unnatural D-amino acids, L-α-hydroxy acids, polyamines, and CoA derivatives of most long-chain fatty acids. Hydrogen peroxide is formed during the oxidation process, and catalase can convert *two* H_2O_2 molecules to O_2 and $2 H_2O$ *(middle reaction)*. Catalase can also catalyze peroxidation *(lower reaction)*, in which one H_2O_2 is converted to H_2O. Substrates ($R'H_2$) for the peroxidation reactions include ethanol, methanol, quinones, and formate. [From C. De Duve and P. Baudhuin, 1966, *Physiol. Rev.* **46**:233.]

In most mammalian cells, and probably in most eukaryotes, the peroxisome is the principal organelle in which fatty acids are degraded. In the liver, mitochondria also oxidize fatty acids (see Figure 17-12), and fatty acids are a major source of ATP in liver tissue; therefore, at one

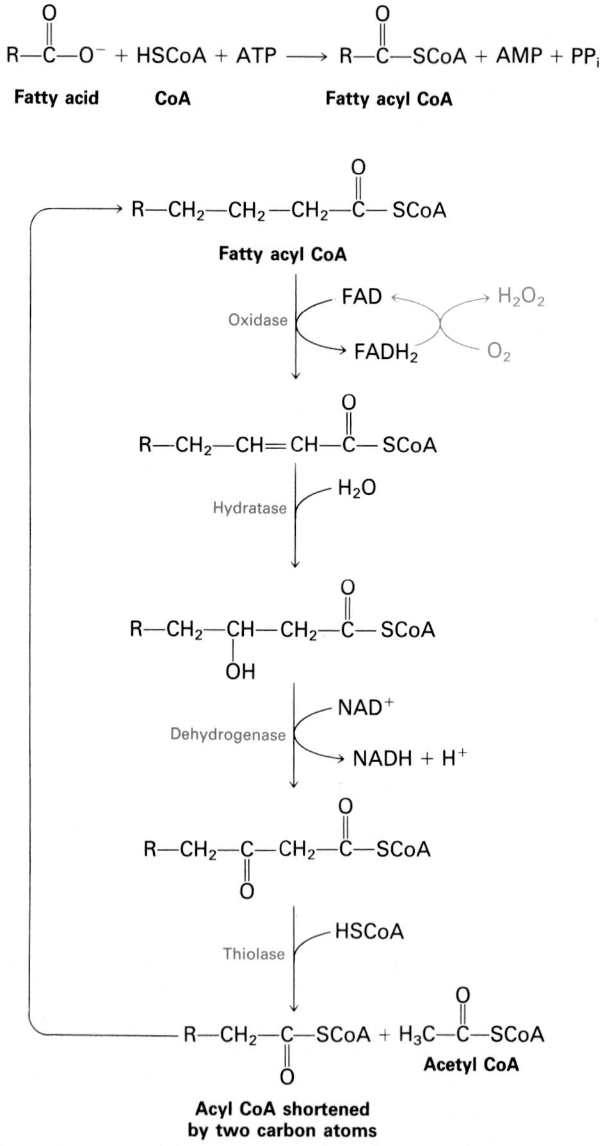

▲ FIGURE 17-33   Oxidation of fatty acids by peroxisomes. Peroxisomes degrade fatty acids with more than 12 carbon atoms by a series of reactions similar to those used by liver mitochondria (see Figure 17-12). Four enzymatically catalyzed reactions convert a fatty acyl CoA molecule to acetyl CoA and a fatty acyl CoA shortened by two carbon atoms. Concomitantly, one NAD^+ molecule is reduced to NADH and one FAD molecule is reduced to $FADH_2$. The cycle is repeated on the shortened acyl CoA until fatty acids with an even number of carbon atoms are completely converted to acetyl CoA. Importantly, the electrons and protons transferred to FAD during the oxidase reaction react with oxygen, forming H_2O_2.

time it was thought that mitochondria are the sites of fatty acid oxidation in all cell types. The recognition that peroxisomes can carry out fatty acid oxidation resulted from studies in which rats were treated with clofibrate, a drug used to reduce the level of blood lipoproteins (page 725) in people with faulty lipid metabolism. The drug was found to cause both an increase in the rate of fatty acid oxidation and a large increase in the number of peroxisomes in liver cells. These findings led to further discoveries: a set of enzymes in peroxisomes that oxidize fatty acids (Figure 17-33) and the presence, in these organelles, of intermediates in fatty acid degradation.

The peroxisome is the only organelle in which very-long-chain fatty acids (those containing over ≈ 20 CH_2 groups) are degraded; fatty acids of ≈ 10–20 CH_2 groups are degraded in both peroxisomes and mitochondria. The pathway of peroxisomal degradation of fatty acids is similar to that used in liver mitochondria (see Figure 17-12).

Peroxisomes lack an electron transport chain, and electrons released during the oxidation of fatty acids are used to form H_2O_2. H_2O_2 is highly toxic to the cell, but it is efficiently decomposed into H_2O within the peroxisome by catalase, an enzyme present in peroxisomes at concentrations as high as 4 percent.

During peroxisomal oxidation of fatty acids, no ATP is formed, and the released energy is converted to heat. In mitochondria, by contrast, electrons released during the oxidation of fatty acids are transferred, via $FADH_2$ and NADH, to the electron transport chain; as a result, mitochondrial fatty acid oxidation is coupled to the generation of ATP. Acetyl CoA is generated during peroxisomal oxidation of fatty acids (see Figure 17-33); its acetyl group is transported into the cytosol, where it is used for various biosynthetic reactions, for instance the synthesis of cholesterol and other steroids.

Before fatty acids can be degraded in the peroxisome, they must first be transported into the organelle from the cytosol. Midlength fatty acids (~ 10–20 CH_2 groups) are esterified to Coenzyme A in the cytosol and the fatty acyl CoA is then transported into the peroxisome by a specific transporter. However, a different transporter is used for very-long-chain fatty acids, and these fatty acids are esterified to CoA inside the peroxisome. In the human genetic disease X-linked adrenoleukodystrophy[*] (ALD), peroxisomal oxidation of very-long-chain fatty acids is specifically defective, while the oxidation of midlength fatty acids is normal. In ALD, very-long-chain fatty acids are transported normally into the peroxisome, but are not esterified to the CoA derivative and so cannot be oxidized. Patients with the severe form of the disease are unaffected until midchildhood, when severe neurological disorders appear, followed by death within a few years. As we discuss in Chapter 19, positional cloning of the ALD gene revealed that it encodes the peroxisome membrane transporter required for the uptake of the enzyme *long-chain fatty acyl CoA synthase* into the peroxisome from the cytosol.

SUMMARY

The immediate energy source for ATP synthesis comes from a combination of a proton concentration (pH) gradient (exoplasmic face > cytosolic face) and an electric potential (negative cytosolic face) across the inner mitochondrial membrane, the chloroplast thylakoid membrane, or the bacterial plasma membrane. The gradient and potential are known collectively as the proton-motive force. The multiprotein F_0F_1 ATPase complex catalyzes ATP synthesis as protons flow back through the membrane down their electrochemical proton gradient. This complex has a very similar structure in all three systems: F_0 is a transmembrane complex that forms a regulated H^+ channel; F_1 is tightly bound to F_0 and contains the site for ATP synthesis from ADP and P_i.

Glucose catabolism begins in the cytosol of eukaryotic cells, where it is converted to pyruvate via the glycolytic pathway, with the net formation of two ATPs and the net reduction of two NAD^+ molecules to NADH. In anaerobic cells, pyruvate can be metabolized further to lactate or ethanol plus CO_2, with the re-oxidation of NADH.

Most of the ATP in mitochondria and aerobic bacteria is generated during the oxidation of pyruvate to CO_2. Pyruvate is first converted to acetyl CoA and CO_2; NADH and $FADH_2$ are formed by the subsequent oxidation of acetyl CoA to CO_2. Acetyl CoA is also a key intermediate in the oxidation of fats and amino acids. The oxidation of acetyl CoA in the citric acid cycle is catalyzed by a set of enzymes localized in the mitochondrial matrix. Acetyl CoA

---

[*]This is the disease of the child in the recent movie *Lorenzo's Oil*, in which the father tried unsuccessfully to cure his son by feeding him a particular concoction of oils.

condenses with the four-carbon molecule oxaloacetate to form the six-carbon citrate. In a series of reactions, citrate is converted to oxaloacetate and to CO_2, concomitant with the reduction of NAD^+ to NADH and FAD to $FADH_2$. The NADH generated in the cytosol during the reactions of the glycolytic pathway is oxidized to NAD^+ by enzymes of an electron shuttle, concomitant with the reduction of NAD^+ to NADH in the matrix. This provides another source of reduced nucleotides for oxidative phosphorylation.

In the mitochondrion, electron flow from NADH or $FADH_2$ to O_2 is coupled to proton transport across the inner membrane from the matrix, generating the proton-motive force. The major components of the electron transport chain are four multiprotein complexes with defined orientations in the inner membrane: succinate-CoQ reductase, NADH-CoQ reductase, $CoQH_2$–cytochrome c reductase, and cytochrome c oxidase; the last complex transfers electrons to O_2 to form H_2O. The movement of electrons through each of the last three complexes is coupled to proton movement. CoQ functions as a lipid-soluble reversible transporter of electrons and protons across the inner membrane. The Q cycle—cycling of electrons through the $CoQH_2$–cytochrome c reductase complex—allows addi-

tional protons to be translocated per pair of electrons moving to cytochrome c. Because the mitochondrial membrane is impermeable to anions, the predominant component of the proton-motive force is the membrane electric potential (~ 160 mV, matrix negative). The proton-motive force powers both the synthesis of mitochondrial ATP from ADP and P_i and the uptake of P_i and ADP from the cytosol in exchange for mitochondrial ATP. This force is also used to generate heat, particularly in brown fat because it contains the natural uncoupler protein thermogenin.

The conversion of glucose to pyruvate is under tight regulation, mainly by the allosteric inhibition or activation of phosphofructokinase-1. The glycolytic pathway is inhibited by an excess of ATP and the production of pyruvate from glucose and the oxidation of pyruvate by mitochondria are tightly coupled.

Peroxisomes are the principal organelles in which fatty acids, especially very-long-chain fatty acids, are oxidized. No ATP is produced in the process. Oxygen is the ultimate electron acceptor in the peroxisomal oxidation of fatty acids and many other organic substances, and H_2O_2 is formed in the process. The H_2O_2 is efficiently decomposed within the organelle by catalase, which is present in peroxisomes at high concentrations.

REVIEW QUESTIONS

1. The metabolism of glucose is of central importance in biochemistry. It can be broken down via glycolysis, synthesized during gluconeogenesis, or converted to glycogen for storage.

Review the pathway for glycolysis that is presented in Figure 17-2. Notice that it is essentially the pathway that oxidizes one molecule of glucose to two molecules of pyruvate. At which points in the pathway are molecules of ATP consumed? How many molecules of ATP are expended per molecule of glucose that enters glycolysis? At which points during glycolysis do ATP molecules form via substrate level phosphorylation? How many ATP molecules are formed by this method per molecule of glucose oxidized? What is the net number of ATP molecules made during glycolysis?

Except for three of the reactions involved in glycolysis, all have values of ΔG that are very near to 0 under physiological conditions. What does that tell you about most of the reactions in this pathway? Are most of them readily reversible? The three reactions that have values of ΔG that are not near 0 are those catalyzed by hexokinase, phosphofructokinase, and pyruvate kinase. Why do you think all these reactions have values of ΔG that are equal to -4.0 kcal/mole or less?

An important feature of any biochemical pathway is its regulation. Review the chief control points in glycolysis. Be sure you understand the relationship between these highly regulated reactions and the ΔG values associated with them.

a. The value of $\Delta G^{\circ\prime}$ for the conversion of glucose to glucose 6-phosphate as catalyzed by hexokinase is -4.0 kcal/mole;

the value of ΔG for this reaction is -8.0 kcal/mole under physiological conditions. (Be sure you understand the differences between the ΔG and $\Delta G^{\circ\prime}$ for a reaction in terms of the general definition of each.) The reaction is described as follows:

glucose + ATP $\rightleftharpoons$ glucose 6-phosphate + ADP.

What is the approximate ratio of glucose 6-phosphate to glucose in the cell? Assume that the intracellular concentrations of ATP and ADP are 5.0 mM and 1.0 mM, respectively; also assume that the temperature is 37°C. Recall that the value of R is 1.98×10^{-3} kcal/mole · °K. What is the advantage of having hexokinase phosphorylate glucose rapidly and efficiently?

b. Review the features of the glycolytic enzyme phosphofructokinase-1 which catalyzes the following reaction

fructose 6-phosphate + ATP $\rightleftharpoons$
fructose 1,6-bisphosphate + ADP.

This enzyme is highly regulated, and the ΔG associated with this reaction under physiological conditions is -5.3 kcal/mole, indicating that the reverse reaction is thermodynamically unfavorable. Gluconeogenesis, which is the synthesis of glucose from pyruvate, has many reactions in common with glycolysis, but it is not the direct reversal of this pathway. This should be clear from the fact that the reaction catalyzed by phosphofructokinase-1 is not readily reversible. In gluconeogenesis, the enzyme fructose 1,6-bisphosphatase catalyzes the following

$$\text{fructose 1,6-bisphosphate} + H_2O \rightleftharpoons$$
$$\text{fructose 6-phosphate} + P_i$$

which is favorable as written under physiological conditions. Why do you think this reaction for the formation of fructose 6-phosphate occurs readily whereas the reverse of the reaction catalyzed by phosphofructokinase is unfavorable? Which of the other enzymes involved in glycolysis would need to be replaced by enzymes catalyzing other reactions in order for gluconeogenesis to occur? Explain your answer.

2. The actual reaction for the oxidation of glucose to CO_2 involves both glycolysis and the citric acid cycle, and it is summarized as

$$C_6H_{12}O_6 + 6O_2 + 32P_i^{2-} + 32ADP^{3-} + 32H^+ \longrightarrow$$
$$6CO_2 + 32ATP^{4-} + 38H_2O.$$

What overall reaction describes the portion of glucose metabolism that is classified as glycolysis? What overall reaction describes the portion of glucose metabolism due to the citric acid cycle? How many ATP molecules are synthesized either directly or indirectly as a consequence of the oxidation reactions that occur during the citric acid cycle?

a. The citric acid cycle is illustrated in Figure 17-13. Notice that it begins and ends with oxaloacetate. The first step involves the reaction of oxaloacetate with acetyl-CoA to form citrate. Glycolysis ends with the formation of pyruvate. What is the reaction that bridges glycolysis and the citric acid cycle? How many molecules of CO_2 are generated during this reaction? This is a redox reaction. What is being oxidized, and what is being reduced?

b. Are any high energy molecules synthesized during the citric acid cycle by way of substrate level phosphorylation? How many are made per turn of the cycle? How many would be made per molecule of glucose? In the reaction describing the overall oxidation of glucose to CO_2 and H_2O, GTP is not present. Do you think this poses a contradiction to the reaction describing the metabolism of glucose as presented above? Explain your answer.

c. What is the net total number of high energy nucleoside triphosphate molecules generated by substrate level phosphorylation for one molecule of glucose being completely oxidized to CO_2 and H_2O? How many ATP molecules remain to be accounted for as products of glucose metabolism?

3. Aerobic respiration involves the use of O_2 as the ultimate agent that accepts the reducing equivalents generated during the oxidation of glucose. Where does this process occur? The electron transport chain is linked to the pumping of protons from the mitochondrial matrix to the intermembrane space. How is the energy stored in this gradient linked to the synthesis of ATP?

Figure 17-9 summarizes several of the processes that occur in the mitochondria. Pay particular attention to the points at which NADH and $FADH_2$ enter the electron transport chain. How do the reducing equivalents from NADH that is made in the cytosol enter the mitochondrial matrix? Oxidation of NADH and $FADH_2$ contributes electrons that are ultimately used to reduce O_2 to water; however, the number of protons pumped for each electron pair donated is different for each. Why is this the case?

Refer to Figure 17-22. Notice that there are three reductase complexes and one oxidase complex in the electron transport system. Which of these are coupled to the extrusion of protons from the mitochondrial matrix? How many protons are pumped by each complex per pair of electrons through the complex?

a. You are investigating a series of inhibitors of the electron transport system in human mitochondria. One of these is a drug called antimycin C which inhibits the transfer of electrons within the $CoQH_2$-cyt c reductase complex. What would be the ultimate consequences of this inhibition? If you added an agent that could accept electrons from $CoQH_2$, and you added it in the presence of antimycin C, would proton pumping resume? How many ATP molecules would be made per molecule of glucose oxidized under these conditions?

b. Cyanide is an extremely potent poison. It has many deleterious effects in biological systems, one of which is to block the transfer of electrons from the cytochrome c oxidase complex to O_2. What would be the effect of cyanide on the electron transport chain? That is, would it shut it down, or only uncouple the last step? Suppose there were a drug available that could accept electrons within reduced cytochrome c and transfer them to O_2 without pumping protons. If this drug were added along with cyanide to isolated mitochondria, what would be the effect on the number of ATP molecules synthesized per molecule of glucose oxidized?

c. You have added a newly developed inhibitor of the electron transport system that works by preventing the succinate-CoQ reductase complex from reducing CoQ. What effect would this have on the total number of ATP molecules synthesized per molecule of glucose oxidized? What would be the effect of this drug on the citric acid cycle?

4. Oxidation of fatty acids is another important source of reducing equivalents used for the synthesis of ATP. Review the steps of fatty acid oxidation that occur in the mitochondrial matrix. How do these contrast with the oxidation of fatty acids that occurs in peroxisomes?

a. Stearic acid is a saturated fatty acid that contains 18 carbon atoms. If one molecule of this fatty acid were completely oxidized in a mitochondrion, how many molecules of ATP would result?

b. Oleic acid is also 18 carbon atoms in length, but it is unsaturated. Refer to Figure 14-4 for its structure. If one molecule of this fatty acid were oxidized in a mitochondrion, what would be the number of ATP molecules produced? In general, do saturated or unsaturated fatty acids yield more ATP per CO_2 produced?

References

Glycolysis and the Citric Acid Cycle

General References

BRIDGER, W. A., and J. F. HENDERSON. 1983. *Cell ATP*. Wiley.

FERSHT, A. 1985. *Enzyme Structure and Mechanism*, 2d ed. W. H. Freeman and Company. Contains an excellent discussion of the reaction mechanisms of key enzymes.

LEHNINGER, A. L., D. L. NELSON, and M. M. COX. 1993. *Principles of Biochemistry*. Worth. Chapters 13–16, 18.

STRYER, L. 1988. *Biochemistry*, 3d ed. W. H. Freeman and Company. Chapters 15 and 16.

Glycolysis

BOITEUX, A., and B. HESS. 1981. Design of glycolysis. *Phil. Trans. R. Soc. Lond.* B293:5–22.

FOTHERGILL-GILMORE, L. A., and P. A. MICHELS. 1993. Evolution of glycolysis. *Prog. Biophys. Mol. Biol.* 59:105–235.

The Citric Acid Cycle

BALDWIN, J. E., and H. A. KREBS. 1981. The evolution of metabolic cycles. *Nature* 291:381–382.

GUEST, J. R., and G. C. RUSSELL. 1992. Complexes and complexities of the citric acid cycle in *Escherichia coli. Curr. Top. Cell Regul.* 33:231–247.

KREBS, H. A. 1970. The history of the tricarboxylic acid cycle. *Perspect. Biol. Med.* 14:154–170.

REED, L. J., Z. DAMUNI, and M. L. MERRYFIELD. 1985. Regulation of mammalian pyruvate and branched-chain α-keto acid dehydrogenase complexes by phosphorylation-dephosphorylation. *Curr. Top. Cell Regul.* 27:41–49.

REMINGTON, S. J. 1992. Structure and mechanism of citrate synthase. *Curr. Top. Cell Regul.* 33:209–229.

SRERE, P. A. 1987. Complexes of sequential metabolic enzymes. *Ann. Rev. Biochem.* 56:89–124.

———. 1992. The molecular physiology of citrate. *Curr. Top. Cell Regul.* 33:261–275.

SUMEGI, B., et al. 1991. Is there tight channelling in the tricarboxylic acid cycle metabolon? *Biochem. Soc. Trans.* 19:1002–1005.

Energy Metabolism and the Chemiosmotic Theory

General References

DEVLIN, T. M., ed. 1992. *Textbook of Biochemistry with Clinical Correlations*, 3d ed. Wiley-Liss. pp. 237–287.

DICKERSON, R. E. 1980. Cytochrome *c* and the evolution of energy metabolism. *Sci. Am.* 242(3):137–153.

ERNESTER, L., ed. 1985. *Bioenergetics*. New Comprehensive Biochemistry, Vol. 9. Elsevier.

HAROLD, F. M. 1986. *The Vital Force: A Study of Bioenergetics*. W. H. Freeman and Company.

HATEFI, Y. 1985. The mitochondrial electron transport and oxidative phosphorylation system. *Ann. Rev. Biochem.* 45:1015–1070.

KELL, D. B. 1992. The proton-motive force as an intermediate in electron transport–linked phosphorylation: problems and prospects. *Curr. Top. Cell Regul.* 33:279–289.

LÄUGER, P. 1991. *Electrogenic Ion Pumps*. Sinauer Associates.

MITCHELL, P. 1979. Keilin's respiratory chain concept and its chemiosmotic consequences. *Science* 206:1148–1159. Mitchell's Nobel Prize lecture.

MONTGOMERY, R., T. W. CONWAY, and A. A. SPECTOR. 1990. *Biochemistry: A Case Oriented Approach*, 5th ed. Mosby. pp. 191–243.

NICHOLLS, D. G., and S. J. FERGUSON. 1992. *Bioenergetics 2*. Academic Press.

RACKER, E. 1980. From Pasteur to Mitchell: a hundred years of bioenergetics. *Fed. Proc.* 39:210–215.

SKULACHEV, V. P. 1992. The laws of cell energetics. *Eur. J. Biochem.* 208:203–209.

YOUVAN, D.C., and F. DALDAL, eds. 1986. *Microbial Energy Transduction: Genetics, Structure, and Function of Membrane Proteins*. Cold Spring Harbor Laboratory.

Synthesis of ATP and the F_0F_1 ATPase

BOYER, P. D. 1989. A perspective of the binding change mechanism for ATP synthesis. *FASEB J.* 3:2164–2178.

———. 1993. The binding change mechanism for ATP synthase—some probabilities and possibilities. *Biochim. Biophys. Acta* 1140:215–250.

CAPALDI, R. A., et al. 1992. Structure of the *Escherichia coli* ATP synthase and role of the gamma and epsilon subunits in coupling catalytic site and proton channeling functions. *J. Bioenerg. Biomembr.* 24:435–439.

FILLINGAME, R. H. 1992. H^+ transport and coupling by the F_0 sector of the ATP synthase: insights into the molecular mechanism of function. *J. Bioenerg. Biomembr.* 24:485–491.

FUTAI, M., T. NOUMI, and M. MAEDA. 1989. ATP synthase (H^+-ATPase): results by combined biochemical and molecular biological approaches. *Ann. Rev. Biochem.* 58:111–136.

LEWIS, M. J., J. A. CHANG, and R. D. SIMONI. 1990. A topological analysis of subunit alpha from *Escherichia coli* F_1F_0-ATP synthase predicts eight transmembrane segments. *J. Biol. Chem.* 265:10541–10550.

NELSON, N. 1992. Evolution of organellar proton-ATPases. *Biochim. Biophys. Acta* 1100:109–124.

PEDERSEN, P. L., and L. M. AMZEL. 1993. ATP synthases: structure, reaction center, mechanism, and regulation of one of nature's most unique machines. *J. Biol. Chem.* 268:9937–9940.

PENEFSKY, H. S., and R. L. CROSS. 1991. Structure and mechanism of F_0F_1-type ATP synthases and ATPases. *Adv. Enzymol. Relat. Areas Mol. Biol.* 64:173–214.

SENIOR, A. E. 1992. Catalytic sites of *Escherichia coli* F_1-ATPase. *J. Bioenerg. Biomembr.* 24:479–484.

Transport of Metabolites into and out of the Mitochondrion

KLINGENBERG, M. 1989. Survey of carrier methodology: strategy for identification, isolation, and characterization of transport systems. *Methods Enzymol.* 171:12–23.

———. 1993. Mitochondrial carrier family: ADP/ATP carrier as a carrier paradigm. *Soc. Gen. Physiol. Ser.* 48:201–212.

KUAN, J., and M. SAIER, JR. 1993. The mitochondrial carrier family of transport proteins: structural, functional, and evolutionary relationships. *Crit. Rev. Biochem. Mol. Biol.* 28:209–233.

SORGATO, M. C., and O. MORAN. 1993. Channels in mitochondrial membranes: knowns, unknowns, and prospects for the future. *Crit. Rev. Biochem. Mol. Biol.* 28:127–171.

The Electron Transport Chain in Mitochondria

BABCOCK, G. T., and C. VAROTSIS. 1993. Discrete steps in dioxygen activation—the cytochrome oxidase/O_2 reaction. *J. Bioenerg. Biomembr.* 25:71–80.

BABCOCK, G. T., and M. WIKSTROM. 1992. Oxygen activation and the conservation of energy in cell respiration. *Nature* 356:301–309.

BRANDT, U. and B. TRUMPOWER. 1994. The Protonmotive Q cycle in mitochondria and bacteria. *Crit. Rev. Biochem. Mol. Biol.* 29:165–197.

CAPALDI, R. A. 1990. Structure and function of cytochrome *c* oxidase. *Ann. Rev. Biochem.* 59:569–596.

CHEN, L. B. 1988. Mitochondrial membrane potential in living cells. *Ann. Rev. Cell Biol.* 4:155–181.

ESPOSTI, M. D., et al. 1993. Mitochondrial cytochrome *b*: evolution and structure of the protein. *Biochim. Biophys. Acta* 1143:243–271.

HILL, B. C. 1993. The sequence of electron carriers in the reaction of cytochrome *c* oxidase with oxygen. *J. Bioenerg. Biomembr.* 25:115–120.

HOPE, A. B. 1993. The chloroplast cytochrome *bf* complex: a critical focus on function. *Biochim. Biophys. Acta* 1143:1–22.

PETTIGREW, G. W., and G. R. MOORE. 1987. *Cytochrome c: Biological Aspects*. Springer-Verlag.

TRUMPOWER, B. L. 1990. Cytochrome *bc*1 complexes of microorganisms. *Microbiol. Rev.* 54:101–129.

———. 1990. The proton-motive Q cycle. Energy transduction by coupling of proton translocation to electron transfer by the cytochrome *bc*1 complex. *J. Biol. Chem.* 265:11409–11412.

VAROTSIS, C., et al. 1993. Resolution of the reaction sequence during the reduction of O_2 by cytochrome oxidase. *Proc. Natl. Acad. Sci. USA* **90**:237–241.

WIKSTROM, M. 1989. Identification of the electron transfers in cytochrome oxidase that are coupled to proton-pumping. *Nature* **338**:776–778.

WOODRUFF, W. H. 1993. Coordination dynamics of heme-copper oxidases. The ligand shuttle and the control and coupling of electron transfer and proton translocation. *J. Bioenerg. Biomembr.* **25**:177–188.

ZHANG, Y. Z., G. EWART, and R. A. CAPALDI. 1991. Topology of subunits of the mammalian cytochrome *c* oxidase: relationship to the assembly of the enzyme complex. *Biochemistry* **30**:3674–3681.

Regulation of Glucose and Fatty Acid Metabolism

BROWN, G. C. 1992. Control of respiration and ATP synthesis in mammalian mitochondria and cells. *Biochem. J.* **284**:1–13.

GUZMAN, M., and M. J. GEELEN. 1993. Regulation of fatty acid oxidation in mammalian liver. *Biochim. Biophys. Acta* **1167**:227–241.

PILKIS, S. J., and D. K. GRANNER. 1992. Molecular physiology of the regulation of hepatic gluconeogenesis and glycolysis. *Ann. Rev. Physiol.* **54**:885–909.

ERICINSKA, A., and D. F. WILSON. 1982. Regulation of cellular energy metabolism. *J. Membr. Biol.* **70**:1–14.

KEMP, R. G., and L. G. FOE. 1983. Allosteric regulatory properties of muscle phosphofructokinase. *Mol. Cell Biochem.* **57**:147–154.

NEWSHOLME, E. A., and C. START. 1975. *Regulation of Metabolism.* Wiley.

OCHS, R. S., R. W. HANSON, and J. HALL, eds. 1985. *Metabolic Regulation.* Elsevier. A collection of articles originally published in *Trends Biochem. Sci.*

Thermogenesis

KLINGENBERG, M. 1990. Mechanism and evolution of the uncoupling protein of brown adipose tissue. *Trends Biochem. Sci.* **15**:108–112.

Oxidation of Fatty Acids in Peroxisomes

MANNAERTS, G. P., and P. P. VAN VELDHOVEN. 1993. Metabolic pathways in mammalian peroxisomes. *Biochimie* **75**:147–158.

MOSSER, J., et al. 1993. Putative X-linked adrenoleukodystrophy gene shares unexpected homology with ABC transporters. *Nature* **361**:726–730.

SINGH, I. 1992. Peroxisomal activation, transport and oxidation of fatty acids: implications to peroxisomal disorders. *Prog. Clin. Biol. Res.* **375**:211–222.

SINGH, I., et al. 1992. Transport of fatty acids into human and rat peroxisomes. Differential transport of palmitic and lignoceric acids and its implication to X-adrenoleukodystrophy. *J. Biol. Chem.* **267**:13306–13313.

VALLE, D., and J. GARTNER. 1993. Human genetics. Penetrating the peroxisome. *Nature* **361**:682–683.

WANDERS, R. J., et al. 1990. The inborn errors of peroxisomal beta-oxidation: a review. *J. Inherit. Metab. Dis.* **13**:4–36.

18 Photosynthesis

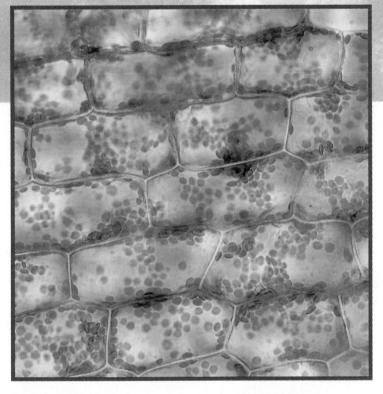

▲ Chloroplasts in leaf cells of the pond weed *Elodea*.

In the previous chapter we learned how animal cells generate phosphoanhydride bonds in ATP by the oxidation of glucose or fats. Here we focus on *photosynthesis,* the process by which light energy is converted to the chemical energy of a phosphoanhydride bond of ATP and stored in the chemical bonds of carbohydrates.

The main products of photosynthesis are polymers of six-carbon sugars, usually sucrose or starch. Oxygen is also formed during photosynthesis in plants, in eukaryotic single-celled algae, and in two groups of photosynthetic prokaryotes—the *cyanobacteria* (formerly called the *blue-green algae*) and the *prochlorophytes.* This oxygen is the source of virtually all of the oxygen in our air. The overall reaction of oxygen-generating photosynthesis, namely

$$6CO_2 + 6H_2O \longrightarrow 6O_2 + C_6H_{12}O_6$$

is the reverse of the overall reaction by which carbohydrates are oxidized to CO_2 and H_2O.

About half of the total photosynthesis in the world is carried out by single-celled algae in the ocean; much of the rest is by plants in tropical rain forests. However, photosynthesis occurs in many bacteria as well as in algae and higher plants. It is essential for almost all life on earth, as the carbohydrates produced during photosynthesis are the ultimate source of energy for virtually all nonphotosynthetic organisms, including all animals.*

Although our emphasis in this chapter is on photosynthesis in plant chloroplasts, we also discuss the simpler type of photosynthesis that occurs in bacteria. The three-

---

* Sunlight, contrary to popular view, is not the ultimate source of energy for *all* organisms on earth. As noted on page 43, in deep ocean vents, where there is no sunlight (and where the temperature of liquid water is well over 100°C), there are bacteria that use, instead, a process known as *chemolithotropy* to obtain energy for converting CO_2 into carbohydrates, amino acids, and other cellular constituents: the energy comes from the oxidation of reduced inorganic compounds in dissolved vent gas.

dimensional structure of the photosynthetic systems in certain bacteria allows us to trace in molecular detail the first stages of photosynthesis—how light energy is converted to a separation of negative and positive charges across the thylakoid membrane, with the simultaneous generation of a strong oxidant and a strong reductant. We also point out some striking parallels between photosynthesis and oxidative phosphorylation; namely, that some components in chloroplasts and mitochondria are virtually identical: the particles that use the proton-motive force to synthesize ATP, some of the nonprotein electron and H^+ carriers (the quinones), and also the electron-transport complex that carries out the Q cycle of proton transport.

► An Overview of Photosynthesis

Photosynthesis in plants occurs in *chloroplasts,* large organelles found mainly in leaf cells (Figure 18-1). The principal end products of photosynthesis are two carbohydrates that are polymers of six-carbon sugars: starch and sucrose (Figure 18-2). Leaf starch, an insoluble polymer of glucose, is stored in the chloroplast. Sucrose, a water-soluble disaccharide, is synthesized in the cytosol from three-carbon precursors generated in the chloroplast. Sucrose is transported from the leaf through the phloem to other parts of the plant; nonphotosynthetic (nongreen) plant tissues like roots and seeds metabolize sucrose for energy.

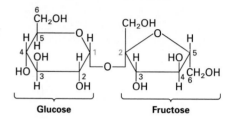

▲ **FIGURE 18-2**   Structure of starch and sucrose. Starch and sucrose are the principal end products of photosynthesis. Both are built of six-carbon sugars.

Photosynthesis Occurs on Thylakoid Membranes

Chloroplasts (Figure 18-3) have three membranes. The two outermost membranes do not contain chlorophyll, the principal light-absorbing pigment, are not green, and do not participate directly in photosynthesis. Of these two membranes, the outer one, like the outer membrane of the mitochondrion, is permeable to metabolites of small molecular weight; it contains proteins that form very large aqueous channels. The inner membrane of the two, conversely, is the permeability barrier of the chloroplast; it contains transporters that regulate the movement of metabolites into and out of the organelle.

Photosynthesis occurs on the third, innermost chloroplast membrane, the *thylakoid membrane.* In each chloroplast the thylakoid membranes are believed to constitute a single, interconnected sheet that encloses a single, continuous space, the *thylakoid lumen.* The thylakoid membrane frequently forms into stacks of small flattened membranes, termed *grana* (singular, *granum;* see Figure 18-3). Thylakoid membranes contain a number of integral membrane proteins to which are bound several important prosthetic groups and light-absorbing pigments, most notably chlorophyll. Carbohydrate synthesis occurs in the *stroma,* the soluble phase between the thylakoid membrane and the inner membrane. In photosynthetic bacteria extensive invaginations of the plasma membrane form a set of internal membranes, also termed *thylakoid membranes,* or simply *thylakoids,* where photosynthesis occurs (Figure 18-4).

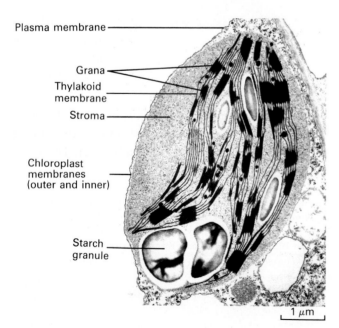

Plasma membrane

Grana
Thylakoid membrane
Stroma

Chloroplast membranes (outer and inner)

Starch granule

1 μm

▲ FIGURE 18-1   Electron micrograph of a chloroplast in part of a *Phleum pratanese* cell. [Courtesy of Biophoto Associates/M. C. Ledbetter/Brookhaven National Laboratory.]

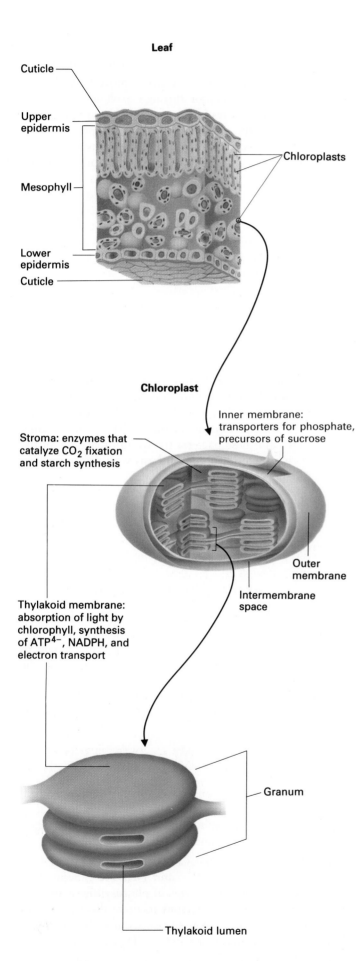

Leaf

Cuticle

Upper epidermis

Mesophyll

Chloroplasts

Lower epidermis

Cuticle

Chloroplast

Inner membrane: transporters for phosphate, precursors of sucrose

Stroma: enzymes that catalyze CO_2 fixation and starch synthesis

Thylakoid membrane: absorption of light by chlorophyll, synthesis of ATP^{4-}, NADPH, and electron transport

Outer membrane

Intermembrane space

Granum

Thylakoid lumen

◄ **FIGURE 18-3** The structure of the chloroplast. Different reactions of photosynthesis occur in different regions of the chloroplast. The chloroplast is bounded by a double membrane: the outer membrane contains proteins that render it permeable to small molecules (MW < 6,000); the inner membrane forms the permeability barrier of the organelle. Photosynthesis occurs on thylakoid membranes—a series of flattened membranes that enclose a single interconnected luminal space. The green color of plants is due to the green color of chlorophyll, all of which is localized to the thylakoids. Stacks of thylakoid vesicles are termed grana. The stroma is the space within the inner membrane surrounding the thylakoid vesicles.

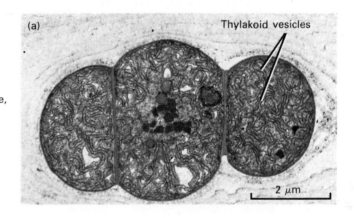

(a)

Thylakoid vesicles

2 μm

(b)

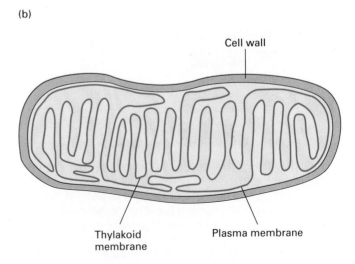

Cell wall

Thylakoid membrane

Plasma membrane

▲ **FIGURE 18-4** (a) Electron micrograph of a thin section through three attached cells of the photosynthetic *Nostoc carneum,* one of the cyanobacteria. The extensive array of thylakoid membranes is characteristic of this prokaryotic group; these internal photosynthetic membranes are formed by invagination of the plasma membrane, as depicted in the diagram in part (b). [Part (a) courtesy of T. E. Jensen and C. C. Bowen.]

Photosynthesis Consists of Both "Light" and "Dark" Reactions

It is convenient to divide the processes of photosynthesis into four stages, each occurring in a defined area of the chloroplast:

1. Absorption of Light   The initial step in photosynthesis is the absorption of light by chlorophyll attached to proteins in the thylakoid membranes. Chlorophyll (Figure 18-5) is a ringed compound similar in structure to heme, except that a magnesium atom, Mg^{2+} (rather than an iron atom Fe^{3+}), is in the center and besides the four central 5-atom rings, there is an additional 5-atom ring. The energy of the absorbed light is used first to remove electrons from an unwilling donor—water, in green plants, forming oxygen:

$$2H_2O \xrightarrow{\text{light}} O_2 + 4H^+ + 4e^-$$

and then to transfer the electrons to a primary accep-

Chlorophyll *a*

Phytol

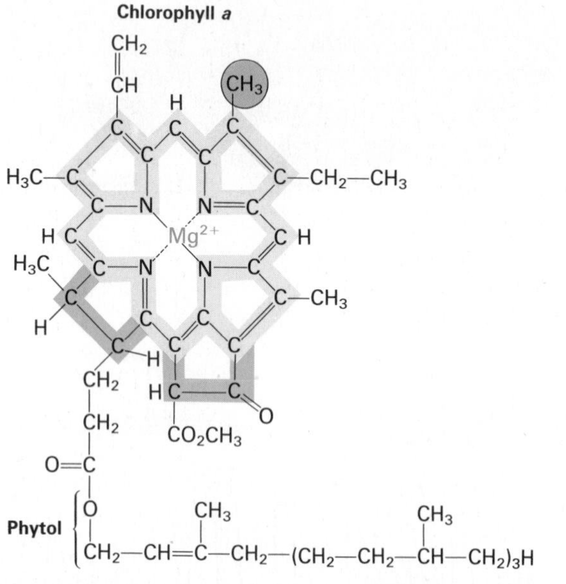

▲ **FIGURE 18-5**   The structure of chlorophyll *a*, the principal pigment that traps light energy. Chlorophyll *b* differs from chlorophyll *a* by having a CHO group in place of the CH_3 group (green). The heme group (four central rings), or porphyrin ring, is a highly conjugated system; electrons are delocalized among three of the four central rings and the atoms that interconnect them in the molecule (blue). A Mg^{2+} atom, rather than an Fe^{3+} atom, is in the center; otherwise, its structure is similar to that of heme found in molecules such as hemoglobin (see Figure 3-15) or cytochromes (see Figure 17-26). The 5-atom ring not found in hemes is colored yellow. The hydrophobic phytol "tail" facilitates the binding of chlorophyll to hydrophobic regions of chlorophyll-binding proteins.

tor. All these reactions occur in a complex of proteins, termed a *photosystem*, in the thylakoid membrane.

2. Electron Transport   Electrons move from the primary electron acceptor through a chain of electron transport molecules in the thylakoid membrane until they reach the ultimate electron acceptor, usually $NADP^+$, reducing it to NADPH. The transport of electrons is coupled to the movement of protons across the membrane from the stroma to the thylakoid lumen, forming a pH gradient across the thylakoid membrane, in much the same way that a proton-motive force is established across the mitochondrial inner membrane during electron transport (see Figure 17-14).

Thus the overall reaction of stages 1 and 2 can be summarized as

$$2H_2O + 2NADP^+ \xrightarrow{\text{light}} 2H^+ + 2NADPH + O_2$$

Many bacteria do not use water as the donor of electrons. Rather, they use molecules such as hydrogen gas (H_2) or hydrogen sulfide (H_2S) as the ultimate source of electrons to reduce NAD^+, rather than $NADP^+$. In these bacteria the overall reactions of the first two stages of photosynthesis would therefore be:

$$H_2S + NAD^+ \xrightarrow{\text{light}} H^+ + NADH + S$$

and:

$$H_2 + NAD^+ \xrightarrow{\text{light}} H^+ + NADH$$

3. Generation of ATP   Protons move down their concentration gradient from the thylakoid lumen to the stroma through a set of transport proteins, the CF_0CF_1 complex (similar to the F_0F_1 complex in mitochondria; see Figure 17-15) which couples proton movement to the synthesis of ATP from ADP and P_i:

$$H^+ + ADP^{3-} + P_i^{2-} \longrightarrow ATP^{4-} + H_2O$$

This use of the proton-motive force to synthesize ATP is very similar to the analogous process occurring during oxidative phosphorylation in the mitochondrion (see Figures 17-15 and 17-17).

4. CO_2 Fixation—the Conversion of CO_2 into Carbohydrates   The ATP^{4-} and NADPH generated by the second and third stages of photosynthesis provide the energy and the electrons to drive the synthesis of polymers of six-carbon sugars from CO_2 and H_2O:

$$6CO_2 + 18ATP^{4-} + 12NADPH + 12H_2O \longrightarrow$$
$$C_6H_{12}O_6 + 18ADP^{3-} + 18P_i^{2-} + 12NADP^+ + 6H^+$$

All four stages of photosynthesis are tightly coupled and controlled so as to produce the amount of carbohydrate required by the plant. All of the reactions involving the absorption of light and the formation of NADPH and ATP (stages 1–3) are catalyzed by proteins in the thylakoid membrane. The enzymes that incorporate CO_2 into chemical intermediates (the process of CO_2 fixation) and then convert it to starch are soluble constituents of the chloroplast stroma (see Figure 18-3). The enzymes that form sucrose from three-carbon intermediates are in the cytosol.

The reactions that generate ATP and NADPH are *directly* dependent on light energy and thus stages 1–3 are called the *light reactions* of photosynthesis. The reactions in stage 4 are *indirectly* dependent on light energy; they are sometimes called the *dark reactions* of photosynthesis because sugar can form in the dark, utilizing the supplies of ATP and NADPH generated by light energy. However, the reactions in stage 4 are not confined to the dark; in fact, they primarily occur during illumination.

► The Light-Absorbing Step of Photosynthesis

We begin our detailed discussion of photosynthesis at the beginning—with a look at some fundamental properties of light waves and the absorption of light by chlorophyll.

Each Photon of Light Has a Defined Amount of Energy

Light is a form of electromagnetic radiation; quantum mechanics established that light has properties of both waves and particles. When it interacts with matter, light can be thought of as discrete packets of energy (*quanta*) called *photons*. The energy of a photon, ϵ, is proportional to the frequency of the light wave:

$$\epsilon = h\gamma$$

where h is Planck's constant (1.58×10^{-34} cal · s, or 6.63×10^{-34} J · s), and γ is the frequency of the light wave. It is customary in biology to refer to the wavelength of the light wave, λ, rather than to its frequency, γ; the two are related by the simple equation

$$\gamma = \frac{c}{\lambda}$$

where c is the velocity of light (3×10^{10} cm/s in a vacuum).

Note that photons of *shorter* wavelength have *higher* energies.

Also, the energy in 1 mol of photons can be denoted by $E = N\epsilon$, where N is Avogadro's number (6.02×10^{23} molecules or photons/mol). Thus

$$E = Nh\gamma = \frac{Nhc}{\lambda}$$

The energy of light is considerable, as we can calculate for light with a wavelength of 550 nm (550×10^{-7} cm), typical of sunlight:

$$E = \frac{(6.02 \times 10^{23} \text{ photons/mol}) \cdot (1.58 \times 10^{-34} \text{ cal} \cdot \text{s}) \cdot (3 \times 10^{10} \text{ cm/s})}{(550 \times 10^{-7} \text{ cm})}$$

$$= 51{,}881 \text{ cal/mol}$$

or about 52 kcal/mol, enough energy to synthesize several moles of ATP from ADP and P_i if all of the energy were used for this purpose.

Chlorophyll *a* Is the Primary Light-Absorbing Pigment

Chlorophyll a (see Figure 18-5), the principal pigment involved in photosynthesis, is found in all photosynthetic organisms, both eukaryotic and prokaryotic. Chlorophyll *a* is found in two types of multiprotein complexes in the thylakoid membrane, *antennas* and *reaction centers*. Each antenna (named by analogy with radio antennas) contains one or more *light-harvesting complexes* (LHCs). Antenna chlorophylls—chlorophyll *a* molecules in the LHCs—absorb light and funnel the energy to the two chlorophylls in the reaction-center; the primary events of photosynthesis occur in the reaction centers (see Figure 18-8).

Vascular plants also contain *chlorophyll b*, a pigment that absorbs light at slightly different wavelengths from those absorbed by chlorophyll *a*. Both plants and photosynthetic bacteria contain *carotenoids*, such as β-carotene (Figure 18-6), which absorb light at still other wavelengths (Figure 18-7). These and other light-absorbing pigments localized to the antennas funnel the energy of absorbed light to the two chlorophyll *a* molecules of the reaction center and greatly extend the range of light that can be absorbed and used for photosynthesis.

One of the strongest pieces of evidence for the involvement of chlorophylls and β-carotene in photosynthesis is that the *absorption spectrum* of these pigments is similar to the *action spectrum* of photosynthesis, which is the ability of light of different wavelengths to support photosynthesis, measured by fixation of CO_2 into carbohydrate (see Figure 18-7).

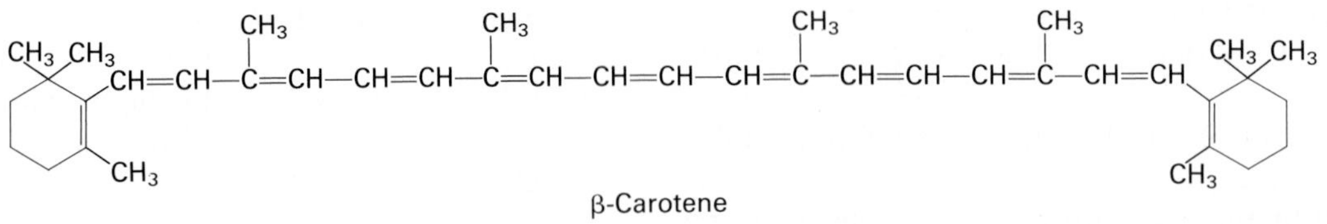

β-Carotene

▲ FIGURE 18-6   The structure of β-carotene, a pigment that assists in light absorption by chloroplasts. β-Carotene, which is related to the visual pigment retinal (see Figure 21-49), is one of a family of carotenoids containing long hydrocarbon chains with alternating single and double bonds.

When chlorophyll *a* (or any other molecule) absorbs visible light, the absorbed light energy raises the chlorophyll *a* to a higher energy state, termed an *excited state*. This differs from the ground (unexcited) state largely in the distribution of electrons around the C and N atoms of the porphyrin (heme) ring (see Figure 18-5). Excited states are unstable, and will return to the ground state by one of several competing processes. For chlorophyll *a* molecules dissolved in organic solvents, such as ethanol, the principal reactions that dissipate the excited-state energy are the emission of light (fluorescence and phosphorescence) and thermal emission (heat). The situation is quite different when the same chlorophyll *a* is bound to the unique protein environment of the reaction center, as we shall see.

The Absorption of Light by Reaction-Center Chlorophylls Causes a Charge Separation across the Thylakoid Membrane

The absorption of a quantum of light of wavelength ≈ 680 nm causes a chlorophyll molecule to enter the *first excited state*. The energy in 1 mol of such photons is considerable: equal, for chlorophyll *a*, to an increase in energy of 42 kcal/mol. In the reaction-center chlorophylls the excited state is used to promote a charge separation across the thylakoid membrane: an electron is transported from the chlorophyll to a *primary electron acceptor* (Q) on the stromal surface of the membrane, leaving a positive charge on the chlorophyll close to the luminal surface (Figure 18-8). The reduced primary electron acceptor becomes a powerful reducing agent: it has the tendency to transfer the electron to another molecule. The positively charged chlorophyll is a strong oxidizing agent, and will attract an electron from an electron donor on the luminal surface. These potent biological reductants and oxidants provide all of the energy needed to drive all the subsequent reactions of photosynthesis: electron transport, ATP synthesis, and CO_2 fixation.

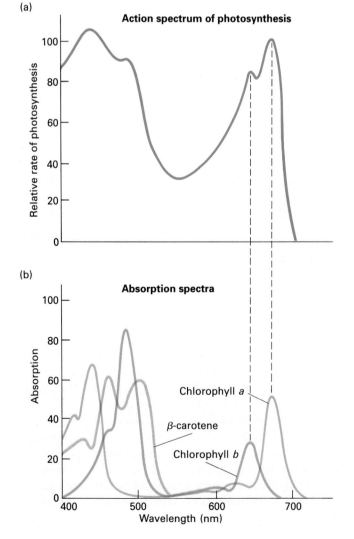

(a)

Action spectrum of photosynthesis

(b)

Absorption spectra

Chlorophyll *a*

β-carotene

Chlorophyll *b*

◀ FIGURE 18-7   Photosynthesis at different wavelengths. (a) The action spectrum of photosynthesis in plants; that is, the ability of light of different wavelengths to support photosynthesis. (b) The absorption spectra for three photosynthetic pigments: chlorophyll *a*, chlorophyll *b*, and β-carotene. Each spectrum shows how well light of different wavelengths is absorbed by one of the pigments. A comparison of the action spectrum with the individual absorption spectra suggests that photosynthesis at 650 nm is primarily due to light absorbed in the antenna complex by chlorophyll *b*; at 680 nm, to light absorbed by chlorophyll *a*; and at shorter wavelengths, to light absorbed by chlorophyll *b* and by carotenoid pigments, including β-carotene.

The following model summarizes the significant features of the primary reactions of photosynthesis, where P represents the chlorophyll *a* in the reaction center, and A the primary electron acceptor:

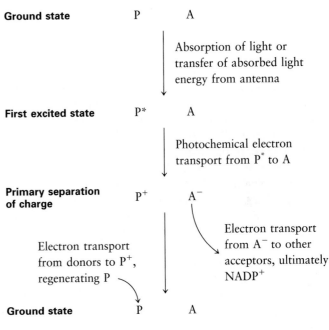

Ground state	P	A

Absorption of light or transfer of absorbed light energy from antenna

First excited state	P*	A

Photochemical electron transport from P* to A

Primary separation of charge	P⁺	A⁻

Electron transport from A⁻ to other acceptors, ultimately NADP⁺

Electron transport from donors to P⁺, regenerating P

Ground state	P	A

The key feature of this model is that an electron will not move spontaneously from P (the ground state of the chlorophyll) to A; P is not a strong enough reductant to reduce A. However, P*, the excited state of the reaction-center chlorophyll, is an excellent reductant, and the formation of P* is followed rapidly (in 10^{-12} s) by the photochemical movement of an electron from P to A, generating P⁺ and A⁻. This photochemical electron movement is a result of the unique environment of both the chlorophylls and the acceptor within the reaction center complex, and

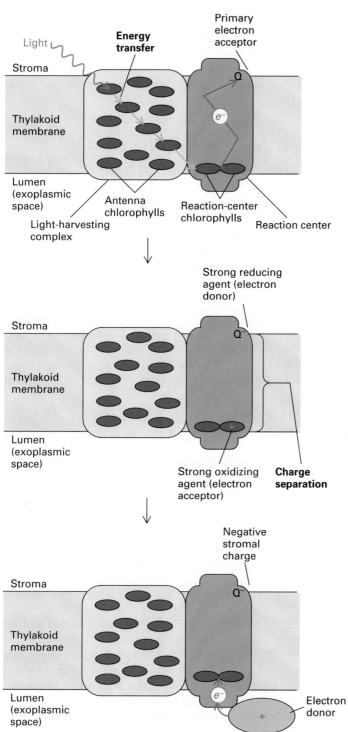

▲ FIGURE 18-8 The primary event in photosynthesis. The absorption of a photon of wavelength ≈ 680 nm results in the formation of a strong oxidizing agent and a strong reducing agent and a charge separation across the thylakoid membrane. The photon of light is absorbed by one of the many chlorophyll molecules in one of the light-harvesting complexes (LHCs) of an antennae (only one is shown), and some of the absorbed energy is transferred to the pair of chlorophyll molecules in the reaction center. The resultant

energized reaction-center chlorophylls donate an electron to an acceptor molecule (Q) on the stromal surface of the thylakoid membrane. The electron acceptor with its extra electron is a powerful reducing agent—it has the tendency to transfer the electron to another molecule. The positively charged chlorophyll is a strong oxidizing agent, and will attract an electron from an electron donor on the luminal surface. The result is a negative charge on the stromal side of the thylakoid membrane and a positive charge on the luminal side. This charge separation is essentially irreversible; the electron cannot easily return through the reaction center to neutralize the positive charge.

occurs nearly every time a photon is absorbed. Acceptor A^- is a powerful reducing agent capable of transferring the electron to still other molecules, ultimately to $NADP^+$. The P^+ is a powerful oxidant that can remove electrons from other molecules to regenerate the original P. In plants, the oxidizing power of four molecules of P^+ is used, by way of intermediates, to remove four electrons from H_2O to form O_2:

$$2H_2O + 4P^+ \longrightarrow 4H^+ + O_2 + 4P$$

Chlorophyll a also absorbs light at discrete wavelengths shorter than 680 nm (see Figure 18-7b); this raises the molecule into one of several higher excited states. Within 1 ps (one picosecond, or 10^{-12} seconds) the chlorophyll a decays to the first excited state P*, with loss of the extra energy as heat. Photochemical charge separation occurs only from the first excited state of the reaction-center chlorophyll a, P*. This means that the quantum yield—the amount of photosynthesis per absorbed photon—is the same for all wavelengths of visible light shorter than 680 nm.

The chlorophyll a pigments in the reaction centers are capable of directly absorbing light and initiating photosynthesis. However, the amount they can absorb is seldom adequate for the plant's requirements. At the maximum light intensity encountered by photosynthetic organisms (tropical noontime sun, $\approx 1.2 \times 10^{20}$ photons/m²/s), each reaction-center chlorophyll will absorb about one photon per second. At more typical light intensities, the reaction-center chlorophylls do not absorb enough light to support photosynthesis sufficient for the needs of the plant.

The *antenna* associated with each reaction center (see Figure 18-8) overcomes this limitation. Each antenna contains several *light-harvesting complexes* (LHCs), packed with chlorophyll a molecules. The LHCs contain chlorophyll b and other pigments as well, so that they promote photosynthesis even further by increasing the range of wavelengths of light that can be absorbed (see Figure 18-7).

As we have seen, photons are absorbed by any of the pigment molecules in each LHC. The absorbed energy is then rapidly transferred (in $<10^{-9}$ s) to one of the two chlorophyll a molecules in the reaction centers (see Figure 18-8), where it promotes the primary photosynthetic charge separation. For this *resonance energy transfer* from antenna pigments to reaction-center chlorophylls to occur efficiently, the antenna chlorophyll molecules must be close together and in a defined position and orientation, so that the LHCs can absorb light energy and funnel it to a reaction center. Within an LHC is one of several transmembrane proteins of 25,000–27,000 MW whose role is to maintain the chlorophyll molecules in the orientation and position that are optimal for energy transfer; Figure 18-9b shows the structure of one LHC.

But what is the structure of the reaction center? Why are the chlorophylls in it able, after absorbing a quantum of light, to release an electron? And what is the primary electron acceptor? For answers to these and other questions, we turn to a study of the photosynthetic purple bacteria.

► Molecular Analysis of Bacterial Photosynthesis

Photosynthetic bacteria* such as the green and purple bacteria have only one type of *photosystem*, in contrast to cyanobacteria and higher plants which have, as we detail later, two. Since the three-dimensional structures of the reaction centers in the photosystems from two purple bacteria (*Rhodopseudomonas viridis* and *Rhodobacter sphaeroides*) have been determined (see Figure 18-9; see also Figure 14-17), scientists can actually trace the detailed paths of electrons during and after the absorption of light. Although these photosynthetic bacteria do not generate O_2, this system provides great insight into the mechanism of photosynthesis in chloroplasts.

Purple Photosynthetic Bacteria Utilize Only One Photosystem and Do Not Evolve O_2

In the photosystem of purple bacteria (Figure 18-10), as in other photosystems, energy from absorbed light is used to transport an electron from a reaction-center chlorophyll to an acceptor quinone on the cytosolic membrane face (this would generate the A^- in the model on page 785). The chlorophyll thereby acquires a positive charge (and thus is converted from P to P^+).

After the quinone accepts a second electron (from the same reaction-center chlorophyll, after a second photon is absorbed), it binds two protons from the cytosol, forming the reduced quinone QH_2, which is then released and replaced by an oxidized Q. In a process very similar to the one that occurs in the inner mitochondrial membrane (see Figure 17-29), the QH_2 diffuses within the bacterial membrane to a *cytochrome bc_1 complex* (analogous to the mitochondrial $CoQH_2$-cytochrome c reductase complex), where it releases its two electrons to a site on the exoplasmic (external) surface. Simultaneously, QH_2 releases two protons into the periplasmic space (the space between the plasma membrane and the bacterial cell wall), thereby gen-

---

* There is a very different type of bacterial photosynthesis that occurs only in certain archaebacteria; we shall not discuss it here because it is unrelated to photosynthesis in higher plants. This type of photosynthesis utilizes bacteriorhodopsin, a small protein with seven membrane-spanning segments that pumps one proton from the cytosol to the extracellular space for every quantum of light absorbed (see Figures 14-18 and 17-19).

(a)

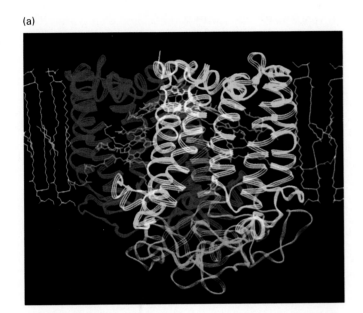

◄ FIGURE 18-9   (a) The three-dimensional structure of the reaction center from *Rhodobacter sphaeroides*, showing all of the pigments in purple. The L, M, and H chains appear in yellow, dark blue, and light blue, respectively. Figure 14-17 shows the conformation of the three polypeptides in greater detail. (b) atomic structure of the light-harvesting complex associated with photosystem II (LHC-II). Each LHC contains a polypeptide (blue) of 232 amino acids with three transmembrane alpha-helices. Attached to it are two carotenoids (orange) and 12 chlorophylls (green). LHC-II contains the most abundant membrane protein in chloroplasts; it absorbs light and transfers the energy to the photosystem reaction center chlorophylls. [Part (a) Courtesy of D. Rees. Part (b) adapted from W. Kühlbrandt, D. N. Wang, and Y. Fujiyoshi, 1994, Atomic Model of Plant Light-harvesting Complex by Electron Crystallography, *Nature*, **367**:614–621.]

(b)

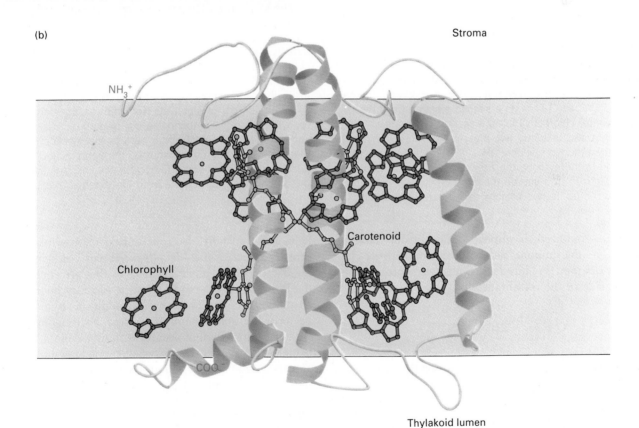

Stroma

NH_3^+

Carotenoid

Chlorophyll

COO⁻

Thylakoid lumen

erating a proton-motive force. Electrons are then transferred from the cytochrome bc_1 oxidoreductase complex to a soluble cytochrome, a one electron carrier, in the periplasmic space; this reduces the cytochrome from an Fe^{3+} to an Fe^{2+} state. The cytochrome then diffuses to a reaction center where it releases its electron to the positively charged (P^+) chlorophyll, returning the chlorophyll to the ground state P.

As in other systems, the proton-motive force in the bacterial photosystem is used by the $F_0F_1ATPase$ to synthesize ATP and also to transport molecules across the membrane. However, this photosynthetic process does not evolve O_2 or split H_2O. Electron flow is cyclic, and there is no reduction of $NADP^+$ to NADPH.

In summary, the *primary photochemical event* in bacterial photosynthesis is the formation of a strong *oxidizing*

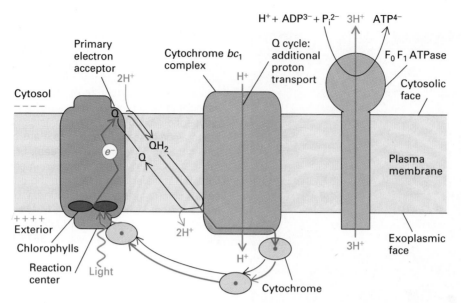

▲ FIGURE 18-10 Photosynthesis in purple bacteria, which utilize a single photosystem. Prior to the steps shown here, the LHCs (not illustrated here) absorb light and funnel the energy to the reaction-center chlorophyll (see Figure 18-8). Light absorption by the reaction-center chlorophyll then results in the transport of one chlorophyll electron to the primary acceptor, a quinone (Q) on the cytosolic face of the plasma membrane; this in turn results in a positively charged chlorophyll. When Q accepts a second electron from the reaction-center chlorophyll (after a second photon is absorbed), it also takes up two protons from the cytosol, forming the reduced quinone QH₂. QH₂ diffuses through the membrane to the cytochrome bc_1 complex, where it donates its two electrons at a site on the exoplasmic face. Simultaneously, it gives up its two protons to the external medium, restoring the oxidized quinone, Q, and generating a proton-motive force. The proton-motive force is used by the F_0F_1ATPase to synthesize ATP and, as in other bacteria, to transport molecules in and out of the cell. The oxidized Q diffuses back through the membrane to a binding site on the cytosolic face of the reaction center. Electrons are transported back to the reaction-center chlorophyll via the heme (blue dot) of a soluble cytochrome. The blue arrows trace the cyclic movement of electrons from the reaction-center chlorophyll to Q to the cytochrome bc_1 oxidoreductase complex and then, via the cytochrome, back to the reaction center. Electrons also may be transported through the bc_1 oxidoreductase complex during a Q cycle that transports additional protons across the membrane to the external medium.

agent (P^+)—oxidized chlorophyll—and a strong *reducing agent* (A^-)—the quinone electron acceptor with its electron. The quinone is reduced and binds two protons on the cytosolic side of the membrane:

$$Q + 2e^- + 2H^+ \longrightarrow QH_2$$

and is oxidized and releases two protons on the exoplasmic side:

$$QH_2 \longrightarrow Q + 2e^- + 2H^+$$

Photoelectron Transport in the Photosynthetic Reaction Center of Purple Bacteria Results in a Charge Separation

The photosynthetic reaction center of purple bacteria has three integral proteins (L, M, and H; see Figures 18-9 and 14-17). These contain a total of 11 transmembrane α helices, to which are bound the prosthetic groups that absorb light and transport electrons during photosynthesis. The prosthetic groups include a "special pair" of light-absorbing bacteriochlorophyll a molecules that are equivalent to the P of our earlier model, the site of the primary photochemical reaction. There are, in addition, two "voyeur" bacteriochlorophyll molecules, two bacteriopheophytins (bacteriochlorophyll molecules without a Mg^{2+}), one nonheme Fe atom, and two quinones, termed Q_A and Q_B, that are structurally similar to mitochondrial ubiquinone. Q_B is the primary acceptor depicted in Figure 18-10. The exact arrangement of all of these elements is known (Figure 18-11; see also Figure 18-9).

The pathway that the electron traverses can be determined by the technique of *optical absorption spectroscopy*. Each photosynthetic pigment absorbs light of only certain wavelengths (see Figure 18-7). The absorption spectrum of each pigment changes when it possesses an extra electron. Thus the pathway of electrons can be determined by monitoring the changes in absorption of the various pigments as a function of time after the absorption of a light photon. Since electron movements are completed in less than 1 millisecond (1 ms = 10^{-3} s), a special technique called *pico-*

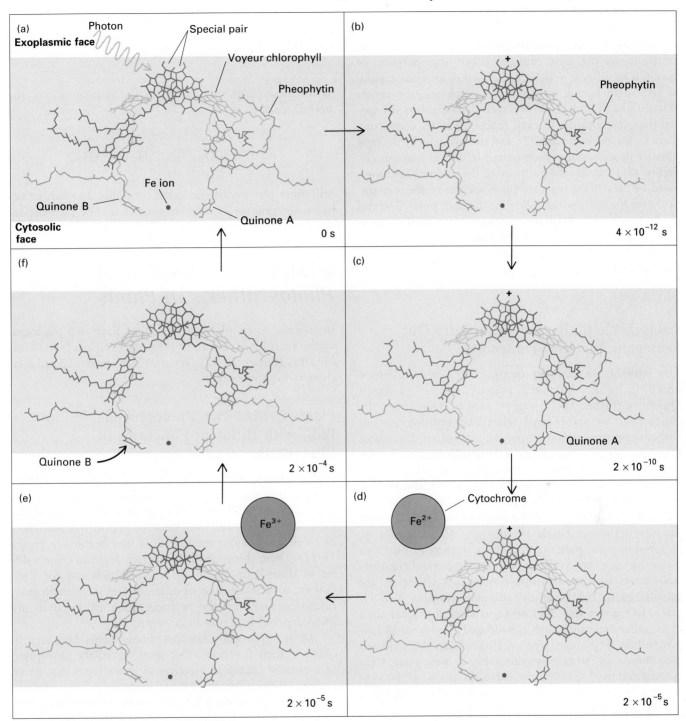

▲ FIGURE 18-11   The steps of electron transport during bacterial photosynthesis. All of the pigments and other electron transporters are bound to the three integral proteins that constitute the bacterial photosynthetic reaction center (see Figure 18-9); the yellow band depicts the approximate limits of the phospholipid bilayer.

In the first step of photosynthesis (a), a photon is absorbed by the "special pair" of reaction-center bacteriochlorophyll molecules (red); the energy is transferred to the chlorophyll electrons. Within 4 picoseconds (ps), an electron moves (b) to one of the pheophytin molecules (red), probably via the right "voyeur" bacteriochlorophyll, leaving a positive charge on the special chlorophyll pair. In ≈200 ps (c), the electron moves to the bound quinone A (Q_A) molecule near the end of the chain of prosthetic groups. Next, a soluble reduced (Fe^{2+}) cytochrome molecule approaches the reaction center (d) and transfers an electron to the special pair (e). The cytochrome thus acquires a Fe^{3+} charge, and the positive charge on the special pair of chlorophylls is neutralized. In (f), lasting 2×10^{-4} s, the cytochrome with its positive charge diffuses away, and the electron is passed from quinone A to the second acceptor, quinone B. As shown in Figure 18-10, quinone B, after accepting a second electron and two protons, diffuses away from the reaction center to the cytochrome bc₁ complex. [From D. C. Youvan and B. L. Marrs, 1987, Molecular Mechanisms of Photosynthesis, *Sci. Am.*, **256**(6):42–48. See W. Holzapfel, et al., 1990, *Proc. Natl. Acad. Sci. USA* **87**:5168–5172.]

second absorption spectroscopy (1 ps = 10^{-12} s) is required to see the early changes. When a preparation of bacterial membrane vesicles is exposed to an intense pulse of laser light lasting less than 1 ps, each reaction center absorbs one photon. Light is directly absorbed by the special-pair chlorophylls in each reaction center, converting them to the excited state, P*, and the subsequent electron transfer processes are synchronized in all reaction centers. Figure 18-11 traces the pathway of the electron, as monitored by measuring the absorption spectra of the preparation over the millisecond following the light pulse. The end result is a *charge separation*: a positively charged cytochrome (electron donor) on the exoplasmic membrane face and a reduced quinone Q_B (primary electron acceptor) on the cytosolic face.

Photosynthetic Bacteria also Carry Out Noncyclic Electron Transport

The pathway of electron transport depicted in Figure 18-10 is *cyclic*. After the primary electron acceptor Quinone$_B$ (Figure 18-11) accepts two electrons from the reaction center chlorophyll (after two protons are absorbed) and two protons (from the cytosol), it dissociates from the reaction center, crosses the plasma membrane, and denotes its electrons to the cytochrome bc_1 oxidoreductase complex. The electrons return, via cytochromes, to the reaction center, and the protons released into the exoplasmic space form part of the proton-motive force across the membrane (see Figure 18-10).

A *noncyclic* pattern of electron transport can also occur during photosynthesis by purple bacteria: here, electrons removed from chlorophyll molecules ultimately are transferred to NAD^+ as the acceptor, forming NADH. Still, H_2O is not split, and no O_2 is formed. Rather than H_2O, other molecules, such as hydrogen gas (H_2) or hydrogen sulfide (H_2S), can give up electrons to the oxidized cytochrome, in this case cytochrome c. In these cases, CO_2 is fixed and used to synthesize polymers of six-carbon sugars and other molecules. In the overall reaction, the light-driven oxidation of H_2S produces sulfur:

$$12H_2S + 6CO_2 \xrightarrow{\text{light}} C_6H_{12}O_6 + 12S + 6H_2O$$

Here, light causes the removal of an electron from the reaction-center chlorophyll that is ultimately transferred to NAD^+ (NAD^+, rather than $NADP^+$ as in plants, being the electron acceptor); then an electron is transferred from cytochrome c to the reaction center (forming oxidized cytochrome c); and finally electrons from H_2S then reduce the oxidized cytochrome c. The overall reaction for light-powered electron transport is

$$H_2S + H^+ + NAD^+ \xrightarrow{\text{light}} 2H^+ + S + NADH$$

When these organisms use H_2 as the electron donor, the overall reaction is

$$12H_2 + 6CO_2 \xrightarrow{\text{light}} C_6H_{12}O_6 + 6H_2O$$

Still other photosynthetic bacteria can use a variety of organic compounds as electron sources for the light-dependent reduction of CO_2.

▶ Molecular Analysis of Photosynthesis in Plants

In contrast to purple bacteria, plants have two photosystems: one splits H_2O into O_2; the other reduces $NADP^+$ to NADPH (Figure 18-12). We now turn to the specifics of these two photosystems.

Plants Utilize Two Photosystems, PSI and PSII, with Different Functions in Photosynthesis

Higher plants and algae, in contrast to purple bacteria, generate oxygen. To do this they utilize two photosystems, termed PSI and PSII. Both PSII and PSI contain two chlorophyll a molecules, and both undergo light-driven transport of an electron across the thylakoid membrane (see Figure 18-12). These chlorophylls in the two reaction centers differ in their light-absorption maxima (680 nm for PSII, 700 nm for PSI) because of differences in the protein environments, and thus the reaction-center chlorophylls are often denoted P_{680} and P_{700}, respectively.

More importantly, the two photosystems differ significantly in their functions. PSII splits water: the absorption of a photon causes an electron to move from P_{680} to an acceptor quinone on the stromal surface, and the resultant positive charge on the P_{680} strips electrons from the highly unwilling donor H_2O, forming O_2 as well as protons that remain in the thylakoid lumen and form part of the proton-motive force. The electrons in the acceptor quinone move, via a series of carriers, to the electron-donor site on the luminal surface of the PSI reaction center; during this movement, additional protons are transported into the thylakoid lumen. PSI uses the energy of an absorbed photon to transfer the electron to ferridoxin, an Fe-S-containing acceptor protein on the stromal surface. From there, electrons are passed to the ultimate acceptor, $NADP^+$, forming NADPH. Associated with both PSI and PSII are light-harvesting complexes that absorb light and transfer the absorbed energy to the reaction center chlorophylls.

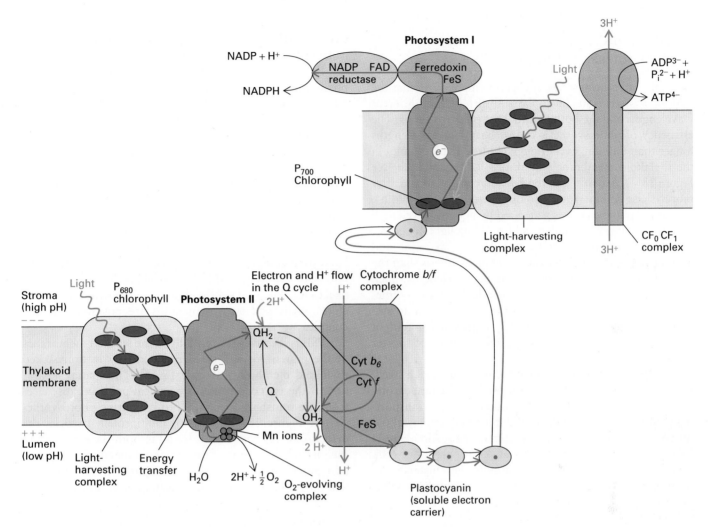

▲ FIGURE 18-12   Photosynthesis in plants, which utilize two photosystems, PSI and PSII. Both photosystems use the energy of absorbed light to move an electron across the thylakoid membrane. The blue lines indicate electron movement.

In PSII, light energy absorbed by the antenna's light-harvesting complex is transferred to P_{680}, the a chlorophylls of the PSII reaction center. Electrons are freed from P_{680}, forming P_{680}^+, which is then reduced by electrons derived from H_2O in the thylakoid lumen: an oxygen-evolving complex removes the electrons from the H_2O, forming O_2, and manganese ions in the complex transport these electrons to the P_{680}^+. The protons from the H_2O molecule remain in the lumen, contributing to the pH gradient that is the major part of the proton-motive force. The electrons freed from P_{680} move to an acceptor quinone, labeled Q in the figure, on the stromal face of the thylakoid membrane. After the antenna absorbs a second photon, Q accepts a second electron from P_{680} and adds two protons from the stroma, forming the reduced quinone QH_2. QH_2 diffuses through the membrane to the cytochrome b/f complex, where it simultaneously releases its two electrons at a site on the luminal face and its two protons into the lumen; the proton release adds to the proton-motive force. Electrons may also be transported through the cytochrome b/f complex during a Q cycle (circular blue arrow); this transports additional protons across the

membrane to the thylakoid lumen. In the lumen, transported protons, together with those released from H_2O, generate a proton concentration (pH) gradient across the thylakoid membrane; since the thylakoid membrane is permeable to Cl^- as well as to Mg^{2+} and K^+, a pH gradient rather than a membrane potential is the principal component of the proton-motive force.

PSI is localized to a different part of the thylakoid membrane (see Figure 18-13), and contains the reaction-center chlorophyll P_{700}. Electrons from the cytochrome b/f complex are carried to PSI by plastocyanin, a soluble electron carrier that diffuses through the thylakoid lumen. The plastocyanin molecule transfers its electron to P_{700}^+, formed when the absorption of a photon by a light-harvesting complex in the PSI antenna causes P_{700} to lose an electron. The electron released from P_{700} is transported, by a series of carriers in the reaction center, to the stromal surface, where soluble ferredoxin (an Fe-S protein) transfers the electron to FAD and finally to $NADP^+$. Two electrons, together with one proton removed from the stromal space, convert each $NADP^+$ to NADPH. As in mitochondria, the movement of 3 protons down their concentration gradient through a CF_0CF_1 complex is coupled to the synthesis of one ATP from ADP and P_i. Thus both NADPH and ATP are generated in the stroma of the chloroplast, where they are utilized for CO_2 fixation.

Both PSI and PSII Are Essential for Photosynthesis in Chloroplasts

A simple and crucial series of experiments showed that the two photosystems in chloroplasts absorb light at different wavelengths and are both essential for photosynthesis. The efficiency of photosynthesis drops sharply at wavelengths longer than 680 nm, even though chlorophyll molecules in the thylakoid membrane still absorb light at 700 nm (see Figure 18-7). According to the *Emerson effect* (named after biophysicist R. Emerson, who discovered it in the 1940s), the rate of photosynthesis generated by light of wavelength 700 nm can be greatly enhanced by adding light of shorter wavelength; a combination of light at, say, 600 and 700 nm supports a greater rate of photosynthesis than the sum of the rates for the two separate wavelengths. Such observations led researchers to conclude that photosynthesis in plants involves the interaction of two separate systems of light-driven reactions: PSI is driven by light of wavelength 700 nm or less; PSII, only by light of shorter wavelength (<680 nm).

The two photosystems are structurally distinct as well. Each is a complex of many chlorophyll molecules, carotenoid pigments, cytochromes, and other electron-transporting proteins that have specific orientations in the thylakoid membrane. Membrane fractions containing predominantly PSI or PSII can be purified, so that the properties of each photosystem can be studied separately. For instance, by using detergents to extract chloroplast membranes selec-tively, particles containing only PSI complexes can be isolated.

Each photosystem contains several hundred chloro-phyll molecules. However, as in the bacterial reaction cen-ter (see Figure 18-11), only a pair of specialized reaction-center chlorophyll *a* molecules—P_{680} in PSII and P_{700} in PSI—are capable of undergoing light-driven electron trans-fer in each complex. Both of these pairs of specialized chlo-rophyll *a* molecules are bound to integral membrane pro-teins: a pair of 82,000-MW proteins contains the P_{700} reaction center for PSI, and a pair of 32,000-MW proteins contains the P_{680} reaction center for PSII. As in photosyn-thetic bacteria (see Figure 18-10), each reaction center has an associated antenna that consists of a group of light-harvesting complexes (LHCs). The LHC's associated with PSII and PSI contain different proteins. The structure of one LHC (Figure 18-19b) illustrates how the 12 bound chlorophylls and two carotenoids are positioned in the complex.

PSII is located preferentially in the stacked membranes of the grana; PSI is located preferentially in nonstacked thylakoid membranes (Figure 18-13). The spatial separa-tion of the two complexes is a critical point because elec-trons are transferred from PSII to PSI during photosynthe-sis (see Figure 18-12). A third multiprotein particle, the *cytochrome b/f complex,* is found in both the stacked and the unstacked regions of the thylakoid membrane. It binds two mobile electron carriers, the quinone plastoquinone and the protein plastocyanin. *Plastoquinone* diffuses in the

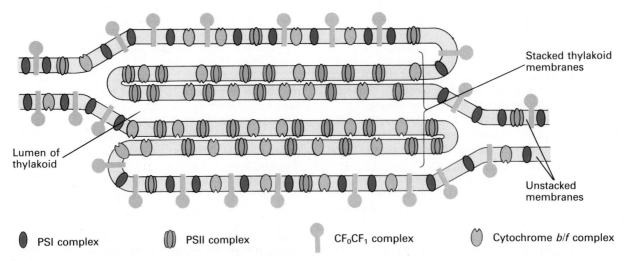

PSI complex   **PSII complex**   CF$_0$CF$_1$ complex   Cytochrome *b/f* complex

▲ **FIGURE 18-13** Distribution of multiprotein complexes in the thylakoid membrane. PSI is located primarily in non-stacked regions; PSII primarily in stacked regions. The cyto-chrome b/f complex transports electrons from PSII to PSI and is found in both stacked and nonstacked regions. The stacking of the thylakoid membranes may be due to the binding properties of the proteins in PSII. Evidence for this distribution came from studies in which thylakoid mem-branes were gently fragmented into vesicles by ultrasound. Stacked and unstacked thylakoid vesicles were then fraction-ated by density-gradient centrifugation to determine their protein and chlorophyll compositions. [After J. M. Anderson and B. Andersson, 1982, *Trends Biochem. Sci.* **7**:288.]

(a)

(b)

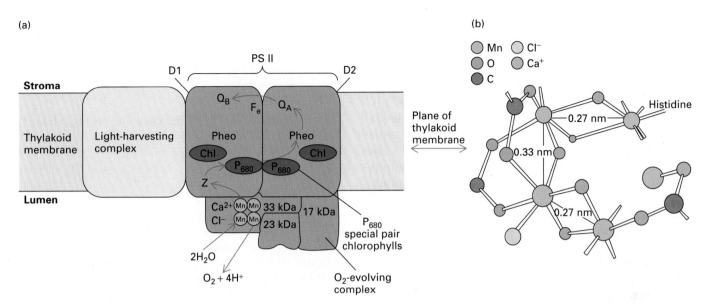

▲ FIGURE 18-14   (a) Molecular model of PSII. Two integral 32,000-MW proteins, D1 and D2, bind the two P_{680} reaction-center "special pair" chlorophylls, two other chlorophylls (Chl), two pheophytins (Pheo), one Fe atom, and two quinones (Q_A and Q_B)—all of which are used for electron transport. Associated with PSII is an antenna (light-harvesting) complex. Three extrinsic proteins (33, 23, and 17 kDa) comprise the oxygen-evolving complex; they bind the four Mn^{2+} ions and the Ca^{2+} and Cl^- ions that function in the splitting of H_2O, and they maintain the environment essential for high rates of O_2 evolution. Z is tyrosine residue 161 of the D1 polypeptide; it conducts electrons from the Mn atoms to the P_{680} reaction-center chlorophyll. (b) Three-dimensional structure of the oxygen-evolving manganese cluster. Each Mn atom has six ligands; among the atoms in the complex are one Cl^- and one Ca^{2+}. [Part (a) After G. Babcock, 1993, *Proc. Natl. Acad. Sci. USA* **90**:10893–10895. Part (b) from V. Yachandra, et al., 1993, *Science* **260**:675–679.]

membrane plane, and transports electrons from PSII to the cytochrome *b/f* complex. *Plastocyanin* is soluble in the thylakoid lumen, and transports electrons from the cytochrome *b/f* complex to PSI. The cytochrome *b/f* complex is similar in structure and function to the cytochrome bc_1 complex in photosynthetic bacteria (see Figure 18-10) and to the $CoQH_2$–cytochrome *c* reductase complex in mitochondria (see Figure 17-29). As the reduced plastoquinone, QH_2, moves electrons from PSII to the cytochrome *b/f* complex, it also transports protons from the stroma to the thylakoid lumen, generating part of the proton-motive force. Additional protons can be transported across the thylakoid membrane by a Q cycle in the cytochrome *b/f* complex (see Figure 18-12), similar to the analogous process in oxidative phosphorylation (see Figure 17-29).

PSII Splits H_2O

PSII removes electrons from H_2O, forming O_2. Somewhat surprisingly, in its structure PSII resembles the reaction center of a photosynthetic bacterium, which, as noted, does not form O_2. Like the bacterial reaction centers, PSII contains two reaction center chlorophylls, two other chlorophylls, two pheophytins, two quinones, called Q_A and

Q_B, and one nonheme iron atom. These molecules are bound to two 32,000-MW PSII proteins, called D1 and D2, whose sequences are remarkably similar to the sequences of the L and M peptides of the bacterial reaction center. When PSII absorbs a photon with a wavelength of <680 nm, it triggers the loss of an electron from a P_{680} chlorophyll *a* molecule, generating P_{680}^+. As in photosynthetic bacteria, the electron is transported via a pheophytin, a quinone (Q_A), and an Fe atom to the primary electron acceptor—another quinone (Q_B)—on the outer (stromal) surface of the thylakoid membrane (Figure 18-14a).

P_{680}^+, the photochemically oxidized reaction-center chlorophyll of PSII, is the strongest biological oxidant known. It is strong enough to oxidize H_2O, which is a more abundant electron donor than H_2S and H_2 used by bacteria; bacteria cannot oxidize water because the P^+ of the bacterial reaction center is not a sufficiently strong oxidant.

Electrons for the reduction of P_{680}^+ are obtained from an H_2O molecule bound to the luminal surface of the thylakoid membrane (see Figure 18-14), where a three-protein complex, the *oxygen-evolving complex*, aids in the splitting of H_2O. The oxygen-evolving complex contains four manganese (Mn) ions as well as bound Cl^- and Ca^{2+} ions (Figure 18-14b); this is one of the very few cases in which Mn plays a role in a biological system. The Mn ions together with three extrinsic proteins can be removed from the reaction center by treatment with solutions of concentrated salts; this abolishes O_2 formation but does not affect light absorption or the initial stages of electron transport.

The oxidation of two molecules of H_2O to form O_2 requires the removal of four electrons. Because the absorption of each photon by PSII results in the transfer of one electron, either several such photosystems must cooperate to oxidize H_2O through the oxygen-evolving complex, or each PSII must lose an electron and then oxidize the oxygen-evolving complex four times in a row. The latter explanation is the correct one; spectroscopic studies indicate that the oxygen-evolving complex can cycle through five different oxidation states (S_0–S_4):

S_0 being the most reduced and S_4 being the most oxidized. S_1–S_4 are thought to represent various oxidation states of the bound Mn ions.

A total of two H_2O molecules are split into four protons, four electrons, and one O_2 molecule. The electrons from H_2O are transferred, via the Mn ions and a tyrosine side chain (Z in Figure 18-14) on the D1 polypeptide, to the P_{680} reaction center, where they regenerate the reduced chlorophyll. The protons released from H_2O remain in the thylakoid lumen; they represent two of the four protons that are released into the lumen by the transport of each pair of electrons.

An electron that is removed from P_{680} is transported to the acceptor plastoquinone (Q_B in Figure 18-14a), bound to a site on the stromal surface of the thylakoid membrane, initially forming the semiquinone $Q_B^{-\cdot}$. The transport of a second electron from P_{680} (after a second photon is absorbed) is followed by the addition of two protons from the stroma, forming the reduced quinone Q_BH_2, abbreviated as QH_2 in Figure 18-12.

In thermodynamic terms, the absorption of a photon of 680 nm by P_{680} is equivalent to an energy change of 42 kcal/mol; this is the difference in free energy of the P_{680} and P_{680}^* states, and is equivalent to a change in reduction potential of ≈ 1.8 volts:

$$P_{680} + light \longrightarrow P_{680}^* \qquad (\text{input of } 1.8 \text{ V})$$

The energy of the photon is conserved in the two subse-

quent reactions. The charge-separation reaction releases 0.7 V:

$$P_{680}^* \longrightarrow P_{680}^+ + e^-$$

The cycle is completed, and the remaining energy released, by the reduction of P_{680}^+:

$$P_{680}^+ + e^- \longrightarrow P_{680}$$

Thus, light absorption converts P_{680} to a strong reductant, P_{680}^* (negative E_o). All of the energy of absorbed light is conserved in the P_{680}^* excited state. As is depicted in Figure 18-15, the transfer of electrons from H_2O ($E_o = +0.8$ V) to P_{680}^+ ($E_o = +1.1$ V) is energetically favored ($\Delta G \simeq -7$ kcal/mol), as is the transfer of electrons from P_{680}^* ($E_o = -0.7$ V) to Q_B ($E_o = 0.0$ V), the electron acceptor of PSII ($\Delta G \simeq -16$ kcal/mol). Thus, of the 42 kcal/mol energy in the excited state of P_{680}^*, only 18 kcal/mol is conserved in the chemical redox products of PSII, O_2 and QH_2. That is, the potential difference of the partial reactions

$$H_2O \longrightarrow \tfrac{1}{2}O_2 + 2e^- + 2H^+ \qquad (E_o = +0.8 \text{ V})$$

and

$$QH_2 \longrightarrow Q + 2e^- + 2H^+ \qquad (E_o = 0.0 \text{ V})$$

is 0.8 V, equivalent to a ΔG of ≈ -18 kcal/mol of electrons for the overall reaction of PSII:

$$H_2O + Q \xrightarrow{\ light\ } \tfrac{1}{2}O_2 + QH_2$$

Electrons Are Transported from PSII to PSI

Recall that the PSI and PSII complexes are spatially separate in the thylakoid membrane (see Figure 18-13). A single structural electron transport system containing both photosystems does not exist. Rather, mobile electron carriers, such as plastoquinone and plastocyanin, transfer electrons between the two photosystems. In mitochondria, similarly, the mobile electron carriers CoQ and cytochrome c are used to transfer electrons between electron transport complexes (see Figure 17-24).

First, as noted, two electrons released from P_{680}, in PSII, combine with a plastoquinone (Q_B) molecule and two protons from the stromal space to generate reduced hydroquinone (QH_2). This QH_2 dissociates from its binding site on the PSII reaction center, and is replaced with an (oxi-

(a)

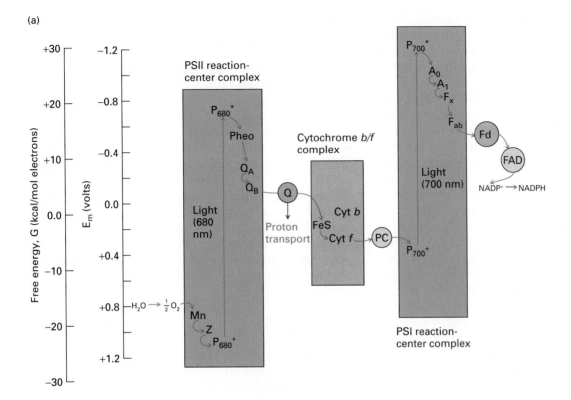

(b)

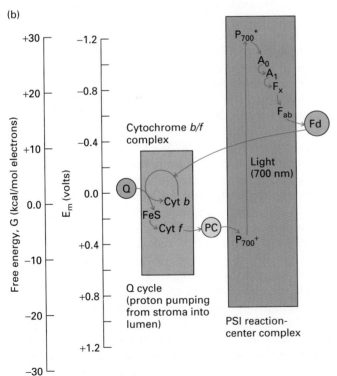

▲ FIGURE 18-15   Energetics of electrons as they flow through the photosynthetic transport system. There are two equivalent scales. The electron volt scale denotes the reduction potential (that is, the tendency to take up an electron) for all of the molecules depicted. The free energy scale is relative: what is important is the difference in free energy (ΔG) of an electron in two molecules. Electrons will tend to flow from a state of higher to lower free energy, or, equivalently, from a more negative to a more positive reduction potential.

(a) Linear electron flow. Photons absorbed by PSII excite electrons from a reduction potential of +1.1 V (the reduction potential of P_{680}) to one of $\approx$ −0.7 V (the reduction potential of P_{680}^*). The transfer of electrons from H_2O to P_{680}^+ is energetically favored, as is the transfer of electrons from P_{680}^* to a pheophytin (pheo) and quinone$_A$ (Q_A) to Q_B, the ultimate electron acceptor of PSII. Of the 42 kcal/mol energy in the excited state of P_{680}^*, (see text) only 18 kcal/mol is conserved in the chemical redox products of PSII, O_2 and Q_BH_2. As electrons move from QH_2 through the cytochrome b/f complex, some of the acquired electron energy is used to transport protons into the thylakoid lumen (see Figure 18-12). Photon absorption by PSI causes an additional increase of 1.6 V in reduction potential. Electrons excited by PSI are transferred, in reactions with negative ΔG values, via several bound electron carriers (A_0, a monomeric chlorophyll a; A_1, a derivative of vitamin K; F_x, a nonheme 4 Fe:4S complex; and F_{AB}, another nonheme 4 Fe:4S complex) to soluble ferredoxin (Fd) and thence, via FAD, to $NADP^+$, forming NADPH. In this process of linear electron flow, PSII and PSI are coupled, in that electrons must move from PSII to PSI via the quinone Q and the protein plastocyanin (PC).

(b) Cyclic electron flow. Only PSI is involved. Electrons excited by PSI are transferred, via ferredoxin and quinones (Q), through the cytochrome b/f complex to plastocyanin (PC) and then back to PSI. During this process, protons are transported from the stroma into the thylakoid lumen, both directly by a reduced quinone, and indirectly via a Q cycle involving the cytochrome b/f complex. [Courtesy of Thomas Owens.]

dized) Q. QH_2 diffuses in the thylakoid membrane to the cytochrome b/f complex (equivalent to the cytochrome bc_1 complex in Figure 18-10), where it releases its two electrons to a site on the luminal (exoplasmic) surface. Simultaneously, the two protons that QH_2 picked up from the stroma are released into the thylakoid lumen, generating a proton-motive force. A Q cycle of proton transport through the cytochrome b/f complex is apparently not involved during *linear* electron transport from PSII to PSI (as determined from the number of protons transported per quantum of light absorbed, and from the finding that antimycin, an inhibitor of a cytochrome in the b/f complex, has no effect on linear electron transport). However, a Q cycle may be involved in *cyclic* electron transport through the b/f complex, as discussed below.

From the cytochrome b/f complex, the electrons move to the carrier *plastocyanin*, a small soluble protein with a single Cu atom. The Cu alternates between the $+1$ and $+2$ states:

$$e^- + Cu^{2+} \rightleftharpoons Cu^+$$

Plastocyanin is a peripheral protein that is soluble in the lumen of the thylakoid vesicle. With its electron, plastocyanin diffuses in the thylakoid lumen from the cytochrome b/f complex to PSI (see Figure 18-12).

PSI Forms NADPH

The absorption of a photon leads to the removal of an electron from P_{700}, the reaction-center pair of chlorophyll a's of PSI. The oxidized chlorophyll P_{700}^+ is reduced by an electron passed from PSII via plastocyanin. The electron given up by P_{700} moves from the luminal surface to the stromal surface of the thylakoid membrane, where it is accepted by ferredoxin, an iron-sulfur (FeS) protein. (Several electron carriers intervene between P_{700} and ferredoxin, including two other iron-sulfur complexes and vitamin K). The net gain in the reduction potential of the electron, moving from plastocyanin, via PSI to ferredoxin, is ≈ 1.0 V, equivalent to a gain in free energy of $\Delta G = 23$ kcal/mol of electrons.

Electrons excited by PSI face one of two fates. They can be transferred from ferredoxin, via the electron carrier FAD, to $NADP^+$, forming, together with one proton picked up from the stroma, the reduced molecule NADPH (see Figure 18-15a). This process is called *linear electron flow*, a process that involves PSII and PSI in an obligate series in which electrons are transferred from H_2O to $NADP^+$.

PSI Can Also Function in Cyclic Electron Flow

Alternatively (Figure 18-15b), electrons in ferredoxin can reduce plastoquinone, forming QH_2 on the stromal sur-

face. Once QH_2 diffuses through the thylakoid membrane to the cytochrome b/f complex, it releases two protons to the lumen and its electrons to the b cytochromes. As in linear electron transport, these electrons return to PSI via plastocyanin. This process is called *cyclic electron flow*. A similar cyclic process occurs during photosynthesis in bacteria (Figure 18-10).

In cyclic electron flow, a *Q cycle* involving plastoquinone and the cytochrome b/f complex, and similar to the analogous process in oxidative phosphorylation (see Figure 17-29), is thought to transport two additional protons into the lumen for each pair of electrons transported. The pH gradient generated permits the synthesis of additional ATP, but no NADPH is produced. PSII is not involved, and no O_2 evolves. In this manner, PSI is used solely to generate a pH gradient and ATP; its function is similar to that of the photosystem in purple bacteria (see Figure 18-10).

PSI and PSII Are Functionally Coupled

Thus in linear (noncyclic) electron flow, the two photosystems have different functions: PSI transfers electrons to $NADP^+$, forming NADPH; PSII removes electrons from H_2O. Evidence supporting the obligatory transport of electrons from PSII to PSI comes from spectroscopic studies measuring the oxidized state of the cytochromes b_6 and f in the b/f complex (which transfer electrons from PSII to PSI) under different conditions of illumination. As discussed earlier, red light of wavelength 700 nm excites only PSI; the wavelength of this light is too long to be absorbed by PSII. Shining only red light on chloroplasts causes the b_6 and f cytochromes to become more oxidized. (This can be determined spectroscopically because the oxidized and reduced states have different absorption maxima.) Under these conditions, electrons are drawn into PSI; none are provided from PSII. The addition of light of shorter wavelength, however, activates PSII; the b_6 and f cytochromes immediately become partly reduced as electrons flow to them.

Many commercially important herbicides inhibit photosynthesis, and studies of their effects have proved useful in dissecting the pathway of photoelectron movement. One such class of herbicides—the s-triazines, such as atrazine—bind specifically to D1, one of the two 32,000-MW proteins that constitute the PSII reaction center (see Figure 18-14). Recall that the plastoquinone Q_B in PSII dissociates after it has been reduced to QH_2. s-Triazines block electron transfer by inhibiting the binding of an oxidized plastoquinone to this plastoquinone binding site. When added to illuminated chloroplasts, these inhibitors cause all downstream electron carriers to accumulate in the oxidized form, since no electrons can be provided to the electron transport system from PSII. Triazine-resistant weeds are prevalent and present a major agricultural problem. The D1 protein is encoded in the chloroplast DNA; in atrazine-resistant mutants, a single amino acid change in D1 renders it unable to bind the herbicide.

Because PSII and PSI act in sequence during linear electron flow, the amount of light energy delivered to the two reaction centers must be controlled so that each center activates the same number of electrons. One control mechanism regulates the rates of phosphorylation and dephosphorylation (the addition and removal, respectively, of phosphate groups) of the proteins associated with the PSII light-harvesting complex (LHC). A membrane-bound protein kinase senses the relative activities of the two photosystems by recognizing the oxidized-reduced state of the plastoquinone pool that transfers electrons from PSII to PSI. If too much plastoquinone is reduced (indicating a higher activity of PSII relative to PSI), the kinase is activated and the PSII LHCs are phosphorylated. Phosphorylation causes the LHCs to dissociate from PSII and migrate to the unstacked thylakoid membranes; this decreases the antenna size of PSII and thus the rate of generation of QH$_2$ by PSII. In this manner the relative rates of PSII and PSI are balanced; the reverse occurs if the plastoquinone pool becomes too oxidized.

The oxidants and reductants formed in PSII can lead to dangerous side reactions. Oxidative damage to the PSII photosystem could occur if PSII had no available electron acceptor, as would occur if all of the quinones were reduced, or if the light intensity were too high for the electron transport systems to handle. Under these conditions protective structural changes occur in the PSII LHCs, induced in part by a large pH gradient across the thylakoid membrane, so that the energy of absorbed light is dissipated as heat and not transferred to P$_{680}$.

The Necessity for Two Photosystems in Plants
The light energy absorbed by reaction-center chlorophylls ($\approx$42 kcal/mol of photons) has to be used in part to stabilize early reactive intermediates, so that not enough is left over to permit the direct movement of electrons from H$_2$O to NADPH. Each of the two photosystems boosts the electrons part of the way (see Figure 18-15).

Cyanobacteria also use H$_2$O as an electron donor and also have two photosystems. Purple (and green) photosynthetic bacteria, in contrast, have only one photosystem. These organisms extract electrons from molecules that have a lower (more negative) reduction potential than that of H$_2$O, so a single photon can boost these electrons to the reduction potential of the NAD$^+$-NADH reaction. For instance, for the partial reaction

$$H_2S \longrightarrow S + 2H^+ + 2e^-$$

that occurs in bacteria with one photosystem, $E_0 = -0.25$ V, compared to $E_0 = +0.8$ V for the analogous reaction for H$_2$O:

$$H_2O \longrightarrow \tfrac{1}{2}O_2 + 2H^+ + 2e^-$$

Less energy is required to boost electrons removed from H$_2$S to a level sufficient to reduce NADP$^+$ (or NAD$^+$) to NADPH (or NADH).

▶ CO$_2$ Metabolism during Photosynthesis

Chloroplasts perform many metabolic reactions in green leaves. In addition to CO$_2$ fixation, chloroplasts are sites for the synthesis of almost all amino acids, all fatty acids and carotenes, all pyrimidines, and probably all purines. However, the synthesis of sugars from CO$_2$ is the most extensively studied biosynthetic reaction in plant cells—and it is certainly a unique one. We now turn to the series of energy-requiring and enzymatically catalyzed reactions known as the *Calvin cycle* (after discoverer Melvin Calvin). These reactions fix CO$_2$ and convert it to hexose sugars; they are powered by energy released by ATP hydrolysis and by the reducing agent NADPH. The enzymes that catalyze the Calvin-cycle reactions are rapidly inactivated in the dark, so that carbohydrate formation generally ceases when light is absent.

CO$_2$ Fixation Is Catalyzed by Ribulose 1,5-Bisphosphate Carboxylase

The reaction that actually fixes CO$_2$ into carbohydrates is catalyzed by the enzyme *ribulose 1,5-bisphosphate carboxylase*, which adds CO$_2$ to the five-carbon sugar ribulose 1,5-bisphosphate to form two molecules of 3-phosphoglycerate (Figure 18-16). Ribulose 1,5-bisphosphate carboxylase makes up almost 50 percent of the chloroplast protein and is believed to be the most abundant protein on earth. It is composed of eight copies each of two subunits, one of which is encoded in chloroplast DNA, the other, in nuclear DNA. When photosynthetic algae are exposed to a brief pulse of ^{14}C-labeled CO$_2$ and the cells are then quickly disrupted, 3-phosphoglycerate is radiolabeled most rapidly, and all the ^{14}C radioactivity is found in the carboxyl group (see Figure 18-16). This establishes that ribulose 1,5-bisphosphate carboxylase fixes the CO$_2$.

The fate of the 3-phosphoglycerate formed by this reaction is complex: some is converted to starch or sucrose, but some is used to regenerate ribulose 1,5-bisphosphate. At least nine enzymes are required to regenerate ribulose 1,5-bisphosphate from 3-phosphoglycerate (Figure 18-17). Quantitatively, for every 12 molecules of 3-phosphoglycerate generated by ribulose 1,5-bisphosphate carboxylase (a total of 36C atoms), two molecules (6C atoms) are converted to two molecules of glyceraldehyde 3-phosphate and 10 molecules (30 C atoms) are converted to six molecules of ribulose 1,5-bisphosphate. The fixation of six CO$_2$

(a)

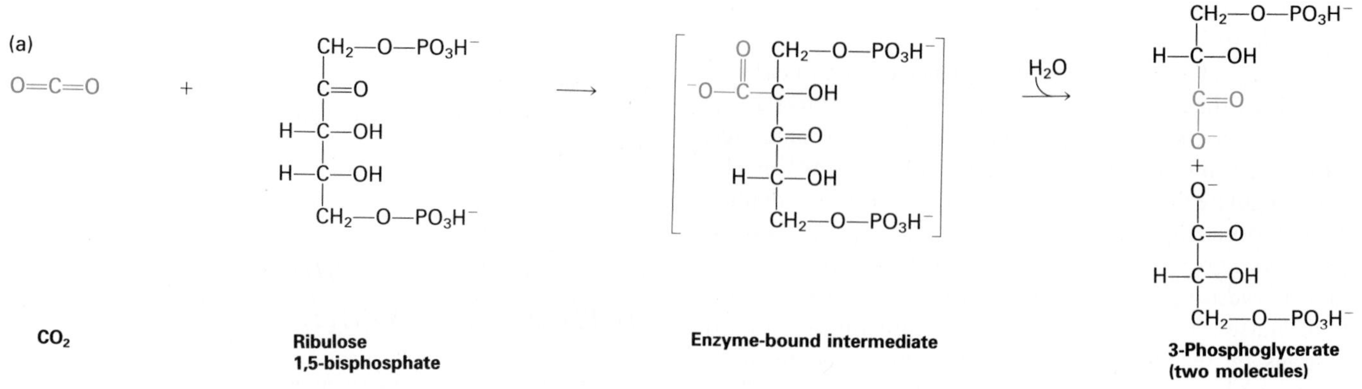

CO₂

Ribulose
1,5-bisphosphate

Enzyme-bound intermediate

3-Phosphoglycerate
(two molecules)

(b)

▲ FIGURE 18-16   CO₂ fixation by ribulose 1,5-bisphos-phate carboxylase. (a) The initial reaction that fixes CO₂ into organic compounds. The reaction, catalyzed by ribulose 1,5-bisphosphate carboxylase, is a condensation of CO₂ with the five-carbon sugar ribulose 1,5-bisphosphate. The products of the reaction are two molecules of 3-phosphoglycerate. If ¹⁴C-labeled CO₂ is used, all the ¹⁴C radioactivity is in the carboxyl carbon atom of 3-phosphoglycerate. (b) Structure of the catalytic domain of ribulose 1,5-bisphosphate carboxyl-ase. The eight parallel β strands are shown in green, and the eight α helices are shown in red. [Part (b) courtesy of J. Richardson.]

molecules and the net formation of two glyceraldehyde 3-phosphate molecules require the consumption of 18ATPs and 12NADPHs, generated by the light-requiring pro-cesses of photosynthesis (Figure 18-17).

Glyceraldehyde 3-phosphate is formed in the stroma; an antiport protein then transports it, in exchange for phosphate, to the cytosol; there, the final steps of sucrose synthesis occur. In these reactions, one molecule of glycer-aldehyde 3-phosphate is isomerized to dihydroxyacetone phosphate. This compound condenses with a second mole-cule of glyceraldehyde 3-phosphate to form fructose 1,6-bisphosphate, an intermediate in both glycolysis (see Fig-ure 17-2) and glucose biosynthesis. In leaf cells, however, most of the fructose 1,6-bisphosphate is converted to su-crose. One-half is converted to fructose 6-phosphate; one-half is converted to glucose 1-phosphate, which then forms uridine-diphosphate glucose (UDP glucose). These two compounds condense to form sucrose 6-phosphate; a final,

irreversible removal of phosphate then generates the ex-portable sucrose.

The antiporter that transports triosephosphate from the chloroplast in exchange for cytosolic inorganic phos-

▶ FIGURE 18-17   The pathway of carbon during photo-synthesis. Six molecules of CO₂ are converted into two molecules of glyceraldehyde 3-phosphate. These reactions, the Calvin cycle, occur in the stroma of the chloroplast. Via the phosphate-triosephosphate antiport, some glyceraldehyde 3-phosphate is transported to the cytosol in exchange for phosphate. In the cytosol, an exergonic series of reactions converts glyceraldehyde 3-phosphate to fructose 1,6-bisphos-phate and, ultimately, to the disaccharide storage form sucrose. Some glyceraldehyde 3-phosphate (not shown here) is also converted to amino acids and fats, compounds essen-tial to plant growth.

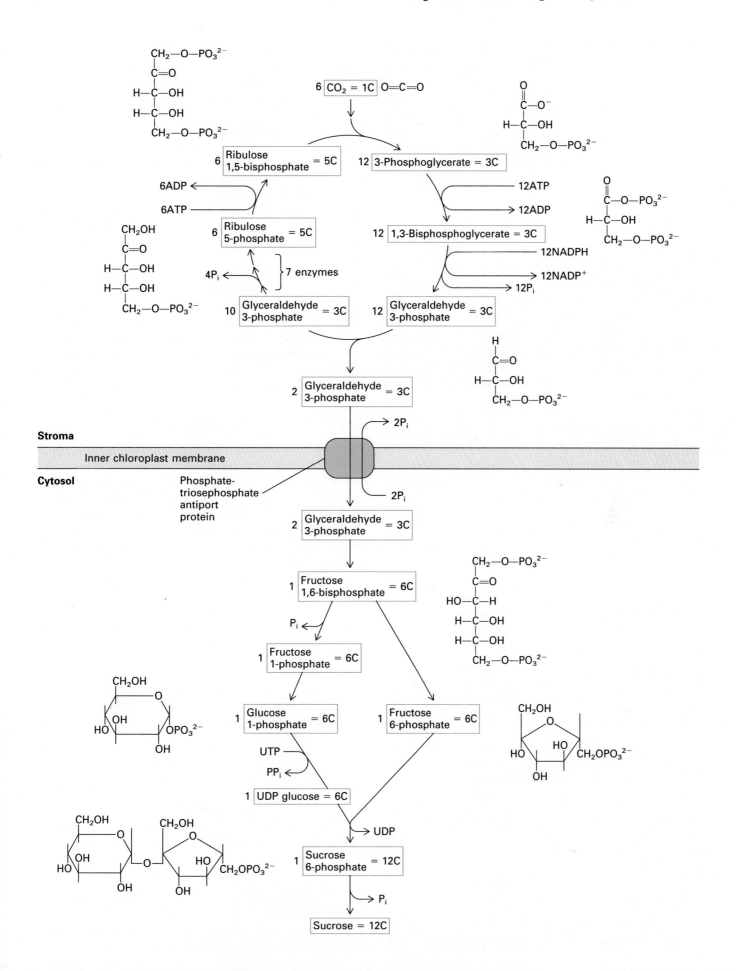

phates brings fixed CO_2 (as glyceraldehyde 3-phosphate) into the cytosol when the cell is exporting sucrose vigorously. This is a strict antiporter: no fixed CO_2 leaves the chloroplast unless phosphate is fed into it. The phosphate is generated in the cytosol, primarily during the formation of sucrose, from phosphorylated three-carbon intermediates (see Figure 18-17). Thus the synthesis of sucrose and its export from the cytosol to other cells encourages the export of additional glyceraldehyde 3-phosphate from the chloroplast.

CO_2 Fixation Is Activated in the Light

The Calvin-cycle enzymes that catalyze CO_2 fixation turn off rapidly in the dark, thereby conserving ATP that is generated in the dark for other synthetic reactions, such as lipid and amino acid biosynthesis. Several mechanisms control this. The pH of the stroma is ≈7 in the dark and ≈8 in the light (due to the light-driven transport of protons from the stroma into the thylakoid lumen), and several of the Calvin-cycle enzymes function better at the higher pH. The stromal concentration of Mg^{2+}, needed for function of the ATP-requiring enzymes, is higher in the light, due to enhanced activity in the light of a thylakoid membrane Mg^{2+} transporter. Additionally, several enzymes are activated in the light by reduction, due to the flow of electrons from PSI, and are inactivated in the dark by oxidation.

The activity of ribulose 1,5-bisphosphate carboxylase (colloquially known as "rubisco," for *ribulose bisphosphate carboxylase*) is increased in the light by the enzyme *rubisco activase*. For ribulose 1,5-bisphosphate carboxylase to be active, CO_2 must carbamylate a lysine residue near the active site:

$$—CH_2—CH_2—CH_2—NH_2 + CO_2 \longrightarrow$$
Lysine side chain $$—CH_2—CH_2—CH_2—NH—COO^-$$

since only the carbamylated enzyme can bind Mg^{2+}, an essential cofactor. In the presence of high CO_2 and Mg^{2+} concentrations, this activating reaction occurs spontaneously. Under normal conditions, however, with ambient levels of CO_2, the reaction requires catalysis by rubisco activase, an enzyme that simultaneously hydrolyzes ATP and uses the energy to attach the CO_2 to the lysine. This enzyme was discovered during the study of a mutant strain of *Arabidopsis thaliana* that required high CO_2 levels to grow and that did not exhibit light activation of ribulose 1,5-bisphosphate carboxylase; the mutant had a defective rubisco activase.

Photorespiration Liberates CO_2 and Consumes O_2

Photosynthesis is always accompanied by *photorespiration*—a process that takes place in light, consumes O_2, and

converts ribulose 1,5-bisphosphate in part to CO_2. As Figure 18-18 shows, ribulose 1,5-bisphosphate carboxylase catalyzes two competing reactions: the addition of CO_2 to ribulose 1,5-bisphosphate to form two molecules of 3-phosphoglycerate, and the addition of O_2 to form one molecule of 3-phosphoglycerate and one molecule of the two-carbon compound phosphoglycolate. Phosphoglycolate is recycled into three-carbon intermediates; this complex process involves peroxisomes and mitochondria as well as chloroplasts (see Figure 18-18). For every two molecules of phosphoglycolate formed by photorespiration (four C atoms), one molecule of 3-phosphoglycerate is ultimately formed and recycled and one molecule of CO_2 is lost.

Photorespiration is wasteful to the energy economy of the plant: it does not generate NADPH, it consumes ATP and O_2, and it generates CO_2. It is surprising, therefore, that no ribulose 1,5-bisphosphate carboxylases have evolved that do not catalyze photorespiration; probably the necessary structure of the active site of ribulose 1,5-bisphosphate carboxylase makes it impossible for plants to evolve such an enzyme.

Peroxisomes Play a Role in Photorespiration

We learned in Chapter 5 that *peroxisomes* are found in all eukaryotic cells, and contain the enzyme catalase, which decomposes hydrogen peroxide into water. In most cells, peroxisomes also contain enzymes that catalyze the oxidation of fatty acids, with the concomitant production of hydrogen peroxide (see Figure 17-33). Peroxisomes in leaf cells and other photosynthetic plant tissues contain the unique enzyme *glycolate oxidase*, which oxidizes glycolate to glyoxylate (see Figure 18-18), an important step in photorespiration. Like other peroxisomal enzymes, glycolate oxidase is a flavin-containing enzyme that uses oxygen as a final electron acceptor and produces hydrogen peroxide as one of its products:

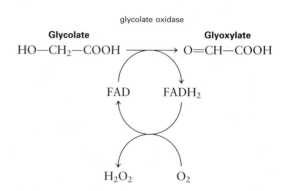

As with all peroxisomes, the hydrogen peroxide is decomposed by catalase.

Leaf peroxisomes also contain another unique enzyme that converts the glyoxylate to glycine. Glycine is an inter-

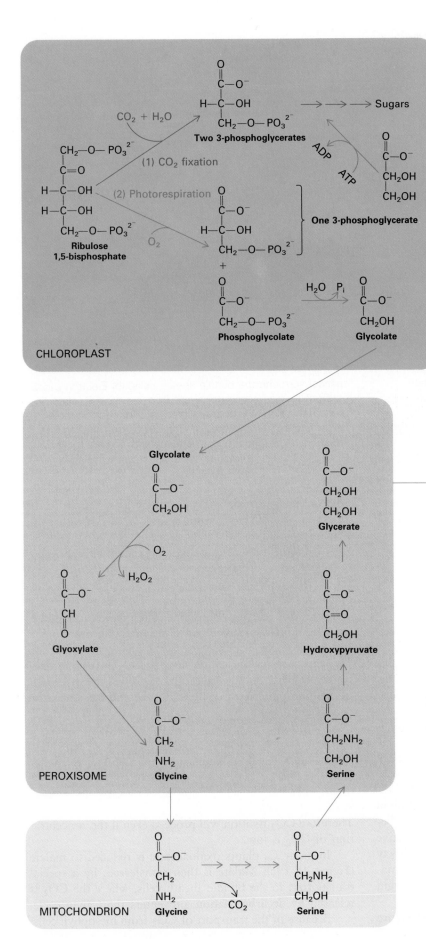

▲ FIGURE 18-18   The relationship between CO_2 fixation and photorespiration. These two competing reactions are catalyzed by ribulose 1,5-bisphosphate carboxylase, and both utilize ribulose 1,5-bisphosphate. CO_2 fixation, reaction (1), is favored by high CO_2 and low O_2 pressures; photorespiration, reaction (2), occurs at low CO_2 and high O_2 pressures (that is, under normal atmospheric conditions). Phosphoglycolate, formed during photorespiration, is converted into three-carbon intermediates of sugar biosynthesis. The phosphoglycolate is first hydrolyzed to glycolate within the chloroplast. The glycolate is transported to the peroxisome, where it is oxidized to glyoxylate and then converted to glycine. The glycine is then transported to a mitochondrion where, in a complex process, two molecules of glycine are converted to one CO_2 molecule and one serine molecule. The serine is transported back to the peroxisome, where it is converted to glycerate, and glycerate re-enters the chloroplast and is converted there to phosphoglycerate. Thus of the four C atoms (from two glycolates) undergoing photorespiration, only one is actually lost as CO_2.

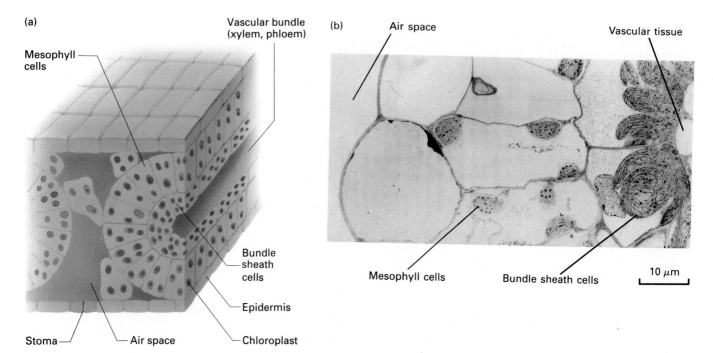

(a)

Vascular bundle (xylem, phloem)

Mesophyll cells

Bundle sheath cells

Epidermis

Stoma

Air space

Chloroplast

(b)

Air space

Vascular tissue

Mesophyll cells

Bundle sheath cells

10 μm

▲ FIGURE 18-19 The C$_4$ pathway for CO$_2$ fixation. (a) The anatomy of a leaf in a C$_4$ plant. Bundle sheath cells line the vascular bundles containing the xylem and phloem. Sucrose is synthesized in bundle sheath cells by photosynthesis and then carried to the rest of the plant via the phloem. Mesophyll cells (adjacent to the substomal air spaces) surround the bundle sheath cells. (b) Electron micrograph of a cross section of a leaf from a typical C$_4$ plant. In mesophyll cells, CO$_2$ is assimilated into four-carbon molecules that are pumped into the interior bundle sheath cells. These cells contain abundant chloroplasts and are the sites of photosynthesis and sucrose synthesis. (c) Diagram

mediate in the conversion of glycolate to Calvin-cycle intermediates, a pathway that involves the mitochondria and chloroplasts as well as peroxisomes (see Figure 18-18).

The C$_4$ Pathway for CO$_2$ Fixation Is Used by Several Tropical Plants

In a hot, dry environment, plants must keep their stomata (the gas-exchange pores in the leaves; see Figure 15-29) closed much of the time to prevent excessive loss of moisture. This causes the CO$_2$ level inside the leaf to fall below the dissociation constant (K_m; see Figure 3-29) of ribulose 1,5-bisphosphate carboxylase. Under these conditions, the rate of photosynthesis is slowed and photorespiration is greatly favored.

Corn, sugar cane, crabgrass, and other plants that can grow in hot, dry environments have evolved a way to avoid this problem by utilizing a two-step pathway of CO$_2$ fixation in which a CO$_2$-hoarding step precedes the Calvin cycle. The pathway has been named the C$_4$ *pathway* because [^{14}C]CO$_2$ labeling showed that the first radioactive molecules formed during photosynthesis in this pathway are four-carbon compounds, such as oxaloacetate and malate, as opposed to the three-carbon molecules that begin the C$_3$ *pathway* of the Calvin cycle.

The C$_4$ pathway involves two types of cells (Figure

18-19a and b): *mesophyll cells*, which are adjacent to the air spaces in the leaf interior, and *bundle sheath cells*, which surround the vascular tissue. In the mesophyll cells, CO$_2$ from the air is assimilated by reacting with phosphoenolpyruvate, a three-carbon molecule (also seen as an intermediate in glycolysis; see Figure 17-2), to generate oxaloacetate, a four-carbon compound. The enzyme that catalyzes this reaction, phosphoenolpyruvate carboxylase, is found almost exclusively in C$_4$ plants: importantly, unlike ribulose 1,5-bisphosphate carboxylase, phosphoenolpyruvate carboxylase is insensitive to O$_2$. The overall reaction from pyruvate to oxaloacetate involves the hydrolysis of one phosphoanhydride bond in ATP and has a negative ΔG value:

$$\text{Pyruvate} + \text{ATP} + P_i + CO_2 \longrightarrow$$
$$\text{oxaloacetate} + \text{AMP} + P_i + PP_i$$

Therefore CO$_2$ fixation will proceed even if the concentration of CO$_2$ is low.

In many C$_4$ plants oxaloacetate is reduced to malate (Figure 18-19c). Malate is then transferred, by a special transporter, to the bundle sheath cells, where the CO$_2$ is released by decarboxylation and enters the Calvin cycle.

Because of the transport of CO$_2$ from mesophyll cells,

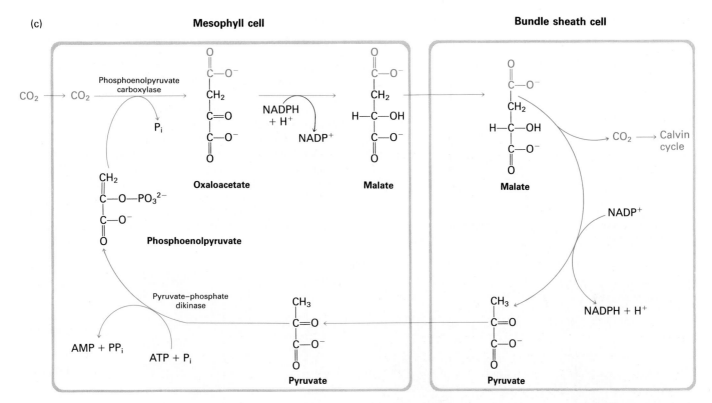

(Continued) showing the C₄ pathway. CO_2 is assimilated by phosphoenolpyruvate carboxylase in mesophyll cells. C₄ molecules, such as malate, are transferred from the mesophyll cells to the bundle sheath cells. CO_2 is then released by decarboxylation for fixation by the standard Calvin cycle. The C₃ compound pyruvate, also generated by decarboxylation, is transported back to the mesophyll cells, where it is reutilized in the C₄ pathway. In certain C₄ plants, oxaloacetate in mesophyll cells is aminated to form aspartate, which is transferred to the bundle sheath cells; it is converted there in several steps to CO_2 and alanine, and the alanine is then recycled to the mesophyll cells. [Photograph in part (b) by S. Craig; courtesy of M. D. Hatch.]

the CO_2 concentration in the bundle sheath cells of C₄ plants is much higher than it is in the normal atmosphere. Bundle sheath cells are also unusual in that they lack PSII and only carry out cyclic electron flow catalyzed by PSI. No oxygen is evolved which might compete with CO_2 fixation by ribulose 1,5-bisphosphate carboxylase. The high CO_2 and reduced O_2 concentrations in the bundle sheath cells favor the fixation of CO_2 to form 3-phosphoglycerate and inhibit the utilization of ribulose 1,5-bisphosphate by photorespiration.

The high O_2 concentration in the atmosphere favors the oxygenation of ribulose 1,5-bisphosphate (reaction 2 in Figure 18-18); in C₃ plants, as much as 50 percent of the photosynthetically fixed carbon may be reoxidized to CO_2 during photorespiration. Compared with C₃ plants, C₄ plants are superior utilizers of available CO_2, since the C₄ enzyme phosphoenolpyruvate carboxylase (see Figure 18-19c) has a lower K_m for CO_2 than does the ribulose 1,5-bisphosphate carboxylase of the Calvin cycle. Since one phosphodiester bond of ATP is consumed in the cyclic C₄ process (to generate phosphoenolpyruvate from pyruvate), the overall efficiency of the photosynthetic produc-

tion of sugars from NADPH and ATP is lower than it is in C₃ plants, which use only the Calvin cycle for CO_2 fixation. However, the net rates of photosynthesis for C₄ grasses, such as corn or sugar cane, can be two to three times the rates for otherwise similar C₃ grasses, such as wheat, rice, or oats, due to the elimination of losses from photorespiration.

Sucrose Is Transported from Leaves through the Phloem to All Plant Tissues

Of the two carbohydrate products of photosynthesis, *starch* remains in the mesophyll cells and, mainly in the dark, is subjected to glycolysis, forming ATP, NADH, and small molecules that are used as building blocks for the synthesis of amino acids, lipids, and other cellular constituents. *Sucrose*, in contrast, is exported from the photosynthetic cells and transported throughout the plant.

There are two components of the vascular system used by higher plants to transport water, ions, sucrose, and other water-soluble substances: the *xylem* and the *phloem*,

(a)

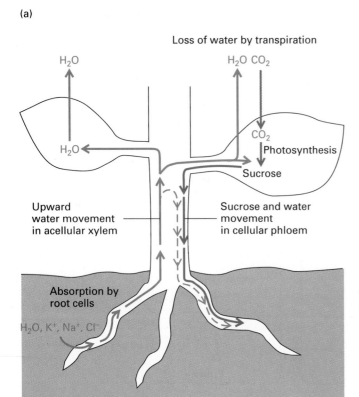

Loss of water by transpiration

H₂O

H₂O CO₂

CO₂

Photosynthesis

H₂O

Sucrose

Upward water movement in acellular xylem

Sucrose and water movement in cellular phloem

Absorption by root cells

H₂O, K⁺, Na⁺, Cl⁻

◄ FIGURE 18-20   The xylem and phloem in higher plants. (a) A schematic diagram of the two vascular systems, xylem and phloem, showing the transport of water (red) and sucrose (green); (b) and (c) micrographs of cross sections of a root from (b) a sunflower (*Helianthus* sp.) and (c) a buttercup (*Ranunculus* sp.), showing vascular bundles containing the xylem and phloem. Water and salts enter the xylem through the roots. Water is lost by evaporation, mainly through the leaves, creating a suction pressure that draws the water and dissolved salts upward through the xylem. The phloem is used to conduct dissolved sucrose, produced in the leaves, to other parts of the plant. Sucrose is actively transported into the phloem by associated companion cells (see Figure 18-21) and actively removed from the phloem by nonphotosynthetic cells in the root and stem. [Parts (b) and (c) courtesy of Runk/Schoenberger, from Grant Heilman.]

(b)

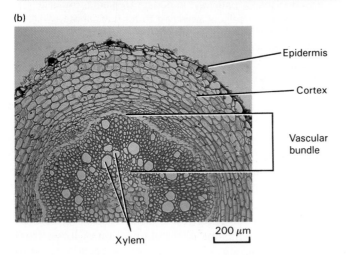

Epidermis

Cortex

Vascular bundle

Xylem

200 μm

(c)

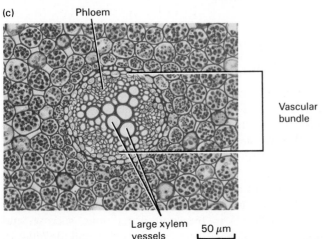

Phloem

Vascular bundle

Large xylem vessels

50 μm

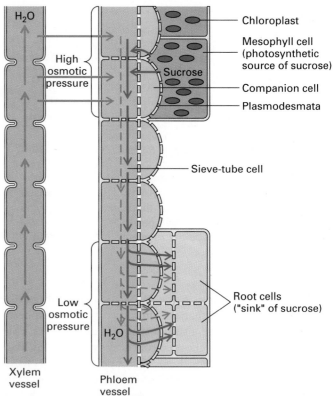

H₂O

High osmotic pressure

Chloroplast

Mesophyll cell (photosynthetic source of sucrose)

Sucrose

Companion cell

Plasmodesmata

Sieve-tube cell

Low osmotic pressure

H₂O

Root cells ("sink" of sucrose)

Xylem vessel

Phloem vessel

▲ FIGURE 18-21   The flow of sucrose in a higher plant, driven by osmotic pressure differences between the source and the sink of sucrose. Sucrose is produced by photosynthesis in leaf mesophyll cells, the "source" of sucrose. The sucrose is actively transported into the companion cells, and then moved through plasmodesmata into the sieve-tube cells that constitute the phloem vessels. The resulting increase in osmotic pressure within the phloem causes water to enter the phloem by osmotic flow. Root cells and other nonphotosynthetic cells constitute the "sink" for sucrose: they actively remove sucrose and metabolize it. This lowers the osmotic pressure in the phloem, causing water to exit the phloem. These differences in osmotic pressure in the phloem between the source and the sink of sucrose provide the force that drives the sucrose solution through the phloem.

generally found grouped together in the *vascular bundle* (Figure 18-20; see also Figure 18-19a). The tubes of the *xylem* conduct salts and water from the roots through the stems to the leaves. Water is absorbed by specialized root hair cells and is then transported upward through the xylem; it is lost from the plant by evaporation, primarily from the leaves. In young plants the xylem is built of cells interconnected by plasmodesmata (see Figures 5-3b and 18-21) but in mature tissues the cell body degenerates, leaving only the cell walls.

The *phloem*, in contrast, transports dissolved sucrose and organic molecules such as amino acids from their sites of origin to tissues throughout the plant. Phloem (Figure 18-21) consists of long, narrow cells, called *sieve-tube* cells, interconnected by *sieve plates*, a type of cell wall that contains many plasmodesmata and is highly perforated. Numerous plasmodesmata also connect the sieve-tube cells to *companion cells*, which line the phloem vessels. Sieve-tube cells have lost their nuclei and most other organelles but retain a water-permeable plasma membrane and cytoplasm, through which the sucrose and water move. In ef-fect, the sieve-tube cells form one continuous tube of cytosol that extends throughout the plant.

Figure 18-21 shows how differences in osmotic strength cause the movement of sucrose from photosynthetic cells in the leaves, through the phloem, to the roots and other nonphotosynthetic tissues. Sucrose produced by photosynthesis in leaf mesophyll cells is actively transported into the companion cells and thence, through the connecting plasmodesmata, into the sieve-tube cells. This increases the osmotic pressure within the sieve-tube cells, and water enters them by osmotic flow. Within the phloem, the sucrose solution moves from sieve-tube cell to sieve-tube cell through the sieve plate. At the far end of the phloem vessels, the sucrose is actively transported into root cells and other nonphotosynthetic cells, which act as "sinks" for sucrose—sites that accumulate sucrose and ultimately metabolize it to CO_2. This active removal of sucrose lowers the osmotic pressure in the phloem, causing water to exit. The resulting difference in osmotic water pressure between the source and the sink drives the sucrose solution from the leaves through the phloem to the roots.

SUMMARY

The mechanism of photosynthetic energy transduction is best understood in purple bacteria, which have a single type of photosynthetic reaction center whose structure is known in molecular detail. The absorption of light by a "special pair" of reaction-center chlorophylls excites a chlorophyll electron. In a few picoseconds, the electron is transferred (via a pheophytin and a quinone) to a second quinone, an electron acceptor, on the cytosolic face of the plasma membrane, and a cytochrome on the exoplasmic membrane face donates an electron to the oxidized reaction-center chlorophylls. The acceptor quinone (Q) picks up two electrons and also two protons from the cytosol, forming QH_2. QH_2 diffuses to the cytochrome bc_1 complex, where it releases its protons to the extracellular space, generating a proton-motive force. The electrons released by QH_2 are transported back to the reaction center by the cytochrome, completing the cycle. As in other systems, the proton-motive force is used mainly to power ATP synthesis through a CF_0CF_1 complex.

Plants contain two photosystems, PSI and PSII; the latter is similar in structure and function to the bacterial photosystem. During photosynthesis in the plant chloroplast, light energy is absorbed by chlorophylls *a* and *b* and other pigments in an antenna, made up of several light-harvesting complexes (LHCs), that is associated with each photosystem. Light energy absorbed by the LHCs is transferred to a specialized reaction-center chlorophyll *a* pair (P_{680} in PSII and P_{700} in PSI). Excitation of P_{680} results in the transport of an electron from it, via a pheophytin and a quinone, Q_A, to an acceptor quinone, Q_B, on the stromal side of the thylakoid membrane. The resulting P_{680}^+ is a powerful oxidant; it removes electrons from H_2O bound to a three-protein complex, the oxygen-evolving complex, on the luminal surface of the thylakoid. O_2 is formed, and the resulting protons remain in the lumen and generate part of the proton-motive force. The quinone Q picks up two electrons and two protons from the stroma, forming QH_2. QH_2 diffuses to the cytochrome *b/f* complex, where it releases its protons to the thylakoid lumen, adding to the proton-motive force. From the cytochrome *b/f* complex, electrons are transferred, by the carrier plastocyanin in the thylakoid lumen, to *PSI*.

Electrons excited by PSI can undergo one of two possible fates. By a process of linear electron transport, they can be transferred via a series of carriers to $NADP^+$ to form NADPH. Alternatively, by a process of cyclic electron flow, electrons can be transferred back to PSI while a Q cycle concomitantly transports additional protons across the thylakoid membrane to the lumen from the stroma. The cyclic process does not involve PSII, and neither NADPH nor O_2 is formed.

Because thylakoid membranes are permeable to anions and cations, a pH gradient (inside pH $\approx$ 5.0 versus stromal pH = 7.8), rather than a membrane electric potential, is the principal component of the proton-motive force. The thylakoid-membrane pH gradient is primarily used in ATP synthesis. Plants use photosynthesis-generated ATP and NADPH in a series of enzymatic reactions (the Calvin cycle) to convert CO_2 to sucrose or starch—the principal products of photosynthesis.

In C_3 plants, CO_2 is fixed in the reaction catalyzed by ribulose 1,5-bisphosphate carboxylase. In C_4 plants, CO_2 is fixed initially in the outer mesophyll cells by reaction with phosphoenolpyruvate. The four-carbon molecules so generated are shuttled to the interior bundle sheath cells, where the CO_2 is released and then used in the Calvin cycle. Sucrose from photosynthetic cells is transported through the phloem to nonphotosynthetic parts of the plant; osmotic pressure differences provide the force that drives the sucrose transport.

REVIEW QUESTIONS

1. The overall reaction of photosynthesis results in the synthesis of six carbon sugars and O_2 from CO_2 and H_2O, and the process can be divided into four principal stages. What are they?

A major event in photosynthesis is the conversion of light energy from photons into chemical energy in the form of ATP. What is the equation that describes the relationship between the energy of a particular wavelength of light and its frequency? Which has a higher energy level per photon, blue light or red light?

What are the light absorbing pigments involved in photosynthesis? Figure 18-7 illustrates an important point about these pigments by showing the absorption spectra for several of them. Where do chlorophyll *a* and chlorophyll *b* absorb most of their energy? Do these spectra explain why most plants are green?

a. Consider two wavelengths of light, 650 nm and 720 nm. What is the total energy associated with a mole of photons at each of these wavelengths?

b. The reaction for the synthesis of ATP and ADP is as follows:

$$ADP + P_i \rightleftharpoons ATP \qquad \Delta G^{\circ\prime} = +7.2 \text{ kcal/mole.}$$

Clearly, energy is stored in ATP in the terminal phosphoanhydride bonds. Is there sufficient energy in one mole of photons at a wavelength of 650 nm to synthesize one mole of ATP from ADP and P_i? Likewise, is there sufficient energy in one mole of photons at a wavelength of 720 nm to make one mole of ATP?

c. In reality, plants would not grow well in light of 720 nm. Why is this the case? Does this contradict your answer to the previous question concerning sufficient energy to synthesize ATP at various wavelengths of light? Justify your answer.

2. Review the principles of photosynthesis in bacteria; use Figure 18-10 to do this. In bacterial photosynthesis, light is used to transport an electron from the chlorophyll to a quinone.

During cyclic electron flow, how do the electrons ultimately reduce the chlorophyll? Is O_2 generated during the process of cyclic electron flow in bacteria?

Noncyclic electron flow can also occur in bacteria. During this process, reduction of the oxidized chlorophyll requires a source of electrons. What are these sources in bacteria? Is H_2O a source? Is O_2 produced? What products are made during noncyclic electron transport?

a. Consider the following two half-reactions.

$$H_2 \longrightarrow 2H^+ + 2e^- \qquad\qquad E_0^{'} = +0.42 \text{ V}$$
$$NADP^+ + 2H^+ + 2e^- \longrightarrow NADPH + H^+ \qquad E_0^{'} = -0.32 \text{ V}$$

When H_2 is used as a source of reducing equivalents in noncyclic electron flow, what is the value of $\Delta G^{\circ\prime}$ for the transfer of electrons from one mole of H_2 to one mole of $NADP^+$?

b. Recall from question 1 that the energy required to synthesize ATP from ADP and P_i has a $\Delta G^{\circ\prime}$ of +7.2 kcal/mole. What would be the minimum number of molecules of H_2 that would need to be oxidized to supply sufficient energy to achieve the synthesis of one molecule of ATP if noncyclic electron flow were used as the photosynthetic pathway?

3. There are two photosystems that are utilized in plants for photosynthesis, PSI and PSII. Review the fundamental properties of PSII. Where is it located in the chloroplast? What wavelength(s) of light is (are) capable of activating this system? When one photon of light possessing the correct minimal energy is absorbed by P_{680}, one electron is released from the P_{680} chlorophyll. What is the source of electrons that will reduce P_{680}^+ back to P_{680}? How many photons of light are required to yield one molecule of O_2?

Ultimately, the electrons generated from PSII are transferred to PSI. Refer to Figure 18-15a to review this process. The PSII and PSI photosystems are not spatially connected. How are the electrons actually transferred between them?

When P_{700} absorbs a photon of light, it becomes oxidized. To where is the electron transferred? During noncyclic photosynthesis, what is the ultimate acceptor of the reducing equivalents in this system? What is the source of the electron that reduces P_{700}^+ to P_{700}? The process by which the electrons from water are transferred to $NADP^+$ through the two photosystems is referred to as linear electron flow. What is cyclic electron flow, and where and how does it occur? Is O_2 generated during cyclic electron flow? Are protons pumped during the process?

a. You are growing plants under different lighting conditions. Which photosystem would operate when only light having a wavelength of 700 nm is used? Would linear electron flow occur? Could cyclic electron flow occur? If cyclic electron flow occurred, would proton pumping accompany it, thus supporting the subsequent synthesis of ATP? Would NADPH be synthesized?

b. You are investigating inhibitors of photosynthesis with the hope of developing new herbicides. You have developed a compound that prevents the transfer of electrons from Q_A in PSII to Q_B (Refer to Figure 18-15a). Could cyclic electron flow occur in the presence of the inhibitor?

c. Another inhibitor that you have prepared prevents the transfer of electrons from plastocyanin to P_{700}^+. What would happen to linear electron flow in this case? Would cyclic electron flow be affected? Explain your answers.

4. The light reactions of photosynthesis are responsible for making ATP and NADPH. The dark reactions fix CO_2 using ATP and NADPH made during the course of the light reactions. The Calvin cycle that is responsible for the first steps in CO_2 fixation is illustrated in Figure 18-17. In what compartment does the Calvin cycle occur? At which points in the cycle is ATP energy consumed? How many molecules of ATP are needed to fix six molecules of CO_2 into one molecule of fructose 1,6-bisphosphate? In which cellular compartment are hexoses ultimately made?

Notice that the first step in the Calvin cycle adds CO_2 to ribulose 1,5-bisphosphate. What is the enzyme responsible for this reaction? How is it regulated?

Another extremely important point in the Calvin cycle is the step whereby 1,3-bisphosphoglycerate is reduced to glyceraldehyde 3-phosphate. What is the source of the reducing equivalents for this step?

The Calvin cycle involves three-carbon molecules. What is C_4 fixation? Which plants use it to fix CO_2, and why? Do C_4 plants still require an active Calvin cycle to synthesize hexoses? Are the energy requirements for fixing CO_2 by the C_4 pathway the same as that for plants that do not use this method? If not, how much more or less ATP is used to fix one molecule of CO_2 in the C_4 pathway?

a. The balance between ATP and NADPH levels in plants is critical. Both are required for synthesizing six carbon sugars. Suppose you exposed a plant to an inhibitor that prevented linear electron flow but permitted cyclic electron flow. What would be the effect on NADPH synthesis? Could this potentially compromise the synthesis of sugars?

b. Do you think cyclic electron flow can replace linear electron flow in plants? Why, or why not?

References

General References on Chloroplast Structure and Photosynthesis

BARBER, J., ed. 1992. *The Photosystems: Structure, Function, and Molecular Biology.* Topics in Photosynthesis, Vol. 11. Elsevier.

DENNIS, D. T., and D. H. TURPIN, eds. 1990. *Plant Physiology and Molecular Biology.* Longman.

DIESENHOFER, J., and J. R. NORRIS, eds. 1993. *The Photosynthetic Reaction Center,* Academic Press. Vols. 1 and 2.

GOVINDJEE, ed. 1982. *Photosynthesis: Energy Conversion by Plants and Bacteria.* Academic Press.

GOVINDJEE and W. J. COLEMAN. 1990. How plants make oxygen. *Sci. Am.* **262**(2):50–58.

HAROLD, F. M. 1986. *The Vital Force: A Study of Bioenergetics.* W. H. Freeman and Company. Chapter 8.

ROCHAIX, J. D. 1992. Genetic analysis of photosynthesis in prokaryotes and lower eukaryotes. *Curr. Opin. Genet. Dev.* **2**:785–791.

STEINBACH, K. E., et al., eds. 1985. *Molecular Biology of the Photosynthetic Apparatus.* Cold Spring Harbor Laboratory.

General Properties of Antennas and Reaction Centers

BABCOCK, G. 1993. Proteins, radicals, isotopes, and mutants in photosynthetic oxygen evolution. *Proc. Natl. Acad. Sci. USA* **90**:10893–10895.

BARBER, J. 1987. Photosynthetic reaction centers: a common link. *Trends Biochem. Sci.* **12**:321–326.

GLAZER, A. N. 1989. Light guides. Directional energy transfer in a photosynthetic antenna. *J. Biol. Chem.* **264**:1–4.

GOLBECK, J. H. 1993. Shared thematic elements in photochemical reaction centers. *Proc. Natl. Acad. Sci. USA* **90**:1642–1646.

GREEN, B. R., E. PICHERSKY, and K. KLOPPSTECH. 1991. Chlorophyll a/b-binding proteins: an extended family. *Trends Biochem. Sci.* **16**:181–186.

HUNTER, C. N., G. R. VAN, and J. D. OLSEN. 1989. Photosynthetic antenna proteins: 100 ps before photochemistry starts. *Trends Biochem. Sci.* **14**:72–76.

KÜHLBRANDT, W., D. N. WANG, and Y. FUJIYOSHI. 1994. Atomic model of plant light-harvesting complex by electron crystallography. *Nature* **367**:614–621.

MARTIN, J. L., and M. H. VOS. 1992. Femtosecond biology. *Ann. Rev. Biophys. Biomol. Struct.* **21**:199–222.

MOSER, C. C., and P. L. DUTTON. 1992. Engineering protein structure for electron transfer function in photosynthetic reaction centers. *Biochim. Biophys. Acta* **1101**:171–176.

NITSCHKE, W., and A. W. RUTHERFORD. 1991. Photosynthetic reaction centres: variations on a common structural theme? *Trends Biochem. Sci.* **16**:241–245.

Structure and Function of the Bacterial Photosynthetic Reaction Center

ALLEN, J. P., et al. 1987. Structure of the reaction center from *Rhodobacter sphaeroides* R-26: The cofactors. *Proc. Natl. Acad. Sci. USA* **84**:5730–5734.

DEISENHOFER, J., and H. MICHEL. 1989. The photosynthetic reaction center from the purple bacterium *Rhodopseudomonas viridis. Science* **245**:1463–1473 (The Nobel Prize Lecture).

DEISENHOFER, J., and H. MICHEL. 1991. Structures of bacterial photosynthetic reaction centers. *Ann. Rev. Cell Biol.* **7**:1–23.

HOLZAPFEL, W., et al. 1990. Initial electron transfer in the reaction center from *Rhodobacter sphaeroides. Proc. Natl. Acad. Sci. USA* **87**:5168–5172.

HUBER, R. 1989. A structural basis of light energy and electron transfer in biology. *EMBO J.* **8**:2125–2147 (The Nobel Prize Lecture).

SINNING, I. 1992. Herbicide binding in the bacterial photosynthetic reaction center. *Trends Biochem. Sci.* **17**:150–154.

Photosystem II and the Oxygen-Evolving Complex

ARO, E. M., I. VIRGIN, and B. ANDERSSON. 1993. Photoinhibition of photosystem II. Inactivation, protein damage, and turnover. *Biochim. Biophys. Acta* **1143**:113–134.

BABCOCK, G. T., et al. 1989. Water oxidation in photosystem II: from radical chemistry to multielectron chemistry. *Biochemistry* **28**:9557–9565.

BARRY, B. A. 1993. The role of redox-active amino acids in the photosynthetic water-oxidizing complex. *Photochem. Photobiol.* **57**:179–188.

GHANOTAKIS, D. and C. YOCUM. 1990. Photosystem II and the oxygen-evolving complex. *Ann. Rev. Plant Physiol. Plant Mol. Biol.* **41**:255–276.

DEBUS, R. J. 1992. The manganese and calcium ions of photosynthetic oxygen evolution. *Biochim. Biophys. Acta* **1102**:269–352.

MATTOO, A. K., J. B. MARDER, and M. EDELMAN. 1989. Dynamics of the photosystem II reaction center. *Cell* **56**:234–246.

MICHEL, H., and J. DEISENHOFER. 1988. Relevance of the photosynthetic reaction center from purple bacteria to the structure of photosystem II. *Biochemistry* **27**:1–7.

RUTHERFORD, A. W. 1989. Photosystem II, the water-splitting enzyme. *Trends Biochem. Sci.* **14**:227–232.

Photosystem I

GOLBECK, J. H. 1992. Structure and function of photosystem I. *Ann. Rev. Plant Physiol. Plant Mol. Biol.* **43**:293–324.

KRAUSS, N., et al. 1993. Three-dimensional structure of system I of photosynthesis at 6 Å resolution. *Nature* **361**:326–331.

Photosynthetic Electron Transport and ATP Synthesis; Regulation of Photosynthesis

ALLEN, J. F. 1992. How does protein phosphorylation regulate photosynthesis? *Trends Biochem. Sci.* **17**:12–17.

ANDERSON, J. M., and B. ANDERSSON. 1988. The dynamic photosynthetic membrane and regulation of solar energy conversion. *Trends Biochem. Sci.* **13**:351–355.

BARBER, J., and B. ANDERSSON. 1992. Too much of a good thing: light can be bad for photosynthesis. *Trends Biochem. Sci.* **17**:61–66.

HOPE, A. B. 1993. The chloroplast cytochrome bf complex: a critical focus on function. *Biochim. Biophys. Acta* **1143**:1–22.

KNAFF, D. 1991. Regulatory phosphorylation of chloroplast antenna proteins. *Trends Biochem. Sci.* **16**:82–83.

MCCARTY, R. E., and G. G. HAMMES. 1987. Molecular architecture of chloroplast coupling factor 1. *Trends Biochem. Sci.* **12**:234–237.

MALMSTROM, B. G. 1989. The mechanism of proton translocation in respiration and photosynthesis. *FEBS Lett.* **250**:9–21.

SCHERER, S. 1990. Do photosynthetic and respiratory electron transport chains share redox proteins? *Trends Biochem. Sci.* **15**:458–462.

CO$_2$ Fixation

BASSHAM, J. A. 1962. The path of carbon in photosynthesis. *Sci. Am.* **206**(6):88–100.

BUCHANAN, B. B. 1991. Regulation of CO$_2$ assimilation in oxygenic photosynthesis: the ferredoxin/thioredoxin system. Perspective on its discovery, present status, and future development. *Arch. Biochem. Biophys.* **288**:1–9.

HUANG, A. H. C., R. N. TRELEASE, and T. S. MOORE, JR. 1983. Plant peroxisomes. *American Society of Plant Physiologists Monograph Series.* Academic Press.

KNAFF, D. B. 1989. Structure and regulation of ribulose 1,5-bisphosphate carboxylase/oxygenase. *Trends Biochem. Sci.* **14**:159–160.

PORTIS, A. 1992. Regulation of ribulose 1,5-bisphosphate carboxylase/oxygenase activity. *Ann. Rev. Plant Physiol. Plant Mol. Biol.* **43**:415–437.

RAWSTHORNE, S. 1992. Towards an understanding of C$_3$-C$_4$ photosynthesis. *Essays Biochem.* **27**:135–146.

RIESMEIER, J., L. WILLMITZER, and W. FROMMER. 1992. Isolation and characterization of a sucrose carrier cDNA from spinach by functional expression in yeast. *EMBO J.* **11**:4705–4713.

SICHER, R. C. 1986. Sucrose biosynthesis in photosynthetic tissue: rate-controlling factors and metabolic pathway. *Physiol. Plant* **67**:118–121.

SCHNEIDER, G., Y. LINDQVIST, and C. I. BRANDEN. 1992. RUBISCO: Structure and Mechanism. *Annu. Rev. Biophys. Biomol. Struct.* **21**:119–143.

WOLOSIUK, R. A., M. A. BALLICORA, and K. HAGELIN. 1993. The reductive pentose phosphate cycle for photosynthetic CO$_2$ assimilation: enzyme modulation. *FASEB J.* **7**:622–637.

19

Organelle Biogenesis: The Mitochondrion, Chloroplast, Peroxisome, and Nucleus

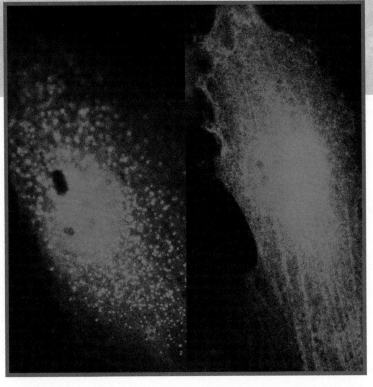

▲ Normal but not mutant human cells incorporate Firefly luciferase into peroxisomes.

We discussed the functions of mitochondria, peroxisomes, and chloroplasts in Chapters 17 and 18, and now return to the question raised in Chapter 16: How is the biogenesis of these organelles accomplished?

Mitochondria, peroxisomes, and chloroplasts grow throughout interphase by incorporating lipids as well as newly made proteins; most of the proteins are synthesized on cytosolic ribosomes. When large enough, the organelles divide by pinching off daughter organelles. The nucleus also incorporates proteins during interphase, in preparation for mitotic division. Thus these membrane-bounded organelles must import most or all of their proteins from the cytosol. In this chapter, therefore, we focus on the importing of newly synthesized proteins from the cytosol into the organelle.

Mitochondria, chloroplasts, and nuclei are bounded by two membranes, and the peroxisome by one. Chloroplasts contain, in addition, an internal membrane compartment—the thylakoid vesicles—on which photosynthesis takes place. Except for the outer nuclear membrane, which is a continuation of the rough endoplasmic reticulum, these organelles are not part of the secretory pathway discussed in Chapter 16; no transport vesicles bud from or fuse with mitochondria, peroxisomes, or chloroplasts. How then are proteins directed to the correct organelle and to the correct membrane or compartment within the organelle? Also, what are the functions of the chloroplast and mitochondrial DNAs, and how are their functions coordinated with that of nuclear DNA in organelle biosynthesis? These are the primary topics we consider in the present chapter.

➤ *An Overview of Organelle Biogenesis Outside the Secretory Pathway*

The growth and division of mitochondria, peroxisomes, and chloroplasts are not coupled to nuclear division. The three organelles grow by the incorporation of proteins (and lipids), a process that occurs continuously during the interphase period of the cell cycle. As the organelles increase in size, one or more daughters pinch off in a manner similar to the way in which bacterial cells grow and divide. Similarly, during interphase the nucleus doubles its content of all proteins, such as RNA and DNA polymerases and transcription factors, in preparation for the division of the nucleus that occurs during mitosis.

Most chloroplast and mitochondrial proteins, and all peroxisomal and nuclear proteins are synthesized outside the organelle on "free" cytosolic ribosomes—that is, on ribosomes not bound to the rough endoplasmic reticulum. The newly made proteins are released into the cytosol and are then taken up specifically into the proper organelle (Figure 19-1). Specific receptor proteins on the organelle surface recognize specific targeting sequences on the new proteins that direct the proteins to the correct organelles (Table 19-1). In general, proteins uptake into each of these organelles is an energy-requiring process that depends on integral proteins in the organellar membranes.

Proteins enter the nucleus through nuclear pores, the conduits that connect the inner and outer nuclear membrane and that also are used for the exit of messenger RNAs and ribosomes from the nucleus into the cytosol. Because nuclear pores are aqueous-filled channels, proteins and particles do not cross phospholipid bilayers as they enter or leave the nucleus. Consequently, large protein or ribonucleoprotein particles can be formed in the cytosol and then imported into the nucleus, and ribosomal sub-

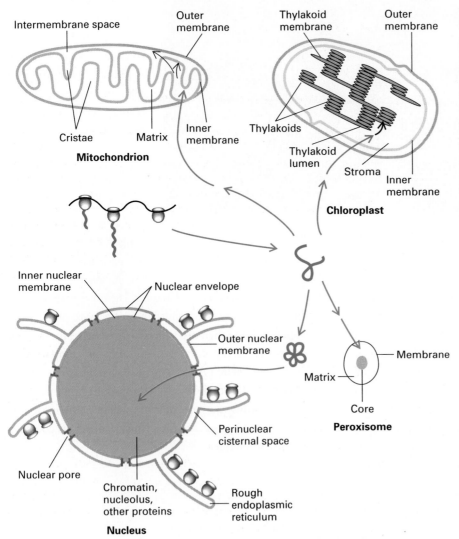

◄ FIGURE 19-1   Import of proteins from the cytosol into the mitochondrion, chloroplast, peroxisome, and nucleus. Immediately after their synthesis on free cytosolic ribosomes (those not bound to membranes), these proteins are released into the cytosol. Mitochondrial proteins insert into the outer membrane, or cross the outer and inner mitochondrial membranes to enter the matrix space. Some remain there; others (black arrows) cross into the intermembrane space or become incorporated into the inner membrane. Similarly, chloroplast proteins cross the outer and inner chloroplast membranes to enter the stroma. Some remain there; others (black arrow) enter the thylakoids. Peroxisomal proteins cross the peroxisome membrane, and nuclear proteins enter through the nuclear pores. Ribosomes and mRNAs exit the nucleus through the pores.

TABLE 19-1   Sequences That Target Proteins from The Cytosol to Organelles

Organelle	Location of Signal within Protein	Signal Generally Removed?	Nature of Signal
Endoplasmic reticulum	Usually N-terminal	Usually	"Core" of 6–12 mostly hydrophobic amino acids, often preceded by one or more basic amino acids
Mitochondrion*	Usually N-terminal	Usually	3–5 nonconsecutive Arg or Lys residues; often contains Ser and Thr; no Glu or Asp residues
Chloroplast*	Usually N-terminal	Usually	No common sequence motifs; generally rich in Ser, Thr, and small hydrophobic amino acids and poor in Glu and Asp residues
Peroxisome	Usually C-terminal	No	Usually Ser-Lys-Leu at extreme C-terminus
Nucleus	Usually internal	No	One cluster of 5 basic amino acids, or 2 smaller clusters of basic residues separated by ≈10 amino acids

* These signals direct the protein from the cytosol into the matrix space of the mitochondrion or the corresponding stroma of the chloroplast; other signals discussed in the text redirect proteins into other subcompartments of these organelles.

units that are formed in the nucleus can be transported into the cytosol.

In contrast, proteins moving from the cytosol into mitochondria, peroxisomes, or chloroplasts must cross one or more phospholipid bilayers. As discussed in Chapter 16, proteins cross membranes only in an unfolded state. Therefore, proteins destined for these organelles must bind to chaperone proteins in the cytosol, which keep them in an unfolded state prior to their incorporation into the proper organelle.

The mitochondrion and chloroplast contain multiple membranes and membrane-limited spaces (Figure 19-1). The transport of proteins from the cytosol into these organelles occurs through protein-lined channels localized to "contact sites" where the inner and outer membranes are close together. In some cases the sequential action of two targeting sequences and two membrane-bound receptor systems is required—one to direct the protein into the organelle, and the other to direct it into the correct organellar compartment or membrane.

Both the chloroplast and the mitochondrion also contain DNA that encodes the synthesis of several proteins within the organelle. Many proteins, such as the F_0F_1 ATPase, contain multiple subunits, some encoded by the nuclear DNA and others by the organellar genome, and the synthesis of these protein subunits must be coordinated.

The presence of DNA in mitochondria and chloroplasts reflects the ancient origin of these organelles as *endosymbionts*. It is thought that the chloroplast began as an ancestral cyanobacterium (a photosynthetic prokaryote) that became endocytosed by an ancestral eukaryotic cell and replicated within the cytoplasm. Over eons of evolution some of the bacterial DNA, encoding some of the proteins of what evolved into a chloroplast, moved to the nucleus, and thus some of the chloroplast proteins are now imported into the organelle after their synthesis in the cytosol. Mitochondria probably originated by a similar process from a prokaryotic cell capable of oxidative phosphorylation: the plasma membrane of the prokaryote, the site of oxidative phosphorylation, became the inner mitochondrial membrane. The outer membranes of the mitochondrion and chloroplast may be evolutionary remnants of the membrane of the endocytic vesicle into which these prokaryotes were internalized. Several types of evidence strongly supporting an endosymbiotic origin of these organelles will be discussed in this chapter.

We begin by looking at mitochondrial DNA, focusing on its protein-coding capacity and its expression. We will also review the inheritance of this organellar DNA, as the genetics of mitochondrial DNA differs from that of nuclear genes in many crucial ways. We will note, en route, some surprising variations among different mitochondrial DNAs—even differences in genetic codes! In the next section we turn to the assembly of mitochondria, focusing on the import and assembly of the multiprotein complexes involved in electron transport and ATP synthesis. We continue with sections on the transport of proteins destined for the chloroplast and the peroxisome, and we round off the chapter with a look at the role of nuclear pores in the importation of nuclear proteins.

➤ *Mitochondrial DNA: Structure, Expression, and Variability*

Individual mitochondria are large enough to be seen under the light microscope and even the mitochondrial DNA (mtDNA) can be detected by fluorescence microscopy (Figure 19-2). The mtDNA is located in the mitochondrial matrix and is sometimes found attached to the inner mitochondrial membrane. As judged by the number of yellow fluorescent "dots" of mtDNA, a *Euglena gracilis* cell contains at least thirty mtDNA molecules.

Since the dyes used in the experiment in Figure 19-2 do not affect cell growth or division, replication of mtDNA and division of the mitochondrial network can be followed in living cells using time-lapse microscopy (see Figure 5-19). Such studies show that mtDNA replicates throughout interphase. At mitosis each daughter cell receives approximately the same amount of mitochondria and mtDNA, but there is no system for apportioning exactly equal numbers of mtDNAs to the daughter cells.

Cytoplasmic Inheritance and DNA Sequencing Have Established the Existence of Mitochondrial Genes

Prior to the isolation and sequencing of mitochondrial DNA, studies of mutants in yeasts and other single-celled organisms indicated that mitochondria exhibit cytoplasmic inheritance and thus must contain their own genetic system (Figure 19-3). *Petite yeast mutants* exhibit structurally abnormal mitochondria and are incapable of oxidative phosphorylation. Petite cells produce ATP by the fermentation of glucose to ethanol, which provides less ATP per mole of glucose than oxidative phosphorylation does; as a result, petite cells grow more slowly than wild-type yeasts and form smaller colonies (hence the name "petite"). Mitochondria in petite cells do contain many Krebs-cycle enzymes and other enzymes that are essential for the biosynthesis of several amino acids and nucleotides. Genetic crosses between different (haploid) yeast strains showed that the petite mutation does not segregate with any known nuclear gene or chromosome. Because the inheritance of the petite or wild-type mitochondrial phenotype is clearly nonchromosomal, researchers postulated that some mutable element in the cytoplasm must be a determinant in mitochondrial synthesis. Later studies have shown that most petite mutations contain deletions of some or all mtDNA.

Mitochondrial inheritance in yeasts is biparental: during the fusion of haploid cells, both parents contribute equally to the cytoplasm of the diploid. In mammals and most other animals, the sperm contributes little (if any) cytoplasm to the zygote, and virtually all of the mitochondria in the embryo are derived from those in the egg, not

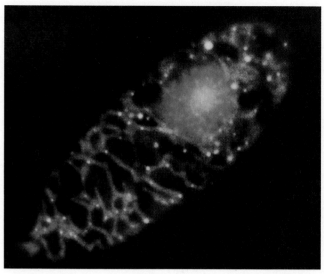

▲ FIGURE 19-2   Visualization of mitochondrial DNA in a growing *Euglena gracilis* cell. Cells were treated with a mixture of two dyes. One, ethidium bromide, binds to DNA and emits a vermilion fluorescence; the other, DiOC6, is incorporated specifically into mitochondria and emits a green fluorescence. Thus the nucleus emits a vermilion fluorescence, and areas rich in mitochondrial DNA fluoresce yellow—a combination of vermilion DNA and green mitochondrial fluorescence. [From Y. Huyashi and K. Veda, 1989, *J. Cell Sci.* **93,** 565].

the sperm. An experiment designed to detect the very low, but significant, paternal contribution to mtDNA took advantage of different strains of mice that have mtDNAs that vary slightly in sequence. When two such strains were mated and the mtDNA in the F_1 animals was analyzed by sequencing, about one part in 10^4 of the mtDNA was found to be the paternal type; the balance was from the mother. Thus 99.99 percent of mtDNA is maternally inherited, but there was a small (0.01 percent) paternal contribution. In higher plants, mtDNA is inherited exclusively in a uniparental fashion through the female parent (egg), not the male (pollen).

The entire mitochondrial genome has now been cloned and sequenced for a number of different organisms, and mtDNAs from all these sources have been found to encode a similar set of rRNAs, tRNAs, and essential mitochondrial proteins. All proteins encoded by mtDNA are synthesized on mitochondrial ribosomes. All mitochondrially synthesized polypeptides identified thus far (with one possible exception) are not complete enzymes but subunits of multimeric complexes used in electron transport or ATP synthesis (see Table 19-2). Most proteins localized in mitochondria, such as the mitochondrial RNA and DNA polymerases, are synthesized on cytoplasmic ribosomes and are imported into the organelle.

(a)

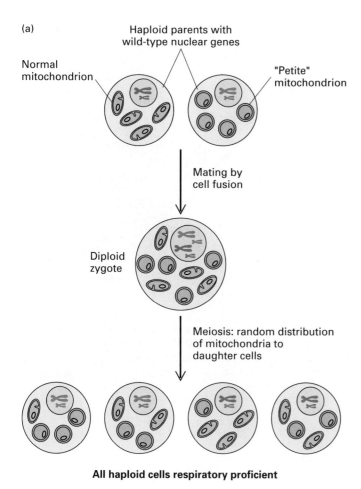

Haploid parents with
wild-type nuclear genes

Normal
mitochondrion

"Petite"
mitochondrion

Mating by
cell fusion

Diploid
zygote

Meiosis: random distribution
of mitochondria to
daughter cells

All haploid cells respiratory proficient

(b)

Mitosis: random
distribution
of mitochondria
to daughter cells

Mitosis   Mitosis

Respiratory proficient   **Petite**   **Respiratory
proficient**

◄ FIGURE 19-3   Cytoplasmic inheritance of the petite
mutation in yeast. Petite-strain mitochondria are defective in
oxidative phosphorylation due to a deletion in mtDNA.
(a) Haploid cells fuse to produce a diploid cell that undergoes
meiosis, during which a random segregation of parental chromosomes and mitochondria containing mtDNA occurs. Since
yeast normally contains ≈50 mtDNA molecules per cell, all
products of meiosis usually contain both normal and petite
mtDNAs and are capable of respiration. (b) As these cells
grow and divide mitotically, the cytoplasm (including the
mitochondria) is randomly distributed to the daughter cells.
Occasionally, a cell is generated that contains only defective
petite mtDNA and yields a petite colony. Thus formation of
such petite cells is independent of any nuclear genetic
marker.

The Size and Coding Capacity of mtDNA Varies in Different Organisms, Reflecting Evolutionary Movement of DNA between Mitochondrion and Nucleus

Surprisingly, the size of mtDNA, the number and nature of
the proteins they encode, and even the mitochondrial genetic code itself, are very different in various organisms.
The human mtDNA, a completely sequenced circular molecule of 16,569 bp, is among the smallest known (Figure
19-4). It encodes the two rRNAs found in mitochondrial
ribosomes and the 22 tRNAs used to translate the mitochondrial mRNAs. Human mtDNA has 13 sequences that
begin with an ATG (methionine) codon, end with a chain-
termination codon, and are long enough to encode a polypeptide of more than 50 amino acids. All of these possible
proteins have been identified (Table 19-2). Three are the
subunits of the cytochrome c oxidase complex that form
the catalytic core of the enzyme (see Figure 17-30a); highly
conserved during evolution, these subunits are similar to
the three subunits that comprise bacterial cytochrome c
oxidases. Two others are subunits of F_0 ATPase (see Figure
17-15a), seven are subunits of the NADH-CoQ reductase
complex (see Figure 17-27), and one is the cytochrome b
subunit of the $CoQH_2$–cytochrome c reductase complex
(see Figure 17-29).

Invertebrate mtDNA is about the same size as human
mtDNA, but yeast mtDNA is almost five times as large
(≈78,000 bp). Yeast mtDNA and mtDNA from other
lower eukaryotes encode many of the same gene products
as mammalian mtDNA does (see Table 19-2). There are
differences, however. Yeast and fungal mtDNAs encode
one ribosomal protein (termed var-1) not present in human
mtDNA. The seven subunits of the NADH-CoQ reductase
complex, encoded by mtDNA in mammals and in the fungus *Neurospora crassa*, are encoded in yeasts by nuclear
DNA. Conversely, subunit 9 of the F_0 ATPase, another

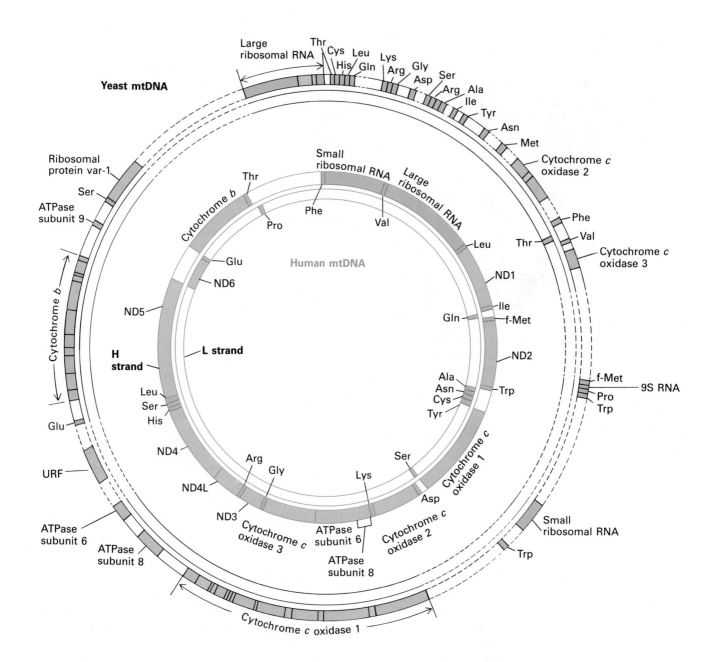

▲ FIGURE 19-4   The organization of human (inner rings) and yeast (outer rings) mtDNA. Yeast mtDNA is five times larger than human mtDNA but is represented here as only two and one-half times larger. Proteins and RNAs encoded by each of the two strands are shown separately. Transcription off the outer (H) strand of each mtDNA is clockwise and off the inner (L) strand is counterclockwise.

The entire human mtDNA has been sequenced. For yeast, the diagram is based on a compilation of partially complete DNA sequence data from several laboratories; the dashed lines represent unsequenced regions. Exons of structural genes and URFs (unidentified or open reading frames) are orange; rRNA and tRNA genes are green. The abbreviations for amino acids denote the tRNA genes. Yeast genes for cytochrome c oxidase subunits 1 and 2, cytochrome b, and 21S rRNA contain introns (blue). No mammalian mtDNA

genes contain introns. The human 207-bp gene encoding F_0 ATPase subunit 8 overlaps, out of frame, with the N-terminal portion of the segment encoding F_0 ATPase subunit 6.

Note that yeast and mammalian mtDNAs encode some different proteins. Human mtDNA encodes seven subunits of the NADH-CoQ reductase complex (ND1, ND2, ND3, ND4L, ND4, ND5, ND6) that have no counterparts in yeast mtDNA; also, F_1 ATPase subunit 9 is encoded by mtDNA in yeast but by nuclear DNA in humans. Most of the nontranscribed (and unsequenced) regions of yeast mtDNA contain adenine and thymine almost exclusively, and their function (if any) is not known. [See P. Borst and L. A. Grivell, 1981, *Nature* **290**:443; L. A. Grivell, 1983, *Sci. Am.* **248**(3):78–89; A. Tzagoloff and A. Myers, 1986, *Ann. Rev. Biochem.* **55**:249–285; A. Chomyn et al., 1985, *Nature* **314**:592; A. Chomyn et al., 1986, *Science* **234**:614–618.]

TABLE 19-2   Mitochondrial Genes and Their Products

	Mammal	Yeast	Fungus (Neurospora)	Plant
Size (thousands of base pairs)	14–18	78	19–108	200–2500
rRNAs				
Large subunit	16S	21S	21S	26S
Small subunit	12S	15S	15S	18S
5S RNA	−	−	−	+
No. of tRNAs	22	23–25	23–25	≈30
Ribosomal protein (var-1)	−	+	+	?
Subunits 1, 2, 3 of cytochrome c oxidase	+	+	+	+
Apocytochrome b subunit of $CoQH_2$–cytochrome c reductase	+	+	+	+
F_0 ATPase				
Subunit 6	+	+	+	+
Subunit 8	+	+	+	+
Subunit 9	−	+	−	+
Subunit α of F_1 ATPase	−	−	−	+
No. of NADH-CoQ reductase subunits	7	0	6	6

Plus (+) = present; minus (−) = absent.
SOURCE: V. K. Eckenrode and C. S. Levings. 1986. *In Vitro Cell Dev. Biol.* **22**:169–176; C. S. Levings and G. G. Brown, 1989, *Cell* **56**:171–179.

mitochondrial protein, is encoded by mtDNA in yeasts and higher plants but by nuclear DNA in *N. crassa* and mammals. Interestingly, in *Neurospora* the mtDNA also contains an apparently nonfunctional gene (a *pseudogene*) that is homologous in sequence to the nuclear gene for subunit 9.

In contrast to the single circular genome of animal and fungal mtDNAs, plant mtDNAs contain multiple circular DNAs that appear to recombine with each other. Plant mtDNAs are much larger and more variable than the mtDNAs of other organisms (see Table 19-2). They range in size from 200,000 to 2,500,000 bp; even in a single family, there can be as much as an eightfold variation in size (watermelon = 330,000 bp; muskmelon = 2,500,000 bp)! Plant mtDNAs contain a few genes not found in other mtDNAs: a 5S mitochondrial rRNA (not found in other mitochondrial ribosomes) and one subunit (α) of the F_1 ATPase. The mitochondrial rRNAs of plants are also considerably larger than those of animals and fungi, but this does not account for even a fraction of the "extra" mtDNA in plants, and details of the coding capacity of plant mtDNA must await DNA sequencing.

Since several proteins are encoded by mtDNA in some species and by nuclear DNA in others, it would appear that entire genes moved from the mitochondrion to the nucleus, or vice versa, during evolution. Most strikingly, the gene *cox II*, which encodes subunit 2 of cytochrome c oxidase, is found in mtDNA in all organisms studied except for one species of legume, the mung bean: in this organism only, the *cox II* gene is nuclear. As described below, many RNA transcripts of plant mitochondrial genes are edited, mainly by the enzyme-catalyzed conversion of selected C residues to U (and occasionally U to C). The nuclear mung bean *cox II* gene more closely resembles edited cox II RNA transcripts than it does the mitochondrial *cox II* genes found in other legumes. These facts are strong evidence that movement of the *cox II* gene from mitochondrion to nucleus occurred during legume evolution and involved an RNA intermediate; this process would be similar to the process by which pseudogenes are formed in the nuclear genome by reverse transcription of nuclear-encoded mRNAs.

Translocations of segments of mtDNA to nuclear DNA may be occurring still. DNA hybridization and sequence studies have identified short ($\approx$50-bp) segments of mtDNA interspersed randomly in the nuclear DNA of all animals and plants studied. These short segments are different in different species, do not appear to encode any proteins, and appear to have moved by an unknown process from mtDNA into nuclear DNA. The mitochondrial genomes of many plant species contain various chloroplast DNA sequences, such as fragments of the ribulose 1,5-bisphosphate carboxylase gene, suggesting the transfer of DNA between these two organelles as well.

Proteins Encoded by Mitochondrial DNA Are Synthesized on Mitochondrial Ribosomes

As far as is known, all RNA transcripts of mtDNA and their translation products remain in the mitochondrion. There is no export of these RNAs or proteins, and all mtDNA-encoded proteins are synthesized on mitochondrial ribosomes. Mitochondria encode the rRNAs that form mitochondrial ribosomes, although all but one or two of the ribosomal proteins (depending on the species) are imported from the cytosol. In most organisms, such as plants, yeasts, fungi, and mammals, all of the tRNAs used for protein synthesis in the mitochondrion are encoded by

mtDNAs. However, in wheat, in the parasitic protozoan *Trypanosoma brucei* (the cause of African sleeping sickness), and in ciliated protozoa, all mitochondrial tRNAs are encoded by the nuclear DNA and imported into the mitochondrion.

Reflecting the bacterial ancestry of mitochondria, mitochondrial ribosomes resemble prokaryotic ribosomes and differ from cytoplasmic ribosomes in their RNA and protein compositions, their size, and their sensitivity to certain antibiotics. For instance, chloramphenicol blocks protein synthesis by bacterial and most mitochondrial ribosomes, but not by cytoplasmic ribosomes. Conversely, cycloheximide inhibits protein synthesis by eukaryotic cytoplasmic ribosomes but does not affect protein synthesis by mitochondrial ribosomes or bacterial ribosomes. In cultured mammalian cells the only proteins synthesized in the presence of cycloheximide are encoded by mtDNA and produced by mitochondrial ribosomes.

Mitochondrial Genetic Codes Are Different from the Standard Nuclear Code, and They Differ among Organisms

The genetic code used in animal and fungal mitochondria is different from the standard code used in all prokaryotic and eukaryotic nuclear genes; remarkably, the code even differs in mitochondria from different species (Table 19-3). Why and how this phenomenon happened during evolution is mysterious. UGA, for example, is normally a stop codon, but is read as tryptophan by human and fungal

	Standard Code: Nuclear-Encoded	Mitochondria				
Codon	Proteins	Mammals	*Drosophila*	*Neurospora*	Yeasts	Plants
UGA	Stop	Trp	Trp	Trp	Trp	Stop
AGA, AGG	Arg	Stop	Ser	Arg	Arg	Arg
AUA	Ile	Met	Met	Ile	Met	Ile
AUU	Ile	Met	Met	Met	Met	Ile
CUU, CUC, CUA, CUG	Leu	Leu	Leu	Leu	Thr	Leu

TABLE 19-3   Alterations in the Standard Genetic Code in Mitochondria

SOURCE: S. Anderson et al., 1981, *Nature* **290**:457; P. Borst, in *International Cell Biology 1980–1981*, H. G. Schweiger, ed., Springer-Verlag, p. 239; C. Breitenberger and U. L. Raj Bhandary, 1985, *Trends Biochem. Sci.* **10**:478–483; V. K. Eckenrode and C. S. Levings, 1986, *In Vitro Cell Dev. Biol.* **22**:169–176; J. M. Gualber et al., 1989, *Nature* **341**:660–662; and P. S. Covello and M. W. Gray, 1989, *Nature* **341**:662–666.

mitochondrial translation systems; however, in plant mitochondria, UGA is still a stop codon. AGA and AGG are standard codons for arginine in fungal and plant mtDNA, but they are stop codons in mammalian mtDNA and serine codons in *Drosophila* mtDNA.

There is a general principle that plant mitochondria utilize the standard genetic code. However, a comparison of the amino acid sequences of plant mitochondrial proteins with the nucleotide sequences of plant mtDNAs originally appeared to indicate, remarkably, that CGG could code for *either* arginine (the "standard" amino acid) or tryptophan. How could this be? Further information clarified this apparent contradiction. As mentioned earlier, cytosine residues in RNA transcripts of mtDNA are often edited (converted) to uracil. In the case of plant mtDNA, if a CGG sequence is edited to UGG, the codon specifies tryptophan, the standard amino acid for UGG, whereas unedited CGG codons encode the standard arginine. Thus in plant mitochondria the translation system of the ribosomes and tRNAs does utilize the standard genetic code.

As noted in Chapter 4 (see Table 4-5), in certain organisms the genetic code is different even for nuclear-encoded proteins. In the prokaryotic organisms known as mycoplasmas, UGA codes for tryptophan, not the normal "stop"; in ciliated protozoa, such as *Tetrahymena*, the normal stop codons UAA and UAG code for glutamine in the mRNAs synthesized from nuclear DNAs. Both RNA editing and the similarity but nonuniversality of the genetic code have profound implications for the evolution of prokaryotic and eukaryotic cells and their organelles, as they challenge the once generally accepted notion that a single genetic code was established early in evolution and then passed on intact to all extant cells. The differences in codon usage between mitochondrial and cytosolic protein-synthesizing systems also make it difficult to explain how genes could have moved from mitochondrial to nuclear DNA and remained functional, since certain codons would then specify different amino acids!

In Animals, Mitochondrial RNAs Undergo Extensive Processing

Like their nuclear counterparts, mitochondrial rRNAs and mRNAs are generated by the enzymatic cleavage of longer precursors (Figure 19-5). However, processing of mammalian mtDNA transcripts shows a number of novel features. Transcription initiates at only two points on the H strand (so called because it bands at a heavier density in cesium chloride). Primary transcript I begins just upstream of the tRNA^Phe gene and terminates just after the 16S rRNA gene; it is cleaved to tRNA^Phe, tRNA^Val, 12S rRNA, and 16S rRNA. Primary transcript II initiates just downstream of primary transcript I, near the 5' end of the 12S rRNA

gene, but it continues past the termination site of transcript I and around the circular DNA to the start point. Transcript II is processed by enzymatic cleavage to yield the remaining tRNAs and ≈13 monocistronic, polyadenylated mRNAs, but apparently not any rRNAs. Very often some of these cleavages occur on nascent chains. The cleavage sites must be very precise to separate the mRNAs from the tRNAs, because the 5' end of each mRNA is immediately adjacent to the 3' end of a tRNA. Transcript I initiates about 10 times more frequently than transcript II, so that rRNAs and, curiously, two tRNAs are synthesized in excess of mRNAs and other tRNAs.

In contrast to mammalian nuclear-synthesized mRNAs, mammalian mitochondrial mRNAs contain no introns and very few untranslated sequences. The first three bases at the 5' end of each mRNA are generally the AUG (or AUA) initiator codon. A UAA terminator codon is usually at or very near the 3' end of the mRNA. In some cases, only the U of this terminator codon is encoded in the mtDNA; the final two A's are part of the polyadenylate sequence, which is added after the mRNA precursor is synthesized, as it is in cytoplasmic mRNAs. Thus human and other mammalian mtDNAs have evolved to contain as few untranslated sequences as possible, and the genome is the minimum required to generate the requisite mitochondrial mRNAs, tRNAs, and rRNAs.

Mutations in Mitochondrial DNA Cause Several Genetic Diseases in Man

Because mammalian mtDNA, in contrast to nuclear DNA, lacks introns and contains no long noncoding sequences, it has a much higher information content than nuclear DNA. For this reason it would seem to be an excellent target for mutations that give rise to human disease. However, because mtDNA is predominantly maternally inherited, and because each cell contains hundreds of mitochondria and thousands of copies of the mtDNA genome, it has been difficult to detect mutations in mtDNA that cause genetic diseases.

The severity of disease caused by a mutation in mtDNA will depend on the nature of the mutation and on the proportion of mutant and wild-type DNAs present in a particular cell type. Generally, cells contain mixtures of wild-type and mutant mtDNAs—a condition known as *heteroplasmy*. Much as in yeast cells (see Figure 19-3), each time a mammalian somatic or germ-line cell divides, the mutant and wild-type mtDNAs will segregate randomly into the daughter cells. Thus, the mtDNA genotype fluctuates from one generation and from one cell division to the next, and can drift toward predominantly wild-type or predominantly mutant mtDNAs. Since all enzymes for the replication and growth of mitochondria, such as DNA

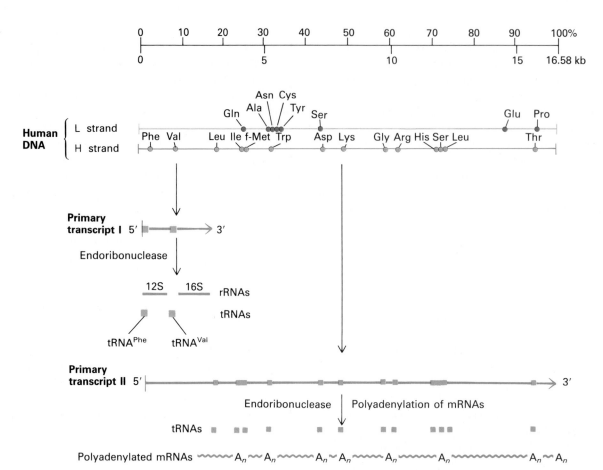

▲ FIGURE 19-5   Transcription map of human mtDNA. As deduced from the DNA sequence and RNA-DNA hybridization studies, the light (L) DNA strand encodes only eight tRNAs (dark green circles); this strand is transcribed right to left. The heavy (H) DNA strand encodes the 12S and 16S rRNAs, 14 tRNAs (light green circles), and all but one of the polyadenylated mRNAs that encode mitochondrial proteins. (The ND6 subunit of the NADH-CoQ reductase complex is encoded by the L strand). Transcription of the H strand is initiated at two sites. Primary transcript I initiates just upstream of the tRNA^{Phe} gene and terminates just after the 16S rRNA; it is processed by endoribonucleases to yield one molecule each of tRNA^{Phe}, tRNA^{Val}, and 12S and 16S rRNA. Primary transcript II initiates near the 5' end of the 12S rRNA and apparently continues completely around the circular mtDNA; it is processed to yield the other tRNAs and mRNAs, which are subsequently polyadenylated (A$_n$). Some of the cleavages of this transcript begin while the chain is nascent. [See D. Ojala et al., 1981, *Nature* **290**:470; J. Montoya, G. L. Gaines, and G. Attardi, 1983, *Cell* **34**:151.]

and RNA polymerases, are imported from the cytosol, a mutant mtDNA should not be at a "replication disadvantage"; mutants that involve large deletions of mtDNA might even be at a selective advantage in replication.

Since all cells have mitochondria, why do mutations in mtDNA only affect some tissues? Those most usually affected are tissues that have a high requirement for ATP produced by oxidative phosphorylation, and tissues that require most or all of the mtDNA in the cell to synthesize functional mitochondrial proteins. Leber's hereditary optic neuropathy (degeneration of the optic nerve, accompanied by increasing blindness), for instance, is caused by a missense mutation in the mtDNA gene encoding subunit 4 of the NADH-CoQ reductase. Another set of diseases

(chronic progressive external ophthalmoplegia and Kearns-Sayre syndrome, characterized by eye defects and, in Kearns-Sayre syndrome, also by abnormal heartbeat and central nervous system degeneration) are caused by any of several large deletions in mtDNA. A third condition, causing "ragged" muscle fibers (with improperly assembled mitochondria) and associated uncontrolled jerky movements, is caused by a single mutation in the TΨCG loop (Figure 4-31) of the mitochondrial lysine tRNA (tRNA^{Lys}); the mutation apparently interferes with the translation of several mitochondrial proteins.

Many of these diseases are hard to diagnose, and require the sequencing of clones of mtDNAs from the affected individuals. Inheritance studies, used to map most types of genetic diseases (Chapter 7), can establish that a disease is maternally inherited and thus is most likely caused by an alteration in mtDNA, but since there is apparently no recombination between mammalian mtDNAs it is not possible to develop a purely genetic (recombination) map of such mutations in mtDNA.

The activity of individual mitochondria normally declines somewhat with age, so that diseases caused by mutations in mtDNA usually become more severe with age. Thus, an individual with, say, 15 percent normal and 85 percent defective mtDNA in his muscle may have enough mitochondrial-encoded proteins to be relatively healthy at age 20, but have a severe disease phenotype at age 60.

➤ Synthesis and Localization of Mitochondrial Proteins

Only a few proteins are encoded by mitochondrial DNA and synthesized on mitochondrial ribosomes. Most such proteins are subunits of integral proteins in the inner mitochondrial membrane. They are incorporated into the inner membrane immediately after synthesis, but little is known of how this happens. The vast majority of mitochondrial proteins are encoded by nuclear genes, synthesized on cytosolic ribosomes, and imported into the mitochondrion. To these proteins we now turn our attention.

Most Mitochondrial Proteins Are Synthesized in the Cytosol as Precursors

Proteins that are synthesized in the cytosol and incorporated into the mitochondrion include the majority of proteins required for oxidative phosphorylation, the mitochondrial DNA and RNA polymerases, and all but one or two of the mitochondrial ribosomal proteins. Some cytosol-synthesized proteins are transported to the intermembrane space, others to the outer membrane, and still others to the inner membrane (Table 19-4). The largest number are transported to the matrix, including the F_1 ATPase subunits, ribosomal proteins, RNA polymerase, and the enzymes of the citric acid cycle. How is a protein imported into the mitochondrion and how is it specifically targeted to its final destination? What drives these specific, unidirectional transport processes?

Various research findings demonstrated that proteins are imported into the mitochondria after they are synthesized in the cytosol. Pulse-chase studies on yeast and *Neurospora* cells established that newly made mitochondrial proteins are initially located in the cytosol outside the mitochondria and accumulate gradually, during the chase period, at their proper destinations in the mitochondrion.

TABLE 19-4 Some Mitochondrial Proteins Synthesized in the Cytosol

Mitochondrial Location	Protein*

Matrix	F_1 ATPase
	  Subunit α (except plants)
	  Subunits β, γ
	  Subunit δ (certain fungi)
	Carbamoyl phosphate synthase (mammals)
	Mn^{2+}-superoxide dismutase
	RNA polymerase
	DNA polymerase
	Ribosomal proteins
	Citrate synthase and other citric acid enzymes
	Ornithine transcarbamoylase (mammals)
	Ornithine aminotransferase (mammals)
	Alcohol dehydrogenase (yeast)

Inner membrane	ADP-ATP antiport phosphate-OH^- antiport Cytochrome c oxidase subunits 4, 5, 6, and 7 Proteolipid of F_0 ATPase $CoQH_2$-cytochrome c reductase complex: subunits 1, 2, 5 (Fe-S protein), 6, 7, and 8 Thermogenin
Intermembrane space	Cytochrome c Cytochrome c peroxidase Cytochrome b_2 $CoQH_2$-cytochrome c reductase complex subunit 4 (cytochrome c_1).
Outer membrane	Mitochondrial porin

* Most proteins (except the ADP-ATP antiport, cytochrome c, and porin) are fabricated as longer precursors.

Furthermore, most mitochondrial proteins destined for the matrix, the intermembrane space, or the inner membrane begin as precursors with additional amino acids at the N-terminus that are not present in the mature protein. These N-terminal residues comprise one or more targeting sequences that direct the protein to its proper destination. The precursor proteins can be synthesized in cell-free systems in the absence of mitochondria. When mitochondria are subsequently added, the precursors are incorporated into the organelle and the N-terminal uptake-targeting sequences are removed (Figure 19-6).

Figure 19-7 presents an overview of the import of proteins from the cytosol into the mitochondrial matrix, the route into the mitochondrion followed by most imported proteins. In the cytosol, precursor proteins are kept in an unfolded or partly folded state by cytosolic chaperone proteins. After binding to receptors on the outer mitochon-

drial membrane, the precursors are transported through the outer and inner membranes via protein-lined channels, in a process that requires a proton-motive force across the inner membrane as well as hydrolysis of ATP. Chaperones in the mitochondrial matrix facilitate the import of the precursors and assist in their folding. We will refer often to Figure 19-7 as we discuss the details of each step in protein transport into the matrix. Then we will consider how proteins are targeted to other compartments of the mitochondrion.

Matrix-Targeting Sequences Direct Imported Proteins to the Mitochondrial Matrix

The precursors of mitochondrial proteins (including hydrophobic integral membrane proteins) are soluble in the cytosol. It is there that they bind to one or more *chaperone*

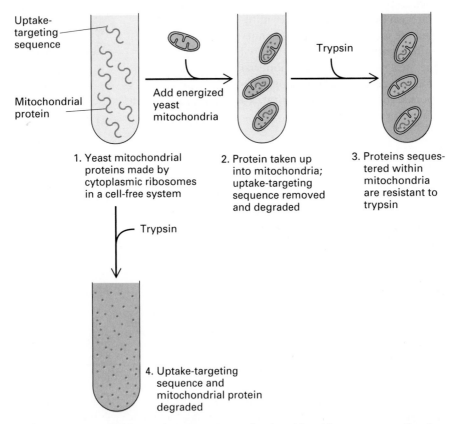

▲ FIGURE 19-6   Uptake of mitochondrial protein precursors. Mitochondrial proteins are imported into the organelle post-translationally, with cleavage of the N-terminal uptake-targeting sequence. Most mitochondrial proteins synthesized on cytosolic ribosomes are in a precursor form with an uptake-targeting sequence at their N-terminus that is not found on the mature protein. Such precursors can be identified in cells following a brief pulse of radioactive amino acid, or they can be synthesized in a cell-free system programmed with cytosolic mRNA, as depicted in step 1. When respiring

mitochondria with a proton-motive force across the inner membrane are added, the protein is taken up into the organelle and the uptake-targeting sequence is removed by a protease in the matrix (step 2). Protein uptake can be demonstrated by adding trypsin or another protease to the reaction medium. Proteins sequestered in the mitochondrion are resistant to the protease (step 3) because it cannot penetrate the mitochondrial membranes. In contrast, the precursor protein in the cytosol is totally destroyed by the protease (step 4).

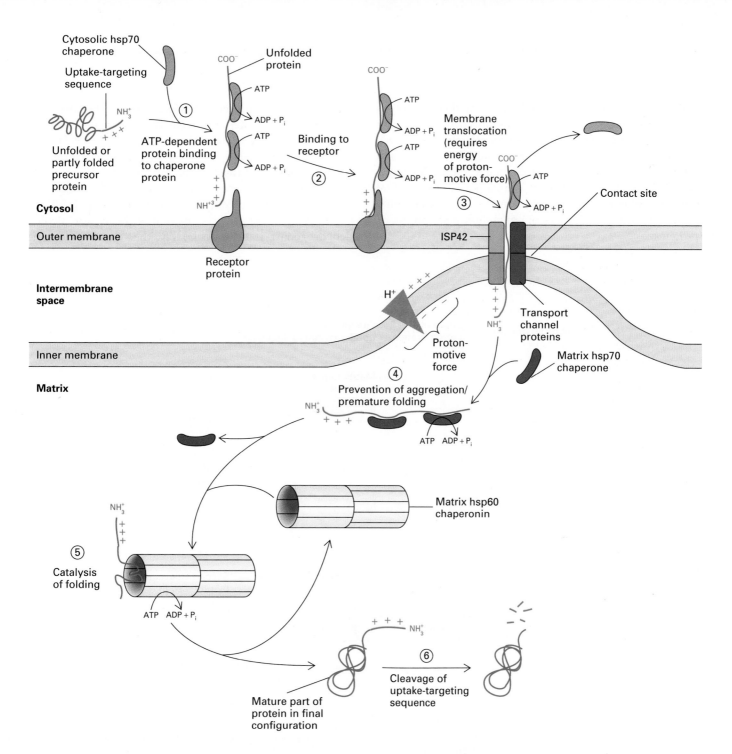

▲ FIGURE 19-7 The import of a polypeptide into the matrix of a mitochondrion. The precursor protein, with its N-terminal matrix-targeting sequence, is synthesized in the cytosol. There (step 1) it binds to a cytosolic hsp70 chaperone protein (similar in structure and function to the endoplasmic reticulum chaperone, Bip), which uses energy released by ATP hydrolysis to keep the precursor unfolded. Step 2: The precursor-chaperone complex binds to one or more receptors on the outer membrane near a site of contact with the inner membrane. Step 3: The protein is then translocated across the outer and inner membranes. through a channel lined with several different types of proteins; one of these, ISP42 (42,000 MW), actually forms part of the import channel. This process requires a proton-motive force (pmf) across the inner membrane; the pmf is a combination of a membrane electric potential and a pH gradient. Note that translo-cation occurs at rare sites at which the inner and outer membranes appear to touch. Step 4: The newly imported protein binds to the matrix chaperone hsp70 (another Bip-like protein), which uses the energy of ATP hydrolysis to assist import into the matrix and to prevent aggregation or premature folding. Step 5: After its release from hsp70, the protein binds to one end of the 14-subunit chaperonin hsp60. Apparently, while the precursor is bound to hsp60, energy released by the hydrolysis of ATP assists it to fold into its final, active configuration. Step 6: The uptake-targeting sequence, having served its function, is removed by a matrix protease. Some imported proteins remain in the matrix space; others insert into the inner membrane or are transported into the intermembrane space. [See K. Baker and G. Schatz, 1991, *Nature* **349**:205–208; N. Pfanner et al., 1992, *Cell* **68**:999–1002.]

proteins, which use the energy released by ATP hydrolysis to keep the precursor proteins in an unfolded state so that they can be taken up by mitochondria (Figure 19-7). One such chaperone, cytosolic hsp70, is similar in sequence and function to Bip, the endoplasmic reticulum (ER) chaperone (see Figure 16-16). Another chaperone, mitochondrial import stimulation factor, binds to the matrix-targeting sequences on a number of mitochondrial precursor proteins.

Most matrix proteins (for example, alcohol dehydrogenase) are fabricated as precursors that have a matrix-targeting sequence at their N-terminus (Figure 19-8). This sequence contains all the information required to target the protein from the cytosol to the mitochondrial matrix. Indeed, this sequence is sufficient to direct a normally cytosolic protein, such as dihydrofolate reductase (DHFR), to the matrix: if the matrix-targeting sequence for alcohol dehydrogenase is fused to DHFR, the resulting chimeric

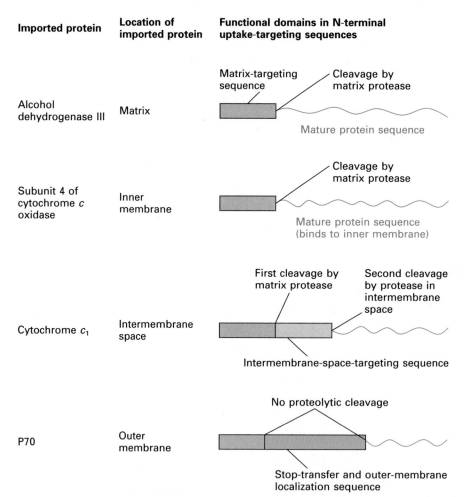

▲ FIGURE 19-8   Uptake-targeting sequences of imported mitochondrial proteins. One or more N-terminal uptake-targeting sequences direct most imported mitochondrial proteins first to the matrix and then to the correct mitochondrial subcompartment. Proteins targeted to the matrix or inner membrane have a single matrix-targeting sequence (red). Following translocation of the protein to the matrix and cleavage of the sequence, the protein either remains in the matrix (e.g., alcohol dehydrogenase) or inserts into the inner membrane (e.g., cytochrome c oxidase subunit 4) by binding to the other subunits of this multiprotein oxidase complex. Proteins such as cytochrome c_1, which are destined for the intermembrane space, carry two sequences: (1) a matrix-targeting sequence (red) that directs the N-terminus of the protein to the matrix and is cleaved, and (2) an intermembrane-space-targeting sequence of hydrophobic amino acids (yellow) that redirects the protein to the intermembrane space, where it is cleaved. As detailed in Figure 19-12, the entire protein may not pass into the matrix space. P70 and other outer-membrane proteins have a typical matrix-targeting sequence (red), followed by a stop-transfer sequence—a long sequence of non-polar amino acids (green). Neither of these sequences is cleaved. The latter sequence appears to function in stopping the transport of P70 across the outer membrane and in anchoring the protein to that membrane. [After E. C. Hurt and A. P. G. M. van Loon, 1986, *Trends Biochem. Sci.* **11**:204–207; F. U. Hartl et al., 1989, *Biochem. Biophys. Acta* **988**:1–45.]

protein is transported to the matrix, and the sequence is cleaved normally (see Figure 19-10a). Conversely, deleting or mutating its matrix-targeting sequence causes a protein normally transported to the matrix to remain in the cytosol.

Proteins destined for the inner mitochondrial membrane include subunits of multiprotein complexes such as the electron transport chain and F_0F_1 ATPase. These proteins are also directed first to the matrix by a matrix-targeting sequence at their N-terminus. They then insert into the inner membrane by binding to other subunits of the multiprotein complex, but the details of how this happens are not known. During their mitochondrial uptake, the precursors of many integral inner-membrane proteins also undergo a major conformational change in order to bind to the hydrophobic core of the membrane. This change may be due in part to the removal of all or a portion of the N-terminal uptake-targeting sequence in the matrix. As detailed later, these conformational changes are facilitated by at least two chaperones in the matrix.

As Table 19-1 makes clear, the proteins that travel from the cytosol to the same organellar destination all have targeting signals that share common motifs, although the signal sequences are generally not identical. Thus, the receptors that recognize such sequences, including the mitochondrial outer-membrane receptors that bind matrix-targeting sequences, are able to bind to a number of different but related sequences. The matrix-targeting sequences of the various mitochondrial proteins are rich in positively charged amino acids—arginine and lysine—and hydroxylated ones—serine and threonine; they also are devoid of the acidic residues aspartate and glutamate. One or more receptors for matrix-targeting sequences on the outer mitochondrial membrane can recognize this large number of related amino acid sequences.

Most proteins imported into the mitochondrion do not have cytosolic counterparts. Some enzymes, such as aminoacyl tRNA synthetases, are found both in the cytosol and in the mitochondrial matrix, but most cytosolic and mitochondrial tRNA synthetases are encoded by different nuclear genes. Occasionally, however, a single gene encodes both the mitochondrial and cytosolic forms of an amino acid tRNA synthetase (Figure 19-9). In this case, two promoters generate two mRNAs that differ at their 5′ ends.

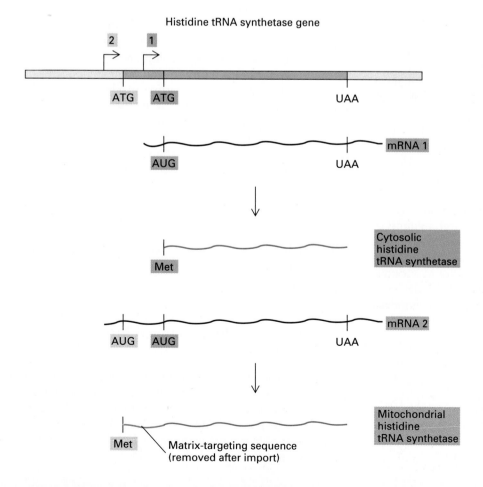

◄ FIGURE 19-9 How a single yeast gene encodes both a cytosolic and a mitochondrial histidine tRNA synthetase. The gene has two in-frame translation start sites, labeled ATG, located 60 base pairs apart. The predominant gene transcript begins from promoter 1 (short mRNA) and initiates translation at the second translation start site, generating the cytosolic form of the synthetase. The minor mRNA begins at promoter 2, and initiates translation at the upstream translation start site, generating a protein with 20 amino acids at its N-terminus that constitute a matrix-targeting sequence. This sequence is removed after this form of the protein is imported into the mitochondria. [Adapted from G. Natsoulis, F. Hilger, and G. Fink, 1986, *Cell* **46**:235–243].

One mRNA encodes the cytosolic protein; the other encodes the same protein, but with a matrix-targeting sequence at its N-terminus that directs it into the mitochondrial matrix.

A matrix protease removes all N-terminal matrix-targeting sequences soon after they arrive in the matrix. During in vitro reactions, this enzyme, a two-subunit, metal-containing protease, specifically cleaves the N-terminal matrix-targeting sequence from several different precursor proteins.

Mitochondrial Receptors Bind Matrix-Targeting Sequences

The import of a protein into the mitochondrial matrix involves two discrete steps: binding to receptors on the outer mitochondrial membrane (Figure 19-7, step 2) and then transport across both the outer and inner membrane (Figure 19-7, step 3). An energized mitochondrion is not required for binding of the matrix-targeting sequence to the receptor but is required for the actual transport. As evidence for this point, when the proton-motive force across the inner mitochondrial membrane is abolished by uncoupling agents, the precursors of the imported proteins still bind tightly to specific receptors on the outer membrane, but they are not transported into the organelle. If the energy block is removed, the bound precursors are rapidly imported and processed.

The outer mitochondrial membrane contains one receptor that binds the matrix-targeting sequence on most imported proteins. In one experiment that showed that this receptor binds many matrix-targeting sequences, the binding of any one precursor protein with a typical matrix-targeting signal was inhibited by the presence of much larger amounts of most other precursors with similar sequences. However, a different receptor is utilized for proteins that lack a typical matrix-targeting signal and that do not undergo proteolytic cleavage during their import into mitochondria. One such protein is the inner-membrane ATP-ADP antiport (see Figure 17-20). Binding of the ATP-ADP antiport to an outer-membrane receptor and its uptake into the organelle are not blocked by large amounts of a precursor protein with a typical matrix-targeting signal. Indeed, the receptor for the ATP-ADP antiport has been identified and cloned. Deletion of this 70,000-MW outer-membrane protein blocks the import of the ATP-ADP antiport but not the import of most other mitochondrial proteins.

Intermediates in Translocation of Proteins into the Mitochondrion Can Be Accumulated and Studied

The experiment depicted in Figure 19-10 makes two crucial points: (1) proteins can be imported into mitochondria only if they are unfolded, and (2) the translocation of precursors to the matrix occurs at the rare sites where the outer and inner membranes are close together.

For the experiment, a chimeric protein was generated; it consisted of an N-terminal matrix-targeting sequence, a spacer sequence of ≈ 50 amino acids, and then, at the C-terminus, the complete sequence of the normally cytosolic enzyme dihydrofolate reductase (DHFR). Normally, chaperones prevent the C-terminal DHFR segment from folding in the cytosol, and the chimera is transported into the mitochondrial matrix, where the matrix-uptake sequence is cleaved. The drug methotrexate, an inhibitor of DHFR, binds in the cytosol to the active site on the DHFR segment of the chimera, causing it to become locked in a folded state and thus unable to cross into the matrix. However, the N-terminal matrix-targeting sequence of the chimera still can enter the matrix space, where it is cleaved, leaving the rest of the chimera stuck in the membrane as a stable *translocation intermediate*.

This experiment provided further evidence that translocated proteins traverse the membrane in an unfolded state: the spacer sequence of 50 amino acids in the stuck chimera clearly spans both membranes, and is long enough to do so only if it is in an extended conformation, stretched to its maximum possible length. If the chimera contained a shorter spacer—say 35 amino acids—no stable translocation intermediate was obtained, because the spacer could not traverse both membranes.

That precursor proteins enter the mitochondrion at sites where the inner and outer membranes are close together was shown by microscopic studies in which the precursors were seen to accumulate at the points of contact between the inner and outer membranes (see Figure 19-10c). Outer-membrane receptors and other components of the mitochondrial protein-importing machinery including ISP42 (Figure 19-7) are believed to be localized at or near these contact sites. Proteins are thought to traverse both the outer and inner mitochondrial membranes in protein-lined channels, since the stable translocation intermediates can be chemically cross-linked (see Figure 20-8) to integral proteins of both the outer and inner membranes. These channel-lining proteins are now being cloned and characterized. Since a typical yeast mitochondrion can contain ≈ 1000 stuck translocation intermediates, it is thought that mitochondria have ≈ 1000 transport channels for the uptake of mitochondrial proteins.

The Uptake of Mitochondrial Proteins Requires Energy

What is the source of the energy that drives protein import into the mitochondrion? Studies on mitochondrial proteins in cell-free systems have shown that three separate inputs of energy are required: ATP hydrolysis in the cytosol, a proton-motive force across the inner membrane, and ATP hydrolysis in the mitochondrial matrix.

(a) Chimeric protein

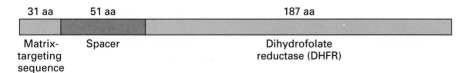

31 aa   51 aa   187 aa

Matrix-targeting sequence   Spacer   Dihydrofolate reductase (DHFR)

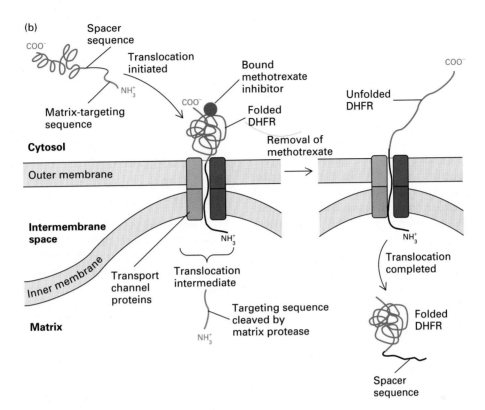

(b)

Spacer sequence

Translocation initiated

Bound methotrexate inhibitor

Folded DHFR

Matrix-targeting sequence

COO⁻

Cytosol

Outer membrane

Removal of methotrexate

Unfolded DHFR

Intermembrane space

Inner membrane

Transport channel proteins

Translocation intermediate

Targeting sequence cleaved by matrix protease

Matrix

Translocation completed

Folded DHFR

Spacer sequence

▲ FIGURE 19-10 Demonstration that mitochondrial proteins are imported in an unfolded state and that they are translocated at contact sites between the inner and outer membranes. (a) The chimeric protein used in the experiment was generated by recombinant DNA. It contained at its N-terminus a matrix-targeting signal, followed by a spacer, ≈50 amino acids long, of no particular function, and then by a complete enzyme, dihydrofolate reductase (DHFR), which is normally cytosolic. After synthesis in a cell-free system, the chimeric protein can be transported to the matrix of energized mitochondria, where the matrix-targeting signal is removed. (b) How a translocation intermediate is accumulated. Methotrexate is an inhibitor of DHFR that binds tightly to the active site DHFR and locks it in the folded state. It prevents translocation of the folded DHFR at the C-terminus of the chimera, which remains on the cytosolic side of the mitochondrion. Thus in the presence of methotrexate the N-terminus of the chimeric protein is translocated across the inner and outer membranes, and the N-terminal matrix-targeting signal is cleaved by the matrix protease, while the folded C-terminus remains in the cytosol. When methotrexate is removed, the DHFR on the cytosolic surface can unfold sufficiently so that the entire chimera is translocated into the matrix space where the DHFR segment again folds into a functional enzyme. (c) Accumulation of matrix-protein precursors at sites where inner and outer membranes are in close contact. The C-terminus of the translocation intermediate in (b) can be detected by incubating the mitochondria with antibodies that bind to the DHFR segment, followed by gold particles coated with a protein (protein A) that binds to the antibody. After sectioning and visualization under the electron microscope, points of close contact between the inner and

(c)

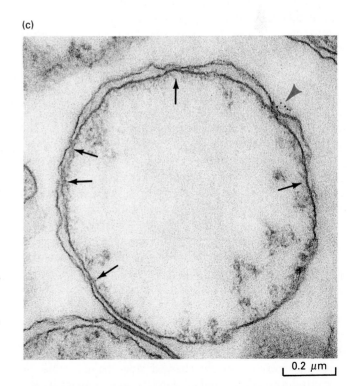

0.2 μm

outer membranes are evident (arrows); some contact sites contain gold particles (arrowhead) bound to the translocation intermediate. [Parts (a) and (b) modified from J. Rassow et al., 1990 *FEBS Letters* **275**:190–194; (c) From M. Schweiger, V. Herzog, and W. Neupert, 1987, *J. Cell Biol.* **105**:235–246; courtesy of W. Neupert.]

ATP hydrolysis in the cytosol is required to keep precursor proteins in an unfolded or partially folded state, so that they can interact with the mitochondrial uptake machinery (see Figure 19-7). As noted earlier, several different cytosolic chaperone proteins have been identified—some related in sequence and function to the ER Bip protein (Figure 16-16)—that bind to precursor mitochondrial proteins and couple the energy released by ATP hydrolysis to the maintenance of the bound proteins in an unfolded state.

The only role of cytosolic ATP in mitochondrial protein transport is to keep the precursor proteins in an unfolded state. One study that demonstrated this used an experimental protocol similar to that in Figure 19-6. A precursor protein was purified and then denatured by urea. When added to a mixture of yeast cytosol and energized mitochondria, the denatured protein was incorporated into the matrix in the absence of cytosolic ATP. In contrast, import of the native, undenatured precursor required ATP for the normal unfolding function of cytosolic chaperones. In a second study, recombinant DNA techniques were used to construct mutant versions of certain matrix or inner-membrane proteins that were unable to fold normally into a functional conformation. These mutant proteins could be incorporated into energized mitochondria without the presence of ATP in the cytosol, presumably because a chaperone was unnecessary to prevent the mutant proteins from folding.

However, whenever a protein is imported from the cytosol into the mitochondrial matrix, a proton-motive force across the inner membrane is always required. To demonstrate this, mitochondria were "poisoned" with inhibitors or uncouplers of oxidative phosphorylation, such as cyanide or dinitrophenol, which dissipated the proton-motive force across the inner membrane. Receptors on the debilitated mitochondria bound precursor proteins (step 2, Figure 19-7), but the proteins could not be imported (step 3), either in the intact cell or in cell-free reactions, even in the presence of ATP and chaperone proteins.

Exactly how the membrane electric potential is used to "pull" a receptor-bound precursor protein into the matrix (see Figure 19-7, step 3) is not clear. Once a protein is partially inserted into the inner membrane, it becomes subjected to a transmembrane potential of 200 mV (matrix space negative), which is equivalent to an electric gradient of about 400,000 V/cm. The electric potential then may alter the conformation of these proteins, in much the same way that a membrane electric potential affects the conformation of voltage-dependent ion channels in nerve cells. The change in protein folding could pull the precursor protein across the energized inner membrane. A related possibility is that the N-terminal matrix-targeting sequence, with its many positively charged side chains, could be "electrophoresed," or pulled, into the matrix space by the inside negative membrane electric potential.

In summary, the outer mitochondrial membrane contains multiple specific receptors for the binding and uptake of different proteins. These receptors recognize sequences that target the proteins to their proper destinations in the organelle. Protein uptake usually requires an expenditure of energy in the form of both cytosolic ATP and a proton-motive force. The proteins are kept in an unfolded conformation to allow their translocation, through protein-lined channels, across the outer and inner membranes.

Matrix Chaperones Are Essential for the Import and Folding of Mitochondrial Proteins

Most imported proteins do not fold spontaneously in the matrix but require protein-folding catalysts. All newly imported proteins first bind to the ATP-hydrolyzing *matrix hsp70 protein*, a chaperone similar in structure and function to the cytosolic hsp70 chaperone and to the ER protein Bip. The matrix hsp70 chaperone is essential for the import of precursor proteins. Using the energy released by ATP hydrolysis, it binds the unfolded form of the protein within the matrix, preventing protein aggregation or precipitation and premature folding (Figure 19-7, step 4). This is particularly important for subunits of multiprotein complexes, since the proper folding of any one subunit requires the presence of all of the protein's subunits.

Usually matrix hsp70 hands over an unfolded protein to a second mitochondrial matrix chaperone, *hsp60*, for final folding (Figure 19-7, step 5). Hsp60 is essential for many monomeric and multimeric proteins to fold properly. Evidence for its importance in protein folding comes from yeast mutants defective in hsp60. In these mutants, proteins such as the β subunit of the F_1 ATPase are imported normally into the matrix, and cleavage of the uptake-targeting sequence is also normal, but the imported polypeptide fails to assemble into a normal multiprotein complex.

Hsp60 is not homologous to any cytosolic or ER chaperones, but is related to a class of bacterial chaperones called GroEL. The hsp60 class of matrix chaperones, often called *chaperonins*, all consist of 14 identical 60,000-MW subunits that are arranged in two stacked rings of 7 subunits each. In electron micrographs these look like a double doughnut with a central cavity ≈ 6 nm in diameter (Figure 19-11). At one end of the double doughnut is a particle called the *co-chaperonin*, called GroES in bacteria, containing seven copies of a 10,000-MW protein; in the cavity at the other end one molecule of the partially folded precursor protein to be folded can be bound.

Initially, all or a part of the partially folded protein binds to the wall of the cavity within the hsp60 chaperonin. The cavity provides a confined space that excludes

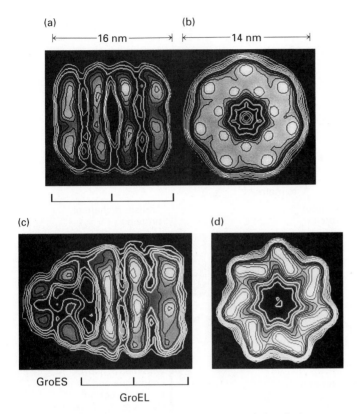

(a) |←——— 16 nm ———→| (b) |←——— 14 nm ———→|

(c)   (d)

GroES

GroEL

▲ FIGURE 19-11   Computer reconstructions of electron micrographs of the bacterial hsp60 chaperonin GroEL, a molecule similar in structure and function to the mitochondrial hsp60 chaperonin. Panel (a) shows a side view and (b) an end view of the symmetric "double-doughnut" containing two rings [brackets in (a)], each with seven subunits, seen best in panel (b). Panel (c) shows a side view and (d) an end view of the functional chaperonin, with one copy of a GroES co-chaperonin, consisting of seven subunits of identical 10,000-MW subunits, attached to one end of the hsp60 complex. In cell-free reactions the hsp60 complex enhances many-fold the rate of refolding of a denatured protein by binding it in the cavity, seen in panel (d), at the end opposite to the end occupied by the co-chaperonin GroES. [From T. Langer et al., 1992, *EMBO J.* **11**:4757–4765. See also K. Braig et al., 1993, *Proc. Natl. Acad. Sci. USA* **90**:3978–3982.]

many matrix proteins that might otherwise bind to a precursor and prevent it from folding. Fourteen molecules of ATP bind to the chaperonin and are then hydrolyzed simultaneously—one by each of the 14 hsp60 subunits—and the precursor is released from the cavity wall and allowed to fold. A properly folded protein then leaves the cavity; one that does not fold is again bound to the wall of the cavity, whereupon the cycle of ATP binding and hydrolysis is repeated. Binding of the 10,000-MW co-chaperonin at the other end of the doughnut also acts to release the bound precursor protein from the chaperonin, and conversely, binding of a partially folded protein causes the release of the co-chaperonin from the other end.

Thus, like Bip, the related hsp70 chaperones utilize the energy released by ATP hydrolysis mainly to prevent the misfolding or aggregation of precursor proteins. The hsp60 chaperonins utilize the energy released by ATP hydrolysis to enable proteins to achieve their mature, folded state.

Proteins Are Targeted to the Correct Submitochondrial Compartment by Multiple Signals and Several Pathways

How are proteins targeted to mitochondrial compartments other than the matrix? Several processes are involved, as we shall now see. Precursors to such intermembrane-space proteins as cytochromes c_1, a subunit of $CoQH_2$ cytochrome c reductase, and b_2, carry two different N-terminal uptake-targeting sequences that get them to the intermembrane space (see Figure 19-8). First, the most N-terminal of the two sequences directs the precursor to the matrix (Figure 19-12), where that sequence is removed by the matrix protease. Exactly what happens next is not yet clear. According to the "conservative sorting mechanism," the entire protein goes into the matrix, and second uptake-targeting sequence directs the precursor, presumably bound to matrix hsp70, across the inner membrane to the intermembrane space. According to the "nonconservative" model, the second uptake-targeting sequence acts as a stop-transfer sequence (see Figure 16-18) that serves to block translocation of the C-terminus of the precursor through the inner membrane. The membrane-anchored intermediate then diffuses, within the inner membrane, away from the translocation site. Finally, cleavage of the second uptake-targeting sequence by a protease in the intermembrane space releases the protein into that space.

Several experiments indicate the importance of the second uptake-targeting sequence. For instance, recombinant DNA techniques allow a precursor cytochrome b_2 protein to be synthesized that lacks only the intermembrane-space-targeting sequence (bright yellow in Figures 19-8 and 19-12); in its absence, the protein accumulates in the matrix space, not in the intermembrane space. This result is consistent with either of the models for sorting proteins to the intermembrane space, since according to either model the intermembrane-space-targeting sequence, if present, would divert the precursor from a destination in the matrix to the intermembrane space.

Cytochrome c is imported directly from the cytosol to the intermembrane space, and utilizes a specific uptake system different from that for other proteins targeted to the intermembrane space. The cytosolic form of cytochrome c, called *apocytochrome c,* has the same amino acid sequence as the mature protein, has no matrix-targeting signal, and lacks the covalently bound heme group found in the mature protein. Apocytochrome c, but not the mature *holo*

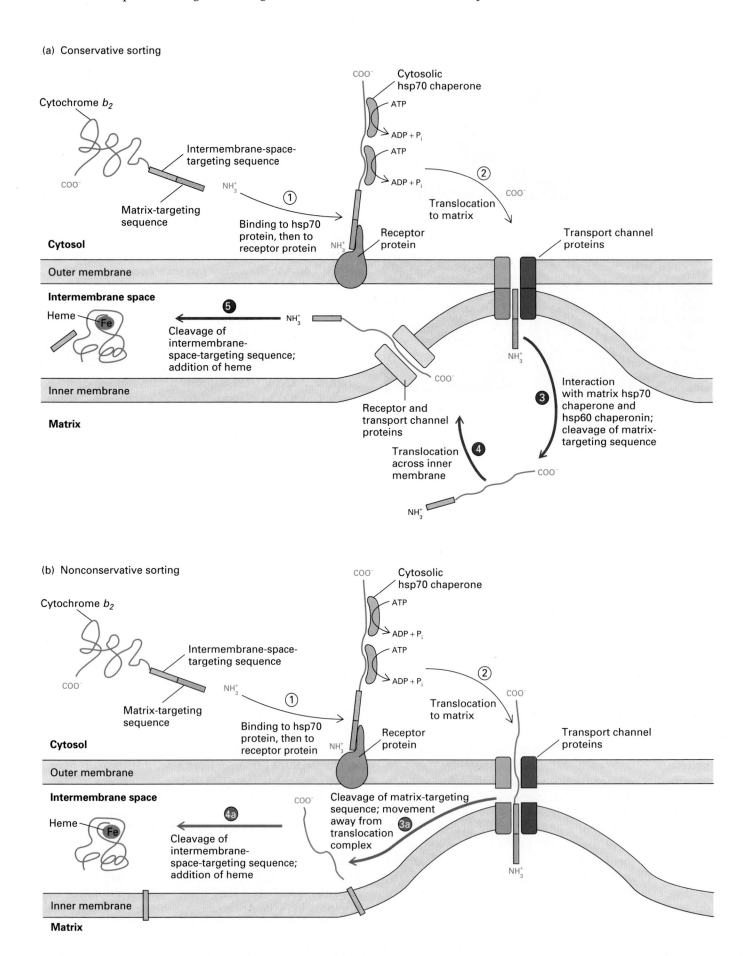

(a) Conservative sorting

(b) Nonconservative sorting

◄ FIGURE 19-12   Transport of proteins from the cytosol to the mitochondrial intermembrane space. There are two proposed models for this process, both of which utilize, but in different ways, the same uptake-targeting sequences at their N-termini. (a) In the first, "conservative sorting," two successive translocations are required. First (steps 1 and 2), a matrix-targeting sequence (red), functions to target the protein (here, cytochrome b_2) to the mitochondrial matrix, exactly as if it were a typical mitochondrial matrix protein (see Figure 19-7). The matrix-targeting sequence is cleaved by the matrix protease (step 3) and the protein remains unfolded, presumably bound to matrix hsp70 (not shown here). Then (step 4) the intermembrane-space-targeting sequence (yellow), targets the protein to the inner membrane, presumably by binding to a receptor on the matrix side of the inner membrane, and initiates translocation of the protein across the inner membrane through a protein-lined channel and into the intermembrane space. (Neither the receptor nor the channel proteins have been characterized.) In the intermembrane space (step 5), the sequence is cleaved by a specific protease that is related to the ER signal peptidase, and heme is added, enabling the cytochrome to fold into its mature configuration.

(b) In the second, "nonconservative," model, the matrix-targeting sequence initiates translocation through both the outer and inner membranes, but the hydrophobic intermembrane-space-targeting sequence functions as a stop-transfer sequence: it remains as an anchor in the inner membrane and prevents translocation of the C-terminus of the protein through it. The protein, anchored in the membrane by the intermembrane-space targeting sequence, apparently causes disassembly of the translocation channel, the rest of the protein moves across the outer membrane into the inter-membrane space, and the protein diffuses, within the inner membrane, away from the translocation site (step 3a). Finally the intermembrane-space-targeting sequence is cleaved by a specific protease, heme is added, and the cytochrome folds into its mature configuration (step 4a).

[Evidence for and against these two models is summarized in B. S. Glick, E. Beasley, and G. Schatz, 1992, *Trends Biochem. Sci.* **17**:453–459 and H. Koll, et al., 1992, *Cell* **68**:1163–1175.]

form, can diffuse freely through the outer mitochondrial membrane, probably by traveling through P70, an abundant mitochondrial outer-membrane protein. Mitochondrial P70 forms channels through the phospholipid bilayer and accounts for the unusual permeability of the outer membrane to small proteins. Once apocytochrome c is in the intermembrane space of the mitochondrion, the heme group is added by the enzyme *cytochrome* c *synthetase* (also called *cytochrome* c *heme lyase*). The addition of heme causes the conformation of the protein to change so that it cannot diffuse through the outer membrane, thus "locking" it into the intermembrane space.

Proteins in the outer mitochondrial membrane are also synthesized on cytosolic ribosomes. How are they targeted

to the outer mitochondrial membrane? The N-terminus of P70, a well-studied outer membrane protein, contains a short matrix-targeting sequence followed by a long stretch of hydrophobic amino acids (green in Figure 19-8); the latter sequence causes the protein to divert from the matrix pathway and accumulate in the outer mitochondrial membrane. Normally, neither sequence is removed. If the hydrophobic sequence is experimentally deleted from P70, the protein accumulates in the matrix space with its matrix-targeting sequence still attached. This suggests that the long hydrophobic sequence functions as a stop-transfer sequence in the outer membrane, preventing the transfer of the protein into the matrix and anchoring it as an integral protein in the outer-membrane.

Certain Mitochondrial Proteins Are Essential for Life

In yeasts, of all the nuclear-encoded mitochondrial proteins only five are essential for life, in that deletion of any of their five genes results in nonviable cells. All five are involved in the import of proteins from the cytosol: two are subunits of the matrix protease that removes matrix-uptake sequences; two are chaperones—the matrix hsp70 and hsp60 (chaperonin) proteins; and one, ISP42, is an outer-membrane protein that is essential for import across the outer membrane; it appears to be a part of the channel through which imported proteins move (see Figure 19-7, step 3). Cytosolic hsp70, though not a mitochondrial protein, is also involved in the import of proteins into the mitochondrial matrix, and it also is essential for life.

These vital proteins are required for the import of nuclear-encoded enzymes into the matrix, such as enzymes that catalyze the biosynthesis of amino acids, nucleosides, and other molecules essential for yeast cell growth. For example, when yeasts grow anaerobically (without oxygen) their mitochondria lack enzymes required for oxidative phosphorylation. However, these mitochondria must still contain enzymes essential for the biosynthesis of amino acids and other molecules. Many of these enzymes cannot function in the cytosol, possibly because chaperonins essential for their folding are found only in the mitochondrial matrix space, and possibly because many of their substrates are generated only in the mitochondrion.

The Synthesis of Mitochondrial Proteins Is Coordinated

As noted earlier, all mitochondrially synthesized proteins are parts of multienzyme complexes, such as cytochrome c oxidase and the F_0F_1 ATPase, that also contain subunits synthesized in the cytosol. Since all of the subunits must be fabricated in appropriate ratios, the close coordination of nuclear and mitochondrial genome expression is required for the assembly of a mitochondrion. As yet, little is known about how the expression of these two genomes is coordinated in animals or plants. This topic has been studied mainly in yeasts because yeast mtDNA can be readily deleted. Petite yeast strains that have no functional mtDNA do contain normal amounts of all mitochondrial proteins encoded by the nucleus, such as cytochrome c and the F_1 ATPase. Thus in yeasts, mitochondrial gene products are not essential for the expression of nuclear genes. Since mtDNA cannot be readily deleted in other organisms, it is impossible to say whether this result is true in all cells.

Yeasts and other eukaryotic microorganisms that can grow anaerobically or aerobically—in the absence or presence of O_2 respectively—provide a striking example of the coordination of nuclear and mitochondrial genes. When grown anaerobically with glucose as a carbon source, these organisms generate ATP solely by the Embden-Meyerhoff pathway (see Figures 17-2 and 17-7). Anaerobically grown yeasts lack a complete respiratory chain, and do not contain cytochromes such as a, a_3, b, or c_1, some of whose subunits are encoded by mitochondrial DNA and some by nuclear DNA. The synthesis of heme is also very low. When viewed under an electron microscope, anaerobically grown cells lack typical mitochondria, although they do contain some petite mitochondria (small organelles with inner and outer membranes but no cristae). These cells also lack enzymes essential to the citric acid cycle (all of which are encoded by nuclear DNA) and the F_0F_1 ATPase complex. The levels of cytosolic mRNAs for proteins such as cytochrome c, subunits of the cytochrome c oxidase complex, and the F_1 ATPase are as much as 100 times lower in anaerobic cells than they are in aerobically grown cells. In anaerobic cells, however, mtDNA is replicated normally, so the organelles must contain sufficient amounts of the imported mitochondrial DNA polymerase.

The addition of oxygen to anaerobic yeast results in induction of synthesis of all mitochondrial components encoded by the nuclear and mitochondrial genomes; moreover, the petite mitochondria expand into true mitochondria with cristae. Heme synthesis is directly activated by oxygen. Heme, in turn, (not O_2 itself) is required for the transcription of the nuclear genes for cytochrome c and other mitochondrial proteins. When heme is abundant, certain transcriptional regulatory proteins are activated and bind to specific enhancer sequences in the genes for cytochrome c and other nuclear-encoded mitochondrial proteins, enhancing their transcription. Since the synthesis of heme requires oxygen, this process may be representative of a general mechanism of down-regulating the expression of multiple nuclear genes that encode mitochondrial proteins during anaerobic fermentation.

Heme also regulates its own synthesis by a novel method: blocking the import into the mitochondrion of a key enzyme in heme biosynthesis. The first enzyme in the pathway of heme biosynthesis, δ-aminolevulinic acid synthase, is localized to the mitochondrial matrix, but it is synthesized in the cytosol and subsequently imported. If excess heme is present in the cell, it binds to newly made δ-aminolevulinic acid synthase in the cytosol and prevents its incorporation into mitochondria. This eventually reduces the rate of heme biosynthesis, thereby inhibiting the transcription of heme-dependent genes.

► Chloroplast DNA and the Biogenesis of Chloroplasts and Other Plastids

The biogenesis of a chloroplast is similar in many respects to that of a mitochondrion, although many key steps are not understood in molecular detail. Some chloroplast proteins are encoded by chloroplast DNA (Table 19-5) and are translated by chloroplast ribosomes in the organelle using the "standard" genetic code. Others are fabricated on cytosolic ribosomes and are incorporated into the organelle after translation; these imported proteins are synthesized with N-terminal uptake-targeting sequences that direct each protein to its correct subcompartment and are subsequently cleaved (Table 19-1). Chloroplasts, like mitochondria, grow by expansion and then fission. This process can be observed easily in unicellular algae, such as *Chlamydomonas*, that contain a single large chloroplast (Figure 19-13).

Chloroplast DNA Contains over 120 Different Genes

Chloroplast DNAs are circular molecules of 120,000–160,000 bp, depending on the species. The complete sequences of several chloroplast DNAs have been determined, including liverwort (121,024 bp) and tobacco (155,844 bp). The liverwort chloroplast genome has two inverted repeats, each consisting of 10,058 bp, that contain the rRNA genes and a few other duplicated genes. These repeats are separated by two single-copy sequences, one small (19,813 bp) and one large, that contain the bulk of the tRNA and protein-coding genes listed in Table 19-5. Despite the difference in size, the overall organization and gene composition of the liverwort and tobacco DNAs are very similar; the size differential is due primarily to the length of the inverted repeat in which some genes are duplicated.

Of the $\approx$120 genes in chloroplast DNA, about 60 are involved in RNA transcription and translation, including genes for rRNAs, tRNAs, RNA polymerase subunits, and ribosomal proteins (see Table 19-5). About 20 genes encode subunits of the chloroplast photosynthetic electron transport complexes and the F_0F_1 ATPase complex; also encoded is the larger of the two subunits of ribulose 1,5-bisphosphate carboxylase.

Reflecting the endosymbiotic origin of chloroplasts, some regions of chloroplast DNA are strikingly similar to those of the DNA of present-day bacteria. Chloroplast

TABLE 19-5   Genes Encoded by Chloroplast DNA from a Liverwort,
Marchantia Polymorpha

Genes	Description of Gene Product
rRNA (duplicated in inverted repeats IR_A and IR_B)	16S, 23S, 4.5S, and 5S RNAs
tRNA	37 tRNAs
RNA polymerase	Homologous to *E. coli* RNA polymerase:
rpo A	subunit α
rpo B	subunit β
rpo C_1	subunit β'
rpo C_2	subunit β'
Ribosomal proteins	
rpl	50S subunit: 8 proteins
rps	30S subunit: 11 proteins
Proteins essential for photosynthesis	
rbcL	Large subunit of ribulose 1,5-bisphosphate carboxylase
psaA, psaB	Chlorophyll *a*–binding proteins in photosystem I
psbA	Photosystem II D1 protein
psbB, psbC	Chlorophyll *a*–binding proteins in photosystem II
psbD	Photosystem II D2 protein
psbE, psbF	Cytochrome b_{559}
psbG	Photosystem II G protein
atpA, atpB, *atpE*	Subunits α, β, and ϵ of F_1 ATPase
atpF, atpH, *atpI*	Subunits 1, 3, and 4 of F_0 ATPase
petA	Cytochrome *f*
petB	Cytochrome b_6
petD	Subunit 4 of cytochrome *b/f* complex
Genes predicted by amino acid sequence homology	
ndh1, ndh2, *ndh3, ndh4,* *ndh4L, ndh5,* *ndh6*	Homologous to subunits of the human mitochondrial NADH-CoQ reductase complex
frxA, frxB, frxC	Homologous to ferredoxin
Others	More than 28 unidentified open reading frames

SOURCE: K. Ohyama et al., 1986, *Nature* **322**:572–575.

DNA encodes four subunits of RNA polymerase that are highly homologous to subunits α, β, and β' of *Escherichia coli* RNA polymerase. One segment of chloroplast DNA encodes eight proteins that are homologous to eight *E. coli* ribosomal proteins; the order of these genes is the same in the two DNAs.

Liverwort chloroplast DNA has some genes that are not detected in the larger tobacco chloroplast DNA, and vice versa. Since the two types of chloroplasts contain virtually the same set of proteins, these data suggest that some genes are present in the chloroplast DNA of one species and in the nuclear DNA of the other, indicating that some exchange of genes between chloroplast and nucleus has occurred during evolution.

The sequencing of chloroplast DNA has revealed some unexpected features. For instance, the DNA contains seven protein-coding sequences that are homologous to the seven mitochondrial-encoded subunits of the NADH-CoQ reductase complex. Such an electron transport system had not been identified previously in chloroplasts of higher plants, and the DNA sequence has prompted investigators to study the role of this unexpected protein complex.

(a)

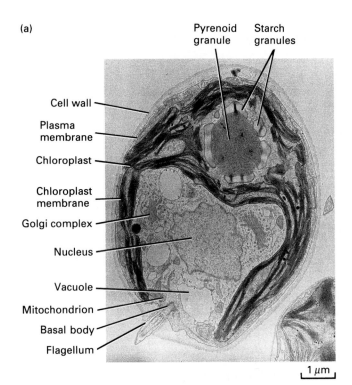

Pyrenoid granule   Starch granules

Cell wall

Plasma membrane

Chloroplast

Chloroplast membrane

Golgi complex

Nucleus

Vacuole

Mitochondrion

Basal body

Flagellum

1 µm

(b)

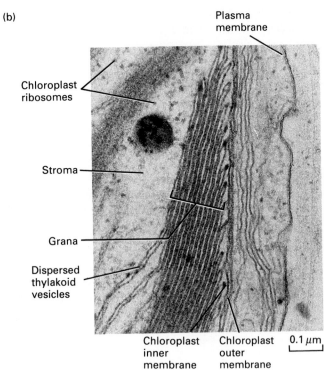

Plasma membrane

Chloroplast ribosomes

Stroma

Grana

Dispersed thylakoid vesicles

Chloroplast inner membrane

Chloroplast outer membrane

0.1 µm

▲ FIGURE 19-13   Electron micrographs of *Chlamydomonas reinhardtii*. (a) The single, large chloroplast. The stacked thylakoid vesicles are major components of the chloroplast, as is a large pyrenoid granule with its associated starch granules. (b) The chloroplast is surrounded by a double-membrane envelope, seen to advantage at this higher magnification in which the chloroplast ribosomes are visible. The thylakoid vesicles are both dispersed and packed into grana; obliquely sectioned thylakoid vesicles appear somewhat fuzzy. [Courtesy of I. Chad, P. Siekevitz, and G. E. Palade.]

Several Uptake-Targeting Sequences Direct Proteins Synthesized in the Cytosol to the Appropriate Chloroplast Compartment

All proteins encoded by chloroplast DNA are synthesized in the stroma and are then directed to their appropriate destinations by mechanisms that are beginning to be uncovered. However, the vast majority of chloroplast proteins are synthesized in the cytosol and then imported into the organelle, although the exact number is unknown. As in mitochondrial biogenesis, different proteins are targeted to different chloroplast subcompartments (Figure 19-1): ferredoxin and the small subunit of ribulose 1,5-bisphosphate carboxylase (rubisco) to the stroma, the binding proteins for chlorophylls *a* and *b* in the light-harvesting complexes (LHC) to the thylakoid membrane, and plastocyanin to the thylakoid lumen. Protein import from the cytosol into the stroma occurs, as in mitochondria, at points where the outer and inner organelle membranes are in close contact (Figure 19-14).

Ribulose 1,5-bisphosphate carboxylase provides an excellent example of how proteins are imported into the chloroplast stroma and how they are assembled there into a multisubunit protein. Rubisco is a component of the Calvin cycle (see Figure 18-16), and is the most abundant protein in chloroplasts. It is located in the stromal compartment of the chloroplast, along with the other enzymes involved in the Calvin cycle.

The rubisco enzyme (MW 550,000) is made up of 16 subunits. Eight are identical large (L) subunits (MW 55,000) that contain the catalytic sites. The other eight are identical small (S) subunits (MW 12,000) that are necessary for full enzyme activity. The L subunit of rubisco is encoded by the chloroplast DNA. The S subunit is encoded in the nucleus and is synthesized on free cytosolic polyribosomes in a precursor form that traverses both the outer and inner chloroplast membranes to reach its final destination (see Figure 19-14).

An experimental protocol similar to that illustrated in Figure 19-6 has shown that the S subunit precursor can be taken up post-translationally by isolated chloroplasts. The precursor polypeptide has an N-terminal stromal-import sequence of about 44 amino acids. After uptake of the precursor, it binds transiently to a stromal hsp70 chaperone, and the N-terminal sequence is cleaved in the stromal space. Eight S subunits combine with the 8 L subunits to yield the active rubisco enzyme.

The L and S subunits of rubisco do not assemble spontaneously to form the mature enzyme; instead, newly made L subunits are bound to a chaperonin that is very similar in structure and function to the mitochondrial matrix hsp60 and that is itself imported into the chloroplast stroma. The chaperonin forms complexes with newly synthesized L subunits and releases them in the presence of ATP. The chaperonin stores excess L subunits in chloroplasts, pending the import of S subunits, and also may facilitate

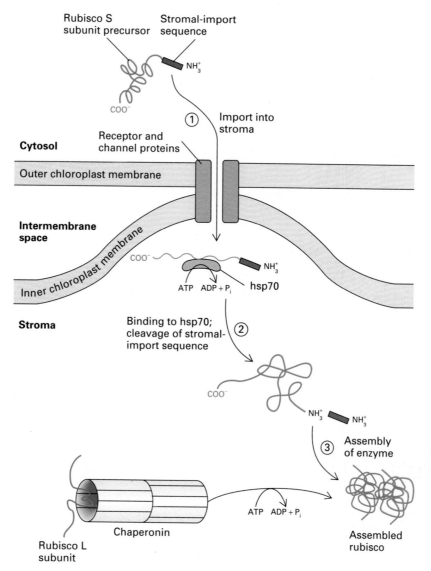

▲ FIGURE 19-14   Import of rubisco small (S) subunits into the chloroplast stroma and the assembly of small and large (L) subunits into the active rubisco enzyme. An S sub-unit is fabricated in the cytosol as a precursor with an N-terminal stromal-import sequence. As with import into mitochondria, a receptor protein located at the point of contact between the outer and inner chloroplast membranes mediates translocation from the cytosol to the stroma. The receptor binds the stromal-import sequence, and the precur-sor then passes through a channel protein into the stroma; neither the receptor nor the channel proteins have been characterized. Within the stroma the precursor transiently binds to an hsp70 protein and the stromal-import sequence is cleaved. L subunits are synthesized in the stroma by chloroplast ribosomes and are stored complexed to hsp60 chaperonins. Eight L subunits, released by the chaperonins, bind to eight imported S subunits, forming the mature rubisco enzyme.

the folding of L subunits and the assembly of S and L polypeptides.

Proteins destined for chloroplast subcompartments other than the stroma have an additional pathway to travel. Proteins targeted to the thylakoid membrane or lumen not only must traverse both the outer and inner chloroplast membranes but also must travel through the stroma and either be inserted into the thylakoid membrane or cross that membrane and enter the thylakoid lumen.

Proteins that are destined for the thylakoid lumen, such as plastocyanin, require the successive action of two uptake-targeting sequences (Figure 19-15). The first, like that of the rubisco S subunit, targets the protein to the stroma; the second targets the protein from the stroma to the thylakoid lumen. The role of these sequences has been shown by cell-free experiments measuring the uptake, into chloroplasts, of mutant proteins that have been generated by recombinant DNA techniques. For instance, when the thylakoid

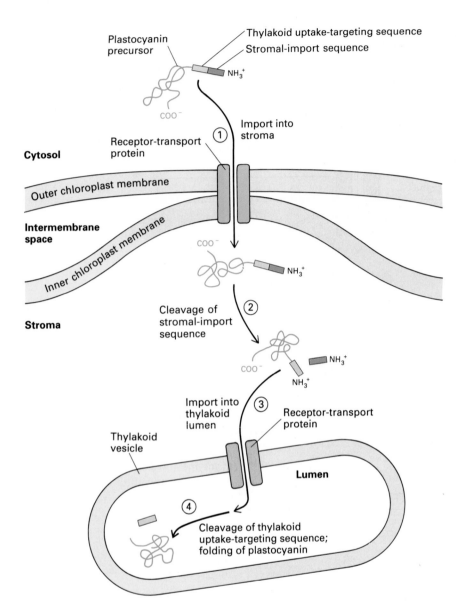

◄ **FIGURE 19-15** Transport of plastocyanin from cytosol to thylakoids. The successive actions of two uptake-targeting sequences direct plastocyanin to the thylakoid lumen. The plastocyanin precursor is synthesized in the cytosol; the 66 amino acids at its N-terminus (purple and blue) are not found on the mature protein in the thylakoid lumen. Step 1: The ≈30 N-terminal residues (purple) signal import into the chloroplast stroma. Step 2: This stromal-import sequence is removed by a stromal signal protease. Step 3: The thylakoid uptake-targeting sequence (blue) of ≈25 residues, with its core of hydrophobic amino acids, is now the N-terminal sequence; it causes the stromal precursor to be imported into the thylakoid lumen, presumably by binding to a distinct receptor-transport protein on the thylakoid membrane. Step 4: This N-terminal sequence is removed in the thylakoid lumen by a separate endoprotease that is similar in its sequence requirements to the analogous enzyme in the endoplasmic reticulum and also in bacterial cells. Not shown here are the chaperones that prevent premature folding of this protein in the stromal space. [After S. Smeekins et al., 1986, *Cell*, **46**:365–375 and J. Shackelton and C. Robinson, 1991, *J. Biol. Chem.* **266**:12152–12156.]

uptake-targeting sequence is deleted, plastocyanin accumulates in the stroma and is not transported into the thylakoid lumen.

Many different precursor proteins appear to share a common chloroplast translocation system, since a synthetic peptide corresponding to the N-terminal stromal-targeting sequence of the S subunit of rubisco will specifically block the uptake of many other precursors, for instance, those of ferredoxin, plastocyanin, and the binding proteins for chlorophylls *a* and *b* in the LHC. However, the N-terminal uptake-targeting sequences on these and other proteins have little resemblance to each other (see Table 19-1), and the presumed receptors on the outer membrane that bind them have not yet been purified or cloned.

As in mitochondria, protein import into chloroplasts requires energy. Experiments suggest that import into the stroma, much like import into the mitochondrial matrix, depends on ATP hydrolysis catalyzed by chaperones in the stroma. Precisely how ATP acts to trigger import is not known, but chaperones within the stroma (see Figure 19-14) similar to the hsp70 in the mitochondrial matrix (see Figure 19-7) are likely candidates. Unlike import into mitochondria, import into chloroplasts does not require an electrochemical potential across the inner chloroplast membrane. However, incorporation of some, but not all, proteins from the stroma into the thylakoid does require an electrochemical potential across the thylakoid membrane.

In many respects, the system for moving proteins from the stroma to the thylakoids resembles the process for the secretion of bacterial proteins across the bacterial plasma membrane, a similarity consistent with the evolution of chloroplasts from ancestral photosynthetic bacteria. For one, thylakoid vesicles form by budding from the inner chloroplast membrane (see Figure 19-16), the membrane that would correspond to the plasma membrane of the ancestral bacterium; thus the chloroplast stroma corresponds to the bacterial cytoplasm. Further, the secretion of bacterial proteins occurs post-translationally and requires binding of the precursor proteins to cytoplasmic chaperones, much as in the (post-translational) uptake of proteins into the thylakoid from the stroma. Additionally, the thylakoid uptake-targeting sequences of plastocyanin and other thylakoid-lumen proteins resemble the signal sequences that target bacterial proteins to cross the bacterial plasma membrane: in both, a sequence of up to 20 amino acids contains a continuous sequence of ≈ 5 hydrophobic amino acids near the N-terminus, usually preceded by a Lys or Arg residue. Also, the bacterial and thylakoid peptidases that remove the targeting sequences are extremely similar with respect to the amino acid sequences at which they bind and cleave: a small amino acid such as glycine or alanine is found just N-terminal to the cleavage site and also at the position two residues toward the N-terminus.

Proplastids Can Differentiate into Chloroplasts or Other Plastids

The chloroplast (Figure 19-16) is one of several types of organelles termed *plastids*, all of which are formed from the same precursor proplastid and all of which contain the same chloroplast DNA (Figure 19-17a). *Proplastids* are composed only of an inner membrane, an outer membrane, and a small stromal space that contains the chloroplast DNA. Proplastids are found in embryonic tissue; similar organelles are found in such tissues as leaves deprived of light, where they are often called *etioplasts* (Figures 19-17a and 19-18). Chloroplast DNA is replicated in proplastids, but a proplastid does not contain LHC proteins, chlorophyll, electron transport systems, or thylakoid vesicles.

Chloroplasts are the only type of plastid to contain internal thylakoid membranes. In leaves and other tissues, the presence of light stimulates the synthesis of chloroplast proteins and the expansion of the inner membrane of the proplastid. Membrane vesicles bud from the inner membrane and arrange themselves into stacks; these vesicles incorporate essential proteins and chlorophyll and are transformed into mature thylakoid vesicles (see Figure 19-16). Chloroplasts do much more for plants besides providing sites for photosynthesis. These plastids are involved in nitrogen assimilation, as well as in the synthesis of some amino acids, many fatty acids and lipids, and even some plant hormones.

Depending on the plant tissue and on environmental cues, proplastids differentiate into several other types of plastids as well as chloroplasts: chromoplasts, amyloplasts, and elaioplasts (see Figure 19-17a). *Chromoplasts*, as their name implies, are pigmented organelles that synthesize and retain yellow, orange, and red carotenoids and other pig-

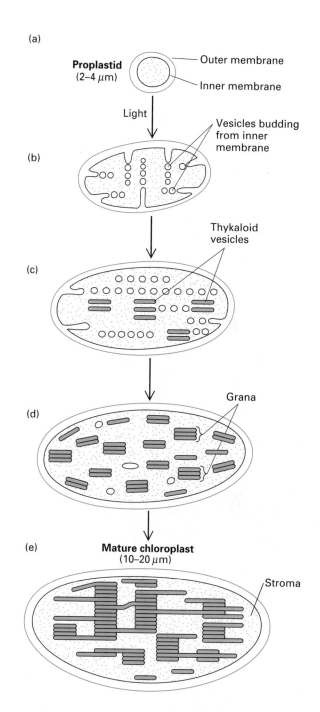

▲ FIGURE 19-16 Steps in the light-induced differentiation of a proplastid into a chloroplast. (a) The proplastid in dark-adapted cells contains only the outer and inner chloroplast membranes. (b) Light triggers the synthesis of chlorophyll, phospholipid, chloroplast stroma and thylakoid proteins, and the budding of small vesicles from the inner chloroplast membrane. (c) As the proplastid enlarges, some of the spherical vesicles fuse, eventually forming one continuous set of flattened thylakoid vesicles. Some thylakoid vesicles stack into grana. The adhesiveness of the vesicles is due to a protein found in the light-harvesting complex (LHC) that is synthesized in abundance during this stage. (d, e) The proplastid enlarges and matures into a chloroplast as more thylakoid vesicles and grana are formed. [See D. von Wettstein, 1959, *J. Ultrastruct. Res.* **3**:235.]

(a)

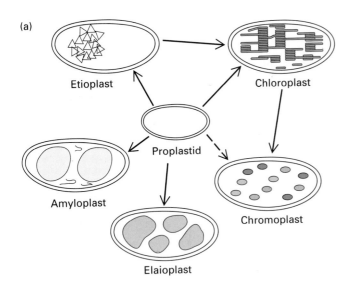

(b)

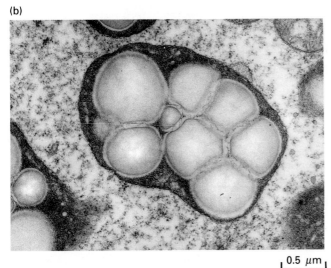

▲ FIGURE 19-17   The proplastid and its progeny. (a) Pro-plastids can differentiate into many organelles, depending on the plant tissue in which they are located and on exposure to light. The etioplast is an intermediate in the conversion of a proplastid to a chloroplast. Only the chloroplast contains internal thylakoid membranes. (b) Electron micrograph of an amyloplast from the root cap of the onion *Allium cepa*. Amy-loplasts contain massive starch granules. In the root cap, these plastids sink to the bottom of the cell, indicating to the root which way is down. [Part (a) courtesy of Barbara Sears; part (b) courtesy of Jeremy Burgess and Science Photoli-brary/Photo Researchers, Inc.]

ments. They are responsible for the colors in flowers, ripening fruits, autumn leaves, and some roots (for example, carrot roots). In most cases, chromoplasts form directly from chloroplasts; this process involves the controlled breakdown of thylakoid membranes and chlorophyll, the appearance of new invaginations from the inner membrane, and the synthesis of new types of carotenoids. *Amyloplasts* (Figure 19-17b), are nonpigmented, starch-containing plastids found in tissues such as potato tubers, and *elaioplasts* store droplets of oils and lipids. Thus these

(a) Chloroplast

(b) Etioplast

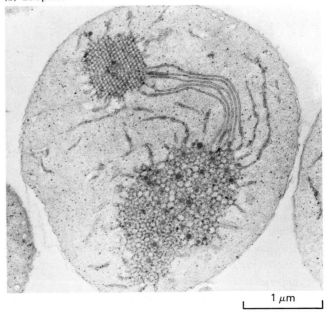

▲ FIGURE 19-18   Electron micrographs of thin sections of plant tissue. (a) A chloroplast from a normal barley seedling leaf. (b) An etioplast from a dark-grown barley seedling. In the absence of light, major polypeptides, such as proteins in the LHC and photosynthetic reaction center, are not synthesized. Etioplasts from such cells contain membranes that take the form of primary lamellae and interconnected vesicles containing some chloroplast pigments. Light triggers the development of the etioplast into the mature chloroplast. [From H. G. Schweiger, ed., *International Cell Biology 1980–1981*. Springer-Verlag. Courtesy of D. von Wettstein.]

plastids in nonphotosynthetic tissues store essential energy sources, such as starch and lipids. What controls the differentiation of these various plastids from a common progenitor plastid is not known but is the subject of much current research.

➤ *Peroxisome Biosynthesis*

Peroxisomes, also called *microbodies*, are small organelles, approximately 0.2–1 μm in diameter. They are lined by a single membrane, and in this respect are different from the other organelles described in this chapter. Peroxisomes are probably present in all eukaryotic cells. Their size and enzyme composition vary considerably, but all peroxisomes contain enzymes that form H_2O_2 and also contain *catalase*, the enzyme that efficiently decomposes H_2O_2 into H_2O (see Figure 17-32). Peroxisomes are most abundant in liver cells, where they constitute about one to two percent of the cell volume; they are present in much lower quantities in other tissues, making their study more difficult. Although peroxisomes lack DNA and ribosomes, they contain an unusually high concentration of protein, visible as a dense core in electron micrographs (see Figure 5-37), and can be conveniently separated from other organelles because of their resulting high equilibrium density in sucrose gradients.

All Peroxisomal Proteins Are Imported from the Cytosol

All peroxisomal proteins, both membrane and luminal, are encoded by nuclear genes and synthesized on free polyribosomes in the cytosol. The newly synthesized proteins are released into the cytosol and then are imported into pre-existing peroxisomes (Figure 19-1). As is the case with mitochondria, this leads to progressive enlargement of the peroxisomes, which then divide into new ones.

The biogenesis of *catalase* has been studied in considerable detail (Figure 19-19). A catalase molecule contains four identical polypeptide chains, each with an attached heme group. Newly made catalase is released from polyribosomes into the cytosol as *apocatalase*, a single-chain polypeptide without a heme group. Apocatalase chains are taken up by peroxisomes some 20 min after their synthesis; once transported, the four chains are assembled and heme is added, forming the mature catalase molecule. The addition of heme may cause a conformational change in the polypeptide that prevents it from moving back across the

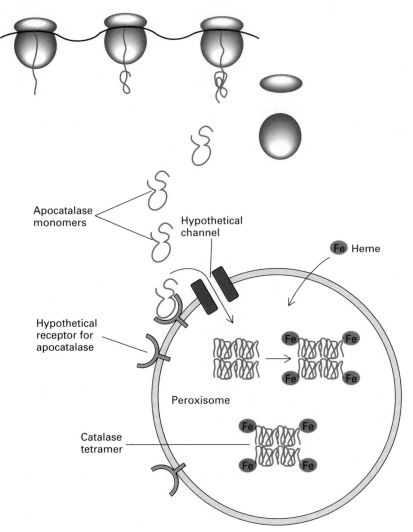

Apocatalase monomers

Hypothetical channel

Fe Heme

Hypothetical receptor for apocatalase

Peroxisome

Catalase tetramer

◄ FIGURE 19-19   Synthesis of catalase and its incorporation into peroxisomes. Catalase monomers without heme (apocatalase) are initially found free in the cytosol, and then are imported into the peroxisome by a process involving a specific receptor on the organelle membrane and a transport protein in the peroxisomal membrane. Inside the peroxisome the polypeptides assemble into tetramers and the heme prosthetic groups are added.

peroxisomal membrane into the cytosol, much in the way that the addition of heme to cytochrome *c* locks it into the intermembrane space of mitochondria.

The import of proteins into rat liver peroxisomes can be studied in a cell-free system similar to that used for studying transport into mitochondria (Figure 19-6). Import into peroxisomes requires ATP hydrolysis but not a membrane potential. Receptor and transport proteins in the peroxisomal surface membrane are also required; these are only now being characterized as the result of analysis of the genetic diseases affecting peroxisome biogenesis, discussed below.

For many peroxisomal proteins the uptake-targeting signal is a Ser-Lys-Leu sequence (SKL in one-letter code) at the very C-terminus; the sequence is not cleaved from the protein (Table 19-1). Examples include fatty acyl CoA oxidase in rat liver peroxisomes, urate oxidase in cucumber, and, in the firefly, luciferase (the enzyme that generates the flashes of light). When luciferase is expressed (by recombinant DNA techniques) in cultured mammalian or plant cells, it is efficiently targeted to peroxisomes, attesting to the universality of peroxisome targeting mechanisms. Furthermore, the SKL sequence is both necessary and sufficient for uptake into peroxisomes: mutant luciferase without the SKL sequence remains in the cytosol, and a normally cytosolic protein becomes targeted to the peroxisomes in cultured monkey cells by the addition of an SKL sequence to its C-terminus. However, sequences other than SKL will also target certain proteins to the peroxisome; an example is thiolase (see Figure 17-33), which is fabricated as a precursor with an N-terminal uptake-targeting sequence of 26 amino acids that is subsequently cleaved.

Peroxisomal membrane proteins are also synthesized on free polyribosomes and incorporated into peroxisomes after their synthesis; little is known about the nature of the targeting sequences except that they do not contain an SKL sequence.

Genetic Diseases Have Helped to Elucidate the Process of Peroxisome Biogenesis

Mutations that cause defective peroxisome assembly occur naturally in the human population. An example is *Zellweger syndrome*, which, like adrenoleukodystrophy (ALD; see Chapter 17), results in severe impairment of many organs and death. In Zellweger syndrome and related disorders, the transport of all proteins into the peroxisomal matrix is impaired; newly synthesized peroxisomal enzymes remain in the cytosol and are eventually degraded. A remarkable feature of the Zellweger syndrome is that cells contain empty peroxisomes; these do contain a normal complement of peroxisomal membrane proteins (Figure 19-20). These findings demonstrate that peroxi-

somes from patients with Zellweger syndrome are defective in the uptake of matrix proteins, but not of peroxisomal membrane proteins, from the cytosol. The Zellweger mutation causes a defect in a peroxisomal receptor or transport protein for peroxisomal matrix proteins but not membrane proteins.

A genetic study of cultured cells from different Zellweger patients has shown that mutations in any of eight different genes can cause this phenotype, so at least eight proteins are involved in the uptake of proteins into the peroxisomal matrix. In this study, cultured fibroblasts from different patients were fused together. In many cases normal peroxisome function was restored—that is, complementation occurred (see Chapter 8)—indicating that the cells from different patients were defective in different proteins required for the uptake of peroxisomal proteins. Similar kinds of peroxisome-assembly mutants have now also been generated in yeast.

Cloning of the wild-type allele of the mutant gene from one Zellweger patient revealed that its product is a MW 70,000 protein that is a major component of peroxisomal membranes. The protein is required for the post-translational peroxisomal uptake of proteins with an SKL sequence, but its exact function is not yet known. Eventually, cloning of the wild-type alleles of the other Zellweger mutations and of the similar yeast mutations should elucidate the identity of all proteins involved in the transport of proteins into the peroxisome.

We noted in Chapter 17 that in patients with ALD the peroxisomal oxidation of very-long-chain fatty acids is defective, due to the absence of long-chain fatty acyl CoA synthase, the enzyme that links coenzyme A to very-long-chain fatty acids within the peroxisome. Positional cloning

➤ FIGURE 19-20 Peroxisomes from normal persons and from patients with Zellweger syndrome. (a) In a section of a cultured fibroblast from a normal individual, an antibody to catalase stains the peroxisomes (P). In Zellweger syndrome the empty peroxisomes cannot be detected with this antibody, or with any antibody to a peroxisomal matrix enzyme, indicating a defect in peroxisome assembly. (b) When an antibody to peroxisomal membrane proteins is added to a section of normal cells, the peroxisomes are stained [arrows in (b)]. (c) Similar but larger vesicles are stained with the antibody to peroxisomal membrane proteins in sections of cells from Zellweger patients [arrows in (c)], indicating that assembly of the peroxisomal membrane is normal. Thus Zellweger cells are defective in the translocation of peroxisomal matrix proteins from the cytosol, but not of peroxisomal membrane proteins. The insets show the stained vesicles at higher magnification. ER = endoplasmic reticulum; M = mitochondrion; N = nucleus; Bar = 0.25 μm. [From M. Santos, et al., 1988, *J. Biol. Chem.* **263**:10502–10509.]

studies (see Chapter 8) revealed that the protein encoded by the ALD gene has a structure similar to that of the CFTR (cystic fibrosis transmembrane regulator) membrane transport protein and of the multidrug resistance protein (see Figure 15-15). Apparently the ALD protein is the peroxisomal membrane transporter specific for the uptake of long-chain fatty acyl CoA synthase from the cytosol.

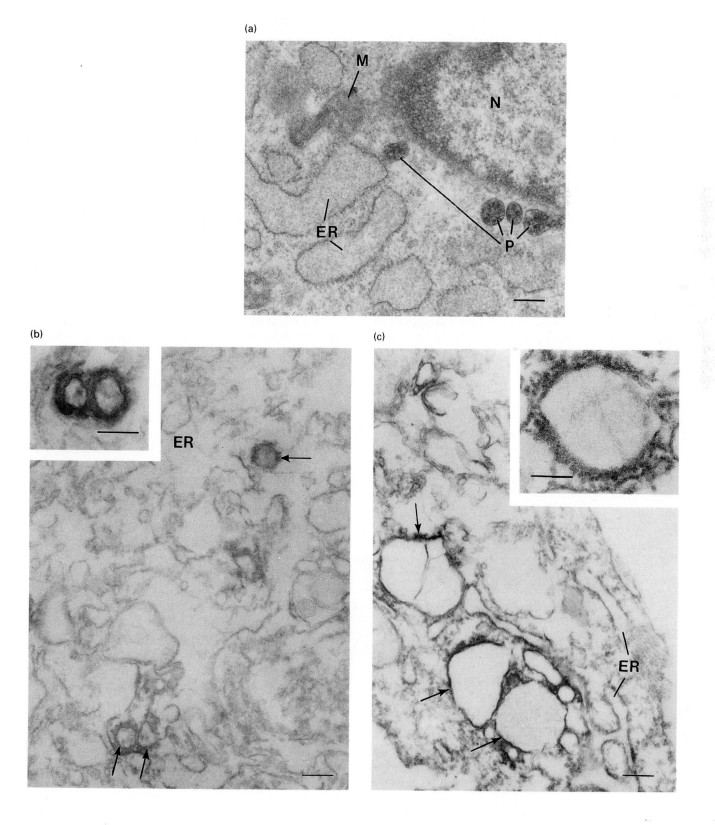

➤ *Protein Traffic into and out of the Nucleus*

The nucleus contains no functional ribosomes and does not carry out protein synthesis (recall that, though assembled in the nucleolus, ribosomes function in the cytosol and attached to the endoplasmic reticulum, including the outer nuclear membrane). Therefore, all nuclear proteins must be synthesized in the cytosol and then taken up into the organelle. These include nuclear proteins such as lamins and histones, ribosomal proteins that are incorporated into ribosomes in the nucleolus, and DNA and RNA polymerases. The list also includes snRNPs (small nuclear ribonucleoprotein particles that participate in RNA splicing); most if not all of the protein components of snRNPs bind to small nuclear RNAs in the cytosol, and the completed snRNP is then taken up into the nucleus.

The movement of proteins from the cytosol into the nucleus is different in several respects from protein translocation into the mitochondrion, chloroplast or peroxisome: fully formed and folded particles (such as snRNPs) are imported, in contrast to the unfolded proteins that are translocated into the mitochondrion, chloroplast or peroxisome, and uptake occurs through a visible pore (Figures 3-1a and 19-21) in the nuclear envelope, in contrast to the much smaller channels in the membranes of the other organelles.

Assembly of the nucleus does present other problems, one being changes in the nuclear envelope during the cell cycle. In higher plant and animal cells, the envelope fragments into a number of small vesicles during the late prophase stage of mitosis, and new envelopes re-form around the daughter nuclei during telophase (see Figure 5-50). Because the control of disassembly and reassembly of the envelope is intricately linked to the control of other steps in the cell cycle, we defer a discussion of this topic until Chapter 25.

In this section we deal mainly with the import of proteins from the cytosol into the nucleus. Our focus here will be on traffic through the nuclear pores, the highly selective passageways both into and out of the nucleus. Perhaps surprisingly, almost nothing is known of the nature or biogenesis of integral proteins in the inner or outer nuclear membrane.

Nuclear Proteins Are Selectively Imported into Nuclei

Pores in the nuclear envelope have many of the properties of a molecular sieve, as was shown in experiments in which scientists injected radioactive substances into the cytosol of a frog oocyte and measured their incorporation into the nucleus. The pores are freely permeable to ions and small molecules; a small molecule, such as radiolabeled sucrose, diffuses into the nucleus as fast as it passes through the cytosol. These injection experiments also showed that virtually any small, globular, nonnuclear protein <9 nm in diameter (corresponding to ≈60,000 MW) was able to diffuse into the nucleus, although smaller proteins or particles enter the organelle at faster rates.

As a result of these studies, it was originally believed that all proteins and particles enter the nucleus by free diffusion through pores and are selectively retained by binding to some nuclear component that is not freely diffusible, such as the lamina or chromatin. More recent work, however, has shown that nuclear proteins and ribonucleoproteins contain one or more *nuclear localization sequences* (NLSs) that cause them to be actively and selectively imported into the nucleus, and that large particles such as snRNPs are actively transported through pores into the nucleus.

0.1 μm

▲ FIGURE 19-21 Scanning electron micrograph of isolated pores in the nuclear membrane of an amphibian oocyte, viewed from the cytoplasmic side. Hand-isolated nuclei were extracted with detergent and then fixed, the water removed, and then shadowed with platinum (Figure 5-34). The eight subunits that form the annulus of each pore are visible, as are cytoskeletal fibers that interconnect the pores. [Courtesy of Hans Ris, University of Wisconsin. See Ris, H., 1991, *Electron Microsc. Soc. Am. Bull.* **21**:54–56.]

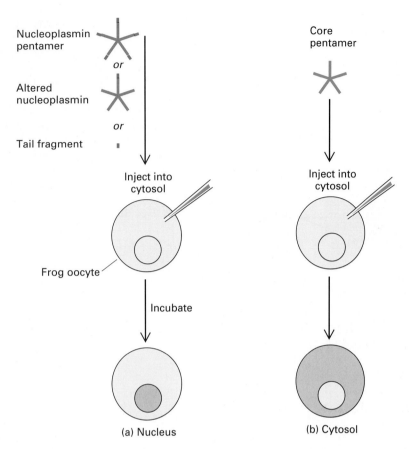

▲ FIGURE 19-22   The role of the nucleoplasmin C-terminal tail sequence in regulating transport into the nucleus.
(a) After its microinjection into the oocyte cytosol, the intact
nucleoplasmin, or a proteolytic pentameric fragment of
nucleoplasmin with at least one intact subunit, or the isolated tail fragment accumulates in the nucleus. (b) The core
pentamer of nucleoplasmin, minus all tails, is not found in
the nucleus after its microinjection into the cytosol. [After
C. Dingwall and R. Laskey, 1986, *Ann. Rev. Cell Biol.* **2**:367–
390.]

Many of these studies have made use of *nucleoplasmin*, a large (MW 165,000) pentameric protein present in
high concentrations in the soluble phase of the frog oocyte
nucleus. Nucleoplasmin assists in the assembly of chromatin during the rapid cell cleavages that occur after fertilization. When microinjected into the oocyte cytosol,
nucleoplasmin rapidly accumulates in the oocyte nucleus
(Figure 19-22a), implying the existence of some sort of specific energy-requiring uptake mechanism. Removal of all
five C-terminal tail regions from a nucleoplasmin pentamer
yields a pentameric core that does not accumulate in the
nucleus (Figure 19-22b). However, if just one tail remains
on a nucleoplasmin pentamer, the pentamer enters into the
nucleus, as does a single, unattached tail (Figure 19-22a).
These studies demonstrate that the C-terminal tail domain
bears a discrete sequence for accumulation in the nucleus—
the NLS.

Nuclear Pores Are the Portals for Protein Transport

When nuclear pores were discovered by electron microscopy, they were assumed to be the portals of entry into and
exit from the nucleus. The definitive experiments that
showed that pores indeed are the routes for protein import
made use of nucleoplasmin. In one key study, small gold
particles coated with nucleoplasmin or with a nonnuclear
protein were microinjected into the cytosol of frog oocytes.
The location of the gold particles was examined by electron microscopy after sectioning the oocytes. Shortly after
injection, the nucleoplasmin-coated gold particles clustered
at the nuclear pores (Figure 19-23); later, they accumulated
in the nucleus after passing through the pore complexes.
Gold particles coated with nonnuclear proteins, by contrast, remained in the cytosol and did not bind to nuclear
pores.

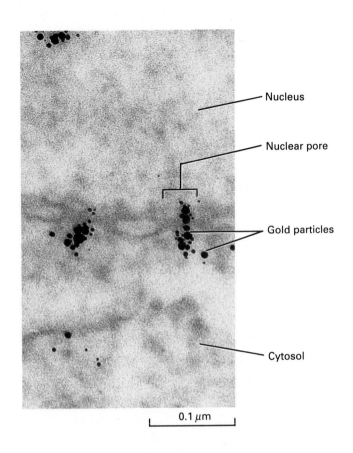

— Nucleus

— Nuclear pore

— Gold particles

— Cytosol

0.1 μm

◄ FIGURE 19-23   Proteins moving through nuclear pores from the cytosol into the nucleus. Gold particles 5 nm in diameter were coated with nucleoplasmin (see Figure 19-22) and then were microinjected into the cytosol of a frog oocyte; 15 min later, the cells were fixed, sectioned, and examined under the electron microscope. The gold particles are clustered at nuclear pore sites along the nuclear membrane, and some appear to be within pores. When the cells were examined at later times (not shown here), many coated gold particles had entered the nucleus. If gold particles are coated with proteins that lack sequences for nuclear entry, they do not accumulate at the pores or enter the nucleus. [After C. M. Feldherr, E. Kallenbach, and M. Schultz, 1984, *J. Cell. Biol.* **99**:2216–2222. Courtesy of C. M. Feldherr.]

These same pores were found to transport RNA out of the nucleus (see Chapter 12). In one experiment, gold particles were coated with tRNA or 5S RNA and injected into the nucleus of frog oocytes; the particles migrated outward into the cytosol. Electron microscopy revealed that RNA-coated gold particles were clustered near all of the pores and that some could be seen inside pores.

Multiple Types of Nuclear Localization Sequences Direct Proteins and Ribonucleoproteins to the Nucleus

A diverse array of proteins are specifically targeted to the nucleus and, as with uptake into other organelles, a variety of sequences can direct proteins from the cytosol into the nucleus. Some examples of proteins with defined nuclear localization sequences (NLSs) for nuclear import are listed in Table 19-6. It is immediately obvious from the table that several different kinds of amino acid sequences serve as signals for the nuclear import of proteins.

The prototype protein for studies of nuclear import was the *T antigen* (also called the *large T antigen*), a nuclear protein of the SV40 virus that binds to the origin of SV40 DNA and induces its replication and transcription. A

variety of mutations in the gene encoding the T antigen were generated by recombinant DNA techniques. The mutant T proteins were synthesized in cultured cells, and the fate of each mutant protein was monitored. The mutant T antigens all accumulated in the nucleus unless the mutation altered the NLS; those mutant proteins remained in the cytosol.

The SV40 NLS was found to be a sequence of seven amino acids that contains five consecutive positively charged amino acids (see Table 19-6). In an experiment to show that the seven amino acids of the NLS are sufficient for nuclear uptake, this sequence was linked to the N-terminus of the normally cytosolic protein pyruvate kinase; virtually all of the protein was redirected into the nucleus (Figure 19-24). Furthermore, if pyruvate kinase was linked to an altered NLS (with threonine instead of lysine at residue 128), which did not direct the SV40 T antigen to the nucleus, the pyruvate kinase remained in the cytosol. Thus these seven amino acids are both necessary and sufficient for nuclear uptake.

The SV40 NLS also functions in yeasts and plant cells to redirect normally cytosolic proteins to the nucleus, results indicating that the function of the SV40-type NLS—containing a core of five consecutive positively charged residues—is general.

Nevertheless, other proteins utilize different types of NLSs (see Table 19-6). Nucleoplasmin, for example, uses a

TABLE 19-6   Amino Acid Sequences That Act as Signals for Translocation to the Cell Nucleus

Type of Signal	Nuclear Protein	Location of Signal in Protein	Deduced Nuclear Localization Sequence*
Single	large T antigen of SV40 virus	Residues 126–132	Pro **Lys Lys Lys Arg Lys** Val
	Adenovirus E1a	C-terminus	**Lys Arg** Pro **Arg** Pro
	Influenza virus nucleo-protein (NP)	Near C-terminus; residues 336–345	Ala Ala Phe Glu Asp Leu **Arg** Val Leu Ser
Bipartite	*Xenopus* nucleoplasmin	Near C-terminus; residues 155–171	**Lys Arg** Pro Ala Ala Thr **Lys Lys** Ala Gly Gln Ala **Lys Lys Lys Lys** Leu
	Xenopus N1 nuclear protein	Internal; residues 534–550	**Lys Arg** Lys Thr Glu Glu Glu Ser Pro Leu **Lys** Asp **Lys** Asp Ala **Lys Lys**
	Yeast SWI 5 tran-scription factor	Internal; residues 636–652	**Lys Lys** Tyr Glu Asn Val Val Ile **Lys Arg** Ser Pro **Arg Lys Arg** Gly **Arg**
	Mouse poly-ADP-ribose polymerase	Internal; residues 208–224	**Arg Lys** Gly Asp Glu Val Asp Gly Thr Asp Glu Val Ala **Lys Lys Lys** Ser
	Chicken estrogen receptor	Internal; residues 250–266	**Arg Lys** Asp **Arg** Arg Gly Gly Glu Met Met **Lys** Gln **Lys Arg** Gln **Arg** Glu

* Boldface = the basic residues thought to be essential for nuclear uptake.
SOURCE: C. Dingwall and R. Laskey, 1986, *Ann. Rev. Cell Biol.* **2**:367–390; and C. Dingwall and R. Laskey, 1991, *Trends Biochem. Sci.* **16**:478–481.

bipartite NLS, a sequence of 17 amino acids with the following general pattern: two positively charged amino acids, a spacer of any ten amino acids, and then a sequence of five amino acids, of which at least three must have a positive charge. Neither group of positively charged amino acids by itself can redirect a cytosolic protein to the nucleus, but the entire 17-aa sequence can.

Some NLSs do not conform to either the SV40 or the bipartite prototype; an example is the NLS of the influenza virus NP protein (see Table 19-6). Also, NLSs can be found

➤ FIGURE 19-24   Demonstration that the nuclear localization sequence (NLS) of the SV40 large T antigen can direct a cytosolic protein to the cell nucleus. (a) Normal pyruvate kinase, visualized by immunofluorescence after treating cultured cells with a specific antibody, is localized to the cytosol. (b) A chimeric pyruvate kinase protein, containing the SV40 NLS at its N-terminus, is directed to the nucleus. [From D. Kalderon, et al., 1984, *Cell* **39**:499–509. Courtesy Dr. Alan Smith.]

(a)
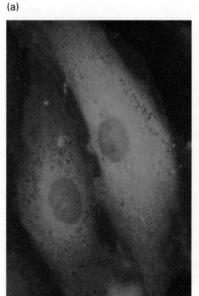

(b)
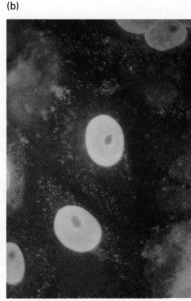

in different parts of a protein: not all are N-terminal or C-terminal. The only general rule is that an NLS must be on the surface of the folded protein in order to direct nuclear uptake. Several proteins that enter the nucleus lack an NLS altogether, but these proteins enter by forming a stable, specific complex with an NLS-bearing protein, a process known as "piggyback transport."

All NLSs with a core of positively charged amino acids enter the nucleus by a common pathway. This was shown by studies in which large amounts of protein containing multiple SV40 NLSs were co-injected into the cytosol of frog oocytes: this blocked the nuclear uptake of other proteins with either an SV40-type NLS or a bipartite NLS. In all known cases the NLS is not removed after nuclear import.

The uptake of snRNPs into the nucleus involves yet another kind of nuclear localization signal. When radiolabeled U2 snRNP (a splicing catalyst; see Figure 12-24) is injected into frog oocytes, it is targeted to the nucleus. Uptake is not blocked by the co-injection of protein containing multiple SV40 NLSs; thus, snRNPs such as U2 utilize a different signal for uptake into the nucleus than do proteins that utilize a series of positively charged amino acids as part of the NLS, though all of these proteins and particles enter via pores. Indeed, the critical determinant for nuclear uptake is the trimethylguanosine cap that is at the 5' end of the small nuclear RNAs for U2 (and other snRNPs), since altered snRNPs without this cap remain in the cytosol. Thus, several types of sequences, some on RNA, some on protein, can direct proteins to the nuclear pores and into the nucleus.

A number of cytosolic proteins have been purified that bind specifically to various NLSs (but not to related import-incompetent sequences). Several are chaperones that are thought to bind to nuclear-destined proteins in the cytosol until these bind to nuclear pores.

Receptor Proteins in Nuclear Pores Bind Nuclear Proteins for Import

The studies that identified receptors for nuclear uptake made use of isolated frog oocyte nuclei. When supplemented with oocyte cytosol. These nuclei bound specifically all proteins with an NLS. Further studies showed that the cytoplasmic face of the pore complex contains Further specific receptors—binding sites for nuclear proteins. When added to preparations of permeabilized cells or isolated nuclei plus cytosol, purified and radiolabeled nuclear proteins are transported into the nuclei. However, ATP hydrolysis is required: in the absence of ATP, the proteins bind specifically at the cytosolic face of the pore complex and remain there, as shown by electron microscopy studies similar to those in Figure 19-23; when ATP is added, the bound proteins are imported into the nucleus. Thus the energy released by ATP hydrolysis is essential to import a nuclear protein across a pore but is not required to bind the protein to the pore receptor.

To sum up the highly specific process of nuclear import, nuclear proteins are synthesized in the cytosol with a signal sequence which causes the nuclear proteins to bind to chaperone proteins in the cytosol and then to receptors on the cytosolic face of nuclear pores. An energy-dependent process then translocates the bound proteins to the nucleus. Precisely how the proteins traverse the pores is not known.

SUMMARY

Both the mitochondrion and chloroplast contain organelle DNA, which encodes organelle rRNAs and tRNAs but only a few organelle proteins. Mitochondrial gene products identified thus far include several inner-membrane components of the electron transport chain. The size, coding capacity, and organization of mtDNA varies widely among different organisms. Certain proteins are encoded by nuclear DNA in one species and by mtDNA in another, suggesting that genes have moved from one DNA to the other during evolution. The mitochondrial genetic code differs from the code used in prokaryotes and in eukaryotic nuclear-encoded proteins; it also differs in different species. Animal mitochondrial RNAs undergo extensive processing, including polyadenylation.

The vast majority of mitochondrial proteins are encoded by nuclear genes, synthesized on cytosolic ribosomes, and imported post-translationally into the mitochondrion. Most such proteins are synthesized as precursors. Proteins destined for the matrix have an uptake-targeting sequence that permits their entry and then is cleaved off within the matrix. The N-terminus of protein destined for the intermembrane space is first imported into the matrix by an N-terminal matrix-targeting sequence and then the entire protein is redirected across the inner membrane by an intermembrane-space-targeting sequence. Other targeting sequences direct proteins to the inner or outer membrane.

Proteins to be imported into the mitochondrion are

kept in the cytosol in a partially unfolded state by the action of a chaperone—an ATP-dependent hsp70 protein related to the ER protein Bip. Import occurs at rate sites where the inner and outer mitochondrial membranes are in close contact. After the unfolded proteins bind to receptors on the outer mitochondrial membrane an energy-driven process, powered by the proton-motive force across the inner membrane, moves these proteins through a protein-lined channel into the organelle. Uptake into the matrix requires ATP hydrolysis catalyzed by a matrix hsp70 protein. The folding of many imported proteins is facilitated by an ATP-hydrolyzing chaperonin related to bacterial hsp60 chaperonins.

Chloroplast DNA encodes more than 120 genes, including rRNAs, tRNAs, RNA polymerase subunits, ribosomal proteins, subunits of the chloroplast photosynthetic electron transport complexes, subunits of the F_0F_1 ATPase complex, and the large (L) subunit of ribulose 1,5-bisphosphate carboxylase (rubisco). Like mitochondrial proteins, most chloroplast proteins are synthesized in the cytosol and then imported into the organelle. Such proteins contain uptake-targeting sequences that direct them first to the stromal space of the chloroplast; other sequences redirect them to the proper subcompartment, such as the thylakoid membrane or thylakoid lumen. Rubisco contains eight copies of the chloroplast-encoded L subunit and eight copies of the imported small (S) subunit; assembly requires a stromal hsp60 chaperonin—an L subunit–binding protein that stores excess L subunits and facilitates the binding of L and S subunits.

The chloroplast is only one type of plastid. Chromo-plasts, amyloplasts, and elaioplasts are plastids without chlorophyll or thylakoid vesicles; they store, respectively, colored pigments, starch, and lipids. All plastids contain the same chloroplast DNA. The formation of chloroplasts from the generic proplastid precursor is triggered by light induction.

All peroxisomal membrane and matrix proteins are incorporated into the organelle post-translationally. Many but not all peroxisomal matrix proteins utilize a C-terminal SKL targeting sequence that is not cleaved after import. Catalase is locked into the matrix by the addition of heme and the formation of tetramers. Patients with Zellweger syndrome are defective in the import of most proteins into the peroxisomal matrix; genetic analysis of patients' cells indicated that at least eight genes are involved in the import process. The gene defective in adrenoleukodystrophy encodes the protein that imports a peroxisomal enzyme required for the oxidation of very-long-chain fatty acids.

Nuclear proteins and small nuclear ribonucleoprotein particles (snRNPs) are imported from the cytosol through the nuclear pores, the same structures that allow completed RNAs and ribosomal subunits to move outward into the cytosol. Proteins to be imported into the nucleus contain nuclear localization sequences that cause them to bind to chaperones in the cytosol and then to receptors localized to the cytosolic side of the pores. Some nuclear localization signals are groups of positively charged amino acids; the trimethylguanosine cap that is at the 5′ end of small nuclear RNAs is the signal for uptake of snRNPs. ATP hydrolysis powers the import of bound proteins across the pores and into the nucleus.

REVIEW QUESTIONS

1. The biogenesis of individual organelles within a cell is an extremely important process. Likewise, the methods employed by the cell for targeting proteins to these individual compartments is vital for organelle function.

Mitochondria have their own genome (mtDNA). How is it structurally different from the nuclear genome? What DNA polymerase is responsible for replicating the mitochondrial genome? Are the growth and division of mitochondria limited to specific periods of the cell cycle? Are there any genetic anomalies that have been identified due to mutations within the mtDNA?

Considering that mitochondria contain their own genetic material, what proteins are encoded by it? Are the general types of proteins encoded within mitochondria always the same from species to species? If proteins are encoded by the mtDNA, transcription and translation must also occur within the mitochondria. Are the ribosomes and other enzymes associated with these processes always encoded by mtDNA, or are they recruited from the cytosol? Is the genetic code within this organelle identical to that found for genes in the nucleus? Review how the mtDNA is transcribed. Where does initiation occur? Do mitochondrial mRNAs contain introns? Recall, too, that in certain species many of the transcripts made following expression of mtDNA are subsequently edited. What is meant by this?

Human mtDNA is slightly larger than 16,000 base pairs; clearly it cannot encode all the proteins found within the mitochondria. How do the mitochondrial proteins encoded by genes in the nucleus enter the mitochondria? What are some of the characteristics of the targeting sequences that direct proteins to the mitochrondria? Figure 19-7 illustrates the process by which proteins enter the mitochondria from the cytosol. What is the function of the ATP needed for the process? How does the cell distinguish between proteins destined for the mitochondrial matrix and those that are to be inserted into the mitochondrial membranes? How are proteins targeted to the intermembrane space?

a. You are studying the uptake of proteins into the matrix of isolated mitochondria from yeast. (How would you purify mitochondria for this purpose?) To do this, you have prepared recombinant plasmids, and used them to synthesize the following derivatives of alcohol dehydrogenase (ADH) using cytoplasmic ribosomes in a cell-free system along with radiolabeled [^{35}S]methionine.

>   I. MTS--------ADH---------------- 38.2 kD
>   II. ADH--------------------- 35.3 kD
>   III. MTS*--------ADH---------------- 38.2 kD

MTS represents the matrix-targeting sequence normally associated with ADH. In I, the entire wild type sequence is present; in II, the MTS has been deleted; and in III, an MTS modified by site-specific mutagenesis to change arginine residues to aspartate amino acids has been added and is represented by MTS*.

Each of the three protein synthesis reactions is divided into two portions, A and B. No mitochondria are added to A, but they are added to B. Following incubation, a portion of each sample is exposed to the protease trypsin. An inhibitor of trypsin is then mixed with each sample, detergent is added, and the proteins are analyzed by gel electrophoresis to elucidate the size of the remaining ADH. The table below summarizes the experiment and the results; the numbers are the sizes of the radiolabeled proteins.

	A (− mitochondria)		B (+ mitochondria)	
	−trypsin	+trypsin	−trypsin	+trypsin
I	38.2 kD	—	38.2 kD	35.3 kD
II	35.3 kD	—	35.3 kD	—
III	38.2 kD	—	38.2 kD	—
The minus sign (−) indicates that no ADH was detected.				

Explain these results.

b. How do you think these data would be affected if an agent were added to the samples that collapsed the proton gradient of the mitochondria prior to incubation? What compound would you use to do this?

2. When yeast are grown anaerobically, which pathways are principally involved in the metabolism of glucose? Is the respiratory chain in anaerobically grown yeast complete? Would the yield of ATP per molecule of glucose metabolized be greater or less in yeast grown anaerobically as opposed to aerobically?

Figure 19-4 illustrates the organization of the yeast and human mitochondrial genomes (mtDNA). Pay particular attention to the yeast mtDNA and the products encoded by it.

How are mitochondria structurally different in yeast that are grown anaerobically as opposed to those grown in the presence of oxygen?

What happens to the mitochondria when anaerobically grown yeast are transferred to an oxygen-containing environment? What is the role of heme in expression of proteins involved in aerobic respiration?

What are *petite* mutant yeast strains?

a. You have isolated several strains of yeast, and you are investigating their growth under anaerobic and aerobic conditions. One criterion that you use is the size of the colonies formed under both conditions, which you label as small or large. A wild type yeast strain grows as small colonies under anaerobic conditions, but it grows as large colonies under aerobic conditions. Two mutant strains, A and B, grow as small colonies under both anaerobic and aerobic conditions, and you label them both as *petite* yeast strains. If strains A and B were unable to use oxygen, how would they be able to grow at all under aerobic conditions?

b. You propose the hypothesis that the mutant strains A and B are not capable of aerobic respiration because their mitochondria lack DNA. You decide to test this by isolating mtDNA from each strain. How would you do this?

You discover that strain A lacks mitochondrial DNA and strain B has it. In order to characterize the mutants further, you decide to perform some genetic crosses with these strains. You cross wild type yeast with strain A, generate diploid yeast, and allow them to sporulate, yielding haploids. Both the diploid yeast and all the haploid yeast grow well under aerobic conditions. You perform a similar cross using wild type yeast and strain B. In this case, the diploid yeast grow well under aerobic conditions, but only 50 percent of the haploid yeast yield large colonies while 50 percent yield small colonies. Explain these results in terms of mutations involving mtDNA, which would produce cytoplasmic *petites*, and those mutations that occur in the nuclear DNA.

Suppose you cross strains A and B. What would be the growth phenotype associated with the diploid yeast? What would be the growth phenotype of the haploid yeast following sporulation?

3. Chloroplasts, which are the sites of photosynthesis in plant cells, also contain their own genomes. What is the progenitor of these organelles? How do chloroplasts multiply? How is the organization of the chloroplast genome different from that of the nuclear genome? What types of proteins do they encode? In general, are the genomes of chloroplasts larger or smaller than those associated with mitochondria?

As with mitochondria, not all proteins found within the chloroplast are encoded by its own DNA. The remaining proteins are encoded by the nucleus and must be transported from the cytosol into the chloroplast. Review Figure 19-14. How do proteins destined for the stroma enter from the cytosol? How are proteins that must enter the thylakoid space targeted to that compartment?

a. Refer to Figure 19-15. Notice that the plastocyanin precursor has two signal sequences: one is designed for import into the stroma, and the other allows for import into the thylakoid space. Suppose you were to construct the plastocyanin precursor protein such that the thylakoid uptake-targeting sequence was deleted from it. You perform experiments using isolated chloroplasts, and you add your modified protein. In which compartment would you expect to find plastocyanin? If you wanted to determine whether proteins were taken up into the chloroplast from the cytosol at sites where the inner/outer chloroplast membranes are in apposition, how would you modify the system? Given that you have developed such a method, how would you identify biochemically the channel proteins and the contact sites that would allow cytosolic proteins to cross the membrane?

b. Suppose you deleted only the stromal-import sequence from the plastocyanin precursor molecule, and once again added this protein to isolated chloroplasts. Where would the protein be found?

c. Finally, you reverse the order of the import sequences, such that the N-terminus of the plastocyanin precursor is the thylakoid uptake-targeting sequence followed by the stromal-import sequence and then plastocyanin. If this protein were added to isolated chloroplasts, where would the plastocyanin be found following incubation?

4. Another important organelle within cells is the peroxisome. Do peroxisomes contain genetic material used to encode peroxisomal proteins? Where do the proteins within these organelles originate? What is the import targeting sequence that guides proteins from the cytosol to the peroxisome? Is this sequence removed once the protein enters the peroxisome?

What is Zellweger syndrome? How many complementation groups exist for this disease? What does this tell you about the number of gene products necessary for importing peroxisomal proteins into this organelle? Recall that the peroxisomes in cells derived from patients with Zellweger syndrome are void of the array of enzymes normally associated with the interior of the organelle. How would this differ from a disorder caused by a single peroxisomal protein having a mutation in the SKL signal uptake sequence?

5. The cell nucleus is also an organelle that must import proteins from the cytoplasm in order to function. Included among these proteins are DNA polymerase, RNA polymerase, and histones. What is the general mechanism for transporting proteins from the cytoplasm to the nucleus? Be sure you understand the generalities concerning the signal sequences. Is the signal sequence located at the N-terminus, the C-terminus, or elsewhere within the protein? What is the difference between a single and bipartite nuclear localization signal sequence? Is ATP necessary for transport to occur?

a. You are investigating the import of chicken estrogen receptor (CER) (see Table 19-6) protein by nuclei isolated from chicken cells. You are using both wild type protein and a mutant protein where you have deleted amino acids 250–266. You have prepared each of these proteins by synthesis in a cell-free systems with [^{35}S]methionine. You incubate each with isolated nuclei in the presence of ATP and then add trypsin. You then add a general protease inhibitor, lyse the nuclei, and resolve the proteins by gel electrophoresis to determine if the receptor protein is present. The results are presented in the table, where a plus sign (+) represents the presence of radioactive chicken estrogen receptor, and a minus sign (−) represents that no protein was detected.

| | −nuclei | | +nuclei | |
	−trypsin	+trypsin	−trypsin	+trypsin
wild type CER	+	−	+	+
mutant CER	+	−	+	−

Explain these results.

b. If you were to exclude ATP from the reaction, what would be the result of the experiment. Why? How could you visualize or isolate the CER bound to a nuclear pore?

c. There are many cytosolic proteins that bind to nuclear localization sequences; they probably behave as chaperones. How would you determine if chaperones were involved in the uptake of a specific protein into the nucleus?

References

The Endosymbiont Hypothesis

GRAY, M. W. 1992. The endosymbiont hypothesis revisited. *Int. Rev. Cytol.* **141**:233–357.

GRAY, M. W. 1993. Origin and evolution of organelle genomes. *Curr. Opin. Genet. Dev.* **3**:884–890.

MARGULIS, L. 1993. *Symbiosis in Cell Evolution: Microbial communities in the Archean and Proterozoic Eons,* 2d ed. W. H. Freeman and Company, New York.

Mitochondria

Mitochondrial DNA

ANDERSON, S., et al. 1981. Sequence and organization of the human mitochondrial genome. *Nature* **290**:457–465.

BONEN, L. 1991. The mitochondrial genome: so simple yet so complex. *Curr. Opin. Genet. Dev.* **1**:515–522.

GRIVALL, L. A. 1989. Mitochondrial DNA: Small, beautiful, and essential. *Nature* **341**:569–571.

LEVINGS, C. S., III, and G. G. BROWN. 1989. Molecular biology of plant mitochondria. *Cell* **56**:171–179.

NEWTON, K. J. 1988. Plant mitochondrial genomes: organization, expression, and variation. *Ann. Rev. Plant Physiol. and Mol. Biol.* **39**:503–532.

Genetic Diseases

BROWN, M. D., et al. 1992. Leber's hereditary optic neuropathy: a model for mitochondrial neurodegenerative diseases. *FASEB J.* **6**:2791–2799.

HARDING, A. E. 1991. Neurological disease and mitochondrial genes. *Trends Neurosci.* **14**:132–138.

SCHAPIRA, A. H. 1993. Mitochondrial disorders. *Curr. Opin. Genet. Dev.* **3**:457–465.

WALLACE, D. C. 1993. Mitochondrial diseases: genotype versus phenotype. *Trends Genet.* **9**:128–33.

Mitochondrial RNA

CLAYTON, D. A. 1991. Replication and transcription of vertebrate mitochondrial DNA. *Ann. Rev. Cell Biol.* **7**:453–478.

GRAY, M. W., and P. S. COVELLO. 1993. RNA editing in plant mitochondria and chloroplasts. *FASEB J.* **7**:64–71.

SHADEL, G. S., and D. A. CLAYTON. 1993. Mitochondrial transcription initiation. Variation and conservation. *J. Biol. Chem.* **268**:16083–16086.

SIMPSON, L., and J. SHAW. 1989. RNA editing and the mitochondrial cryptogenes of kinetoplastid protozoa. *Cell* **57**:355–366.

STUART, K. 1993. The RNA editing process in trypanosoma brucei. *Semin. Cell Biol.* **4**:251–260.

Protein Import

BAKER, K. P., and G. SCHATZ. 1991. Mitochondrial proteins essential for viability mediate protein import into yeast mitochondria. *Nature* **349**:205–208.

GLICK, B. S., C. WACHTER, and G. SCHATZ. 1992. The energetics of protein import into mitochondria. *Biochem. Biophys. Acta* **1101**:249–251.

GLICK, B. S., E. M. BEASLEY, and G. SCHATZ. 1992. Protein sorting in mitochondria. *Trends Biochem. Sci.* **17**:453–459.

PFANNER, N., et al. 1992. A dynamic model of the mitochondrial protein import machinery. *Cell* **68**:999–1002.

PON, L. A., et al. 1989. Interaction between mitochondria and the nucleus. *J. Cell Sci. Suppl.* **11**:1–11.

SEGUI-REAL, B., R. A. STUART, and W. NEUPERT. 1992. Transport of proteins into the various subcompartments of mitochondria. *FEBS LETT.* **313**:2–7.

STUART, R. S., D. M. CYR, E. A. CRAIG, AND W. NEUPERT. 1994. Mitochondrial molecular chaperones: their role in protein translocation. *Trends Biochem Sci.* **19**:87–92.

Chaperones and protein folding in the mitochondrion

GETHING, M. J., and J. SAMBROOK. 1992. Protein folding in the cell. *Nature* **355**:33–45.

HARTL, F. U., J. MARTIN, and W. NEUPERT. 1992. Protein folding in the cell: the role of molecular chaperones hsp70 and hsp60. *Ann. Rev. Biophys. Biomol. Struct.* **21**:293–322.

HARTL, F. U., R. HOLDAN, and T. LAUGER. 1994. Molecular chaperones in protein folding: the art of avoiding sticky situations. *Trends Biochem. Sci.* **19**:20–25.

HENDRICK, J. P., and F. U. HARTL. 1993. Molecular chaperone functions of heat-shock proteins. *Ann. Rev. Biochem.* **62**:349–384.

LANGER, T., et al. 1992. Chaperonin-mediated protein folding: GroES binds to one end of the GroEL cylinder, which accommodates the protein substrate within its central cavity. *EMBO J.* **11**:4757–4765.

STUART, R. A., et al. 1994. Mitochondrial molecular chaperones: their role in protein translocation. *Trends Biochem. Sci.* **19**:87–92.

Chloroplasts

Chloroplast DNA and RNA

OHYAMA, K., et al. 1986. Chloroplast gene organization deduced from complete sequence of liverwort Marchantia polymorpha chloroplast DNA. *Nature* **322**:572–574.

OHYAMA, K., et al. 1988. Newly identified groups of genes in chloroplasts. *Trends Biochem. Sci.* **13**:19–22.

SUGIURA, M. 1992. The chloroplast genome. *Plant Mol. Biol.* **19**:149–168.

WOLFE, K. H., C. W. MORDEN, and J. D. PALMER. 1991. Ins and outs of plastid genome evolution. *Curr. Opin. Genet. Dev.* **1**:523–529.

Chloroplast RNA and Gene Expression

CLEGG, M. T., B. S. GAUT, G. H. LEARN, and B. P. MORTON. 1994. Rates and patterns of chloroplast DNA evolution. *Proc. Natl. Acad. Sci.* **91**:6795–6801.

GRUISSEM, W. 1989. Chloroplast gene expression: how plants turn their plastids on. *Cell* **56**:161–170.

MAYFIELD, S. P. 1990. Chloroplast gene regulation: interaction of the nuclear and chloroplast genomes in the expression of photosynthetic proteins. *Curr. Opin. Cell Biol.* **2**:509–513.

MOSES, P. B., and N.-H. CHUA. 1988. Light switches for plant genes. *Sci. Am.* **258**(4):88–93.

PRING, D., A. BRENNICKE, and W. SCHUSTER. 1993. RNA editing gives a new meaning to the genetic information in mitochondria and chloroplasts. *Plant Mol. Biol.* **21**:1163–1170.

ROCHAIX, J. D. 1992. Post-transcriptional steps in the expression of chloroplast genes. *Ann. Rev. Cell Biol.* **8**:1–28.

Protein Import Into Chloroplasts

ARCHER, E. K., and K. KEEGSTRA. 1990. Current views on chloroplast protein import and hypotheses on the origin of the transport mechanism. *J. Bioenerg. Biomembr.* **22**:789–810.

DOUWE, D. B. A., and P. J. WEISBEEK. 1991. Chloroplast protein topogenesis: import, sorting and assembly. *Biochem. Biophys. Acta* **1071**:221–253.

HEMMINGSEN, S. M. 1990. The plastid chaperonin. *Semin. Cell Biol.* **1**:47–54.

LI, H. M., T. MOORE, and K. KEEGSTRA. 1991. Targeting of proteins to the outer envelope membranes uses a different pathway than transport into chloroplasts. *Plant Cell* **3**:709–717.

PERRY, S. E., H. M. LI, and K. KEEGSTRA. 1991. In vitro reconstitution of protein transport into chloroplasts. *Methods Cell Biol.* **34**:327–344.

ROBINSON, C., et al. 1993. Protein translocation across the thylakoid membrane—a tale of two mechanisms. *FEBS Lett.* **325**:67–69.

SMEEKENS, S., P. WEISBEEK, and C. ROBINSON. 1990. Protein transport into and within chloroplasts. *Trends Biochem. Sci.* **15**:73–76.

Peroxisomes

Biogenesis of Peroxisomes

DE HOOP, M. J., and G. AB. 1992. Import of proteins into peroxisomes and other microbodies. *Biochem. J.* **286**:657–669.

GARTNER, J., and D. VALLE. 1993. The 70 kDa peroxisomal membrane protein: an ATP-binding cassette transporter protein involved in peroxisome biogenesis. *Semin. Cell Biol.* **4**:45–52.

LAZAROW, P. B. 1989. Peroxisome biogenesis. *Curr. Opin. Cell Biol.* **1**:630–634.

SUBRAMANI, S. 1993. Proteins import into peroxisomes and biogenesis of the organelle. *Ann. Rev. Cell Biol.* **9**:445–478.

Genetic Diseases

MOSER, H. W. 1993. Peroxisomal diseases. *Adv. Hum. Genet.* **21**:1–106.

MOSSER, J., et al. 1993. Putative X-linked adrenoleukodystrophy gene shares unexpected homology with ABC transporters. *Nature* **361**:726–730.

RIZZO, W. B. 1993. Lorenzo's oil—hope and disappointment. *N. Engl. J. Med.* **329**:801–802.

SHIMOZAWA, N., et al. 1992. A human gene responsible for Zellweger syndrome that affects peroxisome assembly. *Science* **255**:1132–1134.

SINGH, I., et al. 1992. Transport of fatty acids into human and rat peroxisomes. Differential transport of palmitic and lignoceric acids and its implication to X-linked adrenoleukodystrophy. *J. Biol. Chem.* **267**:13306–13313.

Protein Import Into the Cell Nucleus

DINGWALL, C., and R. A. LASKEY. 1986. Protein import into the cell nucleus. *Ann. Rev. Cell Biol.* **2**:367–390.

DINGWALL, C., and R. A. LASKEY. 1991. Nuclear targeting sequences—a consensus? *Trends Biochem. Sci.* **16**:478–481.

FELDHERR, C. M., and D. AKIN. 1990. EM visualization of nucleocytoplasmic transport processes. *Electron. Microsc. Rev.* **3**:73–86.

FORBES, D. J. 1992. Structure and function of the nuclear pore complex. *Ann. Rev. Cell Biol.* **8**:495–527.

GERACE, L. 1992. Molecular trafficking across the nuclear pore complex. *Curr. Opin. Cell Biol.* **4**:637–645.

HANOVER, J. A. 1992. The nuclear pore: at the crossroads. *FASEB J.* **6**:2288–2295.

NEWMEYER, D. D. 1993. The nuclear pore complex and nucleocytoplasmic transport. *Curr. Opin. Cell Biol.* **5**:395–407.

NIGG, E. A., P. A. BAEUERLE, and R. LUHRMANN. 1991. Nuclear import-export: in search of signals and mechanisms. *Cell* **66**:15–22.

OSBORNE, M. A., and P. A. SILVER. 1993. Nucleocytoplasmic transport in the yeast Saccharomyces cerevisiae. *Ann. Rev. Biochem.* **62**:219–254.

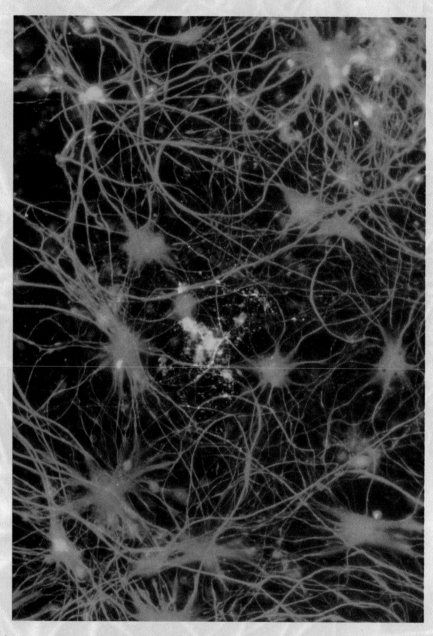

▲ Two types of non-neuronal cells in primary cultures of the rat central nervous system. The oligodendrocyte, a cell in the central nervous system that synthesizes myelin, is revealed by a green-fluorescing antibody specific for myelin basic protein, a principal component of myelin. The astrocytes are visualized by a red-fluorescing antibody to Glial Fibrillary Acidic Protein, a type of intermediate filament protein expressed exclusively in glial cells, almost always astrocytes. *Photograph courtesy of Nancy Kedersha.*

In Part IV we discuss a variety of specific properties of cells. We consider first the integrative functions, the methods by which cells tell each other what they are up to and direct each other to carry out particular functions. This involves sending both chemical and electrical signals, the latter activity being a special capability of nerve cells. We then turn to mechanical issues: the structural properties of cells, their ability to move materials inside themselves, their crawling movements, and finally, the ability of muscle cells to move parts of the body. This is followed by a consideration of how tissues are constructed by specific interactions of cells with each other and with intercellular materials. Next we discuss the control of normal cell growth and the abnormal growth of cells in cancerous tissues. We end Part IV with the most highly integrative section of the book, a presentation of how a number of different cell types cooperate together to allow the body to fight off invading microorganisms.

Chapter 20 deals with cell to cell signaling by hormones and hormone-like proteins. To integrate their activities, cells participate in a network of interactions in which any one cell may either direct the activities of other cells or be directed by them. Secretion of informational molecules by one cell gives directions that are recognized by specific receptors or target cells. This elaborate communications network insures that the body works as a unitary entity. Chemical signals sent throughout the body, however, do not allow for fine control; therefore the nervous system evolved as an electrical conduction system, with chemical relays that can deliver and integrate precise signals, for instance, the signals that allow your eyes to scan this page. Chapter 21 describes how nerve cells carry signals and shows how this is a specialized use of the more general ability of cells to maintain a voltage difference between them-

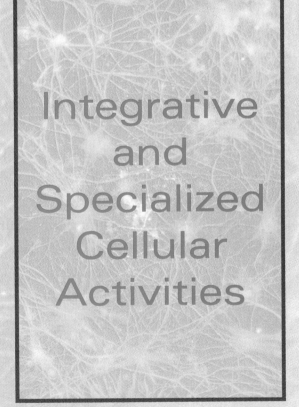

Integrative
and
Specialized
Cellular
Activities

selves and the fluid that bathes them.

Not only must cells integrate activities among themselves, they must orchestrate the activities within themselves. They do this partially through networks of fibers that define a cell's shape and underlie the ability of cells to move and translocate materials to particular intracellular locations. In Chapter 22 we describe the actin-based microfilament system of cells, which allows cells to elongate, to move, and, in muscle cells, to do the body's mechanical work. Chapter 23 describes the two other major intracellular fiber systems: the microtubules and the intermediate filaments. Both microtubules and intermediate filaments provide cells with internal rigidity, but the former also act as highways for sending particles, organelles, and even the chromosomes at mitosis to their appropriate locations in cells. Protein motors attached to the moving substances are the secret to their transport ability.

Having discussed cell shape in Chapter 24 from the perspective of the cell's interior, we turn back to the outside of the cell to consider what holds cells together in tissues. This involves interactions among identical cells—like muscle cells in a muscle fiber or liver cells in liver lobe—and interactions among different cell types, such as the passage of blood vessels through the liver. It involves the important interaction of cells with the extracellular matrix, a complex of fibers and connectors that fills the spaces between cells and that can itself form organized structures like the basement membranes that define tissue spaces. The matrix also provides pathways for cell migration.

One crucial aspect of cellular integration is taken up in Chapter 25: the decision of cells to grow or remain quiescent. To maintain a reasonably constant size, cells divide as they grow, making cell division and the duplication of chromosomes key elements of cell behavior. The control of cell division brings us back to signal transduction mechanisms and protein kinases because kinases are the key to regulating each decision step of the cell cycle. This naturally leads to Chapter 26, where we discuss why cells sometimes grow when more of that cell type is not required by the body, leading ultimately to the uncontrolled growth and division characteristic of cancer cells. Here we learn that the genes whose products are involved in signalling normal cell growth can be mutated to forms, called oncogenes, that signal cancer cell growth.

We end the book with a chapter that describes how events in different cells can be integrated to allow a particular bodily process, immunity. Chapter 27 introduces the B cells that make antibodies and the T cells that both control B cell behavior and provide their own immune function of killing infected or foreign cells. We also learn about the central role of macrophages that ingest and digest foreign materials and then present peptides on their cell surface for recognition by T cells. The chapter also shows how immune cells can specifically reorganize their DNA, allowing a relatively small number of genes to encode millions of different antibodies and T cell receptor molecules. All of the themes in earlier chapters come together in this consideration of a single bodily process.

20

Cell-to-Cell Signaling: Hormones and Receptors

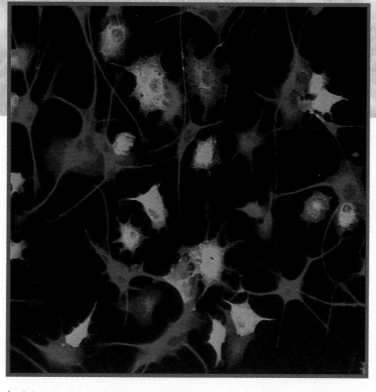

▲ Injection of a dominant-interfering mutation of MAP kinase kinase blocks signal-dependent neurite outgrowth (yellow cells).

No cell lives in isolation. In all multicellular organisms, survival depends on an elaborate intercellular communication network that coordinates the growth, differentiation, and metabolism of the multitude of cells in the diverse tissues and organs. Cells within small groups often communicate by direct cell-cell contact (see Chapter 24). Specialized junctions in the plasma membranes of adjacent cells permit them to exchange small molecules and to coordinate metabolic responses; other junctions between adjacent cells determine the shape and rigidity of many tissues. In addition, the establishment of specific cell-cell interactions between different types of cells is a necessary step in the development of many tissues. In some cases a particular protein on one cell binds to a receptor protein on the surface of an adjacent target cell, triggering its differentiation.

In this chapter, we examine another important mechanism by which cells communicate, principally by means of extracellular *signaling molecules*. These specific substances are synthesized and released by *signaling cells* and produce a specific response only in *target cells* that have receptors for the signaling molecules. An enormous variety of chemicals, including small molecules (e.g., amino acids, acetylcholine), peptides, and proteins, are used in this type of cell-to-cell communication. The extracellular products synthesized by signaling cells can diffuse away or be transported in the blood, thus providing a means for cells to communicate over longer distances than is possible by chains of direct cell-cell contacts.

We begin with a general discussion of signaling molecules, cell-surface receptors, and the role of extracellular signals. We then describe techniques for detecting, purifying, and cloning cell-surface receptors. Next we examine in some detail several signal-transduction pathways and the role of various intracellular molecules, termed *second messengers,* that relay signals from activated receptors and help regulate cellular metabolism. Several mechanisms whereby binding of signaling molecules to cell-surface receptors leads to activation of particular transcription factors and changes in gene expression are discussed in the last section. Obviously, there are too many important signaling molecules and receptors to cover them all in this chapter, and we focus on those whose function in animal cells is reasonably well understood at the cellular and molecular level. Table 20-1 lists many of the abbreviations for signaling molecules and other components of signal pathways that are used throughout this chapter.

▶ Overview of Extracellular Signaling

Communication by extracellular signals usually involves six steps: (1) synthesis and (2) release of the signaling molecule by the signaling cell; (3) transport of the signal to the target cell; (4) detection of the signal by a specific receptor protein; (5) a change in cellular metabolism or gene expression triggered by the receptor-signaling molecule complex; and (6) removal of the signal, often terminating the cellular response.

In many microorganisms (e.g., yeast, slime molds, and

TABLE 20-1   Common Abbreviations Related to Cell-to-Cell Signaling

Abbreviation	Name	Type of Molecule
AC	Adenylate cyclase	Effector enzyme coupled to G_s protein–linked receptors
ACTH	Adrenocorticotropic hormone	Extracellular signaling molecule
cAMP	3',5'-cyclic AMP	Second messenger
cAPK	cAMP-dependent protein kinase	Enzyme that triggers cascade leading to glycogen breakdown in response to rise in intracellular cAMP
CG	Chorionic gonadotropin	Extracellular signaling molecule
cGMP	3',5'-cyclic GMP	Second messenger
DAG	1,2-diacylglycerol	Second messenger
EGF	Epidermal growth factor	Extracellular signaling molecule
FGF	Fibroblast growth factor	Extracellular signaling molecule
FSH	Follicle-stimulating hormone	Extracellular signaling molecule
GAP	GTPase-activating protein	Enzyme in RTK-Ras signaling pathway that converts active Ras · GTP to inactive Ras · GDP
GEF	Guanine nucleotide–exchange factor	Cytosolic protein in RTK-Ras signaling pathway that promotes formation of active Ras · GTP
GPK	Glycogen phosphorylase kinase	Key regulatory enzyme in glycogenolysis that is activated in response to a rise in intracellular cAMP or cytosolic Ca^{2+}
IGF-1	Insulin-like growth factor 1	Extracellular signaling molecule
IP_3	Inositol 1,4,5-trisphosphate	Second messenger
IRS1	Insulin receptor substrate 1	Relay protein that functions in signaling initiated by insulin receptor
LH	Luteinizing hormone	Extracellular signaling molecule
PDGF	Platelet-derived growth factor	Extracellular signaling molecule
PI	Phosphatidylinositol	Phospholipid in plasma membrane
PI-3 kinase	Phosphatidylinositol-3 kinase	Enzyme localized to cell membrane by binding to activated RTKs
PIP_2	Phosphatidylinositol 4,5-bisphosphate	Membrane-bound substrate in synthesis of the second messengers DAG and IP_3
PLC	Phospholipase C	Membrane-associated enzyme that converts PIP_2 to DAG and IP_3
RTK	Receptor tyrosine kinase	Type of cell-surface receptor with tyrosine kinase activity in its cytosolic domain
RYR	Ryanodine receptor	Ca^{2+} channels in membranes of intracellular compartments in muscle and nerve cells

protozoans), secreted molecules coordinate the aggregation of free-living cells for sexual mating or differentiation under certain environmental conditions. Chemicals released by one organism that can alter the behavior or gene expression of other organisms of the same species are called *pheromones.* Yeast mating-type factors discussed in other chapters are a well-understood example of pheromone-mediated cell-to-cell signaling. Some algae and animals also release pheromones, usually dispersing them into the air or water, to attract members of the opposite sex. More important in plants and animals are extracellular signaling molecules that function *within* an organism to control metabolic processes within cells, the growth of tissues, the synthesis and secretion of proteins, and the composition of intracellular and extracellular fluids. To date, no receptor for any plant signaling molecule has been identified with certainty, and the understanding of signaling in plants is much less advanced than that of signaling in animals and microorganisms. For this reason, we focus in this chapter on cell signaling in eukaryotic microorganisms and in a variety of higher eukaryotes, particularly mammals.

Signaling Molecules Operate over Various Distances in Animals

In animals, signaling by extracellular, secreted molecules can be classified into three types—endocrine, paracrine, or autocrine—based on the distance over which the signal acts. In addition, certain membrane-bound proteins on one cell can directly signal an adjacent cell (Figure 20-1).

In *endocrine signaling,* signaling molecules, called *hormones,* act on target cells distant from their site of synthesis by cells of endocrine organs. In animals, an endocrine hormone usually is carried by the blood from its site of release to its target.

In *paracrine signaling,* the signaling molecules released by a cell only affect target cells in close proximity to it. The conduction of an electric impulse from one nerve cell to another or from a nerve cell to a muscle cell (inducing or inhibiting muscle contraction) occurs via paracrine signaling. The role of this type of signaling, mediated by *neurotransmitters* and *neurohormones,* in nerve conduction is discussed in Chapter 21.

In *autocrine signaling,* cells respond to substances that they themselves release. Many growth factors act in this fashion, and cultured cells often secrete growth factors that stimulate their own growth and proliferation. Tumor cells commonly overproduce and release growth factors that stimulate inappropriate, unregulated proliferation of the tumor cells, as well as adjacent nontumor cells; this process may lead to formation of tumor mass.

Some compounds can act in two or even three types of cell-to-cell signaling. Certain small peptides, for example, function both as neurotransmitters (paracrine signaling) and as systemic hormones (endocrine signaling). Some pro-

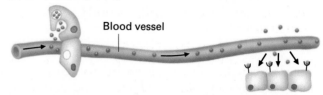

(a) Endocrine signaling

Blood vessel

Hormone secretion into blood by endocrine gland

Distant target cells

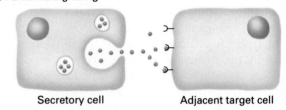

(b) Paracrine signaling

Secretory cell   Adjacent target cell

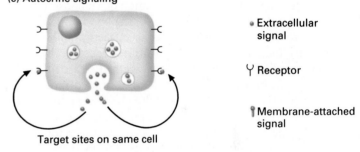

(c) Autocrine signaling

• Extracellular signal

Y Receptor

Membrane-attached signal

Target sites on same cell

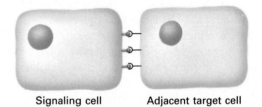

(d) Signaling by plasma membrane–attached proteins

Signaling cell   Adjacent target cell

▲ FIGURE 20-1 General schemes of intercellular signaling in animals. (a–c) Cell-to-cell signaling by extracellular chemicals occurs over distances from a few micrometers in autocrine and paracrine signaling to several meters in endocrine signaling. (d) Proteins attached to the plasma membrane of one cell can interact directly with receptors on an adjacent cell.

tein hormones, such as epidermal growth factor (EGF), are synthesized as the exoplasmic part of a plasma-membrane protein; membrane-bound EGF can bind to and signal an adjacent cell by direct contact. Cleavage by a protease releases secreted EGF, which acts as an endocrine signal on distant cells.

Receptor Proteins Exhibit Ligand-Binding Specificity and Effector Specificity

As noted earlier, the cellular response to a particular extracellular signaling molecule depends on its binding to a specific *receptor protein* located on the surface of a target cell or in its nucleus or cytosol. We can refer to the signaling molecule (a hormone, pheromone, or neurotransmitter) as the *ligand,* which binds to, or "fits," a site on the receptor. Binding of a ligand to its receptor causes a conformational change in the receptor that initiates a sequence of reactions leading to a change in cellular function.

The response of a cell or tissue to specific hormones is dictated by the particular hormone receptors it possesses and by the intracellular reactions initiated by the binding of any one hormone to its receptor. Different cell types may have different sets of receptors for the same ligand, each of which induces a different response. Or the same receptor may occur on various cell types, and binding of the same ligand may trigger a different response in each type of cell. Clearly, different cells respond in a variety of ways to the same ligand. For instance, acetylcholine receptors are found on the surface of striated muscle cells, heart muscle cells, and pancreatic acinar cells. Release of acetylcholine from a neuron adjacent to a striated muscle cell triggers contraction, whereas release adjacent to a heart muscle slows the rate of contraction. Release adjacent to a pancreatic acinar cell triggers exocytosis of secretory granules that contain digestive enzymes. On the other hand, different receptor-ligand complexes can induce the same cellular response in some cell types. In liver cells, for example, the binding of either glucagon to its receptors or of epinephrine to its receptors can induce degradation of glycogen and release of glucose into the blood.

These examples show that a receptor protein is characterized by *binding specificity* for a particular ligand, and the resulting hormone-ligand complex exhibits *effector specificity* (i.e., mediates a specific cellular response). For instance, activation of either epinephrine or glucagon receptors on liver cells by binding of their respective ligands induces synthesis of the same intracellular second messenger (i.e., cyclic AMP); as a result, the effects of both receptors on liver-cell metabolism are the same. Thus, the binding specificity of epinephrine and glucagon receptors differ, but their effector specificity is identical.

In most receptor-ligand systems, the ligand appears to have no function except to bind to the receptor. The ligand is not metabolized to useful products, is not an intermediate in any cellular activity, and has no enzymatic properties. The only function of the ligand appears to be to change the properties of the receptor, which then signals to the cell that a specific product is present in the environment. Target cells often modify or degrade the ligand and, in so doing, can modify or terminate their response or the response of neighboring cells to the signal.

Hormones Can Be Classified Based on Their Solubility and Receptor Location

Hormones fall into three broad categories: (1) small lipophilic molecules that diffuse across the plasma membrane and interact with intracellular receptors; and (2) hydrophilic or (3) lipophilic molecules that bind to cell-surface receptors (Figure 20-2). Here we briefly describe each class of hormone; later we discuss the mechanisms that regulate the synthesis, release, and degradation of hormones.

Lipophilic Hormones with Intracellular Receptors Many lipid-soluble hormones diffuse across the plasma membrane and interact with receptors in the cytosol or nucleus. The resulting hormone-receptor complexes bind to transcription-control regions in DNA thereby affecting expression of specific genes. Hormones of this type include the steroids, thyroxine, and retinoic acid (Table 20-2; see also Figure 11-57).

All steroids are synthesized from cholesterol and have similar chemical skeletons. After crossing the plasma membrane, steroid hormones interact with intracellular receptors, forming complexes that can increase or decrease transcription of specific genes (see Figure 11-62). These receptor-steroid complexes also may affect the stability of specific mRNAs. Steroids are effective for hours or days and often influence the growth and differentiation of specific tissues. For example, estrogen and progesterone, the female sex hormones, stimulate the production of egg-white hormones in chickens and cell proliferation in the hen oviduct. In mammals, estrogens stimulate growth of the uterine wall in preparation for embryo implantation. In insects and crustaceans, α-ecdysone (which is chemically related to steroids) triggers the differentiation and maturation of larvae; like estrogens, it induces the expression of specific gene products.

Thyroxine, including tetraiodothyronine and triiodothyronine—the principal iodinated compounds in the body—is formed in the thyroid by intracellular proteolysis of the iodinated protein thyroglobulin. Thyroxine is produced by thyroid cells and immediately released into the blood. Thyroid hormones stimulate the catabolism of glucose, fats, and proteins by increasing the levels of many enzymes that catalyze these metabolic reactions (e.g., liver hexokinase) and mitochondrial enzymes for oxidative phosphorylation.

Water-Soluble Hormones with Cell-Surface Receptors Water-soluble signaling molecules, which cannot dif-

(a) Intracellular receptors

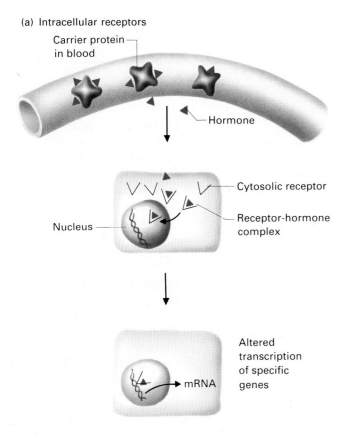

(b) Cell-surface receptors

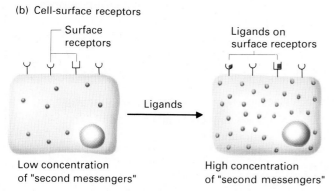

▲ FIGURE 20-2   Some hormones bind to intracellular receptors; others, to cell-surface receptors. (a) Steroid hormones and thyroxine, being lipophilic, are transported by carrier proteins in the blood. After dissociation from these carriers, such hormones diffuse across the cell membrane and bind to specific receptors in the cytosol or nucleus. The receptor-hormone complex then acts on nuclear DNA to alter transcription of specific genes. (b) Polypeptide hormones and catecholamines (e.g., epinephrine), which are water soluble, and prostaglandins, which are lipophilic, all bind to cell-surface receptors. This binding triggers an increase or decrease in the cytosolic concentration of second messengers (e.g., cAMP, Ca^{2+}), activation of a protein kinase, or a change in the membrane potential.

fuse across the plasma membrane, bind to *cell-surface receptors*. This large class of compounds is composed of two groups: (1) peptide hormones, such as insulin, growth hormones, and glucagon, which range in size from a few amino acids to protein-size compounds, and (2) small charged molecules, such as epinephrine and histamine, that are derived from amino acids and function as hormones and neurotransmitters (Table 20-3).

Many water-soluble hormones induce a modification in the activity of one or more enzymes already present in the target cell. In this case, the effects of the surface-bound hormone usually are nearly immediate, but persist for a short period only. The effects of peptide growth factors, however, often extend over days. These factors are required for the growth and differentiation of specific types of cells, and many induce changes in gene expression in a few minutes.

Lipophilic Hormones with Cell-Surface Receptors

The primary lipid-soluble hormones that bind to cell-surface receptors are the *prostaglandins* (see Table 20-3). These hormone-like compounds contain a cyclopentane ring and are synthesized from *arachidonic acid,* a 20-carbon fatty acid with four double bonds (see Table 14-2). There are at least 16 different prostaglandins in nine different chemical classes, designated PGA–PGI. In both vertebrates and invertebrates, prostaglandins are synthesized and secreted continuously by many types of cells and rapidly broken down by enzymes in body fluids.

Many prostaglandins act as local mediators during paracrine signaling and are destroyed near the site of their synthesis. They modulate the responses of other hormones and can have profound effects on many cellular processes. Certain prostaglandins cause blood platelets to aggregate and adhere to the walls of blood vessels. Because platelets play a key role in clotting blood and plugging leaks in blood vessels, these prostaglandins can affect the course of vascular disease and wound healing; aspirin and other anti-inflammatory agents inhibit their synthesis. Other prostaglandins initiate the contraction of smooth muscle cells; they accumulate in the uterus at the time of childbirth and appear to be important in inducing uterine contraction.

Effects of Many Hormones Are Mediated by Second Messengers

The binding of ligands to many cell-surface receptors leads to a short-lived increase (or decrease) in the concentration of intracellular signaling molecules termed *second messengers.* Among the most important second messengers are $3',5'$-cyclic AMP (cAMP); $3',5'$-cyclic GMP (cGMP); 1,2-diacylglycerol (DAG); inositol 1,4,5-trisphosphate (IP_3);

TABLE 20-2   Structure and Function of Some Lipid-Soluble Hormones that Bind to Intracellular Receptors

Hormone	Structure	Origin	Major Effects
STEROIDS			
Progesterone		Ovary, mainly corpus luteum; placenta	Differentiation of the uterus in preparation for implantation of the early embryo; maintenance of early pregnancy; development of the alveolar system in mammary glands
Estradiol (one of three estrogens)		Ovary (stroma, granulosa, theca); placenta	Differentiation of the uterus and other female sex organs; maintenance of secondary female sex characteristics and the normal cyclic function of accessory sex organs; development of the duct system in mammary glands
Testosterone		Testis	Maturation and normal function of accessory male sex organs; development of male sex characteristics
STEROIDLIKE HORMONES			
α-Ecdysone		Endocrine glands (insects, crustaceans)	Differentiation and maturation of larvae
OTHER HORMONES			
Thyroxine Tetraiodo-thyronine (T$_4$)		Thyroid	Increased heat production; maintenance of metabolism of glucose and other fuels; broad effects on gene expression and induction of enzymatic synthesis

and Ca^{2+} (Figure 20-3). The elevated intracellular concentration of one or more such second messengers following hormone binding triggers a rapid alteration in the activity of one or more enzymes or nonenzymatic proteins. The metabolic functions controlled by hormone-induced second messengers include uptake and utilization of glucose, storage and mobilization of fat, and secretion of cellular products. These intracellular molecules also control proliferation, differentiation, and survival of cells, in part by regulating the transcription of specific genes. The mode of action of cAMP and other second messengers is discussed in later sections. Removal (or degradation) of a ligand or second messenger, or inactivation of the ligand-binding receptor, can terminate the cellular response to an extracellular signal.

Cell-Surface Receptors Can Be Categorized into Four Major Classes

The different types of cell-surface receptors that interact with water-soluble ligands are schematically represented in Figure 20-4. Binding of ligand to some of these receptors induces second-messenger formation, whereas ligand binding to others does not. For convenience, we can sort these receptors into four classes:

1. G protein–linked receptors (see Figure 20-4a): Ligand binding activates a G protein, which in turn activates or inhibits an enzyme that generates a specific second messenger or an ion channel, causing a change in membrane potential. The receptors for epinephrine, serotonin, and glucagon are examples.

2. Ion-channel receptors (see Figure 20-4b): Ligand binding changes the conformation of the receptor so that specific ions flow through it; the resultant ion movements alter the electric potential across the cell membrane. The acetylcholine receptor at the nerve-muscle junction is an example.

3. Receptors lacking intrinsic catalytic activity but directly associated with cytosolic protein tyrosine kinases (see Figure 20-4c): Ligand binding causes receptor monomers to dimerize; the dimeric receptor then interacts with and activates one or more cytosolic protein tyrosine kinases. The receptors for many cytokines, the interferons, and for human growth factor are of this type. These tyrosine kinase–linked receptors sometimes are referred to as the *cytokine-receptor superfamily.*

4. Receptors with intrinsic enzymatic activity: Several types of receptors fall into this class. Binding of ligand to some receptors activates guanylate cyclase activity or protein phosphatase activity in the cytosolic domain (see Figure 20-4d,e). The receptors for insulin and many growth factors are ligand-triggered protein kinases; in most cases, ligand binding leads to dimerization and activation of either serine/threonine kinase or tyrosine kinase activity (see Figure 20-4f,g). These receptors—often referred to as *receptor serine/threonine kinases* and *receptor tyrosine kinases* (RTKs)—phosphorylate residues in the cytosolic domain of the receptor itself. Autophosphorylation of RTKs creates binding sites for several cytosolic enzymes, bringing them near their substrates in the plasma membrane; in some cases, these enzymes then generate second messengers. In addition, RTKs can phosphorylate various substrate proteins, thereby altering their activity. Little is yet known about how receptor serine/threonine kinases transduce signals.

The discussion in this chapter focuses primarily on G protein–linked receptors and receptor tyrosine kinases. The general structure and mechanism of action of the intracellular receptors for steroid hormones are discussed in Chapter 11; ion channels are covered in detail in Chapters 15 and 21; and certain receptor serine/threonine kinases are described briefly in Chapter 24.

The Synthesis, Release, and Degradation of Hormones Are Regulated

Because of their potent effects, hormones and neurotransmitters must be carefully regulated. The release and degradation of some signaling compounds are regulated to produce rapid, short-term effects; others to produce slower-acting but longer-lasting effects (Table 20-4). In some cases, complex regulatory networks coordinate the levels of hormones whose effects are interconnected.

Peptide Hormones and Catecholamines Organisms must be able to respond instantly to many changes in their internal or external environment. Such rapid responses are mediated primarily by peptide hormones and catecholamines. The signaling cells that produce these hormones and neurotransmitters store them in secretory vesicles just under the plasma membrane. The supply of stored, preformed signaling molecules is sufficient for 1 day in the case of peptide hormones and for several days in the case of catecholamines. All peptide hormones, including insulin and *adrenocorticotropic hormone* (ACTH), are synthesized as part of a longer propolypeptide, which is cleaved by specific proteases to generate the active molecule just after it is transported to a secretory vesicle (see Figures 16-44 and 16-45).

TABLE 20-3   Some Mammalian Hormones that Interact with Cell-Surface Receptors*

Hormone	Structure	Origin	Major Effects
AMINO ACID DERIVATIVES			
Epinephrine		Adrenal medulla	Increase in pulse rate and blood pressure; contraction of most smooth muscles; glycogenolysis in liver and muscle; lipid hydrolysis in adipose tissue
Histamine		Mast cells	Dilation of blood vessels
DERIVATIVES OF ARACHIDONIC ACID			
Prostaglandins (PGE$_2$)		Most body cells	Contraction of smooth muscle
PEPTIDE HORMONES			
Follicle-stimulating hormone (FSH)	Protein: α chain, 92 aa β chain, 118 aa	Anterior pituitary	Stimulates growth of oocyte and ovarian follicles and estrogen synthesis by follicles
Glucagon	Peptide, 29 aa†	Pancreas α cells	Stimulates glucose synthesis and glycogen degradation in liver; lipid hydrolysis in adipose tissue
Insulin	Polypeptide A chain, 21 aa B chain, 30 aa	β cells of pancreas	Stimulates glucose uptake into fat and muscle cells; carbohydrate catabolism; stimulates lipid synthesis by adipose tissue; general stimulation of protein synthesis and cell proliferation

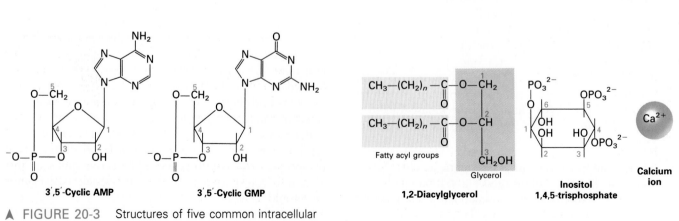

3′,5′-Cyclic AMP   3′,5′-Cyclic GMP   1,2-Diacylglycerol   Inositol 1,4,5-trisphosphate   Calcium ion

▲ FIGURE 20-3   Structures of five common intracellular second messengers.

TABLE 20-3 *(Continued)*

Hormone	Structure	Origin	Major Effects
LH-releasing hormone (LHRH)	α chain, 92 aa	Hypothalamus, neurons	Induces secretion of leuteinizing hormone by anterior pituitary
Luteinizing hormone (LH)	Protein: Polypeptide, 10 aa β chain, 115 aa	Anterior pituitary	Maturation of oocyte; stimulates estrogen and progesterone secretion by ovarian follicles

PEPTIDE GROWTH/DIFFERENTIATION FACTORS

Epidermal growth factor (EGF)	Polypeptide, 53 aa	Salivary and other glands?	Growth of epidermal and other body cells
Insulinlike growth factor 1 (IGF-1)	Polypeptide, 70 aa	Liver and other cells	Autocrine/paracrine growth factor induced by somatotropin; stimulates cells growth and division and glucose and amino acid uptake; increase in liver glycogen synthesis
Nerve growth factor	Two identical chains of 118 aa	All tissues innervated by sympathetic neurons	Growth and differentiation of sensory and sympathetic neurons
Platelet-derived growth factor (PDGF)	Hetero- and homo-dimers of A and B chains: A chain, 125 aa B chain, 109 aa	Platelets and cells in many other tissues	Chemotaxis of connective-tissue cells and inflammatory cells; wound healing; development and survival of cells in the nervous system

* Molecules primarily used as neurotransmitters or neurohormones are discussed in Chapter 21.
† aa = amino acid.

Stimulation of the signaling cell by a neurotransmitter or hormone causes immediate exocytosis of the peptide hormone into the surrounding medium or the blood. Signaling cells also are stimulated to synthesize the hormone and replenish the cell's supply. The released peptide hormones persist in the blood for only seconds or minutes before being degraded by blood and tissue proteases. Released catecholamines are inactivated by different enzymes or taken up by specific cells. The initial actions of these signaling substances on target cells (the activation or inhibition of specific enzymes) also last only seconds or minutes. Thus the catecholamines and some peptide hormones mediate short responses that are terminated by their own degradation.

Steroid Hormones and Thyroxine Steroid hormones are synthesized from cholesterol by pathways involving 10 or more enzymes. Steroid-producing cells, like those in the adrenal cortex, store a small supply of hormone precursor but none of the mature, active hormone. When stimulated, the cells convert the precursor to the active hormone, which then diffuses across the plasma membrane into the blood. Likewise, *thyroglobulin,* the iodinated precursor of thyroxine is stored in thyroid follicles. When cells lining these follicles are exposed to *thyroid-stimulating hormone* (TSH), they take up thyroglobulin; controlled proteolysis of this glycoprotein by lysosomal enzymes yields thyroxine, which is released into the blood.

Because the signaling cells that produce thyroxine and

(a) G protein–linked receptors (epinephrine, glucagon, serotonin)

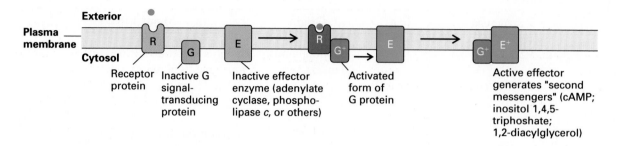

Exterior

Plasma membrane

Cytosol

Receptor protein

Inactive G signal-transducing protein

Inactive effector enzyme (adenylate cyclase, phospholipase *c*, or others)

Activated form of G protein

Active effector generates "second messengers" (cAMP; inositol 1,4,5-triphoshate; 1,2-diacylglycerol)

(b) Ion-channel receptors (acetylcholine)

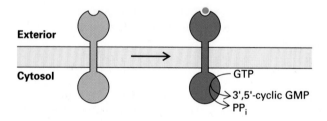

Ligand

Ligand binding-site

Exterior

Ion

Cytosol

Receptor protein

(c) Tyrosine kinase–linked receptors (erythropoietin, interferons)

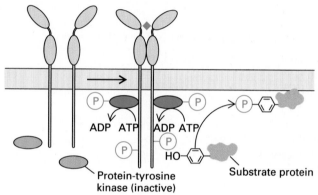

ADP   ATP   ADP ATP

Protein-tyrosine kinase (inactive)

HO

Substrate protein

(d) Receptor guanylate cyclase (atrial naturitic factor)

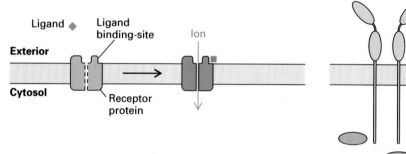

Exterior

Cytosol

GTP

3',5'-cyclic GMP

PP$_i$

(e) Receptor tyrosine phosphatase (leukocyte CD45 protein)

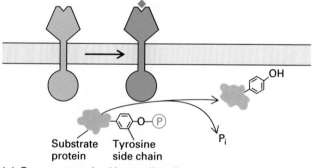

Exterior

Cytosol

OH

Substrate protein

Tyrosine side chain

P$_i$

(f) Receptor serine/threonine kinases (transforming growth factor β)

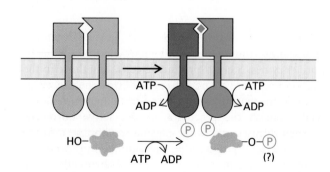

ATP

ADP

ATP

ADP

HO

ATP ADP

O

(?)

(g) Receptor tyrosine kinases (insulin, epidermal growth factor)

Ligand

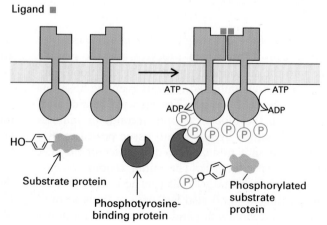

ATP

ADP

ATP

ADP

HO

Substrate protein

Phosphotyrosine-binding protein

Phosphorylated substrate protein

◄ FIGURE 20-4   Ligand-triggered cell-surface receptors. Common ligands for each receptor type are shown in parentheses. (a) G protein–linked receptors. Ligand binding triggers activation of a G protein, which then binds to and activates an enzyme that catalyzes synthesis of a specific second messenger. (b) Ion-channel receptors. A conformational change triggered by ligand binding opens the channel for ion flow. (c) Tyrosine kinase–linked receptors. Ligand binding causes formation of a homodimer or heterodimer, triggering the binding and activation of a cytosolic protein tyrosine kinase. The activated kinase phosphorylates tyrosines in the receptor; substrate proteins then bind to these phosphotyrosine residues and are phosphorylated. (d–g) Receptors with ligand-triggered enzymatic activity in the cytosolic domain. Some activated receptors are monomers with guanine cyclase activity and can generate the second messenger cyclic cGMP (d). Ligand binding to other receptors activates tyrosine phosphatase activity; these receptors can remove phosphate groups from phosphotyrosine residues in substrate proteins, thereby modifying their activity (e). The receptors for many growth factors have intrinsic protein kinase activity. Ligand binding to these receptors causes either identical or nonidentical receptor monomers to dimerize and activates their enzymatic activity. Activated receptors with serine/threonine kinase activity are heterodimers (f), whereas those with tyrosine kinase activity are homodimers (g). In both cases, the activated dimeric receptor phosphorylates several residues in its own cytosolic domain. Receptor tyrosine kinases also can phosphorylate certain substrate proteins, thereby altering the activity of these proteins; it is not known whether receptor serine/threonine kinases phosphorylate specific substrate proteins. [Part (c) see J. E. Darnell, I. M. Kerr, and G. Stark, 1994, *Science* **264**:1415; part (d) see S. Schulz, M. Chinkers, and D. Garbers, 1989, *FASEB J.* **3**:2026, and D. Garbers, 1989, *J. Biol. Chem.* **264**:9103; part (e) see H. Charbonneau et al., 1989, *Proc. Nat'l. Acad. Sci. USA* **86**:5252; part (f) see H. Y. Lin and H. F. Lodish, 1993, *Trends Cell Biol.* **3**:14, and J. Massagué et al., 1994, *Trends Cell Biol.* **4**:172; part (g) see W. J. Fantl, D. E. Johnson, L. T. Williams, 1993, *Annu. Rev. Biochem.* **62**:453.

TABLE 20-4   Characteristic Properties of Principal Types of Mammalian Hormones

Property	Steroids	Thyroxine	Peptides and Proteins	Catecholamines
Feedback regulation of synthesis	Yes	Yes	Yes	Yes
Storage of preformed hormone	Very little	Several weeks	One day	Several days, in adrenal medulla
Mechanism of secretion	Diffusion through plasma membrane	Proteolysis of thyroglobulin	Exocytosis of storage vesicles	Exocytosis of storage vesicles
Binding to plasma proteins	Yes	Yes	Rarely	No
Lifetime in blood plasma	Hours	Days	Minutes	Seconds
Time course of action	Hours to days	Days	Minutes to hours	Seconds or less
Receptors	Cytosolic or nuclear	Nuclear	Plasma membrane	Plasma membrane
Mechanism of action	Receptor-hormone complex controls transcription and stability of mRNAs		Hormone binding triggers synthesis of cytosolic second messengers or protein kinase activity	Hormone binding causes change in membrane potential or triggers synthesis of cytosolic second messengers

SOURCE: Adapted from E. L. Smith et al., 1983, *Principles of Biochemistry: Mammalian Biochemistry*, 6th ed., McGraw-Hill, p. 358. Reproduced by permission of McGraw-Hill.

steroid hormones store little of the active hormone, release of these hormones takes from hours to days (see Table 20-4). These hormones, which are poorly soluble in aqueous solution, are transported in the blood by carrier proteins; the tightly bound active hormones are not rapidly degraded. Thus, cellular responses to thyroxine and steroid hormones take awhile to occur but last much longer (hours to days) than do the effects mediated by peptide hormones and catecholamines.

Feedback Control of Hormone Levels   The synthesis and/or release of many hormones are subject to feedback regulation. This type of regulation is particularly important in coordinating the action of multiple hormones on various cell types during growth and differentiation. Often, the levels of several hormones are interconnected by *feedback circuits,* in which changes in the level of one hormone affect the levels of other hormones. One example involves the regulation of *estrogen* and *progesterone,* steroid hormones that stimulate the growth and differentiation of cells in the *endometrium,* the tissue lining the interior of the uterus. Changes in the endometrium prepare the organ to receive and nourish an embryo. The levels of both hormones are regulated by complex feedback circuits involving several other hormones (Figure 20-5). A key role is played by the *anterior pituitary gland,* an organ separated from but controlled by the brain. The anterior pituitary gland is connected directly to the *hypothalamus,* at the base of the brain, by a special set of blood vessels. Nerve cells in the hypothalamus discharge hypothalamic peptide-releasing factor, which enter these vessels and bind to receptors on anterior pituitary cells. These factors, in turn, cause specific pituitary hormones to be secreted.

Each developing mammalian egg, or oocyte, matures into an ovum inside an ovarian follicle made up of many cells. Under the influence of *follicle-stimulating hormone* (FSH) released by the anterior pituitary gland, the follicle increases in size and number of cells. The follicular cells secrete estrogens that stimulate the growth of the uterine wall and its glands in preparation for embryo implantation, should fertilization occur. Estrogen, in turn, acts on the hypothalamus to reduce the secretion of FSH-releasing factor and on the anterior pituitary gland directly to inhibit the release of FSH. This *negative feedback* by estrogen modulates the level of circulating estrogen in the nonpregnant female. As the follicle matures, pituitary secretion of FSH is reduced, but a different pituitary peptide hormone, *luteinizing hormone* (LH), is secreted under the influence of the hypothalmic LH-releasing factor. LH completes the maturation and release of the ovum into the oviduct and transforms the follicle into an endocrine organ, the *corpus luteum,* which secretes progesterone. The hormone progesterone then, in turn, causes additional growth of the endometrium.

In the absence of fertilization of the ovum and pregnancy, the corpus luteum degenerates. The resulting decrease in the level of circulating estrogen and progesterone causes the uterine wall to break down, resulting in menstruation. If fertilization and implantation occur, progesterone stimulates the placental tissues to produce *chorionic gonadotropin* (CG), a peptide hormone similar in structure and function to LH. CG then acts on the corpus luteum, preventing its degeneration and maintaining progesterone production. Thus *positive feedback* maintains the high level of progesterone required to maintain the endometrial lining of the uterus, so that a good blood supply can form to nourish the implanted embryo.

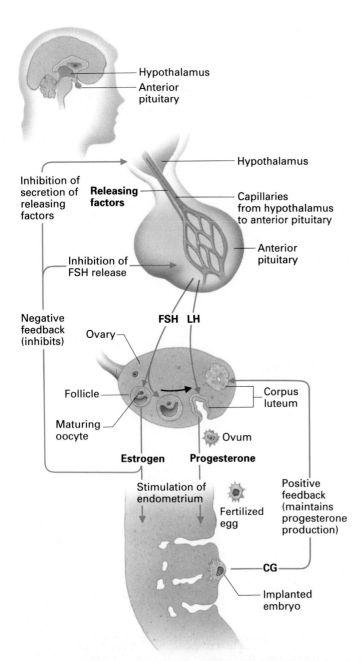

▲ FIGURE 20-5   Feedback control of estrogen and progesterone levels in the blood of female mammals. FSH = follicle-stimulating hormone; LH = luteinizing hormone; CG = chorionic gonadotropin. See text for discussion.

► Identification and Purification of Cell-Surface Receptors

As noted earlier, hormone receptors bind ligands with great specificity and high affinity. Binding of a hormone to a receptor involves the same types of weak interactions—ionic and van der Waals bonds and hydrophobic interactions—that characterize the specific binding of a substrate to an enzyme. The *specificity* of a receptor refers to its ability to distinguish closely related substances; the insulin receptor, for example, binds insulin and a related hormone called insulin-like growth factor 1, but not other peptide hormones.

Hormone binding usually can be viewed as a simple reversible reaction,

$$R + H \rightleftharpoons RH$$

which can be described by the following equation:

$$K_D = \frac{[R][H]}{[RH]} \qquad (1)$$

where [R] and [H] are the concentrations of free receptor and hormone (ligand), respectively, and [RH] is the concentration of the receptor-hormone complex. K_D, the dissociation constant of the receptor-ligand complex, measures the affinity of the receptor for the ligand. This binding equation can be written in a form similar to that of the Michaelis-Menten equation (Chapter 3) used to analyze enzymatic reactions:

$$\frac{[RH]}{R_T} = \frac{1}{1 + K_D/[H]} \qquad (2)$$

where R_T is the sum of free and bound receptors: [R] + [RH].

The lower the K_D value, the higher the affinity of a receptor for its ligand. The K_D value is equivalent to the concentration of ligand at which one-half of the receptors contain bound ligand. If [H] = K_D, then from equation (2) we can see that [RH] = 0.5 R_T.

Hormone Receptors Are Detected by Binding Assays

It is difficult to identify and purify hormone receptors, mainly because they are present in such minute amounts. The surface of a typical cell bears 10,000–20,000 receptors for a particular hormone, but this quantity is only $\approx 10^{-6}$ of the total protein in the cell, or $\approx 10^{-4}$ of the plasma-membrane protein. Purification is also difficult because these integral membrane proteins first must be solubilized with a nonionic detergent.

Usually, receptors are detected and measured by their ability to bind radioactive hormones to a cell or to cell fragments. The development of chemically or enzymatically catalyzed syntheses of radioactive hormones that retain normal hormone activity has been crucial to the identification of many receptors. When increasing amounts of a radiolabeled hormone (e.g., insulin) are added to a cell suspension, the amount that binds to the cells increases at first and then tapers off at higher concentrations (Figure 20-6, curve A). Much of the radiolabeled hormone is bound specifically to its receptor, but some is bound nonspecifically to the multitude of other proteins and phospholipids on the cell surface. Nonspecific binding of a labeled hormone can be measured by conducting the binding assay in the presence of a large excess of unlabeled hormone. Because the specific (high-affinity) binding sites are saturable, they all are filled by unlabeled hormone under these conditions and bind no labeled hormone. Nonspecific sites, however, do not saturate, so that binding of labeled hormone in the

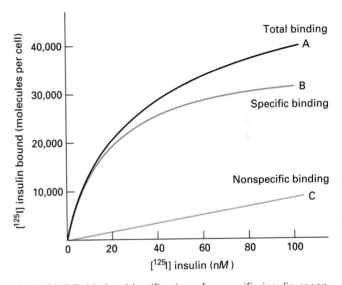

▲ FIGURE 20-6   Identification of a specific insulin receptor on the surface of cells by their binding of radioactive insulin. A suspension of cells is incubated for 1 h at 4°C with increasing concentrations of ^{125}I-labeled insulin; the low temperature is used to prevent endocytosis of the cell-surface receptors (see Figure 16-48). The total binding curve A represents insulin specifically bound to high-affinity receptors as well as insulin nonspecifically bound with low affinity to other molecules on the cell surface. The contribution of nonspecific binding to total binding is determined by repeating the binding assay in the presence of a 100-fold excess of unlabeled insulin, which saturates all the specific high-affinity sites. In this case, all the labeled insulin binds to nonspecific sites, yielding curve C. The specific binding curve B, which fits the simple binding equation (2), is calculated as the difference between curves A and C. For this insulin receptor, K_D = 20 nM (2 × 10^{-8} M), and the number of receptor molecules per cell R_T = $\approx$30,000. [Adapted from A. Ciechanover, A. Schwartz, and H. F. Lodish, 1983, *Cell* **32**:267.]

presence of excess unlabeled hormone represents nonspecific binding (Figure 20-6, curve C). Specific binding is calculated as the difference between total binding and nonspecific binding.

The number of hormone-binding sites per cell can be calculated from the saturation value of the specific binding curve (Figure 20-6, curve B). For example, a single hepatoma cell has $\approx$30,000 insulin-binding sites. The K_D, the hormone concentration at which the receptor is half-saturated, also can be calculated from the specific binding curve, which is described by equation (2). For the hepatoma-cell insulin receptor, K_D is 2×10^{-8} M, so the receptor will be half-saturated at an insulin concentration of 2×10^{-8} M (0.12 μg/ml). Since blood contains $\approx$10 mg of total protein per milliliter, the insulin receptor can specifically bind insulin in the presence of a 100,000-fold excess of unrelated proteins. The binding affinity of the erythropoietin receptor on erythrocyte precursor cells ($K_D = \approx 1 \times 10^{-10}$ M) is even greater than that of the insulin receptor.

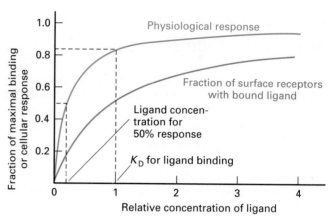

▲ FIGURE 20-7   Comparison of binding curve and response curve for a cell-surface receptor and its ligand. As illustrated here, the maximal physiological response to many hormones occurs when only a fraction of the cell receptors are occupied by ligand. In this example, 50 percent of the maximal response is induced at a ligand concentration at which only 18 percent of the receptors are occupied. Likewise, 80 percent of the maximal response is induced when the ligand concentration equals the K_D value, at which 50 percent of the receptors are occupied.

K_D Values for Cell-Surface Hormone Receptors Approximate the Concentrations of Circulating Hormones

In general, the K_D value of a cell-surface hormone receptor approximates the blood level of its ligand. Changes in hormone concentration are reflected in proportional changes in the fraction of receptors occupied. Suppose, for instance, that the normal (unstimulated) concentration of a hormone in the blood is 10^{-9} M and that the K_D for its receptor is 10^{-7} M; we can calculate the fraction of receptors with bound hormone (RH) from equation (2):

$$\frac{[RH]}{[R_T]} = \frac{1}{1 + 10^{-7}/10^{-9}} = 0.0099$$

Thus about 1 percent of the total receptors will be filled with hormone. If the hormone concentration rises tenfold to 10^{-8} M, the concentration of receptor-hormone complex will rise proportionately, so that about 10 percent of the total receptors would have bound hormone. If the induced cellular response is proportional to the amount of RH, as is often the case, then the cellular responses also will increase tenfold.

In the case of many hormone receptors, a maximal cellular response is induced at a ligand concentration lower than that needed to saturate all the receptor molecules on a cell. In the example of this phenomenon depicted in Figure 20-7, the ligand concentration that induces a 50-percent maximal response is less than 25 percent of the K_D value.

Affinity Techniques Permit Purification of Receptor Proteins

Cell-surface hormone receptors often can be identified and followed through isolation procedures by *affinity labeling*. In this technique, cells are mixed with an excess of a radio-labeled hormone to saturate the hormone-binding sites on its specific receptor. After unbound hormone is washed away, the mixture is treated with a chemical agent that covalently cross-links the bound labeled hormone to the receptor (Figure 20-8). Most cross-linking agents contain two reactive groups that react with free amino groups in proteins. A radiolabeled ligand that is cross-linked to its receptor remains bound even in the presence of detergents and other denaturing agents that are used to solubilize receptor proteins from the cell membrane.

Cell-surface receptors that retain their hormone-binding ability when solubilized can be purified by *affinity chromatography*. In this technique, a ligand of the receptor of interest is chemically linked to polystyrene beads. A crude, detergent-solubilized preparation of membrane proteins is passed through a column containing these beads. Only the receptor binds to the beads; the other proteins are washed through the column by excess fluid. When an excess of the ligand is passed through the column, the bound receptor is displaced from the beads and eluted from the column (see Figure 3-39c). In some cases, a hormone receptor can be purified as much as 100,000-fold in a single affinity chromatographic step.

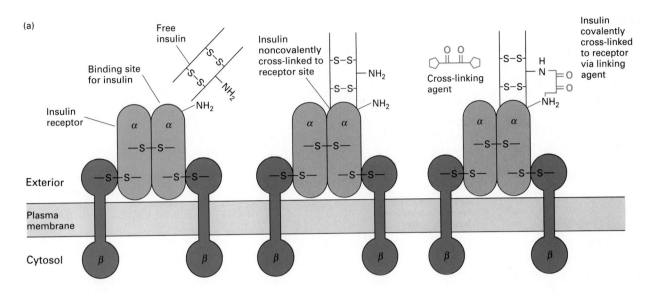

(a)

Free insulin

Binding site for insulin

Insulin receptor

Insulin noncovalently cross-linked to receptor site

Cross-linking agent

Insulin covalently cross-linked to receptor via linking agent

Exterior

Plasma membrane

Cytosol

(b)

Sites of reaction with free amino groups

Receptor protein + 2

▲ **FIGURE 20-8** Affinity labeling of insulin receptor. This tetrameric receptor contains two α subunits, which bind to insulin, and two β subunits, which span the plasma membrane. All the subunits are linked by disulfide bonds. When insulin is bound to the receptor (a), free NH_2 groups on the hormone and receptor proteins are brought close enough together to be covalently joined by a cross-linking agent (b). If the insulin is radiolabeled, then the cross-linked hormone will remain attached to the receptor throughout isolation and purification procedures, providing a means for easily detecting the receptor. [See J. Massagué and M. P. Czech, 1982, *J. Biol. Chem.* **257**:729; A. Ullrich et al., 1985, *Nature* **313**:756.]

Many Receptors Can Be Cloned without Prior Purification

The number of receptors for many hormones on the surface of cells is too small to be purified by affinity chromatography. For example, each nucleated erythrocyte precursor cell possesses only about 1000 cell-surface receptors for erythropoietin, a hormone essential to the growth and differentiation of precursor cells into mature erythrocytes. Because the erythropoietin receptor constitutes only about 1 part per million (10^{-4} percent) of the total cellular protein, it is impossible to purify sufficient amounts of the receptor protein by conventional approaches in order to characterize or sequence it.

With the advent of DNA cloning and other recombinant DNA techniques (Chapter 7), scientists now can obtain key receptor proteins by cDNA cloning without prior purification of the receptor protein. The sequence of a receptor protein can be deduced from sequencing of the cDNA encoding it, and expression of the cDNA in cell lines can be used to produce large amounts of a particular recep-

tor protein, which then can be used to study all aspects of receptor function.

The technique for identifying and cloning a cDNA encoding a desired receptor protein is summarized in Figure 20-9. In this approach, cloned cDNAs are transfected into cells that normally do not synthesize the desired receptor. Generally, when a foreign receptor protein is produced in such transfected cells, the protein is transported to the cell surface, enabling the cells to bind specifically to the appropriate ligand. Those transfected cells containing a cDNA coding for the receptor can be distinguished from those that do not by their ability to bind radiolabeled ligand, which is detected by autoradiography (see Figure 20-9); rescreening of cDNA pools that give rise to receptor-producing cells eventually results in a pure cDNA clone encoding the receptor of interest. Alternatively, transfected cells can be treated with a fluorescent-labeled ligand and passed through a fluorescence-activated cell sorter (see Figure 5-26). Since only cells containing a cDNA encoding the receptor will bind the fluorescent label, they will be separated from other cells in the sorter.

Plasmid expression vector

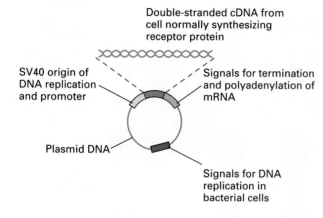

Double-stranded cDNA from cell normally synthesizing receptor protein

SV40 origin of DNA replication and promoter

Signals for termination and polyadenylation of mRNA

Plasmid DNA

Signals for DNA replication in bacterial cells

(a) Initial screening of cDNA pools

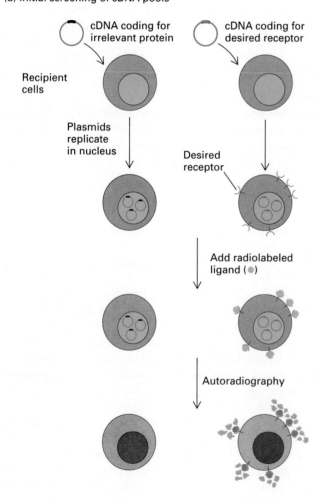

cDNA coding for irrelevant protein

cDNA coding for desired receptor

Recipient cells

Plasmids replicate in nucleus

Desired receptor

Add radiolabeled ligand (●)

Autoradiography

(b) Rescreening of positive cDNA pools

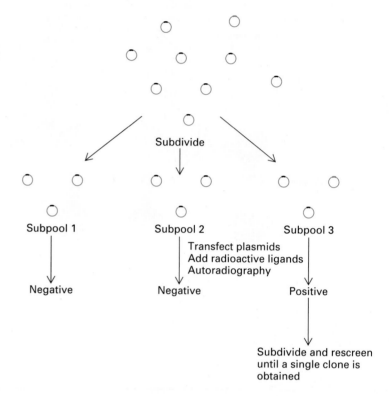

Subdivide

Subpool 1

Subpool 2

Subpool 3

Transfect plasmids
Add radioactive ligands
Autoradiography

Negative

Negative

Positive

Subdivide and rescreen until a single clone is obtained

(c) Transfected COS cells incubated with labeled endothelin 1 and subjected to autoradiography

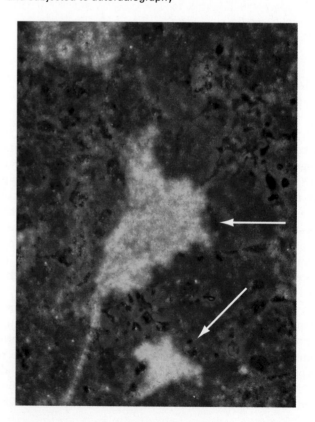

◄ FIGURE 20-9   Identification and isolation of a cDNA encoding a desired cell-surface receptor by plasmid expression cloning. All mRNA is extracted from cells that normally express the receptor and reverse-transcribed into double-stranded cDNA. The entire population of cDNAs is inserted into plasmid expression vectors in between a strong promoter and a terminator of transcription (top). The plasmids are transfected into bacterial cells, and the resulting cDNA library is divided into pools, each containing about 1000 different cDNAs. (a) Plasmids in each pool are transfected into a population of cultured cells (e.g., COS cells) that lack the receptor of interest. Only transfected cells that contain the cDNA encoding the desired receptor synthesize it; other transfected cells produce irrelevant proteins. To detect the few cells producing the desired receptor, a radiolabeled ligand specific for the receptor is added to the culture dishes containing the transfected cells; the cells are fixed and subjected to autoradiography. Positive cells synthesizing the specific receptor will be covered with many grains. (b) Plasmid cDNA pools giving rise to a positive signal are maintained in bacteria and subdivided into smaller pools, each of which is rescreened by transfection into cultured cells. After several cycles of screening and subdividing positive cDNA pools, a pure cDNA clone encoding the desired receptor is obtained. (c) Photomicrograph of a monolayer of COS cells transfected with an expression vector encoding the endothelin A receptor, incubated with radiolabeled endothelin 1, and then subjected to autoradiography. The many grains over the positive cells (arrows) indicate abundant expression of endothelin receptors. These cells are viewed in a dark-field microscope with oblique illumination, so each silver grain acts as a point source of light. [See A. Aruffo and B. Seed, 1987, *Proc. Nat'l. Acad. Sci. USA* **84**:8573; A. D'Andrea, H. F. Lodish, and G. Wong, 1989, *Cell* **57**:277. Part (c) from H. Y. Lin et al., 1992, *Cell* **68**:775 Erratum (1992) **70**:1069; courtesy of H. Y. Lin.]

▶ Seven-spanning G Protein–Linked Receptors

Ligand binding to many different mammalian cell-surface receptors activates a *signal-transducing G protein* that, in turn, activates an *effector enzyme* to generate an intracellular second messenger (see Figure 20-4a). Although these receptors bind different hormones and may mediate different cellular responses, they form a class of receptors that functions similarly and exhibits the following properties:

1. The amino acid sequence contains seven stretches of $\approx$22–24 hydrophobic residues, which are thought to form seven transmembrane α helices (Figure 20-10). The overall structure of these receptors thus is similar to that of bacteriorhodopsin, which is not a receptor (see Figure 14-18).

2. The loop in the receptor between α helices 5 and 6 and the C-terminal segment, which face the cytosol, are important for interactions with a G protein.

3. The signal-transducing G protein associated with the receptor functions as an on-off molecular switch, which is in the off state when it binds GDP. Binding of ligand to the receptor causes the G protein to release its bound GDP and to bind GTP, converting the G protein to the on state,

4. The activated G protein, with a bound GTP, binds to and activates (or inhibits) an effector enzyme, which catalyzes formation of a second messenger.

5. Hydrolysis of GTP, bound to the G protein, switches the G protein back to the inactive (off) state.

▶ FIGURE 20-10   Schematic diagram of the general structure of G protein–linked receptors. Because all receptors of this type contain seven transmembrane loops, they also are called seven-spanning receptors. See text for discussion.

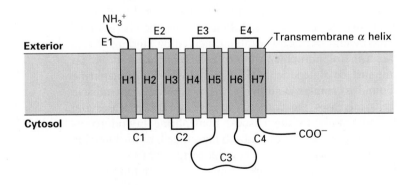

E1 – E4 = extracellular loops

H1 – H7 = transmembrane helices

C1 – C4 = cytosolic loops

To illustrate the operation of this important class of receptors, we discuss the structure and function of epinephrine/norepinephrine receptors and of their associated signal-transducing G proteins. In this receptor system, the effector enzyme is *adenylate cyclase,* which synthesizes the second messenger cAMP. Later we describe how other receptors and other G proteins allow cells to integrate the actions of different types of receptors, to modify other enzymes, and to control essential metabolic processes.

Binding of Epinephrine to β- and α-Adrenergic Receptors Induces Tissue-Specific Responses Mediated by cAMP

Epinephrine and norepinephrine were originally recognized as products of the *medulla,* or core, of the adrenal gland and are also known as *adrenaline* and *noradrenaline.* Embryologically, nerve cells derive from the same tissue as adrenal medulla cells, and norepinephrine is also a secretory product of differentiated nerve cells. Both hormones are charged compounds that belong to the catecholamines, active amines containing the compound *catechol:*

In times of stress, such as fright or heavy exercise, all tissues have an increased need for glucose and fatty acids. These principal metabolic fuels can be supplied to the blood in seconds by the rapid breakdown of glycogen in the liver (*glycogenolysis*) and of triacylglycerol in the adipose storage cells (*lipolysis*). In mammals, the liberation of glucose and fatty acids can be triggered by binding of epinephrine or norepinephrine to *β-adrenergic receptors* on the surface of hepatic (liver) and adipose cells. Epinephrine bound to similar β-adrenergic receptors on heart muscle cells increases the contraction rate, which increases the blood supply to the tissues. Epinephrine bound to β-adrenergic receptors on smooth muscle cells of the intestine causes them to relax. Another type of epinephrine receptor, the *α-adrenergic receptor,* is found on smooth muscle cells lining the blood vessels in the intestinal tract, skin, and kidneys. Epinephrine bound to these α receptors causes the arteries to constrict, cutting off circulation to these peripheral organs. These diverse effects of one hormone are directed to a common end: supplying energy for the rapid movement of major locomotor muscles in response to bodily stress.

All of the very different tissue-specific responses induced by binding of epinephrine to β-adrenergic receptors are mediated by a rise in the intracellular level of cAMP, resulting from activation of adenylate cyclase. As a second messenger, cAMP modifies the rates of different enzyme-catalyzed reactions in specific tissues generating various metabolic responses. Binding of numerous other hormones to their receptors also leads to a rise in intracellular cAMP and characteristic tissue-specific metabolic responses (Table 20-5).

Two types of experiments have been used to establish the identity of the β-adrenergic receptor. The data in Figure 20-11 show that the K_D for binding of epinephrine and other catecholamines to cell-surface β-adrenergic receptors is about the same as the ligand concentration that induces a half-maximal activation of adenylate cyclase. Further evidence that the β-adrenergic receptor mediates induction of epinephrine-initiated cAMP synthesis came from studies with receptors purified by affinity chromatography. The

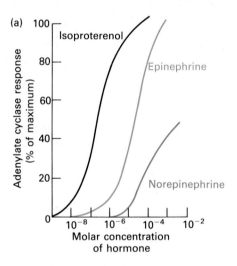

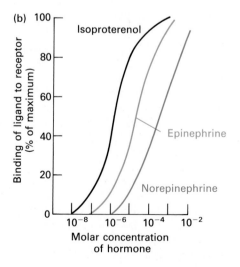

TABLE 20-5 Metabolic Responses to Hormone-Induced Rise in cAMP in Various Tissues

Tissue	Hormone Inducing Rise in cAMP	Metabolic Response
Adipose	Epinephrine; ACTH; glucagon	Increase in hydrolysis of triglyceride; decrease in amino acid uptake
Liver	Epinephrine; norepinephrine; glucagon	Increase in conversion of glycogen to glucose; inhibition of synthesis of glycogen; increase in amino acid uptake; increase in gluconeogenesis (synthesis of glucose from amino acids)
Ovarian follicle	FSH; LH	Increase in synthesis of estrogen, progesterone
Adrenal cortex	ACTH	Increase in synthesis of aldosterone, cortisol
Cardiac muscle cells	Epinephrine	Increase in contraction rate
Thyroid	TSH	Secretion of thyroxine
Bone cells	Parathyroid hormone	Increase in resorption of calcium from bone
Skeletal muscle	Epinephrine	Conversion of glycogen to glucose
Intestine	Epinephrine	Fluid secretion
Kidney	Vasopressin	Resorption of water
Blood platelets	Prostaglandin I	Inhibition of aggregation and secretion

Source: E. W. Sutherland, 1972, *Science* **177**:401.

purified receptors were incorporated into liposomes, which then were fused with cells containing adenylate cyclase and the appropriate G proteins but no β-adrenergic receptors (Figure 20-12). Cells that incorporated receptor-laden liposomes responded to epinephrine by synthesizing high levels of cAMP, proving that the receptor is involved in inducing cAMP synthesis. Similar results have been obtained by transfecting cloned cDNA encoding the β-adrenergic receptor into receptor-negative cells; the transfected cells acquire the ability to activate adenylate cyclase in response to epinephrine.

Analogs Provide Information about Essential Features of Hormone Structure and Are Useful as Drugs

Studies with chemically synthesized analogs of epinephrine and other natural hormones have provided additional evidence that saturable cell-surface receptors are physiologically relevant. These analogs fall into two classes: *agonists*, which mimic the function of a hormone by binding to its receptor and causing the normal response, and *antagonists*, which bind to the receptor but do not activate hormone-induced effects. An antagonist acts as an inhibitor of the natural hormone (or agonist) by competing for binding sites on the receptor, thereby blocking the physiological activity of the hormone.

Comparisons of the molecular structure and activity of various catecholamine agonists and antagonists (Table 20-6) have been used to define the parts of the hormone molecule necessary for binding to β-adrenergic receptors as well as the parts necessary for the subsequent induction of a cellular response. Such studies indicate that the side chain containing the NH group determines the affinity of the

◄ **FIGURE 20-11** Comparison of the abilities of three catecholamines to activate adenylate cyclase, which catalyzes synthesis of cAMP, and to bind to cell-surface β-adrenergic receptors. (a) Different concentrations of norepinephrine, epinephrine, and isoproterenol (an agonist) were incubated with a suspension of frog erythrocytes at 37°C. The cells then were broken and the adenylate cyclase activity determined. (b) Ligand binding to the receptor was measured by an indirect competition assay. In this procedure, binding of the unlabeled hormone or agonist to the receptor inhibits binding of the ^{3}H-labeled antagonist alprenolol, allowing the binding affinities of the unlabeled ligands for the receptor to be estimated. The curves show that each ligand induces adenylate cyclase activity in proportion to its ability to bind to the receptor. Moreover, the concentration required for half-maximal binding of each ligand to the receptor is about the same as that required for activation of adenylate cyclase. Note that the ligand concentration is plotted on a logarithmic scale ranging from 10^{-9} to 10^{-2} M.

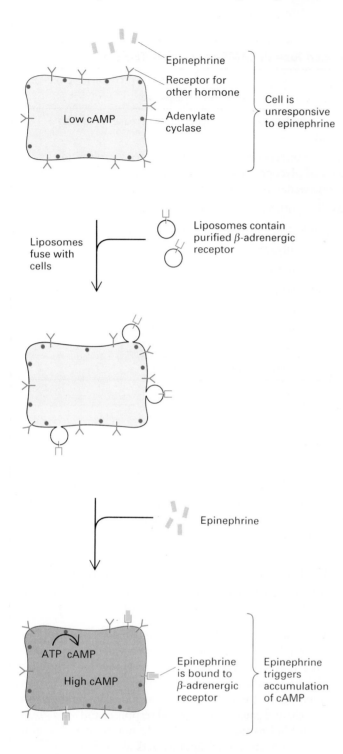

▲ FIGURE 20-12 Experimental demonstration that β-adrenergic receptor mediates the induction of epinephrine-initiated cAMP synthesis. Target cells lacking any receptors for epinephrine but expressing adenylate cyclase and the appropriate signal-transducing G proteins were incubated with liposomes containing purified β-adrenergic receptors. Cells that fused with the liposomes became responsive to epinephrine, producing high levels of cAMP when the hormone was added to the medium. [See R. A. Cerione et al., 1983, *Nature* **306**:562.]

ligand for the receptor, while the catechol ring is required for the ligand-induced increase in cAMP level.

As is true for epinephrine, the K_D for binding of an agonist to β-adrenergic receptors generally is the same as the concentration required for half-maximal elevation of cAMP (see Figure 20-11). This relationship indicates that activation of adenylate cyclase by the epinephrine agonist *isoproterenol* is proportional to the number of β-adrenergic receptors filled with the agonist. Interestingly, the K_D for binding of isoproterenol to β-adrenergic receptors and subsequent induction of cAMP synthesis is 10 times lower than the K_D for epinephrine; other agonists are even more potent with still lower K_D values. The affinity of various antagonists for the β-adrenergic receptor also varies over a wide range (see Table 20-6).

Two types of β-adrenergic receptors have been identified in humans. Cardiac muscle cells possess β_1 *receptors*, which promote increased heart rate and contractility by binding catecholamines with the rank order of affinities isoproterenol > norepinephrine > epinephrine. So-called beta blockers, such as practolol (see Table 20-6), are used to slow heart contractions in the treatment of cardiac arrhythmias and anginas. These β_1-selective antagonists usually have little effect on β-adrenergic receptors on other cell types. The smooth muscle cells lining the bronchial passages possess β_2 *receptors*, which mediate relaxation by binding catecholamines with the rank order of affinities isoproterenol ≫ epinephrine > norepinephrine. Agonists selective for β_2 receptors, such as terbutaline, are used in the treatment of asthma because they specifically mediate opening of the bronchioles, the small airways in the lungs.

Studies with Mutant β-Adrenergic Receptors Identify Residues That Interact with Catecholamines

Mutant forms of the β-adrenergic receptor generated by site-specific mutagenesis have been expressed in cultured cells and their ability to bind the agonist isoproterenol determined. Based on such studies, the model shown in Figure 20-13 has been proposed. Isoproterenol is nestled among the seven transmembrane α helices near the outer (extracellular surface) of the membrane and roughly parallel to the plane of the membrane. Two serine residues in helix 5 of the receptor are hydrogen bonded to the two hydroxyl groups on the catechol ring. Mutation of either of these serines to alanine greatly reduces the ability of the receptor to bind the agonist. The carboxylate group of an aspartate in helix 3 forms an ionic bond with the $-NH_3^+$ group on the catecholamine. Mutation of this aspartate to alanine also inhibits ligand binding by the receptor. Finally, a phenylalanine in helix 6 forms hydrophobic interactions with the catechol ring of the agonist.

TABLE 20-6 Structure of Typical Agonists and Antagonists of the β-Adrenergic Receptor

Structure	Compound	K_D for Binding to the Receptor on Frog Erythrocytes
HO— / HO— benzene —CH(OH)—CH$_2$—NH—CH$_3$	Epinephrine	5×10^{-6} M
AGONIST HO— / HO— benzene —CH(OH)—CH$_2$—NH—CH(CH$_3$)$_2$	Isoproterenol	0.4×10^{-6} M
ANTAGONISTS CH$_2$=CH—CH$_2$— (o-substituted) O—CH$_2$—CH(OH)—CH$_2$—NH—CH(CH$_3$)$_2$	Alprenolol	0.0034×10^{-6} M
naphthalene—O—CH$_2$—CH(OH)—CH$_2$—NH—CH(CH$_3$)$_2$	Propranolol	0.0046×10^{-6} M
CH$_3$—C(O)—NH— benzene —OCH$_2$—CH(OH)—CH$_2$—NH—CH(CH$_3$)$_2$	Practolol	21×10^{-6} M

Source: R. J. Lefkowitz et al., 1976, *Biochim. Biophys. Acta* 457:1.

Experiments described in a later section provide evidence that the loop connecting helices 5 and 6 bind to the signal-transducing G proteins. Binding of ligand to a β-adrenergic receptor is thought to cause several of the transmembrane helices, particularly helices 5 and 6, to move relative to each other. As a result, the conformation of the long cytosolic loop connecting these two helices changes in a way that allows this loop to bind and activate the transducing G proteins.

Trimeric Signal-Transducing G$_s$ Protein Links β-Adrenergic Receptors and Adenylate Cyclase

As explained already, β-adrenergic receptors on different types of mammalian cells mediate distinct tissue-specific responses (see Table 20-5), but the initial response following binding of epinephrine is always the same: an elevation in the intracellular level of cAMP (Figure 20-14). This in-

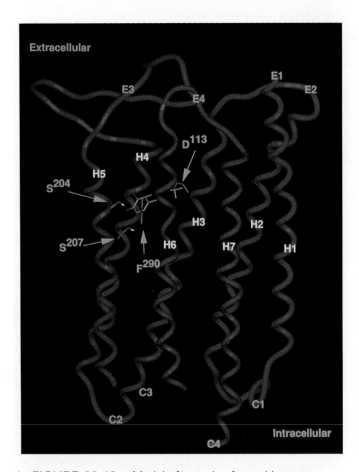

▲ FIGURE 20-13   Model of complex formed between isoproterenol and the β_2-adrenergic receptor. The polypeptide backbone of the receptor is shown in red and green. The transmembrane helices of the receptor are labeled H1–H7; the extracellular domains, E1–E4; and the intracellular domains, C1–C4. Most of the C3 domain (the long cytosolic loop between H5 and H6), C4 domain (the C-terminal tail), and N-terminal extracellular domain E1 are deleted. Isoproterenol (yellow) interacts with several residues in the receptor. Its amino group forms an ionic bond with the carboxylate side chain of aspartate 113 (D^{113}) in H3; the catechol ring engages in an aromatic interaction with phenylalanine 290 (F^{290}) in H6; and two hydroxyl groups on the catechol ring hydrogen-bond to the hydroxyl groups in two serine residues (S^{204} and S^{207}) in H5. [See C. D. Strader et al., 1994, *Annu. Rev. Biochem.* **63**:101; courtesy of C. D. Strader and D. Underwood.]

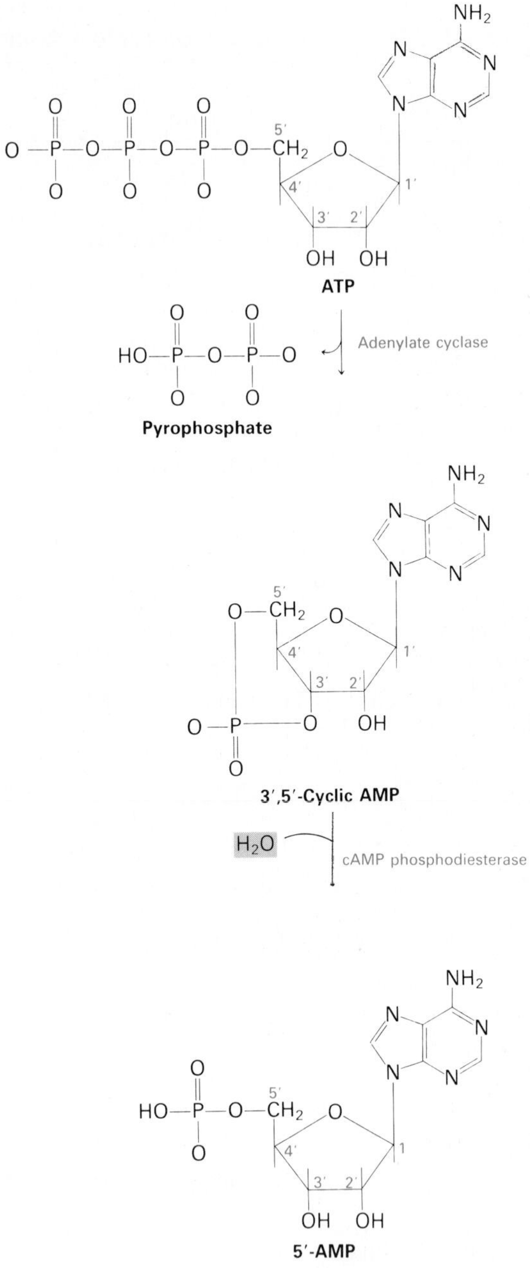

▲ FIGURE 20-14   Synthesis of 3′,5′-cyclic AMP (cAMP) from ATP is catalyzed by adenylate cyclase. This important second messenger is degraded by cAMP phosphodiesterase.

crease in cAMP results from activation of adenylate cyclase, a membrane-bound enzyme with two catalytic domains on the cytoplasmic face of the plasma membrane (Figure 20-15). The link between hormone binding to an exterior domain of the receptor and activation of adenylate cyclase is provided by a trimeric *stimulatory G protein* called G$_s$, which functions as a signal transducer.

Cycling of G$_s$ between Active and Inactive Forms
The G proteins that transduce signals from seven-spanning receptors (see Figure 20-10), such as the β-adrenergic receptor, contain three subunits designated α, β, and γ. When no ligand is bound to a β-adrenergic receptor, the α subunit of G$_s$ protein (called G$_{s\alpha}$) is bound to GDP and complexed with the β and γ subunits (Figure 20-16). Bind-

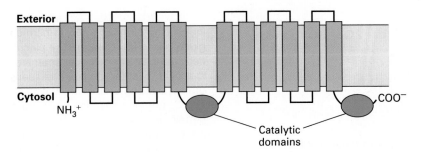

▲ FIGURE 20-15   Schematic diagram of mammalian adenylate cyclases. The membrane-bound enzyme contains two catalytic domains on the cytosolic face of the membrane and two integral membrane domains, each of which is thought to contain six transmembrane α helices. There are six adenylate cyclase isoforms in mammals that are activated or inhibited by transducing G proteins following hormone binding to an appropriate receptor. One isoform found mainly in the brain also is activated by Ca^{2+} ions complexed to the protein calmodulin. [Adapted from W.-J. Tang and A. G. Gilman, 1992, *Cell* **70**:869.]

ing of a hormone or agonist to the receptor changes its conformation, causing it to bind to the trimeric G_s protein in such a way that GDP is displaced from $G_{s\alpha}$ and GTP is bound. The $G_{s\alpha} \cdot$ GTP complex, which dissociates from the $G_{\beta\gamma}$ complex, then binds to and activates adenylate cyclase. This activation is short-lived, however, because GTP bound to $G_{s\alpha}$ hydrolyzes to GDP in seconds, leading to the association of $G_{s\alpha}$ with $G_{\beta\gamma}$ and inactivation of adenylate cyclase. The G_s protein thus alternates between an on state with bound GTP and an off state with bound GDP and shuttles between two membrane proteins—the hormone receptor and adenylate cyclase. In this way, G_s functions as a *signal transducer,* relaying the conformational change in the receptor triggered by hormone binding to adenylate cyclase.

Important evidence supporting this model has come from studies with a nonhydrolyzable analog of GTP called GMPPNP, in which a P–NH–P replaces the terminal phosphodiester bond in GTP:

$$\text{Guanine—ribose—O—} \overset{\overset{\displaystyle O}{\|}}{\underset{\underset{\displaystyle O^-}{|}}{P}} \text{—O—} \overset{\overset{\displaystyle O}{\|}}{\underset{\underset{\displaystyle O^-}{|}}{P}} \text{—NH—} \overset{\overset{\displaystyle O}{\|}}{\underset{\underset{\displaystyle O^-}{|}}{P}} \text{—O}^-$$

Although this analog cannot be hydrolyzed, it binds to $G_{s\alpha}$ as well as GTP does. The addition of GMPPNP and an agonist to an erythrocyte membrane preparation results in a much larger and longer-lived activation of adenylate cyclase than occurs with an agonist and GTP. Once the GDP bound to $G_{s\alpha}$ is displaced by GMPPNP, it remains permanently bound to $G_{s\alpha}$. Because the $G_{s\alpha} \cdot$ GMPPNP complex is as functional as the normal $G_{s\alpha} \cdot$ GTP complex in activating adenylate cyclase, the enzyme is in a permanently active state.

Amplification of Hormone Signal   Because the cellular responses triggered by cAMP may require tens of thousands or even millions of cAMP molecules per cell, the hormone signal must be amplified in order to generate sufficient second messenger from the few thousand β-adrenergic receptors present on a cell. *Amplification* is possible because both receptor and G_s proteins can diffuse rapidly in the plasma membrane. A single receptor-hormone complex causes conversion of up to 100 inactive G_s molecules to the active form. Each active $G_{s\alpha} \cdot$ GTP, in turn, probably activates a single adenylate cyclase molecule, which then catalyzes synthesis of many cAMP molecules during the time $G_s \cdot$ GTP is bound to it. Although the exact extent of this amplification is difficult to measure, binding of a single hormone molecule to one receptor molecule can result in the synthesis of at least several hundred cAMP molecules per receptor-hormone complex before the complex dissociates and activation of adenylate cyclase ceases.

Termination of Cellular Response   Successful cell-to-cell signaling also requires that the response of target cells to a hormone terminate rapidly once the concentration of circulating hormone decreases. *Termination* of the response to hormones recognized by β-adrenergic receptors is facilitated by a decrease in the affinity of the receptor that occurs when G_s is converted from the inactive to active form. When the GDP bound to $G_{s\alpha}$ is replaced with a GTP following hormone binding, the K_D of the receptor-hormone complex increases, shifting the equilibrium toward dissociation. The GTP bound to $G_{s\alpha}$ is quickly hydrolyzed, reversing the activation of adenylate cyclase and terminating the cellular response unless the concentration of hormone remains high enough to form new receptor-hormone complexes. Thus, the continuous presence of hormone is required for continuous activation of adenylate cyclase.

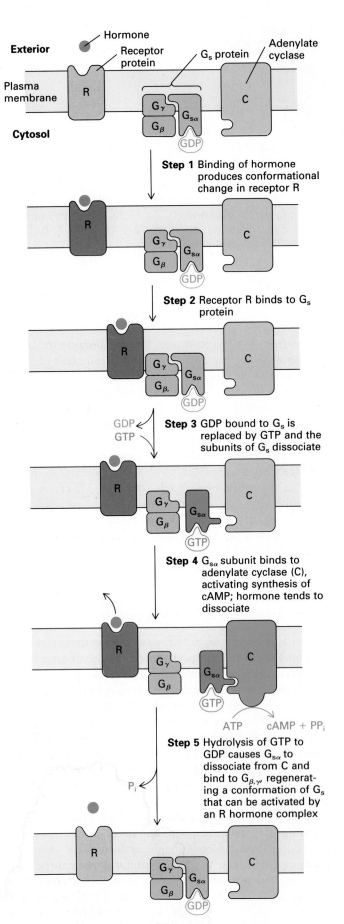

Step 1 Binding of hormone produces conformational change in receptor R

Step 2 Receptor R binds to Gs protein

Step 3 GDP bound to Gs is replaced by GTP and the subunits of Gs dissociate

Step 4 Gsα subunit binds to adenylate cyclase (C), activating synthesis of cAMP; hormone tends to dissociate

Step 5 Hydrolysis of GTP to GDP causes Gsα to dissociate from C and bind to Gβ,γ, regenerating a conformation of Gs that can be activated by an R hormone complex

◄ **FIGURE 20-16**   Activation of adenylate cyclase following binding of an appropriate hormone (e.g., epinephrine, glucagon) to its receptor. The patch of cell membrane depicted contains two transmembrane proteins—a seven-spanning receptor (purple) and adenylate cyclase (blue)—and the transducing protein G_s (green) on its cytosolic surface. The G_s protein relays the hormone signal to the effector protein, in this case adenylate cyclase. G_s cycles between an inactive form with bound GDP and active form with bound GTP. Dissociation of the active form yields the $G_{s\alpha} \cdot$ GTP complex, which directly activates adenylate cyclase. Activation is short lived because GTP is rapidly hydrolyzed (step 5). This terminates the hormone signal and leads to reassembly on the inactive $G_s \cdot$ GDP, returning the system to the resting state. Binding of another hormone molecule causes repetition of the cycle. See text for further discussion.

$G_{s\alpha}$ Belongs to GTPase Superfamily of Intracellular Switch Proteins

Insight into how $G_{s\alpha}$ subunits cycle between the active and inactive forms has come from studies of a related and extremely important intracellular signal-transduction protein called *Ras*. Like $G_{s\alpha}$, Ras alternates between an active on state with a bound GTP and an inactive off state with a bound GDP. As discussed in a later section, in the on state Ras binds to and activates specific effector proteins that control the growth and differentiation of cells. Both $G_{s\alpha}$ and Ras are members of a family of intracellular switch proteins collectively referred to as the *GTPase superfamily*. Other members of this family include the Rab proteins, which regulate fusion of vesicles within cells (see Figure 16-43).

Ras ($\approx$170 amino acids) is smaller than $G_{s\alpha}$ ($\approx$300 amino acids), and its three-dimensional structure is similar to that of the part of $G_{s\alpha}$ that binds GTP. Activation of Ras, which is triggered by binding of many growth hormones to their receptors, occurs in two steps: dissociation of the bound GDP followed by binding of GTP (Figure 20-17). The first reaction is accelerated by a protein called *guanine nucleotide-exchange factor* (GEF), which binds to the Ras · GDP complex, causing dissociation of the bound GDP. Because GTP is present in cells at a higher concentration than GDP, GTP binds spontaneously to "empty" Ras molecules, with release of GEF. GEF thus functions as an activator to promote formation of the active Ras · GTP complex.

The average lifetime of a GTP bound to Ras is about 1 min, indicating that the rate of GTP hydrolysis by Ras is much slower than that by $G_{s\alpha}$ subunits. The deactivation of Ras, like its activation, occurs in two steps: binding of a *GTPase-activating protein* (GAP) to the Ras · GTP complex, and hydrolysis of GTP with release of P_i and GAP (see Figure 20-17). GAP thus functions as an inhibitor by promoting formation of the inactive Ras · GDP complex.

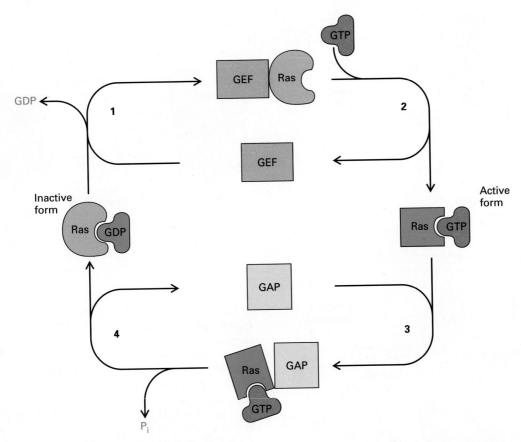

▲ FIGURE 20-17   Cycling of the Ras protein between the inactive form with bound GDP and the active form with bound GTP. By mechanisms discussed later, binding of certain growth factors to their receptors induces formation of the active Ras · GTP complex. (1) Guanine-nucleotide exchange factor (GEF) facilitates dissociation of GDP from Ras. It may do so by stabilizing the nucleotide-free state of Ras. (2) GTP then binds spontaneously, and GEF dissociates yielding the active Ras · GTP form. (3,4) Hydrolysis of the bound GTP to regenerate the inactive Ras · GDP form is accelerated a hundredfold by GTPase-activating protein (GAP). Unlike $G_{s\alpha}$, cycling of Ras thus requires two proteins, GEF and GAP; otherwise, $G_{s\alpha}$ and Ras exhibit many common features.

Based on the similarity in the three-dimensional structures of Ras and the $G_{s\alpha}$ subunit of trimeric G proteins and discovery of the role of GAP in cycling of the Ras protein, a model of $G_{s\alpha}$ has been proposed (Figure 20-18a). According to this model, $G_{s\alpha}$ consists of a Ras-like domain, which includes a GTP-binding site, and a 133-residue domain that functions as a built-in GAP. Experiments with separated fragments of $G_{s\alpha}$ showed that the Ras-like domain binds GTP but hydrolyzes it extremely slowly; addition of the purified GAP domain markedly stimulates GTP hydrolysis (Figure 20-18b). In normal, intact $G_{s\alpha}$, the GAP-like domain is connected to the Ras-like domain precisely where GAP binds to the Ras · GTP complex during cycling of Ras protein.

Unlike Ras, activation of the trimeric G_s protein does not require an activator protein equivalent to GEF. Rather, as noted earlier, interaction of a receptor-ligand complex with G_s causes a conformational change in $G_{s\alpha}$ · GDP that facilitates release of the bound GDP. This change has been proposed to involve a hinge-like motion in which the two domains of $G_{s\alpha}$ pivot relative to each other in a way that opens a cleft between them, leading to release of GDP and binding of GTP (see Figure 20-18c).

The many similarities between the structure and function of Ras and $G_{s\alpha}$, and the identification of both proteins in all eukaryotic cells, indicate that a single type of signal-transducing GTPase originated very early in evolution. The gene encoding this protein subsequently duplicated and evolved to the extent that cells today contain a superfamily of such GTPases, comprising perhaps a hundred different intracellular switch proteins. These related proteins control many aspects of cellular growth and metabolism.

Some Bacterial Toxins Irreversibly Modify G Proteins

So far we have discussed the stimulatory G protein (G_s) that links some seven-spanning cell-surface receptors to adenylate cyclase. Cells, however, contain a number of other trimeric signal-transducing G proteins. Studies with several bacterial toxins that irreversibly modify G proteins

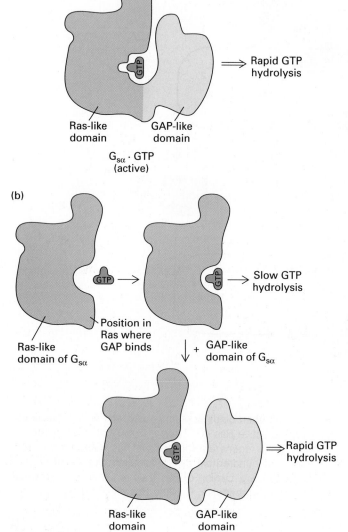

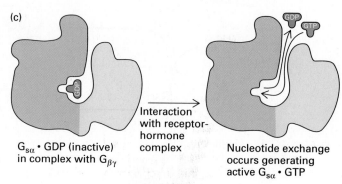

▲ **FIGURE 20-18** A proposed model of GTP hydrolysis of $G_{s\alpha}$ proteins based on analogy with cycling of Ras protein (see Figure 20-17). (a) The α subunit of trimeric G_s proteins comprises two domains. One ≈170-aa domain (blue), similar in structure to Ras protein, contains the nucleotide-binding site. The other domain (yellow), which contains 133 amino acids, is thought to be similar in function to the Ras-associated GAP protein. (b) The Ras-like domain of $G_{s\alpha}$ alone binds GTP but hydrolyzes it very slowly. When the GAP-like domain, produced by recombinant DNA techniques, is added, rapid GTP hydrolysis is restored. These studies provide an explanation for the higher intrinsic GTPase activity of $G_{s\alpha}$ compared with Ras. (c) Interaction of a receptor-hormone complex with the inactive $G_{s\alpha} \cdot GDP \cdot G_{\beta\gamma}$ complex has been proposed to cause the two domains to move relative to each other, opening a cleft through which GDP is released and GTP then has access to the binding site. [See D. W. Markby et al., 1993, *Science* **262**:1895; H. R. Bourne, 1993, *Nature* **366**:628; J. P. Noel et al., 1993, *Nature* **366**:654.]

initially helped to unravel the functions of these other G proteins.

The action of *cholera toxin*, a peptide produced by the bacterium *Vibrio cholerae*, was discovered first. The classic symptom of cholera is massive diarrhea, caused by water flow from the blood through intestinal epithelial cells into the lumen of the intestine; death from dehydration is common. Cholera toxin initially was shown to irreversibly activate adenylate cyclase in the intestinal epithelial cells, causing a sustained high cAMP level. Later studies showed that this toxin has the same effect on many different cell types.

Like diptheria toxin, cholera toxin is a heterodimer. One subunit is an enzyme that penetrates the plasma membrane and enters the cytosol, where it catalyzes the covalent addition of the ADP-ribose moiety from intracellular NAD^+ to $G_{s\alpha}$. ADP-ribose is added to arginine-174, which is located near the GTP-binding site in $G_{s\alpha}$, irreversibly modifying the protein. ADP-ribosylated $G_{s\alpha} \cdot GTP$ can ac-

tivate adenylate cyclase normally but cannot hydrolyze the bound GTP to GDP; thus $G_{s\alpha}$ remains in the active on state, continuously activating adenylate cyclase (Figure 20-19). As a result, the level of cAMP in the cytosol rises 100-fold or more. In intestinal epithelial cells, this rise apparently causes certain membrane proteins to permit a massive flow of water from the blood into the intestinal lumen. The studies with cholera toxin provided additional confirmation of the cycling of G_s described earlier.

Pertussis toxin is secreted by *Bordetella pertussis*, the bacterium causing whooping cough. This toxin catalyzes addition of ADP-ribose to the α subunits of several other G proteins. This prevents release of GDP locking G_α in the GDP-bound state. One of these G proteins, an inhibitory protein called G_i, is linked to adenylate cyclase; another is a stimulatory protein called G_o, which is linked to phospholipase C. The recent identification of several other G proteins is discussed in a later section.

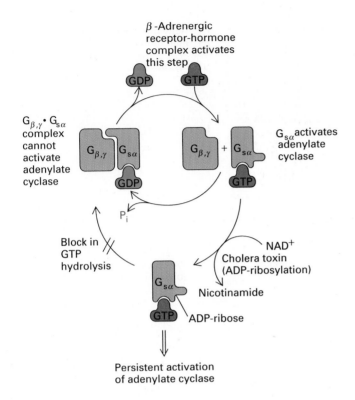

The normal cycling of $G_{s\alpha}$ between the active and inactive forms is coupled to activation and inactivation of adenylate cyclase, so that the rise in cAMP persists only as long as hormone stimulation. Hydrolysis of GTP to GDP is catalyzed by $G_{s\alpha}$ itself. In the presence of cholera toxin, $G_{s\alpha}$ is irreversibly modified by addition of an ADP-ribosyl group; the modified $G_{s\alpha}$ can bind GTP but cannot hydrolyze it. As a result, there is an excessive, nonregulated rise in the intracellular cAMP level.

peptide hormone) and epinephrine bind to different seven-spanning receptors, but both activate adenylate cyclase and thus trigger the same metabolic responses (see Table 20-5). Both types of receptors interact with and activate G_s, converting the inactive $G_{s\alpha} \cdot GDP \cdot G_{\beta\gamma}$ to the active $G_{s\alpha} \cdot GTP$ form. Activation of adenylate cyclase, and thus the cAMP level, is proportional to the total concentration of $G_{s\alpha} \cdot GTP$ resulting from binding of both hormones to their respective receptors.

In some cells, the cAMP level can be both up-regulated and down-regulated by the action of different hormones. In adipose cells, for example, epinephrine, glucagon, and ACTH all stimulate adenylate cyclase, whereas prostaglandin PGE_1 and adenosine inhibit the enzyme. The receptors for PGE_1 and adenosine interact with G_i, an inhibitory G protein that contains the same β and γ subunits as stimulatory G_s but a different α subunit $(G_{i\alpha})$, which also binds GDP or GTP. In response to binding of an inhibitory ligand to its receptor, the associated G_i protein releases GDP and binds GTP; the active $G_{i\alpha} \cdot GTP$ complex then dissociates from $G_{\beta\gamma}$ and inhibits (rather than stimulates) adenylate cyclase (Figure 20-20). As discussed later, $G_{\beta\gamma}$ can directly inhibit the activity of some isoforms of adenylate cyclase.

Adenylate Cyclase Is Stimulated and Inhibited by Different Receptor-Ligand Complexes

The versatile G proteins enable different receptor-hormone complexes to modulate the activity of the same effector protein. In many types of cells, for example, binding of different hormones to their respective receptors induces activation of adenylate cyclase. In the liver, glucagon (a

Hormone-induced activation and stimulation of adenylate cyclase is mediated by $G_{s\alpha}$ (green) and $G_{i\alpha}$ (red), respectively. Binding of $G_{s\alpha} \cdot GTP$ to adenylate cyclase activates the enzyme (see Figure 20-16), whereas binding of $G_{i\alpha}$ inhibits adenylate cyclase. The $G_{\beta\gamma}$ subunit in both stimulatory and inhibitory G proteins is identical; the G_α subunits and the receptors differ. [Adapted from A. G. Gilman, 1984, *Cell* **36**:577.]

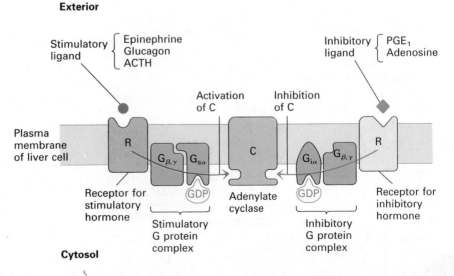

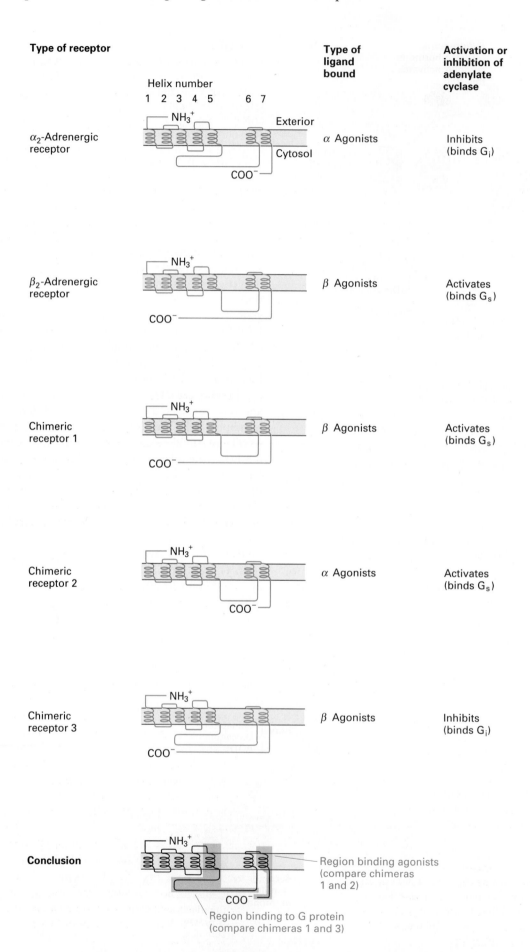

Analogous Regions in All Seven-Spanning Receptors Determine G Protein and Ligand Specificity

Although G protein–linked receptors are thought to span the membrane seven times, and hence their three-dimensional structures are predicted to be similar (see Figures 20-10 and 20-13), their amino acid sequences generally are quite dissimilar. For example, the sequences of the closely related β_1- and β_2-adrenergic receptors are only 50 percent identical; the sequences of the α- and β-adrenergic receptors exhibit even less homology. The specific amino acid sequence of each receptor determines which ligands it binds and which G proteins interact with it.

Studies with recombinant chimeric receptor proteins, containing part of an α_2 receptor and part of a β_2 receptor, localized certain functional domains within specific regions of the receptor sequence (Figure 20-21). In particular, the hydrophilic loop between helices 5 and 6 was shown to determine specificity for a particular G protein. It is thought that specific regions within the loop assume a unique three-dimensional structure in all receptors that bind the same G protein (e.g., G_s or G_i). The chimeric studies also showed that helix 7 functions in determining ligand specificity. As discussed, more recent studies with site-directed mutant β_2 receptors have shown that helix 3, 5, and 6 also participate in ligand binding (see Figure 20-13).

Degradation of cAMP Also Is Regulated

The level of cAMP usually is controlled by the hormone-induced activation of adenylate cyclase. Another point of regulation is the hydrolysis of cAMP to 5'-AMP by *cAMP*

◄ FIGURE 20-21 Demonstration of functional domains in seven-spanning receptors by experiments with chimeric proteins containing portions of the β_2- and α_2-adrenergic receptors. *Xenopus* oocytes microinjected with mRNA encoding the wild-type receptors or chimeric α-β receptors expressed the corresponding receptor protein on the cell surface. *Xenopus* oocytes do not express adrenergic receptors but do express G proteins, which can couple to the foreign receptors. Binding assays were conducted using agonists known to bind to α or β receptors to determine the ligand-binding specificity of the chimeric receptors. The effects of the agonists on adenylate cyclase activity were taken as a measure of whether the receptor protein bound to the stimulatory (G_s) or inhibitory (G_i) type of oocyte G protein. A comparison of chimeric receptor 1, which interacts with G_s, and chimeric receptor 3, which interacts with G_i, shows that the G protein specificity is determined primarily by the source of the cytosol-facing loop between α helices 5 and 6. A comparison of chimeras 1 and 2 indicates that the source of α helix 7 determines the ligand-binding specificity. [See B. Kobilka et al., 1988, *Science* **240**:1310; W. A. Catterall, 1989, *Science* **243**:236.]

phosphodiesterase (see Figure 20-14). This hydrolysis terminates the effect of hormone stimulation.

Many cAMP phosphodiesterases are activated by increases in cytosolic Ca^{2+} (another intracellular second messenger), which are often induced by neuron or hormone stimulation. As discussed later, Ca^{2+} ions bind to *calmodulin*, an ubiquitous Ca^{2+}-binding protein. When a Ca^{2+}-calmodulin complex binds to cAMP phosphodiesterase, the enzyme is activated and increases its rate of cAMP hydrolysis. Some cells also modulate the level of cAMP by secreting it into the extracellular medium.

The synthesis and degradation of cAMP are both subject to complex regulation by multiple hormones, which allows the cell to integrate responses to many types of changes in its internal and external environments.

► *Role of cAMP in the Regulation of Cellular Metabolism*

In the previous section, we saw that hormone stimulation of G_s protein–linked cell-surface receptors leads to an elevation of the level of cAMP. Recall that cAMP is the second messenger for many hormones and that the effects of elevated cAMP differ markedly in various types of cells (see Table 20-5). In this section, we discuss how cAMP affects enzymatic activity, thereby regulating cellular metabolism.

cAMP and Other Second Messengers Activate Specific Protein Kinases

The diverse effects of cAMP are thought to be mediated through the action of *cAMP-dependent protein kinases* (cAPKs; also referred to as protein kinase A or PKA), which modify the activities of specific enzymes in various cell types. Protein kinases transfer the terminal phosphate group in ATP to the hydroxyl group in serine, threonine, and tyrosine residues of substrate proteins (Figure 20-22). Phosphorylation of many enzymes increases their catalytic activity, whereas phosphorylation of others decreases their activity.

The cAMP-dependent protein kinases are tetramers, consisting of two regulatory (R) subunits and two catalytic (C) subunits. In the tetrameric form, cAPK is enzymatically inactive. Binding of cAMP to the R subunits causes dissociation of the two C subunits, which then can phosphorylate specific acceptor proteins (Figure 20-23). Each R subunit binds two cAMP molecules in a cooperative fashion; that is, binding of the first cAMP molecule lowers the K_D for binding of the second. Thus, small changes in the level of cytosolic cAMP can cause proportionately large changes in the amount of dissociated C subunits and hence in the activity of a cAPK.

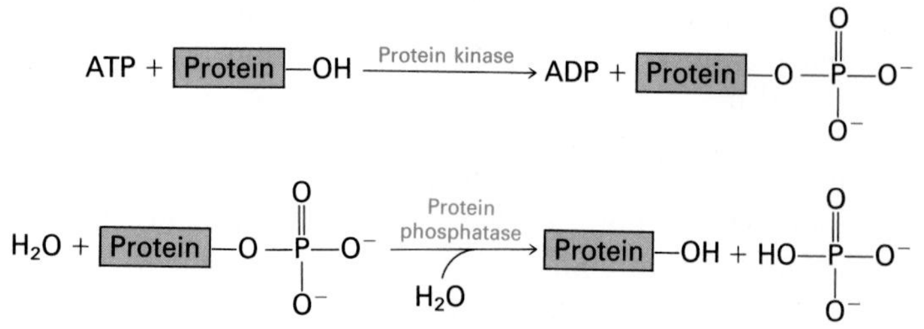

◄ FIGURE 20-22   Reactions catalyzed by protein kinases and protein phosphatases. The phosphorylated and dephosphorylated forms of an enzyme often differ markedly in catalytic activity.

As discussed later, other second messengers also activate protein kinases via a similar mechanism: in the absence of the second messenger, the kinase shows low activity; binding of the second messenger increases the kinase activity. Each type of second messenger–dependent kinase is inhibited by the binding of a peptide sequence to the active site. This peptide sequence can be part of a distinct regulatory subunit, as in cAMP-dependent protein kinases, or it can be part of a *regulatory domain* within the same polypeptide chain that contains the active site. In either case, experimental removal of the regulatory domain—by treating the kinase with an agent that dissociates the subunits or with a proteolytic enzyme—leads to kinase activation. Binding of a second messenger to the inactive kinase induces a conformational change that leads to release of the part of the regulatory domain bound to the active site of the kinase or to dissociation of the regulatory subunit; in both cases, the active sites of the kinase are unmasked and their enzymatic activities activated. Regulatory sequences are often called kinase *pseudosubstrates:* their sequences are similar to those of regions of substrate proteins that are phosphorylated by the kinase. Like substrates, they bind to the active site with high affinity, but often they cannot be phosphorylated, because they lack a serine or threonine residue.

Epinephrine Stimulates Glycogenolysis in Liver and Muscle Cells

The cAMP-dependent protein kinases induce many effects depending on the particular substrate proteins that they phosphorylate. The first cAMP-mediated cellular response to be discovered—the release of glucose from glycogen (*glycogenolysis*)—has been studied the most. This reaction occurs in muscle and liver cells stimulated by epinephrine or agonists of β-adrenergic receptors. Before describing how cAPKs regulate glycogen metabolism, we review the pathways of glycogen synthesis and degradation.

Glycogen, a polymer of glucose linked by $\alpha(1 \rightarrow 4)$ glycosidic bonds, is the major storage form of glucose in the body and is found principally in liver and muscle cells. Like most polymers, glycogen is synthesized by one set of enzymes and degraded by another (Figure 20-24). The primary intermediate in glycogen synthesis is *uridine diphosphoglucose* (UDP-glucose), which is produced from glucose by three enzyme-catalyzed reactions. The glucose residue of UDP-glucose is transferred by glycogen synthase to the free hydroxyl group on carbon 4 of a glucose residue at the end of a growing glycogen chain. Degradation of glycogen involves the stepwise removal of glucose residues from the same end. This is a *phosphorolysis* reaction, catalyzed by glycogen phosphorylase, in which a phosphate is added yielding glucose 1-phospate.

The fate of glucose 1-phosphate resulting from degradation of glycogen differs in liver and muscle cells. In muscle cells, glucose 1-phosphate produced from glycogen is converted by *phosphoglucomutase* to glucose 6-phosphate. This is metabolized via the Embden-Meyerhoff glycolytic pathway (see Figure 17-2) to generate ATP, which is a source of energy for contraction. In contrast, glucose from glycogen is not a major source of ATP in the liver. Rather, the liver stores and releases glucose primarily for use by other tissues. Unlike muscle cells, liver cells contain a glucose 6-phosphatase, which hydrolyzes glucose 6-phospate to glucose. The free glucose is immediately released into the blood and transported to other tissues, particularly the muscles and brain.

▲ FIGURE 20-23   Activation of cAMP-dependent protein kinase by cAMP. This kinase contains two regulatory (R) subunits, with binding sites for cAMP, and two catalytic (C) subunits. The tetrameric protein is enzymatically inactive because the regulatory subunits block the catalytic sites on the C subunits. Binding of cAMP, which occurs cooperatively, causes dissociation of the catalytic subunits, which then can phosphorylate various substrate proteins (see Figure 20-22).

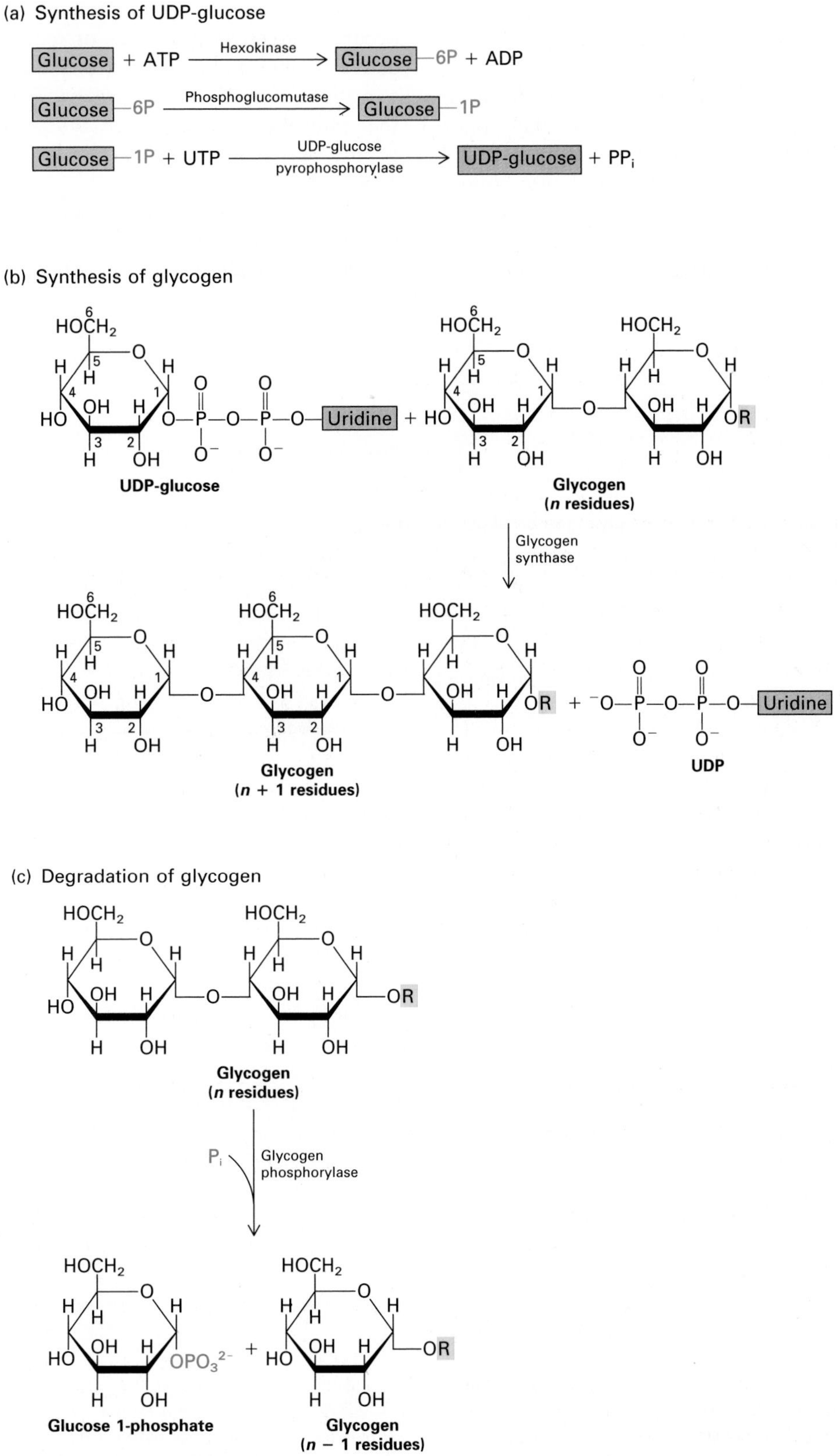

(a) Synthesis of UDP-glucose

$$\boxed{\text{Glucose}} + \text{ATP} \xrightarrow{\text{Hexokinase}} \boxed{\text{Glucose}}\text{—6P} + \text{ADP}$$

$$\boxed{\text{Glucose}}\text{—6P} \xrightarrow{\text{Phosphoglucomutase}} \boxed{\text{Glucose}}\text{—1P}$$

$$\boxed{\text{Glucose}}\text{—1P} + \text{UTP} \xrightarrow[\text{pyrophosphorylase}]{\text{UDP-glucose}} \boxed{\text{UDP-glucose}} + \text{PP}_i$$

(b) Synthesis of glycogen

(c) Degradation of glycogen

▲ FIGURE 20-24   Synthesis and degradation of glycogen. (a) Uridine diphosphoglucose (UDP-glucose), the immediate precursor of glycogen, is synthesized from glucose. (b) Incorporation of glucose from UDP-glucose into glycogen is catalyzed by glycogen synthase (c) Removal of glucose units from glycogen is catalyzed by glycogen phosphorylase. Because two different enzymes catalyze the formation and degradation of glycogen, the two reactions can be independently regulated. R stands for the remainder of the glycogen molecule.

cAMP-Dependent Protein Kinase Regulates the Enzymes of Glycogen Metabolism

In liver and muscle cells, the epinephrine-stimulated elevation in the cAMP level enhances the conversion of glycogen to glucose 1-phosphate in two ways: by inhibiting glycogen synthesis and by stimulating glycogen degradation (Figure 20-25a). In the presence of cAMP, the dissociated C subunit of cAPK phosphorylates *glycogen synthase*, converting it to a much less active enzyme and thus inhibiting glycogen synthesis. Stimulation of glycogen breakdown occurs indirectly via a series of sequential reactions. The active C subunit phosphorylates and activates another protein kinase, *glycogen phosphorylase kinase*. This enzyme then phosphorylates *glycogen phosphorylase*, converting it to the active form that degrades glycogen to glucose 1-phosphate.

The entire process is reversed when epinephrine is removed and the level of cAMP drops (Figure 20-25b). This reversal is mediated by *phosphoprotein phosphatase*,

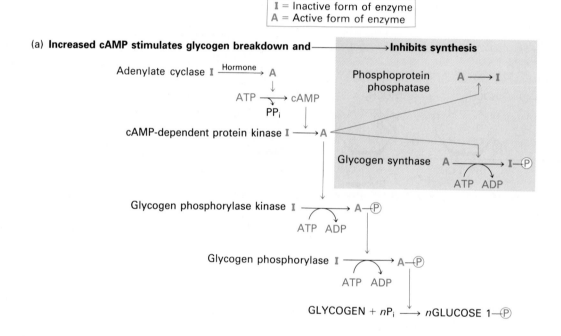

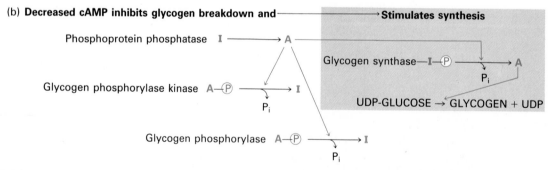

▲ FIGURE 20-25   Regulation of glycogen breakdown and synthesis by cAMP in liver and muscle cells. (a) An increase in cytosolic cAMP activates cAMP-dependent protein kinase, which then triggers a protein kinase cascade that results in breakdown of glycogen. Active cAMP-dependent protein kinase also phosphorylates glycogen synthase and an inhibitor of phosphoprotein phosphatase (see Figure 20-26), inactivating both enzymes. As a result, glycogen synthesis is inhibited, as is dephosphorylation of the activated enzymes in the kinase cascade. (b) A decrease in cAMP inactivates cAMP-dependent protein kinase, leading to formation of the active form of phosphoprotein phosphatase. This enzyme then removes phosphate residues from glycogen phosphorylase kinase and glycogen phosphorylase, thereby inhibiting glycogen degradation. The phosphatase also removes phosphate from inactive glycogen synthase, activating this enzyme and stimulating glycogen synthesis.

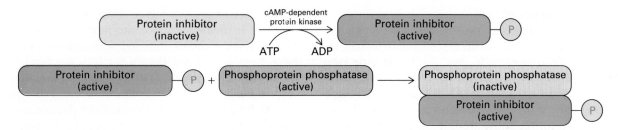

▲ FIGURE 20-26   Regulation of phosphoprotein phosphatase activity by cAMP is mediated by an inhibitor protein. At high levels of cAMP, cAMP-dependent protein kinase phosphorylates an inhibitor protein (blue), which then binds to and inactivates phosphoprotein kinase (orange). When the cAMP level decreases, constitutive phosphatases dephosphorylate the inhibitor, releasing phosphoprotein phosphatase in its active form.

which removes the phosphate residues from glycogen synthase, thereby activating it, and from glycogen phosphorylase kinase and glycogen phosphorylase, thereby inactivating them. The activity of phosphoprotein phosphatase also is regulated by cAMP, although indirectly. At high cAMP levels, cAPK phosphorylates an inhibitor of phosphoprotein phosphatase (Figure 20-26). Phosphorylation of the inhibitor allows it to bind to phosphoprotein phosphatase and inhibit its activity. At low cAMP levels, the inhibitor is not phosphorylated and phosphoprotein phosphatase is active. As a result, the synthesis of glycogen by glycogen synthase is enhanced and the degradation of glycogen by glycogen phosphorylase is inhibited.

Kinase Cascade Permits Multienzyme Regulation and Amplifies Hormone Signal

The set of protein phosphorylations and dephosphorylations just described constitute a *cascade,* a series of reactions in which the protein catalyzing one step is activated (or inhibited) by the product of a previous step. Although such a cascade may seem overcomplicated, it has at least two advantages for the cell.

First, a cascade allows an entire group of enzyme-catalyzed reactions to be regulated by a single type of molecule. As we have seen, the three enzymes in the glycogenolysis cascade—cAMP-dependent protein kinase, glycogen phosphorylase kinase, and glycogen phosphorylase—are all regulated, directly or indirectly, by cAMP (see Figure 20-25a). Other metabolic pathways also are regulated by hormone-induced cascades, some mediated by cAMP and some by other second messengers.

Second, a cascade provides a huge amplification of an initially small signal. For example, blood levels of epinephrine as low as 10^{-10} M can stimulate glycogenolysis and release of glucose, resulting in an increase of blood glucose levels by as much as 50 percent. An epinephrine stimulus of this magnitude generates an intracellular cAMP concentration of 10^{-6} M (Figure 20-27), an amplification of 10^4.

Because three more catalytic steps precede the release of glucose, another 10^4 amplification can occur. In striated muscle, the concentrations of the three successive enzymes in the glycogenolytic cascade are in a 1:10:240 ratio, which dramatically illustrates the amplification of the effects of epinephrine and cAMP.

Cellular Responses to cAMP Vary among Different Cell Types

The effects of cAMP on the synthesis and degradation of glycogen are confined mainly to liver and muscle cells, which store glycogen. However, cAMP also mediates the

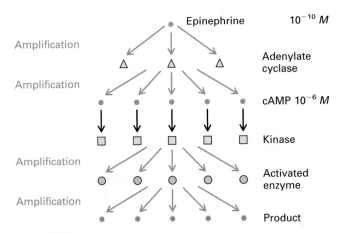

▲ FIGURE 20-27   Intracellular transduction of an extracellular signal via a cascade of sequential reactions produces a large amplification of the signal. In this example, binding of a single epinephrine molecule to one receptor molecule induces synthesis of a large number of cAMP molecules. These in turn activate multiple enzyme molecules, each of which produces multiple product molecules (e.g., active phosphorylated proteins). The more steps in such a cascade, the greater the signal amplification possible.

intracellular responses of many other cells to a variety of hormones that stimulate G_s protein-linked receptors (see Table 20-5). In virtually all eukaryotic cells studied, the action of cAMP appears to be mediated by one or more cAMP-dependent protein kinases. The response of a particular cell type to an elevation of cAMP depends on the substrate specificity of the cAMP-dependent protein kinase(s) it contains and on the substrates for cAPK expressed.

In adipocytes, for example, an epinephrine-induced cascade regulates the synthesis and degradation of triacylglycerols, the storage form of fatty acids (see Figure 2-14). Activation of the β-adrenergic receptors on adipose cells triggers an increase in cytosolic cAMP and activation of a cAMP-dependent protein kinase. The lipase that hydrolyzes triacylglycerols to fatty acids is activated by phosphorylation by this cAPK. The released fatty acids are transferred to the blood, where they bind to albumin, a major serum protein. In this form, fatty acids are transported to other tissues—particularly the heart, muscles, and kidneys—where they are used as a source of ATP.

In some cells, an increase in intracellular cAMP is induced by other hormones: for example, adrenocorticotropic hormone (ACTH) in cells of the adrenal cortex and follicle-stimulating hormone (FSH) in ovarian cells. As with epinephrine stimulation of muscle and liver cells, binding of ACTH or FSH to their G_s-linked receptors activates adenylate cyclase, elevating the level of cAMP, which then activates a cAMP-dependent protein kinase. In adrenal cortex cells, this cAPK activates the enzymes involved in synthesis of several steroid hormones (e.g., cortisone, corticosterone, and aldosterone). In ovarian cells, this kinase enhances synthesis of estradiol and progesterone, two steroid hormones crucial to the development of female sex characteristics (see Figure 20-5).

▶ Receptor Tyrosine Kinases

Receptor tyrosine kinases (RTKs) form a large and important class of cell-surface receptors whose ligands are soluble or membrane-bound peptide/protein hormones including insulin and epidermal growth factor (see Figure 20-4g). Binding of a ligand to this type of receptor stimulates the receptor's protein tyrosine kinase activity, which subsequently stimulates a signal-transduction cascade leading to changes in cellular physiology and/or patterns of gene expression. RTK signaling pathways have a wide spectrum of functions including regulation of cell proliferation and differentiation, promotion of cell survival, and adjustments in cellular metabolism.

Some RTKs have been identified in studies on human cancers associated with mutant forms of growth-factor receptors, which send a proliferative signal to cells even in the absence of growth factor. One such mutant receptor, encoded at the *neu* locus, contributes to the uncontrolled proliferation of certain human breast cancers. Other RTKs have been uncovered during analysis of developmental mutations in *C. elegans, Drosophila,* and the mouse that lead to blocks in differentiation of certain cell types.

All RTKs comprise an extracellular domain containing a ligand-binding site, a single hydrophobic transmembrane α helix, and a cytosolic domain that includes a region with protein tyrosine kinase activity (Figure 20-28). Binding of ligand causes most RTKs to dimerize; the protein kinase of each receptor monomer then phosphorylates a distinct set of tyrosine residues in the cytosolic domain of its dimer partner, a process termed *autophosphorylation.* The subunits of some RTKs, including the tetrameric insulin receptor, are covalently linked (see Figure 20-8). Although these receptors exist as dimers or tetramers even in the absence of ligand, binding of ligand is required for autophosphorylation to occur. The phosphotyrosine residues of activated RTKs play a crucial role in transducing hormone signals to intracellular signaling molecules. As discussed later, some RTKs also phosphorylate substrate proteins that are components of other signaling pathways.

Although RTKs were first recognized during the 1970s, the mechanisms by which these receptors transduce signals have been elucidated only during the past few years. Four different experimental approaches have provided important insights into these mechanisms: (1) identification of substrates phosphorylated by RTKs; (2) identification of proteins that bind to specific phosphotyrosine residues in the cytosolic domain; (c) biochemical analysis of the pathways activated by RTKs; and (d) genetic analysis of RTK signaling pathways in *C. elegans* and *Drosophila.* In this section, we discuss key results obtained with each of these approaches.

SH2-Containing Proteins Bind to Specific Phosphotyrosine Residues in Activated RTKs

Two different classes of proteins associate with the cytosolic domain of activated RTKs: (1) *adapter proteins* (e.g., GRB2) that couple the activated receptors to other signaling molecules but have no intrinsic signaling properties, and (2) enzymes involved in signaling pathways. In the latter class are GAP, a Ras-associated protein discussed earlier (see Figure 20-17); Syp, a protein phosphatase; and enzymes involved in the synthesis of phosphatidylinositol derivatives, including phosphatidylinositol-3 kinase and phospholipase C_γ (Figure 20-29). The roles of these RTK-binding proteins in signal-transduction are explained later.

These proteins bind to different phosphotyrosine residues on RTKs via a conserved polypeptide domain, called

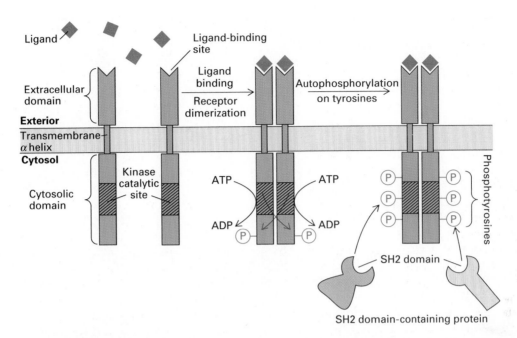

▲ FIGURE 20-28 Dimerization and activation of most receptor tyrosine kinases (RTKs) is induced by ligand binding. These receptors contain an extracellular ligand-binding domain, a single transmembrane α helix, and a cytosolic domain with tyrosine kinase activity. An activated receptor autophosphorylates specific tyrosine residues in the cytosolic domain. The resulting phosphotyrosines function as binding sites for various proteins containing SH2 domains, as described in the text. [Adapted from G. Panayotou and W. D. Waterfield, 1993, *Bioessays* **15**:171.]

the *Src homology 2 (SH2) domain*. The domain derives its name from its homology with a region in the prototypical cytosolic tyrosine kinase encoded by *src*. As we later shall see, the SH3 domain, which exhibits homology with another Src region, also plays a role in mediating protein-protein interactions required in some signal-transduction pathways.

The three-dimensional structures of SH2 domains in different proteins are very similar. Each binds to a distinct sequence of amino acids surrounding a phosphotyrosine residue. The unique amino acid sequence of each SH2 domain determines the specific phosphotyrosine residues it binds. The SH2 domain of the Src tyrosine kinase, for example, binds strongly to any peptide containing the critical core sequence of phosphotyrosine–glutamic acid–glutamic acid–isoleucine (Figure 20-30). These four amino acids make intimate contact with the peptide-binding site in the Src SH2 domain. Binding resembles the insertion of a two-pronged "plug"—the phosphotyrosine and isoleucine residues of the peptide—into a two-pronged "socket" in the SH2 domain. The two glutamic acids fit snugly onto the surface of the SH2 domain between the phosphotyrosine socket and the hydrophobic socket that accepts the isoleucine residue. Variations in the nature of the hydrophobic socket in different SH2 domains allow them to bind to

phosphotyrosines adjacent to different sequences, accounting for differences in their binding specificity.

Ras Protein Is a Key Component of Signaling Pathways in Many Eukaryotes

We saw earlier that Ras proteins belong to the GTPase superfamily of intracellular switch proteins that cycle between active (on) and inactive (off) states (see Figure 20-17). Mammalian Ras proteins have been studied in great detail because mutant Ras proteins are associated with many types of human cancer. These mutant proteins, which bind but cannot hydrolyze GTP, are permanently in the on state and cause neoplastic transformation (Chapter 27).

Ras proteins play a central role in transducing signals from many RTKs to a cascade of serine/threonine kinases that function in controlling cell growth and differentiation. The first indication that Ras functioned downstream from RTKs in a common signaling pathway came from the experiments summarized in Figure 20-31. In these experiments, cultured fibroblast cells (NIH 3T3) were induced to proliferate by treatment with a mixture of platelet-derived growth factor (PDGF) and epidermal growth factor (EFG), two polypeptide growth factors whose receptors are RTKs.

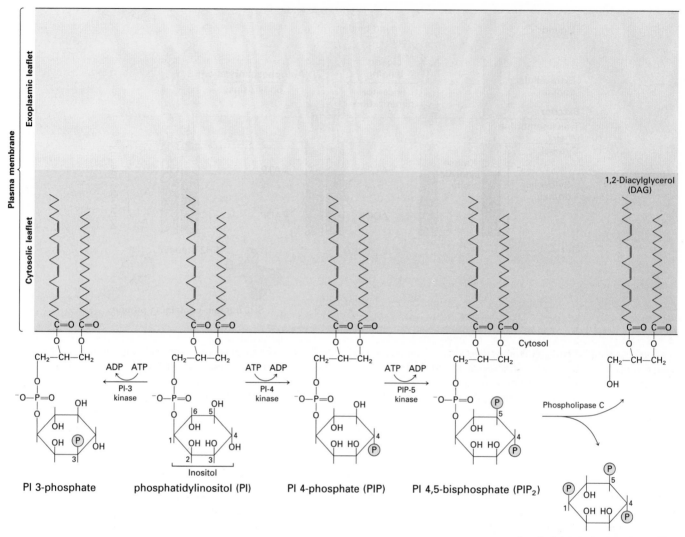

▲ FIGURE 20-29 Synthesis of derivatives of plasma-membrane phosphatidylinositol (PI) by membrane-associated kinases. Each enzyme replaces a hydroxyl group on a specific carbon of the inositol ring with a phosphate group (yellow). The resulting phosphorylated molecules play important roles in several signaling pathways. PI-3 kinase and the γ isoform of phospholipase C bind to the cytosolic domain of many activated RTKs, thereby bringing the enzymes close to their substrates and stimulating their catalytic activity. Ligand binding to seven-spanning receptors induces activation of another phospholipase C. As discussed later, hydrolysis of PI 4,5-bisphosphate by either phospholipase C generates two important second messengers: 1,2-diacylglycerol and inositol 1,4,5-trisphosphate.

Microinjection of anti-Ras antibodies into these cells blocked cell proliferation. Conversely, injection of a constitutively active mutant Ras protein (i.e., Ras^D) caused the cells to proliferate in the absence of polypeptide growth factors. Ras^D hydrolyzes GTP inefficiently and thus remains in the active on state longer than normal (see Figure 20-17). These findings are consistent with studies showing that addition of fibroblast growth factor (FGF) to fibroblasts leads to a rapid increase in the proportion of Ras found in the GTP-bound active form.

But how does binding of a growth factor (e.g., EGF) to an RTK (e.g., the EGF receptor) lead to activation of Ras? Two cytosolic proteins—GRB2 and Sos—provide the key links (Figure 20-32). An SH2 domain in GRB2 binds to a specific phosphotyrosine residue in the activated EGF receptor. GRB2 also contains two SH3 domains, which bind to and activate Sos. GRB2 thus functions as an adapter protein for the EGF receptor. In its active form, Sos functions as a guanine nucleotide–exchange protein (GEF); as noted earlier, binding of a GEF to Ras converts Ras to the

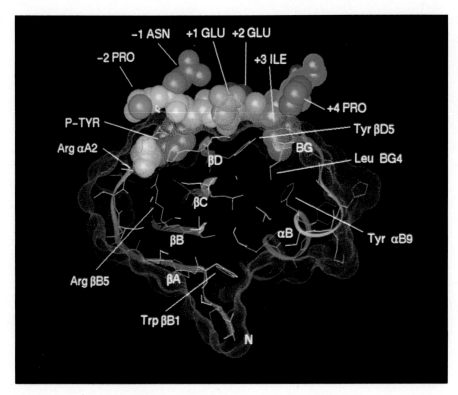

◄ FIGURE 20-30   A cutaway diagram of an SH2 domain bound to a phosphopeptide containing six amino acid residues: PRO–ASN–P-TYR–GLU–GLU–ILE. The space-filling model of the phosphopeptide shows the backbone in yellow, the side chains in green, and the phosphate group covalently attached to the tyrosine residue in white. Like a two-pronged plug, the phosphotyrosine (P-TYR) and isoleucine (+3ILE) fit into a two-pronged socket on the surface of the SH2 domain (red). See text for discussion. [From G. Waksman et al., 1993, *Cell* **72**:779.]

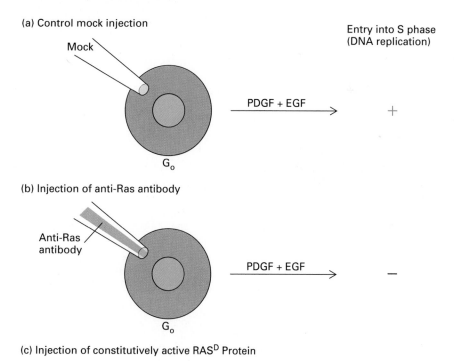

(a) Control mock injection

(b) Injection of anti-Ras antibody

(c) Injection of constitutively active RAS^D Protein

► FIGURE 20-31   Experimental demonstration that activated Ras protein is required and sufficient for induction of DNA synthesis in cultured NIH 3T3 cells (a fibroblast cell line) arrested in G_o. PDGF = platelet-derived growth factor; EGF = epidermal growth factor. See text for discussion. [See L. S. Mulcahy et al., 1985, *Nature* **313**:241; J. R. Feramisco et al., 1984, *Cell* **38**:109.]

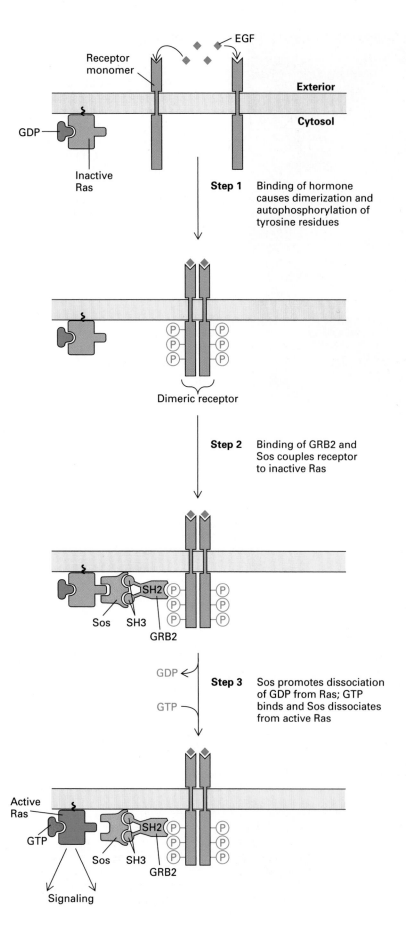

Step 1 Binding of hormone causes dimerization and autophosphorylation of tyrosine residues

Step 2 Binding of GRB2 and Sos couples receptor to inactive Ras

Step 3 Sos promotes dissociation of GDP from Ras; GTP binds and Sos dissociates from active Ras

◄ FIGURE 20-32 Activation of Ras following binding of a hormone (e.g., EGF) to an RTK. Binding of hormone causes the receptor monomers to dimerize and autophosphorylate tyrosine residues in its cytosolic domain (step 1). The SH2 domain in GRB2 binds to a specific phosphotyrosine on the activated RTK; the SH3 domains in GRB2 then bind Sos, which acts as a guanine nucleotide–exchange factor (GEF). Sos then interacts with the inactive Ras · GDP, which is associated with the plasma membrane (step 2). Exchange of GDP for GTP generates the active form of Ras (step 3). [See L. Buday and J. Downward, 1993, *Cell* **73:**611; J. P. Olivier et al., 1993, *Cell* **73:**179; S. E. Egan et al., 1993, *Nature* **363:**45; E. J. Lowenstein et al., 1992, *Cell* **70:**431; M. A. Simon et al., 1993, *Cell* **73:**169.]

active GTP-bound form, which then triggers a protein kinase cascade. As discussed later, activation of this cascade
leads to changes in gene expression.

Because the RTK-Ras signal-transduction pathway
controls differentiation of many cell types as well as cell
proliferation, we examine it in some detail. Genetic analysis of mutants blocked at particular stages of differentiation have proven to be powerful tools for identifying the
proteins that connect RTKs to activation of Ras and the
proteins that, in turn, are activated by Ras. Most of these
genetic studies were done in the worm *C. elegans* and in the
fly *Drosophila*. Mutants in these species in which only one
kind of cell differentiation is blocked have provided a relatively easy means of identifying genes encoding other proteins in the signal-transduction pathway.

Genetic Analysis of *Drosophila* Eye Development Identified Three Proteins That Link RTKs to a Kinase Cascade

Genetic analyses of two developmental systems, one in
Drosophila and one in *C. elegans*, have provided considerable insight into RTK signaling. Because the results with
the *Drosophila* system, which involves development of a
particular type of eye cell, are more complete and detailed,
we describe that system here.

As illustrated in Figure 20-33, the compound eye of the
fly is composed of some 800 individual eyes called *ommatidia*. Each ommatidium consists of 22 cells, eight of which
are photosensitive neurons called retinula, or R cells. An
RTK called Sevenless (Sev) specifically regulates develop-

(a)

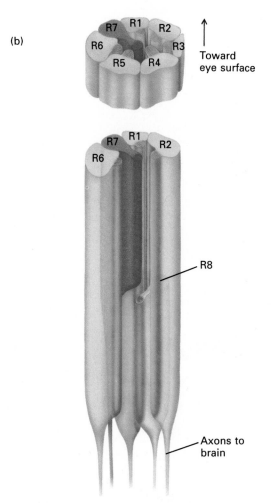

(b)

▲ **FIGURE 20-33** Structure of the compound eye of
Drosophila melanogaster. (a) Scanning electron micrograph
showing individual ommatidia that compose the fruit fly eye.
(b) Longitudinal and cut-away views of a single ommatidium.
Each of these tubular structures contains eight photoreceptors, designated R1–R8, which are long, cylindrically shaped
light-sensitive cells. R1–R6 (yellow) extend throughout the
depth of the retina, whereas R7 (orange) is located toward
the surface of the eye and R8 (blue) toward the backside,
where the axons exit. Each R cell has a tightly packed bundle of photosensitive microvilli that projects toward the center of the ommatidium. [Part (a) from E. Hafen and K. Basler,
1991, *Development* **1**(Suppl.):123; part (b) adapted from
R. Reinke and S. L. Zipursky, 1988, *Cell* **55**:321.]

ment of one of the R cells, designated R7. In flies with a mutant *sevenless* (*sev*) gene, the R7 cell in each ommatidium does not form (see Figure 8-2b, c). Since the R7 photoreceptor is necessary for flies to see in ultraviolet light, mutants that lack functional R7 cells are easily isolated.

During development of each ommatidium, a protein called Boss (Bride of sevenless) is expressed on the surface of the R8 cell. This membrane-bound protein binds to the Sev RTK on the surface of the neighboring R7 precursor cell, signaling it to develop into a photosensitive neuron (Figure 20-34a). In mutant flies that do not express a functional Sev RTK, interaction between the Boss and Sev proteins cannot occur, and no R7 cells develop (Figure 20-34b).

To identify additional proteins in the Sev RTK pathway, investigators produced mutant flies expressing a temperature-sensitive Sev protein. When these flies were al-

lowed to grow at a permissive temperature, all their ommatidia contained R7 cells; but when they were maintained at a nonpermissive temperature, no R7 cells developed (see Figures 20-34a, b). The investigators found that at a particular intermediate temperature just enough of the Sev RTK was functional to mediate normal R7 development. They reasoned that at this intermediate temperature, the signaling pathway would become defective, and thus no R7 cells would develop, if the levels of other proteins involved in the pathway were reduced. A recessive mutation that affected a protein in the Sev signal-transduction pathway would have this effect because, in diploid organisms like *Drosophila*, a heterozygote containing one wild-type and one mutant allele of a gene will produce half the normal amount of the gene product; hence even if such a recessive mutation is in an essential gene, the organism will be viable. However, a fly carrying a temperature-sensitive

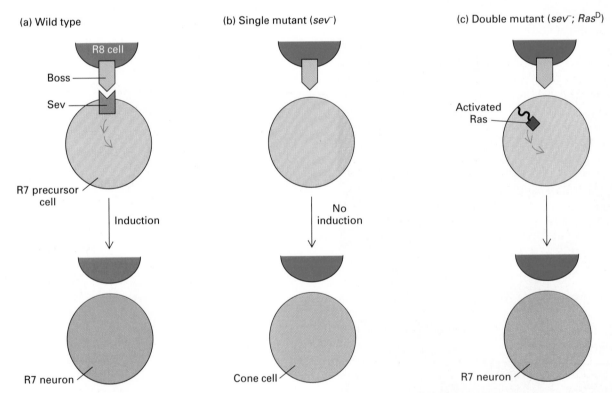

▲ FIGURE 20-34   Genetic analysis of induction of the R7 photoreceptor in the *Drosophila* eye. (a) During larval development of wild-type flies, the R8 cell (blue) in each developing ommatidium expresses a cell-surface protein, called Boss (red), that binds to the Sev RTK (green) on the surface of its neighboring R7 precursor cell (light orange). This interaction induces changes in gene expression that result in differentiation of the precursor cell into a functional R7 cell (dark orange). (b) In fly embryos with a mutation in the *sevenless(sev)* gene, R7 precursor cells cannot bind Boss and

therefore do not differentiate normally into R7 cells. Rather the precursor cell enters an alternative pathway of cellular development and eventually becomes a cone cell (purple). (c) Double-mutant (*sev; Ras*[D]) larvae express a constitutively active Ras (Ras[D]) in the R7 precursor cell, which induces differentiation of R7 precursor cells in the absence of the Boss-mediated signal. This finding shows that activated Ras can mediate induction. [See M. A. Simon et al., 1991, *Cell* **67**:701; M. E. Fortini et al., 1992, *Nature* **355**:559.]

mutation in the *sev* gene and a second mutation affecting another protein in the signaling pathway would be expected to lack R7 cells at the intermediate temperature.

By use of this screen, researchers identified genes encoding three important proteins in the Sev signal-transduction pathway (see Figure 20-32):

1. A Ras protein exhibiting 80 percent identity with its mammalian counterparts

2. A GEF called Sos (Son of sevenless), which is 45 percent identical to its mouse counterpart

3. An SH2-containing adapter protein similar in sequence to human GRB2 (i.e. 64 percent identical)

These three proteins have been found to function in other RTK signal-transduction pathways initiated by ligand binding to different receptors and used at different times and places in the developing fly. Recessive lethal mutations in these essential genes can be identified by the strategy described here much more easily than by the procedure described in Chapter 8 (see Figure 8-11).

Additional studies with *sevenless* flies demonstrated that activation of Ras is sufficient to transmit the signal normally initiated by binding of Boss protein to the Sev RTK. By in vitro mutagenesis, researchers produced a *Ras*^{D} gene encoding a Ras protein with reduced GTPase activity. This mutant Ras is constitutively activated; that is, it is present in the GTP-bound active form even in the absence of a hormone signal. The *Ras*^{D} gene was introduced into fly embryos carrying the *sevenless* mutation and selectively expressed from a cell type–specific promoter in a subset of developing cells in the eye, including R7 precursor cells. Even though no functional Sev RTK was expressed in these double-mutants (*sev-*; *Ras*^{D}), R7 cells formed normally, indicating that activation of Ras is sufficient for induction of R7-cell development (Figure 20-34c). This finding is consistent with the results with cultured fibroblasts described earlier (see Figure 20-31c).

GRB2 Is an Adapter Protein That Binds to Activated RTKs

An understanding of how GRB2 participates in RTK signaling came from biochemical studies in mammalian cells. GRB2 was identified in mammalian cells by an expression cloning method used to identify proteins that associate with specific phosphotyrosine residues on receptor proteins. First, cDNAs are synthesized from mRNAs isolated from a tissue of interest; then a cDNA library is prepared in the λgt11 expression vector and the vectors are plated on a lawn of *E. coli* cells (see Figure 7-21). A fragment of an activated RTK containing the phosphotyrosine residues in the cytosolic domain is used to screen the resulting plaques

to identify those expressing a phosphotyrosine-binding protein. When this method was used to screen a cDNA library prepared from human brain-stem tissue, using a fragment of the human EGF receptor as the probe, two cDNAs were identified. One encoded a subunit of PI-3 kinase, which contains an SH2 domain, and the other encoded GRB2, a homolog of the SH2-containing adapter protein identified in the *Drosophila* Sev pathway. These studies thus identify GRB2 as an adapter protein that functions downstream from RTKs but upstream of Ras in both *Drosophila* and mammalian cells.

Sos Protein Is Localized to the Plasma Membrane by Binding to the SH3 Domains in GRB2

As noted earlier, GRB2 contains one SH2 domain, which binds to phosphotyrosine residues in RTKs, and two SH3 domains, which bind to Sos, a guanine nucleotide–exchange factor (see Figure 20-32). SH3 domains, which contain ≈55–70 residues, are present in a large number of signaling molecules. Although the three-dimensional structure of various SH3 domains is similar, their specific amino acid sequences differ. SH3 domains selectively bind to proline-rich sequences in Sos and other proteins; different SH3 domains bind to different proline-rich sequences.

Prolines play two roles in the interaction between an SH3 domain in an adapter protein (e.g., GRB2) and a proline-rich sequence in another protein (e.g., Sos). First, the proline-rich sequence assumes an extended conformation that permits extensive contacts with the SH3 domain, thereby facilitating interaction. Second, a subset of these prolines fit into binding "pockets" on the surface of the SH3 domain (Figure 20-35a). Several nonproline residues also interact with the SH3 domain and are responsible for determining the binding specificity (Figure 20-35b). Hence the binding of peptides to SH2 and SH3 domains follows a similar strategy: certain residues provide the overall structural motif necessary for binding, and neighboring residues confer specificity to the binding.

Following hormone-induced activation of an RTK (e.g., the EGF receptor), a complex containing the activated RTK, GRB2, and Sos is formed on the cytosolic face of the plasma membrane. Formation of this complex depends on the dual binding ability of GRB2. Receptor activation leads to relocalization of Sos from the cytoplasm to the membrane, bringing Sos near to its substrate, membrane-bound Ras protein. Whether hormone binding stimulates the GEF activity of Sos, or alternatively whether the protein is constitutively active, is not yet clear. In either case, as part of this membrane-associated complex, Sos can mediate conversion of Ras from its inactive to active form (see Figure 20-32).

(a)

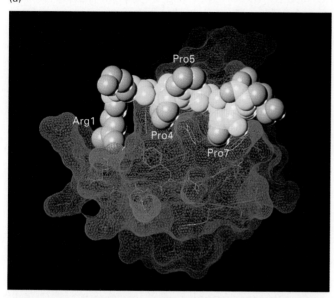

(b)

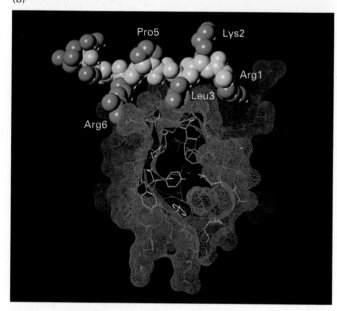

◄ **FIGURE 20-35**   Two views of an SH3 domain bound to a proline-rich sequence in a target peptide. The accessible surface of the SH3 domain is shown in red. The space-filling model of the bound peptide shows the backbone in yellow and the side chains in green. The extended helical conformation of the proline residues permit multiple contacts with the SH3 domain (*top*). Nonproline residues, such as Arg1, Leu3, and Arg6 in this peptide, play an important role in determining the specificity of binding to different SH3-containing proteins (*bottom*). [From H. Yu et al., 1994, *Cell* **76**:933.]

We now examine the functions of Ras and of the signal-transduction proteins that lie downstream from it in RTK signaling pathways. A remarkable convergence of biochemical and genetic studies in yeast, *C. elegans*, *Drosophila*, and mammals has revealed a highly conserved cascade of protein kinases that operate in sequential fashion downstream from activated Ras as follows (Figure 20-36):

1. Activated Ras binds to the N-terminal domain of *Raf*, a serine/threonine kinase.

2. Raf binds to and phosphorylates *MEK*, a dual-specificity protein kinase that phosphorylates both tyrosine and serine residues.

3. MEK phosphorylates and activates *MAP kinase*, another serine/threonine kinase.

4. MAP kinase phosphorylates many different proteins including transcription factors that regulate expression of important cell-cycle and differentiation specific proteins.

In this section we describe several experiments that have established the specific interactions between the different proteins in this important pathway in mammalian cells.

Binding of Ras·GTP to Raf   Several lines of evidence indicate that signaling downstream of Ras is mediated by the serine/threonine kinase Raf. For example, constitutively active mutant Raf proteins induce quiescent cultured cells to proliferate. These mutant Raf proteins, which initially were identified in tumor cells, are encoded by oncogenes and stimulate uncontrolled cell proliferation. Conversely, cultured mammalian cells that express a mutant, defective Raf protein cannot be stimulated to proliferate uncontrollably by a mutant, constitutively active form of Ras. This finding establishes a link between the Raf and Ras proteins.

Various studies have shown that Ras protein binds directly to Raf. For instance, beads coated with covalently bound purified Ras protein selectively bind Raf protein, and few other proteins, in cell homogenates. Only the active Ras·GTP form binds Raf. An interaction between the mammalian Ras and Raf proteins also has been demonstrated in the *yeast two-hybrid system*, a genetic system in

A Highly Conserved Kinase Cascade Transmits RTK-Mediated Signals Downstream from Ras

Thus far we have seen that activation of Ras (i.e., conversion to the form with a bound GTP) is sufficient to induce cell differentiation or, depending on cell type, proliferation. Normally, Ras activation and the subsequent cellular response is induced by ligand binding to an RTK. However, if cells contain a mutant, constitutively active Ras, the cellular response occurs in the absence of ligand binding (see Figures 20-31c and 20-34c).

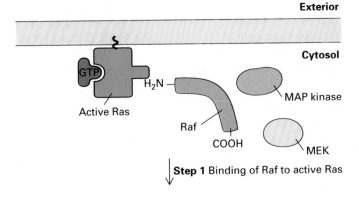

Exterior

Cytosol

Active Ras

MAP kinase

Raf

COOH

MEK

Step 1 Binding of Raf to active Ras

STIMULATED CELL

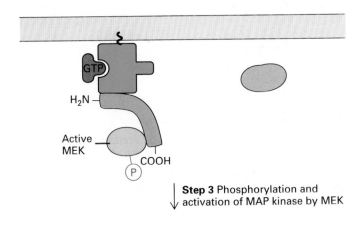

Raf

COOH

Step 2 Binding and phosphorylation of MEK by Raf

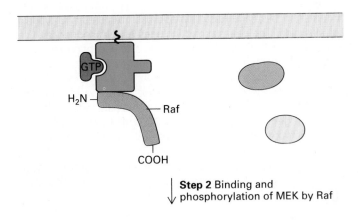

Active MEK

COOH

Step 3 Phosphorylation and activation of MAP kinase by MEK

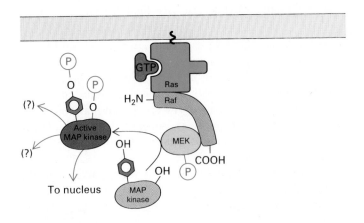

Active MAP kinase

(?)

(?)

To nucleus

MEK

COOH

MAP kinase

◄ FIGURE 20-36   Kinase cascade that transmits signals downstream from activated Ras protein. In unstimulated cells, most Ras is in the inactive form with bound GDP; binding of a growth factor to its RTK leads to formation of the active Ras · GTP (see Figure 20-32). A signaling complex then is assembled downstream of Ras, leading to activation of MAP kinase by phosphorylation of threonine and tyrosine residues separated by a single amino acid. Phosphorylation at both sites is necessary for activation of MAP kinase. See text for discussion. [See A. B. Vojtek et al., 1993, *Cell* **74**:205; L. Van Aelst et al., 1993, *Proc. Nat'l. Acad. Sci. USA* **90**:6213; S. A. Moodie et al., 1993, *Science* **260**:1658; H. Koide et al., 1993, *Proc. Natl. Acad. Sci. USA* **90**:8683; P. W. Warne et al., 1993, *Nature* **364**:352.]

yeast used to select cDNAs encoding proteins that bind to target, or "bait," proteins (Figure 20-37). The binding of Ras and Raf to each other apparently does *not* induce the Raf kinase activity. Perhaps activated Ras acts to relocate catalytically active Raf from the cytosol to the plasma membrane in response to extracellular signals, similar to the relocation of Sos by GRB2.

Activation of MAP Kinase by Activated MEK By analogy with the kinase cascade that regulates glycogen metabolism following stimulation of G_s-linked receptors (see Figure 20-25), the early investigators studying RTKs hypothesized that stimulation of these receptors also would include a kinase cascade. Initial attention focused on a serine/threonine kinase called MAP kinase (for microtubule-associated protein or mitogen-activated protein kinase). The activity of this enzyme was found to be enhanced after induction of cell proliferation by growth factors such as EGF. Activation of MAP kinase was shown to involve phosphorylation of both threonine and tyrosine residues on the enzyme.

Several experiments indicated that MAP kinase lies downstream of Ras and Raf in the RTK signal-transduction cascade. For example, when a constitutively active form of Ras was expressed in quiescent cultured cells, phosphorylated, activated MAP kinase was generated in the absence of hormone stimulation. More importantly, in *Drosophila* mutants that lack a functional Ras or Raf but express a constitutively active MAP kinase specifically in the developing eye, R7 photoreceptors were found to develop normally. This finding indicates that activation of MAP kinase is sufficient to transmit a proliferation or differentiation signal normally initiated by ligand binding to an RTK. The signaling pathway thus appears to be a linear one: activated RTK → Ras → Raf → (?) → MAP kinase.

Determination of how far downstream MAP kinase is came from fractionation of cultured cells that had been stimulated with growth factors. This approach led to identification of MEK, a kinase that specifically phosphorylates threonine and tyrosine residues on MAP kinase, thereby

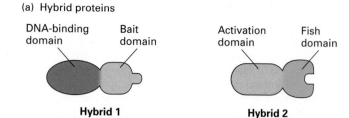

(a) Hybrid proteins

DNA-binding domain | Bait domain

Activation domain | Fish domain

Hybrid 1   **Hybrid 2**

(b) Transcriptional activation by hybrid proteins in yeast

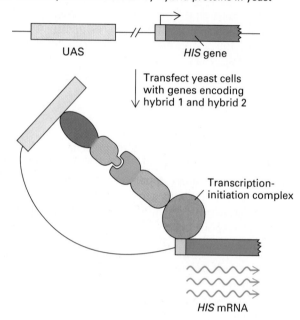

UAS   HIS gene

Transfect yeast cells with genes encoding hybrid 1 and hybrid 2

Transcription-initiation complex

HIS mRNA

(c) Fishing for proteins that interact with Ras

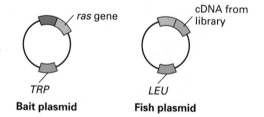

ras gene   cDNA from library

TRP   LEU

Bait plasmid   **Fish plasmid**

1. Transfect into *trp, leu, his* mutant yeast cells
2. Select for cells that grow in absence of tryptophan and leucine
3. Plate selected cells on medium lacking histidine

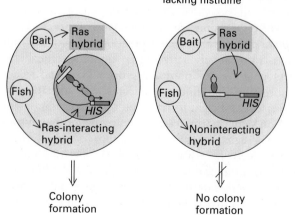

Bait → Ras hybrid

Fish

Ras-interacting hybrid

Bait → Ras hybrid

Fish

HIS

Noninteracting hybrid

Colony formation   No colony formation

◄ **FIGURE 20-37**   Yeast two-hybrid system for detecting proteins that interact. (a) Recombinant DNA techniques can be used to prepare genes that encode hybrid (chimeric) proteins consisting of the DNA-binding domain (purple) or activation domain (orange) of a transcription factor fused to one of two interacting proteins, referred to as the "bait" domain (red) and "fish" domain (green). (b) If yeast cells are transfected with genes encoding both hybrids, the bait and fish portions of the chimeric proteins interact to produce a functional transcriptional activator. One end of this protein complex binds to the upstream activating sequence (UAS) of a test gene (in this example, the *HIS3* gene); the other end, consisting of the activation domain, stimulates assembly of the transcription-initiation complex (gray) at the promoter (yellow). (c) This strategy can be used to screen a cDNA library for clones expressing proteins that interact with a protein of interest, in this case Ras. This approach requires two types of plasmids: The bait plasmid includes a DNA sequence encoding the DNA-binding domain of a transcription factor (purple) connected to the coding sequence for Ras (red). The fish plasmids contain individual cDNAs (green) from a library connected to the coding sequence for the activation domain (orange). Each type of plasmid also contains a wild-type selection gene (e.g., *TRP1* or *LEU2*). Both types of plasmids are transfected into yeast cells with mutations in genes required for tryptophan, leucine, and histidine biosynthesis (*trp1, leu2, his3* cells) and then grown in the absence of tryptophan and leucine. Only cells that contain the bait plasmid and at least one fish plasmid survive under these selection conditions. The cells that survive then are plated on medium lacking histidine; only cells that contain the bait plasmid and a fish plasmid encoding a protein that binds to Ras are able to grow, thus identifying cDNAs encoding Ras-binding proteins. [See A. B. Vojtek et al., 1993, *Cell* **74**:205; S. Fields and O. Song, 1989, *Nature* **340**:245.]

activating its catalytic activity. (The acronym MEK comes from MAP and ERK kinase, where ERK is another acronym for MAP.) Later studies showed that MEK binds to the C-terminal catalytic domain of Raf and is phosphorylated by the Raf serine/threonine kinase activity; this phosphorylation activates the catalytic activity of MEK. Hence, activation of Ras induces a kinase cascade that includes Raf, MEK, and MAP kinase (see Figure 20-36).

Ras-Coupled RTKs Transduce Extracellular Signals by a Common Pathway

Ras-coupled RTKs appear to utilize a highly conserved signal-transduction pathway. We can summarize the sequential steps in this pathway, supported by both genetic and biochemical studies, as follows (Figure 20-38):

   1. Binding of ligand activates a kinase activity in the

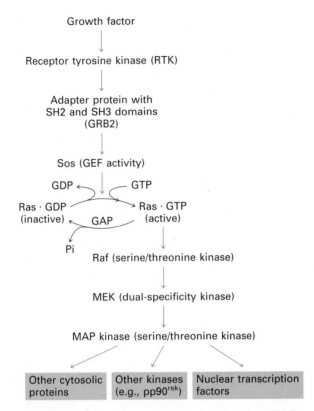

Growth factor

↓

Receptor tyrosine kinase (RTK)

↓

Adapter protein with
SH2 and SH3 domains
(GRB2)

↓

Sos (GEF activity)

↓

GDP ⤺ ⤹ GTP

Ras · GDP → Ras · GTP
(inactive)   (active)
       GAP

Pi

↓

Raf (serine/threonine kinase)

↓

MEK (dual-specificity kinase)

↓

MAP kinase (serine/threonine kinase)

↙   ↓   ↘

| Other cytosolic proteins | Other kinases (e.g., pp90^rsk) | Nuclear transcription factors |

▲ FIGURE 20-38   Flow chart summarizing the RTK-Ras signaling pathway. This pathway is initiated by binding of many growth factors to their receptors. Red arrows indicate activation of a component due to a conformational change or enzymatic modification induced by binding of the prior component in the sequence. Cycling of Ras between its inactive and active forms is facilitated by Sos, which functions as a guanine nucleotide–exchange factor (GEF), and GAP. Although it is clear that GAP can bind to phosphotyrosines on an activated RTK and that GAP can regulate Ras activity, just how GAP binding fits into the signaling pathway is unclear. See text and Figures 20-17, 20-32, and 20-36 for details.

cytosolic domain of the RTK leading to autophosphorylation of tyrosine residues.

2. The SH2 domain of GRB2, an adapter protein, binds to specific phosphotyrosine residues in the activated RTK.

3. The SH3 domains of GRB2 bind specifically to cytosolic Sos, which has guanine nucleotide–exchange activity, thereby localizing Sos to the plasma membrane. It is not known whether GRB2 also regulates the exchange activity of Sos.

4. Sos binds to Ras·GDP, leading to the release of GDP. Ras binds GTP, thus generating the active Ras·GTP form.

5. Activated Ras binds to the N-terminal domain of Raf, localizing this cytosolic protein to the plasma membrane.

6. The C-terminal end of Raf binds to and phosphorylates MEK, a dual-specificity protein kinase, thereby activating its kinase activity.

7. MEK phosphorylates and activates MAP kinase.

8. MAP kinase phosphorylates many substrates, including a number of transcription factors, thereby modulating their activity.

Genetic evidence strongly argues for the physiological importance of this pathway. Although only a small number of RTKs have been studied in detail and clearly shown to be coupled to Ras, no physiologically relevant RTK signal-transduction pathway that is not coupled to Ras has been identified to date. Thus it seems likely that most RTKs utilize this pathway, although some variations may occur. For example, the PDGF receptor is an RTK that lacks a GRB2-binding site. Nevertheless, stimulation of the PDGF receptor leads to Ras activation. In addition, as discussed later, activated RTKs also can induce other signaling pathways. The importance of these alternative RTK signaling pathways and their relationship to the RTK-Ras pathway currently are areas of intense investigation.

Yeast Mating-Factor Receptors Are Linked to G Proteins That Transmit Signals to MAP Kinase

It is likely that most, and perhaps all, eukaryotic cells contain kinase cascades that activate MAP kinase. In multicellular eukaryotes, this MAP kinase pathway is activated by ligand binding to RTKs. However, yeasts and other eukaryotic microorganisms lack RTKs. How, then, is MAP kinase activated in such organisms? And what cell functions does it control?

In *S. cerevisiae*, a MAP kinase pathway is linked to the receptors for two secreted peptide pheromones, the **a** and α factors. These pheromones control mating between haploid yeast cells of the opposite mating type, **a** or α. An **a** haploid cell secretes the **a** mating factor and has cell-surface receptors for the α factor; and an α cell secretes the α factor and has cell-surface receptors for the **a** factor (see Figure 13-12).

The **a** and α receptors are seven-spanning receptors coupled to a trimeric G protein. As with the G protein–linked receptors discussed earlier, ligand binding to these yeast receptors triggers the exchange of GTP for GDP on the α subunit and dissociation of G_α·GTP from the $G_{\beta\gamma}$ complex (Figure 20-39). In contrast to most G protein–linked receptor systems, however, the dissociated $G_{\beta\gamma}$ sub-

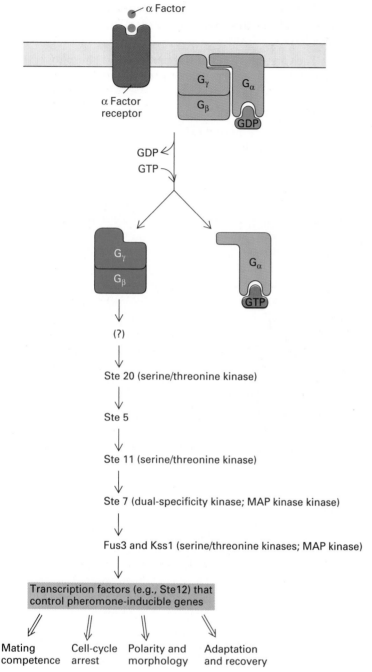

Mating   Cell-cycle   Polarity and   Adaptation
competence   arrest   morphology   and recovery

◄ FIGURE 20-39   Signaling pathway stimulated by binding of yeast **a** or α mating factors to their receptors, which are seven-spanning G protein–linked receptors. Both receptors are linked to the same trimeric G protein. Haploid **a** cells express the α-factor receptor (shown here), and haploid α cells express the **a**-factor receptor. Physiological responses are induced by the dissociated $G_{\beta\gamma}$ complex, not by $G_\alpha \cdot$ GTP as in the G protein–mediated signaling pathways discussed previously. $G_{\beta\gamma}$ activates a protein kinase cascade that is functionally equivalent to the cascade downstream of Ras in the RTK pathway but contains several additional components. The final components, Fus3 and Kss1, are equivalent to MAP kinase in higher eukaryotes (see Figure 20-38), and probably phosphorylate transcription factors (e.g., Ste12) that control expression of proteins involved in mating-specific cellular responses. [See M. Whiteway et al., 1989, *Cell* **56**:467; G. F. Sprague and J. Thorner, 1992, in *The Molecular Biology of the Yeast Saccharomyces*, Cold Spring Harbor Press, p. 657; H. Liu, C. Styles, and G. Fink, 1993, *Science* **262**:1741.]

at least three major biochemical events: (1) cell-surface alterations that enhance the ability of cells to bind strongly and selectively to cells of the opposite mating type and may involve the synthesis of new surface glycoproteins; (2) alterations in the cell-wall and plasma-membrane macromolecules that facilitate fusion of the two mating cells and eventual fusion of the two nuclei; and (3) arrest of cell growth by specifically blocking initiation of new DNA synthesis. This last response results in synchronization of the mating partners in the G_1 stage of the cell cycle.

The kinase cascade that induces all of these responses was uncovered mainly through analyses of yeast mutants that possess functional **a** and α receptors and G proteins but are defective in mating responses. The immediate targets of activated $G_{\beta\gamma}$ have not been conclusively demonstrated, but genetic studies indicate that two proteins—Ste20 and Ste5—couple $G_{\beta\gamma}$ to Ste11, a serine/threonine kinase analogous to Raf. Activated Ste11 then phosphorylates Ste7, a dual-specificity MEK, which in turn activates Fus3 and Kss1, which are equivalent to MAP kinase (see Figure 20-39). Finally, Fus3 and Kss1 are thought to phosphorylate transcription factors that induce synthesis of proteins essential for the mating-specific biochemical events described above.

Thus, the activation of MAP kinase in response to extracellular signals proceeds by a highly conserved kinase cascade in all eukaryotic cells, and the function of activated MAP kinase is likewise similar. Different extracellular hormones can trigger the MAP kinase pathway in different cell types, depending on the receptors they possess. These receptors may be RTKs, seven-spanning receptors coupled to G proteins, or, perhaps, some yet undiscovered type of receptor. Although these different signals all induce the MAP

units (not $G_\alpha \cdot$GTP) mediate all the physiological responses triggered by activation of the yeast pheromone receptors. This unusual feature of the yeast receptors has been demonstrated by genetic analysis. For example, deletion of the gene encoding G_α does not prevent mating but produces the opposite effect: the mating-hormone response is always on in the cell. Deletion of G_β or G_γ, in contrast, abolishes the ability of either **a** or α cells to respond to a mating hormone. Thus the $G_{\beta\gamma}$ subunits—not the G_α subunit—are required to induce all cellular responses to the mating factors.

Binding of mating factors to haploid yeast cells triggers

kinase cascade, they often generate different cellular responses. Several factors could account for the varying responses observed in different cell types: activation of additional pathways in some cells; differences in the transcription factors or other proteins phosphorylated by activated MAP kinase; or differences in the substrates of these phosphorylated proteins. Recent genetic studies in yeast have shown that there are at least three different MAP kinase pathways mediating different signals. This suggests that multiple MAP kinase signaling pathways probably also will be uncovered in animal cells.

➤ Other Important Second Messengers

As we discussed earlier in this chapter, cAMP is widely used as the second messenger in signaling systems in animals. The synthesis and degradation of cAMP are regulated by many seven-spanning G_s or G_i protein–linked receptors but not, so far as is known, by any RTKs. Three other molecules—Ca^{2+}, inositol 1,4,5-trisphosphate, and 1,2-diacylglycerol (see Figure 20-3)—function as second messengers in signaling pathways initiated by both seven-spanning receptors and RTKs.

We first discuss how a rise in cytosolic Ca^{2+} ions induces various metabolic responses and then consider how many hormones acting through inositol 1,4,5-trisphosphate (IP_3) cause this rise in the cytosolic Ca^{2+} level.

We also examine the role of 1,2-diacylglycerol (DAG) in regulating other cellular functions. Both IP_3 and DAG are formed from the same precursor, and signaling pathways involving these second messengers sometimes are referred to as *inositol-lipid pathways*. All of these second messengers interact in complex circuits to regulate crucial aspects of the growth and metabolism of cells.

Cellular Effects of Ca^{2+} Depend on Its Cytosolic Level and Often Are Mediated by Calmodulin

Most intracellular Ca^{2+} ions are sequestered in the mitochondria and endoplasmic reticulum (ER) or other cytoplasmic vesicles. The concentration of Ca^{2+} ions free in the cytosol usually is kept below 0.2 μM. Ca^{2+} ATPases pump cytosolic Ca^{2+} ions across the plasma membrane to the cell exterior or into the lumens of the endoplasmic reticulum or other intracellular vesicles that store Ca^{2+} ions (see Figure 15-11). Localized increases in the cytosolic level of free Ca^{2+} is critical to its function as a second messenger. Local concentrations of Ca^{2+} ions can be monitored with fluorescence dyes; in large cells, different Ca^{2+} concentrations can actually be detected in specific regions of the cytosol (see Figures 5-12 and 5-13).

Small increases in the level of cytosolic Ca^{2+}, which often are mediated by a rise in IP_3, trigger many cellular responses (Table 20-7). In secretory cells, such as the insulin-producing β cells in the pancreatic islets, a rise in

TABLE 20-7   Cellular Responses to Hormone-Induced Rise in Inositol 1,4,5-Trisphosphate (IP_3) and Subsequent Rise in Cytosolic Ca^{2+} in Various Tissues

Tissue	Hormone Inducing a Rise in IP_3	Cellular Response
Pancreas (acinar cells)	Acetylcholine	Secretion of digestive enzymes, such as amylase and trypsinogen
Parotid (salivary gland)	Acetylcholine	Secretion of amylase
Pancreas (β cells of islets)	Acetylcholine	Secretion of insulin
Vascular or stomach smooth muscle	Acetylcholine	Contraction
Liver	Vasopressin	Conversion of glycogen to glucose
Blood platelets	Thrombin	Aggregation, shape change, secretion of hormones
Mast cells	Antigen	Histamine secretion
Fibroblasts	Peptide growth factors, such as bombesin and PDGF	DNA synthesis, cell division
Sea urchin eggs	Spermatozoa	Rise of fertilization membrane

SOURCE: M. J. Berridge, 1987, *Ann. Rev. Biochem.* 56:159–193; M. J. Berridge and R. F. Irvine, 1984, *Nature* 312:315–321.

Ca^{2+} triggers the exocytosis of secretory vesicles and the release of insulin. In smooth or striated muscle cells, a rise in Ca^{2+} triggers contraction; in both liver and muscle cells, an increase in Ca^{2+} activates the degradation of glycogen to glucose 1-phosphate.

A small cytosolic protein called *calmodulin*, which is ubiquitous in eukaryotic cells, mediates many cellular effects of Ca^{2+} ions. Each calmodulin molecule binds four Ca^{2+} ions (Figures 20-40a,b) in a cooperative fashion (i.e., binding of one Ca^{2+} ion facilitates binding of additional Ca^{2+} ions). Binding of Ca^{2+} causes calmodulin to undergo a conformational change that enables the Ca^{2+}-calmodulin complex to bind to and activate many enzymes (Figure 20-40c). Because binding of Ca^{2+} is cooperative, a small change in the level of cytosolic Ca^{2+} leads to a large change in the level of active calmodulin.

(a)

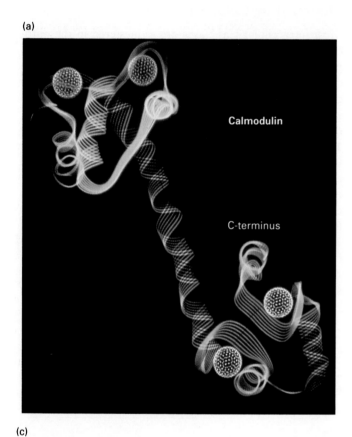

(b)

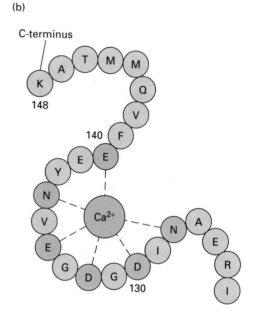

(c)

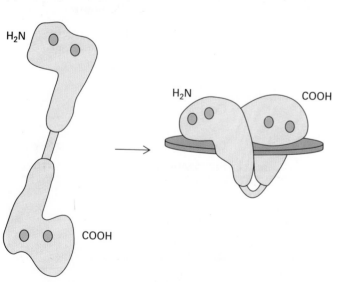

▲ FIGURE 20-40   Structure of calmodulin, a cytoplasmic protein of 148 amino acids that binds Ca^{2+} ions. (a) The backbone of the calmodulin molecule, deduced from crystallographic analysis of the Ca^{2+}-calmodulin complex. The four bound Ca^{2+} ions are represented by a blue sphere. (b) Amino acid sequence of one Ca^{2+}-binding site in calmodulin. Each site contains aspartate, glutamate, and asparagine residues (purple) whose side chains establish ionic bonds with a Ca^{2+} ion, forming a loop in the backbone. In other Ca^{2+} binding sites in calmodulin, oxygen atoms on the side chains of threonine and serine residues also participate in Ca^{2+} binding. (c) On binding Ca^{2+}, calmodulin undergoes a major conformational change, enabling it to bind to a target protein. Shown here is the binding of Ca^{2+}-calmodulin (tan) to a segment of myosin light-chain kinase (green), a protein that regulates the activity of myosin. Binding of Ca^{2+}-calmodulin to other proteins modifies their enzymatic activity. [Part (a) courtesy of Y. S. Babu and W. J. Cook; part (b) adapted from C. Y. Cheung, 1982, *Sci. Am.* **246**(6):62, copyright 1982 by Scientific American, Inc.; part (c) adapted from W. Meador, A. Means, and F. Quiocho, 1992, *Science* **257**:1251.]

One well-studied enzyme that is activated by the Ca^{2+}-calmodulin complex is cAMP phosphodiesterase, which degrades cAMP and terminates its effects (see Figure 20-14). This action thus links Ca^{2+} and cAMP, one of many examples in which these two second messengers interact to fine-tune certain aspects of cell regulation. The Ca^{2+}-calmodulin complex also activates several protein kinases that, in turn, phosphorylate target proteins and alter their activity levels.

Ca^{2+} Ions and cAMP Induce Hydrolysis of Muscle Glycogen

Stimulation of muscle cells by nerve impulses causes the release of Ca^{2+} ions from the sarcoplasmic reticulum and an increase in the cytosolic Ca^{2+} concentration. This rise not only triggers muscle contraction (Chapter 22) but also stimulates degradation of glycogen to glucose 1-phosphate, which fuels prolonged contraction. As we saw earlier, glycogen degradation and synthesis also are modulated by β-adrenergic activation of the cAMP cascade (see Figure 20-25). However, the key regulatory enzyme in glycogenolysis, glycogen phosphorylase kinase (GPK), is activated by a rise in the level of cytosolic Ca^{2+}, as well as by phosphorylation by cAMP-dependent protein kinase (Figure 20-41).

This multiple regulation of GPK results from its multimeric subunit structure $(\alpha\beta\gamma\delta)_4$. The γ subunit is the catalytic protein; the regulatory α and β subunits, which are similar in structure, are phosphorylated by cAMP-dependent protein kinase; and the δ subunit is calmodulin. GPK is maximally active when Ca^{2+} ions are bound to the calmodulin subunit and at least the α subunit is phosphorylated; in fact, binding of Ca^{2+} to the calmodulin subunit may be essential to the enzymatic activity of GPK. Phosphorylation of the α and β subunits increases the affinity of the calmodulin subunit for Ca^{2+}, enabling Ca^{2+} ions to

bind to the enzyme at the submicromolar Ca^{2+} concentrations found in cells not stimulated by nerves. Thus increases in the cytosolic concentration of Ca^{2+} or cAMP, or both, induce incremental increases in the activity of GPK. As a result of the elevated level of cytosolic Ca^{2+} after neuron stimulation of muscle cells, GPK will be active even if it is unphosphorylated; thus glycogen will be hydrolyzed to fuel continued muscle contraction.

When nerve stimulation of muscle cells ceases, Ca^{2+} ions are pumped back into the sarcoplasmic reticulum, the level of cytosolic Ca^{2+} declines, the activity of glycogen phosphorylase kinase is reduced, and less glycogen is converted to glucose phosphate. In several other cells (but not in striated muscle cells), a rise in cytosolic Ca^{2+} increases the rate of Ca^{2+} pumping out of cells and causes a more rapid drop in cytosolic Ca^{2+} to the resting state, terminating all Ca^{2+}-induced signals unless additional Ca^{2+} is released. This rapid decline is due to Ca^{2+}-calmodulin activation of the plasma-membrane Ca^{2+} ATPase.

Inositol 1,4,5-Trisphosphate Causes the Release of Ca^{2+} Ions from the ER

Binding of many hormones to their cell-surface receptors on liver, fat, and other cells induces an elevation in cytosolic Ca^{2+} even when Ca^{2+} ions are absent from the surrounding medium. In this situation, Ca^{2+} is released into the cytosol from the ER and other intracellular vesicles. The mechanism by which a hormone-receptor signal on the cell surface is transduced to the ER became clear in the early 1980s, when it was shown that a rise in the level of cytosolic Ca^{2+} often is preceded by the hydrolysis of an unusual phospholipid, *phosphatidylinositol 4,5-bisphosphate* (PIP$_2$), one of several inositol phospholipids found in the cytosolic leaflet of the plasma membrane.

As shown in Figure 20-29, hydrolysis of PIP$_2$ by the plasma-membrane enzyme *phospholipase C* (PLC) yields

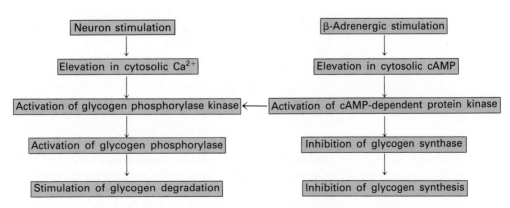

▲ FIGURE 20-41 Glycogenolysis in striated muscle cells is induced by two signaling pathways, one resulting from neuron stimulation and the other by binding of epinephrine to β-adrenergic receptors. The key regulatory enzyme, glycogen phosphorylase kinase (purple), is activated by Ca^{2+} ions and by a cAMP-dependent protein kinase.

two important products: 1,2-diacylglycerol (DAG), which remains in the membrane, and inositol 1,4,5-trisphosphate (IP$_3$), which is water soluble. Binding of hormones to certain G protein–coupled seven-spanning receptors activates PLC. Pertussis toxin ADP ribosylates the α subunit of G proteins called G$_o$, and treatment with pertussis toxin abolishes hormone-induced activation of PLC in some cells. This phenomenon is similar to the effect of cholera toxin on G$_s$ protein–linked receptors coupled to adenylate cyclase discussed earlier (see Figure 20-19). Another G protein that activates PLC, called G$_q$, is not modified by pertussis toxin. As we noted previously, certain activated RTKs also can increase the activity of the γ isoform of PLC. Thus hormone-induced stimulation of PLC activity

and subsequent generation of IP$_3$ is mediated by G protein–linked receptors and receptor tyrosine kinases.

Once formed following hormone stimulation of a target cell, IP$_3$ diffuses to the ER surface, where it binds to a specific IP$_3$ receptor, a Ca^{2+} channel protein composed of four identical subunits, each containing an IP$_3$-binding site. Each subunit contains a C-terminal domain that spans the lipid bilayer twice; together the transmembrane domains of the four monomers form the Ca^{2+} channel. The N-terminal domain is large and contains the IP$_3$-binding site. IP$_3$ binding induces opening of the channel allowing Ca^{2+} ions to exit from the ER into the cytosol. The overall sequence of events from binding of hormones to a G protein–linked receptor to Ca^{2+} release is depicted in Figure 20-42.

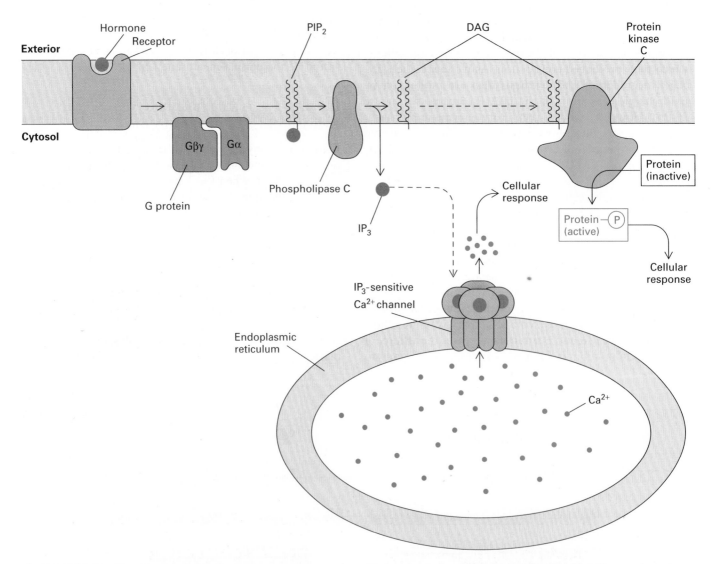

▲ FIGURE 20-42   The inositol-lipid signaling pathway can be coupled to seven-spanning G protein–linked receptors, as illustrated here, and receptor tyrosine kinases. Binding of a hormone to its receptor triggers activation of the G protein (G$_o$ or G$_q$), which in turn activates phospholipase C by a mechanism analogous to activation of adenylate cyclase (see Figure 20-16). Phospholipase C then cleaves phosphatidylinositol 4,5-bisphosphate (PIP$_2$) to inositol 1,4,5-trisphosphate

(IP$_3$) and 1,2-diacylglycerol (DAG). The IP$_3$ diffuses through the cytosol and interacts with IP$_3$-sensitive Ca^{2+} channels in the membrane of the endoplasmic reticulum, causing release of stored Ca^{2+} ions (blue), which mediate various cellular responses. As discussed later, protein kinase C is activated by Ca^{2+} and DAG, which remains in the membrane. The activated kinase then phosphorylates several cellular enzymes and receptors, thereby altering their activity.

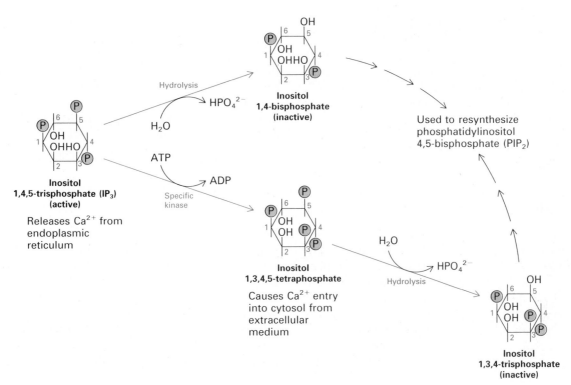

▲ FIGURE 20-43   Fates of cytosolic inositol 1,4,5-tris-phosphate (IP_3). Rapid hydrolysis generates inositol 1,4-bisphosphate, which is inactive as a second messenger and is used to resynthesize phosphatidylinositol 4,5-bisphosphate (PIP_2). Phosphorylation by a specific kinase yields inositol 1,3,4,5-tetraphosphate; in some cells, this compound facili-tates Ca^{2+} entry into the cytosol from the extracellular medium. Eventually, inositol 1,3,4,5-tetraphosphate is hydro-lyzed to inositol 1,3,4-trisphosphate; this trisphosphate, although similar to the active IP_3 is inactive and is recycled to synthesize PIP_2.

Within a second of its formation, most IP_3 is hydrolyzed to *inositol 1,4-bisphosphate* (Figure 20-43), a molecule that cannot induce release of Ca^{2+} ions from the ER. As a result, the release of Ca^{2+} ions terminates quickly unless more IP_3 is generated by action of phospholipase C.

Because continued hormone stimulation can deplete the store of Ca^{2+} ions in the ER within a few minutes, maintenance of elevated cytosolic Ca^{2+} levels requires transport of extracellular Ca^{2+} ions across the plasma membrane into the cell. Some evidence suggests that Ca^{2+} entry into pancreatic cells is mediated by *inositol 1,3,4,5-tetraphosphate,* which is formed by phosphorylation of IP_3 by a specific kinase (see Figure 20-43). In time, this inositol tetraphosphate also is hydrolyzed by a phosphatase, rendering it inactive. In other types of cells, the depletion of Ca^{2+} stores in the ER is thought to trigger entry of extra-cellular Ca^{2+} ions by another mechanism not involving inositol phosphates.

The ability of IP_3 to induce release of Ca^{2+} stores can be demonstrated easily by adding IP_3 to a preparation of ER vesicles. Although cells contain other phosphorylated inositols, only IP_3 causes release of Ca^{2+} ions from ER vesicles. In addition, not all cells respond identically to IP_3. This differential sensitivity may reflect expression of differ-ent isoforms of the IP_3 receptor in the ER membrane and/or variation in the Ca^{2+} content of the ER itself. Because of this variability, cells may exhibit very different responses to the same extracellular signal.

IP_3-sensitive Ca^{2+} channels are present in the ER mem-brane of many cell types. Although cytosolic Ca^{2+} cannot initiate opening of these Ca^{2+} channels in the absence of IP_3, low levels of Ca^{2+} ions potentiate the action of IP_3, resulting in greater release of stored Ca^{2+}. High concentra-tions of cytosolic Ca^{2+}, however, inhibit IP_3-induced re-lease of Ca^{2+} from intracellular stores. (The plasma mem-brane also may contain IP_3-regulated Ca^{2+} channels.) The complex regulation of the IP_3 receptor in ER membranes can have strange consequences. Frequently, rapid oscilla-tions in the cytosolic Ca^{2+} level are seen when the IP_3 path-way in cells is stimulated. Stimulation of hormone-secret-ing cells in the pituitary by LHRH (see Figure 20-5) causes rapid, repeated spikes in the cytosolic Ca^{2+} level; each spike is associated with a burst in secretion of luteinizing hormone (LH). One explanation for these Ca^{2+} spikes is that activation of phospholipase C and generation of IP_3 causes release of a small amount of Ca^{2+} from the ER; the resultant increase in cytosolic Ca^{2+} causes, together with IP_3, more IP_3-sensitive Ca^{2+} channels to open and more Ca^{2+} to be released; this sets into motion an explosive, short-lived release of Ca^{2+} from the ER. Eventually, the rising level of cytosolic Ca^{2+} inhibits these Ca^{2+} channels, no more Ca^{2+} is released, and the level of cytosolic Ca^{2+} drops quickly as Ca^{2+} is pumped back into the ER. Each cycle takes only a few seconds. The purpose of the fluctuations of Ca^{2+}, rather than a sustained rise in cytosolic Ca^{2+}, is not understood. One possibility is that a sustained rise in Ca^{2+} might be toxic to cells.

Release of Intracellular Ca^{2+} Stores Also Is Mediated by Ryanodine Receptors in Muscle Cells and Neurons

In addition to IP_3-sensitive Ca^{2+} channels, muscle cells and neurons possess other Ca^{2+} channels called *ryanodine receptors* (RYRs), because of their sensitivity to the plant alkaloid ryanodine. In skeletal muscle cells, these receptors are located in the membrane of the sarcoplasmic reticulum and associate with the cytoplasmic domain of the dihydropyridine receptor, a voltage-sensing protein in the plasma membrane. A change in potential across the plasma membrane induces a conformational change in the RYR, so that

Ca^{2+} ions are released from the sarcoplasmic reticulum into the cytosol (Figure 20-44a). The ER membranes in cardiac muscle cells and neurons also contain RYRs, but they do not associate directly with a plasma-membrane protein. In these cells, depolarization of the plasma membrane leads to a small influx of extracellular Ca^{2+} ions through voltage-gated channels; binding of these Ca^{2+} ions to RYRs induces a burst of Ca^{2+} release from the ER or other intracellular stores (Figure 20-44b).

1,2-Diacylglycerol Activates Protein Kinase C

As noted previously, hydrolysis of phosphatidylinositol 4,5-bisphosphate following hormone binding to cell-surface receptors generates IP_3 and another second messenger, 1,2-diacylglycerol (DAG), which remains associated with the membrane (see Figures 20-29 and 20-42). The principal function of DAG is to activate a family of plasma-membrane protein kinases collectively termed *protein kinase C*. In the absence of hormone stimulation, protein kinase C is present as a soluble cytosolic protein that is

(a) Voltage-sensitive RYR
(skeletal muscle)

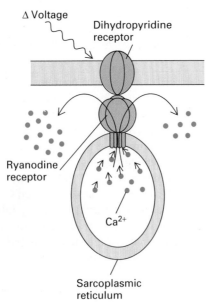

(b) Ca^{2+}-sensitive RYR
(cardiac muscle, neurons)

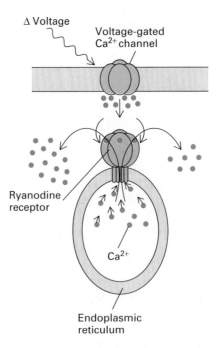

▲ FIGURE 20-44   Release of Ca^{2+} stores mediated by ryanodine receptors (RYRs). (a) In skeletal muscle, voltage-sensing dihydropyridine receptors in the plasma membrane contact ryanodine receptors located in the membrane of the sarcoplasmic reticulum. In response to a change in voltage, the dihydropyridine receptors undergo a conformational change; this produces a conformational change in the associated RYRs, opening them so that Ca^{2+} ions (blue) can exit into the cytosol. (b) In cardiac muscle and probably neurons, membrane depolarization triggers opening of plasma-membrane voltage-gated Ca^{2+} channels, permitting a small influx of extracellular Ca^{2+} ions. Binding of Ca^{2+} to ryanodine receptors in the ER membrane leads to a massive release of intracellular Ca^{2+} stores. Also see Figure 20-42. [Adapted from M. J. Berridge, 1993, *Nature* **361**:315.]

catalytically inactive. A rise in the cytosolic Ca^{2+} level causes protein kinase C to bind to the cytoplasmic leaflet of the plasma membrane, where it can be activated by the membrane-associated DAG. Thus activation of protein kinase C depends on both Ca^{2+} ions and DAG, suggesting an interaction between the two branches of the inositol-lipid signaling pathway.

The activation of protein kinase C in different cells results in a varied array of cellular responses, indicating that it plays a key role in many aspects of cellular growth and metabolism. In liver cells, for instance, protein kinase C phosphorylates glycogen synthase, yielding the phosphorylated inactive form of this enzyme. Thus, hormone-induced activation of phospholipase C, which leads to generation of IP_3 and DAG, mediates glycogen breakdown by the two branches of the inositol-lipid signaling pathway: DAG, by activating protein kinase C, mediates inhibition of glycogen synthesis; IP_3, by inducing an increase in cytosolic Ca^{2+}, enhances the activation of glycogen phosphorylase kinase as discussed earlier (see Figure 20-41). This is another example of the coordination of multiple second messengers to achieve a unified cellular response.

Protein kinase C also phosphorylates various transcription factors; depending on the cell type, these induce or repress synthesis of certain mRNAs. Another substrate is a protein that binds to and inhibits the transcription factor NF-κB in the cytoplasm of unstimulated cells. Phosphorylation of this inhibitor by protein kinase C causes release of NF-κB, which then migrates to the nucleus where it induces transcription of certain genes (see Figure 13-36). These few examples of the function of protein kinase C suggest that it plays a fundamental role in controlling cell growth. The eventual understanding of its regulation and substrates should provide important insights into many aspects of carcinogenesis and cellular metabolism.

► *Multiplex Signaling Pathways*

In earlier sections, we described in detail two classes of cell-surface receptors and two signal-transduction pathways induced by ligand stimulation of these receptors: G_s protein–linked receptors coupled to adenylate cyclase leading to production of cAMP (see Figure 20-16) and receptor tyrosine kinases coupled to Ras protein leading to activation of MAP kinase (see Figure 20-38). The previous discussion of these receptors simplified the actual situation. In fact, activation of each class of receptor generally leads to production of multiple second messengers, and both classes of receptor activate or inhibit production of many of the same second messengers, including Ca^{2+}, inositol phosphates, and 1,2-diacylglycerol. In addition, as mentioned earlier, different G proteins couple seven-spanning receptors to a variety of effector proteins. Here we consider a few examples of these signaling complexities.

Some Activated RTKs Stimulate Activity of Phospholipase C_γ

As discussed already, several enzymes can bind to the cytosolic phosphotyrosine residues of activated RTKs via their SH2 domains (see Figure 20-28). These enzymes include the γ isoform of phospholipase C (PLC$_\gamma$), phosphatidylinositol-3 (PI-3) kinase, a tyrosine phosphatase called Syp, and the Ras-associated GAP protein. Binding of these enzymes to activated RTKs localizes them to the plasma membrane, bringing them into proximity of their substrates.

The substrate for PLC$_\gamma$, for example, is phosphatidylinositol 4,5-bisphosphate (PIP_2), which is associated with the plasma membrane (see Figure 20-29). The SH2 domains of PLC$_\gamma$ bind to specific phosphotyrosines of certain RTKs, thus positioning the enzyme close to PIP_2. In addition, the RTK phosphorylates tyrosine residues on the bound PLC$_\gamma$, enhancing its lipase activity. Thus activated RTKs stimulate PLC$_\gamma$ activity in two ways: by localizing the enzyme to the membrane and by phosphorylating it.

The importance of RTK-mediated stimulation of PLC$_\gamma$ has been demonstrated in the case of the receptor for platelet-derived growth factor (PDGF). Binding of PDGF to its receptor, which is an RTK, induces cell proliferation. Investigators produced an altered PDGF receptor in which the phosphotyrosines that bind PI-3 kinase, Syp, and GAP were mutated. This receptor, capable of binding only to PLC$_\gamma$, was found to mediate PDGF-induced cell proliferation. Interestingly, binding of PDGF to this mutated receptor led to activation of Ras, but the signal-transduction pathway from PLC$_\gamma$ to Ras has not been determined yet.

Multiple G Proteins Transduce Signals from Seven-Spanning Receptors to Different Effector Proteins

We noted previously that studies with bacterial toxins originally led to identification of several G proteins coupled to different effector proteins. More recently, molecular cloning has led to the isolation of a large number of proteins related to the α, β, and γ subunits of these previously characterized G proteins. In mammals, for instance, 16 distinct G_α subunits, 4 G_β subunits, and 5 G_γ subunits have been identified so far.

It is now clear that various trimeric G proteins link seven-spanning receptors to a variety of effector proteins, including ion channels, adenylate cyclase, phospholipase C, and a cGMP-specific phosphodiesterase in photoreceptor cells (Table 20-8). Stimulation of the receptors coupled to these G proteins thus can modulate many cellular functions. The role of trimeric G proteins in these signaling pathways has been demonstrated by inhibition of the pathways with nonhydrolyzable GTP analogs and, in some cases, by their sensitivity to cholera or pertussis toxin. Because the same cell may express diverse G proteins, it is

TABLE 20-8   Properties of Mammalian G Proteins Linked to Seven-Spanning Receptors

G_α Subclass*	Effect	Associated Effector Protein	2nd Messenger
G_s	↑	Adenylate cyclase	cAMP
	↑	Ca^{2+} channel	Ca^{2+}
	↓	Na^+ channel	Change in membrane potential
G_i	↓	Adenylate cyclase	cAMP
	↑	K^+ channel	Change in membrane potential
	↓	Ca^{2+} channel	Ca^{2+}
G_q	↑	Phospholipase C	IP_3, DAG
G_o	↑	Phospholipase C	IP_3, DAG
	↓	Ca^{2+} channel	Ca^{2+}
G_t	↑	cGMP phosphodiesterase	cGMP

* A given G_α may be associated with more than one effector protein. To date, only one major $G_{s\alpha}$ has been identified, but multiple $G_{q\alpha}$ and $G_{i\alpha}$ proteins have been described.
KEY: ↑ = stimulation; ↓ = inhibition. IP_3 = inositol 1,4,5-trisphosphate; DAG = 1,2-diacylglycerol.
SOURCE: See A. C. Dolphin, 1987, *Trends Neurosci.* **10**:53; L. Birnbaumer, 1992, *Cell* **71**:1069.

often difficult to determine which specific G protein mediates the effect of a particular ligand.

The presence of multiple G_α subunits in a single cell raises the possibility that a single ligand could initiate signaling through more than one effector protein. Several examples of such multiplex signaling have been described, although the precise molecular details of which G proteins and which specific subunits mediate these effects are not yet known. In some cells, modulation of different effectors coupled to the same cell-surface receptor is observed at different ligand concentrations or when different concentrations of the receptor are expressed on the surface. The large number of possible combinations of different G protein subunits, seven-spanning receptors, and effector proteins provide cells with the ability to respond in remarkably diverse ways to precisely control their development and function.

$G_{\beta\gamma}$ Acts Directly on Some Effectors in Mammalian Cells

In the signaling pathway stimulated by binding of mating factors to haploid yeast cells, the signal is transduced by the $G_{\beta\gamma}$ subunit complex, not by the G_α·GTP complex (see Figure 20-39). This activation of a downstream signaling pathway by $G_{\beta\gamma}$ initially was thought to reflect an idiosyncrasy of yeast biology. More recent research has shown that in some mammalian cells $G_{\beta\gamma}$ can directly regulate certain effector proteins.

For example, one isoform of adenylate cyclase (AC_I) present in the brain is stimulated by $G_{s\alpha}$·GTP as described previously (see Figure 20-16); $G_{\beta\gamma}$ inhibits the activation of AC_I by $G_{s\alpha}$. Another isoform (AC_{II}), however, is stimulated by binding of $G_{\beta\gamma}$ but only if free $G_{s\alpha}$·GTP also is present. Yet other adenylate cyclase isoforms (e.g., AC_{III}) are insensitive to $G_{\beta\gamma}$. Some brain cells contain two adenylate cyclase isoforms that are regulated differently; in such cells adenylate cyclase activity is subject to dual regulation. Similarly, some K^+ channel proteins in the heart are affected in a complex manner by both free G_α·GTP and $G_{\beta\gamma}$ (see Figure 21-42). In general, considerably higher concentrations of $G_{\beta\gamma}$ than of G_α·GTP are required to modulate the activity of an effector protein. Regulation of certain effector enzymes by $G_{\beta\gamma}$ and various G_α·GTP complexes contributes to the integration of cellular metabolism (Figure 20-45).

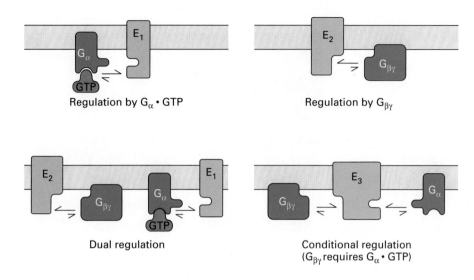

Multiple regulation of effector proteins mediated by G protein–linked receptors. Different isoforms of an effector protein (E), such as adenylate cyclase or phospholipase C, have different binding affinities for the $G_\alpha \cdot GTP$ complex and $G_{\beta\gamma}$, leading to stimulation or inhibition by various G subunits. See text for details.

➤ The Insulin Receptor and Regulation of Blood Glucose

In an animal subjected to stress or exercise, epinephrine induces a wide array of responses, including an increase in the blood glucose level (see Figure 20-25). During normal daily living, two other hormones—insulin and glucagon—regulate blood glucose levels. Both hormones are produced by cells within the *islets of Langerhans,* cell clusters scattered throughout the pancreas (Figure 20-46). Insulin contains two polypeptide chains, A and B, linked by disulfide bonds (see Figure 3-18); synthesized by the β cells in the islets, insulin acts to reduce the level of blood glucose. Glucagon, a single-chain peptide containing 29 amino acids, is produced by the α cells; it has the opposite effect, causing an increase in blood glucose by stimulating glycogenolysis in the liver. Each islet functions as an integrated unit, delivering the appropriate amounts of both hormones to the blood to meet the metabolic needs of the animal. Hormone secretion is regulated by a combination of neuron and hormone signals. Of the hormones that affect blood glucose levels, insulin is the most important. Although the insulin receptor exhibits tyrosine kinase activity (see Figure 20-8), and thus is classified as an RTK, the signal-transduction pathway initiated by insulin binding differs significantly from other RTK-mediated pathways.

Insulin Has Short-Term Effects on Glucose Metabolism and Long-Term Growth-Promoting Effects

Insulin acts on many cells in the body to produce both immediate and long-term effects (Table 20-9). Its immedi-ate effects include an increase in the rate of glucose uptake from the blood into muscle cells and adipocytes and modulation of the activity of various enzymes involved in glucose metabolism. These effects occur within minutes, do not require new protein synthesis, and occur at insulin concentrations of 10^{-9} to 10^{-10} M. Continued exposure to insulin produces longer-lasting effects including increased expression of liver enzymes that synthesize glycogen and of adipocyte enzymes that synthesize triacylglycerols; insulin also functions as a growth factor for many cells (e.g., fibroblasts). These effects are manifested in hours and require continuous exposure to $\approx 10^{-8}$ M insulin.

Within 15 min of the addition of insulin to adipocytes, the rate of glucose transport increases 10–20 times, mainly due to an increase in the number of transporters in the plasma membrane. The membranes of endosome-like vesicles located just under the plasma membranes of unstimulated cells contain abundant GLUT-4 glucose transporters, an isoform found only in adipocytes and muscle cells. Binding of insulin to its receptor induces rapid fusion of these vesicles with the plasma membrane, increasing the number of transporters in the plasma membrane tenfold (Figure 20-47). Exocytosis of vesicles containing transporters and some activation of these transporters accounts for the large increase in glucose transport.

Many of the growth-promoting actions of insulin (e.g., induction of DNA synthesis) may be initiated by its binding to the receptor for *insulin-like growth factor 1* (IGF-1), rather than to the insulin receptor. IGF-1, which is similar in structure and sequence to insulin, is produced mainly in the liver in response to growth hormone secreted by the pituitary gland and appears to be a primary regulator of growth in an organism. Indeed, the ability of exogenous growth hormone to stimulate growth (e.g., in dwarfs who

(a)

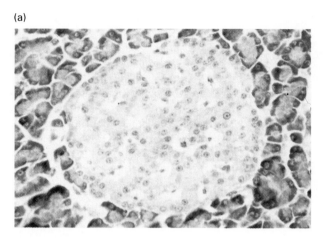

▲ FIGURE 20-46   Structure of islet of Langerhans in rat pancreas. Consecutive serial sections of an islet were stained with hematoxylin-eosin (a) or fluorescent antibodies to insulin (b) or glucagon (c). Insulin-secreting β cells are located in the center of the islet, surrounded by glucagon-secreting α cells. [From L. Orci, 1982, *Diabetes* **31**:538; courtesy of L. Orci.]

(b)

(c)

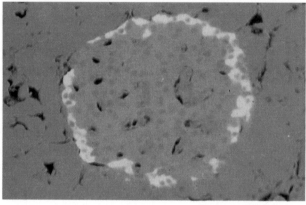

100 µm

TABLE 20-9	Chronology of Insulin Action
Seconds	Binding to the insulin receptor
	Receptor autophosphorylation
	Activation of the receptor protein tyrosine kinase
Minutes	Activation of hexose transport
	Alterations of intracellular enzymatic activities
	Changes in gene regulation
	Insulin-induced receptor internalization and down-regulation
	Phosphorylation of the insulin receptor by other protein kinases
Hours	Induction of DNA, RNA, protein, and lipid synthesis
	Cell growth
	Maximum down-regulation of the insulin receptor

SOURCE: O. M. Rosen, *The Harvey Lectures* (*1986–1987*), Vol. 82, pp. 105–122.

► FIGURE 20-47   Regulation of glucose uptake into adipocytes by insulin. The two isoforms of the glucose transporter, GLUT-1 and GLUT-4, are identical in 60 percent of their 500 amino acid residues. (a) In resting cells, most GLUT-1 transporters (orange) are embedded in the plasma membrane; the more numerous GLUT-4 transporters (blue) are present in the membranes of intracellular endosomelike vesicles. (b) By an unknown mechanism, binding of insulin triggers exocytosis of the GLUT-4-containing vesicles, resulting in a tenfold increase in cell-surface glucose transporters and a corresponding tenfold increase in the rate of glucose uptake. (c) Once the insulin signal is removed, regions of the plasma membrane rich in GLUT-4 undergo endocytosis, depleting the plasma membrane of these transporters. [See A. W. Simpson and S. W. Cushman, 1986, *Annu. Rev. Biochem.* **55**:1059; B. Thorens, M. J. Charron, and H. F. Lodish, 1990, *Diabetes Care* **13**:209; and G. Bell et al., 1993, *J. Biol. Chem.* **268**:19161.]

(a) Resting cell

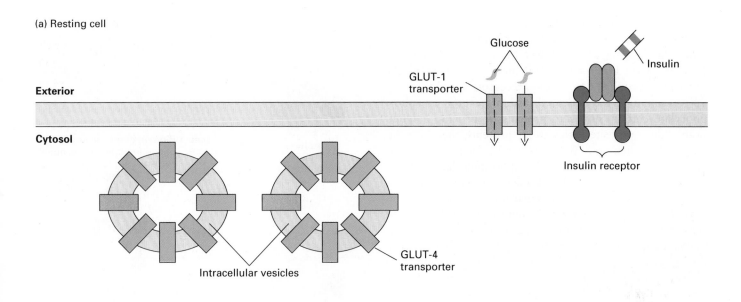

Exterior

Cytosol

Glucose

GLUT-1 transporter

Insulin

Insulin receptor

GLUT-4 transporter

Intracellular vesicles

(b) Insulin-stimulated cell

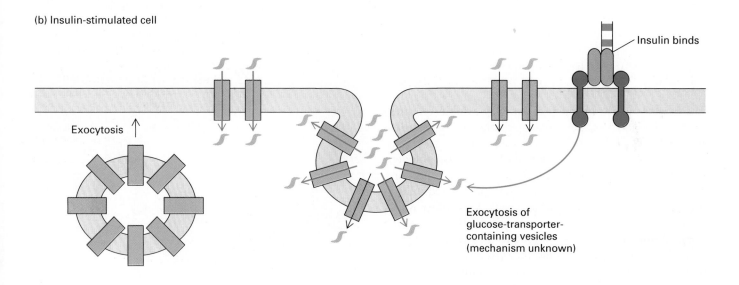

Insulin binds

Exocytosis

Exocytosis of glucose-transporter-containing vesicles (mechanism unknown)

(c) Insulin removed

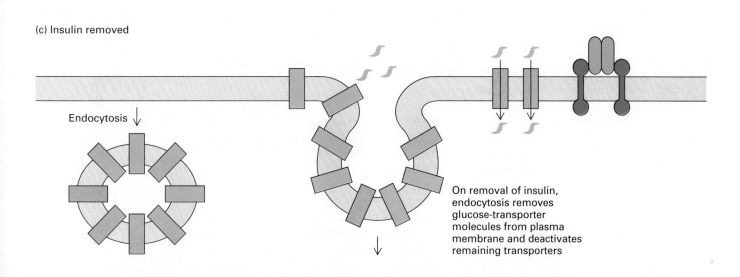

Endocytosis

On removal of insulin, endocytosis removes glucose-transporter molecules from plasma membrane and deactivates remaining transporters

produce insufficient growth hormone) is largely mediated by IGF-1. The insulin and IGF-1 receptors have similar molecular weights, and both have tyrosine kinase activity. Each hormone binds about 100 times less tightly to the other's receptor than to its own receptor.

Presumably, the activated insulin receptor can phosphorylate certain substrate proteins or provide a membrane "docking" site for cytosolic proteins, thereby regulating their activity. We saw earlier that epinephrine stimulation of muscle cells leads to phosphorylation of glycogen synthase by cAMP-dependent protein kinase (see Figure 20-25); the phosphorylated form of glycogen synthase is inactive. In contrast, insulin stimulation of muscle cells induces activation of glycogen synthase and subsequent synthesis of glycogen. Binding of insulin to the insulin receptor is believed to trigger phosphorylation of a phosphoprotein phosphatase, thereby activating it. The active phosphatase then dephosphorylates the inactive form of glycogen synthase, stimulating glycogen synthesis.

Insulin Signaling Pathway Involves a Soluble "Relay" Protein That Does Not Bind to the Receptor

The insulin signal-transduction pathway exhibits an interesting variation on the signaling pathway downstream from activated RTKs that we discussed earlier. Biochemical studies in the mid-1980s identified a 130-kDa polypeptide, called *insulin receptor substrate 1* (IRS1), containing tyrosine residues that were rapidly phosphorylated following insulin stimulation of cultured cells. Most other RTKs do not phosphorylate IRS1, but the related IGF-1 receptor does. Although IRS1 does not bind to the insulin receptor, several lines of evidence suggest that insulin action depends on phosphorylation of IRS1. For instance, injection of antibodies to IRS1 into cultured cells blocks the normal proliferative response induced by insulin. Phosphorylated IRS1, not the insulin receptor, binds to the SH2 domains of several proteins including GRB2, PI-3 kinase, and Syp (a tyrosine phosphatase). In signaling pathways involving most other RTKs, these proteins bind directly to phosphotyrosine residues in the cytosolic domain of the receptor.

Binding of PI-3 kinase to phosphorylated IRS1 leads to a tenfold stimulation in its activity, accounting for the rapid rise in phosphorylated phosphatidylinositols observed in insulin-stimulated cells. Binding of GRB2 to IRS1, and the subsequent binding of GRB2 to Sos protein, may account for the increase in the proportion of the active Ras·GTP complex following insulin stimulation (Figure 20-48). Insulin stimulation of target cells in liver, fat, and muscle also leads to activation of MAP kinase, further indicating that the insulin receptor is linked downstream to Ras, similar to other RTKs. Binding of Syp to IRS1 causes

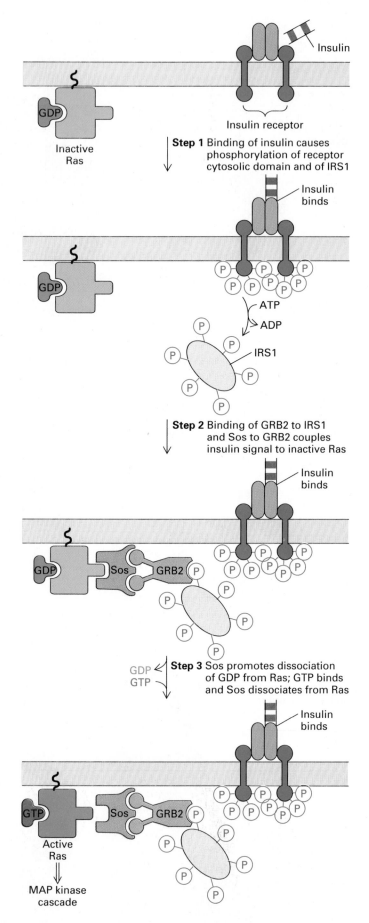

◄ FIGURE 20-48   Activation of Ras following binding of insulin to its receptor, which is an atypical receptor tyrosine kinase. The activated receptor autophosphorylates its own cytosolic domain and a cytosolic "relay" protein called insulin receptor substrate 1 (IRS1). Phosphotyrosine residues on IRS1 bind SH2 domains on many intracellular signaling proteins, including GRB2, which then binds Sos, coupling the insulin signal to Ras activation. Activated Ras then triggers the same protein kinase cascade depicted in Figure 20-36. IRS1 also binds SH2 domains in PI-3 kinase and the protein tyrosine phophatase Syp. [Adapted from M. Myers and M. F. White, 1993, *Diabetes* **42**:643; M. F. White and C. R. Kahn, 1994, *J. Biol. Chem.* **269**:1.]

a marked increase in its tyrosine phosphatase activity. Activated Syp may dephosphorylate IRS1, thereby terminating signaling, but the role of this phosphatase has not been conclusively demonstrated.

The ability of phosphorylated soluble proteins, such as IRS1, to bind SH2-containing proteins may provide a way to "relay" a signal from an RTK anchored in the plasma membrane to other cellular compartments. However, the molecular mechanisms by which IRS1 specifically regulates different physiological responses to insulin are not yet known. Recent studies suggest that similar relay proteins may be involved in pathways coupled to other cell-surface receptors.

Insulin and Glucagon Work Together to Maintain a Stable Blood Glucose Level

Glucagon is released from α cells in the islets of the pancreas in response to low levels of blood glucose and low levels of insulin. The glucagon receptor, found primarily on liver cells, is a seven-spanning G_s protein–linked receptor; like the epinephrine receptor, it is coupled to adenylate cyclase and the cAMP cascade that leads to glycogenolysis and inhibition of glycogen synthesis (see Figure 20-25a).

The availability of glucose for cellular metabolism is regulated during periods of abundance (following a meal) or scarcity (following fasting) by the adjustment of insulin and glucagon concentrations in the blood (Figure 20-49). (Epinephrine is used only under stressful conditions.) When, after a meal, blood glucose rises above its normal level of 80–90 mg/100 ml, the pancreatic β cells of the islets respond to the rise in glucose or amino acids by releasing insulin into the blood, which transports the hormone throughout the body. By binding to muscle and adipocyte cell-surface receptors, insulin causes glucose to be removed from the blood and stored in muscle cells as glycogen. Insulin also affects hepatocytes, primarily by inhibiting glucose synthesis from smaller molecules, such as lactate and acetate, and by enhancing glycogen synthesis from glucose. If the blood glucose level falls below ≈80 mg/100 ml, the α cells of the islets start secreting glucagon. Glucagon binds to glucagon receptors on liver cells, triggering degradation of glycogen and the release of glucose into the blood.

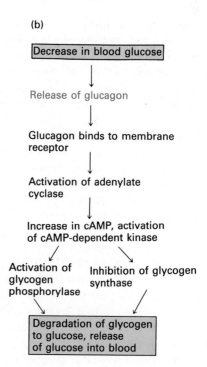

▲ FIGURE 20-49   Regulation of blood glucose level by the opposing effects of insulin and glucagon. (a) Insulin causes an increase in glucose uptake, mainly in muscle cells and adipocytes, and stimulates storage of glucose as glyco-gen, mainly in liver cells. (b) Glucagon acts mainly on liver cells to stimulate glycogen degradation. This effect is mediated by the second messenger cAMP (see Figure 20-25a).

▶ *Regulation of Cell-Surface Receptors*

The number and activity of functional hormone receptors on the surface of cells is not constant. The level of receptors for a particular hormone can be increased (up-regulation) or decreased (down-regulation), permitting cells to respond optimally to small changes in the hormone level. Prolonged exposure of a cell to a high concentration of a hormone usually results in a reduction in the number of its functional receptors, thereby desensitizing the cell to that hormone. Up to a point, an increase in hormone concentration causes typical hormone-induced responses, rather than excessive ones.

Down-regulation of cell-surface hormone receptors can occur in several ways. Receptors can be internalized by endocytosis, thus decreasing the number on the cell surface, and then either destroyed or stored in intracellular vesicles. Alternatively, the number of receptors on the cell surface is not decreased but their activity is modified, so that they either cannot bind ligand, or bind ligand but form a receptor-ligand complex that does not induce the normal cellular response. Receptors for many hormones are regulated by two or more of these mechanisms.

Receptors for Many Peptide Hormones Are Down-Regulated by Endocytosis

Endocytosis is the principal mechanism for down-regulating the receptors for many peptide hormones including insulin, glucagon, EGF, and PDGF. The receptor-hormone complex is brought into the cell by receptor-mediated endocytosis. The internalized hormone is subsequently degraded in lysosomes—a fate similar to that of other endocytosed proteins, such as low-density lipoproteins (see Figure 16-48). Internalization and degradation most likely terminate the hormone signal. Even if the hormone receptors recycle to the cell surface by exocytosis, as they do in many cases, a substantial fraction will be in the internal membrane compartments at any one time. Therefore, fewer receptors will be on the cell surface and thus available to bind extracellular hormone.

Unlike the low-density lipoprotein (LDL) receptor, internalized receptors for many peptide hormones do not recycle efficiently to the cell surface. In the presence of EGF, the average half-life of an EGF receptor on a fibroblast cell is about 30 min; during its lifetime, each receptor mediates the binding, internalization, and degradation of only two EGF molecules. Each time an EGF receptor is internalized with bound EGF, it has a high probability (about 50 percent) of being degraded in an endosome or lysosome. Exposure of a fibroblast cell to high levels of

EGF for 1 h induces several rounds of endocytosis and exocytosis of EGF receptors, resulting in degradation of most receptor molecules. If the concentration of extracellular EGF is then reduced, the number of EGF receptors on the cell surface recovers, but only after 12–24 h. Synthesis of new receptors is needed to replace those degraded by endocytosis, which is a slow process that may take more than a day.

The fewer hormone receptors present on the surface of a cell, the less sensitive the cell is to the hormone; as a consequence, a higher hormone concentration is necessary to induce the usual physiological response. A simple numerical example illustrates this important point. Suppose a cell has 10,000 insulin receptors on its surface with a K_D of 10^{-8} M. As noted earlier, in many cases only a fraction of the available receptors must bind ligand to induce the maximal physiological response (see Figure 20-7). If we assume only 1000 receptors must bind insulin to induce a physiological response (e.g., activation of glucose transport), we can calculate the insulin concentration [H] needed to induce this response from equation (2) rewritten in the following form:

$$[H] = \frac{K_D}{\dfrac{R_T}{[RH]} - 1} \qquad (3)$$

where R_T = 10,000 (the total number of insulin receptors), $K_D = 10^{-8}$ M, and $[RH]$ = 1000 (the number of insulin-occupied receptors). In this example, the necessary insulin concentrations is 1.1×10^{-9} M. If R_T is reduced to 2000/cell, then a ninefold higher insulin concentration (10^{-8} M) is required to occupy 1000 receptors and induce the physiological response. If R_T is further reduced to 1200/cell, an insulin concentration of 5×10^{-8} M, a 50-fold increase, is necessary to generate a response.

Studies with the RTK that binds PDGF suggest that PI-3 kinase plays an important role in the endocytosis and down-regulation of this class of receptors. As with other RTKs, specific phosphotyrosine residues in the cytosolic domain of the activated PDGF receptor serve as binding sites for PI-3 kinase (two sites), PLC$_\gamma$, GAP, and Syp. To determine which, if any, of these phosphotyrosines is important for endocytosis, researchers produced mutant receptors in which one or more of these tyrosines were mutated to phenylalanine. Only the mutations that abolished the binding sites for PI-3 kinase caused a reduction in the rate of internalization of the receptor. Although it is not known how PI-3 kinase controls receptor internalization, it is intriguing that yeasts with mutations in PI-3 kinase are defective in sorting of proteins to the vacuole, suggesting that this enzyme plays an important role in membrane trafficking both in yeasts and mammalian cells.

Phosphorylation of Cell-Surface Receptors Modulates Their Activity

Cultured cells that are exposed to epinephrine for several hours can bind epinephrine, but adenylate cyclase is not activated and the cAMP level does not rise. This receptor desensitization can be reversed by removing epinephrine from the medium; within several hours, the activity of the β-adrenergic receptor is restored to normal.

This desensitization of the β-adrenergic receptor results from phosphorylation of several serine and threonine residues in the cytosolic domain (Figure 20-50a). Phos-

phorylation of four residues is catalyzed by a cAMP-dependent protein kinase (cAPK). The activity of this kinase is enhanced by the high cAMP level induced by epinephrine, explaining the receptor desensitization observed after prolonged exposure to epinephrine. If the phosphorylated receptor is purified and incorporated into cells (see Figure 20-12), it can bind ligand normally but cannot activate G_s protein or stimulate cAMP synthesis in response to epinephrine. Subsequent enzymatic removal of these phosphate groups reactivates the ability of the receptor to stimulate adenylate cyclase. cAMP-dependent protein kinase phosphorylates all G protein–linked cell-surface

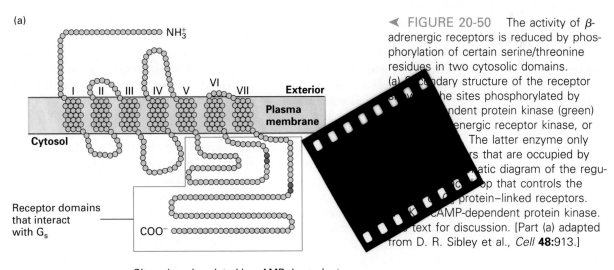

(a)

• = Sites phosphorylated by cAMP-dependent protein kinase

○ = Sites phosphorylated by β-adrenergic receptor kinase (BARK); phosphorylates only agonist-occupied receptors

◄ FIGURE 20-50   The activity of β-adrenergic receptors is reduced by phosphorylation of certain serine/threonine residues in two cytosolic domains.
(a) Secondary structure of the receptor showing the sites phosphorylated by cAMP-dependent protein kinase (green) or β-adrenergic receptor kinase, or BARK. The latter enzyme only phosphorylates that are occupied by ... schematic diagram of the regu... loop that controls the ... protein–linked receptors. ... cAMP-dependent protein kinase. ... text for discussion. [Part (a) adapted from D. R. Sibley et al., *Cell* **48**:913.]

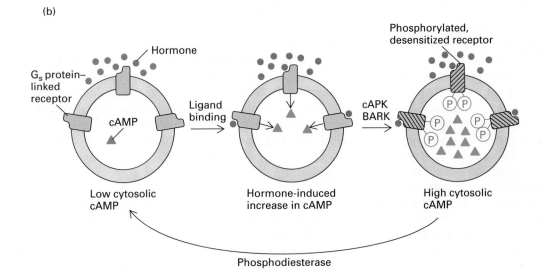

(b)

receptors, not just the β-adrenergic receptor. Since the activity of all such receptors is reduced by phosphorylation by cAPK, this process is called *heterologous desensitization*.

The related process of *homologous desensitization* involves receptor-specific enzymes. *β-adrenergic receptor kinase* (BARK), for example, phosphorylates only the β-adrenergic receptor and only when it is occupied by ligand (see Figure 20-50a). Prolonged treatment of cells with epinephrine or other agonists results in BARK-catalyzed phosphorylation of the β-adrenergic receptor and inhibition of its ability to activate G_s and adenylate cyclase. In this case, a protein called β-arrestin has been shown to bind to the phosphorylated receptor and to sterically block interaction of the receptor with G_s, thus preventing activation of adenylate cyclase following hormone binding. A similar protein has been identified in the visual system (see Figure 21-52).

Figure 20-50b illustrates the feedback loop for modulating the activity of the β-adrenergic and related G_s protein–linked receptors. This loop permits a cell to adjust receptor sensitivity to the constant hormone level at which it is being stimulated, so as to maintain a normal physiological response. Because the phosphorylated receptors are constantly being dephosphorylated by constitutive phosphatases, the number of phosphates per receptor molecule reflects how much ligand has been bound in the recent past (1–10 min). If the hormone level is increased, the resulting rise in the intracellular level of cAMP leads to phosphorylation and desensitization of more receptors, so that production of cAMP and hence the response remain relatively constant. If the hormone is removed, the receptor is completely dephosphorylated and "reset" to a high sensitivity, in which case it can respond to very low levels of hormone.

A similar feedback loop regulates the activity of the EGF receptor, which is an RTK. Phosphorylation of this receptor by protein kinase C decreases its affinity for EGF, thereby moderating the growth-stimulating effect of EGF. As noted earlier, protein kinase C is activated by DAG, which is generated by hormone stimulation of the inositol-lipid signaling pathway. We can summarize the feedback loop modulating the activity of the EGF receptor as follows: binding of EGF to its receptor → activation of PLC_γ → generation of DAG → activation of protein kinase C by DAG → phosphorylation of EGF receptor by protein kinase C leading to down-regulation of the receptor.

Phosphorylation modulates the activity of receptors coupled to many other signaling pathways, including the rhodopsin light receptors in the visual system, and actually may be the basis of short-term memory, as we discuss in the next chapter. Phosphorylation of some receptors leads to an increase in activity, rather than the decrease observed with G_s protein–linked receptors and the EGF receptor described here.

From Plasma Membrane to Nucleus

As noted earlier, the action of steroid hormones, which bind to intracellular receptors, provides the simplest example of the regulation of gene expression by extracellular hormones (see Figure 20-2). The mechanism of action of steroid hormones was discussed in Chapter 11 and is not considered here. Many water-soluble hormones, growth factors, and neurotransmitters, which bind to cell-surface receptors, also induce long-term changes in cellular behavior that invariably involve changes in gene expression. The mechanisms by which binding of a ligand to a cell-surface receptor induces changes in gene expression have been thoroughly studied in several systems. In all cases, binding of a ligand to its receptor stimulates protein kinases that directly or indirectly phosphorylate serine, threonine, or tyrosine residues on specific transcription factors.

The most commonly used experimental approach in studying these mechanisms is to identify a gene whose expression is induced (or repressed) by a hormone; then to define the cis-acting regulatory DNA sequences in that gene (see Figure 11-43a); and finally to identify and clone the cognate proteins that specifically bind to these DNA sequences (see Figure 11-43b). In the remainder of this chapter, we describe activation of several transcription factors by signaling pathways mediated through G protein–linked receptors and receptor tyrosine kinases. Stimulation of gene expression by interferon γ (IFN_γ), whose receptor binds to and activates a cytosolic protein kinase, was discussed in Chapter 11 and is considered in more detail here.

Activation of Some Transcription Factors Occurs via Several Signaling Pathways Coupled to G Protein–Linked Receptors and RTKs

Studies on modulation of gene expression by water-soluble hormones and growth factors have identified the *CRE-binding (CREB) protein* and *serum-response factor (SRF)* as transcription factors that stimulate expression of many genes. Here we describe pathways coupling ligand binding to cell-surface receptors to activation of these two transcription factors. In each case, multiple pathways can lead to transcription-factor activation and subsequent gene expression.

Activation of CREB Protein In mammalian cells, an elevation in the cytosolic cAMP level induces the expression of many genes, including the somatostatin gene in certain endocrine cells and the liver genes encoding several enzymes involved in converting three-carbon compounds to glucose (gluconeogenesis). All such genes contain a cis-acting DNA sequence, called *cAMP-response element* (CRE), that is essential for induction by cAMP. The phosphorylated form of the transcription factor called CRE-binding (CREB) protein binds to this sequence. As discussed previously, binding of neurotransmitters and hormones to G_s protein–linked receptors activates adenylate cyclase, leading to an increase in cAMP and subsequent activation of the catalytic subunit of cAMP-dependent protein kinase (see Figure 20-23). The catalytic subunit then translocates to the nucleus where it phosphorylates serine-133 on CREB protein; phosphorylated CREB protein binds to CRE-containing target genes, stimulating their transcription (Figure 20-51a). Microinjection of different

forms of CREB protein into the nucleus of cultured cells has shown that phosphorylation of serine-133 not only is required for its binding to CRE but also enhances the ability of an adjacent part of the protein to stimulate transcription.

The CREB protein also is activated through signaling pathways that elevate the cytosolic Ca^{2+} level. In certain cells, for instance, depolarization of the plasma membrane results in an increase in cytosolic Ca^{2+} and activation of a Ca^{2+}-calmodulin–stimulated (CaM) kinase. CaM kinase phosphorylates serine-133 of CREB protein, leading to transcriptional activation of target genes. Hence, CREB protein may be a transcription factor associated with several signal-transduction pathways.

(a) G protein – cAMP pathway

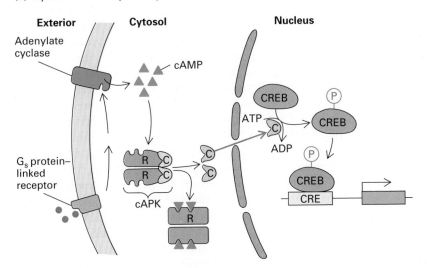

(b) RTK-Ras pathway

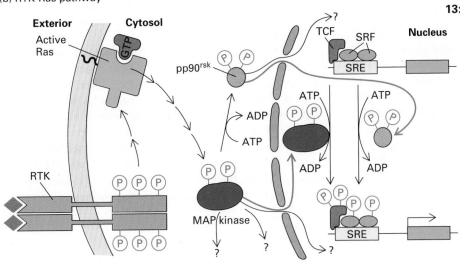

◄ FIGURE 20-51 Signaling pathways leading to activation of transcription factors and modulation of gene expression following ligand binding to certain G_s protein–linked receptors (a) and receptor tyrosine kinases (b). In each case, receptor stimulation leads to activation of protein kinases that translocate to the nucleus (red arrow), where they phosphorylate and activate transcription factors (red), such as CREB, TCF, and SRF. This stimulates transcription of genes controlled by DNA regulating elements (i.e., CRE and SRE) recognized by these transcription factors. cAPK = cAMP-dependent protein kinase. See text for details. [Part (a) see P. K. Brindle and M. R. Montminy, 1992, *Curr. Opin. Genet. Devel.* **2**:199; K. A. Lee and N. Masson, 1993, *Biochim. Biophys. Acta* **1174**:221. Part (b) see R. Marais et al., 1993, *Cell* **73**:381; V. M. Rivera et al., 1993, *Mol. Cell. Biol.* **13**:6260.]

Activation of SRF Addition of a growth factor (e.g., EGF or PDGF) to quiescent cultured mammalian cells in G_0 causes a rapid increase in the expression of as many as 100 different genes. These are called early-response genes because they are induced well before cells enter the S phase and replicate their DNA. One important early-response gene encodes the transcription factor c-Fos; together with other proteins, c-Fos induces expression of many genes encoding proteins necessary for cells to progress through the cell cycle (Chapter 25). Induction of the c-*fos* gene itself by various growth factors is mediated via several intracellular signaling pathways that lead to activation of different protein kinases: those involving cAMP (cAMP-dependent protein kinase), Ca^{2+}/DAG (protein kinase C), and Ras protein (MAP kinase). The regulatory sequence of the c-*fos* gene contains a *serum-response element* (SRE), so named because it is activated by many growth factors in serum. Mutational analysis of the SRE in the c-*fos* gene led to the surprising finding that different signaling pathways act through different sequences in this response element. For instance, cells carrying certain mutations in SRE cannot respond to protein kinase C activation (induced by phorbol esters such as TPA) but can respond to growth factors.

Binding of serum-response factor (SRF) and associated proteins to SREs activates gene expression. As discussed earlier in this chapter, the RTK-Ras signaling pathway activates the MAP kinase cascade (see Figure 20-38). Activated MAP kinase translocates to the nucleus where it phosphorylates specific sites in the C-terminal domain of a protein called *ternary complex factor* (TCF). Phosphorylated TCF associates with two molecules of SRF, forming an active trimeric DNA-binding factor. SRF and TCF may remain bound to promoters in vivo. Mutational studies have shown that if TCF lacks the serine residues phosphorylated by MAP kinase, it does not activate gene expression in transfection experiments. Growth factor-induced phosphorylation of SRF on serine-103 also may stimulate transcription from SREs. Serine-103 of SRF is rapidly phosphorylated in response to growth factors; this modification increases the rate and affinity of SRF binding to the SRE in vitro. A MAP kinase-activated kinase called pp90 has been shown to phosphorylate this serine in SRF in vitro.

Interestingly, the SRF-TCF complex is similar in structure to the yeast transcription factors MCM1 and Ste12 (see Figure 13-13). Transcription of yeast genes with promoters controlled by MCM1 and Ste12 is stimulated by mating pheromones via a signaling pathway that activates MAP kinase (see Figure 20-39). Thus, the later stages of a least one signal-transduction pathway leading from the cell surface to nuclear transcription factors are conserved from yeast to humans.

STATs Are Transcription Factors Activated by Protein Tyrosine Kinases Associated with Cell-Surface Receptors

Interferon α, β, and γ are glycoprotein hormones produced and secreted by animal cells after infection by many viruses. Binding of these hormones to specific cell-surface receptors on other cells inhibits their growth and induces synthesis of a set of proteins that make the target cells resistant to virus infection. By studying the response of cultured mammalian cells to the different interferons (IFNs), investigators have discovered new classes of protein kinases and transcription factors (called STATs for Signal Transducers and Activators of Transcription), that link stimulation of cell-surface receptors to gene expression. These studies on IFNs have identified the components of the signaling pathways initiated by many different polypeptides.

Initial studies focused on the response of cells to IFNα. Within 15–30 min after addition of IFNα, transcription of some genes was shown to increase more than twentyfold. Investigators found that the promoter elements in several of these genes contain a similar sequence, termed the *interferon-stimulated response element* (ISRE), that is both necessary and sufficient for IFNα-stimulated expression. Subsequently, four proteins that bind to the ISRE sequence were identified and cloned. Two of these proteins, Stat1α and Stat1β, 91 and 84K_D respectively, are identical except that Stat1α has 38 extra amino acids at its C-terminus. The two proteins are encoded by the same gene and produced by alternative mRNA splicing. A third protein, Stat2, a 113 K_D protein, exhibits about 59 percent homology with Stat1α and Stat1β. All three of these proteins contain an SH2 domain. The fourth protein, p48, which is a DNA-binding protein from another family unrelated in sequence to the Stats, does not contain an SH2 domain. In cells unstimulated by IFNα, all three Stats are present in the cytosol; following stimulation by IFNα, they move to the nucleus. In contrast, p48 is found in the cytoplasm and nucleus in both unstimulated and IFNα-stimulated cells; it can bind to DNA in the absence of a hormone signal and independent of Stats.

The receptors for all three interferons belong to the *cytokine-receptor superfamily*. These receptors do not have intrinsic protein kinase activity; rather, the cytosolic domain of the activated receptors associates with and activates one or more protein kinases present in the cytosol (see Figure 20-4c). The activated IFNα receptor is thought to interact with two kinases, designated Tyk2 and JAK1, thereby activating them. Activated Tyk2 or JAK1 then phosphorylates tyrosine residues on the cytosolic STATs. The phosphorylated STATs dimerize through interaction

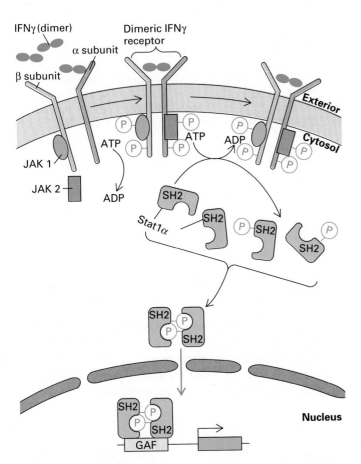

◄ FIGURE 20-52 Signaling pathway initiated by binding of interferon γ (IFNγ) to its receptor, a member of the cytokine-receptor superfamily. The active receptor is a heterodimer; the α subunit probably is identical in different cell types, but the β subunit appears to differ and may confer different properties on the receptor. The interactions and intermediates shown here have been demonstrated experimentally. Two cytosolic protein kinases, JAK1 and JAK2, associate with the cytosolic domain of the receptor. Following IFNγ binding and receptor dimerization, tyrosine residues in the cytosolic domain of the receptor and both JAK1 and JAK2 are phosphorylated. It is not known which kinase is responsible for the various phosphorylation events. Once the phosphorylated receptor/JAK1/JAK2 complex has formed, Stat1α monomers probably bind via their SH2 domain to phosphotyrosine residues on the activated complex. The Stat1α monomers then are phosphorylated by the receptor-associated kinases and released into the cytosol where they dimerize. The Stat1α homodimer then translocates to the nucleus and activates transcription of genes controlled by the GAF regulatory element [See A. C. Greenlund et al., 1993, *J. Biol. Chem.* **269**:14333; J. Soh et al., 1994, *Cell* **76**:793; S. Hemmi et al., 1994, *Cell* **76**:803; K. Shuai et al., 1994, *Cell* **76**:821.]

of the phosphotyrosine in one monomer with the SH2 domain in another, yielding Stat1-Stat2 heterodimers. (The close physical proximity of the phosphotyrosine residue and SH2 domain in each monomer probably prevents intramolecular binding.) The STAT dimers then move to the nucleus and interact with p48 bound to ISRE sequences, thereby activating gene expression.

An important feature of the interferon-STAT signaling pathway is its specificity: each type of interferon induces transcription of a unique subset of genes. The molecular basis of this specificity is becoming clearer as researchers identify other receptor-associated kinases, STAT proteins, and cis-acting DNA response elements. Isolation of mutant cell lines defective in their response to IFNα, IFNγ, or both have been important in uncovering the components of the IFNα and IFNγ signaling pathways. For instance, cell lines missing the Tyk2 kinase are insensitive to IFNα but are sensitive to IFNγ, indicating that Tyk2 is a component of the IFNα pathway but not the IFNγ pathway. Cell lines missing the JAK1 kinase, however, are insensitive to both IFNα and IFNγ, showing that this kinase plays a role in both pathways. The IFNγ receptor also has been shown to associate with a second kinase called JAK2.

Binding of IFNγ to its receptor induces transcription of genes containing a response element called GAF, which binds phosphorylated Stat1 homodimers (Figure 20-52). In cells expressing a mutant Stat1α that cannot be phosphorylated, Stat1α is not translocated to the nucleus following IFNγ stimulation (see Figure 11-63b). Stimulation of cells by IFNγ, in the absence of IFNα, leads to phosphorylation and dimerization of Stat1 and transcription of GAF-containing genes only, because the IFNγ treated cells cannot phosphorylate Stat2. In cells stimulated by IFNα alone, Stat1-Stat2 heterodimers predominate so GAF-containing genes are poorly induced.

Signaling pathways downstream from other members of the cytokine-receptor superfamily are similar to the interferon-JAK-STAT pathway just described. Binding of ligand to any of these receptors causes them to dimerize (Figure 20-53); the cytosolic domain of the activated receptor then interacts with and activates a cytosolic kinase belonging to the JAK family. Binding of ligands to other types of cell-surface receptor also may trigger the activation of STATs. For example, EGF, which binds to an RTK, has been shown to stimulate phosphorylation of a cytosolic STAT, which subsequently binds to a sequence in the c-*fos*

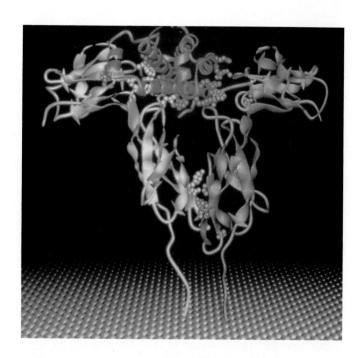

◄ FIGURE 20-53   Model of human growth hormone bound to the extracellular domain of its receptor, a member of the cytokine-receptor superfamily. Binding of a single hormone molecule (red) to the extracellular domains of the receptor monomers (blue and green) induces formation of an activated, dimeric receptor. The hormone signal is transduced to the nucleus via the JAK-STAT pathway illustrated in Figure 20-52, resulting in expression of proteins required for growth and differentiation. [See A. M. DeVos, M. Ultsch, and A. A. Kossiakoff, 1992, *Science* **255**:306. Graphics produced by UCSF MIDAS-plus software; courtesty of A. M. DeVos.]

promoter called *serum-inducible element* (SIE). Investigators are rapidly identifying and cloning new members of the JAK and STAT families. Most recent investigation has revealed that ligands that use transmembrane tyrosine kinases, as well as cytokine receptors (other than IFN recep-

tors), activate various members of the JAK and STAT families. Research in the coming years should clarify the different combinations of receptors, cytosolic kinases, and transcription factors that regulate diverse cellular responses to an array of extracellular signals.

SUMMARY

Three broad groups of molecules mediate cell-to-cell signaling in multicellular animals. Lipid-soluble molecules—principally thyroxine and the steroid hormones—diffuse across the plasma membrane and interact with protein receptors in the cytosol or nucleus; the hormone-receptor complex then induces or represses transcription of specific genes. Peptide hormone (e.g., insulin, glucagon, and various growth factors) and small hydrophilic molecules including the catecholamines (e.g., epinephrine) bind to cell-surface receptors triggering both short-term and long-term cellular responses. Prostaglandins are lipid-soluble derivatives of arachidonic acid that bind to cell-surface receptors; they have numerous effects on cellular processes but primarily mediate close-range paracrine signaling.

Cell-surface receptors can be classified into four classes: seven-spanning receptors coupled via trimeric G proteins to effector enzymes, ion-channel receptors, receptors that lack intrinsic enzymatic activity but associate directly with cytosolic protein kinases, and receptors with intrinsic enzymatic activity. Ligand binding to many receptors results in generation of intracellular signaling com-

pounds called second messengers. Among the most important second messengers are cAMP, cGMP, Ca^{2+} ions, 1,2-diacylglycerol (DAG), and inositol 1,4,5-triphosphate (IP_3). Receptors can be identified and quantified by binding assays with specific agonists and antagonists; characterized by chemical cross-linking to ligands; purified by affinity chromatography; and cloned by recombinant DNA techniques. In this chapter, we described in detail seven-spanning G protein–linked receptors and receptor tyrosine kinases.

Binding of hormones (e.g., epinephrine and glucagon) to some seven-spanning receptors triggers activation of adenylate cyclase and subsequent elevation in the intracellular concentration of the second messenger cAMP. Interaction between these receptors and adenylate cyclase is mediated by the transducing G_s protein, which exchanges bound GDP for GTP in response to hormone binding. The $G_s \cdot GTP$ complex dissociates into $G_{\beta\gamma}$ and the $G_{s\alpha} \cdot GTP$ complex; the latter then activates adenylate cyclase. Binding of other hormones to seven-spanning receptors linked to G_i protein leads to inhibition of adenylate cyclase. Still other seven-spanning receptors are coupled to phospholipase C via the trimeric G_q protein. In recent years, a large number of other G proteins have been identified. In most mammalian receptor systems studied to date, the G_α subunit transduces the signal to the effector protein; however, recent research has shown that in some cases, the $G_{\beta\gamma}$ subunit directly modulates the activity of a coupled effector protein.

The second messengers generated by stimulation of cell-surface receptors have various effects, which often are specific to particular cell types. One well-studied case is cAMP-induced glycogenolysis in epinephrine-stimulated liver and muscle cells. The action of cAMP is mediated by a cAMP-dependent protein kinase, which phosphorylates glycogen synthase, thereby inactivating it. This kinase also phosphorylates, thereby activating, glycogen phosphorylase kinase, which in turn phosphorylates and activates glycogen phosphorylase, the enzyme that degrades glycogen to glucose 1-phosphate. This system is a classic example of a protein kinase cascade leading to hormone-induced activation of an important cellular enzyme.

Ligand binding to certain seven-spanning G protein-linked receptors and receptor tyrosine kinases activates phospholipase C or its γ isoform, which catalyzes formation of the second messengers DAG and IP_3 by hydrolysis of the plasma-membrane lipid phosphatidylinositol 4,5-bisphosphate. DAG activates protein kinase C, which phosphorylates several proteins important in regulating the growth and metabolism of cells. IP_3 binds to Ca^{2+}-channel proteins on ER membranes, triggering the release of stored Ca^{2+} ions into the cytosol. Many of the effects of Ca^{2+} are mediated by calmodulin or related proteins. In muscle cells and neurons, the release of intracellular Ca^{2+} stores also occurs via ryanodine receptors located in the sarcoplasmic reticulum or endoplasmic reticulum. These receptors are stimulated directly or indirectly by depolarization of the plasma membrane. In muscle cells, Ca^{2+} ions trigger contraction and also activate glycogen phosphorylase kinase, thereby stimulating breakdown of glycogen to provide energy for prolonged contraction.

Many peptide hormones, including insulin and various growth factors, bind to receptor tyrosine kinases (RTKs), the second type of receptor examined in detail in this chapter. Following ligand binding, these receptors dimerize and autophosphorylate tyrosine residues on their cytoplasmic domains. Specific phosphotyrosines on activated RTKs bind to various proteins containing SH2 domains. These proteins either are enzymes involved in signaling pathways (e.g., phospholipase C_γ) or adapter proteins (e.g., GRB2) that physically link the receptor to other signaling proteins. Many RTKs are coupled via GRB2 to Ras, a small membrane-bound GTPase structurally related to the G_α subunit of the trimeric G proteins. The active form of Ras has a bound GTP, similar to G_α; but unlike trimeric G proteins, guanine nucleotide binding to Ras is regulated by other proteins: GAPs inhibit Ras by stimulating GTP hydrolysis, and GEFs stimulate Ras by catalyzing the release of GDP. Ligand binding to a receptor tyrosine kinase promotes formation of the active Ras · GTP complex by recruiting Sos protein, which has GEF activity, to the plasma membrane. Sos binds to two SH3 domains in GRB2, which is bound by its SH2 domain to the activated receptor. SH3 domains are present in many proteins that interact in various signaling pathways. The active Ras · GTP complex initiates the sequential activation of three protein kinases: Raf protein kinase → MEK, a dual-specificity kinase → MAP kinase. Activated MAP kinase phosphorylates numerous substrate proteins, including other kinases, cytoskeletal elements, and transcription factors, thereby affecting their activity.

The insulin receptor is an atypical receptor tyrosine kinase. Like other RTKs, the activated insulin receptor autophosphorylates tyrosine residues in its cytosolic domain. Unlike other RTKs, the insulin receptor also phosphorylates multiple sites on a soluble cytosolic protein called IRS1. The phosphotyrosines on IRS1 bind various SH2-containing adapter proteins and signaling enzymes, which bind directly to the cytosolic domain of other RTKs. Insulin has both short-term and long-term effects on cells. In adipocytes and muscle cells, insulin binding triggers exocytosis of vesicles containing a particular type of glucose transporter, thereby stimulating uptake of glucose from the blood. The signaling pathways by which this and other insulin-induced effects occur are not known. The blood glucose level is maintained within a narrow range after meals and during periods of fasting by the antagonistic actions of insulin and glucagon.

The level of functional cell-surface receptors often is regulated, and the continued exposure of a cell to a particular hormone generally results in down-regulation of its receptor. The receptors for insulin and various growth factors can be down-regulated by endocytosis of the receptor-hormone complex. Internalized hormone is degraded by lysosomal enzymes, as are many of the receptors. PI-3 kinase has been shown to be important in this process in the case of the PDGF receptor. The activities of some receptors are down-regulated by phosphorylation. The β-adrenergic receptor is phosphorylated by cAMP-dependent protein kinase (as are other seven-spanning receptors) and by a specific β-adrenergic receptor kinase (BARK), which only is active on ligand-occupied receptors. The EGF receptor, a receptor tyrosine kinase, also is down-regulated by phosphorylation, in this case catalyzed by protein kinase C. Phosphorylation-mediated down-regulation often involves a feedback loop that permits the cell to modulate receptor activity so as to maintain normal physiological responses to a hormone even when it is present for prolonged periods or at levels above or below those necessary to induce normal responses.

Several different pathways for transmitting signals from cell-surface receptors to the nucleus have been described. Ultimately, these signal-transduction pathways lead to activation (or inhibition) of transcription factors and consequent changes in gene expression. Signals originating by ligand binding to seven-spanning G_s protein-linked receptors lead to activation of cAMP-dependent protein kinase; the catalytic subunit of this kinase then translocates to the nucleus where it phosphorylates and thereby activates CREB protein, a transcription factor that regulates many genes in mammalian cells. Stimulation of the RTK-Ras pathway by binding of various growth factors induces activation of MAP kinase, which phosphorylates TCF. Association of TCF with serum-response factor yields an active transcription factor. Studies of interferon-inducible genes led to discovery of the STAT proteins. These proteins are phosphorylated at the plasma membrane by receptor-associated tyrosine kinases and then translocate to the nucleus, where they bind and activate specific interferon-inducible genes. Binding of different interferons leads to activation of different STATs, which in turn activate different sets of genes.

REVIEW QUESTIONS

1. The ability of cells in multicellular species to communicate with one another is vital for survival. In these organisms, there are three general classes of extracellular signals that are used to convey information among cells: They are endocrine, paracrine, and autocrine signaling. Compare and contrast these three signaling mechanisms. What are the six basic steps involved in communication among cells by extracellular signals?

The regulation and duration of hormonal stimulation are very important. Table 20-2 summarizes many of the features of hormones. Be certain you understand the fundamental points presented. What are feedback circuits?

The mere presence of a ligand needed for presenting a signal to a cell is not sufficient in and of itself to perform the task: A receptor for the hormone is a critical link in transducing the signal. What is meant by the binding specificity and the effector specificity of a receptor? What is meant by the K_D value for a receptor? How can such a value be determined experimentally? What is meant by down-regulation and up-regulation of receptors? How are each of these achieved by the cell? What is the effect of each of these events on hormone signaling?

What is a second messenger? What general type of receptor makes use of second messengers?

Chemical analogs of hormones are used as tools in the laboratory to investigate hormone function. Often, they make their way into clinical settings as drugs to treat or diagnose a variety of conditions. What is the difference between an agonist and an antagonist?

Describe techniques for purifying receptor proteins. If the receptor were an integral membrane protein, would you need to consider the lipid environment when performing reconstitution experiments?

a. You are examining the binding of norepinephrine to two different β-adrenergic receptors. The K_D for norepinephrine for each is different. The value of K_D can be described as follows:

$$K_D = k_{off}/k_{on}$$

where k_{on} is the rate constant for the binding of the drug to

its receptor and k_{off} is the rate constant for the release of the drug from its receptor. Both are from the reaction

$$\text{receptor} + \text{hormone} \underset{k_{off}}{\overset{k_{on}}{\rightleftharpoons}} \text{receptor} \cdot \text{hormone}.$$

Suppose that the values of k_{on} for norepinephrine's interaction with each of the β-adrenergic receptors are identical as described in the table.

Adrenergic Receptor	K_D for Norepinephrine (moles · l)	k_{on} for Norepinephrine (liter/mole · s)
β-1	1.2×10^{-9}	2.0×10^7
β-2	2.8×10^{-6}	2.0×10^7

What would be the values of k_{off} for norepinephrine release from each receptor? Be certain to include units. The half-life ($t_{1/2}$) for a receptor · ligand complex is defined as follows:

$$t_{1/2} = \ln(2)/K_D = 0.69/K_D.$$

What are the half-lives for β-1 and β-2? How would this influence your ability to obtain accurate values of K_D for the binding of norepinephrine to each of the receptors?

b. You are developing compounds that will act as agonists or antagonists at β-adrenergic receptors. The ultimate goal is to find new drugs to assist in the treatment of pathological cardiovascular conditions. The agents you are studying are labeled A, B, and C. You do not have radiolabeled forms of the compounds, but you do have radiolabeled epinephrine. Devise a method for establishing the K_D for each of these compounds as it applies to the β-adrenergic receptor.

c. You have determined the K_D values for each of the drugs you are testing; the data are in the following table.

Drug	K_D (M)
A	0.016×10^{-6}
B	8.1×10^{-6}
C	0.31×10^{-6}

Which of these drugs has the greatest affinity for the β-adrenergic receptor? Which has the weakest affinity? Recall that epinephrine is a natural agonist for the β-adrenergic receptor. Which of the drugs have a greater affinity than epinephrine? Refer to Table 20-6 for data needed to answer this.

d. By knowing the K_D value alone for these drugs, do you think you can predict which would be agonists or antagonists towards the responses mediated by the β-adrenergic system? Explain your answer.

2. The receptors for the catecholamines epinephrine and norepinephrine are excellent paradigms for understanding systems that use G proteins during signal transduction. Where are epinephrine and norepinephrine synthesized in the body? What are the α-adrenergic and β-adrenergic receptors? During stressful situations, how does the binding of norepinephrine and epinephrine coordinate glucose and fatty acid levels with blood flow to assist in dealing with the event?

cAMP is the second messenger associated with adrenergic receptors. What is cAMP? What enzyme is responsible for its synthesis? What is cAMP-dependent protein kinase, and how does cAMP activate it? What is the function of activated cAMP protein kinase? How is cAMP degraded?

In general, what are G proteins? Describe their common characteristics. A diagram concerning signal transduction by G proteins is shown in Figure 20-16. Be certain you understand the series of events leading from ligand binding to synthesis of cAMP. How is adenylate cyclase inactivated? How does the formation of a second messenger enhance the signal resulting from the binding of one ligand to its receptor?

a. Liver cells, which contain abundant quantities of glycogen, have receptors that bind the hormone glucagon. You are studying the cellular response to glucagon in cultured hepatocytes, using four cell lines—A, B, C, and D—that respond quite differently to the addition of it to the medium. You are assessing the response by measuring changes in cAMP levels following addition of the hormone. Only cell line A yields significant increases in cAMP; cell lines B, C, and D do not. After significant amounts of arduous research, you compile the table in the opposite column, which summarizes your findings about each cell line. Based on these data, propose a reason for the limited response to glucagon in cell lines B, C, and D. Why would each of these phenotypes interfere with the production of cAMP?

b. If a heterokaryon were prepared between cell lines B and D, would cAMP levels rise in response to normal concentrations of glucagon? Explain your answer.

Cell line	Phenotype
A	A normal response results in increased cAMP.
B	$G_{s\alpha}$ cannot release GDP.
C	Adenylate cyclase activity is absent.
D	The K_D for the binding of glucagon to its receptor is several orders of magnitude higher than the normal dissociation constant.

3. What are receptor tyrosine kinases (RTKs)? Describe their general structure. What is the general mechanism that describes how these kinases work? Once RTKs are activated, what are the two general types of cytoplasmic proteins that interact with them? What is the Src homology 2 (SH2) domain, and what is its function?

What is Ras protein? Describe the pathway for the activation of Ras following the binding of epidermal growth factor to its RTK receptor. What is the Src homology 3 (SH3) domain? Once Ras is activated by the binding of GTP, how is it involved in the pathway that ultimately induces the synthesis of cell-cycle and differentiation-specific proteins? Why do oncogenic forms of Ras signal constitutively?

Describe the mechanisms by which steroids, cAMP, EGF, and interferons can stimulate transcription. (See Chapter 11 for the mechanisms by which steroids regulate transcription.)

4. Where is Ca^{+2} stored within cells? Are Ca^{+2} levels within the cytosol relatively high or low? What is calmodulin? How is it involved in calcium's role as a second messenger? How does it assist in the modulation of cAMP levels in the cell? How does it assist in the degradation of glycogen to glucose 1-phosphate?

What is the role of inositol 1,4,5-triphosphate (IP$_3$) in the regulation of Ca^{+2} levels? List the steps involved in the overall sequence by which IP$_3$ is involved in the influx of Ca^{+2} into the cytosol? How is IP$_3$ inactivated? What are ryanodine receptors?

What is the function of 1,2-diacylglycerol as a second messenger? Where does it come from; what enzyme does it activate? Is an additional second messenger needed for this activation?

a. Phorbol esters have structural similarities to 1,2-diacylglycerol, but they are not readily metabolized to inactive compounds that cannot interact with protein kinase C; therefore they probably maintain the enzyme in an activated state for a prolonged period of time. If cells having the epidermal growth factor (EGF) receptor were exposed to a phorbol ester, what effect would this have on their sensitivity to EGF?

5. The interrelationships between glucagon and insulin in controlling blood glucose levels is critically important in humans. Where is insulin made? What stimulates its release, and what are the target organs for its action? How does insulin affect the metabolism of glucose?

Where is glucagon made, and what stimulates its release? What is its function in regulating glucose levels in the body?

a. *Diabetes mellitus* is the form of diabetes with which most people are familiar. The disease is manifested in two general forms: type 1 where insulin is virtually absent in the afflicted individual, and type 2 where active insulin is present but not effective. In uncontrolled type 1 *diabetes mel-litus,* the patient has elevated blood glucose, and markedly increased levels of fatty acids and glycerol. Based on your knowledge of insulin's functions, why would these symptoms by present?

b. In type 2 *diabetes mellitus,* where active insulin is present in the patient, the cells exhibit a limited response to it. Propose a plausible cause for this form of diabetes.

References

General Properties of Hormone Systems

BAULIEU, E. E., and P. A. KELLY. 1990. *Hormones: From Molecules to Disease.* Chapman and Hall, London.

BARRITT, G. J. 1992. *Communication within Animal Cells.* Oxford Science Publications, Oxford, United Kingdom.

BERRIDGE, M. 1985. The molecular basis of communication within the cell. *Sci. Am.* 253(4):142–150.

WALLIS, M., S. L. HOWELL, and K. W. TAYLOR, eds. 1986. *The Biochemistry of the Polypeptide Hormones.* Wiley.

WILSON, J. D., and D. W. FOSTER. 1985. *Williams Textbook of Endocrinology,* 7th ed. Saunders.

Isolation and Cloning of Receptors

ALLEN, J. M., and B. SEED. 1989. Isolation and expression of functional high-affinity Fc receptor complementary DNAs. *Science* 243:378–381.

GEARING, D. P., et al. 1989. Expression cloning of a receptor for human granulocyte-macrophage colony-stimulating factor. *EMBO J.* 8:3667–3676.

LIN, H. Y., et al. 1992. Expression cloning of the TGF-beta type II receptor, a functional transmembrane serine/threonine kinase. *Cell* 68:775–785, Erratum, 1992, 70:1069.

KLOTZ, I. M. 1982. Number of receptor sites from Scatchard graphs: facts and fantasies. *Science* 217:1247–1249.

SIMONSEN, H., and H. F. LODISH. 1994. *Trends Pharmacol. Sci.* (in the press).

β-Adrenergic Receptors

CARON, M. G., and R. J. LEFKOWITZ. 1993. Catecholamine receptors: structure, function, and regulation. *Recent Prog. Horm. Res.* 48:277–290.

COLLINS, S., M. G. CARON, and R. J. LEFKOWITZ.1992. From ligand binding to gene expression: new insights into the regulation of G-protein-coupled receptors. *Trends Biochem. Sci.* 17:37–39.

LEFKOWITZ, R. J., and M. G. CARON. 1988. Adrenergic receptors: models for the study of receptors coupled to guanine nucleotide regulatory proteins. *J. Biol. Chem.* 263:4993–4996.

LEVITZKI, A. 1988 From epinephrine to cyclic AMP. *Science* 241:800–806.

OSTROWSKI, J., et al. 1992. Mutagenesis of the beta 2-adrenergic receptor: how structure elucidates function. *Annu. Rev. Pharmacol. Toxicol.* 32:167–183.

STRADER, C. D., et al. 1994. Structure and function of G protein-coupled receptors. *Annu. Rev. Biochem.* 63:101–132.

WANG, H.-Y., et al. 1989. Site-directed antipeptide antibodies define the topography of the β-adrenergic receptor. *J. Biol. Chem.* 264:14424–14431.

G_s and G_i Proteins and Activation and Inhibition of Adenylate Cyclase

ALLENDE, J. E. 1988. GTP-mediated macromolecular interactions: the common features of different systems. *FASEB J.* 2:2356–2367.

ARMSTRONG, D. L., and R. E. WHITE. 1992. An enzymatic mechanism for potassium channel stimulation through pertussis-toxin-sensitive G proteins. *Trends Neurosci.* 15:403–408.

BIRNBAUMER, L. 1992. Receptor-to-effector signaling through G proteins: roles for βγ dimers as well as α subunits. *Cell* 71:1069–1072.

BOURNE, H. R., D. A. SANDERS, and F. MCCORMICK. 1990. The GTPase superfamily: a conserved switch for diverse cell functions. *Nature* 348:125–132.

HEPLER, J. R., and A. G. GILMAN. 1992. G proteins. *Trends Biochem. Sci.* 17:383–387.

KRUPINSKI, J., et al. 1989. Adenylyl-cyclase amino acid sequence: possible channel- or transporter-like structure. *Science* 244:1558–1564.

MARKBY, D. W., R. ONRUST, and H. R. BOURNE. 1993. Separate GTP binding and GTPase activating domains of a G_α subunit. *Science* 262:1895–1900.

NOEL, J. P., H. E. HAMM, and P. B. SIGLER. 1993. The 2.2 Å crystal structure of transducin-α complexed with GTPγS. *Nature* 366:654–663.

SIMON, M. I., M. P. STRATHMANN, and N. GAUTAM. 1991. Diversity of G proteins in signal transduction. *Science* 252:802–808.

TAUSSIG R., et al. 1994 Distinct patterns of bidirectional regulation of mammalian adenylyl cyclases. *J. Biol. Chem.* 269:6093–6100.

TANG, W. J., and A. G. GILMAN. 1992. Adenylyl cyclases. *Cell* 70:869–872.

Receptor Tyrosine Kinase

BAR-SAGI, D., et al. 1993. SH3 domains direct cellular localization of signaling molecules. *Cell* 74:83–91.

BOOKER, G. W., et al. 1992. Structure of an SH2 domain of the p85α subunit of phosphatidylinositol-3-OH kinase. *Nature* 358:684–687.

BUDAY, L., and J. DOWNWARD. 1993. Epidermal growth factor regulates p21[ras] through the formation of a complex of receptor, Grb2 adapter protein, and Sos nucleotide exchange factor. *Cell* 73:611–620.

CLARK, S. G., M. J. STERN, and H. R. HORVITZ. 1992. C. elegans cell-signalling gene *sem-5* encodes a protein with SH2 and SH3 domains. *Nature* 356:340–344.

EGAN, S. E., et al. 1993. Association of Sos Ras exchange protein with Grb2 is implicated in tyrosine kinase signal transduction and transformation. *Nature* 363:45–51.

FANTL, W. J., D. E. JOHNSON, and L. T. WILLIAMS. 1993. Signalling by receptor tyrosine kinases. *Annu. Rev. Biochem.* 62:453–481.

HUNTER, T., et al. 1992. Receptor protein tyrosine kinases and phosphatases. *Cold Spring Harbor Symp. Quant. Biol.* 57:25–41.

LOWENSTEIN, E. J., et al. 1992. The SH2 and SH3 domain-containing protein GRB2 links receptor tyrosine kinases to ras signaling. *Cell* 70:431–442.

PANAYOTOU, G., and M. D. WATERFIELD. 1993. The assembly of signalling complexes by receptor tyrosine kinases. *BioEssays* 15:171–177.

PAWSON, T., and G. D. GISH. 1992. SH2 and SH3 domains: from structure to function. *Cell* 71:359–362.

REN, R., et al. 1993. Identification of a ten-amino acid proline-rich SH3 binding site. *Science* 259:1157–1161.

SKOLNIK, E. Y., et al. 1991. Cloning of P13 kinase-associated p85 utilizing a novel method for expression/cloning of target proteins for receptor tyrosine kinases. *Cell* 65:83–90.

VALIUS, M., and A. KAZLAUSKAS. 1993. Phospholipase C-γ1 and phosphatidylinositol-3-kinase are the downstream mediators of the PDGF receptor's mitogenic signal. *Cell* 73:321–334.

YU, H., et al. 1994. Structural basis for the binding of proline-rich peptides to SH3 domains. *Cell* **76**:933–945.

ZIPURSKY, S. L., and G. M. RUBIN. 1994. Determination of neuronal cell fate: lessons from the R7 neuron of *Drosophila. Annu. Rev. Neurosci.* **17**:373–397.

Other Receptor Systems

CUNNINGHAM, B. C., et al. 1991. Dimerization of the extracellular domain of the human growth hormone receptor by a single hormone molecule. *Science* **254**:821–825.

GARBERS, D. L. 1992. Guanylyl cyclase receptors and their endocrine, paracrine, and autocrine ligands. *Cell* **71**:1–4.

KELLY, P. A., et al. 1993. The growth hormone/prolactin receptor family. *Recent Prog. Horm. Res.* **48**:123–164.

MASSAGUÉ, J., L. ATTISANO, and J. L. WRANA. 1994. The TGF-β family and its composite receptors. *Trends Cell Biol.* **4**:172–178.

MIYAJIMA, A., et al. 1992. Cytokine receptors and signal transduction. *Annu. Rev. Immunol.* **10**:295–331.

WEISS, A. 1993. T cell antigen receptor signal transduction: a tale of tails and cytoplasmic protein-tyrosine kinases. *Cell* **73**:209–212.

cAMP-dependent Protein Kinase and Cascades of Protein Phosphorylation/Dephosphorylation

COHEN, P. 1983. *Control of Enzyme Activity,* 2d ed. Chapman & Hall.

COHEN, P. 1989. The structure and regulation of protein phosphatases. *Annu. Rev. Biochem.* **58**:453–508.

HARDIE, G. 1988. Pseudosubstrates turn off protein kinases. *Nature* **335**:592–593.

HUNTER, T. 1987. A thousand and one protein kinases. *Cell* **50**:823–829.

KNIGHTON, D. R., et al. 1991. Structure of a peptide inhibitor bound to the catalytic subunit of cyclic adenosine monophosphate-dependent protein kinase. *Science* **253**:414–420.

TAYLOR, S. S. 1989. cAMP-dependent protein kinase: model for an enzyme family. *J. Biol. Chem.* **264**:8443–8446.

TAYLOR, S. S., et al. 1992. Structural framework for the protein kinase family. *Annu. Rev. Cell Biol.* 429–462.

Ca^{2+} Ions and Calmodulin as Second Messengers

CHEUNG, W. Y. 1982. Calmodulin. *Sci. Am.* **246**(6):48–56.

EVERED, D., and J. WHELAN, eds. 1986. *Calcium and the Cell.* Ciba Foundation Symposium 122. Wiley.

HANSON, P. I., and H. SCHULMAN. 1992. Neuronal Ca^{2+}/calmodulin-dependent protein kinases. *Annu. Rev. Biochem.* **61**:559–601.

PUTNEY, J. W., JR. 1993. Excitement about calcium signaling in inexcitable cells. *Science* **262**:676–678.

RASMUSSEN, H. 1989. The cycling of calcium as an intracellular messenger. *Sci. Am.* **261**(4):66–73.

SCHULMAN, H., and L. L. LOU. 1989. Multifunctional Ca^{2+}/calmodulin-dependent protein kinase: domain structure and regulation. *Trends Biochem.: Sci.* **14**:62–66.

STRYNADKA, N C. J., and M. N. G. JAMES. 1989. Crystal structure of the helix-loop-helix calcium-binding proteins. *Annu. Rev. Biochem.* **58**:951–998.

TSIEN, T. Y., and M. POENIE. 1986. Fluorescence ratio imaging: a new window into intracellular ionic signaling. *Trends Biochem. Sci.* **11**:450–455.

Inositol Triphosphate, 1,2-Diacylglycerol, and Protein Kinase C

BERRIDGE, M. J. 1993. Inositol triphosphate and calcium signaling. *Nature* **361**:315–325.

HANNUN, Y. A., and R. M. BELL. 1993. The sphingomyelin cycle: a prototypic sphingolipid signaling pathway. *Adv. Lipid Res.* **25**:27–41.

HOUSEY, G. M., et al. 1988. Overproduction of protein kinase C causes disordered growth control in rat fibroblasts. *Cell* **52**:343–354.

KIKKAWA, U., A. KISHIMOTO, and Y. NISHIZUKA. 1989. The protein kinase C family: heterogeneity and its implications. *Annu. Rev. Biochem.* **58**:31–44.

MAJERUS, P. W. 1992. Inositol phosphate biochemistry. *Annu. Rev. Biochem.* **61**:225.

NAKAMURA, S., et al. 1993. Protein kinase C for cell signaling: a possible link between phospholipases. *Adv. Second Messenger Phosphoprotein Res.* **28**:171–178.

NISHIZUKA, Y. 1992. Intracellular signaling by hydrolysis of phospholipids and activation of protein kinase C. *Science* **258**:607–614.

Ras Signaling

AHN, N. G., et al. 1991. Multiple components in an epidermal growth factor-stimulated protein kinase cascade: in vitro activation of a myelin basic protein/microtubule-associated protein 2 kinase. *J. Biol. Chem.* **266**:4220–4227.

BOGUSKI, M. S., and F. MCCORMICK. 1993. Proteins regulating Ras and its relatives. *Nature* **366**:643–654.

BRUNNER, D., et al. 1994. A gain-of-function mutation in *Drosophila* MAP kinase activates multiple receptor tyrosine kinase signaling pathways. *Cell* **76**:875–888.

CREWS, C. M., and R. L. ERIKSON. 1993. Extracellular signals and reversible protein phosphorylation: What to Make of it all. *Cell* **74**:215–217.

HAN, M., et al. 1993. The *C. elegans* lin-45 Raf gene participates in let-60-ras-stimulated vulval differentiation. *Nature* **363**:133–140.

HOWE, L. R., et al. 1992. Activation of the MAP kinase pathway by the protein kinase Raf. *Cell* **71**:335–342.

JURNAK, F. 1988. The three-dimensional structure of c-H-ras p21: implications for oncogene and G protein studies. *Trends Biochem. Sci.* **13**:195–198.

MOODIE, S. A., et al. 1993. Complexes of Ras-GTP with Raf-1 and mitogen-activated protein kinase. *Science* **260**:1658–1661.

MULCAHY, L. S., M. R. SMITH, and D. W. STACEY. 1985. Requirement for *ras* protooncogene function during serum-stimulated growth of NIH 3T3 cells. *Nature* **313**:241–243.

NISHIDA, E., and Y. GOTOH. 1993. The MAP kinase cascade is essential for diverse signal transduction pathways. *Trends Biochem. Sci.* **18**:128–131.

SIMON, M. A., et al. 1991. Ras1 and a putative guanine nucleotide exchange factor perform crucial steps in signaling by the sevenless protein tyrosine kinase. *Cell* **67**:701–716.

THOMAS, S. M., et al. 1992. Ras is essential for nerve growth factor- and phorbol ester-induced tyrosine phosphorylation of MAP kinases. *Cell* **68**:1031–1040.

VOJTEK, A. B., S. M. HOLLENBERG, and J. A. COOPER. 1993. Mammalian Ras interacts directly with the serine/threonine kinase Raf. *Cell* **74**:205–214.

WAKSMAN, G., et al. 1993. Binding of high affinity phosphotyrosyl peptide to the Src SH2 domain: crystal structures of the complexed and peptide-free forms. *Cell* **72**:779–790.

WARD, Y., et al. 1994. Control of MAP kinase activation by the mitogen-induced threonine/tyrosine phosphatase PAC1. *Nature* **367**:651–654.

WOOD, K. W., et al. 1992. Ras mediates nerve growth factor receptor modulation of three signal-transducing protein kinases: MAP kinase, Raf-1, and RSK. *Cell* **68**:1041–1050.

The Insulin Receptor and Regulation of Blood Glucose

CZECH, M., et al. 1992. Complex regulation of simple sugar transport in insulin-responsive cells. *Trends Biochem. Sci.* **17**:197–201.

KELLER, S. R., and G. E. LIENHARD. 1994. Insulin signaling: the role of the insulin receptor substrate 1. *Trends in Cell Biol.* **4**:115–119.

ROSEN, O. M. 1987. After insulin binds. *Science* **237**:1452–1458.

SIMPSON, I. A., and S. W. CUSHMAN. 1986. Hormonal regulation of mammalian-glucose transport. *Annu. Rev. Biochem.* **55**:1059–1089.

SLOT, J. W., et al. 1991 Immuno-localization of the insulin regulatable glucose transporter in brown adipose tissue of the rat. *J. Cell. Biol.* **113**:123–135.

WHITE, M., and C. R. KAHN. 1994. The insulin signaling system. *J. Biol. Chem.* **269**:1–4.

Receptor Regulation by Covalent Modification

INGLESE, J., et al. 1993. Structure and mechanism of the G protein-coupled receptor kinases. *J. Biol. Chem.* **268**:23735–23738.

LEFKOWITZ, R. J. 1993. G protein-coupled receptor kinases. *Cell* **74**:409–412.

LEFKOWITZ, R. J., et al. 1992. G.-protein-coupled receptors: regulatory role of receptor kinases and arrestin proteins. *Cold Spring Harbor Symp. Quant. Biol.* **57**:127–133.

Endocytosis and Receptor Degradation

JOLY, M., et al. 1994. Disruption of PDGF receptor trafficking by mutation of its PI-3 kinase binding sites. *Science* **263**:684–687.

KNUTSON, V. P., G. V. RONNETT, and M. D. LANE. 1983. Rapid, reversible internalization of cell-surface insulin receptors. *J. Biol. Chem.* **258**:12139–12142.

RONNET, G. V., et al. 1983. Kinetics of insulin-receptor transit to and removal from the plasma membrane: effect of insulin-induced down regulation in 3T3-L1 adipocytes. *J. Biol. Chem.* **258**:283–290.

YU, S. S., R. J. LEFKOWITZ, and W. P. HAUSDORFF. 1993. β-Adrenergic receptor sequestration—a potential mechanism of receptor resensitization. *J. Biol. Chem.* **268**:337–341.

Hormone Signaling in Yeast

KURJAN, J. 1992. Pheromone response in yeast. *Annu. Rev. Biochem.* **61**:1097–1129.

LEVITZKI, A. 1988. Transmembrane signaling to adenylate cyclase in mammalian cells and in *Saccharomyces cerevisiae*. *Trends Biochem. Sci.* **13**:298–303.

SPRAGUE, G. F., JR. and J. W. THORNER. 1992. Pheromone response and signal transduction during the mating process of *Saccharomyces cerevisiae*. In *The Molecular and Cellular Biology of the Yeast* Saccharomyces: *Gene Expression*, Vol. II, Cold Spring Harbor Laboratory, pp. 657–744.

Signal Transduction to the Nucleus

BRINDLE, P., S. LINKE, and M. MONTMINY. 1993. Protein-kinase-A-dependent activator in transcription factor CREB reveals new role for CREM repressors. *Nature* **364**:821–824.

DARNELL, J. E., I. M. KERR, and G. STARK. 1994. Jak-STAT pathways and transcriptional activation in response to IFN's and other extracellular signaling proteins. *Science* **264**:1415–1420.

FERRERI, K., G. GILL, and M. MONTMINY. 1994. The cAMP-regulated transcription factor CREB interacts with a component of the TFIID complex. *Proc. Nat'l. Acad. Sci. USA* **91**:1210–1213.

HAGIWARA, M., et al. 1992. Transcriptional attenuation following cAMP induction requires PP-1-mediated dephosphorylation of CREB. *Cell* **70**:105–113.

MARAIS, R., J. WYNNE, and R. TREISMAN. 1993. The SRF accessory protein Elk-1 contains a growth factor-regulated transcriptional activation domain. *Cell* **73**:381–393.

RIVERA, V. M., C. K. MIRANTI, R. P. MISRA, D. O. GINTY, R.-H. CHEN, J. BLENIS, and M. GREENBERG. 1993. *Mol. Cell Biol.* **13**:6260.

SHENG, M., M. A. THOMPSON, and M. E. GREENBERG. 1991. CREB: a Ca^{2+}-regulated transcription factor phosphorylated by calmodulin-dependent kinases. *Science* **252**:1427–1430.

NICHOLS, M., et al. 1992. Phosphorylation of CREB affects its binding to high and low affinity sites: implications for cAMP induced gene transcription. *EMBO J.* **11**:3337–3346.

STRUTHERS, R. S., et al. 1991. Somatotroph hypoplasia and dwarfism in transgenic mice expressing a nonphosphorylatable CREB mutant. *Nature* **350**:622–624.

21 Nerve Cells

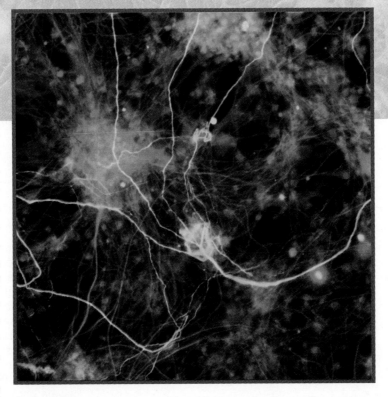

▲ Cultured rat brain nerve cells with long yellow/green-fluorescing axons.

The nervous system regulates all aspects of bodily function and is staggering in its complexity. Millions of specialized *neurons* (nerve cells) sense features of both the external and internal environments and transmit this information to other nerve cells for processing and storage. Millions of other neurons regulate the contraction of muscles and the secretion of hormones. The human brain—the control center that stores, computes, integrates, and transmits information—contains about 10^{12} neurons, each forming as many as a thousand connections with other neurons. The nervous system also contains neuroglial (glial) cells that fill the spaces between neurons, nourishing them and modulating their function.

Despite the complexity of the nervous system as a whole, the structure and function of individual nerve cells is understood in great detail, perhaps in more detail than for any other type of cell. The function of a neuron is to communicate information, which it does by two methods: *Electric signals* process and transmit information within a cell. *Chemical signals* transmit information between cells, utilizing processes similar to those employed by other types of cells to signal each other (Chapter 20). Information from the environment creates special problems because of the diverse types of signals that must be sensed —light, touch, pressure, sound, odorants, the stretching of muscles. *Sensory neurons* have specialized receptors that convert these stimuli into electric signals. These electric signals are then converted into chemical signals that are passed on to other cells—called *interneurons*—that convert the information back into electric signals. Ultimately the information is transmitted to muscle-stimulating *motor neurons* or to other neurons that stimulate other types of cells, such as glands. Thus the output of a nervous system is the result of its circuit properties, that is, the wiring, or interconnections, between neurons, and the strength of these interconnections. At times, the properties of a nervous system must change, for example in the development of new memories. Such changes can be explained as alterations in the number and nature of the interconnections between individual neurons.

In this chapter we focus on how individual nerve cells function and how small groups of cells function together. A great deal of information has been gleaned from simple nervous systems. Squids, sea slugs, and nematodes have large neurons that are relatively easy to identify and manipulate experimentally. Moreover, in these species, only a few identifiable neurons may be involved in a specific task; thus their function can be studied in some detail. Because the principles studied are basic, the findings are applicable to complex nervous systems including that of humans. Indeed, current techniques make it possible to study the role of individual cell types and even of specific proteins in complex functions such as learning in the mammalian brain.

▶ *Neurons, Synapses, and Nerve Circuits*

Since the neuron is the fundamental unit of all nervous systems, we introduce in this section the structural features that are unique to these cells, and the types of electrical signals that they use to process and transmit information. We then introduce *synapses*, the specialized sites where neurons send and receive information from other cells and some of the circuits that allow groups of neurons to coordinate complex processes. Each of these topics will be covered in more detail in later sections of this chapter.

Specialized Regions of Neurons Carry Out Different Functions

Most neurons contain four distinct regions with differing functions: the *cell body*, the *dendrites*, the *axon*, and the *axon terminals* (Figure 21-1).

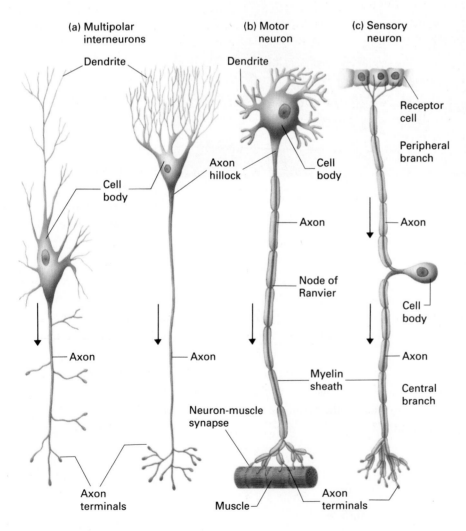

(a) Multipolar interneurons

Dendrite

Cell body

Axon

Axon terminals

(b) Motor neuron

Dendrite

Axon hillock

Cell body

Axon

Node of Ranvier

Myelin sheath

Neuron-muscle synapse

Muscle

Axon terminals

(c) Sensory neuron

Receptor cell

Peripheral branch

Axon

Cell body

Axon

Central branch

◀ FIGURE 21-1   Structure of typical mammalian neurons. Arrows indicate the direction of conduction of action potentials in axons (red). (a) Multipolar interneurons. Each has profusely branched dendrites (which receive signals at synapses with several hundred other neurons) and a single long axon that branches laterally and at its terminus. (b) A motor neuron that innervates a muscle cell. Typically, motor neurons have a single long axon extending from the cell body to the effector cell. In mammalian motor neurons an insulating sheath of myelin usually covers all parts of the axon except at the nodes of Ranvier and the axon terminals. (c) A sensory neuron in which the axon branches just after it leaves the cell body. The peripheral branch carries the nerve impulse from the receptor cell to the cell body, which is located in the dorsal root ganglion near the spinal cord; the central branch carries the impulse from the cell body to the spinal cord or brain. Both branches are structurally and functionally axons, except at their terminal portions, even though the peripheral branch conducts impulses toward, rather than away from, the cell body.

The cell body contains the nucleus and lysosomes, and is where virtually all neuronal proteins and membranes are synthesized. Some proteins are synthesized in dendrites, but axons and axon terminals do not contain ribosomes and no proteins are synthesized there. Proteins and membranes that are required for renewal of the axon and nerve termini are synthesized in the cell body and assembled there into membranous vesicles or multiprotein particles. These are transported along microtubules down the length of the axon—the process called *orthograde axoplasmic transport* (Chapter 23)—to the terminals, where they are inserted into the plasma membrane or other organelles. Axonal microtubules also are the tracks along which damaged membranes and organelles move up the axon toward the cell body, where they are degraded in lysosomes; this process is called *retrograde transport.*

Most neurons have a single axon, whose diameter varies from a micrometer in certain nerves of the human brain to a millimeter in the giant fiber of the squid. Axons are specialized for the conduction of a particular type of electric impulse, called an *action potential,* away from the cell body. An action potential is a series of sudden changes in the electric potential across the plasma membrane (Figure 21-2). In the resting (or nonstimulated) state the potential is ≈ -60 mV (the inside is negative relative to the outside), which is similar to the potential in most nonneu-

ronal cells (see Figure 15-9). At the peak of an action potential the membrane potential can be as much as +50 mV (inside positive), a net change of ≈ 110 mV. This *depolarization* of the membrane is followed by a rapid *repolarization,* or return to the resting potential. These characteristics distinguish an action potential from other types of changes in electric potential across the plasma membrane and allow an action potential to move along an axon without diminution. These and other changes in membrane potential are generated and propagated by the opening and closing of specific ion channel proteins in the neuron plasma membrane, many of which have been cloned and characterized in great detail (Figure 21-3).

Action potentials move rapidly, at speeds up to 100 meters per second. In humans, axons may be more than a meter long, yet it takes only a few milliseconds for an action potential to move along their length. An action potential originates at the *axon hillock,* the junction of the axon and cell body, and is actively conducted down the axon into the axon terminals, small branches of the axon that form the synapses, or connections with other cells. At synapses signals are passed to other neurons, to a muscle cell at a neuromuscular junction, or to any of various other types of cells. A single axon in the central nervous system can synapse with many neurons and induce responses in all of them simultaneously.

► FIGURE 21-2 Action potentials and synaptic transmission in neurons. The membrane potential across the plasma membrane of the presynaptic neuron is measured by a small electrode inserted into it. In this example the neuron is "firing" about every 4 milliseconds, or generating about 250 action potentials a second; the action potentials move down the axon at speeds up to 100 meters per second. Arrival of an action potential at a synapse causes release of neurotransmitters that bind to receptors in the postsynaptic cell, generally inducing an action potential in it. Typical neurons have many axon termini; only one is shown here.

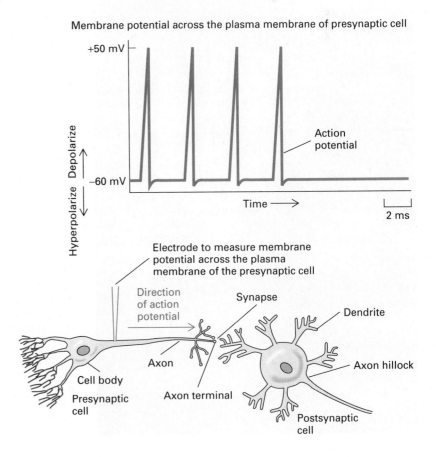

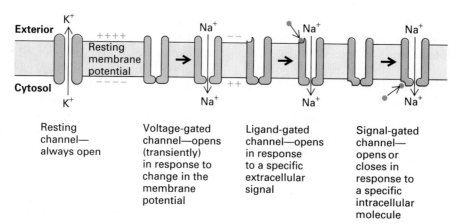

Exterior

K⁺   Na⁺   Na⁺   Na⁺

++++ Resting membrane potential   − −

Cytosol

K⁺   Na⁺   Na⁺   Na⁺

Resting channel— always open

Voltage-gated channel—opens (transiently) in response to change in the membrane potential

Ligand-gated channel—opens in response to a specific extracellular signal

Signal-gated channel— opens or closes in response to a specific intracellular molecule

▲ FIGURE 21-3   The types of ion channels in neuron plasma membranes. Each type of channel protein allows movement only of certain ions. Resting channels are open in unstimulated cells and are responsible for generating the resting membrane potential across the membrane. Voltage-gated channels open, generally only for a fraction of a second, in response to a change in membrane potential; they are responsible for propagation of the electric signals called action potentials along the membrane. Two types of chan-nels that open (or close) in response to chemical signals are responsible for generating electric signals. Ligand-gated chan-nels open in response to a specific extracellular signaling molecule. Some signal-gated channels open (or close) in response to changes in the concentration of specific intracel-lular "second messenger" molecules, such as Ca^{2+} or cyclic GMP; others open or close in response to a G_α or $G_{\beta\gamma}$ sub-unit of a transducing G protein that is activated by a cell-surface receptor.

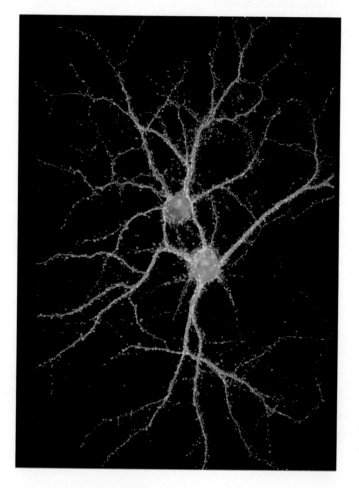

Most neurons have multiple dendrites, which extend outward from the cell body and are specialized to receive chemical signals from the axon termini of other neurons. Dendrites convert these signals into small electric impulses and transmit them to the cell body. Neuronal cell bodies can also form synapses and thus receive signals, as can be seen in the cell depicted in Figure 21-4. Particularly in the central nervous system, neurons have extremely long den-drites with complex branches. This allows them to form synapses with and receive signals from a large number of other neurons, perhaps up to a thousand. Electric distur-bances generated in the dendrites or cell body spread to the axon hillock. If the electric disturbance there is great enough, an action potential will originate and will be ac-tively conducted down the axon.

◄ FIGURE 21-4   A typical interneuron from the hippo-campal region of the brain makes about a thousand syn-apses. The dendrites and cell body of this cultured neuron fluoresce green because the cell was stained with a fluores-cent antibody specific for the microtubule-associated protein MAP2 (see Figure 23-17 and Table 23-1) found only in den-drites and cell bodies. The synapses are stained orange-red by a second fluorescent antibody specific for synaptotagmin, a protein found in presynaptic axon terminals (see Figure 21-37). Thus the orange-red dots indicate presynaptic axon terminals from neurons that are not visible in this field. [Pho-tograph courtesy of O. Mundigl and P. deCamilli.]

Synapses Are Specialized Sites Where Neurons Communicate with Other Cells

Synapses generally conduct signals in only one direction. An axon terminal from the presynaptic cell sends signals that are picked up by the postsynaptic cell. There are two types of synapses, *electric* and *chemical,* which differ in both structure and function. Chemical synapses are much more common than electric synapses.

In a chemical synapse (Figure 21-5a), the axon terminal of the presynaptic cell contains vesicles filled with a particular neurotransmitter substance (Figure 21-5b), such as epinephrine or acetylcholine. The postsynaptic cell can be a dendrite or cell body of another neuron, a muscle or gland cell, or, rarely, even another axon. When the postsynaptic cell is a muscle cell, the synapse is called a *neuromuscular junction* or *motor end plate.* When an action potential in the presynaptic cell reaches an axon terminal,

(a) Chemical synapse

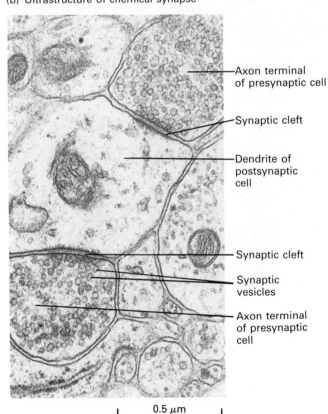

(b) Ultrastructure of chemical synapse

0.5 μm

◄ FIGURE 21-5   Chemical synapses (a) A narrow region—the synaptic cleft—separates the plasma membranes of the presynaptic and postsynaptic cells. Transmission of electric impulses requires release of a neurotransmitter by the presynaptic cell, its diffusion across the synaptic cleft, and its binding by specific receptors on the plasma membrane of the postsynaptic cell. (b) Electron micrograph showing a cross section of a dendrite synapsing with two axon terminals filled with synaptic vesicles. In the synaptic region, the plasma membrane of the presynaptic cell is specialized for vesicle exocytosis; synaptic vesicles, which contain a neurotransmitter, are clustered in these regions. The opposing membrane of the postsynaptic cell (in this case, a neuron) contains receptors for the neurotransmitter. (c) In retrograde signaling, a signal is sent from the postsynaptic to the presynaptic cell. The signal can be a gas, such as nitric oxide or carbon monoxide, or a polypeptide growth factor. Retrograde signals can alter the ability of the presynaptic cell to release neurotransmitter, and are thought to be important in learning. [Part (b) from C. Raine, G. J. Siegel et al., eds., 1981, *Basic Neurochemistry,* 3d ed., Little, Brown, p. 32.]

(c) Retrograde signaling

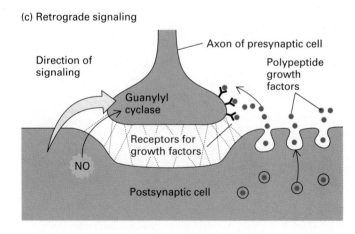

some of the vesicles fuse with the plasma membrane, releasing their contents into the synaptic cleft, the narrow space between the cells. The neurotransmitter diffuses across the synaptic cleft and, after a lag period of about 0.5 millisecond (ms), binds to receptors on the postsynaptic cells. The bound neurotransmitter changes the ion permeability of the postsynaptic plasma membrane, which, in turn, changes the membrane's electric potential at this point. If the postsynaptic cell is a neuron, this electric disturbance may be sufficient to induce an action potential. If the postsynaptic cell is a muscle, the change in membrane potential following binding of the neurotransmitter may induce contraction; if a gland cell, the neurotransmitter may induce hormone secretion. In some cases, enzymes attached to the fibrous network connecting the cells destroy the neurotransmitter after it has functioned; in other cases, the signal is terminated when the neurotransmitter diffuses away or is transported back into the presynaptic cell.

In certain types of synapses, the postsynaptic neuron sends signals to the presynaptic one (Figure 21-5c). Such *retrograde* signals can be gases, such as nitric oxide (NO) and carbon monoxide, or peptide hormones. This type of signaling modifies the ability of the presynaptic cell to signal the postsynaptic one; it is thought to be important in many types of learning. Sometimes an axon terminal of one neuron will synapse with the axon terminal of another neuron (Figure 21-6). Such a synapse may either inhibit or stimulate the ability of an axon terminal to secrete the contents of its synaptic vesicles and signal to a postsynaptic cell. We shall see later how the molecular properties of such synapses permit certain types of learning.

Neurons communicating by an electric synapse are connected by *gap junctions* (Chapter 24) through which electric impulses can pass directly from the presynaptic cell to the postsynaptic one (Figure 21-7). Electric synapses allow a presynaptic cell to induce an action potential in the postsynaptic cell with greater certainty than chemical synapses and without a lag period.

(a) Electric synapse

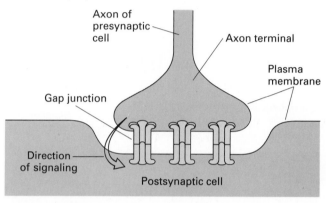

(b)

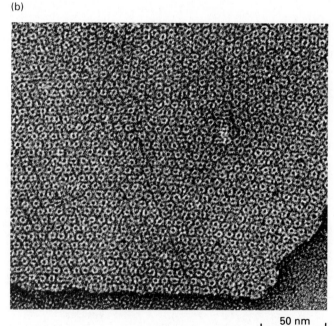

50 nm

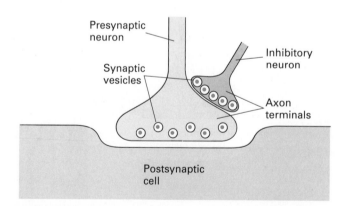

▲ FIGURE 21-6   A modulatory synapse between an inhibitory neuron and the axon terminus of a presynaptic neuron. Stimulation of the inhibitory neuron causes its axon terminal to release neurotransmitter, which reduces the presynaptic axon's ability to transmit a signal to the postsynaptic cell. This type of presynaptic inhibition does not affect the ability of the postsynaptic cell to respond to signals from other neurons.

▲ FIGURE 21-7   An electric synapse. The plasma membranes of the presynaptic and postsynaptic cells are linked by gap junctions; flow of ions through these channels allows electric impulses to be transmitted directly from one cell to the other. (b) Negatively stained, electron microscopic image of the cytosolic face of a region of plasma membrane enriched in gap junctions; each "doughnut" forms a channel connecting two cells. [Part (b) courtesy of N. Gilula.]

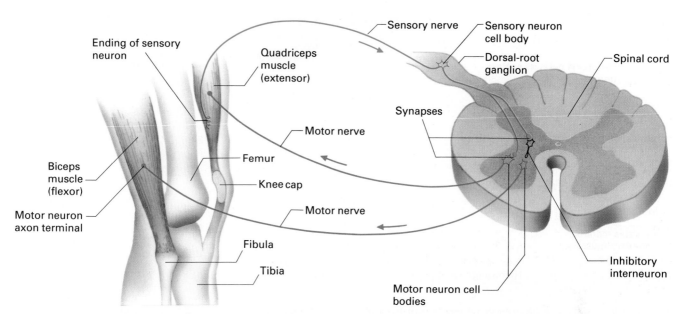

▲ FIGURE 21-8   The knee-jerk reflex arc in the human. Positioning and movement of the knee joint are accomplished by two muscles that have opposite actions: contraction of the quadriceps muscle straightens the leg, whereas contraction of the biceps muscle bends the leg. The knee-jerk response, a sudden extension of the leg, is stimulated by a blow just below the knee cap. The blow directly stimulates sensory neurons (blue) located in the tendon of the quadriceps muscle. The axon of each sensory neuron extends from the tendon to its cell body in a dorsal root ganglion. The sensory axon then continues to the spinal cord, where it branches and synapses with two neurons: (1) a motor neuron (red) that innervates the quadriceps muscle and (2) an inhibitory interneuron (black) that synapses with a motor neuron (red) innervating the biceps muscle. Stimulation of the sensory neuron causes a contraction of the quadriceps and, via the inhibitory neuron, a simultaneous inhibition of contraction of the biceps muscle. The net result is an extension of the leg at the knee joint. (Each cell illustrated here actually represents a nerve, that is, a population of neurons.)

Neurons Are Organized into Circuits

In complex multicellular animals, such as insects and mammals, signaling circuits consist of two or more neurons and, in some cases, highly specialized *sensory receptor cells,* which respond to specific environmental stimuli such as light, heat, stretching, pressure, and concentrations of many chemical substances. In the type of circuit called a reflex arc, interneurons connect multiple sensory and motor neurons, allowing one sensory neuron to affect multiple motor neurons and one motor neuron to be affected by multiple sensory neurons; in this way interneurons integrate and enhance reflexes. For example, the knee-jerk reflex in humans involves a complex reflex arc in which one muscle is stimulated to contract while another is inhibited from contracting (Figure 21-8). Such circuits allow an organism to respond to a sensory input by the coordinated action of sets of muscles that together achieve a single purpose.

The sensory and motor neurons of circuits such as the knee-jerk reflex are contained within the peripheral nervous system. These circuits send information to and receive information from the central nervous system, which comprises the brain and spinal cord (Figure 21-9) and is composed mainly of interneurons. The peripheral nervous system contains two broad classes of motor neurons. The *somatic motor neurons* stimulate voluntary muscles, such as those in the arms, legs, and neck; the cell bodies of these neurons are located inside the central nervous system, in either the brain or the spinal cord. The *autonomic motor neurons* innervate glands, heart muscle, and smooth muscles not under conscious control, such as the muscles that surround the intestine and other organs of the gastrointestinal tract. The two classes of autonomic motor neurons, sympathetic and parasympathetic, generally have opposite effects: one class stimulates the muscle or gland and the other inhibits it. Sensory neurons, which convey information to the central nervous system, have their cell bodies clustered in ganglia, masses of nerve tissue that lie just outside the spinal cord. The cell bodies of the motor neurons of the autonomic nervous system also lie in ganglia. Each peripheral nerve (Figure 21-10) is a bundle of axons; some are parts of motor neurons; others are parts of sensory neurons.

Having surveyed the general features of neuron structure, interactions, and circuits, let us turn to the mechanism by which a neuron generates and conducts electric impulses.

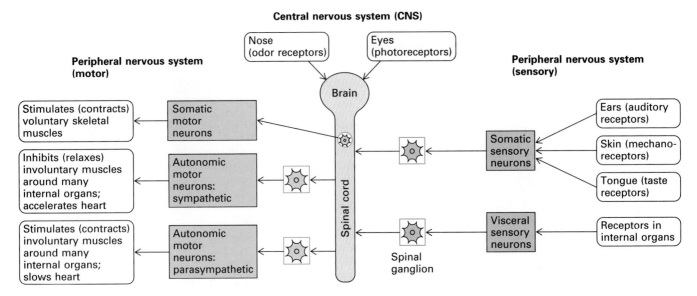

Central nervous system (CNS)

Peripheral nervous system (motor)

Peripheral nervous system (sensory)

Nose (odor receptors)

Eyes (photoreceptors)

Brain

Stimulates (contracts) voluntary skeletal muscles ← Somatic motor neurons

Inhibits (relaxes) involuntary muscles around many internal organs; accelerates heart ← Autonomic motor neurons: sympathetic

Stimulates (contracts) involuntary muscles around many internal organs; slows heart ← Autonomic motor neurons: parasympathetic

Spinal cord

Spinal ganglion

Ears (auditory receptors)

Skin (mechano-receptors)

Tongue (taste receptors)

Somatic sensory neurons

Visceral sensory neurons

Receptors in internal organs

▲ **FIGURE 21-9**   A highly schematic diagram of the vertebrate nervous system. The central nervous system (CNS) comprises the brain and spinal cord. It receives direct sensory input from the eyes and nose. The peripheral nervous system (PNS) comprises three sets of neurons: (1) somatic and visceral sensory neurons, which relay information to the CNS from receptors in somatic and internal organs; (2) somatic motor neurons, which innervate voluntary skeletal muscles; and (3) autonomic motor neurons, which innervate the heart, the smooth involuntary muscles such as those surrounding the stomach and intestine, and glands such as the liver and pancreas. The sympathetic and parasympathetic autonomic motor neurons frequently cause opposite effects on internal organs. The cell bodies of somatic motor neurons are within the CNS; those of sensory neurons and of autonomic motor neurons are in ganglia adjacent to the CNS.

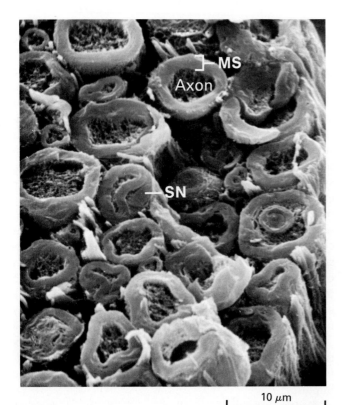

10 μm

The Action Potential and Conduction of Electric Impulses

We saw in Chapter 15 that an electric potential exists across the plasma membrane of all cells. The potential across the plasma membrane of large cells can be measured with a microelectrode inserted inside the cell and a reference electrode placed in the extracellular fluid. The two are connected to a voltmeter capable of measuring small potential differences (Figure 21-11). In virtually all cases the inside of the cell membrane is negative relative to the outside; typical membrane potentials are between −30 and −70 mV. The potential across the surface membrane of most animal cells generally does not vary with time. In contrast, neurons and muscle cells—the principal types of

◄ FIGURE 21-10   Freeze-fracture preparation of a rat sciatic nerve viewed in a scanning electron microscope. The axon of each neuron in the nerve is surrounded by a myelin sheath (MS) formed from the plasma membrane of a Schwann cell (SN). The axonal cytoplasm contains abundant filaments—mostly microtubules and intermediate filaments—that run longitudinally and serve to make the axon rigid. [From R. G. Kessel and R. H. Kardon, 1979, *Tissues and Organs: A Text-Atlas of Scanning Electron Microscopy*, W. H. Freeman and Company, p. 80.]

(a)

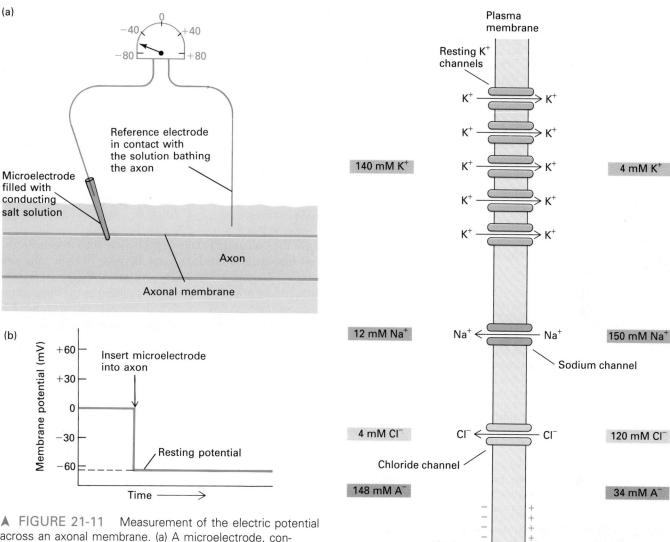

(b)

▲ FIGURE 21-11   Measurement of the electric potential across an axonal membrane. (a) A microelectrode, constructed by filling a glass tube of extremely small diameter with a conducting fluid such as KCl, is inserted into an axon in such a way that the surface membrane seals itself around the electrode. A reference electrode is placed in the bathing medium. A potentiometer connecting the two electrodes registers the potential. The potential difference maintained across the cell membrane in the absence of stimulation is called the *resting potential,* in this case, −60 mV. (b) A potential difference is registered only when the microelectrode is inserted into the axon; no potential is registered if the microelectrode is in the bathing fluid.

▲ FIGURE 21-12   Origin of the resting potential in a typical vertebrate neuron. The ionic compositions of the cytosol and of the surrounding extracellular fluid are different. A^- represents negatively charged proteins, which neutralize the excess positive charges contributed by Na^+ and K^+ ions. In the resting neuron there are more open K^+ channels than Na^+ or Cl^- channels; as a consequence more positively charged K^+ ions exit the cell than Na^+ or Cl^- ions enter, and the outside of the plasma membrane acquires a net positive charge relative to the inside.

electrically active cells—undergo controlled changes in their membrane potential (Figure 21-2); they conduct action potentials along their membrane by sequentially opening and closing ion channels that are specific for Na^+ and K^+ ions (Figure 21-3). Thus, we need to explain how the opening and closing of ion channels and the resultant movement of small numbers of ions from one side of the membrane to the other causes changes in the membrane potential.

The Resting Potential Is Generated Mainly by Open Potassium Channels

The concentration of K^+ ions inside typical metazoan cells is about 10 times that in the extracellular fluid, whereas the concentrations of Na^+ and Cl^- ions are much higher outside the cell than inside (Figure 21-12); these concentration gradients are maintained by Na^+-K^+ ATPases with the expenditure of cellular energy (see Figure 15-13). Another important property of the plasma membrane is that it contains open, so-called "resting", ion channels that allow passage only of Na^+, K^+ or Cl^-, the principal cellular ions. Since the number of open K^+ channels exceeds that of Na^+ or Cl^- channels, more K^+ ions than Na^+ or Cl^- move through the plasma membrane. Each type of ion moves down its concentration gradient. The resting potential—inside negative—is determined mainly by the concentration gradient of K^+ ions: movement of a K^+ ion across the membrane down its concentration gradient leaves an excess negative charge on the cytosolic face and deposits a positive one on the exoplasmic face.

To see this point quantitatively, imagine that the plasma membrane in the neuron depicted in Figure 21-12 has only K^+ channels and thus is impermeable to any other ion; we can use the Nernst equation (see equations 15-7 and 15-9) to determine the membrane potential:

$$E_K = \frac{RT}{Z\mathscr{F}} \ln \frac{K_o}{K_i}$$

$$= 59 \log_{10} \frac{K_o}{K_i} = 59 \log_{10} \frac{4}{140} = -91 \text{ mV} \quad (21\text{-}1)$$

where E_K is the potassium equilibrium potential, K_o and K_i are the potassium concentrations outside and inside the cell, R is the gas constant, T is the absolute temperature, $\mathscr{F}$ is the Faraday constant, and Z is the valency of the ion ($+1$ for K^+). E_K is close to the typical resting potential of -60 mV. If in a resting cell one were to vary the extracellular concentration of K^+ and measure the resultant membrane potential, E would always be close to the calculated value of E_K; this is evidence that the resting potential is due mainly to movements of K^+ through open K^+ channels in the plasma membrane.

The actual situation in cells is complicated because there are also open Na^+ and Cl^- channels in the plasma membranes of the resting cell. Cells, of course, contain other ions, such as HPO_4^{2-}, Ca^{2+}, SO_4^{2-}, and Mg^{2+}, but there are few channels that admit these ions. Furthermore, the membrane potential of electrically active cells such as nerves and muscles is affected only by changes in the channels for K^+, Na^+, and Cl^- (and occasionally Ca^{2+}). Thus these three ions are the only ones we need consider here.

To calculate the membrane potential as a function of concentrations of different ions, it is useful to define a permeability constant P for each ion. P is a measure of the ease with which an ion can cross a unit area (1 cm^2) of membrane driven by a 1 M difference in concentration; it is proportional to the number of open ion channels and to the number of ions each channel can conduct per second (the channel conductivity). Thus P_K, P_{Na}, and P_{Cl} are measures of the numbers of K^+, Na^+, and Cl^- ions that move across a unit area of membrane per second. The units of permeability are expressed in centimeters per second (cm/s). Permeabilities are generally not measured directly;

rather, both the number of open ion channels and the conductivity of each channel are measured by techniques we discuss later in the chapter.

The resultant membrane potential across a cell-surface membrane is given by a more complex version of the Nernst equation in which the concentrations of the ions are weighted in proportion to their permeability constants:

$$E_K = \frac{RT}{Z\mathscr{F}} \ln \frac{P_K \, K_o + P_{Na} \, Na_o + P_{Cl} \, Cl_i}{P_K \, K_i + P_{Na} \, Na_i + P_{Cl} \, Cl_o}$$

$$= 59 \log_{10} \frac{P_K \, K_o + P_{Na} \, Na_o + P_{Cl} \, Cl_i}{P_K \, K_i + P_{Na} \, Na_i + P_{Cl} \, Cl_o} \quad (21\text{-}2)$$

where the "o" and "i" subscripts denote the ion concentrations outside and inside the cell. Because of their opposite charges (Z value in the Nernst equation), K_o and Na_o are placed in the numerator, but Cl_o is placed in the denominator; conversely, K_i and Na_i are in the denominator, but Cl_i is in the numerator. The membrane potential at any time and at any position in the neuron can be calculated with this equation if the relevant ion concentrations and permeabilities are known.

Note that if $P_{Na} = P_{Cl} = 0$, then the membrane is permeable only to K^+ ions and equation 22-2 reduces to equation 22-1. Similarly, if $P_K = P_{Cl} = 0$, then the membrane is permeable only to Na^+ ions and equation 21-2 reduces to the following:

$$E = \frac{RT}{Z\mathscr{F}} \ln \frac{Na_o}{Na_i} = E_{Na} \quad (21\text{-}3)$$

Let us apply equation 21-2 to resting neurons in which the ion concentrations are typically those shown in Figure 21-12. For typical resting neurons (and for most nonexcitable cells), the permeability of the membrane to K^+ ions is much greater than that for Na^+ or Cl^- ions; that is, there are more open channels for K^+ than for Na^+ or Cl^-. The resultant membrane potential is closer to E_K (-91.1 mV) than to E_{Na} ($+64.7$ mV); E_{Cl} -87.2 mV) is close to E_K. We can see this relationship by substituting into equation 21-2 typical values of the three permeability constants ($P_K = 10^{-7}$ cm/s, $P_{Na} = 10^{-8}$ cm/s, and $P_{Cl} = 10^{-8}$ cm/s) and of the ion concentrations:

$$E = 59 \log_{10} \frac{(10^{-7})(0.004) + (10^{-8})(0.15) + (10^{-8})(0.004)}{(10^{-7})(0.14) + (10^{-8})(0.012) + (10^{-8})(0.12)}$$

$$= -52.9 \text{ mV}$$

The potential of -53 mV is close to, but less negative than, the equilibrium K^+ potential E_K. The potential is not equal to E_K because the membrane also contains open Na^+ channels; influx of Na^+ ions adds positive charges to the inside of the cell membrane, making the membrane potential more positive (or less negative).

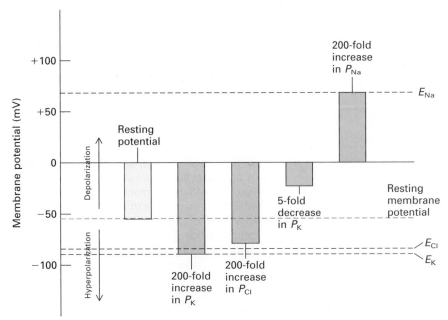

◄ FIGURE 21-13   Effect of changes in ion permeability on membrane potential calculated with equation 21-2 using the permeability constants given in the text and the ion concentrations shown in Figure 21-12. The resting membrane potential is −53 mV; E_{Na}, E_K, and E_{Cl} are the potentials if the membrane contains only channels for Na^+ or K^+ or Cl^-, respectively.

The Opening and Closing of Ion Channels Cause Specific, Predictable Changes in the Membrane Potential

The membrane potential changes predictably if the membrane permeability of an ion changes (Figure 21-13):

1. Opening of K^+ channels (increasing P_K) causes hyperpolarization of the membrane; the membrane potential becomes more negative, approaching E_K. Intuitively, this occurs because more K^+ ions flow outward from the cytosol, leaving excess negative ions on the cytosolic surface of the membrane and putting more positive ones on the outer surface. Conversely, closing K^+, channels, decreasing P_K, causes depolarization of the membrane and a less negative potential.

2. Opening of Na^+ channels (increasing P_{Na}) causes depolarization of the membrane; if the increase is large enough, the potential can become positive inside, approaching E_{Na}. Intuitively, Na^+ ions tend to flow inward from the extracellular medium, leaving excess negative ions on the outer surface of the membrane and putting more positive ions on the cytosolic surface. Conversely, closing Na^+ channels, decreasing P_{Na}, causes membrane hyperpolarization, a more negative potential.

3. Opening of Cl^- channels (increasing P_{Cl}) causes hyperpolarization of the membrane, and the potential approaches E_{Cl}. Intuitively, Cl^- ions tend to flow inward from the extracellular medium, leaving excess positive ions on the outer surface of the membrane and putting more negative ions on the cytosolic surface. Conversely, closing Cl^- channels, decreasing P_{Cl}, causes depolarization and a less negative potential.

Neurons, like all cells, contain "resting ion channels," channels that are open at the resting potential and that admit only one type of ion. As discussed, resting K^+ channels generate the resting potential of ≈−50 to −70 mV. Axons are specialized regions of the neuron that conduct action potentials-sequential depolarizations and repolarizations of the plasma membrane over long distances without diminution (see Figure 21-2). To do this they must have, besides the resting channels, other types of Na^+ and K^+ channels, called *voltage-gated channels*. At the resting potential, voltage-gated channels are closed; no ions move through them. However, when the region of the plasma membrane becomes depolarized (see Figure 21-3) these channels open for a short period, and allow movement of one type of ion (e.g., Na^+ or K^+) through them. In order to appreciate the role of voltage-gated channels in conducting action potentials in one direction down an axon, let us examine how a plasma membrane with only resting ion channels would conduct an electric depolarization.

Membrane Depolarizations Would Spread Only Short Distances without Voltage-Gated Cation Channels

In its electric properties, a nerve cell with only resting ion channels resembles a long underwater telephone cable. It consists of an electrical insulator, the poorly-conducting cell membrane, separating two media—the cell cytosol and the extracellular fluid—that have a high conductivity for ions.

Suppose that a single microelectrode is inserted into the axon and that the electrode is connected to a source of electric current (e.g., a battery) such that the electric poten-

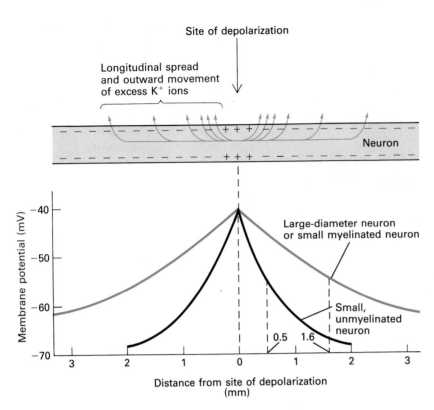

Passive spread of a depolarization of a neuronal plasma membrane with only resting K$^+$ and Na$^+$ ion channels. The neuronal membrane is depolarized from −70 to −40 mV at a single point and clamped at this value. The voltage is then measured at various distances from this site. Because of the outward movement of K$^+$ ions through resting K$^+$ channels, the extent of depolarization falls off with distance from the initial depolarization. Passive spread occurs equally in both directions from the site of depolarization. The length constant is the distance over which the magnitude of the depolarization falls to a value of 1/e (e = 2.718) of the initial depolarization. The length constant for a small neuron with a large number of resting K$^+$ channels (black curve) can be as small as 0.1 mm; in this example it is about 0.5 mm. For a large axon, or one surrounded with a myelin sheath (Fig. 21-10) (blue curve), the length constant can be as large as 5 mm; in this example it is about 1.6 mm.

tial at that point is suddenly depolarized and maintained at this new voltage. At this site the inside of the membrane will have a relative excess of positive charges, principally K$^+$ ions. These ions will tend to move away from the initial depolarization site, thus depolarizing adjacent sections of the membrane. This is called the *passive spread of depolarization*. In contrast to an action potential, passive spread occurs equally in both directions. Also, the magnitude of the depolarization diminishes with distance from the site of initial depolarization, as some of the excess cations leak back across the membrane through resting cation channels (Figure 21-14). Only a small portion of the excess cations are carried longitudinally along the axon for long distances. The extent of this passive spread of depolarization is a function of two properties of the nerve cells: the permeability of the membrane to ions and the conductivity of the cytosol.

The passive spread of a depolarization is greater for neurons of large diameter; this is because the conductivity of the cytosol of a nerve cell depends on its cross-sectional area: the larger the area, the greater the number of ions there will be (per unit length of neuron) to conduct current. Thus K$^+$ ions are able to move, on the average, farther along the axon before they "leak" back across the membrane. As a consequence large-diameter neurons passively conduct a depolarization faster and farther than thin ones. Nonetheless, a membrane depolarization can spread passively for only a short distance, from 0.1 to about 5 mm.

Depolarizations in dendrites and the cell body generally spread in this manner, though some dendrites can conduct an action potential. Neurons with very short axons also conduct axonal depolarizations by passive spread. However, passive spread does not allow propagation of electric signals over long distances.

Opening of Voltage-Gated Sodium Channels Depolarizes the Nerve Membrane during Conductance of an Action Potential

The action potential is a cycle of membrane depolarization, hyperpolarization, and return to the resting value. The cycle lasts 1–2 ms, and an action potential can be generated hundreds of times a second (Figures 21-3 and 21-15a). All of these changes in the membrane potential can be ascribed to transient increases in the conductance of a region of the membrane, first to Na$^+$ ions, then to K$^+$ ions (Figure 21-15b). More specifically, these electric changes are due to voltage-gated Na$^+$ and K$^+$ channels that open and shut in response to changes in the membrane potential. The role of these channels in the generation and conduction of action potentials was elucidated in classic studies done on the giant axon of the squid, in which multiple microelectrodes can be inserted without causing damage to the integrity of the plasma membrane. However, the same basic mechanism is used by all neurons.

(a) Depolarization (↑) and hyperpolarization (↓)

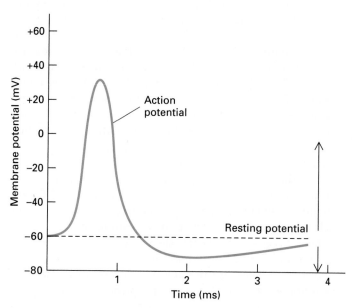

◄ FIGURE 21-15   (a) Changes in membrane potential at a single point on the giant axon of a squid, following stimulation at time 0, as measured by a microelectrode. (b) Changes in permeability (conductance) of the axonal membrane for Na$^+$ and K$^+$ associated with the action potential shown in (a). The transient increase in Na$^+$ influx, resulting from the transient opening of voltage-gated Na$^+$ channels, causes the membrane to become depolarized. This precedes opening of voltage-gated K$^+$ channels and the resultant efflux of K$^+$ ions, which causes the membrane to become hyperpolarized. [See A. L. Hodgkin and A. F. Huxley, 1952, *J. Physiol.* **117**:500.]

(b) Changes in ion permeabilities

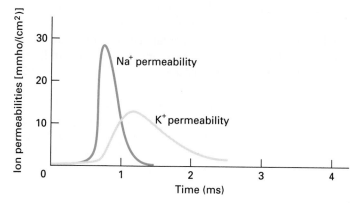

The sudden but short-lived depolarization of a region of the plasma membrane during an action potential (Figures 21-2, 21-15a) is caused by a sudden massive, but transient, influx of Na$^+$ ions through opened voltage-gated Na$^+$ channel proteins in that region (see Figure 21-15b). At the resting membrane potential these voltage-gated channels are closed. The depolarization of the membrane changes the conformation of these proteins, opening the Na$^+$-specific channels and allowing Na$^+$ influx through them.

During conduction of an action potential, the depolarization of a region of membrane spreads passively to the adjacent distal region of membrane. This depolarizes the new region slightly, opening a few of the voltage-dependent Na$^+$ channels in this segment of the membrane and causing an increase in Na$^+$ influx. A combination of two forces acting in the same direction drives Na$^+$ ions into the cell (see Figure 15-9). One is the concentration gradient of Na$^+$ ions. The other is the resting membrane potential—inside negative—which tends to attract Na$^+$ ions into the cell. As more Na$^+$ ions enter the cell, the inside of the cell membrane becomes more positive and thus the membrane becomes depolarized further. This depolarization causes the opening of more voltage-gated Na$^+$ channels, setting into motion an explosive entry of Na$^+$ ions that is completed within a fraction of a millisecond. For a fraction of a millisecond, at the peak of the action potential, the permeability of this region of the membrane to Na$^+$ becomes vastly greater than that for K$^+$ or Cl$^-$, and the membrane potential approaches E_{Na}, the equilibrium potential for a membrane permeable only to Na$^+$ ions (see Figure 21-13).

When the membrane potential almost reaches E_{Na}, further net inward movement of Na$^+$ ions ceases, since the concentration gradient of Na$^+$ ions (outside > inside) is

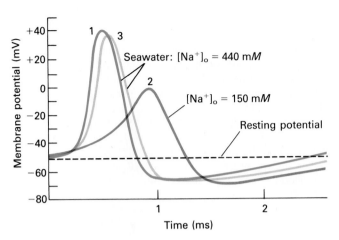

▲ FIGURE 21-16   Effect of changing the external Na$^+$ concentration on the magnitude of the action potential in a squid giant axon. Curves 1 and 3 are membrane potentials measured in normal seawater before and after the same axon was placed in a low-sodium solution and measured to have the membrane potential plotted by curve 2. [See A. L. Hodgkin and B. Katz, 1949, *J. Physiol.* **108**:37.]

balanced by the membrane potential E_{Na} (inside positive). The action potential is at its peak. The measured peak value of the action potential for the squid giant axon is 35 mV (see Figure 21-15a), which is close to the calculated value of E_{Na} (55 mV) based on Na$^+$ concentrations of 440 mM outside and 50 mM inside. The relationship between the magnitude of the action potential and the concentration of Na$^+$ ions inside and outside the cell has been confirmed experimentally. For instance, if the concentration of Na$^+$ ions in the solution bathing the squid axon is reduced to one-third of normal, the magnitude of the depolarization is reduced by 40 mV, exactly as predicted (Figure 21-16).

Voltage-Dependent Sodium Channel Proteins Propagate Action Potentials Unidirectionally without Diminution

Electrophysiological studies on the squid axon and other axons have established some of the remarkable properties of voltage-gated channel proteins and have helped to explain the generation and propagation of an action potential. These studies have been extended by studies of purified channel proteins incorporated into phospholipid vesicles, and by expression of recombinant channel proteins in living cells.

As we discussed, in the resting state most of the voltage-gated Na$^+$ channels are closed but capable of being opened if the membrane is depolarized (Figure 21-17a and b). The greater the depolarization, the greater the chance that any one channel will open. Opening is triggered by movement of *voltage-sensing α helices* in response to the

membrane depolarization, causing a small conformational change in the *channel-blocking segment* that allows ion flow. Once opened, the channels stay open about 1 ms, during which time about 6000 Na$^+$ ions pass through them (Figure 21-17c). Then, further ion movements are prevented by the movement of the *channel-inactivating segment* into the channel opening. As long as the membrane remains depolarized the channel is inactivated and cannot be reopened (Figure 21-17d). A few milliseconds after the inside-negative resting potential is reestablished the channels return to the closed state, capable of being opened by depolarization (Figure 21-17a).

At the peak of the action potential the extent of membrane depolarization is sufficient to depolarize a "downstream" segment of membrane, and to open the Na$^+$ channels in this region. Thus, propagation of the action potential is ensured. The transient inactive state of the Na$^+$ channels (Figure 21-17d) ensures that the action potential is propagated unidirectionally, in contrast to the passive membrane depolarization that spreads bidirectionally (Figure 21-14). Na$^+$ channels are momentarily inactive after opening during passage of the action potential (Figure 21-17d), and thus those Na$^+$ channels "behind" the action potential cannot reopen even though the potential in this segment is still depolarized (Figure 21-18). The inability of Na$^+$ channels to reopen for several milliseconds ensures that action potentials are propagated unidirectionally from the cell body to the axon terminus, and limits the number of action potentials per second that a neuron can conduct. Reopening of Na$^+$ channels "behind" the action potential is also prevented by membrane hyperpolarization caused by the opening of voltage-gated K$^+$ channels, discussed below.

The Opening of Voltage-Gated Potassium Channels Repolarizes the Plasma Membrane during an Action Potential

During the time that the voltage-gated Na$^+$ channels are closing and fewer Na$^+$ ions are entering the cell, voltage-gated K$^+$ channel proteins open. This causes the observed increase in potassium ion permeability (Figure 21-15b), an increase in efflux from the cytosol of K$^+$ ions, and thus, in accord with Figure 21-13, a repolarization of the plasma membrane—a return to its resting potential. Actually, for a brief instant the membrane potential becomes more negative than the resting potential (see Figure 21-15a); it approaches the potassium equilibrium potential E_K, which is more negative than the resting potential (Figure 21-13). Opening of the K$^+$ channels is induced by the membrane depolarization of the action potential. Like the voltage-gated Na$^+$ channels, most types of voltage-gated K$^+$ channels soon close and remain inactivated as long as the membrane is depolarized. Because the K$^+$ channels open a fraction of a millisecond or so after the initial depolarization, they are called *delayed K$^+$ channels*. The increase in P_K

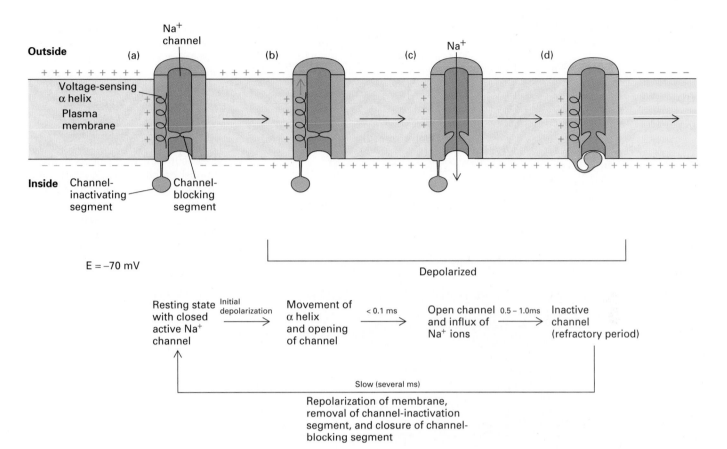

▲ FIGURE 21-17   Operation of voltage-gated Na⁺ channels. In the resting state (a), the *channel-blocking segment* of
the protein closes the channel, and the *channel-inactivating
segment* is free in the cytosol. The channel protein contains
four *voltage-sensing α helices* (discussed later in Figure
21-29), which have positively charged side chains every third
residue. The attraction of these charges for the negative
interior of resting cells helps to ensure that the channel is
closed. The channel-blocking segment inhibits movement of
Na⁺ ions. When the membrane is depolarized (outside
becomes negative), the gating helices move (red arrow)
toward the outer plasma membrane surface (b). Within a

fraction of a millisecond, the channel-blocking segment also
moves, opening the channel for influx of Na⁺ ions (c). Within
a millisecond after opening, the gating helices return to the
resting position and the channel-inactivating segment moves
into the open channel, preventing further ion movements (d).
When the membrane potential is reversed so that the inside
is again negative, the channel-blocking segment moves back
into the blocking position (not shown). After 1–2 ms the
channel-inactivating segment is displaced from the channel
opening and the protein reverts to the closed, resting, state
(a) where it can be opened again by depolarization. [After
C. Miller, 1991, *Science* **252**:1092–1096.]

accounts for the transient hyperpolarization of this region
of the membrane. Eventually all voltage-gated K⁺ and Na⁺
channels close, movement through the membrane of Na⁺
and K⁺ ions returns to the values characteristic of the resting state, and the membrane potential returns to its resting
value.

Movements of Only a Few Sodium and Potassium Ions Generate the Action Potential

The changes in membrane potential characteristic of an
action potential are caused by rearrangements in the balances of ions on either side of the membrane, and not by
changes in the concentrations of ions in the solutions on

either side. The voltage changes are generated by the movements of Na⁺ and K⁺ across the plasma membrane
through voltage-gated channels but the actual number of
ions that move is very small relative to the total number in
the neuronal cytosol. In fact, measurements of the amount
of radioactive sodium entering and leaving single squid
axons and other axons during a single action potential
show that, depending on the size of the neuron, only about
one K⁺ ion per 3000–300,000 in the cytosol (0.0003–0.03
percent) is exchanged for extracellular Na⁺ to generate the
reversals of membrane polarity.

The membrane potential in nerve cells is dependent
primarily on a gradient of Na⁺ and K⁺ ions that is generated and maintained by the Na⁺-K⁺ ATPase. This ATPase
plays no direct role in nerve conduction. If dinitrophenol
or another inhibitor of ATP production is added to cells,

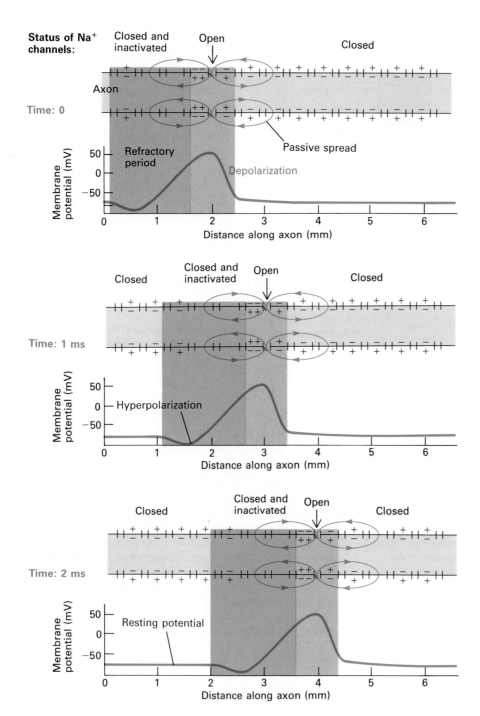

FIGURE 21-18 Because they close and become inactivated shortly after opening, voltage-dependent Na$^+$ channels ensure unidirectional conduction of an action potential. At time 0, an action potential (light purple) is at the 2-mm position on the axon. The membrane depolarization spreads passively (Figure 21-14) in both directions along the axon but the Na$^+$ channels at the 1-mm position are still inactivated (light green; see Figure 21-17d) and cannot yet be reopened. Each region of the membrane is refractory (inactive) for a few milliseconds after an action potential has passed. Thus, the depolarization at the 2-mm site at time 0 triggers action potentials downstream only; at 1 ms an action potential is passing the 3-mm position.

the membrane potential gradually falls to zero as all the ions equilibrate across the membrane. In most nerve cells this equilibration is extremely slow, requiring hours. This and similar experiments indicate that the membrane potential is essentially independent of the supply of ATP over the short time spans required for nerve cells to function. Nerve cells normally can fire thousands of times in the absence of an energy supply because the ion movements during each discharge involve only a minute fraction of the cell's K$^+$ and Na$^+$ ions.

Myelination Increases the Rate of Impulse Conduction

In man, the cell body of a motor neuron that innervates a leg muscle is in the spinal cord and the axon is about a meter in length. Because the axon is coated with a myelin sheath (Figure 21-10) it takes only about 0.01 second for an action potential to travel the length of the axon (velocity ≈100 meters/second) and stimulate muscle contraction. Without a myelin sheath the velocity would be ≈1 meter/

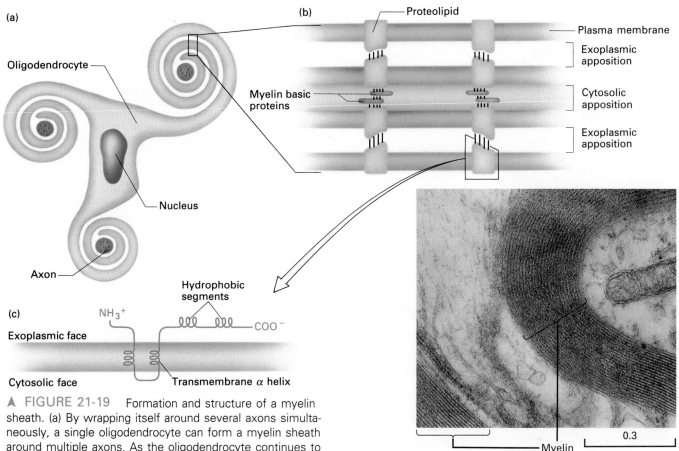

▲ FIGURE 21-19   Formation and structure of a myelin sheath. (a) By wrapping itself around several axons simultaneously, a single oligodendrocyte can form a myelin sheath around multiple axons. As the oligodendrocyte continues to wrap around the axon, all the spaces between its plasma membranes, both cytosolic and exoplasmic, are reduced. Eventually all cytosol is forced out and a structure of compact stacked membranes is formed. This compaction of plasma membranes is generated by proteins that are synthesized only in myelinating cells. (b) Molecular structure of compact myelin. The close apposition of the cytosolic faces of the membrane may result from interactions between myelin basic protein and proteolipid and between myelin basic protein molecules. Apposition of the exoplasmic faces may result from interactions between proteolipid molecules.

(c) Each proteolipid molecule (276 amino acids) has two membrane-spanning α helices and two hydrophobic segments (each containing about 30 amino acids) on its exoplasmic face; the latter are thought to generate proteolipid-proteolipid interactions. (d) Electron micrograph of a cross section of the axon of a myelinated peripheral neuron, surrounded by the Schwann cell that produced the myelin sheath. [Part (c) adapted from L. D. Hudson et al., 1989, *J. Cell Biol.* **109**:717. Part (d) from P. C. Cross and K. L. Mercer, 1993, *Cell and Tissue Ultrastructure, A Functional Perspective*, W. H. Freeman and Company, p. 137.]

second (m/s), and coordination of movements such as running would be impossible.

Myelin is a stack of specialized plasma membrane sheets produced by a glial cell that wraps itself around the axon (Figure 14-11). In the peripheral nervous system these glial cells are called *Schwann cells*; in the central nervous system they are called *oligodendrocytes*. Often several axons are surrounded by a glial cell (Figure 21-19a). In both vertebrates and some invertebrates, axons are accompanied along their length by glial cells, but specialization of these glial cells to form myelin occurs predominantly in vertebrates. Vertebrate glial cells that will later form myelin have on their surface a *myelin-associated glycoprotein* and other proteins that bind to adjacent axons and may trigger the formation of myelin.

A myelin membrane, like all membranes, contains phospholipid bilayers, but unlike many other membranes, it contains only a few types of proteins. *Myelin basic protein* and a *proteolipid* found only in myelin in the central nervous system allow the plasma membranes to stack tightly together (Figure 21-19b and c). Myelin in the peripheral nervous system is constructed of other unique membrane proteins. The myelin surrounding each myelinated axon is formed from many glial cells. Each region of myelin formed by an individual glial cell is separated from the next region by an unmyelinated area called the *node of Ranvier* (or simply, node); only at nodes is the axonal membrane in direct contact with the extracellular fluid (Figure 21-20).

The myelin sheath, which can be 50–100 membranes

(a)

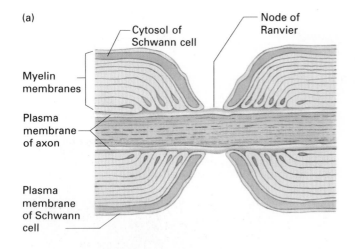

(b)

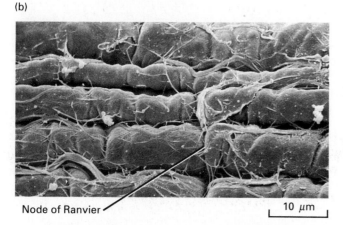

Node of Ranvier ◄ |⎯⎯ 10 μm ⎯⎯|

▲ FIGURE 21-20   (a) Structure of a myelinated axon near a node of Ranvier, the gap that separates the portions of the myelin sheath formed by two adjacent Schwann cells. These nodes are the only regions along the axon where the axonal membrane is in direct contact with the extracellular fluid. (b) A scanning electron micrograph of a peripheral myelinated nerve. The deep folds are the nodes of Ranvier. Numerous strands of collagen surround individual axons and bind them together. [Part (b) from R. G. Kessel and R. H. Kardon, 1979, *Tissues and Organs: A Text-Atlas of Scanning Electron Microscopy*, W. H. Freeman and Company, p. 80.]

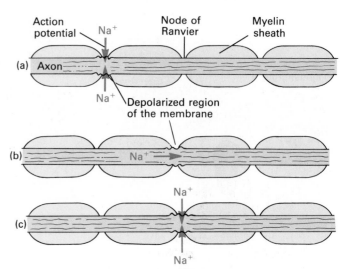

▲ FIGURE 21-21   Regeneration of action potentials at the nodes of Ranvier. (a) The influx of Na⁺ ions associated with an action potential at one node results in depolarization of that region of the axonal membrane. (b) Depolarization moves rapidly down the axon because the excess positive ions cannot move outward across the myelinated portion of the axonal membrane. The buildup of these cations causes depolarization at the next node. (c) This depolarization induces an action potential at that node. By this mechanism the action potential jumps from node to node along the axon.

thick, acts as an electric insulator of the axon by preventing the transfer of ions between the axonal cytosol and the extracellular fluids. Thus all electric activity in axons is confined to the nodes of Ranvier, the sites where ions can flow across the axonal membrane. Node regions contain a high density of voltage-gated Na⁺ channels, about 10,000 per μm^2 of axonal plasma membrane, whereas the regions of axonal membrane between the nodes have few, if any, channels.

The excess cytosolic positive ions generated at a node during the membrane depolarization associated with an action potential diffuse through the axonal cytosol to the next node with very little loss or attenuation because ions are capable of moving across the axonal membrane only at the myelin-free nodes; myelinated nerves have length constants of several millimeters (Figure 21-14). Thus a depolarization at one node spreads rapidly to the next node, and the action potential "jumps" from node to node (Figure 21-21). For this reason, the conduction velocity of myelinated nerves is much greater than that of unmyelinated nerves of the same diameter. For example, a 12-μm-diameter myelinated vertebrate axon and a 600-μm-diameter unmyelinated squid axon both conduct impulses at 12 m/s. Not surprisingly, myelinated nerves are used for signaling in circuits where speed is important.

One of the leading causes of serious neurologic disease among human adults is multiple sclerosis (MS). This disorder, usually characterized by spasms and weakness in one or more limbs, bladder dysfunctions, local sensory losses, and visual disturbances, is caused by patchy loss of myelin in areas of the brain and spinal cord. It is the prototype *demyelinating disease*. In MS patients, conduction of action potentials by the demyelinated neurons is slowed and the Na⁺ channels spread outward from the nodes. The cause of the disease is not known but appears to involve either the body's production of auto-antibodies (antibodies that bind to normal body proteins) that react with myelin basic protein or the secretion of proteases that destroy myelin proteins.

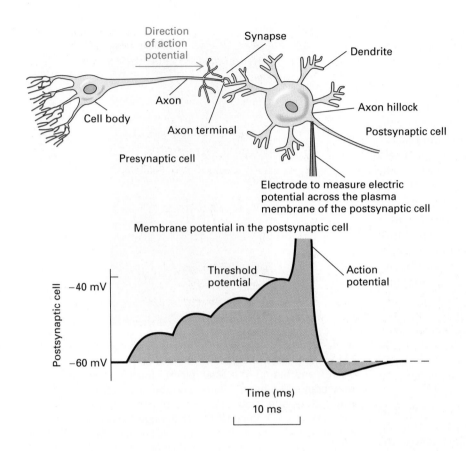

Direction of action potential

▶ FIGURE 21-22   The threshold potential for generation of an action potential. In this example the presynaptic neuron is generating about one action potential every four milliseconds. Arrival of each action potential at the synapse causes a small change in the membrane potential of the postsynaptic cell, in this example a depolarization of ≈5 mV. Note the voltage scale of 10 mV. This synapse is excitatory; when, due to multiple stimuli, the membrane of this postsynaptic cell becomes depolarized to the threshold potential, here ≈−40 mV, an action potential is induced.

Action Potentials Are Generated in an All-or-Nothing Fashion by Summation of Electric Disturbances

A single neuron can be affected simultaneously by synapses with many axons (Figure 21-4). Most synapses are *excitatory*, in that they cause a local depolarization in the membrane of the postsynaptic cell. Others are *inhibitory*, in that they generate a local hyperpolarization of the plasma membrane. The various depolarizations and hyperpolarizations move by passive spread along the dendrite plasma membrane from the synapses to the cell body and then to the axon hillock. Whether a neuron generates an action potential in the axon hillock depends on the balance of the timing, amplitude, and localization of all the various inputs it receives. Action potentials are generated whenever the membrane at the axon hillock is depolarized to a certain voltage called the *threshold potential* (Figure 21-22). The generation of an action potential is thus said to be all-or-nothing. Depolarization to the threshold always leads to an action potential; any depolarization that does not reach the threshold potential never induces it.

In a sense, each neuron is a tiny computer that averages all the electric disturbances on its membrane and makes a decision whether to trigger an action potential and conduct it down the axon. An action potential will always have the same *magnitude* in any particular neuron. The *frequency* with which action potentials are generated in a particular neuron is the important parameter in its ability to signal other cells: the more action potentials in a given neuron in a particular period of time the greater the numbers of signals it sends, via synapses, to neurons or to other target cells.

▶ Molecular Properties of Voltage-Gated Ion Channel Proteins

Voltage-gated ion channel proteins have three remarkable properties that enable nerve cells to conduct an electric impulse: (1) opening in response to changes in the membrane potential (voltage gating); (2) subsequent channel closing and inactivation; and (3) exquisite specificity for ions that will permeate and those that will not. In this section, we describe the molecular analysis of these voltage-dependent ion channel proteins. The technique of *patch-clamping*, or *single-channel recording*, enables workers to investigate the opening, closing, and ion conductance of individual ion channels—that is, of *single* plasma membrane proteins. One of the surprising results to emerge from molecular cloning of ion channel proteins is that all voltage-gated ion channels, be they Na^+, K^+, or Ca^{2+} channels, are related in structure and in function.

Patch Clamps Permit Measurement of Ion Movements through Single Sodium and Potassium Channels

The patch-clamp technique, illustrated in Figure 21-23, measures, across a small patch of isolated membrane, the electric current caused by the movement of ions. In general, the membrane is electrically depolarized or hyperpolarized and maintained (clamped) at that potential by an electronic feedback device. The membrane potential cannot change, in contrast to the situation during an action potential.

The inward or outward movement of ions across a patch of membrane can be quantified from the amount of electric current needed to maintain the membrane potential at the designated "clamped" value. To preserve electroneutrality, the entry of each positive ion (e.g., a Na^+ ion) into the cell across the plasma membrane is balanced by the entry of an electron into the cytosol from the electrode placed in it. Conversely, the movement of each positive ion from the cell (e.g., a K^+ ion) is balanced by the withdrawal of an electron from the cytosol.

In the study depicted in Figure 21-24a, two patches of muscle membrane, each containing one voltage-gated Na^+ channel, were depolarized about 10 mV and clamped at that voltage. The current flow across each patch of membrane was then monitored. Under these circumstances, the transient pulses of current that cross the membrane result from the opening and closing of individual Na^+ channels. Each channel is either open or closed; there are no graded permeability changes for individual channels. From the recording shown in Figure 21-24a, it can be determined that a channel is open for an average of 0.7 ms and that 9900 Na^+ ions per millisecond move through an open channel. We know that these channels conduct only Na^+, since replacement of NaCl with KCl or choline chloride within the patch pipette (corresponding to the outside of the cell) abolishes current through the channels.

Part (b) illustrates the properties of voltage-gated potassium channels. At the depolarizing voltage of -10 mV the channels in the membrane patch open infrequently and remain open for only a few milliseconds, as judged, respectively, by the number and width of the "upward blips" on the electric recording. Further, the ion flux through them is rather small, as measured by the electric current passing through each open channel (the height of the blips). Depolarizing the membrane further to $+20$ mV causes the channels to open about twice as frequently. Also, more K^+ ions move through each open channel (the height of the blips is greater) because the force driving cytosolic K^+ ions outward is greater at a membrane potential of $+20$ mV than at -10 mV. Depolarizing the membrane further to $+50$ mV, such as at the peak of an action potential, causes the opening of more K^+ channels and also increases the flux of K^+ through them. Sometimes two channels in the same patch open simultaneously. Thus, by opening during the peak of the action potential, these K^+ channels cause

(a)

▲ FIGURE 21-23   Outline of the patch-clamping technique. (a) Photomicrograph of the cell body of a cultured neuron and the tip of a patch pipette touching the cell membrane. The bar is 10 μm, and the diameter of the tip of the micropipette is about 0.5 μm. (b) *Facing page:* Arrangement for measuring current flow through individual ion channels in the plasma membrane of a living cell. The patch electrode is filled with a current-conducting saline solution. It is applied, with a slight suction, to the plasma membrane; a region of membrane 0.5 μm in diameter contains only one or a few ion channels. The second electrode is inserted into the cytosol. The recording device measures current flow only through the channels in the patch of plasma membrane. (c) Different ways in which patch clamping can be used. In the on-cell mode (1), pulling the pipette away from the cell results in a piece of membrane in the mouth of the pipette (2). In this configuration, one can study the effects of intracellular second messengers on individual channels. By clamping the potential across the isolated membrane patch and recording the flow of ion current across it, one can measure the effect of the membrane potential on the opening and closing of single channel proteins. The effects of ion composition of the solution on either side of the membrane can also be examined. In the whole-cell mode, (1) gentle suction removes the piece of membrane in the patch (3), allowing measurement of ion movement through all of the plasma membrane ion channels. Pulling the cell away from the pipette results in a piece of plasma membrane being retained in the pore of the pipette, outside-out (4). This allows study of the effects of extracellular substances directly on ion channels within the patch. [Part (a) from B. Sakmann, 1992, *Neuron* **8**:613–629 (Nobel lecture); also published in E. Neher and B. Sakmann, 1992, *Sci. Am.* **266**(3):44. Part (c) after B. Hille, 1992, *Ionic Channels of Excitable Membranes*, Sinauer Associates, p. 89.]

the outward movement of K^+ ions and thus the repolarization of the membrane potential towards the resting state.

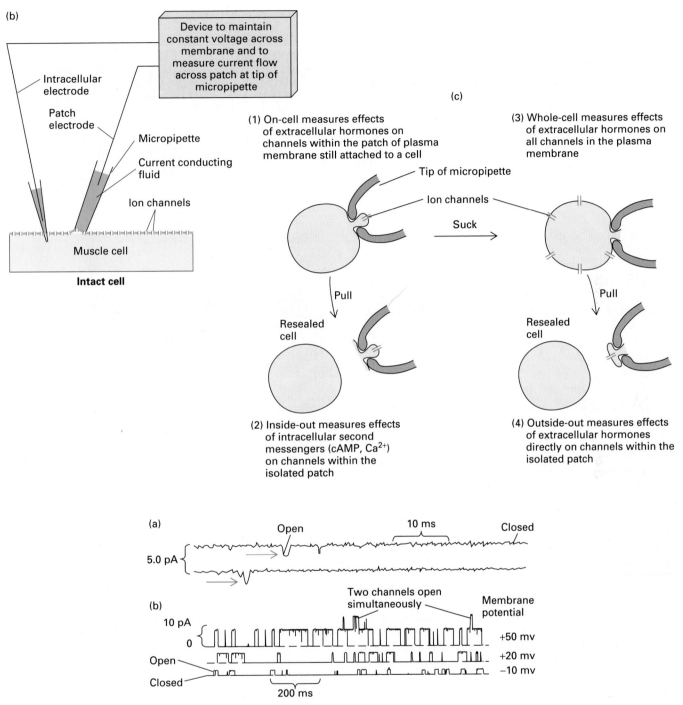

▲ FIGURE 21-24   Use of the patch clamp technique to measure current flux through individual voltage-gated sodium channels (a) or potassium channels (b). In part (a), patches of muscle cell membrane (see Figure 21-23(c), were depolarized by 10 mV and clamped at that value. The transient pulses of electric current (pA = picoamperes), recorded as large downward deviation (arrows), indicate the opening of a Na$^+$ channel and movement of Na$^+$ ions across the membrane. The smaller deviations in current represent background noise. The average current through an open channel is 1.6 picoamperes [1.6×10^{-12} amperes; 1 ampere = 1 coulomb (C) of charge per second]. This is equivalent to the movement of about 9900 Na$^+$ ions per channel per millisecond:

$(1.6 \times 10^{-12}$ C/s$)(10^{-3}$ s/ms$)(6 \times 10^{23}$ molecules/mol$) \div 96,500$ C/mol. (b) Patches of membrane of a cultured muscle cell were clamped at the indicated potentials. Current flux through a patch is recorded as upward deviations since, in contrast to the inward movement of Na$^+$ through the sodium channels in part (a), K$^+$ ions are moving outward across the membrane. Increasing the extent of membrane depolarization from -10 mV to $+50$ mV increases the probability a channel will open, the time it stays open, and the amount of electric current (numbers of ions) that pass through it. [Part (a) see F. J. Sigworth and E. Neher, 1980, *Nature* **287**:447. Part (b) from B. Pallota, K. Magleby, and J. Barrett, 1981, *Nature* **293**:471–474 as modified by B. Hille, 1992, *Ionic Channels of Excitable Membranes*, Sinauer Associates, p. 122.]

(a) Voltage-gated Na⁺ channel protein

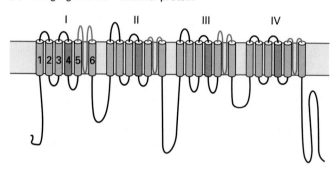

(b) Voltage-gated Ca²⁺ channel protein

(c) Voltage-gated K⁺ channel protein

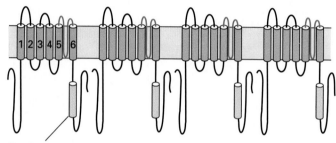

(d) Cyclic nucleotide-gated channel protein

Binding site for
cAMP or cGMP

▲ FIGURE 21-25   Proposed transmembrane structures of three voltage-gated channels and a cyclic GMP-gated channel. Voltage-gated Na⁺ channel proteins (a) contain 1800–2000 amino acids. About 29 percent of the residues are identical in sequence to those in the voltage-gated Ca²⁺ channel protein (b); another 36 percent of the residues in both proteins have similar side chains. Both Na⁺ and Ca²⁺ channel proteins have four homologous domains (indicated by Roman numerals), each of which is thought to contain six transmembrane α helices (indicated by Arabic numerals) and a nonhelical domain, H5 (blue line), that lines the ion pore. One α helix in each domain (no. 4, in red) is thought to function as a voltage sensor. These α helices have multiple arginine and lysine side chains. The *shaker* K⁺ channel protein (c) isolated from *Drosophila* has only 656 amino acids; it is similar in sequence and transmembrane structure to each of the four domains in the Na⁺ and Ca²⁺ channel proteins. The K⁺ channel is a tetramer composed of four identical polypeptides, whereas Na⁺ and Ca²⁺ channels are single polypeptides. Cyclic AMP- or cyclic GMP-gated ion channels (d) are not voltage-gated and helix 4 does not function as a voltage-sensor. However, these do have a pore-lining H5 segment similar to that in the voltage-gated channels. The segment that faces the cytosol that binds cyclic AMP or cyclic GMP, triggering channel opening, is indicated. These channels are found in abundance in the sensing cells in the visual and olfactory systems (see Figure 21-50). [Adapted from W. A. Catterall, 1988, *Science* **242**:50; L. Heginbotham, T. Abramson, and R. MacKinnon, 1992, *Science* **258**:1152; and T. M. Jessell and E. R. Kandel, 1993, in *Cell* vol. 72/*Neuron* vol. 10 (Suppl), pp. 1–3.]

All Voltage-Gated Ion Channels Have a Similar Molecular Structure

The voltage-gated ion channels that generate the action potential have all been cloned, and all have a remarkably similar structure (Figure 21-25). A typical potassium channel contains four copies of a ≈600-amino-acid polypeptide, each of which has six membrane-spanning α-helices.

In contrast, Na⁺ and Ca²⁺ channels are single polypeptides of ≈2000 amino acids that contains four homologous transmembrane domains, each similar in sequence and structure to a K⁺ channel protein. These domains are connected and flanked by shorter stretches of nonhomologous residues. We will explore how these channels were cloned, how their structures were determined, and how they function.

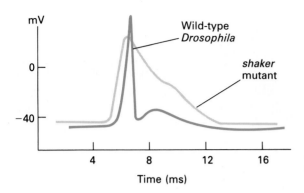

▲ FIGURE 21-26 Action potentials in axons of wild-type *Drosophila* and *shaker* mutants. The *shaker* mutants exhibit an abnormally prolonged action potential because of a defect in the K$^+$ channel protein that is required for normal repolarization. [See L. A. Salkoff and R. Wyman, 1983, *Trends Neurosci.* **6**:128.]

Study of *shaker* Mutants in *Drosophila melanogaster* Led to the Cloning of a Large Family of Voltage-Gated Potassium Channels

The breakthrough in identification and cloning potassium channels came from a genetic analysis of a mutation of the fruit fly *Drosophila melanogaster* that caused a specific abnormality in certain motor neurons. Flies carrying the *shaker* mutation shake vigorously under ether anesthesia, reflecting a loss of motor control and a defect in excitable cells. The X-linked *shaker* mutation was thought to cause a defect in a K$^+$ channel protein. Because the axons of giant nerves in *shaker* mutants have an abnormally prolonged action potential (Figure 21-26), it was proposed that defective K$^+$ channels fail to open normally immediately upon depolarization.

The wild-type *shaker* gene was cloned by chromosome walking (Figure 8-29). To show that the encoded protein indeed was a K$^+$ channel, the *shaker* cDNA was used as a template to produce *shaker* mRNA in a cell-free system. This mRNA was microinjected into frog oocytes (Figures 15-27 and 21-27) and translated into a new voltage-gated K$^+$ channel protein on the oocyte plasma membrane. Patch-clamp studies on the channel in oocyte plasma membrane showed that it had properties identical to those of the K$^+$ channel in the neuronal membrane, demonstrating conclusively that the *shaker* gene encodes that K$^+$ channel.

The 656-amino-acid polypeptide encoded by the *shaker* gene, like most K$^+$ channel polypeptides, contains six membrane-spanning α helixes, labeled S1–S6, a segment, H5, between S5 and S6 that lines the pore of the channel, and an amino-terminal segment—the "ball"—that moves into the open channel, inactivating it. The func-

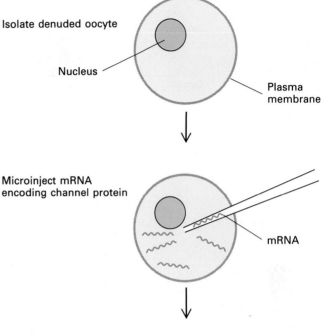

Isolate denuded oocyte

Nucleus

Plasma membrane

Microinject mRNA encoding channel protein

mRNA

Incubate 24–48h for synthesis of channel protein and its movement to plasma membrane

Newly synthesized channel protein

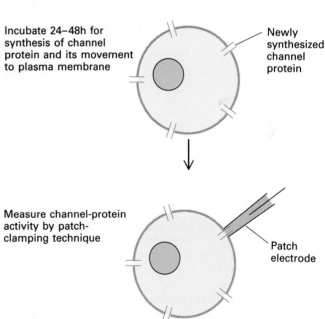

Measure channel-protein activity by patch-clamping technique

Patch electrode

▲ FIGURE 21-27 Expression of a channel protein in a cell that does not normally contain it provides a way to study the activity of normal and mutant channel proteins. A convenient cell for such studies is the oocyte from the frog *Xenopus laevis* (See Figure 15-27). Messenger RNA encoding the channel protein under study is produced in a cell-free transcription reaction using the cloned gene as a template, or a mixture of mRNAs is directly isolated from a tissue. A follicular oocyte is first treated with collagenase to remove the surrounding follicle cells, leaving a denuded oocyte. After microinjection of the mRNA and incubation, the activity of the channel protein is determined by the patch-clamping technique. [Adapted from T. P. Smith, 1988, *Trends Neurosci.* **11**:250.]

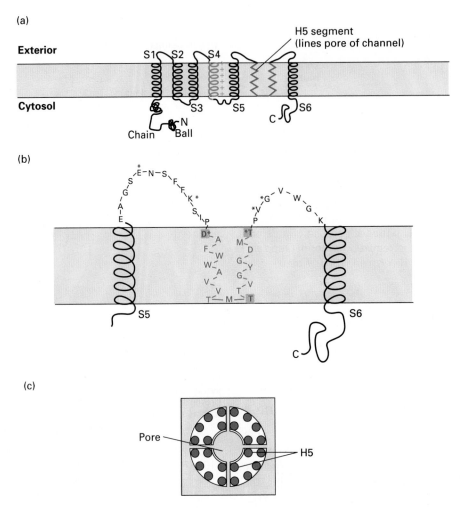

▲ FIGURE 21-28   Molecular structure of voltage-gated potassium channels. Each of the four polypeptides that constitute a channel (a) is thought to contain six membrane-spanning α helixes, S1 through S6. S4 (red) contains several positively charged amino acids and is the voltage-sensing α helix (see Figure 21-17). The N-terminus of the polypeptide is in the cytosol, and contains a globular domain, the ''ball,'' that is linked to the first membrane-spanning segment by a flexible peptide ''chain.'' The ball is essential for inactivation of the open channel. The pore-lining segment H5 (blue), shown in detail in (b), is probably not α-helical. Mutant channels with altered amino acids (one letter code is used) at the sites in the pore indicated by * have altered resistance to the channel-blocking toxin charybdotoxin, and those at **bold** sites (orange boxes) have altered sensitivities to the channel blocker tetraethylammonium ion. Thus the entire H5 sequence is thought to line the pore itself or its entrance. A complete channel, shown in cross section in (c), consists of four subunits. Four H5 segments, one from each subunit, constitute the lining of the ion-selective pore. The position of the voltage-sensing S4 helix is not known. [After C. Miller, 1991, *Science* **252**:1092–1096, and 1992, *Curr. Biol.* **2**:573; and C. Stevens, 1991, *Nature* **349**:657–658.]

tional channel is built of four *shaker* polypeptides (Figure 21-28), but the arrangement of the six α helixes in each segment is not known.

Study of Toxin-Resistant Mutants Led to the Identification of Amino Acids that Line the Pore of the Potassium Channel

To identify the amino acid residues that line the pore of the K⁺ channel, researchers used a neurotoxin from scorpion venom, charybdotoxin. This toxin physically plugs the channel mouth. Site-specific mutation of any of several amino acids (as depicted in part b of Figure 21-28) in the H5 segment inhibited toxin binding, and thus rendered the channel, when expressed in oocytes, resistant to inhibition by the toxin. Similarly, a tetraethylammonium ion was known to plug most kinds of K⁺ channels, and different mutations in the H5 region of the *shaker* protein greatly reduced channel inhibition by tetraethylammonium. Finally, different kinds of K⁺ channels have different conductances (number of ions transported per millisecond), and different sensitivities to tetraethylammonium ions. Chimeric proteins were made in which only the H5 segment from one protein was replaced with that from an-

Voltage-gated Na⁺ channel

Helix 4: *domain I*	S	A	L	R	T	F	R	V	L	R	A	L	K	T	I	S	V	I	P	G	L	K
domain II		G	L	S	V	L	R	S	F	R	L	L	R	V	F	K	L	A	K	S	W	P
domain III	G	A	I	K	S	L	R	T	L	R	A	L	R	P	L	R	A	L	S	R	F	E
domain IV	R	V	I	R	L	A	R	I	G	R	I	L	R	L	I	K	G	A	K	G	I	R

Voltage-gated Ca²⁺ channel

Helix 4: *domain I*	K	A	L	R	T	F	R	V	L	R	P	L	R	V	L	S	G	V	P	S	L	Q
domain II	L	G	I	S	V	L	R	C	I	R	L	L	R	L	F	K	I	T	K	Y	W	T
domain III	S	V	V	K	I	L	R	V	L	R	A	L	R	P	L	R	A	I	N	R	A	K
domain IV	I	S	S	A	F	F	R	L	F	R	V	M	R	L	I	K	L	L	S	R	A	E

Voltage-gated K⁺ channel

Helix 4	R	V	I	R	L	V	R	V	F	R	I	F	K	L	S	R	H	S	K	G	L	O

◄ **FIGURE 21-29** Amino acid sequences of the voltage-sensing α helix (S4) of voltage-gated channel proteins. One S4 α helix is present in each of the four transmembrane domains in the Na⁺ and Ca²⁺ channel polypeptides; one is present in each K⁺ channel polypeptide. These helices have a positively charged lysine (K) or arginine (R) every third or fourth residue (red) whose side chains tend to be localized on one face of an α helix. Amino acids are represented by the single-letter code. [Adapted from W. A. Catterall, 1988, *Science* **242**:50.]

other; in all cases the conductances and sensitivities to tetraethylammonium of the resultant channels correlated with the origin of the H5 segment. Thus, it was concluded that the H5 regions of the *shaker* channel polypeptides constitute the lining of the pore.

A Complete *shaker* Potassium Channel Is Assembled from Four Subunits

Since the pore-lining segment is so small, consisting of about 20–30 amino acids, it would seem that the functional channel would be built of several *shaker* polypeptides, each of which would contribute one segment to line the pore. Indeed, the charybdotoxin-resistant mutant of the *shaker* channel was used to show that the channel comprises four *shaker* polypeptides (Figure 21-28c). In the critical experiment, two kinds of *shaker* mRNA were injected together into frog oocytes; 10 percent coded for the wild-type (toxin-sensitive) channel, and 90 percent for the toxin-resistant mutant. Assuming that the two types of polypeptides mix at random during channel assembly, and that a single copy of the wild-type polypeptide makes a channel toxin sensitive, the fraction of toxin-resistant channels would be 0.9^n, where n is the number of polypeptides that constitute the channel. The determined percentage of toxin-resistant channels was 66 percent, and thus $n = 4$; the channel protein is a tetramer. This result was not surprising, since voltage-gated Na⁺ channels contain, in a single polypeptide, four copies of a sequence resembling the *shaker* polypeptide (Figure 21-25), and a single polypeptide generates a functional Na⁺ channel.

The S4 Segment Is the Voltage Sensor

Since voltage-gated K⁺ (and Na⁺) channels open when the membrane depolarizes, some segment of the protein must "sense" the change in potential. Sensitive electric measurements suggest that the opening of each K⁺ (and Na⁺) channel is accompanied by the movement of four to six protein-

bound positive charges from the cytosolic to the exoplasmic surface of the membrane; alternatively, a larger number of charges may move a shorter distance across the membrane. The movement of these gating charges (or voltage sensors) under the force of the electric field is believed to trigger a conformational change in the protein that opens the channel. The voltage sensor is thought to be the S4 transmembrane α helix present in each *shaker* polypeptide (Figure 21-28a); similar S4 segments are found in voltage-gated Na⁺ and Ca²⁺ channel proteins (Figure 21-25). These voltage-sensing helices, often called *gating helices*, have multiple positively charged side chains (Figure 21-29); when the membrane is depolarized, these amino acids are thought to move toward the exoplasmic surface of the channel (see Figure 21-17).

The role of the S4 helix in voltage sensing was demonstrated in studies with mutant *shaker* K⁺ channel proteins produced by site-specific mutagenesis. In these mutant proteins, one or more arginine or lysine residues in the S4 α helix was replaced with neutral or acidic residues. The ability of such mutant proteins, expressed in frog oocytes, to open in response to membrane depolarization was tested using the patch-clamp technique (Figure 21-24b). As expected, when expressed in oocytes, the mutant proteins exhibited altered opening responses to depolarization voltages. In particular, the fewer the positively charged residues in the gating helices, the larger the depolarization required to open the K⁺ channels.

The N-Terminal Segment of the *shaker* Protein Causes Channel Inactivation

An important characteristic of all voltage-gated channels is inactivation: soon after opening they close spontaneously, forming an inactive channel that will not reopen until the membrane is repolarized (Figure 21-17). The N-terminus of the *shaker* polypeptide forms a globular ball that swings into the open channel, inactivating it. The key experiments in Figure 21-30 support this model. First, deleting the ball

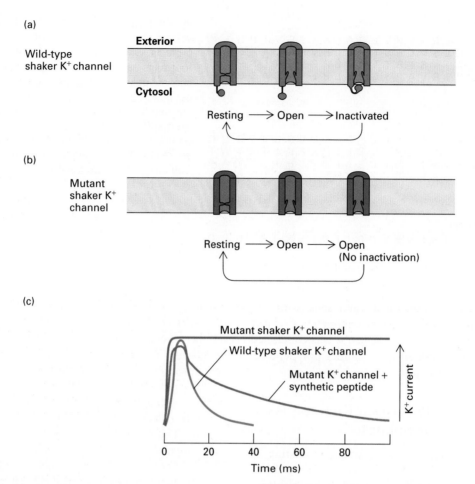

(a)

Wild-type shaker K+ channel

Exterior

Cytosol

Resting $\longrightarrow$ Open $\longrightarrow$ Inactivated

(b)

Mutant shaker K+ channel

Resting $\longrightarrow$ Open $\longrightarrow$ Open (No inactivation)

(c)

Mutant shaker K+ channel

Wild-type shaker K+ channel

Mutant K+ channel + synthetic peptide

K+ current

0   20   40   60   80

Time (ms)

▲ FIGURE 21-30   (a) The "ball and chain" model for inactivation of a voltage-gated potassium channel. In the resting state the ball at the N-terminus of the wild-type channel is free in the cytosol, and remains there during the first milliseconds after the channel is opened by depolarization. The ball then moves into the open channel, inactivating it. After a few milliseconds (depending on the exact type of channel) the ball is displaced from the channel and the protein reverts to the closed, resting state. Evidence for the ball and chain model was obtained by expressing the wild-type and mutant *shaker* potassium channel (b) in *Xenopus* oocytes. When the plasma membrane, in a patch pipette (Figure 21-23), was depolarized from −70 to +30 mV, the wild-type channel opened for ≈5 milliseconds and then closed (red curve in c). In contrast, a mutant channel lacking the ball amino acids 6 through 46 (b) opened normally, but could not close (green curve). When a chemically synthesized ball peptide was added to the cytosolic face of the patch, the mutant channel opened normally and then closed (purple curve). This demonstrated that the added peptide inactivated the channel after it opened. [Parts (a) and (b) after C. Miller, 1991, *Science* **252**:1092–1096; part (c) from W. N. Zagotta, T. Hoshi, and R. W. Aldrich, 1990, *Science* **250**:568–671.]

domain (b) results in a K+ channel that opens normally in response to depolarization and has a normal conductivity, but cannot close (green curve in part c). Second, when a chemically synthesized ball peptide is added to the cytosol, the mutant channel opens normally and then closes (purple curve). This demonstrates that the added peptide inactivates the channel after it opens, and that the ball does not have to be tethered to the protein in order to function. Further experiments suggested that there is a chain as well. Deletion of part of the ≈40-amino-acid segment connecting the ball to the first membrane-spanning segment increases the rate of inactivation—the shorter the chain the more rapid the inactivation—as if a ball attached to a

shorter chain can move into the open channel more readily. Conversely, addition of random amino acids to the normal chain slows channel inactivation.

Potassium Channel Proteins Are Diverse

A great diversity of K+ channel types is necessary to account for the electrical activity of different types of neurons. In *Drosophila*, at least five different *shaker* polypeptides are produced by alternative splicing of the primary transcript of the *shaker* gene. In the oocyte expression assay, these polypeptides exhibit different voltage dependencies and K+ conductivities. Thus differential expression of the *shaker* gene can affect the properties of the action potential in different neurons. Also, if a single K$^+$ channel is constructed from two different kinds of channel polypeptides, the properties of the resultant "hybrid" channel differ from those of both of the channels formed from four identical polypeptides.

Using the *shaker* gene as a hybridization probe, workers isolated genes encoding more than two dozen K$^+$ channel proteins from vertebrates; the encoded channel proteins exhibit different voltage dependencies, conductances, channel open times, and other properties. Many have the property of opening at the strongly depolarizing voltages (as in Figure 21-24b) required for the channel to repolarize the membrane during an action potential. All of the members of this large protein family are identical in essential aspects of mechanism and structure (Figure 21-28).

Thus, starting with the isolation of a mutant fruit fly that shakes under ether anesthesia, scientists have derived a remarkably complete molecular picture of how a voltage-gated potassium channel operates. Using the *shaker* gene, workers have isolated a large number of K$^+$ channel polypeptides expressed in the human brain. As this example illustrates, studies on the nervous systems of invertebrates are indeed directly applicable to an understanding of the human nervous system.

The Sodium Channel Protein Has Four Homologous Transmembrane Domains, Each Similar to a Potassium Channel Polypeptide

Like K$^+$ channels, the density of voltage-gated Na$^+$ channels is very low. Depending on the type of unmyelinated axon, there are only 5–500 voltage-gated Na$^+$ channels per square micrometer of membrane; roughly one membrane protein molecule in a million is a voltage-gated Na$^+$ channel protein. Despite its low concentration, the Na$^+$ channel protein has been purified. The use of neurotoxins that bind tightly and specifically to Na$^+$ channels made this purification possible (Figure 21-31).

Tetrodotoxin, for example, is a powerful poison concentrated in the ovaries, liver, skin, and intestines of the puffer fish. A toxin with related properties, saxitoxin, is produced by certain red marine dinoflagellates. These toxins specifically bind to and inhibit the voltage-gated Na$^+$ channels in neurons, preventing action potentials from forming. One molecule of either toxin binds to one Na$^+$ channel with exquisite affinity and selectivity. Measurements of the amount of radioactive tetrodotoxin or saxitoxin that binds to a typical unmyelinated neuron have shown that an axon contains 5–500 Na$^+$ channels per square micrometer of membrane. This agrees with the numbers of channels estimated from patch-clamp studies. The Na$^+$ channels in these membranes are thus spaced, on average, about 200 nm apart.

The Na$^+$ channel protein was purified by affinity chromatography of detergent-extracted membrane proteins on columns of immobilized toxin that specifically bind Na$^+$ channels. Rat brain and the electroplax of the electric eel

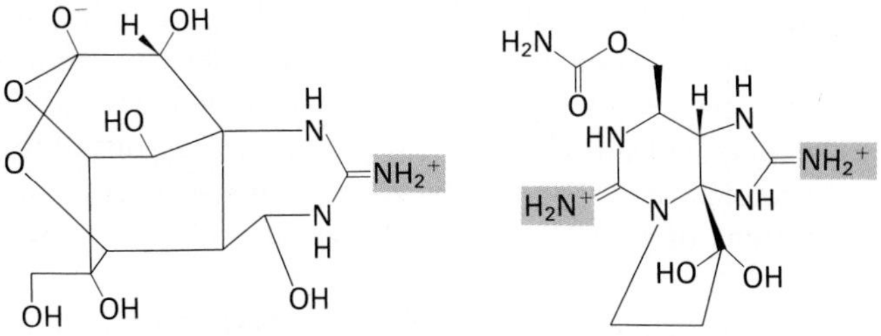

▲ FIGURE 21-31   Structures of two sodium-channel blockers—tetrodotoxin and saxitoxin. The positively charged groups on these neurotoxins may bind to negatively charged carboxylate groups in the Na$^+$ channel protein. The size of these toxins prevents their passage through the channel and blocks the passage of Na$^+$ ions.

are rich and convenient sources of this protein. The major component of the Na$^+$ channels from these sources is a single polypeptide with a molecular weight of 250,000–270,000. Purified Na$^+$ channel protein(s) can be incorporated into artificial lipid membranes, which then exhibit many of the predicted properties of the Na$^+$ channel proteins, such as voltage dependence. From the cloned cDNA, the sequence of the major polypeptide of the Na$^+$ channel has been determined. It contains four homologous transmembrane domains, each similar in sequence and structure to the central part of the *shaker* protein. These domains are connected and flanked by shorter stretches of nonhomologous residues (Figure 21-25a). Other subunits of the Na$^+$ channel are regulatory; they affect the rate at which the channel opens and becomes inactivated, and the voltage dependence of channel inactivation.

All Voltage-Gated Channel Proteins Probably Evolved from a Common Ancestral Gene

The similarities among the voltage-gated Na$^+$, Ca^{2+}, and K$^+$ channels suggest that all three proteins evolved from a common ancestral gene. These similarities include the following:

1. Na$^+$, K$^+$, and Ca^{2+} channels all open when the membrane is depolarized.

2. The voltage-sensing S4 α helices in all three ion channels have a positively charged lysine or arginine every third or fourth residue (see Figure 21-29).

3. The Na$^+$ and Ca^{2+} channel proteins have extensive sequence homology throughout their length. Each contains four transmembrane domains with six α helices and one H5 pore-lining segment per domain. The sequence and transmembrane structure of the much smaller K$^+$ channels, such as the *shaker* protein from *Drosophila*, are similar to those of each domain in the Na$^+$ and Ca^{2+} channels (see Figure 21-25). Each K$^+$ channel comprises four copies of such a polypeptide.

4. All of these channels have an H5 pore-lining sequence in a similar position. The specificity of the ion that passes through the channel is determined by amino acid side chains in the pore-lining regions of the proteins, as was shown by analysis of site-specific mutations of several of these channels.

Voltage-gated K^+ channels have been found in all yeasts and protozoa studied. In contrast, voltage-gated Ca^{2+} channels (important in synaptic transmission, discussed below) are present in only a few of the more complex protozoa, such as *Paramecium,* and only multicellular organisms have voltage-gated Na^+ channels. Thus voltage-gated K^+ channel proteins probably arose first in evolution. The Ca^{2+} and Na^+ channel proteins are believed to have evolved by repeated duplication of an ancestral one-domain K^+ channel gene.

➤ *Synapses and Impulse Transmission*

As noted earlier, synapses are the junctions where neurons pass signals to target cells, which may be other neurons,

muscle cells, or gland cells. In most nerve-to-nerve signaling and all known nerve-to-muscle and nerve-to-gland signaling, the neuron releases chemical neurotransmitters at the synapse that act on the target cell. Much rarer, but simpler in function, are electric synapses in which the action potential is transmitted directly and very rapidly from the presynaptic to the postsynaptic cell.

Impulse Transmission across Electric Synapses Is Nearly Instantaneous

In an electric synapse, ions move directly from one neuron to another via gap junctions (see Figure 21-7). The membrane depolarization associated with an action potential in the presynaptic cell passes through the gap junctions, leading to a depolarization, and thus an action potential, in the postsynaptic cell. Such cells are said to be electrically coupled. Gap junctions are also found in nonneuronal cells where, as discussed in Chapter 24, they enable small molecules such as cAMP and amino acids to pass from cell to cell.

Electric synapses have the advantage of speed; the direct transmission of impulses avoids the delay of about 0.5 ms that is characteristic of chemical synapses (Figure 21-32). In certain circumstances, a fraction of a millisecond can mean the difference between life and death. Elec-

➤ FIGURE 21-32 Transmission of action potentials across electric and chemical synapses. In both cases, the presynaptic neuron was stimulated and the membrane potential was measured in both the presynaptic and postsynaptic cells (see Figure 21-11a). Signals are transmitted across an electric synapse within a few microseconds because ions flow directly from the presynaptic cell to the postsynaptic cell through gap junctions. In contrast, signal transmission across a chemical synapse is delayed about 0.5 ms—the time required for secretion and diffusion of neurotransmitter and the response of the postsynaptic cell to it.

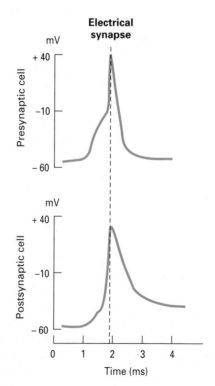

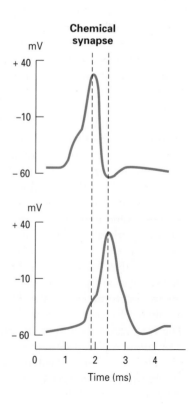

tric synapses in the goldfish brain, for example, mediate a reflex action that flaps the tail, which permits a fish to escape from predators. Examples also exist of electric coupling between groups of cell bodies and dendrites, ensuring simultaneous depolarization of an entire group of coupled cells. The large number of electric synapses in many cold-blooded fishes suggests that they may be an adaptation to low temperatures, as the lowered rate of cellular metabolism in the cold reduces the rate of impulse transmission across chemical synapses.

The efficiency with which an electric signal is transmitted across an electric synapse is proportional to the number of gap junctions that connect the cells. The permeability of the gap junction is regulated by the level of cellular H^+ and Ca^{2+} ions; changes in the concentration of these ions might modulate the efficiency of impulse transmission at electric synapses.

Chemical Synapses Can Be Fast or Slow, Excitatory or Inhibitory, and Can Exhibit Signal Amplification and Computation

In a chemical synapse, neurotransmitters are stored in small membrane-bounded vesicles, called synaptic vesicles, in the axon terminals of the presynaptic cell (see Figure 21-5a and b). As we shall explain in detail, the arrival of an action potential at an axon terminal causes the opening of voltage-gated Ca^{2+} channels. This causes a rise in the cytosolic Ca^{2+} concentration, which triggers exocytosis of the synaptic vesicles and release of neurotransmitter. Synaptic vesicles are uniformly sized organelles, 40–50 nm in diameter, that store one of the "classic" small-molecular-weight neurotransmitters, most of which are depicted in Table 21-1. Except for acetylcholine, they are either amino acids, derivatives of amino acids, or nucleosides or their derivatives such as adenosine and ATP. Each neuron generally contains one type of "classic" neurotransmitter. Synaptic vesicles, which are clustered over the synaptic zone, release their neurotransmitters into the region called the *synaptic cleft* that is adjacent to the postsynaptic cell. The neurotransmitter then binds to specific receptors on the plasma membrane of this cell, causing a change in its permeability to ions.

Many neurons secrete neuropeptides (some of which are listed in Table 21-2) in addition to classic neurotransmitters. The neuropeptides are stored in a different type of vesicle. Exocytosis of neuropeptides is also triggered by a rise in cytosolic Ca^{2+}, but they are released outside the synaptic zone. The effects of the neuropeptide transmitters are very diverse and often long-lived (hours to days), and the following discussions will be confined mainly to the actions of the classic neurotransmitters listed in Table 21-1.

Excitatory and Inhibitory Synapses One way of classifying synapses is whether the action of the neurotransmitter tends to promote or inhibit the generation of an action potential in the postsynaptic cell. Many nerve-nerve and most nerve-muscle chemical synapses are excitatory. The binding of a neurotransmitter to an excitatory receptor changes the permeability of the plasma membrane of the postsynaptic cell, generally by opening Na^+ channels (Figure 21-13), such that the postsynaptic plasma membrane becomes depolarized. This depolarization promotes the generation of an action potential (Figure 21-33a).

The binding of a neurotransmitter to an inhibitory receptor on the postsynaptic cell also causes a localized change in ion permeability—generally by opening K^+ or Cl^- channels (Figure 21-13)—that tends to hyperpolarize the membrane and thus block the generation of an action potential (Figure 21-33b).

Fast Synapses and Ligand-Gated Ion Channels One class of neurotransmitter receptors are ligand-gated ion channels (Figure 21-3). Binding of the neurotransmitter causes an immediate conformational change in the protein that opens the channel and allows ions to cross the membrane. The membrane potential of the postsynaptic cell changes very rapidly—in 0.1–2 milliseconds. Such rapid changes can be excitatory or inhibitory, depending on the neurotransmitter receptor. Certain receptors for acetylcholine—the nicotinic acetylcholine receptors—and for glutamate are ligand-gated (i.e., ligand-opened) cation channels that transduce an excitatory signal (see Figure 21-33a). Binding of the ligand to one of these receptors opens a cation channel that allows passage of both Na^+ and K^+ ions and thus depolarizes the membrane of the postsynaptic cell. In contrast, receptors for the neurotransmitters γ-aminobutyric acid and glycine form ligand-gated Cl^- channels. Binding of the ligand hyperpolarizes the postsynaptic plasma membrane; thus these receptors transduce a rapid inhibitory signal (Table 21-3).

Slow Synapses and Receptors Coupled to G Proteins Many functions of the nervous system operate with time courses of seconds or minutes; regulation of the heart rate, for instance, requires that action of neurotransmitters extend over several beating cycles measured in seconds. In general, the types of neurotransmitter receptors utilized in slow synapses are not ligand-gated ion channels. Rather, these receptors are coupled to G proteins (Table 21-2). The sequence is similar to what happens in nonneuronal cells: Binding of a neurotransmitter to its receptor activates a G protein that, in most cases, directly binds to a separate ion channel protein and causes an increase or decrease in its ion conductance. In other cases, the receptor-activated G protein activates adenylate cyclase or phospholipase C, triggering a rise in cytosolic cAMP or Ca^{2+}, respectively,

TABLE 21-1   Some Small Molecules Identified as Neurotransmitters

Name	Structure	Derivation or Group
Acetylcholine	$CH_3-\overset{\overset{O}{\|\|}}{C}-O-CH_2-CH_2-N^+-(CH_3)_3$	
Glycine	$H_3N^+-CH_2-\overset{\overset{O}{\|\|}}{C}-O^-$	Amino acid
Glutamate	$H_3N^+-CH-CH_2-CH_2-\overset{\overset{O}{\|\|}}{C}-O^-$ with $C=O$, O^- below	Amino acid
Dopamine	(3,4-dihydroxyphenyl)$-CH_2-CH_2-NH_3^+$	Derived from tyrosine
Norepinephrine	(3,4-dihydroxyphenyl)$-CH-CH_2-NH_3^+$, OH	Derived from tyrosine
Epinephrine	(3,4-dihydroxyphenyl)$-CH-CH_2-NH_2^+-CH_3$, OH	Derived from tyrosine
Serotonin (5-hydroxytryptamine)	5-hydroxyindole$-CH_2-CH_2-NH_3^+$	Derived from tryptophan
Histamine	$HC=C-CH_2-CH_2-NH_3^+$ (imidazole ring)	Derived from histidine
γ-Aminobutyric acid (GABA)	$H_3N^+-CH_2-CH_2-CH_2-\overset{\overset{O}{\|\|}}{C}-O^-$	Derived from glutamate

TABLE 21-1   (Continued)

Name	Structure	Derivation or Group
ATP	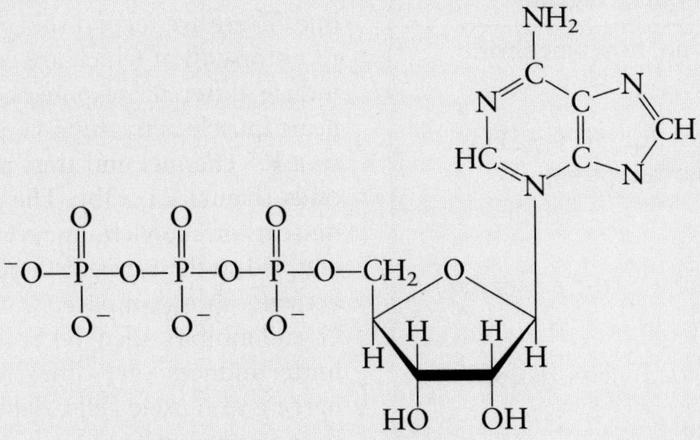	Nucleotide
Adenosine	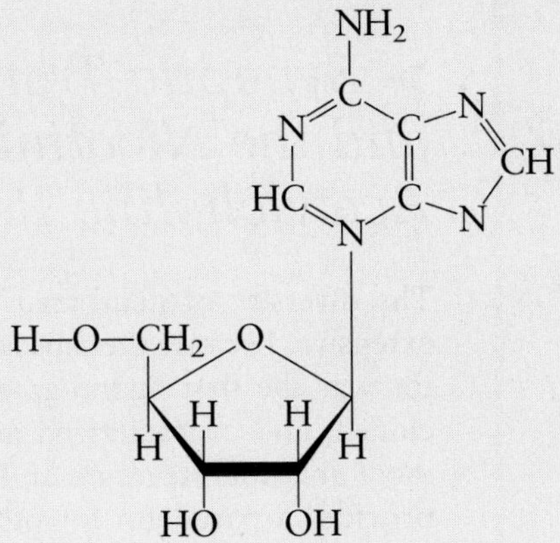	Nucleoside

which in turn affects the ion permeability of a channel protein. G protein-coupled responses are intrinsically slower and longer lasting than those induced by ligand-gated channels; in the heart, for example, the postsynaptic hyperpolarization response to acetylcholine lasts several seconds (Figure 21-33b).

Signal Amplification and Computation Chemical synapses have two important advantages over electric ones. The first is signal amplification, which is common at nerve-muscle synapses. An action potential in a single presynaptic motor neuron can cause contraction of multiple muscle cells because release of relatively few signaling molecules at a synapse is all that is required to stimulate contraction. The second advantage is signal computation, which is common at synapses involving interneurons, especially in the central nervous system. Many interneurons can receive signals at multiple excitatory and inhibitory synapses. Some of the resultant changes in ion permeability

are short-lived—a millisecond or less; others may last several seconds. Whether an action potential is generated depends upon a complex function of all of the incoming signals; this signal computation differs for each type of interneuron.

Many Types of Receptors Bind the Same Neurotransmitter

The diversity of receptors for and responses to a single kind of neurotransmitter is illustrated by acetylcholine. Synapses in which acetylcholine is the neurotransmitter are termed *cholinergic* synapses. The type of acetylcholine receptor that causes excitatory responses lasting only milliseconds are called *nicotinic acetylcholine receptors*. They are so named because nicotine is an agonist of this type of receptors—like acetylcholine, nicotine causes a rapid depolarization (see Figure 21-33a). As noted already, these

TABLE 21-2   Some Neurotransmitter Receptors That Are Coupled to G Proteins

Receptor Class	Typical Ligands
Small neurotransmitters	Epinephrine, norepinephrine Dopamine
	Serotonin* (for 5HT$_1$, 5HT$_2$, 5HT$_4$ class receptors)
	Histamine
	Acetylcholine* (for muscarinic receptors)
	GABA* (for B-class receptors)
	Glutamate*
	ATP
	Adenosine
Neuropeptides	Opiods (e.g. β-endorphin)
	Tachykinins (e.g. substance P)
	Bradykinin
	Luteinizing hormone–releasing hormone (LHRH)
	Thyrotropin-releasing hormone (TRH)
	Vasoactive intestinal peptide (VIP)
	Adrenocorticotropic hormone (ACTH)
	Cholcystokinin (CCK)
	Gastrin
	Endothelin
	Vasopressin
	Oxytocin
Environmental signals	Light (receptor is the visual pigment rhodopsin)
	Odorants (hundreds of different receptors)

*Ligands with an asterisk are also known to have other receptors that are ligand-gated ion channels, as listed in Table 21-3.
SOURCE: B. Hille, 1992, *Neuron* 9:187–195.

receptors are ligand-gated channels for Na$^+$ and K$^+$ ions. The other types of acetycholine receptors are called *muscarinic acetylcholine receptors* because muscarine (a mushroom alkaloid) causes the same response as does acetylcholine. There are several subtypes of muscarinic acetylcholine receptors, all of which are coupled to G proteins but which induce different responses. The M2 receptor present in heart muscle activates a G$_i$ protein that causes the opening of a K$^+$ channel and thus a hyperpolarization lasting seconds (Figure 21-33b). The M1, M3, and M5 subtypes of muscarinic acetylcholine receptors, found in other cells, are coupled to the types of G proteins known as G$_o$ or G$_q$ and activate phospholipase C, while the M4 subtype activates G$_i$ and inhibits adenylyl cyclase. Thus, a single neurotransmitter induces very different responses in different target nerve and muscle cells, depending on the type of receptor found in the cells.

► Synaptic Transmission and the Nicotinic Acetylcholine Receptor

The diversity of neurotransmitters and their receptors is extensive. Because the muscle nicotinic acetylcholine receptor was the first ligand-gated ion channel to be purified, cloned, and characterized at a molecular level, we know more about its structure and mechanism than any other. It provides a paradigm for other ligand-gated ion channels. Thus in this section we discuss in detail the muscle nicotinic acetylcholine receptor and the events at synapses containing this receptor—how synaptic vesicles align with the plasma membrane, how exocytosis of synaptic vesicles is triggered by opening of voltage-gated calcium channels and a rise in cytosolic calcium, and how hydrolysis of acetylcholine terminates the depolarization signal. In the following sections, we briefly consider the properties of other neurotransmitters, their receptors, and their transporters.

Acetylcholine Is Synthesized in the Cytosol and Stored in Synaptic Vesicles

The membrane-bounded vesicles that store acetylcholine are about 40 nm in diameter and accumulate, often in rows, in presynaptic axon terminals near the plasma membrane (Figures 21-34 and 21-35). A single axon terminus of a frog motor neuron may contain a million or more synaptic vesicles, each containing 1000–10,000 molecules of acetylcholine; such a neuron might form synapses with a single skeletal muscle cell at several hundred points.

(a) Excitatory synapse

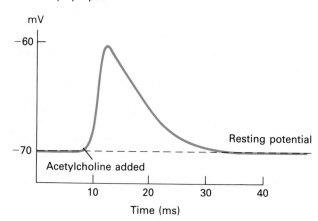

(b) Inhibitory synapse

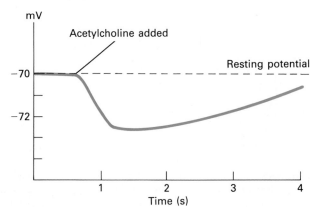

◄ FIGURE 21-33 Excitatory and inhibitory responses in postsynaptic cells. (a) An excitatory response: application of acetylcholine (or nicotine) to frog skeletal muscle produces a rapid postsynaptic depolarization of about 10 mV, which lasts 20 ms. These cells contain nicotinic acetylcholine receptors, a ligand-gated Na^+ and K^+ channel protein. (b) An inhibitory response: application of acetylcholine (or muscarine) to frog heart muscle produces, after a lag period of about 40 ms (not visible in graph), a hyperpolarization of 2–3 mV, which lasts several seconds. These cells contain muscarinic acetylcholine receptors. Thus depending on the type of receptor present in the postsynaptic cell, acetylcholine can either increase muscle contraction (skeletal muscle) or decrease it (cardiac muscle). Note the difference in time scales in the two graphs. [See H. C. Hartzell, 1981, *Nature* **291**:539.]

TABLE 21-3 Neurotransmitter Receptors That Are Ligand-Gated Ion Channels

Functional Type	Ligand[†]	Ion Channel
Excitatory Receptors*	Acetylcholine (for nicotinic receptor)	Na^+/K^+
	Glutamate (for NMDA class receptors)[‡]	Na^+/K^+ and Ca^{2+}
	Glutamate (for non-NMDA class receptors)[§]	Na^+/K^+
	Serotonin (for $5HT_3$ class receptors)	Na^+/K^+
Inhibitory Receptors*	GABA (γ-aminobutyric acid) (for A-class receptors)	Cl^-
	Glycine	Cl^-

* All of these receptor proteins have a very similar structure, as depicted for the nicotinic receptor for acetycholine in Figures 21-39 and 21-40.

† Most of these ligands also bind to receptors that are coupled to G proteins (Table 21-2); thus it is important to define the class of receptors for each ligand.

‡ Glutamate receptors of this class are selectively activated by the nonnatural amino acid N-methyl-D aspartate (NMDA); receptors of this class require both glutamate (or NMDA) and a slightly depolarized membrane in order to open the ion channel. These receptors are thought to be important in long-term potentiation, a form of memory (Figure 21-56).

§ For glutamate receptors of this class, opening of the cation channel only requires binding of glutamate.

SOURCE: H. Lester, 1988, *Science* **241**:1057; E. Barnard, 1988, *Nature* **335**:301; N. Unwin, 1993, *Cell* vol. 72/*Neuron* vol. 10 (Suppl), pp. 31–41; A. Maricq, A. Peterson, A. Brake, R. Myers, and D. Julius, 1991, *Science* **254**:432.

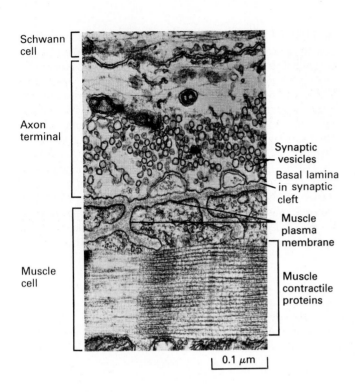

Schwann cell

Axon terminal

Synaptic vesicles

Basal lamina in synaptic cleft

Muscle plasma membrane

Muscle cell

Muscle contractile proteins

0.1 μm

◄ FIGURE 21-34   Longitudinal section through a frog nerve-muscle synapse. The plasma membrane of the muscle cell is extensively folded. The basal lamina lies in the synaptic cleft separating the neuron from the muscle membrane. The axon terminal is surrounded by a Schwann cell, which periodically interdigitates between the terminal and the muscle. Acetylcholine receptors are concentrated in the postsynaptic muscle membrane at the top and part way down the sides of the folds in the membrane. Synaptic vesicles in the axon terminal are clustered near specializations lying just inside the presynaptic plasma membrane where exocytosis occurs. From J. E. Heuser and T. Reese, 1977, in E. R. Kandel, ed., *The Nervous System*, vol. 1, *Handbook of Physiology*, Williams & Wilkins, p. 266.]

Acetylcholine is synthesized in the cytosol of axon terminals from acetyl coenzyme A (CoA) and choline by the enzyme choline acetyltransferase:

$$CH_3-\overset{\overset{\displaystyle O}{\|}}{C}-S-CoA$$
Acetyl CoA

$$+ \; HO-CH_2-CH_2-N^+-(CH_3)_3 \; \xrightarrow{\text{Choline acetyltransferase}}$$
Choline

$$CH_3-\overset{\overset{\displaystyle O}{\|}}{C}-O-CH_2-CH_2-N^+-(CH_3)_3 + CoA-SH$$
Acetylcholine

The synaptic vesicles take up and concentrate acetylcholine from the cytosol against a steep concentration gradient, using a proton-acetylcholine antiport transporter in the vesicle membrane (Figure 21-36).

Exocytosis of Synaptic Vesicles Is Triggered by the Opening of Voltage-Gated Calcium Channels and a Rise in Cytosolic Calcium

The arrival of an action potential in the axon terminal depolarizes the plasma membrane, opening voltage-gated Ca^{2+} channels and thus causing an influx of Ca^{2+} ions into the cytosol from the extracellular medium. The amount of Ca^{2+} that enters an axon terminal through voltage-gated Ca^{2+} channels is sufficient to raise the level of Ca^{2+} from $<0.1 \; \mu M$, characteristic of the cytosol in the resting state, to several micromolar in the region of the cytosol near the synaptic vesicles. The Ca^{2+} ions bind to proteins that connect the synaptic vesicle with the plasma membrane, inducing membrane fusion and thus exocytosis of the neurotransmitter into the synaptic cleft. The extra Ca^{2+} ions are

► FIGURE 21-36   Release of neurotransmitters and the endocytic cycle of synaptic vesicles. Vesicles import neurotransmitters from the cytosol (step 1) using a proton-neurotransmitter antiport. The low intravesicular pH powers neurotransmitter import, and is generated by a V-type ATPase in the vesicle membrane. The vesicles then move to the "active zone" near the plasma membrane (step 2) where they dock at defined sites (step 3). Exocytosis of the docked vesicles (step 4) is triggered by the rise in cytosolic Ca^{2+}, with release of neurotransmitters into the synaptic cleft. The synaptic vesicle membrane proteins are then specifically recovered by endocytosis in clathrin-coated vesicles (step 5) and move away from the active zone (step 6). Non-clathrin-coated vesicles (not shown) may also be used for endocytosis. The clathrin coat is depolymerized, and the vesicles fuse with larger, low pH, endosomes (step 7). New synaptic vesicles then form (step 8) by budding from these endosomes and then are filled with neurotransmitters (step 1), completing the cycle. [From T. Südhof and R. Jahn, 1991, *Neuron* **6**:665–677.]

(a)

(b)

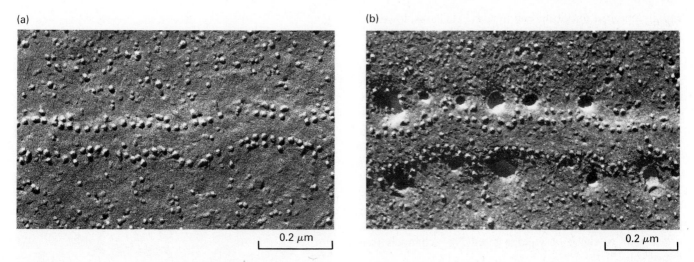

0.2 µm

0.2 µm

▲ FIGURE 21-35   Freeze-fracture image of the external face of the axonal plasma membrane at a neuron-muscle synapse, viewed from the vantage point of the muscle. (a) In the resting state, the membrane contains rows of particles that are aligned with rows of synaptic vesicles. The function of these particles is not known, but it is suspected that they are voltage-gated Ca^{2+} channels. (b) During stimulation, large pits in the membrane, resulting from exocytosis of synaptic vesicles, are visible near the rows of particles. [From J. E. Heuser and T. Reese, 1977, in E. R. Kandel, ed., *The Nervous System,* vol. 1, *Handbook of Physiology,* Williams & Wilkins, p. 268.]

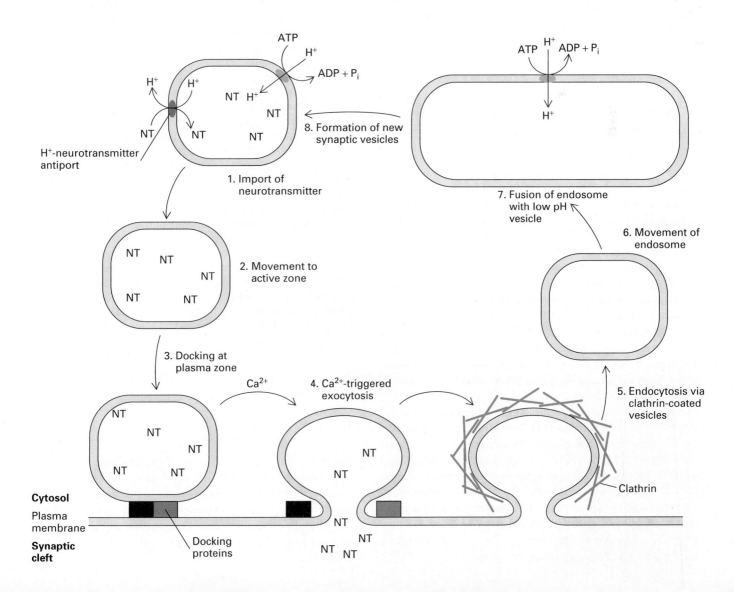

rapidly pumped out of the cell by Ca^{2+} ATPases, lowering the cytosolic Ca^{2+} level and preparing the terminal to respond again to an action potential.

The presence of voltage-gated Ca^{2+} channels in axon terminals has been demonstrated in neurons treated with drugs that block Na^+ channels and thus prevent conduction of action potentials. When the membrane of axon terminals in such treated cells is artificially depolarized, an influx of Ca^{2+} ions into the neurons occurs and exocytosis is triggered. Patch-clamping experiments show that voltage-gated Ca^{2+} channels, like voltage-gated Na^+ channels, open transiently upon depolarization of the membrane. Several voltage-gated Ca^{2+} channels have been purified and cloned. Their sequences and structures are similar to those of voltage-gated Na^+ channels (see Figure 21-25).

More generally, voltage-gated Ca^{2+} channels can be viewed as the transducer of the electric signal in a nerve membrane; that is, the action potential is transduced into a chemical signal. Depolarization of the plasma membrane cannot, in itself, cause synaptic vesicles to fuse with the plasma membrane. The electric signal must be converted into a rise in Ca^{2+} concentration, which, in turn, induces synaptic vesicle fusion.

Multiple Proteins Align Synaptic Vesicles with the Plasma Membrane and Participate in Vesicle Exocytosis and Endocytosis

There are two pools of synaptic vesicles, those docked at the plasma membrane that can be readily exocytosed and those that provide a reservoir near the plasma membrane in the active zone. The triggering of vesicle exocytosis by a rise in Ca^{2+} is but one of a series of steps involved in forming synaptic vesicles, filling them with neurotransmitter, moving them to the active zone near the plasma membrane, docking them at the plasma membrane, and then, after vesicle fusion with the plasma membrane, recycling their membrane components by endocytosis (Figure 21-36). Recycling of synaptic vesicle membrane proteins is rapid, as indicated by the ability of many neurons to fire fifty times a second, and quite specific, in that several membrane proteins unique to the synaptic vesicles are specifically internalized by endocytosis.

The axon terminal exhibits a highly organized arrangement of cytoskeletal fibers that appear essential for localizing the vesicles to the plasma membrane at the synaptic cleft (Figure 21-37a). The vesicles themselves are linked together by the fibrous phosphoprotein synapsin (Figure 21-37b). Synapsin is localized in the cytosolic surface of all synaptic vesicle membranes and constitutes 6 percent of vesicle proteins; it is related in structure to band 4.1 pro-

(a)

Axon terminal   Synaptic vesicles

Postsynaptic cell   0.1 μm

▲ FIGURE 21-37   (a) Micrograph of an axon terminal obtained by the rapid-freezing deep-etch technique. (b) *Facing page:* Structure and function of synaptic vesicle proteins and their interactions with proteins in the plasma membrane of the axon terminal. Synapsins interconnect synaptic vesicles and also connect vesicles to spectrinlike filaments extending inward from the plasma membrane. At least two vesicle proteins, synaptobrevin and synaptotagmin, are important for docking the vesicle to the underside of the plasma membrane; a complex of proteins including NSF, α-SNAP, γ-SNAP, and SNAP-25 binds synaptobrevin to the plasma membrane protein syntaxin, as detailed in (c). Neurexin, Ca^{2+} channels, and other plasma membrane proteins localized to the synaptic region also interact with synaptic vesicle proteins. Synaptophysin contains four membrane-spanning α helices and may form part of the fusion pore between the synaptic vesicle and plasma membrane. Rab3A is important for localizing the synaptic vesicles to the fusion zone, but the proteins with which it interacts are not known. The synaptic vesicle membrane also contains transporters for specific neurotransmitters and, not shown here, V-type proton ATPases. [Part (a) from D. M. D. Landis et al., 1988, *Neuron* **1**:201. Parts (b) and (c) after T. M. Jessell and E. R. Kandel, 1993, *Cell* vol. 72/*Neuron* vol. 10 (Suppl), pp 1–30 and T. Söllner et al., 1993, *Nature* **362**:318–324.]

tein in erythrocyte membranes, a protein that binds both actin and spectrin (see Figure 14-38). Thicker filaments composed of a spectrinlike protein radiate from the plasma membrane and bind to vesicle-associated synapsin. Probably these interactions keep the synaptic vesicles close to

(b)

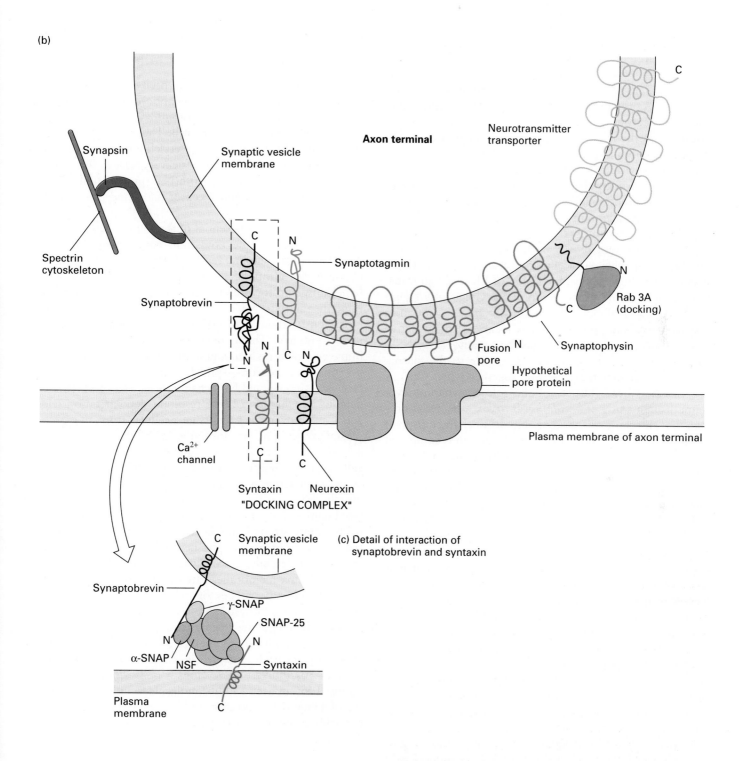

(c) Detail of interaction of
    synaptobrevin and syntaxin

the part of the plasma membrane facing the synapse. Synapsins are substrates of cAMP-dependent and calcium-calmodulin-dependent protein kinases, and the rise in cytosolic Ca^{2+} triggers their phosphorylation. This apparently causes the release of synaptic vesicles from the cytoskeleton and increases the number of vesicles available for fusion with the plasma membrane.

Synaptic vesicles contain several proteins important for their function. Six synaptophysin polypeptides

(MW 38,000) form a complex in the synaptic vesicle membrane. This complex may be part of the fusion pore between the synaptic vesicle and plasma membrane together with as yet unidentified proteins in the plasma membrane. V-type ATPases (see Figure 15-10) generate a low intravesicular pH, and proton-coupled antiporters for neurotransmitters catalyze import of neurotransmitters from the cytosol where they are synthesized (Figure 21-36). Rab3A is a neuron-specific GTP-binding protein similar in sequence and function to the rab proteins that control vesicular traffic in the secretory pathway (see Figure 16-43). It appears to be essential for localization of the synaptic vesicles to the active zone, as indicated by an experiment in which a mutant Rab3A protein that cannot bind GTP was expressed in cultured neurons: the synaptic vesicles became randomly localized in the axon terminals. Synaptotagmin binds the plasma membrane protein neurexin (Figure 21-37b) and may also participate in vesicle docking.

Two experimental observations established that synaptobrevin, another abundant synaptic vesicle protein, is essential for interaction of the vesicle with the plasma membrane. First, synaptobrevin binds a protein complex containing NSF and SNAPs, proteins that are also required for fusion of other intracellular vesicles with their target membranes (for instance, for fusion of ER-to-Golgi transport vesicles with the *cis* region of the Golgi—see Figure 16-42c). The NSF-SNAP complex also binds syntaxin, a plasma membrane protein that is localized to the active zone of axon terminals, and thus the synaptobrevin-NSF-syntaxin complex may anchor the vesicle to the plasma membrane so that they can fuse together when Ca^{2+} is increased. Other evidence for the role of synaptobrevin came from a study of *botulinum-B toxin,* a bacterial protein that can cause the paralysis and death characteristic of botulism, a type of food poisoning. The toxin is composed of two polypeptides, one of which binds to cholinergic motor neurons and facilitates the entry of the other into the cytosol. The second polypeptide is a protease with a remarkable specificity—the only protein it cleaves is synaptobrevin, and by destroying synaptobrevin it blocks acetylcholine release at the neuromuscular synapse.

The synaptic vesicle protein synaptotagmin contains two binding sites for Ca^{2+} in its cytosolic domain; it is thought to be the key Ca^{2+} sensing protein that triggers vesicle exocytosis. In one set of experiments, injection of an antibody specific for the cytosolic domain of synaptotagmin into a neuron inhibited Ca^{2+}-stimulated vesicle exocytosis. Other experiments used a series of synaptotagmin mutants of *Drosophila.* Embryos that completely lacked synaptotagmin failed to hatch and exhibited very reduced, uncoordinated muscle contractions. Larvae with partial lack-of-function mutations of synaptotagmin survived, but their neurons were defective in Ca^{2+}-stimulated vesicle exocytosis. How synaptotagmin functions is beginning to be understood. In resting cells (with low cytosolic Ca^{2+}) synaptotagmin apparently binds to the complex of synaptobrevin and syntaxin; the presence of synaptotagmin blocks the binding of the fusion protein α-SNAP to the complex and thus prevents vesicle fusion. When synaptotagmin binds Ca^{2+} it is displaced from the complex, allowing α-SNAP to bind and thus initiating membrane fusion. Thus synaptotagmin operates as a "clamp" to prevent fusion from proceeding in the absence of a Ca^{2+} signal.

The Nicotinic Acetylcholine Receptor Protein Is a Ligand-Gated Cation Channel

Once acetylcholine molecules have been released from neurons, they diffuse across the synaptic cleft and combine with *nicotinic acetylcholine receptor* molecules in the membrane of the postsynaptic cell. The interaction of acetylcholine with the nicotinic acetylcholine receptor produces within 0.1 ms a large transient increase in the permeability of the membrane to both Na^+ and K^+ ions. Thus the nicotinic acetylcholine receptor is a ligand-gated cation channel.

Since the resting potential of the muscle plasma membrane is near E_K, the potassium equilibrium potential, opening of acetylcholine receptor channels causes little increase in the efflux of K^+ ions; Na^+ ions, on the other hand, flow into the muscle cell. The simultaneous increase in permeability to Na^+ and K^+ ions produces a net depolarization to about -15 mV from the resting potential of about -60 mV. This depolarization of the muscle membrane generates an action potential, which—like an action potential in a neuron—spreads along the membrane surface via voltage-gated Na^+ channels (see Figure 21-18). As detailed in Chapter 22, the resultant depolarization triggers Ca^{2+} movement from its intracellular store, the sarcoplasmic reticulum, into the cytosol, and then contraction.

The snake venom toxin α-bungarotoxin binds specifically and irreversibly to nicotinic acetylcholine receptors. When radioactive α-bungarotoxin is added to sections of muscle cells, the toxin—as detected by autoradiography (see Figure 6-24)—becomes localized in the membranes of postsynaptic striated muscle cells immediately adjacent to the terminals of presynaptic neurons. Thus, the nicotinic acetylcholine receptors are clustered in this domain of the muscle plasma membrane. Patch-clamping studies on isolated outside-out patches of muscle plasma membranes (Figure 21-23c), showed that, in response to acetylcholine, a cation channel in the receptor opened and remained open for several milliseconds before closing spontaneously. When open, each channel is capable of transmitting 15,000–30,000 Na^+ or K^+ ions a millisecond. The time required for an acetylcholine molecule to induce the opening of a channel is too short to be measured directly but is probably a few microseconds.

Spontaneous Exocytosis of Synaptic Vesicles Produces Small Depolarizations in the Postsynaptic Membrane

Careful monitoring of the membrane potential of the muscle membrane at a synapse with a cholinergic motor neuron has demonstrated spontaneous, intermittent depolarizations of about 0.5–1.0 mV in the absence of stimulation

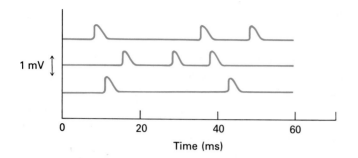

1 mV

0   20   40   60

Time (ms)

▶ FIGURE 21-38 Spontaneous depolarizations in the postsynaptic membrane of a cholinergic synapse. Recordings were taken from intracellular electrodes in an unstimulated frog muscle cell near the neuron-muscle synapse (see Figure 21-11a). Such depolarizations—each generated by the spontaneous release of acetylcholine from a single synaptic vesicle—seem to occur at random intervals.

of the motor neuron (Figure 21-38). Each of these depolarizations is caused by the spontaneous release of acetylcholine from a single synaptic vesicle. Indeed, demonstration of the spontaneous small depolarizations led to the notion of quantal release of acetylcholine (later applied to other neurotransmitters) and thereby led to the hypothesis of vesicle exocytosis at synapses. The release of one vesicle of acetylcholine results in the opening of about 3000 ion channels in the postsynaptic membrane—sufficient to depolarize the region of membrane by about 1 mV but insufficient to reach the threshold for an action potential.

The Nicotinic Acetylcholine Receptor Contains Five Subunits, Each of Which Contributes to the Cation Channel

In most nerve and muscle tissues the nicotinic acetylcholine receptor constitutes a minute fraction of the total membrane protein. Electric eel and electric ray (torpedo) electric organs are particularly rich in this receptor and are the sources used for its purification. Computer-generated averaging of many high-resolution electron microscopic images of purified acetylcholine receptors, viewed from several angles, has led to the pentameric model shown in Figure 21-39a. Each molecule has a diameter of about 9 nm and protrudes from the membrane surfaces about 6 nm into the extracellular space and about 2 nm into the cytosol. A tapered central pore, with a maximum diameter of 2.5 nm, is the cation channel. The channel opens when the receptor cooperatively binds two acetylcholine molecules.

The nicotinic acetylcholine receptor can be solubilized from these membranes by nonionic detergents. The crucial step in its purification is affinity chromatography on columns of immobilized cobra toxin, which binds the receptor but from which it can be subsequently eluted and recovered. The molecular weight of the receptor protein is 250,000–270,000. The receptor consists of four different polypeptides with the composition $\alpha_2\beta\gamma\delta$.

Messenger RNAs encoding all four receptor subunits have been cloned. When all four are microinjected into a single frog oocyte (see Figure 21-27), functional nicotinic acetylcholine receptors form. No channels, or poorly functional ones, are obtained if the mRNA for one subunit is omitted. Thus all four subunit polypeptides—and only these—are needed for receptor function.

The α, β, γ, and δ subunits have considerable sequence homology; on average, about 35–40 percent of the residues in any two subunits are homologous. Each subunit is thought to contain one transmembrane α helix, called M2; the bulk of the receptor consists of multiple β strands. The complete receptor has a fivefold symmetry, and the actual cation channel is thought to be formed by segments from each of the five subunits (Figures 21-39 and 21-40). Studies measuring the permeability of different small cations have suggested that the ion channel is, at its narrowest, about 0.65–0.80 nm in diameter, sufficient to allow passage of both Na^+ and K^+ ions with their tightly bound shell of water molecules (see Figure 2-12).

The structure of the channel is not known in molecular detail. However, much evidence indicates that the channel is lined by five transmembrane M2 α helices, one from each of the five subunits. One reason for thinking that M2 helices form the channel is that certain positively charged organic molecules, such as chlorpromazine, inhibit receptor function by "plugging" the ion channel. Chlorpromazine can be chemically cross-linked to serine residues (number 254 in the β subunit, 262 in δ) in the middle of the M2 helices (Figure 21-39b).

A second line of evidence for the role of M2 helices comes from expression of mutant receptor subunits in frog oocytes. Amino acid residues with negatively charged side chains (glutamate, aspartate) are on both sides of the membrane-spanning M2 helices. If just one of these residues in one subunit is mutated to a lysine, and the mutant mRNA is injected together with mRNAs for the other three wild-type subunits, a functional channel forms but its ion conductivity—the number of ions that can cross it during its open state—is reduced. The greater the number of glutamate or aspartate residues mutated (in one or another subunit), the greater the reduction in conductivity. It is

(a)

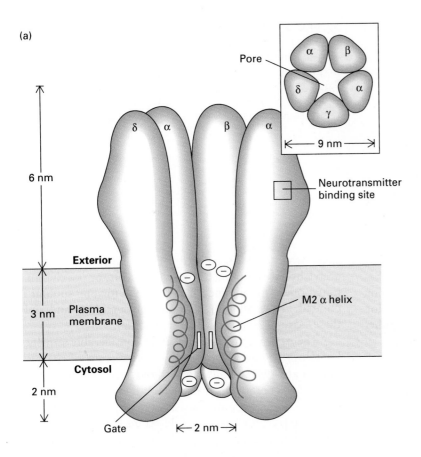

(b) Sequence of M2 helices

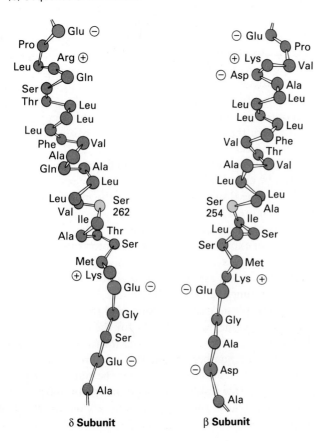

δ Subunit   β Subunit

▲ **FIGURE 21-39** *Proposed structure of the nicotinic acetylcholine receptor.* (a) Schematic model of the pentameric receptor, which contains four different polypeptides and has the composition $\alpha_2\beta\gamma\delta$; for clarity, the γ subunit is not shown. This model is based on amino acid sequence data, information from site-specific mutations, and analysis of electron micrographs. Most of the mass of the protein protrudes from the outer (synaptic) surface of the plasma membrane. Each α subunit contains an acetylcholine binding site and most of the mass of the receptor is composed of β sheets. The M2 α helix (red) in each subunit is part of the lining of the ion channel. The gate, which is opened by binding of the acetylcholine neurotransmitter, lies within the pore. Inset: Cross section of the exoplasmic face of the receptor showing the arrangement of subunits around the central pore. At its narrowest (in the membrane), the ion channel is about 0.65–0.80 nm in diameter. (b) Amino acid sequences of the M2 helix in the β subunit and δ subunit. Negatively charged glutamate or aspartate residues (blue) are present at both ends of the M2 helices, on either side of the pore. They help to screen out anions, and assist in binding Na$^+$ and K$^+$ ions in the channel. Cross-linking of chlorpromazine to serine residues (yellow) in the middle of the M2 helices inhibits receptor function, indicating that these serines are in the pore. [See N. Unwin, 1993, *Cell* vol. 72/*Neuron* vol. 10 (Suppl.), pp. 31–41.]

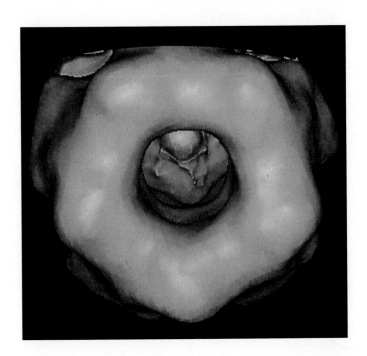

◄ FIGURE 21-40   A view of the three-dimensional structure of the nicotinic acetylcholine receptor from above, looking into the synaptic entrance of the channel. The tunnel made by the synaptic entrance narrows abruptly after a length of about 6 nm, exposing the surfaces proposed by the model in Figure 21-39b to be the negatively charged rings at the ends of the M2 α-helical segments. [From N. Unwin, 1993, *Cell* vol. 72/*Neuron* vol. 10 (Suppl.), pp. 31–41.]

thought that the aspartate and glutamate residues form rings—one residue from each of the five chains—on either side of the channel and participate in attracting Na^+ or K^+ ions as they enter the pore.

The most dramatic evidence came from study of the $\alpha 7$ subunit of a neuronal nicotinic acetylcholine receptor. In neurons, it probably forms a receptor of the composition $(\alpha 7)_2 \beta_3$. However, in *Xenopus* oocytes the $\alpha 7$ mRNA alone directs the expression of a functional acetylcholine-gated cation channel, presumably of composition $(\alpha 7)_5$. The α polypeptide of the GABA receptor is similar in sequence to the acetylcholine receptor α polypeptides; however, the GABA receptor is selective for anions, such as Cl^- (Table 21-3). Replacement of only three amino acids in the M2 helix of the $\alpha 7$ acetylcholine receptor by the analogous amino acids in the GABA receptor converted the protein from an acetylcholine-gated *cation* channel to an acetylcholine-gated *anion* channel. Since the specificity of the channel for anions or cations is determined by the amino acid side chains in the M2 α helix, it is reasonable to conclude that the M2 helices line the ion channel.

Hydrolysis of Acetylcholine Terminates the Depolarization Signal

To restore a synapse to its resting state, the neurotransmitter must be removed or destroyed. There are three main ways to end the signaling: (1) the transmitter may diffuse away from the synaptic cleft; (2) the transmitter may be taken up by the presynaptic neuron; and (3) the transmitter may be enzymatically degraded. Signaling by acetylcholine is terminated by enzymatic degradation of the transmitter, but uptake is used to terminate signaling by most other neurotransmitters.

Acetylcholine is hydrolyzed to acetate and choline by the enzyme *acetylcholinesterase*, which is localized in the synaptic cleft between the neuron and muscle cell membranes. There are several forms of the enzyme (Figure 21-41a), one of which is bound to a network of collagen forming the basal lamina that fills this space (see Figure 21-34).

During hydrolysis of acetylcholine by acetylcholinesterase, a serine at the active site reacts with the acetyl group forming an enzyme-bound intermediate (Figure 21-41b). A large number of nerve gases and other neurotoxins inhibit the activity of acetylcholinesterase by reacting with the active-site serine. Physiologically, these toxins prolong the action of acetylcholine, thus prolonging the period of membrane depolarization. Such inhibitors can be lethal if they prevent relaxation of the muscles necessary for breathing.

► Functions of Other Neurotransmitters, Their Receptors, and Their Transporters

Because it causes a rapid, short-lived, and dramatic excitatory response in skeletal muscle cells, the nicotinic acetylcholine receptor was one of the first neurotransmitter re-

(a)

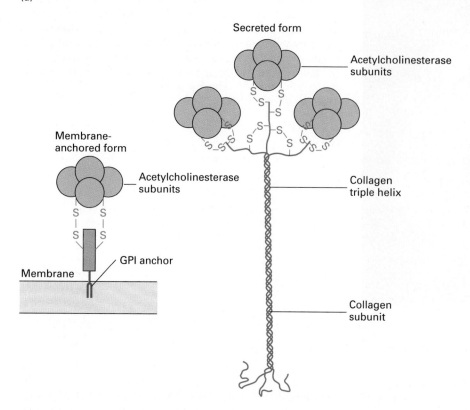

Secreted form

Acetylcholinesterase
subunits

Membrane-
anchored form

Acetylcholinesterase
subunits

Collagen
triple helix

GPI anchor

Membrane

Collagen
subunit

(b) Enzyme—Ser—OH

$$H_3C—\overset{\overset{\displaystyle O}{\|}}{C}—O—CH_2—CH_2—N^+—(CH_3)_3$$

$$HO—CH_2—CH_2—N^+—(CH_3)_3$$

$$Enzyme—Ser—O—\overset{\overset{\displaystyle O}{\|}}{C}—CH_3$$

$$H_2O$$

$$H^+$$

$$Enzyme—Ser—OH + H_3C—\overset{\overset{\displaystyle O}{\|}}{C}\overset{\diagdown}{_{O^-}}$$

▲ FIGURE 21-41   (a) Alternative forms of acetylcholines-
terases. The membrane-anchored form contains one catalytic
domain consisting of four enzymatically active polypeptides,
linked through disulfide bonds to a subunit that is inserted in
the plasma membrane by a glycosylphosphatidylinositol
anchor (see Figure 14-20). In the secreted form three catalytic domains are connected, by disulfide bonds, to a three-
stranded collagenlike subunit that anchors the enzyme to
components of the extracellular matrix in the synaptic cleft.
(b) Mechanism of action of acetylcholinesterase, showing
that acetylserine is an essential enzyme-bound intermediate.
[Adapted from P. Taylor et al., 1987, *Trends Neurosci.* **10**:93
and P. Taylor, 1991. *J. Biol. Chem.* **266**:4025–4028.]

ceptors to be characterized definitively. Many other
receptors are now understood in molecular detail; our
study of these begins with several that trigger inhibitory,
rather than excitatory, responses in postsynaptic cells.

GABA and Glycine Receptors Are Ligand-Gated Anion Channels Used at Many Inhibitory Synapses

Synaptic inhibition in the vertebrate central nervous system
is mediated primarily by two amino acids, glycine and

γ-aminobutyric acid (GABA); the latter is formed from glutamate by loss of a carboxyl group. The concentration of
GABA in the human brain is 200—1000 times higher than
that of other neurotransmitters such as dopamine, norepinephrine, and acetylcholine. Glycine is the major inhibitory neurotransmitter in the spinal cord and brain stem;
GABA predominates elsewhere in the brain. Both glycine
and GABA activate ligand-gated Cl^- channels.

The opening of Cl^- channels tends to drive the membrane potential toward the Cl^- equilibrium potential E_{Cl},
which in general is slightly more negative than the resting

membrane potential (see Figure 21-13). In other words, the membrane becomes slightly hyperpolarized. If many Cl^- channels are opened, the membrane potential will be held near E_{Cl}, and a much larger than normal increase in the sodium permeability will then be rquired to depolarize the membrane. The effect of GABA or glycine on Cl^- permeability is induced rapidly (a fraction of a millisecond) but can last for a second or more, a long time compared with the millisecond required to generate an action potential. Thus GABA or glycine induces a slow, inhibitory postsynaptic response.

GABA and glycine receptors have been purified, cloned, and sequenced. In overall structure and sequence both resemble the nicotinic acetylcholine receptors (see Table 21-3). All are pentamers of similar subunits, although functional GABA and glycine receptors are built of only one or two different types of subunits. As noted already, the M2 helices of nicotinic acetylcholine receptors are thought to line the ion channel and to discriminate cations from anions. The negatively charged glutamate and aspartate side chains at the ends of the M2 helices in acetylcholine receptors may participate in cation binding (see Figure 21-39). Strikingly, the M2 helices of the GABA and glycine receptor subunits have lysine or arginine residues at these positions; positively charged side chains of these residues may attract Cl^- ions specifically.

Cardiac Muscarinic Acetylcholine Receptors Activate a G Protein and Open Potassium Channels

The response of skeletal muscle cells to the release of acetylcholine at the neuron-muscle junction is very rapid—the permeability changes and resulting membrane depolarization are completed within a few milliseconds (see Figure 21-33a). But many other synapses do not work as rapidly, and many functions of the nervous system operate with time courses of seconds or minutes. Responses that begin after a lag period of milliseconds following addition of transmitter are called slow postsynaptic potentials. Binding of acetylcholine to muscarinic acetylcholine receptors in frog aorta cardiac muscle generates a slow inhibitory response. Stimulation of the cholinergic nerves in heart muscle causes a long-lived (several seconds) hyperpolarization of the membrane (see Figure 21-33b) and slows the rate of heart muscle contraction.

Figure 21-42 shows how activation of the cardiac muscarinic receptor leads to opening of K^+ channels and subsequent hyperpolarization of the plasma membrane. Binding of acetycholine to the receptor activates a transducing G protein; the active G_α-GTP subunit then directly binds to and opens a particular K^+ channel protein. That G_α-GTP

directly activates the K^+ channel has been shown by single-channel recording experiments: when purified G_α-GTP was added to the cytosolic face of a patch of heart muscle plasma membrane (see Figure 21-23c), potassium channels opened immediately and in the absence of acetylcholine or other neurotransmitters. The $G_{\beta,\gamma}$ subunits, generated by receptor activation, also bind to and open these K^+ channels.

The cardiac muscarinic receptor illustrates one way in which G protein–coupled receptors affect ion channels: the active G_α-GTP and $G_{\beta,\gamma}$ subunits binds to a channel protein. Receptors for other neurotransmitters activate G proteins that, in turn, affect the activity of enzymes that synthesize or degrade intracellular second messengers. These, in turn, can affect the activity of channel proteins, as several examples will illustrate.

Different Catecholamine Receptors Affect Different Intracellular Second Messengers

Epinephrine and norepinephrine function as both systemic hormones and neurotransmitters. Norepinephrine is the transmitter at synapses with smooth muscles that are innervated by sympathetic autonomic motor neutrons, the neurons of the peripheral nervous system that increase the activity of the heart and internal organs in "fight or flight" reactions. Norepinephrine is also found at synapses in the central nervous system. Epinephrine is synthesized and released into the blood by the adrenal medulla, an endocrine organ that has a common embryologic origin with neurons of the sympathetic system. Unlike neurons, the medulla cells do not develop axons or dendrites.

The neurotransmitters epinephrine, norepinephrine, and dopamine all contain the catechol moiety and are synthesized from tyrosine (Figure 21-43). These transmitters are referred to as catecholamines, and nerves that synthesize and use epinephrine or norepinephrine are termed *adrenergic*. All known receptors for catecholamines are coupled to G proteins, yet different ones affect different G proteins and thus different intracellular second messengers. The binding of agonists to β-adrenergic receptors on nerve cells causes activation of G_s and an increase in cAMP synthesis, the same mechanism by which β-adrenergic receptors function in nonneuronal cells (see Figures 20-14 and 20-15). As happens with the multiple neuronal muscarinic receptors, activation of other neuronal adrenergic receptors inhibits the synthesis of cAMP or increases the level of other intracellular second messengers, such as inositol *tris* phosphate, diacylglycerol, and arachidonic acid. The existence of multiple receptor-signaling pathways for the same neurotransmitter allows for great flexibility in nerve-nerve signaling.

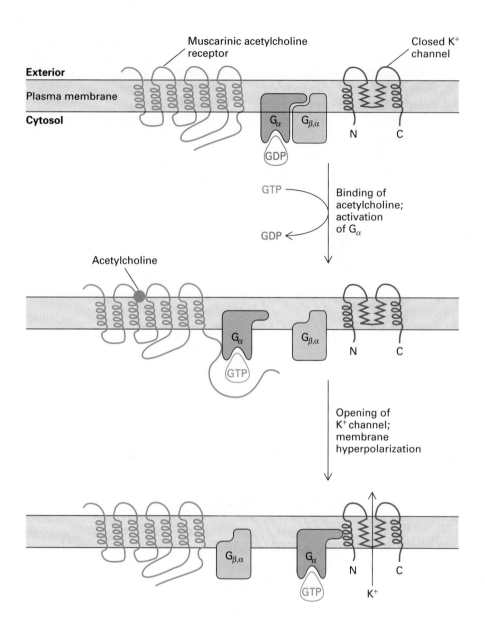

Exterior

Plasma membrane

Cytosol

Muscarinic acetylcholine receptor

Closed K$^+$ channel

G_α $G_{\beta,\alpha}$

GDP

N C

GTP ⟶

Binding of acetylcholine; activation of G_α

GDP ⟵

Acetylcholine

G_α

$G_{\beta,\alpha}$

GTP

N C

Opening of K$^+$ channel; membrane hyperpolarization

$G_{\beta,\alpha}$

G_α

GTP

K$^+$

N C

◄ FIGURE 21-42 Opening of K$^+$ channels in heart muscle plasma membrane. Binding of acetylcholine by muscarinic acetylcholine receptors triggers activation of a transducing G protein by catalyzing exchange of GTP for GDP on the α subunit. The active G_α-GTP and (not shown here) the released $G_{\beta,\gamma}$ subunit bind to and opens a K$^+$ channel. The increase in K$^+$ permeability hyperpolarizes the membrane, which reduces the frequency of heart muscle contraction. Though not shown here, the activation is terminated when the GTP bound to G_α is hydrolyzed to GDP and $G_\alpha \cdot$ GDP recombines with $G_{\beta,\gamma}$. The K$^+$ channel contains an H5 pore-lining segment, but is different from those in the *shaker* family (Figure 21-28) in that it has only two membrane-spanning α helixes. It, like other K$^+$ channels, probably is a tetramer. It is similar in sequence to the resting K$^+$ channels that maintain the resting membrane potential of all cells. [See K. Dunlap, G. Holz, and S. G. Rane, 1987, *Trends Neurosci.* **10**:241; E. Cerbai, U. Klockner, and G. Isenberg, 1988, *Science* **240**:1782; K. Ho et al., 1993, *Nature* **362**:31–37; and Y. Kubo et al., 1993, *Nature* **362**:127–133.]

A Serotonin Receptor Modulates Potassium Channel Function via the Activation of Adenylate Cyclase

The effect of an increase in cAMP on channel function can best be seen in one well-studied synapse of the sea slug *Aplysia* (Figure 21-44). A particular type of interneuron,

called a facilitator neuron (involved in a neural circuit we discuss later, Figure 21-54), forms a synapse with the axon terminal of a sensory neuron that stimulates a motor neuron via an unknown transmitter (Figure 21-45). Stimulation of the facilitator neuron increases the ability of the sensory neuron to stimulate the motor neuron. When the facilitator neuron is stimulated, it secretes serotonin, which

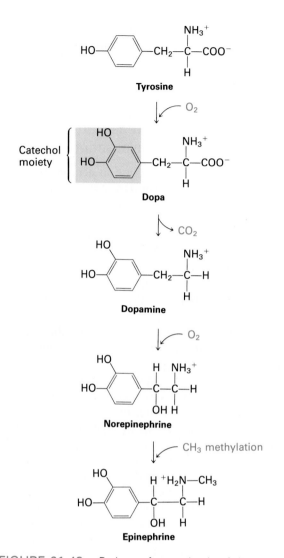

▲ FIGURE 21-43   Pathway for synthesis of the catecholamine neurotransmitters from tyrosine. The first step produces the catechol moiety, which is retained in all three transmitters.

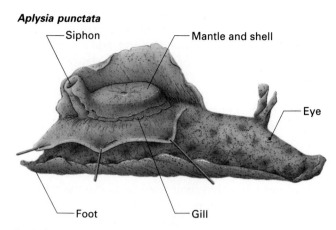

▲ FIGURE 21-44   The sea slug *Aplysia punctata*. The gill is under the protective mantle; it can be seen if the overlying tissue is pulled aside. [Adapted from E. R. Kandel, 1976, *Cellular Basis of Behavior*, W. H. Freeman and Company, p. 76.]

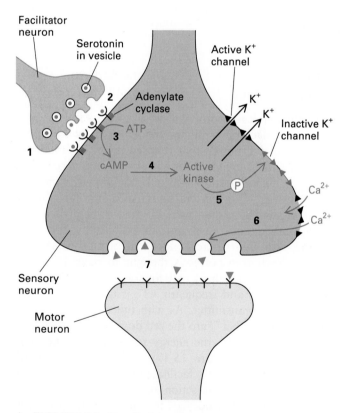

▲ FIGURE 21-45   Pathway by which serotonin, released by stimulation of a facilitator neuron, increases the ability of a sensory neuron to activate a motor neuron in the sea slug *Aplysia*. The effect of serotonin is mediated through adenylate kinase and cAMP. Phosphorylation of the voltage-gated K^+ channel protein or a channel-binding protein, indicated by the circled P, prevents the K^+ channels from opening, leading to prolonged depolarization. See text for discussion. [See E. R. Kandel and J. Schwartz, 1982, *Science* **218**:433; M. B. Boyle et al., 1984, *Proc. Nat'l. Acad. Sci. USA* **81**:7642; and M. J. Schuster et al., 1985, *Nature* **313**:392.]

binds to serotonin receptors on the sensory neuron (Figure 21-45, steps 1 and 2). This binding activates adenylate cyclase, triggering the synthesis of cAMP in the sensory neuron (step 3). cAMP then activates a cAMP-dependent protein kinase, which phosphorylates a voltage-gated K^+ channel protein or an associated protein leading to an inability of the K^+ channels to open during an action potential (steps 4 and 5). This decreases the outward flow of K^+ ions that normally repolarizes the membrane of the sensory neuron after an action potential reaches the axon terminal. The resulting prolonged membrane depolarization increases the influx of Ca^{2+} ions through voltage-gated Ca^{2+} channels (step 6). The increased Ca^{2+} level leads to greater exocytosis of synaptic vesicles in the sensory neuron (step

7), and hence greater activation of the motor neuron each time an action potential reaches the terminal.

As evidence for this model, direct administration of serotonin through a micropipette to the sensory neuron causes decreased efflux of K^+ ions and prolonged depolarization of the membrane. Also, the *Aplysia* sensory neuron is large enough that proteins such as the active catalytic subunit of the cAMP-dependent protein kinase can be injected into it. Such treatment mimics the effect of applying the natural transmitter serotonin to the nerve. Additional supporting evidence that serotonin acts by means of cAMP and a protein kinase has come from patch-clamping studies on isolated inside-out pieces of sensory neuron plasma membrane (see Figure 21-23c). When both ATP and the purified active catalytic subunit of cAMP-dependent protein kinase were added to the cytosolic surface of the patches, the K^+ channels closed. Thus the protein kinase indeed acts on the cytosolic surface of the membrane to phosphorylate the channel protein itself or a membrane protein that regulates channel activity. We shall return to this particular synapse later, as these modifications in synapse efficiency are part of a simple learning response.

Neurotransmitter Transporters Are the Proteins Affected by Drugs Such as Cocaine

After their release into synapses, catecholamines and amino acid neurotransmitters are removed by transport into the axon terminals that released them. Re-uptake is the most widely used mechanism of transmitter removal; enzymatic degradation is used only for acetylcholine and neuropeptides. Transporters for GABA, norepinephrine, dopamine, and serotonin were the first to be cloned and studied. These four transporters are encoded by gene family, and are 60–70 percent identical in amino acid sequences. Each transporter is thought to have 12 membrane-spanning α helices. All four are Na^+-neurotransmitter symports and frequently Cl^- is transported along with the neurotransmitter. As with other Na^+ symports, the movement of Na^+ into the cell down its electrochemical gradient provides the energy for uptake of the neurotransmitter (see Figure 15-17). Study and cloning of these transporters was facilitated by the observation that, following microinjection of mRNA from regions of the brain into frog oocytes, functional transporters for these neurotransmitters were expressed on the oocyte plasma membrane.

The norepinephrine, serotonin, and dopamine transporters are all inhibited by cocaine. Cocaine inhibits dopamine uptake and thus prolongs signaling at key brain synapses; the dopamine transporter is the principle brain "cocaine receptor." Therapeutic agents such as the antidepressant drugs fluoxetine, imipramine, and Prozac, block the serotonin transporter, and the tricyclic antidepressant

desipramine blocks norepinephrine uptake. However, the precise role of transporters in the antidepressant action of these drugs is not yet clear.

Some Peptides Function as Both Neurotransmitters and Neurohormones

Many of the small peptides found in nervous tissue function as synaptic neurotransmitters. Others act in a paracrine fashion (see Figure 20-1c) as "diffusible" hormones that affect many neurons over great distances. Yet other neuropeptides also act as regulators of nerve cell growth and division. Many of the peptides listed in Table 21-2 are found both in the brain and in nonneural tissues. However, in contrast to capillaries in other parts of the body, capillaries in the brain are essentially impermeable to peptides. Thus, any peptide hormones traveling through the body in the blood will be excluded from the brain: this constitutes the blood-brain barrier. Hormones in the blood do not "confuse" the functioning of the central nervous system.

Neurons that secrete peptide hormones, called neurosecretory cells, were first discovered in the hypothalamus. Secretion of peptide hormones by the anterior cells of the pituitary gland is controlled by the hypothalamus, which in turn is regulated by other regions of the brain. The hypothalamus is connected to the anterior pituitary by a special closed system of blood vessels (See Figure 20-5). Hypothalamic neurons secrete hypothalamic peptide hormones into these vessels, and the hormones then bind to receptors on the anterior pituitary cells. One such hypothalamic hormone, thyrotropin-releasing hormone (TRH), stimulates secretion by the anterior pituitary of prolactin and thyrotropin. Another hypothalamic hormone—luteinizing hormone–releasing hormone (LHRH)—causes other cells in the anterior pituitary to secrete follicle-stimulating hormone (FSH) and luteinizing hormone (LH), which are important in regulating the growth and maturation of oocytes in the ovary (see Figure 20-5). The receptors for many neuropeptides have been cloned and all contain seven membrane-spanning α-helical segments and are coupled to G proteins (Table 21-2). Thus, the intracellular signaling pathways induced by neuropeptides are the same as those induced by the classical neurotransmitters.

Neuropeptide hormones, like all neuronal proteins, are synthesized in the cell body, packaged into secretory vesicles, and sent by axonal transport to the axon terminals. Peptide hormones are packaged into different vesicles than those used to store the "classical" neurotransmitters, such as serotonin, produced in the same neuron. In at least one organism, the sea slug *Aplysia*, different peptide hormones produced by proteolysis from the same precursor are packaged into different secretory vesicles; exocytosis of the different types of secretory vesicles is regulated. In contrast to serotonin but like acetylcholine, peptide neurohormones and neurotransmitters are used only once and then degraded by proteases; they are not recycled.

Endorphins and Enkephalins Are Neurohormones That Inhibit Transmission of Pain Impulses

The activity of neurons in both the central and peripheral nervous systems is affected by a large number of neurohormones that act on cells quite distant from their site of release. Neurohormones can modify the ability of nerve cells to respond to synaptic neurotransmitters. Several small polypeptides with profound hormonal effects on the nervous system have been discovered; examples are Met-enkephalin, with the sequence Tyr-Gly-Gly-Phe-Met; Leu-enkephalin (Tyr-Gly-Gly-Phe-Leu) and the 31-amino-acid peptide β-endorphin. These three contain a common tetra-peptide sequence, Tyr-Gly-Gly-Phe, that is essential to their functions. Enkephalins and endorphins function as natural pain killers, or opiates, and decrease the pain responses in the central nervous system.

Enkephalins were discovered during research in the early 1970s on opium addiction. Several groups of researchers discovered that brain plasma membranes contain high-affinity binding sites for purified opiates such as the alkaloid morphine. The sites were presumed to be the receptors that mediated the effects of these narcotic, analgesic drugs. Since such receptors exist in the brains of all vertebrates from shark to man, the question was raised why vertebrates should have highly specific receptors for alkaloids produced by opium poppies. Since none of the neurotransmitters and neuropeptides then known could serve as agonists or antagonists for the binding of opiates to brain receptors, a search was begun for natural compounds that could. This led to the discovery of two penta-peptides, Metenkephalin and Leu-enkephalin, both of which bind to the "opiate" receptors in the brain and have the same effect as morphine (a profound analgesia) when injected into the ventricles (cavities) of brains of experimental animals. Enkephalins and endorphins appear to act by inhibiting neurons that transmit pain impulses to the spinal cord; presumably these neurons contain abundant endorphin or enkephalin receptors.

▶ Sensory Transduction: The Visual and Olfactory Systems

The nervous system receives input from a large number of sensory receptors (see Figure 21-9). Photoreceptors in the eye, taste receptors on the tongue, odorant receptors in the nose, and touch receptors on the skin monitor various aspects of the outside environment. Stretch receptors surround many muscles and fire when the muscle is stretched. Internal receptors monitor the levels of glucose, salt, and water in body fluids. The nervous system, the brain in particular, processes and integrates this vast barrage of information and coordinates the response of the organism.

The "language" of the nervous system is electric signals. Each of the many types of receptor cells must convert, or transduce, its sensory input into an electric signal. A few sensory receptors are themselves neurons that generate action potentials in response to stimulation. However, most are specialized epithelial cells that do not generate action potentials but synapse with and stimulate adjacent neurons that then generate action potentials. The key question that we will consider is how a sensory cell transduces its input into an electric signal.

In some cases, the sensory receptor protein is a Na^+ channel, and activation of the receptor causes an influx of Na^+ ions and thus membrane depolarization. Examples include the stretch and touch receptors that are activated by stretching of the cell membrane, and that have been identified in a wide array of cells, ranging from vertebrate muscle and epithelial cells to yeast, plants, and even bacteria.

The cloning of genes encoding touch receptors began with the isolation of mutant strains of the nematode *Caenorhabditis elegans* that were insensitive to touch. Three of the genes in which mutations were isolated, *MEC4*, *MEC6*, and *MEC10*, encode three similar subunits of a Na^+ channel in the touch receptor cells. These channels are necessary for touch sensitivity, and may open directly in response to mechanical stimulation. Similar kinds of channels are found in prokaryotes and lower eukaryotes; by opening in response to membrane stretching they may play a role in osmoregulation and the control of a constant cell volume. It is thought that such receptors were among the first sensory receptors to evolve. Similarly, of the four taste stimuli used by vertebrates (salty, sweet, bitter, and sour), the "receptors" for salt are the best understood. They are simply Na^+ channels in the apical membrane of taste receptor cells; entry of Na^+ through these channels depolarizes the membrane.

More often, though, the connection between the sensory receptor protein and the ion channel is indirect; the sensory receptor activates a G protein that, in turn, directly or indirectly induces the opening or closing of ion channels. The light receptors in the rod cells in the mammalian retina function in this manner, as do the olfactory (odor) receptors in the nose. We will discuss the light receptor cells in detail, as they are one of the best understood sensory systems, and they illustrate how a sensory system adapts to varying intensities of stimuli—here it is levels of ambient light that vary.

The Light-Triggered Closing of Sodium Channels Hyperpolarizes Red Cells

The human retina contains two types of photoreceptors, rods and cones (Figure 21-46). The cones are involved in color vision, and are discussed on page 977. The rods are stimulated by weak light over a range of wavelengths; in bright light, such as sunlight, the rods become inactive, for reasons that we will discuss. In the outer segment of the rod cell are membrane disks that contain the photoreceptor rhodopsin (Figure 21-47). Rod cells form synapses with neurons that in turn synapse with others that transmit impulses to the brain.

In the dark, the membrane potential of a rod cell is about −30 mV, considerably less than the resting potential (−60 to −90 mV) typical of neurons and other electrically active cells. As a consequence of this depolarization, rod cells in the dark are constantly secreting neurotransmitters, and the bipolar neurons with which they synapse are continually being stimulated. A pulse of light causes the membrane potential in the outer segment of the rod cell to become slightly hyperpolarized—that is, more negative (Figure 21-48). The light-induced hyperpolarization causes a decrease in release of neurotransmitters.

The depolarized state of the plasma membrane of resting, dark-adapted rod cells is due to the presence of a large

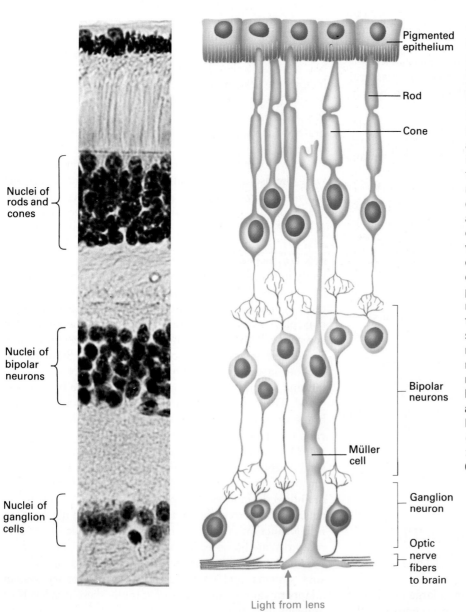

Light from lens

◄ FIGURE 21-46   Some of the cells in the neural layer of the human retina. The outermost layer of cells (in the rear of the eyeball) forms a pigmented epithelium in which the tips of the rod and cone cells are buried. Light focused from the lens passes through all of the cell layers of the retina and is absorbed by the photoreceptor proteins in the rods and cones. The axons of these cells synapse with many bipolar neurons. These, in turn, synapse with cells in the ganglion layer that send axons—optic nerve fibers—through the optic nerve to the brain. By synapsing with multiple rod cells, certain bipolar cells integrate the responses of many cells. They are involved in recognizing patterns of light that fall on the retina—for instance, a band of light that excites a set of rod cells in a straight line. Müller cells are supportive nonneural cells that fill much of the retinal spaces. Other types of cells are not depicted; all of the cells depicted here make many more synapses than are shown. [From R. G. Kessel and R. H. Kardon, 1979, *Tissues and Organs: A Text-Atlas of Scanning Electron Microscopy*, W. H. Freeman and Company, p. 87.]

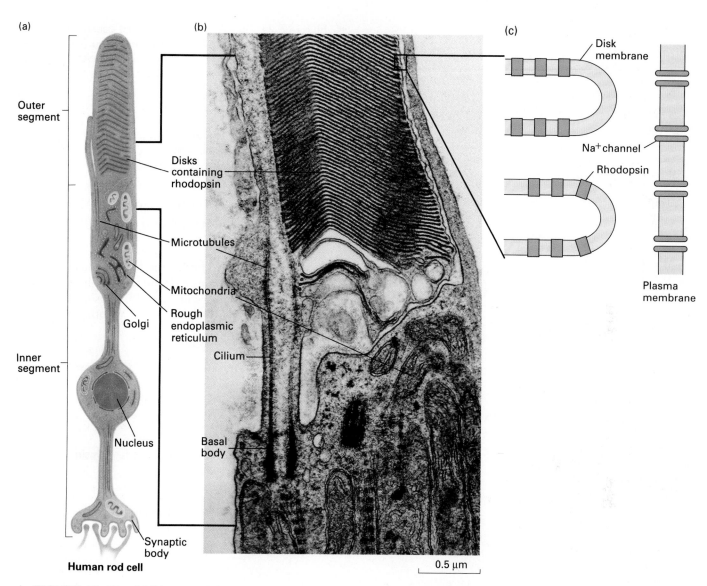

(a) Human rod cell

Outer segment

Disks containing rhodopsin

Microtubules

Mitochondria

Golgi

Rough endoplasmic reticulum

Inner segment

Cilium

Nucleus

Basal body

Synaptic body

(b)

0.5 μm

(c)

Disk membrane

Na⁺ channel

Rhodopsin

Plasma membrane

▲ **FIGURE 21-47** (a) Diagram of the structure of a human rod cell. At the synaptic body, the rod cell forms a synapse with one or more bipolar neurons. (b) Electron micrograph of the region of the rod cell indicated by the bracket in (a); this region includes the junction of the inner and outer segments. (c) A diagram of a small region of the rod cell, showing that the disks, which contain rhodopsin, are not in contact with the Na⁺ channels or other proteins in plasma membrane. [Part (b) from R. G. Kessel and R. H. Kardon, 1979, *Tissues and Organs: A Text-Atlas of Scanning Electron Microscopy*, W. H. Freeman and Company, p. 91.]

number of open Na⁺ channels. The effect of light is to close Na⁺ channels; as shown in Figure 21-13, the closing of Na⁺ channels causes the membrane potential to become more negative. The more photons absorbed, the more Na⁺ channels are closed, the more negative the membrane potential becomes, and the less neurotransmitter is released.

Remarkably, a single photon absorbed by a resting rod cell produces a measurable response, a hyperpolarization of about 1 mV, which lasts a second or two. Humans are able to detect a flash of as few as five photons. A single photon blocks the inflow of about 10 million Na⁺ ions due to the closure of hundreds of Na⁺ channels. Only about 30–50 photons need to be absorbed by a single rod cell in order to cause half-maximal hyperpolarization. The photoreceptors in rod cells, like many other types of receptors, exhibit the phenomenon of adaptation. That is, more photons are required to cause hyperpolarization if the rod cell is continuously exposed to light than if it is kept in the dark.

Let us now turn to three key questions: how is light

◄ FIGURE 21-48   A brief pulse of light causes a transient hyperpolarization of the rod-cell membrane. The membrane potential is measured by an intracellular microelectrode (see Figure 21-11a).

absorbed; how is the signal transduced into the closing of Na^+ channels; and how does the rod cell adapt to 100,000-fold or greater variations in light intensity?

Absorption of a Photon Triggers Isomerization of Retinal and Activation of Opsin

The photoreceptor in rod cells, rhodopsin, consists of the transmembrane protein opsin bound to the light-absorbing pigment 11-*cis*-retinal (Figure 21-49). Opsin has seven membrane-spanning α helices, similar to other receptors

➤ FIGURE 21-49   The photoreceptor in rod cells is rhodopsin, which is formed from 11-*cis*-retinal and opsin, a transmembrane protein. Absorption of light causes rapid photoisomerization of the *cis*-retinal to the *trans* isomer, forming the unstable intermediate *meta*-rhodopsin II, or activated opsin. The latter dissociates spontaneously to give all-*trans*-retinal and opsin. [See J. Nathans, 1992, *Biochemistry* **31**:4923–4931.]

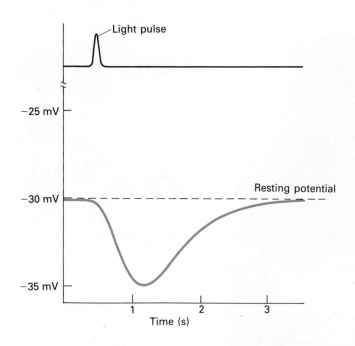

that interact with transducing G proteins. Rhodopsin is localized to the thousand or so flattened membrane disks that make up the rod's outer segment; a human rod contains about 4×10^7 rhodopsin molecules.

The pigment 11-*cis*-retinal absorbs light in the visible range (400–600 nm). The primary photochemical event is isomerization of the 11-*cis*-retinal moiety in rhodopsin to all-trans-retinal, which has a different conformation than the *cis* isomer; thus the energy of light is converted into atomic motion. The stable intermediate in which opsin is bound to all-*trans*-retinal is called *meta*-rhodopsin II, or *activated opsin*. The light-induced formation of activated opsin is both extremely efficient and rapid. Approximately 20 percent of photons at a wavelength of 500 nm, that of maximum rhodopsin absorption, that strike the retina lead to a signal transduction event, an efficiency comparable to that of the best photomultiplier tubes. Of the 57 kcal/mole of energy of photons of 500 nm, 27 kcal/mole, or 47 percent is stored in the conformationally activated *meta*-rhodopsin II intermediate, making it an effective and reliable trigger of the next signaling step. An absorbed photon triggers opsin activation in less than 10 ms. In contrast, the spontaneous isomerization of 11-*cis*-retinal is extremely slow—about once per thousand years. This means that there is very little spontaneous activation of opsin, and that the system has a very good ratio of signal to noise.

Activated opsin is unstable and spontaneously dissociates, releasing opsin and all-*trans*-retinal. In the dark, all-*trans*-retinal is isomerized back to 11-*cis*-retinal in a reaction catalyzed by enzymes in rod-cell membranes; the *cis* isomer can then rebind to opsin, re-forming rhodopsin.

Cyclic GMP is a Key Transducing Molecule

The key transducer molecule 3',5' cyclic GMP links activated opsin to the closing of Na^+ channels. Rod outer segments contain an unusually high concentration of 3',5'-cGMP, about 0.07 mM, and its concentration *drops* upon illumination. Rod outer segments contain a specific phosphodiesterase for cGMP that is activated by a cascade triggered by light:

$$3',5'\text{-cGMP} + H_2O \xrightarrow[\text{phosphodiesterase}]{\text{cGMP}} 5'\text{-GMP}$$

However, light appears to have no immediate effect on the synthesis of cGMP from GTP:

$$GTP \xrightarrow[\text{cyclase}]{\text{guanylate}} 3',5'\text{-cGMP} + PP_i$$

Injection of cGMP into a rod cell depolarizes the cell membrane, and the effect is potentiated if an analog of cGMP that cannot be hydrolyzed is injected. Thus, in the dark the high level of cGMP acts to keep the Na^+ channels open.

Several hundred molecules of cGMP phosphodiesterase are activated by a single photon. Activation is coupled to light absorption, which generates activated opsin, by the rod protein transducin (Figure 21-50). Transducin (T), a member of the family of signal-transducing G proteins that is found only in rods, has three subunits: T_α, T_β, and T_γ. The β and γ subunits are similar to those in other G proteins. In the resting state, the α subunit has a tightly bound GDP (T_α-GDP) and is incapable of affecting cGMP phosphodiesterase. Light-activated opsin catalyzes the exchange of free GTP for a GDP on the α subunit of transducin and the subsequent dissociation of T_α-GTP from the β and γ subunits. Free T_α-GTP then activates cGMP phosphodiesterase. A single molecule of activated opsin in the disk membrane can activate 500 transducin molecules; this is the primary stage of signal amplification in the visual system.

Biochemical studies have shown that T_α-GTP activates cGMP phosphodiesterase by binding to and removing the inhibitory γ subunit, thus releasing the catalytic α and β subunits in an active form (Figure 21-50). As with other α subunits of G proteins, a GTPase activity is intrinsic to the α subunit of transducin, which slowly converts light-induced T_α-GTP back to T_α-GDP. Once re-formed, T_α-GDP combines with T_β and T_γ, thus regenerating trimeric transducin. As a result, cGMP phosphodiesterase is again

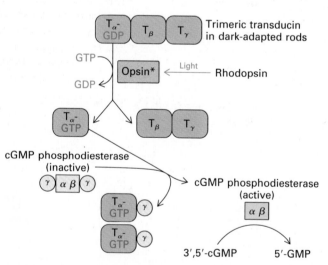

▲ FIGURE 21-50 Coupling of light absorption to activation of cGMP phosphodiesterase via light-activated opsin (opsin*) and transducin in rod cells. In dark-adapted rod cells, a high level of cGMP acts to keep Na^+ channels open and the membrane depolarized compared with the resting potential of other cell types. Light absorption leads to a decrease in cGMP by the pathway shown; as result, many Na^+ channels close and the membrane becomes transiently hyperpolarized (see Figure 21-48). [Adapted from M. Applebury, 1987, *Nature* **326**:546; and L. Stryer, 1991, *J. Biol. Chem.* **266**:10711–10714.]

inactivated and the 3',5'-cGMP level gradually returns to its dark-adapted level.

Direct support for the role of cGMP in rod-cell activity has been obtained in patch-clamping studies using isolated patches of rod outer segment plasma membrane, which contains abundant Na^+ channels. When cGMP is added to the cytosolic surface of these patches, there is a rapid increase in the number of open Na^+ channels. The effect occurs in the absence of protein kinases or phosphatases, and it appears that cGMP acts directly on the channels to keep them open. Three cGMP molecules must bind per channel in order to open it; this allosteric interaction makes channel opening very sensitive to changes in cGMP levels. In sequence the cGMP-gated Na^+ channel protein resembles voltage-gated K^+ channels, but contains a binding site for the cyclic nucleotide in its cytosol-facing domain (Figure 21-25). Presumably, as with voltage-gated K^+ channels, the cGMP-gated channel contains four subunits. Light closes the channels by activating the cGMP phosphodiesterase and lowering the level of cGMP.

Rod Cells Adapt to Varying Levels of Ambient Light

Cone cells are insensitive to low levels of illumination, and the activity of rod cells is inhibited at high light levels. Thus when we move from daylight into a dimly lighted room, we are initially blinded. As the rod cells slowly become sensitive to the dim light, we gradually are able to see and distinguish objects. This process, called adaptation, is mediated by negative feedback acting on the light receptor rhodopsin and on the levels of the second messenger cGMP.

A rod cell is able to adapt to a more than 100,000-fold variation in the ambient light level so that differences in light levels, rather than the absolute amount of absorbed light, are used to form visual images. One process contributing to this adaptation involves Ca^{2+} ions. As shown in Figure 21-51, the activity of guanylate cyclase, the enzyme that synthesizes cGMP, is relatively low in 0.5 μM Ca^{2+}, the concentration characteristic of resting rod cells. Cyclic GMP, besides opening Na^+ channels, also opens Ca^{2+} channels; the level of Ca^{2+} in the cell is balanced by Ca^{2+} pumps that export Ca^{2+} from the cytosol into the extracellular space. Light, as we noted, causes a reduction in cGMP levels; this leads to a closing of both Na^+ channels and Ca^{2+} channels. The resultant drop in Ca^{2+} concentration (due to the continual export of Ca^{2+} ions) causes activation of a Ca^{2+}-sensing protein that, in turn, binds to and activates guanylate cyclase, causing synthesis of more cGMP. This "resets" the system to a new baseline level, so that a greater change in light level will be necessary to hy-

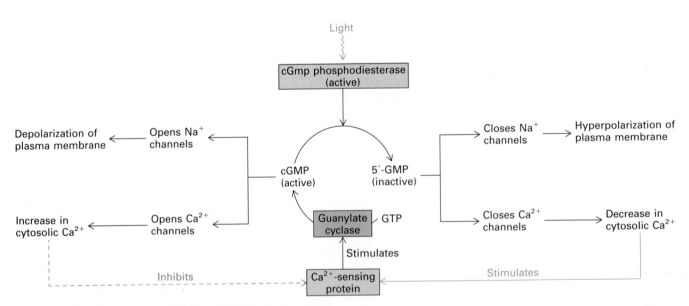

▲ FIGURE 21-51 Role of a Ca^{2+}-sensing protein, guanylate cyclase, and Ca^{2+} in adaptation of rod cells to changes in ambient light levels. In dark-adapted cells, the high level of cGMP opens both Na^+ and Ca^{2+} channels; the relatively high level of cytosolic Ca^{2+} blocks the Ca^{2+}-sensing protein from activating guanylate cyclase. A reduction in cGMP, triggered by light activation of cGMP phosphodiesterase, causes a decrease in the cytosolic Ca^{2+} level as well as hyperpolariza-tion of the plasma membrane. The reduction in the Ca^{2+} level causes activation of the Ca^{2+}-sensing protein and thus of guanylate cyclase, which catalyzes synthesis of more cGMP, restoring the cells to a new baseline state in which they are less sensitive to small changes in light level. [Adapted from E. Pugh and J. Altman, 1988, *Nature* **334**:16 and L. Stryer, 1991, *J. Biol. Chem.* **266**:10711–10714.]

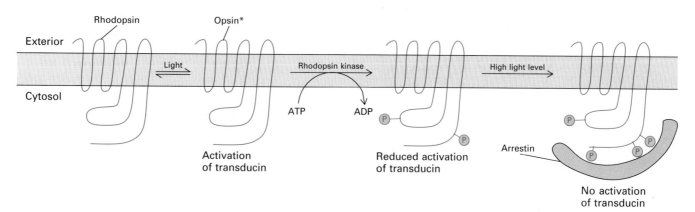

▲ FIGURE 21-52 Role of opsin phosphorylation in adaptation of rod cells to changes in ambient light levels. Light-activated opsin (opsin*), but not dark-adapted rhodopsin, is a substrate for rhodopsin kinase. The extent of opsin* phosphorylation is directly proportional to the ambient light level, and the ability of an opsin* molecule to catalyze activation of transducin (see Figure 21-50) is inversely proportional to the number of sites phosphorylated. Thus the higher the ambient light level, the larger the increase in light level needed to activate the same number of transducin molecules. At very high light levels, arrestin binds to the completely phosphorylated opsin, forming a complex that cannot activate transducin at all. [See L. Lagnado and D. Baylor, 1992, *Neuron* **8**:995–1002.]

drolyze cGMP, to close the same number of Na^+ channels, and to generate the same visual signal than if the cells had not been exposed to light. In other words, the cells become less sensitive to small changes in levels of illumination.

A second process, affecting the protein opsin itself, participates in adaptation of rod cells to ambient levels of light and also prevents overstimulation of the rod cell in very high ambient light (Figure 21-52). A rod-cell enzyme, rhodopsin kinase, phosphorylates light-activated opsin (O) but not dark-adapted rhodopsin. Phosphorylated opsin is less able to activate transducin than is nonphosphorylated opsin. Each O molecule has seven phosphorylation sites; the more sites that are phosphorylated, the less able O^* is to activate transducin. Because the extent of O^* phosphorylation is proportional to the amount of time each opsin molecule spends in the light-activated form, it is a measure of the background level of light. Under high light conditions, phosphorylated opsin is abundant. Then a greater increase in light level will be necessary to generate a visual signal. When the level of ambient light is reduced, most of the opsins become dephosphorylated, and thus activation of transducin increases. Then fewer additional photons will be necessary to generate a visual signal. At high ambient light (such as noontime outdoors), the level of opsin phosphorylation is such that the protein *arrestin* binds to opsin; arrestin binds to the same site on opsin as does transducin, totally blocking activation of transducin and causing a shutdown of all rod-cell activity. The mechanism by which rod-cell activity is controlled by rhodopsin kinase is similar to adaptation of the β-adrenergic receptor to high levels of hormone (see Figure 20-50). Indeed, rhodopsin kinase is very similar to β-adrenergic receptor kinase—the enzyme that phosphorylates and inactivates only the ligand-occupied β-adrenergic receptor—and each protein can phosphorylate the other's substrate.

Color Vision Utilizes Three Opsin Pigments

There are three classes of cone cells in the human retina. Each absorbs light at a different wavelength (Figure 21-53), and each contains a different rhodopsin photopigment. One absorbs mainly blue light, one green, and one red. As in rods, the relative amount of light absorbed by each class of cones is translated into electrical signals that are transmitted to the brain. There the overall pattern of absorption of light of different wavelengths is converted into what we perceive as *color*. All cone opsins bind the same retinal as found in rods, and the three cone opsins are similar to the rod opsin and to each other. The unique absorption spectra of the three cone rhodopsins are due to different amino acid side chains that contact the retinal on the inside of rhodopsin and that affect its ability to absorb light of different wavelengths.

The most interesting results to emerge from the study of cone opsins were molecular explanations of the different types of color blindness in humans. The "blue" opsin is encoded on human chromosome 7, while the red and green opsin genes are located next to each other, head-to-tail, on the X chromosome. The red and green opsin genes are the product of an evolutionary recent gene duplication because they are 98 percent identical in sequence. Furthermore, new world monkeys have only a single opsin gene on their X chromosome, while old world monkeys, which are more closely related to humans, have two. Two adjacent and almost identical genes can be expected to recombine un-

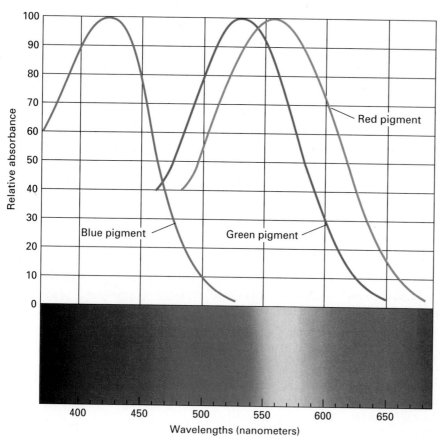

◄ FIGURE 21-53   The absorption spectra of the three human opsins responsible for color vision. What is called the blue pigment is maximally responsive to the short-wavelength region of the visible spectrum; the green and red pigments absorb primarily in the intermediate and long-wavelength regions. Individual cone cells express one of the three opsins. The spectra were determined by measuring in a microspectrophotometer the light absorbed by individual cone cells, obtained from cadavers. [From J. Nathans, 1989, *Sci. Am.* **260**(2):44.]

equally during gamete formation rather frequently, resulting in X chromosomes with only a green or only a red opsin gene. This results in red-green color blindness, a phenotype not uncommon in males, because they have a single X chromosome, but very rare in females.

Remarkably, because of polymorphisms in the red opsin genes, even individuals with "normal" color vision see colored objects differently. Many individuals have an alanine at position 180 (in the middle of the fourth membrane-spanning α helix, a region that contacts the retinal) of the red opsin; the absorption maximum of that pigment is ≈530 nm. Others have a serine in that position, resulting in an absorbance maximum of ≈560 nm. Thus, the subjects with serine at position 180 have a higher sensitivity to red light than the others; they "see" colors differently due to the change in a single nucleotide.

More Than a Thousand Different G Protein–Coupled Receptors Detect Odors

The visual system functions efficiently with only four types of photoreceptors, three in cones and one in rods. In contrast, the olfactory system utilizes at least several hundred homologous olfactory receptors in responding to the millions of different chemicals we can smell.

Signal transduction in the olfactory system is different from that in the visual system. Each receptor cell in the olfactory epithelium in the nose probably has only one specific odorant receptor on its apical (outward-facing) plasma membrane, and "senses" only one or a few odorants. Most of these receptors are coupled to a G protein, G_{olf}, unique to olfactory epithelia; G_{olf}, like G_s, activates adenylyl cyclase, and the level of cAMP increases as a result. In olfactory cells cAMP binds to and opens a cAMP-gated Na^+ channel, unique to olfactory epithelia, that is similar in structure to the cGMP-gated Na^+ channel in the visual system (Figure 21-25). This leads to a depolarization of the cell membrane (rather than the hyperpolarization induced by activation of rhodopsin), initiating the electrical signal that is sent to the brain.

Several hundred genes encoding odorant receptors were isolated by a novel cloning strategy. First, workers identified sequences of amino acids that were conserved in many other known G protein–coupled receptors. Assuming that odorant receptors were also coupled to G proteins, the workers devised primers for the polymerase chain reaction that would allow amplification of cDNA sequences that encoded novel G protein–coupled receptors, and indeed hundreds were cloned using that approach from a cDNA library made from olfactory epithelia. The diversity of these receptors is entirely encoded in the nuclear genome, and there is no evidence for somatic recombination, as is found in the immune system, for generation of odorant receptors. In situ hybridization showed that each of these receptor genes is expressed in only a few of the millions of olfactory epithelial cells, as might be expected for a receptor that binds a specific kind of odorant. However, it has not yet been possible to identify which of the thousands of known odorants binds to any one of the cloned receptors. It is striking that, during evolution, there was selection for so many different odorant receptors.

► *Memory and Neurotransmitters*

In its most general sense, learning is a process by which humans and other animals modify their behavior as a result of experience or as a result of acquisition of information about the environment. Memory is the process by which this information is stored and retrieved.

Psychologists have defined two types of memory, depending on how long it persists: short term (minutes to hours) and long term (days to years). It is generally accepted that memory results from changes in the structure or function of particular synapses, but until recently learning and memory could not be studied with the tools of cell biology or genetics. Most researchers believe that long-term memory involves the formation or elimination of specific synapses in the brain and the synthesis of new mRNAs and proteins. Because short-term memory is too rapid to be attributed to such gross alterations, some have suggested that changes in the release and function of neurotransmitters at particular synapses are the basis of short-term memory. Indeed, recent work has identified several types of proteins that function to modify synaptic activity. These proteins integrate two different signaling pathways; they respond to two (or more) different but coincident signals by generating an output that is different from that produced by either signal acting separately. We shall see how two such proteins are involved in elemental forms of learning. We begin with a molecular analysis of elemental forms of learning in the fruit fly *Drosophila* and the sea slug *Aplysia,* and then turn to long-term potentiation, a form of learning exhibited by many synapses in the mammalian brain.

Mutations in *Drosophila* Affect Learning and Memory

Remarkable as it sounds, fruit flies can be trained to avoid certain noxious stimuli. During the training period, a population of flies is exposed to two different stimuli, either two odoriferous chemicals or two colors of light. One of the two is associated with an electric shock. The flies are then removed and placed in a new apparatus, and the two stimuli are repeated but without the electric shock. The flies are tested for their avoidance of the stimulus associated with the shock. About half the flies learn to avoid the stimulus associated with the shock, and this memory persists for at least 24 h. Painstaking observation of mutagenized flies has led to the identification of six different genes in which mutations cause defects in this learning process.

Two mutations that disrupt memory affect cAMP levels. One, *dunce,* is due to a mutation in one of two isoforms of the enzyme cAMP phosphodiesterase. A second, *rutabaga,* is caused by a mutation in the gene encoding one of four types of adenylate cyclase—the one activated by both Ca^{2+}-calmodulin and an activated $G_{s,\alpha}$ signal-transducing protein. It is not yet known which synapses are involved in the learning response or how the high levels of cAMP that result from the phosphodiesterase deficiency might affect learning. However, a dual-regulated adenylate cyclase also plays a key role in learning at a particular synapse in the sea slug *Aplysia.*

Gill-Withdrawal Reflex in *Aplysia* Exhibits Three Elementary Forms of Learning

Sea slugs exhibit three of the most elementary forms of learning familiar in vertebrates: habituation, sensitization, and classical conditioning. Habituation is a decrease in behavioral response to a stimulus following repeated exposure to the stimulus with no adverse effect. For example, an animal that is startled by a loud noise may show decreasing responses on prolonged repetition of the noise. Sensitization, in contrast, is an increase in behavioral response to a stimulus that does have an adverse effect.

Classical conditioning is one of the simplest types of associative learning—the recognition of predictive events within an animal's environment. The animal learns that one event, termed the conditioned stimulus (CS), always precedes, by a critical and defined period, a second or reinforcing stimulus or event, the unconditioned stimulus (US). A well known example is the Pavlovian response in dogs: a bell (CS) is rung a few seconds before food (US) is presented; the dogs soon learn to associate the two stimuli and to salivate in response to the bell alone. In such a learning process, be it in mammals or sea slugs, it is essential that the conditioning stimulus always precede the unconditioned stimulus by a small and critical time interval.

Habituation   When a sea slug (Figure 21-44) is touched gently on its siphon, the gill muscles contract vigorously and the gill retracts into the mantle cavity. The gill-withdrawal reflex is mediated by a simple reflex arc (Figure 21-54). Sensory neurons in the siphon synapse with motor neurons that innervate the gill muscles. However, if the siphon is touched 10–15 times in rapid sequence, the gill response decreases to only about one-third of its initial in-

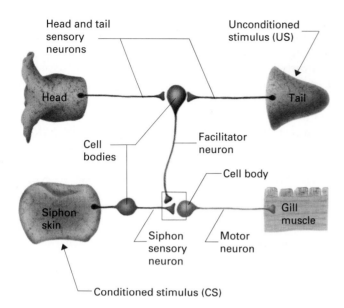

▲ FIGURE 21-54   Neural circuits in the gill-withdrawal reflex of the sea slug *Aplysia*. For simplicity, certain of the interneurons are omitted. This reflex exhibits habituation, sensitization, and classical conditioning. The details of the synapses (boxed) between the sensory, facilitator, and motor neurons are shown in Figure 21-45). [See E. R. Kandel and J. H. Schwartz, 1982, *Science* **218**:433 and T. W. Abrams and E. R. Kandel, 1988, *Trends Neurosci.* **11**:128.]

tensity. By recording the electric changes in the motor neurons to the gill, researchers discovered that this habituative response is due to a progressive decrease in the amount of neurotransmitter released at the synapses between the siphon sensory neurons and the motor neurons. In other words, repeated stimulation of the siphon leads to a decrease in the magnitude of the excitatory postsynaptic potential.

We have noted that release of neurotransmitters is triggered by a rise in the intracellular Ca^{2+} concentration following opening of voltage-gated Ca^{2+} channels. Measurements of Ca^{2+} movements in the *Aplysia* siphon sensory neuron have shown that habituation results from a decrease in the number of voltage-gated Ca^{2+} channels that open in response to the arrival of the action potential at the terminal, thus reducing the amount of neurotransmitter released. Habituation does not affect the generation of action potentials in the siphon sensory neuron or the response of the receptors in the postsynaptic cells.

Sensitization

If a habituated sea slug is given a strong, noxious stimulus, such as a blow on the head or tail, it will respond to the next weak stimulus to the siphon by a rapid withdrawal of the gill. The noxious stimulation is said to sensitize the animal so that it exhibits an enhanced response to touching of the siphon. *Aplysia* sensitization is mediated by interneurons called facilitator neurons (Fig-

ures 21-45 and 21-54) that are activated by shocks to the head or tail. Electron microscopy shows that the axon of a facilitator neuron synapses with the terminal of a siphon sensory neuron near the site where the siphon sensory neuron synapses with a motor neuron (Figure 21-54). Stimulation of the facilitator neuron causes the siphon sensory neuron to release more transmitter in its synapse with the motor neuron, thus increasing the magnitude of the gill-withdrawal reaction.

As illustrated in Figure 21-45, stimulation of the *Aplysia* facilitator neurons leads to inhibition of voltage-gated K^+ channels in the siphon sensory neurons, which normally participate in repolarizing the membrane after an action potential. As a result, action potentials reaching the nerve terminals decay more slowly. This prolonged depolarization causes a longer and larger than usual influx of Ca^{2+} ions via the voltage-gated Ca^{2+} channels. The increased cytosolic Ca^{2+} level leads to (1) more extensive exocytosis of neurotransmitter by siphon sensory neurons at their synapses with motor neurons; (2) enhanced activity of motor neurons; and (3) enhanced contraction of the gill muscle. The effect of facilitator neuron stimulation is mediated by cAMP and a cAMP-dependent protein kinase in the siphon sensory neuron terminal. Short-term sensitization persists as long as the concentration of cAMP is elevated and the kinase is activated, about 1 h after each sensitizing stimulus.

Classical Conditioning

The gill-withdrawal reflex also exhibits classical conditioning. In the "training" process, a weak touch to the siphon—the conditioned stimulus (CS)—is followed immediately by a sharp blow to the tail or head—the unconditioned stimulus (US)—which, of course, evokes a marked gill-withdrawal response. After a series of such trials, the gill-withdrawal response to the CS alone is substantially enhanced, as if the animal "learns" that a weak siphon touch (CS) is followed by a noxious, sharp blow (US). As in conditioning in other animals, the CS must precede the US by a short and definite interval, in this case 1–2 s.

Sensitization occurs when the facilitator neuron is activated by the US (a blow to the head or tail) in the absence of the CS; it triggers activation of adenylate cyclase and closing of K^+ channels in the siphon sensory neuron. During conditioning, the rise in cAMP in the siphon sensory neuron terminal is much greater when the sensory neuron is triggered by the CS to fire an action potential just before the US arrives.

Figure 21-55 outlines how an adenylate cyclase activated both by $G_{s,\alpha}\bullet GTP$ and by Ca^{2+}-calmodulin is the probable molecular site for convergence of the US and CS in the axon terminal of the siphon sensory neuron. The brief Ca^{2+} influx triggered by the CS action potential and resulting in the siphon sensory neuron activates this adenylate cyclase. Serotonin released by the facilitator neuron triggered by the US also stimulates this cyclase. However, activation of the cyclase, and hence the increase in cAMP, is greatest when the cyclase is first "primed" by Ca^{2+} influx and then, within 1–2 s, activated by binding of serotonin. In this way, the enhancement of adenylate cyclase activity triggered by the unconditioned stimulus makes the sensory neuron more sensitive to a conditioning stimulus; the animal learns to associate the CS with the US and to

respond to the CS alone with an enhanced response. Indeed, in isolated membranes prepared from these neurons, adenylate cyclase activity was greater if the membranes were exposed to elevated Ca^{2+} *before* exposure to $G_{s,\alpha} \cdot GTP$ than vice versa. This biochemical asymmetry mirrors the key temporal requirement for conditioning in the intact animal; conditioning is produced only when the CS precedes the US.

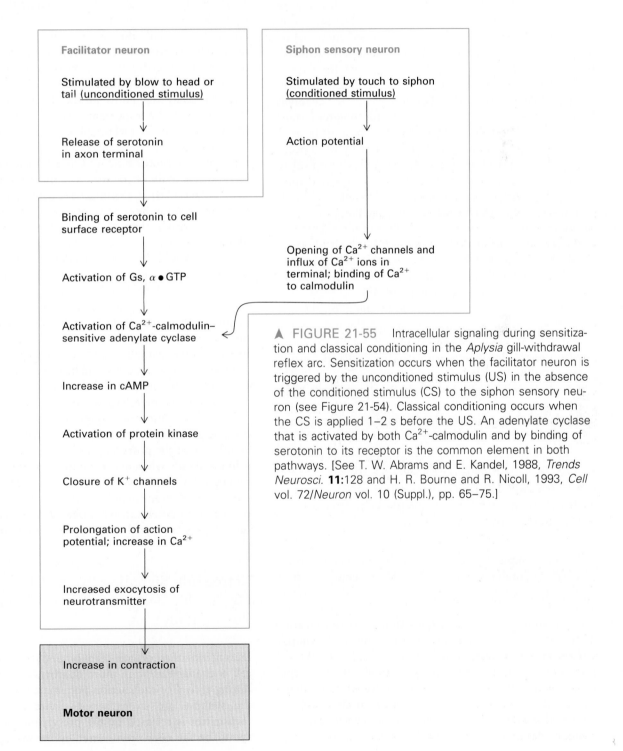

▲ FIGURE 21-55   Intracellular signaling during sensitization and classical conditioning in the *Aplysia* gill-withdrawal reflex arc. Sensitization occurs when the facilitator neuron is triggered by the unconditioned stimulus (US) in the absence of the conditioned stimulus (CS) to the siphon sensory neuron (see Figure 21-54). Classical conditioning occurs when the CS is applied 1–2 s before the US. An adenylate cyclase that is activated by both Ca^{2+}-calmodulin and by binding of serotonin to its receptor is the common element in both pathways. [See T. W. Abrams and E. Kandel, 1988, *Trends Neurosci.* **11**:128 and H. R. Bourne and R. Nicoll, 1993, *Cell* vol. 72/*Neuron* vol. 10 (Suppl.), pp. 65–75.]

Thus, the adenylate cyclase activated both by $G_{s,\alpha}\bullet GTP$ and by Ca^{2+}-calmodulin functions as a *molecular coincidence detector;* it responds to two different but coincident signals by generating an output that is different from that produced by either signal acting separately.

Long-Term Memory The short-term sensitization and conditioning responses in *Aplysia* can occur in the presence of inhibitors of protein synthesis, suggesting that no new proteins (or cells) are required for short-term learning responses (short-term memory). On the other hand, a series of closely spaced tail shocks (unconditioned stimulus) delivered over a few hours will produce a long-term sensitization (long-term memory), which can persist for days or even weeks. Both long-term and short-term sensitizations affect the same synapses, and even the same K^+ channels. However, protein synthesis is essential for long-term sensitization, suggesting that certain new proteins must be made in these synapses in order for long-term memory to occur. Long-term sensitization of the *Aplysia* gill-withdrawal reflex can be induced in cells in culture by the repeated application of serotonin, the hormone normally released by the facilitator neuron (Figure 21-45). Repeated application of serotonin leads to induction of synthesis of several proteins in the sensory neuron and inhibition of synthesis of others, among which are several plasma membrane proteins that function in neural cell-cell adhesion (N-CAM, or neural cell adhesion proteins—see Figure 24-32).

Sensitization and classical conditioning of the gill-withdrawal reflex of *Aplysia* are among the few cases in which short-term changes in synaptic function are understood in molecular detail. These simple forms of learning have served as models for more complex forms of behavior, such as short-term and long-term memory in vertebrates. Increasingly, neuroscientists are identifying molecules that may function in the memory of mammals.

A Novel Glutamate Receptor Is the Coincidence Detector for Long-Term Potentiation at Many Synapses in the Mammalian Brain

The hippocampus is the region of the mammalian brain associated with many types of short-term memory. Certain types of hippocampal neurons, here simply called *postsynaptic cells*, receive inputs from hundreds of presynaptic cells. In *long-term potentiation*—similar to sensitization discussed above—continual stimulation of a postsynaptic neuron makes it more responsive to subsequent stimulation by presynaptic neurons. For example, in the hippocampus stimulation of a presynaptic nerve with 100 depolarizations acting over only 200 milliseconds causes an increased sensitivity of the postsynaptic neuron that lasts hours to days.

Figure 21-56 shows how two types of glutamate receptors in the postsynaptic neuron combine to generate long-term potentiation. Both receptors are glutamate-gated cation channels that are similar in structure and amino acid sequence to the nicotinic acetylcholine receptors (Table 21-3 and Figure 21-39); both receptors depolarize the plasma membrane when activated. Because the two receptors were initially distinguished by their ability to be activated by the nonnatural amino acid *N*-methyl-D aspartate (NMDA), they are called NMDA glutamate receptors and non-NMDA glutamate receptors. Non-NMDA receptors are "conventional"; the ion channels open whenever glutamate, released from the presynaptic cell, binds to the receptors on the postsynaptic neuron (Figure 21-56).

NMDA glutamate receptors are different in two key respects. First, they allow influx of Ca^{2+} as well as Na^+. Second, and more important, two conditions must be fulfilled for the ion channel to open: glutamate must be bound *and* the membrane must be partly depolarized. In this way, the NMDA receptor functions as a coincidence detector; it integrates activity of the postsynaptic cell—reflected in its depolarized plasma membrane—with release of neurotransmitter from the presynaptic cell, generating a cellular response greater than that caused by glutamate release alone.

NMDA receptors are dependent upon membrane voltage because of the voltage-sensitive blocking of the ion channel by a Mg^{2+} ion from the extracellular solution. Depolarization of the membrane causes the Mg^{2+} ion to dissociate from the receptor, making it possible for glutamate binding to open the channel. Indeed, the Mg^{2+} ion is binding in the channel itself, since mutagenesis of a single asparagine residue in the M2 helix of the NMDA receptor—the segment thought to line the pore itself (see Figure 21-39)—abolishes the effect of Mg^{2+}.

Since activation of a single synapse, even at high frequency, generally causes only a small depolarization of the membrane of the postsynaptic cell, long-term potentiation is induced only when many synapses on a single postsynaptic neuron are activated simultaneously. Thus the requirement for membrane depolarization explains a key property of long-term potentiation: *cooperativity*—a large number of synapses on a cell must be activated simultaneously.

Retrograde Signaling by the Gas Nitric Oxide May be a Part of Long-Term Potentiation

During induction of long-term potentiation the presynaptic cell is changed as well; more neurotransmitter is released during arrival of each action potential. Thus, both sides of the synapse are strengthened—more glutamate neurotransmitter is released by the presynaptic cell and the response to the neurotransmitter by the postsynaptic cell is

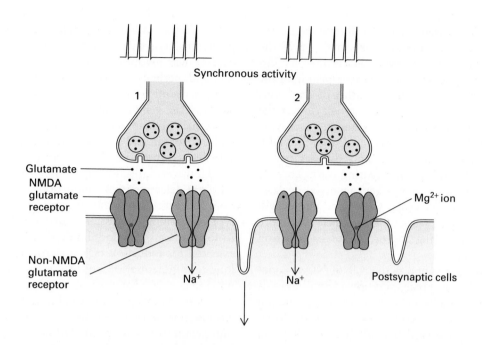

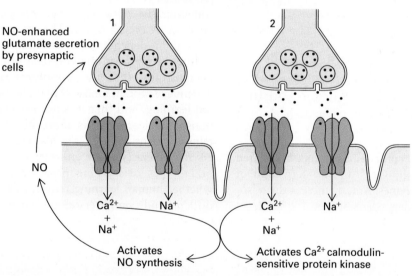

▲ FIGURE 21-56   Involvement of two types of glutamate receptors and nitric oxide in long-term potentiation—a type of short-term memory. This postsynaptic cell in the hippocampus region in the brain has two types of glutamate receptors, NMDA (green) and non-NMDA (pink). The ion channel in the NMDA receptor is normally blocked by a Mg^{2+} ion, and thus the glutamate released by firing of presynaptic neurons leads, at first, to opening of only the non-NMDA glutamate receptors. The resultant influx of Na^+ partially depolarizes the membrane. If many presynaptic neurons (here two are shown) fire in synchrony, the membrane of the postsynaptic cell becomes sufficiently depolarized so that the Mg^{2+} ions blocking the NMDA receptors are removed and thus the NMDA as well as the non-NMDA glutamate receptors open in response to glutamate. Ca^{2+} ions as well as Na^+ ions enter through the open NMDA receptors, causing an enhanced response in the postsynaptic cells. Among the responses induced by Ca^{2+} is induction of NO synthesis. NO diffuses to the presynaptic cell—retrograde signaling—and there activates a protein that enhances the secretion of glutamate in response to the depolarization induced by arrival of an action potential. As a result, the same response in the postsynaptic neuron can be induced by fewer action potentials in the presynaptic neurons or, in other words, the synapse "learns" to have an enhanced response to the electrical signals in the presynaptic cells. [After T. M. Jessell and E. R. Kandel (1993) *Cell* vol. 72/*Neuron* vol. 10 (Suppl.), pp. 1–30 and C. F. Stevens (1993) *Cell* vol. 72/*Neuron* vol. 10 (Suppl.), pp. 55–63.]

also increased. The increased activity of the presnaptic cell is thought to be induced by a retrograde signal sent by the postsynaptic cell; the gas nitric oxide (NO) is one such signal. Figure 21-56 traces how this happens. Opening of the NMDA glutamate receptors causes an influx of Ca^{2+} ions and an activation of many Ca^{2+}-dependent enzymes in the postsynaptic cell. Among these is NO synthase, the enzyme that synthesizes NO from the amino acid arginine. NO, unlike classical neurotransmitters, simply diffuses out of the postsynaptic cell and into neighboring cells. There it activates several enzymes, among them guanylyl cyclase, an enzyme that synthesizes cyclic GMP, and the cGMP is thought to change the presynaptic cell such that more glutamate is secreted during the arrival of each action potential.

Substantial evidence supports the role of NO in long-term potentiation. For instance, addition of inhibitors of NO synthesis to cultured brain slices blocks the induction of long-term potentiation, as does addition of hemoglobin to the brain slices, a protein that binds released NO and prevents it from reaching neighboring cells. Moreover, addition of NO to hippocampal neurons in culture causes an increase in neurotransmitter release, as expected from its role as a retrograde signaling molecule involved in the induction of long-term potentiation.

Mice Defective in the Hippocampal α-Ca^{2+}-Calmodulin-Activated Protein Kinase Are Impaired in Long-Term Potentiation and in Spatial Learning—the Beginnings of a Molecular Psychology

Opening of the NMDA receptors is, as noted, associated with an increase in cytosolic Ca^{2+} and an activation of several Ca^{2+}-dependent enzymes. Among these is the α isoform of Ca^{2+}-calmodulin-dependent protein kinase II, an enzyme found in abundance mainly in neurons in the hippocampus. For this reason, workers suspected that this enzyme was involved in induction of long-term potentiation and, perhaps, in certain types of learning. To test this hypothesis, mice were genetically engineered to have a deletion in the gene encoding this enzyme. In most respects these mice grow and behave normally; for instance, their abilities to eat and mate are normal, and they have the coordinated motor skills to swim normally in water.

However, cultured hippocampal neurons from these mice are defective in the induction of long-term potentiation. More strikingly, the mice are impaired in a particular type of learning process, supporting the notion that the α isoform of Ca^{2+}-calmodulin-dependent protein kinase II is essential for induction of long-term potentiation and that long-term potentiation, in turn, is the electrophysiological basis of a particular type of learning.

In the critical psychological test, the mice are placed in a round pool of opaque water; to escape the water the mice must swim to a submerged platform. The mutant mice can find the platform normally if it is made visible by a flag, indicating that they can learn to associate the flag with the submerged platform. In contrast, when the platform is hidden, the mice must learn to find it from integrating multiple spatial relationships between objects in the room surrounding the pool (e.g., pictures on the wall) and the position of the platform. The mice with defective Ca^{2+}-calmodulin-dependent protein kinase take longer to learn to find the platform than do their normal littermates. This demonstrates that the α isoform of Ca^{2+}-calmodulin-dependent protein kinase II plays an important role in spatial learning, and that it is not essential for some types of nonspatial learning. As more hippocampal-specific proteins are identified by molecular cloning techniques, workers may identify other enzymes or receptors essential for other types of learning processes in mice. Determining whether human learning requires these enzymes will be a difficult and controversial area of research.

SUMMARY

An electric potential exists across the plasma membrane of all eukaryotic cells because the ion compositions of the cytosol and extracellular fluid differ, as do the permeabilities of the plasma membrane to the principal cellular ions—Na^+, K^+, Cl^-, and Ca^{2+}. In most nerve and muscle cells, the resting membrane potential is about 60 mV, negative on the inside; the potential is due mainly to the relatively large number of open K^+ channels in the membrane.

Impulses are conducted along a nerve axon by action potentials. An action potential consists of a sudden (less than a millisecond) depolarization of the membrane followed by a rapid hyperpolarization and a gradual return to the resting potential. The initial depolarization in the membrane potential is caused by a sudden and transient opening of transmembrane voltage-gated Na^+ channels that admit Na^+ ions into the cytosol. Voltage-gated K^+ channels also open in response to membrane depolarization;

their opening repolarizes the membrane by permitting the efflux of K^+ ions. K^+ channels are assembled from four similar subunits; particular segments of the channel protein have been identified that line the pore, that cause channel inactivation, and that sense voltage changes. Voltage-gated Na^+ and Ca^{2+} channel proteins are single polypeptides containing four homologous domains each similar to a K^+ channel protein; they, too, have a similar positively charged "gating" helix that moves in response to a voltage change of a sufficient magnitude.

Neurons only generate action potentials when the plasma membrane in the region of the axon hillock is depolarized to the threshold value. An action potential generated at one point along an axon will lead to depolarization of the adjacent segment and thus to propagation of the action potential along its length. The speed of impulse conduction depends on the diameter of the axon and conductivity of the neuronal cytosol. Thick neurons conduct faster than thin ones, and myelinated nerves conduct faster than unmyelinated nerves of similar diameter because of insulation of the neuron by the myelin sheath.

Impulses are transmitted from neurons to other cells at specialized junctions called synapses. In electric synapses, ions pass from the presynaptic cell to the postsynaptic cell through gap junctions, and an action potential is generated in the postsynaptic cell with no time delay. In the more common chemical synapses, the arrival of an action potential in the presynaptic axon triggers the release of neurotransmitters into the synaptic cleft; from there the transmitters bind to receptors on the postsynaptic cell. Transmitters are stored in membrane-bounded vesicles, and exocytosis of these vesicles is triggered by a rise in the cytosolic Ca^{2+} level induced by the opening of voltage-gated Ca^{2+} channels.

At excitatory synapses, the neurotransmitter acts to depolarize the postsynaptic cell and generate an action potential. At the synapse of a motor neuron and striated muscle cell, binding of acetylcholine to the well-studied nicotinic acetylcholine receptor triggers a rapid increase in permeability of the membrane to both Na^+ and K^+ ions, leading to depolarization. Much information is available on the structure of the receptor protein and of the segments that line its ion channel. In other postsynaptic cells, the depolarization of the postsynaptic membrane is less extensive but longer lived, on the order of seconds. In many such synapses, receptors are coupled to G transducing proteins that directly or indirectly open or close ion channel proteins. At inhibitory synapses, the release of neurotransmitter triggers a hyperpolarization of the postsynaptic membrane, making it more difficult for the cell to generate an action potential. Depending on the specific receptor in the postsynaptic cell, the same neurotransmitter can induce either an excitatory or inhibitory response.

The inhibitory actions of GABA and glycine are mediated through ligand-gated Cl^- channels whose structures are similar to that of the nicotinic acetylcholine receptor. In some postsynaptic cells, receptors for epinephrine and serotonin modulate the activity of adenylate cyclase. The electric response of these cells is believed to be caused by phosphorylation of Na^+ or K^+ channel proteins by the cAMP-dependent protein kinase. In cardiac muscle, binding of ligand to the muscarinic acetylcholine receptor activates a transducing G protein; this in turn opens a K^+ channel and causes hyperpolarization of the membrane and a decrease in muscle contraction.

Chemical synapses allow a single postsynaptic cell to amplify, modify, and integrate signals from multiple presynaptic neurons. Especially in the central nervous system, many neurons must integrate excitatory and inhibitory stimuli from dozens or hundreds of other neurons, if not a thousand or more. Whether a threshold potential is induced at the axon hillock depends on the timing and magnitude of these stimuli, the localization and duration of the resultant local hyperpolarizations and depolarizations, and the ability of the localized changes in potential to be conducted along the plasma membrane.

Many compounds released by neurons are systemic hormones as well as neurotransmitters, affecting both distant secretory cells and adjacent neurons. Recent work suggests that small peptides such as endorphins, enkephalins, and hypothalamic releasing factors function as neurotransmitters in particular synapses in the brain and also act as hormones.

Removal of neurotransmitter from the synapse is essential for ensuring its repeated functioning. The action of acetylcholine is terminated by the enzyme acetylcholinesterase. Peptide neurotransmitters are hydrolyzed to amino acids. Other neurotransmitters are removed by reuptake into the presynaptic cell; antidepressant drugs and narcotics such as cocaine inhibit certain catecholamine transporters.

Many sensory transduction systems convert signals from the environment—light, taste, sound, touch—into electric signals. These signals are collected, integrated, and processed by the central nervous system. The sensory system understood in the most molecular detail is that of the photoreceptor rod cells. Absorption of even a single photon results in hyperpolarization of the rod-cell plasma membrane and reduces the release of chemical transmitters to adjacent nerve cells. Light causes isomerization of the 11-*cis*-retinal moiety in rhodopsin and formation of activated opsin, which then activates the signal-transducing G protein transducin (T) by catalyzing exchange of free GTP for bound GDP on the T_α subunit. Activated T_α-GTP, in turn, activates cGMP phosphodiesterase. This enzyme lowers the cGMP level, which leads to closing of the membrane Na^+ channels, hyperpolarization of the membrane, and release of less neurotransmitter. Modifications in the activity of guanylate cyclase by Ca^{2+}, and also phosphorylation of the light-activated form of opsin, result in adaptation of rod-cell activity to more than a 100,000-fold range of illumination.

Modifications in the activity of certain synapses are associated with short-term memory, at least in some invertebrate and mammalian systems. Certain *Drosophila* mutants that cannot learn are defective in cAMP metabolism, most notably in a Ca^{2+}-calmodulin-activated adenylate cyclase. In the sea slug *Aplysia*, the gill-withdrawal reflex exhibits habituation, sensitization, and classical conditioning—three forms of simple learning. Habituation is linked to the closing of Ca^{2+} channels in the presynaptic axon terminals of siphon sensory neurons originating in the siphon; this alters the flux of Ca^{2+} in the terminals and the amount of transmitter released to the motor neurons. Sensitization and classical conditioning are mediated by facilitator neurons that synapse with the siphon sensory neurons. Serotonin released by stimulation of the facilitator neurons causes an increase in activity of a Ca^{2+}-calmodulin-activated adenylate cyclase in the siphon sensory neurons, which in turn leads to an elevation in the level of cAMP and to closure of voltage-gated K^+ channels. This prolongs depolarization and increases exocytosis of neurotransmitter. In classical conditioning, a conditioned stimulus, triggering the siphon sensory neurons and increasing Ca^{2+} levels, and an unconditioned stimulus, triggering the facilitator neurons, converge on activation of adenylate cyclase in the terminals of the siphon sensory neurons.

Long-term potentiation in the hippocampus region of the brain is a form of synaptic plasticity associated with learning. Activation of the NMDA glutamate receptors in these cells requires both glutamate and depolarization of the postsynaptic membrane; thus continuous activation of the postsynaptic cell makes it more sensitive to additional stimulation. Activation of the NMDA receptors causes an influx of Ca^{2+}; among the responses induced by Ca^{2+} is induction of NO synthesis. NO diffuses to the presynaptic cell—retrograde signaling—and there activates a protein that enhances glutamate secretion in response to the depolarization induced by arrival of an action potential. As a result, fewer action potentials in the presynaptic neurons can induce the same response in the postsynaptic neuron; the synapse "learns" to have an enhanced response to the electric signals in the presynaptic cells.

REVIEW QUESTIONS

1. The vast majority of cells have a resting voltage potential across their membranes, where the cytoplasm is negative with regard to the surrounding milieu. What is the origin of the resting membrane potential, and how is it maintained?

Neurons are the specialized cells within the nervous system that conduct electrical impulses. Describe the basic structure of a neuron. What is meant by orthograde and retrograde transport? Recall that each neuron has only one axon. In which direction along an axon does an action potential move? How is an impulse conducted down an axon? What guarantees the unidirectionality of the impulse? Which conducts action potentials faster, an axon with a small diameter or an axon with a large diameter? Why?

What happens to the membrane potential of a neuron when the permeability of the plasma membrane to K^+ is increased? What happens when the permeability to Na^+ is increased? Likewise, what happens when the permeability to Cl^- is increased?

For an impulse to be transmitted between two neurons, it must cross a synapse. What are the differences between a chemical and an electrical synapse? Electrical synapses conduct action potentials with little delay. Why, then, are chemical synapses much more common than electrical synapses if they transmit action potentials relatively slowly?

The nervous system is often divided into two principal components, the central nervous system and the peripheral nervous system. What comprises each of these systems, and what are their functions?

a. Refer to Figure 21-15 which shows the values of the membrane potential over time following excitation of a nerve cell. Of great importance is the relationship of the action potential to the membrane permeabilities of the ions involved.

When an action potential occurs, is the change in the internal concentration of Na^+ ions significant, or is it negligible? Explain your answer.

b. Refer to equation 21-2 on page 934, which presents a complex version of the Nernst equation that approximates the membrane potential based on the concentrations of relevant ions and their corresponding membrane permeabilities. These values are summarized in the table.

Ion	Concentration (mM)		Permeability (cm/s)
	Cytosol	Medium	
K^+	140	4	10^{-7}
Na^+	12	150	10^{-8}
Cl^-	4	120	10^{-7}

At the peak of an action potential, the $E_{membrane}$ can reach $+50$ mV. The opening of the voltage-dependent sodium channels is the initial occurrence that permits the formation of an action potential; the changes in K^+ and Cl^- permeabilities are negligible at the onset of the event. Assuming that the depolarization during an action potential peaks at $+50$ mV, and assuming that the initial permeabilities of K^+ and Cl^- do not change, what is the membrane permeability to Na^+ at the peak of an action potential? How much larger or smaller is this than the permeability of the membrane to Na^+ when the cell is at rest?

2. The rate of conductance of nerve impulses along long neurons is enhanced by myelination. How does this happen?

Review the technique of patch clamping. How is it used to investigate the flow of ions through ion channels?

What is meant by the term threshold potential? How does this contribute to the all-or-none phenomenon as it relates to voltage-gated sodium channels?

What is known about the comparative structures of voltage-gated ion channels? Be certain you are aware of how the *Drosophila shaker* mutant was used to isolate the voltage dependent K^+ channels.

a. You are using the technique of patch clamping to study the characteristics of a voltage-dependent Na^+ channel from human cells. How would you set up the patch clamp to ensure that you measure sodium current alone?

b. In experiment I, you maintain the membrane potential at -55 mV and determine the ability of Na^+ to move through the channel. You do the same in experiments II, III, and IV, but you maintain the membrane potentials at -40 mV, -20 mV, and $+20$ mV, respectively. The results in terms of Na^+ permeability are presented in the following table.

Experiment	Clamped Membrane Potential (mV)	Na^+ Permeability
I	-55	Absent
II	-40	Absent
III	-20	Present
IV	$+20$	Present

Explain these results. Can you state anything about the approximate magnitude of the threshold potential for these cells?

c. If you performed the same experiments in the presence of tetrodotoxin, what would occur?

3. Where are neurotransmitters stored in chemical synapses? What causes secretion of a neurotransmitter into the synaptic cleft?

What are the differences between excitatory and inhibitory synapses? What are ligand-gated ion channels? How do neurotransmitter receptors coupled to G proteins function? What is signal computation?

What are cholinergic synapses? How is acetylcholine made and stored in the presynaptic vesicles? What is the underlying mechanism governing the fusion of these vesicles with the presynaptic membrane? What is the difference between a nicotinic and a muscarinic receptor, and what types of channels are controlled by each?

What are the ways of inactivating a neurotransmitter?

a. You are investigating two neuronal cell lines derived from two strains of mice that exhibit neurological defects. In cells from the strain marked A, you have discovered that the class V ATPase in endosomes is defective, resulting in an internal pH of 7.0 in the organelle. Where would the neurotransmitter be found in the cell, and how would this affect the ability of the cell to transmit an impulse across the synapse?

b. In cells obtained from the strain marked B, you have determined that the K_m for the binding of the neurotransmitter for the H^+/neurotransmitter antiport is approximately one order of magnitude higher than that found in the normal mouse. How would this affect the storage of the neurotransmitter and its subsequent function at the synapse?

4. The visual system involving rods and cones presents a good illustration of a sensory system. How do rod cells respond to increased levels of light? Rhodopsin is the photoreceptor in rod cells. What are the components of rhodopsin? What happens to rhodopsin following absorption of a photon? What happens to internal cGMP levels following the absorption of light? How do rod cells adjust to increased or decreased levels of light? What is rhodopsin kinase? What is arrestin? Suppose that a patient had a visual defect resulting from partial lack of the phosphatase that removes phosphate residue from phosphorylated opsin. How would this affect the ability of the patient to adapt to dim light after being out in the sun?

What are the three classes of cone cells in humans? How do we achieve a sense of color from these cells? Why is red/green color blindness more prevalent in males? Why is loss of the red pigment gene more common than the loss of the blue pigment gene?

References

General Properties of Neurons and Nervous Systems

Cell vol. 72/*Neuron* vol. 10. 1993 Review Supplement: Signaling at the synapse. Several excellent review articles.

HILLE, B. 1991. *Ionic Channels of Excitable Membranes*. 2d ed. Sinauer Associates.

KANDEL, E. R., J. H. SCHWARTZ, and T. M. JESSELL. 1991. *Principles of Neural Science*, 3d ed. Elsevier.

KATZ, B. 1966. *Nerve, Muscle and Synapse*, 2d ed. McGraw-Hill.

KELNER, K. L., and D. E. KOSHLAND, JR., eds. 1989. *Molecules to Models: Advances in Neuroscience*. American Association for the Advancement of Science. A collection of research articles and reviews in the neurosciences.

LEVITAN, I. B., and L. K. KACZMAREK. 1991. *The Neuron: Cell and Molecular Biology*. Oxford University Press.

Molecular Biology of Signal Transduction. 1988. Cold Spring Harbor Symp. Quant. Biol., vol. 53.

NICHOLLS, J. G., A. R. MARTIN, and B. G. WALLACE. 1992. *From Neuron to Brain.* Sinauer Associates.

WHEAL, H. V., and A. M. THOMSON. 1991. *Excitatory Amino Acids and Synaptic Transmission.* Academic Press.

The Action Potential: Ionic Movements

HODGKIN, A. L. 1964. *The Conduction of the Nervous Impulse.* Liverpool University Press, Liverpool, U.K.

KEYNES, R. D. 1979. Ion channels in the nerve-cell membrane. *Sci. Am.* **240**(3):126–135.

General Properties of Ion Channels

ARMSTRONG, C. M. 1992. Voltage-dependent ion channels and their gating. *Physiol. Rev.* **72**:S5–13.

BETZ, H. 1990. Homology and analogy in transmembrane channel design: lessons from synaptic membrane proteins. *Biochemistry* **29**:3591–3599.

CATTERALL, W. A., et al. 1991. Structure and modulation of voltage-gated ion channels. *Ann. NY Acad. Sci.* **625**:174–180.

HILLE, B. 1992. *Ionic Channels of Excitable Membranes,* 2d ed. Sinauer Associates.

ISUM, L. L., K. S. DEJONGH, and W. A. CATTERALL. 1994. Auxiliary Subunits of Voltage-Gated Ion Channels. *Neuron* **12**: 1183–1194.

JAN, L. Y., and Y. N. JAN, 1992. Tracing the roots of ion channels. *Cell* **69**:715–718.

MILLER, M. 1989. Genetic manipulation of ion channels: a new approach to structure and mechanism. *Neuron.* **2**:1195–1205.

MONTAL, M. 1990. Molecular anatomy and molecular design of channel proteins. *FASEB J.* **4**:2623–2635.

NEHER, E. 1992. Nobel lecture. Ion channels for communication between and within cells. *Neuron.* **8**:605–12 and *Science* **256**:498–502.

NEHER, E., and B. SAKMANN. 1992. The patch clamp technique *Sci. Am.* (3):28–35.

RANGANATHAN, R. 1994. Evolutionary Origins of Ion Channels. *Proc. Nat'l Acad. Sci.* **91**:3484–3486.

SAKMANN, B. 1992. Nobel Lecture. Elementary steps in synaptic transmission revealed by currents through single ion channels. *EMBO J.* **11**:2002–2016 and *Science* **256**:503–512.

UNWIN, N. 1989. The structure of ion channels in membranes of excitable cells. *Neuron.* **3**:665–676.

YAU, K-W. 1994. Cyclic Nucleotide-gated Channels; an expanding new family of ion channels. *Proc. Nat'l Acad. Sci.* **91**:3481–3483.

Voltage-Gated Sodium Channels

CATTERALL, W. A. 1992. Cellular and molecular biology of voltage-gated sodium channels. *Physiol. Rev.* **72**:S15–48.

HARTSHORNE, R., M. TAMKUN, and M. MONTAL. 1986. The reconstituted sodium channel from brain. In C. Miller, ed., *Ion Channel Reconstitution,* Plenum.

VASSILEV, P. M., T. SCHEUER, and W. A. CATTERALL. 1988. Identification of an intracellular peptide segment involved in sodium channel inactivation. *Science* **241**:1658–1660.

Voltage-Gated Potassium Channels

ALDRICH, R. W. 1990. Potassium channels. Mixing and matching. *Nature* **345**:475–476.

BALDWIN, T. J., et al. 1992. Elucidation of biophysical and biological properties of voltage-gated potassium channels. *Cold Spring Harbor Symp. Quant. Biol.* **57**:491–499.

JAN, L. Y., and Y. N. JAN. 1990. How might the diversity of potassium channels be generated? *Trends Neurosci.* **13**:415–419.

JAN, L. Y., and Y. N. JAN. 1992. Structural elements involved in specific K$^+$ channel functions. *Ann. Rev. Physiol.* **54**:537–555.

MACKINNON, R. 1991. New insights into the structure and function of potassium channels. *Curr. Opin. Neurobiol.* **1**:14–19.

MILLER, C. 1991. 1990: Annus mirabilis of potassium channels. *Science* **252**:1092–1096.

Impulse Transmission and Myelin

COMPSTON, A., et al. 1991. The pathogenesis of demyelinating disease: insights from cell biology. *Trends Neurosci.* **14**:175–182.

DAVISON, A. N. 1991. Myelin structure and demyelination in multiple sclerosis. *Ann. NY Acad. Sci.* **633**:174–177.

LEMKE, G. 1993. The molecular genetics of myelination: an update. *Glia* **7**:263–271.

LI, C. et al. 1994. Myelination in the absence of myelin-associated glycoprotein. *Nature* **369**:747–750.

MARTIN, R., H. F. MCFARLAND, and D. E. MCFARLIN. 1992. Immunological aspects of demyelinating diseases. *Ann. Rev. Immunol.* **10**:153–187.

RAINE, C. D. 1984. Morphology of myelin and myelination. In P. Morell, ed., *Myelin,* 2d ed., Plenum.

READHEAD, C. et al. 1990. Role of myelin basic protein in the formation of central nervous system myelin. *Ann. NY Acad. Sci.* **605**:280–285.

SALZER, J. L., et al. 1990. Structure and function of the myelin-associated glycoproteins. *Ann. NY Acad. Sci.* **605**:302–312.

WAXMAN, S. G., and J. M. RITCHIE. 1985. Organization of ion channels in the myelinated nerve fiber. *Science* **228**:1502–1507.

Electric Synapses

BENNETT, M. V., et al. 1991. Gap junctions: new tools, new answers, new questions. *Neuron.* **6**:305–320.

HANNA, R. B., J. S. KEETER, and G. P. PAPPAS. 1978. The fine structure of a rectifying electrotonic synapse. *J. Cell Biol.* **79**:764–773.

STAUFFER, K. A., and N. UNWIN. 1992. Structure of gap junction channels. *Semin. Cell Biol.* **3**:17–20.

Chemical Synapses: Structure and Function of Synapses and Synaptic Vesicles

BENNETT, M. K., and R. H. SCHELLER. 1993. The molecular machinery for secretion is conserved from yeast to neurons. *Proc. Nat'l. Acad. Sci. USA* **90**:2559–2563.

DI ANTONIO, A. and T. L. SCHWARZ. 1994. The effect on synaptic physiology of synaptotagmin mutations in Drosophila. *Neuron.* **12**:909–920.

DE CAMILLI, P., et al. 1990. The synapsins. *Ann. Rev. Cell Biol.* **6**:433–460.

FERRO-NOVICK, S. and R. JAHN. 1994. Vesicle fusion from yeast to man. *Nature* **370**:191–193.

GEPPERT, M., et al. 1992. Neurexins. *Cold Spring Harbor Symp. Quant. Biol.* **57**:483–490.

GREENGARD, P., et al. 1993. Synaptic vesicle phosphoproteins and regulation of synaptic function. *Science* **259**:780–785.

JESSELL, T. M., and E. R. KANDEL. 1993. Synaptic transmission: a bidirectional and self-modifiable form of cell-cell communication. *Cell* **72**:1–30.

NEIMANN, H., J. BLASI, and R. JAHN. 1994. Clostridial Neurotoxins: New tools for dissecting exocytosis. *Trends in Cell Biology* **4**: 179–185.

O'CONNOR, V., G. J. AUGUSTINE, and H. BETZ. 1994. Synaptic Vesicle Exocytosis: Molecules and Models. *Cell* **76**:785–787.

SUDHOF, T. C., et al. 1993. Membrane fusion machinery: insights from synaptic proteins. *Cell* **75**:1–4.

THOMAS P., and W. ALMERS. 1992. Exocytosis and its control at the synapse. *Curr. Opin. Neurobiol.* **2**:308–311.

VALTORTA, F., F. BENFENATI, and P. GREENGARD. 1992. Structure and function of the synapsins. *J. Biol. Chem.* **267**:7195–7198.

WALCH-SOLIMENA, C., R. JAHN, and T. C. SUDHOF. 1993. Synaptic vesicle proteins in exocytosis: what do we know? *Curr. Opin. Neurobiol.* **3**:329–336.

Voltage-gated Calcium Channels and Exocytosis

ALSOBROOK, J. P., II, and C. F. STEVENS. 1988. Cloning the calcium channel. *Trends Neurosci.* **11**:1–3.

BEAM, K. G., T. TANABE, and S. NUMA. 1989. Structure, function, and regulation of the skeletal muscle dihydropyridine receptor. *Ann. NY Acad. Sci.* **560**:127–137.

CATTERALL, W. A. 1991. Functional subunit structure of voltage-gated calcium channels. *Science* **253**:1499–1500.

CATTERALL, W. A., and J. STRIESSNIG. 1992. Receptor sites for Ca^{2+} channel antagonists. *Trends Pharmacol. Sci.* **13**:256–262.

HESS, P. 1990. Calcium channels in vertebrate cells. *Ann. Rev. Neurosci.* 13:337–356.

MONCK, J. R., and J. FERNANDEZ. 1994. The Exocytic Fusion Pore and Neurotransmitter Release. *Neuron.* 12:707–716.

Nicotinic Acetylcholine Receptors

BREHM, P. 1989. Resolving the structural basis for developmental changes in muscle ACh receptor function: it takes nerve. *Trends Neurosci.* 5:174–177.

CHANGEUX, J. P. 1990. The TIPS lecture. The nicotinic acetylcholine receptor: an allosteric protein prototype of ligand-gated ion channels. *Trends Pharmacol. Sci.* 11:485–492.

CHANGEUX, J. P., et al. 1992. New mutants to explore nicotinic receptor functions. *Trends Pharmacol. Sci.* 13:299–301.

FATT, P., and B. KATZ. 1951. An analysis of the end-plate potential recorded with an intracellular electrode. *J. Physiol.* 115:320–370.

KARLIN, A. 1993. Structure of nicotinic acetylcholine receptors. *Curr. Opin. Neurobiol.* 3:299–309.

LENA, C., and J. P. CHANGEUX. 1993. Allosteric modulations of the nicotinic acetylcholine receptor. *Trends Neurosci.* 16:181–186.

SARGENT, P. B. 1993. The diversity of neuronal nicotinic acetylcholine receptors. *Ann. Rev. Neurosci.* 16:403–434.

TOYOSHIMA, C., and N. UNWIN, 1988. Ion channel of acetylcholine receptor reconstructed from images of postsynaptic membranes. *Nature* 336:247–250.

UNWIN, N. 1993. Neurotransmitter action: opening of ligand-gated ion channels. *Cell* 72:31–41.

The Neuromuscular Junction and Acetylcholinesterase

HALL, Z. W., and J. R. SANES. 1993. Synaptic structure and development: the neuromuscular junction. *Cell* 72:99–121.

SALPETER, M. M., ed. 1987. *The Vertebrate Neuromuscular Junction.* Neurology and Neurobiology Series, vol. 23. Alan R. Liss.

TAYLOR, P. 1991. The cholinesterases. *J. Biol. Chem.* 266:4025–4028.

GABA, Glycine, and Inhibitory Synapses

BARNARD, E. A., M. G. DARLISON, and P. SEEBURG. 1987. Molecular biology of the GABA$_A$ receptor: the receptor/channel superfamily. *Trends Neurosci.* 10:502–509.

BECKER, C. M. 1990. Disorders of the inhibitory glycine receptor: the spastic mouse. *FASEB J.* 4:2767–2774.

BETZ, H. 1991. Glycine receptors: heterogeneous and widespread in the mammalian brain. *Trends Neurosci.* 14:458–461.

GOTTLIEB, D. I. 1988. GABAergic neurons. *Sci. Am.* 258(2):38–45.

LANGOSCH, D., C. M. BECKER, and H. BETZ. 1990. The inhibitory glycine receptor: a ligand-gated chloride channel of the central nervous system. *Eur. J. Biochem.* 194:1–8.

G Proteins and Direct Coupling of Neurotransmitter Receptors to Ion Channels

BONNER, T. I. 1989. The molecular basis of muscarinic receptor diversity. *Trends Neurosci.* 12:148–151.

BROWN, A. M., and L. BIRNBAUMER. 1990. Ionic channels and their regulation by G protein subunits. *Ann. Rev. Physiol.* 52:197–213.

BROWNSTEIN, M. J. 1993. A brief history of opiates, opioid peptides, and opioid receptors. *Proc. Nat'l. Acad. Sci. USA* 90:5391–5393.

HEMMINGS, H. C., JR., et al. 1989. Role of protein phosphorylation in neuronal signal transduction. *FASEB J.* 3:1583–1592.

HILLE, B. 1992. G protein-coupled mechanisms and nervous signaling. *Neuron.* 9:187–195.

KACZMAREK, L. K., and I. B. LEVITAN, eds. 1987. *Neuromodulation: The Biochemical Control of Neuronal Excitability.* Oxford University Press.

KUBO, Y., et al. 1993. Primary structure and functional expression of a rat G-protein-coupled muscarinic potassium channel. *Nature* 364:802–806.

ROBISHAW, J. D., and K. A. FOSTER. 1989. Role of G proteins in the regulation of the cardiovascular system. *Ann. Rev. Physiol.* 51:229–244.

Neurotransmitter Transporters

AMARA, S. G., and J. L. ARRIZA. 1993. Neurotransmitter transporters: three distinct gene families. *Curr. Opin. Neurobiol.* 3:337–344.

AMARA, S. G., and M. J. KUHAR. 1993. Neurotransmitter transporters: recent progress. *Ann. Rev. Neurosci.* 16:73–93.

BARONDES, S. 1994. Thinking about Prozac. *Science* 263:1102–1103.

COOPER, J. R., F. E. BLOOM, and R. H. ROTH. 1991. *The Biochemical Basis of Neuropharmacology,* 6th ed., Oxford University Press.

KANAI, T., C. P. SMITH, and M. A. HEDIGER. 1994. A new family of neurotransmitter transporters: the high-affinity glutamate transporters. *FASEB J.* 8:1450–1459.

LEONARD, B. E. 1992. *Elements of Psychopharmacology,* Wiley, Chichester, U.K.

Sensory Transduction: Olfactory and Touch Receptors

CHESS, A., et al. 1992. Molecular biology of smell: expression of the multigene family encoding putative odorant receptors. *Cold Spring Harbor Symp. Quant. Biol.* 57:505–516.

DIONNE, V. E. 1994. Emerging complexity at odor transduction. *Proc. Nat'l. Acad. Sci.* 91:6253–6254.

GOULDING, E. H., et al. 1992. Molecular cloning and single-channel properties of the cyclic nucleotide-gated channel from catfish olfactory neurons. *Neuron.* 8:45–58.

JENTSCH, T. J. 1994. Trinity of ion channels (a review of touch receptors) *Nature* 367:412–413.

NGAI, J., et al. 1993. Coding of olfactory information: topography of odorant receptor expression in the catfish olfactory epithelium. *Cell* 72:667–680.

RONNETT, G. V., and S. H. SNYDER. 1992. Molecular messengers of olfaction. *Trends Neurosci.* 15:508–513.

Sensory Transduction: The Visual System

DE VRIES, S. H., and D. A. BAYLOR. 1993. Synaptic circuitry of the retina and olfactory bulb. *Cell* 72:139–149.

DOWLING, J. 1987. *The Retina: An Approachable Part of the Brain.* Harvard University Press.

HUBEL, D. H. 1988. *Eye, Brain, and Vision.* Scientific American Library.

KAUPP, U. B., and K. W. KOCH. 1992. Role of cGMP and Ca^{2+} in vertebrate photoreceptor excitation and adaptation. *Ann. Rev. Physiol.* 54:153–175.

KHORANA, H. G. 1992. Rhodopsin, photoreceptor of the rod cell. An emerging pattern for structure and function. *J. Biol. Chem.* 267:1–4.

LAGNADO, L., and D. BAYLOR. 1992. Signal flow in visual transduction. *Neuron.* 8:995–1002.

NATHANS, J. 1994. In the eye of the beholder: visual pigments and inherited variation in human vision. *Cell* 78:357–360.

NATHANS, J. 1992. Rhodopsin: structure, function, and genetics. *Biochemistry* 31:4923–4931.

NATHANS, J., et al. 1992. Molecular genetics of human visual pigments. *Ann. Rev. Genet.* 26:403–424.

RANDO, R. R. 1991. Membrane phospholipids as an energy source in the operation of the visual cycle. *Biochemistry* 30:595–602.

YARFITZ, S. and J. B. HURLEY. 1994. Transduction mechanisms of vertebrate and invertebrate photoreceptors. *J. Biol. Chem.* 269:14329–14332.

Memory and Neurotransmitters; *Drosophila* Mutations Affecting Learning

DUDAI, Y. 1988. Neurogenetic dissection of learning and short-term memory in *Drosophila*. *Ann. Rev. Neurosci.* **11**:537–563.

HALL, J. C. 1994. The mating of a fly. *Science* **264**:1702–1714.

HOTTA, Y., and S. BENZER. 1972. Mapping of behavior in *Drosophila* mosaics. *Nature* **240**:527–535.

QUINN, W. G., W. A. HARRIS, and S. BENZER. 1974. Conditioned behavior in *Drosophila melanogaster*. *Proc. Nat'l. Acad. Sci. USA* **71**:708–712.

TULLY, T. 1987. *Drosophila* learning and memory revisted. *Trends Neurosci.* **10**:330–335.

Learning in the Sea Slug *Aplysia*

ABRAMS, T. W., and E. R. KANDEL. 1988. Is contiguity detection in classical conditioning a system or a cellular property? Learning in *Aplysia* suggests a possible molecular site. *Trends Neurosci.* **11**:128–135.

HAWKINS, R. D., E. R. KANDEL, and S. A. SIEGELBAUM. 1993. Learning to modulate transmitter release: themes and variations in synaptic plasticity. *Ann. Rev. Neurosci.* **16**:625–665.

JESSELL, T. M., and E. R. KANDEL. 1993. Synaptic transmission: a bidirectional and self-modifiable form of cell-cell communication. *Cell* **72**:1–30.

SCHACHER, S., et al. 1990. Long-term facilitation in *Aplysia*: persistent phosphorylation and structural changes. *Cold Spring Harbor Symp. Quant. Biol.* **55**:187–202.

SCHWARTZ, J. H. 1993. Cognitive kinases. *Proc. Nat'l. Acad. Sci. USA* **90**:8310–8313.

SCHWARTZ, J. H., and S. M. GREENBERG. 1987. Molecular mechanisms for memory: second-messenger-induced modifications of protein kinases in nerve cells. *Ann. Rev. Neurosci.* **10**:459–476.

Long-term Potentiation and Learning in the Mammalian Brain

BLISS, T. V., and G. L. COLLINGRIDGE. 1993. A synaptic model of memory: long-term potentiation in the hippocampus. *Nature* **361**:31–39.

BREDT, D. S., and S. H. SNYDER. 1992. Nitric oxide, a novel neuronal messenger. *Neuron.* **8**:3–11.

DAVIS, G. W., and R. K. MURPHEY. 1994. Long-term regulation of short-term transmitter release properties: retrograde signaling and synaptic development. *Trends Neurosci.* **17**:9–13.

GRANT, S., and A. J. SILVA. 1994. Targeting learning. *Trends Neurosci.* **17**:71–75.

LOWENSTEIN, C. J., and S. H. SNYDER. 1992. Nitric oxide, a novel biologic messenger. *Cell* **70**:705–707.

SNYDER, S. H. 1992. Nitric oxide: first in a new class of neurotransmitters. *Science* **257**:494–496.

Microfilaments: Cell Motility and Control of Cell Shape

Cell motility is one of the crowning achievements of evolution. Primitive cells were probably immobile, carried by currents in the primordial milieu. Diffusion within the cell was sufficient to distribute metabolites, but as cells grew larger and acquired more sophisticated functions, transport systems evolved to move materials within the cell. These systems became components of a motility apparatus which enabled cells to disperse to locations more favorable for growth. With the evolution of multicellular organisms, primitive organs were formed by migrations of single cells and groups of cells from distant parts of the embryo. In adult organisms, movements of single cells in search of foreign organisms are integral to the host's defenses against infection; on the other hand, we also know that uncontrolled cell motility is an ominous sign of a cancerous cell.

Not all cells migrate. In fact, most cells in the body are stationary. Nevertheless they also exhibit dramatic changes in their morphology— the contraction of muscle cells, the elongation of nerve axons, the formation of cell surface protrusions, such as microvilli and filopodia, the constriction of a dividing cell during mitosis. Perhaps the most subtle movements are those within cells—the active separation of chromosomes, the streaming of cytosol, the transport of membrane vesicles. From our discussions in other chapters, we recognize these internal movements as essential elements in the growth and differentiation of cells. In all of these examples, the movements are not random events but instead are carefully controlled by the cell to take place at specified times and in particular locations.

Several elements contribute to cell movements. Movement is a manifestation of mechanical work, which requires a fuel (ATP) and proteins that convert the energy stored in ATP into motion. Also, all movements involve the *cytoskeleton*, a cytoplasmic system of fibers. Like steel girders supporting the shell of a building, the cytoskeleton

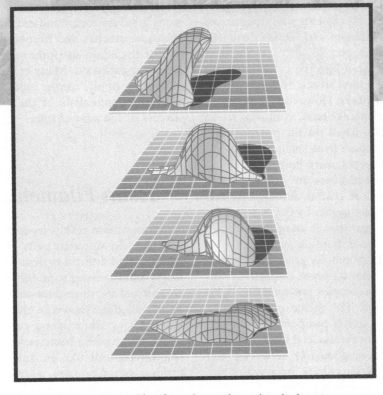

▲ Two minutes in the life of a polymorphonuclear leukocyte.

plays a structural role by supporting the cell membrane or by forming tracks along which organelles and other elements move in the cytosol. Unlike the passive framework of a building, though, the cytoskeleton is dynamic, undergoing constant rearrangements, which also produce movements.

The cytoskeleton is an internal network of three types of cytosol fibers: 7- to 9-nm-diameter *microfilaments*, 10-nm-diameter *intermediate filaments,* and 24-nm-diameter *microtubules.* Only microfilaments and microtubules are involved in motility; intermediate filaments are not known to generate any cell movements. Instead, they structurally brace the inside of a cell. Microfilaments will be discussed in this chapter, while microtubules and intermediate filaments will be discussed in Chapter 23.

When biologists first peered into the cytoplasm with the electron microscope, they were struck by the dense and seemingly random array of fibers. However, we now recognize that this array, the cytoskeleton, conforms to an architecture, an organization based on the principle that large, complex structures are built from small, simple components. Just as proteins are polymers built from amino acid monomers, so cytoskeletal fibers are polymers built from small protein subunits. A crucial difference is that these protein subunits are held together by the same noncovalent bonds that hold other proteins in complexes instead of by the covalent peptide bonds that link amino acid residues. Microfilaments and microtubules grow by the polymerization of actin and tubulin subunits, respectively; intermediate filaments are more complex polymers built with a common α-helical subunit. The theme of assembling complicated structures from simple components is carried further by the cross-linkages of cytoskeletal fibers into such structures as bundles, geodesic-dome-like networks, and gel-like lattices. The cell takes its shape by bonding the plasma membrane to these protein supports.

Cells have evolved two basic mechanisms for motility. One mechanism involves a special class of enzymes called *motor proteins.* These proteins use energy gained from the hydrolysis of ATP and the release of ADP and phosphate to walk or slide along a microfilament or a microtubule. Some motor proteins carry membrane-bound organelles and vesicles along the cytoskeletal fiber tracks; other motor proteins cause the fibers to slide past each other. The other mechanism for motility, which is responsible for many of the changes in the shape of a cell, results from the polymerization of tubulin and actin and their assembly into bundles and networks. A few movements involve both the action of motor proteins and cytoskeleton rearrangements.

We begin our discussion of the cytoskeleton with actin for two reasons: first because we know more about actin than we do about the other cytoskeletal proteins, and second because actin can play a role in every type of motility, from cell migration to cytosol transport. To provide a background for understanding the material in the subsequent sections, we will begin with a brief description of the movements in a crawling animal cell. Then we will cover the biochemistry of actin (the structure and function of actin and proteins which interact with it) and discuss examples in which actin polymerization is the major mechanism for cell motility and changes in shape. We next introduce myosin, the actin motor protein, examining its structure and assembly and how it transduces the energy of ATP into a sliding movement along actin filaments. With a basic understanding of the key components of the cytoskeleton, we can examine the mechanisms responsible for motility by studying model cell systems, starting with muscle (the paradigm for understanding how actomyosin interactions generate force) and ending with the crawling movements of amebas and fibroblasts. The chapter will finish with a discussion about the regulation of movement by cell signaling pathways. Many of the principles that we discuss here also apply to the microtubule and intermediate-filament components of the cytoskeleton, structures we discuss in the next chapter.

➤ *Actin Filaments*

Cell locomotion results from the coordination of motions generated by different parts of a cell. These motions are complex and difficult to describe, but we can pick out their major features using sophisticated computer programs that reconstruct the three-dimensional shapes of a moving cell. In the diagram shown in the opening to this chapter, a leukocyte, a white blood cell that patrols the body in search of invading bacteria, displays motions typical of a fast-moving cell like an ameba. Initially, the leukocyte forms a *pseudopodium,* a large, broad membrane protrusion, which arches forward from the cell. After contacting the substratum, the pseudopodium quickly fills with cytosol and the rear of the cell retracts forward toward the body of the cell.

In a slow-moving cell like a fibroblast (Figure 22-1; see also Figure 5-19), the motions involve different structures than those formed in a moving leukocyte and their dynamics are much slower. As a fibroblast moves, it extends either slender "fingers" of membrane, called *filopodia,* or thin membrane sheets, *lamellipodia,* from the leading edge (Figure 22-1a). Where these structures contact the substratum, the cell membrane quickly assembles special structures, called *adhesion plaques,* on the ventral surface of the cell. These plaques serve to hold the cell to the substratum. As the cell continues to travel, its tail is pulled forward, often leaving behind small patches of the cell still firmly attached to the substratum through the adhesion plaques. Some areas of the cell membrane do not form stable adhesions at the leading edge of the cell; these areas project upward as a thin veil, or *ruffle,* that moves as an undulating ridge back along the dorsal cell surface toward the cell

(a)

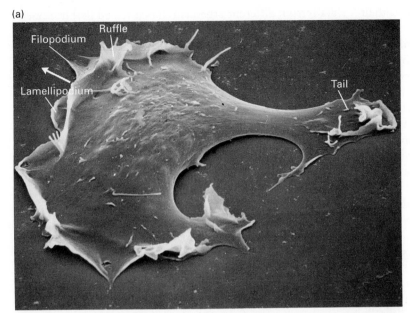

Filopodium
Ruffle
Lamellipodium
Tail

(b)

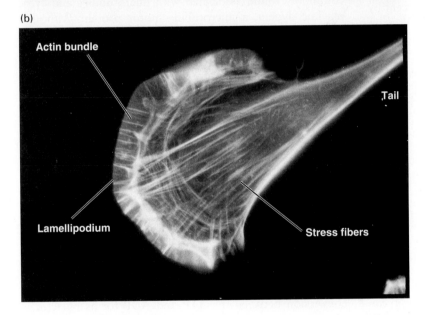

Actin bundle
Tail
Lamellipodium
Stress fibers

◄ FIGURE 22-1   Actin structures in a fibroblast. (a) Scanning electron micrograph of a cultured fibroblast. At the front of the cell, filopodia, lamellipodia, and ruffles project from the cell membrane. At the rear of the cell, the tail is firmly attached to the surface. The arrow indicates the direction of movement. (b) Fluorescence micrograph of a fan-shaped fibroblast, stained with rhodamine phalloidin. Visible are numerous actin bundles in the lamella and stress fibers in the cell body. [Part (a) Courtesy of J. Heath; part (b) courtesy of B. Hollifield.]

body. From this brief description of moving cells we can take note of one other feature—all moving cells display *polarity*; that is, certain structures are always formed at the front of the cell, while other structures are found at the rear.

The machinery that powers these movements is built from the *actin cytoskeleton*. Visualized in a fibroblast by fluorescence microscopy after the actin filaments are stained with a fluorescent dye (Figure 22-1b), the actin cytoskeleton is huge. It is larger than any organelle, filling the cytosol with actin filaments. Radially oriented actin filament bundles lie at the leading edge, and axial bundles, called *stress fibers*, run along the entire length of the cell. In

addition, a network of actin filaments fills the rest of the cell, but the individual filaments of this network are difficult to resolve in the light microscope.

Much of the discussion in this section will focus on the ability of huge actin filament structures to control the shape of a cell. Because the actin cytoskeleton is so big, it can easily change cell morphology just by assembling or disassembling itself. Therefore, we will discuss in detail the special properties of the proteins that comprise the cytoskeleton and control its organization.

In previous chapters, we have seen examples of large protein complexes in which the number and positions of the subunits are fixed. For example, all ribosomes have the same number of protein and RNA components and their three-dimensional geometry is invariant. However, the actin cytoskeleton is different—the lengths of filaments vary greatly, the filaments are cross-linked into imperfect bundles and networks, and the ratio of cytoskeletal proteins is not rigidly maintained. This lack of regularity and order is a property needed for flexibility of function because a cell can have many shapes and can vary them easily. Thus the cytoskeleton must assemble rapidly and does not have the chance to form well-organized, highly ordered structures. Keeping this in mind, we will examine actin as a model for understanding how polymeric proteins form the structural framework of a cell and how the cell tailors this framework to carry out various tasks.

All Eukaryotic Cells Contain Abundant Amounts of Actin

Actin is the most abundant intracellular protein in a eukaryotic cell. In muscle cells, actin comprises 10 percent by weight of the total cell protein, and in nonmuscle cells actin makes up 1–5 percent of the cell's protein. A typical concentration of actin in the cytosol of human platelets is 0.5 mM. Even higher concentrations of actin exist in special structures like microvilli, where the actin concentration is 5 mM. To put this in perspective, compare the number of actin molecules with the 10,000 insulin receptor molecules. In a liver cell having a volume of 3.4 pl (3.4 picoliters, or 3.4×10^{-12} l), there are approximately half a billion (0.5×10^9) actin molecules. The number emphasizes a fact about cytoskeletal proteins in general: because they form structures that must cover large spaces in a cell, these proteins are the most abundant proteins in a cell.

The Actin Sequence Has Changed Little during Evolution

Actin, a moderate-sized protein consisting of approximately 375 residues, is encoded by a large, highly conserved gene family. Some single-celled organisms like yeasts and amebas have a single actin gene. Many multicellular eukaryotic organisms, contain multiple actin genes—the sea urchin has 11, humans have six, the slime mold *Dictyostelium* has 17, and some plants have as many as 60. In the sea urchin, some actin genes are expressed during embryonic development, whereas others are expressed in adult tissues. In some organisms, several genes encode identical actin protein sequences; in other cases, some of the genes are pseudogenes, which do not produce an actin protein.

At least six different actins have been identified in birds and mammals. Three are called *α-actins:* each one is unique to a different type of muscle. Two other actins, termed nonmuscle *β-actin* and *γ-actin,* are found in nearly all nonmuscle cells. The sixth actin, another *γ-actin* occurs in smooth muscles that line the intestine. The sequencing of actin from different organisms has revealed that it is one of the most conserved proteins in a cell, comparable with histones, the structural proteins of chromatin (see Chapter 9). The various *α-actins* differ in only four to six amino acids; nonmuscle *β-* and *γ-actin* differ from each other at just a single position and differ from striated muscle *α-actin* in only 25 residues. Moreover, actin residues from amebas and from animals are identical at 80 percent of the positions. Inferring phylogeny from the similarities in actin sequences, we suspect that muscle is a recent evolutionary adaptation. Perhaps, the contractile function of muscle requires a specialized muscle-specific actin to carry out contraction more efficiently, whereas actin in nonmuscle actin cells must be involved in more basic and varied cellular functions.

Recently, a family of actin-related proteins sharing 50 percent similarity with actin has been identified in many eukaryotic organisms. Paradoxically, actin-related proteins are not part of the actin cytoskeleton; instead, they are commonly found associated with microtubules and cytoplasmic dynein, a microtubule motor protein. Although their exact function is unknown, we will discuss actin-related proteins in the next chapter.

ATP Holds Together the Two Lobes of the Actin Monomer

Actin exists as a *globular* monomer called *G-actin* and as a *filamentous* polymer called *F-actin,* that is a string of G-actin subunits. (In a cell seen by electron microscopy, an F-actin filament, along with any proteins bound to it, is the *microfilament* of the cytoskeleton.) Each actin molecule contains a Mg^{2+} cation complexed with either ATP or ADP. Thus there are four states of actin: ATP–G-actin, ADP–G-actin, ATP–F-actin, and ADP–F-actin. Two of these forms, ATP–G-actin and ADP–F-actin, predominate in a cell. In a later section, we will discuss how the interconversion between the ATP and ADP forms of actin is important in the assembly of the cytoskeleton.

Although G-actin appears globular in the electron microscope, x-ray crystallographic analysis of G-actin (Figure 22-2a) reveals that it is separated into two lobes by a deep cleft. The lobes and the cleft comprise an *ATPase fold,* the site where ATP and Mg^{2+} are bound by ionic and hydrogen bonds to amino acid side chains.

A clue to the function of ATP and Mg^{2+} came from the surprising discovery that the ATPase fold is also found in other ATP-binding proteins, including hsp70 (a molecular chaperone (see Figure 19-7), sugar kinases such as hexokinase (see Figure 3-28), and some bacterial phosphatases. By comparing the structures, x-ray crystallographers noted that the floor of the cleft acts as a hinge that allows the lobes of the proteins to flex relative to one another. When bound to the cleft, ATP becomes a latch that holds the lobes together. From these observations, we can easily understand how the presence of ATP or ADP in the cleft will

(a)

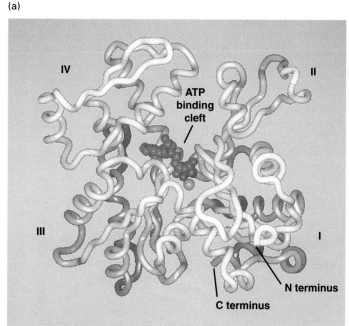

(b)

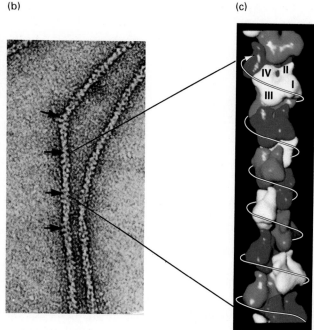

(c)

▲ FIGURE 22-2   Structures of an actin monomer and filament. (a) The atomic structure of a β-actin monomer from a nonmuscle cell shows a platelike molecule (measuring 5.5 × 5.5 × 3.5 nm) which is divided by a central cleft into two approximately equal-sized lobes and four subdomains, numbered I–IV. ATP, colored red, is at the bottom of the cleft and contacts both lobes (the yellow ball represents Mg^{2+}). The N- and C-termini lie in subdomain I. (b) In the electron microscope, negatively stained actin filaments appear as long, flexible, and twisted strands of beaded subunits. The helical arrangement of the subunits is most easily seen if you lift the page and "sight" along the filament.

Successive narrow (7 nm diameter) and wide (9 nm diameter) views (arrows) of the helix repeat every 36 nm. (c) One model of the arrangement of subunits in an actin filament, showing 14 subunits. The subunits lie in a tight helix along the filament, as indicated by the arrow. One repeating unit consists of 28 subunits (13 turns of the helix), covering a distance of 72 nm. The ATP-binding cleft is oriented in the same direction (*top*) in all actin subunits in the filament. [Part (a) adapted from C. E. Schutt, et al., 1993, *Nature* **365**:810; courtesy of M. Rozycki. Part (b) courtesy of R. Craig. Part (c) see M. F. Schmid, et al., 1994, *J. Cell Biol.* **124**:341; courtesy of M. Schmid.]

affect the conformation of the actin molecule; in fact, without a bound nucleotide, actin denatures very quickly.

G-Actin Assembles into Long F-Actin Polymers

The addition of ions—Mg^{2+}, K^+, or Na^+—to a solution of G-actin will induce the polymerization of G-actin into F-actin filaments. The process is also reversible: F-actin depolymerizes into G-actin when the ionic strength of the solution is lowered. The F-actin filaments that form in vitro are indistinguishable from microfilaments isolated from cells. This indicates that other factors such as accessory proteins are not required for polymerization. The assembly of G-actin into F-actin is accompanied by the hydrolysis of ATP to ADP and P_i; however, as we will discuss later, ATP hydrolysis affects the kinetics of polymerization but is not necessary for polymerization to occur.

The ability of G-actin to polymerize into F-actin and of F-actin to depolymerize into G-actin are perhaps the most important properties of actin. In this chapter, we will repeatedly discuss how the reversible assembly of actin lies at the core of many forms of motility.

F-Actin Is a Helical Polymer of Identical Subunits

When negatively stained by uranyl acetate for electron microscopy, F-actin is a long, flexible filament whose diameter varies between 7 and 9 nm. At higher magnifications, the globular subunits in the filament can be recognized, so that the filaments appear as twisted strings of beads (Figure 22-2b). From x-ray diffraction studies of actin filaments and the actin monomer structure shown in Figure 22-2a, scientists have produced a model of an actin filament which shows the subunits organized as a helix (Figure 22-2c). With this arrangement, each subunit is surrounded by four other subunits, one above, one below, and two to one side. Each subunit corresponds to a bead seen in the electron micrographs of a filament.

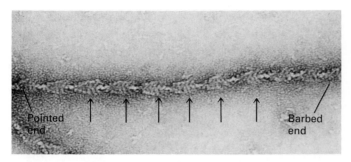

Pointed end

Barbed end

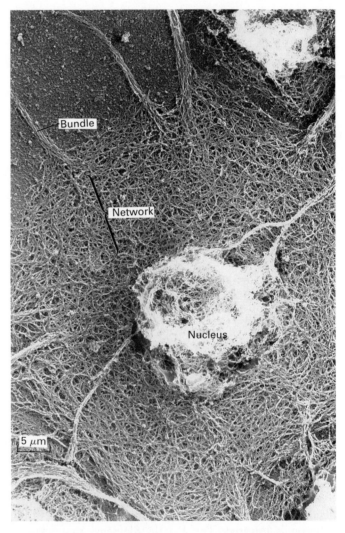

FIGURE 22-3   Polarity of subunits in an actin filament as detected by myosin S1 head binding. When all actin subunits are bound by myosin S1, the filament becomes thicker and the wider and narrower views of the filament become more exaggerated. The coating of myosin heads produces a series of arrowhead-like decorations, most easily seen at the wide views of the filament. This polarity in decoration defines a pointed end and a barbed end; the former corresponds to the top of the model in Figure 22-2c. [Courtesy of R. Craig.]

F-Actin Has Structural and Functional Polarity

All subunits in a filament have the same *polarity;* that is, they all point toward the same filament end. Consequently, at one end of the filament (the top end in Figure 22-2c), the ATP-binding cleft of an actin subunit is exposed to the surrounding solution, while at the opposite end the cleft contacts the neighboring actin subunit. Without the atomic vision of x-ray crystallography, we would find it impossible to see the cleft in a subunit and therefore the polarity of a filament; but using electron microscopy, we can deduce the polarity from *"decoration" experiments* that exploit the ability of the protein myosin to bind specifically to actin filaments. In this type of experiment an excess of myosin S1, the globular head domain of myosin, is mixed with actin filaments and binding is permitted to occur. Myosin attaches to the sides of a filament with a slight tilt (Figure 22-3). When all subunits are bound by myosin the filament appears coated, or "decorated," with arrowheads that all point toward one end of the filament, the "pointed" end, and away from the other, "barbed," end (Figure 22-3). Because myosin binds to actin filaments and not to microtubules or intermediate filaments, the arrowhead decoration was one criterion by which actin filaments were identified among the other cytoskeletal fibers in electron micrographs of thin-sectioned cells.

▶ *Actin Architectures*

So far we have discussed the organization of the actin cytoskeleton at its most basic level—the structure of the actin subunit and the helical arrangement of subunits in a filament. This knowledge of the filament gives us the background to explore how actin filaments themselves are organized in a cell. In this section, we will examine the two major structures of actin filaments in the cytosol, actin bundles and actin networks, and study their functions in cells.

Bundle

Network

Nucleus

5 μm

▲ FIGURE 22-4   Actin bundles and networks. Bundles and networks of actin filaments fill the cytosol of a spreading platelet treated with detergent to remove the plasma membrane. Actin bundles project from the cell to form the spike-like filopodia. In the lamellar region of the cell, the actin filaments form a network that fills the cytosol. In contrast to the roughly parallel alignment of bundled filaments, the filaments in networks lie at various angles approaching 90°. [Courtesy of J. Hartwig.]

The Actin Cytoskeleton Is Organized into Bundles and Networks of Filaments

On first looking at an electron micrograph or immunofluorescence micrograph of a cell, one is struck by the dense, seemingly disorganized mat of filaments present in the cytosol (Figure 22-4). However, a keen eye will start to pick out areas, generally where the membrane protrudes from the cell surface, in which the filaments are concentrated into bundles. From these bundles the filaments continue into the cell interior, where they fan out and become part of a network of filaments. These two structures, bundles and networks, are the most common arrangements of actin filaments in a cell. Functionally, bundles and networks have identical roles in a cell: both provide a framework which supports the plasma membrane and, therefore, determines a cell's shape. Structurally, bundles differ from networks mainly in the organization of actin filaments. In bundles the actin filaments are closely packed in parallel arrays, whereas in a network the actin filaments crisscross, often at right angles, and are loosely packed. We find two types of actin networks in a cell. A network associated with the plasma membrane is planar or two-dimensional, like a net or a web. Within the cell, the network of actin filaments is three-dimensional, giving the cytosol gel-like properties.

In all bundles and networks, the filaments are held together by *actin–cross-linking proteins* (Table 22-1). To connect a pair of filaments, a cross-linking protein must have two actin-binding sites, one site for each filament.

TABLE 22-1   Actin–Cross-Linking Proteins

Protein*	MW	Domain Organization†	Location
GROUP I			
30 Kd	33,000		Filopodia, lamellipodia, stress fibers
EF-1a	50,000		Pseudopodia
Fascin	55,000		Filopodia, lamellipodia, stress fibers, microvilli, acrosomal process
Scruin	102,000		Acrosomal process
GROUP II			
Villin	92,000		Intestinal and kidney brush border microvilli
Dematin	48,000		Erythrocyte cortical network
GROUP III			
Fimbrin	68,000		Microvilli, stereocilia, adhesion plaques, yeast actin cables
α-Actinin	102,000		Filopodia, lamellipodia, stress fibers, adhesion plaques
Spectrin	α: 280,000 β: 246,000– 275,000		Cortical networks
Dystrophin	427,000		Muscle cortical networks
ABP 120	92,000		Pseudopodia
Filamin	280,000		Filopodia, pseudopodia, stress fibers

* Cross-linking proteins are placed into three groups. Group I proteins have unique actin-binding domains; Group II have a 7,000-MW actin-binding domain; and Group III have pairs of a 26,000-MW actin-binding domain.
† Calmodulin-like calcium-binding domains (red), actin-binding domains (blue), α-helical repeats (purple), β-sheet repeats (green), uncharacterized domains (tan or yellow).

(a)

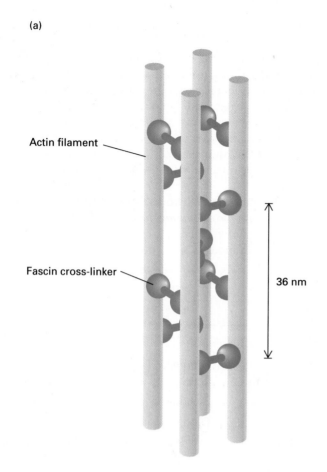

Actin filament

Fascin cross-linker

36 nm

(b)

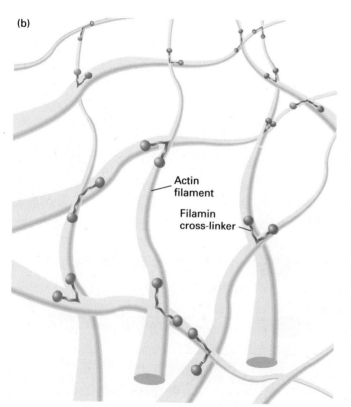

Actin filament

Filamin cross-linker

◄ FIGURE 22-5   Actin cross-linking proteins bridging pairs of actin filaments. (a) When cross-linked by the short, stiff fascin, actin filaments form a bundle. (b) Long cross-linking proteins are flexible and thus can cross-link pairs of filaments lying at various angles. Filamnin can adopt a 90° bend and thus can cross-link orthogonally oriented filaments. Generally, there are fewer cross-links per filament in a network than in a bundle.

The length and flexibility of a cross-linking protein critically determine whether bundles or networks are formed. Short cross-linking proteins will hold actin filaments closely together; this will tend to force the filaments into a parallel alignment, a *bundle* (Figure 22-5a). In contrast, long, flexible cross-linking proteins are able to adapt to any arrangement of actin filaments and thus will tether orthogonally oriented actin filaments in *networks* (Figure 22-5b). We can see this dependence on length in one group of cross-linking proteins (group III in Table 22-1). The proteins in this superfamily share a common actin-binding domain, but their lengths are determined by α-helical or β-sheet motifs which are used as spacers between actin-binding domains. The shortest members of this group, fimbrin and α-actinin, are found in actin bundles in a cell, while the longest members of the group, filamin, spectrin, and dystrophin, are found in actin networks. Many cross-linking proteins contain a camodulin-like calcium binding domain. The role of calcium is regulatory; it prevents these proteins from cross-linking actin filaments.

Actin Bundles and Networks Are Connected to the Membrane

The distinctive shape of a cell is dependent not only on the organization of actin filaments but also on proteins that connect the filaments to the membrane. These proteins, called *membrane-microfilament binding proteins,* act as spot welds which tack the sheet of membrane to the underlying cytoskeleton framework. When attached to a bundle of filaments, the membrane acquires a fingerlike shape; when attached to a planar network of filaments, the membrane is held flat. Membrane-microfilament binding proteins vary greatly in their structures. The simplest connections are made by integral membrane proteins that bind directly to actin filaments. In one example, the actin cytoskeleton of the ameba *Dictyostelium* is linked to *ponticulin,* an integral membrane protein. More commonly found are complex linkages that connect actin filaments to integral membrane proteins through peripheral membrane proteins.

Cortical Networks of Actin Filaments Stiffen Cell Membranes and Immobilize Integral Membrane Proteins

The richest area of actin filaments in a cell lies in the *cortex*, a narrow zone just beneath the plasma membrane. In this region, most actin filaments are arranged into a network which is seen in electron micrographs to exclude most organelles from the cortical cytoplasm. The cell has developed several ways to organize the cortical actin cytoskeleton (Figure 22-6). Perhaps the simplest cytoskeleton is the two-dimensional network of actin filaments adjacent to the erythrocyte plasma membrane. In more complicated cortical cytoskeletons, such as those in platelets, epithelial cells, and muscle, actin filaments are part of a three-dimensional network that fills the cytosol and that anchors the cell to the substratum. We will briefly review the most important lessons learned about the cytoskeleton in the erythrocyte and then compare it with the more complex platelet cytoskeleton.

The Erythrocyte Cytoskeleton   A red blood cell must squeeze through narrow blood capillaries and yet retain its

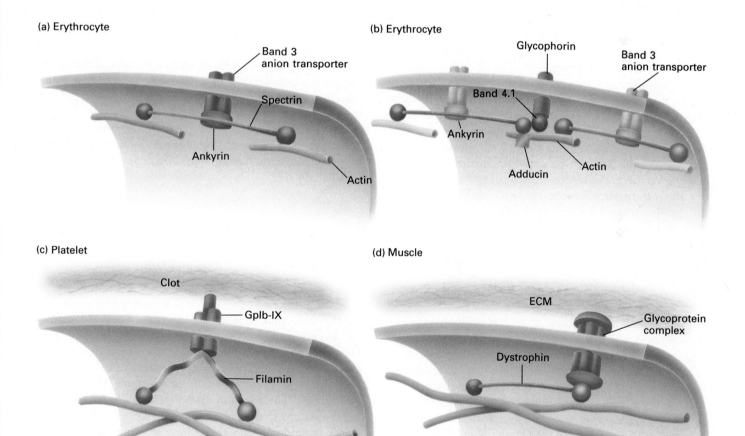

▲ **FIGURE 22-6**   The network of filaments that supports the inner surface of the plasma membrane. The network is indirectly linked to integral membrane proteins by different peripheral membrane proteins. (a) At the erythrocyte membrane, spectrin and ankyrin link short actin filaments to an integral membrane protein, the band 3 anion transporter. (b) A second set of connections composed of adducin and band 4.1 connects actin to another integral membrane protein, glycophorin. (The location of ankyrin linkages is shown in light green, for reference.) (c) At the platelet membrane, a three-dimensional network of actin filaments is attached to the integral membrane glycoprotein complex GpIb-IX by filamin. GpIb/IX binds to proteins in a blood clot outside the platelet. In addition, there is a two-dimensional network of actin and spectrin, like that depicted in part (a) (not shown; see Figure 22-7a). (d) In the muscle cell membrane, dystrophin attaches actin filaments to an integral membrane glycoprotein complex. This complex binds to laminin and agrin in the extracellular matrix (ECM). In addition to the linkages shown in parts (a)–(d), actin filaments can also connect to the membrane by direct binding to integral membrane proteins (not shown).

biconcave shape. As discussed in Chapter 14, the strength and flexibility of the erythrocyte plasma membrane depend on a planar network of actin filaments located along the inner surface of the membrane. The erythrocyte cytoskeleton is based on a proven engineering structure, a geodesic sphere. In this structure, the actin filaments are arranged in a spoke-and-hub network (see Figures 14-38 and 14-39). A single spectrin tetramer comprises one spoke, which cross-links a pair of hubs. Each hub is composed of a short (14-subunit) actin filament plus tropomyosin and tropomodulin; the latter two strengthen the network by preventing the actin filament from depolymerizing. (Tropomyosin is a short fibrous protein that runs the length of the actin filament; tropomodulin, which, as we shall see later, binds to the end of the actin filament, provides additional stability by preventing subunits at the end of the filament from dissociating.) Because several spectrin molecules can bind the same actin filament, several spectrin spokes radiate from a single hub to form the network organization of the cytoskeleton.

To ensure that the erythrocyte retains its characteristic shape, the spectrin-actin cytoskeleton is firmly attached to the overlying erythrocyte membrane by three different peripheral membrane proteins, ankyrin, band 4.1 protein, and adducin. Each protein cross-links a particular part of the cytoskeleton to a particular integral membrane protein: ankyrin connects the center of spectrin to band 3 protein, the anion-transporter protein in the membrane (Figure 22-6a), while band 4.1 protein and adducin connect the integral membrane protein glycophorin to the actin-spectrin complex (Figure 22-6b). In this pattern of binding ensures that the membrane is connected to both the spokes and the hubs of the cytoskeleton.

Platelet Cortical Cytoskeleton It is likely that the cortical network of actin filaments in the erythrocyte is similarly present in other cells, because non-erythroid isoforms of spectrin and ankyrin have been found at the membranes of various cells. It is clear, however, that actin filaments are arranged in a more complicated structure, especially where it attaches to the membrane. An example of this more complicated structure is seen in the platelet, a small, nonnucleated cell that is important in blood clotting and wound repair. The platelet cytoskeleton (Figure 22-7; see also Figure 22-6c) must undergo complicated rearrangements that are responsible for a repertoire of changes in cell shape during a blood clotting reaction (Figure 22-7b). These platelet changes could not be generated by the simple cytoskeleton seen in the erythrocyte; thus additional components are needed in the platelet cytoskeleton.

The cytoskeleton of an unactivated platelet (Figure 22-7a) consists of three components: a rim of microtubules (the *marginal band*), a cortical actin network, and a cytosolic actin network. [A marginal band of microtubules is

also present in most erythrocytes (human red cells are one exception)].

The cortical network of actin in platelets shows similarities to the erythrocyte actin cytoskeleton (see Figure 22-6a). In platelets this network consists of actin filaments cross-linked into a two-dimensional network (Figure 22-7a) by a non-erythroid isoform of spectrin and linked through ankyrin to an anion transporter, the Na/K ATPase, in the membrane. A critical difference between the erythrocyte and platelet cytoskeletons is the presence in the platelet of a second network of actin filaments, organized by filamin cross-links into a three-dimensional gel (Figure 22-6c). The gel fills the cytosol of a platelet and is anchored by filamin to the glycoprotein Ib-IX complex in the platelet membrane. (Glycoprotein Ib-IX is the membrane receptor for thrombin and von Willebrand's factor; two molecules that are products of blood clotting reactions.)

In order to close a wound, a platelet must be able to transmit cytoskeletal changes within the cell to the blood clot outside the cell. This is effectively accomplished by linking the cytoskeleton to the same proteins in the membrane that also bind to the clot. This principle is repeated over and over in other cells: the cytoskeleton is directly connected to the extracellular matrix through shared membrane proteins. We will discuss other examples of membrane-cytoskeleton linkages in this chapter and also in Chapter 24.

Dystrophin Anchors a Cortical Actin Network Directly to the Extracellular Matrix

Membrane-cytoskeleton linkages play a critical role in muscle contraction. This role was uncovered by studies on degenerative muscle diseases which led to the discovery of a very large protein, dystrophin (MW 426,000). Dystrophin was discovered in studies on Duchenne muscular dystrophy (DMD), a fatal, degenerative sex-linked genetic disease of muscle that affects about one of every 3500 males born. The gene associated with DMD is the largest known human gene, containing more than 2 million bases. Many individuals with DMD have deletions in this locus, and DNA sequences corresponding to these missing X-chromosome segments were used to isolate the cDNA for the DMD gene product, the protein missing in DMD (see Figure 8-24). Subsequent cloning of this cDNA led to the determination that its product is *dystrophin*, which is localized to the cytosolic face of the plasma membrane in striated and cardiac muscle cells and is missing in muscles from DMD patients.

While it is only 0.002 percent of the total protein in muscle, dystrophin comprises 5 percent of the membrane-associated cytoskeleton. Its membrane association and its homology with other actin cross-linking proteins (see

(a)

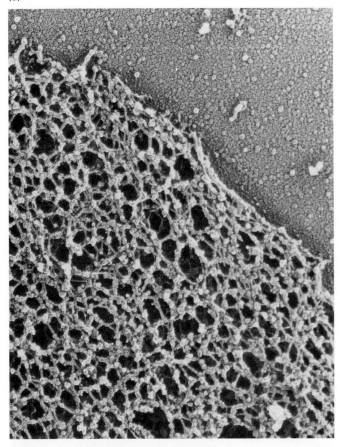

(b)

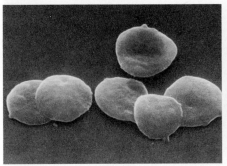 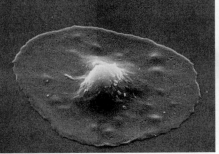 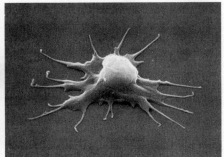

Table 22-1) suggested that dystrophin might be involved in anchoring actin filaments to the plasma membrane, and this proved to be the case. On the basis of dystrophin's similarity to spectrin and filamin, scientists hypothesized and demonstrated that dystrophin is a membrane-microfilament linkage which anchors an integral membrane complex of glycoproteins to the cortical actin network (see Figure 22-6d).

What is the function of dystrophin in the cortex? One answer is that without the support of a cortical cytoskeleton, the membrane of a muscle cell would be easily damaged by the stress of repeated muscle contraction. The exact organization of actin filaments at the muscle membrane is unknown, but we assume that it is similar to that in the platelet cytoskeleton. Like filamin in the platelet cytoskeleton, dystrophin has two functions: cross-linking actin filaments into a network and attaching this supportive network to a glycoprotein complex in the muscle cell membrane. This membrane complex is also bound to laminin and agrin, proteins in the extracellular matrix (ECM). Thus the dystrophin-glycoprotein complex in the membrane links the internal cytoskeleton to the external ECM.

The identification of dystrophin and its functions was an outstanding example of "reverse genetics"—identifying the gene for a protein whose absence causes a genetic disease, even when the function of the protein is not known. More importantly, such studies on diseases often uncover basic mechanisms important for the normal function of a cell or tissue.

Actin Bundles Support Projecting Fingers of Membrane

The surface of a cell that is exposed to the surrounding medium is not smooth. Instead, as shown in an electron micrograph of a dividing sea urchin embryo, the cell surface is studded with numerous membrane projections, the most common being microvilli (Figure 22-8). A phospho-lipid membrane is fragile, and without support such projections would be unable to withstand the stresses encountered by cells in an organism. However, the underlying cytoskeleton supports these cell-surface projections by providing a bundle of actin filaments to which the membrane is anchored. Because it is held together by protein cross-links, the actin bundle is stiff and provides a rigid structure that reinforces the projecting membrane, enabling it to maintain its long, slender shape.

Two types of fingerlike membrane projections, micro-villi and filopodia, are supported by an internal actin bundle. A *microvillus* ranges in length from 0.5 to 10 μm and is found where the cell membrane faces the fluid environment. Because microvilli are bathed in the medium surrounding the cell, they increase the surface area available for the absorption of molecules into a cell. Hence a dense

(a)

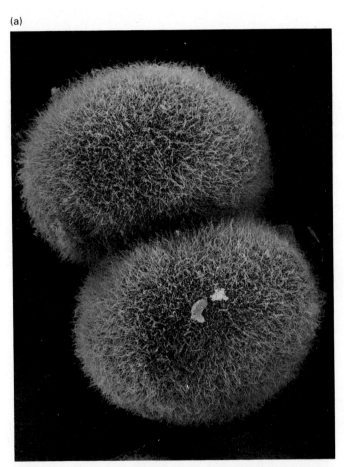

(b)

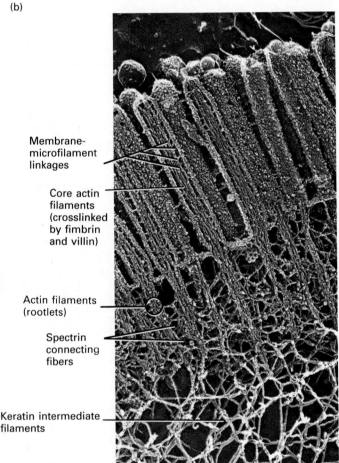

Membrane-microfilament linkages

Core actin filaments (crosslinked by fimbrin and villin)

Actin filaments (rootlets)

Spectrin connecting fibers

Keratin intermediate filaments

▲ FIGURE 22-8   Microvilli. The surface of all cells is covered by microvilli. (a) The surface of a sea urchin embryo at the 2-cell stage has a hairy appearance from 10-μm-long microvilli. The microvillar membrane contains receptors, ion channels and pumps, and transport systems. (b) The absorptive surface of intestinal epithelial cells is lined with 2-μm-long microvilli. At the core of each microvillus, a bundle of actin filaments, cross-linked by fimbrin and villin, stabilizes the finger-like shape. The microvillar membrane is attached to the sides of the bundle by evenly spaced membrane-microfilament linkages. Each bundle continues into the cell as a 0.5-μm-long rootlet. The rootlets are cross-braced by connecting fibers composed of an intestinal isoform of spectrin, and the bases of the rootlets form attachment sites for keratin filaments. These numerous connections anchor the rootlets in a meshwork of filaments and thereby support the upright orientation of the microvilli. [Part (a) courtesy of T. Schroeder; part (b) courtesy of N. Hirokawa.]

carpet or *brush border* of microvilli covers the surface of a cell whose primary function is to transport nutrients, such as an intestinal epithelial cell in the lining of the gut lumen (see Figures 14-40 and 14-42). Microvilli are relatively stable and change little in length or shape during the lifetime of the cell.

The bundle of actin filaments in the core of a microvillus (Figure 22-8b) can be formed by different cross-linking proteins. In most microvilli, including those on the surface of fibroblasts and sea urchin embryos, actin filaments are cross-linked by fascin, while fimbrin cross-links the filaments that support intestinal microvilli and stereocilia, the fingers of membrane that sense motion in the inner ear. It is not known why one cross-linking protein is found in one actin bundle but not in another.

The second type of membrane extension, a *filopodium*, attaches cells to a solid surface. Typically found at the edge of moving or spreading cells (see Figure 22-1), filopodia are transient structures, present only during the time required to establish a stable contact with the underlying substratum. As with the actin bundles found in microvilli, several different cross-linking proteins, including filamin and fascin, organize the actin filaments in filopodia.

▶ The Dynamics of Actin Assembly

Thus far we have treated the actin cytoskeleton as if it were an unchanging structure consisting of bundles and networks of filaments. However, in a cell the cytoskeleton is dynamic—microfilaments are constantly shrinking or growing in length, and bundles and meshworks of microfilaments are continually forming and dissolving. From our discussions of the cytoskeleton so far, you would agree that changes in the organization of actin filaments must cause equally large changes in the shape of a cell—that the dynamic nature of the cytoskeleton is due in large part to changes in actin. In this section, we will discuss how actin polymerizes into filaments and how polymerization is controlled by several small actin-binding proteins.

Actin Polymerization in Vitro Proceeds in Three Steps

As we mentioned earlier, we can initiate actin polymerization by adding salts to a solution of G-actin, creating F-actin filaments. We can follow the polymerization process with various assays, including viscometry, sedimentation, and fluorescence spectroscopy. Viscometry detects when actin filaments become long enough to become entangled: this causes the solution to become viscous and is identified as a decrease in the flow rate of the solution. In the second assay, sedimentation, F-actin but not G-actin will pellet at high g-forces (i.e., centrifugation at 100,000 g for 30 min). In a fluorescence assay, a fluorescent dye is covalently attached to G-actin; the fluorescence spectrum

of the modified G-actin monomer changes when the G-actin becomes part of a filament. These assays are useful both when we are studying the kinetics of actin polymerization and when we are purifying actin-binding proteins, which cross-link or depolymerize actin filaments.

The polymerization of actin filaments proceeds in three sequential phases (Figure 22-9). The first phase is marked by a *lag period* in which G-actin aggregates slowly into short, unstable oligomers. Once the oligomer reaches a certain length (three or four subunits) it can act as a stable *seed*, or *nucleus*, which in the second phase rapidly *elongates* into a filament by the addition of actin monomers onto both of its ends. This growth phase is enhanced by the spontaneous and random breakage of the growing filaments, producing additional filament ends that also act as nuclei for elongation. The lag period can be eliminated by adding a small number of F-actin nuclei to the solution of G-actin (Figure 22-9b). As F-actin filaments grow, the concentration of G-actin monomers decreases until it is in equilibrium with the filament. This equilibrium condition, the third phase, is called *steady state* because G-actin monomers exchange with subunits at the filament ends but there is no net change in the mass of filaments. The equilibrium concentration of monomers is called the *critical concentration* (C_c). This value is important because it measures the ability of a solution of G-actin to polymerize. Under typical in vitro conditions, the C_c is 0.1 μM. Above this value, a solution of G-actin will polymerize; below this value a solution of F-actin will depolymerize.

Actin polymerization is accompanied by ATP hydrolysis. ATP–G-actin monomers add to the ends of a filament; once they are incorporated into the filament, the ATP bound to the actin subunit is slowly hydrolyzed to ADP. As a result of ATP hydrolysis, most of the filament consists of ADP–F-actin, but ATP–F-actin is found at the ends. However, we know that hydrolysis is not essential for polymerization to occur, because actin containing ADP or a nonhydrolysable ATP analog is able to form filaments.

ATP Enhances Assembly from One End of a Filament

Recall that the inherent structural polarity of F-actin ensures that the two ends of the filament are different. This structural difference is also detected in polymerization experiments because one end of the filament, called the (+) end, elongates five to ten times faster than does the opposite, or (−), end. The unequal growth rates are demonstrated by a simple experiment exploiting the ability of F-actin to nucleate polymerization of a G-actin solution. If actin filaments decorated by myosin subfragments (see Figure 22-3) are used as the nuclei, electron microscopy will show bare actin filaments that have elongated from both ends of the decorated actin. The newly polymerized (undecorated) actin is five to ten times longer at the (+) end than at the (−) end of the filaments (Figure 22-10a). The

(a)

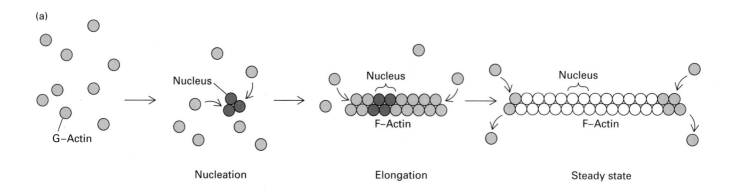

Nucleation   Elongation   Steady state

(b)

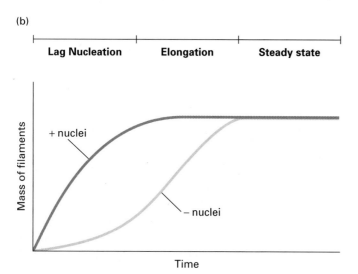

▲ FIGURE 22-9 The three phases of actin polymerization. (a) During the polymerization of G-actin in vitro, the mass of actin filaments increases after an initial lag period and eventually reaches a steady state. In the first phase, G-actin monomers bound with ATP (pink subunits), form nuclei—stable complexes of actin (purple subunits) which in the second phase are elongated by the addition of subunits to both ends of the filament. In the third phase of assembly, steady state, the ends of actin filaments are in equilibrium with monomeric ATP-G-actin. After their incorporation into a filament, subunits slowly hydrolyze ATP and become stable ADP-F-actin (white subunits). (b) Polymerization can be speeded by adding actin filament fragments (nuclei) at the start of assembly. The lag period is thus bypassed because the actin nuclei promote immediate elongation into filaments.

arrowhead decoration permits us to conclude that the more rapidly elongating (+) end corresponds to the barbed end of a filament and the more slowly growing (−) end corresponds to the pointed end of a filament.

The difference in elongation rates at the opposite ends of a filament is caused by a difference in C_cs at the filament ends. This difference can be measured by blocking one or the other end with proteins that "cap" the ends of actin filaments. (We will discuss (+) end and (−) end capping proteins in a following section.) Shown in the schematic in Figure 22-10b, a capping protein on the (+) end of actin causes the filament to elongate from its (−) end. Conversely, elongation occurs at the (+) end when the (−) end of a actin filament is blocked. In either case, the C_c of each end of a filament is measured in actin polymerization assays (Figure 22-10c)—0.1 μM for addition at the (+) end of the filament and 0.8 μM for addition at the (−) end. Based on the C_c values for the (+) and (−) ends of a filament, we would observe that: (1) no filament growth at G-actin concentrations below 0.1 μM, (2) growth only from the (+) end at G-actin concentrations between 0.1 and 0.8 μM, and (3) growth on both ends at G-actin concentrations above 0.8 μM [remember, growth will still be faster at the (+) end than at the (−) end].

This difference in the C_c of filament ends leads to an interesting phenomenon called *treadmilling* (Figure 22-10d). At G-actin concentrations intermediate between the C_cs for the (+) and (−) ends, subunits dissociate from the (−) end but add to the (+) end. The newly added subunits would travel through the filament, as if on a treadmill, until they reach the (−) end, where they dissociate. A treadmill is interesting because it can perform work; theoretically, objects attached to the filaments could be carried along by the transit of subunits through the filament. In the actual cell, actin filaments probably do not treadmill, because their ends are usually bound by *capping proteins*; as we shall see, these proteins block the addition or loss of G-actin monomers. A similar process of treadmilling occurs during microtubule assembly and disassembly.

ATP is the controlling parameter of actin polymerization because it affects the C_c for polymerization. This role of ATP is demonstrated by an experiment in which actin filaments are polymerized with ADP-actin monomers. The result is that both ends of a filament elongate at the slower rate characteristic of the (−) end. Thus, despite the inherent structural polarity of a filament, the C_c of the (+) end becomes equal to the (−) end of the filament when ADP-actin monomers are incorporated into filaments.

(a)

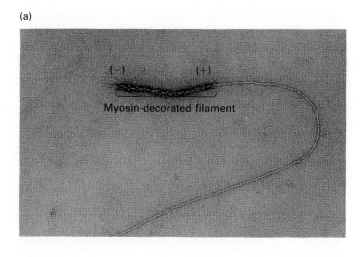

(b)

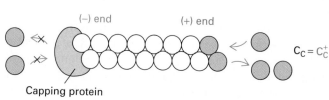

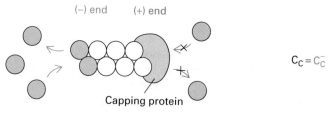

(c)

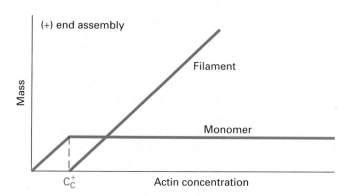

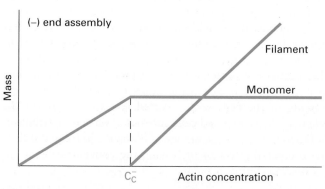

(d)

$$C_C^+ < \text{G-actin concentration} < C_C^-$$

▲ **FIGURE 22-10** Uneven growth at the two ends of an actin filament. The two ends of a filament have different critical concentrations. (a) When short myosin-decorated filaments are the nuclei for actin polymerization, ATP-G-actin monomers add unequally to the two ends. Because of the polarity of the myosin heads, the barbed, or (+), end of the filament grows measurably faster than the pointed, or (−), end. The (−) end corresponds to the clefted (top) end of the actin filament model shown in Figure 22-2c. (b) Simple experiments blocking the (+) or (−) ends of a filament with actin-capping proteins permit growth only at the opposite

end. Under these conditions, the critical concentration is determined by the sole growing end. (c) The critical concentration (C_c) is the concentration of monomers which support actin polymerization. Below the C_c, actin remains monomeric, while at concentrations greater than C_c, actin filaments assemble and the monomer concentration remains invariant. The C_c for the (+) end (blue trace) is less than the C_c for the (−) end (red trace). (d) At concentrations intermediate between the C_cs for the (−) and (+) ends, actin subunits can flow through the filaments by attaching preferentially to the (+) end (follow the colored subunits) and dissociating preferentially from the (−) end of the filament, a phenomenon known as "treadmilling." The oldest subunits in a treadmilling filament lie at the C − D end. [Part (a) courtesy of T. Pollard.]

Fungal Toxins Disrupt the Monomer-Polymer Equilibrium

In earlier chapters we learned how mutations or inhibitory antibodies can be employed to interfere with a particular cellular process. In the 1970s, before these techniques were perfected for cell biology experiments, drugs were used to disrupt the actin cytoskeleton. The first of such drugs studied was cytochalasin B (or its more potent derivative, cytochalasin D), a fungal alkaloid. Cytochalasin causes actin filaments in a cell to depolymerize by binding the (+) end of F-actin, where it blocks further addition of subunits. The critical concentration then shifts to that of the (−) end, which causes subunits to dissociate from the filament. When cytochalasin is added to live cells, the actin cytoskeleton disappears and cell movements like locomotion and cytokinesis are inhibited. These observations were among the first that implicated actin filaments in cell motility.

A second toxin, phalloidin, has the opposite effect on actin: it poisons a cell by preventing actin filaments from depolymerizing. Isolated from *Amanita phalloides* (the "Angel of Death" mushroom), phalloidin binds at the interface between subunits in F-actin, where it is in an ideal position to lock subunits together. Even when actin is diluted below its critical concentration, the phalloidin-stabilized filaments will not depolymerize. Phalloidin binds only F-actin and its most common use is as an F-actin–specific stain for light microscopy. Tagged with fluorescent dyes, phalloidin binds any actin filament in a cell (see Figure 22-1b), and the distribution of F-actin is detected in a fluorescence microscope.

Actin-Binding Proteins Control the Lengths of Actin Filaments

In the perfect world of a test tube, we can start the polymerization process by adding salts to G-actin or we can depolymerize F-actin by simply diluting the filaments. Let us compare this with the real world of the cell. In the actual cell the ionic composition of the cytosol is relatively constant; any change in salt concentration is a few millimolar at best, and most of the salt ions are adsorbed to proteins and other macromolecules. In addition, the cell cannot tolerate any dilution of its cytosol, because the membrane would swell and possibly burst open. Thus we would conclude that the cell cannot regulate actin polymerization by methods used in test-tube experiments but instead must employ a different mechanism. We know that the cellular mechanism involves regulatory proteins, because biochemical studies have isolated and characterized several actin-binding proteins that either promote or inhibit actin polymerization.

Sequestration of Actin Monomers by Small Actin-Binding Proteins Considering the C_c of actin (0.1 μM)

and the ionic conditions of the cell, we can calculate that nearly all cellular actin (0.5 mM) should exist as filaments; there should be very little G-actin. Actual measurements, however, show that as much as 40 percent of actin in a cell is unpolymerized. What keeps the cellular concentration of G-actin above its C_c? The most likely explanation is that cytosol proteins *sequester* actin, holding it in a form that is unable to polymerize. Two candidate proteins, *thymosin β_4* (Tβ_4) and *profilin,* have emerged, based on biochemical studies of their actin-binding properties. Both proteins bind actin monomers into a 1:1 complex and are abundant in the cytosol of a cell. A third protein, hsc70, sequesters actin monomers in the cytosol of *Dictyostelium* ameba, but it is not certain whether its interactions with actin are related to protein folding or to the regulation of actin polymerization.

Thymosin β_4 Thymosin β_4 (MW 5000) binds an actin monomer in a 1:1 complex but is unable to bind F-actin. As a complex with thymosin β_4, actin cannot polymerize. In platelets, the concentration of thymosin β_4 is 0.55 mM, approximately twice the concentration of unpolymerized actin (0.25 mM). On the basis of these numbers, we would calculate that approximately 70 percent of the monomeric actin in a platelet should be sequestered by thymosin β_4.

Thymosin β_4 functions like a buffer for monomeric actin. The 1:1 complex of actin and thymosin β_4 is in equilibrium with unbound thymosin β_4, monomeric G-actin, and F-actin. In a simple equilibrium, an increase in the cytosolic concentration of thymosin β_4 would increase the concentration of sequestered actin subunits and correspondingly decrease F-actin (remember actin filaments are in equilibrium with actin monomers). Two complementary experiments confirmed this prediction for live cells. In both, the cytosolic concentration of thymosin β_4 was artificially increased: in one experiment, by expression of a transfected plasmid containing the thymosin β_4 gene and in the other experiment (Figure 22-11) by microinjection of chemically synthesized thymosin β_4 peptides. The outcome of both experiments was the same. F-actin filaments were missing in the cells with elevated levels of thymosin β_4, whereas the unaffected cells showed a normal distribution of F-actin. The disappearance of filaments in treated cells is the predicted effect of increasing the concentration of an actin-sequestering protein.

Profilin Profilin is a 15,000-MW protein that was first found, in high concentrations and complexed with actin, in unactivated sperm from the sea cucumber, a marine invertebrate. Biochemical analysis of profilin isolated later from calf spleen or brain showed that profilin preferentially binds ATP-actin monomers as a stable 1:1 complex. On the basis of its binding constant and cellular concentration, we would predict that profilin can buffer up to 20 percent of the unpolymerized actin in the cytosol, a level too low for it to act as an effective sequestering protein. As

(a)

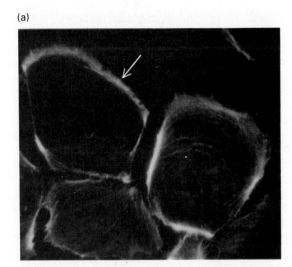

(b)

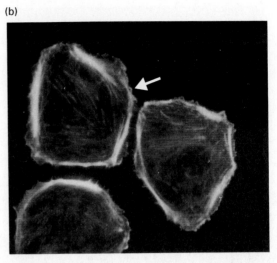

▲ FIGURE 22-11   Effect of thymosin β_4 on actin. The pool of unpolymerized actin in live kidney epithelial cells can be experimentally changed by microinjection of thymosin β_4. (a) In the cell injected with thymosin β_4 (arrow), the bulk of the actin stress fibers are missing, suggesting that thymosin β_4 has depolymerized actin filaments. The surrounding unin-jected cells retain their stress fibers. The actin filaments were stained with fluorescent phalloidin. (b) The control cell injected with an unrelated peptide (arrow), shows a normal organization of stress fibers, like those in the uninjected cells around it. [See M. C. Sanders, A. L. Goldstein, and Y.-L. Wang, 1992, *Proc. Natl. Acad. Sci. USA* **89**:4678.]

a result, most of the unpolymerized actin in a cell is probably complexed with thymosin β_4.

Profilin has several interesting properties. First, it is the only actin-binding protein that allows the exchange of ATP for ADP. Usually, when actin is complexed with other proteins, ATP or ADP is trapped in an ATP-binding cleft of actin. In the three-dimensional structure of the profilin-actin complex (Figure 22-12a), a profilin is bound to a site on the opposite side from the ATP-binding site of actin. In this position, the nucleotide associates or dissociates easily from actin. With profilin acting to recharge actin with ATP, any ADP-actin subunit released from a filament would be rapidly converted to ATP-actin. This activity of profilin would replenish the pool of ATP-actin.

A second intriguing feature of profilin is that it binds to phosphoinositol 4,5 bisphosphate (PIP$_2$), a membrane lipid, which inhibits its actin-binding activity. In Chapter 20, we discussed how the hydrolysis of PIP$_2$ by phospholipase C is the first step in the inositol 1,4,5-trisphosphate/diacylglycerol second-messenger signaling pathway. As both a PIP$_2$-binding and actin-binding protein, profilin could play a central role in stimulating actin polymerization in response to cell-cell signals. Perhaps because of its PIP$_2$-binding activity, profilin as seen in immunofluorescence micrographs is located at cell membranes. This location is interesting because the cell membrane is one region where actin filaments polymerize.

All of these observations suggest that profilin is involved in actin polymerization. A current model hypothesizes that profilin shuttles actin between G-actin and F-actin pools. Because its binding affinity for actin is intermediate between that of thymosin β_4 and the (+) end of actin filaments, profilin may regulate the store of unpolymerized actin by drawing actin from thymosin β_4 and releasing it to actin filaments. Thus profilin's actin-sequestering properties may be part of a complex membrane-based actin assembly system (Figure 22-12b).

A Family of Actin-Severing Proteins Generates New Filament Ends by Breaking Actin Filaments

Cell biologists have long been fascinated by the ability of the cytosol to flow in the center of an ameba and then turn into a gel when it reaches the front of the cell. This "sol-to-gel" transformation is most certainly dependent on the assembly of actin filaments at new sites in the cytosol and the disassembly of old actin filaments. Because the actin concentration in a cell favors the formation of filaments, there must be additional proteins to break down existing actin filaments and filament networks. We have just dealt with proteins that regulate the assembly of actin by binding actin monomers; now we will discuss a family of proteins (generically called *severing proteins*) that regulate the lengths of actin filaments by breaking them into short pieces or blocking their ends (Table 22-2).

The details of the severing mechanism are not well understood, but it must involve binding to an actin filament and breaking the filament by disrupting the bond

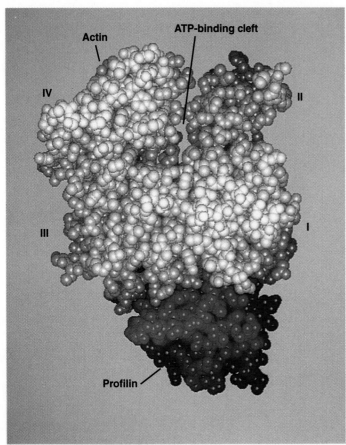

◄ FIGURE 22-12   Effect of profilin on actin. Profilin forms transient complexes with actin which are involved in actin assembly. (a) The three-dimensional structure of the profilin-actin complex. Profilin (green) binds to the edge of subdomains I and III of the actin monomer (white) at the end opposite to the ATP-binding cleft. Profilin is unique among actin-binding proteins because it permits ATP exchange with the nucleotide-binding site of actin: when actin is complexed with any other actin-binding protein, the nucleotide in the ATP-binding cleft of actin is nonexchangeable. (b) Profilin is hypothesized to play a central role in modulating the levels of unpolymerized actin. In cells (*left panel*), profilin is bound with PIP$_2$, a membrane lipid, while most actin is complexed with thymosin β_4. The actin–thymosin β_4 pool is in equilibrium with a smaller pool of profilin-actin complexes. In response to an extracellular signal such as chemotactic molecules (see Figure 22-45), which stimulate actin assembly (*middle panel*), profilin is released from the membrane by hydrolysis of PIP$_2$, causing a recruitment of actin monomers from thymosin β_4-actin complexes into profilin-actin complexes. Because of their relative affinities [K_d (+) end < K_d profilin < K_d thymosin β_4], the (+) end of actin filaments (located close to the plasma membrane) will polymerize by binding monomeric actin (*right panel*). The incorporation of actin monomers into filaments eventually depletes the pools of profilin-actin and thymosin β_4–actin complexes. ADP-G-actin subunits that have dissociated from a filament are converted into ATP-G-actin by profilin, thus helping to restore the cytoplasmic pool of ATP-G-actin. [Part (a) see C. E. Schutt, et al., 1993, *Nature* **365**:810; courtesy of M. Rozycki and C. E. Schutt.]

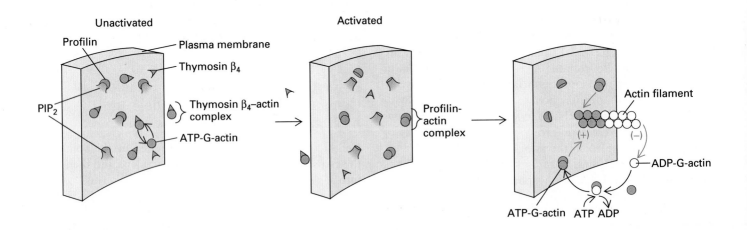

between two subunits. This breakage produces a pair of shorter actin filaments. The protein *gelsolin* is capable of instantaneously severing all of the filaments in a network into short fragments (Figure 22-13). As a consequence of the severing event, gelsolin remains bound at the (+) end of one piece where it prevents the exchange of actin subunits, an activity called *capping*. The (−) end of the same filament remains uncapped. The half-life of a gelsolin-actin

complex is six weeks; thus once gelsolin severs a filament, it effectively remains bound to the (+) end of the filament and is unable to sever additional filaments.

The capping and severing proteins are regulated by two signaling pathways, one mediated by calcium and one by PIP$_2$. Normally, in the low calcium concentrations of the cytosol (<0.1 μM), capping and severing proteins are inactive. However, when a cell's calcium concentration is

TABLE 22-2 Actin-Capping and Actin-Severing Proteins

Protein	MW	Domain Organization*	Activity
gCAP39	40,000		Capping
Severin (fragmin)†	40,000		Capping, severing
Gelsolin	87,000		Capping, severing
Villin	92,000		Capping, severing, cross-linking

*Actin monomer binding domains are colored yellow; F-actin binding domains are colored orange.
†Severin and fragmin are synonyms for the same protein.

increased from resting levels, then the severing protein binds the calcium and this in turn activates the capping and severing activities of the protein. Like that of profilin, the actin-binding activity of gelsolin is inhibited by PIP$_2$. The counteracting influence of two second messengers, calcium and PIP$_2$, from separate signaling pathways provides a mechanism for the reciprocal regulation of these proteins. At the end of this chapter, we will discuss how cell signals coordinate the activities of different actin-binding proteins, including severing proteins, during cell movements.

Actin Filaments Are Stabilized by Actin-Capping Proteins

A second group of proteins can cap the ends of actin filaments but, unlike severing proteins, cannot break filaments to create new ends. One such protein, *CapZ*, binds the (+) ends of actin filaments independently of calcium. Its affinity for actin is so high ($K_m > 10^{-10}$) that one molecule of CapZ will dissociate from a filament once a month. As a result CapZ prevents actin subunits from dissociating from filaments. As its name indicates, CapZ was first discovered in the sarcomere Z disk of muscle, a weblike matrix to which the (+) ends of actin filaments are tethered (see Fig-

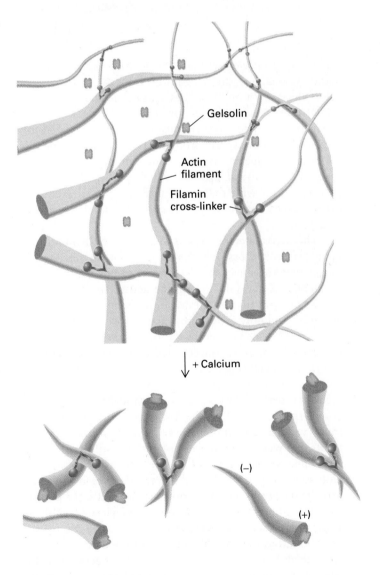

► FIGURE 22-13 Actin-severing proteins. In the absence of calcium, severing proteins such as gelsolin are not bound to actin, but in the presence of calcium, a severing protein binds an actin filament, breaking it into pieces. The severing protein remains tightly bound to the newly created (+) end of the filament, where it blocks monomer growth. The action of severing proteins can convert a gel of long actin filaments cross-linked by a protein such as filamin into a solution (sol) of short actin filaments.

ure 22-26). Subsequently, CapZ was found in nonmuscle cells at regions where the ends of actin filaments are anchored to the membrane. Another protein in this group, *tropomodulin*, is unrelated to CapZ in sequence. Tropomodulin caps the (−) end of actin filaments in both the muscle sarcomere and the erythrocyte membrane network. Its capping activity is enhanced in the presence of tropomyosin, which suggests that they function as a complex to stabilize a filament. The ability to cap the (+) and (−) ends of the same filament prevents an actin filament from depolymerizing. Capped actin filaments are needed in places where the organization of the cytoskeleton is unchanging, as in a muscle sarcomere or at the erythrocyte membrane.

Many Movements Are Driven by Actin Polymerization

By manipulating actin polymerization and depolymerization, the cell creates forces which results in several types of movements.

An Actin Bridge from Sperm to Egg

Actin polymerization solves an engineering problem for the sperm cells of echinoderms (sea urchins, starfish, or sea cucumbers) during fertilization. The oocyte, or unfertilized egg, of echinoderms is surrounded by a viscous coat of jelly that is 50 μm thick, which bars sperm from contact. To surmount this jelly barrier, a sperm extrudes a finger of membrane 80 μm long, the *acrosomal process*, which pushes through the jelly and fuses with the egg membrane. This sequence of events, called the *acrosome reaction*, is completed within 1–2 min.

The acrosomal process is supported by a bundle of actin filaments cross-linked by fascin, but the remarkable fact is that the unactivated sperm does not contain actin filaments. As a result, during the acrosome reaction, the bundle of actin filaments must form de novo. How is this achieved, and how is the acrosomal process extruded from the cell? The answers lie in the observation that actin in unactivated sperm is concentrated in a pocket at the head of the sperm cell (Figure 22-14) and is kept in an unpolymerized state as a complex with profilin. At the start of the acrosome reaction, two events are thought to trigger the polymerization of actin. The first is exocytosis of the acrosomal vesicle; this exposes the (+) ends of short actin filaments that were covered by a bed of the profilin-actin complex. The second event is the release of profilin from monomeric actin, which could be triggered by a rise in either the intracellular pH or the amount of PIP_2, the plasma membrane lipid that binds profilin (see Figure 22-12b). As a result, actin monomers elongate from the (+) ends of the actin seed and push the membrane of the acrosomal process ahead. Following the polymerization of actin microfilaments and the fusion of the sperm and egg plasma mem-

► FIGURE 22-14   The acrosome reaction in echinoderm sperm. (a) Thin section of an unactivated sea urchin spermatozoon. Lying within an indentation of the nucleus (N) is the spherical, membrane-bounded acrosomal vesicle (A). Beneath and lateral to it is the area of cytoplasm called the periacrosomal region (P), which contains unpolymerized profilin-actin complexes. Note that there are no filaments visible in this region. (b) Electron micrograph of a sea cucumber sperm cell after the activation that occurs upon encountering an egg. *Inset*: Cross section of the acrosomal process, showing the many actin fibers cut in transverse section. (c) Steps in the formation of an acrosomal process. Step 1: An unactivated sperm cell binds to the egg jelly coat by means of specific receptor proteins on its plasma membrane. This binding triggers exocytosis of the acrosomal vesicle, releasing enzymes from the vesicle that digest both the jelly coat and the vitelline layer covering the egg plasma membrane and exposing the (+) ends of short actin filaments that had been blocked by the vesicle membrane. Step 2: Explosive polymerization of actin, stored as profilin-actin complexes in the periacrosomal region of the sperm cytoplasm, then occurs at the (+) ends of the actin filaments. The polymerization causes outgrowth of the acrosomal process, whose plasma membrane is derived from the membrane of the acrosomal vesicle. Step 3: When the acrosomal process reaches the egg plasma membrane, the sperm and egg membranes fuse. Fusion is triggered by surface proteins on the acrosomal process that were originally part of the membrane of the acrosomal vesicle. The sperm nucleus then enters the cell. [Parts (a) and (b) courtesy of L. Tilney.]

branes, the sperm nucleus enters the egg and nuclear fusion occurs.

Bacterial Movement on a Growing Tail of Actin

When *Listeria monocytogenes* infects mammalian cells, it moves through the cytosol at a rate of 11 μm/min. Various fluorescent stains showed that a meshwork of short actin filaments followed the moving bacterium like the plume of a rocket (Figure 22-15) but the protein myosin, assumed to be necessary for movement, was missing. Does actin alone move the bacterium? And if so, how? The answers were provided by a microinjection experiment in which fluorescent actin was injected into *Listeria*-infected cells. In the microscope, the fluorescent actin could be seen incorporating into the tail-like meshwork at the end nearest the bacterium, with a simultaneous loss of actin throughout the tail. This result shows that actin polymerizes into filaments at the base of the bacterium and suggests that as the tail-like meshwork assembles it pushes the bacterium ahead. However, bacterial motility requires not only actin polymerization but also factors which determine the site of polymerization. One protein, profilin, is implicated in guiding the movement, because it binds to the area of the bacterium where the active actin polymerization takes place. To test whether profilin is involved in bacterial motility, mutations

(a)

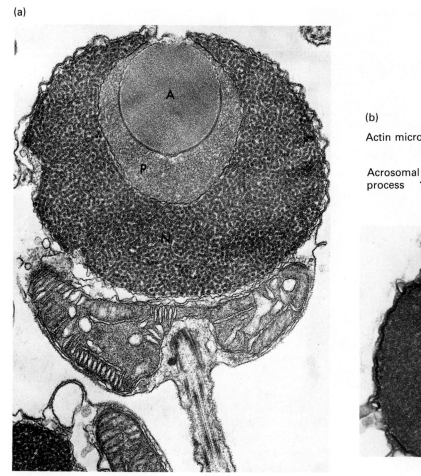

(b)

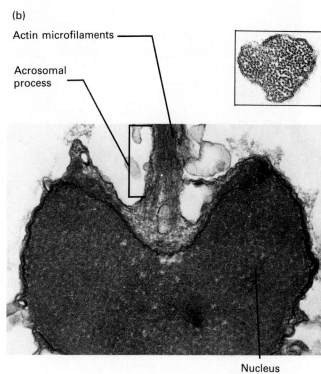

Actin microfilaments

Acrosomal process

Nucleus

(c)

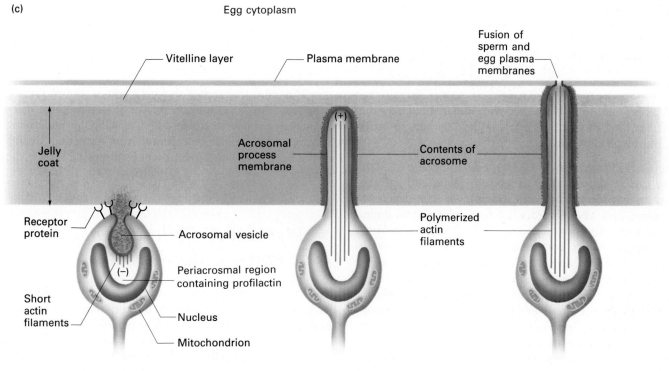

Egg cytoplasm

Vitelline layer

Plasma membrane

Fusion of sperm and egg plasma membranes

Jelly coat

Acrosomal process membrane

Contents of acrosome

(+)

Receptor protein

Acrosomal vesicle

Periacrosmal region containing profilactin

Polymerized actin filaments

(−)

Short actin filaments

Nucleus

Mitochondrion

Step 1: Binding and exocytosis

Step 2: Polymerization of actin

Step 3: Fusion of sperm and egg

(a)

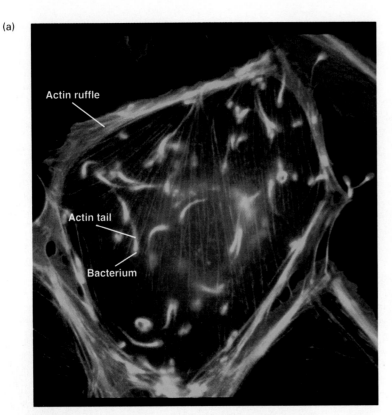

(b)

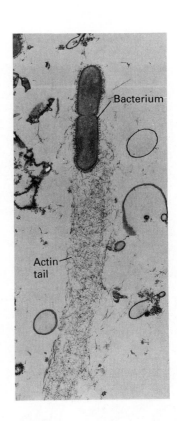

▲ **FIGURE 22-15** *Listeria* motility in infected fibroblasts. (a) Numerous bacterial cells move independently within the cytosol of an infected mammalian cell. Infection is transmitted to other cells when a spike of cell membrane, generated by a bacterium, protrudes into a neighboring cell and is engulfed by a phagocytotic event. Bacteria (red) are stained with an antibody raised against ActA, a bacterial membrane protein that binds cellular profilin and is essential for infectivity and motility. Behind each bacterium is a "tail" of actin (green) stained with fluorescent phalloidin. (b) An electron micrograph shows that the tail is composed of a meshwork of short actin filaments. As the bacterium moves, the length of the tail remains invariant but there is constant addition of actin subunits at the profilin-associated end of the bacterium. Dissociation of subunits occurs throughout the tail. [Part (a) courtesy of J. Theriot and T. Mitchison; part (b) courtesy of J. Theriot, M. Siebert, and T. Mitchison.]

were made in ActA, the bacterial membrane protein that binds profilin. In cells infected with mutated bacteria that did not bind profilin, the bacteria did continue to move, but at much slower rates.

Actin Polymerization at the Leading Edge of Moving Cells   Although the acrosome reaction and bacterial rockets may seem to be exotic examples of actin-dependent motility, they have played a key role in helping us understand how cells crawl forward. As in the extrusion of the acrosomal process and in bacterial motility, profilin is thought to play a central role because it is located at the leading edge of the cell, where polymerization occurs. Cell movement is led by changes in the position of the plasma membrane at the front of the cell, and video microscopy reveals that a major feature of this movement is the polymerization of actin at the membrane (Figure 22-16). In addition, at the leading edge, the actin filaments are rapidly cross-linked into bundles and networks in the projecting filopodia and lamellipodia. As we shall see later, these structures then form stable contacts with the underlying surface and prevent the membrane from retracting. Details of the steps in cell movements will be discussed later in the chapter.

► *Myosin: A Cellular Engine that Powers Motility*

We have seen that actin filaments provide a lattice that supports the plasma membrane and organizes the cytosol, and also that the ability of actin to polymerize is harnessed by the cell to generate some forms of motility. However, for many types of cellular movements, actin does not func-

(a)

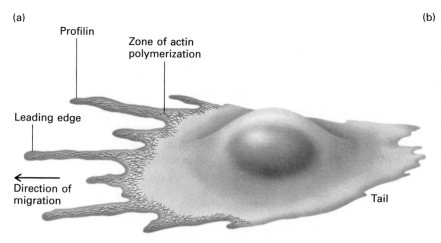

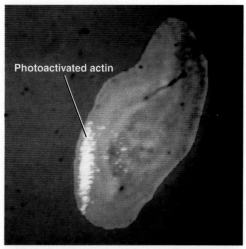

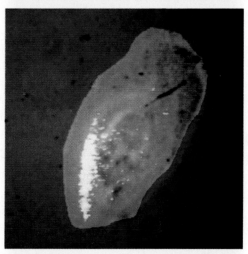

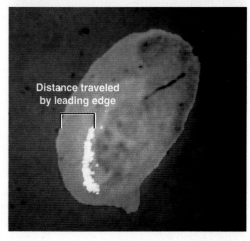

▲ FIGURE 22-16   Assembly of actin filaments at the leading edge of migrating cells. (a) As shown in this diagram of a fibroblast, profilin is located at the leading edge of the cell, the site where actin filaments are assembled. (b) In a microinjection experiment, the location of F-actin relative to the front of the cell is monitored by video microscopy. G-Actin, modified with a caged fluorescent dye, is microinjected into a cell. (A caged dye is initially nonfluorescent, but UV irradiation induces a chemical change which renders the dye fluorescent.) When a narrow band of the cell at the leading edge is irradiated with UV light (first panel), the actin in that zone is made fluorescent (white band), while the remainder of the cell remains nonfluorescent. The G-actin has assembled into F-actin at the leading edge, and by 48 sec (second panel) and 81 sec (third panel) after UV irradiation, this zone moves away from the leading edge as the cell moves forward. The band becomes less intense, suggesting that F-actin dissociates from the zone as it moves back from the leading edge. [Part (b) see J. A. Theriot and T. J. Mitchison, 1991, *Nature* **352**:126. Courtesy of J. Theriot.]

tion alone. These types of motility depend upon interactions between actin filaments and the motor protein myosin, an ATPase that moves along actin filaments.

Myosin is a unique enzyme because it can move along an actin filament by coupling the hydrolysis of ATP to conformational changes. We call this type of enzyme, one that converts chemical energy into mechanical energy, a *mechanochemical enzyme* or, colloquially, a *motor protein*. Myosin is the motor, actin filaments are the tracks along which myosin moves, and ATP is the fuel that powers the motility.

In previous chapters, we have discussed enzymes that exhibit nucleotide-dependent movement; for example, polymerases hydrolyze NTPs as they move along DNA or RNA strands, and GTP is hydrolyzed as ribosomes move along mRNA during translation. However, we normally do not consider these to be mechanochemical enzymes.

The following sections on myosin will describe the structure and functions of three major types of myosin—myosin I, myosin II, and myosin V—and will provide examples of movements that result from actin-myosin (actomyosin) interactions. Then we will investigate the mechanism of myosin-dependent movement by examining the interactions between a myosin head and an actin filament. Afterwards, we will discuss the role of myosin in muscle. One special form of movement, contraction, is caused when actin filaments slide past filaments composed of myosin II. Because contraction is highly evolved in muscle cells, we will study muscle as a model for understanding contractile events involving less organized actomyosin systems in nonmuscle cells.

In the next chapter, we will discuss the microtubule motors kinesin and dynein. Because they exhibit many of the same properties as myosin, we use myosin as a general model for understanding mechanochemical processes.

Myosin Is a Diverse Family of Proteins Characterized by Distinct Head, Neck, and Tail Domains

Myosin (Figure 22-17) is present in all eucaryotic cells. Fifteen years ago, we only knew of two types of myosin, now named myosin I and myosin II. Myosin II, the first one discovered, was found to be the motor protein in muscle that powers muscle contraction. It was soon realized that nonmuscle cells also contain myosin II, and this finding suggested that movements could result from a muscle-type mechanism. Later, myosin I was discovered; it had many of the properties of myosin II, such as actin binding, but it also had the unusual properties of being a monomer (myosin II is a dimer) and of being associated with membranes.

By now, we have identified, from gene cloning and DNA sequencing studies, more than 10 classes of the myosin gene family. Myosin I and myosin II proteins are the most abundant and most thoroughly studied and are found in most cells. Scientists are studying a third protein, myosin V; the functions of the remaining myosin genes are unknown. From the biochemical characterization of myosin proteins, the genetic analysis of mutations in myosin genes, and the immunolocalization of myosin proteins in cells, we have learned that the three best studied myosins, I, II, and V, have completely different functions, yet they share the same basic property of being a myosin motor. Myosin II powers muscle contraction and cytokinesis, while myosins I and V are involved in cytoskeleton-membrane interactions such as the transport of membrane vesicles.

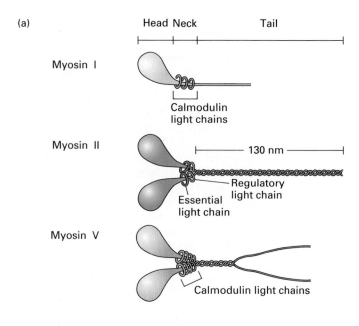

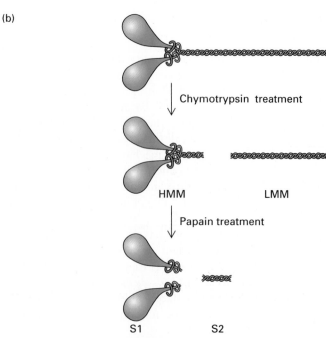

▶ FIGURE 22-17   Structure of various myosin molecules. All myosins are composed of an actin- and ATP-binding heavy chain and two or more calcium-binding regulatory and essential light chains. (a) Proteins in the three major myosin families are organized into head, neck, and tail domains. Myosin I is single-headed; myosin II and myosin V are two-headed molecules. Myosin II and myosin V are dimeric because α-helical segments in the tail induce coiled-coil interactions; myosin I lacks any oligomerization region in its tail domain, and so remains a monomer. The head, neck, and tail domains each carry out separate functions. The head domain binds actin and hydrolyzes ATP. The light chains bind to the neck domain and regulate myosin activity. The tail domain controls the unique function of each myosin type. The tail domain of myosin II polymerizes dimers into thick filaments, but myosin V tails only form dimers. (b) The head-neck-tail domain structure of myosin is easily dissected apart by proteolysis. For example, most proteases cleave myosin II in the tail domain to generate a rodlike LMM fragment and a two-headed HMM fragment. Further cleavage at the neck separates the single-headed S1 fragments from the S2 fragment.

An explanation for the diversity in the functions of a motor protein is shown in Figure 22-17a. All myosins are composed of a heavy chain and several light-chain subunits, and they are organized into three structurally and functionally different domains. The *head domain* is the most conserved region among the myosins; it contains the actin- and ATP-binding sites and is responsible for generating force. The mechanics of the head domain will be discussed in the following sections. Adjacent to the head domain lies the highly α-helical *neck region*. It regulates the activity of the head domain by binding either calmodulin or calmodulin-like regulatory light chain subunits. The *tail domain* contains the binding sites that determine whether the tail binds to the membrane or binds to other tails, and thus whether a myosin dimer or a myosin filament will be formed.

The differences and similarities among myosins I, II, and V are summarized in Figure 22-17. Myosin II and myosin V are dimers because highly α-helical sequences in the tail associate to form a coiled-coil structure. Myosin I, a monomer, lacks this sequence but like myosin V it contains a membrane-binding site in the tail. The three myosins differ in the number and type of light chains bound in the neck region. The light chains of myosin I and myosin V are calmodulin, a calcium-binding regulatory subunit in many intracellular enzymes. However, myosin II contains two different light chains (called essential and regulatory light chains); both are calcium-binding proteins like calmodulin, but differ in their calcium-binding properties. The similarities among light chains suggest that all myosins will be regulated in some way by calcium. The differences in the light chains ensure that the activities of different myosins will differ in their response to calcium signals in the cell.

Electron micrographs of myosin clearly show the head, neck, and tail organization of the myosins. The biochemical properties of these domains were determined by examining proteolytic fragments of myosin II. Various proteases can cleave the heads of myosin I or myosin II away from their tails. For example, chymotrypsin cleaves myosin II in the tail at the base of the forked neck domain, leaving the pair of heads connected (Figure 22-17b). The head and neck fragment is called *heavy meromyosin* (HMM), while the tail is called *light meromyosin* (LMM). (*Mero*, Greek, means "part.") Further digestion of HMM with papain, a protease isolated from papaya, cleaves the part of the neck that connects each head to the tail. As a result, single myosin head fragments, called subfragment 1 or S1, are produced in addition to a second fragment, subfragment 2 or S2, which originates from the distal portion of the neck (Figure 22-17b). The proteolytic studies confirmed that myosin is organized into head, neck, and tail domains; later experiments showed that the head domain corresponds to the N-terminal half of the heavy chain while the tail domain consists of the C-terminal half of the heavy chain.

The Myosin Tail Domain Regulates Binding to Membranes or the Assembly of Thick Filaments

All myosin heads exert force on actin but the particular function of a myosin molecule is related to its tail domain (Figure 22-18). For example, the function of myosin I and myosin V is membrane-related because their tail domains bind the plasma membrane or the membranes of intracellular organelles (Figure 22-18a). Because of its membrane-binding activity, myosin I serves as a linkage between membranes and the microfilament bundles in brush-border microvilli and in filopodia. Also, myosin I is associated with the Golgi membrane, suggesting that it may transport Golgi vesicles along the actin filament in the cell.

In contrast to the membrane-binding properties of myosin I, the rodlike tail domains of myosin II dimers oligomerize together into a *thick filament* (Figure 22-18b). In electron micrographs, we can see that a thick filament from skeletal muscle has a bipolar organization—the heads are located at both ends of the filament and are separated by a central bare zone which is devoid of heads (Figure 22-18c). When packed tightly together in a thick filament, many myosin head domains can interact simultaneously with actin filaments.

The Myosin Head Domain Is an Actin-Activated ATPase

Myosin is a specialized enzyme, an ATPase that is able to couple the hydrolysis of ATP with motion. A critical feature of the myosin ATPase activity is that it is *actin-activated*. In the absence of actin, solutions of myosin will catalyze the conversion of ATP into ADP and phosphate; this reaction proceeds slowly—four ATP molecules per hour—and the rate is identical in single myosin molecules and in myosin thick filaments. In a complex of actin and myosin, however, the myosin ATPase activity is stimulated to 20 ATPs per second. The actin-activation step ensures that the myosin ATPase is at its maximum activity only when the myosin head domain is bound to actin.

Myosin Heads Walk along Actin Filaments

Studies of muscle contraction provided the first evidence that myosin heads slide or walk along actin filaments. The mechanism of muscle contraction was greatly clarified when in vitro motility assays were developed, because these allowed the movement of a single myosin molecule to be studied. The first indications that actin-myosin (actomyosin) interactions could be mimicked in vitro came from observations that mixtures of purified actin and myosin form a gel in solution and then precipitate if ATP is added. The precipitation represented the contraction of the gel into a dense mass of actin and myosin.

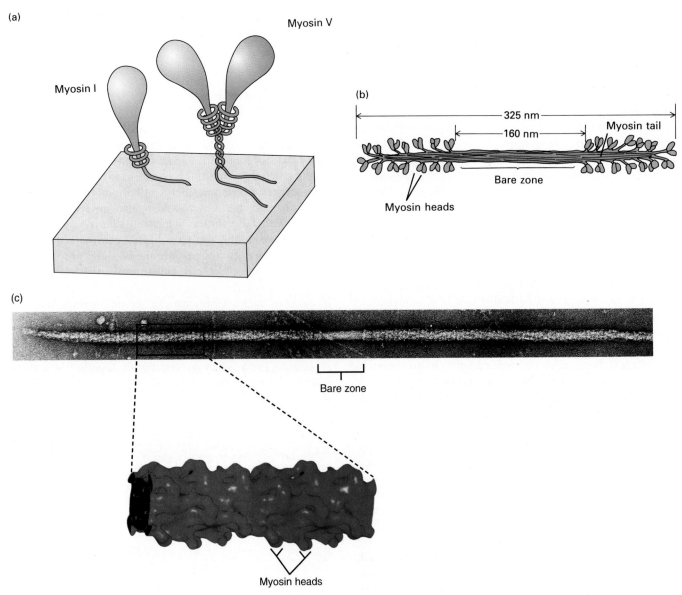

▲ FIGURE 22-18   Functions of the myosin tail domain. Thick-filament formation and membrane-binding properties of myosin are determined by its tail domain. (a) Myosin I and myosin V are localized to the membrane by undetermined sites in their tails. As a result, these myosins are associated with intracellular membrane vesicles or the cytoplasmic face of the plasma membrane. (b) In contrast, the coiled-coil tail domains of myosin II molecules pack side-by-side, forming a thick filament from which the heads project. In a skeletal muscle, the thick filament is bipolar. Heads are at the ends of the thick filament and are separated by a bare zone, which consists of the side-by-side tails. (c) An electron micrograph (*top panel*) of a myosin thick filament that was isolated from tarantula muscle shows a long rod-like structure. The rows of myosin heads generate a cross-banding pattern from the edge of the bare zone to the tip of the thick filament. A three-dimensional reconstruction (*bottom panel*) of the head region shows the pair of heads derived from each myosin II dimer projecting from the backbone of the rod; two pairs of heads are indicated. [Part (c) adapted from R. Padron and R. Craig, 1992, *J. Mol. Biol.* **33**:384; courtesy of R. Craig.]

More powerful assays allow us to study the movements of myosin and actin in cell extracts from amebas and algal cells, using the light microscope. One such assay, the *sliding-filament assay*, monitors the movement of actin filaments, along a bed of myosin molecules (Figure 22-19). In this assay, the myosin molecules are tethered to a coverslip so that they cannot move; therefore the actin filaments are forced to move along the myosin instead. The moving actin filaments are easily seen with a fluorescence microscope if they are labeled with phalloidin bound to a fluorescent dye.

When myosin is adsorbed onto the surface of a glass coverslip, many molecules are bound by their heads and thus are prevented from binding actin. However, many

(a)

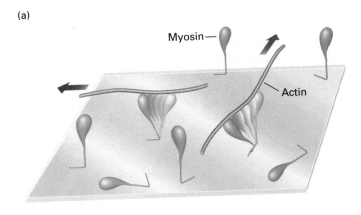

◀ **FIGURE 22-19** The sliding-filament assay. Sliding movements of fluorescent actin filaments generated by the myosin head can be quantified by video microscopy. (a) A solution of myosin molecules is adsorbed onto the surface of a glass coverslip. After removal of excess myosin, the coverslip is placed myosin-side down on a glass slide to form a chamber through which solutions can flow. A solution of actin filaments, made visible by staining with rhodamine-labeled phalloidin, is allowed to flow into the chamber, and individual filaments are observed under a fluorescence light microscope. (The coverslip in the diagram is shown inverted from its orientation on the flow chamber to make it easier to see the positions of the molecules.) (b) Photographs showing the positions of three actin filaments (numbered 1, 2, 3) at 15-s intervals. In the presence of ATP, the actin filaments move at a velocity which can vary widely depending on the myosin tested (see Table 22-3). The effects of changes in the assay conditions (ionic strength, temperature, ATP concentration, calcium concentration, etc.) on filament velocity are easily measured. [Part (b) courtesy of M. Footer and S. Kron.]

(b)

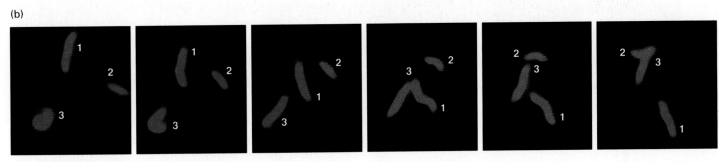

other myosin molecules are bound by their tails, leaving their heads free to bind actin filaments. In the fluorescence microscope, the actin filaments are seen to adhere to the coverslip and not to move, if ATP is absent, but if ATP is present the actin filaments glide along the surface of the coverslip. The movement is always with the (−) end of the actin filament in the lead. This polarity of movement must be caused by a myosin head (bound to the coverslip) that is "walking" toward the (+) end of the filament.

Using a video camera to capture the movements, scientists can calculate the velocity at which myosin moves an actin filament. Table 22-3 shows the actin velocities measured for different types of myosin using this assay. The values vary greatly, showing that some myosins are faster than others.

With this assay, we can identify the role of myosin domains, determine the effects of mutations on myosin, and study how myosin is regulated by other proteins. For example, the sliding-filament assay is able to show that the myosin head by itself is sufficient for movement to occur. When the tails of myosin molecules are removed proteolytically, the head fragments that remain are able to generate movement in the actin filaments. Indeed, the two-headed

myosin HMM fragment (but not the single-headed S1 fragment) can generate sliding velocities that are comparable with the velocity of intact myosin or myosin filaments.

A Myosin Head Takes an 11-nm Step Each Time an ATP Molecule Is Hydrolyzed

Given that the head domain is sufficient to cause movement, what is the force generated by a myosin head and how far does it travel when an ATP molecule is hydrolyzed? To determine these measurements, we must deal with the properties of a single myosin molecule. The forces generated by single myosin molecules have been measured using a device called an *optical trap* (Figure 22-20a), which was originally developed by physicists to trap single atoms in focused laser beams. As part of a light microscope, the laser beam of the optical trap can hold a polystyrene bead (or any other object that refracts light) in the center of the beam just as fictitious "tractor beams" can trap spaceships in science fiction stories. The strength of the force holding the bead is adjusted by increasing or decreasing the intensity of the laser beam.

TABLE 22-3 Rate of Actin Filament Sliding by Different Myosins

Myosin	Actin Sliding Speed (μm/s)*
Chicken brush-border myosin I	0.04
Human platelet myosin II	0.04
Turkey gizzard smooth-muscle myosin II	0.22
Rabbit cardiac myosin II	0.49
Rabbit slow contracting skeletal muscle myosin II	0.81
Rabbit fast contracting skeletal muscle myosin II	4.5
Rabbit Heavy meromyosin (chymotryptic fragment)	4.6
Rabbit Subfragment 1 (generated by various proteases)	0.9–1.8

* Measured at 25°C.
SOURCE: J. R. Sellers, et al., 1993, Myosin-specific adaptations of the motility assay, in *Methods in Cell Biology*, Vol. 39, J. M. Scholey, ed., Academic Press, pp. 23–49.

➤ FIGURE 22-20   Use of the optical trap to determine myosin-generated force. The light from a single laser beam is used to measure the force exerted on a refractive polystyrene bead. (a) The optical trap captures refractive objects—polystyrene beads, bacteria, or cell organelles—in the beam of an infrared laser focused by the objective of a light microscope. (Biological materials do not absorb infrared light and hence do not generate damaging heat.) (b) In a motility assay, a polystyrene bead is fixed to the end of an actin filament. With the optical trap turned off, the actin filament and its attached bead move in response to the force generated by myosin adsorbed on the coverslip, as in Figure 22-19. When the optical trap is turned on and captures a bead, the trap will hold the bead immobile if the trapping force is great enough to resist the force generated by myosin. The minimum force needed to hold the bead in the trap is a measure of the force generated by myosin. The distance traveled by a myosin head during one cycle of ATP hydrolysis can be measured by the movement of the bead. The forces normally used in this type of experiment are much too low to pull an actin filament apart; they are in the range of force exerted on a bacterium by gravity.

X-Ray Crystallography Reveals the Atomic Structure of the Motor Domain

To understand how hydrolysis of ATP is coupled to the movement of a myosin head along actin filaments, we need to know the three-dimensional structure of the head domain. Fortunately, x-ray crystallography of the myosin S1 fragment has recently produced a model which reveals three key pieces of information: the shape of the myosin head domain, the positions of the essential and regulatory light chains, and the locations of the ATP- and actin-binding sites. The myosin head is an elongated domain, measuring 16.5 × 6.5 × 4 nm, on the end of an α-helical neck (Figure 22-21). Both light chains lie at the base of the head where they wrap around the neck, like C-clamps. In this

If a bead is attached to the end of an actin filament (Figure 22-20b), then an optical trap can capture the filament, via the bead, and hold the filament to the surface of a myosin-coated coverslip. To measure the movement and force generated by a single myosin head, the density of myosin on the coverslip must be low so that an actin filament interacts with only one myosin molecule.

The force exerted by a single myosin molecule on an actin filament is measured from the movement of the bead held in the optical trap. A computer-controlled electronic feedback system keeps the bead centered in the trap, and movement of the bead caused by myosin is counteracted by an opposing movement of the trap. The distance traveled by the actin filaments is measured from the displacement of the bead in the trap.

Such studies show that myosin II moves in discrete steps, approximately 11–15 nm long, and generates 3–4 picoNewtons (pN) of force—the same force as that exerted by gravity on a single bacterium. This force is sufficient to cause myosin thick filaments to slide past actin thin filaments during muscle contraction or to transport a membrane bounded vesicle through the cytoplasm. A step size of 11–15 nm suggests that as myosin walks along an actin filament, it binds to every second or third actin subunit on one strand of the filament.

➤ FIGURE 22-21   Three-dimensional structure of the myosin head (S1 fragment; see Figure 22-17b). X-ray crystallography reveals a curved, elongated domain (16.5 × 6.5 × 4 nm) which is marked by two clefts. In a cleft on one side of the head is the ATP-binding site. It lies 3.5 nm across from the second cleft, which bisects the actin-binding site near the end of the domain. A second prominent feature of myosin is a long α-helix which extends from the head to form the neck. Wrapped around the shaft of the helix are a pair of calmodulin-like light chains, named essential and regulatory light chains. They act to stiffen the neck, which can then act as a lever arm for the head. [Adapted from I. Rayment, et al., 1993, *Science* **261**:50; photograph courtesy of I. Rayment.]

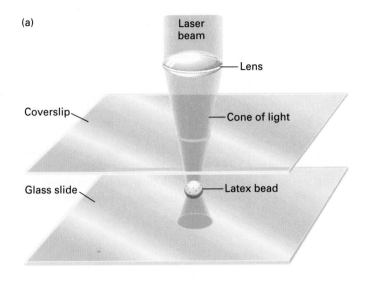

(a)

Laser beam

Lens

Coverslip

Cone of light

Glass slide

Latex bead

(b)

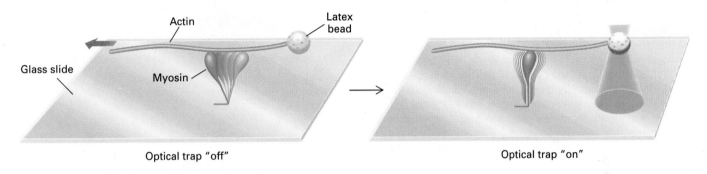

Actin

Latex bead

Glass slide

Myosin

Optical trap "off"

Optical trap "on"

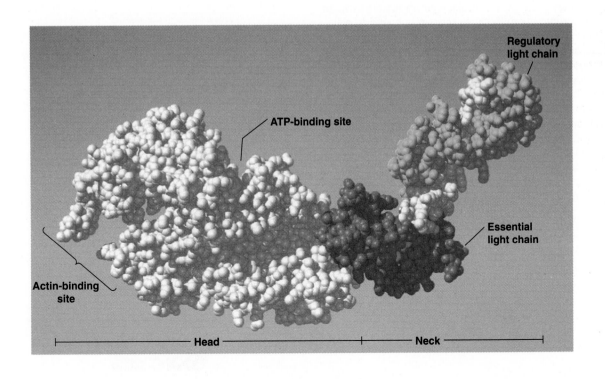

Regulatory light chain

ATP-binding site

Actin-binding site

Essential light chain

Head

Neck

position, the light chains stiffen the neck region and therefore are able to regulate the activity of the head domain.

An important feature in the structure of the myosin head is the presence of two clefts on its surface. One cleft, at the tip of the head, divides the actin-binding site in two while the other cleft, on the opposite side of the head, contains the ATP binding site. (The clefts are separated by 3.5 nm, a long distance in a protein.) The presence of surface clefts provides an obvious mechanism for generating large movements of the head domain. We can imagine how opening or closing of a cleft in the head domain, by binding or releasing actin or ATP, causes the head domain to pivot about the neck region. As discussed in detail in the next section, the ATP- and actin-binding sites are most likely coupled by large changes in the conformation of the head domain.

Conformational Changes in the Head Couple ATP Hydrolysis to Movement

Armed with the three-dimensional structure of the myosin head (see Figure 22-21) and the knowledge of the kinetics of the myosin ATPase, we can ask how myosin couples the energy from ATP to generate the force for movement. In

▶ FIGURE 22-22   The coupling of ATP hydrolysis to myosin motility. Myosin transduces chemical energy from ATP into motion by coupling actin binding with cycles of ATP-dependent conformational changes in the head. Movement results from the cyclic attachment and detachment of the myosin head from the actin filament and the movement of the head. In the absence of bound nucleotide, myosin binds actin tightly in a "rigor" state. When ATP binds (step 1), it closes the nucleotide-binding cleft in the head; this weakens actin binding by opening the actin-binding cleft. Freed of actin, the myosin head hydrolyzes ATP (step 2), causing a conformational change in the head, moving it to a new position before rebinding to the filament. Note that rebinding occurs at a different subunit than in step 1. Force is generated during the so-called power stroke (step 3). Myosin rebinds to the filament, releasing phosphate and then ADP (step 4). As phosphate dissociates from the ATP-binding cleft, the myosin head undergoes a second conformational change—the power stroke—which restores myosin to its rigor conformation. Because myosin is bound to actin, this conformational change exerts a force which causes myosin to move along the actin filament. Movement is not only dependent on ATP hydrolysis and conformational changes in the head but also on the fact that myosin binds to a different subunit (the purple one) than at the beginning of the cycle (where it was bound to the green one): rebinding to the same actin subunit would not result in movement. The diagram shows the cycle for a myosin II head attached to a thick filament but we can assume that other myosins attached to a membrane operate according to the same mechanism. [Adapted from I. Rayment and H. M. Holden, 1994, *Trends in Biochem. Sci.* **19**:129.]

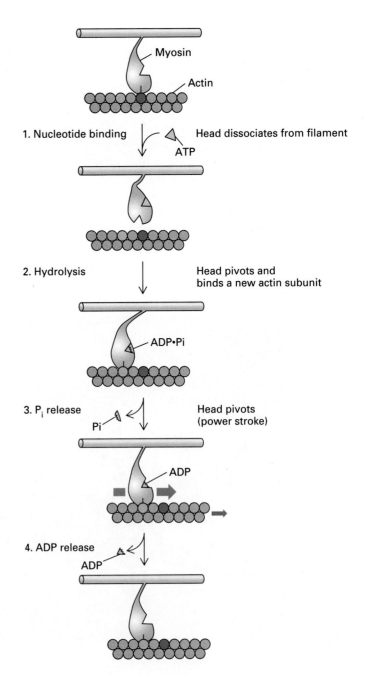

1. Nucleotide binding — Head dissociates from filament

ATP

2. Hydrolysis — Head pivots and binds a new actin subunit

ADP·Pi

3. P$_i$ release — Head pivots (power stroke)

Pi

ADP

4. ADP release

ADP

the following discussion, we will consider the general case in which a myosin molecule walks along an actin filament. Because we believe that all myosins must move using the same mechanism, we will ignore for the moment whether the myosin is bound to a vesicle or is part of a thick filament. One assumption in the model that we will discuss is that the hydrolysis of a single ATP molecule is coupled to one cycle of myosin movement.

Consider the interactions between one myosin head and an actin filament as diagrammed in Figure 22-22. We begin with an initial state in which the ATP-binding site on myosin is empty but myosin is strongly bound to actin (the actin-binding cleft on the myosin head is closed). In muscle, this state corresponds to *rigor mortis*. When muscle is depleted of ATP, at death, it becomes stiff and rigid because without ATP, myosin heads become firmly attached to thin (actin) filaments.

At the first step in the cycle, a molecule of ATP binds in the ATP-binding cleft of myosin, which causes the actin-binding cleft on the myosin head to open. This change disrupts actin-binding at the cleft and the myosin head dissociates from the actin filament. After the myosin head has detached from actin, ATP is hydrolysed to ADP and P_i. During this step, the ATP-binding cleft closes, causing the myosin head to bend. In this new conformation, the myosin head is ready to bind at a new subunit in the actin filament.

The next steps in the cycle occur when myosin is bound to actin. Sequential release of P_i and then ADP is coupled with conformation changes in the myosin head to produce the force for movement. Initially, the binding of the myosin head to actin is weak, but as P_i is released from the ATP-binding cleft, actin binding becomes strong (probably because the actin binding cleft closes and the actin-binding site is restored). The cycle is completed when ADP dissociates from the ATP-binding cleft and the myosin head reorients to its rigor-like position on actin. During this step, the head domain changes its conformation which in turn causes the neck and tail domains of myosin to shift along the actin filament toward its (+) end. This step is called the *power stroke* because all the force generated is thought to occur at this step.

In this model, myosin walks along an actin filament; however, the type of movement that results from this interaction depends on how myosin or actin is anchored. For example, in a bipolar thick filament, myosin heads are firmly anchored to the thick filament backbone. Because the heads in the two halves of the thick filament have opposite polarities, their interactions would cause actin filaments to slide toward the middle of the thick filament while the thick filament remains immobile. In contrast, a single myosin I molecule, bound to a membrane-bounded vesicle, will move along the actin filament because the actin filament part of a massive structure, the cytoskeleton. Thus the frame of reference for movement changes depending on whether actin or myosin is immobile.

Muscle, A Specialized Contractile Machine

Muscle cells have evolved to carry out one highly specialized function—to contract. Muscle contraction is a firsthand experience for everyone: just the act of picking up this book and turning its pages invokes the use of many different muscles in your arms and hands.

In this section, we will build on our discussion of myosin as a motor by describing how myosin and actin are organized in muscle. Once the structure of muscle has been reviewed, we will discuss how the motor properties of myosin are exploited to contract a muscle. Throughout this section we will mainly discuss the structure and function of skeletal muscle because it is the best understood muscle. However, sometimes we will describe important details of smooth muscle because its structure and activity are very similar to actomyosin systems in nonmuscle cells. Thus when we later compare actomyosin structures in muscle with those in nonmuscle cells, we will understand why muscle contraction is so efficient, and how actin and myosin in nonmuscle cells cause movement.

Muscle contractions must occur quickly, repetitively, through long distances, and with enough force to move large loads. To carry out this function, muscle actin and myosin are organized into the highly ordered structure of muscle, having the ability to do work very efficiently. Like any mechanical engine, muscles can be characterized by their power output, the rate at which they can work. Muscle power depends on several parameters: velocity of contraction, ability to contract repetitively, and force of contraction. Compared with that of other cellular systems (Table 22-4), the power output of muscle is like that of a flagellum but is 330,000-fold higher than the power output

TABLE 22-4 Power of Cellular and Mechanical Engines		
System	Specific Power Output (erg-sec^{-1}/cm^3)	Power/Engine-Weight Ratio (erg-sec^{-1}/g)
Striated muscle	2×10^6	2×10^6
Flagellum	3×10^6	3×10^5
Cytokinetic furrow	3×10^2	
Mitotic spindle	6	
Passenger car engine		3×10^6
Racing car engine		2×10^7
Aircraft gas turbine engine		8×10^7

SOURCE: R. B. Nicklas, 1984, A quantitative comparison of cellular motile systems, *Cell Motil.* 4:1–5.

of the mitotic spindle, the microtubule machinery that separates chromosomes. Perhaps more instructive is to compare the power output of muscle with that of mechanical engines (Table 22-4). We find when corrected for weight differences that a racing car engine and an aircraft engine are only tenfold to fortyfold more powerful than a muscle. A passenger car engine is only $1\frac{1}{2}$ times more powerful than muscle. In muscle, nature has obviously designed a very efficient and powerful engine.

Some Muscles Contract, Others Generate Tension

Vertebrates and many invertebrates have two classes of muscle—skeletal and smooth; in vertebrates, cardiac (heart) muscle forms a third class. Skeletal and cardiac muscles have a striped appearance in the light microscope

and are therefore called *striated*. Skeletal muscles connect the bones in the arms, legs, and spine and are used in complex coordinated activities, such as walking or positioning of the head; they generate rapid movements by contracting quickly. Another major function of skeletal muscle is to hold objects immobile. Clenching of fists or tensing of muscles are common examples where our muscles undergo *isometric contraction*, in which pairs of contracting muscles work to oppose each other and thus cancel out any movements. Cardiac muscle resembles skeletal muscle in many respects (Figure 22-23a), but it is specialized for the continuous, involuntary rhythmic contractions needed in pumping of the blood. Smooth muscles lack striations (Figure 22-23b); they surround internal organs such as the large and small intestines, the uterus, and large blood vessels. The contraction and relaxation of smooth muscles controls the diameter of blood vessels and also propels

(a)

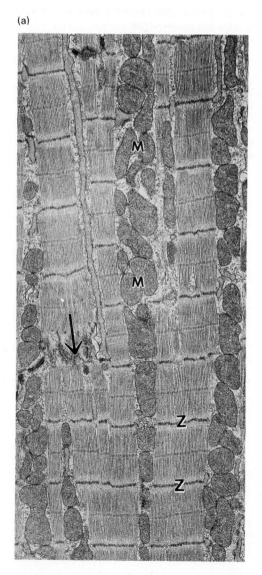

(b)

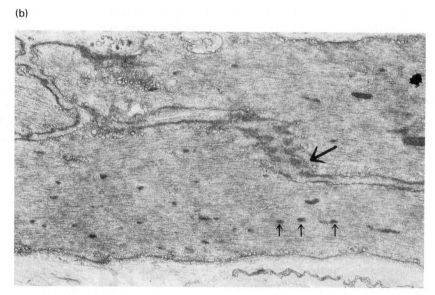

◄ FIGURE 22-23   Comparison of cardiac muscle and smooth muscle. (a) Electron micrograph of cardiac striated muscle. Myofibrils (separated by mitochondria, M) are in exact alignment, producing characteristic striations. Within a myofibril, transverse lines (Z) mark the Z-disk boundaries of the sarcomere. Large arrow marks the intercalated disk, a site of cell-cell contact and attachment of the sarcomere to the plasma membrane. A skeletal muscle cell appears very similar to a cardiac muscle cell. One major difference is that a cardiac cell contains one nucleus, while a skeletal cell is multinucleated. (b) Electron micrograph of a typical smooth muscle cell lining the stomach. The fibers are not organized or parallel to each other and the actin and myosin filaments are loosely packed together. Numerous dense bodies (small arrows) are scattered throughout the cyosol. Regions where one cell contacts a neighboring cell are marked by special adhesion sites (large arrow) called attachment plaques. [Part (a) adapted from J. R. Sommer and E. A. Johnson 1979, in: *Handbook of Physiology*, Section 2. The Cardiovascular System. American Physiological Society. Vol. 1, page 122. Part (b) courtesy of J-Q. Xu and R. Craig.]

food along the gastrointestinal tract. Cardiac muscle and smooth muscles are typically under the involuntary (unconscious) control of the central nervous system. Compared with striated muscles, smooth muscle cells contract and relax slowly, and they can create and maintain tension for long periods of time.

Striated Muscles Contain a Regular Array of Actin and Myosin

A typical striated muscle cell, or *myofiber* (Figure 22-24), is cylindrical, large (measuring 1–40 mm in length and 10–50 μm in width), and multinucleated (containing as many as 100 nuclei). A muscle fiber is packed with bundles of filaments that extend the length of the cell. Each bundle, called a *myofibril*, is further subdivided into alternating

| Muscle |
| Bundle of myofibers |
| Myofibril |
| One myofiber |
| Enlarged myofibril showing sarcomere and adjacent bands |

Plasma membrane

50 μm

Nucleus

Sarcomere

30 μm

I band   M line   Z disk

A band

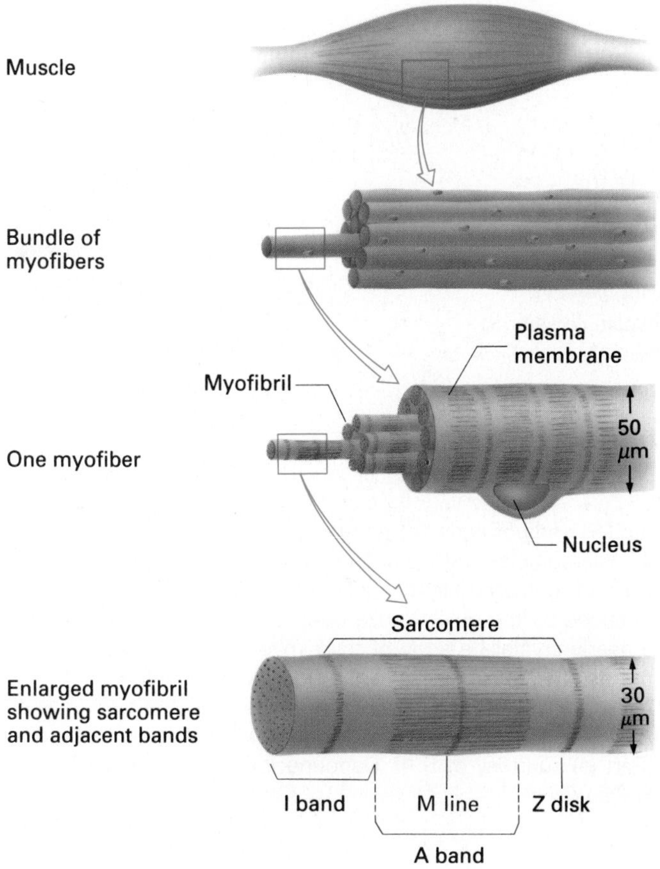

▲ FIGURE 22-24   Structure of muscle. Muscle tissue is composed of bundles of multinucleated muscle cells, or myofibers. Each muscle cell, or fiber, is packed with bundles of actin and myosin filaments, or myofibrils, which extend the length of the cell. Packed end to end in a myofibril is a chain of sarcomeres. The sarcomere is the functional unit of contraction; it is about 2 μm long in resting muscle. Contraction of the sarcomeres shortens the length of a myofibril.

light and dark bands that run along the length of the muscle cell. Closer examination reveals that the dark bands, called *A bands,* are bisected by a dark line, the *M line,* while the light bands, called *I bands,* are bisected by a different dark line, the Z disk (also called the *Z line* because when seen in profile on electron micrographs, the Z disk appears as a line). The segment from one Z disk to the next, consisting of two halves of an I band and an A band, is termed a *sarcomere.* A myofibril comprises a chain of sarcomeres, each about 2 μm long in resting muscle. In contracted muscle, the sarcomeres become shortened to 70 percent of their uncontracted, resting lengths. A sarcomere is both the structural and the functional unit of skeletal muscle. Electron microscopy and biochemical analysis have shown that each sarcomere contains two types of filaments: *thick filaments,* composed of myosin, and *thin filaments,* containing actin (Figure 22-25). Near the center of the sarcomere, thin filaments overlap with the thick filaments in the *AI zone.*

Thick Filaments Composed of Myosin   A thick filament of the A band is bipolar, with its myosin heads lying at the distal tips of the filament and its tails at the center (Figure 22-25b, c; see also Figure 22-18b, c). The bare zones of the thick filaments correspond to the M line, a darkly staining region of the A band. Salt extraction of muscle dissolves the thick filaments and causes the loss of the A bands, but the thin filaments and I bands remain. Biochemical analysis shows that myosin but not actin is extracted, and dialysis of the extracted myosin into low-salt solutions reconstitutes thick filaments. Thus myosin is the primary structural component of the thick filaments.

Thin Filaments Containing Actin   The I band is a bundle of thin filaments. All the filaments in an I band are the same length, but can vary among different muscles. Biochemical studies show that a thin filament is basically an actin filament plus two additional proteins that are involved in regulating actomyosin interactions. These proteins, tropomyosin and troponin, will be discussed later. Evidence that actin is a main component of the I bands is drawn from findings in three experiments. First, thin filaments appear identical to actin filaments reconstituted from purified G-actin monomers. Second, the thin filaments can be decorated with myosin S1 or HMM. Third, thin filaments but not thick filaments can be removed with the actin-severing protein gelsolin.

In electron micrographs, one end of a thin filament is associated with the Z disk, while the other end is near the center of the sarcomere (see Figure 22-25a, b). Myosin decoration experiments show that all thin filaments have the same polarity with respect to the Z disk—the barbed or (+) end of the filament is always closest to the Z disk. The heads of the myosin molecules protrude from the thick filaments, in the AI zone, forming cross-bridges with adjacent actin thin filaments (Figure 22-25c).

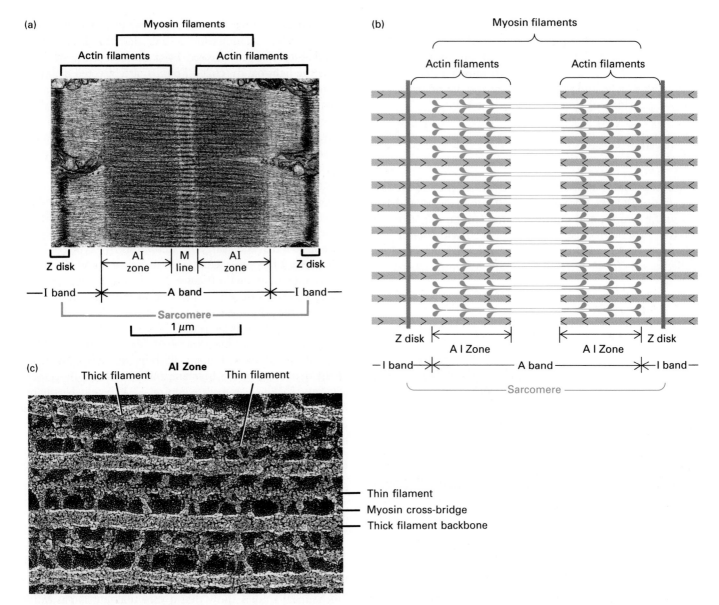

▲ FIGURE 22-25   Structure of the sarcomere. (a) Electron micrograph of mouse striated muscle in longitudinal section, showing one sarcomere. On either side of the Z disks are the lightly stained I bands, composed of actin filaments. These thin filaments extend from both sides of the Z disk to interdigitate with the dark-stained myosin thick filaments that make up the A band. The region containing both thick and thin filaments (the AI zone) is darker than the area containing only myosin thick filaments. The M line is a dark-staining stripe. (b) Schematic diagram of a sarcomere. The barbed ends of actin filaments are attached to the Z disks. (c) Micrograph showing actin-myosin cross-bridges in the AI zone of a striated flight muscle of an insect. This image, obtained by the quick-freeze deep-etching technique, shows a nearly crystalline array of thick myosin and thin actin filaments. The muscle was in the rigor state at preparation. Note that the myosin heads protruding from the thick filaments connect with the actin filaments at regular intervals. [Part (a) courtesy of S. P. Dadoune; part (c) courtesy of J. Heuser.]

Capping Proteins That May Stabilize Thin Filaments   The ends of a thin filament are capped by two proteins, CapZ and tropomodulin. CapZ is present in the Z disk of skeletal muscle (Figure 22-26), where it may possibly prevent actin filaments from depolymerizing at their (+) end. CapZ probably also anchors the (+) ends of actin filaments to other Z-disk proteins. At the opposite end of the thin filament lies tropomodulin, which, as in the erythrocyte skeleton, may cap the (−) end and protect it from depolymerization. Capping at both ends would cause the thin filaments to be very stable.

The Z Disk   The Z disk is a lattice of fibers whose major function is to anchor the (+) ends of actin filaments. How

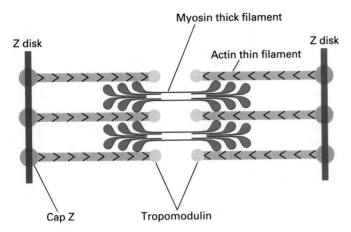

▲ FIGURE 22-26   Capping proteins stabilizing the ends of thin filaments. The diagram shows the location of CapZ (green) at the (+), or barbed, ends of filaments at the Z disk. The (−), or pointed, ends of thin filaments are stabilized by tropomodulin (yellow). The presence of these two proteins at opposite ends of a thin filament prevents actin subunits from dissociating during muscle contraction.

the filaments are attached is not certain, but scientists believe that the attachment must involve the actin-capping protein CapZ, as just discussed above, and α-actinin. α-Actinin is a major component of isolated Z disks and its actin cross-linking activity, localized at the Z disk, can serve to organize thin filaments in the I band into a bundle of filaments.

In Smooth Muscle, Thick and Thin Filaments Are Not in Regular Arrays

A smooth muscle consists of spindle-shaped or tapered cells which are packed with thick and thin filaments. As we have seen (Figure 22-23b), the filaments are not organized into well-ordered sarcomeres, but instead are gathered into loose bundles of thick and thin filaments at dense bodies in the cytosol. (*Dense body* is a term used by microscopists to describe an electron-dense material seen in electron micrographs of smooth muscle cells. Dense bodies apparently serve the same function as Z disks in skeletal muscle.) Additionally, in many smooth (and striated) muscle cells, actin filaments at the ends of the muscle fibers are connected to *attachment plaques*, which are similar to dense bodies but are located at the *sarcolemma*, the plasma membrane of a muscle cell. Like a Z disk, an attachment plaque contains a high density of α-actinin; it also contains a second protein, *vinculin* (MW 130,000), not found in Z disks. Vinculin binds tightly to α-actinin in cell-free experiments. Most likely it binds directly to an integral membrane protein in the plaque and serves a general role of attaching actin filaments to membrane adhesion sites (see Figure 22-35).

Scientists study smooth muscle cells not only to learn about muscle contraction but also because they believe the organization and regulation of actin and myosin in smooth muscle more closely resembles that in a nonmuscle cell.

Thick and Thin Filaments Slide Past Each Other during Contraction

Previously, we examined the movement of myosin along actin filaments (see Figures 22-19a and 22-22b). The ideas that guided those experiments were based on studies of muscle contraction that began in the 1950s. Because the muscle sarcomere is so well organized, scientists quickly saw that thick and thin filaments did not change in length during contraction, yet the sarcomere shortened. This observation could be most easily explained if the two filament systems slid past one another. Thus a model, called the *sliding-filament model*, was proposed to explain how muscle contraction occurs (Figure 22-27).

The central tenet of the model is that ATP-dependent interactions between thick filaments (myosin) and thin filaments (actin) generate a force which causes thin filaments to slide past thick filaments. The force is generated by *myosin heads*, which make cross-bridge attachments with actin; conformational changes in a cross-bridge cause the myosin heads to walk along an actin filament. The overlap

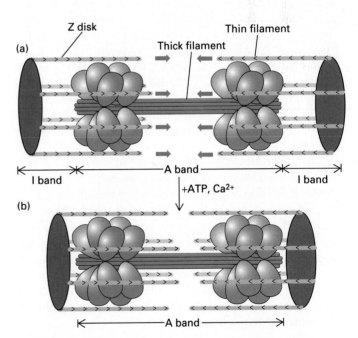

▲ FIGURE 22-27   The sliding-filament model. The diagrams show the arrangement of thick myosin and thin actin filaments in striated muscle in (a) the relaxed and (b) the contracted state. Note that the (+) ends of actin filaments are anchored at the Z disks; pivoting of the myosin heads pulls the actin filaments toward the center of the thick filament, thus reducing sarcomere length.

region, the *AI zone,* is the only part of the sarcomere where myosin heads can bind actin filaments. The sliding-filament model predicted that the force of contraction should be proportional to the overlap between the two filament systems.

Several technical difficulties lay in the path of proving the model. Electron microscopy and x-ray diffraction studies have shown that myosin and actin interact, but no one has yet seen a myosin cross-bridge make the type of conformational changes demanded by the model. The heads move too quickly and asynchronously for scientists to be confident that a particular head orientation seen in electron micrographs corresponds to a particular step in the model. However, two recent developments, determining the X-ray structure of the myosin head (Figure 22-21) and measuring the force and step size during the movement of single myosin molecules (Figure 22-20), have enabled us to envisage how a myosin head might move every time it hydrolyses in ATP molecule. Now, forty years after the sliding filament model was first stated, we believe that it best describes the mechanics of muscle contraction.

To understand how a muscle contracts, consider the interactions between one myosin head (among the hundreds in a thick filament) and a thin filament as diagrammed in Figure 22-22b. During these steps, the myosin head binds ATP, detaches from the thin filament, and undergoes a conformation change to its new position. After hydrolysis of ATP, the head rebinds to a new position on the thin filament and executes the conformational changes that are associated with the power stroke. At the completion of the cycle, also called the *cross bridge cycle,* a myosin head has moved two subunits closer to the Z disk or the (+) end of the filament. Because the heads are physically connected to the rigid rodlike backbone of the thick filament (see Figure 22-25c), myosin remains near actin and probably dissociates very briefly from the filament during contraction. Thus the heads, once they dissociate from a thin filament, are always available for rebinding to a thin filament.

Muscle contraction is the product of hundreds of myosin heads on a single thick filament, amplified by the hundreds of thick filaments in a sarcomere and thousands of sarcomeres in a muscle fiber. Because the thick filament is bipolar, the action of the myosin heads at opposite ends of the thick filament draws the thin filaments toward the center of the thick filament and therefore toward the center of the sarcomere (Figure 22-27). This movement shortens the sarcomere until the ends of the thick filaments abut the Z disk or the (−) ends of the thin filaments jam together at the center of the A band.

A Third Filament System of Long Proteins Organizes the Sarcomere

Muscle is elastic like a rubber band. Resting muscle can be stretched until the thick and thin filaments no longer overlap. However, it soon develops a resisting force or passive tension which is greater than the force normally developed by contracting muscle. If the stretching forces are removed, the muscle resumes its normal resting length and the regular arrangement of thick and thin filaments is restored. The source of this inherent elasticity was an enigma until scientists isolated some very long proteins from muscle. They found that a third filament system, composed of a distinctive set of extremely large proteins, organizes the thick and thin filaments in their three-dimensional arrays and gives muscle much of its elastic properties.

One function of the third system is to keep thick filaments centered during contraction. This function is carried out by the gigantic fibrous protein *titin* (also called *connectin*), which connects the ends of myosin thick filaments to Z disks and extends along the thick filament to the M line (Figure 22-28). Titin appears to function like an elastic band, keeping the myosin filaments centered in the sarcomere when muscle contracts or is stretched. As its name suggests, titin is huge (MW ≈ 2,700,000; that is, over 25,000 amino acids); the protein is about 1 μm long, a length that spans half of a sarcomere.

Thin filaments retain their regular organization after thick filaments are removed with salt treatment. This suggests that other structures maintain the lattice of thin filaments. Another gigantic protein, *nebulin* (MW ≈ 700,000), is implicated in performing this role. An abundant protein in skeletal muscle, constituting about 3 percent of muscle protein, nebulin forms long nonelastic filaments that extend from each side of the Z disk and along the thin filaments (Figure 22-28). Each nebulin filament is as long as its adjacent actin filament; thus the nebulin also may act as a ruler which regulates the number of actin monomers that polymerize into each thin filament during the formation of mature muscle fibers. Nebulin is missing in cardiac muscle; perhaps this explains why cardiac thin filaments are not uniformly aligned.

Calcium from the Sarcoplasmic Reticulum Triggers Contraction

Having examined how a muscle contracts, we now turn to the question of how muscle contraction is regulated. With the exception of cardiac muscle cells, which must contract rhythmically throughout the lifetime of the animal, all other muscles exhibit periods of activity followed by inactivity. In Chapters 20 and 21, we discussed how different signals—electrical, hormonal, and chemical—act to excite or intercellular responses, including muscle contraction. However, these stimuli are external to the cell. How do they produce signals within the cell that trigger the sliding interactions between actin and myosin filaments? In this section, starting with skeletal muscle and ending with smooth muscle, we will discuss the various cytosolic pathways through which a signal at the plasma membrane activates muscle contraction.

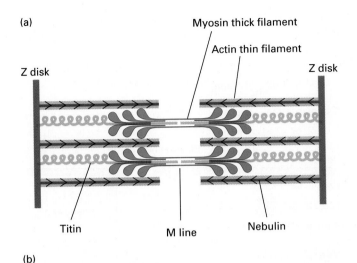

(a)

Z disk   Myosin thick filament   Z disk

Actin thin filament

Titin   M line   Nebulin

(b)

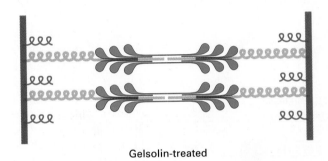

Gelsolin-treated

▲ FIGURE 22-28   The titin-nebulin filament system. A third filament system, composed of the extremely large proteins titin and nebulin, maintains the alignment of thick and thin filaments. (a) A titin filament attaches at one end to the Z disk and spans the distance to the M line. Thick filaments are thus connected at both ends to Z disks through titin. Nebulin is associated with the entire length of a thin filament, from the Z disk to the thin filament (−) end. This system of large filaments maintains its connections during muscle contraction and maintains a passive tension when muscle is stretched. (b) To visualize the titin filaments in a sarcomere, muscle is treated with the actin-severing protein gelsolin which removes the thin filaments. Without a supporting thin filament, nebulin condenses at the Z disk, leaving titin still attached to the Z disk and thick filament.

Like many processes in a cell, muscle contraction is triggered by an increase in the cytosolic calcium concentration. As described in Chapter 15, the calcium concentration of the cytosol is normally kept low, below 0.1 μM. Nonmuscle cells maintain this low concentration by Ca^{2+} ATPases in the plasma membrane, which pump calcium out of the cell. The calcium concentration in a muscle cell is also kept at the same low level, but through a different mechanism. Here, a unique Ca^{2+} ATPase continually pumps calcium from the cytosol into the sarcoplasmic reticulum (SR), a network of tubules in the muscle-cell cytosol. This activity establishes a reservoir of calcium in the SR.

Because it needs to respond instantaneously; a skeletal muscle cell has evolved a signaling pathway that is quick and efficient (Figure 22-29). The major anatomic feature of this pathway is that fingers of the plasma membrane, the *T (transverse) tubules,* invaginate into the cytosol and terminate next to the SR, at a structure called a *triad* (Figures 22-29a,b). This system is designed to bring the membrane depolarization signal into the cytosol at a triad, where the signal then stimulates the SR to release its calcium into the cytosol through Ca^{2+} channels in the SR membrane (Figure 22-29c). The T tubules and SR membranes are separated by 16 nm; this close apposition physically ensures that depolarization, an electrical signal, is converted rapidly, within milliseconds, into a cytosolic calcium signal. Muscle contraction is initiated when the calcium is released from the SR into the cytosol. Conversely, a muscle stops contracting when the channels close and calcium is pumped back into the SR.

Tropomyosin, Troponin, and Skeletal Muscle Contraction   Once the cytosolic Ca^{2+} concentration rises, it activates actomyosin interactions. So long as the Ca^{2+} concentration is sufficiently high and ATP is present, the myosin-actin bridges will cycle continuously, and the muscle will contract. In skeletal muscle, contraction is regulated by four accessory proteins on the actin thin filaments: tropomyosin and troponins C, I, and T, (Figure 22-30). Calcium changes the position of these proteins on the thin filament, and this in turn controls myosin-actin interactions. *Tropomyosin* (Tm) is a rope like molecule, about 40 nm in length; its subunits bind together head to tail, forming a continuous chain along the actin thin filament. Each TM molecule has seven actin-binding sites and binds to seven actin monomers in the thin filament. Associated with tropomyosin is *troponin* (TN), a complex of the three proteins TN-T, TN-I, and TN-C. TN-T (MW 37,000) is an elongated protein that binds along the C-terminal third of tropomyosin and links both TN-I and TN-C to tropomyosin. TN-I (MW 22,000) binds to actin as well as to TN-T. TN-C (MW 18,000) is the calcium-binding subunit of troponin. Similar in sequence to calmodulin and the myosin light chains, TN-C controls the position of TM on the surface of an actin filament.

Scientists currently think that, under the control of calcium and TN, TM can occupy two positions on a thin filament—an "off" state and an "on" state (Figure 22-30b). In the absence of calcium (the off state), myosin can bind to a thin filament but the TM-TN complex prevents myosin from sliding along the thin filament. When Ca^{2+} ions bind the Ca^{2+}-binding sites on TN-C (the on state), this inhibition is suppressed, thus activating contraction(summarized in Figure 22-31a). As revealed by electron-microscopic reconstructions of thin filaments (Figure

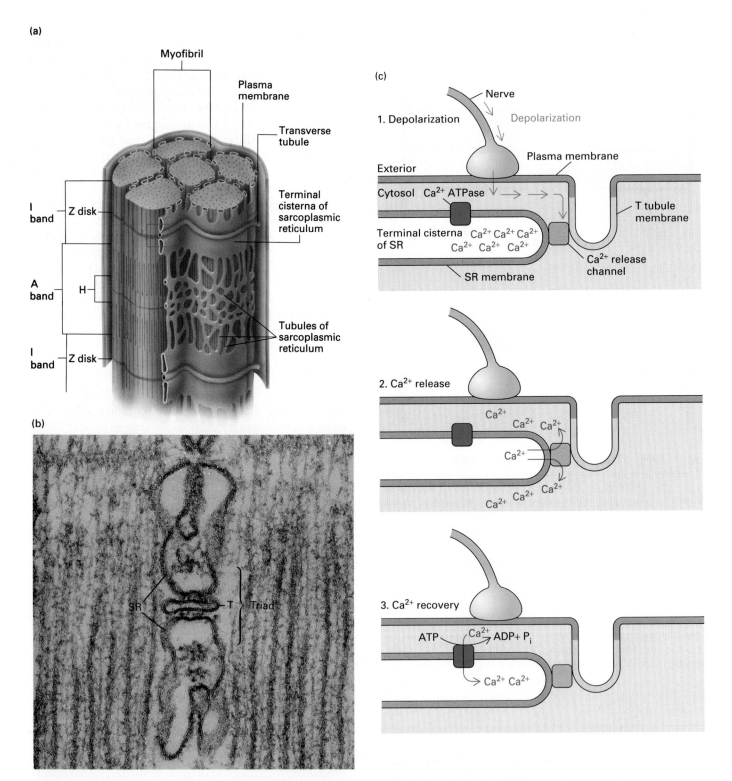

▲ **FIGURE 22-29** Regulation of cytosol Ca^{2+} in skeletal muscle by the sarcoplasmic reticulum. (a) Three-dimensional structure of six myofibrils in a muscle fiber. The transverse (T) tubules, which are invaginations of the plasma membrane, enter muscle fibers at the Z disks, where they come in close contact with the terminal cisternae of the sarcoplasmic reticulum (SR). The terminal cisternae store Ca^{2+} ions and connect with the lacelike network of SR tubules that are abundant over the H zone of the A band. (b) Electron micrograph of cardiac muscle showing a triad, consisting of a T tubule flanked by the terminal cisternae of the SR. (c) Release and recovery of Ca^{2+} by the SR. When a muscle is depolarized (step 1), Ca^{2+} ions stored in the SR are released into the cytosol through the Ca^{2+} release channel protein in the SR membrane (step 2). The calcium concentration of the cytosol returns to resting levels when Ca^{2+} is recovered from the cytosol by Ca^{2+} ATPases in the SR membrane (step 3). [Part (b) courtesy of C. Franzini-Armstrong.]

(a) Thin filament

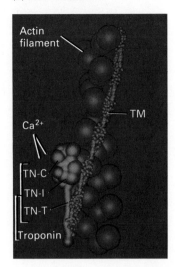

(b)

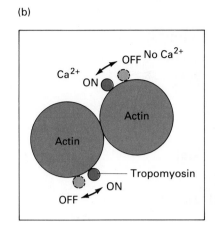

(c)

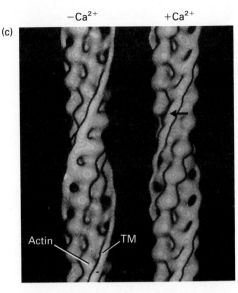

▲ **FIGURE 22-30**   Effect of Ca²⁺ ions on tropomyosin binding to actin filaments. (a) Model of the tropomyosin-troponin (TM-TN) regulatory complex on a thin filament. TN, a club-like complex of TN-C, TN-I, and TN-T subunits, is bound to the α-helical TM molecule. (b) Diagram of the position of TN in the "off" state (−Ca²⁺) and the "on" state (+Ca²⁺). The binding of Ca²⁺ to TN causes TM to shift toward the center of the filament. (c) Three-dimensional electron-microscopic reconstructions of the TM helix on a thin filament from scallop muscle. TM in its "off" state (left) shifts to its new position (arrow) in the "on" state (right) when Ca²⁺ is present. (TN is not shown in this representation.) [Part (a) from G. N. Phillips, Jr., J. P. Filliers, and C. Cohen, 1986, *J. Mol. Biol.* **222**:111; courtesy of C. Cohen. Part (c) adapted from W. Lehman, R. Craig, and P. Vibert, 1993, *Nature* **123**:313; courtesy of P. Vibert.]

22-30c), the binding of Ca²⁺ to TN-C triggers a slight movement of TM toward the center of the actin filament, which exposes the myosin-binding sites on actin. Thus the role of calcium is to relieve the inhibition exerted by the TM-TN complex.

Caldesmon Dissociation and Smooth Muscle Contraction   Changes in the cytosolic Ca²⁺ level are much slower in smooth muscle than in skeletal muscle—on the order of seconds to minutes—thereby allowing the slow, steady response in contractile tension that is required by vertebrate smooth muscle. There are several reasons why the contractile response of smooth muscle is slow. Recall that vertebrate smooth muscle does not have a sarcomere organization (see Figure 22-23b), and the actomyosin network is more disordered than in striated muscle. In addition, the SR membrane network is poorly developed and sparse. Much of the calcium enters the cell through the plasma membrane. Most importantly, smooth muscle contains TM but lacks TN. Therefore, the mechanism for controlling muscle contraction in smooth muscle must differ from the TM-TN system of skeletal muscle (Figure 22-31).

There are, in fact, several pathways that increase or decrease smooth muscle contractions. One pathway is through *caldesmon*, a calmodulin-binding protein in the smooth-muscle thin filament (Figure 22-31b). Smooth-muscle caldesmon (MW 150,000) is an elongated protein, about 75 nm in length. When Ca²⁺ ions are in short supply, caldesmon forms a complex with TM and actin, thereby restricting the ability of myosin to bind to actin. Caldesmon-actin binding is also regulated by phosphorylation by various kinases, including protein kinase C (PK-C). When phosphorylated by PK-C, caldesmon is unable to bind to a thin filament and therefore is unable to inhibit smooth-muscle myosin from binding to actin. As a result, PK-C activates smooth muscle contraction.

(a) Skeletal muscle

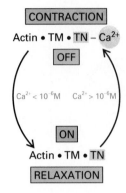

(b) Smooth muscle

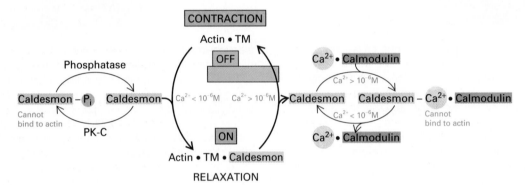

▲ FIGURE 22-31   Comparison of regulation of contraction in skeletal and smooth muscle. (a) Regulation of skeletal muscle contraction by Ca^{2+} binding to TN. Note that the TM-TN complex remains bound to the thin filament whether muscle is relaxed or contracting. (b) Regulation of smooth muscle contraction by caldesmon. At low Ca^{2+} concentrations ($<10^{-6}$ M), caldesmon binds to TM and actin, reducing the binding of myosin to actin and keeping muscle in the relaxed state. At higher Ca^{2+} concentrations, a Ca^{2+}-calmodulin complex binds to caldesmon, releasing it from actin; thus myosin can bind to actin and the muscle can contract. Phosphorylation by several kinases, including protein kinase C (PK-C), also regulates caldesmon's actin-binding activity. When phosphorylated, caldesmon is unable to bind actin and therefore cannot block actin-myosin interaction. This inhibition is relieved by phosphatases.

Calcium Activation of Myosin Light Chains Regulates Contraction in Smooth Muscle and Invertebrate Muscle

So far we have examined examples where proteins on actin filaments control actomyosin interactions. However, smooth muscle and invertebrate skeletal muscle are also regulated by a different mechanism, one that is directed toward myosin instead of toward actin. In these muscles, calcium, either by binding to the regulatory light chains of myosin or by activating calcium-dependent phosphorylation of those light chains, regulates myosin binding to actin (Figure 22-32).

Calcium-Binding to the Regulatory Light Chain

The simplest example of myosin-linked regulation is found in invertebrate muscle. Typically, in mollusks, such as the scallop, the interaction between myosin heads and actin filaments is inhibited at low Ca^{2+} concentrations (Figure 22-32a) by the regulatory light chain (LC), one of the two pairs of LCs in the myosin neck region (see Figures 22-17 and 22-21). The binding of Ca^{2+} ions to the regulatory LC in the neck region induces a conformational change in the myosin head that allows it to bind to actin; this, in turn, permits activation of the myosin ATPase and contraction of the muscle.

Myosin Light-Chain Kinase–Dependent Activation of Myosin

Contractions in the smooth muscle of vertebrates are regulated primarily by a complex program of myosin LC phosphorylations and dephosphorylations (Figure 22-32b). As in mollusks, one of the two myosin LC pairs in smooth muscle inhibits actin stimulation of the myosin ATPase activity at low Ca^{2+} concentrations. This

(a) Ca^{2+} binding to light chains

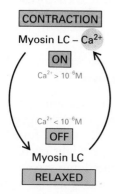

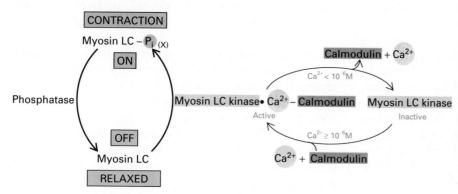

(b) Phosphorylation of light chains

(c) Regulation by protein kinase C

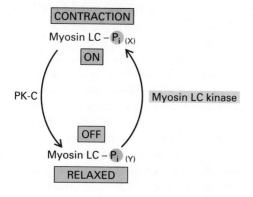

◄ FIGURE 22-32   Three myosin-dependent mechanisms for regulating contraction and relaxation in muscle. (a) Ca^{2+} binding to the regulatory light chain (LC) of scallop myosin activates the muscle to contract. (b) Smooth muscle myosin activity is regulated by phosphorylation of the LC by myosin light-chain kinase. In its active form, this enzyme phosphorylates the myosin regulatory LC (see Figure 22-21) on site X; as a result, myosin can bind to actin and the muscle contracts. The activity of myosin LC kinase itself is controlled by Ca^{2+} and calmodulin. When the Ca^{2+} concentration is $\geq 10^{-6}$ M, Ca^{2+} binds to calmodulin and this complex binds to myosin LC kinase, activating the enzyme. A decrease in the Ca^{2+} concentration to 10^{-7} M leads to dissociation of Ca^{2+} and calmodulin from the kinase, thereby inactivating the kinase. Also under these conditions, a myosin LC phosphatase, which is not dependent on Ca^{2+} for activity, dephosphorylates the myosin LC, causing muscle relaxation. (c) Activation of PK-C by diacylglycerol and Ca^{2+} leads to phosphorylation of the myosin regulatory LC at site Y. Modification of this site hinders phosphorylation at site X by myosin LC kinase and causes a decrease in the activity of myosin.

inhibition is relieved and the smooth muscle contracts when the regulatory LC is phosphorylated by the enzyme myosin LC kinase. Because Ca^{2+} is required for activation of myosin LC kinase, the Ca^{2+} level regulates the extent of LC phosphorylation and hence regulates contraction. This mode of regulation relies on the diffusion of Ca^{2+} and the action of protein kinases, and thus muscle contraction is much slower in smooth muscle than in skeletal muscle.

Myosin LC kinase activity is regulated by calcium through calmodulin. Calcium first binds to calmodulin, and the Ca^{2+}-calmodulin complex then binds to myosin LC kinase and activates it (Figure 22-32b). Evidence for the role of activated myosin LC kinase can be obtained by microinjecting into smooth muscle cells, an inhibitor that blocks the activity of the kinase. The inhibitor also blocks muscle contraction, even though it does not block the rise in the cytosolic Ca^{2+} level associated with membrane depolarization. The effect of the inhibitor can be overcome by microinjecting a proteolytic fragment of myosin LC kinase that is active even in the absence of Ca^{2+}-calmodulin (this treatment also does not affect Ca^{2+} levels).

Protein Kinase C–Dependent Inhibition of Myosin
The regulation of smooth muscle contraction utilizes the same signal transduction pathways found in nonmuscle cells (see Chapter 20). Unlike skeletal muscle, which has one source for stimulation—nerves—smooth muscle is regulated in addition by many humoral factors; some are activators of contraction while others are inhibitors of contraction. Whatever the stimulant, all funnel into the same regulatory mechanism by modulating cytosolic Ca^{2+} levels and the activities of myosin light-chain kinases and phosphatases. For example, some hormones, acting through the phospholipase C-PIP_2 pathway, cause the relaxation of smooth muscle through actions of kinases and phosphatases. In one example, when protein kinase C(PK-C) is activated by Ca^{2+} and 1,2-diacylglycerol, it phosphorylates two target proteins. As we discussed, one target is caldesmon. When caldesmon is phosphorylated, it cannot bind actin filaments and therefore inhibit myosin activity. The second substrate of PK-C is the myosin regulatory light chain. Smooth muscle myosin activity decreases when the regulatory light chain is phosphorylated on a site different from that phosphorylated by myosin light chain kinase (Figure 22-32c). The counteracting effects of PK-C illustrates that the regulation of smooth muscle is complex because various cellular factors modulate the activity of the muscle through PK-C and other kinases and phosphatases.

➤ *Actin and Myosin in Nonmuscle Cells*

On the basis of our discussion of actin and myosin in a muscle cell, we can now examine the function of actomyosin structures in a nonmuscle cell. At first, scientists thought that most cell movements were caused by a muscle-like mechanism involving a contraction of actin and myosin filaments. This idea was based on observations of contractile-like movements of cytosolic extracts and the purification of actin and myosin II from several types of nonmuscle cells. In addition, in nonmuscle cells myosin II and actin were found to be organized into bundles similar to muscle sarcomeres both in their organization and in having their ends anchored to the plasma membrane by proteins also found at the muscle Z disk.

Later, biochemical studies extracted a second myosin, myosin I, which did not form thick filaments. By now, as discussed earlier in the chapter, we have identified more than ten members of the myosin family but have only studied three members: myosin I, myosin II, and myosin V. Because it was the first myosin to be purified from a nonmuscle cell and also because it is the most abundant and most easily purified, we know more about myosin II than

about myosins I and V. In this section, we will present features of the functions of these three myosins in various nonmuscle cells.

Actin and Myosin II Are Arranged in Sarcomere-Like Structures

In a nonmuscle cell, myosin II is an abundant protein (although the ratio of myosin to actin in a nonmuscle cell is much less than in a muscle cell), and it forms thick filaments like those in a sarcomere. Light micrographs show actin and myosin to be organized into sarcomere-like contractile bundles, which are usually found along the cell membrane at sites where the cell contacts the substratum or another cell. The bundles are classified as contractile because, when isolated from the cell, they contract upon addition of ATP. Contractile bundles also called circumferential belts or stress fibers, differ from the noncontractile bundles of actin described earlier in this chapter (see Figure 22-5a) in two respects. First, contractile bundles contain myosin II interspersed among actin filaments; the myosin II is responsible for the contractility. In the noncontractile actin bundles present in a microvillus or a filopodium, myosin I is sometimes found along the sides or tips of the actin bundles, but myosin II is not usually present as a major component. Second, a noncontractile bundle comprises the core of a finger of membrane (see Figure 22-6b), while a contractile bundle always wraps around the inner surface of the plasma membrane like a belt.

Actin Stress Fibers in Attachment of Cells to Surfaces One example of a contractile bundle is the long bundles of microfilaments, called *stress fibers*, which lie along the ventral surfaces of cells (Figure 22-33; see also Figure 22-1b) cultured on artificial (glass or plastic) surfaces. The ends of the bundles terminate at *adhesion plaques*, special structures that attach a cell to the underlying substratum. (The structure of an adhesion plaque will be discussed later). Fluorescent-antibody techniques reveal that myosin and α-actinin are distributed in alternating patches in stress fibers, much like the pattern of alternating thick filaments and Z bands in muscle sarcomeres. Stress fibers also contain tropomyosin, caldesmon, and the regulatory protein myosin LC kinase. The contractility of stress fibers has been demonstrated: after the ends of stress fibers are cut with a laser microbeam (thereby separating the stress fibers from the adhesion plaques), the addition of ATP causes the stress fibers to shorten. This ability to contract suggests that stress fibers are functionally similar to sarcomeres.

Although stress fibers are contractile, they probably function in cell adhesion rather than in movement. This interpretation of stress fiber function is supported by several observations. For one, fibroblasts that are migrating have few stress fibers during the time of rapid movement;

(a)

Adhesion plaque

Stress fiber

(b)

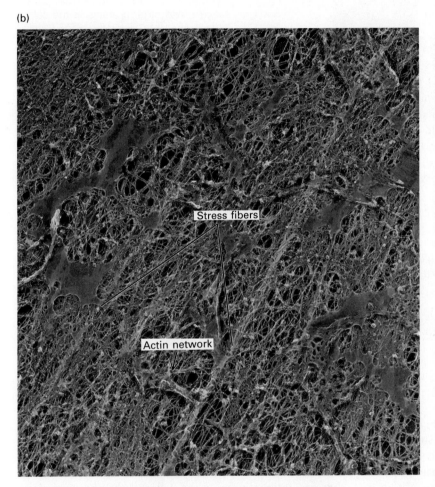

Stress fibers

Actin network

(c)

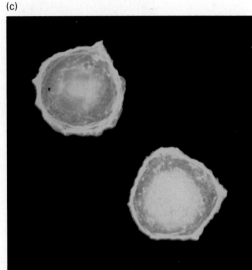

(d)

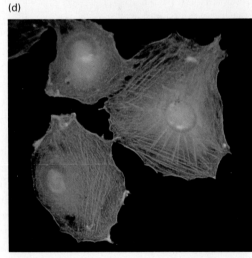

▲ FIGURE 22-33   Role of stress fibers and cell adhesion plaques in cell attachment to the underlying substratum. (a) A stress fiber lying along the ventral surface of a cell has its ends attached to adhesion plaques (purple). Myosin (blue) is interspersed with the actin filaments (red) in the stress fiber. (b) Electron micrograph of the cytoskeleton attached to the cell membrane shows the long stress fibers embedded in a network of actin filaments. (c) Cells become spherical immediately after treatment with trypsin has detached them from the surface of a culture dish. As detected by fluorescent antibodies, actin is found in ruffles at the perimeter of the cell and diffusely throughout the body of the cell. Microfilaments are not present at this stage. (d) After 4 h of spreading on a culture dish, the trypsin-treated cell has flattened and contains polygonal arrays of actin filaments at the periphery; stress fibers also crisscross the cell. [Part (b) courtesy of J. Hartwig. Parts (c) and (d) from R. Hynes and A. T. Destree, 1978, *Cell* **15**:875; courtesy of R. Hynes.]

(a)

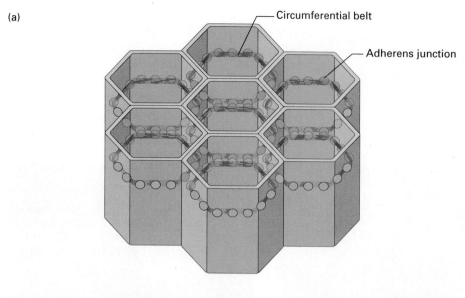

Circumferential belt

Adherens junction

◄ FIGURE 22-34   The circumferential belt is located near the apical surface of epithelial cells. (a) In epithelial tissue, a belt of actin (red) and myosin (blue) filaments rings the inner surface of the cell adjacent to the adherens junctions (purple), where cell-cell contacts are maintained. This circumferential belt is attached to E-cadherin by vinculin and other proteins at the cell membrane. (b) Contraction of the circumferential belt can fold an epithelial sheet. If contraction occurs in a row of cells, the resulting constriction at the apical portion of the cells causes the epithelial sheet to pucker. If there is contraction at several zones the sheet will form a groove.

(b)

Zone of apical contraction

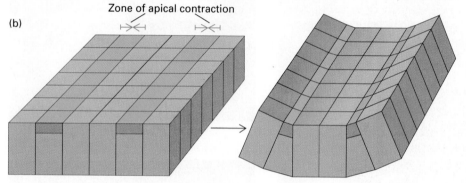

however, when these cells stop migrating, the stress fibers increase in number. Also, when cultured fibroblasts are removed from their substratum, the cell becomes spherical and the stress fibers disappear (Figure 22-33c). Within a few hours after the cell settles back on its substratum, the stress fibers reappear. Apparently, the adhesion of cells to a substratum, perhaps through the formation of cell-substratum attachments and stresses on the cytoskeleton, induce the random assortment of actin microfilaments to collect into microfilament bundles. Because the ends of stress fibers are firmly anchored to adhesion plaques, the bundles are prevented from contracting and instead are only able to generate tension.

Epithelial Cell-Cell Adhesion by Belts of Actin and Myosin II   In epithelial cells, a bundle of actin and myosin filaments form a *circumferential belt* which encircles the inner surface of the cell at the level of the adherens junction (Figure 22-34; see Figure 14-42). The circumferential belt, like a stress fiber, resembles a primitive sarcomere in its organization and contains many proteins found in stress

fibers and smooth muscle, including vinculin, tropomyosin, and α-actinin. As a complex with the adherens junction, the circumferential belt functions as a tension cable that can internally brace the cell and thereby control its shape.

Contraction of the circumferential belt may play a role in generating folds in a sheet of cells (Figure 22-34b), a characteristic occurrence in the developing embryo. The early stages of embryonic development are marked by morphogenetic changes during which sheets of cells fold and move to new positions. During neurulation, for example, the primitive brain and spinal cord form when the *neural plate,* a sheet of column-shaped cells on the dorsal surface of the embryo, folds into a tube. The folding process is initiated when the apical region of each neural plate cell becomes constricted by a contractile bundle of actin which encircles the apical rim of the cell. In a manner analogous to drawing closed a purse string, the actin bundle tightens around the neck of the cell, giving it a bottle-shaped appearance. Within a cell sheet, this type of change in cell shape deforms the sheet and causes a groove to form.

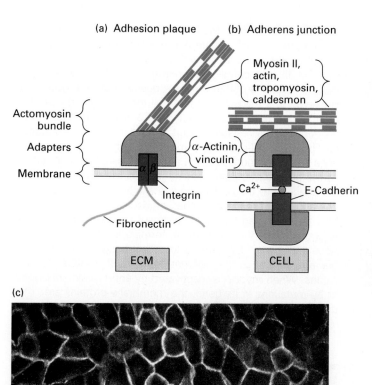

(a) Adhesion plaque   (b) Adherens junction

Myosin II, actin, tropomyosin, caldesmon

Actomyosin bundle

Adapters

Membrane

α-Actinin, vinculin

Integrin

Ca²⁺   E-Cadherin

Fibronectin

ECM   CELL

(c)

◄ FIGURE 22-35   Actin-myosin attachments at cell junctions. (a) The end of a stress fiber is attached to the cell membrane at adhesion plaques. An adhesion plaque is organized around integrin, the membrane protein receptor for the extracellular matrix (ECM). Attachment between the stress fiber and integrin is mediated through adapter proteins, a complex of peripheral membrane proteins which include actin-binding proteins such as vinculin and α-actinin. (b) The circumferential belt is attached to the adherens junction through the same set of adapter proteins. However, instead of integrin, the belt is connected to E-cadherin, an integral membrane protein that makes Ca²⁺-dependent cell-cell contacts. (c) A view through the top of a monolayer of cultured kidney epithelial cells labeled with a fluorescent anti–E-cadherin antibody shows the E-cadherin surrounding each cell in the regions of cell-to-cell contact. [Part (c) from B. Gumbiner, B. Stevenson, and A. Grimaldi, 1988, *J. Cell Biol.* **107**:1575; courtesy of B. Gumbiner.]

Contractile Actin Bundles Are Attached to Specialized Sites at the Plasma Membrane

We now turn to the question of how contractile bundles like stress fibers and circumferential belts are attached to the plasma membrane. This is of interest because their role in cell adhesion is an important one in cell-cell and cell-matrix interactions (discussed in Chapter 24). We have already discussed two examples, the Z disk of skeletal muscle and the membrane attachment site in smooth muscle, where the ends of a contractile bundle are attached to a specialized structure. In nonmuscle cells, the distinctive feature of a contractile bundle is that they are found where the cell contacts the substratum (adhesion plaque) or contacts another cell (adherens junction). The attachment site for a contractile bundle has a complicated structure. As one indication of this complexity, the number of different proteins found just at an adhesion plaque is now known to exceed 20; an equally large number of proteins will probably be found at an adherens junction. Despite their com-

plexity, we can see that the attachment sites are organized into three parts (Figure 22-35): integral membrane proteins which connect the cell to the extracellular matrix or to another cell; adapter proteins which connect the integral membrane protein to the contractile actin bundle; and lastly the contractile bundle itself. As we will see in Chapters 23 and 24, when intermediate filaments make similar contacts with the cell membrane, they use different adapter molecules and similar integral membrane proteins.

Integrins at Cell-Substratum Attachments   The ends of stress fibers are anchored to the *adhesion plaques*, the specialized sites that attach the ventral plasma membrane to the extracellular matrix (Figure 22-35a; see also Figure 24-29). Clustered within an adhesion plaque are the *integrins*, the integral membrane proteins that bind to fibronectin molecules in the extracellular matrix. Actin filaments of the stress fibers are attached to integrins through *adapter proteins*; these peripheral membrane proteins, in-

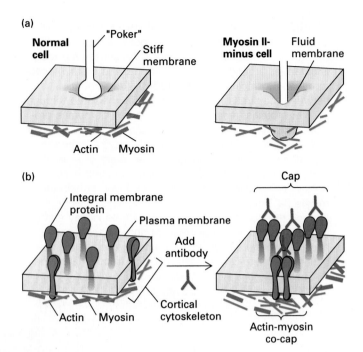

◄ FIGURE 22-36   Myosin stiffening of the cortex.
(a) The presence of myosin II in the cortex can be detected by poking the cell surface with a small glass ball at the end of a rod. In cells containing a normal cortical cytoskeleton, the membrane is stiff and resists pushes from the glass ball. However, in a myosin II-mutant cell that lacks all myosin II protein, the membrane is fluid-like. Consequently, the poker encounters less resistance as it pushes into the cell surface. (b) The myosin II-dependent movement of cell-surface proteins can be demonstrated in a cell "capping" experiment. At the plasma membrane, many integral membrane proteins are attached to the underlying cytoskeleton in the cortex. The addition of reagents, either antibodies or lectins which bind the membrane proteins, will cause membrane proteins to aggregate at the cell surface, forming a patch or "cap." Actin and myosin filaments also collect beneath the membrane even though they are not bound by the antibodies or lectins. When myosin is inactivated by anti-myosin antibodies or by mutations in its gene, the membrane proteins are unable to cap.

cluding α-actinin and vinculin, are generally located in the region of the cytosol just inside the plasma membrane. Many other proteins—actin-binding proteins, kinases, and membrane-binding proteins—are also localized in cell adhesion plaques but their precise functions are still unknown.

Cadherins at Cell-Cell Junctions   In contrast to the adhesion plaque, which joins the ventral surface of a cell to its substratum, a cell-to-cell contact, as in a sheet of epithelial cells, is held together by an *adherens junction* (Figure 22-35b). Here, the plasma membranes of the adjacent cells are parallel and only 15–20 nm apart. Concentrated in this region, the transmembrane protein *E-cadherin* links the plasma membranes of adjacent cells (Figure 22-35; see also Figure 24-31). In a similar fashion to the attachment of a stress fiber to an adhesion plaque, the circumferential belt of actin and myosin is attached along its side to E-cadherin through vinculin and α-actinin.

Myosin II Stiffens Cortical Membranes

Not all of myosin II is organized into contractile bundles; myosin II is also part of the cortical cytoskeleton, along with the network of actin filaments, where it contributes to the stiffness of the plasma membrane. Stiffness is measured by poking the membrane with a small glass ball at the end of a rod (Figure 22-36a) and calculating the force required to deform the membrane. The plasma membrane is a phos-pholipid bilayer, which is very fluidlike and therefore not stiff. However, if the membrane is supported by a highly cross-linked cytoskeleton it becomes very stiff. The membranes of mutant cells lacking myosin II are deformed more easily by the glass ball than the membranes of normal cells. These experiments indicate that myosin II molecules in the cortex act as small tension rods which "tighten up" the cortical actin cytoskeleton.

Cortical myosin II is also responsible for the movements of some cell-surface proteins (Figure 22-36b). Although cell-surface proteins are normally immobile in the membrane, we can induce them to cluster, or "cap," at one region in the membrane by adding antibodies or lectins that bind them. Capping is inhibited in cells lacking myosin II, suggesting that myosin II in the cortex provides the force that aggregates the membrane proteins.

Actin and Myosin II Have Essential Roles in Cytokinesis

During cell division, or *cytokinesis,* the last step in mitosis (see Chapter 23), a *contractile ring* assembles at the equator of the dividing cell. This ring is a contractile bundle of actin filaments that is similar to a stress fiber or a circumferential belt. Fluorescence microscopy shows that during mitosis actin and myosin accumulate at the equator of a cell, midway between the poles of the spindle (Figure 22-37). There they align into a bundle which encircles the cell.

(a)

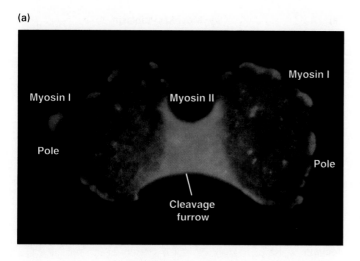

(b)

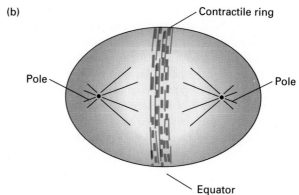

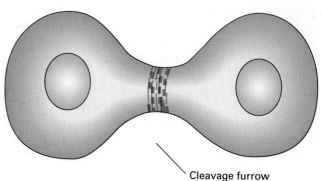

▲ **FIGURE 22-37** The role of myosin II in cytokinesis. (a) During cell division of an *Dictyostelium* ameba, myosin II (red) is concentrated in the cleavage furrow while myosin I (green) is localized at the poles of the cell. (b) During mitosis (see Chapter 23) when the spindle poles and aster microtubules separate, myosin (blue) and actin (red) in the cortex of the cell assemble into an equatorial contractile ring around the cell. As the nuclei in the daughter cells start to reform, the contractile ring constricts, causing the membrane to form a cleavage furrow. In the last step of mitosis, the cell is pinched into two parts. [Part (a) Courtesy of Y. Fukui.]

As cytokinesis proceeds, the diameter of the contractile ring decreases and the cell is pinched into two parts by a deepening cleavage furrow. Dividing cells stained with antibodies against myosin I and myosin II show that myosin II is localized to the contractile ring while myosin I is at the cell poles (Figure 22-37a). This localization indicates that myosin II but not myosin I is involved in cytokinesis.

Two types of experiments have demonstrated that cytokinesis is indeed dependent on myosin II (Figure 22-38). In both kinds of experiments, active myosin II was eliminated from the cell. In one case, this was done by microinjecting anti–myosin II antibodies into one blastomere of a

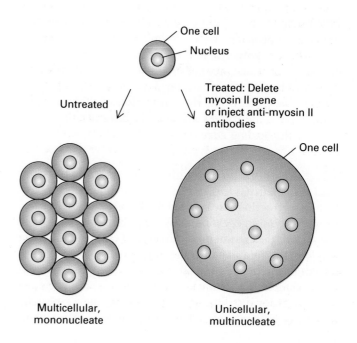

➤ **FIGURE 22-38** The role of myosin II in cytokinesis. The need for myosin II was shown when its activity was inhibited, either by deleting its gene or by microinjecting anti–myosin II antibodies into a cell. A cell that lacked myosin II was able to replicate its DNA and nucleus, but it failed to divide; this defect caused the cell to become large and multinucleate. In comparison, an untreated cell during the same period continued to divide and formed a multicellular ball of cells in which each cell contained a single nucleus.

sea urchin embryo at the two-cell stage. In the other case, expression of myosin II was inhibited in the *Dictyostelium* ameba by genetic deletion of the myosin gene or by antisense inhibition of myosin mRNA expression. The results were identical in all cases: cells lacking myosin II became multinucleated because cytokinesis (cell separation) but not karyokinesis (chromosome separation) was inhibited. Without myosin II, the cells failed to assemble a contractile ring although other events in the cell cycle proceeded.

Myosins I and V Move Membrane-Bounded Cargoes along Actin Filaments

Among the many movements exhibited by cells, vesicle translocation has been one of the most fascinating to cell biologists. In early studies of the cytoplasm, researchers found that certain particles, now known to be membrane vesicles, moved in straight lines within the cytosol, sometimes stopping only to resume movement again, at times after changing direction. This type of behavior could not be caused by diffusion because the movement was clearly not random. Therefore, researchers reasoned, there must be tracks along which the particles travel as well as some type of motor which powers the movement. At first scientists thought myosin II caused the movements, but this theory proved incorrect.

Although the most dramatic result of genetic deletion of the myosin II gene was the inability of a cell to divide, an equally significant finding was that other motile processes, such as vesicle transport and membrane movements at the leading edge, proceeded although somewhat abnormally in these cells. This finding suggested that other proteins, possibly other myosins, must be involved in these forms of cell motility. In the next chapter, we will discuss how microtubule motor proteins are responsible for the movement of some vesicles; however, in this section, we will discuss the evidence that some myosins are motors which move along an actin filament while carrying a membrane vesicle as its cargo. We will leave for the last section of this chapter a discussion of the role of myosin I in directing the movement of the leading edge of cells.

Myosin-Generated Movements of the Endoplasmic Reticulum in Cytoplasmic Streaming Not all actomyosin interactions are sarcomere-like. One example is *cell streaming*. In large, cylindrical green algae such as *Nitella* and *Chara*, the cytosol flows rapidly, at a rate approaching 4.5 mm/min, in an endless loop around the inner circumference of the cell (Figure 22-39a). The rapid flow of cytosol, sustained over millimeter-long distances, is a principal mechanism for distributing cellular metabolites, especially in large cells such as plant cells and amebas. This type of

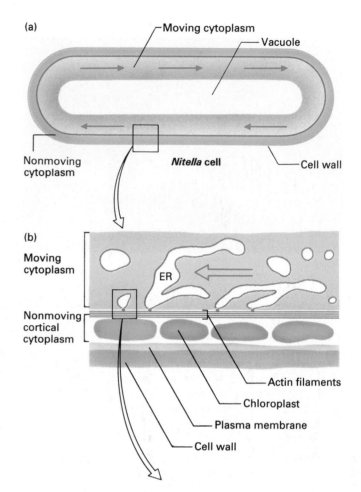

(a)

Moving cytoplasm
Vacuole
Nonmoving cytoplasm
Nitella cell
Cell wall

(b)

Moving cytoplasm
Nonmoving cortical cytoplasm
ER
Actin filaments
Chloroplast
Plasma membrane
Cell wall

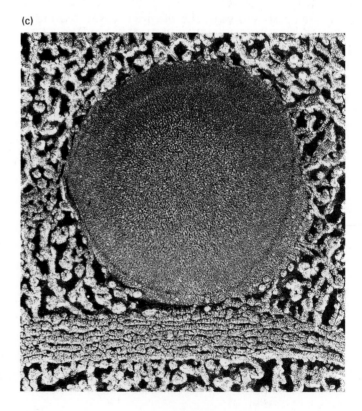

(c)

◄ FIGURE 22-39 Cytoplasmic streaming in cylindrical giant algae. (a) Movement of cytoplasm in a *Nitella* cell is indicated by arrows. The cell center is filled with a single large water-filled vacuole. (b) Expanded diagram of region boxed in (a). A nonmoving layer of cortical cytoplasm filled with chloroplasts lies just under the plasma membrane. On the inner side of this layer are bundles of stationary actin filaments, all oriented with the same polarity. (c) An electron micrograph of the cortical cytoplasm shows a large vesicle connected to an underlying bundle of actin filaments. This vesicle is part of the endoplasmic reticulum (ER) network, which contacts the stationary actin filaments and moves along them by a myosin-like motor [shown in blue in part (b)]. The sliding of the ER network propels the entire viscous cytoplasm, including organelles that are enmeshed in the ER network. [Part (c) from B. Kachar.]

movement probably represents an exaggerated version of the smaller-scale movements exhibited during the transport of membrane vesicles.

Close inspection of objects caught in the flowing cytosol, such as the endoplasmic reticulum (ER) and other membrane-bounded vesicles, showed that the velocity of streaming increases from the cell center (zero velocity) to the cell wall (maximum velocity). This gradient in the rate of flow is most easily explained if the motor generating the flow lies at the membrane. In electron micrographs, bundles of actin filaments can be seen aligned along the length of the cell, lying across chloroplasts embedded at the membrane. Attached to the actin bundles are vesicles of the ER network (Figure 22-39b). The bulk cytosol is propelled by myosin attached to the ER along the bundles (Figure 22-39c). Although the *Nitella* myosin has not been isolated, it must be one of the fastest known, because the flow rate of the cytosol in *Nitella* is at least 15 times faster than the rate produced by any other myosin.

Membrane-Bound Myosin in Vesicle Movements

Both myosin I and myosin V appear to be important in vesicle transport. The role of myosin I in vesicle transport has been deduced from studies of amebas. Amebic myosin I was the first myosin I molecule to be identified and characterized. The cDNA sequences of three myosin I genes have now been identified in *Acanthameba,* a common soil ameba. Detected by antibodies specific for each myosin I gene, each myosin I isoform is localized to different membrane structures in the cell. For example, myosin IA is associated with small cytoplasmic vesicles. Myosin IC, by contrast, is found at the plasma membrane and at the

contractile vacuole (Figure 22-40a), a vesicle that regulates the osmolarity of the cytosol by fusing with the plasma membrane. The introduction of antibodies against myosin IC into a living ameba inhibits water expulsion, causing the vacuole to expand uncontrollably, eventually bursting the cell.

Myosin I in vertebrate cells is also implicated in vesicle transport. For example, in intestinal epithelial cells, myosin I is purified with vesicles derived from Golgi membranes. The presence of this motor on Golgi membranes suggests myosin I moves membrane vesicles between membrane compartments in the cytoplasm.

Myosin V also could be involved in the intracellular transport of membrane-bounded vesicles, as indicated by studies on the localization of myosin V and genetic analysis of myosin V mutants in mice (*dilute*) and in yeast (*myo2/myo4*). Myosin V is highly expressed in brain tissue, where specific antibodies show it to be concentrated on Golgi stacks (Figure 22-40b). This association with membranes is consistent with the effects of myosin V mutations on intracellular vesicle transport. Mutations in the mouse myosin V gene cause death from seizures and are implicated in defects in synaptic transmission, while myosin V mutations in yeast disrupt protein secretion and cause vesicles to accumulate in the cytoplasm.

Myosin I and Myosin II Are Not Essentially Required for Cell Migration

While we believe that the myosins provide the motive force for many cell movements, we are still uncertain of the roles played by myosin I and myosin II in cell migration. Immunofluorescence micrographs show that in a migrating ameba, myosin I is localized to the leading edge of the cell while myosin II is in the rear (Figure 22-41a). Myosin I is in a position to guide the cell, while myosin II appears to be in a position to push the cell, although in early studies of *Dictyostelium*, the deletion of either myosin I or myosin II did not inhibit cell migration, indicating that neither of the myosins is essential for this type of movement. After careful analysis, however, later experiments demonstrate that myosin I or myosin II has some function. Video micrographs Figure 22-41b show that a migrating cell normally sends out single pseudopodium in the direction of movement. In contrast, the *Dictyostelium* cells lacking myosin I send out many pseudopodia, as if they were trying to move in all directions at once. This behavior makes it difficult to start sustained or persistent movement in one direction. Although cells lacking myosin I move, cells lacking myosin II do not form pseudopodia. As a result, myosin II-mutants are non-motile. Thus while we are unable to describe a particular role of myosin in cell migration, these results suggest that myosin I and myosin II participate in coordinating cell movements at opposite parts of a cell.

(a)

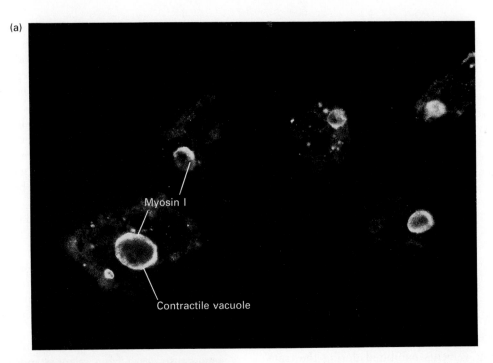

(b)

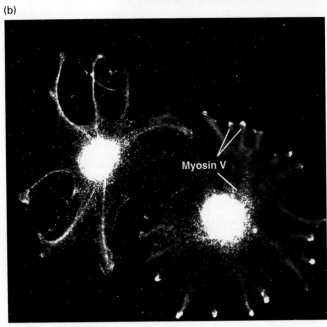

▲ **FIGURE 22-40**   Myosin I and myosin V associations with cell membranes. (a) Myosin IC is one of three myosin I isoforms in *ameba*. This isoform is associated with the membrane of contractile vacuoles, where it regulates the volume of water. Other myosin I isoforms (not shown) are located at the plasma membrane or around small cytoplasmic vesicles. (b) Intracellular vesicle sorting may also involve myosin V because immunofluorescence microscopy localizes the protein in the Golgi region of astrocytes. Myosin V is also located at the tips of membrane processes that extend from the cell. This distribution indicates that myosin V is associated with membranes and suggests the protein could be involved in membrane transport from the Golgi to the cell periphery. [Part (a) from I. Baines and E. Korn, 1990, *J. Cell Biol.* **111**:1895; courtesy of I. Baines. part (b) from E. M. Espreafico et al., 1992, *J. Cell Biol.* **119**:1541; courtesy of R. E. Cheney and M. Mooseker.]

➤ *Cell Motility*

A time-lapse series of micrographs of a moving leukocyte (see chapter opening figure) or fibroblast (see Figure 5-19) shows that when moving, both cells display the same sequence of changes in cell morphology—extension of the cell membrane and attachment to the substratum, forward flow of cytosol, and retraction of the rear of the cell. These steps are diagrammed in Figure 22-42. We have now exam-

ined the different mechanisms used by a cell to create movement—from the assembly of actin filaments and the formation of actin-filament bundles and networks to the contraction of bundles of actin and myosin and the sliding of single myosin molecules along an actin filament. Our belief is that most of the major mechanisms for generating force have been discovered. However, scientists still do not agree on the overall mechanism that moves a cell forward because we do not know how the various types of cell motility are integrated or coordinated.

(a)

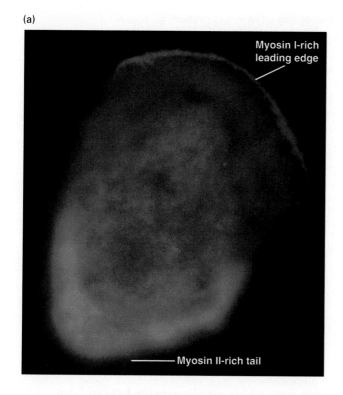

Myosin I-rich leading edge

———— Myosin II-rich tail

(b)

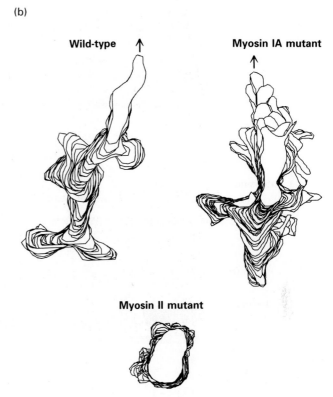

Wild-type

Myosin IA mutant

Myosin II mutant

▲ **FIGURE 22-41** The role of myosin I and myosin II in locomoting cells. (a) In a moving ameba stained with antibodies to myosin I and myosin II, the myosin I (green) is located at the front, where active forward movement takes place. In the same cell, myosin II (red) is limited to regions in the rear. (b) Comparisons between the movement of a normal cell with myosin IA or myosin II mutants reveal abnormal behaviors in cell motility. Each panel shows the outline of the cell, recorded at 4 second intervals on videotape. By superimposing the outlines, we see the changes in cell shape and position with time. A normal wild-type cell exhibits a classic elongated shape and, led by a single pseudopodium at its anterior, moves in a polar direction with few pauses or turns in the process. In contrast, a myosin IA-deficient mutant turns in many directions during its migration and as a result travels a shorter distance. From the outlines of this mutant, we see that it generates many pseudopodia simultaneously and therefore has difficulty in moving persistently in one direction. A myosin II deficient cell is unable to locomote under these conditions. It does not extend pseudopodia and exhibits a rounded cell shape. [Part (a) courtesy of Y. Fukui; part (b) courtesy of D. Soll].

Movements of Fibroblasts Involve Controlled Polymerization and Rearrangements of Actin Filaments

In a slow-moving cell like a fibroblast or in a neuron extending its growth cone, we can observe how a cell takes its first step—the protrusion of the leading lamella. As depicted in Figure 22-43a, the process is accompanied by the controlled polymerization of actin filaments from sites at the leading edge and the subsequent cross-linking of those filaments into the bundles and networks of the cytoskeleton. But what is propelling the membrane forward? One hypothesis (Figure 22-43b) is that the membrane is extended forward by the pushing action of the polymerizing actin filaments, as in the extension of an acrosomal process

or the movement of *Listeria* in the cytosol of an infected cell. Another hypothesis (Figure 22-43c) proposes that myosin I with its cargo of plasma membrane crawls forward along the actin cytoskeleton at the leading edge. A third model (not shown) suggests that the protein-rich cytoskeleton in the lamellipodium swells by an osmotic mechanism, causing the membrane to balloon forward.

Whatever propels it, once the membrane is extended and the cytoskeleton is assembled, then the membrane becomes firmly attached to the substratum. Time lapse microscopy shows that actin bundles in the leading edge become anchored to the attachment site, which quickly develops into the adhesion plaque. The attachment serves two purposes: it anchors the cell to the substratum and it prevents the leading lamella from retracting.

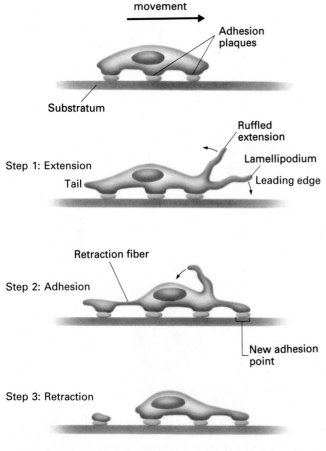

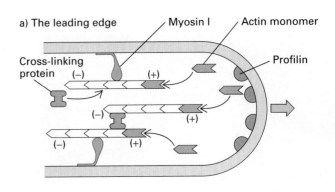

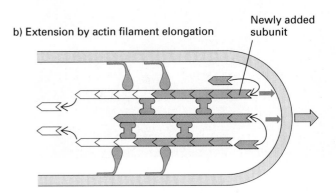

▲ FIGURE 22-42   Steps in cell motility. Stationary fibroblasts adhere to the substratum through cell adhesion plaques on their undersurface. During movement, lamellipodia that are rich in actin bundles are extended from the leading edge of the cell (step 1); some lamellipodia adhere to the substratum, while others move backward over the ventral surface of the cell (step 2). As the fibroblast moves forward, the trailing edge of the cell remains attached to the substratum, and the tail—called the retraction fiber—becomes greatly elongated under the resulting tension. At the same time, the ruffled extensions on the dorsal surface of the cell move backward and collapse. Eventually, the tail ruptures, leaving a bit of itself attached to the substratum; the major part of the tail retracts into the cell body (step 3).

▲ FIGURE 22-43   Models for the roles of actin polymerization and myosin I in the extension of the cell membrane. (a) At the leading edge, the polymerization of actin filaments is stimulated by profilin, located at the leading edge, membrane. Cross-linking proteins stabilize the actin filaments into networks and bundles. In addition, myosin I is thought to link actin filaments to the leading edge plasma membrane. (b) According to one model, attachment of actin monomers to the (+) ends of the actin filaments and their loss from the (−) ends pushes the membrane outward (to the right in the diagram). (c) According to a second model, extension and retraction are powered by movement of myosin I. If the actin filaments were anchored to other cytoskeletal fibers, then, because myosin I is attached to the plasma membrane, the movement of myosin I toward the (+) ends of the actin filaments would cause extension of the plasma membrane. The space between the (+) ends of the filament and the membrane is then filled by newly polymerized actin.

After the forward attachments have been made, the bulk contents of the cell are brought forward. How this is accomplished is also unknown; one speculation is that the nucleus and the other organelles are embedded in the cytoskeleton and that the entire cytoskeletal framework is pulled or pushed forward. Finally, in the last step of movement, the rear of the cell, the tail, is brought forward. In the light microscope, the tail is seen to "snap" loose from its connections—perhaps by contraction of stress fibers in the tail—but it leaves a little bit of its membrane behind,

(a)

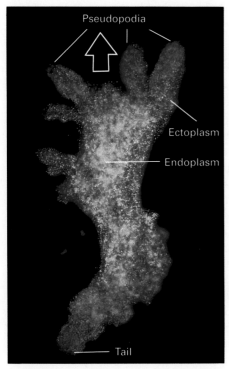

(b)

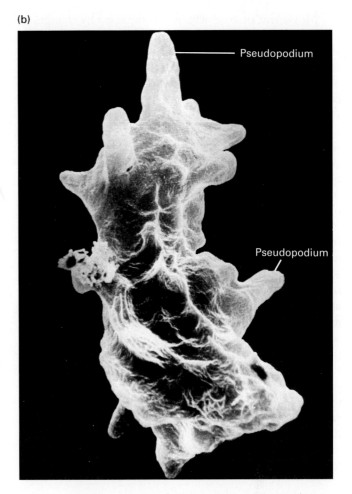

(c)

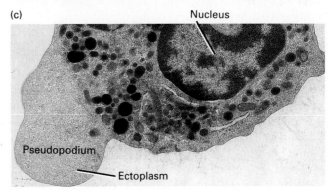

▲ **FIGURE 22-44** Movements of amebas. Ameboid movement is characterized by two dynamic movements—extension of large pseudopodia and cytosol streaming. (a) In a light micrograph of a moving ameba, numerous pseudopodia extend from the cell front. Internally, the ameba's cytosol is organized into the inner flowing endoplasm and the outer gel-like ectoplasm. The endoplasm flows so quickly that it is blurred in the photo. (b) A scanning electron micrograph of the ameba *Proteus* shows the numerous pseudopodia that project from the cell. (c) A transmission electron micrograph of a thin section of a guinea pig leukocyte shows clearly that organelles and vesicles found elsewhere in the cytoplasm are not present in pseudopodia. [Part (a) courtesy of M. Abbey/Photo Researchers. Part (b) courtesy of G. Antipa. Part (c) from D. Fawcett, 1981, *The Cell,* Saunders; courtesy of Photo Researchers, Inc.]

still firmly attached to the substratum (see Figure 22-42, step 3).

Ameboid Movement Involves Reversible Gel-Sol Transitions of an Actin Network

A fast-moving cell like an ameba or a leukocyte must have more vigorous mechanisms for moving the membrane and cytoplasm forward. Thus the motility of a fast-moving cell is characterized by pseudopodial extension and cytoplas-mic streaming (Figure 22-44), accompanied by fluctuation of the cytoskeleton between "sol" and "gel" states. In large soil amebas, these cellular changes are especially dramatic; nevertheless, the ameba exhibits the same basic steps in motility seen with all moving cells. In the ameba, movement is initiated when the plasma membrane balloons forward to form a *pseudopodium*, or "false foot." As the pseudopodium attaches to the substratum, it fills with cytosol that is flowing forward through the cell. In the last step in movement, the rear of the ameba is pulled forward, breaking its attachments to the substratum.

Movement of an ameba is accompanied by changes in the viscosity of its cytosol, which cycles between sol and gel states. Perhaps as a mechanism to propel the cytoplasm forward, the central substance of the cell, the *endoplasm,* is a fluid, or sol, which flows rapidly toward the front of the cell, filling the pseudopodium. Here, the endoplasm is converted into the *ectoplasm,* a gel which forms the cortex, just beneath the plasma membrane. As the cell crawls forward, the ectoplasmic gel at the tail end of the cell is converted back into endoplasmic sol, only to be converted once again into ectoplasm when it again reaches the front of the cell. This cycling between sol and gel states continues only when the cell migrates.

The transformation between sol and gel states results from the disassembly and reassembly of microfilament networks in the cytosol. Several actin-binding proteins probably control the viscosity of the cytosol: profilin at the front of the cell promotes actin polymerization and α-actinin and filamin form gel-like actin networks in the more viscous ectoplasm, while proteins like gelsolin sever actin filaments to form the more fluid endoplasm. The actions of these proteins can explain how differences in the viscosity of the cytosol are generated, but in a large cell such as an ameba, what forces produce the movement of the cytosol? One scenario envisions that myosin II at the rear of the cell causes cortical contractions which squeeze the tail and thus push the cytosol forward. Meanwhile, myosin I at the front of the cell cross-links the cytoskeleton to the membrane in the pseudopodia.

Cell Movements Are Coordinated by Various Second Messengers and Signal Transduction Pathways

One striking feature of a moving cell is its polarity: a cell has a front and a back. When a cell makes a turn, a new leading lamella or pseudopodium forms in the new direction. If lamellae form in all directions, as in the myosin I mutants, then the cell is unable to pick a new direction of movement. To sustain movement in a particular direction, a cell requires signals to coordinate events at the front of the cell with events at the back and, indeed, signals to tell the cell where its front is. In Chapter 20, we discussed the various mechanisms for cell signaling. In this section, we will examine the roles of signal transduction pathways and external factors in activating movement and controlling the direction of movement, and the role of calcium in the regulation of forward cell movement.

Signal Transduction: Activation of Membrane Ruffles and Stress Fibers by Small G Proteins In response to stimulation by a growth factor, such as platelet-derived growth factor, a quiescent, nongrowing fibroblast will begin to grow and divide: the fibroblast immediately polymerizes actin filaments and later forms tight adhesions to the substratum. These cytoskeletal activities lead to the active ruffling of the cell's leading membranes and later to the formation of stress fibers. Recall that in Chapter 20 we examined how growth factors, acting through membrane receptors, activate several signal transduction pathways, including G protein–induced hydrolysis of PIP_2. Similarly, two Ras-related G proteins, Rac and Rho, were tested for their signal-transduction roles in cell movement by examining the effects on the actin cytoskeleton when either protein was microinjected into a cell.

When Rac was injected, the fibroblast membrane immediately started to ruffle, and adhesion plaques and stress fibers formed 5–10 minutes later. The injection of an inactive form of Rac inhibited all reorganization of the actin cytoskeleton when growth factors were added to the cell. When Rho, a related small GTP-binding protein, was injected, it mimicked the mitogenic effects of lysophospholipid (also called lysophosphatidic acid, or LPA), a growth factor in serum and a potent stimulator of platelet aggregation. Both Rho and LPA induced the assembly of stress fibers and cell adhesion plaques within 2 minutes but did not induce membrane ruffling.

These findings led to the hypothesis that growth factors activate a signal transduction pathway involving G proteins, in which actin polymerization at the leading edge membrane is an early event and formation of adhesion plaques is a later event. If the model is correct, then the inhibition of stress-fiber assembly should not affect membrane ruffling. To test the model, Rac and ADP-ribosylase, an enzyme which inactivates Rho by covalently attaching ADP onto it, were co-injected into a fibroblast. As predicted by the hypothesis, membrane ruffles were formed but the assembly of stress fibers was blocked. These observations suggest that Rho-dependent events like stress-fiber formation are "downstream" of control by Rac (Figure 22-45a). The intermediate steps in the activation pathway have not been identified, but Rac probably activates the metabolism of PIP_2 whereas Rho activates a tyrosine kinase.

External Factors: Steering of Cell Movements by Chemotactic Molecules Under certain conditions, extracellular chemical cues guide the movement of a cell in a particular direction. In some cases, the movement is guided by insoluble molecules in the underlying substratum. In other cases, the cell senses soluble molecules and follows them, along a concentration gradient, to their source. This latter response is called *chemotaxis.* One of the best-studied examples of chemotaxis is the migration of *Dictyostelium* amebas along an increasing concentration of cAMP: following cAMP to its source, the amebas aggregate into a slug and then differentiate into a fruiting body. Many other cells also display chemotactic movements. For example,

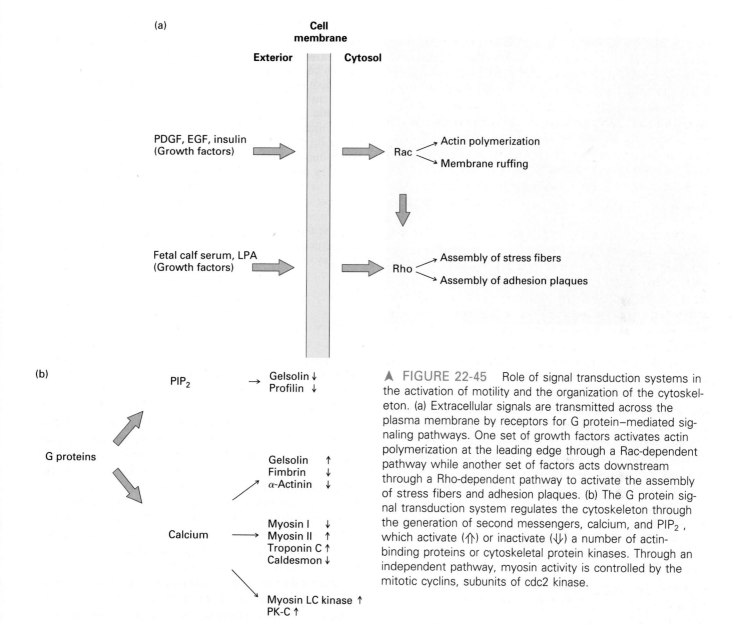

▲ FIGURE 22-45 Role of signal transduction systems in the activation of motility and the organization of the cytoskeleton. (a) Extracellular signals are transmitted across the plasma membrane by receptors for G protein–mediated signaling pathways. One set of growth factors activates actin polymerization at the leading edge through a Rac-dependent pathway while another set of factors acts downstream through a Rho-dependent pathway to activate the assembly of stress fibers and adhesion plaques. (b) The G protein signal transduction system regulates the cytoskeleton through the generation of second messengers, calcium, and PIP$_2$, which activate (⇑) or inactivate (⇓) a number of actin-binding proteins or cytoskeletal protein kinases. Through an independent pathway, myosin activity is controlled by the mitotic cyclins, subunits of cdc2 kinase.

leukocytes are guided by f-Met-Leu-Phe (*N*-formylmethio-nyl-leucyl-phenylalanine), a tripeptide secreted by bacterial cells.

Despite the variety of different chemotactic molecules—sugars, peptides, cell metabolites, cell-wall or membrane lipids—they all work through a common and familiar mechanism: binding to cell-surface receptors, activation of G protein–mediated signal transduction pathways, and remodeling of the cytoskeleton through the activation or inhibition of various actin-binding proteins (Figure 22-45b).

Calcium as Regulator: Coincident Gradients of Calcium and Chemotactic Molecules When a cell's movement is directed by chemotaxis, how is the cell able to sense the difference between the concentrations of chemo-

(a)   (b)

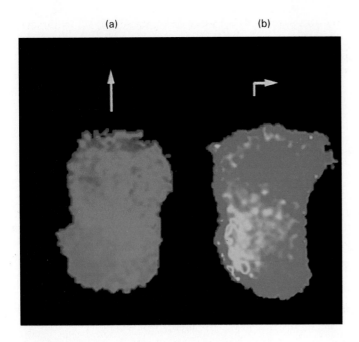

◄ FIGURE 22-46   Changes in cytoplasmic calcium that accompany motility. Fura-2, a fluorescent dye whose intensity changes at different calcium levels, shows the relative calcium concentrations in a moving eosinophil. (a) A gradient of calcium is established in a moving cell. The highest levels are at the rear of the cell where cortical contractions take place and the lowest levels are at the cell front where actin polymerization occurs. (b) When the cell is induced to turn by placing a pipette filled with chemotactic molecules to the side of the cell, the calcium concentration momentarily increases throughout the cytoplasm and a new gradient of calcium is established. The gradient is oriented such that the region of lowest calcium (blue) lies in the direction the cell will turn, whereas a region of high calcium (yellow) always forms at the site that will become the rear of the cell. [From R. A. Brundage et al., 1991, *Science* **254**:703–706; courtesy of F. Fay.]

tactic molecules at the front and the back of the cell, a difference of only a few molecules? Optical microscopy using fluorescent dyes that act as internal calcium sensors (see Figure 5-13) suggests that calcium may play a role (Figure 22-46).

In a gradient of chemotactic molecules, a moving cell displays a cytosolic gradient of calcium, with the lowest concentration at the front of the cell and the highest concentration at the rear. This internal calcium concentration can be disrupted by placing the cell in a new external chemotactic gradient. In one experiment, a pipette containing f-Met-Leu-Phe was placed to the side of a migrating neutrophil. The peptide diffusing from the pipette established a new chemotactic gradient which immediately caused a general increase in the overall concentration of cytosolic calcium. However, the calcium gradient then reoriented, with the lowest concentrations on the side of the cell closest to the pipette, causing the cell to turn toward the pi-

pette. After the pipette was removed, the cell continued to move in the direction of its newly established calcium gradient.

We have seen that many actin-binding proteins, including myosins I and II, gelsolin, α-actinin, and fimbrin, are regulated by calcium. Hence the cytosolic Ca^{2+} gradient may regulate the sol-to-gel transitions that occur during cell movement. The low Ca^{2+} concentration at the front of the cell would favor the formation of actin networks by activating myosin I, inactivating actin-severing proteins, and reversing the inactivation of Ca^{2+}-regulated actin–cross-linking proteins. The high calcium concentration at the rear of the cell would cause actin networks to disassemble and a sol to form, by activating gelsolin or cause cortical actin networks to contract by activating myosin II. Thus an internal gradient of calcium would contribute to the cycling of the ectoplasm and endoplasm.

SUMMARY

A network of fibers, the cytoskeleton, fills the cytosol of all cells. Composed of three fiber systems (microfilaments, microtubules, and intermediate filaments), the cytoskeleton is responsible for giving a cell its shape, for generating

the forces required for cell migration, and for providing a framework to which cellular organelles are attached.

One of the most conserved proteins in a cell, actin is an ATP-binding protein which exists in two states—G- and

F-actin. The globular monomers, called G-actin, polymerize into filaments 7–9 nm in diameter, called F-actin. F-actin has a polarity, as demonstrated by its asymmetric decoration with myosin head fragments and by the unequal rates of assembly at the two ends of the actin filament. The polymerization of actin in vitro is accompanied by the slow hydrolysis of ATP to ADP.

In the cell, the actin cytoskeleton is a network of filaments and bundles in equilibrium with a pool of actin monomers. Bundles of filaments support fingerlike projections of membrane, while networks of filaments stiffen plasma membranes. Actin bundles and networks are held together by actin–cross-linking proteins. The filaments form a framework to which the plasma membrane is attached through membrane-microfilament binding proteins. All attachments of actin to the membrane are through integral membrane proteins. The integral membrane protein in many cases is a receptor for extracellular matrix proteins as well as a binding site for an actin–cross-linking protein. The polymerization of actin filaments and their cross-linking into bundles and networks is one of the mechanisms by which the cell generates force to push a membrane forward during cell movement. The assembly of actin filaments starts with the formation of short actin oligomers, known as seeds or nuclei. Nuclei rapidly elongate into filaments, which form a steady-state equilibrium with actin monomers. An actin filament has two nonequivalent ends; assembly at the ($+$) end occurs faster than at the ($-$) end. The concentration of actin monomers required for assembly into filaments is called the critical concentration. The critical concentration differs at each end in the presence of ATP-G-actin but not ADP-G-actin.

In the cell, actin-binding proteins control the lengths of actin filaments. One class of small actin-binding proteins form a 1:1 complex with actin monomers that are unable to polymerize. A second class of proteins bind F-actin and, in the presence of calcium, break or sever the filaments into shorter fragments. In the process, these proteins remain bound to the ($+$) end of the filament, where they act as a cap which blocks polymerization. A third class of proteins also cap the ends of filaments. When both ends of a filament are capped by proteins, the filament is prevented from depolymerizing.

A mechanism for generating movements is the ATP-dependent walk of myosin along an actin filament. Myosin is a family of mechanochemical enzymes which convert energy derived from ATP hydrolysis into conformational changes which allow myosin to walk along an actin filament. A myosin molecule is organized into head, neck, and tail domains. The actin-binding and ATP-binding activities lie in the head domain; these activities are controlled by calmodulin-like subunits bound to the neck domain. The tail domain organizes myosin into a monomer or dimer through coiled-coil interactions. The tail domain is also a critical determinant of the type of function carried out by a

myosin. Myosin I (a monomer) and myosin V (a dimer) bind to phospholipid membranes, while myosin II (a dimer) oligomerizes into thick filaments. The membrane-bound myosins are involved in vesicle transport and movements of the plasma membrane; myosin II participates in contractions of large bundles of actin filaments.

Muscle is a bundle of actin filaments that is specially organized to carry out contraction efficiently. A skeletal muscle cell contains long bundles of actin filaments; each bundle is further subdivided into repeated arrays of sarcomeres, the fundamental contractile units. Actin filaments attached to the Z disk form the boundaries of a sarcomere, while myosin thick filaments lie in its center. Contraction of a sarcomere is caused by the sliding of actin filaments past the myosin heads on a thick filament. The sarcomere-like arrangement is maintained by titin and nebulin, long proteins which link thick myosin and thin actin filaments to the Z disk. In a smooth muscle cell, actin and myosin are not as highly organized. The contractile units of smooth muscle are attached at the plasma membrane to dense bodies that appear to function like the Z disk.

Muscle contraction is activated by calcium. The cytosolic calcium levels in muscle are maintained by pumps in the sarcoplasmic reticulum (SR); release of calcium from the SR causes an increase in cytosolic calcium. In skeletal muscle, contraction is activated when calcium binds the tropomyosin/troponin complex on the thin filament and causes the complex to shift. This allows myosin to walk along a thin filament and generate the force for muscle contraction. In smooth muscle and nonmuscle cells, calcium triggers contraction by activating kinases which phosphorylate myosin light chains, or by binding to the regulatory thin-filament protein caldesmon.

In nonmuscle cells, actin and myosin II are organized into sarcomere-like contractile bundles. These bundles are involved in cell adhesion or cell division; they are associated with the plasma membrane at sites where cells are bound to one another at adherens junctions or to the substratum at adhesion plaques. Myosin I in nonmuscle cells is the membranes of intracellular vesicles and the plasma membrane to actin filaments and generates force for vesicle and membrane movements.

Movements of nonmuscle cells are characterized by actin polymerization and the assembly of actin filaments into bundles and networks at the leading edge, the formation of adhesion plaques, at the cell-substratum junction, the flow of cytosol to the front of the cell, and the retraction of the rear of the cell. The organization of the cytoskeleton and the activity of different myosins at the front and back of the cell are controlled by an internal calcium gradient. Using signal transduction pathways that are mediated by small GTP-binding proteins, external factors such as chemotactic molecules are able to guide cell movement in a particular direction.

1. The cytoskeleton is an extremely important component of cells; it supplies an internal foundation that supports the cell membrane. What are the three types of fibers that constitute the cytoskeleton? Which are necessary for cell motility? What are the two principal mechanisms underlying cell motility?

What are pseudopodia, filopodia, and lamellipodia? What are adhesive plaques? Actin is the protein that forms a major component of the machinery for mobility. Describe actin molecules in birds and mammals. Specifically, which are found in muscle? Which are associated with nonmuscle tissue? Are the different forms of actin similar, or are they quite different in amino acid sequence? When actin molecules are compared among species, are the proteins similar or different? What are G-actin and F-actin? Are these forms of actin associated with nucleotides?

2. Actin filaments do have common arrangements in cells. What are bundles and networks? What is the function of actin-cross-linking proteins? Refer to Table 22-1 which presents a list of actin-cross-linking proteins. What is the basic purpose of membrane-microfilament binding proteins? Contrast the properties of microvilli and filopodia. What is the richest area of actin filaments in the cell, and where is it located? What are spectrin, filamin, and dystrophin?

   a. Listed in the following table are a series of situations where actin–cross-linking molecules have undergone mutations at the sites described. For each, characterize the phenotypic change in the cytoskeleton that would accompany each mutation by describing how the mutation would affect actin cross-linking.

Mutation	Phenotypic Changes
Heterozygous for the filamin gene; one allele is wild type and the other has a mutation at the site of dimer formation.	
Heterozygous for the filamin gene; one allele is wild type and the other has a mutation at the actin binding site. The protein can still form dimers.	
Homozygous for the filamin gene such that it prevents the dimer from binding to the membrane. How would this be similar to phenotypes caused by mutations in the dystrophin gene?	
Mutant dystrophic gene; Is there a gender aspect to this?	

3. What are the three sequential steps in the formation of F-actin from G-actin? What is meant by the critical concentration (C_c) of G-actin? What is the function of ATP in F-actin assembly? Is ATP's presence required for polymerization? Is ATP hydrolysis required for polymerization? The structural polarity of F-actin is quite important. Which end grows faster? What is treadmilling, and how does it relate to the C_c value for each end of an F-actin filament? What is the function of sequester proteins like profilin and thymosin $\beta4$, and how do they affect the C_c of G-actin? What are severing proteins? What are capping proteins?

   a. The formation of F-actin from G-actin can be described by the following equation

$$G\text{-actin} \rightleftharpoons F\text{-actin}.$$

   The critical parameter dictating whether polymerization or depolymerization will occur is the C_c value. Suppose that you are investigating the formation of F-actin in several mutant cell lines labeled I through IV derived from a wild type cell where the C_c for actin is 0.1 μM; the mutants are described below. Would you expect the mutants to have more F-actin, less F-actin, or the same amount of F-actin when compared to the wild type cell? Explain your answer for each.

   I. This mutant overproduces thymosin $\beta4$ at a level ten times higher than the wild type cell.
   II. This mutant produces profilin at a level that is one-half that found in the wild type cell.
   III. This mutant overproduces G-actin.
   IV. This mutant overexpresses gelsolin, but the normal intracellular Ca^{+2} concentrations are less than 0.1 μM. (What would happen if you stimulated Ca^{+2} influx into the cytoplasm of this mutant?)

4. How many different members of the myosin family of proteins have been identified? Which three are well characterized, and what are their functions? The three structural domains of myosin are the head, neck, and tail regions. What is the basic characteristic of each? Specifically, which region interacts with actin? What proteins bind to the neck regions? Which domain confers functional specificity to myosin?

Myosin is a specialized enzyme having ATPase activity. What is the effect of actin on myosin ATPase activity? How, ultimately, does myosin act as a protein that converts energy from ATP into motion? What domain of myosin is essential for motility?

Describe a sarcomere present in striated muscle; specifically, what are A bands, Z disks, I bands, and the M line; what are the structural components associated with each? What are some of the structural differences between smooth and striated muscle? What is vinculin? What are titin and nebulin; what are their functions in striated muscle? What are troponin and tropomyosin? What is the function of Ca^{+2} in muscle contraction? Where is Ca^{+2} stored; what signals its release into the sarcoplasm? What is MLCK; how does it regulate smooth muscle contraction?

   a. Recall that the neuromuscular junction, which is the junction between a nerve and striated muscle, uses acetylcholine (ACh) as the neurotransmitter. Outline the processes that occur at the molecular level from the moment of ACh release at the neuromuscular junction until the muscle contracts, placing particular emphasis on the role of Ca^{+2} in the process.

b. The poison known as curare binds to ACh receptors but is an antagonist of ACh. At the molecular level, outline how this drug would influence muscle contraction.

c. Another drug which binds to the ACh receptor is succinylcholine. This particular agent is an ACh agonist, but it is very poorly hydrolyzed by acetylcholinesterase. How would this drug affect muscle contraction at the molecular level?

References

Actin: General Reviews

POLLARD, T. D., and J. A. COOPER. 1986. Actin and actin-binding proteins. A critical evaluation of mechanisms and functions. *Ann. Rev. Biochem.* 55:987–1035.

SHETERLINE, P., and J. C. SPARROW. 1994. *Protein profile: Actin.* Academic Press. Vol. 1.

Actin Structure and Membrane Interactions

BENNETT, V., and D. M. GILLIAN. 1993. The spectrin-based membrane skeleton and micron-scale organization of the plasma membrane. *Ann. Rev. Cell Biol.* 9:27–66.

COLUCCIO, L. M., and A. BRETSCHER. 1989. Reassociation of microvillar core proteins: making a microvillar core in vitro. *J. Cell Biol.* 108:495–502.

DOOLITTLE, R. F. 1992. Reconstructing history with amino acid sequences. *Prot. Sci.* 1:191–200.

HARTWIG, J. H. 1992. Mechanisms of actin rearrangements mediating platelet activation. *J. Cell Biol.* 188:1421–1442.

KABSCH, W., et al. 1990. Atomic structure of the actin:DNase I complex. *Nature* 347:37–44.

LEES-MILLER, J. P., G. HENRY, and D. M. HELFMAN. 1992. Identification of *act2*, an essential gene in the fission yeast *Schizosaccharomyces pombe* that encodes a protein related to actin. *Proc. Natl. Acad. Sci. USA* 89:80–83.

LEWIS, A. K., and P. C. BRIDGMAN. 1992. Nerve growth cone lamellipodia contain two populations of actin filaments that differ in organization and polarity. *J. Cell Biol.* 199:1219–1243.

LORENZ, M., D. POPP, and K. C. HOLMES. 1993. Refinement of the F-actin model against x-ray fiber diffraction data by the use of a directed mutation algorithm. *J. Mol. Biol.* 234:826–836.

LUNA, E. J., and A. L. HITT. 1992. Cytoskeleton–plasma membrane interactions. *Science* 258:955–964.

MATSUDAIRA, P. 1991. Modular organization of actin cross-linking proteins. *Trends Biochem. Sci.* 16:87–92.

MOOSEKER, M. S. 1985. Organization, chemistry, and assembly of the cytoskeletal apparatus of the intestinal brush border. *Ann Rev. Cell Biol.* 1:209–241.

ZUCKER, M. B., and V. T. NACHMIAS. 1985. Platelet Activation. *Arteriosclerosis* 5:2–18.

Actin Filament Dynamics

BUSS, F., et al. 1992. Distribution of profilin in fibroblasts correlates with the presence of highly dynamic actin filaments. *Cell Motil. Cytoskeleton* 22:51–61.

CAO, L.-G., et al. 1992. Effects of profilin and profilactin on actin structure and function in living cells. *J. Cell Biol.* 117:1023–1029.

FECHHEIMER, M., and S. H. ZIGMOND. 1993. Focusing on unpolymerized actin. *J. Cell Biol.* 123:1–5.

KIRSCHNER, M. W. 1980. Implications of treadmilling for the stability and polarity of actin and tubulin polymers in vivo. *J. Cell Biol.* 86:330–334.

KORN, E. D., M.-F. CARLIER, and D. PANTALONI. 1987. Actin polymerization and ATP hydrolysis. *Science* 238:638–644.

PANTALONI, D., and M.-F. CARLIER. 1993. How profilin promotes actin filament assembly in the presence of thymosin β4. *Cell* 75:1007–1014.

SANDERS, M. C., A. L. GOLDSTEIN, and Y.-L. WANG. 1992. Thymosin β4 (Fx peptide) is potent regulator of actin polymerization in living cells. *Proc. Nat'l. Acad. Sci. USA* 89:4678–4682.

SCHUTT, C. E., et al. 1993. The structure of crystalline profilin-β-actin. *Nature* 365:810–816.

THERIOT, J. A., and T. J. MITCHISON. 1991. Actin microfilament dynamics in locomoting cells. *Nature* 352:126–131.

TILNEY, L. G., et al. 1983. Actin from *Thyone* sperm assembles on only one end of an actin filament; a behavior regulated by profilin. *J. Cell Biol.* 97:112–124.

TILNEY, L. G., D. J. DE ROSIER, and M. S. TILNEY. 1992. How *Listeria* exploits host cell actin to form its own cytoskeleton. I. Formation of a tail and how that tail might be involved in movement. *J. Cell Biol.* 118:71–81.

Myosin Structure and Regulation

EMERSON, C. P., and S. I. BERNSTEIN. 1987. Molecular genetics of myosin. *Ann. Rev. Biochem.* 56:695–726.

FINER, J. T., R. M. SIMMONS, and J. A. SPUDICH. 1994. Single myosin molecule mechanics: piconewton forces and nanometre steps. *Nature* 368:113–119.

GOODSON, H. V., and J. A. SPUDICH. 1993. Molecular evolution of the myosin family: relationships derived from comparisons of amino acid sequences. *Proc. Nat'l. Acad. Sci. USA* 90:659–663.

LOZANNE, A. D., and J. A. SPUDICH. 1987. Disruption of the *Dictyostelium* myosin heavy chain gene by homologous recombination. *Science* 236:1086–1091.

MILLIGAN, R. A., and P. F. FLICKER. 1987. Structural relationships of actin, myosin, and tropomyosin revealed by cryo-electron microscopy. *J. Cell Biol.* 105:29–39.

RAYMENT, I., et al. 1993. Three-dimensional structure of myosin subfragment-1: a molecular motor. *Science* 261:50–57.

RAYMENT, I., et al. 1993. Structure of the actin-myosin complex and its implications for muscle contraction. *Science* 261:58–65.

WESSELS, D., and D. R. SOLL. 1990. Myosin II heavy chain null mutant of *Dictyostelium* exhibits defective intracellular particular movement. *J. Cell Biol.* 111:1137–1148.

Membrane Myosins

ADAMS, R. J., and T. D. POLLARD. 1986. Propulsion of organelles isolated from *Acanthamoeba* along actin filaments by myosin-I. *Nature* 322:754–756.

ADAMS, R. J., and T. D. POLLARD. 1989. Binding of myosin I to membrane lipids. *Nature* 340:565–568.

BAINES, I. C., H. BRZESKA, and E. D. KORN. 1992. Differential localization of *Acanthamoeba* myosin I isoforms. *J. Cell Biol.* 119:1193–1203.

FUKUI, Y., et al. 1989. Myosin I is located at the leading edges of locomoting *Dictyostelium* amoebae. *Nature* 341:328–331.

KACHAR, B., and T. S. REESE. 1988. The mechanism of cytoplasmic streaming in characean algal cells: sliding of endoplasmic reticulum along actin filaments. *J. Cell Biol.* 106:1545–1552.

In Vitro Motility Assays

SCHOLEY, J. M. 1993. Motility assays for motor proteins. *Methods in Cell Biology,* Volume 39. Academic Press.

SELLERS, J. R., and B. KACHAR. 1990. Polarity and velocity of sliding filaments: control of direction by actin and of speed by myosin. *Science* 249:406–408.

SHEETZ, M. P., and J. A. SPUDICH. 1983. Movement of myosin-coated fluorescent beads on actin cables in vitro. *Nature* 303:31–35.

TOYOSHIMA, Y., et al. 1987. Myosin subfragment-1 is sufficient to move actin filaments in vitro. *Nature* 328:536–539.

WARRICK, H. M., et al. 1993. In vitro methods for measuring force and velocity of the actin-myosin interaction using purified proteins. *Methods Cell Biol.* **39**:1–21.

Muscle Sarcomere Organization

BOND, M., and A. V. SOMLYO. 1982. Dense bodies and actin polarity in vertebrate smooth muscle. *J. Cell Biol.* **95**:403–413.

COHEN, C., and D. A. D. PERRY. 1986. α-Helical coiled coils—a widespread motif in proteins. *Trends Biochem. Sci.* **11**:245–248.

ERVASTI, J. M., and K. P. CAMPBELL. 1993. A role for the dystrophin-glycoprotein complex as a transmembrane linker between laminin and actin. *J. Cell Biol.* **122**:809–823.

FULTON, A. B., and W. B. ISSACS. 1991. Titin, a huge, elastic sarcomeric protein with a probable role in morphogenesis. *Bioessays* **13**:157–161.

FUNATSU, T., H. HIGUCHI, and S. ISHIWATA. 1990. Elastic filaments in skeletal muscle revealed by selective removal of thin filaments with plasma gelsolin. *J. Cell Biol.* **110**:53–62.

WANG, K. 1984. Cytoskeletal matrix in striated muscle: the role of titin, nebulin and intermediate filaments. *Adv. Exp. Med. Biol.* **170**:285–305.

Mechanisms of Muscle Contraction

EISENBERG, E., and T. L. HILL. 1985. Muscle contraction and free energy transduction in biological systems. *Science* **227**:999–1006.

HUXLEY, A. F. 1980. *Reflections on Muscle.* Princeton University Press.

NICKLAS, R. B. 1984. A quantitative comparison of cellular motile systems. *Cell Motil.* **4**:1–5.

POLLARD, T. D. 1987. The myosin crossbridge problem. *Cell* **48**:909–910.

Regulation of Calcium Release

BLOCK, B. A. 1988. Structural evidence for direct interaction between the molecular components of the transverse tubule/sarcoplasmic reticulum junction in skeletal muscle. *J. Cell Biol.* **107**:2587–2600.

CATTERALL, W. A., M. J. SEAGER, and M. TAKAHASKI. 1988. Molecular properties of dihydropyridine-sensitive calcium channels in skeletal muscle. *J. Biol. Chem.* **263**:3535–3538.

FLEISCHER, S., and M. INUI. 1989. Biochemistry and biophysics of excitation-contraction coupling. *Ann. Rev. Biophys. Biochem.* **18**:333–364.

MILLER-HANCE, W. C., et al. 1988. Biochemical events associated with activation of smooth muscle contraction. *J. Biol. Chem.* **263**:13979–13982.

PAYNE, M. R., and S. E. RUDNICK. 1989. Regulation of vertebrate striated muscle contraction. *Trends Biochem. Sci.* **14**:357–360.

WAGENKNECHT, T., et al. 1989. Three-dimensional architecture of the calcium channel/foot structure of sarcoplasmic reticulum. *Nature* **338**:167–170.

Thin Filament Regulation

LEES-MILLER, J. P., and D. M. HELFMAN. 1991. The molecular basis for tropomyosin isoform diversity. *Bioessays* **13**:429–437.

VIBERT, P., R. CRAIG, and W. LEHMAN. 1993. Three-dimensional reconstruction of caldesmon-containing smooth muscle thin filaments. *J. Cell Biol.* **123**:313–321.

WHITE, S. P., C. COHEN, and G. N. PHILLIPS, JR. 1987. Structure of co-crystals of tropomyosin and troponin. *Nature* **325**:826–828.

YAMAKITA, Y., S. YAMASHIRO, and F. MATSUMURA. 1992. Characterization of mitotically phosphorylated caldesmon. *J. Biol. Chem.* **267**:12022–12029.

ZOT, A. S., and J. D. POTTER. 1987. Structural aspects of troponin-tropomyosin regulation of skeletal muscle contraction. *Ann. Rev. Biophys. Biochem.* **16**:535–559.

Myosin Regulation

EGELHOFF, T. T., R. J. LEE, and J. A. SPUDICH. 1993. *Dictyostelium* myosin heavy chain phosphorylation sites regulate myosin filament assembly and localization in vivo. *Cell* **75**:363–371.

ITOH, T., et al. 1989. Effects of modulators of myosin light-chain kinase activity in single smooth muscle cells. *Nature* **338**:164–167.

KOLODNEY, M. S., and E. L. ELSON. 1993. Correlation of myosin light chain phosphorylation with isometric contraction of fibroblasts. *J. Biol. Chem.* **268**:23850–23855.

RASMUSSEN, H., Y. TAKUWA, and S. PARK. 1987. Protein kinase C in regulation of smooth muscle contraction. *FASEB J.* **1**:177–185.

TAN, J. L., S. RAVID, and J. A. SPUDICH. 1992. Control of nonmuscle myosins by phosphorylation. *Ann. Rev. Biochem.* **61**:721–759.

YAMAKITA, Y., S. YAMASHIRO, and F. MATSUMURA. 1994. In vivo phosphorylation of regulatory light chain of myosin II during mitosis of cultured cells. *J. Cell Biol.* **124**:129–137.

Muscle-Like Structures

BYERS, H. R., and K. FUJIWARA. 1982. Stress fibers in situ: immunofluorescence visualization with anti-actin, anti-myosin and anti-alpha-actinin. *J. Cell Biol.* **93**:804–811.

HYNES, R. O., and A. D. LANDER. 1992. Contact and adhesive specificities in the associations, migrations, and targeting of cells and axons. *Cell* **68**:303–322.

SANGER, J. W., J. M. SANGER, and B. M. JOCKUSCH. 1983. Differences in the stress fibers between fibroblasts and epithelial cells. *J. Cell Biol.* **96**:961–969.

SINGER, I. I. 1979. The fibronexus: a transmembrane association of fibronectin-containing fibers and bundles of 5-nm microfilaments in hamster and human fibroblasts. *Cell* **16**:675–685.

Cell Motility

BRAY, D., and J. G. WHITE. 1988. Cortical flow in animal cells. *Science* **239**:883–888.

CHEN, W.-T. 1981. Mechanism of retraction of the trailing edge during fibroblast movement. *J. Cell Biol.* **90**:187–200.

CONDEELIS, J. 1993. Life at the leading edge: The formation of cell protrusions. *Ann. Rev. Cell Biol.* **9**:411–444.

MITCHISON, T., and M. KIRSCHNER. 1988. Cytoskeletal dynamics and nerve growth. *Neuron* **1**:761–772.

MURRAY, J., et al. 1992. Three-dimensional motility cycle in leukocytes. *Cell Motil. Cytoskeleton* **22**:211–223.

PASTERNAK, C., J. A. SPUDICH, and E. L. ELSON. 1989. Capping of surface receptors and concomitant cortical tension are generated by conventional myosin. *Nature* **341**:549–551.

SMITH, S. J. 1988. Neuronal cytomechanics: the actin-based motility of growth cones. *Science* **242**:708–715.

STOSSEL, T. P. 1993. On the crawling of animal cells. *Science* **260**:1086–1094.

THERIOT, J. A., and T. J. MITCHISON. 1992. The nucleation-release model of actin filament dynamics in cell motility. *Trends Cell Biol.* **2**:219–222.

WANG, Y. L. 1985. Exchange of actin subunits at the leading edge of living fibroblasts: possible role of treadmilling. *J. Cell Biol.* **101**:597–602.

ZAHALAK, G. I., W. B. MCCONNAUGHEY, and E. L. ELSON. 1990. Determination of cellular mechanical properties by cell poking, with an application to leukocytes. *J. Biomech. Eng.* **112**:283–294.

Signaling the Cytoskeleton

BRUNDAGE, R. A., et al. 1991. Calcium gradients underlying polarization and chemotaxis of eosinophils. *Science* **254**:703–706.

RIDLEY, A. J., et al. 1992. The small GTP-binding protein Rac regulates growth factor-induced membrane ruffling. *Cell* **70**:401–410.

RIDLEY, A. J., and A. HALL. 1992. The small GTP-binding protein Rho regulates the assembly of focal adhesions and actin stress fibers in response to growth factors. *Cell* **70**:389–399.

23 Microtubules and Intermediate Filaments

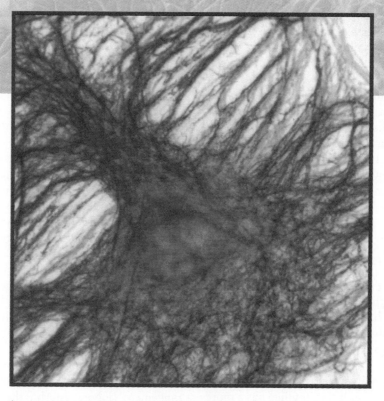

▲ Microtubule and vimentin cytoskeletons in a fibroblast.

In Chapter 22, we looked at microfilaments and their associated proteins. These represent only one of the types of filamentous structures involved in cellular and subcellular movements and in the determination of cell shape. This chapter focuses on microtubules and intermediate filaments (IFs)—the other two cytoskeletal systems of a cell. Both are polymers: microtubules are composed of tubulin subunits, and intermediate filaments are constructed from IF protein subunits. In the cell, long, wavy microtubules and IFs fill the cytosol, spanning the distance between the nucleus and the cell membrane. In many places, the two types of cytoskeletal fibers overlap and follow one another through the cytosol, but in spite of their similar distributions, they carry out different functions in the cell.

Microtubules are responsible for many cell movements—examples include the beating of cilia and flagella, the transport of membrane vesicles in the cytoplasm, and,

in some protists, the capture of prey by spiny extensions of the surface membrane. All these movements result from the polymerization and depolymerization of microtubules or the actions of the microtubule motor proteins dynein and kinesin. Some other cell movements, such as the alignment and separation of chromosomes during meiosis and mitosis, result from both mechanisms. Microtubules also direct cell movement—the ends of axonal microtubules promote the extension of the neuronal growth cone. But in some cases, microtubules serve purely structural functions. For example, a band of microtubules in the outer rim of certain disk-shaped blood cells give the cell its resilience, so that it can squeeze through a narrow blood capillary by folding in half, and can then spring back to its disk shape when released from the confining space.

In contrast to the motile functions of a microtubule, the function of IFs is structural, and, in fact, we do not know of any movements that are dependent on IFs. IFs are usually attached at one end to membranes through junctions such as desmosomes or hemidesmosomes, or to integral membrane proteins through IF-binding proteins. Whereas actin-myosin bundles act as tension cables in the cytosol, IFs are like the fiber threads in adhesive tape: they reinforce cellular membranes while the cell changes its shape during cell movements.

Thus there are many similarities and differences between the microfilament, microtubule, and IF cytoskeletal systems. In this chapter, we will build on the general principles learned in the last chapter about the structure and function of the microfilament cytoskeleton and show how many of the same concepts also apply to microtubules and IFs. We will begin the chapter by examining the structure and assembly of microtubules and then discuss how microtubule assembly and microtubule motors can power cell movements. The microtubule section will end with a detailed study of two important cell movements, flagellar and ciliary beating and chromosome translocation during mitosis. The last part of the chapter will discuss intermediate filaments, concentrating on their structures and the recent experiments which demonstrate that IFs are dynamic polymers in a cell. The chapter will close with a brief discussion of the role of IFs in human disease.

► *Microtubule Structures*

A microtubule is a polymer of globular *tubulin* subunits which are arranged in a cylindrical tube measuring 24 nm in diameter—more than twice the width of an IF and three times the width of a microfilament. Varying in length from a fraction of a micrometer to hundreds of micrometers, microtubules are much stiffer than either microfilaments or IFs because of their tubelike construction. As suggested from its cylindrical construction, a microtubule has a more complex structure than a microfilament.

Tubulin Subunits Comprise the Wall of a Microtubule

Two microtubules are shown in Figure 23-1a. A microtubule is composed of subunits that are heterodimers of α-tubulin and β-tubulin monomers. Each tubulin monomer is a globular protein 4 nm in diameter; the heterodimer subunit is therefore 8 nm long. Each heterodimer binds two molecules of GTP nucleotide. One GTP-binding site, located in α-tubulin, binds GTP irreversibly and does not hydrolyze it, whereas the second site, located on β-tubulin, binds GTP reversibly and hydrolyzes it to GDP. The second site is also called the *exchangeable* site because GDP can be displaced by GTP. As we will discuss, the hydrolysis of β-tubulin-bound GTP is linked to the addition of tubulin subunits at the ends of a microtubule. A third tubulin, γ-

tubulin, does not itself polymerize into microtubules but instead is somehow involved in serving as a nucleus for the polymerization of αβ microtubules.

In a microtubule, lateral and longitudinal interactions between the tubulin heterodimer subunits are responsible for maintaining the tubular form (Figure 23-1b). One set of interactions is responsible for holding α-tubulin and β-tubulin in a heterodimeric complex which is so stable that a heterodimer rarely dissociates into individual α-tubulin and β-tubulin monomers. A second set of interactions, longitudinal contacts between heterodimeric subunits, links the subunits head to tail into a straight column called a *protofilament*. Within each protofilament, the dimer subunits are spaced 8 nm apart. The tubulin dimer is the building block of a microtubule; through lateral interactions, protofilaments associate side by side into a sheet or cylinder—a microtubule.

Virtually every microtubule in a cell is a simple tube, a singlet microtubule, built from 13 protofilaments (Figure 23-2); however, in rare cases, other numbers of protofilaments are possible: for example, certain microtubules in the neurons of nematode worms contain 11 or 15 protofilaments. In addition to providing the simple structure of a singlet microtubule, protofilaments often associate into doublet or triplet microtubules (Figure 23-2), which are found in specialized structures such as cilia and flagella (doublet microtubules) and centrioles and basal bodies (triplet microtubules). A doublet microtubule consists of an A and a B tubule. The A tubule is a complete 13-protofilament microtubule, while the B tubule consists of 10 protofilaments that form a tube by fusing to the wall of the A tubule. A triplet microtubule has an additional 10-protofilament tube fused to the B tubule of a doublet microtubule.

The exact arrangement of protofilaments in the wall of a microtubule is currently debated. In one possible arrangement (Figure 23-1b), the heterodimers in adjacent protofilaments are slightly staggered, forming spiralling rows of α- and β-tubulin monomers in the microtubule wall. In the other arrangement (not shown), the α-tubulin and β-tubulin subunits are staggered enough to give the microtubule wall a checkerboard pattern. The arrangement is uncertain because, unlike the one for actin filaments, no model for the atomic structure of tubulin has yet been determined. Without this missing piece of information, scientists are unable to establish the precise details of the bonds that form a heterodimer, protofilament, or microtubule wall.

Like an actin filament, a microtubule exhibits a polarity in its structure. The polarity arises from the head-to-tail arrangement of the α- and β-tubulin dimers in a protofilament. Because all protofilaments in a microtubule have the same polarity, one end of a microtubule is ringed by α-tubulin, while the opposite end is ringed by β-tubulin. The polarity of a microtubule was revealed by experiments designed like those which revealed the polarity of actin fila-

(a)

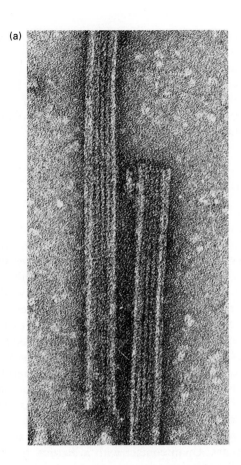

(b)

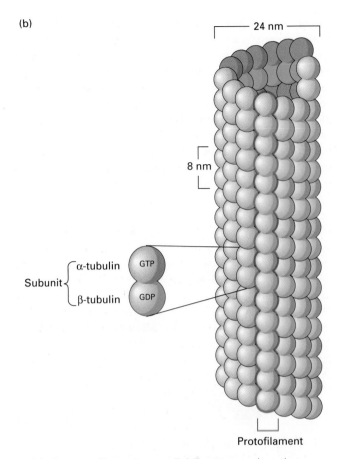

▲ FIGURE 23-1 Microtubule structure. (a) Electron micrograph of a negatively stained microtubule. Globular tubulin subunits form the tubular walls of a microtubule. (b) The organization of subunits in a microtubule. The 8-nm tubulin subunit is a heterodimer of an α-tubulin and β-tubulin molecule. The subunits are aligned end to end into a protofilament (outlined with solid dark green line in the diagram). The columns of protofilaments are packed side by side to form the wall of the microtubule. There are two possible ways that protofilaments pack in a microtubule. In this model, the protofilaments are slightly staggered so that α-tubulin in one protofilament is in contact with α-tubulin in the neighboring protofilaments. In a different model, the protofilaments are staggered by one half subunit, forming a checkerboard pattern. In the microtubule wall α-tubulin binds a single GTP nucleotide while GDP is bound to β-tubulin. [Part (a) courtesy of E. M. Mandelkow; part (b) modified from Y. H. Song and E. Mandelkow, 1993, *Proc. Nat'l. Acad. Sci. USA* **90**:1671.]

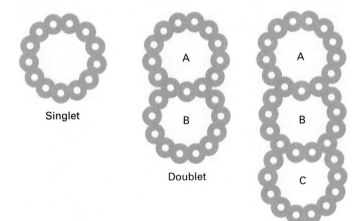

◄ FIGURE 23-2 Arrangement of protofilaments in singlet, doublet, and triplet microtubules. In cross section, a typical microtubule, a singlet, is a simple tube built from 13 protofilaments. Doublet and triplet microtubules are more complex microtubules in which additional sets of 10 protofilaments form microtubules ("B" and "C" in the figure) by attaching to the wall of a singlet ("A") microtubule.

ments. Microtubule assembly experiments (discussed later in this chapter) show, as in the case of actin filaments, that a microtubule has a (+) and a (−) end, which differ in their rates of assembly. In a second type of experiment, a decoration experiment, the polarity of a microtubule was revealed by the way curved sheets of protofilaments assemble along the wall of a microtubule (Figure 23-3). (In solutions of high salt concentration, exogenously added tubulin subunits do not polymerize into new microtubules but instead grow from the wall of an existing microtubule). In a cross

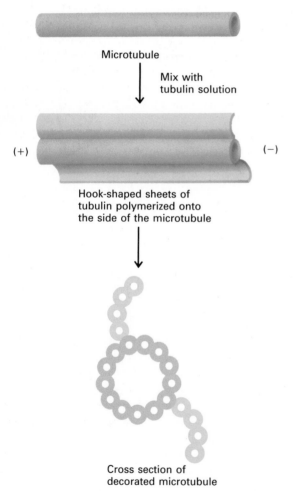

Microtubule

Mix with
tubulin solution

(+)   (−)

Hook-shaped sheets of
tubulin polymerized onto
the side of the microtubule

Cross section of
decorated microtubule

▲ FIGURE 23-3   The polarity of microtubules. This is identified by incubating microtubules in a solution containing a high concentration of soluble tubulin. Because the salt concentration of the solution is abnormally high, tubulin polymerizes in curved sheets or incomplete microtubule walls onto the wall of preexisting microtubules. In cross section, the microtubule appears to be decorated by hook-shaped arms. The clockwise curvature of the arms indicates that the microtubule is being viewed from its (−) end toward the (+) end.

section through such a decorated microtubule, the sheets appear as hooks which curve around the microtubule. When viewed from the (−) end, all the hooks curve in a clockwise direction. Just as the tilt of a myosin head identifies a barbed (+) end and a pointed (−) end of an actin filament, the direction of the hooks reveals a polarity in microtubules.

Microtubules Form a Diverse Array of Both Permanent and Transient Structures

As with microfilaments, there are two populations of microtubules: stable, long-lived microtubules and dynamic, short-lived microtubules. Dynamic microtubules are found when the microtubule structures need to assemble and disassemble quickly. For example, during mitosis, the cytosolic microtubule network characteristic of interphase cells disappears, and the tubulin from it is used to form the spindle apparatus (Figure 23-4a), which partitions chromosomes equally to the daughter cells. When mitosis is complete, the spindle disassembles and the interphase microtubule network reforms. Dynamic microtubules are also used by certain protozoa in capturing prey. These organisms are covered by long slender extensions, called tentacles (Figure 23-4b), which are supported by a core of microtubules. Upon contact with a diatom, for example, the bundle of microtubules undergoes a dramatic rearrangement which causes the tentacle to constrict. In other organisms, the tentacles retract by depolymerizing their microtubules.

In contrast to these short-lived, transient structures, some cells, usually nonreplicating cells, contain stable microtubule-based structures such as the bundle of microtubules in a cilium (Figure 23-4c) or the marginal band of microtubules in red blood cells and platelets. A neuronal cell, for example, is long-lived and seldom needs to establish new connections in an adult. However, the cell must maintain its long nerve processes and it does so with an internal core of stable microtubules (Figure 23-4d). The disassembly of stable structures would have catastrophic consequences—sperm would be unable to swim, a red blood cell would lose its springlike pliability, and axons would retract.

In some cases, the stability of cytoplasmic microtubules changes in response to a change in the environment surrounding a cell. For example, in a confluent culture of cells, if some of the cells are scraped off the culture dish, the cells remaining at the edge of the "wound" begin to move into the open area and eventually fill in the wound. This movement is accompanied by a rearrangement of the microtubule cytoskeleton, which changes from the symmetric arrangement in the previously "resting" cells at the border of the wound into an array in which most of the microtubules radiate toward the open wound. This restructuring of the microtubule network precedes a more general polarization of the cell cytosol and directed cell migration into the wound.

Microtubules Grow from Microtubule-Organizing Centers

In an interphase cell, a seemingly haphazard and random network of microtubules permeates the entire cytosol. However, upon closer analysis, we can see that the cytosolic microtubules are arranged in a hub-and-spoke array centered near the nucleus (Figure 23-5a on page 1058). The microtubule spokes radiate from a central site occupied by a *centrosome*, or *microtubule-organizing center* (*MTOC*), which appears in electron micrographs as a darkly staining, amorphous material in the cytosol (Figure 23-5b). We will use the term MTOC in describing the structures used by interphase cells to organize microtubules. In many cases, an MTOC is a centrosome: however, some epithelial cells or newly fertilized eggs have many MTOCs that do not look like centrosomes. Later, we will use the term centrosome to describe how microtubules are organized in a mitotic cell.

The major role of the MTOC is to nucleate the assembly of most cytosolic microtubules. The presence of a master control site for microtubule assembly is a feature unique to the microtubule cytoskeleton. In contrast, actin filaments are nucleated from numerous sites throughout the cytosol of the cell.

This central role of the MTOC was deduced from several microtubule assembly studies. In cells treated with colchicine, a microtubule-depolymerizing drug, almost all of the cytosolic microtubules, except those in the centrosome, are depolymerized (Figure 23-6a on page 1059). When colchicine is removed by washing the cells, with a colchicine-free culture medium, tubulin repolymerizes to form new microtubules, which radiate from the MTOC (Figure 23-6b). This result could arise by either of two mechanisms: the MTOC could nucleate microtubule polymerization or it could gather together the ends of microtubules that independently polymerized in the cytosol. To identify the correct mechanism, centrosomes were purified and tested for their interaction with microtubules. The addition of purified centrosomes to a solution of tubulin dimers nucleated the assembly of microtubules whose (−) ends remained associated with the centrosome. In the absence of centrosomes, the concentration of dimers was too low to permit spontaneous polymerization of microtubules.

In many animal cells but not in plant cells or fungi, the MTOC may contain a pair of centrioles at the center of the amorphous cytosol (see Figure 23-5c). Each centriole is a pinwheel array of triplet microtubules. The centriole lies in the center of the MTOC but does not make direct contact with the ends of the cytosolic microtubules.

Despite its amorphous appearance, the pericentriolar material of an MTOC contains two proteins that are necessary for initiating the growth of microtubules. The first protein, γ-tubulin, was identified by genetic studies designed to discover proteins which interact with β-tubulin. Subsequent studies demonstrated that γ-tubulin is bound in the pericentriolar material of centrosomes. The second protein, pericentrin, corresponds to an antigen demonstrated in MTOCs by autoimmune antibodies. The introduction of antibodies against pericentrin or γ-tubulin into cells was found to block microtubule assembly. It is still unclear how these two proteins are organized in the MTOC and how they act to direct microtubule assembly.

Besides organizing cytosolic microtubules in an interphase cell, MTOCs control the assembly of two other structures: the mitotic spindle of dividing cells and cilia and flagella. During mitosis, the centrosome duplicates, sheds its cytosolic microtubules, and migrates to new positions flanking the nucleus. There the centrosome becomes the organizing center for astral and spindle microtubules which will separate the chromosomes into the daughter cells during mitosis. An organizing function is also carried out by the *basal body*, which lies at the base of a cilium or flagellum. In contrast to a centrosome, a basal body contains a single centriole.

The Microtubule-Organizing Center Determines the Polarity of Cellular Microtubules

Figure 23-7 (page 1060) illustrates a key property of most cellular microtubules: they are oriented with respect to the MTOC or basal body such that the (−) end of each microtubule is closest to the MTOC or basal body while the (+) end of a microtubule is distal, often near the plasma membrane. In flagella and cilia, the (+) end of microtubules lie at the distal tips of the whiplike motile structures. At the poles of the mitotic spindle, all of the spindle microtubules have their (−) ends pointing to one of the poles, and their (+) ends pointing to the equator at the center of the cell. It is the (+) ends of spindle microtubules that make contact with chromosomes.

Neurons provide exceptions to the rule that the (+) end of a microtubule points toward the plasma membrane. For example, no obvious MTOC is present in dendrites, and dendritic microtubules can have either their (+) or their (−) ends pointing toward the neuron's cell body (see Figure 23-7). In contrast, all the microtubules within an axon have the same polarity, and those that end near the cell body radiate from an MTOC. However, many microtubules within an axon do not appear to emerge from any MTOC; nevertheless they maintain the same orientation as other axonal microtubules, with their (+) ends distal to the cell body.

(a)

(b)

Microtubules

8 μm

1 μm

1 μm

(c)

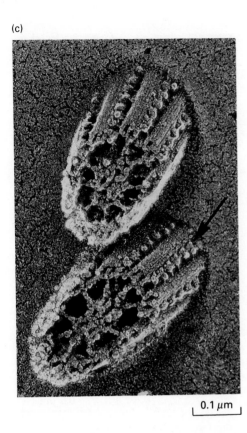

0.1 μm

◀ FIGURE 23-4   Diverse microtubule structures in cells.
(a) In cultured animal cells at interphase, microtubules are
arranged into long, wavy fibers which fill the entire cytosol.
In a cell undergoing mitosis, the network of microtubules
disappears and is replaced with the spindle-shaped arrangement of microtubules in the mitotic apparatus (inset). (b) On
the left, a differential interference micrograph of the suc-
torian protozoan *Heliophrya erhardi* shows tentacles radiating
from the spherical body. The tentacles trap prey, such as
other microorganisms. The tentacles then contract, moving
the prey toward the cell body, where it is ingested. At right
in a cross section electron micrograph through the midregion
of an expanded tentacle, we can see that tentacles derive
their strength and rigidity not only from the large number of
microtubules but also from the extensive cross-bridges
between them. As the tentacles contract (*bottom*), the rows
of microtubules spiral inward, and the microtubule network
becomes thicker with a smaller overall diameter. (c) Electron
micrograph of two *Tetrahymena* ciliary axonemes freeze-
fractured at an oblique angle. Arrow indicates the row of
outer doublet microtubules at the periphery of the axoneme.
(d) Microtubules and intermediate filaments in a quick-frozen
frog axon visualized by the deep-etching technique. There
are several 24-nm-diameter microtubules running longitudinally. Thinner, 10-nm-diameter intermediate filaments also
run longitudinally and form occasional connections with
microtubules. [Part (a) courtesy of B. R. Brinkley and
B. Scott, Baylor College of Medicine; part (b) light micrograph
courtesy of M. Hauser and G. Hanke; electron micrographs
courtesy of M. Hauser and H. Van Eys; part (c) from U. W.
Goodenough and J. E. Heuser, 1982, *J. Cell Biol.* **95**:798;
courtesy of U. W. Goodenough; part (d) from N. Hirokawa,
1982, *J. Cell Biol.* **94**:129; courtesy of N. Hirokawa.]

(d)

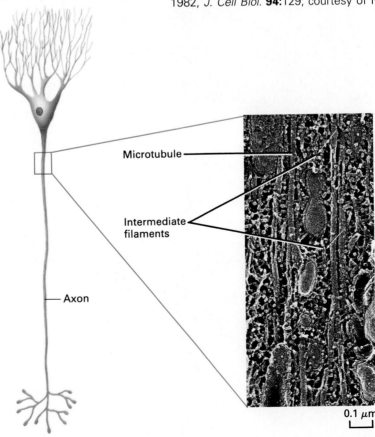

Microtubule

Intermediate
filaments

Axon

0.1 μm

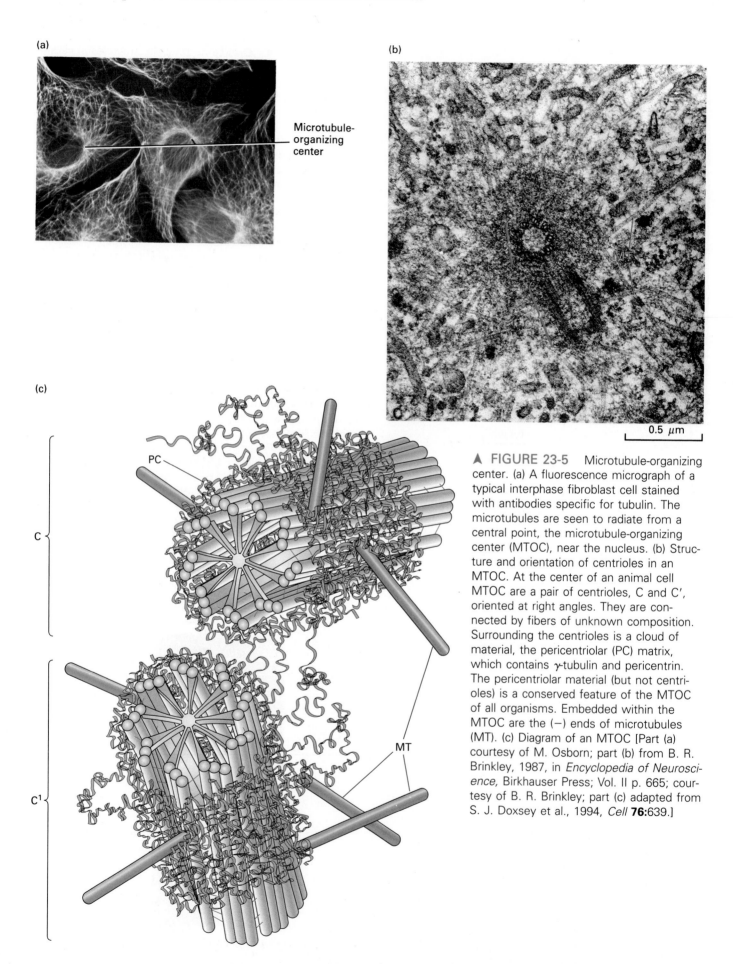

(a)

Microtubule-organizing center

(b)

0.5 μm

(c)

PC

C

C¹

MT

▲ FIGURE 23-5   Microtubule-organizing center. (a) A fluorescence micrograph of a typical interphase fibroblast cell stained with antibodies specific for tubulin. The microtubules are seen to radiate from a central point, the microtubule-organizing center (MTOC), near the nucleus. (b) Structure and orientation of centrioles in an MTOC. At the center of an animal cell MTOC are a pair of centrioles, C and C′, oriented at right angles. They are connected by fibers of unknown composition. Surrounding the centrioles is a cloud of material, the pericentriolar (PC) matrix, which contains γ-tubulin and pericentrin. The pericentriolar material (but not centrioles) is a conserved feature of the MTOC of all organisms. Embedded within the MTOC are the (−) ends of microtubules (MT). (c) Diagram of an MTOC [Part (a) courtesy of M. Osborn; part (b) from B. R. Brinkley, 1987, in *Encyclopedia of Neuroscience,* Birkhauser Press; Vol. II p. 665; courtesy of B. R. Brinkley; part (c) adapted from S. J. Doxsey et al., 1994, *Cell* **76**:639.]

(a)

(b)

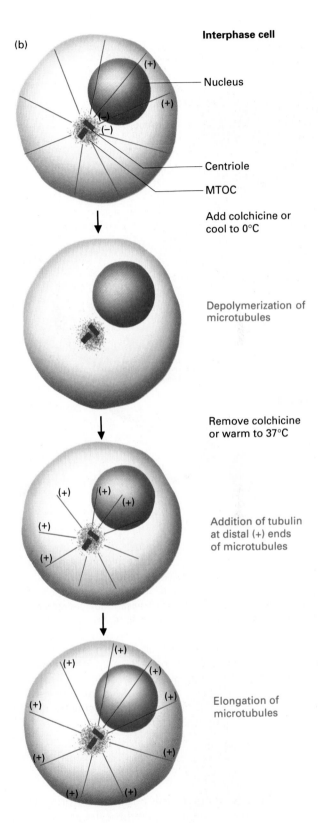

▲ FIGURE 23-6   (a) Micrograph of colchicine-treated Indian muntjac fibroblasts exposed to a fluorescent antibody specific for tubulin (nuclear DNA is stained red with propidium iodide). Comparison with Figure 23-5 shows that most cytoplasmic microtubules are lost after incubation with high concentrations of colchicine, although the centriole and MTOC are retained. The microtubules in centrioles, cilia, and flagella are not affected by colchicine. (b) The disassembly and reassembly of microtubules in interphase cultured animal cells can be induced either by adding and subsequently removing colchicine or by cooling to 0°C and subsequently rewarming to 37°C. Both the addition and the removal of tubulin occurs at the (+) ends of the microtubules. [Part (a) courtesy of B. R. Brinkley and B. Scott, Baylor College of Medicine.]

Multiple Tubulin Genes and Chemical Modification Leads to Tubulin Diversity

From the sequences of many cloned α and β-tubulin genes, it is clear that even simple eukaryotes such as fungi express several different α- and β-tubulins. For example, yeast cells contain two α-tubulin genes whose protein products are 80 percent identical, and only a single β-tubulin gene. In contrast to the few tubulin isoforms in microorganisms, vertebrates and plants produce multiple isoforms of both α- and β-tubulins. For example, chickens and mice express at least six α-tubulin genes and six β-tubulin genes, and plants express seven α-tubulin genes and nine β-tubulin genes. For each tubulin isoform, the amino acid sequences are almost identical. Five of the β-tubulin isoforms typically differ from one another in only 2–8 percent of their 450

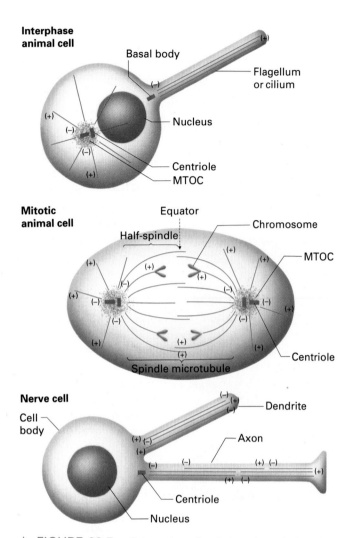

Interphase animal cell

Basal body
Flagellum or cilium
Nucleus
Centriole
MTOC

Mitotic animal cell

Equator
Chromosome
Half-spindle
MTOC
Centriole
Spindle microtubule

Nerve cell

Cell body
Dendrite
Axon
Centriole
Nucleus

▲ FIGURE 23-7   Orientation of cellular microtubules. In interphase cells, the (−) ends of most microtubules are proximal to the MTOC. (Microtubules extend around the nucleus during interphase.) Similarly, the microtubules in flagella and cilia have their (−) ends facing the basal body, which acts as the MTOC in these structures. As cells enter mitosis, the microtubule network rearranges, forming a mitotic spindle; the (−) ends of all spindle microtubules point toward one of the two MTOCs, or poles, as they are called in mitotic cells. In nerve cells, the (−) ends of axonal microtubules are oriented toward the MTOC at the base of the axon. However, a dendrite lacks an MTOC at its base; thus dendritic microtubules have mixed polarities.

amino acid residues, but a sixth β-tubulin isoform, which is found in the marginal band of erythrocytes and platelets, is unusual because its amino acid sequence is only about 75 percent identical to that of the other five β-tubulin isoforms.

Although sequence diversity often indicates that a protein has become specialized in its function, much of the evidence suggests that most α-tubulins or β-tubulins are functionally identical. For example, genetic studies in yeast

cells show that if one of the two α-tubulin genes is disrupted, then yeast cells are still able to grow and divide normally as long as the remaining α-tubulin gene can synthesize sufficient amounts of protein. In contrast, if the single β-tubulin gene is disrupted, the cells die. A similar pattern is observed for algal cells such as *Chlamydomonas*. From these observations we would conclude that a single isoform each of α- and β-tubulin is sufficient for constructing all microtubular structures: mitotic and meiotic spindles, cytoplasmic microtubules, MTOCs, and flagella. Therefore, in these organisms tubulin isoforms are not limited to special tasks.

In contrast with fungi and algae, genetic studies demonstrate that during spermatogenesis in insects, one β-tubulin isoform cannot be substituted with another β-tubulin isoform. The mutant sperm cells lack the microtubular structures in their flagella. This result suggests that isoform diversity in insects arose because of a need for special tubulins to carry out particular functions efficiently.

In addition to having genetic diversity, tubulin is further diversified by post-translational modification. Two types of modifications of α-tubulin have been studied very closely because the chemical changes may influence microtubule structure, assembly, or interactions with other proteins. One type of modification is acetylation. In swimming microorganisms such as *Chlamydomonas*, α-tubulin in flagellar microtubules but not in cytoplasmic microtubules contains an acetylated lysine residue. Acetylation occurs during or just after the incorporation of α-tubulin into these flagellar microtubules. Although the enzyme that incorporates the acetyl group is found both in flagella and in the cytosol, a second enzyme that removes the acetyl group is localized to the cytosol. Thus the steady-state level of microtubule acetylation is shifted completely to the unacetylated state in the cytosol. So far, there is no evidence that acetylation alters the function of microtubules: in *Chlamydomonas*, mutation of the lysine acetylation site has no phenotypic effects on the flagellar motion.

A second type of covalent modification of α-tubulin involves removal and readdition of the carboxyl-terminal tyrosine residue. The enzyme that removes tyrosine, a detyrosinase, acts on α-tubulin that is incorporated into a microtubule. The readdition of tyrosine requires the hydrolysis of ATP and is catalyzed by a second enzyme, tubulin-tyrosine ligase. Cells stained with an antibody that recognizes only the tyrosinated form of α-tubulin shows that tyrosinated α-tubulin is present in all classes of cellular microtubules, whereas detyrosinated α-tubulin is found preferentially in the more stable cytosolic microtubules (e.g., centrioles, cilia, and the few microtubules that remain after colchicine treatment has depolymerized most microtubules). This observation could mean that detyrosination makes a microtubule more resistant to disassembly. Alternatively, a trivial explanation is that other factors cause a microtubule to be stable and a long-lived microtubule is more likely to be detyrosinated merely because it is exposed to the detyrosinase for a longer time.

➤ Microtubule Dynamics

The ability of microtubules to grow and shorten rapidly allows the rapid re-formation of microtubule-containing structures (e.g., conversion of interphase microtubules into mitotic spindles). Before proceeding further in our discussion of microtubule-containing structures and microtubule-based movements, we will take a close look at the assembly, disassembly, and polarity of microtubules. Compared with microfilament assembly, the dynamics of a microtubule are more complex because a microtubule can oscillate between growing and shortening phases. This complex behavior is used by the cell to quickly assemble or disassemble microtubule structures. In this section, we will examine these complexities in detail because they represent key steps in the function of the microtubule cytoskeleton.

Microtubule Assembly and Disassembly Occur by Preferential Addition and Loss of $\alpha\beta$ Dimers at the (+) End

Microtubules assemble by polymerization of tubulin $\alpha\beta$ dimers. Once assembled the stability of most microtubules is temperature-dependent. If microtubules purified from sperm, brain, or certain protozoan flagella are cooled to 4°C, the microtubules depolymerize into stable $\alpha\beta$-tubulin dimers (Figure 23-8a). When warmed to 37°C in the presence of GTP, the soluble tubulin dimers polymerize into microtubules, which can be recovered by centrifugation at low *g*-forces.

From the kinetics of tubulin polymerization and the structural intermediates observed during microtubule assembly or disassembly, scientists have been able to show that in many respects, microtubule assembly is similar to actin filament assembly. For example, microtubule assembly is concentration-dependent; at tubulin dimer concentrations above the critical concentration (C_c), microtubules polymerize, while at concentrations below the C_c, microtubules depolymerize (Figure 23-8b). The addition of nuclei, in the form of fragments of flagellar or other microtubules, to a solution of $\alpha\beta$-tubulin dimers greatly accelerates the initial rate at which tubulin dimers polymerize into microtubules. As in actin polymerization, at tubulin dimer concentrations higher than the C_c for microtubule polymerization, tubulin subunits add to both ends of a growing microtubule. Also as in actin polymerization, however, the addition (or loss) of tubulin subunits occurs preferentially at one end of a microtubule (Figure 23-9a). This difference between the two ends of a growing microtubule is easily demonstrated by examining electron micrographs of microtubules which have assembled from the ends of nucleating flagellar fragments (Figure 23-9b). The electron micrographs show a tuft of microtubules sprouting from opposite ends of the nucleus, but one tuft is much longer than the other. Using the same terminology as in actin assembly, the preferred assembly end is designated the (+) end, while the end that assembled more slowly is the (−) end. When the samples are diluted below the C_c, then the microtubules disassemble twice as rapidly at the (+) end as at the (−) end. Thus both assembly and disassembly occur preferentially at the (+) end.

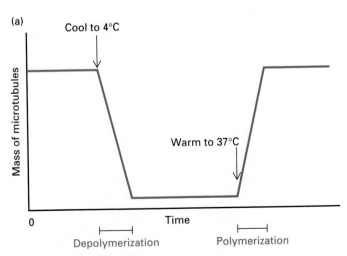

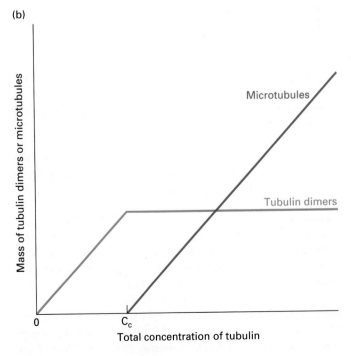

▲ FIGURE 23-8   Microtubule depolymerization and repolymerization. (a) Cycles of cooling and warming of tubulin solutions result in the depolymerization and repolymerization of microtubules. (b) Effect of tubulin concentration on microtubule assembly and disassembly. Above the critical tubulin concentration, C_c, tubulin polymerizes into microtubules; below the C_c, only tubulin dimers are present.

(a)

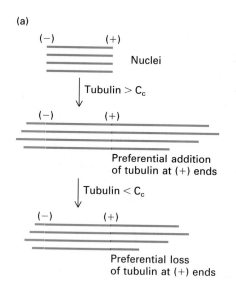

(b)

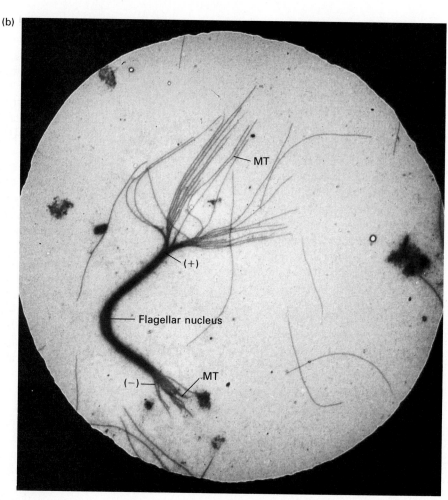

▲ FIGURE 23-9   Polarity of tubulin polymerization. A cell-free experiment demonstrates that microtubules grow and shrink preferentially at one end. Fragments of flagellar microtubules are used as nuclei for the in vitro addition of tubulin. (a) Above the C_c, the growth of microtubules (red) occurs primarily at one end of the flagellar fragment, the (+) end. When the concentration of tubulin is diluted less than the C_c, the nucleated microtubules depolymerize, again primarily from their (+) ends. (b) Because of its highly organized structure, the nucleating flagellar fragment can be distinguished in the electron microscope from the new-formed microtubules (MT), seen radiating from the ends of the flagellar fragment. Note that the microtubules at one end are longer than at the other end. [Part (b) courtesy of G. Borisy.]

One major difference between the assembly of microtubules and microfilaments is a consequence of the more complicated protofilament-based organization of a microtubule. Microtubule assembly must first involve the formation of protofilaments from tubulin subunits, followed by the association of protofilaments to form the wall of the microtubule. During elongation, GTP-containing tubulin dimers add to the ends of microtubules. After the tubulin dimer is incorporated onto an end of the microtubule, GTP bound to β-tubulin is hydrolyzed to GDP. Consequently, the end of a microtubule will consist of GTP-containing tubulin subunits while the bulk of the microtubule will contain GDP in the exchangeable site on β-tubulin (Figure 23-10). In other words, a microtubule has a cap of GTP-subunits at each of its ends.

In the electron microscope, the ends of growing microtubules appear jagged, as if the protofilaments elongate unevenly (Figure 23-11a). However, the mechanism of dis-

▶ FIGURE 23-10   Assembly of microtubules. (a) The subunits first associate longitudinally to form short protofilaments. The protofilaments are probably unstable and (b) quickly associate laterally into curved sheets which are more stable. (c) Eventually a sheet wraps around into a microtubule with 13 protofilaments. The microtubule grows by the addition of dimers to protofilament ends in the microtubule wall. The free tubulin dimers have GTP bound to the exchangeable nucleotide binding site on the β-tubulin subunit. After incorporation of the dimer into a microtubule, the GTP on the β tubulin (but not on the α-tubulin) is hydrolyzed to GDP. If polymerization is faster than the rate of GTP hydrolysis, then a cap of GTP-bound subunits is generated. The rate of polymerization is twice as fast at the (+) end as at the (−).

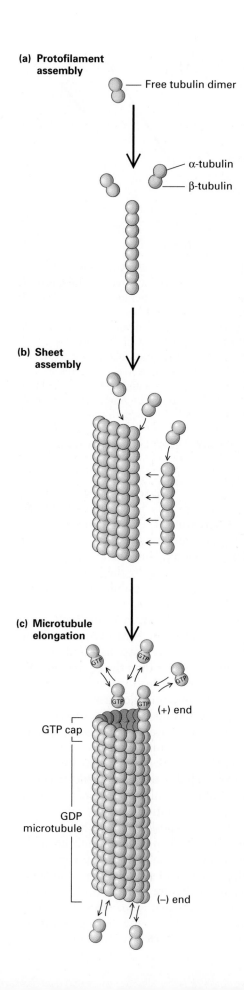

(a) Protofilament assembly

Free tubulin dimer

α-tubulin

β-tubulin

(b) Sheet assembly

(c) Microtubule elongation

GTP

GTP   GTP   (+) end

GTP cap

GDP microtubule

(−) end

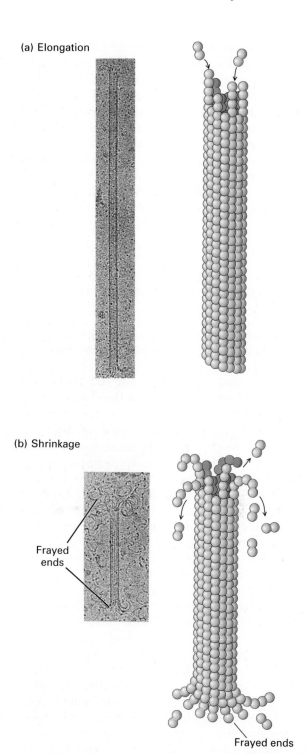

(a) Elongation

(b) Shrinkage

Frayed ends

Frayed ends

▲ FIGURE 23-11   Smooth and frayed ends of microtubules correspond to microtubule assembly and disassembly. Microtubules undergoing assembly or disassembly have been quickly frozen in liquid ethane and examined in the frozen state in a cryoelectron microscope. The contrast in the image is generated by the difference in density of protein (dark) and the surrounding water (light). (a) In assembly conditions, microtubule ends are relatively smooth; occasionally a short protofilament is seen to extend from one end. (b) In contrast, in disassembly conditions, the protofilaments splay apart at microtubule ends, giving the ends a frayed appearance. [Micrographs courtesy of E. Mandelkow and E. M. Mandelkow.]

assembly must be fundamentally different from that of assembly, because in the electron microscope, microtubules undergoing shortening have curled protofilaments at their ends (Figure 23-11b). The microtubule end has splayed apart as if the lateral interactions between protofilaments have been broken. Once frayed apart and freed from lateral stabilizing interactions, the protofilaments depolymerize by endwise dissociation of tubulin subunits. The splayed appearance of a shortening microtubule was unexpected, but as we will discuss in the next section, the images provided clues about the potential instability of a microtubule.

Dynamic Instability Is an Intrinsic Property of Microtubules

So far in this discussion we have painted a simple picture of the dynamic behavior of microtubules; that is, above the C_c, tubulin subunits polymerize into microtubules, while below the C_c, microtubules depolymerize into tubulin subunits. However, this simple picture of the behavior of an individual microtubule is misleading. A variety of experiments on different types of cells have shown that some microtubules will be growing, while simultaneously other nearby microtubules are shortening. Furthermore, a single microtubule can oscillate between growth and shortening phases (Figure 23-12). In all cases, microtubule growth is slow and microtubule shortening is rapid. This more complex behavior of microtubules, termed *dynamic instability*, was unexpected because we had thought that under any condition all the microtubules in a solution or the same cytosol would behave identically.

One clue that microtubules exhibit complex dynamics came from observing the after-effects of the amputation of flagella from *Chlamydomonas*. After deflagellation, a *Chlamydomonas* cell can regenerate both flagella. In the absence of protein synthesis, the new flagella grow to only one-half their original length, because the pool of precursor proteins in the cytosol is limited and cannot be replenished by newly synthesized material. What would happen if only one of the two flagella were amputated? The answer is that a new flagellum would grow, but at the expense of the remaining flagellum. Initially, the surviving flagellum shortens at the same rate as the new one is regenerated. Once both flagella reach half of normal length, both begin to grow to normal length. This observation establishes that flagellar microtubule assembly and disassembly can occur simultaneously in a single cell.

Individual cytosolic microtubules also display this opposing dynamic behavior in live cultured cells. After fluorescent tubulin subunits are microinjected into a cell and allowed to incorporate into cytoplasmic microtubules, video records of a small region in the cell show that some

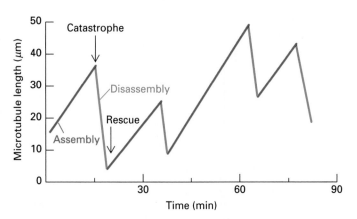

▲ FIGURE 23-12   Dynamic instability of microtubules in vitro. Individual microtubules can be observed in the light microscope and their lengths can be plotted during stages of assembly and disassembly. Assembly and disassembly each proceed at uniform rates, but there is a large difference between the rate of assembly and that of disassembly, as seen in the different slopes of the lines. During periods of growth, the microtubule elongates at a rate of 1 μm/min. Notice the abrupt transitions, at "catastrophe", to the shrinkage stage and (rescue) to the elongation stage. The microtubule shortens much more rapidly (7 μm/min) than it elongates [Adapted from P. M. Bayley, K. K. Sharma, and S. R. Martin, 1994, in *Microtubules*, Wiley-Liss, Inc., p. 119.]

microtubules are growing while others are shortening (Figure 23-13). Also, within a few minutes, one microtubule can alternately grow and shrink. Since we know that most microtubules in a cell are anchored by the (−) ends to the MTOCs, we can conclude that the instability is largely limited to the (+) ends of microtubules.

Two conditions favor instability. First, the oscillations between growth and shrinkage in vitro occur at tubulin concentrations near the C_c. As we have discussed, at concentrations either higher or lower than the C_c, the population of microtubules will either grow or shrink. At concentrations near the C_c, some microtubules grow while other microtubules shrink. The second condition affecting dynamics is whether GTP or GDP occupies the exchangeable nucleotide-binding site on the β-tubulin at the (+) end of a microtubule (Figure 23-14). A microtubule becomes unstable and depolymerizes rapidly if the (+) end becomes capped with GDP-tubulin subunits instead of the usual GTP-tubulin subunits. This situation can arise when a microtubule shrinks rapidly, exposing GDP-tubulin in the walls of the microtubule, or when a microtubule grows so slowly that hydrolysis of the bound GTP cap converts it into a GDP cap before additional subunits can be added to the end of the microtubule. Before it vanishes, the shortening microtubule can be "rescued" and start to grow if GTP-bound tubulin subunits add to the (+) end before the

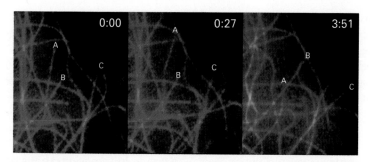

▲ FIGURE 23-13   In vivo growth and shrinkage of individual microtubules. Fluorescent-labeled tubulin was microinjected into cultured human fibroblasts. The cells were chilled to deploymerize preexisting microtubules into tubulin dimers and were then incubated at 37°C to allow repolymerization, thus incorporating the fluorescent tubulin into all the cell's microtubules. A region of the cell periphery was viewed in the fluorescence microscope at 0 s, 27 s later, and 3 min 51 s later (left to right panels). During this period several microtubules elongate and shorten. The letters ''mark'' the position of ends of 3 microtubules. [From P. J. Sammak and G. Borisy, 1988, *Nature* **332**:724.]

➤ FIGURE 23-14   Dynamic instability model of microtubule growth and shortening tubulin. Tubulin dimers with two bound GTP molecules (red) add preferentially to the (+) end of a preexisting microtubule. After incorporation of a tubulin dimer, one bound GTP is hydrolyzed to GDP. This hydrolysis is apparently catalyzed by the microtubule itself but may be facilitated by cytosolic proteins. Only microtubules whose terminal tubulin is associated with GTP (those with a GTP cap) are stable and can serve as primers for polymerization of additional tubulin. Microtubules with tubulin bound to GDP (blue) at the end (those with a GDP cap) are rapidly depolymerized and may disappear within 1 min. At high concentrations of unpolymerized, GTP-bound tubulin, the rate of addition of tubulin is faster than the rate of hydrolysis of the GTP bound in the microtubule or the rate of dissociation of GTP-bound tubulin from the end; thus the microtubule grows. At low concentrations of unpolymerized, GTP-bound tubulin, the rate of addition of tubulin is decreased; consequently, the rate of GTP hydrolysis exceeds the rate of addition of tubulin dimers and a GDP cap forms. Because the GDP cap is unstable, the microtubule end peels apart to release tubulin subunits. [See T. Mitchison and M. Kirschner, 1984, *Nature* **312**:237; M. Kirschner and T. Mitchison, 1986, *Cell* **45**:329; and R. A. Walker et al., 1988, *J. Cell Biol.* **107**:1437.]

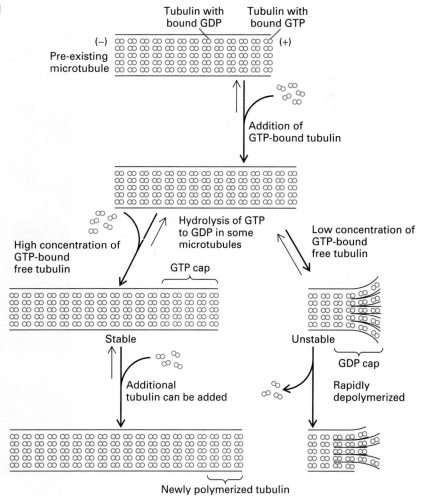

Newly polymerized tubulin

bound GTP hydrolyzes. Thus the one parameter that determines the stability of a microtubule is the rate at which tubulin-GTP is added to the (+) end. The specific factors that switch a microtubule between growth and shortening are unknown, but they must act by modulating GTP hydrolysis and tubulin subunit concentrations.

Colchicine and Other Anti-Cancer Drugs Poison Microtubule Assembly or Disassembly

Before antibodies or gene disruption methods were available for cell biologists to use as tools in dissecting apart the functions of microtubules, drugs that inhibited mitosis provided the first means to manipulate the microtubules in a cell. Three drugs, colchicine, vinblastine, and taxol, all purified from plants, have proved to be very powerful probes

of microtubule function, partly because they bind only to tubulin or microtubules and not to other proteins, and also because their concentrations in cells can be easily controlled. Let us consider colchicine (Figure 23-15a) as an example, because it is the best understood of the microtubule poisons.

Colchicine has two effects on a cell. As we have said, high concentrations of colchicine depolymerize cytosolic microtubules, leaving an MTOC (see Figure 23-6). However, in low concentrations of colchicine, microtubules do not depolymerize but instead remain intact. Consequently, a plant or animal cell becomes "blocked" at metaphase; that is, it is unable to proceed past metaphase in the cell cycle, and condensed chromosomes are frozen in their metaphase configuration (Figure 23-15b). Scientists speculate that at low concentrations colchicine inhibits microtubule dynamics, so that microtubules are unable to depolymerize and tubulin subunits are unable to polymerize. When the treated cells are washed with a colchicine-free solution,

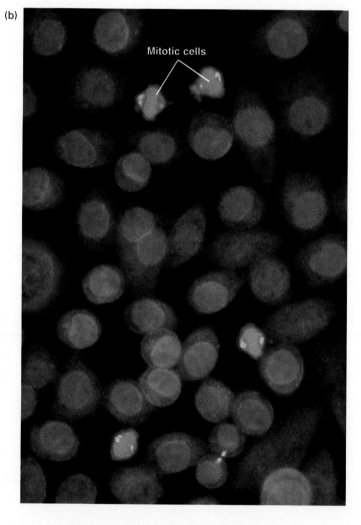

(a)

Colchicine

(b)

Mitotic cells

▲ **FIGURE 23-15**   (a) Structure of colchicine. This water-soluble drug blocks polymerization of tubulin subunits into microtubules. (b) Micrograph of colchicine-treated mouse fibroblasts stained with tubulin-specific antibodies (green) and a DNA-specific dye (blue). At low concentrations of colchicine, such as the one used here, the interphase organization of microtubules is preserved but, in a thick cell, the individual microtubules are not easily seen. Note, however, that the four mitotic cells are in a metaphase configuration. The cells have proceeded through the earlier steps of mitosis but have stopped or become "blocked" at metaphase. [Part (b) courtesy of M. A. Jordan and L. Wilson.]

colchicine diffuses from the cell and mitosis resumes normally. Thus colchicine can be used for cytogenetic studies to accumulate cells at metaphase and, after removal of the colchicine, to provide a population of cells whose cell cycle is in synchrony.

Nearly thirty years ago, colchicine was found to bind one high-affinity binding site on a tubulin dimer. In fact, the binding of radiolabeled colchicine was used as an assay for tubulin in the purifying of microtubules. More detailed biochemical studies showed that a tubulin-colchicine complex, at concentrations much less than the concentration of free tubulin subunits, can add to the end of a growing microtubule and block microtubule assembly. The presence of one or two colchicine-bearing tubulins at the end prevents tubulin subunits from dissociating from a microtubule or assembling onto it. Thus colchicine works by "poisoning" the end of a microtubule and alters the steady-state balance between assembly and disassembly. At very high colchicine concentrations, microtubules depolymerize and accumulate as free tubulin dimers.

Other microtubule drugs nocodazole, taxol, and vinblastine bind to other sites on tubulin or microtubules and therefore affect microtubule stability through different mechanisms. Low concentrations of taxol or vinblastine stabilize microtubules by inhibiting microtubule dynamics— the lengthening and shortening of microtubules. High concentrations of vinblastine are able to depolymerize microtubules and to assemble tubulin dimers into nearly crystalline arrays called vinblastine paracrystals. In contrast, nocodazole causes microtubules to depolymerize.

Because of their effects on mitosis, microtubule inhibitors have been widely used to treat illness and more recently as anticancer agents, since blockage of spindle formation will preferentially inhibit rapidly dividing cells like cancer cells. A highly effective anti-ovarian cancer agent is taxol. In ovarian cancer cells, which undergo rapid cell divisions, mitosis is blocked by taxol treatment while other functions carried out by intact microtubules are not affected. Other microtubule inhibitors are used to treat various illnesses. In fact, the ancient Egyptians used colchicine more than 2,500 years ago to treat heart problems.

Besides increasing the concentration of tubulin dimers by binding to microtubules and triggering their depolymerization, colchicine also modulates tubulin concentrations in a cell by affecting tubulin protein synthesis. A cell carefully controls its microtubule-tubulin content at the level of tubulin protein synthesis by monitoring the concentration of the cytosolic pool of unpolymerized tubulin. A simple feedback mechanism autoregulates tubulin synthesis: the nascent tubulin polypeptide chain binds to the ribosome during tubulin synthesis, and this triggers degradation of tubulin mRNA. Since colchicine increases the level of unpolymerized tubulin, this autoregulation mechanism inhibits tubulin synthesis; synthesis resumes after colchicine is removed.

► *Microtubule-Associated Proteins*

Tubulin is typically isolated by pelleting microtubules from a cell lysate, depolymerizing the microtubules by cooling them to 4°C, centrifuging the cooled solution to remove the insoluble material, and then polymerizing the tubulin-containing supernatant by warming to 37°C. After several cycles of these assembly and disassembly steps, tubulin becomes highly enriched but the preparation still contains other proteins. These proteins are present in lower amounts than α- and β-tubulin, and they maintain their quantitative ratio to α- and β-tubulin through successive cycles of polymerization-depolymerization. Their co-purification with microtubules suggests that these proteins, called *microtubule-associated proteins (MAPs)*, are not nonspecific contaminants but rather have specific associations with α- and β-tubulin. One piece of evidence that the MAPs bind specifically to microtubules is seen in the microtubule-like organization of a MAP displayed in immunofluorescence micrographs of cultured cells. This pattern of staining, seen in Figure 23-16 using MAP 4 antibodies, is highly suggestive of a microtubule-binding activity.

MAPs Organize Bundles of Microtubules

One major family of MAPs, called *assembly MAPs,* are responsible for cross-linking microtubules in the cytosol. These MAPs are organized into two domains (Table 23-1), a basic microtubule-binding domain and an acidic projection domain (so called because in the electron microscope it appears as a filamentous arm that projects from the wall of the microtubule). The projection domain can bind to membranes, IFs, or other microtubules, and its length controls how far apart microtubules are spaced. For example, transfection of insect cells with DNA expressing either short-armed Tau protein or long-armed MAP2 protein causes the cells to grow long axonlike processes. In the *tau*-expressing cells, the microtubules in these processes are closely spaced, while in *MAP2*-expressing cells, the microtubules are widely spaced (Figure 23-17).

Based on sequence analysis, assembly MAPs can be grouped into two types (see Table 23-1). Type I MAPs, MAP1A and MAP1B, contain several repeats of the amino acid sequence KKEX, which is implicated as a binding site for negatively charged tubulin. This sequence is postulated to neutralize the charge repulsion between tubulin subunits and stabilize the polymer. MAP1A and MAP1B are large, filamentous molecules found in axons and dendrites of neurons and also in non-neuronal cells. One interesting feature of both MAP1A and MAP1B is that each is derived from a single precursor polypeptide which is proteolyti-

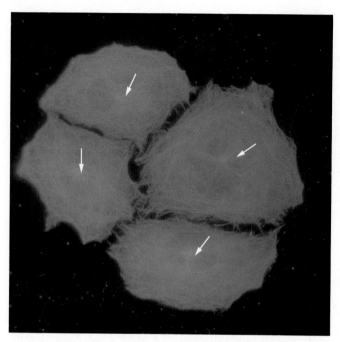

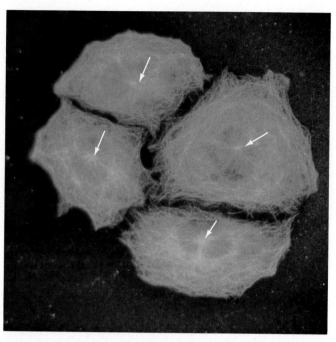

▲ FIGURE 23-16   Comparison between the organization of microtubules (left) and MAP4 (right) in the same interphase HeLa cells. The colinear arrangement of MAP4 and microtubules, even at the MTOC (arrow), is suggestive of a binding interaction. [Courtesy of J. Chloe-Bulinski.]

TABLE 23-1   Microtubule-Associated Proteins

Protein	MW	Domain Organization*	Location
TYPE I			
MAP1A	300,000 heavy chain 34,000 light chain 30,000 light chain 18,000 light chain		Dendrites and axons
MAP1B	255,000		Dendrites and axons
TYPE II			
MAP2a	280,000		Dendrites
MAP2b	200,000		Dendrites
MAP2c	42,000		Embryonic dendrites
MAP4	210,000		Non-neuronal cells
Tau	55,000–62,000		Dendrites and axons

* Yellow, microtubule-binding domain; pink, projection domain; blue, 18 amino acid repeats.

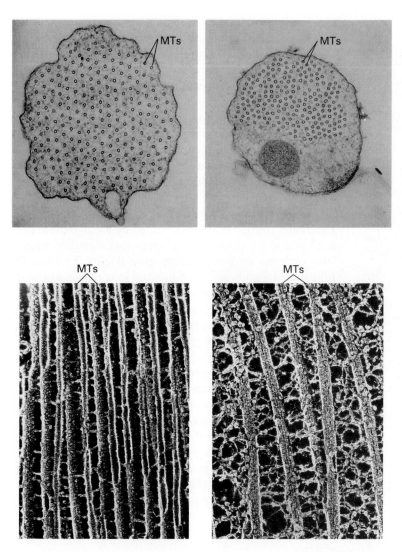

◄ **FIGURE 23-17**   Spacing of microtubules (MTs). Above are electron micrographs of cross sections through the axonlike extensions of transfected insect cells that are induced when long-armed *MAP2* (left) or short-armed *tau* (right) is expressed. Note that the spacing between microtubules in *MAP2*-containing cells is larger than in *tau*-containing cells. Both cell types contain approximately the same number of microtubules but the effect of *MAP2* expression is to enlarge the caliber of the extension. Below, micrographs of quick-frozen, deep-etched insect cell cytosol show microtubules cross-linked with MAP2 protein (left) or Tau protein (right). [(*top*) From J. Chen et al., 1992, *Nature* **360**:674; (*bottom*) from *Neuronal cytoskeleton: Morphogenesis, Transport, and Synaptic Transmission*, N. Hirokawa, ed., *CRC Press*, 1993, pp. 3–32. Photographs courtesy of N. Hirokawa.]

cally processed in a cell to generate one light chain and one heavy chain. Although the light-chain sequence is located at the C-terminus of the precursor, it eventually reassociates with the N-terminal domain of the heavy chain. MAP1B binds microtubules through a spherical domain and forms cross-bridges between microtubules through its projection domain. MAP1A and MAP1B have similar microtubule-binding and projection domains, but MAP1A lacks a spherical domain. This difference in appearance is attributed to the absence of light-chain subunits.

Type II MAPs include MAP2, MAP4, and Tau. These proteins are characterized by three or four repeats of an 18-residue sequence in the microtubule-binding domain. MAP2 is found only in dendrites, where it forms fibrous cross-bridges between microtubules and also links microtubules to IFs. MAP4 is a non-neuronal microtubule-binding protein that co-localizes with microtubules during interphase and mitosis.

The axons and dendrites of nerve cells contain Tau.

This protein exists in four or five forms (MW 55,000–62,000) derived from alternative splicing of a *tau* mRNA. The Tau protein has two effects on microtubules: it accelerates the polymerization of tubulin and it cross-links microtubules through its 18-nm-long projection domain. The ability of Tau to cross-link microtubules into thick bundles may contribute to the great stability of axonal microtubules.

In a cell, the cross-linking function of Tau is easily detected in cells transfected with a *tau* gene. When the recombinant Tau protein is synthesized in high levels in a cultured insect cell, the cell produces several axonlike projections which are filled with bundles of microtubules. The role of Tau in promoting axon growth was tested in complementary studies, in which tau anti-sense RNA was expressed in neuronal cells. With Tau protein synthesis blocked, neurons were unable to grow an axon. A similar effect was also seen for dendrites when the levels of MAP2 were reduced.

MAPs Stabilize Microtubules

You can easily imagine how an assembly MAP, bound to the side of a microtubule, can alter the dynamics of assembly and disassembly. When MAPs coat the outer wall of a microtubule, tubulin subunits are unable to dissociate from the middle or ends of a microtubule. Although the depolymerization of microtubules is generally dampened by MAPs, the assembly of microtubules is affected to varying degrees: some MAPs promote polymerization while other MAPs have no effect.

One mechanism to control the length of microtubules is the regulation of the binding of MAPs to a microtubule. In most cases, this is accomplished by phosphorylating or dephosphorylating the projection domain of a MAP. When phosphorylated on its projection domain, a MAP is unable to bind a microtubule. One major kinase involved is named *MAP kinase*; it is also called the *mitogen-activated protein kinase* (with the same acronym), a name chosen by scientists who study cell signalling by phosphorylation. The multiple names for this kinase emphasize that MAPs are targets of signalling pathways. MAPs are also targets of a second kinase, *cdc2 kinase*, which plays a major role in controlling the activities of various proteins during the cell cycle.

► *Kinesin, Dynein, and Intracellular Transport*

Within cells, membrane-bounded vesicles and proteins are frequently transported many micrometers along well-defined routes in the cytosol and delivered to particular addresses. Diffusion alone cannot account for the rate, directionality, or destinations of such transport processes. Cell biologists have long suspected that transport involves some kind of tracks, because video light microscopy showed that the long distance movements follow straight paths in the cytosol, frequently along cytosolic fibers. Two well-studied experimental systems—nerve cells and fish-scale pigment cells—have provided the evidence showing the tracks are microtubules and the intracellular transport of membrane-bounded vesicles and organelles is dependent on microtubule motor proteins.

Fast Axonal Transport Occurs along Microtubules

Nerve cells are the longest cells in the body, often achieving lengths of several meters. In Chapter 21, we discussed how the growth cone at the end of an axon eventually becomes

the nerve terminal, where electrical connections are established with neighboring cells. The neuron must constantly supply new materials—proteins and membranes—to the terminal to replenish those lost by exocytosis at the synapse. Where do these new materials originate? Ribosomes are present only in the cell body and dendrites of nerve cells, so no protein synthesis can occur in the axons and synaptic terminals. Therefore, proteins and membranes must be synthesized in the cell body and then transported down the axon to the synaptic regions. This process, first described in 1948, is called *axonal transport*. Only in the last ten years have we come to understand that axonal transport occurs on microtubules.

Axonal transport can be detected by a pulse-chase experiment and the transported components identified by gel electrophoresis and autoradiography (Figure 23-18a). If radiolabeled amino acids are injected into the ganglion of the sciatic nerve, they become incorporated into proteins made in the cell body. (Neurons in the mammalian sciatic nerve are studied because their cell bodies are conveniently located in the dorsal root ganglion near the spinal cord and their nerve axons are very long.) At various times after injection, the nerve is removed and cut into sections which are analyzed for radioactivity. Autoradiography of SDS-polyacrylamide gels of proteins solubilized from the nerve segments detects the radioactive proteins. Soon after injection, these are found in the nerve segments closest to the cell body. With increasingly longer chase periods, radioactive protein is detected in nerve segments more and more distal to the cell body (Figure 23-18b). Further analysis shows that different types of proteins are transported, including proteins found in the synaptic vesicles, organelles, and plasma membrane, cytoskeletal proteins, and soluble proteins.

Studies on axonal transport determined that materials are transported in both directions. *Anterograde transport*, which we have just discussed, proceeds from the cell body to the synaptic junctions and is associated with axonal growth and the renewal of synaptic vesicles. In the opposite, *retrograde*, direction other substances move along the axon rapidly toward the cell body. These substances mainly consist of "old" membrane from the synaptic terminals, and are destined to be degraded in lysosomes in the cell body.

Furthermore, axonal transport is divided into three groups (Table 23-2) according to the speed of movement. The fastest-moving material, consisting of membrane-bounded vesicles, has a velocity of about 250 mm/day, or about 3 μm/s. The slowest material, comprising mostly polymerized cytoskeletal proteins, moves only a fraction of a millimeter per day. Organelles such as mitochondria move down the axon at an intermediate rate.

Having characterized the materials transported along the axon, scientists suspected that transport occurs along microtubules when, using video light microscopy, they saw

(a)

(b)

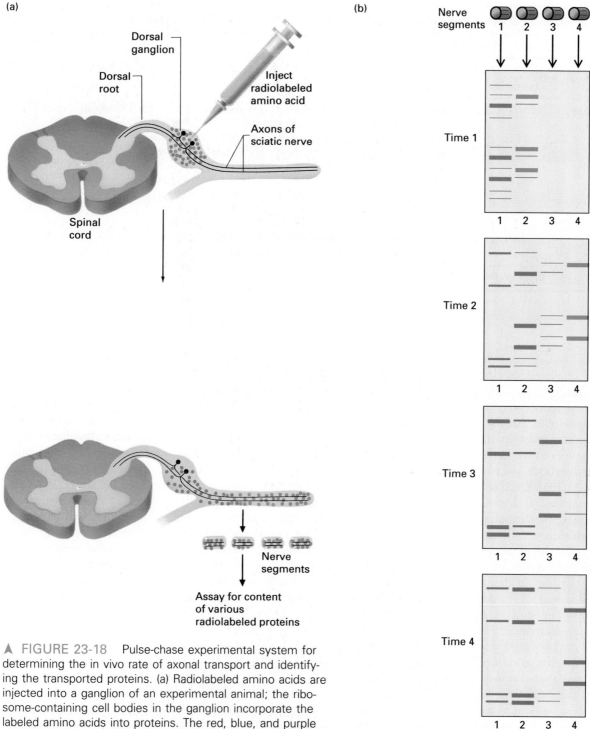

▲ FIGURE 23-18   Pulse-chase experimental system for determining the in vivo rate of axonal transport and identifying the transported proteins. (a) Radiolabeled amino acids are injected into a ganglion of an experimental animal; the ribosome-containing cell bodies in the ganglion incorporate the labeled amino acids into proteins. The red, blue, and purple dots in the ganglion and axon represent groups of proteins that are transported down the axon at different rates, red most rapidly, purple least rapidly. Animals are killed at different times after injection, and the sciatic nerve is dissected and cut into 5-mm segments. (b) The amount of radiolabeled protein in each fragment is measured, and then the various proteins are resolved by gel electrophoresis. One set of polypeptides (red bands), characterized by molecular weight, are detected first in the segments proximal to the injection site and later in more distal segments. Lagging behind this fast-traveling component are other polypeptides whose rate of transport is slower (blue and purple bands). The distribution of labeled proteins in part (a) corresponds to the gel pattern at time 2. [See O. Ochs, 1981, in *Basic Neurochemistry*, 3d ed., G. J. Sigel et al., eds. Little, Brown, p. 425.]

TABLE 23-2   Rate of Axonal Transport of Various Materials in Mammalian Neurons

Component	Transport Rate (mm/day)	Transport Rate (μm/s)	Materials Transported	Cellular Structures Transported
Fast anterograde	250–400	3.0–4.5	Glycoproteins, glycolipids, acetylcholinesterase	Vesicles, smooth ER, small granules
retrograde	100–200	1.5–2.5	Nerve growth factor, lysosomal enzymes	Prelysosomal structures, multivesicular bodies
Intermediate	50	0.6	F1-ATPase	Mitochondria
Slow	2–4	0.03	Actin, clathrin, enolase, calmodulin	Microfilaments
	0.2–1	0.002–0.01	Neurofilament proteins tubulin, MAP	Microtubule–intermediate filament network

SOURCE: Adapted from S. Brady and R. Lasek, 1982, *Meth. in Cell Biol.* 25:365.

the movement of small vesicles along cytosolic fibers in the giant axon of squid (Figure 23-19a). Video microscopy revealed this movement occurred at rates corresponding to fast axonal transport and that the movements occurred in both the anterograde and retrograde directions. In some cases, two organelles can be seen to move along the same fiber in opposite directions and to pass each other without colliding (Figure 23-19a). Electron microscopy of the same region of the axoplasm has demonstrated the transport fibers are individual microtubules (Figure 23-19b). The movements do not require an intact cell; in fact many studies on fast axonal transport are conducted with extruded axoplasm. In this preparation, the cytosol is squeezed from the axon with a roller onto a glass coverslip. When such an extract is provided with ATP, the movement of vesicles along microtubules can be observed by video microscopy. The rate of vesicle movement (1–2 μs) in this cell-free system is similar to that of fast axonal transport in intact cells. These experiments, conducted in the last ten years, established that fast axonal transport occurs along microtubules and that the movement requires ATP. As we will discuss shortly, these two observations led to the identification of microtubule motor proteins which generate the movements.

Microtubules Provide Tracks for the Movement of Pigment Granules

Another system for observing rapid transport along microtubules is provided by the specialized pigment cells—called melanophores—that are found in the skin of many amphibians and on the scales of many fish. Nerves and hor-

mones control the color of the skin by triggering the transport of membrane-enclosed pigment granules either to the cell periphery, to darken the color of the skin, or inward, toward the center of the cell, to lighten the color (Figure 23-20). In this way, an animal can adjust its color to blend better with its surroundings in order to escape predators.

The role of microtubules in color adjustment is easily studied if the pigment cells are placed in culture. If a melanophore-covered fish scale is placed in a culture medium, the pigment granules in the melanophores will be seen to move inward and outward spontaneously. In cases when individual granules could be followed, it has been observed that after movement to the cell periphery, each granule always returns to its prior location in the center of the cell. During this movement, microtubules serve as tracks along which the pigment granules can move in either direction.

Intracellular Membrane Vesicles Travel along Microtubules

Having seen that membrane vesicles in specialized cells are transported between the cell body and the cell periphery, we can next ask whether membrane vesicles are transported along microtubules in every eukaryotic cell. The preliminary answer is a qualified yes: some types of vesicle transport are dependent on microtubules, though microfilaments may also be involved in some cases. The best studied system is the intracellular membrane transport between the Golgi complex and other membrane compartments. In cultured fibroblasts, the Golgi complex is concentrated near the MTOC (Figure 23-21). During mitosis (or after the depolymerization of microtubules by colchicine), the

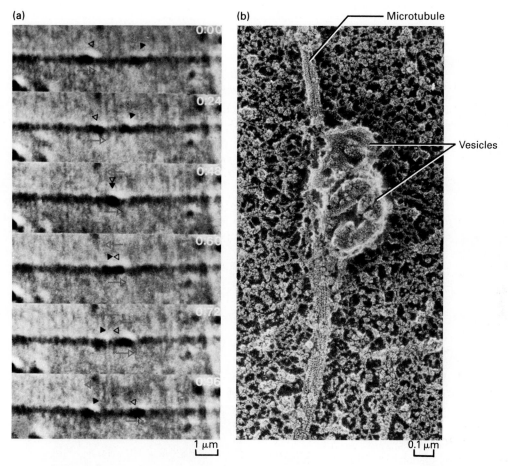

▲ FIGURE 23-19   Video micrographs showing bidirectional movement of two vesicular organelles on a single transport microtubule filament. (a) A piece of squid giant axon was dissected, the cytoplasm was extruded, and a buffer containing ATP was added. The preparation was then viewed in a differential interference contrast microscope, and the images were recorded on videotape. The two organelles (located at positions indicated by open and solid triangles) move in opposite directions (indicated by colored arrows) along the same filament, pass each other, and continue in their original directions. Elapsed time in seconds appears at the top right corner of each video frame. (b) A region of cytoplasm similar to that shown in (a) was freeze-dried, rotary-shadowed with platinum, and viewed in the electron microscope. Two large structures attached to one microtubule are visible; these presumably are small vesicles that were moving along the microtubule when the preparation was frozen. [See B. J. Schnapp et al., 1985, *Cell* **40**:455. Courtesy of B. J. Schnapp, R. D. Valle, M. P. Sheetz, and T. S. Reese.]

Golgi complex breaks into small vesicles that are dispersed throughout the cytosol. When the cytosolic microtubules re-form during interphase (or after removal of the colchicine), the Golgi vesicles move along these microtubule tracks toward the MTOC, where they reaggregate to form large membrane complexes.

In addition to the Golgi complex, microtubules are also associated with the endoplasmic reticulum (ER). Fluorescence microscopy, using anti-tubulin antibodies and $DiOC_6$, a fluorescent dye specific for the ER, reveals an anastomosing network of tubular membranes in the cytosol which co-localizes with microtubules (Figure 23-22). If microtubules are destroyed by drugs such as nocodazole or colchicine, then the ER loses its network-like organization.

After the drug is washed from the cell, tubular fingers of ER grow as new microtubules polymerize. In cell-free systems, the ER can be reconstituted with microtubules and an ER-rich cell extract. Even under this simple condition, ER membranes elongate along a microtubule. This close association between ER and intact microtubules suggests there are proteins which are involved in binding ER membranes to microtubules.

A key feature of the in vitro assembly of the ER is that the tubular membranes seem to elongate along microtubules. This outgrowth of the ER membrane is blocked by microtubule inhibitors like colchicine or nocodazole, suggesting that intact microtubules are required for membrane elongation.

(a)

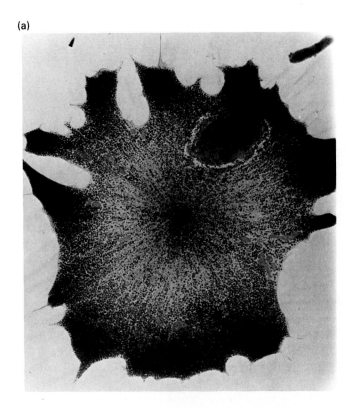

(b)

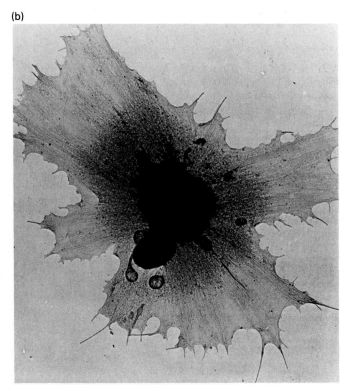

(c)

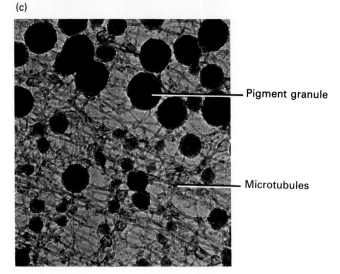

Pigment granule

Microtubules

▲ FIGURE 23-20   High-voltage electron micrographs showing movement of pigment granules in a melanophore, or red-pigment cell, of the squirrelfish, *Holocentrus ascensionis*. (a) The pigment granules are dispersed to the cell periphery. (b) The granules are condensed around the nucleus. (c) A portion of a dispersing melanophore showing the pigment granules associated with tracks of microtubules. [Courtesy of K. Porter.]

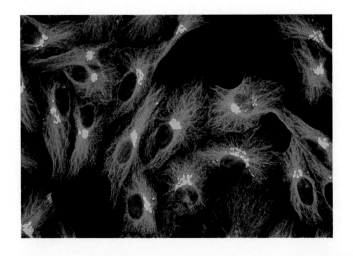

◄ FIGURE 23-21   Distribution of the Golgi complex (yellow) and microtubules (green) in cultured monkey fibroblasts. Microtubules were visualized with an anti-tubulin antibody and the Golgi complex with an antibody specific for galactosyltransferase (see Figure 16-35). The Golgi complex is concentrated near the MTOC. [Courtesy of T. Kreis.]

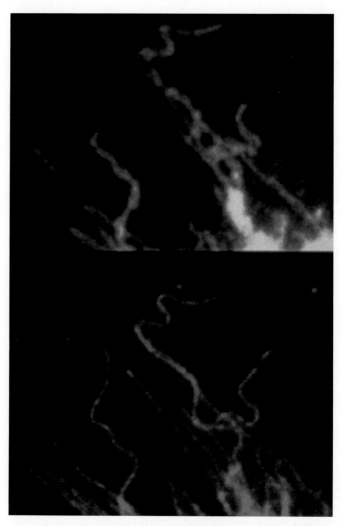

▲ FIGURE 23-22   In a frog fibroblast, the striking alignment of ER and microtubules is easily seen because the cell has sparse microtubules. In many but not all regions of the cytoplasm, the endoplasmic reticulum stained with DiOC$_6$ (green) colocalizes with cytoplasmic microtubules (red). [Courtesy of M. Terasaki.]

Microtubule Motor Proteins Promote Vesicle Translocation along Microtubules

Knowing that axonal transport occurs along microtubules, cell biologists turned to biochemistry to isolate the proteins that caused the movements. The plan was to identify an active component in neuronal cytosolic extracts which caused membrane vesicles to move along MAP-free microtubules assembled from purified tubulin subunits. The drug taxol was used to prevent these microtubules from depolymerizing. When synaptic vesicles were added with ATP to these MAP-free microtubules, the vesicles neither bound to the microtubules nor moved along them. However, the addition of squid nerve axoplasm (free of tubulin) caused the vesicles to bind to the microtubules and to move

along them, indicating that a soluble protein in the nerve cytosol is required for translocation.

Vesicles and nerve cytosol (squid axoplasm) were mixed with microtubules in the presence of an analog of ATP that could not be hydrolyzed, AMPPNP:

$$\text{Adenine—ribose—O—}\overset{\displaystyle O}{\underset{\displaystyle O^-}{\overset{\|}{\underset{|}{P}}}}\text{—O—}\overset{\displaystyle O}{\underset{\displaystyle O^-}{\overset{\|}{\underset{|}{P}}}}\text{—NH—}\overset{\displaystyle O}{\underset{\displaystyle O^-}{\overset{\|}{\underset{|}{P}}}}\text{—OH}$$

Under these conditions, the vesicles bound tightly to the microtubules but did not move. However, they did move when ATP was added. These results suggest that a motor protein in the cytosol binds to microtubules in the presence of ATP or AMPPNP but movement requires hydrolysis of the terminal phosphoanhydride bond of ATP.

To purify the soluble motor protein, scientists used the trick of substituting AMPPNP for ATP so that the motor protein would bind tightly to microtubules, which were used as an affinity matrix. A mixture of microtubules, brain extract, and AMPPNP was incubated and then proteins in the extract which had bound to the microtubules were collected by centrifugation to sediment the microtubules. Treatment of the microtubule-rich material in the pellet with ATP released one predominant protein back into solution; this protein was named *kinesin*.

Kinesin Is a (+) End–Directed Motor Protein

Kinesin (MW 380,000), isolated from squid axoplasm, consists of a dimer of two heavy chains (MW of each, 124,000), each heavy chain complexed with a light chain (MW 64,000). The electron microscope reveals that kinesin is organized into three domains (Figure 23-23), a pair of large globular head domains connected by a long central stalk to a pair of small globular tail domains which contain the light chains.

Each domain carries out a particular function. The head domain is responsible for the motor activity of kinesin. The force-generating motor activity of the head domain was demonstrated by a chimeric protein created by combining the kinesin motor domain with α-spectrin, a subunit of an actin-crosslinking protein. In in vitro motility experiments (see following discussion for a description of the assays), the kinesin head was sufficient to cause microtubule gliding as long as it was a part of the spectrin. The head alone did not cause movement presumably because it was too close to the glass surface. These results confirm that the head domain is responsible for the motor function of kinesin but also suggest the kinesin stalk region spaces the head from the surface of a membrane. In contrast to the head domain, which binds microtubules and ATP, the tail domain is responsible for binding to membranes. In light of

(a) **Kinesin**

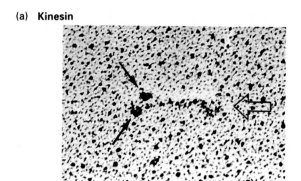

(b)

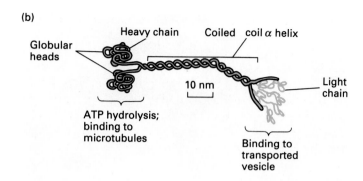

▲ FIGURE 23-23   Kinesin. (a) Electron micrograph of rotary-shadowed brain kinesin. The two globular heads (solid arrows) that bind to tubulin are visible on the left side of the molecule, and the tail (open arrow) on the right. (b) Sche-

matic model of kinesin showing the arrangement of the two heavy chains and two light chains. [Part (a) from N. Hirokawa et al., 1989, *Cell* **56**:867; courtesy of N. Hirokawa.]

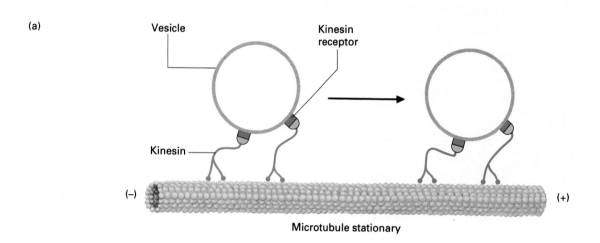

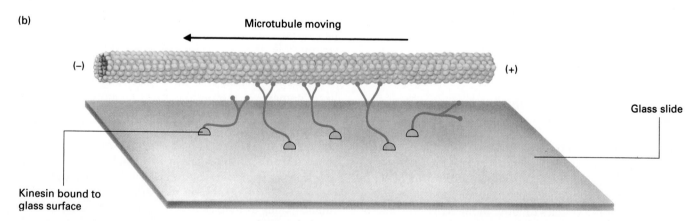

▲ FIGURE 23-24   Model of kinesin-catalyzed anterograde transport. (a) Kinesin-motored transport of vesicles along immobile microtubules. The kinesin molecules, attached to unidentified receptors on the vesicle surface, transport the vesicles from the (−) to the (+) end of a stationary microtubule. (b) Kinesin-catalyzed movement of microtubules. The kinesin molecules bound to the glass surface move toward

the (+) end of the microtubule. Because the kinesin molecules are immobilized onto the coverslip, the sliding force is transmitted to the microtubule, which then moves in the direction of its (−) end. ATP is required for movement in both cases. [Adapted from R. D. Vale, et al., 1985, *Cell* **40**:559; and T. Schroer, et al., 1988, *J. Cell Biol.* **107**:1785.]

the transport function of kinesin, the membrane vesicle is often referred to as kinesin's "cargo."

Kinesin movements have polarity; this was established by in vitro motility assay using microtubules nucleated from isolated centrosomes. In this assay, a vesicle or a plastic bead coated with kinesin is added to a glass slide along with a preparation of centrosome-nucleated microtubules. The kinesin-coated beads bind to the microtubules, and, in the presence of ATP, kinesin carries the beads along a microtubule from the (−) to the (+) end (Figure 23-24a). Alternatively, kinesin alone, adsorbed onto a glass coverslip, causes the microtubules to glide across the surface, with their (−) ends in the lead (Figure 23-24b); this direction of movement results from kinesin walking toward the (+) end of the microtubule. Thus kinesin is a (+) end-directed microtubule motor protein. Because this direction corresponds to anterograde transport, kinesin is implicated by these studies as the motor protein responsible for anterograde movements and for other (+) end-directed movements, such as the transport of secretory vesicles to the plasma membrane and the radial movements of ER membranes and pigment granules.

Recent in vitro motility experiments using an optical trap have determined two fundamental characteristics of the kinesin motor, its step size and force. In a manner similar to that of the experiments performed on myosin (see Figure 22-20), a two-headed kinesin molecule was found to move in 8-nm steps and exert a force of 6 pN. The step size matches the distance between successive α- or β-tubulin monomers in a protofilament, suggesting that kinesin binds only to one or the other monomer. A force of 6 pN is similar to the force generated by a myosin molecule. Unlike the case with myosin, we do not have a three-dimensional model of kinesin and therefore we are unable to speculate on the mechanism of force transduction. However, kinetic studies indicate that in many of its features kinesin will be similar to myosin.

Like myosin, kinesin also belongs to a superfamily of motor proteins. The latest estimate counts over twelve different kinesin family members; all contain the kinesin motor domain but differ in two important aspects. In some kinesin family members, the motor domain is at the N-terminus, while in other members, the motor domain is central or at the C-terminus. In the latter case, these kinesins are (−) end–directed motor proteins, that is, they move toward the (−) end of a microtubule and not (+) end like the other kinesin members. Kinesins can be divided into two groups (Table 23-3): cytosolic kinesins involved in vesicle transport and kinesin-related proteins (KRPs) involved in spindle assembly and chromosome segregation at

TABLE 23-3 Microtubule Motor Proteins

Motor	Subunit Composition (MW)			Cargo	Direction of Movement[‖]
	Heavy Chain (No./molecule)	Intermediate Chain (No./heavy chain)	Light Chain (No./heavy chain)		
Cytosolic kinesin	110,000–135,000 (2)	60,000–70,000 (2)		Cytosolic vesicles	+
KRP (+) motors*	(?)			Nuclear spindle	+
KRP (−) motors[†]	(?)			Nuclear spindle	−
Cytosolic dynein[‡]	400,000 (2)	74,000 (3)	50,000–58,000 (4)	Cytosolic vesicles, kinetochore	−
Axonemal outer-arm dynein	500,000 (3)			A tubule	−
Axonemal inner-arm dynein[§]	500,000 (2)	Actin-related protein	p28 (28,000)	A tubule	−

* KRP = kinesin-related protein. Includes bimC-like family (cut7, Kip1p, Klp61F, Eg5, and Cin8p) and nod family (nod, KIPA, MLPA, MKPL1, CENPE); exact cargo not identified.
[†] Includes Kar3p, klpA, and ncd.
[‡] Also named MAP1C.
[§] Shown is inner-arm dynein I2 family; not listed are families I1 and I3.
[‖] Movement toward microtubule (+) or (−) end.

meiosis and mitosis. We will discuss the latter group, including one member, ncd, a (−) end–directed motor protein, in the mitosis section of this chapter.

Dynein Is a (−) End–Directed Motor Protein

The discovery of (+) end–directed kinesin only partially explained how membrane vesicles are transported in cells. Some movements, such as retrograde axonal transport or the transit of endocytotic vesicles of the plasma membrane to lysosomes, are in the direction opposite to this kinesin-dependent movement. Thus there must be a second group of motor proteins responsible for (−) end–directed motility. Such a group exists; it is composed of the dyneins, a superfamily of exceptionally huge multimeric proteins. Dyneins are divided into two classes, cytosolic and axonemal (flagellar) (see Table 23-3). These proteins have native molecular weights exceeding 1,000,000 and are composed of two to three heavy chains (MW 470,000–540,000) complexed with a poorly determined number of intermediate and light chains. The number of heavy-chain proteins corresponds to the number of globular heads in the dynein complex. Under the most favorable conditions, dynein can be partially denatured and examined by electron microscopy. This procedure reveals that cytosolic dynein is a two-headed molecule (Figure 23-25), consisting of two identical or nearly identical heavy chains. (Axonemal, or flagellar, dynein will be examined later, in the section on ciliary and flagellar motility.)

The activity of cytosolic dynein may be regulated by a 20S multiprotein complex which co-purifies with it. Called the *dynactin complex,* it consists of four polypeptides, including an actin-related protein and dynactin, a 150,000-MW protein. The function of the dynactin complex is unknown, but scientists speculate that it may mediate interactions with membranes, because electron micrographs show that dynactin (visualized by staining with antibodies), is localized at the tail of cytosolic dynein.

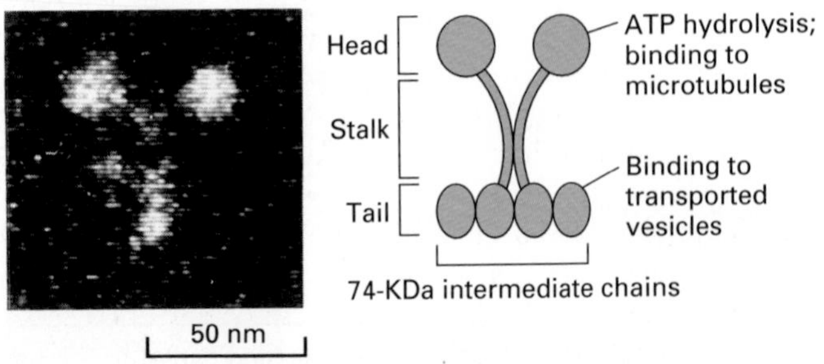

MAP1C

▲ FIGURE 23-25   Electron micrograph and schematic diagram of cytosolic dynein (MAP1C), showing the two globular heads that bind to microtubules. The intermediate chains are located at the base of the molecule. [Photograph from R. B. Vallee et al., 1988, *Nature* **332**:561.]

Multiple Motor Proteins Are Associated with Membrane Vesicles

The identification of kinesin and cytosolic dynein motor proteins explains not only how cell organelle movement is powered but also how the direction of movement is controlled (remember also that the direction of vesicle transport is dependent on the polarity of microtubules, which is fixed by the MTOC). However, there are several instances where movement of a vesicle must involve more than one motor protein. For example, because pigment granules can alternate their direction of movement along a single microtubule, both anterograde and retrograde microtubule motor proteins must be present; but at any one time, only one motor protein is active or, alternatively, only one motor protein is bound to the vesicle.

However, there is increasing evidence that a membrane vesicle is the cargo of both a microtubule motor protein and a microfilament motor protein. As discussed in Chapter 16, endocytosis carries the vesicles from the plasma membrane inward, while secretion carries vesicles from the ER and Golgi outward. In both instances, a vesicle must traverse microtubule-poor but microfilament-rich regions in the cell. Several complementary experiments suggest that microtubule and microfilament motor proteins are bound to the same membrane vesicles and cooperate in their transport. One piece of evidence was obtained from microscopy of vesicle movements in extruded squid axoplasm. As observed many times before, membrane vesicles traveled along microtubule tracks; however, at the periphery of the extruded axoplasm, movement continued even though no microtubules were present. This region contained microfilaments, and subsequent experiments demonstrated that a vesicle could move on a microtubule or a microfilament. Thus at least two motor proteins, myosin and either kinesin or dynein, must be bound to the same vesicle. This capability of movement along both cytoskeletal systems suggests a model for the delivery of a synaptic vesicle to the nerve terminal. In a neuron, a synaptic vesicle could be transported at a fast rate by a microtubule motor in the microtubule-rich axon and then travel through the actin-rich cortex at the synapse on a myosin motor (Figure 23-26).

Two other independent approaches also suggest that microtubule and microfilament motor proteins are bound to the same vesicle. Biochemical studies have shown that Golgi-derived vesicles, purified from extracts of intestinal epithelial cells, contain both dynein and myosin I. In intestinal epithelial cells, microtubules are oriented with their (−) ends near the apical membrane and their (+) ends pointed toward the basal membrane. The presence of both motor proteins explains how Golgi-derived vesicles can be transported to the apical membrane along microtubules and then through the actin-rich terminal web in the apical cytosol of the cell. Genetic studies in yeast add support to the biochemical evidence. In yeast, mutations in myosin V disrupt secretion. A genetic search was conducted for genes

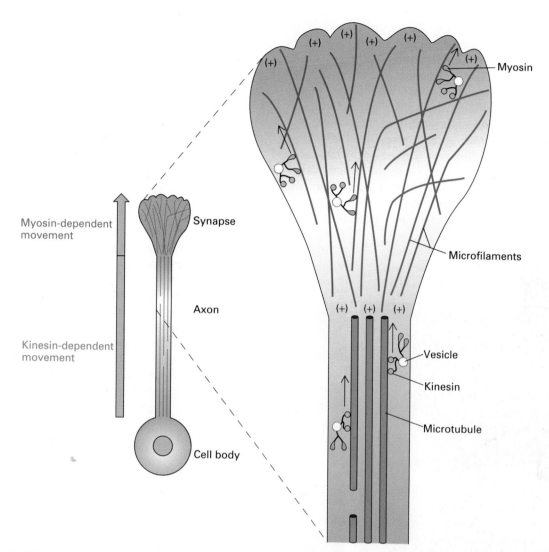

▲ FIGURE 23-26 Anterograde transport probably involves kinesin (pink) and myosin (blue) bound to the same vesicle. In the axon, vesicles powered by kinesin travel along microtubules toward the (+) ends, located in the nerve ter-minal. At the nerve terminal, the vesicles must traverse the actin-rich cortex (microfilaments) at the synapse. The vesicle attaches to the microfilaments via its myosin motor and then translocates to the synapse.

that suppress the secretion defect by encoding for proteins which can replace the function of myosin V. Surprisingly, the search discovered a KRP and not another myosin family member. The genetic result suggests that two motor proteins, a KRP and myosin V, participate in membrane transport in yeast.

➤ *Cilia and Flagella: Structure and Movement*

Swimming is the major form of movement exhibited by sperm and by many protozoa. Some cells are propelled at velocities approaching 1 mm/s by the beating of *cilia* and *flagella*, whiplike membrane extensions of the cell. Some cilia and flagella are no more than a few microns in length, whereas some insect sperm flagella are longer than 2 mm!

Some cells possess only one or two flagella, whereas other cells have many cilia. For example, the mammalian spermatozoon has a single flagellum, the unicellular green alga *Chlamydomonas* has two flagella, and the unicellular ciliated protozoan *Paramecium* has a few thousand cilia over its surface, which are used both for locomotion and to trap food particles. In mammals, many epithelial cells are ciliated in order to sweep materials across the tissue surface. For instance, the mammalian oviduct is ciliated in order to propel the ovum from the ovary to the uterus, and huge numbers of cilia (more than $10^7/mm^2$) cover the surfaces of mammalian respiratory passages (the nose, pharynx, and trachea), where they dislodge and expel particulate matter that collects in the mucus secretions of these tissues.

Ciliary and flagellar beating is a series of bends, originating at the base of the structure and propagated toward the tip (Figure 23-27). When frozen by high-speed strobe microscopy, the waveform of the beat is easily seen. Beating can be planar or three-dimensional and, like waves that you have studied in physics, the beating can be described by its amplitude, wavelength, and frequency. The bends push against the surrounding fluid, propelling the cell forward or moving the fluid across a fixed epithelium.

The cilium or flagellum is bent in two directions, called the *principal* and *reverse bends* (Figure 23-28). When the principal and reverse bends both have the same amplitude, the waveform is symmetric. Many sperm beat with a nearly symmetric pattern (Figure 23-28a); consequently they swim along a curved path. In contrast, when the principal and reverse bends are not equivalent, an asymmetric beating pattern is produced (Figure 23-28b,c). The cilia of epithelial cells typically beat asymmetrically, producing a beating pattern in two parts: a stiff, whip-like *effective stroke*, in which a low-amplitude bend is propagated so that the cilium moves perpendicular to the cell surface, and a "limp-noodle"-like *recovery stroke*, in which a high-amplitude bend sweeps the cilium near the cell surface (Figure 23-28b). In this way, the epithelial cilia produce a large net movement of material in one direction by the effective stroke and move very little material in the "wrong" direction during the recovery stroke.

The two flagella of *Chlamydomonas* provide another example of the importance of bend symmetry. These flagella are able to beat either symmetrically or asymmetrically. The symmetric beating of the two flagella pushes the cell through the liquid medium in one direction like a swimmer's butterfly kick, whereas the asymmetric beating pulls the cell in the opposite direction by a breast-stroke-like motion. Regardless of the pattern of the beat, ciliary and flagellar movement, like muscle contraction and cell crawling, requires ATP hydrolysis.

All Eukaryotic Cilia and Flagella Contain Bundles of Doublet Microtubules

Virtually all eukaryotic cilia and flagella are remarkably similar in their organization (Figure 23-29), possessing a central bundle of microtubules, called the *axoneme,* in

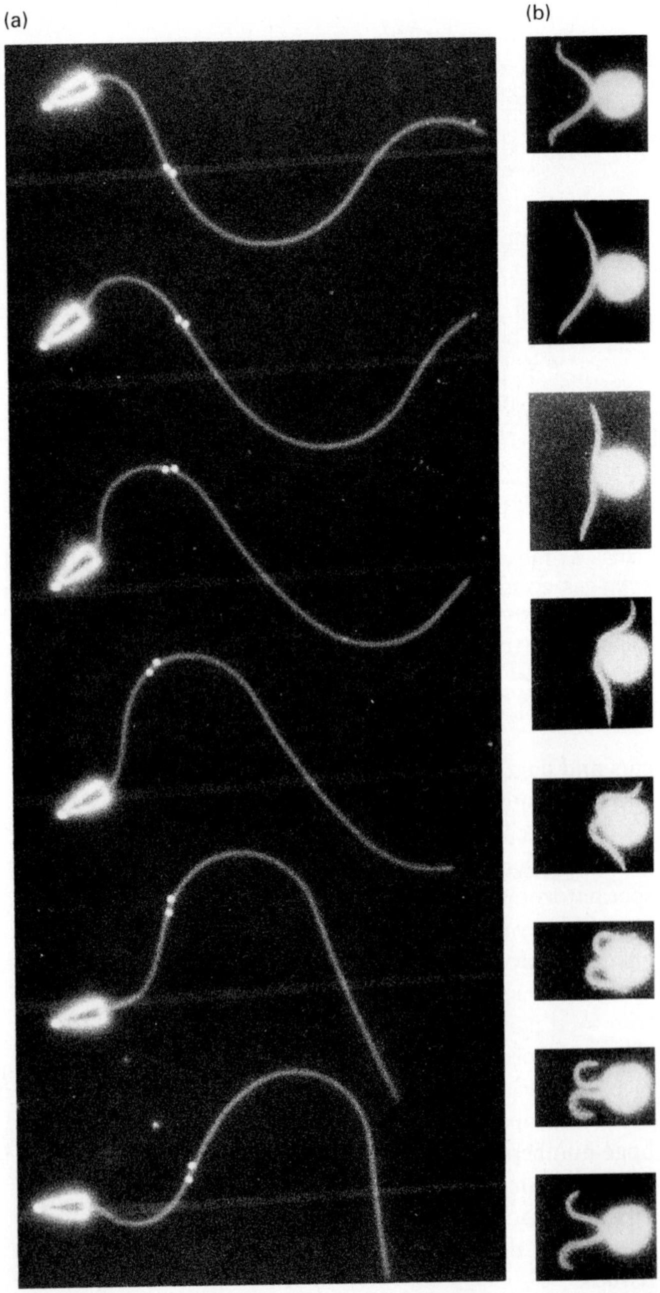

(a)

(b)

◀ FIGURE 23-27   Characteristic motions of cilia and flagella. (a) In the typical sperm flagellum, successive waves of bending originate at the base and are propagated out toward the tip; these waves push against the water and propel the cell forward. Captured in this multiple-exposure sequence, a bend at the base of the sperm in the first (top) frame has moved distally halfway along the flagellum by the last frame. The sinusoidal waveform of the flagellum is easily seen, and from such pictures a frequency of 30–40 waves per second is measured. A pair of beads on the flagellum are seen to slide apart as the bend moves through their region. (b) Beating of a cilium, which occurs 5–10 times per second, has two stages, called the effective stroke and the recovery stroke. In a motion similar to the breast stroke of a swimmer, the cilium is stiff during the effective stroke, and this movement pulls the organism through the water. During the recovery stroke, a different wave of bending moves outward along the cilium from its base, pushing the limp cilium along the surface of the cell until it reaches the position to initiate another effective stroke. ATP is hydrolyzed during both the effective stroke and the recovery stroke. [Part (a) from C. Brokaw, 1991, *J. Cell Biol.* **114** (6): cover photo; courtesy of C. Brokaw; part (b) courtesy of S. Goldstein.]

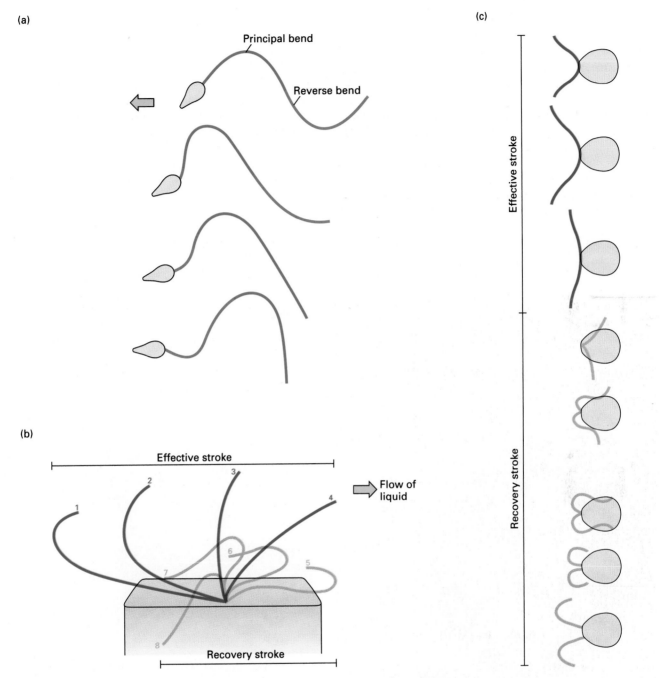

▲ FIGURE 23-28   Diagram of the principal (blue) and reverse (red) bends in (a) a beating flagellum, and the effective (green) and recovery strokes (orange) in (b) an epithelial cell cilium, and (c) *Chlamydomonas* cilia. In a and c, the cells swim toward the left. [Courtesy of D. Asai.]

which nine outer doublet microtubules surround a central pair of singlet microtubules. This characteristic "9 + 2" arrangement of microtubules is easily seen when the axoneme is viewed in cross-section with the electron microscope. As we have seen (Figure 23-2), each doublet microtubule consists of A and B tubules, or subfibers: the A tubule is a complete microtubule with 13 protofilaments, while the B tubule contains 10 protofilaments. The bundle of microtubules comprising the axoneme is surrounded by

a cylinder of plasma membrane. Regardless of the organism or cell type, the axoneme is about 0.25 μm in diameter, but it varies greatly in length, from a few microns to more than 2 mm.

At its point of attachment to the cell, the axoneme connects with the basal body (Figure 23-30a). Like centrioles, basal bodies are cylindrical structures, about 0.4 μm long and 0.2 μm wide, which contain nine triplet microtubules (Figure 23-30b). Each triplet contains one complete

(a)

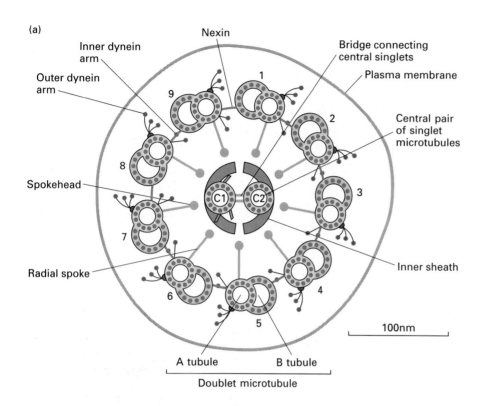

(b)

(c)

(d)

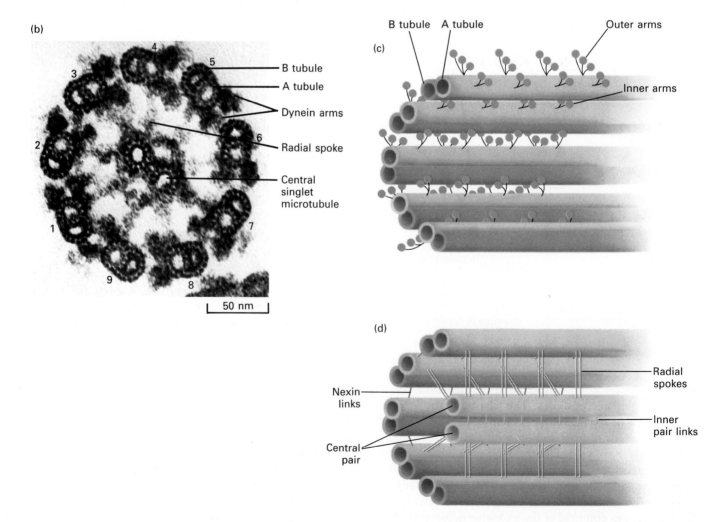

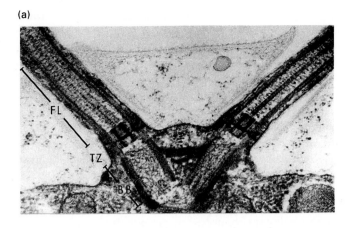
(a)

◄ FIGURE 23-29   Structure of ciliary and flagellar axonemes. (a) Cross-sectional view of an idealized flagellum showing all major structures. The dynein arms and radial spokes with attached heads occur only at intervals along the longitudinal axis. The central microtubules, C1 and C2, are distinguished by fibers bound only to C1. (b) Micrograph of a transverse section through an isolated demembranated cilium. The two central singlet microtubules are surrounded by nine outer doublets, each composed of an A and a B subfiber. (c) A cut-away longitudinal view of the axoneme. The organization of inner and outer dynein arms along the A tubule shows their different spacings. The inner dynein arms lie close to the radial spokes, which connect the outer doublets to the central pair of microtubules. The axoneme is oriented with its basal body to the left, and the central pair of microtubules are not shown. (d) The organization of the three elastic links which hold the axoneme together. Nexin links cross-link pairs of outer doublets, radial spokes connect the central pair of microtubules to the outer doublets, and the central pair of microtubules are held together by inner bridges. [Part (b) courtesy of L. Tilney. See U. W. Goodenough and J. E. Heuser, 1985, *J. Cell Biol.* **100**:2008.]

13-protofilament microtubule, the A tubule, fused to the incomplete B tubule, which in turn is fused to the incomplete C tubule. The A and B tubules of basal bodies continue into the axonemal shaft, whereas the C tubule terminates within the transition zone between the basal body and the shaft (Figure 23-30c). The two central tubules in a flagellum or a cilium also end in the transition zone, above the basal body. As we shall see later, the basal body plays an important role in initiating the growth of the axoneme.

Within the axoneme, the two central singlet and nine outer doublet microtubules are continuous for the entire length of the structure (Figure 23-30c). The outer doublet represents a specialized polymer of tubulin that is found

► FIGURE 23-30   (a) Electron micrograph of the basal regions of the two flagella in *Chlamydomonas reinhardtii*. The bundles of microtubules and some fibers connecting them are visible in the flagella (FL). The two basal bodies (BB) form the point of a "V"; a transition zone (TZ) between the basal body and flagellum proper contains two dense-staining cylinders of unknown structure. (b) Cross sections of flagellar basal bodies in the protozoan *Trichonympha*. The proximal region contains a single central tubule and radial spokes; these are missing in the distal region. Each of the nine outer fibers is built of three tubules, labeled A, B, and C. (c) A schematic view of a basal body. The nine sets of outer doublet microtubules are continuous with the nine sets of triplet microtubules. [Part (a) From B. Huang, et al., 1982, *Cell* **29**:745; part (b) reproduced from the *Journal of Cell Biology*, 1981, by copyright permission of the Rockefeller University Press; courtesy of I. Gibbons.]

(b)
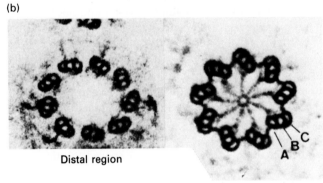
Distal region

Proximal region

(c)
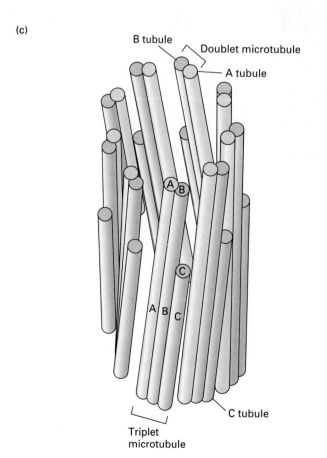

only in the axoneme. Permanently attached to the A tubule of each doublet microtubule is an inner and an outer row of dynein arms (see Figure 23-29c). These dyneins reach out to the B tubule of the neighboring doublet. The junction between A and B tubules of one doublet are probably strengthened by the protein *tektin*, a highly α-helical protein that is similar in structure to intermediate-filament proteins. Each tektin filament, which is 2 nm in diameter and approximately 48 nm long, runs longitudinally along the wall of the outer doublet where the A tubule is joined to the B tubule.

The axoneme is held together by three sets of protein cross-links (see Figure 23-29d). The central pair of singlet microtubules are connected by periodic bridges, like rungs on a ladder, and are surrounded by a fibrous structure termed the *inner sheath*. A second set of linkers, composed of the protein *nexin*, joins adjacent outer doublet microtubules. Spaced every 86 nm along the axoneme, nexin links are probably elastic, since, as we shall see, they must allow the outer doublets to slide past each other longitudinally during beating. Comprising the third linkage system are the *radial spokes*, which radiate from the central singlets to each A tubule of the outer doublets. Detailed analyses of electron micrographs have shown that the radial spokes are not evenly spaced along the axoneme; instead they are grouped into pairs, and each pair is spaced 96 nm apart.

While the 9 + 2 pattern is the fundamental pattern of virtually all cilia and flagella, the axonemes of certain protozoa and some insect sperm show some interesting exceptions. Some axonemes, though motile, lack the central pair of singlet microtubules. The simplest such axoneme is the 3 + 0 arrangement in *Daplius*, a parasitic protozoan. Its flagellum beats slowly (1.5 beats/s) in a helical pattern. Other axonemes consist of 6 + 0 or 9 + 0 arrangements of microtubules. Since these cilia and flagella are all motile, the central pair of microtubules is evidently not necessary for axonemal beating. Numbers of outer doublets less than nine also can sustain motility, but at a lower frequency.

Clearly, at the level of resolution afforded by the electron microscope, flagella and cilia are seen to be quite complex structures which contain many protein components in addition to microtubules. The roles these various structures play in accomplishing the characteristic motion of flagella and cilia are discussed next.

Ciliary and Flagellar Beating Is Produced by Controlled Sliding of Outer Doublet Microtubules

Isolated cilia and flagella, from which the plasma membrane has been removed by nonionic detergents, can beat when ATP is added; the reactivated movement is indistinguishable from the flagellar beat of a living cell. Thus the forces that generate movement must reside within the axo-

neme and are not located in the plasma membrane or elsewhere in the cell body.

As in the movement of muscle during contraction, the basis for axonemal movement is the sliding of protein filaments relative to one another. In cilia and flagella, the filaments are the doublet microtubules, arranged all in parallel with the same structural polarity: the (+) ends of the microtubules are at the outer tip of the axoneme. Axonemal bending is produced by forces that cause sliding between pairs of doublet microtubules. The active sliding occurs all along the axoneme, so that the resulting bends are propagated without damping.

Three kinds of experiments demonstrate that ciliary and flagellar bending is caused by outer doublet sliding. In the first (Figure 23-31a,b), electron micrographs of the tip of a cilium were taken at the end of the effective and the recovery strokes. These showed that the doublet microtubules located on the inside of the bend in the cilium protruded beyond those on the outside of the bend. This arrangement of microtubule ends would be expected if the microtubules slide past each other. (You can simulate the sliding motion if you place your hands together as in prayer, and then bend your fingers from side to side. You should immediately see that the fingers on one hand protrude past the fingers on the other hand.) Since the microtubules do not change in length, their relative displacement was exactly the distance expected from the degree of bending.

In the second kind of experiment, the sliding between doublet microtubules was also visualized in a beating flagellum by letting small gold particles attach to the surface of the axoneme (see Figure 23-27a). As a bend passes through the tagged region, the distance between the gold particles increases and then decreases. Because the length of the doublet microtubules is fixed (the doublet microtubules do not stretch or compress), then the change in distance must be caused by sliding between outer doublets.

In a third kind of experiment (Figure 23-31c), demembranated axonemes were briefly treated with proteolytic enzymes such as trypsin or elastase to digest the nexin linkages and the radial spokes. Without internal connections to hold it together, the axoneme then disintegrates upon the addition of ATP. The digested axonemes telescoped apart, with the outer doublets sliding out from the tip of the axoneme (Figure 23-31c,d). The sliding was often nearly complete, so that the resulting structure was as much as nine times longer than the original length of the axoneme.

Dynein Arms Generate the Sliding Forces

The sliding exhibited by the digested axonemes suggested that force-generating proteins were localized on the doublet microtubules and the inner and outer-arm dyneins

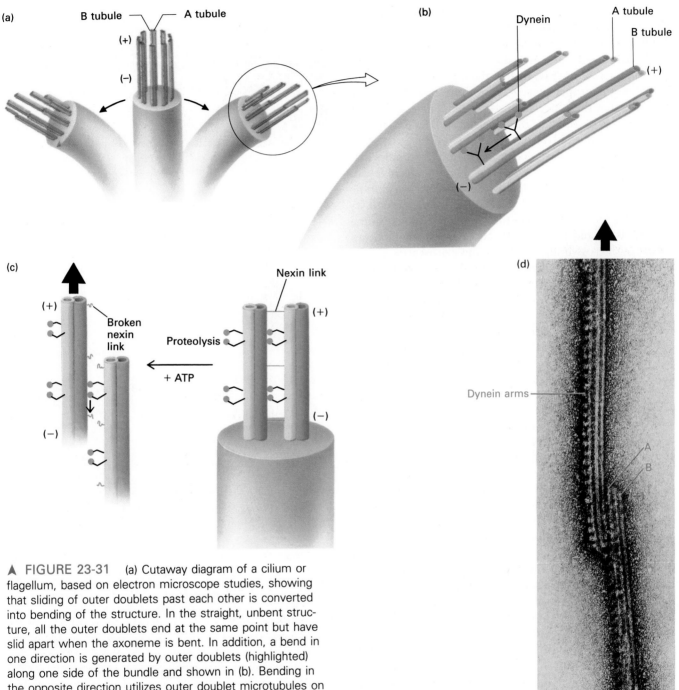

▲ FIGURE 23-31 (a) Cutaway diagram of a cilium or flagellum, based on electron microscope studies, showing that sliding of outer doublets past each other is converted into bending of the structure. In the straight, unbent structure, all the outer doublets end at the same point but have slid apart when the axoneme is bent. In addition, a bend in one direction is generated by outer doublets (highlighted) along one side of the bundle and shown in (b). Bending in the opposite direction utilizes outer doublet microtubules on the other side of the bundle. (c) The role of dynein arms in the movement of cilia and flagella is shown in this diagram of the dynein-walking model. Axonemes are treated with proteases to destroy the nexin links between doublets. Without the nexin links, the doublets slide apart freely when incubated with ATP. In the model, sliding movements of the doublet microtubules are generated when the dynein arms that are attached to the A subfiber of one axonemal microtubule, walk along the B subfiber of the adjacent doublet toward its (−) end. (d) An electron micrograph, taken after a disrupted isolated axoneme was incubated with ATP, shows that one doublet has moved relative to its neighbor. [Part (d) courtesy of P. Satir.]

which bridge between the doublet microtubules were the best candidates. The identity of the motor proteins was revealed by biochemical studies which showed that cilia and flagella possess a potent ATP-hydrolyzing activity that is associated with the dynein arms. The treatment of isolated axonemes with solutions containing high concentrations of salt removes the outer-arm dyneins. With the loss of the outer-arm dyneins, the rate of ATP hydrolysis, microtubule sliding, and beat frequency of the axoneme are all reduced by 50 percent. If the extracted outer-arm dyneins are added back to the salt-stripped axonemes, both the ATPase activity and the beat frequency are restored and electron microscopy reveals that the outer arms have reattached to the proper places. From these experiments, we conclude that the active sliding between a pair of doublet microtubules is caused by the dynein arms bound to each A tubule (see Figure 23-31c).

Based on the polarity of the doublet microtubules and the direction of sliding by doublet microtubules, we can propose a model in which the dynein arms on the A tubule "walk" along the adjacent doublet's B tubule toward its base, the (−) end. The force producing the active sliding requires ATP and is caused by successive formation and breakage of cross-bridges between the dynein arm and the B tubule. Successive binding and hydrolysis of ATP causes the dynein arms to successively break and make bonds with the adjacent doublet. We believe this model to be correct but many important details such as the mechanism of force transduction by dynein are still unknown.

Axonemal Dyneins Are Multi-Headed Motor Proteins

Axonemal dyneins are complex multimers of heavy chains, intermediate chains, and light chains (see Table 23-3). Isolated axonemal dyneins, when slightly denatured and spread out on an electron microscope grid, are seen as a bouquet of two or three "blossoms" (Figure 23-32a). Each blossom consists of a large globular domain attached to a small globular domain (the "head") through a short stalk and another stalk connects one or more blossoms to a base (Figure 23-32b)). The interpretation of these images is that the base is the site where the dynein arm attaches to the A tubule, and the small globular head is the portion of the dynein that binds the adjacent B tubule (Figure 23-32c).

Each globular head and its stalk is formed from a single long polypeptide, called the dynein heavy chain. The dynein heavy chain is enormous, approximately 4,500 amino acids in length with a molecular weight exceeding 540,000. Each heavy chain is capable of hydrolyzing ATP; based on sequences commonly found at an ATP-binding site, the ATP-binding domain is predicted to lie in the globular head portion of the heavy chain. The intermediate and light chains are thought to cluster at the base of the dynein arm. These proteins help mediate the attachment of the dynein arm to the A tubule and may also participate in the regulation of dynein activity.

The axoneme contains at least eight or nine different dynein heavy chains. The outer dynein arms are two or three-headed structures (e.g., the outer-arm dynein of the sea urchin sperm flagellum has two heavy chains; the outer-arm dynein of *Chlamydomonas* flagella has three), and inner-arm dyneins are two-headed structures.

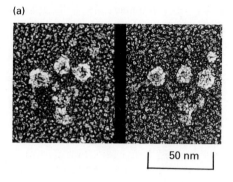

(a)

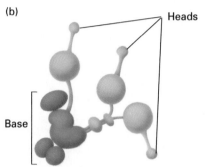

(b)

Heads

Base

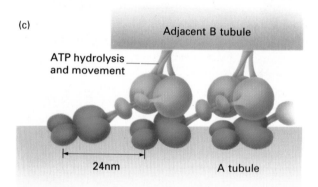

(c)

Adjacent B tubule

ATP hydrolysis and movement

24nm

A tubule

▲ FIGURE 23-32   (a) Electron micrographs of freeze-etched outer-arm dynein from *Tetrahymena* cilia, showing the three globular heads connected by stems to a common base. (b) An artist's interpretation of the electron micrographs shows the arrangement of globular domains and short stalks. Doublet microtubules are thought to bind the small heads at the tips of the terminal stalks. (c) Model showing the attachment of the outer dynein arm to the A tubule of one doublet and the formation of cross-bridges to the B tubule of an adjacent doublet. The attachment to the A tubule is stable. In the presence of ATP, the successive formation and breakage of cross-bridges to the adjacent B tubule leads to movement of one doublet relative to the other. [Part (a) from U. W. Goodenough, and J. E. Heuser, 1984, *J. Mol. Biol.* **18**:1083.]

Flagellar Beat Requires Conversion of Sliding to Bending

The flagellar beat is characterized by the propagation of bends that originate from the base or tip of the axoneme. An important point to discuss is how the active sliding between doublet microtubules is converted to bending. Formally, a bend will be generated at any point in the axoneme when there is a difference in the velocity of sliding at two points along the pair of microtubules. The difference in sliding velocity occurs through a combination of several factors. A major contribution comes from the resistance to free sliding created by some of the cross-linking structures that hold the axoneme together. Thus when the nexin and radial spokes are digested away, the internal resistances normally produced by these structures are eliminated and the axoneme, instead of bending, disintegrates when the microtubules slide. A second contribution to the difference in velocity is that at a given place along the axoneme, only some of the dynein arms are active. A close examination of the axoneme cross-section reveals that the nine outer doublets and their dynein arms are arranged in a circle so that, when viewed from the base of the axoneme, the arms all point clockwise. Thus in Figure 23-29a, the dynein arms on the left side of the cross-section (on doublets 6, 7, and 8) are facing toward the top of the page while the dynein arms on the opposite side (on doublets 2, 3, and 4) are facing toward the bottom of the page. Since the dynein arms walk in only one direction, toward the (−) end, and each doublet slides down only one of its two neighboring doublets, active sliding on one half of the axoneme produces bending toward one side and active sliding on the other half produces bending toward the opposite side. Thus by regulating the timing and location in which dynein arms are active, the axoneme propagates bends in both directions from base to tip.

Genetic studies of mutations in *Chlamydomonas* flagella which generate a non-motile phenotype reveal that the inner- and outer-arm dyneins contribute differently to the waveform and beat frequency of an axoneme. For example, the absence of one set of inner-arms affects the waveform of the beat and the absence of all of the inner arms renders the flagellum immotile. In contrast, mutants lacking outer arms have normal waveform but slower beat frequencies. Thus the outer-arm dyneins speed the active sliding of the outer doublets but do not produce the bends. In contrast, the inner-arm dyneins are responsible for producing the sliding forces that are converted to bending; this suggests that inner-arm dyneins are essential for bending.

Genetic Studies Provide Information about the Roles of the Central Microtubules and the Radial Spokes

The biflagellated, unicellular alga *Chlamydomonas reinhardtii* has proved especially amenable to biochemical and genetic studies on the function, structure, and assembly of flagella. A population of cells, shorn of their flagella by mechanical or chemical methods, provide flagella in good purity and high yield, and the deflagellated cells quickly regenerate new flagella. Analysis of the sheared flagella by two-dimensional gel electrophoresis reveals approximately 200 discrete polypeptides, in addition to the α and β-tubulins. To identify the functions of these polypeptides, scientists use standard genetic methods for isolating many viable mutants that are nonmotile and defective in flagellar function. Microscopic analysis of flagella from some of these nonmotile mutants reveals they are missing an entire substructure, such as the radial spokes (Figure 23-33a) or inner dynein arms. Many of the mutants that are missing a flagellar substructure have been found to lack certain specific proteins, so that individual proteins can be assigned to a specific substructure and associated with specific genes. For instance, 17 of the axonemal polypeptides are missing in nonmotile mutants that lack the radial spokes and spokeheads (Figure 23-33b), and 6 of these 17 proteins are also missing in mutants that lack just a spokehead. Therefore, the 6 proteins are components of the spokehead, and the other 11 contribute to the radial spoke. A similar mutational analysis has identified many of the proteins comprising the inner and outer dynein arms.

Other studies have dealt with the role of the central microtubules. In one interesting class of poorly motile cells, the C1 central microtubule and 10 nontubulin polypeptides bound to it are missing but the C2 microtubule and its 7 associated polypeptides are present. A second class of mutants missing the radial spokes or both central microtubules have paralyzed flagella and is nonmotile. Motility can be restored in these mutants by making genetic recombinants with other mutants. In some of these recombinants, the flagella are able to beat, albeit abnormally, even though the central pair or the radial spokes remain absent. Thus, the central pair of microtubules and the radial spokes are not necessary for movement, although they may play a role in regulating the activity of dynein and the switching on and off of sets of dyneins during beating. One clue to their possible function in regulating dynein comes from electron microscopic analysis of beating cilia in some organisms, such as *Paramecium*. Micrographs of different regions along the length of a cilium show that the central pair of microtubules rotates within the axoneme. One model proposes their orientation defines the direction of the bending by signalling different dynein arms to be active.

Calcium Regulates the Direction of Swimming

Swimming organisms, like *Paramecium*, *Tetrahymena*, and *Chlamydomonas*, display a very revealing behavior when they bump into something. Contact at the rear causes the organism to swim faster, but contact at the front activates an avoidance response: the organism immediately reverses direction and swims backward away from the site of contact. Analysis of the waveform of beating flagella or cilia

(a)

Wild-type —

Mutant —

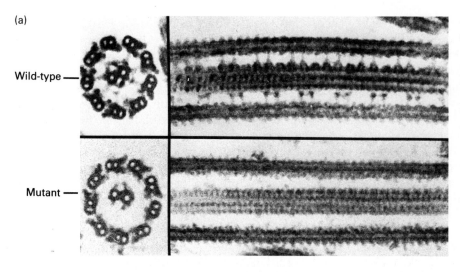

◄ FIGURE 23-33   Analysis of flagellar axonemes from a wild-type *Chlamydomonas reinhardtii* and from a paralyzed mutant. (a) Electron micrographs of isolated wild-type and mutant flagellar axonemes cut in transverse (left) and longitudinal (right) sections. Note that the mutant axonemes lack radial spokes and spokeheads. (b) Autoradiographs of two-dimensional gels of radioactive axonemal proteins. In the gel prepared from wild-type flagella, solid arrows point to 17 polypeptides that are missing in the gel prepared from the mutant flagella; these missing polypeptides are indicated by the open arrows on the gel of the mutant proteins. Tubulin subunits not resolved in the gels form the central dark streaks. [Part (a) courtesy of B. Huang and D. Luck. Reproduced from the *Journal of Cell Biology,* 1981, vol. 88, p. 73, by copyright permission of the Rockefeller University Press; part (b) courtesy of G. Piperno, B. Huang, Z. Ramanis, and D. Luck.]

(b)   Wild-type   Mutant

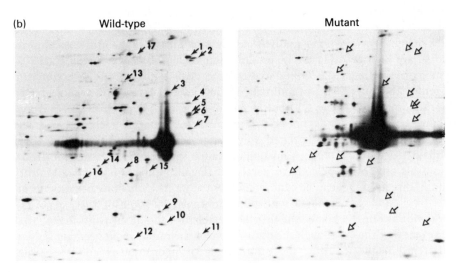

shows that the change in swimming direction can be caused by a reversal in the direction of the stroke, as in the case of the surface cilia on *Paramecium,* or by a switch in the symmetry of beating, as in the two flagella of *Chlamydomonas.* Electrophysiological recordings from swimming *Paramecium* cells show that an influx of K^+ increases the beat frequency, while an increase in the cytosolic Ca^{2+} concentration reverses the direction of the stroke. Contact induces voltage-regulated Ca^{2+} channels in the plasma membrane of the axoneme to open and extracellular Ca^{2+} crosses the membrane, raising the internal concentration of Ca^{2+}. In *Chlamydomonas,* changes in Ca^{2+} concentration mediate the switching between a symmetric waveform, which pushes the cell through the medium, and an asymmetric waveform, which pulls the cell along. How Ca^{2+} changes are coupled to changes in axonemal bending is not well understood, but the mechanism must lie within the axoneme itself, because the movement of flagella is sensitive to very small changes in Ca^{2+} concentration in the reactivation solution.

Axonemes Assemble from Basal Bodies

For microtubules to participate in movement or to act as a structural framework, they must be anchored by at least one end. As noted already, a cilium or flagellum is anchored at its cytosolic end to a basal body. In addition to its anchoring role, the basal body serves as a nucleus for the assembly of flagellar microtubules. Recall that the basal body has nine triplet microtubules (see Figure 23-30). The nine A and B tubules of these triplets appear to initiate assembly of the nine outer doublet microtubules of the cilium or flagellum by growing outward from the basal body during elongation of the axonemal shaft.

Studies on the assembly of flagella and cilia provided the first evidence that a microtubule elongates by incorporating tubulin subunits at its tip. This model was originally proven by autoradiography of cells that were regenerating their flagella in the presence of radioactive tubulin subunits (and other axonemal components). Recent studies using a recombinant *Chlamydomonas* cell (Figure 23-34) have

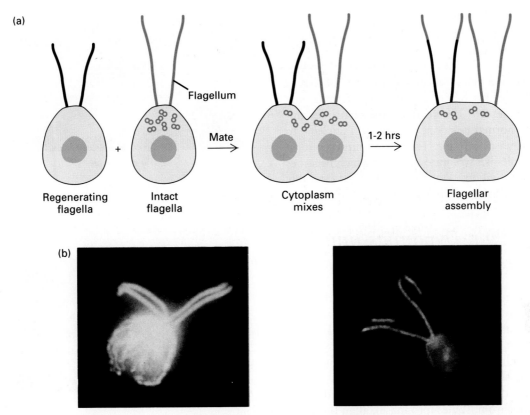

▲ FIGURE 23-34   Assembly of flagellar microtubules.
(a) Schematic diagram of an experiment in which two *Chlamydomonas* cells were mated. One cell had its flagella amputated and is regenerating them, while the other cell contains a soluble pool of tubulin tagged with a small epitope. After mating, the cytoplasm mixes in the dikaryon and the flagella continue to regenerate, now using tubulin subunits from both parental cells. (b) The first immunofluorescence micrograph shows the location of all tubulin (identified with a tubulin-specific antibody) in the mated cell. Note all four flagella are full-length. However, when similar cells are stained with the antibody specific for the epitope tag on the tubulin then the two flagella from the parental cell in which the epitope tag was expressed are labeled along their full length, but only the tips of the two regenerated flagella are labeled. This shows that the epitope-tagged tubulin subunits have added to the tips, the (+) end, of the regenerating microtubules. [Part (b) courtesy of K. Johnson and J. Rosenbaum.]

confirmed that elongation occurs at the tip not only for the tubulin subunits but also for other axonemal components. In these experiments, a *Chlamydomonas* cell expressing an epitope-tagged tubulin subunit was mated with a wild-type cell whose flagella had been amputated. After mating, which involves fusion of the two cells and mixing of their cytoplasms, the diploid cell regenerated full-length flagella by incorporating the tagged tubulin subunits. Antibodies to the epitope tag showed that the recombinant tubulin was localized within the distal tips of the regenerated flagella. This pattern could only arise if elongation occurs at the tip and not the base.

Basal Bodies Closely Resemble Centrioles

In cells that contain both a centriole and a basal body, axonemes are assembled only from the basal body. The basal body does not act as an MTOC for cytosolic microtubules, suggesting that centrioles and basal bodies are functionally different. However, the close morphologic resemblance between centrioles and basal bodies suggests that the two structures are closely related despite their functional differences. For example, just as with centriole replication, a "daughter" basal body is formed at right angles to a "parent" basal body, so that at this stage the replicating basal body resembles the paired structure of a centriole. Eventually the daughter basal body separates from the parent and then generates a new flagellum or cilium.

In some cases, basal bodies develop into centrioles with microtubule-organizing activity. In many species, a set of microtubules, called the aster, is nucleated in the fertilized egg from the basal body of the sperm flagellum a few minutes after it enters the egg. Apparently the egg in this case, although lacking the centrioles itself, contains some factor that confers microtubule-organizing activity on the sperm basal body.

Conversely, centrioles can convert into basal bodies. For instance, the precursors of the ciliated epithelial cells that line the human trachea and esophagus contain only centrioles. As these cells differentiate, the centrioles migrate from their normal position, near the cell nucleus, to the luminal plasma membrane. There the centrioles form numerous centriole-like structures, each of which becomes a basal body for a cilium. During this process, cytosolic microtubule–organizing activity is lost.

➤ Microtubule Dynamics and Motor Proteins during Mitosis

Mitosis is the process that takes newly replicated chromosomes and partitions them equally into separate parts of a cell. This is the last step in the cell cycle. In a typical animal cell, mitosis spans a one-hour period in the cell cycle. In that period, the cell builds and then disassembles a specialized microtubule structure, the *mitotic apparatus*. Larger than the nucleus, the mitotic apparatus is designed to attach and capture chromosomes (prophase), align the chromosomes (metaphase), and then separate them (anaphase) (Figure 23-35; see also Figure 5-50). Fifteen hours later, the whole process is repeated by the two daughter cells. In the life span of an organism, there are billions of cell divisions, and remarkably few errors occur. Mistakes in mitosis would lead to missing or extra chromosomes and this re-

➤ FIGURE 23-35 The organization of chromosomes and microtubules during mitosis. Fluorescence micrographs of mitotic cultured PtK2 fibroblasts stained with a fluorescent anti-tubulin antibody (green) and the DNA-binding dye ethidium homodimer (blue); left and right photographs in each panel show the same cell. (a) During early prophase, the nucleus is surrounded by interphase microtubules. Note that the centrosomes have replicated (most easily visible in the right photograph as two dots near the center of the nucleus). (b) By late prophase, the nuclear membrane has broken down and the chromosomes have condensed. The centrosomes have migrated to the poles of the developing spindle. The microtubules radiate from the poles. (c) At metaphase, the chromosomes have aligned midway between the poles to form the metaphase plate. The dense bundles of microtubules connects the chromosomes to the poles. (d) During late anaphase the chromosomes are pulled to the poles. Afterwards during telophase (not shown), the cell is pinched in half to form two daughter cells, each with a complete set of chromosomes. [From J. C. Waters, R. W. Cole, and C. L. Rieder, 1993, *J. Cell Biol.* **122**:361; courtesy of C. L. Rieder.]

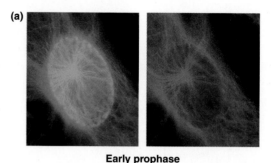

Early prophase

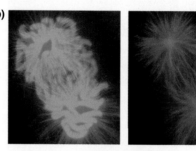

Late prophase

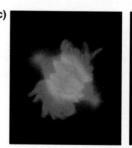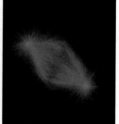

Metaphase

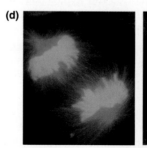

Anaphase

sults in abnormal patterns of development when it occurs during embryogenesis and causes pathologies when it occurs after birth. To ensure that mitosis is fail-safe, a highly redundant mechanism has evolved in which each crucial step is carried out concurrently by microtubule motor proteins and microtubule assembly dynamics.

The mitotic cycle of chromosome separation is linked to two other parts of the cell cycle: the replication of centrosomes and the synthesis of DNA. In Chapter 5, we were introduced to the cell cycle (see Figure 5-48) and the major steps of mitosis (see Figure 5-50). Later, in Chapter 25, we will return to discuss the details of the cell cycle, especially the regulatory steps which control its timing. In this section, we will focus on the mechanism by which microtubules separate chromosomes during mitosis in a "typical" animal cell. Studies of mitosis in animal cells have provided most of the information about the kinematics and force generation of mitosis, the actual motions and the locations where forces are generated to cause movements.

The Mitotic Apparatus Is a Microtubule Machine for Separating Chromosomes

The mitotic apparatus has no fixed structure: it is constantly changing during mitosis. For one brief moment at metaphase, however, when the chromosomes are aligned at the equator of the cell, the mitotic apparatus remains static. We will begin our discussion by examining the structure of the mitotic apparatus at metaphase and then discuss how it first organizes chromosomes during prophase, how it separates chromosomes during anaphase, and how it determines where cells are cleaved during telophase.

The mitotic apparatus at metaphase (Figure 23-36; see also Figure 23-35c) is organized into two parts: (1) a central *spindle*—a bilaterally symmetric bundle of microtubules with the overall shape of a football, but divided into opposing halves at the equator of the cell by a plate of metaphase chromosomes—and (2) a pair of *asters*—a tuft of microtubules at each pole of the spindle.

In each half of the spindle, a single *centrosome* at the pole organizes three distinct sets of microtubules, whose (−) ends all point toward the centrosome. One set, the *astral microtubules,* forms the aster; they radiate outward from the centrosome toward the cortex of the cell, where they help to position the mitotic apparatus and later help to determine the cleavage plane during cytokinesis. The other two sets of microtubules comprise the spindle. They also originate from the centrosome but extend inward toward the equator of the cell. A second set, the *kinetochore microtubules,* attach to chromosomes at specialized attachment sites on the chromosomes called *kinetochores.* The third set, *polar microtubules,* do not interact with chromosomes but instead interdigitate with polar microtubules from the opposite pole.

In an overlap zone at the equator, two types of interactions hold the spindle halves together to form the bilaterally symmetric mitotic apparatus: (1) lateral interactions between the (+) ends of the interdigitating polar microtubules from each pole, and (2) end-on interactions between the kinetochore microtubules from each pole and the kinetochores of the sister chromatids.

The spindle-pole-aster organization of the mitotic apparatus is basic to mitosis in all organisms, but the appearance of the mitotic apparatus can vary widely. The number of microtubules in a spindle, the overall size of the mitotic apparatus, and the timing and duration of mitotic movements all vary among different organisms (Figure 23-37). In addition, organisms differ in the length or number of their astral microtubules—in some organisms the asters are absent altogether (anastral spindles).

In a simple organism like yeast, mitosis is carried out by a structurally simple mitotic apparatus which lacks centriole-based centrosomes and asters (Figure 23-37b; see also Figure 5-52). Instead of a centrosome, the microtubules are organized around a *spindle pole body,* a trilaminated structure in the nuclear membrane. Furthermore, because a yeast cell is small, it does not require well-developed asters to assist in mitosis. Thus, the yeast mitotic apparatus is simplified to just a spindle which itself is constructed from a minimal number of kinetochore and polar microtubules. The common bakers yeast, *S. cerevisiae,* contains 16 chromosomes. Its spindle contains 32 kinetochore microtubules plus a few polar microtubules. (At metaphase, each kinetochore is attached to one microtubule and so each chromosome is attached to two kinetochore microtubules.) In contrast, in an animal cell mitotic apparatus, each kinetochore is connected to many microtubules and hundreds of astral microtubules radiate from a centrosome.

The Kinetochore Is a Specialized Attachment Site at the Chromosome Centromere

The sister chromatids of a metaphase chromosome are transported to each pole along the kinetochore microtubules. In terms of cargos delivered on microtubules, a chromosome is a much different piece of baggage than the membrane vesicles and organelles we have discussed elsewhere. The chromosome is a highly condensed nucleoprotein particle containing two coiled sister chromatids. The kinetochore microtubules attach to the surface of a chromatid at a *kinetochore,* a plate-like structure lying within a small, highly specialized region of the chromosome called the *centromere* (Figure 23-38). (Do not confuse the centromere with a centrosome: the centromere is a region of the chromosome, while the centrosome is a microtubule orga-

(a)

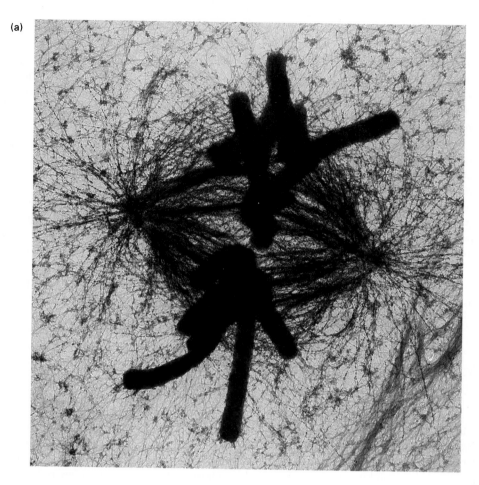

(b)

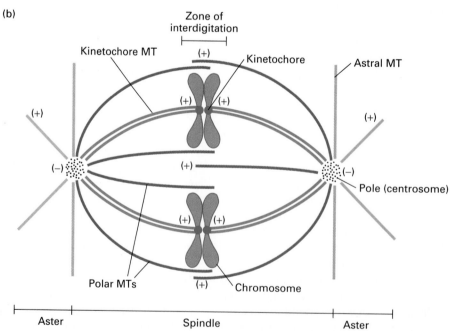

▲ FIGURE 23-36   Metaphase spindle. (a) A high-voltage electron micrograph of a mammalian mitotic apparatus. To visualize the spindle microtubules more clearly, biotin-tagged anti-tubulin antibodies were added to make microtubules more massive. (b) Diagram showing the three sets of microtubules (MTs) in a mitotic apparatus. Centered around the poles are astral microtubules, kinetochore microtubules, which are connected to chromosomes (blue), and polar microtubules. The (+) ends of these microtubules all point away from the centrosome at each pole. [Part (a) courtesy of J. R. McIntosh.]

(a)

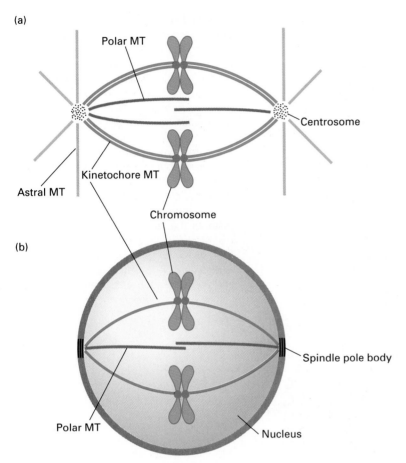

Polar MT

Centrosome

Kinetochore MT

Astral MT

Chromosome

(b)

Spindle pole body

Polar MT

Nucleus

◄ FIGURE 23-37   Variations of mitosis. (a) The most common variation in the organization of the mitotic spindle is in the size of the asters, which vary from nonexistent (anastral spindles) to small asters. (b) In fungi, the nucleus remains intact during mitosis and instead of centrosomes the poles contain spindle pole bodies, which are attached to the nuclear membrane. Thus the chromosomes are isolated from direct interaction with the cytosol during mitosis.

nizing center.) The centromere is easily recognized in the microscope as a constriction in the chromatin where the sister chromatids are most closely associated and where kinetochore microtubules are attached (Figure 23-38). The location of the centromere and hence that of the kinetochore is directly controlled by a unique segment of chromosomal DNA termed *centromeric DNA*. In animal cells the importance of the centromere is seen in the fragments of a chromosome that are occasionally generated by drugs or x-irradiation. If the fragment lacks a centromere, then it cannot be incorporated into a metaphase spindle and is lost during mitosis.

The kinetochore is first recognizable during late prophase, after the chromosomes have condensed but well before the mitotic apparatus has assembled. The structure of the kinetochore is complex but is analogous to an endplate to which a post (a microtubule) is attached. This is most clearly seen by electron microscopy. In ultra-thin sections of chromosomes, the kinetochore is seen as a stack of disc-like plates (see inset of Figure 23-38). The innermost disk probably consists of chromatin that is condensed differently than the surrounding heterochromatic chromatin. The outer disk is a fibrous structure where kinetochore

microtubules attach by their (+) ends. A 25- to 30-nm layer, the middle layer, separates the inner and outer disk. A series of fine filaments which bridge the middle layer may help hold the two disks together. Finally, covering the surface of the kinetochore are fibers in which (−) end-directed microtubule motors are thought to reside.

We do not yet know how the end of a microtubule attaches to the kinetochore but because the kinetochore is the physical interface between microtubules and DNA, some of the components of the kinetochore must bind DNA or microtubules. The identities of some of these components have been uncovered by their reaction with antibodies that specifically recognize kinetochores. For unknown reasons, these antibodies are frequently produced by patients suffering from scleroderma (an autoimmune disease of unknown origin that causes fibrosis of connective tissue). Such antibodies identify four proteins in the mammalian centromere which are localized to the inner kinetochore disk. No proteins in the outer disk have been identified, but other antibodies have detected cytoplasmic dynein in the fibers on the surface of the kinetochore.

The anti-kinetochore antibodies are also used as immunofluorescence reagents for visualizing the movements

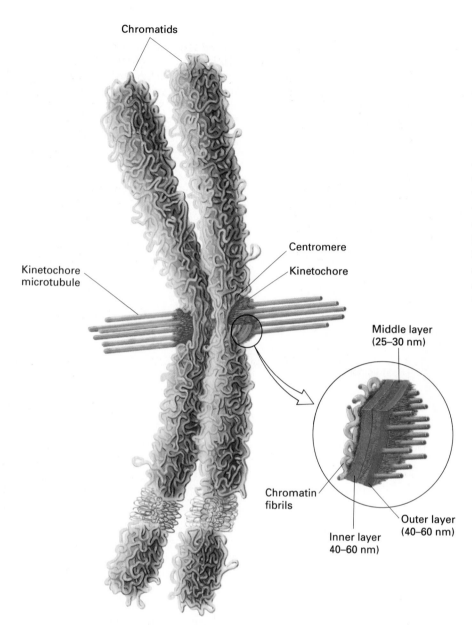

Chromatids

Kinetochore
microtubule

Centromere

Kinetochore

Middle layer
(25–30 nm)

Chromatin
fibrils

Inner layer
40–60 nm)

Outer layer
(40–60 nm)

◀ FIGURE 23-38   Schematic diagram of a metaphase chromosome, showing the kinetochore microtubules and the three-layer kinetochore characteristic of animals and lower plants (inset); each kinetochore fiber contains many microtubules. Note that the sister chromatids have not yet separated. The (+) ends of microtubules insert into the outer layer of each kinetochore and the microtubules extend toward one of the two poles of the cell. At anaphase, the sister chromatids separate, and the chromosomes are pulled to opposite poles of the cell by the kinetochore microtubules. [From B. R. Brinkley, A. Tousson, and M. M. Valdivia, 1985, in *Aneuploidy*, V. L. Dellarco, P. E. Voyter, and A. Hollaender, eds. Plenum, p. 243.]

of kinetochores in mitotic cells. For example, in metaphase cells, kinetochores stained with such an antibody appear as dots (Figure 23-39a). The treatment of interphase cells with caffeine causes kinetochores to detach from the rest of the chromosome. During mitosis, the free kinetochores in the caffeine-treated cells can be seen to move poleward along the kinetochore microtubules (Figure 23-39b,c), suggesting that only kinetochores are necessary for the mitotic movements of chromosomes.

Yeast Centromeres Bind a Single Microtubule

Much of the information about the structure of centromeres comes from genetic studies of the simple centro-

meres in yeast. Unlike the well-organized platelike mammalian kinetochore, a yeast centromere is too small to be studied by electron microscope. We have previously seen that centromeric (CEN) DNA sequences from yeast chromosomes can be identified by their ability to make self-replicating plasmids into artificial chromosomes that can be passed from mother to daughter at mitosis (see Figure 9-58b). Sequence analysis of cloned CEN DNAs from yeast chromosomes reveals that they are generally organized into three regions, denoted CDEs, or *centromere DNA elements*, I, II, and III (Figure 23-40). Of the three regions, mutational analysis implicates CDE II and CDE III as the most critical for centromere function; CDE I is conserved in sequence but is not required. CDE II and CDE III appear to mediate interactions with microtubules through

(a)   (b)   (c)

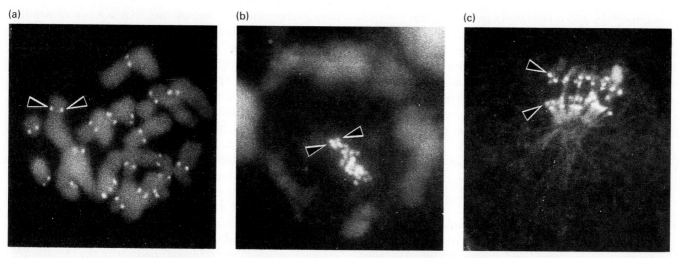

▲ FIGURE 23-39   Kinetochores detached experimentally from chromosomes to show normal movement during mitosis. Kinetochores can be detached from the rest of the chromosome by treatment of cultured mammalian cells with caffeine while they are in the G_1, G_2, or S stages of the cell cycle. (a) A preparation of untreated metaphase Chinese hamster ovary (CHO) cell chromosomes stained with an anti-kinetochore antibody (arrowheads) and with a DNA-binding dye that stains the chromosomes. (b) CHO cells after caffeine treatment; kinetochores (arrowheads) have become detached from the chromosomes. (c) An anaphase caffeine-treated cell exhibiting kinetochores (arrowheads) moving poleward, detached from the rest of the chromosome. [See B. R. Brinkley et al., 1988, *Nature* **336**:251; courtesy of B. R. Brinkley.]

CBF2 and CBF3 (centromere binding factors 2 and 3), respectively, and through the kinesin-related protein Kar3p and other identified proteins. Several lines of study suggest that these proteins are an important link between the kinetochore microtubules and the centromere. In vitro binding studies show that CBF3 and Kar3p can bind centromeres

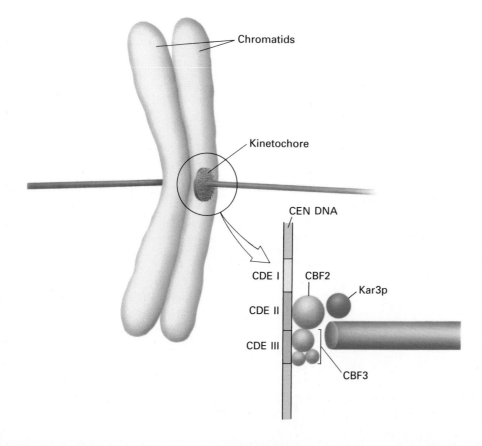

◄ FIGURE 23-40   The structure of a yeast centromere and kinetochore. The centromeric DNA (CEN DNA) is divided into three contiguous segments (CDE I–III). Two groups of centromere-binding factors, CBF2 and CBF3, are proteins associated with CDE II and CDE III, respectively. The CBFs mediate the attachment of a single microtubule to the centromere. Other proteins, including a putative (−) end–directed motor protein, kinesin-related protein (Kar3p), form the rest of the kinetochore. During anaphase, two events at the kinetochore, depolymerization at the (+) end of the microtubule and (−) end movement, by KAR3p combine to pull the chromosome to the pole.

to microtubules and then translocate toward the (−) ends of the microtubules. Conversely, mutations in CBF3 proteins affect the ability of chromosomes to sort properly during mitosis. These observations suggest that CBF3, possibly in association with microtubule motor proteins, binds and directs the movement of chromosomes during metaphase or anaphase. Later in the section, we will discuss the role of kinesin-related motor proteins in this process.

Centrosome Duplication and Migration During Interphase and Prophase Initiate the Assembly of the Mitotic Apparatus

The assembly of a mitotic apparatus is linked with the *centriole cycle* (Figure 23-41), a series of steps in which the centrosomes are duplicated and then redistributed to opposite halves of the cell. In some respects, mitosis can be understood as a migration of duplicated centrosomes, which along their journey pick up chromosomes, pause in metaphase, and recontinue at anaphase to new locations in the daughter cells, where they release the chromosomes and organize the cytoplasmic microtubules.

The centriole cycle begins during interphase. The centrosome duplicates just before S phase, but the two resulting daughter centrosomes remain paired until mitosis. We can see the major stages in centrosome duplication in electron micrographs of centrosomes that contain centrioles (Figure 23-42a). Recall that in a centrosome, a pair of barrel-like centrioles are oriented at right angles and are connected by thin fibrils. This orthogonal orientation is lost during S phase and a small bud develops on the side of each centriole. As the cell progresses through S phase, the bud elongates; by G2, two pairs of full-sized centrioles are present (see Figure 23-41).

The next stage in the centriole cycle (see Figure 23-41), centrosome migration, occurs during prophase: the two pairs of centrioles separate and each pair migrates to the opposite side of the nucleus, establishing the bipolarity of the dividing cell. Centrosome migration begins with the assembly of astral and polar microtubules. Some of the microtubules extend from one centrosome to the other centrosome, while others terminate midway in the spindle. We will discuss the mechanism for centrosome (or spindle pole body) migration in the next section.

At the end of the migration stage, the nuclear membrane has broken down but the condensed chromosomes have not yet attached to spindle microtubules. Located at opposite ends of the cell, the centrosomes organize the chromosomes into the mitotic apparatus and execute the steps in mitosis. When mitosis is completed, the centrosome becomes the MTOC and initiates the assembly of cytosolic microtubules in the interphase cell.

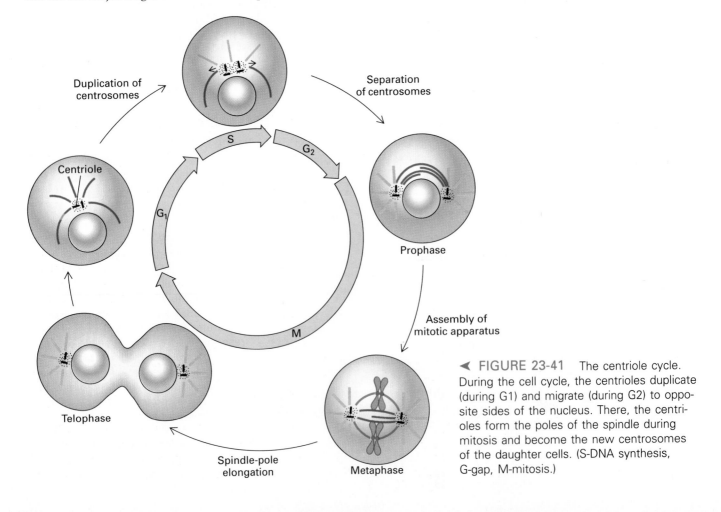

◀ FIGURE 23-41   The centriole cycle. During the cell cycle, the centrioles duplicate (during G1) and migrate (during G2) to opposite sides of the nucleus. There, the centrioles form the poles of the spindle during mitosis and become the new centrosomes of the daughter cells. (S-DNA synthesis, G-gap, M-mitosis.)

(a)

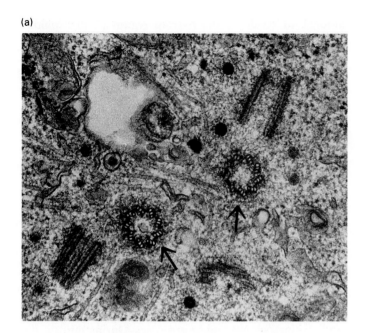

(b)

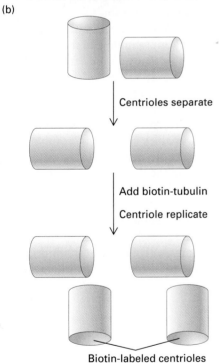

Centrioles separate

Add biotin-tubulin

Centriole replicate

Biotin-labeled centrioles

◄ FIGURE 23-42 (a) Electron micrograph of a pair of newly replicated centrioles (arrows) in a Chinese hamster ovary cell. In this family portrait, both parental centrioles and daughter centrioles are shown. (b) The replication of centrioles begins when the pair of centrioles separate. In the next step, a daughter centriole buds from the side of a parent. In the resulting pair of centrioles, one centriole is old (the parent) and the other centriole is new (the daughter). This model of replication is supported by experiments in which replication proceeds in the presence of biotin-labeled tubulin subunits. The biotin-labeled tubulin is found in one centriole of the pair indicating that this centriole assembles de novo rather than from fission of an old centriole and regrowth. [Part (a) from M. McGill et al., 1976, *J. Ultrastruct. Res.* **57**:43; courtesy of W. Brinkley.]

rated the modified tubulin subunits into only one centriole of each pair of centrioles. As the cell proceeded through mitosis, the labeled centrioles separated and migrated to opposite ends of the cell to form the poles of the mitotic apparatus. When replication was completed, only two and not four biotin-labeled centrioles were detected in the cell.

During Prophase, Kinesin-Related Proteins and Cytoplasmic Dynein Participate in the Movements of Kinetochores and Centrosomes

In the early stages of mitosis that lead to metaphase, both microtubule dynamics and microtubule motor proteins participate in the assembly of the mitotic apparatus—reorganization of the spindle microtubules, separation of centrosomes, and capture and alignment of chromosomes. Mitosis researchers are actively studying the location of the forces involved in centering chromosomes at the equator because the same forces are probably involved in the subsequent chromosome separation that brings chromosomes toward opposite poles. We have already discussed how some microtubule-dependent movements in the cytosol are generated by microtubule dynamics or by microtubule motor proteins. As we will discuss in the following sections, movements during mitosis involve both mechanisms—perhaps as a way to ensure a fail-safe mechanism for separating chromosomes. We will start with the role of microtubule motor proteins in mitosis and then discuss the involvement of microtubule dynamic instability.

Mitosis uses its own special versions of the motor protein kinesin, called *kinesin-related proteins (KRPs)*. Genetic and cell biology studies, primarily in yeast and flies, have implicated several KRPs in the separation and migration of centrosomes or spindle pole bodies and the poleward movement of chromosomes. Within the family of KRPs, some members probably power movements of cen-

A budding type of mechanism predicts that centrioles can be distinguished by their age: of the pair of centrioles, one is newly synthesized while the other is one cell cycle older. This suggests that centrosomes replicate by a semiconservative process. This type of replication mechanism was confirmed by experiments (Figure 23-42b) in which biotin-labeled tubulin was injected into a cell before centriole duplication. As the centrioles replicated, they incorpo-

trosomes (Figure 23-43), while others power movements of kinetochores (or chromosomes). The list of KRPs is large and still growing, as counterparts are discovered in model organisms such as the yeast, worm, fly, frog, and mouse. We will not attempt to list the KRPs except by broad classes according to their role in mitosis. Like kinesin, most KRPs are (+) end-directed motors; however, one unusual class of KRPs consists of (−) end-directed motors that are not dynein (Table 23-3).

Antibody microinjection experiments suggest that the (+) end-directed KRPs, are involved in the separation and migration of centrosomes and spindle pole bodies during mitosis (and meiosis). This is supported by preliminary studies of animal cells which show that antibodies against a (+) end-directed KRP will inhibit the formation of a bipolar spindle when microinjected into a cell before but not after prophase. A (+) end-directed motor protein is

thought to reside in the overlap region of the spindle, where it cross-links antiparallel polar microtubules (Figure 23-43b). When bound to a microtubule by a microtubule-binding domain, the motor domain is free to bind a neighboring microtubule, where it would stabilize the spindle microtubules.

Other motor proteins are also implicated in centrosome migration: antibodies detect cytosolic dynein at the centrosome and cortex of animal cells; when these antibodies are microinjected into animal cells, mitosis is blocked.

Assembly of the Mitotic Apparatus Involves Dynamic Microtubules

From our earlier discussions, we learned that dynamic instability is a characteristic of microtubules in vivo: they undergo rapid fluctuations in length (see Figure 23-12). Microtubules display the greatest instability during mitosis, and microtubule dynamics could be one mechanism for moving kinetochores and centrosomes. During mitosis, the first indication of a change in microtubule stability occurs at prophase, when long interphase microtubules disappear and are replaced by the mitotic spindle and asters (Figure 23-44). Mitotic microtubules that are nucleated from the newly replicated centrosomes are more numerous, shorter, and less stable than interphase microtubules. The average lifetime of a microtubule decreases from 10 minutes in interphase cells to 30 seconds in the mitotic spindle. This increase in the turnover enables spindle microtubules to assemble and disassemble more quickly during mitosis.

The rapid fluctuation in the length of a microtubule is an unusual mechanism employed by a kinetochore microtubule to capture a chromosome during prophase. By quickly lengthening and shortening at its (+) end, the dynamic microtubule acts like a poker which can probe into a chromosome-rich environment (Figure 23-45). If the growing microtubule does not encounter a kinetochore, it re-

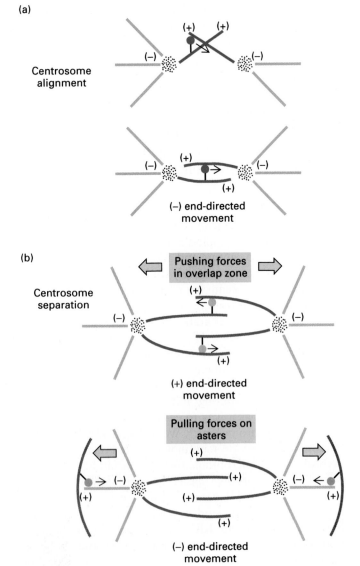

(a) Centrosome alignment

(+) (+)
(−) (−)

(−) (−)
(+)
(+)
(−) end-directed movement

(b) Centrosome separation

Pushing forces in overlap zone

(+)
(−) (−)
(+)
(+) end-directed movement

Pulling forces on asters

(+)
(−) (+) (−)
(+) (+)
(+)
(−) end-directed movement

◀ FIGURE 23-43 Roles for microtubule motor proteins in centrosome movements during mitosis. (a) During late prophase centrosomes are aligned by (−)end–directed motors (dark green) which pull on polar microtubules and thus align them. (b) The family of (+) end–directed KRPs (pink), located between the spindle poles, is involved in the separation of poles—beginning after centrosome duplication at prophase, continuing through the formation of the spindle poles, and ending, as we shall see later, with spindle pole separation at anaphase (see Figure 23-50). In addition, a (−) end–directed force exerted by cytoplasmic dynein (light green) located at the cortex may pull asters toward the poles in combination with the (+) end–directed motors in the spindle.

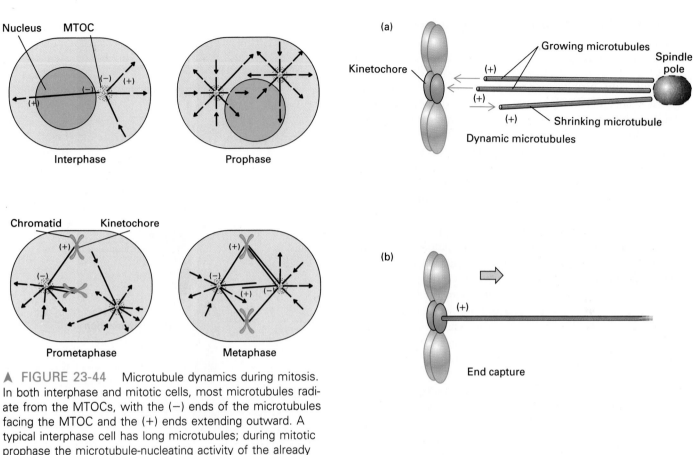

▲ FIGURE 23-44 Microtubule dynamics during mitosis. In both interphase and mitotic cells, most microtubules radiate from the MTOCs, with the (−) ends of the microtubules facing the MTOC and the (+) ends extending outward. A typical interphase cell has long microtubules; during mitotic prophase the microtubule-nucleating activity of the already replicated MTOCs increases; and this leads to a larger number of shorter, more dynamic microtubules. In late prophase, some of the microtubules interact with kinetochores (dark blue), causing the microtubules to be partially stabilized. In the metaphase mitotic apparatus, astral microtubules elongate and shorten.

tracts by shortening, only to lengthen again but in a different direction. In this way, a dynamic microtubule pokes around, sampling a large volume of the cytoplasm, until it locates a kinetochore. Sometimes the microtubule scores a "bull's-eye" and its end contacts the kinetochore, as in Figure 23-45b, but usually binding will instead occur along the side of a microtubule and the chromosome will slide along the microtubule to the (+) end, as in Figure 22-45c. When the chromosome reaches the (+) end, its kinetochore "caps" the microtubule, preventing it from depolymerizing: studies have shown that if a kinetochore is detached from its kinetochore microtubule by a fine needle, the microtubule rapidly depolymerizes from the (+) end. This process of kinetochore capture is unusual because we would normally expect the connection between a kinetochore and microtubule to arise from microtubule polymerization. Instead, the kinetochore is grabbed by a microtubule. This mechanism requires cross-linking proteins on the microtubule, on the kinetochore, or on both.

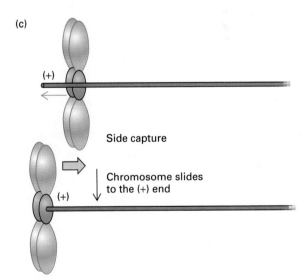

▲ FIGURE 23-45 Dynamic instability and the capture of chromosomes. (a) During mitotic prophase, some kinetochore microtubules are growing at their distal (+) end, while others are shrinking rapidly from their distal end. (b) In late prophase some of the microtubules interact with kinetochores (dark blue), causing those microtubules to be stabilized and to be pulled to the metaphase plate. (c) In addition, some microtubules just miss the kinetochore but the kinetochore binds to the side of the microtubule and then slides to the (+) end.

At Metaphase Forces at the Kinetochore Move Chromosomes to the Equator of the Spindle

During late prophase (prometaphase), the newly condensed chromosomes, having attached to their kinetochore microtubules, *congress* or move to the equator of the spindle. Along the way, the chromosomes exhibit *salatory behavior*, oscillating between movements toward and then away from the equator.

The microtubule linkage between kinetochores and poles can be modified experimentally. The prometaphase chromosomes in certain large cells can be micromanipulated with fine glass needles. When the chromosomes are rotated by 180° with respect to the axis of the spindle, the attachment between the pole and the chromosomes is broken. Now oriented toward the opposite pole, the pair of chromatids, however, soon reattach to microtubules from the new pole and later, at anaphase, both are pulled into the same daughter cell. These and other studies indicate that some force pulls the two kinetochores on sister chromatids toward opposite poles (Figure 23-46). Micromanipulation experiments suggest that the strength of the force is proportional to the distance from the chromosome to the pole; if a metaphase chromosome is displaced toward one pole by micromanipulation, then the force exerted from the opposite pole momentarily increases and quickly pulls the displaced chromosome back to the equator. An alternative explanation is that the pole closest to the chromosome exerts a pushing force that restores the chromosome to the equator. Whatever mechanism is in operation, these opposing forces are balanced when a chromosome is at the equator of the spindle, so the chromosome remains stationary there.

By metaphase, the chromosomes are aligned at the equator, their position fixed midway between each pole of the spindle. Although the lengths of kinetochore and polar microtubules have stabilized (Figure 23-44), there continues to be a flow, or treadmilling, of subunits through the microtubules toward the poles (Figure 23-46b), but the loss of subunits at the (−) end is balanced by the addition of subunits at the (+) end.

During Anaphase Chromosomes Separate and the Spindle Elongates

The same forces which form the spindle during prophase and metaphase also direct the separation of chromosomes toward opposite poles at anaphase. Anaphase is divided into two distinct stages, *anaphase A* and *anaphase B*. Anaphase A is characterized by the shortening of kinetochore microtubules, which pulls the chromosomes toward the poles. During the second stage, anaphase B, the two poles move further apart, bringing the chromosomes with them into what will become the two daughter cells.

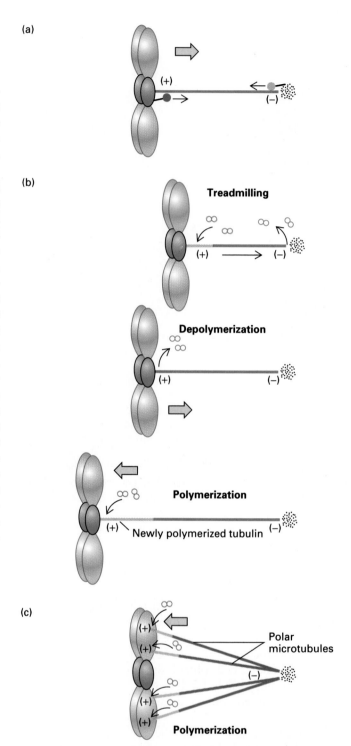

▲ FIGURE 23-46   Suggested alternative mechanisms for chromosome movements. (a) A (−) end–directed motor (dark green) attached to the kinetochore or a (+) end–directed motor (pink) attached to the pole could both generate forces that pull (arrow) the chromosome poleward. (b) The flow of tubulin subunits through kinetochore microtubules can be used to pull or push the chromosome toward the pole, especially if rapid polymerization or depolymerization occurs at the (+) end of a microtubule. (c) Non-kinetochore microtubules can push on the chromatid arms of a chromosome by polymerization at their (+) ends.

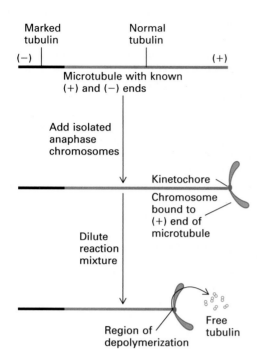

Marked tubulin
Normal tubulin

Microtubule with known (+) and (−) ends

Add isolated anaphase chromosomes

Kinetochore

Chromosome bound to (+) end of microtubule

Dilute reaction mixture

Region of depolymerization

Free tubulin

◄ **FIGURE 23-47** In vitro experiment demonstrating poleward movement of anaphase chromosomes. A population of microtubules is prepared in vitro, using both "marked" and normal tubulin, in such a way that their (−) and (+) ends can be distinguished. When these microtubules are mixed with purified anaphase chromosomes, the chromosomes bind only to the (+) ends of the microtubules, which in cells would face the equator. When the reaction mixture is diluted, to lower the concentration of free (unpolymerized) tubulin, the chromosomes move toward the (−) ends of the microtubules, which face the pole of the cell. Simultaneously, only the length of the unmarked portion of the microtubules decreased, indicating that the microtubules had depolymerized from the (+) ends, just behind the kinetochores. Since the reaction mixture contains no source of energy, such as ATP, it is depolymerization of microtubules at the kinetochore that both regulates and powers poleward chromosome movement during anaphase. [Adapted from D. E. Koshland, T. J. Mitchison, and M. Kirschner, 1988, *Nature* **331**:499.]

In part because anaphase A and anaphase B can be studied in certain cell-free extracts, we understand a good deal about their molecular mechanisms; we shall discuss the key experiments and concepts in the next two sections.

Microtubule Shortening during Anaphase A Chromosomes move toward the poles on shortening kinetochore microtubules. In vitro studies have indicated that the depolymerizing of microtubules can generate sufficient force to move chromosomes. In one such study (Figure 23-47), purified microtubules were mixed with purified anaphase chromosomes, and as expected, the kinetochores bound preferentially to the (+) ends of the microtubules. To induce depolymerization of the microtubules, the reaction mixture was diluted, thus lowering the concentration of free tubulin dimers. Video microscopy analysis then showed that the chromosome moved along the depolymerizing microtubule toward the (−) end, at a rate similar to that of chromosome movement in intact cells. Since no ATP (or any other energy source) was present in these experiments, chromosome movement toward the (−) end must be powered, in some way, by microtubule disassembly at the kinetochore and must not involve microtubule motor proteins. In vivo fluorescence tagging experiments provide additional evidence that microtubules do depolymerize when chromosomes are moving toward the pole. As shown in the experiment in Figure 23-48, when fluorescently-labeled microtubules in a live cell are photobleached during anaphase in the region between the poles and the chromosomes, the kinetochores move toward the bleached

region of the microtubules. The flux toward the poles is faster than the flux measured during metaphase. Because the kinetochore microtubules remain stationary, they must disassemble mainly from their (+) ends but also to a lesser extent from the poles.

Spindle Elongation during Anaphase B Anaphase B is an elaborate process that involves sliding between polar microtubules, elongation of the polar microtubules, and pulling forces exerted by the cellular cortex on astral microtubules. During anaphase B, microtubule motor proteins are clearly involved in the separation of the spindle poles (see Figure 23-43b), because anaphase B movements can only be reactivated in detergent-permeabilized cells when ATP is present. Detergent treatment of mitotic cells, which causes ATP to leak out, does not affect poleward chromosome movement (anaphase A). The best example demonstrating anaphase B movements in vitro comes from studies of mitotic spindles isolated from diatoms (Figure 23-49). Unlike the loose bundle of spindle microtubules found in most animal and plant cells, the spindles of diatoms are almost crystalline in arrangement and are stable to the rigors of isolation. In the presence of ATP, isolated diatom spindles elongate, simulating anaphase B; the zone of overlap between the two half-spindles decreases in length as the spindle elongates by a distance similar to the original length of the overlap. These and other studies suggest that overlap zone interactions between microtubules from opposite half-spindles are what generate the force of anaphase B.

Early anaphase

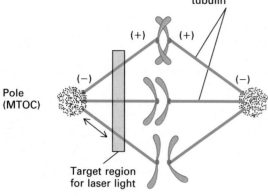

Early anaphase

Kinetochore microtubules containing fluorescent tubulin

(+)   (+)

Pole (MTOC)   (−)   (−)

Target region for laser light

◄ FIGURE 23-48 Experiment demonstrating that chromosomes move poleward along stationary kinetochore microtubules, which coordinately disassemble from their kinetochore ends during anaphase A. Fibroblasts are injected with fluorescent tubulin and then allowed to enter metaphase, so that all of the microtubules are fluorescent. Only the kinetochore microtubules are shown. In early anaphase, a band of microtubules (yellow box) is subjected to a laser light, which bleaches the fluorescence but leaves the microtubules continuous and functional across the bleached region. The bleached segment of each microtubule thus provides a marker for the fate of that part of the polar microtubule. During anaphase the distance of the bleached zone from the poles (measured in diagrams by the black double-headed arrows) does not change, indicating that no depolymerization of the microtubules occurs at the poles. Rather, the kinetochore microtubules disassemble just behind the kinetochore, and the kinetochores move poleward along the microtubules. [Adapted from G. J. Gorbsky, P. J. Sammak, and G. Borisy, 1987, *J. Cell Biol.* **104**:9; and G. J. Gorbsky, P. J. Sammak, and G. Borisy, 1988, *J. Cell Biol.* **106**:1185.]

Expose to laser light

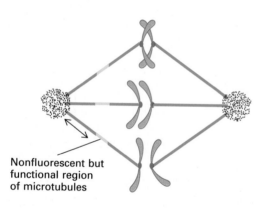

Nonfluorescent but functional region of microtubules

Kinetochore microtubules depolymerize at (+) ends and kinetochores move toward poles

Anaphase

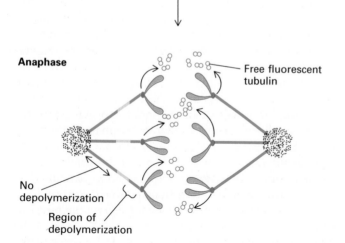

Free fluorescent tubulin

No depolymerization

Region of depolymerization

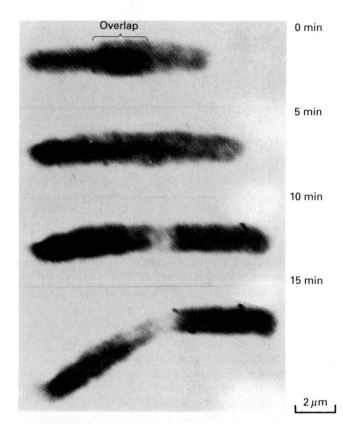

▲ FIGURE 23-49 Anaphase B movements in vitro. An isolated diatom spindle was incubated with a solution containing ATP and was viewed under the polarization microscope before (0 min) and after 5, 10, and 15 min of incubation. Note the decrease in birefringence of the central overlap segment as the two half-spindles slide apart. Sliding of the half-spindles requires ATP hydrolysis and is thought to involve a kinesin-like molecule motor. [See W. Z. Cande and K. L. McDonald, 1985, *Nature* **316**:168; courtesy of W. Z. Cande.]

If we were to analyze the direction of microtubule movement during anaphase B, we would find that adjacent microtubules migrate in the direction of their pole-facing (−) ends. This polarity of movement suggests that a (+) end-directed KRP can be responsible for generating the force for centrosome separation during anaphase B. In one model, a KRP attached to a microtubule in the overlap region would walk toward the (+) end of a neighboring but anti-parallel microtubule and thus push adjacent microtubules in the direction of their (−) ends (Figure 23-50). This model is supported by experiments in which antibodies raised against a conserved region of the kinesin superfamily inhibit ATP-induced elongation of diatom spindles in vitro. Control experiments show that normal spindle elongation is observed when the isolated spindles are preincubated with both the antibody and the kinesin antigen peptide. The involvement of a kinesin-like protein and the requirement for ATP hydrolysis are two strong pieces of evidence that a KRP could be responsible for anaphase B movements.

In addition to demonstrating the sliding forces at work between spindle microtubules, experiments on isolated diatom spindles also suggest that microtubule polymerization at the (+) end in the overlap zone also occurs during anaphase B. During anaphase-B movement of the isolated spindle from a diatom, elongation is limited to the length of the microtubule overlap zone. If tubulin is added to the reaction mixture, however, spindle elongation is several times the length of the original overlap zone. This result indicates that during anaphase B the microtubules attached to opposite poles not only slide by each other but also can lengthen by addition of tubulin at their (+) ends (see Figure 23-50).

A third component of anaphase B is the interaction between astral microtubules and the cortex. In cells which assemble an aster, scientists have shown that a pulling force is experienced by the asters. During anaphase, if the spindle is cut in half with a microneedle, the resulting half-spindles move quickly to the poles, at a rate faster than usual during anaphase. The results suggest that a (−) end directed motor, maybe cytoplasmic dynein or a KRP, is associated at the cortex, where it pulls the asters to the poles of the daughter cells (see Figure 23-50).

Astral Microtubules Determine Where Cytokinesis Takes Place

Once the chromosomes have been separated to the poles of the cell during anaphase, the cell divides in two, capturing a complete set of chromosomes in each daughter cell. This is the last step in mitosis and is the beginning step in the life of a daughter cell. In Chapter 22, we discussed how the contractile ring of actin and myosin constricts the cell during cytokinesis, but not the mechanism that determines the plane of cleavage through the cell. It is clear that the contractile ring, and hence the cleavage furrow, always develops where the chromosomes lined up during metaphase, but it is not obvious which component of the mitotic apparatus dictates where the contractile ring will assemble.

Using dividing sea urchin embryos, ingenious micromanipulation studies, rather than biochemical or genetic studies, were able to obtain evidence that asters are the guiding structures for determining where the cleavage furrow forms. First, scientists showed that it was possible to produce multiple cleavage furrows by using a glass microneedle to move the mitotic apparatus to different parts of the egg, leaving it in each location just long enough for the furrow to form before they push the mitotic apparatus to a new position (Figure 23-51a). This established that the mitotic apparatus fixes the position of the cleavage plane and that it must send some sort of signal to the membrane.

A second set of experiments established that only the presence of two asters, not the spindle itself, is necessary to determine the cleavage plane (Figure 23-51b). Before the mitotic apparatus has started to assemble, a hole is poked

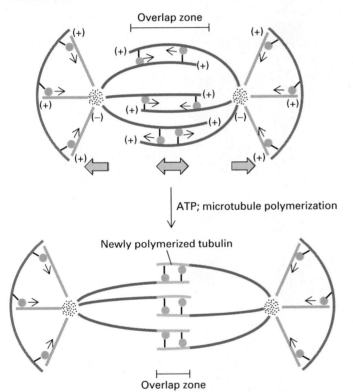

▲ FIGURE 23-50   Model of spindle elongation and movement during anaphase B. Tubulin (light purple) adds to the (+) ends of all polar microtubules, lengthening these fibers. Simultaneously, (+) end–directed KRPs (pink) bind to the polar microtubules in the overlap region. Each KRP, bound to a microtubule in one half-spindle, "walks" along a microtubule in the other half-spindle, toward its (+) end, utilizing the energy from ATP hydrolysis. In addition, cytoplasmic dynein (light green) located in the cortex of the plasma membrane could pull on the microtubules in the aster. These actions of KRPs and cytoplasmic dynein result in a sliding of the two half-spindles toward their respective poles. [Adapted from H. Masuda and W. Z. Cande, 1987, *Cell* **49**:193.]

(a)

(b)

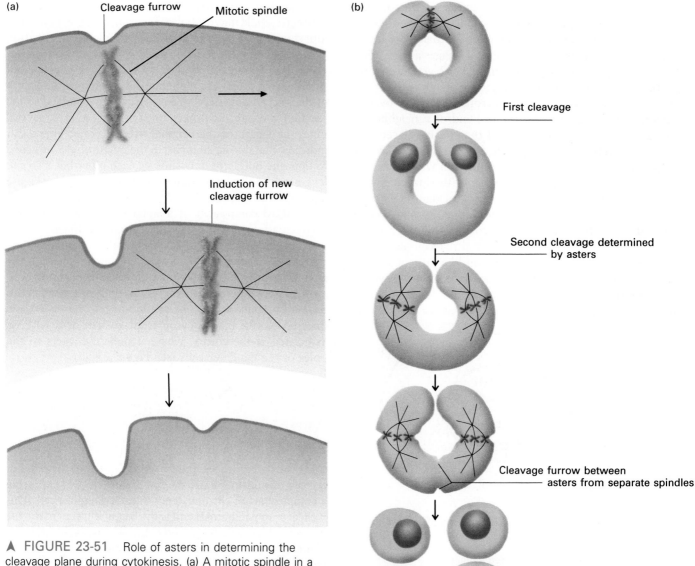

▲ FIGURE 23-51   Role of asters in determining the cleavage plane during cytokinesis. (a) A mitotic spindle in a sea urchin egg is allowed to induce the formation of a cleavage furrow. Before furrowing is completed, the spindle is moved with a glass microneedle to a new position near the cortex. After a brief period, a second furrow appears. This process can be repeated several times, resulting in the formation of many cleavage furrows. (b) A small glass ball is pressed against a fertilized sea urchin egg until membranes from opposite sides of the cell touch and fuse. This changes the spherical egg into a doughnut shape. During the first cell division, a normal spindle develops and the doughnut-shaped cell divides between the two asters. The result is a single C-shaped cell with two normal nuclei. At the next cell division, the two nuclei each produce a normal spindle. Two cleavage furrows bisect at spindles, in the usual manner. But because the asters from the two separate spindles are close enough together, a third cleavage furrow also forms and divides the cell in a region not occupied by either spindle but lying between asters from the two separate spindles. Thus the cell divides in three planes, producing four daughter cells. This experiment provides conclusive evidence that asters and not spindles determine the site of cell cleavage.

into a one-cell-stage sea urchin embryo to transform it into a doughnut-shaped cell. Now pushed to one side of the cell, a spindle assembles, and the cell undergoes cell division between the asters to form a single, binucleated, C-shaped cell. Normally, the embryo would now be at the two-cell stage. But still at the one-cell stage, the embryo's two nuclei undergo another round of mitosis and set up a pair of mitotic apparatuses. If the spindle determines the cleavage plane, then we would expect the C-shaped cell to form two cleavage furrows and divide into three cells, but if cleavage occurs between any two asters, then three cleavage furrows form and hence four cells should be generated. The "extra" cell in the second case is generated by cleavage in the region between asters connected to separate spindles; that is, in a region where there is no spindle. The result is that four cells are produced, demonstrating the role of asters and not spindles in determining the site where the contractile ring is assembled. These results also suggest that astral microtubules send a signal to the region of the cortex midway between asters. This signal activates the assembly of actin and myosin, resulting in the formation of the contractile ring followed by the development of the cleavage furrow. The signal is unidentified as yet, but an alluring candidate is one of the protein kinases that are activated in

the cell cycle (see Chapter 25). Caldesmon and myosin are known substrates for one of these, cdc2 kinase. As discussed in the previous chapter, phosphorylation of these two proteins will activate the assembly of myosin filaments and cause caldesmon to relieve the inhibition of actin-myosin interactions.

Plant Cells Build a New Cell Wall During Cell Division

Because plant cells are bound by a rigid, undeformable cell wall, cytokinesis in plants involves quite different processes (Figure 23-52). At cytokinesis, plant cells must construct both a cell membrane and a cell wall to separate the daugh-

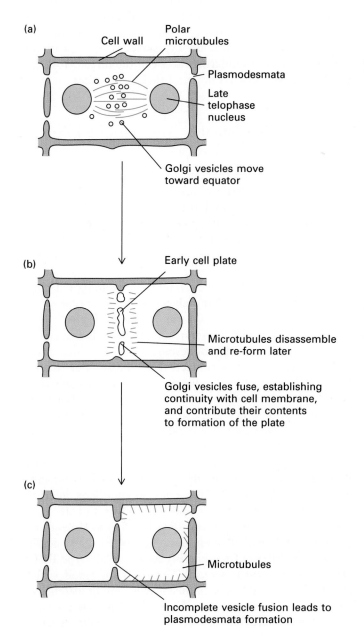

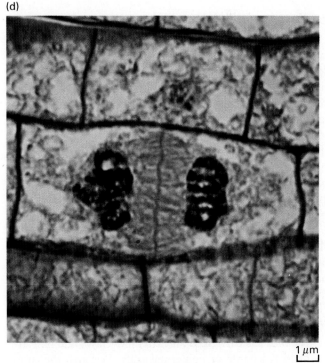

1 μm

▲ FIGURE 23-52 Cytokinesis in a higher plant cell. (a) In late telophase, the nuclear membrane has reformed, and the polar microtubules have not yet dispersed. A set of small vesicles derived from the Golgi complex, which contain cellulose and other precursors of the cell wall, accumulate at the equatorial plate. (b) These vesicles fuse with each other to form the phragmoplast—a large membrane-limited vesicle in the center of the cell. Additional vesicles fuse with the phragmoplast, extending it outward. (c) Eventually the phragmoplast fuses with the plasma membrane, and the two cells separate. The daughter cells remain connected by thin membrane-lined passages, called plasmodesmata, that penetrate the separating cell wall. (d) Light micrograph of an onion root cell in telophase, showing the new cell wall being laid down between the daughter nuclei.

ter cells. In place of a contractile ring, a membrane structure, the *phragmoplast*, is assembled from vesicles in the interzone between daughter nuclei. The vesicles originate from the Golgi complex and are first observed during metaphase, when they extend into the mitotic apparatus and in some cases even appear to contact the kinetochores. Later in anaphase, by travelling along microtubules that radiate from the nucleus, the vesicles line up near the center of the dividing cell, where they fuse to form the phragmoplast. The membranes of the vesicles become the plasma membranes of the daughter cells, and their contents form the intervening immature cell wall. The vesicles also contain material for the future cell wall, such as polysaccharide precursors of cellulose and pectin.

➤ *Intermediate Filaments*

Intermediate filaments (IFs) comprise a third set of cytoskeletal fibers in cells. The name derives from their 10-nm diameter—smaller than microtubules (24 nm) but larger than microfilaments (7 nm). IFs are found in nearly all eukaryotic cells (they have not been detected in fungi and other lower eukaryotes). In epidermal cells and the axons of neurons, IFs are at least ten times more abundant than microfilaments or microtubules. In immunofluorescence micrographs, we can see that a network of IFs fills the entire cytosol of a cell and that the pattern is distinct from that of microfilaments but similar to that of microtubules (Figure 23-53; see also the opening figure of this chapter). The organization of IFs and their association with plasma membranes suggest that the principal function of IFs is primarily structural—to reinforce cells and to organize cells into tissues. This organization indicates that the supportive role of IFs is similar in many respects to the structural roles of microfilaments and microtubules. Indeed, the supportive role of IFs is carried to an extreme—claws and hair are the dead remnants of epidermal cells. Perhaps, the most important function of IFs is to provide mechanical support for the plasma membrane where it comes into contact with other cells or with the extracellular matrix. Unlike microfilaments and microtubules, IFs do not share a role in motility. There are no known examples of IF-dependent cell movements or of motor proteins that move along IFs.

Several physical and biochemical characteristics distinguish IFs from microfilaments and microtubules. To begin with, the most notable property of IFs is their extreme stability. Even after extraction with solutions containing detergents and high concentrations of salts, most IFs in a cell remain polymerized while microfilaments and microtubules are solubilized. In fact, most IF purification methods employ these treatments to free IFs from other proteins. The IFs are then solubilized in urea, a denaturing solvent, and purified by ion-exchange chromatography (see Figure 3-39b); the IF subunits then repolymerize into filaments when the urea is removed by dialysis. Second, in contrast to the beaded filaments and the hollow tubules formed by the globular actin and tubulin subunits, respectively, IF subunits are α-helical rods that assemble into ropelike filaments. A third difference between IFs and microtubules or microfilaments is that IF subunits do not bind nucleotides and therefore IF assembly does not require nucleotide (GTP or ATP) hydrolysis, as does actin filament or microtubule polymerization. However, many of the details concerning the assembly of IFs in cells remain speculative.

In this section, we will discuss the structure of IFs, the assemblies formed by IFs, and the functions of IFs. Much of our discussions about IFs will seem familiar because IFs are organized in ways similar to that of the microtubule and microfilament components of the cytoskeleton, and in some cases IFs perform functions analogous to those of actin filaments. Using our previous discussions of the microfilament and microtubule cytoskeleton as a basis, we will point out the similarities between the three cytoskeletal systems and discuss in detail their differences.

Intermediate Filaments Are Classified into Five Types

In higher vertebrates, IFs constitute a superfamily of highly α-helical proteins that is divided into five major classes or types on the basis of similarities in sequence (Table 23-4). The expression of each IF protein is characteristic of a certain tissue or cell type. Unlike the actin and tubulin isoforms, IF protein classes are widely divergent in sequence and vary greatly in molecular weight.

Keratins are obligate heterodimers containing 1:1 mixtures of acidic and basic IF polypeptides; either type alone cannot assemble into a keratin filament. Typically expressed in epithelial cells, keratin IFs are usually associated with desmosomes and hemidesmosomes—sites that form cell-cell and cell-matrix contacts. Like the actin cytoskeleton, which is attached to adherens junctions and adhesion plaques, keratins in epithelial cells form a flexible but resilient framework which gives structural support to the epithelium. A large number of keratin isoforms are expressed but they can be divided into two groups. About 10 keratins are specific for "hard" epithelial tissues, which give rise to nails, hair, and wool; about 20, called *cytokeratins,* are more generally found in the epithelia that line internal body cavities. Each type of epithelium always expresses a characteristic combination of type I and type II keratins.

Four proteins, vimentin, desmin, glial fibrillary acidic protein (GFAP), and peripherin, comprise the type III IF proteins. *Vimentin* is typically expressed in blood vessel endothelial cells, in some epithelial cells, and in mesenchymal cells such as fibroblasts. Vimentin fibers often terminate at the nuclear membrane and at cell-surface desmosomes or hemidesmosomes. At these locations, vimentin

(a) Intermediate filaments (vimentin)

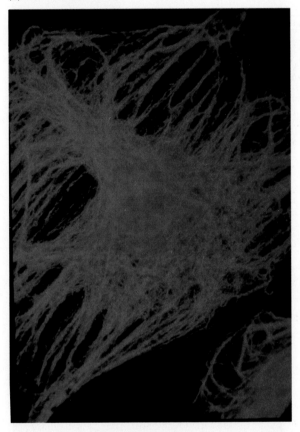

◄ FIGURE 23-53   Intermediate filaments, microtubules, and microfilaments in the same cell. The distribution of the intermediate filament protein vimentin (a), tubulin (b), and actin (c) in the same cultured fibroblasts, as detected by fluorescent vimentin-specific (green) and tubulin-specific (blue) and rhodamine phalloidin (red) antibodies. Many vimentin fibers are colinear with microtubules, suggesting a close association between the two filament networks. Actin filaments in stress fibers traverse the cell but do not show patterns similar to those of IFs or microtubules. Images (a) and (b) were superimposed in a computer and printed in a different color scheme in the chapter opening figure [Courtesy of V. Small.]

(b) Microtubules (tubulin)

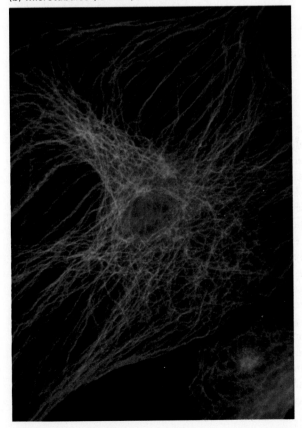

(c) Microfilaments (actin)

TABLE 23-4   Mammalian Intermediate-Filament Proteins

IF Protein	MW (10^{-3})*	No. of Polypeptides	Tissue Distribution
TYPE I			
Acidic keratins	40–57	>15	Epithelia
TYPE II			
Basic keratins	53–67	>15	Epithelia
TYPE III			
Desmin	53	1	Muscle
Glial fibrillary acidic protein	50	1	Glial cells and astrocytes
Vimentin	57	1	Mesenchyme
Peripherin	57	1	Peripheral and CNS neurons α-Internexin
TYPE IV			
NF-L	62	1 ⎫	Mature peripheral and central
NF-M	102	1 ⎬	neurons
NF-H	110	1 ⎭	
Internexin	66	1	Developing central nervous system
Nestin	240	1	Neuro-epithelial stem cells
TYPE V⁺			
Lamin A	70	1 ⎫	All cells
Lamin B	67	1 ⎬	
Lamin C	67	1 ⎭	

NF = neurofilament; L, M, and H = low, medium, and high molecular weight.
*IFs show species-dependent variations in MW.

IFs may keep the nucleus or other organelles in a defined place within the cell. For example, in the adipocyte (fat cell), vimentin filaments form a cage around lipid droplets, most likely preventing them from fusing with one another or with cellular membranes (Figure 23-54). Vimentin is frequently associated with microtubules (Figure 23-53); the alignment of the two cytoskeletal systems is seen clearly in the chapter opening figure, in which the pattern of vimentin filaments (purple) and microtubules (light orange) are superimposed. (These two patterns are identical to those in Figure 23-53a and b, but they are printed in different colors.) Later in this section, we will discuss the proteins that may be responsible for cross-linking IFs to microtubules.

In muscle cells, *desmin* filaments link myofibrils into bundles. Found specifically in striated muscle, desmin filaments encircle the Z disk and make connections between neighboring Z disks or with the overlying plasma membrane. Through these intermyofibrillar connections, desmin filaments align Z disks in neighboring cells. This inte-

grating role of desmin is responsible for stabilizing sarcomeres in contracting muscle.

Glial fibrillary acidic protein forms IFs in the glial cells that surround neurons and in astrocytes. *Peripherin* is found in neurons of the peripheral nervous system, but little is known about it.

The core of neuronal axons is filled with *neurofilaments (NFs)*, each a heteropolymer composed of three type IV polypeptides—NF-L, NF-M, and NF-H—which differ greatly in their molecular weights (L, M, and H indicate low, medium, and high molecular weight size classes). NFs are frequently associated with the axonal microtubules. A fourth protein, nestin, is also included in this class because its exons are organized similarly to the exons in the genes of other type IV IF proteins. It is expressed in developing neurons and muscle. As we have previously discussed in this chapter, axonal microtubules are responsible for the transport of material to the nerve terminal and for directing the elongation of an axon. In contrast, neurofilaments

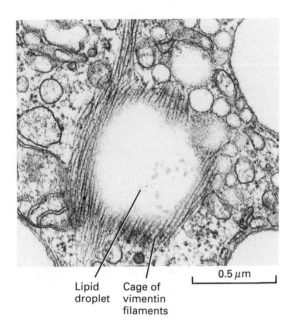

Lipid   Cage of
droplet   vimentin
             filaments

$\blacktriangle$ FIGURE 23-54   Electron micrograph of a section of an adipocyte, showing a cage of parallel vimentin filaments surrounding the "hollow" lipid droplet. [From W. W. Franke, M. Hergt, and C. Grund, 1987, *Cell* **46**:131; courtesy of W. W. Franke.]

are responsible for the radial growth of an axon and the diameter of an axon, and hence the speed of impulse transmission is directly dependent on the number of neurofilaments present. A mutation named *quiver* blocks the assembly of neurofilaments in quails. As a result, the velocity of nerve conduction is severely reduced.

In contrast with the cytosolic location of the other four IF types, type V IFs are found exclusively in the nucleus, where they support the nuclear membrane. Type V IF proteins consist of the nuclear *lamins*, a set of three polypeptides, lamin A and lamin C, which are alternatively spliced products from a common gene, and lamin B, which is encoded by a separate lamin gene. The nuclear lamins form the nuclear lamina, a web of filaments adherent to the inner surface of the nuclear membrane, and are attached by lamin B to the inner nuclear membrane through a lamin B receptor in the nuclear membrane. The role of lamins in the assembly of the nucleus during the cell cycle will be discussed in Chapter 25 (see Figures 25-10 to 25-13); later in this chapter, we will introduce you to the structure of the nuclear lamina and the regulation of its assembly by cell-cycle kinases.

Compiling a list of IFs is not only important in comparing and contrasting their structural and functional properties; it is also helpful in the diagnosis of cancer, especially when a diagnosis depends on whether the cell is epithelial or mesenchymal in origin. The type of IFs found in a cell can often identify the origin (i.e., epithelial, mesenchy-

mal, or neuronal) of the cell. In a tumor, cells lose their normal appearance (see Chapter 26) and thus cannot be identified by their morphology. However, because tumor cells retain many of the differentiated properties of the cells from which they are derived, including expression of particular IF proteins, their tagging by antibodies specific for those IF proteins allows the identification of cell type. For example, the most common carcinomas of the breast and gastrointestinal tract contain keratins and lack vimentin; thus they are derived from epithelial cells (which contain keratins) rather than from the underlying stromal mesenchymal cells (which contain vimentin but not keratins). Because epithelial cancers and mesenchymal cancers are sensitive to different treatments, identifying the IF proteins in a cancer cell helps a physician in selecting the correct treatment to destroy the cancer.

All Subunit Proteins of Intermediate Filaments Have a Similar Structure

Besides sharing an ability to form filaments 10 nm in diameter, all IF subunit proteins have in common their organization into three domains—a central α-helical core and flanking globular N- and C- terminal domains (Figure 22-55a). The core domain is conserved among all IF proteins. It consists of four long α-helices which are separated by three nonhelical regions. The positions of these "spacer" elements are also highly conserved among all IF subunit proteins. The α-helical segments pair to form a coiled-coil dimer.

In electron micrographs, the dimer appears as a rodlike molecule with globular domains at the ends. Theoretically, the two IF polypeptides in the dimer could be paired in a parallel or an antiparallel orientation. The orientation can be determined by labeling the dimer molecule with antibodies specific for the N- or C-terminal globular domains. When bound with an antibody, the globular domain is much larger than a globular domain in an unlabeled dimer. The difference in size is easily detected by electron microscopy. Separately, an antibody to the N-terminus or the C-terminus labels only one or the other end of the dimer but mixtures of the two antibodies label both ends. From this pattern of binding we can conclude that the polypeptide chains are parallel in a dimer.

When an IF is partially dissociated with denaturants or is assembled in the presence of denaturants that slow assembly, substructures representing different oligomeric forms of IFs (and possibly intermediates in assembly) are seen (Figure 23-55a). Using denaturants in their studies, scientists propose that a pair of dimers associate laterally into a tetramer about 70 nm long. An antibody specific for the globular C-terminal domain labels both ends of the tetramer, showing that the dimers are oriented antiparallel in a tetramer. As we will discuss later, some evidence suggests that the tetramer and not the dimer is the subunit for

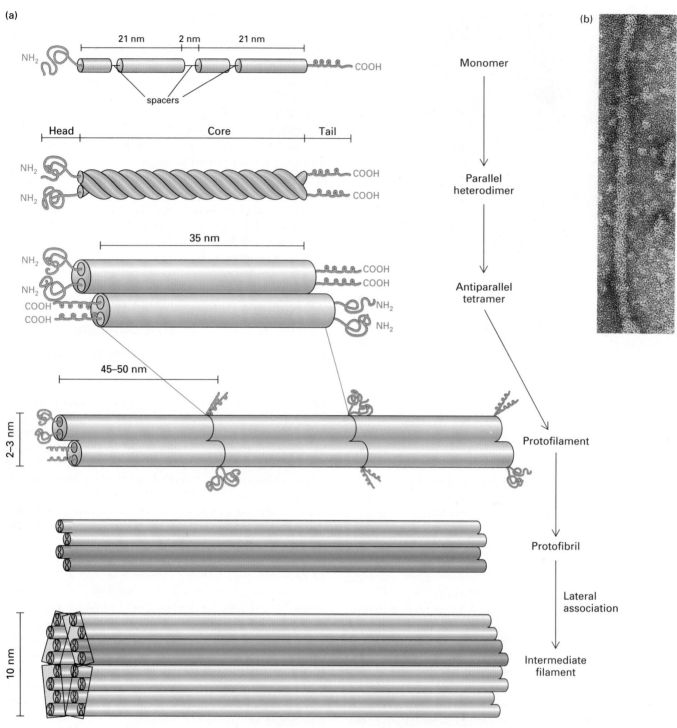

▲ **FIGURE 23-55** Levels of organization and assembly of intermediate filaments. (a) A common structural model for IF-subunit proteins. These dimeric proteins have a central α-helical coiled-coil core domain (green), which is highly conserved, and nonhelical tails and heads (blue), which are variable in length and sequence. The central core domain contains three nonhelical spacer elements. A tetramer is formed by antiparallel, staggered side-by-side aggregation of two dimers. Tetramers aggregate end to end, forming a protofilament, pairs of protofilaments laterally associate into a protofibril, and four protofibrils form a cylindrical filament 10 nm thick. (b) Electron micrograph of a negatively stained preparation of neurofilament intermediate filaments. [Part (b) courtesy of E. M. Mandelkow.]

assembly of IFs. Tetramers bind end to end, forming protofilaments 2 nm thick, which pair together into 4-nm-thick protofibrils. Finally, four protofibrils form a single IF that is 10 nm in diameter (Figure 23-55b). Interestingly, because the tetramer is symmetric, an IF may not have a po-larity like an actin filament or a microtubule. However, none of the kinds of tests for polarity of actin filaments or microtubules have been successfully applied to IFs. There-fore, the polarity of an IF is unresolved.

While the α-helical core is common to all IF proteins,

peptides. However, keratins do not form heteropolymers with other IF proteins, whereas some IFs, vimentin for example, can assemble into a heteropolymer with desmin or with neurofilaments. NF-L self-associates to form the core of a neurofilament, but NF-H and NF-M also co-assemble with the NF-L backbone.

Whether a heteropolymer or a homopolymer forms is probably not dictated by simple hydrophobic interactions in the α-helical segments but probably involves spacer sequences in the coiled coil or interactions with diverse N- or C-terminal domains. In fact, mutations in these regions generate extremely useful mutated IF polypeptides which are able to form hetero-oligomers with normal proteins. These hybrid molecules often "poison" IF polymerization by blocking assembly at an intermediate stage. For example, when the mutated IF protein is introduced into a cell in a transfection or microinjection experiment, the assembly of protofilament intermediates is inhibited. This property of mutated IF proteins is used in experiments to reveal the function of IFs in a cell; at the end of the chapter, we will discuss how such mutations in keratins reveal the role of keratins in the epidermis.

▲ FIGURE 23-56   Electron micrograph of a neurofilament purified from spinal cord. After the filament is rotary-shadowed with metals, long whiskers are seen to project from its walls. These projections correspond to the N- and C-terminal domains of the neurofilament protein. [Courtesy of U. Aebi.]

Intermediate Filaments Are Dynamic Polymers in the Cell

Compared with our knowledge of microfilaments and microtubules, we understand very little about the assembly of IFs, either in vitro or in the cell. Scientists are still discussing, for example, whether a tetramer is the subunit for assembly of an IF, analogous to the actin monomer and tubulin dimer. The main supporting evidence for involvement of the tetramer comes from cell fractionation experiments which show that although most vimentin in cultured fibroblasts is polymerized into filaments, 1–5 percent of the protein exists as a soluble pool of tetramers. The presence of a tetramer pool suggests that vimentin monomers are rapidly converted into dimers which rapidly form tetramers.

That the tetramer pool is soluble suggests that IF subunits can exchange with the IF cytoskeleton. Exchange between soluble IF subunits and insoluble IF polymer is detected in a cell which is injected with a biotin-labeled or fluorescent IF protein. In one experiment, a biotin-labeled type I keratin is injected into an epithelial cell (Figure 23-57); within four hours after injection, the labeled protein has become incorporated into the already existing keratin cytoskeleton. In other experiments, the vimentin cytoskeleton is made fluorescent by the incorporation of rhodamine-labeled vimentin, and small patches of the fluorescent IF cytoskeleton are then photobleached (see Figure 14-34 for a discussion of the technique). Shortly after exposure to ultraviolet light, the bleached patches of filaments rapidly recover their fluorescence. These experiments demonstrate that IF subunits in a soluble pool are able to add themselves to preexisting filaments and that subunits are also able to dissociate from intact filaments

the N- and C-terminal domains of different types of IF proteins vary greatly in molecular weight and sequence. This lack of sequence conservation, and the fact that electron micrographs (Figure 23-56) show these domains as short whiskers projecting out from the walls of the IF, led scientists initially to speculate that the N- and C-terminal domains are not involved in IF assembly. However, this idea was tested by several experiments and found to be incorrect. In one series of experiments, if the N- or C-terminal domains are shortened, either by proteolysis or by deletion mutagenesis, then the resulting protein cannot assemble into an IF. (Keratins are the exception; they form filaments even if both terminal domains are absent.) Hence most scientists would now agree that the N- and C-terminal domains play an important role in most IF assembly. In addition, these domains may control the stability of IFs and interactions between IFs and other cellular components.

Sequences within the N- and C-terminal domains and the core domain are also important in determining whether an IF protein dimerizes with itself to form a homopolymeric IF, or with another IF protein to form a heteropolymeric IF. Vimentin, desmin, glial fibrillary acidic protein, and NF-L are able to form homodimers. In contrast, keratins are obligate heteropolymers of type I and type II poly-

(a)

(b)

(c)

(d)

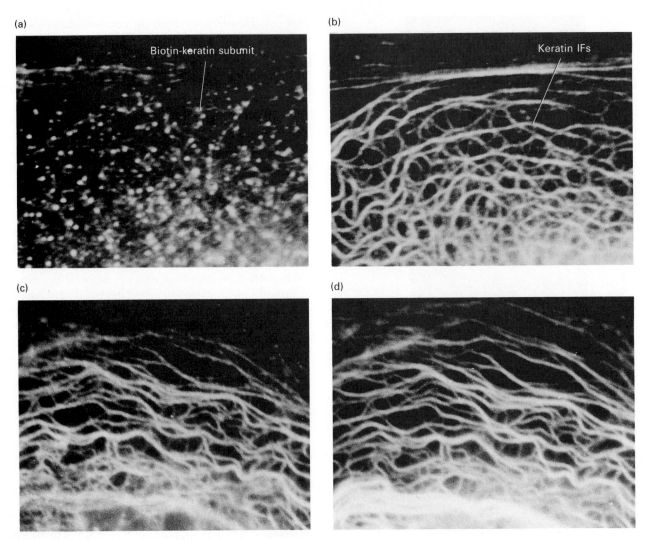

▲ FIGURE 23-57   Assembly of biotin-labeled type I keratin in PtK2 cells. Type I keratin was purified, chemically modified with biotin, and injected into living PtK2 cells. The cells were then fixed at different times after injection and the distribution of biotin-labeled *keratin* was revealed with a fluorescent antibody to biotin (a) and (c). To visualize the organization of all the keratin filaments, the same cells were also stained with antibodies to keratin (b) and (d). At 20 min after injection, the microinjected biotin-labeled keratin is concentrated into small foci scattered through the cytoplasm (a); the injected protein has not disturbed the organization of the endogenous keratin cytoskeleton (b). By four hours, the biotin-labeled subunits (c) and the keratin filaments (d), display identical patterns, indicating that the microinjected protein has become incorporated into the existing cytoskeleton. [From R. K. Miller, K. Vistrom, and R. D. Goldman, 1991, *J. Cell Biol.* **113**:843; courtesy of R. D. Goldman.]

(since the bleached patches disappeared), but neither experiment is able to discern the oligomeric state of the subunit or whether subunits exchange at the ends or along the length of filaments.

Phosphorylation of the N-Terminal Domain Regulates Polymerization of Intermediate Filaments during Mitosis

The stability of IFs presents special problems in mitotic cells, which must reorganize IF networks during the cell cycle and disassemble IFs during cytokinesis. We have discussed previously how calcium and various kinases control the organization of the microfilament and microtubule cytoskeleton during the cell cycle. Similarly, under the control of cdc2 kinase, filaments of vimentin, desmin, and lamins disassemble prior to mitosis and reassemble after cell division. In lamin A and vimentin, the phosphorylation of serine residues in the N-terminal domain by cdc2 kinase induces the disassembly of intact filaments and prevents reassembly. The importance of phosphorylating this site for filament stability was demonstrated by mutating the target serine residues in lamin A, changing them to residues that cannot be phosphorylated. Without these sites, the lamin filaments were prevented from depolymerizing. The causal relationship between lamin phosphorylation and lamin filament disassembly is also demonstrated by inactivating protein phosphatases with drugs. This treatment causes rapid collapse of the network-like nuclear lamina. The opposing action of kinases and phosphatases thus provides a rapid mechanism for controlling the assembly state of IFs. (See Figures 25-11 to 25-14 for a more detailed discussion on the role of nuclear lamins during the assembly of the nucleus.)

Not all IFs are disassembled by kinases. In neurons, which are postmitotic cells, neurofilaments also are phosphorylated, but the modification does not affect the stability of the filaments. Instead, the phosphorylated neurofilaments are spaced farther apart in the axon than are unphosphorylated proteins, leading to differences in the diameters of IF bundles and, consequently, of the axon. Since the diameter of the axon affects the velocity of nerve transmission, phosphorylation-dependent changes in the packing of neurofilaments may alter the physiologic properties of nerves.

Intermediate Filament-Associated Proteins Cross-Link Intermediate Filaments to Membranes and Microtubules

The interactions of IFs with one another and with other cell structures are mediated by intermediate filament–associated proteins (IFAPs), which cross-link IFs into a bundle (also called a *tonofilament*), into a network, or to the plasma membrane. Unlike actin-binding proteins or MAPs, no IFAP has been discovered which severs or caps IFs, sequesters IF proteins in a soluble pool, or is an IFAP motor protein. Although only few IFAPs have been identi-

fied (Table 23-5), many more will undoubtedly be added to the list because current research is now focused on the proteins that control IF organization and assembly.

One possible role of IFAPs is to cross-link IFs to the microfilament and microtubule cytoskeleton and the existence of such linkages has been deduced from light and electron micrographs. For example, in a nerve axon (see Figure 23-4d), numerous filamentous connections, similar in appearance to MAPs, bridge between microtubules and neurofilaments. Although the identity of these connections is unknown, they may represent IFAPs whose function is to cross-link neurofilaments and microtubules into a stable cytoskeleton.

The close association between microtubules and IFs in a specialized cell like a neuron is a general feature of all cells. In cultured fibroblasts, microtubules and vimentin filaments have similar organization (see Figure 23-53). In some parts of the cytoplasm, both sets of filaments display the same curvy, looping pattern. In these regions of the cell, the microtubules and vimentin filaments are coincident, suggesting there must be IFAPs which cross-link the two cytoskeletons together. In other regions of the cell, vimentin filaments and microtubules do not overlap and show no evidence of association. (In contrast, the actin cytoskeleton is represented by straight, parallel stress fibers which are arranged completely differently than IFs or microtubules.) This difference in organization between the three types of cytoskeletal fibers would be difficult to detect in a long cylindrical process such as the axon.

The physical linkage between IFs and microtubules that is detected in micrographs can also be detected by treating cells with drugs. Recall that the microtubule inhibitor colchicine causes the complete dissolution of microtubules after a period of several hours. In the same colchicine-treated cells, vimentin filaments remain intact but they clump into disorganized bundles near the nucleus. These

TABLE 23-5	**Intermediate Filament-Associated Proteins**		
Name	MW	Location	Function
BPAG1*	230,000	Hemidesmosome	IF cross-link to dense plaque
Plakoglobin	83,000	Desmosome	IF cross-link to dense plaque
Desmoplakin I	250,000	Desmosome	IF cross-link to dense plaque
Desmoplakin II	215,000	Desmosome	IF cross-link to dense plaque
Plectin	300,000	Cortex	Vimentin cross-linker; cross-link to MAP1, MAP2, & spectrin
Ankyrin	140,000	Cortex	Vimentin cross-link to plasma membrane
Filaggrin	30,000	Cytoplasm	Keratin cross-link
Lamin B receptor	58,000	Nucleus	Lamin cross-link to inner surface of nuclear membrane

* Bullous pemphigoid antigen 1.

findings demonstrate that the organization of vimentin filaments is dependent on intact microtubules and suggest the presence of protein connectors between the two types of filaments. Based on its ability to bind both IFs and microtubules, the IFAP plectin is one candidate for the protein cross-link between IFs and microtubules.

A major function of IFs is to provide a flexible structural support for the plasma membrane. In our discussions in this section, you will see immediately many similarities between the organization of the IF cytoskeleton and the actin cytoskeleton; indeed, in several cases the similarities extend to the use of shared proteins.

Support of Nuclear Membranes by Nuclear and Cytosolic IF Networks   A network of IFs is often found as a laminating layer adjacent to a membrane, where it provides mechanical support. The best example is the *nuclear lamina* along the inner surface of the nuclear membrane (see Figure 25-10). This supporting network is composed of lamin A and lamin C filaments which are cross-linked into an orthogonal lattice that adheres to the inner nuclear surface. The nuclear lamina is attached to the membrane through polyisoprenoid lipids covalently attached to the C-terminus of lamin B and through interactions with a lamin B receptor.

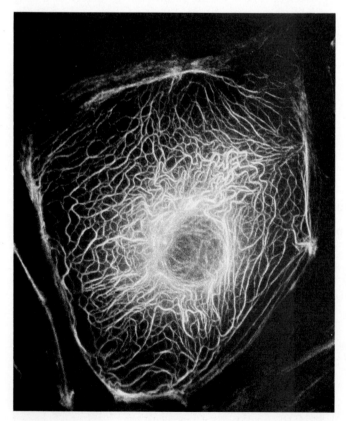

▲ FIGURE 23-58   Keratin filament network in PtK2 cells that reaches from the cell surface to the nucleus. [Courtesy of R. D. Goldman.]

In addition to forming the nuclear lamina, IFs are typically organized in the cytosol as an extended system that stretches from the nuclear envelope to the plasma membrane (Figure 23-58). At the plasma membrane, vimentin binds two proteins: ankyrin, the actin-binding protein associated with the Na^+/K^+ ATPase in non-erythroid cells, and plectin, a 300,000 MW IFAP. Plectin has been shown to interact with other cytoskeletal proteins, including spectrin, MAPs, and lamin B. Through these two IFAPs, vimentin is attached to both the plasma membrane and nuclear membrane in a cell.

Anchorage of IFs to Desmosomes and Hemidesmosomes   The epithelial cells of organs and skin are held together by desmosome and hemidesmosome membrane junctions (Figure 23-59; see also Figures 14-42, 24-36, and 24-37). As we shall see in Chapter 24, desmosomes mediate cell-cell adhesions, while hemidesmosomes are responsible for attaching cells to the underlying basement membrane. In the electron microscope, both junctions appear as darkly staining proteinaceous plaques bound to the cytosolic face of the plasma membrane and are attached to bundles of IFs. Some IFs run parallel to the cell surface, while others traverse the cytosol; together they form an internal framework which supports the shape and resilience of the cell. As shown in Figure 23-59, IFs in one cell are directly connected to the IFs in the neighboring cell through desmosomes. Thus through desmosomes and hemidesmosomes, IFs can distribute shearing forces from one region of a cell layer to the entire sheet of epithelial cells, providing strength and rigidity to the entire epithelium. Without this supporting network of IFs, the cells remain connected by desmosomes but are easily damaged by abrasive forces.

The detailed structure of desmosomes and hemidesmosomes will be discussed in Chapter 24 (see Figures 24-36 and 24-37), but we can see that the overall organization of both junctions is very similar to the actin-associated adherens junctions and adhesion plaques. All four junctions share a common structural plan; the cytoskeleton is attached to integral membrane proteins through adapter proteins. This similarity results from their use of common proteins. For example, cell–cell contacts mediated through desmosomes are made through E-cadherin, the same protein that mediates cell–cell contacts in adherens junctions. Similarly, cell-matrix interactions in both hemidesmosomes and adhesion plaques are dependent on integrins. Although IFs and actin filaments are different polymers, they are attached to the membrane through common adapter or *plaque proteins*. Many of the plaque proteins in a desmosome are IFAPs but one in particular, plakoglobin (see Table 23-5), is similar to vinculin, the actin-binding protein found in adherens junctions and adhesion plaques (see Figure 22-35).

Anchorage of the Sarcomere to the Plasma Membrane by Desmin   In muscle, a lattice composed of a band of desmin filaments surrounds the sarcomere (Figure

(a)

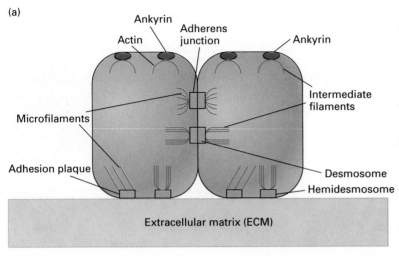

◄ FIGURE 23-59   Intermediate filaments anchored to desmosomes and hemidesmosomes. (a) Schematic diagram comparing IF-membrane and microfilament-membrane connections in a pair of typical epithelial cells. IFs (green) and microfilaments (red) both are attached to the membrane at three sites: (1) between cells (light purple), (2) at cell-matrix surfaces (pink), and (3) at ankyrin-containing sites (dark purple) at the plasma membrane. These sites represent, respectively, cell-cell connections by desmosomes and adherens junctions, cell-matrix connections by hemidesmosomes and adhesion plaques, and general microfilament-membrane or IF-membrane connections. (b) A schematic diagram of desmosomes and hemidesmosomes showing that the IFs loop through the junctions. The IFs are cross-linked to the plasma membrane through proteins in the darkly stained plaque. Desmosomes maintain cell-cell contacts through the integral membrane protein E-cadherin, while hemidesmosomes are connected to the extracellular matrix through integrins. (see Figure 22-35).

(b)

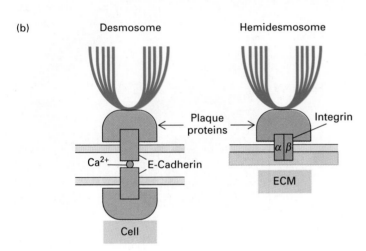

23-60). The desmin filaments encircle the Z disk and are cross-linked to the plasma membrane by several IFAPs, including paranemin, synemin, and ankyrin. Additional desmin filaments cross to neighboring Z disks within the myofibril and serve to cross-link myofibrils into bundles within a muscle cell. The lattice is also attached to the sarcomere through interactions with myosin thick filaments, possibly by skelemin at the M line. Because the desmin filaments lie outside the sarcomere, they do not actively participate in generating contractile forces. Instead, through their interactions with the muscle plasma membrane and Z disks, desmin filaments ensure that muscle fibers maintain their regular architecture and contract as a unit.

Skin Diseases Caused by Disruption of Keratin Networks The epidermis is a tough outer layer of tissue which acts as a water-tight barrier to prevent desiccation and serves as a protection against abrasion. In the epidermal cells, bundles of keratin filaments are cross-linked by filaggrin, an IFAP, and are anchored at their ends to desmosomes. As epidermal cells differentiate, the cells condense and die, but the keratin filaments remain intact,

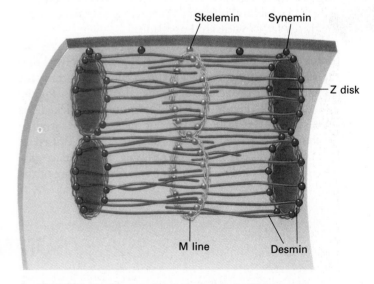

▲ FIGURE 23-60   A diagram of desmin filaments in muscle. Desmin filaments encircle the Z disk and make additional connections to neighboring Z disks. Longitudinally oriented desmin filaments bridge between Z disks in the same myofibril. This alignment of IFs with the muscle sarcomere is held in place by numerous IFAPs, including skelemin at the M line and synemin at the Z disk.

(a)

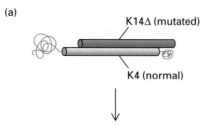

Protofilaments unable to assemble

▲ FIGURE 23-61   Keratins and skin disease. (a) Normally in humans and mice, K4 and K14 keratin polypeptides form heterodimers which assemble into protofilaments. With a mutated K14 polypeptide whose N- and C-terminal domains are missing, heterodimers form in vitro but cannot assemble into protofilaments. In a cell, the mutated K14 polypeptide causes the IF network to break down into aggregates. Expression of the same mutant polypeptide in mice carrying the mutant transgene causes increased death in utero, and pups that are born develop skin blisters easily. The gene is targeted to basal stem cells deep in the epidermis. (b) Histologic sections through the skin of a normal and a transgenic mouse. In a normal mouse (top), the epidermis consists of a hard outer epidermal layer covering the soft inner dermal layer. In contrast, the skin of mice expressing a mutant K14 gene (bottom) resembles the skin of humans suffering from epidermolysis bullosa simplex, a blistering disease. The mutation of keratin weakens the cells at the base of the epidermis, thus causing the two layers to separate (arrow). [Part (b) from Coulombe et al., 1991, *Cell* **66**:1301; courtesy of E. Fuchs.]

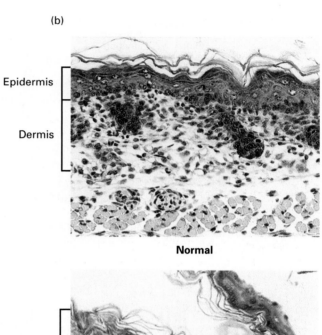

forming the structural core of the dead, keratinized layer of skin. The structural integrity of keratin is essential in order for this layer to withstand abrasion.

Studies of transgenic mice who express mutant keratin proteins in the basal stem cells of the epidermis reveal that the supportive function of keratin is more important for the mechanical properties of epithelial tissue than for the individual cell (Figure 23-61). The mutations picked for these studies, generally deletions of the keratin C-terminal domain and parts of the α-helical domains, all caused IF networks to disassemble. The resulting mutant mice displayed gross skin abnormalities, primarily blistering of the epidermis which resembled the human skin disease epidermolysis bullosa simplex (EBS). Histologic examination of the blistered area detected a high incidence of dead basal cells. Death of these cells appeared to be caused by mechanical trauma from rubbing of the skin during movement of the limbs. Without their normal bundles of keratin filaments, the mutant basal cells became fragile and easily damaged, causing the overlying epidermal layers to delaminate and blister. Like the role of desmin filaments in supporting muscle tissue, keratin filaments appear to play a general role of maintaining the structural integrity of tissues by mechanically reinforcing the cell connections.

SUMMARY

Microtubules are built of tubulin, a heterodimer of one α and one β subunit. Despite a multiplicity of genes encoding α- and β-tubulin in most eukaryotes, all tubulins are functionally equivalent. Microtubules form a diverse array of structures. Some, such as axonal microtubules, are more or less permanent; others, such as interphase microtubules, mitotic spindle microtubules, and, in some cases, flagellar and ciliary microtubules, assemble and disassemble.

Microtubules grow by the addition of GTP-bound tubulin, preferentially to the (+) end. The GTP is then hydrolyzed to GDP, or the tubulin bound to GTP dissociates. According to the current dynamic instability model, microtubules that retain a GTP cap at both ends will, depending on the concentration of tubulin, continue to grow; those with a GDP cap are rapidly depolymerized. The microtubule-organizing center (MTOC)—an amorphous region surrounding the pair of centrioles nucleates the growth of new microtubules. All MTOCs contain γ-tubulin, a non-polymerizing isotype of tubulin; in some cases, as in animal cells, the MTOC also contains a centriole. The (+), growing, end of a microtubule is pointed away from the MTOC. Similarly, in flagellar microtubules the (+) end is distal to a centriole-based basal body, and the growth and shrinkage of flagellar microtubules also occurs at the (+) end. In cultured cells, interphase microtubules rapidly grow and shrink. The assembled microtubules are cross-linked into bundles by microtubule associated proteins (MAPs). MAPs are organized into two domains, a microtubule binding domain and a projection domain. The length of the projection domain determines the distance between microtubules in a bundle.

Microtubules act as tracks along which Golgi vesicles, other vesicles, and small particles move. Movement is rapid (2 μm/s) and can be bidirectional along a single microtubule. Two molecular motor proteins that power axonal transport in neurons and vesicle movement in other cells have been purified. One, kinesin, uses energy released by ATP hydrolysis to move vesicles toward the (+) end of the axonal microtubule. The other, cytosolic dynein, powers movement in the opposite direction, toward the (−) end.

Together with other proteins, microtubules form the axonemes of cilia and flagella. An axoneme contains two central singlet microtubules surrounded by an outer ring of nine doublet microtubules, all interconnected by several types of fibrous proteins. Inner and outer arms attached to the outer doublet microtubules are dynein ATPases, which are crucial for ciliary or flagellar movement. Like cytosolic dynein, flagellar dyneins make and then break bonds with adjacent microtubules, causing each doublet to slide relative to its neighbor. Because of the way the doublets are interconnected circumferentially, this motion is converted into the bending of the cilium or flagellum.

Microtubules are the principal structural elements of the mitotic spindle. Bundles of microtubules radiate from the two poles, or microtubule-organizing centers (MTOCs), of the spindle. In animal cells a centriole is at the center of the polar region. Centrioles and flagellar basal bodies are barrel-shaped structures composed of a rim of

nine triple microtubules. In some circumstances a centriole can differentiate into a basal body, and vice versa; both have properties of self-replicating organelles in which new centrioles usually grow during the S phase from the side of a "parent" centriole.

Kinetochore microtubules run from the pole of the mitotic spindle to the kinetochore, an attachment site for microtubules at the centromere of each chromosome. Polar microtubules run from the poles to the equator of the spindle. Forces of an unknown nature align the chromosomes at the equatorial metaphase plate. Polar microtubules grow outward from the pole, and kinetochores bind to the (+) ends of these microtubules. The movement of chromosomes poleward (anaphase A) is driven by depolymerization of the kinetochore microtubules at or near the kinetochore. In contrast, separation of the poles (anaphase B) involves growth of the polar microtubules at the equatorial (+) end; concomitantly, the polar microtubules from each half-spindle slide past each other. ATP hydrolysis is required for anaphase B, but not for anaphase A.

Intermediate filaments (IFs) are, in many cases, key determinants of cellular structure. There are at least five classes of IFs. Desmin filaments connect adjacent Z disks in muscle cells. Keratin filaments in epithelial cells interact with desmosomes that interconnect adjacent cells; these junctions give strength and rigidity to the epithelium, and the keratin fibers allow stresses to be transmitted from cell to cell. Other, related keratin proteins are the major structural proteins of skin, hair, and wool. Long neurofilaments form the core of nerve cell axons and appear to be a major determinant of their elongated shape. Other classes of IF proteins are found in other cell types—glial filaments in glial cells and vimentin in fibroblasts and adipocytes.

All IF proteins have a similar structure. A high conserved, 40-nm-long central rodlike domain is a coiled-coil dimeric structure, formed by coiling of the α-helical sections of two polypeptides around each other. The N and C-terminal domains of an IF protein are globular and vary widely among IF proteins. Some IFs like keratin are obligate heterodimers of type I and type II keratin proteins while other IFs like desmin and lamin can assemble from homodimers. Dimers of IF proteins form antiparallel tetramers, which, in turn, form long protofilaments; eight of these form an IF 10 nm in diameter. In the cytosol, an IF is a dynamic polymer which can be rapidly depolymerized by phosphorylation. IFs are cross-linked to the plasma membrane by Intermediate Filament Associated Proteins (IFAPs) especially at sites of cell-cell (desmosomes) and cell-matric (hemidesmosomes) adhesions. IFs often are found in association with microtubles. Studies of human and mouse skin diseases demonstrate that IFs are like reinforcing threads: through their connections to desmosomes and hemidesmosomes, the criss-cross network of filaments protects the cells from damage by abrasion.

REVIEW QUESTIONS

1. Microtubules are responsible for cell movements; they are polymers of the protein tubulin. Discuss the fundamental structure of a microtubule, including polarity and the role of GTP in its organization. What is a protofilament? What are doublet and triplet microtubules?

What are the differences between short- and long-lived microtubules? Cite examples of each.

What is the microtubule organizing center (MTOC), and what are several of its functions? What proteins are found in the pericentriolar material of the MTOC? Discuss microtubule polarity in relation to the MTOC.

Discuss the isoforms of α-tubulin and β-tubulin. What post-translational modifications are associated with tubulin proteins?

Discuss microtubule assembly and disassembly, being certain to include microtubule polarity in the discussion. What does the phrase microtubule dynamic instability refer to? What two conditions favor microtubule instability; in particular, what is the effect of GTP on the process?

What are the basic functions of microtubule associated proteins (MAPs)? What are the assembly MAPs, and how can they alter the assembly and disassembly of microtubules?

a. You are studying microtubule assembly in vitro. Specifically, you are investigating the effect of mutations in several tubulin monomers on the synthesis of microtubules. The table below describes the properties of several α-tubulin and β-tubulin molecules. Answer each question with a brief explanation.

Tubulin	Phenotype
α-tubulin	Wild type
β-tubulin	Wild type
αI-tubulin	Binds GTP five-times better than wild type
βI-tubulin	Unable to hydrolyze GTP to GDP

(1) If you were to place α-tubulin alone into a test tube with GTP, would polymerization occur?

(2) If you were to place β-tubulin alone into a test tube with GTP, would polymerization occur?

(3) If you were to place α-tubulin and β-tubulin along with GTP in a test tube, would polymerization occur? What

would happen in the presence of colchicine? What would happen in the presence of taxol?

(4) If you were to mix β-tubulin and αI-tubulin in the presence of GTP, would polymerization occur? What about depolymerization?

(5) If you were to mix α-tubulin and βI-tubulin in the presence of GTP, would polymerization occur? What about depolymerization?

2. In neurons, what is meant by anterograde and retrograde transport? What is kinesin, and how does it assist microtubule dependent transport, particularly that associated with anterograde motion in neurons? What are dyneins; how are they involved in retrograde transport?

Describe the axoneme, emphasizing the 9 + 2 arrangement of microtubules and how they are linked? How are dynein and nexin involved in axonemic structure? What is the basal body? This arrangement is characteristically found in cilia and flagella. What are ciliary beat and flagellar beat, and how is axonemal sliding involved in motility by these mechanisms?

3. What is the mitotic apparatus, and how is it organized? Be sure you describe both asters and spindles. What are kinetochores? What is the centromere? Describe the centriole cycle, emphasizing the role of microtubules and kinesin-related proteins in the process.

a. During the course of mitosis, the dynamics of microtubule assembly and disassembly are extraordinarily important. As mentioned in this chapter, several drugs can interfere with microtubule dynamics. Colchicine binds to tubulin monomers, thus preventing polymerization. Where during mitosis would this drug exert its effect?

b. The drug taxol stabilizes microtubule structure, preventing disassembly. Explain at the molecular level how this would interfere with mitosis.

c. Suppose you were investigating GTP analogues that were not hydrolyzable to GDP. How would these interfere with mitosis?

4. Intermediate filaments (IFs) comprise another set of cytoskeletal fibers; describe their characteristics. How are they different from microtubules and microfilaments? What is the fundamental organization of IF subunit proteins? Which regions dictate if IF proteins will form heteropolymeric or homopolymeric dimers? Discuss what is known about polymerization and depolymerization of IFs. What is the function of IF associated proteins? How do IFs stabilize cellular membranes?

a. Fill in the following table concerning the five types of IFs.

IF Type	Examples	Intracellular Location	Cell Types
I			
II			
III			
IV			
V			

References

General References on Microtubules and the Cytoskeleton

BRAY, D. 1993. *Cell Movements.* Garland Press.

HYAMS, J. S., and C. W. LLOYD. 1994. *Microtubules. Wiley-Liss.*

Structure of Microtubules

COHEN, W. D., et al. 1982. The cytoskeletal system of nucleated erythrocytes. I. Composition and function of major elements. *J. Cell Biol.* 93:828–838.

HAYS, T. S., et al. 1989. Interacting proteins identified by genetic interactions: a missense mutation in α-tubulin fails to complement alleles of the testis-specific β-tubulin gene of *Drosophila melanogaster. Mol. Cell Biol.* 9:975–984.

WEBER, K., and M. OSBORN. 1979. Intracellular display of microtubular structures revealed by indirect immunofluorescent microscopy. In *Microtubules,* K. Robert and J. S. Hyams, eds. Academic Press, pp. 279–313.

Assembly and Disassembly of Microtubules in Vitro

BINDER, L. I., W. L. DENTLER, and J. L. ROSENBAUM. 1975. Assembly of chick brain tubulin onto flagellar microtubules from *Chlamydomonas* and sea urchin sperm. *Proc. Nat'l. Acad. Sci. USA* 72:1122–1126.

HORIO, H., and H. HOTANI. 1986. Visualization of the dynamic instability of individual microtubules by dark field microscopy. *Nature* 321:605–607.

KEATES, R. A., and F. R. HALLET. 1988. Dynamic instability of sheared microtubules observed by quasi-elastic light scattering. *Science* 241:1642–1645.

MITCHISON, T., and M. KIRSCHNER. 1984. Dynamic instability of microtubule growth. *Nature* 312:237–242.

TOSO, R. J., et al. 1993. Kinetic stabilization of microtubule dynamic instability in vitro by vinblastine. *Biochemistry* 32:1283–1293.

WALKER, R. A., S. INOUE, and E. D. SALMON. 1989. Asymmetric behavior of severed microtubule ends after ultraviolet-microbeam irradiation of individual microtubules in vitro. *J. Cell Biol.* 108:931–937.

WALKER, R. A., et al. 1988. Dynamic instability of individual microtubules analyzed by video light microscopy: rate constants and transition frequencies. *J. Cell Biol.* 107:1437–1448.

PRYER, N. K., et al. 1992. Brain microtubule-associated proteins modulate microtubule dynamic instability in vitro. *J. Cell Sci.* 103:965–976.

Assembly and Disassembly of Microtubules in Cells

CASSIMERIS, L., N. K. PRYER, and E. D. SALMON. 1988. Real-time observations of microtubule dynamic instability in living cells. *J. Cell Biol.* 107:2223–2231.

GUNDERSEN, G. G., and J. C. BULINSKI. 1988. Selective stabilization of microtubules oriented toward the direction of cell migration. *Proc. Nat'l. Acad. Sci. USA* 85:5946–5950.

KIM, S., 1987. Development of a differentiated microtubule structure: formation of the chicken erythrocyte marginal band in vivo. *J. Cell Biol.* 104:51–59.

KIRSCHNER, M., and T. MITCHISON. 1986. Beyond self-assembly: from microtubules to morphogenesis. *Cell* 45:329–342.

OKABE, S., and N. HIROKAWA. 1988. Microtubule dynamics in nerve cells: analysis using microinjection of biotinylated tubulin into PC12 cells. *J. Cell Biol.* 107:651–664.

SAMMAK, P. J., G. J. GORBSKY, and G. G. BORISY. 1987. Microtubule dynamics in vivo: a test of mechanisms of turnover. *J. Cell Biol.* 104:395–405.

SCHULZE, E., and M. KIRSCHNER. 1987. Dynamic and stable populations of microtubules in cells. *J. Cell Biol.* 104:277–288.

Microtubule-Associated Proteins

BLACK, M. M., and P. W. BAAS. 1989. The basis of polarity in neurons. *Trends Neurosci.* 12:211–214.

ENNULAT, D. J., et al. 1989. Two separate 18-amino acid domains of Tau promote the polymerization of tubulin. *J. Biol. Chem.* 264:5327–5330.

HIMMLER, A., et al. 1989. Tau consists of a set of proteins with repeated C-terminal microtubule-binding domains and variable N-terminal domains. *Mol. Cell Biol.* 9:1381–1388.

HIROKAWA, N., S-I. HISANAGA, and Y. SHIOMURA. 1988. MAP2 is a component of crossbridges between microtubules and neurofilaments in the neuronal cytoskeleton: quick-freeze, deep-etch immunoelectron microscopy and reconstitution studies. *J. Neurosci.* 8:2769–2779.

HIROKAWA, N., Y. SHIOMURA, and S. OKABE. 1988. Tau proteins: the molecular structure and mode of binding on microtubules. *J. Cell Biol.* 107:1449–1459.

MCNALLY, F. J., and R. D. VALE. 1993. Identification of katanine, an ATPase that severs and disassembles stable microtubules. *Cell* 75:419–430.

OKABE, S., and N. HIROKAWA. 1989. Rapid turnover of microtubule-associated protein MAP2 in the axon revealed by microinjection of biotinylated MAP2 into cultured neurons. *Proc. Nat'l. Acad. Sci. USA* 86:4127–4131.

SHIOMURA, Y., and N. HIROKAWA. 1987. The molecular structure of microtubule-associated protein 1A (MAP1A) in vivo and in vitro. An immunoelectron microscopy and quick-freeze, deep-etch study. *J. Neurosci.* 7:1461–1469.

SONG, Y-H., and E. MANDELKOW. 1993. Recombinant kinesin motor domain binds to β-tubulin and decorates microtubules with a B surface lattice. *Proc. Nat'l. Acad Sci. USA* 90:1671–1675.

Multiplicity and Modification of Tubulins

BARRA, H. S., C. A. ARCE, and C. E. ARGARANA. 1988. Posttranslational tyrosination and detyrosination of tubulin. *Mol. Neurobiol.* 2:133–153.

BRÉ, M.-H., T. E. KREIS, and E. KARSENTI. 1987. Control of microtubule nucleation and stability in Madin-Darby canine kidney cells: the occurrence of noncentrosomal, stable detyrosinated microtubules. *J. Cell Biol.* 105:1283–1296.

GU, W., S. A. LEWIS, and N. J. COWAN. 1988. Generation of antisera that discriminate among mammalian α-tubulins: introduction of specialized isotypes into cultured cells results in their coassembly without disruption of normal microtubule function. *J. Cell Biol.* **106**:2011–2022.

KOZMINSKI, K. G., D. R. DIENER, and J. L. ROSENBAUM. 1993. High level expression of nonacylatable α-tubulin in *Chlamydomonas reinhardtii. Cell Motil. Cytoskel.* **25**:158–170.

MARUTA, H., K. GREER, and J. L. ROSENBAUM. 1986. The aceylation of α-tubulin and its relationship to the assembly and disassembly of microtubules. *J. Cell Biol.* **103**:571–579.

SCHATZ, P. J., F. SOLOMON, and D. BOTSTEIN. 1986. Genetically essential and nonessential α-tubulin genes specify functionally interchangeable proteins. *Mol. Cell Biol.* **6**:3722–3733.

SCHULZE, E., et al. 1987. Posttranslational modification and microtubule stability. *J. Cell Biol.* **105**:2167–2177.

WEBSTER, D. R., et al. 1987. Differential turnover of tyrosinated and detyrosinated microtubules. *Proc. Nat'l. Acad. Sci. USA* **84**:9040–9044.

Microtubules and Intracellular Transport

ACHLER, C., et al. 1989. Role of microtubules in polarized delivery of apical membrane proteins to the brush border of the intestinal epithelium. *J. Cell Biol.* **109**:179–189.

ALLEN, R. D., et al. 1985. Gliding movement of and bidirectional transport along single native microtubules from squid axoplasm: evidence for an active role of microtubules in cytoplasmic transport. *J. Cell Biol.* **100**:1736–1752.

DABORA, S. L., and M. P. SHEETZ. 1988. The microtubule-dependent formation of a tubulovesicular network with characteristics of the ER from cultured cell extracts. *Cell* **54**:27–35.

GELLES, J., B. J. SCHNAPP, and M. P. SHEETZ. 1988. Tracking kinesin-driven movements with nanometre-scale precision. *Nature* **331**:450–453.

HIROKAWA, N., et al. 1989. Submolecular domains of bovine brain kinesin identified by electron microscopy and monoclonal antibody decoration. *Cell* **56**:867–878.

HOWARD, J., A. J. HUDSPETH, and R. D. VALE. 1989. Movement of microtubules by single kinesin molecules. *Nature* **342**:154–158.

LEES-MILLER, J. P., D. M. HELFMAN and T. A. SCHROER. 1992. A vertebrate actin-related protein is a component of a multisubunit complex involved in microtubule-based vesicle motility. *Nature* **359**:244–246.

LILLIE, S. H., and S. S. BROWN. 1992. Suppression of a myosin defect by a kinesin-related gene. *Nature* **356**:358–361.

RINDLER, M. J., I. E. IVANOV and D. SABATINI. 1987. Microtubule-acting drugs lead to the nonpolarized delivery of the influenza hemagglutinin to the cell surface of polarized Madin-Darby canine kidney cells. *J. Cell Biol.* **104**:231–241.

ROGALSKI, A. A. and S. J. SINGER. 1984. Association of the Golgi apparatus with microtubules. *J. Cell Biol.* **99**:1101–1109.

ROZDZIAL, M. M., and L. T. HAIMO. 1986. Bidirectional pigment granule movements of melanophores and regulated by protein phosphorylation and dephosphorylation. *Cell* **47**:1061–1070.

SCHNAPP, B. J., and T. S. REESE. 1989. Dynein is the motor for retrograde axonal transport of organelles. *Proc. Nat'l. Acad. Sci. USA* **86**:1548–1552.

SCHROER, T. A., E. R. STEUER, and M. P. SCHEETZ. 1989. Cytoplasmic dynein is a minus end-directed motor for membranous organelles. *Cell* **56**:937–946.

VALLEE, R. B., et al. 1988. Microtubule-associated protein 1C from brain is a two-headed cytosolic dynein. *Nature* **332**:561–563.

YANG, J. T., R. A. LAYMON, and L. S. B. GOLDSTEIN. 1989. A three-domain structure of kinesin heavy chain revealed by DNA sequence and microtubule binding analyses. *Cell* **56**:879–889.

Structure and Function of Cilia and Flagella

AMOS, W. G., L. A. AMOS, and R. W. LINCK. 1986. Studies of tektin filaments from flagella microtubules by immunoelectron microscopy. *J. Cell Sci.* Suppl. 5:55–68.

ASAI, D. J., and C. J. BROKAW. 1993. Dynein heavy chain isoforms and axonemal motility. *Trends Cell Biol.* **3**:398–402.

BROKAW, C. J. 1991. Microtubule sliding in swimming sperm flagella: direct and indirect measurements on sea urchin and tunicate spermatozoa *J. Cell Biol.* **114**:1201–1215.

GIBBONS, I. R. 1981. Cilia and flagella of eukaryotes. *J. Cell Biol.* **91**:107s–124s.

GIBBONS, I. R. 1988. Dynein ATPases as microtubule motors. *J. Biol. Chem.* **263**:15837–15840.

GOODENOUGH, U. W., and J. E. HEUSER. 1985. Outer and inner dynein arms of cilia and flagella. *Cell* **41**:341–342.

HUANG, B. 1986. *Chlamydomonas reinhardtii*: a model system for genetic analysis of flagellar structure and motility. *Int. Rev. Cytol.* **99**:181–215.

KAMIYA, R. 1988. Mutations at twelve independent loci result in absence of outer dynein arms in *Chlamydomonas reinhardtii. J. Cell Biol.* **107**:2253–2258.

LEFEBVRE, P. A., and J. L. ROSENBAUM. 1986. Regulation of the synthesis and assembly of ciliary and flagellar proteins during regeneration. *Ann. Rev. Cell Biol.* **2**:517–546.

LUCK, D. J. L. 1984. Genetic and biochemical dissection of the eukaryotic flagellum. *J. Cell Biol.* **98**:789–794.

SALE, W. S., and P. SATIR. 1977. Direction of active sliding of microtubules in *Tetrahymena* cilia. *Proc. Nat'l. Acad. Sci. USA* **74**:2045–2049.

SALE, W. S., U. W. GOODENOUGH, and J. E. HEUSER. 1985. The substructure of isolated and in situ outer dynein arms of sea urchin sperm flagella. *J. Cell Biol.* **101**:1400–1412.

SATIR P. 1968. Studies on cilia. 3. Further studies on the cilium tip and a "sliding filament" model of ciliary motility. *J. Cell Biol.* **39**:77–94.

SUMMERS, K. E., and I. R. GIBBONS 1971. Adenosine triphosphate-induced sliding of tubules in trypsin-treated flagella of sea-urchin sperm. *Proc. Nat'l. Acad. Sci. USA* **68**:3092–3096.

WARNER, F. D., and P. SATIR. 1974. The structural basis of ciliary bend formation. *J. Cell Biol.* **63**:35–63.

Basal Bodies and Centrioles

BORNENS, M., et al. 1987. Structural and chemical characterization of isolated centrosomes. *Cell Motil. Cytoskel.* **8**:238–249.

BRINKLEY, B. R. 1985. Microtubule organizing centers. *Ann. Rev. Cell Biol.* **1**:145–172.

DOXSEY, S. J., et al. 1994. Pericentrin, a highly conserved centrosome protein involved in microtubule organization. *Cell* **76**:639–650.

JOHNSON, K. A., and J. E. ROSENBAUM. 1992. Replication of basal bodies and centrioles. *Curr. Opin. Cell Biol.* **4**:80–85.

JOSHI, H. C., et al. 1992. γ-Tubulin is a centrosomal protein required for cell-dependent microtubule nucleation. *Nature* **356**:80–83.

KARSENTI, E., and B. MARO. 1986. Centrosomes and the spatial distribution of microtubules in animal cells. *Trends Biochem. Sci.* **11**:460–463.

MCINTOSH, J. R. 1983. The centrosome as an organizer of the cytoskeleton. *Mod. Cell Biol.* **2**:115–142.

MITCHISON, T., and M. W. KIRSCHNER. 1984. Microtubule assembly nucleated by isolated centrosomes. *Nature* **312**:232–237.

STEARNS, T., and M. KIRSCHNER. 1994. In vitro reconstitution of centrosome assembly and function: the central role of γ-tubulin. *Cell* **76**:623–637.

Structure and Function of the Mitotic Spindle

BEGG, D. A., and G. W. ELLIS. 1979. Micromanipulation studies of chromosome movement. *J. Cell Biol.* **82**:528–541.

DING, R., K. L. MCDONALD and J. R. MCINTOSH. 1993. Three-dimensional reconstruction and analysis of mitotic spindles from the yeast *Schizosaccharomyces pombe. J. Cell Biol.* **120**:141–151.

FUNABIKI, H., et al. 1993. Cell cycle-dependent specific positioning and clustering of centromeres and telomeres in fission yeast. *J. Cell Biol.* **121**:961–976.

GORBSKY, G. J., P. J. SAMMAK, and G. G. BORISY. 1988. Microtubule dynamics and chromosome motion visualized in living anaphase cells. *J. Cell Biol.* **106**:1185–1192.

HIUTOREL, P., and M. W. KIRSCHNER. 1988. The polarity and stability of microtubule capture by the kinetochore. *J. Cell Biol.* **106**:151–159.

KIRSCHNER, M., and T. MITCHISON. 1986. Beyond self-assembly: from microtubules to morphogenesis. *Cell* **45**:329–342.

KOSHLAND, D. E., T. J. MITCHISON, and M. W. KIRSCHNER. 1988. Polewards chromosome movement driven by microtubule depolymerization in vitro. *Nature* **331**:499–504.

LI, Y.-Y., et al. 1993. Disruption of mitotic spindle orientation in a yeast dynein mutant. *Proc. Nat'l. Acad. Sci. USA* **90**:10096–10100.

MASUDA, H., K. L. MCDONALD, and W. Z. CANDE. 1988. The mechanism of anaphase spindle elongation: uncoupling of tubulin incorporation and microtubule sliding during in vitro spindle reactivation. *J. Cell Biol.* **107**:623–633.

MCNEILL, P. A., and M. W. BERNS. 1981. Chromosome behavior after laser microirradiation of a single kinetochore in mitotic PtK2 cells. *J. Cell Biol.* **88**:543–553.

MITCHISON, T. J. 1989. Polewards microtubule flux in the mitotic spindle: evidence from photoactivation of fluorescence. *J. Cell Biol.* **109**:637–652.

MITCHISON, T., and E. SALMON. 1992. Poleward kinetochore fiber movement occurs during both metaphase and anaphase-A in newt lung cell mitosis. *J. Cell Biol.* **119**:569–582.

MURRAY, A. W., and J. W. SZOSTAK. 1985. Chromosome segregation in mitosis and meiosis. *Ann. Rev. Cell Biol.* **1**:289–315.

NICKLAS, R. B. 1988. The forces that move chromosomes in mitosis. *Ann. Rev. Biophys. Biophys. Chem.* **17**:431–450.

NICKLAS, R. B. 1989. The motor for poleward chromosome movement in anaphase is in or near the kinetochore. *J. Cell Biol.* **109**:2245–2255.

PICKETT-HEAPS, J. D. 1986. Mitotic mechanisms: an alternative view. *Trends Biochem. Sci.* **11**:504–507.

RIEDER, C. L., and E. D. SALMON. 1994. Motile kinetochores and polar ejection forces dictate chromosome position on the vertebrate mitotic spindle. *J. Cell Biol.* **124**:223–233.

RIEDER, C. L., et al. 1986. Oscillatory movements of mono-oriented chromosomes and their position relative to the spindle pole result from the ejection properties of the aster and half-spindle. *J. Cell Biol.* **103**:581–591.

ROOF, D. M., P. B. MELUH and M. D. ROSE. 1992. Kinesin-related proteins required for assembly of the mitotic spindle. *J. Cell Biol.* **118**:95–108.

ROOS, U.-P. 1976. Light and electron microscopy of rat kangaroo cells in mitosis. III. Patterns of chromosome behavior during prometaphase. *Chromosoma* **54**:363–385.

SAWIN, K. E., et al. 1992. Mitotic spindle organization by a plus end-directed microtubule motor. *Nature* **359**:540–543.

SCHLEGEL, R. A., M. S. HALLECK, and P. N. RAO, eds. 1987. *Molecular Regulation of Nuclear Events in Mitosis and Meiosis.* Academic Press.

SPURCK, T. P., and J. D. PICKETT-HEAPS. 1987. On the mechanism of anaphase A: evidence that ATP is needed for microtubule disassembly and not generation of polewards force. *J. Cell Biol.* **105**:1691–1705.

SULLIVAN, D. S., and T. C. HUFFAKER. 1992. Astral microtubules are not required for anaphase B in *Saccharomyces cerevisiae. J. Cell Biol.* **119**:379–388.

TELZER, B. L., and L. T. HAIMO. 1983. Decoration of spindle microtubules with dynein: evidence for uniform polarity. *J. Cell Biol.* **89**:373–378.

WATERS, J. C., R. W. COLE and C. L. RIEDER. 1993. The force-producing mechanism for centrosome separation during spindle formation in vertebrates is intrinsic to each aster. *J. Cell Biol.* **122**:361–372.

Intermediate Filaments: General References

ALBERS, K., and E. FUCHS. 1992. The molecular biology of intermediate filament proteins. *Int. Rev. Cytol.* **134**:243–279.

STEWART, M. 1993. Intermediate filament structure and assembly. *Curr. Opin. Cell Biol.* **5**:3–11.

Structure of Intermediate Filaments

AEBI, U., et al. 1986. The nuclear lamina is a meshwork of intermediate filament-type filaments. *Nature* **323**:560–564.

AEBI, U., et al. 1983. The fibrillar substructure of keratin filaments unraveled. *J. Cell Biol.* **97**:1131–43.

DONG, D.L.-Y., et al. 1993. Glycosylation of mammalian neurofilaments: Localization of multiple O-linked N-acetylglucosamine moieties on neurofilament polypeptides L and M. *J. Biol. Chem.* **268**:16679–16687.

GEISLER, N., E. KAUFMANN, and K. WEBER. 1985. Antiparallel orientation of the two double-stranded coiled-coils in the tetrameric protofilament unit of intermediate filaments. *J. Mol. Biol.* **182**:173–177.

GEISLER, N., J. SHUNEMANN, and K. WEBER. 1992. Chemical crosslinking indicates a staggered and antiparallel protofilament of desmin intermediate filaments and characterizes one higher-level complex between protofilaments. *Eur. J. Biochem.* **206**:841–852.

HATZFELD, M., and W. W. FRANKE. 1985. Pair formation and promiscuity of cytokeratins: formation in vitro of heterotypic complexes and intermediate-sized filaments by homologous and heterologous recombination of purified polypeptides. *J. Cell Biol.* **101**:1826–1841.

HEINS, S., et al. 1993. The rod domain of NF-L determines neurofilament architecture, whereas the end domains specify filament assembly and network formation. *J. Cell Biol.* **123**:1517–1533.

LEE, M., et al. 1993. Neurofilaments are obligate heteropolymers in vivo. *J. Cell Biol.* **122**:1337–1350.

MCCORMICK, M. B., et al. 1993. The roles of the rod end and the tail in vimentin IF assembly and IF network formation. *J. Cell Biol.* **122**:395–407.

MCKEON, F. D., M. W. KIRSCHNER, and D. CAPUT. 1986. Homologies in both primary and secondary structure between nuclear envelope and intermediate filament proteins. *Nature* **319**:463–468.

SOELNER, P., R. A. QUINLAN, and W. W. FRANKE. 1985. Identification of a distinct soluble subunit of an intermediate filament protein: tetrameric vimentin from living cells. *Proc. Nat'l. Acad. Sci. USA* **82**:7929–7933.

STEVEN, A. C., et al. 1983. Epidermal keratin filaments assembled in vitro have masses-per-unit-length that scale according to average subunit mass: structural basis for homologous packing of subunits in intermediate filaments. *J. Cell Biol.* **97**:1939–44.

Assembly of Intermediate Filaments

BUXTON, R. S., and A. I. MAGEE. 1992. Structure and interactions of desmosomal and other cadherins. *Semin. Cell Biol.* **3**:157–167.

CHOU, Y. H., et al. 1990. Intermediate filament reorganization during mitosis is mediated by p34cdc2 phosphorylation of vimentin. *Cell* **62**:1063–1071.

ERICKSON, J. E., et al. 1992. Cytoskeletal integrity in interphase cells requires protein phosphatase activity. *Proc. Nat'l. Acad. Sci. USA* **89**:11093–11097.

PASDAR, M., and W. J. NELSON. 1988. Kinetics of desmosome assembly in Madin-Darby canine kidney epithelial cells: temporal and spatial regulation of desmoplakin organization and stabilization upon cell-cell contact. II. Morphological analysis. *J. Cell Biol.* **106**:687–695.

PETER, M., et al. 1990. In vitro disassembly of the nuclear lamina and M phase-specific phosphorylation of lamins by cdc2 kinase. *Cell* **61**:591–602.

SHETTY, K. T., W. T. LINK and H. C. PANT. 1993. Cdc2-like kinase from rat spinal cord specifically phoshorylates KSPXK motifs in neurofilament proteins: Isolation and characterization. *Proc. Nat'l. Acad. Sci. USA* **90**:6844–6848.

VIKSTROM, K. L., S-S. LIM, and R. D. GOLDMAN. 1992. Steady state dynamics of intermediate filament networks. *J. Cell Biol.* **118**:121–129.

VIKSTROM, K. L., G. G. BORISY, and R. D. GOLDMAN. 1989. Dynamic aspects of intermediate filament networks in BHK-21 cells. *Proc. Nat'l. Acad. Sci. USA* **86**:549–553.

Functions of Intermediate Filaments

COULOMBE, P. A., et al. 1991. Point mutations in human keratin 14 genes of epidermolysis bullosa simplex patients: genetic and functional analysis. *Cell* **66**:1301–1311.

COULOMBE, P. A., et al. 1991. A function for keratins and a common thread among different types of epidermolysis bullosa simplex disease. *J. Cell Biol.* **115**:1661–1674.

FRANKE, W. W., M. HERGT, and C. GRUND. 1987. Rearrangement of the vimentin cytoskeleton during adipose conversion: formation of an intermediate filament cage around lipid globules. *Cell* **49**:131–141.

GEROGATOS, S. D., and G. BLOBEL. 1987. Two distinct attachment sites for vimentin along the plasma membrane and nuclear envelope in avian erthrocytes: a basis for vectorial assembly of intermediate filaments. *J. Cell Biol.* **105**:105–115.

GEORGATOS, S. D., J. MEIER and G. SIMOS. 1994. Lamins and lamin-associated proteins. *Curr. Opin. Cell Biol.* **6**:347–353.

HOFFMAN, P. N., et al. 1987. Role of neurofilaments in the control of axonal caliber in myelinated nerve fibers. In eds., R. Lasek and M. Black, *Intrinsic Determinants of Neuronal Form and Function.* Alan R. Liss, pp. 389–402.

JONES, J. C. R., et al. 1985. The organizational fate of intermediate filament networks in two epithelial cell types during mitosis. *J. Cell Biol.* **100**:3–102.

MILLER, R. K., K. VIKSTROM and R. D. GOLDMAN. 1991. Keratin incorporation into intermediate filament networks is a rapid process. *J. Cell Biol.* **113**:843–855.

OHARA, O., et al. 1993. Neurofilament deficiency in quail caused by non-sense mutation in neurofilament-L gene. *J. Cell Biol.* **121**:387–395.

VASSAR, R., et al. 1991. Mutant keratin expression in transgenic mice causes marked abnormalities resembling a human genetic skin disease. *Cell* **64**:365–380.

Multicellularity: Cell-Cell and Cell-Matrix Interactions

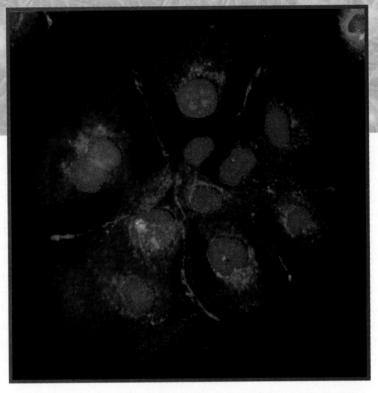

▲ Human skin carcinoma cells forming tight cell–cell contacts in culture; an as yet unidentified cell adhesion protein is stained with a green-fluorescing monoclonal antibody.

The evolution of multicellular organisms permitted specialized cells and tissues to form; a flowering plant has at least 15 types of cells, and a vertebrate hundreds. In both plants and animals, cells specialized to carry out a particular task are found together in tissues in which the task is performed: a xylem or meristem; a liver, a muscle, or a nerve ganglion. Different types of cells in a tissue are precisely arranged in patterns often exhibiting staggering complexity. In the human brain, for instance, there are hundreds of different types of neurons interconnected to one another through a network of some 10^{15} synaptic connections! The coordinated functioning of many types of cells within tissues, and of multiple specialized tissues, permits the organism as a whole to move, metabolize, reproduce, and carry out other essential activities.

A key step in the evolution of multicellularity must have been the ability of cells to contact tightly and interact specifically with other cells. Cell-surface proteins enable many animal cells to adhere tightly and specifically with cells of the same, or a similar, type; these interactions allow populations of cells to segregate into distinct tissues. Following aggregation, cells elaborate specialized junctions that stabilize these interactions and promote local communication between adjacent cells. Animal cells also secrete a complex network of secreted proteins and carbohydrates, the *extracellular matrix,* that fills the spaces between cells. The matrix helps bind the cells in tissues together and is a reservoir for many hormones controlling cell growth and differentiation. The matrix also provides a lattice through which cells can move, particularly during the early stages of differentiation.

In contrast, the cell walls that surround plant cells are thicker, more impermeable, and more rigid than the extracellular matrix around animal cells. Most importantly, plant cells can grow larger but do not move in relation to their neighbors. Because of these differences, the plant cell wall and its interactions with plant cells will be treated separately at the end of this chapter. Our main focus is on animal cells. In the first part of the chapter we describe the components of the extracellular matrix and the cell-surface proteins that mediate adhesion of cells to the matrix and to other cells.

Local interactions between cells often lead to changes in cell shape and movement, and to changes in gene expression leading to cellular differentiation. For instance, wounding of the adult animal leads to invasion of leukocytes (white blood cells) into tissue at the sites of infection; this and subsequent steps of wound healing are mediated by local interactions among different types of cells. During development, local interactions between cells and their environment induce the formation of different cell types and their organization in tissues; they also guide the migration of cells and direct the formation of precise patterns of synaptic connections among developing nerve cells. Principally, these interactions include locally active secreted proteins, direct contact between surface proteins on adjacent cells, and specific contacts between cells and constituents of the extracellular matrix. We explore the role of these types of interactions in animal development in the middle part of this chapter. In earlier chapters, we discussed the regulation of development at the level of mRNA synthesis (Chapters 11, 12 and 13) and the mechanisms by which cells interpret extracellular signals that lead to changes in cellular behavior and transcriptional activity (Chapter 20). It is through the integration of these three basic cellular processes—signaling interactions between cells, intracellular transduction of these signals, and activation of specific

transcriptional programs—that multicellular organisms develop.

The Extracellular Matrix: Primary Components and Functions

Animals contain many types of extracellular matrices, each specialized for a particular function such as strength (in a tendon), filtration (in the kidney glomerulus where urine is formed), or adhesion. For instance, the smooth muscle cells that surround an artery are connected by an extracellular matrix that provides strong but flexible connections. Extracellular matrices consist of different combinations of *collagen* proteins, which form either long fibers or porous sheets; the long polysaccharide *hyaluronan;* and *proteoglycans,* macromolecules consisting of a core protein covalently attached to highly charged polysaccharides called *glycosaminoglycans.* Multiadhesive proteins also are important matrix constituents; they bind to receptor proteins on the surfaces of cells and to other matrix components, imparting strength and rigidity to the matrix (Figure 24-1).

In many animal tissues such as the epidermis of the skin, there is little extracellular space between the cells (Figure 24-2). Most epithelia and other organized groups of cells (such as muscle) rest upon, and are tightly bound to, a thin matrix called the *basal lamina,* which is linked to the plasma membrane by specific receptor proteins. The basal lamina is tightly connected to fibrous collagens and other materials in the underlying loose connective tissue

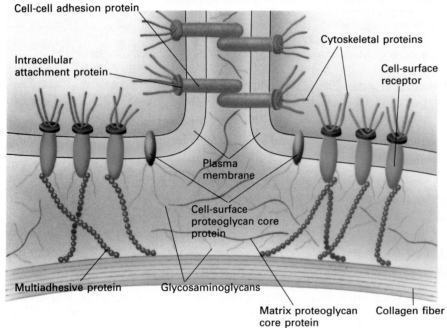

Cell-cell adhesion protein

Intracellular attachment protein

Cytoskeletal proteins

Cell-surface receptor

Plasma membrane

Cell-surface proteoglycan core protein

Multiadhesive protein

Glycosaminoglycans

Matrix proteoglycan core protein

Collagen fiber

◄ FIGURE 24-1 Cell-adhesion molecules bind cells together and, *via* intracellular attachment proteins, form anchors for cytoskeletal proteins. Other plasma-membrane receptor proteins form connections with components of the extracellular matrix and also, *via* attachment proteins, with cytoskeletal proteins. Multiadhesive proteins bind to cell-surface receptor proteins and to other matrix components. Proteoglycans, consisting of a core protein, to which glycosaminoglycan chains are attached, also participate in adhesion of cells to each other and to the protein components of the matrix. Together, these interactions allow cells to adhere to each other, interconnect the cytoskeletons of adjacent cells, and give tissues their strength and resistance to shear forces.

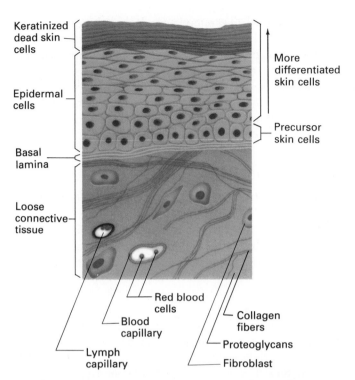

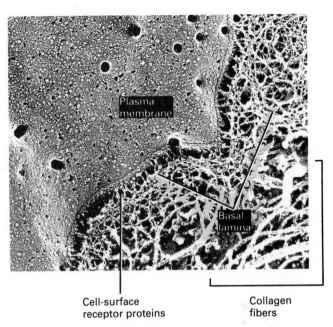

▲ FIGURE 24-2   A schematic drawing of a section through the outer part of the skin of a pig. The cellular epidermis rests on the thin basal lamina, which in turn contacts the thick layer of loose connective tissue consisting of abundant collagen fibers and cells—mostly fibroblasts—that synthesize the connective-tissue proteins, hyaluronan, and proteoglycans. The regions between the collagen fibers are filled with proteoglycans and multiadhesive proteins. Blood and lymph capillaries course through the loose connective tissue. As the epidermal precursors of skin cells differentiate into mature keratinized cells they lose receptors that bind them to the basal lamina and synthesize other proteins that form the desmosome junctions that bind adjacent cells tightly together.

▲ FIGURE 24-3   Association of the plasma membrane of skeletal muscle with the basal lamina. In this quick-freeze deep-etch preparation, the basal lamina is seen as a meshwork of filaments. Some of the basal lamina filaments contact receptor proteins in the plasma membrane; others bind to the thick collagen fibers that form the connective tissue around the muscle. [From D. W. Fawcett, 1981, *The Cell*, 2d ed., Saunders/Photo Researchers; courtesy John Heuser.]

(Figure 24-3). In the glomerulus of the kidney, the basal lamina forms a porous filter that allows water, ions, and small molecules in blood to cross into the urinary space while retaining proteins and cells in the blood.

Animals contain abundant connective tissue in which much of the tissue volume consists of spaces between cells. *Loose connective tissue* forms the bedding on which most small glands and epithelia lie and connects to the basal lamina around the cells. Proteins such as fibrous collagens and the rubber-like flexible protein *elastin* give tissues their shape, rigidity, and flexibility. Most of the space is filled with the highly hydrated hyaluronan and proteoglycans, giving tissues a gel-like consistency. Loose connective tissue is highly cellular and contains numerous fibroblasts, which synthesize much of the extracellular matrix, and also blood-borne cells such as lymphocytes. Most blood

capillaries that bring O_2 and nutrients to the cells in all parts of the body are confined to loose connective tissue. A principal function of the spaces between cells in this tissue is to allow O_2 and nutrients to diffuse freely into the cells of the epithelia and glands (see Figure 24-2).

Dense connective tissue, including bone, cartilage, and tendon, is a major component of organs in which strength or flexibility or both are essential attributes—the skeletal system, for instance. Unlike loose connective tissue, it consists almost entirely of fibrous extracellular matrix materials and contains few cells. The extracellular matrices of dense connective tissue consist of densely packed collagens and other fibrous proteins surrounded by glycoproteins, proteoglycans, and other substances (e.g., elastin) that are produced and secreted by the relatively few cells present.

The extracellular matrix is not just an inert framework or cage that supports or surrounds cells. The matrix is required for many cells to carry out specialized functions. For example, hepatocytes—the principal cell in the liver—must contact the basal lamina or other extracellular matrix in order to synthesize cell-type specific proteins. When liver cells are dissociated and plated on plastic tissue-culture dishes or even on certain types of fibrous collagens, they stop translating mRNAs encoding liver-specific proteins (e.g., albumin) and stop synthesizing these mRNAs; production of these mRNAs and proteins is maintained only if the cultured cells are surrounded by a matrix of appropriate composition. Hepatocytes, like most other cells, also require contact with an appropriate matrix to maintain their proper cytoskeletal architecture.

The extracellular matrix has other functions as well. It binds many growth factors and other hormones and can either sequester these signals from the cells that contact it or, conversely, present the hormonal signal to these cells. Specific matrix components are needed for cells to differentiate, that is, to acquire the functions of a specific cell type. Morphogenesis—the later stage of embryonic development during which form is achieved by cell movements and rearrangements—also is critically dependent on matrix molecules. In developing organisms, matrix components are constantly being remodeled, degraded, and resynthesized locally. Even in adults—in areas of wounding, for example—degradation and resynthesis of matrix components occurs.

We begin our discussion of the extracellular matrix with its most abundant structural component, collagen, and then discuss, in turn, the other principal matrix proteoglycans and proteins.

► Collagen: A Class of Multifunctional Fibrous Proteins

Collagen is the major class of insoluble fibrous protein in the extracellular matrix and in connective tissue. In fact, it is the single most abundant protein in the animal kingdom.

TABLE 24-1   The Collagens

Type	Chains	Molecule Composition	Length of Triple Helix; Structural Details	Representative Tissues
COLLAGENS THAT FORM LONG FIBRILS WITH STRIATIONS EVERY 67 NM				
I	$\alpha1(I)$; $\alpha2(I)$	$[\alpha1(I)]_2[\alpha2(I)]$	300 nm; 67-nm banded fibrils	Skin, tendon, bone, dentin
		$[\alpha1(I)]_3$		Fetal form in skin, other tissues
II	$\alpha1(II)$	$[\alpha1(II)]_3$	300 nm; small 67-nm banded fibrils	Cartilage, vitreous humor
III	$\alpha1(III)$	$[\alpha1(III)]_3$	300 nm; 67-nm banded fibrils	Skin, muscle, blood vessels, frequently together with type I collagen
V	$\alpha1(V)$; $\alpha2(V)$; $\alpha3(V)$	$[\alpha1(V)]_3$ $[\alpha1(V)]_2[\alpha2(V)]$ $[\alpha1(V)][\alpha2(V)][\alpha3(V)]$	390 nm; N-terminal globular domain; frequently together with type I collagen	Cell cultures, fetal tissues, fetal membranes, skin, bones, placenta, most interstitial tissues
XI	$\alpha1(XI)$; $\alpha2(XI)$; $\alpha3(XI)$	$[\alpha1(XI)][\alpha2(XI)][\alpha3(XI)]$	300 nm; small fibers; frequently together with type II collagen	Cartilage
FIBRIL-ASSOCIATED COLLAGENS WITH INTERRUPTED TRIPLE HELIXES				
IX	$\alpha1(IX)$; $\alpha2(IX)$; $\alpha3(IX)$	$[\alpha1(IX)][\alpha2(IX)]$ $[\alpha3(IX)]$	200 nm; N-terminal globular domain; bound glycosamino-glycan; associated with type II collagen	Cartilage, vitreous humor
XII	$\alpha1(XII)$	$[\alpha1(XII)]_3$	Large N-terminal domain; cross-shaped molecule; interacts with type I collagen	Embryonic tendon and skin
XIV	$\alpha1(XIV)$	$[\alpha1(XIV)]_3$	Large N-terminal domain; cross-shaped molecule	Fetal skin and tendon

There are at least fourteen types of collagen (Table 24-1). Types I, II, and III are the most abundant; these collagen molecules pack together to form long thin *fibrils* of similar structure (Figure 5-24). Type IV, in contrast, forms a two-dimensional reticulum and is a principal component of the basal lamina. The structural features of each type of collagen make it suitable for a particular function. At one time it was thought that all collagens were secreted by connective-tissue cells called fibroblasts, but we now know that numerous epithelial cells make certain types of collagens.

The Basic Structural Unit of Collagen Is a Triple Helix

Because its abundance in tendon-rich tissue such as rat tail makes the fibrous type I collagen easy to isolate, it was the first to be characterized. Its fundamental structural unit is a long (300 nm), thin (1.5-nm diameter) protein that consists of three coiled subunits: two $\alpha1(I)$ chains and one $\alpha2(I)$.[1] Each chain contains precisely 1050 amino acids wound around each other in a characteristic right-handed triple helix (Figure 24-4a). All collagens were eventually shown to contain three-stranded helical segments of similar structure; the unique properties of each type of collagen are due mainly to segments that interrupt the triple helix, and that fold into other kinds of three-dimensional structures.

---

[1] In collagen nomenclature, the collagen type is in Roman numerals and is placed in parentheses.

TABLE 24-1 *(Continued)*

Type	Chains	Molecule Composition	Length of Triple Helix; Structural Details	Representative Tissues
FIBRIL-ASSOCIATED COLLAGENS THAT FORM BEADED FILAMENTS				
VI	$\alpha1(VI)$; $\alpha2(VI)$; $\alpha3(VI)$	$[\alpha1(VI)][\alpha2(VI)]$ $[\alpha3(VI)]$	150 nm; N- and C-terminal globular domains; microfibrils banded every 100 nm; associated with type I	Most interstitial tissues
COLLAGENS THAT FORM SHEETS				
IV	$\alpha1(IV)$; $\alpha2(IV)$; $\alpha3(IV)$; $\alpha4(IV)$; $\alpha5(IV)$	$[\alpha1(IV)]_2[\alpha2(IV)]$ other forms	Two-dimensional cross-linked network	All basal laminas
VIII	$\alpha1(VIII)$; $\alpha2(VIII)$	Not known	Regular triangular lattice	Endothelial cells; Descement's membrane separating corneal epithelial cells from the stroma
X	$\alpha1(X)$	$[\alpha1(X)]_3$	150 nm; C-terminal globular domain	Growth plate (mineralizing cartilage forming bone)
COLLAGENS THAT FORM ANCHORING FIBRILS				
VII	$\alpha1(VII)$	$[\alpha1(VII)]_3$	450 nm; dimeric; globular domains at each end	Epithelia (anchors skin basal lamina to underlying stroma)
COLLAGENS KNOWN FROM cDNA CLONING ONLY				
XIII	$\alpha1(XIII)$	Not known	Not known	Endothelial cells

SOURCE: K. Kuhn, 1987, in R. Mayne and R. Burgeson, eds., *Structure and Function of Collagen Types,* Academic Press, p. 2; M. van der Rest and R. Garrone, 1991, *FASEB J.* 5:2814–2823.

(a)

α Chain

Portion of a collagen
molecule (a right-hand
triple helix)

(b)

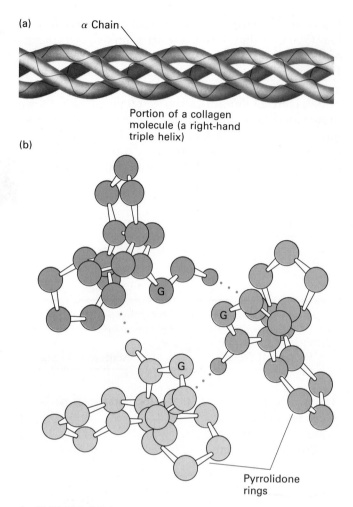

G

G

G

Pyrrolidone
rings

▲ FIGURE 24-4   The structure of collagen. (a) The basic structural unit is a triple-stranded helical molecule. Hydrogen bonds (not shown here) link residues in each chain to the other two chains, which makes the helix rigid. (b) Ball-and-stick model of a collagen triple helix viewed down the helix axis. The α carbon of glycine is labeled G; every third residue must be glycine because the space in the center of the helix will not accommodate a side-chain larger than H. The five-membered pyrrolidone rings from the proline residues are on the outside of the helix. The dots indicate the hydrogen bonds between the NH of a glycine residue on one chain with a peptide C=O residue on another.

There are three amino acids per turn of the collagen helix and every third amino acid is glycine, the smallest amino acid. Collagen is also rich in proline and hydroxyproline (see Figure 3-17). The rigid five-membered pyrrolidone rings in these amino acids are on the outside of the triple helix; the fixed angle of the C–N peptidyl-proline or peptidyl-hydroxyproline bond enables each polypeptide chain to fold into a helix with a geometry such that three polypeptide chains can twist together to form a three-stranded helix. The side chain of glycine, an H atom, is the only one that can fit into the crowded center of a three-stranded helix. Many regions of collagen chains are com-

posed of the repeating motif Gly-Pro-X, where X can be any amino acid. Hydrogen bonds linking the peptide bond NH of a glycine residue with a peptide carbonyl (C=O) group in an adjacent polypeptide help hold the three chains together (see Figure 24-4b). Interestingly, although the rigid peptidyl-proline linkages disrupt the packing of amino acids in an α helix, they stabilize the rigid three-stranded collagen helix.

Most Exons in Fibrous Collagen Genes Encode Multiple Gly-X-Y Sequences

The unusual amino acid sequence of the fibrous collagens (i.e., types I, II, III, and V) reflects their gene organization. For example, the gene encoding the human α1(I) chain, called the pro-α1(I) gene, contains 51 exons; 46 of these exons are quite small and encode portions of the triple-helical segment of the collagen molecule (Figure 24-5). About half of these 46 exons consist of 54 base pairs and encode one so-called *primordial unit* containing six repeats of a Gly-X-Y sequence. Since three amino acids form one turn of the triple-stranded helix, each primordial unit corresponds to six turns of the helix.

It is thought that the ancestral collagen gene encoded one primordial unit and that the ancestral collagen protein contained six Gly-X-Y sequences. Present-day fibrous collagen genes evolved by multiple duplications of this ancestral primordial-unit exon. Some exons in the human pro-α1(I) gene encode two or three of these primodial units (e.g., exon 36 with 108 bp and exon 39 with 162 bp). Other exons are 45 base pairs long and appear to have lost 9 base pairs (encoding one Gly-X-Y) from the ancestral primordial-unit exon during evolution (see Figure 24-5). The genes encoding the chains of other fibrous collagens have a similar structure.

Collagen Fibrils Form by Lateral Interactions of Triple Helices

Many three-stranded type I collagen molecules pack together side-by-side, forming fibrils of 50 to 200 nm diameter. In fibrils, adjacent collagen molecules are displaced from one another by about one-quarter their length, 67 nm, as shown in Figure 24-6a. This staggered array produces a striated effect that can be seen in electron micrographs of stained collagen fibrils; the characteristic pattern of bands is repeated about every 67 nm (Figure 24-6b; see also Figure 5-24). The bands are caused by binding of a metal or a stain to certain repetitive sequences of amino acids in collagen; the stain also fills the space between the ends of collagen molecules. The unique properties of the fibrous collagens—types I, II, III, and V—are due to the ability of the rodlike triple helices to form such side-by-side interactions.

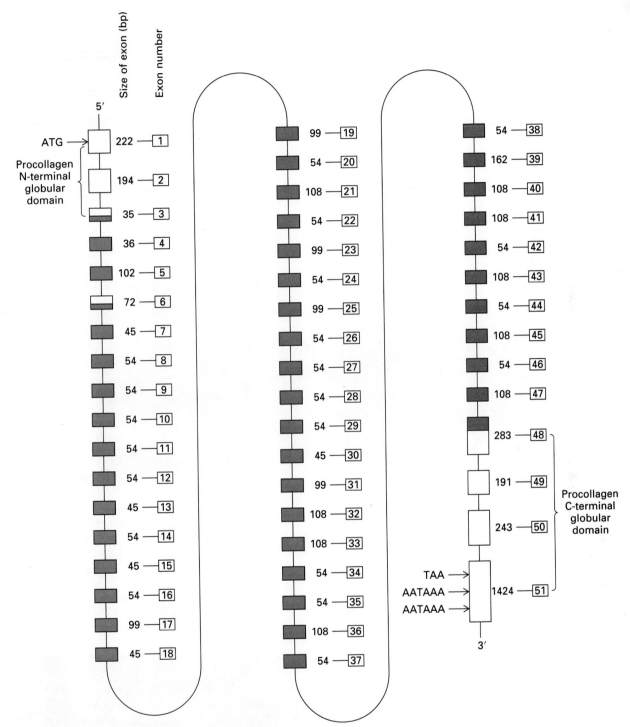

▲ FIGURE 24-5   Intron-exon structure of the entire human pro-α1(I) collagen gene. The exons colored red encode Gly-X-Y sequences that form the triple helices. One primordial unit equals (Gly-X-Y)$_6$ and is encoded by 54 bp.

[Adapted from L. Sandell and C. Boyd, 1990, in L. Sandell and C. Boyd, eds., *Extracellular Matrix Genes*, Academic Press, San Diego, pp. 1–56.]

Type I collagen fibrils have enormous tensile strength; that is, such collagen can be stretched without being broken. These fibrils, roughly 50 nm in diameter and several micrometers long, are packed side-by-side in parallel bundles, called *collagen fibers*, in tendons, where they connect

muscles with bones and must withstand enormous forces (Figure 24-7). Gram for gram, type I collagen is stronger than steel.

Short segments at either end of the collagen chains are of particular importance in the formation of collagen

(a)

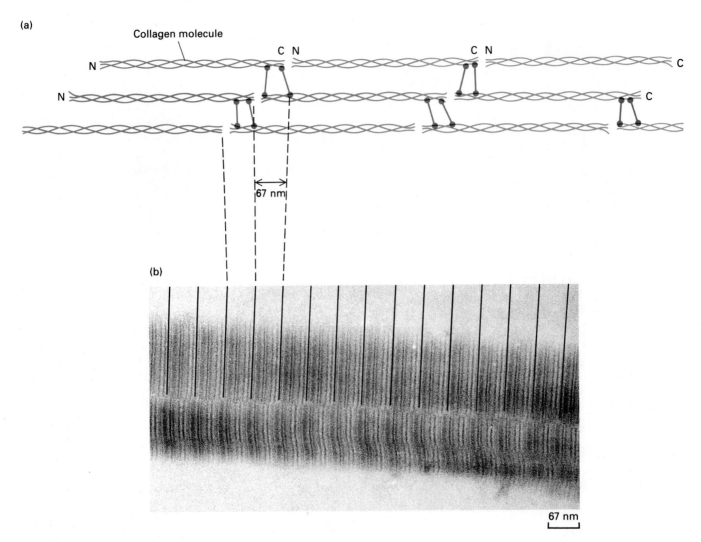

(b)

67 nm

▲ FIGURE 24-6   The structure of type I collagen fibrils.
(a) Each triple-stranded collagen molecule is 300 nm long. In
fibrous collagen, collagen molecules pack together side by
side. Adjacent molecules are displaced 67 nm, or slightly less
than one-fourth the length of a single molecule. A small gap
separates the "head" of one collagen from the "tail" of the
next. The side-by-side interactions are stabilized by covalent
bonds (red) between the N-terminus of one molecule and
the C-terminus of an adjacent one. (b) An electron micrograph of calfskin collagen fibrils stained with phosphotungstic
acid. As indicated by the leaders, the striations created by
the 67-nm periodic pattern of the packing are clearly visible.
[Part (b) courtesy of R. Bruns.]

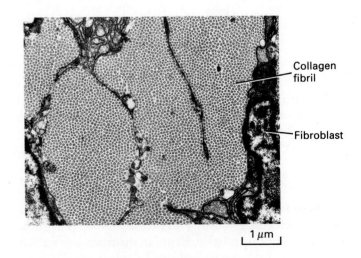

▶ FIGURE 24-7   Electron micrograph of the dense connective tissue of a chick tendon. Most of the tissue is occupied by parallel type I collagen fibrils, about 50 nm in diameter, seen here in cross section. The cellular content of the
tissue is very low. [From D. A. D. Parry, 1988, *Biophys.
Chem.* **29**:195.]

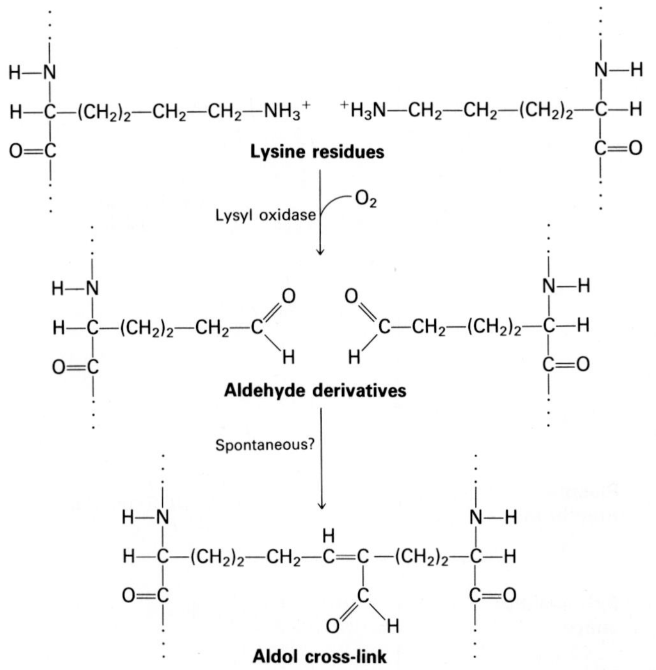

▲ FIGURE 24-8 The side-by-side interactions of collagen helices are stabilized by an aldol cross-link between two lysine side chains. The extracellular enzyme lysyl oxidase catalyzes formation of the aldehyde groups.

fibrils (see Figure 24-6). These segments do not assume the triple-helical conformation and contain the unusual amino acid *hydroxylysine* (see Figure 3-17). Covalent aldol cross-links form between two lysine or hydroxylysine residues at the C-terminus of one collagen molecule with two similar residues at the N-terminus of an adjacent molecule (Figure 24-8). These cross-links stabilize the side-by-side packing of collagen molecules and generate a strong fibril. Figure 5-24 shows a micrograph of collagen fibrils.

Denatured Collagen Polypeptides Cannot Renature to Form a Triple Helix

Largely due to the covalent cross-links between adjacent molecules, type I collagen purified from tissues of mature animals is insoluble in aqueous solutions. In type I collagen from young animals, however, many of these cross-links have not yet formed; thus this collagen is soluble. In type I collagen from young animals, the three-dimensional structure of the triple helix is stabilized mostly by multiple, weak interactions including hydrogen bonds between the three chains. When a solution of type I collagen from a young animal is heated above 40°C, the rodlike helix becomes denatured, and the three chains separate from each other. Such denatured collagen cannot spontaneously renature to form a completely normal collagen triple helix, be-

cause the polypeptides may associate at many places other than the ends, forming triple helices that are out of register, with single-stranded segments at each end. During biosynthesis, in-register alignment of the three polypeptides is facilitated by special propeptide sequences at each end, as explained in the next section.

Procollagen Chains Assemble into Triple Helices in the Rough ER and Are Modified in the Golgi Complex

Collagen chains are synthesized as longer precursors called *procollagens;* the growing peptide chains are co-translationally transported into the lumen of the rough endoplasmic reticulum (ER). Within the ER lumen, the procollagen chains are modified and then assembled into triple helices (Figure 24-9). All three chains in type I procollagen contain 150 more amino acids at the N-terminus and 250 more at the C-terminus than are present in mature type I collagen. These additional amino acids constitute the *propeptides*. Except for a short collagen-like helical region in the N propeptide, both propeptides have a globular conformation. Although type I collagen triple helices themselves contain no disulfide bonds (some are present in other collagens), both propeptides contain several intrachain disulfide bonds, and five interchain disulfide bonds connect the C-terminal propeptides of the three procollagen chains. The interchain disulfide bonds are important in producing in-register alignment of the three chains before formation of the triple helix. These bonds are believed to form first; then, the three chains can zip together in the C → N direction to form the triple helix. This hypothesis has been supported by the results of in vitro studies. For example, in vitro renaturation of thermally denatured procollagen is accelerated substantially if the proper interchain disulfide bonds form first, and the chains in the resulting triple helices are in register.

As outlined in Figure 24-9, the signal sequence is cleaved from procollagen chains and several other modifications occur before the triple helix is assembled. Specific proline residues in the middle of the chains are hydroxylated by the membrane-bound enzymes prolyl 4-hydroxylase and prolyl 3-hydroxylase (Figure 24-10), and certain lysine residues are hydroxylated by lysyl hydroxylase. Ascorbic acid (vitamin C) is an essential cofactor for both prolyl hydroxylases. In the absence of ascorbate, the procollagen chains are not hydroxylated sufficiently to form stable triple helices at 37°C (Figure 24-11), nor can they form normal fibrils. Nonhydroxylated procollagen chains are degraded within the cell. Consequently, fragility of blood vessels, tendons, and skin is characteristic of the disease scurvy, which results from a deficiency of vitamin C in the diet.

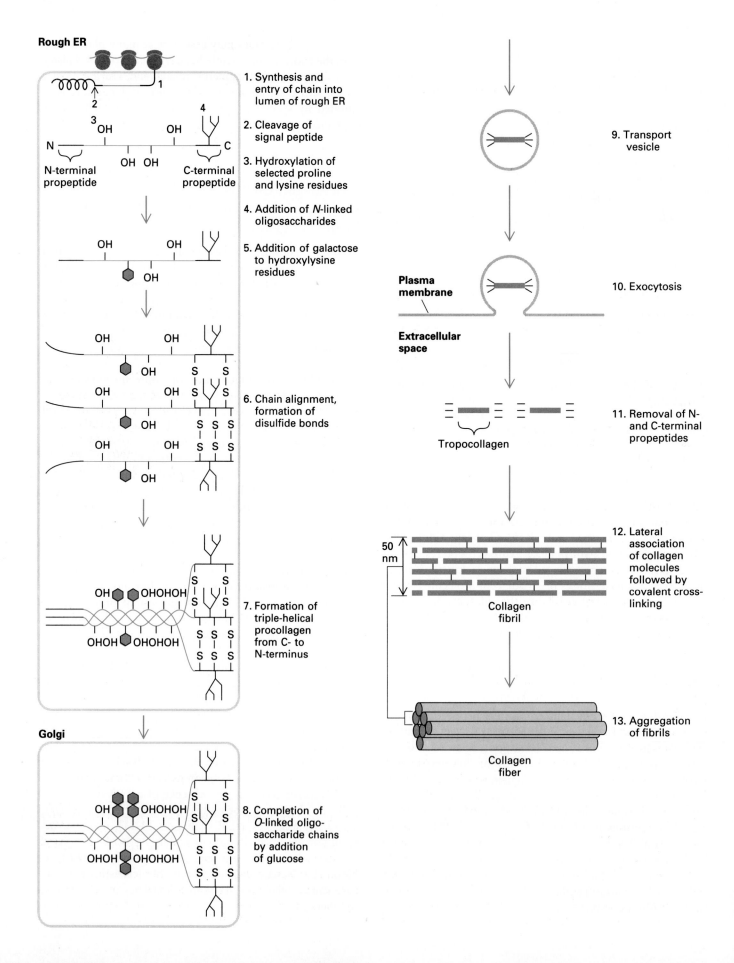

Rough ER

1. Synthesis and entry of chain into lumen of rough ER

2. Cleavage of signal peptide

N-terminal propeptide

C-terminal propeptide

3. Hydroxylation of selected proline and lysine residues

4. Addition of N-linked oligosaccharides

5. Addition of galactose to hydroxylysine residues

6. Chain alignment, formation of disulfide bonds

7. Formation of triple-helical procollagen from C- to N-terminus

Golgi

8. Completion of O-linked oligosaccharide chains by addition of glucose

9. Transport vesicle

Plasma membrane

10. Exocytosis

Extracellular space

Tropocollagen

11. Removal of N- and C-terminal propeptides

50 nm

Collagen fibril

12. Lateral association of collagen molecules followed by covalent cross-linking

Collagen fiber

13. Aggregation of fibrils

◄ FIGURE 24-9   Major events in the biosynthesis of fibrous collagens. Modifications of the collagen polypeptide in the endoplasmic reticulum include hydroxylation, glycosylation, and disulfide-bond formation. Interchain disulfide bonds between the C-terminal propeptides of three procollagens align the chains in register and initiate formation of the triple helix. The process continues, zipperlike, toward the N-terminus. All modifications occur in a precise sequence in the rough ER, Golgi complex, and the extracellular space, and allow lateral alignment and formation of the covalent cross-linkers that enable helices to pack into 50-nm-diameter fibrils. [After M. E. Nimni, 1993, in M. Zern and L. Reid, eds., *Extracellular Matrix*, Marcel Dekker, pp. 121–148.]

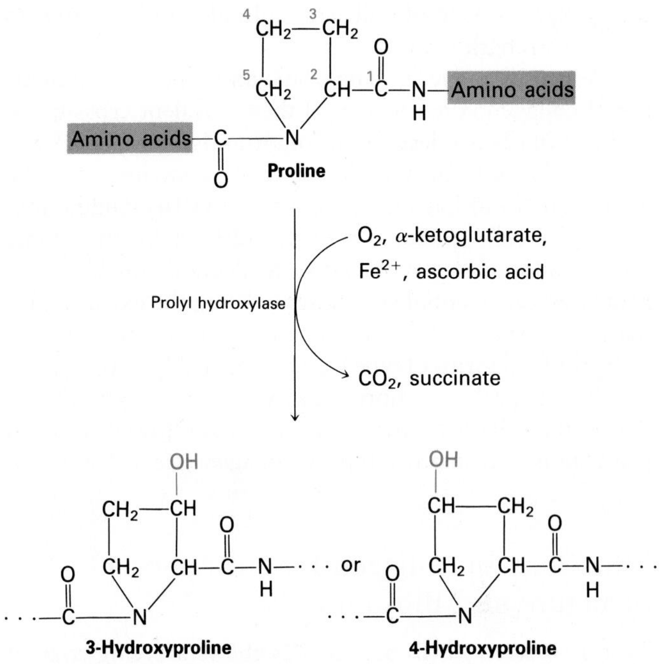

As with other secreted proteins, glycosylation of procollagen occurs in the rough ER and Golgi complex. Galactose and glucose residues are added to hydroxylysine residues, and long oligosaccharides are added to certain asparagine residues in the propeptides.

Collagen Is Assembled into Fibrils after Secretion

During or following exocytosis, type I procollagen molecules undergo proteolytic cleavage by extracellular enzymes, the procollagen peptidases, that remove the N-terminal and C-terminal propeptides. The resulting protein, often called *tropocollagen* (or simply collagen), consists almost entirely of a triple-stranded helix (see Figure 24-9). Excision of both propeptides allows the collagen molecules to polymerize into normal fibrils. The potentially catastrophic assembly of fibrils within the cell does not occur because the propeptides are removed only after secretion. The inability to remove the propeptides from tropocollagen results in formation of abnormal collagen fibrils with detrimental consequences. For example, dermatosparaxis in cattle is caused by a deficiency in one of the procollagen peptidases. In this disorder, the skin and tendons are very deformable, because the type I collagen forms mostly disorganized bundles rather than the normal highly organized and strong fibrils.

Once the propeptides are excised, type I collagen fibrils form by spontaneous association of the 300-nm-long collagen triple helices, which are staggered by about 67 nm, as discussed previously. Fibrils average about 50 nm in diameter and can be up to several micrometers in length. Collagen types I and V are often found together in the same fibril, and these collagen fibrils can be viewed as molecular alloys. It is thought that the diameter of the fibril is determined by the amount of type V present. Accompanying fibril formation is the oxidation of certain lysine and hydroxylysine residues into reactive aldehydes by the extracellular enzyme lysyl oxidase. Such aldehydes spontaneously form specific covalent cross-links between two chains (see Figure 24-8), which stabilizes the staggered

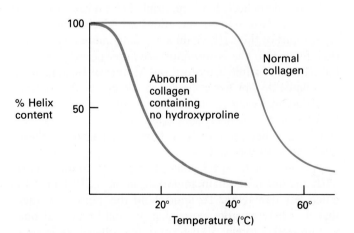

▲ FIGURE 24-10   Certain proline residues in procollagen are hydroxylated in the rough ER before the triple helices form. Oxygen, Fe^{2+}, ascorbic acid, and α-ketoglutarate are all required in this complex reaction. Proline can be hydroxylated on position 4 only if it is in the amino acid sequence Gly-X-Pro, where X is any amino acid. Hydroxylated prolines, especially the predominant 4-hydroxyproline, are important for stability of the triple helix.

▲ FIGURE 24-11   Denaturation of collagen containing a normal content of hydroxyproline and of abnormal collagen containing no hydroxyproline. Without hydrogen bonds between hydroxyproline residues, the collagen helix is unstable and loses most of its helical content at temperatures above 20°C. Such collagens are formed by experimental animals (or man) in the absence of ascorbic acid (vitamin C). Normal collagen is more stable and resists thermal denaturation until a temperature of 40°C is reached.

array characteristic of collagen molecules and contributes to fibril strength.

As noted above, in embryonic and immature animals type I collagen contains few of these covalent cross-links, and the fibrils are less rigid. Importantly, type I collagen from embryonic or growing animals contains only the $\alpha 1(I)$ chain, and has the composition $[\alpha 1(I)]_3$. Unlike mature type I collagen containing the $\alpha 2(I)$ chain, embryonic type I collagen is very sensitive to degradation by tissue proteases, an essential step in remodeling the extracellular matrix in tendons and bones as the animal grows. Similarly, the fetal forms of type V collagen, mainly $[\alpha 1(V)]_3$ and $[\alpha 1(V)]_2[\alpha 2(V)]$, are more sensitive to proteolytic degradation than the heterotrimeric form $[\alpha 1(V)][\alpha 2(V)][\alpha 3(V)]$ characteristic of mature type V collagen (see Table 24-1).

Mutations in Collagen Reveal Aspects of Its Structure and Biosynthesis

Type I collagen fibrils are used as the reinforcing rods in construction of bone. Certain mutations in the $\alpha 1(I)$ or $\alpha 2(I)$ genes lead to *osteogenesis imperfecta,* or brittle-bone disease. The most severe type is an autosomal dominant, lethal disease, resulting in death in utero or shortly after birth. Milder forms generate a severe crippling disease.

As might be expected, many cases of osteogenesis imperfecta are due to deletions of all or part of the very long $\alpha 1(I)$ gene. However, a single amino acid change is sufficient to cause certain forms of this disease. As we have seen, a glycine must be at every third position for the collagen triple helix to form; mutations of glycine to almost any other amino acid are deleterious, producing poorly formed and unstable helices. Since the triple helix forms from the C- to the N-terminus, mutations of glycine near the C-terminus of the $\alpha 1(I)$ chain are usually more deleterious than those near the N-terminus; the latter permit substantial regions of triple helix to form. Mutant unfolded collagen chains do not leave the rough ER of fibroblasts (the cells that make most of the type I collagen) or leave it slowly. As the ER becomes dilated and expanded, the secretion of other proteins (e.g., type III collagen) by these cells also is slowed down.

Because each type I collagen molecule contains two $\alpha 1(I)$ and one $\alpha 2(I)$ chain, mutations in the $\alpha 2(I)$ chains are much less damaging. To understand this point, consider that in a heterozygote expressing one wild type and one mutant $\alpha 2(I)$ protein, 50 percent of the collagen molecules will have the abnormal $\alpha 2(I)$ chain. In contrast, if the mutation is in the $\alpha 1(I)$ chain, 75 percent of the collagen molecules will have one or two mutant $\alpha 1(I)$ chains. In fact, even low expression of a mutant $\alpha 1(I)$ gene can be deleterious, because the mutant chains can disrupt the function of wild-type $\alpha 1(I)$ chains when combined with them. To study such mutations, experimenters constructed a mutant $\alpha 1(I)$ collagen gene with a glycine-to-cysteine substitution near the C-terminus. This mutant gene was used to create lines of transgenic mice with otherwise normal collagen genes. High-level expression of the mutant transgene was lethal, and expression at a rate 10 percent that of the normal $\alpha 1(I)$ genes caused severe growth abnormalities.

In summary, fibrous collagen has rigid structural requirements and is very susceptible to mutation, especially in glycine residues. Because mutant collagen chains can affect the function of wild-type ones, such mutations have a dominant phenotype.

Collagens Form Diverse Structures

Type II is the major collagen in cartilage. Its fibrils are smaller in diameter than type I and are oriented randomly in the viscous proteoglycan matrix. Such rigid macromolecules impart a stiffness and compressibility to the matrix and allow it to resist large deformations in shape. This property allows joints to absorb shocks.

Bound to the surface of type II fibrils at regular intervals is a collagen of a different structure—type IX (Figure 24-12a). Type IX collagen consists of two long triple helices connected by a flexible kink. A glycosaminoglycan chain is linked to the $\alpha 2(IX)$ chain at the flexible kink and protrudes from the composite fibrils, as does the globular N-terminal domain of the type IX molecule. These nonhelical domains are thought to anchor the fibril to proteoglycans and other components of the matrix. The interrupted triple-helical structure of type IX collagen, which also is characteristic of types XII and XIV, prevents these collagens from assembling into fibrils; instead, these three colla-

▶ FIGURE 24-12   Interactions of fibrous and nonfibrous collagens. (a) Types II and IX in a cartilage matrix. Type II forms fibrils similar in structure to type I, with a similar 67-nm periodicity, though smaller in diameter. Type IX contains two long triple helices connected at a flexible kink. At this point a chondroitin sulfate chain (see Figure 24-17) is linked to the $\alpha 2(IX)$ chain. Type IX collagens are bound at regular intervals along type II fibrils, with an N-terminal nonhelical domain of type IX projecting outward. It is thought that these domains bind the collagen fibrils to the proteoglycan-rich matrix. (b) Organization of the major fibrous components in the extracellular matrix of tendons. Type I fibrils, with their characteristic 67-nm period, are all oriented longitudinally, that is, in the direction of the stress applied to the tendon. The fibrils are coated with an array of proteoglycans, as shown in red on the right-hand fibril. Type VI fibrils bind to and link together the type I fibrils. Type VI collagen consists of thin triple helices, about 60 nm long, with globular domains at either end (blue). The globular domains of several type VI molecules bind together, giving a "beads-on-a-string" appearance to the type VI fibril. [Part (a) after L. M. Shaw and B. Olson, 1991, *Trends Biochem. Sci.* **18**:191; part (b) after R. R. Bruns, et al., 1986, *J. Cell Biol.* **103**:393.]

gens associate with fibrils formed from other collagen types and thus are called *fibril-associated collagens* (see Table 24-1).

Type VI collagen is found in many connective tissues, where it often is bound to the sides of type I fibrils and may bind them together to form thicker collagen fibers (Figure 24-12b; see also Figure 24-9). Type VI collagen is unusual in that the molecule consists of relatively short triple-helical regions about 60 nm long separated by globular domains about 40 nm long. Fibrils of pure type VI collagen thus give the impression of beads on a string.

In summary, fibrous collagen is built of different kinds of collagen chains and is organized in different ways in different tissues to fulfill specific needs. One of the most remarkable organizations is seen in both mature bone and in the transparent cornea of the eye. Here collagen fibers are arranged in layers; fibers in each layer are parallel to each other but perpendicular to fibers in the adjacent layers (Figure 24-13). Like plywood, this arrangement creates a thick, nondeformable material.

Type IV Collagen Forms the Two-Dimensional Reticulum of the Basal Lamina

The basal lamina is the thin sheetlike network of extracellular matrix components lining the basal surface of most epithelial and endothelial cells and surrounding muscle cells and adipocytes (see Figures 24-2 and 24-3). Usually no more than 60–100 nm thick, the basal lamina connects cells to the underlying connective tissue (Figure 24-14). A principal constituent of the basal lamina is type IV collagen, which forms a two-dimensional reticulum, a chicken-wire–like network that gives the basal lamina its shape and rigidity.

Three type IV collagen chains form a 400-nm-long triple helix with large globular domains at the C-termini and smaller ones of unknown structure at the N-termini. The helical segment is unusual in that the Gly-X-Y sequences

(a)

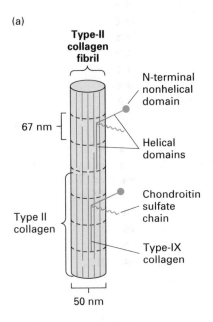

(b)

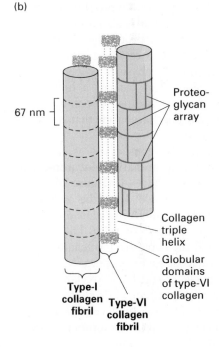

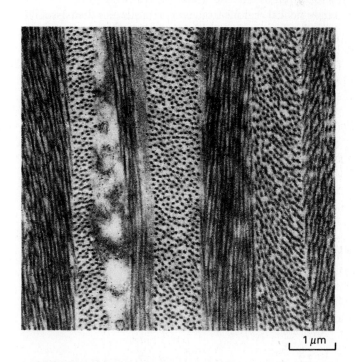

▲ FIGURE 24-13   Electron micrograph of a section from a rabbit cornea, showing the layered arrangement of fibrous collagen. In each layer fibers are parallel to each other and perpendicular to those in adjacent layers. [Courtesy D. Parry.]

Plasma membrane
of epithelial cell

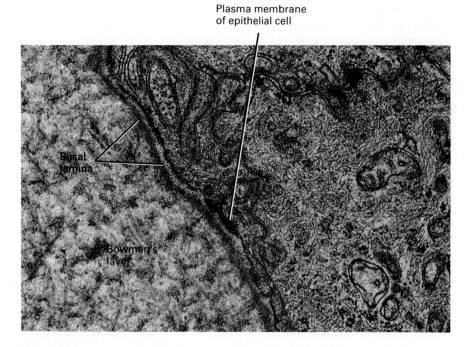

◄ FIGURE 24-14   Electron micrograph of a section of human corneal epithelial cells, showing the basal lamina lining the cell. The basal lamina is anchored to the plasma membrane by receptor proteins and is also connected, probably by fibers of type VII collagen, to the thick collagen-filled connective tissue, called Bowman's layer, that supports the epithelium. [From D. Fawcett, 1981, *The Cell*, 2d ed., Saunders, p. 47, Photo Researchers, Inc.; courtesy T. Kuwahara and D. Fawcett.]

are interrupted about 24 times with segments that cannot form a triple helix (Figure 24-15a); these nonhelical regions introduce flexibility into the molecule. Lateral association of the N-terminal regions of four type IV molecules yields a characteristic tetrameric unit (Figure 24-15b). Triple-helical regions from several molecules then associate laterally, in a manner similar to fibril formation among fibrous collagens (see Figure 24-6a), to form branching strands of variable but thin diameters. These interactions, together with those between the C-terminal globular domains and the triple helices in adjacent type IV molecules, generate an irregular two-dimensional fibrous network (Figure 24-15c). Such cross-linked type IV collagens assemble spontaneously and form the predominant fibers in the basal lamina (see Figure 24-3).

In summary, collagens are a family of matrix proteins, all of whose members have segments that fold into a characteristic triple-helical structure. The varied structures and binding properties of the different collagens are due in part to the differing abilities of collagen triple helices to interact with each other, and in part to the distinguishing features of the segments of the chains that do not form triple helices. Examples of the latter are the N- and C-terminal domains of the type IV collagen chains, which enable the two-dimensional lattice to form.

➤ *Hyaluronan and Proteoglycans*

Several times in this chapter we have made reference to various polysaccharides and the proteoglycans, a set of large molecules that contain both protein and carbohy-

drate and that are ubiquitous in all matrices. We turn now to a discussion of their structure and function, beginning with hyaluronan.

Hyaluronan Is an Immensely Long, Negatively Charged Polysaccharide That Forms Hydrated Gels

Hyaluronan (HA) is a major component of the extracellular matrix that surrounds migrating and proliferating cells, particularly in embryonic tissues. It is also a major structural component of the complex proteoglycans that are found in many extracellular matrices, particularly cartilage. Because of its remarkable physical properties, HA imparts stiffness and resilience as well as a lubricating quality to many types of connective tissue such as joints. HA is the only extracellular oligosaccharide that is not covalently linked to a protein.

Each hyaluronan molecule consists of as many as 50,000 repeats of the simple disaccharide glucuronic acid $\beta(1 \rightarrow 3)$ N-acetylglucosamine $\beta(1 \rightarrow 4)$ (Figure 24-16). If stretched end-to-end, one molecule would be as long as 20 μm! Individual segments of an HA molecule fold into a stiff rodlike conformation because of the β linkages between the saccharides and the extensive intrachain hydrogen bonding between adjacent sugar residues. Mutual repulsion between negatively charged carboxylate groups that protrude outward at regular intervals contributes to these local rigid structures. Overall, however, HA is not a long, rigid rod as is collagen; rather, in solution it behaves as a random coil about 500 nm in diameter.

(a) Type IV collagen molecule

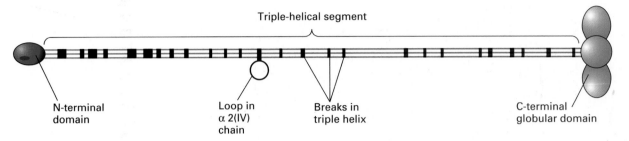

Triple-helical segment

N-terminal
domain

Loop in
α 2(IV)
chain

Breaks in
triple helix

C-terminal
globular domain

(b) Tetrameric type IV complex

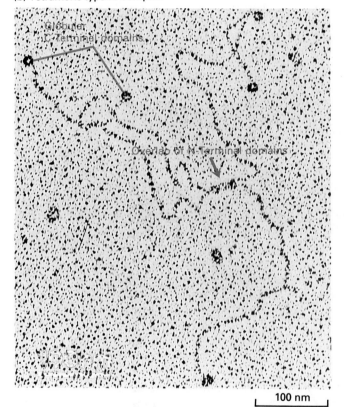

Globular
C-terminal domains

Overlap of N-terminal domains

100 nm

▲ **FIGURE 24-15**   Structure and assembly of type IV collagen. (a) Schematic diagram of 400-nm-long triple-helical
molecule of type IV collagen. This molecule has a noncolla-

(c) Assembly of type IV collagen network

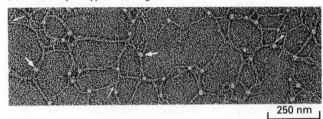

250 nm

gen domain at the N-terminus (green) and a large globular
domain at the C-terminus (blue); the triple helix is interrupted
by many nonhelical segments (black bands), including a loop
of unknown structure formed from part of the α2(IV) chain.
(b) An electron micrograph of one of the oligomeric forms of
type IV collagen. In this complex formed from four type IV
triple helices in vitro, the N-terminal domains of four molecules are laterally associated, and the C-terminal globular
domains are at the ends of the "arms" of the complex.
(c) Assembly of type IV collagen network from triple-stranded
molecules in vitro. The network is stabilized by lateral interactions between the N-terminal domains of several molecules, lateral interactions of triple-helical segments, and interactions between the globular C-terminal domains (arrows)
and the sides of triple-helical segments. The resulting network has a chicken-wire appearance similar to that of the
basal lamina (see Figure 24-2). [Part (a) adapted from
H. Furthmayr, 1993, in M. Zern and L. Reids, eds., *Extracellular Matrix*, Marcel Dekker, pp. 149–185; parts (b) and (c)
courtesy of P. Yurchenco; see P. Yurchenco and G. C.
Ruben, 1987, *J. Cell Biol.* **105**:2559.]

Because of the large number of hydrophilic residues on
its surface, HA binds a large amount of water and forms,
even at low concentrations, a viscous hydrated gel. Given
no constraints, an HA molecule will occupy a volume
about 1000 times the space of the HA molecule itself.
When placed in a confining space, such as in a matrix between two cells, the long HA molecules will tend to push
outward. This creates a swelling, or turgor pressure, within
the space; the HAs push against any fibers or cells that
block their motion. Importantly, by binding cations, the
COO− groups on the surface increase the concentration of
ions and thus the osmotic pressure in the HA gel. Large

amounts of water are taken up into the matrix, contributing to the turgor pressure within the HA matrix. These
swelling forces give connective tissues their ability to resist
compression forces, in contrast to collagen fibers, which
are able to resist stretching forces.

Hyaluronan Inhibits Cell-Cell Adhesion and Facilitates Cell Migration

Hyaluronan is bound to the surface of many migrating
cells by a 34-kDa receptor protein termed CD44, or by a
homologous protein in the CD44 family. The domain in

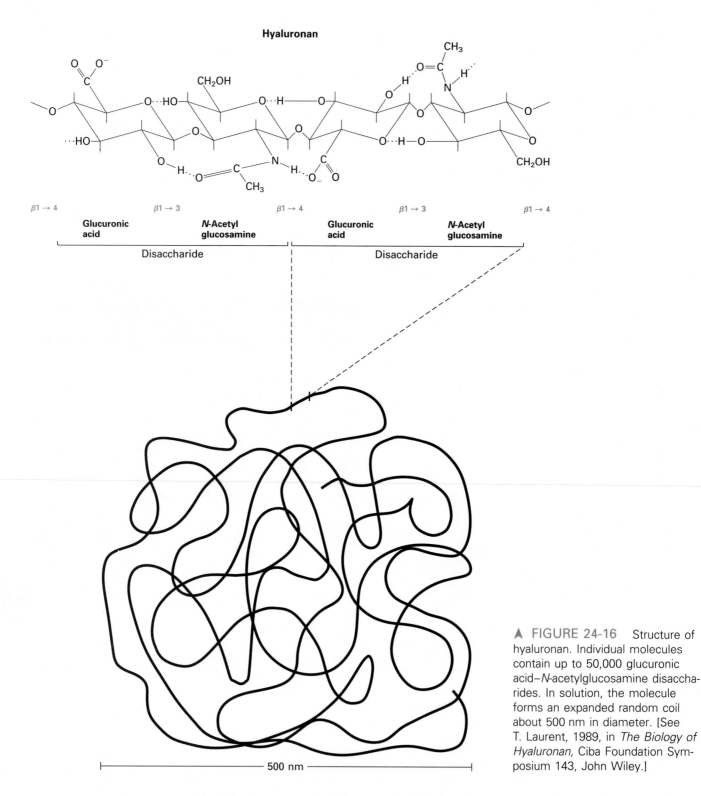

Hyaluronan

$\beta 1 \to 4$ $\beta 1 \to 3$ $\beta 1 \to 4$ $\beta 1 \to 3$ $\beta 1 \to 4$

Glucuronic acid   *N*-Acetyl glucosamine   Glucuronic acid   *N*-Acetyl glucosamine

Disaccharide   Disaccharide

500 nm

▲ FIGURE 24-16   Structure of hyaluronan. Individual molecules contain up to 50,000 glucuronic acid–*N*-acetylglucosamine disaccharides. In solution, the molecule forms an expanded random coil about 500 nm in diameter. [See T. Laurent, 1989, in *The Biology of Hyaluronan,* Ciba Foundation Symposium 143, John Wiley.]

CD44 that binds HA is similar in sequence and structure to ones found in various extracellular-matrix proteoglycans that bind HA. This is one of many examples we shall encounter where a number of different matrix and cell-surface proteins contain domains, or "modules," of similar structure and function (see Figure 3-14). Almost certainly these arose during evolution from a single ancestral gene that encoded just this domain.

Because of its loose, hydrated, porous nature, the HA coat appears to keep cells apart from each other and gives cells the freedom to move about and proliferate. Cessation of cell movement and initiation of cell-cell attachments are frequently correlated with a decrease in HA, a decrease in the cell-surface molecules that bind HA, and an increase in the extracellular enzyme hyaluronidase, which degrades the matrix HA. These functions of HA are particularly important during the many cell migrations that facilitate differentiation.

HA plays a key role in formation of striated muscle cells, such as in the differentiation of the myotome. Undifferentiated muscle-cell precursors, called myoblasts, express no muscle-specific proteins such as α-actin or troponin C. Hundreds of myoblasts fuse together to form a multinucleate syncytium that will induce muscle-specific proteins and become a multinucleated, nondividing muscle fiber (see Figure 6-6). Myoblasts bear an HA-rich coat that helps to prevent premature cell fusion. At the time of fusion, the HA-rich cell coat and the receptor proteins that bind HA are lost. The fused cells begin to synthesize a different, HA-free extracellular matrix, the basal lamina, that, as we have seen, allows the cells to attach to the surrounding collagen matrix.

Proteoglycans Comprise a Diverse Family of Cell-surface and Extracellular-matrix Macromolecules

Proteoglycans are macromolecules found in all connective tissues and extracellular matrices and on the surface of many cells. They consist of a *core protein* covalently attached to one or more polysaccharides called *glycosaminoglycans*. Glycosaminoglycans are long repeating linear polymers of specific disaccharides; usually one sugar is a uronic acid (either D-glucuronic acid or L-iduronic acid) and the other is either N-acetylglucosamine or N-acetylgalactosamine (Figure 24-17). One or both of the sugars contain one or two sulfate residues. Thus each glycosaminoglycan chain bears many negative charges. Proteoglycans are remarkable for their diversity. A given matrix may contain several different types of core proteins, and each may contain different numbers of glycosaminoglycans with differing lengths and compositions. Thus, the molecular weight and charge density of a population of proteoglycans can only be expressed as an average; individual molecules can differ considerably. Nonetheless, a good deal is known of the structure and function of certain proteoglycans.

Proteoglycans are named according to the structure of their principal repeating disaccharide. Frequently some of the residues in an oligosaccharide chain are modified after synthesis; dermatan sulfate is formed from chrondroitin sulfate, for instance. Similarly, heparin (used medicinally as an anticlotting drug) is formed, only in mast cells, as a result of enzymatic addition of sulfate groups at specific sites in heparan sulfate.

In all proteoglycans, heparan sulfate and chondroitin sulfate chains are covalently connected, via a three-sugar "linker," to serine residues in the core protein molecule (Figure 24-18). One of the "signal sequences' in a core protein that specifies addition of this linker sugar is Ser-Gly-X-Gly, where X is any amino acid. However, not all such sites in the core protein become substituted, and glycosaminoglycans are attached to serines in other sequences. Thus the conformation of the core protein may be more important than localized primary sequences in determining where the glycosaminoglycan chains attach. In addition, the mechanisms determining how long the chains are or which glycosaminoglycans (e.g., chondroitin sulfate, heparan sulfate, keratan sulfate) are attached to which serine residues are unknown.

Extracellular Matrix Proteoglycans   One of the most important extracellular proteoglycans is *aggrecan,* the predominant proteoglycan in cartilage. As its name implies, aggrecan forms very large aggregates, termed proteoglycan aggregates. A single aggregate, one of the largest macromolecules known, can be more than 4 μm long and have a volume larger than that of a bacterial cell! These aggregates give cartilage its unique gel-like properties and its resistance to deformation, essential for distributing the load in weight-bearing joints.

The central component of the cartilage proteoglycan aggregate is a long molecule of hyaluronan. Bound to it, tightly but noncovalently at 40-nm intervals, are aggrecan molecules (Figure 24-19a). The aggrecan core protein ($\approx$250,000 MW) has one N-terminal globular domain that binds with high affinity to a decasaccharide sequence in hyaluronan; this binding is facilitated by a *link protein* that binds to the aggrecan core protein and hyaluronan (Figure 24-19b). Covalently attached to each aggrecan core protein, via the trisaccharide linker, are multiple chains of chondroitin sulfate and keratan sulfate. The molecular weight of an aggrecan monomer—that is, the core protein plus the bound glycosaminoglycans—averages 2×10^6. The entire proteoglycan aggregate, which may contain upward of 100 aggrecan monomers, has a molecular weight in excess of 2×10^8.

Not all matrix proteoglycans form large aggregates like aggrecan. A class of proteoglycans present in the basal lamina, for example, consist of a core protein (20,000–40,000 MW) to which are attached several heparan sulfate chains. Such proteoglycans bind to type IV collagen and other structural proteins discussed later, thereby imparting structure to the basal lamina.

The importance of the glycosaminoglycan chains that are part of various matrix proteoglycans is illustrated by the rare humans who have a genetic defect in one of the enzymes required for synthesis of dermatan sulfate. These individuals have many defects in their bones, joints, and muscles; do not grow to normal height; and have wrinkled skin, giving them a prematurely aged appearance.

(a) Chondroitin sulfate class formed from [D-glucuronic acid-*N*-acetyl-D-galactosamine]$_n$

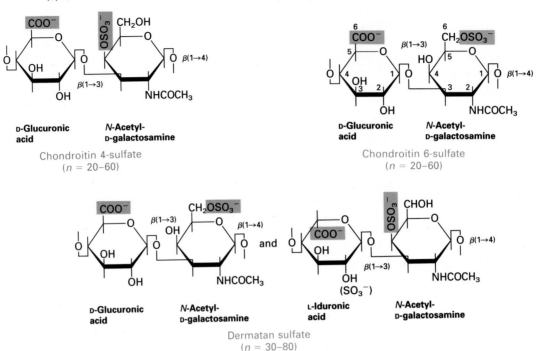

(b) Heparan sulfate class formed from [D-glucuronic acid-*N*-acetyl-D-glucosamine]$_n$

(c) Keratan sulfate class formed from [D-galactose-*N*-acetyl-D-glucosamine]$_n$

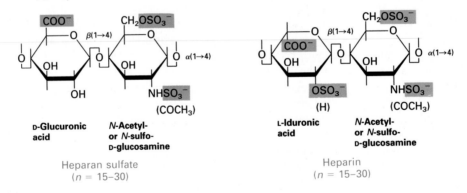

▲ FIGURE 24-17   Structures of various glycosaminogly-cans, the polysaccharide components of proteoglycans. Each of the three classes of glycosaminoglycans are formed by polymerization of a specific disaccharide and subsequent modifications including addition of sulfate groups and inversion (epimerization) of the carboxyl group on carbon-5 of D-glucuronic acid to yield L-iduronic acid. Heparan sulfate, which is ubiquitous, and its derivative heparin, found mostly in mast cells, are actually complex mixtures resulting from different modifications indicated in parentheses. The number *(n)* of disaccharides typically found in each glycos-aminoglycan chain is given. Green boxes highlight the nega-tively charged residues.

$$(GlcUA \xrightarrow{\beta(1\rightarrow3)} GalNAc)_n \xrightarrow{\beta(1\rightarrow4)} GlcUA \xrightarrow{\beta(1\rightarrow3)} Gal \xrightarrow{\beta(1\rightarrow3)} Gal \xrightarrow{\beta(1\rightarrow4)} Xyl \longrightarrow Ser$$

Chondroitin sulfate repeats

Linking sugars

Core protein

Gal = Galactose
GalNAc = N-Acetylgalactosamine

GlcUA = Glucuronic acid
Xyl = Xylose

▲ **FIGURE 24-18** Linkage of chondroitin sulfate to core proteins in proteoglycans occurs via a trisaccharide linker. Synthesis of the polysaccharide is initiated by transfer of a xylose residue to a serine residue in the protein, most likely in the Golgi complex, followed by sequential addition of two galactose residues. Glucuronic acid and N-acetylgalactosamine residues then are added sequentially, forming the chondroitin sulfate chain. In the Golgi, sulfate groups are added to sugars after their incorporation into the oligosaccharide. Heparan sulfate is connected to core proteins by the same three-sugar linker.

(a)

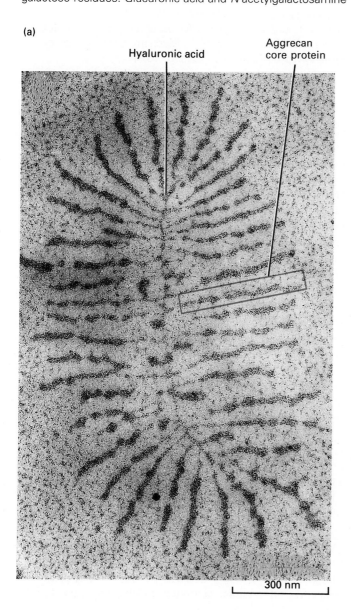

Hyaluronic acid

Aggrecan core protein

300 nm

(b)

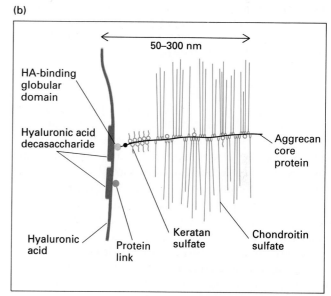

50–300 nm

HA-binding globular domain

Hyaluronic acid decasaccharide

Hyaluronic acid

Protein link

Keratan sulfate

Chondroitin sulfate

Aggrecan core protein

▲ **FIGURE 24-19** Structure of cartilage proteoglycan aggregate. (a) Electron micrograph of a proteoglycan aggregate from fetal bovine epiphyseal cartilage. Aggrecan core proteins are bound at ≈40-nm intervals to a molecule of hyaluronan. Numerous keratan sulfate and chondroitin sulfate chains are attached to the aggrecan core proteins. (b) Diagram of detailed structure of an aggrecan monomer. Each core protein contains 2124 amino acids and has one N-terminal globular domain (yellow), which binds with high affinity to a decasaccharide (green) portion of the central hyaluronan molecule. Binding is facilitated by a link protein (red), which binds to both the hyaluronan disaccharide and to the aggrecan core protein. Each aggrecan core protein has 127 Ser-Gly sequences at which the glycosaminoglycan chains are added; 30 short keratan sulfate chains and 97 longer chondroitin sulfate chains are added to each core protein molecule in aggrecan. [Part (a) from J. A. Buckwalter and L. Rosenberg, 1983, *Coll. Rel. Res.* **3**:489; courtesy of L. Rosenberg. Part (b) adapted from D. Heinegard and A. Oldberg, 1989, *FASEB J.* **3**:2042, and T. Hardingham and A. Fosang, 1992, *FASEB J.* **6**:861.]

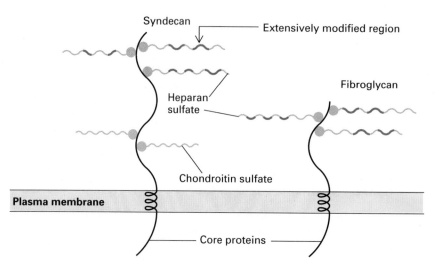

▲ **FIGURE 24-20**   Schematic diagrams of two cell-surface proteoglycans, syndecan and fibroglycan. The core protein in all cell-surface proteoglycans spans the plasma membrane. The syndecan core protein (56,000 MW) is considerably larger than the fibroglycan core protein, but the sequences of their membrane-spanning and cytosolic domains are similar. Both proteoglycans contain three hepa-ran sulfate chains, which are extensively modified. Syndecan also contains two chondroitin sulfate chains. [Adapted from M. Yanagashita and V. Hascall, 1992, *J. Biol. Chem.* **267**:9451, and A. Rapraeger, M. Jakanen, and M. Bernfeld, 1987, in T. N. Wright and R. P. Mecham, eds., *Biology of Proteoglycans,* Academic Press, p. 137.]

Cell-surface Proteoglycans Proteoglycans are attached to the surface of many types of cells, particularly epithelial cells. Two examples are *syndecan* and *fibroglycan,* whose general structures are illustrated in Figure 24-20. The core protein of cell-surface proteoglycans spans the plasma membrane and contains a short cytosolic domain, as well as a long external domain to which a small number of heparan sulfate chains are attached; some cell-surface proteoglycans also contain chondroitin sulfate. The glycosaminoglycan chains are bound to serine residues in the core protein as they are in extracellular proteoglycans (see Figure 24-18). The attached heparan sulfate chains bind to fibrous collagens (type I, III, and V) and to the large glycoprotein fibronectin, which are present in the interstitial matrix surrounding the basal lamina. In this way, cell-surface proteoglycans are thought to anchor cells to matrix fibers.

Proteoglycans Can Bind Many Growth Factors

Besides acting as structural components of the extracellular matrix and anchoring cells to the matrix, both extracellular and cell-surface proteoglycans, particularly those containing heparan sulfate, also bind many protein growth factors (Chapter 20). One example is fibroblast growth factor (FGF), which despite its name stimulates proliferation of many types of cells other than fibroblasts. FGF binds tightly to the heparan sulfate chains in extracellular proteoglycans. Since the bound growth factor is resistant to degradation by extracellular proteases, it serves as a reservoir of matrix-bound FGF (Figure 24-21). Active hormone is released by proteolysis of the proteoglycan core protein or by partial degradation of the heparan sulfate chains, processes that occur during tissue growth and remodeling or after infection. FGF also binds to cell-surface heparan sulfate proteoglycans such as syndecan, which then "present" the bound FGF to its receptor (an RTK, see Figure 20-4g), inducing proliferation. Free FGF cannot interact with the cell-surface FGF receptor, and cells that cannot synthesize heparan sulfate proteoglycans do not respond to FGF.

Another example is transforming growth factor β (TGFβ) whose role in embryonic development is discussed later. TGFβ binds to the core protein of another important cell-surface proteoglycan called *beta-glycan,* also called the type III TGFβ receptor. TGFβ bound to beta-glycan is presented; or "handed off," to types I and II receptors, which are serine/thronine kinases (see Figure 20-4f); binding of TGFβ to these receptors generates intracellular signals. The ligand-binding extracellular domain of beta-glycan often is cleaved off by tissue proteases; TGFβ bound to this domain serves as an extracellular reservoir of this hormone. These examples illustrate how proteoglycans can function as extracellular hormone reservoirs and facilitate binding of hormones to their cell-surface receptors, thus triggering intracellular signaling pathways.

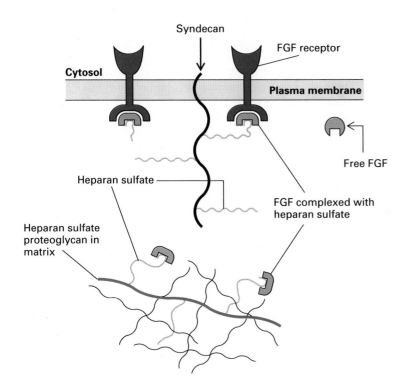

◀ FIGURE 24-21 Modulation of activity of fibroblast growth factor (FGF) by heparan sulfate proteoglycans. Free FGF cannot bind to FGF receptors on cell membranes. Binding of FGF to heparan sulfate chains such as those in syndecan, a cell-surface proteoglycan (Figure 24-20), induces a conformational change that enables FGF to bind to its receptors. Binding of FGF to extracellular heparan sulfate proteoglycan protects the growth factor from degradation and forms a reservoir of active FGF. [Adapted from E. Ruoslahti and T. Yamaguchi, 1991, *Cell* **64**:867.]

▶ *Multiadhesive Matrix Proteins and Their Cell-Surface Receptors*

Collagens and proteoglycans are two important components of all extracellular matrices. We turn now to another major component—a group of proteins that bind not only to collagens and proteoglycans but also to specific cell-surface receptors. These *multiadhesive proteins* are important for organizing the other components of the matrix and also for regulating cell attachment to the matrix, cell migration, and cell shape.

Laminin and Nidogen Are Principal Structural Proteins of All Basal Laminae

All basal laminae contain a common set of proteins and glycosaminoglycans: type IV collagen, heparan sulfate proteoglycans, nidogen, and laminin. The basal lamina is often called the type IV matrix, named after its collagen component. It supports all epithelial cells, regenerating hepatocytes (liver cells), and endothelial (blood vessel–lining) cells. All of the matrix components are synthesized by cells that rest on the basal lamina.

Laminin is a family of cross-shaped molecules almost 70 nm long. (Figure 24-22a); it is as long as the basal lamina is thick. In adult animals, laminin contains three polypeptides of total molecular weight 820,000 and has high-affinity binding sites for other components of the basal lamina. Nidogen, a 158,000 molecular weight rod-shaped molecule with three globular domains (Figure 24-22b), binds tightly to laminin and is often considered a laminin subunit. Nidogen also binds to type IV collagen and to proteoglycans, and thus could mediate the assembly of all of the components of the basal lamina into a network.

The basal lamina is structured differently in different tissues. Different types of cells that interact with a basal lamina, such as epithelial cells, fat cells, and smooth and striated muscle may utilize different cell-surface receptors that bind to matrix components. For instance, capillary cells are polarized, with one surface facing the blood. The surface not facing the blood is surrounded by a basal lamina that forms a filter that regulates passage of proteins and other molecules from the blood into the tissues; similarly, the double-thickness basal lamina in the kidney glomerulus acts as a filter in forming the urine (Figure 24-23). In smooth muscle, on the other hand (Figure 5-32), the basal lamina connects adjacent cells and maintains the integrity of the tissue. Laminin and other type IV matrix components are also found in mammalian four- and eight-celled embryos; the basal lamina helps these cells adhere together in a ball.

During development of the nervous system, axon and dendrite projections of nerve cells (neurites) often elongate to form connections with specific target cells. The growth cones that direct this neuronal outgrowth migrate along

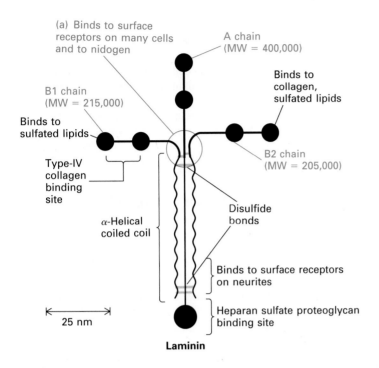

◄ **FIGURE 24-22**   Structure of laminin, a large heterotrimeric multiadhesive protein found in all basal laminae, and nidogen, a monomeric multiadhesive protein frequently associated with laminin. (a) Laminin is a cross-shaped molecule containing globular domains (black circles) and a coiled-coil region in which the three chains are covalently linked via several disulfide bonds (yellow). Different regions of laminin bind to cell-surface receptors and various matrix components. (b) The three globular domains of nidogen—G1, G2, and G3—are connected by flexible linker segments. Nigoden binds to three other components of the basal lamina: type IV collagen, proteoglycan, and laminin. [Part (a) adapted from G. R. Martin and R. Timpl, 1987, *Ann. Rev. Cell Biol.* **3**:57, and K. Yamada, 1991, *J. Biol. Chem.* **266**:12809. Part (b) adapted from D. Reinhardt et al., 1993, *J. Biol. Chem.* **268**:10881.]

extracellular-matrix pathways that contain laminin and other matrix components. The mechanism by which matrix proteins mediate movement of growth cones is discussed in a later section.

Integrins Are Cell-Surface Receptors That Mediate Adhesion to the Extracellular Matrix and Cell-Cell Interactions

The integrins are a large class of cell-surface receptors that bind different components of the extracellular matrix; some of these transmembrane proteins also mediate cell-cell interactions. Most integrins are expressed on a variety of cells, and most cells express several integrins, enabling them to bind to several matrix molecules. Furthermore, individual integrins often bind more than one ligand, and individual matrix molecules are often bound by more than one integrin. Because they play multiple roles in differentiation and in adult organisms, integrins are mentioned several times in this chapter. Here we describe their general

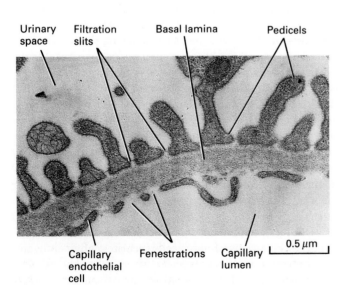

◄ **FIGURE 24-23**   Electron micrograph of a section through a rat kidney glomerulus, showing how the basal lamina acts to filter capillary blood, forming a filtrate in the urinary space that ultimately becomes urine. The endothelial cells that line the capillaries have many gaps within them, as do the epithelial cells that line the urinary space, so that the basal lamina is exposed to both the blood and urine spaces. The basal lamina in this region is a fusion of two basal laminae, one formed by the endothelial and one by the epithelial cells, and is twice as thick as basal lamina in most other tissues. [From R. Kessel and R. Kardon, 1979, *Tissues and Organs: A Text-Atlas of Scanning Electron Microscopy*, Freeman, p. 233.]

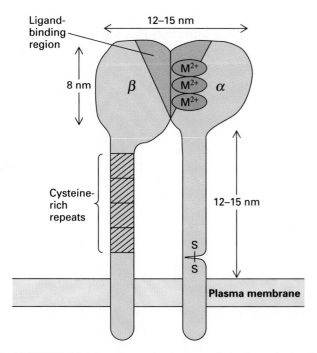

▲ FIGURE 24-24   General structure of the integrin family of cell-surface receptors. Each receptor consists of two transmembrane polypeptides, α and β, whose molecular weight varies from about 100,000 to 140,000 in different molecules. α chains usually undergo a proteolytic cleavage in the extracellular domain, and the two fragments remain linked by a disulfide bond. Each β chain has a four cysteine-rich repeated sequences (cross-hatched), and each α chain binds several divalent cations. The dark orange area represents the ligand-binding region, which contains sequences from both subunits. [Adapted from R. O. Hynes, 1992, *Cell* **69**:11.]

features, focusing on their binding to components of the type IV matrix.

Integrin Structure and Diversity   Integrins are heterodimers of α and β subunits, and the ligand-binding site is composed of parts of both chains (Figure 24-24). In mammals, at least 20 integrin heterodimers, comprised of 14 types of α subunits and 8 types of β subunits, are known (Table 24-2). A single β chain can interact with multiple α chains, forming integrins that bind different ligands. In general, integrins containing the β_1 subunit bind components of the extracellular subunit. For example, the $\alpha_1\beta_1$ and $\alpha_2\beta_1$ integrins both bind a segment within the C-terminal domain of type IV collagen; these integrins, as well as the widely expressed $\alpha_6\beta_1$, also bind at least two different regions of laminin (see Figure 24-22a); and $\alpha_5\beta_1$ binds fibronectin, as discussed below. In contrast, the β_2 subunit is expressed exclusively on leukocytes (white blood cells) along with three leukocyte-specific α subunits (see Table 24-2). These integrins mediate cell-cell interactions, rather than cell-matrix interactions. We will see later how one such integrin participates in the binding of leukocytes to specific ligands on endothelial cells at sites of infection or inflammation.

Several integrin subunits in mammals have alternatively spliced cytosolic domains. This adds a further level of complexity, since these different integrins probably bind different intracellular attachment proteins, and thus interact differently with components of the cytoskeleton (See Figures 22-35 and 24-1).

Regulation of Integrin Activity   Even though integrins are present on the cell surface, they may require activation in order to bind their ligand and thus to anchor the cell to the extracellular matrix or to another cell. One important example concerns platelets, the small cell fragments that circulate in blood and that are important for blood clotting. The $\alpha_{IIb}\beta_3$ integrin normally is present on the plasma membrane of platelets but is unable to bind the blood protein fibrinogen or the other protein ligands listed in Table 24-2, all of which participate in formation of a blood clot. Only after a platelet becomes "activated," (see Figure 22-7), by binding collagen or thrombin in a forming clot, can $\alpha_{IIb}\beta_3$ integrin bind fibrinogen; this interaction accelerates the formation of the clot. Platelet activation is accompanied by a conformational change in the $\alpha_{IIb}\beta_3$ integrin. The nature of this change is unknown, but as platelet activation also involves a major change in the platelet cytoskeleton (Figure 22-7), this change probably involves binding of a cytoskeletal protein to the integrin cytosolic domain. Patients with genetic defects in the β_3 integrin subunit are prone to excessive bleeding, attesting to the role of this integrin in formation of blood clots.

Integrins typically exhibit relatively low affinities for their ligands (dissociation constants K_D between 10^{-6} and 10^{-8} moles/liter) compared with the high affinities (K_D values of 10^{-9} to 10^{-11} moles/liter) of typical cell-surface hormone receptors. However, the multiple weak interactions generated by binding of hundreds or thousands of integrin molecules to extracellular matrix proteins allows a cell to remain firmly anchored to the matrix. Alternatively, in situations where cells are migrating it is essential that they be able to make and break specific contacts with the extracellular matrix; this is facilitated if individual contacts are weak. One of the reasons cells express several different integrins that bind the same ligand is that, at any one time, one class may be activated (e.g., by binding to a specific cytosolic protein) and thus able to bind to a matrix component, while another class of integrins is inactivated.

Later in this chapter we examine other examples where the binding activities of integrins are modulated to serve

TABLE 24-2   Vertebrate Integrins Grouped in Subfamilies Sharing a Common β Subunit

Subunits		Ligands	Minimal Sequence of Integrin-Binding Site*
$\beta_1{}^\dagger$	α_1	Collagens, laminin	—
	α_2	Collagens, laminin	DGEA
	$\alpha_3{}^\dagger$	Fibronectin, laminin, collagens	RGD
	α_4	Fibronectin, VCAM-1	EILDV
	α_5	Fibronectin	RGD
	$\alpha_6{}^\dagger$	Laminin	—
	α_7	Laminin	—
	α_8	?	—
	α_V	Vitronectin, fibronectin	RGD
β_2	α_L	ICAM-1, ICAM-2	—
	α_M	C3b component of complement (inactivated), fibrinogen, factor X, ICAM-1	—
	α_X	Fibrinogen, C3b component of complement	GPRP
$\beta_3{}^\dagger$	α_{IIb}	Fibrinogen, fibronectin, von Willebrand factor, vitronectin, thrombospondin	RGD, KQAGDV
	α_V	Vitronectin, fibrinogen, von Willebrand factor, thrombospondin, fibronectin, osteopontin, collagen	RGD
$\beta_4{}^\dagger$	$\alpha_6{}^\dagger$	Laminin	—
β_5	α_V	Vitronectin	RGD
β_6	α_V	Fibronectin	RGD
β_7	α_4	Fibronectin, VCAM-1	EILDV
	α_{IEL}	?	—
β_8	α_V	?	—

* Amino acid sequence (one-letter code) of integrin-binding site on ligand proteins. The ability of an integrin to interact with its ligand(s) can be modified by the environment of the binding site in the ligand protein and by the state of activation of the integrin. A (—) means the sequence of the binding site is not known.
† These subunits can have multiply spliced isoforms with different cytosolic domains.
SOURCE: R. O. Hynes, 1992, *Cell* **69**:11.

important physiological purposes. First, however, we need to complete the discussion of the multiadhesive proteins in the extracellular matrix.

Fibronectins Bind Many Cells to Fibrous Collagens and Other Matrix Components

Fibronectins are another important class of multiadhesive matrix proteins. Their primary role is attaching cells to a variety of extracellular matrices—all matrices, in fact, that contain the fibrous collagens (types I, II, III, and V). The presence of fibronectin on the surface of nontransformed cultured cells, and its absence on transformed (or tumorigenic) cells, first led to the mdentification of fibronectin as an adhesive protein. By their attachments, fibronectins reg-

ulate the shape of cells and the organization of the cytoskeleton; they are essential for migration and cellular differentiation of many cell types during embryogenesis. Fibronectins also are important for wound healing because they facilitate migration of macrophages and other immune cells into the affected area.

Fibronectins are dimers of two similar polypeptides linked at their C-terminii by two disulfide bonds; each chain is about 60–70 nm long and 2–3 nm thick. At least 20 different fibronectin chains have been identified, all of which are generated by alternative splicing of the RNA transcript of a single fibronectin gene. Analysis of digests of fibronectin with low amounts of proteases shows that the polypeptides consist of six tightly folded domains. Each domain, in turn, contains small repeated sequences whose

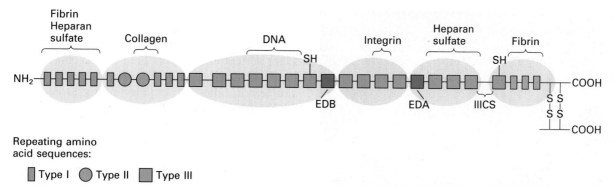

Repeating amino acid sequences:

▮ Type I ● Type II ▢ Type III

▲ FIGURE 24-25 Structure of fibronectin chains. Only one of the two chains present in the dimeric fibronectin molecule is shown; both chains have very similar sequences. Each chain contains about 2446 amino acids and is composed of three types of repeating amino acid sequences. Each repetitive sequence is encoded by one or two exons. Circulating fibronectin lacks one or both of the type III repeats designated EDA and EDB due to alternative mRNA splicing. At least five different sequences may occur in the IIICS region as the result of alternative splicing. Each chain contains six domains (tan ovals) containing specific binding sites for heparan sulfate, fibrin (a major constituent of blood clots), denatured forms of collagen, DNA, and cell-surface integrins. Binding to integrins is dependent on an Arg-Gly-Asp (RGD) sequence. Two domains have binding sites for heparan sulfate and fibrin; these differ in their affinity for the ligand. [Adapted from G. Paolella, M. Barone, and F. Baralle, 1993, in M. Zern and L. Reid, eds., *Extracellular Matrix*, Marcel Dekker, pp. 3–24.]

similarities in amino acid sequence allow them to be classified into three types (Figure 24-25). Different domains of fibronectin possess specific high-affinity binding sites for cell-surface integrins, denatured forms of collagen, fibrin, and sulfated proteoglycans. Fibronectin secreted by cultured cells does not spontaneously form fibrils, but will do so when in contact with a cell expressing certain integrins, usually $\alpha_5\beta_1$. Figure 24-26 depicts these insoluble fibronectin filaments surrounding a cultured fibroblast.

Proteolytic digestion of the cell-binding domain in one of the type III repeats showed that a segment of about 100 amino acids could bind to integrins. Studies with synthetic peptides corresponding to parts of this segment identified the tripeptide sequence Arg-Gly-Asp (usually abbreviated RGD) as the minimal structure required for recognition by cell-surface receptors. For example, when this tripeptide was covalently linked to an inert protein (e.g., albumin) and dried on a culture dish, it stimulated adhesion of fibroblasts to the surface of the dish similar to the effect of intact fibronectin. In the three-dimensional structure of this fibronectin type III repeat, the RGD sequence is at the apex of a loop that protrudes outward from the molecule, in a position to bind to an integrin or other protein (Figure 24-27).

Although the RGD peptide is the minimal structure required for binding to several integrins [see Table 24-2], its affinity for integrins is substantially less than that of intact fibronectin or of the entire cell-binding domain. Thus, sequences surrounding the RGD sequence in fibronectin and other proteins apparently enhance binding to integrins. Further, the affinity of the RGD sequence for integrins is dependent on the absorbed state of fibronectin. Fibronectin in solution or circulating in blood binds to in-

tegrins on fibroblasts poorly. Absorption of fibronectin to a surface—in animals to a collagen matrix or the basal lamina surrounding an endothelial cell or, experimentally, to a tissue-culture dish—enhances its ability to bind to cells, probably because the segment of the protein containing the RGD sequence becomes more exposed.

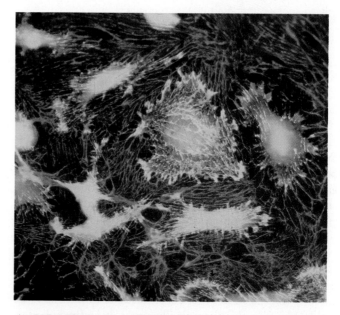

▲ FIGURE 24-26 Fibronectin (red) secreted by cultured fibroblasts (yellow) forms a fibrillar array that is densest where fibronectin contacts the cells. Fluorescent-labeled antibodies were used to stain fibronectin and the cells. [From R. Hynes, 1986, *Sci. Am.* **254**(6):42; courtesy of R. Hynes.]

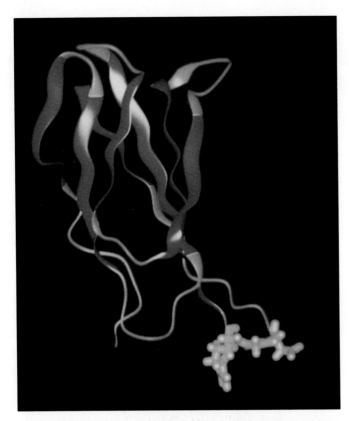

▲ FIGURE 24-27   Three-dimensional structure of the type III repeat of fibronectin that contains the RGD integrin-binding sequence. This domain has a structure similar to that of immunoglobulin domains (see Figure 27-16), consisting of a core of seven β strands (red). The side chains of the Arg[78]-Gly[79]-Asp[80] sequence of the integrin-binding site are shown in yellow. [From A. L. Main et al., 1993, *Cell* **71**:671; courtesy of A. Main and I. D. Campbell.]

Fibronectin circulating in the blood, which is secreted by the liver, lacks one or both of the type III repeats designated EDA and EDB (for extra domain A and B) present in fibronectin secreted by cultured fibroblasts (see Figure 24-25). Circulating fibronectin forms insoluble matrices somewhat less readily than fibronectin within tissues but does bind to fibrin, a constituent of blood clots. Following binding to fibrin, the immobilized fibronectin binds, via its exposed RGD-containing cell-binding domain, to integrins expressed on passing, activated platelets. As a result, the platelets are localized to damaged regions of blood vessels and can participate in expansion of blood clots. As noted already, the $\alpha_{IIb}\beta_3$ integrin on activated platelets also binds directly to fibrin and other constituents of blood clots (see Table 24-2). As this example illustrates, both the polymerization of fibronectin into filaments and adhesion of cells to fibronectin are closely controlled to meet the organism's needs.

Fibronectins Promote Cell Adhesion to the Substratum

Eight vertebrate integrins bind fibronectin; most of these bind specifically to the RGD sequence found in fibronectin and certain other extracellular matrix proteins (see Table 24-2). Many cultured animal cells secrete fibronectin, which binds to the culture dish, forming a substratum along with other matrix components. Fibronectin-binding integrins on the cells then interact with the fibronectin in the substratum, causing the cells to adhere to it.

In such cultures, fibronectin tends to concentrate at the small points where nonmigrating cells adhere to the substratum. As discussed in Chapter 22, nonmigrating cells contain abundant actin-rich stress fibers that maintain the cell's shape and attach to the plasma membrane where it adheres to substratum (Figure 24-28a). Several actin-binding proteins, such as vinculin and α-actinin, also are enriched in these regions. Electron micrographs occasionally reveal exterior fibronectin fiber bundles that appear continuous with bundles of actin fibers within the cell (Figure 24-28b). Integrin $\alpha_5\beta_1$, a fibronectin receptor, is localized to these focal contacts and may anchor both fibers to the opposite sides of the plasma membrane, as depicted in Figure 22-35. The apparent continuity of matrix and cytoplasmic fibers is probably not coincidental, as the connection through the integrin allows the cell to attach rigidly to the substratum.

Many tumor cells growing in culture do not stick to plastic culture dishes. Such cells synthesize little fibronectin and contain few stress fibers. Adsorption of purified fibronectin to the culture dish frequently results in tight adhesion and, concomitantly, to the development of stress fibers.

Regulation of the number of integrin molecules on cells can modulate cell attachment to the substratum. An important example is the $\alpha_4\beta_1$ integrin, a fibronectin receptor, found on many hematopoietic cells (precursors of red and white blood cells). In order for these cells to proliferate and differentiate, they must be attached to extracellular-matrix fibronectin synthesized by supportive ("stromal") cells in the bone marrow. The $\alpha_4\beta_1$ integrin on hematopoietic cells binds to an EILDV sequence present in the IIICS domain of stromal-cell fibronectin (see Figure 24-25), thereby anchoring the cells to the matrix. This integrin also binds to an EILDV sequence in VCAM-1, an integral membrane protein present on stromal cells of the bone marrow. Thus hematopoietic cells directly contact the stromal cells, as well as attach to the matrix. A decrease in the number of $\alpha_4\beta_1$ integrin molecules present on hematopoietic cells at a late stage in their differentiation is thought to allow mature blood cells to detach from the matrix and stromal cells in the bone marrow and subsequently enter the circulation.

(a)

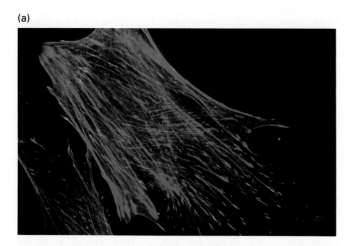

▲ FIGURE 24-28   Connections between extracellular fibronectin and intracellular actin filaments may be mediated by the $\alpha_5\beta_1$ integrin, a fibronectin receptor. (a) Immunofluorescence of a fixed, stationary cultured fibroblast showing colocalization of the $\alpha_5\beta_1$ integrin (green) and actin-containing stress fibers (red). At the ends of the stress fibers, where the cells contact the substratum, there is a coincidence of actin and the fibronectin-binding integrin (yellow). (b) Electron micrograph of the junction of fibronectin and actin fibers in a cultured fibroblast. Individual actin-containing 7-nm microfilaments, components of a stress fiber, end at the obliquely sectioned cell membrane. The microfilaments appear in close proximity to the thicker, densely stained fibronectin fibers on the outside of the cell. [Part (a) from J. Duband et al., 1988, *J. Cell Biol.* **107**:1385. Part (b) from I. J. Singer, 1979, *Cell* **16**:675; courtesy of I. J. Singer; copyright, 1979, M.I.T.]

(b)

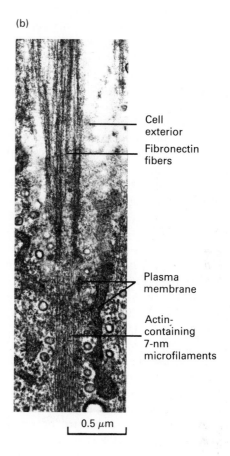

Cell exterior

Fibronectin fibers

Plasma membrane

Actin-containing 7-nm microfilaments

0.5 μm

Fibronectins Promote Cell Migration

Fibronectins are abundant constituents of several types of extracellular matrices, generally together with fibrous collagens and specific proteoglycans. The six-armed protein tenascin (Figure 24-29) is frequently associated with fibronectin, particularly in connective tissues in many developing organs and in the mesenchyme that underlies many epithelia. In many cases, these matrices are loosely packed with several fibrillar components secreted by various interstitial cells and provide trails along which other cells can migrate.

➤ FIGURE 24-29   Electron micrograph of tenascin, a multimeric protein frequently found together with fibronectin in mesenchymal extracellular matrices. Each of the six arms is formed by an identical ≈230-kDa subunit. The six subunits are bound together by the globular domains at their N-terminii. The distal C-terminal region of each subunit contains a number of fibronectin-like type III repeats (see Figure 24-25). [From R. Chiquet-Ehrismann, 1990, *FASEB J.* **4**:2598.]

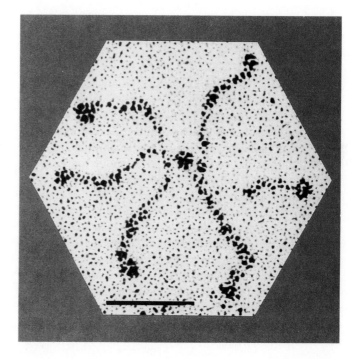

In contrast to embryonic cells, discussed later in this chapter, cells in an adult organism generally do not migrate. Cells involved in wound healing are an exception. These cells—including fibroblasts, macrophages, and other immune-system cells—migrate into the affected area and move across blood clots, which are composed primarily of the fibrous protein fibrin. Fibronectin is incorporated into the clot as it forms by virtue of its fibrin-binding domains (see Figure 24-25). Integrins on the surface of fibroblasts, in turn, adhere to the fibronectin as they migrate across the clot. Fibroblasts migrating on fibronectin make and break contacts with the matrix. In such cells, the fibronectin receptors are laterally mobile in the plasma membrane, as determined by fluorescence recovery from photobleach experiments (see Figure 14-34). As the fibroblasts stop migrating and form stress fibers, integrin $\alpha_5\beta_1$, which binds to the RGD sequence of fibronectin, becomes immobilized at the focal contacts in contact with the substratum and keeps the cells tightly bound to the matrix; immobilization may be due to binding of the receptor, via α-actinin and vinculin, to the actin cytoskeleton (see Figure 24-28). Interestingly, fibroblasts also express integrin $\alpha_3\beta_1$ that, like $\alpha_5\beta_1$, binds to the RGD sequence of fibronectin; however, unlike $\alpha_5\beta_1$, it is not localized to focal contacts but is dispersed on the plasma membrane. Possibly $\alpha_3\beta_1$ participates in cell migration on fibronectin.

Emerging from studies of the cytoskeleton, plasma-membrane receptors for matrix proteins, and extracellular-matrix components is a picture of multiple and changing interactions among the elements that allows coordinated activities to occur as different as migration of cells along extracellular-matrix fibers and rigid adhesion of cells to the matrix.

▶ Cell-Cell Adhesion: Adhesive Proteins

Adhesion of like cells is a primary feature of the architecture of many tissues. A sheet of absorptive epithelial cells, for instance, forms the lining of the small intestine, and sheets of hepatocytes two cells thick make up much of the liver. A number of cell-surface adhesive molecules are now known; a single cell type may use more than one such molecule to mediate such *homophilic* (like-binds-like) adhesion. Most such proteins are uniformly distributed along the regions of plasma membranes that contact other cells, and the cytosol-facing domains of these proteins are usually connected to elements of the cytoskeleton.

Tissues like the liver, pancreatic acinus, and the intestinal epithelium also are stabilized by very localized, specialized surface junctions that permit or restrict the passage of ions and molecules between cells—the gap junctions—and

the desmosomes that strengthen the adhesion of cells to each other (see Figure 14-42). Since formation of these junctions occurs after formation of the initial cell contacts mediated by cell-surface adhesive proteins, we discuss the adhesive proteins first, then the junctions.

Adhesive Proteins Mediate Cell-Cell Interactions

Many features of cell-surface interactions were shown in now classical experiments performed beginning in the 1940s. Various types of embryonic tissues (which dissociate more easily than adult tissues) were dissociated into single cells and allowed to reassociate in the presence of other cell types. Invariably, it was found that cells of a particular type, such as liver, kidney, or retina, would reassociate preferentially with cells of a similar origin. Mixtures, say, of dissociated liver or kidney cells would form small aggregates consisting only of liver or only of kidney cells. If a single large mass was formed, one cell type was localized to the center, surrounded by cells of the other type. Adhesion of similar cells even extended across species boundaries. Mouse liver cells, for instance, adhere tightly to chicken liver cells and form aggregates; however, they do not aggregate with mouse kidney cells.

To analyze such cell-adhesion phenomena, experimenters prepared monoclonal antibodies directed against proteins on the plasma membrane of a specific cell type, say chick liver. Some of these antibodies bound to cell-surface adhesive proteins and blocked the ability of dissociated chick-liver cells to reaggregate. The monoclonal antibodies subsequently were used to characterize, purify, and clone the corresponding cell-surface adhesive protein. Cell-adhesive proteins are now known to fall into three principal classes (Figure 24-30): the *cadherins*, *selectins*, and *immunglobulin (Ig) superfamily*. Cell-cell adhesion involving cadherins and selectins depends on Ca^{2+} ions, whereas interactions involving Ig-superfamily proteins do not. Many cells use several different adhesive proteins to mediate cell-cell adhesion. As noted already, certain integrins also mediate specific cell-cell interactions.

E-Cadherin Is a Key Adhesive Protein Expressed by Epithelial Cells

Cadherins are a family of Ca^{2+}-dependent cell-cell adhesion molecules that are important for tissue differentiation and structure. The most widely expressed, particularly during early differentiation, are the E-, P-, and N-cadherins. Over 30 different cadherins are known; some are summarized in Table 24-3. The brain expresses the largest number of different cadherins, presumably due to the necessity of forming very specific cell-cell contacts.

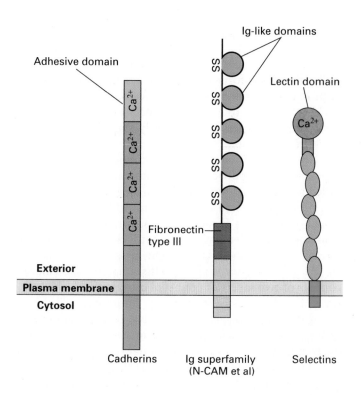

◄ FIGURE 24-30   Major families of cell-adhesion proteins; all are built of multiple domains. Cadherins are homophilic adhesion molecules. Four domains in the extracellular segment bind Ca²⁺; the domain farthest from the membrane (blue) is thought to mediate cell-cell adhesion. Receptors of the immunoglobulin (Ig) superfamily are Ca²⁺-independent and participate in heterophilic and homophilic interactions. They contain multiple domains similar in structure to immunoglobulins (green) and frequently contain type III fibronectin repeats (purple). Selectins bind carbohydrate groups on adjacent cells only in the presence of Ca²⁺. The lectin (carbohydrate-binding) domain (gray), found at the tip of the molecule, also binds Ca²⁺. Selectins also contain a different series of repeated sequences (pink). [Adapted from R. O. Hynes and A. Lander, 1992, *Cell* **68**:303.]

TABLE 24-3   Major Cadherin Molecules on Mammalian Cells

Molecule	Predominant Cellular Distribution
E-cadherin (uvomorulin)	Preimplantation embryos, epithelial cells (particularly at belt desmosomes)
P-cadherin	Trophoblast, heart, lung, intestine
N-cadherin	Nervous system, lung, heart, lens, embryonic mesoderm and neural ectoderm
R-cadherin	Retinal nerve and glial cells
M-cadherin	Myoblasts (muscle cell precursors) and mature skeletal muscle cells

SOURCE: M. Takeichi, 1988, *Development* **102**:639 and, 1991, *Science* **251**:1451; H. Inuzuka et al., 1991, *Neuron* **7**:69; and M. Donalies et al., 1991, *Proc. Nat'l. Acad. Sci.* **88**:8024.

Each cadherin is an integral membrane glycoprotein of 720–750 amino acids. On average, 50–60 percent of the sequence is identical among different cadherins. Importantly, each cadherin has a characteristic tissue distribution. During differentiation, the amount or nature of the cell-surface cadherins change, affecting many aspects of cell-cell adhesion and cell migration.

In adult vertebrates, *E-cadherin*, also known as *uvomorulin*, is the intercellular glue that holds most epithelial sheets together. Sheets of polarized epithelial cells, such as those that line the small intestine or kidney tubules, contain abundant E-cadherin at the sites of cell-cell contact along their lateral surfaces. When a monoclonal antibody to E-cadherin is added to a monolayer of cultured epithelial cells (see Figure 6-9), the cells detach from each other, directly demonstrating the requirement for E-cadherin in cell-cell adhesion. The removal of Ca²⁺ from the medium also disrupts cell-cell adhesion, showing that E-cadherin–mediated interactions require Ca²⁺. If E-cadherin–mediated adhesion is blocked during cell aggregation none of the other junctions between epithelial cells (e.g., spot desmosomes and gap junctions) are generated.

E-cadherin, like the other cadherins, preferentially mediates homophilic interactions. This phenomenon was demonstrated in experiments with L cells, a line of cultured transformed mouse fibroblasts that express no cadherins and adhere poorly to themselves or to other cultured cells. Lines of transfected L cells that express either E-cadherin or P-cadherin were generated; such cells were found to adhere preferentially to cells expressing the same class of cadherin molecules. For instance, E-cadherin–expressing L cells adhere tightly to each other and to epithelial cells from embryonic lung that express E-cadherin; they do not attach to untransfected L cells or to L cells (or other cell types) expressing P-cadherin. L cells expressing P-cadherin adhere to each other, and to other types of cells that express this cadherin. Thus, cadherins directly cause homophilic interactions among cells.

The molecular sites responsible for the binding specificity of cadherins are localized to the most N-terminal ≈100 amino acids, the domain at the "tip" of the molecule (see Figure 24-30). One experiment that established this point made use of a chimeric cadherin in which the N-terminal ≈100 amino acids was from P-cadherin and the remainder of the molecule from E-cadherin; fibroblasts expressing this chimera had the binding properties of P-cadherin, not of E-cadherin. Moreover, site-specific mutagenesis of E-cadherin showed that a change in only two amino acids in this N-terminal domain converted its binding specificity to that of P-cadherin. Clearly, all of the cadherins evolved from a common progenitor, which acquired different homophilic binding specificities, permitting similar kinds of cells in a tissue to adhere specifically to each other.

Cadherins Influence Morphogenesis and Differentiation

Cell-cell contacts form and break as essential parts of many steps in cell differentiation and tissue morphogenesis. Cadherins play a key role in these processes, as we will illustrate with examples from the early stages of mouse and chick differentiation.

After fertilization and fusion of mouse sperm and egg pronuclei, the egg divides several times into a mass of cells called a morula. At this stage the major cell-surface adhesive protein expressed is E-cadherin. Addition of a monoclonal antibody to E-cadherin causes the cells to detach from each other, demonstrating that E-cadherin is essential for the cells to adhere into a tight mass.

At a much later stage in differentiation, the embryo consists of two layers of epithelial cells, both of which express E-cadherin: the outermost embryonic cell layer, called the *ectoderm,* which gives rise to skin, the nervous system, and other tissues; and the inner *endoderm,* which differentiates into the gastrointestinal tract and other tissues. First, some of the ectoderm cells lose their E-cadherin, releasing them from the epithelium; these cells move to the middle of the embryo and form the mesoderm (Figure 24-31a). After a number of additional cell divisions, a region of the ectoderm along the midline of the embryo folds in-

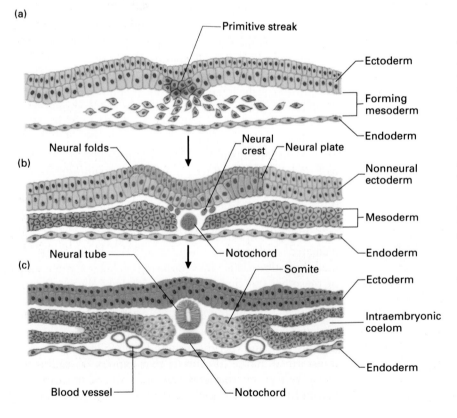

◄ FIGURE 24-31   Changes in cell-surface cadherins during early differentiation in the chick. (a) The embryo at this stage consists of two layers of cells, the endoderm and ectoderm, each of which expresses E-cadherin, or uvomorulin (yellow). Near the midsection of the embryo, ectodermal cells stop producing E-cadherin (gray) and migrate inward to form the mesoderm. (b) After many cell divisions, another segment of surface ectodermal cells along the midline—the neural plate—loses E-cadherin and folds inward. Eventually these cells synthesize N-cadherin (blue) and form the neural crest. (c) Still later the ectodermal cells express both E- and P-cadherins (red) and the notochord N- and P-cadherins (purple). As mesodermal cells condense into somites (precursors of muscles), they express N-cadherin. [After M. Takeichi, 1988, *Trends Genet.* **3**:213.]

ward to form the neural tube that runs the length of the embryo (Figure 24-31b). As these ectodermal cells invaginate, they lose E-cadherin and synthesize a different but related cadherin, N-cadherin. During further differentiation of the central nervous system from the neural tube, N-cadherin becomes a major adhesive protein; P cadherin also is expressed by several types of cells, allowing for selective cell-cell interactions.

At the time the neural tube is formed, the neural crest is also produced. These cells form a group on the dorsal (top) side of the neural tube. Some neural-crest cells migrate along paths to sites near the neural tube where they differentiate into aggregates of nerve cells called ganglia. The region of the ectoderm that generates neural-crest cells also loses E-cadherin and expresses N-cadherin. However, expression of N-cadherin by neural-crest cells is only temporary. As the neural-crest cells begin migrating, they lose all cadherins, and thus are distinguished from the overlying ectoderm (expressing E-cadherin) and the neural tube (expressing N-cadherin). Presumably the loss of adhesive proteins is essential for cells to migrate. Once the neural-crest cells reach their destinations next to the neural tube, they differentiate into neurons, simultaneously re-express N-cadherin, and form ganglia. This adhesive protein helps maintain the integrity of neuronal aggregates in the various ganglia.

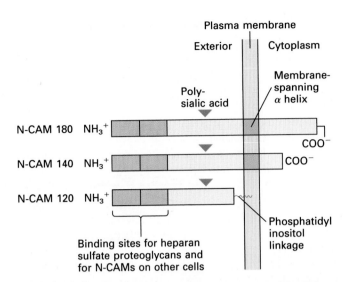

▲ FIGURE 24-32   Three of the forms of N-CAM produced by alternative splicing. N-CAM 180 (180,000 MW) and N-CAM 140 are anchored in the membrane by a single hydrophobic α helix and differ in the length of their cytoplasmic domains. N-CAM 120 is anchored by a complex inositol-phosphate-containing phospholipid (see Figure 14-20). Each of the three forms can also vary in the length of the poly α $(2 \rightarrow 8)$ sialic acid chain, whose attachment site is indicated. [After G. M. Edelman, 1988, *Biochemistry* **27**:3533, and T. M. Jessell, 1988, *Neuron* **1**:3.]

N-CAMs Mediate Ca²⁺-Independent Adhesion of Cells in Nervous Tissue and Muscle

N-CAMs, a group of Ca²⁺-independent cell-adhesion proteins in vertebrates, belong to the Ig superfamily (see Figure 24-30). Their full name—*nerve-cell adhesion molecule*—reflects their particular importance in nervous tissue. Like cadherins, N-CAMs primarily mediate homophilic interactions, binding together cells that express similar N-CAM molecules. Unlike cadherins, N-CAMs are encoded by a single gene; their diversity is generated by alternative mRNA splicing and by differences in glycosylation (Figure 24-32).

Like N-cadherin, N-CAMs appear during morphogenesis as neural epithelial cells fold inward to form the neural crest and neural tube. They are uniformly distributed along the neural tube and maintain neuron-neuron adhesion. N-CAMs, like N-cadherin, disappear from the neural-crest cells during migration but are expressed again once the cells stop migrating and form ganglia. Differentiating muscle, glial, and nerve cells all express N-CAMs. Their role in cell adhesion has been directly demonstrated by use of specific antibodies. For instance, adhesion of cultured retinal neurons is inhibited by addition of antibodies to N-CAMs.

The adhesive properties of N-CAMs are modulated by long chains of sialic acid, a negatively charged sugar.

N-CAMs that are heavily sialylated form weaker homophilic interactions than do less sialylated forms, possibly due to repulsion between the negatively charged sialic acid residues. In embryonic tissues such as brain, polysialic acid constitutes as much as 25 percent of the mass of N-CAMs; in contrast, N-CAMs from adult tissues contain only one-third as much sialic acid. The lower adhesive properties of embryonic N-CAMs enable cell-cell contacts to be made and then broken, a property necessary for specific cell contacts to form in the developing nervous system. The higher adhesive properties of the adult forms of N-CAM stabilize these contacts. Thus, the strength of cell-cell adhesions is modified during differentiation by differential glycosylation of the N-CAMs. Additionally, binding of heparan sulfate proteoglycans to a site near the N-terminus of N-CAMs also can modify cell-cell adhesion during neuronal differentiation.

Movement of Leukocytes into Tissues Requires Sequential Interaction of Specific Adhesive Proteins

Thus far we have considered cell interactions in solid tissues, such as epithelia and neuronal tissue. Once these interactions form during differentiation, they generally are

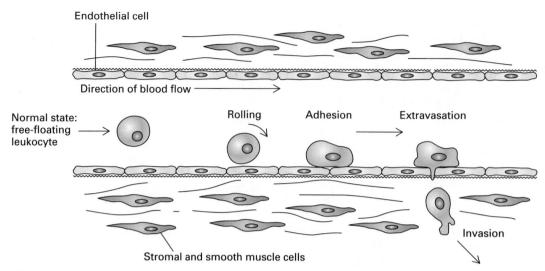

▲ FIGURE 24-33   General process by which free-floating leukocytes (white blood cells) interact with activated endothelial cells in regions of infection or inflammation and then migrate (extravasate) into the underlying tissue. The specific interactions involved in this process are depicted in Figure 24-34. [Adapted from R. O. Hynes and A. Lander, 1992, *Cell* **68**:303.]

stable for the life of the cells. In adult organisms, many types of cells that participate in defense against foreign invaders (e.g., bacteria and viruses) must move from the blood, where they circulate as unattached cells, into the underlying tissue at sites of infection or inflammation (Figure 24-33). Movement into tissue, termed *extravasation,* of four types of leukocytes (white blood cells) is particularly important: monocytes, the precursors of macrophages, which can ingest foreign particles; neutrophils, which release several antibacterial proteins; and T and B lymphocytes, the antigen-specific cells of the immune system (Chapter 27).

Extravasation requires the successive formation and breaking of cell-cell contacts between leukocytes in the blood and endothelial cells lining the vessels. These contacts are mediated by specific cell-adhesion proteins and receptors. A key protein in this process, *P-selectin,* is localized to the blood-facing surface of endothelial cells. Like other members of the selectin family of cell-adhesion proteins, P-selectin is a *lectin,* a protein that binds to carbohydrates. Each type of selectin binds to specific oligosaccharide sequences in glycoproteins or glycolipids. As with cadherins, binding of selectins to their ligands is Ca^{2+} dependent. The sugar-binding lectin domain in selectins generally is at the end of the extracellular region of the molecule (see Figure 24-30). The ligand for P-selectin is a specific oligosaccharide sequence, called the *sialyl Lewis-x antigen,* that is part of longer oligosaccharides present in abundance on leukocyte glycoproteins and glycolipids. The sialyl Lewis-x antigen contains four sugars in very specific linkages:

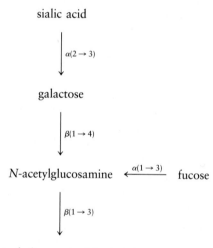

As illustrated in Figure 24-34, in normal endothelial cells P-selectin is localized to intracellular vesicles and is not present on the plasma membrane. These cells are activated by various inflammatory signals released by surrounding cells in areas of infection or inflammation. Once endothelial cells are activated, the vesicles containing P-selectin undergo exocytosis within seconds, and P-selectin appears on the plasma membrane. As a consequence, passing leukocytes adhere weakly to the endothelium; because of the force of the blood flow, these "trapped" leukocytes are slowed but not stopped and seem to roll along the surface of the endothelium (see Figure 24-33).

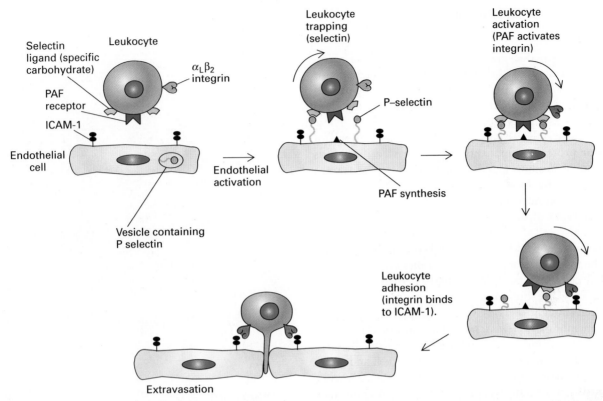

▲ FIGURE 24-34   Interactions between cell-adhesion proteins during the initial binding and tight binding of T cells, a kind of leukocyte, to endothelial cells. PAF = platelet-activating factor. See text for discussion. [Adapted from R. O. Hynes and A. Lander, 1992, *Cell* **68**:303.]

In order for tight adhesion to occur between activated endothelial cells and leukocytes, integrins containing the β_2 subunit present on the surface of leukocytes also must be activated. For example, activation of the $\alpha_L\beta_2$ integrin, which is expressed by T lymphocytes, is induced by *platelet-activating factor* (PAF), a phospholipid released by activated endothelial cells at the same time that P-selectin is exocytosed. Binding of PAF to its receptor on T lymphocytes apparently induces phosphorylation of the cytosolic domain of integrin $\alpha_L\beta_2$, triggering a conformational change in its extracellular domain. The activated integrin then binds to *ICAM-1* and *ICAM-2* (intracellular adhesion molecules 1 and 2), which are expressed constitutively on the surface of endothelial cells. Both ICAM-1 and ICAM-2 are members of the Ig superfamily of adhesive proteins, and their binding to integrin $\alpha_L\beta_2$ does not require Ca^{2+} ions. The tight adhesion mediated by interaction of $\alpha_L\beta_2$ and the ICAMs leads to spreading of T lymphocytes on the surface of the endothelium; soon the adhered T lymphocytes move between adjacent endothelial cells and into the underlying tissue (see Figure 24-34).

Thus, the selective adhesion of T lymphocytes to the endothelium near sites of infection or inflammation depends on the sequential appearance and activation of several different adhesive proteins on the surface of the inter-

acting cells. Other leukocytes, which express specific integrins containing the β_2 subunit, move into tissues by a similar mechanism; $\alpha_M\beta_2$, for instance, is found primarily on macrophages. As might be expected, humans with a genetic defect in synthesis of the integrin β_2 subunit, termed *leukocyte-adhesion deficiency*, are susceptible to repeated bacterial infections.

▶ Cell-Cell Adhesion: Cell Junctions

Adhesion of cells to each other generally is initiated by one or more of the types of cell-surface molecules just discussed. In order for cells in tissues to function in an integrated manner, specialized junctions are essential. There are three major classes of junctions: the *tight junction, desmosome,* and *gap junction.* In Chapter 14, we discussed the structure and function of tight junctions, which connect epithelial cells (e.g., those lining the intestine) and prevent passage of fluids through the cell layer. Here we consider desmosomes and gap junctions and their specialized functions (see Figure 14-42).

Three Types of Desmosomes Impart Rigidity to Tissues

Desmosomes are thickened regions of the plasma membrane where cells are tightly attached to their neighbors or to the extracellular matrix. They increase the rigidity of tissues, such as epithelia, by holding cells securely together, and they act as anchorages for intracellular fibers. These junctions have only recently been purified, and several of the major protein components have been identified. This discussion of desmosomes is based mainly on what has been revealed by localizing these specific proteins with immunoelectron microscopy.

Three types of desmosomes are recognized. *Adherens junctions,* also called belt desmosomes, are found primarily in epithelial cells, and form a belt of cell-cell adhesion just under the tight junction. *Spot desmosomes* are found in all epithelial cells and many other tissues, such as smooth muscle. They are buttonlike points of contact between cells, often thought of as a "spot-weld" between adjacent plasma membranes. *Hemidesmosomes* anchor the plasma membrane to regions of the extracellular matrix. Bundles of intermediate filaments course through the cell and interconnect spot desmosomes and hemidesmosomes; likewise, bundles of actin filaments interconnect belt desmosomes and adhesion plaques.

Intermediate Filaments Stabilize Epithelia by Connecting Spot Desmosomes

A spot desmosome consists of proteinaceous cytoplasmic adhesion plaques (15–20 nm thick) attached to the cytosolic face of the plasma membranes of adjacent cells and connected by transmembrane linker proteins (Figure 24-35). *Plakoglobin* is a major constituent of the plaques; it is closely related in sequence to vinculin, a protein that underlies the plasma membrane at adherens junctions and interconnects transmembrane proteins to the actin cytoskeleton (see Figure 22-35). Fibrous transmembrane linker proteins, called *desmoglein* and *desmocollin,* bind to plakoglobin and other proteins in the plaques, and extend into the intercellular space, where the fibers from the two cells form an interlocking network that binds the two together. Desmoglein is related in sequence and in structure to the cadherins (see Figure 24-30), in particular to E-cadherin, a transmembrane protein found in adherens junctions. Thus, adherens junctions (belt desmosomes), where the actin cytoskeleton is linked to the plasma membrane at points of cell-cell contact, are related in structure to spot desmosomes, where intermediate filaments are linked to the plasma membrane at points of cell-cell contact.

Desmoglein was first identified by an unusual but revealing skin disease called *pemphigus vulgaris,* an autoimmune disease. Patients with autoimmune disorders synthesize antibodies that bind to a normal body protein. In this

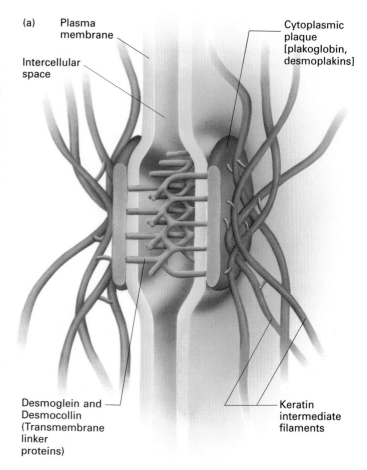

(a)
Plasma membrane
Intercellular space
Cytoplasmic plaque [plakoglobin, desmoplakins]
Desmoglein and Desmocollin (Transmembrane linker proteins)
Keratin intermediate filaments

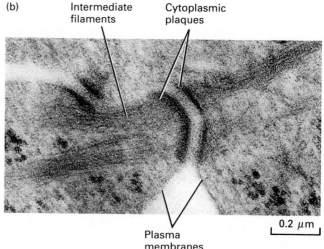

(b)
Intermediate filaments
Cytoplasmic plaques
Plasma membranes
0.2 μm

▲ **FIGURE 24-35** Spot desmosomes. (a) Schematic model showing components of a spot desmosome between epithelial cells and attachments to the sides of keratin intermediate filaments, which crisscross the interior of cells. (b) Electron micrograph of a thin section of a spot desmosome connecting two cultured differentiated human keratinocytes. Bundles of intermediate filaments radiate from the two darkly staining cytoplasmic plaques that line the inner surface of the adjacent plasma membranes. See text for discussion. [Part (a) see B. M. Gumbiner, 1993, *Neuron* **11**:551; D. R. Garrod, 1993, *Curr. Opin. Cell Biol.* **5**:30. Part (b) courtesy of R. van Buskirk.]

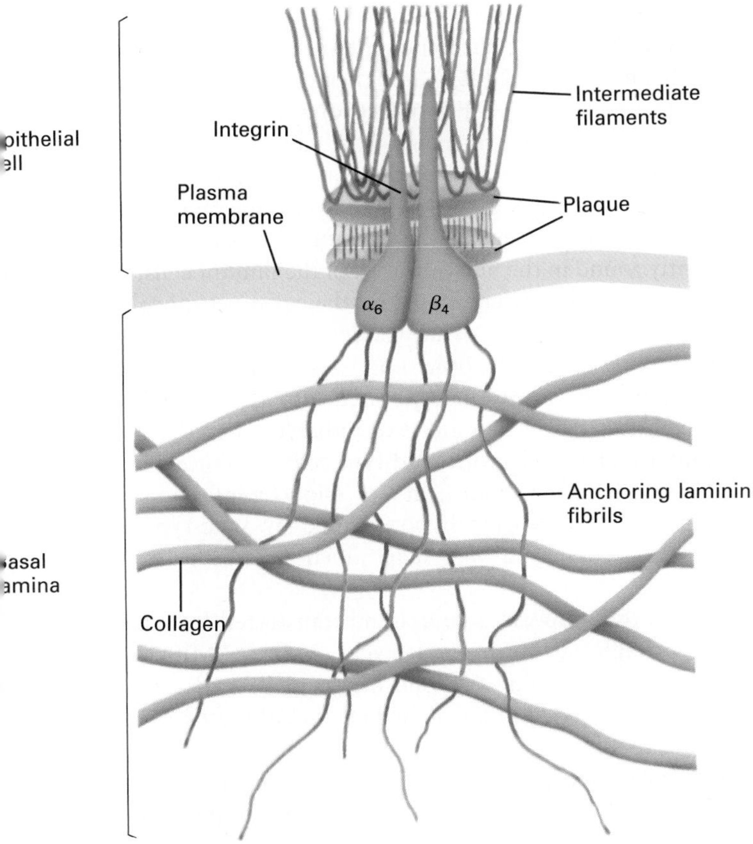

▲ FIGURE 24-36   Schematic model of a hemidesmosome connecting an epithelial cell to the basal lamina. The integrin $\alpha_6\beta_4$, which is localized to hemidesmosomes, binds to proteins in the plaques and to laminin in the extracellular matrix. The nature of the proteins comprising the cytosolic plaques is not known. [Adapted from V. Quaranta and J. Jones, 1991, *Trends Cell Biol.* **1**:2.]

case, the autoantibodies disrupt adhesion between epithelial cells, causing blisters of the skin and mucous membranes. The predominant autoantibody was shown to be specific for desmoglein, a major protein in the skin desmosomes; indeed, addition of such antibodies to normal skin induces formation of blisters and disruption of cell adhesion.

In epithelial cells, keratin intermediate filaments course near the cytoplasmic plaques of spot desmosomes and apparently are linked to them by the *desmoplakin proteins*. Some of these filaments run parallel to the cell surface, and others penetrate and traverse the cytoplasm. They are thought to be part of the internal structural framework of the cell, giving it shape and rigidity. If so, spot desmosomes also could transmit shearing forces from one region of a cell layer to the epithelium as a whole; they thus provide strength and rigidity to the entire epithelial cell layer.

Cardiac (heart) muscle cells are also interconnected by

spot desmosomes. Desmin intermediate filaments interconnect these desmosomes, allowing the stress and strain of the contractile force of one muscle cell to be transmitted to the others. In smooth muscle, desmin filaments are anchored to the "dense bodies," regions of the plasma membrane that anchor together adjacent cells.

Hemidesmosomes Connect Epithelial Cells to the Basal Lamina

Hemidesmosomes are found mainly on the basal surface of epithelial cells where they form tight points of contact with the basal lamina. By interconnecting the intermediate filaments of the cytoskeleton with the fibers of the basal lamina, they increase the overall rigidity of epithelial tissues. Hemidesmosomes consist of two proteinaceous plaques, on the cytosolic face of the plasma membrane, which are attached to the ends of intermediate filaments (Figure 24-36). The proteins composing hemidesmosomes differ from those in spot desmosomes. Plakoglobin, for instance, is not found in hemidesmosomes. Rather, integrin $\alpha_6\beta_4$ is localized to hemidesmosomes and is thought to bind to proteins within the plaques and to the extracellular-matrix protein laminin. In this way, the cell is anchored to the extracellular network.

The β_4 chain in integrin $\alpha_6\beta_4$ has more than 1000 amino acids in its cytosolic domain, dwarfing the $\approx$50-residue cytosolic "tails" of other integrins that associate with intracellular actin filaments. The very long cytosolic domain of integrin $\alpha_6\beta_4$ could extend deep into the cytoskeleton, enabling it to bind to multiple proteins, perhaps including intermediate filaments.

Gap Junctions Allow Small Molecules to Pass between Adjacent Cells

Gap junctions are distributed along the lateral surfaces of adjacent cells and allow them to exchange small molecules. As we discussed in Chapter 21, gap junctions allow ions, and thus electric current, to pass from one nerve cell to the next, thereby allowing a presynaptic cell to induce an action potential in the postsynaptic cell without a lag period. But gap junctions also are present in many non-neuronal tissues, where they help to integrate the metabolic activities of all the cells in a tissue by permitting exchange of ions and small molecules (e.g., cyclic AMP and precursors of DNA and RNA).

Early electron micrographs showed that almost all animal cells that come in contact with each other have regions of junctional specialization characterized by an intercellular gap, which is filled by a well-defined set of cylindrical particles (Figure 24-37; see also Figure 21-7b). Morphologists named these regions *gap junctions*, but in retrospect the gap is not their most important feature. The cylindrical particles, which are tiny water-filled channels, are the key

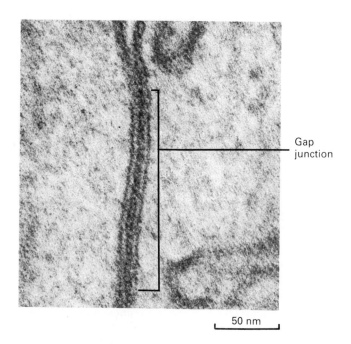

Gap
junction

50 nm

▲ **FIGURE 24-37**   Electron micrograph of a thin section through a gap junction connecting two mouse liver cells. The two plasma membranes are closely associated for a distance of several hundred nanometers, separated by a "gap" of 2–3 nm. [Courtesy of D. Goodenough.]

to the function of gap junctions. These channels directly link the cytosol of one cell with that of an adjacent cell, providing a passageway for movement of very small molecules and ions between the cells.

The size of these intercellular channels can be measured by injecting a cell with a fluorescent dye covalently linked to molecules of various sizes and using a fluorescence microscope to observe whether they pass into neighboring cells. Gap junctions between mammalian cells permit the passage of molecules as large as 1.2 nm in diameter. In insects, these junctions are permeable to molecules as large as 2 nm in diameter. Generally speaking, molecules with a molecular weight of lower than 1200 pass freely and those of 2000 or more do not pass; the passage of intermediate-sized molecules is variable and limited. Thus ions and many low-molecular-weight building blocks of cellular macromolecules, such as amino acids and nucleoside phosphates, can pass from cell to cell.

A vivid example of this cell-cell transfer is the phenomenon of *metabolic coupling*, or *metabolic cooperation*, in which a cell can transfer molecules to a neighboring cell that is unable to synthesize them. For example, AMP, ADP, or ATP can pass through gap junctions. In normal fibroblasts, hypoxanthine (adenine, with the C—NH$_2$ group replaced by C = O) can serve as a precursor of DNA. By the pathway depicted in Figure 6-11, hypoxanthine is con-

verted first to inosine 5'-phosphate (IMP), in a reaction catalyzed by the enzyme hypoxanthine-guanine phosphoribosyltransferase (HGPRT), and then to dATP, the immediate precursor of DNA. Fibroblasts of a mutant line lacking HGPRT are unable to incorporate radioactive hypoxanthine into their DNA. But if cells lacking HGPRT are co-cultured with cells that have this enzyme and labeled hypoxanthine is added to the medium, radioactivity is frequently found in the nuclear DNA of the mutant cells. (The two cells can be differentiated by their distinct morphologies or by feeding one of the cell lines carbon particles before mixing it with the other line.) The dATP derived from hypoxanthine is only incorporated into the DNA of the mutant cells that are in direct or indirect contact (through an intermediate cell) with wild-type cells. It is thought that labeled adenosine mono-, di-, or triphosphate is synthesized from the labeled hypoxanthine by wild-type cells and then passed through gap junctions to the mutant cells.

Another important compound transferred from cell to cell through gap junctions is cyclic AMP (cAMP), which acts as an intracellular second messenger. As discussed in Chapter 20, the amount of cellular cAMP increases in response to stimulation of cells by binding of many different hormones. The fact that cAMP can pass through gap junctions means that the hormonal stimulation of just one or a few cells in an epithelium initiates a metabolic reaction in many of them. For instance, secretory hormones, such as secretin, bind to receptors on the basal plasma membranes of pancreatic acinar cells and increase the intracellular concentration of either cAMP or Ca^{2+} ions, both of which trigger secretion of the contents of secretory vesicles. Because Ca^{2+} and cAMP can pass through the gap junctions, hormonal stimulation of one cell triggers secretion by many. As we saw in Chapter 22, an elevation in cytosolic Ca^{2+} in smooth muscle cells induces contraction. Gap junction–mediated transfer of Ca^{2+} ions between adjacent smooth muscle cells thus allows the coordinated contractile waves in the intestine during peristalsis and in the uterus during birth.

An important aspect of gap-junction physiology is that the channels close in the presence of very high concentrations of Ca^{2+} in the cytosol. Recall that the Ca^{2+} concentration in extracellular fluids in quite high (from 1×10^{-3} M to 2×10^{-3} M), whereas normally the concentration of Ca^{2+} free in the cytosol is lower than 10^{-6} M (see Table 15-2). If the membrane of one cell in an epithelium is ruptured, Ca^{2+} enters the cell, closing the channels that connect the cell with its neighbors and thus preventing leakage of the low-molecular-weight substances that are present in the cytoplasm of all epithelium cells. Even slight increases in the level of cytosolic Ca^{2+} ions or decreases in cytosolic pH can decrease the permeability of gap junctions. Thus cells may modulate the degree of coupling with their neighbors, but precisely why and how they accomplish this is a matter of debate.

Connexin, a Transmembrane Protein, Forms Cylindrical Channels in Gap Junctions

In livers and many other tissues, large numbers of individual gap junctions cluster together in an area about 0.3 μm in diameter (see Figure 21-7b). This property has enabled researchers to separate gap junctions from other components of the plasma membrane by shearing the purified plasma membrane into small fragments. Due to their relatively high protein content, fragments containing gap junctions have a higher density than the bulk of the plasma membrane and can be purified on an equilibrium density gradient (see Figure 5-30). Electron micrographs of stained, isolated gap junctions reveal a lattice of hexagonal particles with hollow cores as intercellular channels (see Figure 21-7). A model of the structure of the gap junction, based on these and other studies, is shown in Figure 24-38a. The transmembrane particles from purified liver gap junctions are composed of *connexin 32*, a protein with a molecular weight of 32,000. Each hexagonal particle consists of 12 connexin molecules: six of the molecules are arranged in a hexagonal cylinder in one plasma membrane and joined to six arranged in the same array in the adjacent cell membrane.

The sequences of several connexin proteins, expressed in different tissues, have been determined from their cDNAs. All connexins have related amino acid sequences. Experiments suggest that each connexin polypeptide spans the plasma membrane four times (see Figure 24-38b) and that one conserved transmembrane α helix lines the aqueous channel. The connexins differ mainly in the length and sequence of their most C-terminal segment, which faces the cytosol. At least 11 different genes in the connexin family have been cloned; many are expressed in specific types of cells.

In order to study the properties of various connexins, researchers have microinjected connexin cDNAs into cells that normally do not express connexin and thus do not form gap junctions (e.g., frog oocytes and L cells, a line of mouse tumor cells). These experiments showed that different types of connexins form gap junctions that differ in their channel size and regulation. Interestingly, a cell expressing a particular type of connexin generally can form a gap junction only with another cell expressing the same type of connexin; that is, all twelve of the connexin subunits in a junction must be of the same type. This property of connexins insures that most cells in a tissue will metabolically couple only to cells of the same type. For instance, the connexin expressed in endothelial cells cannot form gap junctions with the type of connexin expressed on the smooth muscle cells that surround the endothelium, but endothelial cells couple to each other as do smooth muscle cells. In a few cases, however, cells expressing different but related connexin proteins can form functional gap junctions, allowing coupling between different types of adjacent cells.

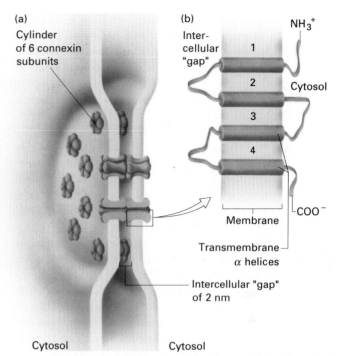

▲ FIGURE 24-38 (a) Model of a gap junction, based on electron microscopic and x-ray diffraction analyses. Two plasma membranes are separated by a gap of about 2–3 nm. Both membranes contain cylinders of six dumbbell-shaped connexin subunits. Two such cylinders join in the gap between the cells to form a channel, about 1.5–2.0 nm in diameter, that connects the cytoplasm of the two cells. (b) Arrangement of a connexin subunit within the plasma membrane. Transmembrane α helix 3 is amphipathic: one face of the helix contains mostly hydrophilic side chains; the other face is mostly hydrophobic. The sequence of α helix 3 is highly conserved among gap junction proteins in different cell types and species; it is thought that the aqueous channel is lined by transmembrane α helices from the 12 connexin subunits. [See M. Bennett et al., 1991, *Neuron* **6**:305; Part (b) courtesy of Bruce Nicholsen.]

▶ *Dorsoventral Patterning during Embryogenesis*

Local interactions between cells and between cells and the extracellular matrix play a prominent and widespread role in regulating the development of multicellular organisms. In early vertebrate embryogenesis, local cellular interactions determine regionalization along the major axes of the embryo. This contrasts with what we learned in Chapter 13 about regionalization in the early *Drosophila* embryo, where the developmental determinants, for the most part transcription factors, diffuse through the common cytoplasm of the syncytial blastoderm to influence transcription from different spatially distributed nuclei. Local interactions between different cell populations, however, is the primary mechanism regulating the formation of internal organs such as the kidney, lung, and pancreas. The vast number of highly specialized cells and their stereotyped arrangement in different tissues also is a consequence of locally acting signals. And finally, within the developing nervous system, local interactions guide the developing neuronal processes to their targets. In Chapter 20, we primarily considered the mechanisms by which different types of receptors transduce signals from the cell surface. In this and the next three sections, we focus on several different types of locally acting signals that specify different cell types and determine their precise organization into tissues. We begin by considering establishment of the dorsoventral axis in invertebrate and vertebrate embryos.

As discussed in Chapter 13, during early development the embryo is divided progressively into domains through the action of patterning genes. The early *Drosophila* embryo is a syncitium (see Figure 13-29), and many of the important patterning events are controlled by the spatial distribution of transcription factors. Two mechanisms operate to subdivide the embryo along the two axes: (1) gradients of transcription factors, acting as morphogens, activate different sets of genes at different threshold concentrations, and (2) the action of one patterning gene may induce expression of another patterning gene, which in turn functions to control the expression of other genes (see Figure 13-37). During later stages of *Drosophila* embryogenesis and in early vertebrate embryos, whose cells are separated by membranes, extracellular molecules act as morphogens and can induce the expression of other extracellular signals.

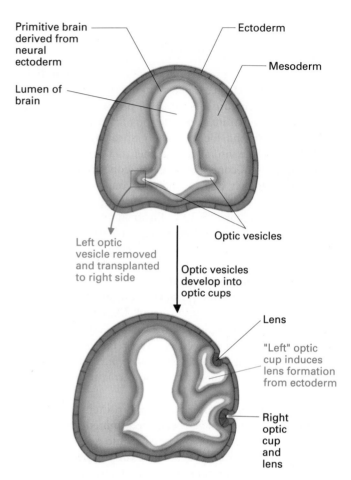

▲ FIGURE 24-39   Diagrams of a cross section through the head of a vertebrate embryo illustrating the interaction of the optic cup with the overlying ectodermal cells that become the optic lens. When the left optic vesicle is transplanted to another site in the head, it induces a lens in that inappropriate (ectopic) site. [After N. K. Wessels, 1977, *Tissue Interaction in Development*, Benjamin-Cummings, p. 46.]

Embryologic Development Is Directed by Induction

In embryology, the term *induction* refers to any mechanism whereby one cell population influences the development of neighboring cells. The phenomenon of induction was discovered in the early 1900s through two complementary experiments illustrated in Figure 24-39. In one, destruction of an optic-vesicle primordium in developing frogs prevented formation of the lens from the overlying ectodermal cells. Conversely, transplantation of an optic-vesicle primordium to a region of ectoderm that normally does not give rise to a lens induced formation of a lens in an abnormal (ectopic) site.

Inductive mechanisms can be divided into two broad categories. In *permissive induction,* a responding tissue (or cell) differentiates in the presence of an inducing signal, but fails to develop at all in the absence of the signal. In *instructive induction,* a tissue (or cell) responds to a signal by differentiating into one cell type, whereas it differentiates into another cell type in the absence of the signal.

Inducing signals fall into three classes. Some are tethered to the cell surface and are available only to immediate neighboring cells. Others may be highly localized by their tight binding to the extracellular matrix, and yet others are freely diffusible and hence can act at a distance. Early embryologists noted that cells differed in their ability to respond to inducing signals. Cells that can respond to such signals are referred to as *competent.* Competence may reflect the expression of receptors specific for a given signaling molecule, the ability of the receptors to activate specific intracellular signaling pathways, or the presence of the transcription factors necessary to stimulate expression of the genes required to implement the developmental program induced. The following sections describe inductive mechanisms in several different developmental systems.

Transforming Growth Factor β (TGFβ) Has Numerous Inductive Effects in Invertebrates and Vertebrates

A number of related extracellular signaling molecules that play widespread roles in regulating development in both invertebrates and vertebrates constitute the *transforming growth factor β (TGFβ) superfamily*. The first members of the TGFβ superfamily were identified on the basis of their ability to induce a transformed phenotype of certain cells in culture, but these secreted growth factors are now known to have a remarkable spectrum of effects on growth and development. Some TGFβ-superfamily members have antiproliferative effects on certain tissues and proliferative effects on others. These growth factors also promote the production of cell-adhesion molecules, other growth factors, and extracellular matrix molecules. Here we describe the general structure of TGFβ proteins and the cell-surface receptors that bind them.

Structure of TGFβ   A schematic diagram of TGFβ monomers is shown in Figure 24-40a. The mature, active

(a) Formation of mature TGFβ-super family growth factors

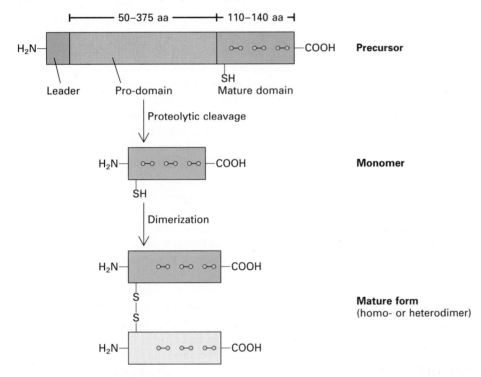

(b) Structure of mature TGF β

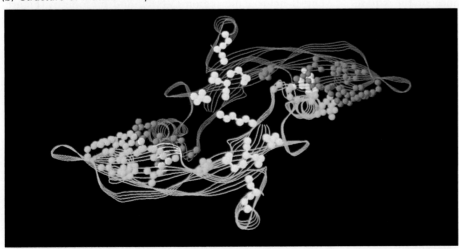

◄ FIGURE 24-40   TGFβ-superfamily proteins, which function as inducers during vertebrate and invertebrate development. (a) Schematic diagram of formation of mature dimeric TGFβ proteins from secreted monomeric TGFβ precursors. The mature domain contains six cysteine residues (yellow dots), which form three intrachain disulfide bonds. Another N-terminal cysteine in the mature domain forms an interchain disulfide bond that links monomers into the active homo- or heterodimeric proteins. (b) Ribbon diagram of mature TGFβ dimer based on x-ray crystallographic analysis. The two subunits are shown in yellow and blue, with the hydrophobic residues at the interfaces indicated by balls. The white balls represent disulfide-linked cysteine residues. The three intrachain disulfide linkages in each monomer form a cystine-knot domain, which is resistant to degradation. [Part (b) from S. Daopin et al., 1992, *Science* **257**:369.]

form of these growth factors, containing 110–140 amino acids, is derived from the C-terminal region of a precursor protein through proteolytic processing. The precursor contains an N-terminal signal peptide followed by a pro-domain containing 50–375 amino acids. X-ray crystallographic studies reveal a compact structure with four antiparallel β strands and three intramolecular disulfide linkages forming a structure called a cystine knot (Figure 24-40b). The cystine-knot domain, which is relatively resistant to denaturation, may be well suited to an extracellular role. An additional N-terminal cysteine in each monomer then links two TGFβ monomers into a dimer. Much of the sequence variation among different TGFβ proteins is observed in the N-terminal regions, the loops joining the β strands, and the α helices. TGFβ monomers can form both homodimers and heterodimers; different heterodimeric combinations may increase the functional diversity of these proteins beyond that generated by differences in the primary sequence of the monomer.

The primary sequence of TGFβ monomers exhibits less than 10 percent homology with nerve growth factor and platelet-derived growth factor (PDGF). Nonetheless, the remarkable similarity in the three-dimensional structures of the monomers of these protein growth factors suggest a common ancestral origin with much sequence drift during evolution. The organization of the subunits in the dimeric proteins, however, varies among the three growth factors.

Cell-Surface Receptors for TGFβ By cross-linking radio-iodinated TGFβ molecules bound to the surface of cells, investigators identified three different polypeptides with apparent molecular weights of 55, 85, and 280 kDa referred to as types I, II, and III TGFβ receptors, respectively. The type I and type II receptors are both transmembrane serine/threonine kinases; binding of TGFβ to a heterodimer containing these subunits is required for signal transduction (see Figure 20-4f). Although mutagenesis studies have demonstrated the importance of the kinase domains of the type I/II heterodimer, the mechanism by which signals are transduced is not yet known. As discussed earlier, the type III receptor is a cell-surface proteoglycan called β-glycan, which appears to regulate the accessability of TGFβ to the signal-transducing heterodimer of the type I and type II receptor.

The function of TGFβ receptors can be modified in several different ways. Alternative splicing has been shown to generate different forms of the receptors with different affinities for ligand. The ligand affinity of a type II receptor is modulated by its dimerization with different type I receptors. The type III receptor (β-glycan) though not essential for signaling activity, plays an important modulatory role. In some cases, the type I/II receptor complex binds much more strongly to a ligand complexed to β-glycan. Finally, soluble proteins or components of the extracellular matrix can bind to TGFβ ligand and sequester it in a form that is not recognized by the receptors.

TGFβ Homolog Encoded by the *decapentaplegic* Gene Controls Dorsoventral Patterning in *Drosophila* Embryos

In *Drosophila*, the *decapentaplegic* (*dpp*) gene encodes a TGFβ-superfamily member, the Dpp protein. The role of Dpp in embryonic induction is understood in some detail, and studies in vertebrate systems suggest that other members of the TGFβ superfamily play similar roles. In both *Drosophila* and vertebrates, TGFβ proteins have been shown to control cell fate along the dorsoventral axis during embryogenesis. In this section, we discuss the mechanism of this patterning in flies.

As explained in Chapter 13, initial dorsoventral patterning in *Drosophila* is controlled by a set of extracellular signaling molecules encoded by maternal mRNAs, particularly the *Toll* and *dorsal* mRNAs (see Figure 13-37d). Generation of a signal in the most ventral part of the embryo activates the Toll protein, a cell-surface receptor, in the most ventral region of the embryo. The resulting Toll-activation gradient leads to nuclear localization of Dorsal in the ventral region. In lateral regions, corresponding to lower levels of the Toll signal, Dorsal protein is evenly distributed between the cytoplasm and the nucleus. In the dorsal region of the embryo, the Dorsal protein is found only in the cytoplasm (see Figure 13-34a). As a consequence of this gradient in the nuclear concentration of Dorsal, a protein that both activates and represses gene transcription, the *dpp* gene is expressed only in the most-dorsal 40 percent of the embryo.

At this early stage of embryonic development, four distinct cell types are found in bands along the dorso-ventral axis: the extraembryonic amnioserosa, dorsal ectoderm, ventral neurogenic region (VNR), and mesoderm (Figure 24-41a). These different cell identities can be assessed by examining the pattern of cuticular structures, as well as by the expression of specific marker genes. A combination of both genetic and molecular genetic evidence suggests that Dpp acts as a morphogen to induce establishment of different cell types at different threshold concentrations. For instance, complete removal of Dpp function leads to the loss of all dorsal structures and their conversion into more ventral ones (Figure 24-41b). Embryos carrying only one wild-type *dpp* allele show an increase in the number of cells assuming a ventral fate, whereas embryos with three copies of *dpp* form more dorsal cells. By injecting in vitro synthesized *dpp* mRNA into mutant embryos genetically altered to give rise only to lateral structures (i.e., ventral neurogenic ectoderm), investigators have demonstrated that dif-

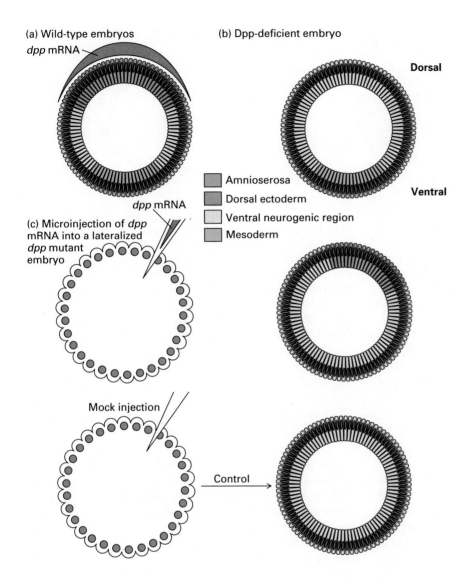

(a) Wild-type embryos
dpp mRNA

(b) Dpp-deficient embryo

Dorsal

Ventral

☐ Amnioserosa
☐ Dorsal ectoderm
☐ Ventral neurogenic region
☐ Mesoderm

dpp mRNA

(c) Microinjection of dpp mRNA into a lateralized dpp mutant embryo

Mock injection

Control →

◄ FIGURE 24-41   Effect of Dpp, a TGFβ-superfamily protein, on dorsoventral patterning in the *Drosophila* embryo. Dorsal is toward the top; ventral, toward the bottom. (a) Four identifiable cell types are present in wild-type fly embryos soon after cellularization, shown schematically in this cross-sectional diagram. The Dpp (blue) concentration is a gradient that is greatest in the most-dorsal region. (b) In mutant embryos lacking Dpp, cells in the dorsal and lateral region of the embryo all assume a ventral neurogenic fate. (c) Microinjection of *dpp* mRNA into mutant embryos in which the entire dorsoventral axis develops into ventral neurogenic region tissue demonstrates the inductive effect of Dpp. The area closest to the injection site, where the Dpp level is highest, develops into amnioserosa; as the distance from the injection site increases, more-ventral structures characteristic of dorsal ectoderm are induced. Even farther away, where the Dpp level falls below the threshold, cells differentiate along a ventral neurogenic pathway. In the control mock injection, no dorsal structures develop. [Adapted from J. M. W. Slack, 1993, *Mech. Devel.* **41**:91; see E. L. Ferguson and K. V. Anderson, 1992, *Cell* **71**:451.]

ferent levels of Dpp specify different cell fates. Close to the injection site, where the Dpp concentration is highest, amnioserosa tissue forms; somewhat farther from the injection site, the intermediate Dpp concentration induces formation of dorsal ectoderm. In regions where the Dpp concentration is below the threshold concentration for inducing dorsal ectoderm, ventral ectoderm develops (Figure 24-41c).

Recent studies suggest cell fate along the dorsoventral axis may not simply reflect the graded activity of Dpp. Several other genes have been identified that regulate dorsoventral patterning and may do so by directly modulating Dpp or acting in combination with Dpp to specify different fates. For instance, in *screw* mutants dorsal-most structures are lost. *Screw* encodes a TGF-β homolog and raises the possibility that different cell fates along the dorsoventral axis may be specified by different homodimeric and heterodimeric combinations of Dpp and Screw. The activity of Dpp also may be modulated by posttranslational

modification. In the dorsal region of the embryo, for example, the *tolloid* gene produces a metalloprotease that promotes Dpp function. Interestingly, a vertebrate homolog of Tolloid has been proposed to activate TGF-β-like ligands.

Sequential Inductive Events Regulate Early *Xenopus* Development

The large size of the frog embryo and its ease of experimental manipulation make it a favored system for investigating the role of extracellular signaling molecules in regulating early pattern formation in vertebrate embryos. Most recent research, particularly at the molecular level, has concentrated on development of the frog *Xenopus laevis*.

Establishment of Vertebrate Body Plan   As illustrated in Figure 24-42a, the unfertilized *Xenopus* egg has an inherent asymmetry with a pigmented *animal* hemisphere (pole or cap) and an unpigmented *vegetal*

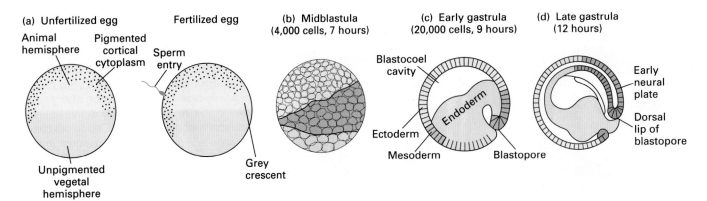

▲ FIGURE 24-42   Early embryogenesis of the frog *Xenopus laevis*. (a) The vegetal pole is the yolk-rich hemisphere. The site of sperm entry defines the ventral side of the embryo. Fertilization leads to rapid cell divisions and formation of the blastula (not shown), a hollow ball of 32 cells called blastomeres. (b) Signals from the vegetal pole induce the formation of mesodermal cells in the marginal zone (blue) separating the vegetal and animal caps. (c, d) During gastrulation, mesodermal cells fold into the embryo. Signals from the invaginating mesoderm induce the development of both the underlying endoderm and neural tissue from the overlying ectoderm. The anterioposterior axis is determined by the mesoderm; cells that invaginate first induce anterior structures. Subsequent interactions between different cell populations play an important role in organogenesis. [Adapted from E. M. De Robertis, G. Oliver, and C. V. E. Wright, *Sci. Am.* **263**(1):46.]

hemisphere. Entry of sperm into the animal pole induces the actin-rich cortical region to rotate about 30 degrees relative to the yolky cytoplasm of the vegetal pole. Cortical rotation results in a second asymmetry that leads to the future dorsoventral axis with the point of sperm entry defining the ventral side of the embryo. If cortical rotation is prevented—for instance, by UV irradiation—the establishment of dorsoventral polarity is blocked. A stereotyped pattern of rapid synchronous cell divisions then ensues. As these divisions occur, a cavity, called the *blastocoel*, forms in the animal hemisphere; the developing embryo at this stage is referred to as a *blastula*. After the first 12 cell divisions, the cell-division rate slows and becomes asynchronous as zygotic transcription begins. During this stage, the three germ layers become arranged in a simple layered pattern along the animal/vegetal axis (Figure 24-42b). The animal pole will give rise to the ectoderm (e.g., skin and neural tissue); the vegetal pole, to endoderm (e.g., gut); and the marginal zone between the two poles, to mesoderm (e.g., notochord, muscle, blood). During gastrulation, the next stage, the mesodermal and endodermal cells move inside the embryo. The movement of cells into the embryo is initiated at a structure on the dorsal side of the embryo called the *dorsal lip of the blastopore*. At the completion of gastrulation, virtually the entire surface of the embryo is covered by ectoderm (Figure 24-42c,d). The endoderm is found towards the center of the embryo, and a layer of mesodermal cells separates the endoderm and ectoderm.

Sequence of Inductive Events   Early *Xenopus* development progresses as a series of inductive events (Figure 24-43). A popular model for induction in this system invokes three signals. In brief, during the blastula stage a signal from the ventral vegetal pole induces ventral mesodermal tissue. A second signal arising in the dorsal vegetal pole (a region called the Nieuwkoop center) induces a region of the overlying marginal zone to form dorsal mesoderm. This region of the dorsal mesoderm called the *organizer* (or the Spemann organizer) specifies the organization of the mesoderm along the dorsoventral axis through the production of a third signal called the *horizontal dorsalizing signal*. During gastrulation, the mesoderm induces the overlying ectoderm to form neural tissue and the endoderm to form various endodermal derivatives such as the pharynx and intestine. In addition to specifying cell types, the mesoderm also regionalizes both the ectoderm and endoderm. Anterior regions of the mesoderm (regions of the blastula that fold into the gastrula first) induce anterior neural tissue (e.g., the forebrain) in the overlying ectoderm, and more posterior regions induce more posteriorly located neural structures (e.g., the spinal cord). These inductive interactions have been demonstrated by transplanting explants from one developing embryo into an ectopic location in a recipient embryo.

Role of Growth Factors in Mesodermal Induction
Study of the induction of mesoderm derivatives has progressed rapidly in recent years, largely due to development

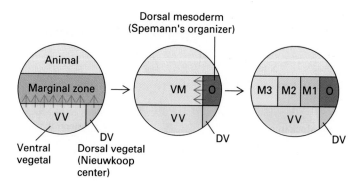

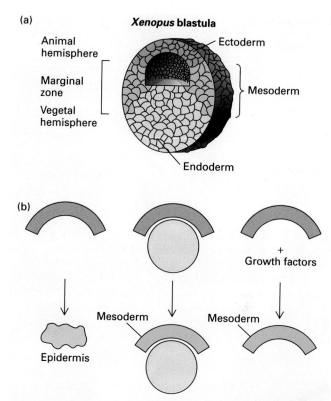

▲ FIGURE 24-43 Model for induction at blastula stage in *Xenopus* embryogenesis. Following cleavage, three signals (indicated by red arrows) are proposed to induce mesoderm formation in the marginal zone. Signals from the ventral vegetal (VV) pole induce ventral mesoderm (VM) tissue in the marginal zone. A signal from the dorsal vegetal (DV) pole, called the Nieuwkoop center, induces formation of the most-dorsal mesodermal tissue, called Spemann's organizer (O). A horizontal dorsalizing signal originating in the organizer further regionalizes the mesodermal tissue (M1, M2, M3) with more-dorsal cell fates specified in tissue closer to the organizer. [Adapted from S. F. Gilbert, 1991, *Developmental Biology*, Sinauer.]

▲ FIGURE 24-44 The animal-cap assay for mesodermal inductive activity. (a) The top-most region of the animal hemisphere (tan), referred to as the animal cap, will give rise to ectoderm and the vegetal hemisphere to endoderm (yellow). The region between them, the marginal zone, will give rise to mesoderm (blue). (b) Cultured animal-cap fragments give rise to epidermal-like ectodermal tissue *(left)*, but addition of vegetal tissue fragments induces mesoderm formation *(center)*. Addition of certain growth factors also induces mesoderm *(right)*.

of an in vitro assay for inductive activity. Although early observations indicated that isolated animal caps develop into ectodermal derivatives (i.e., skinlike epidermal cells), later studies revealed that during normal development, some animal-cap cells contribute to formation of mesodermal structures. This finding suggested that the fate of animal cells may be regulated in part by other regions of the blastula. Indeed, when explants of isolated animal caps were combined with vegetal cells, mesodermal derivatives (e.g., muscle) developed from the animal caps (Figure 24-44). This *animal-cap assay* has been used to measure the ability of various growth factors to induce development of animal caps into mesoderm. Other assays permit the inductive role of specific growth factors to be assessed. For example, injection of mRNAs encoding certain growth factors into a cell (called a *blastomere*) on the ventral, vegetal side of a 32-cell blastula will induce a second embryonic axis (i.e., a two-headed embryo).

These assays have identified four different types of secreted proteins that induce mesodermal tissue. These include members of the TGFβ superfamily (particularly activins and TGFβ), fibroblast growth factor (FGF), the wingless (Wnt) family, and noggin protein. Experiments with the receptors for activin and FGF have provided additional evidence for the role of these growth factors during normal development. When mRNAs encoding altered forms of these receptors lacking the C-terminal cytosolic signaling domain were injected into frog embryos, meso-

dermal induction was blocked. In these experiments, the truncated receptor monomers are expressed and then dimerize with endogenous receptors, thereby inactivating them. Because the receptors for noggin and the Wnt proteins have not been identified, similar experiments have not been conducted as yet with these receptors.

Many different growth factors have been shown to induce mesoderm in the *Xenopus* experimental system. However, definitive proof that any one of these functions to regulate mesodermal fates along the dorsoventral axis during normal development is lacking. As we saw in the last section, *Drosophila* Dpp protein, a TGFβ homolog, acts as a morphogen along this axis. A *Xenopus* TGFβ homolog, called Vg1, may play a role in inducing Spemann's organizer, which in turn specifies mesodermal cell fates along the dorsoventral axis (see Figure 24-43). The

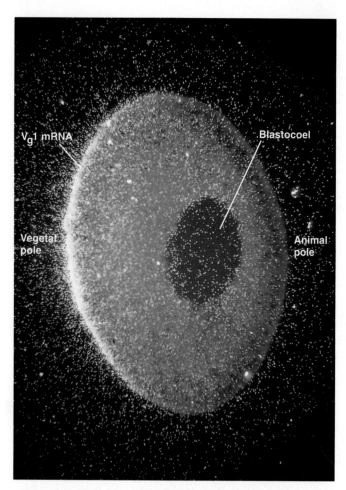

▲ **FIGURE 24-45**   Localization of maternally derived mRNA encoding Vg1, a TGFβ homolog, in *Xenopus* blastula. Tissue sections were incubated with a radiolabeled RNA probe that hybridizes to Vg1 mRNA and then subjected to autoradiography. In this section, visualized by dark-field illumination, regions of hybridization are white. The Vg1 mRNA clearly is localized to the vegetal cortical region of the embryo. [From D. L. Weeks and D. A. Melton, 1987, *Cell* **51**:861.]

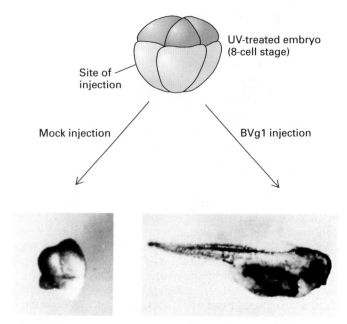

▲ **FIGURE 24-46**   Experimental demonstration of Vg1 function in *Xenopus* embryo. UV irradiation of an early embryo prevents cortical rotation; as a result, the dorsoventral axis is not formed *(lower left)*. Injection of an mRNA encoding a chimeric Vg1 precursor, designated BVg1 (see text), into a vegetal cell rescued UV-irradiated embryos, permitting development of normal tadpoles *(lower right)*. β-Galactosidase mRNA was coinjected with the BVg1 mRNA, and the tadpoles that developed were stained for β-galactosidase activity (blue). This tracer technique showed that the injected cells gave rise to endoderm. During normal development, the Nieuwkoop center, which induces Spemann's organizer, gives rise to endoderm. These results suggest that Vg1 normally functions as the inducer in the Nieuwkoop center. [Adapted from G. H. Thomsen and D. A. Melton, 1993, *Cell* **74**:433; photographs courtesy of D. A. Melton.]

finding that maternally derived Vg1 mRNA is localized specifically to the vegetal pole suggested that it might play an important role in mesodermal induction (Figure 24-45). However, injection of mRNA encoding the full-length Vg1 precursor had no inductive effect, in contrast to mRNAs encoding other TGFβ homologs. Analysis of uninjected and injected *Xenopus* embryos revealed an accumulation of the uncleaved precursor of Vg1 and sparse amounts of the cleaved mature form. Researchers then constructed a chimeric gene that encoded the pro-domain from another TGFβ homolog fused to the mature domain of Vg1 (see Figure 24-40a). Injection of the mRNA encoding this chimeric precursor, called BVg1, led to expression of mature Vg1 protein and mesodermal induction. Remarkably, injection of BVg1 mRNA induces a normal dorsoventral axis even in UV-irradiated embryos, which otherwise do not establish dorsoventral polarity (Figure 24-46).

These findings make Vg1 an excellent candidate as the primary inducer of the organizer, a function previously ascribed to the dorsal vegetal pole, or Nieuwkoop center. The finding that mRNA encoding the full-length Vg1 precursor has no inductive effect has led to the proposal that induction requires a specific proteolytic processing of the Vg1 precursor into the mature form. Some have suggested that cortical rotation, normally induced by sperm entry, may lead to local activation of Vg1 in the Nieuwkoop center, which is located just beneath the organizer. This intriguing notion suggests an analogy with the posttranslational modification of Dpp that, as described earlier, is likely to play an important role in regulating Dpp activity in specifying the dorsoventral axis in the *Drosophila* embryo. Proof that Vg1 indeed induces the organizer will require disruption of its function during normal development.

► *Formation of Internal Organs and Organization of Tissues*

Formation of internal organs such as the kidney, gut, pancreas, and lung is regulated by interactions between epithelial and mesenchymal cells. As discussed in Chapter 14 and mentioned earlier in this chapter, epithelial cells form continuous sheets of polarized cells with the apical and basal regions separated by tight junctions. In contrast, the mesenchyme comprises loosely associated nonpolarized cells. Epithelial cells in different organs are derived from one of the three germ layers: the ectoderm, mesoderm, or endoderm. Mesenchyme is derived from either the mesoderm or the ectoderm. Extracellular-matrix material separates the epithelial and mesenchymal cell populations. Underlying the epithelium is a basal lamina containing laminin, type IV collagen, nidogen, and certain proteoglycans. Several integrins on the surface of epithelial cells bind to components of the basal lamina, anchoring them to the extracellular matrix (see Table 24-2). The basal lamina is linked to the mesenchyme through a reticular network of extracellular components containing collagen (types I and III) and fibronectin.

In this section, we consider two examples of epithelial-mesenchymal interactions that control organ development: the first, formation of the salivary gland, highlights the importance of the extracellular matrix; the second, kidney development, requires direct contact between epithelial and mesenchymal cells. We then discuss the role of secreted molecules in determining the organization of cells within a particular tissue.

The Basal Lamina Is Essential for Differentiation of Many Epithelial Cells

In the developing salivary gland, groups of epithelial cells, derived from the endoderm, secrete proteins and proteoglycans that form a basal lamina, grow, and form clusters of cells that become the secreting cells of the mature gland. Mesodermal cells are attached under the basal lamina. These groups of epithelial cells together with the basal lamina can be separated from the mesodermal cells by microdissection of embryos. The experiment depicted in Figure 24-47 demonstrates that two cell types—epithelial and mesodermal—and the basal lamina all interact and are necessary for the endodermal cells to differentiate into the mature secreting cells of the salivary gland.

Although the extracellular matrix components required for salivary-gland differentiation have not been identified, dramatic changes in extracellular matrix components have been found to occur during other inductive interactions. For instance, syndecan is expressed in a dynamic fashion in the developing kidney. However, since syndecan is expressed by mesenchymal cells after induction, it is likely to play a role in subsequent morphogenesis

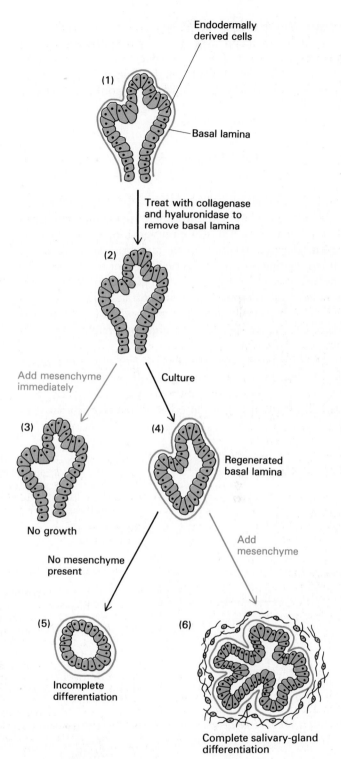

▲ FIGURE 24-47   Elements necessary for salivary-gland differentiation. Salivary-gland explants can be dissected away from cells (1) and treated with enzymes to remove the basal lamina (2). If the epithelial cells are then mixed immediately with mesenchymal cells, there is no growth or differentiation (3). If, on the other hand, the epithelial cells are cultured, they regenerate the basal lamina (4), but do not differentiate further (5) unless they are allowed to interact with mesenchymal cells (6). [After N. K. Wessels, 1977, *Tissue Interactions in Development*, Benjamin-Cummings, p. 225.]

rather than in induction. Tenascin, a large matrix protein containing six identical subunits of 210,000 MW, also is expressed during the period of inductive interactions in the developing kidney, gut, and tooth. Despite its appearance at the right time and place to play an important role during epithelial-mesenchymal interactions, knockout mice lacking tenascin develop normally!

Direct Cell-Cell Contact Regulates Kidney Induction

Both the epithelium, called the *ureteric bud,* and the mesenchyme that give rise to the kidney are mesodermal derivatives (Figure 24-48). As embryonic development proceeds, the ureteric bud grows and branches, eventually differentiating into the collecting ducts of the kidney. The mesenchyme is transformed into a distinct epithelium that will form the proximal and distal tubules and the glomeruli. Both the mesenchyme and the ureteric bud send and receive inductive signals, a phenomenon referred to as *reciprocal induction.*

Several types of experiments have demonstrated reciprocal induction in the developing kidney. For example, if the developing kidney is removed from an 11- to 12-day-old mouse and placed in culture, the ureteric bud continues to grow and branch. Branching ceases, however, when the mesenchyme is separated from the epithelium by mild trypsin treatment; addition of kidney mesenchyme to the trypsin-treated culture restores branching and differentiation. Other in vitro and in vivo experiments involving disruption of kidney development have demonstrated that the epithelium is necessary for conversion of the mesenchyme into tubules. Interestingly, although kidney epithelium shows a specific requirement for kidney mesenchyme for induction of branching, kidney mesenchyme can be induced to form tubules by many different tissues. In other developing organs, however, the signal regulating epithelial development can be derived from other mesenchymal derivatives. The molecular basis for these differences is not known.

Little is yet known about the signaling molecules that mediate specific epithelial-mesenchymal interactions, but in vitro experiments suggest direct cell-cell contact is necessary for kidney induction. Moreover, recent genetic studies have shown that c-*ret*, a proto-oncogene, is critical for kidney morphogenesis. This proto-oncogene encodes a transmembrane receptor tyrosine kinase, c-Ret, whose ligand is unknown. In situ hybridization has shown the c-Ret is expressed in the ureteric bud but not the mesenchyme of the developing kidney. Newborn mice carrying homozygous

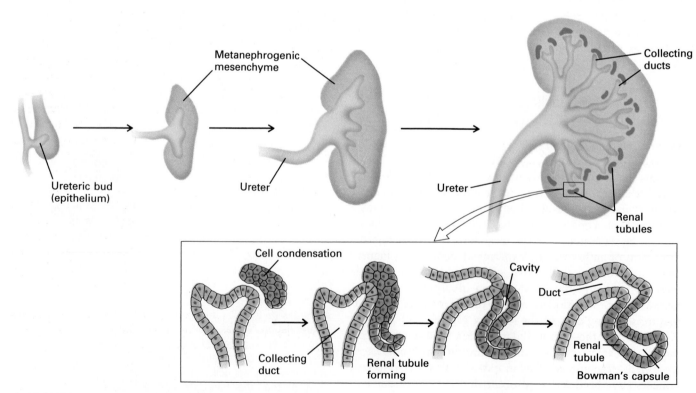

▲ FIGURE 24-48   Embryonic development of the kidney. The ureteric-bud epithelium is induced to branch by interactions with the metanephrogenic mesenchyme. A reciprocal induction drives the mesenchyme to form an epithelium that gives rise to the renal tubules.

(a) Wild type

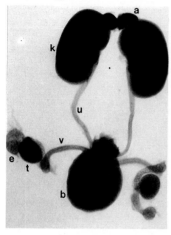

(b) c-ret knockout

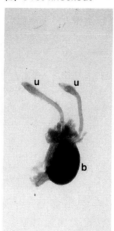

◄ FIGURE 24-49   Knockout mutations in c-ret produce severe defects in kidney morphogenesis in mice. (a, b) Urogenital systems dissected from wild-type and mutant newborn mice. In this mutant, blind-ended ureters formed but no kidney development occurred. (c) Diagrammatic comparison of normal urogenital system and mutant phenotypes resulting from knockout of c-ret. a = adrenal; b = bladder; e = epididymis; k = kidney; u = ureter; v = vas deferens. [From A. Schuchardt et al., 1994, *Nature* **367**:382.]

(c) Spectrum of mutant phenotypes

Normal

Homozygous mutant phenotypes

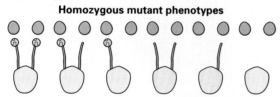

knockout mutations in c-ret exhibit a variety of phenotypes, all involving highly abnormal rudimentary kidneys and in some cases blind-ended ureters (Figure 24-49). Developmental studies have shown that the ureteric bud often fails to form during development in c-ret knockout mice. This finding has led some researchers to suggest that formation of the ureteric bud itself, not only its branching, may depend on a signal from the mesenchyme. These findings are particularly intriguing since c-ret mRNA is specifically localized to the tip of the ureteric bud as well as to the tips of the branching epithelium. This is consistent with the notion that c-ret is a receptor for a cell-surface signaling molecule produced by mesenchyme that induces budding of the kidney epithelium.

Hedgehog Organizes Pattern in the Chick Limb and *Drosophila* Wing

Inductive processes not only are important in the cell-type specialization required for organ development but also play a critical role in determining the organization of cells within a particular tissue. Two well-studied examples—tissue patterning in the chick limb and *Drosophila* wing—appear to involve an inductive protein called Hedgehog.

Embryonic patterning in the vertebrate limb occurs along three axes: *proximodistal* (shoulders to finger tips), *anterioposterior* (thumb to little finger), and *dorsoventral* (upper surface of the hand to the palm). During early em-

bryogenesis, mesodermal cells migrate from the lateral plate to specific regions beneath the epidermis causing a local bulge in the tissue called the limb bud. Mesodermal cells induce a specialized structure within the epidermis (the apical epidermal ridge) that controls patterning along the proximodistal axis. Mesodermal cells at the posterior edge of the limb bud form a region called the *zone of polarizing activity* (ZPA), which acts as an organizer, patterning cells along the anterioposterior axis (Figure 24-50a). This organizer activity was demonstrated by transplanting the ZPA from one limb bud to the anterior edge of a recipient limb bud, thus producing a bud with two ZPAs. Such transplantation resulted in formation of a limb with a mirror symmetric duplication of the digits (Figure 24-50b). When a nonpermeable filter was inserted between the ectopic ZPA and recipient tissue, no response occurred, suggesting that the ZPA produces a diffusible signal. The activity of this diffusible substance appeared graded, since transplantation of fewer cells induced progressively more posterior structures.

Until recently the molecular basis of patterning in the vertebrate limb was poorly understood. Reasoning that other secreted inductive proteins found in *Drosophila* (e.g., Dpp) are homologous to proteins in vertebrates with inductive capacities, investigators isolated vertebrate homologs of another *Drosophila* protein called Hedgehog. The *hedgehog* gene initially was studied as a segment-polarity gene (Chapter 13). A chick homolog of *hedgehog*, called

(a) Wild-type limb development

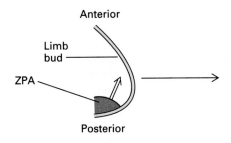

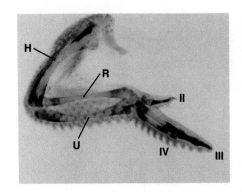

(b) ZPA transplant

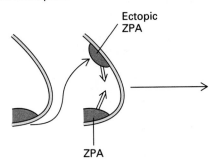

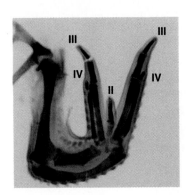

(c) Sonic hedgehog implant

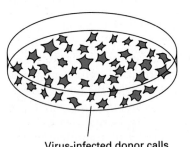

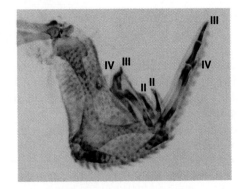

▲ **FIGURE 24-50** Patterning of limbs along the antero-posterior axis in the chick. Diagrams of the embryonic limb bud are to the left; photographs of the adult structure are to the right. (a) Normal limb development is induced by the zone of polarizing activity (ZPA), a group of mesodermal cells at the posterior edge of the limb bud. (b) Transplantation of the ZPA to the anterior region of a normal limb bud results in limb duplication. Considerable variation in the nature of the duplication is observed among limb-bud preparations. In this case, digit II was not duplicated. (c) When donor cells infected with a virus carrying the *sonic hedgehog* gene are implanted in a limb bud from a virus-resistant chick embryo, limb duplication occurs. These results suggest that Sonic Hedgehog, normally expressed in the ZPA, controls limb patterning in the chick. [Adapted from R. D. Riddle et al., 1993, *Cell* **75**:1401; photographs courtesy of C. Tabin.]

sonic hedgehog, was found to be expressed in the ZPA. Insertion of chick fibroblasts infected with a retrovirus carrying a recombinant *sonic hedgehog* gene into the anterior region of a limb bud led to a mirror-image duplication of the limb (Figure 24-50c). Experiments with animals in which Sonic Hedgehog function is knocked out are needed to conclusively confirm its role in limb patterning. Hedgehog also has been shown to induce patterning in other tissues in developing vertebrate embryos.

In *Drosophila*, Hedgehog not only functions in establishing segment polarity early in embryogenesis but also is critical for patterning adult appendages. The wings and

(a)

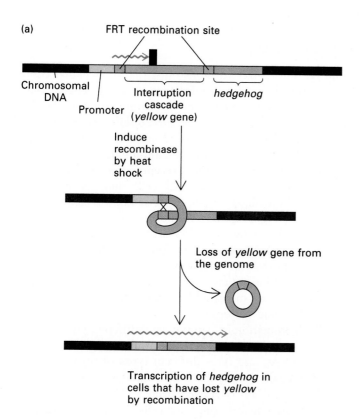

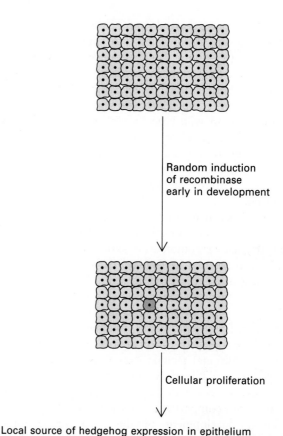

▲ FIGURE 24-51 Effect of ectopic expression of Hedgehog on *Drosophila* wing development. (a) During normal development, Hedgehog is expressed in the posterior compartment of the wing disc, but a yeast recombination system introduced into *Drosophila* can be used to express Hedgehog in the anterior compartment of the wing disc. *(Left)* Diagram of the yeast DNA construct, which contains two cis-acting FRT recombination sites and a constitutive promoter inserted into a *Drosophila* chromosome. Expression of the *hedgehog* gene (orange) is disrupted by the intervening *yellow* gene (blue). The recombinase enzyme, encoded by a gene with a heat-shock–inducible promoter, also inserted into a *Drosophila* chromosome, can be induced by brief exposure to heat shock. Recombination at the FRTs leads to excision of the *yellow* gene and fusion of the constitutive promoter to the *hedgehog* gene, leading to its expression. *(Right)* In fly embryos containing the yeast construct and exposed to heat shock early in development, the rare cells (orange) in which recombination occurs will begin to express Hedgehog. In some cases, these will be in the anterior compartment of the wing disc, and their proliferation leads to ectopic clones of Hedgehog-expressing cells. In the adult wing, loss of the *yellow* gene is easily detected by the lighter color of the tissue. These light patches are derived from cells that underwent recombination and thus expressed Hedgehog constitutively earlier in development. (b) In the developing wing of a normal fly, Hedgehog is expressed in the posterior (P) compartment (orange). Expression of Hedgehog in a localized sector in the anterior (A) compartment during development leads to mirror image duplication of anterior structures. This effect is similar to that resulting from ectopic expression of Sonic Hedgehog in the anterior region of the chick limb bud (see Figure 24-50). [See K. Basler and G. Struhl, 1994, *Nature* **368**:208.]

(b) Effect of ectopic Hedgehog on fly wing

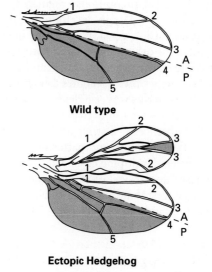

legs of *Drosophila* are derived from epithelia called imaginal discs, which are transformed into appendages during metamorphosis (see Figure 13-27b). Since *hedgehog* is an essential gene for early embryogenesis, its role in later development has been studied using temperature-sensitive mutations and by ectopically expressing the gene during development. In wild-type fly embryos, Hedgehog is expressed in the posterior compartment of the wing disc. Loss-of-function mutations in the *hedgehog* gene cause a dramatic decrease in the size of both the anterior and posterior portions of the wing and severe pattern abnormalities.

Grafting experiments similar to those in chick embryos (see Figures 24-50) cannot be performed in *Drosophila*. However, the yeast recombination system outlined in Figure 24-51a can be used in *Drosophila* to ectopically express specific genes. Ectopic expression of Hedgehog in the anterior compartment of the wing disc, where it normally is not expressed, leads to marked pattern duplication (Figure 24-51b) in *Drosophila* wings, qualitatively similar to that caused by ectopic expression of Sonic Hedgehog in the chick limb. Additional genetic experiments suggest that Hedgehog does not directly pattern tissue but rather induces Dpp expression at the boundary between the anterior and posterior compartments of the wing disc. Interestingly, during development of the chick limb, Sonic Hedgehog has been shown to induce expression of a Dpp homolog called BMP-2 in cells adjacent to the ZPA.

These various studies suggest that structurally similar molecules can function as organizers for limb patterning in both vertebrates and invertebrates. The precise details of how Hedgehog works in these different systems is not known.

▶ *Developmental Regulation by Direct Cell-Cell Contact*

Lineage studies in the vertebrate nervous system have shown that cellular interactions also regulate the fine detail of cellular organization late in development. In such studies, investigators mark a developing cell by injection of an intracellular dye (or other marker) and then follow the developmental fate of the progeny of the marked cell. In virtually every region of the vertebrate nervous system that has been analyzed in this way, single precursor cells labeled late in development were found to give rise to a diversity of cell types. The absence of a strict lineage restriction indicates that cell-type diversity and patterning result from interactions between developing cells. The specific molecules regulating these interactions are not yet known, but studies in *Drosophila* suggest that direct cell-cell contacts also may play an important role in patterning.

Boss Is a Cell-Surface Inductive Ligand for the Sev Receptor

Lineage studies have demonstrated that intercellular communication is involved in determining the precise cellular composition and organization of the ommatidium, the elementary neuronal unit in the *Drosophila* compound eye (see Figure 20-33). As discussed in Chapter 20, development of the R7 cell in each ommatidium depends on interaction between Sev protein, a receptor tyrosine kinase expressed on the precursor R7 cell, and its inductive transmembrane ligand Boss, expressed on the neighboring R8 cell (see Figure 20-34). Additional genetic studies with temperature-sensitive *sevenless* (*sev*) mutants identified the proteins that functioned in the signal-transduction cascade downstream from Sev. Here we describe how this system has provided insight into the strategies by which local cellular interactions mediated by direct cell-cell contact specify cell fate and patterning.

Each ommatidium contains eight different photoreceptor neurons designated R1–R8; four cone cells, which make the lens; and three different types of pigment cells, which optically isolate the R cells in adjacent ommatidia. Each ommatidium is derived from a cluster of epithelial cells; interactions between these cells specify the cell types and their organization during development. The only ligand-receptor pair known to function in controlling patterning in the developing ommatidium is Boss and Sev. Although Boss is expressed only on the R8 cell, Sev is expressed on five R-cell precursors contacting the R8 cell and four cone-cell precursors located only one-cell diameter away from the R8 cell (Figure 24-52). Nevertheless, only one R7 cell is induced, and the Sev protein is not required for any of the other cells to develop normally.

Two mechanisms restrict the number and position of R7 cells that form in each ommatidium. First, since the inductive cell-surface ligand Boss is expressed exclusively on R8, only adjacent Sev-expressing R-cell precursors can bind the ligand. Thus, cone-cell precursors normally cannot interact with R8 and do not develop into R7 cells. However, in transgenic flies misexpression of Boss on R cells contacting cone-cell precursors induces the cone-cell precursors to develop into R7 cells. And second, four of the five Sev-expressing R-cell precursors that contact R8 fail to respond to Boss because they have already been committed to alternative fates. If the ability of these cells to assume their normal fates is disrupted by mutation, they become competent to respond to Boss and develop into R7 cells (see Figure 24-52). Thus strict spatial expression of the inductive ligand Boss and the competence of the responding R7 precursor cell specifically control induction in the *Drosophila* ommatidium. Similar mechanisms control induction in other systems.

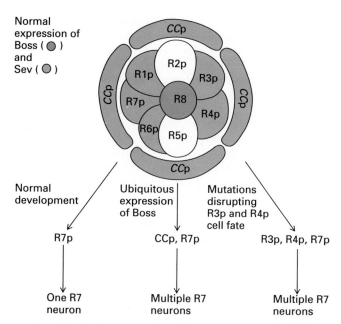

Normal expression of Boss (●) and Sev (○)

CCp

R1p   R2p   R3p

R7p   R8   R4p

R6p   R5p

CCp

CCp

Normal development

Ubiquitous expression of Boss

Mutations disrupting R3p and R4p cell fate

R7p

CCp, R7p

R3p, R4p, R7p

One R7 neuron

Multiple R7 neurons

Multiple R7 neurons

▲ FIGURE 24-52   Specificity of Boss-mediated induction of one R7 neuron in each ommatidium in the *Drosophila* compound eye. Nine precursor cells in the developing ommatidium express Sev (green), a receptor tyrosine kinase. Binding of Boss (red), a transmembrane ligand expressed by the R8 cell, to the Sev receptor induces the R7 cell fate. Cone-cell precursors normally do not respond to Boss because they do not contact the R8 cell. Expression of Boss throughout the developing ommatidium drives many CCp cells to become neurons. The commitment of Sev-expressing cells contacting R8 (other than R7p) to alternative fates prevents them from responding to Boss. For instance, mutations disrupting the fates of R3p and R4p result in their conversion to R7 neurons. The position of the precursor cells in the developing ommatidium is highly stereotyped. [See D. L. Van Vactor, Jr et al., 1991, *Cell* **67**:1145.]

Cell-Surface Notch and Delta Proteins Control Signaling between Many Different Cell Types

In addition to highly specific cell-cell interactions mediated by proteins such as Boss and Sev, direct cellular contact mediated by other cell-surface proteins regulates many different cellular interactions. Other ligand-receptor pairs play more widespread roles in development. One well-studied example is the *Drosophila* Notch receptor and one of its ligands, Delta. The mechanism whereby interaction of these proteins regulates development is not yet clear, but recent findings suggest that their binding may modulate the competence of cells to respond to extracellular signals.

Notch is a transmembrane protein with a large extracellular domain containing 36 tandemly arranged EGF-like repeats, each containing 40 amino acids (Figure 24-53a). The cytosolic domain of Notch contains a so-called *opa repeat*, consisting of some 30 glutamines. Although opa repeats are found in many *Drosophila* proteins, their functional significance is unknown. Notch also contains a cytosolic motif called PEST (Pro-Glu-Ser-Thr); this motif targets proteins for rapid degradation and is present in many developmentally important proteins. Although much smaller than Notch, the transmembrane Delta protein also contains EGF-like repeats in its extracellular domain. Biochemical experiments have shown the Delta and Notch bind to each other (Figure 24-53b). Because both proteins are expressed widely and often on the same cells, determining which acts as the ligand and which as the receptor has been difficult. However, genetic studies suggest that Delta acts as the ligand and Notch functions as the receptor; that is, the cytosolic domain of Notch transduces the signal initiated by Notch-Delta binding.

Loss-of-function mutations in the *Notch* or *Delta* genes produce a wide spectrum of phenotypes. The hallmark of such mutations in either gene is neural hypertrophy, that is, an increase in the number of neuroblasts in the central nervous system or of sensory organ precursors (SOPs) in the peripheral nervous system of the developing fly. As discussed in Chapter 13, during neurogenesis of the *Drosophila* peripheral nervous system, a single cell within a group of cells called a *proneural cluster* develops into an SOP. During normal development, cells not selected to become SOPs subsequently give rise to epidermis. Studies with temperature-sensitive loss-of-function mutations in either the *Notch* or *Delta* gene have shown that interactions between cells within a proneural cluster play an important role in this selection process. Disruption of these interactions by functional loss of either protein leads to development of additional SOPs (or neuroblasts in the CNS) from a proneural cluster (Figure 24-54). In contrast, in developing flies that express a constitutively active (i.e., active in the absence of a normal ligand), truncated Notch protein lacking the extracellular domain, all the cells in a proneural cluster develop into epidermal cells.

The ability of Notch to regulate a highly specific cellular interaction was tested in the developing ommatidium. As we saw in the previous section, development of an R7 precursor cell into a mature R7 neuron requires interaction between Sev receptors on the precursor cell with Boss on the adjacent R8 cell. Expression of the truncated, constitutively active form of Notch by the R7 precursor cell prevents its development into an R7 neuron. Likewise, transient expression of activated Notch in a Sev-expressing cell that normally does not respond to Boss prevents the cell from responding to its normal cue. This leaves the cell competent to respond to Boss at a later time in development when activated Notch is not expressed. These studies support the view that one function of Notch is to regulate

(a)

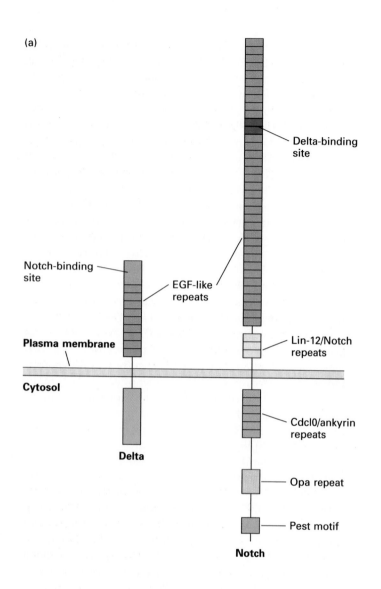

Delta-binding site

Notch-binding site

EGF-like repeats

Plasma membrane

Cytosol

Lin-12/Notch repeats

Delta

Cdcl0/ankyrin repeats

Opa repeat

Pest motif

Notch

◄ FIGURE 24-53   Notch and Delta are transmembrane proteins that mediate a wide spectrum of cellular interactions in *Drosophila*. (a) Both Notch and Delta contain numerous EGF-like repeats. The much larger Notch protein contains several other repeats and motifs found in other transmembrane proteins. See text for discussion. (b) Biochemical studies have shown that Notch and Delta interact. Genetic studies have shown that Notch is a receptor and Delta is one of its ligands. Activation of Notch by ligand binding initiates signal transduction via its cytosolic domain.

(b)

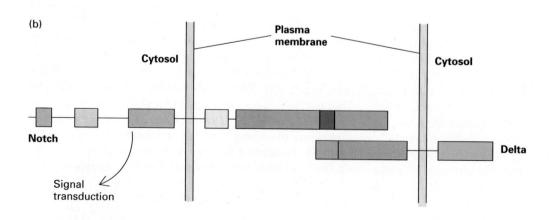

Plasma membrane

Cytosol

Cytosol

Notch

Delta

Signal transduction

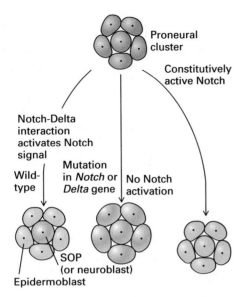

▲ FIGURE 24-54   Role of Notch and Delta during neurogenesis in the *Drosophila* embryo. In the developing peripheral nervous system of wild-type embryos, a single sensory organ precursor (SOP) develops from a proneural cluster, a group of neural precursor cells each of which is competent to assume the SOP cell fate. The remaining cells become epidermoblasts. Interactions between cells within a cluster play a role in restricting SOP selection to a single cell. Mutations in either the *Notch* or *Delta* gene, which disrupt this interaction, lead to development of additional SOPs. Truncated forms of Notch, which are constitutively active (i.e., signal in the absence of Delta binding), prevent SOP selection and all cells become epidermoblasts. In the central nervous system, a similar selection of a single neuroblast from each proneural cluster occurs.

the competence of cells to respond to extracellular signals. Studies of activated forms of the *Xenopus* Notch introduced into the *Xenopus* embryo also suggest that it is involved in regulating competence.

► *Regulation of Neuronal Outgrowth*

Proper function of the nervous system depends on the intricate array of synapses that are formed during development. Within the vertebrate central nervous system, a highly specific network of synapses must be made among thousands of billions of interneurons, motor neurons, and sensory neurons; in the peripheral nervous system, motor neurons must innervate the proper muscles in order to control body activities. Invertebrates, such as insects, have fewer neurons but face similar developmental tasks in constructing patterns of neuronal connectivity.

As illustrated in Figure 21-2, neurons form synapses with other cells through their axons and dendrites. During differentiation dendrites and axons, which may be very long, grow out from the cell body of each neuron. Axons grow outward toward those target cells with which they will form synapses. To fathom how the nervous system is constructed, we thus need to ask: How does an axon select the correct pathway along which to grow? How does it choose a specific target region within which to terminate? And how does it recognize certain cells with which to synapse? The process of neuronal wiring can be divided into two different stages: one dependent on neuronal activity (i.e., firing of an action potential) and the other independent of it. In general, the early stages of wiring, in which the *growth cone* (see Figure 24-57) at the leading edge of an axon navigates to the specific region within which it will form synapses, do not require neuronal activity; the fine tuning of the projection pattern, particularly in vertebrate systems, does. Because little is known about the molecular mechanisms by which neuronal activity regulates synaptic specificity, we will restrict our discussion to activity-independent processes.

Individual Neurons Can Be Identified Reproducibly and Studied

A prerequisite for understanding how specific neural contacts form is identifying specific neurons in the developing embryo and observing them as their axons elongate and form contacts. Embryos of grasshoppers, *Drosophila*, and other insects in which the central nervous system is formed by a string of ganglia, one per body segment, provide excellent experimental systems. A single ganglion may contain a thousand or so nerve-cell bodies; some of these have a characteristic size and position, permitting precise identification of the same cell in different embryos. When a nerve cell is sufficiently large, it can be microinjected with a fluorescent dye (e.g., lucifer yellow) that spreads throughout its cytosol. The cell's projections then can be visualized in the fluorescence microscope (Figure 24-55). Analogous techniques can be used to identify vertebrate motor neurons. For example, the enzyme peroxidase injected into a nerve near its terminus is transported back to the cell body by retrograde axonal transport (Chapter 23). The enzyme can then be detected in fixed tissue by histochemical staining. In this way, it is possible to localize in the spinal cord the cell body of a particular neuron that innervates a specific muscle. Lipophilic dyes (e.g., DiO and DiI) have proven to be particularly important for studying neuronal development in vertebrates. They diffuse within the membrane highlighting the entire extent of the axon and neuritic processes. In contrast to peroxidase labeling, lipophilic dyes can be used to stain fixed material or to follow events in living tissue using confocal microscopy.

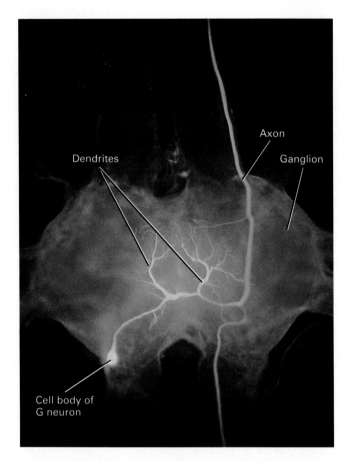

▲ FIGURE 24-55   Identification of a single neuron in a grasshopper embryo. The fluorescent dye Lucifer yellow was microinjected into the cell body of the G neuron, one of 2000 neurons in the second thoracic ganglion. The axon extends from the cell body, crosses the ganglion, then extends forward (upward in this picture); a smaller axon branch extends rearward. [From C. S. Goodman et al., 1984, *Sci. Am.* **251**(6):58.]

The simple soil worm, *Caenorhabditis elegans*, which exists in two forms—a self-fertilizing hermaphrodite and a male that facilitates cross-fertilization—has provided important insights into neuronal wiring (see Figure 5-16d). The adult hermaphrodite contains 959 somatic cells; 301 of these are neurons. Because of the small number of neurons in *C. elegans*, researchers have been able to reconstruct the complete neuronal wiring pattern from thin sections visualized by electron microscopy. Moreover, the short generation time of *C. elegans* (52 h at 25°C) and the ease of genetic analysis in this system have permitted identification of many different mutations affecting neuronal wiring. In the past, it has been difficult to detect the effects of mutations on developing neurons. However, new labeling methods, such as that illustrated in Figure 24-56, allow analysis of the effects of mutations on neuronal outgrowth in living worms.

Growth Cones Guide the Migration and Elongation of Developing Axons

During differentiation of the nervous system, precursors of neurons, called neuroblasts, arise at particular locations at specific times. Such cells lack axonal or dendritic projections. Certain neuroblasts migrate to specific destinations where they form ganglia. One example, mentioned earlier in this chapter, is the migration of neural-crest cells to form sympathetic ganglia. Vertebrate cells migrate extensively in the central nervous system. As a newly born neuron begins to differentiate, it grows one or more axonal projections. At the leading edge of the elongating axon is the highly motile growth cone, which possesses cell-surface receptors for extracellular signals. Both secreted and membrane-bound molecules have been identified that attract and inhibit (or repel) growth-cone progression. Components

➤ FIGURE 24-56   Individual neuronal cell bodies and axons can be visualized in *C. elegans* (see Figure 5-16d). In the example shown here, a jellyfish gene encoding a fluorescent protein was introduced into a living worm. This gene was linked to a worm neuron-specific promoter that is highly active in the touch-sensitive neurons designated ALMR and PLMR. Expression of the fluorescent protein and observation by fluorescent microscopy reveals the cell bodies (light spots) and processes of these two neurons (arrows). In this photograph, the worm is bent in a U shape. [From M. Chalfie et al., 1994, *Science* **263**:802.]

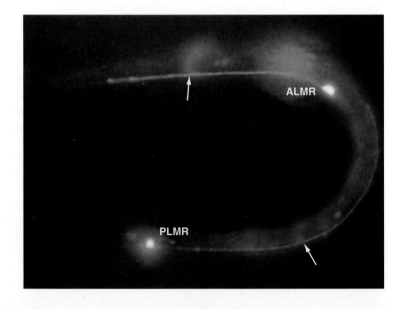

bound to the extracellular matrix also have been shown to profoundly affect axon outgrowth. As the growth cone moves outward, the cell body stays put and the axon elongates due, in part, to the polymerization of tubulin into microtubules that gives the axon its rigidity. Ligand binding to receptors on growth cones is thought to trigger intracellular signaling that regulates their motility, in part, by controlling actin polymerization in the filopodia and lamellapodia at the distal end. Recognition of signaling molecules by receptors located at different positions along the periphery of advancing growth cones is thought to steer them toward the appropriate target cells.

Different Neurons Navigate along Different Outgrowth Pathways

Although the nervous systems of vertebrates and invertebrates are vastly different in their structure and complexity, it is thought that similar principles of axon guidance are used: growth cones of elongating axons use a changing set of cell-surface receptors to move along specific matrix fibers and also along specific cells. The notion that specific molecules guide growth cones to their targets was first proposed nearly a century ago and in this section we describe a few examples that argue persuasively for the importance of specific guidance cues directing growth cones to their targets.

In vertebrates, the cell bodies of motor neurons are located in the spinal cord, and axons extend out of the central nervous system by ventral roots. Axons that innervate a specific muscle are bound together to form a peripheral nerve. Obviously it is essential for each motor neuron to innervate only the appropriate muscle, and indeed this high degree of specificity is found throughout the nervous system.

In certain favorable experimental systems, scientists can observe individual motor neurons as their axons exit the spinal cord and elongate. During embryogenesis in the zebra fish, for example, the axons of three pioneer motor neurons are the first to emerge from the spinal cord in each segment of the embryo (Figure 24-57). Investigators have identified the cell bodies of these three neurons and followed the trajectories of their axons by microinjection with a fluorescent dye. These studies have shown that the axons of all three pioneer neurons initially grow out of the spinal cord along a common pathway. They then diverge: one axon grows upward to innervate dorsal muscles; one downward to innervate ventral muscles; and one laterally to innervate both muscles. These pioneer axons are never seen to send branches or growth cones off in an inappropriate direction.

At a later stage of development, additional axons grow out of the central nervous system to innervate the same muscles. Growth cones of these axons migrate along the surfaces of the three pioneer axons, which guide the secondary axons to the correct muscles. However, experimen-

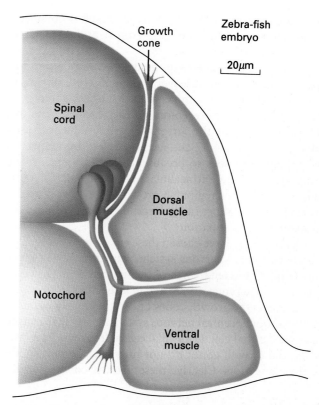

▲ **FIGURE 24-57**   During development of the zebra-fish embryo, growth cones of pioneer motor neurons follow different pathways. A cross-sectional view of a trunk section of a 19-h-old zebra-fish embryo shows that axons of three adjacent motor neurons extend outward from the developing spinal cord at the same ventral root. They follow the same pathway out of the spinal cord, but then follow different pathways. One axon extends downward, innervating the ventral muscles; one upward, innervating the dorsal; and one laterally, innervating both. [From M. Westerfield and J. S. Eisen, 1988, *Trends Neurosci.* **11**:18.]

tal destruction of one of the pioneer neurons by an intense laser beam does not affect the ability of the secondary motor neurons to select the pathway that led to the appropriate target muscle. This finding suggests that by the time a growth cone exits the spinal cord, the neuron is already programmed to follow a specific pathway. Experiments with the chick embryo—in which nerve transplantation experiments are easier—support this contention. Here, motor neurons from four adjacent segments of the spinal cord innervate various muscles of the hindlimb. Multiple axons destined for different muscles grow out of the spinal cord together and follow the same path to plexuses; from there the individual neurons follow different paths to the appropriate muscle. If, before nerve outgrowth, a piece of spinal cord containing motor-neuron cell bodies is transplanted from one segment to an adjacent one, these motor-neuron axons will exit the spinal cord at the "wrong" site. However, they will grow into the proper plexus and still innervate the correct muscle. A similar result is obtained if a segment is inverted, so that anterior neurons are moved to the posterior.

Clearly, the specificity in guidance of any axon does not depend on where it leaves the spinal cord, nor on its transplanted position within a segment. It seems to depend on an inherent set of receptors on each growth cone that allow it to recognize the surrounding environment. For instance, in Figure 24-57 the neuron that grows along the ventral pathway does so because it specifically recognizes components of the extracellular matrix or glial or other cells located in this region. A growth cone most probably moves from one short-range target to another as it guides an axon to its destination.

Although we have chosen these examples from the vertebrate peripheral nervous system, similar principles probably apply to the outgrowth of the first axons in the vertebrate central nervous system and in insect nervous systems. For instance, the growth cones of certain pioneer axons in the grasshopper central nervous system always make a turn at a specific glial cell. If the glial cell is destroyed by a laser beam, the growth cones do not turn but continue in the original direction. Clearly, these growth cones use glial cells as landmarks or guides, as do those of other pioneer neurons in both vertebrate and invertebrate central nervous systems. Evidence suggests very intimate contact between pioneer neurons and guidepost glial cells; gap junctions, for instance, form transiently between the two contacting cells.

Different Extracellular-Matrix Components Are Permissive for Neuronal Outgrowth

The extracellular matrix and particularly laminin has been shown to regulate neuronal outgrowth. By layering different extracellular matrix materials onto tissue-culture dishes in specific patterns (i.e., some regions of the plate are covered with one matrix component and other regions covered with a different one), investigators showed that growth cones select specific substrates over others for outgrowth (Figure 24-58). Some concluded from this finding that the pathway taken by a growth cone to its target is determined by a series of choice points between substrates of different adhesivity. Recent studies, however, have shown there is no simple relationship between the degree of adhesivity and growth-cone motility on a variety of sub-

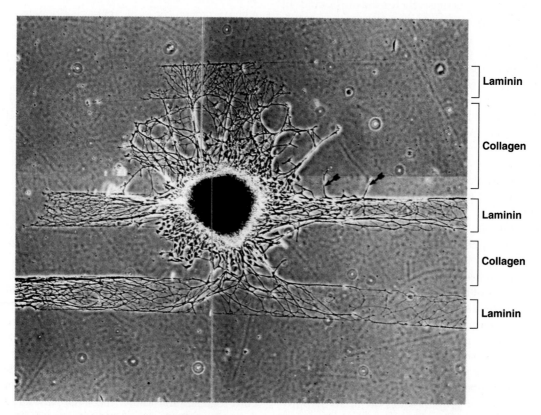

Laminin

Collagen

Laminin

Collagen

Laminin

▲ FIGURE 24-58 Experimental demonstration that growth cones preferentially migrate on laminin rather than collagen IV. Alternating stripes of laminin and collagen IV were affixed to a tissue-culture dish. When sensory neurites were cultured on the matrix materials, outgrowth was observed predominantly along the laminin stripes. The arrows indicate the infrequent outgrowth of small fascicles into a collagen stripe. [From R. W. Gundersen, 1987, *Dev. Biol.* **121**:423; Courtesy of R. Gundersen.]

strates. Indeed, under certain conditions, laminin has anti-adhesive properties. Hence, it is probably more accurate to view outgrowth as reflecting the ability of different extra-cellular matrix components (or components expressed on the surfaces of cells along the pathway) to promote motility through intracellular signaling pathways regulating cytoskeletal dynamics in the growth cone (e.g., integrin-laminin interactions).

Although many common extracellular matrix components support growth-cone movement, they do not appear capable of determining the directionality of outgrowth. For instance, neurons in culture do not show directed outgrowth even on a steep gradient of laminin. How is directionality determined? The convergence of genetic studies in *C. elegans* and elegant biochemical studies in vertebrate systems have led to the identification of an evolutionarily conserved class of proteins that provide directional information to the growth core. We discuss these studies in the next two sections.

Three Genes Control Dorsoventral Outgrowth of Neurons in *C. elegans*

Numerous mutations affecting movement of *C. elegans* have been identified. Worms carrying these mutations, called *unc* (for uncoordinated body movement), are easily identified under a dissecting microscope. Many *unc* mutants have been shown to have specific defects in neuronal organization and wiring. These mutations fall into three classes: (1) those affecting basic processes of axonogenesis and outgrowth; (2) those affecting the ability of axons to form bundles called *fascicles*; and (3) those specifically affecting the directionality of neuronal outgrowth. Here we discuss the last class of mutations.

The initial outgrowth of axons in the developing worm occurs between the epidermis and the basal lamina secreted by the epidermal cells. Some axons extend dorsally and others grow ventrally. Upon reaching the dorsal or ventral side of the body, these axons turn and extend along the anterioposterior axis. Three genes—*unc5*, *unc6*, and *unc40*—have been shown to be required for dorsoventral guidance. The *unc5* gene is required for dorsal outgrowth; *unc40*, for ventrally directed growth, and *unc6*, for both ventral and dorsal outgrowth (Figure 24-59). Growth along the anterioposterior axis is unaffected in mutations in these genes. Comprehensive phenotypic analysis of outgrowth and cell migration (these genes also regulate migration) show that these three genes are part of a common genetic pathway controlling these processes. A remarkable feature of phenotypic analysis in *C. elegans* is that the effects on any given mutation can be assessed on the wiring patterns and cell migrations of every cell in the organism! This provides an unprecedented degree of precision in phenotypic analysis.

The *unc5* gene has been cloned and sequenced. The

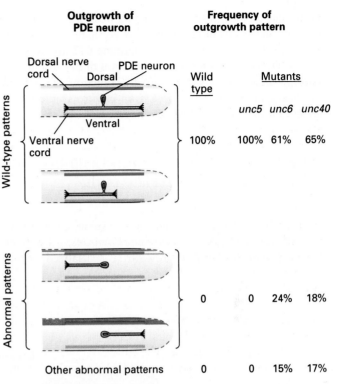

▲ **FIGURE 24-59** Mutational analysis of directional outgrowth of the PDE neuron in *C. elegans*. In wild-type worms, the PDE neuron grows ventrally, branches at the nerve cord and grows both anteriorly and posteriorly; in 5 percent of worms, the posterior branch is shorter. Mutations in *unc6* and *unc40*, but not in *unc5*, cause errors in ventral outgrowth. In the case of *unc6* and *unc40* mutants, additional projection abnormalities differing from the two shown here were observed in some worms. Similar analyses with other identified neurons have shown that *unc5* is required for dorsal outgrowth; *unc6* for both dorsal and ventral outgrowth; and *unc40* for ventral outgrowth. [Adapted from E. M. Hedgecock et al., 1990, *Neuron* **2**:61.]

deduced sequence of its product, Unc5, suggests that it is a transmembrane protein with an extracellular domain containing two immunoglobulin repeats and two different repeats found in the protein thrombospondin and with a large intracellular domain containing a sequence similar to an SH3 domain (see Figure 20-37). If the Unc5 protein is misexpressed in a neuron that normally grows ventrally (e.g., the PVM neuron), then outgrowth is respecified to the dorsal direction. Interestingly, respecification of PVM outgrowth is dependent on the protein encoded by *unc6*, suggesting that Unc5 and Unc6 function along the same genetic pathway specifying dorsoventral outgrowth in *C. elegans*. The structure of Unc5 and its ability to respecify the directionality of neuronal outgrowth suggests that it functions as a receptor for directional cues.

The *unc6* gene also has been cloned and sequenced.

Molecular studies raise the intriguing possibility that *unc6* encodes an extracellular signal that regulates directionality. Unc6 is a secreted protein containing 612 amino acids with a significant homology to a portion of the laminin B2 subunit. Important insights into the function of Unc6 in regulating guidance have come from in vitro studies of its vertebrate homolog, called netrin.

A Chemoattractant Related to Unc6 Is Produced in the Floor Plate of the Vertebrate Neural Tube

During development of the embryonic spinal cord, a subclass of neurons, called *commissural neurons*, grow from dorsal positions ventrally toward a specialized region called the *floor plate* found at the ventral midline (Figure 24-60). By co-culturing floor-plate cells with explants of embryonic spinal cord in collagen gels, investigators showed that commissural neurons, which were distinguished from other axons by their expression of a specific cell-surface marker, grew out of the explants and turned towards the floor-plate tissue. This finding suggested that the oriented growth of commissural neurons ventrally was due to secretion of a chemoattractant by the floor-plate cells. The floor-plate activity was also found in embryonic brain, which provided a more abundant source of material for purification.

Using outgrowth of commissural neurons as an assay, researchers purified two proteins called netrin 1 and netrin 2 (from Sanskrit *netr*, "one who guides"). The genes encoding them then were cloned and sequenced. Remark-

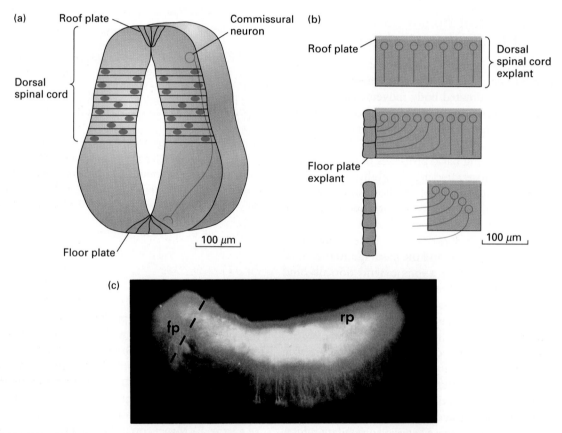

▲ **FIGURE 24-60** Experimental demonstration that floor-plate tissue attracts outgrowth of commissural neurons in embryonic spinal cord. (a) Schematic cross-section of spinal cord. Cell bodies of commissural neurons are located near the dorsal roof plate. During normal development, they grow ventrally toward the floor plate. (b) Explants of the dorsal spinal cord and floor plate were arranged in a collagen gel as diagrammed and co-cultured for 40 h. Staining for a specific cell-surface marker on commissural neurons showed that outgrowth was directed toward the floor-plate tissue. (c) Photograph of embryonic explant injected with the fluorescent tracer dye DiI. Cell bodies in the roof plate (rp) were injected, leading to the filling of commisural axons. Axons as far away as 300 μm were attracted towards the floor plate (fp). [Part (c) from M. Tessier-Lavigne and M. Placzek, 1991, *Trends Neurosic.* **14**:303; Courtesy of M. Tessier-Lavigne.]

ably, the vertebrate netrins are highly related in structure to Unc6 from *C. elegans* and to a region of the laminin B2 subunit. Expression of netrin cDNAs in tissue-culture cells promoted directed outgrowth of commissural axons toward these cells, supporting the idea that netrins are chemoattractants. Netrin 1 was shown to be expressed in the floor plate and netrin 2 more weakly in the ventral and lateral regions of the embryonic spinal cord. Similarly, Unc6 is expressed in the ventral region of the worm. Since Unc6 and the netrins have similar structures and act to promote directional outgrowth, they most likely are functional homologs. Whereas genetic evidence indicated that Unc6 can function as a chemoattractant *in vivo*, the *in vitro* studies on the netrins provide a direct demonstration that these proteins can act as chemoattractants.

The genetic studies in *C. elegans* show that Unc6 controls neuronal outgrowth in the ventral and dorsal directions, presumably by acting as a chemoattractant for some neurons and a chemorepellent for others. The simplest interpretation is that an Unc6 protein gradient forms with the highest concentration on the ventral side. Growth towards the ventral side, by analogy to that of commissural neurons in vertebrates, is directed by Unc6 acting as a chemoattractant. Conversely, Unc6 may act as a chemorepellent for neurons extending dorsally. Interestingly, recent studies have led to the identification of several molecules that act to repel axonal outgrowth. Recent studies on the netrins indicate that they also can act as chemorepellants is not yet known.

Extracellular Signals Can Repel Growth Cones

Dramatic evidence that repulsive mechanisms can influence the direction of growth-cone movement first came from experiments with cultured sympathetic and retinal growth cones isolated from chick embryos. When retinal growth cones contact retinal axons or sympathetic growth cones contact sympathetic axons from dorsal root ganglia, their morphology remains largely unchanged. In marked contrast, retinal growth cones collapse upon contacting sympathetic axons, and sympathetic growth cones collapse upon contact with retinal axons. Extracts from a variety of neural and non-neural tissue can induce collapse of cultured growth cones (Figure 24-61). The active factors within these extracts are referred to as *collapsing factors*. In the presence of collapsing factors, growth-cone lamellar protrusion abruptly ceases, and growth-cone paralysis ensues. This is followed by a loss of filamentous actin from the leading edge of the growth cone and collapse of filopodia and lamellapodia. Although the signal-transduction pathways leading from the binding of ligand to growth-cone collapse are not known in detail, both G protein and Ca^{2+} signaling have been implicated in some cases.

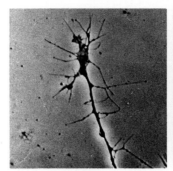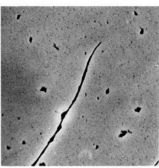

▲ FIGURE 24-61   (a) A typical growth cone from the chick dorsal root ganglion. (b) Addition of a membrane fraction from embryonic chick brain induced complete collapse of the growth cone within 15 min. [From J. A. Raper and J. P. Kapfhammer, 1990, *Neuron* **4**:21.]

Collapsing factors are thought to influence neuronal outgrowth in several different systems. In the chick embryo, for example, motor axons selectively grow out through the anterior half of each somite, giving rise to the segmental arrangement of spinal nerves (Figure 24-62). When investigators rotated a somite, so that the anterior half was in the posterior position, the motor neurons preferentially grew over the portion of the somite in the posterior position. Removal of a somite altogether led to continuous neuronal outgrowth along the anterioposterior axis (i.e., loss of the segmented pattern). This result suggests that selective outgrowth through the anterior part of each somite is due to a repellent expressed in the posterior half rather than an attractant in the anterior half. Two glycoproteins expressed selectively in the posterior half of somites exhibit collapsing activity in vitro, but whether these proteins are sufficient to confer the selectivity of motor neuron outgrowth in vivo is not yet known.

Growth-cone collapsing activity has been observed and studied in several other system. In the chick visual system, for instance, collapsing factors appear to play a role in the formation of connections between the retina and its target, the tectum. Growth-cone collapsing factors also have been isolated from the central nervous system (CNS) of adult vertebrates, raising the intriguing possibility that the presence of such inhibitory factors is what prevents regeneration of injured nerves within the CNS.

Investigators have isolated several collapsing factors using growth-cone collapse as an assay to follow purification. A chick gene encoding one of these factors, called *collapsin*, has been cloned and sequenced. Collapsin expressed from a cDNA in tissue-culture cells induces collapse activity in vitro. A secreted glycoprotein containing 772 residues, it contains an immunoglobulin-type C2 domain and shows extensive homology to a class of proteins

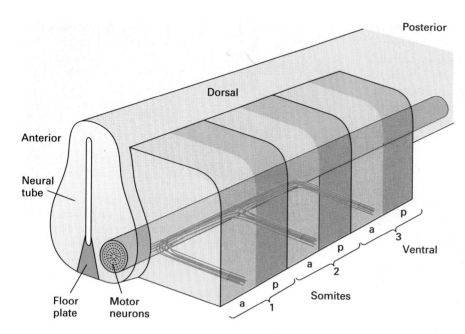

◄ **FIGURE 24-62**   Dorsal view of the developing spinal cord. Motor neurons (red) grow out from a ventral position in the spinal cord into the anterior region (yellow) of each somite (numbered), producing the segmental pattern of spinal nerves. This selective outgrowth may result from a repellent expressed in the posterior region of each somite (orange).

called *semaphorins,* which have been isolated from both insects and mammals including humans. Both transmembrane and secreted semaphorins have been identified.

A semaphorin has been proposed to play a role in guiding growth cones in the developing limb bud of the grasshopper. During normal limb-bud development in grasshopper embryos, semaphorin is expressed selectively along a stripe of epithelial cells. A pair of Ti pioneer neurons extend their growth cones together from distal positions toward the CNS with their axons forming a fascicle. Upon encountering the semaphorin stripe, the growth cones abruptly change direction and extend ventrally along this boundary. When they encounter a specific guidepost cell, called a Cx1 cell, they change direction again and continue on in the original direction (Figure 24-63a). In the presence of anti-semaphorin antibody, the Ti growth cones do not make an abrupt turn at the stripe of semaphorin-expressing cells; instead, they defasciculate and branch extensively as they cross the stripe of cells (Figure 24-63b). Although the precise mechanism by which semaphorin mediates this abrupt change in growth-cone behavior is not known, it most likely functions in conjunction with other guidance cues in the developing limb. For instance, semaphorin may make growth cones less sensitive to an attractant in the proximal region of the limb bud and more sensitive to a cue directing ventral growth.

Different Growth Cones Navigate along Different Axons

As pioneer axons extend using extracellular-matrix and cell-surface cues, more and more of the space in the central

nervous system becomes occupied by axonal processes. As secondary neurons differentiate, their growth cones navigate on the surfaces of other axons and their axons eventually bundle together. Different growth cones select different axonal surfaces in different fascicles on which to migrate. This can be seen vividly in the development of the grasshopper central nervous system.

Grasshopper and *Drosophila* embryos are divided into segments, most of which have a similar pattern of neurons. Each segment has two ganglia, one on each side of the embryo, which contain the neuron cell bodies. Bundles of axons (*longitudinal fascicles*) run along either side of the embryo and bundles of axons (*transverse fascicles*) cross the embryo, connecting the segmental ganglia in both directions. Of the hundreds of neurons in each segmental ganglion, many sprout axons whose growth cones migrate to the opposite side of the embryo by growing along a transverse fascicle. Six such identifiable neurons are depicted in Figure 24-64a. When these six growth cones reach the opposite side, each chooses a different longitudinal fascicle to follow or a different direction in which to migrate.

For instance, the G neuron growth cone initially makes contact with about 100 axons in 25 longitudinal fascicles, but it only follows the A-P fascicle, which is composed of the axons of four neurons called A1, A2, P1, and P2. More detailed studies have shown that the G growth cone actually moves along only two of the four axons in the fascicle, P1 and P2 (Figure 24-69b). If the A1 and A2 neurons are experimentally destroyed by a laser beam, the differentiation of the G neuron as well as most of the other neurons is unaffected; the G growth cone moves normally along the P1 and P2 axons. However, if the P1 and P2 neurons are

(a)

(b)

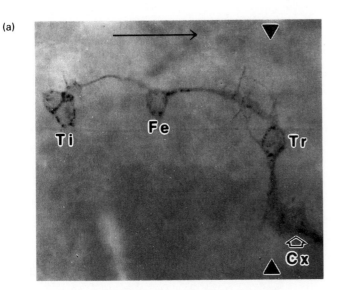
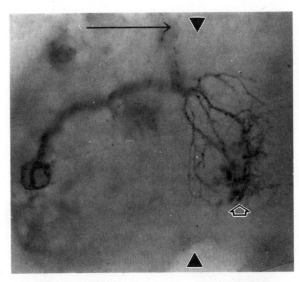

▲ **FIGURE 24-63** Experimental demonstration that semaphorin guides outgrowth of Ti pioneer neurons in the grasshopper limb bud. Arrows indicate direction of outgrowth. (a) During normal development, the growth cones of a pair of Ti neurons make an abrupt ventral turn at a boundary corresponding to a stripe of epithelial cells (arrowheads) that express the cell-surface protein semaphorin. After proceeding ventrally, the Ti neurons change direction again when they encounter a Cx cell. (b) Addition of antibody against semaphorin disrupts guidance of Ti outgrowth. Fe and Tr are other cells in the Ti pathway. Cells and their processes are visualized by immunohistochemistry. See text for discussion. [From A. Kolodkin et al., 1992, *Neuron* **9**:831.]

destroyed, the G neuron grows abnormally, and its growth cone behaves as if it were undirected and does not bind to any other axon. Thus, migration of the G growth cone relies absolutely on binding to the P axons.

These observations suggest that the P1 and P2 axons bear a unique surface marker that is recognized by a receptor present on the G growth cone but not on growth cones of other neurons that do not follow the P axons. To identify these guide molecules, investigators prepared monoclonal antibodies to preparations of grasshopper neuronal membranes and screened them for their ability to stain specific subsets of fascicles. With these anti-fascicle antibodies, several cell-surface fascicle proteins were identified and purified; the genes encoding these proteins then were identified and cloned using techniques described in Chapter 7. Remarkably, when four of these proteins—*fasciclin I, II, III* and *neuroglian*—were individually expressed in transfected nonadherent tissue-culture cells, they each promoted aggregation of cells expressing the protein, suggesting that they function as cell-adhesion molecules.

To further characterize these proteins and study their role in neuronal outgrowth, their *Drosophila* homologs were identified. All are transmembrane proteins with a large extracellular domain; fly neuroglian and fasciclin II contain immunoglobulin-type repeats and fibronectin type III repeats, whereas fasciclin III contains only immunoglobulin repeats. Researchers then introduced various mutations that disrupted the structure of these proteins. Surprisingly, these mutations had little or no effect on the

pattern of axonal wiring in mutant flies. Similar results have been obtained in recent experiments on N-CAM, the vertebrate homolog of fasciclin II, which has long been viewed as a critical determinant of neuronal development, including formation of fascicles. For instance, knockout mice that produce no functional N-CAM are viable and fertile, and their neuronal organization exhibits few abnormalities.

In view of the multitude of adhesive proteins identified on fascicles and the minimal effects of mutations in any one protein, it is likely that considerable redundancy exists in the mechanisms guiding movement of growth cones. Indeed, such redundancy has been directly demonstrated in antibody-disruption experiments with cultured chick embryonic cells. In this system, the interactions between three different adhesive proteins and their receptors on growth cones must be neutralized to prevent outgrowth of ciliary ganglion cells over Schwann-cell membranes. The emerging view is that selection of specific fascicles by a particular growth cone involves recognition of a unique combination of widely expressed cell-surface molecules rather than unique fascicle-specific molecules.

In summary, we find that studies from a variety of systems suggest that the route growth cones navigate is determined by a complex pattern of different soluble and membrane-bound attractants and repellents. Nothing is known about the mechanisms by which a growth cone selects its ultimate target region and the specific cell(s) with which it synapses.

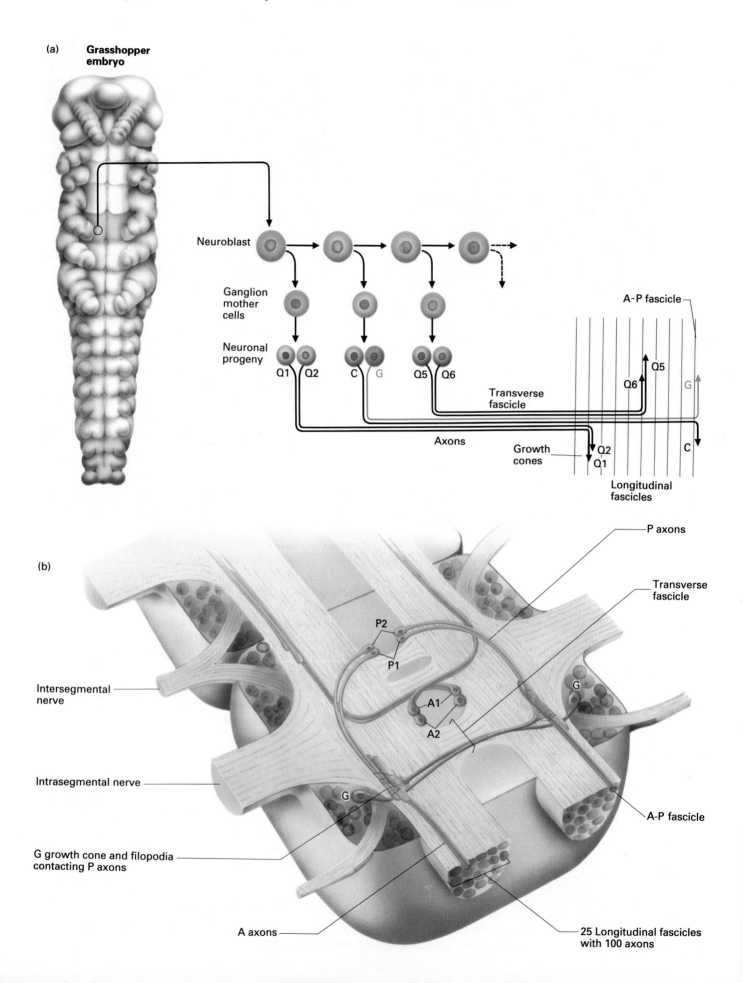

(a) **Grasshopper embryo**

Neuroblast

Ganglion mother cells

Neuronal progeny

Q1 Q2   C   G   Q5 Q6

A-P fascicle

Q5
Q6

Transverse fascicle

G

Axons

Growth cones

Q2
Q1

C

Longitudinal fascicles

(b)

P axons

Transverse fascicle

Intersegmental nerve

P2
P1

A1
A2

G

Intrasegmental nerve

G

A-P fascicle

G growth cone and filopodia contacting P axons

A axons

25 Longitudinal fascicles with 100 axons

◄ FIGURE 24-64   Different stereotyped pathways are taken by the axons of sister neurons. (a) Each of the 17 segments of the grasshopper embryo have two segmental ganglia; these ganglia have a virtually identical pattern of nerve differentiation. One identifiable neuroblast in each half of each segment divides repeatedly to form about 50 ganglion mother cells, each of which divides to yield two sister neurons. The first six neurons formed, shown here, extend axons across the ganglion, forming part of a transverse fascicle. The growth cone of each neuron then recognizes a different longitudinal fascicle and moves along it, elongating the axon in a specific direction. (b) Details of the selective fasiculation of the growth cone of the identifiable G neurons. Each half segment has one G neuron, whose axon grows along a transverse fascicle to the opposite side of the embryo. There the G growth cones explore the surfaces of 25 fascicles with a total of 100 axons. The G growth cones adhere to and migrate along only the axons of the P1 and P2 neurons (blue) in the A-P fascicle (green and blue axons). [After C. S. Goodman and M. J. Bastiani, 1984, *Sci. Am.* **251**(6):58.]

The Basal Lamina at the Neuromuscular Junction Directs Differentiation of Regenerating Nerve and Muscle

Once the axon of a motor neuron reaches its target muscle, a specialized neuromuscular junction is established. The neuromuscular junction is a chemical synapse, an intricate arrangement of cellular and molecular architecture designed for rapid focal transmission of the nerve impulse to the muscle (see Figure 21-5). The highly specialized pre- and postsynaptic components of the synapse must be concentrated and then assembled to lie in precise apposition from one another across the synaptic cleft. This requires communication between the nerve and muscle cells. Evidence from frogs and other amphibians suggests that a specialized region of the basal lamina, which is present in the synaptic cleft, contains molecules that induce both nerve and muscle to form the specialized structures of the neuromuscular junction.

For example, if both the nerve and muscle are damaged, the nerve terminals and the muscle fibers degenerate and are phagocytosed. All that remains are the muscle basal lamina and the Schwann cells that form the myelin around the motor neuron. With time, the axon of the motor neuron regenerates, and a new muscle fiber forms within the original basal lamina. Strikingly, a new neuromuscular junction is formed precisely at the original site (Figure 24-65). Even if muscle regeneration is prevented, axons still return to precisely the original site on the basal lamina and form an axon terminus with synaptic vesicles. Clearly the muscle cell is unnecessary for precise innervation. Moreover, if myofibers regenerate—but the axon does not—specialized regions of the plasma membrane with concentrations of acetylcholine receptors form at the original synaptic site on the basal lamina.

Such experiments show that the basal lamina in the original synapse contains signaling molecules that attract

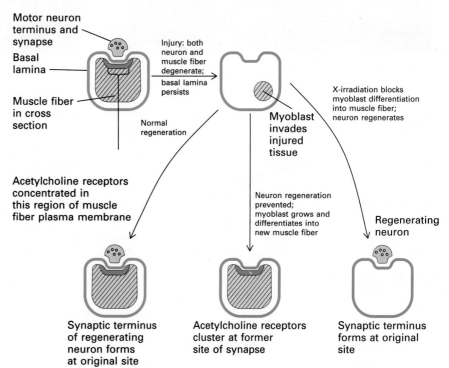

► FIGURE 24-65   A specialized region of the basal lamina determines the site of the neuromuscular junction. When a striated frog limb muscle and innervating motor neurons are damaged, both the muscle and the nerve axon degenerate. All that remains are the muscle basal lamina and some surrounding Schwann cells. With time, myoblasts invade the space and differentiate into muscle fibers, and the motor neuron regenerates a new axon; the regenerated axons form synapses with the regenerating myofiber precisely at the sites of the original synapses. Even if muscle regeneration is prevented by x-irradiation, the axon regenerates synaptic termini at the original site. If nerve regeneration is prevented, the regenerating muscle fiber will form a synaptic specialization, with concentrations of acetylcholine receptors, again at the original site. [After S. J. Burden, 1987, in *The Vertebrate Neuromuscular Junction*, Liss, pp. 163–186.]

the growth cone of the regenerating axon and regulate the distribution of muscle plasma-membrane proteins. Observations such as these led investigators to search for molecules in the basal lamina that direct where on the muscle cell a postsynaptic specialization will form. From the time of initial nerve-muscle contact to when a functional synapse is formed, more than 40 known molecules become concentrated specifically at the synapse. One of the earliest to become concentrated, and one of obvious physiological significance, is the acetylcholine receptor (AChR) on the muscle-cell membrane. The localization of AChRs is easily monitored using fluorescent-labeled bungarotoxin, which directly binds to these receptors. In muscle cells cultured in the absence of motor neurons, the AChRs are distributed randomly over the surface of the muscle cell. Contact with a co-cultured motor neuron results in rapid clustering of AChRs via lateral diffusion of receptor molecules in the muscle-cell plasma membrane to the site of nerve-muscle contact.

Cultured chick myotubes were used to assay the AChR–clustering activity of protein fractions of basal lamina prepared from a particularly rich source of cholinergic synapses, the electric ray *Torpedo californica* (Figure 24-66a). From this source, investigators isolated and partially sequenced a protein they named *agrin*. Antibodies prepared against agrin show it to be stably associated with the basal lamina at the neuromuscular junction. Furthermore, inclusion of anti-agrin antibodies in nerve-muscle co-cultures prevents the aggregation of AChRs (Figure 24-66b). Agrin thus mediates aggregation of the AChR, as well as a dozen other components, at the neuromuscular junction, and is thought to be a centrally important signaling molecule in the formation of the synapse.

Sequencing of the agrin cDNA revealed that it encodes a 200 kDa protein related to several known proteins, including laminin, another component of the basal lamina. The N-terminal half of the agrin molecule contains matrix-binding sites; the C-terminal half contains membrane-

(a)

−Basal lamina

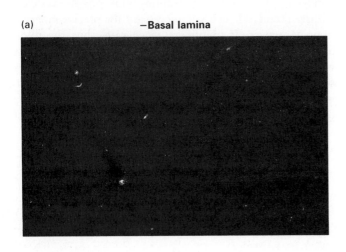

+Basal lamina

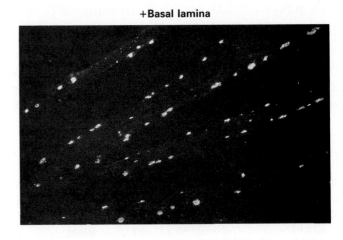

(b)

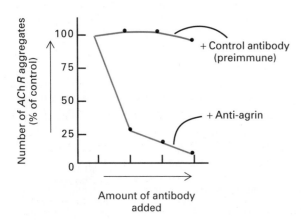

▲ **FIGURE 24-66** Experimental demonstration that agrin, which is associated with the basal lamina, promotes aggregation of acetylcholine receptors (AChRs). (a) Cultured chick myotubes were incubated in the presence or absence of basal-lamina extracts from the electric ray of *Torpedo californica*. Staining with fluorescent-labeled bungarotoxin, which binds specifically to acetylcholine receptors, showed that aggregation depended on the basal lamina. This system was used to assay for AChR-clustering activity during purification of agrin. (b) Addition of antibodies against agrin to co-cultures of chick motor neurons and myotubes prevented formation of AChR aggregates. [Part (a) from U. J. McMahan, 1989, *Dev. Neurosci.* **11**:227; part (b) adapted from N. E. Reist, M. J. Werle, and U. J. McMahan, 1992, *Neuron* **8**:865.]

binding sites and is required for the AChR-clustering activity of agrin. The protein has several repeats found in other extracellular-matrix proteins or signaling molecules. Alternative splicing of agrin transcripts results in a number of agrin isoforms with differing properties; some forms are inactive in AChR-clustering assays. The significance and function of the different isoforms is not understood. Localization studies indicate that both muscle and motorneurons produce agrin, that motorneurons transport agrin to the nerve terminus, and that agrin from both cellular sources is localized at the neuromuscular junction.

Although neuron-derived agrin induces clustering of AChR in myotubes in vitro, it is unclear whether neuron-derived agrin does so during normal development in vivo. Agrin causes AChR clustering at very low concentrations (0.1–1 ppm) and exhibits saturable, high-affinity binding to muscle-cell membranes. These data and the finding that one molecule of agrin mediates clustering of ≈200 AChRs suggest that agrin does not directly bind to the AChR. Rather, it most likely functions as a signaling molecule controlling the assembly of postsynaptic specializations at the site of motor neuron contact.

Recent studies indicate that α-dystroglycan, a peripheral membrane glycoprotein, binds to agrin and may mediate binding of agrin to the muscle-cell membrane. At the synapse, α-dystroglycan binds to a transmembrane complex that is linked to the membrane cytoskeleton by its association with utropin, a homolog of dystrophin found at synapses. These protein complexes may stabilize the binding of the muscle cell to the synapse. Extrajunctional contacts between the muscle and the extracellular matrix are stabilized in an analogous fashion by the interaction of α-dystroglycan with laminin and through a transmembrane complex to dystrophin and the cytoskeleton (Chapter 22). Although α-dystroglycan clearly is an agrin-binding protein, conclusive evidence that it is required for agrin-induced AChR clustering has not been obtained as yet.

► Structure and Function of the Plant Cell Wall

Depending on its composition of proteins and proteoglycans, the extracellular matrix in animals can fulfill several types of functions: it can facilitate cell-cell adhesion (as in epithelia); it can provide a network (for instance, the basal lamina) on which cells can grow; and it can provide strength, rigidity, and compressibility to tissues, particularly mature connective tissue such as tendon or bone where few cells are present. Plant cell walls serve many of the same functions, even though they are composed of entirely different macromolecules from those that make up animal extracellular matrices.

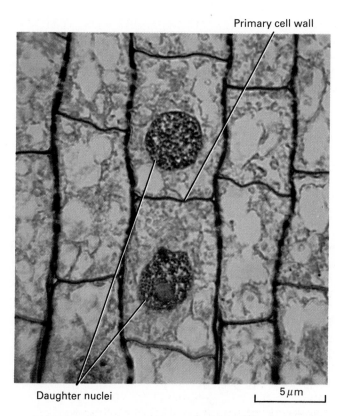

Primary cell wall

Daughter nuclei

5 μm

▲ FIGURE 24-67   Light micrograph of young root tip cells of an onion, showing the thin primary cell wall separating two recently separated cells. [Courtesy of Jim Solliday and Biological Photo Service.]

Cell division in plants is restricted to specific regions called *meristems*, including root tips and the tip of the stem. Young cells are connected by thin primary cell walls (Figure 24-67), which can be loosened and stretched to allow subsequent cell elongation. After cell elongation ceases, the cell wall generally is thickened, either by secretion of additional macromolecules into the primary wall or, more usually, by formation of a secondary cell wall composed of several layers. In mature tissues such as the xylem—the tubes that conduct salts and water from the roots through the stems to the leaves (see Figure 18-21)—the cell body degenerates, leaving only the cell wall. The unique properties of wood and of plant fibers such as cotton are due to the molecular properties of the cell walls in the tissues of origin.

Cell walls are constructed of only a few types of macromolecules. All plant walls contain fibers of *cellulose,* a polysaccharide made of glucose units. Like collagen in tendons, cellulose provides the tensile strength of the wall. Cellulose fibers are imbedded in and linked to a matrix containing two other types of polysaccharides, *hemicellulose* and *pectin,* and also a group of proteins the most prominent of which is a family of hydroxyproline-rich

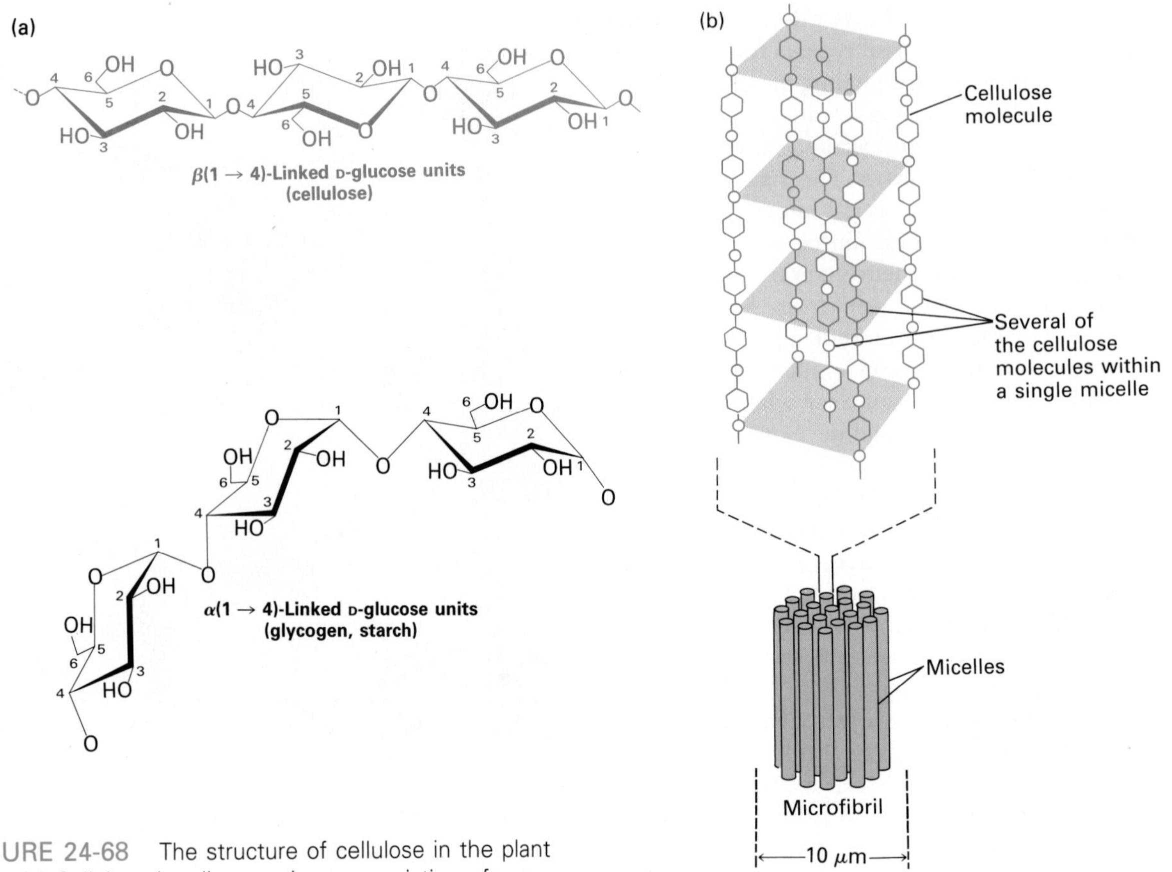

(a)

$\beta(1 \rightarrow 4)$-Linked D-glucose units
(cellulose)

$\alpha(1 \rightarrow 4)$-Linked D-glucose units
(glycogen, starch)

(b)

Cellulose molecule

Several of the cellulose molecules within a single micelle

Micelles

Microfibril

$\longleftarrow 10 \ \mu m \longrightarrow$

▲ FIGURE 24-68   The structure of cellulose in the plant cell wall. (a) Cellulose is a linear polymer consisting of more than 500 glucose residues linked together by $\beta(1 \rightarrow 4)$ glycosidic bonds. The $\beta(1 \rightarrow 4)$ linkages cause the molecule to form straight chains. In contrast, the $\alpha(1 \rightarrow 4)$ linkage in polyglucose molecules such as glycogen and starch causes a turning of the chain; such molecules adopt a coiled helical conformation. (b) The linear cellulose molecules pack together to form rodlike structures termed micelles. These are stabilized by hydrogen bonds between the cellulose molecules. Not all of the chains in a micelle are illustrated. Micelles are packed into microfibrils that are usually several micrometers in length and 3–10 μm in diameter. In certain algae they can be up to 30 μm in diameter. Each cellulose molecule is polar because its two ends are distinct, and all cellulose molecules in a microfibril have the same polarity.

fibrous glycoproteins termed *extensins*. The composition and structure of the cell wall varies in different parts of a plant and in different types of plants. We begin our discussion of the structure of the cell wall with cellulose, the most abundant cell-wall macromolecule.

Cellulose Molecules Form Long, Rigid Microfibrils

Cellulose molecules are linear polymers of glucose residues linked together by $\beta(1 \rightarrow 4)$ glycosidic bonds (Figure 24-68). The $\beta(1 \rightarrow 4)$ linkage causes the polysaccharide to adopt a fully extended conformation; this is accomplished by the flipping of each glucose unit 180° relative to the preceding one. In contrast, glycogen and starch are $\alpha(1 \rightarrow 4)$-linked glucose polymers; the $\alpha(1 \rightarrow 4)$ linkage causes the chain to adopt a coiled helical conformation. Cellulose molecules bundle together into fibers termed microfibrils that can be many micrometers in length. Ex-

tensive hydrogen bonding within cellulose molecules and between adjacent molecules makes the microfibril an almost crystalline aggregate.

Other Polysaccharides Bind to Cellulose to Generate a Complex Wall Matrix

The plant cell wall contains a number of polysaccharides whose compositions are more complex and variable than the composition of cellulose. These polymers contain several different saccharides including arabinose, xylose, fucose, rhamnose, galactose, and galacturonic acid. Biochemical study of these polysaccharides is difficult because they are heterogeneous and frequently are insoluble, being cross-linked to other polysaccharides. The best-studied constituents are those of the primary cell wall produced by cultured sycamore cells, but those in other cell walls are similar (Figure 24-69).

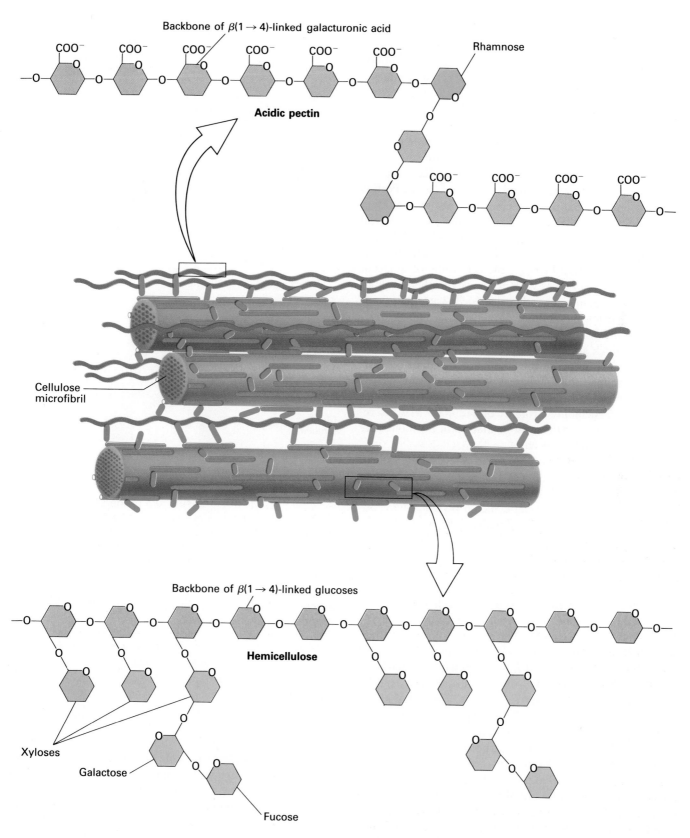

▲ FIGURE 24-69   The structure and interconnections of the major polymers in the primary cell wall. Hemicellulose molecules are hydrogen bonded to the surface of cellulose microfibrils. The backbone of hemicellulose (red) is similar to that of cellulose, but there are multiple branches (blue) of xylose, galactose, and fucose residues. These may link the microfibrils to pectins (purple). Pectins contain a backbone of short $\beta(1 \rightarrow 4)$-linked galacturonic acid (purple) chains, bound at their ends to short chains of rhamnose (green) and galacturonic acid. [See P. Albersheim, 1980, in N. E. Tolbert, ed., *Biochemistry of Plants*, Vol. I, Academic Press, pp. 91–162.]

Hemicelluloses are highly branched polysaccharides with a backbone of about 50 $\beta(1 \rightarrow 4)$-linked sugars of a single type. Hemicelluloses are linked by hydrogen bonds to the surface of cellulose microfibrils. The hemicellulose branches help bind the microfibrils to each other and to other matrix components, particularly the pectins.

Pectins, like hyaluronan (see Figure 24-16), contain multiple negatively charged saccharides. Pectins bind cations such as Ca^{2+} and, like hyaluronan, are highly hydrated. When purified, they bind water and form a gel—hence the use of pectins in many processed foods. Pectins are often cross-linked to hemicelluloses, thus participating in forming a complex network of all the principal wall components. This interlinked network helps bind adjacent cells together. Pectins are particularly abundant in the middle lamella, the layer between the cell walls of adjacent cells. Treatment of tissues with pectinase or other enzymes that degrade pectin frequently causes cells with their walls to separate from one another.

The primary cell wall can be thought of as a gel-like matrix in which are imbedded cellulose microfibrils. Such walls are more impermeable than are the matrices surrounding animal cells. Whereas water and ions diffuse freely in cell walls, diffusion of particles of diameter greater than ≈ 4 nm, including proteins of >20,000 MW, is reduced. This is one of the reasons that plant hormones are small, water-soluble molecules.

As illustrated in Figure 24-70, the secondary wall of the mature cell may have several layers; within each layer the cellulose fibrils are parallel to each other, but the orientation differs in adjacent layers. Such a plywood-like construction adds considerable strength to the wall.

Cell Walls Contain Lignin and an Extended Hydroxyproline-Rich Glycoprotein

As much as 15 percent of the primary cell wall may be composed of *extensin*, a glycoprotein made up of roughly 300 amino acids. Extensin, like collagen, contains abundant hydroxyproline (Hyp) and about half of its length represents variations of the four-residue sequence Ser-Hyp-Hyp-Hyp. Most of the hydroxyprolines are glycosylated with chains of three or four arabinose residues, and galactose is linked to the serines. Thus extensin is about 65 percent carbohydrate, and its protein backbone forms an extended rodlike helix with carbohydrates protruding outward. Extensins, like other cell-wall proteins, are incorporated into the insoluble polysaccharide network and are believed to have a structural role, forming the scaffolding upon which the cell-wall architecture is formed.

Lignin—a complex, insoluble polymer of phenolic residues—is a strengthening material in all cell walls. It is particularly abundant in wood, where it accumulates in primary cell walls and in the secondary walls of the xylem. Lignin associates with cellulose; like cartilage proteoglycans, it resists compression forces on the matrix. Particu-

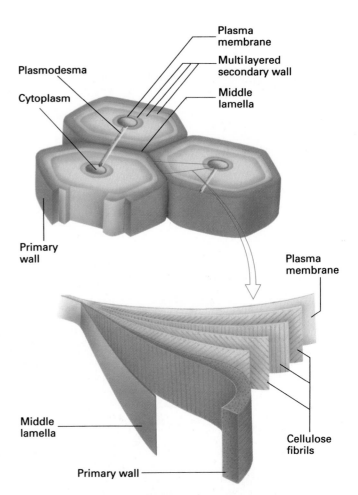

▲ FIGURE 24-70   The structure of the secondary cell wall, built up of a series of layers of cellulose. In each layer the fibers run more or less in the same direction, but the direction varies in different layers. As plant cells grow, they deposit new layers of cellulose adjacent to the plasma membrane. Thus the oldest layers are in the primary wall (the outer wall) and in the middle lamella (the pectin-rich part of the cell wall laid down between two daughter cells as they cleave during cell division.) Younger regions of the wall—collectively the secondary cell wall—are laid down as successive layers, adjacent to the plasma membrane. The cytoplasms of adjacent cells are usually connected by plasmodesmata that run through the layers of the cell walls.

larly for soil-grown plants, lignin is essential for strengthening the xylem tubes to enable them to conduct water and salts over long distances. Lignin also protects the plant against invasion by pathogens and against predation by insects or other animals.

Plants Grow Primarily by Auxin-Induced Cell Enlargement

Cell growth in higher plants frequently occurs without an increase in the volume of the cytosol. Because of the low

(a)

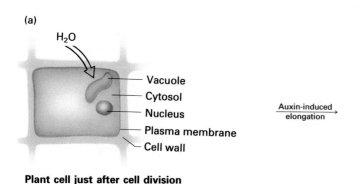

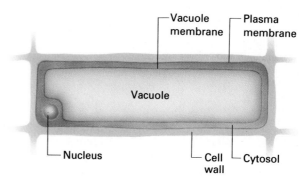

Plant cell just after cell division

(b)

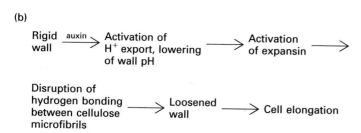

▲ FIGURE 24-71   Elongation of plant cells. (a) Change in structure of a plant cell during elongation. Uptake of water causes an internal pressure (turgor); in the presence of auxin, the cell wall is loosened and the turgor pressure against the loosened wall leads to elongation. (b) Proposed mechanism of cell-wall loosening in plant cells. [Part (b) adapted from L. Taiz, 1994, *Proc. Nat'l. Acad. Sci. USA* **91**:7387.]

ionic strength of the cell wall, water tends to enter the cytosol and vacuole, causing the cell to expand. Because the tensile strength of the surrounding cell wall resists this expansion, plant cells develop considerable internal pressure, or *turgor.* A localized loosening of the primary cell wall, induced by *auxin,* allows the cell to expand in a particular direction (Figure 24-71a); the size and shape of a plant are determined primarily by the amount and direction of this enlargement. Individual plant cells can increase in size very rapidly by loosening the wall and pushing the cytosol and plasma membrane outward against it. During this elongation, the amount of cytosol remains constant; the increase in cell volume is due only to the expansion of the intracellular vacuole by uptake of water. We can appreciate the magnitude of this phenomenon by considering that if all cells in a redwood tree were reduced to the size of a typical liver cell ($\approx 20\ \mu$m in diameter), then the tree would have a maximum height of only 1 meter!

Elongation of plant cells is induced by auxin (indole-3-acetic acid), a principal plant growth hormone. The rapid effects of auxin on cell elongation were first observed in classical experiments on *coleoptiles* from grasses and oats. Coleoptiles are the protective sheaths that surround the primary shoots and leaves; like many plant parts, they are *phototropic* (i.e., they bend toward a source of light). The growing meristematic cells at the tip of a coleoptile synthesize auxin, which is then transported down the coleoptile, where it causes many effects. Light inhibits auxin production, so that more auxin and preferential cell elongation occur on the side away from the light, causing the coleoptile to bend toward light (Figure 27-72a). Indeed, if the coleoptile is decapitated, removing its endogenous source

of auxin, and a small amount of auxin is placed on one side of the cut, the seedling bends away from the side where auxin was applied (Figure 24-72b).

Auxin is thought to induce cell elongation by causing the cell wall to soften at the "growing" end of the cell. In some plants, this is believed to be due to the induction by auxin of localized proton secretion at the ends of the cells; auxin apparently activates (directly or indirectly) a proton pump bound to the plasma membrane, which lowers the pH of the cell wall region near the plasma membrane (possibly to as low as pH 4.5 from the normal pH 7.0). The low pH activates a class of wall proteins termed *expansins* which disrupt the hydrogen bonding between cellulose microfibrils. As a consequence, the rigidity of the wall is reduced, and the cell can elongate (Figure 24-71b).

Expansins were discovered and purified using a novel biochemical assay on pure cellulose paper, since paper, like the plant cell walls from which it is made, derives its mechanical strength from hydrogen bonding between cellulose microfibrils. Extracts of plant cell walls were tested for their ability to mechanically weaken paper at pH values between 3.0 and 5.0, but not at pH 7. Importantly, expansins are not enzymes that degrade cellulose, and the expansin-triggered loosening of the wall is reversed when the pH is raised back to 7.0. Additional evidence for the *acid-growth hypothesis* stems from studies of the fungal compound *fusicoccin.* Like auxin, fusicoccin induces rapid cell elongation, and it triggers proton pumping out of sensitive cells and localized wall loosening. The action of fusicoccin or auxin can be blocked by permeating the cell wall with buffers that prevent the extracellular pH from being lowered.

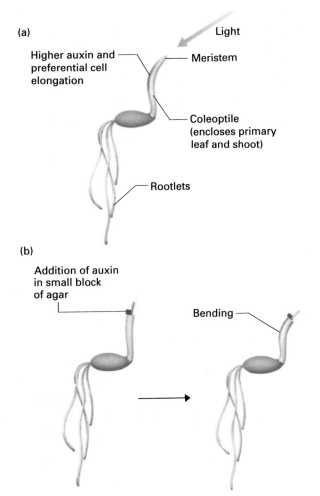

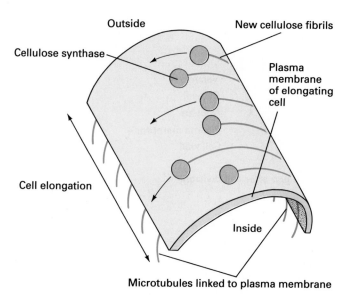

▲ FIGURE 24-73 Microtubules and cellulose synthesis in an elongating root tip cell. Circumferential rings of microtubules lie just inside the plasma membrane, perpendicular to the direction of cell elongation. Cellulose synthase, a large integral membrane protein, synthesizes cellulose fibrils on the outer face of the plasma membrane. As it spins out insoluble cellulose fibrils, the synthase moves in the plasma membrane (arrows) parallel to the underlying microtubule network. Thus in a growing cell, the new fibrils are arranged in circumferential rings perpendicular to the direction of elongation.

▲ FIGURE 24-72 Auxin-induced cell elongation causes phototropism in oat coleoptiles. (a) The release of auxin by the growing tip (meristem) on the side of the coleoptile away from the light causes the cells on the far *(left)* side to elongate preferentially and bend the tip toward the light. (b) Phototropism can be demonstrated experimentally by removing the tips from coleoptiles grown in the dark, keeping the decapitated seedlings in the dark, and applying an agar block containing auxin to one side. The coleoptiles subsequently bend away from the side at which auxin is applied.

The Orientation of Newly Made Cellulose Microfibrils Is Affected by the Microtubule Network

As noted already, all the microfibrils in each layer of the secondary wall are oriented in the same direction (see Figure 24-70). In the primary cell wall in elongating cells, newly made cellulose microfibrils encircle the cell like a belt perpendicular to the axis. Determining the mechanism whereby cellulose microfibrils are precisely oriented has been hampered because of difficulties in studying cellulose biosynthesis. Other cell-wall polysaccharides, like animal

proteoglycans, are fabricated in the Golgi complex and then secreted, but cellulose is synthesized on the external surface of the plasma membrane from UDP-glucose and ADP-glucose formed in the cytosol. The polymerizing enzyme, called cellulose synthase, is thought to be a large complex of many identical subunits, each of which "spins out" cellulose molecules that spontaneously form microfibrils. The long microfibrils are insoluble, which probably explains why they are not formed within the cell.

Experiments with elongating root tip cells suggest that in the primary wall, at least, microtubules influence the direction of cellulose deposition. These cells have oriented bands or rings of microtubules located just under the plasma membrane; these microtubules are perpendicular to the direction of elongation but parallel to many of the cellulose microfibrils in the primary cell wall of the elongating cell (Figure 24-73). Moreover, disruption of the microtubular network by drugs eventually disrupts the pattern of cellulose disposition. Thus, many investigators believe that cellulose synthase complexes move within the plane of the plasma membrane, as cellulose is formed, in directions determined by the underlying microtubule cytoskeleton. Any linkage, however, between the microtubules and cellulose synthase remains to be determined.

(a)

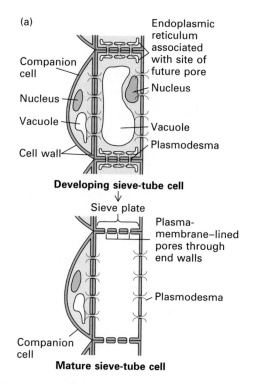

Companion cell

Nucleus

Vacuole

Cell wall

Endoplasmic reticulum associated with site of future pore

Nucleus

Vacuole

Plasmodesma

Developing sieve-tube cell

Sieve plate

Plasma-membrane–lined pores through end walls

Plasmodesma

Companion cell

Mature sieve-tube cell

(b)   Companion cell   Sieve plate

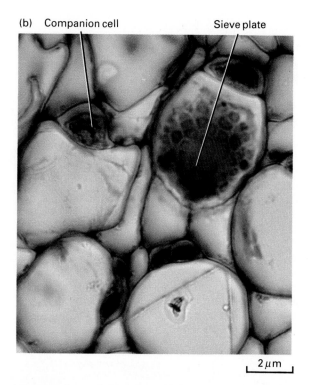

2 μm

▲ FIGURE 24-74   (a) Modifications of the cell wall during formation of the phloem sieve-tube cells. The developing sieve-tube cell and its companion cell are derived from the same mother cell. The primary cell wall of the sieve cell thickens, and the nucleus, vacuole, and other internal organelles are lost, but the plasma membrane is retained. The end walls, called the sieve plate, become extensively perforated by expansions of the plasmodesmata. The companion cell is connected to the sieve-tube cell by many plasmodesmata and actively secretes substances such as sucrose into the sieve-tube cell for transport, or actively resorbs such molecules. (b) Cross section of the phloem from *Curcurbita*, showing a sieve plate and companion cells. [Courtesy of J. R. Waaland and Biological Photo Service.]

Plasmodesmata Interconnect the Cytoplasms of Adjacent Cells in Higher Plants

All cells in higher plants are separated from each other by walls containing cellulose. However, the cytoplasms of adjacent cells are interconnected by tubelike *plasmodesmata* that penetrate the wall. Like gap junctions, plasmodesmata provide intercellular channels for molecules of up to 1000 molecular weight, including a variety of metabolic and signaling compounds. In the formation of the phloem, the plasmodesmata in the end cell walls that connect adjacent cells are enlarged, forming large pores that facilitate fluid movement. The phloem cells lose their nuclei and most of their organelles and depend on plasmodesmata with their companion cells for provision of nutrients (see Figure 24-74).

Electron micrographs reveal that in plasmodesmata the plasma membranes of the adjacent cells extend continuously through the cell wall (Figure 24-75a). The diameter of the channel is about 60 nm, and plasmodesmata can traverse cell walls up to 90 nm thick. In some cases, an extension of the endoplasmic reticulum called a *desmotubule* passes through the channel (Figure 24-75b,c). Membrane-bound molecules may pass from cell to cell via the desmotubule. Depending on the plant type, the density of plasmodesmata varies from 1 to 10 per μm^2, and even the smallest meristematic cells (the growing cells at the tips of roots or stems) have more than 1000 interconnections with their neighbors.

Much evidence establishes that plasmodesmata are in fact used in cell-cell communication. Fluorescent water-soluble chemicals microinjected into plant cells spread to the cytoplasm of adjacent cells but not into the cell wall. Many normal metabolic products, such as sucrose, spread from cell to cell; transport of such substances is proportional to the number of plasmodesmata and does not occur between cells not connected by such junctions. As with gap junctions, transport through plasmodesmata is reversibly inhibited by an elevation in cytosolic Ca^{2+}. Thus, despite considerable structural differences between gap junctions and plasmodesmata, many of their functional aspects are remarkably similar.

(a)

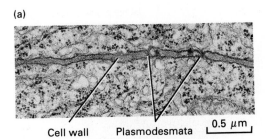

Cell wall   Plasmodesmata   0.5 μm

▲ FIGURE 24-75   The structure of plasmodesmata. (a) Electron micrograph of the
cell wall between two root tip cells of the
bean *Phaseolus vulgaris*. Plasmodesmata
connecting the cytoplasms of the two cells
are lined by extensions of the plasma membrane. (b) Components of plasmodesmata.
Note the desmotubule, an extension of the
endoplasmic reticulum, and the annulus, a
ring of cytosol that interconnects the cytosol
of adjacent cells. [Part (a) by W. P. Wergin;
courtesy of E. H. Newcomb/BPS.]

(b)

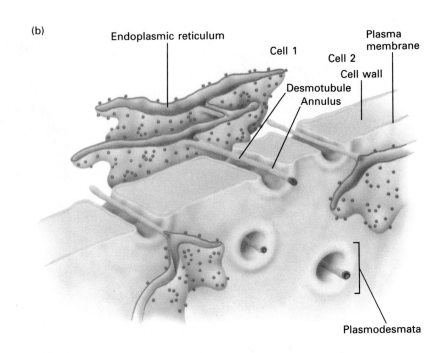

Endoplasmic reticulum   Cell 1   Plasma
                                           membrane
                                    Cell 2
                                Cell wall
                          Desmotubule
                          Annulus

Plasmodesmata

SUMMARY

Extracellular matrices in animals are composed of different combinations of collagens, proteoglycans, hyaluronan, laminin, fibronectin, and other glycoproteins.

Collagens are characteristically fibrous three-stranded helices with glycine as every third amino acid and abundant proline and hydroxyproline. Fibrous collagens appear to have evolved from an ancestral exon encoding six Gly-Pro-X motifs. Fibrous collagen, such as type I in tendons, is formed by parallel aggregates of triple helices stabilized by lateral lysine cross-links. Fibrous collagens are fabricated with the aid of N- and C-terminal propeptides that enable the three polypeptides to interact in register. In the rough ER, formation of disulfide bonds in the propeptides facilitates these interactions; other essential modifications such as proline hydroxylation also occur in the ER. Generally, the propeptides are removed after secretion, and then collagen fibrils form. There are 14 known collagens. Some retain their globular domains and do not form fibrils but rather interact with other fibrous collagens that do. Type IV collagen is unique to the basal lamina. It forms a two-dimensional fibrous network, stabilized by interactions with heparan sulfate proteoglycan and laminin.

Hyaluronan is an extremely long, negatively charged polysaccharide. It forms viscous, hydrated gels, and in several systems inhibits cell-cell adhesion and facilitates cell migration. HA forms the central component of cartilage proteoglycans; bound to the HA at regular intervals are core proteins to which are bound glycosaminoglycans such as chondroitin sulfate and dermatan sulfate. Smaller proteoglycans are found in other matrices and also attached to cell surfaces, where they facilitate cell-matrix interactions.

Laminin, a major protein in basal lamina, binds heparan sulfate, type IV collagen, and specific cell-surface receptor proteins. Nidogen interacts with type IV collagen, laminin, and heparan sulfate proteoglycans, and may mediate the assembly of all components of the basal lamina into a network. Fibronectins serve analogous functions in other matrices; they find fibrous collagens, proteoglycans, fibrin in blood clots, and other substances. Fibronectin promotes migration and adhesion of many cells. A variety of cell-surface receptors of the integrin family bind fibronectin, laminin, or other multiadhesive matrix proteins. In stationary fibroblasts, integrins anchor cytoskeletal actin filaments to extracellular fibronectin.

A variety of plasma-membrane proteins promote homophilic adhesion between similar types of cells. Examples include E-cadherin, which is essential for formation and stabilization of sheets of epithelial cells. During differentiation, cells can change their set of surface adhesion molecules; migrating cells such as neuroblasts lose adhesive proteins and then resynthesize them after migration ceases. Similarly, binding of leukocytes to endothelial cells and their subsequent movement into underlying tissues is orchestrated by changes in several adhesion molecules, including integrins and selectins, on both types of cells.

In order for tissues to function in an integrated manner, specialized junctions interconnecting similar kinds of cells are essential. Tight junctions seal together adjacent epithelial cells and prevent the passage of fluids through the cell layer. Belt desmosomes and spot desmosomes are junctions that bind the membranes of adjacent cells in a way that gives strength and rigidity to the entire tissue. On the cytosolic face of a cell, desmosomal complexes connect with intermediate filaments, contributing rigidity to the cell layer. Hemidesmosomes contain a specific integrin that binds both the intermediate filament network and elements of the basal lamina. Gap junctions are constructed of 12 copies of a single protein formed into a transmembrane channel that interconnects the cytoplasm of two adjacent cells and allows small molecules to pass between them.

During embryonic development, interactions between cells play a central role in determining the organization of cells and tissues in multicellular organisms. In many cases, one cell population induces the development of neighbouring cells. The TGFβ superfamily of secreted glycoproteins and their receptors (transmembrane serine-threonine kinases) play a widespread inductive role in development both in vertebrates and invertebrates. For instance, a Drosophila TGFβ homolog called Dpp acts as morphogen to control different cell fates along the dorsoventral axis. Similarly, several different TGFβ homologs in Xenopus (e.g., Vg1) induce mesodermal tissue; in some cases TGFβ family members induce different mesodermal cell fates as a function of their concentration. Inductive interactions between epithelial and mesenchymal cells regulates the formation of internal tissues such as the kidney, lung, pancreas, and gut. Both direct cell-cell contacts and cell-matrix interactions have been shown to be important regulators of induction.

Interactions between cells not only control cell-type specification, they also can pattern tissue. During chick-limb development, a zone of polarizing activity (ZPA) located at the posterior edge of the limb bud regulates the formation of limb pattern along the anterioposterior axis (i.e., thumb to little finger). The secreted protein, Sonic Hedgehog, is likely to be the inducer in the ZPA. Expression of Sonic Hedgehog protein in the anterior edge of the limb bud produces a mirror image duplication, mimicking the effects of ZPA transplantation. Similarly, the Drosophila Hedgehog protein regulates patterning in appendages such as the wing and leg. Both the fly and vertebrate Hedgehog proteins have other inductive functions.

Direct cell-cell contact mediated by cell-surface proteins on the interacting cells also can generate inductive signals. During development of the Drosophila compound eye, the R7 neuron is induced by direct contact with a specific neighboring cell, the R8 neuron. R7 precursors express Sev, a receptor tyrosine kinase, which interacts with Boss, an inductive ligand present on the surface of R8 cells. The precise spatial localization of Boss plays a central role in regulating the number of Sev-expressing cells induced to become a R7 neuron. Whereas the Boss-Sev interaction triggers a highly specific response, the interaction of two other transmembrane proteins—Notch (the receptor) and Delta (the ligand)—affects numerous cellular interactions in Drosophila. These proteins may function to regulate the competence of cells to respond to more specific intercellular cues.

During neuronal development, specific interactions between a developing growth cone and its environment guide it to its synaptic targets. Growth cones navigate through a constantly changing array of extracellular cues. Different extracellular-matrix components are more or less permissive for outgrowth. Generally, ubiquitous matrix components such as laminin support outgrowth but do not appear to be able to confer directionality. Conserved secreted proteins, the product of the unc6 locus in the worm C. elegans, and the netrins purified from chick spinal cord, act as directional cues for outgrowth along the dorsoventral axis in both organisms. The netrins, produced by the floor plate in the ventral-most region of the neural tube, are chemoattractants for ventral outgrowth of the dorsally located commissural neurons. Genetic evidence in C. elegans indicates that Unc6 is required for neurons that grow both dorsally and ventrally, suggesting that it may act as a repulsive factor for some neurons and an attractant for others. The unc5 gene encodes a transmembrane protein necessary for dorsally directed growth; misexpression of Unc5 causes ventrally destined growth cones to be redirected dorsally.

Other extracellular and cell-surface molecules also influence growth-cone navigation. Repellent molecules that mediate rapid growth-cone collapse in vitro have been identified. One of these, called collapsin, has been isolated from the chick and is homologous to proteins called semaphorins found in grasshoppers, flies, and humans. Antibody-disruption experiments in the grasshopper limb indicate that semaphorin, expressed in a stripe of epithelial cells along the limb bud, plays a role in guiding the growth cone to its target. Cell-surface adhesion proteins are often distributed on different subclasses of axons and have long been considered important guidance cues. Knockout experiments in Drosophila and in the mouse reveal surprisingly mild phenotypes, suggesting there may be a redundancy of recognition mechanisms regulating selective growth along different axonal fascicles.

Although it is not known how a growth cone ultimately selects its specific synaptic target once a target has been identified, a complex interaction between the two cells engaged in forming the synapse takes place. This interaction has been best characterized at the neuromuscular junction. The glycoprotein agrin is an important constituent in the specialized basal lamina at the synapse. During development it is synthesized both by the neuron and the muscle. It catalyzes the aggregation of acetylcholine receptors and other junctional proteins, thereby promoting the formation of the neuromuscular junction.

Microfibrils of cellulose are crystalline aggregates of $\beta(1 \rightarrow 4)$-linked glucose polymers. They are ubiquitous constituents of plant cell walls and are responsible for much of its tensile strength. Cellulose microfibrils are imbedded in and linked to a matrix that contains hemicellulose, negatively charged pectins, and the fibrous hydroxyproline-rich glycoprotein extensin. By binding cations and water, pectins give the wall a gel-like consistency. Extensin and lignin provide reinforcements; lignin is responsible for much of the strength of xylem tubes and also of wood.

The primary cell wall between growing plant cells is thin and extensible. Cells grow by localized loosening of the cell walls and expansion of the vacuole resulting from osmotic influx of water. Auxin induces this cell elongation by causing a reduction in the wall pH and activation of expansins, proteins in the cell wall that disrupt the hydrogen bonding between cellulose microfibrils. Circumferential bands of cellulose fibers are frequently laid down perpendicular to the direction of cell elongation; circumferential bands of microtubules just under the plasma membrane determine the direction in which the cell-surface cellulose synthase synthesizes new cellulose microfibrils. The secondary cell wall often has several layers; in each layer the cellulose microfibrils are parallel but different in orientation from those in adjacent layers. Adjacent cells in higher plants are interconnected by plasmodesmata, which allow small molecules such as sucrose to pass between the cells.

REVIEW QUESTIONS

1. The extracellular matrices found in animals are quite varied depending upon their specific functions. What are the chief proteins that constitute extracellular matrices? What are the basal lamina, loose connective tissue, and dense connective tissue?

Describe the fundamental structural unit of collagen; be certain to discuss the importance of having a glycine as every third amino acid, the requirement for having a large proportion of the amino acids proline and hydroxyproline, and the need for hydroxylysine at the chain termini. How is the gene encoding $\alpha 1(I)$ collagen organized?

What are collagen fibrils, and how are they formed? What is a collagen fiber?

What is procollagen? In terms of collagen fibril formation, what is the importance of the interchain disulfide bonds present in the C-terminal propeptide? What are the steps of collagen synthesis that occur in the rough endoplasmic reticulum? What are the steps that occur in the Golgi apparatus? Where are collagen fibrils and collagen fibers formed; what are the stages involved in the processes?

a. Scurvy is an illness caused by a deficiency of vitamin C (ascorbic acid). Which step in the pathway of collagen synthesis is affected in this disease? In the advanced stages of scurvy, two of its many symptoms are the inability to stand and the presence of spots of blood on the skin. Based on your knowledge of the source of the disease, why would patients present with these disorders?

b. Suppose a heterozygous mouse expressed one wild type $\alpha 2(IV)$ allele and one mutant $\alpha 2(IV)$ allele. Which types of collagen would be affected? Would all the collagen containing the $\alpha 2(IV)$ chains be affected? Why? Would this condition be more or less severe than that for a heterozygous individual having one wild type and one mutant allele for $\alpha 1(IV)$ chains? Explain your answer.

2. What is hyaluronan (HA), and what are its unique properties? What classes of cells is HA associated with? Do cell surface receptor specific for HA exist? What enzyme degrades HA?

What are proteoglycans, and what are some of their functions? What is a glycosaminoglycan? Ser-gly-X-gly, where X is any amino acid, is one of the signal sequences in a protein that specifies the addition of an oligosaccharide. Is this site always substituted with carbohydrate at the ser residue? What other aspect of protein structure is probably involved in the attachment of an oligosaccharide?

What are laminin and nidogen, and what are some of their functions? What are integrins? What is the common general structure of integrin molecules? What is the structure and function of fibronectin?

Cell adhesion molecules include cadherins, selectins, members of the immunoglobulin superfamily, and integrins. What are the cadherins, and what are some of their functions? What are N-CAMs? Discuss the roles of cadherins and N-CAMs in neuronal development.

What is a lectin? What are selectins, and how are they involved in the localization of neutrophils at sites of infection?

What is the function of spot desmosomes, and what are their protein constituents? What are hemidesmosomes, and what proteins constitute these contact points?

Gap junctions permit cell-cell communications to occur. What is the structural protein found in gap junctions, and how is it arranged in the junction? What size molecules can pass through gap junctions? What effect does Ca^{2+} have on gap junction function?

a. You are investigating the ability of four cell lines, A, B, A1, and B1 to associate with one another. Cell line A is derived from nerve cells, cell line B is from glial cells. Cell line A1 was derived from A by transfection with an expression vector encoding E-cadherin; cell line B1 was derived from cell line B transfected with an expression vector encoding N-cadherin. When each of the cell lines is cultured individually, the cells associate readily. When cells of one line are mixed with cells of another, and the combination is cultured, mixed associated aggregates sometimes form. The results of several of these types of mixing experiments are presented in the table, where a (+) sign indicates the presence of mixed colonies due to association of both cell types. A (−) sign indicates that all colonies formed were only one cell type. Explain these results.

Cells Lines Grown Together	Presence of Mixed Associated Aggregates
A/B	−
A/A1	+
A/B1	+
B/A1	−
B/B1	+
A1/B1	+

b. Part of your studies with cell lines A and B involves uptake of a fluorescein derivative. Cell lines A and A1 are able to take up the indicator which is confined to the cytosol and colors the cells red under a fluorescence microscope, but cell lines B and B1 do not incorporate it and are white. Furthermore, when cell lines A and B are mixed and subsequently cultured, colonies from cell line A are red under fluorescent microscopy, but colonies from cell line B remain white. Of the mixed associated aggregates that form as indicated in the table, all but B/B1 are red. Also, within all the mixed associated aggregates that take up dye, all the cells are red. Offer an explanation for these results.

3. Cell-cell interactions are extremely important during the devel-

opment of multicellular organizations. What is meant by induction? What is permissive induction; what is instructive induction? What are competent cells, and what are some of the probable mechanisms for conferring competence?

The superfamily of proteins known as TGFβ has important functions in development. Describe the basic characteristics of the receptors for these growth factors. What mechanism can modulate the response of these receptors?

What are the ZPA and the Spemann's organizer? Compare and contrast these two organizers.

Compare the proposed role for the *Drosophila* Hedgehog protein to its vertebrate analog in regulating tissue patterning.

4. Development of the nervous system is a highly complex process in that the axons must travel along the correct paths and terminate in the proper vicinity. What is the growth cone? As a growth cone moves, does the cell body move as well, or does it remain stationary? What are pioneer axons and secondary axons? What are netrin 1 and netrin 2?

What are collapsing factors?

What evidence supports the notion that the basal lamina at the neuromuscular junction is essential for differentiation of regenerating nerve and muscle? What are some of the signaling molecules in the basal lamina? What is agrin?

What genes control outgrowth along the dorsoventral axis in *C. elegans*? What proteins do they encode?

5. What macromolecules are components of the plant cell wall? Cellulose is the predominant macromolecule in the plant cell wall. Discuss the structure of cellulose. What are hemicellulose and pectin? What is extensin? What is the abundant modified amino acid in extensin? What is lignin?

a. Plasmodesmata are analogous to gap junctions. Devise an experiment that would allow you to determine the maximum size molecules that are able to pass through plasmodesmata.

References

General References on the Extracellular Matrix and its Roles in Differentiation

ADAIR, W. S., and R. P. MECHAM, eds., 1990. *Organization and Assembly of Plant and Animal Extracellular Matrix.* Academic Press.

ENGEL, J. 1991. Common structural motifs in proteins of the extracellular matrix. *Curr. Opin. Cell Biol.* 3:779–785.

GILBERT, S. F. 1991. *Developmental Biology,* 3d ed. Sinauer Associates.

HAY, E. D., ed., 1991. *Cell Biology of Extracellular Matrix,* 2d ed. Plenum.

HYNES, R. O., and A. D. LANDER. 1992. Contact and adhesive specificities in the associations, migrations, and targeting of cells and axons. *Cell.* 68:303–322.

LAWRENCE, P. A. 1992. *The Making of a Fly.* Blackwell Scientific Publications.

PAULSSON, M. 1992. Basement membrane proteins: structure, assembly, and cellular interactions. *Crit. Rev. Biochem. Mol. Biol.* 27:93–127.

PIEZ, K. A., and A. H. REDDI, eds. 1984. *Extracellular Matrix Biochemistry.* Elsevier.

SANDELL, L. J., and C. D. BOYD. 1990. *Extracellular Matrix Genes.* Academic Press.

SLACK, J. M. W. 1991. From egg to embryo: regional specification in early development. In P. W Barlow et al., eds., *Developmental and Cell Biology Series,* vol. 26, 2d ed., Cambridge Univ. Press.

YAMADA, K. M. 1991. Adhesive recognition sequences. *J. Biol. Chem.* 266:12809–12812.

ZERN, M. A., and L. REID, eds. 1993. *Extracellular Matrix: Chemistry, Biology, and Pathobiology with Emphasis on the Liver.* Marcell Dekker.

Fibrous Collagens: Structure and Synthesis

BURGESON, R. E. 1988. New collagens, new concepts. *Ann. Rev. Cell Biol.* 4:551–577.

DAVIDSON, J. M., and R. A. BERG. 1981. Posttranslational events in collagen synthesis. *Methods Cell Biol.* 23:199–136.

ENGEL, J., and D. J. PROCKOP. 1991. The zipper-like folding of collagen triple helices and the effects of mutations that disrupt the zipper. *Ann. Rev. Biophys. Chem.* 20:137–152.

EYRE, D. R., M. A. PAZ, and P. M. GALLOP. 1984. Cross-linking in collagen and elastin. *Ann. Rev. Biochem.* 53:717–748.

KUIVANIEMI, H., G. TROMP, and D. J. PROCKOP. 1991. Mutations in collagen genes: causes of rare and some common diseases in humans. *FASEB J.* 5:2052–2060.

MARTIN, G. R., et al. 1985. The genetically distinct collagens. *Trends Biochem. Sci.* 10:285–287.

PARRY, D. A. D. 1988. The molecular and fibrillar structure of collagen and its relationship to the mechanical properties of connective tissue. *Biophys. Chem.* 29:195–209.

SHAW, L. M., and B. R. OLSEN. 1991. FACIT collagens: diverse molecular bridges in extracellular matrices. *Trends Biochem. Sci.* 16:191–194.

STACEY, A., et al. 1988. Perinatal lethal osteogenesis imperfecta in transgenic mice bearing an engineered mutant pro-α1(I) collagen gene. *Nature* 332:131–136.

Type IV Collagen and the Basal Lamina

YURCHENCO, P. D., and G. C. RUBEN. 1987. Basement membrane structure in situ: evidence for lateral associations in the type IV collagen network. *J. Cell Biol.* 105:2559–2568.

YURCHENCO, P. D., and J. C. SCHITTNY. 1990. Molecular architecture of basement membranes. *FASEB J.* 4:1577–1590.

Matrix Proteoglycans

EVERED, D., and J. WHELAN, eds. 1986. *Functions of the Proteoglycans.* Ciba Foundation Symposium 124. Wiley.

HASSELL, J. R., J. H. KIMURA, and V. C. HASCALL. 1986. Proteoglycan core protein families. *Ann. Rev. Biochem.* 55:539–567.

HEINEGÅRD, D., and Å. OLDBERG. 1989. Structure and biology of cartilage and bone matrix noncollagenous macromolecules. *FASEB J.* 3:2042–2045.

JACKSON, R. L., S. J. BUSCH, and A. D. CARDIN. 1991. Glycosaminoglycans: molecular properties, protein interactions, and role in physiological processes. *Physiol. Rev.* 71:481–539.

KJELLÉN, L., and LINDAHL, U. 1991. Proteoglycans: structures and interactions. *Ann. Rev. Biochem.* 60:443–475.

KNUDSON, C. B., and W. KNUDSON. 1993. Hyaluronan-binding proteins in development, tissue homeostasis, and disease. *FASEB J.* 7:1233–1241.

LANDER, A. D. 1993. Proteoglycans in the nervous system. *Curr. Opin. Neurobiol.* 3:716–723.

RUOSLAHTI, E. 1988. Structure and biology of proteoglycans. *Ann. Rev. Cell Biol.* 4:229–255.

RUOSLAHTI, E. 1989. Proteoglycans in cell regulation. *J. Biol. Chem.* 264:13369–13372.

RUOSLAHTI, E., and Y. YAMAGUCHI. 1991. Proteoglycans as modulators of growth factor activities. *Cell* 64:867–869.

TIMPL, R. 1993. Proteoglycans of basement membranes. *Experientia* 49:417–428.

TOOLE, B. P. 1990. Hyaluronan and its binding proteins, the hyaladherins. *Curr. Opin. Cell Biol.* 2:839–844.

WIGHT, T. N., and R. P. MECHAM, eds. 1987. *Biology of Proteoglycans.* Academic Press.

Cell-Surface Proteoglycans

BERNFIELD, M., et al. 1992. Biology of the syndecans: a family of transmembrane heparan sulfate proteoglycans. *Ann. Rev. Cell Biol.* 8:365–393.

DAVID, G. 1993. Integral membrane heparan sulfate proteoglycans. *FASEB J.* 7:1023–1030.

YANAGISHITA, M., and V. C. HASCALL. 1992. Cell surface heparan sulfate proteoglycans. *J. Biol. Chem.* 267:9451–9454.

Laminin, Nidogen, and Tenascin

CHIQUET-EHRISMANN, R. 1990. What distinguishes tenascin from fibronection? *FASEB J.* 4:2598–2604.

CHIQUET-EHRISMANN, R. 1991. Anti-adhesive molecules of the extracellular matrix. *Curr. Opin. Cell Biol.* 3:800–804.

ENGEL, J. 1992. Laminins and other strange proteins. *Biochemistry* 31:10643–10651.

MARTIN, G. R., and R. TIMPL. 1987. Laminin and other basement membrane components. *Ann. Rev. Cell Biol.* 35:57–85.

SASAKI, M., et al. 1988. Laminin, a multidomain protein. *J. Biol. Chem.* 263:16536–16544.

Fibronectin and Other Multiadhesive Proteins

HYNES, R. O. 1989. *Fibronectins.* Springer-Verlag.

MOSHER, D. F., ed. 1989. *Fibronectin.* Academic Press.

MOSHER, D. F., et al. 1992. Assembly of extracellular matrix. *Curr. Opin. Cell Biol.* 4:810–818.

RUOSLAHTI, E. 1988. Fibronectin and its receptors. *Ann. Rev. Biochem.* 57:375–414.

Integrins

BURRIDGE, K., et al. 1988. Focal adhesions: transmembrane junctions between the extracellular matrix and the cytoskeleton. *Ann. Rev. Cell Biol.* 4:487–526.

HYNES, R. O. 1987. Integrins: a family of cell surface receptors. *Cell* 48:549–554.

HYNES, R. O. 1992. Integrins: versatility, modulation, and signaling in cell adhesion. *Cell* 69:11–25.

HYNES, R., and A. LANDER. 1992. Contact and adhesive specificities in the associations, migrations, and targeting of cells and axons. *Cell* 68:303–322.

MECHAM, R. P. 1991. Receptors for laminin on mammalian cells. *FASEB J.* 5:2538–2546.

PHILLIPS, D. R., I. F. CHARO, and R. M. SCARBOROUGH. 1991. GPIIb-IIIa: the responsive integrin. *Cell* 65:339–362.

QUARANTA, V., and J. JONES. 1991. The internal affairs of an integrin. *Trends Cell Biol.* 1:2–5.

RUOSLAHTI, E. 1991. Integrins. *J. Clin. Invest.* 87:1–5.

RUOSLAHTI, E., and M. D. PIERSCHBACHER. 1987. New perspectives in cell adhesion: RGD and integrins. *Science* 238:491–497.

SINGER, I. I. 1979. The fibronexus: a transmembrane association of fibronectin-containing fibers and bundles of 5 nm microfilaments in hamster and human fibroblasts. *Cell* 16:675–685.

Cadherins

BUXTON, R. S., et al. 1993. Nomenclature of the desmosomal cadherins. *J. Cell Biol.* 121:481–483.

BUXTON, R. S., and A. I. MAGEE. 1992. Structure and interactions of desmosomal and other cadherins. *Semin. Cell Biol.* 3:157–167.

GEIGER, B., and O. AYALON. 1992. Cadherins. *Ann. Rev. Cell Biol.* 8:307–332.

GUMBINER, B. 1988. Cadherins: a family of Ca^{2+}-dependent adhesion molecules. *Trends Biochem. Sci.* 13:75–76.

GUMBINER, B. M., and P. D. MCCREA. 1993. Catenins as mediators of the cytoplasmic functions of cadherins. *J. Cell. Sci. Suppl.* 17:155–158.

HYNES, R. O. 1992. Specificity of cell adhesion in development: the cadherin superfamily. *Curr. Opin. Genet. Dev.* 2:621–624.

KEMLER, R. 1992. Classical cadherins. *Semin. Cell Biol.* 3:149–155.

KEMLER, R. 1993. From cadherins to catenins: cytoplasmic protein interactions and regulation of cell adhesion. *Trends Genet.* 9:317–321.

NATHKE, I. S., L. E. HINCK, and W. J. NELSON. 1993. Epithelial cell adhesion and development of cell surface polarity: possible mechanisms for modulation of cadherin function, organization and distribution. *J. Cell Sci. Suppl.* 17:139–145.

TAKEICHI, M. 1988. The cadherins: cell-cell adhesion molecules controlling animal morphogenesis. *Development* 102:639–655.

TAKEICHI, M. 1991. Cadherin cell adhesion receptors as a morphogenetic regulator. *Science* 251:1451–1455.

N-CAMs and Related Cell-Adhesion Molecules

CUNNINGHAM, B. A., et al. 1987. Neural cell adhesion molecule: structure, immunoglobulin-like domains, cell surface modulation, and alternative RNA splicing. *Science* 236:799–806.

RUTISHAUSER, U., et al. 1988. The neural cell adhesion molecule (NCAM) as a regulator of cell-cell interactions. *Science* 240:53–57.

RUTISHAUSER, U. 1993. Adhesion molecules of the nervous system. *Curr. Opin. Neurobiol.* 3:709–715.

WILLIAMS, A. F., and A. N. BARCLAY. 1988. The immunoglobulin superfamily—domains for cell surface recognition. *Ann. Rev. Immunol.* 6:381–406.

Selectins and Movement of Leukocytes from the Blood into Underlying Tissues

BEVILACQUA, M. P. 1993. Endothelial-leukocyte adhesion molecules. *Ann. Rev. Immunol.* 11:767–804.

BEVILACQUA, M. P., and R. M. NELSON. 1993. Selectins. *J. Clin. Invest.* 91:379–387.

BRANDLEY, B. K., S. J. SWIEDLER, and P. W. ROBBINS. 1990. Carbohydrate lectins of the LEC cell adhesion molecules. *Cell* 63:861–863.

BUTCHER, E. C. 1991. Leukocyte-endothelial cell recognition: three (or more) steps to specificity and diversity. *Cell* 67:1033–1036.

GRAVES, R. J., et al. 1994. Insight into E-selectin/ligand interactions from the crystal structure and mutagenesis of the Lec/EGF Domains. *Nature* 367:532–538.

HOGG, N., and R. C. LANDIS. 1993. Adhesion molecules in cell interactions. *Curr. Opin. Immunol.* 5:383–390.

LASKY, K. A. 1992. Selectins: interpreters of cell-specific carbohydrate information during inflammation. *Science* 258:964–969.

SPRINGER, T. A. 1994. Traffic signals for lymphocyte recirculation and leukocyte emigration: the multistep paradigm. *Cell* 76:301–314.

VARKI, A. 1994. Selectin ligands. *Proc. Natl. Acad. Sci.* 91:7390–7397.

VESTWEBER, D. 1993. The selectins and their ligands. *Curr. Topics Microbiol. Immunol.* 184:65–75.

Cell Junctions

BOCK, G., and S. CLARK. 1987. *Junctional Complexes of Epithelial Cells.* Ciba Foundation Symposium 125. Wiley.

STAEHELIN, L. A., and B. E. HULL. 1978. Junctions between cells. *Sci. Am.* 238(5):140–152.

Gap Junctions and Plasmodesmata

BENNETT, M. V., et al. 1991. Gap junctions: new tools, new answers, new questions. *Neuron* 6:305–320.

BENNETT, M. V., and V. K. VERSELIS. 1992. Biophysics of gap junctions. *Semin. Cell Biol.* 3:29–47.

BENNETT, M. V. L. 1994 Connexins in disease. *Nature* 368:18–19.

BENNETT, M., and D. SPRAY, eds. 1985. *Gap Junctions.* Cold Spring Harbor Laboratory.

GUTHRIE, S. C., and N. B. GILULA. 1989. Gap junctional communication and development. *Trends Neurosci.* 12:12–16.

KUMAR, N. M., and N. B. GILULA. 1992. Molecular biology and genetics of gap junction channels. *Semin. Cell Biol.* 3:3–16.

MEINERS, S., O. BARON-EPEL, and M. SCHINDLER. 1988. Intercellular communication—filling in the gap. *Plant Physiol.* 88:791–793.

ROBARDS, A. W., and W. J. LUCAS. 1991. Plasmodesmata. *Ann. Rev. Plant Physiol. Plant Mol. Biol.* 41:369–419.

SAEZ, J. C., et al. 1993. Gap junctions. Multiplicity of controls in differentiated and undifferentiated cells and possible functional implications. *Adv. Second Messenger Phosphoprotein Res.* 27:163–198.

SCHWARZMANN, G., et al. 1981. Diameter of the cell-to-cell junctional membrane channels as probed with neutral molecules. *Science* 213:551–553.

TILNEY, L. G., et al. 1991. The structure of plasmodesmata as revealed by plasmolysis, detergent extraction, and protease digestion. *J. Cell. Biol.* 112:739–747.

Desmosomes

GARROD, D. R. 1993. Desmosomes and hemidesmosomes. *Curr. Opin. Cell Biol.* 5:30–40.

SCHWARZ, M. A., et al. 1990. Desmosomes and hemidesmosomes: constitutive molecular components. *Ann. Rev. Cell Biol.* 6:461–491.

Embryonic Induction: Early Patterning in Invertebrates and Vertebrates

ARORA, K., M. S. LEVINE, and M. B. O'CONNOR. 1994. The *screw* gene encodes a ubiquitously expressed member of the TGF-β family required for specification of dorsal cell fates in the *Drosophila* embryo. *Genes & Dev.* 8:2588–2601.

ATTISANO, L., et al. 1993. Identification of human activin and TGFβ type I receptors that form heteromeric kinase complexes with type II receptors. *Cell* 75:671–680.

FERGUSON, E. L., and K. V. ANDERSON. 1992. Localized enhancement and repression of the activity of the TGF-β family member, *decantaplegic*, is necessary for dorsal-ventral pattern formation in the *Drosophila* embryo. *Development* 114:583–597.

FRANÇOIS, V., M. SOLLOWAY, J. W. O'NEILL, J. EMERY, and E. BIER. 1994. Dorsal-ventral patterning of the *Drosophila* embryo depends on a putative negative growth factor encoded by the *short gastrulation* gene. *Genes & Dev.* 8:2602–2616.

GREEN, J. B. A., H. V. NEW, and J. C. SMITH. 1992. Responses of embryonic *Xenopus* cells to activin and FGF are separated by multiple dose thresholds and correspond to distinct axes of the mesoderm. *Cell* 71:731–739.

HEMMATI-BRIVANLOU, A., and D. A. MELTON. 1992. A truncated activin receptor inhibits mesoderm induction and formation of axial structures in *Xenopus* embryos. *Nature* 359:609–614.

HEMMATI-BRIVANLOU, A., O. G. KELLY, and D. A. MELTON. 1994. Follistatin, an antagonist of activin, is expressed in the Spemann-organizer and displays direct neuralizing activity. *Cell* 77:283–295.

MASON, E. D., K. D. KONRAD, C. D. WEBB, and J. L. MARSH. 1994. Dorsal midline fate in *Drosophila* embryos requires *twisted* gastrulation, a gene encoding a secreted protein related to human connective tissue growth factor. *Gene & Dev.* 8:1489–1501.

MASSAGUÉ, J., L. ATTISANO, and J. L. WRANA. 1994. The TGF-β family and its composite receptors. *Trends Cell Biol.* 4:172–178.

MATHEWS, L. S., and W. W. VALE. 1991. Expression cloning on activin receptor, a predicted transmembrane serine kinase. *Cell* 65:973–982.

NELLEN, D. M., M. AFFOLTER, and K. BASKER. 1994. Receptor serine/threonine kinases implicated in the control of *Drosophila* body pattern by *decapentaplegic*. *Cell* 78:225–237.

SLACK, J. M. W. 1994. Inducing factors in *Xenopus* early embryos. *Curr. Biol.* 4:116–126.

SMITH, W. C., et al. 1993. Secreted *noggin* protein mimics the Spemann organizer in dorsalizing *Xenopus* mesoderm. *Nature* 361:547–549.

THOMSEN, G. H., and D. A. MELTON. 1993. Processed Vg1 protein is an axial mesoderm inducer in *Xenopus*. *Cell* 74:433–441.

WANG, X. F., et al. 1991. Expression cloning and characterization of the TFG-β type III receptor. *Cell* 67:797–805.

Epithelial-Mesenchymal Interactions

AUFDERHEIDE, E., and P. EKBLOM. 1988. Tenascin during gut development: appearance in the mesenchyme, shift in molecular forms, and dependence on epithelial-mesenchymal interactions. *J. Cell Biol.* 107:2341–2349.

BIRCHMEIER, C., and W. BIRCHMEIER. 1993. Molecular aspects of mesenchymal-epithelial interactions. *Ann. Rev. Cell Biol.* 9:511–540.

EKBLOM, P. 1989. Developmentally regulated conversion of mesenchyme to epithelium. *FASEB J.* 3:2141–2150.

PACHNIS, V., B. MANKOO, and F. CONSTANTINI. 1993. Expression of c-ret proto-oncogene during mouse embryogenesis. *Development* 119:1005–1017.

SAGA, Y., et al. 1992. Mice develop normally without tenascin. *Genes & Dev.* 6:1821–1831.

SCHUCHARDT, A., et al. 1994. Defects in the kidney and enteric nervous system of mice lacking the c-Ret tyrosine kinase receptor. *Nature* 367:380–383.

Pattern Formation in Limbs

BASLER, K., and G. STRUHL. 1994. Compartment boundaries and the control of *Drosophila* limb pattern by *hedgehog* protein. *Nature* 368:208–214.

LEE, J. J., et al. 1992. Secretion and localized transcription suggest a role inpositional signaling for products of the segmentation gene *hedgehog*. *Cell* 71:33–50.

RIDDLE, R. D., et al. 1993. Sonic hedgehog mediates the polarizing activity of the ZPA. *Cell* 75:1401–1416.

TABIN, C. 1991. Retinoids, homeoboxes and growth factors: toward molecular models for limb development. *Cell* 66:199–217.

Local Interactions Directed by Cell-Cell Contact

COFFMAN, C. R., et al. 1993. Expression of an extracellular deletion of *xotch* diverts cell fate in *Xenopus* embryos. *Cell* 73:659–671.

ELLISEN, L. W., et al. 1991. *Tan-1*, the human homolog of the Drosophila *Notch* gene, is broken by chromosomal translocations in T lymphoblastic neoplasms. *Cell* 66:649–661.

FEHON, R. G., et al. 1990. Molecular interactions between the protein products of the neurogenic loci *Notch* and *Delta*, two EGF-homologous genes in *Drosophila*. *Cell* 61:523–534.

FORTINI, M. E., et al. 1993. An activated Notch receptor blocks cell-fate commitment in the developing *Drosophila* eye. *Nature* 365:555–557.

HAFEN, E., et al. 1987. Sevenless, a cell-specific homeotic gene of *Drosophila*, encodes a putative transmembrane receptor with a tyrosine kinase domain. *Science* 236:55–63.

KRÄMER, H., R. L. CAGAN, and S. L. ZIPURSKY. 1991. Interaction of bride of sevenless membrane-bound ligand and the sevenless tyrosine-kinase receptor. *Nature* 352:207–212.

STRUHL, G., K. FITZGERALD, and I. GREENWALD. 1993. Intrinsic activity of the Lin-12 and Notch intracellular domains in vivo. *Cell* 74:331–345.

VAN VACTOR, D. L. JR., et al. 1991. Induction in the developing compound eye of *Drosophila*: multiple mechanisms restrict R7 induction to a single retinal precursor cell. *Cell* 67:1145–1155.

Regulation of Neuronal Outgrowth

BIEBER, A. J., et al. 1989. *Drosophila* neuroglian: a member of the immunoglobulin superfamily with extensive homology to the vertebrate neural adhesion molecule L1. *Cell* 59:447–460.

CALOF, A. L., and A. D. LANDER. 1991. Relationship between neuronal migration and cell-substratum adhesion: laminin and merosin promote olfactory neuronal migration but are anti-adhesive. *J. Cell Biol.* 115:779–794.

EISEN, J. S., S. H. PIKE, and B. DEBU. 1989. The growth cones of identified motorneurons in embryonic zebrafish select appropriate pathways in the absence of specific cellular interactions. *Neuron* 2:1097–1104.

GARRIGA, G., C. DESAI, and H. R. HORVITZ. 1993. Cell interactions control the direction of outgrowth, branching and fasciculation of the HSN axons of *Caenorhabditis elegans*. *Development* 117:1071–1087.

GOODMAN, C. S., and C. J. SHATZ. 1993. Developmental mechanisms that generate precise patterns of neuronal connectivity. *Cell* 72/*Neuron* 10 (Suppl.):77–98.

HYNES, R. O., and A. D. LANDER. 1992. Contact and adhesive specificities in the associations, migrations, and targeting of cells and axons. *Cell* 68:303–322.

LEMMON, V., et al., 1992. Neurite growth on different substrates: permissive versus instructive influences and the role of adhesive strength. *J. Neurosci.* 12:818–826.

MCINTIRE, S. L., et al. 1992. Genes necessary for directed axonal elongation and fasciculation in *C. elegans*. *Neuron* 8:307–322.

NEUGEBAUER, K. M., et al. 1988. N-cadherin, NCAM, and integrins promote retinal neurite outgrowth on astrocytes in vitro. *J. Cell Biol.* 107:1177–1187.

PATEL, N. H., P. M. SNOW, and C. S. GOODMAN. 1987. Characterization and cloning of fasciclin III: a glycoprotein expressed on a subset of neurons and axon pathways in *Drosophila*. *Cell* 48:975–988.

RAPER, J. A., M. J. BASTIANI, and C. S. GOODMAN. 1984. Pathfinding by neuronal growth cones in grasshopper embryos. *J. Neurosci.* 4:2329–2345.

REICHARDT, L. F., and K. J. TOMASELLI. 1991. Extracellular matrix molecules and their receptors: functions in neural development. *Ann. Rev. Neurosci.* 14:531–570.

Chemoattractants, Repellents, and Directional Outgrowth

BAIER, H., and F. BONHOEFFER 1992. Axon guidance by gradients of a target-drived component. *Science* 255:472–475.

CARONI, P., and M. E. SCHWAB. 1988. Two membrane protein fractions from rat central myelin with inhibitory properties for neurite growth and fibroblast spreading. *J. Cell Biol.* 106:1281–1288.

DAVIES, J. A., et al. 1990. Isolation from chick somites of a glycoprotein fraction that causes collapse of dorsal root ganglion growth cones. *Neuron* 2:11–20.

HEDGECOCK. E. M., J. G. CULOTTI, and D. H. HALL. 1990. The *unc-5*, *unc-6*, and *unc-40* genes guide circumferential migrations of pio-

neer axons and mesodermal cells on epidermis in *C. elegans*. *Neuron* 2:61–85.

KAPFHAMMER, J. P., and J. A. RAPER. 1987. Collapse of growth cone structure on contact with specific neurites in culture. *J. Neurosci.* 7:201–212.

KENNEDY, T. E., et al. 1994. Netrins are chemotropic factors for commissural axons in the embryonic spinal cord. *Cell* 78:425–435.

KOLODKIN, A. L., D. J. MATTHES, and C. S. GOODMAN. 1993. The *semaphorin* genes encode a family of transmembrane and secreted growth cone guidance molecules. *Cell* 75:1389–1399.

LUO, Y., D. RAIBLE, and J. A. RAPER. 1993. Collapsin: a protein in brain that induces the collapse and paralysis of neuronal growth cones. *Cell* 75:217–227.

SCHWAB, M. E., J. P. KAPFHAMMER, and C. E. BANDTLOW. 1993. Inhibitors of neurite growth. *Ann. Rev. Neurosci.* 16:565–595.

SERAFINI, T., et al. 1994. The netrins define a family of axon outgrowth-promoting proteins homologous to *C. elegans* Unc-6. *Cell* 78:409–424.

TESSIER-LAVIGNE, M., and M. PLACZEK. 1991. Target attraction: are developing axons guided by chemotropism? *Trends Neurosci.* 14:303–310.

Formation of the Neuromuscular Junction

ANGLISTER, L., U. J. MCMAHAN, and R. M. MARSHALL. 1985. Basal lamina directs acetylcholinesterase accumulation at synaptic sites in regenerating muscle. *J. Cell Biol.* 101:735–743.

CAMPENELLI, J. T., et al. 1994. A role for dystrophin-associated glycoproteins and utrophin in agrin-induced AChR clustering. *Cell* 77:663–674.

GEE, S. H., et al. 1994. Dystroglycan-α, a dystrophin-associated glycoprotein, is a functional agrin receptor. *Cell* 77:675–686.

HUNTER, D. D., et al. 1989. A laminin-like adhesive protein concentrated in the synaptic cleft of the neuromuscular junction. *Nature* 338:229–234.

RUEGG, M. A., et al. 1992. The agrin gene codes for a family of basal lamina proteins that differ in function and distribution. *Neuron* 8:691–699.

SANES, J. R., L. M. MARSHALL, and U. J. MCMAHAN. 1978. Reinnervation of muscle fiber basal lamina after removal of myofibers. *J. Cell Biol.* 78:176–198.

SEALOCK. R., and S. C. FROEHNER. 1994. Dystrophin-associated proteins and synapse formation: is α-dystroglycan the agrin receptor? *Cell* 77:617–619.

Structure and Function of the Plant Cell Wall

ALONI, R. 1987. Differentiation of vascular tissue. *Ann. Rev. Plant Physiol.* 38:179–204.

CASSAB, G. I., and J. E. VARANER. 1988. Cell wall proteins. *Ann. Rev. Plant Physiol.* 39:321–353.

CRONSHAW, J., W. J. LUCAS, and R. T. GIAQUINTA, eds. 1986. *Phloem Transport*. Liss.

DELMER, D. P. 1987. Cellulose biosynthesis. *Ann. Rev. Plant Physiol.* 38:259–290.

FRY, S. 1986. Cross-linking of matrix polymers in the growing cell walls of angiosperms. *Ann. Rev. Plant Physiol.* 37:165–186.

FRY, S. C. 1988. *The Growing Plant Cell Wall: Chemical and Metabolic Analysis*. Wiley.

LEVY, S., and L. A. STAEHELIN. 1992. Synthesis, assembly, and function of plant cell wall macromolecules. *Curr. Opin. Cell Biol.* 5:856–862.

LEWIS, N. G., and E. YAMAMOTO. 1990. Lignin: occurrence, biogenesis, and biodegradation. *Ann. Rev. Plant Physiol. Plant Mol. Biol.* 41:455–496.

LLOYD, C. W., and R. W. SEAGULL. 1985. A new spring for plant cell biology: microtubules as dynamic helices. *Trends Biochem. Sci.* 10:476–478.

MCNEIL, M., et al. 1984. Structure and function of the primary cell walls of plants. *Ann. Rev. Biochem.* 53:625–663.

TAIZ, L. 1994. Expansins: proteins that promote cell wall loosening in plants. *Proc. Nat'l. Acad. Sci.* 91:7387–7389.

VARNER, J. E., and L.-S. LIN. 1989. Plant cell wall architecture. *Cell* 56:231–239.

WICK, S. M., et al. 1981. Immunofluorescence microscopy of organized microtubule arrays in structurally stabilized meristematic plant cells. *J. Cell Biol.* 89:685–690.

25

Regulation of the Eukaryotic Cell Cycle

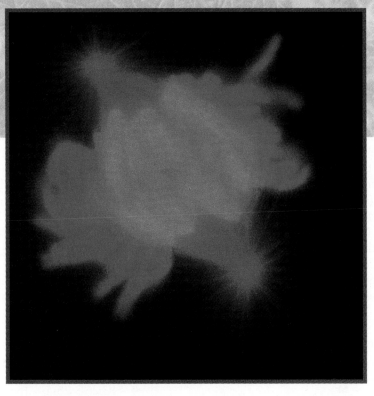

▲ A newt lung cell in metaphase. Microtubules were detected by indirect immunofluorescence (green) and chromosomes were visualized by staining DNA with an intercalating dye (blue).

Cells divide by duplicating their chromosomes and segregating one copy of each duplicated chromosome, as well as providing essential organelles, to each of two daughter cells. Regulation of cell division is critical for the normal development of multicellular organisms. Loss of control ultimately leads to cancer, an all-too-familiar disease that kills one in every six people in the developed world (see Chapter 26). In the late 1980s, it became clear that the molecular processes regulating chromosome replication and cell division are fundamentally similar in all eukaryotic cells. Because of this similarity, research with diverse organisms, each with its own particular experimental advantages, has contributed to a growing understanding of how the molecular events required for cell division are coordinated and controlled. Biochemical and genetic techniques, as well as recombinant DNA technology, have been employed in studying various aspects of the eukaryotic cell cycle.

These studies have revealed that cell replication is primarily controlled by regulating the timing of two critical events in the cell cycle: the initiation of nuclear DNA synthesis and the initiation of mitosis. A small number of protein kinases are the master controllers of these events. They regulate the activities of multiple proteins involved in these complex processes by phosphorylating them at specific regulatory sites, activating some and inhibiting others to coordinate their activities. These protein kinases are heterodimeric proteins. The regulatory subunits are called *cyclins* because their concentrations cycle in phase with the cell cycle. The catalytic subunits are called *cyclin-dependent kinases* (Cdks) because they have no kinase activity unless they are associated with a cyclin. Each Cdk catalytic subunit can associate with different cyclins, and the associated cyclin determines which proteins are phosphorylated by the Cdk-cyclin complex. Cyclins were first discovered in marine invertebrates, while Cdks were first recognized through genetic analysis of the cell cycle in yeasts, and were named after the genes encoding them, *cdc2* in *Schizosaccharomyces pombe* and *CDC28* in *Sacharomyces cerevisiae*.

In this chapter, we first review the stages of the cell cycle and briefly describe the experimental systems that have provided revealing information about cell-cycle regulation in eukaryotes. Following these preliminary sections, we discuss how the major molecular mechanisms for controlling the cell cycle have been unraveled in the past decade or so.

➤ *Phases of the Cell Cycle*

An overview of the cell cycle was presented in Chapter 5, and the times required for each phase in a typical mammalian cell with a generation time of 16 h are shown in Figure 5-48. In Figure 25-1, the cell cycle is depicted in a way that emphasizes the fate of chromosomes during the different phases. In cycling (replicating) somatic cells, chromosomes are replicated during the *S* (synthesis) phase. After a period called G_2, cells begin the complicated process of mitosis, also called the *M phase* (see Figures 5-50 and 23-35). Chromosomes condense during the prophase period of mitosis, probably by successive coiling of loops of the 30-nm chromatin fiber attached to the chromosome scaffold (see Figure 9-54). Sister chromatids, produced by

DNA replication during the S phase, remain attached at the centromere and several other points along their length. During the anaphase portion of mitosis, sister chromatids are segregated to opposite poles of the *mitotic spindle* (see Figure 23-36), segregating one of the two sister chromatids to each daughter cell. In most higher cells, the nuclear envelope breaks down into multiple small vesicles early in mitosis and re-forms around the segregated chromosomes as they decondense at the end of mitosis. The Golgi complex and endoplasmic reticulum also vesiculate during mitosis and re-form in the two daughter cells after cell division. In yeasts and other fungi, the nuclear envelope does not break down. In these organisms, the mitotic spindle forms within the nuclear envelope, which then pinches off, forming two nuclei at the time of *cytokinesis*, the physical process of cell division. Following mitosis, cycling cells enter G_1, the period before DNA synthesis is reinitiated in the S phase.

In vertebrates and diploid yeasts, cells in G_1 have a diploid number of chromosomes ($2n$), one inherited from each parent. In haploid yeasts, cells in G_1 have one of each chromosome ($1n$). Rapidly replicating human cells progress through the full cell cycle in about 24 h: mitosis takes ≈ 30 min; G_1, 9 h; the S phase, 10 h; and G_2, 4.5 h. In contrast, the full cycle takes only ≈ 90 min in rapidly growing yeast cells.

Postmitotic cells in multicellular organisms can "exit" the cell cycle and remain for days, weeks, or in some cases

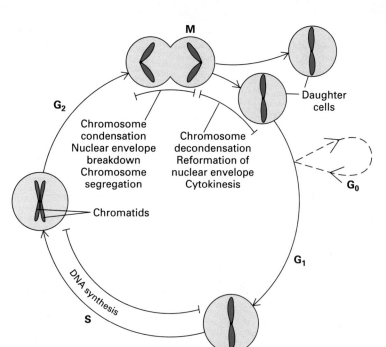

◄ FIGURE 25-1   The fate of a single parental chromosome is depicted throughout the eukaryotic cell cycle. Although chromosomes condense only during mitosis, they are shown in condensed form to emphasize the number of chromosomes at different cell-cycle stages. The nuclear envelope is not depicted. Following mitosis (M), daughter cells contain $2n$ chromosomes in diploid organisms and $1n$ chromosomes in haploid organisms including yeasts maintained in the haploid state. In proliferating cells, G_1 is the period between "birth" of a cell following mitosis and the initiation of DNA synthesis, which marks the beginning of the S phase. Most nonproliferating cells in vertebrates leave the cell cycle in G_1, entering the G_0 state. At the end of the S phase, cells enter G_2 containing twice the number of chromosomes as G_1 cells ($4n$ in diploid organisms). The end of G_2 is marked by the onset of mitosis, during which numerous events leading to cell division occur.

(e.g., nerve cells and cells of the eye lens) even the lifetime of the organism without proliferating further. Most postmitotic cells in vertebrates exit the cell cycle in G_1, entering a phase called G_0 (see 25-1). Postmitotic cells returning to the cell cycle enter into the S phase; this reentry is regulated, thereby providing control of cell proliferation.

➤ Experimental Systems in Cell-Cycle Research

For biochemical studies of the cell cycle, the eggs and early embryos of amphibians and marine invertebrates are particularly suitable. In these organisms, multiple synchronous cell cycles follow fertilization of a large egg. By isolating large numbers of eggs from females and fertilizing them simultaneously by addition of sperm (or treating them in ways that mimic fertilization), researchers can obtain extracts for analysis of proteins and enzymatic activities that occur at specific points in the cell cycle.

The budding yeast *S. cerevisiae* and the distantly related fission yeast *Schizosaccharomyces pombe* have been especially useful for isolation of mutants that are blocked at specific steps in the cell cycle or that exhibit altered regulation of the cycle. In both of these yeasts, temperature- and cold-sensitive mutants with defects in specific proteins required to progress through the cell cycle are readily recognized microscopically and therefore easily isolated. *S. cerevisiae* daughter cells form from a growing bud, whose size relative to the parental cell increases during the cell cycle (see Figure 5-52). Mutant *S. cerevisiae* cells with a cell-cycle defect are easily identified because at the nonpermissive temperature they are arrested in the budding process (Figure 25-2a). Such cells are called *cdc* (cell division cycle) mutants. *S. pombe* cells, in contrast, increase in length and then divide in the middle to form daughter cells. In this yeast, cdc mutants grow without dividing and form enormously elongated cells at the nonpermissive temperature. Other *S. pombe* mutants, called *wee* (from the Scottish word for small), divide before the parental cell has grown to the normal length, forming cells that are shorter than normal.

Temperature- and cold-sensitive mutations that block progression through the cell cycle at the nonpermissive temperature obviously prevent colony formation from a single haploid yeast cell. The wild-type alleles of recessive temperature- and cold-sensitive mutant alleles can be isolated readily by transforming haploid mutant cells with a plasmid library prepared from wild-type cells and then plating at the nonpermissive temperature (Figure 25-2b). Complementation of the recessive mutation by the wild-type allele on one of the plasmids in the library allows a transformed mutant cell to grow into a colony; the plasmid bearing the wild-type allele can then be recovered from those cells. Because many of the proteins that regulate the cell cycle are highly conserved, human cDNAs cloned into yeast expression vectors often can complement yeast cell-cycle mutants, leading to the rapid isolation of human genes encoding cell-cycle control proteins.

➤ Control of Entry into and Exit from Mitosis

The first evidence that trans-acting factors regulate the cell cycle came from cell-fusion experiments with cultured mammalian cells. When interphase cells in the G_1, S, or G_2 stage of the cell cycle were fused to cells in mitosis, their nuclear envelopes broke down and their chromosomes condensed (Figure 25-3). This finding indicates that some trans-acting component or components in the cytoplasm of the mitotic cells forced interphase nuclei to undergo many of the processes associated with mitosis. Similarly, when cells in G_1 were fused to cells in the S phase and the fused cells exposed to radiolabeled thymidine, the label was incorporated into the DNA of the G_1 nucleus, indicating that DNA synthesis began in the G_1 nucleus shortly after fusion. However, when cells in G_2 were fused to S-phase cells, no incorporation of labeled thymidine occurred in the G_2 nuclei. Thus trans-acting factors in an S-phase cell can enter the nucleus of a G_1 cell and stimulate DNA synthesis, but these factors cannot induce DNA synthesis in a G_2 nucleus. Although these cell-fusion experiments demonstrated that diffusable, trans-acting factors control entry into the S and M phases of the cell cycle, other types of experiments were needed to identify these factors. Important progress came from biochemical studies of the cell cycle in developing frog oocytes and early embryos.

The Same Factor Promotes Oocyte Maturation and Mitosis in Somatic Cells

A breakthrough in identification of the factor that induces mitosis came from studies of oocyte development in the frog *Xenopus laevis*. To understand these experiments, we must first lay out the events of oocyte maturation. As oocytes develop in the frog ovary, they replicate their DNA and become arrested in G_2 for 8 months as they grow in size to a diameter of 1 mm, stockpiling all the materials needed for the multiple cell divisions required to generate a swimming, feeding tadpole. When stimulated by a male, an adult female's ovarian cells secrete the steroid hormone progesterone, which induces the G_2-arrested oocytes to enter meiosis I, the first cell division of meiosis (see Figure 5-53b–e). Following this exposure to progesterone, frog oocytes continue through meiosis I, the succeeding inter-

(a) Identification of cdc mutants

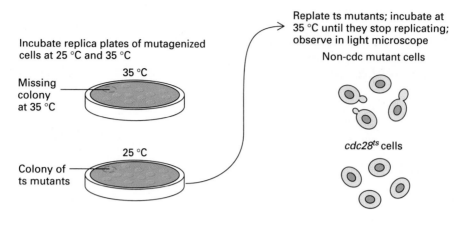

Incubate replica plates of mutagenized cells at 25 °C and 35 °C

35 °C

Missing colony at 35 °C

25 °C

Colony of ts mutants

Replate ts mutants; incubate at 35 °C until they stop replicating; observe in light microscope

Non-cdc mutant cells

$cdc28^{ts}$ cells

(b) Isolation of wild-type *CDC28* gene

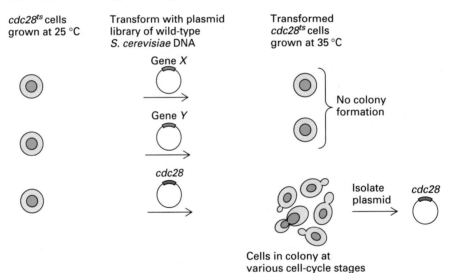

$cdc28^{ts}$ cells grown at 25 °C

Transform with plasmid library of wild-type *S. cerevisiae* DNA

Gene *X*

Gene *Y*

cdc28

Transformed $cdc28^{ts}$ cells grown at 35 °C

No colony formation

Isolate plasmid

cdc28

Cells in colony at various cell-cycle stages

▲ FIGURE 25-2   Identification of *S. cerevisiae* cdc mutants and isolation of the corresponding wild-type *CDC* gene. (a) Temperature-sensitive (ts) and cold-sensitive (cs) mutants are first identified by replica plating mutagenized yeast cells at high (35°C) and low (25°C) temperatures. Wild-type cells form colonies at both temperatures; a ts mutant (shown here) forms colonies at the low temperature, but not at the high one; a cs mutant does the reverse. The ts and cs mutants then are replated and incubated at the nonpermissive temperature. Most of the mutants will not have mutations in cdc genes; at the nonpermissive temperature, such non-cdc mutants stop growing a various points in the cell cycle, yielding arrested cells with buds of varying sizes. Each type of cdc mutant arrests at a specific point in the cell cycle, so that all cells have the same bud phenotype. For example, *cdc28* mutants, shown here, cannot initiate bud formation at the nonpermissive temperature. Figure 8-12 shows a photomicrograph of *cdc13* cells arrested in G_2 with large buds. *S. pombe* cdc mutants can be similarly identified, but at the nonpermissive temperature these mutants form greatly elongated cells. (b) The wild-type version of a *cdc* gene (e.g., *CDC28*) can be isolated by transforming mutant cells cultured at the permissive temperature with a genomic library prepared from wild-type cells and then plating the transformed cells at the nonpermissive temperature. Each transformed cell takes up a single plasmid containing one genomic DNA fragment. Most such fragments include genes (e.g., *X* and *Y*) that do not encode the defective Cdc protein; transformed cells that take up such fragments do not form colonies at the nonpermissive temperature. The rare cell that takes up a plasmid containing the wild-type version of the mutant gene is complemented, allowing the cell to replicate and form a colony at the nonpermissive temperature. Plasmid DNA isolated from this colony carries the wild-type *CDC* gene. The same procedure is used to isolate wild-type cdc genes in *S. pombe*.

(a)

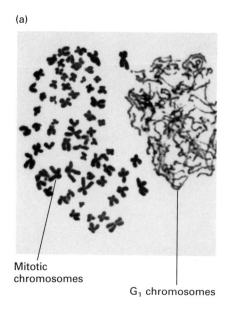

Mitotic
chromosomes

G₁ chromosomes

(b)

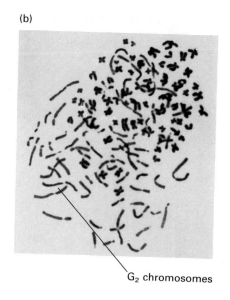

G₂ chromosomes

▲ FIGURE 25-3   Fusion of mitotic cells with interphase cells in G₁ or G₂. In unfused interphase cells, the nuclear envelope is intact and the chromosomes are not condensed, so individual chromosomes cannot be distinguished (see Figure 5-2). In both fused cells shown here, the nuclear envelope of the interphase cells has broken down and is not visible. Chromosomes from the mitotic cells are highly condensed, and the two sister chromatids, joined at the centro- mere, are distinct. (a) Chromosomes from G₁ cells fused to mitotic cells partially condense but homologous chromo- somes do not associate. (b) Chromosomes from G₂ cells fused to mitotic cells also condense, although again not to the same extent as the mitotic chromosomes. Sister chro- matids of the replicated G₂ chromosomes associate along their length and are constricted at the centromere. [From R. T. Johnson and P. N. Rao, 1970, *Biol. Rev.* **46**:97.]

phase, and then arrest during the second meiotic meta- phase. At this stage the cells are called eggs. When fertilized by sperm, the egg nucleus is released from its metaphase arrest and completes meiosis (see Figure 5-53f–j). The re- sulting haploid egg nucleus then fuses with the haploid sperm nucleus, producing a diploid zygote, and the mitotic divisions of early embryogenesis begin.

The process of *oocyte maturation,* from G₂-arrested oocyte to the egg arrested in metaphase of meiosis II, can be studied in vitro by surgically removing G₂-arrested oo- cytes from the ovary of an adult female frog and treating them with progesterone (Figure 25-4a). The microinjection experiment outlined in Figure 25-4b permitted identifica- tion of a factor in egg cytoplasm, named *maturation- promoting factor* (MPF), that can stimulate maturation of oocytes in vitro in the absence of progesterone. As we will see shortly, MPF turned out to be the key factor that regu- lates the initiation of mitosis in all eukaryotic cells.

The microinjection system can be used to assay MPF activity in the cytoplasm of oocytes and eggs at different times during maturation in vitro. As shown in Figure 25-5, untreated G₂-arrested oocytes have low levels of MPF ac-

tivity; treatment with progesterone induces MPF activity as the cells enter meiosis I. As the cells enter the interphase between meiosis I and II, MPF activity falls; it then rises as the cells enter meiosis II and are arrested. Following fertil- ization, MPF activity falls again until the zygote (fertilized egg) enters the first mitosis of embryonic development. All the cells in early frog embryos undergo 12 synchronous cycles of mitosis. MPF assays throughout these cycles showed that MPF activity is low in the interphase periods between mitoses and then rises as the cells enter mitosis.

MPF activity is not unique to frogs but has been found in mitotic cells from all species assayed. For example, cultured mammalian cells can be arrested in mitosis by treatment with compounds (e.g., colchicine) that inhibit assembly of microtubules. When cytoplasm from such mitotically arrested mammalian cells was injected into G₂- arrested *Xenopus* oocytes, it stimulated oocyte maturation; that is, the mammalian somatic mitotic cells exhibited frog MPF activity. This finding suggested that MPF controls the entry of mammalian somatic cells into mitosis as well as the entry of frog oocytes into meiosis. When cytoplasm from mitotically arrested mammalian somatic cells was

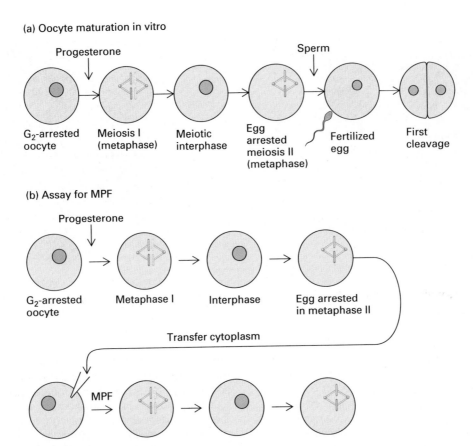

(a) Oocyte maturation in vitro

(b) Assay for MPF

▲ FIGURE 25-4   In vitro maturation of *Xenopus* oocytes and assay of maturation-promoting factor (MPF). (a) Treatment of G$_2$-arrested *Xenopus* oocytes with progesterone stimulates them to proceed through meiosis I, interphase, and the first half of mieosis II before arresting in the metaphase of meiosis II. Two pairs of duplicated homologous chromosomes connected to kinetochore fibers (red) are shown schematically to represent metaphase cells; individual sister chromatids are not shown. (See Figure 5-53 for more detailed illustrations of meiotic stages in mammalian cells; frog oocytes contain much more cytoplasm relative to the nuclear material than do mammalian gametes.) After addition of sperm and fertilization, fertilized eggs complete meiosis II. The resulting haploid egg nucleus fuses with the haploid sperm nucleus to produce a diploid zygote, which undergoes the first of 12 synchronous early embryonic cleavages. (b) When cytoplasm from unfertilized eggs arrested in metaphase of meiosis II is injected into G$_2$-arrested oocytes, the oocytes mature into eggs in the absence of progesterone. This process can be repeated multiple times without further addition of progesterone. [See Y. Masui and C. L. Markert, 1971, *Exp. Zool.* **177**:129.]

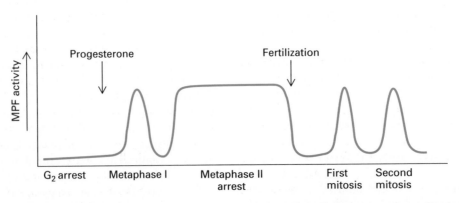

▲ FIGURE 25-5   Oscillation of MPF activity during meiotic and mitotic cell cycles of *Xenopus* oocytes and early frog embryos. Diagrams of the cell structures corresponding to each stage are shown in Figure 25-4a. The MPF activity rises at the beginning of mitosis and falls between mitosis for each of the 12 synchronous cycles that early frog embryos undergo. See text for discussion. [See J. Gerhart, M. Wu, and M. W. Kirschner, 1984, *J. Cell Biol.* **98**:1247; adapted from A. Murray and M. W. Kirschner, 1989, *Nature* **339**:275.]

injected into interphase cells, the interphase cells entered mitosis; that is, their nuclear membranes broke down into small vesicles and their chromosomes condensed. Thus MPF is the trans-acting factor, first revealed in cell-fusion experiments (see Figure 25-3), that promotes entry of cells into mitosis. Conveniently, the acronym MPF also can stand for *mitosis-promoting factor*, a name that denotes the more general activity of this factor.

Because the assay for MPF is cumbersome, several years passed before MPF was purified by column chromatography and the MPF proteins were characterized. As discussed in more detail later, MPF functions as a protein kinase and regulates a number of proteins that mediate mitotic events by phosphorylating them. MPF thus provides yet another example of regulation of the biochemical activity of proteins via phosphorylation. Although many of its significant substrates are still unidentified, MPF phosphorylates histone H1, which is a particularly convenient substrate for assaying MPF protein kinase activity. Years of research have shown that MPF is a heterodimer composed of a catalytic protein kinase subunit and a regulatory subunit that controls which proteins are phosphorylated by the catalytic subunit. Each MPF subunit was recognized through different experimental approaches. First we discuss how the regulatory subunit, called cyclin, was found, and then we show how yeast genetic experiments led to the discovery of the catalytic subunit.

Mitotic Cycling in Early Embryos Depends on Synthesis and Degradation of Cyclin B

Experiments with inhibitors showed that new protein synthesis is required for the increase in MPF during the mitotic phase of each cell cycle in early frog embryos (see Figure 25-5). Following this clue, researchers looked for a protein whose synthesis followed the time course of MPF activity in early embryos, reasoning that such a protein might be a component of MPF. Sea urchin embryos proved to be a convenient experimental system for this search. Female sea urchins produce prodigious numbers of eggs, which can be synchronously fertilized by the addition of sea urchin sperm. As in early frog embryos, the initial cell cycles in the early sea urchin embryo occur synchronously, with all the embryonic cells entering mitosis simultaneously. Because of this property, large numbers of embryos can be cultured, each containing cells progressing through several cell cycles simultaneously. This is an ideal situation for biochemical studies.

In these studies, synchronously fertilized sea urchin eggs were incubated with a radiolabeled amino acid and samples were removed every 10 min. Protein was isolated from each sample and analyzed by gel electrophoresis followed by autoradiography. The amount of radiolabel in the vast majority of proteins increased steadily through several cell cycles. However, one protein peaked in intensity at each mitosis, fell abruptly during anaphase of each mitosis and then slowly accumulated during the next interphase to peak at the beginning of the following mitosis. Careful analysis of the synthesis of this protein, called *cyclin*, showed that it is synthesized continuously during the embryonic cell cycles and abruptly destroyed at the onset of anaphase. Similar *mitotic cyclins* have now been found in all eukaryotic cells analyzed. As we shall see in a later section, several types of cyclins are found in eukaryotic cells. The critical cyclin abruptly destroyed at anaphase in early sea urchin embryos is called *cyclin B*.

In subsequent experiments, a cDNA clone for sea urchin cyclin B was used to isolate a homologous cyclin B cDNA from *Xenopus laevis*. Western blotting of MPF purified from *Xenopus* eggs (see Figure 3-41), using antibody prepared against the protein encoded by cyclin B cDNA, showed that one subunit of MPF is indeed cyclin B. The other subunit is the protein kinase catalytic subunit, first identified in genetic experiments with yeasts discussed in a later section.

Cycling Extracts from Xenopus Eggs Some unusual aspects of the rapid cell cycles in early animal embryos provided a way to study the role of mitotic cyclin in controlling MPF activity. Of particular importance, in the 12 rapid, synchronous cell cycles that occur following fertilization of *Xenopus* eggs, the G_1 and G_2 periods are minimized, and the cell cycle consists of alternating M and S phases. Once mitosis is complete, the early embryonic cells proceed immediately into the S phase, and once DNA replication is complete, the cells progress almost immediately into the next mitosis.

Remarkably, the oscillation in MPF activity that occurs as early frog embryos enter and exit mitosis (see Figure 25-5) is observed even when the nucleus is removed from a fertilized egg. This finding shows that a *cell-cycle clock* operates in the cytoplasm of early frog embryos completely independently of nuclear events. This phenomenon occurs only in synchronously dividing cells of early animal embryos. No transcription occurs during these rapid cell cycles, and all the cellular components required are stored in the unfertilized egg. In somatic cells generated later in development and in yeasts considered in later sections, specific mRNAs must be produced at particular points in the cell cycle for progress through the cycle to proceed. But in early animal embryos, all the mRNAs necessary for the early cell divisions are present in the unfertilized egg.

Extracts of frog eggs that can support many cell cycle events are prepared as outlined in Figure 25-6. These extracts contain all the materials required for multiple cell cycles, including the enzymes and precursors needed for DNA replication, the histones and other chromatin proteins involved in assembling the replicated DNA into chromosomes, and the proteins and lipids required in formation of the nuclear envelope. These egg extracts also synthesize proteins encoded by mRNAs in the extract, including cyclin B.

To test the function of mitotic cyclin, researchers added chromatin prepared from interphase frog sperm to a *Xenopus* egg extract. After addition of sperm chromatin, a nuclear envelope develops around it, forming a haploid nucleus. Following formation of a nuclear envelope, the sperm DNA replicates one time. Following DNA replication, the sperm chromosomes condense and the nuclear envelope breaks down into vesicles, just as it does in intact cells entering mitosis. About 10 min after the nuclear envelope breaks down, cyclin B suddenly is degraded, as it is in intact cells during anaphase. Following cyclin B degradation, the sperm chromosomes decondense and a nuclear envelope re-forms around them, as in an intact cell at the end of mitosis. After about 20 min the cycle begins again. DNA within the nuclei formed after the first mitotic period (now 2*n*) replicates, forming 4*n* nuclei. Cyclin B, synthesized from the cyclin B mRNA present in the extract, accumulates. As cyclin B approaches peak levels, the chromosomes condense once again, the nuclear envelopes break down, and about 10 min later cyclin B is once again suddenly destroyed. These remarkable extracts can mediate several of these cycles, which mimic the rapid synchronous cycles of an early frog embryo.

Dependence of MPF Activity on Cyclin B In the experimental system described above, MPF protein kinase activity, assayed by its phosphorylation of histone H1, was found to rise and fall in synchrony with the concentration of cyclin B (Figure 25-7a). The early events of mitosis—chromosome condensation and nuclear envelope breakdown—occurred when MPF protein kinase activity reached its highest levels in parallel with the rise in cyclin B concentration. Addition of cycloheximide, an inhibitor of protein synthesis, prevented cyclin B synthesis and also prevented the rise in MPF activity, chromosome condensation, and nuclear envelope breakdown. These observations, although consistent with the model that cyclin B controls MPF activity and early mitotic events, were not conclusive because many proteins are synthesized in the cycling egg extracts. Proof that cyclin B is the key protein whose synthesis is required for cycling came from a bold extension of the initial experiment.

All mRNAs in the egg extract were degraded by digestion with a low concentration of RNase, which then was inactivated by addition of a specific inhibitor. This treatment destroys mRNAs without affecting the tRNAs and rRNAs required for protein synthesis. When sperm chromatin was added to the RNase-treated extracts, nuclear envelopes assembled around the sperm chromatin and the resulting 1*n* nuclei replicated their DNA, but the increase in MPF activity and the early mitotic events (chromosome condensation and nuclear envelope breakdown), which the untreated extract supports, did not occur (Figure 25-7b). In the next part of the experiment, cyclin B mRNA, tran-

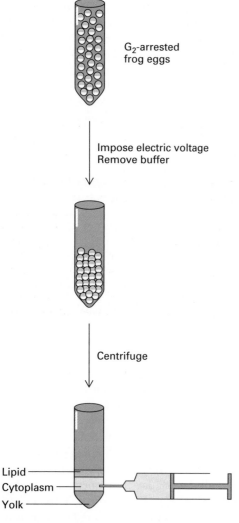

Preparation of cycling extracts from activated frog eggs

G₂-arrested frog eggs

Impose electric voltage
Remove buffer

Centrifuge

Lipid
Cytoplasm
Yolk

▲ FIGURE 25-6 Large numbers of frog eggs are prepared by treating isolated G₂-arrested oocytes with progesterone (see 25-4a). Fertilization normally starts the arrested eggs cycling again by inducing an influx of calcium ions. In this procedure, the eggs are activated with a pulse of electrical voltage, which drives calcium ions into the cells. *Xenopus* eggs are asymmetric with one hemisphere (light orange) having more yolk granules than the other. The eggs are crushed by centrifugation, which separates the contents into three layers: lipid droplets on top; yolk particles on the bottom; and cytoplasm in the middle. The cytoplasmic layer is recovered. [Adapted from A. Murray and T. Hunt, 1993, *The Cell Cycle: An Introduction*, W. H. Freeman and Company.]

(a) Untreated extract

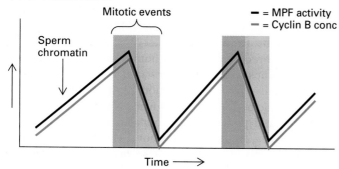

(b) RNase-treated extract

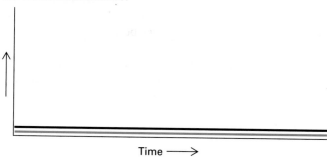

(c) RNase-treated extract + wild-type cyclin B mRNA

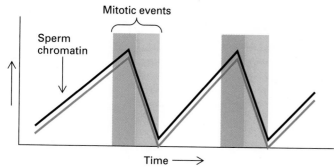

(d) Nondegradable cyclin B

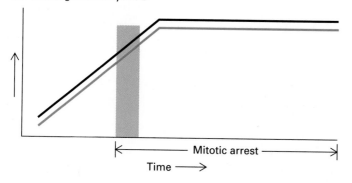

◄ FIGURE 25-7 Experimental demonstration that cyclin B is the only protein whose synthesis is required for the rise in MPF activity that induces entry into mitosis and that its degradation is required for the decrease in MPF activity leading to exit from mitosis. In all cases, MPF activity and cyclin B concentration were determined at various times after addition of sperm chromatin to a *Xenopus* egg extract treated as indicated. Microscopic observations determined the occurrence of early mitotic events (blue shading), including chromosome condensation and nuclear envelope breakdown, and of late events (yellow shading), including chromosome decondensation and nuclear envelope reformation. (a) With untreated egg extracts, cyclin B levels rose and fell in parallel with MPF activity. Early mitotic events coincided with the rise in MPF activity, and late events coincided with the precipitous fall in MPF activity. (b) In an RNase-treated extract depleted of all mRNAs, no cyclin B was synthesized, and no oscillation in MPF activity or mitotic events occurred. (c) Addition of cyclin B mRNA to the RNase-treated extract restored the oscillations in MPF activity and mitotic events. Since cyclin B was the only protein synthesized under these conditions, it clearly is the key protein whose synthesis is required for the rise in MPF activity. (d) When a mutant mRNA encoding a nondegradable cyclin B was added to the RNase-treated extract, the concentration of cyclin B rose continuously with MPF activity. The added sperm nuclei underwent chromosome condensation and nuclear envelope breakdown but did not proceed to late mitotic events, a situation referred to as mitotic arrest. This finding shows that exit from mitosis depends on degradation of cyclin B. [See A. W. Murray, M. J. Solomon, and M. W. Kirschner, 1989, *Nature* **339**:275–280; adapted from A. Murray and T. Hunt, 1993, *The Cell Cycle: An Introduction,* W. H. Freeman and Company.]

scribed in vitro from cloned cyclin B cDNA with bacteriophage RNA polymerase (see Figure 7-36), was added to the RNase-treated egg extract and sperm chromatin. This restored the oscillations in MPF activity in parallel with the rise and fall in the cyclin B level and the characteristic early and late mitotic events as observed with the untreated egg extract (Figure 25-7c). Since cyclin B is the only protein synthesized under these conditions, these results demonstrate that it is the crucial protein whose synthesis is required to regulate MPF activity and the cycles of chromosome condensation and nuclear envelope breakdown mediated by cycling egg extracts.

Proteosome Degradation of Polyubiquitinated Cyclin In the experiments described above, chromosome decondensation and nuclear envelope formation (late mitotic events) coincided with decreases in MPF activity and the cyclin B level. Before considering whether degradation of cyclin B is required for exit from mitosis, we describe the mechanism of cyclin degradation. Animal cells actually contain three cyclins that can function like cyclin B to stim-

ulate *Xenopus* oocyte maturation, cyclin A (which was the first cyclin shown to have this function), and two closely related cyclin Bs. Sequencing of cDNAs encoding several mitotic cyclins from various eukaryotes has shown that all the encoded proteins contain a homologous sequence near the N-terminus called the *destruction box* (Figure 25-8a). In intact cells, cyclin degradation begins at the onset of anaphase, the period of mitosis when sister chromatids are separated and pulled toward opposite spindle poles (see Figure 5-50e). As mentioned above, the cyclin B synthesized in *Xenopus* egg extracts also is periodically degraded. However, mutant forms of cyclin B that lack the destruction box are not degraded.

Further biochemical studies with *Xenopus* egg extracts have shown that wild-type mitotic cyclins are posttranscriptionally modified in a way that marks them for degradation, whereas mutant cyclins lacking the destruction box are not modified. This modification involves the covalent attachment of multiple copies of a highly conserved, 76-residue protein called *ubiquitin*. Mitotic cyclins provide just one example of *polyubiquitination*, a widely occurring

(a) Mitotic cyclin destruction box

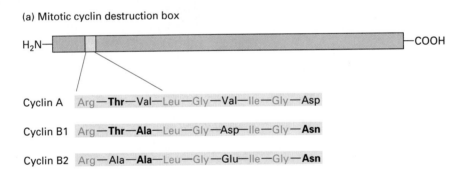

(b) Polyubiquitination of mitotic cyclin

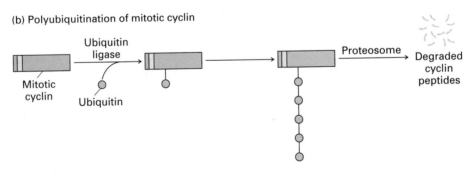

▲ **FIGURE 25-8** (a) Diagram of mitotic cyclin structure showing position of the destruction box (yellow) near the N terminus. The destruction-box sequences of *Xenopus* mitotic cyclins are shown with amino acid residues conserved in all three proteins in red; residues conserved in two of the three proteins in boldface. *Xenopus* and other vertebrates contain two closely related, functionally equivalent B-type cyclins called B1 and B2. Cyclin A, a third mitotic cyclin discussed later, also is present at low levels during early mitosis. The mitotic cyclins are 60–63 kDa and contain ≈550–580 residues. (b) At the onset of anaphase, mitotic cyclins are polyubiquitinated on lysine residues C-terminal to the destruction box. The polyubiquitinated protein then is rapidly degraded by a multiprotein proteosome complex. [See M. Glotzer, A. W. Murray, and M. W. Kirschner, 1991, *Nature* **349**:132; adapted from A. Murray and T. Hunt, 1993, *The Cell Cycle: An Introduction*, W. H. Freeman and Company.]

process for marking proteins for rapid degradation in eukaryotic cells. Polyubiquitination requires two proteins; a *recognition protein* that binds the substrate protein to be degraded and the enzyme *ubiquitin ligase*, which adds ubiquitin to the side-chain amino group of lysine residues in the substrate protein. Each ubiquitin covalently attached to a lysine residue of the substrate protein is further ubiquitinated at a lysine residue in the ubiquitin sequence itself.

Polyubiquitinated proteins are degraded rapidly by multiprotein complexes of proteolytic enzymes called *proteosomes*. This degradation mechanism is regulated by controlling the protein that initially recognizes a specific region in the substrate protein and directs its polyubiquitination by ubiquitin ligase. Different recognition proteins regulate the degradation of different classes of proteins, but the same type of ubiquitin ligase catalyzes addition of ubiquitin to all substrates. In the mitotic cyclins, the recognition protein (as yet unidentified) interacts with the destruction box, and ubiquitin is added to lysine residues on the C-terminal side of the destruction box (Figure 25-8b). Since a cyclin mutant with a deletion of the destruction box is not polyubiquitinated at anaphase, it is not rapidly degraded.

Now we can return to the question of whether degradation of cyclin B is required for exit from mitosis. To answer this question, RNase-treated *Xenopus* egg extracts and sperm chromatin were mixed, as described earlier, but in this case a mutant mRNA encoding cyclin B that lacked the destruction box was added. As shown in Figure 25-7d, MPF activity increased in parallel with the level of the mutant cyclin B, triggering condensation of the sperm chromatin and nuclear envelope breakdown. However, the mutant cyclin B synthesized in this reaction never was degraded as in the reaction with wild-type cyclin B mRNA (see Figure 25-7c). As a consequence, MPF activity continued to increase and the late mitotic events of chromosome decondensation and nuclear envelope re-formation were both blocked. This experiment clearly demonstrates that the fall in MPF activity and exit from mitosis depends on degradation of cyclin B.

The signal that triggers degradation of cyclin B during anaphase is not yet known. According to the favored current model (Figure 25-9), when MPF protein kinase activity reaches a high level, it phosphorylates and thereby activates the cyclin destruction-box recognition protein. Polyubiquitination of cyclin B then occurs, leading to the degradation of cyclin B. Since cyclin B is an essential subunit of MPF, its degradation causes inactivation of the MPF protein kinase activity. The phosphorylated recognition protein is thought to be dephosphorylated by a phosphatase that acts continuously at a slow rate. As a result, the level of phosphorylated, active recognition protein depends on the MPF activity. When MPF activity is high, it

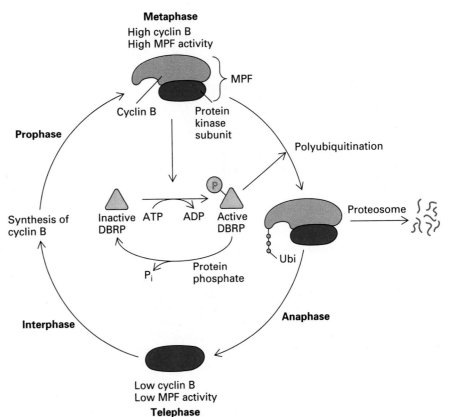

➤ FIGURE 25-9 Proposed model for regulation of mitotic cyclin levels in cycling cells. The cyclin destruction-box recognition protein (DBRP) is activated only when MPF activity is high. The active DBRP binds to cyclin B, and ubiquitin ligase then adds multiple ubiquitin (Ubi) molecules. As the polyubiquitinated cyclin B is degraded, MPF activity declines, triggering the onset of anaphase and telophase. A constitutive phosphatase then dephosphorylates the active DBRP. Following cytokinesis, synthesis of cyclin B occurs in the interphase daughter cells; when the MPF activity is high enough, another mitotic cycle ensues.

can rapidly rephosphorylate the recognition protein as soon as the phosphatase inactivates it, but when MPF activity falls, the activity of the recognition protein also falls because of the unopposed action of the phosphatase. Since cyclin B is synthesized continuously during the cell cycle, this model would account for the rise in the cyclin B levels following mitosis (during interphase) and the sudden fall in cyclin B levels during mitosis.

Mathematical modeling of the overall model depicted in Figure 25-9 shows that it could account for the oscillations in MPF activity that control entry into mitosis in the early *Xenopus* embryo. This model predicts that if the action of the recognition protein that directs polyubiquitination of cyclin B is blocked, mitotic arrest would result, exactly what is observed in *Xenopus* eggs before fertilization (see Figure 25-5).

MPF-Catalyzed Phosphorylation of Nuclear Lamins and Other Proteins Induces Early Mitotic Events

So far, we have seen that a rise in cyclin B levels accompanied by an increase in MPF activity induces entry into mitosis; likewise, the degradation of cyclin B accompanied by a decrease in MPF activity induces exit from mitosis. Presumably, the entry into mitosis is a consequence of the phosphorylation of specific proteins by the protein kinase activity of MPF. For the most part, however, the proteins phosphorylated by MPF in vivo have not been identified. The active phosphorylated forms of these proteins are thought to mediate the many remarkable events of mitosis including chromosome condensation (see Figure 9-54) and formation of the mitotic spindle (see Figures 23-35 and 23-36). The molecular mechanisms of these events have not yet been elucidated. In contrast, the mechanism by which MPF-catalyzed phosphorylation induces breakdown of the nuclear lamina is understood and provides some insight into how other mitotic events may be regulated by MPF.

Depolymerization of Nuclear Lamins As discussed in earlier chapters, the nuclear envelope is a double-membrane extension of the rough endoplasmic reticulum containing many nuclear pore complexes. The lipid bilayer of the inner nuclear membrane is supported by the *nuclear lamina*, a meshwork of lamin filaments located adjacent to the inside face of the nuclear envelope (Figure 25-10). The three nuclear lamins (A, B, and C) present in vertebrate cells belong to a class of proteins called intermediate filaments, which are discussed in Chapter 23. Lamins A and C, which are encoded by the same transcription unit and produced by alternative splicing of a single pre-mRNA, are identical except for a 133-residue region at the C-terminus of lamin A, which is absent in lamin C. Lamin B, which is

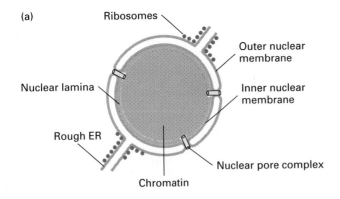

(a)

Ribosomes

Outer nuclear membrane

Inner nuclear membrane

Nuclear lamina

Rough ER

Nuclear pore complex

Chromatin

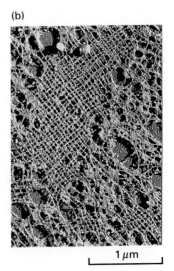

(b)

1 μm

▲ FIGURE 25-10   The nuclear envelope and underlying nuclear lamina. (a) Diagram of the double-membrane nuclear envelope of an interphase cell. The inner nuclear membrane is bound to a protein meshwork, the nuclear lamina (red). (b) Electron micrograph of the nuclear lamina. A nuclear membrane from a hand-dissected *Xenopus* oocyte was fixed to an electron microscope grid and then extracted with a nonionic detergent to remove the lipid membranes and non-polymerized proteins. The nuclear lamina consists of two orthogonal sets of 10-nm-diameter filaments built of lamins A, B, and C. [Part (b) from U. Aebi et al., 1986, *Nature* **323**:560; courtesy of U. Aebi.]

encoded by a different transcription unit, associates with the inner leaflet of the inner nuclear membrane by a hydrophobic isoprenyl anchor. As illustrated in Figure 25-11, the isoprenyl group is covalently attached to the sulfhydryl group of a cysteine near the C-terminus of lamin B. All three nuclear lamins form dimers containing a rodlike α-helical coiled-coil central section and globular head and tail domains; polymerization of these dimers generates the intermediate filaments (see Figure 22-55) that compose the nuclear lamina.

Early in mitosis, MPF phosphorylates specific serine residues in all three nuclear lamins, causing depolymerization of the lamin intermediate filaments. The resulting phosphorylated lamin A and C dimers are released into solution (Figure 25-12), whereas the phosphorylated lamin B dimers remain associated with the nuclear membrane via their isoprenyl anchor. Depolymerization of the nuclear lamins leads to disintegration of the nuclear lamina meshwork and contributes to the breakdown of the nuclear envelope into small vesicles.

To analyze the importance of phosphorylation to the depolymerization of lamin, researchers first subjected the human lamin A gene to site-directed mutagenesis that changed the serines phosphorylated by MPF to alanines, which cannot be phosphorylated. Expression vectors that produced either the wild-type human lamin A protein or the mutant protein then were constructed and separately transfected into cultured hamster cells. The human lamin A produced in these cells was visualized with a fluorescent-labeled monoclonal antibody specific for the human protein. When the transfected cells entered mitosis, the wild-type lamin A depolymerized normally beginning in prophase, whereas the mutant lamin A did not (Figure 25-13).

Phosphorylation of Other Nuclear Proteins The demonstration that nuclear envelope breakdown depends on phosphorylation of nuclear lamins suggests that phosphorylation of other proteins may play a role in other early mitotic events. For example, phosphorylation of histone H1, other chromatin proteins, and the components of the chromosome scaffold may regulate the higher order folding that results in condensation of chromosomes at the beginning of mitosis. Similarly, phosphorylation of microtubule-associated proteins by MPF probably is required for the dramatic changes in microtubule dynamics that result in the formation of the mitotic spindle and asters (Chapter 23). In addition, all vesicular traffic in the cell ceases during mitosis, and the endoplasmic reticulum and Golgi complex break down into small vesicles as the nuclear membrane does. Phosphorylation of proteins associated with these membranous organelles, by MPF or other protein kinases activated by MPF-catalyzed phosphorylation, likely is responsible for these mitotic events as well.

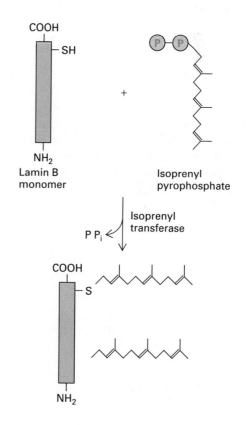

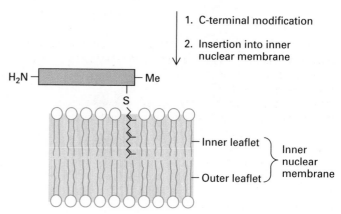

➤ FIGURE 25-11 Modification of lamin B and association with inner leaflet of the inner nuclear membrane. A hydrophobic isoprenyl group, which contains 15 carbon atoms, is covalently attached to the -SH group of a cysteine residue near the carboxyl end of lamin B. Three residues are removed from the C-terminus, which then is modified to a methyl ester. For simplicity, a lamin B monomer is depicted; however, lamin B exists as a homodimer (see Figure 25-12 *bottom*). Each subunit undergoes the reactions shown here, and the modified dimer then associates with the inner nuclear membrane via its hydrophobic anchors. Polymerization into filaments then occurs. [See S. Clarke, 1992, *Ann. Rev. Biochem.* **61**:355.]

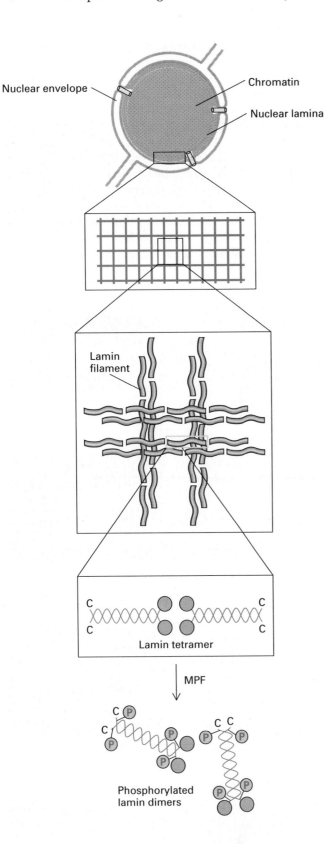

Nuclear envelope

Chromatin

Nuclear lamina

Lamin filament

Lamin tetramer

C   C

C   C

MPF

Phosphorylated lamin dimers

C

C

P

P

P

P

P

C   C

P

P

◄ FIGURE 25-12   Schematic diagram showing depolymerization of the nuclear lamina resulting from phosphorylation of lamin proteins by MPF. Individual lamin filaments are formed by end-to-end polymerization of lamin tetramers, which consist of two lamin dimers. The red circles represent the globular N-terminal domains. Phosphorylation near the ends of the coiled-coil rodlike central section of lamin dimers causes the filaments and tetramers to depolymerize. In the case of lamins A and C, the resulting phosphorylated dimers are not associated with the nuclear envelope, but the lamin B dimers remain attached via their isoprenyl anchor. [Adapted from A. Murray and T. Hunt, 1993, *The Cell Cycle: An Introduction*, W. H. Freeman and Company.]

► FIGURE 25-13   Experimental demonstration that phosphorylation of human nuclear lamin A is required for lamin depolymerization. By site-directed mutagenesis, a mutant lamin A gene was prepared encoding alanines, which cannot be phosphoryated, rather than serines at the sites that are phosphorylated in wild-type lamin A (see Figure 25-12). Expression vectors carrying the wild-type or mutant gene then were separately transfected into cultured hamster cells. In each case, transfected cells at various stages in the cell cycle were stained with a fluorescent-labeled monoclonal antibody specific for human lamin A and with a fluorescent dye that binds to DNA. (a) Interphase: In cells expressing wild-type or mutant lamin A, the bright fluorescence around the periphery of the nucleus indicates that lamin A is assembled into the nuclear lamina. The chromosomes are decondensed as revealed by diffuse DNA staining. (b) Prophase: In a cell expressing wild-type lamin A, the diffuse staining of the cytoplasm shows that depolymerization of lamin A is beginning; DNA staining reveals chromosomes in the early stages of condensation. A cell with mutant lamin A reveals no depolymerization, but the chromosomes are beginning to condense. (c) Metaphase: In a cell with wild-type lamin A, depolymerization is complete, as revealed by diffuse staining throughout the cell. The chromosomes are fully condensed and aligned on the mitotic spindle. In a cell with mutant lamin A, the chromosomes have condensed, but no lamin A depolymerization has occurred. The mutant lamin A continues to form a meshwork, producing a bright edge of fluorescence around the chromosomes caged within. (d) Anaphase: Wild-type lamin A remains depolymerized, and chromosomes are segregating. In a cell with mutant lamin A, the chromosomes are separating but are constrained by the cage of polymerized mutant lamin A. (e) Telophase: Cytokinesis has occurred in a cell with wild-type lamin A. Polymerized lamin A is revealed on the inner surface of the nuclear envelope that has formed around the decondensing chromosomes of the two separated daughter cells. In a cell expressing mutant lamin A, the nuclear lamina still is intact, but separate clusters of decondensing chromosomes are visible. [From R. Heald and F. McKeon, 1990, *Cell* **61**:579.]

Wild-type lamin A

Mutant lamin A

Lamin A stain   DNA stain   Lamin A stain   DNA stain

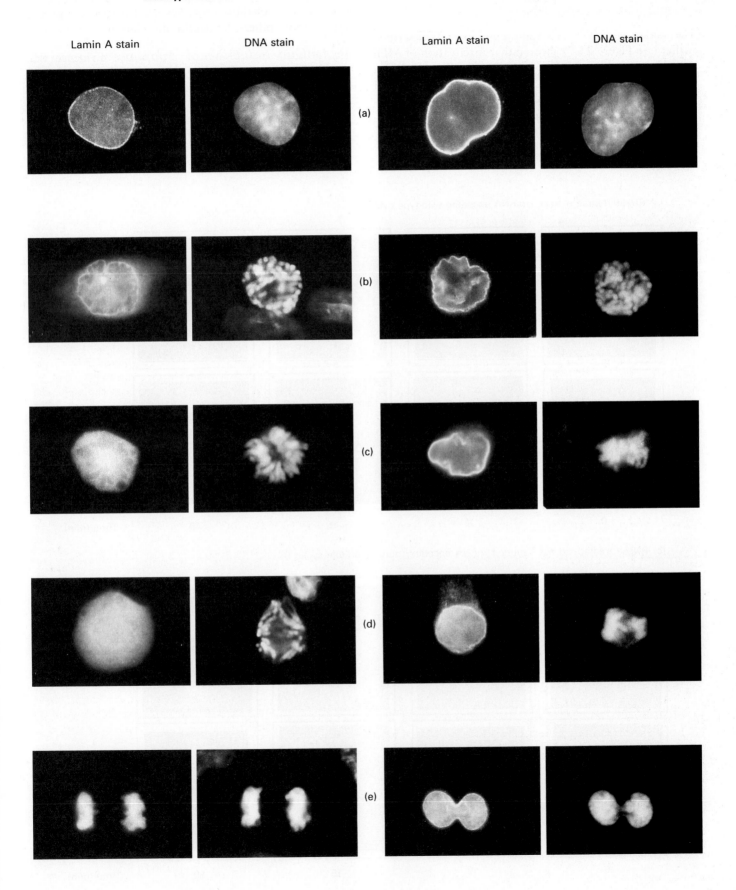

(a)

(b)

(c)

(d)

(e)

Wild-type lamin A

Mutant lamin A

Protein Degradation and Dephosphorylation Trigger Late Mitotic Stages

The studies with *Xenopus* egg cycling extracts described earlier (see Figure 25-7) showed that inactivation of MPF coincides with the late stages of mitosis (anaphase and telophase). During these stages, sister chromatids separate and move to opposite spindle poles, microtubule dynamics return to interphase conditions, the chromosomes decondense, the nuclear envelope re-forms, the endoplasmic re-

ticulum and Golgi complex are remodeled, and cytokinesis occurs. Some of these processes are triggered by protein degradation; others, by dephosphorylation.

Proteolytic Induction of Anaphase   Evidence that anaphase may be induced by the same polyubiquitinating enzymes that target cyclin B for destruction came from additional experiments with RNase-treated *Xenopus* egg extracts. Preparation of the extracts and incubation conditions were such that when frog sperm nuclei and cyclin

(a) RNase treated extract + mRNA encoding wild-type cyclin B

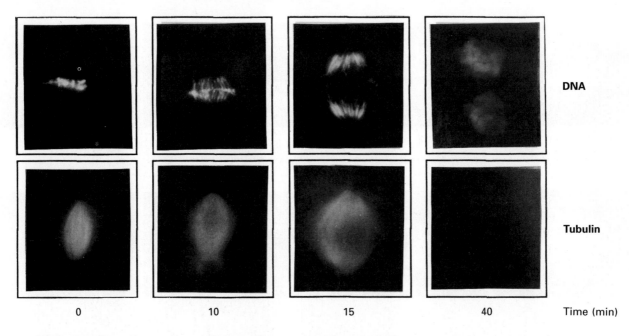

(b) RNase treated extract + mutant mRNA encoding nondegradable cyclin B

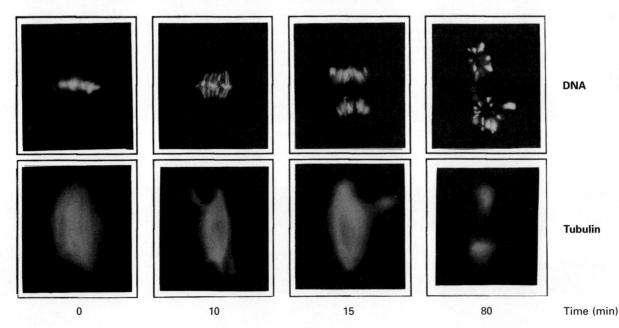

B mRNA were added to these extracts, the sperm chromosomes replicated and condensed and the nuclear envelope broke down as described earlier (see Figure 25-7), but in addition the mitotic spindle apparatus also assembled. Kinetochores of condensed sperm chromosomes attached to microtubules of these spindles and aligned between the spindle poles as they do during metaphase in intact cells (Figure 25-14a, time 0). As the level of cyclin B fell in the reactions, the captured chromosomes moved toward the spindle poles, just as they do during anaphase in intact cells (Figure 25-14a, time 10 and 15 min). Over the next half hour, the spindle depolymerized and the chromosomes decondensed (Figure 25-14a, 40 min). When mRNA encoding a nondegradable cyclin B was substituted for wild-type mRNA, MPF activity remained high as in the experiments described earlier (see Figure 25-7d). As before, chromosome decondensation did not occur, but chromosome segregation was observed to occur normally, indicating that chromosome segregation during anaphase does not require MPF inactivation (Figure 25-14b).

(c) Untreated extract + cyclin B destruction-box peptide

15-min reaction time

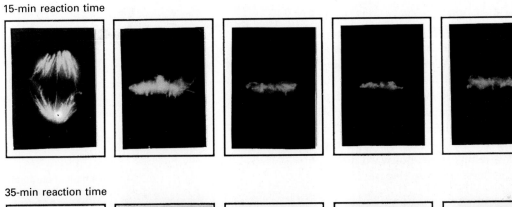

DNA

35-min reaction time

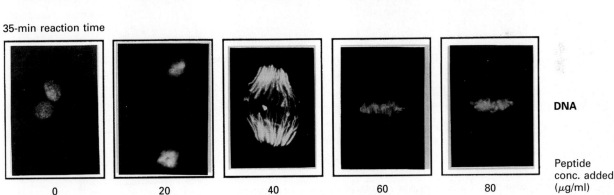

DNA

0	20	40	60	80

Peptide conc. added (μg/ml)

◄ **FIGURE 25-14** Experimental evidence that onset of anaphase depends on polyubiquitination of proteins other than cyclin B. The reaction mixtures contained an untreated or RNase-treated *Xenopus* egg extract and isolated *Xenopus* sperm nuclei, plus other components indicated below. Chromosomes were visualized with a fluorescent DNA-binding dye. Addition of fluorescent rhodamine-labeled tubulin to the reactions permitted observation of the microtubules forming the spindle apparatus. (a,b) The egg extract was treated with RNase to destroy endogenous mRNAs, and an RNase inhibitor was added before addition of mRNA encoding wild-type cyclin B or nondegradable cyclin B. In both samples, the condensed chromosomes and assembled spindle apparatus are visible at 0 min. In the presence of wild-type cyclin B (a), condensed chromosomes attached to the spindle microtubules and segregated toward the poles of the spindle as cyclin B was degraded. By 40 min, the spindle had depolymerized (thus is not visible), and the chromosomes had decondensed. In the presence of nondegradable cyclin B (b), chromosomes segregated to the spindle poles (15 min), as in (a), although the spindle microtubules did not depolymerize and the chromosomes did not decondense even after 80 min. These observations indicate that degradation of cyclin B is not required for chromosome segregation during anaphase, although it is required for depolymerization of spindle microtubules and chromosome decondensation during telophase. (c) Various concentrations of a cyclin B peptide containing the destruction box were added to extracts that had not been treated with RNase; the samples were stained for DNA 15 min and 35 min after formation of the spindle apparatus. The two lowest peptide concentrations delayed chromosome segregation, and the higher concentrations completely inhibited chromosome segregation. In this experiment, the added cyclin B peptide is thought to function as a competitive inhibitor for polyubiquitination of target proteins whose degradation is required for onset of anaphase. [From S. L. Holloway et al., 1993. *Cell* **73**:1393; courtesy of A. W. Murray.]

Researchers then prepared a peptide corresponding to residues 13–110 of cyclin B, which contains the destruction-box sequence. When this peptide was added to a reaction mixture containing untreated egg extract and sperm chromatin, movement of chromosomes toward the spindle poles was greatly delayed at peptide concentrations of 20–40 µg/ml and blocked altogether at higher concentrations (Figure 25-14c). The added excess destruction-box peptide is thought to act as a substrate for the polyubiquitination system and to compete with the normal endogenous target proteins, thereby delaying or preventing their degradation by proteosomes.

The implication of these results is that polyubiquitination and subsequent degradation of proteins other than cyclin B is required to initiate anaphase. Quite possibly these substrates include the proteins that link sister chromatids at the centromere. As discussed in Chapter 23, each sister chromatid is attached through its kinetochore to microtubules (kinetochore fibers), whose opposite ends associate with one of the spindle poles. At metaphase, the spindle is in a state of tension with forces pulling the two kinetochores towards the opposite spindle poles balanced by forces pushing the spindle poles apart. Degradation of the protein cross-bridges that link the sister chromatids at the centromeres would leave the polewards forces on the kinetochores unopposed, resulting in the poleward movement of chromatids that defines anaphase (Figure 25-15).

Phosphatase-Mediated Events Earlier we saw that phosphorylation of nuclear lamins results in their depolymerization, leading to breakdown of the nuclear envelope, a crucial event of early mitosis. Removal of these phosphates coincides with lamin repolymerization and reformation of the nuclear lamina associated with the daughter-cell nuclei later in mitosis (see Figure 25-13e, *left*). Studies with *Xenopus* egg extracts and analyses of various organisms with temperature-sensitive mutations in protein phosphatases indicate that reassembly of the nuclear envelope occurs as depicted in Figure 25-16. Vesicles derived from the breakdown of the nuclear envelope during prophase associate with the surface of the decondensing chromosomes during telophase. These vesicles fuse to form continuous double membranes around each chromosome. Nuclear pore complexes, which are broken down into subpore complexes during prophase, reassemble into the nuclear membrane around each chromosome, forming individual mininuclei called *karyomeres*. Subsequent fusion of the karyomeres associated with each spindle pole generates the two daughter-cell nuclei, each one containing a full set of chromosomes. Lamins A and C appear to be imported through the reassembled nuclear pore complexes during this period and reassemble into a new nuclear lamina. Reassembly of the nuclear lamina in the daughter nuclei probably is initiated on lamin B molecules, which re-

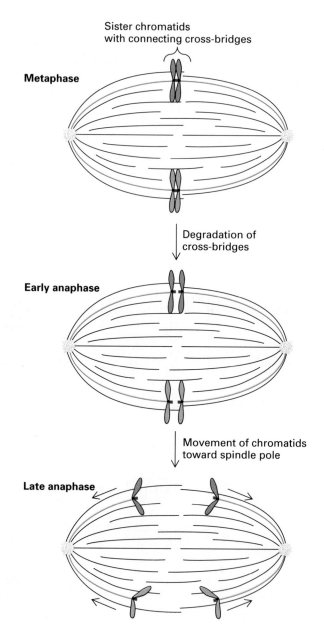

▲ **FIGURE 25-15** Model for induction of anaphase by degradation of protein cross-bridges connecting the centromeres of sister chromatids. Arrows indicate direction of the forces acting on the kinetochores. As long as the cross-bridges (green) are intact, the linked chromatids remain at the equator. Breakdown of the cross-bridge proteins permits the polewise movement of the chromatids. See text for discussion. [See S. L. Holloway et al., 1993, *Cell* **73**:1993.]

main associated with the nuclear envelope membrane throughout mitosis (Figure 25-17).

Cytokinesis, the final step in mitosis, also is dependent on phosphatase action. In previous chapters, we described the role of cortical actin and myosin filaments in forming the mitotic cleavage furrow (see Figure 22-37) and how the mitotic asters appear to control the location of the cleavage

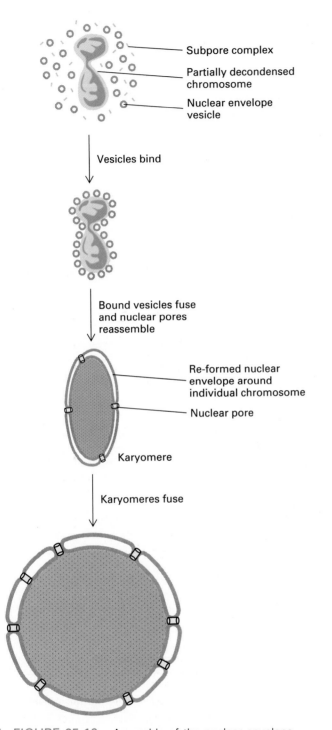

Subpore complex

Partially decondensed chromosome

Nuclear envelope vesicle

Vesicles bind

Bound vesicles fuse and nuclear pores reassemble

Re-formed nuclear envelope around individual chromosome

Nuclear pore

Karyomere

Karyomeres fuse

▲ FIGURE 25-16 Assembly of the nuclear envelope during telophase. Nuclear envelope vesicles, generated by the breakdown of the envelope during prophase, associate with decondensing chromosomes and then fuse. Subpore complexes reassemble into nuclear pores, forming individual mininuclei called karyomeres. The enclosed chromosome further decondenses, and subsequent fusion of the nuclear envelopes of all the karyomeres at each spindle pole forms a single nucleus containing a full set of chromosomes. Reassembly of the nuclear lamina is not shown. [See G. P. Vigers and M. J. Lohka, 1991, *J. Cell Biol.* **112**:545; adapted from A. Murray and T. Hunt, 1993, *The Cell Cycle: An Introduction*, W. H. Freeman and Company.]

furrow (see Figure 23-51). As MPF activity rises early in mitosis, it phosphorylates the myosin light chain, thereby inhibiting the ATPase activity of myosin and its association with actin filaments (Figure 25-18). As MPF is inactivated toward the end of anaphase due to the degradation of cyclin B, protein phosphatases dephosphorylate myosin light chain. As a result, the contractile machinery is activated, the cleavage furrow can form, and cytokinesis proceeds. This regulatory mechanism assures that cytokinesis does not occur until anaphase is completed.

Biochemical Studies with *Xenopus* Egg Extracts Identified Central Role of MPF in Regulating Entry into Mitosis

These studies of MPF in *Xenopus* egg extracts demonstrated that MPF protein kinase activity induces the onset of mitosis, and that inactivation of MPF protein kinase activity by the degradation of cyclin B is required for the late stages of mitosis. Many of the critical proteins regulated by MPF that are required for mitotic processes remain to be identified. However, some MPF substrates have been characterized, and analysis of how their activities are regulated by phosphorylation gives insight into how MPF can coordinate mitotic events. The nuclear lamina breaks down early in mitosis because MPF phosphorylates regulatory sites in the nuclear lamin proteins, causing them to depolymerize. Cytokinesis is blocked early in mitosis because MPF phosphorylates a site in myosin light chain that inhibits its ATPase activity and association with actin filaments. When MPF is inactivated by the degradation of cyclin B at the onset of anaphase, the action of continuously acting phosphatases that remove these regulatory phosphates is unopposed, and the proteins are dephosphorylated, allowing nuclear lamina to re-form in the two daughter cell nuclei and permitting myosin light chain to interact with cortical actin, forming the mitotic cleavage furrow. Further insights into the structure and regulation of MPF, and the control of entry into the S phase that operates in most cells came from genetic studies of the cell cycle in the yeasts *S. cerevisiae* and *S. pombe*.

► Regulation of MPF Activity

The studies with *Xenopus* egg extracts described in the previous section clearly show that continuous synthesis of cyclin B followed by its periodic degradation at anaphase is required for the rapid cycles of mitosis observed in early animal embryos. Identification of the catalytic subunit of MPF and further insight into its regulation came from genetic analysis of the cell cycle in the fission yeast *S. pombe*.

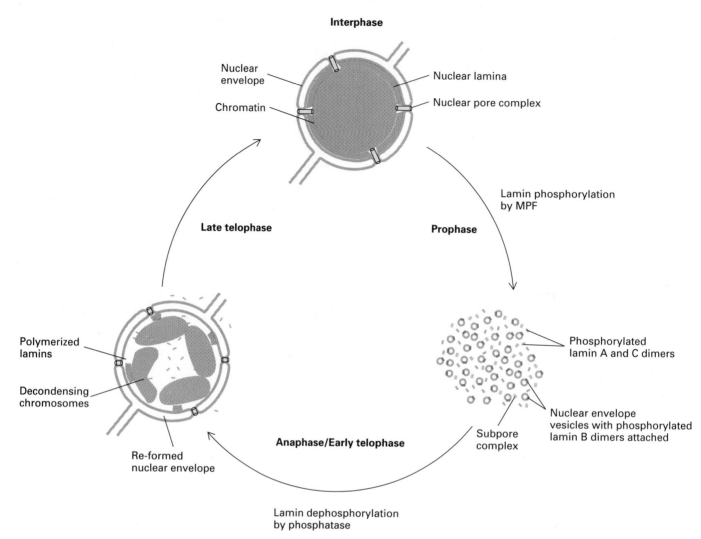

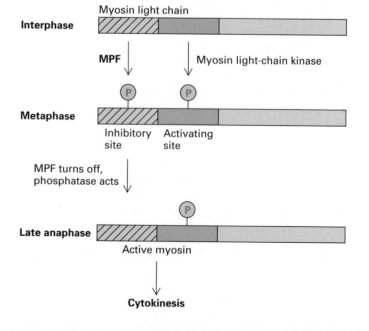

▲ FIGURE 25-17 Overview of depolymerization and polymerization of nuclear lamins during the cell cycle. Phosphorylation of nuclear lamins (red) by MPF during prophase causes lamin depolymerization. Soluble lamin A and C dimers are released, and the nuclear envelope breaks down into small vesicles that remain associated with phosphorylated lamin B dimers. As MPF activity drops during anaphase, the lamin dimers are dephosphorylated. After the nuclear envelope re-forms during telophase (see 25-16), lamins A and C are transported through nuclear pores into the nucleus, where they polymerize with lamin B associated with the inner nuclear membrane, forming a new nuclear lamina around each daughter nucleus.

▶ FIGURE 25-18 MPF phosphorylates inhibitory sites on the myosin light chain early in mitosis. This prevents active myosin light chains, resulting from phosphorylation of the activating site, from interacting with actin filaments and inhibits myosin ATPase activity. When MPF activity falls during anaphase, a constitutive phosphatase dephosphorylates the inhibitory sites, permitting cytokinesis to proceed. (See L. L. Satterwhite et al., 1992, *J. Cell Biol.* **118**:595; adapted from A. Murray and T. Hunt, 1993, *The Cell Cycle: An Introduction*, W. H. Freeman and Company.]

This yeast grows as a rod-shaped cell that divides in the middle during mitosis to produce two daughter cells of equal size (Figure 25-19).

Many temperature- and cold-sensitive mutants of *S. pombe* with conditional defects in the ability to progress through the cell cycle have been isolated (see Figure 25-2). The vast majority of standard temperature- and cold-sensitive mutations cause individual *S. pombe* cells to stop growing, at random lengths within the normal range, when some essential function unrelated to control of the cell cycle becomes limiting. In contrast, cdc and wee mutants have defects in the mechanisms regulating the cell cycle. As a result, cdc mutants form extremely long cells because they fail to divide, and wee mutants form smaller-than-normal cells because they are defective in the proteins that normally prevent cells from dividing when they are too small.

Complementation and recombination analyses (Chapter 8) of *S. pombe* mutants have identified a number of different *cdc* genes, which are designated with individual numbers. Wild-type genes are indicated in italics with a superscript plus sign (e.g., $cdc2^+$); genes with a recessive mutation, in italics with a superscript minus sign (e.g., $cdc2^-$). The protein encoded by a particular gene is desig-

nated by the gene symbol in Roman type with an initial capital letter (e.g., Cdc2).

MPF Catalytic Subunit is Encoded by $cdc2^+$ Gene in *S. pombe*

S. pombe cells with a temperature-sensitive recessive mutation in *cdc2* duplicate their DNA at the nonpermissive temperatures but fail to enter mitosis. These $cdc2^-$ mutants were distinguished from other cdc mutants because dominant mutations at this locus designated $cdc2^D$, gave rise to the wee phenotype (Figure 25-20). Generally, recessive phenotypes result from the absence of wild-type protein function, whereas dominant phenotypes are due to increased protein function, either because of overproduction or lack of regulation. Isolation of these mutants indicates that an absence of Cdc2 activity prevents entry into mitosis, and an excess of Cdc2 activity brings on mitosis earlier than normal. Thus, Cdc2 was identified as a key regulator of entry into mitosis in *S. pombe*. The wild-type $cdc2^+$ gene contained in a *S. pombe* plasmid library was identified by its ability to complement $cdc2^-$ mutants and isolated (see Figure 25-2). Sequencing of the isolated $cdc2^+$

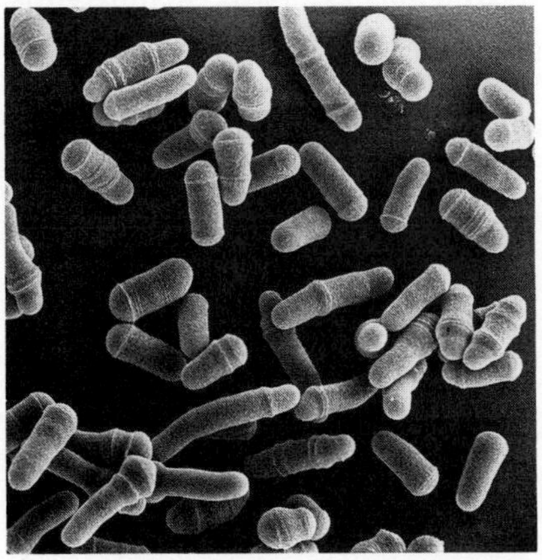

(b)

(a)

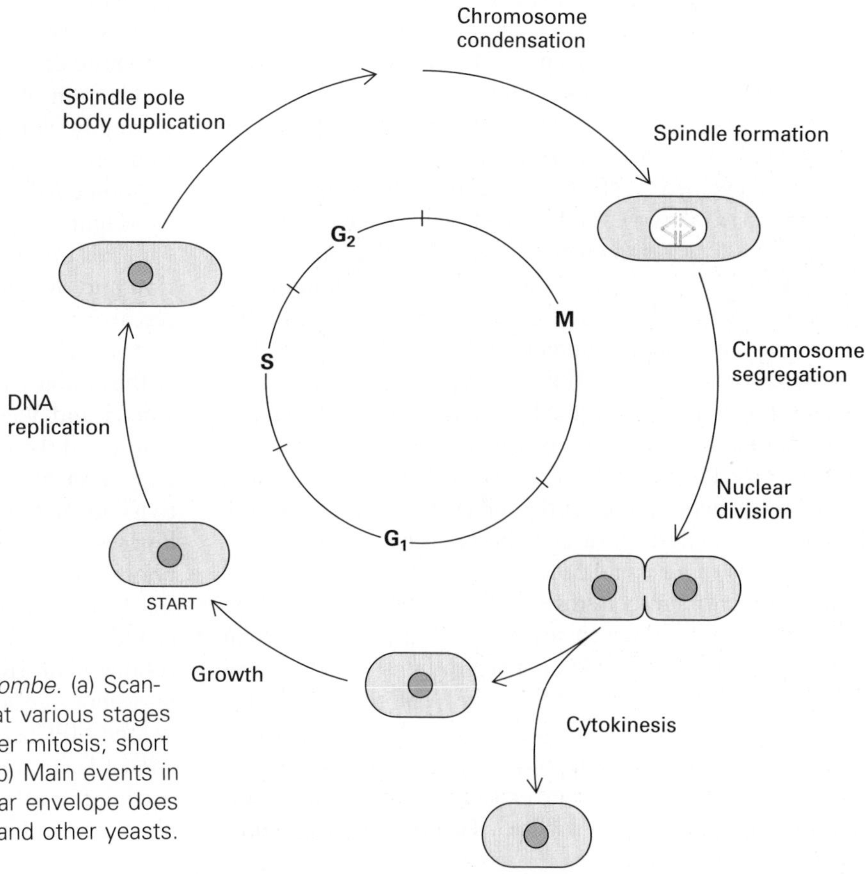

▲ **FIGURE 25-19** The fission yeast *S. pombe*. (a) Scanning electron micrograph of *S. pombe* cells at various stages of the cell cycle. Long cells are about to enter mitosis; short cells have just passed through cytokinesis. (b) Main events in the *S. pombe* cell cycle. Note that the nuclear envelope does not break down during mitosis in *S. pombe* and other yeasts. [Part (a) courtesy of N. Hajibagheri.]

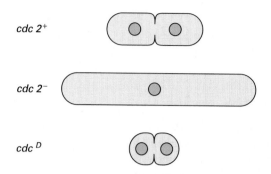

$cdc\,2^+$

$cdc\,2^-$

$cdc\,^D$

◀ FIGURE 25-20   Schematic diagrams of phenotypes of
S. pombe cdc2 mutants. Wild-type cell ($cdc2^+$) is depicted
just before cytokinesis with two normal size daughter cells.
A recessive $cdc2^-$ mutant cannot enter mitosis and appears
as an elongated cell with a single nucleus containing duplicated chromosomes at the nonpermissive temperature. A
dominant $cdc2^D$ mutant enters mitosis prematurely before
reaching normal size in G_2; thus, the two daughter cells just
before cytokinesis are smaller than normal.

gene showed that it encodes a 34-kDa protein with homology to eukaryotic protein kinases.

The finding that Cdc2 is homologous to protein kinases immediately suggested that it might regulate the
function of multiple proteins by phosphorylating them,
and in this way stimulate entry into mitosis. To search for
genes related to *S. pombe cdc2$^+$* in other eukaryotes,
cDNA clones from other organisms were tested for their
abilities to complement *cdc2$^-$ S. pombe* mutants. Remarkably, a cDNA from human cells was isolated encoding a
protein identical to *S. pombe* Cdc2 in 63 percent of its
residues. This was one of the first demonstrations that
complicated functions, such as induction of mitosis, often
are carried out by proteins that are highly conserved during
evolution. The ability of the human protein to perform all
the functions of *S. pombe* Cdc2 also indicates that the
mechanism for controlling entry into mitosis is fundamentally similar among all eukaryotes.

The discovery that Cdc2 from *S. pombe* is a protein
kinase was one of the important clues that led researchers
who later purified MPF from *Xenopus* eggs to test it for
protein kinase activity. As discussed earlier, one of the subunits of MPF was shown to be cyclin B. The other MPF
subunit from *Xenopus* was shown to react with antibody
prepared against the region of Cdc2 that is most highly
conserved between the human and yeast Cdc2 proteins;
moreover the *Xenopus* MPF catalytic subunit is the same
size as the Cdc2 protein. This exciting finding demonstrated that *Xenopus* MPF is a heterodimer composed of
cyclin B and a protein kinase similar to *S. pombe* Cdc2 and
linked the genetic studies of the cell cycle in yeasts with the
biochemical analysis of early embryonic cell cycles in frogs.

Analysis of a second *S. pombe* gene (*cdc13$^+$*), also required for entry into mitosis, revealed that it encodes a
protein with homology to sea urchin and *Xenopus* cyclin
B. Further studies showed that a heterodimer of Cdc13 and
Cdc2 form the *S. pombe* MPF; like *Xenopus* MPF, this
heterodimer has protein kinase activity. Moreover, Cdc2
protein kinase activity was found to rise as *S. pombe* cells
enter mitosis and fall as they exit mitosis in parallel with
the rise and fall in the Cdc13 level. These findings are com-

pletely analogous to the results obtained with *Xenopus*
eggs (see Figure 25-7a).

MPF Catalytic Subunit Contains Activating and Inhibitory Sites That Are Phosphorylated

Analysis of other cdc mutants suggested that proteins encoded by other genes might influence the protein kinase
activity of *S. pombe* MPF (the Cdc2–Cdc13 heterodimer).
For example, temperature-sensitive *cdc25$^-$* mutants do not
enter mitosis at a nonpermissive temperature. On the other
hand, overexpression of *cdc25$^+$* from a plasmid present in
multiple copies per cell decreases the length of G_2 causing
premature entry into mitosis and small cells (Figure 25-
21a). Conversely, *wee1$^-$* mutants exhibit premature entry
into mitosis indicated by their small cell size, whereas overproduction of Wee1 protein increases the length of G_2 resulting in elongated cells. A logical interpretation of these
findings is that Cdc25 protein stimulates the activity of
S. pombe MPF, whereas Wee1 protein inhibits MPF activity (Figure 25-21b).

The interesting phenotypes associated with inactivation and overproduction of Cdc25 and Wee1 proteins led
researchers to isolate and sequence the wild-type *cdc25$^+$*
and *wee1$^+$* genes and then produce the encoded proteins
with suitable expression vectors. The deduced sequences of
Cdc25 and Wee1 and biochemical studies of the proteins
confirmed the inferences from the genetic analyses in spec-
tacular fashion. These studies demonstrated that the activity of *S. pombe* MPF, like that of many other protein
kinases, is regulated by phosphorylation and dephosphorylation at specific regulatory sites.

Cdc2 is active as a protein kinase only when it is associated with a cyclin such as Cdc13. Phosphorylation at one
residue (Thr-161) in the Cdc2 subunit activates MPF and
phosphorylation at another residue (Tyr-15) in Cdc2 inactivates MPF, even when the activating site is phosphorylated. These regulatory phosphorylations of Cdc2 only
occur after it is bound by the mitotic cyclin Cdc13. As
illustrated in Figure 25-22, Wee1 is the protein kinase that

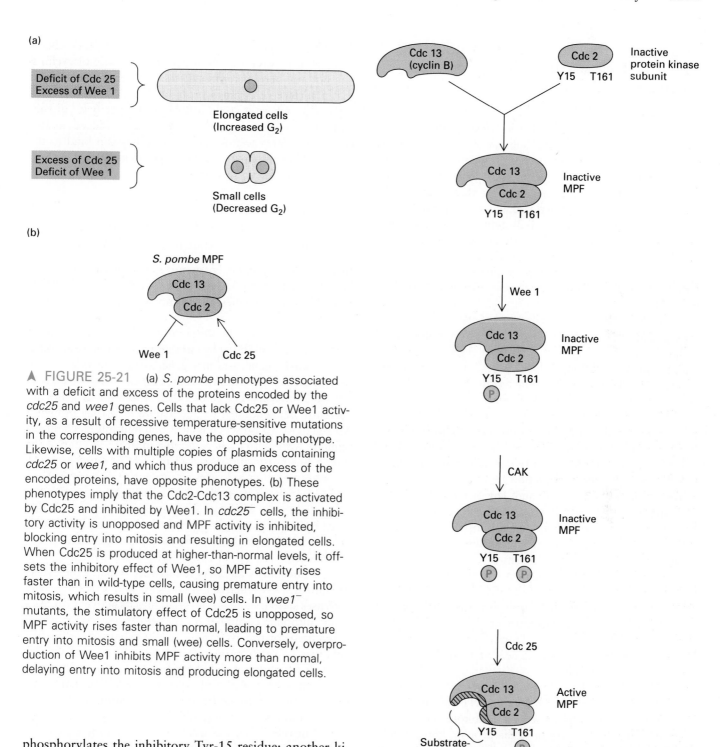

(a)

Deficit of Cdc 25
Excess of Wee 1

Elongated cells
(Increased G₂)

Excess of Cdc 25
Deficit of Wee 1

Small cells
(Decreased G₂)

(b)

S. pombe MPF

Cdc 13 / Cdc 2

Wee 1   Cdc 25

▲ FIGURE 25-21   (a) *S. pombe* phenotypes associated with a deficit and excess of the proteins encoded by the *cdc25* and *wee1* genes. Cells that lack Cdc25 or Wee1 activity, as a result of recessive temperature-sensitive mutations in the corresponding genes, have the opposite phenotype. Likewise, cells with multiple copies of plasmids containing *cdc25* or *wee1*, and which thus produce an excess of the encoded proteins, have opposite phenotypes. (b) These phenotypes imply that the Cdc2-Cdc13 complex is activated by Cdc25 and inhibited by Wee1. In *cdc25⁻* cells, the inhibitory activity is unopposed and MPF activity is inhibited, blocking entry into mitosis and resulting in elongated cells. When Cdc25 is produced at higher-than-normal levels, it offsets the inhibitory effect of Wee1, so MPF activity rises faster than in wild-type cells, causing premature entry into mitosis, which results in small (wee) cells. In *wee1⁻* mutants, the stimulatory effect of Cdc25 is unopposed, so MPF activity rises faster than normal, leading to premature entry into mitosis and small (wee) cells. Conversely, overproduction of Wee1 inhibits MPF activity more than normal, delaying entry into mitosis and producing elongated cells.

phosphorylates the inhibitory Tyr-15 residue; another kinase, designated Cdc2-activating kinase (CAK), phosphorylates the activating Thr-161 residue. The resulting diphosphorylated MPF is still inactive. Finally, Cdc25, which has protein phosphatase activity, removes the phosphate from Tyr-15, yielding an active MPF. Site-specific mutagenesis that changed the Tyr-15 in Cdc2 to a phenylalanine, which cannot be phosphorylated, produced mutants with the wee phenotype, similar to that of *wee1⁻* mutants. Both mutations prevent the inhibitory phosphorylation at Tyr-15, leading to increased MPF activity, resulting in premature entry into mitosis.

▲ FIGURE 25-22   Model for regulation of MPF protein kinase activity in *S. pombe* by Cdc13 (cyclin B), Cdc2-activating kinase (CAK), Wee1, and Cdc25. Wee1 and CAK are protein kinases, and Cdc25 is a protein phosphatase. Once bound by Cdc13, the catalytic Cdc2 subunit can be phosphorylated at two regulatory sites, tyrosine-15 (Y15) and threonine-161(T161). Only when Cdc2 is monophosphorylated at T161 is MPF active. Cdc13 contributes to the specificity of substrate binding, probably by forming part of the substrate-binding surface (cross-hatch), which also includes the inhibitory Y15 residue.

Structure of Human Cyclin-Dependent Kinase 2 Suggests How Phosphorylation Regulates MPF Activity

The experiments described so far demonstrate that the protein kinase activity of the Cdc2 subunit of *S. pombe* MPF is controlled by phosphorylation but provide no information about the regulatory mechanism. Some insight about how Cdc2 may be regulated comes from the three-dimensional structure of the unphosphorylated, inactive form of human cyclin-dependent kinase 2 (Cdk2), which is discussed in a later section. Although the three-dimensional structures of Cdc2 and most other cyclin-dependent kinases have not been determined, their extensive sequence homology with Cdk2 suggests that all these cyclin-dependent kinases have a similar structure.

The three-dimensional structure of inactive Cdk2 complexed with ATP, as determined by x-ray crystallography, is shown in Figure 25-23a. A flexible region of inactive Cdk2, called the T loop, blocks access to the active site where ATP is bound. A threonine residue located at the top of this loop corresponds to the activating site in Cdc2 that is phosphorylated by CAK (see Figure 25-22). The structure of activated Cdk2 phosphorylated at Thr-161 is not available, but comparison to the known structure of the active catalytic subunit of cAMP-dependent protein kinase (cAPK) provides the information needed to suggest how phosphorylations at Tyr-15 and Thr-161 regulate Cdc2. The structures of inactive Cdk2 and the active catalytic subunit of cAPK are shown superimposed in Figure 25-23b. Activation of cAPK occurs by autophosphorylation of a threonine residue whose position in the sequence is equivalent to that of the regulatory Thr-161 in Cdk2. As a result of this phosphorylation, the T loop in cAPK is bent out of the way, exposing the enzyme's active site to protein substrates. By analogy, phosphorylation of Cdc2 Thr-161 is thought to cause the Cdc2 T loop to bend away from the active site, thereby allowing Cdc2 to bind protein substrates at the active site.

The inhibitory Tyr-15 of Cdc2 is in the middle of the region of the protein that binds the ATP phosphates. In vertebrate Cdk2 proteins, a second inhibitory site (Thr-14) is located in the same region of the protein. Phosphorylation of Tyr-15 and Thr-14 in these proteins prevents binding of ATP because of electrostatic repulsions between the phosphates linked to the protein and the phosphates of ATP. Thus, these phosphorylations inhibit protein kinase activity even when the activating site is phosphorylated.

As noted previously, phosphorylation of the Cdc2 regulatory sites occurs only when the protein is bound to cyclin B (Cdc13). In the space-filling model of Cdk2 shown in Figure 25-23c, the residues in the upper right of the molecule, shown in red, are critical for binding cyclin B. The cyclin B–binding site is ideally positioned to influence the conformation of the T loop, thereby stimulating phosphorylation of the activating threonine (yellow) by CAK.

The phosphorylation of Tyr-15 in Cdc2 by Wee1 also is stimulated by cyclin B. Although the structure of cyclin B has not yet been determined, these observations suggest that a surface of cyclin B probably lies close to the Cdc2 active site, where it could influence the specificity of substrate binding (see Figure 25-22, *bottom*). Indeed, as discussed in a later section, different cyclins can bind to the same Cdk, changing its substrate specificity so that it phosphorylates different sets of proteins.

Entry into Mitosis is Controlled by Multiple Mechanisms That Regulate MPF Activity

In wild-type *S. pombe*, entry into mitosis is carefully regulated in order to properly coordinate cell division with cell growth. Further studies of mutants with altered cell cycles have revealed additional regulatory components that influence MPF activity. The Wee1 protein kinase that inhibits MPF is in turn inhibited by another protein kinase encoded by the *nim1*+ gene. Another gene, *mik1*+, encodes a protein kinase, very similar to Wee1, that can also phosphorylate the inhibitory Tyr-15 of Cdc2. Several other genes are currently being studied to understand how they influence MPF activity. At present it is clear that multiple mechanisms regulate MPF activity in *S. pombe* in order to control the timing of mitosis and therefore the size of daughter cells.

Enzymes with activities equivalent to *S. pombe* Wee1 and Cdc25 have been found in cycling *Xenopus* egg extracts. *Xenopus* Wee1 tyrosine kinase activity is high and Cdc25 phosphatase activity is low during interphase, holding the MPF assembled from *Xenopus* Cdc2 and newly synthesized cyclin B in the inactive state with the Cdc2 Tyr-15 phosphorylated. As the extract initiates the events of mitosis, Wee1 activity diminishes and Cdc25 activity increases so that MPF is converted into its active form. As a result, although cyclin B is the only protein whose *synthesis* is required for the cycling of early *Xenopus* embryos, the *activities* of other proteins, including Wee1 and Cdc25, must be properly regulated for cycling to occur. In its active form, Cdc25 is phosphorylated, suggesting that its activity is also controlled by one or more additional protein kinases and phosphatases, additional steps in a kinase cascade, but these enzymes have not yet been identified.

MPF activity also can be regulated by controlling transcription of the genes encoding the proteins that regulate MPF activity. For example, after the initial rapid synchronous cell divisions of the early *Drosophila* embryo, all the mRNAs are degraded, and the cells become arrested in G$_2$. This arrest occurs because the *Drosophila* homolog of Cdc25, called String, is degraded fairly rapidly. Since *string* mRNA also is degraded, synthesis of String ceases. Because of the resulting decrease in String phosphatase activity, MPF is maintained in its inhibited state. The subsequent regulated entry into mitosis by specific groups of cells is triggered by transcription of the *string* gene at specific times in development of the embryo.

(a)

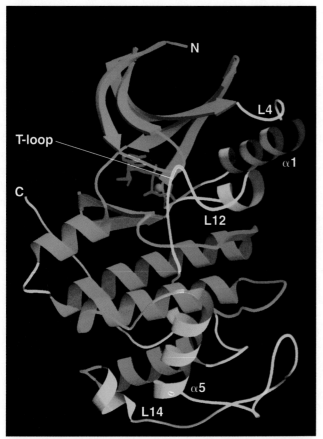

(b)

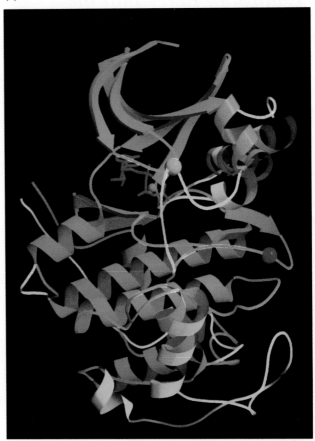

(c)

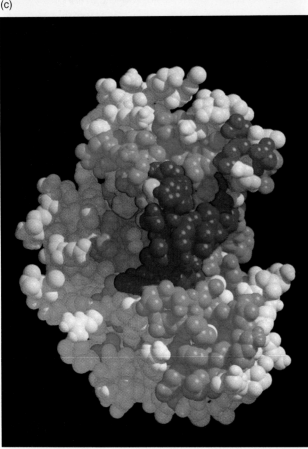

▲ **FIGURE 25-23**   Structural models of inactive, unphosphorylated human Cdk2 and the active catalytic subunit of cAMP-dependent protein kinase (cAPK) based on x-ray crystallographic analysis. (a) A ribbon diagram of Cdk2 in which regions similar to those in cAPK are blue and regions unique to Cdk2 are yellow. ATP is red, and the Mg^{2+} ion bound to the ATP phosphates is green. The T-loop region of Cdk2 is positioned in front of the active site where it blocks access of protein substrates to the γ-phosphate of the bound ATP. (b) Superimposed ribbon diagrams of Cdk2 and cAPK. Common regions are in blue; unique structures in Cdk2 are yellow; and unique structures in cAPK are pink. (c) A space-filling model of Cdk2 viewed from the same angle as the ribbon diagram in (a). Residues that are highly conserved in different cyclin-dependent kinases are pink; the T loop is green with threonine-161 highlighted in yellow. Mutations in the red residues in the upper right corner of the molecule interfere with binding to cyclin B. The visible portion of the bound ATP is shown in dark blue, just to the left of the yellow threonine-161. The γ-phosphate of ATP is hidden from view by the T loop. See text for discussion. [From H. L. DeBondt et al., 1993, *Nature* **363**:595.]

➤ *Control of Entry into the S Phase*

As mentioned earlier, in most vertebrate cells the key decision determining whether or not a cell will divide is the decision to enter the S phase. In most cases, once a vertebrate cell has become committed to enter S phase, it enters S phase several hours later and progresses through the remainder of the cell cycle until it completes mitosis. *S. cerevisiae* cells regulate their proliferation similarly, and much of our current understanding of molecular mechanisms controlling entry into the S phase originates with genetic studies of *S. cerevisiae*.

S. cerevisiae cells replicate by budding (Figure 25-24). The daughter cell initially is smaller than the mother cell

and must grow in size considerably before it attempts to divide. Both mother and daughter cells remain in the G_1 period of the cell cycle while growing, although it takes mother cells a shorter time to reach a size compatible with cell division. By unknown mechanisms, when *S. cerevisiae* cells in G_1 have grown sufficiently, they begin a program of gene expression that leads to entry into the S phase. If G_1 cells are shifted from a rich medium to a medium low in nutrients before they reach a critical size, they remain in G_1 and grow slowly until they are large enough to enter the S phase. However, once G_1 cells reach the critical size, they become committed to completing the cell cycle, entering the S phase and proceeding through G_2 and mitosis, even if they are shifted to a medium low in nutrients. This point in late G_1 of growing *S. cerevisiae* cells when they become irrevocably committed to entering the S phase and traversing the cell cycle is called START (Figure 25-25).

(a)

(b)

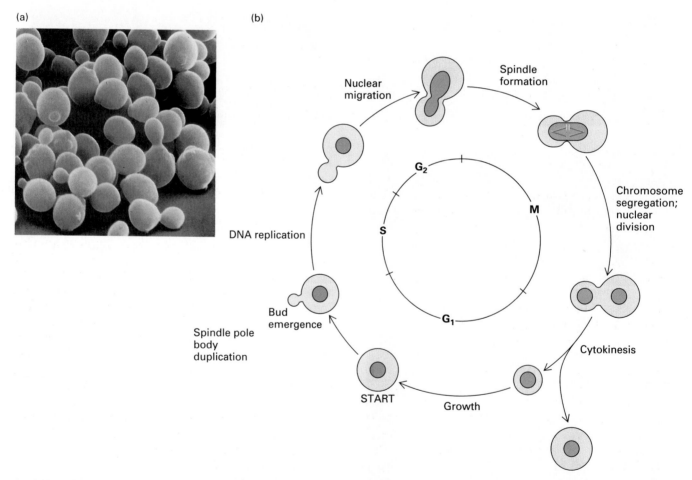

▲ FIGURE 25-24 The budding yeast *S. cerevisiae*. (a) Scanning electron micrograph of *S. cerevisiae* cells at various stages of the cell cycle. The larger the bud, which emerges at the end of the G_1 period, the further along in the cycle the cell is. (b) Main events in *S. cerevisiae* cell cycle. Daughter cells are born smaller than mother cells and must

grow longer in G_1 before they are large enough to enter the S phase. As in *S. pombe*, the nuclear envelope does not break down during mitosis. Unlike *S. pombe* chromosomes, *S. cerevisiae* chromosomes do not condense sufficiently to be visible by light microscopy. [Part (a) courtesy of E. Schachtbach and I. Herskowitz.]

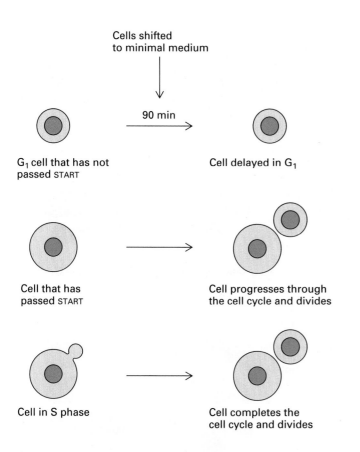

Cells shifted
to minimal medium

90 min

G₁ cell that has not
passed START

Cell delayed in G₁

Cell that has
passed START

Cell progresses through
the cell cycle and divides

Cell in S phase

Cell completes the
cell cycle and divides

◄ FIGURE 25-25   Experimental demonstration of START by shifting *S. cerevisiae* cells from a rich medium to a minimal medium. Cells that have not passed START when shifted to minimal medium are delayed in G₁ until they grow large enough to support production of a daughter cell, which may take several hours. Cells that have passed START when nutrients are reduced experience no delay and progress through the remainder of the cell cycle at the same speed as they would in a rich medium. The entire cycle takes about 90 min in rich medium and about 8 h in minimal medium because of the lengthening of G₁.

S. cerevisiae Cdc28 Is Functionally Equivalent to *S. pombe* Cdc2

The first conditional temperature-sensitive cdc mutants to be isolated were *S. cerevisiae* cells blocked at specific points in the cell cycle. All *S. cerevisiae* cells carrying a particular *cdc* mutation arrest with the same size bud at the nonpermissive temperature (see Figures 25-2 and 8-12). Each type of mutant has a terminal phenotype with a particular bud size: no bud (*cdc28*), intermediate-sized buds, or large buds (*cdc13*). Note that in *S. cerevisiae* wild-type genes are indicated in italic capital letters (e.g., *CDC28*) and recessive mutant genes in italic lowercase letters (e.g., *cdc28*); the corresponding wild-type protein is written in Roman letters with an initial capital (e.g., Cdc28), similar to *S. pombe* proteins.

The phenotype of temperature-sensitive *cdc28* mutants indicates that Cdc28 function is critical for entry into the S phase. When these mutants are shifted to the nonpermissive temperature, they behave like wild-type cells suddenly deprived of nutrients. That is, *cdc28* cells that have grown large enough to pass START at the time of the temperature shift continue through the cell cycle normally and undergo mitosis, whereas those that are too small to have passed START when shifted to the nonpermissive temperature do not enter the S phase even though nutrients are plentiful. Although *cdc28* cells blocked in G₁ continue to grow in

size at the nonpermissive temperature, they do not initiate formation of a bud, synthesize DNA, or duplicate their spindle pole body. In other words, at a temperature that inactivates their Cdc28, these cells cannot pass START and enter the S phase.

The wild-type *CDC28* gene was isolated by its ability to complement mutant *cdc28* cells at the nonpermissive temperature (see Figure 25-2b). Sequencing of *CDC28* showed that the encoded protein is homologous to known protein kinases, and when Cdc28 protein was expressed in *E. coli*, it exhibited protein kinase activity. Actually, Cdc28 from *S. cerevisiae* was the first cell-cycle protein shown to be a protein kinase. When the *S. pombe cdc2⁺* gene was cloned shortly afterwards, it was found to be highly homologous to the *S. cerevisiae CDC28* gene. In fact, the Cdc2 and Cdc28 proteins are functionally analogous, and *S. cerevisiae CDC28* can complement a *S. pombe cdc2⁻* mutant. Each type of yeast contains a single cyclin-dependent kinase, which can substitute for each other. Cdc2 in *S. pombe* and Cdc28 in *S. cerevisiae*.

Even though *cdc2⁺* and *CDC28* encode cyclin-dependent protein kinases, the mutant phenotypes at the nonpermissive temperature differ: *cdc2⁻ S. pombe* cells are arrested in G₂ (see Figure 25-20), whereas *cdc28 S. cerevisiae* cells are arrested in G₁. This difference can be explained in terms of the physiology of the two yeasts. In *S. pombe* cells growing in rich media, cell-cycle control is exerted primarily at the $G_2 \rightarrow M$ transition (i.e., entry to mitosis). In many $cdc2^-$ mutants, including those isolated first, enough Cdc2 activity is maintained at the nonpermissive temperature to permit cells to enter the S phase, but not enough to permit entry into mitosis. Such mutant cells are observed to be arrested in G_2. Completely defective $cdc2^-$ mutants are arrested in both G_1 and G_2. Conversely, as noted earlier, cell-cycle regulation in *S. cerevisiae* is exerted primarily at the $G_1 \rightarrow S$ transition (i.e., entry to the S phase). Therefore, partially defective $cdc28$ cells are arrested in G_1, but completely defective $cdc28$ cells are arrested in both G_1 and G_2. These observations demonstrate that Cdc2 and Cdc28 are required for entry into both the S phase and mitosis. These proteins are now often referred to as *cyclin-dependent kinases*, a more general term.

S Phase–Promoting Factor Consists of a Catalytic Subunit and G_1 Cyclin

By the late 1980s, it was clear that mitosis-promoting factor (MPF) is composed of two subunits: a cyclin-dependent protein kinase (Cdk) and a mitotic cyclin required to activate the catalytic subunit. By analogy, it seemed likely that *S. cerevisiae* contains an *S phase–promoting factor* (SPF) that phosphorylates and regulates proteins required for DNA synthesis. Like MPF, SPF was proposed to be a heterodimer composed of Cdc28 and a cyclin that acts in G_1.

To identify a G_1 cyclin, researchers looked for genes that, when expressed at high levels, could complement certain temperature-sensitive $cdc28$ mutations. The rational of this approach was that some $cdc28$ mutants might be temperature sensitive because of decreased affinity of their Cdc28 for a G_1 cyclin at the nonpermissive temperature. In this case, if the G_1 cyclin was present at high enough levels, it might drive formation of enough SPF, containing G_1 cyclin and the mutant Cdc28, to promote entry into the S phase at the nonpermissive temperature. By using a library of *S. cerevisiae* genomic DNA cloned in a high-copy plasmid vector, researchers isolated two genes, *CLN1* and *CLN2*, that complemented some $cdc28$ mutations in this way (Figure 25-26). *CLN1* and *CLN2* each encodes a protein with an ≈ 100-residue domain exhibiting significant homology with mitotic cyclins from sea urchin, clam, and *S. pombe*.

A third *S. cerevisiae* G_1 cyclin gene was isolated by a different approach that depends on the ability of mating-type pheromones secreted by haploid yeast cells of one mating type (e.g., α) to arrest cells of the other mating type (e.g., **a**) in the G_1 period (see Figure 13-15). In this case, researchers looked for a dominant mutation that would allow haploid **a** cells to grow in the presence of the α factor secreted by α cells, or α cells to grow in the presence of **a** factor. Recessive mutations were avoided in this search because they could simply inactivate any of the numerous proteins involved in the response to pheromone (see Figure 20-38). However, the researchers reasoned that a domi-

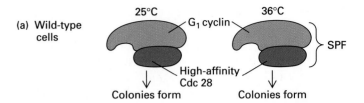

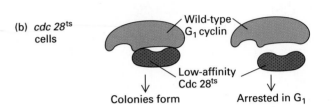

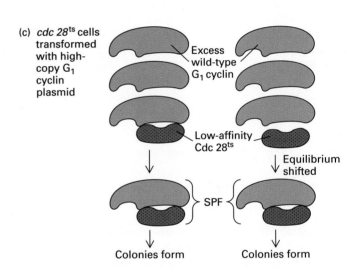

▲ FIGURE 25-26   Proposed interactions between Cdc28 and G_1 cyclins in wild-type and temperature-sensitive (ts) *S. cerevisiae* cells. (a) Wild-type cells produce a high-affinity Cdc28 that associates with G_1 cyclins, forming SPF, at 25 and 36°C. (b) Some $cdc28^{ts}$ mutants are thought to express a Cdc28^{ts} protein with low affinity for G_1 cyclin. At the permissive temperature, enough Cdc28^{ts}–G_1 cyclin (SPF) forms to support growth and colony development, but at the nonpermissive temperature SPF does not form, and consequently no colonies develop. (c) When $cdc28^{ts}$ cells were transformed with a *S. cerevisiae* genomic library cloned in high-copy plasmids, three types of colonies formed at 36°C: one contained a plasmid carrying the wild-type *CDC28* gene; the other two contained plasmids carrying two genes, *CLN1* and *CLN2*, which encode proteins with homology to mitotic cyclins. In transformed cells carrying the *CLN1* or *CLN2* gene, the concentration of G_1 cyclin is thought to be high enough to offset the low affinity of the altered Cdc28^{ts} at 36°C, so enough SPF forms to support entry into the S phase and subsequent mitosis. Untransformed $cdc28$ mutant cells and cells transformed with plasmids carrying other genes are arrested in G_1 and do not form colonies. [See J. A. Hadwiger et al., 1989, *Proc. Nat'l. Acad. Sci. USA* **86**:6255.]

nant mutation might alter a protein normally required to drive cells into the S phase so that it no longer could be controlled by the pheromone signal-transduction pathway. This search led to identification of a mutation ($cln3^D$) that allows **a** cells to grow and form colonies in the presence of the α factor. When the wild-type *CLN3* gene was isolated and sequenced, it was found to encode a protein similar to Cln1 and Cln2.

Gene knockout experiments showed that *S. cerevisiae* cells can grow if they carry any one of the three *CLN* genes, but knockout of all three is lethal. As the data presented in Figure 25-27 show, overproduction of one Cln protein increases the fraction of cells in the S phase and G_2, demonstrating that high levels of the Cdc28–G_1 cyclin complex (SPF) drives cells through START. Moreover, in the absence of any Cln proteins, cells become arrested in G_1, indicating that SPF is required for *S. cerevisiae* cells to enter the S phase. The three related *CLN* genes probably evolved in order to optimize regulation of the G_1 period in response to changing conditions in the environment.

Cln1 and Cln2 increase in concentration during G_1 and fall as cells enter the S phase, in good analogy with the way mitotic cyclins increase in concentration as cells approach mitosis and fall during mitosis. The concentration

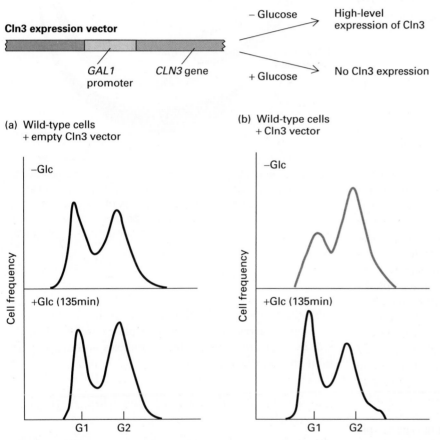

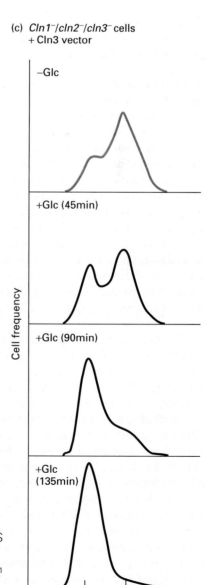

▲ FIGURE 25-27   Experimental demonstration that G_1 cyclins (Cln1, Cln2, and Cln3) regulate entry of *S. cerevisiae* cells into the S phase. The yeast expression vector used in these experiments (*top*) carried the *CLN3* gene linked to the strong *GAL1* promoter, which is turned off when glucose is present in the medium. The distribution of cells in G_1 and G_2 was determined by analysis of their DNA content with a fluorescence-activated cell sorter (see Figure 5-49). (a) Wild-type cells transformed with an empty expression vector displayed the normal distribution of cells in G_1 and G_2 in the absence of glucose (Glc) and after addition of glucose. (b) Wild-type cells transformed with the Cln3 expression vector displayed a higher-than-normal percentage of cells in the S phase and G_2 because overexpression of Cln3 decreased the G_1 period (red curve). When expression of Cln3 from the vector was shut off by addition of glucose, the cell distribution returned to normal (bottom curve). (c) Cells with mutations in all three *CLN* genes and transformed with the Cln3 expression vector also showed a high percentage of cells in S and G_2 in the absence of glucose (red curve). When expression of Cln3 from the vector was shut off by addition of glucose, the cells completed the cell cycle and arrested in G_1 (bottom curve) because cells contained no functional Cln proteins. [Adapted from H. E. Richardson et al., 1989, *Cell* **59**:1127.]

of Cln3, however, does not oscillate greatly as cells progress through the cell cycle. Consequently, the activity of the Cdc28-Cln3 complex is thought to be regulated by posttranslational modifications (e.g., phosphorylation/dephosphorylation at specific sites as occurs with Cdc2) or by binding of a specific inhibitor.

The complexes formed between Cdc28 and the three G_1 cyclins (Cln1, Cln2, and Cln3) in *S. cerevisiae* have protein kinase activity and constitute the hypothesized S phase–promoting factors (SPFs). Although the substrates of these SPFs are not yet known, they must differ from those phosphorylated by Cdc28-mitotic cyclin complexes in *S. cerevisiae,* since that combination of subunits induces mitosis, not the S phase. Because Cdc28 forms the catalytic subunit in both MPF and SPF in *S. cerevisiae,* the cyclin subunit is thought to determine substrate specificity.

Various Cdc28-Cyclin Heterodimers Regulate Progress through the Cell Cycle in *S. cerevisiae*

In addition to the three *CLN* genes described in the previous section, several other cyclin genes have been cloned from *S. cerevisiae.* A current model for how the various cyclin proteins associate with Cdc28 through the course of the cell cycle is depicted in Figure 25-28. According to this model, beginning in G_1 the Cdc28-Cln3 protein kinase stimulates transcription of the *CLN1* and *CLN2* genes. As the levels of Cln1 and Cln2 increase, they form complexes with Cdc28 that promote passage through START and entry into the S phase. In the S phase, two cyclins with homology to sea urchin cyclin B, called Clb5 and Clb6, replace Cln1 and Cln2 as the partners of Cdc28; in the G_2 period, Cdc28 associates with four other B-type cyclins, Clbs 1, 2, 3, and 4. Current research is ongoing to determine the substrates for each of these complexes and how cyclins affect the substrate specificity of Cdc28.

Studies of the G_1 arrest induced by mating-type factors revealed another type of control of Cdc28 activity in *S. cerevisiae.* When **a** cells with a recessive mutation in either the *FAR1* or *FUS3* gene were treated with α factor, they were not arrested in G_1 like wild-type cells. Neither *far1* nor *fus3* mutations, however, interfered with the pheromone signal-transduction pathway that induces transcription of various genes involved in mating. These findings indicate that the Far1 and Fus3 proteins are specifically required for pheromone-induced G_1 arrest but not for any of the multiple steps involved in pheromone-induced transcription. As illustrated in Figure 25-29, Fus3 is a protein kinase that lies at the end of the pheromone signal-transduction cascade. Biochemical studies showed that Fus3 phosphorylates Far1 in response to α factor. Phosphorylated Far1 can bind to the Cdc28-Cln2 complex. This binding is thought to in-

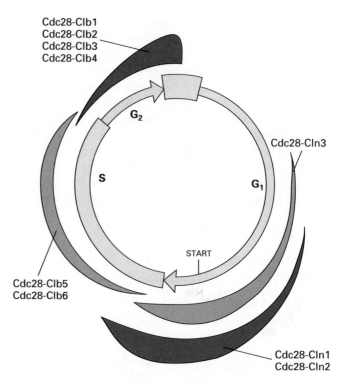

▲ FIGURE 25-28   Model for action of *S. cerevisiae* Cdc28-cyclin complexes through the course of the cell cycle. The green and purple bands indicate the stages when various complexes are present. The band width is approximately proportional to the demonstrated or proposed protein kinase activity of the indicated Cdc28-cyclin complexes.

hibit the S phase–promoting activity of the complex, thus causing G_1 arrest. Binding of phosphorylated Far1 to the Cdc28-Cln1 and Cdc28-Cln3 complexes is thought to inhibit their activity as well.

Cdc28–G_1 Cyclin Complexes May Activate Transcription Factors at START

We saw earlier that MPF regulated various proteins involved in early mitotic events by phosphorylating them. By analogy, SPFs probably induce entry into the S phase by phosphorylating specific proteins required for DNA synthesis. None of these substrates has yet been identified, but they most likely include transcription factors, as well as DNA replication factors discussed in the next section.

Two related transcription factors are activated as *S. cerevisiae* cells pass through START. One of these activates transcription from all *S. cerevisiae* genes encoding proteins required for DNA replication and deoxynucleotide synthesis that have been examined (Table 25-1). This transcription factor acts by binding to a cis-acting control sequence that includes a site for the restriction enzyme *Mlu*I and therefore is known as the *Mlu*I *cell-cycle box*

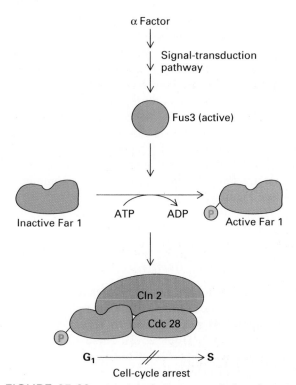

α Factor

Signal-transduction pathway

Fus3 (active)

ATP   ADP

Inactive Far 1   Active Far 1

Cln 2

Cdc 28

G₁ ——————⫽——————→ S

Cell-cycle arrest

▲ FIGURE 25-29   Model of pheromone-induced arrest of *S. cerevisiae* cell cycle by activation of a Cdc28–G₁ Cln complex inhibitor. Binding of active phosphorylated Far1 to an S phase–promoting factor (e.g., Cdc28-Cln2) is thought to inhibit its protein kinase activity, leading to arrest in G₁. See text for discussion. The signal-transduction cascade triggered by yeast mating factors is diagrammed in Figure 20-39. [Adapted from M. Peter, et al., 1993, *Cell* **73**:747.]

TABLE 25-1   *S. cerevisiae* Genes under Control of *Mlu*I Cell-Cycle Box That Are Induced as Cells Enter S Phase

Gene	Gene product
POL1	DNA polymerase I
POL2	DNA polymerase II
POL3	DNA polymerase III
POL30	PCNA
DPB2	DNA polymerase II subunit B
DPB3	DNA polymerase II subunit C
PRI1	DNA primase I
PRI2	DNA primase II
CDC8	Thymidylate kinase
CDC9	DNA ligase
CDC21	Thymidylate synthase
RNR1	Large subunit of ribonucleotide reductase
RFA1 *RFA2* *RFA3*	Replication factor A
CLB5	Cyclin B5
CLB6	Cyclin B6

SOURCE: L. H. Johnston and N. F. Lowndes, 1992, *Nucl. Acids Res.* 20:2403; C. B. Epstein and F. R. Cross, 1992, *Genes & Dev.* 6:1695; E. Schwob and K. Nasmyth, 1993, *Genes & Dev.* 7:1160.

(MCB). This MCB-binding factor, as it is called, is composed of two subunits designated Swi6 and Mbp1. The second transcription factor activated at START is CCBF, which binds to a sequence called the cell-cycle box. CCBF also is made up of two subunits: Swi6, also present in MCB-binding factor, and Swi4, which is closely related to the second subunit in MCB-binding factor. Thus the two transcription factors activated at START are closely related.

As discussed in Chapter 13, cell-cycle boxes are found in the upstream control region of the *HO* gene, and binding of CCBF in this region is required for the G₁-specific expression of the HO endonuclease that initiates mating-type switching (see Figure 13-17). Cell-cycle boxes also are found in the upstream activating sequences of the *CLN1* and *CLN2* genes. As a consequence, once Cln1 and Cln2 are expressed, they further stimulate their own expression by activating CCBF. Although it has not yet been experimentally proved, the most likely hypothesis is that MCB-binding factor and CCBF are activated by phosphorylation catalyzed by a Cdc28–Cln complex.

A Cdk-Cyclin Complex Regulates a DNA Initiation Factor in *Xenopus*

Although there is little direct evidence that a Cdc28–G₁ cyclin complex activates a DNA replication factor in *S. cerevisiae,* experiments with *Xenopus* egg extracts suggest that a Cdk-cyclin complex directly phosphorylates and activates a factor required to initiate DNA replication. As noted earlier, when sperm chromatin is added to an RNase treated *Xenopus* egg extract, a nuclear envelope forms and the sperm DNA subsequently replicates. Thus these egg extracts contain all the enzymes required to initiate and complete DNA replication, and no new protein synthesis is required. In later experiments, Cdc2 and Cdc2-related proteins were removed from the egg extract with an anti-Cdc2 antibody. When frog sperm chromatin was added to this Cdc2-depleted extract, a nuclear envelope formed around the chromatin, but DNA replication was decreased substantially. Similar results were obtained with egg extracts depleted by passage over a matrix-bound *S. pombe* protein called Suc1, which tightly binds Cdc2. Further analysis showed that depletion of Cdc2 inhibited initiation of DNA replication but did not interfere with elongation of DNA strands whose synthesis was initiated before treatment with the anti-Cdc2 antibody. Addition of proteins eluted from the Suc1 matrix to a Cdc2-depleted egg extract restored its ability to initiate DNA replication. These results imply that Cdc2 or a protein closely related to Cdc2 phosphorylates and activates a factor required to initiate DNA replication. Identification of this regulated DNA initiation factor is an important goal of current cell-cycle research.

► Cell-Cycle Control in Mammalian Cells

As described in previous sections, genetic studies with *S. pombe* and *S. cerevisiae* and biochemical studies with eggs and early embryos of *Xenopus* and marine invertebrates have provided considerable information about regulation of the cell cycle in eukaryotes. In all of these systems, cyclin-dependent protein kinases (Cdks) associated with various cyclins mediate passage through the various stages of the cell cycle, in particular entry into the S phase and into mitosis (M phase). The substrate specificity of Cdks is determined by the associated cyclin, and the activity of Cdk-cyclin complexes is regulated by phosphorylation at specific activating sites and inhibitory sites and by binding of inhibitory proteins. As we discuss in this section, the cell-cycle regulatory mechanisms uncovered in yeasts and *Xenopus* also operate in higher eukaryotes including humans and other mammals.

Mammalian Restriction Point is Analogous to START in Yeast Cells

Most studies of mammalian cell-cycle control have been done with cultured cells that require certain growth factors to stimulate cell proliferation. Binding of these polypeptide growth factors to cell-surface receptors initiates a signal-transduction cascade (see Figure 20-38) that stimulates mammalian cells to *progress through* G_1 into the S phase and around the cell cycle. This response is the opposite of that exhibited by *S. cerevisiae* cells that bind mating-type factors; this binding sets off a signal-transduction cascade that *arrests* yeast cells in G_1.

Mammalian cells cultured in the absence of growth factors are arrested with a diploid complement of chromosomes in a period of the cell cycle referred to as G_0 (see Figure 25-1). If growth factors are added to the culture medium, the cells pass through the *restriction point* 14–16 h later, enter the S phase 6–8 h after that, and traverse the remainder of the cell cycle. Like START in yeast cells, the restriction point is the point in the cell cycle at which mammalian cells become committed to entering the S phase and completing the cell cycle. If mammalian cells are moved from a medium containing growth factors to one lacking growth factors before they have passed the restriction point, the cells do not enter the S phase. But once cells have passed the restriction point, they are committed to entering the S phase and progressing through the entire cell cycle, which takes about 24 h for most cultured mammalian cells. This behavior is analogous to that of *S. cerevisiae* cells that are shifted from a nutrient-rich to a minimal medium (see Figure 25-25).

Multiple Cdks and Cyclins Regulate Passage of Mammalian Cells through the Cell Cycle

Unlike *S. pombe* and *S. cerevisiae,* which each have a single Cdk (Cdc2 and Cdc28, respectively), mammalian cells use a small family of related Cdks to regulate progression through the cell cycle (Table 25-2). These have been named Cdk1, 2, 3, and 4 in order of their discovery. As noted earlier, a human cDNA clone was found to complement *S. pombe cdc2⁻* mutants, and the encoded protein initially was called human Cdc2 or $p34^{cdc2}$. This was the first human cyclin-dependent kinase to be identified and thus is designated Cdk1, but the term Cdc2 continues to be commonly used. cDNAs encoding mammalian Cdk2 and Cdk3 were isolated by their ability to complement *S. cerevisiae cdc28* mutants. Cdk4 was isolated based on its homology with other protein kinases, and Cdk5 based on its ability to bind cyclin D. Cdk3 is not expressed at significant levels in most mammalian cell lines, and its function is unclear. Several human cyclins also have been identified including cyclins D and E, which function during G_1 and collectively are referred to as *G_1 cyclins,* and cyclins A and B, which function later in the cell cycle (see Table 25-2).

cDNAs encoding human cyclins D and E were isolated based on their ability to complement *S. cerevisiae* cells mutant in all three *CLN* genes encoding G_1 cyclins. The ability of these human cyclins to perform the S phase–promoting functions of yeast Cln proteins indicates that the critical substrates of G_1 Cdk-cyclin complexes required to promote entry into the S phase are conserved between humans and yeast. These conserved substrates probably include transcription factors and DNA replication-initiation factors. A small gene family encodes three different D-type cyclins called D1, D2, and D3. The relative amounts of the three D-type cyclins expressed in various cell types (e.g., fibroblasts, hematopoetic cells) differs. Here we refer to them collectively as cyclin D.

The two cyclins (A and B) that function in the S phase and G_2 were the first to be discovered. These cyclins initially were detected as proteins whose concentration oscillates in experiments with synchronously cycling early clam embryos, and a cDNA encoding clam cyclin A was the first

TABLE 25-2   Cyclin-Dependent Protein Kinases (Cdks) in Yeast and Mammalian Cells

Cdk	Associated Cyclins	Cell-Cycle Phase of Action
*S. POMBE**		
Cdc2	Cdc13	$G_2 \rightarrow M$ transition; early M
S. CEREVISIAE		
Cdc28	Cln1, Cln2, Cln3	$G_1 \rightarrow S$ transition
	Clb5, Clb6	S
	Clb1, Clb2, Clb3, Clb4	$G_2 \rightarrow M$ transition; early M
MAMMALIAN[†]		
Cdk2	Cyclin D1, D2, D3	G_1
	Cyclin E	$G_1 \rightarrow S$ transition
	Cyclin A	S
Cdk4	Cyclin D1, D2, D3	G_1
Cdk5	Cyclin D1, D2, D3	G_1
Cdc2 (Cdk1)	Cyclins A and B	$G_2 \rightarrow M$ transition; early M

* *S. pombe* contains other cyclins, which act at different stages, that are not discussed here.
[†]Cdk3 is present at low levels and its function is unclear. Two other cyclins (C and F) have been cloned, but their functions are not yet clear. Additional mammalian Cdk proteins and cyclins are likely to be identified.

cyclin cDNA cloned. Homologous cyclin A and B proteins have been found in all higher eukaryotes examined.

Entry into S Phase Figure 25-30 presents a current model for the periods of the cell cycle in which different Cdk-cyclin complexes act in G_0-arrested mammalian cells stimulated to divide by the addition of growth factors. First we describe the entry of G_0-arrested cells into the S phase, and then their subsequent passage through the S phase and G_2.

When mammalian cells in G_0 are stimulated with growth factors, cyclin D appears before cyclin E. Since cyclin D and its mRNA are unstable, the cyclin D level falls precipitously when growth factors are removed. However, in cells that are continuously proliferating because of constant exposure to growth factors, the cyclin D level remains high, since cyclin D mRNA is synthesized and translated into protein continuously. The experiment illustrated in Figure 25-31 indicates that a Cdk–cyclin D complex functions to drive mammalian cells through the restriction point.

As shown in Figure 25-30, Cdk2 associates with cyclins D and E, whereas Cdk4 and Cdk5 associate only with cyclin D. Like Cyclins D and E, Cdk2 and Cdk4 are not expressed in G_0 cells, but are induced by growth factors. The Cdk2 and Cdk4 proteins and the mRNAs encoding them also are unstable. Consequently, when growth

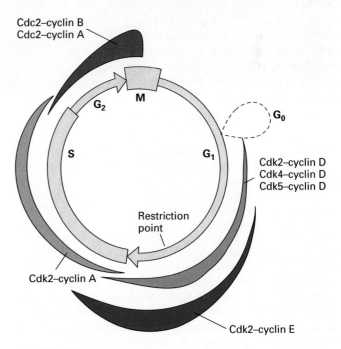

▲ FIGURE 25-30   A current model for the action of mammalian Cdk-cyclin complexes during cell cycle induced by treatment of G_0 cells with growth factors. The width of the green and purple bands is approximately proportional to the activity of the indicated complexes. Cyclin D refers to all three D-type cyclins. Little is yet known about Cdk5. New mammalian cyclins and Cdks continue to be identified, and this model is likely to be modified. [Adapted from C. J. Sherr, 1993, *Cell* **73**:1059.]

(a)

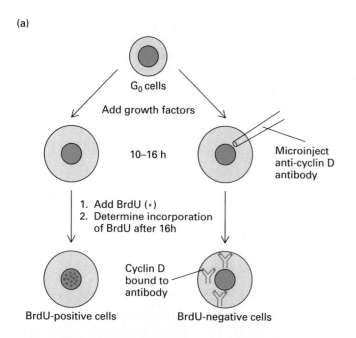

BrdU-positive cells   BrdU-negative cells

(b)

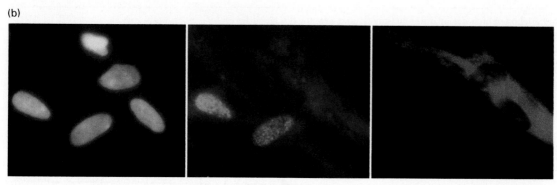

(c)

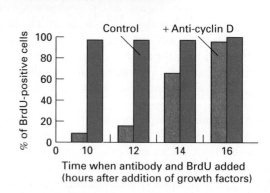

Time when antibody and BrdU added
(hours after addition of growth factors)

▲ **FIGURE 25-31** Experimental demonstration that cyclin D is required for passage through the restriction point in the mammalian cell cycle. The G_0-arrested cells used in these experiments pass the restriction point 14–16 h after addition of growth factors, and enter S phase 6–8 h later. (a) Outline of experimental protocol. At various times 10–16 h after addition of growth factors, some cells were microinjected with rabbit antibodies against cyclin D. Bromodeoxyuridine (BrdU), a thymidine analog, was then added to the medium, and the control and microinjected cells were incubated separately for an additional 16 h. Each sample then was analyzed to determine the percentage of cells that had incorporated BrdU, indicating that they had entered the S phase. (b) Analysis of cells microinjected with rabbit anti-cyclin D antibody eight hours after addition of growth factors. The three panels show the same field of cells stained sixteen hours after microinjection and addition of BrdU to the medium. Cells were stained with a DNA stain that fluoresces blue (*left*), a fluorescein-conjugated mouse monoclonal antibody specific for BrdU (*middle*), and Texas red-conjugated goat antibodies against rabbit antibodies to identify the cells that had been microinjected with the rabbit anti-cyclin D antibody (*right*). Note that the two cells injected with anti-cyclin D antibody (the red cells in the right panel) did not incorporate BrdU into nuclear DNA. (c) Percentage of control cells (blue) and cells injected with anti-cyclin D antibodies (red) that had incorporated BrdU. When cells were injected with anti-cyclin D antibodies 10 h, or 12 h after addition of growth factors, most failed to enter the S phase. In contrast, anti-cyclin D antibodies had little effect on entry into the S phase and DNA synthesis when injected at 14 h or 16 h, after cells had passed the restriction point. These results indicate that cyclin D is required to pass the restriction point, but once cells have passed the restriction point, they do not require cyclin D to enter S-phase six to eight hours later. [Parts (b) and (c) adapted from V. Baldin et al., *Genes & Dev.* **7**:812.]

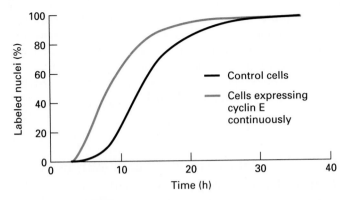

▲ FIGURE 25-32   Experimental demonstration that continuous expression of cyclin E shortens G_1 mammalian cells. Cultured fibroblasts were transformed with a vector that continuously expressed cyclin E. Mitotic cells from the transformed culture and a control (nontransformed) culture were isolated and placed separately in medium containing bromodeoxyuridine (BrdU), which is incorporated into DNA during the S phase. The percentage of cells that had entered the S phase at various times was determined by analyzing the incorporation of BrdU with a specific fluorescent-labeled antibody against BrdU. The half-time for entry into the S phase was ≈8 h for the cells continuously expressing cyclin E (red curve) and ≈12 h for the control cells (black curve). [Adapted from M. Ohtsubo and J. M. Roberts, 1993, *Science* **259**:1908.]

factors are removed, and transcription of the *cdk2* and *cdk4* genes ceases, the protein levels fall.

In continuously proliferating mammalian cells maintained in media containing growth factors, cyclin E levels oscillate, reaching peak levels just before the S phase is initiated. When cyclin E is expressed continuously at high levels from a constitutive promoter, the length of G_1 is shortened (Figure 25-32). This effect is very similar to that resulting from high-level expression of Cln3 in *S. cerevisiae* cells (see Figure 25-27). Experiments with *Xenopus* egg extracts in which Cdk2 was removed using a specific antibody showed that it is essential for DNA replication, but not for mitosis. These findings suggest that the Cdk2–cyclin E complex is the protein kinase principally responsible for activating DNA initiation factors and inducing entry into the S phase. When Cdk2 is associated with cyclin E, threonine and tyrosine residues in Cdk2 can be phosphorylated, suggesting that Cdk2 is regulated by the same mechanism uncovered for *S. pombe* Cdc2, that is, phosphorylation of activating and inhibitory sites by specific kinases and dephosphorylation by one or more activating phosphatases (see Figure 25-22).

As noted above, when growth factors are removed from cultured cells, transcription of the genes encoding Cdk2, Cdk4, and cyclins D and E ceases, and the levels of all these proteins fall rapidly. Consequently, in G_0 cells, the

proteins currently thought to promote passage through the restriction point and to compose the S phase–promoting factors are not present. This would certainly explain why cells in G_0 are arrested and do not enter the S phase. Before G_0 cells can enter the S phase, growth factors must induce the expression of these proteins.

Passage through S Phase and G_2   Soon after mammalian cells enter the S phase, Cdk2 becomes associated with cyclin A (see Figure 25-30). Synthesis of cyclin A begins as cells approach the G_1 → S transition, and the protein is immediately transported into the nucleus. Disruption of cyclin A function inhibits DNA synthesis in mammalian cells, suggesting that the Cdk2–cyclin A complex is crucial for progression through the S phase.

In cycling mammalian cells, cyclin B is first synthesized during the S phase and increases in concentration as cells proceed through G_2, peaking in early mitosis and dropping after anaphase, as in *Xenopus* cycling egg extracts (see Figure 25-7). In human cells, cyclin B first accumulates in the cytoplasm and then enters the nucleus just before the nuclear envelope breaks down early in mitosis. Consequently, MPF function may be controlled in part by regulation of the nuclear transport of cyclin B. The principle mammalian Cdk in G_2 is Cdc2 (Cdk1); as mentioned previously, human Cdc2 is highly homologous with *S. pombe* Cdc2. Mammalian Cdc2 associates with cyclins A and B, and mRNAs encoding these mammalian cyclins can promote oocyte maturation when injected into *Xenopus* oocytes. In mammalian cells, the function of Cdc2–cyclin A and Cdc2–cyclin B in inducing entry into mitosis appears to be regulated by proteins analogous to the *S. pombe* CAK activating enzyme, Wee1 inhibitory kinase, and Cdc25 activating phosphatase (see Figure 25-22).

Growth Factor–Induced Expression of Two Classes of Genes Returns G_0 Mammalian Cells to the Cell Cycle

Addition of growth factors to G_0-arrested mammalian cells induces transcription of multiple genes, most of which fall into one of two classes—early-response or delayed-response genes—depending on how soon their encoded mRNAs appear (Figure 25-33a). Transcription of early-response genes is induced within a few minutes after addition of growth factors by a signal-transduction cascade that leads to activation of transcription factors (see Figure 20-38). Induction of early-response genes is not blocked by inhibitors of protein synthesis (Figure 25-33b), because the transcription factors involved are present in G_0 cells and are activated by posttranslational modifications such as phosphorylation. Many of the early-response genes encode transcription factors such as c-Fos and c-Jun. Both of these transcription factors are leucine-zipper proteins (Figure 11-

(a)

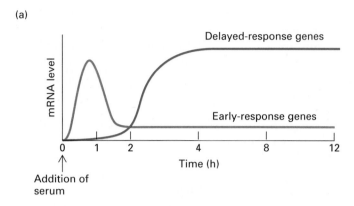

(b)

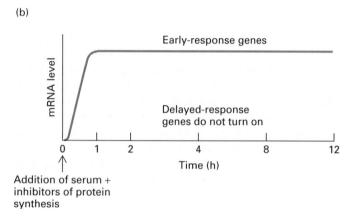

▲ FIGURE 25-33   Time course of expression of early- and delayed-response genes in G_0-arrested mammalian cells after addition of serum containing growth factors in the absence (a) and presence (b) of inhibitors of protein synthesis. See text for discussion. [Adapted from A. Murray and T. Hunt, 1993, *The Cell Cycle: An Introduction*, W. H. Freeman and Company.]

48g) that form heterodimers with each other and with other leucine-zipper proteins; the heterodimeric proteins then bind specific DNA sequences. As discussed in Chapter 26, mutant, unregulated forms of both c-Fos and c-Jun are expressed by oncogenic retroviruses; the discovery that the viral forms of these proteins (v-Fos and v-Jun) can transform normal cells into cancer cells led to identification of the regulated cellular forms of these transcription factors. After peaking at about 30 min following addition of growth factors, the concentrations of the early-response mRNAs fall to a lower level that is maintained as long as growth factors are present in the medium. This drop in transcription is blocked by inhibitors of protein synthesis, indicating that it depends on production of one or more of the early-response proteins.

The proteins encoded by the early-response genes induce transcription of the delayed-response genes. Some of these genes encode additional transcription factors such as E2F-1, discussed in the next section. Also included in the delayed-response genes are those encoding the D-type cyclins, cyclin E, Cdk2, and Cdk4.

Activity of Transcription Factor E2F Is Required for Entry into S Phase

As in *S. cerevisiae* (see Table 25-1), mammalian genes encoding many of the proteins involved in DNA and deoxynucleotide synthesis are induced as cells pass through the $G_1 \rightarrow S$ transition. A family of transcription factors, together called *E2F*, is required for transcription of several of these genes. E2F was first characterized as an activator of the adenovirus early region 2 transcription unit, which encodes the viral DNA polymerase and primase. Binding of two proteins, one called RB and a related protein called p107, to E2F inhibits its ability to activate transcription. RB was initially identified as the product of the prototype *tumor suppressor* gene *Rb*. As discussed in Chapter 8, a child with hereditary retinoblastoma receives one normal Rb^+ allele and one mutant Rb^- allele. If the Rb^+ allele in a retinal cell is somatically mutated to a Rb^- allele, then no functional protein is expressed and the cell or one its descendants is likely to become cancerous, leading to the retinal tumors that characterize this disease (see Figure 8-10).

Experiments with RB indicate that regulation of its ability to inhibit the activity of E2F is an important mechanism for controlling entry into the S phase. For example, cultured cells derived from a human tumor defective in both *Rb* alleles can cycle in the presence of very low levels of growth factors. Transfection of these cells with an RB expression vector causes the cells to arrest in G_1. This finding suggests that inhibition of E2F activity by RB can block entry into the S phase. Mammalian G_1 Cdk-cyclin complexes and Cdk2-cyclin A can phosphorylate RB in vitro; the resulting phosphorylated RB cannot bind to and inhibit E2F. Moreover, when G_1 cyclin expression vectors are cotransfected into Rb^- tumor-derived cells along with an RB expression vector, the cells are not arrested in G_1. These findings fit the model shown in Figure 25-34. The activity of E2F is repressed by binding of RB, but phosphorylation by G_1 Cdk-cyclin complexes and Cdk2-cyclin A releases and derepresses E2F, which then is free to activate transcription of genes encoding proteins required for DNA synthesis. This activation of E2F is similar to the activation of MCB-binding factor and CCBF in *S. cerevisiae* cells by Cdc28-Cln complexes discussed earlier. In agreement with this model, in normal cycling mammalian cells, RB is dephosphorylated at the end of mitosis and phosphorylated in late G_1.

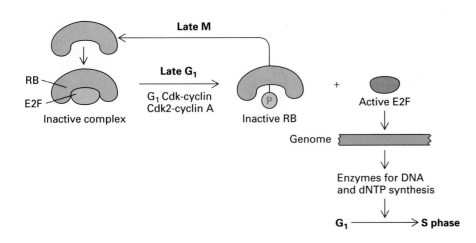

◄ FIGURE 25-34   Model for role of RB protein in regulating the $G_1 \rightarrow S$ transition in mammalian cells. E2F is a transcription factor that is required for expression of various enzymes involved in synthesis of dNTPs and DNA. RB is encoded by a tumor suppressor gene. See text for discussion.

➤ *Role of Checkpoints in Cell-Cycle Regulation*

Catastrophic genetic damage can occur if cells progress to the next phase of the cell cycle before the previous phase is properly completed. For example, when S-phase cells are induced to enter mitosis by fusion to a cell in mitosis, the MPF present in the mitotic cell forces the chromosomes of the S-phase cell to condense. However, since the replicating chromosomes are fragmented by the process (Figure 25-35), such premature entry into mitosis is disastrous for a cell. Another example concerns attachment of kinetochores to microtubules of the mitotic spindle during metaphase. If anaphase is initiated before both kinetochores of a replicated chromosome become attached to microtubules from opposite spindle poles, daughter cells are produced that have a missing or extra chromosome (Figure 25-36). When this process, called *nondisjunction*, occurs during the meiosis that generates a human egg, Down's syndrome can occur from trisomy of chromosome 23, resulting in

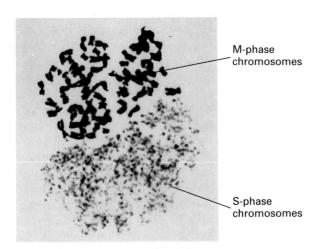

▲ FIGURE 25-35   Micrograph of chromosomes from a hybrid cell resulting from fusion of an S-phase cell and a M-phase cell. Chromosomes from the mitotic cell are highly condensed with distinct sister chromatids visible, as expected. The chromosomes from the S-phase cell are fragmented into pieces of condensed chromosomes. [From R. T. Johnson and P. N. Rao, 1971, *Biol. Rev.* **46**:97.]

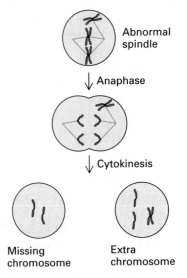

▲ FIGURE 25-36   Nondisjunction occurs when chromosomes segregate in anaphase before the kinetochore of each sister chromatid has attached to microtubules (red lines) from the opposite spindle poles. As a result, one daughter cell contains two copies of one chromosome, while the other daughter cell lacks that chromosome. [Adapted from A. Murray and T. Hunt, 1993, *The Cell Cycle: An Introduction*, W. H. Freeman and Company.]

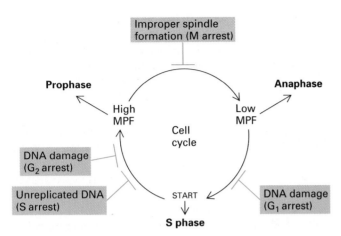

▲ FIGURE 25-37   Stages at which checkpoint controls can arrest passage through the cell cycle. DNA damage due to irradiation or chemical modification prevents G_1 cells from entering the S phase and G_2 cells from entering mitosis. Unreplicated DNA prevents entry into mitosis. Defects in assembly of the mitotic spindle or the attachment of kineto-chores to spindle microtubules prevent activation of the polyubiquitination system that leads to degradation of mitotic cyclins. Consequently, MPF activity remains high and cells do not enter anaphase. [Adapted from A. Murray and T. Hunt, 1993, *The Cell Cycle: An Introduction*, W. H. Freeman and Company.]

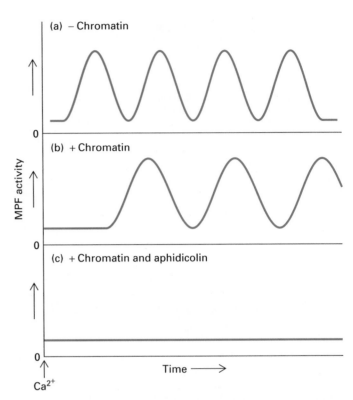

▲ FIGURE 25-38   Experimental demonstration that unreplicated DNA inhibits formation of active MPF in *Xenopus* egg extracts. (a) Addition of Ca^{2+} ions causes oscillation of MPF activity in the absence of chromatin. (b) Addition of a high concentration of *Xenopus* sperm chromatin to Ca^{2+}-stimulated extracts delays the rise in MPF activity. (c) Addition of aphidicolin, an inhibitor of DNA synthesis, plus sperm chromatin completely inhibits the rise in MPF activity. [Adapted from M. Dasso and J. Newport, 1990, *Cell* **61**:811.]

developmental abnormalities and mental retardation. Such mistakes in cell-cycle events are extremely rare because of *checkpoint controls* that operate to ensure that chromosomes are intact and that each stage of the cell cycle is completed before the following stage is initiated. The primary cell cycle checkpoints are illustrated in Figure 25-37 and discussed in the remainder of this chapter.

Presence of Unreplicated DNA Prevents Entry into Mitosis

It would be disastrous for a cell to enter mitosis before replicating all of its chromosomal DNA. A cell cycle checkpoint prevents this from happening. The ability of unreplicated DNA to inhibit entry into mitosis has been studied in *Xenopus* egg extracts. Addition of Ca^{2+} ions to egg extracts induces cycling, even in the absence of DNA, with the characteristic oscillation in MPF protein kinase activity. However, if high concentrations of sperm chromatin are added, the rise in MPF activity is delayed until DNA replication is complete. If aphidicolin, a specific inhibitor of DNA polymerase α, also is added, the rise in MPF activity is blocked entirely (Figure 25-38).

How the presence of unreplicated DNA is recognized and how it prevents MPF activation is not yet understood.

But the recent identification of yeast mutants defective in this checkpoint control will make the characterization of proteins involved in the process possible. Such mutants can be recognized using cells with a temperature-sensitive mutation affecting an enzyme required for DNA synthesis, such as DNA polymerase. When such cells are shifted from the permissive to the nonpermissive temperature during the S phase, DNA synthesis is inhibited before chromosome replication is complete. Mutant cells with intact checkpoint controls do not proceed into mitosis at the nonpermissive temperature; when shifted back to the permissive temperature, such cells complete DNA synthesis and proceed through the cell cycle normally, eventually forming a colony at the permissive temperature. Cells with a second mutation that inactivates a step in this checkpoint control mechanism proceed into mitosis, damage their chromosomes, and die at the nonpermissive temperature; they do not form colonies when shifted back to the permissive temperature.

Defects in Assembly of the Mitotic Spindle Prevent Exit from Mitosis

Inhibitors of microtubule polymerization, which cause cells to arrest during mitosis, commonly are used to prepare cells for karyotyping. When a microtubule inhibitor (e.g., colchicine) is added to cultured cells, the cells enter mitosis and arrest with condensed chromosomes. With increasing time, a large fraction of the cells in a culture become arrested, thus permitting determination of the size, shape, and number of mitotic chromosomes—that is, karyotyping—in multiple cells. A checkpoint control somehow senses when the mitotic spindle has not assembled properly and prevents activation of the polyubiquitination system that normally leads to degradation of mitotic cyclins and onset of anaphase (see Figure 25-9). As a result, MPF activity remains high, chromosomes remain condensed, and the nuclear envelope does not reform.

A microtubule-depolymerizing drug called benomyl has been used to isolate yeast mutants defective in this mitotic checkpoint. Low concentrations of benomyl increase the time required for yeast cells to assemble the mitotic spindle and attach kinetochores to microtubules. Wild-type cells exposed to benomyl delay in anaphase until these processes are completed and then proceed on through mitosis, producing normal daughter cells. In contrast, mutants defective in this cell-cycle checkpoint proceed through anaphase before assembly of the spindle and attachment of kinetochores is complete; consequently, they missegregate their chromosomes, producing abnormal daughter cells that die. Analysis of these mutants, called *bub* (for budding uninhibited by benomyl) and *mad* (for mitotic arrest deficient), should shed light on the mechanism by which this interesting checkpoint operates. The sequence of one of the *BUB* genes indicates that it encodes a protein kinase, which may influence the activities of multiple proteins.

DNA Damage Prevents Entry into the S Phase and Mitosis

Cells whose DNA is damaged by irradiation with UV light or γ-rays or by chemical modification become arrested in G_1 and G_2 until the damage is repaired. Arrest in G_1 prevents copying of damaged bases, which would fix mutations in the genome. Replication of damaged DNA also causes chromosomal rearrangements at high frequency by as-yet unknown mechanisms. Arrest in G_2 allows DNA double-stranded breaks to be repaired (Chapter 10) before

mitosis. If a double-stranded break is not repaired, the broken distal portion of the damaged chromosome is not properly segregated because it is not physically linked to a centromere, which is pulled toward a spindle pole during anaphase.

G_2 Checkpoint Control In *S. pombe,* mutation of the Cdc2 inhibitory Tyr-15 (Figure 25-22) to a phenylalanine prevents G_2 arrest subsequent to DNA damage, indicating that phosphorylation of the inhibitory site is necessary for G_2 arrest. In *S. cerevisiae,* however, DNA damage induces G_2 arrest even when the inhibitory site in Cdc28 is mutated so that it cannot be phosphorylated. Yeast mutants defective in DNA damage–induced G_2 arrest have been isolated as a subset of mutants identified as particularly sensitive to UV irradiation. As discussed in Chapter 10, many such *rad* mutants are defective in the repair of lesions introduced into DNA. Some of these mutants have been shown to repair UV-induced DNA lesions normally (see Figure 10-27), but they proceed through mitosis before the repair of double-stranded DNA breaks is complete, and therefore suffer a lethal loss of genetic information. Analysis of these mutants should provide insights into how the DNA damage activates the checkpoint leading to arrest in G_2.

G_1 Checkpoint Control The protein encoded by the *p53* tumor suppressor gene functions in the checkpoint control that arrests human cells with damaged DNA in G_1. Cells with functional p53 arrest in G_1 when exposed to γ-irradiation, whereas cells lacking functional p53 do not (Figure 25-39). The p53 protein, which functions as a transcription factor, causes G_1 arrest by a mechanism similar to the way mating-type pheromones arrest *S. cerevisiae* cells in G_1 (see Figure 25-29). Although p53 normally is extremely unstable, DNA damage somehow stabilizes it, leading to an increase in the concentration of p53, which then stimulates transcription of a number of genes. One of these encodes a cyclin-dependent kinase inhibitor, which binds to and inhibits mammalian G_1 Cdk-cyclin complexes; as a result, cells are arrested in G_1 until the DNA damage is repaired and p53 and the cyclin-dependent kinase inhibitor levels fall.

Both alleles of the *p53* gene are mutated in a high percentage of human metastatic cancers. Inactivation of p53 by mutations in both *p53* alleles prevents the delay of entry into the S phase that normally is induced by DNA damage (see Figure 25-39). When this checkpoint control does not operate properly, damaged DNA can replicate, producing mutations and DNA rearrangements that contribute to the development of a highly transformed, metastatic cell (Chapter 26). The consequences of mutations in *p53* provide a dramatic example of the significance of cell-cycle checkpoints to the health of a multicellular organism.

(a) Wild-type cells

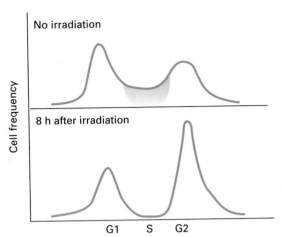

(b) p53⁻ mutant cells

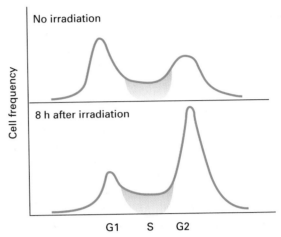

▲ FIGURE 25-39   Effect of mutation of *p53* tumor suppressor gene on G_1 checkpoint control. The distribution of cultured human cells in G_1, S, and G_2 was determined by analysis of their DNA content with a fluorescence-activated cell sorter (see Figure 5-49). (a) By 8 h following exposure of wild-type cells to γ-radiation, cells that were in the S phase (red shading) had completed DNA synthesis, entered G_2, and then arrested, accounting for the rise in the G_2 peak. The absence of S-phase cells indicates that irradiation prevented new cells from entering the S phase, causing them to arrest in G_1. (b) The presence of an S peak 8 h after irradiation of *p53⁻* mutant cells indicates that the G_1 checkpoint does not operate in these cells. The increase in the G_2 peak indicates that the checkpoint blocking entry of irradiated cells into mitosis still operates in these mutant cells. [See S. J. Kuerbitz et al., 1992, *Proc. Nat'l. Acad. Sci. USA* **89**:7491; adapted from A. Murray and T. Hunt, 1993, *The Cell Cycle: An Introduction*, W. H. Freeman and Company.]

SUMMARY

All eukaryotic cells grow and divide by fundamentally similar mechanisms. This has made it possible to study cell-cycle control in diverse organisms that each have their experimental advantages. In particular genetic studies in yeasts and biochemical analyses of frog eggs have complemented one another to give us considerable understanding of the molecular mechanisms that regulate DNA replication and cell division. A cell that is destined to grow and divide passes through specific phases of a cell cycle: G_1, S (the period of DNA synthesis), G_2, and M (mitosis). Most molecular mechanisms that control a cell's progress through the cell cycle regulate initiation of the S phase or initiation of mitosis. Both of these transitions are regulated by heterodimeric protein kinases composed of a catalytic subunit and a regulatory subunit that contributes to substrate specificity. The regulatory subunits are called *cyclins* because the concentrations of most of them cycle in phase with the cell cycle; each reaches peak levels during the specific cell-cycle phase(s) in which they function and then are rapidly degraded as the cell enters the next phase of the cycle. The protein kinase activity of the catalytic subunits depends on their association with a cyclin, leading to the name cyclin-dependent kinase, or Cdk. When activated, these Cdk-cyclin heterodimers phosphorylate multiple different proteins, activating some and inhibiting others, to control the many molecular events associated with DNA replication and mitosis.

In keeping with their significant functions in controlling cell replication, the Cdk-cyclin complexes are themselves extensively regulated. The activity of the Cdk catalytic subunit is regulated by other protein kinases and by protein phosphatases that are also regulated so as to coordinate cell division in concert with the requirements of the organism. Regulated inhibitor proteins also can control Cdk-cyclin catalytic activity. In addition, the concentrations of most of the cyclin subunits and of the Cdk subunits in mammalian cells are controlled by regulating transcription of the genes encoding them and by regulating their degradation.

Replicating mammalian cells contain a few different cyclins that function in G_1 to promote entry into the S phase. Most of these are not expressed in nonproliferating cells, which exited the cell cycle while in G_1 to enter an arrested phase called G_0. When growth factors are added to G_0 mammalian cells a group of genes called early-response genes are induced. Among the growth factor–induced proteins encoded by these genes are several transcription factors that induce the expression of a second set of genes, the delayed-response genes. G_1 cyclins and Cdks are included among the proteins encoded by the delayed-response genes. Transcription of the genes encoding G_1 cyclins also is induced in haploid yeast cells as they prepare to enter the S phase. Both S. pombe and S. cerevisiae each contain a single Cdk, whose concentration does not vary greatly during the cell cycle. In contrast, mammalian cells possess three principle Cdks, whose levels rise and fall during the course of the cell cycle.

In S. cerevisiae cells and most mammalian cells, induction of the genes encoding the G_1 cyclins and Cdks commits cells to entering the S phase, a process called passing START in yeast and passing the restriction point in mammalian cells. In both yeasts and mammals, G_1 Cdk-cyclin complexes activate transcription factors already present in the cell; the activated transcription factors then stimulate transcription of multiple genes encoding the enzymes required for DNA replication and synthesis of deoxynucleotides. DNA replication-initiation factors also are activated by G_1 Cdk-cyclin complexes, but the critical regulated proteins that control initiation of DNA synthesis at replication origins have not yet been identified. The recent characterization of origin-binding proteins in S. cerevisiae may lead to rapid progress in this area. G_1 Cdk-cyclin complexes also activate duplication of the microtubule-organizing center into the two centrosomes that organize the poles of the mitotic spindle apparatus during mitosis (Chapter 23).

Cells enter the G_2 period after completion of DNA replication. Entry into mitosis following G_2 depends on activation of mitosis-promoting factor (MPF), which originally was named maturation-promoting factor because of its ability to induce maturation of Xenopus oocytes. The MPF catalytic subunit, also a cyclin-dependent protein kinase, was originally identified in S. pombe $cdc2^-$ mutants and generally is called Cdc2. The MPF regulatory subunit is cyclin B, although cyclin A of higher eukaryotes, which is expressed earlier in the cell cycle than cyclin B, also can promote maturation of Xenopus oocytes. Regulation of MPF activity by phosphorylation of the Cdc2 subunit is understood in more detail than regulation of other Cdk-cyclin kinases. All phosphorylations require that Cdc2 be bound by a cyclin. Phosphorylation by Cdk-activating kinase (CAK) is required for MPF activity, probably by causing a large conformational change in the Cdc2 subunit that exposes the active site to potential protein substrates. Phosphorylation by Wee1 protein kinase inactivates MPF by introducing a phosphate into the ATP-binding site of the Cdc2 subunit. MPF doubly phosphorylated on the Cdc2 subunit is finally converted into an active protein kinase by removal of the inhibitory phosphate by the Cdc25 protein phosphatase. This dephosphorylation is the critical step that regulates MPF activity in cells of the early Drosophila embryo.

In most eukaryotic cells, breakdown of the nuclear envelope into small vesicles is an early event of mitosis. This process requires depolymerization of the nuclear lamina, a protein meshwork attached to the inner membrane of the nuclear envelope. Depolymerization of the nuclear lamina is a direct consequence of phosphorylation of nuclear lamin proteins by MPF. MPF also phosphorylates an inhibitory site in cortical myosin molecules, preventing contraction of the cortical ring and cytokinesis until the end of mitosis. Phosphorylation of chromatin proteins by MPF, or possibly a related kinase activated by MPF phosphorylation in a kinase cascade, is thought to cause the condensation of mitotic chromosomes. Similarly, MPF-induced phosphorylation of microtubule-associated proteins is thought to cause the dramatic reorganization of microtubules into the mitotic spindle apparatus.

The onset of anaphase is induced by degradation of mitotic cyclins, which contain a specific sequence, the destruction box, recognized by an enzyme system that marks proteins for degradation. Multiple copies of a small protein called ubiquitin are added to mitotic cyclins at the onset of anaphase, and the polyubiquitinated cyclins are degraded by large protease complexes called proteosomes. The enzymes that polyubiquitinate mitotic cyclins are thought to be activated by high levels of MPF kinase activity, so that the cyclins are degraded only after MPF activity has induced all the events of mitosis through metaphase. The segregation of sister chromatids that occurs during anaphase also may be initiated by polyubiquitination of the protein cross-links between the sister chromatids, thus marking these proteins for degradation. If these cross-links were degraded by proteosomes, then the microtubule forces acting on the chromatid kinetochores could pull the separated chromosomes to the opposite spindle poles.

Degradation of the mitotic cyclins inactivates MPF, leaving the activity of constitutive protein phosphatases unopposed. As a consequence, the sites in various proteins that were phosphorylated by MPF are dephosphorylated, leading to the late mitotic events. For example, dephosphorylated nuclear lamins repolymerize into the nuclear lamina of the two daughter nuclei during late telophase. Chromosome decondensation is thought to result from dephosphorylation of chromatin proteins. Similarly, dephosphorylation of microtubule-associated proteins is thought to be responsible for disassembly of the mitotic spindle by returning microtubule dynamics to those of an interphase cell (Chapter 23). Finally, dephosphorylation of the inhibitory site on cortical myosin initiates contraction of the cortical ring, causing cytokinesis and cell separation.

Multiple levels of regulation are required for the complex processes of the cell cycle to be properly coordinated. Checkpoint controls operate to ensure that each phase of the cell cycle is completed before the ensuing phase is begun. Unreplicated DNA blocks entry into mitosis. Anaphase is not initiated until all chromosomal kinetochores are attached to a properly assembled mitotic spindle. Analysis of yeast mutants defective in these checkpoints is just beginning to reveal how they operate. But already it is clear that (surprise!) protein kinases are involved; these probably function to regulate the activities of Cdk-cyclin complexes and other proteins such as those involved in specific polyubiquitination during anaphase.

DNA damage induces checkpoints that prevent entry into the S phase and the M phase until the damage is repaired. Blocking entry into the S phase prevents the replication of damaged DNA, which otherwise would cause mutations in daughter strands synthesized on damaged templates and lead to DNA rearrangements through poorly understood mechanisms. Blocking cells with double-stranded DNA breaks from entering mitosis prevents segregation of broken chromosomes, which otherwise would cause missegregation of genes distal to the break. In mammalian cells DNA damage leads to activation of the p53 transcription factor, which induces transcription of the gene encoding Cki1, a small protein that binds to and inhibits G_1 Cdk-cyclin complexes, blocking entry into the S phase. In humans, mutations that inactivate p53 predispose cells to develop into metastatic cancer cells, an aspect of tumor progression that is considered in more detail in Chapter 26.

REVIEW QUESTIONS

1. Regulation of the cell cycle is vital for normal growth and development of organisms. What evidence supports the notion that trans-acting factors regulate the cell cycle? Mitosis promoting factor (MPF) is the trans-acting factor that promotes cells to enter mitosis. What type of enzyme is MPF? What are the two subunits of the protein?

Are nuclear events required for the operation of the cell cycle clock in early embryos? Are they involved in cells born later in development?

Cyclin B synthesis must be controlled to regulate MPF activity. Must cyclin B be destroyed for MPF activity to fall? What is a destruction box? What is the function of a destruction box? What is ubiquitin? What is meant by polyubiquitination, and what is its purpose? How are proteosomes involved in protein degradation?

What are nuclear lamins? How is MPF involved in the depolymerization of the nuclear lamina? What chromatin-associated-proteins are substrates for phosphorylation by MPF? What is the role of protein phosphatases in nuclear envelope assembly and cytokinesis?

a. You have isolated the recognition protein for the destruction box in mitotic cyclins and prepared antibodies against it. You microinject some cells with these antibodies. At what point in the cell cycle would they arrest?

2. What are temperature sensitive yeast mutants? How are they used to detect Cdc proteins? Which Cdc protein is the key regulator of entry into M-phase in *S. pombe*? What is the catalytic activity exhibited by Cdc2? Which protein in *Xenopus* is the analog to *S. pombe* Cdc2? What is *S. pombe* Cdc13? What does Cdc25 do to the *S. pombe* MPF activity? What does Wee1 appear to do to *S. pombe* MPF activity?

Which of the MPF subunits in *S. pombe* is phosphorylated as a means of regulating the MPF activity? What enzyme activates it? What is the relationship between Wee1 and Cdc25 in the phosphorylation state and activity of MPF?

In regard to *S. cerevisiae*, what is meant by START? What is CDC28? What is the homolog that is found in *S. pombe*? What are CLN1, CLN2, and CLN3? Do cyclins or the kinase catalytic subunit direct substrate specificity in an MPF or SPF? What is the function of Far1 protein?

What two transcription factors are activated at START, and why are they important?

a. The following table lists a series of temperature sensitive mutations in several genes. Give the resulting point in the cell cycle where arrest would occur when the yeast mutants are grown at the nonpermissive temperature.

Temperature Sensitive Mutant	Cell Cycle Arrest Point
pol1	
cdc9	
cdc28	
clb5	
clb1	

b. It is possible for two different *cdc28*[ts] mutants to arrest in two different stages of the cell cycle. Discuss how this is possible.

c. The cdc mutants isolated were obtained as described in Figure 25-2. Recall that the mutagen ethyl methanesulfonate causes point mutations in DNA, but ionizing radiation creates large deletions. Which would you choose for selecting temperature sensitive mutants for identifying *cdc* genes? Explain your answer.

3. What is the restriction point in the mammalian cell cycle? Which complex is required to drive cells through the restriction point? What are the human cyclins that have been identified?

Which cyclins are sensitive to the presence or absence of growth factors?

a. The need for checkpoint control is emphasized as a means to prevent cells from entering a phase of the cycle prematurely. An important protein that is involved in checkpoint control in human cells is p53. Recall that ultraviolet radiation induces the formation of pyrimidine dimers that impede replication and are mutagenic. Why would it be advantageous to halt the onset of DNA synthesis if a cell were exposed to ultraviolet light?

b. Suppose an individual had one mutant allele for the *p53* gene such that its ability to control the entry into S-phase was compromised. What would be the consequences to a cell if the remaining *wild-type p53* allele were inactivated?

References

General

MURRAY, A., and T. HUNT. 1993. *The Cell Cycle: An Introduction.* W. H. Freeman and Company.

Maturation-Promoting Factor

DUNPHY, W. G., et al. 1988. The *Xenopus* cdc2 protein is a component of MPF, a cytoplasmic regulator of mitosis. *Cell* 54:423–431.

EVANS, T., et al. 1983. Cyclin: a protein specified by maternal mRNA in sea urchin eggs that is destroyed at each cleavage division. *Cell* 33:389–396.

GAUTIER, J., et al. 1990. Cyclin is a component of maturation-promoting factor from *Xenopus. Cell* 60:487–494.

GAUTIER, J., et al. 1988. Purification of maturation-promoting factor from *Xenopus* eggs: the factor contains the product of a *Xenopus* homolog of the fission yeast cell cycle control gene *cdc2+. Cell* 54:433–439.

GERHART, J., M. WU, and M. J. KIRSCHNER. 1984. Cell cycle dynamics of an M-phase specific cytoplasmic factor in *Xenopus laevis* oocytes and eggs. *J. Cell Biol.* 98:1247–1255.

JOHNSON, R. T., and P. N. RAO. 1970. Mammalian cell fusion: induction of premature chromosome condensation in interphase nuclei. *Nature* 226:717–722.

MASUI, Y., and C. L. MARKERT. 1971. Cytoplasmic control of nuclear behavior during meiotic maturation of frog oocytes. *Exp. Zool.* 177:129–145.

MURRAY, A. W., and M. W. KIRSCHNER. 1989. Cyclin synthesis drives the early embryonic cell cycle. *Nature* 339:275–280.

RAO, P. N., and R. T. JOHNSON. 1970. Mammalian cell fusion studies on the regulation of DNA synthesis and mitosis. *Nature* 225:159–164.

SWENSON, K. I., K. M. FARRELL, and J. V. RUDERMAN. 1986. The clam embryo cyclin A induces entry into M phase and the resumption of meiosis in *Xenopus* oocytes. *Cell* 47:861–870.

Nuclear Envelope Breakdown

AEBI, U., et al. 1986. The nuclear lamina is a meshwork of intermediate-type filaments. *Nature* 323:560–564.

HEALD, R., and F. MCKEON. 1990. Mutations of phosphorylation sites in lamin A that prevent nuclear lamina disassembly in mitosis. *Cell* 61:579–589.

NIGG, E. 1992. Assembly-disassembly of the nuclear lamina. *Curr. Opin. Cell Biol.* 4:105–109.

NIGG, E. A. 1993. Targets of cyclin-dependent protein kinases. *Curr. Opin. Cell Biol.* 5:187–193.

Polyubiquitination and Proteosome Degradation

FINLEY, D., and V. CHAU. 1991. Ubiquitination. *Ann. Rev. Cell Biol.* 7:25–69.

GOLDBERG, A. L., and K. L. ROCK. 1992. Proteolysis, proteosomes and antigen presentation. *Nature* 357:375–379.

HOCHSTRASSER, M. 1992. Ubiquitin and intracellular protein degradation. *Curr. Opin. Cell Biol.* 4:1024–1031.

Regulation of Cyclin-Dependent Kinases by Association with Cyclins and Phosphorylation

DESAI, D., Y. GU, and D. O. MORGAN. 1992. Activation of human cyclin-dependent kinases in vitro. *Mol. Biol. Cell* 3:571–582.

EDGAR, B. A., et al. 1994. Distinct molecular mechanisms regulate cell cycle timing at successive stages of *Drosophila* embryogenesis. *Genes & Dev.* 8:440–452.

FEATHERSTONE, C., and P. RUSSEL. 1991. Fission yeast p107^{wee1} mitotic inhibitor is a tyrosine/serine kinase. *Nature* 349:808–811.

FELIX, M. A., P. COHEN, and E. KARSENTI. 1990. cdc2 H1 kinase is negatively regulated by a type 2A phosphatase in *Xenopus* early embryonic cell cycle: evidence from the effects of okadaic acid. *EMBO J.* 9:675–683.

GAUTIER, J., et al. 1989. Dephosphorylation and activation of *Xenopus* p34^{cdc2} protein kinase during the cell cycle. *Nature* 339:626–629.

GAUTIER, J., et al. 1991. Cdc25 is a specific tyrosine phosphatase that directly activates p34-cdc2. *Cell* 67:197–211.

GOULD, K. L., and P. NURSE. 1989. Tyrosine phosphorylation of the fission yeast *cdc2+* protein kinase regulates entry into mitosis. *Nature* 342:39–45.

LEE, T. H., et al. 1991. INH, a negative regulator of MPF, is a form of protein phosphatase 2A. *Cell* 64:415–423.

SOLOMON, M. J., T. LEE, and M. W. KIRSCHNER. 1992. Role of phosphorylation in p34-cdc2 activation: identification of an activating kinase. *Mol. Biol. Cell* 3:13–27.

Genetic Analysis of Cyclin-Dependent Kinases in Yeast

BEACH, D., B. DURKACZ, and P. NURSE, 1982. Functionally analogous cell cycle control genes in budding and fission yeast. *Nature* 300:706–709.

CROSS, F. 1988. DAF1, a mutant gene affecting size control, pheromone arrest and cell cycle kinetics of *Saccharomyces cerevisiae. Mol. Biol. Cell* 8:4675–4684.

FOSBERG, S. L., and P. NURSE. 1991. Cell cycle regulation in the yeasts *Saccharomyces cerevisiae* and *Schizosaccharomyces pombe. Ann. Rev. Cell Biol.* 7:227–256.

HADWIGER, J. A., et al. 1989. A family of cyclin homologs that control G$_1$ phase in yeast. *Proc. Nat'l. Acad. Sci. USA* 86:6255–6259.

HARTWELL. 1991. Twenty-five years of cell cycle genetics. *Genetics* 129:975–980.

HARTWELL, L. H., J. CULOTTI, and B. REID. 1970. Genetic control of cell division in yeast I. Detection of mutants. *Proc. Nat'l. Acad. Sci. USA* 66:352–359.

HINDLEY, J., and G. PHEAR. 1984. Sequence of the cell division gene *cdc2* from *Schizosaccharomyces pombe;* patterns of intron splicing and homology to protein kinases. *Gene* 31:129–134.

LORINCZ, A., and S. I. REED. 1984. Primary structure homology between the product of yeast cell cycle control gene *CDC28* and vertebrate oncogenes. *Nature* 307:183–185.

NASH, R., et al. 1988. The *WHI1+* gene of *Saccharomyces cerevisiae* tethers cell division to cell size and is a cyclin homolog. *EMBO J.* 7:4335–4346.

NASMYTH, K., and T. HUNT. 1993. Dams and sluices. *Nature* 366:634–635.

REED, S. I. 1992. The role of p34 kinases in the G_1 to S-phase transition. *Ann. Rev. Cell Biol.* 8:529–561.

REED, S. I., J. A. HADWIGER, and A. T. LORINCZ. 1985. Protein kinase activity associated with the product of the yeast cell division cycle gene CDC28. *Proc. Nat'l. Acad. Sci. USA* 82:4055–4059.

RICHARDSON, H. E., et al. 1989. An essential G_1 function for cyclin-like proteins in yeast. *Cell* 59:1127–1133.

Regulation of Cdk-Cyclin Complexes by Binding of Inhibitory Proteins

DUNPHY, W. G., and J. W. NEWPORT. 1989. Fission yeast p13 blocks mitotic activation and tyrosine dephosphorylation of the *Xenopus* cdc2 protein kinase. *Cell* 58:181–191.

GU, Y., C. W. TURCK, and D. O. MORGAN. 1993. Inhibition of Cdk2 activity *in vivo* by an associated 20K regulatory subunit. *Nature* 366:707–710.

HARPER, J. W., et al. 1993. The p21 Cdk-interacting protein Cip1 is a potent inhibitor of G_1 cyclin-dependent kinases. *Cell* 75:805–816.

MENDENHALL, D. 1993. An inhibitor of p34^{cdc28} protein kinase activity from *Saccharomyces cerevisiae*. *Science* 259:216–219.

NOBORI, T., et al. 1994. Deletions of the cyclin-dependent-4 inhibitor gene in multiple human cancers. *Nature* 368:753–756.

PETER, M., et al. 1993. FAR1 links the signal transduction pathway to the cell cycle machinery in yeast. *Cell* 73:747–760.

SERRANO, M., G. J. HANNON, and D. BEACH. 1993. A new regulatory motif in cell-cycle control causing specific inhibition of cyclin D/Cdk4. *Nature* 366:704–707.

Cdk Control of Late G_1 and S-Phase Events in Yeasts

ANDREWS, B. J., and I. HERSKOWITZ. 1989. The yeast SWI4 protein contains a motif present in developmental regulators and is part of a complex involved in cell-cycle-dependent transcription. *Nature* 342:830–833.

BLOW, J. J., and P. NURSE. 1990. A cdc2-like protein is involved in the initiation of DNA replication in *Xenopus* egg extracts. *Cell* 62:855–862.

BREEDEN, L., and K. NASMYTH. 1987. Cell cycle control of the yeast HO gene: *cis*- and *trans*-acting regulators. *Cell* 48:389–397.

JOHNSTON, L. H., and N. F. LOWNDES. 1992. Cell cycle control of DNA synthesis in budding yeast. *Nucl. Acids Res.* 20:2403–2410.

KOCH, C., et al. 1993. A role for the transcription factors Mbp1 and Swi4 in progression from G_1 to S phase. *Science* 261:1551–1557.

Mammalian Cyclins, Cdks, and G_1-S Control

BALDIN, V., et al. 1993. Cyclin D1 is a nuclear protein required for cell cycle progression in G_1. *Genes & Dev.* 7:812–821.

COBRINIK, D., et al. 1992. The retinoblastoma protein and the regulation of cell cycling. *Trends Biochem. Sci.* 17:312–315.

DE BONDT, H. L., et al. 1993. Crystal structure of cyclin-dependent kinase 2. *Nature* 363:595–602.

DULIC, V., E. LEES, and S. I. REED. 1992. Association of human cyclin E with a periodic G_1-S phase protein kinase. *Science* 257:1958–1961.

D'URSO, G., et al. 1990. Cell cycle control of DNA replication by a homologue from human cells of the p34^{cdc2} protein kinase. *Science* 250:786–791.

EWEN, M. E., et al. 1993. Functional interactions of the retinoblastoma protein with mammalian D-type cyclins. *Cell* 73:487–497.

FANG, F., and J. W. NEWPORT. 1991. Evidence that the G_1-S and G_2-M transitions are controlled by different cdc2 proteins in higher eukaryotes. *Cell* 66:731–742.

GIRARD, F., et al. 1991. Cyclin A is required for S phase progression in mammalian fibroblasts. *Cell* 67:1169–1179.

HERSCHMAN, H. R. 1991. Primary response genes induced by growth factors and tumor promoters. *Ann. Rev. Biochem.* 60:281–319.

HIEBERT, S. W., et al. 1992. The interaction of RB and E2F coincides with an inhibition of the transcriptional activity of E2F. *Genes & Dev.* 6:177–185.

HINDS, P. W., et al. 1992. Regulation of retinoblastoma protein functions by ectopic expression of human cyclins. *Cell* 70:993–1006.

HOLLINGSWORTH, R. E., JR., C. E. HENSEY, and W. H. LEE. 1993. Retinoblastoma protein and the cell cycle. *Curr. Opin. Genet. Dev.* 3:55–62.

KOFF, A., et al. 1991. Human cyclin E, a new cyclin that interacts with two members of the CDC2 gene family. *Cell* 66:1217–1228.

LEES, E., et al. 1992. Cyclin E/Cdk2 and cyclin A/Cdk2 kinases associate with p107 and E2F in a temporally distinct manner. *Genes & Dev.* 6:1874–1885.

LEW, D. J., V. DULIC, and S. I. REED. 1991. Isolation of three novel human cyclins by rescue of G_1 cyclin (Cln) function in yeast. *Cell* 66:1197–1206.

LEW, D., and S. REED. 1992. A proliferation of cyclins. *Trends Cell Biol.* 2:77–81.

NEVINS, J. R. 1992. E2F: a link between the Rb tumor suppressor protein and viral oncoproteins. *Science* 258:424–429.

NORBURY, C., and P. NURSE. 1992. Animal cell cycles and their control. *Ann. Rev. Biochem.* 61:441–470.

OHTSUBO, M., and J. M. ROBERTS. 1993. Cyclin-dependent regulation of G_1 in mammalian fibroblasts. *Science* 259:1908–1912.

PARDEE, A. B. 1989. G_1 events and regulation of cell proliferation. *Science* 246:603–608.

PINES, J. 1993. Cyclins and cyclin-dependent kinases: take your partners. *Trends Biochem. Sci.* 18:195–197.

POLYAK, K., et al. 1994. p27Kipl, a cyclin-Cdk inhibitor, links transforming growth factor-beta and contact inhibition to cell cycle arrest. *Genes & Dev.* 8:9–22.

SHERR, C. J. 1993. Mammalian G_1 cyclins. *Cell* 73:1059–1065.

SOLOMON, M. J. 1993. Activation of the various cyclin/Cdc2 protein kinases. *Curr. Opin. Cell Biol.* 5:180–186.

TSAI, L.-H., E. HARLOW, and M. MEYERSON. 1991. Isolation of the human cdk2 gene that encodes the cyclin A- and adenovirus EIA-associated p33 kinase. *Nature* 353:174–177.

WEINTRAUB, S. J., C. A. PRATER, and D. C. DEAN. 1992. Retinoblastoma protein switches the E2F site from a positive to negative element. *Nature* 358:259–261.

XIONG, Y., et al. 1991. Human D-type cyclin. *Cell* 65:691–699.

XIONG, Y., H. ZHANG, and D. BEACH. 1992. D-type cyclins associate with multiple protein kinases and the DNA replication and repair factor PCNA. *Cell* 71:505–514.

ZETTERBERG, A., and O. LARSSON. 1985. Kinetic analysis of regulatory events in G_1 leading to proliferation or quiescence of Swiss 3T3 cells. *Proc. Nat'l. Acad. Sci. USA* 82:5365–5369.

Checkpoint Controls

AL-KHODAIRY, F., and A. M. CARR. 1992. DNA repair mutants defining G_2 checkpoint pathways in *Schizosaccharomyces pombe*. *EMBO J.* 11:1143–1150.

DASSO, M., and J. W. NEWPORT. 1990. Completion of DNA replication is monitored by a feedback system that controls the initiation of mitosis *in vitro*: studies in *Xenopus*. *Cell* 61:811–823.

DULIC, V., et al. 1994. p53-dependent inhibition of cyclin-dependent kinase activities in human fibroblasts during radiation-induced G_1 arrest. *Cell* 76:1013–1023.

EL-DEIRY, W. S., et al. 1993. WAF1, a potential mediator of p53 tumor suppression. *Cell* 75:817–825.

ENOCH, Y., and P. NURSE. 1990. Mutation of fission yeast cell cycle control genes abolishes dependence of mitosis on DNA replication. *Cell* 60:665–673.

HARTWELL, L. H. 1992. Defects in a cell cycle checkpoint may be responsible for the genomic instability of cancer cells. *Cell* 71:543–546.

HARTWELL, L. H., and T. A. WEINERT. 1989. Checkpoint controls that ensure the order of cell cycle events. *Science* 246:629–634.

HOYT, M. A., L. TOTIS, and B. T. ROBERTS. 1991. *S. cerevisiae* genes required for cell cycle arrest in response to loss of microtubule function. *Cell* 66:507–517.

LI, R., and A. W. MURRAY. 1991. Feedback control of mitosis in budding yeast. *Cell* 66:519–531.

MURRAY, A. W. 1992. Creative blocks: checkpoints and feedback controls. *Nature* 359:599–604.

PETER, M., AND I. HERSKOWITZ. 1994. Joining the complex: cyclin-dependent kinase inhibitory proteins and the cell cycle. *Cell* 79:181–184.

ROWLEY, R., J. HUDSON, and P. G. YOUNG. 1992. The *wee1* protein kinase is required for radiation-induced mitotic delay. *Nature* **356**:353–355.

WALKER, D. H., and J. L. MALLER. 1991. Role for cyclin A in the dependence of mitosis on completion of DNA replication. *Nature* **354**:314–317.

WEINERT, T. A. 1992. Dual cell cycle checkpoints sensitive to chromosome replication and DNA damage in the budding yeast *Saccharomyces cerevisiae*. *Radiation Res.* **132**:141–143.

WEINERT, T. A., and L. H. HARTWELL. 1988. The *RAD9* gene controls the cell cycle response to DNA damage in *Saccharomyces cerevisiae*. *Science* **241**:317–322.

26 Cancer

The multiplication of cells is carefully regulated and responsive to specific needs of the body. In a young animal, cell multiplication exceeds cell death, so the animal increases in size; in an adult, the processes of cell birth and death are balanced to produce a steady state. For some adult cell types, renewal is rapid: intestinal cells have a half-life of a few days before they die and are replaced; certain white blood cells are replaced as rapidly. In contrast, human red blood cells have approximately a 100-day half-life, healthy liver cells rarely die, and, in adults, there is a slow loss of brain cells with little or no replacement.

Very occasionally, the exquisite controls that regulate cell multiplication break down and, although the body has no need for further cells of its type, a cell begins to grow and divide. When such a cell has descendants that inherit

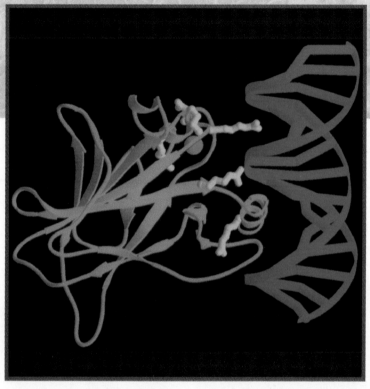

▲ Diagram of p53 protein (light blue) binding to DNA (dark blue) with key amino acids in yellow.

the propensity to grow without responding to regulation, the result is a clone of cells able to expand indefinitely. Ultimately, a mass called a tumor may be formed by this clone of unwanted cells. Because tumors may have devastating effects on animals that harbor them, much research has gone into understanding how they form.

In this chapter we describe current knowledge about the genetic and physiological events that transform a normally regulated cell into one that grows without responding to controls. These genetic events are generally not inherited through the gametes; rather, they are changes in the DNA of somatic cells. The principal type of change is the alteration of preexisting genes to oncogenes, whose products cause the inappropriate cell growth. Thus DNA alteration is at the heart of cancer induction.

➤ *Characteristics of Tumor Cells*

Although most research into the molecular basis of cancer utilizes cells growing in culture, we need first to consider tumors as they occur in experimental animals and in humans. In this way we can see the gross properties of the disease—the properties that ultimately must be explained by analysis of cells and molecules.

Malignant Tumor Cells Are Invasive and Can Spread

Tumors arise with great frequency, especially in older animals and humans, but most pose little risk to their host because they are localized. We call such tumors *benign;* an example is warts. Tumors become life-threatening if they spread throughout the body. Such tumors are called *malignant* and are the cause of cancer.

It is usually apparent when a tumor is benign because it contains cells that closely resemble normal cells and that may function like normal cells. The surface interaction molecules that hold tissues together keep benign tumor cells, like normal cells, localized to appropriate tissues. Benign liver tumors stay in the liver, and benign intestinal tumors stay in the intestine. A fibrous capsule usually delineates the extent of a benign tumor and makes it an easy target for a surgeon. Benign tumors become serious medical problems only if their sheer bulk interferes with normal functions or if they secrete excess amounts of biologically active substances like hormones (Figure 26-1).

The major characteristics that differentiate malignant tumors from benign ones are their invasiveness and spread. Malignant tumors do not remain localized and encapsulated; instead they invade surrounding tissues, get into the body's circulatory system, and set up areas of proliferation away from the site of their original appearance. The spread of tumor cells and establishment of secondary areas of growth is called *metastasis;* malignant cells have the ability to *metastasize* (Figure 26-2).

Malignant cells are usually less well differentiated than benign tumor cells. Furthermore, their properties often vary over time. For example, liver cancer cells may lack certain enzymes characteristic of liver cells and may ulti-

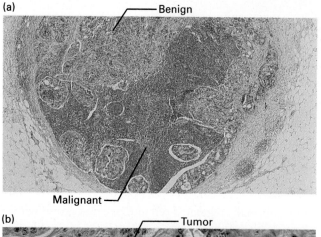

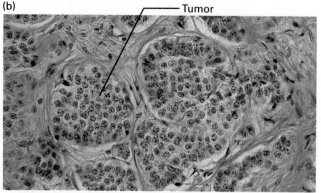

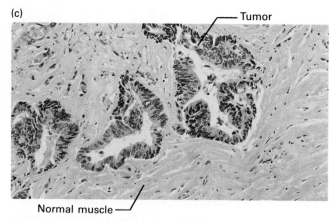

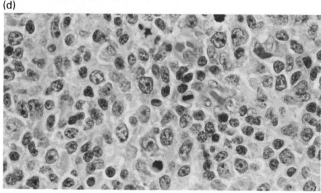

▲ FIGURE 26-1   Sections of various types of tumors. (a) A benign polyp of the lower intestine that has within it malignant tissue. (b) A tumor derived from cells that secrete neuroendocrine hormones. It is organized like a little gland in the midst of normal tissue. (c) A rectal epithelial tumor seen here as invaginations into the normal smooth muscle tissue of the rectum. (d) Lymphoma cells. These cells are leukemic cells that adhere together to form a solid mass. [Photographs courtesy of Dr. J. Aster.]

(a)

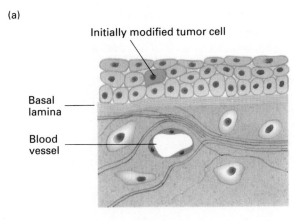

Initially modified tumor cell

Basal lamina

Blood vessel

(b)

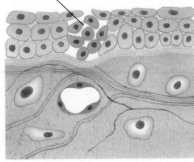

Mass of tumor cells (localized-benign tumor)

(c)

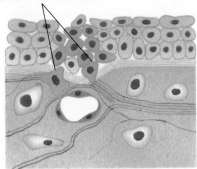

Invasive tumor cells

(d)

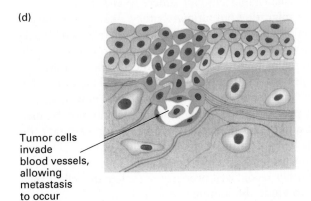

Tumor cells invade blood vessels, allowing metastasis to occur

◄ FIGURE 26-2   Stages in tumor growth and metastasis. (a) A single modified cell appears in a tissue. (b) The modified cell divides, although surrounding cells do not, and a mass of localized tumor cells forms. This tumor is still benign. (c) As it progresses to malignancy, the tumor invades the basal lamina that surrounds the tissue. (d) The tumor cells spread into blood vessels that will distribute them to other sites in the body. If the tumor cells can exit from the blood vessels and grow at distant sites, they are considered malignant, and a patient with such a tumor is said to have cancer.

mately evolve to a state in which they lack most liver-specific functions. This variability of phenotype is often correlated to a variability of genotype; cancer cells have abnormal and unstable numbers of chromosomes, as well as many chromosomal abnormalities. Because apparently benign tumors may progress to malignancy and the earliest stages of malignant tumors are hard to identify, we are rarely sure how a malignancy began.

Cancer cells can be distinguished from normal cells by microscopic examination. In a specific tissue, cancer cells usually exhibit the characteristics of rapidly growing cells, that is, a high nucleus-to-cytoplasm ratio, prominent nucleoli, many mitoses, and relatively little specialized structure. The presence of invading cells in an otherwise normal tissue section is the most diagnostic indication of a malignancy (Figure 26-3).

Malignant cells usually retain enough resemblance to the normal cell type from which they derived that it is possible to classify them by their relationship to normal tissue. Normal animal cells are often classified according to their embryonic tissue of origin, and the naming of tumors has followed suit. Malignant tumors are classified as *carcinomas* if they derive from endoderm or ectoderm and *sarcomas* if they derive from mesoderm. The *leukemias*, a class of the sarcomas, grow as individual cells in the blood, whereas most other tumors are solid masses. (The name leukemia is derived from the Latin for "white blood": the massive proliferation of leukemic cells can cause a patient's blood to appear milky.)

Alterations in Cell-to-Cell Interactions Are Associated with Malignancy

The restriction of a normal cell type to a given organ and/or tissue is maintained by cell-to-cell recognition and by physical barriers. Cells recognize one another by surface interactions, largely by specific protein-protein binding, which dictate that cells of the same type bind together and that certain cells interact within tissues. Primary among the physical barriers that keep tissues separated is the basal lamina, which underlies layers of epithelial cells as well as the endothelial cells of blood vessels (see Figure 23-3).

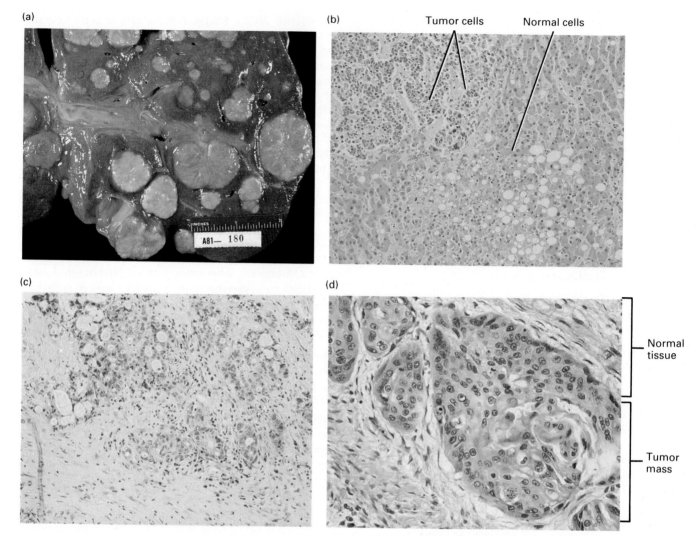

▲ FIGURE 26-3 Gross and microscopic views of tumors invading normal tissue. (a) The gross morphology of a human liver in which a metastatic lung tumor is growing. The white protrusions on the surface of the liver are the tumor masses. (b) A light micrograph of a section of the tumor in (a) showing areas of small, dark-staining tumor cells invading a region of larger, light-staining, normal, liver cells. (c) A section of the adrenal gland from the same patient. Again, invading metastatic tumor cells are evident. (d) Squamous cell carcinoma invading the normal connective tissue of the dermis. [Courtesy of J. Braun.]

Basal laminas define the surfaces of external and internal epithelia and the structure of blood vessels.

Malignant cells overcome the restriction provided by basal laminas and other barriers. Metastatic tumor cells break their contacts within their tissue of origin, make their way into the circulation, and set up a nidus of growth in a foreign environment distant from their normal location. In the process of metastasizing, they may invade adjoining tissue before spreading to distant sites through the circulation (see Figure 26-2c). They often produce elevated levels of both receptors for the basal lamina proteins and enzymes that digest collagen and other basal lamina proteins, such as proteoglycans and glycosaminoglycans. As the proteins disintegrate, the tumor cells penetrate the lamina. The metastatic cells can adhere to endothelial cells to aid their migration but ultimately they escape from the circulation and begin growth in a foreign environment. To set up a metastasis, tumor cells must be able to multiply without a mass of surrounding identical cells and to adhere to new types of cells. The wide range of altered behaviors that underlie malignancy may have their basis in new or variant surface proteins made by malignant cells. In one case, metastatic ability has been conferred on a nonmetastatic rat cell line by expression of an altered form of the CD44 cell-adhesion protein.

Another recognized property of malignant cells is their ability to elude the immune system. It might be expected that cell-surface alterations associated with malignancy would lead the immune system to recognize tumor cells as foreign. The blood indeed contains a *natural killer cell* that can recognize and kill many types of tumor cells while sparing normal cells. Despite this, tumor cells find ways to elude immune detection. They may cover up antigens that would otherwise mark them for destruction, or they may rid themselves of the cell-surface proteins that lymphocytes use to recognize foreign cells (major histocompatibility antigens). In fact, natural killer cells apparently recognize and kill experimental tumor cells injected into an animal precisely because of their low content of major histocompatibility antigens, thus countering one strategy by which tumor cells avoid attack. It remains to be understood just how tumor cells manage it, but they can develop defenses that make the immune system ineffective at eliminating them.

Tumor Cells Lack Normal Controls on Cell Growth

In an animal or in culture, cells are either growing or quiescent. Cell growth involves an increase in cellular mass that leads a cell to divide; cell growth and cell division are thus interrelated phenomena. A quiescent cell is one that is not increasing its mass or passing through the cell cycle. Quiescent cells carry out the characteristic functions of tissues, such as the synthesis of export proteins in the liver or the transmission of impulses in a nerve. Cell growth is a regulated process, so the fraction of growing cells in a given tissue is a function of both the age of the organism and the properties of the tissue. In adults, certain tissues (e.g., the intestine) maintain a constant size by the continued growth of new cells and the death of older cells; such tissues may have many dividing cells. Embryos and expanding tissues of young animals also contain a large fraction of growing cells.

Control of cell division is one of the most important aspects of an animal's physiology. Cells of an adult must divide frequently enough to allow tissues to remain in a steady state, and division must be stimulated at wounds or when special requirements are placed on a tissue. A myriad of cell-specific cytokines signal individual cell types whether to divide or not. Some of these proteins are known—notably ones controlling the growth of blood cells—but many more await discovery.

▶ *Use of Cell Cultures in Cancer Research*

Although questions about cell growth and the induction of cancer are ultimately questions about the behavior of individual cells in a living organism, the study of cells in an animal is impractical because of the difficulty in identifying

the relevant cells, in manipulating their behavior in a controlled manner, and in separating the effects due to the intrinsic properties of the cells from the effects due to the interactions among the many cell types present in the organism. Scientists avoid these difficulties by studying cells growing in culture. The environment of cultured cells can be manipulated by the investigator, the type of target cell can be well defined, the changes in cells following treatment with a cancer-causing agent (a *carcinogen*) can be examined, and the fate of the carcinogen can be determined. Furthermore, cultured cells can be quiescent or growing; they can have, in fact, precisely defined growth parameters. They can also be manipulated genetically. For these reasons, studies of normal cell growth as well as of cancer induction depend heavily on the use of cultured cells.

Fibroblastic, Epithelial, and Nonadherent Cells Grow Readily in Culture

Cell cultures are established by removing cells from an organism and placing them at 37°C in a medium with nutrients plus 5–20 percent blood serum. For many years, most cell types were difficult, if not impossible, to culture. Recent modifications in culture methods have allowed experimenters to grow many specialized cells, and the utility of cell culture as a method of examining cell behavior has increased enormously. Most published studies, however, describe work with the few cell types that grow readily in culture. These are not cells of a defined type; rather, they represent whatever grows when a tissue or an embryo is placed in culture.

The cell type that usually predominates in such cultures is called a *fibroblast* because it secretes the types of proteins associated with the fibroblasts that form fibrous connective tissue in animals. Cultured fibroblasts have the morphology of tissue fibroblasts, but they retain the ability to differentiate into other cell types; thus they are not as differentiated as tissue fibroblasts. They have the properties of mesodermal stem cells because, with appropriate stimulation, they differentiate into mesodermal cells such as fat, connective tissue, muscle, and others. In most studies, however, cultured "fibroblasts" are considered simply as convenient prototypical cells for study. Many studies also have been conducted with cultured *epithelial* cells. Although these do not necessarily correspond to normal tissue cells, they are representative of the type of cell that comes from the ectodermal or endodermal embryonic cell layers.

Cultured fibroblasts and epithelial cells are grown on glass or plastic dishes to which they adhere tightly due to their secretion of matrix proteins such as laminin, fibronectin, and collagen. Neither cell type will ordinarily grow if it is not adhering to a substratum. To prepare tissue cells for culture or to remove adherent cells from a culture dish for biochemical studies, trypsin or another protease is used.

The process of putting cells into culture or of transferring cells to a new culture is often called *plating*.

Certain cells cultured from blood, spleen, or bone marrow adhere poorly, if at all, to a culture dish but nonetheless grow well. In the body, such nonadherent cells are held in suspension (in the blood) or they are loosely adherent (in the bone marrow and spleen). Because these cells often come from immature stages in the blood cell lineages, they are very useful for studying the development of leukemias.

Some Cell Cultures Give Rise to Immortal Cell Lines

When cells are removed from an embryo or an adult animal, most of the adherent ones grow continuously in culture for only a limited time before they spontaneously cease growing. Such a culture dies out even if it is provided with fresh supplies of all of the known nutrients that cells need to grow, including blood serum. This can be seen when human fetal cells are explanted into cell culture. The

majority of cells die and the "fibroblasts" become the predominant cell type. The fibroblasts then divide about 50 times before they cease growth. Starting with 10^6 cells, 50 doublings can produce $10^6 \times 2^{50}$, or more than 10^{20} cells, which is equivalent to the weight of about 10^5 people. Thus, even though its lifetime is limited, a single culture, if carefully maintained, can be studied through many generations. Such a lineage of cells originating from one initial culture is called a *cell strain* (Figure 26-4a).

To be able to clone individual cells, modify cell behavior, or select mutants, scientists often want to maintain cell cultures for many more than 100 doublings. This is possible with cells from some animal species because these cells undergo a change that endows them with the ability to grow indefinitely. A culture of cells with an indefinite life span is considered immortal; such a culture is called a *cell line* to distinguish it from an impermanent *cell strain*. The ability of cultured cells to grow indefinitely varies depending on the animal species from which the cells originate. Among human cells, only tumor cells grow indefinitely, and therefore the HeLa tumor cell has been invaluable for

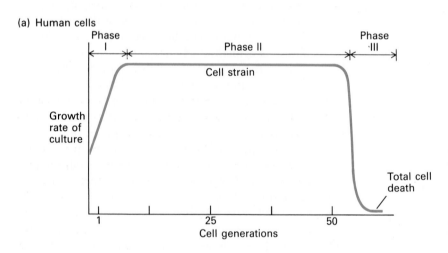

(a) Human cells

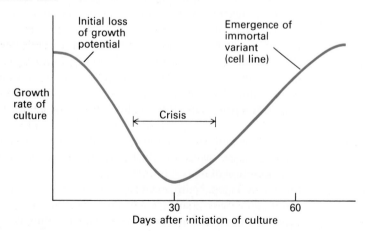

(b) Mouse cells

◄ FIGURE 26-4   Stages in the establishment of a cell culture. (a) Human cells. When an initial explant is made (e.g., from foreskin), some cells die and others (mainly fibroblasts) start to grow; overall, the growth rate increases (*phase I*). If the remaining cells are continually diluted, the cell strain grows at a constant rate for about 50 cell generations (*phase II*), after which growth begins to slow. Ultimately, during the ensuing period of increasing cell death (*phase III*), all the cells in the culture die. (b) Mouse or other rodent cells. In a culture prepared from mouse embryo cells, there is initial cell death coupled with the emergence of healthy growing cells. As these are diluted and allowed to continue growth, they soon begin to lose growth potential and most cells die (the culture goes into crisis). Very rare cells do not die but continue growing until their progeny overgrow the culture. These cells constitute a cell line, which will grow forever if it is appropriately diluted and fed with nutrients: the cells are immortal.

research on human cells. Chicken cells die out after only a few doublings, and even tumor cells from chickens almost never become immortal. Cultures of embryonic adherent cells from rodents however, routinely give rise to cell lines.

When adherent rodent cells are first explanted, they grow well, but after a number of serial replatings they lose growth potential and the culture goes into *crisis*. During crisis most of the cells die, but often a rapidly dividing variant cell arises spontaneously and takes over the culture. A culture founded by such a variant will grow forever if it is provided with the necessary nutrients (Figure 26-4b). Cells in an established line usually have more chromosomes than the normal cell from which they arose, and their chromosome complement continually expands and contracts in culture. The culture is said to be *aneuploid* (i.e., to have an inappropriate number of chromosomes), and the cells of such a culture are obviously mutants.

If rodent cell cultures are maintained at a low cell density until a cell line emerges, the line will consist of flat cells that adhere tightly to the dish in which they are grown. A number of mouse cell lines derived in this fashion have been used extensively in cancer research. These lines are called 3T3 cells because they were derived according to a schedule whereby 3×10^5 cells were *t*ransferred every *3* days into petri dishes with a 50-mm diameter to maintain the appropriate cell density. As is true for other cultured fibroblasts, the exact cell type that gives rise to 3T3 cells is uncertain, but they can differentiate into a range of mesodermally derived cell types, especially endothelial cells (those that line blood vessels). The ability to derive lines of flat cells such as 3T3 set the stage for studying the transition to malignancy in cell culture because cancer cells differ dramatically from 3T3 cells in their growth properties. Before we describe the use of 3T3 cells in cancer research, however, we will consider the control of 3T3 cell growth.

Certain Factors in Serum Are Required for Long-Term Growth of Cultured Cells

If a culture of 3T3 cells is plated at 3×10^5 cells per dish in a medium with 10 percent blood serum, the cells will grow for a few days and then cease growth when there are about 10^6 cells per dish. The culture is said to have reached *saturation density* and the cells are considered to be in the G_0 phase of the cell cycle. Although the quiescent cells in a saturated culture have stopped growing, they can remain viable for a long time and can resume growth if supplied with fresh medium.

Among the treatments that will reinitiate growth in a quiescent 3T3 cell culture is the addition of extra serum to the medium. In fact, the density at which the cells stop growing is in direct proportion to the amount of serum with which they are initially provided (Figure 26-5). Although this result appears to indicate that serum factors are the primary determinants of whether cells remain qui-

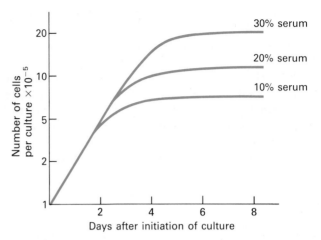

▲ FIGURE 26-5   The dependence of cell growth on serum concentration. The same number of 3T3 cells was used to initiate multiple cultures, each of which was fed with a medium containing the indicated percentages of fetal calf serum. The number of cells per culture was determined daily. The initial growth rates were indistinguishable, but the final number of cells was proportional to the amount of added serum. The experiment shows that serum factors rather than cell contacts control cell growth because cells are already touching one another in 10 percent serum. [See R. W. Holley and J. A. Kiernan, 1968, *Proc. Nat'l. Acad. Sci. USA* **60**:300.]

escent or initiate growth, other results show that they are not the whole story. For example, if a strip of cells is removed from a quiescent cell culture (the culture is said to be *wounded*), the cells at the border of the wound will begin growing and will divide a few times to fill the gap. Because the cell medium is not altered in such an experiment and because the cells that are not adjacent to the wound do not initiate growth, it is clear that local effects (cell-to-cell contacts) also control cell growth. In all probability, the degree to which crucial serum growth factors are available to cells is limited by close contacts.

In recent years, long-term growth of certain cell lines has been achieved with *defined media* containing known components and no serum (see Table 6-3). In such media, serum is replaced with purified proteins present in serum. For example, *epidermal growth factor* (EGF) is needed for growth by almost all cells in culture; many cells also need insulin or an insulinlike factor for growth. These two substances plus transferrin (which makes iron available to cells; Figure 16-51) can largely satisfy the serum requirement for most types of cells. Individual cell types, however, often have exotic requirements, and no universal defined serum substitute has been concocted.

Growth factors and hormones are signaling agents that direct the cell to carry out whatever steps are needed for growth. The factors and hormones themselves neither provide nutrient value to the cell nor play any known role in metabolic pathways. Only their ability to bind to specific cell-surface receptors enables them to control cellular events.

These characteristics of cell growth in vitro provide the background for a consideration of how cancer cells differ from normal cells. Quiescent normal cells are ones that have not been stimulated to grow by factors or hormones. Because unrestricted growth is a characteristic of cancer cells, overcoming the growth inhibition in quiescent cells is a key aspect of cancer induction.

Malignant Transformation Leads to Many Changes in Cultured Cells

Treatment of adherent cells with various agents (e.g., viruses, certain chemicals, radiation) can dramatically change their growth properties in culture. Furthermore, such treated cells can form tumors after they are injected into susceptible animals. Such changes in the growth properties of cultured cells and their subsequent development of tumor-forming capacity are collectively referred to as *malignant transformation*, or just *transformation*. Because

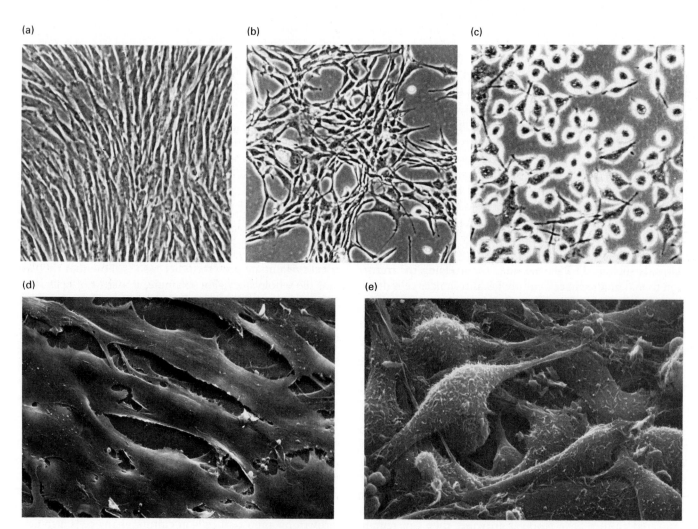

▲ FIGURE 26-6   Normal and transformed rat embryo fibroblasts as viewed in the phase-contrast light microscope. (a) Cultured normal rat embryo fibroblasts. Note that the cells are aligned and closely packed in an orderly fashion. (b) Rat embryo fibroblasts that have been transformed by integration of the polyoma virus gene encoding the mid-T antigen. The cells are crisscrossed and chaotic in their growth. (c) Rat embryo fibroblasts transformed by the Abelson murine leukemia virus. The cells have lost adherence to the dish so completely that they appear almost round. Each has a white halo because rounded cells refract light.

Scanning electron micrographs of normal and transformed 3T3 cells. (d) Normal 3T3 cells. Each straplike image is a cell. Like the normal rat embryo fibroblasts in part (a), the cells are lined up in one direction and are spread out into thin lamellae with bulges representing the nuclei. (e) 3T3 cells transformed by Rous sarcoma virus. The cells are much more rounded, and they are covered with small hairlike processes and bulbous projections. The cells grow one atop the other and they have lost the side-by-side organization of the normal cells. These transformed cells truly appear malignant. [Courtesy of L.-B. Chen.]

transformation can be carried out entirely in cell culture, it has been widely studied as a model of cancer induction in animals, although the two processes are not precisely the same.

Transformed adherent cells usually can be identified by changes in their cellular morphology and growth habit. If, for instance, a growing culture of 3T3 fibroblasts is exposed to SV40 virus, a small proportion of the cells will be infected, and these will become more rounded. The virus-infected cells will be less adherent to one another and to the dish than are the normal surrounding cells, and they will continue to grow when the normal cells have become quiescent. A group, or *focus*, of these loosely adherent, transformed cells can be recognized under the microscope (Figure 26-6). If a focus of transformed cells is recovered and a culture is grown from it, the result will be a line of transformed cells (called, e.g., SV-3T3 cells).

Many properties of a transformed cell line differ from those of its parental line. These differences relate to aspects of growth control, morphology, cell-to-cell interactions, membrane properties, cytoskeletal structure, protein secretion, and gene expression. As will become evident soon, these alterations arise from the multiple actions of the genes that the virus brings into the cell. Some of the changes caused by transformation are probably interrelated, but some seem independent of one another. Not all transformed cells show all of the changes that can be induced by transformation, but the changes described below are common.

Alterations in Growth Parameters and Cell Behavior A key characteristic of transformed cells is that they continue to divide when normal cells cease dividing (Figure 26-7). This increase in saturation density, considered an important analog of malignant cell growth in animals, results from numerous alterations in transformed cells.

Decreased Growth Factor Requirements Because transformed cells have apparently lost some of the hormone and growth factor requirements of normal cells, they grow in initial serum concentrations that are much lower than those required by normal cells. Some transformed cells produce both growth factors and their cognate receptors, providing themselves an autostimulatory growth impetus (called *autocrine* stimulation; see Figure 20-1).

Loss of Capacity for Growth Arrest When the concentration of isoleucine, phosphate, EGF, or any other substance that regulates growth falls below a certain threshold level, normal cells go into quiescence because cells will not initiate growth unless they are nutritionally satisfied. Transformed cells are deficient in their ability to arrest growth in response to lowered nutrient or factor concentrations. Tumor cells may even kill themselves trying to continue growth in an impossible environment.

Loss of Dependence on Anchorage for Growth Normal adherent cells require firm contact with the substratum for growth. If they are plated onto a surface to which they cannot adhere, they will not grow, although they can remain viable for long periods of time. Similarly, if normal cells are suspended in a semisolid medium such as agar, they will metabolize but they will not grow. Most transformed cell lines have lost the requirement for adherence; they grow without attachment to a substratum, as indi-

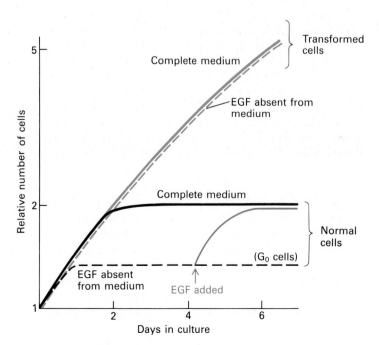

► FIGURE 26-7   Differences in growth of cultures of normal and transformed cells revealed by studies with epidermal growth factor (EGF). Equal numbers of normal and transformed cells were plated into a defined medium with or without EGF at day 0. Cell numbers were determined daily. Transformed cells in the complete medium grew after normal cells had ceased growth. Transformed cells grew even in the medium without EGF.

At day 4, EGF was added to some dishes of normal cells that were initially lacking it. Because the cultures responded to the addition of EGF by growing, it is evident that without EGF, normal cells can remain viable, in a G_0 state.

cated by their ability to form colonies when suspended as single cells in agar. This characteristic correlates extremely well with the ability of transformed cells to form tumors: cells that have lost anchorage dependence generally form tumors with high efficiency when they are injected into animals that cannot immunologically reject them. How the cell senses shape and why transformation abrogates the need for anchorage are both unanswered questions.

Changed Cell Morphology and Growth Habits   The individual transformed cell is very different in shape and appearance from its normal counterpart (see Figure 26-6). It adheres much less firmly to the substratum and therefore is more rounded with fewer processes. Furthermore, transformed cells adhere poorly to each other and do not appear to sense their neighbors. As normal cells in a culture dish become more crowded, they form ordered patterns; transformed cells form chaotic masses. The low mutual adherence of transformed cells coupled with their loss of anchorage to the substratum allows them to grow in multiple layers, whereas normal cells grow in monolayers with some overlap at the cell borders.

Loss of Contact Inhibition of Movement   In culture, normal fibroblastic cells are motile. When two normal cells moving around in a culture dish come into contact, one or both of them will stop and then take off in another direction. This ensures that the cells do not overlap. When a normal cell is surrounded by others in such a way that it has nowhere to go, it ceases movement and forms gap junctions with the surrounding cells (see Figures 24-37 and 24-38). This phenomenon is referred to as *contact inhibition of movement*. Transformed cells lack contact inhibition of movement: they pass over or under one another, they grow on top of one another, and they infrequently form gap junctions.

Cell-Surface Alterations   Many of the foregoing properties of transformed cells relating to growth and behavior are probably consequences of cell-surface events. Some of the differences in the surface properties of normal and transformed cells are described here.

Increased Mobility of Surface Proteins   The glycolipids and glycoproteins that are abundant on the surfaces of normal cells are modified in transformed cells. For example, protein-linked N-acetylneuraminic (sialic) acid is decreased (see Figure 16-27), as is the ganglioside content of all lipids. The general structure of the plasma membrane is not altered, however, and the lateral mobility of surface lipids remains the same. Probably the most important difference is that cell-surface proteins are much more mobile in transformed cells than in normal cells; for example, antibodies can more easily agglutinate a given surface protein on a transformed cell than on a normal cell. Also, plant lectins that bind to sugars on surface proteins can more readily agglutinate tumor cells than normal cells (Figure 26-8).

Although the lipids are not intrinsically more mobile, alteration of the linkages between surface proteins and the underlying cytoskeletal elements are thought to make the proteins more mobile. Modification of the links between the surface proteins and the underlying cytoplasm is probably the basis for the altered morphology of transformed cells.

Increased Glucose Transport   Transformed cells transport glucose more efficiently than normal cells. They have in their membrane a glucose transporter with a higher affinity [lower K_M] for glucose than that ordinarily expressed mainly in brain and erythrocytes. The rapid and high-affinity uptake of glucose due to this transporter correlates to the high glycolytic activity of tumor cells.

Reduced or Absent Surface Fibronectin   Normal quiescent cells in monolayer culture become covered with a dense fibrillar network containing fibronectin as a major protein component (see Figure 24-26). Even growing cells have a diffuse covering of fibronectin. Transformed cells either totally lack fibronectin or have greatly reduced amounts. They may either make little fibronectin or lack the fibronectin receptor. The addition of high concentrations of pure fibronectin derived from normal cells can cause many tumor cells to flatten out and take on a fairly normal appearance; thus the loss of surface fibronectin may be an important determinant of the transformed state.

Loss of Actin Microfilaments   Not only are the surfaces of transformed cells different from those of normal cells, but some cytoskeletal elements also are different. For example, the actin microfilaments that extend the length of normal cells (see Figure 23-53c) are either diffusely distributed or concentrated beneath the cell surface in transformed cells. The loss of cytoskeletal elements has been considered a possible cause of the increased mobility of cell-surface proteins.

Release of Transforming Growth Factors   Transforming growth factors α and β (TGFs) are proteins secreted by transformed cells that can stimulate growth of normal cells. They are factors or hormones that control normal embryonic cell growth but are inappropriately expressed by many tumor cells. Other growth factors can also be made by tumor cells.

Protease Secretion   Transformed cells often secrete a protease called *plasminogen activator*, which cleaves a peptide bond in the serum protein *plasminogen*, converting it to the protease *plasmin*. Thus the secretion of a small amount of plasminogen activator causes a large increase in protease concentration by catalytically activating the abundant plasminogen in normal serum. Normal cells treated

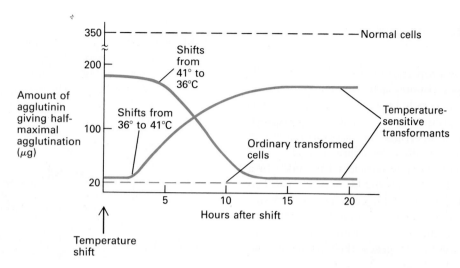

▲ FIGURE 26-8   The agglutination of normal, transformed, and temperature-sensitive transformed cells by wheat germ agglutinin. The chicken embryo cells used for this experiment were of three types: normal fibroblasts, fibroblasts transformed by Rous sarcoma virus, and fibroblasts transformed by a viral mutant that produces transformed cells at 36°C and nearly normal cells at 41°C. The temperature of cells was raised or lowered, and at various times after the temperature shift the amount of wheat germ agglutinin that would give half of the maximal agglutination was determined. For normal cells, 350 μg was required; for the ordinary transformed cells, only 20 μg was required. The temperature-sensitive transformants began to change their properties a few hours after the temperature was shifted. At the higher temperature they became quasinormal, whereas at the lower temperature they took on the characteristics of transformed cells. [See M. M. Burger and G. S. Martin, 1972, *Nature New Biol.* **237**:356.]

with protease exhibit some characteristics of transformation (loss of actin microfilaments, growth stimulation, etc.), suggesting that plasminogen activator secretion may help maintain the transformed state of certain cell lines.

Secretion of plasminogen activator by transformed cells may be related to their tumor-forming capacity because the resulting increase in plasmin may help the cells penetrate the basal lamina. The normally invasive extraembryonic cells of the fetus secrete plasminogen activator when they are implanting in the uterine wall; this provides a compelling analogy to invasion by tumor cells.

Altered Gene Transcription   All of the foregoing characteristics of transformed cells are cytoplasmic or cell-surface activities but changes in proteins encoded in the nucleus are the key aspect of the induction of malignancy. The extraordinary range of differences between normal and transformed cells are largely due to alterations in the transcription of specific genes and in the relative stability of the transcripts. Because tumor cells carry out the same basic processes as normal cells, the mRNA populations of normal cells and transformants derived from them are quite similar. The concentrations of some mRNAs are increased and those of others are decreased, but only about 3 percent of the total mRNA is specific to transformed cells. The transformation specific mRNA, however, includes many different low-abundance species that together can profoundly alter cell behavior in the many ways outlined

above. The proteins translated from these transformation-specific mRNAs, although low in concentration, have diverse effects on cell growth and morphology. Some transformation-specific mRNAs also appear in embryonic cells, and tumor cells have many proteins that are characteristic of embryonic cells; transformation thus alters protein composition toward that characteristic of embryos.

Immortalization of Cell Strains   When cell strains, with their limited growth potential in culture, are used as targets for cell transformation, then one measurable characteristic of transformation is the induction of unlimited growth potential—that is, the conversion of a cell strain into an immortal cell line. (Obviously, immortalization cannot be used as an indicator of transformation in cell lines because they are immortal before exposure to transforming stimuli.) The ease with which transforming stimuli can generate immortal cell lines from cell strains depends on the underlying propensity of the cells to spontaneously acquire immortality: nonadherent (blood) cells from many animals are routinely immortalized by transformation; adherent human cells are rarely immortalized. Adherent chicken cells are almost never immortalized, but adherent rodent cells are readily immortalized. The other characteristics of transformation described above are as relevant to adherent human and chicken cells as they are to mouse cells, which implies that immortalization is quite distinct from other transformation parameters.

Transcription of Oncogenes Can Trigger Transformation

The differences in the properties of normal cultured cells and their transformed counterparts, summarized in the previous section, provide only limited clues to the mechanism or cause of transformation. Despite the variety of these properties, the events triggering transformation can be comparatively simple; often the expression of one or two genes, called *oncogenes,* is sufficient to transform a cell. An oncogene is any gene that encodes a protein able to transform cells in culture or to induce cancer in animals. The word derives from the Greek *onkos,* meaning a bulk or mass; thus *oncology* is the scientific study of tumors. We turn our attention now to oncogenes, their products, and their effects.

➤ *Oncogenes and Their Proteins: Classification and Characteristics*

Of the many oncogenes known, all but a few are derivatives of normal cellular genes. Cellular genes known to be progenitors of oncogenes are called *proto-oncogenes.* They are important for normal cellular processes, but they can be altered, often in very simple ways, to become oncogenes. Because most proto-oncogenes are basic to animal life, they have been highly conserved over eons of evolutionary time.

In this section we discuss how *oncoproteins,* the protein products encoded by oncogenes, can transform cells to malignancy. Most oncogenes derive from genes whose products act along normal cellular growth-controlling pathways. Thus normal growth control and malignancy are two faces of the same coin. Some oncoproteins are mutant forms of the protein products of the corresponding proto-oncogenes. In other cases, the protein products of an oncogene and its proto-oncogene are the same; the oncoprotein causes cancer either by being present where it normally is absent or by being expressed at a level much higher than normal. In these cases the mutation creating the oncogene changes transcriptional control rather than protein structure.

A proto-oncogene and its oncogene are named with a three-letter designation (e.g., *src*). The same code—but not italicized and starting with a capital letter—identifies their protein products (e.g., Src).

Oncogenes Were Initially Identified in Viruses and Tumor Cell DNA

The first oncogenes were found as the transforming genetic material of oncogenic viruses. It was then found that the DNA from many tumors, both human and experimental, can transform 3T3 cells. Using the ubiquitous *Alu* human-specific sequences as a handle, biologists cloned the active

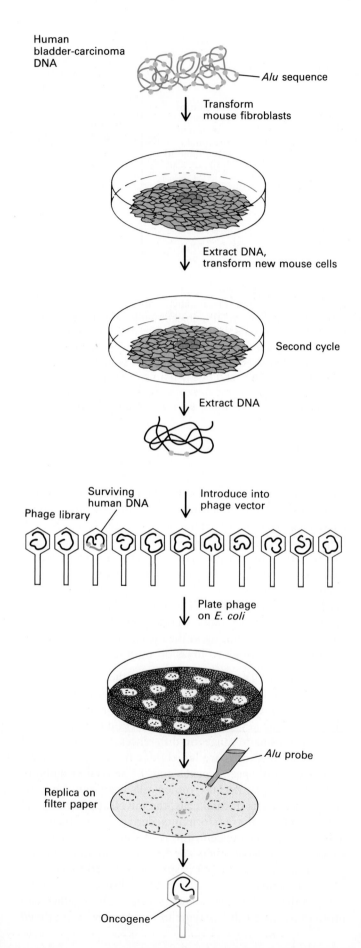

◄ FIGURE 26-9   The isolation of a cellular oncogene from a human tumor by molecular cloning. Like almost all human genes, oncogenes have nearby repetitive DNA sequences called *Alu* sequences. Cloning of an oncogene is achieved by first isolating mouse cells transformed by the oncogene and separating the gene from adventitious human DNA by secondary transfer to mouse cells (see Figure 26-31). The total DNA from a secondary transfected mouse cell is then cloned into bacteriophage λ, and the phage receiving human DNA is identified by hybridization to an *Alu* probe. The hybridizing phage should contain part or all of the oncogene. This expected result can be proved by showing either that the phage DNA can transform cells (if the oncogene has been completely cloned) or that the cloned piece of DNA is always present in cells transformed by DNA transfer from the original donor cell.

genes from tumor cell DNA (Figure 26-9). Many of these genes turned out to be the same ones found in viruses. In particular, the *ras* family of genes was identified as common oncogenes in human and animal tumors.

The oncogene in a virus is denoted with the prefix v (e.g., v-*src*). The corresponding cellular proto-oncogene is given the prefix c (e.g., c-*src*).

Five Types of Proteins Participate in Control of Cell Growth

Although growth control in mammalian cells is only understood in outline, we can differentiate five types of proteins, all discussed elsewhere in this book, that participate in the process: growth factors, growth factor receptors, intracellular signal transducers, nuclear transcription factors, and cell-cycle control proteins (Figure 26-10). Genes for each type of growth-controlling protein have given rise to one or more oncogenes.

Our understanding of growth control starts from the observation that cells in culture can be induced to grow by adding specific *growth factors* to the medium, epidermal growth factor (EGF) being a paradigm. Certain hormones

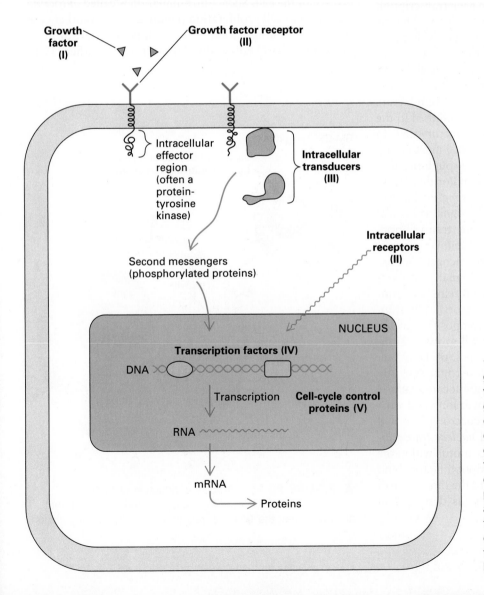

◄ FIGURE 26-10   Control of cell growth involves five types of proteins whose mutant forms lead to cancer: growth factors (I), receptors (II), transducers (III), transcription factors (IV) and cell-cycle control proteins (V). Mutations changing the structure or expression of the first four classes can give rise to dominantly active oncogenes. The class V proteins mainly act as tumor suppressors and mutations in the genes encoding them act recessively to release cells from control and surveillance, greatly increasing the probability that the mutant cells will become tumor cells.

also induce the cultures to grow. Such growth factors and hormones are pure signals; they serve no metabolic purpose except to be ligands for cognate receptors. They are one cell's way to send a message to another cell. Such growth factors and hormones induce many responses in cells—such as mobilization of energy stores and differentiation—as well as entry into the growth cycle. The responding cell has a specific receptor (e.g., the EGF receptor) that is sensitive to the signal. Because different cells have different receptors, specific signals can produce a response in some cell types but not others. For example, EGF stimulates growth of epithelial cells not other cells.

Initiating cell division commits the cell to leave G_0, traverse G_1 and then proceed through the rest of the cell cycle (G_2, S, and M; see Figure 25-1). Occupancy of growth factor receptors is sensed by the cell and, if metabolic conditions are propitious, the cell responds by commiting itself to the growth cycle.

The chain of events that leads to growth is started by ligand-receptor interactions either at the cell surface or inside the cell. Occupancy of a receptor generates many alterations in the cell, particularly changes in gene transcription in the cell's nucleus. Intracellular receptors, typified by the steroid receptors, interact with lipophilic ligands that passively cross through the cell membrane to reach the receptor in the cytoplasm or the nucleus. Those intracellular receptors in the cytoplasm translocate to the nucleus and there the intracellular receptors act as transcription factors (Chapter 11). However, most receptors are tethered to the plasma membrane because they recognize factors that do not permeate into the cell but whose presence must be signaled to the nucleus. The signals are sent through the activation of intracellular *signal transducers* and/or the release of diffusible substances called *second messengers* that can transmit signals throughout the cell. The signal transduction pathways can lead to changes in gene transcription, cell shape, cell metabolism, or any other cellular process.

For the purposes of this discussion of oncogene action, the transcriptional response to a growth signal is of paramount importance because many oncoproteins are transcriptional modifiers (activators or repressors). The transcriptional response alters the protein composition of the cell, providing the proteins critical for cell growth. Transcription is controlled through two types of DNA sequence: promoters, which are located close to the start site of transcription, and enhancers, which are located farther from the start site and are able to act over long and variable distances (see Chapter 11). The control is exerted through specific binding proteins called *nuclear transcription factors* that recognize short sequence motifs within the promoters and enhancers. These transcription factors bind to the motifs and then accelerate or retard the rate at which RNA polymerase II initiates transcripts (as described in Chapter 11).

Oncoproteins Affect the Cell's Growth-Control Systems in Various Ways

We have described five types of gene products that participate in growth control. Most known oncoproteins appear to be identical or similar to one or another of these five types of growth controllers (see Figure 26-9). Table 26-1 on page 1262 lists some representative oncogenes, classified in terms of the five types of growth-controlling gene products.

Growth Factors (Class I) Oncogenes rarely arise from genes encoding growth factors. In fact, only one naturally occurring growth factor oncogene—*sis*—has been discovered. The *sis* oncogene, which encodes a type of the platelet-derived growth factor (PDGF), can transform cells that naturally have the PDGF receptor. Artificial class I oncogenes have been created. For example, if the gene encoding the granulocyte-macrophage colony stimulating factor (GM-CSF) is inserted into a cell that has the GM-CSF receptor, the GM-CSF protein continually stimulates growth of the cell. Such autostimulation is called *autocrine* induction of cell proliferation (Figure 26-11). Another case of autocrine stimulation is the release of transforming growth factors (TGFs), which we mentioned earlier. TGF-

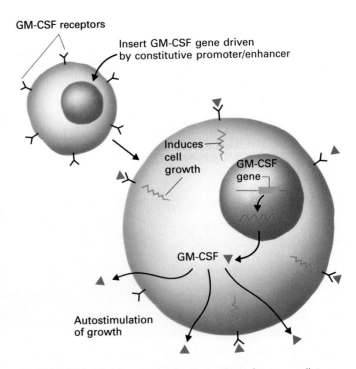

GM-CSF receptors

Insert GM-CSF gene driven by constitutive promoter/enhancer

Induces cell growth

GM-CSF gene

GM-CSF

Autostimulation of growth

▲ FIGURE 26-11 Autocrine induction of tumor-cell growth. A factor (GM-CSF) gene is inserted into a cell that already carries the GM-CSF receptor, causing autostimulation of growth.

α, for example, is an EGF analog that is released by many tumor cells and binds to EGF receptors, stimulating cell growth.

Receptors for Growth Factors and Hormones (Class II) A receptor binds a specific factor and then the receptor, recognizing that it has bound the factor, sends a growth signal to the cell through one or more signal transduction pathways. Some cell-surface receptors have integral protein-tyrosine kinases in their cytoplasmic domains; these receptors transmit the growth signal by phosphorylating tyrosine residues on themselves as well as on one or more target proteins, thus initiating a cascade of events (see Figure 20-3). The genes for such receptors become oncogenes when they are mutated in such a way that the receptor remains active even in the absence of its bound ligand. For example, a change in one codon specifying an amino acid in the transmembrane domain of the receptor protein encoded by the *neu* gene produces the *neu* oncogene (also called *her-2*) (Figure 26-12). In most cases, much of the extracellular ligand-binding domain is deleted during the formation of the oncogene (as in ErbB). Thus it is believed that the signal for growth given by the oncoprotein is the usual signal, but the altered receptor has lost ligand control and thus is constitutively active. A cell containing a constitutively active oncogene grows independently of the factors that are supposed to regulate its growth.

Another example is the product of the *erbA* oncogene, whose proto-oncogene codes for the intracellular receptor for thyroid hormone. Binding of the hormone transforms the receptor into a transcription factor. In this case, the oncogene codes for a modified receptor, ErbA, which appears to act as a negative factor, competing with the endogenous thyroid hormone receptor and somehow facilitating cell growth. ErbA does not fully transform cells; it rather works coordinately with ErbB, the EGF-receptor derivative. Both the *erbA* and *erbB* oncogenes were first recognized in the avian erythroblastosis virus in which they act synergistically to give full oncogenicity.

Intracellular Signal Transducers (Class III) The largest class of oncogenes are those whose proto-oncogenes encode proteins that act as intracellular transducers, proteins that transmit signals from a receptor to a cellular target. Among the best understood transducers are the G proteins, one of which, G_S, controls cAMP synthesis. G_S protein senses that a ligand has occupied a cell-surface receptor, binds GTP, and activates the adenyl cyclase enzyme that forms cAMP. It then hydrolyzes the GTP and returns to an inactive state. A mutation in the gene for G_S can make it an oncogene by eliminating the GTPase activity, thus causing it to signal perpetually, producing a constitutive rise in cAMP. A high constitutive cAMP concentration can cause unregulated proliferation of pituitary cells and therefore pituitary tumors.

Other class III oncoproteins are variant forms of other transducing proteins. Many proto-oncogenes of class III oncogenes, such as *src* and *abl*, encode nonreceptor protein-tyrosine kinase (PTK). These differ from the cell-surface receptors produced by the class II oncogenes in that they are cytoplasmic or nuclear proteins that have no transmembrane or extracellular domains. Many such PTK proteins have myristate, a long-chain fatty acid, bound to

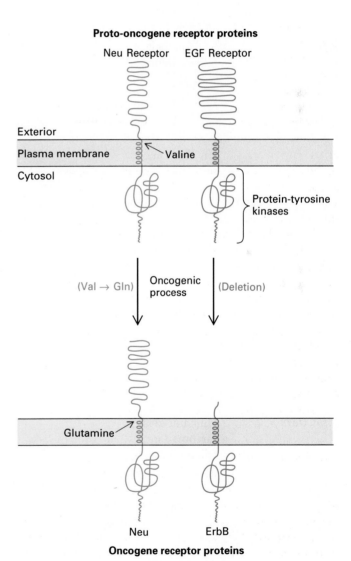

▲ FIGURE 26-12 Creation of oncogenes from proto-oncogenes that encode cell-surface receptors. In the example diagrammed in red, the *neu* oncogene arises from a mutation that alters a single amino acid (valine to glutamine) in the transmembrane region, somehow making the protein constitutively active as a kinase. In the other example, the receptor is for EGF; the oncogene arises by loss of the coding region for the EGF-binding domain.

TABLE 26-1   Selected Oncogenes and the Proteins of Their Proto-Oncogenes

Oncogene	Oncogene Found in: Animal Retrovirus	Nonviral Tumor	Subcellular Location of Protein	Protein Encoded by Proto-Oncogene
CLASS I: GROWTH FACTORS				
sis	Simian sarcoma		Secreted	Platelet-derived growth factor
CLASS II: RECEPTORS				
A. Cell-surface receptors with protein-tyrosine kinase activity				
fms	McDonough feline sarcoma		Plasma membrane	CSF-1 receptor
erbB	Avian erythroblastosis		Plasma membrane	Epidermal growth factor receptor
neu (or *erbB-2*)		Neuroblastoma	Plasma membrane	Related to epidermal growth factor receptor
ros	UR II avian sarcoma		Plasma membrane	Related to insulin receptor
B. Intracellular receptors				
erbA	Avian erythroblastosis		Nuclear	Thyroid hormone receptor
CLASS III: INTRACELLULAR TRANSDUCERS				
A. Protein-tyrosine kinase				
src	Rous avian sarcoma		Cytoplasm	⎫
yes	Yamaguchi avian sarcoma		Cytoplasm	
fps (*fes*)	Fujinami avian sarcoma (and feline sarcoma)		Cytoplasm	Protein kinases that phosphorylate tyrosine residues
abl	Abelson murine leukemia	Chronic myelogenous leukemia	Cytoplasm and nucleus	
met		Murine osteosarcoma		⎭
B. Protein-serine/threonine kinases				
mos	Moloney murine sarcoma		Cytoplasm	⎫ Protein kinases specific for serine or threonine
raf (*mil*)	3611 murine sarcoma		Cytoplasm	⎭

TABLE 26-1 *(Continued)*

Oncogene	Oncogene Found in: Animal Retrovirus	Oncogene Found in: Nonviral Tumor	Subcellular Location of Protein	Protein Encoded by Proto-Oncogene
C. *Ras* proteins				
Ha-*ras*	Harvey murine sarcoma	Bladder, mammary, and skin carcinomas	Plasma membrane	
Ki-*ras*	Kirsten murine sarcoma	Lung and colon carcinomas	Plasma membrane	Guanine nucleotide-binding proteins with GTPase activity
N-*ras*		Neuroblastoma and leukemias	Plasma membrane	
D. Adaptors				
crk	Avian sarcoma virus		Cytoplasm	Contains protein with SH2 and SH3 domains but no catalytic domain
CLASS IV: NUCLEAR TRANSCRIPTION FACTORS				
jun	Avian sarcoma virus 17		Nucleus	Transcription factor AP1
fos	FBJ osteo-sarcoma		Nucleus	
myc	Avian MC29 myelocytomatosis		Nucleus	
N-*myc*		Neuroblastoma	Nucleus	
myb	Avian myelo-blastosis	Leukemia	Nucleus	Proteins that regulate transcription
ski	Avian SKV770		Nucleus	
rel	Avian reticuloendo-theliosis		Nucleus and cytoplasm	
CLASS V: CELL CYCLE CONTROL PROTEINS				
RB		Retinoblastoma and osteosarcoma	Nucleus	Tumor suppressors
p53		Most types of human cancer	Nucleus	

their N-terminal glycine. This causes them to be partially bound to the plasma membrane and puts their kinase domain in the same perimembrane region as that of the receptor kinases (see Figure 14-20). For this reason, it is assumed that the cytoplasmic kinases serve to transduce signals in the same way as do the receptor kinases. If the N-terminal glycine, which is required for myristoylation, is removed from proteins encoded by the *src* and *abl* oncogenes, they can no longer bind to the plasma membrane. They then do not transform fibroblasts, supporting this hypothesis.

Tyrosine-specific protein phosphorylation occurs only rarely in normal cells. Actually, protein-tyrosine kinases were first recognized as oncoproteins; later it was shown that about 0.1 percent of protein-linked phosphate in normal cells is bound to tyrosine residues. Cells transformed by the viruses that encode tyrosine-specific kinases have levels of phosphotyrosine in their proteins 10-fold or more higher than normal cells. The excess phosphotyrosine is distributed over many proteins, which suggests that either the viral kinases phosphorylate a number of different protein substrates or that the phosphorylation of certain key proteins sets off a spate of further phosphorylations. Even nonkinase oncoproteins can increase phosphotyrosine levels in cells by activation of cellular kinases.

Direct evidence that the viral kinases have a very low substrate specificity has come from experiments in which oncoproteins have been produced by bacteria. To accomplish this, oncogenes removed from cloned retroviral DNA were placed into vectors that allowed their expression in bacteria. The kinase encoded by Abelson virus, for example, has been expressed in *Escherichia coli* with this procedure. When a bacterial cell contains the Abelson kinase, many tyrosine residues in bacterial proteins become phosphorylated. Because ordinarily there is no phosphotyrosine in *E. coli* proteins, the kinase encoded by v-*abl* must phosphorylate many substrate proteins. These experiments and others have shown that kinases encoded by oncogenes recognize and phosphorylate a broad range of target proteins. Thus phosphorylation of one or more crucial target proteins appears to be the mechanism by which some oncogenes induce the transformed state. Multiple tyrosine-phosphorylated proteins appear to conspire to cause transformation.

Vinculin, a protein that helps link the actin cytoskeleton to the cell membrane (see Figure 22-35), is one of the proteins known to have an elevated phosphotyrosine level in transformed cells. The alteration of such a protein could easily be imagined to cause some of the hallmarks of the transformed state: changed cell morphology, reduced concentrations of actin microfilaments, loss of the anchorage dependence of growth, and changed mobility of cell-surface proteins. In this regard it may be pertinent that the *src* and *abl* oncogenes encode protein-tyrosine kinases that are localized to adhesion plaques on the plasma membrane, sites of bonding between cells and their under-lying support where vinculin is also concentrated (see Figure 23-27).

The alteration by which c-*src* becomes an oncogene has been studied in great detail. The c-*src* product, a 60-kDa protein denoted $pp60^{c-src}$ (or c-Src), has multiple phosphorylation sites through which it is regulated. Phosphorylation of a tyrosine residue at position 527, six from the C-terminus, causes a great reduction in its kinase activity and is thought to negatively regulate c-Src. This site is often altered in Src oncoproteins that have constitutive kinase activity. In Rous sarcoma virus, for instance, the *src* gene has suffered a deletion that eliminates the C-terminal 18 amino acids of c-Src. The kinase activity of c-Src can also be augmented by its binding to polyoma mid-T protein, and this may be part of the mechanism by which polyoma transforms cells. Both c-Src and c-Abl can also be activated by changes in the N-terminal region, which is not part of the kinase domain. This region, which must regulate the kinase, is considered later in the discussion of the *crk* oncogene.

Another set of class III oncogenes are the various *ras* genes (see Figure 20-17). Their protein products bind GTP and slowly hydrolyze it, as do the G proteins described above. Like the *src*-related proteins, they have a covalently attached fatty acid (a farnesyl group) and are found at the inner side of the cell's plasma membrane (see Figure 14-22).

The *ras* oncogenes were the first nonviral oncogenes to be recognized. One mutation by which a *ras* proto-oncogene becomes an oncogene causes only one change in the protein, substitution of valine for glycine at position 12 of the sequence. This simple mutation reduces the protein's GTPase activity, thus linking GTP hydrolysis to the maintenance of normal, controlled Ras function (as in the G_s protein described earlier). The structures of the Ras proteins are known from crystallography; the oncogenic change causes only a slight alteration in the structure, but this is sufficient to change the proto-oncogene into an oncogene. This tiny change initiates the transformation of cells from normal to malignant. Many human tumors have activated *ras* genes.

The product of the *crk* (pronounced "crack") oncogene has a particularly intriguing structure. Crk protein has no known biochemical activity but has domains called SH2 and SH3 in common with other oncogenes (Figure 26-13). These two regions of the protein interact with other proteins. They are called SH domains because they were first recognized in *src* and are therefore Src Homologies. They are found in many signalling proteins such as protein kinases and phospholipases (Figure 26-13). SH2 domains interact with short peptides at sites in proteins that have a phosphorylated tyrosine; SH2 domains thus bind Crk and the many other proteins that have them onto the target proteins of protein-tyrosine kinases. SH3 domains interact with proline-rich regions of targets. In both cases, the SH domains bind selectively to particular se-

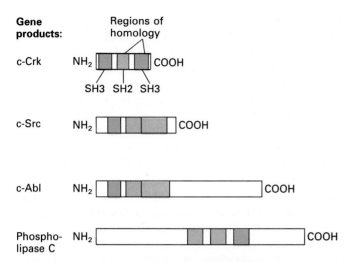

FIGURE 26-13 Proteins with SH2 and SH3 domains. Numerous signal transduction and cytoskeleton proteins have domains of 50 or 100 amino acids that act as specific binding regions for other proteins, a sort of molecular Velcro but with great specificity of interaction. Two of these are SH2—which binds to short peptides containing phosphorylated tyrosine residues—and SH3—which binds to stretches of about 10 amino acids containing a proline-X-X-proline motif (where X is any amino acid). Shown are four proteins that have these domains: an adaptor protein (Crk), two protein-tyrosine kinases (Src and Abl) and a form of phospholipase C. The orange region represents the kinase domains while the SH2 and SH3 regions are shown in blue and green, respectively.

quences around the phosphotyrosine or prolines and mediate highly specific interactions. SH domains are found in many oncoproteins and can be key elements in their transforming activity. Thus, a central event of oncogenesis is the formation of specific protein aggregates that normally serve as signalling units for cellular events but that inappropriately signal the growth and metastatic abilities of cancer cells. This is particularly apparent from the structure of Crk because it causes cancer without having any catalytic domain and is therefore a pure *adaptor,* a molecule that can only glue other proteins together.

Nuclear Transcription Factors (Class IV) By one mechanism or another, all oncogenes eventually cause changes in the proportions of different mRNAs in cells both because growing cells make many proteins at different rates than they are made in quiescent cells and because tumor cells have properties different from their normal counterparts. The proteins encoded by the class IV oncogenes exert direct effects on transcriptional rates by actually binding to DNA and modifying rates of transcriptional initiation. A good example is the *jun* oncogene. The c-*jun* proto-oncogene encodes part of the AP-1 transcription factor, which binds to a sequence found in promoters and enhancers of many genes. AP-1 often consists of a dimeric complex of

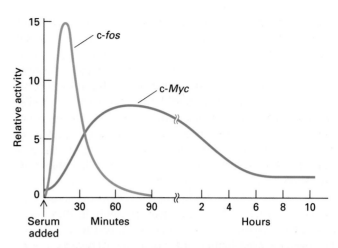

FIGURE 26-14 The response of two proto-oncogenes, *fos* and *myc*, to serum stimulation of the growth of quiescent 3T3 cells. Serum contains factors like platelet-derived growth factor (PDGF) that stimulate the growth of quiescent cells. [See M. E. Greenberg and E. B. Ziff, 1984, *Nature* **311**:433.]

Jun with Fos; *fos*, too, is a known oncogene. Presumably Fos and Jun act as oncoproteins by activating transcription of key genes that encode growth-promoting proteins—or by inhibiting transcription of growth-repressing genes—but the targets and pathways have yet to be elucidated.

Many nuclear proto-oncogene proteins are induced when normal cells are stimulated to grow, indicating their direct role in growth control. For example, PDGF treatment of quiescent 3T3 cells induces an increase (up to 50-fold) in the production of c-Fos and c-Myc, the normal products of the *fos* and *myc* proto-oncogenes. Initially there is a transient rise of c-Fos and later a more prolonged rise of c-Myc (Figure 26-14). Although c-Fos rapidly disappears, c-Myc stays at a somewhat elevated level. In both cases, the proteins are maintained at high levels only briefly, probably because prolonged expression would be oncogenic. In fact, the oncogenic forms of these proteins are expressed at high and unregulated levels. To assure the rapid loss of c-Fos after its induction in normal cells, both the protein and its mRNA are intrinsically unstable and some of the changes that turn *fos* from a normal gene to an oncogene involve loss of sequences in the gene that lead to the instability.

Cell-Cycle Control Proteins (Class V) The entry of cells into the cell cycle from a quiescent state and the progression of cells around the cycle are precisely controlled events. This assures that cellular growth and the coordination of DNA synthesis with cell-size increase and cytokinesis are monitored and do not fall out of regulated synchrony. Cyclins, cyclin-dependent kinases, and the p53 and RB check-point regulators are all elements of this control system (see Chapter 25). Altered regulation of expression of at least one cyclin, as well as mutation of *p53* and *RB*, can be oncogenic.

RB and p53 are the prototypes of a class of proteins encoded by *tumor-suppressor genes*. Their ability to check cell-cycle progression, and hold cells in quiescence or even lead cells to commit suicide unless conditions are appropriate for cell-cycle progression, means that they can prevent cells from becoming cancerous (see Chapter 25). This provides a strong selective advantage to cells that lose both copies of a tumor suppressor gene and therefore can grow even when conditions are not optimal. Certain cancer cells lack both copies of RB because of multiple mutations, chromosome losses or gene conversion. p53 functions as a tetramer, or even higher order oligomer. This means that the mutation of even one *p53* allele in a cell can abrogate almost all p53 activity because virtually all of the oligomers will contain at least one defective subunit: the *p53* mutations act as dominant negatives. In contrast, RB protein functions as a momomer and mutation of a single *RB* allele has little consequence. Mutations in *RB* are recessive to wild type, making them quite different from the dominant oncogenes or the dominant negative mutations in *p53*. Like *RB*, most tumor-suppressor genes are recessive to wild type.

Tumor-suppressor genes have often been described as *anti-oncogenes* because of their ability to prevent cancer. In cells that lack p53, the loss of the ability to check cell-cycle progression leads to increased rates of DNA alteration—mutations and translocations—speeding up the evolution of new oncogenes and thus making these cells particularly dangerous. About half of all human tumors have mutations in *p53*, most of them missense mutations that can act as dominant negatives (Figure 26-15).

Other genes that are linked to p53 are also mutated in human cancer cells. The Cip1 protein, whose synthesis is controlled by p53 (see Chapter 25), has been shown to be mutated, and a relative, called p16, may also be mutated. p53 is controlled by a protein encoded by the *mdm2* gene, and tumors that do not lack functional p53 may have over-expression of *mdm2*. Therefore, many human cancers have mutations in genes that exert control over the cell cycle, such as *RB*, *p53*, *mdm2*, *Cip1*, *p16*, and *cyclin D*, showing that cell-cycle control is a key element that must be neutralized for cancer to develop.

There are other types of recessive tumor-suppressor genes. Many were first identified in humans as inherited propensities to develop cancer. Inherited mutation of one copy of these genes increases the probability that a tumor will develop because now one further mutation is sufficient to ablate the function of the gene. *RB* was first identified in this way as a mutation that caused retinoblastoma as was

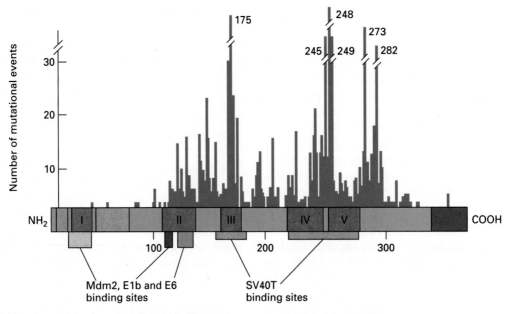

▲ FIGURE 26-15 Mutations in human tumors that inactivate the function of p53 protein. The upper bar graph shows the sites of 1361 mutations in p53 that have been found in human tumors. Mutational hot spots are evident at sites such as amino acid 248 where 112 of the mutations occurred. The central part of the protein is required for DNA binding and most of the mutations eliminate the ability of the protein to bind to DNA. The structural map of the protein shows its complexity. It includes five evolutionarily conserved regions (red), transactivation regions (orange) and an oligomerization region (green). The Mdm2 protein binds at the indicated site and inactivates p53 function as part of the normal control of p53 function. Viral proteins can also inactivate it by binding to the indicated sites (E1b is from adenovirus, E6 is from a papilloma virus). [From C. C. Harris, 1993, *Science* **262**:1980.]

WT1 for Wilm's tumor, *NF1* for Schwannomas and *APC* for colon cancer. *p53* is mutated in the Li-Fraumeni syndrome of multiple inherited cancers. Genes involved in the differentiation of particular cell types can be cell-specific suppressors because their loss prevents differentiation and therefore predisposes to development of unregulated growth in the cell. Loss of genes for excising damaged DNA is oncogenic because of the possibility that oncogenes will be formed if damage is not repaired. Such a situation has been uncovered in an inherited propensity to develop colon cancer.

Apoptosis, or Induced Cell Suicide, Is One Mechanism of Protection Against Cancer

One of the most important protections we have against cancer is the death of cells that are not wanted by the body. This death mechanism, called, *apoptosis,* is induced in cells when they are damaged, unnecessary to the body, or dangerous to it—as are autoimmune cells (see Chapter 27). Many cells, maybe all cells, can be triggered to kill themselves and, in the process, to signal macrophages to engulf and degrade their corpses. Apoptosis is a key mechanism in developing organisms, used to rid the body of excess cells. For instance, our fingers are separated because embryonic cells that joined them died during development; in aquatic birds these cells live and form the webbing of the hind legs. Apoptosis is also a mechanism used in the thymus to eliminate self-reactive T lymphocytes and avoid autoimmunity (see Chapter 27). When T cells kill other cells, they activate the latent apoptosis pathway in their targets.

P53 can induce apoptosis as one its modes of protecting the body against cells that are behaving in a discoordinated fashion or that have damaged DNA. The Bcl-2 protein is one that can overcome the apoptosis mechanism. Its normal function is apparently to protect valuable cells against apoptosis and it is generally down-regulated when cells become sensitive to apoptotic mechanisms. Inappropriate expression of *bcl-2*, however, induces cell overgrowth. In *bcl-2* transgenic mice, cells that would ordinarily die can live; conversely, animals lacking *bcl-2* experience loss of cells that would ordinarily live. The gene *bcl-2* is activated in a particular form of human B cell cancer where it acts not as a direct oncogene but rather to allow the continued proliferation of cells that should die, thus increasing the chance that they will transform into lethal tumors.

Oncoproteins Act Cooperatively in Transformation and Tumor Induction

When primary rat embryo cells are transfected with a *ras* oncogene, they show the morphological changes associated with transformation but are not immortalized. Transfection with *ras* plus *myc*, however, leads to fully trans-

formed, immortal, tumorigenic cell lines. Various other nuclear oncogenes can replace *myc*. Thus it appears that two classes of transforming function can be defined: *ras*-like and *myc*-like. Because the Ras proteins are cytoplasmic and the Myc protein is nuclear, transformation would appear to be a two-step process, a cytoplasmic alteration of cell behavior plus a change in gene transcription.

This neat picture of transformation as a two-step process is clouded by observations that if *ras* or *myc* is provided with retroviral LTRs, which act as strong promoters and enhancers, either gene will morphologically transform cell lines, and *ras* will also transform primary cells to immortality. Thus high levels of the oncoproteins may have different effects from lower levels, and the transformation of a given cell may be affected both by which oncogenes the cell expresses and by the amount of the oncoproteins in the cell.

The two-step model of transformation, however, sheds some light on a phenomenon described early in this chapter. We noted that embryonic, adherent, rodent cells have a low but reproducible probability of spontaneously becoming cell lines. We can now hypothesize that the event responsible is the activation of an immortalizing oncogene or loss of a tumor suppressor gene. Thus 3T3 cell lines—as opposed to embryonic rodent cells—may provide such a good substrate for assaying *ras* oncogenes because they already have an active *myc*-like oncogene.

The ability of an oncogene to immortalize cells is consistent with our earlier portrayal of immortalization as a parameter of transformation quite distinct from the others. Cell "mortality" is probably a consequence of differentiation events that change a stem cell into an end-stage, nondividing cell. Thus immortalization may actually be the outcome of a blockade of differentiation caused by the action of oncogenes. In fact, in myeloid or erythroid cells, *myc* does act as an inhibitor of differentiation.

One especially powerful technique for analyzing oncogene action is to insert an oncogene with a specific promoter/enhancer into the genome of a mouse strain, forming a transgenic mouse (see Figure 8-39). In a prototype of such experiments, the SV40 T oncogene, a DNA virus oncogene described in the next section, has been coupled to an insulin promoter/enhancer and inserted as a transgene. Consistent with the known fact that insulin is made only in the β cells of pancreatic islets of Langerhans, such transgenic mice get tumors of islet β cells. Although many cells express the gene, only a rare cell gives rise to a malignant clone. Thus T-protein expression is not sufficient to cause transformation to malignancy; another event must occur and it is rare. Likely possibilities are the activation of a second, cellular oncogene or the inactivation of a tumor suppressor gene.

The cooperativity of multiple oncogenes in mouse tumor formation has been shown most dramatically using the *ras* and *myc* genes as transgenes driven by a mammary cell-specific promoter/enhancer from a virus (Figure 26-

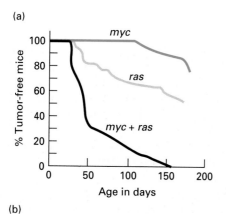

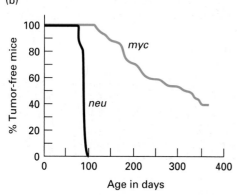

▲ FIGURE 26-16   Tumors in transgenic mice. (a) Kinetics of tumor appearance in female transgenic mice carrying transgenes driven by the mouse mammary tumor virus (MMTV) LTR. Shown are results for mice carrying *myc* and *ras* transgenes as well as for the progeny of a cross of *myc* carriers by *ras* carriers. The percentage of tumor-free mice graphically depicts the time-course of tumorigenesis. Females were studied because the hormonal stimulation of pregnancy activates the MMTV-driven oncogenes. (b) Comparison of the tumor incidence in *myc* transgenics and *neu* transgenics. Note that the appearance of *neu*-induced tumors is almost simultaneous. [Part (a) see E. Sinn et al., 1987, *Cell* **49**:465. Part (b) see W. J. Muller et al., 1988, *Cell* **54**:105.]

16a). By itself, *myc* causes tumors only after 100 days and then in only a few mice; *ras* causes tumors earlier but still slowly and with about 50 percent efficiency over 150 days. When the *myc*- and *ras*-transgenics are crossed, however, so that all mammary cells express both genes, tumors arise much more rapidly and all animals succumb to cancer. Such experiments emphasize the synergistic effects of multiple oncogenes. Not all oncogenes, however, need partners. The *neu* oncogene in certain transgenic animals causes tumors so efficiently (Figure 26-16b) that little normal mammary epithelium is present in pregnant *neu*-transgenic mice because so many cells have been transformed to malignancy.

In humans, the kinetics of appearance of tumors suggests that many different events conspire together to cause cancer. This is generally considered as the reason that cancer is a disease of old age—it takes a long time for the causative mutations to accumulate in any single cell. Such mutations may be either dominantly acting oncogenes or recessive mutations of tumor suppressor genes. Our longevity relative to that of mice, for instance, suggests that the human species has evolved multiple ways to counter the tendency of cells to accumulate mutations, so that human cells have protections that rodent cells either lack or have in a less efficient form. Like the continual battle that we carry out with the infectious agents that surround us, we also are continually battling the tendency of cells to become transformed into cancer cells.

Consistent Chromosomal Anomalies Associated with Tumors Point to the Presence of Oncogenes

It has long been known that chromosomal abnormalities abound in tumor cells. Human cells ordinarily have 23 pairs of chromosomes, recognized by their well-defined substructure, but tumor cells are usually *aneuploid* (i.e., they have an abnormal number of chromosomes—generally too many), and they often contain *translocations* (fused elements from different chromosomes). As a rule, these unusual characteristics are not the same from tumor to tumor: each tumor has its own set of anomalies. Certain anomalies recur, however, and they point to the presence of oncogenes. The first to be discovered, the Philadelphia chromosome, is found in the cells of virtually all patients with the disease chronic myelogenous leukemia. This chromosome is a fusion of most of chromosome 9 to a piece of chromosome 22. The reciprocal translocation—in which a tiny piece of chromosome 9 is fused to the broken end of chromosome 22—is also present, although it is not easily recognized. *Abl* was first identified as an oncogene in human cancer when it was discovered that the c-*abl* proto-oncogene lies at the break point on chromosome 9, and the translocation causes the formation of an mRNA derived partly from chromosome 22 and partly from c-*abl*.

In cells of the tumor known as Burkitt's lymphoma, antibody genes are translocated to chromosome 8, just at the site of the *myc* gene. The analogous site in the mouse genome is also involved in mouse myelomas. In both cases, the tumor cells are antibody-producing cells, and such cells carry out DNA rearrangements during their maturation (see Chapter 27). c-*myc* translocation is a rare abberance of the normal rearrangement events, bringing it from its normal, distant chromosomal location into juxtaposition with the antibody genes. The translocated gene comes under the transcriptional regulation of the antibody genes

(a)

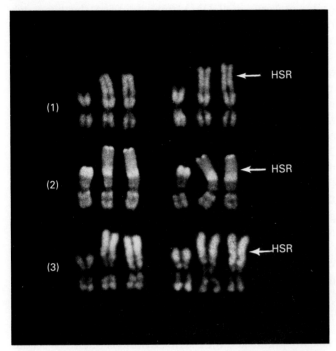

(b)

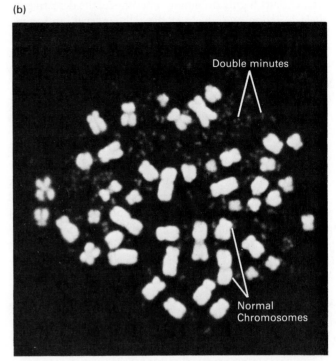

 FIGURE 26-17 Visible DNA amplifications. (a) Homogeneously staining regions (HSRs) in chromosomes from two neuroblastoma cells. In each set of three chromosomes, the left-most one is a normal chromosome 1 and the other two are HSR-containing chromosomes. The three lines (1, 2, and 3) represent three different methods of staining the chromosomes. Method 1 is quinacrine staining, which highlights AT-rich regions; method 2 is staining with chromomycin A3 plus methyl green, which highlights GC-rich areas; and method 3 is 33258 Hoechst staining after a pulse of bromodeoxyuridine late during the S phase, which highlights the early repli- cating regions. In all three cases the HSRs stain homogeneously whereas the rest of the chromosomes are somewhat banded. (b) Quinacrine-stained double minute chromosomes from a human neuroblastoma cell. The normal chromosomes are the large white structures; the double minute chromosomes are the many small paired dots. Both the HSRs and the double minute chromosomes shown here contain the N-*myc* oncogene. [Part (a) see S. Latt et al., 1975, *Biopolymers* **24**:77. Part (b) see N. Kohl et al., 1983, *Cell* **35**:359. Photographs courtesy of Dr. S. Latt.]

and is continually expressed, causing the cell to become cancerous.

Abl and *myc* are only two of many oncogenes that have been found at points of chromosomal translocation in human tumor cells. The translocations are fingers pointing into the genome at potential oncogenes.

Another common chromosomal anomaly in tumor cells is the localized reduplication of DNA to produce as many as 100 copies of a given region (usually a region spanning hundreds of kilobases). This anomaly may take either of two forms: the duplicated DNA may be tandemly organized at a single site on a chromosome, or it may exist as small, independent chromosomelike structures. The former case leads to a homogeneously staining region (HSR) that is visible in the light microscope at the site of the duplication; the latter case causes double minute chromosomes to pepper a stained chromosomal preparation (Figure 26-17). Again, oncogenes have been found in the duplicated

regions. Most strikingly, the *myc*-related gene called N-*myc* has been identified in both HSRs and double minute chromosomes of human nervous system tumors.

Inherited Human Propensities to Develop Cancer Point to Tumor-Suppressor Genes

Inheriting certain genes increases to almost 100 percent the probability that a person will develop a specific tumor. A classic case is retinoblastoma, which is, like many inherited tumors, a disease of childhood. This disease led to the identification of the first tumor-suppressor gene, *RB*. Children who inherit a single defective copy of the *RB* gene, often seen as a small deletion on chromosome 13, develop, on the average, three retinoblastoma tumors, each derived from a single transformed cell. The developing retina contains about 4×10^6 cells, but only about 1 in 10^6 cells actually becomes a tumor cell. This finding shows that even with its highly dominant inheritance, the defective *RB* allele is acting recessively at the cellular level, and that a second event is needed to bring on the transformed state. The second event is the deletion or mutation of the normal *RB* gene on the other chromosome. Rare somatic events can cause this deletion either by loss of all, or a segment, of the chromosome or by small alterations in the remaining *RB* gene. When chromosomal loss occurs, it is balanced by duplication of the affected chromosome.

A variety of rare childhood tumors are associated with inherited dominant mutations and they have turned out to be inactivating mutations of tumor-suppressor genes. Some families have a strong propensity to the development of cancers in adults and, again, mutant tumor-suppressor genes have been implicated. Although all human tumors involve some somatic mutation, inherited mutant genes are being recognized as an important contributor to the probability that any one individual will develop cancer.

➤ DNA Viruses as Transforming Agents

Some animal viruses have RNA and some have DNA as their genetic material (see Table 6-5). One group of RNA viruses, the *retroviruses,* and many types of DNA viruses can be transforming agents and are called *tumor viruses.* Tumor viruses cause transformation as a consequence of their ability to integrate their genetic information into the host cell's DNA; most often they cause the chronic production of one or more oncoproteins that maintain an infected cell in the transformed state. Most oncoproteins encoded by oncogenes in viral genomes are intracellular products. The known oncogenes of DNA viruses are integral parts of the viral genome required for viral replication. The oncogenes of retroviruses are normal or slightly modified cellular genes that are either appropriated from or hyperactivated in the host cell; these genes play no part in viral replication. Because the oncogenes in retroviruses and DNA viruses arise in different ways, we shall consider these two classes of tumor viruses separately.

DNA Viruses Can Transform Nonpermissive Cells by Random Integration of the Viral Genome into the Host-Cell Genome

Among the many different types of animal DNA viruses that exist, the genome size varies from 5 to 200 kb. The viruses with the smallest genomes carry genes for very few proteins and rely mainly on host-cell functions for their replication; the larger genomes of other DNA viruses encode many enzymes and provide many of their own replication functions. Although all types of DNA viruses can be tumor viruses, we shall focus on one group of very small DNA tumor viruses, the *papovaviruses,* of which the best-known representatives are SV40 and polyoma.

Infection of cells by a papovavirus is divided into an early phase, during which it makes one set of viral proteins and induces cellular proteins, and a late phase, during which viral DNA is replicated, coat protein is made, and new virions mature. At the end of the early phase, a quiescent infected cell is activated to synthesize cellular DNA. The small papovaviral genome encodes a few key proteins that induce the cell to make the enzymes necessary for DNA replication. The virus accomplishes this by causing the cell to progress from the G_0 or G_1 phase into the S phase. Thus both cellular and viral DNA are replicated during the infection cycle. It is believed that the induction of the S phase underlies the transforming ability of papovaviruses. Approximately half of the genetic information in the 5-kb circular genomes of SV40 and polyoma specifies their coat proteins; the other half encodes two or three early proteins.

Cells in which the late phase of viral infection follows the early phase, and the massive synthesis of virions is coupled with cell death, are said to be *permissive;* such cells experience a *lytic* infection. Monkey cells are permissive for SV40, but mouse cells are not (Figure 26-18); mouse cells are, however, permissive for polyoma.

In mouse cells infected with SV40 and papovavirus-infected cells from many other animal species, the induction of the S phase does not ordinarily lead to viral DNA replication, apparently because of an incompatibility between the cellular enzymes and a crucial viral sequence or protein. In such *nonpermissive* cells, there is no switch to the late phase and no cell death. As long as the expression of the early viral functions continues, however, infected cells cannot rest in G_0, because they are continually induced to proceed through the cell cycle. In most cases, the viral genome is degraded or lost by dilution during cell proliferation; once the viral genome is lost, the cells revert to normal. Thus most such cells experience an *abortive transformation.* In a small percentage of infected, nonpermissive cells, the viral genome is *integrated* into the cell's genome. These cells thereafter continuously display most of the parameters of transformation described previously; thus they have been permanently transformed into cancer cells by the continual expression of the early papovavirus genes.

Cultures of permissive cells also can be transformed by certain viral mutants that are unable to replicate their DNA and thus can express only early proteins even in permissive cells. Such mutants appear frequently enough that by polyoma virus can cause cancer after it is injected into baby mice.

In theory, viral DNA could integrate into cellular DNA by either a site-specific or a random mechanism. *Site-specific* integration has a defined target sequence in the

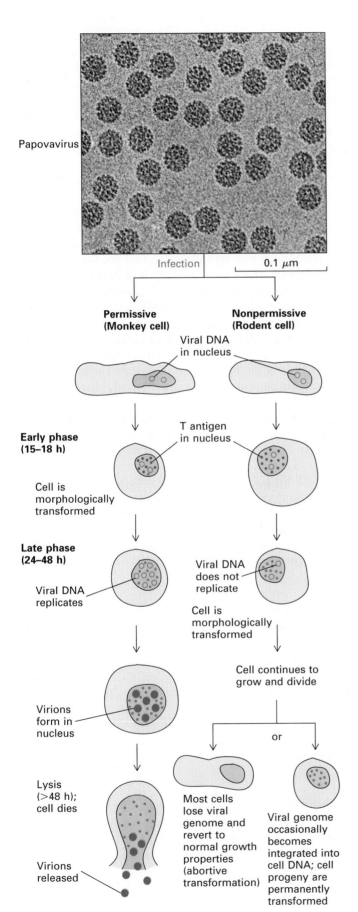

Papovavirus

Infection   0.1 μm

Permissive (Monkey cell)   **Nonpermissive (Rodent cell)**

Viral DNA in nucleus

T antigen in nucleus

Early phase (15–18 h)

Cell is morphologically transformed

Late phase (24–48 h)

Viral DNA replicates

Viral DNA does not replicate

Cell is morphologically transformed

Cell continues to grow and divide

Virions form in nucleus

or

Lysis (>48 h); cell dies

Most cells lose viral genome and revert to normal growth properties (abortive transformation)

Viral genome occasionally becomes integrated into cell DNA; cell progeny are permanently transformed

Virions released

◄ FIGURE 26-18   Responses of permissive and nonpermissive cells to infection by SV40, a papovavirus. In permissive cells, viral infection leads to cell death and the production of progeny virions. In nonpermissive cells, the synthesis of the early viral proteins (T antigens) causes cells to become transformed. If the viral DNA becomes permanently integrated into the cell DNA, permanent transformation of the cell results. The electron micrograph shows an unstained virus preparation embedded in a thin layer of ice. [Photograph courtesy of T. Baker and M. Bina.]

cellular DNA and/or a defined integration site in the viral DNA. One or more proteins (probably viral) would bind to these sequences and direct integration. Papovaviruses do not integrate site-specifically; they apparently use *random* sites both in the viral and in the host-cell DNA. The viral proteins do not appear to play an active role in the integration process. Papovavirus DNA integration is probably no different from the rare integration of any DNA that has been transfected into cells. Integration of viral DNA into the cellular DNA of nonpermissive cells can have no role in propagating the virus because no viral progeny are made in such cells.

Transformation appears here to be simply the consequence of three features of the viral replication: (1) the occurrence in papovavirus genomes of transforming genes used by the virus for the early phase of its lytic cycle, (2) the occasional integration of viral DNA into the DNA of nonpermissive cells, and (3) the lack of cell death after infection of nonpermissive cells. These three circumstances allow a virus to permanently transform nonpermissive cells whenever the viral DNA is integrated in such a way that the expression of early genes is not impeded. This event happens frequently enough that some cells are transformed whenever a culture of nonpermissive cells is infected by SV40 or polyoma virus. Transformation by these viruses is a laboratory phenomenon, however, and neither virus is known to induce cancer in wild animal populations.

Transformation by DNA Viruses Requires Interaction of a Few Independently Acting Viral Proteins

In permanently transformed nonpermissive cells, the viral genome is integrated into the host-cell genome in such a way that the early viral genes are expressed constituitively. That the continual activity of two or three early viral gene products maintains the transformed phenotype has been shown in experiments with various temperature-sensitive viral mutants. The morphology and growth properties of cells infected with these mutants can be varied from normal to transformed by shifting the temperature. The genes encoding these early proteins are, by definition, oncogenes—

that is, they transform cells and induce cancer in experimental animals. The papovavirus oncogenes were, in fact, the first ones to be recognized. Because it was evident that an understanding of how these genes can cause cancer would illuminate the more general issue of cancer induction, much attention has been focused on the products of papovavirus early genes.

SV40 makes two early proteins called T (large T) and t (small t); polyoma makes three early proteins: T, middle-T (mid-T), and t. T protein was originally called *T antigen* because it was first demonstrated by immunofluorescence using serum from animals bearing virus-induced tumors. The T antigens are 90-kilodalton (kDa) proteins found in the cell nucleus. They bind tightly to DNA and play an important role in viral DNA transcription and replication during the lytic cycle. The mid-T of polyoma is a 45-kDa plasma membrane–bound protein that has no counterpart in SV40. A fraction of the cell's SV40 T antigen, however, is bound to the plasma membrane and appears to serve the same function as the mid-T protein of polyoma. The 20-kDa t proteins are cytoplasmic. The three polyoma proteins have independent roles in the transformation process.

The early region of the polyoma genome gives rise to one complex transcript that can be spliced in three different ways to produce mRNAs for the three early proteins (Figure 26-19). By deletion of specific nucleotides it has been possible to construct three different DNA molecules, each of which encodes an RNA that specifies only one of the three polyoma early proteins. The DNA molecule encoding only mid-T protein can transform a rat 3T3 cell line as defined by many of the criteria described earlier, but such cells cannot grow in a low-serum medium. Thus mid-T apparently cannot eliminate the requirement for all growth factors, although it can do almost everything else

associated with transformation. In contrast, a cell that expresses both the T and the mid-T proteins can grow in low serum or even in no serum. Cells transfected with polyoma DNA encoding only T protein can grow in low-serum media, but such cells are not morphologically transformed and they do not grow to high saturation densities. This dramatic separation of transformation parameters was the first recognized case of the general principle of cell transformation enunciated earlier that transformation often involves the interaction of a small number of independently acting proteins.

In the experiments described in the previous paragraph, rat 3T3 cells were used as the targets of transformation. When plasmids encoding the T and mid-T proteins are introduced into recently explanted rat embryo fibroblasts (cells with a limited life span in culture), the division of labor among the polyoma early proteins is even more obvious. A plasmid encoding mid-T produces no transformed colonies in these cells, suggesting that the mid-T protein cannot immortalize cells although it can transform 3T3 cells, which are already immortal. Consistent with this explanation is the finding that a plasmid encoding T protein confers immortality on rat embryo fibroblasts without changing them morphologically. The subsequent incorporation into the cell of a plasmid encoding mid-T morphologically transforms the cells already immortalized by the action of T protein, and the resulting transformants behave almost like wild-type polyoma-transformed cells. In other experiments, polyoma t protein has been shown to have a role in producing complete transformation.

For SV40, recombinant DNA methods have shown that different parts of the T protein have different effects. Thus this single protein is responsible for several different aspects of transformation. Papovaviruses have apparently

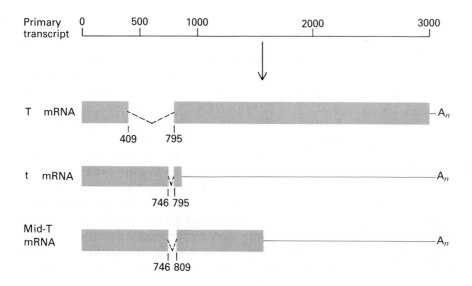

◄ FIGURE 26-19 The early region of polyoma virus is expressed in one 3000-nucleotide RNA that can be spliced in three ways to yield three different mRNAs and therefore three early proteins. The thick orange bars denote protein-coding regions; the thin red lines, noncoding regions. [See M. Rassoulzadegan et al., 1982, *Nature* **300**:713.]

evolved a number of distinct ways of changing cell behavior. When these changes occur independently, they do not produce a complete tumor cell. When they occur together, however, a full-blown tumor cell results.

We noted earlier that both activation of oncogenes and inactivation of tumor-suppressor genes are involved in cancer induction. Similarly, the DNA tumor viruses, along with their ability to induce quiescent cells into the cell cycle, also evolved methods of neutralizing RB and p53. The polyoma and SV40 early proteins, as well as proteins of the oncogenic adenoviruses and papilloma viruses, can bind these proteins and either prevent their action or induce their degradation. These abilities are crucial ones for the transforming potential of the viruses.

► RNA-Containing Retroviruses as Transforming Agents

Transformation by retroviruses is more complicated than transformation by DNA viruses because the basic retroviral genome does not encode proteins with transforming activity. To understand the transforming ability of retroviruses, we must first examine in some detail their life cycle, which was outlined in Figure 6-23.

Virion-Producing Infection Cycle of Retroviruses Requires Integration into the Host-Cell Genome

Retroviruses have an RNA genome consisting of two apparently identical, ≈8500-nucleotide RNA molecules noncovalently bonded to one another; each has a cellular tRNA molecule bound to it. A virion also contains about 50 molecules of a DNA polymerase that is capable of copying the genomic RNA into DNA; this enzyme is commonly known as *reverse transcriptase* because it reverses the normal flow of information in biological systems. The infection process brings both the RNAs and the reverse transcriptase into the host cell's cytoplasm. Retroviruses are similar in their genetic organization and their mechanism of duplication to the retrotransposons described earlier (Chapter 9).

Once in the cytoplasm, reverse transcriptase uses the tRNA as a primer to initiate DNA synthesis. In a complicated series of events, a complete complementary (or minus) strand of DNA is formed; this minus strand is used as a template for the formation of a plus strand of DNA. The ultimate product of reverse transcription is thus a double-stranded DNA copy of the information in the viral RNA genome (see Figure 9-24). The double-stranded DNA *reverse transcript* is actually longer than its template. Because of the "jump" made by the initial minus-strand DNA

and a second "jump" made by the initial plus-strand DNA, the short repeat at either end of the viral RNA is extended to form a long terminal repeat (LTR) at either end of the reverse transcript. The LTR is a very important component of the reverse transcript because, as we shall see, it contains many of the signals that allow retroviruses to function.

After their formation, the double-stranded DNA reverse transcripts migrate into the cell's nucleus where they integrate more or less at random into the cell's chromosomes. In contrast to the variable integration sites on genomes of DNA tumor viruses, the integration sites on retroviral DNA are located specifically at the ends of the linear molecule. Integration is accomplished largely through the activity of a viral protein called *integrase*. The site specificity of the integration site within the retroviral genome, the approximate randomness of the chromosomal target site, and the duplication of the target site make reverse transcript integration similar to retrotransposon movement in bacteria or yeast (Chapter 9).

Once the reverse transcript is integrated into the host-cell DNA, the transcript—which is called *proviral* DNA in its integrated form—becomes a template for RNA synthesis (Figure 26-20). Provirus-directed transcription, like most other eukaryotic transcription, involves a *promoter*—a sequence that directs the RNA polymerase to a specific initiation site—and an *enhancer*—a sequence that facilitates transcription although it need not be located near the initiation site. The promoter and enhancer sequences, as well as the polyadenylation signals, are located in the LTR. The two LTRs at either end of proviral DNA are identical in sequence but have different functions: the upstream LTR acts to promote transcription, whereas the downstream LTR specifies the polyadenylylation site. The RNA made from the proviral DNA serves both as mRNA in the synthesis of viral proteins and as genomic RNA for the next generation of virus particles.

Because they lack most metabolic machinery, all viruses are obligate intracellular parasites. Retroviruses are perhaps the ultimate intracellular parasites because they can link their genome stably to that of their host cell without seriously damaging the cell. Their genetic information is then replicated as part of the cellular DNA and distributed to all daughter cells, which actively transcribe it so that new virus particles are made perpetually. Thus the life cycle of retroviruses differs from that of DNA tumor viruses in two fundamental respects: integration is site-specific on the retroviral genome and integration of the retroviral genome into the host genome is part of the virion-producing infection cycle. Because the proviral DNA can continually direct synthesis of the proteins for new viral progeny by using only a small fraction of the infected cell's metabolic machinery, it does not kill the host cell, which can continue to grow at a virtually normal rate.

Retroviruses occasionally are integrated into an animal's germ-line cells; they are then inherited by all offspring of that animal and can even become part of the

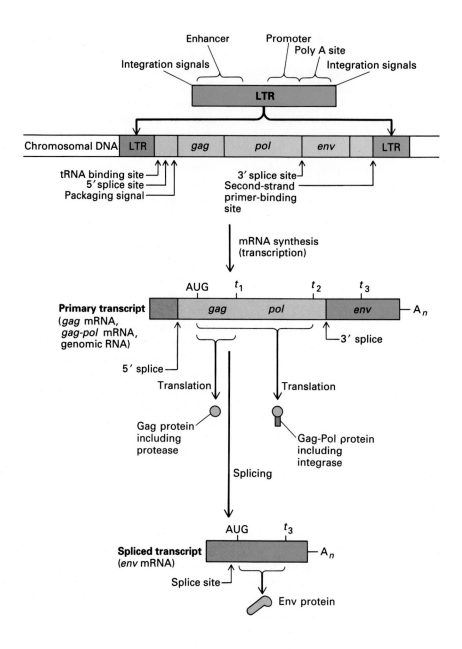

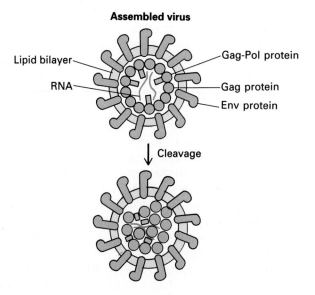

◀ FIGURE 26-20 Genetic elements of proviral DNA and the corresponding gene products. A standard retroviral provirus is shown integrated into cellular DNA. The genetic structure of its flanking LTRs is indicated. Although the two LTRs are identical, their positions somehow allow the left-hand LTR to be a promoter for transcription and the right-hand LTR to specify the poly A site for the transcript. The primary RNA transcript has three functional roles. It is an mRNA for the synthesis of both the Gag and the Gag-Pol proteins because its first translation termination signal (t_1, the termination signal for *gag*) is suppressed occasionally (5–10 percent of the time) to give rise to some Gag-Pol protein. By virtue of its packaging signal, the transcript is also the genome of the virus. About 50 percent of the transcripts are spliced; why the 5′ and 3′ splice sites are only partially utilized remains an unanswered question. The spliced transcript, which is not packaged because the packaging signal has been spliced out, serves as the mRNA for the Env protein. All three of the translation products are virion proteins: the Env protein is inserted into the lipid bilayer that surrounds the virus, the Gag protein forms an inner shell, and the Gag-Pol protein (which contains the reverse transcriptase) is also inside the particle. After particles are formed, both the Gag and the Gag-Pol proteins are extensively processed by proteolysis.

inheritance of a species. In mice, for example, between 0.1 and 1 percent of the genome consists of proviral DNA. Most such inherited retroviruses, as opposed to the ones that are integrated during somatic cell growth, are transcriptionally inactive. Integrated retroviruses are similar to retrotransposns; the main difference is that retroviruses generally do not transpose in a cell but rather must be produced by one cell and then infect another cell—they have an obligate extracellular phase in their life cycle.

The genome of most retroviruses has three protein-coding regions. They are *gag*, which encodes a polyprotein that is cleaved to form four internal virion structural proteins; *pol*, which encodes the reverse transcriptase and the integrase; and *env*, which encodes a glycoprotein that covers the virion surface (see Figure 26-20). A gene segment encoding a protease that cleaves the polyprotein precursors is part of *gag* in some viruses and of *pol* in others. None of the corresponding retroviral proteins, all of which are found in virions, changes the growth properties of fibroblasts. Thus the basic retrovirus (in contrast to DNA tumor viruses) is not a transforming virus; in fact, infected cells are difficult to distinguish from uninfected cells by any parameter of cellular life.

Although the basic retroviral life cycle does not include a transforming event, many retroviruses can cause cell transformation and cancers in animals and even in humans. In fact, the first tumor virus to be recognized was the Rous sarcoma virus, a chicken retrovirus described by Peyton Rous in 1911. Many other tumor-inducing retroviruses have been found, including the mouse mammary tumor virus (MMTV); leukemia viruses of chickens, mice, cats, apes, and, most recently, humans; sarcoma viruses of many species; and a few carcinoma-inducing viruses. Retroviruses can transform cells and induce tumors by two quite distinct mechanisms. One is utilized by transducing retroviruses; the other, by slow-acting retroviruses.

Oncogenic Transducing Retroviruses Convert Cellular Proto-Oncogenes into Oncogenes

The key to our present understanding of the transforming and tumor-inducing ability of some retroviruses came from the realization that such viruses contain genetic information other than *gag*, *pol*, and *env*. The additional genetic material contained by these viruses are oncogenes appropriated from normal cellular DNA. Because phages that acquire cellular genes are said to *transduce* those genes, retroviruses that have acquired cellular sequences are called *transducing retroviruses*.

The first indication of the existence of transducing retroviruses came with the discovery that Rous sarcoma virus contains nucleotide sequences that are not found in closely related but nontransforming retroviruses. The extra nucleic acid sequence, called *src*, was then identified in the DNA of normal chickens as well as in the DNA of all vertebrate and even invertebrate species. This landmark discovery that an oncogene was derived from a cellular proto-oncogene fundamentally reoriented thinking in cancer research because it showed that cancer may be induced by the action of normal, or nearly normal, genes. The discovery of many other transduced cellular genes in other retroviruses implies that the normal vertebrate genome contains many potentially cancer-causing genes.

Transducing retroviruses arise because of complex, apparently spontaneous, rearrangements following the integration of a retrovirus near a cellular proto-oncogene (Figure 26-21). Most commonly, the acquired genetic information in a transducing retrovirus replaces part or all of *gag*, *pol*, and *env*, making the new retrovirus defective. However, the products of *gag*, *pol*, and *env* can be provided by a *helper virus* (a wild-type retrovirus infecting the same cell), so propagation of the defective virus is possible. The genome structures of three transducing retroviruses are shown in Figure 26-22.

Many different oncogenes, derived from cellular proto-oncogenes, have been identified in retroviruses (see Table 26-1). The characteristics of different transducing retroviruses are determined in part by which oncogenes they contain. For instance, some transform only adherent cells, whereas others transform both adherent and nonadherent cells. In animals, the ones that transform nonadherent cells cause leukemia; the remainder cause sarcomas or carcinomas. Abelson murine leukemia virus transforms only cells of the B lymphocyte lineage in cultures of bone

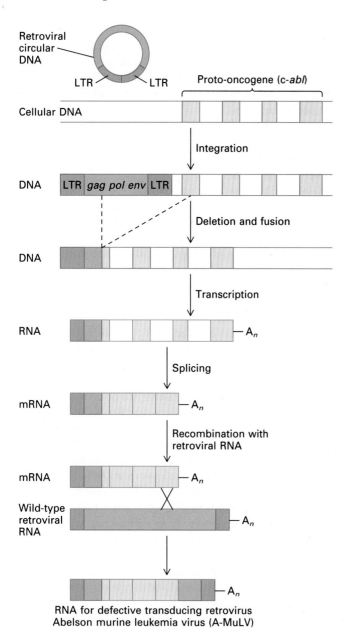

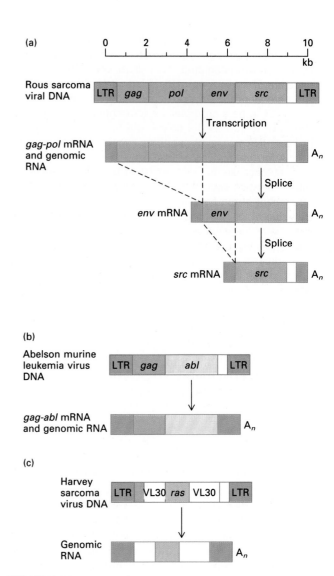

▲ FIGURE 26-21   The formation of a transducing retrovirus. Although the events that generate a transducing retrovirus occur at such a low frequency that they have not been studied in the laboratory, the structure of the transducing viruses strongly suggests that they are formed as depicted here. Formation of the Abelson virus is depicted because it has many features found in other retroviruses. The c-abl gene is shown as a series of exons in cellular DNA. A wild-type retrovirus probably integrates upstream of c-abl randomly once in about 10^6 integrations. A subsequent deletion-fusion event fuses the c-abl exons into the gag region of the retroviral DNA. Transcription of the fused DNA and subsequent splicing produces a gag-abl mRNA. If the cell is also infected by a wild-type retrovirus, it will package the gag-abl mRNA along with wild-type RNA. In the next generation, the two RNAs can recombine (presumably by a switch of templates during reverse transcription) to generate the RNA of the transducing retrovirus (now called Abelson murine leukemia virus). A retrovirus formed as shown here would be defective because it lacks pol, env, and part of gag.

▲ FIGURE 26-22   The structure and expression of the genes in three transducing retroviruses. The src, abl, and ras genes were derived from host-cell genomes. (a) Rous sarcoma virus, the only known nondefective transducing retrovirus, produces three different mRNAs: one unspliced mRNA (carrying gag and pol), which is also the genomic RNA, and two spliced mRNAs (one carrying env, the other carrying only src). (b) Abelson murine leukemia virus, a highly defective virus, yields one RNA molecule that serves as the mRNA for the Gag-Abl fusion protein and acts as the viral genome. (c) Harvey sarcoma virus has a complicated origin because it is a mouse retrovirus that acquired its oncogene during passage in a rat. Two recombination events must have occurred in the rat. One was a recombination of the mouse retrovirus with an endogenous retrovirus of the rat, called VL30. The second recombination was a fusion of a mutated form of the rat c-Ha-ras proto-oncogene into the mouse/rat recombinant retrovirus within the VL30 sequences. The only gene expressed by the Harvey virus is the v-Ha-ras gene, which encodes a 21-kDa protein that is probably synthesized from the genomic RNA. The v-Ha-ras gene differs from the c-Ha-ras gene by one crucial point mutation in the codon for amino acid 12; this mutation converts the proto-oncogene into the oncogene.

marrow cells; in animals, however, Abelson virus causes other types of tumors, and it can even affect the differentiation of fetal red blood cells. Harvey virus, which primarily causes sarcomas, also can produce tumors of the red blood cell lineage in mice.

Although the transforming activity of a transducing retrovirus is provided by its acquired cellular genetic information, the genetic elements that allow the oncogenes to be expressed and transferred from cell to cell are all inherited from the retroviral parent. As noted already, the LTRs of the viral genome contain most of the control elements including the promoter and the enhancer, the polyadenylylation signal, and the integration signals (see Figure 26-20). The genome also contains a sequence that binds the tRNA just at the border of the left-hand LTR, a sequence for initiating plus-strand DNA synthesis at the border of the right-hand LTR, and a sequence that allows for packaging into virus particles (located between the left-hand LTR and the AUG that initiates Gag protein synthesis). These vital control elements are preserved in the hybrid genomes of transducing retroviruses, even highly defective ones such as the Abelson virus (see Figure 26-21).

The demonstration in transducing retroviruses that cellular genetic information can be an agent of transformation clearly implies that human genomes harbor genes that can, in certain circumstances, cause cancer. Furthermore, this notion also suggests that other agents, like chemicals and radiation, could change normal genes into cancer-inducing genes by causing relatively simple mutations. Both of these implications have had a profound impact on cancer research.

Three kinds of alterations, affecting either the coding region of the gene or its expression, are involved in the conversion of a proto-oncogene into an oncogene. First, transcription of a cellular gene acquired by a retrovirus is determined by the viral control elements, which can promote high transcription rates in many different cell types.

Thus a gene transduced by a retrovirus may be both expressed at inappropriately high levels and or in inappropriate cells. This alone appears to explain why certain normal genes become transforming genes when they are captured by a virus. Indeed, proto-oncogenes cloned from cellular DNA can become transforming genes after being inserted into a retrovirus or otherwise placed in an unregulated, high-transcription environment. A second kind of alteration is the loss of parts of the normal cellular gene that may leave the protein product with an unregulated activity that allows it to transform cells. Third, a single point mutation or other small change may produce a crucial sequence difference between a cellular gene and its transforming counterpart. These alterations in a proto-oncogene may result in a quantitative change in the amount of the encoded protein or a qualitative change in the protein itself. A single change or multiple changes may convert a given proto-oncogene into an oncogene.

Nononcogenic Transducing Retroviruses Have Been Constructed Experimentally

All of the transducing retroviruses identified to date have been oncogenic; recognizing such viruses is relatively easy because they produce an obvious disease. It seems probable that retroviruses containing acquired genes that are not oncogenes also arise naturally; recognition of such nononcogenic transducing retroviruses, however, would be difficult if they produce no obvious effects on infected cells.

Nononcogenic transducing retroviruses have been constructed in the laboratory by in vitro genetic manipulation. An example of an artifically constructed virus that can transduce both a globin cDNA and a neomycin-resistance gene is shown in Figure 26-23. Such artificial retroviruses have been used to insert new or altered genes into cells and animals in a process known as *gene therapy*. It is hoped that such constructed genetic elements may be used in

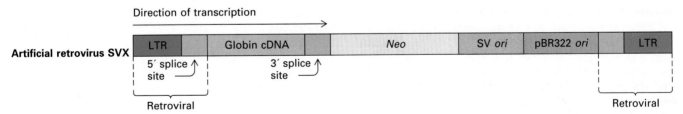

▲ FIGURE 26-23   The artificial retrovirus SVX. It was built to be a shuttle vector, a recombinant molecule that is able to carry genes back and forth between mammalian cells and bacteria. In this vector the crucial LTRs from a mouse retrovirus are coupled to the following elements: a globin cDNA that will direct globin synthesis when SVX is present in a mammalian cell; a 3' splice site that can be joined to the 5' splice site near the left-hand LTR to form an mRNA for the expression of the next gene (downstream); a gene (*Neo*) that encodes a protein conferring resistance to an analog of the antibiotic neomycin; an SV40 virus origin of replication; and a pBR322 bacterial plasmid origin of replication. This shuttle vector is used to introduce the globin gene into cells (other cDNAs can be substituted for globin). Cells integrating the retrovirus can be selected because of the presence of the *Neo* gene, which is expressed both in bacteria and in mammalian cells. To rescue the provirus from cells, the cells can be fused to other cells that make SV40 T antigen. This fusion activates replication from the SV40 origin, and the resulting circular DNA molecules can be purified and reinserted into bacteria. Their multiplication in bacteria is caused by the pBR322 origin of replication. Bacteria containing the recombinant retrovirus can be selected through the *Neo* marker. [See C. Cepko et al., 1984, *Cell* **37**:1053.]

the future to treat many human genetic diseases, such as lipid storage diseases, immunodeficiency due to lack of adenosine deaminase, or diseases that affect hemoglobin synthesis.

Slow-Acting Carcinogenic Retroviruses Can Activate Nearby Cellular Proto-Oncogenes after Integration into the Host-Cell Genome

The second type of carcinogenic retrovirus—the slow-acting retrovirus—induces cancer over a period of months or years, rather than the days or weeks required for cells to respond to a transducing virus. The genomes of slow-acting retroviruses differ from those of transducing viruses in one crucial respect: they lack an oncogene. The slow-acting retroviruses not only have no affect on growth of cells in culture, but they are also completely proficient to multiply themselves because they have suffered no debilitating deletions or insertions during acquisition of an oncogene.

The mechanism by which avian leukemia viruses cause cancer appears to operate in all slow-acting retroviruses.

Like other retroviruses, avian leukemia viruses generally integrate into cellular chromosomes more or less at random. However, the site of integration in the cells from tumors caused by these viruses was found to be near a gene called c-*myc*, known to be a proto-oncogene. This finding suggested that the slow-acting avian leukemia viruses cause disease by activating c-*myc* to become an oncogene. The viruses act slowly both because integration near c-*myc* is a random, rare event and because secondary events probably have to occur before a full-fledged tumor becomes evident.

An integrated viral genome can act in a number of ways to cause a cellular gene to become an oncogene. In some tumors, the 5′ end of the *myc* gene transcript includes a sequence from a retroviral LTR. In such cases the right-hand LTR of the integrated retrovirus—which usually serves as a terminator—is believed to act as a promoter, initiating synthesis of RNA transcripts from the c-*myc* gene (Figure 26-24a). Such c-*myc* transcripts apparently encode a perfectly normal c-*myc* product. The enhanced level of c-*myc* RNA resulting from the strong promoting activity of the retroviral LTR appears to be part of the explanation of oncogene activation in such cases. A second aspect is that

(a) Promoter insertion

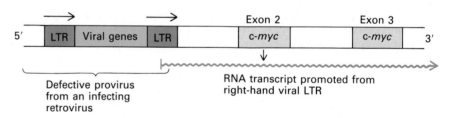

(b) Enhancer insertions

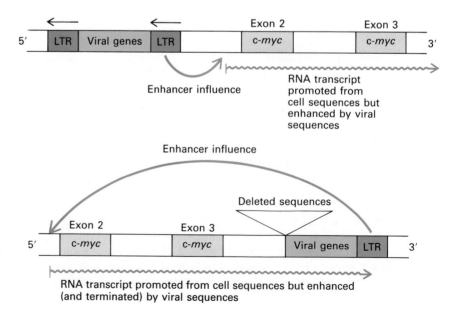

> FIGURE 26-24 Activation of proto-oncogene c-*myc* by retroviral promoter and enhancer insertions. (a) The promoter can be activated when the retrovirus inserts upstream (5′) of the c-*myc* exons. The right-hand LTR may then act as a promoter if the provirus has a defect preventing transcription through to the right-hand LTR. The c-*myc* gene is shown as two exons; there is a further upstream exon but it has no coding sequences. (b) The enhancer can be activated when a retrovirus inserts either upstream of the c-*myc* gene in the opposite transcriptional direction or downstream of the gene. In both cases, a viral LTR acts on a c-*myc* promoter sequence. In the downstream case, the LTR also terminates the transcript. [Modified from actual cases of retroviral insertion described in G. G. Payne et al., 1982, *Nature* **295**:209.]

c-*myc* expression is usually down-regulated when cells are induced to differentiate, but the LTR-driven c-*myc* will not respond to such signals. This mechanism of oncogene activation, which is called *promoter insertion,* has been implicated in many tumors induced by avian leukemia viruses.

In a few tumors induced by avian leukemia viruses, the proviral DNA is located near the c-*myc* gene but at the 3' end of the gene; in others, it is at the 5' end of c-*myc* but oriented in the opposite transcriptional direction (Figure 26-24b). In such cases the promoter-insertion model cannot be applicable, and the proviral DNA is thought to exert an indirect enhancer activity that apparently activates c-*myc* by increasing its level of transcription and making it insensitive to normal regulation. This mechanism is called *enhancer insertion.*

Promoter insertion and enhancer insertion probably explain many tumors caused by retroviruses that do not carry oncogenes. The slow-acting retroviruses include the mouse mammary tumor virus and a number of mouse leukemia viruses. In some cases, activation of c-*myc* has been demonstrated; other proto-oncogenes have been identified near other insertion points. Slow-acting viruses act as fingers pointing to the genomic location of potential oncogenes and many oncogenes have first been discovered as sites of viral integration in tumor cells.

The activation of a proto-oncogene following integration of a slow-acting retrovirus should be clearly distinguished from the acquisition of a proto-oncogene by a virus that converts it into an oncogene-containing transmissible virus. As is evident from Figure 26-24a, the first step in the formation of a transducing virus may well be promoter insertion, but several additional steps must occur before a defective, oncogene-containing retrovirus emerges. Those steps are all low-probability events, meaning that very few promoter insertions will progress to the formation of a transmissible, transducing virus. In fact, because defective transducing retroviruses depend on a helper virus for their cell-to-cell transmission, they are not maintained in natural animal populations. Most oncogene-containing retroviruses have arisen in laboratories or in domesticated animals and have been maintained for experimental purposes; the defective viruses are not readily spread from animal to animal. In natural populations, insertional oncogene activation is probably the major cause of retrovirus-induced cancer.

➤ *Human Tumor Viruses*

The discovery of many RNA and DNA tumor viruses in animals has raised the question of whether humans also have tumor viruses, and if so, what fraction of human cancer might be caused by them. In certain specific situations, human tumor viruses have been found, but there is little

evidence that viruses cause the major types of human tumors that are prevalent in developed countries (e.g., breast, prostate, lung, and colon). A lack of evidence does not constitute proof of noninvolvement, however, and it remains an open question whether viruses might have a wider role in human cancer than is now suspected. In this section, we describe briefly the various viruses that are presently associated with human cancers.

In 1980, after many years of fruitless search for a human retrovirus, oncogenic or nononcogenic, the first plausible candidate was reported: *human T-cell leukemia virus* (HTLV). Infection with this virus is clearly associated with a type of T-lymphocytic human tumor that is especially prevalent in southern Japan, the Caribbean, and parts of Africa. How the virus causes the tumor is being actively studied; oddly, the process appears to involve neither a cell-derived oncogene within the virus nor a reproducible site of integration in tumor-cell DNA. Rather, the basis of the carcinogenicity of HTLV is a unique viral genetic region, totally distinct from *gag, pol,* and *env* and more like the oncogenes of DNA tumor viruses.

Epstein-Barr virus, a herpeslike DNA virus, probably plays a role in at least two human tumors: Burkitt's lymphoma and nasopharyngeal carcinoma (Figure 26-25). In certain hot, wet sections of Africa, Burkitt's lymphoma is the predominant cancer in children, and in southern China nasopharyngeal carcinoma is a major cancer. In the United States and other developed countries, Epstein-Barr virus mainly causes mononucleosis in adolescents and young adults; it leads to cancer very rarely, and then usually in

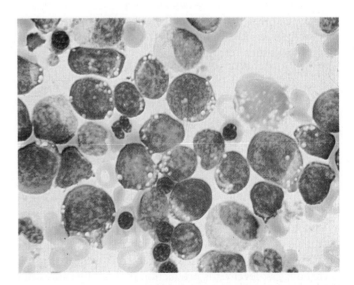

▲ FIGURE 26-25 Stained tumor cells of Burkitt's lymphoma. These cells carry the genome of the Epstein-Barr virus. [Courtesy of Dr. R. Van Etten.]

people with a malfunctioning immune system. The cancer-inducing ability of the virus in Africa may be due to the presence of a second endemic factor, thought to be malaria, which works in concert with the virus.

Infection with *hepatitis B virus*—a very small DNA virus that is mainly responsible for human serum hepatitis—is correlated with liver cancers, especially in underdeveloped areas of the world. Although this virus has a DNA genome, it multiplies through an RNA intermediate like a retrovirus. Its genome has been found integrated into the DNA of liver tumor cells.

The *papilloma viruses,* which are small DNA viruses belonging to the papovavirus group, not only are responsible for warts and other benign human tumors but also for malignancies. Currently over 30 distinct human papilloma viruses are recognized. Two types of papilloma virus are associated with human cervical carcinoma and appear to be the cause of this sexually transmitted disease. Papilloma viruses, which also cause a variety of warts, have multiple oncogenes that together transform cells. They are probably the most significant viruses for inducing cancers in developed countries.

Two viruses, BK and JC, that are direct human counterparts of SV40 have been identified. BK can transform cells in culture, but neither virus has been associated with any specific human cancer.

Any discussion of viruses and human cancer today must include consideration of the *acquired immunodeficiency syndrome (AIDS)*, a retroviral induced disease that, among other symptoms, makes its victims susceptible to numerous cancers. AIDS is caused by the *human immunodeficiency virus (HIV)*, a retrovirus different from all of the others discussed thus far (Figure 26-26a). The HIV genome includes five or six genes in addition to the *gag, pol,* and *env* genes of other retroviruses (Figure 26-26b). These extra genes, which are all small and expressed from a battery of spliced mRNAs, allow HIV to grow more vigorously than other retroviruses in infected cells. The exact function of the HIV-specific genes is still debated but involves activation of transcription, regulation of splicing or transport of mRNA, and possibly other alterations of macromolecular metabolism. HIV kills cells rather than transforming them; because its prime, and perhaps sole, targets are cells that mount the immune response, HIV causes a profound immunodeficiency. HIV infection has two consequences, one short-term, one long-term. During the first six weeks or so following exposure, the HIV-infected person develops flu-like symptoms and there is abundant virus in the blood. Thereafter, the virus largely, but not entirely, disappears from the blood, probably because of the actions of T killer cells (described in Chapter 27). Over years there is chronic low-level virus production that finally causes the immune system to collapse and AIDS to develop. One type of immune system cells, T helper cells, slowly disappear from the blood and the AIDS patient begins to suffer from

infectious diseases that his or her body cannot fight off. AIDS patients often suffer from cancers, probably caused by viruses like Epstein-Barr that are released from their usual control by the immune system. A common tumor of AIDS patients is Kaposi's sarcoma, a tumor of endothelial (blood vessel) origin, which may be induced by growth factors released by HIV-infected cells (Figure 26-27).

HIV-infected people generally die of AIDS at an average of about 10 years after infection. Some 5 percent or so of HIV-infected people may never develop AIDS, but since the epidemic only started about 1979, insufficient time has passed to be certain that any infected people will survive. AIDS is one of the most devastating diseases ever to spread within the human population. Already in 1995, perhaps 20 million people worldwide are HIV-infected.

Development of treatments for AIDS has been slow. Even though HIV provides good targets for drug action—like the reverse transcriptase and the viral protease—the viral genome can mutate so rapidly, making its proteins resistant to the drugs that have been developed, that no drug has been able to eliminate the virus or even slow its growth for very long. Vaccines are the protection we have against most viruses, but thus far no effective vaccine against HIV infection has been developed. Scientists in universities, research laboratories and companies continue with efforts to make effective drugs or vaccines, but HIV is a very elusive target.

▶ *Chemical Carcinogens*

Chemicals, which are thought to be the cause of many human cancers, were originally associated with cancer through experimental studies in animals. The classic experiment is to repeatedly paint a test substance on the back of a mouse and look for development of both local and systemic tumors in the animal. In this way, many substances have been shown to be *chemical carcinogens.*

Most Chemical Carcinogens Must Undergo Metabolic Conversion to Become Active

Chemical carcinogens have a very broad range of structures with no obvious unifying chemistry. Because many carcinogens have structures that appear to be highly unreactive, early workers were puzzled about how such unreactive and water-insoluble compounds could be potent cancer inducers. The situation was clarified when it was realized that there are two broad categories of carcinogens, *direct-acting* and *indirect-acting* (Figure 26-28); the latter require metabolic activation to become carcinogens. The

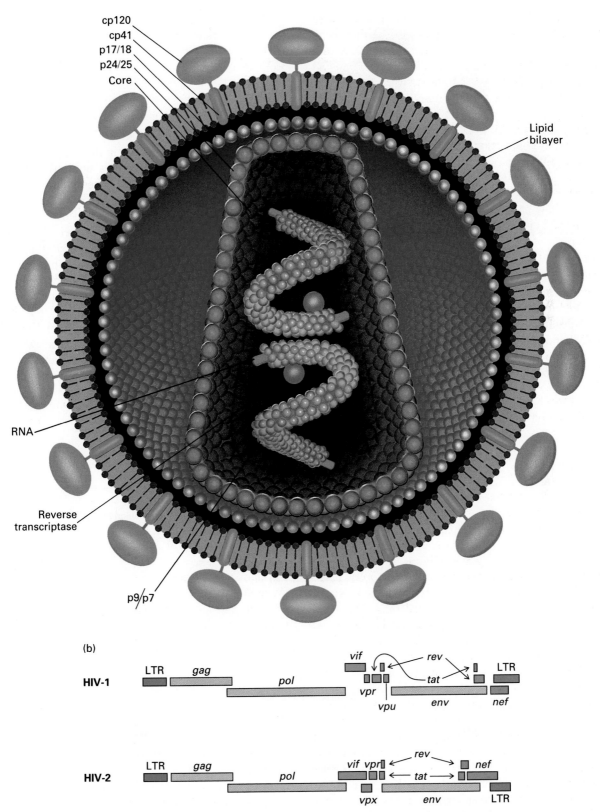

▲ FIGURE 26-26   (a) Schematic structure of the human immunodeficiency virus (HIV) particle showing the location of various components. (b) Structures of the genomes of HIV-1 and HIV-2, two strains of human immunodeficiency virus, which causes AIDS. The standard retroviral genes (purple)— *gag*, *pol*, and *env*—are present along with five or six other genes (red). The full length of the genomes is about 9000 nucleotides. Each strain has one gene not present in the other (*vpu* or *vpx*). [See R. Gallo et al., 1988, *Nature* **333**:504.]

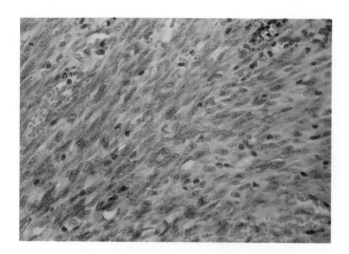

◀ FIGURE 26-27 Section of a Kaposi's sarcoma, a cancer common in AIDS patients. The red spots are red blood cells; otherwise the field is filled with tumor cells. [Courtesy of Dr. J. Aster.]

direct-acting carcinogens, of which there are only a few, are reactive electrophiles (compounds that seek out and react with negatively charged centers in other compounds). Indirect carcinogens are converted to *ultimate carcinogens* by introduction of electrophilic centers.

The metabolic activation of carcinogens is carried out by enzymes that are normal body constituents. Animals have such enzymes, especially in the liver, because they are part of a system that detoxifies noxious chemicals that make their way into the body. Therapeutic drugs, insecticides, polycyclic hydrocarbons, and some natural products are often so fat-soluble and water-insoluble that they would accumulate continually in fat cells and lipid membranes if there were no way for the animal to excrete them. The detoxification system works by solubilization: it adds hydrophilic groups to water-insoluble compounds, thus allowing the body to rid itself of noxious or simply insoluble materials.

The detoxification process begins with a powerful series of oxidation reactions catalyzed by a set of proteins called cytochrome P-450s. These enzymes, which are bound to endoplasmic reticulum membranes, can oxidize

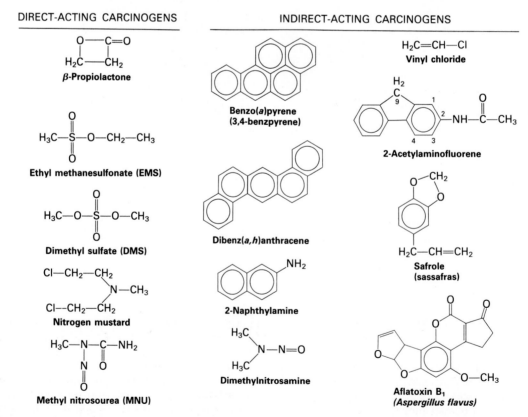

▲ FIGURE 26-28   Structures of some chemical carcinogens. Direct-acting carcinogens are highly electrophilic compounds that can react with DNA. Indirect-acting carcinogens must be metabolized before they can react with DNA.

even highly unreactive compounds such as polycyclic aromatic hydrocarbons (Figure 26-29). Oxidation of polycyclic aromatics produces an epoxide, a very reactive electrophilic group. Usually these epoxides are rapidly hydrolyzed into hydroxyl groups, which are then coupled to glucuronic acid or other groups, producing compounds soluble enough in water to be excreted. Some intermediate epoxides, however, are only slowly hydrolyzed to hydroxyl groups, probably because the relevant enzyme (epoxide hydratase) cannot get to the epoxide to act on it. Such compounds are the highly reactive electrophiles whose precursors are referred to as carcinogens. Other types of carcinogens are activated by different oxidative pathways, which also involve P-450 enzymes.

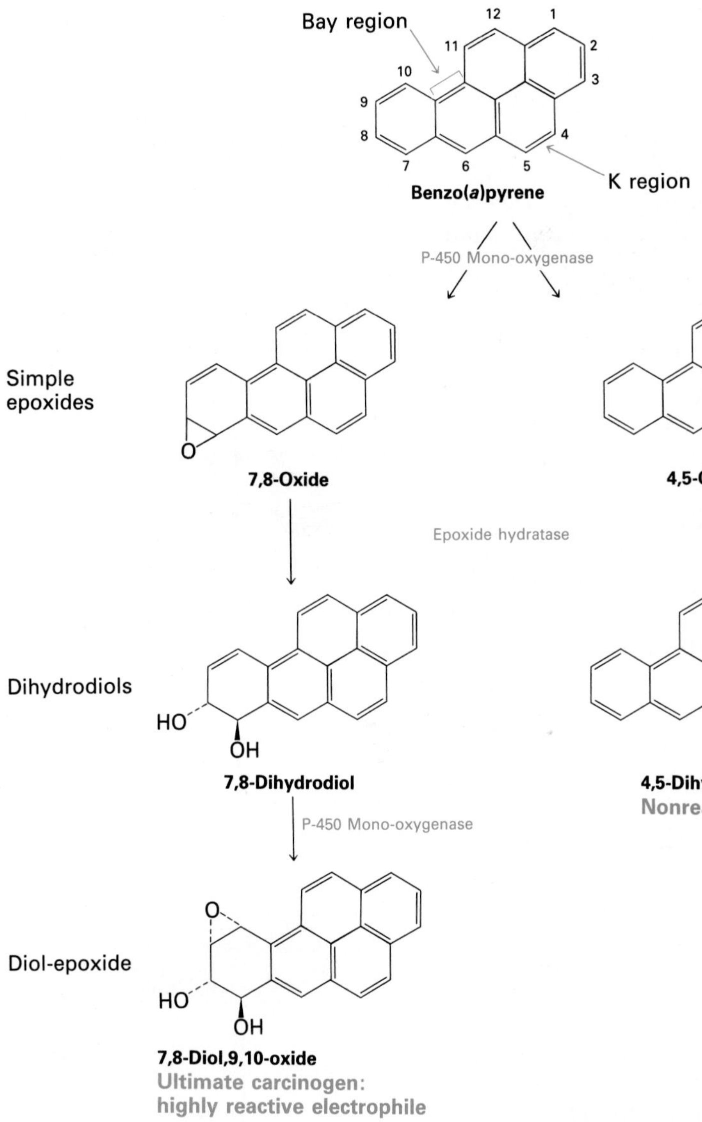

▲ FIGURE 26-29 The metabolic activation of benzo(a)pyrene, a polycyclic aromatic hydrocarbon that is a powerful carcinogen. Although chemically almost inert, it is metabolically converted into a highly reactive electrophile. Two metabolic pathways are shown. In the right-hand pathway, an intermediate epoxide forms by the attack of the cytochrome P-450 system on what is called the "K region" of the molecule. An epoxide in this region is rapidly hydrolyzed to a nonreactive dihydrodiol. In the left-hand pathway, the molecule is initially oxidized at the 7,8 double bond, leading to a 7,8-oxide that is rapidly converted to a 7,8-dihydrodiol. This compound is still a good substrate for the P-450 system and is again epoxidated, now near the "bay region," at the 9,10 double bond. The 7,8-diol,9,10-oxide (or diol-epoxide) is not a good substrate for epoxide hydratase, so it is released into the cell as a highly reactive electrophile. This form is carcinogenic because it can readily react with negatively charged centers in DNA.

The Carcinogenic Effect of Chemicals Depends on Their Interaction with DNA

Once inside cells, electrophiles can react with negatively charged centers on many different molecules—protein, RNA, and DNA, to name the most obvious. It is only the reaction with DNA, however, that leads to cancer. Depending on their size and structure, electrophiles react with particular positions on individual DNA bases, causing changes in the DNA sequence. Thus the electrophiles are mutagens.

The mutagenicity of chemicals is easily assayed in bacteria and by this means most compounds identified as carcinogens for experimental animals have been shown to be mutagenic. In fact, the mutagenic potential of compounds is roughly proportional to their carcinogenic potential. For this reason, bacterial mutagenesis has become a test for carcinogens. The first and most popular of these tests is the Ames test, named for its developer Bruce Ames, a bacterial geneticist. This test, however, only shows if a compound has the potential to be carcinogenic; the actual danger posed by any chemical is often assessed in animal studies, and even these are not a definitive indication of the danger to humans.

The strongest evidence that carcinogens act as mutagens comes from the observation that the cellular DNA altered by exposure of cells to carcinogens can transform cultured cells. This very important result was first obtained by extracting DNA from chemically transformed mouse cells and treating 3T3 cells with the DNA. A small fraction of the treated 3T3 cells became transformed, and their DNA acquired a new gene—an oncogene—that could be passed on to other cells (Figure 26-30). This type of experiment has been performed also with DNA from animal and human tumor cells and has revealed many oncogenes.

In summary, exposure of cells to a chemical carcinogen often induces a permanently altered state in cellular DNA. In other words, a carcinogen causes cancer by acting as a mutagen. In theory, a carcinogen could do this by binding to DNA and causing a mistake during the replication of the DNA, but the evidence suggests that in reality DNA sequence changes often are induced by the very repair processes cells use to rid themselves of DNA damage.

➤ *The Role of Radiation and DNA Repair in Carcinogenesis*

The living world is constantly being bombarded by radiation of many different kinds. Two types of radiation are especially dangerous because they can modify DNA; ultraviolet (UV) radiation and the ionizing radiations (x-rays and atomic particles). UV radiation of the appropriate wavelength can be absorbed by the DNA bases and can produce chemical changes in them. The most common

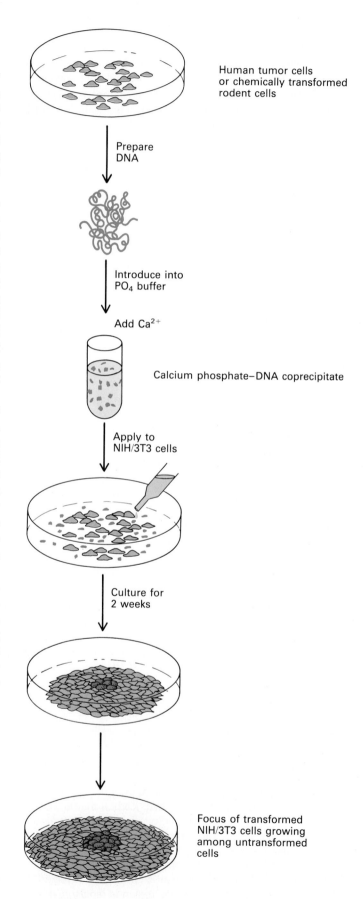

Human tumor cells or chemically transformed rodent cells

Prepare DNA

Introduce into PO$_4$ buffer

Add Ca^{2+}

Calcium phosphate–DNA coprecipitate

Apply to NIH/3T3 cells

Culture for 2 weeks

Focus of transformed NIH/3T3 cells growing among untransformed cells

▲ FIGURE 26-30   Standard assay for a cellular oncogene in tumor cells or chemically transformed cells.

damage is the production of dimers between adjacent pyrimidine residues in one strand of DNA (see Table 10-3 and Figure 10-26). These dimers interfere with both transcription and replication of DNA. Ionizing radiation mainly causes breaks in DNA chains. Both types of radiation can cause cancer in animals and can transform cells in culture. The ability of ionizing radiation to cause human cancer, especially leukemia, was dramatically shown by the increased rates of leukemia among survivors of the atomic bombs dropped in World War II.

UV radiation has been widely used in research on mutations and cancer because it is an easily manipulable and measurable agent that directly damages DNA. However, the UV-induced DNA lesions, including the dimers, are probably not primary consequences of the radiation. This has been shown directly in bacteria with a defective *recA*, a gene involved in DNA repair. Mutants which lack *recA*'s product are easily killed by UV radiation because DNA repair is prevented; there are virtually no mutants among the survivors. This result indicates that the mutational effect of UV radiation is caused during repair and not as a primary consequence of the radiation.

Ineffective or Error-Prone Repair of Damaged DNA Perpetuates Mutations

Because organisms are continually being exposed to chemicals that can react with DNA and to radiation that can damage DNA, very effective methods of repairing DNA have evolved. As more complex organisms with larger genomes and longer generation times evolved, the effectiveness of DNA repair processes increased. If repair processes were 100 percent effective, chemicals and radiation would pose no threat to cellular DNA. Unfortunately, repair of some lesions is relatively inefficient, and these can become progenitors of mutations.

One hard-to-repair lesion is a double-strand break of the DNA backbone. Such breaks, caused either by ionizing radiation or by chemicals, can be correctly repaired only if the free ends of the DNA rejoin exactly; however, without overlapping single-stranded regions there is no base-pair homology to catalyze the joining. Because cells will not tolerate free DNA ends, broken ends of molecules are generally joined to other broken ends. A cell that has suffered a particular double-strand break usually contains other breaks; thus a broken end has a number of possible segments to which it can join. The joining of broken ends on different chromosomes leads to translocation of pieces of DNA from one chromosome to another. Such a translocation may activate a protooncogene. Consequently, the carcinogenic effect of chemicals or radiation can result from their production of double-strand DNA breaks followed by translocations that occasionally activate oncogenes.

Proper repair of DNA also is difficult when cells attempt to replicate damaged DNA before repair processes

have had a chance to act on the lesion. This situation can arise when a cell suffers so much damage over a short time that its repair systems are saturated. It then runs the danger of having repair fall so far behind that extensive replication of unrepaired lesion occurs. In such situations, both bacteria and animal cells use inducible reserve systems for repair. Such systems are not expressed in undamaged cells, but some aspect of the accumulated damage causes their derepression (induction) and expression.

One such inducible system is the SOS repair system of bacteria. In contrast to other repair systems, this one makes errors in the DNA as it repairs lesions, so it is referred to as error-prone. The SOS system is responsible for UV-induced mutations, and its activity is dependent on the *recA*-gene product. The errors induced by the SOS system are at the site of lesions, suggesting that the mechanism of repair is insertion of random nucleotides in place of the damaged ones in the DNA. This inducible rather than constitutive system is only a last resort used when error-free mechanisms of repair cannot cope with damage. SOS repair may accelerate evolution at a time when the organism is under stress. Bacteria lacking an SOS system have greatly reduced mutation rates, which supports the idea that most of the mutations produced by treating bacteria with radiation or chemicals are caused by the error-prone SOS repair system.

Where animal cells have an error-prone repair system is not known, but they certainly have inducible repair systems. If one or more of these systems are error-prone, inducible repair is likely to play a role in mutagenesis and therefore in carcinogenesis. In any case, many investigators believe that in animal cells, as in bacteria, most mutation is an indirect consequence of DNA damage and not a direct result.

Because radiation- or carcinogen-induced DNA damage must be repaired before the DNA is replicated, cells have sensing mechanisms that react to DNA damage and stop DNA replication. This is part of the way that the p53 protein suppresses tumors. In cells that have lost both copies of their *p53* gene, there are increased rates of mutation in response to damage, accelerating the production of oncogenes.

Some Defects in DNA-Repair Systems Are Associated with High Cancer Rates in Humans

A link between DNA-repair systems and carcinogenesis also is suggested by the finding that humans with inherited genetic defects that make specific repair systems nonfunctional have an enormously increased probability of developing certain cancers. One such disease is xeroderma pigmentosum, an autosomal recessive disease. Cells of affected patients are unable to repair UV damage or to remove bulky chemical substitutents on DNA bases. The patients get skin cancers very easily if their skin is exposed to the UV rays in sunlight. The complexity of mammalian repair systems is shown by the fact that there are many different genes in which defects lead to xeroderma pigmentosum lesions, all having the same phenotype and the same consequences.

Another human disease that affects DNA repair and increases cancer risks is ataxia telangiectasia. This disease sensitizes cells to x-rays but not to UV. These two forms of radiation cause different types of DNA lesions, which are handled by different repair systems. Ataxia telangiectasia results in the loss of a system that repairs x-ray-induced cross-links in the DNA duplex, whereas xeroderma pigmentosum results in the loss of a pathway that repairs pyrimidine dimers. Because most chemical carcinogens mimic the effects of either x-rays or UV irradiation, cells from patients with one of these two diseases are hypersensitive to treatment with one or another set of chemicals.

▶ The Multicausal, Multistep Nature of Carcinogenesis

In this chapter we have stressed the role of oncogenes and tumor suppressor genes in the development of cancer and ignored most other potential mechanisms. The rationale for this emphasis is simply that we know much more about the mutational origin of cancer than about other possible cancer-inducing mechanisms. We must, however, maintain an open mind on the problem. In this section we describe several possible nongenetic mechanisms of carcinogenesis and review the evidence that cancer results from multiple interacting causes.

Epigenetic Alterations May Occur in Teratocarcinomas

The largest conceivable class of nongenetic cancer-inducing mechanisms involve *epigenetic* inheritance. An epigenetic condition or process is one that is passed from a specific cell to its progeny without any alteration in the germ line coding sequence of the DNA. The differentiated state of cells, which is passed on to progeny cells, is generally determined epigenetically. How epigenetic states are maintained is still a matter for speculation; it clearly involves protein regulatory molecules interacting with DNA. DNA methylation may also be involved. Whatever its basis, a process determined epigenetically can be inherited with the same fidelity as one determined by a genetic alteration.

Epigenetic mechanisms may underlie formation of malignant mouse teratocarcinomas. These tumors of very early embryonic cells maintain the ability to differentiate into the full range of body cell types. In fact, if certain teratocarcinoma cells are incorporated into a preimplantation embryo, they contribute as normal cells to all of the tissues of the adult, including the germ-line cells, without forming tumors. Conversely, if teratocarcinoma cells are injected into adult mice, they form lethal tumors. Thus their malignant state is conditional on their environment. Mouse teratocarcinomas can be formed by the simple procedure of transferring early embryos into older mice at sites such as the testis; these tumors can also be formed by explanting very early embryos into cell culture. Induction of these tumors thus requires no mutagenic treatment and no mutation is evident; in the right environment the cells appear absolutely normal. We presently believe that the teratocarcinomas are a unique case, but their occurrence certainly alerts us to the possible importance of nongenetic events in carcinogenesis.

Some Cancer-Inducing Chemicals Act Synergistically

The phenomenon of *tumor promotion* illustrates both the possible involvement of a nongenetic mechanism in cancer induction and the synergistic interaction of cancer-inducing agents. The initial discovery of tumor promotion in the early 1940s was based on the finding that if the skin of an animal is treated with certain chemicals (now recognized as DNA-damaging carcinogens), the odds of a tumor appearing could be greatly augmented by further treatment with a second compound having a very different structure and chemistry. The first compound is called an *initiator*, and the second is called a *promoter*. In contrast to initiators, promoters do not need to be metabolized to be active, they have no tendency to react as electrophiles, and they rarely induce tumors by themselves. Therefore promoters must act on cells through a mechanism quite different from that of initiators. To be effective, a promoter must be applied after an initiator and it must be applied repeatedly over many weeks or months, whereas the initiator need be applied only once. Although treatment with an initiator often produces a tumor (i.e., the initiator may act both to initiate and to promote), promoters require previous initiation to be tumorigenic.

One class of promoters, the phorbol esters (Figure 26-31), has been intensively studied and provides a model for the action of all promoters. Many types of cells, when treated with phorbol esters, take on a transformed morphology and growth habit but revert to normal when the compound is removed. Thus phorbol esters can produce a phenocopy of the transformed state (i.e., they alter the phenotype but not the genotype). If cells that have the potential to differentiate are treated with phorbol esters, differentiation will often be either blocked or facilitated, which again attests to the powerful physiological effects of phorbol esters.

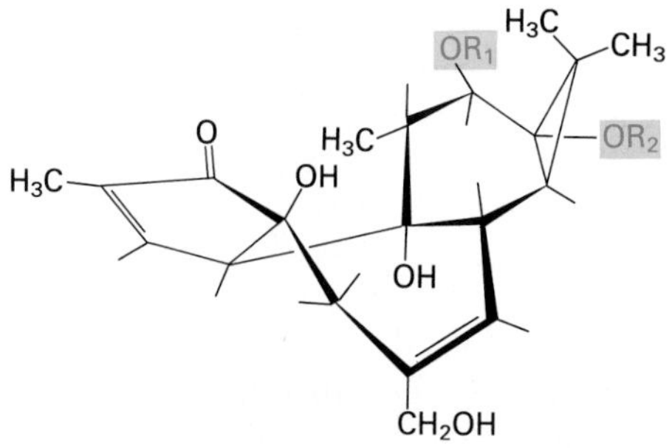

Phorbol ester

▲ FIGURE 26-31   The structure of phorbol esters, the prototypical tumor promoters. The molecule has four rings, one each containing seven, six, five, and three carbons. Crucial components are the two long-chain fatty acids (OR_1 and OR_2) esterified to the phorbol backbone. The molecule apparently acts as an analog to diacylglycerol, a molecule with two esterified fatty acid chains and a hydroxyl group.

The site of action of phorbol esters became known from studies of protein kinase C, an enzyme that phosphorylates serine or threonine residues on other proteins. The activation of protein kinase C requires the presence of a number of cofactors: Ca^{2+}, phospholipids, and diacylglycerol. This latter compound is produced when phospholipids are degraded as part of a signal transduction process. A specific class of progenitors of diacylglycerol is the inositol-containing phospholipids, which are degraded by a complex pathway involving phosphorylation of the inositol moiety before its cleavage (see Figure 20-44).

Phorbol esters can substitute for diacylglycerol as a cofactor for protein kinase C and, in association with Ca^{2+} and phospholipids, strongly activate protein kinase C for phosphorylation of substrate proteins. The mechanism of action of phorbol esters appears to be a facilitation of Ca^{2+} binding. The stereochemistry of phorbol esters makes them somewhat analogous in shape to diacylglycerol; this analogy is presumably the basis for their ability to substitute for diacylglycerol and activate protein kinase C.

In the absence of prior initiation, tumor promoters have no permanent effects on cell metabolism, whereas following initiation they can set in motion events that lead to an irreversible alteration in the cell. The end result could be a DNA alteration, or it could be a changed state of cellular differentiation. Whatever the mechanism, a promoter must produce a second permanent alteration in a cell that has already undergone an initiator-induced mutational change. Thus tumor promoters act to amplify the effects of initiators.

Evidence of synergism between various cancer-inducing agents in human carcinogenesis has been discovered, and the initiator-promoter model may be just as applicable to humans as it is to animal model systems. If so, control of cancer could be achieved through control of either initiators or promoters. Short-term mutagenesis tests can only identify initiators; no reliable tests for promoters have yet been devised. Some investigators believe that a major aspect of the cancer-inducing capacity of cigarettes may lie in a promoterlike activity released from the burning of the tobacco.

The initiator-promoter model of carcinogenesis describes just one of a variety of ways in which chemicals may interact to cause tumors. For example, low concentrations of potentially carcinogenic compounds may interact when the compounds are applied together (cocarcinogenesis). Certain compounds, notably antioxidants, can counteract the carcinogenic effects of other compounds (anticarcinogenesis).

Natural Cancers Result from the Interaction of Multiple Events over Time

The evidence supporting the concept that carcinogenesis is a multicausal, multistep process is powerful. First is epidemiologic evidence showing that cancer generally occurs late in life in a pattern that follows multihit kinetics. Second is the initiator-promoter phenomenon in which the involvement of at least two events is clear. Third is the synergism between oncogenes, particularly *myc* and *ras,* in cultured cells and in transgenic animals. Fourth is the common observation that tumors are monoclonal—even in transgenic animals in which all cells contain a single dominant oncogene—implying that a second, rare event was part of the causation chain that led to the tumor.

► Human Cancer

A central motivation of basic cancer research is to understand human cancer well enough to be able to alter its course. Although what we have learned in the laboratory is certainly relevant to human cancer, the understanding we seek has proved elusive.

To assess the role of a substance in cancer induction, it is not useful to think in terms of a single cause. As we emphasized above, naturally occurring cancer is most often a consequence of multiple factors that interact over long periods of time. Because each of these factors increases the possibility that a cancer will occur, they are called *risk factors.* For example, Epstein-Barr virus is a risk factor for Burkitt's lymphoma, although it is clearly not the sole cause. Identifying risk factors for human cancer has been a slow and frustrating activity. One successful endeavor stands out: the identification of cigarette smoking as a crucial risk factor in lung cancer. A risk factor of this potency gives a clear indication of how to act to avoid lung cancer: avoid cigarettes.

Unfortunately, lung cancer is the only major human cancer for which a clear-cut risk factor has been identified. A diet high in animal fat is thought to be a risk factor for colon and breast cancer, and many viruses and chemicals have been correlated with minor cancers; however, hard evidence that would help us avoid breast cancer, colon cancer, prostate cancer, leukemias, and others is generally lacking.

A corollary to the belief that multiple interacting events in our environment play the major roles in inducing cancer is the belief that genetic inheritance plays only a small role. This proposition is supported by the finding that people who migrate to a new environment take on the risk profile for cancers in their new environment within a generation. For instance, when Japanese citizens move to

California, they rapidly lose the oriental propensity toward stomach cancer and soon show the occidental propensity toward breast cancer. Genetic inheritance does, however, play some role in human cancer. We described earlier the knowledge that mutated tumor suppressor genes, such as *RB*, can be inherited and increase the propensity to develop particular tumors to almost 100 percent. Inherited mutations at least partially explain why people in some families get cancers relatively early in their lives. It may also explain why some people seem resistant to carcinogens in cigarette smoke while many others succumb to the cancer-inducing effects in smoke.

SUMMARY

Cancer is a fundamental aberration in cellular behavior touching on many aspects of molecular cell biology. To become a cancer cell, a normal cell undergoes many significant changes. It continues to multiply when normal cells would be quiescent; it invades surrounding tissues, often breaking through the basal laminas that define the boundaries of tissues; and it spreads through the body and sets up secondary areas of growth in a process called metastasis. The various cell types of the body can all give rise to cancer cells. Cancer cells are usually closer in their properties to immature normal cells than to more mature cell types. Because cancer cells grown in culture retain malignant properties, we know that the alteration from a normal cell to a cancer cell is a change within the cancer cell itself.

Experimenters can freely manipulate cells grown in culture to learn about the properties that distinguish normal cells from cancer cells. Carcinogens, cancer-causing agents, can be used to alter cultured cells to produce full-fledged cancer cells in a process called transformation. Transformed cells differ from normal cells in many ways, including cell-growth control, cell morphology, cell-to-cell interactions, membrane properties, cytoskeletal structure, protein secretion, gene expression, and mortality (cultures of transformed cells can grow indefinitely).

A normal cell can be transformed to malignancy by the expression of one or a few oncogenes. Most oncogenes derive from growth-controlling genes that encode one of five types of proteins: growth factors, receptors, intracellular signal transducers, nuclear transcription factors, and cell-cycle control proteins. Signaling division in a cell begins with an interaction between a growth factor and a cellular receptor and continues with one or more steps of signal transduction, leading, among other events, to changes in the genes transcribed in the cell. The overall process is co-ordinated and monitored by controlling proteins like RB and p53. The majority of oncogenes encode intracellular transducing proteins including protein-tyrosine kinases

(e.g., the historic *src* of Rous sarcoma virus) and GTP-binding and hydrolysis proteins (the *ras* oncogenes). The other major class is the nuclear transcription factor oncogenes. Certain cell-cycle control proteins act as tumor-suppressor genes that counter the action of oncogenes. Inactivation of these genes on both chromosomes makes cells much more susceptible to oncogene action and also leads to increased mutability. They are often inactivated by mutation in human tumors.

Oncogenes often act collaboratively, no one oncogene being sufficient to induce malignant growth. This has been demonstrated in transgenic mice whose inheritance is modified by insertion of one or more oncogenes into their genome. Also, human tumor cells may harbor more than one oncogene, suggesting that multiple genetic changes contribute to the malignancy of a given cell. The accumulation of multiple alterations may explain the long latency period between the initial exposure to a carcinogen and the development of a full-fledged malignancy.

Oncogenes created in situ in the genome of human or other animal cells are called cellular oncogenes. Cellular oncogenes are activated from their normal, benign, proto-oncogenic state by local alterations in DNA structure (point mutations, deletions, etc.), by chromosomal translocations, or by extensive reduplication of DNA regions including an oncogene. They are often revealed when transfer of DNA from a malignant to a normal cell confers malignant properties on the recipient, implying the transfer of an oncogene.

Oncogenes have generally first been identified either as the oncogenic principles in viruses, as the active genes in malignant cells, or as genes at translocation breakpoints that recur in specific tumors. Mutated tumor-suppressor genes have generally been found as the active genes in human families with a high inherited propensity to get certain forms of cancer.

Oncogenes are formed in cells or carried into cells by

transforming agents, of which we recognize three main types: viruses, chemical carcinogens, and radiation. Certain DNA viruses and RNA-containing retroviruses can transform cells by permanently integrating new genes into the DNA of infected cells. The DNA-containing papovaviruses carry oncogenes that transform cells by inducing a growing state that facilitates virus multiplication. Permanent cell transformation results when virus multiplication is not possible.

Retroviruses cause cancer in two ways, both of which depend on cellular genes. These viruses, which contain RNA as their genetic material, employ reverse transcriptase to make an intracellular DNA copy that can integrate into cellular DNA. Such integrated proviruses, by recombination with cellular genes, can acquire oncogenes that turn them from relatively benign viruses into cancer-inducing agents. Retroviruses can also cause cancer by integrating near a proto-oncogene in such a way that the gene is activated to become an oncogene. The difference between these two mechanisms of retroviral-induced cancer is that in the former case the oncogene becomes part of a transmissible virus (i.e., the gene is transduced), whereas in the latter case the proto-oncogene is activated in situ to become an oncogene that cannot be transmitted to other cells or animals.

Viruses cause certain human tumors. These include retroviruses that cause leukemia; the Epstein-Barr virus, a herpeslike virus that causes leukemia; hepatitis B virus causing liver tumors; and papilloma viruses causing cervical tumors. AIDS is caused by a unique human retrovirus that does not directly cause cancer.

Chemical carcinogens have a variety of structures with one unifying characteristic: electrophilic reactivity (either they are electrophiles or they are metabolized in the body to become electrophiles). Metabolic activation occurs via the cytochrome P-450 system, a pathway generally used by cells to rid themselves of noxious chemicals. The reactive electrophiles combine with many parts of the cell, but their reaction with DNA is the primary carcinogenic event. The reaction of a carcinogen and DNA leads to a DNA alteration that ultimately produces an oncogene from a proto-oncogene.

Ultraviolet, x-ray, and atomic particle radiations can all cause cancer. All affect DNA and presumably alter proto-oncogenes. Cellular DNA-repair processes have been implicated both in protecting against radiation-induced carcinogenesis and in causing it. The protective effect is most vividly demonstrated by the association of certain human diseases that cause DNA-repair deficiencies with high rates of cancer; the deficiencies in these diseases inactivate error-free repair systems. The inductive effect is seen most clearly in bacterial strains that lack secondary, error-prone repair systems; such strains have much lower mutation rates than wild-type bacteria.

Nongenetic mechanisms may play some role in cancer induction. For example, chemicals called promoters can potentiate the activity of electrophilic carcinogens. The best-understood promoters are the phorbol esters, which cause nongenetic cellular changes that often mimic transformation. These substances activate a cellular protein kinase. Long-term treatment with phorbol esters leads to permanent cellular alterations that may or may not be genetic. An apparently clear-cut case of a nongenetic change that causes cancer is the epigenetic alteration leading to a teratocarcinoma. The cells in these tumors revert to normal when they are implanted into early embryos.

REVIEW QUESTIONS

1. What is a benign tumor? Can benign tumors be harmful? Explain your answer.

   a. What are malignant tumors, and what are some of their characteristics? What is meant by metastasis, how does it apply to malignant tumors, and how does it occur? What is the difference between a carcinoma and a sarcoma?

   b. Why are cultured cells useful in understanding cancer? Describe the characteristics of fibroblasts and epithelial cells. What is a cell strain? Are cells isolated from an embryo or adult animal immortal, or do they have a finite life span? What is a cell line? Are cell lines immortal? What is meant by aneuploid?

   c. What is defined medium, and what are some of its essential components?

   d. What is meant by malignant transformation, and what

agents can cause it to occur? Outline the changes that can be induced in cells following transformation.

2. What are oncogenes? What are the five general types of proteins that control cell growth?

   a. What are tumor suppressor genes? What is *p53*, and what is its basic function in a cell? Why does *p53* behave as a dominant negative when it exists as one wild type allele and one mutant allele in a cell?

   b. What is RB, and how is it involved in cell cycle progression? Why does RB not behave as a dominant negative when there is one mutant allele and one wild type allele?

   c. What is apoptosis? How does it protect an organism against cancer? What is Bcl-2, and how is it involved in apoptosis?

d. What is the two step model of transformation? Discuss the relationship between *ras* and *myc* as an example of this model.

e. What are chromosomal translocations, and give some examples of how oncogenes are influenced by the phenomenon? What physical or chemical agent is particularly notorious for inducing translocations? Why?

f. Refer to Figure 26-10, which illustrates the five types of growth control in cells. Fill in the following table by giving at least one example of a gene and its encoded protein's function. Remember that mutations that affect the first four classes of genes can result in active oncogenes, whereas mutations in class V proteins result in changes in tumor suppressor genes.

Protein Class	Gene	Protein Product Function
I		
II		
III		
IV		
V		

g. You are part of a team involved in studies of the *p53* gene. One member of the group has been taking histories of families exhibiting inherited forms of cancer, and has also assembled pedigrees from several families where no evidence of inherited cancers are present. You have been assigned the task of using restriction enzymes to map the *p53* loci from several members of these families and to sequence the region when necessary. After much arduous work, three very interesting observations are made concerning three patients; they are presented in the table below.

	p53 Characteristics		Presence of Inherited Cancer in Family
Patient	Restriction Map	Sequence	
A	Wild type	Wild type	No
B	Wild type	Several point mutations	Yes
C	85% of one allele is deleted	One allele is wild type	No

Explain the results presented in this table in terms of the susceptibility of each family to cancer. Why in particular might family C not exhibit an inherited form of cancer in spite of the fact that one allele of the p53 gene is virtually deleted?

3. Discuss briefly the stages of papovaviral infection. What is meant by permissive and nonpermissive cells, and which class is more easily transformed? Explain your answer.

a. What is the difference between site-specific and random integration of viral genomes? How do papovaviruses integrate?

b. What are the early proteins of SV40 and of polyoma? Which are able to transform immortalized cells?

c. Review the life cycle of retroviruses. What is the reverse transcript and how is it made? What are the LTR regions, and how does each affect transcription of proviral DNA? What are transducing retroviruses, and how do they transform cells? What are slow-acting retroviruses, and how are they fundamentally different from transducing retroviruses? How do they transform cells?

4. What is the difference between a direct acting and an indirect acting carcinogen? What is the function of the P-450 enzymes in chemical carcinogenesis?

a. Both prokaryotes and eukaryotes have evolved elaborate mechanisms for maintaining the genome, including DNA repair proteins for removing a vast array of covalent modifications to the genome. What is the SOS response?

b. You are involved in testing new drugs for their potential ability to cause cancer. You must establish the carcinogenicity for each of two new compounds that have been developed. One set of experiments involves testing the compounds on rodent skin for their ability to induce tumors. In experiment I, you shave the skin from some rodents, and apply a solution of either drug A or B to the animals. After several months, no tumors formed in an amount greater than that found for rodents where only water was applied to the skin. You conclude that these agents do not cause skin tumors in rodents.

You are then directed to perform another experiment, which you designate II, where you first apply a low dose of methyl nitrosourea (MNU) (see Figure 26-29) to the skin. After two weeks go by, you apply either drug A or B to these animals. After several months, tumors form in animals exposed to drug A. The data are presented in the table.

Experiment	Drug Applied	Pretreatment with MNU	Number of Skin Tumors Per 50 Rodents
I	A	–	0
	B	–	1
	H$_2$O	–	1
II	A	+	24
	B	+	0
	H$_2$O	+	3

Explain these results.

c. You are interested in sequencing the *p53* gene from some of the skin tumors that formed, but the amount of tissue is very limited. Propose a method for doing this such that the amount of tissue is not necessarily a restricting factor.

References

General References

Advances in Cancer Research (yearly volumes). Academic Press.

BECKER, F. F., ed. 1982. *Cancer: A Comprehensive Treatise*. Plenum. (Multiple volumes are released under this title.)

BOCK, G., and J. MARSH, eds. 1989. *Genetic Analysis of Tumour Suppression* (Ciba Foundation symposium; 142). Wiley, London.

BOICE, J. D., JR., and J. F. FRAUMENI, eds. 1984. Radiation carcinogenesis: epidemiology and biological significance. In (same editors) *Progress in Cancer Research and Therapy*, vol. 26. Raven.

BRADSHAW, R. A., and S. PRENTIS, eds. 1987. *Oncogenes and Growth Factors*. Elsevier.

BURCK, K. B., E. T. LIU, and J. W. LARRICK. 1988. *Oncogenes: An Introduction to the Concept of Cancer Genes*. Springer-Verlag.

BUYSE, M. E., M. J. STAQUET, and R. J. SYLVESTER, eds. 1984. *Cancer Clinical Trials: Methods and Practice*. Oxford University Press.

CAIRNS, J., ed. 1974. *Cancer: Science and Society*. W. H. Freeman and Company.

CAIRNS, J. 1975. The cancer problem. *Sci. Am.* 233(5):64–72, 77–78.

DEVITA, V. T., JR., S. HELLMAN, and S. A. ROSENBERG, eds. 1982. *Cancer: Principles and Practice of Oncology*. Lippincott.

HOLLAND, J. F., ed. 1982. *Cancer Medicine*. Lea & Febiger.

KAHN, P., and T. GRAF, eds. 1986. *Oncogenes and Growth Control*. Springer-Verlag.

KAISER, H. E. 1981. *Neoplasms: Comparative Pathology of Growth in Animals, Plants, and Man*. Williams & Wilkins.

KLEIN, G., ed. 1988. *Cellular Oncogene Activation*. Dekker.

LEVINE, A. J., W. C. TOPP, and J. D. WATSON, eds. 1984. *Cancer Cells*, vol. 1: *The Transformed Phenotype*. Cold Spring Harbor Laboratory.

MELNICK, J. L., ed. 1985. *Viruses, Oncogenes and Cancer*, vol. 2. Karger.

PIMENTEL, E. 1987. *Hormones, Growth Factors, and Oncogenes*. CRC.

VAN DE WOUDE, G. F., et al., eds. 1984. *Cancer Cells*, vol. 2: *Oncogenes and Viral Genes*. Cold Spring Harbor Laboratory.

VARMUS, H., and J. M. BISHOP, eds. 1986. *Cancer Surveys: Advances and Prospects in Clinical, Epidemiological and Laboratory Oncology*, vol. 5, no. 2. *Biochemical mechanisms of oncogene activity: proteins encoded by oncogenes*. Oxford University Press, Oxford.

VARMUS, H., and R. A. WEINBERG. 1993. *Genes and the Biology of Cancer*. Scientific American Library.

Characteristics of Tumor Cells

BEHRENS, J., et al. 1992. Cell adhesion in invasion and metastasis. *Semin. Cell Biol.* 3:169–178.

CAIRNS, J. 1975. Mutational selection and the natural history of cancer. *Nature* 255:197–200.

CIFONE, M. A., and I. J. FIDLER. 1981. Increasing metastatic potential is associated with increasing instability of clones isolated from murine neoplasms. *Proc. Nat'l. Acad. Sci. USA* 78:6949–6952.

FIDLER, I. J., and I. R. HART. 1982. Biological diversity in metastatic neoplasms: origins and implications. *Science* 217:998–1003.

ILLMENSEE, K., and B. MINTZ. 1976. Totipotency and normal differentiation of single teratocarcinoma cells cloned by injection into blastocysts. *Proc. Nat'l. Acad. Sci. USA* 73:549–553.

NICHOLSON, G. L. 1979. Cancer metastasis. *Sci. Am.* 240(3):66–76.

OSSOWSKI, L., and E. REICH. 1983. Antibodies to plasminogen activator inhibit human tumor metastasis. *Cell* 35:611–619.

Cell Culture and Transformation

ABERCROMBIE, M. 1970. Contact inhibition in tissue culture. *In Vitro* 6:128–142.

ABERCROMBIE, M., and J. E. M. HEAYSMAN. 1954. Observations on the social behaviour of cells in tissue culture. *Exp. Cell Res.* 6:293–306.

BURGER, M. M., and G. S. MARTIN. 1972. Agglutination of cells transformed by Rous sarcoma virus by wheat germ agglutinin and concanavalin A. *Nature New Biol.* 237:356–359.

CARPENTER, G., and S. COHEN. 1975. Human epidermal growth factor and the proliferation of human fibroblasts. *J. Cell Physiol.* 88:227–238.

CARREL, A. 1912. On the permanent life of tissues outside of the organism. *J. Exp. Med.* 15:516–528.

CHERINGTON, P. V., B. L. SMITH, and A. B. PARDEE. 1979. Loss of epidermal growth factor requirement and malignant transformation. *Proc. Nat'l. Acad. Sci. USA* 76:3937–3941.

DULAK, N. D., and H. M. TEMIN. 1973. A partially purified polypeptide fraction from rat liver cell conditioned medium with multiplication-stimulating activity for embryo fibroblasts. *J. Cell Physiol.* 81:153–160.

FOLKMAN, J., and A. MOSCONA. 1978. Role of cell shape in growth control. *Nature* 273:345–349.

GAFFNEY, B. J. 1975. Fatty acid chain flexibility in the membranes of normal and transformed fibroblasts. *Proc. Nat'l. Acad. Sci. USA* 72:510–516.

GEY, G. O., W. D. COFFMAN, and M. T. KUBICEK. 1952. Tissue culture studies of the proliferative capacity of cervical carcinoma and normal epithelium. *Cancer Res.* 12:264–265.

GOSPODAROWICZ, D., and J. S. MORAN. 1976. Growth factors in mammalian cell culture. *Ann. Rev. Biochem.* 45:531–558.

HAKOMORI, S. 1975. Structures and organization of cell surface glycolipid: dependency on cell growth and malignant transformation. *Biochim. Biophys. Acta* 417:58–80.

HATANAKA, M. 1974. Transport of sugars in tumor cell membranes. *Biochim. Biophys. Acta* 355:77–104.

HAYFLICK, L., and P. S. MOREHEAD. 1961. The serial cultivation of human diploid cell strains. *Exp. Cell Res.* 25:585–621.

HOLLEY, R. W. 1975. Factors that control the growth of 3T3 cells and transformed 3T3 cells. In D. B. Rifkin and E. Shaw, eds., *Proteases and Biological Control*. Cold Spring Harbor Laboratory.

HOLLEY, R. W., and J. A. KIERNAN. 1968. "Contact inhibition" of cell division in 3T3 cells. *Proc. Nat'l. Acad. Sci. USA* 60:300–304.

HYNES, R. O. 1973. Alteration of cell-surface proteins by viral transformation and by proteolysis. *Proc. Nat'l. Acad. Sci. USA* 70:3170–3174.

HYNES, R. O., ed. 1979. *Surfaces of Normal and Malignant Cells*. Wiley.

MACPHERSON, I., and L. MONTAGNIER. 1964. Agar suspension culture for the selective assay of cells transformed by polyoma virus. *Virology* 23:291–294.

OSSOWSKI, L., et al. 1973. An enzymatic function associated with transformation of fibroblasts by oncogenic viruses, II: Mammalian fibroblast cultures transformed by DNA and RNA tumor viruses. *J. Exp. Med.* 137:112–126.

SATO, G., ed. 1979. *Hormones and Cell Culture*. Cold Spring Harbor Laboratory.

SHIN, S. I., et al. 1975. Tumorigenicity of virus-transformed cells in nude mice is correlated specifically with anchorage independent growth in vitro. *Proc. Nat'l. Acad. Sci. USA* 72:4435–4439.

STOKER, M. G. P. 1967. Transfer of growth inhibition between normal and virus-transformed cells: Autoradiographic studies using marked cells. *J. Cell Sci.* 2:239–304.

TODARO, G. J. 1963. Quantitative studies of the growth of mouse embryo cells in culture and their development into established lines. *J. Cell Biol.* 17:299–313.

TODARO, G. J., J. E. DELARCO, and S. COHEN. 1976. Transformation by murine and feline sarcoma viruses specifically blocks binding of epidermal growth factor to cells. *Nature* 264:26–31.

TODARO, G. J., and H. GREEN. 1964. An assay for cellular transformation by SV40. *Virology* 23:117–119.

UNKELESS, J. C., et al. 1973. An enzymatic function associated with transformation of fibroblasts by oncogenic viruses, I: Chick embryo fibroblast cultures transformed by avian RNA tumor viruses. *J. Exp. Med.* 137:85–111.

WEBER, K., et al. 1974. Localization and distribution of actin fibers in normal, transformed and revertant cells. *Cold Spring Harbor Symp. Quant. Biol.* 39:363–369.

Oncogenes and Their Proteins

ALITALO, K., et al. 1983. Homogeneously staining chromosomal regions contain amplified copies of an abundantly expressed cellular oncogene (c-*myc*) in malignant neuroendocrine cells from a human colon carcinoma. *Proc. Nat'l. Acad. Sci. USA* 80:1707–1711.

BALMAIN, A., and I. B. PRAGNELL. 1983. Mouse skin carcinomas induced *in vivo* by chemical carcinogens have a transforming Harvey-*ras* oncogene. *Nature* 303:72–74.

BARBACID, M. 1987. *ras* Genes. *Ann. Rev. Biochem.* 56:779–827.

BARGMANN, C. I., M.-C. HUNG, and R. A. WEINBERG. 1986. Multiple independent activations of the *neu* oncogene by a point mutation altering the transmembrane domain of 649. *Cell* 46:185.

BISHOP, J. M. 1983. Cellular oncogenes and retroviruses. *Ann. Rev. Biochem.* 52:301–354.

BOGUSKI, M. S., and F. MCCORMICK. 1993. Proteins regulating Ras and its relatives. *Nature* 266:643–654.

BRODEUR, G., et al. 1984. Amplification of N-*myc* in untreated human neuroblastomas correlates with advanced disease stage. *Science* 224:1121–1124.

CANAANI, E., et al. 1983. Activation of the c-*mos* oncogene in a mouse plasmacytoma by insertion of an endogenous intracisternal A-particle genome. *Proc. Nat'l. Acad. Sci. USA* 80:7118–7122.

COLLETT, M. S., and R. L. ERIKSON. 1978. Protein kinase activity associated with the avian sarcoma virus *src* gene product. *Proc. Nat'l. Acad. Sci. USA* 75:2021–2024.

COWELL, J. K. 1982. Double minutes and homogeneously staining regions: gene amplification in mammalian cells. *Ann. Rev. Genet.* 16:21–59.

CROCE, C. 1987. Role of chromosomal translocations in human neoplasia. *Cell* 49:155.

CROCE, C. M., et al. 1984. Translocated c-*myc* oncogene of Burkitt lymphoma is transcribed in plasma cells and repressed in lymphoblastoid cells. *Proc. Nat'l. Acad. Sci. USA* 81:3170–3174.

DEKLEIN, A., et al. 1982. A cellular oncogene is translocated to the Philadelphia chromosome in chronic myelocytic leukaemia. *Nature* 300:765–767.

DOOLITTLE, R. F., et al. 1983. Simian sarcoma virus *onc* gene, v-*sis*, is derived from the gene (or genes) encoding a platelet-derived growth factor. *Science* 221:275–276.

DOWNWARD, J., et al. 1984. Close similarity of epidermal growth factor receptor and v-*erb*-B oncogene protein sequences. *Nature* 307:521–527.

ERIKSON, J., et al. 1982. Translocation of immunoglobulin V$_H$ genes in Burkitt lymphoma. *Proc. Nat'l. Acad. Sci. USA* 79:5611–5615.

FANIFI, A., E. A. HARRINGTON, and G. I. EVAN. 1992. Cooperative interaction between c-*myc* and *bcl*-2 proto-oncogenes. *Nature* 359:554–556.

FIEG, L. A. 1993. The many roads that lead to Ras. *Science* 260:767–768.

GERONDAKIS, S., S. CORY, and J. M. ADAMS. 1984. Translocation of the *myc* cellular oncogene to the immunoglobulin heavy chain locus in murine plasmacytomas is an imprecise reciprocal exchange. *Cell* 36:973–982.

HAYWARD, W. S., B. G. NEEL, and S. M. ASTRIN. 1981. Activation of a cellular *onc* gene by promoter insertion in ALV-induced lymphoid leukosis. *Nature* 209:475–479.

HEISTERKAMP, N., et al. 1983. Localization of the c-*abl* oncogene adjacent to a translocation breakpoint in chronic myelocytic leukaemia. *Nature* 306:239–242.

HERMANS A., et al. 1987. Unique fusion of *bcr* and c-*abl* genes in Philadelphia chromosome positive acute lymphoblastic leukemia. *Cell* 51:33.

HOFFMAN-FALK, H., et al. 1983. *Drosophila melanogaster* DNA clones homologous to vertebrate oncogenes: evidence for a common ancestor to *src* and *abl* cellular genes. *Cell* 32:589–598.

HUNTER, T. 1987. A thousand and one protein kinases. *Cell* 50:823.

HUNTER, T. 1993. Editorial overview: oncogenes and the cell cycle. *Curr. Opin. Genet. Dev.* 3:1.

HUNTER, T., and B. M. SEFTON. 1980. Transforming gene product of Rous sarcoma virus phosphorylates tyrosine. *Proc. Nat'l. Acad. Sci. USA* 77:1311–1315.

KAMPS, M. M., J. E. BUSS, and B. M. SEFTON. 1985. Mutation of NH$_2$-terminal glycine of p60^{src} prevents both myristoylation and morphological transformation *Proc. Nat'l. Acad. Sci. USA* 82:4625.

KRONTIRIS, T., and G. M. COOPER. 1981. Transforming activity of human tumor DNAs. *Proc. Nat'l. Acad. Sci. USA* 78:1181–1184.

LAND, H., L. F. PARADA, and R. A. WEINBERG. 1983. Tumorigenic conver-

sion of primary embryo fibroblasts requires at least two cooperating oncogenes. *Nature* 304:596–601.

LEVINSON, A. D., et al. 1978. Evidence that the transforming gene of avian sarcoma virus encodes a protein kinase associated with a phosphoprotein. *Cell* 15:561–572.

MOTOKURA, T., and A. ARNOLD. 1993. Cyclin D and oncogenesis. *Curr. Opin. Genet. Dev.* 3:5.

NISHIDA, E., and Y. GOTOH. 1993. The MAP kinase cascade is essential for diverse signal transduction pathways. *Trends Biochem. Sci.* 18:128–131.

PARADA, L. F., et al. 1982. Human EJ bladder carcinoma oncogene is a homologue of Harvey sarcoma virus *ras* gene. *Nature* 297:474–478.

PURCHIO, A. F., et al. 1978. Identification of a polypeptide encoded by the avian sarcoma virus *src* gene. *Proc. Nat'l. Acad. Sci. USA* 75:1567–1571.

SCHWAB, M., et al. 1984. Enhanced expression of the human gene N-*myc* consequent to amplification of DNA may contribute to malignant progression of neuroblastoma. *Proc. Nat'l. Acad. Sci. USA* 81:4940–4944.

SHIH, C., et al. 1979. Passage of phenotypes of chemically transformed cells via transfection of DNA and chromatin. *Proc. Nat'l. Acad. Sci. USA* 76:5714–5718.

STEHELIN, D., et al. 1976. DNA related to the transforming gene(s) of avian sarcoma viruses is present in normal avian DNA. *Nature* 260:170–173.

STEWART, T. A., P. K. PATTENGALE, and P. LEDER. 1984. Spontaneous mammary adenocarcinomas in transgenic mice that carry and express MTV/*myc* fusion genes. *Cell* 38:627–637.

SUKUMAR, S., et al. 1983. Induction of mammary carcinomas in rats by nitroso-methylurea involves malignant activation of H-*ras*-1 locus by single point mutations. *Nature* 306:658–661.

TABIN, C. J., et al. 1982. Mechanism of activation of a human oncogene. *Nature* 300:143–149.

TRAHEY, M., and F. MCCORMICK. 1987. A cytoplasmic protein stimulates normal n-*ras* p21 GTPase but does not affect oncogenic mutants. *Science* 238:542–545.

USHIRO, H., and S. COHEN. 1980. Identification of phosphotyrosine as a product of epidermal growth factor-activated protein kinase in A431 cell membranes. *J. Biol. Chem.* 255:8363–8365.

VARMUS, H., and A. J. LEVINE, eds. 1983. *Readings in Tumor Virology.* Cold Spring Harbor Laboratory.

VAN DE WOUDE, G. F., et al. 1984. *Cancer Cells*, vol. 2: *Oncogenes and Viral Genes.* Cold Spring Harbor Laboratory.

VOGELSTEIN, B., and K. W. KINZLER. 1993. The multistep nature of cancer. *Trends Genet.* 9:138–141.

WANG, J. Y. J. 1993. Abl tyrosine kinase in signal transduction and cell-cycle regulation. *Curr. Opin. Genet. Dev.* 3:35.

WATERFIELD, M. D., et al. 1983. Platelet-derived growth factor is structurally related to the putative transforming protein p28^{sis} of simian sarcoma virus. *Nature* 304:35–39.

WEINBERG, R. A. 1982. Oncogenes of spontaneous and chemically induced tumors. *Adv. Cancer Res.* 36:149–164.

WEINBERG, R. A. 1983. A molecular basis of cancer. *Sci. Am.* 249(5):126–143.

YUNIS, J. J. 1983. The chromosomal basis of human neoplasia. *Science* 221:227–235.

Tumor Suppressor Genes

CHELLAPPAN, S. P., et al. 1991. The E2F transcription factor is a cellular target for the RB protein. *Cell* 65:1053–1061.

CULOTTA, E., and D. E. KOSHLAND, JR. 1993. Molecule of the year. p53 sweeps through cancer research. *Science* 262:1958–1961.

DECAPRIO, J. A., et al. 1988. SV40 large tumor antigen forms a specific complex with the product of the retinoblastoma susceptibility gene. *Cell* 54:275–283.

FRIEND, S. H., et al. 1986. A human DNA segment with properties of the gene that predisposes to retinoblastoma and osteosarcoma. *Nature* 323:643–646.

FRIEND, S. H., et al. 1987. Deletions of a DNA sequence in retinoblastomas and mesenchymal tumors: organization of the sequence and its encoded protein. *Proc. Nat'l. Acad. Sci., USA.* 84:9059–9063.

HARLOW, E., et al. 1986. Association of adenovirus early-region 1A proteins with cellular polypeptides. *Mol. Cell Biol.* **6**:1579–1589.

HARRIS, C. C. 1993. p53: At the crossroads of molecular carcinogenesis and risk assessment. *Science* **262**:1980–1981.

HARTWELL, I. 1992. Defects in a cell cycle checkpoint may be responsible for the genomic instability of cancer cells. *Cell* **71**:543–546.

HUANG, H. J., et al. 1988. Suppression of the neoplastic phenotype by replacement of the Rb gene in human cancer cells. *Science* **242**:1563–1566.

LIVINGSTONE, L. R., et al. 1992. Altered cell cycle arrest and gene amplification potential accompany loss of wild-type p53. *Cell* **70**:923–935.

MARX, J. 1994. New tumor suppressor may rival p53. *Science* **264**:344–345.

MOMAND, J., et al. 1992. The *mem-2* oncogene product forms a complex with the p53 protein and inhibits p53-mediated transactivation. *Cell* **69**:1237–1245.

MURRAY, A. W. 1992. Creative blocks: cell-cycle checkpoints and feedback controls. *Nature* **359**:599–604.

WHYTE, P., et al. 1988. Association between an oncogene and an antioncogene: the adenovirus E1A proteins bind to the retinoblastoma gene product. *Nature* **334**:124–129.

Viruses as Agents of Transformation

BALTIMORE, D. 1970. RNA-dependent DNA polymerase in virions of RNA tumour viruses. *Nature* 226:1209–1211.

BISHOP, J. M., and H. E. VARMUS. 1984. Functions and origins of retroviral transforming genes. In R. Weiss et al., eds., *Molecular Biology of Tumor Viruses: RNA Tumor Viruses.* Cold Spring Harbor Laboratory.

COOK, P. J., and D. P. BURKITT. 1971. Cancer in Africa. *Brit. Med. Bull.* 27:14–20.

DESGROSEILLERS, L., E. RASSART, and P. JOLICOEUR. 1983. Thymotropism of murine leukemia virus is conferred by its long terminal repeat. *Proc. Nat'l. Acad. Sci. USA* 80:4203–4207.

DUESBERG, P. H., and P. K. VOGT. 1970. Differences between the ribonucleic acids of transforming and non-transforming avian tumor viruses. *Proc. Nat'l. Acad. Sci. USA* 67:1673–1680.

EPSTEIN, M. A., B. G. ACHONG, and Y. M. BARR. 1964. Virus particles in cultured lymphoblasts from Burkitt's lymphoma. *Lancet* 1:702–703.

FIELDS, B. N., and D. M. KNIPE. 1990. *Virology,* 2d ed., 2 vols. Raven.

GROSS, L., ed. 1970. *Oncogenic Viruses,* 2d ed. Pergamon.

GROSS, L., ed. 1983. *Oncogenic Viruses,* 3d ed. Pergamon.

HUEBNER, R. J., and G. J. TODARO. 1969. Oncogenes of RNA tumor viruses as determinants of cancer. *Proc. Nat'l. Acad. Sci. USA* 64:1087–1094.

NUSSE, R., and H. E. VARMUS. 1982. Many tumors induced by the mouse mammary tumor virus contain a provirus integrated in the same region of the host genome. *Cell* 31:99–109.

RASSOULZADEGAN, M., et al. 1982. The roles of individual polyoma virus early proteins in oncogenic transformation. *Nature* 300:713–718.

RASSOULZADEGAN, M., et al. 1983. Expression of the large T protein of polyoma virus promotes the establishment in culture of "normal" rodent fibroblast cell lines. *Proc. Nat'l. Acad. Sci. USA* 80:4354–4358.

ROUS, P. 1911. A sarcoma of the fowl transmissible by an agent separable from the tumor cells. *J. Exp. Med.* 13:397–411.

STEFFEN, D. 1984. Proviruses are adjacent to c-*myc* in some murine leukemia virus-induced lymphomas. *Proc. Nat'l. Acad. Sci. USA* 81:2097–2101.

TAKEYA, T., and H. HANAFUSA. 1983. Structure and sequence of the cellular gene homologous to the RSV *src* gene and the mechanism for generating the transforming virus. *Cell* 32:881–890.

TEMIN, H. 1964. Nature of the provirus of Rous sarcoma. *Nat'l. Cancer Inst. Monogr.* 17:557–570.

TEMIN, H., and S. MIZUTANI. 1970. RNA-dependent DNA polymerase in virions of Rous sarcoma virus. *Nature* 226:1211–1213.

TIOLLAIS, P., C. POURCEL, and A. DJEAN. 1985. The hepatitis B virus. *Nature* 317:489.

TOOZE, J., ed. 1980. *DNA Tumor Viruses,* 2d ed. Cold Spring Harbor Laboratory.

WHYTE, P., et al. 1988. Association between an oncogene and an antioncogene: the adenovirus E1A proteins bind to the retinoblastoma gene product. *Nature* 334:124–129.

Chemical Carcinogenesis

AMES, B. N. 1979. Identifying environmental chemicals causing mutations and cancer. *Science* 204:587–593.

BROWN, K., et al. 1986. v-*ras* genes from Harvey and Balb murine sarcoma viruses can act as initiators of two-stage mouse skin carcinogenesis. *Cell* 46:447–456.

BROWN, P. C., T. D. TLSTY, and R. T. SCHIMKE. 1983. Enhancement of methotrexate resistance and dihydrofolate reductase gene amplification by treatment of mouse 3T6 cells with hydroxyurea. *Mol. Cell Biol.* 3:1097–1107.

CHEN, T. T., and C. HEIDELBERGER. 1969. Quantitative studies on the malignant transformation of mouse prostate cells by carcinogenic hydrocarbons *in vitro. Int. J. Cancer* 4:166–178.

MCCANN, J., and B. N. AMES. 1977. The Salmonella/microsome mutagenicity test: predictive value for animal carcinogenicity. In H. H. Hiatt, J. D. Watson, and J. A. Winsten, eds., *Origins of Human Cancer.* Cold Spring Harbor Laboratory.

MILLER, J. H. 1982. Carcinogens induce targeted mutations in *Escherichia coli. Cell* 31:5–7.

OLSSON, M., and T. LINDAHL. 1980. Repair of alkylated DNA in *Escherichia coli. J. Biol. Chem.* 255:10569–10571.

POLAND, A., and A. KENDE. 1977. The genetic expression of aryl hydrocarbon hydroxylase activity: evidence for a receptor mutation in nonresponsive mice. In H. H. Hiatt, J. D. Watson, and J. A. Winsten, eds., *Origins of Human Cancer.* Cold Spring Harbor Laboratory.

QUINTANILLA, M., et al. 1986. Carcinogen-specific mutation in mouse skin tumors; *ras* gene involvement in both initiation and progression. *Nature* 322:78–80.

SIMS, P. 1980. The metabolic activation of chemical carcinogens. *Brit. Med. Bull.* 36:11–18.

Radiation and DNA Repair in Carcinogenesis

KENNEDY, A. R., J. CAIRNS, and J. B. LITTLE. 1984. Timing of the steps in transformation of C3H 10T1/2 cells by X-irradiation. *Nature* 307:85–86.

OLSSON, M., and T. LINDAHL. 1980. Repair of alkylated DNA in *Escherichia coli. J. Biol. Chem.* 255:10569–10571.

Promoters of Carcinogenesis

BERENBLUM, I. 1974. *Carcinogenesis as a Biological Problem.* Elsevier/North-Holland.

DAVIS, R. J., and M. P. CZECH. 1985. Tumor-promoting phorbol diesters cause the phosphorylation of epidermal growth factor receptors in normal human fibroblasts at threonine-654. *Proc. Nat'l. Acad. Sci. USA* 82:1974–1978.

DELCLOS, K. B., D. S. NAGLE, and P. M. BLUMBERG. 1980. Specific binding of phorbol ester tumor promoters to mouse skin. *Cell* 19:1025–1032.

EBELING, J. G., et al. 1985. Diacylglycerols mimic phorbol diester induction of leukemic cell differentiation. *Proc. Nat'l. Acad. Sci. USA* 82:815–819.

KIKKAWA, U., et al. 1983. Protein kinase C as a possible receptor protein of tumor-promoting phorbol esters. *J. Biol. Chem.* 258:11442–11445.

WEINSTEIN, I. B. 1983. Protein kinase, phospholipid and control of growth. *Nature* 302:750.

Human Cancer

BAKHSHI, A., et al. 1983. Lymphoid blast crises of chronic myelogenous leukemia represent stages in the development of B-cell precursors. *New England J. Med.* 309:826–831.

BOS, J. L., et al. 1987. Prevalence of *ras* gene mutations in human colorectal cancers. *Nature* 327:293–297.

CAVENEE, W. K., et al. 1983. Expression of recessive alleles by chromosomal mechanisms in retinoblastoma. *Nature* 305:779–784.

COMMITTEE ON DIET, NUTRITION AND CANCER, ASSEMBLY OF LIFE SCIENCES, NATIONAL RESEARCH COUNCIL. 1982. *Diet, Nutrition and Cancer.* National Academy Press.

DOLL, R., and R. PETO, eds. 1981. *The Causes of Cancer.* Oxford University Press.

FRIEND, S. H., et al. 1986. A human DNA segment with properties of the gene that disposes to retinoblastoma and osteosarcoma. *Nature* 323:643.

HANSEN, M. F., and W. K. CAVENEE. 1987. Genetics of cancer predisposition. *Cancer Res.* 47:5518–5527.

KNUDSON, A. G., JR. 1977. Genetic predisposition to cancer. In H. H. Hiatt, J. D. Watson, and J. A. Winsten, eds., *Origins of Human Cancer.* Cold Spring Harbor Laboratory.

KRONTIRIS, T. G., et al. 1985. Unique allelic restriction fragments of the human Ha-*ras* locus in leukocyte and tumor DNAs of cancer patients. *Nature* 313:369–374.

SLAMON, D. J., et al. 1987. Human breast cancer: correlation of relapse and survival with amplification of HER/*neu* oncogene. *Science* 235:177–182.

VOGELSTEIN, B., et al. 1988. Genetic alterations during colorectal-tumor development. *New England J. Med.* 319:525–532.

YOSHIDA, M., I. MIYOSHI, and Y. HINUMA. 1982. Isolation and characterization of retrovirus from cell lines of human adult T-cell leukemia and its implication in the disease. *Proc. Nat'l. Acad. Sci. USA* 79:2031–2035.

Immunity

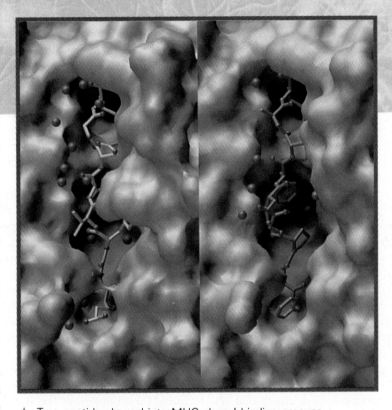

▲ Two peptides bound into MHC class I binding grooves.

The life of every organism is constantly threatened by other organisms—this is the nature of the living world. In response, each species has evolved protective mechanisms, varying from camouflage colors, to poisons, to effective running muscles. From their continual battle with microorganisms, vertebrates have evolved an elaborate set of protective measures called, collectively, the *immune system*. The word "immune" (Latin *immunis,* meaning "exempt") implies freedom from a burden: an animal that is immune to a specific infecting agent will remain free of infection by that agent. The study of the immune system constitutes the discipline of *immunology.*

The immune system works by a learning process. Our first encounter with a bacterial, fungal, protozoan or viral pathogen leads to an infection, often accompanied by disease symptoms. The immune system aids in recovery from the infection, and after recovery, we usually remain free of that disease forever. Our immune system has learned to recognize this specific pathogen as a foreign infecting agent; should it attack us again, it will be rapidly killed.

A key process carried out by the immune system is *recognition.* The system must recognize the presence of an invader. It must also be able to discriminate between foreign invaders and the natural constituents of the body: we call this *discrimination between nonself and self.* The importance of this is underscored by disorders in which such discrimination fails. Such disorders are known as *autoimmune diseases,* and the most severe ones can be fatal. The recognition of a foreign invader is only the first step of the immune system's attack; it must be followed by steps that will kill and eliminate the invader. Thus the immune system carries out two types of activities: *recognition processes* directed against individual, discrete aspects of a target, and *destructive responses* that follow from the recognition and allow the system to mount an attack against the invader. Precise recognition is the function of cells called *lymphocytes* while destruction is carried out both by lymphocytes and by cells called *macrophages* and *neutrophils*.

The immune system also has other abilities besides the recognition and killing of invading pathogens. It can kill cancer cells and, in experimental animals at least, it can protect the body against certain tumors. (How active the immune system is against human tumors is still a debated question.) The immune system also prevents tissue transplantation between individuals. This is because every vertebrate individual has a unique set of molecules on the surface of its cells. Its immune system recognizes those cells as self, but grafted cells from other individuals, even of the same species, are seen as foreign and are killed, or rejected. Inhibition of this *rejection reaction* is necessary whenever a surgeon attempts to graft a patch of skin or transplant a kidney from one person to another. Without special treatment, the only transplants that are accepted without a rejection reaction are those from an identical twin.

If we are to understand the immune system, we must answer some key questions. How does the system recognize the individuality of a specific pathogen? How does it discriminate foreign matter from endogenous material, nonself from self? How does it translate the recognition of a foreign invader into a killing reaction? How does it learn, so that the second attack by an invader is repulsed so much faster than the first? The answers to these questions are not fully known. While we understand, for instance, a great deal about how recognition develops, we know less about how the system avoids reacting with self-components.

This chapter concentrates on aspects of the immune system that are reasonably well known. Especially highlighted are the molecular biological mechanisms that underlie the recognition process because they have a remarkable aspect—the reorganization of DNA as a mechanism in cellular differentiation. We begin with an overview of the system, followed by a discussion of antibodies, the best-understood molecules of the immune system, and how they develop their vast diversity and hence their specificity for foreign molecules. We then turn to the T lymphocytes and describe their particular properties, after which we consider the interactions of lymphocytes in generating an immune response.

► *Overview*

The immune system works in three fundamentally different ways (Figure 27-1) by *humoral immunity*, by *cellular immunity*, and by secretion of stimulatory proteins, called *lymphokines*. The term "humor" refers to a fluid, and *humoral immunity* relies on molecules in solution in the body. These molecules are proteins collectively called *immunoglobulin* (abbreviated *Ig*). They constitute 20 percent of the proteins in the blood. A single immunoglobulin molecule is called an *antibody*, but "antibody" is also used to mean many individually different molecules all directed against the same target molecule. Humoral immunity also involves *complement*, a set of proteins that are activated to kill bacteria both nonspecifically and in conjunction with antibody.

In *cellular immunity*, intact cells are responsible for recognition and elimination reactions. The body's first line of defense is the recognition and killing of microorganisms by *phagocytes*, cells specialized for the ingestion and digestion of unwanted materials. These cells include neutrophils and macrophages. A key role of antibodies is to help phagocytes recognize foreign materials. There is also an important class of cells that carry antibody-like recognition molecules on their surface and can directly kill cells infected by microorganisms. Another class of cells carrying recognition molecules responds by secretion of *lymphokines* that stimulate the activity of other immune system cells and therefore these cells act indirectly in the clearance of an infection.

The injection of almost any foreign macromolecular substance into an animal elicits the formation of both antibodies and immune cells that specifically bind to that substance. The cells that carry specific recognition molecules are called *lymphocytes* (see Figure 27-1), while a substance that provokes antibody or lymphocyte formation, or is recognized by an antibody or lymphocyte, is called an *antigen*. This chapter concentrates on the formation and activities of lymphocytes because lymphocytes display well how complex molecular biological events can generate specialized cellular capabilities.

The invasion of the body by a potential pathogen that has not been previously encountered is potentially devastating for an organism and it cannot wait for an antibody response to develop, which takes days. The body has a number of lines of defense that come into play instantaneously. *Phagocytic cells* are activated by an encounter with a bacterial cell wall; they engulf the bacterium and kill it. *Complement* is also activated by the cell walls of pathogens, and it kills many invading pathogens. Cells that are infected with viruses make *interferons*; these proteins interact with receptors on neighboring cells, triggering a response that makes the cells poor hosts for further infection. These *nonspecific reactions* are critical: a mouse unable to respond to bacterial cell walls is super-sensitive to bacterial infections; a mouse without interferon receptors on its cells is rapidly killed by many viruses.

The nonspecific initial responses to infection are superseded by *specific responses*, mediated by activated lymphocytes (Figure 27-1). Lymphocytes are activated either directly by the surface of a pathogen or indirectly by macrophages that digest a pathogen, carry its peptide fragments to their surface, and then present the peptides to the lymphocytes. Lymphocytes also respond to peptide frag-

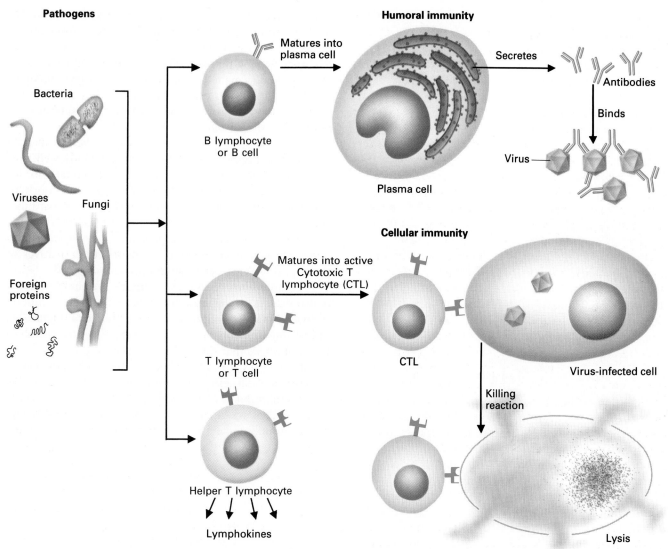

▲ FIGURE 27-1 Humoral immunity and cellular immunity, specified by lymphocytes. When a pathogen invades the body, lymphocytes respond with three types of reaction. The lymphocytes of the humoral system (B cells) secrete antibodies that can bind to the pathogen, signalling its degradation by macrophages and other cells. The lymphocytes of the cellular system (T cells) carry out two major types of functions. Cytotoxic T lymphocytes (CTLs) develop the ability to directly recognize and kill cells infected by the pathogen. Helper T cells (T_H cells) independently recognize the pathogen and secrete protein factors (lymphokines) that stimulate growth and responsiveness of B cells, T cells, and macrophages, thus greatly strengthening the power of the immune response.

ments from viruses when these are presented on the surface of infected cells. The lymphocytes recognize the foreign nature of these materials and initiate antibody formation.

Antibodies Bind to Epitopes and Have Two Functional Domains

The major function performed by an antibody is high-affinity binding to an antigen. Hydrophobic forces, ionic forces, and van der Waals forces bind antibodies and antigens, but covalent bonds are not formed between them.

The site on an antigen at which a given antibody molecule binds is called a determinant or *epitope* (Figure 27-2a). The surface of a protein molecule provides many epitopes to which antibodies might bind. Some antigens, such as virus particles (Figure 27-2b), have a repeating structure, with the same epitope on them many times. Such an antigen is said to be *multivalent*.

New epitopes can be created on a protein by attaching a small molecule onto its surface. If, for instance, bovine serum albumin, a fairly big protein, is coupled to the simple molecule dinitrophenol, and the conjugate is injected

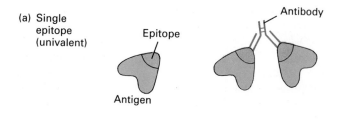

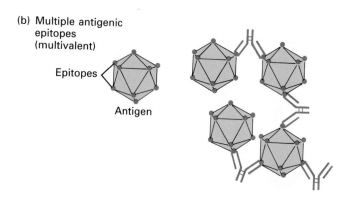

▲ FIGURE 27-2   Antibodies binding to single or multiple epitopes on an antigen. (a) When antibody binds to a protein with a single epitope, the antibody's two sites can both be filled but the complex cannot grow further. (b) Some antigens, notably viruses, have multiple identical epitopes on a single particle. Antibodies can form large aggregates when reacting with such antigens.

into a mouse, the mouse will make antibody molecules that bind to the dinitrophenyl group. This mechanism can be demonstrated by coupling dinitrophenol to another protein and showing that this second protein conjugate reacts with antibodies made to the original dinitrophenol-conjugated protein. In this context, dinitrophenyl, or any small substituent on a protein that can elicit an antibody response, is called a *hapten* (Figure 27-3).

A key property of antibodies, which is best probed by haptens, is *specificity*. An antibody that binds well to a given hapten may bind poorly to a similar hapten. For instance, many anti-dinitrophenyl antibodies bind poorly to trinitrophenyl. Simple changes in short peptides that function as haptens show the specificity of antibodies: a peptide of 7–15 residues, linked to a protein, can elicit antibodies that bind well to the peptide but poorly to derivatives of the peptide having even one amino acid changed.

We noted before that the immune system carries out two processes: recognition and destruction. The antibody is divided into two regions for these functions (Figure 27-4): *binding domains* that interact with the antigen and *effector domains* that signal the initiation of processes (such as phagocytosis) to rid the body of the antigen bound to the antibody. The effector region also provides signals that distribute antibodies to various bodily fluids. Some types of antibodies are directed to secretions, such as saliva, mucus, or milk; others go across the placenta to protect the fetus.

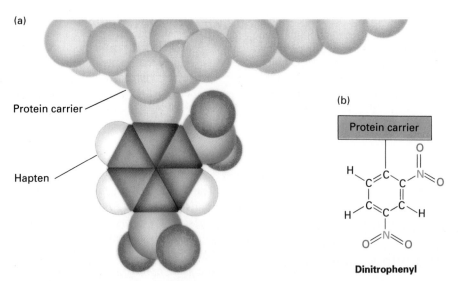

▲ FIGURE 27-3   Hapten linked to a protein. The hapten shown is dinitrophenyl, one commonly used in experimental immunology (see Figure 3-34). (a) The space-filling representation shows that the hapten's varying surface presents many different determinants to which antibodies can bind. (b) Chemical formula of dinitrophenyl. [From G. M. Edelman, 1970, The Structure and Function of Antibodies, *Sci. Am.* **223**(2). Copyright 1970 by Scientific American, Inc.]

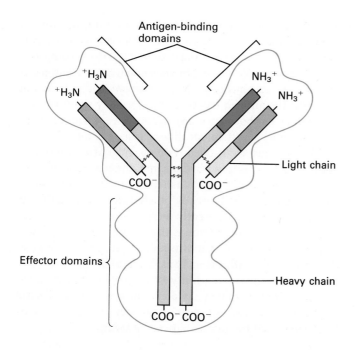

Antigen-binding domains

^+H_3N NH_3^+

^+H_3N NH_3^+

Light chain

COO^- COO^-

Effector domains

Heavy chain

COO^- COO^-

◄ FIGURE 27-4   Schematic representation of an antibody molecule. Antigen binding occurs at the two upper ends of the Y-shaped molecule, where the H and L chains are in intimate contact. The lower portion of the Y contains the effector region that is made only from heavy chains. The N-terminal (NH_3^+) and C-terminal (COO^-) ends of the chains are indicated. Disulfide bridges (S–S) link the chains (see also Figures 3-32, 27-15, and 27-16).

A given binding domain is found on only a very small set of antibodies, whereas the effector domains are common to many antibodies. This dichotomy of a highly specific portion and a common portion in the antibody molecule has its counterpart in the genes that encode antibodies. These genes are bipartite, being constructed by selection of one DNA segment from a locus with a vast array of variable regions and another segment from a very small number of constant regions.

Each antibody molecule consists of two classes of polypeptide chains, *light (L) chains* and *heavy (H) chains*. A single antibody molecule has two identical copies of the L chain and two of the H chain. Therefore the basic antibody structure is a four-chain molecule (Figure 27-4). The N-terminal ends of one H chain and L chain together form one binding domain. The antibody molecule is thus *bivalent*: it has two binding sites for antigen.

Antibody Reaction with Antigen Is Reversible

The binding of an antibody to its antigen is a simple bimolecular, reversible reaction capable of analysis by standard kinetic theory. Looking only at a single binding site we write

$$Ag + Ab \underset{k_2}{\overset{k_1}{\rightleftharpoons}} Ag\text{-}Ab$$

where Ag is the antigen, Ab the uncomplexed antibody, and Ag-Ab the bound complex. The forward and reverse binding reactions are characterized by k_1 and k_2 rate constants, respectively. The *affinity* of the antibody's binding site is measured by the ratio of complexed to free reactants at equilibrium. An *affinity constant, K,* is defined by

$$K = \frac{[Ag\text{-}Ab]}{[Ag][Ab]} = \frac{k_1}{k_2}$$

where the brackets denote concentration. Typical values of K are 10^5–10^{11} liters/mole. K can be determined by measuring the concentration of free antigen needed to fill half the binding sites of an antibody (Figure 27-5). At this point [Ag-Ab] must equal [Ab], and the above formula resolves to

$$K = \frac{1}{[Ag]}$$

Many proteins may have weak affinities for other proteins; consequently we need to define an affinity below which binding is considered to be nonspecific and above which binding is considered to be a specific antibody-antigen reaction. From experience, any K less than 10^4 liters/mole is considered nonspecific.

But antibody-antigen reactions are not necessarily the simple bimolecular events treated in the above equation.

(a)

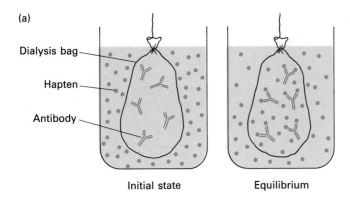

Dialysis bag

Hapten

Antibody

Initial state   Equilibrium

(b)

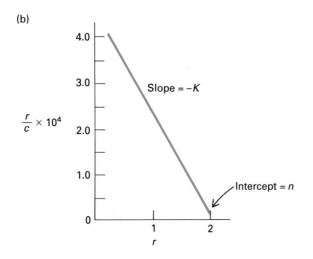

Slope = $-K$

Intercept = n

Scatchard equation $\dfrac{r}{c} = Kn - Kr$

where $r = \dfrac{\text{Moles of hapten bound to antibody}}{\text{Moles of antibody in bag}}$

c = Concentration of hapten in bulk solution

n = Number of binding sites on antibody molecule

K = Equilibrium association constant of antibody

▲ FIGURE 27-5   Determining the affinity of an antibody for a hapten. (a) Equilibrium dialysis. A large volume of a solution of hapten is equilibrated through a semipermeable dialysis membrane with a solution of antibody that can bind to the hapten. (b) Typical data are plotted in a fashion that allows application of the Scatchard binding equation. In the experiment shown, dinitrophenol hapten bound to lysine was used, and a monoclonal antidinitrophenol antibody from a myeloma was placed inside the dialysis bag. The Scatchard equation gives two values, number of binding sites per antibody molecule (n) and the equilibrium binding constant (K). The number of binding sites per molecule is given by the intercept on the x axis. It is 2, as expected, because of the two binding arms on an antibody molecule. [See H. N. Eisen, 1980, *Immunology*, 2d ed., Harper & Row, pp. 299–301.]

Such a treatment is appropriate for an antigen with one potential epitope for reaction and for a homogeneous antibody population, even though there are two binding sites per antibody molecule. If, however, binding of antigen to one site on an antibody blocks reaction with the second site, or if the antigen has two appropriately spaced identical sites so that both antibody sites can react with the same antigen molecule, then ideality is lost and more complicated equations are needed. For instance, if the two antibody sites can react with the same antigen molecule, the second site will react much faster than the first because the antigen molecule will already be very close by and the apparent affinity will be greatly increased, usually by at least 10^4. A further complication is that natural antibody populations can react with a multitude of epitopes on antigens, with a variety of affinities. Thus the term *avidity* is used to describe the apparent affinity of an antibody population reacting with an antigen. Avidity is measured by half-saturation of antibody, but we must recognize that this measure has no simple mathematical meaning.

Antibodies Come in Many Classes

Because antibodies have to be distributed to various parts of the body and serve multiple effector functions, animals make multiple *classes* of antibodies. Each class has a different set of effector domains with its own specific set of properties. There are five major classes and a number of subclasses. The classes are IgM, IgD, IgG, IgE, and IgA (Table 27-1; Figures 27-4 and 27-6). The differences among them derive from the different H chains of which they are constituted.

When an animal begins its response to antigenic challenge (for instance, following inoculation of an antigen), the first antibody class it makes is *IgM*. The IgM molecule (Figure 27-6) is a pentamer of the basic four-chain antibody units; the units are held together by disulfide bonds and by a single copy of a polypeptide known as the *J chain* (because it *joins* the units together). Because of its multiple, closely spaced binding sites, IgM has a very high binding avidity for microorganisms that are covered with identical subunits, such as viruses. One major effector function of IgM is *complement fixation*—that is, activation of the complement system, a group of proteins able to kill cells to which antibody is bound. IgM also activates phagocytic *macrophages*.

The *IgD* molecule, a monomer, is an enigma. Mature B cells have IgD on their surface but very little is found in bodily fluids and no special function is yet known for it.

IgG is the main serum antibody. A four-chain monomer, it is made in copious amounts in response to an antigenic challenge, especially after multiple stimulations with antigen. Like IgM, it stimulates both complement and macrophages. IgG antibodies can pass through the placenta to the fetus. Their effector regions are specialized for binding

TABLE 27-1 Properties of the Five Antibody Classes

Properties	IgM	IgD	IgG	IgE	IgA
Heavy chain	μ	δ	γ	ϵ	α
Mean human adult serum level (mg/ml)	1	0.03	12	0.0003	2
Half-life in serum (days)	5	3	25	2	6
Number of four-chain monomers	5	1	1	1	1, 2, or 3
Special properties	Early appearance; fixes complement; activates macrophages	Found mainly on B cell surfaces; traces in serum	Activates complement; crosses placenta; binds to macrophages and granulocytes	Stimulates mast cells to release histamines; found mainly in tissues	Found mainly in secretions

to a receptor on the maternal side of placental cells; the receptor then carries the IgG through the cell into the fetal circulation.

IgE is one of the most bothersome types of antibody. It is found mainly in tissues, where, in complex with an antigen, it activates the release of histamine from specialized, blood-derived cells called *mast cells*. Histamine release is the cause of such allergic reactions as hives, asthma, and hay fever. The positive role of IgE is as a defense against parasitic infections.

IgA, the fifth major class of antibody, is perhaps the most important. It exists as a monomer and is also polymerized into a J-chain-linked dimer or trimer. Polymerized IgA binds to the *poly-Ig receptor* on the blood-facing sur-

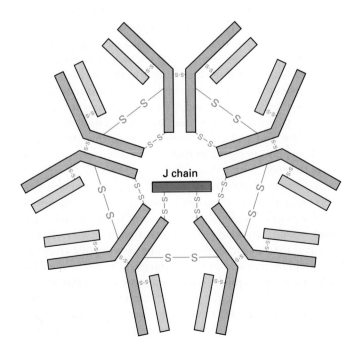

◀ FIGURE 27-6 Schematic representation of a pentameric IgM molecule. The antibodies of most classes are monomeric, formed of only four chains as depicted in Figure 27-4. Circulating IgM is a pentamer whose polymerization is initiated by an ancillary protein called the J chain. Holding the various polypeptides together are disulfide (S–S) bonds. The IgM found on the surface of B cells is a monomer, as in Figure 27-4.

J chain

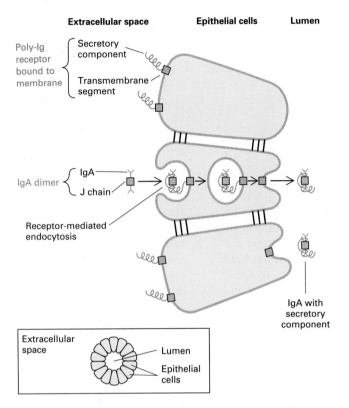

Extracellular space   Epithelial cells   Lumen

Poly-Ig receptor bound to membrane

Secretory component

Transmembrane segment

IgA dimer

IgA

J chain

Receptor-mediated endocytosis

IgA with secretory component

Extracellular space

Lumen

Epithelial cells

▲ FIGURE 27-7   Transfer of IgA across epithelial cells by the poly-Ig receptor. IgA molecules are shown being transported across epithelial cells from the blood-facing surface into a lumen (for instance, from the intestinal lining into the intestinal space). The other class of polymerized immunoglobulin, IgM, can be transported in a similar way. The receptor that carries out the transport is found on the blood-facing surface of the intestinal cell. There it binds to an IgA dimer and is carried into the cell by endocytosis. Rather than staying in the cell, the fate of most transported molecules, the poly-Ig receptor complex with IgA is carried through the cell to the luminal face of the cell, where the IgA separates from the cell. This separation is achieved by a cleavage of the poly-Ig receptor, leaving the "secretory component" bound to the IgA that is released into the lumen. [See K. E. Mostov, M. Friedlander, and G. Blodel, 1984, *Nature* **308**:37.]

face of many epithelia and is transported through the epithelial cells into whatever compartment the epithelial layer surrounds, for instance, into the intestinal lumen (Figure 27-7). During the passage, most of the receptor is cut away from its membrane-binding region and remains bound to the antibody. When IgA emerges from the epithelial cells, a piece of the receptor, called the *secretory component*, is still bound to it. The poly-Ig receptor will also bind to polymerized IgM; persons unable to make IgA will secrete IgM, which also protects them. Once secreted IgA (or IgM) is present in saliva, tears, the lungs, and the intestines it becomes the body's first line of defense against infection by microorganisms.

Antibodies Are Made by B Lymphocytes

Antibodies are made by several cell types, but these are all of a single cell lineage: the B lymphocytes, or B cells. (They are designated "B" because in birds, where they were first studied, these cells mature in an outpocketing at the end of the cloaca called the *bursa of Fabricius*. This organ has no counterpart in mammals, but most mammalian B lymphocytes originate from the bone marrow so the "B" remains appropriate.) B lymphocytes arise as cells having IgM or IgM plus IgD on their surface. When the surface-bound antibody reacts with an antigen, this activates the B lymphocyte to mature into a cell type, known as a *plasma cell*, that is highly specialized for antibody production and secretion (see Figure 27-1).

The Immune System Has Extraordinary Versatility

Many haptens, such as the dinitrophenyl group described earlier (see Figure 27-3), are the products of synthetic chemists. During the long evolution of vertebrates, no animal outside a laboratory ever encountered a dinitrophenylated protein. Yet at the first encounter with such a protein, most vertebrates produce dinitrophenol-specific antibodies. This response is the most remarkable aspect of immune function—the ability to recognize the whole universe of potential determinants whether the species has encountered them at any time in evolution or not. The plasticity of the immune system—its ability to extend itself to meet unprecedented challenges—makes the system's molecular events especially intriguing. All the other body systems, except the nervous system, have evolved their properties in response to specific environmental pressures encountered by their ancestors; that is, by natural selection. Although the immune system itself is certainly a product of evolution, not all of the genes encoding individual antibody proteins could have been selected during evolution because not all of the determinants recognized by antibodies could have been previously encountered. The genetic basis for this extraordinary plasticity of protein structure is now fairly well understood and will be described later in this chapter. At this point, however, we might ask: by what cellular strategy is this plasticity achieved?

When the plasticity of the immune system was first appreciated, many suggestions were proposed concerning how the system might operate. Most of the proposals were based on an approach called *instructive theory*: it was imagined that all antibodies had one polypeptide structure that could be induced to fold in different ways by combining with antigen, which served as a structural template; the binding specificity of the antibody would then arise from this previous encounter with the antigen (Figure 27-8). This notion has now been disproved by the observation

Instructive theory

Virgin antibody; no specificity   +   Antigen   Antigen-antibody complex   →   Antigen-specific antibodies now created

Selective theory

Specific preexisting antibodies   +   Antigen   →   Antigen-antibody complex

◄ FIGURE 27-8 Two theories of the origin of antibody specificity. The instructive theory, now discarded, suggested that the antigen is a template for determining the structure of the antibody. With today's knowledge of genetic encoding of protein structure, such a notion is hard to imagine, but before the 1970s so much less was known about the determination of protein structure that template formation of antibodies was considered possible. Selective theory, now considered the only likely approach, suggests that the immune system generates a very large number of specific molecules and that antigens are bound to the ones that fit.

that antibodies can be denatured and renatured in the absence of antigen and still retain their specificity. Thus the one-dimensional sequence of the polypeptides must already contain the information for the structure of the binding site. Furthermore, antigen-specific antibodies can be detected on lymphocyte surfaces before the exposure of an animal to the antigen.

Even before instructive-theory concepts had been disproved, a competing concept based on preexisting antigen-specific antibodies was developed. It provided a consistent explanation for the cellular events that produce specificity and plasticity. This latter theory is now the basis of all thinking in immunology.

Clonal Selection Theory Underlies All Modern Immunology

The class of theories that replaced the instructive models is called *selective theory* because an antigen is thought to select the appropriate antibody from a preexisting pool of antibody molecules (Figure 27-8). The notion of selection was brought into the field by Niels Jerne, a brilliant theorist and experimentalist of immunology, and was then incorporated by Sir MacFarlane Burnet into one of the great intellectual constructs of modern biology, the *clonal selection theory* of antibody production (Figure 27-9). This the-

ory supposes that (1) the body is continually elaborating B lymphocytes that have immunoglobulin molecules on their surface, (2) all of the surface immunoglobulin molecules on any one B cell have the same binding specificity, and (3) for any given antigenic determinant, only a very small subset of the entire pool of B cells will have surface antibody with which it can bind.

When a vertebrate encounters a foreign antigen in its circulation, this antigen combines with preexisting immunoglobulins only on those B cells that have molecules with the appropriate specificity. The interaction of antigen and antibody on the B-cell surface then triggers that B cell to multiply, producing daughter cells that synthesize and secrete its specific antibody (Figure 27-10). The antigen has *selected* from a preexisting pool those cells with appropriate specificity and has caused their proliferation into a large population of cells all making the same antibody. Such a population of identical cells, all coming from the same ancestral cell, is called a *clone;* thus the term *clonal selection.* Clonal selection theory is, like natural selection, a Darwinian theory because the antigen selects the cells that will multiply from a set of variants that have arisen independently of the selective force. Clonal selection has received such complete verification that it no longer is considered a theory but rather is accorded the status of an accepted explanation of the functioning of the immune system.

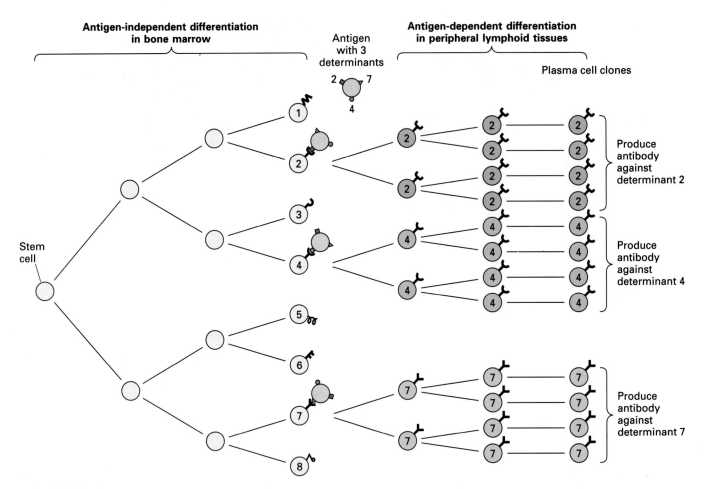

▲ FIGURE 27-9   The clonal selection process by which the immune system develops specificity. From a large number of cells having different antibodies, cells carrying appropriate antibodies bind to an antigen's determinants. Then clones of these cells are produced, with all cells of each clone carrying the same surface antibody.

Clonal selection proceeds through two distinct phases in the life of the B cell (see Figure 27-9). In the *antigen-independent phase,* before any encounter with an antigen, the bone marrow produces a large number of B cells, each with a different antigen-binding specificity. Although the bone marrow elaborates B cells continually throughout the life of the animal, each newly arising cell has a lifetime of only a few days as a circulating cell of the blood or lymph. During this time, if a B cell encounters an antigen to which its surface antibody can bind, then the cell enters an *antigen-dependent phase* in which it is activated to grow, divide, and secrete antibody. Also, long-lived progeny cells called *memory B cells* result from this activation so that what was a transient cell becomes permanent (see Figure 27-10).

If clonal selection is to operate, there must be a mechanism for the *generation of antibody diversity:* an enormous number of structurally different variable regions in immunoglobulin molecules must be generated in the absence of any selective force. As we shall see later, a variety of mechanisms creates this antibody diversity. In part, it arises from multiple alternative possibilities encoded in germ-line DNA—that is, from choices among multiple germ-line gene segments. In part, it results from events that take place in somatic tissue, including *somatic mutation.*

The Immune System Has a Memory

Among the remarkable capabilities of the immune system is its ability to learn. A first encounter of a B cell with an antigen leads to a slowly rising synthesis of antibody, chiefly IgM (this is known as the *primary immune response*). A second encounter with the same antigen leads to a more rapid and greater response, chiefly IgG production (the *secondary immune response*). Only a previously encountered antigen provokes the secondary response: the system has learned to recognize the antigen to which it was previously exposed (Figure 27-11). The basis of learning in the immune system is the formation of long-lived *memory cells* during the primary response (see Figure 27-10). The memory cells persist in the circulation for the life of the organism, even without further antigenic stimulation.

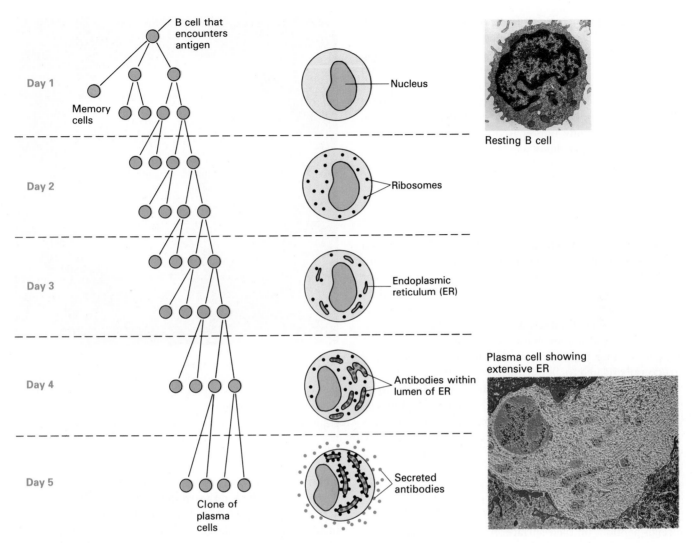

▲ FIGURE 27-10   The activation of B lymphocytes, leading to both growth and secretion. After an encounter with antigen and stimulation by factors produced by helper T cells, the B cell initiates a series of divisions, during which it develops the apparatus for secreting copious amounts of antibody. The end result is a large clone of antibody-secreting plasma cells with extensive endoplasmic reticulum. [After L. Hood et al., 1984, *Immunology*, 2d ed., Benjamin, p. 11. Upper photograph from *Atlas of Blood Cells: Function and Pathology*, 1981, Lea & Febiger; courtesy of D. Zucker-Franklin. Lower photograph from Science Photo Library/Photo Researchers, Inc.]

► FIGURE 27-11   Responses to initial and later injections of an antigen. At day 0, antigen was injected into a mouse and the response was followed by measuring levels of serum antibody to the antigen. At day 28, when the initial response had subsided, a second immunization was done with both antigen A and a new antigen B. The secondary response to antigen A was faster and greater than the initial response to antigen A. The response to antigen B followed the course of an initial encounter with antigen. Thus the secondary response to antigen A is a specific, not a general, response of the system.

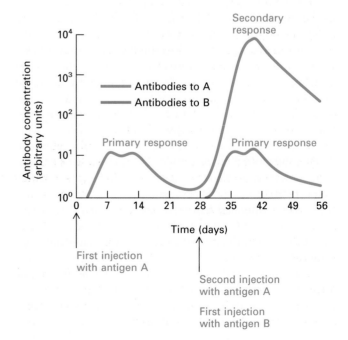

These circulating memory cells each carry on their surface the particular immunoglobulin that will avidly bind to a particular reinvading antigen. The memory cells mainly make IgG or IgA and respond so rapidly that a second encounter with an antigen leads to a much faster and more effective response than did the first encounter. Hence once a person has been infected by a given virus, that virus can never catch the body unprepared again. The presence of memory cells also explains why adults are less susceptible to infectious diseases than are children.

Other Parts of the Immune Response Are Carried Out by T Lymphocytes

While the B lymphocyte carries surface immunoglobulin and can be activated to secrete immunoglobulin by an encounter with antigen, the immune system contains another cell population that also carries surface antigen-binding molecules, but these cells never secrete antibody molecules. They are known collectively as the *T lymphocytes*, or simply *T cells*, because they mature in the thymus gland. The antigen-binding molecules on the surface of T cells are structurally related to, but distinct from, antibody; they are called *T-cell receptors*. These receptors do not react with intact or soluble antigens but rather with peptide antigens bound to the surfaces of macrophages and other cells. This complex reaction of surface molecules on one cell with surface molecules on another cell is at the heart of T-cell biology.

Two fundamentally different types of T lymphocytes exist: cytotoxic and helper. The *cytotoxic T lymphocyte* (*CTL*) or *killer T cell* recognizes and kills cells that display foreign peptides on their surface. A classic CTL target is a virus-infected cell that displays fragments of viral proteins on its surface.

Precursors to CTLs display antibody-like surface receptors that recognize the foreign peptides on cells. The recognition process apparently activates the growth of the CTL precursors, providing the animal with a clone of killer T cells able to destroy target cells displaying the foreign peptide. The mature CTLs do their killing by secreting toxic molecules that induce the target cells to kill themselves by *apoptosis* or *programmed cell death* (see page 1267). CTLs rid the body of infected cells by activating this latent suicide mechanism. The killing process can be demonstrated by incubating CTLs with target cells that have previously taken up the radioactive isotope chromium-51 (^{51}Cr). If the CTLs come from an animal that has previously responded to the target cells, those CTLs will kill the target cells. The ^{51}Cr leaks out of the dead cells, so that cell destruction by CTLs can be measured by ^{51}Cr release.

The second type of T cell is a *helper T lymphocyte*, or T_H *cell*. The function of the T_H cell is to recognize foreign peptides on cell surfaces and to secrete protein factors that

▲ FIGURE 27-12   False-color scanning electron micrograph of a T_H lymphocyte infected with human immunodeficiency virus (HIV), the causative agent of AIDS. The small spherical virus particles (green) on the surface of the T_H cell are in the process of budding away from the cell membrane. From NIBSC/Science Photo Library/Photo Researchers, Inc.

stimulate the various other cells involved in the immune response. B cells need this stimulation: without it, the interaction of antigen with antibodies on the surface of B cells is insufficient to stimulate B-cell growth and secretion of soluble antibody. In fact, without T-cell help, a B-cell interaction with antigen may lead to paralysis of the B cell. T_H cells also secrete factors to stimulate other cells that participate in immune reactions, such as CTLs and macrophages. The T_H cells, like CTLs, cannot recognize free antigen; they recognize only cell-bound peptides derived from antigens that macrophages and B cells have taken up from the fluids that surround them. T_H cells are central players in most immune reactions, as we shall learn later. They are also the targets of human immunodeficiency virus (HIV), the retrovirus that causes AIDS (Figure 27-12). Loss of T_H cells because of HIV infection causes failure of the immune system, leading to the symptoms of AIDS.

B Cells and T Cells Have Identifying Surface Markers

The B and T lymphocytes are indistinguishable in size and general morphology. To distinguish among them, immunologists use differences in their surface proteins (Figure 27-13a) which have therefore come to be called *surface markers*. B cells not only have surface antibody but

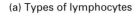

(a) Types of lymphocytes

T cells

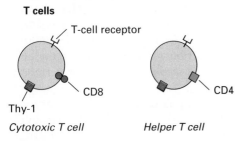

Cytotoxic T cell   *Helper T cell*

B cells

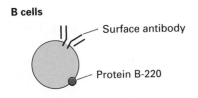

(b) Panning for T-cells

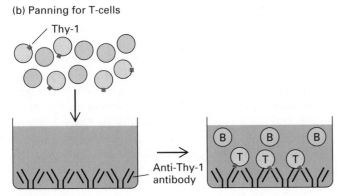

▲ FIGURE 27-13 Surface markers on lymphocytes. Each class of lymphocyte has a distinct set of surface proteins. (a) Some of these are indicated for various lymphocyte subsets. The T cells fall into two categories. All have surface protein Thy-1 and the T-cell receptor, but CTLs have CD8 protein while T_H cells have CD4. B cells have none of these T-cell surface proteins but have two distinctive surface markers: surface antibody and protein B-220. All of the cell-surface proteins can be specifically recognized with the use of antibodies that bind selectively to them. (b) An anti-Thy-1 antibody is used to select T cells from a mixture of T and B cells in the process called panning. The anti-Thy-1 antibody is bound to the surface of a dish. When the mixture of T and B cells is poured over the bound antibody, the T cells are bound to the antibody while the B cells remain free. The B cells are poured off, leaving a pure population of T cells on the dish. [See L. J. Wysocki and V. L. Sato, 1978, *Proc. Nat'l. Acad. Sci. USA* **75**:2844.]

also other molecules, such as B-220, that are not found on T cells. Similarly, T cells carry the antibody-like T-cell receptors and a protein called Thy-1. T cells also have subtype-specific surface markers: CD4 on helper cells and CD8 on CTLs (therefore, T_H cells are referred to as $CD4^+$, and CTLs as $CD8^+$).

Operationally, immunologists distinguish the cell types by using antibodies that recognize one or another of the characteristic surface proteins. Reasonably pure populations of cells can be made by using these antibodies in one of two ways. In the first method, a form of negative selection, cells with a given surface antigen are killed with specific antibody and the remaining cells are retained. A convenient way to make pure T cells or pure B cells is to use anti-immunoglobulin or anti-Thy-1, respectively. In the second positive-selection method, a dish is coated with the antibody and cells that bind to the dish are retained—the procedure is called *panning* (Figure 27-13b). A more elegant positive-selection method is to use an antibody tagged with a fluorescent compound and a fluorescence-activated cell sorter (FACS) to select the cells to which the antibody has bound (see Figure 5-26).

Macrophages Play a Central Role in Stimulating Immune Responses

A cell type quite distinct from the lymphocyte, the *macrophage,* is very important to the immune response. These cells are found in all tissues and also circulate in the blood, where they are called *monocytes.* They are generally the first cells to encounter a foreign substance in the body. They nonspecifically engulf such materials, as well as scavenge normal cellular debris, and degrade them, using powerful hydrolytic enzymes and oxidative attack. Peptides from the degraded proteins are then carried to the macrophage cell surface where they can be recognized by T lymphocytes. Peptides from self-proteins are ignored but foreign peptides activate precursors of helper T cells to mature into fully active, lymphokine-secreting T_H cells.

Macrophages, B cells, and T_H cells collaborate in two separate processes by which antigen generates antibody. The antigen activates B cells by direct interaction with antibody on the B-cell surface, and it activates T_H cells through peptides displayed on the macrophage cell surface that interact with receptors on the T_H cell. Although this understanding emerged only in the late 1980s, it had earlier been recognized that antigen has two independent roles, a notion enshrined in the concept of an antigen as made of a hapten for B-cell response and a "carrier" for T-cell response. In other words, it was formerly evident that B cells and T_H cells respond to different determinants; it is now clear that some parts of a protein are most easily recognized by B cells while other parts, in the form of degraded but stable peptides bound to a supporting protein, are recognized by T cells.

Cells Responsible for the Immune Response Circulate throughout the Body

We have briefly described some of the cells that together provide an immune response. How are these cells organized and coordinated in the body? How does antigen reach these cells? These questions pertain to the organization of vertebrate lymphoid tissues.

The immune system has two compartments (Figure 27-14). B lymphocytes arise in the bone marrow, and T lym-

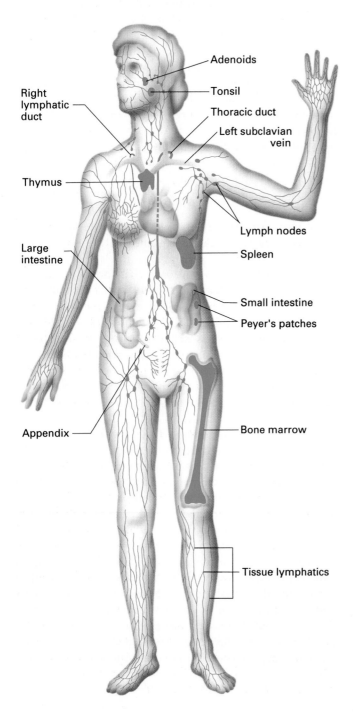

phocytes in the thymus; hence these structures are known as the *primary organs* of the immune system. Here the lymphocytes go through their antigen-independent phase of establishing surface receptors. The lymphocytes then leave these organs and are distributed to many sites in the body that together make up the *secondary organs*, or *peripheral lymphoid tissue*.

The lymphoid system guards all portals of entry into the body. The tonsils in the throat and the adenoids in the nose are lymph nodes. There are lymph nodes in the armpit that filter fluid from the arms, and lymph nodes in the groin that filter fluid from the legs. A special set of lymphoid organs, the *Peyer's patches*, are found in the intestinal wall where they filter out antigens that enter in food or come from the bacteria growing in the intestines. The lymphocytes in Peyer's patches can be activated by an antigen after which they migrate out of the node into the blood. They are finally captured back in the tissue spaces just inside the intestinal lining where they secrete IgA. As already described, the IgA is then rapidly transported across the epithelial cells into the gut lumen.

The peripheral lymphoid tissue is organized around the two fluid systems of the body, the blood and the lymph. Lymph consists of fluid transported from the blood to the spaces within and around tissues. From these extracellular spaces, lymph flows into thin-walled *lymphatic vessels*, where it is slowly moved to larger central collecting vessels. Ultimately the lymph is returned to veins, where it reenters the blood.

The lymphocytes in human blood constitute 20–30 percent of the nucleated blood cells; in lymph they constitute 99 percent. Until they become activated by antigen, lymphocytes continually cycle from blood to lymph to blood, making a complete round once or twice a day. A lymphocyte recognizes a particular site through *homing receptors* that bind to specific *addresins* in individual tissues and lymph nodes. The lymphocytes enter and leave the blood circulation by squeezing between the endothelial cells that line the blood vessels. The lymph circulation is filtered through the lymph nodes, in which all types of lymphocytes take up temporary residence. Antigens that enter the body find their way into the blood or lymph and are filtered out by lymph nodes or by the spleen, which is the blood's filter. In the lymph nodes, macrophages that have engulfed and degraded an antigen display its peptides on their surface. The rare lymphocyte in the lymph node or

◀ FIGURE 27-14   The human lymphoid system. The primary organs (thymus and bone marrow) are colored blue; secondary organs and lymph vessels are shown in red. The whole system consists of the circulating lymphocytes, the vessels that carry the cells (both blood vessels and lymphatic vessels), and the lymphoid organs. Only one bone is shown, but all major bones contain marrow.

spleen that carries an antibody or T-cell receptor with the appropriate specificity will bind either to unprocessed antigen or to a macrophage-bound peptide; it will then become activated to multiply and differentiate into either secreting B lymphocytes or mature T lymphocytes. The binding of the receptor on a previously unactivated lymphocyte is the moment of clonal selection. In the lymph nodes or spleen, helper T lymphocytes and B lymphocytes collaborate; here the concentration of cells facilitates the interaction and magnifies the clonal selection. The antibody made by the secreting B lymphocytes and their mature progeny, the plasma cells, leaves the node and is transported by the blood.

Tolerance Is a Central Concept of Immunology

A key attribute of the immune system is its ability to discriminate between self and nonself. Surface recognition molecules on lymphocytes are sufficiently diverse for any antigen to find some cells with a surface antibody or T-cell receptor whose binding constant for the antigen is high enough to lead to activation. But what prevents self-antigens from activating any of the body's own cells that may carry antibody able to react with them?

Immunologists call the inability to react with oneself *tolerance,* or *self-tolerance,* and they say that the immune system is tolerant to self-antigens. The mechanisms that make for tolerance either eliminate or render unreactive the clones of both B and T cells that would otherwise carry out antiself reactions.

Most relevant evidence comes from the experimental production of a tolerant state in mice. The immune system in mice becomes responsive to foreign antigens within days of birth. All antigens present in the body at the time the system matures are considered by the immune system to be self. Thus if foreign antigens are incorporated into a newborn mouse, it will not be able to mount an immune response when later presented with those antigens. Tolerance to a specific antigen can also be induced in mature animals. Inoculation with large amounts of a protein, for instance, can make a mouse nonresponsive to a later challenge with a dose and form of the protein that would otherwise stimulate a high production of antibody.

The consequences of a failure of tolerance can be severe. Especially in their older years, people may begin to lose self-tolerance and make antibodies to their own proteins. This can lead to many kinds of disorders, including autoimmune hemolysis (an attack on red blood cells), rheumatoid arthritis, and systemic lupus erythematosus. When a body reacts to a major self-protein of the serum, high levels of antigen-antibody complexes appear in the blood. The deposition of these autoimmune complexes in the kidney results in glomerulonephritis, an inflammation of the filtering apparatus of the kidney. The deposits can be severe enough to cause kidney failure.

Immunopathology Is Disease Caused by the Immune System

Autoimmunity is only one type of disease caused by a faulty immune system. So many immune-system problems exist that the term *immunopathology* has been coined. Some disorders are due to inherited deficiencies of important components of the immune system. A person may even lack all B cells or T cells, a condition known as severe combined immunodeficiency disease (SCID). One form of this disease is caused by inactivation of the gene that encodes the enzyme adenosine deaminase. In the absence of this enzyme, excess adenosine, which is toxic to lymphocytes, accumulates. Patients with SCID lack lymphocytes and die at a young age from viral or bacterial infection. They previously could be kept alive only in a sterile environment, but now can be cured by bone marrow transplantation. Patients lacking a suitable marrow donor are being treated experimentally by gene therapy.

Many less severe immunodeficiencies exist. Some persons lack components of the complement system, some lack any antibodies, some lack T cells. In certain cases, medical therapies for diseases not related to the immune system can affect immune function; this is especially true of treatment with corticosteroids, which depresses immune function.

Several types of cancer result from a loss of growth control in lymphocytes and macrophages. Hodgkin's disease is one example; others are acute lymphocytic leukemia, chronic lymphocytic leukemia, and many mature T-cell leukemias. Human T-lymphotropic virus (HTLV) is a retrovirus able to transform helper T cells into tumor cells, causing adult T-cell leukemia. As mentioned earlier, another human retrovirus, HIV, kills helper cells and is responsible for the disease AIDS (acquired immune deficiency syndrome).

Along with the deficiency states and malignancies of the immune system there are hyperreactivity problems. Autoimmunity is one such problem, but more common are the allergies, which represent a hyperreactivity to foreign antigens. Allergies result from the hyperproduction of IgE in reaction to environmental substances such as pollen and dust. Other types of IgE-induced hyperreactivity include asthma, food sensitivities, and reactions to toxins injected by stinging insects. In all cases, it is the local release of mast-cell products, induced by the binding of IgE-antigen complexes, that causes the often violent symptoms. These released products include heparin as well as substances that cause constriction of smooth muscles, such as histamine (see Figure 3-26). Such reactions are treated with antihistamines and with epinephrine (adrenalin), which acts by raising the cyclic AMP levels of the mast cell and preventing the release of its components.

▶ *Antibodies, B Cells, and the Generation of Diversity*

The previous section introduced antibodies as protein molecules with an antigen-binding region and an effector region. It also described the different classes of antibody: IgM, IgG, IgD, IgE, and IgA (see Table 27-1). We will now examine the structure of antibodies in much greater detail, and will consider how a population of molecules with so many common properties can also have such extensive variability. The answers to this question lie in understanding the mechanisms behind antibody diversity.

Because a population of antibodies comprises molecules with various structures, it is difficult to chemically characterize antibody that comes from immunized animals. Tumors, however, initially arise as single transformed cells; therefore the antibody made by such B-cell tumors is homogeneous. The *plasmacytoma*, or *myeloma*, is especially useful because such tumors are analogues of plasma cells, the cells that secrete large amounts of antibody. Up to 10 percent of the protein made by a myeloma cell population can be a single homogeneous type of antibody with one light chain and one heavy chain. Myelomas arise spontaneously in humans and can be induced at will in mice of certain strains by injecting mineral oil into their peritoneum. Most early analysis of protein sequences of specific antibody chains came from the study of myeloma proteins.

Heavy-Chain Structure Differentiates the Classes of Antibodies

Examination of the antibodies made by myelomas, and also the study of normal immunoglobulins, has shown that all five classes of antibody have the same general organization. Each antibody molecule has two kinds of chains, *light (L) chains* (MW ≈23,000) and *heavy (H) chains* (MW 53,000 or more). There are two types of L chains, called kappa (κ) and lambda (λ) chains; each antibody molecule can have either κ chains or λ chains, but not both types.

The differences between the five classes arise from the H chains. Each antibody class has a different type of H chain, the name of which is the Greek equivalent of the class name. Thus IgM molecules have μ chains, IgGs have γ chains, and the others have δ, ϵ, or α chains (see Table 27-1). There are actually four subclasses of IgG proteins, each with somewhat different properties, and each with a different H chain—making a total of eight types of immunoglobulin. Attached to all H chains are asparagine-linked carbohydrate chains.

Antibodies Have a Domain Structure

Every individual antibody molecule has two identical L chains and two identical H chains. The four chains are held

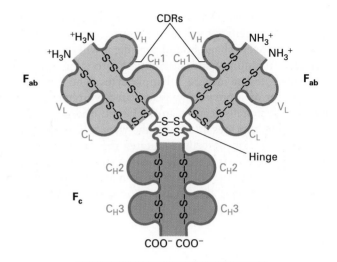

▲ FIGURE 27-15   Domain structure and the complementarity-determining regions (CDRs) of antibody. The molecule is organized in disulfide-bonded 110-amino-acid domains, four in the H chain and two in the L chain. The domain closest to the N-terminal of each chain is variable in sequence (V_H and V_L regions): the other domains are constant in sequence (regions C_H1, C_H2, and C_H3 in the H chain and C_L in the L chain). When the molecule is protease-digested, the enzyme cuts at the hinge region connecting C_H1 and C_H2, the most sensitive spot on the molecule. This splits the molecule into two parts, F_{ab} (the antigen-binding domain) (yellow) and F_c (the effector region) (green). Within the V regions are segments of highly variable sequence constituting the CDRs (red), the amino acids that actually contact the antigen.

together by disulfide bonds to form the basic structure of the molecule (see Figure 27-4). Within each chain, units made of about 110 amino acids fold up to form compact *domains* (Figure 27-15). Each domain is held together by a single internal disulfide bond. An L chain has two domains; H chains have four or five domains. The two L-chain domains interact with the first two N-terminal domains of the H chains, producing a compact unit that acts as the antigen-binding region of the antibody. In most H chains a *hinge* region, consisting of a small number of amino acids, is found after the first two domains. The hinge is flexible and allows the binding regions to move freely relative to the rest of the molecule. Located at the hinge region are cysteine residues whose sulfhydryl groups are linked to form disulfide bridges between the two identical halves of the antibody's basic structure. The hinge region is the place on the molecule most susceptible to the

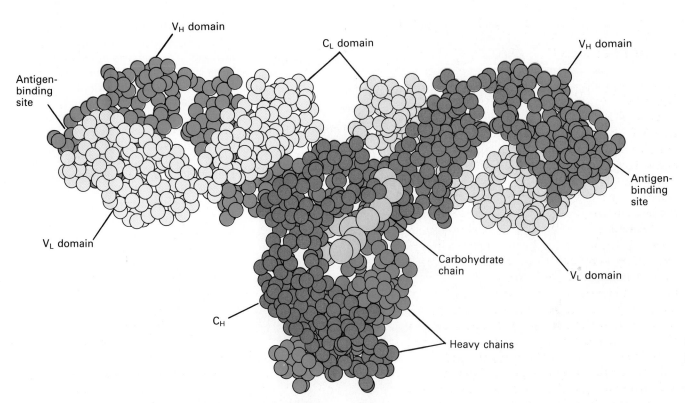

▲ FIGURE 27-16   Model of antibody molecule derived from x-ray crystallographic analysis. The model shows all of the atoms as solid balls and thus describes the outer contours of the molecule. Its Y shape is evident. The individual chains are color-coded so that they can be distinguished: the two L chains are light red and light blue; one H chain is dark red and the other is dark blue. [After E. W. Silverton, M. A. Navia, and D. R. Davies, 1977, *Proc. Nat'l. Acad. Sci. USA* **74**:5140.]

action of protease, which can split an antibody into two pieces, called F_{ab} and F_c fragments. The F_{ab} portion has the antigen-binding site, the F_c portion has the effector regions.

The N-Terminal Domains of H and L Chains Have Highly Variable Structures That Constitute the Antigen-Binding Site

Amino acid sequence comparisons of H and L chains of various immunoglobulins have shown that the N-terminal domains have variable structures while the C-terminal domains are quite constant in structure. The N-terminal domain is called the *variable region* and the C-terminal domain is called the *constant region* (Figure 27-15). The variable domains of L and H chains interact closely to form a single compact unit (Figure 27-16). An antigen that contains a reactive chemical group will form a covalent bond to the variable domain of the H or L chain, showing that the variable domains form the antigen-binding segment of the molecule.

Antibodies of the same specificity are likely to have similar variable-region sequences. Conversely, antibodies that have different specificities have different sequences. This observation illustrates a central principle of immunology, that binding specificity is determined by the amino acid sequence of the variable regions. The rule can be shown quite directly, using a myeloma protein that binds the dinitrophenyl group. If the L and H chains of this anti-dinitrophenyl antibody are separated, denatured, and then renatured together, they regain their specificity for binding dinitrophenyl but not other haptens. The experiment is then varied by substituting either the L or H chains of this antidinitrophenyl antibody with other L or H chains. Such an experiment shows that both the correct L chain and the correct H chain are needed to get dinitrophenyl-binding specificity. Thus it is the combined structure formed by two variable regions that produces a binding site, and not either variable region alone. For antidinitrophenyl we do not have an x-ray crystallographic structure to demonstrate this, but for antiphosphorylcholine, which happens to use mainly H-chain residues for its binding, the binding region is known with great precision from x-ray analysis (Figure 27-17).

X-ray crystallographers have determined the high-resolution structure of a few antibodies. They see a vari-

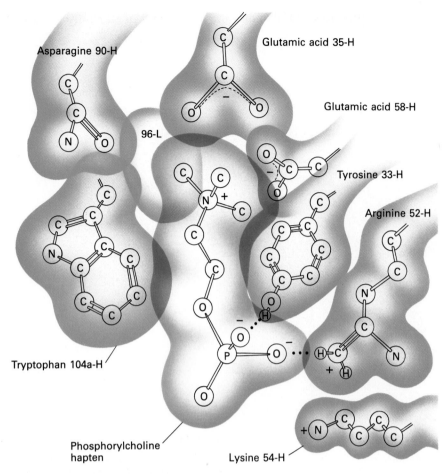

Asparagine 90-H

96-L

Glutamic acid 35-H

Glutamic acid 58-H

Tyrosine 33-H

Arginine 52-H

Tryptophan 104a-H

Phosphorylcholine hapten

Lysine 54-H

◄ FIGURE 27-17 Amino acids in the CDR regions forming the binding site for the hapten phosphorylcholine. Each of these amino acids is part of one of the three CDRs of either the L or H chain (here, mostly H-chain amino acids are involved). The numbers of the amino acids indicate their positions in the chains (see Figure 27-20). The CDR amino acids form a cleft or pocket into which the phosphorylcholine hapten fits. A combination of electrostatic, hydrogen-bonding, and van der Waals forces hold the hapten in the cleft. [From J. D. Capra and A. B. Edmundson, 1977, The Antibody Combining Site, *Sci. Am.* **236**(1). Copyright 1977 by Scientific American, Inc.]

(a)

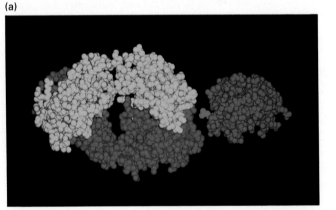

(b)

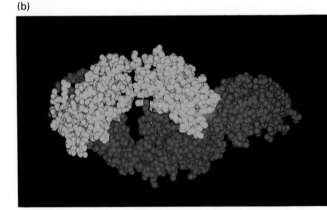

(c)

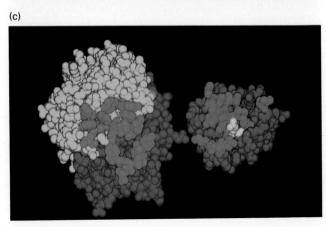

▲ FIGURE 27-18 Binding of a protein antigen to an antibody. The antigen (green) is lysozyme; the antibody (yellow for L chain, blue for H chain) is a monoclonal antilysozyme antibody. The amino acid glutamine on lysozyme (red) fits into a pocket on the antibody. Otherwise the two proteins contact each other through surface interactions. (a) The antigen and antibody separated. (b) The antigen and antibody joined into a complex. (c) The separated proteins rotated about 90 degrees around the vertical axis from part (b) to show the interacting surfaces. The amino acid side chains that interact are shown in red in (c), with the glutamine now shown in pink. [From A. G. Amit et al., 1986, *Science* **233**:747; courtesy of R. J. Poljak.]

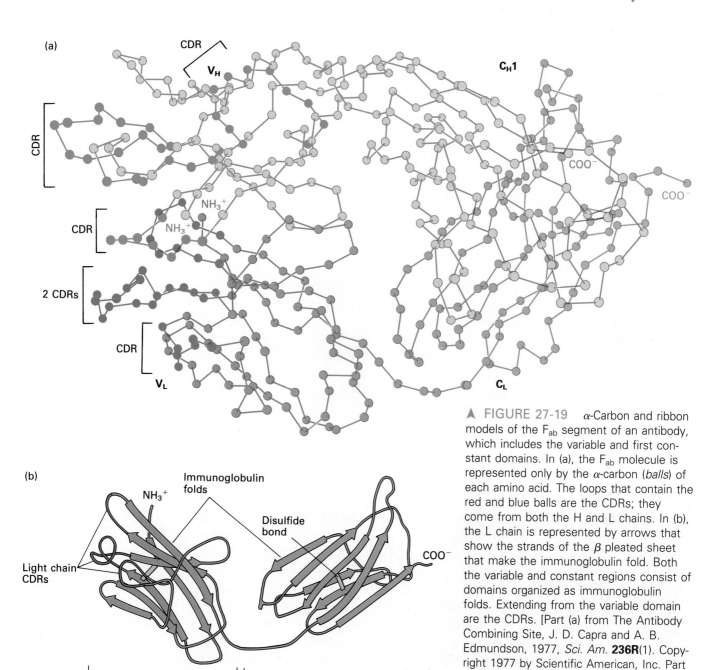

▲ FIGURE 27-19   α-Carbon and ribbon models of the F_ab segment of an antibody, which includes the variable and first constant domains. In (a), the F_ab molecule is represented only by the α-carbon (*balls*) of each amino acid. The loops that contain the red and blue balls are the CDRs; they come from both the H and L chains. In (b), the L chain is represented by arrows that show the strands of the β pleated sheet that make the immunoglobulin fold. Both the variable and constant regions consist of domains organized as immunoglobulin folds. Extending from the variable domain are the CDRs. [Part (a) from The Antibody Combining Site, J. D. Capra and A. B. Edmundson, 1977, *Sci. Am.* **236R**(1). Copyright 1977 by Scientific American, Inc. Part (b) after M. Schiffer et al., 1973, *Biochemistry* **12**:4620.]

able pocket, cleft, or plane on the surface of the antibody formed by three short polypeptide segments from each chain. From crystals of hen egg lysozyme bound to antilysozyme we have learned that in a complex of antibody and a protein antigen, a surface on the antibody interacts with a surface on the antigen in such a way that many hydrophobic, hydrogen-bonding, and charge interactions hold the complex together (Figure 27-18). Different antibodies react with different parts of lysozyme, and it is likely that the whole surface is potentially antigenic. There are no simple rules that relate the amino acid sequence of the antibody to that of the antigen: the surface complemen-

tarity of the two molecules is a consequence of many individual local interactions summing to give a very high-affinity interaction.

The large portions of the variable region that do not directly contact the antigen interact among themselves to produce a stable domain structure by forming planes of protein known as β pleated sheets (see Figure 3-12). The sheet structure of the variable domains is so characteristic that it has been called the *immunoglobulin fold*. The three amino acid segments of the variable region that bind antigens are loops that extend from the immunoglobulin fold (Figure 27-19). Because the binding site is complementary

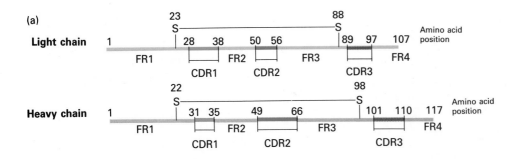

(a)

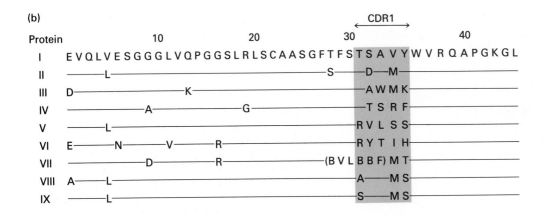

(b)

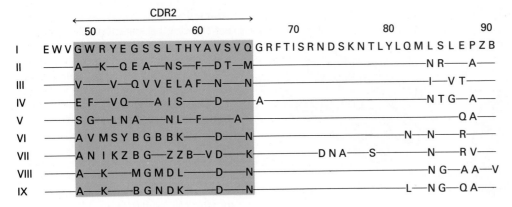

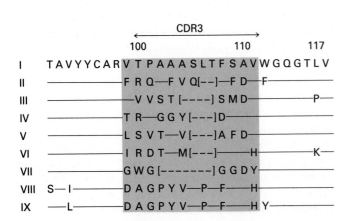

▲ FIGURE 27-20 Location of the CDRs on the variable regions of the L and H chains. (a) Schematic representation of the CDRs interspersed among framework regions (FRs). The numbers start with amino acid 1 at the N-terminus. (b) Some actual sequences of human H-chain variable regions. This collection demonstrates the extent of variation found in the CDR sequences. In the top line, the sequence of protein I is shown using the single-letter amino acid code (see Figure 3-3). For proteins II–IX, the sequences are shown using the convention that a solid line indicates identity with protein I, a letter signifies the location of an amino acid different from that in protein I, parentheses indicate an uncertainty, and broken lines represent deletion of a sequence relative to protein I. One region of variability between CDR2 and CDR3, at positions 84 through 88, is not in the antigen-binding site. [Part (b), see J. D. Capra and J. M. Kehoe, 1974, *Proc. Nat'l. Acad. Sci. USA* **71**:4032.]

in structure to the antigen, these three segments are called *complementarity-determining regions*, or CDRs. Because the CDR structure is much more variable than that of the rest of the variable region, the CDRs are often called *hypervariable regions*. The variable regions therefore consist of two types of sequence: highly variable CDRs embedded in less variable *framework regions* (FRs) that form the immunoglobulin fold. The variable region has four FRs and three CDRs (Figure 27-20).

Several Mechanisms Generate Antibody Diversity

Only the variable regions of an antibody contribute to the diversity of antigen-binding specificities. During the generation of diversity, how does one portion of a molecule come to vary enormously while the other portions remain essentially invariant? There are two distinct ways to imagine a solution to the problem. One is to have a single germ-line gene that would undergo extensive mutation in lymphoid cells, but only at its variable-region end. Then each mutation would make a protein with a unique variable-region sequence, but all proteins would have the same constant region. The other way to solve the problem is to have many variable-region gene segments in the DNA, any one of them able to join to a single constant-region gene segment. These two different approaches to the diversity problem are generically described as the *somatic mutation hypothesis* and the *somatic recombination hypothesis*, respectively. The immune system actually uses both mechanisms. In fact, there are many independent mechanisms that contribute to diversity. The recombination hypothesis was the first to gain experimental support, and is discussed here. Our discussion of these events will focus, for convenience, on the κ genes, but similar processes characterize the formation of both λ-chain and H-chain genes. We save our discussion of somatic mutation for later, since this mechanism takes place after B-cell activation by an antigen.

DNA Rearrangement Generates Antibody Diversity

The key experiment that convinced immunologists that DNA rearrangement is central to the immune response relied on myeloma cells (Figure 27-21). The κ light-chain mRNAs were extracted from a mouse myeloma, and some were broken in half. The 3′ halves of broken molecules were purified and used as molecular probes for the constant region; the intact molecules served as a probe for both variable and constant regions. Mouse myeloma-cell DNA and mouse embryo DNA were probed with the two mRNAs. The results showed that the κ-chain coding region was organized differently in the myeloma-cell DNA than in the DNA from the rest of the animal. DNA rearrangement was thus shown to be a central aspect of B-cell differentiation.

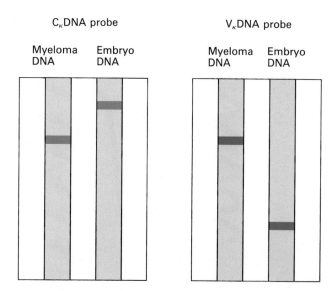

▲ **FIGURE 27-21**   Demonstration of κ-gene rearrangement during B-cell maturation. In the experiment, DNA was extracted from two sources: a clonal mouse myeloma cell and a mouse embryo. The two DNA samples were cleaved with restriction enzymes, the DNA was fractionated by electrophoresis, and the separate DNA fragments were visualized by hybridization with specific radioactive DNA probes. A C_κ constant-region probe was used for the left panel, a V_κ variable-region probe for the right. In both panels, the hybridizing bands in the myeloma DNA are different from those in embryo DNA. All cells from adult mice except B cells have the same patterns as embryo DNA. The myeloma cell has one band that hybridizes with both the C_κ and V_κ probes, indicating that the V_κ and C_κ regions, which are on separate fragments in germ-line DNA, have been so closely juxtaposed in myeloma DNA that they are on the same DNA fragment (within 10 kb of each other). [See N. Hozumi and S. Tonegawa, 1976, *Proc. Nat'l. Acad. Sci. USA* **73**:3628 for the related experiment described in the text.]

Complete analysis of immunoglobulin coding regions has shown that the H chains, κ L chains, and λ L chains are encoded in three genetic loci on three different chromosomes (on human chromosomes 14, 2, and 22, respectively). Each locus has multiple variable regions and one or at most a few constant regions. We can call each group of variable regions a *library*. In a cell that makes an antibody consisting of a κ chain and an H chain, DNA is rearranged from its germ-line configuration at both the κ locus and the H-chain locus to form the two genes that encode the two chains.

A Single Recombination Event Generates Diversity in L Chains

To appreciate in detail how L-chain genes are constructed by recombination, we need first to recall that mRNAs are formed by the splicing out of regions from a nuclear precursor RNA. In the recombining of cellular DNA to form

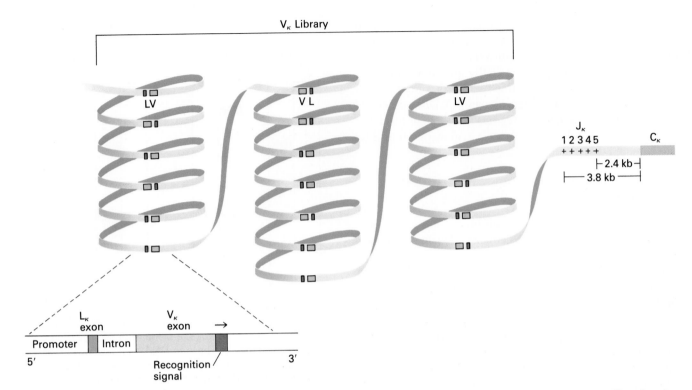

▲ FIGURE 27-22   Organization of the κ locus. In a library spread over thousands of kilobases of DNA, there are hundreds of V_κ regions. They are organized in either transcriptional orientation relative to the J_κ and C_κ segments. Both J_κ and C_κ are in the 5'-to-3' orientation (left to right in the illustration). One L_κV_κ unit is shown expanded. It consists of a promoter at the 5' end (beginning of transcription), an exon encoding a leader peptide (L_κ), an intron, an exon encoding the V_κ region, and a recognition signal that specifies the site at which the V_κ region is to join to the J_κ region. The recognition signal is actually made of two pieces, one seven nucleotides long (7-mer) and the other nine nucleotides long (9-mer). The arrow shows the orientation of the 7-mer followed by the 9-mer. [After Paul D. Gottleib, *Mol. Immunol.* **17**:1423.]

an immunoglobulin gene, the recombination has to bring together the relevant DNA into one transcriptional unit, but splicing can then eliminate any parts of the RNA from that transcriptional unit that have no role in the ultimate mRNA (introns are eliminated and exons are maintained).

The mRNA for a κ chain is made from three exons. At the 5' end is the L_κ exon; it encodes a *leader*, or *signal*, *peptide* that directs newly made κ protein into the endoplasmic reticulum. The second exon is the variable region proper and the third is the constant region.

As shown in Figure 27-22, in the germ line, the leader peptide and most of the variable region are encoded in one library consisting of a few hundred units. Each unit consists of one L_κ exon and one V_κ region. (The V_κ region makes up most, but not all, of the final variable region; we denote it to V_κ to distinguish it from the complete variable region.) The L_κ + V_κ units are tandemly arrayed along one long stretch of DNA. Each unit is about 400 nucleotides long, and they are separated by about 7 kilobases (kb); thus 100 L_κ + V_κ units would cover about 740 kb of DNA. The variable region is followed by five *joining re-gions*, or J_κ regions (not related to the J chain of the IgM molecule), and then by the one *constant-region* exon in the cell's DNA, the C_κ region. The five J_κ units are tandemly arranged and are separated by about 20 kb from the V_κ regions. Each is about 30 nucleotides long, and they are spread over 1.4 kb of DNA. Between the J_κ units and the C_κ region lies an intron of 2.4 kb of DNA.

When DNA reorganizes to make a functional κ gene, one V_κ region joins to one J_κ region, with the deletion or inversion of the intervening sequence (Figure 27-23). This forms the completed variable region. So far as is known, any V_κ can join to any J_κ, and the choice is random. Once V_κ and J_κ are joined, the variable and constant regions are transcribed together into a nuclear RNA and the intervening sequences between L_κ and V_κ and between J_κ and C_κ are removed by RNA splicing to produce the mature mRNA for the κ protein (see Figure 27-23).

The V_κ-J_κ joining combines the V_κ region diversity and the J_κ region diversity. The combination of 300 V_κ regions and 5 J_κ regions can produce, obviously, 1500 different possible chains. But V_κ-J_κ joining generates even more se-

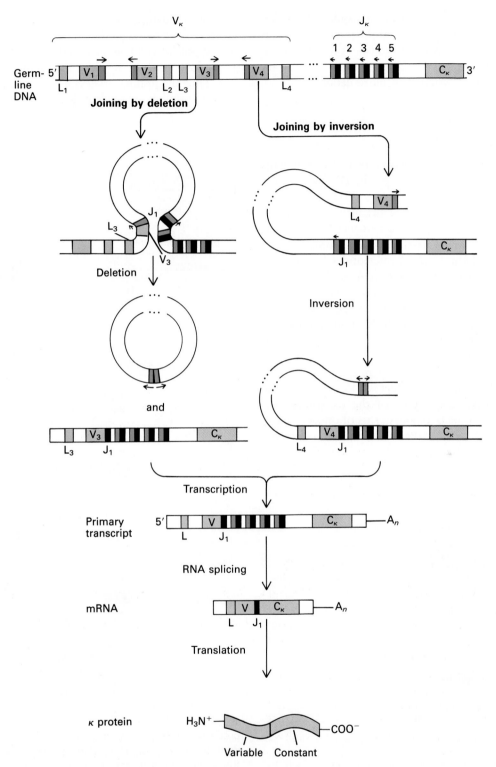

▲ FIGURE 27-23   Joining of V_κ to J_κ and formation of a κ light chain. The $L_\kappa V_\kappa$ segments are oriented in either transcriptional direction relative to C_κ in germ-line DNA. Each V_κ and J_κ region has a recognition signal (red) to specify the points at which joining should take place. The little arrows indicate the heptamer-to-nonamer orientation as defined in Figure 27-22. The joining process either deletes the intervening DNA or inverts it, depending on the relative orientations of V_κ and J_κ. In deletional joinings, the deleted DNA is lost from the cell. In inversional joinings, all of the DNA is conserved. Once the V_κ-J_κ joining has occurred, the now complete gene can be transcribed to produce a nuclear RNA that is spliced to remove all unwanted segments (including the J_κ regions that were not joined to V_κ). The spliced mRNA then encodes a complete κ light chain.

quence variability than this calculation would suggest because in the vicinity of the V_κ-J_κ joint the joining process is imprecise and can generate many additional combinations. To see the consequences of this imprecision, it is necessary to examine the joining reaction in more detail.

Imprecision of Joining Makes an Important Contribution to Diversity

At the 3' end of the coding sequence of the V_κ segments and at the 5' edge of the J_κ segments lie DNA sequences that signal the joining process (Figure 27-24). They are called *recognition signals* and are organized as follows: abutting each V_κ and J_κ is a 7-base sequence (a heptamer, or 7-mer) that forms an approximate palindrome; beyond this is a spacer of about 11 or 23 nucleotides (one or two turns of the DNA helix) and then an AT-rich 9-base sequence (a nonamer, or 9-mer). As we shall see, these recognition sequences allow an as-yet-uncharacterized enzymatic system to carry out an orderly though imprecise joining reaction. The imprecision of this joining contributes greatly to diversity.

When two pieces of DNA join, there are two products. In the joining of V_κ to J_κ, one product is a $V_\kappa J_\kappa$ unit and the second is a back-to-back joining of the recognition sequences. When many such joining events of one V_κ and J_κ were studied, it became clear that the recognition elements are joined identically in almost all cases, with the heptamers linked to each other. The $V_\kappa J_\kappa$ unit, however, is not joined precisely: a few nucleotides from V_κ and a few from J_κ are lost from the DNA at the joining point (Figure 27-25). Thus the joining process has the unique property that a small number of nucleotides at the joint are lost from

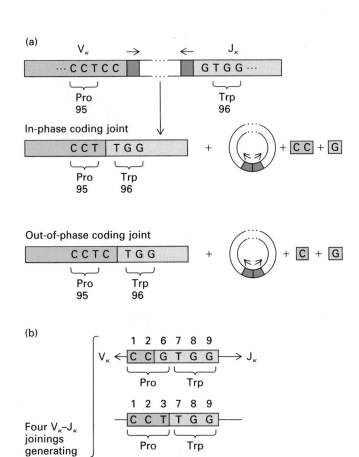

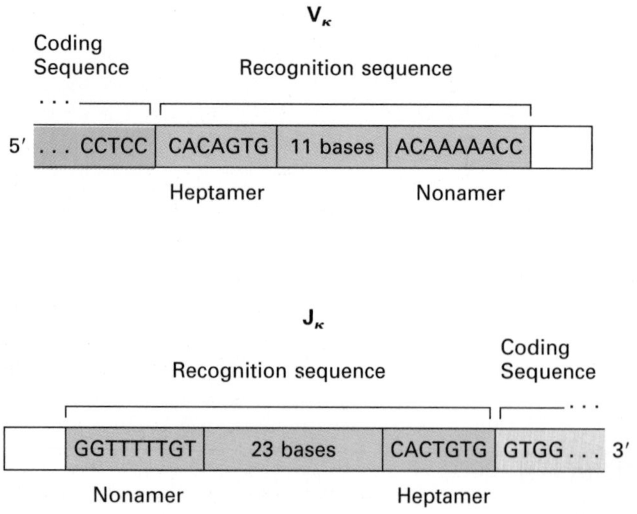

FIGURE 27-24 V_κ and J_κ recognition signals. Each J_κ and V_κ region has a short characteristic DNA sequence (recognition signal) preceding or following its coding sequence (see organization in Figure 27-23). This signal is recognized by DNA joining enzymes and directs those enzymes to join a V_κ to a J_κ. The recognition signals consist of a heptamer and a nonamer, separated by a spacer of about 11 nucleotides in a V_κ and 23 nucleotides in a J_κ. The joining system always joins DNA with an 11-base spacer to DNA with a 23-base spacer, so that a V_κ joins to a J_κ but not to another V_κ.

FIGURE 27-25 The joining of V_κ to J_κ. (a) The joining process can give rise to in-phase or out-of-phase joints. During the joining, a small, random number of bases is lost from the ends of both V_κ and J_κ. This can leave J_κ joined in the appropriate reading frame (Trp at position 96 encoded by TGG just following the CCT codon at position 95) or joined out of frame (for instance, if one more C is left on V_κ, the CCT at 95 will be followed by CTG and all of J_κ will be out of frame with V_κ). (b) The joining process can give rise to many in-phase joints with various amino acids at the joining point if variable ends of V_κ are joined to variable ends of J_κ. In the example, four different in-phase joinings encode Pro-Trp, Pro-Arg, and Pro-Pro.

DNA entirely; they appear in neither product. The random loss of nucleotides at the joining site generates significant diversity at that point, but the system pays for its diversity. The cost is evident if we remember the constraints on a coding sequence.

Recall that a coding sequence in DNA must be read in 3-base codons, and that the initiation codon, AUG (methionine), defines a *reading frame* that groups the rest of the coding region into codons. Reading DNA in one of the two other reading frames would generate a meaningless string of amino acids. Thus, when two pieces of coding DNA join, as in the joining of V_κ to J_κ, it can be an *in-phase* joint that maintains a sensible reading frame or it can be an *out-of-phase* joint that encodes a nonsense protein (Figure 27-25a). The V_κ-J_κ joining process is a random one, and two out of every three random joints make no sense. Thus the system pays for its diversity in making two *nonproductive* joints for each *productive* joint.

The diversity gained by the imprecise joining process is significant. Figure 27-25b shows how four different in-phase joinings can be made between one V_κ sequence and one J_κ sequence. All of these joints have actually been found in sequenced κ proteins.

We now have three sources of diversity: variability in the structure of the many V_κ regions in its library of sequences, variability in the structure of the five J_κ regions, and variability in the number of nucleotides deleted at V_κ-J_κ joints. Recall that antibody chains have three CDRs (see Figure 27-20). The diversity within the V_κ library contributes to all three, whereas the diversity at the recombination joint contributes only to CDR3 because the site of joining is within CDR3. The diversity in both CDR1 and CDR2 has developed over evolutionary time as the individual V_κ regions have evolved. By contrast, the diversity in CDR3 is a somatic process: it happens in cells within the body of the individual animal, not over evolutionary time.

As we said, the λ-chain and H-chain genes form by similar recombinational processes. However, they show some interesting differences, which we now consider.

Lambda Diversity Derives from Multiple Constant Regions

The ratio of κ to λ light chains differs radically from species to species. Mice have 95 percent κ, humans about 60 percent κ, and chickens virtually only λ. In most species, λ proteins have not one but multiple C_λ regions, each with its own J_λ region. In mice, probably because of the paucity of λ protein, there are only a few V_λ regions, each associated with a J_λ-C_λ cluster. Humans have even more C_λ regions and many V_λ. Mouse λ chains have much less diversity than mouse κ chains because V_λ and J_λ diversities are so minimal. Moreover, very little recombinational diversity has been found at the V_λ-J_λ joining site. The situation in chickens is remarkable: most chicken antibody has λ light chains, but there is only one functional V_λ segment. Diversity in chicken λ chains arises from a string of pseudogene segments upstream of the one intact V_λ. By a process of gene conversion, these pseudogenes diversify the intact V_λ after it has joined to J_λ. A similar process occurs in rabbits.

H-Chain Variable Regions Derive from Three Libraries

Analysis of the antigen-binding sites of numerous antibodies suggests that the H-chain contribution to diversity is even greater than the L-chain contribution (see, for example, Figure 27-17). Consistent with a need for greater diversity, H-chain structure is determined by three libraries of sequence elements instead of just the two (V and J) that make up L chains. The third library consists of the *D regions* (D for diversity), which constitute the bulk of the H-chain's CDR3. The D-region segments are found between the other two libraries, whose segments are called V_H and J_H (Figure 27-26). Thus two joining reactions, V_H-D and D-J_H form the variable region for the H chain. Clearly, having three segments, rather than two, greatly increases the combinational diversity. There are hundreds of V_H segments, perhaps 20 D segments, and four J_H segments. Recombinational diversity at the V_H-D and D-J_H joints, like that at the V_κ-J_κ joint, is also created by the removal of small numbers of nucleotides.

The third CDR of H chains is diversified by yet another mechanism. When a D segment joins to a J_H, or when a V_H joins to a D, not only are nucleotides removed but a few nucleotides not found in either parental sequence are added at the joint. These extra added nucleotides comprise an *N region*; the enzyme responsible for their addition is *terminal deoxynucleotidyl transferase* (TdT), a template-independent DNA polymerase known to be present in cells making H-chain joints (but absent from cells when they are making L-chain joints). N regions are absent in a cell whose TdT gene has been disrupted.

With a complete H-chain variable region being a $V_H NDNJ_H$ unit, maintaining the reading frame presents a problem similar to that for L chains. The nucleotides of the NDN region between V_H and J_H must keep the reading frame correct and, as in L chains, two out of three joinings will be out of phase. A danger not usually present in the simpler V-J joint of L chains is that the formation of the NDN sequence could put a termination codon in the reading frame. The possibility of this is minimized by the high G + C content of many N regions (terminators are UAG, UGA, or UAA) and by the absence of terminators from most D regions.

The potential diversity generated by the NDN region is truly vast because the N regions have wholly random sequences (with some bias toward G + C). NDN regions up to 30 nucleotides long have been found, and this highly variable segment can encode 0–10 amino acids.

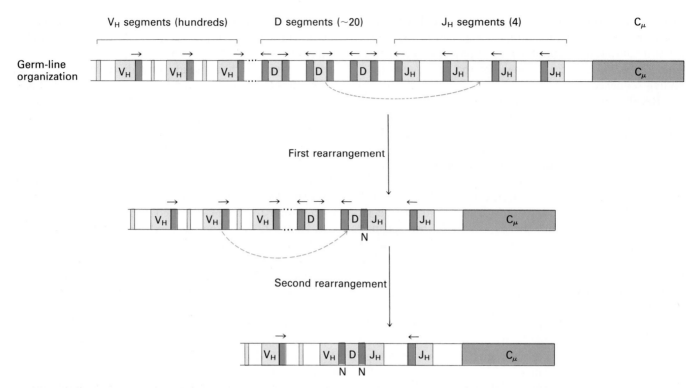

▲ FIGURE 27-26   Organization and rearrangement of H-chain genes. In the germ line, the H-chain-related DNA is organized with a large library of V_H regions all in the same 5′-to-3′ orientation, followed by a library of D regions, a library of J_H regions, and the constant region for the μ chain (C_μ). The V_H, D, and J_H gene segments have recognition signals like those found at the ends of V_κ and J_κ (overlined with arrows). In the first stage of rearrangement, a D segment joins to a J_H segment, deleting the intervening DNA. In the second stage a V_H joins to the preformed DJ_H unit, forming a $V_H DJ_H$ H-chain variable region. At each joint, a few random nucleotides may be inserted (N regions).

Recognition Sequences for All Joining Reactions Are Indistinguishable

We have described here a number of different joining events: V_κ to J_κ, V_λ to J_λ, V_H to D, and D to J_H. All of these reactions follow one inviolable rule that relates to the structure of their recognition sequences. Before we state the rule, a bit of background information is needed. Each sequence element that participates in these joining reactions has an adjacent recognition sequence. Just as we described for κ gene segments, these recognition sequences are of two types: one has a characteristic heptamer followed by an 11- or 12-base spacer of random sequence and an AT-rich nonamer; the other has a similar heptamer and nonamer but separated by a 21- to 23-base spacer (see Figure 27-24). The length of the spacer makes the difference with the shorter spacer being approximately one turn of a double-stranded DNA helix (10.5 bp), and the other spacer about two turns. The inviolable rule is that in all unions a one-turn recognition sequence always combines with a two-turn recognition sequence. Thus V_κ segments have one-turn elements, and J_κ segments have two-turn; D segments are flanked by one-turn elements, and V_H and J_H

segments have two-turn elements. Why the spacers fall into this pattern is not totally clear, but very likely there are proteins that bind to these recognition sequences, and they may have evolved to recognize signals arrayed together on one side of a DNA double helix.

The enzymology of joining remains to be worked out. We imagine that recognition proteins bind to the recognition elements and specify that one-turn and two-turn elements join to each other. Endonucleases must cut the strands, exonucleases might degrade free ends, TdT may add nucleotides, polymerases must provide flush ends, and then ligases must seal the gap. Many enzymes must participate, and the events must be highly regulated so that joining occurs only when needed and then only between the correct sequence elements. As yet, only two lymphocyte-specific proteins are known to be needed for recombination: the products of the two recombination-activating genes, *RAG-1* and *RAG-2*. While these proteins are known to be involved, their roles remain to be determined.

We have considered joining processes that create diversity by combining regions from different libraries in the germ line. We also mentioned the insertion of N regions, a de novo synthesis of DNA, which is a form of somatic mutation that is coupled to joining. Another form of somatic mutation, separate from joining, involves the changing of individual base pairs throughout a fully assembled variable region to generate diversity across the whole region. This mutational process occurs later than the other mechanisms generating diversity and will be described later.

The Synthesis of Immunoglobulins Is Like That of Other Extracellular Proteins

All immunoglobulin synthesis takes place on the rough endoplasmic reticulum membranes of the B cell. This would be expected because immunoglobulins are either cell-surface molecules or secreted molecules, and the pathways to both end points begin with newly made protein entering the cisternae of the endoplasmic reticulum. Newly made polypeptides are usually targeted to the endoplasmic reticulum by a hydrophobic signal sequence. In immunoglobulins, such a sequence is found at the N-terminus of newly made H and L chains. It is encoded by the L exon located on the 5' side of each V region (see Figure 27-22). The signal sequence is cut away from the chains after its synthesis, so that mature antibodies do not contain it.

➤ *The Antigen-Independent Phase of B-Lymphocyte Maturation*

Our discussion of clonal selection touched upon the two phases in the development of a B lymphocyte. In the *antigen-independent phase,* diversity is generated. By the end of this phase a B cell has acquired rearranged H- and L-chain genes and is synthesizing IgM antibody as a surface receptor protein. At this time the cell is a *virgin B cell,* not having undergone any selection for the reactivity of its receptor with an antigen. During the second or *antigen-dependent phase,* B cells have three possible fates: (1) to react with a foreign antigen and thus be positively selected for further growth, maturation, and synthesis of secreted antibody; (2) to react with a self-antigen and be paralyzed or killed, thus ensuring self-tolerance; or (3) not to react at all and die within a few days. In this section we consider the ordering of events in the antigen-independent phase of B-cell maturation. Later, the antigen-dependent phase and the key role that T cells play in its progression will be discussed.

Before discussing the individual events of the antigen-independent phase, the overall strategy of the developmental process is worth emphasizing. Cellular development can be driven in two ways, by internal events of a cell or by external influences. As far as is known, the antigen-independent phase is a good example of internally driven dif-

ferentiation. Although it is probably initiated by an external inducing agent acting on an uncommitted cell, that event has yet to be characterized. After the initial commitment, however, the events from that point through the production of the virgin B lymphocyte seem to be internally determined. Progression through a variety of stages comes about through internal signals, mainly generated by the successful rearrangement of first H-chain genes and then L-chain genes. It is reasonable that the antigen-independent phase should need no extracellular stimulation: it is the working out of a differentiation program that has a single goal, the production of virgin B cells. Once the virgin cell is produced, many external influences come into play, making the antigen-dependent phase a dialogue between the cell and its surroundings.

Cellular development in general relies on both internal and external signaling. B-cell differentiation happens to be a convenient system in which to see the two styles of developmental decision-making at work.

B Lymphocytes Go Through an Orderly Process of Gene Rearrangement

B lymphocytes arise continually throughout life by differentiation from bone-marrow stem cells (see Figure 27-9). Many of the intermediate stages of B-cell maturation have been defined by examining the structure of immunoglobulin-related DNA in the cells. Tumors of early-stage cells have been very helpful in defining the intermediate stages, just as myelomas have helped our understanding of end-stage cells.

The earliest recognizable B cell has cell-surface proteins (defined by antibody binding) that mark its B-cell lineage, but its DNA is still germ-line in organization (Figure 27-27). At this point, the cell is known as a *pro-B cell.* Next, this differentiating cell begins H-chain gene rearrangement; its L-chain genes are not yet actively rearranging. It first joins a D to a J_H, and then joins a V_H to the preformed DNJ_H to generate the complete V_HNDNJ_H. The cell can then make an H chain. Because the H-chain constant region nearest to the V_HNDNJ_H unit encodes a μ chain, the cell makes μ and is recognized by immunofluorescence as a μ-positive, L-chain-negative bone marrow cell, now called a *pre-B lymphocyte.*

The next stage of B-cell differentiation is rearrangement of the κ L-chain genes (Figure 27-27). Most mature mouse B cells make a κ chain; however, a few, those that generally rearrange κ genes nonproductively, rearrange λ genes and become λ producers. Once a cell has constructed a complete in-phase κ or λ gene, it makes an L chain. It also produces two accessory proteins, called Ig-α and Ig-β. The L chain binds to an H chain, along with Ig-α and Ig-β, and this unit is then transported to the cell surface. The newly arisen cell now has surface antibody and is a *virgin B lymphocyte.*

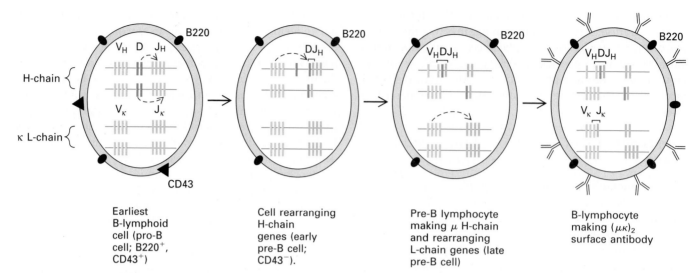

Earliest
B-lymphoid
cell (pro-B
cell; B220$^+$,
CD43$^+$)

Cell rearranging
H-chain
genes (early
pre-B cell;
CD43$^-$).

Pre-B lymphocyte
making μ H-chain
and rearranging
L-chain genes (late
pre-B cell)

B-lymphocyte
making $(\mu\kappa)_2$
surface antibody

▲ FIGURE 27-27   Ordered rearrangement of variable
regions of antibody genes during the antigen-independent
phase of B-cell differentiation. The earliest cell has B-lymphoid surface markers but has not started rearrangements. It
begins maturation by first making D-to-J$_H$ rearrangements at
both heavy-chain loci. It then begins V$_H$-to-DJ$_H$ rearrangements, probably sequentially on the chromosomes. Production of an in-phase V$_H$NDNJ$_H$ unit leads to H-chain (μ) synthesis, and this (or possibly some other signal) prevents
further rearrangement. The example shown here would be
an in-phase joint preventing V$_H$-to-DJ$_H$ rearrangement on the

second chromosome. After μ-chain synthesis begins, the cell
is designated a pre-B lymphocyte. After an unknown number
of further divisions, κ-gene rearrangement begins and continues only long enough to make one κ gene that can encode a
complete κ protein able to bind to μ. The $(\mu\kappa)_2$ dimers then
appear on the cell surface, the cell stops division, and it
becomes a mature B lymphocyte. If no in-phase κ join is
produced, the cell can still rearrange its λ L-chain genes.
About 5 percent of mouse antibody and almost 50 percent
of human antibody have λ L chains.

We noted earlier that the signal for rearranging immunoglobulin genes is always the same: a heptamer, then a
one- or two-turn spacer, then a nonamer. How then is the
patterned rearrangement of the various immunoglobulin
gene segments regulated to coordinate the events in time?
The answer is thought to lie in transcriptional control:
those segments that are being transcribed are the ones that
rearrange. Thus the patterned unfolding of a transcriptional control program, described later, is thought to underlie the stages in the rearrangement process.

The Antigen-Independent Phase Can Generate B Cells with 10^{11} Different Specificities

The antigen-independent phase of B-lymphocyte development generates cells containing surface antibodies with a
wide range of specificities. If we assume a random association of the units that make up antibody genes, we can estimate the potential number of different antibody molecules
that could be produced by the multiple diversification
mechanisms available. Actually, our calculation will be an
overestimate because some V$_H$ regions, such as those most
proximal to the constant regions, are favored for rear-

rangement. In addition, one reading frame of each D region is highly favored.

Starting with the H chain, if we assume that in the
mouse there are 300 V$_H$ regions, 20 D regions, and 4 J$_H$
regions, we calculate that $300 \times 20 \times 4 = 2.4 \times 10^4$ different combinations of these units could exist. Because
nucleotides are lost at each D-J$_H$ joining and each V$_H$-D
joining, all three constituents can be truncated to various
extents, and this increases the diversity at least tenfold. In
addition, the D-J$_H$ and V$_H$-D joinings have N regions that
are random in length and sequence, generating at least 100
different possibilities. Thus about $2.4 \times 10^4 \times 10 \times
100 = 2.4 \times 10^7$ different H chains are possible. L chains
are less diverse. Ignoring λ chains because they are rare in
mice, if there are 100 V$_\kappa$ regions, 4 functional J$_\kappa$ regions
(one of the five is not functional), and 10 ways of putting
them together, then $100 \times 4 \times 10 = 4 \times 10^3$ possibilities
exist. Finally assuming that any H chain could combine
with any L chain, we see that $2.4 \times 10^7 \times 4 \times 10^3 = 10^{11}$
physically different antibodies can be produced by the
immune system of a mouse during the antigen-independent
phase. The number 10^{11} is huge—a mouse makes only
about 10^8 lymphocytes per day, so that all possible combinations might not even be expressed over the lifetime of the
animal.

Another characteristic of the antigen-independent phase is that it is organized hierarchically, as Figure 27-9 indicates, to produce a highly diverse population of single cells. Through the early stages of their lineage, B cells are continually growing and dividing about once every 8 hours. How long the early stages last is not clear, but a cell just past the DJ_H rearrangement stage can probably give rise to many progeny, each of which will have its own V_HDJ_H rearrangement. Similarly, a cell with a V_HDJ_H rearrangement can give rise to many cells with independent $V_\kappa J_\kappa$ rearrangements. But when a virgin B lymphocyte appears in the marrow, it rapidly ceases growth and moves out of the marrow to the peripheral blood-lymph circulation system. Because a cell appears to cease growth very soon after it has acquired surface antibody, each cell produced by the bone marrow could be virtually unique. In fact, however, certain gene structures appear to form preferentially. How this is accomplished remains unclear, but because all the individual mice of a certain inbred strain respond to a given hapten by making virtually identical antibodies (called a *shared idiotype*), many with specific N regions, some gene configurations seem to be preferred during the antigen-independent phase of maturation. ,

The Immune System Requires Allelic Exclusion

A third aspect of B-lymphocyte maturation is that cells are diploid and therefore could, in principle, make two different H chains if both chromosomes rearranged productively, that is, if each made an in-phase V_HNDNJ_H unit. In addition, two κ L chains plus two or more λ L chains can conceivably be made. One cell would therefore be able to make multiple types of antibody. But such a situation would violate the precondition for clonal selection to operate. Clonal selection depends on the manufacture of only one antibody type per cell so that when the cell is activated it will go on to secrete only one type of antibody. If multiple chains were to be made, antibodies could have two different binding specificities on their bivalent molecules and would lose the advantage of high avidity gained from multiple interactions with a single antigen. Mechanisms have evolved, however, that ensure that B lymphocytes make antibody with one, and only one specificity. Because this means that only one of two alleles carried by a cell will be expressed on any one cell, the process has been called *allelic exclusion.*

Allelic exclusion apparently works by shutting down a particular step in the rearrangement process after there has been one productive rearrangement. The mechanisms are not known, but one type of experiment clearly indicates that the exclusion process is at work. An already rearranged H-chain or L-chain gene can be introduced into a mouse's germ line by the microinjection of a plasmid into a fertilized mouse egg (see Figures 8-33 and 8-39). The animals are called *transgenic,* denoting their acquisition of a foreign gene. When a mouse is transgenic for a rearranged H-chain gene, the rearrangement of endogenous H-chain genes is suppressed. Transgenics that have acquired a rearranged κ chain suppress the rearrangement of endogenous κ chains. In both cases, the expression of a rearranged gene suppresses the rearrangement of other similar genes—the result we would expect if allelic exclusion is a consequence of the suppression of secondary rearrangements.

Antibody Gene Expression and Rearrangement Are Controlled by Transcription Factors

We have thus far considered only the construction of immunoglobulin genes and not their transcription. The control of their transcription, unlike that of many other genes in animals, should be simple because they are expressed in only one cell lineage in the body, the B lymphocyte. The transcriptional control regions for H chains and κ L chains are easily identified; they consist of a few promoter elements that are encoded in the V segments upstream of the coding region, and enhancer elements that are located either between the J segments and the C-region exons or 3' to the constant regions (Figure 27-28a). The position of the enhancers ensures that nonrearranged V regions are not under their influence and that all rearranged genes are influenced by the same enhancer. Each V region, however, has its own promoter, but all contain common OCTA and TATA motifs.

The TATA motif (see Figure 11-42) is found in many genes and does not display cell specificity. By contrast, OCTA, an 8-bp sequence, is regulatory: if it is artificially appended to another gene, say the globin gene, it causes that gene to be preferentially expressed in cells of the B-lymphocyte lineage. Thus the promoters of immunoglobulin genes are cell-specific regulatory elements.

The enhancers are also regulatory; they function only in B-lineage cells. The H-chain enhancer (Figure 27-28b) contains many motifs that bind individual proteins. The OCTA motif is the same one as found in the promoter. It, along with the π, μB, and the various μE motifs, give the enhancer many of its cell-specific regulatory properties. There is also negative regulation of the enhancer in non-B lymphoid cells, probably carried out partly by the binding of a ubiquitous protein called ZEB to the μE5 site (Figure 27-28b).

For the κ L-chain enhancer, the κB and E2 sites are positive regulatory elements (Figure 27-28c). The κB site binds a protein called NF-κB that is activated by the external stimulation of many cells of the body and presumably is involved in the antigen-dependent activation of the gene (see Figure 11-34). Whether NF-κB or another factor acts on this site during development remains to be determined. Another element within the κ enhancer, the silencer motif,

(a) H-chain and κ L-chain transcriptional control elements

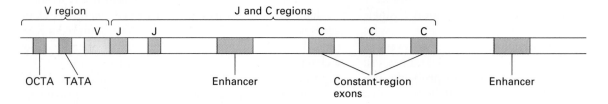

(b) The H-chain intronic enhancer

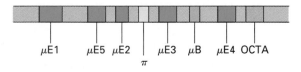

μE1 μE5 μE2 μE3 μB μE4 OCTA
                    π

(c) The κ L-chain intronic enhancer

Silencer   κB   κE1   κE2   κE3

▲ FIGURE 27-28   Transcriptional control regions for H chains and κ L chains. The control elements for immunoglobulin gene expression consist of DNA sequence motifs, 6- to 12-bp segments that bind specific proteins. The promoter regions contain common prominent motifs, OCTA and TATA; the enhancers are different for H and L chains. Both H and L chain genes have multiple enhancers, one in the intron between the J and C regions, the other 3′ to all coding regions. The protein-binding sites for the intronic enhancers are shown in (b) and (c).

acts negatively to prevent the enhancer from acting in non-B cells.

These transcriptional control elements work together to ensure that H chains and κ L chains are synthesized in B lymphocytes and not elsewhere in the body (λ L chains are under similar control). As mentioned earlier, it is the ordered transcription pattern of the H-chain and L-chain segments that signals their orderly rearrangement. Because H chains and L chains are under separate regulatory control by different constellations of factors, regulation of these factors determines the order of gene rearrangement.

Many other transcription factors are also regulated during B-cell development, and they, in turn, regulate a collection of genes whose products are needed during the B-cell differentiation process. Much remains to be learned, but a schematic representation of some present knowledge is given in Figure 27-29. Also included in the figure is an indication of how T-cell differentiation may be orchestrated by transcriptional control.

► *T Lymphocytes*

Before considering the antigen-dependent phase of B-cell maturation, we must switch to a discussion of T cells because T-cell–B-cell collaboration plays a crucial role once

virgin B cells leave the bone marrow and start their antigen-dependent activities.

The T lymphocytes are equal partners with B lymphocytes in generating the immune response. They too show very high specificity, recognizing one particular antigen but not its close relatives. T cells have on their surface a recognition molecule able to make discriminations as fine as those made by antibody. This molecule, the *T-cell receptor,* is structurally related to antibody but has numerous distinguishing properties. One key difference is that antibodies are mainly produced as soluble secreted molecules, whereas the T-cell receptor exists only at the cell surface. All T-cell functions, therefore, involve reactions on the surface of the cell which is why T cell activity is called cell-mediated immunity. The molecules to which T-cell receptors bind are sometimes designated antigens although they have radically different properties from the antigens to which antibodies bind. Again, a key difference is that the antigens recognized by T cells are always cell-bound molecules. Hence interactions between the T-cell receptor and its antigen are actually cell-to-cell interactions.

T cells, like B cells, mature through a series of stages before they are released to the periphery to be activated by interacting with antigen. Early T-cell development occurs in the thymus gland; here T-cell receptor gene rearrangements and the steps to T cell self-tolerance take place.

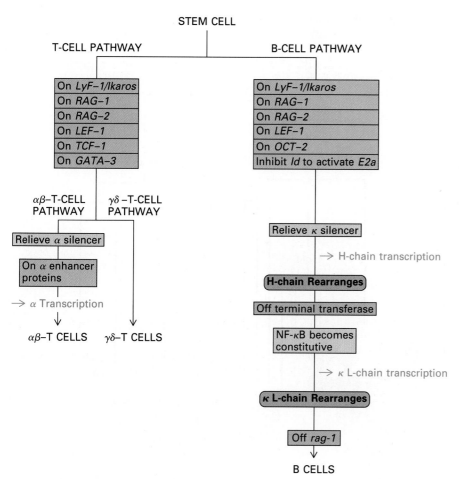

◄ FIGURE 27-29   Landmark events in the differentiation of lymphoid cells. Some of the known events during lymphocyte differentiation are placed here on a time line. Some represent activation of the genes for specific transcription factors (E2A, which is sequestered by Id proteins in undifferentiated cells, OCT-2, LyF-1/Ikaros, LEF-1, TCF-1, GATA-3, NF-κB). Many genes for such proteins as receptors, adherence molecules, and signalling molecules are also activated in the individual pathways.

There Are Two T-Cell Receptor Molecules

We introduced earlier the two types of well-characterized T cells, T_H (helper) cells and CTLs (killer cells, or cytotoxic T lymphocytes) (see Figures 27-1 and 27-13). The T-cell receptors on both these types of cells are two-chain molecules built similarly to antibodies. The chains of the most common T-cell receptor are called α and β; they are 40,000-MW glycoproteins comprising two domains, one variable and one constant. A second T-cell receptor, found on a minor population of T cells, is made of γ and δ chains. Although its function is poorly understood, this receptor appears to be a key player in immunity to certain bacteria. The γ-δ T cells are formed selectively in the fetal thymus, and different subsets of them home to different tissues, such as the skin or the gut.

The T-cell receptor proteins, the immunoglobulins, and many other cell-surface proteins are all built with the immunoglobulin fold (see Figure 27-19b) as their backbone structure; they presumably all evolved from a single precursor protein and are considered members of the immunoglobulin superfamily (Figure 27-30). The genes for the α, β, γ, and δ chains of the T-cell receptors have organi-

zations similar to that of antibody genes: there are libraries of V, D, and J regions from which members are joined to form entire genes. The joining process is identical to that used for antibody gene segments, being directed by heptamer-spacer-nonamer recognition sequences. The libraries for α, β, and γ are separate entities found on human chromosomes 14, 7q, and 7p, respectively. The δ-chain segments are on chromosome 14, interspersed among the α-chain segments (Figure 27-31).

T-Cell Receptors Recognize a Foreign Antigen in Combination with a Self-Molecule

Like antibodies, the T-cell receptor is highly specific in its recognition of a foreign antigen. It also displays tolerance to self-antigens. The antigen, however, must be presented to the T-cell receptor as a short peptide held on the surface of a cell in a complex with a cellular protein called an MHC molecule. This requirement became clear in a classic experiment done in 1974 to examine how CTLs recognize

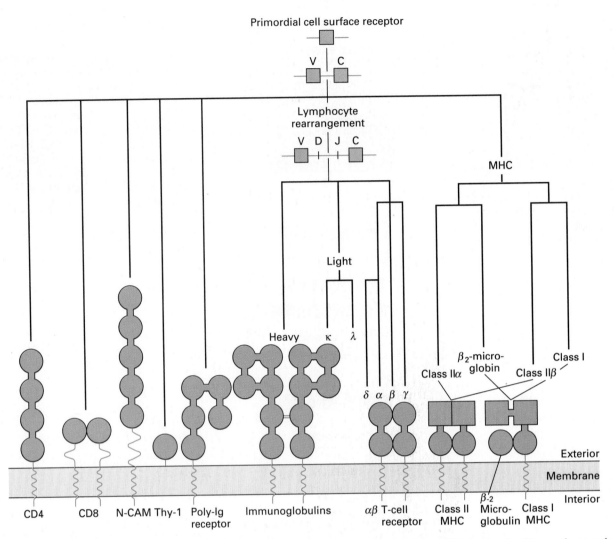

▲ FIGURE 27-30   The superfamily of immunoglobulin-related proteins. For each protein, the individual balls represent domains of protein structure that are derivatives of the immunoglobulin fold described in Figure 27-19. Regions of MHC proteins denoted by boxes are not immunoglobulin folds. The genes encoding these proteins all have homologies and are considered a superfamily of related genes that must have evolved over hundreds of millions of years from a common ancestor. Many other proteins are part of the family, including *Drosophila* proteins that guide nerve growth, so the family predates the evolution of the immune system. [After L. Hood, M. Kronenberg, and T. Hunkapiller, 1985, *Cell* **40**:225.]

T-cell receptor-chain genes

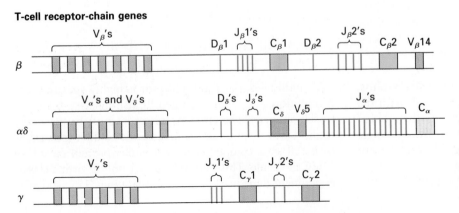

◄ FIGURE 27-31   Structure of mouse or human T-cell receptor-chain genes. There are three gene clusters on three separate chromosomes, unrelated to the antibody gene clusters. The αδ cluster is the only case in which genes for two chains are interspersed.

and kill infected target cells. The investigators injected a virus into an inbred mouse, producing in the mouse anti-viral CTLs. When studied in culture these CTLs were found to recognize and kill cells from that mouse which had been infected with that virus. Just as is seen with antibodies, the CTLs reacted only with cells infected by that particular virus. The experimenters then asked an apparently innocuous question: would the CTLs kill cells taken from a different inbred strain of mice infected by the same virus? To almost everyone's surprise, they would not. Apparently the CTL receptor recognized both the identity of the infected cell and the particular type of virus: it recognized self- and foreign determinants together, but neither one separately.

What is this surface self-molecule that is so important to the T-cell receptor? As we shall now see, they are proteins encoded in a region of the genome called the major histocompatibility complex (MHC)—CTL killing is restricted by self-MHC molecules (Figure 27-32).

How MHC Genes Got Their Names Although the MHC molecules are now known to serve as scaffolds for presenting peptide antigens to T cells, their importance was first recognized in tissue transplantation experiments, which, as we shall see, accounts for their unwieldy names. It happened in this way. If skin from a mouse of one inbred strain is transplanted to the back of a mouse of another strain, the graft is recognized as foreign and soon dies; we say that the graft has been *rejected*. Grafts are not rejected when the donor and recipient mice are both of the same inbred strain. To find out which genes encoded the proteins causing the rejection, a series of graft experiments were performed, during which the genetic differences between the donor and recipient animals were progressively minimized by inbreeding. Finally, a region containing a small number of genes could be identified as causing the rejection. Further graft experiments using different combinations of mice showed that a fair number of cell-surface proteins can cause rejection. But one gene complex stood out in all these experiments; when donor and host genes showed differences in this complex, rejection was very rapid. Because the genes in this region were so important for graft acceptance or rejection, they were called *histocompatibility genes*, and the especially potent gene complex was called the *major histocompatibility complex*, or *MHC*.

Many individual genes have been found within the MHC, and most have been molecularly cloned. A map of the MHC (Figure 27-33) was found to include numerous genes involved in peptide presentation. It also includes many other genes whose products relate to T cell function, such as peptidases, peptide transporters, and even cytokines; still other MHC genes have no obvious immune function.

MHC proteins are able to elicit graft rejection because each individual animal of a given species has some MHC

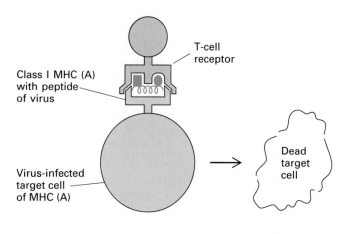

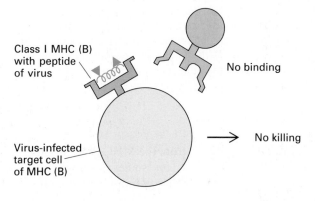

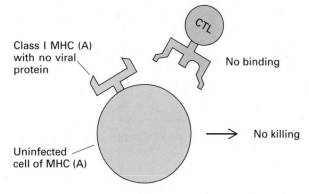

▲ FIGURE 27-32   Regulation of CTL killing by class I MHC molecules. For the experiment outlined, CTLs were prepared from animals of MHC type A inoculated with a virus. These CTLs will kill cultured cells from type A mice infected with the virus. They will not kill infected cells of type B mice because the type B MHC molecule is not recognized by the T-cell receptor on the CTLs. Also, type B MHC will probably bind a different viral peptide. The CTLs will also not kill uninfected type A cells because recognition by the CTL receptor requires that the MHC protein have the viral peptide bound to it.

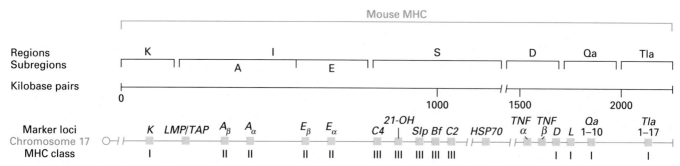

▲ FIGURE 27-33   The MHC gene complex of the mouse. This complex of genes, called the H-2 region in the mouse and the HLA region in humans, was first identified by tumor rejection and later shown to be responsible for skin graft rejection. Genetic mapping uncovered the predominant MHC class I genes at the K and D loci. Later work on T_H cell reactions defined the I region, which contains the immune-response-associated A and E genes that encode the MHC class II molecules. Mapped between K and A are genes for proteases that participate in degrading proteins to peptides in cells as well as genes for proteins that transport the peptides into the endoplasmic reticulum for union with the class

I proteins (LMP = Large Multifunctional Proteasome genes; TAP = Transporter associated with Antigen Processing genes). The S region was found to encode blood proteins (mainly of the complement system). Following the S region the heat shock gene, HSP70, was found. In the D region are genes for tumor necrosis factor (TNF)α and lymphotoxin (TNFβ). Following the D and L class I genes are multiple Qa and Tla loci that encode class I-related proteins. [See L. Hood et al., 1984, Immunology, 2d ed., Benjamin; E. Lai, R. Wilson, and L. Hood, 1989, Adv. Immunol. 46:1–59; and W. E. Paul, 1993, Fundamental Immunology, 3rd Ed., Raven.]

genes that differ in sequence from the MHC genes of other individuals (except, perhaps, for some close relatives). When genes have multiple alleles without a predominant one, as the MHC genes do, they are said to be *polymorphic.* MHC polymorphism is so extensive that each individual has an identity, or "self," defined by his or her particular set of MHC genes. CTLs are tolerant toward self-MHC and therefore did not kill cells grafted between inbred strains. However, the T-cell receptor on CTLs can bind to foreign MHC molecules, allowing the CTLs to kill foreign cells and thus reject grafts.

T-Cell Subsets and MHC Classes   The MHC gene products that provide CTL self-recognition are called *class I MHC genes,* typified by a series of genes called *H-2D* and *H-2K* genes in the mouse and *HLA-A, -B,* and *-C* genes in humans. Class I MHC gene products are two-chain proteins with one highly polymorphic chain and one constant chain called *β₂-microglobulin* (Table 27-2). These MHC proteins are part of the immunoglobulin gene superfamily (see Figure 27-30). Most cells in a vertebrate's body have class I MHC proteins on their surface. The T-cell receptors on CTLs do not mediate killing of self-cells because the

TABLE 27-2   Properties of Class I and Class II MHC Molecules

	Name in Mice	Name in Humans	Chain Structure	Cells Found on	Peptide Antigens Bound		Type of T cell Recognized by
					Source of Protein	Type of Organism	
MHC Class I	H-2 and H-2K	HLA-A HLA-B and HLA-C	One polymorphic chain plus one β₂-microglobulin chain	Most cell types	Intracellular	Viruses	CTL (CD8⁺)
MHC Class II	I	HLA-D	Two polymorphic chains	Macrophages B cells	Extracellular	Bacteria	T_H (CD4⁺)

MHC = major histocompatibility complex; HLA = histocompatibility locus A; CTL = cytotoxic T lymphocyte; T_H = helper T cell.

CTLs have been selected during their thymic maturation to avoid reaction with the body's own cells, as we will discuss below. However, the CTLs will attack a self-cell if the cell's surface carries a class I self-MHC protein with a peptide from a foreign protein bound to it (see Figure 27-32). We call this a *joint recognition process*: both self-MHC and the foreign peptide must be jointly recognized on the same cell at the same time. Again, the T-cell receptor molecule on CTLs recognizes MHC and the foreign peptide together; however, neither one separately can trigger a CTL's attack response.

The T-cell receptor on T_H cells must also jointly recognize self-MHC molecules complexed with foreign cell-surface molecules. The self-molecule in this case, however, is not a class I MHC molecule but a class II MHC molecule (Table 27-2). Class II molecules also have two chains, but both are polymorphic and both are smaller than the polymorphic chain of class I molecules (see Figure 27-30).

There is a fundamental difference between the peptide-MHC complexes formed by class I and class II MHC molecules. Class I molecules bind peptides derived from *intracellular* proteins, while class II molecules bind peptides derived from *extracellular* proteins (Table 27-2). Thus peptides from viral proteins are made inside a virus-infected cell and are presented on the cell's surface in association with class I molecules (as in Figure 27-32). By contrast, when macrophages digest free-living bacteria, they present the bacterial peptides in association with class II molecules. Because most CTLs recognize class I–associated peptides, CTLs kill virus-infected cells. T_H cells, by contrast, generally recognize class II molecules and therefore help B cells make antibacterial antibodies. Also, class I molecules are found on the surface of most cells, allowing those cells to signal to CTLs when they have been invaded. Class II molecules are found mainly on macrophages, the professional digesters of foreign materials in the body, and on B cells, allowing macrophages to recruit help from T_H cells and B cells to receive their help.

The T Cell's True Antigen The true antigen for a T-cell receptor is a tight complex composed of a foreign peptide and an MHC molecule. The complex does not include an entire foreign protein, but rather a peptide derived from such a protein, which is about nine amino acids long for Class I MHC and up to 25 amino acids long for Class II. To become a T-cell antigen, a foreign protein first enters a cell and is degraded there to peptides, which are transported into the cell's endoplasmic reticulum. Here some of the peptides bind to newly-made MHC molecules; these complexes are then transported to the cell's surface. When peptides bound to MHC molecules are displayed on a cell's surface, we say that the MHC molecules are *presenting* the peptide antigen to a T-cell receptor.

The MHC molecule has a cleft into which certain peptides fit (Figure 27-34), and the MHC molecule clasps a matching peptide so tightly that the complex takes days to dissociate. Of all the peptides that could be derived from an antigenic protein, only a few can be antigenic substances for a given individual's T cells, namely those that fit well into that individual's own particular MHC molecules. Because MHC molecules vary in structure so much among members of a species, the antigenic peptides recognized by individuals also vary. Because a given peptide can be presented by the MHC of one person but not by the MHC of another, the full range of immunologic potential is carried by the species as a whole with any one individual utilizing a limited subset.

T Cells Are Educated in the Thymus to React with Foreign Proteins but Not Self-Proteins

The CTLs and T_H cells develop their receptor specificity in the thymus gland. There, the α-chain and β-chain genes rearrange—or, more rarely, γ and δ—and the T-cell receptor is expressed on the cell surface. Soon thereafter, *positive* and *negative selection* events ensue (Figure 27-35), guaranteeing that the cells that leave the thymus are poised to recognize peptides in association with the animal's own specific class I or class II MHC molecules. In fact, the specificity of T-cell receptors represents a balancing act: the T cells must recognize self-MHC well enough to react strongly when foreign peptides are in the self-MHC cleft but not when self-peptides are there.

This process of selection is very exacting: 95 percent of newborn T cells die in the thymus through apoptosis, apparently because their T-cell receptor does not pass muster. [Apoptosis, or programmed cell death, is a suicide process that can be induced in cells (see Chapter 26).] Many cells die because their T-cell receptor does not effectively recognize either class I or class II molecules of that particular animal (positive selection). Other cells are eliminated because they recognize the individual's class I or class II molecules with a self-peptide in its cleft, so that they would react with normal body cells if allowed out into the body (negative selection). The latter case is especially important because it lies at the heart of tolerance—the T cells, like the B cells, must be tolerant of self-molecules. The positive and negative selection events result from intracellular signals that regulate whether the T cell will commit suicide through apoptosis or will live on to exit from the thymus and become a peripheral, functional T cell.

Once T cells leave the thymus, their receptors are fixed; T-cell receptors do not undergo somatic diversification as happens in B cells. This makes sense because once intrathymic selection is completed, any change in the T-cell receptor would run the risk of creating uncontrollable self-reactivity.

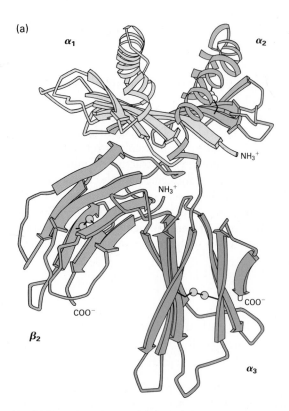

(a)

α_1 α_2

NH_3^+

NH_3^+

COO⁻   COO⁻

β_2

α_3

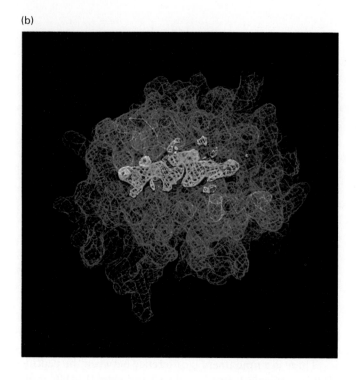

(b)

(c)

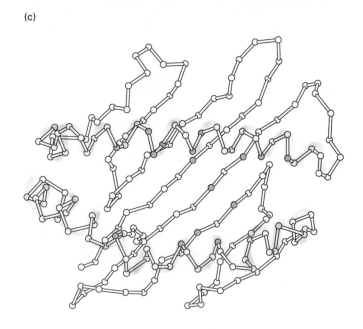

▲ FIGURE 27-34   Structure of the peptide-binding cleft of the class I MHC molecule. (a) Structure of the extracellular domain of an MHC class I molecule derived from x-ray diffraction data. The molecule is a dimer of β_2-microglobulin (green) and a three-domain α chain (blue). The α_1 and α_2 domains together form a peptide-binding channel lined on the bottom by β pleated sheets and on the sides by α helices, as shown at the top of the picture. (b) Computer-generated model of the peptide-binding site at the top of the MHC class I molecule. The pink region represents a peptide bound into the cleft formed by the α_1 and α_2 domains. (c) Stick diagram showing the cleft without a peptide, emphasizing that the walls are formed by α helices and the floor consists of a β pleated sheet. Amino acids (orange) that affect the molecule's antigen-binding capacity are clustered in the cleft. [Part (a) after original drawing by M. Silver, from P. J. Bjorkman et al., 1987, *Nature* **329**:506; part (b) courtesy of D. C. Wiley et al.; part (c) adapted from H. M. Gray et al., 1989, *Sci. Am.* **261**(5):64.]

The selection process is carried out at a stage when the intrathymic T cell has already committed itself to the CTL or T_H lineage. Initially, all T cells carry both CD4 and CD8 surface markers, and part of a T cell's commitment involves the decision to keep one marker type and downregulate the other, thus ensuring that CTLs are CD8⁺ and T_H cells are CD4⁺ (see Figure 27-13).

T Cells Respond to Antigen by Either Killing Cells or Secreting Protein Factors

The helper and killing responses of T_H cells and CTLs are quite similar: both rely on the secretion of specific proteins.

T_H cells respond to antigen by secreting *lymphokines*, factors that stimulate both T-cell and B-cell growth and maturation. Lymphokines function through specific receptors on the surface of target cells. Not all T_H cells secrete all factors: different subsets (T_H1 and T_H2) orchestrate responses to different types of infections by the particular lymphokines they secrete. As shown in Table 27-3, individual lymphokines can have multiple actions, and a given response can be elicited by more than one lymphokine. There are many lymphokines known and more to be discovered. Many of the lymphokines are called *interleukins* (abbreviated IL); one of the first to be discovered was IL-2, known then as T-cell growth factor.

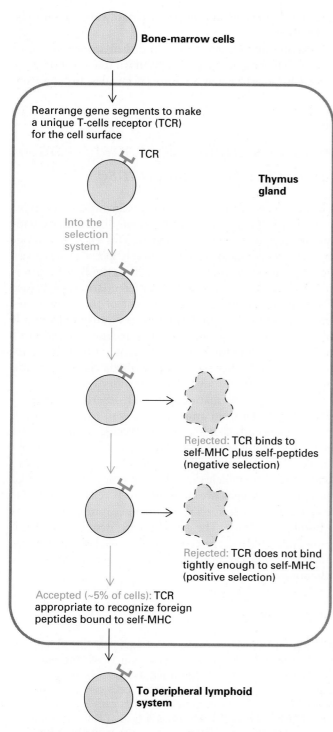

Bone-marrow cells

Rearrange gene segments to make a unique T-cells receptor (TCR) for the cell surface

TCR

Thymus gland

Into the selection system

Rejected: TCR binds to self-MHC plus self-peptides (negative selection)

Rejected: TCR does not bind tightly enough to self-MHC (positive selection)

Accepted (~5% of cells): TCR appropriate to recognize foreign peptides bound to self-MHC

To peripheral lymphoid system

▲ FIGURE 27-35   Schematic representation of the role of the thymus gland in selecting T cells able to bind foreign peptides but not self-proteins. The cells enter the selection process after they have rearranged their T-cell receptor genes to generate a single T-cell receptor (TCR) on their surface. The selection system then rejects T cells with TCRs that react too well with self-proteins (negative selection) and T cells that cannot recognize a self-MHC molecule well enough for it to be a useful carrier of foreign peptides (positive selection). The rejected cells die by apoptosis and only the accepted cells exit the thymus.

TABLE 27-3   Some Lymphokines Secreted by T_H Cells in Response to Antigen*

Lymphokine	Function
IL-2	T-cell growth stimulation Stimulation of antibody secretion by B cells
Interferon γ (IFNγ)	Macrophage activation Ig class switching by B cells
IL-4	B-cell activation Ig class switching by B cells T-cell growth stimulation Macrophage activation
IL-3	Stimulation of blood-cell differentiation in many lineages
GM-CSF	Stimulation of growth and differentiation of granulocytes and macrophages B cells Many other effects on nonblood cells
IL-7	Stimulation of pre-B and pre-T cell growth

*This is a very partial list of factors and their effects; IL-16 is already known.

The lymphokines have multiple effects on B cells, as described later. They also stimulate T cells, particularly T_H cells. Actually, the activation of a T_H cell by an encounter with a peptide-MHC complex induces the formation of both IL-2 receptor and secreted IL-2; this leads to a self-stimulation known as an *autocrine growth response*. Lymphokines also stimulate other blood cells, particularly phagocytes such as macrophages and granulocytes. Some skin-hypersensitivity reactions, such as poison ivy and the skin test for reactivity to tuberculosis antigens, are due to the mobilization of phagocytes by T_H cells. Such reactions are called *delayed-type hypersensitivity* (DTH) because they take hours or days to develop, in contrast to allergic reactions (to insect stings, for instance), which develop very quickly.

CTLs respond to antigen in two ways: one is by secreting proteases and proteins able to form channels in the target-cell membrane. These proteins probably activate apoptosis in the target cell. The second mode of CTL killing is by activation of the FAS receptor on a target cell, which signals the activation of the apoptosis pathway. When a cell is attacked by a CTL, it appears to burst and then shrink. One CTL can kill many targets. CTLs also secrete certain lymphokines along with their killer proteins.

The surface CD4 and CD8 molecules play important roles in the T cell's helper or killing functions as demonstrated by the ability of antibodies to CD4 and CD8 to prevent T cells from responding to antigen. The interaction between antigen and T-cell receptor requires the intimate participation of either a class I MHC–CD8 interaction or a class II MHC–CD4 interaction (Figure 27-36).

When a T cell interacts with a cell containing an MHC-peptide complex recognized by the T-cell receptor, other surface molecules on the T cell and the cell it recognizes also interact, tightening the association and strengthening the intracellular signals that result from a T cell's encounter with antigen. These signals are mainly the consequence of activities of accessory transmembrane proteins tightly bound to the T-cell receptor. One, called the CD3 complex, consists of γ, δ, and ϵ components; there is also a critical ζ chain (see Figure 27-36). The signalling process first becomes evident as a transfer of phosphoryl groups from ATP to tyrosine residues in the cytosolic tails of the accessory proteins. The kinase involved in this phosphorylation on tyrosine is either the protein-tyrosine kinase, encoded by *lck* and bound to both CD4 and CD8, or one encoded by *fyn*. The accessory proteins' cytosolic tails have dual sites for the binding of SH2 regions after phosphorylation on tyrosine, and they recruit from the cytoplasm another tyrosine kinase, Zap70, which has dual SH2 regions. The phosphorylation of other proteins and the activation of Ras ensue, but the detailed pathways are yet to be understood.

Having introduced T cells and their unique mode of recognizing antigen, we can now return to considering the events that ensue when a pathogen invades an organism.

► *The Antigen-Dependent Phase of the Immune Response*

A key event of the immune response is an antigen's encounter with a B lymphocyte that bears surface antibody with binding specificity for that antigen. This encounter, coupled to a coordinated response from helper T cells, causes the virgin B cell to mature further; thus the B cell that resulted from antigen-independent differentiation (see Figure 27-27) becomes an active component of the immune response system. Many molecular events transpire from a B cell's encounter with antigen. They include activation of the B cell's proliferation, its secretion of antibody, production of IgG and other secondary antibody classes, somatic mutation, and the production of memory cells.

The activation, maturation, and differentiation processes that ensue from an encounter with antigen, as well as the collaboration of T cells and B cells, all occur in the secondary lymphoid tissues: the lymph nodes, Peyer's patches, and spleen (see Figure 27-14). These tissues are subdivided into follicles in which immune reactions take place. Here antigens are presented to B cells on *follicular dendritic cells,* cells specialized for the capture of antigen-antibody complexes. The actual events, described in detail below, take place in the center of the follicles in structures called *germinal centers,* where the various cell types can collaborate.

◄ FIGURE 27-36   Interaction of a peptide-MHC complex on a presenting cell with a receptor on a T cell. The presenting cell can be a macrophage using a class II MHC or any tissue cell using a class I MHC. The T cells can be T$_H$ cells carrying the CD4 surface protein or CTLs carrying CD8; the class I MHC binds CD8, and class II binds CD4. The T-cell receptor not only has CD8 or CD4 as components of its interaction system, it also has four accessory proteins—three CD3 components (γ, ϵ, δ) and ζ—that are involved in sending signals that modify the cell's behavior and growth following the T cell receptor's binding to MHC plus peptide. The accessory proteins can bind to the protein-tyrosine kinase Zap70 when their cytosolic tails are phosphorylated. The CD4 or CD8 protein is bound to the *lck*-encoded protein-tyrosine kinase, which is thought to phosphorylate a CD3 component on tyrosine when the T-cell receptor interacts with the MHC-peptide complex, thus helping to initiate the signal transduction process.

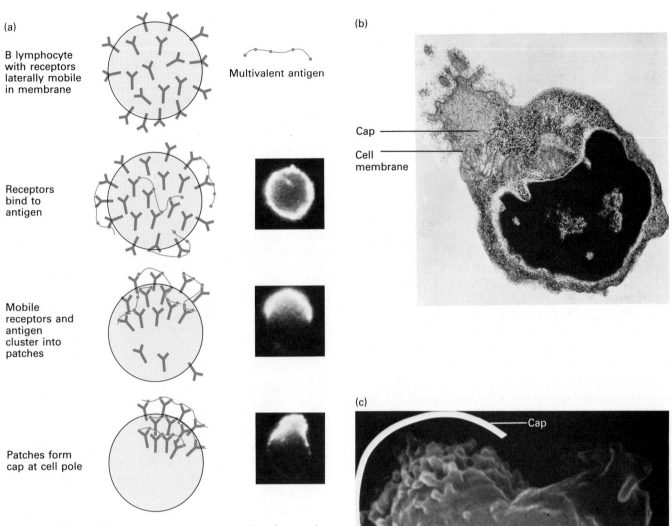

(a)

B lymphocyte with receptors laterally mobile in membrane

Multivalent antigen

Receptors bind to antigen

Mobile receptors and antigen cluster into patches

Patches form cap at cell pole

(b)

Cap

Cell membrane

(c)

Cap

▲ FIGURE 27-37   Patching and capping of surface antibody molecules. (a) Schematic representation of the aggregation of surface antibody and experimental demonstration by immunofluorescence. The diagram illustrates a multivalent antigen linking together surface antibody molecules on a B lymphocyte. The end result is a cap at one place on the surface of the cell. With the use of fluorescence-labeled antibody that can react with the surface antibody, the process of patch and cap formation is visually demonstrated. (b) Transmission micrograph of a sectioned B lymphocyte with its cap region evident at the left. Within the cell, the cap region is clear of organelles because it is filled with actin and myosin that move to the cap region along with the surface antibody. (c) The cap is shown in a scanning electron micrograph of a B lymphocyte. [Photographs courtesy of J. Braun.]

B-Cell Activation Involves B Cell–T$_H$ Cell Collaboration

After the B lymphocyte has developed into a virgin B cell and has been expelled from the bone marrow into the periphery, it will live for only a few days as a circulating cell unless an antigen interacts with its surface antibody, triggering further growth and maturation. That antigen will generally be a soluble protein or a microorganism. The immune response is most effective against *multivalent antigens*—those with multiple antigenic determinants on one molecule—so that many antibody molecules on one B cell can be bound together to form a tight patch, or aggregate, on the B cell surface (Figure 27-37). The surface antibody can be aggregated by multivalent antigens because of the fluidity of the lipid-bilayer plasma membrane in which the

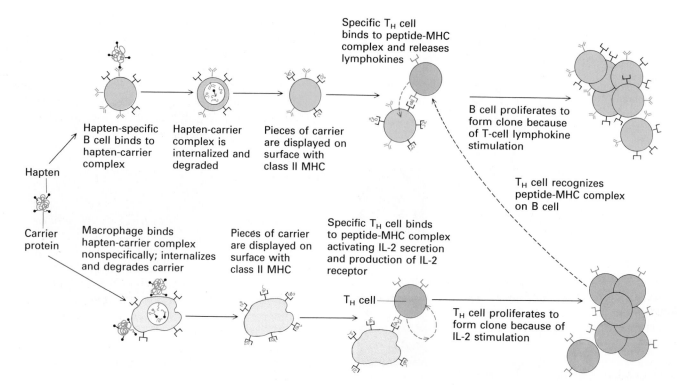

▲ FIGURE 27-38   B-lymphocyte activation. A hapten-carrier complex activates B cells by first stimulating T_H cells. The hapten-carrier complex is an intact protein to which several hapten molecules have been covalently coupled. The hapten portion binds to a B cell but cannot by itself initiate B-cell proliferation. The B cell internalizes the hapten-carrier complex and digests it, and peptide portions of the carrier are displayed on the B-cell surface as a complex with class II MHC protein. A macrophage cell also internalizes carrier protein and displays peptide fragments on its surface in association with class II MHC. The T_H cells with peptide-specific receptors bind to the peptide-MHC complex displayed on the macrophage surface. This binding stimulates the T_H cells, which then proliferate, recognize the identical peptide-MHC complex on the B cell, and secrete factors that stimulate the B cell to grow.

antibody is embedded. After islands of patches form on the cell surface, these coalesce into a cap. Ultimately, the cap is either shed into the surrounding fluid or internalized by endocytosis and degraded. It is generally said that a competent antigen *cross-links* surface antibody; this cross-linking, or aggregation, appears to be a key to the activation of a B cell. Hence a polymeric antigen, such as a carbohydrate with repeating sugars, can be a very good activator.

The achievement of B-cell activation requires the help of T_H cells. The collaboration of specific T_H cells and specific B cells induces the B cells to proliferate and secrete specific antibody (Figure 27-38). The key to this process is that a B cell can degrade a protein to peptides and display them on its surface in association with MHC class II molecules, just as a macrophage can. However, the macrophage processes all foreign proteins indiscriminately, whereas the B cell can endocytose and process specifically only those proteins that bind to its surface receptor. Thus when the B cell binds a hapten-carrier complex, it internalizes the complex by endocytosis and then expresses the carrier peptides on its surface along with class II MHC molecules. The T_H cells previously stimulated by macrophage-bound carrier

peptides will then recognize the peptides on the B-cell surface and become further stimulated. In this interaction, the lymphokines that the T_H cell produces are *B-cell growth and differentiation factors* (see Table 27-2), which activate the B cell to proliferate.

This mechanism explains why haptens must be bound to carriers to be antigenic. Free hapten could bind to B cells, but it cannot activate them for two reasons: haptens lack multivalency, and they cannot stimulate T_H cells.

Hapten-protein conjugates made in the laboratory, such as dinitrophenol coupled to bovine serum albumin, are surrogates for natural proteins, which are constructed solely of amino acids. Natural proteins can act as both haptens and carriers because they possess a variety of antigenic determinants. In proteins, the "haptens" are the antigenic regions of the protein that interact with the surface antibody on B cells; the "carriers" are those regions whose peptides interact with T-cell receptors on T_H cells. Self-proteins will fail to activate T cells, so that if a B cell that can interact with a self-protein should arise, say by somatic mutation, it will not be dangerous because it cannot stimulate T_H cells and therefore cannot develop into a clone of self-reactive cells. In fact, a B cell may be rendered entirely unreactive if it is stimulated by antigen in the absence of T_H-generated help.

Activation Is the Result of Cellular Stimulation

It is evident that antibody secretion requires the activation of two kinds of cells, B cells and T_H cells. Activation is the result of the intracellular signalling pathways that are stimulated by encounters of lymphoid cells with antigens or lymphokines. The word *activation,* as used here, is actually a technical term: it means the transformation from a small, nondividing lymphocyte to a larger cell called a *blast.* The blast is metabolically much more active than the small lymphocyte: it enters the division cycle and divides every 8–24 h, it actively secretes important substances, it has receptors for lymphokines, and the B cell blast can mature to become an antibody-secreting plasma cell.

Activated T_H cells secrete many substances, the most important for this discussion being the lymphokines IL-2, IL-4, IL-5, and IL-6. These factors find receptors on activated B cells and stimulate B-cell growth and maturation to plasma cells; the receptors are not found on resting B cells. In the joint activation of B cells and T_H cells by antigen, only B cells and T cells specific for the antigen are activated because each cell must recognize the antigen in order to develop the receptors for the growth factors. Bystander cells that have not seen antigen are not activated.

The stimulation of T cells through their T-cell receptors and accessory molecules was described earlier (see Figure 27-36). B-cell activation through the stimulation of its surface antibody receptor is a similar process. In this receptor, surface antibody is noncovalently linked to a complex similar to the T-cell receptor's CD3 complex. The B-cell complex consists of two accessory proteins, known as Ig-α and Ig-β that have tails very like those on the T-cell accessory molecules. Signalling involves a bound protein-tyrosine kinase and a recruited one called Syk that is a close structural relative of Zap70. Another protein-tyrosine kinase, called Btk, is also involved in signalling, but its function is still uncharacterized. Btk is known to be needed for the development of B cells in humans because its mutation leads to a serious immunodeficiency disease called Bruton's agammaglobulinemia.

Antibody Secretion by Activated B Cells Entails Many Cellular Changes

A massive increase in the synthesis of immunoglobulins accompanies the activation of B cells. This increase is a consequence of an increased synthesis rate of immunoglobulin-specific transcripts, more efficient processing to mRNA, and stabilization of the mRNAs. To accommodate the increased rate of synthesis of secreted proteins, the activated B cell develops a larger cytoplasm with more endoplasmic reticulum. The end stage of this maturation is a plasma cell that secretes, as antibody, up to 10 percent of the protein it makes. The plasma cell is highly specialized for secretion, with a large Golgi region as well as a large, layered endoplasmic reticulum (see Figure 27-10). Since the plasma cell is the final stage of B-lymphocyte development, it loses its proliferative ability entirely and dies after some days of antibody production.

The Activation Process Produces Both Plasma Cells and Memory Cells

Thus two consequences of activation are the proliferative response and the cellular maturation response. Maturation, characterized by the increased synthesis of immunoglobulins and a switch to secretion, not only requires the just-discussed changes in cell architecture (see Figure 27-10), but also involves the synthesis of many new proteins and the inhibition of synthesis of others. For instance, J-chain synthesis (needed for the polymerizing of secreted IgM) is turned on by activation. Cell surface markers also change after activation. The end result of B-cell activation is a *plasma cell.*

Not all progeny of an activated B lymphocyte are plasma cells. The other important product is the *memory B cell.* These are cells that retain, often for the life of the animal, a record of the antigens that it has previously encountered. Memory cells preserve variable regions that have previously proved useful, so that a second encounter with an antigen can elicit a rapid, highly avid response. Like virgin B lymphocytes, memory cells carry on their surfaces the antibody they are programmed to make. Unlike virgin B cells, which live for about a week, memory cells last for years, circulating as relatively quiescent cells. They may be maintained by residual antigen stably bound to follicular dendritic cells. A new exposure to antigen activates them just as it does virgin cells. Since memory cells become most active in such a secondary immune response, we will meet up with these cells again later in our discussion. There are also *memory T cells* that circulate, and memory B cell responses are helped by specific memory T_H cells which also persist after a primary immune response. Memory cells are differentiable from virgin cells by their surface antigens; CD44 is a particular cell-surface marker for memory T and B cells.

The Change from Surface Antibody to Secreted Antibody Requires the Synthesis of an Altered H Chain

One of the necessities of clonal selection is that a virgin B lymphocyte, after its activation by an encounter with antigen, must secrete antibody with exactly the binding specificity of the antibody it carries on its surface. This means that the variable regions of the L and H chains must be the same in the virgin cell and the secreting cell, and therefore the same variable-region DNA must be used for creating both forms of antibody. To switch to the secreted form requires a change in the membrane-binding portion of the surface antibody's H-chain constant region without any alteration of the variable region. The constant region of the L chain need not be transformed because it binds to the H chain, not directly to the B-cell membrane. The alteration in the membrane-binding segment at the C-terminal end of the H chain is accomplished by a change in the type of RNA transcript made by the cell's DNA. An exact understanding of this requires a closer look at the H-chain's constant region.

The form of immunoglobulin on the surface of a virgin B cell is *membrane-bound IgM*. The IgM previously described as a pentamer of the basic four-chain immunoglobulin molecule (see Figure 27-6) is *secreted IgM*. The membrane-bound form is a monomer, a single four-chain unit; it is held to the membrane by a hydrophobic sequence at the C-terminal end of each of its H chains. Eight exons contribute to specifying the structure of the μ H chain of membrane-bound IgM (the μ_m chain): a leader exon, a variable-region exon, four exons that encode the four domains of the constant region, and two exons that encode the membrane-binding segment of the molecule (Figure 27-39). The mRNA for the μ_m chain is polyadenylated just beyond the exon specifying its C-terminus. Splicing brings together the exons to generate the mature mRNA for μ_m.

The C-terminus of μ_s, the secreted form of the μ chain,

is encoded by a segment of DNA contiguous with the sixth exon (see Figure 27-39). The mRNA for μ_s is polyadenylated just after the coding region for μ_s, so that the two terminal exons specifying the μ_m C-terminus are excluded. Consequently, an intron sequence of μ_m becomes an exon sequence of μ_s; thus the two mRNAs have different 3' structures, generating different C-terminal ends on the proteins.

The detailed structures of the μ_s and μ_m C-terminal ends are just what would be predicted from the behaviors of the two proteins. The μ_m terminus includes a string of 26 uncharged amino acids preceded and followed by clusters of charged residues—a classic membrane-spanning region (see Figure 14-14). The secreted form lacks a membrane-spanning region but includes a cysteine residue that becomes the disulfide bridge linking the four-chain monomers into the pentamer of the secreted molecule. It also has a site for carbohydrate addition, presumably as a way of making the molecule more soluble.

Because the mRNAs for μ_m and μ_s differ in their site of polyadenylation, it is thought that control over polyadenylation is the basis for the switch from μ_m mRNA to μ_s mRNA when a B lymphocyte is activated (see Figure 27-39). This is the most obvious hypothesis but not necessarily the correct one. Some evidence exists for premature termination of transcription as a mechanism of control, and some control may be a consequence of differential protein stability.

The synthesis of μ_m and μ_s by alternative use of the same DNA sequence ensures the identity of the variable

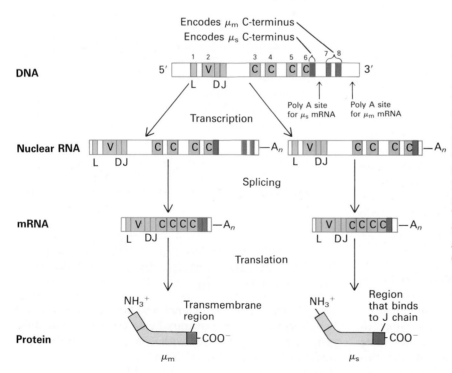

◄ FIGURE 27-39   Production of μ_m and μ_s chains by alternative polyadenylation of mRNAs. The genome organization in the C_μ region includes two polyadenylation sites. When one is used, the mRNA formed by splicing the nuclear RNA precursor has the coding region for the μ_s C-terminus. The other polyadenylation site is farther toward the 3' end and includes two additional exons that encode the μ_m C-terminus. When this 3' site of polyadenylation is used, the μ_s coding region is spliced out but the μ_m region is included. Thus, the C_μ region can encode both surface antibodies (with the μ_m C-terminus) and secreted antibodies (with the μ_s C-terminus).

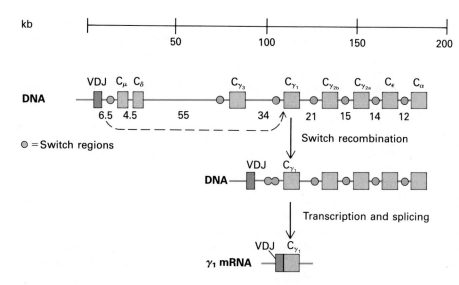

Organization of the H-chain region of DNA and switch recombination. The positions of the H-chain constant regions are indicated. A permanent change in H-chain synthesis from μ to a downstream constant region is produced by recombination between switch regions (green) located upstream of each of the constant regions. The constant regions are composed of multiple exons, and each region has alternative secreted and membrane-binding C-termini, but these details have been omitted. The numbers between units indicate the distances in kilobases.

region on the two proteins and thus is a central mechanism for allowing the clonal selection process to operate in the immune system. This pattern of alternative transcripts derived from one region of DNA is widespread in the immune system. For instance, there are two types of virgin B lymphocyte, a less-mature one with IgM on its surface and a more mature one with both IgM and IgD. IgM and IgD molecules have identical variable regions and their L chains are the same. They differ in their H chain constant regions as a consequence of a different splicing and polyadenylation pattern that utilizes for IgD a set of constant regions (C_δ) located 3′ to the constant regions of IgM (Figure 27-40).

Antibody Class Switching Also Requires an Altered H Chain

The antibody classes IgG, IgE, and IgA play no role in the earliest stages of an immune response because they are not found on the surface of virgin B cells. For this reason they are referred to as the *secondary antibody classes*. Activated cells *switch* from IgM/IgD synthesis to the synthesis of these secondary classes. Since switching is an event that follows the activation of a B lymphocyte, the progeny cells of a single activated B cell can differ in the classes of antibody they make. When a B cell encounters an antigen, it initially makes IgM, but then secondary antibody classes, such as IgG, begin to predominate (Figure 27-41).

Because the various classes differ in their H chains and not their L chains, it is an *H-chain switch* that underlies the ability to change antibody-class synthesis. Analogous events described earlier, where cells switch from μ_m to μ_s synthesis (see Figure 27-39) and from synthesis of IgM

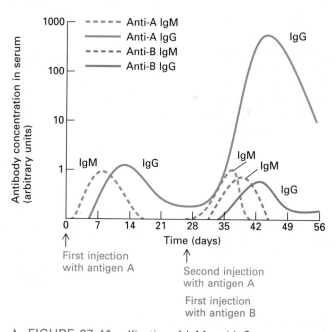

▲ FIGURE 27-41   Kinetics of IgM and IgG responses after primary and secondary immunization. The graph expands upon that in Figure 27-11, which indicated only total antibody concentrations. Here the initial immune response to an antigen injection is seen to change with time from an IgM-dominated response to an IgG-dominated response. Once the initial response has subsided, a second injection with the same antigen leads to a prompt and massive IgG response because of the large number of memory cells prepared to make IgG. The secondary response also includes an IgM response, probably emanating from virgin B cells that are always present and prepared to be activated. Antigen B, introduced during the second injection, gives the IgM-to-IgG pattern of a primary response.

alone to IgM-plus-IgD synthesis, are due to changes in the processing of RNA transcripts. In contrast, synthesis of the secondary antibody classes requires DNA rearrangements called *switch recombinations*. In a switch recombination, regions of DNA recombine with each other, deleting an intervening segment.

The mechanisms of switching are inherent in the organization of the H-chain locus (see Figure 27-40). Previously we focused on the variable-region components and the C_μ region (the region encoding the six C_μ exons). Downstream of C_μ and C_δ are segments that encode the other constant regions: the C_γ's, C_ϵ, and C_α, in that order. Because there are four IgG subclasses and thus four C_γ's, there are a total of eight different C regions. Each C region is made of multiple exons that encode the domains of the individual H chains. Every C region (except C_δ) has an associated switch region somewhat 5' of its coding exons. The switch regions are constructed from internally reiterated short sequences that can recombine with each other. As a result of switch recombination, the VDJ unit is brought close to a new C region (see Figure 27-40). The DNA is then polyadenylated just beyond that C region, establishing that the cell will make only a single class of antibody.

The mechanism of switching involves two steps. First a switch region is activated by transcription from a local promoter. Second, the switch recombination occurs through DNA scissions in the switch regions and physical recombination of the DNA. The signals that activate and direct the switching process in B cells appear to be the lymphokines secreted by T cells. For instance, IL-4 can induce cells to switch to γ_1 or ϵ synthesis, IL-5 induces a switch to α synthesis, and γ-interferon induces a switch to γ_{2a}. The lymphokines probably control switching by inducing localized transcription within the H-chain constant-region DNA. The specificity of switching (to γ, ϵ, or α) is ultimately a function of the antigen: skin parasites, for example, induce IgE (ϵ chain) while intestinal microorganisms induce IgA (α chain). Which lymphokines are induced by a particular pathogen is a consequence of whether T_H1 or T_H2 cells are activated by the pathogen because each T_H subset makes a separate constellation of lymphokines.

Memory Cells Participate in the Secondary Immune Response

Memory cells usually are programmed to carry on their surface one of the secondary antibody types. When memory cells respond to antigen, they make antibodies of the class represented on their surface. Because they have already been through many of the maturation stages of an antibody response, a second encounter with antigen elicits a very rapid response of synthesis of a secondary antibody (Figure 27-41). This response is also of much greater magnitude than the initial response. If, during a second immu-

nization with an antigen, a new antigen is also introduced, the animal undergoes a memory-dominated (secondary) response to the previously encountered antigen and an IgM-dominated "naive" (primary) response to the new antigen (see Figure 27-41). It is thus clear that immunologic memory is specific for only those antigens previously encountered.

Somatic Mutation of Variable Regions Follows from Activation

We have described the variable regions of antibodies as arising from combinatorial joining events among V, D, and J regions. We have then followed the fate of the joined VJ and VDJ regions through transcriptional alterations and switching events without discussing how the variable regions themselves are altered. After a B cell encounters antigen, a mechanism comes into play that increases the diversity of variable regions; it is called *somatic mutation*. This mechanism involves the replacement of individual bases in a joined VDJ segment with alternative bases, causing apparently random variation throughout the VDJ segment (and also in flanking DNA for some distance on either side of the VDJ). Somatic mutation has been documented in both L-chain and H-chain variable regions by comparing germ-line DNA sequences with expressed DNA gene sequences (Figure 27-42).

The changes in amino acid sequence caused by somatic mutation have the consequence of varying the fit between the antibody and the antigen, changing the affinity of the antibody for the antigen. Somatic mutation throughout the variable region should have its greatest effect in the CDRs, where mutation can lead to higher affinity of binding. In fact, somatic mutations that cause amino acid replacements are especially prevalent in the regions encoding the three CDRs (see Figure 27-42).

It is thought that the higher the affinity of the surface antibody for antigen, the easier is activation of B lymphocytes by antigen, and that therefore there is a continual selection among the somatically mutated variable regions in the germinal centers for cells bearing higher-affinity antibody. Because antigen is continually selecting for affinity, these events represent a continuation of the process of clonal selection.

The rate of somatic mutation in germinal center B cells has been estimated to be as high as 10^{-3} per nucleotide per cell generation, a rate 10^6-fold higher than the spontaneous rate of mutation in other genes. Such a high rate of mutation over the whole genome would be insupportable. There must exist mechanisms that direct mutational activity to the variable-region sequences. How these might work is not known; probably a sequence in the area of the variable region directs a special enzyme system to carry out point replacements of nucleotides independently of template specification.

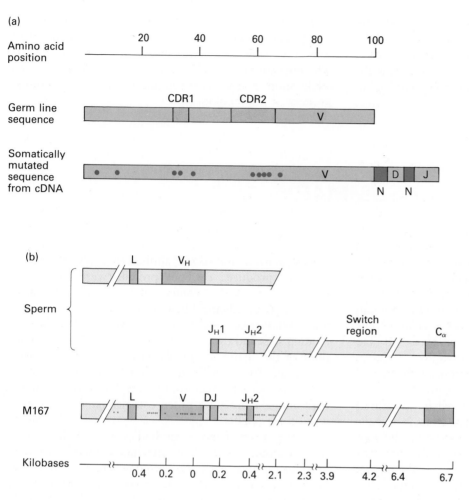

◄ FIGURE 27-42   Somatic muta-
tion of rearranged V_H regions. (a) Comparison of the nucleotide sequence of
a germ-line V_H region with the
sequence of the cDNA for an H-chain
variable region that has undergone
somatic mutation. The scale is given as
amino acid positions, but 10 nucleotide
changes are indicated (dots). [See A.
Bothwell et al., 1981, *Cell* **24**:625–
637.] (b) Comparison of germ-line
(sperm) DNA sequences and somatic
mutations found in H-chain genomic
DNA from a myeloma (M167). The
consequences of somatic mutation are
evident as alterations in nucleotide
sequence, indicated as dots. Somatic
mutation is localized to the variable
region and a small region around it.
[After S. Kim et al., 1981, *Cell* **27**:573.]

The consequence of somatic mutation is an increase in the kinds of potential antibodies an animal can make, from the 10^{11} we calculated earlier to a number many orders of magnitude higher. The greatly increased diversity will include not only better-fitting antibodies but also, and more frequently, poorly designed antibodies. Somatic mutation occurs during the proliferative response following an encounter with antigen, and we assume that if a cell making a poor antibody is produced, it stops dividing and dies, so that only cells with appropriately designed surface antibody are continually stimulated to grow and mature. Darwinian selection processes continue from the first moment of antigen encounter until the cell ceases somatic mutation.

Tolerance Is Achieved Partly by Making B Cells Unresponsive

Nothing thus far has explained how B cells avoid making antiself antibodies. In fact, the somatic mutational process could easily produce self-reactive clones. Part of the tolerance to self antigens is that T_H cells are tolerized to the carrier portion of antigens and therefore do not provide the help needed for antibody synthesis. However, B cells are also directly tolerized in two different ways. B cells can be tolerized by deletion in a process akin to that which tolerizes T cells. For example, a mouse transgenic for synthesis of an antibody to a particular MHC molecule will have all of its transgenic B cells eliminated if it also carries that particular MHC molecule.

The second form of B cell tolerance is called B cell *anergy*. A cell is said to be *anergic* if it gives no activation response when exposed to an antigen that binds to its surface immunoglobulin. B cell anergy as a mechanism of tolerance was demonstrated by an experiment that used transgenic technology in a particularly elegant manner. For that experiment, two types of transgenic mice were created, one making the enzyme hen-egg lysozyme and the other making high-affinity antibodies to hen-egg lysozyme. Progeny of crosses between these mice have both the antigen and the ability to make an antibody to it, but they display tolerance. Because they have been exposed from early in life to the hen-egg lysozyme, it is treated as self, and despite the presence of the genes for the antibody in all of the animal's B cells, little antibody is made. B cells are present but not secreting antibody in these mice. Thus, the autoreactive B cells were not eliminated but rather were prevented from reacting by induction of anergy. How the activation of B cells is prevented remains to be learned, but the nonreactivity is a property of the B cells themselves since, if they are transferred to a normal recipient mouse, they remain nonresponsive to antigenic stimulation for some time even if effective T cells are provided. Exactly why one situation leads to deletion of self-reactive B cells while another leads to B cell anergy is another unsolved mystery of immunology.

SUMMARY

The immune system evolved as the body's protective mechanism against invasion by pathogens such as viruses, bacteria, and fungi. There are two arms to the system: humoral immunity mediated by soluble protein antibodies in bodily fluids, and cellular immunity mediated by circulating cells. B lymphocytes make antibodies; T lymphocytes and macrophages carry out cellular immune reactions.

Each antibody molecule has two identical binding sites that can bind to a specific antigen. The binding reaction is a simple, reversible binding characterized by an affinity constant. Antibody molecules also have effector domains that allow the body to rid itself of the antigen. There are five major antibody classes with different effector activities: IgM, IgD, IgG, IgE, and IgA.

The mechanism by which antibodies are formed is called clonal selection. The system works by producing an enormous variety of B lymphocytes, each one with a homogeneous population of cell-surface antibody molecules. An antigen selects, from the population of antibody-bearing B cells, those that carry molecules able to bind to it. Such B cells are activated to multiply and mature into antibody-secreting cells. Some activated B cells mature into plasma cells, which are specialized for immediate antibody production; others mature into long-lived memory B cells, which respond rapidly in further encounters with the antigen.

There are two major types of T cells: cytotoxic T lymphocytes (CTLs) and helper T lymphocytes (T_H cells). The CTLs (also known as killer T cells) recognize target cells with their surface antibody-like receptors and cause those to die. The T_H cells assist B cells in their reaction to antigens.

The ability of an animal's immune system to avoid reacting to the molecules in its own body is called tolerance. A failure of tolerance leads to autoimmune disease, one of a number of disease types that can be caused by the immune system. Other diseases come from failures of immune function, tumors of the immune system, and hyperreactive conditions such as allergies.

Our knowledge of the details of the structure and synthesis of antibodies is heavily dependent on studies of myelomas, tumors that secrete antibodies. From the study of myeloma products, antibody proteins were found to be four-chain molecules, composed of two identical heavy chains and two identical light chains. Each chain is folded into a number of compact domains that form the antigen-binding sites and the effector regions. The N-terminal portions of the chains are highly variable in amino acid sequence, producing the binding variability of the molecules. Within the variable regions, three regions directly interact with antigen; these are called complementarity-determining regions (CDRs).

The genes that encode antibodies are bipartite in structure: constant segments are attached to members of a library of variable segments. Antibody genes are not carried intact in the genome but rather are carried as gene segments that come together during lymphocyte differentiation to form the variable regions. The joining process is signaled by a DNA recognition element consisting of a heptamer and a nonamer separated by about 11 or 23 nucleotides—that is, by either one or two turns of the DNA helix. The joining rule is that a one-turn element always joins to a two-turn element. The rearrangement of DNA to make antibody genes is an orderly process; first heavy-chain genes, then light-chain genes are rearranged.

The diversity of antibody structure is partly a consequence of the large size of the variable-segment libraries and partly a consequence of combinatorial diversity in the joining of members of the libraries. Imprecision at the joints between segments further increases diversity, as does the insertion of random-sequence elements into the third CDR during heavy-chain gene formation. The variability inherent in the system can make as many as 10^{11} different molecules. Thus the immunodifferentiation process produces a vast array of cells from which antigens can choose those that fit best.

The T-cell classes carry out their function using an antibody-like T-cell receptor that is similarly encoded, but by a set of genes quite separate from those that encode antibodies. The receptor is a two-chain molecule on the T-cell surface, each chain having a variable and a constant region. Because T-cell receptors remain membrane-bound and T cells recognize antigens on the surface of other cells, T-cell recognition is a cell-to-cell recognition process. The structure on a target cell recognized by the T-cell receptor is not the foreign antigen itself but peptides derived from the foreign protein. These peptides are approximately nine amino acids in length and are bound to self-proteins called MHC (major histocompatibility complex) proteins; the T-cell receptor binds to the MHC-peptide complex. Thus T-cell recognition is a joint process of binding to both self- and foreign protein molecules.

The activation of a B cell by an encounter with an antigen is the key process in successful clonal selection. Most antigens are unable to activate B cells directly and activation requires help from T helper (T_H) cells. The antigens are first degraded by macrophages to peptides which are transported into the endoplasmic reticulum. Within the endoplasmic reticulum of these phagocytic cells are MHC class II proteins, which bind peptide fragments of the anti-

gen and carry the bound peptides to the cell surface for presentation to T_H cells.

Because surface antibody and secreted antibody differ, respectively, by having or lacking membrane-binding domains in their heavy chains, alternative exons encode the C-terminal ends of the heavy chains. Part of the activation of antibody synthesis involves the increased use by the cell of the secretion exon by a process of differential polyadenylation and splicing.

The process of B-cell activation often leads also to a switching process by which progeny cells make antibodies with different constant domains than those of the parental cells, and therefore with different properties. Furthermore, activation involves an extensive process of specific somatic mutation of the heavy- and light-chain variable regions. Thus the clone of cells resulting from activation actually varies extensively from the initially virgin cell. Continual Darwinian selection acts to maintain those cells that make antibodies with the highest affinity for antigen.

REVIEW QUESTIONS

1. The immune system is responsible for recognizing invaders in an organism. What are the three general ways in which the immune system accomplishes this task?

What is an antigen? What is a determinant? What is meant by the phrase multivalent antigen? What is a hapten? What forces allow antibodies and antigens to bind together? What are the two binding domains of an antibody, and what is the function of each? All antibodies consist of two classes of polypeptide chains. What are they? What is meant by the word bivalent as it applies to an antibody?

What is the difference between affinity and avidity? What is Scatchard analysis, and how is it used to determine the number of binding sites on an antibody molecule?

What are the five classes of antibodies? Describe the structural and functional characteristics of each.

B lymphocytes are responsible for the synthesis of antibodies? What causes a B cell to mature into a plasma cell?

a. Refer to Figure 27-2, which presents an important point about antigenic determinants. Often, precipitation with an antibody is used as an experimental procedure to remove or isolate a molecule. Which would be more easily precipitated with an antibody, a univalent or multivalent antigenic molecule? Explain your answer.

b. You are trying to characterize an antibody you have produced. The data you have obtained are in the table; use them to determine the number of binding sites and the association constant for the antibody.

r	c (μmole/liter)
0.10	0.250
0.50	1.32
0.75	2.03
1.00	2.78
1.50	4.41

c. Based on your results from question b, what class of antibody do you think you have obtained?

2. What is meant by the statement "The immune system has extraordinary plasticity"? What are the suppositions behind the clonal selection theory? Briefly describe how an antigen selects cells with the correct specificity; be certain to emphasize the two phases in the life of a B cell.

An extremely important capability of the immune system is memory. What does this refer to? What are the differences between the primary and secondary responses of the immune system to an antigen?

a. You are using a rabbit to produce antibodies, and you inject it with the hapten-protein conjugate dinitrophenyl-albumin (DNP-albumin), where the albumin is rabbit albumin. You test for anti-DNP antibodies every week for four weeks. At the beginning of the fifth week (day 28), you inject the rabbit with more DNP-albumin conjugate, and you measure the antibody response for DNP every week for four weeks. The results are presented in the table, where the relative antibody response is presented in terms of the number of + signs. (As a control, you cleverly obtained serum from the rabbit prior to any immunizations.)

Days Following Immunization	Presence of Anti-DNP Antibodies
0	−
7	++
14	++
21	+
28	+
35	+++++++
42	++++++
49	+++++

Explain these results. Specifically, why is the antibody response to the dinitrophenyl group low after day 28 and high after 35 days? Be certain to mention how T cells are involved in the process.

b. If you were to inject DNP-albumin into the rabbit 100 days after the first DNP-albumin injection, what do you think the relative antibody response would be like for DNP at day 110?

c. You decide to inject the rabbit with a DNP-protein conjugate, where the protein is not related to albumin. You are surprised because the response produces a very low titer of antibodies against DNP; it looks more typical of a primary response to an immunogen. Why do you think this might happen?

3. A second major class of cells other than B lymphocytes comprises the immune system: T lymphocytes. What are the two different types of T cells that are found? Give a brief description of each.

What are macrophages, and how are they involved in the immune response? Do they have an effect on T-cell maturation?

What are the primary and secondary organs of the immune system?

What is meant by tolerance?

4. Describe the basic structure of antibodies. Be certain to include the interactions of the different domains. What is the hinge region? What are the F_{ab} and F_c portions of an antibody? What are the variable and constant regions of an antibody? What is the immunoglobulin fold? What are the hypervariable and framework regions?

The diversity of antibodies is obtained by several mechanisms. Be certain you understand the mechanism behind κ chain diversity; pay particular attention to Figure 27-23 which illustrates this point. What are the V_κ, J_κ, and C_κ regions? At the joining point between V_κ and J_κ, a few nucleotides can be lost. How does this enhance antibody diversity? What is the difference between a productive and nonproductve joint?

H chain diversity also involves rejoining of DNA. How many libraries determine H chain structure? What is the N region of an H chain, and how does it contribute to diversity? What is the order in which gene rearrangements occur that marks the conversion of a B-lymphoid cell to a pre-B lymphocyte and finally to a virgin B lymphocyte? What is allelic exclusion, and why is it important for clonal selection to operate properly? What are the transcriptional control elements that regulate immunoglobulin expression in B lymphocytes?

5. What are histocompatibility genes? What is the major histocompatibility complex? Specifically, what are the class I and class II MHC genes? What is meant by an MHC presenting a peptide? Which type of T lymphocyte is mostly responsible for killing virus infected cells; why? Which type of T lymphocyte is mostly responsible for helping B cells produce antibacterial antibodies; why? How do CTLs ultimately kill a target cell? What factors do T cells secrete?

Once a virgin B lymphocyte encounters antigen, what change occurs that allows secretion of antibody that has an identical variable region to that found on the cell surface? Recall that an activated B lymphocyte is a plasma cell that secretes antibody. What is a memory B cell? What is switch recombination, and how does it allow a B cell that secretes IgM to convert to a cell that secretes IgG? What is somatic mutation, and how does it assist in antibody diversity?

References

General References

Advances in Immunology. Yearly volumes.
Annual Review of Immunology. Yearly volumes.
Immunological Reviews. Bimonthly volumes on topics in immunology. Edited by G. MÖLLER. Munksgard, Copenhagen.
Immunology Today. Monthly issues covering a wide range of immunology.
Life, death, and the immune system. 1993. *Sci. Am.* **269**(3). Special issue (Sept. 1993) on the immune system.
JERNE, N. K. 1973. The immune system. *Sci. Am.* **229**(1):52.
KABAT, E. A. 1976. *Structural Concepts in Immunology and Immunochemistry.* 2d ed., Holt.
LANDSTEINER, K. 1945. *The Specificity of Serologic Reactions.* Harvard.
PAUL, W. E. 1993. *Fundamental Immunology.* 3d ed., Raven.

Clonal Selection

BURNET, F. M. 1957. A modification of Jerne's theory of antibody production using the concept of clonal selection. *Austral. J. Sci.* **20**:67.
MANSER, T., S. Y. HUANG, and M. L. GEFTER. 1985. The influence of clonal selection on the expression of immunoglobulin variable region genes. *Science* **226**:1283.

Antibodies

AMIT, A. G., et al. 1986. Three-dimensional structure of an antigen-antibody complex at 2.8 Å resolution. *Science* **233**:747.
AMZEL, L., et al. 1974. The three-dimensional structure of a combining region–ligand complex of immunoglobulin NEW at 3.5 Å resolution. *Proc. Nat'l. Acad. Sci. USA* **71**:1427.
CARAYANNOPOULOS, L., and J. D. CAPRA. 1993. Immunoglobulins: struc-

ture and function. In W. PAUL, ed. *Fundamental Immunology*. 3d ed., Raven, p. 283.

DAVIES, D. R., and H. METZGER. 1983. Structural basis of antibody function. *Ann. Rev. Immunol.* 1:87.

DAVIES, D. R., E. A. PADLAN, and D. M. SEGEL. 1975. Three-dimensional structure of immunoglobulins. *Ann. Rev. Biochem.* 44:639.

EDELMAN, G. M. 1970. The structure and function of antibodies. *Sci. Am.* 223(2):34.

LERNER, R. A., S. J. BENKOVIC and P. G. SCHULTZ. 1991. At the crossroads of chemistry and immunology: catalytic antibodies. *Science* 252:659.

LOH, D. Y., et al. 1983. Molecular basis of a mouse strain–specific anti-hapten response. *Cell* 33:153.

SIEKEVITZ, M., S. Y. HUANG, and M. L. GEFTER. 1983. The genetic basis of antibody production: a single heavy chain variable region gene encodes all molecules bearing the dominant anti-arsonate idiotype in the strain A mouse. *Eur. J. Immunol.* 13:123.

Diversity Generated from Joining Immunoglobulin Gene Segments

CAPRA, J. D., and A. B. EDMUNDSON. 1977. The antibody combining site. *Sci. Am.* 236:50.

EARLY, P., et al. 1980. An immunoglobulin heavy-chain variable region gene is generated from three segments of DNA: V_H, D, and J_H. *Cell* 19:981.

HILSCHMANN, H., and L. C. CRAIG. 1965. Amino acid sequence studies with Bence-Jones proteins. *Proc. Nat'l. Acad. Sci. USA* 53:1403.

HOZUMI, N., and S. TONEGAWA. 1976. Evidence for somatic rearrangement of immunoglobulin genes coding for variable and constant regions. *Proc. Nat'l. Acad. Sci. USA* 73:3628.

KUROSAWA, Y., and S. TONEGAWA. 1982. Organization, structure, and assembly of immunoglobulin heavy-chain diversity DNA segments. *J. Exp. Med.* 155:201.

LEDER, P. 1982. The genetics of antibody diversity. *Sci. Am.* 246(5):102.

LEWIS S., et al. 1982. Continuing κ gene rearrangement in an Abelson murine leukemia virus transformed cell line. *Cell* 30:807.

MAX, E. 1993. Immunoglobulins: molecular genetics. in W. PAUL, ed. *Fundamental Immunology*. 3d ed., Raven, p. 315.

SAKANO, H., et al. 1980. Two types of somatic recombination are necessary for the generation of complete immunoglobulin heavy-chain genes. *Nature* 286:676.

TONEGAWA, S. 1983. Somatic generation of antibody diversity. *Nature* 302:575.

WEIGERT, M., et al. 1980. The joining of V and J gene segments creates antibody diversity. *Nature* 283:497.

Allelic Exclusion

ALT, F., et al. 1984. Ordered rearrangement of immunoglobulin heavy chain variable region segments. *EMBO J.* 3:1209.

Immunoglobulin Synthesis

ALT, F., and D. BALTIMORE. 1982. Joining of immunoglobulin heavy-chain gene segments: implications from a chromosome with evidence of three D-J_H fusions. *Proc. Nat'l. Acad. Sci. USA* 79:4118.

CORY, S., J. JACKSON, and J. M. ADAMS. 1980. Deletions in the constant region locus can account for switches in immunoglobulin heavy-chain expression. *Nature* 285:450.

DUTTON, R. W., and R. I. MISHELL. 1967. Cellular events in the immune response. The in vitro response of normal spleen cells to erythrocyte antigens. *Cold Spring Harbor Symp. Quant. Biol.* 32:407.

EARLY, P., et al. 1980. Two mRNAs can be produced from a single immunoglobulin E gene by alternative RNA processing pathways. *Cell* 20:3131.

KOHLER, G., and C. MILSTEIN. 1975. Continuous cultures of fused cells secreting antibody of predefined specificity. *Nature* 256:495.

MILSTEIN, C. 1980. Monoclonal antibodies. *Sci. Am.* 243(4):66.

MOORE, K. W., et al. 1981. Expression of IgD may use both DNA rearrangement and RNA splicing mechanisms. *Proc. Nat'l. Acad. Sci. USA* 78:1800.

POTTER, M. 1972. Immunoglobulin-producing tumors and myeloma proteins of mice. *Physiol. Rev.* 62:631.

RUSCONI, S., and G. KOHLER. 1985. Transmission and expression of a specific pair of rearranged immunoglobulin μ and κ genes in a transgenic mouse line. *Nature* 314:330.

SHIMIZU, T., et al. 1981. Ordering of mouse immunoglobulin heavy-chain genes by molecular cloning. *Nature* 289:149.

WEISSMAN, I. L. 1975. Development and distribution of immunoglobulin-bearing cells in mice. *Transplant. Rev.* 24:159.

WHITLOCK, C. A., and O. N. WITTE. 1982. Long-term culture of B lymphocytes and their precursors from murine bone marrow. *Proc. Nat'l. Acad. Sci. USA* 79:3608.

Somatic Mutation

BALTIMORE, D. 1981. Somatic mutation gains its place among the generators of diversity. *Cell* 26:295.

CREWS, S., et al. 1981. A single V_H gene segment encodes the immune response to phosphorylcholine: somatic mutation is correlated with the class of antibody. *Cell* 25:59.

GEARHART, P., and D. BOGENHAGEN. 1983. Clusters of point mutation are found exclusively around rearranged antibody variable region genes. *Proc. Nat'l. Acad. Sci. USA* 80:3439.

KIM, S., et al. 1981. Antibody diversity: somatic hypermutation of rearranged V_H genes. *Cell* 27:573.

T Lymphocytes

BARTH, R. K., et al. 1985. The murine T-cell receptor uses a limited repertoire of expressed V_β gene segments. *Nature* 316:517.

CANTOR, H., and E. A. BOYSE. 1975. Functional subclasses of T lymphocytes bearing different Ly antigens. I. The generation of functionally distinct T cell subclasses is a differentiative process independent of antigen. *J. Exp. Med.* 141:1375.

FORD, C. E., et al. 1966. The inflow of bone-marrow cells to the thymus. Studies with part body-irradiated mice injected with chromosome-marked bone marrow and subjected to antigenic stimulation. *Ann. N.Y. Acad. Sci.* 129:238.

LEDOUARIN, N. M., and F. JOTEREAU. 1975. Tracing of cells of the avian thymus through embryonic life in interspecific chimeras. *J. Exp. Med.* 142:17.

MILLER, J. F. A. P. 1961. Immunological function of the thymus. *Lancet* 2:748.

MOORE, M. A. S., and J. J. T. OWEN. 1967. Experimental studies on the development of the thymus. *J. Exp. Med.* 126:715.

WEISMAN, I. L. 1967. Thymus cell migration. *J. Exp. Med.* 126:291.

ZINKERNAGEL, R., and P. DOHERTY. 1976. The concept that surveillance of self is mediated via the same set of genes that determines recognition of allogenic cells. *Cold Spring Harbor Symp. Quant. Biol.* 41:505.

T-Cell Receptor

BRENNER, M., et al. 1986. Identification of a putative second T-cell receptor. *Nature* 322:145.

CHIEN, Y.-L., et al. 1987. A new T-cell receptor gene located within the alpha locus and expressed early in T-cell differentiation. *Nature* 327:677.

GARMAN, R. D., P. J. DOHERTY, and D. RAULET. 1986. Diversity, rearrangement and expression of murine T cell gamma genes. *Cell* 45:733.

HEDRICK, S., et al. 1984. Isolation of cDNA clones encoding T cell–specific membrane-associated proteins. *Nature* 308:149.

MCINTYRE, B., and J. ALLISON. 1983. The mouse T cell receptor: structural heterogeneity of molecules of normal T cells defined by Xenoantiserum. *Cell* 34:739.

YAGUE, J., et al. 1985. The T cell receptor: the α and β chains define idiotype, and antigen and MHC specificity. *Cell* 42:81.

MHC Genes

BILLINGHAM, R., and W. SILVERS. 1971. *The Immunobiology of Transplantation*. Prentice-Hall.

BJORKMAN, P., et al. 1987. Structure of the human class I histocompatibility antigen, MLA-A2. *Nature* 329:506.

DOHERTY, P. C., and R. M. ZINKERNAGEL. 1975. H-2 compatibility is required for T-cell-mediated lysis of target cells infected with lymphocytic choriomeningitis virus. *J. Exp. Med.* 141:502.

GERMAIN, R. N. 1994. MHC-dependent antigen processing and peptide presentation: providing ligands for T lymphocyte activation. *Cell* **76**:287.

HOOD, L., M. STEINMETZ, and B. MALISSEN. 1983. Genes of the major histocompatibility complex of the mouse. *Ann. Rev. Immunol.* **1**:529.

KLEIN, J. 1975. *Biology of the Mouse Histocompatibility-2 Complex.* Springer-Verlag.

KLEIN, J. 1979. The major histocompatibility complex of the mouse. *Science* **203**:516.

MADDEN, D. R., et al. 1991. The structure of HLA-B27 reveals nonamer self-peptides bound in an extended conformation. *Nature* **353**:321.

SNELL, G. D. 1981. Studies in histocompatibility. *Science* **213**:172.

SNELL, G. D., J. DAUSSET, and S. NATHENSON. 1976. *Histocompatibility.* Academic Press.

STEINMETZ, M., et al. 1982. A molecular map of the immune response region from the major histocompatibility complex of the mouse. *Nature* **300**:35.

STEINMETZ, M., A. WINOTO, K. MINARD, and L. HOOD. 1982. Clusters of genes encoding mouse transplantation antigens. *Cell* **28**:489.

YANAGI, Y., et al. 1984. A human T cell–specific cDNA clone encodes proteins having extensive homology to immunoglobulin chains. *Nature* **308**:145.

Lymphokines

CLARK, E. A. and J. A. LEDBETTER. 1994. How B and T cells talk to each other. *Nature* **367**:425.

KISHIMOTO, T., T. TAGA, and S. AKIRA. 1994. Cytokine signal transduction. *Cell* **76**:253.

PAUL, W. E. 1989. Pleiotropy and redundancy: T cell–derived lymphokines in the immune response. *Cell* **57**:521.

Tolerance

GOODNOW, C. C., S. ADELSTEIN and A. BASTAN. 1990. The need for central and peripheral tolerance in the B cell repertoire. *Science* **248**:1373.

NOSSAL, G. J. V. 1994. Negative selection of lymphocytes. *Cell* **76**:229.

VON BOEHMER, H. 1994. Positive selection of lymphocytes. *Cell* **76**:219.

SCHWARTZ, R. M. 1989. Acquisition of immunologic self-tolerance. *Cell* **57**:1073.

Glossary

Terms that are used in these definitions and that are also defined in this glossary are **boldfaced.** Alternative names for main entries that appear within the definition are in *italics*.

α helix (pl. α helices) Common secondary structure of proteins in which the linear sequence of amino acids is folded into a right-handed spiral that is stabilized by hydrogen bonds between the carboxyl oxygen of each **peptide bond** and the amide group three residues away in the sequence. (See Figure 3-9.)

acetyl CoA Small, water-soluble metabolite that consists of an acetyl group linked to **coenzyme A** (CoA). Pyruvate from glycolysis, fatty acids, and amino acids are all converted to acetyl CoA, which then transfers its two-carbon acetyl group to citrate at the beginning of the **citric acid cycle.** (See Figure 17-11.)

actin Abundant structural protein in eukaryotic cells that interacts with many other proteins. The monomeric globular form (G actin) polymerizes to form actin filaments (F actin). In muscle cells, F actin interacts with **myosin** during contraction. See also **microfilaments.**

action potential Rapid, all-or-none electrical activity that is propagated in the plasma membrane of neurons and muscle cells. It may be initiated by electrical, chemical, or mechanical stimuli. The action potential spreads down the length of neuronal **axons** as the result of the opening and closing of voltage-sensitive ion channels and the influx of Na^+ and K^+ ions. Action potentials allow long-distance signaling in the nervous system. (See Figure 21-18.)

activation energy The input of energy required to overcome the barrier to initiate a chemical reaction. (See Figure 2-26.)

active site Region on the surface of an **enzyme** where the substrate binds and undergoes a catalyzed reaction. The active site contains residues involved in substrate binding and others involved in catalysis. (See Figure 3-27.)

active transport Movement of an ion or small molecule across a membrane against the concentration gradient or electrochemical gradient of the transported molecule by a process directly coupled to ATP hydrolysis.

adenylate cyclase Enzyme that catalyzes formation of **cyclic AMP** (cAMP), a **second messenger.** Binding of ligands to certain **seven-spanning** and other **receptors** leads to activation of membrane-bound adenylate cyclase and a rise in intracellular cAMP. (See Figure 20-15.)

aerobic Referring to a cell, organism, or metabolic process that utilizes O_2 or that can grow in the presence of O_2.

allele One of two or more alternative forms of a gene located at the corresponding site (**locus**) on homologous chromosomes.

allosteric site Region on the surface of an enzyme (or other protein) that is distinct from the **active site** and to which activators or inhibitors (effectors) bind; also called the *regulatory site*. Effector binding induces a conformational change that modulates enzymatic activity. (See Figure 3-30.)

amino acids Class of organic compounds in which an amino group and carboxyl group are linked to a central carbon atom, called the α carbon, to which a variable side chain is bound. Twenty common amino acids are the monomers that polymerize to form proteins. Some of these amino acids can then become modified (See Figure 3-3.)

aminoacyl-tRNA Activated form of an amino acid that is used in protein synthesis. It consists of an amino acid linked via a high energy ester bond to the 3'-hydroxyl group of a tRNA molecule. (See Figures 4-31 and 4-34.)

amphipathic Referring to a molecule or structure that has both a **hydrophobic** and a **hydrophilic** part. Both proteins and lipids may be amphipathic.

anabolism Cellular processes whereby energy is used to synthesize complex molecules from simpler ones. See also **catabolism**.

anaerobic Referring to a cell, organism, or metabolic process that functions in the absence of O_2.

anaphase Mitotic stage during which the sister **chromatids** separate and move apart (segregate) toward the spindle poles. (See Figure 5-50.)

antibody A protein that interacts with a particular site (epitope) on an **antigen** and facilitates clearance of that antigen by various mechanisms. See also **immunoglobulin**. (See Figure 27-4.)

anticodon Sequence of three nucleotides in a transfer RNA (**tRNA**) that is complementary to a **codon** in a messenger RNA (**mRNA**). During protein synthesis, base pairing between a codon and anticodon aligns the tRNA carrying the corresponding amino acid for addition to the growing peptide chain.

antigen Any material (usually foreign) that elicits and/or is specifically bound by an **antibody**.

antigenic determinant The part of an **antigen** molecule that binds to the antigen-binding region of an **antibody**; also called *epitope*.

antiport Cotransport process in which the movement of a molecule or ion across a membrane against its concentration gradient is driven by the movement in the opposite direction of a second ion down its concentration gradient. Transport is mediated by specific membrane-bound proteins called antiporters. See also **symport**.

apoptosis Programmed cell death; a specific suicide process in animal cells that includes fragmentation of nuclear DNA. Apoptosis occurs normally in development, as in the resorption of the tadpole tail during metamorphosis into a frog; or it can be induced, for example by DNA damage that exceeds the capacity of repair mechanisms.

archaebacteria Prokaryotic organisms, often found in extreme environments, phylogenetically distant from the so-called true bacteria (**eubacteria**).

aster Cellular structure seen during mitosis composed of **microtubules** (called *astral fibers*) radiating outward from a **centrosome** or mitotic pole. (See Figure 5-50.)

asymmetric carbon atom A carbon atom bonded to four different atoms; also called *chiral carbon atom*. The bonds can be arranged in two different ways producing **stereoisomers**, which are mirror images of each other. (See Figure 2-4.)

ATP (adenosine 5'-triphosphate) A **nucleotide** containing two high-energy phosphoanhydride bonds whose hydrolysis or transfer is each accompanied by a large **free-energy change** (ΔG) of ≈ 7 kCal/mole. It is the most important molecule for capturing and transferring free energy in all cells. (See Figure 2-22.)

ATPase One of a large group of enzymes that catalyze hydrolysis of **ATP** to yield ADP and inorganic phosphate with release of free energy. These enzymes normally are associated with cellular membranes or cytoskeletal structures and the energy released is utilized in some coupled activity such as active transport, muscle contraction, unwinding of DNA, or movement of cilia and flagella.

ATP synthase, mitochondrial Multimeric protein complex bound to inner mitochondrial membranes that catalyzes synthesis of ATP coupled to proton movement down its electrochemical gradient; also called F_0F_1 *complex*. Similar enzymes are present in bacterial plasma membranes and the thylakoid membranes of chloroplasts (the CF_0CF_1 complex). (See Figure 17-15.)

autonomously replicating sequence (ARS) Sequence that permits a DNA molecule to replicate in yeast; a yeast DNA replication origin. (See Figure 9-58.)

autoradiography Technique for visualizing radioactive molecules in a sample (e.g., a tissue section or electrophoretic gel) by exposing a photographic film or emulsion to the sample. The exposed film is called an autoradiogram or autoradiograph. Individual silver grains deposited on a tissue are visualized by light microscopy. (See Figure 6-24.)

autosome Any chromosome other than a sex chromosome.

auxotroph A cell or microorganism that grows only when the medium contains a specific nutrient or metabolite that is not required by the **wild type**; commonly, a microbial mutant strain that cannot synthesize an essential metabolite as judged by its inability to grow on a minimal medium.

axon Process extending from the cell body of a neuron that is capable of conducting an electric impulse (**action potential**) generated at the junction with the cell body (called the axon hillock) toward its distal, branching end (called the axon terminal). (See Figure 21-1.)

axoneme Bundle of **microtubules** present in **cilia** and **flagella** and responsible for their movement. It is enveloped by an extension of the plasma membrane and connects with the **basal body**. (See Figure 23-29.)

β pleated sheet A planar secondary structure element of proteins that is created by hydrogen bonding between the backbone atoms in two different polypeptide chains or segments of a single folded chain. (See Figure 3-12.)

bacteriophage (phage) Any virus that infects bacterial cells. Some bacteriophage are widely used as **cloning vectors**.

basal body Structure at the base of **cilia** and **flagella** from which microtubules forming the **axoneme** radiate. The (−) ends of microtubules are associated with the basal body, and the (+) ends point away from it.

basal lamina (pl. basal laminae) A thin sheetlike network of extracellular-matrix components that underlies most animal epithelial layers and other organized groups of cells (e.g., muscle), separating them from connective tissue. It usually contain type IV **collagen, laminin, and heparan sulfate proteoglycans**.

base A compound, usually containing nitrogen, that can accept a H^+. Commonly used to describe the non-sugar components of nucleotides in DNA and RNA.

base composition Frequency of adenine (A), guanosine (G), cytosine (C), thymine (T), or uracil (U) in a **nucleic acid**.

base pair Association of two complementary **nucleotides** in a DNA or RNA molecule stabilized by hydrogen bonding. Adenine pairs with thymine or uracil (A-T, A-U) and guanine pairs with cytosine (G-C). (See Figure 4-7.)

B cell See **B lymphocyte**.

benign Referring to a **tumor** containing cells that closely resemble normal cells. Benign tumors stay in the tissue where they originate. See also **malignant**.

bilayer See **phospholipid bilayer**.

biomembrane Permeability barrier, surrounding cells or organelles, that consists of a **phospholipid bilayer**, associated membrane proteins, and in some cases **cholesterol** and **glycolipids**. See also **integral membrane protein** and **peripheral membrane protein**.

blastula An early embryonic form produced by cleavage of a fertilized ovum and usually consisting of a single layer of cells surrounding a fluid-filled spherical cavity.

blotting (blot analysis) Widely used biochemical technique for detecting the presence of specific macromolecules (proteins, mRNAs, or DNA sequences) in a mixture. A sample first is separated on an agarose or polyacrylamide gel usually under denaturing conditions; the separated components are transferred (blotted) to a nitrocellulose sheet, which is exposed to a radiolabeled molecule that specifically binds to the macromolecule of interest, and then subjected to **autoradiography**. In Northern blotting, mRNAs are detected with a complementary DNA; in Southern blotting, DNA **restriction fragments** are detected with complementary nucleotide sequences; and in Western blotting, proteins are detected by specific antibodies.

B lymphocyte A small white blood cell that matures in the bone marrow and expresses membrane-bound **antibody**; also called *B cell*. When B cells interact with **antigen**, they give rise both to plasma cells, which secrete circulating antibody, and to long-lived memory cells. They mediate humoral immunity.

buffer A solution of the acid (HA) and base (A$^-$) form of a compound that undergoes little change in pH when small quantities of strong acid or base are added.

cadherins A family of Ca^2-dependent **cell-adhesion molecules** that play roles in tissue differentiation and structure. (See Table 24-3.)

calmodulin A small cytosolic protein that binds four Ca^{2+} ions; the Ca^{2+}-calmodulin complex binds to and activates many enzymes. (See Figure 20-42.)

Calvin cycle See **dark reactions.**

capsid The outer proteinaceous coat of a **virus**, formed by self-assembly of one or more protein subunits and enclosing the viral nucleic acid. The capsid plus enclosed nucleic acid is called a nucleocapsid.

carbohydrate Certain polyhydroxyaldehydes, polyhydroxyketones, or compounds derived from these with the formula $(CH_2O)_n$. They are classified as **monosaccharides, oligosaccharides,** and **polysaccharides.**

carbon fixation See **dark reactions.**

carcinogen Any chemical or physical agent that can cause cancer when cells or organisms are exposed to it.

carcinoma A malignant **tumor** derived from epithelial cells.

catabolism Cellular processes whereby complex molecules are degraded to simpler ones and energy is released. See also **anabolism.**

catabolite repression Depression by glucose or a glucose metabolite of **transcription** of genes encoding a wide variety of sugar-metabolizing enzymes in prokaryotes. It is mediated by CAP (catabolite activator protein). (See Figure 11-20.)

catalyst A substance that increases the rate of a chemical reaction without undergoing a permanent change in its structure. See also **enzyme.**

cDNA (complementary DNA) DNA molecule copied from an mRNA molecule by reverse transcriptase and therefore lacking the **introns** present in **genomic DNA**. Sequencing of a cDNA permits the amino acid sequence of the encoded protein to be deduced; expression of cDNAs in recombinant cells can be used to produce large quantities of given proteins in vitro.

cDNA library Collection of cloned DNA molecules consisting of DNA copies of all the mRNAs produced by a cell type inserted into a suitable **cloning vector**. See also **gene cloning** and **DNA library.**

cell-adhesion molecules (CAMs) Integral membrane proteins that mediate cell-cell binding. These adhesive proteins primarily fall into three major classes: the Ca^2-dependent **cadherins** and selectins, and the Ca^2-independent **immunoglobulin superfamily** including N-CAM. (See Figure 24-32.)

cell cycle Ordered sequence of events in which a cell duplicates its chromosomes and divides into two. Most eukaryotic cell cycles can be commonly divided into four phases: G$_1$ before DNA synthesis occurs; S when DNA replication occurs; G$_2$ after DNA synthesis; and M when **cell division** occurs, yielding two daughter cells. Under certain conditions, cells exit the cell cycle during G$_1$ and remain in the G$_0$ state as nongrowing, nondividing (quiescent) cells. Appropriate stimulation of such cells induces them to return to G$_1$ and resume growth and division. (See Figure 5-48.) See also **checkpoint.**

cell division Separation of a cell into two daughter cells. In higher eukaryotes, it involves division of the nucleus (**mitosis**) and of the cytoplasm (cytokinesis); mitosis often is used as a synonym, referring to both nuclear and cytoplasmic division.

cell fusion Production of a hybrid cell (**heterokaryon**) containing two or more nuclei by various techniques that stimulate the fusion of the plasma membranes of two cells at the point of contact and intermingling of their cytoplasms. See also **hybridoma.**

cell junctions Specialized regions on the cell surface through which cells are joined to each other: tight junctions are ribbon like bands connecting adjacent epithelial cells that prevent leakage of fluid across the cell layer. Gap junctions are protein-lined channels between adjacent cells that allow passage of ions and small molecules between the cells. Desmosomes and *adherens junctions* consist of dense protein plaques connected to intermediate filaments that mediate adhesion between adjacent cells and between cells and the extracellular matrix. (See Figure 14-42.)

cell line A population of cultured cells, of plant or animal origin, that has undergone a change allowing the cells to grow indefinitely. Cell lines can result from chemical or viral **transformation.**

cell strain A population of cultured cells, of plant or animal origin, that has a finite lifespan.

cellular immunity Immune responses mediated by cells including **macrophages** and **cytotoxic T lymphocytes**; also called *cell-mediated immunity.* (See Figure 27-1.)

cellulose A structural **polysaccharide** made of glucose units linked together by $\beta(1 \rightarrow 4)$ **glycosidic bonds**. It forms long microfibrils, which are the major component of plant cell walls. (See Figure 24-68.)

cell wall A specialized, rigid extracellular matrix that lies next to the plasma membrane, protecting a cell and maintaining its shape. It is prominent in most fungi, plants, and prokaryotes, but is not present in most multicellular animals.

central nervous system (CNS) The part of the vertebrate nervous system comprising the brain and spinal cord; the main information-processing organ.

centrifugation Technique involving application of a large centrifugal force for collecting or separating macromolecules, particles, or cells from a suspension based on differences in mass, shape, or density. (See Figure 3-36.)

centriole Either of two cylindrical structures within the **centrosome** of animal cells and containing nine sets of triplet **microtubules**; structurally similar to a **basal body**. (See Figure 23-5.)

centromere Region of a chromosome required for proper chromosome segregation during mitosis and meiosis; the region where sister **chromatids** in mitotic chromosomes are attached and from which **kinetochore** fibers extend toward a spindle pole.

centrosome (cell center) Organelle located near the nucleus of animal cells that is the primary **microtubule-organizing center** and contains a pair of **centrioles.** It divides during mitosis, forming the spindle poles.

CF$_0$CF$_1$ complex See **ATP synthase, mitochondrial.**

chaperone Any protein that binds to an unfolded or partially folded target protein, thus preventing misfolding, aggregation, and/or degradation of the target protein and facilitating its proper folding.

checkpoint Any of several points in the eukaryotic **cell cycle** at which progression of a cell to the next stage can be halted until conditions are suitable. These regulatory mechanisms are important in preventing formation of defective daughter cells. (See Figure 25-37.)

chemiosmosis Process whereby an electrochemical proton gradient (pH plus electric potential) across a membrane is used to drive ATP synthesis, to pump metabolites across the membrane against their concentration gradient, or to power the rotation of bacterial flagella. (See Figure 17-1.)

chiasma (pl. chiasmata) Point at which non sister **chromatids** of homologous chromosomes cross each other during **meiosis.** (See Figure 5-55.)

chimera An animal or tissue composed of elements derived from genetically distinct individuals; also a protein molecule containing segments derived from different proteins.

chlorophylls A group of light-absorbing porphyrin pigments, critical in **photosynthesis,** that contain a central Mg^{2+} ion surrounded by five rings, one of which has a long hydrocarbon tail. (See Figure 18-5.)

chloroplast A specialized organelle in plant cells that is surrounded by a double membrane and functions in photosynthesis, the light-absorbing step of which occurs in a group of internal membranes collectively called the thylakoid membrane.

cholesterol An **amphipathic** lipid containing the four-ring **steroid** structure with a hydroxyl group on one ring; a major component of many eukaryotic membranes and precursor of steroid hormones. (See Figure 14-5.)

chromatid One of a pair of **homologus chromosomes,** formed during the S phase of the **cell cycle,** that is still joined at the **centromere** to its homolog. During mitosis, the two **chromatids** separate, each becoming a chromosome of one of the two daughter cells.

chromatin Complex of **DNA, histones,** and nonhistone proteins present in the nucleus of eukaryotic cells. It exists in two forms: a less condensed form which can be transcribed (**euchromatin**) and a highly condensed form which is generally not transcribed (**heterochromatin**). Condensation of chromatin during mitosis yields the visible metaphase chromosomes. (See Figure 9-47.)

chromatography, liquid Group of biochemical techniques for separating mixtures of molecules in solution in which a sample is passed through a column filled with a solid matrix with which the dissolved molecules interact more or less frequently depending on their charge (ion exchange), polarity (hydrophobic interaction chromatography) or ability to bind specifically to other molecules covalently attached to the matrix (affinity). Gel filtration columns separate molecules on the basis of size. Commonly used technique for separating and purifying proteins. (See Figure 3-39.)

chromosome In eukaryotes, the structural unit of the genetic material consisting of a single, linear double-stranded DNA molecule and associated proteins. During mitosis, the **chromatin** composing each chromosome condenses into a compact structure visible in the light microscope. Each member of a species normally possesses the same number of chromosomes in its somatic cells, characteristic of the species. In prokaryotes, a single, circular double-stranded DNA molecule constitutes the bulk of the genetic material. See also **karyotype.**

cilium (pl. cilia) Membrane-enclosed motile structure extending from the surface of eukaryotic cells and composed of a specific arrangement of **microtubules** forming the **axoneme.** Cilia usually occur in groups and beat rhythmically to move a cell (e.g., single-celled organism) or to move small particles or fluid along the surface (e.g., trachea cells). See also **flagellum.** (See Figure 23-27.)

cis-active Referring to a DNA regulatory sequence (e.g., **enhancer, promoter,** or **operator**) that can control a gene only on the same chromosome. In bacteria, cis-active elements are adjacent or proximal to the gene(s) they control, whereas in eukaryotes they may also be far away. See also **trans-active.**

cisterna (pl. cisternae) Flattened membrane-bounded compartment present in the **Golgi complex** and **endoplasmic reticulum** in various tissues.

cistron A genetic unit that encodes a single polypeptide.

citric acid cycle A set of nine coupled reactions occurring in the matrix of the **mitochondrion** in which acetyl groups from food molecules are oxidized, generating CO_2 and reduced intermediates used to produce ATP; also called *Krebs cycle* and *tricarboxylic acid cycle* (TCA). (See Figure 17-13.)

class switching The process by which a B lymphocyte producing IgM and IgD switches to the production of a secondary immunoglobulin class (an IgG, IgE, or IgA) with the same antigen binding specificity. The process involves site-specific DNA recombination which brings exons for alternative heavy chain constant regions into the heavy chain transcription unit (See Figure 27-40).

clonal selection Mechanism whereby antigen binds to membrane-bound antibody on a B lymphocyte, thereby stimulating the cell to proliferate and develop into a **clone** of plasma cells all of which secrete the same antibody, with the same antigen specificity, as made in the original B lymphocyte. A similar process results in production of clones of T lymphocytes. (See Figure 27-9.)

clone A population of identical cells or DNA molecules descended from a single progenitor. Also viruses or organisms that are genetically identical and descended from a single progenitor.

cloning vector An autonomously replicating genetic element used to carry a **cDNA** or fragment of **genomic DNA** into a host cell for the purpose of **gene cloning.** The most commonly used vectors are bacterial **plasmids** and modified bacteriophage genomes; yeast artificial chromosomes (YACs) have been used as vectors for very large fragments of genomic DNA. (See Figures 7-1 and 7-12.)

cloning See **gene cloning** and **expression cloning.**

coated pit See **endocytosis.**

coding region Sequences in DNA that encode the amino acid sequence of all or part of a protein, as distinct from regulatory sequences, which control transcription and translation, and other non-transcribed or non-translated regions of a gene such as **introns** and spacers.

codon Sequence of three nucleotides in DNA or mRNA that specifies a particular amino acid during protein synthesis; also called *triplet.* Of the 64 possible codons, three are stop codons, which do not specify amino acids. The amino acid corresponding to each codon is the same in nearly all organisms; the set of all codons constitutes the **genetic code.** (See Table 4-3.)

coenzyme Tightly bound small molecule or metal that is required for catalytic activity of some enzymes, often by forming a transient covalent bond to the substrate; also called *prosthetic group* or *cofactor*. Examples include pyridoxal phosphate (see Figure 3-26), coenzyme A, and NAD.

coenzyme A (CoA) See **acetyl CoA**.

coiled coil Stable rodlike quaternary protein structure formed by two or three α helices interacting with each other along their length; commonly found in fibrous proteins and basic-zipper transcription factors (See Figures 3-11d and 11-48g and h.) See also **leucine zippers**.

collagen A triple-helical protein that forms fibrils of great tensile strength; a major component of the extracellular matrix and connective tissues. The numerous collagen subtypes differ in their tissue distribution and the extracellular components and cell-surface proteins with which they associate. (See Table 24-1.)

complement A group of cytotoxic serum proteins involved in the mediation of immune responses. The complement cascade, a series of enzymatic reactions, is activated by antibody-antigen complexes or microbial cell walls and generates specific complement complexes that directly lyse microorganisms or infected cells or promote their **phagocytosis**.

complementary Referring to two nucleic acid sequences or strands that can form a perfect base-paired double helix with each other; also describing regions on two interacting molecules (e.g., an enzyme and its substrate) that fit together in a lock-and-key fashion.

complementary DNA (cDNA) See **cDNA**.

complementation In genetics, the restoration of a wild-type function (e.g., ability to grow on galactose) in diploid **heterozygotes** generated from **haploids** each of which carries a mutation in a different gene whose encoded protein is required for the same biochemical pathway. Complementation analysis of a set of mutants exhibiting the same mutant phenotype (e.g., inability to grow on galactose) can be used to determine if mutations are in the same or different genes. (See Figure 8-14.)

conformation The precise shape of a protein or other macromolecule in three dimensions resulting from the spatial location of the atoms in the molecule. The conformation of proteins is most commonly determined by **x-ray crystallography**. Small changes in the conformations of some proteins affects their activity considerably.

consensus sequence The nucleotides or amino acids most commonly found at each position in the sequences of related DNAs, RNAs, or proteins. See also **homology**.

constitutive Referring to cellular production of a molecule at a constant rate, which is not regulated by internal or external stimuli.

constitutive mutant (1) A mutant in which a protein is produced at a constant level, as if continuously induced; (2) a bacterial regulatory mutant in which an operon is transcribed in the absence of inducer. (3) A mutation in which a regulated enzyme is in a continuously active form.

cooperativity Property exhibited by some proteins with multiple ligand-binding sites whereby binding of one ligand molecule increases (positive cooperativity) or decreases (negative cooperativity) the binding affinity of successive ligand molecules. (See Figure 3-31.)

cosmid A type of **cloning vector** used to clone large DNA fragments. (See Figure 7-14.)

cotransport Protein-mediated transport of an ion or small molecule across a membrane against a concentration gradient driven by coupling to movement of a second molecule down its concentration gradient. See also **antiport** and **symport**.

coupled reaction Two linked chemical reactions in which the **free energy** released by one of the reactions is used to drive the other.

covalent bond A chemical force that holds the atoms in molecules together by sharing of one or more pairs of electrons. Such a bond has a strength of 50–200 kcal/mol. (See Table 2-1.)

crossing over Exchange of genetic material between nonsister **chromatids** of **homologous chromosomes** during meiosis to produce recombined chromosomes. (See Figure 5-55.) See also **recombination**.

cyclic AMP (cAMP) A **second messenger**, produced in response to hormonal stimulation of certain **seven-spanning receptors**, that activates cAMP-dependent protein kinase.

cyclin Any of several related proteins many of whose concentrations rise and fall during the course of the eukaryotic cell cycle. Cyclins form complexes with **cyclin-dependent protein kinases**, thereby activating and determining the substrate specificity of these crucial enzymes which regulate passage through the cell cycle.

cyclin-dependent protein kinase (Cdk) A protein kinase that is catalytically active only when bound to a **cyclin**. Various Cdk-cyclin complexes trigger progression through various stages of the eukaryotic **cell cycle** by phosphorylating specific target proteins. (See Figure 25-30.)

cytochrome Any of a group of colored, heme-containing proteins that transfer electrons during cellular respiration and photosynthesis. (See Figure 17-26.)

cytokine Any of numerous secreted, small proteins (e.g., interferons, interleukins) that bind to cell-surface receptors on certain cells to trigger their differentiation or proliferation. Some cytokines, also called *lymphokines*, function to regulate the intensity and duration of the **immune response**. (See Table 27-3.)

cytokinesis See **cell division**.

cytoplasm Viscous contents of a cell that are contained within the plasma membrane but, in eukaryotic cells, outside the nucleus. The part of the cytoplasm not contained in any **organelle** is called the cytosol.

cytoskeleton Network of fibrous elements, consisting of **microtubules**, actin **microfilaments**, and **intermediate filaments**, found in the cytoplasm of eukaryotic cells. The cytoskeleton provides structural support for the cell and permits directed movement of organelles, chromosomes, and the cell itself.

cytosol See **cytoplasm**.

dalton Unit of molecular mass approximately equal to the mass of a hydrogen atom (1.66×10^{-24} g).

dark reactions All the reactions and processes that fix CO_2 into sugar during **photosynthesis**; also called *Calvin cycle* and *carbon fixation*. These processes are indirectly dependent on the **light reactions** and occur both in the dark and light.

degenerate In reference to the **genetic code**, having more than one **codon** specifying a particular amino acid.

denaturation In proteins, disruption of various **noncovalent bonds** resulting in unfolding of the polypeptide chain; in nucleic acids, disruption of **hydrogen bonds** between nucleotides converting double-stranded molecules or portions of molecules into single-stranded ones. Heating or exposure to certain chemicals can cause denaturation, which usually results in loss of biological function.

dendrite Process extending from the cell body of a neuron that is relatively short and typically branched and receives signals from **axons** of other neurons. (See Figure 21-1.)

deoxyribonucleic acid See **DNA.**

dephosphorylation Removal of a phosphate group from a molecule by **hydrolysis.** The activity of many phosphorylated proteins is modulated by removal of the phosphate groups by various phosphoprotein phosphatases.

depolarization Change in membrane potential, resulting in a less negative membrane potential, that typically is induced by an excitatory **neurotransmitter.**

desmosome See **cell junctions.**

determination In **embryogenesis,** a change in a cell that commits the cell to a particular developmental pathway.

development Process involving growth and **differentiation** by which a fertilized egg gives rise to an adult plant or animal.

differentiation Process usually involving changes in gene expression by which a precursor cell becomes a distinct specialized cell type.

diploid Referring to an organism or cell having two full sets of **homologous chromosomes** and hence two copies (**alleles**) of each gene or genetic locus. Somatic cells contain the diploid number of chromosomes ($2n$) characteristic of a species. See also **haploid.**

disulfide bond (–S–S–) A common **covalent** linkage between two cysteine residues in different proteins or in different parts of the same protein; formed in an **oxidation** reaction involving two **sulfhydryl groups.** Generally found only in extracellular proteins or protein domains.

DNA (deoxyribonucleic acid) Long linear polymer, composed of four kinds of deoxyribose **nucleotides** linked by **phosphodiester bonds,** that is the carrier of genetic information. In its native state, DNA is a **double helix** of two antiparallel strands. (See Figure 4-6.)

DNA library Collection of cloned DNA molecules consisting of fragments of the entire genome (genomic library) or of DNA copies of all the mRNAs produced by a cell type (cDNA library) inserted into a suitable **cloning vector.** See also **gene cloning.**

DNA polymerase An enzyme that copies one strand of DNA (the template strand) to make the complementary strand. Using rules of Watson-Crick base pairing, deoxyribonucleotides are added one-at-a-time, using as substrates deoxyribonucleotide triphosphates and releasing pyrophosphate.

DNAse (deoxyribonuclease) An enzyme that cuts a DNA strand or completely hydrolyzes a DNA, forming deoxyribonucleotides.

domain Region of a protein with a distinct **tertiary structure** (e.g., globular or rodlike) and characteristic activity; homologous domains may occur in different proteins.

dominant In genetics, referring to that **allele** of a gene expressed in the **phenotype** of a **heterozygote;** the nonexpressed allele is **recessive.** Also referring to the phenotype associated with a dominant allele.

dorsal Relating to the back of an animal or structure (e.g., leaf, wing).

double helix, DNA The most common three-dimensional structure for cellular DNA in which the two polynucleotide strands are antiparallel and wound around each other with complementary bases hydrogen bonded. Most DNA double helices are "right-handed." (See Figure 4-9.)

downstream For a gene, in the direction RNA polymerase moves during transcription and ribosomes move during translation, which is towards the end of a DNA strand with a 3′ hydroxyl group. By convention, the +1 position of a gene is the first transcribed nucleotide; nucleotides downstream from the +1 position are designated +2, +3, etc. See also **upstream.**

duplex DNA Double-stranded DNA.

duplicated gene The result of a non-homologous recombination or other event by which two copies of a gene come to lie near each other on a chromosome. Frequently the sequence of duplicated protein-coding genes diverges during evolution, generating a gene family encoding a class of related homologous proteins.

dynein An ATP powered motor protein in **cilia** and **flagella** that accounts for the movement of these organelles. ATP binding to dynein and its hydrolysis causes the dynein bridges to sequentially break and form new bonds with adjacent microtubule doublets. (See Figure 23-32.)

electron carrier A molecule that accepts electrons from donor molecules and transfers them to acceptor molecules. Most are prosthetic groups (e.g., heme, copper, iron-sulfur clusters) associated with membrane-bound proteins. (See Table 17-3.)

electron transport Flow of electrons from reduced electron donors (e.g., NADH) to O_2 via a series of **electron carriers** located in the inner mitochondrial membrane, or from H_2O to NADP in the thylakoid membrane of plant chloroplasts. In mitochondria and chloroplasts, electron movement is coupled to pumping of H ions across the membrane, setting up the **proton-motive force** that drives the formation of ATP. (See Figures 17-24 and 18-12.)

electrophoresis Any of several techniques for separating **macromolecules** based on their migration in a gel or other medium subjected to a strong electric field. (See Figures 3-37 and 3-38.)

elongation factor One of a group of non-ribosomal proteins involved in **translation** of mRNA following initiation. Functions include regulation of binding of an aminoacyl-tRNA to a ribosome and release of tRNA after addition of an amino acid to the growing peptide chain. See also **initiation factor.** (See Figure 4-42.)

embryogenesis Early development of an individual from a fertilized egg (**zygote**). Following cleavage of the zygote, the major axes are established during the **blastula** stage; in the subsequent **gastrula** stage, the early embryo invaginates and acquires three cell layers. (See Figure 24-42.)

endergonic Referring to a chemical reaction that absorbs energy (i.e., has a positive **free-energy change**).

endocytosis Uptake of extracellular materials by invagination of the plasma membrane to form a small membrane-bounded vesicle (endosome). Receptor-mediated endocytosis involves the specific uptake of a receptor-bound ligand typically by invagination of clathrin-coated pits on the plasma membrane of animal cells. See also **exocytosis** and **phagocytosis.** (See Figure 16-48.)

endoplasmic reticulum (ER) Network of interconnected membranous structures within the cytoplasm of eukaryotic cells. Two forms can be distinguished cytologically: rough ER, which is associated with **ribosomes** and functions in the synthesis and processing of secretory and membrane proteins, and smooth ER, which lacks ribosomes and is involved in synthesis of lipids.

endothelium Layer of highly flattened cells that forms the lining of all blood vessels and regulates exchange of materials between the bloodstream and surrounding tissues; it usually is underlain by a **basal lamina.**

endothermic Referring to a chemical reaction that absorbs heat (i.e., has a positive change in **enthalpy**).

enhancer A type of regulatory sequence in eukaryotic DNA (rarely in prokaryotic DNA) that may be located at a great distance either upstream or downstream from the transcription start site of the gene it controls. Binding of specific proteins to an enhancer either stimulates or decreases the rate of transcription of the associated gene. See also **promoter-proximal element**. (See Figure 11-42.)

enthalpy *(H)* Heat; in a chemical reaction, the enthalpy of the reactants or products is equal to their total bond energies.

entropy *(S)* A measure of the degree of disorder or randomness in a system; the higher the entropy, the greater the disorder.

enzyme A biological macromolecule that acts as a **catalyst**. Most enzymes are proteins, but certain RNAs, called ribozymes, also have catalytic activity.

epithelium Coherent sheet covering an external body surface or lining an internal cavity that is formed from one or more layers of cells. (See Figure 6-7.)

epitope See **antigenic determinant**.

equilibrium The state of a chemical reaction in which the concentration of all products and reactants is constant and the rates of the forward and reverse reactions are equal.

equilibrium constant *(K)* Ratio of forward and reverse rate constants for a reaction; equals the association constant.

erythrocyte Small biconcave blood cell that contains hemoglobin, transports O_2 and CO_2 to and from the tissues of vertebrates and (in mammals) lacks a nucleus.

eubacteria Prokaryotic organisms including the so-called true bacteria such as *Escherichia coli* and *Salmonella typhimurium*. Phylogenetically distinct from **archaebacteria**.

euchromatin Less condensed portions of **chromatin**, including most transcribed regions.

eukaryotes Class of organisms, including all plants, animals, fungi, yeast, protozoa, and most algae, that are composed of one or more cells containing a membrane-enclosed **nucleus** and **organelles**. See also **prokaryotes**.

exergonic Referring to a chemical reaction that releases energy (i.e., has a negative **free-energy change**).

exocytosis Release of intracellular molecules (e.g., hormones, matrix proteins) contained within a membrane-bounded vesicle by fusion of the vesicle with the plasma membrane of a cell. This is the process whereby most molecules are secreted from eukaryotic cells. See also **endocytosis**.

exon Segments of a eukaryotic gene (or of the **primary transcript** of that gene) that reaches the cytoplasm as part of a functional, mature, mRNA, rRNA, or tRNA molecule. See also **intron**.

exoplasmic face The face of a membrane directed away from the cytosol. The exoplasmic face of the plasma membrane faces the cell exterior.

exothermic Referring to a chemical reaction that releases heat (i.e., has a negative change in **enthalpy**).

expression cloning Recombinant DNA techniques for isolating a cDNA or genomic segment based on functional properties of the encoded protein and without prior purification of the protein. (See Figures 7-21 and 20-9.)

expression vector A modified **plasmid** or virus that carries a gene or cDNA into a suitable host cell and there directs abundant synthesis of the encoded protein. (See Figures 7-35 and 7-36.)

expression See **gene expression**.

extracellular matrix A usually insoluble network consisting of **glycosaminoglycans**, **collagen**, and various adhesive proteins (e.g., **laminin**, **fibronectin**), which are secreted by animal cells. It provides structural support in tissues and can affect the development and biochemical functions of cells.

extrinsic protein See **peripheral membrane protein**.

F_0F_1 complex See **ATP synthase, mitochondrial**.

facilitated diffusion Protein-aided transport of an ion or molecule across a membrane down its concentration gradient. This type of transport is characterized by three properties: (1) a rate faster than that predicted by **passive diffusion** through a **phospholipid bilayer** according to Fick's law; (2) ligand specificity (expressed by K_m); and (3) saturation (V_{max}). The glucose transporter is a well-studied example of a protein that mediates facilitated diffusion. (See Figure 15-7.)

fatty acid Any hydrocarbon chain that has a carboxyl group at one end; a major source of energy during metabolism and precursors for synthesis of **phospholipids**.

feedback inhibition Decrease in the catalytic activity of one of the enzymes in a metabolic pathway caused by binding of the ultimate product of the pathway. Usually, the inhibited step is the first step in the pathway that does not lead to other products.

fertilization Fusion of a female and male gamete (both **haploid**) to form a **diploid** zygote, which develops into a new individual.

fibroblast A type of connective-tissue cell, found in almost all vertebrate organs, that secretes collagen and other components of the extracellular matrix; it migrates and proliferates during wound healing and in tissue culture.

fibronectins An extracellular multiadhesive protein that binds to other matrix components, fibrin, and cell-surface receptors of the **integrin** family. It functions to attach cells to the **extracellular matrix** and is important in wound healing. (See Figure 24-25.)

flagellum (pl. **flagella**) Membrane-enclosed locomotory structure extending from the surface of eukaryotic cells and composed of a specific arrangement of microtubules, called an **axoneme**. Usually there is only one flagellum per cell (as in sperm cells), and its bending propels the cell forwards or backwards. See also **cilium**. (See Figures 23-27 to 23-29.) Bacterial flagella are much simpler structures containing a single predominant type of protein.

fluorescein See **fluorescent staining**.

fluorescent dye A molecule that absorbs light at one wavelength and emits it at a specific longer wavelength within the visible spectrum. Such dyes are used by themselves or linked to other molecules in **fluorescent staining**.

fluorescent staining General technique for visualizing cellular components by treating cells with a fluorescent-labeled agent that binds specifically to a component of interest and then observing the cells by fluorescence microscopy. Commonly, a **fluorescent dye** is chemically linked to an antibody to detect a specific protein; two commonly used dyes for this purpose are fluorescein, which emits green light, and rhodamine, which emits red light. Various fluorescent dyes that bind specifically to DNA are used to detect chromosomes or specific chromosomal regions.

footprinting, DNase Technique for identifying DNA regions that bind protein by digesting a radiolabeled DNA sample with DNase in the presence or absence of a DNA-binding protein and then subjecting the samples to gel **electrophoresis**. Because regions of DNA with bound protein are protected from digestion, the patterns of fragment bands from protected and unprotected samples will differ, permitting identification of the protein-binding regions of the DNA. (See Figure 11-10.)

free energy *(G)* A measure of the potential energy of a system that is a function of the **enthalpy** *(H)* and **entropy** *(S)*.

free-energy change *(ΔG)* The difference in the **free energy** of the product molecules and of the starting molecules in a chemical reaction, which can be used to predict the **equilibrium** of a reaction. A large negative value of ΔG indicates that a reaction has a strong tendency to occur from left to right, as written.

G₀, G₁, G₂ phase See **cell cycle.**

gamete Specialized **haploid** cell (in animals either a sperm or an egg) produced by **meiosis** of germ cells; in sexual reproduction, union of a sperm and an egg (**fertilization**) initiates the development of a new individual.

ganglioside Any glycolipid containing one or more *N*-acetylneuraminic acid (sialic acid) residues in its structure. Gangliosides are found in the plasma membrane of eukaryotic cells and are especially abundant in neurons; they confer a net negative charge on most animal cells.

gap junction See **cell junctions.**

gastrula An early embryonic form subsequent to the **blastula** characterized by invagination of the cells to form a rudimentary gut cavity and development of three cell layers.

gene Physical and functional unit of heredity, which carries information from one generation to the next. In molecular terms, a gene is the entire **DNA** sequence necessary for the synthesis of a functional polypeptide or RNA molecule. In addition to **coding regions** most genes also contain noncoding intervening sequences (**introns**) and **transcription-control regions.** See also **cistron** and **transcription unit.**

gene cloning Recombinant DNA technique in which specific **cDNAs** or fragments of **genomic DNA** are inserted into a **cloning vector,** which then is incorporated into cultured **host cells** (e.g., *E. coli* cells) and maintained during growth of the host cells; also called *DNA cloning.* (See Figures 7-3 and 7-16.)

gene control All of the mechanisms involved in regulating **gene expression.** These mechanisms can operate at several molecular steps including **transcription, RNA processing,** RNA stabilization, and **translation.** In a few cases, amplification or rearrangement of chromosomal segments affect gene expression.

gene conversion Phenomenon in which one allele of a gene is converted to another during meiotic **recombination.** (See Figure 10-36.)

gene expression Overall process by which the information encoded in a gene is converted into an observable **phenotype** (most commonly production of a protein).

genetic code The set of rules whereby nucleotide triplets (**codons**) in RNA specify amino acids in proteins. (See Table 4-3.)

genetic drift Spontaneous variation in the sequence of duplicated genes that developed during evolution; also called *sequence drift.*

genome Total genetic information carried by a cell or organism.

genomic DNA All the DNA sequences composing the **genome** of a cell or organism. See also **cDNA.**

genotype Entire genetic constitution of an individual cell or organism; also, the **alleles** at one or more specific loci.

germ cell Any precursor cell that can give rise to **gametes.**

germ line Genetic material transmitted from one generation to the next through the **gametes.**

glucagon A peptide **hormone** produced in the α cells of the pancreas that triggers the conversion of **glycogen** to glucose by the liver and with **insulin** controls blood glucose levels.

glycocalyx Carbohydrate-rich layer covering the outer surface of the plasma membrane of eukaryotic cells; composed of membrane glycolipids and oligosaccharide side chains of integral membrane proteins and absorbed peripheral membrane proteins.

glycogen A very long, branched polymer composed exclusively of glucose units, which is the primary storage carbohydrate in animal cells. It is found primarily in the liver and muscle.

glycogenolysis Breakdown of glycogen to glucose, which occurs primarily in liver and muscle cells. The process is stimulated by a rise in **cAMP** following epinephrine stimulation of cells and, in muscle, by a rise in Ca^{2+} following neuronal stimulation. (See Figure 20-25a.)

glycolipid A class of molecules, frequently found in the plasma membrane, in which a short carbohydrate chain is covalently linked to a lipid.

glycolysis Anaerobic conversion of sugars to lactate or pyruvate in the cytosol with the production of ATP. Further degradation via the **citric acid cycle** occurs in the mitochondrion and is coupled to **oxidative phosphorylation.** (See Figure 17-2.)

glycoprotein A class of molecules in which one or more oligosaccharide chains are covalently linked to a protein; frequently found in the plasma membrane or secreted from the cell.

glycosaminoglycan A long, linear polymer of a repeating disaccharide in which one member of the pair usually is a sugar acid (uronic acid) and the other is an amino sugar and many residues are sulfated. These highly charged polysaccharides are covalently bound to core proteins forming **proteoglycans,** which are major components of the **extracellular matrix.** (See Figure 24-17.)

glycosidic bond The covalent linkage between two **monosaccharides** formed by a condensation reaction in which one carbon, usually carbon #1, of one sugar reacts with a hydroxyl group on a second sugar with the loss of a water molecule. (See Figure 2-24.)

glycosyl transferase An enzyme that forms a **glycosidic bond** between a sugar residue (monosaccharide) and an amino acid side chain of a protein or a residue in an existing carbohydrate chain.

Golgi complex Stacks of membranous structures in eukaryotic cells that function in processing and sorting of proteins and lipids destined for other cellular compartments or for secretion; also called *Golgi apparatus.*

G protein Any of numerous heterotrimeric guanine nucleotide-binding proteins that usually are linked to a **seven-spanning receptor** on the cell surface. Binding of hormone to the receptor converts the inactive form with GDP bound to the α subunit to the active $G_\alpha \cdot$ GTP form, which in turn stimulates or inhibits an effector protein that generates a **second messenger** or functions as an ion channel.

growing fork Site in double-stranded DNA at which the **template** strands are separated and addition of deoxyribonucleotides to each newly formed chain occurs; also called *replication fork.* As DNA synthesis proceeds, the growing fork continuously moves in the direction of synthesis of the **leading strand.** (See Figure 10-12.)

growth factor An extracellular polypeptide molecule that binds to a cell-surface receptor triggering a **signal-transduction pathway** leading to cell proliferation or, in other cases, to specific differentiation responses. The receptors for many growth factors (e.g., epidermal growth factor, platelet-derived growth factor, insulin) are **receptor tyrosine kinases.**

GTP (guanosine 5′-triphosphate) A **nucleotide** that is a precursor in RNA synthesis and also plays a special role in protein synthesis, signal-transduction pathways, and microtubule assembly. See also **GTPase superfamily.**

GTPase superfamily Group of guanine nucleotide-binding proteins that cycle between an inactive state with bound GDP and an active state with bound GTP. These proteins—including **G proteins**, **Ras proteins**, and certain polypeptide **elongation factors**—function as intracellular switch proteins.

haploid Referring to an organism or cell having only one member of each pair of **homologous chromosomes** and hence only one copy (**allele**) of each gene or genetic locus. Gametes and bacterial cells are haploid. See also **diploid**.

heat-shock response Increased expression of a specific group of genes (*hsp* genes) in response to elevated temperature or other stressful treatment accompanied by decreased transcription of other genes and decreased translation of other mRNAs. This response is very widespread among both prokaryotic and eukaryotic organisms and helps the organism survive the stress.

HeLa cell Line of human epithelial cells, derived from a human cervical carcinoma, that grows readily in culture and is widely used in research.

helix See **α helix** and **double helix, DNA**.

helix-loop-helix A conserved protein dimerization **motif** characterizing one class of eukaryotic **transcription factors**. (See Figure 11-48g.)

helix-turn-helix A DNA binding **motif** found in most bacterial DNA binding proteins (See Figure 11-15).

heterochromatin Regions of **chromatin** that remain highly condensed and transcriptionally inactive during **interphase**.

heteroduplex A **duplex DNA** containing one or more mispaired bases.

heterokaryon Cell with more than one functional nucleus produced by the fusion of two or more different cells. See also **cell fusion**.

heterozygote Referring to a diploid cell or organism having two different **alleles** of a particular gene.

hexose A six-carbon **monosaccharide**.

high-energy bond Covalent bond that releases a large amount of energy when hydrolyzed under the usual intracellular conditions. Examples include the **phosphoanhydride bonds** in ATP, **thioester bond** in acetyl CoA, and various phosphate ester bonds. (See Table 2-9.)

histones A family of small, highly conserved basic proteins, found in the chromatin of all eukaryotic cells, that associate with DNA in the **nucleosome**. The five major types are H1, H2A, H2B, H3, and H4 histone. Histone-like proteins also are present in some prokaryotes.

Holliday structure Intermediate structure in DNA **recombination** whose resolution can result in **recombination** and/or **heteroduplex** formation. (See Figures 10-28 and 10-30.)

homeobox Conserved protein sequence which forms a DNA-binding domain (**homeodomain**) in a class of transcription factors encoded by certain **homeotic** genes. (See Figure 11-48a.)

homeodomain A conserved DNA-binding **motif** found in many developmentally important transcription factors. See **homeobox**.

homeotic gene A gene in which mutations cause cells in one region of the body to act as though they were located in another, giving rise to conversions of one cell, tissue, or body region into another. Most but not all, homeotic genes encode **homeodomain**-containing proteins.

homologous chromosome One of two copies of a particular chromosome in a **diploid** cell; also called *homologue*. Each is derived from a different parent.

homology Similarity in the sequence of a protein or nucleic acid or in the structure of an organ that reflects a common evolutionary origin. In contrast, analogy is a similarity in structure or function that does not reflect a common evolutionary origin.

homozygote Referring to a diploid cell or organism having two identical **alleles** of a particular gene.

hormone General term for any extracellular substance that induces specific responses in target cells. Hormones coordinate the growth, differentiation, and metabolic activities of various cells, tissues, and organs in multicellular organisms.

host cell A cell in which a virus or **cloning vector** can survive and replicate.

humoral immunity Immunity conferred by circulating antibodies produced by **B lymphocytes** and plasma cells. (See Figure 27-1.)

hybridization Association of two **complementary** nucleic acid strands to form double-stranded molecules. Hybrids can contain two DNA strands, two RNA strands, or one DNA and one RNA strand. In situ hybridization is a technique for determining the location of a specific RNA sequence within a tissue or cell by treatment with a labeled (e.g., radiolabeled) single-stranded nucleic acid **probe** followed by detection (e.g., **autoradiography**). In situ hybridization is also used to map the location of genes to specific chromosomal locations.

hybridoma A **clone** of hybrid cells, formed by fusion of normal B or T **lymphocytes** with myeloma cells, which are immortal and produce antibodies or T-cell receptors. Hybridomas commonly are used to produce **monoclonal antibodies**.

hydrogen bond A noncovalent association between an electronegative atom (commonly oxygen or nitrogen) and a hydrogen atom covalently bonded to another electronegative atom. Although relatively weak, hydrogen bonds are numerous in macromolecules; they are particularly important in stabilizing the three-dimensional structure of proteins and are responsible for formation of **base pairs** in nucleic acids.

hydrolysis Reaction in which a covalent bond is cleaved with addition of an H from water to one product of the cleavage and of an OH from water to the other.

hydrophilic Interacting effectively with water. See also **polar**.

hydrophobic Not interacting effectively with water; in general, poorly soluble or insoluble in water. See also **nonpolar**.

hydrophobic interaction The force that drives association of **nonpolar** molecules or parts of molecules with each other in aqueous solution. A type of **noncovalent bond** that is particularly important in stabilization of the **phospholipid bilayer**.

hydroxyl group (–OH) A hydrogen atom covalently bonded to an oxygen atom. A common substituent group in sugars and in the side group of several amino acids. Hydroxyl groups often participate in formation of intra- and intermolecular **hydrogen bonds** in biological molecules.

hypertonic A hypertonic solution is one having an osmotic strength greater than that of a cell ($\approx$300 mOsm). Such a solution causes water to move out of a cell due to **osmosis**.

hypotonic A hypotonic solution has an osmotic strength lower than that of a cell ($\approx$300 mOsm). Such a solution causes water to move into a cell due to **osmosis**.

immortality Property of a **cell line** that permits it to undergo an unlimited number of cell divisions.

immune response Variety of host defenses mediated by **B** and **T lymphocytes**, and other white blood cells that result in the lysis of microorganisms or foreign cells (cell-mediated immunity) and elimination of foreign molecules (**antigens**) by interaction with specific **antibody** (humoral immunity). See also **inflammation**. (See Figure 27-1.)

immunogen A substance capable of eliciting an **immune response**. See also **antigen**.

immunoglobulin (Ig) Any protein that functions as an **antibody**. An Ig molecule contains two or more identical heavy chains and two or more identical light chains. The five major classes of vertebrate immunoglobulins (IgA, IgD, IgE, IgG, and IgM) differ in the heavy chains they contain, their average serum levels, and their specific functions in the immune response. (See Table 27-1.)

immunoglobulin (Ig) superfamily Group of proteins that contain immunoglobulin-fold domains, or structurally related domains, including **immunoglobulins**, **T-cell receptors**, **MHC** molecules, and some **cell-adhesion molecules**. (See Figure 27-30.)

induction In embryogenesis, a change in the developmental fate of one cell or tissue caused by direct interaction with another cell or tissue or with an extracellular **signaling molecule**; in metabolism, an increase in the synthesis of an enzyme or series of enzymes mediated by a specific molecule (inducer).

inflammation Localized tissue response to injury or infection involving an influx of phagocytic and other white blood cells, which mediate removal of antigens and microorganisms, and healing of the damaged tissue.

initiation factor One of a group of proteins that promote the proper association of **ribosomes** and **mRNA** and are required for initiation of protein synthesis. See also **elongation factor**. (See Figure 4-42.)

initiator A eukaryotic promoter sequence for RNA polymerase II that specifies transcription initiation within the sequence. (See Figure 11-35.)

inositol phospholipids A family of lipids containing phosphorylated inositol derivatives that are important in **signal-transduction pathways** in eukaryotic cells. (See Figure 20-29.)

insulin A protein **hormone** produced in the β cells of the pancreas that stimulates uptake of glucose into fat and muscle cells and with **glucagon** helps to regulate blood glucose levels. Insulin also functions as a growth factor for many cells.

integral membrane protein Any membrane-bound protein all or part of which interacts with the hydrophobic core of the **phospholipid bilayer** and can be removed from the membrane only by extraction with detergent; also called *intrinsic membrane protein*. Most are transmembrane proteins, which span the bilayer and have domains extending on each side of the membrane. (See Figure 14-1.)

integration The insertion of one DNA molecule into another. Integration occurs during the **lysogenic** cycle of bacteriophage λ (see Figure 6-19), infection by **retroviruses** (see Figure 6-23) and the transposition of mobile DNA elements, including insertion sequences, **transposons**, and **retrotransposons** (see Figure 9-14).

integrins A large family of heterodimeric cell-surface receptors that promote adhesion of cells to the **extracellular matrix** by binding to **fibronectin**, **laminin**, and other matrix components, or to the surface of other cells by interacting with members of the Ig superfamily. (See Table 24-2.)

interferon (IFN) A group of small proteins released from macrophages following stimulation or from many cells after virus infection that bind to **tyrosine kinase-linked receptors** on target cells inducing changes in gene expression leading to an antiviral state (IFNα, IFNβ, and IFNγ) or other cellular responses important in the immune response (IFNγ).

intermediate filaments Cytoskeletal fibers (10 nm in diameter) formed by polymerization of several classes of cell-specific subunit proteins. They attach to spot **desmosomes** and **hemidesmosomes**; form the major structural proteins of skin and hair; form the scaffold that holds Z disks and myofibrils in place in muscle; and generally function as important structural determinants in many animal cells and tissues.

intermediate-repeat DNA The eukaryotic DNA fraction that reassociates at an intermediate rate and is derived from **mobile DNA elements**. This fraction constitutes 25–40 percent of total mammalian DNA. See also **simple-sequence DNA** and **single-copy DNA**.

interphase Long period of the **cell cycle** (G$_1$, S phase, and G$_2$) between one **mitosis** and the next.

intervening sequence See **intron**.

intrinsic protein See **integral membrane protein**.

intron Part of a **primary transcript** (or the DNA encoding it) that is removed by **splicing** during **RNA processing** and is not included in the mature, functional mRNA, rRNA, or tRNA; also called *intervening sequence*.

inverted repeat Self-complementary region of a nucleic acid chain containing an axis of symmetry and capable of folding back on itself and base-pairing about this axis of symmetry (e.g., CGCAT . . . . . ATGCG).

in vitro Denoting a reaction or process taking place in an isolated cell-free extract; sometimes used to distinguish cells growing in culture from those in an organism (**in vivo**).

in vivo In an intact cell or organism.

ion channel Any transmembrane protein complex that forms a water-filled channel across the phospholipid bilayer allowing selective ion transport down its electrochemical gradient. See also **ion pump**.

ionic bond A **noncovalent bond** between a positively charged ion (cation) and negatively charged ion (anion).

ionophore Any compound that increases the permeability of membranes to a specific ion.

ion pump Any transmembrane ATPase that couples hydrolysis of ATP to the transport of a specific ion across a membrane against its electrochemical gradient. The three principal classes of ion pumps—P, V, and F—differ in structure, location, and the ions transported. (See Table 15-3 and Figure 15-10.)

isoelectric focusing Technique for separation of molecules based on their charge by gel **electrophoresis** in a pH gradient subjected to an electric field. A protein migrates to the pH at which its overall net charge is zero; this pH is called the isoelectric point.

isoforms Multiple forms of the same protein whose amino acid sequences differ slightly but whose general activity is identical. They may be produced by alternative splicing of RNA transcripts from the same gene or be encoded by different genes.

isotonic Referring to a solution whose solute concentration and osmotic strength equals that of a cell so that there is not net movement of water in or out of the cell.

karyotype Number, sizes, and shapes of the entire set of **metaphase** chromosomes of a eukaryotic cell. (See Figure 9-51.)

kinase An enzyme that adds the γ phosphate group from ATP to a substrate. (See Figure 7-20 and **phosphorylation**.)

kinesin A motor protein that uses the energy of ATP to move along a microtubule, transporting vesicles or particles in the process. (See Figure 23-23.)

kinetochore A three-layer protein structure located at or near the **centromere** of each mitotic chromosome from which microtubules (kinetochore fibers) extend toward the spindle poles of the cell; plays an active role in movement of chromosomes toward the poles during **anaphase**. (See Figure 23-38.)

K_m A parameter that describes the affinity of an enzyme for its substrate and equals the substrate concentration that yields the half-maximal reaction rate; also called the *Michaelis constant*. A similar parameter describes the affinity of a transport protein for the transported molecule or the affinity of a receptor for its ligand, since the kinetics of transport and ligand binding are similar to those of enzymatic reactions. (See Figure 3-29.)

knockout, gene Technique for selectively inactivating a gene by replacing it with a mutant allele in an otherwise normal organism. (See Figure 8-38.)

Krebs cycle See **citric acid cycle.**

label A fluorescent or radioactive chemical group (or atom) added to a molecule in order to spatially locate the molecule or follow it through a reaction or purification scheme. As a verb, to add such a group or atom to a cell or molecule. See also **radioisotope.**

lagging strand Newly synthesized DNA strand formed at the **growing fork** as short, discontinuous segments, called **Okazaki fragments,** which are later joined by DNA ligase. Although overall lagging strand synthesis occurs in the 3′ to 5′ direction, each Okazaki fragment is synthesized in the 5′ to 3′ direction. See also **leading strand.** (See Figure 10-12.)

laminin A component of the **extracellular matrix** that is found in all basal laminae and has binding sites for cell-surface receptors, collagen, and heparan sulfate proteoglycans. (See Figure 24-22.)

lamins A group of **intermediate filament** proteins that form the fibrous network (**nuclear lamina**) on the inner surface of the nuclear envelope. Phosphorylation and dephosphorylation of lamins play an important role in nuclear-envelope decondensation and condensation during **mitosis.** (See Figures 25-12 and 25-17.)

leading strand Newly synthesized DNA strand formed by continuous synthesis in the 5′ → 3′ direction at the **growing fork.** The direction of leading-strand synthesis is the same as movement of the growing fork. See also **lagging strand.** (See Figure 10-12.)

lectin Any protein that binds tightly to specific sugars. Lectins can be used in affinity chromatography to purify glycoproteins or as reagents to detect them in situ.

leucine zipper A coiled-coil of two α-helices that connect the polypeptide chains of one class of dimeric eukaryotic **transcription factors.** (See Figure 11-48g and h.)

leukocyte A nucleated blood cell that does not contain hemoglobin; also called *white blood cell.* These cells, including **lymphocytes,** macrophages, and neutrophils, are important in the **immune response.**

library See **DNA library.**

ligand Any molecule, other than an enzyme **substrate,** that binds tightly and specifically to a macromolecule, usually a protein, forming a macromolecule-ligand complex, which usually signals some cellular response.

ligase An enzyme that links together the 3′ end of one nucleic acid strand with the 5′ end of another, forming a continuous strand.

light reactions The sum total of the light-dependent reactions and processes that generate ATP and NADPH during **photosynthesis.** These reactions are not dependent on the **dark reactions,** but the dark reactions are indirectly dependent on the light reactions.

LINES (long interspersed elements) One of the abundant **intermediate repeat DNA** sequences in mammals, a non-viral **retrotransposon** about 6 to 7 kb long. (See Figure 9-30.)

linkage In genetics, the tendency of two different loci on the same chromosome to be inherited together. The closer two loci are, the greater their linkage and the lower the frequency of **recombination** between them.

lipid A molecule that is insoluble in water but is soluble in organic solvents. Lipids contain covalently-linked fatty acids and are found in fat droplets and, as **phospholipids,** in biomembranes.

lipophilic Soluble in hydrocarbons and other non-polar solvents. See **hydrophobic.**

liposome Spherical phospholipid bilayer structure with an aqueous interior that forms in vitro from **phospholipids** and may contain protein. (See Figure 15-4.)

locus In genetics, the specific site of a gene on a chromosome. All the **alleles** of a particular gene occupy the same locus.

lymphocyte See **B lymphocyte** and **T lymphocyte.**

lymphokine See **cytokine.**

lysogenic cycle Series of events in which a bacterial **virus** of a certain type enters a host cell and its DNA is incorporated into the host-cell genome in such a way that the virus (the prophage) lays dormant. The association of a prophage with the host-cell genome is called lysogeny. By various mechanisms, the prophage can be activated so that it replicates, produces new viral particles, and lyses the bacterium. (See Figure 6-19.)

lysogeny See **lysogenic cycle.**

lysosome Small membrane-bounded organelle having an internal pH of 4–5 and containing hydrolytic enzymes, which aid in the digestion of material ingested by **phagocytosis** and **endocytosis.**

lytic cycle Series of events in which a **virus** enters and replicates within a host cell to produce new viral particles eventually causing lysis of the cell. See also **lysogenic cycle.** (See Figure 6-17.)

M phase See **cell cycle.**

macromolecule Any large, usually polymeric molecule (e.g., a protein, nucleic acid, polysaccharide) with a molecular mass greater than a few thousand daltons.

major histocompatibility complex See **MHC.**

malignant Referring to a **tumor** or tumor cells that can invade surrounding normal tissue and/or undergo **metastasis.** See also **benign.**

mapping, DNA Various techniques for determining the relative order of genes on a chromosome (genetic map), the absolute position of genes (physical map), or the relative position of restriction sites (restriction map).

meiosis In eukaryotes, a special type of cell division comprising two successive nuclear and cellular divisions with only one round of DNA replication resulting in production of four genetically nonequivalent **haploid** cells (**gametes**) from an initial **diploid** cell. See also **mitosis.** (See Figure 5-53.)

membrane potential Voltage difference across a membrane due to the slight excess of positive ions (cations) on one side and negative ions (anions) on the other resulting from the selective permeability of the membrane to different ions or from pumping of ions across the membrane by **active transport.** (See Figure 15-8.)

messenger RNA See **mRNA.**

metabolism The sum of the chemical processes that occur in living cells; includes **anabolism** and **catabolism.**

metaphase Mitotic stage at which chromosomes are fully condensed and attached to the mitotic spindle at its equator but have not yet started to segregate toward the opposite spindle poles. (See Figure 5-50.)

metastasis Spread of **tumor** cells from their site of origin and establishment of areas of secondary growth.

MHC (major histocompatability complex). A number of genes, located in close proximity, most of which are important in T cell-mediated immune responses. They include genes that encode the key class I and II MHC molecules that bind peptides from foreign proteins and present them to the T cell receptors. Also present are genes encoding proteases that degrade foreign proteins and transporters that bring the peptides to the class I and II MHC molecules in the endoplasmic reticulum or other organelles. The MHC was originally identified as being responsible for rapid graft rejection between individuals of a species because of their high polymorphism (multiple alleles at the locus). (See Figures 27-33 and 27-34.)

microfilaments Cytoskeletal fibers ($\approx$7 nm in diameter) that are formed by polymerization of monomeric globular (G) **actin**, exhibit **polarity**, and associate with a variety of actin-binding proteins; also called *actin filaments*. Microfilaments play an important role in muscle contraction, cytoplasmic streaming, cytokinesis, cell movement, platelet activation, and other cellular functions and structures. (See Figure 22-2.)

microtubule-associated protein (MAP) Any protein that binds to **microtubules** in a constant ratio and determines the unique properties of different types of microtubules. Numerous MAPs have been identified including the motor proteins **dynein** and **kinesin**.

microtubule-organizing center (MTOC) Origin of the radial array of **microtubules** seen in non-mitotic (interphase) cells. See also **centrosome**.

microtubules Cytoskeletal fibers (24 nm in diameter) that are formed by polymerization of α,β-tubulin monomers and exhibit structural and functional **polarity**. They are important components of the **cilium**, **flagellum**, mitotic spindle, and other cellular structures. Vesicles and protein particles often are transported along microtubules in a process mediated by **kinesin** or **dynein**. (See Figure 23-1.)

microvillus (pl. microvilli) Small, membrane-covered projection on the surface of an animal cell containing a core of actin filaments and various associated proteins. Numerous microvilli are present on the absorptive surface of intestinal epithelial cells, increasing the surface area for transport of nutrients and forming the brush border. (See Figure 22-8.)

missense mutation Any mutation that alters a **codon** so that it codes for a different amino acid. See also **nonsense mutation**.

mitochondrion (pl. mitochondria) Large **organelle** that is surrounded by two phospholipid bilayer membranes, contains DNA, and carries out **oxidative phosphorylation**, thereby producing most of the ATP in eukaryotic cells.

mitosis In eukaryotes, the process whereby a cell divides to produce two genetically equivalent daughter cells. Mitosis refers specifically to nuclear division, whereas cytokinesis refers to division of the cytoplasm. Both processes occur during the M phase of the **cell cycle**, which is conventionally divided into four stages: prophase, metaphase, anaphase, and telophase. See also **meiosis**. (See Figure 5-50.)

mobile DNA element Any DNA sequence that is not present in the same chromosomal location in all individuals of a species. Examples include Tn elements in bacteria, *copia* and P elements in *Drosophila*, Ty elements in yeast, Ac and Ds in maize, and LINES and SINES in mammals. See also **retrotransposons** and **transposons**.

monoclonal antibody Antibody produced by the progeny of a single B cell and thus a homogenous protein exhibiting a single antigen specificity. Experimentally, it is produced by use of a **hybridoma**. (See Figure 6-12.)

monomer Any small molecule that can be linked with others of the same type to form a **polymer**. Examples include **amino acids**, **nucleotides**, and **monosaccharides**.

monomeric For proteins, consisting of a single polypeptide chain.

monosaccharide Any simple sugar with the formula $(CH_2O)_n$ where $n = 3$–7.

motif In proteins, a unit exhibiting a particular three-dimensional architecture that is found in a variety of proteins and usually is associated with a particular function. Many DNA-binding proteins contain one of a small number of DNA-binding motifs including the **helix-loop-helix, homeodomain, leucine zipper,** and **zinc finger**. (See Figure 11-48.)

MPF (mitosis-promoting factor) A heterodimeric protein that triggers entrance of a cell into the M phase by inducing chromatin condensation and nuclear-envelope breakdown; originally called *maturation-promoting factor* because it supports maturation of G_2-arrested frog oocytes into mature eggs. It consists of an A or B type **cyclin** and a **cyclin-dependent protein kinase** (cdc2, also called cdk1).

mRNA (messenger RNA) Any **RNA** that specifies the order of amino acids in protein synthesis. It is produced by **transcription** of DNA by RNA polymerase and, in RNA viruses, by transcription of viral RNA. In eukaryotes, the initial RNA product (primary transcript) undergoes **RNA processing** to yield functional mRNA, which is transported to the cytoplasm. See also **translation**.

multimeric For proteins, containing several polypeptide chains (or subunits).

mutation In genetics, a permanent, transmissable change in the nucleotide sequence of a chromosome, usually in a single **gene**, which leads to a change in or loss of its normal function.

myelin sheath Stacked specialized cell membrane that forms an insulating layer around vertebrate **axons** and increases the speed of impulse conduction. It is produced by Schwann cells in the peripheral nervous system and oligodendrocytes in the central nervous system. (See Figure 14-11.)

myofiber A long (1–40 nm), multinucleate cell that contains **actin** and **myosin** organized in **myofibrils** and capable of contraction; a muscle cell. (See Figure 22-24.)

myofibril Long, highly organized bundle of thin (**actin**) filaments, thick (**myosin**) filaments, and other proteins that constitute the basic structural unit of muscle cells (myofibers) and contract by a sliding-filament mechanism. A myofibril is divided into **sarcomeres**. (See Figure 22-24.)

myosin A protein with a globular head region and coiled-coil tail region that has actin-stimulated ATPase activity, which drives movement along actin filaments during muscle contraction and cell movement. (See Figure 22-17.)

NAD$^+$ (nicotinic adenine dinucleotide) A widely used **coenzyme** that participates in oxidation reactions by accepting two electrons from a donor molecule and one H$^+$ from the solution to form NADH, which is an important electron carrier in **oxidative phosphorylation**. (See Figure 17-3.)

NADP$^+$ (nicotinic adenine dinucleotide phosphate) Phosphorylated form of **NAD$^+$**, which is used extensively as an electron carrier in biosynthetic pathways. (See Figure 17-3.) NADP$^+$ is reduced, forming NADPH, during photosynthesis. (See Figure 18-12).

Nernst equation Mathematical expression that defines the electric potential E across a membrane as directly proportional to the logarithm of the ratio of the ion concentrations on either side of the membrane and inversely proportional to the **valency** of the ions.

neuron (nerve cell) Any of the impulse-conducting cells of the nervous system. A typical neuron contains a cell body; several short, branched processes (**dendrites**); and one long process (**axon**). (See Figure 21-1.)

neuropeptide A peptide secreted by neurons that functions as a **signaling molecule** either at a synapse or elsewhere. These molecules have diverse, often long-lived effects in contrast to **neurotransmitters**. (See Table 21-2.)

neurotransmitter Extracellular **signaling molecule** that is released by the presynaptic neuron at a chemical **synapse** and relays the signal to the postsynaptic cell. A neurotransmitter can elicit either an excitatory or inhibitory response, but the response is determined by the **receptor** activated by the neurotransmitter not by the chemical nature of the neurotransmitter. Examples include acetylcholine, dopamine, GABA (γ-aminobutyric acid), and serotonin. (See Table 21-1.)

noncovalent bond Any chemical bond between atoms that does not involve an intimate sharing of electrons. Although covalent bonds are relatively weak, large numbers of these bonds can stabilize the three-dimensional structure of biological molecules, particularly proteins and nucleic acids. The four main types are the **hydrogen bond, hydrophobic interaction, ionic bond,** and **van der Waals interaction.**

nonpermissive temperature See **temperature-sensitive mutant.**

nonpolar Referring to a molecule or structure that lacks any net electric charge or asymmetric distribution of positive and negative charges. Nonpolar molecules generally are insoluble in water.

nonsense mutation Any mutation that alters a **codon** so that it codes for a stop (termination) codon and thus results in production of a truncated polypeptide which is nonfunctional or has an altered function. See also **missense mutation.**

Northern blotting Technique for detecting specific mRNAs. See also **blotting.** (See Figure 7-32.)

nuclear envelope Double-membrane structure surrounding the nucleus; the outer membrane is continuous with the endoplasmic reticulum and the two membranes are perforated by **nuclear pores.** (See Figure 25-10.)

nuclear lamina Fibrous network on the inner surface of the inner nuclear membrane composed of **lamin** filaments. (See Figure 25-12.)

nuclear pore A protein-lined channel through the **nuclear envelope** that permits passage of RNA from the nucleus to the cytoplasm and of proteins from the cytoplasm to the nucleus. (See Figure 12-35.)

nucleic acid A polymer of **nucleotides** linked by phosphodiester bonds. **DNA** and **RNA** are the primary nucleic acids in cells.

nucleocapsid See **capsid.**

nucleolus Large structure in the nucleus of eukaryotic cells where **rRNA** synthesis and processing occurs and ribosome subunits are assembled.

nucleoside A small molecule composed of a **purine** or **pyrimidine** base linked to a **pentose** (either ribose or deoxyribose). (See Table 4-1.)

nucleosome Small structural unit of **chromatin** consisting of a disk-shaped core of **histone** proteins around which a $\approx$146-bp segment of DNA is wrapped. (See Figure 9-48.)

nucleotide A **nucleoside** with one or more phosphate groups linked via an ester bond to the sugar moiety. DNA and RNA are polymers of nucleotides. (See Figure 4-4 and Table 4-1.)

nucleus Large membrane-bounded organelle that contains DNA organized into chromosomes; synthesis and processing of RNA and ribosome assembly occur in the nucleus. Nearly all eukaryotic cells contain a nucleus, whereas prokaryotic cells do not.

Okazaki fragments Short (<1000 bases), single-stranded DNA fragments base paired to the DNA template and covalently linked at the 5' end to a short RNA **primer.** They are formed during discontinuous synthesis of the **lagging strand** in DNA replication. (See Figure 10-12.)

oligodendrocyte See **myelin sheath.**

oligomer General term for a short polymer most commonly consisting of amino acids (oligopeptides), nucleic acids (oligonucleotides), or sugars (oligosaccharides).

oncogene A gene whose product is involved either in transforming cells in culture or in inducing cancer in animals. Most oncogenes are mutant forms of normal genes (**proto-oncogene**) involved in the control of cell growth or division.

operator Short DNA sequence in a bacterial or viral genome that binds a repressor protein and controls transcription of an adjacent gene.

operon In bacterial DNA, a cluster of contiguous genes transcribed from one **promoter** that gives rise to a **polycistronic** mRNA.

organelle Any membrane-limited structure found in the cytoplasm of eukaryotic cells.

osmosis Movement of water across a semipermeable membrane from a solution of lesser to one of greater solute concentration. The membrane must be permeable to water but not to solute molecules.

osmotic pressure Hydrostatic pressure that must be applied to the more concentrated solution to stop the net flow of water across a semipermeable membrane separating solutions of different concentrations. (Figure 15-25.)

oxidation Loss of electrons from an atom or molecule as occurs when hydrogen is removed from a molecule or oxygen is added. The opposite of reduction.

oxidative phosphorylation The phosphorylation of ADP to form ATP driven by the transfer of electrons to oxygen (O_2) in bacteria and mitochondria. This process involves generation of a **proton-motive force** during electron transport, and use of the proton-motive force to power ATP synthesis. (See Figure 17-9.)

palindromic sequence Nucleotide sequence that is identical to its complementary sequence when each is read in the same direction (e.g., $5' \rightarrow 3'$). Many sites recognized by **restriction enzymes** are palindromes.

passive diffusion, also called **simple diffusion** Movement of a molecule across a membrane at a rate proportional to its concentration gradient across the membrane and the permeability of the membrane. Only water, gases (O_2, CO_2, N_2), small hydrophobic molecules, and small uncharged polar molecules such as urea and ethanol can move across biomembranes by simple diffusion. See also **active transport** and **facilitated diffusion.**

pathogen Any disease-producing agent (e.g. a bacterium or virus).

PCR (polymerase chain reaction) Technique for amplifying a specific DNA segment in a complex mixture by multiple cycles of DNA synthesis from short oligonucleotide **primers** followed by brief heat treatment to separate the complementary strands. (See Figure 7-37.)

pentose A five-carbon **monosaccharide.** The pentoses ribose and deoxyribose are present in RNA and DNA, respectively.

peptide A small polymer usually containing fewer than 30 **amino acids** connected by peptide bonds.

peptide bond Covalent bond that links adjacent amino acid residues in proteins; formed by a condensation reaction between the amino group of one amino acid and the carboxyl group of another with release of a water molecule. (See Figure 3-4.)

permissive temperature See **temperature-sensitive mutant.**

peroxisome Small **organelle** in eukaryotic cells whose functions include degradation of fatty acids and amino acids by means of reactions that generate hydrogen peroxide, which is converted to water and oxygen by catalase.

pH A measure of the acidity or alkalinity of a solution defined as the negative logarithm of the hydrogen ion concentration in moles per liter: $pH = -\log [H^+]$. Neutrality is equivalent to a pH of 7; values below this are acidic and those above are alkaline.

phage See **bacteriophage.**

phagocytosis Process by which relatively large particles (e.g., bacterial cells) are internalized by certain eukaryotic cells. Phagocytosis of **pathogens** by macrophages and neutrophils is an important part of the **immune response** in vertebrates.

phenotype The observable characteristics of a cell or organism as distinct from its **genotype.**

phosphatase An enzyme that removes a phosphate group from a substrate by hydrolysis. See **dephosphorylation.**

phosphoanhydride bond A **phosphodiester bond**, a type of **high-energy bond**, present in **ATP**, that releases considerable free energy when hydrolyzed. (See Figure 2-22.)

phosphodiester bond A set of covalent bonds in which two hydroxyl groups form ester linkages to the same phosphate group. Adjacent nucleotides in DNA and RNA are joined by phosphodiester bonds. (See Figure 4-5.)

phospholipid The major class of lipids present in **biomembranes**, usually composed of two fatty acid chains esterified to two of the carbons of glycerol phosphate, with the phosphate esterified to one of various polar groups. (See Figure 14-2.)

phospholipid bilayer A symmetrical two-layer structure formed by **phospholipids** in aqueous solution; the basic structure of all **biomembranes**. In a bilayer, the **polar** head groups are exposed to the aqueous medium, while the **nonpolar** hydrocarbon chains of the fatty acids are in the center. (See Figures 14-7 and 14-8.)

phosphorylation Reaction in which a phosphate group becomes covalently linked to another molecule. The activity of many proteins is regulated by phosphorylation of hydroxyl-containing residues (serine, threonine, tyrosine) by various protein kinases.

photosynthesis Complex series of **light reactions** and **dark reactions** in some bacteria and the **chloroplasts** of plants that uses light energy to generate carbohydrates from CO_2, usually with the consumption of H_2O and evolution of O_2.

phylogeny The evolutionary history of an organism or group of organisms; often represented in chart form as a phylogenetic "tree." (See Figure 9-4.)

pinocytosis The nonspecific uptake of small droplets of extracellular fluid into endocytic vesicles. See **endocytosis.**

plaque assay Technique for determining the number of infectious viral particles in a sample by culturing a diluted sample on a layer of susceptible host cells and then counting the clear areas of lysed cells (plaques) that develop. (See Figure 6-16.)

plasma membrane The membrane surrounding a cell that separates the cell from its external environment, consisting of a **phospholipid bilayer** and associated proteins.

plasmid Small, circular extrachromosomal DNA molecule capable of autonomous replication in a cell. Commonly used as a **cloning vector.**

pluripotent Referring to a **stem cell** capable of giving rise to multiple types of differentiated cells. See also **unipotent.** (See Figure 13-16.)

point mutation Change of a single nucleotide in DNA, especially in a region coding for protein.

polar Referring to a molecule or structure with a net electric charge or asymmetric distribution of positive and negative charges. Polar molecules are usually soluble in water.

polar bond A **covalent bond** in which the electrons are unequally shared due to a difference in the ability of the linked atoms to attract electrons, a property called electronegativity. One end of polar bond carries a small negative charge; the other end, a small positive charge.

polarity Used to denote functional and/or structural differences in distinct regions of a cell or cellular component. For example, the two ends of a microtubule or actin filament—the (+) and (−) ends—exhibit distinct properties, and the plasma membrane of many epithelial cells is organized into two discrete regions—the apical and basolateral—with different functions.

polymer Any large molecule composed of multiple identical or similar units (**monomers**) linked by covalent bonds. Proteins, nucleic acids, and polysaccharides are the primary polymers in cells.

polymerase chain reaction See **PCR.**

polypeptide Linear polymer of **amino acids** connected by **peptide bonds**. Proteins are large polypeptides, and the two terms commonly are used interchangeably.

polyribosome (polysome) A complex containing several ribosomes all translating a single messenger RNA.

polysaccharide Linear or branched polymer of **monosaccharides**, linked by **glycosidic bonds**, usually containing more than 15 residues. Examples include **glycogen, cellulose,** and **glycosaminoglycans.**

pre-mRNA Precursor mRNA; the **primary transcript** and intermediates in **RNA processing** that yield functional mRNA.

primary structure In proteins, the linear arrangement (sequence) of **amino acids** and the location of covalent (mostly disulfide) bonds within a polypeptide chain. (See Figure 3-5.)

primary transcript Initial RNA product, containing **introns** and **exons**, produced by **transcription** of DNA. Many primary transcripts must undergo **RNA processing** to form the physiologically active RNA species.

primase A specialized RNA polymerase that synthesizes short stretches of RNA used as primers for DNA synthesis. (See Figures 10-11 and 10-14.)

primer A short nucleic acid sequence containing a free 3′ hydroxyl group that forms **base pairs** with a complementary **template** strand and functions as the starting point for addition of nucleotides to copy the template strand.

probe Defined RNA or DNA fragment, radioactively or chemically labeled, that is used to locate specific nucleic acid sequences by **hybridization.**

prokaryotes Class of organisms, including the **eubacteria** and **archaebacteria**, that lack a true membrane-limited nucleus and other organelles. See also **eukaryotes.**

promoter DNA sequence which determines the site of **transcription** initiation for an RNA polymerase.

promoter-proximal element Any regulatory sequence in eukaryotic DNA that is located close to (within 200 base pairs of) a **promoter** and binds a specific protein thereby modulating transcription of the associated protein-coding gene. Many genes are controlled by multiple promoter-proximal elements. (See Figure 11-42.)

prophase Earliest stage in **mitosis** during which the chromosomes condense and the centrioles begin moving toward the spindle poles. (See Figure 5-50.)

prosthetic group See **coenzyme.**

protein A linear polymer of **amino acids** linked together in a specific sequence and usually containing more than 50 residues. Proteins—the most abundant macromolecules in cells—serve as **en-**

zymes, structural elements, **antibodies, hormones, electron carriers,** etc., and are involved in nearly all cellular activities. X-ray crystallographic analysis of proteins has revealed four levels of structure: primary (the sequence of amino acids), secondary, tertiary, and quaternary. (See Figure 3-5.)

proteoglycans A group of glycoproteins that contain a core protein to which is attached one or more **glycosaminoglycans.** They are found in nearly all extracellular matrices, and some are attached to the plasma membrane. (See Figures 24-19 and 24-20.)

proton-motive force The energy equivalent of the proton concentration gradient and electric potential gradient across a membrane. In chloroplasts and mitochondria, these gradients are generated by **electron transport,** maintained by the thylakoid or inner mitochondrial membrane, and drive ATP synthesis by **ATP synthase.**

proto-oncogene A normal cellular gene that encodes a protein usually involved in regulation of cell growth or proliferation and that can be mutated into a cancer-promoting **oncogene,** either by changing the protein-coding segment or by altering the regulation of the protein. (See Table 26-1.)

pseudogene A duplicated gene copy that has become nonfunctional because of **genetic drift.**

pulse-chase A type of experiment in which a radioactive small molecule is added to a cell for a brief period (the pulse) during which it is incorporated into macromolecules. The fate of the newly-synthesized, radioactive, molecules is examined during the subsequent period when the radioactive small molecule is removed from the cell medium and an excess of the same small molecule, but unlabeled, is added (the chase).

pump Any transmembrane protein that mediates the **active transport** of an ion or small molecule across a biomembrane. See **ion pump.**

purine A basic compound, containing two fused heterocyclic rings, that occurs in nucleic acids. The purines commonly found in DNA and RNA are adenine and guanine. (See Figure 4-3.)

pyrimidine A basic compound, containing one heterocyclic ring, that occurs in nucleic acids. The purines commonly found in DNA are cytosine and thymine; in RNA, uracil replaces thymine. (See Figure 4-3.)

quaternary structure The number and relative positions of the polypeptide chains (subunits) in **multimeric** proteins. Subunits are held together by noncovalent bonds, may be identical or different, and frequently associate in a regular (symmetrical) arrangement. (See Figure 3-5.)

quiescent Referring to a cell that has exited the **cell cycle** and is in the G_0 state.

radioisotope Unstable form of an atom that emits radiation as it decays. Several radioisotopes are commonly used as **labels** in biological molecules. Such radiolabeled molecules can be detected by **autoradiography** or by measurement of the radiation emitted. (See Table 6-6.)

Ras protein A monomeric guanine nucleotide-binding protein that occurs in **signal-transduction pathways** and is activated by binding of ligand to **receptor tyrosine kinases** and other receptors on the cell surface. Ligand binding leads to conversion of the inactive Ras · GDP form to the active Ras · GTP. **Constitutively** active forms of Ras are **oncogenes.** (See Figure 20-32.)

reading frame The sequence of nucleotide triplets (**codons**) that runs from a specific **translation** start codon in an mRNA to a stop codon. Some mRNAs can be translated into different polypeptides by reading in two different reading frames. (See Figure 4-27.)

receptor Any protein that binds a specific extracellular **signaling molecule** (ligand) that induces a cellular response. Receptors for steroid hormones, which diffuse across the plasma membrane, are located within the cell; receptors for water-soluble hormones (e.g., epinephrine) and peptide growth factors, as well as lipid-soluble prostaglandins, are located in the plasma membrane with their ligand-binding domain exposed to the external medium. Receptors for **neurotransmitters,** located in postsynaptic membranes, can regulate ion flow through the cell membrane or activate other signal transduction pathways.

receptor tyrosine kinases (RTKs) An important class of cell-surface **receptors** whose cytosolic domain has tyrosine-specific protein kinase activity. This activity is activated by ligand binding and leads both to receptor autophosphorylation and to phosphorylation of target proteins. Both types of phosphorylations activate particular intracellular signal transduction pathways.

recessive In genetics, referring to that **allele** of a gene that is not expressed in the **phenotype** when the **dominant** allele is present; also refers to the phenotype of an individual (homozygote) carrying two recessive alleles.

recombinant DNA Any DNA molecule formed by joining DNA fragments from different sources. They are commonly produced by cutting DNA molecules with **restriction enzymes** and then joining the resulting fragments from different sources with DNA ligase.

recombination In genetics, the exchange (crossing over) of DNA sequences on maternal and paternal **chromatids** during meiosis resulting in new combinations of genes. The farther apart two genes on the same chromosome are, the more frequently recombination between them occurs; thus analysis of recombination frequencies is a way to determine the position of genes relative to each other on a chromosome. (See Figures 5-55 and 8-21.)

reduction Gain of electrons by an atom or molecule as occurs when hydrogen is added to a molecule or oxygen is removed. The opposite of oxidation.

replication fork See **growing fork.**

residue That part of a **monomer** (amino acid, sugar, nucleotide) that is retained as part of a **polymer** following chemical linkage of the monomers.

resolution The minimum distance that can be distinguished by an optical apparatus; also called *resolving power*. For example, a microscope with a resolution of 0.5 μm cannot distinguish between two objects that are positioned less than 0.5 μm apart.

respiration General term for any cellular process involving the uptake of O_2 coupled to production of CO_2.

resting potential The electric potential across the membrane of an excitable cell when the cell is not stimulated by a **neurotransmitter** or other signaling molecule and no **action potential** is being propagated. It ranges from -30 to -70 mV (inside negative). (See Figures 21-11 and 21-12.)

restriction enzyme (endonuclease) Any enzyme that recognizes and cleaves a specific short sequence, the restriction site, in double-stranded DNA molecules. These enzymes are widespread in bacteria and are used extensively in recombinant DNA technology. (See Table 7-1 and Figure 7-5.)

restriction fragment A defined DNA fragment resulting from cleavage with a restriction enzyme. These fragments are used in the production of recombinant DNA molecules and **gene cloning.**

retrotransposon Type of eukaryotic **mobile DNA element** whose movement in the genome is mediated by an RNA intermediate and involves a step of reverse transcription. See also **transposon.**

retrovirus A type of eukaryotic virus containing an RNA genome that replicates in cells by first making a DNA copy of the RNA, a process termed **reverse transcription**. This proviral DNA is inserted into cellular chromosomal DNA, and gives rise to further genomic RNA as well as the mRNAs for viral proteins.

reverse transcription Process catalyzed by reverse transcriptase in which a DNA strand is copied from an RNA template. (See Figure 9-24.)

rhodamine See **fluorescent staining.**

ribosomal RNA See **rRNA.**

ribosome A complex comprising several different **rRNA** molecules and more than 50 proteins, organized into a large subunit and small subunit, that is the site of protein synthesis. (See Figure 4-39.)

RNA (ribonucleic acid) Long, linear, single-stranded polymer, composed of ribose **nucleotides** linked by **phosphodiester bonds;** is formed by transcription of DNA or, in some viruses, by copying of RNA. The three types of cellular RNA—mRNA, rRNA, and tRNA—play different roles in protein synthesis.

RNA polymerase An enzyme that copies one strand of DNA or RNA (the template strand) to make the complementary RNA strand. Using rules of Watson-Crick base pairing, ribonucleotides are added one-at-a-time, using as substrates ribonucleotide triphosphates and releasing pyrophosphate.

RNA processing Various modifications that occur to the initial RNA molecules produced by **transcription.** Many but not all RNAs undergo processing to yield functional molecules. RNA processing is more complex in eukaryotes than in prokaryotes. See also **RNA splicing.** (See Figure 12-9.)

RNAse (ribonuclease) An enzyme that cuts an RNA strand or completely hydrolyzes an RNA, forming ribonucleotides.

RNA splicing A process that results in removal of **introns** and joining of **exons** in RNAs. Splicing in the pre-mRNA of higher eukaryotes occurs in large ribonucleoprotein complexes called spliceosomes. (See Figures 12-20 and 12-24.)

rRNA (ribosomal RNA) Any one of several large RNA molecules that are structural and functional components of **ribosomes.** Often designated by their sedimentation coefficient: 28S, 18S, 5.8S, and 5S rRNA in higher eukaryotes. (See Figure 12-50.)

S phase See **cell cycle.**

sarcomere Repeating unit of a **myofibril** in striated muscle that extends from one Z disk to an adjacent one and shortens during contraction; it is composed of overlapping thick (**myosin**) filaments and thin (**actin**) filaments. (See Figure 22-25.)

sarcoplasmic reticulum Network of membranes, derived from the smooth endoplasmic reticulum, that surrounds each **myofibril** in a muscle cell and sequesters Ca^2 ions. Depolarization (stimulation) of a muscle cell induces release of Ca^2 ions into the cytosol, triggering coordinated contraction along the length of the cell. (See Figure 22-29.)

satellite DNA See **simple-sequence DNA.**

saturated Referring to a molecule containing only single bonds between adjacent carbon atoms (C–C); commonly used to describe fatty acids. See also **unsaturated.**

Schwann cell See **myelin sheath.**

second messenger An intracellular **signaling molecule** whose concentration increases (or decreases) in response to binding of an extracellular **ligand** to a cell-surface receptor and participates in mediating the cellular response to the ligand. Examples include cAMP, Ca^2, diacylglycerol, and inositol 1,4,5-trisphosphate.

secondary structure In proteins, local folding of a polypeptide chain into regular structures including the α **helix,** β **pleated sheet,** and U-shaped turns and loops. (See Figure 3-5.)

sequence The linear order of **monomers** in polymeric molecules, especially proteins and nucleic acids. Various techniques, collectively termed sequencing, are used to determine the identity and position of each monomer in a polymer.

seven-spanning receptors An important class of cell-surface **receptors** that have seven transmembrane α helices and are coupled to **G proteins.**

signaling molecule General term for any extracellular or intracellular molecule involved in mediating the response of a cell to its external environment or other cells. Extracellular signaling molecules include **hormones, growth factors,** and **neurotransmitters.** Intracellular signaling molecules include **second messengers** and various protein components of **signal-transduction pathways** (e.g., G proteins and Ras protein).

signal sequence A relatively short amino acid sequence that directs a protein to a specific location within the cell; also called *signal peptide.* One example is the N-terminal sequence of a nascent secretory or membrane protein that causes the growing polypeptide to cross the membrane of the endoplasmic reticulum into the lumen; the signal sequence is cleaved in the process. (See Table 16-3.)

signal-transduction pathway A series of coupled intracellular events, triggered by binding of a **signaling molecule** to a **receptor,** that occur in a sequential fashion to convert an extracellular signal into a cellular response. Examples include pathways initiated by binding of ligand to G protein-linked receptors (see Figure 20-16), receptor tyrosine kinases (see Figure 20-38), and tyrosine kinase-linked receptors (see Figure 20-52).

silencer sequence A sequence in eukaryotic DNA where **repressors** bind and inhibit transcription from **promoters** within several hundred base pairs.

SINES Short interspersed elements. An abundant intermediate DNA sequence in mammals about 300 bp long. (See Figure 9-32.)

simple-sequence DNA The eukaryotic DNA fraction that reassociates most rapidly and consists of short, tandemly repeated sequences; also called *satellite DNA.* This fraction constitutes 10–15 percent of total mammalian DNA. Simple-sequence DNA is found in **centromeres** and **telomeres** as well as at other chromosomal locations and is not transcribed. See also **intermediate-repeat DNA** and **single-copy DNA.**

single-copy DNA The eukaryotic DNA fraction that reassociates most slowly and contains most of the protein-coding genes. This fraction constitutes 50–60 percent of total mammalian DNA. See also **intermediate-repeat DNA** and **simple-sequence DNA.**

somatic cell Any plant or animal cell other than a **germ cell** or germ-cell precursor.

Southern blotting Technique for detecting specific DNA sequences. See also **blotting.** (See Figure 7-31.)

START A point in the G_1 stage of the yeast **cell cycle** that controls entry of cells into the S phase. Passage of a cell through START commits a cell to proceed through the remainder of the cell cycle. (See Figure 25-25.)

stem cell A self-renewing cell that divides to give rise to a cell with an identical developmental potential and/or one with a more restricted developmental potential. (See Figure 13-16.)

stereoisomers (D and L) Two compounds that have identical molecular formulas and atoms linked in the same order, but which have a different arrangement of atoms about an **asymmetric carbon atom.** D and L refer to particular configurations related to the configurations of D- and L-glyceraldehyde. All naturally occurring amino acids are L-isomers, and naturally occurring sugars generally are D-isomers.

steroid A group of four-ring hydrocarbons including **cholesterol** and related compounds. Many important hormones (e.g., estrogen and progesterone) are steroids.

substrate Molecule that undergoes a change in a reaction catalyzed by an **enzyme.**

substrate-level phosphorylation Formation of ATP from ADP and P$_i$ catalyzed by cytosolic enzymes in reactions that do not depend on a **proton-motive force.** During glycolysis, two molecules of ATP are produced by this process. (See Figure 17-4.)

sulfhydryl group (–SH) A hydrogen atom covalently bonded to a sulfur atom; also called a *thiol group.* A substituent group present in the amino acid cysteine and other molecules. A disulfide bond is formed by an oxidation reaction between two sulfhydryl groups.

supercoiled DNA Region of DNA in which the double helix is twisted on itself. (See Figure 4-14.)

suppressor mutation A mutation that reverses the phenotypic effect of a second mutation. Suppressor mutations are frequently used to identify genes encoding interacting proteins. (See Figure 8-17.)

symport A type of **cotransport** in which two different molecules or ions move across a membrane in the same direction. Transport is mediated by specific membrane-bound proteins called symporters. See also **antiport.**

synapse Region between an axon terminus of a neuron and an adjacent neuron or other excitable cell (e.g., muscle cell) across which impulses are transmitted. At a chemical synapse, the impulse is conducted by a **neurotransmitter;** at an electric synapse, impulse conduction occurs via gap junctions connecting the cytoplasms of the pre- and postsynaptic cells. (See Figures 21-5 and 21-7.)

syncytium A multinucleated mass of cytoplasm enclosed by a single plasma membrane. The early *Drosophila* embryo is a syncytium resulting from nuclear division in the absence of cytokinesis; cell fusion also produces a syncytium (e.g. formation of a muscle fiber from myoblasts).

TATA box A conserved sequence in the **promoter** of many eukaryotic protein-coding genes that binds the general **transcription factor** TFIID, thereby beginning formation of the transcription-initiation complex, which contains RNA polymerase. The TATA box is typically 25–35 base pairs **upstream** from the transcription start site. (See Figure 11-53.)

T cell See **T lymphocyte.**

T-cell receptor (TCR) Heterodimeric membrane proteins expressed by **T lymphocytes** that specifically recognize peptide antigens associated with class I or class II **MHC molecules.**

telomere End regions of a eukaryotic chromosome containing characteristic telomeric (TEL) sequences that are replicated by a special process catalyzed by telomerase. This process counteracts the tendency of the chromosome to be shortened during each round of replication. (See Figure 10-17.)

telophase Final mitotic stage during which the nuclear-envelope reforms around the two sets of separated chromosomes; the chromosomes decondense; and division of the cytoplasm (cytokinesis) is completed. (See Figure 5-50.)

temperature-sensitive (ts) mutant A cell or organism with a mutant gene encoding an altered protein that functions normally at one temperature (the permissive temperature) but is nonfunctional at another temperature (the nonpermissive temperature).

template A molecular "mold" that dictates the structure of another molecule; most commonly, one strand of DNA that directs synthesis of a **complementary** DNA strand during DNA replication or of an RNA during **transcription.**

tertiary structure In proteins, overall three-dimensional form of a polypeptide chain, which is stabilized by multiple non-covalent interactions between side chains. (See Figure 3-5.)

thioester bond A **high-energy bond** present in acetyl CoA and many enzyme-substrate complexes formed by a condensation reaction between a carboxyl group and a **sulfhydryl group.** (See Figure 17-5.)

tight junction See **cell junctions.**

T lymphocyte A small white blood cell that matures in the thymus and expresses an antigen-specific **T-cell receptor;** also called *T cell.* There are several distinct types of T cells, including cytotoxic T lymphocytes (CTLs), which are responsible for cell-mediated immunity, and helper T lymphocytes (T$_H$ cells), which release **cytokines** (lymphokines) that have a variety of specific and nonspecific effects on other cells of the immune system.

tolerance Diminished or nonexistent capacity to mount an **immune response** to a particular antigen. Tolerance is produced by contact with an antigen during development of the immune system (self-tolerance) or by introduction of antigen under nonimmunizing conditions (induced tolerance).

tracer A molecule containing a **label.**

trans-active Referring to DNA sequences that encode diffusible proteins (e.g. repressors and transcription factors) that control genes on the same or different chromosomes. See also **cis-active.**

transcript See **primary transcript.**

transcription Process whereby one strand of a DNA molecule is used as a **template** for synthesis of a **complementary** RNA. RNA polymerase and various accessory proteins called **transcription factors** form a complex that initiates transcription. (See Figure 4-20.)

transcription-control region Collective term for all the cis-acting DNA regulatory sequences that regulate transcription of a particular gene.

transcription factor (TF) General term for any protein, other than RNA polymerase, required to initiate or regulate **transcription** in eukaryotic cells. General factors, required for transcription of all genes, participate in formation of the transcription-initiation complex near the start site. Specific factors stimulate (or repress) transcription of particular genes by binding to their regulatory sequences (e.g., **enhancers, promoter-proximal elements**).

transcription unit A region in DNA, bounded by an initiation (start) site and termination site, that is transcribed into a single **primary transcript.** (See Figures 9-1 and 9-2.)

transfection Experimental introduction of foreign DNA into cells in culture, usually followed by expression of genes in the introduced DNA.

transfer RNA See **tRNA.**

transformation Permanent, heritable alteration in a cell resulting from the uptake and incorporation of a foreign DNA. Also, conversion of a "normal" mammalian cell in tissue culture to a cell characterized by **immortality** and/or uncontrolled cell division usually induced by treatment with a virus or other cancer-causing agent.

transgene A cloned gene that is introduced and stably incorporated into a plant or animal and is passed on to successive generations. (See Figures 8-39 and 8-40.)

transgenic Referring to any plant or animal carrying a **transgene**.

translation The **ribosome**-mediated production of a polypeptide whose amino acid sequence is specified by the **codon** sequence in an mRNA. (See Figure 4-42.)

transposable element A mobile DNA element.

transposon A relatively long **mobile DNA element**, present in prokaryotes and eukaryotes, that moves in the genome by a mechanism involving DNA synthesis and transposition. See also **retrotransposon**.

tRNA (transfer RNA) A group of small RNA molecules that function as amino acid donors during protein synthesis. Each tRNA becomes covalently linked to a particular amino acid, forming an **aminoacyl-tRNA**; this reaction is catalyzed by a specific enzyme that recognizes one amino acid and one tRNA species. In addition, each tRNA contains a three-nucleotide **anticodon**, which base pairs to the corresponding codon in mRNA. (See Figure 4-31.)

tubulin A family of cytoskeletal proteins consisting of three highly conserved GTP-binding proteins. Dimers of α and β tubulin polymerize into **microtubules** which are necessary for movement of **flagella** and intracellular vesicles. A third tubulin, γ tubulin organizes microtubules at **centrosomes**.

tumor A mass of cells, generally derived from a single cell, that is not controlled by normal regulators of cell growth.

tyrosine kinase-linked receptors Class of cell-surface **receptors** that are linked to cytosolic tyrosine kinases; also called *cytokine-receptor superfamily*. Ligand binding to these receptors leads to cytosolic activation of STAT proteins, which translocate to the nucleus and stimulate transcription of specific genes.

ubiquitin A small protein that becomes covalently linked to a cellular protein targeted for degradation; the ubiquitin-protein conjugate is usually degraded by complex of proteases.

uncoupler An agent that dissipates the **proton-motive force** across the inner mitochondrial membrane and thylakoid membrane of chloroplasts, thereby inhibiting ATP synthesis.

unipotent Referring to a **stem cell** capable of giving rise to only a single type of differentiated cell. See also **pluripotent**. (See Figure 14-16.)

unsaturated Referring to a molecule containing one or more double (C=C) or triple (C≡C) bonds between adjacent carbon atoms; generally used to describe fatty acids. See also **saturated**.

upstream The direction on a DNA opposite to the direction RNA polymerase moves during transcription. By convention, the +1 position in a gene is the first transcribed base; nucleotides upstream from the +1 position are designated −1, −2, etc. See also **downstream**.

upstream activating sequence (UAS) Any protein-binding regulatory sequence in the DNA of yeast and other simple eukaryotes that is necessary for maximal gene expression and is equivalent to an enhancer or **promoter-proximal element** in higher eukaryotes. (See Figure 11-42.)

V_{max} Parameter that describes the maximal velocity of an enzyme-catalyzed reaction or other process such as protein-mediated transport of molecules across a membrane. (See Figure 3-29.)

valency Numerical measure of the capacity of an atom, ion, or molecule to combine. In proteins that bind specific ligands, the number of binding sites is equivalent to the valency.

van der Waals interaction A weak **noncovalent** attraction due to small, transient asymmetric electron distributions around atoms (dipoles). These interactions are particularly important in stabilizing the binding between sites on different molecules with complementary shapes (e.g., enzyme and its substrate; antigen and antibody).

vector See **cloning vector** and **expression vector**.

ventral Relating to the belly, or anterior, surface of an animal.

virion An individual viral particle. See **virus**.

virus A small parasite consisting of nucleic acid (RNA or DNA) enclosed in a protein coat (**capsid**) that can replicate only within a susceptible host cell. Bacterial viruses exhibit either a **lytic cycle** or **lysogenic cycle** of growth. Viruses are widely used in cell biology research. (See Table 6-5.)

Western blotting Technique for detecting specific proteins in a mixture. After separation of the proteins by gel electrophoresis, proteins are **blotted** to a filter paper and specific ones detected by labeled antibodies. (See Figure 3-41.)

wild type Normal, nonmutant form of a macromolecule, cell, or organism.

x-ray crystallography Most commonly used technique for determining the three-dimensional structure of macromolecules (particularly proteins and nucleic acids) by passing x-rays through a crystal of the purified molecules and analyzing the diffraction pattern of discrete spots that results. Analysis of the diffraction pattern, a complex and tedious process, can reveal the complete structure.

zinc finger Several types of conserved DNA-binding **motifs** composed of protein domains folded around a zinc ion; present in several types of eukaryotic **transcription factors**. (See Figure 11-48.)

zygote A fertilized egg; diploid cell resulting from fusion of a male and female **gamete**.

Index

Note: Page numbers in *italics* indicate illustrations; page numbers followed by t indicate tables.

A antigen, 615, 615t
A band, 1023, *1024*
AB antigen, 615, 615t
Abelson murine leukemia virus, 1275–1277, *1276*
A box, *471, 471, 472,* 490–491, *491*
Absorption spectrum, 783
a/α cell switching, in yeast, 551–556
Acetylation, 71, *71*
Acetylcholine
  hydrolysis of, 965
  muscle contraction and, *957*
  structure of, 954t
  synthesis of, 956–958, *958*
Acetylcholine receptor(s), 955–956, *957*
  aggregation of, *1186,* 1186–1187
  muscarinic, 956, *957*
    cardiac, *967, 968*
  nicotinic, 956–965, *957, 962. See also* Nicotinic acetylcholine receptor
Acetylcholinesterase, 965, *966*
Acetyl CoA
  in citric acid cycle, 749–750, *750*

in fatty acid metabolism, *748,* 748–749, *749*
  in pyruvate metabolism, *748,* 748–749
N-Acetylgalactosamine, 612–613, *613*
N-Acetylglucosamine, 612–613, *613, 614,* 614
N-Acetylglycosamine phosphotransferase, 709
N-Acetylneuraminic acid, 612–613, *613, 614, 614*
achaete gene, 561–562, *563, 564*
Acid(s)
  amino. *See* Amino acid(s)
  definition of, 31
  dissociation of, 32–34
  fatty. *See* Fatty acid(s)
  pH of, 31t, 31–32
  polyamino, 56
Acid-growth hypothesis, for plant cell growth, 1191
Acid hydrolase, 173
Acidic late endosome/CURL organelle, 726–727, *727, 728*
Acidification, gastric, *657,* 657–658
Ac mobile element, *327,* 327–328, *328*
Acquired immunodeficiency syndrome (AIDS), 212, 1280, *1281*

human immunodeficiency virus in, 212, 492, *492*
Acrocentric chromosomes, *1350*
Acrosome process, 1010
Actin, 604
  ADP forms of, 994
  assembly of, 1003–1012
  ATP forms of, 994
  concentrations of, 994
  in cytokinesis, 1036–1038, *1037*
  depolymerization of, fungal toxins and, 1006
  filamentous. *See* Actin filaments
  globular, 994–995, *995*
    in actin filaments, *995, 995, 996*
    concentration of, 1006
  in nonmuscle cells, 1032–1039
  polymerization of, 1003–1007, *1004, 1061*
    in acrosome reaction, 1010, *1011*
  actin-binding proteins in, 1006–1007, *1007, 1008*
  ATP in, 1003, *1004*
  bacterial movement and, 1010–1012, *1012*
  capping proteins in, *1004,* 1009t, 1009–1010
  cell movement and, 1010–1012

Actin (*continued*)
  polarity in, 1003–1004, *1005*
    steps in, 1003, *1004*
    treadmilling in, 1004
  reversible assembly of, 995
  sequestration of, 1006–1007
  states of, 994
  structure of, 994–995, *995*
  types of, 994
α-Actin, 994
β-Actin, 994
γ-Actin, 994
Actin-binding proteins, 1006–1007, *1007*
Actin bundles, 997–999, *998*, 1002–1003
  contractile, 1032–1036, *1033*
  in filopodia, 1003
  in microvilli, *1002*, 1002–1003
Actin-capping proteins, 1004, *1005*, 1008–1010, 1009t
Actin–cross-linking proteins, 997t, 997–998, *998*
Actin cytoskeleton. *See* Cytoskeleton
Actin filaments, 169, *170*, 992–996
  in actin bundles, 997–999, *998*, 1002–1003
  in actin networks, 997–1002, *998*
  anchoring of to plasma membrane, *999*, 1000–1002
  capping of, 1004, *1005*, 1008–1010, 1009t
  in cell streaming, *1038*, 1038–1039
  in circumferential belt, 1034, *1034*
  depolymerization of into G-actin, 995
  fibronectin and, 1148
  formation of, 1003–1007, *1004*. *See also* Actin, polymerization of
  globular subunits of, 995, *995*, *996*
  length of, 1006–1007
    actin-severing proteins and, 1007–1009, 1009t
  in muscle, 1023, *1024*, *1025*, 1025–1026
  myosin movement on, 1015–1021. *See also* Myosin
    in muscle contraction, *1025*, 1025–1026
    in vesicle translocation, 1038–1039
  polarity of, 996, *996*, 1003–1004, *1005*
  polymerization of, in cell motility, 1041–1043, *1042*
  severence of, 1007–1009, *1009*, 1009t
  structure of, 52, *53*, *152*, 169, *170*, *995*, 995–996, *996*
  synthesis of, 1061
  in transformed cells, 1256
  treadmilling in, 1004, *1005*
  tropomyosin binding to, 1027–1029, *1029*

α-Actinin, 1025, 1035–1036
Actin mRNA, cytoplasmic localization of, 521, *522*
Actin networks, 997–1002, *998*
Actin-related proteins, 994
Actin-severing proteins, 1007–1009, 1009t
Action potential, 927, *927*, 932–943
  conduction of, by voltage-gated ion channels, 935–939
  definition of, 927, 936
  function of, 927
  generation of
    all-or-nothing nature of, 943
    frequency of, 943
    ion concentraton gradient and, 939–940
  magnitude of, 943
  membrane depolarization during, 927, 936–940
  membrane repolarization during, 938–939
  movement of on axon, 927, *927*
  potassium channels and, 933–934
  propagation of, 938, *939*
    myelination and, 940–942
    synapses in, 952–965. *See also* Synapse(s)
  in signal transduction, 971–979
  threshold, 943, *943*
Action spectrum, photosynthetic, 783, *784*
Activated complexes, 18
Activation-binding sites, 435–442
  TATA box as, *435*, 435–436, *436*
Activation domains, transcription factor, 445, 447, 452–453, 456
  amino acid sequences in, 452, *452*
Activation energy, 18, 18t, 21–22, *44–46*, 44–46
  enzymes and, 76
Activator
  enzymatic, 84, *84*
  transcription, 425, 425–426
Activator-binding sites
  enhancers as, 439–442
  proximal-promoters as, 436–439, 441–442
Activator element, 327, 327–328, *328*
Active site, 75–76, *76*
Active transport, *634*, 634–635
Activin, 1165
Actomyosin, 1015–1021. *See also* Actin; Myosin
  in cell streaming, *1038*, 1038–1039
  in cytokinesis, 1036–1038, *1037*
  in muscle contraction, 1023–1026, *1025*, 1030–1032, *1031*
  in nonmuscle cells, 1032–1039
Acylation, fatty acid, 71, *71*
Acyl CoA, in fatty acid metabolism, 748–749, *749*

Acylenzyme, hydrolysis of, 78, *79*
Adaptation, receptor, 973
Adapter proteins, *1035*, 1035–1036
  in signaling, 886, 893
Adaptor, 1265
Addressins, 1308
Adenine, *102*, 102, 103t
  pairing of with thymine, 104–105, *105*
Adenosine, structure of, 954t
Adenosine deaminase deficiency, gene therapy for, 299
Adenosine diphosphate. *See* ADP
Adenosine monophosphate. *See* AMP; cAMP
Adenosine triphosphate. *See* ATP
Adenovirus(es), 208, 210t, *211*
Adenylate cyclase, 870
  activation of
    cAMP synthesis and, 870, 870–881, *872*, *874*, *876*
    G protein cycling and, 879, *879*
    hormonal regulation of, 879, *879*
  in classical conditioning, 980–981, *981*
  potassium ion channels and, 968–970
  structure of, *875*
Adherens junctions, 628, 1035, *1035*, 1036, 1156
Adhesion plaque, *1022*, 1025
Adhesion plaques, 992, 1032, *1033*, *1035*, 1035–1036
  in spot desmosome, 1156, *1156*
Adhesive proteins, 1150–1155. *See also* Multiadhesive matrix proteins
  cadherins as, 1035, *1035*, 1036, *1115*, 1115, 1150–1153, *1151*, 1151t, *1152*
  in leukocyte extravasation, *1154*, 1154–1155, *1155*
  N-CAMs as, 1153, *1153*
  in neuronal outgrowth, 1182–1183, *1185*
ADL. *See* Adrenoleukodystrophy (ADL)
ADP
  in ATP synthesis, 741, *741*, 742, 746, 747, 751–756, *754*
  in respiratory control, 770
ADP-F-actin, 994
ADP-G-actin, 994
Adrenergic nerves, 967
α-Adrenergic receptor(s), 870
β-Adrenergic receptor(s), 870, 870–871, *872*
  agonists/antagonists of, 871, 873t
  binding to, 872–873
  cardiac muscle, 872
  smooth muscle, 872
β-Adrenergic receptor kinase (BARK), 914
Adrenocorticotropic hormone (ACTH), 859

Adrenoleukodystrophy (ADL), 773, 838–839
Adsorption, 205
Aerobe, obligate, 744
Aerobic metabolism, 43
Affinity chromatography, 87, 88, 96, 96
oligo-dT column, 235, 235
in receptor protein purification, 866, 867
sequence-specific DNA, 442
Affinity labeling, in receptor protein purification, 866, 867
Agar, 190
Agarose gel electrophoresis, 242, 242–243, 243
Aggrecan, 1139, 1141
Agonists, hormone, 871, 873t
Agrin, 1186, 1186–1187
Agrobacterium tumefaciens, gene transfer in, 297, 298
AIDS, 212, 1280, 1281
Alanine, 54, 55. *See also* Amino acid(s)
stereoisomers of, 20, 20
ALA synthase mRNA, translation of, 527, 527
Algae
blue-green, 142
flagella in. *See* Flagella
Aliphatic side chains, amino acid, 54, 55
Alkalinity, 31t
Allele
definition of, 266
dominant, 266
recessive, 266
stubble, 266
Allelic exclusion, 1323
Allergies, 1309
Allostery, enzymatic, 84, 84
α helix, 57, 63–65, 63–65
in domains, 67–68
in Golgi enzmyes, 719
in membrane proteins, 605, 606–608
as topogenic sequence, 690–692, 691, 692
in motif, 66–67, 67
in voltage-gated ion channels, 949, 949, 951
voltage-sensing, 938
α strand, in domains, 67–68
α subunit
of Ca²⁺ ATPase, 647, 648, 648t
of Na²⁺/K⁺ ATPase, 648, 649
RNA polymerase, 416
Alu sequences, 336, 337
Ameba, movement of, 1043, 1043–1044
Ames test, 1284, 1284
Amino acid(s)
activated, 126
in α helix, 57, 63–65, 63–65

amphipathic, 65, 65
analysis of, 59, 60
in β sheet, 57, 66, 66
in catalysis, 77, 77–78, 79
codons for. *See* Codon(s)
in collagen, 1128, 1128
C-terminal, 56
distribution of, 56
essential, for culture, 193, 193t
function of, 51–52
in genetic code, 120–122, 121, 121t
histone, 347, 348
hydrophilic, 54, 55
in membrane proteins, 607
hydrophobic, 54, 55
in membrane proteins, 607–608
isomers of, 20–21
membrane transport of, 652–654
mitochondrial metabolism of, 749
molecular weight of, 56
N-terminal, 56
peptide bonds of, 56, 57
polarity of, 54
in probes, 238–239
protein structure and, 52, 53, 56–59, 58
in protein synthesis, 133–138, 134–135, 137. *See also* Protein(s), synthesis of
radiolabeling of, 214–215
tRNA linkage to, 126–128, 127
sequence of, 56
in activation domains, 452, 452
analysis of, 59, 60
homology of, 68–70, 69, 70
immunoglobulin structure and, 86, 86–87
protein structure and, 68–70, 69, 70
in signal peptides, 682t, 683
topogenic, 670–671
side chains of, 52–54, 54, 55
solubility of, 52–54
structure of, 32, 52–56, 54, 55
variable-region, 1311–1315, 1312–1314
Amino acid acceptor stem, 125
Amino acid residue, 56, 57
Aminoacylation, of tRNA, 126–128, 127
Aminoacyl-tRNA synthetase (ARS), 124, 126–128, 127t, 127–129, 133
γ-Aminobutyric acid (GABA), 966–967
structure of, 954t
AMP. *See also* cAMP
hydrolysis of, 40
Amphipathic helix, 64–65, 65
Amphipathic lipids, 27, 27, 597
Ampholytes, 94
Amplification, by polymerase chain reaction, 254–256, 255
Amyloplasts, 836, 836

Anaerobe
facultative, 744
obligate, 744
Anaerobic fermentation, 744
Anaerobic metabolism, 43
of glucose, 744, 744–745
Anaphase, 179–180, 180, 182
chromosome separation in, 1100–1103, 1102, 1103
proteolytic induction of, 1216–1218, 1217
stages of, 1100–1103
Anergy, B-cell, 1339–1340
Aneuploid cell culture, 1253
Anfinsen, Christian, 74
Angular aperture, 149
Animal, transgenic, 1323
Animal-cap assay, 1165, 1165
Animal cell, 208–212, 209, 210t, 211, 212
Animal pole, in *Xenopus*, 1163–1164, 1164
Animal viruses, 204
Anion, 24. *See also under* Ion(s)
Anion antiporter(s). *See also* Antiporter(s)
in cell volume regulation, 663–664, 664
in gastric acidification, 657, 657–658
Anion exchange, transmembrane, 655–656
Ankyrin, 621, 622, 623, 1113t, 1114
Annealing. *See* Renaturation
Antagonists, hormone, 871, 873t
ANT-C genes. *See* Antennapedia (ANT-C) complex genes
Antenna chlorophyll, 783, 786
Antennapedia (ANT-C) complex genes, in *Drosophila*, 581–583, 581–584
mammalian homologs of, 584, 584–586, 585
Antennapedia gene, 579, 580, 581
Anterior pituitary
in feedback control, 864, 864
in peptide secretion, 970
Anterograde transport, 1070
Anthocyanin, 327, 328
Antibiotic-resistance plasmid, in cloning, 222, 223–224, 224
Antibodies, 86–88. *See also* Immunoglobulin(s)
antigen binding of, 1297–1298, 1298
reversibility of, 1299–1300, 1300
sites of, 1310–1314, 1311–1315
specificity in, 1298, 1311
structural aspects of, 1297–1299, 1298, 1299
anti-kinetochore, 1093–1094
anti-Sm, 503
binding affinity of, 1299, 1300
binding avidity of, 1300
binding sites on, 87, 87

Antibodies (*continued*)
  bivalency in, 1299
  capping of, *1333*, 1334
  catalytic, 87–88, *89*
  classes of, 1300–1302, 1301t
    differentiation of, 1310
    secondary, 1337
    switching of, 1335–1338, *1336*, *1337*
  complementarity-determining regions in, 87, *87*, *1310*, *1312–1314*, 1315
    diversity in, 1319
  constant regions in, *1310*, 1311, 1316–1318, *1336*, 1336–1337, *1337*
  cross-linkage of, 1334
  definition of, 1296
  diversity of, 1315–1319
    DNA rearrangements and, 344, *1315–1318*, 1315–1319
    generation of, 1304
    joining imprecision in, *1318*, 1318–1319
    lambda, 1319
    somatic mutation hypothesis for, 1315, 1338–1339
    somatic recombination hypothesis for, 1315–1318
  domains of, 1298–1299, *1299*, *1310–1314*, 1310–1315
  D regions in, 1319, *1320*
  framework regions in, *1314*, 1315
  functions of, 86, 1297–1298
  gene rearrangement in, 344, *1315–1318*, 1315–1319
    transcriptional control of, 1323–1324, *1324*
  heavy chains in, 1299, *1310–1314*, 1310–1315
    in antibody class switching, 1335–1338
    diversity in, 1319, *1320*
    synthesis of, 1319–1321, 1323–1324
  hinge region in, 1310–1311
  in immunization response, *1337*
  joining reactions in, 1315–1321
    recognition sequences for, 1320–1321
  joining regions in, 1316–1318, *1316–1318*
  kappa chains in, 1315–1318
  lambda chains in, 1319
  light chains in, 1299, *1310–1314*, 1310–1315
    diversity in, *1315–1317*, 1315–1318
    gene rearrangement in, 1321–1322, *1322*
    recombinational construction of, 1315–1318, *1316–1318*

synthesis of, 1315–1319, 1320–1321, 1323–1324
  monoclonal
    in protein purification, 202
    synthesis of, *201*, 201–202
    uses of, 202
  N regions in, 1319, *1320*
  in organelle purification, 165, *166*
  peptide synthesis and, *59*
  polyclonal, 201
  recognition signals in, 1318, *1318*
  secretion of
    B-cell, 1335–1337
    T-cell, 1330–1331
  shared-idiotype, 1323
  specificity of, 86, 1298, 1311
  structure of, *86*, 86–87, *1298*, 1298–1299, *1299*, 1310–1321, *1311–1314*, 1311–1315
  synthesis of
    from B lymphocytes, 1302
    clonal selection in, *1303*, 1303–1304, *1304*, 1309, 1323, 1337
    by memory cells, 1338
    switching in, 1335–1338
  variable regions in
    heavy-chain, *1310*, 1311–1315, 1319, *1320*
    light-chain, *1310*, 1311–1318, *1315–1318*
Antibody-affinity chromatography, 87, *88*
Antibody gene
  DNA rearrangement in, 344, 1315–1319
  joining process for, 1315–1321
  transcriptional control of, 1323–1324, *1324*
Anticodon, 125
  base pairing of with codons, 124–126, *125*, *128*, 136
  in tRNA identification, 128
Antifolates, salvage pathways and, 200, *200*
Antigen(s), 86, 1296
  antibody binding to, 87, 1297–1298, *1298*
    reversibility of, 1299–1300, *1300*
    sites of, *1310–1314*, 1311–1315
    specificity of, 1298, 1311
    structural aspects of, 1297–1299, *1298*, *1299*
  B, 615, 615t
  blood-group, 615, 615t
  epitopes on, 1297–1298, *1298*
  Forssman, 628
  in lymphocyte activation, 1296–1297, 1304, *1305*, 1307
  multivalent, 1297, *1298*
  polyclonal, 201

  proliferating cell nuclear, 379, *379*
  self, tolerance to, 1295, 1309, 1339–1340
  sialyl Lewis-x, 1154
  synthetic peptides as, *59*
  T, 379, 1272, 1329
    in nuclear protein transport, 842
    T-cell receptor interaction with, 1325–1327, *1327*, 1329–1332
Antigen binding sites, *1310–1314*, 1311–1315
Antigenic determinants, 87, *87*
Anti-kinetochore antibodies, 1093–1094
Antiport, 652, 654–659
Antiporter(s), *634*, 635, 654–659. *See also* Membrane transport proteins
  anion, 655–656
    in cell volume regulation, 663–664, *664*
    in gastric acidification, *657*, 657–658
  ATP-ADP, *758*, 758–759, 824
  cation, 654–655
  in cotransport, 654–659, *655–657*
  in cytosolic pH regulation, *658*, 658–659
  Na^+/H^+, 658–659, 664
  oligosaccharide synthesis, 703, *703*
  phosphate, 758–759, *759*
  proton, in plants, 659–660
  proton-sucrose, 660
  sodium-calcium, 654–655
Antisense RNA, in translational control, 527–528, *528*
Antisense vector, 528
Anti-Sm antibody, 503
Antitermination, *490*, 490–491, *491*, 492, *492*
AI zone, 1026
Apical membrane, 623–625
Aplysia punctata. *See* Sea slug
Apo-B
  in low-density lipoproteins, 724
  regulation of, 535–536
Apocatalase, 837
Apocytochrome (c), 827–829
Apoptosis, 195, 1267, *1297*, 1306
Aporepressor, *trp*, 419, *420*
Apotransferrin, 728, *728*
Aquaporin, 663, *663*
Aqueous solution
  hydrogen ion concentration in, 30–34
  phospholipid bilayer in, 27–28. *See also* Phospholipid bilayer
Arabidopsis thaliana, gene transfer in, 298
Arachidonic acid, 598t, 857
Arboviruses, 211
Archaebacteria, 475
  transcription in, 475
ARF, in transport vesicle formation, 715, *717*

Arginine, 54, 55. *See also* Amino acid(s)
  biochemical pathway of, 274, *275*
Arrestin, 977
ARS. *See* Aminoacyl-tRNA synthetase (ARS); Autonomously replicating sequence (ARS)
Ascorbic acid, in collagen synthesis, 1131, *1131*
asense gene, 561–562, *563*, *564*
Asialoglycoprotein receptor
  in endocytosis, 726–727
  membrane orientation of, 690, *690*
Asparagine, 54, 55. *See also* Amino acid(s)
Aspartate
  in malate shuttle, *760*
  in protein hydrolysis, 76, 76–77
Aspartic acid, 54, 55. *See also* Amino acid(s)
Assay, protein, 88, 96–98, *97*
Assembly particles, in clathrin-coated vesicles, 712, *712*
Asters, 179, 1089–1090, 1091, *1092*, *1093*
  in cytokinesis, 1103–1105
Astral microtubules, 1091, *1092*, 1103–1105, *1104*, *1105*
Ataxia telangiectasia, 1286
Atom(s)
  asymmetric, 20
  bonding of. *See* Bond(s)
  electronegativity of, 19, 19t
  electrons in. *See under* Electron
  models of, *18*
ATP
  in actin, 994–995
    in polymerization, 1003, 1004
  hydrolysis of
    in anaphase B, 1103, *1103*
    free energy from, 40–43, 41t, *42*
    in ion transport, *634*, 634–635, 644–654
    in mitochondrial protein uptake, 824–826
    in muscle contraction, 1–26
    in myosin movement, 1015, 1017–1021, 1018t, *1020*
    in protein transport, *687*, 687–688
    in chloroplasts, 834
    in nucleus, 844
    in transcription, 425–426
  in metabolic regulation, 770–773
  mitochondrial synthesis of, 174–175
  radiolabeled, 213
  in respiratory control, 770
  structure of, *40*, 954t
  synthesis of
    ATP-ADP antiport protein in, *758*, 758–759
    binding-change mechanism in, 755, *757*

chemiosmosis in, 740, *740*
citric acid cycle in, 749–750, *750*
closed vesicles in, 752–753, 756–758
electron transfer in, 750–751
electron transport in, 750–752, 759–770
energy sources for, 42–43
fatty acid metabolism in, *748*, 748–749
F_0F_1 ATPase complex in, 751, 753–756, *754*
in glucose metabolism, 42, 740–745
metabolic regulation of, 770–772
metabolite transport in, 752–759
in mitochondria, 746–750, *747–752*
phosphate transporter in, 758–759
phosphorylation in, 742–744, *743*, *744*
in photosynthesis, 42, 43, 782. *See also* Photosynthesis
in plants. *See* Photosynthesis
proton concentration gradient in, 740, *740*, 750–752
proton-motive force in, 740, *740*, 750, 750–752, 753–759
proton transport in, 766–770
pyruvate metabolism in, *748*, 748–749
in respiration, 42, 43
ATP-ADP antiporter, *758*, 758–759, 824
ATP-ADP transporter, 674
ATPase(s)
  F_1F_0, mitochondrial, 815t
  Ca^{2+}, *645*, 645–648, *647*
  H^+, *650*, 650–651
  H^+/K^+, 657–658
  in ion transport, 644–652
  myosin as, 1015
  Na^+/K^+, 648–650, *649*
  P-class, 644–650, 650–651
  in proton transport, *650*, 650–651
  V-class, *645*, 659–660
ATPase complex, in ATP synthesis, 752, 753–756, *754*
ATPase fold, 994
ATP-binding cleft
  in actin, 994–995, *995*, *996*
  in myosin, 1020
ATP cycle, *42*, 42–43
ATP-F-actin, 994
ATP-G-actin, 994
ATP-powered pumps, 634–635
ATP synthase. *See* ATPase complex
Attachment plaque. *See* Adhesion plaque
Attenuation, transcriptional, 486–488, *487*, *489*

Attenuator, 488, *489*
Autocrine growth response, 1331
Autocrine induction, of tumor cell growth, 1260, *1260*
Autocrine signaling, 855. *See also* Signaling
Autoimmune disorders, 1295, 1309
Autonomously replicating sequence (ARS), 355, *356*, 371
Autophosphorylation, 886
Autoradiography, *214*, 214–215
Autoregulation, of transcription, 547, *548*
Autosomal dominant mutation, *278*, 278–279
Autosomal recessive mutation, 277–278, *278*
Auxin, in plant cell growth, 1190–1191, *1191*, *1192*
Auxotroph, 274
  nutritional, isolation of, 192
Avery, O.T., 223
Avian leukemia viruses, 1278–1279
Avogadro's number, 16n
Axial rise per residue, 63
Axon(s)
  action potential movement along, 927, *927*
  functions of, 927, 935
  growth cone of. *See* Growth cones
  migration and elongation of, 1176–1184. *See also* Neuronal outgrowth
  myelinated, 940–942, *941*, *942*
  pioneer, 1177–1178
  synaptic vesicles in, 956. *See also* Synaptic vesicle(s)
Axonal transport, 1070–1072, *1071*, 1072t, *1073*
Axonemes
  elongation of, 1088–1089, *1089*
  movement of, 1084–1089, *1085*, *1086*
  mutant, 1087–1088, *1088*
  structure of, 1080–1081, *1082*, *1083*, 1083–1084, 1088–1089, *1089*
Axon hillock, 927
Axon potential, generation of, threshold potential of, 943, *943*
Axon terminal, *958*
  calcium ion channels in, 960
  structure of, *960*
5-Azacytidine, in myogenesis, 557, 557–558
Azamyoblast, 558, *558*, *559*

Bacillus subtilis, sigma factors for, 425
Bacteria. *See also Escherichia coli*; Prokaryote
  ATP synthesis in, 751–752, *752*

Bacteria (*continued*)
classification of, 142
culture of, 190t, 190–193, *191. See also* Cell culture
gram-negative, 143
gram-positive, 143
movement of, actin polymerization and, 1010–1012, *1012*
nutrient uptake in, *660,* 660–661
photosynthesis in, 779–780, 786–790, *787–789*
Bacterial infections, endocytosis in, 731–732
Bacterial insertion sequences, in transposition, 323–326, *323–326*
Bacterial mutagenesis, carcinogenesis and, 1284
Bacterial photosynthesis, 786–790, *787–789*
Bacterial photosynthetic reaction center, 607–608, *608*
Bacterial repressors, 418, *418–420*
Bacterial ribosome, 132, *132*
Bacterial transposons, 323–326, *323–326*
Bacterial viruses, in research, 206–208
Bacteriophage(s), 204
DNA, 206–208, *207*
in research, 206–208
RNA, 208
T, 206
temperate, 206–207
transcription in, 473
Bacteriophage lambda. *See* Lambda phage
Bacteriophage P1 vector, 233t, 234
Bacteriorhodopsin, 608–609, *609,* 786n
in proton-motive force experiments, 757–758
Balbiani ring, 466, *467*
Balbiani ring mRNPs, 516, *516*
Banding, chromosomal, *354,* 354–355
Banding analysis, *281–283,* 281–284, *282, 823*
Band 3 protein, 621, *622,* 623
in anion transport, *655,* 655–656, *656*
Band 4.1 protein, 621, *622,* 623
Band-shift assay, 419–421, *421*
B antigen, 615, 615t
BARK. *See* β-Adrenergic receptor kinase (BARK)
Barr, M.L., 465
Barr body, 465
Basal body, 1055, 1081–1082, *1083,* 1088–1090
centriole and, 1089–1090
Basal lamina, 197, 1124–1125, *1125,* 1167
in cancer, 1249–1250
collagen in, 1135–1136, *1136, 1137*

components of, 1143
in epithelial cell differentiation, *1167,* 1167–1168
functions of, 1143, *1144*
laminin in, 1143–1144, *1144*
neuromuscular junction and, *1185,* 1185–1187, *1186*
nidogen in, 1143–1144, *1144*
structure of, 1143, *1144*
Base
definition of, 31
dissociation of, 32
nucleic acid, *102,* 102–103, 103t
pH of, 31t, 31–32
Base pairing, 104–108, *105–108*
of codons and anticodons, 124–126, *125*
complementarity in, 105
in recombination, 392–393
RNA, 112
in splicing, *503,* 503–504, *505*
of sticky ends, 226
wobble, *125,* 125–126, *126*
Basic copy (BC), 342, *342*
Basic-zipper proteins, 452, *452*
Basolateral membrane, 623–625
of intestinal epithelial cells, 624
of pancreatic acinar cells, 624
B box, 471, *471, 472,* 490–491, *491*
BC. *See* Basic copy (BC)
B cell(s)
activation of, 1296–1297, 1304, *1305,* 1307, 1309, *1333,* 1333–1335, *1334*
cellular response and, 1335
prevention of, 1339–1340
proliferative response and, 1335
T cells and, 1333–1335
antibody secretion by, 1335–1337
in antibody synthesis, 1302
antigen interaction with, 1332–1340
circulation of, 1308
in clonal selection, *1303,* 1303–1304, *1304*
differentiation of, 1323, *1324*
functions of, 1296, 1303–1307
gene rearrangement in, 1321–1322, *1322*
identification of, 1306–1307, *1307*
maturation of
antigen-dependent phase of, 1321, 1324–1332
antigen-independent phase of, 1321–1324
memory, 1304–1306, 1335, 1338
plasma, 1335
somatic mutations in, 1304, 1315, 1338–1339, *1339*
surface markers on, 1306–1307, *1307*
switching and, 1335–1338

T helper cells and, 1306, 1333–1335, *1334*
tolerance of, 1295, 1309, 1339–1340
virgin, 1321, 1337
B-cell anergy, 1339–1340
B-cell growth and differentiation factors, 1334
Bcl-2 protein, apoptosis and, 1267
Beadle, George W., 11
Beads-on-a-string chromatin, 347, *347*
Belt desmosomes, 628, 1035, 1036, *1036,* 1156
Benign tumor, 1248, *1248. See also* Tumor(s)
Bent DNA, 106, *107*
Benzidine stain, 150
Beta-glycan, 1142
β pleated sheet, 57, *58,* 65–66, *66*
β strand, 57, *58*
in domains, 67–68
in motif, 66–67, *67*
porin, 609, *609*
β subunit
of DNA polymerase III, 376, *377*
of Na^{2+}/K^+ ATPase, 648, *649*
Bicarbonate ions. *See also under* Ion(s)
in CO_2 transport, 656
bicoid gene/Bicoid protein, in *Drosophila* regionalization, 568–572, *569–571, 577, 580*
Binding. *See also* Bond(s); Ligand(s); Receptor(s)
antigen-antibody. *See* Antigen(s), antibody binding to
hormone, 856, 865, 874–875
Binding assays, in receptor identification, *865,* 865–866, *866*
Binding specificity
antigen-antibody, 86, 1298, 1311
noncovalent bonding and, 26–27
in signaling, 856, 865
Biochemical pathways, genetic dissection of, 274, *275*
Biochemical reaction. *See* Chemical reaction
Biology, as historical science, 4–5
Biomembranes. *See* Membrane(s)
Bipartite nuclear localization sequence, 843
Bip protein, *687,* 687–688
mutant protein binding to, 697
in protein folding, 696, 697–698
retention/retrieval of, 698–699, *699*
Bithorax complex (BX-C) genes, in *Drosophila,* 581–583, 581–584
mammalian homologs of, *584,* 584–586, *585*
Bithorax mutation, 583, *583*
BK virus, 1280
Blast, lymphocyte, 1335
Blastocoel, 1164

Blastoderm, syncytial, in *Drosophila*, 567, *567*
Blastomere, 1165
Blastopore, dorsal lip of, 1164
Blastula
  in *Drosophila*, 567, *567*
  in *Xenopus*, 1164, *1165*
Blood, ion concentrations of, 640, 641t
Blood-brain barrier, 970
Blood clot formation, 1150
Blood-group antigens, 615, 615t
Blood proteins, physical characteristics
  of, 89t
Blue-green algae, 142
  flagella in. *See* Flagella
Blunt ends, ligation of, 227–228, 236
B lymphocytes. *See* B cell(s)
Bond(s)
  breaking of, 18, 18t
    energy for, 18, 18t, 21–22, 44–46,
    44–46
  covalent, 16–20, *18*, 18t, *19*, 19t
    double, 19
    nonpolar, 19
    orientation of, *18*, 18–19, *19*
    polar, *19*, 19–20
    single, 18
    triple, 19
  disulfide
    formation of, 694–697, *695*, *696*
    rearrangement of, *695*, *696*
    in recombinant DNA technology,
    695
  enzyme-substrate, 75–76, *76*, 80–81,
    84
  glycosidic, 43, *43*
  hydrogen, 17, *18*, 18t, 22, 22–23, *23*,
    23t
    in DNA helix, 104–105, *105*, *106*
  ionic, 23–24, *24*
  noncovalent, 16–17, 21–27, *22*
    binding specificity and, 26–27
    hydrogen, 22, 22–23, *23*, 23t
    hydrophilic, 24
    hydrophobic, 25–26, *26*
    ionic, 23–24, *24*
    membrane structure and, 27–28
    van der Waals, 24–25, *25*, 25t, *26*,
    26
  peptide, 56, *57*
    hydrolysis of, 76–78, *77*, 79
  phosphoanhydride, in ATP, 40, *40*
    hydrolysis of, 739
  phosphodiester, 104, *104*
Bone marrow, 1308
Boss protein, 891–893, *892*, 1172,
  1172
Botulinum-B toxin, 962
Box(es), cell-cycle, 553, 1231
Box A, *471*, 471, *472*, 490–491, *491*
Box B, *471*, 471, *472*, 490–491, *491*

Box C, 471–472, *472*
Box P, 553
BPAG1, 1113t
Branch point, 501
BRF, 471
Bright-field microscopy, 148–150, *149*,
  151, 156
Brown-fat tissue, 770–771
Brush border, 197, *197*, 623, *624*
  actin bundles in, *1002*, 1003
bub mutant, 1239
Budding
  yeast, 181, *183*, 1226, *1226*
  in yeast, 1226, *1226*
Buffer, 32–34
Bundle sheath cells, in C₄ pathway,
  802–803, *802–803*
α-Bungarotoxin, 963
Burkitt's lymphoma
  chromosomal translocations in,
    1268–1269
  Epstein-Barr virus and, *1279*,
    1279–1280
Burnet, MacFarlane, 1303
Bursa of Fabricus, 1302
BX-C genes, in *Drosophila*, 581–583,
  581–584
  mammalian homologs of, *584*,
    584–586, *585*

Ca²⁺. *See also* Calcium ion(s)
Ca²⁺ ATPase
  catalytic subunit of, 647, *648*, 648t
  in ion transport, *645*, 645–646
    in sarcoplasmic reticulum,
      645–646, *647*, 647–648
Ca²⁺ binding protein, high-affinity, 646
Ca²⁺-calmodulin, in classical
    conditioning, 980–982, *981*
Ca²⁺-calmodulin–activated protein
    kinase, in long-term potentiation,
    984
Cactus protein, *575*, 575
Cadherins, *1035*, 1036, *1115*, 1115,
  1150–1153, *1151*, 1151t, *1152*
Caenorhabditis elegans
  genomic mapping for, 284
  neuronal outgrowth in, *1176*, *1176*,
    1179, 1179–1180
Calcium
  in actin filament capping and
    severence, 1008–1009
  in muscle contraction, 1026–1032
Calcium ion(s). *See also under* Ion(s)
  antiport movement of, 654–655
  ATPase and, 645–648
  in cell movement, 1045–1046, *1046*
  in classical conditioning, 980–982,
    981

concentration of, in cell vs. blood,
  640, 641t
  in flagella/cilia beating, 1087–1088
  fluorescent microscopy of, *153*,
    153–154
  gap junctions and, 1158
  in glycogenolysis, 901, *901*
  in inositol-lipid pathway, 901–904,
    902
  in light adaptation, *977*, 976–977
  in long-term potentiation, 982, *983*
  regulation of
    by inositol 1,4,5-trisphosphate, 899,
      899t, 901–904
    by ryanodine receptors, *904*, *904*
  as second messenger, 857–859, *860*,
    899–905
  tissue responses to, 899t, 899–901
Calcium ion channel(s)
  ligand-gated, GABA/glycine activation
    of, 966–967
  voltage-gated
    evolution of, 951–952
    in habituation, 980
    properties of, 943–952
    as signal transducers, 960
    in synaptic vesicle exocytosis,
      958–960
Calcium ion channel proteins. *See* Ion
    channel proteins
Calcium pump, *645*, 645–646
  muscle, 645–646, *647*, 647–648,
    1027–1029, *1029*
  plasma-membrane, 646
Caldesmon, in muscle contraction,
  1029, *1030*
Callus, 298
Calmodulin, 646, 881, *900*, 900–901
  in classical conditioning, 980–982,
    981
  in long-term potentiation, 984
  in muscle contraction, *1031*,
    1031–1032
Calorie, 16, 16n
Calsequestrin, 646
Calvin cycle, 797–805, *798*
Cambrian Period, 4, *6*
cAMP
  calcium ions and, 901
  cAMP phosphodiesterase regulation
    of, 881
  cellular response to, 885–886
  gap junctions and, 1158
  in gene expression, 914–915
  glucose metabolism and, 421–423,
    422
  hormonal regulation of, 879, *879*
  hormone-induced rise in, tissue
    response to, 870, 871t
  in metabolic regulation, 881–886
  in olfaction, 978

cAMP (*continued*)
  potassium ion channel function and, 968–970, *969*
  as second messenger, 857–859, *860*, 870, 871t, 968–970, *969*
  structure of, *103*, *421*
  synthesis of
    adenylate cyclase in, *870*, 870–881, *874*
    epinephrine-induced, *870*, 870–871, 873–875
  in transcription, 421–423, *422*
cAMP-dependent kinases, 881–882, *882*, 886
cAMP phosphodiesterase, 881, 901
cAMP receptor protein, 422–423, *423*
cAMP response element, 915
Cancer, 1247–1289. *See also under* Carcinogen(s); Carcinogenesis; Tumor(s)
  in AIDS, 1280, *1282*
  apoptosis and, 1267
  genetic factors in, 1269–1270, 1288
  human, viruses and, 1279–1280
  immune system and, 1250–1251, 1296, 1309
  malignant transformation in, 196, *196*, 1254–1257, *1255*, *1257*
    abortive, 1270
    oncoproteins in, 1267–1268
    retroviruses and, 1273–1279
    T proteins and, 1272–1273
    tumor viruses and, 1270–1273
    two-step model of, 1267–1268, *1268*
  metastasis of, 1248, *1249*, *1250*, 1250
  mutations in, 1266–1267, *1267*, 1284
  oncogenes in, 1258–1270. *See also* Oncogene(s)
  risk factors for, 1287–1288
Cancer research
  cell cultures in, 1251–1258
  3T3 cells in, 1253–1255, *1254*
CAP. *See* Catabolite activator protein (CAP)
Capacitor, 641
Capping
  in actin filaments, 1004, *1005*, 1009–1010
  in muscle, 1024, *1025*
  in antibodies, *1333*, 1334
CAP protein, 412
Capsid, viral, 202–203, *203*
CapZ, 1009–1010
  in muscle, 1024, *1025*
Carbohydrates
  blood-group antigen, 615
  membrane, 612–615, *613–615*
  metabolism of, mitochondrial, 747–750
  structure of, 20–21, *21*

Carbon
  bonding of, 17, *18*, 18t
  chiral, 20–21, *21*
  α Carbon, 20, *21*
Carbon-14, 213t
Carbon cycle, 43
Carbon dioxide
  in citric acid cycle, 749–750, *750*
  in glucose metabolism, 744, 745
  transport of, 655–656
Carbon dioxide fixation, in photosynthesis, 175, 782–783, 797–805, *798–804*
  C₄ pathway for, 802–803, *802–803*
Carbonic anhydrase, in CO₂ transport, 655–656
Carboxyl-terminal domain (CTD), 432, *432*
Carcinogen(s), 1251
  Ames test for, 1284, *1284*
  chemical
    activation of, 1283, *1283*
    DNA damage from, 1284
    synergistic, 1286–1287, *1287*
  direct-acting, 1280–1282, *1282*
  indirect-acting, 1280–1282, *1282*
  ultimate, 1282
Carcinogenesis
  chemical, 1280–1284
  DNA repair and, 1284–1286
  DNA viruses in, 1270–1273
  epigenetic alterations in, 1286
  initiator-promoter model of, 1286–1287
  as multifactorial multistep process, 1268, 1286–1287
  radiation and, 1284–1286
  synergy in, 1286–1287, *1287*
Carcinoma. *See also* Cancer
  definition of, 1249
Cardiac muscarinic acetylcholine receptor, 967, *968*
Cardiac muscle, 1022, *1022*
Cardiolipin, 599, 674
  mitochondrial, 747
Carotenoids, 783, *784*
Carrier, 277
Cartilage. *See also* Connective tissue; Extracellular matrix
  collagen in, 1134
  proteoglycan aggregate in, 1139, *1141*
Cascade, 885
Catabolism, aerobic, 43
Catabolite activator protein (CAP), *422*, 422–423
Catabolite repression, 421
Catalase, 174
  synthesis of, *837*, 837–838
Catalysis, interfacial, *611*, 612
Catalyst, 28, 29. *See also* Enzyme(s)
  action of, 46, *46*
  definition of, 27

RNA as, 112, *113*, 508, 530–532, *532*
Catalytic antibody, 87–88, *89*
Catecholamine(s). *See also* Dopamine; Epinephrine; Norepinephrine
  α-adrenergic receptor for, 870
  β-adrenergic receptor for, 870–878
  degradation of, 859–861
  properties of, 863t
  structure of, 954t
  synthesis of, 859–861, 967, *969*
  transporters for, 970
Catecholamine receptors, 967
Catenane, 383
Catenation, 382, *383*
Cation, 24. *See also under* Ion(s)
Cation antiporter, 654–655
Cation channels, ligand-gated, 953
CBF. *See* Centromere-binding factor (CBF)
C box, 471–472, *472*
CCBF transcription factor, 1231
CD44, 1137–1138
Cdc2, 1070, 1232, 1233t
  in intermediate filament disassembly, 1112–1113
  in mitosis, 1221–1224, 1227–1228
  phosphorylation of, 1221–1224
Cdc25, in mitosis, 1222–1224, *1223*
Cdc28, in S phase regulation, 1227–1231, 1233t
Cdc2-activating kinase, 1223, *1233*
Cdc28-cyclin complexes, *1230*, 1230–1231
cdc mutants, 1203, *1204*
  in mitosis, 1221–1224
*cdc*28 mutants, 273
CD3 complex, 1332
Cdk(s)
  in mammals, 1232–1235, 1233t
  in mitosis, 1127–1128, 1201–1202, 1222–1224
  in S phase, 1227–1232
  in yeast, 1226–1232, *1227*, 1233t
Cdk1. *See* Cdc2
Cdk2, 1232, 1233t
  phosphorylation of, 1224
Cdk3, 1232, 1233t
Cdk4, 1232, 1233t
Cdk5, 1232, 1233t
Cdk-cyclin complex(es)
  in DNA replication, 1231–1232
  in mammalian cell cycle, 1233–1235, *1233–1235*
CD4 surface protein, T helper cells and, 1331–1332, *1332*
CD8 surface protein, T helper cells and, 1331–1332, *1332*
Cell(s)
  cancer. *See* Tumor cells
  connections between, 626–628. *See also* Junction

cytoskeleton of. *See* Cytoskeleton
diploid, 178
DNA in. *See also* DNA
   organization of, 147
   variation in, *312,* 312–313
   volume of, 144–146, 145t, 146t
electrically coupled, 953
eukaryotic
   DNA in, 144–147, 146t
   macromolecular content of,
      143–144, 145t, 146t
   structure of, 7–8, *8,* 143, *144–147*
functional adaptation of, 13
growth of, 13–14. *See also* Cell
     division; Growth
   autocrine induction of, 1260–1261
   steps in, 1260
haploid, 181, 264
historical view of, 4
hybrid. *See also* Cell fusion
   culture of, 198–202
   gene mapping with, 199
immortal, 196, 1252–1253
   oncogenes and, 1267–1268
ion concentrations of, 640, 641t
meiotic, 181–185, *184*
mitotic, 179–181, *180*
movement of. *See* Cell movement
nonpermissive, 1270
nucleus of. *See* Nucleus
organelles of, 143, *144, 145.* *See also*
     Organelle(s)
parietal, 657, *657*
permissive, 1270
plant. *See* Plant cells
polarized, 196, 623, 993
   plasma membrane of, 623–625
polar regions of, 179
postmitotic, 1202–1203
postsynaptic, 929
programmed death of, 195, 1267,
   1297, 1306
prokaryotic
   DNA in, 144–147, 146t
   macromolecular content of,
      143–144, 145t, 146t
   structure of, *142,* 142–143
protein content of, 143
quiescent, 1251
rupturing methods for, 162–164
sensory receptor, 931
shape of, myosin II and, 1036, *1036*
somatic, 199
structure of, 5–9, 7–8, *8, 142,*
     142–143, *144–147*
transformation of, 196, *196*
tumor. *See* Tumor cells
turgor of, 174, 1191
Cell adhesion, 1150–1159
   adhesive proteins in, 1150–1155
   circumferential belts in, 1034, *1034*
   homophilic, 1150, 1151–1152

hyaluronan in, 1138–1139
junctions in. *See* Cell junctions
multiadhesive matrix proteins in,
   1143–1150. *See also*
     Multiadhesive matrix proteins
stress fibers in, 1032–1034, *1033*
Cell adhesion plaques, 992, 1032, *1033,*
   1035, 1035–1036
   in spot desmosomes, 1156, *1156*
Cell cloning. *See* Cloning
Cell colony, 190, *191*
Cell communication, plasma membrane
   in, 167. *See also* Signaling
Cell cortex, actin filaments in, 999, *999*
Cell culture, 190–198
   amino acids for, 193, 193t
   aneuploid, 1253
   of animal cells, 193–198, 194t, *195*
   bacterial, 190t, 190–193, *191*
   in cancer research, 1251–1258
   cell fusion in, 198–199, *199*
   cell line in. *See* Cell line
   cell strain in, 194–196, 1252, *1252*
      immortalization of, 196, 1257
   colony in, 190, *191*
   crisis in, *1252,* 1253
   epidermal growth factor in, 1253,
      1255
   epithelial cells in, 1251–1252
   fibroblasts in, 194–196, 1251–1252
   of hybrid cells, 198–202
   malignant transformation in,
      1254–1257, *1255, 1257*
   media for
      for animal cells, 193, 194t
      for bacteria and yeasts, 190, 190t
      defined, 1253
      HAT, 201
   microorganisms in
      isolation of, 190–193, *191, 192*
      types of, 190
   mutation isolation in, 190–193, *193*
   plating in, 190–193, *191,* 1252
   primary, 194–196
   salvage pathways in, *200,* 200–201
   saturation density in, 1253
   serum factors in, 1253
   stages in, *1252*
   transformed cells in, 196, *196*
   wounded, 1253
   yeast, 190t, 190–193, *191*
Cell cycle, 178, *178. See also* Cell
     division
   anaphase in, 179–180, *180, 182.* See
     also Anaphase
   cdc28-cyclin complexes in, *1230,*
     1230–1231
   G_1 phase in, *178, 178, 179,* 1202,
     1202
   G_2 phase in, *178, 178, 179,* 1202,
     1202
   interphase in, 178, *178,* 178, 179,
     180, 181

meiosis in, *178,* 181–185, *184, 185.*
     See also Meiosis
mitosis in, *178,* 178–181, *180, 182,*
     183. See also Mitosis
M phase in, 1202, *1202*
mutations in, 1203
phases of, 178–185, 1202–1202
regulation of, 1201–1241
   checkpoints in, 1237–1239, *1238,*
     1240
   in eukaryotes, 1201–1232
   in mammalian cells, 1232–1236
   maturation-promoting factor in,
     1205–1225. *See also*
     Mitosis-promoting factor (MPF)
   trans-acting factors in, 1203, *1205*
   in yeast, 1226–1232
research on, experimental systems in,
   1203, *1204*
S phase in, 1202, *1202*
   control of entry into, 1226–1232
Cell-cycle boxes, 553, 1231
Cell-cycle clock, 1207
Cell-cycle proteins, oncogenes and,
   1265–1267, *1266*
Cell death, programmed, 195, 1267,
   1297, 1306
Cell differentiation
   cadherins in, *1152,* 1152–1153
   internally driven, 1321
Cell division, 177–185. *See also*
     Cytokinesis; Meiosis; Mitosis
   cleavage furrow in, 1103
   DNA synthesis and, 177–178, *179*
   eukaryotic, 177–185, *178–180,*
     182–185
   in plants, 180–181, *182, 1105,*
     1105–1106, 1187
   prokaryotic, 177, *177*
   rate of, 177
   in yeast, 181, *183,* 1226–1232
Cell fractionation techniques, 161–165,
   163–166
Cell fusion, 198–199, *199*
   in monoclonal antibody production,
     201, 201–202
   salvage pathways in, *200,* 200–201
Cell fusion experiments, in cell cycle
     regulation studies, 1203, *1205*
Cell interactions, 1123–1195
   in cancer, 1249–1251
   in cell adhesion, 1150–1159
   cell-cell contact in, 1172–1175
   in embryogenesis, 1159–1166
   with extracellular matrix, 1124–1149
   integrin in, 1144–1146
   overview of, 1123–1124
   signaling in. *See* Signaling
Cell junctions
   adherens, 628, *1035,* 1035, *1036,*
     1036, 1156
   desmosome, 628, 1035, *1036,* 1036,
     1114, *1156,* 1156–1157, *1157*

Cell junctions (*continued*)
  gap, 628, *930*, 930, 1157–1159,
    1158, 1159
  neuromuscular, 929, *1185,*
    1185–1187, *1186*
  tight, 625–628, *626, 627*
  types of, 1155–1156
Cell line, *1252,* 1252–1253
  definition of, 196
  differentiated, 196t, 196–197, *197*
  immortal, 196, 1252–1253
    oncogenes and, 1267–1268
  transformed, 1254–1257, *1255*
  undifferentiated, 197–198
Cell membrane. *See* Membrane(s)
Cell metabolism. *See* Metabolism
Cell migration. *See also* Cell movement
  cadherins and, 1153
  fibronectins in, *1149,* 1149–1150
  myosin in, 1039, *1041*
  in wound healing, 1150
Cell motility. *See* Cell movement
Cell movement, 991–1046. *See also* Cell
    migration
  actin polymerization in, 1010–1012,
    1011–1013, 1041–1043, *1042*
  ameboid, *1043,* 1043–1044
  calcium in, 1045–1046, *1046*
  chemotactic, 1044–1046
  cilia in, 1079–1090
  contact inhibition of, 1256
  cytoplasmic streaming in, *1043,*
    1043–1044
  flagella in, 1079–1090
  hyaluronan in, 1138–1139
  microtubule rearrangement in,
    1054–1055
  motor proteins in, 992
  in muscle cells. *See* Muscle; Muscle
    contraction
  myosin in, *1036,* 1036–1043, *1039,*
    1041, 1042
  polarity in, 993, 1044
  pseudopodial extension in, *1043,*
    1043–1044
  second messengers in, 1044–1046
  signal tranduction pathways in,
    1044–1046, *1045*
  steps in, 1040, *1042*
Cell signaling. *See* Signaling
Cell sorting techniques, 161–166
Cell strain, 194–196, 1252, *1252*
  immortalization of, 1257
Cell streaming, *1038,* 1038–1039
  in cell movement, *1043,* 1043–1044
Cell surface markers, lymphocyte,
    1306–1307, *1307*
Cell-surface proteins, in transformed
    cells, 1256
Cell-surface proteoglycans, 1142, *1142*
Cell-surface receptors. *See* Receptor(s),
    cell-surface

Cell theory, 4
Cell transformation, 1254–1257
  phosphorylation and, 1264
Cell-type specification, in yeast,
    550–556
Cellular immunity, 1296, *1297*
Cellulose, 43, 176–177
  microfibrils in, 1188, *1188*
    orientation of, *1190,* 1190, *1192,*
    1192
  polysaccharide binding to,
    1188–1190, *1189*
  structure of, 1187–1188, *1188*
  synthesis of, 1192
Cellulose synthase, 1192
Cell volume, regulation of, 661–664
Cell wall, 143, *146*
  prokaryotic, 143
CEN sequence, 355, 357, *357*
Centimorgan, 279
Central nervous system, structure and
    function of, 931, *932*
Centrifugation
  in cell fractionation, 164–165, *165*
  differential, 90, *91,* 164–165, *165*
  equilibrium density-gradient, 92,
    164–165, *166*
  in protein purification, 88–92, *91*
  rate-zonal, 90–92, *91,* 164–165, *165*
  sedimentation constant for, 89t, 90
Centriole
  basal body and, 1089–1090
  in microtubule-organizing center,
    1055, *1058, 1059*
  in mitosis, 179, *180,* 1096,
    1096–1097, *1097*
Centriole cycle, 1096, 1096–1097
Centromere, 349, 1091–1093
  kinetochore in, 1091–1094, *1094*
  in meiosis, 181, *184*
  in mitosis, 179, 1090–1096
  in mitotic segregation, 355–357, *356*
  structure of, 1094–1096, *1095*
  yeast, microtubule binding by,
    1094–1096, *1095*
Centromere-binding factor (CBF), 357,
    357, 1094–1096, *1095,*
    1095–1096
Centromere DNA elements (CDE), 1094
Centromeric DNA, 355–357, 1093
Centrosome(s), 1055, *1057, 1058,*
    1091, *1092. See also*
    Microtubule-organizing center
  migration of, 1096–1098, *1096–1098*
  in mitosis, 1055, *1090*
  replication of, *1096,* 1096–1097,
    1097
C/EPB, 458, *458*
Ceramide, membrane, 597, *597*
Cervical cancer, papilloma viruses and,
    1280

c-fos gene, 916
Channel(s)
  ion. *See* Ion channel(s)
  membrane, 143
  water, 663, *663*
Channel proteins
  ion. *See* Ion channel proteins
  water, 663, *663*
Chaperone proteins, 74–75, 687,
    687, 687–688, 695–696
  in clathrin-coated vesicle
    depolymerization, 713
  mitochondrial, 820–822, *821,*
    826–827, *827*
Chaperonins, 826–827, 829
Chemical bonds. *See* Bond(s)
Chemical carcinogens, 1280–1284
  activation of, 1283, *1283*
  as mutagens, 1284
  synergistic, 1286–1287, *1287*
Chemical concepts, 15–47
Chemical equilibrium, 28–30
Chemical modification, of proteins, 71,
    71–72
Chemical reaction
  activation energy for, 18, 44–46,
    44–46
  coupling of, 39–41
  direction of, free energy and, 34–35
  endothermic, 34
  entropy and, 26, 34–35, *35*
  equilibrium in, 28–30
  exothermic, 34
  free energy in, 34–43
  oxidation-reduction, 37–39, 39t
  rate of, 28–30
    enzymes and, 46, *46*
    pH and, 45
  transition state in, 45
Chemical signaling, 925. *See also*
    Signaling
  cessation of, 965
  neurotransmitters in, 929, 929–930,
    930, 953–955
Chemical synapse(s), 929, 929–930. *See*
    also Synapse(s)
  excitatory, 953
  fast, 953, *957*
  inhibitory, 953, *957*
  signal amplification in, 955
  signal computation in, 955
  slow, 953–955
Chemiosmosis, 740, *740*
Chemiosmotic theory, 740
Chemorepellents, in neuronal
    outgrowth, *1181,* 1181–1182,
    1182
Chemotaxis, 1044–1045
Chimera, 226, 294, 683
CHIP (Aquaporin), 663, *663*
Chirality, 20

Chironomous tentans, Balbiani rings in, 466, *467*
CHI site, 392
Chlamydomonas reinhardtii, flagella in, 1087
Chloride-bicarbonate antiporter, 659
Chloride channels. *See also* Ion channel(s)
  in cystic fibrosis, 651–652
  in gastric acidification, 657, *657–658*
  membrane polarization and, 935
  multidrug-transport protein as, 651–652
Chloride ions. *See also under* Ion(s)
  concentration of, in cell vs. blood, 640, 641t
  in CO₂ transport, 656
Chlorophyll, 175, 783–784
  antenna, 783
  excited-state, 784
  in PSI, 792, *792*
  in PSII, 792, *793, 793*
  reaction-center, 784–786, *785, 792*
  types of, 783
Chloroplast(s), 143, *146,* 175, *175,* 780
  ATP synthesis in, 751–752, *752*
  biogenesis of, *810,* 810–811, 811t, *835,* 835–836, *836*
  division of, 180
  functions of, 797
  photosynthesis in. *See* Photosynthesis
  protein targeting to, *810,* 810–811, 811t
  structure of, *780, 781, 832*
  synthesis of, 830–836
  thylakoid membrane of, 780, *781*
  transcription in, 474–475
Chloroplast DNA, 830–836
  gene content of, 830–831, 831t
  sequencing of, 831–832
Chloroplast proteins
  synthesis of, 832–833, *833*
  targeting and uptake of, 832–835, *834*
Cholera toxin, 878, *879,* 902
Cholesterol
  in familial hypercholesterolemia, 724–726
  in low-density lipoprotein endocytosis, *723, 724, 725*
  membrane, 597–599
    membrane fluidity and, 601
  structure of, *599, 599*
Cholinergic synapses, 955
Chondroitin sulfate, 1139, *1140, 1141*
Chorionic gonadotropin, 864
Chromatid(s)
  daughter, segregation of, topoisomerase II in, 383–384, *385*
  in meiosis, 181, *184, 185,* 1202, *1202*

in mitosis, 179–180, *180, 1202, 1202*
  recombination of, 184–185, *185*
  sister, 349
    segregation of, 355–359, *356–358*
Chromatin, 179, 346, *347*
  in chromosome scaffold, 351–353, *352*
  condensed, *347, 348, 352, 355*
  decondensed, transcription and, 466–467, *467*
  extended, *347, 347*
  interphase, nonhistone proteins in, 352
  nucleosomes in, 347, *348*
Chromatography
  affinity, *96, 96*
    oligo-dT column, *235, 235*
    in receptor protein purification, 866, *867*
  antibody-affinity, 87, *88*
  gel filtration, 94, *95*
  ion exchange, 94–96, *95*
  liquid, 59, 94–96, *95*
  sequence-specific DNA, 442
Chromatosome, 348
Chromogenic enzyme reactions, 96–97
Chromoplasts, 835–836, *836*
Chromosomal puffs, *432, 433,* 466
Chromosomal replication. *See* DNA replication
Chromosome(s)
  abnormalities of, 267, *267. See also* Mutation(s)
    in tumors, 1268–1269
  acrocentric, *350*
  alignment of, 1100, *1100*
    on mitotic spindle, 1100, *1100*
  banding patterns in, *354,* 354–355
  condensation of, 349, 1203, *1205*
  congress of, 1100
  crossing over of, 185, *185*
  decondensation of, *1216, 1217*
    cyclin B and, *1216, 1217*
  definition of, 349
  deletion of, 267, *267*
  division of, 179–180, *180,* 181–185, *184, 185*
  DNA in. *See also* DNA
    organization of, 344–348, *345–349*
    rearrangement of, 344, *344*
    single linear molecule of, 354, 359
  in eukaryotes vs. prokaryotes, 147
  fate of, in cell cycle, 178–185, *1202, 1202–1203*
  fragmentation of, *1237, 1237*
  genes in. *See* Gene(s)
  homologous, 181
  interphase, 351–353
  kinetochore microtubule binding to, 1091–1094, *1094*
  mapping of. *See* Mapping

  in meiosis, 181–185, *184, 185,* 1202, *1202*
  metacentric, *350*
  metaphase, 349
  microtubule capture of, 1098–1099, *1099*
  in mitosis, 179–180, *180,* 1090–1103, 1202, *1202*
  morphology and functional characteristics of, 349–359
  mutant. *See* Mutation(s)
  mutation sites on, identification of, 277
  nondisjunction of, *1237,* 1237–1238
  numbers of, 349, 351t
  organization of, mitotic apparatus and, 1091
  Philadelphia, 1268
  polytene, *283,* 283–284, 333, *333, 432,* 432–433
    banding in, 355
    formation of, 343, *343*
  prokaryotic, 147, 344–346, *345, 346*
  puffs on, *432, 433,* 466
  replication of, 355–359
  retroviral integration into, 1273
  salutory behavior of, 1100
  segregation of, *1216, 1217*
  separation of, 1100–1103, *1102, 1103*
  specificity of, 349
  SV40, replication in, 378–380, *379*
  in synapsis, 184–185, *185*
  telocentric, *350*
  translocation of, 267, *267*
    in tumor cells, 1268
  yeast artificial, 355
    in cloning, 359
Chromosome scaffold, 349–353, *352, 353*
Chromosome walking, 287, *288*
Chronic progressive external ophthalmoplegia, 818
Chymotrypsin, 76, *76–77,* 79
  hydrolysis by, 76, *76–78, 77, 79*
  vs. trypsin, 78
Chymotrypsinogen, 76, *76–78*
C₂H₂ zinc finger, *448,* 451
Cilia, 175–176, *176,* 1079–1090
  axoneme in, 1080–1089, *1082*
  basal bodies in, 1081–1082, *1083,* 1088–1090, *1089*
  beating of, 1080, *1080, 1081,* 1084–1087, *1085*
  microtubules in, 1080–1089
  structure of, 1080–1084, *1082, 1083*
  swimming direction in, 1087–1088
Cip1 protein, 1266
Circular DNA, *109,* 109–110, *110, 345,* 345–346, *346*
Circumferential belt, 1034, *1034*

Cis-acting mutation, 410, 411
Cis-acting regulatory sequences, DNA, 435–442
Citrate, in citric acid cycle, 750, 750
Citric acid cycle, 749–750, 750
  glycolysis and, 771
Cl⁻. See also under Chloride
Cl⁻/HCO₃⁻ antiporter, 659
Classical conditioning, 979, 980–982, 981
Clathrin, 709, 710
  in organelle purification, 165, 166
Clathrin-coated pits, 711, 713
  in endocytosis, 722, 723, 724–726
Clathrin-coated vesicles, 172, 711–713, 712, 713
  in endocytosis, 722–724, 727
Clathrin triskelions, 712, 713
Cleavage, 114, 114
  proteolytic, 72, 72–73
Cleavage and polyadenylation, of pre-mRNA, 498, 499
  on-off regulation of, 509, 510
Cleavage and polyadenylation specificity factor (CPSF), 498, 498
Cleavage furrow, 1103
Cln proteins, 1228–1232, 1229
Clonal selection, 1303, 1303–1304, 1304, 1309, 1323, 1337
Clone(s)
  definition of, 190, 221, 224, 1303
  cDNA, 234, 235–236, 236
  in genomic library, 232, 232–233, 233
  identification of, 236–240
    by expression cloning, 240, 241, 252–254, 253
    by membrane hybridization, 237–240, 237–240
Cloning, 193, 221
  DNA, 222–228, 234–236, 252–254
    with bacteriophage P1 vector, 233t, 234
    with cosmid vectors, 233t, 233–234, 234
    DNA ligase in, 227–228, 228
    expression, 240, 241, 252–254, 253
    with lambda-phage vectors, 230–233, 231–233, 233t
    with plasmid vectors, 222–225, 224, 225
    polymerase chain reaction in, 256
    restriction enzymes in, 225–226, 226, 227t
    restriction fragments in. See Restriction fragments
    with YAC vector, 233t, 234
  expression, 240, 241, 252–254, 253
  of mutation-defined genes, 284–290. See also Mutation-defined genes

physical mapping in, 284–287, 285–287
  positional, 284–290. See also Mutation-defined genes
  receptor, 867, 868
  viral, 205
  yeast artificial chromosomes in, 359
Cloning vectors
  bacteriophage lambda, 230–233, 230–233, 233t
  bacteriophage P1, 233t, 234
  cosmid, 233t, 233–234, 234
  DNA separation from, 240–242
  plasmid, 222, 223
    in expression systems, 252–254, 253
    restriction fragment insertion into, 228, 228
  YAC, 233t, 234
Closed complex, 415–416
Closed vesicles, in ATP synthesis, 752–753, 756–758
Clot formation, 1150
c-myc gene
  as proto-oncogene, 492, 1278, 1278–1279
  transcription in, 492
c-myc translocation, 1268
Coated vesicles
  clathrin-coated, 172, 711–173, 712, 713
    in endocytosis, 722–724, 727
  non–clathrin-coated, 172, 713–716, 714
Coatomer, 714, 715, 716, 717
Co-chaperonin, 826
Coding region, 308
Codon(s), 120–122, 121t, 121–123, 122t
  base pairing of with anticodons, 124–126, 125, 128
  iniatiator (start), 121
  synonymous, 121
  terminator (stop), 121
Coenzyme, 78–80, 80
Coenzyme Q, 761, 761, 763–764, 764. See also Electron carriers
Coil
  coiled, 65, 65, 451
  random, 56
    in domains, 67–68
Coiled coil, 65, 65, 451
Cointegrate, 326
Colchicine, microtubule stability and, 1066, 1066–1067
Coleoptiles, 1191, 1192
Collagen, 194, 195
  amino acids in, 1128, 1128
  banding in, 1128, 1130
  in basal lamina, 1135–1136, 1136, 1137

in cartilage, 1134
  in cornea, 1135, 1135, 1136
  cross-linkage in, 1131, 1131, 1133
  denatured, 1131, 1133
  fibril-associated, 1126t–1127t, 1135
  fibrous and nonfibrous, interaction of, 1134, 1134–1135
  formation of, 1131–1134, 1132, 1133
  gene organization in, 1128, 1129
  mutations in, 1134
  primoridal unit for, 1128
  strength of, 1129
  structure of, 614, 1127–1128, 1128, 1130, 1131, 1134–1135, 1135, 1137
  triple helix in, 1127–1131, 1128
    formation of, 1131–1133, 1132
  types of, 1126t–1127t, 1127, 1134–1136
Collagen fibers, 1129
Collagen fibrils, 1127
  formation of, 1128–1131, 1130, 1131, 1133–1134
  structure of, 1128–1129, 1130
Collagen proteins, 194
Collapsin, 1181–1182
Collapsing factors, 1181, 1181–1182
Color vision, 977–978, 978
Commissural neurons, 1180, 1180–1181
Companion cells, 805
Competence, in induction, 1160
Complement, 1296
Complementarity-determining regions (CDRs), 87, 87, 1310, 1312–1314, 1315
  diversity in, 1319
Complementary DNA, 118, 234
  cloning of, 234, 235–236, 236, 256. See also DNA cloning
  sequencing of, in identification of mutation-defined genes, 289–290
Complementary mutation, 274, 275
Complementation analysis, 274, 275
Complex transcription unit, 309–310, 310, 509
Compound microscope, 148, 149
Concatomer, 231, 231
Concentration gradient
  ion. See Ion concentration gradient
  proton. See Proton pump
Conditional mutation, 269–270
Conditioned stimulus, 979, 980
Conditioning, classical, 979, 980–982, 981
Cone cells, 977–978, 978
cI protein
  in lysogenic cycle, 544, 545–548, 546–548
  in lytic cycle, 548–550, 549, 550

Confocal scanning microscopy, 154–155, *155*
Connectin, 1026, *1027*
Connective tissue, 176
  collagen in. *See* Collagen
  dense, 1125, *1130*
  hyaluronan in, 1136–1139, *1138*
  loose, 1125
Connexin, 1159
Consensus sequence, in transcription, 414, *415*
Constant region, antibody, *1310*, 1311, 1316–1318, *1336*, 1336–1337, *1337*
Constitutive transcription, 408
Contact inhibition of movement, 1256
Contractile belts. *See* Contractile bundles
Contractile bundles, 1032–1036, *1033*
  membrane adhesion of, *1035*, 1035–1036
Contractile ring, 1036–1038, *1037*
  formation of, 1103–1105
Contractile vacuole, 662, *662*
Contraction. *See* Muscle contraction
Cooperative binding, 423, 545–547
COP-coated transport vesicles, 713–714, *714*
Copia elements, *329*, 333, 333–334
Copying errors, repair of, 385, *386*
CoQ, 761, *761*, 763–764, *764*. *See also* Electron carriers
CoQH$_2$-cytochrome *c* reductase. *See also* Electron carriers
  electron-proton transport by, *767*, 767–768
CoQH$_2$-cytochrome *c* reductase complex, 761, 762, 762t, 764–765
Core polymerase, 376, 415
Corey, Robert B., 63
Corn, mobile elements in, *327*, 327–328, *328*
Cornea, collagen in, 1135, *1135*, 1136
Corpus luteum, 864
Cortex, cell, actin filaments in, 999, *999*
Cosmid cloning, 233–234, *234*
COS sites, 231, *231*
Cotransport, membrane, 652–659
C$_o$t$_{1/2}$ (reassociation rate), 316
Covalent bonds, 16–20, *18*, 18t, *19*, 19t. *See also* Bond(s), covalent
coxII, 815
C$_4$ pathway, for CO$_2$ fixation, 802–803, *802–803*
C peptide, 721
CpG island, 436
CRE-binding (CREB) protein, 914–915
c-ret, in kidney development, 1168–1169
Crick, Francis H.C., 65, 104, 114, 320

Cristae, mitochondrial, 175, *175*, 747
Critical micelle concentration, 605–606
crk oncogene, 1263t, 1264–1265
cro gene
  in lysogenic cycle, 545–547, *546*, 548–549
  in lytic cycle, 548–549, *549*
Cro protein, 490, *490*
  in lysogenic cycle, 545–547, *546*
  in lytic cycle, 548–550, *549*, *550*
Cross-bridge cycle, 1026
Crossing over. *See also* Recombination
  in meiosis, 185, *185*
  unequal
    DNA fingerprinting and, 319–320, *321*, *322*
    gene duplication and, 314, *314*
Cross-linking
  antibody, 1334
  protein, *114*, 114
Cryptic splice sites, 518
Crystallography, x-ray, 59–61
c-src proto-oncogene, 1264
CTD. *See* Carboxyl-terminal domain (CTD)
cIII protein, in lysogeny, 547–548, *548*
CTLs. *See* Cytotoxic T lymphocytes (CTLs)
cII protein, in lysogeny, 547–548, *548*
Culture. *See* Cell culture
CURL organelles, 709, 727, *727*–728
C value, 312
C-value paradox, 312–313
Cyanobacteria, 142
Cyclic AMP. *See* cAMP
Cyclic electron flow, *795*, 796
Cyclic GMP
  photon absorption and, *975*, 975–976
  as second messenger, 857–859, *860*
Cyclin(s), 1201, 1207–1212, *1209–1211*
  degradation of, *1210*, 1210–1212, *1211*
  in DNA replication, 1231–1232
  G$_1$, 1228–1232
  in mammalian cell cycle, 1232–1235, 1233t, *1233–1235*
  in yeast cell cycle, 1228–1232, 1233t
Cyclin B, 1208–1212, *1209–1211*, 1233t
Cyclin D, 1232, 1233t
Cyclin-dependent kinases (Cdks)
  in mammalian cell cycle, 1232–1236, 1233t
  in mitosis, 1201–1202, 1222–1224, 1227–1228
  in S phase, in yeast, 1227–1232, 1233t
Cyclin destruction box, 1210, *1210*, *1211*

Cysteine, 54, *55*. *See also* Amino acid(s)
Cystic fibrosis
  gene knockout and, 295–296
  gene therapy for, 202
  inheritance patterns for, 277–278, *278*
  mutant protein in, 698
  prevention of, 254–256
Cystic fibrosis transmembrane regulator protein, 651–652
Cystine knot, *1161*, 1162
Cystosine, pairing of with guanine, 104–105, *105*
Cytidine residues, methylation of, 468, 558
Cytochalasin, 1006
Cytochemical staining, 150, *151*
Cytochrome, 607
  in electron transport, *761*, 763, 764–770
  in bacterial photosynthesis, 788–790, *789*
Cytochrome *b$_2$*, 827
Cytochrome *bc$_1$* complex, 786
Cytochrome *b/f* complex, 792, *792*, 793
Cytochrome *c*, 761, 762, 762t, 765, 827
Cytochrome *c* heme lyase, 829
Cytochrome *c* oxidase complex, 755t, 761, 762, 762t, 765. *See also* Electron carriers
  electron-proton transport by, 766, *766*, 768–770, *769*
Cytochrome *c* reductase, in ATP synthesis, 755t
Cytochrome *c* synthetase, 829
Cytochrome P-450, in detoxification, 1282–1283
Cytokeratins, 1106
Cytokine-receptor superfamily, 859, *862*, 916–917
Cytokinesis, 180, 1202
  actomyosin in, 1036–1038, *1037*
  astral microtubules in, 1103–1105, *1104*, *1105*
  cleavage plane in, 1103–1105, *1104*
  phosphorylation and, 1218–1219
  in plants, *1105*, 1105–1106
Cytometry, flow, 161, *162*
Cytoplasm, 143
  RNA transport to, 513–520, 534
  sol-gel transformation in, in cell movement, 1044
Cytoplasmic streaming, *1038*, 1038–1039
  in cell movement, *1043*, 1043–1044
Cytosine, 102, *102*, 103t
Cytosine methylation, gene inactivation and, 468
Cytoskeletal proteins, families of, 314

Cytoskeleton, 143, 168–169
    attachment to, in protein sorting, 731
    in cell movement, 991
    cytosol fibers in, 992
    erythrocyte, 620–623, *621–623, 999,*
        999–1000
    intermediate filaments in. *See*
        Intermediate filament(s) (IF)
    marginal band in, 1000
    membrane binding to, 618–623
        dystrophin in, *999,* 1000–1002
    microfilaments in. *See* Actin filaments;
        Myosin filaments
    microtubules in. *See* Microtubule(s)
    myosin II stiffening of, 1036, *1036*
    platelet, 1000, *1001*
    poly-A tail and, 521
    mRNA and, 521
    structure of, 992, *993*
Cytosol, 143, 164
    glycose metabolism in, 740–745
    intermediate filaments in, 1114, *1114*
    pH of, 640–641
    structure and function of, 168–170,
        169
Cytosolic dynein, 1077t, 1078, *1078*
Cytosolic hsp70, 829
Cytosolic pH
    in cell volume regulation, 663–664
    transporter regulation of, *658,*
        658–659
Cytosolic proteins, 169–170
    in vesicle-membrane fusion, 716, *717*
Cytotoxic drugs, microtubule stability
        and, *1066,* 1066–1067
Cytotoxic T lymphocytes (CTLs), *1297,*
        1306
    activity of, 1325–1327, 1331–1332
    self-recognition of, 1328–1330
C_4 zinc finger, *448, 450, 451,* 452
C_6 zinc finger, *448, 451*

DAG. *See* 1,2-Diacylglycerol (DAG)
Dalton, 56
Dark reactions, in photosynthesis, 782,
        783
Darwin, Charles, 4, *7*
Daughter cells
    in meiosis, 181, *184*
    in mitosis, 179–180, *180*
    separation of, 179
Daughter chromatids. *See also*
        Chromatid(s)
    segregation of, topoisomerase II in,
        383–384, *385*
daughterless gene, 561–562, *564*
DBA-binding proteins, in looping out,
        425
Decatenation, 382, *383*
Decoration experiments, 996

Deep etching, 169, *170*
    in membrane studies, 616, *617*
Defined media, 1253
Degenerate probe, 239
Delayed-type hypersensitivity, 1331
Deletion, chromosomal, 267, *267*
Delta protein, 1173–1175, *1174, 1175*
Demyelinating diseases, 942
Denaturation, of DNA, *108,* 108–110,
        109
Dendrites, 928, *928*
Dense body, in smooth muscle, *1022,*
        1025
Dense connective tissue, 1125, *1130*
Density, definition of, 88
Density gradient, 90
Density-gradient centrifugation,
        164–165, *166*
Deoxyribonucleic acid. *See* DNA
Deoxyribonucleosides, 103, 103t
    radiolabeled, 215
Deoxyribose, structure of, *102*
Dephosphorylation
    cytokinesis and, 1219
    in late mitosis, 1216–1219
Depolarization. *See* Membrane
        depolarization
Dermatan sulfate, *1140*
Dermomyotome, *558*
Desmin, 1108, 1108t, 1114–1115,
        1115, 1157. *See also* Intermediate
        filament(s) (IF)
    Z disks and, 1114–1115, *1115*
Desmocollin, 1156
Desmoglein, 1156
Desmoplakin, 1113t
Desmoplakin proteins, 1157
Desmosomes, 628, *1156,* 1156–1157,
        1157
    belt, 628, 1035, 1036, *1036,* 1156
    intermediate filament anchorage to,
        1114, *1115*
    spot, 628, *1156,* 1156–1157
    structure of, 1114
Desmotubule, 1193, *1194*
Destruction box, cyclin, 1210, *1210,*
        1211
Detergents, 604–606, *606, 607*
Detoxification, cellular, 1282–1283
Dextro isomer, 20
1,2-Diacylglycerol (DAG)
    in protein kinase C activation,
        904–905
    as second messenger, 857–859, *860*
Dicot, 297
Dideoxy sequencing, 245–246, *247*
Differential centrifugation, 90, *91,*
        164–165, *165*
Differential interference microscopy,
        156–158, *156–158*
Diffraction pattern, 60

Diffusion
    facilitated, 636–640
    of membrane lipids, 616–618
    of membrane proteins, 616–619, *618,*
        619
    passive, *635,* 635–636
    prevention of, by tight junctions,
        625–628, *626, 627*
    rate of, calculation of, 636
Diffusion coefficient, 618
Digoxin, *655*
2,4-Dinitrophenol (DNP), 770
Dintzis, Howard, 216–218
Diploid, 264
Diploid cell, 178
Dipole, 24–25
Dipole moment, *19,* 19–20
Direct repeats, in insertion sequences,
        324, *324*
Disaccharide, formation of, 43, *43*
Diseases, genetic. *See* Genetic diseases
Dissociation, acidic, 32–34
Dissociation constant, 32
Dissociation element, *327,* 327–328,
        328
Distributive enzyme, 376
Disulfide bonds
    formation of, 694–697
    rearrangement of, 694–695, *696*
    in recombinant DNA technology, 695
DNA, 9–10, 104–111. *See also* Nucleic
        acid(s)
    A form of, 106
    amplification of
        generalized, 343, *343*
        localized, 343–344, *344*
    analysis of, 10–11
    base pairing in, 104–108, *105–108*
        of codons and anticodons,
            124–126, *125*
        complementarity of, 105
        in recombination, 392–393
        of sticky ends on, 226
        wobble, 125, *125, 126*
    bent, 106, *107*
    B form of, 106
    cellular
        organization of, 147
        volume of, 144–146, 145t, 146t
    centromeric, 355–357, 1093
    chloroplast, 830–836, 831t
    chromosomal
        organization of, 344–348, *345–349*
        rearrangement of, 344, *344*
        as single linear molecule, 354, 359
    circular, *109,* 109–110, *110,*
        344–345, *345*
    cis-acting regulatory sequences in,
        435–442
    cloned, 224. *See also* Cloning, DNA
        analysis and sequencing of,
            240–248

mapping of
    by DNA walking, 287, *288*
    by in situ hybridization,
        285–287, *286*
complementary, *118*, 234
    cloning of, 234, 235–236, *236*,
        256. *See also* DNA cloning
    sequencing of, in identification of
        mutation-defined genes, 289–290
cooperative protein binding to,
    545–547
damage to, cell cycle arrest and, 1239
deletion of, 344
denaturation of, *108*, 108–110, *109*
duplex
    synthesis of, 115–116
    unwinding of, 372–373
eukaryotic
    classification of, 312t
    organization of, 346–347, *347*
    vs. prokaryotic, 144–147, 145t,
        146t
form I, 110, *110*
form II, 110, *110*
functional rearrangements of,
    338–344
growing fork in, 115–116, *117*,
    366–368, *368*. *See also* Growing
    fork(s)
heteroduplex. *See* Heteroduplex
intermediate-repeat, 317, 320–338.
    See also Insertion sequences;
    Mobile elements;
    Retrotransposon(s); Transposon(s)
junk, 307n
lagging strand of, 374–378, *375*, *377*,
    378
leading strand of, 374, *375*, 375–378,
    377, *378*
length of, 308
linking number of, 110, *111*
mapping of. *See* Mapping
melting temperature of, 108, *109*
minisatellite, 320, *321*
minus strand of, 208
mitochondrial, 473–474, *474*,
    811–819. *See also under* DNA
mobile elements in. *See* Mobile
    elements
modification of, 113, *114*
native state of, 104–108
nicking of, 110, *110*, 115
noncoding, 311. *See also* DNA,
    repetitious
nuclear, mitochondrial DNA in,
    815–816
nucleotides of, *102*, 102–104, 103t
operator. *See* Operator
plasmid, 222
prokaryotic
    vs. eukaryotic, 144–147, 145t, 146t
    organization of, 344–346, *345*, *346*

promoter. *See* Promoter(s)
proviral, 1273, *1274*
rearrangement of, 344
reassociation rates for, 316–317, *317*
recombinant. *See* Recombinant DNA
    technology
reduplication of, in tumor cells, 1269,
    1269
renaturation of, *108*, 108–109
    rate of, 316–317, *317*
repair of, 385–389, *387*, 387t, *388*
    defects in, cancer and, 1285–1286
    mutations and, 385–389
    RecA protein in, 393
repetitious, 312t, 316–317
    interspersed, 316
    tandem, 316
repressor. *See* Repressor(s)
retroviral, *329*, 1273–1275, *1274*
satellite, 318, *319*
selfish, 320
simple-sequence, 317, 318t, 318–320
single-copy, 317
spacer, 317
structure of, 104–111, *105–111*
supercoiling of, 110–111, *111*, 346,
    346, 372
    negative, 381
    positive, 381
    topoisomerases and, 381–382
synthesis of, 114–119. *See also* DNA
    replication
    cell divsion and, 177–178, *179*
    definition of, 365
    in eukaryote, 177–178, *179*
    primer in, 115
    in prokaryote, 177
    rules of, 113–114, 115–116
synthetic. *See also* Recombinant DNA
    technology
    in probes, 238–240, *239*
    production of, 229, *229*
    uses of, 229, *229*, 238–240, *239*
telomeric, 358
transcription-control regions in,
    435–442
transcription of. *See* Transcription
transfer of, 292, *292*
underwound, 110, 111
unwinding of, 372–373, *373*
variation in, *312*, 312–313
writhe in, 111, *111*
Z, 106, *107*
cDNA, *118*, 234
    cloning of, 234, 235–236, *236*, 256.
        See also Cloning, DNA
    polymerase chain reaction in, 256
    sequencing of, in identification of
        mutation-defined genes, 289–290
mtDNA, 811–819
    coding capacity of, 813–816
    disease-causing mutations in, 817

isolation and sequencing of, 812, *813*
    in nuclear DNA in, 815–816
    organization of, 813, *814*
    plant, 815
    processing of, 817
    size of, 813, *814*
    visualization of, 812, *812*
dnaA gene, in DNA replication,
    372–373
DNA amplification, by polymerase
    chain reaction, 254–256, *255*
DnaA protein, in DNA replication,
    372–373, *373*
dnaB gene, in DNA replication,
    372–373
DNA-binding domains, transcription
    factor, 445, 447–452, *448–452*,
    456
DNA-binding proteins, 445, 447–452,
    452, 454–455, 457–458
    histone, 346–347
    nonhistone, 354
    in transcription. *See* Transcription
        factor(s)
DNA binding site, half-site, 552
DnaB protein, in DNA replication,
    372–373, *373*, 374, *374*
DNA catenane, 383
DNA fingerprinting, 319–320, *321*, *322*
DNA gyrase, 382
DNA helix, 104–108, *105–108*
    base pairing in. *See* Base pairing
    circular, 109–110, *109–111*
    hybrid, 111–112
    linking number of, 110, *111*
    right vs. left handed, 106, *106*, *107*
    supercoiled, *110*, 110–111, *111*
    twist in, 110, *111*
    underwound, 110–111, *111*
    unwound, *108*, 108–109, *109*
    writhe in, 111, *111*
DNA initiation factors. *See* Initiation
    factor(s) (IF)
DNA inversions, 338–339, *339*
DNA library
    cDNA, 234–236, *235*, *236*
    genomic, 229–233
        construction of, 231–233, *232*, *233*
        size of, 232–233
cDNA library, 234–236, *235*, *236*
    screening of, by membrane
        hybridization, 237–240, *238*
DNA ligase, 375, *375*
    in DNA cloning, 227, 227–228
    in replication, 116, 227
DNA mapping. *See* Mapping
DNA-mediated transposition, 323–328
DNA phage
    small, 207–208
    T, 206

DNA polymerase(s), 115, *116*
  in DNA replication
    in *E. coli*, 372, 374–376, *375,*
      376t, 378
    in eukaryotes, *379,* 379–380
  proofreading by, 385, *386*
  β subunit of
    in replication, 376, *377*
    structure of, 52, *53*
DNA polymorphisms, 280
  restriction fragment length, 279–281,
    280
  single-stranded conformation, 289,
    290
DNA probes, 238–240, *239, 240*
  in expression cloning, 240
  in nuclease-protection method,
    249–251, *250*
  in Southern blotting, 248
cDNA probes, 238–240, *239, 240*
DNA recombinant, 221
DNA replication, 103–104, 114–119
  bidirectionality of, 366–369,
    368–370
  catenation in, 382, *383*
  of circular molecules, 383, *384*
  copying errors in, correction of, 385,
    386
  decatenation in, 382, *383*
  definition of, 365
  directionality of, 113, *114,* 115
  DNA ligase in, 116, 227
  DNA unwinding in, 372–373, *373,*
    379–380
  duplex, 115
  in *E. coli*, 372–378
  eukaryotic, 378–381
  growing fork in, 115–116, *117,*
    366–368, *368. See also* Growing
    fork(s)
  helicase in, 372–373, *373, 378*
  incomplete, mitosis and, *1238,*
    1238–1239
  initiation factor for, 1231–1232
  initiation of, 223, *223,* 372, *373*
  lagging strand in, 115–116, *117,*
    358–359, 374–378, *375,*
    377–380, 379–381
  leading strand in, 115–116, *117,* 374,
    375, 375–378, *377–380,*
    379–380
  in lytic pathway vs. lysogenic
    pathway, 544–550
  methylation in, 385
  mitochondrial, 180
  Okazaki fragments in, 116
  plasmid, *223,* 223–224
  polymerases in, 115, *116,* 372,
    374–376, *375, 376t, 378*
  primase in, 374, *374, 374t, 378*
  prokaryotic, 372–378

proteins in, 372–378, *378*
    prokaryotic, 372–373, *373,* 374,
      374, 378
  replication origin in, 223, *223*
  in retroviruses, 115
  RNA primer in, 115, *374, 374t,*
    374–375
  rolling circle, 344
  rules for, 115–116
  semiconservative, 366, *367, 368*
  sites of, 370–372, *371*
  SV40 chromosome, 378–380, *379*
  telomerase in, 380–381, *381*
  template in, 115
  termination of, 378
  topoisomerases in, 381–384, *381–385*
  viral, 205–206, *206, 207,* 1273
DNA replication bubble, 368–369, *369,*
    374
DNA replication initiation factor,
    1231–1232
DNA replication origin, 370–371, *371*
  in *E. coli*, 370–371
  SV40, 370–371
  in yeast, 355, *356,* 371
DNA replication termination sites, 378
DNA rescue, in positional cloning, 284
DNA restriction fragments. *See*
    Restriction fragments
DNase footprinting, 412, *413, 414*
  in vivo, 416, *417*
DNase I
  hypersensitivity to, 467, *467*
  resistance to, transcription inhibition
    and, 466–467
DNA sequences, homologous, 248
DNA sequencing, 221–222, 245–248
  data storage, distribution and analysis
    of, 246–248
  Maxam-Gilbert method of, 245, *246*
  Sanger (dideoxy) method of,
    245–246, *247*
DNA viruses
  classification of, 208–209, *209,* 210t
  tumor, 1270–1273
DNA walking, 287, *288*
Dolichol, 704, *705*
Domain(s)
  antibody, 1298–1299, *1299,*
    1310–1314, 1310–1315
  protein, 67–68, *68*
  regulatory, 882
  RNA, 112, *112*
  steroid hormone receptor, 459–461,
    460, 461
  transcription factor, 445–452, 456
Dominant gene, 226, *226*
Dominant mutation, 266, *266*
Dopamine. *See also* Catecholamine(s)
  function of, 967
  structure of, 954t

synthesis of, 967, *969*
  transporter for, 970
Dopamine receptors, 967
Dorsal lip of blastopore, 1164
Dorsal protein, 573–575, *574,* 1162
Dorsal vegetal pole, *1165,* 1166
Dorsoventral patterning. *See* Patterning
  in embryogenesis, 1159–1166
Double helix. *See* Helix, DNA
Double-sex gene, 513
Drosophila melanogaster
  compartment in, 577
  copia elements in, *329, 333,* 333–334
  DNA in, 146t
  embryogenesis in
    regionalization in, 565–584. *See*
      also Regionalization, in
      Drosophila
    steps in, *567, 567*
  embryonic mutants in, *273,* 273–274
  eye development in
    cell signaling and, *891,* 891–893,
      892, 1172, 1173
    mutations in, 264, *265,* 891–893,
      892
  heat-shock genes in, transcription of,
    492–493, *493*
  HOM-C genes in, 584
  learning and memory in, 979
  life cycle of, 565, *566*
  mutations in, 264, *265,* 891–893
    bobbed, 316
    genetic screen for, *273,* 273–274
    induction of, 271–272, *272*
  neurogenesis in, 561–562, *563, 564*
  oogenesis in, 565–566, *566*
  parasegments in, 576–577, *578, 580*
    BX-C genes and, *582,* 582–583
  P element of, 326–327, 327t
  polytene chromosomes in, *283,*
    283–284, *343, 343*
    banding in, 355
  position effect variegation in, *464,*
    464–465
  mRNA localization in, 522
  segments in, 576–577, *578*
  sex determination in, 511–513
  shaker mutants in, *947,* 947–950
  signal-dependent translocation in,
    573–575, *575*
  tandemly repeated arrays in, 316t
  transcription control in, *429,*
    429–430
  transgenic, 297, *298*
  wing development in, 1169–1172,
    1170, 1171
Drug resistance, multidrug-transport
  protein in, *651,* 651–652
Drug-resistance plasmid, in cloning,
    222, *223*–224, *224*
Ds mobile element, *327,* 327–328, *328*

dsx gene, 513
D-type cyclins, 1232
Duchenne muscular dystrophy (DMD)
  dystrophin deficiency in, 1000–1002
  inheritance patterns for, 277, *278*
  mutations in, 334
    mapping of, *282*, 283
dunce mutation, 979
Duplex, replication of, 115–116
Duplicated genes, 313–315, *314, 315*
  nonfunctional, 314–315
Dynactin complex, 1078
Dynamin, 711–712
Dynein
  cytosolic, 1077t, 1078, *1078*
  flagellar, *1082*, 1084–1087, *1085,*
    1086
α-Dystroglycan, 1187
Dystrophin, *999*, 1000–1002

E2A protein, in myogenesis, 560
Early endosome, 717, 724
Early-response gene, 916
  transcription of, 1235–1236
E bos, 560
E-cadherin, *1035*, 1036, 1115, *1115,*
  1150–1153, *1151*, 1151t, *1152*
αEcdysone, structure and function of,
  858t
Echinoderms, acrosome reaction in,
  1010, *1011*
Ectoderm
  cadherins and, *1152*, 1152–1153
  embryonic development of, 1164
Ectoplasm, 1044
Edman degradation, *59, 60*
EF. *See* Elongation factor (EF)
E2F, in S phase regulation, 1236, *1237*
Effector, enzymatic, 84, *84*
Effector specificity, in signaling, 856
EGF. *See* Epidermal growth factor
  (EGF)
Elaioplasts, 836, *836*
Elastin, 1125
ELC. *See* Expression-linked copy (ELC)
Electrical signaling. *See also* Action
  potential
  in sensory transduction, 971–979
Electrical synapse, 930, *930*, 952–953.
  See also Synapse(s)
Electric energy, 16
Electric potential, 16
  transmembrane. *See* Membrane
  potential
Electron(s), arrangement of, 17
Electron acceptor, primary, 784
Electron carriers, *761–764*, 762t,
  762–766
  in chloroplasts, 794–796, *795*
  lateral mobility of, 765
  orientation of, 765

proton transport and, 766–767, *767*
  reduction potential of, 765
Electronegativity, 19, 19t
Electron flow
  cyclic, *795, 796*
  linear, *795, 796*
Electron micrography, *148*
Electron microscopy, 158–161,
  159–161
  cell isolation and, 161–162, *163*
  freeze-fracture, 168, *168, 169*
    in membrane studies, 616, *617,*
    626, 627
  of membrane, 616, *617*
  scanning, *148*, 161, *161*
  transmission, *148*, 158–161, *159–161*
    quick freeze technique for, 169,
    170
Electron shuttles, 759, *760*
  in ATP synthesis, 759
Electron-spin spectroscopy, 599
Electron transfer, in oxidation-reduction
  reactions, 37–39
Electron transport
  in ATP synthesis, 750–752, 759–770
  carriers in. *See* Electron carriers
  in photosynthesis, 782, 784–786,
    790, 793–797
    in bacteria, 788–790, *789*
  proton-motive force and, 759–770
Electron transport complexes, *761–764*,
  762t, *762–766*
Electrophoresis, *92*, 92–94, *93, 96*
  gel, *92*, 92–93, *96*
    in Maxam-Gilbert method, 245,
    246
    pulsed-field, 244–245, *245*
    in restriction fragment separation,
    242, 242–243, *243*
    in restriction-site mapping,
    243–244, *244*
    in Sanger method, 245–246, *247*
    in Southern blotting, 248–249, *249*
Electrophoretic mobility shift assay,
  419–421, *421*
  in transcription factor identification,
  442, *443*
Electroporation, 292
Elliptocytosis, 623
Elongation factor (EF), *134–135*, 490
Embryogenesis
  basal lamina in, 1167–1168
  dorsoventral patterning in,
    1159–1166
  in *Drosophila*
    regionalization in, 565–584. *See*
    also Regionalization, in
    Drosophila
    steps in, *567, 567*
  epithelial-mesenchymal interactions in,
    1167–1169
  induction in, 1160–1166

internal organ formation in,
  1167–1172
  kidney development in, 1168–1169
  salivary-gland development in,
    1167–1168
  tissue organization in, 1167–1172
  in *Xenopus*, 1163–1166, *1164–1166*
Embryonic stem (ES) cells, in gene
  knockout, 294, *294–296*
Emerson effect, 792
Emphysema, mutant protein in, 698
EMS. *See* Ethylmethane sulfonate (EMS)
Endergonic reaction, 739
Endocrine signaling, 855. *See also*
  Hormone(s); Signaling
Endocytosis
  pinocytotic, 722
  receptor-mediated, 710, 722–729,
    723, 723t. *See also*
    Receptor-mediated endocytosis
Endoderm
  cadherins and, *1152*, 1152–1153
  embryonic development of, 1164
Endonuclease(s), restriction, 225–226,
  226, 227t
Endoplasm, 1044
Endoplasmic reticulum (ER), *153*,
  170–172
  collagen synthesis in, 1131, *1133*
  in cytoplasmic streaming, *1038*, 1039
  disulfide bonding in, 694–697
  in glycosylation, 703–704
  lipid synthesis in, 671–674, *672, 673*
  microtubules and, 1073, *1075*
  nuclear membrane and, 167–168
  phospholipid synthesis in, 671
  post-translational protein modification
    in, 694–697
  protein degradation in, 698
  protein glycosylation in, 699–711
  protein orientation in, 670–671,
    688–694
  protein retention in, 698–699, *699*
  protein synthesis in, 670, *670*,
    694–697
  protein targeting to, *810*, 810–811,
    811t
  protein transport and, *681*, 681–688,
    711–718, *717*
  quality control in, 697–699
  resident proteins in. *See also* Bip
    protein; Protein disulfide
    isomerase (PDI)
  ribosomal binding to, *675, 675*
  rough, 143, *144, 145*
    lumen of, *674, 676*
    structure and function of, *171*,
    171–172, *676, 700*
  secretory proteins in, 676–677, *677,*
    681
  smooth, 143, *144, 145*, 170–171,
    171

Endoproteases, in proteolytic processing, *721*, 721–722

Endorphins, 971

Endosome
early, 717, 724
late, 726–727, *727*

Endosymbionts, 811

Endothermic reaction, 34

Energy
activation, 18, 18t, 21–22, *44–46*, 44–46
enzymes and, 76
cellular storage of, 8–9
of chemical reactions, 18, 18t, 44–46, *44–46*
definition of, 16
electric, 16
free. *See* Free energy
hydration, 24
interconvertibility of, 16
kinetic, 16
lattice, 24
potential, 16
in chemiosmotic theory, 740
radiant, 16
thermal, 16

engrailed gene, *577, 578, 579, 580*

Enhancer, 425, 439–442, *440, 442*
in provirus-directed transcription, 1273

Enhancer insertion, in oncogene activation, 1279

Enkephalins, 971

Enthalpy, 34, 35

Entropy, 26
free energy and, 34–35, *35*

env gene, 331
retroviral, *1274, 1275*

Enzyme(s), 75–86. *See also specific enzymes*
active site of, 75–76, *76*
allosteric regulation of, 84, *84*
catalytic activity of, 44, *44–46*, 45–46, 75
quantification of, 81–83, *83*
regulation of, 83–86, *84, 85*
subunit interaction and, 84–85, *85*
coenzymes and, 78–80, *80*
definition of, 75
distributive, 376
effectors and, 84, *84*
feedback inhibition of, 83–84
gloxisomal, 174
Golgi, 706
retention of, 707–708, 719
inhibitors and, 84, *84*
lock-and-key mechanism of, 80–81, *81*
lysosomal, 173
mechanochemical, 1013
mitochondrial, 829

modification, 225
multimeric, *57*, 84–85, *85*
peroxisomal, 174
processive, 376
proteolytic, *76*, 76–78, *77, 79*
restriction, 221
in DNA cloning, 225–226, *226*, 227t
in separation of DNA from vector, 240–242
RNA, 112, *113*
secretory protein, 677
specificity of, 75
quantification of, 81–83, *83*
substrate binding to, 75–76, *76*
conformational change and, 80–81, *81*, 84
induced-fit, 80, *81*
lock-and-key, 80–81, *81*
substrate for, 46
variety of, 75

Ephrussi, Boris, 11

Epidermal growth factor (EGF)
in cell culture, 1253
as protein domain, 68, *68*
receptor for, down-regulation of, 912
structure and function of, 861t

Epi-fluorescent microscope, *152*

Epigenetic alterations
cancer and, 1286
in position effect variegation, 464

Epinephrine. *See also* Catecholamine(s)
in glycogenolysis, 882, *883*
receptor binding of, 870–871
structure and function of, 860t, 954t, 967
synthesis of, *967, 969*

Epinephrine receptors, 870–871, 967

Episome, 292

Epithelial cell(s)
circumferential belt in, 1034, *1034*
cultured, 196–197, *198*, 1251–1252
desmosomes and, 1156–1157
differentiation of, 1167–1168
intermediate filaments in, 1114, *1115*
intestinal, plasma membrane of, 623–624, *624*
mesenchymal cells and, 1167–1172
structure of, 196–197, *197*
tight junctions of, 625–628, *626, 627*
types of, 196–197, *197*

Epithelial cell lines, 196–197, *198*

Epitopes, 87, *87*, 1297–1298, *1298*

Epoxide, 1283, *1283*

Epstein-Barr virus, *1279*, 1279–1280

Equatorial plane, 179, *180*

Equilibrium, chemical, 28–30

Equilibrium constant, 28
free-energy change and, 36–37
pH and, 32

Equilibrium density-gradient centrifugation, 92, 164–165, *166*

ER. *See* Endoplasmic reticulum (ER)

erbA oncogene, 261

Erythrocyte(s)
ctyoskeleton of, *999*, 999–1000
hereditary defects in, 623
plasma membrane of, 620–623, *621–623*
anion transport across, *655*, 655–656

Erythrocyte ghost, *621*, 621–623

Erythrocyte glucose transporter, 638–640, *639, 640*, 640t

ES cells. *See* Embryonic stem (ES) cells

Escherichia coli. See also Bacteria; Prokaryote
cellular macromolecules in, 144, 145t
chromosomal DNA in, *345*, 345–346, *346*
culture of, 190t, 190–193, *191. See also* Cell culture
in DNA cloning, 223–225, *224, 225*
DNA in, 144–146, 145t, 146t
gene mapping of, *192*, 193
glucose metabolism in, 421–423, *422*
lac operon of, 406–411
lambda phage infection in, lysogeny vs. lysis in, 544–550
phages of, 206–208
proton-lactose symporter in, *660*, 660–661, *661*
recombination in, 391–396, *392–395*
replication origin in, 370–371, *371*
structure of, *142*

Escherichia coli expression systems, 252–254, *253*

Escherichia coli plasmid vectors, 222–225, *223–225*

Estradiol, structure and function of, 858t

Estrogen, feedback control of, 864, *864*

Ethylmethane sulfonate (EMS), 268, *270*

Etioplasts, 835, *836*

Eubacteria, 142, *142*
types of, 142

Euchromatin, 355

Eukaryote
cell content in, 143–147, *145*, 145t, 146t
cell division in, 177–185. *See also* Cell division
cell structure in, 7–8, *8*, 143, *144–147*
chromosomes in, 147
definition of, 142
DNA in, 312, 312t
DNA replication in, 378–381
gene expression in, 406, 426–473
gene organization in, 116–119, 117t, *118, 119*, 311–315, 311–316
organelles of, 8, 9

protein synthesis in, *118*, 133–138
transcription in, 116–119, *118–120*, 406, 426–473
translation in, 116–119, *118–120*
Eukaryotic expression vectors, 253
eve gene, *579*, 587–579
Even-skipped gene expression, *429*, 429–430
Evolution, 4–5
  Darwinian theory of, 4
  duplicated genes in, 314–315
  mobile elements in, 321–322
  sequence drift in, 314–315
Exchange proteins, phospholipid, *673*, 674–674
Exergonic reaction, 742
Exocytosis, 172, 670
  in transcytosis, 729, *729*
Exon, 119
  splicing of, 119, *120. See also* Splicing
Exon shuffling, 322
Exoplasm, 1044
Exothermic reaction, 34
Expansins, 1191
Expression cloning, 240, *241*, 252–254, *253*
Expression-linked copy (ELC), 342, *342*
Expression vectors, 240, *241*, 252–254, *253*
  eukaryotic, 253
  lambda, 240, *241*
  plasmid, 252
Extensin, 1188, 1190
Extracellular matrix, 1124–1149
  cell attachment to, by multiadhesive proteins, 1143–1150
  collagen in, 1126–1136. *See also* Collagen(s)
  composition of, 194, *195*, *1124*, 1124–1125
  in culture, 194
  definition of, 193, 1123
  functions of, 1125–1126
  hyaluronan in, 1136–1139, *1138*
  membrane binding to, 176
  in neuronal outgrowth, *1178*, 1178–1179
  proteins in, 194, *195*
    multiadhesive, 1124, *1124*, 1143–1150. *See also* Multiadhesive matrix proteins
  proteoglycans in, 1139–1142
Extracellular matrix proteoglycans, 1139, *1141*
extramachrochaete gene, 561–562, *564*
Eye development, in *Drosophila*, 891–893, *892*, 1172, *1173*
  mutations in, 264, *265*, 891, 891–893, *892*

Facilitator neuron, 968, 980
FACS. *See* Fluorescence-activated cell sorter (FACS)
F actin. *See* Actin filaments
Facultative anaerobe, 744
FAD/FADH
  in citric acid cycle, 749–751, *750*, 751t
  in fatty acid metabolism, 749, *749*, 751
  in glycolysis, 747, *747*, 751t
$FADH_2$, electron transport from, 759–770
Familial hypercholesterolemia, 724
Faraday constant, 38
Farnesyl, in membrane-protein binding, *610*, *611*, 611–612
Fascicle, 1179
  longitudinal, 1182
  transverse, 1182
Fasciclin, 1183
Fascin, 997, *998*
Fatty acid(s), 27
  membrane, 597, *598*, 598t
  metabolism of
    mitochondrial, 746, 746–750
    peroxisomal, 772, 772–773
  saturated, 27
  synthesis of, 671
    cAMP-dependent protein kinases in, 886
  transport of, 886
  unsaturated, 27
Fatty acid acylation, 71, *71*
Fatty acyl chains, 27, *27*
  membrane, 597, *597*
Feedback control, hormonal, 864, *864*
Feedback inhibition, 83–84
Feed-forward activation, 771
Fermentation, anaerobic, 744
Ferredoxin, *795*, 796
Ferritin mRNA, translation of, *527*, 527
Ferrotransferrin, 728, *728*
Fe_2S_2, 762, *763*
Fe_4S_4, 762, *763*
F_0F_1 ATPase complex. *See* ATPase complex
Fibrils, collagen, 1127
  formation of, 1128–1131, *1130*
Fibroblast(s), 194, *195*
  cultured, 194–196, 1251–1252
  movement of, 992
Fibroblast growth factor
  in mesodermal induction, 1165
  proteoglycan binding of, 1142, *1143*
Fibroglycan, 1142, *1142*
Fibronectin, 1146–1150
  alternative, cell-type–specific expression of, 509–511, *510*
  in cell adhesion, *1147*, 1147–1148, *1148*

  in cell migration, *1149*, 1149–1150
  integrin binding to, 1147–1148, *1148*, *1149*
  RNA processing in, 513–515, *514*
  structure of, 1146–1147, *1147*
  surface, in transformed cells, 1256
  tenascin and, 1149, *1149*
  in wound healing, 1150
Fick's law, 636
Filaggrin, 1113t
Filaments. *See* Actin filaments; Intermediate filaments; Microfilaments
Filamin, 997, *998*
  in cytoplasmic viscosity, 1044
  in platelet cytoskeleton, 1000, *1001*
Filopodia, 992, *993*
  actin bundles in, 1003
F ion pump, 644–645, *645*, 646t
5′ cap, of mRNA, 119, *119*, 133–136
  in nuclear transport, 516–518, *517*
5′ deletions, in transcription-control region mapping, 437, *437*
5′ end, of nucleic acid, 104, *104*
5S-rRNA, synthesis of, by tandemly repeated genes, *315*, 315–316, 316t
Flagella, 175–176, 1079–1090
  axoneme in, 1080–1089, *1082*
  basal bodies in, 1081–1082, *1083*, 1088–1090, *1089*
  beating of, 1080, *1080*, *1081*, 1084–1087, *1085*
  microtubules in, 1080–1089
  structure of, 1080–1084, *1082*, *1083*
  swimming direction in, 1087–1088
Flavin adenine dinucleotide. *See* FAD
Flippase proteins, 671–673, *673*
Flow cytometry, 161, *162*
Fluid mosaic model, 617
Fluorescein staining, 152
Fluorescence-activated cell sorter (FACS), 161, *162*
Fluorescence microscopy, 150–154, *152–154*
Fluorescence recovery after bleaching (FRAP) studies, 618, *619*
Follicle cells, in *Drosophila*, 566
Follicle-stimulating hormone (FSH), 864
Follicular dendritic cells, 332
Footprinting
  DNase, 412, *413*, *414*
  in vivo, 416, *417*
  RNA, 132
Forkhead proteins, 448, *451*
Forssman antigen, 628
fos oncogene, 1265
Fossils, Cambrian, 4, *6*
Fraenkel-Conrat, H., 208
Fragmin, 1009t
Frame shifting, 121

Frameshift mutation, 267, *267*

Framework regions, antibody, *1314,* 1315

FRAP studies, 618, *619*

Free energy, 34–43

  from ATP hydrolysis, 40–43, 41t, *42*

  change in

    calculation of, 36–37

    determinants of, 35–36

    direction of chemical reaction and, 34

    enthalpy and, 34–35

    entropy and, 34–35, *35*

    in generation of concentration gradient, 37

    reduction potential and, 38–39

    values for, 36t, 37t, 39t

  in coupled reaction, 39–41

  definition of, 34

  in oxidation-reduction reaction, 37–39, 39t

Freeze-fracture technique, 168, *168, 169,* 616, *617,* 626, *627*

Frog. *See Xenopus laevis*

Fructose 2,6-biphosphate, allosteric control of, 771, *771*

Fructose 6-phosphate, 798, *799*

Fruit fly. *See Drosophila melanogaster*

Fura-2, *153,* 153–154

Furin, 721, *722*

Fuschin stain, 150

Fusicoccin, 1191

Fusion peptide, 731

Fusion protein, in expression cloning, 240, *241*

G_1 cyclins, 1228–1232, *1229*

G_1 phase, 178, *178,* 1202, *1202*

  arrest in, 1239

G_2 phase, 178, *178,* 1202, *1202*

  arrest in, 1239

GABA, 966–967

  transporter for, 970

G-actin, 994–995, *995*

  concentration of, 1006

  in F-actin, *995, 995, 996,* 1003, *1004*

gag gene, 331

  retroviral, *1274, 1275*

Galactolipids, 599, *599*

  in chloroplasts, 603

Galactose, 21, *21*

  in oligosaccharide, 612, *614,* 614

β-Galactosidase, synthesis of, 406, *406, 407, 408*

Galactosyltransferase, 700–703, *702*

  localization of, 707–708, 719

GAL4 gene, identification of, 445

Gall, 297

Gal4 protein, functional domains in, 445–447, *446*

Gamete, 181

Ganglioside GM_2, in Tay-Sachs disease, 173

Gangliosides, 614, *614*

GAP. *See* GTPase activating protein (GAP)

Gap genes, in *Drosophila* regionalization, *573, 573,* 577–578

Gap junctions, 628, 930, *930,* 1157–1159, *1158, 1159*

  connexin in, 1159, *1159*

  in metabolic coupling, 1158

Garrod, Archibald, 11

Gastric acidification, *657,* 657–658

Gastric pH, regulation of, 657, *657*

Gastrulation, 1164

Gated channels. *See* Ion channel(s), voltage-gated

Gating helix, 949, *949*

G band, 354

gCAP39, 1009t

GCSF. *See* Granulocyte colony-stimulating factor (GCSF)

GDP. *See* Guanosine diphosphate (GDP)

  in cell signaling, 869, 874–877, *876–878*

  in microtuble assembly and disassembly, 1062, *1062,* 1064–1066, *1065*

  photon absorption and, *975,* 975–976

GDP binding sites, tubulin, 1052, *1053*

Geiger counter, 215

Gel electrophoresis, 92, 92–943, *93,* 96

  in Maxam-Gilbert method, 245, *246*

  pulsed-field, 244–245, *245*

  in restriction fragment separation, *242,* 242–243, *243*

  in restriction-site mapping, 243–244, *244*

  in Sanger method, 245–246, *247*

  in Southern blotting, 248–249, *249*

Gel filtration chromatography, 94, *95*

Gel-shift assay, 419–421, *421*

Gelsolin, 1008, 1009t

Gene(s), 4

  antibody

    DNA rearrangement in, 344, 1315–1319

    joining process for, 1315–1321

    transcriptional control of, 1323–1324, *1324*

  cloned. *See also* Clone(s); Cloning

    in transgenic organisms, 296–298, *297–299*

    in vitro mutagenesis of, 291, *291*

  coding region of, 308

  conservation of, 4

  DNA repair, 388–389

  dominant, 266, *266*

double-sex, 513

  duplicated, 313–315, *314, 315*

    nonfunctional, 314–315

  early-response, 916

    transcription of, 1235–1236

  eukaryotic, 116–119, 117t, *118, 119,* 308–311

  gap, in *Drosophila* regionalization, *573, 573,* 577–578

  heat shock, 492–493, *493*

  histocompatibility, 1327

  homeotic, 447

  isolation of, by physical mapping, 284–287, *285–287*

  late-response, transcription of, 1235–1236

  mitochondrial, 813, 815t

    coordination of with nuclear genes, 830

  mitochondrial-nuclear movement of, in evolution, 815

  molecular, definition of, 308–311

  mutation-defined. *See* Mutation-defined genes

  mutations in. *See* Mutation(s)

  noncoding, 308, 310–311

  pair-rule, 576, 578–579, *579, 580*

  polymorphic, 1328

  prokaryotic, 116–118, 117t, *118,* 308

  recessive, 266, *266*

  reporter, 429, 437

  segmentation of, 5

  segment-polarity, 576, 579, *579, 580*

  selector, *576,* 579–584, *580–583*

  solitary, 313

  tandemly repeated, *315,* 315–316

  trans-acting, 411

  transcription factor-encoding, identification of, 445

  transcription of. *See* Transcription

  transcription units and, 308–311, *309, 310*

  transfer, 222

  tumor-suppressor, *1266,* 1266–1267, 1269–1270

Gene activation. *See* Transcription, initiation of

Gene amplification

  generalized, 343, *343*

  localized, 343–344, *344*

Gene control. *See also* Transcription; Translation

  by antisense RNA, 527–528

  by cell-specific transcription, 551–556

  in development, 543–588

  differential splicing in, 509–513

  in embryogenic regionalization, 565–584

  in eukaryotes, 406, 426–473

  lysogenic vs. lytic, 544–550

  mating-type, 550–556

mRNA cytoplasmic localization in, 520–522
mRNA stability in, 522–524
in myogenesis, 556–561
   vs. in neurogenesis, 562
in neurogenesis, 561–565
by on-off regulation of polyadenylation, 509
posttranscriptional, 485, 509–513
in prokaryotes, 406–426
RNA-binding proteins in, 524–527
by RNA editing, 535–537
by RNA processing regulation, 509–513
signal transduction pathways in, 914–918
silencers in, 465, 556
by splicing. See Splicing
translation initiation in, 525–528
transport regulation in, 519–520
Gene conversion, 339–343, 396–397, 397–399
in trypanosomes, 341–343, 342
in yeast, 55, 339–341, 340, 341, 554–556, 555
Gene duplication, 313–315, 314, 315
Gene expression. See also Transcription; Translation
cAMP and, 914–915
cell-type-specific
   in mammals, 556–561
   in yeast, 550–556
definition of, 405
in eukaryotes, 406, 426–473
even-skipped, 429, 429–430
posttranscriptional control in, 485, 509–513
in prokaryotes, 406–426, 426
regulation of. See Gene control
steps in, 405
Gene families, 313–314
Gene inactivation. See Transcription, inhibition of
Gene knockout, 291, 292–296
in mice, 293–296, 294–296
in myogenesis studies, 560–561, 561t
in yeast, 292–293, 293
Gene mapping. See Mapping
Gene regulation. See Gene control
Gene replacement, 291–299
by gene knockout, 292–296
by transgenic techniques, 296–298
Gene therapy, 299, 1277–1278
for cystic fibrosis, 202
viruses in, 202
Genetic analysis, 264–300
Genetic code, 120–124
breaking of, 122, 122–124, 123
codons in, 120–122, 121t, 121–123
degeneracy of, 121, 121t, 238–239
mitochondrial, 816t, 816–817

reading frame for, 121, 121
translation of. See Translation
as triplet code, 120
universality of, 817
Genetic diseases, 268–269, 270
with DNA repair defects, 389t
gene knockout and, 295–296
gene therapy for, 202, 299, 1277–1278
inheritance patterns for, 277–279, 278
mutations and, 268–269, 270, 817–819, 838, 838
X chromosome inactivation and, 465–466
Genetic dissection, of metabolic pathways, 274, 275, 276
Genetic engineering. See Recombinant DNA technology
Genetic map, 12, 13, 279. See also Mapping
definition of, 279
of E. coli, 192
Genetic-map unit, 279, 279
Genetic marker, 192, 271, 279
Genetic recombination. See Recombination
Genetics
evolution of, 10–12
reverse, 1002
Genetic screens, for mutations, 269–274
Genetic suppression, 274–277, 276
Genetic traits
inheritance patterns for, 277, 277–279, 278
linked, 185, 277
Genome, 12
mapping of. See Mapping
retroviral, 1273, 1274, 1275
sequencing of, 245–248
size of, phylogenetic complexity and, 312–313
viral, 202
Genomic library, 231, 232
construction of, 231–233, 232, 233
screening of, by membrane hybridization, 237–240, 238
size of, 232–233
Genotype, 264
Germinal centers, 332
$G_{\beta\gamma}$ complex
effector protein regulation by, 906, 907
in yeast, 897–898, 898
G_α-GTP subunit, 967
Ghost, erythrocyte, 621, 621–623
giant gene/Giant protein, 570, 573, 573, 578, 579
Gibbs, Josiah Willard, 34
Gibb's law, 34
Gierer, A., 208

Gill-withdrawal reflex, 979–982, 980, 981
Glial cells, 941, 941
Glial fibrillary acidic protein, 1108, 1108t. See also Intermediate filament(s) (IF)
Globin fold, 64, 64
α-Globin gene, as pseudogene, 314–315
β-Globin-like gene family, 313–315, 314
Gloxisome, 174
Gloxylate, 800, 801
Glucagon, 907
insulin and, 911, 911
structure and function of, 860t
Glucocorticoid receptor, 459–460, 460, 461, 462. See also Receptor(s)
D-Glucofuranose, 20, 21
D-Glucopyranose, 20, 21
Glucose
metabolism of. See also Glycolysis
   anaerobic, 744, 744–745
   ATP synthesis and, 42, 740–745
   in cytosol, 740–745
   in E. coli, 421–423, 422
   glycolysis in, 740–745
   insulin regulation of, 907–911, 908–911
storage of, 43, 43
structure of, 20, 21
transport of
   into erythrocyte, 637, 638, 638–640, 640t
   kinetics of, 639–640
   symporter-catalyzed, 652–654, 653, 654
   topogenic sequences in, 692
   in transformed cells, 1256
   uniporter-catalyzed, 638–640
Glucose-1-phosphate, 882, 883
Glucose-6-phosphate, 882
Glucose transporter (GLUT), 907, 909
Glucosylcerebroside, 614, 614
Glucuronic acid, 1136
GLUT. See Glucose transporter (GLUT)
Glutamate, structure of, 954t
Glutamate receptors
in long-term potentiation, 982, 983
RNA editing regulation of, 536
Glutamic acid, 54, 55. See also Amino acid(s)
Glutamine, 54, 55. See also Amino acid(s)
Glutathione, structure and function of, 694, 695
β-Glycan, 1162
Glyceraldehyde 3-phosphate, 797–800, 798, 799
Glyceraldehyde 3-phosphate dehydrogenase, in glycolysis, 743, 743

Glycine, 54, 55, 800–802, 966–967
    in collagen, 1128, 1128
    structure of, 954t
Glycocalyx, 197, 197, 624, 624
Glycogen, 43
    cytosolic granules of, 169, 171
    metabolism of, cAMP in, 882–885,
        883, 884
    synthesis of, insulin and, 910
Glycogenolysis, 870
    calcium ion–induced, 901, 901
    epinephrine-induced, 882–885, 883,
        884
Glycogen phosphorylase, 884
Glycogen phosphorylase kinase (GPK),
    884, 901, 901
Glycogen synthase, 884
Glycolate oxidase, 800, 801
Glycolipids, membrane, 614, 614–615,
    615. See also Lipid(s), membrane
    leaflet, 616, 616, 627–628
    in protein sorting, 730–731
Glycolysis. See also Glucose,
        metabolism of
    in ATP synthesis, 740–745
    enzymatic control of, 771, 771–772
    in leaf starch, 803
    rate of, regulation of, 771, 771–772
Glycolytic pathway, 741, 741
Glycophorin, 614, 621–623
    membrane interaction of, 604, 605
Glycoprotein(s)
    membrane, 612–614, 613, 614. See
        also Membrane proteins
        maturation of, 680–681
        orientation of, 615–616, 688–694
    myelin-associated, 941
    oligosaccharides of. See
        Oligosaccharide(s)
    synthesis of, 71
    variable surface, 341–343, 342
    viral, 206, 207
Glycosaminoglycans, 1124
    in proteoglycans, 1139, 1140
Glycosidic bonds, 43, 43
Glycosylation, 71, 71, 700–711
Glycosylphosphatidylinositol
    in membrane-protein binding, 610,
        610
    in protein sorting, 730
Glycosylphosphatidylinositol membrane
    anchor, 692–694, 693
Glycosyltransferase, 700–703, 702
Glyoxisome, 143
GMP, photon absorption and, 975,
    975–976
Golgi complex
    collagen synthesis in, 1131, 1133
    disruption of, 1073, 1074
    oligosaccharide-processing enzymes
        sites in, 707–708

protein glycosylation in, 699–711
    protein retention in, 707–708, 719
    protein sorting and transport in, 172,
        172–173, 714–722, 717
    structure of, 700
Golgi enzymes, 706
    retention of, 707–708, 719
Golgi membrane
    acceptor, 714
    donor, 714
    vesicular fusion with, 716–718, 717,
        718
Golgi reticulum, 677, 678
    protein retrieval from, 698–699, 699
Golgi vesicles, 143, 144, 145
    dispersion and reaggregation of, 1073
    glycosylation in, 703–704, 706,
        706–707
    protein sorting and transport to/from,
        172, 172–173, 677–678,
        718–722
    transport of, 1073, 1074
Gortner, E., 602
G₁ phase, of cell cycle, 178, 178, 179
G₂ phase, of cell cycle, 178, 178, 179
GPI. See Glycosylphosphatidylinositol
GPK. See Glycogen phosphorylase
        kinase (GPK)
G protein(s)
    bacterial toxin effects on, 877–878
    in cell movement, 1044, 1045
    cycling of, 874–875, 876–877, 877
        adenylate cyclase activation and,
            879, 879
    in inositol-lipid pathway, 901–904,
        902
    as oncogenes, 1261
    receptor specificity for, 880, 881
    signal-transducing, 869–881, 875
        in multiplex signaling, 905–906
        types of, 877–878
    stimulatory, 874–877, 875, 876
    vesicular transport of, 714–715, 715
G-protein-linked receptors, 859, 860,
        862, 869–881. See also
        β-Adrenergic receptor(s);
        Receptor(s)
    binding to, 870–873
    effector protein regulation by, 906,
        907
    in gene regulation, 914–918
    ion channels and, 967
    MAP kinase activation and, 894–899
        in yeast, 897–899, 898
    in olfaction, 978–979
    neurotransmitter, 953–955, 956t
    properties of, 869–870, 906t. See also
        Epinephrine
    specificity of, 880, 881
    structure of, 869, 869
Graft rejection, 1327–1328

Gram-negative bacteria, 143
Gram-positive bacteria, 143
Grana, 175, 780, 781
Granulocyte colony-stimulating factor
    (GCSF), synthesis of, 252
Granulocyte-macrophage
    colony-stimulating factor
    (GMCSF), 1331t
GRB2 protein, 888, 890, 893, 897, 897
Grendel, F., 602
GroEl chaperone, 826–827, 827
Group I and II introns, 508. See also
    Intron(s)
Growing fork(s), 115–116, 117
    lagging strand of, 374–378, 375, 377,
        378
    leading strand of, 374, 375, 375–378,
        377, 378
    number of, 369
    rate of movement of, 369
Growth cones, 1175, 1176–1177
    collapse of, 1181, 1181–1182
    guidance of, 1177, 1177–1178
        adhesive proteins in, 1182–1183,
            1185
        semaphorins in, 1182, 1183
    repulsion of, 1181, 1181–1182
Growth control, proteins in, 1259,
    1259–1260, 1262t–1263t
Growth factor(s), 1259, 1259–1260,
    1262t. See also specific factors
    in cell movement, 1044, 1045
    in mesodermal induction, 1164–1166,
        1165, 1166
    oncogenes and, 1260–1261, 1262t
    proteoglycan binding of, 1142, 1143
Growth factor receptors, 1259–1260,
    1262t
    oncogenes and, 1261, 1261, 1262t
Growth media
    for animal cells, 193, 194t
    for bacteria and yeasts, 190, 190t
    defined, 1253
    HAT, 1253
GTP
    in cell signaling, 869, 875–877,
        876–878
    in cGMP synthesis, 975
    in microtuble assembly and
        disassembly, 1062, 1062,
        1064–1066, 1065
    in protein synthesis, 133, 134–135,
        136
    in vesicle fusion, 716–718, 718
GTPase activating protein (GAP),
    876–877, 877, 886
GTPase superfamily, 876–877
GTP binding sites, tubulin, 1052, 1053
Guanine, 102, 102, 103t
    pairing of with cytosine, 104–105,
        105

Guanine-nucleotide exchange factor (GEF), 876, 888–891, *890*
Guanosine diphosphate. *See* GDP
Guanosine triphosphate. *See* GTP
Guanylate cyclase, in light adaptation, 976, *976*
Guard cells, stomata, 664
Guide RNA, 537, *537*
Gyrase, 382, *383*

Habituation, 979–980
Hairpin configuration, in RNA, 112, *112*
Half-life, 213
Half-site, 552
  operator, 418, *418*
Halophiles, 142
Haltere mimic mutation, 583, *583*
Haploid cell, 181, 264
HA protein
  influenza, 731–732, *732*
  synthesis of, 696–697, *697*
Hapten, 1298, *1298*
Hapten-carrier complexes, *1334*, 1334
Hartwell, L.C., 272
Harvey sarcoma virus, *1276*, 1277
HAT medium, 201
H⁺ ATPase
  in noneukaryotes, 660
  in proton transport, 650–651, *651*
Haworth projection, 20, *21*
HCO₃⁻. *See* Bicarbonate
Heat, 16
Heat-shock genes, transcription of, 492–493, *493*
Heavy chains, antibody, 1299, *1310–1314*, 1310–1315
  in antibody class switching, 1335–1338
  diversity in, 1319, *1320*
  synthesis of, 1319–1321, *1323–1324*
Heavy meromyosin (HMM), 1015
Hedgehog, 1169–1172
HeLa cell
  culture of, *195*, 198
  macromolecular contents of, 145t
  protein synthesis in, 144
Helicase, 372–373, *373*
Helix
  in actin filament, 995, *995*
  α, 57, 63–65, *63–65*
    in domains, 67–68
    in Golgi enzymes, 719
    in membrane proteins, *605*, 606–608, 690–692
    in motif, 66–67, *67*
    in voltage-gated ion channels, 949, *949*, 951
    voltage-sensing, 938
    amphipathic, 64–65, *65*

coiled-coil, 65, *65*
DNA, 104–108, *105–108*
  base pairing in. *See* Base pairing
  circular, 109–110, *109–111*
  hybrid, 111–112
  linking number of, 110, *111*, 308
  right- vs. left-handed, 106, *106, 107*
  supercoiled, *110*, 110–111, *111*
  twist in, 110, *111*
  underwound, 110–111, *111*
  unwound, *108*, 108–109, *109*
  writhe in, 111, *111*
gating, 949, *949*
hybrid, 111–112
M2
  in GABA/glycine receptor, 967
  in nicotinic acetylcholine receptor, 963–965, *964*
random-coil, 56, 67–68
recognition, 418, *419*, 549, *550*
triple, in collagen, 1127–1133, *1128, 1132*
Helix bundle, *64*, 64
Helix-loop-helix proteins, 67, *67*, 448, *452*, 452
  in myogenesis, 560, 562
  in neurogenesis, 561–565
Helix-turn-helix motif, 418
Helper T cells, *1297, 1306*, 1306
  activity of, 1330–1331
  B cell activation and, 1333–1335, *1334*
  lymphokine secretion by, 1330–1331, 1331t
Helper virus, 1275
Hemagglutinin (HA), structure of, 56–57, *58*
Hemagglutinin (HA) protein
  influenza, 731–732, *732*
  synthesis of, 696–697, *697*
Hematoxylin and eosin stain, 150, *151*
Heme
  structure of, *70*, 70–71
  synthesis of, 830
Hemicellulose, 1187, *1189*, 1190
Hemidesmosomes, 628, 1156, 1157, *1157*
  intermediate filament anchorage to, 1114, *1115*
  structure of, 1114
Hemoglobin
  homology of with myoglobin, 68–70, *69, 70*
  oxygen binding by, 84–85, *85*
  types of, 313–314
Hemophilia, mutations in, 334
Henderson-Hasselbalch equation, 32
Heparan sulfate, 1139, *1140*
  fibroblast growth factor binding of, 1142

Heparin, *1140*
Hepatitis B virus, liver cancer and, 1280
Hepatocyte(s)
  insulin effects on, 911
  protein sorting in, 731
  transcription factors in, 457–458, *458*
Hepatocyte nuclear factor (HNF), 458, *458*
Hepatoma cell lines, 196, 196t
Hereditary spherocytosis, 623
Herpesviruses, 209, 210t, *211*
Heterochromatin, *347, 348, 352*, 355
  in transcription inhibition, *464*, 464–468
Heterocyclic bases, 103
Heterodimers, formation of, 452
Heteroduplex, 323, *323*
  gene conversion and, 396–397
  in Holliday recombination, 390, *390, 391*
Heterogeneous nuclear RNA, 494–496, *495, 496*
Heterogeneous ribonucleoprotein particle proteins, 494–498, *495–497*
Heterokaryon, 199, *199*
Heterologous desensitization, receptor, 913–914
Heteroplasmy, 817–818
Heterozygosity, 266
Hexokinase
  allosteric control of, 771
  substrate binding to, 80–81, *81*
Hfl protease, 548, *548*
High-affinity Ca²⁺ binding protein, 646
High-mannose oligosaccharide, 704, *706, 707*
High mobility group proteins, 354
Hippocampus, in long-term potentiation, 982
Histamine
  structure and function of, 860t, 954t
  synthesis of, 79–80, *80*
Histidine, 54, *55*. *See also* Amino acid(s)
  codon for, 122–123, *123*
  conversion of to histamine, 79–80, *80*
  in protein hydrolysis, 76, 76–77
Histocompatibility genes, 1327
Histone(s), 346, 346–347, *348*
  synthesis of, by tandemly repeated genes, *315*, 315–316
Histone genes, tandemly repeated, 315–316
HIV infection, 212, *492*, 492, 1280, *1281*
HIV Rev protein, in mRNA transport, 519–520, *520*
H⁺/K⁺ ATPase, 657–658
HML locus, in yeast, 339–341, *340, 341*, 554

HMR locus, in yeast, 339–341, *340, 341, 554*

HNF. *See* Hepatocyte nuclear factor (HNF)

HO endonuclease, 340, 554–556

Holliday recombination, 389–391, *390, 391*

Holliday structure, 390, *390, 391*
crossed-strand, 390
formation of, RecA protein in, 393
resolution of, 393–395, *394*

Holliday-type structure, heteroduplex, 392, *393*

HO locus, in yeast, 554–556

Holoenzyme, 415

HOM-C genes, in *Drosophila,* *581–583, 581–584*
mammalian homologs of, *584, 584–586, 585*

Homeobox, 447
in *Drosophila,* 570–571

Homeodomain proteins, 447, *448*

Homeosis, 581

Homeotic gene, 447

Homeotic mutation, 581

Homing receptors, 1308

Homologous desensitization, receptor, 913–914

Homologous proteins, 68–70, *69, 70*

Homologous recombination, 292, 389–397

Homologous sequences, 248

Homozygosity, 266

Hooke, Robert, 4

Horizontal dorsalizing signal, 1164

Hormone(s). *See also specific types*
analogs of, 871–872
binding of, 856, 865, *865*
adenylate cyclase activation and, 874–875, *876*
classification of, 856–857, *857, 858t*
concentration of
calculation of, 866
receptor numbers and, 912
degradation of, 859–864
lipophilic, 856–857
with intracellular receptors, 856, *857*
peptide, 857
structure and function of, 860t–861t
synthesis, release, and degradation of, 859–861
properties of, 863t
proteolytic processing of, *721,* 721–722
receptors for. *See* Receptor(s), cell-surface
regulation of, 859–864
feedback control of, 864, *864*
release of, 859–864

second messengers and, 857–859, *860*
synthesis of, 859–864
in transcription regulation, 458–462
water-soluble, 856–857, *857,* 858t

Hormone agonists, 871–872, 873t

Hormone antagonists, 871–872, 873t

Host range, viral, 204–205

Hox complex, 583–586, *585, 586*

H^+ pumps, in plants, 659–660

Hsc70 protein, 1006

Hsp60 protein, 826–827, 829

Hsp70 protein, 826, 827, 829

Human cyclin-dependent kinase, phosphorylation of, 1224

Human Genome Project, 12

Human immunodeficiency virus (HIV), 212
transcription termination in, 492, *492*

Human immunodeficiency virus infection, 1280, *1281*

Human T-cell leukemia virus (HTLV), 1279

Human T-cell lymphotrophic virus (HTLV), 212, 1309

Humoral immunity, 1296, *1297*

hunchback gene/Hunchback protein, in *Drosophila* regionalization, *570–573, 571–573, 577–580, 578*

Huntington's disease, inheritance patterns for, *278,* 278–279

Hyaluronan (HA), 1136–1139, *1138*

Hyaluronic acid, 194

Hybrid cells. *See also* Cell fusion gene mapping with, 199

Hybridization
membrane, 237–240, *237–240*
probes in, 238–240, *239, 240*
molecular, 237
nucleic acid, 109

Hybridoma, in monoclonal antibody production, *201,* 201–202

Hydrocarbons, 25

Hydrochloric acid, in gastric acidification, *657,* 657–658

Hydrogen bonds, 17, *18,* 18t, *22,* 22–23, *23,* 23t
in DNA helix, 104–105, *105, 106*

Hydrogen ion concentration
in aqueous solution, 30–34
pH and, 30–34

Hydrogen pump, in noneukaryotes, 660

Hydrolysis
ATP. *See* ATP, hydrolysis of
protein, 76–78, *77, 79*

Hydronium ion, 30

Hydrophilic amino acids, 54, *55*
in membrane proteins, 607

Hydrophilic bonds, 24

Hydrophobic amino acids, 54, *55*
in membrane proteins, 607–608

Hydrophobic bonds, 25–26, *26*

Hydrophobic cleft, 77

Hydrophobicity, measurement of, 636

Hydroxyl ion concentration, pH and, 30–34

Hydroxylysine, 1131

Hydroxyproline, in collagen, 1128, *1128*

Hypersensitivity reactions, 1309
delayed-type, 1331

Hypertonic solution, 164
osmotic flow and, 662

Hypothalamus
in feedback control, 864, *864*
in peptide secretion, 970

Hypotonic solution, 164
osmotic flow and, 662

Hypoxanthine, 1158

I band, 1023, *1024*

I-cell disease, 711

Icosahedra, 202–203, *203*

Id protein, in myogenesis, 560

IEF. *See* Isoelectric focusing (IEF)

IF. *See* Initiation factor(s) (IF); Intermediate filament(s)

IFAP. *See* Intermediate filament–associated proteins (IFAPs)

IFN. *See* Interferon

IgA, 1301t, 1301–1302, *1302*

IgD, 1300, 1301t

IgE, 1300–1301, 1301t

IGF-1. *See* Insulin-like growth factor-1

IgG, 89t, 1300–1301, 1301t

IgM, 89t, 1300, *1301,* 1301t
membrane-bound, 1336
secreted, 1336

Imaginal discs, *Drosophila,* 565, *566*

Immersion oil, 149

Immortal cells, 196, 1252–1253, 1257
oncogenes and, 1267–1268

Immune disorders
autoimmune, 1309
immunopathologic, 1309

Immune response
antigen-dependent phase of, 1332–1340
antigen-independent phase of, 1321–1324
deficient, 1309
DNA rearrangement and, 1315–1318, *1315–1318*
hyperreactive, 1309
nonspecific, 1296
primary, 1304
secondary, 1304, 1337–1338
memory cells in, 1338
specific, 1296–1297, *1297*
tolerance and, 1295, 1309, 1339–1340

Immune system. *See also* Immunity
  allelic exclusion in, 1323
  antibodies in. *See* Antibodies
  antigens in. *See* Antigen(s)
  cancer and, 1250–1251, 1296, 1309
  complement in, 1296
  definition of, 1295
  functions of, 1295–1296
  homing receptors in, 1308
  lymphocytes in. *See* B cell(s); T cell(s)
  lymphoid tissue in, 1308, *1308*
  lymphokines in, 1296
  macrophages in, 1296, 1307
  memory cells in, 1304–1306, *1305,*
    1335, 1338
  organization of, 1308, *1308*
  overview of, 1296–1309, *1297*
  phagocytes in, 1296
  plasticity of, 1302–1303
  primary organs of, 1308, *1308*
  secondary organs of, 1308, *1308*
  T-cell receptor in. *See* T-cell
    receptor(s)
  versatility of, 1302–1303
Immunity, 1295–1241
  cell-mediated, 1324
  cellular, 1296, *1297*
  destructive responses in, 1296
  humoral, 1296, *1297*
  instructive theory of, 1302–1303,
    1303
  recognition in, 1295–1296
  rejection in, 1296
  selective theory of, *1303,* 1303–1304,
    1304
  tolerance in, 1295, 1309
Immunization, antibody response in,
  1337
Immunoblotting, 97, *97*
Immunodeficiency
  in AIDS, 212, 1280, *1281*
  in immune disorders, 1309
Immunofluorescence microscopy,
  150–154, *152–154*
Immunoglobulin(s), 1300–1302, 1301t.
  See also Antibodies
  definition of, 1296
  membrane-bound, 1336
  secondary, 1337
  secreted, 1336
  structure of, *86,* 86–87
  synthesis of, 1321
    switching in, 1335–1338
  in transcytosis, 729, *729*
Immunoglobulin fold, 1313, *1313*
Immunoglobulin superfamily, 1325,
  1326
Immunological techniques, for organelle
  purification, 165, *166*
Immunology, 1295
Immunopathology, 1309

Inclusion bodies, 169
Induced fit, 80, *81*
Inducer, 407
Induction, 1160–1166
  competence and, 1160
  instructive, 1160
  mesodermal, 1164–1166, *1165, 1166*
  in organ development, 1167–1169
  permissive, 1160
  reciprocal, 1169
  signaling in, 1160
  in tissue organization, 1169–1172
  transforming growth factor *a* and,
    1161, 1161–1163, *1163*
  in *Xenopus,* 1163–1166, *1164–1166*
Infection
  receptor-mediated endocytosis in,
    731–732, *732, 733*
  viral, lysis vs. lysogeny in, 544–545
Influenza HA protein, synthesis of,
  696–697, *697*
Influenza virus, 210t, *211,* 211
  endocytosis of, 731–732, *732, 733*
Informatics, 246–248
Inhibitor
  enzymatic, 84, *84*
  in hormone-controlled transcription,
    461, *461*
Initiation complex, 133–136, *134*
Initiation factor(s) (IF), 133–136, *134,*
  415
  activation of, by Cdk2-cyclin
    complex, 1235
  for 5S RNA transcription, 471–472,
    472
  for tRNA transcription, 471, *472*
Initiator
  transcription, 435, *436*
  in tumor promotion, 1286–1287
Initiator codon, 121
Inosine, 126, *126*
Inositol 1,4-bisphosphate, 903, *903*
Inositol-lipid pathway, 899, 901–904,
  902, 903
Inositol 1,3,4,5-tetraphosphate, 903,
  903
Inositol 1,4,5-trisphosphate (IP$_3$)
  in calcium regulation, 899, 899t,
    901–904
  as second messenger, 857–859, *860*
Insertion, chromosomal, 267, *267*
Insertion sequences, in transposition,
  323–326, *323–326*
In situ hybridization, in isolation of
    mutation-defined genes, 285–287,
    287, 289, 289
Instars, *Drosophila,* 565, *566*
Instructive induction, 1160
Instructive theory, of immunity,
  1302–1303, *1303*

Insulin
  in glucose metabolism, 907–911,
    908–911
  proinsulin conversion to, 721, 722
  structure and function of, 860t
  synthesis of, 72
Insulin-like growth factor 1 (IGF-1),
  907
  structure and function of, 861t
Insulin receptor, 907–911
  synthesis and membrane insertion of,
    689, 689
Insulin receptor substrate 1 (IRS-1), 910
Integral membrane proteins. *See*
    Membrane proteins, integral
    (intrinsic)
Integrase, 331, 395–396, 548, 1273
Integrin(s), 1035, *1035,* 1144–1146,
  1145
  cell binding by, 1145–1146
  diversity of, 1145, 1146t
  fibronectin binding to, 1147–1148,
    1148, 1149
  hemidesmosomes and, 1157
  in leukocyte adhesion, 1155
  RDG binding sequence for, 1147,
    1148
  regulation of, 1145–1146
  structure of, 1145, *1145*
Interactions, 21
  ionic, 24
  Van der Waals, 24–25, *25,* 25t, *26,* 26
Intercalated disk, *1022*
Interfacial catalysis, 612, *613*
Interferon(s), 1296, 1331t
  in cell signaling, 916–918
Interferon gamma, in transcription, 462,
  463
Interferon-stimulated response element,
  916–917
Interleukins, 1330–1331, 1331t
Intermediate filament(s) (IF), 143,
  1106–1116
  abundance of, 1106
  anchorage of to demosomes and
    hemidesmosomes, 1114, *1115*
  assembly and disassembly of,
    1111–1112, *1112*
  in cell identification, 1109
  cross-linkage of, 1113–1116
  cytosolic, 1114, *1114*
  dimerization of, 1109–1111, *1110*
  as dynamic polymers, 1111–1112,
    1112
  in epithelial cells, 1114, *1115*
  function of, 1051–1052, 1106
  microtubules and, 1113
  nuclear, 1114, *1114*
  polymerization of, in mitosis,
    1112–1113
  properties of, 1106

Intermediate filament(s) (*continued*)
  spot desmosomes and, *1156,*
    1156–1157
  structure of, 1109–1111, *1110*
  subunit proteins of, 1109–1111, *1110*
  types of, 1106–1109, 1108t
Intermediate filament–associated
    proteins (IFAPs), 1113t,
    1113–1116
Intermediate-repeat DNA, 317,
    320–338. *See also* Insertion
    sequences; Mobile elements;
    Retrotransposon(s); Transposon(s)
Internal control regions, 471
Interneurons, 925
Interphase
  centrosome replication and migration
    in, 1096–1097, *1097*
  meiotic, 181
  mitotic, 178, *178, 179, 180*
Interphase chromosomes, 351–353
Intestinal epithelial cells, plasma
    membrane of, 623–624, *624*
int gene, 331, 548
Intracellular switch proteins, 876–877
Intrinsic membrane proteins. *See*
    Membrane proteins, integral
    (intrinsic)
Intron(s), 119
  excision of. *See* Splicing
  group I, 508
  group II, 508
  as lariat structure, 501
  pre-rRNA, 531–532, *532*
  pre-tRNA, 533, *533*
  self-splicing, *508,* 508–509, 531–532,
    532
Inversion(s)
  chromosomal, 267, *267*
  DNA, 338–339, *339*
Inverted repeat, in insertion sequence,
    323, *324*
In vitro fertilization, for congenital
    disease prevention, 254–256
Iodine-125, 213t, 214
Iodine-131, 213t
Ion(s). *See also specific ions*
  in cell vs. blood, 640–641, 641t
  dipolar, 52
  movement of, measurement of, 944,
    944, 945
  permeability constant for, 934
  properties of, 24
Ion channel(s), 24
  ion movement through, 944, *945,*
    946
  ligand-gated, *928,* 953
    nicotinic acetylcholine receptor as,
    956–965. *See also* Nicotinic
    acetylcholine receptor
  membrane potential and, 935, *935*
  in neurons, 927, *928, 928, 928*

resting, 643, *928, 935*
  as sensory receptors, 971
  signal-gated, *928*
  voltage-gated, *928*
    action potential and, 935–939
    cloning of, 946–947, *947*
    evolution of, 951–952
    inactivation of, 938, 949–950, *950*
    membrane potential and, 935–939
    patch-clamping studies of, *944,*
      944–945, *945*
    properties of, 943–952
    similarity of, 951–952
    structure of, 946, *946,* 947–950,
      948
    voltage sensor in, 949, *949*
Ion channel–blocking segment, 938, *939*
Ion channel–inactivating segment, 938,
    939
Ion channel proteins, *634, 635, 686,*
    686–687. *See also* Ion channel(s);
    Membrane transport proteins
  in patch-clamping studies
  similarities among, 951–952
  structure of, 947–948, *948*
  voltage-gated, 943–952
    properties of, 943
Ion channel receptors, 859, *860, 862*
Ion concentration gradient, 16,
    640–641, 641t
  action potential generation and,
    939–940
  functions of, 641
  generation of, energy for, 37
  ion neurons, 939–940
  in ion transport, 643–644
  membrane potential and, 641–644,
    642
  in plants, *659,* 659–660
Ion exchange chromatography, 94–96,
    95
Ionic bonds, 23–24, *24*
Ionizing radiation, as carcinogen,
    1284–1286
Ion permeability, 934
  membrane potential and, 934
Ion pump(s), 42, *634,* 634–635
  calcium, *645,* 645–646, 647–648,
    648
  classes of, 644–645, *645,* 646t
  sodium/potassium, 648–650, *649*
Ion transport. *See also* Membrane
    transport
  ATP hydrolysis and, 644–652
  concentration gradient in, 643–644
  membrane potential in, 641–644
IP₃ (inositol 1,4,5-trisphosphate)
  in calcium regulation, 899, 899t,
    901–904
  as second messenger, 857–859, *860*
IPTG, *lac* operon and, 407, *407*
IRE (iron-response element), 525

IRE-binding protein, in translation, 525,
    526–527
Iron-response elements (IRE), 525
Iron-sulfur clusters, 762, *763*
Iron transport, by transferrin, 727–728,
    728
IS elements, in transposition, 323–326,
    323–326
Islets of Langerhans, 907, *908*
Isoelectric focusing (IEF), 94
Isoelectric point, 94
Isoleucine, 54, *55. See also* Amino
    acid(s)
  synthesis of, *83,* 83–84
Isomer
  chirality of, 20
  optical, 20
Isometric contraction, 1022
Isoproterenol, 872
  receptor binding of, 872–873, *874*
Isotonic solution, 164
IS*10* transposase, regulation of,
    527–528, *528*

Jacob, Francis, 406, *406,* 408
JAK kinases, 916–917
J chain, 1300
JC virus, 1280
Jerne, Niels, 1303
Joining regions, in antibodies,
    1316–1318, 1316–1319
Joint recognition process, 1329
Joule, 16n
Junction(s), 625–628, 1155–1159
  adherens, 115–116, 628, *1035,* 1035,
    1036, *1036*
  desmosome, 628, 1035, *1036,* 1036,
    1114, *1156,* 1156–1157, *1157*
  gap, 628, *930,* 930, *930,* 1157–1159,
    1158, 1159
  neuromuscular, 929, *1185,*
    1185–1187, *1186*
  tight, 625–628, *626,* 627
Junctional complex, 621, *622, 623*
Junk DNA, 307n
jun oncogene, 1265

K⁺. *See* Potassium ion(s)
Kaposi's sarcoma, 1280, *1282*
κ gene, in antibody diversity, *1316,*
    1316–1318
Karyomere, 1218
Karyotype, 349, *350*
KDEL receptor, 698–699, *699*
Kearns-Sayre syndrome, 818
Kendrew, John, 59
Keratan sulfate, 1139, *1140, 1141*
Keratin(s), 1106, 1108t. *See also*
    Intermediate filament(s) (IF)
  in skin diseases, 1115–1116, *1116*

spot desmosomes and, *1156*, 1156–1157
KH motif, 496, *496*
Kidney, embryonic development of, *1168*, 1168–1169
Killer cells, *1297*, 1306
Kilocalorie, 16, 16n
Kinase(s)
  cdc2, 1070
    in intermediate filament disassembly, 1112–1113
    in mitosis, 1221–1224
  Cdc2-activating, 1223, *1233*
  cyclin-dependent, 1201–1202, 1222–1224, 1227–1228
    in mammalian cells, 1232–1235
    phosphorylation of, 1224
    in S phase, 1227–1232
    in yeast, 1226–1232, 1233t
  in phosphorylation, 72, 1224
  receptor tyrosine. *See* Receptor tyrosine kinases (RTKs)
Kinase cascade, 885, *885*, 886–899
  Ras protein and, 891–896, *892*, *895*, *896*
  in yeast, 897–899, *898*
Kinase pseudosubstrates, 882
Kinesin, 1075–1078, *1076*, 1077t
Kinesin-related proteins (KRPs), 1077t, 1077–1078
  in centrosome separation, 1103
  in mitosis, 1097–1098, *1098*
Kinetic energy, 16
Kinetochore, 179–180, 1091–1094, *1094*, *1095*
  alignment of, 1100
Kinetochore microtubules, 1091, *1092*, 1093, *1094*
  chromosome capture by, 1098–1099, *1099*
  depolymerization of, 1101, *1101*, *1102*
Kinetoplast, RNA editing in, 536–537, *537*
Knee-jerk reflex, 931, *931*
knirps gene/Knirps protein, 570, 573, *573*, *580*
Knockout. *See* Gene knockout
Kozak sequence, 133
Krebs cycle, 749–750, *750*
  glycolysis and, 771
KRPs. *See* Kinesin-related proteins (KRPs)
Krüpple gene/Krüpple protein, 570, 573, *573*, *578*, *579*, *580*

lac control region, 412, *412*, 413, *417*
lac operon
  cAMP-CAP regulation of, 421–423, *421–423*

protein-binding sites on, 412, *412*, *413*
  in transcription, 406–426
    control region for, 412, *412*, 413, *417*
lac promoter, 252, 409
  RNA polymerase and, 413–414, *414*, *415*
  strength of, 413–414, *415*
lac repressor, 408, *408*, 416–418, *417*, *422*
  binding site for, 408–409
Lactose permease, synthesis of, 406, 407, *408*
Lactose transporter, *660*, 660–661
lacZ gene, 437
  in *Drosophila* regionalization, 579
Lagging strand, in DNA replication, 374–378, *375*, *377–380*, *379–381*
Lambda chains, antibody, 1319
Lambda expression vector, 240, *241*
Lambda operators
  in lysogenic pathway, 545–547, *548*
  in lytic pathway, *549*, 549–550, *550*
Lambda phage, in recombination, *395*, 395–396
Lambda-phage cloning, 230–233, *230–234*
Lambda phage infection
  lysogeny vs. lysis in, 544–550
  in recombination, *395*, 395–396
  transcription termination in, 490, *490*
Lambda phage vector, *230–232*, 230–233, 233t
Lambda repressor
  in lysogeny, 545–547, *546*, *547*
  in lytic pathway, *549*, 549–550
Lambda vector arms, 232
Lamellipodia, 992, *993*
Lamin(s), 168, 1108t, 1109. *See also* Intermediate filament(s) (IF)
  basal. *See* Basal lamina
  nuclear, 1108t, 1109, 1113, 1114, *1114*
    depolymerization of, 1212–1213, *1213–1216*, 1218, *1219*
    polymerization of, 1218, *1219*
Lamin B receptor, 1113t
Laminin, 1143–1144, *1144*
  growth cone migration on, *1178*, 1178–1179
Lariat structure, 501
Late endosome, 709, 726–727, *727*
Late-response genes, transcription of, 1235–1236
Lattice energy, 24
Leader sequence, 487, 488, *489*
Leading strand, in DNA replication, 374, *375*, *377–380*, 379–380
Leaflets, membrane, 599
  lipids of, 616, *616*, 627–628

Leaf starch
  fate of, 803
  structure of, 780, *780*
Leaf stomata, opening of, osmotic pressure and, 664
Learning. *See also* Memory
  classical conditioning in, 979, 980–982, *981*
  habituation in, 979–980
  neurotransmitters and, 979–984
  sensitization in, 980, *980*, *981*
  synapses in, 930
Leber's hereditary optic neuropathy, 818
Lectin, 1154
Lederberg, Esther, 191
Lederberg, Joshua, 191
Leeuwenhoek, Antonie van, 4
L1 elements, *334*, 334–335, *335*
lethal of scute gene, 561–562, *563*, *564*
Leucine, 54, *55*. *See also* Amino acid(s)
Leucine-zipper proteins, 448, 451–452
Leukemia, 1249
  Philadelphia chromosome in, 1268
Leukocyte
  extravasation of, adhesive proteins and, *1154*, 1154–1155, *1155*
  movement of, 992
  structure of, *148*
Leukocyte-adhesion deficiency, 1155
Levo isomer, 20
LexA protein, in DNA repair, 393
LHC. *See* Light-harvesting complex (LHC)
Library
  cDNA, 234–236, *235*, *236*
  genomic, 229–233
    construction of, 231–233, *232*, *233*
    size of, 232–233
  screening of, by membrane hybridization, 237–240, *238*
Ligand(s), 86
  function of, 856
  receptor binding to, 856, 865
    adenylate cyclase activation and, 874–875, *876*
    in receptor-mediated endocytosis, 722–727, 723t
  types of, 723t, 862
Ligand-gated ion channel, 928, 953
  neurotransmitter receptors as, 953t, 957t
  nicotinic acetylcholine receptor as, 956–965. *See also* Nicotinic acetylcholine receptor
Ligase. *See* DNA ligase
Ligation, 114, *114*
  of blunt ends, 227–228, 236
  of sticky ends, 227, 227–228
Light
  adaptation to, 976, 976–977
  photons of, 783. *See also* Photons

Light (*continued*)
  properties of, 783
  ultraviolet. *See* Ultraviolet radiation
Light absorption
  opsin activation and, 974–975
  in photosynthesis, 782, 783–788
  retinal isomerization and, 974–975
Light chains
  antibody, 1299, *1310–1314*,
    1310–1315
    diversity in, *1315–1317*,
      1315–1318
      gene rearrangement in, 1321–1322,
        1322
      synthesis of, 1315–1319,
        1320–1324, *1323–1324*
  muscle, 1030–1031
Light-harvesting complex (LHC), 783,
    786, 787
Light meromyosin (LMM), 1015
Light microscopy, 148–158. *See also*
    Microscopy
Light reactions, in photosynthesis, 782,
    783
Light receptors. *See* Photoreceptors
Lignin, 177, 1190
LINE, 317, *334*, 334–335, *335*
Linear electron flow, *795*, 796
Linkage disequilibrium, 281
Linked traits, 185, 277
Linker, restriction-site, 236
Linker-scanning mutation, 437–439,
    438, 439
Linking number, 110, *111*
Lipid(s). *See also* Phospholipid(s)
  amphipathic, 27, *27*, 597
  membrane, 6–7, *7*, 167, 596t,
    596–599
    glycolipid, *614*, 614–615, *615*
    leaflet, 616, *616*, 627–628
    mobility of, 616–618
    organization of, 615–618
    in protein sorting, 730–731
    synthesis of, 671–674, *672, 673*
  mitochondrial metabolism of,
    746–750
Lipid binding, of membrane proteins,
    610, 610–612, *611*
Lipolysis, 870
Lipoproteins, low-density, endocytosis
    of, 723, 724–727, *725, 726*
Liposomes, 27, *28*, 600
Liquid chromatography, 59, 94–96, *95*
  affinity, *95*, 96, *96*
  gel filtration, 94, *95*
  ion exchange, 94–96, *95*
Liver, glycogen metabolism in, 882, *884*
Liver cancer, hepatitis B virus and, 1280
Lock-and-key mechanism, 80–81, *81*
Long-chain fatty acyl CoA synthase,
    773

Long interspersed element (LINE), 317
Long terminal repeat (LTR), 328, *329*
  in retrovirus, 1273, *1274*, 1278
  in viral retrotransposons, *328–330*,
    329–330
Long-term memory, 979, 982
Long-term potentiation, 982–984, *983*
Loop
  in motif, 66–67, *67*
  protein, 66
Looping out, 425
LTR. *See* Long terminal repeat (LTR)
Luciferase, 838
Luteinizing hormone (LH), 864
  structure and function of, 861t
Luteinizing hormone–releasing hormone
    (LHRH), structure and function
    of, 861t
Lwoff, Andre, 406, *406*
Lymph, circulation of, 1308, *1308*
Lymphatic vessels, 1308
Lymph nodes, 1308, *1308*
Lymphocyte(s). *See* B cell(s); T cell(s)
Lymphoid system, *1308*, 1308–1309.
    See also B cell(s); T cell(s)
Lymphoid tissue, secondary, 1308, 1332
Lymphokines, 1296, 1330–1331, 1331t
Lymphoma
  Burkitt's, Epstein-Barr virus and,
    1279, 1279–1280
  chromosomal translocations in,
    1268–1269
  human T-cell lymphotropic virus and,
    1279
Lyonization, 465–466
Lyons, Mary, 465
Lysine, 54, *55*. *See also* Amino acid(s)
Lysogen, 544
Lysogenic pathway, in lambda-phage
    infection, 544–550
Lysogeny, in lambda-phage infection,
    544–550
Lysosomal proenzymes, 709
Lysosomes, 143, *145*, 173, *173*
  protein targeting to, 709–710
Lysozyme
  antibody binding to, 87, *87*
  physical characteristics of, 89t
Lytic pathway, in lambda-phage
    infection, 544–550

———————

Macleod, C. M., 223
Macromolecular assemblies, 57–58
Macromolecules, 9–10
  eukaryotic vs. prokaryotic, 143–144,
    145t, 146t
  radiolabeling of. *See* Radiolabeled
    compounds
Macrophages, 1307
  functions of, 1296

Madin-Darby canine kidney (MDCK)
    cell line, 197, *198*
mad mutant, 1239
Magnesium ions. *See also under* Ion(s)
  in actin, 994, 995, *995*
  concentration of, in cell vs. blood,
    640, 641t
  in long-term potentiation, 982, *983*
Maize, mobile elements in, *327*,
    327–328, *328*
Major histocompatibility complex
    (MHC), 1327, 1327–1328, *1328*,
    1328t, *1330*
  in B-cell activation, 1334, *1334*
Malate shuttle, 759, *760*
  in C4 pathway, 802
Malignant transformation, 196, *196*,
    1254–1257, *1255, 1257*
  abortive, 1270
  mutations in, 1264
  oncogenes in, 1260–1267,
    1262t–1263t
  oncoproteins in, 1260–1267,
    1262t–1263t, 1267–1268
  phosphorylation and, 1264
  retroviruses and, 1273–1279
  T proteins and, 1272–1273
  tumor viruses in, 1270–1273
  two-step model of, 1267–1268
Malignant tumors, 1248, *1248*. *See also*
    Cancer
Mammalian restriction point, passage
    through, 1232, 1233–1235, *1234*
Manganese, in photosynthesis, 792,
    793–794
Mannose, 21, *21*
  in oligosaccharide, 612, *614*, 614
Mannose-6-phosphate receptor, 709,
    710
Mannose phosphorylation
  in I-cell disease, 711
  in protein targeting to lysosomes,
    709–711
MAP kinase, 1070
  insulin and, 910, *910*
  in kinase cascade, 894, *895*,
    895–896, 897–899
  in RTK-Ras protein signaling
    pathway, 897, *897*
Mapping
  of *E. coli*, 192, *193*
  genetic, 12, *13*, 192, 279
  with hybrid cells, 199
  of mutations, 277–284. *See also*
    Mutation(s), mapping of
  physical, 284–287, *286, 287*
  restriction-site, 243–244, *244*
    by nuclease-protection method, 251
    by Southern blotting, 248–249, *249*
  of scaffold-associated regions, 353,
    353

of transcription-control regions, 435–442

of transcription start sites, 251, *433*, 433–435, *434*

MAPs. *See* Microtubule-associated proteins (MAPs)

Markers
    genetic, *192*
    phenotypic, 271, 279
    surface, on lymphocytes, 1306–1307, *1307*

MASH1, in neurogenesis, 562–565, *564*

Mass, definition of, 88

Mass spectrometry, 59, *60*

Mast cells, 1301

Mating factors, yeast, 553, 553–554

Mating-type conversion, in yeast
    cell-type–specific, 551–556
    gene conversion and, 339–341, *340, 341*

MAT locus, in yeast, 339–341, *340, 341*, 554

Matrix
    extracellular. *See* Extracellular matrix
    mitochondrial, 175, *175*

Matrix proteins, 826, 827, 829
    multiadhesive matrix, 1124, *1124*, 1143–1150

Maturases, in splicing, 508

Maturation pathways, genetic dissection of, 274, *276*

Maturation-promoting factor. *See* Mitosis-promoting factor (MPF)

Maxam-Gilbert method, of restriction fragment sequencing, 245, *246*

MCB-binding factor, 1231, 1231t

McCarty, M., 223

McClintock, Barbara, 11, *11*, 322, 328

MCM1, in cell-specific transcription, in yeast, *551*, 551–554, *552*

MDCK cell(s)
    protein sorting in, 729, *730*
    tight junctions of, 627–628

MDCK cell line, 197, *198*

Mechanochemical enzyme, 1013

Media, culture
    for animal cells, 193, 194t
    for bacteria and yeasts, 190, 190t
    defined, 1253
    HAT, 201

Meiosis, *178*, 181–185, *184, 185*, 1202, *1202*

Meiotic recombination, 279, *279*
    gene conversion in, 396–397, *397–399*
    in yeast, 396

MEK protein
    in kinase cascade, 894, 895–896
    in RTK-Ras protein signaling pathway, 897, *897*

Melanophores, transport of, 1072

Melting temperature, DNA, 108–109, *109*

Membrane(s), 27–28
    asymmetry of, 615–616, *616*
    bilayer organization of, 602–603
    chloroplast, *176*, 176–177, *780, 781*
    compartmentalization in, 603, *603*
    composition of, *596*, 596t, *596–599*
    cortical, myosin II in, 1036, *1036*
    cytoskeletal binding of, 618–623
    cytosolic face of, 603, *603*
    degradation of, phospholipases in, *611*, 612
    diffusion coefficient for, 618
    electron microscopy of, 616, *617*
    eukaryotic, 143, *144, 145*
    exoplasmic face of, 603, *603*
    fatty acids in, 597, *598*, 598t
    fluid mosaic model of, 617
    functions of, 8–9
    glycolipids in, *614*, 614–615, *615*
    glycoproteins in, 612–614, *613, 614*
    Golgi, 714
    leaflets of, *599*
    lipids in. *See* Lipid(s), membrane
    lumenal surface of, 616
    mitochondrial, 174–175, *175*, 745, 745–747
    nuclear, 167–168, *168*
        intermediate filaments in, 1114, *1114*
    permeability of, 633–634
        osmotic pressure and, 662
    phospholipid bilayer of. *See* Phospholipid bilayer
    plasma, 6, *7*, 623–628
        actin network in, *999*, 999–1000
        apical region of, 623–625
        basolateral region of, 623–625
        calcium pump in, 646
        as capacitor, 641
        cilia of. *See* Cilia
        cytoskeleton binding of, dystrophin and, *999*, 1000–1002
        electric potential across. *See* Membrane potential
        erythrocyte, 620–623, *621–623*
            anion transport across, *655*, 655–656
        eukaryotic, 143
        flagella of. *See* Flagella
        functions of, 167, *167*
        of intestinal epithelial cells, 623–624, *624*
        ionic gradient in, 640–641
        junctional regions of, 626–628
        matrix binding of, 176
        of pancreatic acinar cells, 624–625, *625*
        permeability of, 633–634, 662

of plant cells, *176*, 176–177, *780, 781*
    in polarized cells, 623–625
    prokaryotic, 6, *7*
    rupturing methods for, 162–164
    specialization in, 623–628
    tight junctions in, 625–628, *626, 627*
    postsynaptic, depolarization of, 962–963, *964*
    prokaryotic, *142*, 142–143
    protein interactions with, 604, *605*. *See also* Membrane proteins
    proteins in. *See* Membrane proteins
    structure of, *596*
    thylakoid, 603
        antennas in, 783, 786
        charge separation across, 784, *785*
        photosynthesis on, 780, *781*, 783–786
        reaction centers in, 783, 786

Membrane anchor
    glycophosphatidylinositol, 692–694, *693*
    stop transfer, 689, *689*

Membrane-binding proteins, 8–9, 998

Membrane carbohydrates, 612–615, *613–615*

Membrane channel, 143. *See also* Ion channel(s)

Membrane depolarization, 935
    acetylcholine and, 962–963, *964*
    during action potential, 927, 936–940, 953, 957, 962–963
    in olfaction, 978
    passive spread of, 936, *936*
    in postsynaptic membrane, 962–963, *964*
    in rod cells, 972
    synaptic vesicle exocytosis and, 962–963
    voltage-gated ion channels and, 935–938

Membrane diffusion. *See* Diffusion

Membrane fusion, in viral infections, 731–732, *732, 733*

Membrane hybridization, 237–240, *237–240*
    probes in, 238–240, *239, 240*

Membrane hyperpolarization, 935
    action potential inhibition and, 953, 957
    photon absorption and, 972–973

Membrane leaflets, 599
    lipids of, 616, *616*, 627–628

Membrane lipids. *See* Lipid(s), membrane; Phospholipid(s), membrane

Membrane-microfilament binding proteins, 998

Membrane potential, 641–643, 927, 927
  in ATP synthesis, 740, 750, 750–752, 753–759
  calculation of, 753, 933–934
  functions of, 641
  generation of, 641–643, 642
  in ion transport, 643–644
  magnitude of, 643
  in neurons, 932–943. See also Action potential
  in noneukaryotes, 660
  in plants, 659, 659–660
  resting, 933–934
Membrane projections, actin bundles in, 1002, 1002–1003
Membrane proteins, 596, 596t, 604–612. See also Protein(s)
  attachment mechanisms of, 604, 605, 610, 610–612, 611
  cytoskeletal binding of, 618–622
  diffusion of, 616–618, 618, 619
  disulfide bond formation in, 694–695
  erythrocyte, 620–623
  flippase, 671–673, 673
  glycoprotein, 612–614, 613, 614
    maturation of, 680–681
    orientation of, 616
  glycosylation of, 699–711
  integral (intrinsic), 604
    α helix in, 605, 606–607
    lipid binding of, 610, 610–612, 611
    porins as, 609, 609
    sorting of, 670, 670–671
    transmembrane, 604
    in vesicle-membrane fusion, 716, 717
  ion channel. See Ion channel proteins
  isolation and purification of, 604–606, 606
  maturation of, 680–681
  membrane interactions of, 604, 605, 610, 610–612, 611
  membrane-spanning domains in, 604
  mobility of, 616–618, 618, 619
  multispanning, 688, 690–692, 692
  oligosaccharide stabilization of, 708–709
  orientation of, 671
    asymmetric, 615–616
    determination of, 612
    topogenic sequences and, 688–694
  peripheral (extrinsic), 604
  photosynthetic reaction center in, 607–608, 608
  posttranslational modification of, 694–697
  proteolytic processing of, 720–722, 721
  seven-spanning, 608–609, 609

  solubilization of, 604–606, 606, 607
  sorting of. See Protein sorting
  synthesis of, sites of, 674t, 674–676, 675
  in tight junctions, 626, 626–628, 627
  transport. See Membrane-transport proteins
  transport of. See Protein transport
  uniporter, 636–640, 637–639
Membrane repolarization, 927
Membrane transport, 171, 172–173, 633–666. See also Diffusion
  active, 634, 634–635
  antiporter-catalyzed, 654–659
  ATP hydrolysis in, 644–652
  ATP-powered pumps in, 634, 634–635
  concentration gradient in, 640–641, 643–644
  coupled, 652–659
  facilitated, 636–640
  ion pumps in, 634, 644–652
  membrane potential in, 641–644
  symporter-catalyzed, 652–654, 658
  uniporter-catalyzed, 636–640, 637–639
  of vesicles
    axonal, 1070–1072
    microtubular, 1072–1075, 1074, 1075
    myosin V in, 1038, 1038–1039, 1078–1079
  via passive diffusion, 635, 635–636
Membrane transport proteins, 624, 634, 634–635
  ATP-powered, 644–652
  channel
    ion. See Ion channel proteins
    water, 663, 663
  cystic fibrosis transmembrane regulator, 651–652
  glucose. See Glucose, transport of
  multidrug, 651, 651–652
  in plants, 659, 659–660
  in prokaryotes, 659, 660–661
  properties of, 636–640
  pump. See Ion pump(s); Proton pump(s)
  transporter. See Transporter(s)
  types of, 634, 634–635
Membrane vesicles. See Vesicle(s)
Memory. See also Learning
  long-term, 979, 982
  neurotransmitters and, 979–984
  short-term, 979–982
Memory cells, 1304–1306, 1305, 1335
  in secondary immune response, 1338
  T-cell, 1335
Mendel, Gregor, 11, 11, 264, 266, 266
Mendel's second law, 277

Mensenchyme, in embryogenesis, 1167–1169
Meristem, 1187
Merrifield, Bruce, 59
Meselsohn-Stahl experiment, 366, 367
Meselson, Matthew, 391
Meselson-Radding model, of recombination, 391, 391, 392
Mesodermal induction, growth factors in, 1164–1166, 1165, 1166
Mesophyll cells
  in C4 pathway, 802–803, 802–803
  sucrose synthesis in, 804
Mesosome, 177
Messenger ribonucleoproteins, transport of, 515, 515–516, 516
Messenger RNA. See under RNA
Metabolic control
  of respiration, 770
  of thermogenesis, 770–771
Metabolic coupling (cooperation), 1158
Metabolic pathways, genetic dissection of, 274, 275
Metabolism
  aerobic, 43
  anaerobic, 43, 744–745
  glucose. See Glucose, metabolism of
  rate of, cytosolic pH and, 658–659
Metacentric chromosomes, 350
Metal shadowing, 159–160, 160
Metaphase, mitotic, 179, 180, 182
Metaphase chromosomes, 349
Metastasis, 1248, 1249, 1250, 1250
Methanogens, 142
Methionine, 54, 55. See also Amino acid(s)
  in protein synthesis, 133, 134–135
Methionine-tRNA, in protein synthesis, 133, 134–135
Methylation
  cytosine, gene inactivation and, 468
  in DNA replication, 385
  in pre-rRNA processing, 529
MHC. See Major histocompatibility complex (MHC)
M2 helix
  in GABA/glycine receptor, 967
  in nicotininc acetylcholine receptor, 963–965, 964
Mice. See Mouse
  gene knockout in, 293–296, 294–296
  transgenic, 297, 297
Micelle, 27, 28, 671
  critical concentration for, 605–606
Microbodies. See Peroxisomes
Microfibrils, cellulose, 1188, 1188
  orientation of, 1190, 1190, 1192, 1192
Microfilament motor proteins, 1038–1039, 1078–1079

Microfilaments, 143, 991–1047, 992. See also Actin filaments; Myosin filaments
Microscope
    compound, 148, 149
    electron, 158–159, 159
    light, sample preparation for, 150, 150, 152, 152
    magnification of, 148
    resolution of, 148–149
        limit of, 149–150
Microscopy
    bright-field, 148–150, 149, 151, 156
    confocal scanning, 154–155, 155
    electron. See Electron microscopy
    fluorescence, 150–154, 152–154
    Nomarski interference, 156–158, 156–158
    phase-contrast, 156–158, 156–158
    time-lapse, 156–158, 158
Microsomes, rough, 676
Microtubule(s), 143
    assembly and disassembly of, 1059, 1061–1067
        drug manipulation of, 1066, 1066–1067
        dynamic instability and, 1054, 1064, 1064–1066, 1065
        in mitosis, 1098–1099, 1099
        tubulin polymerization in, 1061–1063, 1061–1064
    astral, 1091, 1092, 1103–1105, 1104, 1105
    bundles of, 1054, 1057
        in axoneme, 1080–1081, 1082, 1083, 1083–1089
        MAP organization of, 1067–1069, 1068
    in cellulose deposition, 1192, 1192
    in cilia, 1080–1089
    diversity of, 1054–1055, 1056–1057
    doublet, 1052, 1053
        in axoneme, 1081–1087
    endoplasmic reticulum and, 1073, 1075
    in flagella, 1080–1089
    function of, 1051
    growth of, 1055, 1057, 1058, 1088–1089, 1089
    intermediate filaments and, 1113
    in intracellular transport, 1070–1079
        axonal, 1070–1072, 1071, 1072t, 1073
        motor proteins for, 1075–1079, 1076, 1077t, 1078
        vesicular, 1072–1075, 1074, 1075
    kinetochore, 1091, 1092, 1093, 1094
        chromosome capture by, 1098–1099, 1099
        depolymerization of, 1101, 1101, 1102
    in mitosis, 179, 1090–1116

    in mitotic spindle, 1054, 1056, 1091, 1092
    in orthograde axoplasmic transport, 927
    in pigment granule transport, 1072, 1074
    polar, 1091, 1092
        elongation of, 1103, 1103
        sliding of, 1101–1103, 1102
    polarity in, 1052–1054, 1054, 1055, 1059
        in assembly and disassembly, 1061–1064, 1061, 1062
    protofilaments in, 1052–1054, 1053
    rearrangement of, in cell movement, 1054–1055
    stable, 1054
    structure of, 1051, 1052–1060, 1053, 1054, 1056–1060
    in tentacles, 1054, 1056
    triplet, 1052, 1053
    tubulin subunits in, 1052–1054, 1053
    visualization of, 169, 170
Microtubule-associated proteins (MAPs), 1067–1070, 1068, 1068t, 1069
Microtubule dynamics, 1054, 1064, 1064–1066, 1065
    in mitosis, 1098–1099, 1099
Microtubule motor proteins, 1075–1079, 1076, 1078, 1079
    in centromere migration, 1097–1098, 1098
    in cilia/flagella, 1086, 1086
Microtubule-organizing center (MTOC), 1055, 1057, 1058. See also Centrosome
Microvilli, 197, 197, 623, 624
    actin bundles in, 1002, 1002–1003
Mini-exon, 507–508
Minisatellite DNA, 320, 321
Minus strand, 208
Missense mutation, 267, 267
Mitchell, Peter, 756
Mitochondria, 143, 144, 145
    amino acid metabolism in, 749
    ATP synthesis in, 174–175, 740, 740, 751–752, 752
    biogenesis of, 810–830
    brown-fat, 770–771
    cytoplasmic inheritance in, 812
    division of, 180
    electron transport in, 750–752, 759–770
    fatty acid metabolism in, 746, 747, 748–750
    functions of, 174–175
    genetic code in, 816t, 816–817
    intermembrane particles in, 746, 746–747
    lipid metabolism in, 746–750
    metabolite transport in, 752–759

    in plants, 747
    protein targeting to, 810, 810–811, 811t
    proton transport in, 759–770
    pyruvate metabolism in, 746, 747, 748–750
    structure of, 174–175, 175, 745, 745–747
    transcription in, 473–474, 474
Mitochondrial DNA, 811–819. See also under DNA
Mitochondrial enzymes, 829
Mitochondrial genes, 813, 815t
    coordination of with nuclear genes, 830
Mitochondrial membranes, 174–175, 175, 745, 745–747
Mitochondrial porin, 746
Mitochondrial proteins, 813, 815t, 819–824
    essential, 829
    folding of, 826–827
    human vs. yeast, 813–815, 814
    matrix, 819t, 820–824
    in multienzyme complexes, 813, 815t, 829
    plant, 815
    precursor, 819–820, 820, 821
    receptors for, 824, 826
    synthesis of, 810, 816, 819–820, 829–830
    targeting of, 811t, 819–824, 821, 827–829, 828
    translocation intermediates for, 824
    transport of, 811, 819–824, 821, 827–829, 828
    uptake of, 820–822
        energy for, 824–826
Mitochondrial RNA, 816–817
    editing of, 817
    processing of, 817
Mitogen-activated protein kinase, 1070
Mitosis, 178, 178–181, 180, 182, 183, 1202, 1202
    anaphase in, 179–180, 180, 182, 1100–1103, 1216–1218
    centriole cycle in, 1096, 1096–1097
    centromere duplication and migration in, 1096, 1096–1097, 1097
    centrosome in, 1055, 1090
    chromosome condensation in, 349, 1203, 1205
    chromosome organization in, 179–180, 1090, 1090–1103
    cyclin B and, 209–211, 1207–1212
    entry into
        control of, 1203–1219
        in mammalian cells, 1235
        in yeast, 1224
        premature, 1237
        prevention of, 1237–1239

Mitosis (*continued*)
  exit from
    control of, 1203–1219
    cyclin degradation and, 1210–1211,
      1211–1212
    prevention of, 1239
  induction of, 1203–1207
    mitosis-promoting factor in,
      1205–1225. *See also*
      Mitosis-promoting factor (MPF)
  intermediate filament polymerization
    in, 1112–1113
  kinesin-related proteins in,
    1097–1098, *1098*
  lamin depolymerization in,
    1212–1213, *1213–1216*
  metaphase in, 179, *180, 182*
  microtubules in, 179, 1090–1116
  in plants, 180–181, *182, 1105,*
    1105–1106
  prophase in, 179, *180, 182,*
    1096–1098
  protein degradation and
    dephosphorylation in, 1216–1219
  protein phosphorylation in,
    1212–1213, *1213–1216*
  stability of, 1106, 1112–1113
  telophase in, 180, *180, 182*
  variations in, 1091, *1093*
  in yeast, 181, *183*
Mitosis-promoting factor (MPF), 1205,
  1207
  catalytic subunit in. *See* Cdc2
  during cell cycle, 1205–1207, *1206*
  cyclin regulation of, 1207–1212,
    1209–1211
  in oocyte maturation, 1205–1207,
    1206
  purification of, 1207
  regulation of, 1219–1224
    Cdc2 phosphorylation and,
      1221–1224
  regulatory subunit in, 1208–1212,
    1209–1211
  structure of, 1207
Mitotic apparatus, *1090,* 1090–1091
  assembly of, 1096–1099
    dynamic microtubules in,
      1098–1099, *1099*
  structure of, 1091, *1092, 1093*
Mitotic cyclin. *See* Cyclin
Mitotic segregation, 355–359, *356–358*
Mitotic spindle, 179, *181,* 1091, *1092,*
  1202
  assembly of, *1090*
    microtubule-organizing center in,
      1055
  chromosome alignment on, 1100,
    1100
  defects in, mitosis and, 1239
  disassembly of, *1216,* 1217

elongation of, 1101–1103, *1102*
  metaphase, 1091, *1092*
  microtubules in, 1054, *1056*
Mitotic spindle pole body, 181, *183,*
  357, 1091, *1093*
M line, 1023, *1024*
Mlu*I* cell-cycle box (MCB), 1230–1231,
  1231t
Mobile elements, 320
  Alu, 336, *337*
  copia, 329, 333, 333–334
  in corn, *327,* 327–328, *328*
  discovery of, 322
  in *Drosophila,* 326–327, *329, 333,*
    333–334
  elimination of, 321
  function of, 321
  group II introns as, 509
  LINES as, 317, *334,* 334–335, *335*
  P, 326–327
  retrotransposons as, 328–338. *See*
    also Retrotransposon(s)
  SINES as, 317, 335–336, *336*
  transposition of, *322–325,* 322–326
    DNA-mediated, *322–325,* 323–325
    in eukaryotes, 326–328, *327, 328*
    in prokaryotes, 323–326, *323–326*
    RNA-mediated, 328–332
  transposons as, *323–326,* 323–326,
    323–328. *See also* Transposon(s)
  Ty, 328–333, *329, 332*
  in yeast, 328–333, *329, 332*
Moderately repeated sequence, 317. *See*
  also Intermediate-repeat DNA
Modification enzyme, 225
Module. *See* Domain(s)
Molarity, 30
Mold, arginine biosynthetic pathway in,
  274, *275*
Mole, 16n
Molecular cell biology
  central dogma of, 113
  goals of, 15
  scope of, 14
Molecular hybridization, 237–240,
  237–240
  probes in, 238–240, *239, 240*
Molecular weight, 16n
Molecule(s), 9–12
  bonding of. *See* Bond(s)
  nonpolar, 25–26
  polar, 24
  radiolabeled, 212–218, 213t. *See also*
    Radiolabeled compounds
  separation of
    by centrifugation, 90–92, *91*
    by electrophoresis, 92, 92–94, *93,*
      96
    by liquid chromatography, 94–96,
      95
  small, functions of, 9–10

Monocistrionic mRNA, 308–309
Monoclonal antibodies
  in protein purification, 87, *88,* 202
  synthesis of, *201,* 201–202
  uses of, 202
Monocot, 297
Monocytes, 1307
Monod, Jacques, 406, *406,* 408
Monomeric protein, 57
Monosaccharides
  linkage of, 43, *43*
  structure of, 20
Morgan, Thomas Hunt, 11, 279
Morphogen, 568
Motif, 66–67, *67*
Motor end plate, 929
Motor neuron(s). *See also* Neuron(s)
  autonomic, 930
  interaction of with sensory neuron,
    968–970, *969*
  somatic, 930
  structure of, 925, *926*
Motor proteins, 992, 1013
  microfilament, 1038–1039,
    1078–1079, *1079. See also*
    Myosin
  microtubule, 1075–1079, *1076,*
    1077t, *1078, 1079*
    in centromere migration,
      1097–1098, *1098*
    in cilia/flagella, 1086, *1086*
Mouse
  Hox-complex genes in, 584–586
  neurogenesis in, 562–565, *564*
  transgenic, 297, *297,* 1323
MPF. *See* Mitosis-promoting factor
  (MPF)
M phase, in cell cycle, 1202, *1202*
mrf4 gene, 560–561, 561t
Mrf4 protein, 560–561, *561*
mRNA. *See under* RNA
M1 RNA, 534
MRP, in pre-rRNA processing, 529
MTOC. *See* Microtubule-organizing
  center (MTOC)
M2 helix
  in GABA/glycine receptor, 967
  in nicotinic acetylcholine receptor,
    963–965, *964*
Müller cells, 972
Multiadhesive matrix proteins, 1124,
  1124, 1143–1150
  firbonectins as, 1146–1150
  integrins as, 1143–1145. *See also*
    Integrin(s)
  receptors for, 1143–1146
Multicellularity, evolution of, 1123
Multidrug transport protein (P170),
  651, 651–652
Multimeric protein, 57, 84–85, *85*
Multimeric proteins, 84–85, *85*

Multiple cloning-site sequence, 228, *228*
Multiple sclerosis, demyelination in, 942
Muscarinic acetylcholine receptor, 956, *957*
  cardiac, 967, *968*
Muscle
  actin filaments in, 1023–1026, *1025*
  ATP in, 1021
  cardiac, 1022, *1022*
  development of, in mammals, 554–561
  dystrophin in, *999*, 1000–1002
  elasticity of, 1026
  glycogen metabolism in, 882, *884*
  myosin in. *See* Myosin
  nebulin in, 1026, *1027*
  power of, 1021–1022, 1022t
  skeletal, *1022*, 1022–1023
  smooth, *1022*, 1022–1023
  striated, *1022*, 1022–1023
    contraction of. *See* Muscle contraction
    structure of, 1022–1024, *1023, 1024*
  titin in, 129, 1026, *1027*
Muscle calcium pump, 645–646, *647*, 647–648, 1027–1029, *1029*
Muscle cells, 1023, *1023*
  hyaluronan in, 1139
Muscle contraction, 1021–1032
  acetylcholine and, *957*
  actin filaments in, 1023–1026, *1025*
  calcium in, 645–646, 647–648, 1026–1032
  caldesmon in, 1029, *1030*
  cross-bridge cycle in, 1026
  in invertebrate muscle, 1030–1032
  isometric, 1022
  mechanism of, 1021–1026, *1025*
  myosin filaments in, 1023–1026, *1025*, 1030–1032, *1031*
  neurotransmitters and, 930
  power of, 1021–1022, 1022t
  regulation of, 1026–1032
  regulatory light chain in, 1030–1031
  sarcomere in, 1023, *1024*
  signaling in, 1026–1209, *1028–1030*
  sliding-filament model of, *1025*, 1025–1026
  in smooth muscle, 1022–1023, 1029–1032, *1030, 1031*
  tropomyosin in, 1027–1029, *1029*
  troponin in, 1027–1029, *1029*
Muscular dystrophy
  dystrophin deficiency in, 1000–1002
  inheritance patterns in, *278*, 278–279
  mutations in, 334
    mapping of, *282*, 283
Mutagen, 268
Mutagenesis, bacterial, carcinogenesis and, 1284

Mutant(s). *See* aso Mutations
  complementation analysis of, 274, *275*
  constitutive, 408
  definition of, 263
  genetic screens for, 269–274
  isolation and characterization of, 264–277
  petite yeast, 812
Mutant proteins
  degradation of, 698
  transport of, 697–698
Mutation, somatic, 1304
Mutation(s), 267, *267. See also specific mutations*
  autosomal dominant, *278*, 278–279
  autosomal recessive, *278*, 278
  bacterial insertion sequences and, 323–326, *323–326*
  in cancer, 1266–1267, *1267*, 1284
  carrier of, 277
  cis-acting, 410, 411
  complementary, 274, *275*
  conditional, 269–270
  definition of, 263
  detection of
    polymerase chain reaction in, 254–256
    by replica plating, 190–193, *191*
    screening for, 269–274
  disease-causing, 268–269, *270*
  DNA alterations in, 266–267
  DNA repair and, 385–389, 465–466, 817–819, 838, *838*, 1285
  dominant, 266, *266*
  Drosophila regionalization and, 568, *569*
  frameshift, 267, *267*
  homeotic, 581
  induced, 268
    in *Drosophila*, 271–272, *272*
  knockout, 292–296
  linker-scanning, 437–439, *438, 439*
  mapping of, 277–284
    to autosomes, 279
    banding analysis in, *281–283*, 281–284
    to chromosomes, 277–279, *278*
    linkage disequilibrium in, 281
    recombination, 279, *279*
    restriction fragment length polymorphisms in, 279–281, *280*
    segregation patterns in, 277, *277*
  missense, 267, *267*
  mobile elements and, 321. *See also* Mobile elements
  nonsense, 267, *267*
  operator constitutive, 409
  point, 266–267, *267*
    identification of, 289, *290*
    in prokaryotic genes, 308, *309*
    spontaneous, 267–268, *268*

postbithorax, 583, *583*
  in prokaryotic genes, 308, *309*
  promoter, 409
  recessive, 266, *266*
    X-linked, 277, *278*
  recombination errors and, 396
  second-site, 414
  segregation patterns for, 277, *277–279, 279*
  significance of, 264
  in simple vs. complex transcription unit, 309–310, *310*
  site-specific, production of, 291, *291*
  somatic, 1304, 1315, 1338–1339, *1339*
  spontaneous, 267–269, *268*, 280
  suppressor, 274–277, *276*
  switch-independent, 556
  transcription and, 414
  types of, 266–267, *267*
  up, 414
Mutation-defined genes
  complex transcription unit and, 310
  isolation of, 284–290
    chromosome walking in, 287, *288*
    cDNA sequencing in, 289–290
    mapping in, 285–287, *285–287*
    point mutation identification in, 289, *290*
    mRNA analysis in, 288–289
    in situ hybdridization in, 285–287, *287, 289*, 289
    Southern blotting in, 287–288
MYB gene, *13*
myc oncogene, 1267–1268, *1268*
Myelin, structure of, 941, *941*
Myelin-associated glycoprotein, 941
Myelin basic protein, 941
Myelin sheath, 940–942, *941, 942*
  bilayer organization of, *602*, 602–603
Myeloma, antibody studies in, 1310
Myeloma cells, in monoclonal antibody production, *201*, 201–202
myf5 gene, 560–561, 561t
Myf5, 560–561, *561*
Myoblasts
  in cell culture, 196, *196*
  hyaluronan and, 1139
  mRNA localization in, 521, *522*
  in myogenesis, *557*, 557–558, 561, *561*
MyoD, 560–561, *561*
myoD gene, 558–561, *559*, 561t
Myofibers, 1023, *1023*
Myofibrils, 1023, *1023*
  actin filaments in, 1023–1026, *1024, 1025*
  myosin filaments in, 1023–1026, *1024, 1025*
Myogenesis, gene control in, 556–561
  vs. in neurogenesis, 562

Myogenin, 560–561, *561*
myogenin gene, 559, 560–561, 561t
Myoglobin
  homology of with hemoglobin, 68–70, *69, 70*
  oxygen binding by, 85, *85*
Myosin, 1012–1021
  actin and, 1016–1021
  in cell streaming, *1038,* 1038–1039
  filamentous. *See* Myosin filaments
  functions of, 1014–1015
  head domain of, 1015
    movement of, 1016–1018, *1017, 1019*
    in muscle contraction, 1025–1026
    structure of, *1014,* 1018–1020, *1019*
  light-chain kinase-dependent activation of, 1030–1031, *1031*
  membrane binding of, 1015, *1016*
  movement of, 1016–1021
    actin binding in, 1016–1017, *1017,* 1018
    ATP and, 1017–1021, 1018t, *1020*
    measurement of, 1017–1018, *1019*
    optical trap studies of, 1017–1018, *1019*
    power stroke in, 1021
    sliding filament assay for, 1016–1017, 1018t
    sliding filament model of, 1025–1026
  neck domain in, *1014,* 1015
  in nonmuscle cells, 1032–1039
  protein-kinase inhibition of, 1032
  structure of, *1014,* 1015
  tail domain in, *1014,* 1015
  type I, *1014,* 1014, 1032
    in cell motility, 1041–1043, *1042*
    in vesicle transport, 1039, *1040*
  type II
    in cell motility, 1036, *1036*
    in contractile bundles, 1032–1038
    cortical, 1036, *1036*
    in cytokinesis, 1036–1308, *1037*
  type V, 1014, *1014*
    in vesicle transport, 1039, *1040,* 1078–1079
Myosin ATPase, 1015
Myosin bundles, contractile, 1032–1036, *1033*
Myosin filaments
  in circumferential belt, 1034, *1034*
  in smooth muscle, 1025, 1025–1026, 1030–1032, *1031*
  in striated muscle, 1023–1025, *1024*
    sliding-filament model of, *1025,* 1025–1026
  thick, 1015, *1016*
Myotube, 196, *196, 557, 558, 561, 561*
Myristate, in membrane-protein binding, *610,* 610–611

Na$^+$. *See under* Sodium
NAD$^+$, in glucose metabolism, 742, *742, 744, 744*
NADH
  in citric acid cycle, 749–751, 751t
  electron transport from, 759–770
  in glucose metabolism, 742, *742, 744, 744,* 750–751, 751t
NADH-CoQ reductase complex, 761, *762,* 762t, 764–766. *See also* Electron carriers
  electron-proton transport by, 766
NADPH, in photosynthesis, *795, 796*
Na$^+$/H$^+$ antiporter, 658–659, 664
NaHCO$_3$-Cl$^-$ transporter, 658–659
Na$^+$/K$^+$ ATPase, in ion transport, 648–650, *649*
nanos gene/Nanos protein, in *Drosophila* regionalization, *572,* 572–573, *576*
Na$^+$/potassium$^+$-2Cl$^-$ symporter, 664
Nascent-chain analysis, 427, *427, 428*
Nascent transcript analysis, 433, *433, 434*
Nasopharyngeal cancer, Epstein-Barr virus and, 1279–1280
Native state, 73
Natural killer cells, 1251
N-cadherin, 1151t, *1152, 1153*
N-CAMs, 1153, *1153*
Ndn2 protein, 1266, *1266*
Nebulin, 1026, *1027*
Nernst equation, 642
Nerve(s)
  adrenergic, 967
  demyelination of, 942
  myelinated, 940–942, *941*
Nerve-cell adhesion molecules (N-CAMs), 1153, *1153*
Nerve cells. *See* Neuron(s)
Nerve growth factor, structure and function of, 861t
Netrins, 1180–1181
neu oncogene, 1268, *1268*
Neural circuits, 931, *932*
Neural plate, 1034
Neural precursor cell, 562
Neural tube, formation of
  cadherins and, *1152,* 1153
  N-CAMs in, *1152,* 1153
Neural tube floor plate, chemoattractant in, *1180,* 1180–1181
Neurofibromatosis, 248
  mutations in, 336
Neurofilament(s) (NF), 1108t, 1108–1109. *See also* Intermediate filaments (IF)
  cross-linkage of, 1113
  phosphorylation of, 1113
Neurogenesis
  gene control in, 561–565
  in mouse, 562–565, *564*

Notch-Delta binding in, 1173–1175, *1174, 1175*
Neuroglian, 1183
Neurohormones, 855. *See also* Hormone(s)
  peptide, 953, 954t, 970–971
Neuromuscular junction, 929, *1185,* 1185–1187, *1186*
Neuron(s), 925–986
  action potential of. *See* Action potential
  adhesion of, 1153, *1153*
  cell body in, 926, 927
  commissural, *1180,* 1180–1181
  dendrites of, 928, *928*
  facilitator, 968, 980
  functions of, 925
  identification of, 1175–1176, *1176*
  ion channels in, 927, *928. See also* Ion channel(s)
  ion concentration gradient in, 939–940
  membrane depolarization/repolarization in, 927, 935–940
  membrane potential in, 927, *927,* 933–940
  motor, 925
    autonomic, 930
    sensory neurons and, 968–970, *969*
    somatic, 930
    structure of, *926*
  outgrowth pathways for. *See* Neuronal outgrowth
  pioneer, 1177–1178
  regions of, 926–928, *926–928*
  sensory, 925
    interaction of with motor neuron, 968–970, *969*
    structure of, *926*
  structure of, 926–928, *926–928*
Neuronal circuits, 931, *931*
Neuronal outgrowth
  adhesive proteins in, 1182–1183, *1185*
  chemoattractants in, *1180,* 1180–1181
  chemorepellents in, 1181–1182, *1181–1183*
  collapsing factors in, *1181,* 1181–1182
  directionality in, 1179–1180
  extracellular matrix in, *1178,* 1178–1179
  growth cone in, *1176,* 1176–1177, 1181–1183
  laminin in, *1178,* 1178–1179
  pathways for, *1177,* 1177–1178
  regulation of, 1175–1187
  semaphorins in, 1182, *1183*
Neuronal pathways, development of. *See* Neuronal outgrowth

Neuronal specification, 562
Neuron circuits, 931, *932*
Neuropeptide(s), 953, 954t
   degradation of, 970
   as neurohormones, 970–971
   as neurotransmitters, 970–971
   secretion of, 970
Neuropeptide receptors,
      G-protein–linked, 956t
Neurosecretory cells, 970
Neurospora crassa, arginine biosynthetic
      pathway in, 274, *275*
Neurotransmitter(s), 855. *See also*
      specific neurotransmitters
   in chemical signaling, 855, 925, *929,*
      929–930, *930,* 953–955
   drug inhibition of, 970
   endocytosis of, 958, *959,* 960–962
   exocytosis of, *958,* 958–963, *959*
   memory and, 979–984
   receptor binding of, 953, 955–956
   removal/destruction of, *965,* 970
   secretion of, calcium ion channels in,
      958–960
   secreton of, 953
   in synaptic vesicles, 953, 958–963
   types of, 954t–955t, 957t
   in vision, 972
Neurotransmitter receptors
   acetylcholine, 955–956, *957*
   diversity of, 955–956
   in excitatory synapses, 953
   in fast synapses, 953
   G-protein–linked, 953–955, 956t
   in inhibitory synapses, 953
   as ligand-gated ion channels, 953t,
      957t
   in slow synapses, 953–955
   types of, 953–955
Neurotransmitter transporters, 970
Neutrophils, functions of, 1296
Nexin, 1084
NF1 gene, 248
NF$_k$B transcription factor, 574, *575*
Nicotinamide adenine dinucleotide
      (NAD). *See* NAD$^+$; NADH
Nicotinic acetylcholine receptor,
      955–956, 962
   as ligand-gated ion channel, 962
   structure of, 963–965, *964, 965*
   subunits of, 963–965, *964*
Nidogen, 1143–1144, *1144*
Nieuwkoop center, *1165,* 1166
NinaA, 696
Nirenberg, Marshall, *123*
Nitric oxide, in long-term potentiation,
      982–984, *983*
Nitrogen bonding, 17, 18t
Nitrogen fixation, 174
Nitrogen regulatory protein C (NRPC),
      425, 425–426

NMDA glutamate receptors, in
      long-term potentiation, 982, *983*
NMR spectroscopy, 59–61
Nocodazole, microtubule stability and,
      1067
Node of Ranvier, 941, *942, 942*
Noggin protein, 1165
Nomarski interference microscopy,
      156–158, *156–158*
Noncovalent bonds, 16–17, 21–27. *See
      also* Bond(s), noncovalent
Nondisjunction, chromosomal, *1237,*
      1237–1238
Nonhistone proteins
   in chromosome scaffold, 349–351,
      352, 353
   in transcription, 354
Nonmuscle β-actin, 994
Nonmuscle γ-actin, 994
Non-NMDA gluatamte receptors, in
      long-term potentiation, 982, *983*
Nonpermissive cells, 1270
Nonpermissive temperature, 271
Nonpolar bond, *19,* 19–20
Nonpolar molecule, 25–26
Nonreplicative transposition, 324, *324*
Nonself vs. self discrimination, 1295,
      1309, 1339–1340
Nonsense mutation, 267, *267*
Norepinephrine. *See also*
      Catecholamine(s)
   structure and function of, 954t, 967
   synthesis of, 967, *969*
   transporter for, 970
Norepinephrine receptors, 967
Northern blotting, 249, *249*
   of mutation-defined genes, 289
   in transcription start site mapping,
      251
Notch protein, 1173–1175, *1174, 1175*
N protein, 490, *490,* 492
N1 protein, 347, *348*
NSF protein, in vesicular fusion, 716,
      717
NSF-SNAP complex, 962
NTRB, in transcription, *425, 425*
NTRC protein, *425,* 425–426
Nuclear DNA, mitochondrial DNA in,
      815–816
Nuclear envelope, 167–168, *168, 515,*
      1212
   degradation of, 1212–1213, 1218
Nuclear lamins, 1108t, 1109, 1113,
      1114, *1114*
   depolymerization of, 1212–1213,
      1213–1216, 1218, *1219*
Nuclear localization, 574, *574*
Nuclear localization sequences, 840,
      842–844, *843,* 843t
   bipartite, 843
Nuclear magnetic resonance
      spectroscopy, 59–61

Nuclear matrix, 513–515, *515*
Nuclear membrane, 167–168, *168*
   intermediate filaments in, 1114, *1114*
Nuclear pore(s), 52, *53,* 168, *168*
   in protein transport, *515,* 515–516,
      516, 840–842, *842*
   structure of, *840*
Nuclear pore complex, *515,* 515–516,
      516
Nuclear proteins
   receptors for, 844
   transport and uptake of, 840–844
Nuclear skeleton, 513–515, *515*
Nuclear transcription factors, 1260
   oncogenes and, 1263t, *1265, 1265*
Nuclease-protection method, 249–251,
      250
   in transcription start site mapping,
      251
Nucleic acid(s), 102–104. *See also*
      DNA; RNA
   base components of, *102,* 102–103,
      103t
   definition of, 102
   5′ end of, 104, *104*
   modification of, 113, *114*
   orientation of, 104, *104*
   radiolabeling of, 214–215
   structure of, 102–104, *102–104*
   synthesis of, 103–104, 114–119. *See
      also* DNA replication;
      Transcription
      directionality of, 113, *114,* 115
      polymerases in, 115, *116*
      primer in, 115
      protein synthesis and, 113–114,
         114
      rules for, 113–114, *114,* 115–116
      template for, 115
   3′ end of, 104, *104*
   viral, 206–208
Nucleic acid hybridization, 109
Nucleocapsid, 202
Nucleolar organizer, 168
   pre-rRNA gene as, 530, *531*
Nucleolus, 168
Nucleoplasm, 168
Nucleoplasmin, 347, *348,* 841, *841*
Nucleoside(s), 103, 103t
   radiolabeled, in pulse-chase
      experiments, 215–216, *216*
Nucleoside diphosphates, 103, 103t
   in oligosaccharide synthesis, 700–703,
      702
Nucleoside monophosphates, 103, 103t
Nucleoside phosphates, 103, 103t. *See
      also* Nucleotides
Nucleoside triphosphates, 103, 103t
Nucleosomes
   assembly of, 347, *348*
   structure of, 347, *348*
   transcription inhibition by, 467–468

Nucleotides, 102. *See also* DNA; RNA
  in genetic code, 120–122, *121*, 121t
  nucleic acid synthesis from, 103–104,
    115, *116*
  sequencing of, 245–248, *246, 247*
  structure of, *102*, 102, *103*
  synthesis of, salvage pathways for,
    200, 200–201
Nucleotide sugars, in oligosaccharide
  synthesis, 700–704, *702, 703*
Nucleus, 143, *144, 145. See also under*
  Nuclear
  biogenesis of, *810*, 810–811, 811t
  functions of, 168
  protein import and export in,
    840–844
  protein targeting to, *810*, 810–811,
    811t
  substructure of, 513–515, *514, 515*
  transcription domains in, 513–515,
    514
Nuclear lamins, polymerization of,
  1218, 1219
Nurse cells, in *Drosophila*, 566
Nüsslein-Volhard, C., 273
Nut, *547*
Nutritional auxotroph, isolation of, 192

O antigen, 615, 615t
Obligate aerobe, 744
Obligate anaerobe, 744
OCTA motif, 1323, *1324*
Odorant receptors, 978–979
Okazaki fragments, 116, 375, *375*
Oleic acid, *598*
Olfaction, 978–979
Oligodendrocyte, 941, *941*
Oligo-dT column affinity
  chromatography, *235*, 235
Oligomeric proteins, formation of,
  696–697, *697*
Oligonucleotide probe, synthetic, *239*,
  239–240
  in in vitro mutagenesis, 291, *291*
Oligosaccharide(s), 612–615
  constituents of, 612–613, *613, 614*
  N-linked, 614
    classification of, 704
    complex, 704, 706, *707*
    high-mannose, 704, 706, *707*
    hybrid, 704, *707, 707*
    precursors of, 704, *704*
    structure of, 699–700, *701*, 704,
      704
    synthesis of, 704–706, *705, 706*
  O-linked, 613–614
    structure of, 699–700, *701*
    synthesis of, *702, 703*
  precursor, 704
  processing of, in Golgi vesicles,
    706–707, *707*
  in protein stabilization, 708–709
  structure of, 700, *701*
  synthesis of, 700–708
Oligosaccharide-protein transferase, 706
Omega protein, 381
Ommatidia, *Drosophila, 891*, 891–893,
  1172, 1173
Oncogene(s), 212, 1258–1270
  assay for, 1284, *1284*
  cell-cycle protein, 1265–1267, *1266*
  cellular, 1267–1269
  chromosomal abnormalities and,
    1268–1269
  collaborative, 1267–1268
  definition of, 1258
  discovery of, 1258–1259
  growth factor, 1260–1261, 1262t
  growth factor–receptor, 1261, *1261,*
    1262t
  isolation of, *1258*, 1258–1259
  nuclear transcription factor, 1263t,
    1265, *1265*
  retroviral, 1275–1277, *1276*
    activation of, *1278*, 1278–1279
  signal transducer, 1261–1265,
    1262t–1263t, *1265*
  terminology for, 1258, *1259*
  transcription of, 1258
  viral, 1270
    terminology for, 1259
Oncoproteins, 1258–1270
  in malignant transformation,
    1267–1268
  protein-tyrosine kinases as,
    1261–1264, 1262t
  as transcriptional modifiers, 1260
  types of, 1260–1267
One-sodium/one-glucose symporter, 654
Oocytes
  in *Drosophila*, 566
  endocytosis in, 728–729
  maturation of, in *Xenopus*,
    1203–1207
Oogenesis, in *Drosophila*, 565–566,
  566
Opa repeat, 1173
Open-complex formation, 425–426
Operator(s), 252, 408–409, 412, *412,*
  413
  half-site in, 418, *418*
  lambda
    in lysogenic pathway, 545–547,
      546
    in lytic pathway, 548–550, *549,*
      550
  repressor and, 416–419
Operator constitutive mutations, 409
Operon, 117, 308
  attenuation in, 486–488, *487, 489*
  lac, in transcription control. *See lac*
    operon
  in regulon, 424

trp, 424
  transcription of, 486–488, *487,*
    489
Opsin(s), 608–609, *974*, 974–975, *975*
  in color vision, 977–978, *978*
  in light adaptation, 977, *977*
Optical electron spectroscopy, 788
Optical isomer, 20
Optical sectioning, *155*, 155
Optical trap, 1017, *1019*
Orbital, 17
ORC (origin-recognition complex), 371
Organelle(s), 8, 9, 57, 143, *144, 145*
  biogenesis of, *810*, 810–811
  purification methods for, 162–165,
    164–166
Organizer, 1164, *1165, 1165*
ORI. *See* Replication origin (ORI)
oriC, 370–371
Origin-recognition complex (ORC), 371
Orphan receptors, 462
Orthograde axoplasmic transport, 927
Orthophosphate, radiolabeled, 213
Osmosis, 661–664
  in animal cells, 662, 663–664
  in plant cells, 664
  in protozoan cells, 662
Osmotic flow, 662
  in cell separation, 164, *164*
  in plants, 803–805, *804*
Osmotic pressure, *661*, 662
  leaf stomata opening and, 664, *665*
  turgor, 664
Osteoclasts, 650
Osteogenesis imperfecta, 1134
Ouabain, 649, 655
Ovariole, in *Drosophila*, 565–566, *566*
Oxaloacetate
  in citric acid cycle, *750*, 750
  in malate shuttle, *760*
Oxidase, definition of, 772
Oxidation
  in carcinogen activation, 1283, *1283*
  definition of, 37
Oxidation potential, 38
Oxidation-reduction reaction, 37–39,
  39t
Oxidative phosphorylation. *See also*
  ATP, synthesis of
  in mitochondria, 745–752
  proton-motive force and, 750–752,
    753–759
Oxygen
  bonding of, 17, 18t
  photosynthesis of, 793–794
Oxygen-evolving complex, 793

PABII, 498
PABP. *See* Poly-A–binding protein
  (PABP)
Pain, enkephalins/endorphins and, 971

Pair-rule genes, 576, 578–579, *579*, *580*
Palmitic acid, *598*
Panacrine signaling, prostaglandins in, 857
Pancreatic acinar cells, plasma membrane of, 624–625, *625*
Panning, 1307, *1307*
Papilloma viruses, 1280
Papovaviruses, 1270–1273
Paracrine signaling, 855. *See also* Neurotransmitter(s); Signaling
Paralog, 584, *584*
Parasegments, in *Drosophila*, 576–577, *578*, *580*
    BX-C genes and, *582*, 582–583
Parietal cells, 657, *657*
Partition coefficient K, 636
Parvoviruses, 209, 210t
Passing START, 1226, *1227*, 1230
Passive diffusion, *635*, 635–636
Patch-clamp technique, 944, *944*, *945*
Patterning. *See also* Regionalization
    in embryogenesis, 1159–1166
    in neuronal outgrowth, 1179–1181
    in vertebrate limb, 1169–1172
    in *Xenopus*, 1163–1166, *1164–1166*
Patterning genes, in *Drosophila*, 568–579
Pauling, Linus, 63, 88
P box, 553
PC2, in proteolytic processing, 722
PC3, in proteolytic processing, 722
PCNA (proliferating cell nuclear antigen), 379, *379*
PCR. *See* Polymerase chain reaction (PCR)
PDI. *See* Protein disulfide isomerase (PDI)
Pectin, 1187, *1189*, 1190
P element, of *Drosophila*, 326–327, 327t
P-element transformation, 297, *298*
Pelle protein, 575, *575*
Pemphigus vulgaris, 1156–1157
Peptide(s). *See also* Neuropeptide(s); Protein(s)
    definition of, 56
    fusion of, 731
    synthesis of, 59, *61*
Peptide bonds, 56, *57*
    hydrolysis of, 76–78, *77*
Peptide chain, formation of, *134–135*, 136–138, *137*. *See also* Protein(s), synthesis of
Peptide hormones, 857. *See also* Hormone(s)
    properties of, 863t
    structure and function of, 860t–861t
    synthesis, release, and degradation of, 859–861
Peptidoglycan, 143
Peptidyl-prolyl isomerases, 696

Peptidyltransferase reaction, 136
Pericentrin, 1055
Peripherin, 1108, 1108t. *See also* Intermediate filament(s) (IF)
Periplasm, 143
Permeability, ion, 934
Permeability coefficient P, 636
Permissive cells, 1270
Permissive induction, 1160
Permissive temperature, 271
Peroxisomal enzyme, 174
Peroxisomes, 143, *144*, *145*, 174
    biogenesis of, *810*, 810–811, 811t, *837*, 837–839
    defects in
        in adrenoleukodystrophy, 838–839
        in Zellweger syndrome, 838, *838*
    fatty acid oxidation in, *772*, 772–773
    in photorespiration, 800–802, *801*
    protein targeting to, *810*, 810–811, 811t
    structure of, 837
Pertussis toxin, 878, *879*, 902
Perutz, Max, 59, 68
PEST motif, 1173
Peyer's patches, 1308, *1308*
p53 mutation, in tumors, 1266, *1266*, 1267
pH, 30–34, 31t
    buffers and, 32–34
    cytosolic, 640–641
        in cell volume regulation, 663–664
        transporter regulation of, *658*, 658–659
    equilibrium constant and, 32
    gastric, regulation of, 657, *657*
    intracellular, fluorescent microscopy of, *153*, 153–154
    in protein sorting, 727–728
    as rate regulator, 45–46
Phage, 204
    lambda. *See* Lambda phage
    research in, 206–208
    RNA, 208
    T, 206
    temperate, 206–207
    transcription in, 473
Phage vector, P1, 233t, 234
Phagocytes, 1296
Phalloidin, 1006
Phase-contrast microscopy, 156–158, *156–158*
Phase variation, 338–339, *339*
Phenotype, 264, *266*
Phenotypic marker, 271, 279
Phenotypic suppression, 583
Phenylalanine, 54, *55*. *See also* Amino acid(s)
Pheromone, 855
    yeast, 553, *553*, 1230

Pheromone-responsive element, 553–554
Philadelphia chromosome, 1268
Phloem, 803–805, *804*
Phloem sieve tubes, 1193, *1193*
Phorbol esters, as tumor promoters, 1286–1287, *1287*
Phosphatase, in phosphorylation, 72
Phosphate
    as buffer, 33–34
    inorganic, in ATP synthesis, 741, *741*, 742, 746, 751–756, *754*
Phosphate transporter, 758, *758*, 758–759
Phosphatidic acid, 27
Phosphatidylcholine, 628
Phosphatidylethanolamine, synthesis of, 671, *672*
Phosphatidylinositol 4,5-bisphosphate (PIP$_2$), 901–902
    in actin filament capping and severence, 1008–1009
    prolifin binding of, 1007
Phosphatidylinositol derivatives, synthesis of, 886, *888*
Phosphatidyl-3 kinase, synthesis of, 886, *888*
Phosphoanhydride bonds,
    in ATP, 40, *40*
    hydrolysis of, 739
Phosphodiester bonds, 104, *104*
Phosphoenolpyruvate carboxylase, 802, *802–803*
Phosphofructokinase-1, allosteric control of, 771, *771*
Phosphoglucomutase, 882, *883*
3-Phosphoglycerate, 797, *798*
Phosphoglycerides, 27
    membrane, 597, *597*
Phosphoinositol 4,5-bisphosphate. *See* Phosphatidylinositol 4,5-bisphosphate (PIP$_2$)
Phospholipase(s), 611, *612*
Phospholipase C, 901–902
Phospholipase C$_\gamma$
    receptor tyrosine kinases and, 905
    synthesis of, 886, *888*
Phospholipid(s)
    amphipathic, 597
    cleavage of, *611*, 612
    membrane, 596–599
        leaflet, 599, 616, *616*
        movement of, 671–674
        properties of, 27–28, 596t, 597–599
        structure of, 27, *27*, 597, 597–599, 598t
        synthesis of, 671–674, *672*, *673*
Phospholipid bilayer, 6–7, *7*, 27–28, *28*, 167. *See also* Membrane(s)
    in aqueous solution, 30–34
    cell transport across. *See* Membrane transport

Phospholipid bilayer (*continued*)
  compartmentalization in, 603, *603*
  definition of, 599
  experimental studies of, 602–603
  external face of, 603, *603*
  fluidity of, 599–601
  internal face of, 603, *603*
  liposome, 600, *601*
  in myelin sheath, 602, 602–603
  phase transition in, 600, *601*
  planar, 600, *601*
  structure of, *596, 599, 600*
  universality of, 602–603
  in viral envelope, 203, *203*
Phospholipid exchange proteins, *673*,
      673–674
Phospholipid flippase, 671–673, *673*
Phospholipid leaflet, 599, 616, *616*
Phosphoprotein phosphatase, 884–885,
      885
Phosphoric acid, dissociation of, 33–34
Phosphorolysis, 882
Phosphorus, bonding of, 17, 18t
Phosphorus-32, 213t, 214, 218
Phosphorus-33, 213t
Phosphorylation, *71*, 71–72, 1207
  in cell transformation, 1264
  cytokinesis and, 1218–1219, *1220*
  lamin depolymerization and, 1213,
      1214–1216
  mannose
      in I-cell disease, 711
      in protein targeting to lysosomes,
          709–711
  in membrane transport, 645, 647
  in mitosis-promoting factor
      regulation, 1221–1224
  of nuclear lamins, 1212–1213,
      1218–1219, *1220*
  of nuclear proteins, 1212–1213,
      1214–1216
  oxidative. *See also* ATP, synthesis of
      in mitochondria, 745–752
      proton-motive force and, 750–752,
          753–759
  in photosynthesis, 797
  of proteins, 71–72, 1207,
      1212–1213, 1221–1224
  of receptors, *913*, 913–914
  substrate-level, in ATP synthesis,
      742–744, *743, 744*, 752
  in transcription initiation, 462, *463*
Phosphotyrosines, in receptor
      endocytosis, 912
Photons, 783
  absorption of
      opsin activation and, 974–975
      retinal isomerization and, 974–975
  in photosynthesis, 782, 783–788
Photoreceptors, 972–978
  in rods, 972–974, *972–974*

Photorespiration, 800, *801*
Photosynthesis, 16, 779–805
  action spectrum of, 783, *784*
  ATP synthesis in, 42, 43, 782
  bacterial, 607–608, *608*, 786–790,
      787–789
  in carbon cycle, 43
  carbon dioxide fixation in, 175,
      782–783, 797–805, *798–804*
  charge separation in, 784–786
      in bacteria, 788–790, *789*
  chemical reaction in, 779
  chemiosmosis in, 740, *740*
  chlorophyll in, 175, *782*, 782,
      783–784, *784*
  dark reactions of, 782, 783
  electron transport in, 782, 784–786,
      790, 793–797
      in bacteria, 788–790, *789*
  Emerson effect in, 792
  light absorption in, 782, 783–786
  light-harvesting complexes in, 783,
      786, *787*
  light reactions of, 782, 783
  molecular analysis of, 790–797
  overview of, 780–783
  photorespiration and, 800, *801*
  photosystems in, 790–797
  primary electron acceptor in, 784,
      785
  primary event in, 784, *785*
  proton-motive force in, 740, *740*,
      782–790, *789*, 793–797
  resonance energy transfer in, 786
  ribulose 1,5-bisphosphate carboxylase
      in, 797–800, *798*
  starch production in, 780, *780*, 803
  steps in, 782–783, *785*
  thylakoid membrane in, 780, *781*,
      783–786
Photosynthetic reaction center,
      607–608, *608*
Photosystems
  in bacteria, 786
  in plants, 790–797, *791*
Phototropism, 1191
Phragmoplast, *1105*, 1106
Physical map, 284–287, *286, 287*
Picornaviruses, structure of, *204*
Picosecond absorption spectroscopy,
      788–790
Piggyback transport, 844
Pigment granules, transport of, 1072,
      1074
PI-3 kinase, in receptor endocytosis, 912
Pilus, 222
Pinocytosis, 722
P ion pump, 644–645, *645*, 646t
  Ca^{2+} ATPase, *645*, 645–646,
      647–648
  Na^+/K^+ ATPase, 648–650, *649*

PIP$_2$. *See* Phosphatidylinositol
      4,5-bisphosphate (PIP$_2$)
Pituitary, anterior
  in feedback control, 864, *864*
  in peptide secretion, 970
pK_a, 32
Plakoglobin, 1113t, 1114, 1156
Plant(s)
  C_3, 803
  C_4, 802–803
  cell division in, 180–181, *182, 1105*,
      1105–1106, 1187
  cellulose in. *See* Cellulose
  growth of, 1190–1191, *1191, 1192*
  membrane transport proteins in, *659*,
      659–660
  mitochondria in, 747
  mitochondrial DNA in, 815
  osmotic flow in, 803–805, *804*
  photorespiration in, 800, *801*
  photosynthesis in. *See* Photosynthesis
  plasmodesmata in, 1193, *1193, 1194*
  sucrose transport in, 803–805, *804*
  transgenic, 297–298, *299*
  turgor pressure in, 664
Plant cell(s)
  chloroplasts in, 175, *175*
  connection of, 1193, *1193, 1194*
  division of, 180–181, *182, 1105*,
      1105–1106, 1187
  growth of, 1190–1191, *1191, 1192*
  mitosis in, 180–181, *182*
  plasma membrane of, 176, 176–177
  structure of, 143, *146–147*
  turgor in, 174, 1191
Plant cell wall, structure of, 1187–1190,
      1187–1190
Plant vacuoles, 143, *146*, 174
  sucrose in, 659–660
  transport across, *659*, 659–660
Plant viruses, 204, 208
Plaque(s), 181
  adhesion, 992, 1032, *1033, 1035*,
      1035–1306
  in spot desmosome, 1156, *1156*
Plaque assay, 205, *205*
Plaque proteins, 1114
Plasma cells, 1335
Plasmacytoma, antibody studies in,
      1310
Plasma membrane. *See* Membrane(s),
      plasma
Plasmid
  definition of, 222
  drug resistance, 222
Plasmid vectors
  cloning with, 222–225, *223–225*
      in expression systems, 252–254,
          253
      restriction fragment insertion into,
          228, *228*

Plasminogen, in transformed cells, 1256–1257
Plasmodesmata, 628, 1193, *1193*, *1194*
Plastids, differentiation of, *835*, 835–836
Plastocyanin, *791*, *793*, *794*, *796*
  transport of, 833–834, *834*
Plastoquinone, 792–793, *794*, *796*
Platelet(s)
  activation of, 1145, 1148
  cytoskeleton of, 1000, *1001*
Platelet-activating factor, 1155, *1155*
Platelet-derived growth factor (PDGF)
  activation of, 905
  structure and function of, 861t
Plating, 190–193, *191*, 1252
PLC. *See* Phospholipase C
Pleated β sheet, 57, *58*, 65–66, *66*
  antibody, 1313, *1313*
Plectin, 1113t, 1114
  cytosolic, 1114
Plus strand, 208
Point mutation, 266–267, *267*
  identification of, 289, *290*
  in prokaryotic genes, 308, *309*
  spontaneous, 267–268, *268*
Polar bond, *19*, 19–20
Polarized cells, 196, 623, 993
  plasma membrane of, 623–625
Polar microtubules, 1091, *1092*
  elongation of, 1103, *1103*
  sliding of, 1101–1103, *1102*
Polar molecules, 24
Pole, 179
pol gene, 331
  retroviral, *1274*, *1275*
Poliovirus, 210t, *211*, 211, *211*
Poly-A-binding protein (PABP), 498, *499*
  in mRNA processing, 520
Polyacrylamide gel electrophoresis, *92*, 92–93, *96*, 242, 242–243, *243*
Polyadenylation
  on-off regulation of, 509, *510*
  of pre-mRNA, 498, *499*, 509, 510
Polyamino acid, 56
Poly-A polymerase, 119
Poly-A site, 494, 498
Poly-A tail, 235, *235*
  cytoskeleton and, 521
  synthesis of, 498, *499*
Polycistronic mRNA, 308
Polyclonal antibodies, 201
Polyclonal antigens, 201
Polycyclic aromatic hydrocarbon, as carcinogen, 1283, *1283*
Poly-Ig receptor, 1301–1302
Polylinker, 228, *228*
Polymerase(s), in nucleic acid synthesis, 115
Polymerase α, *379*, 379

Polymerase chain reaction (PCR), 254–256, *255*
  in physical mapping, 284
Polymerase κ, *379*, 379
Polymorphic genes, 1328
Polymorphisms
  DNA, 280
  restriction fragment length, 279–281, *280*
  single-stranded conformation, 289, *290*
Polyoma virus, malignant transformation by, 1272, *1272*
Polypeptides. *See also* Protein(s)
  amino acid sequence in, 56
  arrangement of, 56–57
  definition of, 56
  structure of, 56, *57*
Polyprotein, 211
Polytene chromosomes, *283*, 283–284, *333*, 333, *432*, 432–433
  banding in, 355
  formation of, 343, *343*
Polytenization, 343, *343*
Polyubiquitination, 1210–1212
  anaphase induction and, 1216–1218, *1217*
  of cyclin, 1211–1212
P170 (multidrug-transport protein), *651*, 651–652
P1 phage vector, 233t, 234
Pores, nuclear. *See* Nuclear pore(s)
Porins, 143, 609, *609*
  mitochondrial, 746
Positional cloning, 284–290. *See also* Mutation-defined genes
Position effect variegation, *464*, 464–465
Postbithorax mutation, 583, *583*
Postmitotic cells, 1202–1203
Postsynaptic cell, 929
Postsynaptic membrane, depolarization of, 962–963, *964*
Postsynaptic potential, slow, 967
Posttranscriptional control, 509–513
  definition of, 485
Potassium ion(s). *See also under* Ion(s)
  concentration of, in cell vs. blood, 640, 641t
Potassium ion channel(s). *See also* Ion channel(s)
  cardiac muscarinic acetylcholine receptor and, 967
  delayed, 938–939
  in gastric acidification, 657, 657–658
  ion movement through, 944, *945*, *946*
  membrane polarization and, 935
  membrane potential and, 641–643, *642*
  resting, 643

  resting potential and, 933–934
  voltage-gated
    amino acids in, 948–949
    cloning of, *947*, 947–948
    evolution of, 951–952
    inactivation of, 949–950, *950*
    membrane potential and, 936–940
    in membrane repolarization, 939–940
    properties of, 943–952
    serotonin and, 968–970, *969*
    structure of, *946*, 946, 947–950, *948*
    voltage sensor in, *949*, 949
Potassium ion channel proteins. *See also* Ion channel proteins
  diversity of, 950–951
  identification of, 948–949
Potential energy, 16
  in chemiosmotic theory, 740
Poxviruses, 209, 210t
PP$_i$. *See* Pyrophosphate, inorganic
PP$_i$-hydrolyzing proton pump, 650, 659–660
p16 protein, 1266
p53 protein, 1266
  in tumor suppression, 1285
p53 tumor suppressor gene, in G$_1$ checkpoint control, 1239, *1240*
Pre-B cell, 1321
pre-mRNA. *See under* RNA
Preproinsulin, 72, *73*
Pre-RNA, splicing of, *500*, 500–501
Pre-rRNA, synthesis and processing of, *315*, 315–316
Pressure, in equilibrium, 28
Pribnow box, 414
Primary electron acceptor, 784
Primary transcript, 494
Primase, 374, *374*
Primer, 115
  in polymerase chain reaction, 256
Primer extension, 251, *251*
Primosome, 374
Pro-B cell, 1321
Probes
  in expression cloning, 240, *241*
  in membrane hybridization, 238–240, *239*, *240*
Processed pseudogene, 338, *338*
Processive enzyme, 376
Procollagen, 1131–1133, *1132*
Proenzymes, lysosomal, 709
Profilin, 1006–1007, *1008*
  in acrosome reaction, 1010, *1011*
  in bacterial movement, 1010–1012, *1012*
  in cytoplasmic viscosity, 1044
Progesterone
  feedback control of, 864, *864*
  structure and function of, 858t

Programmed cell death, 195, 1267, *1297*, 1306

Prohormone, proteolytic processing of, *721*, 721–722

Proinsulin, 72
  proteolytic processing of, 721–722

Prokaryote, 7–8, *8*. *See also* Bacteria
  cell contents in, 143–146, 145t, 146t
  cell division in, 177, *177*
  cell structure in, 7–8, *8*, 142–143
  chromosomes in, 147, 344–346
  definition of, 142
  DNA replication in, 372–378
  gene organization in, 116–118, 117t, *118*, 308
  membrane transport proteins in, *659*, 660–661
  protein synthesis in, 116–118, *118*
  transcription in
    initiation of, 406–426
    termination of, 486–491

Prokaryotic genes, 116–118, 117t, *118*, 308

Proliferating cell nuclear antigen (PCNA), 379, *379*

Proline(s), 54, *55*. *See also* Amino acid(s)
  in collagen, 1128, *1128*
  in signaling, 893

Promoter(s)
  transcription, 252, 409, 413–416, 423–424
    A box, 471, *471*, *472*
    adenovirus major late, 453, *454*, *455*
    in archaebacteria, 475
    B box, 471, *471*, *472*
    in chloroplasts, 475
    in eukaryotes, 453–454, *454*, *455*
    in mitochondria, 474
    in prokaryotes, 252, 409, 413–416, 423–424
    RNA polymerase and, 413–414, *414*, *415*
    strength of, 413–414, *415*, 415
    in provirus-directed transcription, 1273
    in regulon, 424
    strength of, 413–414
    TATA-box, 453–454, *454*, *455*
    tumor, 1286–1287
      in oncogene activation, 1279
      in proto-oncogene activation, *1278*, 1278–1279

Promoter mutations, 409

Promoter-proximal elements, 436–439, *437–439*, 441–442, *442*

Proneural cluster, 562, 1173, *1175*

Proofreading, by DNA polymerase, 385, *386*

Propeptides, in collagen synthesis, 1131, *1132*, 1133

Prophage, 544

Prophase
  centrosome replication and migration in, 1096–1097, *1097*
  kinetochore/centromere movement in, 1097–1098, *1098*
  mitotic, 179, *180*, *182*

Proplastids, *835*, 835–836

Proprotein, proteolytic processing of, *721*, 721–722

Prostaglandins, 857, 960t
  structure and function of, 860t

Prosthetic group, 70–71

Protease, 76, 76–78, *77*
  Hfl, 548, *548*

Protein(s), 51–98. *See also specific proteins*
  acetylation of, 71, *71*
  actin-binding, 1006–1007, *1007*
  actin-capping, 1004, *1005*, 1008–1010, 1009t
    in muscle, 1024, *1025*
  actin–cross-linking, 997t, 997–998, *998*
  actin-related, 994
  adapter, *1035*, 1035–1036
    in signaling, 886, 893
  adhesive. *See* Adhesive proteins
  α helix of, 57, 63–65, *63–65*
  amino acids in. *See* Amino acid(s)
  assay for, 88, 96–98, *97*
  β sheet of, 57, 66, *66*
  blood, physical characteristics of, 89t
  cell-surface, in transformed cells, 1256
  cellular volume of, calculation of, 143–144
  channel, *634*, *635*, *686*, 686–687. *See also* Ion channel proteins
  chaperone, 74–75, *687*, 687–688, 695–696
    in clathrin-coated vesicle depolymerization, 713
    mitochondrial, 820–822, *821*, *822*, 826–827
  chemical modifications of, *71*, 71–72
  chloroplast, 175, 832–835
  collagen, 194
  cytosolic, 169–170
    in vesicle-membrane fusion, 716, *717*
  definition of, 56
  denaturation of, *73*, 73–74
  disulfide bond formation in, 694–695
  domains of, 67–68, *68*
  endoplasmic reticulum, 670, *670*, 687–688, 695, 696, 698–699, *699*
  ER-resident, 698–699, *699*
  extracellular matrix, 194, *195*
  fatty acid acylation in, 71, *71*

fluorescent microscopy of, 150–152
folding of, *73*, 74–75, 826–827
  abnormalities in, 697–698
  chaperones in, 695–696
  oligosaccharides in, 708–709
functions of, 9
fusion, in expression cloning, 240, *241*
genetic suppression in, 274–277, *276*
glycosylation of, 71, *71*, 700–711
growth control, *1259*, 1259–1260, 1262t–1263t
high mobility group, 354
histone, *315*, 315–316, *346*, 346–347, *348*
homologous, 68–70, *69*, *70*
hydrolysis of, 76–78, *77*, *79*
intermediate filament–associated, 1113t, 1113–1116
ion channel, 943–951
isolation of. *See* Protein(s), purification of
kinesin-related, 1077t, 1077–1078
  in centrosome separation, 1103
  in mitosis, 1097–1098, *1098*
loops in, 66
in macromolecular assemblies, 57–58
mass of, 56
membrane. *See* Membrane proteins
membrane-binding, 8–9, 998
microtubule-associated, 1067–1070, *1068*, 1068t, *1069*
mitochondrial. *See* Mitochondrial proteins
modification of, 113, *114*
modules in. *See* Protein(s), domains of
molecular weight of, 56
monomeric, 57
motifs in, 66–67, *67*
motor. *See* Motor proteins
multiadhesive matrix, 1124, *1124*, 1143–1150
multimeric, 57, 84–85, *85*
mutant
  degradation of, 698
  transport of, 697–698
native state of, 73
nonhistone
  as chromosome scaffold, 349–351, *352*, *353*
  in transcription, 354
nuclear
  receptors for, 844
  transport and uptake of, 840–844
oligomeric, formation of, 696–697, *697*
oligosaccharide chains on. *See* Oligosaccharide(s)
peroxisomal, *837*, 837–838
phosphorylation of, *71*, 71–72, 1207. *See also* Phosphorylation

plaque, 1114
polyubiquitination of, 1210–1211, 1210–1212, 1216–1218
processing of, 72, 72–73, 720–722, 721
promoter. *See* Promoter(s)
prosthetic groups in, 70–71
purification of
  by antibody-affinity chromatography, 87, 88, 202
  assays for, 96, 96–97, 97
  by centrifugation, 88–92, 91
  by chromogenic enzyme reactions, 96–97
  by electrophoresis, 92, 92–94, 93, 96
  by immunoblotting, 97, 97
  by liquid chromatography, 94–96, 95
receptor. *See* Receptor(s)
regulation of, by RNA editing, 535–537, 536, 537
renaturation of, 73, 73–74
replication, 372–378, 378
repressor. *See* Repressor(s)
RNA binding to, 112
secretion of, 677. *See also* Secretory proteins
  continuous, 677
  regulated, 677, 719
  steps in, 678–680
secretory. *See* Secretory proteins
self-splicing of, 72, 72–73
sequence of, 56
  analysis of, 55, 60
  homology of, 68–70, 69, 70
signal transduction, 9
size of, 56
structure of, 9, 10, 10, 52–75
  analysis of, 59–61, 60
  coiled coil, 65, 65
  functional implications of, 52, 53
  graphical representations of, 61–63, 62
  hierarchy of, 56–59, 58
  primary, 56, 58
  quarternary, 57, 58
  random-coil, 56
  secondary, 56–57, 58
  in motifs, 66–67, 67
  sequence homology and, 68–70, 69, 70
  subunit interaction and, 84–85, 85
  tertiary, 57, 58
  domains of, 67–68, 68
synthesis of, 119–138
  aminoacyl-tRNA synthetases in, 126–128, 127t, 127–129
  in bursts, 522–524
  codon-anticodon pairing in, 124–126, 125, 128

directionality of, 113, 114, 216–218, 217
elongation in, 134–135, 136–138
in eukaryotes vs. prokaryotes, 116–119, 118–120
by expression cloning, 240, 241, 252–254, 253
expression systems for, 252–254, 253
initiation of, 133–136, 134–135
Kozak sequence in, 133
nucleic acid synthesis and, 113–114, 114
range of, 146
rate of, 144, 216, 217
ribosomes in, 128–132, 130–132, 136–138, 674t, 674–676
RNA in, 12
mRNA in, 119, 120, 120–124, 133–138, 134–135
rRNA in, 120, 120, 131–132, 133–138, 137
tRNA in, 119, 120, 124–128, 133–138, 134–135, 137
rules for, 113–114, 114
secretory pathway for, 670, 670, 676–678
Shine-Dalgarno sequence in, 133, 308
sites of, 674t, 674–676, 675
steps in, 133–138, 134–135
termination of, 134–135, 138
translation and, 120
translocation in, 134–135, 136–138
taxonomy of, 70
transcriptional. *See* Transcription factor(s)
transforming, 610
transport of. *See* Protein transport
turns in, 66
unfolding of, 73–74, 74
variety of, 146
vesicular, 714–716, 717
viral, 205–206
  arrangement of, 202–203, 203
  in yeast two-hybrid system, 894–895, 896
Protein binding, cooperative, 545–547
Protein carrier, haptan, 1298, 1298
Protein disulfide isomerase (PDI), 695, 696
  in protein folding, 696
  retention/retrieval of, 698–699, 699
Protein family, 313, 313t
Protein hormones. *See also* Hormone(s)
  properties of, 863t
Protein kinase(s)
  Ca²⁺-calmodulin–activated, in long-term potentiation, 984

cAMP-dependent, 881–882, 882, 886
  in glycogen metabolism, 884, 884–885
catalytic subunits in. *See* Cdk(s)
in cell cycle regulation, 1201
mitogen-activated, 1070
mitosis-promoting factor as, 1207
in muscle contraction regulation, 1030–1032, 1031
regulatory subunits in. *See* Cyclin(s)
second messenger activation of, 881–882, 882
serine/threonine, as oncoproteins, 1262t
tyrosine, as oncoproteins, 1261–1264, 1262t
Protein kinase C, 604
  1,2-diacylglycerol activation of, 904–905
  functions of, 905
  phorbol esters and, 1287
Protein-serine/threonine kinases, as oncoproteins, 1262t
Protein sorting, 172–173, 669–671, 722–723. *See also* Protein targeting; Protein transport
  to apical or basolateral membrane domains, 729, 729–731, 730
  cytoskeletal attachment in, 731
  definition of, 669
  glycosylation and, 699–711
  Golgi and post-Golgi, 172, 172–173, 716–722
  hierarchy in, 670
  membrane lipids and, 730–731
  of membrane proteins, 670, 670–671, 722–732
  monitoring of, 707–708
  pH in, 727–728
  principles of, 710
  receptor-mediated endocytosis in, 722–732
  retention signal in, 731
  secretory pathway in, 670, 670
  of secretory proteins, 716–722
  to secretory vesicles, 719–720
  signal recognition particle in, 683–685, 684, 685
  signal sequence in, 682t, 682–683, 719
  transcytosis in, 728–729, 729, 731
  in *trans*-Golgi reticulum, 729, 729–730, 730
Protein targeting, 669. *See also* Protein sorting; Protein transport
  of chloroplast proteins, 832–835
  to lysosomes, 709–710
  of mitochondrial proteins, 811t, 819t, 819–824, 820–822, 827–829, 828

Protein targeting (*continued*)
nuclear localization sequences in, 840, 842–844, *843*, 843t
in organelle biogenesis, *810*, 810–811, 811t
Rab proteins in, 716–718, *718*
of secretory proteins, *681*, 681–688
signal recognition particle in, 683–685, *684*, *685*
signal sequence in, 682t, 682–683
Protein translocation, 824
Protein transport. *See also* Protein sorting; Protein targeting; Transport
channel proteins in. *See* Ion channel proteins
chaperone proteins in. *See* Chaperone proteins
of chloroplast proteins, 832–835
cytosol-to-organelle, *810*, 810–811, 811t
glycosylation in, 699–711
of membrane proteins, 688–694
microtubules in, 1070–1079
of mitochondrial proteins, 811, 819–824, *821*, 827–829, *828*
monitoring of, by oligosaccharide processing, 707–708
of mutant proteins, 697–698
of nuclear proteins, 840–844
piggyback, 844
process of, 669–670, 677–678
protein folding and, 697–698
protein unfolding in, 686
of secretory proteins, *681*, 681–688
to/from endoplasmic reticulum, 711–718, *717*
transcellular, 729, *729*
vesicular, 172–173, 711–718. *See also* Transport vesicles
Protein-tyrosine kinases, as oncoproteins, 1261–1264, 1262t
Proteoglycan(s), 194, 1139–1142
cell-surface, 1142, *1142*
extracellular matrix, 1139, *1141*
growth factor binding by, 1142, *1143*
hyaluronan in, 1136–1139, *1138*
types of, 1139–1142
Proteoglycan aggregates, 1139, *1141*
Proteolytic enzymes, *76*, 76–78, *77*, *79*
Proteolytic processing, of proteins, *72*, 72–73, 720–722, *721*
Protofilament, 1052–1054, *1053*
Proton antiporter, in plants, 659–660
Proton-lactose symporter, *660*, 660–661, *661*
Proton-motive force, 740, *740*. *See also* Membrane potential; Proton pump
calculation of, 753
definition of, 740

electron transport and, 759–770
experimental demonstration of, 756–758, *757*, *758*
in photosynthesis, 740, *740*, 782, 784–786, 790, 793–797
in bacteria, 788–790, *789*
Proton-motive force in, in ATP synthesis, *750*, 750–752, 753–759
Proton-motive Q cycle, 767–770
Proton pump(s)
in ATP synthesis, 740, 750–752, 753–759
in bacteria, 660
in fungi, 660
H+ ATPase, 650–651, *651*
in plant vacuole, 659–660
PPi-hydrolyzing, 650, 659–660
Proton-sucrose antiporter, 660
Proton transport
by electron carriers, 766–770
mitochondrial, 759–770
Proto-oncogenes, 492, 1258
oncogenes from, 1259–1267, 1262t–1263t
in retroviruses, 1275–1277, *1276*
retroviral activation of, *1278*, 1278–1279
Proviral DNA, 1273, *1274*
P-selectin, 1154
Pseudogene, *311*, 314–315
processed, 338, *338*
Pseudoknot configuration, in RNA, 112, *112*
Pseudopodium, 992
in cell movement, *1043*, 1043–1044
PSI, in photosynthesis, 790–793, *791*, 794–797
PSII, in photosynthesis, 790–797, *791*
Puffs, chromosomal, *432*, *433*, 466
Pulse-chase experiments, 215–216, *216*
Pulsed-field gel electrophoresis, 244–245, *245*
Pumps
ion. *See* Ion pump(s)
proton. *See* Proton pump(s)
Pupation, *Drosophila*, 565, *566*
Purines
salvage pathway for, *200*, 200–201
Purines, structure of, *102*, 102–103
Purple bacteria, photosynthesis in, 786–790, *787–789*
Pyridoxal phosphate, as coenzyme, 79–80, *80*
Pyrimidines, structure of, *102*, 102–103
Pyrophosphate, inorganic, in fatty acid metabolism, 748
Pyrophosphate-hydrolyzing proton pump, 650, 659–660
Pyruvate
allosteric regulation of, 771, *771*
in glycolysis, 740–741, *741*, 743, *744*

mitochondrial metabolism of, *746*, *747*, 748–750
oxidation of, 745, 748–752
Pyruvate dehydrogenase, 748
Pyruvate kinase, allosteric control of, 771

Q band, 354
Q box, *553*
Q cycle, 767–770
Quality control, in endoplasmic reticulum, 697–699
Quick freeze technique, 169, *170*
Quiescent cell, 1251
Quinone, in photosynthesis
in bacteria, 786, 788, *789*
in plants, *791*, 792–796, *795*
Quiver mutation, 1109

Rab3A, *961*, 961–962
RAB proteins, in vesicular fusion, 716–718, *718*
Rac protein, in cell movement, 1044
Radding, Charles, 391
Radial spokes, in axoneme, 1084, 1087
Radiant energy, 16
Radiation
as carcinogen, 1284–1286
ultraviolet, DNA damage by, repair of, *387*, 387–388
Radioisotopes, 212–218, 213t
choice of, 213–214
half-life of, 213
types of, 213t
Radiolabeled compounds, 212–218, 213t
detection of
by autoradiography, *214*, 214–215
quantitative, 215
pulse-chase experiments with, 215–216, *216*
specific activity of, 213
synthesis time of, 216, *217*
Radius, van der Waals, 25, 25t
Raf protein
in kinase cascade, 894–895
in RTK-Ras protein signaling pathway, 897, *897*
Random coil, 56
in domains, 67–68
Random-coil polypeptide, 56
ras oncogene, 1263t, 1264, 1267–1268
Ras protein, 876, *877*
insulin receptor and, *910*, 910–911
kinase cascade and, 894–896, *895*, *896*
MAP kinase and, 894, 895–896, 897–899
MEK protein and, 894, 895–896

Raf protein and, 894–895
in signaling, 887–897
structure of, 62
Rate-zonal centrifugation, 90–92, 91, 164–165, 165
R bands, 354
RB protein, 1266
in S phase regulation, 1236, 1237
RB tumor suppressor gene, 1269–1270
Reading frame, 121, 121
Reassociation rate, 316–317, 317
RecA protein
in DNA repair, 393
in recombination, 391–393, 392, 393
RecBCD enzyme, in recombination, 392, 392
Receptor(s)
acetylcholine, aggregation of, 1186, 1186–1187
adaptation of, 973
binding to, adenylate cyclase activation and, 874–875, 876
catecholamine, 967
cell growth and, 1259, 1260
cell-surface
α-adrenergic, 870
β-adrenergic, 870–878
binding specificity of, 856
classification of, 857, 859, 862
cloning of, 867, 868
cytokinase, 916–917
desensitization of, 913–914
in endocytosis, 722–727, 912
glucagon, 911
G protein–linked, 859, 860, 862, 869–881. See also G protein–linked receptors
hormone concentration and, 912
identification of, 865–867, 865–868
insulin, 907–911
ion-channel, 859, 860, 862
ligand-triggered, 862
of lipophilic hormones, 857, 857, 859t
phosphorylation of, 913, 913–914
purification of, 866, 867
receptor tyrosine kinase, 886–899
regulation of, 912–914
ryanodine, 904, 904
seven-spanning. See G-protein–linked receptors
in signal transduction, 462
specificity of, 856, 865
of water-soluble hormones, 857, 857
cloning of, 867, 867
domain structure of, 459, 459–461, 460
growth factor, 1259–1260, 1262t
oncogenes and, 1261, 1261, 1262t

homing, 1308
insulin, synthesis and membrane insertion of, 689, 689
intracellular, 856, 857
as transcription factors, 1260. See also Transcription factor(s)
ligand binding to, 722–724, 723, 723t, 856, 862, 865
light, 972–978. See also Photoreceptors
lymphocyte, homing, 1308
mitochondrial protein, 824, 826
for multiadhesive matrix proteins, 1143–1146
neurotransmitter. See Neurotransmitter receptors
nuclear protein, 844
olfactory, 974–975
orphan, 462
response elements and, 460, 460
sensory, 91, 971–979
T-cell. See T-cell receptors
touch, 971
in transcription regulation, 461, 461–462
transforming growth factor β, 1162
zinc fingers in, 449–450
Receptor-mediated endocytosis, 710, 722–727, 722–729, 723, 723t
acidic late endosome/CURL organelle in, 726–727, 727, 728
clathrin-coated pits and vesicles in, 722–724, 723
in hormone down-regulation, 912
in infections, 731–732, 732
ligand in, 722, 723
in low-density lipoproteins, 724–726, 725, 726
in oocytes, 728–729
pH changes in, 727–728
in transcytosis, 729, 729
in transferrin cycle, 727–728, 728
Receptor serine/threonine kinases, 859, 862
Receptor tyrosine kinase–Ras protein signaling pathway, 896–897, 897
Receptor tyrosine kinases (RTKs), 859, 862, 886–899
activation and dimerization of, 886, 887
adapter protein binding to, 886–887
autophosphorylation of, 886, 887
components of, 886
in Drosophila eye development, 891–893, 892
in gene regulation, 914–918
GRB2 protein and, 893
insulin receptor as, 907
phospholipase C$_\gamma$ and, 905
in Ras protein activation, 887–897
Sos protein and, 893

Receptor tyrosine phosphatase, 862
Recessive genes, 266, 266, 266, 266
Recessive mutation, 266, 266
X-linked, 277, 278
Reciprocal gene conversion, 396–397, 397–399
Reciprocal induction, 1169
Reciprocal recombination, 389
Recognition, in immunity, 1295–1296
Recognition helix, 418, 419, 549, 550
Recognition signals, antibody, 1318, 1318
Recognition site, 80
Recombinant DNA technology, 221–258
cloned DNA cleavage in, 240–242
clone identification in, 236–240, 237–241
cloning in, 222–225, 240, 241, 252–254, 253. See also Cloning, DNA
complementary DNA in, 234–236, 235, 236
definition of, 222
DNA ligases in, 227–228, 228
gel electrophoresis in, 242, 242–243, 243
pulsed-field, 244–245, 245
gene-to-protein strategy in, 237–240, 257, 257, 264
genomic library in, 229–233, 232
informatics in, 247–249, 248
in isolation of mutation-defined genes, 284
lambda-phage vectors in, 229–233, 230–233
Maxam-Gilbert method in, 245, 246
membrane hybridization in, 237–240, 237–240
Northern blotting in, 249, 249, 289
nuclease protection in, 249–251, 250
plasmid vectors in, 222–225, 223–225, 228, 228
polymerase chain reaction in, 254–256, 255
primer extension in, 251, 251
probes in, 238–240, 239–241
protein synthesis in, 695
protein-to-gene strategy in, 256–257, 257, 264
restriction enzymes in, 225–226, 226, 227t
restriction fragment sequencing in, 244–248, 246, 247
restriction-site mapping in, 243–244, 244, 248–249, 251
Sanger (dideoxy) method in, 245–246, 247
Southern blotting in, 248–249, 249
S1 protection in, 251, 251
Recombinases, 395
site-specific, 326, 339

Recombination
  antibody diversity and, 1315–1319
  base pairing in, 392–393
  branch migration in, 393, 393–395
  chromatid, 184–185, 185
  in E. coli, 391–396, 392–395
  errors in, mutations and, 396
  Holliday model of, 389–391, 390, 391
  Holliday structure in, 390–391, 391
    crossed-strand, 390
    formation of, RecA protein in, 393
    heteroduplex, 392, 392
    resolution of, 393–395, 394
  homologous, 292, 389–397
  meiotic, 279, 279
    gene conversion in, 396–397, 397–399
    in yeast, 396
  phage integration in, 395, 395–396
  RecA protein in, 392–393, 393
  RecBCD enzyme in, 392, 392
  reciprocal, 389
    gene conversion in, 396–397, 397–399
    resolution in, 390, 390
  Ruv proteins in, 393–395, 394
  site-specific, 395, 395–396
  strand invasion in, 392–393
  switch, 1338
Recombination frequency, 279, 279
Recombination mapping, of mutations, 279, 279
Red blood cell. See Erythrocyte(s)
Red-green color blindness, 977–978
Reduction
  definition of, 37
  oxidation and, 37–39
Reduction potential, 38–39, 39t
  of electron carriers, 765
Reflex
  gill-withdrawal, 979–982, 980, 981
  knee-jerk, 931, 931
Reflex arc, 931, 931
  in habituation, 979–982, 981
Refractive index, 149
Regionalization. See also Patterning
  in Drosophila, 565–584
    ANT-C genes in, 581–583, 581–584
    bicoid gene in, 568–572, 569–571
    BX-C genes in, 581–583, 581–584
    Dorsal protein in, 573–575, 1162
    Dpp protein in, 1162
    epidermal, 581–584, 582, 583
    gap genes in, 573, 573, 577–578, 580
    hedgehog gene in, 1169–1172, 1170, 1171
    hunchback gene in, 570–573, 571–573, 580

    maternal genes in, 568–579, 577, 580
    mutations and, 568, 569
    nanos gene in, 572–573, 573–574
    nuclear localization in, 574, 574
    pair-rule genes in, 576, 578–579, 579, 580
    regulation of by cell-cell contact, 1172–1175
    segment-polarity genes in, 576, 578, 579, 580
    selector genes in, 576, 579–584, 580–583
    signal-dependent, 573–575, 574, 575
    Toll protein in, 575–576, 578, 1162
    torso gene in, 573–574
    zygotic genes in, 572–573, 576–579, 580
  in embryogenesis, 1159–1166
  in neuronal outgrowth, 1179–1181
  in vertebrate limb, 1169–1172
  in Xenopus, 1163–1166, 1164–1166
Regulated protein secretion, 677, 719–720
Regulatory domain, 882
Regulatory switching, cooperative protein binding in, 545–547, 546
Regulon, 424
Rejection, graft, 1327–1328
Renaturation, of DNA, 108, 108–109, 237
  rate of, 316–317, 317
  with repeats, 316–317
Reoviruses, 210t, 211
Repetitious DNA, 312t, 316–317
  interspersed, 316
  tandem, 316
Replica plating, 190–193, 191, 1252
Replication. See DNA replication
Replication origin (ORI), 223, 223, 370–371, 371
  in E. coli, 370–371
  SV40, 370–371
  in yeast, 355, 356, 371
Replication termination sites, 378
Replicative transposition, 324, 325, 325–326
Replicon, 368, 369
Repolarization, membrane, 927
Reporter genes, 429, 437
Repressor(s)
  eukaryotic, 456
  prokaryotic, 408–409, 416–419, 424
    lac, 408, 408, 416–418, 417, 422
    binding site on, 408–409
    operator binding of, 416–419, 419
    structure of, 418, 418–420
    trp, 418–419, 420, 424
Resolution, microscopic, 148–149

Resolvase, 326
Resonance energy transfer, 786
Resonance hybrid, 17
Respiration
  ATP synthesis in, 42
  in carbon cycle, 43
  peroxisomal, 772, 772
Respiratory control, 770
Response elements, 460, 460
Resting ion channels, 643, 928, 935
Resting potential, 933–934
Restriction endonucleases, 225–226, 226, 227t
Restriction enzymes, 221
  in DNA cloning, 225–226, 226, 227t
  in separation of DNA from vector, 240–242
Restriction fragment length polymorphisms (RFLPs), in mutation mapping, 279–281, 280
Restriction fragments
  amplification of, by polymerase chain reaction, 256
  formation of, 225, 226
  homologous sequences in, 248
  identification of, by Southern blotting, 248, 248–249
  insertion of into plasmid vector, 228, 228
  linkage of, 227, 227–228
  measurement of, 243–244, 244
  production of, 225, 226
  separation of
    by gel electrophoresis, 240–242, 241, 242
    by pulsed-field electrophoresis, 244–245, 245
  sequencing of, 245–248, 246, 247
    data storage, distribution and analysis of, 246–248
    Maxam-Gilbert method of, 245, 246
    Sanger (dideoxy) method of, 245–246
  sticky ends of, 226
Restriction-modification system, 225, 227t
Restriction point, mammalian, passage through, 1232, 1233–1235, 1234
Restriction site, 225, 226
Restriction-site linker, 236
Restriction-site mapping, 243–244, 244
  by nuclease-protection method, 251
  by Southern blotting, 248–249, 249
Retention signal, in protein sorting, 731
Retina
  cone cells in, 977–978, 978
  photoreceptors in, 972–974, 972–974
Retinal, isomerization of, 974–975
Retinoblastoma, 268–269, 270, 1269–1270

Retrograde signaling, *929, 930*
  in long-term potentiation, 982–984
Retrograde transport, 1070
Retrotransposed RNA, 336–338, *338*
Retrotransposon(s). *See also* Mobile
    elements
  copia, 329, 333, 333–334
  definition of, *322,* 323
  nonviral, *328, 334, 334–336, 335*
  viral, *323, 328, 328–334, 329*
Retroviral DNA, *329, 1273–1275, 1274*
Retroviral oncogenes, *1275–1277, 1276*
  activation of, *1278, 1278–1279*
Retroviruses, 210t, *211, 211–212*
  artificial, *1277, 1277–1278*
  DNA replication in, 115
  genome of, *1273, 1274, 1275*
  genomic integration of, 1273–1275,
    1274
  human cancer and, 1279
  as intracellular parasites, 1273
  life cycle of, *212,* 1273–1275
  long terminal repeats in, 329–330,
    330, 1273, *1274*
  in malignant transformation, 1270,
    1273–1279
  replication of, 1273
  retrotransposons and, *323, 328,*
    328–334, 329
  slow-acting carcinogenic, *1278,*
    1278–1279
  transducing
    nonocogenic, *1277, 1277–1278*
    oncogenic, *1275–1277, 1276*
Reverse genetics, 1002
Reverse transcriptase, 211–212, 1273
  in cDNA production, *235, 235*
  in long terminal repeat production,
    329–331, *330*
  structure of, *52, 53*
Reverse transcription, in mobile element
    transposition, *331, 331–333, 332*
Revertant, *274, 275*
RFLP. *See* Restriction fragment length
    polymorphisms (RFLPs)
RGD-binding sequence, in fibronectin,
    1147, 1148, 1148
RGG box, *495, 496*
Rhodamine, 152
Rhodopseudomonas viridis,
    photosynthetic reaction center in,
    607, 608
Rhodopsin, *974, 974–975*
Rhodopsin kinase, 977
Rho factor, 486
  in transcription termination, 488–490
Rho protein, in cell movement, 1044
Ribonuclease, denaturation and
    renaturation of, 74
Ribonuclease P (RNase P), 534
Ribonucleic acid. *See* RNA

Ribonucleoprotein particles (RNPs). *See*
    RNPs
Ribonucleosides, 103, 103t
  radiolabeled, 215
Ribophorin, 706
Ribose, structure of, *102*
Ribosomal protein, bacterial synthesis
    of, translational control of,
    525–526, 526
Ribosomal RNA. *See under* RNA
Ribosomal subunits
  in translation, 308, 309, *529, 530*
  transport of, 529
Ribosome(s)
  binding sites on, 136–138, *137*
  cytosolic
    functional equivalence of, 675
    organelle biogenesis on, 810
  membrane-bound, *675, 675*
  membrane-unattached, *675, 675*
  in protein synthesis, 128–132,
    130–132, 136–138, *137,* 674t,
    674–676
  in mitochondria, 816–817
Ribosome-binding site, 133, 308
Ribozyme, 112, *113,* 532
Ribulose 1,5-bisphosphate carboxylase,
    797–800, 798
  synthesis of, *832–833, 833*
Rigor mortis, 1021
Risk factors, cancer, 1287–1288
RNA. *See also* Nucleic acid(s)
  antisense, *527–528, 528*
  catalytic activity of, 112, 508,
    530–532
  5' cap of, in nuclear transport,
    516–518, *517*
  functions of, 12, 112
  guide, *537, 537*
  hairpin configuration in, 112, *112*
  identification of, by Northern
    blotting, *248,* 248–249
  mapping of, by nuclease-protection
    method, 249–251, *250*
  messenger. *See* mRNA
  minus strand of, 208
  mitochondrial. *See* mtRNA
  modification of, 113, *114*
  noncoding, 310–311
  nucleotides of, *102,* 102–104, 103t
  plant virus, 208
  in protein synthesis, 12, 119–138,
    120
  pseudoknot configuration in, 112,
    112
  quantification of, by
    nuclease-protection method,
    249–251, *250*
  retrotransposed, 336–338, *338*
  ribosomal. *See* rRNA
  as ribozyme, 112, *113,* 532

small, nonfunctional copies of,
    336–338, *338*
small nuclear, in splicing, *503,*
    503–509, 505–508
small nucleolar, 529
splicing of, 498–509. *See also*
    Splicing
stem-loop configuration in, 112, *112,*
    124–125, 131–132, 486, *486,*
    488, 489
structured domains of, 112, *112*
structure of, 111–112, *112*
synthesis of. *See also* Transcription
  directionality of, 113, *114,* 115
  nucleus in, 168
  polymerases in, 115, *116*
  rules for, 113–114, *115–116*
transcribed spacer, 468
transfer. *See* tRNA
transport of, 513–522, 534
5S rRNA, 312t, 316. *See also* rRNA
  synthesis of, 430, 470–473, *472*
gRNA, *537, 537*
hnRNA, 494–496, *495, 496. See also*
    pre-mRNA
M1 RNA, 534
mRNA
  adenovirus, preferential transport of,
    519
  antibody light chain, 1316
  cytoplasmic
    localization of, 520–522, *521–523*
    stability of, 522–524, *524*
  cytoskeleton and, 521
  degradation of, regulation of,
    524–525
  cDNA copies of, 234–235, *235*
  eukaryotic, synthesis of, 116–119,
    118–120
  5' cap of, 119, *119,* 133–136,
    516–518, *517*
    in transcription initiation, *434, 435*
  function of, 119
  half-lives of, 524t
  isolation of, *235, 235*
  leader sequence of, 487, *488, 489*
  monocistrionic, 308–309
  in mutation-defined genes,
    identification of, 288–289, *289,*
    290
  nuclear transport of, 513–520
    5' cap in, 516–518, *517*
    HIV Rev protein in, 519–520, *520*
    nuclear pore complexes in, *515,*
      515–516, *516*
    preferential, 519
    proteins in, 518–520, *519, 520*
  plus strand of, 208
  polycistronic, 308
  precursor. *See* pre-mRNA

mRNA (continued)
  processing of, 494–513
    hnRNP proteins in, 494–498
    nuclear domains for, 513–515,
      514, 515
    posttranscriptional, 509–513
    pre-mRNA cleavage and
      polyadenylation in, 498
    splicing in, 498–509. See also
      Splicing
    steps in, 494, 494
  in protein synthesis, 119, 120,
    120–124, 133–138, 134–135
  reverse transcription of, in mobile
    element transposition, 331,
    331–333, 332
  stability of, 522–524, 524, 524t
    regulation of, 524–525, 525
  synthesis of, RNA polymerase II in,
    430
  transferrin-receptor, stability of, 525,
    526
  transformation-specific, 1257
  translation of. See Translation
  transposase, antisense regulation of,
    527–528, 528
mtRNA, 816–817
  editing of, 817
  processing of, 817
pre-mRNA
  cleavage and polyadenylation of, 498,
    499
    on-off regulation of, 509, 510
  processing of, 494–496, 494–496
    by RNA editing, 535–537, 536,
      537
  spliceosome-associated, transport of,
    518
  synthesis of, proteins in, 518, 519
pre-rRNA
  processing of, 528–532
  splicing of, 500, 500–501
  synthesis of, 315, 315–316, 430,
    468–470, 469, 470
    by tandemly repeated genes, 315,
      315–316, 316t
  transcription-control regions in,
    469–470, 470
pre-tRNA, processing of, 532–534
rRNA. See also 5S rRNA
  function of, 120
  gene amplification in, 343–344, 344
  mitochondrial, 815t
  processing of, 528–532
  in protein synthesis, 119, 120,
    131–132, 133–138, 137
  structure of, 131, 131–132
  synthesis of, 430, 470–473, 472
    by tandemly repeated genes, 315,
      315–316, 316t
  transcription unit in, 469, 470t
  transport of, 534

7S RNA, synthesis of, 470–473
7SL RNA, 336, 337
snRNA, in splicing, 501–507, 503,
    505–507
snoRNA, 529
Tfr mRNA, stability of, 525, 526
tRNA
  amino acid linkage to, 126–128, 127
  aminoacyl-tRNA synthetase activation
    of, 124, 126–128, 127t, 127–129
  binding sites for, 136–138, 137
  in codon-anticodon pairing, 124–126,
    125, 128
  cognate, 126
  function of, 119, 124
  identity elements in, 128, 129
  mitochondrial, 815t
  processing of, 532–534
  in protein synthesis, 119, 120,
    124–128, 125–127, 133–138,
    134–135, 137
  structure of, 124, 124–126
  synthesis of, 430, 470–473, 472
    by tandemly repeated genes, 315,
      315–316, 316t
  transport of, 534
tRNA^Met, 133, 134–135
tRNA synthetase,
    cytosolic/mitochondrial,
    single-gene encoding of, 823,
    823–824
RNA-binding motifs, 495–496, 496
RNA domains, 112, 112
RNA editing, 535–537, 536, 537
  mRNA editing, 535–537, 536, 537
RNA enzymes, 112, 113
RNA footprinting, 132
pre-rRNA genes, as nucleolar
    organizers, 530, 531
rRNA genes, tandemly repeated,
    315–316, 316t
tRNA genes, tandemly repeated,
    315–316, 316t
RNA-mediated transposition, 328–338
RNA phage, 208
RNA polymerase(s), 115, 116
  α-subunit of, 416
  bacteriophage, 473, 473
  binding sites for. See Promoter(s)
  core, 415
  mitochondrial, 474
  structure of, 473, 473
  subunits of
    in eukaryotes, 430–433, 431, 431t
    in prokaryotes, 414–416, 424t,
      424–426, 425
  T7, 473, 473
  in transcription, 334–335
    in eukaryotes, 430–434, 430–435,
      468–473
    in prokaryotes, 411–412, 413–416,
      423–426

  in transcription termination,
    491–493, 492, 493
RNA polymerase I, in pre-rRNA
    synthesis, 315, 315–316, 430,
    468–470, 469, 470
RNA polymerase II
  carboxyl-terminal domain in, 431,
    432
  start site for, 432–435, 434
  transcription by, 430–435, 434, 441
  in transcription-initiation complex,
    453–456, 454, 455
RNA polymerase III, transcription by,
    430, 470–473
RNA primers, 115, 374, 374t, 374–375
RNA processing, 119, 119
RNA transcript, processing of, 119, 119
RNA viruses
  classification of, 209, 209–212, 210t
  tumor, 1273–1279. See also
    Retroviruses
RNP motif, 495, 496
RNPs, 494
  in mRNA transport, 513–518
  hnRNPs, 494–498, 495–497
    functions of, 496–498, 497
    RNA-binding motifs in, 495–496,
      496
    structure of, 494–495, 495
  mRNPs
    Balbiani ring, 516, 516
    nuclear, transport of, 515, 515–516,
      516, 842–844
  snRNPs
    evolution of, 503–509
    in splicing, 503, 503–509, 505–508
    transport and uptake of, 840–844
  snoRNPs, 529
    in pre-rRNA processing, 529
Rod cells
  light adaptation of, 976, 976–977
  photoreceptors in, 972–974, 972–974
Rolling circle replication, 344
Rough endoplasmic reticulum. See
    Endoplasmic reticulum (ER),
    rough
Rough microsomes, 676
Rous sarcoma virus, 1275, 1276
R7 protein, 891–893, 892, 1172
RTKs. See Receptor tyrosine kinases
    (RTKs)
Rubisco, synthesis of, 832–833, 833
Rubisco activase, 800
Ruffles, 992, 993
  in cell movement, 1044
Run-off assay, 433, 433–435, 434
Run-on analysis, 427, 427, 428
rutabaga mutation, 979
Ruv proteins, in recombination,
    393–395, 394
Ryanodine receptors, 904, 904

Saccharomyces cerevisiae. See Yeast
Salivary-gland differentiation, *1167,* 1167–1168
Salmonella typhimurium, phase variation in, 338–339, *339*
Salvage pathways, *200,* 200–201
Sample preparation
  for electron microscopy, 159, *159, 160*
  for light microscopy, 150, *150*
Sanger, Fred, 68
Sanger method, of restriction fragment sequencing, 245–246, *247*
SAR. *See* Scaffold-associated region (SAR)
Sarcolemma, 1025
Sarcoma, 1249
  Kaposi's, 1280, *1282*
Sarcomere, 1023, *1024*
  intermediate filaments in, 1114–1115, *1115*
  thick filaments in, 1023, *1024, 1025,* 1025–1026
  thin filaments in, 1023, *1024, 1025,* 1025–1026
Sarcoplasmic reticulum (SR)
  calcium pump of, 644, 645–646, *647,* 647–648, 1027
  in muscle contraction, 1027, *1028*
Satellite DNA, 318, *319*
Scaffold, chromosome, 349–353, *352, 353*
Scaffold-associated region (SAR), 353, *353*
Scanning electron microscopy, *148, 161,* 161. *See also* Electron microscopy
Scatchard binding equation, *1300*
Schiff base, 79–80, *80*
Schizosaccharomyces pombe. See Yeast
Schleiden, Matthias, 4
Schramm, G., 208
Schwann, Theodor, 4
Schwann cell, 941, *958*
Scintillation counter, 215
Scurvy, 1131
scute gene, 561–562, *563, 564*
SDS-polyacrylamide gel electrophoresis, *92,* 92–93, *96*
Sea slug
  learning and memory in, 979–980, *980*
  neuropeptides in, 970–971
  sensory neuron–motor neuron interaction in, 968–970, *969*
sec mutants, 274
Second messenger(s), 857–859, *860*
  calcium ions as, 857–859, *860,* 899–905
  cAMP as, 857–859, *860,* 870, *871t*
  catecholamines and, 967
  in cell growth, 1260

  in cell movement, 1044–1046, *1045*
  definition of, 854
  1,2-diacylglycerol as, 904–905
  inositol phosphates as, 901–904, *902, 903*
  multiple, coordination of, 905–906
  in protein kinase activation, 881–882
Second-site mutation, 414
Second-site revertant, 414
Sec63 protein, *686,* 687
Sec proteins, *686,* 686–687
  in Golgi vesicle fusion, *686,* 716
Secretory pathway, for protein synthesis, *670, 670,* 676–681, *680*
Secretory proteins
  aggregation of, 719–720
  classification of, *676t*
  disulfide bond formation in, 694–695
  enzymatic modification of, 677
  glycosylation of, 699–711
  maturation of, *678, 680*
  oligosaccharide stabilization of, 708–809
  posttranslational modification of, 694–697
  proteolytic processing of, 720–722, *721*
  in rough endoplasmic reticulum, 676–677, *677*
  signal sequence on, *682t,* 682–683
  sorting of, *171, 172,* 172–173, 677, 718–722. *See also* Protein sorting
    regulated, 677, 719–720
  synthesis of, 676–678, *679,* 681
  targeting of, 682–683. *See also* Protein targeting
  transport of, *171,* 172–173, 676–688, *677*–688. *See also* Protein transport
Secretory vesicles, *172,* 172–173
  maturation of, 678
  protein sorting to, 719–720
  protein transport to, 677–678
Sedimentation constant, *89t,* 90
Segment-polarity genes, 576, *578, 579, 580*
Segregation patterns, for mutations, 277, 277–279, *279*
Selective theory, of immunity, *1303,* 1303–1304, *1304*
Selectivity factor 1 (SL1), *470, 470*
Selector genes, 576, 579–584, *580–583*
Selfish DNA, 320
Self-splicing
  protein, 72, 72–73
  RNA, 112, *113, 508,* 508–509, 530–532, *532*
    catalytic activity and, 530–532, *531*
Self-tolerance, 1295, 1309, 1339–1340
Semaphorins, 1182, *1183*
Sensitization, 980, *980, 981*

Sensory neuron(s), 925. *See also* Neuron(s)
  interaction of with motor neuron, 968–970, *969*
  structure of, *926*
Sensory organ precursor (SOP), 562, *563,* 1173, *1175*
Sensory receptors, 931, 971–979
Sensory transduction, 971–979
Sequence drift, 314
Sequence homology, 68–70, *69, 70,* 248
Sequence-specific DNA affinity chromatography, 442
Sequence-tagged site (STS), 284, *285*
Sequencing
  DNA. *See* DNA sequencing
  restriction fragment, 245–248, *246, 247*
Serine, 54, *55*
  in protein hydrolysis, 76, 76–77
Serine/threonine kinases, as oncoproteins, *1262t*
Ser-Lys-Leu sequence, 838
Serotonin, 968–970, *969*
  in classical conditioning, 980–982, *981*
  structure of, *954t*
  transporter for, 970
Serotonin receptor, 968–970, *969*
Serum factors
  in cell culture, 1253
  in culture media, 193
Serum-inducible element, 918
Serum-response element (SRE), 916
Serum-response factor (SRF), 916
Sevenless mutant gene, *265*
Sevenless receptor tyrosine kinase, 891–893, *892,* 1172, *1173*
Seven-spanning receptors. *See* G protein–linked receptors
Severe combined immunodeficiency disease (SCID), 1309
Severin, *1009t*
Severing proteins, 1007–1009, *1010t*
Sev receptor, 891–893, *892,* 1172, *1173*
Sex lethal gene, 511–512
Sex-linked segregation pattern, 278, *279*
shaker mutants, in *Drosophila,* 947, 947–950
SH2 domain, 887, *889, 897, 897*
SH3 domain, 893, *894, 897, 897*
SH domains, in cell transformation, 1264–1265
S4 α helix, 949, *949,* 951
Shell, atomic, 17
shi gene, cloning of, 284–285, *286*
Shine-Dalgarno sequence, 133, 308
Short interspersed element (SINE), 317
Short-term memory, 979–982
  long-term potentiation in, 982–984, *983*

I-48   Index

Sialic acid, 612–613, *613, 614, 614*
Sialyl Lewis-x antigen, 1154
Sialyltransferase, retention of, 719
Sickle cell disease, 268
Sieve plates, 805
Sieve-tube cells, *804, 805*
σ factors, 424t, 424–426, *425, 426*
σ⁷⁰ subunit, of RNA polymerase, 414–416
Signal
  horizontal dorsalizing, 1164
  recognition, in antibodies, 1318, *1318*
Signal amplification, in chemical synapses, *955*
Signal computation, in chemical synapses, *955*
Signal-dependent regionalization, in *Drosophila*, 573–575, *575*
Signaling
  abbreviations in, 854t
  in actin filament capping and severence, 1008–1009
  adapter proteins in, 886, 893
  amplification of, 875, 885, *885*
  autocrine, 855
  in axonal migration, 1177, 1186
  in B-cell activation, 1335
  binding specificity in, 856
  calcium ions in, 857–859, *860,* 899–905
  cAMP in, 873–874
  cascade in, 885, *885*
  in cell growth, 1259–1260
  in cell movement, 1044–1046, *1045*
  cell-to-cell contact in, 1172
  chemical, *925, 929, 929–930, 930, 953–955*
    cessation of, *965*
    neurotransmitters in, *929, 929–930, 930, 953–955*
  Delta protein in, 1173–1175, *1174, 1175*
  Drosophila eye development and, *891, 891–893, 892, 1172, 1173*
  effector specificity in, 856
  electrical. *See also* Action potential
    in sensory transduction, 971–979
  in embryogenesis, 1160–1166
  endocrine, 855. *See also* Hormone(s)
  gap junctions and, 1158
  in gene regulation, 914–918
  G proteins in, 869–881
  in immune response, 1298
  induction, 1160–1166, 1168–1169. *See also* Induction
  inositol-lipid pathway in, 901–904, *902, 903*
  insulin pathway in, 907–911, *910*
  kinase pathway in, 885, 886–899
  in learning and memory, 980–981, *981*

ligands in, 856
in metabolic regulation, 881–886
multiplex pathways in, 905–906
in muscle contraction, 1026–1029, *1028–1030*
neuromuscular junction and, 1185–1187
neurotransmitters in, 855–925. *See also* Neurotransmitter(s), chemical
Notch protein in, 1173–1175, *1174, 1175*
overview of, 854–864
panacrine, prostaglandins in, 857
paracrine, 855
pheromones in, 855
in plants, 855, 1193, *1194*
plasma membrane in, 167
plasmodesmata in, 1193, *1194*
proteoglycans in, 1142
Ras proteins in, 887–897
receptors in. *See* Receptor(s)
receptor tyrosine kinases in, 886–899
retrograde, *929, 930*
  in long-term potentiation, 982–984
second messengers in. *See* Second messenger(s)
steps in, 854–855
synaptic, *929, 929–930, 930,* 1185–1187
in T cell activation, 1332, *1332*
termination of, 875
Signaling circuits, 931
Signaling molecules, 855
Signal peptidase, 682
Signal recognition particle (SRP), 336, 683–685, *684, 685*
  receptor for, *684, 685*
Signal sequence, 682t, 682–683, 719
Signal sequence binding protein, *681, 685*
Signal transducer(s)
  in cell growth, 1260, 1262t–1263t
  G protein as, 875
  oncogenes and, 1261–1265, 1262t–1263t, *1265*
Signal transduction, 462
  in cell growth, 1260
  in olfaction, 978–979
  in vision, 971–978
Signal-transduction cascade, in *Drosophila*, 576–579
Signal-transduction proteins, 9
Silencer motif, 1323–1324
Silencers, 465, 556
Silent mating-type loci, in yeast, 465
Simian virus 40. *See* SV40 virus
Simple-sequence DNA, 317, 318t, 318–320
Simple transcription unit, 309, *309*
Sindbis virus, 210t, *211,* 211

SINE, 317, 335–336, *336*
Single-channel recording, *943, 943–944, 944*
Single-copy DNA, 317
Single-strand binding protein, 373
Single-stranded conformation polymorphism (SSCP), 289, *290*
sis oncogene, 1260, 1262t
Sister chromatids, 349
  in meiosis, 181, *184, 185,* 1202, *1202*
  in mitosis, 179–180, *180,* 1202, *1202*
  recombination of, 184–185, *185*
  segregation of, 355–359, *356–358*
Site-specific recombinase, 339
Site-specific viral integration, 1270–1271
Skeletal muscle, *1022, 1022–1023. See also* Muscle
Skin diseases, keratin in, 1115–1116, *1116*
SKL sequence, 838
Sl 1. *See* Selectivity factor 1
Sliding filament assay, 1016, *1017*
Sliding filament model, *1025,* 1025–1026
Slow postsynaptic potential, 967
Small molecule, functions of, 9–10
Small nuclear ribonucleoprotein particles (snRNPs)
  evolution of, 503–509
  in splicing, *503,* 503–509, *505–508*
  transport and uptake of, 840–844
Small nuclear ribonucleoprotein (snurp), 112
Small nuclear RNA, in splicing, 501–507, *503, 505–507*
Small nucleolar ribonucleoprotein particles (snoRNP), in pre-rRNA processing, 529
Small nucleolar RNA, 529
Smell, 978–979
Smooth endoplasmic reticulum, 143, *144, 145,* 170–171, *171*
Smooth muscle, *1022, 1022–1023. See also* Muscle
SNAP proteins
  in vesicular fusion, 716, *717,* 962
snRNA, nonfunctional copies of, 336–338, *337*
snRNPs. *See* Small nuclear ribonucleoprotein particles (snRNPs)
Snurp, 112
Sodium-bicarbonate-chloride transporter, 658
Sodium-calcium antiporter, 654–655
Sodium-glucose symporter, 652–654, *653, 654*
Sodium-hydrogen antiporter, 658, 664

Sodium ion(s). *See also under* Ion(s)
  concentration of
      action potential and, 938–939, *939*
      in cell vs. blood, 640, 641t
Sodium ion channel(s)
  action potential and, 938, *939*
  ion movement through, 944, *945, 946*
  membrane polarization and, 935
  membrane potential and, 641–643, *642, 643*
  photon absorption and, 972–973
  as sensory receptors, 971–978
  voltage-gated
      action potential and, 936–940
      evolution of, 951–952
      properties of, 943–952
      structure of, 946, 951, *951*
Sodium ion channel proteins. *See also* Ion channel proteins
  domains of, 951, *951*
Sodium ion pumps, in plants, 659–660
Sodium-potassium-chloride symporter, 664
Sodium/potassium pump, in ion transport, 648–650, *649*
Solenoid chromatin fibers, 348, *349*
Sol-gel cycling, in cell movement, 1044
Solitary gene, 313
Solubility, binding specificity and, 25–26
Solution
  hypertonic, 164
      osmotic flow and, 662
  hypotonic, 164
      osmotic flow and, 662
  isotonic, 164
Somatic cell genetics, 199
Somatic mutations, 1304, 1315, 1338–1339, *1339*
Somatic recombination hypothesis, 1315–1318
Somites, in myogenesis, *557*, 557–558
Sonication, 162
sonic hedgehog gene, 1170
SOP. *See* Sensory organ precursor (SOP)
Sos protein, 888, *890*, 893, 897, *897*
SOS repair system, 1285
SOS response, 375, 393
Southern blotting, 248–249, *249*
  of mutation-defined genes, 288, *288*
  in physical mapping, 285–287
  in Ty element identification, 328–329, *329*
Spacer DNA, 317
Spectrin, 604, 621, *622, 623*, 623
Spectroscopy
  nuclear magnetic resonance, 59–61
  optical electron, 788
  picosecond absorption, 788–790
Spemann organizer, 1164, 1165, *1165*

Sperm, in acrosome reaction, 1010, *1011*
Spermidine, 345, *345*
Spermine, 345, *345*
SPF. *See* S-phase promoting factor (SPF)
S phase, 1202, *1202*
  entry into
      in eukaryotes, 1226–1232
      in mammalian cells, 1226, *1227*, 1230, 1233–1235, *1233–1235*
S-phase promoting factor (SPF), *1228*, 1228–1230, *1229*
Spheroplast, 292
Sphingomyelin, membrane, 597, *598*
Sphingosine, *614*
  membrane, 597
Spindle, mitotic. *See* Mitotic spindle
Spindle pole body, 181, *183*, 357, 1091, *1093*
Spliceosomal splicing cycle, 504, *506, 532*, 532
Spliceosome, 504, *507*
  in nuclear retention of pre-mRNA, 518
Splicing
  at cryptic splice sites, 518
  mechanisms of, *532*
  protein, 72, 72–73
      self-splicing in, 72, 72–73
  RNA, *114*, 114, 119, *120*
      alternative, 509–513, *510*
      base pairing in, *503*, 503–504, *505*
      differential, 509–513
      in pre-rRNA processing, 530–532, *532*
      in pre-tRNA processing, 532–534, *533, 534*
      self-splicing in, 112, *113, 508*, 508–509, 530–532, *532*
      catalytic activity and, 530–532, *531*
      snRNA in, 501–507, *503, 505–507*
      spliceosomal model of, 504, *506*
      tissue-specific, 509–513
      transesterification reactions in, 501, *501, 507*
      trans-splicing and, 507–508
  spliceosomal, 504, *507*, 532, *532*
Spot desmosomes, 628, *1156*, 1156–1157
S1 protection, 251, *251*
SP1 transcription factor, 436
  assay for, *444*
SR. *See* Sarcoplasmic reticulum (SR)
Src homology 2 (SH2) domain, 887, *889*
src oncogenes, 1262t, 1264
SRP. *See* Signal recognition particle (SRP)
Ssb protein, 373
SSCP. *See* Single-stranded conformation polymorphism (SSCP)

Staining
  for electron microscopy, 159, *159*
  for fluorescence microscopy, 152, *152*
  for light microscopy, 150, *150, 151*
Standard free energy change
  calculation of, 35–37
  determinants of, 35–36
  values for, 36t, 37t
Standard reduction potential, 38, 39t
Stanley, Wendell, 208
Starch, leaf
  fate of, 803
  structure of, *780*
START, passing, 1226, *1227*, 1230
Start codon, 121
Stat1α, 462, *463*
STATS, 916–918
Steady-state reaction, 82
Stem cells
  embryonic, in gene knockout, 294, *294–296*
  yeast, 555, *555*
Stem-loop configuration
  in RNA, 112, *112*
  in mRNA, 486, *486*, 488, *489*, 508
  in rRNA, *131*, 131–132
  in tRNA, *124*, 124–125
STE12 protein, in yeast mating-type conversion, 554, *554*
Stereoisomer, 20, *20*
Steroid(s), 856, *857*, 858t. *See also* Hormone(s)
  properties of, 863t
  structure of, 597–599, *599*
  synthesis, release, and degradation of, 861–864
Steroid hormone receptor(s)
  domain structure of, *459*, 459–461, *460*
  zinc fingers in, 449–450
Steroid hormone receptor superfamily, *459*, 460. *See also* Receptor(s)
Sticky ends
  base pairing of, 226
  ligation of, *227*, 227–228
Stimulatory G protein, 874–877, *875, 876*
Stimulus, in conditioning, 979, 980
Stokes' law, 90
Stomach. *See under* Gastric
Stomata, opening of, osmotic pressure and, 664, 665
Stop codon, 121
Stop-transfer membrane-anchor sequence, 689, *689*
Stress fibers, 993, 1032–1036, *1033*
  adhesion plaques and, 992, 1032, *1033, 1035*, 1035–1036
  in cell movement, 1044
  matrix fibers and, 1148
  visualization of, 169, *170*

Striated muscle. *See* Muscle, striated
Stroma, in chloroplast, 175, *175*
STS. *See* Sequence-tagged site (STS)
Stubble allele, 266
Sturtevant, A., 279
Substrate, 46
Substrate binding, 75–76, *76*
    conformational change and, 80–81, *81*
    induced-fit, 80, *81*
    lock-and-key, 80, *81*
Succinate-CoQ reductase complex. *See also* Electron carriers
Sucrase-isomaltase, 624, *624*
Sucrose
    photosynthesis of, 797–800, *798, 799*
    in plant vacuole, 659–660
    structure of, *780*
    transport of, 803–805, *804*
Sugars. *See also specific types*
    N-linked, 614
    nucleotide, in oligosaccharide synthesis, 700–704, *702, 703*
    O-linked, 613–614
Sulfur, bonding of, 17, 18t
Sulfur-35, 213t
Supercoiling, DNA, 346, *346*
    negative, 381
    positive, 381
Suppressor mutation, 274–277, *276*
Surface markers, lymphocyte, 1306–1307, *1307*
Svedberg, 90, 131
SV40 chromosome, replication in, 378–380, *379*
SV40 enhancer, 439–441, *440*
SV40 nuclear localization sequence, 842, *843, 843t*
SV40 virus, 208–209, 210t
    replication origin in, 371–372
    T proteins of, *1272*, 1272–1273
Swimming movements, calcium ions in, 1087–1088
Switch genes, *556*
Switch-independent mutation, 556
Switching
    antibody, 1335–1338
    regulatory, cooperative protein binding in, 545–547, *546*
Switch recombination, 1338
sxl gene, 511–512
Symport, 652–654
Symporter(s), *634, 635. See also* Membrane transport proteins
    in cotransport, 652–654, *653, 654*
    in cytosolic pH regulation, *658,* 658–659
    Na$^+$/K$^+$-2Cl$^-$, 664
    one-sodium/one-glucose, 654
    proton-lactose, *660,* 660–661, *661*
    sodium-glucose, 652–654, *653, 654*
    two-sodium/one-glucose, 652

Synapse(s), *929,* 929–930, *930*
    chemical, *929,* 929–930
        excitatory, 953, *957*
            fast, *953*
            inhibitory, 953, *957*
            slow, 953–955
        cholinergic, 955
        definition of, 926
        dendrites and, 928, *928*
        electrical, 930, *930,* 952–953
        excitatory, 943
        impulse transmission and, 952–965
        inhibitory, 943
            GABA and, 966–967
            glycine and, 966–967
        neuromuscular, 1185–1187
        signal amplification in, 955
        signal computation in, 955
Synapsin, 960–961, *961*
Synapsis, 184–185, *185*
Synaptic cleft, *929,* 953
Synaptic vesicle(s), *929,* 953, *958, 959*
    acetylcholine in, 956–958, *958*
    endoytosis of, 958, *959,* 960–962
    exocytosis of, *958,* 958–962, *959*
Synaptic vesicle proteins, 960–962, *961*
Synaptobrevin, *961,* 962
Synaptophysin, *961,* 962
Syncytial blastoderm, in *Drosophila,* 567, *567*
Syndecan, 1142, *1142,* 1167
Synaptotagmin, *961,* 962
Syntaxin, *961,* 962
Synthetic DNA
    in probes, 238–240, *239*
    production of, 229, *229*
    uses of, 229, *229,* 238–240, *239*

TAFs, 453t, *454,* 456, 470
tailless gene, 576
Tandemly repeated array, 315–316
T antigen, 379, *1272,* 1329
    in nuclear protein transport, 842
Taq polymerase, 254
Targeting proteins, 716
TATA box, 435, *435, 436,* 442, *442*
TATA-box binding protein (TBP), 453, *455,* 470, 471, 472–473
TATA-box promoter, 453–454, *454, 455*
TATA motif, 1323, *1324*
Tat protein, antitermination, 492, *492*
Tatum, Edward L., 11
Tau protein, 1067–1069, *1068t, 1069*
Taxol, microtubule stability and, 1067
Tay-Sachs disease, 173
T cell(s), 1324–1332
    activation of, 1296–1297, 1307, 1309
    adhesion of, 1155
    antigen recognition by, 1324, 1329–1332, *1332*

    in cell-mediated immunity, 1324
    circulation of, 1308
    cytotoxic, *1297,* 1306
        activity of, 1325–1327, 1330–1331
        self-recognition of, 1328–1330
    differentiation of, 1323, *1324*
    functions of, 1296, 1306–1307, 1330–1332
    helper, *1297,* 1306, *1306*
        activity of, 1330–1331
        B cell activation and, 1304, 1333–1335, *1334*
        lymphokine secretion by, 1330–1331, 1331t
    identification of, 1306–1307, *1307*
    joint recognition process and, 1329
    killer. *See* T cell(s), cytotoxic
    major histocompatibility complex and, *1327,* 1327–1329, *1328, 1330*
    maturation of, 1324
    memory, 1335
    surface markers on, 1306–1307, *1307*
    thymic selection of, 1329–1330, *1331*
    true antigen for, 1329
    volume regulation of, 663–664, *664*
T-cell receptors, 1306, 1324–1332
    antigen interaction with, 1325–1327, *1327,* 1329–1332
    DNA rearrangement in, 344
    specificity of, 1329–1330, *1331*
    structure of, 1324–1325, *1326*
TdT. *See* Terminal deoxynucleotidyl transferase (TdT)
Tektin, 1084
Telomerase, 358
    in DNA replication, 380–381, *381*
Telomere, 357–359, 380
Telomere terminal transferase, 358
Telophase, mitotic, 180, *180, 182*
TEL sequence, 358
Temperate phage, 206–207
Temperature, permissive, 271
Template, 115
Tenascin, 1149, *1150,* 1168
Tentacles, microtubules in, 1054, *1056*
Teratocarcinoma, epigenetic alterations and, 1286
Terminal deoxynucleotidyl transferase (TdT), 1319
Terminal transferase, 235
Termination factor (TF), 138
Termination sites, 486
    Rho-independent, 486, *486, 487*
Terminator codon, 121
Ternary complex factor (TCF), 916
Testosterone, structure and function of, 858t
Tetrahymena pyriformis
    chromosome replication in, 358
    tandemly repeated arrays in, 316t
Tetrahymena thermophila, self-splicing in, 530–532

Tetraiodothyronine, structure and function of, 856, 858t
TF. *See* Termination factor (TF)
TFIID, 453–456
TGFβ. *See* Transforming growth factor β (TGFβ)
TGFβ receptor, type III, 1142
Thermal energy, 16
Thermoacidophiles, 142
Thermodynamics, first law of, 16
Thermogenesis, 770–771
Thermogenin, 770
3′ end, of nucleic acid, 104, *104*
3T3 cells, 1265, *1265*
    in cancer research, 1253–1255, *1254*
Threonine, 54, *55*
    codon for, 122–123, *123*, 123
Threshold potential, 943, *943*
Thylakoid lumen, 780
    protein uptake by, 832–835, *834*
Thylakoid membrane, 603
    antennas in, 783, 786
    charge separation across, 784, *785*
    photosynthesis on, 780, *781*, 784–786
    reaction centers in, 783, 786
Thylakoid vesicle, 175, *175*
Thymine, 102, *102*, 103t
    pairing of with adenine, 104–105, *105*
Thymosin β₄, 1006, *1007*
Thymus, T-cell selection in, 1329–1330, *1331*
Thyroglobulin, 861
Thyroid hormone receptor, 462
Thyroid hormones, 856, *857*, 858t, 861–864
Thyroxine, 856, *857*, 858t, 861–862. *See also* Hormone(s)
    properties of, 863t
    structure and function of, 856, 858t
    synthesis, release, and degradation of, 861–864
Tight junctions, 625–628, *626, 627*
Time-lapse microscopy, 156–158, *158*
Ti plasmid, 297–298
Tissue morphogenesis, cadherins in, *1152*, 1152–1153
Tissue staining
    for electron microscopy, 159, *159, 160*
    for fluorescence microscopy, 152, *152*
    for light microscopy, 150, *150*
Titin, 129, 1026, *1027*
Titration curve, 32–34, *33*
T7 late promoter, 252–253, *253, 254*
T lymphocytes. *See* T cell(s)
Tobacco mosaic virus (TMV), 202
    nucleic acid studies in, 208
    self-assembly of, 203
Togaviruses, 211
Tolerance, immunologic, 1295, 1309, 1339–1340

toll gene/Toll protein, 573–574, *574, 576, 577*, 1162
Tonofilament, 1113
Topogenic sequences, 670–671, 688–694
    α helix as, 690–692, *691, 692*
    glycosylphosphatidylinositol membrane anchor, 692–694, *693*
    multiple, 690–692, *691*
    stop-transfer membrane-anchor, 689, *689*
    uncleaved internal signal-anchor, 689–690, *691*
Topoisomerase(s), 111
    in DNA replication, 381–384, *381–385*
Topoisomerase II (topo II), in protein scaffold, 351, *353*
torso gene/Torso protein, 575–576, *577*
Touch receptors, 971
TPB (TATA-box binding protein), 453, *455*, 470, 471, 472–473
TPB-associated factors (TAFs), 453t, *454*, 456, 470
T phage, 206
T proteins, malignant transformation and, *1272*, 1272–1273
tra gene, 512–513
TRAM protein, 685, 686–687
Trans-acting gene, 411
Transcribed spacer RNA, 468
Transcript, primary, 494
Transcription
    in archaebacteria, 475
    attenuation in, 486–488, *487, 489*
    autoregulation of, 547, *548*
    in bacteriophages, 473
    basal, nucleosome repression of, 467–468
    cell growth and, 1260
    cell-type–specific
        in mammals, 556–561
        in yeast, 551–556
    in chloroplasts, 474–475
    cI protein in, 545–547, *546*
    closed complex in, 415
    constitutive, 408
    control of, 1260
    definition of, 115
    domains for, 445–452, 513–515, *514*
    in *Drosophila* regionalization, 565–584. *See also* Regionalization, in *Drosophila*
    of early-response genes, 1235–1236
    elongation in, regulation of, 486–493
    enhancers in, 425, 439–442, *440, 442*
    in eukaryotes, vs. in prokaryotes, 116–119, *118*
    holoenzyme in, 415–416
    of immunoglobulin genes, 1323–1324, *1324*

    inhibition of
        cytosine methylation and, 468
        DNase resistance and, *466*, 466–467
        in eukaryotes, 456
        heterochromatin in, *464*, 464–468
        by methylation, 468
        by nucleosomes, 467–468
        in prokaryotes, 408–409, 416–419, 424
        silencers in, 465, 556
    initiation of
        activators in, *425*, 425–426
        ATP hydrolysis in, 425–426
        blocking of, 416–418, *417*
        catabolite activator protein in, 421–423, *422–423*
        chromatin structure and, 464–468
        consensus sequences in, 414, *415*
        control elements in, 439–442
        cooperative binding in, 423
        cyclic AMP in, 421–423, *421–423*
        enhancers in, 425, 439–442, *440, 442*, 1273
        in eukaryotes, 426–473
        hormone-controlled, 458–462
        initiation factor in, 415
        inititator in, 435, *435, 436*
        lac operon in, 406–426. *See also* *lac* operon
        operators in. *See* Operator(s)
        phosphorylation in, 462, *463*
        in prokaryotes, 406–426
        promoter-proximal elements in, 436–439, *437–439*
        promoters in, 252, 409, 413–416, 423–424. *See* Promoter(s), transcription
        strength of, 413–414, *415*
        protein-binding sites for, 412, *412, 413*
        regulon in, 424
        repressors in
            lac, 408, *408*, 414–419, *417–420*, 422
            trp, 418–419, *420*, 424
        RNA polymerase in, 115, *116*. *See also* RNA polymerase
        in eukaryotes, *430–434*, 430–435, 468–473
        in prokaryotes, 411–412, 413–416, 423–426
        σ factors in, 414–416, 424t, 424–426, *425, 426*
        signal transduction in, 462
        start site for, 412, *433*, 433–435, *434*, 436
        mapping of, 251
        Rho-dependent, 488–490
        TATA box in, 435, *435, 436*, 442, *442*
        TPB in, 453, *455*, 470, 471, 472–473

Transcription (*continued*)
  of late-response genes, 1235–1236
  leader sequence in, 487, 488, *489*
  in lysogenic pathway, 545–547,
    546–548
  in lytic pathway, 548–550, *549, 550*
  in mitochondria, 473–474, *474, 817*
  of oncogenes, 1258
  in prokaryotes, vs. in eukaryotes,
    116–119, *118*
  promoter in, 252
  rate of, measurement of, *427,*
    427–428, *428*
  regulation of. *See also* Gene control
    chromatin in, 464–467
    cytosine methylation in, 468
    DNase in, *466,* 466–467, *467*
    hormonal, 458–462
    nucleosomes in, 467–468
    by transcription factors, 428–430,
      442–453. *See also* Transcription
      factor(s)
  of tRNA, 430, 470–473, *472*
  of 5S rRNA, 430, 470–473, *472*
  rules of, 113–114, 115–116
  by tandemly repeated genes, *315,*
    315–316, 316t
  termination of
    antitermination and, *490,* 490–491,
      491
    in eukaryotes, 491–493
    premature, 486–488, *487, 489*
    prevention of, 490–491
    in prokaryotes, 486–491
    Rho-dependent, 488–490
    Rho-independent, 486
    stem-loop structure in, *486, 486,*
      488, 489
  in transformed cells, 1257
Transcription-control elements, 439–442
  enhancers as, *425,* 439–442, *440,*
    442, 1273
  multiple, 441–442, *442*
  promoter-proximal elements as,
    436–439, *437–439,* 441–442,
    442
  TATA box as, *435, 435, 436, 442,*
    442
  upstream activating sequences as,
    442, 442
Transcription-control regions, 308,
    435–442
  identification of, 435–439
  in pre-rRNA, 469–470, *470*
  spacing of, *439, 440*
Transcription factor(s), 354, 428–430,
    442–453
  activation domains in, 445, *447, 447,*
    452–453, 456
    amino acid sequences in, 452–453

  activation of
    cdc-28-G_1 complexes in,
      1230–1231, 1231t
    signal transduction pathways in,
      914–918
  activator, 453–456
  activity of, assay for, 444
  in antibody gene rearrangement,
    1323–1324, *1324*
  basic-zipper domain in, *452, 452*
  binding sites for, 435–442
  dimerization of, 560
  distribution of, 458
  DNA-binding domains of, 445,
    447–452, *448–452*
  DNase-hypersensitive sites and, *467,*
    467
  domains of, 445–452, 456
  flexible protein domains in, *447, 447*
  general, 453t, 453–456
  genes encoding, identification of, 445
  heat-shock, *493, 493*
  helix-loop-helix, *448, 452, 452*
    in myogenesis, 560–561, *562*
    in neurogenesis, 561–565
  heterodimeric, *452, 452*
  homeodomain, *447, 448*
  hormonal regulation of, 456–462
  identification of, 442–445, *443, 444*
  inhibitor-binding of, *461,* 461–462
  interaction of, *452, 452*
  lambda-encoded, 545–547
  leucine-zipper domain in, 451–452
  modular nature of, *447, 447*
  morphogens and, 568
  nuclear, 1260
    oncogenes and, 1263t, *1265, 1265*
  repression domains in, 456
  repressor, 456
  for RNA polymerase II, 453t,
    453–456
  for RNA polymerase III, 471–472,
    472
  specificity of, 457–458
    in myogenesis, 556–561
    in yeast, 552–556
  in transcription-initiation complex,
    453–456, *454, 455*
  translocation of. *See* Translocation
  winged-helix (forkhead) domains in,
    448, 451
  zinc finger domains in, *448–452,*
    449–452
Transcription factor E2F, in S phase
    regulation, 1236, *1237*
Transcription-initiation complex,
    453–456, *454, 455*
Transcription loops, 466–467, *467*
Transcription start site, 412, *433,*
    433–435, *434, 436*
  mapping of, 251
  Rho-dependent, 488–490

Transcription unit. *See also* Exon;
    Intron
  eukaryotic
    complex, 309–310, *310, 509*
    simple, 309, *309*
  prokaryotic, 308
Transcytosis, 729, *729,* 731
Transducin, *975, 975*
Transducing retroviruses
  nononcogenic, *1277,* 1277–1278
  oncogenic, 1275–1277, *1276*
Transduction, signal. *See* Signal
    transduction
Transesterification reactions, in splicing,
    112, *113,* 501, *502,* 508
Transfection experiments, 355–359
  for transcription factor activity, 444,
    444
Transfer gene, 222
Transferrin, iron transport by, 727–728,
    728
Transfer RNA. *See under* RNA
Transformation
  abortive, 1270
  in culture, *196, 196,* 1254–1257,
    1255, 1257
  in DNA replication, 223
  oncoproteins and, 1267–1268
  phosphorylation and, 1264
  T proteins and, 1272–1273
  tumor viruses in, 1270–1273
  two-step model of, 1267–1268
Transformed cell, 196
Transformed cell line, 1254–1257,
    1255, 1257
Transformer gene, 512–513
Transforming growth factor(s), in
    transformed cells, 1256,
    1260–1261
Transforming growth factor β (TGFβ)
  in induction, *1161,* 1161–1163, *1163*
  in mesodermal induction, 1165
  proteoglycan binding of, 1142
  receptors for, 1162
  structure of, *1161,* 1161–1162
Transforming protein, 610
Transgenes, 290
  in gene therapy, 299
Transgenic organisms, 291, 1323
  production of, 296–298
Transition state, 18, 45
Transition-state intermediate, 76, 78
  antibody binding of, 88, *89*
Translation. *See also* Protein(s),
    synthesis of
  aminoacyl-tRNA synthetases in,
    126–128, 127t, *127–129*
  codon-anticodon pairing and,
    124–126, *125, 128*
  in control of ribosomal protein
    synthesis, 525–526, *526*

definition of, 120
elongation in, *134–135, 136–138*
in eukaryotes vs. prokaryotes,
    116–119, *118–120*
initiation of, 133, *134–135*
    sites of, 133, 308–309
in prokaryotes vs. eukaryotes,
    308–309
regulation of
    antisense, 527–528, *528*
    in eukaryotes, 526–527, *527*
    in prokaryotes, 525–526, *526,*
        527–528, 528
    by RNA-binding proteins, 525–527
ribosomes in, 128–132, *130–132*
mRNA in, 120–122
rRNA in, 128–132
tRNA in, 124–128
termination of, *135, 136–138*
viral, 205–206, *206, 207*
Translocation
chromosomal, 267, *267*
    in tumor cells, 1268
of mitochondrial proteins, 824
in protein synthesis, *134–135,*
    136–138
signal-dependent, in *Drosophila,*
    573–575, 575
Translocation intermediates, 824
Transmembrane proteins, 604. *See also*
    Membrane proteins
porins as, 609, *609*
Transmission electron microscopy,
    158–161, *159–161*
Transplicing, variable surface
    glycoprotein switching and, 342
Transport
anterograde, 1070
axonal, 1070–1072, *1071,* 1072t,
    1073
intracellular
    microtubules in, 1070–1079
    motor proteins in, 1075–1079
membrane. *See* Membrane transport
orthograde axoplasmic, 927
protein. *See* Protein transport
proton
    by electron carriers, 786–770
    mitochondrial, 759–770
retrograde, 1070
Transporter(s), 143, *634, 635. See also*
    Membrane transport proteins
antiporter, *634, 635,* 654–660,
    655–658. See also Antiporter(s)
ATP-ADP, 674
carrier model of, 638, *638*
in cotransport, 654–659
erythrocyte glucose, 638–640, *640,*
    640t
lactose, *660,* 660–661, *661*
neurotransmitter, 970

in oligosaccharide synthesis, 703, *703*
phosphate, *758,* 758–759
symporter, *634, 635,* 652–654, *653,*
    654, 658, 658–659. See also
    Symporter(s)
uniporter, *634, 635,* 636–640,
    637–639
Transport proteins. *See* Membrane
    transport proteins
Transport vesicles, *172, 172–173,*
    711–718
clathrin-coated, *172,* 711–713, *712,*
    713
depolymerization of, 713
in endocytosis, 722–724, 727
formation of
    by budding, 711–716, *717*
    by endocytosis, 722–724, *723*
lysosomal, 709, *710*
membrane fusion with, 716–718,
    717, 718
non–clathrin-coated, *172,* 713–716,
    714
for phospholipids, 673
protein integration into, 712, *713*
in receptor-mediated endocytosis,
    722–732
secretory. *See* Secretory vesicles
targeting of, 716–718, *718*
types of, 711
Transposase, *324,* 324–325
Transposase mRNA, antisense
    regulation of, 527–528, *528*
Transposition
of *Drosophila* P element, 326–327
of mobile elements, 322–338. *See*
    also Mobile elements
    DNA-mediated
        in eukaryotes, 326–328, *327, 328*
        in prokaryotes, 323–326, *323–326*
        nonreplicative, 324, *324*
        replicative, 324, *325,* 325–326
    RNA-mediated, 328–338
Transposon(s). *See also* Mobile elements
bacterial, 323–326, *323–326*
definition of, 323
eukaryotic, 326–328, *327, 328*
Trans-splicing, 507–508
Transthyretin, 457–458, *458*
Transverse tubules, 1027, *1028*
Treadmilling, 1004, *1005*
Triacylglycerol
hydrolysis of, 748
metabolism of, 886
Triad, 1027, *1028*
Tricarboxylic acid cycle, 749–750, *750*
    glycolysis and, 771
Trinucleotides, in genetic code studies,
    123, 123–124
Triplet microtubule, 1052, *1053*
Tristearin, 25, *26*

Tritium, 213t, 218
tRNA. *See under* RNA
Tropocollagen, *1132,* 1133
Tropomodulin, 1000, 1010
    in muscle, 1024, *1025*
Tropomyosin (TM), 1000, 1027–1029,
    1029
Troponin (TN), 1027–1029, *1029*
trp aporepressor, 418–419, *420,* 424
trp operon, 424
    transcription of, 486–488, *487, 489*
trp repressor, 418–419, *420*
Trypanosomes
    gene conversion in, 341–343, *342*
    mitochondria of, RNA editing in,
        536–537, *537*
Trypsin, 76, *77*
    vs. chymotrypsin, 78
Trypsinogen, 76
Tryptophan, 54, *55. See also* Amino
    acid(s)
    synthesis of, 486–488, *487, 489*
T tubules, 1027, *1028*
Tube protein, 575, *575*
Tubulin
    aceylation of, 1059
    carboxyl-terminal tyrosine residue in,
        1060
    colchicine and, *1066,* 1066–1067
    diversity of, 1059–1060
    in microtubule-organizing center,
        1055
    polymerization of, in microtubule
        assembly, *1061–1063,* 1061–1064
    post-translational modification of,
        1059–1060
Tubulin subunits, in microtubules,
    1052–1054, *1053*
Tumor(s)
    benign, 1248, *1248*
    classification of, 1249
    malignant, 1248, *1248. See also*
        Cancer
Tumor cells
    aneuploidy in, 1268
    characteristics of, 1248–1251,
        1255–1257
    chromosomal abnormalities in,
        1268–1269, *1269*
    culture of, in cancer research,
        1251–1258
    DNA reduplication in, 1269, *1269*
    growth of, 1251
        autocrine induction of, *1260,*
            1260–1261
    interaction of, 1249–1251
    transformation of. *See* Transformation
Tumor promotion, 1286–1287
Tumor-suppressor genes, *1266,*
    1266–1267
    mutational inactivation of,
        1269–1270

Tumor viruses
 DNA, 1270–1273
 human, *1279*, 1279–1280
 RNA, 1273–1279. *See also*
  Retroviruses
Turgor, 174
Turgor pressure, 664
Turn
 in domains, 67–68
 protein, 57
Tus protein, 378
Twist, in DNA, 110, *111*
twist gene, 574–575
Two-dimensional gel electrophoresis, *93*,
  93–94
Two-sodium/one-glucose symporter, 652
Ty elements, 328–333, *329*, *332*
Tyk2 kinase, 916–917
Tyrosine, 54, *55*. *See also* Amino acid(s)
 in tubulin, 1060
Tyrosine kinase–linked receptors, 862
Tyrosine-specific kinases, as
  oncoproteins, 1261–1264, 1262t

U1A protein, autoregulation of, 509
UBF. *See* Upstream binding factor (UBF)
Ubiquinone, 761, *761*, 763–764, *764*.
  See also Electron carriers
Ubiquitin, 1210
UCE. *See* Upstream control element
  (UCE)
UDP-glucose, 882, *883*
Ultrabithorax gene, 579, *580–583*,
  581–583
Ultraviolet radiation
 as carcinogen, 1284–1286
 DNA damage by, repair of, *387*,
  387–388, *388*
Uncleaved internal signal-anchor
  sequences, 689–690, *691*
unc mutations, in *C. elegans*, *1179*,
  1179–1181
Unconditioned stimulus, 979, *980*
Uncoupler, 770
Unc6 protein, as
  chemoattractant/chemorepellent,
  1179–1181
Unequal crossing over
 DNA fingerprinting and, 319–320,
  321, *322*
 gene duplication and, 314, *314*
Uniporters, *634*, 635, 636–640,
  637–639. *See also* Membrane
  transport proteins
Up mutation, 414
Upstream activating sequence, 442, *442*
Upstream binding factor (UBF),
  469–470, *470*
Upstream control element (UCE),
  469–470, *470*

Upstream regulatory sequence (URS), in
  cell type-specific gene expression,
  551–552, *552*, 555, *555*
Uracil, 102, *102*, 103t
Ureteric bud, 1168, *1168*, 1169
Uridine diphosphoglucose, 882, *883*
Uvomorulin (E-cadherin), 1036, 1115,
  1150–1153
UvrABC repair, *387*, 387–388, *388*

Vaccinia virus, 209, 210t
Vacuole, 143, *146*, 174
 contractile, 662, *662*
 sucrose in, 659–660
Valine, 54, *55*. *See also* Amino acid(s)
Van der Waals contact, 25, *25*
Van der Waals interactions, 24–25, *25*,
  25t, *26*, 26
Van der Waals radius, 25, 25t
Van't Hoff equation, 662
Variable regions, antibody
 heavy chain, *1310*, 1311–1315,
  1312–1314, 1319, *1320*
 light chain, *1310*, 1311–1318,
  1315–1318
Variable surface glycoprotein (VSG),
  341–343, *342*
Variola virus, 209
Vascular bundle, 805
V-class ATPase, *645*, 659–660
Vector(s)
 antisense, 528
 cloning
  bacteriophage lambda, 230–233,
   231–233, 233t
  bacteriophage P1, 233t, 234
  cosmid, 233t, 233–234, *234*
  DNA separation from, 240–242
  plasmid, 222–225, *223–225*
  in expression systems, 252–254,
   253
  restriction fragment insertion into,
   228, *228*
  YAC, 233t, 234
 expression, 240, *241*, 252–254, *253*
 viral, 204
Vegetal pole, in *Xenopus*, 1163–1164,
  1164
Vesicle(s)
 clathrin-coated, 172, 711–713, *712*,
  713
  in endocytosis, 722–724, 727
 closed, in ATP synthesis, 752–753,
  756–758
 coated, 172
 CURL, 726–727, *727*
 Golgi. *See* Golgi vesicles
 secretory, *172*, 172–173
  maturation of, 678
  protein sorting to, 719–720

 protein transport to, 677–678,
  719–720
 synaptic. *See* Synaptic vesicles
 thylakoid, 175, *175*
 transport, 172, 172–173, 711–718.
  See also Transport vesicles
 transport of
  axonal, 1070–1072
  microtubular, 1072–1075, *1074*,
   1075
  myosin V in, *1038*, 1038–1039,
   1078–1079
Vesicles, non-clathrin-coated, 172,
  713–716, *714*
Vesicular stomatitis virus (VSV), 205,
  714–718
VG1, in mesodermal induction,
  1165–1166, *1166*
Villin, 1009t
Vimentin, 1106–1108, *1107*, 1108t,
  1109. *See also* Intermediate
  filament(s) (IF)
 cytosolic, 1114
 microtubules and, 1113
Vinblastine, microtubule stability and,
  1067
Vinculin, 1025, *1034*, 1035–1036, 1114
 in cell transformation, 1264
V ion pump, 644–645, *645*, 646t
Viral capsid, 202–203, *203*
Viral envelope, 203, *204*
Viral infection, lysis vs. lysogeny in,
  544–545
Viral oncogenes, 1270
 terminology for, 1259
Viral proteins, 205–206
 arrangement of, 202–203, *203*
Viral retrotransposons, 328, 328–334,
  329
Virchow, Rudolf, 4
Virgin B cell, 1321, 1337
Virion, 202
 lambda-phage, construction of,
  230–231, *231*
Virology, 205
Virus(es), 202–212
 animal, 204, 208–212, *209*, 210t,
  211, *212*
 assay of, 205, *205*
 bacterial, in research, 206–208
 cell integration of, 1270–1271
 classification of, 208–212, *209*, 210t
 cloning of, 205
 definition of, 202
 DNA, classification of, 208–209, *209*,
  210t
 endocytosis of, 731–732
 Epstein-Barr, *1279*, 1279–1280
 in experiments, 202
 in gene therapy, 202
 helical, 202, *203*

helper, 1275
host range of, 204–205
as intracellular parasites, 1273
isocahedral, 202–203, *203*
nucleic acids in, 206–212
plant, 204, 208
quasi-spherical, 202–203, *203*
replication of
   lysogenic cycle of, 206, *207*, 230
   lytic cycle of, 206, *206*, *207*, 230
RNA, classification of, *209*, 209–212,
   210t
self-assembly of, 203
shapes of, 202–203, *203*, *204*
structure of, *203*, 203–205, *204*
tumor
   DNA, 1270–1273
   human, 1279–1280
   RNA, 1273–1279. *See also*
     Retroviruses
vectors of, 204
Viscometry, 1003
Vision, 971–978. *See also* Photons,
   absorption of
color, 977–978, *978*
Visual system, 971–978
Vitamin C, in collagen synthesis, 1131,
   1131
Voltage change, in oxidation-reduction
   reaction, 38–39
Voltage-gated ion channels. *See* Ion
   channel(s), voltage-gated
Voltage-sensing α helixes, 938
V proton pump, H+ ATPase, 650–651,
   651
VSG. *See* Variable surface glycoprotein
   (VSG)

Water
   bulk flow of, 661–664
   chemical bonding in, *22*, 22–23, *23*
   dissociation products of, 30–31
   molecular structure of, *22*, 23
Water channel proteins, 663, *663*
Watson, James D., 104, 114
wee mutants, 1203, 1222–1224
Wee1 protein, in mitosis, 1222–1224

Western blotting, 97, *97*
White blood cells, migration of,
   adhesive proteins and, *1154*,
   1154–1155, *1155*
White-fat tissue, 770
Wiechaus, E., 273
Wild-type organism, 116, 1203
Williams, R., 208
Wilms' tumor (WT1) gene, 456, *457*
Winged-helix proteins, *448*, 451
Wing imaginal disc, 562
Wingless protein, 1165
Wobble pairing, *125*, 125–126, *126*
Wound healing, cell migration in, 1150
Writhe, in DNA, 111, *111*
WT1 gene, 456, *457*

X chromosome, inactivation of,
   465–466
Xenopus laevis
   DNA replication in, 1231–1232
   embryogenesis in, 1163–1166
   oocyte maturation in, 1203–1207
Xeroderma pigmentosa, 388, 389t,
   1285–1286
X-linked recessive mutation, 277, *278*
X-ray crystallography, 59–61
Xylem, 661, 803–805, *804*

YAC. *See* Yeast artificial chromosome
   (YAC)
YAC vector, 233t, 234
Yeast
   autonomously replicating sequences
     in, 355, *356*, 371
   budding in, 181, *183*, *1226*, 1226
   in cell cycle experiments, 1203, *1204*
   cell cycle mutants in, genetic screen
     for, *272*, 272–273
   cell division in, 181, *183*, 1226–1232
   centromere in, 1094–1096, *1095*
   culture of, 190. *See also* Cell culture
   DNA in, 146t
   gene conversion in, 339–341, *340*,
     341, 554, 554–556, *555*

gene knockout in, 292–293, *293*
gene organization in, 311, *311*
MAP kinase activation in, 897–899,
   898
mating factors in, *553*, 553–554
   in signaling, 855, 897–899, *898*
mating-type conversion in, 550–556
   gene conversion and, 339–341,
     340, *341*
maturation-promoting factor in, *1221*,
   1221–1223, *1222*
meiotic recombination in, 396
   gene conversion in, 396–397,
     397–399
mitotic apparatus in, 1091, *1093*
mutations in, complementation
   analysis of, 274, *275*
pheromones in, *553*, *553*, 1230
protein secretion in, 678–680, *680*
silent mating-type loci in, 465
tandemly repeated arrays in, 316t
transcription in, cell-type–specific,
   550–556
viral retrotransposons in, 328
Yeast artificial chromosome (YAC), 355
   in cloning, 359
   in physical mapping, 284–285, *286*
Yeast artificial chromosome (YAC)
   vector, 233t, 234
Yeast chromosome III, 293
Yeast two-hybrid system, 894–895, *896*

Z disk, 1023, *1024*, 1024–1025
   desmin and, 1108, 1114–1115, *1115*
Z DNA, 106, *107*
Zellweger syndrome, 838, *838*
zen gene, 574–575
Zinc finger, 67, *67*, 448–452, *449–452*
Z line, 1023, *1024*
Zone of polarizing activity (ZPA), 1169,
   1170
Zwitterion, 32, 52
Zygotic genes, in *Drosophila*
   regionalization, 572–573,
   576–579, 580
Zymogen, 76